Lecture Notes in Computer Science 4491

Commenced Publication in 1973
Founding and Former Series Editors:
Gerhard Goos, Juris Hartmanis, and Jan van Leeuwen

Derong Liu   Shumin Fei
Zeng-Guang Hou   Huaguang Zhang
Changyin Sun (Eds.)

Advances in Neural Networks – ISNN 2007

4th International Symposium
on Neural Networks, ISNN 2007
Nanjing, China, June 3-7, 2007
Proceedings, Part I

Springer

Volume Editors

Derong Liu
University of Illinois at Chicago, IL 60607-7053, USA
E-mail: dliu@ece.uic.edu

Shumin Fei
Southeast University, School of Automation, Nanjing 210096, China
E-mail: smfei@seu.edu.cn

Zeng-Guang Hou
Chinese Adacemy of Sciences, Institute of Automation, Beijing 100080, China
E-mail: hou@compsys.ia.ac.cn

Huaguang Zhang
Northeastern University, Shenyang 110004, China
E-mail: zhanghuaguang@ise.neu.edu.cn

Changyin Sun
Hohai University, School of Electrical Engineering, Nanjing 210098, China
E-mail: cysun@hhu.edu.cn

Library of Congress Control Number: 2007926816

CR Subject Classification (1998): F.1, F.2, D.1, G.2, I.2, C.2, I.4-5, J.1-4

LNCS Sublibrary: SL 1 – Theoretical Computer Science and General Issues

ISSN 0302-9743
ISBN-10 3-540-72382-X Springer Berlin Heidelberg New York
ISBN-13 978-3-540-72382-0 Springer Berlin Heidelberg New York

Springer is a part of Springer Science+Business Media

springer.com

Printed in Germany

Typesetting: Camera-ready by author, data conversion by Scientific Publishing Services, Chennai, India
Printed on acid-free paper SPIN: 12060757 06/3180 5 4 3 2 1 0

Preface

ISNN 2007 – the Fourth International Symposium on Neural Networks—was held in Nanjing, China, as a sequel of ISNN 2004/ISNN 2005/ISNN 2006. ISNN has now become a well-established conference series on neural networks in the region and around the world, with growing popularity and increasing quality. Nanjing is an old capital of China, a modern metropolis with a 2470-year history and rich cultural heritage. All participants of ISNN 2007 had a technically rewarding experience as well as memorable experiences in this great city.

A neural network is an information processing structure inspired by biological nervous systems, such as the brain. It consists of a large number of highly interconnected processing elements, called neurons. It has the capability of learning from example. The field of neural networks has evolved rapidly in recent years. It has become a fusion of a number of research areas in engineering, computer science, mathematics, artificial intelligence, operations research, systems theory, biology, and neuroscience. Neural networks have been widely applied for control, optimization, pattern recognition, image processing, signal processing, etc.

ISNN 2007 aimed to provide a high-level international forum for scientists, engineers, and educators to present the state of the art of neural network research and applications in diverse fields. The symposium featured plenary lectures given by worldwide renowned scholars, regular sessions with broad coverage, and some special sessions focusing on popular topics.

The symposium received a total of 1975 submissions from 55 countries and regions across all six continents. The symposium proceedings consists of 454 papers among which 262 were accepted as long papers and 192 were accepted as short papers. We would like to express our sincere gratitude to all reviewers of ISNN 2007 for the time and effort they generously gave to the symposium. We are very grateful to the National Natural Science Foundation of China, K. C. Wong Education Foundation of Hong Kong, the Southeast University of China, the Chinese University of Hong Kong, and the University of Illinois at Chicago for their financial support. We would also like to thank the publisher, Springer, for cooperation in publishing the proceedings in the prestigious series of *Lecture Notes in Computer Science*.

Derong Liu
Shumin Fei
Zeng-Guang Hou
Huaguang Zhang
Changyin Sun

ISNN 2007 Organization

General Chair

Derong Liu, University of Illinois at Chicago, USA, and Yanshan University, China

General Co-chair

Marios M. Polycarpou, University of Cyprus

Organization Chair

Shumin Fei, Southeast University, China

Advisory Committee Chairs

Shun-Ichi Amari, RIKEN Brain Science Institute, Japan
Chunbo Feng, Southeast University, China
Zhenya He, Southeast University, China

Advisory Committee Members

Hojjat Adeli, Ohio State University, USA
Moonis Ali, Texas State University-San Marcos, USA
Zheng Bao, Xidian University, China
Tamer Basar, University of Illinois at Urbana-Champaign, USA
Tianyou Chai, Northeastern University, China
Guoliang Chen, University of Science and Technology of China, China
Ruwei Dai, Chinese Academy of Sciences, China
Dominique M. Durand, Case Western Reserve University, USA
Russ Eberhart, Indiana University Purdue University Indianapolis, USA
David Fogel, Natural Selection, Inc., USA
Walter J. Freeman, University of California-Berkeley, USA
Toshio Fukuda, Nagoya University, Japan
Kunihiko Fukushima, Kansai University, Japan
Tom Heskes, University of Nijmegen, The Netherlands
Okyay Kaynak, Bogazici University, Turkey
Frank L. Lewis, University of Texas at Arlington, USA
Deyi Li, National Natural Science Foundation of China, China
Yanda Li, Tsinghua University, China
Ruqian Lu, Chinese Academy of Sciences, China

Steering Committee Chair

Steering Committee Co-chair

Steering Committee Members

Local Arrangements Chairs

Enrong Wang, Nanjing Normal University, China
Shengyuan Xu, Nanjing University of Science and Technology, China
Junyong Zhai, Southeast University, China

Electronic Review Chair

Xiaofeng Liao, Chongqing University, China

Symposium Secretariats

Ting Huang, University of Illinois at Chicago, USA
Jinya Song, Hohai University, China

ISNN 2007 International Program Committee

Shigeo Abe, Kobe University, Japan
Ajith Abraham, Chung Ang University, Korea
Khurshid Ahmad, University of Surrey, UK
Angelo Alessandri, University of Genoa, Italy
Sabri Arik, Istanbul University, Turkey
K. Vijayan Asari, Old Dominion University, USA
Amit Bhaya, Federal University of Rio de Janeiro, Brazil
Abdesselam Bouzerdoum, University of Wollongong, Australia
Martin Brown, University of Manchester, UK
Ivo Bukovsky, Czech Technical University, Czech Republic
Jinde Cao, Southeast University, China
Matthew Casey, Surrey University, UK
Luonan Chen, Osaka-Sandai University, Japan
Songcan Chen, Nanjing University of Aeronautics and Astronautics, China
Xiao-Hu Chen, Nanjing Institute of Technology, China
Xinkai Chen, Shibaura Institute of Technology, Japan
Yuehui Chen, Jinan University, Shandong, China
Xiaochun Cheng, University of Reading, UK
Zheru Chi, Hong Kong Polytechnic University, China
Sungzoon Cho, Seoul National University, Korea
Seungjin Choi, Pohang University of Science and Technology, Korea
Tommy W. S. Chow, City University of Hong Kong, China
Emilio Corchado, University of Burgos, Spain
Jose Alfredo F. Costa, Federal University, UFRN, Brazil
Mingcong Deng, Okayama University, Japan
Shuxue Ding, University of Aizu, Japan
Meng Joo Er, Nanyang Technological University, Singapore
Deniz Erdogmus, Oregon Health & Science University, USA

Shuxian Lun, Bohai University, China
Fa-Long Luo, Anyka, Inc., USA
Jinwen Ma, Peking University, China
Xiangping Meng, Changchun Institute of Technology, China
Kevin L. Moore, Colorado School of Mines, USA
Ikuko Nishikawa, Ritsumeikan University, Japan
Stanislaw Osowski, Warsaw University of Technology, Poland
Seiichi Ozawa, Kobe University, Japan
Hector D. Patino, Universidad Nacional de San Juan, Argentina
Yi Shen, Huazhong University of Science and Technology, China
Daming Shi, Nanyang Technological University, Singapore
Yang Shi, University of Saskatchewan, Canada
Michael Small, Hong Kong Polytechnic University
Ashu MG Solo, Maverick Technologies America Inc., USA
Stefano Squartini, Universita Politecnica delle Marche, Italy
Ponnuthurai Nagaratnam Suganthan, Nanyang Technological University, Singapore
Fuchun Sun, Tsinghua University, China
Johan A. K. Suykens, Katholieke Universiteit Leuven, Belgium
Norikazu Takahashi, Kyushu University, Japan
Ying Tan, Peking University, China
Yonghong Tan, Guilin University of Electronic Technology, China
Peter Tino, Birmingham University, UK
Christos Tjortjis, University of Manchester, UK
Antonios Tsourdos, Cranfield University, UK
Marc van Hulle, Katholieke Universiteit Leuven, Belgium
Dan Ventura, Brigham Young University, USA
Michel Verleysen, Universite Catholique de Louvain, Belgium
Bing Wang, University of Hull, UK
Dan Wang, Dalian Maritime University, China
Pei-Fang Wang, SPAWAR Systems Center-San Diego, USA
Zhiliang Wang, Northeastern University, China
Si Wu, University of Sussex, UK
Wei Wu, Dalian University of Technology, China
Shunren Xia, Zhejiang University, China
Yousheng Xia, University of Waterloo, Canada
Cheng Xiang, National University of Singapore, Singapore
Daoyi Xu, Sichuan University, China
Xiaosong Yang, Huazhong University of Science and Technology, China
Yingjie Yang, De Montfort University, UK
Zi-Jiang Yang, Kyushu University, Japan
Mao Ye, University of Electronic Science and Technology of China, China
Jianqiang Yi, Chinese Academy of Sciences, China
Dingli Yu, Liverpool John Moores University, UK
Zhigang Zeng, Wuhan University of Technology, China

Table of Contents – Part I

Neural Fuzzy Control

Neural Networks for Control Applications

Adaptive Dynamic Programming and Reinforcement Learning

Neural Networks for Nonlinear Systems Modeling

Robotics

Stability Analysis of Neural Networks

Learning and Approximation

Data Mining and Feature Extraction

Direct Adaptive Fuzzy-Neural Control for MIMO Nonlinear Systems Via Backstepping

Shaocheng Tong and Yongming Li

Department of Basic Mathematics of Liaoning Institute of Technology, Jinzhou, Liaoning .121001, China
jztsc@sohu.com

Abstract. In this paper, an adaptive fuzzy-neural network control problem is discussed for some uncertain MIMO nonlinear systems with the block-triangular structure. The fuzzy-neural networks are utilized to approximate the virtual controllers, and by using backstepping technique, the direct adaptive FNN control scheme is developed. The proposed control method guarantees the closed-loop signals to be semiglobally uniformly ultimately bounded.

1 Introduction

The very rapid developments described in adaptive and robust control techniques are accompanied by an increasing in the use of neural networks or fuzzy logic systems for system identification or identification-based control. With the help of neural networks or fuzzy logic systems, a large number of backstepping design schemes are reported that combine the backstepping technique with adaptive neural networks or fuzzy logic systems [2,3]. In these backstepping design schemes, most of them are centralized on SISO nonlinear systems, and belong to the indirect adaptive control methodology. So far, there are a few of the results on MIMO nonlinear systems. Recently, indirect adaptive backstepping-based neural and backstepping-based fuzzy control approaches were proposed for a class of MIMO nonlinear systems with triangular structure [4,5], respectively. In these methods, neural networks and fuzzy logic systems are utilized to approximate the unknown functions in every recursively design, and the stability of control systems was given by using Lyapunov functions. However, as far as we know, the direct adaptive backstepping-based neural or backstepping-based fuzzy control approaches has not been discussed yet.

In this paper, we focus on developing a direct adaptive fuzzy-neural control of a class of MIMO nonlinear systems. In recursively backstepping designs, fuzzy-neural networks are employed to approximate the optimal virtue controllers, not to approximate the unknown functions of the systems. The adjusting parameter vectors are derived on the Lyapunov functions. The proposed control method can guarantee the closed-loop signals to be semiglobally uniformly ultimately bounded and the tracking error converging to a residual set.

D. Liu et al. (Eds.): ISNN 2007, Part I, LNCS 4491, pp. 1–7, 2007.

2 Problem Formulation

Consider a class of uncertain MIMO nonlinear systems described by the following differential equation:

$$\begin{cases} \dot{x}_{1,i_1} = f_{1,i_1}(\bar{x}_{1,(i_1-\rho_{11})},\cdots,\bar{x}_{m,(i_1-\rho_{1m})}) + g_{1,i_1}(\bar{x}_{1,(i_1-\rho_{11})},\cdots,\bar{x}_{m,(i_1-\rho_{1m})})x_{1,i_1+1}, \ 1\le i_1 \le \rho_1 - 1 \\ \dot{x}_{1,\rho_1} = f_{1,\rho_1}(X) + g_{1,\rho_1}(\bar{x}_{1,(i_1-\rho_{11})},\cdots,\bar{x}_{m,(i_1-\rho_{1m})})u_1 \\ \cdots \\ \dot{x}_{j,i_j} = f_{j,i_j}(\bar{x}_{1,(i_j-\rho_{j1})},\cdots,\bar{x}_{m,(i_j-\rho_{jm})}) + g_{j,i_j}(\bar{x}_{1,(i_j-\rho_{j1})},\cdots,\bar{x}_{m,(i_j-\rho_{jm})})x_{j,i_j+1}, \ 1\le i_j \le \rho_j - 1 \\ \dot{x}_{j,\rho_j} = f_{j,\rho_j}(X,u_1,\cdots,u_{j-1}) + g_{j,\rho_j}(\bar{x}_{1,(\rho_1-1)},\cdots,\bar{x}_{m,(\rho_m-1)})u_j \\ y_j = x_{j,1}, \quad 1\le j\le m \end{cases} \tag{1}$$

where x_{j,i_j}, $i_j = 1,\cdots,\rho_j$ are the state variables of the jth subsystem, $u_j \in R$ and $y_j \in R$ are the control input and output of the jth subsystem, respectively. $\bar{x}_{j,i_j} = [x_{j,1},\cdots,x_{j,i_j}]^T \in R^{i_j}$ and $X = [\bar{x}_{1,\rho_1}^T \ \cdots \ \bar{x}_{m,\rho_m}^T]^T$ denote the state variables of the complete system. $f_{j,i_j}(\cdot)$ and $g_{j,i_j}(\cdot)$ $(i_j = 1,\cdots,\rho_j,\ j=1,\cdots,m)$ are unknown smooth nonlinear functions. j, i_j, ρ_j and m are positive integers. ρ_j denotes the order of the jth subsystem, $\rho_{j,l} = \rho_j - \rho_l$ is the order difference between the jth and lth subsystems, and i_j denotes the subscription of the i_j component of the corresponding items in the jth subsystem.

Assumption 2.1 The signs of $g_{j,i_j}(\cdot)$ are known, and there exist constants $\bar{g}_{j,i_j} > \underline{g}_{j,i_j} > 0$, such that $\forall \bar{x}_{j,i_j} \in \Omega \in R^n$, $\bar{g}_{j,i_j} \ge g_{j,i_j}(\cdot) \ge \underline{g}_{j,i_j}$.

The derivatives of $g_{j,i_j}(\cdot)$ are given by

$$\dot{g}_{j,i_j}(\bar{x}_{1,(i_j-\rho_{j1})},\cdots,\bar{x}_{m,(i_j-\rho_{jm})}) = \sum_{l=1}^{m}\sum_{k=1}^{i_j-\rho_{jl}} \frac{\partial g_{j,i_j}(\cdot)}{\partial x_{l,k}} \times [g_{l,k}(\cdot)x_{l,k+1} + f_{l,k}(\cdot)] \tag{2}$$

Assumption 2.2 There exist constants $g_{j,i_j}^d > 0, i_j = 1,\cdots,\rho_j,\ j = 1,\cdots,m$, such that $\forall \bar{x}_{j,i_j} \in \Omega_{j,i_j}, \left|\dot{g}_{j,i_j}(\cdot)\right| \le g_{j,i_j}^d$.

3 Direct Adaptive Fuzzy-Neural Control Design

In this section, we will give the procedure of the backstepping design for the jth subsystem. For clarity and conciseness, Step1 is described with detailed explanations, while Step i_j and ρ_j are simplified, with the relevant equations and the explanations being omitted.

Step 1. Let $x_{d1} = y_{id}$ and define the tracking error variable $e_{j,1} = x_{j,1} - x_{d1}$. Its derivative is

$$\begin{aligned}\dot{e}_{j,1} &= \dot{x}_{j,1} - \dot{x}_{d1} \\ &= f_{j,1}(\overline{x}_{1,(1-\rho_{j1})},\cdots,\overline{x}_{m,(1-\rho_{jm})}) + g_{j,1}(\overline{x}_{1,(1-\rho_{j1})},\cdots,\overline{x}_{m,(1-\rho_{jm})})x_{j,2} - \dot{x}_{d1}\end{aligned} \tag{3}$$

By viewing $x_{j,2}$ as a virtual control input, apparently there exists a desired feedback control

$$\alpha^{*}_{j,1} = x_{j,2} = -k_{j,1}e_{j,1} - \frac{1}{g_{j,1}}(f_{j,1} - \dot{x}_{d1}) \tag{4}$$

where $k_{j,1}$ is a positive design constant to be specified later. A fuzzy-neural network is utilized to approximate the desired controller $\alpha^{*}_{j,1}$, and a fuzzy–neural virtual controller can be used as follows

$$\hat{u}(\overline{X}_{j,1}|\theta_{j,1}) = \alpha_{j,1} \tag{5}$$

where $\overline{X}_{j,1} = \left[\overline{x}^{T}_{1,(1-\rho_{j1})} \ \cdots \ \overline{x}^{T}_{j-1,(1-\rho_{jj-1})} \ x_{j,1} \ \overline{x}^{T}_{j+1,(1-\rho_{jj+1})} \ \cdots \ \overline{x}^{T}_{m,(1-\rho_{jm})}\right]^{T}$,and $\hat{u}(\overline{X}_{j,1}|\theta_{j,1})$ is a fuzzy-neural networks taken in the form of reference[5].
Define $e_{j,2} = x_{j,2} - \alpha_{j,1}$, and we have

$$\begin{aligned}\dot{e}_{j,1} &= f_{j,1} + g_{j,1}(e_{j,2} + \alpha_{j,1}) - \dot{x}_{d1} \\ &= f_{j,1} + g_{j,1}e_{j,2} + g_{j,1}\hat{u}(\overline{X}_{j,1}|\theta_{j,1}) - g_{j,1}x_{j,2} + g_{j,1}x_{j,2} - \dot{x}_{d1}\end{aligned} \tag{6}$$

Substituting (4) into (6) yields

$$\begin{aligned}\dot{e}_{j,1} &= g_{j,1}e_{j,2} + g_{j,1}(\hat{u}(\overline{X}_{j,1}|\theta_{j,1}) - x_{j,2}) - g_{j,1}k_{j,1}e_{j,1} \\ &= g_{j,1}e_{j,2} + g_{j,1}(\tilde{\theta}^{T}_{j,1}\xi_{j,1}(\overline{X}_{j,1})) + g_{j,1}\omega_{j,1} - g_{j,1}k_{j,1}e_{j,1}\end{aligned} \tag{7}$$

where $\omega_{j,1} = \hat{u}(\overline{X}_{j,1}|\theta^{*}_{j,1}) - x_{j,2}$ is called the minimum fuzzy approximation error and $\tilde{\theta}^{T}_{j,1} = \theta^{T}_{j,1} - \theta^{*T}_{j,1}$ the parameter vector error.

Consider the following Lyapunov function candidate as

$$V_{j,1} = \frac{1}{2g_{j,1}(x_{j,1})}e^{2}_{j,1} + \frac{1}{2\gamma_{j,1}}\tilde{\theta}^{T}_{j,1}\tilde{\theta}_{j,1} \tag{8}$$

The derivative of $V_{j,1}$ is

$$\begin{aligned}\dot{V}_{j,1} &= \frac{e_{j,1}\dot{e}_{j,1}}{g_{j,1}(x_{j,1})} - \frac{\dot{g}_{j,1}(x_{j,1})}{2g_{j,1}^2(x_{j,1})}e_{j,1}^2 + \frac{1}{\gamma_{j,1}}\tilde{\theta}_{j,1}^T\dot{\theta}_{j,1} \\ &= e_{j,1}e_{j,2} - k_{j,1}e_{j,1}^2 + e_{j,1}\tilde{\theta}_{j,1}^T\xi_{j,1}(\overline{X}_{j,1}) + e_{j,1}\omega_{j,1}(\overline{X}_{j,1}) - \frac{\dot{g}_{j,1}(x_{j,1})}{2g_{j,1}^2(x_{j,1})}e_{j,1}^2 + \frac{1}{\gamma_{j,1}}\tilde{\theta}_{j,1}^T\dot{\theta}_{j,1} \\ &= e_{j,1}e_{j,2} - k_{j,1}e_{j,1}^2 - \frac{\dot{g}_{j,1}(x_{j,1})}{2g_{j,1}^2(x_{j,1})}e_{j,1}^2 + e_{j,1}\omega_{j,1}(\overline{X}_{j,1}) + \tilde{\theta}_{j,1}^T[e_{j,1}\xi_{j,1}(\overline{X}_{j,1}) + \frac{1}{\gamma_{j,1}}\dot{\theta}_{j,1}]\end{aligned} \tag{9}$$

Choosing the following adaptive law

$$\dot{\theta}_{j,1} = -\gamma_{j,1}e_{j,1}\xi_{j,1}(\overline{X}_{j,1}) - c_{j,1}\theta_{j,1} \tag{10}$$

where $\gamma_{j,1} > 0$ is a small constant.
Then we have

$$\dot{V}_{j,1} = e_{j,1}e_{j,2} + e_{j,1}\omega_{j,1}(\overline{X}_{j,1}) - \left(k_{j,1} + \frac{\dot{g}_{j,1}(x_{j,1})}{2g_{j,1}^2(x_{j,1})}\right)e_{j,1}^2 - c_{j,1}\tilde{\theta}_{j,1}^T\theta_{j,1} \tag{11}$$

Let $k_{j,1} = k_{j,10} + k_{j,11}$ with $k_{j,10} > 0$ and $k_{j,11} > 0$.
where

$$-c_{j,1}\tilde{\theta}_{j,1}^T\theta_{j,1} = -c_{j,1}\tilde{\theta}_{j,1}^T(\tilde{\theta}_{j,1} + \theta_{j,1}^*) \le -c_{j,1}\left\|\tilde{\theta}_{j,1}\right\|^2 + c_{j,1}\left\|\tilde{\theta}_{j,1}\right\|\left\|\theta_{j,1}^*\right\| \le -\frac{c_{j,1}\left\|\tilde{\theta}_{j,1}\right\|^2}{2} + \frac{c_{j,1}\left\|\theta_{j,1}^*\right\|^2}{2}$$

by completion of the squares, we have

$$e_{j,1}\omega_{j,1} - k_{j,1}e_{j,1}^2 \le e_{j,1}\left|\omega_{j,1}\right| - k_{j,11}e_{j,1}^2 \le \frac{\omega_{j,1}^2}{4k_{j,11}} \le \frac{\varepsilon_{j,1}^2}{4k_{j,11}}$$

Because $-(k_{j,10} + \dot{g}_{j,1}(x_{j,1})/2g_{j,1}^2(x_{j,1}))e_{j,1}^2 \le -(k_{j,10} - g_{j,1}^d/2\underline{g}_{j,1}^2(x_{j,1}))e_{j,1}^2$, by choosing $k_{j,10}$ such that $(k_{j,10}^* = k_{j,10} - g_{j,1}^d/2\underline{g}_{j,1}^2(x_{j,1})) > 0$, we have the following inequality

$$\dot{V}_{j,1} \le e_{j,1}e_{j,2} - k_{j,10}^*e_{j,1}^2 + \frac{\varepsilon_{j,1}^2}{4k_{j,11}} - \frac{c_{j,1}\left\|\tilde{\theta}_{j,1}\right\|^2}{2} + \frac{c_{j,1}\left\|\theta_{j,1}^*\right\|^2}{2} \tag{12}$$

Step i_j $(2 \le i_j \le \rho_j - 1)$ In the similar fashion, we can design a virtual controller α_{j,i_j} to make the error $e_{j,i_j} = x_{j,i_j} - \alpha_{j,i_j-1}$ as small as possible.

$$\alpha_{j,i_j}^* = -e_{j,i_j-1} - k_{j,i_j}e_{j,i_j} - \frac{1}{g_{j,i_j}}(f_{j,i_j} - \dot{\alpha}_{j,i_j-1}) \tag{13}$$

A fuzzy virtual controller is

$$\alpha_{j,i_j} = \hat{u}(\bar{X}_{j,i_j}) = \theta^T_{j,i_j} \xi_{j,i_j} (\bar{X}_{j,i_j}) \tag{14}$$

Differentiating e_{j,i_j} gives

$$\begin{aligned} \dot{e}_{j,i_j} &= \dot{x}_{j,i_j} - \dot{\alpha}_{j,i_j-1} \\ &= f_{j,i_j}(\bar{x}_{1,(i_j-\rho_{j1})},\cdots,\bar{x}_{m,(i_j-\rho_{jm})}) + g_{j,i_j}(\bar{x}_{1,(i_j-\rho_{j1})},\cdots,\bar{x}_{m,(i_j-\rho_{jm})}) x_{j,i_j+1} - \dot{\alpha}_{j,i_j-1} \end{aligned} \tag{15}$$

define $e_{j,i_j+1} = x_{j,i_j+1} - \alpha_{j,i_j}$, and we have

$$\dot{e}_{j,i_j} = g_{j,i_j}[e_{j,i_j+1} - e_{j,i_j-1} - k_{j,i_j} e_{j,i_j} + \tilde{\theta}^T_{j,i_j} \xi_{j,i_j} (\bar{X}_{j,i_j}) + \omega_{j,i_j}] \tag{16}$$

Consider the Lyapunov function candidate

$$V_{j,i_j} = V_{j,i_j-1} + \frac{1}{2g_{j,i_j}(x_{j,i_j})} e^2_{j,i_j} + \frac{1}{2\gamma_{j,i_j}} \tilde{\theta}^T_{j,i_j} \tilde{\theta}_{j,i_j} \tag{17}$$

Choosing the following adaptive law

$$\dot{\theta}_{j,i_j} = -\gamma_{j,i_j} e_{j,i_j} \xi_{j,i_j} (\bar{X}_{j,i_j}) - c_{j,i_j} \theta_{j,i_j} \tag{18}$$

Let $k_{j,i_j} = k_{j,i_j0} + k_{j,i_j1}$. By using (18), (12), and straightforward derivation similar to those employed in the former steps, the derivative of V_{j,i_j} becomes

$$\dot{V}_{j,i_j} \le e_{j,i_j} e_{j,i_j+1} - \sum_{k=1}^{i_j} k^*_{j,k0} e^2_{j,k} + \sum_{k=1}^{i_j} \frac{\varepsilon^2_{j,k}}{4k_{j,k1}} - \sum_{k=1}^{i_j} \frac{c_{j,k} \left\|\tilde{\theta}_{j,k}\right\|^2}{2} + \sum_{k=1}^{i_j} \frac{c_{j,k} \left\|\theta^*_{j,k}\right\|^2}{2} \tag{19}$$

where k_{j,i_j0} is chosen such that $k^*_{j,i_j0} = k_{j,i_j0} - g^d_{j,i_j} / 2\underline{g}^2_{j,i_j} > 0$

Step ρ_j This is the final step, and we define $e_{j,\rho_j} = x_{j,\rho_j} - \alpha_{j,\rho_j-1}$. Its derivative is

$$\begin{aligned} \dot{e}_{j,\rho_j} &= \dot{x}_{j,\rho_j} - \dot{\alpha}_{j,\rho_j-1} \\ &= f_{j,\rho_j}(X, u_1, \cdots, u_{j-1}) + g_{j,\rho_j}(\bar{x}_{1,(\rho_1-1)}, \cdots, \bar{x}_{j,(\rho_j-1)}) u_j - \dot{\alpha}_{j,\rho_j-1} \end{aligned} \tag{20}$$

and exists a desired feedback control

$$u^*_j = -e_{j,\rho_j-1} - k_{j,\rho_j} e_{j,\rho_j} - \frac{1}{g_{j,\rho_j}} (f_{j,\rho_j} - \dot{\alpha}_{j,\rho_j-1}) \tag{21}$$

A fuzzy virtual controller is

$$u_j = \hat{u}_j(\bar{X}_{j,\rho_j} \big| \theta_{j,\rho_j}) = \theta^T_{j,\rho_j} \xi_{j,\rho_j} (\bar{X}) \tag{22}$$

Then

$$\begin{aligned}\dot{e}_{j,\rho_j} &= f_{j,\rho_j} + g_{j,\rho_j}u_j - \dot{\alpha}_{j,\rho_j-1} \\ &= g_{j,\rho_j}[-e_{j,\rho_j-1} - k_{j,\rho_j}e_{j,\rho_j} + \tilde{\theta}^T_{j,\rho_j}\xi_{j,\rho_j}(\bar{X}) + \omega_{j,\rho_j}(\bar{X})]\end{aligned} \tag{23}$$

Consider the overall Layapunov function candidate

$$V_{j,\rho_j} = V_{j,\rho_j-1} + \frac{1}{2g_{j,\rho_j}(x_{j,\rho_j})}e^2_{j,\rho_j} + \frac{1}{2\gamma_{j,\rho_j}}\tilde{\theta}^T_{j,\rho_j}\tilde{\theta}_{j,\rho_j} \tag{24}$$

Choosing the following adaptive law

$$\dot{\theta}_{j,\rho_j} = -\gamma_{j,\rho_j}e_{j,\rho_j}\xi_{j,\rho_j}(\bar{X}) - c_{j,\rho_j}\theta_{j,\rho_j} \tag{25}$$

Let $k_{j,i_j} = k_{j,\rho_j0} + k_{j,\rho_j1}$. Similar to those employed in the former steps, the derivative of V_{j,ρ_j} becomes

$$\begin{aligned}\dot{V}_{j,\rho_j} &\le \dot{V}_{j,\rho_j-1} - k^*_{j,\rho_j0}e^2_{j,\rho_j} + \frac{\varepsilon^2_{j,\rho_j}}{4k_{j,\rho_j1}} - \frac{c_{j,\rho_j}\left\|\tilde{\theta}_{j,\rho_j}\right\|^2}{2} + \frac{c_{j,\rho_j}\left\|\theta^*_{j,\rho_j}\right\|^2}{2} \\ &\le -\sum_{k=1}^{\rho_j}k^*_{j,k0}e^2_{j,k} + \sum_{k=1}^{\rho_j}\frac{\varepsilon^2_{j,k}}{4k_{j,k1}} - \sum_{k=1}^{\rho_j}\frac{c_{j,k}\left\|\tilde{\theta}_{j,k}\right\|^2}{2} + \sum_{k=1}^{\rho_j}\frac{c_{j,k}\left\|\theta^*_{j,k}\right\|^2}{2}\end{aligned} \tag{26}$$

Choosing such that $k^*_{j,k0} > (\mu_j/2g_{j,k}) + (g^d_{j,k}/2\underline{g}^2_{j,k}), k = 1,\cdots,\rho_j$ and μ_j is a positive constant, and letting $\delta_j \le \sum_{k=1}^{\rho_j}\varepsilon^2_{j,k}/4k_{j,k1} + \sum_{k=1}^{\rho_j}c_{j,k}\left\|\theta^*_{j,k}\right\|^2/2$, then from (26), we have the following inequality

$$\begin{aligned}\dot{V}_{j,\rho_j} &\le -\sum_{k=1}^{\rho_j}k_{j,k0}e^2_{j,k} + \delta_j - \sum_{k=1}^{\rho_j}\frac{c_{j,k}\left\|\tilde{\theta}_{j,k_j}\right\|^2}{2} \\ &< -\mu_j V_{j,\rho_j} + \delta_j\end{aligned} \tag{27}$$

and we have

$$\dot{V}_{j,\rho_j} \le -k^*_{j\min}\left\|e_j\right\|^2 - \sigma_{j\min}\left\|\tilde{\theta}_j\right\|^2/2 + \delta_j$$

where $k^*_{j\min}$ and $\sigma_{j\min}$ are the minimum of k_{j,i_j} and σ_{j,i_j}, respectively. Therefore, the derivative of global Lyapunov function is negative as long as

$$e_j \in \Omega_j = \{e_j \left\| \left\|e_j\right\| \le \sqrt{\delta_j/k^*_{j\min}}\}\right. \tag{28}$$

or

$$\tilde{\theta}_j \in \Omega_{\theta j} = \{\tilde{\theta}_j \left\| \left\|\tilde{\theta}_j\right\| \le \sqrt{\delta_j/\sigma_{j\min}}\}\right. \tag{29}$$

According to a standard Lyapunov theorem, it is easily seen that all the signals in the closed-loop system remain bounded. Let $\psi_j = \delta_j / \mu_j$, and we have

$$\sum_{j=1}^{m}\sum_{k=1}^{\rho_j}\frac{1}{2g_{j,k}}e_{j,k}^2 < \sum_{j=1}^{m}\psi_j + (\sum_{j=1}^{m}V_j(0) - \sum_{j=1}^{m}\psi_j)\exp(-\mu t) \tag{30}$$

Let $g_{\max} = \max_{1\le k\le \rho_j}\{g_{k1}\}$ .Then, we have

$$\sum_{j=1}^{m}\sum_{k=1}^{\rho_j}e_{j,k}^2 < 2g_{\max}\sum_{j=1}^{m}\psi_j + 2g_{\max}\sum_{j=1}^{m}V_j(0)\exp(-\mu t) \tag{31}$$

which implies that given $\lambda_j > \sqrt{2g_{\max}\psi_j}$ .there exists T_j for all $\max\{T_j\} \ge t$, the tracking error satisfies

$$\left|e_{j,1}\right| = \left|y_{j,1}(t) - y_{jd}(t)\right| < \lambda_j,\ j = 1,\cdots,m \tag{32}$$

It is easily seen that the increase in the control gain k_{j,i_j} and the number of fuzzy inference rules will result in a better tracking performance.

4 Conclusion

In this paper, the direct adaptive fuzzy-neural network control problem is discussed for a class of uncertain multi-input/multi-output nonlinear systems via backstepping . By theoretical analysis, the closed-loop control system is proven to be semiglobally uniformly ultimately bounded with tracking error converging to a residual set.

Acknowledgements. This paper is supported by NNSF of China (60674056), National Key Basic Research and Development Program of China (2002CB312200) and Outstanding Youth Funds of Liaoning Province (2005219001).

References

1. Kristic, M., Kanellakopoulos, I., Kokotovic, P.V.: Nonlinear and Adaptive Control Design. New York: Wiley (1995)
2. Wan, C. K., Lewis, F. L.: Robust Backstepping Control of Nonlinear Systems using Neural Networks. *IEEE Trans. Syst. Man Cybern: A* **30** (2000) 753–766
3. Zhang, Y., Peng, P. Y., Jiang, Z. P.: Stable Neural Controller Design for Unknown Nonlinear Systems using Backstepping. *IEEE Trans. Neural Networks* **11** (2000) 1347–1359
4. Ge, S. S., Wang, C.: Adaptive Neural Control of Uncertain MIMO Nonlinear Systems. *IEEE Trans. Neural Networks* **15(3)** (2004) 674–692
5. Chen, B., Liu, X.P., Tong, S.C.: Fuzzy Output Tracking Control of MIMO Nonlinear Uncertain Systems by Backstepping Approach. *IEEE Trans. Fuzzy Systems* (2006) (to be appeared)

An Improved Fuzzy Neural Network for Ultrasonic Motors Control

Xu Xu[1], Yuxiao Zhang[2], Yanchun Liang[2,*], Xiaowei Yang[3], and Zhifeng Hao[3]

[1] College of Mathematics, Jilin University, Key Laboratory of Symbol Computation and Knowledge Engineering of the Ministry of Education, Changchun 130012, China
[2] College of Computer Science and Technology, Jilin University, Key Laboratory of Symbol Computation and Knowledge Engineering of the Ministry of Education, Changchun, China
[3] Department of Applied Mathematics, South China University of Technology, Guangzhou 510640, P. R. China
ycliang@jlu.edu.cn

Abstract. A newly developed non-symmetric sinusoidal membership function (NSSMF) is constructed. An improved fuzzy neural network controller using NSSMF is constructed to control the speed of ultrasonic motors. A dynamic algorithm with adaptive learning rate is used to train FNNC online. The global convergence of the FNNC systems could be guaranteed by adjusting the adaptive learning rate. The validity of the proposed scheme is examined by simulated experiments.

1 Introduction

An ultrasonic motor (USM) is a newly developed motor that has many excellent performances, and has been extensively used in many fields [1, 2]. However, USM has complicated nonlinear characteristic and its mathematical models are usually complex. Precise identification and control for ultrasonic motor is difficult. Recently, the combination intelligent methods have become one of the main approaches to control USM [3, 4]. Among these methods, the fuzzy neural network (FNN) is one of the effective methods [5-7]. In FNN system, the most importance is the fuzzy membership function and the training algorithm of connective weights. There are some shortcomings for the conventional membership functions such as hardly breaking symmetry and having no zeros for the Gauss's function; hardly differentiating them at the end of each line for the piecewise linear functions. In order to avoid these, we propose a novel non-symmetric sinusoidal membership function

(NSSMF) which contains many good virtues, such as the flexible form, adaptive parameter, zero point, smooth and comparatively small calculation. And a FNN controller of USM based on the NSSMF are presented. In this method, the connective weights and the membership functions of the FNN can be trained during online

* The authors are grateful to the support of the National Natural Science Foundation of China (10501017, 60673023, 60433020), the science-technology development project of Jilin Province of China (20050705-2), "985" project of Jilin University, and the European Commission under grant No. TH/Asia Link/010 (111084).

D. Liu et al. (Eds.): ISNN 2007, Part I, LNCS 4491, pp. 8–13, 2007.

identification and control, so that the errors between the actual system output and the reference output are decreased in every iteration step.

2 Non-Symmetric Sinusoidal Membership Function (NSSMF)

Usually membership functions are composed of piecewise linear functions. Therefore, one cannot differentiate them at the end of each line. One way of preventing this is to use Gauss' membership function. But this function is very inflexible so we can hardly break symmetry, change the shape, or even reach a given shape. Moreover it is never zero. If we do not shift the Gaussian shape, we have to introduce a cut-off line to obtain a finite set. In both cases we have a lack of completeness, which must be considered by the application of reasoning methods. If we choose, for example, sinusoidal membership functions we have no lack of completeness but likewise two points of discontinuity and also a symmetrical shape as before. To overcome these shortcomings, we propose a non-symmetric sinusoidal membership function shown in Fig. 1 as follows.

$$S(x)=\begin{cases}0 & if\,(x<a)\\ L(x) & if\,(a\le x<c)\\ R(x) & if\,(c\le x<b)\\ 0 & if\,(x\ge b)\end{cases} \tag{1}$$

where a,b,c present the left point, the right point and the center point of NSSMF, $a<c<b$ and

$$L(x)=1/2\sin[\pi(x-a)/(c-a)-\pi/2]+1/2 \tag{2}$$

$$R(x)=1/2\sin[\pi(b-x)/(b-c)-\pi/2]+1/2 \tag{3}$$

Eq. (1) can be written as:

$$S(x)=L(x)\mathrm{sgn}(x-a)(c-x)+R(x)\mathrm{sgn}(x-c)(b-x) \tag{4}$$

where $\mathrm{sgn}(x)=1$ for $x\ge 0$ and $\mathrm{sgn}(x)=0$ for $x<0$.

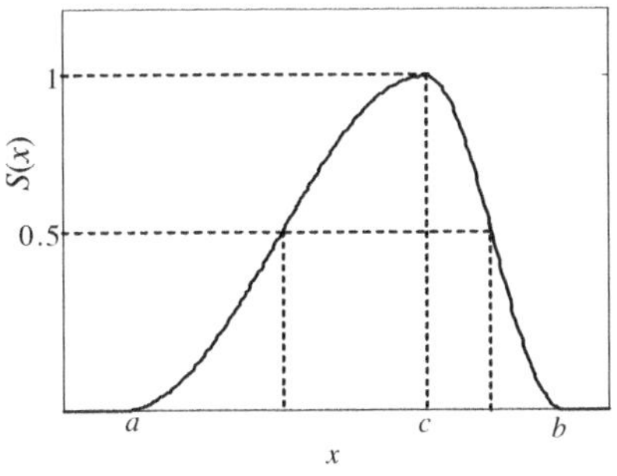

Fig. 1. Figure of the NSSMMF

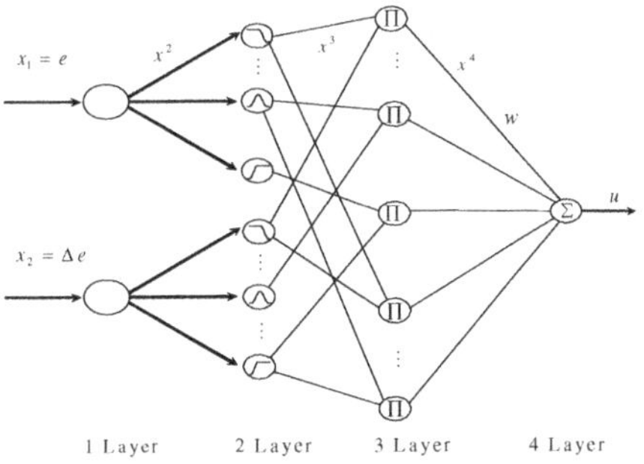

Fig. 2. Architecture of the FNN

In Fig. 1, the NSSMF possesses a very flexible form. When $(c-a)=(b-c)$, the membership function is symmetric, that is similar to the Gauss membership function; when $(a \to -\infty)$ or $(b \to +\infty)$, the membership function is similar to the sigmoid function. Therefore, NSSMF has very good virtues: the flexible form, adaptive parameter, zero point and smooth. The NSSMF has infinite times differentiable, and it can be used to determine the partial derivatives. Through this one can compute the maximum change of the fuzzy controller's output when changing the input by a given value. It even might be possible to determine the stability features by means of this function.

The structure of FNN proposed in this paper is shown in Fig. 2, which is comprised of four layers: an input layer (1 layer), a membership layer (2 layer), a rule layer (3 layer) and an output layer (4 layer). In Fig. 2, we use x_i^m to represent the input of the ith node in layer m, where, m=1,2,3,4.

3 Fuzzy Neural Network (FNN) Controller

Using the NSSMF, we construct the FNN control system for the ultrasonic motors. The overall structure of the FNN control system shown in Fig. 3.

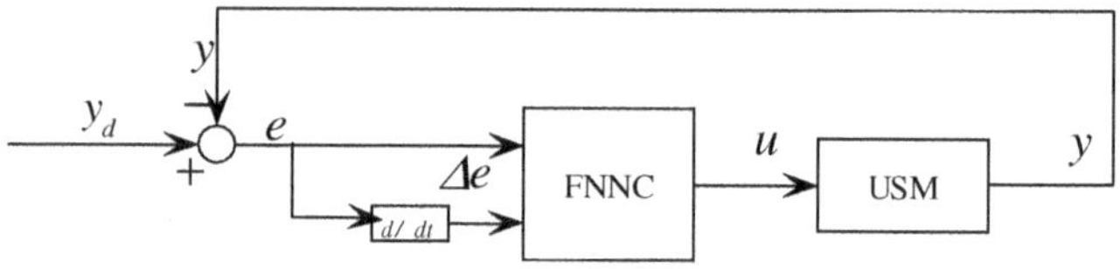

Fig. 3. FNN control system

Where y_d is the reference speed, y is the output speed of USM, u is the control variable of the USM, and Δe is the change of e respectively. Firstly, define $E = 1/2(y_d - y)^2 \equiv 1/2e^2$. Then we have the online learning algorithm of the FNNC

$$\frac{\partial E}{\partial a_j} = \frac{\partial E}{\partial y}\frac{\partial y}{\partial u}\frac{\partial u}{\partial a_j} = -e\frac{\partial y}{\partial u}\frac{\partial u}{\partial a_j}, \tag{5}$$

where $\partial y/\partial u$ can approximately be equal to $\Delta y/\Delta u$. Then (5) can be written as

$$\frac{\partial E}{\partial a_j} \approx -e\frac{\Delta y}{\Delta u}\sum_k\left[\frac{\partial u}{\partial x_k^4}\frac{\partial x_k^4}{\partial x_j^3}\frac{\partial x_j^3}{\partial a_j}\right] = -e\frac{\Delta y}{\Delta u}\sum_k\left[w_k x_i^3\frac{\partial L_j}{\partial a_j}\operatorname{sgn}(x_j^2 - a_j)(c_j - x_j^2)\right], \tag{6}$$

In a similar way, we can obtain

$$\frac{\partial E}{\partial b_j} \approx -e\frac{\Delta y}{\Delta u}\sum_k\left[w_k x_i^3\frac{\partial R_j}{\partial b_j}\operatorname{sgn}(x_j^2 - c_j)(b_j - x_j^2)\right], \tag{7}$$

$$\frac{\partial E}{\partial c_j} \approx -e\frac{\Delta y}{\Delta u}\sum_k \{w_k x_i^3 \frac{\partial L_j}{\partial c_j}[\text{sgn}(x_j^2 - a_j)(c_j - x_j^2) + \frac{\partial R_j}{\partial c_j}\text{sgn}(x_j^2 - c_j)(b_j - x_j^2)]\}, \tag{8}$$

$$\frac{\partial E}{\partial w_k} = - - e\frac{\Delta y}{\Delta u}\sum_k x_k^4, \tag{9}$$

where, $\partial L_j/\partial a_j$, $\partial R_j/\partial c_j$ and $\partial R_j/\partial b_j$ can be obtained from (2) and (3). According to the back-propagation algorithm [8, 9], we have

$$q_j(t) = q_j(t-1) - \eta_q \, \partial E/\partial q_j \tag{10}$$

where q denotes a, b, c or w.

To improve the learning performance, we have the following theorem.

Theorem 1: Suppose that the modification of the weights of the FNNC is determined by Eq.(10). If learning rate η_q is adopted as $\eta_q = \lambda_q E(1+E)/\|\partial E/\partial \mathbf{q}\|^2$, then global convergence of the update rules (10) is guaranteed. Where $\lambda_q > 0$ is the minimal learning rate of weights **q**, $\|\cdot\|$ represents the Euclidean norm.

Proof: Because control error is determined by the weights η_a, η_b, η_c and η_w. Thus the error during the learning process can be represented as $E = E(\mathbf{a}, \mathbf{b}, \mathbf{c}, \boldsymbol{\omega})$, and then we have

$$\frac{da}{dt} = -\eta_a \frac{\partial E}{\partial a}, \quad \frac{db}{dt} = -\eta_b \frac{\partial E}{\partial b}, \quad \frac{dc}{dt} = -\eta_c \frac{\partial E}{\partial c}, \quad \frac{dw}{dt} = -\eta_w \frac{\partial E}{\partial w}, \tag{11}$$

then

$$dE/dt = -E(1+E)(\lambda_{a0} + \lambda_{b0} + \lambda_{c0} + \lambda_{w0}) = -E(1+E)\lambda_0 . \tag{12}$$

Eq. (12) can be written as $dE/(E(1+E)) = -\lambda_0 dt$. Let E_0 is the identified error at initial time, then integrating it we obtain that $E/(1+E) = E_0 \exp(-\lambda_0 t)$. It can be seen that when $t \to \infty$, E is such that $E \to 0$. According to the Lyapunov stability theory, we have shown that the control error converges to the zero as $t \to \infty$. This completes the proof of Theorem 1.

4 Numerical Simulation Results

Numerical simulations are performed for a longitudinal oscillation USM [1, 10]. Some parameters on the USM model are taken as: driving frequency 27.8kHZ, amplitude of driving voltage 300V, allowed output moment 2.5kg·cm, rotation speed 3.8m/s.

Fig. 4 shows the speed control curves using the proposed control method based on the NSSMF when the reference speeds vary as cosine and step types respectively.

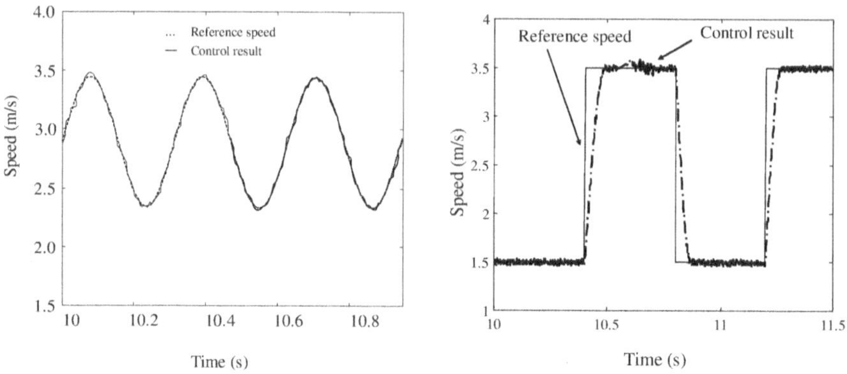

Fig. 4. Speed control curves for different reference speeds

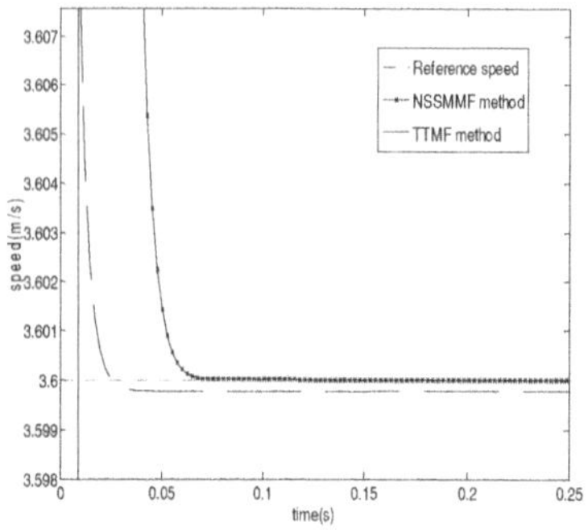

Fig. 5. Comparison of control errors using different Membership Function

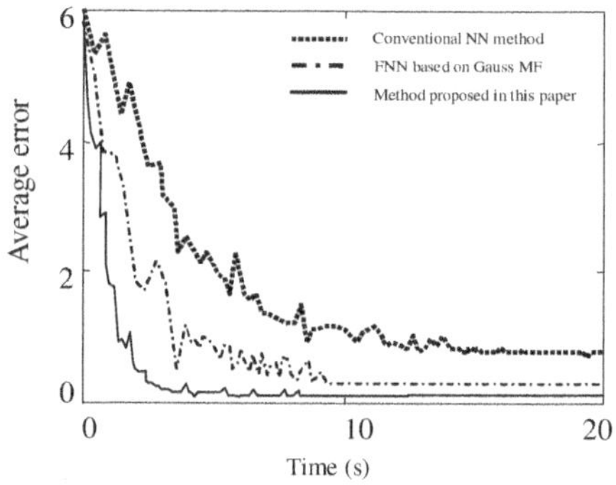

Fig. 6. Comparison of speed control curves using different schemes

From this figure, it can be seen that the method proposed has fairly adaptation for the different reference speeds.

Fig. 5 shows the USM speed control curves using FNN controller based on the NSSMF and the triangle type Membership function (TTMF) when the control speed is constant. It is easy to see that the control errors using the method proposed in this paper are smaller than that of TTMF method.

Fig. 6 illustrates the control errors for the different control schemes. It is easy to see that control performance based on the proposed method is much better than the conventional neural network method [8] and the FNN method discussed in [6] based on Gaussian shape Membership Function (GSMF).

5 Conclusions

We proposed a fuzzy neural network method with non-symmetric sinusoidal membership functions for ultrasonic motors speed control. Numerical experiments show the proposed control method has good performance and favorable adaptation.

And the FNN scheme with NSSMF is more efficient than the FNN with TTMF, FNN with GSMF and conventional NN control scheme for the USM control.

References

1. Sashida, T., Kenjo, T.: An Introduction to Ultrasonic Motors, Oxford: Clarendon, London (1993)
2. Senjyu, T., Yokoda, S., Uezato, K.: A Study on High-efficiency Drive of Ultrasonic Motors. Electric Power Components and Systems **29** (2001) 179-189
3. Xu, X., Liang, Y.C., Lee, H.P., Lin, W.Z., Lim, S.P., Lee, K.H., Shi, X.H.: Identification and Speed Control of Ultrasonic Motors Based on Neural Networks. Journal of Micromechanics and Microengineering **13** (2003) 104–114
4. Xu, X., Liang, Y.C., Lee, H.P., Lin, W.Z., Lim, S.P., Lee, K.H., Shi, X.H., A Stable Adaptive Neural-network-based Scheme for Dynamical System Control. Journal of Sound and Vibration **285** (2005) 653-667
5. Lin, F.J., Wai, R.J., Duan, R.Y.: Fuzzy Neural Networks for Identification and Control of Ultrasonic Motor Drive with LLCC Resonant Technique. IEEE Trans. Industrial Electronics **46** (1999) 999-1011
6. Senjyu, T., Yokoda, S., Uezato, K.: Speed Control of Ultrasonic Motors Using Fuzzy Neural Network. Journal of Intelligent and Fuzzy Systems **8** (2000) 135-146
7. Chou, K.T., Chung, S.W., Chan, C.C.: Neuro-fuzzy Speed Tracking Control of Traveling-Wave Ultrasonic Motor Drives Using Direct Pulsewidth Modulation. IEEE Trans. Industry Applications **39** (2003) 1061-1069
8. Wu, W., Feng, G., Li, Z., Xu, Y.: Deterministic Convergence of an Online Gradient Method for BP Neural Networks. IEEE Trans. Neural Networks **16** (2005) 533- 540
9. Zhang, N., Wu, W., Zheng, G.: Convergence of Gradient Method with Momentum for Two-Layer Feedforward Neural Networks. IEEE Trans. Neural Networks **17** (2006) 522-525.
10. Xu, X., Liang, Y. C., Lee, H. P., Lin, W.Z., Lim, S.P., Lee, K.H.: Mechanical Modeling of a Longitudinal Oscillation Ultrasonic Motor and Temperature Effect Analysis. Smart Materials and Structures **12** (2003) 514-523

Adaptive Neuro-Fuzzy Inference System Based Autonomous Flight Control of Unmanned Air Vehicles

Sefer Kurnaz[1], Okyay Kaynak[1,2], and Ekrem Konakoğlu[1]

[1] Aeronautics and Space Technologies Institute, Air Force Academy, Yesilkoy, Istanbul, Turkey
seferkurnaz1@yahoo.com
[2] BogaziciUniversity, Bebek, 34342 Istanbul, Turkey
o.kaynak@ieee.org, ekremkon@yahoo.com

Abstract. This paper proposes ANFIS logic based autonomous flight controller for UAVs (unmanned aerial vehicles). Three fuzzy logic modules are developed for the control of the altitude, the speed, and the roll angle, through which the altitude and the latitude-longitude of the air vehicle is controlled. The implementation framework utilizes MATLAB's standard configuration and the Aerosim Aeronautical Simulation Block Set which provides a complete set of tools for rapid development of detailed 6 degree-of-freedom nonlinear generic manned/unmanned aerial vehicle models. The Aerosonde UAV model is used in the simulations in order to demonstrate the performance and the potential of the controllers. Additionally, Microsoft Flight Simulator and FlightGear Flight Simulator are deployed in order to get visual outputs that aid the designer in the evaluation of the controllers. Despite the simple design procedure, the simulated test flights indicate the capability of the approach in achieving the desired performance.

1 Introduction

This paper addresses the design of an ANFIS (Adaptive Neuro-Fuzzy Inference System) based controller to autopilot an Unmanned Aerial Vehicles (UAV). UAVs are remotely piloted or self-piloted aircraft that can carry many different types of accessories such as cameras, sensors and communications equipment. They have a very wide range of applications including both civil and military areas. Some important features that make them very popular are their low cost, smaller size and their extended maneuver capability because of absence of a human pilot.

In literature, there can be found many different approaches related to the autonomous control of UAVs; some of the techniques proposed include fuzzy control [1-2], adaptive control [3-4], neural networks [5-7], genetic algorithms [8] and Lyapunov theory [9]. In addition to the autonomous control of a single UAV, research on other UAV related areas such as formation flight [10] and flight path generation [11] are also popular.

The approach proposed in this paper is neuro-fuzzy logic based. Three fuzzy modules are designed, one module is used for adjusting the bank angle value to control the

D. Liu et al. (Eds.): ISNN 2007, Part I, LNCS 4491, pp. 14–21, 2007.

latitude and the longitude coordinates, and the other two are used for adjusting elevator and throttle controls to obtain the desired altitude value.

The performance of the proposed system is evaluated by simulating a number of test flights, using the standard configuration of MATLAB and the Aerosim Aeronautical Simulation Block Set [12]. The latter provides a complete set of tools for rapid development of detailed 6 degree-of-freedom nonlinear generic manned/unmanned aerial vehicle models. As a test air vehicle, a model which is called Aerosonde UAV [13] is utilized. The great flexibility of the Aerosonde, combined with a sophisticated command and control system, enables deployment and command from virtually any location.

The paper is organized as follows. Section 2 starts with a basic introduction ANFIS and then explains the design of the controllers which are used for the autonomous control of the UAV. The inputs and the outputs of each controller are described and the membership functions used are given. The hybrid learning algorithm adopted in this work is described in Section 3 and some representative simulation results are presented in Section 4. In the final sections of the paper some concluding remarks and suggestions for future work are made.

Fig. 1. Aerosonde UAV

Table 1. UAV Specifications

Weight	27-30 lb,
Wing Span	10 ft
Engine	24 cc, 1.2 kw
Flight	Fully Autonomous / Base Command
Speed	18 – 32 m/s
Range	>1800 miles
Altitude Range	Up to 20,000 ft
Payload	Maximum 5 lb with full fuel

2 Adaptive Neuro-Fuzzy Inference System (ANFIS)

ANFIS is a 5 layered feed-forward neural network structure, as shown in Fig.2. The functions of the various layers are well explained in the literature [14] together with its merits over the other types of neuro-fuzzy approaches and therefore will not be dwelled upon here. The only remark that is worth making is the fact that its special architecture based on Sugeno type of inference system enables the use of hybrid learning algorithms (explained below) that are faster and more efficient as compared to the classical algorithms such as the error back propagation technique.

2.1 Hybrid Learning Algorithm

The approach used in this work for updating the ANFIS network parameters is a hybrid learning algorithm which is a two level learning algorithm. In this approach, the parameters of ANFIS network are evaluated in two parts as input and output parameters. Let us express the total parameter set as S=S1+S2, where S1 is the set of input parameters (the parameters of the membership functions) and S2 is the set of output parameters (weights). During the forward pass of the hybrid learning algorithm, the parameters of the membership functions in the input stage (S1) are kept constant. In this manner, the output of the network becomes a linear combination of output parameters of the parameter set S2 and the well known Least Square Error (LSE) based training can be used.

During the backward pass of the hybrid learning algorithm, the parameter set S2 is kept constant and the error is back propagated. The parameter set S1 can now be updated using the well known gradient descent method.

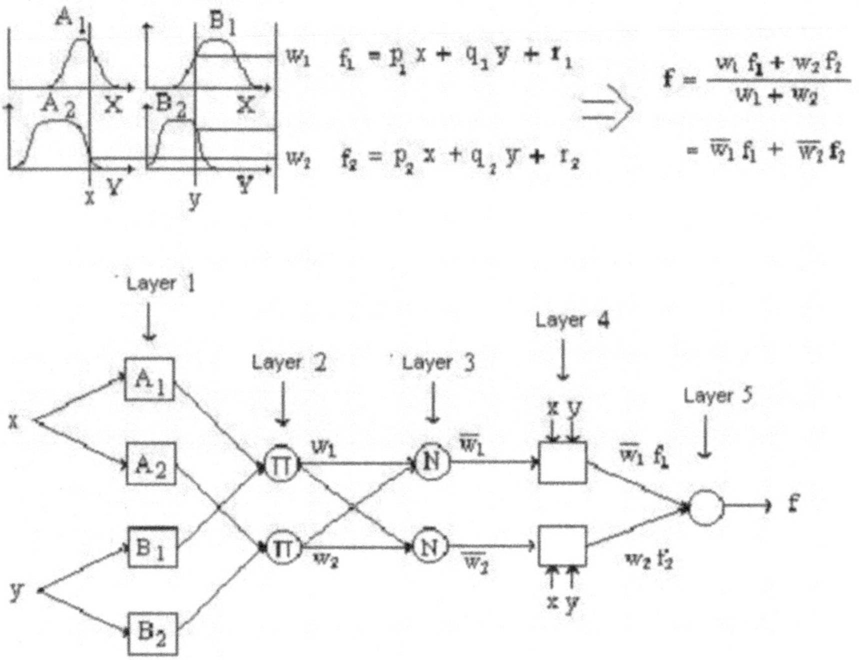

Fig. 2. Upper: 2-input, 2-rule Sugeno inference system. Lower: Equivalent ANFIS architecture.

3 Simulation Studies

In order to evaluate the potential and the performance of the proposed controller extensive simulation studies have been carried out on an Aerosonde UAV mode using MATLAB functions and Aeronautical Simulation Block Set (Aerosim). Additionally, to ease the design process, the Flight Gear Simulator is used to get visual outputs and to see the physical response of air vehicle. Despite the simple design procedure adopted, the simulated test flights indicate the capability of the approach in achieving the desired performance. Figure 3 depicts the Simulink model used for the simulation studies. As can be seen, 3 different ANFIS controllers are used for bank angle control, speed control and altitude control.

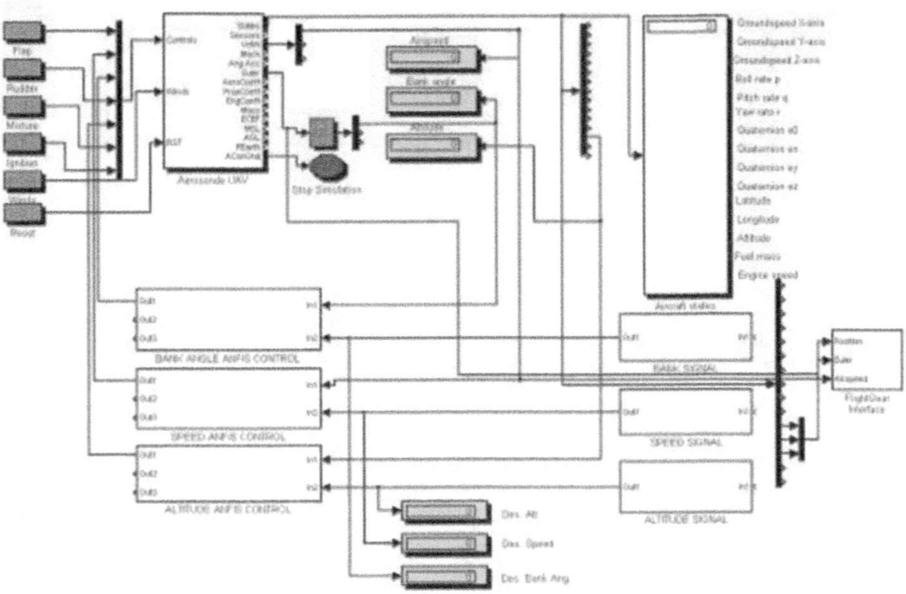

Fig. 3. The SIMULINK block diagram used in simulation studies

3.1 Reference Trajectories

The reference trajectories that are used for the simulation studies are given below, where $X_{d1}(t)$, $X_{d2}(t)$, $X_{d3}(t)$ are the required bank angle, speed and altitude.

$$[X_d(t)] = \begin{bmatrix} X_{d1}(t) \\ X_{d2}(t) \\ X_{d3}(t) \end{bmatrix} = \begin{bmatrix} 20\sin(0.001\pi t) \\ 23 + 5\sin(0.001\pi t) \\ 1000 + 50\sin(0.001\pi t) \end{bmatrix}.$$

3.2 Simulation Results

Extensive simulation studies are carried out on the Aerosonde model for the reference trajectories given above with different atmospheric conditions. A typical result is shown in Fig. 4.

3.3 Flight Gear Visual Interface

In order to be able to visualize the flight of the air vehicle, the software Flight Gear v.9 was used In this way, it was possible to see the effects of the even very small changes of the flight parameters in the flight conditions (that may not apparent from a study of the graphical simulation results). During the simulation studies, some valuable feedbacks were obtained from an F14 pilot who studied the visual outputs.

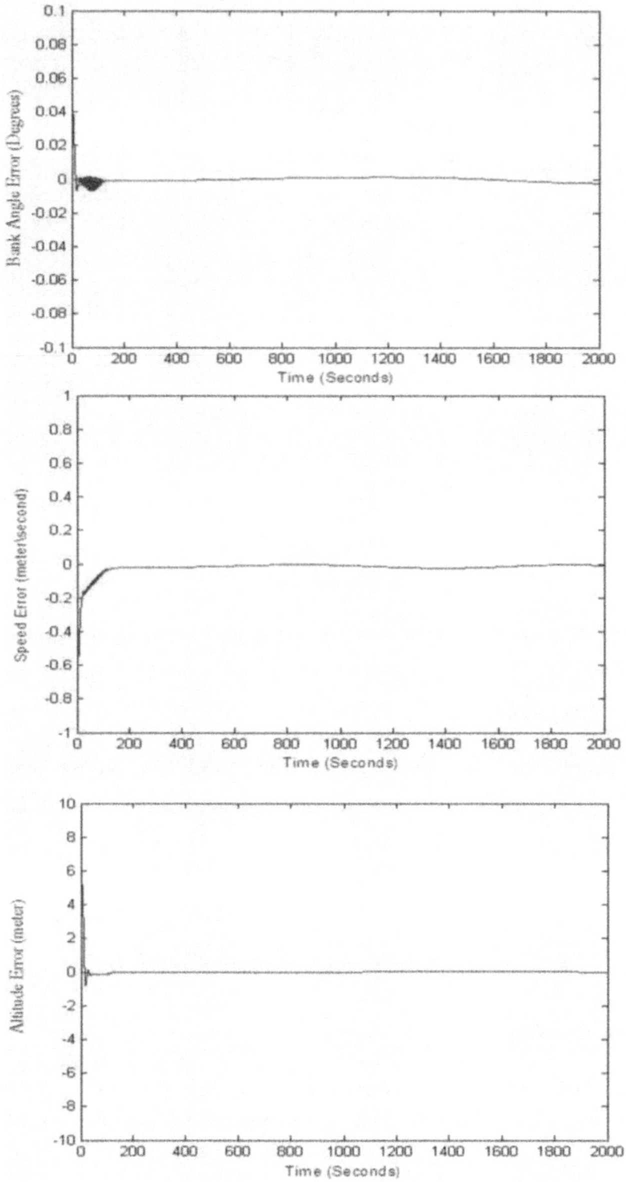

Fig. 4. A typical simulation result

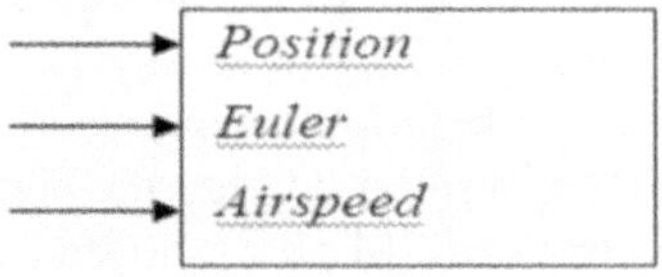

Fig. 5. The Flight Gear interface block

Fig. 6. The Flight Gear cockpit window

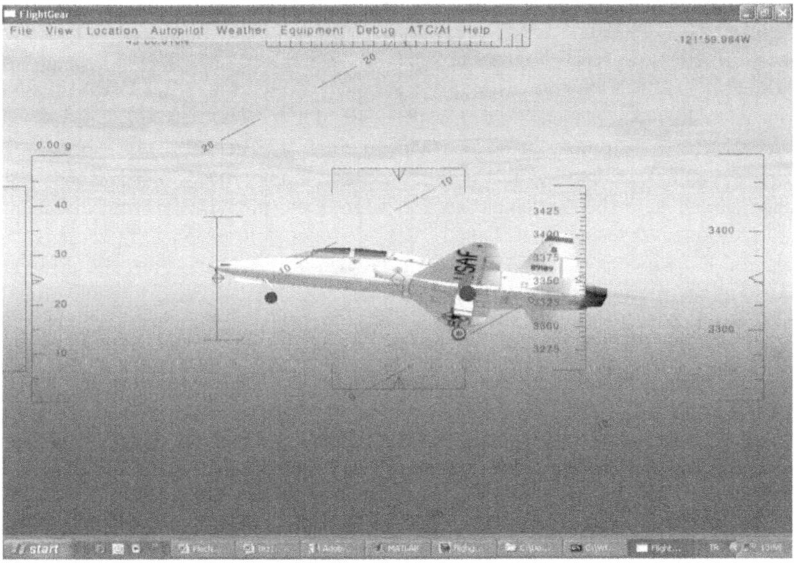

Fig. 7. The Flight Gear Flight conditions and HUD (Head-Up Display) window

Figure 5 depicts the block diagram of the interface between the Simulink and the Flight Gear, the inputs to the block being the aircraft states and the other information available as the outputs of the Aerosim block. UDP is used for communication between the two softwares. In Figs. 6 and 7, two snapshots of the Flight Gear windows are shown. Due to the unavailability of Aerosonde, T-38 TALON aircraft is visualized.

4 Conclusions

The simulation results presented demonstrates the feasibility of ANFIS based controllers for autonomous flight control of UAVs. In order to be able to have a basis for comparison, well-tuned PID type and fuzzy logic type controllers are also designed. It is seen that the performance of ANFIS type of controller is comparable to those obtained from the bank angle controller with a PI type controller and from PID type controller for speed controller despite the model-free approach of the ANFIS approach. However, PI and fuzzy type of altitude controllers have demonstrated superior performance. For some flight conditions, the ANFIS type controller has resulted in unstable performance. This has demonstrated that more stable learning algorithms need to be adopted. One possible solution could be the use of Variable Structure Systems theory based algorithms that are known for their stability [15].

References

1. Kumon, M., Udo, Y., Michihira H., et al: Autopilot System for Kiteplane. IEEE-ASME Transactions on Mechatronics **11** (2006) 615-624
2. Doitsidis, L., Valavanis, K.P., Tsourveloudis, N.C., Kontitsis, M.: A Framework for Fuzzy Logic Based UAV Navigation and Control. IEEE International Conference on Robotics and Automation ICRA '04. **4** (2004) 4041-4046
3. Schumacher, C.J., Kumar, R.: Adaptive control of UAVs in Close-coupled Formation Flight. American Control Conference **2** (2000) 849-853
4. Andrievsky, B., Fradkov, A.: Combined Adaptive Autopilot for an UAV Flight Control, International Conference on Control Applications **1** (2002) 290-291
5. Dufrene, W.R.Jr.: Application of Artificial Intelligence Techniques in Uninhabited Aerial Vehicle Flight. The 22nd Digital Avionics Systems Conference **2** (2003) 8.C.3 - 8.1-6
6. Sundararajan, Y.Li, N., Sratchandran, P.: Neuro-Controller Design for Nonlinear Fighter Aircraft Maneuver Using Fully Tuned RBF Networks. Automatica **37** (2001) 1293-1301
7. Borrelli, F., Keviczky, T., Balas, G.J.: Collision-free UAV Formation Flight Using Decentralized Optimization and Invariant Sets. The 43rd IEEE Conference on Decision and Control CDC. **1** (2004) 1099-1104
8. Marin, J.A., Radtke, R., Innis, D., Barr, D.R., Schultz, A.C.: Using A Genetic Algorithm to Develop Rules to Guide Unmanned Aerial Vehicles. IEEE International Conference on Systems, Man, and Cybernetics, SMC '99 . **1** (1999) 1055-1060
9. Ren, W., Beard, R.W.: CLF-based Tracking Control for UAV Kinematic Models with Saturation Constraints. 42nd IEEE Conference on Decision and Control **4** (2003) 3924-3929
10. Schiller, I., Draper, J.S.: Mission Adaptable Autonomous Vehicles. IEEE Conference on Neural Networks for Ocean Engineering (1991) 143–150
11. Dathbun, D., Kragelund, S., Pongpunwattana, A., Capozzi, B.: An Evolution Based Path Planning Algorithm for Autonomous Motion of a UAV through Uncertain Environments. 21st Digital Avionics Systems Conference **2** (2002) 8D2-1 - 8D2-12
12. Aerosim, Aeronautical Simulation Block Set v1.1, Users Guide, *www.u-dynamics.com*, Baldonado, M., Chang, C.-C.K., Gravano, L., Paepcke, A.: The Stanford Digital Library Metadata Architecture. Int. J. Digit. Libr. (1997) 108–121

13. Aerosonde – Global Robotic Observation System, *www.aeronde.com*
14. Jang, J.-S.R., Sun, C.-T.: Neuro-Fuzzy And Soft Computing:A Computational Approach to Learning and Machine Intelligence. Printice Hall (1997)
15. Topalov, A., Kaynak, O.: Neural Network Modeling and Control of Cement Mills Using a Variable Structure Systems Theory Based On-Line Learning Mechanism. Journal of Process Control **14** (2004) 581-589

A Novel Cross Layer Power Control Game Algorithm Based on Neural Fuzzy Connection Admission Controller in Cellular Ad Hoc Networks

Yong Wang[1], Dong-Feng Yuan[1,2], and Ying-Ji Zhong[1]

[1] School of Information Science and Engineering, Shandong University, Jinan, Shandong, Postfach 250100, Shandong, P.R. China
zhongyingji32@sdu.edu.cn
[2] State Key Lab. on Mobile Communications, Southeast University, Nanjing, Jiangsu, Postfach 210096, Jiangsu, P.R. China

Abstract. The special scenario of the topology in the cellular Ad Hoc networks was analyzed and a novel cross layer power control game algorithm based on Neural Fuzzy Connection Admission Controller (NFCAC) was proposed in this paper. NFCAC has been successfully applied in the control-related problems of neural networks. However, there is no discussion about the power control game algorithm and location recognition based on NFCAC in cellular Ad Hoc networks. The proposed algorithm integrated the attributes both of NFCAC and the topology space in special scenario. The topology and the power consumption of each node can all be optimized due to the minimum link occupation with the help of the algorithm. Simulation results show that the novel algorithm can give more power control guarantee to cellular Ad Hoc networks in the variable node loads and transmitting powers, and make the node more stable to support multi-hops at the same time.

1 Introduction

[1]The cellular Ad Hoc network[1] is the hybrid network which combines cellular network with Ad Hoc[2,3] mechanisms. As for hybrid networks, it should be a trade-off between cellular networks and Ad Hoc networks. We believe that the application of the NFCAC and the topology space analysis in special scenario should be benefit for the modification of the power control game algorithm. In this paper, we base on the special scenario of the network topology to explore the relationship between attributes of topology space and the topology scenario. To this end, we propose a novel cross layer power control game algorithm to effectively utilize location marking information and address the performance issues.

---

[1] The authors thank the following foundations: Outstanding Youth Scientist Awards Foundation of Shaodong (No.2006BS01009), National Scientific Foundation of China (No. 60372030), China Education Ministry Foundation for Visiting Scholars (No. [2003]406), Key Project of the Provincial Scientific Foundation of Shandong (No. Z2003G02), China Education Ministry Foundation for State key Lab. on Mobile Communications (No. A2005010) and China Education Ministry Foundation for State key Lab. on Mobile Communications (No. A0205).

D. Liu et al. (Eds.): ISNN 2007, Part I, LNCS 4491, pp. 22–28, 2007.

The rest of this paper is organized as follows. In Section 2, the architecture of NFCAC and its attributes are given. In Section 3, we give the evaluation models and do the dimensionality analysis of the topology in special scenario. In Section 4, we propose the novel cross layer power control game algorithm and analyze the topology control performance. In Section 5, we evaluate the performance of the proposed algorithm and analyze the improvement of the power control guarantee via simulation. Finally we give the conclusion in Section 6.

2 Architecture of NFCAC and Its Attributes

CHENG [4] has given the architecture of NFCAC. In the first layer of NFCAC, three input nodes with respective input linguistic variables are defined and

$$f_i^{(l_1)}(u_{ij}^{(l_1)}) = u_{ii}^{(l_1)}. \tag{1}$$

In the second layer of NFCAC, there are six nodes in the controller and each performs a bell shaped function, shown in Eq.(1).

$$f_i^{(l_2)}(u_{ij}^{(l_2)}) = -(u_{ij}^{(l_2)} - m_{jn}^{(input)})^2 / \sigma_{jn}^{(input)^2}. \tag{2}$$

The precondition matching of fuzzy control is taken in the third layer of NFCAC, and each node in the network controller performs the fuzzy operation [5,6] defined as

$$f_i^{(l_3)}(u_{ij}^{(l_3)}) = \min(u_{ij}^{(l_3)}; \forall j \in P_i). \tag{3}$$

The nodes in the fourth layer of NFCAC perform down-up mode and up-down mode contemporary, and each node performs the fuzzy operation to integrate the fired strength of the rules defined as

$$f_i^{(l_4)}(u_{ij}^{(l_4)}) = \max(u_{ij}^{(l_4)}; \forall j \in C_i). \tag{4}$$

In the fifth layer of NFCAC, the feedback was given to the controller to adjust the link weights optimally, and

$$f_i^{(l_5)}(u_{ij}^{(l_5)}) = \sum_{j=1}^{4} \sigma_j^{output} u_{ij}^{(l_5)} m_j^{output}. \tag{5}$$

For the optimal nodes,

$$f_i^{(l_5)}(u_{ij}^{(l_5)}) = u_{ii}^{(l_5)}. \tag{6}$$

The attributes of NFCAC can be employed to perform the topology analysis of the cellular Ad Hoc networks, which will be discussed in Section 3.

3 Dimensionality Analysis of Topology and Evaluation Models

The evaluation model is multiple cell environments with seven cells, in which the Mobile Hosts (MHs)[7] are in point wise uniformity. Analysis is based on two-dimension scenario, that is to say, the MHs and the base stations are on a Dual Ring Topology.

The Dual Ring Topology of the MHs is shown in Fig.1. The network topology used in this scenario consists of concentric circles, represented by $\{C_i\}_{i\in[1,2]}$, and the base station of Cell 1 is situated at the center of C_1. Let $\mathcal{E}= C_1 \bigcup C_2$, and C_1 is the compact subset of $\mathcal{E}$, then for the open covering, U, which is composed of the neighborhood basis of $\mathcal{E}$, the finite subset U' of U can cover C_1, then $\mathcal{E}/\bigcup U'$ is the finite set, and U has finite sub-covering. $\{\{e\}: e\in C_2\}$ is the disjoint uncountable open set family of $\mathcal{E}$, so that $\mathcal{E}$ is not a metrizable compact space. Let $K_1\neq\Phi$, then if p has a countable basis in the compact metric space, C_1, and $\{F\}_{n\in N}$ is the finite covering of K_1, and for each $n\in N$ and subset $K_n\subset N$, we have $F_n=\{V(e_{n,i}):i\leq k_n\}$, where $V(e_{n,i})$ is an open arc at the center of C_1. Let

$$U_n=(\bigcup_{i\leq k_n} V(e_{n,i})\bigcup p(V(e_{n,i})/\{e_{n,i}\}))\bigcup K_2, \tag{7}$$

then each U_n is the open set of $\mathcal{E}$ and $K_n\subset U_n$.

For each neighborhood of K in E, when $e\in K_1$, there should be an arc, $V(e)$, of C_1, which has its center at e, and

$$V(e)\bigcup p(V(e)/\{e\})\subset U. \tag{8}$$

Therefore, due to the compactification of K, we have $n\in N$ such that

$$K_1\subset(\bigcup_{i\leq k_n} V(e_{n,i})\bigcup p(V(e_{n,i})/\{e_{n,i}\})\subset U. \tag{9}$$

and $K\subset U_n\subset U$, so that each compact sub-set has a countable neighborhood basis in E. Therefore, it can be seen that the Dual Ring Topology belongs to the Alexandroff Dual Ring Space [5]. The attributes of Alexandroff Dual Ring Space are potentially worthy for the Location Information (LI) application and power control game.

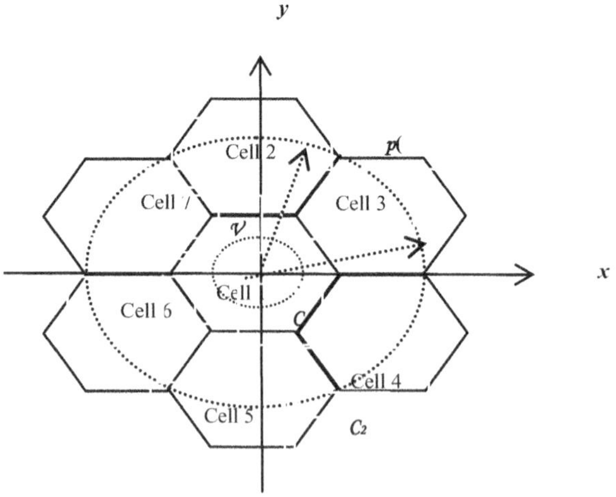

Fig. 1. Dual ring topology scenario with multiple cells

4 Cross Layer Power Control Game Algorithm

The physical layer provides to the upper layers a convex set of capacity graphs supported by a finite set or basis of elementary capacity graphs. The physical layer subproblem addresses the transmission interference among nearby nodes. In this paper, we explore ways of approximating the optimal solution using game theory. Inspired by the work of Saraydar [8], we use a tax mechanism and assume each link player maximizes its own payoff function

$$\max_{p_l} \ Q_l^{PHY} = \mu_l \log(1+\frac{G_{ll}q_l}{\sum_{j\neq l} G_{lj}q_j+\sigma_l^2}) - t_l q_l, \tag{10}$$

$$0 \le q_l \le q_{l,\max} \qquad \forall l,$$

where t_l is the tax rate for link l, and q_l is the action for link l. More power link l uses, more interference it will produce to others.

In general, not every game has a Nash equilibrium, so we propose the power control game algorithm to ensure the game for converging the stable Nash equilibrium.

Cross Layer Power Control Game Algorithm:

1) Initialize $t^{(0)}, q^{(0)}$. *Set* $s = 0$.

2) Set $q^{(\tau_0)} = q^{(s)}$. *Set* $i = 0$, *iteratively update* $q^{(\tau_i)}$ *as follows:*

$$q_l^{(\tau_i+1)} = \frac{\mu_l}{t_l^{(s)}} - \frac{1}{M_{ll}}(M_{lj}q_j^{(\tau_i)} + \sigma_l^2),$$

project $q_l^{(\tau_i+1)}$ *into power constraint interval* $[0, q_{l,\max}]$.

Repeat until $q^{(\tau_i)}$ *converges. Set* $q^{(s+1)} = q^{(\tau_i)}$.

3) Update tax rate s_l *and let*

$$bcm_f = \frac{\mu_f SINR_f^{(s+1)}}{1+SINR_f^{(s+1)}} \frac{SINR_f^{(s+1)}}{M_{ff}q_f^{(s+1)}}.$$

4) Return 2 until convergence.

The power update in step 2) is the best response of link player l given the tax rate and his assessment of others' action. As the tax rates and the bcm_f converge, the power control game Algorithm converges to a stable Nash equilibrium. Such power allocation equilibrium strikes a balance between minimizing interference and maximizing rate.

5 Simulations and Discussion

The terrain model we used is a 1000m×1000m square area with seven cells in it, on which 1000 MHs are pseudo-randomly moving along the dual ring topology. All the MHs are presented by $\{u_i\}_{i\in[1,2000]}$, and all the links between MHs are bi-directional, presented by $\{v_j\}_{j\in[1,Y]}$. Each cell has a base station with omni-directional antenna at the center point and its radius is 250m. Each base station has 256 available data channels. The NFCAC of the neural networks was used in the simulation model and the cellular network combines the MAC protocol together with the DCF in the cellular Ad Hoc networks. We use the modified DSR protocol with location information [9,10] as the routing protocol for the Ad Hoc mode. Assume the power consumption is based on the distance from the transmitting MHs to the base stations. As for handoff mechanism, hard handoff was used in the evaluation model and connectivity is considered under Poisson Boolean Model [11-13] in this kind of sparse network. We use 512 TCP flows in dual ring topology and the simulation time for each point is 3600s.

Employing the proposed algorithm, the traffic requirement η_{td} and the maximal amount of the permitted hops Ψ_{td} are examined in different load and transmitting power of the nodes with or without LI, shown separately in Fig.2, Fig.3, Fig.4 and Fig.5.

Fig.2 shows the different values of η_{td} along with the load of the nodes with or without LI. From this figure we can see that the traffic requirement depend deeply on the load when use no LI, but released by LI. The maximal optimization is about 9.72%. the maximal amounts of the permitted hops in different loads are shown in Fig.3, Ψ_{td} is in fixed value when the load is 0, that is to say, the default value of the hops is 1 when the nodes have no load. The network can tolerate more hops to support reliable transportation with the help of LI. The maximal optimization is 8.31% when the load reached 20b/s.

Fig.4 shows the different values of η_{td} along with the transmitting powers with or without LI. The addressing success ratio can be enhanced with the help of the LI, so the LI can give 12.27% improvement to reform the capacity of the data flows.

Fig.5 shows the maximal amount of the permitted hops of the network in different transmitting powers, in the condition of the unchanged parameters and external information, the LI can make the network more stable and can support more hops. The maximal optimization is about 10.19% and the merits of the proposed algorithm are obviously.

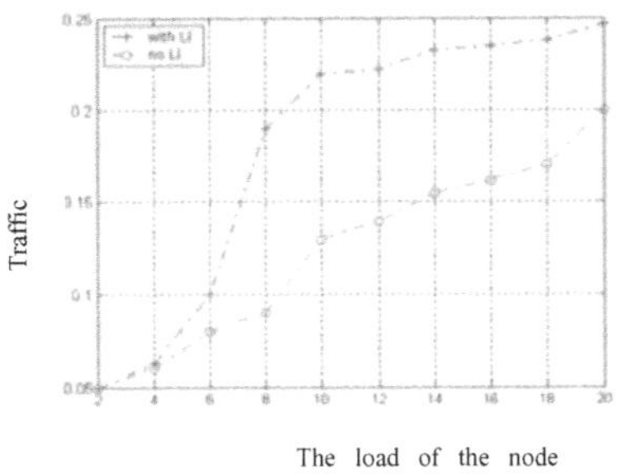

Fig. 2. The traffic requirements in different loads

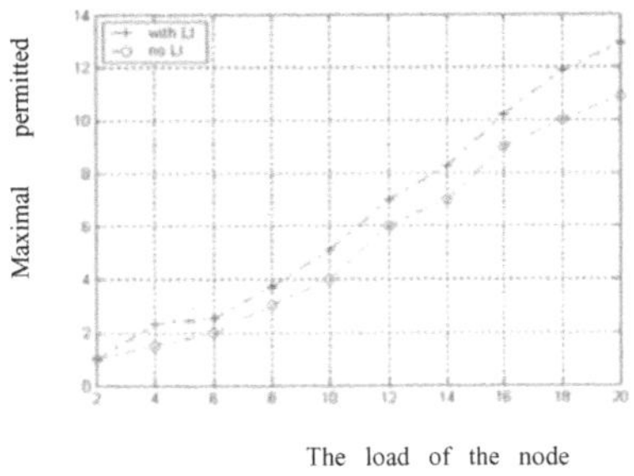

Fig. 3. The maximal amount of the permitted hops in different loads

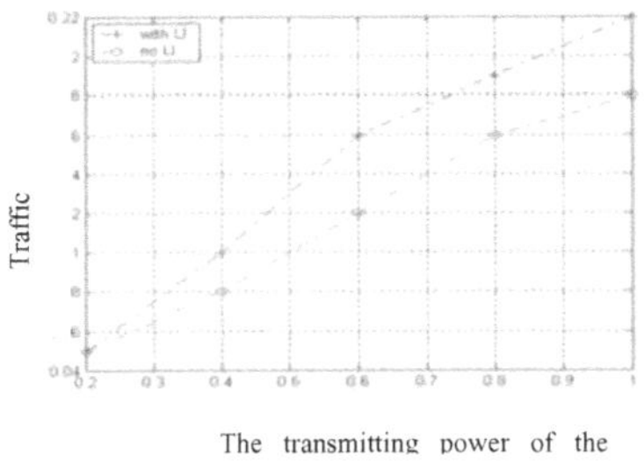

Fig. 4. The traffic requirements in different transmitting powers

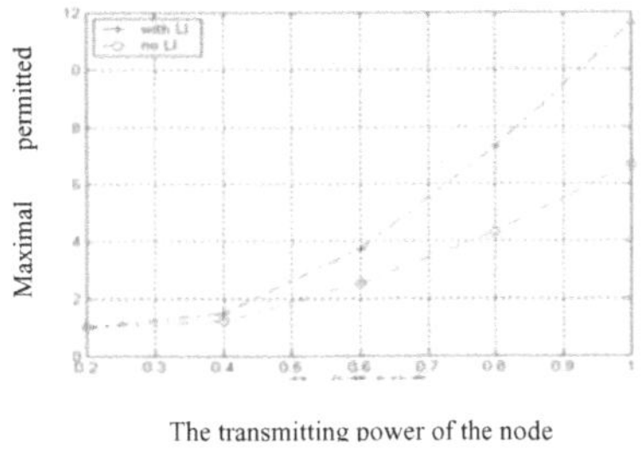

Fig. 5. The maximal amount of the permitted hops in different transmitting powers

6 Conclusions

Based on the analysis of the topology space and cross layer constraints, we induced the special scenario of the topology in the cellular Ad Hoc networks and proposed a novel power control game algorithm based on NFCAC in this paper. The proposed algorithm integrated the attributes both of NFCAC and the topology space in special scenario. The topology and the power consumption of each node can all be optimized due to the minimum link occupation with the help of the algorithm. Simulation results show that the novel algorithm can give more power control guarantee to cellular Ad Hoc networks in the variable node loads and transmitting powers, and the employing of the location recognition can make the node more stable to support multi-hops in the cellular Ad Hoc networks.

Acknowledgement

This work is supported by the foundations in the footnote of the first page. The authors would like to thank all professors concerned the topic.

References

1. Fu, Z., Luo, H., Zerfos, P., Lu, S., Zhang, L., Gerla, M.: The Impact of Multihop Wireless Channel on TCP Performance. IEEE Trans. Mobile Computing (2003)
2. Bettstetter, C.: On the Connectivity of Ad Hoc Networks. The Computer Journal, British Computer Society, OXFORD University Press **47** (4) (2004) 432-447
3. Wattenhofer, R., Li, L., Bahl, P., Wang,Y.: Distributed Topology Control for Wireless Multihop Ad Hoc Networks. IEEE INFOCOM (2001)
4. Cheng, R., Chang, C.: A QoS-Provisioning Neural Fuzzy Connection Admission Controller for Multimedia High-Speed Networks. IEEE/ACM Trans. Networking **17** (1) 1999
5. Liao, X., Wang, J., Zeng, Z.: Global Asymptotic Stability and Global Exponential Stability of Delayed Cellular Neural Networks. IEEE Trans. Circuits and Systems - Part II Express Briefs **52** (7) (2005) 403-409
6. Chen, H.V.: A Neural Architecture for Syntax Analysis. IEEE Trans. Neural Networks **10** (1999) 94-114
7. Hu, L.: Topology Control for Multihop Packet Radio Networks. IEEE Trans. Communication **41** (1993) 1424-1481
8. Saraydar, C. etc.: Efficient Power Control via Pricing in Wireless Data Networks. IEEE Trans. Communication **50** (2) (2002) 291-303
9. Zhong, Y., Yuan, D., Kyung, S.: A New Low Delay Marking Algorithm Based on Topology Space Analysis for Mobile Multimedia Hybrid Networks. International Transaction on Computer Science and Engineering **12** (1) (2005) 211-223
10. Zhong, Y., Yuan, D.: Dynamic Source Routing Protocol for Wireless Ad Hoc Networks in Special Scenario Using Location Information. IEEE ICCT'2003, Beijing **2** (2003) 1587-1592
11. Liu, J., Yuan, D., Ci, S., Zhong, Y.: A New QoS Routing Optimal Algorithm in Mobile Ad Hoc Networks Based on Hopfield Neural Network. IEEE ISNN'2005 **3** (2005) 343-348
12. Zhong, Y., Yuan, D.: A novel Low Delay Marking Algorithm in Multihop Cellular Networks Based on Topology Space Analysis. Chinese Journal of Electronics **15** (3) (2006) 516-520
13. Wu, H., Qiao, C., Swades, D., Ozan, T.: An Integrated Cellular and Ad Hoc Relaying System: Icar. IEEE Journal on Selected Areas in Communications **19** (2001)

A Model Predictive Control of a Grain Dryer with Four Stages Based on Recurrent Fuzzy Neural Network

Chunyu Zhao[1], Qinglei Chi[1], Lei Wang[2], and Bangchun Wen[1]

[1] School of Mechanical Engineering and Automation, Northeastern University, Shenyang 110004, P.R. China
{chyzhao,qlchi,bcwen}@mail.enu.edu.cn
[2] Shenyang Neusoft Software Co.ltd, Shenyang 110179, P.R. China
wang-lei@springer.com

Abstract. This paper proposes a model predictive control scheme with recurrent fuzzy neural network (RFNN) by using the temperature of the drying process for grain dryers. In this scheme, there are two RFNNs and two PI controllers. One RFNN with feedforeward and feedback connections of grain layer history position states predicts outlet moisture content (MPRFNN), and the other predicts the discharge rate of the dryer (RPRFNN). One PI controller adjusts the objective of the discharge rate by using MPRFNN, and the other adjusts the given frequency of the discharge motor to control the discharge rate of the grain dryer to reach its objective by using RPRFNN. The experiment is carried out by applying the proposed scheme on the control of a gain dryer with four stages to confirm its effectiveness.

1 Introduction

Classical feedback control is inadequate for controlling grain dryers because of the long delay and nonlinearity intrinsic to the grain drying process [1,2,3,4,5]. Forbes et al. [6] designed model-based dryer controllers in which the control action is based upon a process model and a so-called pseudo-inlet grain moisture content. The drying-rate parameter is updated intermittently according to the difference between the model-predicted and the sensor-measured outlet moisture contents. Zhang and Litchfield [7] investigated to use fuzzy control on a laboratory dryer for control of both the grain moisture content and breakage susceptibility; both the drying-air temperature and the discharge-Auger speed were adjusted. Liu and Bakker-Arkema [2,4,5] developed a distributed-parameter process model based on the fundamental laws of simultaneous heat and mass transfer and used it to establish a model predictive controller for dryer control. The moisture controller is able to operate in conjunction with a grain quality controller. Jover and Alastruey [8] compared the multivariable and monovariable control scheme for an industrial rotary dryer and concluded that the settling time of the multivariable control scheme is shorter than that of the multivariable control scheme. However, to our knowledge, the online sensor-measured accuracy

D. Liu et al. (Eds.): ISNN 2007, Part I, LNCS 4491, pp. 29–37, 2007.

of inlet and outlet moisture content is poor, especially when the environmental temperature is below $0^{o}C$. All above-mentioned control strategies of grain dryers rely on the sensor-measured moisture content of inlet and outlet grain. This may be the reason that few automatic controllers are widely found in use on commercial grain dryers.

In fact, a gain dryer is an open system of thermodynamics, in which complex heat and mass transfer is carried out [9]. The temperature and its variation of grain in the dryer can be considered as the measurement of the degree of heat and mass transfer. Zhao et al. [10] investigated experimentally the relation between the discharge moisture of dried maize and its temperature in a concurrent drying process with four stages and gave a predictive method of outlet moisture content.

Fuzzy neural network is a successful method for modeling and identification of nonlinear system [11,12,13,14]. Zhang and Julian Morris [11] proposed a type of recurrent neuro-fuzzy network to build long-term prediction models for nonlinear processes and it have been successfully applied to the modeling and control of a neutralization process. The state parameters of a dryer are associated greatly with his history and the outlet moisture content of a particular grain is affected by adjacent inlet grains [10]. This relation is too complex to be described by mathematical model. It is possible to train a RFNN by using the temperature data to describe the drying process in terms of linguistic forms. The objective of this paper is to develop a new predictive control scheme of dryers based recurrent fuzzy neural network to overcome the control difficulty problem of grain dryers. The remainder of this paper is organized as follows. Section 2 is devoted to describe the experiment setup. The predictive model of outlet moisture content based RFNN is presented in Section 3. The control scheme of grain dryers and the formulation of training RFNNs are addressed in Section 4. The experimental process is described in Section 5 together with experimental results. Finally, conclusions are provided in Section 6.

2 Experimental Setup

The drying setup is showed in Fig.1 (a), which is a concurrent grain dryer that consists of four drying stages and a cooling stage. There are two rows of ducts in every drying stage. One row is air inlet ducts, marked as I and another air exhaust ducts marked as E. There are four rows of ducts in the cooling stage. During drying process, drying stage 1 and 2 are supplied with the same temperature hot air by a fan, and drying stage 3 and 4 lower temperature hot air by another fan. The cooling stage is provided with ambient air. The dried material is maize. There are two switches that are installed in Position 1 and 2 in the dryer, respectively. The dryer is supplied discontinuously with inlet grain by an automation supply system according to the states of these switches. When the grain position in the dryer is below Position 2, Switch 2 is off and the supply system begins to supply the dryer with the grain. When the grain position reaches Position 1, Switch 1 is on and the supply system stops. If the end time of the ith time supply grain is $t_{(e,i)}$ and its beginning time of the $(i+1)$th supply grain is

$t_{(b,i+1)}$, the average volume discharge rate of the dryer between $t_{(e,i)}$ and $t_{(b,i+1)}$ can be obtained as [10]

$$V_i = \frac{Ah_0}{t_{(b,i+1)} - t_{(e,i)}}, \tag{1}$$

where A is the area of the dryer section, and h_0 is the distance between Position 1 and 2.

The volume discharge rate between $t_{(b,i)}$ and $t_{(e,i)}$ is assumed to be $0.5(V_{i-1} + V_i)$. Five temperature sensors are installed in position ①, ②, ③, ④, ⑤ of the dryer to measure the temperature of inlet grain, the temperature of dried grain in the end of each stage, respectively. Two other temperature sensors measure the temperatures of the drying high and low hot air, respectively.

3 The Structure of MPRFNN

The structure of the fuzzy neural network for predicting outlet moisture content (MPRFNN) is shown in Fig.1 (b). The temperature variation differences of maize kernels between the drying and tempering stages represent their drying characteristics [10]. During the drying process with multi-stage, the energy of hot air in the first drying stage mainly heats up the dried maize kernels and vaporizes their surface water. From the second drying stage, the moisture within maize

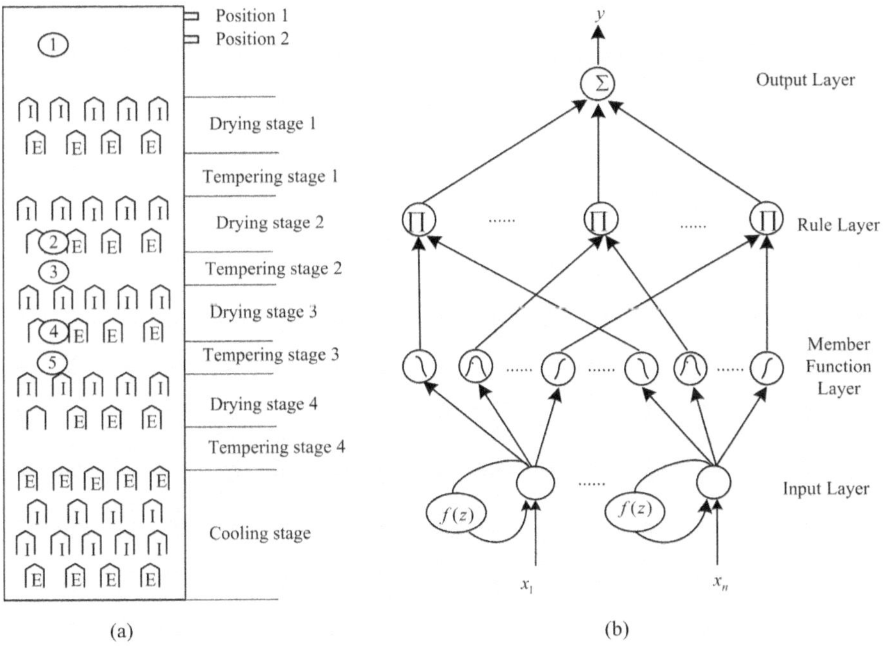

Fig. 1. The experimental setup and its structure of MPRFNN; (a)the experimental setup; (b) the structure of MPRFNN

kernels gradually is removed from outer to inner layer. Therefore, the temperature variations in the ends of tempering stage 2 and 3 can represent the drying characteristic of particular maize kernels. Here, a four-layer network realizes a fuzzy predictive model of outlet moisture content in the following form:

$$R_j : \text{IF } x_1 \text{ is } A_{1j} \text{ and } x_2 \text{ is } A_{2j} \cdots \text{ and } x_{14} \text{ is } A_{14j} \text{ THEN } y_j \text{ is } B_j, \quad (2)$$

where x_1 is the maize inlet temperature, x_2, x_3, x_4 and x_5 are the inlet hot air temperatures when the sampled maize layer passes through drying stage 1, 2, 3, and 4, respectively, x_6, x_7, x_8, x_9 and x_{10} are the discharge rates when the sampled maize layer passes through the corresponding stage, respectively, x_{11} and x_{12} are the sensor-measured temperature in ② and ④, respectively, x_{13} and x_{14} are the error of the measured temperature in ③ and ②, and that in ⑤and ④, respectively, y_j represents the outlet moisture content of the sampled maize measured by the oven box. A_{nj} is the linguistic term of precondition part, B_j is the constant consequent part.

The functions of the nodes in each layer of the RFNN model are described as follows.

Input layer: This layer accepts the input variable. Its nodes transmit the input values to the membership layer. Feedforward and feedback connections of grain layer position are added in this layer to embed temporal relations in the network. As mentioned above, the drying characteristics of the dried maize layers are different, they affect mutually during the drying process. The drying stage is divided into $(L+1)$ layers and the characteristics of the maize kernels in each layer are considered to be the same. Then the characteristic of maize layer that can affect mutually will include $2L$ layers.

$$f(z) = \sum_{j=1}^{L} \left(w_{i,-j}^1 x(k-j) + w_{i,j}^1 x(k+j) \right) . \quad (3)$$

For i^{th} node in this layer, the input and output are represented as

$$u_i^1(k) = x_i(k) + \sum_{j=1}^{L} \left(w_{i,-j}^1 x_i(k-j) + w_{i,j}^1 x_i(k+j) \right) , \quad (4)$$

where k is the number of the sampled maize layer, $w_{i,\pm j}^1$ is the recurrent weight of ahead and behead layer characteristic effect of maize kernels.

By adding feedback connections in input layer of network, the mutual effect of the different drying characteristic maize kernels in the same drying stage is introduced into network to realize fuzzy inference. This is the attribute that distinguishes our RFNN from the others.

Member function layer: Nodes in this layer represent the terms of respective linguistic variables. Each node performs a Gaussian membership function

$$u_{ij}^2 = \exp\left[-\frac{\left(u_i^1 - m_{ij}\right)^2}{\sigma_{ij}^2} \right] , \quad (5)$$

where m_{ij} and σ_{ij} are, respectively, the mean and variance of the Gaussian membership function of the jth term of the ith input variable x_i.

Ruler Layer: Nodes in this layer represent the precondition part of one fuzzy logic rule. They receive the membership degrees of the associated rule from the nodes in member function layer. The input and output of the nodes in this layer can be described as [14]

$$u_j^3 = \prod_{i=1}^{n} u_{ij}^2, i = 1, 2, 3, .., n; j = 1, 2, ..., q, \tag{6}$$

where j represents the jth rule, and q represents the sum of the rule in the rule layer.

Output Layer: There is only one node in this layer, which represents the outlet moisture content. The node performs the defuzzification operation. The input and output of this node can be calculated by

$$y = \sum_{j=1}^{q} \left(u_j^3 w_j^4\right), \tag{7}$$

where the weight w_j^4 is the output action strength associated with the jth rule.

4 The Predictive Control of the Dryer

The block diagram of the predictive control system is shown in Fig. 2. There are the two RFNNs in this scheme: one, called RPRFNN, is used to predicte discharge rate and the other, called MPRFNN, is used to predict outlet moisture content. During the drying process, the discharge rate can be measured on line at intervals of about 20 to 40 minutes by using switches in Position 1 and 2

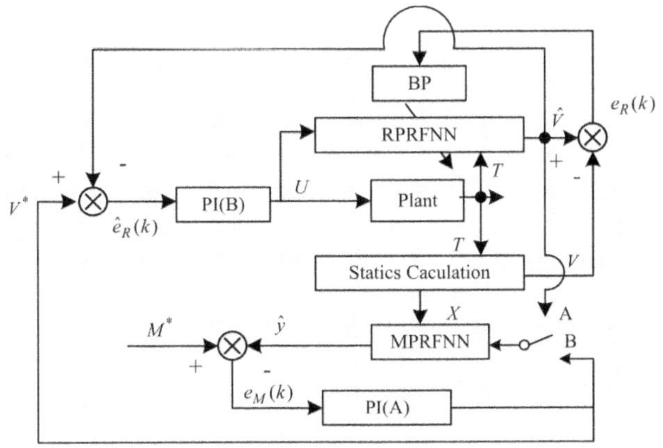

Fig. 2. Structure of the predictive control of the dryer

in Fig.1 (a). The outlet moisture content can be only measured offline by an oven box and the measured results will be delayed about two hours. Therefore, MPRFNN is trained offline and RPRFNN is intermittently trained online by BP algorithm.

The samples of offline training MPRFNN are obtained artificially at the bottom of the dryer and their moisture contents are measured by oven box. The data of the temperature and the discharge rate can be calculated from the history records in the system [10]. For training MPRFNN, the cost function is defined as follows:

$$J_M(k) = \frac{1}{2}\sum_{i=1}^{m}(e_{mi}(k))^2 = \frac{1}{2}\sum_{i=1}^{m}(\hat{y_i} - M_i)^2, \quad (8)$$

where k is the training epoch, $\hat{y_i}$ and M_i are, respectively, the output of MPRFNN for the drying process data of sample i and its moisture content, $e_{mi}(k)$ is the error between $\hat{y_i}$ and M_i at the training epoch k, and m is the number of the samples.

Then, the parameters of MPRFNN can be modified by

$$\theta_M(k+1) = \theta_M(k) + \eta_M \frac{\partial J_M(k)}{\partial \theta_M(k)}, \quad (9)$$

where θ_M includes $w^1_{i,-j}$, $w^1_{i,j}$, m_{ij}, σ_{ij} and w^4_j in MPRFNN, and η_M is the training rate of MPRFNN.

The input variables of RPRFNN are the maize inlet temperature, z_1, the inlet hot air temperatures of the high and lower drying layers at the current time, z_2, z_3, z_4, z_5, the temperatures of drying and tempering stage 2 and 3, z_6, z_7, z_8, z_9, and the given frequency of the discharge motor, z_{10}. 6 feedback connections of time are added to the input nodes of RPRFNN. The output variable of RPRFNN is the volume discharge rate of the dryer, $\hat{V}$. The input variables are sampled at intervals of 5 minutes and saved in the system. For training RPRFNN, the cost function is defined as follows:

$$J_R(k) = \frac{1}{2}\sum_{i=1}^{p}(e_{Ri}(k))^2 = \frac{1}{2}\sum_{i=1}^{m}\left(\overline{\hat{V_i}} - V_i\right)^2, \quad (10)$$

where $\overline{\hat{V_i}}$ and V_i are, respectively, the average output of RPRFNN between two measured times and the calculated discharge rate by using the measured switch time in Position 1 and 2 [10], $e_{Ri}(k)$ is the error between $\overline{\hat{V_i}}$ and V_i at the training epoch k, and p is the number of the samples. The method of modifying weights of RPRFNN is the same as that of MPRFNN, Eq.9.

During the drying process, the number of the data set of training RPRFNN is ensured to be a given value and continuously modified by new data. When the $e_{Ri}(k)$ is over the given error ϵ_R, the system will automatically train RPRFNN. The system will forecast the outlet moisture content of the maize layer in the end of tempering stage 3 (the position of the sensor ⑤) with MPRFNN. The discharge rates of this maize layer that passes through drying stage 4 and the cooling stage will be assumed to be the output of RPRFNN, i.e., the switch K is connected to A, the temperature of hot air in drying stage 4 to be the current sensor-measured value, and other variables will be obtained by calculating the history data [10], see the block of Statistic Calculation in Fig. 2. When the error $e_M(k)$ between the output $\hat{y}$ of MPRFNN and the set-point M^* is over the given value ϵ_M, the system will connect the switch K to B and start PI controller A to adjust the discharge rate till $e_M(k) = 0$. The output V^* of PI controller A is assumed to be the given volume discharge rate of the dryer. The error $e_V(k)$ between V^* and $\hat{V}(k)$ will drive PI controller B to adjust the volume discharge rate to reach V^*.

5 Experiment Results

When the dryer was controlled artificially by the operators, the samples of discharge grain were artificially taken at the bottom of dryer and their moisture contents were measured by an oven box. The number of samples was 300 [10].The temperature and discharge rate parameters of the samples during the drying process are calculated statistically by the method proposed in the literature [10]. The number of input variable domain is assumed to be 5, i.e., there are 5 nodes in the membership layer that are corresponding to a single input variable for the two RFNNs. ϵ_R and ϵ_M are assumed to be 0.2 and 0.3, respectively. During the training process of MPRFNN, L is assumed to be 8, 7, 5, 4, 2, respectively. The training results show that the predictive accuracy of MPRFNN is improved unconspicuously when $L > 4$. By comparing the weights w of the input variables of inlet hot air and discharge rate, the outputs of their nodes in the input layer are close to their average values. Therefore, the recurrent function $f(z)$ of the above-mentioned variables is eliminated and their input variables are assumed to their average value when the grain layer passes through the drying stages. The result of training MPRFNN is shown in Fig.3. From Fig. 3, it can be seen that the predictive errors of MPRFNN range only from -0.4% to +0.4% for the most samples. The errors of a few samples are over 0.5%. The reason is that the moisture content of these samples is away from the average moisture content of the samples and their training errors do not have conspicuous effect on the cost function $J_M(k)$. RPRFNN is trained online during the drying process. The experimental results show that RPRFNN trained by 50 continuously taken samples is enough to predict the discharge rate. When inlet grain temperature changes greatly, RPRFNN is trained frequently and its training time is longer. The first training time is about 20 minutes. The current weights of RPRFNN are assumed to be their initial values for next training, i. e., training RPRFNN is the successive process. The successive training time ranges from 2 to 10 minutes. This is enough for the drying process.

To verify the effectiveness of the proposed scheme, the control experiment was carried out for the maize drying. The objective of outlet moisture content was assumed to be 14%(w. b.). The samples were taken artificially at intervals of one hour and their moisture content was measured with an oven box. The

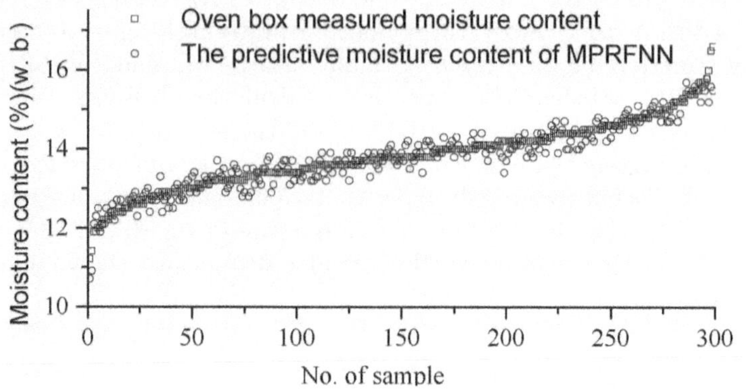

Fig. 3. The results of training MPFNN (Moisture contents are re-arranged from small to big)

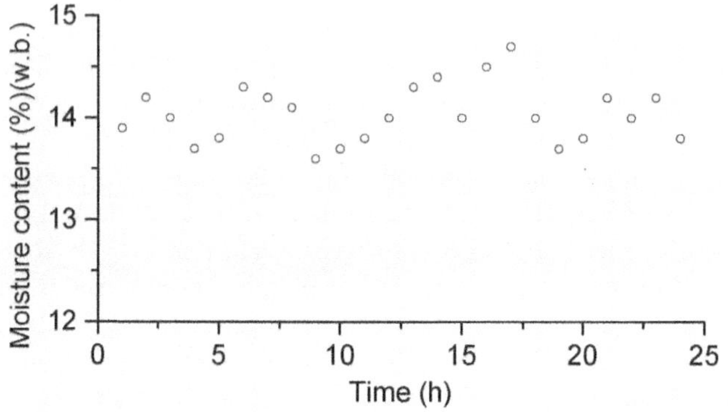

Fig. 4. The experimental results of predictive control of a grain dryer

experimental result of 24 hours is shown in Fig. 4. From Fig. 4, it can be seen that the minimum and maximum of outlet grain moisture content are 13.6% and 14.6%, respectively, and the moisture content average value of the dried samples is 14.04%(w. b.). The safe moisture level for long-term storage ranges from 13.6% to 14.5% (w. b.) in China. There is only one sample in the experiment that its moisture content is over 14.5%. Therefore, the control scheme proposed in this paper can meet the needs of commercial grain dryers.

6 Conclusion

This paper proposes a model predictive control scheme of grain dryers by using the temperature and its variation based recurrent fuzzy neural network. This scheme overcomes the difficult control problem of grain dryers because of long delay, nonlinearity intrinsic to the grain drying process, and the lack of the online sensor-measured accuracy of inlet and outlet moisture content. The experimental results in the predictive control of a maize dryer with four stages show that the proposed scheme can meet the needs of commercial grain dryers.

Acknowledgments. This research is supported by National Natural Science Foundation of China (Grant No.50535010).

References

1. Giner, S.A., Bruce, D.M., Mortimore, S.: Two-Dimensional Simulation Model of Steady-state Mixed-flow Grain Drying, Part 1: The Model. J. agric. Engng. Res. **71** (1998) 37-30
2. Liu, Q., Bakker-Arkema, F.W.: Automatic Control of Crossflow Grain Dryers, Part 1: Development of a Process Model. J. agric. Engng Res. **80** (2001) 81-86
3. Courtois, F., Nouafo, J.L., Trystram, G.: Control Strategies for Corn Mixed-Flow Dryers. Drying Technology **13** (1995) 1153-1165
4. Liu, Q., Bakker-Arkema, F.W.: Automatic Control of Crossflow Grain Dryers, Part 2: Design of a Model-Predictive Controller. J. agric. Engng Res. **80** (2001) 173-181
5. Liu, Q., Bakker-Arkema, F.W.: A Model-Predictive Controller for Grain Drying. J. Food Engineering **49** (2001) 321-326
6. Forbes, J.F., Jacobson, B.A., et al.: Model-Based Control Strategies for Commercial Grain Drying Systems. Canadian Journal of Chemical Engineering **62** (1984) 773-779
7. Zhang, Q., Litchfield , J.B.: Fuzzy Logic Control for a Continuous Crossflow Grain Dryer. Food Process Engineering **16** (1993) 59-77
8. Jover, C., Alastruey, C.F.: Multivarable Control for an Industrial Rotary Dryer. Food Control **17** (2006) 653-659
9. Shi, M.H., Wang, X.: Investigation of Moisture Transfer Mechanism in Porous Media During a Rapid Drying Process. Heat Transfer-Asian Research **30** (2001) 22-27
10. Zhao, C.Y., Zhao, X.G., Chi, Q.L., Wen, B.C.: Experimental Investigation of the Relation between the Moisture Content of Discharge Grain and the Drying Temperatures of the Maize. Journal of the Chinese Cereal and Oil Association **21** (2006) 358-365
11. Zhang, J., Julian Morris, A.: Recurrent Neuro-Fuzzy Networks for Nonlinear Process Modeling. IEEE Trans. Neural Networks **10** (1999) 313-326
12. Yu, Y.L., Xu, L.H., Wu, Q.D.: Generalized Fuzzy Networks. Acta Automation Sinca **29** (2003) 867-875
13. Yi, F.Z., Hu, Z., Zhou, D.: Fuzzy Controller Parameters Optimization by Using Symbiotic Evolution Algorithm. Electric Machines and Control **7** (2003) 54-58
14. Wang, K., Ong, Y.S.: An Adaptive Control for AC Servo System Using Recurrent Fuzzy Neural Network. ICNC 2005, LNCS **3611** (2005) 190-196

Adaptive Nonlinear Control Using TSK-Type Recurrent Fuzzy Neural Network System*

Ching-Hung Lee and Ming-Hui Chiu

Department of Electrical Engineering, Yuan Ze University
Chung-li, Taoyuan 320, Taiwan
chlee@saturn.yzu.edu.tw

Abstract. This paper presents a TSK-type recurrent fuzzy neural network (TRFNN) system and hybrid algorithm to control nonlinear uncertain systems. The TRFNN is modified from the RFNN to obtain generalization and fast convergence rate. The consequent part is replaced by linear combination of input variables and the internal variable- fire strength is feedforward to output to increase the network ability. Besides, a hybrid learning algorithm (GA_BPPSO) is proposed to increase the convergence, which combines the genetic algorithm (GA), back-propagation (BP), and particle swarm optimization (PSO). Several simulation results are proposed to show the effectiveness of TRFNN system and GA_BPPSO algorithm.

1 Introduction

In recent years, fuzzy systems and neural networks are being used successfully in an increasing number of application areas [1-6]. In [4], a recurrent fuzzy neural network (RFNN) is proposed to identify and control nonlinear systems. For TSK-type fuzzy model, the consequent part of each rule is a function input linguistic variable. The general used function is a linear combination of input variable [2, 4]. In this paper, a modified RFNN based on TSK-type fuzzy model (called TRFNN) is presented to generalize and increase the ability of RFNN systems.

Recently, several algorithms are proposed by the observation of real-world, such as, genetic algorithm (GA), DNA computation, particle swarm optimization (PSO) [3, 7-13], etc. GAs are stochastic search procedures based on the mechanics of natural selection, genetics, and evolution [3, 9, 12]. It presumes that the potential solution of a problem is an individual and can be represented by a set of parameters. Particle swarm optimization (PSO): a new evolutionary computation technique is proposed by Kennedy and Eberhart [11]. PSO, similar to GA, a population of random solution is initialized. It was developed by the research of the social behavior of animals, e.g., bird flocking. Compared with GA, PSO has no evolution operators such as crossover and mutation, and moreover, PSO has lesser parameters. It has been applied on optimization problems, neural network learning, fuzzy control, and evolution algorithm [3, 8, 13].

* This work was support partially by National Science Council, Taiwan, R.O.C. under NSC-94-2213-E-155- 039.

D. Liu et al. (Eds.): ISNN 2007, Part I, LNCS 4491, pp. 38–44, 2007.

In this paper, we present the TSK-type recurrent fuzzy neural network (TRFNN) system and hybrid learning algorithm (GA-BPPSO) to control nonlinear uncertain systems. The TRFNN is modified from the previous result RFNN: the consequent part is replaced by linear combination of input variables to obtain generalization and fast convergence rate; the internal variable- fire strength is feedforward to output to increase the network ability.

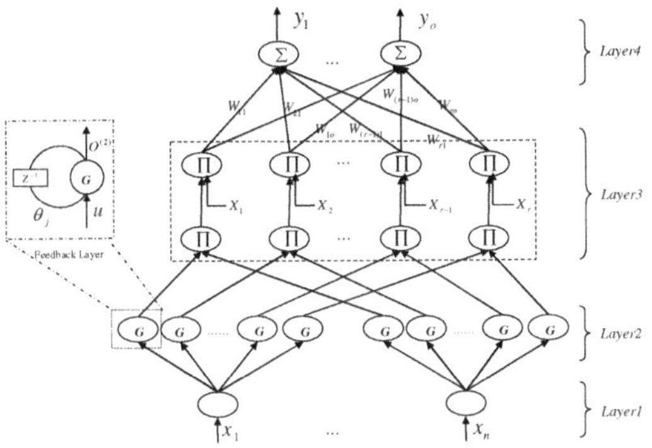

Fig. 1. The TSK-type recurrent fuzzy neural network (TRFNN) system

2 TSK-Type Recurrent Fuzzy Neural Network (TRFNN) System

The structure of the TRFNN system is shown in Fig. 1. Two major modifications from the RFNN are: the consequent part is replaced by linear combination of input variables; the internal variable- fire strength is feedforward to output to obtain generalization and fast convergence rate. Herein, we indicate the signal propagation and the basic function of every node in each layer, $net_i^{(l)}$ denotes the net output, the superscript $^{(l)}$ indicates the lth layer and the subscript i indicates the ith input variable.

Layer 1- input layer: For the ith node of layer 1, the net input and the net output are:

$$O_i^{(1)} = x_i^{(1)}(t) . \tag{1}$$

Layer 2- membership layer: Each node performs a membership function, the representation of each node is

$$O_{ij}^{(2)}(k) = \exp\left\{ -\frac{(u_{ij}^{(2)} - m_{ij})^2}{(\sigma_{ij})^2} \right\} \tag{2}$$

where m_{ij} and σ_{ij} denote center and width of the Gaussian membership function, respectively. Note that $u_{ij}^{(2)} = (O_{ij}^{(2)}(k-1) \cdot \theta_{ij} + O_i^{(1)}(k)$, where θ_{ij} is the self-feedback weight. That is, the TRFNN is a dynamic system.

Layer 3- rule layer: The links in this layer are used to implement the antecedent matching which is determined by fuzzy AND operation.

$$O_j^{(3)}(k) = \prod_i \exp\left\{ -\left[D_i(u_i^{(2)} - m_i)\right]^T \left[D_i(u_i^{(2)} - m_i)\right] \right\} \tag{3}$$

where $D_i = diag\left[\frac{1}{\sigma_{1j}}, \frac{1}{\sigma_{2j}}, \ldots, \frac{1}{\sigma_{nj}}\right]$, $m_i = [m_{1j}, m_{2j}, \cdots, m_{nj}]^T$, and $u_i^{(2)} = [u_{1j}, u_{2j}, \cdots, u_{nj}]^T$.

Layer 4- output layer: The output of TRFNN system is,

$$y_q = O_q^{(4)} = \sum_{j=1}^{r} [W_{jq} \times X_j] \cdot O_j^{(3)} \tag{4}$$

where $W_j = [w_{0jq}, w_{1jq} \cdots, w_{n+1,jq}]$ and $X = [1, x_1, x_2 \cdots, x_n, O_j^{(3)}]^T$.

Therefore, the fuzzy inference of TRFNN can be introduced as IF-THEN rules [4]

$$\begin{aligned} &Rule\ j: IF\ u_{1j}(k)\ is\ A_{1j}\ and\ u_{2j}(k)\ is\ A_{2j} \ldots and\ u_{nj}\ is\ A_{nj} \\ &THEN\ \ y_q(k+1)\ is\ w_{0jq} + x_1 \cdot w_{1jq} + x_2 \cdot w_{2jq} + \cdots + x_n \cdot w_{njq} + O_j^{(3)} \cdot w_{n+1,jq} \end{aligned} \tag{5}$$

Note that the network is not fully connected for each node to avoid the exponential growth of fuzzy rules number and parameters [6].

3 Hybrid Learning Algorithms

3.1 Learning Algorithm- BP, GA, and PSO

Herein, a hybrid learning algorithm (GA_BPPSO) combines the GA, BP, and PSO to have a fast convergent rate. Firstly, the BP algorithm is introduced [4, 6],

$$W(k+1) = W(k) + \Delta W(k) = W(k) + \eta\left(-\frac{\partial E(k)}{\partial W}\right) \tag{6}$$

where $W = [m, \sigma, \theta, w]^T$, and the cost function is defined as

$$E(k) = \frac{1}{2}(y(k) - \hat{y}(k))^2 = \frac{1}{2}\sum_j (y(k) - O^{(4)}(k))^2 \tag{7}$$

$y(k)$ and $\hat{y}(k)$ are the desired output and TRFNN output, respectively.

Secondly, the GA is briefly introduced. The GA uses three basic operators to manipulate the genetic composition of a population: reproduct, crossover, and mutation [12, 15]. The chromosomes consist of a set of genes. Herein, the real-coded GA is used to tune the parameters. It is more natural to represent the genes directly as real numbers since the representations of the solutions are very close to the natural formulation. Therefore, a chromosome here is a vector of floating point numbers. The chromosome is chosen by TRFNN's adjustable parameters as below.

m	…	σ	…	θ	…	w	…

In operation process, an initial population $P(0)$ is given, and then the GA generates a new generation $P(t)$ based on the previous generation $P(t\text{-}1)$ as follows [9, 12]:

```
Initialize P(t)→ P(0)                                  : P(t) Population at time t
Evaluate P(0)
While (not terminate-condition) do
Begin
    T← t+1                  : Increment generation
    Select P(t) from P(t-1)
    Recombine P(t)               : apply genetic operators
    (crossover, mutation)
    evaluate P(t)
```

Particle swarm optimization (PSO): a new evolutionary computation technique is proposed by Kennedy and Eberhart [3, 11]. It was developed by the research of the social behavior of animals, e.g., bird flocking. Compared with GA, PSO has no evolution operators such as crossover and mutation, and moreover, PSO has lesser parameters. It has been applied on optimization problems, neural network learning, fuzzy control, and evolution algorithm [3, 8, 13].

The optimization process of PSO is nonlinear and complicated. The system initially has a population of random solutions. Each particle has a position $\vec{x}_i$ which represents a potential solution, and it is given a random velocity and is flown through the problem space. Each particle has memory and keeps tracking its best position $\vec{p}_i$ and its corresponding fitness. There exist a number of $\vec{p}_i$ for the respective particles in the swarm and the particle with greatest fitness is the best of the swarm $\vec{p}_g$. Thus, we have

$$\vec{v}_i(t+1) = \chi(\vec{v}_i(t) + c_1\phi_1(\vec{p}_i(t) - \vec{x}_i(t)) + c_2\phi_2(\vec{p}_g(t) - \vec{x}_i(t))) \tag{8}$$

where χ is control parameter of $\vec{v}$, c_1, c_2>0, and ϕ_1 , ϕ_2 are uniformly distributed random numbers in [0,1]. In addition, each particle changes its position by

$$\vec{x}_i(t+1) = \vec{x}_i(t) + \vec{v}_i(t+1) . \tag{9}$$

3.2 The Hybrid Algorithm GA_BPPSO

In [5], we combined the advantages of BP and GA to obtain faster convergence of network parameters. As above, the HGAPSO with the concept of elites has the effective performance for recurrent neural network [3]. The GA_BPPSO learning algorithm for TRFNN combines the concept of literature [3, 5]. An initialized population (P_s individuals) is selected randomly. Then all the fitness values of individuals are calculated and ranked to find the "elites". The optimal individual p^* with the highest fitness value is chosen. Subsequently, new population is created by BP and GA [5]. m individuals ($p_1^*, p_2^*, \cdots, p_m^*$) are obtained through the BP with different learning rates $\eta_1(k), \eta_2(k), \cdots, \eta_m(k)$, respectively, where the proper learning rates are determined by PSO. Besides, the others (P_s-m) are generated by GA operations- reproduct, crossover, and mutation. The description can be summarized as bellows and Fig. 2.

initialize P(t) → P(0) : P(t) population at time t
Evaluate P(0)
While (NOT terminate-condition) do
Begin
T←t+1
Select elites p^*
Select $\eta_1, \ldots, \eta_m$ by PSO
Create $p_1^*, p_2^*, \cdots, p_m^*$ by BP
Create $p_{m+1}^*, p_{m+2}^*, \cdots, p_s^*$ by GA (apply GA operators)
Evaluate P(t)
Ranking by P(t)
End

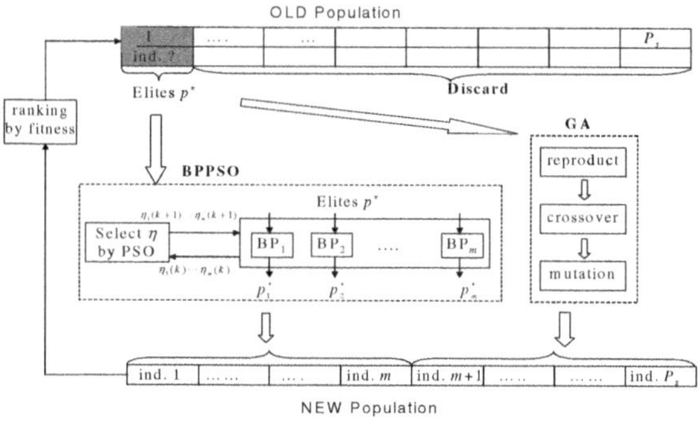

Fig. 2. The flow description of GA_BPPSO algorithm

4 Simulation Result-Adaptive Control for MIMO Nonlinear System

Consider the tracking control of system [4]

$$Y_{p1}(k+1) = 0.5 \cdot \left[\frac{Y_{p1}(k)}{1 + Y_{p2}^2(k)} + u_1(k-1) \right] \tag{10}$$

$$Y_{p2}(k+1) = 0.5 \cdot \left[\frac{Y_{p1}(k) \cdot Y_{p2}(k)}{1 + Y_{p2}^2(k)} + u_2(k-1) \right]. \tag{11}$$

The tracking trajectories are $Y_{r1}(k) = \sin(k\pi / 45)$ and $Y_{r2}(k) = \cos(k\pi / 45)$. The control scheme is shown in Fig. 3.

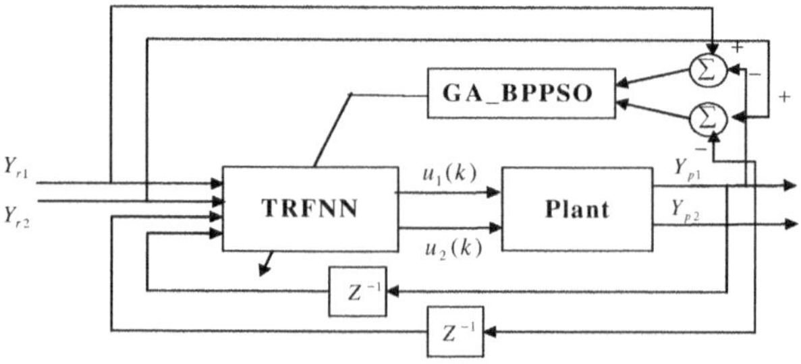

Fig. 3. Adaptive Control Scheme using TRFNN

The learning parameters of the GA_BPPSO algorithm is selected as- input #: 4; output #:2, rule #: 5, population size P_S: 50, # of BP m: 10, crossover probability: 0.8; reproduction probability: 0.03; $c_1 = c_2 = 1$; χ =0.8, max. $\vec{v}_i$: 1; # of generation: 20.

Simulation results are shown in Tables 1, 2, and Fig. 4. Figure 4(a) shows the testing results- solid: desired output; dotted: TRFNN output. Figure 4(b) shows the comparison results of RMSE- solid: TRFNN (RMSE after 20 generations: 0.0254); dotted: RFNN (RMSE after 20 generations: 0.0264). Figure 4(c) shows the RMSE comparison of algorithms for TRFNN. Obviously, we can conclude that the TRFNN has the ability of high speed convergence and small network structure, i.e., increase the ability of network.

Table 1. RMSE comparison results of different algorithms after 20 generations

RMSE / Algorithm	TRFNN	RFNN
BP	0.0694	0.0859
GA	0.4723	0.7447
PSO	0.3729	0.6550
HGAPSO	0.3714	0.6304
GA_BPPSO	**0.0254**	**0.0264**

Table 2. Comparison of network structure for TRFNN and RFNN

	TRFNN	RFNN
Rule number	**5**	10
Node number	**31**	56
Parameters number	**80**	140

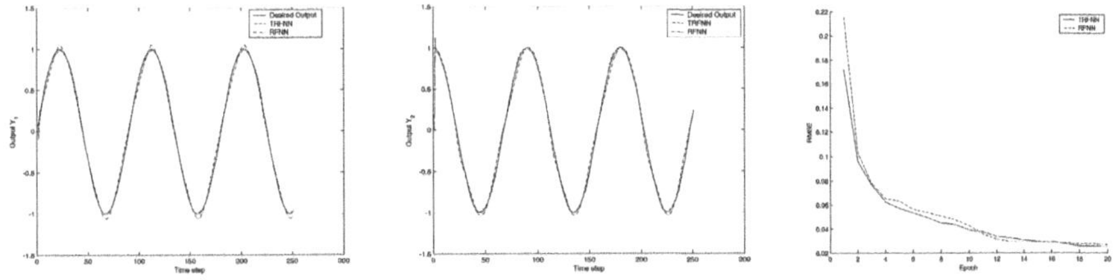

4(a) 4(b) 4(c)

Fig. 4. Simulation results of nonlinear MIMO system tracking control using TRFNN and RFNN

5 Conclusion

In this paper, we have presented a new TSK-type recurrent fuzzy neural network (TRFNN) system to control nonlinear uncertain systems. The TRFNN is modified from the previous result RFNN to increase the network ability. The hybrid learning algorithm GA_BPPSO was proposed to have a fast convergence. Simulation results are proposed to show the effectiveness of TRFNN system and GA-BPPSO algorithm, e.g., the TRFNN has the ability of high speed convergence and small network structure and the GA_BPPSO increases the approximation accuracy.

References

1. Chen, Y.C., Teng, C.C.: A Model Reference Control Structure Using A Fuzzy Neural Network. Fuzzy Sets and Systems **73** (1995) 291–312
2. Juang, C.F.: A TSK-type Recurrent Fuzzy Network for Dynamic Systems Processing by Neural Network and Genetic Algorithms. IEEE Trans. Fuzzy Systems **10** (2002) 155-170
3. Juang, C.F.: A Hybrid of Genetic Algorithm and Particle Swarm Optimization for Recurrent Network Design. IEEE Trans. Systems, Man and Cybernetics Part: B **34**(2) (2004) 997-1006
4. Lee, C.H., Teng, C.C.: Identification and Control of Dynamic Systems Using Recurrent Fuzzy Neural Networks. IEEE Trans. Fuzzy Systems **8**(4) (2000) 349-366
5. Lee, C.H., Lin, Y.C.: Hybrid Learning Algorithm for Fuzzy Neuro Systems Fuzz-IEEE 2004 (2004) 691-696
6. Lin, C.T., Lee, C.S.G.: Neural Fuzzy Systems. Prentice Hall: Englewood Cliff (1996)
7. Adleman, L.M.: Molecular Computation of Solutioins to Combinatorial Problems Science **266** (1994) 1021-1023
8. Clerc, M., Kenney, J.: The Particle Swarm-explosion, Stability, and Convergence in Multi-dimensional Complex Space. IEEE Trans. Evol. Comput. **6** (2002) 58-73
9. Goldberg, D.E.: Genetic Algorithms in Search, Optimization and Machine Learning. Addison-Wesley, Reading (1989)
10. Gudise, V.G., Venayagamoorthy, G.K.: Comparison of Particle Swarm Optimization and Backpropagation as Training Algorithm for Neural Networks. IEEE Swarm Intelligence Symp. USA, April 24 -26 (2003) 110 – 117
11. Kennedy, J., Eberhart, R.: Particle Swarn Optimization. Proc. IEEE Int. Conf. Neural Networks, Perth, Australia (1995) 1942-1948
12. Michalewicz, Z.: Genetic Algorithms+Data Structure=Evolutionary Programs. Springer-Verlag, Berlin, 3rd edition (1997)
13. Zhang, C., Shao, H., Li, Y.: Particle Swarm Optimization for Evolving Artificial Network IEEE Int. Conf. Syst. Man, Cyber. **4** (2000) 2487-2490

GA-Based Adaptive Fuzzy-Neural Control for a Class of MIMO Systems

Yih-Guang Leu, Chin-Ming Hong, and Hong-Jian Zhon

Department of Industrial Education, National Taiwan Normal University
162, Ho-Oing E. Road, Sec 1, Taipei, Taiwan
leuyk@ntnu.edu.tw

Abstract. A GA-based adaptive fuzzy-neural controller for a class of multi-input multi-output nonlinear systems, such as robotic systems, is developed for using observers to estimate time derivatives of the system outputs. The weighting parameters of the fuzzy-neural controller are tuned on-line via a genetic algorithm (GA). For the purpose of on-line tuning the weighting parameters of the fuzzy-neural controller, a Lyapunov-based fitness function of the GA is obtained. Besides, stability of the closed-loop system is proven by using strictly-positive-real (SPR) Lyapunov theory. The proposed overall scheme guarantees that all signals involved are bounded and the outputs of the closed-loop system track the desired output trajectories. Finally, simulation results are provided to demonstrate robustness and applicability of the proposed method.

1 Introduction

Since neural networks [1] and fuzzy systems [2] are universal approximators, some adaptive control schemes of nonlinear systems via fuzzy systems [3][4] or neural networks [5][6][7][8] have been proposed. Likewise, for a class of nonlinear discrete-time systems, adaptive control using neural networks by feedback linearization has been proposed in [9]. Also, a dynamic recurrent neural-network- based adaptive observer for single-input single-output (SISO) nonlinear systems has been presented in [10]. In [11], an output feedback controller has been developed for using a high-gain observer to estimate the time derivatives of the system output. Moreover, fuzzy logic incorporated into neural networks and its applications in function approximation, decision systems and nonlinear control systems have been proposed in [12][13][14][15][16][17]. An on-line tuning approach of fuzzy-neural networks for adaptive control of a SISO nonlinear system has been proposed in [15][16].

Theoretical justification development presented in [15][16] is valid only for SISO nonlinear systems and so is hardly practical in real applications such as the trajectory control of robot manipulators and space vehicles. Although, Hwang and Hu [17] have proposed a robust neural learning controller for multi-input multi-output (MIMO) manipulators, the state feedback control scheme does not always hold in practical applications, because those system states are not always available. An estimation of state from the system output for output feedback control is required. That is to say, we

D. Liu et al. (Eds.): ISNN 2007, Part I, LNCS 4491, pp. 45–53, 2007.

need to design an observer that estimates the states of the system for output feedback control. Therefore, the problem as to how an output feedback adaptive fuzzy controller for MIMO systems is designed remains to be solved.

Besides, because of the capability of genetic algorithms (GAs) in directed random search for global optimization, one is used to evolutionarily obtain the optimal weighting parameters for the fuzzy neural network [18] [19]. Therefore, the objective of this paper is to develop a GA-based algorithm for designing an output feedback adaptive fuzzy-neural controller for a class of MIMO nonlinear systems, such as robotic systems. The weighting parameters of the fuzzy-neural controller are tuned on-line via a GA. The overall adaptive scheme guarantees that all signals involved are bounded and the outputs of the closed-loop system track the desired output trajectories.

2 Description of Fuzzy-Neural Networks

The basic configuration of fuzzy logic systems consists of some fuzzy IF-THEN rules and a fuzzy inference engine. The fuzzy inference engine uses the fuzzy IF-THEN rules to perform a mapping from input linguistic variables to output linguistic variables. Given the input data $x_q, q=1,2,\cdots,n$, and the output data y_p, $p=1,2,\cdots,m$, the ith fuzzy rule has the following form

$$\begin{aligned} R^i: IF \quad x_1 \quad is \quad A_1^i \quad and \cdots x_n \quad is \quad A_n^i \\ THEN \quad y_1 \quad is \quad w_1^i \quad and \cdots y_m \quad is \quad w_m^i \end{aligned} \tag{1}$$

where i is a rule number, A_q^i's are the fuzzy sets of the antecedent part, and w_p^i are real numbers of the consequent part. When the inputs $x=[x_1 x_2 \cdots x_n]^T$ are given, the output y_p of the fuzzy inference can be derived from the following equations

$$y_p(x|w_p) = \sum_{i=1}^{h} w_p^i \left(\prod_{q=1}^{n} \mu_{A_q^i}(x_q)\right) \Big/ \sum_{i=1}^{h} \left(\prod_{q=1}^{n} \mu_{A_q^i}(x_q)\right) = w_p^T \psi \tag{2}$$

where $\mu_{A_q^i}(x_q)$ is the membership function of A_q^i, h is the number of the fuzzy rules. $w_p = [w_p^1 \; w_p^2 \cdots w_p^h]^T$ is a weighting vector related to the pth output $y_p(x)$. Fuzzy-neural networks are generally a fuzzy inference system constructed from structure of neural networks [15][16].

3 GA-Based On-Line Tuning of Weighting Parameters of the Fuzzy-Neural Controller

In this paper, the reduced-form genetic algorithm (RGA) [20] [21] is used to tune on-line the weighting parameters of the fuzzy-neural controller. By using the supervisory control [3], the stability of the closed-loop system can be verified. Moreover, the proposed overall control scheme guarantees that all signals are bounded and the outputs of the closed-loop system track the desired output trajectory well.

First, consider a class of MIMO systems of the form

$$\dot{\mathbf{x}}_i = \mathbf{A}_i\mathbf{x}_i + \mathbf{B}_i(\mathrm{f}_i(\mathbf{x}) + \sum_{j=1}^{n} \mathrm{g}_{ij}(\mathbf{x})u_j)$$
$$y_i = \mathbf{C}_i^T\mathbf{x}_i \qquad i = 1,2,\ldots,n \tag{3}$$

where $\mathbf{x}_i \in \Re^{r_i}$, $\mathbf{A}_i = \begin{bmatrix} 0 & 1 & \cdots & 0 \\ 0 & 0 & \cdots & 0 \\ \cdots & \cdots & \ddots & \cdots \\ 0 & 0 & \cdots & 0 \end{bmatrix}_{r_i \times r_i}$, $\mathbf{B}_i = \begin{bmatrix} 0 \\ 0 \\ \vdots \\ 1 \end{bmatrix}_{r_i \times 1}$, $\mathbf{C}_i = \begin{bmatrix} 1 \\ 0 \\ \vdots \\ 0 \end{bmatrix}_{r_i \times 1}$. Define the reference vectors $\mathbf{y}_{mi} = [y_{mi}, y'_{mi}, \ldots, y_{mi}^{(r_i-1)}]^T$, the state vectors $\mathbf{x}_i = [y_i, y'_i, \ldots, y_i^{(r_i-1)}]^T$, the control input $\mathbf{u} = [u_1, u_2, \ldots, u_n]^T$, the tracking error vectors $\mathbf{e}_i = \mathbf{y}_{mi} - \mathbf{x}_i$, and the estimated tracking error vectors $\hat{\mathbf{e}}_i = \mathbf{y}_{mi} - \hat{\mathbf{x}}_i$, where $\hat{\mathbf{x}}_i$ and $\hat{\mathbf{e}}_i$ denote the estimates of $\mathbf{x}_i$ and $\mathbf{e}_i$, respectively. Note that $\mathbf{x} = [\mathbf{x}_1^T, \mathbf{x}_2^T, \ldots, \mathbf{x}_n^T]^T$ and $\hat{\mathbf{x}} = [\hat{\mathbf{x}}_1^T, \hat{\mathbf{x}}_2^T, \ldots, \hat{\mathbf{x}}_n^T]^T$.

Based on the certainty equivalence approach, the control law is

$$\mathbf{u} = \hat{\mathbf{G}}^{-1}(\hat{\mathbf{x}})\left\{-\hat{\mathbf{f}}(\hat{\mathbf{x}}) + [y_{m1}^{(r_1)}\, y_{m2}^{(r_2)} \cdots y_{mn}^{(r_p)}]^T + [\mathbf{K}_{c1}^T\hat{\mathbf{e}}_1 \mathbf{K}_{c2}^T\hat{\mathbf{e}}_2 \cdots \mathbf{K}_{cn}^T\hat{\mathbf{e}}_n]^T - \mathbf{u}_c - \mathbf{u}_s\right\}. \tag{4}$$

where $\mathbf{K}_{ci} = [k_{r_i}^{ci}, k_{r_i-1}^{ci}, \ldots, k_1^{ci}]^T$ are the feedback gain vectors, chosen such that the characteristic polynomials of $\mathbf{A}_i - \mathbf{B}_i\mathbf{K}_{ci}^T$ are Hurwitz, $\hat{\mathbf{f}}(\hat{\mathbf{x}}) = [\hat{\mathrm{f}}_1(\hat{\mathbf{x}}), \mathrm{f}_2(\hat{\mathbf{x}}), \ldots, \mathrm{f}_n(\hat{\mathbf{x}})]^T$,
$\hat{\mathbf{G}}(\hat{\mathbf{x}}) = \begin{bmatrix} \hat{g}_{11}(\hat{\mathbf{x}}) & \hat{g}_{12}(\hat{\mathbf{x}}) & \cdots & \hat{g}_{1n}(\hat{\mathbf{x}}) \\ \hat{g}_{21}(\hat{\mathbf{x}}) & \hat{g}_{22}(\hat{\mathbf{x}}) & \cdots & \hat{g}_{2n}(\hat{\mathbf{x}}) \\ \vdots & \vdots & \ddots & \vdots \\ \hat{g}_{n1}(\hat{\mathbf{x}}) & \hat{g}_{n2}(\hat{\mathbf{x}}) & \cdots & \hat{g}_{nn}(\hat{\mathbf{x}}) \end{bmatrix}$, $\hat{\mathbf{g}}_i(\hat{\mathbf{x}}) = \left[\hat{g}_{i1}(\hat{\mathbf{x}}) \;\; \hat{g}_{i2}(\hat{\mathbf{x}}) \;\; \cdots \;\; \hat{g}_{in}(\hat{\mathbf{x}})\right]^T$, and the terms $\hat{\mathrm{f}}_i, \hat{g}_{ij}$ are denote the estimates of the uncertain terms f_i, g_{ij}, respectively. Besides, the control term $\mathbf{u}_c$ is employed to compensate for the modeling error, and the control term $\mathbf{u}_s$ is a supervisory control [2]. From equations (3) and (4), we have

$$\dot{\mathbf{e}}_i = \mathbf{A}_i\mathbf{e}_i - \mathbf{B}_i\mathbf{K}_{ci}^T\hat{\mathbf{e}}_i + \mathbf{B}_i(u_{ci} + u_{si}) + \mathbf{B}_i[\hat{\mathrm{f}}_i(\hat{\mathbf{x}}) - \mathrm{f}_i(\mathbf{x}) + (\sum_{j=1}^{n}(\hat{\mathrm{g}}_{ij}(\hat{\mathbf{x}}) - \mathrm{g}_{ij}(\mathbf{x}))u_j]]$$
$$e_{oi} = \mathbf{C}_i^T\mathbf{e}_i \tag{5}$$

where $e_{oi} = y_{mi} - y_i$ denote the output tracking errors. Thus, the tracking problem is converted into the regulation problem of designing observers for estimating the vectors $\mathbf{e}_i$ in (5) in order to regulate e_{oi} to zero.

Consider the following observers that estimate the state vectors $\mathbf{e}_i$

$$\dot{\hat{\mathbf{e}}}_i = \mathbf{A}_i\hat{\mathbf{e}}_i - \mathbf{B}_i\mathbf{K}_{ci}^T\hat{\mathbf{e}}_i + \mathbf{B}_i u_{ci} - \mathbf{B}_i u_{vi} + \mathbf{K}_{oi}(e_{oi} - \hat{e}_{oi})$$
$$\hat{e}_{oi} = \mathbf{C}_i^T\hat{\mathbf{e}}_i \tag{6}$$

where $\mathbf{K}_{oi}=[k_1^{oi},k_2^{oi},...,k_{r_i}^{oi}]^T$ are the observer gain vectors, chosen such that the characteristic polynomials of $\mathbf{A}_i-\mathbf{K}_{oi}\mathbf{C}_i^T$ are strictly Hurwitz, and the term u_{vi} is employed to compensate for the modeling error. We define the observation errors $\tilde{\mathbf{e}}_i=\mathbf{e}_i-\hat{\mathbf{e}}_i$ and $\tilde{e}_{oi}=e_{oi}-\hat{e}_{oi}$. Subtracting (6) from (5), we have

$$\begin{aligned}\dot{\tilde{\mathbf{e}}}_i &= (\mathbf{A}_i-\mathbf{K}_{oi}\mathbf{C}_i^T)\tilde{\mathbf{e}}_i+\mathbf{B}_i[\hat{\mathrm{f}}_i(\hat{\mathbf{x}})-\mathrm{f}_i(\mathbf{x})+\sum_{j=1}^{n}(\hat{\mathrm{g}}_{ij}(\hat{\mathbf{x}})-\mathrm{g}_{ij}(\mathbf{x}))u_j+u_{vi}-u_{si}]\\ \tilde{e}_{oi} &= \mathbf{C}_i^T\tilde{\mathbf{e}}_i\end{aligned} \tag{7}$$

We replace the estimation functions $\hat{\mathrm{f}}_i(\hat{\mathbf{x}})$ and $\hat{\mathrm{g}}_{ij}(\hat{\mathbf{x}})$ in (7) by the fuzzy-neural systems $\hat{\mathrm{f}}_i(\hat{\mathbf{x}}\,|\,\theta_{\mathrm{f}i})=\theta_{\mathrm{f}i}^T\psi(\hat{\mathbf{x}})$ and $\hat{\mathrm{g}}_{ij}(\hat{\mathbf{x}}\,|\,\theta_{\mathrm{g}ij})=\theta_{\mathrm{g}ij}^T\psi(\hat{\mathbf{x}})$, respectively. In order to derive the control method, the following assumptions must be required.

Assumption 1. The parameter vector $\theta_{\mathrm{g}_{ij}}$ is such that $\hat{\mathbf{G}}(\hat{\mathbf{x}}\,|\,\theta_{\mathrm{g}_{ij}})$ is bounded away from singularity. □

Assumption 2 [22]. Let $\mathbf{x}$ and $\hat{\mathbf{x}}$ belong to compact sets $\mathrm{U}_{\mathbf{x}}=\{\mathbf{x}\in\Re^n:\|\mathbf{x}\|\le m_{\mathbf{x}}<\infty\}$, $\mathrm{U}_{\hat{\mathbf{x}}}=\{\hat{\mathbf{x}}\in\Re^n:\|\hat{\mathbf{x}}\|\le m_{\hat{\mathbf{x}}}<\infty\}$. It is known a prior that the optimal parameter vectors $\theta_{\mathrm{f}i}^*$ and $\theta_{\mathrm{g}ij}^*$ lie are defined as

$$\theta_{\mathrm{f}_i}^*=\arg\min_{\theta_{\mathrm{f}_i}\in\mathrm{M}_{\theta_{\mathrm{f}_i}}}[\sup_{\mathbf{x}\in\mathrm{U}_{\mathbf{x}},\hat{\mathbf{x}}\in\mathrm{U}_{\hat{\mathbf{x}}}}\left|\mathrm{f}_i(\mathbf{x})-\hat{\mathrm{f}}_i(\hat{\mathbf{x}}\,|\,\theta_{\mathrm{f}_i})\right|] \text{ and } \theta_{\mathrm{g}_{ij}}^*=\arg\min_{\theta_{\mathrm{g}_{ij}}\in\mathrm{M}_{\theta_{\mathrm{g}_{ij}}}}[\sup_{\mathbf{x}\in\mathrm{U}_{\mathbf{x}},\hat{\mathbf{x}}\in\mathrm{U}_{\hat{\mathbf{x}}}}\left|\mathrm{g}_{ij}(\mathbf{x})-\hat{\mathrm{g}}_{ij}(\hat{\mathbf{x}}\,|\,\theta_{\mathrm{g}_{ij}})\right|]$$

where $\mathrm{M}_{\theta_{\mathrm{f}i}}=\{\theta_{\mathrm{f}i}\in\Re^{h}:\|\theta_{\mathrm{f}i}\|\le m_{\theta_{\mathrm{f}i}}\}$, $\mathrm{M}_{\theta_{\mathrm{g}ij}}=\{\theta_{\mathrm{g}ij}\in\Re^{h}:\|\theta_{\mathrm{g}ij}\|\le m_{\theta_{\mathrm{g}ij}}\}$. □

Now, define the approximation observation errors $w_{oi}=(\hat{\mathrm{f}}_i(\hat{\mathbf{x}}\,|\,\theta_{\mathrm{f}i}^*)-\mathrm{f}_i(\mathbf{x}))+\sum_{j=1}^{n}[(\hat{\mathrm{g}}_{ij}(\hat{\mathbf{x}}\,|\,\theta_{\mathrm{g}ij}^*)-\mathrm{g}_{ij}(\mathbf{x}))]u_j$. The observation error dynamic equation (7) with can be rewritten as

$$\begin{aligned}\dot{\tilde{\mathbf{e}}}_i &= (\mathbf{A}_i-\mathbf{K}_{oi}\mathbf{C}_i^T)\tilde{\mathbf{e}}_i+\mathbf{B}_i[\tilde{\theta}_{\mathrm{f}i}^T\psi(\hat{\mathbf{x}})+\sum_{j=1}^{n}\tilde{\theta}_{\mathrm{g}ij}^T\psi(\hat{\mathbf{x}})u_j+u_{vi}-u_{si}+w_{oi}]\\ \tilde{e}_{oi} &= \mathbf{C}_i^T\tilde{\mathbf{e}}_i\end{aligned} \tag{8}$$

The output error dynamics of (8) can be given as

$$\tilde{e}_{oi}=H_i(s)[\tilde{\theta}_{\mathrm{f}i}^T\psi(\hat{\mathbf{x}})+\sum_{j=1}^{n}\tilde{\theta}_{\mathrm{g}ij}^T\psi(\hat{\mathbf{x}})u_j+u_{vi}-u_{si}+w_{oi}]. \tag{9}$$

where $H_i(s)=\mathbf{C}_i^T(s\mathbf{I}-(\mathbf{A}_i-\mathbf{K}_{oi}\mathbf{C}_i^T))^{-1}\mathbf{B}_i=1/(s^{r_i}+k_1^{oi}s^{r_i-1}+\cdots+k_{r_i}^{oi})$. The transfer function $H_i(s)$ is a known stable transfer function. In order to be able to use the SPR-Lyapunov design approach, equation (9) can be rewritten as

$$\tilde{e}_{oi} = H_i(s)L_i(s)[\tilde{\theta}_{fi}^T\psi(\hat{\mathbf{x}}) + \sum_{j=1}^{n}\tilde{\theta}_{gij}^T\psi(\hat{\mathbf{x}})u_j + z_i - v_i + \delta_i)]. \tag{10}$$

where $\varepsilon_i = [\tilde{\theta}_{fi}^T\psi(\hat{\mathbf{x}}) - L_i\tilde{\theta}_{fi}^T\psi(\hat{\mathbf{x}})] + [\sum_{j=1}^{n}\tilde{\theta}_{gij}^T\psi(\hat{\mathbf{x}})u_j - L_i\sum_{j=1}^{n}\tilde{\theta}_{gij}^T\psi(\hat{\mathbf{x}})u_j)] + w_{oi}$, $\delta_i = L_i^{-1}(s)\varepsilon_i$, $z_i = L_i^{-1}(s)u_{vi}$, $v_i = L_i^{-1}(s)u_{si}$ and $L_i(s)$ is chosen so that $L_i^{-1}(s)$ is a proper stable transfer function and $H_i(s)L_i(s)$ is a proper SPR transfer function. Suppose that $L_i(s) = s^{m_i} + b_{1i}s^{m_i-1} + b_{2i}s^{m_i-2} + \ldots + b_{m_i i}$, where $m_i = r_i - 1$, such that $H_i(s)L_i(s)$ is a proper SPR transfer function. Then, the state-space realization of (10) can be written as

$$\begin{aligned}\dot{\tilde{\mathbf{e}}}_{ci} &= \mathbf{A}_{ci}\tilde{\mathbf{e}}_{ci} + \mathbf{B}_{ci}[\tilde{\theta}_{fi}^T\psi(\hat{\mathbf{x}}) + \sum_{j=1}^{n}\tilde{\theta}_{gij}^T\psi(\hat{\mathbf{x}})u_j + z_i - v_i + \delta_i]\\ \tilde{e}_{oi} &= \mathbf{C}_{ci}^T\tilde{\mathbf{e}}_{ci}.\end{aligned} \tag{11}$$

where $\mathbf{A}_{ci} = (\mathbf{A}_i - \mathbf{K}_{oi}\mathbf{C}_i^T), \mathbf{B}_{ci} = [1, b_{1i}, \ldots, b_{m_i i}]^T$, $\mathbf{C}_{ci} = [1,0,\ldots,0]^T$. For the purpose of stability analysis, the following assumptions and lemma are required.

Assumption 3. ε_i is assumed to satisfy $\|\varepsilon_i\| \le \eta_i$, where η_i is a positive constant. Moreover, the uncertain nonlinear functions $g_{ij}(\mathbf{x})$ is bounded by $|g_{ij}(\mathbf{x})| \le g^U{}_{ij}(\hat{\mathbf{x}})$. The uncertain nonlinear functions $\mathrm{f}_i(\mathbf{x})$ is bounded by $\mathrm{f}_i(\mathbf{x}) \le \mathrm{f}_i^U(\hat{\mathbf{x}})$. □

Consider the Lyapunov-like function candidate $V = \sum_{i=1}^{n} V_i$, where $V_i = \frac{1}{2}\tilde{\mathbf{e}}_{ci}{}^T\mathbf{P}_i\tilde{\mathbf{e}}_{ci}$ with $\mathbf{P}_i = \mathbf{P}_i^T > 0$. Then, we have

$$\dot{V}_i = \frac{1}{2}\tilde{\mathbf{e}}_{ci}{}^T(\mathbf{A}_{ci}{}^T\mathbf{P}_i + \mathbf{P}_i\mathbf{A}_{ci})\tilde{\mathbf{e}}_{ci} + \tilde{\mathbf{e}}_{ci}{}^T\mathbf{P}_i\mathbf{B}_{ci}[\tilde{\theta}_{fi}^T\psi(\hat{\mathbf{x}}) + \sum_{j=1}^{n}\tilde{\theta}_{gij}^T\psi(\hat{\mathbf{x}})u_j + z_i - v_i + \delta_i]. \tag{12}$$

Because $H_i(s)L_i(s)$ is SPR, there exists $\mathbf{P}_i = \mathbf{P}_i^T > 0$ such that

$$\begin{aligned}\mathbf{A}_{ci}{}^T\mathbf{P}_i + \mathbf{P}_i\mathbf{A}_{ci} &= -\mathbf{Q}_i\\ \mathbf{P}_i\mathbf{B}_{ci} &= \mathbf{C}_{ci},\end{aligned} \tag{13}$$

with $\mathbf{Q}_i = \mathbf{Q}_i^T > 0$. Therefore, we get $\dot{V}_i \le -\frac{1}{2}\lambda_{\min}(\mathbf{Q}_i)|\tilde{e}_{oi}|^2 + \tilde{e}_{oi}[\tilde{\theta}_{fi}^T\psi(\hat{\mathbf{x}}) + \sum_{j=1}^{n}\tilde{\theta}_{gij}^T\psi(\hat{\mathbf{x}})u_j + z_i - v_i + \delta_i]$. Let $u_{vi} = -\eta_i \mathrm{sign}(\tilde{e}_{oi})$ we have $\dot{V}_i \le -\frac{1}{2}\lambda_{\min}(\mathbf{Q}_i)|\tilde{e}_{oi}|^2 + \tilde{e}_{oi}[\tilde{\theta}_{fi}^T\psi(\hat{\mathbf{x}}) + \sum_{j=1}^{n}\tilde{\theta}_{gij}^T\psi(\hat{\mathbf{x}})u_j - v_i]$. Then, from (7), we have $\dot{V}_i \le 0$ when the supervisory control is chosen as $u_{si} = I_i * \mathrm{sign}(\tilde{e}_{oi})\left[|\mathrm{f}_i^u(\hat{\mathbf{x}})| + |\mathrm{f}_i(\hat{\mathbf{x}}|\theta_{fi})| + \sum_{j=1}^{n}[|\mathrm{g}_{ij}^u(\hat{\mathbf{x}})| + |\mathrm{g}_{ij}(\hat{\mathbf{x}}|\theta_{g_{ij}})|]\right]$ where $I_i = 1$ if

$V_i > \overline{V} > 0$, $I_i = 0$ if $V_i < \overline{V}$. More specifically, when $V_i < \overline{V}$, the weighting parameters $\theta_{fi}, \theta_{g_{ij}}$ of $\hat{f}_i(\hat{\mathbf{x}} | \theta_{fi})$ and $\hat{g}_{ij}(\hat{\mathbf{x}} | \theta_{gij})$ are tuned on-line by using GA-based algorithm, and here a fitness function of the GA-based algorithm is chosen as

$$\text{fitness} = -\frac{1}{2}\xi \sum_{i=1}^{n} |\tilde{e}_{oi}|^2 + \sum_{i=1}^{n} e_{oi} \{ |f_i^u(\hat{\mathbf{x}})| + |f_i(\hat{\mathbf{x}} | \theta_{fi}) + \sum_{j=1}^{n} [|g_{ij}^u(\hat{\mathbf{x}})| + |g_{ij}(\hat{\mathbf{x}} | \theta_{g_{ij}})|] \} \tag{14}$$

where $\xi = \min_{i=1,2,\dots,n} \{\lambda_{\min}(Q_i)\}$. Besides, when $V_i > \overline{V}$, the supervisory control u_{si} is added to force $V_i < \overline{V}$.

From the above discussion and [15], all signals in the closed-loop system are bounded, and $e_{oi} \to 0$ for i=1,2,...n as $t \to \infty$. The overall scheme of the proposed controller is shown in Fig. 1.

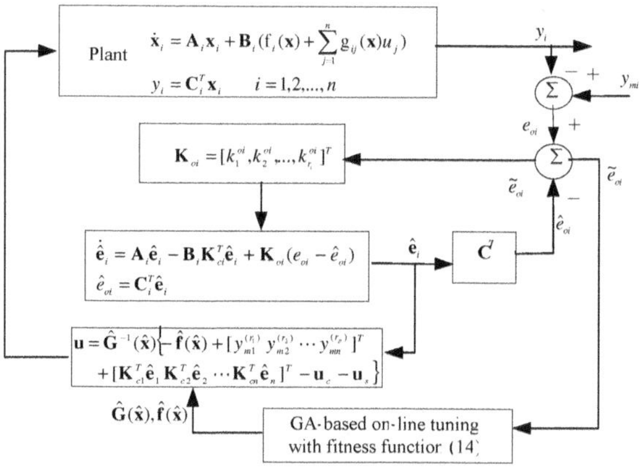

Fig. 1. The overall scheme of the proposed controller

3 Simulation Example

Consider the two-link robot for illustrating the proposed methods. The dynamic equations are given by

$$\mathbf{H}(\mathbf{q})\ddot{\mathbf{q}} + \mathbf{C}(\mathbf{q}, \dot{\mathbf{q}})\dot{\mathbf{q}} + \mathbf{g}(\mathbf{q}) = \mathbf{u}$$

where $\mathbf{H}(\mathbf{q}) = \begin{bmatrix} (m_1 + m_2)l_1^2 + m_2 l_2^2 + 2m_2 l_1 l_2 \cos q_2 & m_2 l_2^2 + m_2 l_1 l_2 \cos q_2 \\ m_2 l_2^2 + m_2 l_1 l_2 \cos q_2 & m_2 l_2^2 \end{bmatrix}$, (15)

$$\mathbf{C}(\mathbf{q}, \dot{\mathbf{q}}) = \begin{bmatrix} -m_2 l_1 l_2 \dot{q}_2 \sin q_2 & -m_2 l_1 l_2 (\dot{q}_1 + \dot{q}_2) \sin q_2 \\ m_2 l_1 l_2 \dot{q}_1 \sin q_2 & 0 \end{bmatrix}, \text{ and}$$

$$\mathbf{g}(\mathbf{q}) = [(m_1 + m_2) l_1 g_e \cos q_1 + m_2 l_2 g_e \cos(q_1 + q_2) \;\; m_2 l_2 g_e \cos(q_1 + q_2)]^T.$$

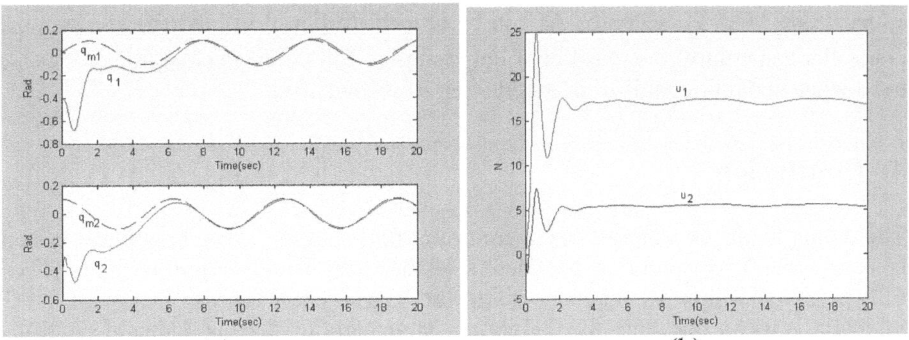

(a) (b)

Fig. 2. (a) The states $q_1(t)$ and $q_2(t)$ (solid line) and the reference outputs $q_{m1}(t)$ and $q_{m2}(t)$ (dashed line). (b) The control inputs u_1 and u_2.

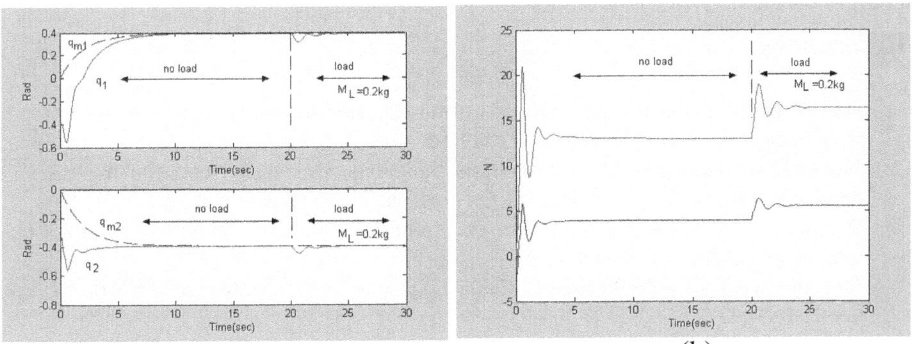

(a) (b)

Fig. 3. (a) The states $q_1(t)$ and $q_2(t)$ (solid line) and the reference outputs $q_{m1}(t)$ and $q_{m2}(t)$ (dashed line). (b) The control inputs u_1 and u_2.

The parameter values are $m_1 = 0.5kg$, $m_2 = 0.5kg$, $l_1 = 1m$, $l_2 = 0.8m$ and $g_e = 9.8 \, m/\mathrm{sec}^2$. The feedback and observer gain vectors are given as $\mathbf{K}_{ci} = [40040]^T$ and $\mathbf{K}_{oi} = [200\ 2000]^T$, respectively. The filter $L_i^{-1}(s)$ is given as $L_i^{-1}(s) = 1/(s+2)$. The initial states and estimation states of the system are assumed as $\mathbf{x}(0) = [-0.4, 0.4, -0.4, 0.4]^T$ and $\hat{\mathbf{x}}(0) = [0.1, 0.1, 0.1, 0.1]^T$, respectively. Two cases corresponding to two different desired trajectories are simulated. In case 1, our objective is to control the outputs q_1 and q_2 of the two-link robot to track the desired trajectories $q_{m1} = (\pi/30)\sin(t)$ and $q_{m2} = (\pi/30)\cos(t)$, respectively. As can be seen from Fig. 2 (a) and (b), the proposed controller can control the two-link robot system to follow the desired trajectories well. In case 2, our objective is to control the outputs q_1 and q_2 of the system to track the desired trajectories $q_{m1} = \pi[1 - \exp(-t/2)]/8$ and $q_{m2} = -\pi[1 - \exp(-t/2)]/8$, respectively. In addition, an additional load ($m_L = 0.2kg$) is added to m_2 after 20 seconds. As can be seen from Fig. 3 (a) and (b), the proposed controller can control the two-link robot system to follow the desired trajectories well even when an additional load is added after 20 seconds.

4 Conclusion

The output feedback adaptive fuzzy controller for a class of MIMO nonlinear systems is developed. The weighting parameters of the fuzzy-neural controller can successfully be tuned on-line via a genetic algorithm (GA). A new fitness function of the GA is obtained to on-line tune the weighting parameters of the fuzzy-neural controller. The proposed overall scheme guarantees that all signals involved are bounded and the outputs of the closed-loop system track the desired output trajectories. The robustness and applicability of the proposed control scheme are demonstrated by the simulation results.

References

1. Hornik, K., Stinchcombe, M., White, H.: Multilayer Feedforward Networks Are Universal Approximators. Neural Networks **2** (1989) 359-366
2. Wang, L.X., Mendel, J.M.: Fuzzy Basis Functions, Universal Approximation, and Orthogonal Least Squares Learning. IEEE Trans. on Neural Networks **3** (1992) 807-814
3. Wang, L.X.: Adaptive Fuzzy Systems and Control: Design and Stability Analysis. Englewood Cliffs, NJ: Prentice-Hall (1994)
4. Jamshidi M., Vadiee N., Ress T. J.: Fuzzy Logic and Control. Englewood Cliffs, NJ: Prentice-Hall (1993)
5. Polycarpou, M., Ioannou, P.A.: Modelling, Identification and Stable Adaptive control of Continuous-Time Nonlinear Dynamical Systems Using Neural Networks. Proc. American Control Conf (1992) 36-40
6. Kosmatopoulos, E.B., Ioannou, P.A., Christodoulou, M.A.: Identification of nonlinear systems using new dynamic neural network structures. In Proc. IEEE Conf. Decision and Control (1992) 20-25
7. Rovithakis, G.A., Christodoulou, M.A.: Adaptive control of unknown plants using dynamical neural networks. IEEE Trans. Syst. Man, Cyber. 24 (1995) 400-411
8. Chen, F.C., Khalil, H.K.: Adaptive control of nonlinear systems using neural networks. Int.J.Contr. **55** (1992) 1299-1317
9. Sanner, R.M., Slontine, J.J.E.: Guassian networks for direct adaptive control. IEEE Trans. on Neural Networks **3** (1992) 837-863
10. Kim, Y.H., Lewis, F.L., Abdallah, C.T.: A dynamic recurrent neural-network-based adaptive observer for a class of nonlinear systems. Automatica **33** (1997) 1539-1543
11. Zhang, T., Ge, S.S., Hang, C.C.: Adaptive output feedback control for general nonlinear systems using multilayer neural networks. Proceedings of the 1998 American Control Conference 520-524
12. Horikawa, S., Furuhashi, T., Uchikawa, Y.: On fuzzy modeling using fuzzy neural networks with the back-propagation algorithm. IEEE Trans. on Neural Networks, **3** (1992)
13. Lin, C.T., Lee, C. S. G.: Neural-network-based fuzzy logic control and decision system. IEEE Trans. on Computer **40** (1991) 1320-1336

14. Wang, C.H., Wang, W.Y., Lee, T.T., Tseng P.S.: Fuzzy B-spline membership function and its Applications in fuzzy-neural control. IEEE Trans. Syst. Man, Cyber **25** (1995) 841-851
15. Wang, W.Y., Leu, Y.G., Lee, T.T.:Output-feedback control of nonlinear systems using direct adaptive fuzzy-neural controller. Fuzzy Sets and Systems **140** (2003) 341-358
16. Leu, Y.G. Lee, T.T. Wang, W.Y.: Observer-based Adaptive Fuzzy-Neural Control for Unknown Nonlinear Dynamical Systems. IEEE Trans. Syst. Man, Cyber.-Part B: Cybernetics **29** (1999)
17. Hwang, M. C., Hu, X.: A Robust Position/Force Learning Controller of Manipulators via Nonlinear H∞ Control and Neural Networks. IEEE Trans. Syst. Man, Cyber.-Part B: Cybernetics **30** (2000) 310-321
18. Yuan, Y., Zhuang H.: A genetic algorithm for generating fuzzy classification rules, Fuzzy Sets and Systems **84** (1996) 1-19
19. Seng, T.L. Khalid, M.B. Yusof, R.: Tuning of a neuro-fuzzy controller by genetic algorithm, IEEE Trans. Syst. Man, Cyber. Part B **29** (1999) 226-236
20. Wang, W.Y., Li, Y.H.: Evolutionary learning of BMF fuzzy-neural networks using a reduced-form genetic algorithm. IEEE Trans. Syst. Man, Cyber.-Part B: Cybernetics **33** (1999) 966-976
21. Wang, W.Y., Cheng C.Y., Leu, Y.G.: An online GA-based output-feedback direct adaptive fuzzy-neural controller for uncertain nonlinear systems. IEEE Trans. Syst. Man, Cyber.-Part B: Cybernetics **34** (2004)
22. Tsakalis, K. S., Ioannou, P. A.: Linear time-varying systems. Englewood Cliffs, NJ: Prentice-Hall (1993)

Filtered-X Adaptive Neuro-Fuzzy Inference Systems for Nonlinear Active Noise Control

Riyanto T. Bambang

School of Electrical Engineering and Informatics
Bandung Institute of Technology, Jalan Ganesha 10, Bandung 40132, Indonesia
briyanto@lskk.ee.itb.ac.id

Abstract. A new method for active noise control is proposed and experimentally demonstrated. The method is based on Adaptive Neuro-Fuzzy Inference Systems (ANFIS), which is introduced to overcome nonlinearity inherent in active noise control. A new algorithm referred to as Filtered-X ANFIS algorithm suitable for active noise control is proposed. Real-time experiment of Filtered-X ANFIS is performed using floating point Texas Instruments C6701 DSP. In contrast to previous work on ANC using computational intelligence approaches which concentrate on single channel and off-line adaptation, this research addresses multichannel and employs online adaptation, which is feasible due to the computing power of the DSP.

1 Introduction

Noise is unwanted or unpleasant sound that needs to be attenuated. Basically, there are two approaches in noise control : active methods and passive methods. Active noise control (ANC) methods have recently attracted much attention from engineers and scientists[3-5,7-11,13,14]. This is due to the fact that it offers advantages in terms of bulk and expenditure over the conventional utilization of passive dampers for attenuating low frequency acoustic noise. The ANC typically employs linear transversal filter algorithm for both identification and control, such as the well-known FX-LMS[4,8]. However, due to nonlinear nature of ANC, such linear filter is not effective in attenuating the noise. Adaptive-Neuro Fuzzy Inference Systems (ANFIS)[1,12] is one of nonlinear adaptive structures that is widely employed in modeling and control. In this paper, the nonlinear mapping capability of ANFIS, together with its structured knowledge representation are employed to model secondary acoustic path and to implement adaptive nonlinear controller for ANC. Back Propagation algorithm [2] is employed to adaptively tune the ANFIS parameters for ANC control task, taking into account tapped delay lines inserted into the ANC control structure. This results in a new algorithm referred to as Filtered-X ANFIS. In this paper, ANFIS is chosen as a system modeling and control mechanism because conventional approaches of system modeling and control perform poorly in dealing with complex and uncertain systems, such as acoustic noise. In acoustic noise environment, the parameters of the acoustic system may change significantly, particularly due to the variations in air temperature, geometry, noise characteristics, and moving noise sources.

D. Liu et al. (Eds.): ISNN 2007, Part I, LNCS 4491, pp. 54–63, 2007.

In this paper, design and implementation of a multi-channel active noise control system employing Filtered-X ANFIS on a Texas Instruments C6701 DSP are addressed. In contrast to previous work on ANC using computational intelligence approaches which concentrate on single channel and off-line adaptation, this paper addresses multichannel and employs online adaptation, which are possible because of the computing power of the DSP.

2 ANFIS Structure

ANFIS is a class of adaptive networks constructed from multilayer feedforward networks which each node performs a particular function (node function) based on incoming signals and a set of parameters pertaining to this node[1,12]. In ANFIS, each node at layer 1 (membership) represents a fuzzy set membership function. Parameters of the membership function are tuned with back propagation in the learning process based on a given training data set. Each node at layer 2 (conjunction) multiplies the incoming signals and sends the product out. The output signal corresponds to the firing strength of a fuzzy rule. The *i*th node at layer 3 (normalization) calculates the ratio of the ith rule's firing strength to the sum of the firing strengths of all the rules, i.e. the relative portion of the *i*th rule in the final result. A node at layer 4 calculates a linear combination of input signals, and multiplies the result with the weight coming from layer 3. During the learning process coefficients of the linear combination are adjusted using a particular learning method to minimize the mean square error between the calculated output and the desired output. Finally the node at layer 5 (summation) produces the weighted sum of the output signals coming from invoked rules.

As one of adaptive network structure, ANFIS is constructed by using fuzzy inference system [1,12]. In the following, we address ANFIS architecture based on Sugeno fuzzy model. For simplicity, assume that fuzzy inference system has two input x and y and one output z. Using first order Sugeno fuzzy model, fuzzy if-then rules are given as follows :

Rule 1 : if u_1 is A_1 and u_3 is A_3, then $f_1 = q_{11}u_1 + q_{12}u_2 + q_{13}$

Rule 2 : if u_2 is A_2 and u_4 is A_4, then $f_2 = q_{21}u_1 + q_{22}u_2 + q_{23}$

From the above rules, the corresponding adaptive networks is constructed as in Figure 1. Denoting $O_{l,i}$ as the output of each layer, computation of each layer is perform as follows :

- **Layer 1**

  Each node in this layer is adaptive and its output is the value of membership function of its input

$$O_{1,i} = \mu A_i(u_1),\ \text{for } i = 1, 2, \quad (1)$$

$$O_{1,i} = \mu A_i(u_2),\ \text{for } i = 3, 4. \quad (2)$$

Parameter of membership function is adaptive, and is called premise parameter.

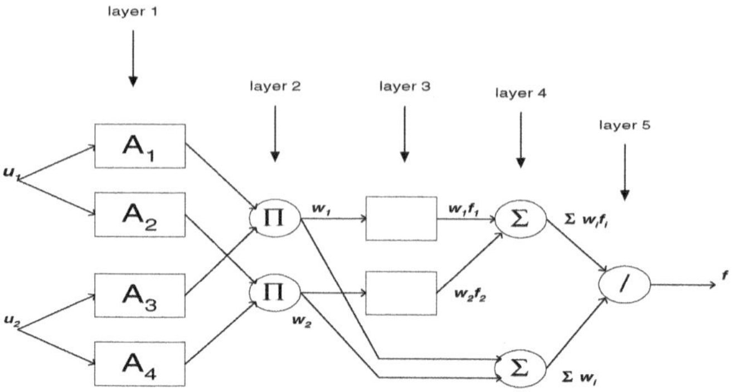

Fig. 1. ANFIS Architecture Based on Sugeno Fuzzy Model

- **Layer 2**

  Each node in this layer performs *T-norm* operation (such as product operation) of its input. Its output is the result of this operation.

$$O_{2,i} = w_i = \mu A_i(x) \times \mu A_{(i+2)}(x) \quad i = 1,2. \tag{3}$$

  Output of each node represents firing strength of the associated fuzzy rule.

- **Layer 3**

  Each node in this layer adaptive. Output of node in this layer is the result of rule inference in fuzzy system.

$$O_{3,i} = w_i f_i = w_i(q_{i1}u_1 + q_{i2}u_2 + q_{i3}), \qquad i = 1,2, \tag{4}$$

  $\{q_{i1}, q_{i2}, q_{i3}\}$ are adaptive parameters, called consequence parameter.

- **Layer 4**

  This layer performs summation.

$$O_{4,1} = \alpha = \textstyle\sum_i w_i f_i, \qquad \text{i = 1,2,} \tag{5}$$

$$O_{4,2} = \beta = \textstyle\sum_i w_i, \qquad \text{i = 1,2,} \tag{6}$$

- **Layer 5**

  The last layer is output layer which computes final result

$$O_5 = f = \frac{\alpha}{\beta} \tag{7}$$

ANFIS learns form environments by adjusting its parameters[1,12]. One of the learning method is based on gradient descent which minimizes cost function

$$\hat{E} = \frac{1}{2}e^2(n)$$

where $e(n) = d(n) - f(n)$. Signal $e(n)$ is error at n^{th} iteration, i.e. the difference between the desired target . $(d(n))$ and actual output of ANFIS $f(n)$. During the learning process ANFIS parameter is adjusted according to

$$parameter(n+1) = parameter(n) - \Delta parameter(n)$$

$$parameter = \{p_{ij}, q_{ij}\}$$

$$\Delta p_{ij}(n) = \eta \frac{\partial \hat{E}}{\partial p_{ij}} = -\eta e(n) \frac{\partial f(n)}{\partial p_{ij}}, \tag{8}$$

$$\Delta q_{ij}(n) = \eta \frac{\partial \hat{E}}{\partial q_{ij}} = -\eta e(n) \frac{\partial f(n)}{\partial q_{ij}}, \tag{9}$$

where η is networks learning rate.

3 Filtered-X ANFIS for Active Noise Control

In an active noise control ANFIS can be used as controller and as secondary path model. Motivated by [3,15] for ANC using neural networks, the structure of ANC based on ANFIS is shown in Figure 2. To capture dynamic behavior of the controller as well as the model, tapped delay lines are introduced before signals are applied into each ANFIS input. The ANFIS model is constructed by using input output information of the secondary path of ANC. This process can be performed off-line. While learning mechanism of model ANFIS is straightforward, the standard gradient descent of ANFIS learning mechanism can not be employed directly due to the tapped delay lines between the controller and the model. Motivated by conventional Filtered-X LMS algorithm in linear ANC problem, the modified ANFIS learning mechanism for controller is based on the fact that instantaneous quadratic error depends at time n depends on controller ANFIS parameter at the last $L + 1$ samples (parameter values at time n, $n - 1$, ..., $n - L$), where L is the number of delay in tapped delay lines. Therefore, instantaneous gradient is given by [3,11,13,15]

$$\frac{1}{2}\sum_{i=0}^{L} \frac{\partial e^2(n)}{\partial parameter_k(n-i)} \tag{10}$$

where

$e^2(n)$ = instantaneous square error

$parameter_k(n-i)$ = values of ANFIS parameter $\{p_{ij}, q_{ij}\}$ at time $(n-i)$.

The resulting algorithm is called FX-ANFIS to represent that it is filtered-X version of standard ANFIS learning algorithm. FX-ANFIS is computed as follows : First, instantaneous gradient is rewritten as

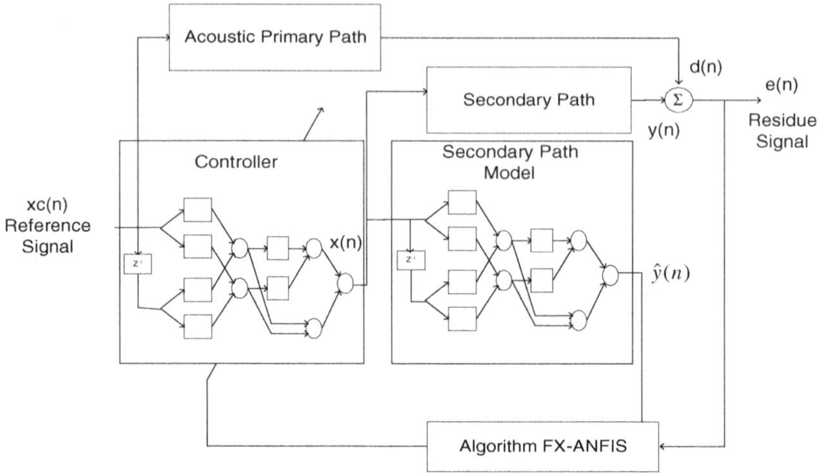

Fig. 2. Block Diagram of ANC Using ANFIS

$$\Delta = \frac{1}{2}\sum_{i=0}^{L} \frac{\partial e^2(n)}{\partial parameter(n-i)} = \sum_{i=0}^{L} \Delta(n-i) \tag{11}$$

with $\Delta(n-i) = \frac{1}{2}\frac{\partial e^2(n)}{\partial parameter(n-i)}$. Using derivative chain rule, it follows that

$$\Delta(n-i) = e(n).\frac{\partial e(n)}{\partial x(n-i)}\frac{\partial x(n-i)}{\partial parameter(n-i)} \tag{12}$$

We find that

$$\frac{\partial \hat{y}(n)}{\partial x(n-i)} = \sum_{j \in P_{(i+1)}} \left(\sum_{k \in M_j} \left(\left(\frac{f_k(n)}{\beta(n)} - \frac{\alpha(n)}{\beta^2(n)} \right) \cdot \varphi(x(n-i)) \cdot \prod_{l \in N_k} \mu A_l(n) \right) \right) \tag{13}$$

The error signal is given by

$$e(n) = d(n) + y(n)$$

where $y(n)$ is output of secondary path, and $d(n)$ is noise response of primary path. Then, by computing $\frac{\partial x(n-i)}{\partial parameter(n-i)}$ and using equation (12), it follows that

$$\Delta(n-i) = e(n) \sum_{j \in P_{(i+1)}} \left(\sum_{k \in M_j} \left(\left(\frac{f_k(n)}{\beta(n)} - \frac{\alpha(n)}{\beta^2(n)} \right) \cdot \varphi(x(n-i)) \cdot \prod_{l \in N_k} \mu A_l(n) \right) \right) \frac{\partial x(n-i)}{\partial parameter \quad (n-i)} \tag{14}$$

Detailed derivation is omitted for lack of space.

4 Experiment Results

ANFIS structure and its learning algorithm is implemented in real-time using DSP. A series of experiments are performed in ANC in free (open) space, to include single channel and multichannel ANC, varying membership function, ANC for saturated signal, etc. For lack of space, however, we only present results of some of the experiments. All experiments are performed with sampling rate of 2030 Hz.

Block diagram of identification process is shown in Figure 3. The goal is to minimize the error. Number of ANFIS input is 3 and for each input fuzzification process is carried out through 2 fuzzy partition. Two types of membership function are employed : triangular and trapezoidal. Thus, in layer 1 there are 6 nodes. Number of consequence parameters in layer 1 depends on type of membership function employed. The number of nodes in layer 2 and layer 3 corresponds to the number of fuzzy rules contained in the fuzzy inference system. For a fuzzy inference system with 3 input and 2 partition, maximum number of rules are 8. To reduce DSP computational load, however, we use only 4 rules, resulting in 16 consequence parameters.

Figure 4 shows error of ANFIS identification using triangular membership function. Using η=0.3 , steady state error is achieved within 0.5 second. This value of learning rate is used as it gives smallest MSE. Figure 5 shows identification error when trapezoidal membership function is employed.

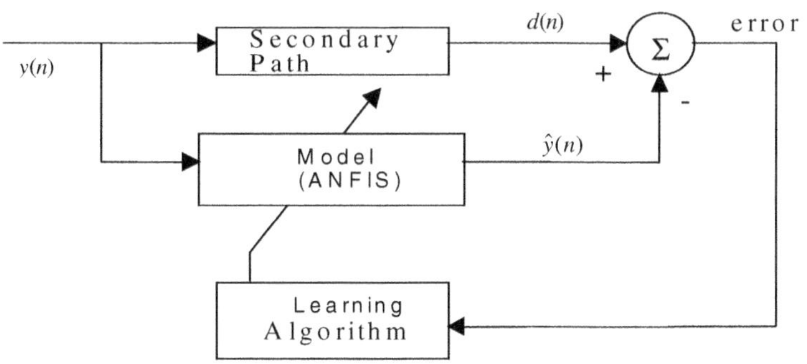

Fig. 3. Identification Using ANFIS

The performance of identification process is also measured using SNR (Signal to Noise Ratio) expressed by $SNR = 10\log(\mathrm{var}(d(n))) - 10\log(\mathrm{var}(e(n)))$.

Table 1 shows performance of ANFIS identification results in terms of MSE, SNR and transient. Observe that the resulting MSE is quite small, while SNR is sufficiently large, indicating that ANFIS identification performs well.

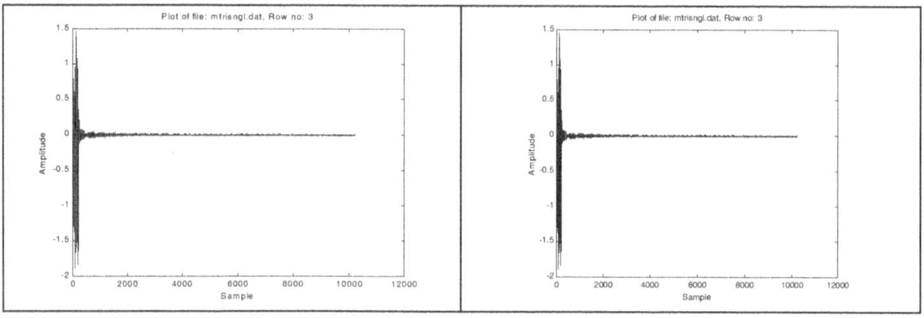

Fig. 4. ANFIS Identification Error with Triangular (left) and Trapezoidal (right) Membership Functions

Table 1. Single Channel Identification

MF	Excitation Signal	Number of Input	MSE	SNR(dB)	Transient (samples)
Triangular I	170 Hz	3	1.0724E-05	39.5593	1000
Triangular II	170 Hz	3	1.8804E-05	37.1261	600
Trapezoidal I	170 Hz	3	1.2200E-05	39.0159	1000
Trapezoidal II	170 Hz	3	1.9654E-06	46.8637	1000

Once secondary path model has been obtained, ANC is ready to be implemented. Figure 5 shows noise signal residue, measured by error microphone, and its power spectral density. The noise residue reaches steady state in relatively short time (less than 1.5 second), and the resulting MSE is 6.7595E-05. Attenuation level of 25 dB is obtained at main noise frequency (170 Hz).

Results of single channel ANC with different membership function and noise frequency are shown in Table 2. Total attenuation is computed through

$$Total\_attenuation = 10\log(\mathrm{var}(residue1)) - 10\log(\mathrm{var}(residue2))$$

where *residue1* is noise measured without ANC, and *residue2* is noise measured with ANC. Note from this table that attenuation level obtained for noise frequency of 210 Hz is worst than that achieved with noise frequency of 170 Hz.

In subsequent discussion, we present experiment results of decentralized multichannel ANC. The similar learning algorithm as single channel ANC is adopted in this configuration (Figure 6), that is the standard gradient descent ANFIS learning algorithm is employed to model the secondary path, while Filtered-X ANFIS algorithm is applied to controller ANFIS structure. Note that coupling between channel is ignored in decentralized configuration. The purpose is to reduce computational load of the DSP. The experiment setup is shown in Figure 7.

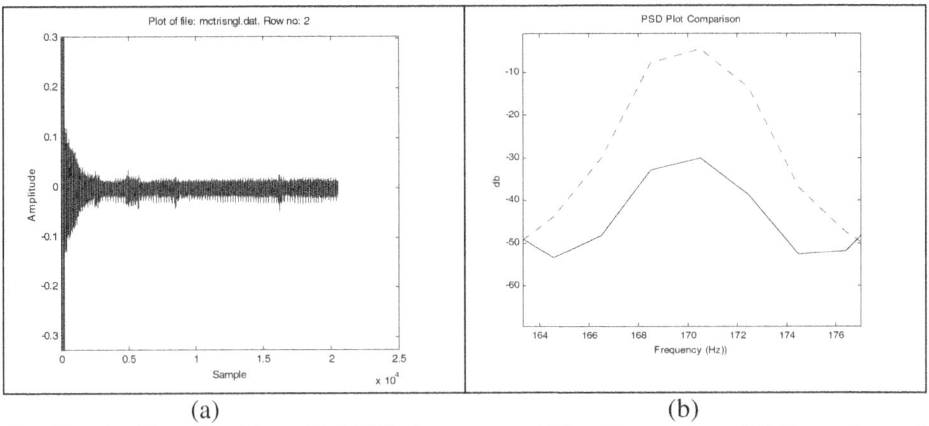

Fig. 5. Noise Signal Residue with 170Hz frequency : a) Time Response, and b) Power Spectral Density (dashed-line : without ANC, solid line : with ANC)

Table 2. Results of Filtered-X ANFIS with Different Membership Function and Frequency

MF	Excitation Signal	Number of Input	MSE	Attenuation at main frequency (dB)	Total Attenuation (dB)	Transient (samples)
Triangular I	170 Hz	3	6.7595E-05	25.5752	12.1524	3000
Triangular II	170 Hz	3	4.2006E-05	25.1602	13.9581	2000
Trapezoidal I	170 Hz	3	5.4084E-05	35.3972	12.8517	4000
Trapezoidal II	170 Hz	3	5.8623E-05	30.7572	12.5058	2000
Triangular I	210 Hz	3	3.8692E-05	16.5243	11.6602	18000
Trapezoidal I	210 Hz	3	2.5370E-05	13.7293	13.7293	16000

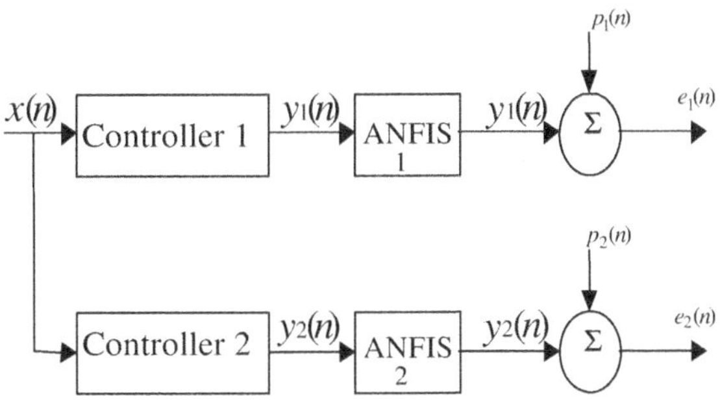

Fig. 6. Decentralized Two Channel ANC

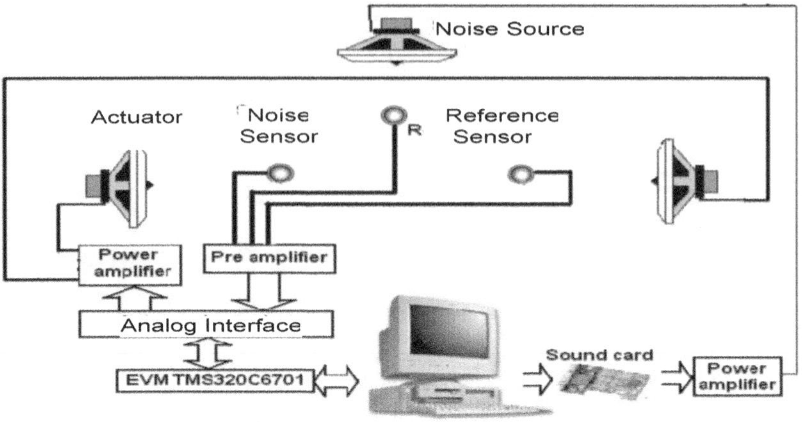

Fig. 7. Decentralized Experiment Setup

The results are shown in Figure 8 for channel 2. The ANC performance is shown in Table 3. While the resulting attenuation level is quite large, the performance of channel 1 is worst than that of channel 2. This could be caused by asymmetry of the geometry of multichannel ANC and by the difference of microphone sensitivity. We also performed ANC experiment by incorporating nonlinearity. We found that the performance of ANFIS is better than the conventional FX-LMS. This is due to the fact that ANFIS can perform arbitrary nonlinear mapping for both controller and the model [1,12].

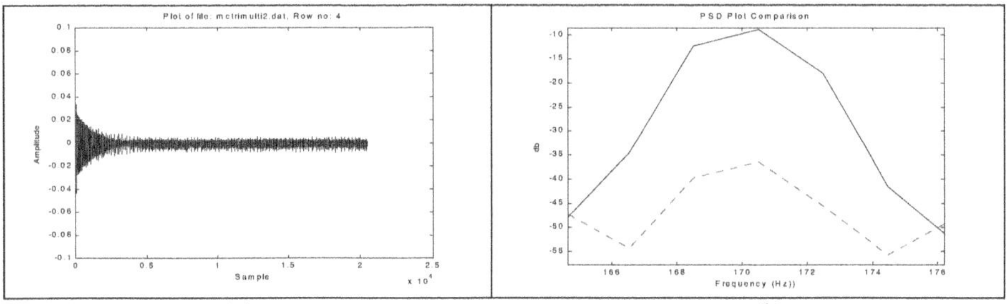

(a) (b)

Fig. 8. Noise Signal Residue Measured by Microphone 2 with 170Hz frequency : a) Time Response, and b) Power Spectral Density (dashed-line : without ANC, solid line : with ANC)

Table 3. Performance of Multichannel ANC

Channel	Excitation Signal	Number of Input	MSE	Attenuation at noise main freq. (dB)	Transient (samples)
1	170 Hz	3	3.6879E-05	21	10000
2	170 Hz	3	4.8874E-06	27.5	5000

5 Conclusion

In this paper a new method for ANC based on neuro-fuzzy inference system was proposed and experimentally demonstrated. Filtered-X ANFIS learning algorithm was developed to cope with nonlinear phenomena arising in ANC. The experiment was performed in ANC in free space where ANFIS was implemented on DSP. The results show that ANFIS is a viable alternative to FX-BP proposed by Bouchard [3] and provides better performance than the conventional FX-LMS algorithm.

References

1. Jang, J.S.R., Sun, C.T.: Neuro Fuzzy Modeling and Control, Proceedings of the IEEE **83** **(3)** (1995)
2. Haykin, S.: Neural Networks: A Comprehensive Foundation, Macmillan College Publishing Company, Inc., New York (1998)
3. Bouchard, M., Paillard, B., Le Dinh, C.T.: Improved Training of Neural Networks for the Nonlinear Active Control of Sound and Vibration, IEEE Trans. on Neural Networks **10** **(2)** (1999) 391-401
4. Elliot, S.J.: Down with Noise, Proceedings of the IEEE (1999)
5. Elliot, S.J., Nelson P.A.: Active Noise Control, IEEE Signal Processing Magazine **10** **(4)** (1993) 12-35
6. Haykin, S.: Adaptive Filter Theory, Englewood Cliffs, NJ: Prentice-Hall (1997)
7. Hong J. et al.: Modeling, identification, and feedback control of noise in an acoustic duct, IEEE Transactions on Control Systems Technology (1996) 283-291
8. Kuo, M.S., Morgan, D.R.: Active Noise Control Systems: Algorithms and DSP Implementations, New York: John Wiley & Sons, Inc. (1996)
9. Bambang, R.: Decentralized Active Noise Control Using U-Filtered Algorithm: An Experimental Study, International Conf. On Modeling, Identification and Control, Innsbruck, Austria (2000)
10. Bambang, R.: On-Line Secondary Path Identification of Active Noise Control Using Neural Networks, Int. Conf. Modeling and Simulation, Pittsburgh , USA (2000)
11. Bambang, R., Uchida, K., Jayawardana, B.: Active Noise Control in 3D Space Using Recurrent Neural Networks, International Congress and Exposition on Noise Control Engineering, Korea (2003)
12. Azeem, M.Z. et al. : Generalization of Adaptive Neuro-Fuzzy Inference Systems, IEEE Trans. Neural Networks **11** **(6)** (2000)
13. Bambang, R., Anggono, L., Uchida, K.: DSP Based Modeling and Control for Active Noise Cancellation Using Radial Basis Function Networks, IEEE Symposium on Intelligent Systems and Control, Vancouver, Canada (2002).
14. Bambang, R., Yacoub, R., Uchida, K.: Identification of Secondary Path in ANC Using Diagonal Recurrent Neural Networks with EKF Algorithm, Proc. 5th Asian Control Conference, Melbourne (2004)
15. Bouchard, M.: New Recursive-Least-Squares Algorithms for Nonlinear Active Control of Sound and Vibration using Neural Networks, IEEE Trans. Neural Networks **12** (2001) 135-147

Neural Network Based Multiple Model Adaptive Predictive Control for Teleoperation System

Qihong Chen[1], Jin Quan[2], and Jianjun Xia[1]

[1]School of Automation, Wuhan University of Technology, Wuhan 430070, China
chenqh@whut.edu.cn
[2]School of Electronics and Information Engineering, Tongji University, Shanghai 200092, China

Abstract. Environment model and communication time delays of a teleoperation system are variant usually, which will induce bad performance, even instability of the system. In this paper, neural network based multiple model adaptive predictive control method is proposed to solve this problem. The whole control system is composed of predictive controller and decision controller. First of all, neural network model set of any possible environment is built up, and time forward state observer based predictive controllers are designed for all models. In succession, decision controller is designed to adaptive switch among all predictive controllers according to performance target. This method can ensure stability and performance of the system. Finally, simulation results show effectiveness of the proposed method.

1 Introduction

Because of variable environment dynamics and time delay between the master and the slave in an Internet based teleoperation, it's difficult to control such a system. Anderson et al. in [1] presented the ideal response for a time-delayed master-slave system by scattering theory. Casavola et al. in [2] presented a predictive strategy. The significance of the method is that stability is preserved and no constraints violation occurs regardless of any possible time delay. Guan et al. in [3] investigated a class of hybrid multi-rate control models with time-delay and switching controllers. Shahdi et al. proposed a multiple model adaptive control scheme for bilateral teleoperation in unknown environments in [4]. But the method didn't address the issue of time delay. Smith et al. used two neural networks to predictive dynamics of the environment in [5]. Nevertheless, when environment model varies rapidly, environment dynamics can't be predicted exactly. Brady et al. in [6] described the time-variant nature of the delay and developed a time forward observer for supervisory control over the Internet. Narendra et al. in [7] presented a general methodology for such adaptive control using multiple models, switching, and tuning. However, all of the mentioned methods can't be directly used to teleoperation systems with time-variant delay and environment.

In order to solve the problem, this paper proposes a neural network based multiple model adaptive predictive control method. The adaptive predictive

D. Liu et al. (Eds.): ISNN 2007, Part I, LNCS 4491, pp. 64–69, 2007.

controller consists of some predictive controllers and a decision controller. In the following sections, neural network model base of any possible environment is built up firstly, then time forward observer based predictive controllers are designed according to every model. In succession, decision controller is designed. Finally, simulation results are given to show effectiveness of the proposed method.

2 System Model

A teloperation system can be represented by five subsystems: the human operator, the master, the communication channel, the slave and environment,as shown in Fig. 1.

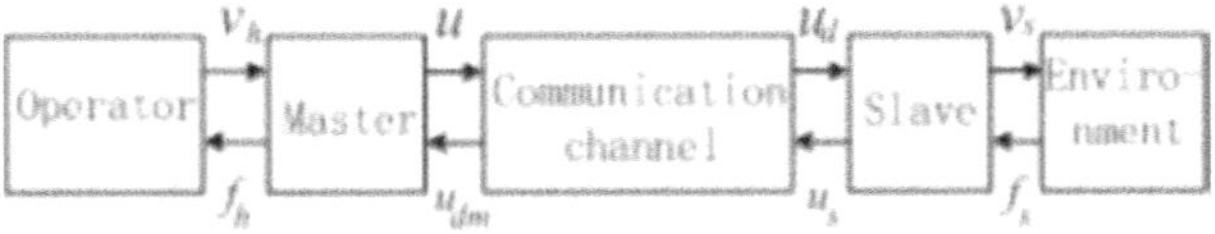

Fig. 1. Teleoperation system configuration

Under the assumption that each degree of freedom (DOF) has linear decoupled from the other, the analysis and design will be focused on a one DOF linear system hereafter. In this paper, the communication channel is Internet. There exist random time delay between the master and the slave. Let forward and backward time delays are $T_r(t)$ and $T_l(t)$, sampling period is T_c.Then

$$d_r(k) = T_r(k)/T_c \ , \ d_l(k) = T_l(k)/T$$

$d_r(k)$, $d_l(k)$ is simplified as d_r , d_l hereafter. Writing out the discrete equations of the master and slave yields

$$x_m(k+1) = a_{m1}x_m(k) + a_{m2}x_m(k-1) + b_{m1}f_h(k) - b_{m1}u_{dm}(k) \ , \qquad (1)$$

$$x_s(k+1) = a_{s1}x_s(k) + a_{s2}x_s(k-1) + b_{s1}u_d(k) - b_{s1}f_e(k) \ , \qquad (2)$$

where subscript m denotes the master and s denotes the slave. x represents position. a_{mi}, a_{si}, b_{m1}, $b_{s1}(i = 1, 2)$ are model parameters. f_h, f_e are operator force and environment force. u_{dm}, u_d are control signal acting on the master and slave. Similarly the environment model can be described as the following equation

$$f_e(k) = a_{e0}x_s(k+1) + a_{e1}x_s(k) + a_{e2}x_s(k-1) \ , \qquad (3)$$

where $a_{mi}(i = 0, 1, 2)$ are the model parameters. The parameters are usually unknown and time-variant. It is assumed that the dynamics of the environment are governed by a model from a finite set of environment models at any given time.

Master controller is designed though predicted states as the follow

$$\begin{aligned} u_{dm}(k) = {} & f_{11}\, x_m(k) + f_{12}x_m(k-1) + f_{13}\hat{x}_s(k+d) + \\ & f_{14}\, \hat{x}_s(k+d-1) + c_1 f_h(k) + c_2\hat{f}_e(k+d_r) \ , \end{aligned} \qquad (4)$$

where $F_1 \triangleq \begin{pmatrix} f_{11} & f_{12} & f_{13} & f_{14} \end{pmatrix}$, c_1, c_2 are feedback coefficients. $\hat{x}_s(k+d_r)$, $\hat{x}_s(k+d_r-1)$, and $\hat{f}_e(k+d_r)$ represent predictive values of $x_s(k+d_r)$, $x_s(k+d_r-1)$, $f_e(k+d_r)$. Slave controller is designed as

$$\begin{aligned} u_d(k) = f_{21}\, x_m(k-d_r) + f_{22} x_m(k-d_r-1) + f_{23} x_s(k) + \\ f_{24}\, x_s(k-1) + c_3 f_h(k-d_r) + c_4 f_e(k) \,, \end{aligned} \tag{5}$$

where $F_2 \triangleq \begin{pmatrix} f_{21} & f_{22} & f_{23} & f_{24} \end{pmatrix}$, c_3, c_4 are control parameters to be designed. In this way, it seems that no time delay exists for the master. Only the slave can feel about forward time delay.

3 Neural Network Based Multiple Model Adaptive Predictive Control

3.1 Neural Network Model Base

Environment dynamics of a teleoperation system is usually unknown and time variant, while predicting states of the slave needs environment model. Neural network is an effective modeling method. Radial basis function network is a typical local approximation neural Network. It's easy to approximate local performance of a function and the training speed is fast. Therefore, RBF network is used to model environment dynamics in this system. Environment model base is built up though store model parameters of all environments at any time.

3.2 Slave State Predict

In the slave station, environment model which fits best to current environment dynamics is selected and sent to the master. The master controller uses the received environment model parameters and the slave model to predict the slave state. In this way time delay won't effect on modeling and prediction precision.

If the current environment model parameters are a_{e0}, a_{e1}, a_{e2} the force between slave and environment is denoted as

$$f_e(k) = a_{e0} x_s(k+1) + a_{e1} x_s(k) + a_{e2} x_s(k-1) \,. \tag{6}$$

To ensure robustness of environment parameters, (6) can be rewritten as the uncertainty expression

$$f_e(k) = (a_{e0} + \Delta a_{e0}) x_s(k+1) + (a_{e1} + \Delta a_{e1}) x_s(k) + (a_{e2} + \Delta a_{e2}) x_s(k-1) \,, \tag{7}$$

where Δa_{e0}, Δa_{e1}, Δa_{e2} are corresponding uncertainty parameters.

For the sake of predicting the slave state, substituting (7) to the slave model yields the following state equation

$$\begin{aligned} \bar{x}_s(k+1) &= (A_s + b_s F_{22} + \Delta A_s(k)) \bar{x}_s(k) + \\ &\quad (b_s + \Delta b_s(k))(F_{21} \bar{x}_m(k-d_r) + c_3 f_h(k-d_r)) \\ y_s(k) &= c_s \bar{x}_s(k) \,, \end{aligned} \tag{8}$$

where $\bar{x}_s(k) = \begin{pmatrix} x_s(k) & x_s(k-1) \end{pmatrix}^T$, $\bar{x}_m(k) = \begin{pmatrix} x_m(k) & x_m(k-1) \end{pmatrix}^T$, A_s, b_s, c_s are parameter matrices obtained by slave-manipulator and environment model, $\Delta A_s(k)$, $\Delta b_s(k)$ are uncertain parameters with corresponding dimensions.

In order to predict $\bar{x}_s(k+d_r)$, $f_e(k+d_r)$, state equation of the slave at $k+d_r$ should be achieved. Shifting (8) into the future by d_r times period, and state observer is used to predict $\bar{x}_s(k+d_r)$ as

$$\begin{aligned} z(k+1) &= (A_s + b_s F_{22})z(k) + b_s(F_{21}\bar{x}_m(k) + c_3 f_h(k)) + \\ &\quad L(y_s(k-d_l) - \bar{y}_s(k-d_r-d_l)) \\ \bar{y}_s(k) &= c_s z(k) , \end{aligned} \tag{9}$$

where L is the observer gain. Let observing error $e(k) = \bar{x}_s(k+d_r) - z(k)$, then

$$\begin{aligned} e(k+1) = (\ & A_s + b_s F_{22})e(k) + \Delta A_s(k)\bar{x}_s((k+d_r) + \Delta b_s(F_{21}\bar{x}_m(k) + c_3 f_h(k)) \\ & - L(y_s(k-d_l) - \bar{y}_s(k-d_r-d_l)) . \end{aligned} \tag{10}$$

3.3 Stability Analysis

State equation of the whole system is depicted as

$$\begin{aligned} \begin{pmatrix} \tilde{x}(k+1) \\ e(k+1) \end{pmatrix} &= \begin{pmatrix} \tilde{A} + \tilde{B}_1 F + \Delta\tilde{A}(k) & \tilde{B}_2 F + \Delta\tilde{B}_2(k) \\ \Delta E(k) & A_s + b_s F_{22} \end{pmatrix} \begin{pmatrix} \tilde{x}(k) \\ e(k) \end{pmatrix} \\ &\quad + \begin{pmatrix} B_3 + \Delta B_3(k) \\ L c_s \end{pmatrix} e(k-d) + (B_4 + \Delta B_4(k)) f_h(k) \\ y(k) &= \tilde{C}\tilde{x}(k) , \end{aligned} \tag{11}$$

where $\tilde{x}^T(k) = \begin{pmatrix} \bar{x}_m^T(k) & \bar{x}_s^T(k+d_r) \end{pmatrix}$, $d = d_r + d_l$, (11) can be simplified as

$$\begin{aligned} x(k+1) &= (A + BF + \Delta A(k))x(k) + (B_1 + \Delta B_1(k))x(k-d(k)) + \\ &\quad (B_2 + \Delta B_2(k)) f_h(k) \\ y(k) &= Cx(k) , \end{aligned} \tag{12}$$

where $x^T(k) = \begin{pmatrix} \tilde{x}^T(k) & e^T(k) \end{pmatrix}$, uncertainty parameters meet

$$\begin{pmatrix} \Delta A(k) & \Delta B_1(k) & \Delta B_2(k) \end{pmatrix} = DF(k) \begin{pmatrix} E & E_1 & E_2 \end{pmatrix} , \quad F^T(k)F(k) \le I$$

Theorems 1. If there exist the matrices $P_1 > 0$, P_2, P_3, $S_1 > 0$, and feedback coefficients c_1, c_2, c_3, c_4, F, such that (13) holds, then, teleoperation system (12) is robustly asymptotically stable under the control law (4) and (5).

$$\begin{pmatrix} \phi & \Omega & (P_2 B_1 - M_1)/2 & P_2 D & 0 \\ * & P_1 + d(k)S_1 + d(k)W_3 - 2P_3 & (PB_1 - M_2)/2 & 0 & P_3 D \\ * & * & E_1^T E_1 - S_2 & 0 & 0 \\ * & * & * & -I/3 & 0 \\ * & * & * & * & -I/3 \end{pmatrix} < 0 , \tag{13}$$

where $\phi = d(k)W_1 + M_1 - P_2B_1 + P_2(A+BF+B_1-I) + (A+BF+B_1-I)^T P_2^T + S_2$, $\Omega = d(k)W_2 + (M_2 - P_3B_1)^T/2 + P_1 - P_2 + P_3(A+BF+B_1-I)$, $\begin{pmatrix} W & M \\ * & S_1 \end{pmatrix} > 0$, $M = \begin{pmatrix} M_1 \\ M_2 \end{pmatrix}$.

3.4 Switch Controller Design

Assuming the number of environment model is n. Predictive controllers are designed for every model. When the system operates, each controller runs online. Switch controller compares output of each model with the factual system output, and selects the model which has the least error as the current model, then sends to the master. Consequently, master controller is switched to the corresponding controller. At the slave station, let modeling error of the ith environment model

$$e_{ie}(k) = f_{ie}(k) - f_e(k) \ , \tag{14}$$

where $f_{ie}(k)$ is the interactive force between the slave and environment computed though the ith environment model, $f_e(k)$ is the factual force between slave and environment. Performance index function for each environment model has the form

$$J_i = \alpha e_{ie}^2(k) + \beta \sum_{j=k-l+1}^{k} e^{-\tau(k-j)} e_{ie}^2(j) \ , \tag{15}$$

where α, β are proportion factors of current error, and accumulated error, τ is a forgetting factor and α, $\beta > 0$, $\tau > 0$, $i \in \{1, 2, \cdots, n\}$.

When the system operates, performance index is monitored at every instant. A natural way to decide when, and to which controller, one should switch, is to determine performance index for each controller and switch to the one with the minimum value.

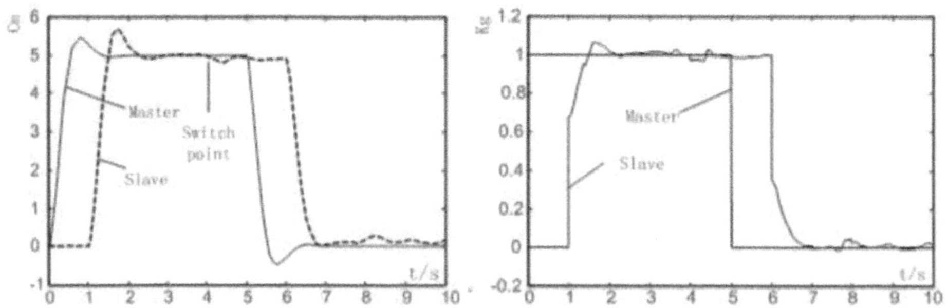

(a) Position tracking curve (b) Force tracking curve

Fig. 2. Simulation result

4 Simulation

Parameters for simulation are: $M_m = M_s = 1.532\,kg$, $B_m = B_s = 0.051$, $M_{e1} = 1\,kg$, $B_{e1} = 0.12$, $K_{e1} = 0.1$, $M_{e2} = 1.5\,kg$, $B_{e2} = 0.22$, $K_{e2} = 0.3$.

Time delay is a random number between $1\,s$ and $1.2\,s$. The simulation result is shown in Fig. 2.

The result suggests that the master and the slave are stable under switch control. The master curves are consistent with the slave's about T_r units later. It's seemed that there is no backward time delay. We can conclude that the adaptive predictive control is exact and the performance is good.

5 Conclusion

In order to ensure stability and performance of a teleoperation system in case of time-variant environment dynamics and time delay, on the basis of building neural network model base and observer based predictive controller base, this paper proposed a kind of multiple model adaptive predictive control method. Simulation result shows that the control system is effectiveness.

Acknowledgments. We would like to thank Natural Science Foundation of Hubei, China for the support to this work under the grant No. 2005ABA226.

References

1. Anderson, R.J. and Spong, M.W.: Bilateral Control of Teleoperators with Time Delay, IEEE Transactions on Automation Control **34**(4) (1989) 494–501.
2. Casavola, A., Mosca, E. and Papini, M.: Predictive Teleoperation of Constrained Dynamic Systems via Internet-like Channels, IEEE Transactions On Control Systems Technology **14**(4) (2006) 681–694.
3. Guan, Z.H., Zhang, H. and Yang S.H.: Robust Passive Control for Internet-based Switching Systems with Time-delay, Chaos & Solitons and Fractals (in press).
4. Shahdi, S.A. and Sirouspou, S: Multiple Model Control for Teleoperation in Unknown Environments, Proceedings of the 2005 IEEE International Conference on Robotics and Automation, Barcelona, Spain (2005) 703–708.
5. Smith, A.C. and Hashtrudi-Zaad, K.: Adaptive Teleoperation Using Neural Network-based Predictive Control, Proceedings of the 2005 IEEE Conference on Control Applications Toronto, Canada (2005) 1269–1274.
6. Brady, K. and Tarn, T.J.: Internet-based Remote Teleoperation, Proceedings of the IEEE International Conference on Robotics and Automation (1998) 65–70.
7. Narendra, K.S. and Balakrishnan, J.: Adaptive Control Using Multiple Models, IEEE Trans. Automat. Contr. **42**(2) (1997) 171–187.

Neural-Memory Based Control of Micro Air Vehicles (MAVs) with Flapping Wings

Liguo Weng, Wenchuan Cai, M.J. Zhang, X.H. Liao, and David Y. Song

Center for Cooperative Systems
Department of Electrical and Computer Engineering
North Carolina A&T State University,
1601 East Market St. Greensboro, NC, USA, 27411

Abstract. This paper addresses the problem of wing motion control of flapping wing Micro Air Vehicles (MAVs). Inspired by hummingbird's wing structure as well as the construction of its skeletal and muscular components, a dynamic model for flapping wing is developed. As the model is highly nonlinear and coupled with unmeasurable disturbances and uncertainties, traditional strategies are not applicable for flapping wing motion control. A new approach called neural-memory based control is proposed in this work. It is shown that this method is able to learn from past control experience and current/past system behavior to improve its performance during system operation. Furthermore, much less information about the system dynamics is needed in construction such a control scheme as compared with traditional NN based methods. Both theoretical analysis and computer simulation verify its effectiveness.

1 Introduction

The development of MAVs has been spear headed by the demand of DoD in developing autonomous, lightweight, small-scale flying machines that are appropriate for a variety of missions including reconnaissance over land, in buildings and tunnels, and other confined spaces. Of particular interest is the ability of these vehicles to operate in the urban environment and perch on buildings to provide situational awareness to the war fighter. Following DoD's lead, numerous national and international government agencies have initiated activities to develop small autonomous flying vehicles.

As a new class of air vehicle, these systems face many unique challenges that make their design and development difficult. Although successful fixed-wing MAV designs have been reported [1]-[2], the potential applications of current fixed-wing designs are essentially limited due to maneuver constraints, they do not possess the flight agility and versatility that would enable missions such as rapid flight beneath a forest canopy or within the confines.

There are numerous examples of highly successful flapping flies exist in the nature, which could provide us with another perspective to design MAVs. During the past few years, a number of flapping mechanisms have been developed and demonstrated in a limited fashion, for example, Aerovironment's Microbat and University of

D. Liu et al. (Eds.): ISNN 2007, Part I, LNCS 4491, pp. 70–80, 2007.

California (UC) Berkeley's Micromechanical flying insect [3]-[4]. Evidently, the creation of a practical MAV remains an elusive goal and has attracted increasing attention.

In consideration of the similarities between the characteristics of numerous flapping fliers exist in the nature and the requirements to the future flapping MAVs, perhaps the one that best demonstrates the characteristics we wish to possess in an agile MAV is the hummingbird, as shown in Figure 1. Hummingbird species bracket the size range of 6 in. and speed range of 25 mph used to define MAV-class vehicles. Wing lengths range from about 33 mm to 135 mm. Wind-tunnel tests have revealed their maximum flight speeds as high as 27 mph. This study explores the vibratory flapping dynamics inspired by the biomechanical system of hummingbirds. In particular, we are interested in the motion of the wing mounted on a MAV. Note that micro air vehicles operate in a very sensitive Reynolds number regime in which many complex flow phenomena take place within the boundary layer. Due to the lack of knowledge about the fundamental flow physics in this regime, many dynamic effects might not be reflected with current flapping wing model, which could causes failures during practical implementation. To neutralize the conflict a novel control approach named neural-memory network control is proposed in this work.

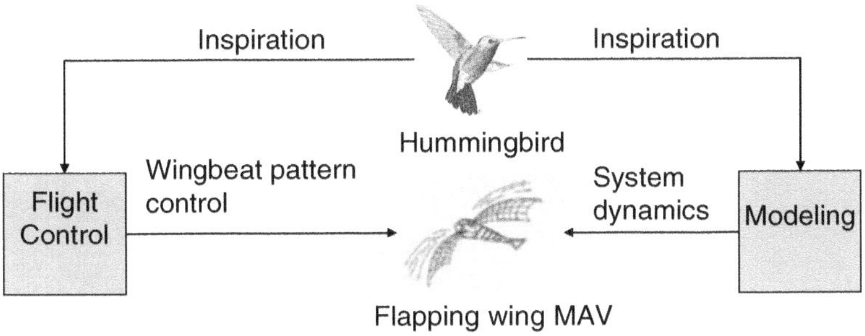

Fig. 1. Flight Control Inspired by Hummingbirds

Neural network based control has been widely applied in various systems, see, for instance, [5]-[9], however, most of NN-based controls more or less suffer the following shortages: 1) Large number of training data is needed to pre-construct the network, although lots of improved neural networks are able to update themselves online, their accuracy heavily rely on the selected training data. 2) There is no theory existing that tell people how to built a neural network, such as how many layers there should be or what kind of base function should be used. As a matter of fact, constructing a neural network is always time consuming and empirical. 3) A practical neural network contains lager quantities of neurons, which brings heavy computational burden and requires a great deal of memory space. 4) Usually, network reconstruction is a must when the system dynamics vary even slightly. 5) No theoretical proof to guarantee the stability of the control system.

Inspired by human memory system, in this work we investigate the neural-memory based control approach jn which the aforementioned disadvantages associated with most NN-based control methods do not exist. More specifically, it does not rely on precise system model, and demands less computation as compared with most other methods. It learns from both past experience and current observed information to improve its performance. There is no need for network reconstruction or consistent weights update even of the system dynamics change significantly. These features have been verified via both theoretical analysis and simulation study.

2 Neural-Memory Based Control Approach

Human response to the occurrences in the real world is based on both the so called Natural Response (NR) – a natural born instinctive response and the Acquired Response (AR) - the learnt response. At the very beginning, the behaviors/responses are dominated by immature and imprecise natural responses, because neither much memory nor experience has been gained at this point. However, as time goes by, the brain is able to retrieve and analyze certain accumulated knowledge and experience (as conceptually illustrated in Figure 2), such as the most recent behaviors and corresponding results (feedback), and combine the memorized information with current behavior to generate more reliable responses (actions), leading to the AR, which will gradually takes domination. In this process, the response to the world becomes more rational and more accountable.

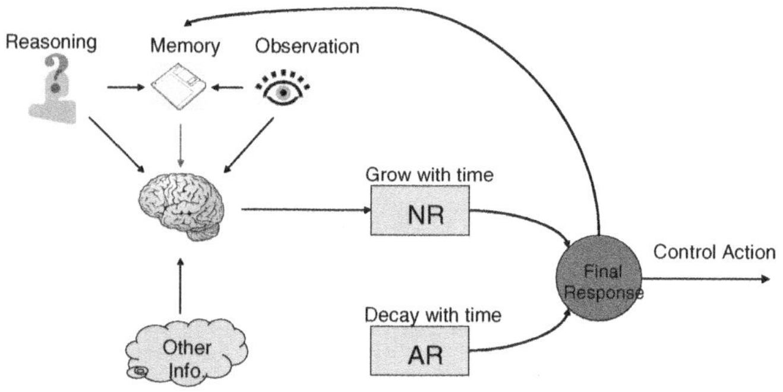

Fig. 2. Process of human learning

2.1 Analogies Between Human Memory System and Neural-Memory Network

The proposed neural-memory based approach utilizes concepts and mimics mechanisms of the Human Memory System (HMS) as described above. To begin with the introduction of the Neural-Memory Network (NMN), some of the analogies between HMS and NMN is presented in Table 1.

2.2 Structure of Neural-Memory Network

Inspired by the process of human memory and learning system, we propose the neural-memory based control scheme, which consists of two sub-networks, namely, NR network and AR network. Moreover, a time-varying function $\phi(t) \in [0,1]$, called

Table 1. Analogies between HMS and NMN

HMS	NMN
Past experience	Past control signal
Current (final) response	Current control output
Observations	system state information
Objective	Desired system response
Feedbacks(Comparison)	Control errors
NR	NR sub-network
AR	AR sub-network

trust-factor, is introduced to credit both NR-network (with factor $1-\phi(t)$) and AR-network (with factor $\phi(t)$), as shown in Figure 3. Fundamentally, $\phi(t)$ represents the gained creditability of the acquired response. When $\phi(t)=0$, only NR network is functioning, whereas $\phi(t)=1$ corresponds to pure AR action. If $0<\phi(t)<1$, both NR and AR contribute to the action of the neural-memory network. In Figure 3, u_{k-m} to u_{k-1} represent the past control experience up to m steps back, e_0 to e_k denote the past control errors (behaviors), x^*_{k-m} to x^*_k represent the desired objectives, z_{k-m} to z_k represent other useful neurons, the subscription m here means m steps back stored information used (retrieved). Correspondingly, the network is called m^{th} order neural-memory network control. u_k is the network output, which acts as the current control signal. w_b, w_o, w_f and w_d are weighting factors or memory coefficients. As creadibility usually gains with time, the trust-factor $\phi(t)$ is a function increasing with time. One of the choices for such factor is, $\phi(t,e)=1-\exp\left(\frac{-\mu t}{\|e_k\|+\varepsilon}\right)$, $\mu>0,\ \varepsilon>0$, in which the magnitude of error is also a determinant. It is seen that when $\|e_k\|$ is large, which means the memory information is imprecise and learning is incomplete, leading to small $\phi(t,e_k)$ and large $1-\phi(t,e_k)$, thus the system is basically under the control of NR network, as usually the case in nature systems. If, however, $\|e\|$ is small, implying that better behaviors are acquired thus AR action should be given more credibility. As $\phi(t,e_k)$ goes to 1, the trained AR completely takes the role. This process is quite similar to what happens during human learning.

2.3 Stability Analysis

For easy description and later development, we consider a second order system dynamics,

$$\ddot{x}=f(x)+g(x)u+\Delta f(.) \tag{1}$$

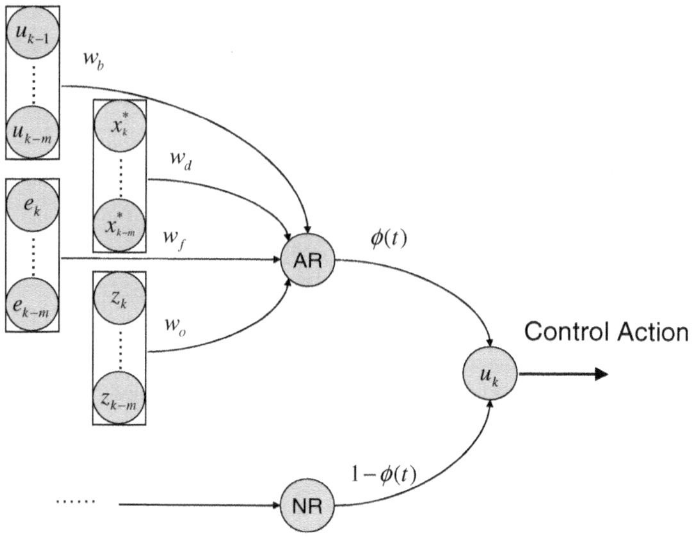

Fig. 3. Conceptual structure of neural-memory network

where *f*(.) and *g*(.) are nonlinear functions and $\Delta f(.)$ is the uncertain term in the system. Define a new variable $s = \dot{e} + \beta e$ with $e = x - x^*$ being the tracking error and $\beta > 0$ a design number. The proposed m^{th} order neural-memory network is of the form:

$$u_k = (1-\phi(t))u_{N,k} + \phi(t)u_{A,k}$$
$$\text{with } u_{A,k} = w_b U + w_f S + \eta(.) \text{ and } u_{N,k} = g^{-1}(-ks_k + \ddot{x}^* - \beta\dot{e}) \tag{2}$$

where u_N stands for the natural response network, u_A stands for the acquired response network. g is the control gain, T is the sampling period, U is a vector of $[u_{k-1}, u_{k-2}, \cdots, u_{k-m}]^T$ storing the control experiences, S is used to store $[s_k, s_{k-1}, \cdots, s_{k-m}]^T$, the system history behaviors, η accounts for available nonlinear information of the system (which could be zero if no such information is available), $w_b \in \mathbb{R}^m$ and $w_f \in \mathbb{R}^{m+1}$ are weight vectors.

For simplicity, we only describe the detailed structure of the *1*st (i.e., $m = 1$) order neural-memory network and its stability here. For the *1*st-order neural-memory network control, we have

$$w_b = [1]^T \quad w_f = g^{-1}/T[-2,1]^T$$
$$\eta(.) = g^{-1}[\ddot{x}_k^* - \ddot{x}_{k-1}^* - \beta(\dot{e}_k - \dot{e}_{k-1}) - (f(.)_k - f(.)_{k-1})] \tag{3}$$

To show the stability, we express (1) in terms of s and use Euler approximation to get

$$s_{k+1} = s_k + T\{(f(x) + g(x)u + \Delta f(.))_k - \ddot{x}_k^* + \beta \dot{e}_k\} \quad (4)$$

$$s_k = s_{k-1} + T\{(f(x) + g(x)u + \Delta f(.))_{k-1} - \ddot{x}_{k-1}^* + \beta \dot{e}_{k-1}\} \quad (5)$$

Eq (5) is obtained by one step back time-shift to (4). From (4)-(5), with u_k being defined as in (2) in which the memory coefficients are given in (3), it can be readily shown that $s_{k+1} = T(\Delta f_k - \Delta f_{k-1})$, Therefore $\|s_{k+1}\| \leq T^2 c_0 < \infty$, where $c_0 = \max \| \frac{d\Delta f(.)}{dt} \|$, donates the maximum variation rate of the disturbances and uncertainties, which is assumed to be bounded due to the fact that in general such varation cannot be infinitely fast. As a result, since the sampling interval T is a very small number, tracking error is confined within a narrow envelop defined by $T^2 c_0$.

The above analysis is based on the 1st order neural-memory network. Similar analysis can be made for higher order case. Presumably, higher order neural-memory network leads to better control precision because more previous (longer-term) memory is incorporated in the control scheme, though more computations are involved.

3 Application to Flapping Wing Motion Control

In this section, we apply the proposed NMN method to flapping wing motion control. The model for the flapping wing motion is derived with inspiration form Hummingbird's flight characters. The control issue is addressed in the section that follows.

3.1 Wing Motion Dynamics

Hummingbirds are well known flying masters with flapping wings. The agility, precision, and flight mode variability exhibited by hummingbirds is astonishing, therefore, in this paper, based on the study of the skeleton structure of hummingbirds and existing literature [10]-[12], an artificial vibratory flapping system, as shown in Figure 4, is developed to describe the wing flapping motion.

Two reference frames consisting of body-fixed axes $\{e_1, e_2, e_3\}$ and wing-fixed axes $\{x, y, z\}$, are defined, both having their origin at the shoulder joint. An ideal column with length l and radius r is used to represent the humerus. Located at distance l_1 from the shoulder joint, a pair of exogenous forces is introduced to represent the depressor muscle, F_1 and F_2 each composes an opposite angle δ to z axis. At the same location, a vertical spring with stiffness k donating the elevator muscle is placed, the jurisdiction for representing this muscle with a single spring is that this muscle travels around and over the top of the coracoid, reversing direction and attaching to the sternum, therefore, this constrains the line of action of the elevator muscle to pass through a point at the top of the coracoid, much as the force generated by the

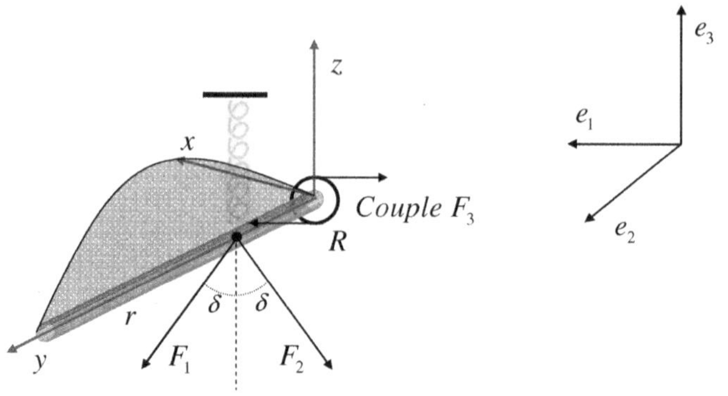

Fig. 4. Schematic representation of the vibratory flapping system

spring. Lastly, a rolling plate with radius R, at which a couple F_3 is acted, is attached to the column at the shoulder joint, and it is assumed that the plate is always perpendicular to the column. The wing motion dynamics is therefore given as:

$$J\begin{bmatrix}\dot{p} & \dot{q} & \dot{r}\end{bmatrix}^T = G\begin{bmatrix}F_1 & F_2 & F_3\end{bmatrix}^T + Nkl_1^2 - vC\begin{bmatrix}p & q & r\end{bmatrix}^T + \Delta_b \tag{6}$$

$$\begin{bmatrix}\dot{\phi} & \dot{\theta} & \dot{\psi}\end{bmatrix}^T = Y\begin{bmatrix}p & q & r\end{bmatrix}^T$$

where $G = \begin{bmatrix} -l_1\cos\delta\cos\phi & -l_1\cos\delta\cos\phi & 0 \\ l_1\sin\delta\cos\theta & -l_1\sin\delta\cos\theta & 0 \\ 0 & 0 & \dfrac{r}{R} \end{bmatrix}$, $N = \begin{bmatrix} -\sin\phi\cos\phi \\ -\sin\theta \\ 0 \end{bmatrix}$,

$C = diag(c_1 \quad c_2 \quad c_3)$, v is the free stream speed, therefore, $G[F_1 \quad F_2 \quad F_3]^T$, Nkl_1^2 and $vC\begin{bmatrix}p & q & r\end{bmatrix}^T$ correspondingly denote moments due to actuating forces, restoring forces, and damping forces. Δ_b represents the model constructing error, J is the moment of moment of inertia. $\begin{bmatrix}p & q & r\end{bmatrix}^T$ represents the angular velocity of the wing- fixed frame with respect to the body-fixed frame, ϕ, ψ and θ are Euler angles respectively referred to flapping, feathering and folding motion. Y is the transformation matrix.

3.2 Flapping Wing Motion Control with NMN

To derive the control network for wing motion adjustment, we rewrite the system dynamics as follows:

$$\ddot{\xi} = f(\dot{\xi}, \xi) + g(\xi)F + \Delta_b \tag{7}$$

$$f = \dot{Y}Y^{-1}\xi + YJ^{-1}(Nkl_1^2 - vCY^{-1}\xi) \quad g = YJ^{-1}G \tag{8}$$

where $\xi = [\phi \quad \theta \quad \psi]^T$, $F = [F_1 \quad F_2 \quad F_3]^T$ and note that: because none of the Euler angles, ϕ, θ and ψ, could physically reach $\pm\frac{\pi}{2}$ in reality, matrix G and Y are always invertible. The 1st order neural-memory network is constructed as in Figure 5, which ensures that $s_{k+1} = T(\Delta_{b,k} - \Delta_{b,k-1})$, thus $\|s_{k+1}\| \le T^2 c_0 < \infty$, where $c_0 = \max \| \frac{d\Delta_b}{dt} \|$, in consideration of that the variation rate of Δ_b can not be infinite. Thus it can be concluded that $\|s\|$ is bounded, and in view of the relationship of e and s as defined earlier, it is readily shown that $\|e\|$ and $\|\dot{e}\|$ are bounded.

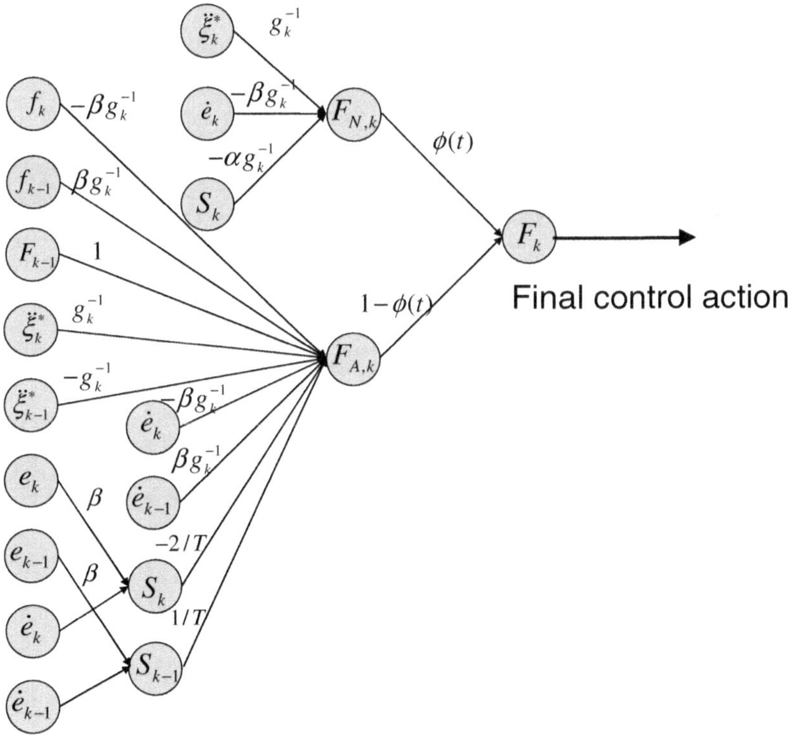

Fig. 5. 1st order neural-memory network for the wing motion control

4 Simulation Results

To verify the effectiveness of the proposed neural-memory network, we conduct numerical simulation on flapping wing motion control using the first-order neural-memory network. The parameters used for simulation are chosen as:

$$T = 0.02 \quad \mu = 5 \quad \varepsilon = 0.01 \quad \alpha = 10 \quad \beta = 5$$

$$k = 10N/m \quad v = 10m/s \quad \delta = 40^{\circ} \quad C = diag(0.1\, 0.06\, 0.04)$$

and the uncertainty $\Delta_b = \left[20\sin(t) + t^2 \quad 30\cos(t) + \sqrt{t} \quad e^{0.01t} \right]^T$. The desired Euler angles are following the trajectory: $\left[60\cos 2t \quad 30\cos 2t \quad 10 \right]^T$. The initial angles are $\left[0 \quad 0 \quad 0 \right]$. Figure 6 presents the tracking trajectories and the control action generated by neural-memory network.

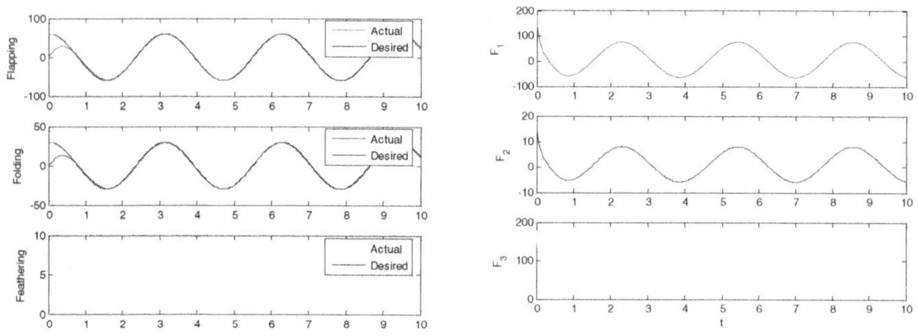

Fig. 6. Tracking trajectories and neural-memory control action (left to right)

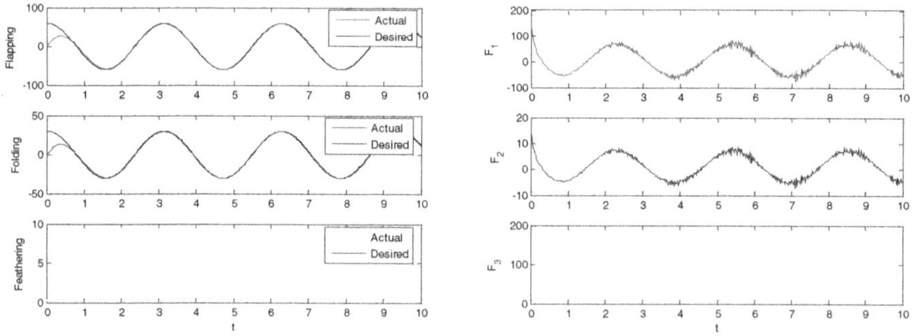

Fig. 7. Tracking trajectories and neural-memory control action with free stream speed randomly changing (left to right)

In order to test the stability and adaptivity of the proposed approach, 30% random variation is appended to the free stream speed by the equation: $v = v \pm rand \times 0.3v$, where ***rand*** is a function generating random number from 0 to 1. Not a minor modification is made to the established network, the corresponding simulation results is shown in Figure 7, which shows good control performance in the presence of significant parameter variations.

5 Conclusions

This work investigated the skeleton and muscle structure of hummingbird, an artificial vibratory flapping system was model is established and a novel neural-memory concept was proposed to design a highly robust and adaptive control scheme for flapping wing motion of MAVs. It was shown that neural-memory based control method removes the shortcomes of traditional NN based approaches and no lengthy training is needed. Furthermore, stability was always insured. Both analysis and simulation confirmed then efficiency of the method.

References

1. Grasmeyer, J.M., and Keennon, M.T.: Development of the Black Widow Micro Air Vehicle. AIAA Paper (2000) 2001-0127
2. Morris, S., and Holden, M.: Design of Micro Air Vehicles and Flight Test Validation. Proceedings of the Conference on Fixed, Flapping and Rotary Wing Vehicles at Very Low Reynolds Numbers, Dept. of Aerospace and Mechanical Engineering, Notre Dame Univ., Notre Dame, Indiana (2000)
3. Pornsin-Sisirak, T.N., Lee, S.W., Nassef, H., Grasmeyer, J.: Tai, Y. C., Ho, C.M., and Keenon, M.: MEMS Wing Technology for a Battery-Powered Ornithopter. Proceedings of the 13th Annual IEEE International Conference on Micro Electro Mechanical Systems, Miyazaki, Japan (2000) 799–804
4. Dickinson, M.H., Lehmann, F., and Sane, S.P.: Wing Rotation and the Aerodynamic Basis of Insect Flight. Science **284** (1999) 1954–1960.
5. Karayiannis, N.B., Xiong, Y.: Training Reformulated Radial Basis Function Neural Networks Capable of Identifying Uncertainty in Data Classification. IEEE Transactions on Neural Networks **17** (2006) 1222 – 1234
6. Yao L., Xu L.: Improving Signal Prediction Performance of Neural Networks Through Multiresolution Learning Approach. IEEE Transactions on Systems, Man and Cybernetics, **36**(2006) 341-352
7. Le Callet, P., Viard-Gaudin, C., Barba, D.: Convolutional Neural Network Approach for Objective Video Quality Assessment. IEEE Transactions on Neural Networks **17** (2006) 1316-1327
8. Liu, S., Wang, J.: A Simplified Dual Neural Network for Quadratic Programming with Its KWTA Application. IEEE Transactions on Neural Network **17** (2006) 1500 – 1510
9. Abdollahi, F., Talebi, H.A., Patel, R.V.: A Stable Neural Network-Based Observer With Application to Flexible-Joint Manipulators. IEEE Transactions on Neural Networks **17** (2006) 118-129
10. David L.R., Eric C. S.: Mechanization and Control Concepts for Biologically Inspired Micro Air Vehicles. Journal of Aircraft **41** (2004) 1257-1265

11. Grant D., Abdulrahim M., Lind R.: Flight Dynamics of a Morphing Aircraft Utilizing Independent Multiple-Joint Wing Sweep. AIAA Atmospheric Flight Mechanics Conference and Exhibit (2006) AIAA-2006-6505.
12. Sibilski K., Loroch L., Buler W., Zyluk A.: Modeling and Simulation of the Nonlinear Dynamic Behavior of a Flapping Wings Micro-Aerial-Vehicle. 42nd AIAA Aerospace Sciences Meeting and Exhibit (2004) AIAA-2004-541

Robust Neural Networks Control for Uncertain Systems with Time-Varying Delays and Sector Bounded Perturbations

Qing Zhu[1], Shumin Fei[1], Tao Li[1], and Tianping Zhang[2]

[1] Department of Automatic Control, Southeast University, Nanjing, Jiangsu, China
[2] College of Information Engineering, Yangzhou University, Yangzhou, Jiangsu, China

Abstract. In this paper, a robust neural networks adaptive control scheme is proposed for the stabilization of uncertain linear systems with time-varying delay and bounded perturbations. The uncertainty is assumed to be unknown continuous function without norm-bounded restriction. The perturbation is sector-bounded. Combined with liner matrix inequality method, neural networks and adaptive control, the control scheme ensures the stability of the close-loop system for any admissible uncertainty.

1 Introduction

Time-delays are frequently encountered in many real control systems. The existence of these delays may be the source of instability of serious deterioration in the performance of the closed-loop systems. Meanwhile, perturbations ,measure errors and modeling errors cause the uncertainty of systems. So the problem of controlling uncertain time delay systems has been widely investigated in recent years[1]-[6].

In [2], a class of uncertain time-varying delay system H_∞ control problem is considered, and corresponding state feedback controller using Linear Matrix Inequalities (LMI) is proposed. In [4], a state feedback control scheme is proposed for a class of uncertain system with time-varying delay input. In [5], an improved global robust asymptotic stability criteria is introduced for delayed cellular neural networks. But [1],[4],[5] need to assume that uncertainty matrices of system satisfy a particular decomposition condition, which is difficult to implement in real control systems. [2],[5] are focused on adaptive control of linear systems with multiple delays. In [3], an absolute stability criteria of time-delay systems with sector-bounded nonlinearity is proposed by using LMI method. But [1],[2] both have the assumption that system matrices are exactly known. Also only unknown constant delays are considered in [1],[2] and [5].

In this paper, we will deal with the problem of a robust neural networks control scheme for the stabilization of uncertain linear systems with time-varying delay and bounded perturbations. The assumption that uncertain matrices of system satisfy a particular decomposition condition ([1],[4],[5]) is canceled. One part of uncertainties are assumed to be norm-bounded, but the bounds are not necessarily known. The other part of uncertainties are unknown continuous functions about state vectors. The perturbation is sector-bounded. Utilizing liner matrix inequality method, we propose a state feedback

D. Liu et al. (Eds.): ISNN 2007, Part I, LNCS 4491, pp. 81–86, 2007.

control for the constant part of the system. Neural network and adaptive control are employed to estimate the unknown continuous functions. The control scheme ensures the stability of the close-loop system for all admissible uncertainties.

Notation. $\|\cdot\|$ denotes the Frobenius norm of matrix, i.e.$\| P \| = \left[tr(P^T P)\right]^{1/2}$. For an arbitrary matrix B and two symmetric matrices A and D, the symmetric term in a symmetric matrix is denoted by $*$, i.e. $\begin{bmatrix} A & B \\ * & D \end{bmatrix} = \begin{bmatrix} A & B \\ B^T & D \end{bmatrix}$.

2 Preliminaries

Lemma 1. [7] *For any $X, Y \in R^n$ and any positive symmetric definite matrix $P \in R^{n\times n}$,*

$$2X^T Y \le X^T P^{-1} X + Y^T PY.$$

3 Problem Statements

Consider the time delay system in the following form:

$$\begin{cases} \dot{x}(t) = (A_1 + \triangle A_1(x(t), t))x(t) + (A_2 + \triangle A_2(x(t), t))x(t - h(t)) + B_1 u(t) \\ \qquad +(B_2 + \triangle B_2(x(t), t))\omega(t) + f(x(t), x(t - h(t)), t), \\ z(t) = Cx(t) + Dx(t - h(t)), \\ x(t) = \phi(t), \qquad \forall t \in (-\infty, 0], \\ \omega(t) = -\varphi(t, z(t)), \end{cases} \tag{1}$$

where $x(t) \in R^n$, $u(t) \in R^m$, $z(t) \in R^q$ are state variables, system input and system output, respectively; A_1, A_2, B_1, B_2 are known parameter matrices of appropriate dimensions, $h(t)$ denotes unknown time-varying delay which satisfies the following conditions: $h(t) \ge 0$, $\dot{h}(t) \le h_1 < 1$, where h_1 is a positive constant. $\phi(t)$ is a continuous vector valued initial function. $\triangle A_1(x(t), t)$, $\triangle A_2(x(t), t)$, $\triangle B_2(x(t), t)$ are time-varying parameter uncertainties which satisfy the following conditions:

$$\begin{aligned} \triangle A_1(x(t), t) &= B_1 \left[\triangle A_{11}(t) + \triangle A_{12}(x(t))\right], \\ \triangle A_2(x(t), t) &= B_1 \left[\triangle A_{21}(t) + \triangle A_{22}(x(t))\right], \\ \triangle B_2(x(t), t) &= B_1 \left[\triangle B_{21}(t) + \triangle B_{22}(x(t))\right], \end{aligned} \tag{2}$$

with $\| \triangle A_{11}(t) \|, \| \triangle A_{21}(t) \|, \| \triangle B_{21}(t) \|$ existing but unknown and $\triangle A_{12}(x(t))$, $\triangle A_{22}(x(t))$, $\triangle B_{22}(x(t))$ are unknown continuous functions. $\omega(t) \in R^p$ denotes the external disturbance. We have $\omega(t) = -\varphi(t, z(t))$, with $\varphi(t, z(t)) : [0, \infty) \times R^q \to R^q$ is a memory, time-varying, nonlinear vector valued function which is piecewise continuous in t, globally Lipchitz in $z(t)$, $\varphi(t, 0) = 0$, and satisfies the following sector condition for $\forall t \ge 0, \forall z(t) \in R^q$:

$$\varphi^T(t, z(t))[\varphi(t, z(t)) - K_1 z(t)] \le 0, \tag{3}$$

where K_1 is constant real matrix. $f(x(t), x(t - h(t)), t) \in R^n$ denotes unknown dynamics which satisfies the following conditions:

$$\begin{aligned} & f^T(x(t), x(t - h(t)), t) \cdot f(x(t), x(t - h(t)), t) \\ & \le x^T(t) G_f^T G_f x(t) + x^T(t - h(t)) H_f^T H_f x(t - h(t)), \\ & \qquad t \in [0, +\infty), \end{aligned} \tag{4}$$

where G_f, H_f are constant real matrices. In the following, we use f to denote $f(x(t), x(t-h(t)), t)$. The target is to find a control scheme to stabilize the close-loop system.

4 Main Results

Now we are ready to present the robust adaptive control scheme for the system with all uncertainty and perturbations. Linear matrix inequality is employed to demonstrate the stability of constant part of the uncertain delay system with state feedback. Since the upper bounds of $\| \triangle A_{11}(t) \|$, $\| \triangle A_{21}(t) \|$,$\| \triangle B_{21}(t) \|$ exist but unknown, we use adaptive control to estimate the unknown values. Furthermore, neural networks are used to estimate the unknown continuous functions. The following theorem shows the result.

Theorem 1. *Consider the time delay system (1)-(4).* $\triangle A_1(x(t),t) \neq 0$, $\triangle A_2(x(t),t) \neq 0$, $\triangle B_2(x(t),t) \neq 0$, *system is asymptotically stable for any time delay* $h(t) \geq 0$, $\dot{h}(t) < h_1$, *if there exist matrices of appropriate dimensions* W, $J > 0$, $M > 0$, $P_3 > 0$, $P_6 > 0$, $N_i > 0, i \in \{1, \ldots, 4\}$ *such that*

$$\Sigma_1 = \begin{bmatrix} \Sigma_{11} & A_2 J & \Sigma_{13} & I & J\varepsilon_2 G_f^T & J & J & 0 & 0 & 0 \\ * & -(1-h_1)M & -J\frac{\varepsilon_1}{2}D^T K_1^T & 0 & 0 & 0 & 0 & J\varepsilon_2 H_f^T & J & J \\ * & * & \Sigma_{33} & 0 & 0 & 0 & 0 & 0 & 0 & 0 \\ * & * & * & -\varepsilon_2 I & 0 & 0 & 0 & 0 & 0 & 0 \\ * & * & * & * & -\varepsilon_2 I & 0 & 0 & 0 & 0 & 0 \\ * & * & * & * & * & -N_1 & 0 & 0 & 0 & 0 \\ * & * & * & * & * & * & -N_2 & 0 & 0 & 0 \\ * & * & * & * & * & * & * & -\varepsilon_2 I & 0 & 0 \\ * & * & * & * & * & * & * & * & -N_3 & 0 \\ * & * & * & * & * & * & * & * & * & -N_4 \end{bmatrix} < 0, \tag{5}$$

where

$$\begin{aligned} \Sigma_{11} &= (A_1 J + B_1 W) + (A_1 J + B_1 W)^T + M, \\ \Sigma_{13} &= B_2 - J\frac{\varepsilon_1}{2}C^T K_1^T, \\ \Sigma_{33} &= -\varepsilon_1 I + P_3 + P_6, \\ N_1 &= P_1^{-1},\ N_2 = P_4^{-1},\ N_3 = P_2^{-1},\ N_4 = P_5^{-1}, \\ J &= P^{-1},\ W = KP^{-1},\ M = P^{-1}QP^{-1}, \end{aligned}$$

and $\varepsilon_1, \varepsilon_2$ *are positive constants chosen beforehand and the control law is chosen as :*

$$u(t) = Kx(t) - \frac{\alpha}{2}B_1^T Px(t) + \Psi(t)\hat{\theta}(t) - \sum_{i=1}^{3}\frac{1}{2} \| P_{i+3}^{-1} \| \left[\hat{W}_i^T \varphi_i(x(t)) + \hat{\sigma}_i\right] B_1^T Px(t), \tag{6}$$

where $P = J^{-1}$, $K = WP$. $\hat{\theta}(t)$ *denotes the estimate of unknown parameters* θ, *and with*

$$\theta = [s_1^2 \quad s_2^2 \quad s_3^2]^T, \tag{7}$$

$$s_1 = \sup_{t\in[0,+\infty)} \| \triangle A_{11}(t) \|, \quad s_2 = \sup_{t\in[0,+\infty)} \| \triangle A_{21}(t) \|, \quad s_3 = \sup_{t\in[0,+\infty)} \| \triangle B_{21}(t) \|, \tag{8}$$

$$\Psi(t) = -\frac{1}{2} B_1^T P x(t) \left[\| P_1^{-1} \| \quad \| P_2^{-1} \| \quad \| P_3^{-1} \| \right], \tag{9}$$

and adaptive law is chosen as

$$\dot{\hat{\theta}}(t) = -R\Psi^T(t) B_1^T P x(t), \tag{10}$$

with R is a constant positive matrix chosen beforehand. The tuning law of neural networks parameters are chosen as

$$\dot{\hat{W}}_i(t) = \frac{1}{2} \| P_{i+3}^{-1} \| \| B_1^T P x(t) \|^2 \Gamma_i \varphi_i(x(t)), \ i \in \{1,2,3\}, \tag{11}$$

$$\dot{\hat{\sigma}}_i(t) = \frac{1}{2} \| P_{i+3}^{-1} \| \| B_1^T P x(t) \|^2 \eta_i, \ i \in \{1,2,3\}. \tag{12}$$

Proof. Firstly, by (1), (3) and (4), it is easy to see that

$$L_1 \triangleq \varepsilon_1 \omega^T(t)[\omega(t) + K_1 C x(t) + K_1 D x(t-h(t))] \le 0, \tag{13}$$

$$L_2 \triangleq \varepsilon_2 \left[f^T f - x^T(t) G_f^T G_f x(t) - x^T(t-h(t)) H_f^T H_f x(t-h(t)) \right] \le 0, \tag{14}$$

where $\varepsilon_1 > 0, \varepsilon_2 > 0$. By Lemma 1, we can deduce

$$2x^T(t) P B_1 \triangle A_{11}(t) x(t) \le x^T(t) P_1 x(t) + \| P_1^{-1} \| \| \triangle A_{11}(t) \|^2 \| B_1^T P x(t) \|^2, \tag{15}$$

$$\begin{aligned} 2x^T(t) P B_1 \triangle A_{21}(t) x(t-h(t)) &\le x^T(t-h(t)) P_2 x(t-h(t)) \\ &+ \| P_2^{-1} \| \| \triangle A_{21}(t) \|^2 \| B_1^T P x(t) \|^2, \end{aligned} \tag{16}$$

$$2x^T(t) P B_1 \triangle B_{21}(t) \omega(t) \le \omega^T(t) P_3 \omega(t) + \| P_3^{-1} \| \| \triangle B_{21}(t) \|^2 \| B_1^T P x(t) \|^2, \tag{17}$$

$$2x^T(t) P B_1 \triangle A_{12}(x(t)) x(t) \le x^T(t) P_4 x(t) + \| P_4^{-1} \| \| \triangle A_{12}(x(t)) \|^2 \| B_1^T P x(t) \|^2, \tag{18}$$

$$\begin{aligned} 2x^T(t) P B_1 \triangle A_{22}(x(t)) x(t-h(t)) &\le x^T(t-h(t)) P_5 x(t-h(t)) \\ &+ \| P_5^{-1} \| \| \triangle A_{22}(x(t)) \|^2 \| B_1^T P x(t) \|^2, \end{aligned} \tag{19}$$

$$2x^T(t) P B_1 \triangle B_{22}(x(t)) \omega(t) \le \omega^T(t) P_6 \omega(t) + \| P_6^{-1} \| \| \triangle B_{22}(x(t)) \|^2 \| B_1^T P x(t) \|^2, \tag{20}$$

where $P_i > 0, i \in \{1, \ldots, 6\}$. Since the upper bounds of $\| \triangle A_{11}(t) \|$,$\| \triangle A_{21}(t) \|$, $\| \triangle B_{21}(t) \|$ exist, we estimate these upper bounds by adaptive method. Define

$$\tilde{\theta}(t) = \hat{\theta}(t) - \theta.$$

Furthermore, we utilize three Radius Basis Function Neural Networks (RBFNN) to estimate three unknown continuous functions as following:

$$\| \triangle A_{12}(x(t)) \|^2 = W_1^{*T} \varphi_1(x(t)) + \sigma_1, \tag{21}$$

$$\| \triangle A_{22}(x(t)) \|^2 = W_2^{*T} \varphi_2(x(t)) + \sigma_2, \tag{22}$$

$$\| \triangle B_{22}(x(t)) \|^2 = W_3^{*T} \varphi_3(x(t)) + \sigma_3. \tag{23}$$

Define $W_1^* = \arg\min_{W_1 \in S} \left\{ \sup_{x(t)\in T} \|\|\| \triangle A_{12}(x(t)) \|^2 - W_1^T \varphi_1(x(t)) \| \right\}$, here S, T are compact sets about $W_1, x(t)$ respectively. W_2^*, W_3^* are defined in a similar way.

$$W_i^* \in R^{m_i},\ \varphi_i(x(t)) = [\varphi_{i1}(x(t)), \varphi_{i2}(x(t)), \dots \varphi_{im_i}(x(t))]^T \in R^{m_i},\ i \in \{1,2,3\}, \quad (24)$$

$$\varphi_{ij}(x(t)) = \exp\left[\frac{-(x(t)-\mu_{ij})^T(x(t)-\mu_{ij})}{\delta_{ij}^2}\right],\ i \in \{1,2,3\}, j \in \{1,2,\dots,m_i\}, \quad (25)$$

where $\mu_{ij} \in R^n$, $\delta_{ij} > 0$. $|\sigma_i| \le \bar{\sigma}_i$, $i \in \{1,2,3\}$, $j \in \{1,2,\dots,m_i\}$, $\bar{\sigma}_i (i \in \{1,2,3\})$ are known positive constants. It is easy to see that $\varphi_{ij}(x(t)) > 0$, therefore

$$W_i^{*T}\varphi_i(x(t)) \ge 0,\ i \in \{1,2,3\}.$$

Define

$$\tilde{W}_i(t) = \hat{W}_i(t) - W_i^*,\ \tilde{\sigma}_i(t) = \hat{\sigma}_i(t) - \bar{\sigma}_i,\ i \in \{1,2,3\},$$

then we choose a Lyapunov-Krasovskii functional candidate for system (1)-(4) as

$$V(t) = x^T(t)Px(t) + \int_{t-h(t)}^{t} x^T(s)Qx(s)ds + \tilde{\theta}^T R^{-1}\tilde{\theta} + \sum_{i=1}^{3} \tilde{W}_i^T \Gamma_i^{-1}\tilde{W}_i + \sum_{i=1}^{3} \eta_i^{-1}\tilde{\sigma}_i^2. \quad (26)$$

By (1),(13),(14), the time-derivative of $V(t)$ gives

$$\begin{aligned}\dot{V}(t) \le\ & 2x^T(t)P[(A_1 + \triangle A_1(x(t),t))x(t) + (A_2 + \triangle A_2(x(t),t))x(t-h(t)) \\ & + B_1u(t) + (B_2 + \triangle B_2(x(t),t))\omega(t) + f] + x^T(t)Qx(t) \\ & -(1-\dot{h}(t))x^T(t-h(t))Qx(t-h(t))) + 2\tilde{\theta}^T(t)R^{-1}\dot{\hat{\theta}}(t) \\ & + \sum_{i=1}^{3} 2\tilde{W}_i^T\Gamma_i^{-1}\dot{\hat{W}}_i + \sum_{i=1}^{3} 2\eta_i^{-1}\tilde{\sigma}_i\dot{\hat{\sigma}}_i - L_1 - L_2, \end{aligned} \quad (27)$$

Noting inequalities (15)-(20), neural networks (21)-(23), control input (6) and adaptive law (10)-(12), by some mathematical calculations, we have

$$\dot{V}(t) \le -\alpha \| B_1^T Px(t) \|^2 + y^T \Sigma_2 y, \quad (28)$$

where

$$y = \begin{bmatrix} x^T(t) & x^T(t-h(t)) & \omega^T(t) & f^T \end{bmatrix}^T, \quad (29)$$

and

$$\Sigma_2 = \begin{bmatrix} \begin{array}{c} P(A_1 + B_1K) \\ +(A_1+B_1K)^TP \\ +Q + \varepsilon_2 G_f^T G_f \\ +P_1 + P_4 \end{array} & PA_2 & \begin{array}{c} PB_2 \\ -\frac{\varepsilon_1}{2}C^TK_1^T \end{array} & P \\ * & \begin{array}{c} -(1-h_1)Q \\ +\varepsilon_2 H_f^T H_f \\ +P_2 + P_5 \end{array} & -\frac{\varepsilon_1}{2}D^TK_1^T & 0 \\ * & * & \begin{array}{c} -\varepsilon_1 I \\ +P_3 + P_6 \end{array} & 0 \\ * & * & * & -\varepsilon_2 I \end{bmatrix} < 0. \quad (30)$$

Pre- and post-multiplying (30) by $diag(P^{-1}, P^{-1}, I, I)$, together with Schur complement, we get that (30) is equivalent to (5). Therefore $\dot{V}(t) < 0$. (only if $y = 0$, $\dot{V}(t) = 0$.) It follows that the close-loop system is globally asymptotically stable. This completes the proof. □

Remark 1. The proof of theorem 1 shows that the state of system is bounded, so it must be in a compact set. The condition of applicating neural networks is met [8],[9].

Remark 2. If several (or both) items of A_1, A_2, B_2 are zeros, the control scheme is still applicable without any modification.

Remark 3. For theorem 1, if several items of $\triangle A_{12}(x(t))$, $\triangle A_{22}(x(t))$, $\triangle B_{22}(x(t))$ are zeros, the control scheme needs some modifications. i.e.if $\triangle A_{12}(x(t)) = 0$, then (21) no longer exists. We need to delete one item from control law (6) which is:
$\frac{-1}{2} \parallel P_4^{-1} \parallel \left[\hat{W}_1^T \varphi_1(x(t)) + \hat{\sigma_1}\right] B_1^T P x(t)$.

5 Conclusions

The control problem of a class of uncertain time-delay systems with sector-bounded perturbations has been addressed in this paper. The uncertainty is not necessary to be norm-bounded. A control scheme combining state feedback, neural networks and adaptive control is present. The feedback controller is used for the constant part of the system and neural networks, adaptive control are used for the uncertain part of the system. The close-loop system is proved to be asymptotically stable for any admissible uncertainty.

References

1. Xu, S., Lam, J., Zou, Y.: New Results on Delay-dependent Robust H_∞ Control for Systems with Time-varying Delays. Automatica **42** (2006) 343–348
2. Wu, H.: Adaptive Stabilizing State Feedback Controllers of Uncertain Dynamical Systems with Multiple Time Delays. IEEE Trans. Automat. Contr. **45** (2000) 1697–1701
3. Han, Q.: Absolute Stability of Time-delay Systems with Sector-bounded Nonlinearity. Automatica **41** (2005) 2171–2176
4. Yue, D., Han, Q.: Delayed Feedback Control of Uncertain Systems with Time-varying Input Delay. Automatica **41** (2005) 233–240
5. Xu, S., Lam, J., Daniel, W.C.Ho., Zou, Yun.: Improved Global Robust Asymptotic Stability Criteria for Delayed Cellular Neural Networks. IEEE Trans. Sys., Man & Cyber.-Part B **35** (2005) 1317–1321
6. Zheng, F., Wang, Q., Lee, T.: Adaptive Robust Control of Uncertain Time Delay Systems. Automatica **41** (2005) 1375–1383
7. Yue, D.: Robust Stabilization of Uncertain Systems with Unkown Input Delay. Automatica **40** (2004) 331–336
8. Ge, S., Hang, C., Zhang, T.: A Direct Method for Robust Adaptive Nonlinear Control with Guranteed Transient Performance. Systems & Control Letters **37** (1999) 275–284
9. Zhang, T., Ge, S., Hang, C.: Design and Performance Analysis of a Direct Adaptive Controller for Nonlinear Systems. Automatica **35** (1999) 1809–1817

Switching Set-Point Control of Nonlinear System Based on RBF Neural Network

Xiao-Li Li

Department of Automation, Information and Engineering School, University of Science and Technology Beijing, Beijing, 100083

Abstract. Multiple controllers based on multiple radial based function neural network(RBFNN) models are used to control a nonlinear system to trace a set-point. Considering the nonlinearity of the system, when the set-point value is time variant, a controller based on a fixed structure RBFNN can not give a good control performance. A switching controller which switches among different controller based on different RBFNN is used to adapt the varing set-point value and improve the output reponse and control performance of the nonlinear system.

1 Introduction

It is well known that the conventional adaptive control system based on a fixed or slowly adaptive model can get good performance. But when the parameters or structure or the set-point value of the system change abruptly from one context to another, the conventional adaptive control will react slowly, the output of the system will change abruptly and may be out of control at this time. One way to solve this problem is to use multi-model adaptive control (MMAC). From mid 1990's to now, a lot of MMAC algorithms combined with a switching index function have been given successively, and this kind of MMAC can guarantee the stability of closed-loop system. In recent ten years, the papers about switching MMAC have covered continuous time system [1], discrete time system [2,3,4], stochastic system [5,6], etc, and there are also some practical implementations in this field.

For linear system, a fixed controller or a slowly adaptive controller can always force the output of the system to a set-point value. But for nonlinear system, the dynamics of the system will change greatly at different equilibrium point. If the set-point is time variant , due to the changing dynamics of the nonlinear system, a controller based on a fixed approximate model of the nonlinear system will always give a worse control performance.

In this paper, multiple RBFNN are set up according to the dynamics character of nonlinear system at different equilibrium point.Then multiple controllers according to these RBFNN models can be obtained. A switching controller based on these controllers will control the nonlinear system to trace a time variant set-point. From the simulation, it can be seen that by using this kind of switching controller, the control performance can be improved greatly.

D. Liu et al. (Eds.): ISNN 2007, Part I, LNCS 4491, pp. 87–92, 2007.

2 Description of Plant

The system to be controlled is a single-input, single-output, discrete time system and described by the following equation.

$$y(t+1) = f(y(t), y(t-1), \cdots, y(t-n)) + B(z^{-1})u(t) \tag{1}$$

where z^{-1} is a back-shift operator.

$$B(z^{-1}) = b_0 + b_1 z^{-1} + \cdots + b_m z^{-m} \tag{2}$$

is known *a priori*.

When $f(y(t), y(t-1), \cdots, y(t-n))$ is a linear function, such as

$$f(y(t), y(t-1), \cdots, y(t-n)) = (a_1 + a_2 z^{-1} + \cdots + a_n z^{-n})y(t) \tag{3}$$

the system (1) can be rewritten in the following regression form

$$y(t+1) = \phi^T(t)\theta \tag{4}$$

$$\phi^T(t) = [y(t), \cdots, y(t-n), u(t), \cdots, u(t-m)] \tag{5}$$

$$\theta^T = [a_1, \cdots, a_n, b_0, \cdots, b_m] \tag{6}$$

$$y^*(t+1) = \phi^T(t)\theta \tag{7}$$

where $y^*(t)$ is the set-point value of output, and the control input can be obtained by (7).

When $f(y(t), y(t-1), \cdots, y(t-n))$ is a nonlinear function, the method above can not be used again, and the controller based on (4)-(7) can not be used again. Compared the weights of RBF neural network with the parameter θ in linear system (4), it can be found that nonlinear system can have the same regression form as linear system if the nonlinear part of system is identified by RBF neural network. So the result in linear controller can be extended to nonlinear system by using the RBF neural network.

3 Switching Control of Nonlinear System by Using Multiple RBF Neural Network Models

When $f(y(t), y(t-1), \cdots, y(t-n))$ is a nonlinear function, RBF neural network can be used to design a controller. In this case, the system (1) should satisfy the following assumption.

$\mathcal{A}1$: Polynomial $B(z^{-1})$ is known *a priori*, the roots of $B(z^{-1})$ lie inside the unit circle in complex plane (i.e. the system is minimum phase).

3.1 RBF Neural Network

Here a three layers RBF neural network with N input $x_i(i = 1, \cdots, N)$, L hidden units and M output $y_i(i = 1, \cdots, M)$ is used to approximate the nonlinear function $f(\cdot)$.

$$q_i(t) = K(|X(t) - S_i(t)|) = e^{-\frac{\sum_{j=1}^{N}(x_j(t) - s_{i,j}(t))^2}{2\alpha_i^2(t)}}, \quad 1 \le i \le L$$

$$y_j(t) = \sum_{i=1}^{L} w_{j,i}(t)q_i(t) + \lambda_j(t) = \sum_{i=0}^{L} w_{j,i}q_i(t) = Q^T(t)W_j(t); \quad 1 \le j \le M \tag{8}$$

$$w_{j,0}(t) = \lambda_j(t); \quad q_0(t) = 1 \tag{9}$$

$$Q(t) = [q_0(t), \cdots, q_L(t)]^T, \quad W_j = [w_{j,0}, \cdots, w_{j,L}]^T \tag{10}$$

where $q_i(t)$ is the output of the hidden unit (i.e. the output of radial based function). $X(t)$ is the input vector. $S_i(t)$ is the transforming center vector of the unit in hidden layer. $\alpha_i(t)$ is the control parameter according to the central vector.

The training step of the weights of network is as follows

$$G(t) = \frac{P(t-1)Q(t)}{\delta + Q^T(t)P(t-1)Q(t)} \tag{11}$$

$$P(t) = P(t-1) - G(t)Q^T(t)P(t-1) \tag{12}$$

$$\hat{W}_i(t) = \hat{W}_i(t-1) + G(t)[d_i(t) - Q^T(t)\hat{W}_i(t-1)] \tag{13}$$

where $P(t)$ is called inverse correlation matrix, δ is a forgetting factor, $d_i(t)$ $(i = 1, 2, \cdots, M)$ is desired output of the unit of output layer.

3.2 Controller Based on RBF Neural Network

Consider nonlinear system (1) satisfies the following assumption:
$\mathcal{A}2$: The nonlinear function $f(\cdot)$ is also bounded, if $y(t), \cdots, y(t-n)$, $(0 \le t \le \infty)$ are bounded.

The objective of the adaptive control problem is to determine a bounded control input $u(t)$ such that the output $y(t)$ of the system asymptotically tracks a specified arbitrary bounded reference output $y^*(t)$, i.e.

$$\lim_{t\to\infty} | y(t) - y^*(t) |= 0 \tag{14}$$

A three layers RBF neural network with $(n + 1)$ input and one output can be used to identify the nonlinear part $f(\cdot)$, and the assumption below should be satisfied

$\mathcal{A}3$: Over a certain compact set, $f(\cdot)$ can be approximated by a RBF neural network with proper choice of the structure and weights, i.e.

$$|f(\cdot) - \hat{f}(\cdot)| < \varepsilon \tag{15}$$

where $\hat{f}(\cdot) = Q^T(t)\hat{W}(t)$, and ε can be any specified positive number.

As $y^*(t+1)$ and $\hat{W}(t)$ are known, $u(t)$ can be got from the equation below

$$y^*(t+1) = Q^T(t)\hat{W}(t) + B(z^{-1})u(t) \tag{16}$$

3.3 Switching Set-Point Controller Based on Neural Network

To solve the problem mentioned in the first part, multiple controllers based on different neural network according to the different dynamics of the nonlinear system around different equilibrium point will be set up, and a switching controller will be given based on these controllers.

Consider the varying scope of set-point value y^* is $[\underline{y},\ \overline{y}]$,this varying range of y^* will be divided into several small subrange as below

$$[\underline{y},\ \underline{y}+a)\,,[\underline{y}+a,\ \underline{y}+2a)\,,\cdots,[\underline{y}+(m-1)a,\ \overline{y}] \tag{17}$$

where

$$a = \frac{\overline{y}-\underline{y}}{m} \tag{18}$$

With same structure as (11)-(13), several RBFNN with different transforming center vector of the units in hidden layer according the difference subrange of set value will be set up as $RBFNN_i,\ i \in \{1,2,...,m\}$. Different controller as (16) will be set up based on these RBFNN models, and a switching controller based on these controllers can be obtained. When the set-point value changes from subrange i to subrange j , the corresponding controller based on $RBFNN_i$ will be switched to the controller based on $RBFNN_j$ to adapt the changing dynamics of nonlinear system, and the output performance of the system can be improved greatly by the way of switching.

4 Simulation Analysis

Consider the following nonlinear system with time variant set-point value

$$y(t+1) = f(y(t), y(t-1)) + 0.5u(t) \tag{19}$$

$$f(t) = \frac{y(t)y(t-1)}{1+y^2(t)+y^2(t-1)} \tag{20}$$

$$y^*(t) = \begin{cases} -1 & ; \quad 0 \le t < 50 \\ 1 & ; \quad 50 \le t < 100 \\ -1 & ; \quad 100 \le t < 150 \\ 1 & ; \quad 150 \le t < 200 \end{cases} \tag{21}$$

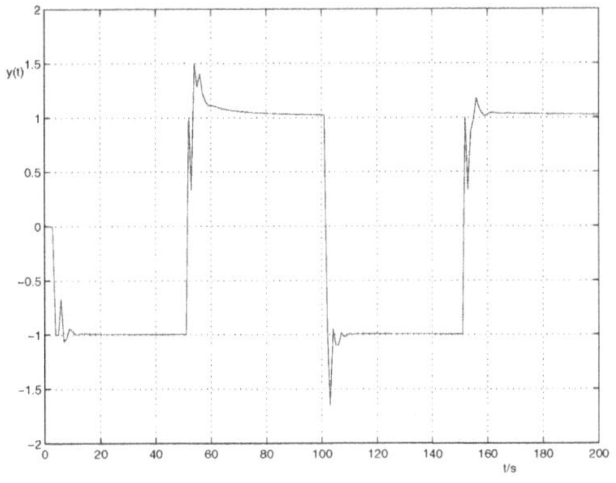

Fig. 1. Output of System by using C_1

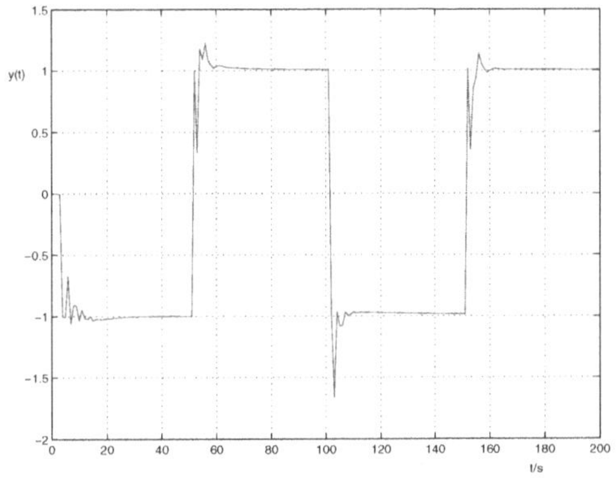

Fig. 2. Output of system by using C_2

Two RBFNN, $RBFNN_1$ and $RBFNN_2$ are all with three layers structure, two hidden units, but they are set up according the two different transforming center vector of the two hidden units, i.e. $\{[-1,-1],\ [0,0]\}$ and $\{[1,1],\ [0,0]\}$. Two controller C_1 and C_2 will be given based on $RBFNN_1$ and $RBFNN_2$ as (16). The output of the nonlinear system can be seen in figure 1 and figure 2 when controller C_1 and C_2 is used to control the system. Figure 3 is the output of the system when a switching controller is used, this controller switches between C_1 and C_2 after each 50 sample time. If the figures can be enlarged enough, from figure 1 and 2, it can be seen very clearly that the steady errors always exist by using C_1 and C_2 and the transient responses are not very well compared with result of figure 3.

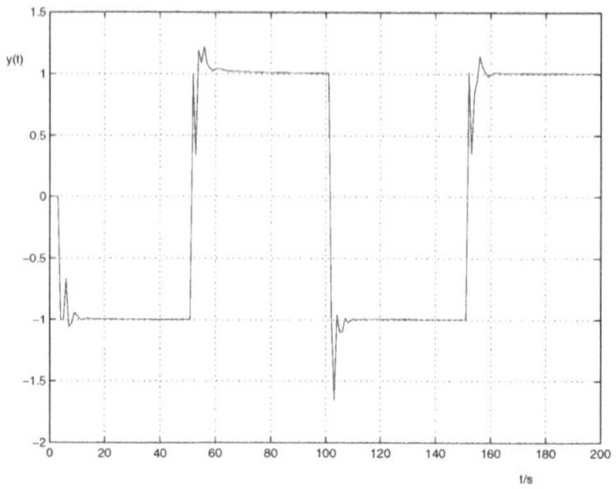

Fig. 3. Output of system by using switching controller

5 Conclusion

A nonlinear system switching controller based on RBF neural network is proposed in this paper. By using this kind of switching controller, the control performance of the nonlinear system can be improved greatly especially for time variant set-point value. From the simulation the effectiveness of the method proposed in this paper can be tested easily. This kind of switching controller can also gives a way for better control of nonlinear system.

Acknowledgements

This work is partially supported by the Fund of National Natural Science Foundation of P.R. China (60604002), Beijing Nova Programme (2006B23), Innovation Talent Project of University of Science and Technology Beijing and Key Discipline Project of Beijing Municipal Commission of Education

References

1. Narendra, K.S., Balakrishnan, J.: Adaptive Control Using Multiple Models. IEEE Trans. Automatic Control **42(2)** (1997) 171-187
2. Narendra, K.S., Xiang, C.: Adaptive Control of Discrete-time Systems Using Multiple Models. IEEE Trans. Automatic Control **45(9)** (2000) 1669-1685
3. Li, X.L., Wang, W.: Minimum Variance Based Multi-model Adaptive Control. Proc. IFAC World Congress. Beijing. China (1999) 325-329
4. Li, X.L., Wang, W., Wang, S.N.: Multiple Model Adaptive Control for Discrete Time Systems. American Control Conference. Arlington. Virginia. USA (2001) 4820-4825
5. Chen, L.J., Narendra, K.S.: Nonlinear Adaptive Control Using Neural Networks and Multiple Models. Automatica **37(8)** (2001) 1245-1255
6. Narendra, K.S., Driollet, O.: Stochastic Adaptive Control Using Multiple Estimation Models. Int. J. Adapt. Control Signal Process **15(3)** (2001) 287-317

Adaptive Tracking Control for the Output PDFs Based on Dynamic Neural Networks

Yang Yi[1], Tao Li[1], Lei Guo[2], and Hong Wang[3]

[1] Research Institute of Automation
Southeast University, Nanjing 210096, China
[2] The School of Instrument Science and Opto-Electronics Engineering
Beihang University, Beijing 100083, China
l.guo@seu.edu.cn
[3] Control Systems Centre
The University of Manchester, Manchester, UK

Abstract. In this paper, a novel adaptive tracking control strategy is established for general non-Gaussian stochastic systems based on two-step neural network models. The objective is to control the conditional PDF of the system output to follow a given target function by using dynamic neural network models. B-spline neural networks are used to model the dynamic output probability density functions (PDFs), then the concerned problem is transferred into the tracking of given weights corresponding to the desired PDF. The dynamic neural networks with undetermined parameters are employed to identify the nonlinear relationships between the control input and the weights. To achieve control objective, an adaptive state feedback controller is given to estimate the unknown parameters and control the nonlinear dynamics.

1 Introduction

Stochastic control has been an important research subject over the past decades, especially for the industrial processes which posses complex nonlinear dynamics. Recently, motivated by some typical examples in practical systems such as paper and board making, a group of new strategies that control the shape of output probability density function (PDF) for general stochastic systems have been developed (see [1, 2, 3, 4]). This novel control framework has been called as stochastic distribution control (SDC) [1]. Stochastic distribution control has two obstacles from any other previous stochastic control approaches. The first obstacle is to characterize the output PDF using analytic methods. In order to obtain some feasible design algorithms, B-spline neural network expansions have been introduced to model the output PDF so that the problem can be reduced to a tracking problem for the weighting systems (see [1, 3, 4]). The second obstacle is to establish the dynamic model of the weight vectors. For convenience, most of papers only concerned linear dynamic models between the control input and the weights related to the PDF. It is noted that a linear mapping cannot change the PDF shape of the stochastic input, which confines the practical applications.

D. Liu et al. (Eds.): ISNN 2007, Part I, LNCS 4491, pp. 93–101, 2007.

In [4], nonlinear models have been considered for the weighting dynamics of the B-spline models. However, it is noted that such nonlinear models are difficult to be provided via classical identification approaches.

It is well known that neural networks are powerful tools for learning dynamic highly nonlinear systems because of their massive parallelism, very fast adaptability, and inherent approximation capabilities [5, 6]. Recently, dynamic neural networks, which are also known as recurrent neural networks, have been shown as the successful techniques in system identification (see [7]-[10]). Dynamic neural network identification is the model estimation process of capturing dynamics using measured data. Compare with the static neural networks, dynamic neural networks incorporate feedback which make them more suitable for identifying complex nonlinear system.

In this paper, we will apply a novel two-step neural network to study the SDC problem for the non-Gaussian systems. Firstly, the B-spline neural networks are used to approximate the probability density function of the system output directly. If the basis functions are fixed, the weights of the approximation can characterize the shape of the output PDFs. Thus, the stochastic distribution control problem can be transformed to a weight tracking problem. Secondly, the dynamic neural networks are applied to identify the nonlinear dynamic relationships between the control input and the weight vectors. Dynamic adaptive controller is developed so that the weight dynamics can follow the outputs of a reference model. Both stability and robustness of the closed loop system can be guaranteed by using the proposed adaptive control strategies.

2 Output PDFs Model Using B-Spline Neural Network

For a dynamic stochastic system, denote $u(t) \in R^m$ as the input, $\eta(t) \in [a, b]$ as the stochastic output and the probability of output $\eta(t)$ lying inside $[a, \sigma]$ can be described as

$$P(a \leq \eta(t) < \sigma, u(t)) = \int_a^\sigma \gamma(y, u(t))dy \tag{1}$$

where $\gamma(y, u(t))$ is the PDF of the stochastic variable $\eta(t)$ under control input $u(t)$. As in [3, 4], it is supposed that the output PDF $\gamma(y, u(t))$, as the control objective, can be measured or estimated. In this paper, the following B-spline model will be adopted

$$\sqrt{\gamma(y, u(t))} = \sum_{i=1}^{n} v_i(u(t))B_i(y) \tag{2}$$

where $B_i(y)(i = 1, 2, \cdots, n)$ are pre-specified basis functions and $v_i(t) := v_i(u(t))$, $(i = 1, 2, \cdots, n)$ are the corresponding weights. Due to $\int_a^b \gamma(y, u(t))dy = 1$, only $n - 1$ weights are independent. (2) can be rewritten as

$$\sqrt{\gamma(y, u(t))} = C_0(y)V(t) + v_n(t)B_n(y) \tag{3}$$

where

$$C_0(y) = [B_1(y) \ \ B_2(y) \ \ \cdots \ \ B_{n-1}(y)] \quad V(t) = [v_1(t) \ \ v_2(t) \ \ \cdots \ \ v_{n-1}(t)]^T$$

Denote

$$\Lambda_1 = \int_a^b C_0^T(y)C_0(t)dy \quad \Lambda_2 = \int_a^b C_0(y)B_n(y)dy \quad \Lambda_3 = \int_a^b B_n^2(y)dy \tag{4}$$

To guarantee $\int_a^b \gamma(y, u(t))dy = 1$, we assume that $\{v_i(t) : i = 1, 2, \cdots, n-1\}$ are independent. In this paper, we consider the following expansion with the approximation error $\epsilon(y, t)$

$$\sqrt{\gamma(y, u(t))} = C_0(y)V(t) + h(V(t))B_n(y) + \epsilon(y, t) \tag{5}$$

where $h(V(t)) = \frac{\sqrt{\Lambda_3 - V^T(t)\Lambda_0 V(t)} - \Lambda_2 B_n(y)}{\Lambda_3}$ and $\Lambda_0 = \Lambda_1\Lambda_3 - \Lambda_2^2$. For $h(V(t))$, it is supposed that Lipschitz condition can be satisfied (see [4] for the detail).

3 Dynamic Neural Network Identification

Once B-spline expansions have been made for the PDFs, the next step is to find the dynamic relationships between the input and the weights related to the PDFs corresponds to a further modeling procedure. However, most published results only concerned linear precise models, while practically the relationships from control input $u(t)$ to weight vectors $V(t)$ should be nonlinear dynamics. The nonlinear models used in [4] actually are difficult to obtain through traditional identification approaches.

Dynamic neural network identifier can be employed to perform black box identification. In the following, we will provide a dynamic neural network model to characterize the weighting dynamics, with a learning strategy for the model parameters. Then a novel adaptive tracking control law will be given for the target weights.

It is assumed that there exists optimal model parameters (also can be seen weight matrices related to the dynamic neural networks) W_1^*, W_2^* such that the nonlinear dynamics between the input and the weights related to the PDFs can be described by the following neural network model

$$\begin{cases} \dot{x}(t) = Ax(t) + BW_1^*\sigma(x) + BW_2^*\phi(x)u(t) - F(t) \\ V(t) = Cx(t) \end{cases} \tag{6}$$

where $x \in R^m$ is the state vector, $A \in R^{m\times m}$ is a stable matrix, $B \in R^{m\times m}$ is a diagonal matrix, of the form $B = diag[b_1, b_2, \cdots, b_m]$. $C \in R^{(n-1)\times m}$ is a known matrix. $F(t)$ represents the error term and there exist unknown positive constants d such that

$$\|F(t)\| \leq d \tag{7}$$

The vector functions $\sigma(x) \in R^m$ is assumed to be m-dimension with the elements increasing monotonically and the matrix function $\phi(x)$ is assumed to be $m \times m$ diagonal matrix. The typical presentation of the elements $\sigma_i(.), \phi_i(.)$ are as sigmoid functions, i.e.

$$\sigma_i(x_i) = \frac{a}{1 + e^{-bx_i}} - c \tag{8}$$

The next step is to construct the following dynamic neural network for identification

$$\begin{cases} \dot{\hat{x}}(t) = A\hat{x}(t) + BW_1\sigma(\hat{x}) + BW_2\phi(\hat{x})u(t) + u_f(t) \\ V(t) = C\hat{x}(t) \end{cases} \tag{9}$$

where $\hat{x}(t) \in R^m$ is the state of the dynamic neural network, W_2 is an $m \times m$ diagonal matrix of synaptic weights, of the form $W_2 = diag[w_{21}, w_{22}, \cdots w_{2m}]$. $u_f(t)$ is the compensation term for the model error and is defined later.

Denoting $\tilde{W}_1 = W_1 - W_1^*, \tilde{W}_2 = W_1 - W_2^*, \tilde{\sigma} = \sigma(\hat{x}) - \sigma(x), \tilde{\phi} = \phi(\hat{x}) - \phi(x)$ and identification error $e(t) = \hat{x}(t) - x(t)$. Because $\sigma(.)$ and $\phi(.)$ are chosen as sigmoid functions, they satisfy the following *Lipshitz* property(see [7, 8, 9])

$$\tilde{\sigma}^T\tilde{\sigma} \le e^T(t)D_\sigma e(t), \quad (\tilde{\phi}u(t))^T(\tilde{\phi}u(t)) \le \bar{u}e^T(t)D_\phi e(t) \tag{10}$$

where D_σ, D_ϕ are known positive-definite matrices and $u(t)$ satisfy $u^T(t)u(t) \le \bar{u}$, $\bar{u}$ is a known constant. From (6) and (9), We can get the error equation

$$\begin{aligned} \dot{e}(t) = Ae(t) + B\tilde{W}_1\sigma(\hat{x}) + B\tilde{W}_2\phi(\hat{x})u(t) + u_f(t) \\ + BW_1^*\tilde{\sigma} + BW_2^*\tilde{\phi}u(t) + F(t) \end{aligned} \tag{11}$$

If we define $Q = D_\sigma + \bar{u}D_\phi + Q_0$, there exist a stable matrix A and a strictly positive definite matrix Q_0 such that the matrix Lyapunov equation

$$A^TP + PA = -Q \tag{12}$$

has a positive definite solution P.

Theorem 1. If the compensation term $u_f = u_{f1} + u_{f2}$, where $u_{f1}(t) = -Pe\|Pe\|^{-1}\hat{d}$, $u_{f2}(t) = -\frac{1}{2}\hat{K}BB^TPe(t)$ and K is a unknown constant and will be defined later. The weights W_1, W_2, $\hat{K}$, $\hat{d}$ are updated as

$$\begin{aligned} \dot{W}_1 &= -\gamma_1 BPe(t)\sigma^T(\hat{x}) \\ \dot{W}_2 &= -\gamma_2\Theta[BPe(t)u^T(t)\phi(\hat{x}(t))] \\ \dot{\hat{K}} &= \frac{\gamma_3}{2}\|BPe(t)\|^2 \qquad \dot{\hat{d}} = \gamma_4\|Pe(t)\| \end{aligned} \tag{13}$$

where $\hat{K}$, $\hat{d}$ are estimation values of unknown constants K, d respectively, $\gamma_i(i = 1, 2, 3, 4)$ are defined positive constants, $\Theta[.]$ represent a kind of transformation that make the common matrix into diagonal matrix, and P is the solution of the Lyapunov equation (12), then the error dynamics of the identification scheme described by (11) satisfy $\lim_{t\to\infty} e(t) = 0$.

Remark 1. Compared with dynamic neural network models in [7, 8], there are two improvements in this paper. Firstly we consider the identification error $F(t)$ and construct the error compensation term $u_{f1}(t)$ to guarantee the $e(t)$ convergence to zero. However, there is a disadvantage when using the compensation term in practical applications. Since from theoretical point of view, $e(t)$ will exactly convergent to zero in finite time which cause singularity in the compensation term. A simple way to overcome the defect is to modify the compensation term as

$$u_{f1}(t) = \begin{cases} -Pe\|Pe\|^{-1}\hat{d} & when \quad \|e(t)\| \geq \kappa \\ 0 & when \quad \|e(t)\| \leq \kappa \end{cases}$$

where κ is a very small positive constant. Secondly the optimal model parameters W_1^*, W_2^* in (6) are exist and bounded, i.e. $W_1^* W_1^{*T} \leq \bar{W}_1$, $W_2^* W_2^{*T} \leq \bar{W}_2$ (see [8]), but their boundaries $\bar{W}_1$, $\bar{W}_2$ are unknown and different to be utility in practical process. In this paper, the compensation term $u_{f2}(t)$ is designed to eliminate the influence of the unknown boundary, where $K = \|\bar{W}_1^* + \bar{W}_2^*\|$, $W_1^* W_1^{*T} = \bar{W}_1^*$, $W_2^* W_2^{*T} = \bar{W}_2^*$. The condition about Riccati equation (see [8] for detail) can be avoided and the identification error can be guaranteed to converge to zero.

4 Adaptive Tracking Control for the Reference Model

In this section, we investigate the tracking problem. Corresponding to (3), a desired (known) PDF to be tracked can be described as

$$\sqrt{g(y)} = C_0(y)V_g + h(V_g)B_n(y) \tag{14}$$

where V_g is the desired weighting vector corresponding to $B_i(y)$. The tracking objective is to find $u(t)$ such that $\gamma(y, u(t))$ can follow $g(y)$. The error is formulated by $\Delta_e = \sqrt{g(y)} - \sqrt{\gamma(y, u(t))}$, i.e.

$$\Delta_e = C_0 V_e + [h(V_g) - h(V(t))]B_n(y) \tag{15}$$

where $V_e = V_g - V(t)$. Due to continuity of $h(V(t))$, $\Delta_e \longrightarrow 0$ holds as long as $V_e \longrightarrow 0$. The considered PDF control problem can be formulated into the tracking problem for the above nonlinear weighting systems, and the control objective is to find $u(t)$ such that the tracking performance, stability are guaranteed simultaneously.

At this stage, a dynamic reference model is considered here as a part of the target model. Indeed, the dynamic reference model has advantages in adjusting the closed-loop dynamic transient behavior and has been used widely in the past for model following control and model reference adaptive control. The desired weight vector $V_g \in R^{n-1}$ can be get by the following dynamic reference model

$$\begin{cases} \dot{x}_m = A_m x_m + B_m r \\ V_m(t) = C_m x_m(t) \end{cases} \tag{16}$$

At this stage, the problem is transformed into a nonlinear dynamical control problem for error vector $e_v = V(t) - V_m(t)$. From (9) and (16), the error $e_v(t)$ can be expressed as

$$e_v(t) = C\hat{x} - C_m x_m \tag{17}$$

Define $\bar{C} = \begin{bmatrix} C \\ C_1 \end{bmatrix}$, $\bar{C}_m = \begin{bmatrix} C_m \\ C_{m1} \end{bmatrix}$, where $C_1 \in R^{(m-n+1)\times m}, C_{m1} \in R^{(m-n+1)\times m}$ are artribary matrices which satisfy $|\bar{C}| \neq 0$ and $|\bar{C}_m| \neq 0$. So we can get

$$\bar{e}_v(t) = \bar{C}\hat{x} - \bar{C}_m x_m \tag{18}$$

$$\begin{aligned}\dot{\bar{e}}_v(t) = \bar{C}A\hat{x} + \bar{C}BW_1\sigma(\hat{x}) + \bar{C}BW_2\phi(\hat{x})u(t) + \bar{C}u_f(t) \\ -\bar{C}_m A_m x_m - \bar{C}_m B_m r\end{aligned} \tag{19}$$

Taking u to be equal to

$$\begin{aligned}u(t) = -[\bar{C}BW_2\phi(\hat{x})]^{-1}[\bar{C}A\bar{C}^{-1}\bar{C}_m x_m + \bar{C}BW_1\sigma(\hat{x}) \\ +\bar{C}u_f(t) - \bar{C}_m A_m x_m - \bar{C}_m B_m r]\end{aligned} \tag{20}$$

and substituting it to (19) we can get $\dot{\bar{e}}_v(t) = \bar{C}A\bar{C}^{-1}\bar{e}_v(t) = \bar{A}\bar{e}_v(t)$

In order to assure the existence of $[\bar{C}BW_2\phi(\hat{x})]^{-1}$, we need to establish is $w_{2i} \neq 0$. In particular, the standard adaptive laws are modified to(see [11])

$$\dot{W}_1 = \begin{cases} -\gamma_1 BPe(t)\sigma^T(\hat{x}) & when \quad \|W_1\| < M_1 \quad or \quad \|W_1\| = M_1 \\ & and \quad tr\{\sigma(\hat{x})e^T(t)PBW_1\} \geq 0 \\ -\gamma_1 BPe(t)\sigma^T(\hat{x}) + \gamma_1 tr\{\sigma(\hat{x})e^T(t)PBW_1\}\frac{W_1}{\|W_1\|^2} & when \quad \|W_1\| = M_1 \\ & and \quad tr\{\sigma(\hat{x})e^T(t)PBW_1\} < 0 \end{cases} \tag{21}$$

when $w_{2i} = \varepsilon$, we adopt

$$\dot{w}_{2i} = \begin{cases} -\gamma_2 b_i u_i \phi_i(\hat{x})e^T(t)P_i & when \quad b_i u_i \phi_i(\hat{x})e^T(t)P_i < 0 \\ 0 & when \quad b_i u_i \phi_i(\hat{x})e^T(t)P_i \geq 0 \end{cases} \tag{22}$$

where u_i is the ith element of $u(t)$ and P_i is the ith column of P. Otherwise

$$\dot{W}_2 = \begin{cases} -\gamma_2 \Theta[BPe(t)u^T(t)\phi(\hat{x}(t)))] & when \quad \|W_2\| < M_2 \quad or \quad \|W_2\| = M_2 \\ & and \quad tr\{BPe(t)u^T(t)\phi(\hat{x}(t))W_2\} \geq 0 \\ -\gamma_2 \Theta[BPe(t)u^T(t)\phi(\hat{x}(t))] + \gamma_2 tr\{BPe(t)u^T(t)\phi(\hat{x}(t))W_2\}\frac{W_2^T}{\|W_2\|^2} & \\ when \quad \|W_2\| = M_2 \quad and \quad tr\{BPe(t)u^T(t)\phi(\hat{x}(t))W_2\} < 0 & \end{cases} \tag{23}$$

$$\dot{\hat{K}} = \frac{\gamma_3}{2}\|BPe(t)\|^2 \qquad \dot{\hat{d}} = \gamma_4\|Pe(t)\| \tag{24}$$

Theorem 2. Consider the control scheme (9), the reference model (16), the control law (20) and the adaptive law (21)–(24), we have the following properties

(1) $\|W_1\| \leq M_1; \|W_2\| \leq M_2, \ \ w_{2i} \geq \varepsilon$, where M_1, M_2, ε are known constants.

(2) $\lim_{t\to\infty} e(t) = \lim_{t\to\infty} e_v(t) = \lim_{t\to\infty} \bar{e}_v(t) = 0$, $\lim_{t\to\infty} \dot{\tilde{W}}_1 = \lim_{t\to\infty} \dot{\tilde{W}}_2 = 0$, $\lim_{t\to\infty} V(t) = \lim_{t\to\infty} V_m(t) = V_g$.

Remark 2: Compared with [3, 4], this paper can not only accomplish the dynamic tracking control problem in designed control input, but also identify the dynamic trajectory of the weights related to the PDFs through dynamic neural network identifier. The modified adaptive laws (21-24) can satisfy our demands than those of [11]. (21) and (23) can guarantee the bound for weights W_1 and W_2. Define Φ_1, Φ_2 are constraint sets for W_1 and W_2 respectively, that is

$$\Phi_1 = \{W_1 : \ tr(W_1 W_1^T) \le M_1, M_1 > 0\}$$

$$\Phi_2 = \{W_2 : \ tr(W_2 W_2^T) \le M_2, M_2 > 0\}$$

In (23), $\Theta[.]$ is applied because W_2 is a diagonal matrix. While the initial value of w_{2i} is chosen to be larger than ε so that $w_{2i} \ge \varepsilon$ can be ensured through (22).

The proof of Theorem is omitted here to save space.

5 An Illustrative Example

In many practical processes such as the particle distribution control problems, the shapes of measured output PDF normally have 2 or 3 peaks. Suppose that the output PDFs can be approximated using the square root B-spline models described by (3) with $n = 3$, $y \in [0, 1.5]$ and $i = 1, 2, 3$

$$B_i(y) = \begin{cases} |\sin 2\pi y| & y \in [0.5(i-1); 0.5i] \\ 0 & y \in [0.5(j-1); 0.5j] \quad i \ne j \end{cases} \tag{25}$$

From the notations in (4), it can be seen that $\Lambda_1 = diag\{0.25, 0.25\}, \Lambda_2 = [0, 0], \Lambda_3 = 0.25$. The desired PDF $g(y)$ is supposed to be described by (16) with $V_g = [\frac{\pi}{3}, \frac{\pi}{6}]^T$.

In this paper, the nonlinear dynamic relationships between the input and the weights related to the PDF are assumed by the difference equation

$$\dot{x}(t) = A_1 x(t) + B_1 f(x) + C_1 u(t) + d_1 \tag{26}$$

where

$$A_1 = \begin{bmatrix} -3 & 0 & -1 \\ 2 & -4 & -1 \\ 2 & 0 & -3 \end{bmatrix}, C_1 = \begin{bmatrix} 1 & 0 & 0 \\ 0 & 1 & 0 \\ 0 & 0 & 1 \end{bmatrix}, x_0 = \begin{bmatrix} 2 \\ 3 \\ 0 \end{bmatrix}, B_1 = \begin{bmatrix} 1 \\ -1 \\ -1 \end{bmatrix}, d_1 = \begin{bmatrix} 0.5 \\ 0.5 \\ -0.5 \end{bmatrix}$$

$$f(x) = 2sinx_1 - 6cosx_2 + 2sinx_3$$

Let us select dynamic neural network as

$$\begin{cases} \dot{\hat{x}}(t) = A\hat{x}(t) + W_1\sigma(\hat{x}) + W_2\phi(\hat{x})u(t) + u_f(t) \\ V(t) = C\hat{x}(t) \end{cases} \tag{27}$$

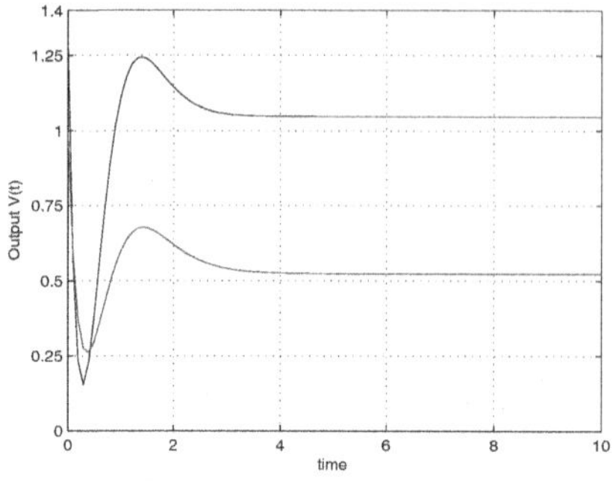

Fig. 1. Outputs of the DNN

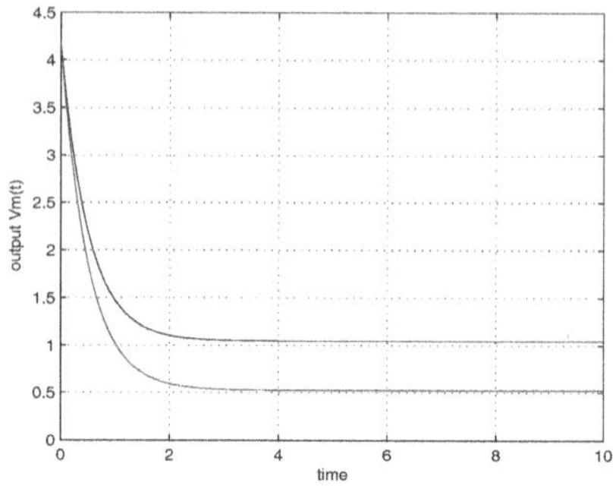

Fig. 2. Outputs of the reference model

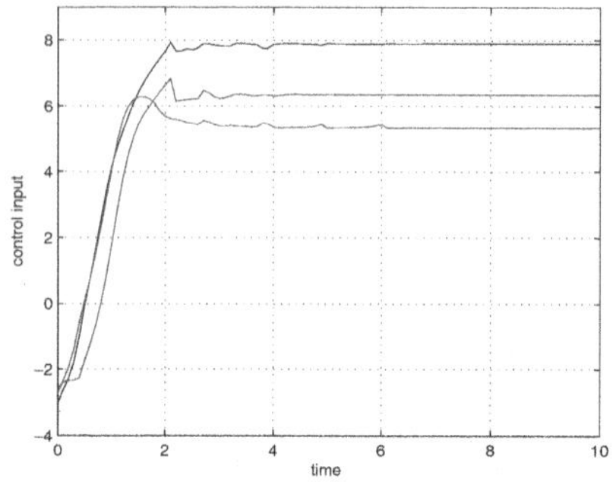

Fig. 3. The control input

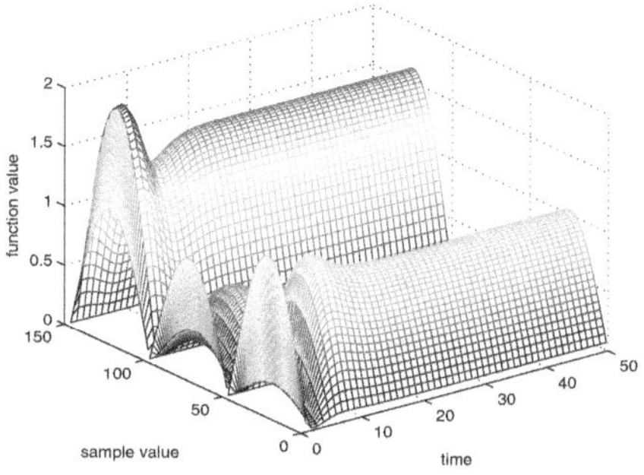

Fig. 4. 3D mesh plot of the output mesh

where

$$\sigma(x_i) = \phi(x_i) = \frac{2}{1 + e^{-0.5x_i}}, \hat{x}_0 = \begin{bmatrix} 2 \\ 3 \\ 0 \end{bmatrix}, A = \begin{bmatrix} -3 & 0 & -2 \\ 0 & -2 & -2 \\ 2 & 0 & -2 \end{bmatrix}, C = \begin{bmatrix} \frac{2}{3} & 0 & 0 \\ 0 & \frac{1}{3} & 0 \end{bmatrix}$$

In reference model (16),

$$A_m = \begin{bmatrix} -2 & 0 & 0 \\ 0 & -2 & 0 \\ 0 & 0 & -2 \end{bmatrix}, B_m = \begin{bmatrix} 1 \\ -0.5 \\ 1 \end{bmatrix}, x_{m,0} = \begin{bmatrix} -2 \\ -2 \\ -1 \end{bmatrix}, C_m = \begin{bmatrix} \frac{2\pi}{3} & 0 & 0 \\ 0 & -\frac{2\pi}{3} & 0 \end{bmatrix}$$

In adaptive laws (21–24),

$$\gamma_i = 3, i = 1, 2, 3, 4, \quad \hat{d}(0) = 2, P = W_{1,0} = W_{2,0} = \begin{bmatrix} 1 & 0 & 0 \\ 0 & 1 & 0 \\ 0 & 0 & 1 \end{bmatrix}$$

Fig. 1 and Fig. 2 are the output trajectories of the dynamic neural network and reference model respectively. The control law is showed in Fig. 3. Fig. 4 shows the 3-D mesh plot of the output PDFs.

6 Conclusion

In this paper, two step neural networks are employed to solve the tracking control problem for general non-Gaussian stochastic systems. After B-spline approximation to the measured output PDFs, the control objective is transferred into the tracking of given weights that correspond to the desired PDF. Dynamic neural networks describe the complex nonlinear relationships between the control input and the weights. An adaptive state feedback controller based on the dynamic neural networks guarantee the tracking performance.

References

1. Wang, H.: Bounded Dynamic Stochastic Systems: Modelling and Control. Springer-Verlag, London (2000)
2. Forbes, M.J., Forbes, J.F., Guay, M.: Regulatory Control Design for Stochastic Processes: Shaping the Probability Density Function. In: Proc. ACC. Denver, USA (2003) 3998-4003
3. Guo, L., Wang, H.: PID Controller Design for Output PDFs of Stochastic Systems Using Linear Matrix Inequalities. IEEE Trans. Systems, Man and Cybernetics-Part B **35** (2005) 65-71
4. Guo, L., Wang, H.: Fault Detection and Diagnosis for General Stochastic Systems Using B-spline Expansions and Nonlinear Filters. IEEE Trans. Circuits and Systems-I **52** (2005) 1644-1652
5. Narendra, K.S., Parthasarath, K.: Identification and Control of Dynamical Systems Using Neural Networks. IEEE Trans. Neural Networks **1** (1990) 4-27
6. Brown, M., Harris, C.J.: Neurofuzzy Adaptive Modeling and Control. Prentice-Hall, Englewood Cliffs, NJ (1994)
7. Poznyak, A.S., Yu, W., Sanchez, E.N., Perez, J.P.: Nonlinear Adaptive Trajectory Tracking Using Dynamic Neural Networks. IEEE Trans. Neural Networks **6** (1999) 402-1411
8. Yu, W., Li, X.O.: Some New Results on System Identification with Dynamic Neural Networks. IEEE Trans. Neural Networks **12** (2002) 412-417
9. Ren, X.M., Rad, A.B., Chan, P.T., W, L.L.: Identification and Control of Continuous-time Nonlinear Systems via Dynamic Neural Networks. IEEE Trans. Industrial Electronics **50** (2003) 478-486
10. Lin, C.M., Hsu, C.F.: Recurrent Neural Network Based Adaptive Backstepping Control for Induction Servomotors. IEEE Trans. Industrial Electronics **52** (2005) 1677-1684
11. Zhang, T.P.: Stable Direct Adaptive Fuzzy Control for a Class of MIMO Nonlinear System. Int. J. Systems Science **34** (2003) 375-388

Adaptive Global Integral Neuro-sliding Mode Control for a Class of Nonlinear System*

Yuelong Hao, Jinggang Zhang, and Zhimei Chen

Institute of Electronic Information Engineering,
Taiyuan University of Science and Technology, 030024, China
haoyuelong@126.com
http://www.springer.com/lncs

Abstract. An scheme of composite sliding control is proposed for a class of uncertainty nonlinear system, which is based on fuzzy neural networks (FNN) and simple neural networks (SNN). The SNN is uniquely determined by the design of the global integral sliding mode surface, the output of which replaces the corrective control, and FNN is applied to mimic the equivalent control. In this scheme, the bounds of the uncertainties and the extern disturbance are not required to be known in advance, and the stability of systems is analyzed based on Lyapunov function. Simulation results are given to demonstrate the effectiveness of this scheme.

1 Introduction

Sliding-mode control is one of the effective nonlinear robust control approaches since it provides fast system dynamic responses with an invariance property to uncertainties once the system dynamics are controlled in the sliding mode [1-2]. The design procedure of sliding-mode control is first to select a sliding surface that models the desired closed-loop performance in the state variable space and then to design the control such that the system state trajectories are forced toward the sliding surface and stay on it. Generally, sliding mode surfaces are in linear forms and the control consists of equivalent control and corrective control.

In the design of the SMC law, it is assumed that the control can be switched from one value to another infinitely fast. However, this is impossible to achieve in practical systems because finite time delays are present for control computation, and limitations exist in the physical actuators. This non-ideal switching results in a major problem, i.e., the chattering phenomenon which is the first disadvantage. This phenomenon not only highly undesirable by itself but it may also excite the high-frequency unmodelled dynamics, neglected in the course of modelling, which could result in unforeseen instability, and can also cause damage to actuators or the plant. To alleviate these difficulties, several modified sliding control laws are proposed [3-4]. The most popular solution is the boundary-layer approach which

* Sponsored by Shanxi Nature Science Foundation(20041049).

D. Liu et al. (Eds.): ISNN 2007, Part I, LNCS 4491, pp. 102–111, 2007.

uses a high-gain feedback when the system motion reaches the ϕ-vicinity of the sliding manifold [5].

The second disadvantage is the difficulty involved in the calculation of what is known as the equivalent control. A thorough knowledge of the plant dynamics is required for this purpose [6]. In order to avoid the computational burden, an estimation technique can be used to calculate the value of equivalent control. More recently, the use of intelligent techniques based on fuzzy logic and neural networks (NNs) have been suggested [7-8]. In [9], the radial basis function (RBF) neural networks (NNs) combined with sliding mode are used to design the adaptive control architecture for continuous-time dynamic nonlinear systems. A fuzzy NNs sliding mode controller was also developed for a class of lager-scale system with unknown bounds of high-order interconnections and disturbance [10]. The whole system has one FNN for control and another FNN for identification.

In this paper, an improved neural control structure is proposed. This neural controller is based on the FNN and the SNN, which is determined by the design of the global integral sliding mode surface, and results in smoothed control performance without requiring complex calculation of equivalent control term. In addition, online adaptive updating of the switching gain in the SNN controller can eliminate the need for larger control signal and avoid the requirement for estimating the bounds on system uncertainties and extern disturbance.

2 Design of General Sliding Mode Control Law

Consider the SISO (single-input and single-output) affine nonlinear dynamical system

$$x_{(n)} = f(X) + g(X)u(t) + d(t), \tag{1}$$

where $X = [x_1, x_2, \cdots, x_n]^T = [x, \dot{x}, \cdots, x^{(n-1)}]^T \in \mathbb{R}^n$ is the system state vector which is assumed to be available for measurement, the scalar x is the variable of interest (for instance the positing of mechanical system), $x^{(i)}$ being the i order time derivative of x, $u(t) \in \mathbb{R}$ is the control input (for example, the motor torque). $f(X)$ and $g(X)$ are nonlinear system functions representing the dynamic system behavior and known control gain respectively, and $d(t)$ is unknown extern disturbance. It is assumed that $f(X) = \hat{f}(x) + \Delta f(X)$, where is the estimation of $f(X)$, and $\Delta f(X)$ is the model uncertainty. Let F and D be the upper bound functions of $\Delta f(X)$ and $d(t)$ i.e, $|\Delta f(X)| < F$ and $|d(t)| < D$. The control problem is to obtain the state X for tracking a desired state X_d in the presence of model uncertainties and external disturbance with tracking error

$$e = X - X_d = [e, \dot{e}, \cdots, e^{(n-1)}]. \tag{2}$$

Then (1) can be rewritten as

$$\begin{cases} \dot{e_1} = e_2, \\ \dot{e_2} = e_3, \\ \cdots \\ \dot{e_n} = -f(X) - g(X) \cdot u - d(t) + x^{(n-1)}. \end{cases} \tag{3}$$

Define a sliding surface in the space of state error as

$$\begin{aligned} s(e) &= c_1 e + c_2 \dot{e} + \cdots + c_{n-1} e^{(n-2)} + e^{(n-1)} \\ &= c^T e. \end{aligned} \tag{4}$$

Where $c = [c_1, c_2, \cdots, c_{n-1}, 1]^T$are the coefficient of the Hurwitiz polynominal $h(\lambda) = \lambda^{n-1} + c_{n-1}\lambda^{n-2} + \cdots + c_1$, the tracking problem $X = X_d$ can be considered as the state error vector remaining on the sliding surface $s(e) = 0$, for all $t \geq 0$. A sufficient condition to achieve this behavior is to select a control strategy such that

$$\frac{1}{2}\frac{d(s^2(t))}{dt} \leq -\eta |s|, \quad \eta > 0. \tag{5}$$

Consider the control problem of nonlinear system (1). We take the derivative of (4) and set $\dot{s}(t) = 0$; then, we have the equivalent control

$$u_{eq} = \frac{1}{g(X)}(-\sum_{i=1}^{n-1} c_i e_{i+1} - \hat{f}(X) - \Delta f(X) - d(t) + x_d^{(n)}). \tag{6}$$

From (6), the control law is taken as

$$u = \frac{1}{g(X)}(-\sum_{i=1}^{n-1} c_i e_{i+1} - \hat{f}(X) + x_d^{(n)} - (F + D + \eta) sgn(s)). \tag{7}$$

Where $\eta > 0$ is constant and

$$sgn(s) = \begin{cases} 1 & for \quad s > 0, \\ 0 & for \quad s = 0, \\ -1 & for \quad s < 0. \end{cases}$$

Now set

$$\hat{u}_{eq} = \frac{1}{g(X)}(-\sum_{i=1}^{n-1} c_i e_{i+1} - \hat{f}(X) + x_d^{(n)}), \tag{8}$$

$$u_c = \frac{1}{g(X)}(-(F + D + \eta)) sgn(s). \tag{9}$$

Where $\hat{u}_{eq}$is the estimation of the desired equivalent control, and u_c is the corrective control, then

$$u = \hat{u}_{eq} + u_c. \tag{10}$$

From the analysis above, we get

$$s \cdot \dot{s} \leq -\eta |s|. \tag{11}$$

The control guarantees the sliding condition of (5). Therefore, under the control law (7), the sliding surface exists and is reachable.

3 Design of Neuro-sliding Mode Controller(NSMC)

In this study, we use neural network to generate u_{eq} and u_c in SMC. u_{eq} is generated by FNN and u_c is generated by SNN. The structure of NSMC is shown in Fig.1 The presentation of FNN and SNN will be discussed in following sections.:

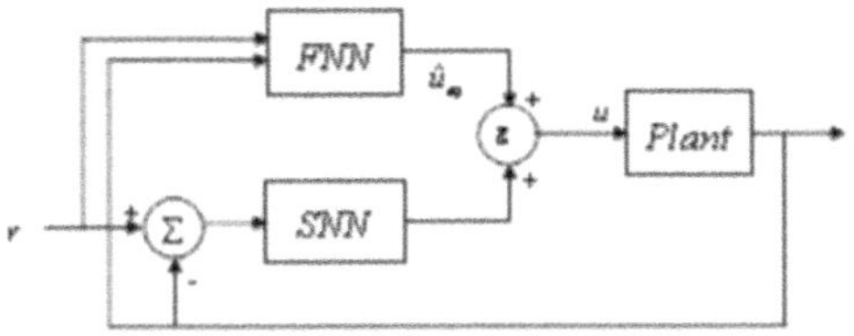

Fig. 1. the structure of NSMC

3.1 Design of Neuro-sliding Mode Equivalent Controller

A four-layer FNN as show in Fig. 2, which comprises the input layer, membership layer, rule layer, and output layer, is adopted to implement the FNN controller in this study. It is assumed that each control rule has two input variable (r_1, r_2) and the output is $\hat{u}_{eq}$. It is also assumed that each input variable has seven membership functions. The term set of each fuzzy variable is $\{NB, NM, NS, Z, PS, PM, PB\}$ are abbreviation for the commonly used name "Negative Big" "Negative Medium" and so on.

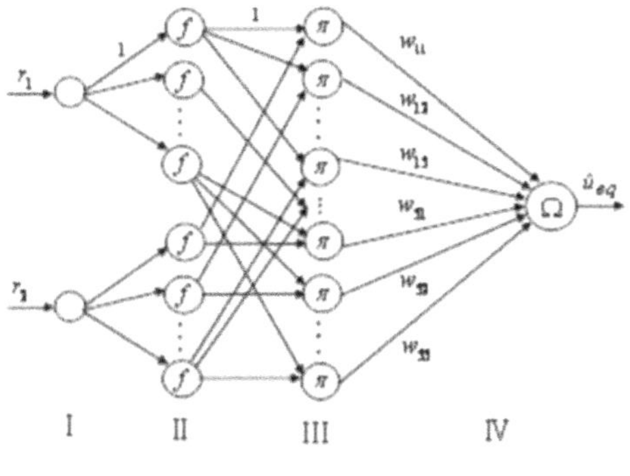

Fig. 2. Architecture of FNN

The layer (I) and (II) in Fig. 2 correspond to the antecedent part of the fuzzy control rules, and the layer (III) and (IV) correspond to the conclusion part. The input-output relationship of units in the FNN are define as follows: (I and O stand for input and output respectively; k takes 1or 2; $i = 1, \cdots, 7$; $j = 1, \cdots, 7$)

$$(I) \qquad I_k^{(1)} = r_k\,; \qquad O_k^{(1)} = I_k^{(1)}. \tag{12}$$

$$(II) \quad I_{ki}^{(2)} = O_k^{(1)}; \quad O_{ki}^{(2)} = exp(-\frac{(I_{ki}^{(2)} - a_{ki})^2}{(b_{ki})^2}). \tag{13}$$

In this layer, the Gaussian function is adopted as the membership function. Where a_{ki} and b_{ki} are, respectively, the mean and the standard deviation of the Gaussian functions

$$(III) \quad I_{ij}^{(3)} = O_{1i}^{(2)} O_{2j}^{(2)}; \quad O_{ij}^{(3)} = I_{ij}^{(3)}. \tag{14}$$

$$(IV) \quad \hat{u}_{eq} = \frac{\sum_i \sum_j O_{ij}^{(3)} w_{ij}}{\sum_i \sum_j O_{ij}^{(3)}}. \tag{15}$$

The online learning algorithm of the FNN is the supervised gradient descent method. The energy function is define as:

$$E = \frac{1}{2}(u_{eq} - \hat{u}_{eq})^2, \tag{16}$$

where $\hat{u}_{eq}$ is the output of FNN and u_{eq} is the desired equivalent control output. The weight$w$ is changed as followed:

$$w_{ij}(t+1) = w_{ij}(t) - \alpha \cdot \frac{\partial E}{\partial w_{ij}} + \beta \cdot \Delta w_{ij}(t), \tag{17}$$

where $\Delta w_{ij} = w_{ij}(t) - w_{ij}(t-1)$, α is the learning rate, and β is a momentum constant which can diminish the oscillation occurring in the learning process. The parameters (a_{ik}, b_{ik}) of the membership function can be modified as:

$$a_{ki}(t+1) = a_{ki}(t) - \alpha \cdot \frac{\partial E}{\partial a_{ki}} + \beta \cdot \Delta a_{ki}(t), \tag{18}$$

$$b_{ki}(t+1) = b_{ki}(t) - \alpha \cdot \frac{\partial E}{\partial b_{ki}} + \beta \cdot \Delta b_{ki}(t). \tag{19}$$

Equation (18) can be derived as follows:

$$\begin{aligned}
\frac{\partial E}{\partial w_{ij}} &= \frac{\partial E}{\partial y(t)} \frac{\partial y(t)}{\partial w_{ij}} \\
&= -(u_{eq}(t) - \hat{u}_{eq}(t)) \cdot \frac{\partial}{\partial w_{ij}} \frac{\sum_i \sum_j O_{ij}^{(3)} w_{ij}}{\sum_i \sum_j O_{ij}^{(3)}} \\
&= -(u_{eq}(t) - \hat{u}_{eq}(t)) \cdot \frac{O_{ij}^{(3)}}{\sum_i \sum_j O_{ij}^{(3)}},
\end{aligned} \tag{20}$$

$$\frac{\partial E}{\partial a_{1i}} = \frac{\partial E}{\partial \hat{u}_{eq}} \frac{\partial \hat{u}_{eq}}{\partial O_{ij}^{(3)}} \frac{\partial O_{ij}^{(3)}}{\partial O_{1i}^{(2)}} \frac{\partial O_{1i}^{(2)}}{\partial a_{1i}}, \tag{21}$$

$$\frac{\partial E}{\partial b_{1i}} = \frac{\partial E}{\partial \hat{u}_{eq}} \frac{\partial \hat{u}_{eq}}{\partial O_{ij}^{(3)}} \frac{\partial O_{ij}^{(3)}}{\partial O_{1i}^{(2)}} \frac{\partial O_{1i}^{(2)}}{\partial b_{1i}}, \tag{22}$$

$$\frac{\partial E}{\partial \hat{u}_{eq}} = -(u_{eq} - \hat{u}_{eq}), \tag{23}$$

$$\frac{\partial \hat{u}_{eq}}{\partial O_{ij}^{(3)}} = \frac{w_{ij} \sum_i \sum_j O_{ij}^{(3)} - \sum_i \sum_j (O_{ij}^{(3)} w_{ij})}{(\sum_i \sum_j O_{ij}^{(3)} w_{ij})^2}, \tag{24}$$

$$\frac{\partial O_{ij}^{(3)}}{\partial O_{1i}^{(2)}} = O_{2j}^{(2)}, \tag{25}$$

$$\frac{\partial O_{1i}^{(2)}}{\partial a_{1i}} = -\frac{2(I_{1i}^{(2)} - a_{1i})}{b_{1i}^2}, \tag{26}$$

$$\frac{\partial O_{1i}^{(2)}}{\partial b_{1i}} = -\frac{2(I_{1i}^{(2)} - a_{1i})^2}{b_{1i}^3}. \tag{27}$$

The a_{2j} and b_{2j} can be obtained in the same manner. The actual equivalent control $u_{eq}(t)$ in equation (16) is unknown. Since the characteristic of $u_{eq}(t) - \hat{u}_{eq}(t)$ and $s(t)$ are similar, the value of function $s(t)$ can be utilized to replace the $u_{eq}(t) - \hat{u}_{eq}(t)$. Then,

$$\frac{\partial E}{\partial w_{ij}} = \frac{\partial E}{\partial y(t)} \frac{\partial y(t)}{\partial w_{ij}} = -s(t) \cdot \frac{O_{ij}^{(3)}}{\sum_i \sum_j O_{ij}^{(3)}}. \tag{28}$$

3.2 Computation of the Global Integral Corrective Controller

In the second NN structure, the corrective control u_c is obtained by a simple neural network (SNN), thus the structure of SNN is easy to determine via designing SMC. The structure of SNN is also a feed-forward network, which has one input layer and one output layer. The structure of SNN for the manipulator is presented in Fig.3

From Fig.3 it can be found that the inputs of neuron are the state error and the error integral, and the threshold function is an exponential one. The output neuron is a full connection structure. The advantage is that the controller possesses the self-tuning characteristic. As a general neural network, the output of the output neuron also passes through an activation function, sign function

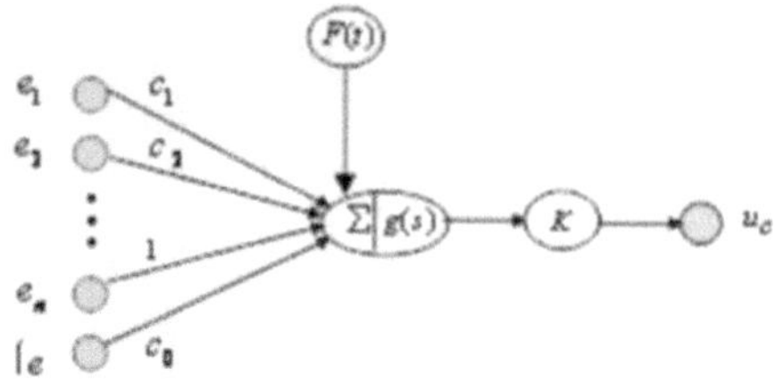

Fig. 3. Architecture of SNN

or sign-like continuous functions. Here, we use a sigmoid transfer function as the expression. The output of the output layer is the corrective control Δu.

In general, the u_c is defined as follows:

$$u_c = K \cdot g(S). \tag{29}$$

Weight Adaptation of SNN: In SNN, for the control objective that is $s \longrightarrow 0$, and for minimizing the value of function, hence, we define a cost function below:

$$J = \frac{1}{2}s^2. \tag{30}$$

The purpose is that we need to minimize J the cost function to minimize . It is necessary to change the parameters in the direction of the negative gradient

$$\Delta K = -\mu \cdot \frac{\partial J}{\partial K}, \tag{31}$$

$$\Delta c_i = -\mu \cdot \frac{\partial J}{\partial c_i}. \tag{32}$$

where μ is the learning rate parameter of the back propagation algorithm and it is a constant. Defining the global integral sliding surface s

$$s = c_1e_1 + c_2e_2 + \cdots + c_{n-1}e_{n-1} + e_n + c_0 \int e - F(t). \tag{33}$$

Where $F(t) = s(0) \cdot exp(-\lambda t)$[12], $\lambda > 0$, $s(0)$ is the initial value of the with the time $t = 0$. $c = [c_1, c_2, \cdots, c_{n-1}]^T$ are the coefficients of the Hurwitiz polynominal $h(\lambda) = \lambda^{n-1} + c_{n-1}\lambda^{n-2} + \cdots + c_1$. The gradient descent for K is

$$\Delta K = -\mu \cdot \frac{\partial J}{\partial K} = -\mu \cdot s \cdot \frac{\partial s}{\partial K}. \tag{34}$$

$$\begin{aligned} s(e) &= c_1e_1 + c_2e_2 + \cdots + c_{n-1}e_{n-1} + e_n \\ &= c_1e_1 + c_2e_2 + \cdots + c_{n-1}e_{n-1} + x_d^{(n-1)} - x_n \\ &= c_1e_1 + c_2e_2 + \cdots + c_{n-1}e_{n-1} + x_d^{(n-1)} - \int [f(X) + b(X)u + d(t)]dt. \end{aligned} \tag{35}$$

That is,

$$\frac{\partial s}{\partial K} = -\int g(s) \cdot b(X)dt. \tag{36}$$

The last form of K-adaptation is obtained as

$$\Delta K = \mu \cdot s \cdot \int g(s) \cdot b(X)dt. \tag{37}$$

Similar to the derivation of (34-37), the gradient decent for can be derived as

$$\begin{aligned}\Delta c_i &= -\mu \cdot \frac{\partial J}{\partial c_i} = -\mu \cdot \frac{\partial J}{\partial s}\frac{\partial s}{\partial c_i} \\ &= -\mu \cdot s \cdot \frac{\partial s}{\partial c_i} = -\mu \cdot s \cdot e_i,\end{aligned} \tag{38}$$

$$\begin{aligned}\Delta c_0 &= -\mu \cdot \frac{\partial J}{\partial c_0} = -\mu \cdot \frac{\partial J}{\partial s}\frac{\partial s}{\partial c_0} \\ &= -\mu \cdot s \cdot \frac{\partial s}{\partial c_0} = -\mu \cdot s \cdot \int e_0.\end{aligned} \tag{39}$$

Noted that the adaptation process K of and c_i should be stopped when the state error is acceptable. They may be sensitive to system perturbations. Additionally, the bounded concept should be considered when designing, ensuring that $K > 0$ and $c_i > 0$. The overall neural network algorithm and the weight adaptation are described in the statements above.

4 Simulation and Experimental Results

In this paper, we consider the inverted pendulum for the simulation studies. By setting $x_1 = \theta$ and $x_2 = \dot{\theta}$, the dynamic equation of the inverted pendulum system [11] are given :

$$\begin{aligned}\dot{x_1} &= x_2, \\ x_2 &= \frac{gsinx_1 - \frac{mlx_2^2 cosx_1 sinx_1}{m_c+m}}{l(\frac{4}{3} - \frac{mcos^2 x_1}{m_c+m})} + \frac{\frac{cosx_1}{m_c+m}}{l(\frac{4}{3} - \frac{mcosx_1^2}{m_c+m})}u + d(t) + \Delta f,\end{aligned} \tag{40}$$

where $g = 9.8m/s^2$ is the acceleration due to gravity, $m_c = 1kg$ is the mass of car, $m = 0.1kg$ is the mass of pole, and $l = 0.5m$ is the half-length of the pole, u is the control input, $d(t)$ is extern disturbance, and Δf represents the uncertainty. We assume, $d(t) = 20e^{-\frac{(t-10)^2}{2}}$ and $\Delta f = sint$, the desired angle trajectory is $y_d = 0.1sint$.

The structure of the applied fuzzy neural network is showed Fig.2. We used a four-layer fuzzy neural network to estimate the equivalent control. The inputs of FNN are set as $[r_1, r_2] = [x, \dot{x}]$, and the initial value of the states are

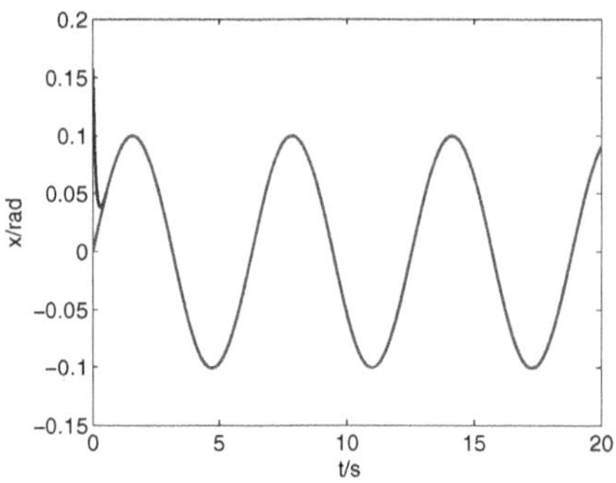

Fig. 4. Desired and observed angular response

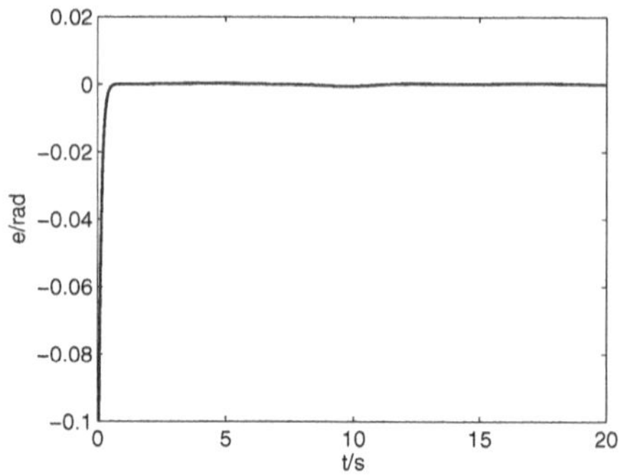

Fig. 5. Tracking error

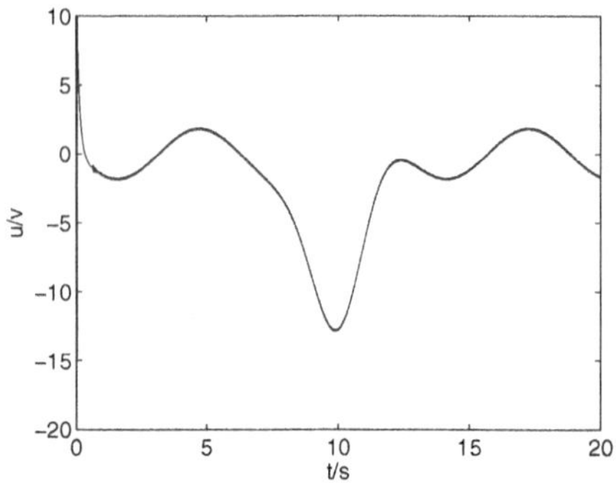

Fig. 6. Control effort

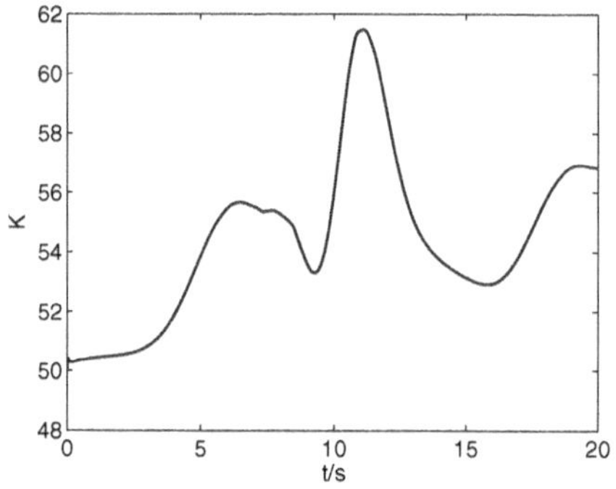

Fig. 7. The adapted Parameter K

$[x, \dot{x}] = [\frac{\pi}{20}, 0]$. The parameters of the Gaussian function in the second layer of FNN are given by experience, in this paper, we assume, the mean of the Gaussian functions is $[a_1, a_2, \cdots, a_{k7}] = [-3, -2, 0, 1, 2, 3]$ and the standard deviation b_{ki} is 1. Let the learning rates $\alpha = 0.5$ and $\beta = 0.1$. Moreover, all the network weights are initialized to random values between $[-1, 1]$.

The simple neural network of the corrective controller is showed Fig.3 . The activated function of neuron is selected as sigmoid function $g(s(t)) = \dfrac{2\alpha}{1 + e^{-\gamma s(t)}} - \alpha$, in simulation, we choose $\alpha = 1$ and $\gamma = 0.4$.

The given sliding surface $s(e) = c_2 e_2 + c_1 e_1 + c_0 \int e - s(0) e^{(-\lambda t)}$ where $\lambda = 10$, choosing the initial values of the coefficients $[c_0, c_1, c_2] = [1, 50, 1]$, and setting the learning rate $\mu = 0.3$. The simulation results show in Fig.4, Fig.5, Fig.6, Fig.7. It can be seen that the tracking performance is good even presence of uncertainty and disturbance, and the chattering of traditional sliding mode control is eliminated. This is because that the proposed controller replaces the discontinuous controller in convention sliding control system. The parameter K is adjusted adaptively to reject the external disturbance, and the static tracking error is minimized by the integral effect.

5 Conclusions

In this paperan approach of composite sliding mode control is proposed for a class of uncertainty nonlinear system. We use two parallel neural networks to implement the NSMC. The SNN structure is simple, is uniquely determined by the design of the global integral sliding mode surface. The FNN is applied to estimate the equivalent control and the SNN is used to compute the corrective control. The learning process is online. Learning and the calculation of the control signal are carried out simultaneously. The bounds of the uncertainties and the extern disturbance are not required to be known in advance. The drawback of chattering in sliding mode control is avoided and the zero steady tracking error can be ensured. The simulation results demonstrate the effectiveness of the proposed control schemes.

References

[1] Astrom, K.J., Wittenmark, B.: Adaptive control. New York: Addison-Wesley (1995)
[2] Utkin, V., Guldner, J., Shi, J.: Sliding mode control in electromechanical systems. New York: Taylor & Francis (1999)
[3] Wong, L.K., Leung, F.H.F., Tam, P.K.S.: A chattering elimination algorithm for sliding mode control of uncertain nonlinear system. Mechatronics **8** (1998) 765-775
[4] Ha, Q.P., Rye, D.C., Durrant-Whyte, H.F.: Fuzzy moving sliding mode control with application to robotic manipulators. Automatica **35** (1999) 607-616
[5] Slotine, J.J.E., Saltry, S.S.: Tracking control of nonlinear systems using sliding surface with application to robot manipulator. Int. J. Control **38** (1983) 465-492
[6] Fujimoto, H., Hori, Y., Kawamura, A.: Prefect tracking control based on multi-rate feedforword control with generalized sampling periods. IEEE Trans. Industrial Electronics **3** (2001) 636-644
[7] Tong, S., Li, H.X.: Fuzzy adaptive sliding-mode control for MIMO nonlinear systems. IEEE Trans. Fuzzy Syst. **11** (2003) 354-360
[8] Tsai, C.H., Chung, H.Y., Yu, F.M.: Neuro-sliding mode control with its applications to seesaw systems. IEEE Trans. Neural Networks **15** (2004) 124-134
[9] Onder Efe, M., Kaynak, O., Yu, X.H., Wilamowski, B.M.: Sliding mode control of nonlinear systems using gaussian radial basis function neural networks. INNS-IEEE Int.Joint Conf.on Neural Networks (2001) 474-479
[10] Da, F.: Decentralized sliding mode adaptive controller design based on fuzzy neural networks for interconnected uncertain nonlinear systems. IEEE Trans. Neural Networks **11** (2000) 1471-1480
[11] Su, J.P., Chen, T.M.: Adaptive fuzzy sliding mode control with GA-based reaching laws. J. Fuzzy and System **120** (2001) 145-158
[12] Liu, J.K., Sun, F.C.: Fuzzy global sliding mode control for a servo system with Lugre friction model. Proceedings of the 6th World Congress on Control and Automation, DaLian China (2006)

Backstepping Control of Uncertain Time Delay Systems Based on Neural Network

Mou Chen[1], Chang-sheng Jiang[1], Qing-xian Wu[1], and Wen-hua Chen[2]

[1]Automation College, Nanjing University of Aeronautics and Astronautics
Nanjing 210016, China
[2]Department of Aeronautical and Automotive Engineering, Loughborough University
Loughborough, Leicestershire LE11 3TU, UK
chenmou@nuaa.edu.cn

Abstract. In this paper, a robust adaptive control scheme is proposed for a class of uncertain MIMO time delay systems based on backstepping method with Radical basis function(RBF) neural network. The system uncertainty is approximated by RBF neural networks, and a parameter update law is presented for approximating the system uncertainty. In each step, the control scheme is derived in terms of linear matrix inequalities (LMI's). A robust adaptive controller is designed using backstepping and LMI method based on the output of the RBF neural networks. Finally, an example is given to illustrate the availability of the proposed control scheme.

1 Introduction

Backstepping control is an effective control method in the control of uncertain systems due to its robust performance with respect to uncertainty not satisfying the matching conditions [1,2].The backstepping design can also be extended to handle uncertain nonlinear systems by combining intelligent control technology such as neural networks [3]-[5]. Ref.[3] has studied two different backstepping neural network control approaches for a class of affine nonlinear systems with unknown nonlinearities. A direct adaptive backstepping neural-network control has been proposed for similar systems in [4]. Recently Ref.[5] studied the problem of controlling uncertain nonlinear systems by combining backstepping design with neural networks, however, only Single-Input-Single-Output (SISO) nonlinear systems without time delay were considered.

Stabilization of nonlinear systems with time delays is receiving much attention [6-8]. The main significance of this paper is that we design the backstepping control scheme for uncertain Multiple-Input-Multiple-Output (MIMO) delayed systems. A robust adaptive controller is designed with RBF neural networks. The structure of this paper is as follows. Section 2 describes the robust adaptive control problem formulation for a class of uncertain systems with time delay. Section 3 develops a robust adaptive control scheme based on RBF neural network. Finally, the simulation example is given in Section 4.

D. Liu et al. (Eds.): ISNN 2007, Part I, LNCS 4491, pp. 112–121, 2007.

2 Problem Formulation

Consider a uncertain delayed system in the form of

$$\begin{aligned} \dot{x_i} &= A_i x_i + B_i x_{i+1} + \sum_{j=1}^{i} H_{ij} x_j(t-\tau_i) + d_i(\overline{x}_i), 1 \leqslant i \leqslant n-1 \\ \dot{x_n} &= A_n x_n + B_n u + \sum_{j=1}^{n} H_{nj} x_j(t-\tau_n) + d_n(\overline{x}_n). \\ y &= x_1 \end{aligned} \tag{1}$$

where $x_i = [x_{i1}, x_{i2}, \ldots, x_{in}]^T \in R^n$, $\overline{x}_i = [x_1, x_2, \ldots, x_i]^T$. $x = [x_1, \ldots, x_n]^T \in R^{n \times n}$is the state vectors of uncertain time delay system (1). $u \in R^n$ is the control input, and is the output of the nonlinear system. $A_i, B_i, H_{ij} (1 \le i \le n, 1 \le j \le n)$ are matrices with corresponding dimensions where B_i is a row full rank matrix. $d_i(\overline{x}_i), 1 \le i \le n$, are the uncertainty with unknown upper bound. $\tau_i (1 \le i \le n)$ are unknown time delays of the states. The sizes of the unknown time delays are bounded by a known constant, i.e., $\tau_i \le \tau_{max}, i = 1, \ldots, n$.

RBF neural networks are chosen to approximate the influence of the uncertainties in this paper. Therefore, the approximations of compound uncertainty $\Psi_i (1 \le i \le n)$ of i-th subsystem can be expressed as

$$\hat{\Psi}_i(z_i, t) = W_i^T \phi_i(z_i). \tag{2}$$

where Ψ_i and z_i are to be defined in Section 3. $\phi_i(z_i), 1 \le i \le n$ is the base function of the corresponding RBF neural networks. $\hat{W}_i = [\hat{W}_{i1}, \hat{W}_{i2}, \ldots, \hat{W}_{in}] \in R^{n \times n}$, $\phi_i(z_i) = [\phi_{i1}, \phi_{i2}, \ldots, \phi_{in}]^T \in R^{n \times 1}$, $\hat{\Psi}_{ji} = \hat{W}_{ji}^T \phi_{ji}$, $\phi_{ji} \in R^{n \times 1}, 1 \le j \le n$.

The optimization weight value in the RBF neural networks is defined as

$$W_i^* = arg \min_{W_i \in \Omega_{\Psi_i}} \left[\sup_{z_i \in S_{z_i}} |\hat{\Psi}_i(z_i/\hat{W}_i) - \Psi_i(z_i)| \right]. \tag{3}$$

where $\Omega_{\Psi_i} = \{\hat{W}_{\Psi_i} : \|\hat{W}_{\Psi_i}\| \le M_{\Psi_i}\}$is the valid field of the parameter and $M_{\Psi_i}$is the designed parameter. $S_{z_i} \subset R^{n_i}$ is the variable space of the state vector.

Under the optimization weight value, the unknown uncertainty can be expressed as

$$\Psi_i = W_i^{*T} \phi_i(z_i) + \varepsilon_i. \tag{4}$$

where $\varepsilon_i = [\varepsilon_{i1}, \varepsilon_{i2}, \ldots, \varepsilon_{in}]^T \in R^{n \times 1}$ is the smallest approximation error of RBF neural networks. Suppose that

$$\|\varepsilon_i\| \le \varepsilon_i^*. \tag{5}$$

where $\varepsilon_i^* > 0$ is the upper bound of the approximation error of $\Psi_i(z_i)$using RBF neural networks.

3 Robust Control for Uncertain Systems Based on Backstepping Method

The task of this section is to design a robust controller based on backstepping method. Before the control design, the following projective operator is introduced.

$$E(e_i) = \begin{cases} 1, \|e_i\| \geq \kappa_i \\ 0, \|e_i\| < \kappa_i \end{cases} \tag{6}$$

where $\kappa_i > 0$ is the design parameter. The design procedure is described below.

Step 1: Let $e_1 = y_1 - y_{1d}$. It follows from Eq.(1) that

$$\dot{e}_1 = A_1 x_1 + B_1 x_2 + H_{11} x_1(t - \tau_1) + d_1(x_1) - \dot{y}_{1d}. \tag{7}$$

Let $V_{z1} = e_1^T P_1 e_1$. Its derivative is given by

$$\dot{V}_{z1} = x_1^T A_1^T P_1 e_1 + e_1^T P_1 A_1 x_1 + x_2^T B_1^T P_1 e_1 + e_1^T P_1 B_1 x_2 \\ + x_1^T(t-\tau_1) H_{11}^T P_1 e_1 + e_1^T P_1 H_{11} x_1(t-\tau_1) + 2e_1^T P_1 \Big(d_1(x_1) - \dot{y}_{1d} \Big). \tag{8}$$

Since the uncertainty $\Psi_1 = d_1(x_1) - \dot{y}_{1d}$ is unknown, the RBF neural network is used to approximate it. Eq.(8) can be rewritten as

$$\dot{V}_{z1} = x_1^T A_1^T P_1 e_1 + e_1^T P_1 A_1 x_1 + x_2^T B_1^T P_1 e_1 + e_1^T P_1 B_1 x_2 \\ + x_1^T(t-\tau_1) H_{11}^T P_1 e_1 + e_1^T P_1 H_{11} x_1(t-\tau_1) \\ + 2e_1^T P_1 \Big(W_1^{*T} \phi_1(Z_1) + \varepsilon_1 \Big). \tag{9}$$

where $z_1 = [x_1, e_1, \dot{x}_{1d}]^T \in \Omega_{z_1} \subset R^3$.

Suppose that there exist positive definite matrices Ω_1, T_1, Y_1 and $K_1 = Y_1 \Omega_1^{-1}$ which satisfy the following linear matrix inequality

$$\begin{bmatrix} \bar{A}_{11} & H_{11}\Omega_1 \\ \Omega_1 H_{11}^T & -T_1 \end{bmatrix} < 0. \tag{10}$$

where $\bar{A}_{11} = \Omega_1 K_1^T B_1^T + B_1 K_1 \Omega_1 + \Omega_1 A_1^T + A_1 \Omega_1$. By viewing x_2 as a virtual control input, let us choose a virtual controller x_2^* as follows:

$$x_2^* = P_1 B_1^T (B_1 P_1 B_1^T)^{-1} \Big(- e_1 - A_1 e_1 - P_1 e_1 - A_1 x_1 \\ - \frac{1}{e_1^T e_1} P_1^{-1} e_1 x_1^T \omega_1 x_1 + B_1 K_1 e_1 - \hat{W}_1^T \phi_1(z_1) \Big). \tag{11}$$

where $\hat{W}_1 = [\hat{W}_{11}, \hat{W}_{12}, \ldots, \hat{W}_{1n}]$, $\phi_1(z_1) = [\phi_{11}, \phi_{12}, \ldots, \phi_{1n}]^T$, $\hat{d}_{1i} = \hat{W}_{i1}^T \phi_{1i}$, $\hat{W}_{1i} \in R^{n \times 1}$, $\phi_{1i} \in R^{n \times 1}$. $\omega_1 > 0$is a positive matrix, $P_1 > 0$ will be defined in following.

Inspecting Eq.(11) reveals that x_2^* is not well-defined when $e_1 = 0$as $\lim_{e_1 \to 0} e_1^T e_1 = 0$. It is noted that point $e_1 = 0$ is not only an isolated point in Ω_{z_1}, but also the case that the system reaches the origin at this point. To facililate the discussion, let us define sets $\Omega_{z_{10}} \subset \Omega_{z_1}$ and $\Omega_{z_1}^0$ as follows:

$$\Omega_{z_1}^0 := \{e_1 | \|e_1\| < \kappa_1\}. \tag{12}$$

$$\Omega_{z_{10}} := \Omega_{z_1} - \Omega_{z_1}^0. \tag{13}$$

The control is only activated when $\Omega_{z_{10}}$. If Ω_{z_1} is a compact set, we can conclude that $\Omega_{z_{10}}$ is a compact set. So Ψ_1 can be also approximated using RBF neural networks in $\Omega_{z_{10}}$. Accordingly, the following practical control law is given

$$\begin{aligned} x_2^* = E(e_1^T e_1) P_1 B_1^T (B_1 P_1 B_1^T)^{-1} \Big(& - e_1 - A_1 e_1 - P_1 e_1 - A_1 x_1 \\ & - \frac{1}{e_1^T e_1} P_1^{-1} e_1 x_1^T \omega_1 x_1 + B_1 K_1 e_1 - \hat{W}_1^T \phi_1(z_1) \Big). \end{aligned} \tag{14}$$

Defining $e_2 = x_2 - x_2^*$ and invoking (14) in Eq.(9) yields

$$\begin{aligned} \dot{V_{z1}} = & e_1^T (K_1^T B_1^T P_1 + A_1^T P_1) e_1 + e_1^T (P_1 B_1 K_1 + P_1 A_1) e_1 - 2 e_1^T P_1 e_1 \\ & - 2 e_1^T P_1 P_1 e_1 + x_1^T (t - \tau_1) H_{11}^T P_1 e_1 + e_1^T P_1 H_{11} x_1 (t - \tau_1) \\ & + 2 e_1^T P_1 (-\tilde{W}_1^T \phi_1(z_1) + \varepsilon_1) - x_1^T \omega_1 x_1. \end{aligned} \tag{15}$$

where $\tilde{W}_1 = \hat{W}_1 - W_1^*$.

Consider the following Lyapunov function candidate:

$$V_1 = e_1^T P_1 e_1 + \sum_{j=1}^{n} \tilde{W_{1j}}^T \Gamma_{1j}^{-1} \tilde{W}_{1j} + \int_{t-\tau_1}^{t} x_1^T \omega_1 x_1 dt. \tag{16}$$

where $\Gamma_{1j} = \Gamma_{1j}^{-1} > 0$.

Substituting Eq.(15) into Eq.(16) gives

$$\begin{aligned} \dot{V}_1 \le & -2 e_1^T P_1 e_1 + e_1^T (K_1^T B_1^T P_1 + A_1^T P_1) e_1 + e_1^T (P_1 B_1 K_1 + P_1 A_1) e_1 \\ & + x_1^T (t - \tau_1) H_{11}^T P_1 e_1 + e_1^T P_1 H_{11} x_1 (t - \tau_1) + \sum_{j=1}^{n} \tilde{W}_{1j}^T \Gamma_{1j}^{-1} \dot{\hat{W}}_{1j} \\ & + 2 e_1^T P_1 (-\tilde{W}_1^T \phi_1(z_1) + \varepsilon_1) - 2 e_1^T P_1 P_1 e_1 - x_1^T (t - \tau_1) \omega_1 x_1 (t - \tau_1). \end{aligned} \tag{17}$$

Choose the following adaptive law

$$\dot{\hat{W}}_{1j} = \dot{\tilde{W}}_{1j} = 2 \Gamma_{1j} \phi_{1j}(z_1) e_1^T P_{1j}. \tag{18}$$

where P_{1j} is the j-th row of P_1 .

Invoking (18), Eq.(17) becomes

$$\dot{V}_1 = -2 e_1^T P_1 e_1 + 2 e_1^T P_1 \varepsilon_1 - 2 e_1^T P_1 P_1 e_1 + X_1^T \Lambda_1 X_1. \tag{19}$$

where $X_1 = [e_1, x_1(t-\tau_1)]^T$, $A_1 = \begin{bmatrix} \bar{A}_{11} & P_1 H_{11} \\ H_{11}^T P_1 & -\omega_1 \end{bmatrix}$, $\bar{A}_1 = K_1^T B_1^T P_1 + P_1 B_1 K_1 + A_1^T P_1 + P_1 A_1$. Defined $\Omega = P_1^{-1}$, $T_1 = P_1 \omega_1 P_1$, $Y_1 = K_1 P_1^{-1}$. It can be shown that$A_1 < 0$ by left and right multiplication of $diag(P_1, P_1)$on both sides of Eq.(10).

By completion of squares, we have

$$e_1^T P_1 \varepsilon_1 + \varepsilon_1^T P_1 e_1 \leq 2 e_1^T P_1 P_1 + \frac{1}{2} \varepsilon_1^T \varepsilon_1 \leq 2 e_1^T P_1 P_1 e_1 + \frac{1}{2} |\varepsilon_1^*|^2. \tag{20}$$

Substituting (20) into (19) yields

$$\dot{V}_1 \leq -2 e_1^T P_1 e_1 + \frac{1}{2} |\varepsilon_1^*|^2. \tag{21}$$

Step i $(2 \leq i \leq n-1)$: This step is to make the error betweenx_i and x_i^*as small as possible. Defining $e_i = x_i - x_i^*$ and differentiating it yields

$$\dot{e}_i = \dot{x}_i - \dot{x}_i^* = A_i x_i + B_i x_{i+1} + \sum_{j=1}^{i} H_{ij} x_j (t - \tau_i) + d_i(\bar{x}_i) - \dot{x}_i^*. \tag{22}$$

where $d_i = d_i(\bar{x}_i) - \dot{x}_i^*$ will be approximated an by RBF neural networks. Let $V_{zi} = V_{i-1} + \frac{1}{2} e_i^T P_i e_i$. One obtainds

$$\begin{aligned} \dot{V}_{zi} \leq & -2 \sum_{j=1}^{i} (e_j^T P_j e_j - \frac{1}{2} |\varepsilon_j^*|^2) + x_i^T A_i^T P_i e_i + e_i^T P_i A_i x_i + x_{i+1}^T B_i^T P_i e_i \\ & + e_i^T P_i B_i x_{i+1} + \sum_{j=1}^{i} x_j^T (t - \tau_i) H_{ij}^T P_i e_i + \sum_{j=1}^{i} e_i^T P_i H_{ij} x_j (t - \tau_i) \\ & + 2 e_i^T P_i (W_i^{*T} \phi_i(z_i) + \varepsilon_i). \end{aligned} \tag{23}$$

where $z_i = [x_1, x_2, \ldots, x_i, e_1, e_2, \ldots, e_i, \dot{x}_i^*] \in \Omega_{z_i} \subset R^{2i+1}$. $P_i > 0$ is to be determined.

Now suppose that there exist positive definite matrices $\Omega_i, T_i, Y_i, K_i = Y_i \Omega_i^{-1}$ satisfying the following linear matrix inequality

$$\begin{bmatrix} \bar{A}_{ii} & H_{i1}\Omega_i & \ldots & H_{ii}\Omega_i \\ \Omega_i H_{i1}^T & -T_{i1} & \ldots & 0 \\ \vdots & \vdots & \ddots & \vdots \\ \Omega_i H_{ii}^T & 0 & \ldots & -T_{ii} \end{bmatrix} < 0. \tag{24}$$

where $\bar{A}_{ii} = \Omega_i K_i^T B_i^T + B_i K_i \Omega_i + \Omega_i A_i^T + A_i \Omega_i$ .By viewing x_{i+1}^* as a virtual control input, let us choose a virtual practical controller x_{i+1}^* as follows:

$$\begin{aligned} x_{i+1}^* = & E(e_i^T e_i) P_i B_i^T (B_i P_i B_i^T)^{-1} \Big(- e_i - A_i e_i - P_i e_i - A_i x_i \\ & - \frac{P_i^{-1} e_i}{e_i^T e_i} \sum_{j=1}^{i} (x_j^T \omega_{ij} x_j) + B_i K_i e_i - \hat{W}_i^T \phi_i(z_i) \Big). \end{aligned} \tag{25}$$

where $z_i = [x_1, x_2, \dots, x_i, e_1, e_2, \dots, e_i, \dot{x}_i^*]^T \in \Omega_{zi0} \subset R^{2i+1}$. $\omega_{ij} > 0$ are positive definite matrices.

In the similar fashion as in Step 1, defining $e_{i+1} = x_{i+1} - x_{i+1}^*$. By substituting Eq.(25), Eq.(23) becomes

$$\begin{aligned}
\dot{V}_{zi} \le & -2\sum_{j=1}^{i-1}(e_j^T P_j e_j - \frac{1}{2}|\varepsilon_j^*|^2) - 2e_i^T P_i e_i - 2e_i^T P_i P_i e_i \\
& + e_i^T (K_i^T B_i^T P_i + A_i^T P_i) e_i + e_i^T (P_i B_i K_i + P_i A_i) e_i \\
& + \sum_{j=1}^{i} x_j^T (t-\tau_i) H_{ij}^T P_i e_i + \sum_{j=1}^{i} e_i^T P_i H_{ij} x_j (t-\tau_i) \\
& - \sum_{j=1}^{i} x_i^T \omega_{ij} x_i + 2e_i^T P_i (\tilde{W}_i^T \phi_i(z_i) + \varepsilon_i).
\end{aligned} \tag{26}$$

where $\tilde{W}_i = \hat{W}_i - W_i^*$.

Consider a Lyapunov function candidate

$$V_i = e_i^T P_i e_i + \sum_{j=1}^{n} \tilde{W}_{ij}^T \Gamma_{ij}^T \tilde{W}_{ij} + \int_{t-\tau_i}^{t} \sum_{j=1}^{i} x_i^T \omega_{ij} x_i dt. \tag{27}$$

where$\Gamma_{ij} = \Gamma_{ij}^{-1} > 0$.

Considering (25), the derivative of V_i is

$$\begin{aligned}
\dot{V}_i \le & -2\sum_{j=1}^{i-1}(e_j^T P_j e_j - \frac{1}{2}|\varepsilon_j^*|^2) - 2e_i^T P_i e_i - 2e_i^T P_i P_i e_i + e_i^T (K_i^T B_i^T P_i + A_i^T P_i) e_i \\
& + e_i^T (P_i B_i K_i + P_i A_i) e_i + \sum_{j=1}^{i} x_j^T (t-\tau_i) H_{ij}^T P_i e_i + \sum_{j=1}^{i} e_i^T P_i H_{ij} x_j (t-\tau_i) \\
& + 2e_i^T P_i (-\tilde{W}_i^T \phi_i(z_i) + \varepsilon_i) + \sum_{j=1}^{n} \tilde{W}_{ij}^T \Gamma_{ij}^{-1} \dot{\hat{W}}_{ij} \\
& - \sum_{j=1}^{i} x_j^T (t-\tau_i) \omega_{ij} x_j (t-\tau_i).
\end{aligned} \tag{28}$$

Choose the following adaptive law

$$\dot{\hat{W}}_{ij} = \dot{\tilde{W}}_{ij} = 2\Gamma_{ij} \phi_i(z_i) e_i^T P_{ij}. \tag{29}$$

where P_{ij} is the j-th row of P_i .

Substituting (29) into (28) yields

$$\dot{V}_i \le -2\sum_{j=1}^{i} \left(e_j^T P_j e_j - \frac{1}{2}|\varepsilon_j^*| \right) + 2e_i^T P_i \varepsilon_i - 2e_i^T P_i P_i e_i + X_i^T A_i X_i. \tag{30}$$

where$X_i = [e_i, x_1(t-\tau_i), x_2(t-\tau_i), \ldots, x_i(t-\tau_i)]^T$, $\bar{A}_i = K_i^T B_i^T P_i + P_i B_i K_i + A_i^T P_i + P_i A_i$, $A_i = \begin{bmatrix} \bar{A}_i & P_i H_{i1} \ldots & P_i H_{ii} \\ H_{i1}^T P_i & -\omega_{i1} \ldots & 0 \\ \vdots & \vdots \quad \ddots & \vdots \\ H_{ii}^T P_i & 0 \quad \ldots & -\omega_{ii} \end{bmatrix}$ .Defined $\Omega_i = P_i^{-1}$, $T_{i1} = P_i \omega_{i1} P_i$, ..., $T_{ii} = P_i \omega_{ii} P_i$, $Y_i = K_i P_i^{-1}$. Similarly, left and right multiplication of $diag(P_i, \ldots, P_i)$ on both sides of Eq.(25), it can be shown that $A_i < 0$.

Similar to Eq.(20), one concludes

$$\dot{V}_i \leq -2 \sum_{j=1}^{i} \left(e_j^T P_j e_j - \frac{1}{2} |\varepsilon_j^*|^2 \right). \tag{31}$$

Step n: This step is to design the robust adaptive controller for the overall uncertain system. Defining $e_n = x_n - x_n^*$ and differentiating it yields

$$\dot{e}_n = \dot{x}_n - \dot{x}_n^* = A_n x_n + B_n u + \sum_{j=1}^{n} H_{nj} x_j(t-\tau_n) + d_n(\bar{x}_n) - \dot{x}_n^*. \tag{32}$$

Let $V_{zn} = V_{n-1} + \frac{1}{2} e_n^T P_n e_n$. Then its derivative is given by

$$\begin{aligned} \dot{V}_{zn} \leq & -2 \sum_{j=1}^{n} (e_j^T P_j e_j - \frac{1}{2} |\varepsilon_j^*|^2) + x_n^T A_n^T P_n e_n + e_n^T P_n A_n x_n + u^T B_n^T P_n e_n \\ & + e_n^T P_n B_n u + \sum_{j=1}^{n} x_j^T (t-\tau_n) H_{nj}^T P_n e_n + \sum_{j=1}^{n} e_n^T P_n H_{nj} x_j (t-\tau_n) \\ & + 2 e_n^T P_n (W_n^{*T} \phi_n)(z_n) + \varepsilon_n). \end{aligned} \tag{33}$$

where$z_n = [x_1, x_2, \ldots, x_n, e_1, e_2, \ldots, e_n, \dot{x}_n^*] \in \Omega_{z_n} \subset R^{2n+1}$ and$P_n > 0$.

Similarly, suppose that there have positive matrix $\Omega_n, T_n, Y_n, K_n = Y_n \Omega_n^{-1}$ which satisfy the following linear matrix inequality

$$\begin{bmatrix} \bar{A}_{nn} & H_{n1} \Omega_n \ldots & H_{nn} \Omega_n \\ \Omega_n H_{n1}^T & -T_{n1} \quad \ldots & 0 \\ \vdots & \vdots \quad \ddots & \vdots \\ \Omega_n H_{nn}^T & 0 \quad \ldots & -T_{nn} \end{bmatrix} < 0. \tag{34}$$

where$\bar{A}_{nn} = \Omega_n K_n^T B_n^T + B_n K_n \Omega_n + \Omega_n A_n^T + A_n \Omega_n$. Choose the control law of the uncertain time delay system as

$$\begin{aligned} u = & E(e_n^T e_n) P_n B_n^T (B_n P_n B_n^T)^{-1} \Big(- e_n - A_n e_n - P_n e_n - A_n x_n \\ & - \frac{P_n^{-1} e_n}{e_n^T e_n} \sum_{j=1}^{n} (x_j^T \omega_{nj} x_j) + B_n K_n e_n - \hat{W}_n^T \phi_n(z_n) \Big). \end{aligned} \tag{35}$$

where$z_n = [x_1, x_2, \ldots, x_n, e_1, e_2, \ldots, e_n, \dot{x}_n^*]^T \in \Omega_{z_{n0}} \subset R^{2n+1}$. $\omega_{nj} > 0$ are positive definite matrices.

Substituting (35) into (33) yields

$$\dot{V}_{zn} \le -2\sum_{j=1}^{n}(e_j^T P_j e_j - \frac{1}{2}|\varepsilon_j^*|^2) - 2e_n^T P_n e_n - 2e_n^T P_n P_n e_n$$
$$+e_n^T(K_n^T B_n^T P_n + A_n^T P_n)e_n + e_n^T(P_n B_n K_n + P_n A_n)e_n$$
$$+\sum_{j=1}^{n} x_j^T(t-\tau_n)H_{nj}^T P_n e_n + \sum_{j=1}^{n} e_n^T P_n H_{nj} x_j(t-\tau_n)$$
$$+2e_n^T P_n(W_n^{*T}\phi_n)(z_n) + \varepsilon_n). \qquad (36)$$

where $\tilde{W}_n = \hat{W}_n - W_n^*$

Consider a Lyapunov function candidate:

$$V_n = e_n^T P_n e_n + \sum_{j=1}^{n} \tilde{W}_{nj}^T \Gamma_{nj}^T \tilde{W}_{nj} + \int_{t-\tau_n}^{t} \sum_{j=1}^{n} x_n^T \omega_{nj} x_n dt. \qquad (37)$$

where $\Gamma_{nj} = \Gamma_{nj}^{-1} > 0$.

Considering (36), the derivative of V_n becomes

$$\dot{V}_n \le -2\sum_{j=1}^{n}(e_j^T P_j e_j - \frac{1}{2}|\varepsilon_j^*|^2) - 2e_n^T P_n e_n - 2e_n^T P_n P_n e_n$$
$$+e_n^T(K_n^T B_n^T P_n + A_n^T P_n)e_n + e_n^T(P_n B_n K_n + P_n A_n)e_n$$
$$+\sum_{j=1}^{n} x_j^T(t-\tau_n)H_{nj}^T P_n e_n + \sum_{j=1}^{n} e_j^T P_n H_{nj} x_j(t-\tau_n) + \sum_{j=1}^{n} \tilde{W}_{nj}^T \Gamma_{nj}^{-1} \dot{\hat{W}}_{nj}$$
$$+2e_n^T P_n(W_n^{*T}\phi_n)(z_n) + \varepsilon_n) - \sum_{j=1}^{n} x_j^T(t-\tau_n)\omega_{nj} x_j(t-\tau_n). \qquad (38)$$

Choose the following adaptive law

$$\dot{\hat{W}}_{nj} = \dot{\tilde{W}}_{nj} = 2\Gamma_{nj}\phi_n(z_n)e_n^T P_{nj}. \qquad (39)$$

where P_{nj} is the j-th row of P_n .

Substituting (39) into (38) yields

$$\dot{V}_n \le -2\sum_{j=1}^{n}(e_j^T P_j e_j - \frac{1}{2}|\varepsilon_j^*|) + 2e_n^T P_n \varepsilon_n - 2e_n^T P_n P_n e_n + X_n^T A_n X_n. \qquad (40)$$

where$X_n = [e_n, x_1(t-\tau_n), x_2(t-\tau_n), \ldots, x_n(t-\tau_n)]^T$, $\bar{A}_n = K_n^T B_n^T P_n + P_n B_n K_n + A_n^T P_n + P_n A_n$, $A_n = \begin{bmatrix} \bar{A}_n & P_n H_{n1} & \ldots & P_n H_{nn} \\ H_{n1}^T P_n & -\omega_{n1} & \ldots & 0 \\ \vdots & \vdots & \ddots & \vdots \\ H_{nn}^T P_n & 0 & \ldots & -\omega_{nn} \end{bmatrix}$.Defined $\Omega_n =$

P_n^{-1},$T_{n1} = P_n\omega_{n1}P_n, \ldots, T_{nn} = P_n\omega_{nn}P_n, Y_n = K_nP_n^{-1}$. It follows from Eq.(34) that $A_n < 0$.

Similar to (20), one proves

$$\dot{V}_n \leq -2\sum_{j=1}^{n}(e_j^T P_j e_j - \frac{1}{2}|\varepsilon_j^*|^2). \tag{41}$$

when condition (34) holds.

The above design procedure and the property of the adaptive controller can be summarized in the following Theorem.

Theorem 1. Considering the closed-loop system consisting of the uncertain time delay system (1), the controller can be designed as (35), and the RBF neural network parameter updating laws can be designed as (18), (29) and (39), Ω_1, T_1, Y_1 satisfy (10), Ω_i, T_i, Y_i satisfy (24) and Ω_n, T_n, Y_n satisfy (34), then the uncertain time delay closed loop system is uniformly ultimately bounded stable.
Proof. The uniform ultimate boundedness stability can be established by Lyapunov theory. Choose a Lyapunov function candidate for the overall closed-loop system as in Eq. (33). It follows from Eq.(41) that $\dot{V}_n < 0$ if there exist appropriate$P_j (1 \leq j \leq n)$ parameters satisfying (10), (24), (34). Thus the closed loop uncertain time delay system is uniformly ultimately bound. This point shows thate_i, $\tilde{W}_i$ and u are bound.

4 Simulation Example

Consider the uncertain nonlinear systems in the form of

$$\begin{aligned}
\dot{x}_1 &= A_1x_1 + B_1x_2 + H_{11}x_1(t-1) + d_1(x_1) \\
\dot{x}_2 &= A_2x_2 + B_2u + \sum_{j=1}^{2} H_{2j}x_j(t-1) + d_2(x_1, x_2) \\
y &= x_1
\end{aligned}$$

where $A_1 = \begin{bmatrix} -45 & -7 \\ -0.9 & -20 \end{bmatrix}$, $B_1 = \begin{bmatrix} -2 & 10 \\ 20 & -8 \end{bmatrix}$, $H_{11} = \begin{bmatrix} 15 & 1 \\ -4 & 10 \end{bmatrix}$, $A_2 = \begin{bmatrix} -0.1 & -0.5 \\ 0 & -1 \end{bmatrix}$, $B_2 = \begin{bmatrix} -0.5 & -0.5 \\ 0 & 5 \end{bmatrix}$, $H_{21} = \begin{bmatrix} -0.5 & -51 \\ 0 & 1 \end{bmatrix}$, $H_{22} = \begin{bmatrix} 0.8 & 0.8 \\ 0 & 0.8 \end{bmatrix}$,$d_1 = 0.2x_{12}sin(x_{11})$, $d_2 = 0.1x_{11}x_{21}cos(x_{22})$.

Choosing $\Omega_1 = \begin{bmatrix} 0.15 & 0.1 \\ 0.1 & 0.15 \end{bmatrix}$, $T_1 = \begin{bmatrix} 2 & 0 \\ 0 & 2 \end{bmatrix}$, $\Omega_2 = \begin{bmatrix} 4.7 & 0 \\ 0 & 1 \end{bmatrix}$, $T_{21} = \begin{bmatrix} 8 & 0 \\ 0 & 100 \end{bmatrix}$, $T_{22} = \begin{bmatrix} 0.5 & 0 \\ 0 & 0.3 \end{bmatrix}$. Parameter update law can be chosen as

$$\dot{\hat{W}}_1 = \dot{\tilde{W}}_1 = \Gamma_1\phi_1(z_1)e_1; \dot{\hat{W}}_2 = \dot{\tilde{W}}_2 = \Gamma_2\phi_2(z_2)e_2 \tag{42}$$

where$\Gamma_1 = \begin{bmatrix} 1 & 0 \\ 0 & 1 \end{bmatrix}$,$\Gamma_2 = \begin{bmatrix} 1.5 & 0 \\ 0 & 1.5 \end{bmatrix}$.$\phi_1$ and ϕ_2 are Radical basis functions.

Choosing design parameter $\alpha_1 = 2, \alpha_2 = 1, \beta_1 = 1$, then the robust adaptive controller is designed according to (35).The closed loop state respondences under the designed controller are shown in Fig.1.

From these simulation results, we can see the closed loop system is uniformly ultimately bounded stable. So the proposed robust adaptive controller is effective for the uncertain time delay system.

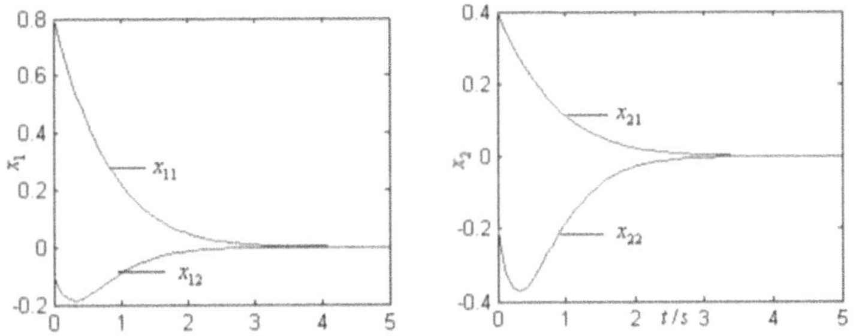

Fig. 1. The state response plots of close-loop system

References

1. Jiang, Z.P., David, J.H.: A Robust Adaptive Backstepping Scheme for Nonlinear Systems with Unmodeled Dynamics. IEEE Transactions on Automatic Control **9** (1999) 1705-1711
2. Zhou, J., Wen, C.Y., Zhang, Y.: Adaptive Backstepping Control of a Class of Uncertain Nonlinear Systems with Unknown Backlash-like Hysteresis. IEEE Transactions on Automatic Control **10** (2004) 1751-1757
3. Li, Y.H., Qiang, S., Zhuang, X.Y., Okyay, K.: Robust and Adaptive Backstepping Control for Nonlinear Systems Using RBF Neural Networks. IEEE Transaction on Neural Networks **3** (2004) 693-701
4. Shuzhi, S.G., Wang, C.: Direct Adaptive NN Control of a Class of Nonlinear Systems. IEEE Transaction on Neural Networks **3** (2002) 214-221
5. Zhang, Y.P., Peng, P.Y., Jiang, Z.P.: Stable Neural Controller Design for Unknown Nonlinear Systems Using Backstepping. IEEE Transaction on Neural Networks **6** (2000) 1347-1360
6. Shuzhi, S.G., Fan, H., Tong, H.L.: Adaptive Neural Network Control of Nonlinear Systems with Unknown Time Delays. IEEE Transactions on Automatic Control **11** (2003) 2004-2010
7. Hsiao, F.H., Huang, J.D.: Stabilization of Nonlinear Singularly Perturbed Multiple Time-delay Systems by Dither. J. Dyna. Syst. Measure. Control **1** (1996) 176-181
8. Nguang, S.K.: Robust Stabilization of a Class of Time-delay Nonlinear Systems. IEEE Transactions on Automatic Control **4** (2000) 756-762

Neural Network in Stable Adaptive Control Law for Automotive Engines

Shiwei Wang and Ding-Li Yu

Control Systems Research Group, School of Engineering Liverpool John Moores University, Byrom Street, Liverpool L3 3AF, UK
D.Yu@ljmu.ac.uk
http://www.ljmu.ac.uk/ENG/72971.htm

Abstract. This paper proposes to use a radial basis function (RBF) neural network in realising an adaptive control law for air/fuel ratio (AFR) regulation of automotive engines. The sliding mode control (SMC) structure is used and a new sliding surface is developed in the paper. The RBF network adaptation and the control law are derived using the Lyapunov method so that the entire system stability and the network convergence are guaranteed. The developed method is evaluated by computer simulation using the well-known mean value engine model (MVEM) and the effectiveness of the method is proved.

1 Introduction

Car emission is a major source for air pollution in urban area. Reducing the harmful gases emitted by cars, such as CO, HC, SO_2, NO_x etc. becomes a main concern of governments and car manufacturers. Experimental results showed that air-fuel ratio (AFR) is a key feature. Maintaining AFR at the stoichiometric value (AFR=14.7) will generate proper ratio between power output and fuel consumption. AFR also influences the effect of emission control because its stoichiometric value ensures the maximum efficiency of three-way catalysts (TWC). Variations of greater than 1% below the stoichiometric value in AFR will result in significant increase of CO and HC emissions. An increase of more than 1% over the stoichiometric value will produce more up to 50% [1][2].

The current production, electric control unit (ECU), uses look-up tables with compensation of a PI controller to control the air-fuel ratio. This method cannot produce desirable accurate control performance due to that the engine dynamics are highly non-linear [1][2][3]. Some researches on air-fuel ratio control have been conducted in recent years. Choi and Hedrick developed an observer based sliding mode control (SMC) method [4] to achieve a fast response. But the chattering in the air-fuel ratio still needs to be improved. Yoon and Sunwoo [5] realized the adaptation of model parameters for fuel delivery and measurement bias of air mass flow to deal with the problem caused by engine uncertainties. However, the major problem is the large amplitude chattering caused by the system uncertainty.

D. Liu et al. (Eds.): ISNN 2007, Part I, LNCS 4491, pp. 122–131, 2007.

In this paper, a new sliding surface is proposed and it is steered to zero in finite time by using a discontinuous first derivative of the control variable $\dot{\alpha}(t)$ and the corresponding actual control variable $\alpha(t)$ turns to be continuous, which can reduce the undesired chattering. By using a RBF neural network based adaptation method, a simple and robust control law is derived. The configuration of the network is fixed while the weight is updated with the input/output data. The system stability is guaranteed using Lyapunov function. This new scheme simplifies the conventional ideal sliding control law and avoids involving immeasurable parameters and system uncertainties. The developed scheme is applied to a well-developed and widely adopted model, the general mean value engine model, developed by Hendricks [6] for evaluation. The simulation results indicate the effectiveness of the proposed control scheme when the engine is subjected to a sudden change of throttle angle.

2 RBF Network-Based Adaptive SMC

Consider an nth order single-input single-output (SISO) nonlinear system:

$$\begin{aligned} \dot{x}_1(t) &= x_2(t) \\ \dot{x}_2(t) &= x_3(t) \\ &\cdots \\ \dot{x}_{n-1}(t) &= x_n(t) \\ \dot{x}_n(t) &= f[x(t)] + g[x(t)]\alpha(t) \end{aligned} \tag{1}$$

where $x(t) = [x_1(t), x_2(t), \cdots, x_n(t)]^T$ is the vector of the available states and $f[x(t)]$ and $g[x(t)]$ are uncertain smooth functions. A sliding surface is proposed as

$$S[x(t)] = \dot{x}_n(t) + \sum_{i=1}^{n} c_i x_i(t) = 0 \tag{2}$$

where c_i is real positive constants such that the characteristic equation $z^n + \sum_{i=1}^{n} c_i z^{i-1} = 0$ is Hurwitz. Its autonomous closed loop dynamics is asymptotically stable.

The first derivative of the sliding mode is

$$\begin{aligned} \dot{S}[x(t)] &= \dot{x}_{n+1}(t) + \sum_{i=1}^{n} c_i x_{i+1}(t) \\ &= \frac{d}{dt} f[x(t)] + \frac{d}{dt} g[x(t)]\alpha(t) + g[x(t)]\dot{\alpha}(t) + \sum_{i=1}^{n} c_i x_{i+1}(t) \end{aligned} \tag{3}$$

Substitute $\dot{S}[x(t)]$ into the well-known sliding condition $\dot{S} = -\eta sgn(S)$ (η is a positive constant). Then the following ideal control law can be obtained

$$\alpha(t_2) = \alpha(t_1) - \int_{t_1}^{t_2} g[x(t)]^{-1} \cdot \left\{ \frac{d}{dt} f[x(t)] + \frac{d}{dt} g[x(t)]\alpha(t) + \sum_{i=1}^{n} c_i x_{i+1}(t) + \eta sgn(S) \right\} \tag{4}$$

In this control law, the actual control variable $\alpha(t)$ turns to be continuous which can reduce the undesired chattering.

However, it is difficult to be applied directly in practical applications because the control law involves many system states and variables. We propose to use a RBF network to approximate a part of the function in the idea control law in Equation 4. Firstly, the RBF network is briefly introduced. The Gaussian activation function is chosen,

$$\phi_j(t) = exp\left(-\frac{\|x(t) - c_j(t)\|^2}{\sigma_j^2}\right), \quad j = 1, 2, \cdots, n_h \tag{5}$$

where σ_j is a positive scalar called a width and n_h is the number of centers. The output is then given by

$$\hat{y}_i(t) = \sum_{j=1}^{n_h} \phi_j(t) w_{ji}, \quad i = 1, 2, \cdots, q \tag{6}$$

where w are the output layer weights and q is the number of outputs.

Thus, the adaptive control law with RBF network is given in Theorem 1.

Theorem 1. *If a sliding controller is designed as*

$$\alpha(t_2) = \alpha(t_1) - \int_{t_1}^{t_2} \hat{V}_M(t) sgn(S) \tag{7}$$

where the sliding gain is estimated by a RBF network

$$\hat{V}_M(t) = \sum_{i=1}^{n_h} \hat{w}_i(t) exp\left(-\frac{\|x(t) - c_i(t)\|^2}{\sigma^2}\right) \tag{8}$$

with its adaptation law

$$\hat{w}_i(t_2) = \hat{w}_i(t_1) + \int_{t_1}^{t_2} sgn(S)\rho \cdot g[x(t)] exp\left(-\frac{\|x(t) - c_i(t)\|^2}{\sigma^2}\right) \tag{9}$$

then it can maintain the SISO nonlinear system in Equation 1 on the sliding surface defined in Equation 2.

Proof. **Step 1:** Define a desirable value of the sliding gain $V_M(t)$ as follows

$$V_M(t) = \sum_{i=1}^{n_h} w_i(t) exp\left(-\frac{\|x(t) - c_i(t)\|^2}{\sigma^2}\right) \tag{10}$$

It should generate the ideal control law in Equation 4, that is

$$\begin{aligned}\alpha(t_2) &= \alpha(t_1) - \int_{t_1}^{t_2} g[x(t)]^{-1} \\ &\cdot \left\{\frac{d}{dt} f[x(t)] + \frac{d}{dt} g[x(t)]\alpha(t) + \sum_{i=1}^{n} c_i x_{i+1}(t) + \eta sgn(S)\right\} \\ &= \alpha(t_1) - \int_{t_1}^{t_2} \hat{V}_M(t) sgn(S)\end{aligned} \tag{11}$$

The estimated $\hat{V}_M(t)$ approximates the desirable sliding gain $V_M(t)$ and the corresponding estimation error $\tilde{V}_M(t)$ can be written as

$$\tilde{V}_M(t) = \sum_{i=1}^{n_h} \tilde{w}_i(t) exp\left(-\frac{\|x(t) - c_i(t)\|^2}{\sigma^2}\right) = V_M(t) - \tilde{V}_M(t) \tag{12}$$

where $\tilde{w}_i = w_i - \hat{w}_i \Longrightarrow \dot{\tilde{w}}_i = -\dot{\hat{w}}_i$

Step 2: According to Equation 11, it can be defined that

$$\begin{aligned}H(t) \equiv &- \int_{t_1}^{t_2} \left\{\frac{d}{dt} f[x(t)] + \frac{d}{dt} g[x(t)]\alpha(t) + \sum_{i=1}^{n} c_i x_{i+1}(t) + \eta sgn(S)\right\} \\ &+ \int_{t_1}^{t_2} g[x(t)] V_M(t) sgn(S) = 0\end{aligned} \tag{13}$$

and its first derivative is

$$\begin{aligned}\dot{H}(t) \equiv &- \left\{\frac{d}{dt} f[x(t)] + \frac{d}{dt} g[x(t)]\alpha(t) + \sum_{i=1}^{n} c_i x_{i+1}(t) + \eta sgn(S)\right\} \\ &+ g[x(t)] V_M(t) sgn(S) = 0\end{aligned} \tag{14}$$

Step 3: Design a Lyapunov function as follows.

$$V = |S| + |H| + \frac{1}{2\rho} \sum_{i=1}^{n_h} [\tilde{w}_i(t) sgn(S)]^2 \tag{15}$$

Here, ρ is a positive constant. If an adaptation law of the network weights $\hat{w}_i$ can be found such that the above Lyapunov function V can satisfy that $V > 0$ and $\dot{V} < 0$, then both the defined sliding mode $S[x(t)]$ and the difference between the adapted sliding gain $\hat{V}_M(t)$ and the desired value $V_M(t)$ will be driven to zero.

When S is not zero, the differentiation of the Lyapunov function is

$$\dot{V} = sgn(S)\dot{S} + \dot{H} - \frac{1}{\rho}\sum_{i=1}^{n_h}\left[\tilde{w}_i(t)\dot{\hat{w}}_i(t)sgn(S)\right] \tag{16}$$

Because $\dot{H}$ is equal to zero, $sgn(S)\dot{H}$ is also equal to zero. Then Equation 16 becomes

$$\dot{V} = sgn(S)\dot{S} - \frac{1}{\rho}\sum_{i=1}^{n_h}\left[\tilde{w}_i(t)\dot{\hat{w}}_i(t)sgn(S)\right] \tag{17}$$

Substitute $\dot{S}$ and $\dot{H}$ into $\dot{V}$ to yield

$$\begin{aligned}\dot{V} &= sgn(S)g[x(t)]\dot{\alpha}(t) - \eta + g[x(t)]V_M(t) - \frac{1}{\rho}\sum_{i=1}^{n_h}\left[\tilde{w}_i(t)\dot{\hat{w}}_i(t)sgn(S)\right] \\ &= -\eta + g[x(t)]\tilde{V}_M(t) - \frac{1}{\rho}\sum_{i=1}^{n_h}\left[\tilde{w}_i(t)\dot{\hat{w}}_i(t)sgn(S)\right] \\ &= -\eta + \sum_{i=1}^{n_h}\tilde{w}_i(t)\left\{g[x(t)]exp\left(-\frac{\|x(t)-c_i(t)\|^2}{\sigma^2}\right) - \frac{1}{\rho}\dot{\hat{w}}_i(t)sgn(S)\right\}\end{aligned} \tag{18}$$

Choosing the adaptation law

$$\hat{w}_i(t_2) = \hat{w}_i(t_1) + \int_{t_1}^{t_2} sgn(S)\rho \cdot g[x(t)]exp\left(-\frac{\|x(t)-c_i(t)\|^2}{\sigma^2}\right) \tag{19}$$

yields $\dot{V} = -\eta < 0$, which ends the prove.

3 Application to AFR Control

The engine dynamics concerned with air/fuel ratio control include air intake manifold, fuel injection, crankshaft speed, and exhaust oxygen measurement. A schematic diagram of the engine dynamics is shown in Fig.1. The system has one input, the injected fuel mass flow rate $\dot{m}_{fi}$ one output, air/fuel ratio AFR. Besides, the system is subjected to a significant disturbance, the throttle angle u. Due to the space limitation, the dynamics of each of the four sub-systems, a number of differential and algebraic equations, are not included. The interested reader can refer to [6].

A sliding surface is chosen as

$$S[x(t)] = x_1(t) = \dot{m}_{ap} - \beta\dot{m}_f = 0 \tag{20}$$

The first derivative of $S[x(t)]$ is

$$\dot{S}[x(t)] = \ddot{m}_{ap} - \beta\ddot{m}_f \tag{21}$$

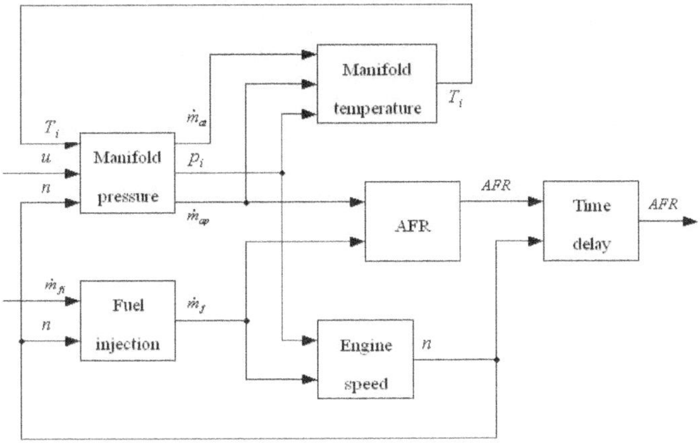

Fig. 1. Schematic diagram of engine dynamics

Using the engine dynamics, it is derived that

$$\begin{aligned}\dot{S} &= \left[\frac{\partial \dot{m}_{ap}}{\partial p_i} \cdot \frac{\kappa R}{V_i}\left(\dot{m}_{at}T_a + \dot{m}_{EGR}T_{EGR} - (S + \beta\dot{m}_f)T_i\right) + \frac{\partial \dot{m}_{ap}}{\partial n} \cdot \dot{n}\right] \\ &\quad - \beta\left[\ddot{m}_{fv} + \frac{1}{\tau_f}(\dot{m}_{fi} - \dot{m}_f)\right] \\ &= -K_pS + K_p\left(\dot{m}_{at}\frac{T_a}{T_i} + \dot{m}_{EGR}\frac{T_{EGR}}{T_i} - \beta\dot{m}_f\right) \\ &\quad + K_n\dot{n} - \beta\ddot{m}_{fv} + \frac{\beta}{\tau_f}\dot{m}_f - \frac{\beta}{\tau_f}\dot{m}_{fi}\end{aligned} \tag{22}$$

According to the sliding condition $S = -\eta sgn(S)$ (η is a positive constant), an ideal control law can be obtained as follows.

$$\begin{aligned}\dot{m}_{fi}(t_2) = \dot{m}_{fi}(t_1) + \frac{1}{\beta(1-X)} \cdot \Bigg\{\int_{t_1}^{t_2}\Bigg[K_p\left(\dot{m}_{at}\frac{T_a}{T_i} + \dot{m}_{EGR}\frac{T_{EGR}}{T_i} - \beta\dot{m}_f\right) \\ + K_n\dot{n} - \frac{\beta}{\tau_f}(\dot{m}_{fi} - \dot{m}_f) + \eta \cdot sgn(S)\Bigg]\Bigg\}\end{aligned} \tag{23}$$

In ideal conditions, that is all of the parameters and variables in Equation 23 can be precisely measured, the obtained control variable $\dot{m}_{fi}$ can guarantee satisfactory control results. Unfortunately, it is very difficult to obtain the precise values of all the parameters and variables. To simplify the implementation of the ideal control law 23, the designed sliding controller is implemented.

$$\dot{m}_{fi} = \dot{m}_{fi}(t_1) - \int_{t_1}^{t_2} \hat{V}_M(t)sgn\{S[x(t)]\} \tag{24}$$

Notice that the sliding gain $V_M(t)$ determines the slope of the sliding trajectory. The value of $V_M(t)$ should cover all of uncertain dynamics of the engine under different operating conditions. Therefore, a suitable choice of the sliding gain $V_M(t)$ is essential in this control scheme and it is necessary to develop a sliding controller using the method proposed before.
Define

$$
\begin{aligned}
H(t) \equiv & \frac{1}{\beta(1-X)} \cdot \left\{ \int_{t_1}^{t_2} \left[K_p \left(\dot{m}_{at} \frac{T_a}{T_i} + \dot{m}_{EGR} \frac{T_{EGR}}{T_i} - \beta \dot{m}_f \right) + K_n \dot{n} \right.\right. \\
& \left.\left. - \frac{\beta}{\tau_f} (\dot{m}_{fi} - \dot{m}_f) + \eta \cdot sgn(S) \right] \right\} - \int_{t_1}^{t_2} V_M sgn(S) = 0
\end{aligned} \tag{25}
$$

Then, choose a Lyapunov function

$$
V \equiv |S| + |H| + \frac{1}{2\rho} \sum_{i=1}^{n_h} [\tilde{w}_i(t) sgn(S)]^2 \tag{26}
$$

ρ is a positive constant. When S is not zero, the derivative of the Lyapunov function is

$$
\dot{V} = sgn(S)\dot{S} + \dot{H} - \frac{1}{\rho} \sum_{i=1}^{n_h} [\tilde{w}_i(t) \dot{\hat{w}}_i(t) sgn(S)] \tag{27}
$$

Substitute $\dot{S}$ and $\dot{H}$ into $\dot{V}$ to yield

$$
\begin{aligned}
\dot{V} = & sgn(S) \left\{ -K_p S + K_p \left(\dot{m}_{at} \frac{T_a}{T_i} + \dot{m}_{EGR} \frac{T_{EGR}}{T_i} - \beta \dot{m}_f \right) + K_n \dot{n} \right. \\
& \left. - \beta \left[(1-X) \ddot{m}_{fi} + \frac{1}{\tau_f} (\dot{m}_{fi} - \dot{m}_f) \right] \right\} \\
& + sgn(S) \left[K_p \left(\dot{m}_{at} \frac{T_a}{T_i} + \dot{m}_{EGR} \frac{T_{EGR}}{T_i} - \beta \dot{m}_f \right) + K_n \dot{n} \right. \\
& \left. - \frac{\beta}{\tau_f} (\dot{m}_{fi} - \dot{m}_f) + \eta \cdot sgn(S) \right] \\
& - sgn(S) \beta (1-X) V_M sgn(S) - \frac{1}{\rho} \sum_{i=1}^{n_h} [\tilde{w}_i(t) \dot{\hat{w}}_i(t) sgn(S)] \\
= & -K_p |S| - \eta - \beta(1-X) \tilde{V}_M - \frac{1}{\rho} \sum_{i=1}^{n_h} [\tilde{w}_i(t) \dot{\hat{w}}_i(t) sgn(S)] \\
= & -K_p |S| - \eta - \sum_{i=1}^{n_h} \tilde{w}_i(t) \left\{ \beta(1-X) \right. \\
& \left. \cdot exp \left(- \frac{\|x(t) - c_i(t)\|^2}{\sigma^2} \right) + \frac{1}{\rho} \dot{\hat{w}}_i(t) sgn(S) \right\}
\end{aligned} \tag{28}
$$

Choose the adaptation law

$$\dot{\hat{w}}_i(t) = -sgn(S) \cdot \beta(1 - X)exp\left(-\frac{\|x(t) - c_i(t)\|^2}{\sigma^2}\right) \tag{29}$$

which yields $\dot{V} = -K_p|S| - \eta < 0$.
Integrate Equation 29 and the adaptation law becomes

$$\hat{w}_i(t) = \hat{w}_i(t_1) - \int_{t_1}^{t_2} sgn(S) \cdot \rho \cdot \beta(1 - X)exp\left(-\frac{\|x(t) - c_i(t)\|^2}{\sigma^2}\right) \tag{30}$$

The RBF network based adaptation method compensates the model-plant mismatch caused by part weariness and batch error in production. It also avoids handling many unavailable parameters and variables in the conventional sliding control. The estimated sliding gain $\hat{V}_M(t)$ generated by the updated weights drives the actual control variable $\dot{m}_{fi}(t)$ to be as close as possible to the desirable value in Equation 23 which can achieve satisfactory control results for air fuel ratio.

4 Simulation Results

In order to illustrate the effectiveness of the proposed control scheme, numerical simulations are executed by using the mean value engine model operating as an economic IC engine with 1.2 L displacement. The EGR mass flow is set to be about 20% of the total mass flow. The two fuel injection parameters are chosen as $X = 0.3$ and $\tau_f = 0.6$. The sampling time is chosen as $10ms$. The air-fuel ratio is to be controlled between the ($\pm 1\%$ bounds of the stoichiometric value (14.7).

As shown in Fig.2, a throttle angle is designed to rapidly change from 26° to 40° with 0.3° uncertainty. This throttle angle signal drives the engine to operate between 3600 and 4300 rpm.

In the simulation, the input vector and the output for the RBF network are chosen as

$$x(t) = \begin{bmatrix} u(t) \\ n(t) \\ p_i(t) \\ \hat{V}_M(t-1) \end{bmatrix}, \quad \hat{y} = \hat{V}_M(t)$$

The input data are scaled to the range of $(0, 1)$ before they are fed into the network. The network centers and widths are chosen to be constant using the K-means clustering method and P-nearest center rule. Thus, only the weights will be updated on-line to adapt the sliding gain.

Fig.3 shows the air fuel ratio control results with the throttle angle changing from 26° to 40°. The corresponding adaptive sliding gain is shown in Fig.4. The sliding gain is automatically adapted to drive the air fuel ratio to be maintained in the required region.

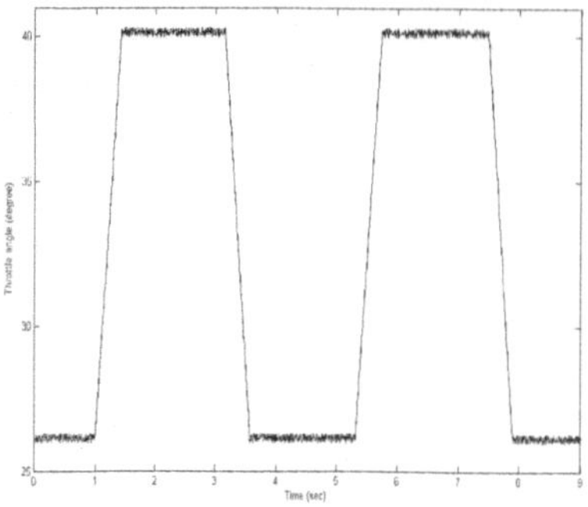

Fig. 2. Throttle angle change

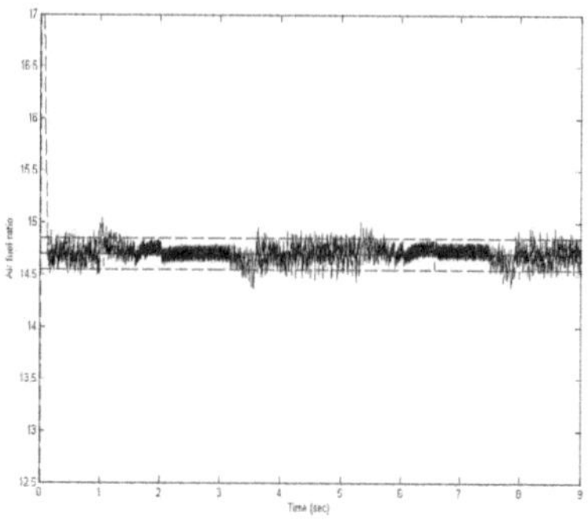

Fig. 3. Air fuel ratio control performance

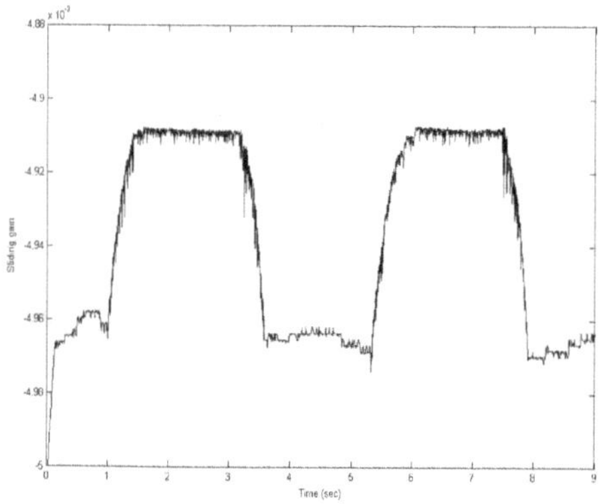

Fig. 4. Sliding gain adaptation

5 Conclusions

A new solution, an adaptive SMC scheme based on a RBF network, is proposed for engine air fuel ratio control. Instead of the actual injected fuel mass flow, the discontinuous first derivative of the control variable is first obtained and the corresponding actual control variable becomes continuous. Using Lyapunov function, the method is guaranteed for system stability and network convergence. By adjusting the sliding gain and driving the control variable to approximate the desirable one as precisely as possible, the scheme reduces the chattering problem and simplifies the conventional SMC. Simulation results show that the adaptive control law can achieve satisfactory control performance for air fuel ratio even with engine parameter uncertainty and modelling errors. Additionally, it avoids involving immeasurable parameters in the design and is robust to system uncertainties.

References

1. Manzie, C., Palaniswami M. and Watson H.: Gaussian networks for fuel injection control, Proceedings of the Institution of Mechanical Engineers, Part D (Journal of Automobile Engineering) **215(D10)** (2001) 1053-1068.
2. Manzie, C., Palaniswami, M., Ralph, D., Watson, H. and Yi X.: Model predictive control of a fuel injection system with a radial basis function network observer, Journal of Dynamic Systems Measurement and Control Transactions of the ASME **124**(4) (2002) 648-658.
3. Choi, S.B. and Hendrick, J.K.: An observer-based controller design method for improving air/fuel characteristics of spark ignition engines, IEEE Transactions on Control Systems Technology **6**(3) (1998) 325-334.
4. De Nicolao, G., Scattolini, R. and Siviero, C.: Modelling the volumetric efficiency of IC engines: parametric, non-parametric and neural techniques, Control Eng. Practice **4**(10) (1996) 1405-1415.
5. Yoon, P. and Sunwoo, M.: An adaptive sliding mode controller for air-fuel ratio control of spark ignition engines, Proceedings of the Institution of Mechanical Engineers, Part D (Journal of Automobile Engineering) **215** (2001) 305-315.
6. Hendricks, E.: A generic mean value engine model for spark ignition engines, 41st Simulation Conference, SIMS 2000, DTU, Lyngby, Denmark, 2000.

Neuro-fuzzy Adaptive Control of Nonlinear Singularly Perturbed Systems and Its Application to a Spacecraft*

Li Li and Fuchun Sun

Dept. of Computer Science and Technology, State Key Lab of Intelligent Technology & Systems, Tsinghua University, Beijing 100084, P.R. China
ll-03@mails.tsinghua.edu.cn

Abstract. In this paper, we first present a series of dynamic TS fuzzy subsystems to approximate a nonlinear singularly perturbed system. Then the reference model with same fuzzy sets is established. To make the states of the closed-loop system follow those of the reference model, a controller including of neuro-fuzzy adaptive and linear feedback term is designed. The linear feedback parameters can be solved by LMI approach. Adaptive term is used to compensate the uncertainty and alleviate the external disturbance. Lyapunov constitute techniques can prove the stability of the closed loop systems. The simulations results illustrate the effectiveness of this approach.

1 Introduction

Recently, TS type fuzzy controller has been successfully applied to the stabilization control design of nonlinear systems and it has been extended to the control of Singularly Perturbed (SP) system too [1-3]. Liu [1] proposes fuzzy SP models by extending the ordinary TS fuzzy model, then H_∞ and H_2 [2] controllers are developed based on fuzzy SP models. The stabilization of closed-loop systems are proved in [3].

However, most of TS model-based controller design approaches assume TS linear model can approximate the nonlinear system exactly and there isn't external disturbance. Only few results are given concerning of robust stability for the nonlinear system with parameter uncertainties [4] or external disturbance [5] or both of them existing. It is still an open problem, although some effort is taken. Several authors adopt LMI approach to get feedback sub-controllers, which need uncertainties and external disturbance satisfying match conditions and the results are conservative since more constraints are added. Adaptive control can maintain the consistent performance of a system in the presence of the uncertainties and neuro-fuzzy control can realize superior control performance even with some parameters unknown, neuro-fuzzy adaptive control might solve the problem above-mentioned.

In this paper, we will combine the advantage of fuzzy SP model and adaptive control and propose a new controller. The linear feedback control gain can be

* This work was jointly supported by the National Natural Science Foundation of China(Grant No: 60625304, 60474025, 60504003, 60321002 and 90405017), the National Key Project for Basic Research of China (Grant No: G2002cb312205), and the Specialized Research Fund for the Doctoral Program of Higher Education (Grant No: 20050003049).

D. Liu et al. (Eds.): ISNN 2007, Part I, LNCS 4491, pp. 132–137, 2007.

achieved by LMI approach. The adaptive control term can be adjusted on line. Lyapunov constitute techniques prove the stability of the closed loop systems. The simulations results illustrate the effectiveness of the approach.

2 Problem Formulations

Considering a continuous-time nonlinear SP system described by TS fuzzy model [1] with the ith rule:

$$\begin{aligned} &R^i : \text{if } z_1 \text{ is } M_1^i \text{ and}\ldots\text{and } z_N \text{ is } M_N^i \text{ , then} \\ &E_\varepsilon \dot{x}(t) = (A^i + \Delta A^i)x(t) + B^i u(t) + D^i \omega(t) \end{aligned} \tag{1}$$

Given a pair of input $x(t)$ and $u(t)$, the TS fuzzy model is inferred as follows:

$$E_\varepsilon \dot{x}(t) = \sum_{i=1}^{p} \xi_i(z)(A^i x(t) + B^i u(t) + \Delta A^i x(t) + D^i \omega(t)) \tag{2}$$

where $x(t) = [x_1(t), x_2(t)] \in R^n$ is as the plant state vector; $x_1(t) \in R^{n_1}$ as the plant slow state vector; $x_2(t) \in R^{n_2}$ as the plant fast state vector; $u(t) \in R^m$ as the control input. $0 < \varepsilon << 1$ is a small constant; $\omega(t)$ the external disturbance; z_{i_1} fuzzy premise variables; $M_{i_1}^i$ a fuzzy set with membership functions $\xi_{i_1}^i(z_{i_1})$. ΔA^i represents the parameters perturbation. $i_1 = 1, \cdots, N$, $n = n_1 + n_2$.

$$\xi_i(z) = \frac{\prod_{j=1}^{n} \xi_j^i(z_j)}{\sum_{i=1}^{p} \prod_{j=1}^{n} \xi_j^i(z_j)}, \qquad E_\varepsilon = \begin{bmatrix} I_{n_1 \times n_1} & 0 \\ 0 & \varepsilon I_{n_2 \times n_2} \end{bmatrix}, \qquad A^i = \begin{bmatrix} 0_{(n-m)\times m} & I_{(n-m)} \\ \bar{A}_1^i & \bar{A}_2^i \end{bmatrix},$$

$$\Delta A^i = \begin{bmatrix} 0_{(n-m)\times m} & 0_{(n-m)\times(n-m)} \\ \Delta \bar{A}_1^i & \Delta \bar{A}_2^i \end{bmatrix}, \quad B^i = \begin{bmatrix} 0_{(n-m)\times m} \\ b_2^i \end{bmatrix}, \qquad D^i = \begin{bmatrix} 0_{(n-m)} \\ d_2^i \end{bmatrix}$$

Suppose the reference model shares the same fuzzy sets with the fuzzy system (1):

$$R_m^i : \text{if } z_1 \text{ is } M_1^i \text{ and}\ldots\text{and } z_N \text{ is } M_N^i \text{, then } E_\varepsilon \dot{x}_m(t) = A_m^i x_m(t) + B_m^i r(t) \tag{3}$$

Where $x_m(t) \in R^n$ denotes reference states; $r(t) \in R^m$ bounded reference input.

$$A_m^i = \begin{bmatrix} 0_{(n-m)\times m} & I_{(n-m)} \\ \bar{A}_{m1}^i & \bar{A}_{m2}^i \end{bmatrix}, \qquad B_m^i = \begin{bmatrix} 0_{(n-m)\times m} \\ b_{m2}^i \end{bmatrix}, \qquad b_{m2}^i \in R^{m\times m}$$

Define the tracking error is:

$$e(t) = x(t) - x_m(t) \tag{4}$$

The control objective is to design a control law $u(t)$ to make the states $x(t)$ follow those of a stable reference model, that is when $t \to \infty$, $e(t) \to 0$.

3 The Design of Neuro-fuzzy Adaptive Controller

In this section, we will present a neuro-fuzzy adaptive controller to realize the objective. The following assumptions are required for our derivations:

Assumption 1: The external disturbance is bounded and $\|\omega(t)\|_\infty < k_\omega$

Assumption 2: Assumption $g = \sum_{i=1}^{p} \xi_i(z) b_2^i$ is nonsingular for any z .

Take derivative of (4) is:

$$E_\varepsilon \dot{e}(t) = \sum_{i=1}^{p} \xi_i(z)[A^i e(t) + (A^i - A_m^i) x_m(t) - B_m^i r(t) + \Delta A^i x(t) + B^i u(t) + D^i \omega(t)]$$

$$= \sum_{i=1}^{p} \xi_i(z)[A^i e(t) + (A^i - A_m^i) x_m(t) - B_m^i r(t) + B^i u(t)] + [0 \quad f(x(t)) + d(t)]^T \tag{5}$$

Where $f(x(t)) = \sum_{i=1}^{p} \xi_i(z)\left[\Delta \bar{A}_1^i, \Delta \bar{A}_2^i\right] x(t)$, $d(t) = \sum_{i=1}^{p} \xi_i(z) \bar{d}_2^i \omega(t)$ .Since $f(x(t))$ is unknown, it can be approximated by RBF network.

Assumption 3: Suppose $f(x(t))$ can be approximated by the output of RBF network in a big enough space Ω , that is, $f(x(t)) = \hat{f}(x(t)) + w^*, x(t) \in \Omega$ .And suppose $\|w^*(x(t))\| = \|f(x(t)) - \hat{f}(x(t))\| < k_{w^*}, x(t) \in \Omega$.

where $\hat{f}(x(t)) = W_0 + \sum_{l}^{N_u} W_l \Psi_l(x) = W\Psi(x)$, $W = [W_0, \cdots, W_{N_u}] \in R^{m\times(N_u+1)}$ is weight matrix and $\Psi(x) = [1, \Psi_1(x), \cdots, \Psi_{N_u}(x)]^T \in R^{N_u+1}$, $\Psi_l(x) (l = 1, \cdots, N_u)$ is the hidden function of RBF network. Here we choose Guassian function.

Then the following controller and adaptive law are given:

$$u(t) = u_1(t) + u_2(t) \tag{6}$$

$$u_1(t) = \sum_{i=1}^{p} \xi_i(z) K_i e(t) \tag{7}$$

$$u_2(t) = -g^{-1} \sum_{i=1}^{p} \xi_i([\bar{A}_1^i - \bar{A}_{m1}^i \quad \bar{A}_2^i - \bar{A}_{m2}^i] x_m - b_{m2}^i r(t)) - g^{-1} W\Psi(x) + u_s(t) \tag{8}$$

$$u_s(t) = -K_D g^{-1} \operatorname{sgn}(\bar{P}_{e2}^T) \tag{9}$$

$$\dot{W}_l^T = \tau_l \Psi_l(x) \bar{P}_{e2} \quad l = 1, 2, \cdots, N_u \tag{10}$$

Where $\boldsymbol{P}_e = \boldsymbol{e}^T \boldsymbol{P}_\varepsilon^T = \begin{bmatrix} \bar{\boldsymbol{P}}_{e1} & \bar{\boldsymbol{P}}_{e2} \end{bmatrix}$, $\boldsymbol{P}_\varepsilon = \begin{bmatrix} \boldsymbol{P}_{11} & \varepsilon \boldsymbol{P}_{21}^T \\ \boldsymbol{P}_{21} & \boldsymbol{P}_{22} \end{bmatrix}$ is the matrix designed in the following proof. $\bar{\boldsymbol{P}}_{e1} \in \boldsymbol{R}^{1\times(n-m)}$, $\bar{\boldsymbol{P}}_{e2} \in \boldsymbol{R}^{1\times m}$. K_D is a design positive parameter satisfying $K_D > k_w + k_{w^*}$. $\boldsymbol{\Gamma} = diag[\tau_0, \cdots \tau_{N_u}] > 0$ is design parameter.

Theorem: If there exist common positive define matrix $\boldsymbol{P}_{11}$, $\boldsymbol{P}_{22}$ and common matrix $\boldsymbol{P}_{21}$ satisfying $\boldsymbol{P}(\boldsymbol{A}^i + \boldsymbol{B}^i \boldsymbol{K}_j) + (\boldsymbol{A}^i + \boldsymbol{B}^i \boldsymbol{K}_j)^T \boldsymbol{P} < -\bar{\boldsymbol{Q}}$ for all $i, j = 1, 2, \cdots, p$, where $\boldsymbol{P} = \begin{bmatrix} \boldsymbol{P}_{11} & \boldsymbol{0} \\ \boldsymbol{P}_{21} & \boldsymbol{P}_{22} \end{bmatrix}$, $\bar{\boldsymbol{Q}}$ is a design positive define matrix, then there exists an $\varepsilon^* > 0$, for $\varepsilon \in (0, \varepsilon^*]$, when we adopt controller (6-9) with $\boldsymbol{K}_j = \boldsymbol{Y}_j \boldsymbol{Q}^{-1}$ and adaptive law (10), the states $\boldsymbol{x}(t)$ can follow those of a stable reference model, that is when $t \to \infty$, $\boldsymbol{e}(t) \to \boldsymbol{0}$.

Proof:
Substitute (6-10) into (5), we get:

$$\boldsymbol{E}_\varepsilon \dot{\boldsymbol{e}}(t) = \sum_{i=1}^{p} \sum_{j=1}^{p} \xi_i \xi_j (\boldsymbol{A}^i + \boldsymbol{B}^i \boldsymbol{K}_j) \boldsymbol{e}(t) + \begin{bmatrix} \boldsymbol{0} \\ \boldsymbol{f}(\boldsymbol{x}(t)) + \boldsymbol{d}(t) \end{bmatrix} - \begin{bmatrix} \boldsymbol{0} \\ \boldsymbol{I}_m \end{bmatrix} [\boldsymbol{W\Psi}(\boldsymbol{x}) + K_D \operatorname{sgn}(\bar{\boldsymbol{P}}_e^T)] \tag{11}$$

We choose Lyapunov function

$$V(t) = 1/2 \boldsymbol{e}^T(t) \boldsymbol{E}_\varepsilon \boldsymbol{P}_\varepsilon \boldsymbol{e}(t) + tr[(\boldsymbol{W} - \boldsymbol{W}^*) \boldsymbol{\Gamma}^{-1} (\boldsymbol{W} - \boldsymbol{W}^*)^T] \tag{12}$$

Where $\boldsymbol{W}^*$ is the optimal approximation parameters of $\boldsymbol{f}(\boldsymbol{x}(t))$.
Take the time derivative of $V(t)$ is:

$$\begin{aligned} \dot{V}(t) &= 1/2 \boldsymbol{e}^T(t) \boldsymbol{E}_\varepsilon \boldsymbol{P}_\varepsilon \dot{\boldsymbol{e}}(t) + 1/2 \dot{\boldsymbol{e}}^T(t) \boldsymbol{P}_\varepsilon^T \boldsymbol{E}_\varepsilon \boldsymbol{e}(t) + tr[(\boldsymbol{W} - \boldsymbol{W}^*) \boldsymbol{\Gamma}^{-1} \dot{\boldsymbol{W}}^T] \\ &= 1/2 \boldsymbol{e}^T(t) [\sum_{i=1}^{p} \sum_{j=1}^{p} \xi_i(z) \xi_j(z) [\boldsymbol{P}_\varepsilon^T (\boldsymbol{A}^i + \boldsymbol{B}^i \boldsymbol{K}_j) + (\boldsymbol{A}^i + \boldsymbol{B}^i \boldsymbol{K}_j)^T \boldsymbol{P}_\varepsilon] \boldsymbol{e}(t) \\ &\quad + \bar{\boldsymbol{P}}_{e2} (\boldsymbol{w}^* + (\boldsymbol{W}^* - \boldsymbol{W}) \boldsymbol{\Psi}(\boldsymbol{x}) + \boldsymbol{d}(t) - K_D \operatorname{sgn}(\bar{\boldsymbol{P}}_{e2}^T)) + tr[(\boldsymbol{W} - \boldsymbol{W}^*) \boldsymbol{\Gamma}^{-1} \dot{\boldsymbol{W}}^T] \end{aligned} \tag{13}$$

Considering assumptions 1, 3 and using the fact $K_D > k_w + k_{w^*}$, (13) leads to

$$\dot{V}(t) < 1/2 \boldsymbol{e}^T(t) [\sum_{i=1}^{p} \sum_{j=1}^{p} \xi^i \xi^j [\boldsymbol{P}_\varepsilon^T (\boldsymbol{A}^i + \boldsymbol{B}^i \boldsymbol{K}_j) + (\boldsymbol{A}^i + \boldsymbol{B}^i \boldsymbol{K}_j)^T \boldsymbol{P}_\varepsilon] \boldsymbol{e}(t) \tag{14}$$

Since

$$\begin{aligned} &\boldsymbol{P}_\varepsilon^T (\boldsymbol{A}^i + \boldsymbol{B}^i \boldsymbol{K}_j) + (\boldsymbol{A}^i + \boldsymbol{B}^i \boldsymbol{K}_j)^T \boldsymbol{P}_\varepsilon + \bar{\boldsymbol{Q}} \\ &= \boldsymbol{P}^T (\boldsymbol{A}^i + \boldsymbol{B}^i \boldsymbol{K}_j) + (\boldsymbol{A}^i + \boldsymbol{B}^i \boldsymbol{K}_j)^T \boldsymbol{P} + \bar{\boldsymbol{Q}} + \boldsymbol{o}(\varepsilon) \end{aligned} \tag{15}$$

Where $\boldsymbol{o}(\varepsilon)$ is the higher order of small parameter ε, from [6,7], we know if we can guarantee

$$P^T(A^i + B^i K_j) + (A^i + B^i K_j)^T P + \bar{Q} < 0,\ i, j = 1, 2, \cdots, p \tag{16}$$

there exists an $\varepsilon^* > 0$, for $\varepsilon \in (0, \varepsilon^*]$

$$P_\varepsilon^T(A^i + B^i K_j) + (A^i + B^i K_j)^T P_\varepsilon + \bar{Q} < 0\ i, j = 1, 2, \cdots, p \tag{17}$$

Then (14) is:

$$\dot{V}(t) < -1/2 e^T(t)\bar{Q}e(t) < -1/2\lambda_{\min} e^T(t)e(t) \tag{18}$$

And

$$\int_0^t e^T(t)e(t)dt < -2/\lambda_{\min}(V(t) - V(0)) < 2/\lambda_{\min} V(0) \tag{19}$$

Using Barbalat lemma, we have $t \to \infty$, $e(t) \to 0$.

Let $Q = P^{-1}$, $Y_j = K_j Q$ and multiply both sides of (16) by Q and use Schur theorem, we can get:

$$\begin{bmatrix} A^i Q + B^i Y_j + (A^i Q)^T + (B^i Y_j)^T & Q \\ Q & -\bar{Q} \end{bmatrix} < 0, \quad i, j = 1, 2, \cdots, p \tag{20}$$

Then the controller gain K_i is achieved and the proof is completed.

4 Simulations

In this section, we will design a neuro-fuzzy adaptive controller for a flexible spacecraft [8]. Sector nonlinearity approach is used to model the system. Let $L_1(t) = \dot{\theta} q$, We have $\max(L_1) = k_1^1$, $\min(L_1) = k_1^2$, the fuzzy rules are as following:

$$R^i : \text{if } L_1(t) \text{ is } M_1^i, \text{ then } E_\varepsilon \dot{x}(t) = (A^i + \Delta A^i)x(t) + B^i u(t) + D^i \omega(t) \tag{21}$$

The membership functions are:

$$M_1 = \frac{L_1 - \min(L_1)}{\max(L_1) - \min(L_1)} \qquad M_2 = \frac{\max(L_1) - L_1}{\max(L_1) - \min(L_1)} \tag{22}$$

Then model parameters are achieved, which is omitted in this paper. The rule premises of the stable reference models are the same as those of (21) and its parameters are as follows:

$$A_m^1 = A_m^2 = \begin{bmatrix} 0 & 1 & 0 & 0 \\ -2 & -2 & 0 & 0 \\ 0 & 0 & 0 & 0 \\ 0 & 0 & 0 & 0 \end{bmatrix} \quad B_m^1 = B_m^2 = \begin{bmatrix} 0 & 0 \\ 1 & 2 \\ 0 & 0 \\ 0 & 0 \end{bmatrix} \quad r(t) = \begin{bmatrix} 0.1 & 0 \end{bmatrix}$$

The simulation results as Fig.1 and Fig.2, we can see that the actually states can track the reference ones successfully.

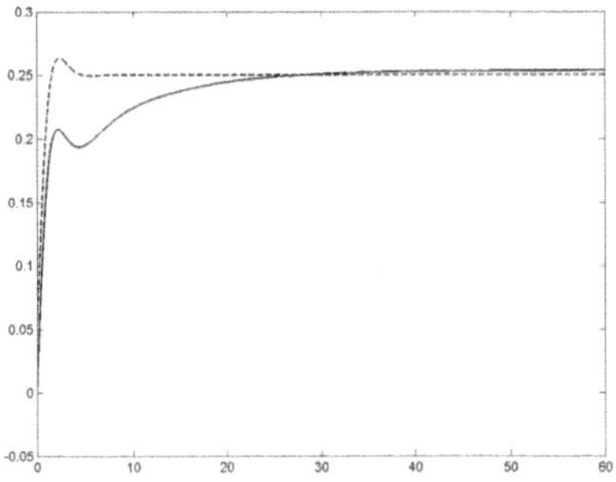

Fig. 1. The state x_1 (solid line) and the reference states x_{m1} (dash line)

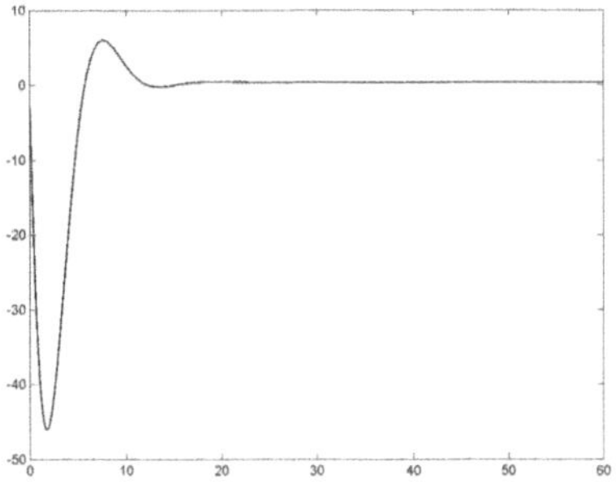

Fig. 2. The fast state z_1 of the actual system

5 Conclusion

In this paper, the nonlinear SP system is modeled as the composition of TS fuzzy linear systems. To alleviate the approximation error of modeling and external disturbance, we present a neuro-fuzzy adaptive controller based on reference model, which does not require the parameter perturbation and external disturbance satisfy matching conditions. Simulations illustrate the effectiveness of this approach.

References

1. Liu, H.P., Sun, F.C. et al: Controller Design and Stability Analysis for Fuzzy Singularly Perturbed Systems. Acta Automatic Sinica **29**(4) (2003) 494-500
2. Liu, H.P., Sun, F.C. et al: H_2 State Feedback Control for Fuzzy Singularly Perturbed Systems. Proceedings of the 42nd IEEE conference on decision and control. Maul, Hawali USA (2003) 5239-5243
3. Liu, H.P., Sun, F.C. et al: Simultaneous Stabilization for Singularly Perturbed Systems via Linear Matrix Inequalities. Acta Automatic Sinica **30**(1) (2004) 1-7
4. Kiriakidis, K.: Non-linear Control System Design via Fuzzy Modeling and LMIs. Int. J. Control **72**(7) (1999) 676–685
5. Teixeira, M.C.M., Zak, S.H.: Stabilizing Controller Design for Uncertain Nonlinear Systems Using Fuzzy Models. IEEE Trans. Fuzzy Syst. **7**(2) (1999) 133–142
6. Garcia, G., Daafouz, J., Bernussou, J.: A LMI Solution in the H_2 Optimal Problem for Singularly Perturbed Systems. Proceedings of the American Control Conference **1** (1998) 550-554
7. Fridman, E.: Effects of Small Delays on Stability of Singularly Perturbed Systems. Automatica **38**(5) (2002) 897-902
8. Nayeri, M.R.D., Alasty, A., Daneshjou, K: Neural Optimal Control of Flexible Spacecraft Slew Maneuver. Acta Astronautica **55**(10) (2004) 817- 827

Self-tuning PID Temperature Controller Based on Flexible Neural Network

Le Chen[1], Baoming Ge[1], and Aníbal T. de Almeida[2]

[1] School of Electrical Engineering, Beijing Jiaotong University, Beijing 100044, China
bm-ge@263.net
[2] Institute of System and Robotics, University of Coimbra, 3030 Coimbra, Portugal
adealmeida@isr.uc.pt

Abstract. A temperature control solution is proposed in this paper, which uses a self-tuning PID controller based on flexible neural network (FNN). The learning algorithm of FNN can adjust not only the connection weights but also the sigmoid function parameters. This makes FNN characterized with online learning and high learning speed. The FNN has the following advantages when applied to temperature control problems: high learning ability, which considerably reduces the controller training time; the mathematical model of the plant is not required, which eases the design process; high control performance. These advantages are verified by its application to a practical temperature controlled box, which is used in medicinal inspection. The proposed system presents better behavior than that when using traditional back-propagation neural network.

1 Introduction

Temperature control in some medicinal inspection is very important. If the temperature is too high or too low, the final result is seriously affected. Therefore, it is necessary to reach some desired temperature points quickly and avoid large overshoot. The PID control method has been used in practical control problems, especially in the field of process control [1]. Classical PID control theory usually requires a mathematical model for designing the controller and the inaccuracy of mathematical modeling of the plant usually degrades the performance of the controller, especially for nonlinear and complex control problems [2]. And the PID controller consists of three parts: proportional, integral, and derivative controllers. In this case, we must determine the three parameters such as k_p, k_i and k_d [3]. These gains were determined by operators based on their experience and knowledge in the process. Therefore, it is difficult to determine which values are the best to produce the desired output [1].

In order to overcome these disadvantages, in [1], [4], and [5], the authors have applied neural controller based on multilayered back-propagation neural network (BPNN) to temperature control problems. BPNN has advantages over traditional control systems, because it does not need to know the mathematical models of the plants and it can tune the three parameters (k_p, k_i and k_d) on-line by itself. Besides these, for a BPNN, its nonlinear mapping and self-learning abilities have been the motivating factors for its application in different control problems, one of which is the temperature control.

D. Liu et al. (Eds.): ISNN 2007, Part I, LNCS 4491, pp. 138–147, 2007.

However, for BPNN, slow convergence speed is the major disadvantage and its long training time usually discourages its practical applications in industry [2]. Owing to these problems, in this paper, we propose the flexible neural network (FNN) controller, which is a better choice in temperature control problems.

Compare to BPNN, the flexibility of FNN is improved, because the sigmoid function (SF) in FNN is different from the SF in BPNN. The learning algorithms of FNN adjust not only connection weights but also the parameters of SFs. The increased flexibility of the structure induces a more efficient learning ability in the system which in turn causes less iteration and better error minimization [6].

We apply the FNN controller to a temperature controlled box system which is used in medicinal inspection. Simulated results verify that the FNN controller gives an excellent temperature control performance.

2 Temperature Controlled Box

As shown in Fig. 1 the temperature controlled box system can be divided into five main components: (1) the temperature box and the prober; (2) the heater module; (3) the sensor module; (4) the controller; (5) the motor and the luggage carrier in the box.

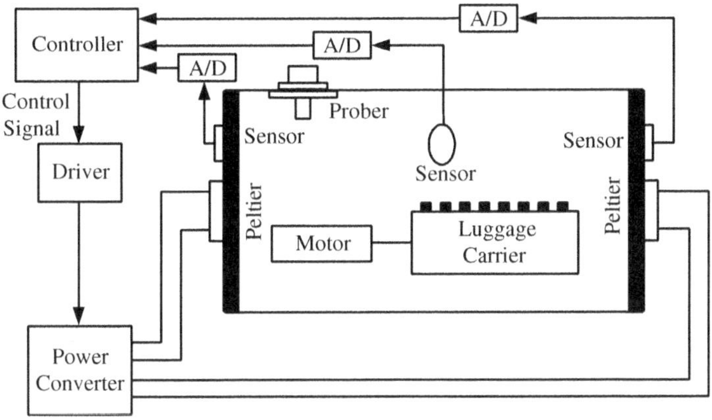

Fig. 1. Schematic diagram of the temperature controlled box system

The two side-walls of the box are made of aluminous, which has good heat dissipation. The rest walls and the outside side of the side-walls are heat insulation. The prober is used to inspect the object, which is on the luggage carrier.

There are two peltiers with a power circuit in the heater module. The power circuit is used as an actuator for the peltiers. The peltiers provide heat to two side-walls of the box, and the two side-walls send heat to the inside of the box.

The sensor module has been developed by using PT100 and amplifiers. It can transform the measured temperatures into corresponding voltages. There are three sensors, the one in the center of the box is used to measure the inner temperature of the box; and the others are used to measure the temperature of two aluminous side-walls of the box.

The motor and the luggage carrier in the box are disturbances to the system. When the motor works, it gives out heat and its temperature can be higher than that of the box, therefore the motor can be considered as an unstable heater. The luggage carrier can go out of the box from the door of the box to bring the inspected objects into the box. Therefore, it will lose heat when it is in the outside of the box, as a result that its temperature will reduce a bit. Because of these disturbances and the difficulty to get the mathematical thermodynamic model, we design the control system by using a double closed-loops control. The inner loop is a hysteresis-band control of temperature in two side-walls, and the outer loop is a self-tuning PID control of temperature in the center of box.

The approximate thermodynamic model of the box is shown in Fig. 2, where u denotes the temperature of the side-walls, u_1 denotes the temperature of the motor, u_2 denotes the temperature of luggage carrier, and y denotes the center temperature of the box. The values of k, k_0, k_1, k_2, and k_3 are determined by the size and structure of the temperature box.

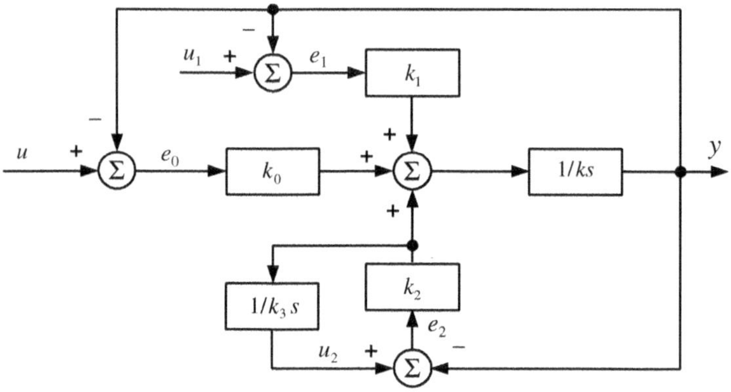

Fig. 2. Temperature model of the box

The temperature model of the box can be described as

$$y = \frac{e_0 k_0 + e_1 k_1 + e_2 k_2}{ks} , \tag{1}$$

$$e_0 = u - y \ , \ e_1 = u_1 - y \ , \ e_2 = u_2 - y \ . \tag{2}$$

The temperature u_2 of the luggage carrier is a variable, because at the moment the luggage carrier comes into the box from the outside its temperature is lower than y, then the carrier will be heated, and u_2 can be described as

$$u_2 = u_2^0 + \frac{e_2 k_2}{k_3 s} , \tag{3}$$

where u_2^0 means the initial value of u_2.

3 FNN-Based Self-tuning PID Controller

3.1 Structure of Controller

In the FNN-based controller the output of the proposed FNN is used as an input to the conventional controller, the output of controller is used as an input to the plant, and the output of the plant is fed back to the FNN, as illustrated in Fig. 3. The FNN can be trained on-line, so that the connection weights and the SF parameters are adjusted at every sampling time.

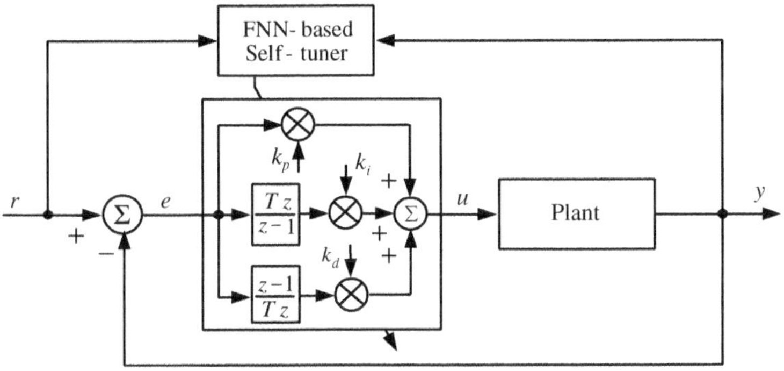

Fig. 3. Block diagram of self-tuning temperature control system

In this controller, the FNN is used as a self-tuner to automatically tune the parameters of the conventional PID controller.

The FNN-based self-tuner has two neural units in the input-layer, five neural units in the hidden-layer, and three neural units in the output-layer. The functions used in the hidden-layer are flexible bipolar sigmoid functions (FBSF), and in the output-layer are flexible unipolar sigmoid functions (FUSF).

From Fig. 3, the plant input signal u is expressed as

$$u = (k_p + \frac{k_i Tz}{z-1} + \frac{k_d(z-1)}{Tz})e \ , \tag{4}$$

where T is a sampling interval, r denotes the desired temperature. k_p, k_i and k_d are the proportional, integral, and derivative gains which are obtained from the outputs of the established FNN.

3.2 Flexible Sigmoid Functions

FUSF, which is used in output-layer, is described by

$$g(x,a) = \frac{2|a|}{1+e^{-2|a|x}} \ . \tag{5}$$

The activation derivative of FUSF with respect to variable x, $g'(x, a)$, is obtained as

$$g'(x,a) = \frac{\partial g(x,a)}{\partial x} = g(x,a)[2|a| - g(x,a)] \ , \tag{6}$$

and the derivative of FUSF with respect to the parameter a, $g^*(x, a)$, is obtained as

$$g^*(x,a) = \frac{\partial g(x,a)}{\partial a} = \frac{1}{|a|}[g(x,a) + g'(x,a)x] \ . \tag{7}$$

FBSF, which is employed in hidden-layer, is described by

$$f(x,a) = \frac{1 - e^{-2xa}}{a(1 + e^{-2xa})} \ . \tag{8}$$

The activation derivative of FBSF with respect to variable x, $f'(x, a)$, is obtained as

$$f'(x,a) = \frac{\partial f(x,a)}{\partial x} = 1 - a^2 f^2(x,a) \ , \tag{9}$$

and the derivative of FBSF with respect to the parameter a, $f^*(x, a)$, is obtained as

$$f^*(x,a) = \frac{\partial f(x,a)}{\partial a} = \frac{1}{a}[f'(x,a)x - f(x,a)] \ . \tag{10}$$

3.3 Learning Algorithm of Connection Weights

The classical incremental digital PID algorithm is

$$\begin{aligned} u(k) &= u(k-1) + \Delta u(k) \ , \\ \Delta u(k) &= k_p[e(k) - e(k-1)] + k_i e(k) + k_d[e(k) - 2e(k-1) + e(k-2)] \ . \end{aligned} \tag{11}$$

The inputs and the outputs of the hidden-layer are

$$\begin{aligned} net_i^{(2)}(k) &= w_{1i} r + w_{2i} y(k) \ , \\ o_i^{(2)}(k) &= f(net_i^{(2)}(k), a_i^{(2)}(k)), \quad i = 1, \ 2, \ \cdots, \ 5. \end{aligned} \tag{12}$$

where w_{ij} is the connection weights between the input- and hidden-layer, the superscript (1), (2), and (3) denote input-, hidden-, and output-layer respectively.

The inputs and the outputs of the output-layer respectively are

$$net_l^{(3)}(k) = \sum_{i=1}^{5} v_{il} o_i^{(2)}(k), \quad o_l^{(3)}(k) = g(net_l^{(3)}(k), a_l^{(3)}(k)), \quad l = 1, \ 2, \ 3. \tag{13}$$

where v_{il} are the connection weights between the hidden- and output-layer and $o_1^{(3)} = k_p(k)$, $o_2^{(3)} = k_i(k)$, $o_3^{(3)} = k_d(k)$.

The cost function is expressed as

$$E(k) = \frac{1}{2}[r(k) - y(k)]^2 \ . \tag{14}$$

By employing the gradient descent method, the increment of v_{il}, denoted by Δv_{il} , can be obtained as

$$\Delta v_{il}(k) = -\eta \frac{\partial E(k)}{\partial v_{il}(k)} + \alpha \Delta v_{il}(k-1) \ , \tag{15}$$

where η is a learning rate given by a small positive constant, α is a stabilizing coefficient in the range [0, 1). And

$$\begin{aligned} &\frac{\partial E(k)}{\partial v_{il}} = \frac{\partial E(k)}{\partial y(k)} \cdot \frac{\partial y(k)}{\partial u(k)} \cdot \frac{\partial u(k)}{\partial o_l^{(3)}(k)} \cdot \frac{\partial o_l^{(3)}(k)}{\partial net_l^{(3)}(k)} \cdot \frac{\partial net_l^{(3)}(k)}{\partial v_{il}(k)} \ , \\ &\frac{\partial net_l^{(3)}(k)}{\partial v_{il}(k)} = o_i^{(2)}(k) \ . \end{aligned} \tag{16}$$

From (11) and (13) we get

$$\begin{aligned} &\frac{\partial u(k)}{\partial o_1^{(3)}(k)} = e(k) - e(k-1), \quad \frac{\partial u(k)}{\partial o_2^{(3)}(k)} = e(k) \\ &\frac{\partial u(k)}{\partial o_3^{(3)}(k)} = e(k) - 2e(k-1) + e(k-2) \ . \end{aligned} \tag{17}$$

Because $\partial y(k)/\ \partial u(k)$ is unknown we replace it by $\mathrm{sgn}(\partial y(k)/\ \partial u(k))$. Therefore, the learning equations for the connection weights at the output-layer are obtained by

$$\begin{aligned} &\Delta v_{il}(k) = \alpha \Delta v_{il}(k-1) + \eta \delta_l^{(3)} o_i^{(2)}(k), \quad i = 1, \ 2, \ \cdots, \ 5. \\ &\delta_l^{(3)} = e(k) \cdot \mathrm{sgn}(\frac{\partial y(k)}{\partial u(k)}) \cdot \frac{\partial u(k)}{\partial o_l^{(3)}} \cdot g'(net_l^{(3)}(k), a_l^{(3)}(k)), \quad l = 1, \ 2, \ 3. \end{aligned} \tag{18}$$

The learning equations for the connection weights at the hidden-layer are obtained by

$$\begin{aligned} &\Delta w_{ji}(k) = \alpha \Delta w_{ji}(k-1) + \eta \delta_i^{(2)} o_j^{(1)}(k), \quad j = 1, \ 2. \\ &\delta_i^{(2)} = f'(net_i^{(2)}(k), a_i^{(2)}(k)) \cdot \sum_{l=1}^{3} \delta_l^{(3)} v_{il}(k), \quad i = 1, \ 2, \ \cdots, \ 5. \end{aligned} \tag{19}$$

where $o_1^{(1)}(k) = r, \ o_2^{(1)}(k) = y(k)$.

3.4 Learning Algorithm of SF Parameters

By employing the gradient descent method, the increment of $a_l^{(3)}$, denoted by $\Delta a_l^{(3)}$, can be obtained as

$$\Delta a_l^{(3)}(k) = -\eta \frac{\partial E(k)}{\partial a_l^{(3)}(k)} + \alpha \Delta a_l^{(3)}(k-1), \quad l = 1, \ 2, \ 3. \tag{20}$$

The partial derivative of $E(k)$ with respect to a is described as

$$\frac{\partial E(k)}{\partial a_l^{(3)}(k)} = \frac{\partial E(k)}{\partial o_l^{(3)}(k)} \cdot \frac{\partial o_l^{(3)}(k)}{\partial a_l^{(3)}(k)} = \frac{\partial E(k)}{\partial o_l^{(3)}(k)} \cdot g^*(net_l^{(3)}(k), a_l^{(3)}(k)) . \quad (21)$$

Here, we define

$$\sigma_l^{(3)} = -\frac{\partial E(k)}{\partial o_l^{(3)}(k)} = -\frac{\partial E(k)}{\partial y(k)} \cdot \frac{\partial y(k)}{\partial u(k)} \cdot \frac{\partial u(k)}{\partial o_l^{(3)}} . \quad (22)$$

The next step is to calculate a at the hidden-layer, namely

$$\Delta a_i^{(2)}(k) = -\eta \frac{\partial E(k)}{\partial a_i^{(2)}(k)} + \alpha \Delta a_i^{(2)}(k-1), \quad i = 1, \ 2, \ \cdots, \ 5.$$

$$\frac{\partial E(k)}{\partial a_i^{(2)}(k)} = \frac{\partial E(k)}{\partial o_i^{(2)}(k)} \cdot \frac{\partial o_i^{(2)}(k)}{\partial a_i^{(2)}(k)} = \frac{\partial E(k)}{\partial o_i^{(2)}(k)} \cdot f^*(net_i^{(2)}(k), a_i^{(2)}(k)) . \quad (23)$$

Using the definition

$$\sigma_i^{(2)} = -\frac{\partial E(k)}{\partial o_i^{(2)}(k)} = \sum_{l=1}^{3} \{\sigma_l^{(3)} g'(net_l^{(3)}(k), a_l^{(3)}(k)) v_{il}(k)\} . \quad (24)$$

Therefore, the learning update equations for parameter a at the output- and hidden-layer are obtained, respectively, by

$$\Delta a_l^{(3)}(k) = \eta \sigma_l^{(3)} g^*(net_l^{(3)}(k), a_l^{(3)}(k)) + \alpha \Delta a_l^{(3)}(k-1), \quad l = 1, \ 2, \ 3. \quad (25)$$

$$\Delta a_i^{(2)}(k) = \eta \sigma_i^{(2)} f^*(net_i^{(2)}(k), a_i^{(2)}(k)) + \alpha \Delta a_i^{(2)}(k-1), \quad i = 1, \ 2, \ \cdots, \ 5. \quad (26)$$

4 Simulation Tests and Results

We apply FNN controller and BPNN controller to the temperature controlled box system via a simulation to verify that the FNN controller gives a better control performance. The approximate thermodynamic model of the box is shown in Fig. 2, where parameters k, k_0, k_1, k_2, and k_3 are respectively obtained as k=0.657 (J/°C), $k_0=1.1\times10^{-2}$ (W/°C), $k_1=5.5\times10^{-3}$ (W/°C), $k_2=1.1\times10^{-3}$ (W/°C), $k_3=2.33\times10^{3}$ (W/°C). We do not need to know these values during the actual control process.

In our simulation the BPNN has the same structure with the FNN, that is, two nodes in input-layer, five nodes in hidden-layer, and three nodes in output-layer. We need to determine some parameters and the initial states of system in the first step. The physical parameters are set as follows: sampling interval is 30s and the initial temperature of the box is 15°C. The temperature of motor will stabilize at 44°C when it operates. The final temperature of luggage carrier will be 30°C when it stays outside of box. Therefore, the initial temperature of luggage carrier can be 30°C. The desired temperature of the box is 36°C.

Fig. 4 shows the simulation results of the controlled temperature of the box using two learning algorithms. It can be observed that the self-tuning PID controller based on the BPNN is slower in achieving the desired temperature, compared to the FNN.

The self-tuning PID controller using the FNN provides a betterment of implementation over the BPNN. Fig. 5 shows the tuning gains of PID controller, which are exactly the outputs of FNN. The PID gains show the dynamic convergence to their steady values during the control operations.

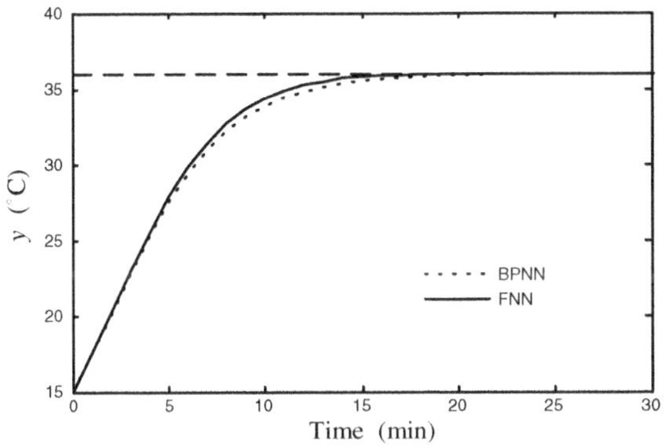

Fig. 4. Simulation results of self-tuning PID control for temperature of the box

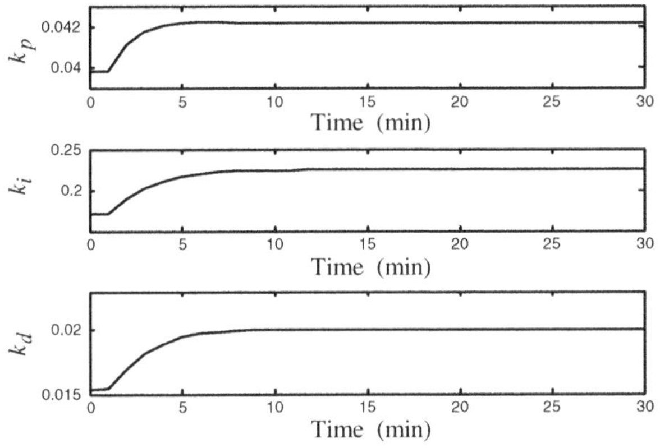

Fig. 5. Outputs of the proposed FNN

Figs. 6 and 7 show the simulation results at the disturbances from the motor and the carrier. The motor begins to operate at 35 second, as a result that its temperature stabilizes at 44°C, namely u_1 =44°C, which will make the box temperature increase, as shown in Fig. 6. However, the tuning PID controller will force the box temperature to the desired value of 36°C, shown in Fig. 6.

Fig. 7 shows the response of box temperature when the luggage carrier has an initial temperature of 30°C at 60 second because the carrier goes out of the box. The result shows that the controller can correct the errors due to the disturbances, accurately and rapidly.

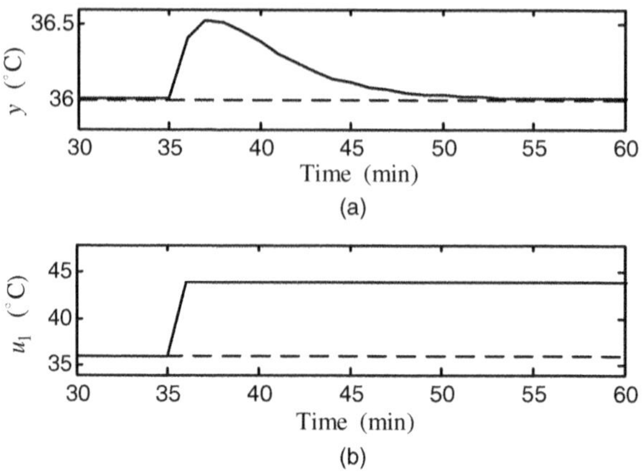

Fig. 6. Simulation results when the motor operates, (a) shows the changes of the box temperature, (b) shows the changes of the motor temperature

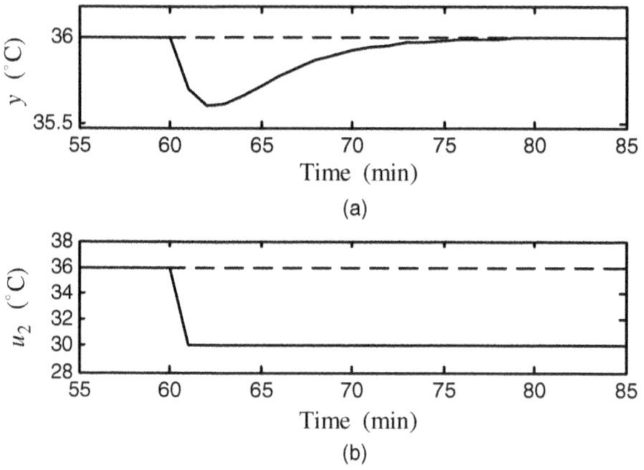

Fig. 7. Simulation results when the carrier has an initial temperature of 30°C, (a) shows the changes of the box temperature, (b) shows the changes of the carrier temperature

5 Conclusions

In this paper we proposed an on-line self-tuning PID temperature controller which is on basis of FNN. The learning algorithms of FNN can adjust not only the connection weights but also the SF parameters, which makes FNN characterized with online learning and high learning speed. We apply the FNN self-tuning PID controller to a practical medicinal box to control its temperature. The proposed temperature controller has shown superior performance both in controller design effort and control performance. These advantages will motivate us further applications of FNN to other temperature control problems in industry.

References

1. Tanomaru J., Omatu S.: Process Control by On-line Trained Neural Controllers. IEEE Transactions on Industrial Electronics **39** (1992) 511-521
2. Lin, C.T., Juang, C.F., Li, C.P.: Temperature Control with a Neural Fuzzy Inference Network. IEEE Transaction on Systems Man and Cybernetics **29** (1999) 440-451
3. Guo, C.Y., Song, Q., Cai, W.J.: Supply Air Temperature Control of AHU with a Cascade Control Strategy and a SPSA Based Neural Controller. Proceedings of the 2005 International Joint Conference on Neural Networks **4** (2005) 2243-2248
4. Omatu S., Iwasa T., Yoshioka M.: Skill-based PID Control by Using Neural Networks. Proceedings of the 1998 IEEE International Conference on System Man and Cybernetics **2** (1998) 1972-1977
5. Hu, Q.H., So A.T.P., Tes W.L., Dong, A.: Use of Adaline PID Control for a Real MVAC System. Proceedings of the 2005 International Conference on Wireless Communications, Networking and Mobile Computing **2** (2005) 1374 – 1378
6. Wang, H.Y., Shi, G.D., Xia, D.S.: Flexible Neural Network and Its Application. Pattern Recognision and Artificial Intelligence **15** (2002) 373-376

Hybrid Neural Network Controller Using Adaptation Algorithm*

ManJun Cai, JinCun Liu, GuangJun Tian, XueJian Zhang, and TiHua Wu

College of Electrical Engineering, YanShan University,
QinHuangDao 066004, China
{liujincun01,zxj20777,infobase1}@163.com

Abstract. Neural network controller using adaptation algorithm is a new and simple controller, in which a feedback network propagating the error is not required. So it can be applied to hardware easily. Nevertheless, our simulations show that while the order of controlled plant is high, some unstable phenomenon appear and we also find that sometimes the error is far from being satisfactory, although when the order of controlled plant is low. Moreover, the present adaptation algorithm can not solve this problem. In this paper we will give our derivation of adaptation algorithm used in the neural network controller and configuration of an adaptive neural network controller. Then give some simulation figures to illustrate defect for the new controller. Finally we will develop a hybrid neural network to solve the problem and improve the accuracy as well as reduce the cost to the least in the practical application.

1 Introduction

In this paper, we will apply the approach of adaptive interaction to neural networks. Before Robert D. Brandt and Feng Lin raised this approach, back propagation algorithm was to adapt synapses in a neural network [1],[2],[3]. In the back propagation algorithm, a dedicated companion (feedback) network to propagate the error back is required. This may complicate implementations [4],[5],[6]. On the other hand, using adaptive interaction, we can eliminate the need for such a feedback network, and hence significantly reduce the complexity of adaptation for complex neural networks. This is particularly important in VLSI implementations of neural networks [7],[8],[9]. The absence of the feedback network means that adding trainability to a chip design does not involve additional wiring-layout complexity between neurons [10],[11]. These trainable neurons can be connected in any way the designer wants, in that increases the potential for designing net works with dynamically reconfigurable topologies [12],[13],[14],[15].

* This paper is supported by the National Natural Science Foundation of China (20577038).

D. Liu et al. (Eds.): ISNN 2007, Part I, LNCS 4491, pp. 148–157, 2007.

2 Adaptive Neural Networks

As mentioned above an adaptive neural network doesn't need feedback network that mean the special train time is not required. When we give the adaptive neural network input the weights that connect neurons will change directly base on the output we +expect. Now we give a neural network as Fig. 1.

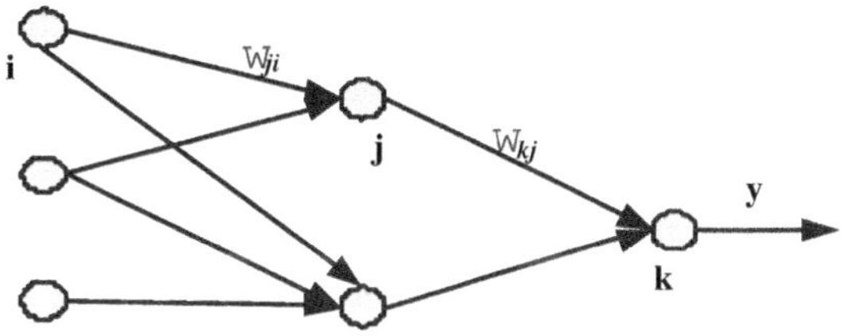

Fig. 1. A part of normal neural network. Here we make notations in neural networks. $s, t \in n$, y_s-the output of neuron s; x_s-the input of neuron s; ξ_s-the external input of neuron s; ζ_s-the desired output of neuron s; ω_{st}-the weight of the connection from neuron t to neuron s. i, j, k stand for the neuron i, neuron j. Neuron k(an output neuron) respectively. $i, j, k \in n$.

2.1 Theorem

Here we give out a theorem which will be used later.

$$\dot{\omega}_{st} = -\gamma \frac{dE}{d\omega_{st}} \ . \tag{1}$$

Here $\gamma > 0$ is a coefficient, and the explanation of this theorem will be given in latter part [16],[17].

2.2 Derivation of Adaptation Algorithm

We denote by $f(x)$ the activation function of a neuron. Ignoring the dynamics, we can describe the neural network by

$$y_k = f(x_k) = f(\sum_{j \in n} \omega_{kj} y_j + \xi_k) \ . \tag{2}$$

Our goal is to minimize the following error

$$E = \frac{1}{2} \sum_{k \in n} e_k^2 \ . \tag{3}$$

Where

$$e_k = y_k - \zeta_k \ . \tag{4}$$

Because E is a functional of y_k, so we can get the following derivative process

$$\frac{dE}{d\omega_{kj}} = \frac{dE}{dy_k}\frac{dy_k}{dx_k}\frac{dx_k}{d\omega_{kj}} = e_k f'(x_k) y_j \ . \tag{5}$$

Now we denote the equation

$$y_j = f_1(x_j) = f_1(\sum_{i \in n} \omega_{ji} y_i + \xi_j) \ . \tag{6}$$

Then there will be

$$y_k = f(x_k) = f(\sum_{j \in n} \omega_{kj} f_1(x_j) + \xi_k) \ . \tag{7}$$

Therefore

$$\frac{dE}{d\omega_{ji}} = \frac{dE}{dy_k}\frac{dy_k}{dx_k}\frac{dx_k}{df_1}\frac{df_1}{dx_j}\frac{dx_j}{d\omega_{ji}} = e_k f'(x_k)\omega_{kj} f_1'(x_j) y_i \ . \tag{8}$$

From the equation (5) we can get equation

$$f'(x_k) = \frac{dE}{d\omega_{kj}}\frac{1}{e_k y_j} \ . \tag{9}$$

Therefore

$$\frac{dE}{d\omega_{ji}} = e_k f'(x_k)\omega_{kj} f_1'(x_j) y_i = e_k \frac{dE}{d\omega_{kj}}\frac{1}{e_k y_j}\omega_{kj} f_1'(x_j) y_i \ . \tag{10}$$

Because the following equation is always satisfied

$$\dot{\omega}_{st} = -\gamma \frac{dE}{d\omega_{st}} \ . \tag{11}$$

So, according to this equation we can substitute dE/ω_{st} by $-\dot{\omega}_{st}/\gamma$ then we have

$$\frac{dE}{d\omega_{ji}} = -\alpha\dot{\omega}_{ji}, \frac{dE}{d\omega_{kj}} = -\beta\dot{\omega}_{kj} \ . \tag{12}$$

Therefore we have

$$\dot{\omega}_{ji} = \frac{\beta}{\alpha} e_k \dot{\omega}_{kj} \frac{1}{e_k y_j} \omega_{kj} f_1'(x_j) y_i = \mu f_1'(x_j) \frac{y_i}{y_j} \omega_{kj} \dot{\omega}_{kj} \ . \tag{13}$$

And

$$\dot{\omega}_{kj} = -\lambda e_k f'(x_k) y_j \ . \tag{14}$$

We need indicate that in this neural network there is only one output neuron k, when more than one output using above derivative process we can easily have

$$\dot{\omega}_{ji} = \mu f_1'(x_j) \frac{y_i}{y_j} \sum_{k \in n} \omega_{kj} \dot{\omega}_{kj} \ . \tag{15}$$

Finally we can come to the conclusion when the weight connecting neuron i to neuron j (neuron i and neuron j are not output neuron) change as above equation the neuron network will have adaptive property.

3 Adaptive Neural Network Controller

Now we apply the adaptation algorithm to the following neural network controller showed in Fig. 2. This controller was designed by Robert D.Brandit and Feng Lin [18].

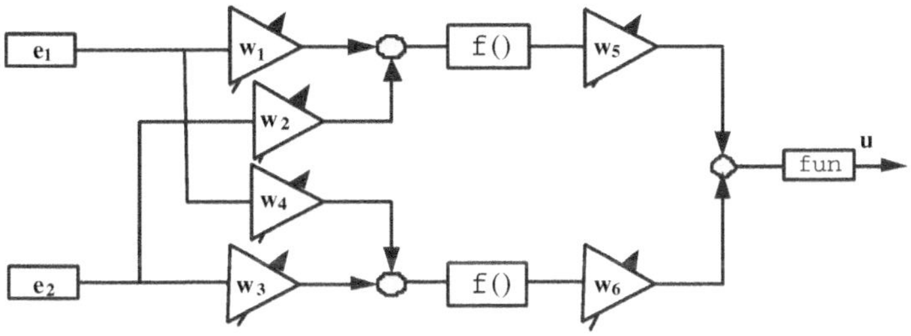

Fig. 2. Adaptive neural network controllers configuration

In neural network the input are error e_1 and its delay e_2, the output is control single. This is a special controller, its configuration very simple and powerful. Our experiments proved when applying adaptation algorithm to the neural network, the effect of this neural network is better than any others by far.

From the paper wrote by Robert D.Brandit and Feng Lin we learn that this controller do a good job when the order of controlled plant is low, input frequency is low and learning rate not big [18],[19],[20]. However, through simulate test we found that the error of output will become very big when the order of controlled plant is higher than third or the input frequency is higher. And we also found the system steady time will be long when the learning rate not high enough, the system will become unstable if the leaning rate is too big. Moreover the output error will oscillate around zero but the amplitude never become smaller as time goes on, in other word the error can not become small as we expect by the controller self. So sometimes the error is far from being satisfactory.

We use MATLAB to make some simulation to show the problems. The input is sine signal, the frequency is 0.1Hz and amplitude is 10. The learning rate is set at 1000. The controller output function is tan-sig, $w_1 - w_4$ changed according to function (15), and w_5, w_6 according to function (14). Our controlled plant is

$$G(S) = \frac{100}{S(S + 21.526)(S + 2.474)} . \tag{16}$$

And then the output error result as followed Fig. 3.

From the figure we can see the biggest error is about 0.2. Because our learning rate is so high that the system responses quickly. Now we change the third order plant to a higher one, keep the leaning rate unchanged. Then the output error figure was showed in Fig. 4.

$$G(S) = \frac{5000}{(S + 1)(S + 5)(S + 10)(S + 100)} . \tag{17}$$

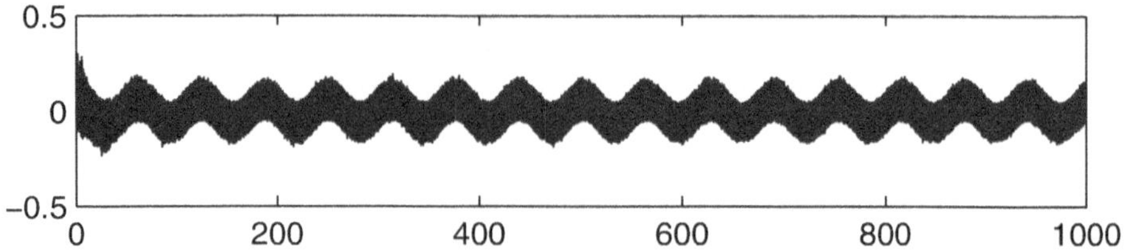

Fig. 3. Third order plant output error

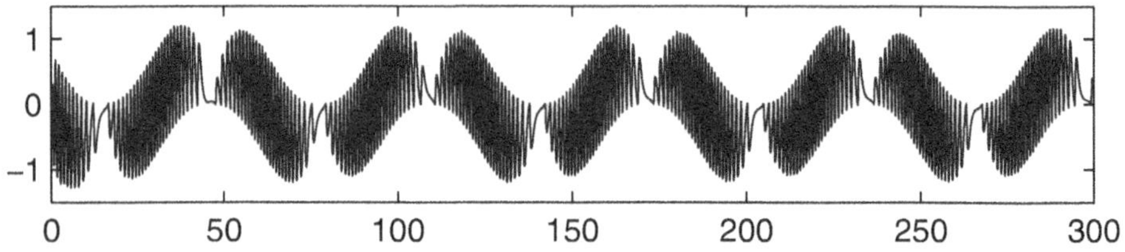

Fig. 4. Forth order plant output error

From the above two figures we can easily find that the output error is so disappointing. In fact when using a smaller learning rate we can not get a better result either, and after repeated experiments we found using a constant gain instead of the original neural network output function the result was not better either.

4 Hybrid Neural Network Controller

From the above section we learn that there are two ways to improve the adaptive neural network controller: One is to change the learning rate and the other is to change neural network output function. At the same time we also learn that these two ways do not take effect obviously. So it is necessary to find out a new way to solve these problems. Let us look the Fig. 5 showing the output from the neural controller, the system error, and the system output. They are all come from forth order plant.

Enlarging the picture to see part of it clearly in the Fig. 6 we can discover the error oscillate around zero with the amplitude never become smaller as time goes on, and the control signal is switching between -1 and 1. Here we need point out that when the error become bigger the time of controller output 1 will become longer. From these phenomena it is not difficult to find out the control signal is too weak that makes error very bad. According to this conclusion we insert a constant gain directly without dropping original output function of the controller to amplify the control signal, because without original output function the system will become unstable in most cases. This may be the new controller's particular property.

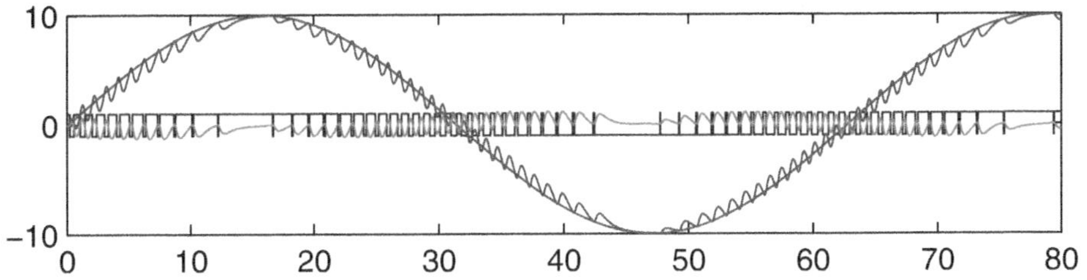

Fig. 5. Control signal, error, input and output. The sine line is input, around which is output; curve line is error, and polyline is control signal.

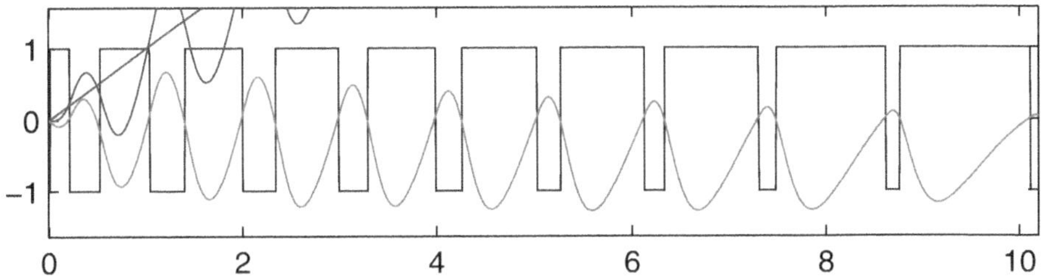

Fig. 6. Control signal(square wave), error(curve line around zero), input and output

Through some simulation we learnt that after inserting a constant-gain the result get better, but a little. If the constant-gain is too big the system become unstable. So we try to design a hybrid neural network to amplify the controller signal according to its need.

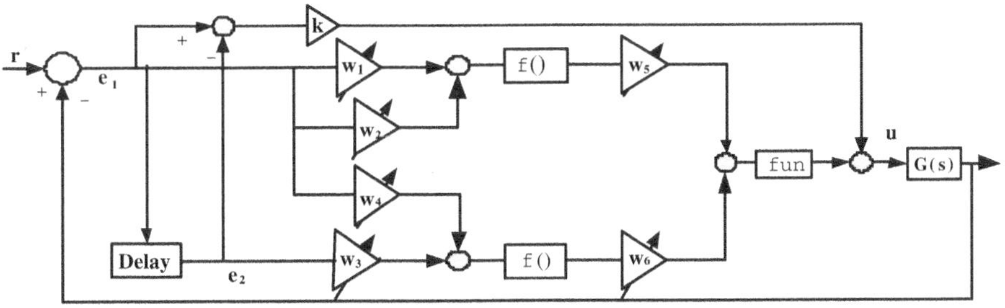

Fig. 7. The hybrid neural network controller

In Fig. 7 we add a new neural network (where the weights are not changed) into the original one to form a hybrid neural network controller. The new part of the neural network just be described as the function $k \cdot \Delta e$ (k is a constant), and at first it is used in the followed method and then we change it to another one.

$$\begin{cases} u = \delta(\cdot) + k \cdot \Delta e & if(e \cdot \Delta e > 0) \\ u = \delta(\cdot) - k \cdot \Delta e & if(e \cdot \Delta e < 0) \end{cases} . \tag{18}$$

Here $\delta(\cdot)$ stand for original function of controller, $k > 0$ is a coefficient, and$\Delta e = e_1 - e_2$, and $e = e_1$. It is declared that we must ensure the symbol of u and e must be the same when use it, if not we need to change the symbol of k. The function and its method mean when the error e is increase ($\Delta e > 0$) and bigger than zero ($u > 0$,$e > 0$) then equation $u = \delta(\cdot) + k \cdot \Delta e$ will enhance the positive controller signal; as the same if $e > 0$,$u > 0$and$\Delta e < 0$ then equation is $u = \delta(\cdot) - k \cdot \Delta e$; if $e < 0$,$u < 0$and$\Delta e < 0$ then $u = \delta(\cdot) + k \cdot \Delta e$. From above all we can get the method using the function.

Applying this method and at the same time inserting a constant-gain randomly to do some tests we can find the speed of system response becomes quick, but accuracy is not improved very obviously sometimes. However, it will do a good job when we use it in a special way(we do not give it out in this paper).

Because the above method is over powerful for the function $k \cdot \Delta e$ when our system input is a sine signal so that we change it in another simple way, and also insert the constant-gain then

$$u = a \cdot \delta(\cdot) - k \cdot \Delta e . \tag{19}$$

In this way we do not need any condition unlike the above one, through simulation we can get the result showed in Fig. 8.

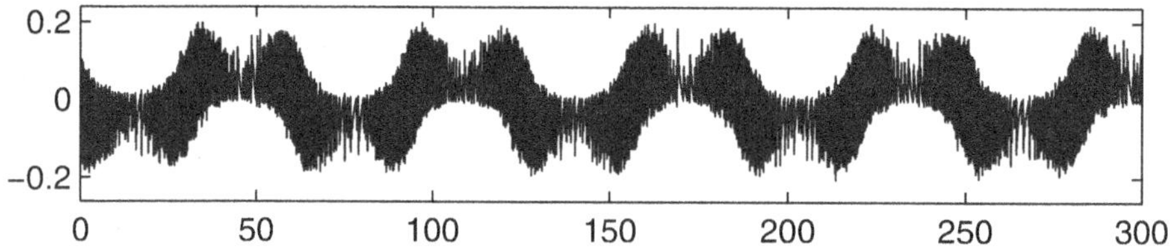

Fig. 8. Forth order plant output error, using simple function in hybrid neural network controller

Comparing with Fig. 4 the accuracy is improved very perfectly, and results of simulation prove that the degree of accuracy will be more higher when the simple function be applied in a second or third order plant. That is because the simple function working in a very special way, which can restrain the oscillation of error, can not the complicated one. In fact we can explain it further more by the theorem which mentioned in this paper. It is just the equation(1), in which E is a function of ω_{st}, and both E and ω_{st} are function of t. This equation is a descent algorithm according to the gradient, if we want to solve out the maximum or minimum value of function E, we must let ω_{st} change in the opposite gradient direction as time goes on. Now we can easily understand that the function (19) just changes in this way when the neural network controller work. The following Fig. 9 can show how the hybrid neural network controller work.

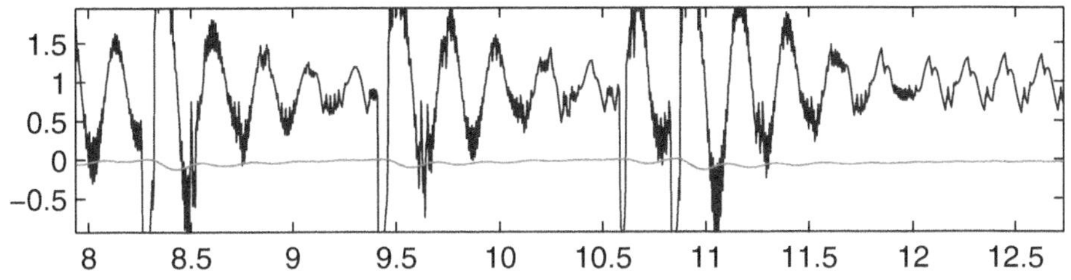

Fig. 9. Control signal, error(around zero) with using simple function

When we apply function(19) to the controller we must pay attention to that the original outputs function of the neural network must play a major role. Only on the premise of that our improvement can be achieved. Considering the learning rate can affect the error of output we try use function (20) to make the output better. Assume the learning rate is γ.

$$\gamma(k) = \begin{cases} 0.7\gamma(k-1) & e(k) \geq 1.04e(k-1) \\ 1.05\gamma(k-1) & e(k) < 0.96e(k-1) \\ \gamma(k) & else \end{cases} . \tag{20}$$

This function is usually used in the back-propagation algorithm, and can change the learning rate in real time to make the system response quickly and the output error more smoothly.

To sum up the above conclusion, we use all the mentioned method into the hybrid neural network controller and choose some constants then get the error as showed in Fig. 10.

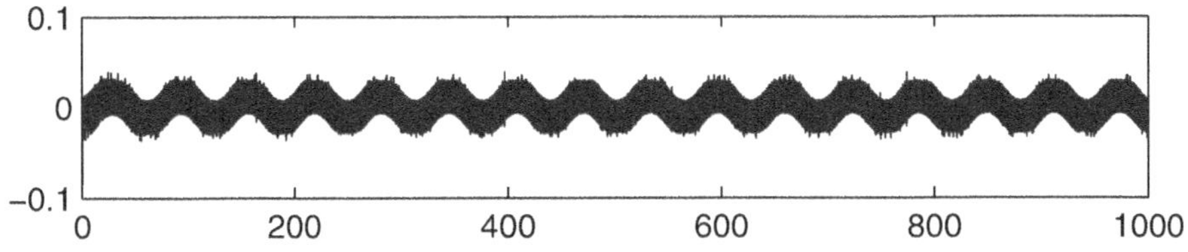

Fig. 10. Third older plant with $,\gamma = 1000, k = 200$

5 Conclusion

In this paper, an approach to system adaptation was proposed and proved, and apply this algorithm to a special neural network controller which has so many its own properties (if we just add a new neuron with adaptive algorithm the controller will do a bad job), then according to its drawback we developed a hybrid network to make its degree of accuracy higher. In the hybrid neural network some weight do not change and some weight change in the adaptation way. And this improvement is far different from adding a new neuron with adaptive algorithm, and more better than it.

References

1. Brandt, R.D., Lin, F.: Supervised Learning in Neural Network Without Explicit Error Back-propagation. Proceeding of the 32nd Annual Allenton Conference on Communication, Control and Computing (1994) 294–303
2. Brandt, R.D., Lin, F.: Can Supervised Learning be Achieved without Explicit Error Back-propagation. Proceeding of International conference on Neural Networks (1996) 300–305
3. Sakelaris, G., Lin, F.: A neural Network Controller by Adaptive Interaction. Proceedings of the American Control Conference Arrington, VA June (2001) 25–27
4. Cabrera, J.B.D., Narendra, K.S.: Issues in the Application of Neural Networks for Tracking Based on Inverse Control. IEEE Trans. Automatic Control **44** (1999) 2007–2027
5. Brandt, R.D., Lin, F.: Adaptive Interaction and Its Application to Neural Networks. Information Sciences **121** (1999) 201–205
6. Brandt, R.D., Lin, F.: Optimal Layering of Neurons. IEEE International Symposium on Intelligent Control (1996) 497–501
7. Narendra, K.S., Parthasarathy, K.: Identification and Control of Dynamical Systems Using Neural Networks. IEEE Trans. Neural Networks **1** (1990) 1–27
8. Kuschewski, J.G., Hui, S.: Application of Feedforward Neural Networks to Dynamical System Identificaion and Control. IEEE Trans. Control systems Technology **1** (1993) 37–49
9. Levin, A.U., Narendra, K.S.: Control of Nonlinear Dynamical Systems Using Neural Networks. IEEE Trans. Neural Networks **7** (1996) 30–42
10. Chen, F.C., Khalil, H.K.: Adaptive Control of Nonlinear Systems Using Neural Networks. IEEE Proceedings on the 29th Conference on Decision and Control **44** (1990) TA-12-1-8:40
11. Narendra, K.S., Parthasarathy, K.: Gradient Methods for Optimization of Dynamical systems Containg Neural Networks. IEEE Trans. Neural Networks **2** (1991) 252–262 e
12. Yamada, T., Yabuta, T.: Neural Network Controller Using Autoturning Method for Nonlinear Functions. IEEE Trans. Neural Networks **3** (1992) 595–601
13. Chen, F.C., Khalil, H.K.: Adaptive Control of a Class of Nonlinear Discrete-Time Systems Using Neural Networks. IEEE Trans. Automatic Control **40** (1995) 791–801
14. Brdys, M.A., Kulawski, G.L.: Dynamic Neural for Induction Motor. IEEE Trans. Neural Networks **10** (1999) 340–355
15. Narendra, K.S., Mukhopadhyay, S.: Adaptive Control Using Neural Networks and Approximate Models. IEEE Trans. Neural Networks **8** (1999) 475–485
16. Park, Y.M., Choi, M.S., Lee, K.Y.: An Optimal Tracking Neuro-Controller for Nonlinear Dynamic Systems. IEEE Trans. Neural Networks **7** (1999) 1099–1110
17. Sakelaris, G., Lin, F.: A neural Network Controller by Adaptive Interaction. Proceedings of the American Control Conference Arrington VA June (2001) 25–27

18. Brandt, R.D., Lin, F.: Supervised Learning in Neural Networks Without Feedback Network. IEEE International Symposium on Intelligent Control (1996) 86–90
19. Shukla, D., Dawson, D.M., Paul, F.W.: Multiple Neural-Network Based Adaptive Controller Using Orthonomal Activation Function Neural Networks. IEEE Trans. Neural Networks **10** (1999) 1494–1051
20. Spooner, J.T., Passino, K.M.: Decentralized Adaptive Control of Nonlinear Systems Using Radial Basis Neural Networks. IEEE Trans. Neural Networks **44** (1999) 2025–2057

Adaptive Output-Feedback Stochastic Nonlinear Stabilization Using Neural Network

Jun Yang[1], Junchao Ni[2], and Weisheng Chen[3]

[1]Department of Mathematics, Linyi Normal College, Linyi 276005, China
[2]Department of Chemistry, Shannxi Institute of Education, Xi'an 710061, China
[3]Department of Applied Mathematics, Xidian University, Xi'an 710071, China
{cwsl113@tom.com,jchni@126.com,wshchen@126.com}

Abstract. This letter extends adaptive neural network control method to a class of stochastic nonlinear output-feedback systems . Differently from the existing results, the nonlinear terms are assumed to be completely unknown and only a neural network is employed to compensate all unknown nonlinear functions. Based on stochastic LaSalle theorem, the resulting closed-loop system is proved to be globally asymptotically stable in probability.

1 Introduction

After a success of constructive control design for deterministic nonlinear systems such as backstepping technique [1], how to extend these techniques to stochastic nonlinear systems is an open research area. Recently, some interesting results were obtained [2-5], Pan and Basar [2] were first to solve the stochastic stabilization problem for the class of strict-feedback systems. By employing a quartic Lyapunov function, Deng and Krstić [3] gave a backstepping design for stochastic output-feedback system. Based on stochastic LaSalle-Yoshizawa theorem, Ji and Xi [4] extended the idea proposed in [3] to the parametric-output-feedback system. Fu *et al.* [5] extended the results in [3,4] to the time-delay stochastic nonlinear system with output-feedback form.

On the other hand, adaptive neural network control (ANNC) method has been successfully applied to some classes of unknown nonlinear systems, such as strict-feedback systems [6], output-feedback systems [7] and so on. Recently, ANNC was extended to nonlinear time-delay systems with strict-feedback form [8] and output-feedback form[9-10].

Motivated by the aforementioned discussion, when the systems states are unavailable , How to apply ANNC method to the output-feedback control of unknown nonlinear stochastic systems is a challenging subject. In this paper, we try to solve this open problem. The main contributions are listed as follows. Firstly, only a NN is employed to approximate all unknown functions of system, so that nonlinear functions are assumed to be completely unknown. Secondly, by constructing a Lyapunov function, the closed-loop system is proved to be asymptotically stable in probability.

D. Liu et al. (Eds.): ISNN 2007, Part I, LNCS 4491, pp. 158–165, 2007.

2 Preliminaries and Systems Description

2.1 Stochastic Stability

To establish stochastic stability as a preliminary, consider the nonlinear stochastic system

$$dx = f(x,t)dt + g(x,t)dw \tag{1}$$

where $x \in R^n$ is the state, w is an independent r-dimensional Wiener process defined on the complete probability space $(\Omega, \mathcal{F}, P)$, with the incremental covariance $E\{dw \cdot dw^{\mathrm{T}}\} = \sigma(t)\sigma(t)^{\mathrm{T}}dt$, the Borel measurable functions $f : R^n \times R_+ \to R^n$ and $g : R^n \times R_+ \to R^{n\times r}$ are locally Lipschitz continuous in $x \in R^n$ (uniformly in $t \in R_+$)with $f(0,t) = 0$ and $g(0,t) = 0$ for all $t \geq 0$.

Lemma 1 (*Stochastic LaSalle Theorem*[11]): Consider system (1) and suppose there exists a twice continuously differentiable function $V(x,t)$, which is positive definite, decrescent and radially unbounded, and another nonnegative continuous function $W(x) \geq 0$, such that the infinitesimal generator of $V(x,t)$ along (1) satisfies

$$LV(x,t) = \frac{\partial V}{\partial x} f(x,t) + \frac{1}{2}\mathrm{Tr}\left\{\sigma^{\mathrm{T}} g^{\mathrm{T}} \frac{\partial^2 V}{\partial x^2} g\sigma\right\} \leq -W(x),\ \forall x \in R^n, t \geq 0 \tag{2}$$

where Tr denotes the matrix trace. Then, the equilibrium $x = 0$ is globally stable in probability and

$$P\left\{\lim_{t\to\infty} W(x(t)) = 0\right\} = 1,\ \forall x(0) \in R^n. \tag{3}$$

2.2 Neural Network Approximation

In this paper, an unknown smooth nonlinear function $G(y) : R \to R$ will be approximated on a compact set D by the following RBF neural network [6].

$$G(y) = W^{\mathrm{T}} S(y) + \delta(y) \tag{4}$$

where $S(y) = [s_1(y), \cdots, s_l(y)]^{\mathrm{T}} : D \to R^l$, is a known smooth vector function with the NN node number $l > 1$. Basis function $s_i(y)$, $1 \leq i \leq l$, is chosen as the commonly used Gaussian function with the form $s_i(y) = \exp[-(y-\mu_i)^2/\eta^2]$, where $\mu_i \in D$ and $\eta > 0$, are the center and the width of basis function $s_i(y)$, respectively. The optimal weight vector $W_=[w_1, \cdots, w_l]^{\mathrm{T}}$ is defined as

$$W := \arg\min_{\hat{W}\in R^l}\left\{\sup_{y\in D}\left|G(y) - \hat{W}^{\mathrm{T}} S(y)\right|\right\} \tag{5}$$

and $\delta(y)$ denotes the NN inherent approximation error. In many previous published works, the approximation error is assumed to be bounded by a fixed constant. However, this may not be true in many case since there is no guarantee the compact set D can easily identified before the stability of the closed-loop system is established. Hence we instead make the following assumption on the approximation error $\delta(y)$.

Assumption 1. There exist a known positive function $\psi(y)$ and an unknown positive constant θ such that

$$|\delta(y)| \leq \psi(y)\theta. \tag{6}$$

Remark 1. If (6) holds only on the compact set D, the results obtained in this paper are semi-global. However, in the special case (6) holds for all $y \in R$, then the stability become global. To simplify the analysis, in this paper we assume that the bounding condition (6) holds globally.

2.3 System Description

In this paper, we consider nonlinear output-feedback system driven by white noise. Its structure is given by the following nonlinear stochastic differential equation:

$$\begin{cases} dx_i = (x_{i+1} + f_i(y))dt + \varphi_i^{\mathrm{T}}(y)d\omega, & i = 1, \cdots, n-1 \\ dx_n = (u + f_n(y))dt + \varphi_n^{\mathrm{T}}(y)d\omega \\ y = x_1 \end{cases} \tag{7}$$

where $x = [x_1, \cdots, x_n]^{\mathrm{T}} \in R^n$, $u \in R$ and $y \in R$ represent the state, control input and output of system, respectively; $f_i(y) : R \to R$ is an unknown function with $f_i(0) = 0$, and $\varphi_i(y) : R \to R^r$ is an unknown vector-valued function with $\varphi_i(0) = 0$; ω is an independent r-dimensional Wiener process with unknown incremental covariance matrix $\sigma(t)$; Only the output y is available.

Remark 2. $f_i(0) = 0$ and $\varphi_i(0) = 0$ imply that $x = 0$ is the equilibrium of system (1). According to the mean value theorem, the following equalities hold

$$f_i(y) = y\bar{f}_i(y) \tag{8}$$

$$\varphi_i(y) = y\bar{\varphi}_i(y) \tag{9}$$

where $\bar{f}_i(\cdot)$ and $\bar{\varphi}_i(\cdot)$ are completely unknown nonlinear functions that will be compensated only by a neural network in this paper.

Assumption 2 [3]**.** The incremental covariance matrix $\sigma(t)$ is bounded with $\sigma(t)^{\mathrm{T}}\sigma(t) \leq \varrho$, where ϱ is an unknown positive constant.

3 Adaptive NN Output Feedback Controller Design

Since the state x is not measured, we first design an observer for x as follows

$$\begin{cases} \dot{\hat{x}}_i = \hat{x}_{i+1} + k_i(y - \hat{x}_1), & i = 1, \cdots, n-1 \\ \dot{\hat{x}}_n = u + k_n(y - \hat{x}_1) \end{cases} \tag{10}$$

where $\hat{x} = [\hat{x}_1, \cdots, \hat{x}_n]^{\mathrm{T}}$ is the observer state. The vector $k = [k_1, \cdots, k_n]^{\mathrm{T}}$ is chosen such that the matrix

$$A = \begin{bmatrix} -k_1 & & & \\ \vdots & & I_{(n-1)\times(n-1)} & \\ -k_n & 0 & \cdots & 0 \end{bmatrix} \tag{11}$$

is Hurwitz, and thus there exists a positive definite matrix P such that

$$A^{\mathrm{T}} P + PA = -I_{n\times n}. \tag{12}$$

Define the observer error $\varepsilon = x - \hat{x}$, which is governed by

$$d\varepsilon = (A\varepsilon + f(y))dt + \varphi^{\mathrm{T}}(y)dw \tag{13}$$

where

$$f(y) = y \begin{bmatrix} \bar{f}_1(y) \\ \vdots \\ \bar{f}_n(y) \end{bmatrix} = y\bar{f}(y), \quad \varphi^{\mathrm{T}}(y) = y \begin{bmatrix} \bar{\varphi}_1^{\mathrm{T}}(y) \\ \vdots \\ \bar{\varphi}_n^{\mathrm{T}}(y) \end{bmatrix} = y\bar{\varphi}^{\mathrm{T}}(y). \tag{14}$$

Based on system (7) and observer (10), we can get the following system

$$\begin{cases} dy = (\hat{x}_2 + \varepsilon_2 + f_1(y))dt + \varphi_1^{\mathrm{T}}(y)dw \\ \dot{\hat{x}}_i = \hat{x}_{i+1} + k_i(y - \hat{x}_1), \quad i = 2, \cdots, n-1 \\ \dot{\hat{x}}_n = u + k_n(y - \hat{x}_1). \end{cases} \tag{15}$$

Clearly, the controller for system (15) can be designed by backstepping technique. We define the following coordinate transformation

$$\begin{cases} z_1 = y = x_1, \\ z_i = \hat{x}_i - \alpha_{i-1}(y, \hat{x}_1, \cdots, \hat{x}_{i-1}, \hat{\theta}, \hat{W}), \quad i = 2, \cdots, n, \end{cases} \tag{16}$$

where $\hat{\theta}$ and $\hat{W}$ are the estimates of the unknown constant θ in (16) and the NN weight W, respectively (see (19)) , α_i is the stabilizing function to be designed later. Under transformation (16), system (15) is changed into (17)

$$\begin{cases} dy = \left[\alpha_1 + z_2 + \varepsilon_2 + f_1\right]dt + \varphi_1^{\mathrm{T}} dw \\ dz_i = \left[\alpha_i + z_{i+1} + \Pi_i - \frac{\partial \alpha_{i-1}}{\partial y}(\varepsilon_2 + f_1) - \frac{1}{2}\left(\frac{\partial^2 \alpha_{i-1}}{\partial y^2}\right)\varphi_1^{\mathrm{T}} \sigma\sigma^{\mathrm{T}} \varphi_1\right]dt \\ \quad - \frac{\partial \alpha_{i-1}}{\partial y}\varphi_1^{\mathrm{T}} dw, \qquad i = 2, \cdots, n-1 \\ \dot{z}_n = \left[u + \Pi_n - \frac{\partial \alpha_{n-1}}{\partial y}(\varepsilon_2 + f_1) - \frac{1}{2}\left(\frac{\partial^2 \alpha_{n-1}}{\partial y^2}\right)\varphi_1^{\mathrm{T}} \sigma\sigma^{\mathrm{T}} \varphi_1\right]dt \\ \quad - \frac{\partial \alpha_{n-1}}{\partial y}\varphi_1^{\mathrm{T}} dw \end{cases} \tag{17}$$

where for $i = 2, \cdots, n$, we define Π_i as follows

$$\Pi_i = k_i(y - \hat{x}_1) - \frac{\partial \alpha_{i-1}}{\partial y}\hat{x}_2 - \sum_{j=1}^{i-1} \frac{\partial \alpha_{i-1}}{\partial \hat{x}_j}\Big(\hat{x}_{j+1} - k_j(y - \hat{x}_1)\Big) - \frac{\partial \alpha_{i-1}}{\partial \hat{\theta}}\dot{\hat{\theta}} - \frac{\partial \alpha_{i-1}}{\partial \hat{W}}\dot{\hat{W}}. \tag{18}$$

From (17), we design the stabilizing functions and the control law as

$$\alpha_1 = -c_1 y - y\left(\hat{W}^{\mathrm{T}} S(y) + \psi(y)\hat{\theta}\right) \tag{19}$$

$$\alpha_i = -c_i z_i - \Xi_i z_i - \frac{1}{4}\lambda_i^2 \left(\frac{\partial^2 \alpha_{i-1}}{\partial y^2}\right)^2 z_i^3 - \frac{3}{4}\delta_i^{4/3} z_i - \Pi_i,\ i=2,\cdots,n-1 \tag{20}$$

$$u = -c_n z_n - \Xi_n z_n - \frac{1}{4}\lambda_n^2 \left(\frac{\partial^2 \alpha_{n-1}}{\partial y^2}\right)^2 z_n^3 - \Pi_n \tag{21}$$

where Ξ_i is defined as

$$\Xi_i = \frac{3}{2}\eta_i^{4/3}\left(\frac{\partial \alpha_{i-1}}{\partial y}\right)^{4/3} + \frac{3}{4\xi_i^2}\left(\frac{\partial \alpha_{i-1}}{\partial y}\right)^4 + \frac{1}{4\delta_{i-1}^4}, i = 2, \cdots, n \tag{22}$$

c_i, $\lambda_i, \delta_i, \eta_i$ and ξ_i are positive design parameters, and $S(y)$ is the vector basis function , and the adaptive laws are designed as

$$\dot{\hat{\theta}} = \gamma\psi(y)y^4, \qquad \dot{\hat{W}} = \Gamma S(y)y^4 \tag{23}$$

where $\gamma > 0$ and $\Gamma > 0$, are the adaptive gains. Substituting (19)-(21) into (17), the error system is given by

$$\begin{cases} dy = \left[-c_1 y - y\left(\hat{W}^{\mathrm{T}} S(y) + \psi(y)\hat{\theta}\right) + z_2 + \varepsilon_2 + f_1 \right] dt + \varphi_1^{\mathrm{T}} dw \\ dz_i = \left[-c_i z_i - \Xi_i z_i - \frac{1}{4}\lambda_i^2 \left(\frac{\partial^2 \alpha_{i-1}}{\partial y^2}\right)^2 z_i^3 - \frac{3}{4}\delta_i^{4/3} z_i + z_{i+1} - \frac{\partial \alpha_{i-1}}{\partial y}(\varepsilon_2 + f_1) \right. \\ \qquad \left. -\frac{1}{2}\left(\frac{\partial^2 \alpha_{i-1}}{\partial y^2}\right)\varphi_1^{\mathrm{T}}\sigma\sigma^{\mathrm{T}}\varphi_1 \right] dt - \frac{\partial \alpha_{i-1}}{\partial y}\varphi_1^{\mathrm{T}} dw, \quad i = 2, \cdots, n-1 \\ dz_n = \left[-c_n z_n - \Xi_n z_n - \frac{1}{4}\lambda_n^2 \left(\frac{\partial^2 \alpha_{n-1}}{\partial y^2}\right)^2 z_n^3 - \frac{\partial \alpha_{n-1}}{\partial y}(\varepsilon_2 + f_1) \right. \\ \qquad \left. -\frac{1}{2}\left(\frac{\partial^2 \alpha_{n-1}}{\partial y^2}\right)\varphi_1^{\mathrm{T}}\sigma\sigma^{\mathrm{T}}\varphi_1 \right] dt - \frac{\partial \alpha_{n-1}}{\partial y}\varphi_1^{\mathrm{T}} dw. \end{cases} \tag{24}$$

Consider the following Lyapunov function

$$V = \frac{b}{2}(\varepsilon^{\mathrm{T}} P\varepsilon)^2 + \frac{1}{4}y^4 + \frac{1}{4}\sum_{i=2}^{n} z_i^4 + \frac{1}{2}\tilde{W}^{\mathrm{T}}\Gamma^{-1}\tilde{W} + \frac{1}{2}\gamma^{-1}\tilde{\theta}^2 \tag{25}$$

where b is a positive designed constant, $\tilde{W} = W - \hat{W}$ and $\tilde{\theta} = \theta - \hat{\theta}$, denote the estimates of W and θ, respectively. Along the solutions of (13), (23) and (24), we have

$$LV = -b(\varepsilon^{\mathrm{T}} P\varepsilon)|\varepsilon|^2 + \underbrace{b(\varepsilon^{\mathrm{T}} P\varepsilon)2\varepsilon^{\mathrm{T}} Pf}_{\text{(Eq.I)}} + \underbrace{b\mathrm{Tr}\{\sigma^{\mathrm{T}}\varphi(2P\varepsilon\varepsilon^{\mathrm{T}} P + \varepsilon^{\mathrm{T}} P\varepsilon P)\varphi^{\mathrm{T}}\sigma\}}_{\text{(Eq.II)}}$$

$$-c_1y^4 - y^4\Big(\hat{W}^{\mathrm{T}}S(y) + \psi(y)\hat{\theta}\Big) + \underbrace{y^3(z_2+\varepsilon_2+f_1)}_{\text{Eq.III}} + \underbrace{\frac{3}{2}y^2\varphi_1^{\mathrm{T}}\sigma\sigma^{\mathrm{T}}\varphi_1}_{\text{Eq.IV}}$$

$$+\sum_{i=2}^{n}\left[-c_iz_i^4 - \varXi_iz_i^4 - \frac{1}{4}\lambda_i^2\Big(\frac{\partial^2\alpha_{i-1}}{\partial y^2}\Big)^2 z_i^6\right] - \sum_{i=2}^{n-1}\frac{3}{4}\delta_i^{4/3}z_i^4 + \underbrace{\sum_{i=2}^{n-1}z_i^3z_{i+1}}_{\text{Eq.V}}$$

$$\underbrace{-\sum_{i=2}^{n}z_i^3\frac{\partial\alpha_{i-1}}{\partial y}(\varepsilon_2+f_1)}_{\text{Eq.VI}}\underbrace{-\frac{1}{2}\sum_{i=2}^{n}z_i^3\Big(\frac{\partial^2\alpha_{i-1}}{\partial y^2}\Big)\varphi_1^{\mathrm{T}}\sigma\sigma^{\mathrm{T}}\varphi_1}_{\text{Eq.VII}}$$

$$+\underbrace{\frac{3}{2}\sum_{i=2}^{n}z_i^2\left(\frac{\partial\alpha_{i-1}}{\partial y}\right)^2\varphi_1^{\mathrm{T}}\sigma\sigma^{\mathrm{T}}\varphi_1}_{\text{Eq.VIII}} - \tilde{W}^{\mathrm{T}}S(y)y^4 - \tilde{\theta}\psi(y)y^4. \tag{26}$$

By using Young's inequality, each underlined term in (26) satisfy

$$\text{Eq.(I)}: b(\varepsilon^{\mathrm{T}}P\varepsilon)2\varepsilon^{\mathrm{T}}Pf \le \frac{3b}{2}\epsilon_1^{4/3}|P|^{8/3}|\varepsilon|^4 + \frac{b}{2\epsilon_1^4}y^4|\bar{f}|^4 \tag{27}$$

$$\begin{aligned}\text{Eq.(II)}: & \; b\mathrm{Tr}\{\sigma^{\mathrm{T}}\varphi(2P\varepsilon\varepsilon^{\mathrm{T}}P + \varepsilon^{\mathrm{T}}P\varepsilon P)\varphi^{\mathrm{T}}\sigma\} \\ & \le \frac{3bn\sqrt{n}}{2\epsilon_2^2}\varrho^2y^4|\bar{\varphi}|^4 + \frac{3bn\sqrt{n}}{2}\epsilon_2^2|P|^4|\varepsilon|^4\end{aligned} \tag{28}$$

$$\begin{aligned}\text{Eq.(III)}: & \; y^3(z_2+\varepsilon_2+f_1) \\ & \le \frac{3}{4}\delta_1^{4/3}y^4 + \frac{3}{4}\epsilon_3^{4/3}y^4 + \frac{1}{4\delta_1^4}z_2^4 + \frac{1}{4\epsilon_3^4}|\varepsilon|^4 + y^4\bar{f}_1\end{aligned} \tag{29}$$

$$\text{Eq.(IV)}: \frac{3}{2}y^2\varphi_1^{\mathrm{T}}\sigma\sigma^{\mathrm{T}}\varphi_1 \le \frac{3}{2}\varrho y^4|\bar{\varphi}_1|^2 \tag{30}$$

$$\text{Eq.(V)}: \sum_{i=2}^{n-1}z_i^3z_{i+1} \le \frac{3}{4}\sum_{i=2}^{n-1}\delta_i^{4/3}z_i^4 + \frac{1}{4}\sum_{i=3}^{n}\frac{1}{\delta_{i-1}^4}z_i^4 \tag{31}$$

$$\begin{aligned}\text{Eq.(VI)}: & -\sum_{i=2}^{n}z_i^3\frac{\partial\alpha_{i-1}}{\partial y}(\varepsilon_2+f_1) \\ & \le \frac{3}{2}\sum_{i=2}^{n}\eta_i^{4/3}\left(\frac{\partial\alpha_{i-1}}{\partial y}\right)^{4/3}z_i^4 + \frac{1}{4}\sum_{i=2}^{n}\frac{1}{\eta_i^4}|\varepsilon|^4 + \frac{1}{4}y^4\sum_{i=2}^{n}\frac{1}{\eta_i^4}|\bar{f}_1|^4\end{aligned} \tag{32}$$

$$\begin{aligned}\text{Eq.(VII)}: & -\frac{1}{2}\sum_{i=2}^{n}z_i^3\Big(\frac{\partial^2\alpha_{i-1}}{\partial y^2}\Big)\varphi_1^{\mathrm{T}}\sigma\sigma^{\mathrm{T}}\varphi_1 \\ & \le \frac{1}{4}\sum_{i=2}^{n}\lambda_i^2\Big(\frac{\partial^2\alpha_{i-1}}{\partial y^2}\Big)^2z_i^6 + \frac{1}{4}y^4\sum_{i=2}^{n}\frac{\varrho^2}{\lambda_i^2}|\bar{\varphi}_1|^4\end{aligned} \tag{33}$$

$$\text{Eq.(VIII)}: \frac{3}{2}\sum_{i=2}^{n}z_i^2\left(\frac{\partial\alpha_{i-1}}{\partial y}\right)^2\varphi_1^{\mathrm{T}}\sigma\sigma^{\mathrm{T}}\varphi_1$$

$$\leq \frac{3}{4}\sum_{i=2}^{n}\frac{1}{\xi_i^2}\left(\frac{\partial\alpha_{i-1}}{\partial y}\right)^4 z_i^4 + \frac{3}{4}y^4\sum_{i=2}^{n}\xi_i^2|\bar{\varphi}_1|^4 \tag{34}$$

Substituting (27)-(34) into (26), we have

$$LV \leq \left(-b\lambda + \frac{3bn\sqrt{n}}{2}\epsilon_2^2|P|^4 + \frac{3b}{2}\epsilon_1^{4/3}|P|^{8/3} + \frac{1}{4}\sum_{i=2}^{n}\frac{1}{\eta_i^4} + \frac{1}{4\epsilon_3^4}\right)|\varepsilon|^4 - \sum_{i=1}^{n} c_i z_i^4$$
$$-y^4\Big(W^{\mathrm{T}}S(y) + \psi(y)\theta\Big) + y^4\Bigg(\underline{\frac{b}{2\epsilon_1^4}|\bar{f}|^4 + \frac{3bn\sqrt{n}}{2\epsilon_2^2}\varrho^2|\bar{\varphi}|^4 + \frac{3}{4}\delta_1^{4/3} + \frac{3}{4}\epsilon_3^{4/3}}$$
$$\underline{+\bar{f}_1 + \frac{3}{2}\varrho|\bar{\varphi}_1|^2 + \frac{1}{4}\sum_{i=2}^{n}\frac{1}{\eta_i^4}|\bar{f}_1|^4 + \frac{1}{4}\sum_{i=2}^{n}\frac{\varrho^2}{\lambda_i^2}|\bar{\varphi}_1|^4 + \frac{3}{4}\sum_{i=2}^{n}\xi_i^2|\bar{\varphi}_1|^4}\Bigg) \tag{35}$$

where $\lambda > 0$, denotes the smallest eigenvalue of P. The unknown function underlined in (35)is denoted by $G(y)$, which can be approximated as following

$$G(y) = W^{\mathrm{T}}S(y) + \delta(y) \tag{36}$$

with the approximation error $|\delta(y)| \leq \psi(y)\theta$. Substituting (36) into (35), we have

$$LV \leq \left(-b\lambda + 3bn\sqrt{n}\epsilon_2^2|P|^4 + \frac{3b}{2}\epsilon_1^{4/3}|P|^{8/3} + \frac{1}{4}\sum_{i=2}^{n}\frac{1}{\eta_i^4} + \frac{1}{4\epsilon_3^4}\right)|\varepsilon|^4 - \sum_{i=1}^{n} c_i z_i^4. \tag{37}$$

For given $0 < \nu < 1$, the parameters $\epsilon_1, \epsilon_2, \epsilon_3, \eta_i$ and b are selected to render the following inequality holds:

$$-b\lambda + \frac{3bn\sqrt{n}}{2}\epsilon_2^2|P|^4 + \frac{3b}{2}\epsilon_1^{4/3}|P|^{8/3} + \frac{1}{4}\sum_{i=2}^{n}\frac{1}{\eta_i^4} - \frac{1}{4\epsilon_3^4} \leq -\nu \tag{38}$$

and we have

$$LV \leq -\nu|\varepsilon|^4 - \sum_{i=1}^{n} c_i z_i^4. \tag{39}$$

The main results are stated as follows

Theorem 1: Under Assumption 1-2, consider the closed-loop system consisting of the system (7), the control law (21) and the adaptive laws (23), for bounded initial conditions, the following properties hold.

(I) The closed-loop system is globally asymptotically stable in probability.

(II) System state $x(t)$ and the parameter estimates $\hat{\theta}(t), \hat{W}(t)$ satisfy

$$P\left\{\lim_{t\to\infty} x(t) = 0\right\} = 1 \tag{40}$$

$$P\left\{\lim_{t\to\infty}\hat{\theta}(t) \text{ and } \lim_{t\to\infty}\hat{W}(t) \text{ exist and are finite}\right\} = 1 \tag{41}$$

Proof. Omitted.

4 Conclusion

In this letter, ANNC method is extended to a class of unknown stochastic nonlinear system. Only a NN is used to compensate all unknown functions of system, and therefore the assumption on the nonlinear terms is relaxed. The globally asymptotically stability of closed-loop system is guaranteed.

References

1. Krstić M., Kanellakopulos I., Kocotovic P.V.: Nonlinear and Adaptive Control Design. New York: Wiley (1995)
2. Pan Z., Basar T.: Backstepping Controller Design for Noninear Stochastic Systems under A Risk-sensitive Cost Criterion. SIAM J. Control and Optimization **37** (1999) 957-995
3. Deng H., Krstić M.: Output-feedback Stabilization of Stochastic Nonlinear s Systems Driven by Noise of Unknown Covariance. Systems & Control Letters **39** (2000) 173-182
4. Ji H.B., Xi H.S.: Adaptive Output-feedback Tracking of Stochastic Nonlinear Systems. IEEE Transactions on Automatica Control **51** (2006) 355-360
5. Fu Y.S., Tian Z.H., Shi S.J.: Output Feedback Stabilization for A Class of Stochastic Time-delay Nonlinear Systems. IEEE Transactions on Automatica Control **50** (2005) 847-851
6. Wang D., Huang J.: Neural Network-based Adaptive Dynamic Surface Control for A Class of Uncertain Nonlinear Sytems in Strict-feedback Form. IEEE Transactions on Neural Networks **16** (2005)195-202
7. Choi J.Y., Farrell J.A.: Adaptive Observer Backstepping Control Using Neural Networks. IEEE Transactions on Neural Networks **12**(2001) 1103-1113
8. Ho D.W.C., Li J.M., Hong Y.G.: Adaptive Neural Control for A Class of Nonlinear Parametric Time Delay Systems. IEEE Transactions on Neural Networks **16** (2005) 625-635
9. Chen W.S., Li J.M.: Adaptive Neural Tracking Control for Unknown Output Feedback Nonlinear Time-delay Systems. ACTA Automatica Sinica **31** (2005) 799-803
10. Chen W.S., Li J.M.: Adaptive Output Feedback Control for Nonlinear Time-delay Systems Using Neural Network. Journal of Control Theory and Application **4** (2006) 313-320
11. Krstić M., Deng H.: Stabilization of Nonlinear Uncertain Systems. London, Spring-Verlag (1998)

Adaptive Control for a Class of Nonlinear Time-Delay Systems Using RBF Neural Networks

Geng Ji[1] and Qi Luo[2]

[1] School of Mathematics and Information Engineering,
Taizhou University, Linhai 317000, P.R. China
gjiwust@yahoo.com.cn
[2] College of Information and Control,
Nanjing University of Information Science and Technology,
Nanjing 210044, P.R. China
hb2-102@nuist.edu.cn

Abstract. In this paper, adaptive neural network control is proposed for a class of strict-feedback nonlinear time-delay systems. Unknown smooth function vectors and unknown time-delay functions are approximated by two neural networks, respectively, such that the requirement on the unknown time-delay functions is relaxed. In addition, the proposed systematic backstepping design method has been proven to be able to guarantee semiglobally uniformly ultimately bounded of closed loop signals, and the output of the system has been proven to converge to a small neighborhood of the desired trajectory. Finally, simulation result is presented to demonstrate the effectiveness of the approach.

1 Introduction

In recent years, adaptive neural network control (ANNC) has received considerable attention and become an active research area [1]. ANNC is a nonlinear control methodology which is particularly useful for the control of highly uncertain, nonlinear and complex systems. In adaptive neural control design, neural networks are mostly used as approximators for unknown nonlinear functions in system models.

By using the idea of backstepping design [2], several adaptive neural controllers [3-7] have been proposed for strict-feedback nonlinear systems. In [3], an indirect adaptive NN control scheme was presented for a class of nonlinear systems. The unknown smooth functions were first approximated on-line by neural networks, and a stabilizing controller was constructed based on the approximation. In [4], a neural controller was proposed for a class of unknown, minimum phase, feedback linearizable nonlinear system with known relative degree. By combining adaptive neural design with backstepping methodology, a smooth adaptive neural controller was proposed in [5], where integral-type Lyapunov functions are introduced and play an important role in overcoming singularity problem. In [6], by utilizing a special property of the affine term, direct

D. Liu et al. (Eds.): ISNN 2007, Part I, LNCS 4491, pp. 166–175, 2007.

adaptive neural-network control scheme was developed to avoid the controller singularity problem completely. Two different backstepping neural network control approaches were presented for a class of affine nonlinear systems in [7]. The proposed controller made the neural network approximation computationally feasible.

However, all these works study the nonlinear systems without time delay. Practically, time delays are often appeared in practical systems. Stabilization of nonlinear systems with time delay has received much attention, and many approaches to this issue have been developed (see [8-12]). In [11], an adaptive neural network control design approach was proposed for a class of nonlinear time-delay systems. The time-delay exists in output variable. The unknown time-delay functions were approximated by neural networks, such that the requirement on the nonlinear time-delay functions was relaxed. In [12], adaptive neural control was presented for a class of strict-feedback nonlinear systems with unknown time delays. By constructing proper Lyapunov-Krasovskii functionals, the uncertainties of unknown time delays were compensated. The unknown time-delay functions were not approximated by neural networks.

In this paper, an adaptive neural network backstepping control design approach is proposed for a class of nonlinear time delay systems. Motivated by references [7, 11], we adopt to two neural networks approximate unknown smooth function vectors and unknown time delay functions. Simulation study is conducted to verify the effectiveness of the approach.

The rest of the paper is organized as follows: The problem formulation is given in section 2. In section 3, an adaptive neural network control design scheme is presented. Simulation result is shown in section 4. Finally, conclusion is given in section 5.

2 Problem Formulation

Consider a class of single-input-single-output (SISO) nonlinear time-delay systems

$$\begin{aligned} \dot{x}_i(t) &= x_{i+1}(t) + f_i\left(\bar{x}_i(t)\right) + h_i\left(\bar{x}_i(t-\tau)\right), \ (1 \le i \le n-1) \\ \dot{x}_n(t) &= u(t) + f_n\left(\bar{x}_n(t)\right) + h_n\left(\bar{x}_n(t-\tau)\right), \\ y &= x_1, \end{aligned} \tag{1}$$

where $\bar{x}_i = [x_1, x_2, \cdots, x_i]^T \in R^i$, $i = 1, 2 \cdots, n$, $u \in R$, $y \in R$ are state variables, system input and output, respectively. $f_i(\cdot)$, $h_i(\cdot)$ $(i = 1, 2, \cdots, n)$ are unknown smooth functions, and τ is known time delay constant of the states. The control objective is to design an adaptive NN controller for system (1) such that 1) all the signals in the closed-loop remain semiglobally uniformly ultimately bounded and 2) the output y follows a desired trajectory y_d, which and whose derivatives up to the nth order are bounded.

Note that in the following derivation of the adaptive neural controller, NN approximation is only guaranteed with some compact sets. Accordingly, the stability result obtained in this work is semiglobal in the sense that, as long as desired, there exists controller with sufficient large number of NN nodes such that all the signals in the closed-loop remain bounded.

In control engineering, radial basis function (RBF) NN is usually used as a tool for modeling nonlinear functions because of their good capabilities in function approximation. In this paper, the following RBF NN [6, 14] is used to approximate the continuous function $h(Z):\ R^q \to R$

$$h_{nn}(Z) = W^T S(Z), \tag{2}$$

where the input vector $Z \in \Omega_Z \subset R^q$, weight vector $W = [w_1, w_2, \cdots, w_l]^T \in R^l$, the NN node number $l > 1$; and $S(Z) = [s_1(Z), s_2(Z), \cdots, s_l(Z)]^T$, with $s_i(Z)$ being chosen as the commonly used Gaussian functions, which have the form

$$s_i(Z) = \exp\left[\frac{-(Z-\mu_i)^T(Z-\mu_i)}{\eta_i^2}\right],\ i = 1, 2, \cdots, l \tag{3}$$

where $\mu_i = [\mu_{i1}, \mu_{i2}, \cdots, \mu_{iq}]^T$ is the center of the receptive field and η_i is the width of the Gaussian function.

It has been proven that network (2) can approximate any continuous function over a compact set $\Omega_Z \subset R^q$ to arbitrary any accuracy as

$$h(Z) = W^{*T} S(Z) + \varepsilon,\ \forall Z \in \Omega_Z \tag{4}$$

where W^* is ideal constant weight vector, and ε is the approximation error which is bounded over the compact set, i.e. $|\varepsilon| \le \varepsilon^*$, where $\varepsilon^* > 0$ is unknown constant. The ideal weight vector W^* is an "artificial" quantity required for analytical purposes. W^* is defined as the value of W that minimizes $|\varepsilon|$ over Ω_Z, that is

$$W^* \triangleq \arg\min_{W \in R^l}\left\{\sup_{Z \in \Omega_Z}\left|h(Z) - W^T S(Z)\right|\right\}. \tag{5}$$

In the following, we let $\|\cdot\|$ denote that 2-norm, $\lambda_{\max}(B)$ and $\lambda_{\min}(B)$ denote the largest and smallest eigenvalues of a square matrix B, respectively.

3 Adaptive Neural Network Control

The detailed design procedure is described in the following steps. For clarity and conciseness, Step 1 is described with detailed explanations, while Step i and Step n are simplified, with the relevant equations and the explanations being omitted.

Step 1: Defined $z_1 = x_1 - y_d$. Its derivative is

$$\dot{z}_1(t) = x_2(t) + f_1(x_1(t)) + h_1(x_1(t-\tau)) - \dot{y}_d(t). \tag{6}$$

By viewing $x_2(t)$ as a virtual control input, if we choose $\alpha_1^* \triangleq x_2$ as the control input for the z_1-subsystem in the above equation, and consider the Lyapunov function candidate $V_{z_1} = \frac{1}{2}z_1^2$, whose derivative is

$$\dot{V}_{z_1} = z_1 \cdot \dot{z}_1 = z_1\left[\alpha_1^* + f_1(x_1(t)) + h_1(x_1(t-\tau)) - \dot{y}_d(t)\right]. \tag{7}$$

Let us choose feedback controller α_1^* as follows:

$$\alpha_1^* = -c_1 z_1 - \left[f_1\left(x_1(t)\right) + h_1\left(x_1(t-\tau)\right) - \dot{y}_d(t)\right], \tag{8}$$

where $c_1 > 0$ is constant. Substituting (8) into (7), We can give $\dot{V}_{z_1} = -c_1 z_1^2 \leq 0$. Therefore, z_1 is asymptotically stable.

Since $f_1\left(x_1(t)\right)$ and $h_1\left(x_1(t-\tau)\right)$ are unknown smooth functions, the desired feedback control α_1^* cannot be implemented in practice. Instead, two neural networks are adopted to approximate the unknown smooth functions $f_1\left(x_1(t)\right)$ and $h_1\left(x_1(t-\tau)\right)$, i.e.

$$f_1\left(x_1\right) = W_{11}^{*\,T} S_{11}(Z_{11}) + \varepsilon_{11},\ \ h_1\left(x_1(t-\tau)\right) = W_{12}^{*\,T} S_{12}\left(Z_{12}\right) + \varepsilon_{12}, \tag{9}$$

where $Z_{11} = \left[x_1\right]^T \subset R^1$, $Z_{12} = \left[x_1(t-\tau)\right]^T \subset R^1$. W_{11}^* and W_{12}^* are the optimal weight vectors of $f_1\left(x_1\right)$ and $h_1\left(x_1(t-\tau)\right)$, respectively. The neural reconstruction error $e_1 = \varepsilon_{11} + \varepsilon_{12}$ is bounded, i.e., there exists a constant $\varepsilon_1^* > 0$ such that $|e_1| < \varepsilon_1^*$. Throughout the paper, we shall define the reconstruction error as $e_i = \varepsilon_{i1} + \varepsilon_{i2}$, where $i = 1, 2, \cdots, n$. Like in the case of e_1, e_i is bounded, i.e., $|e_i| < \varepsilon_i^*$. Since W_{11}^* and W_{12}^* are unknown, let $\hat{W}_{11}$ and $\hat{W}_{12}$ be the estimate of W_{11}^* and W_{12}^*, respectively.

Defining error variable $z_2 = x_2 - \alpha_1$ and choosing the virtual control

$$\alpha_1 = -c_1 z_1 - \hat{W}_{11}^T S_{11}(Z_{11}) - \hat{W}_{12}^T S_{12}\left(Z_{12}\right) + \dot{y}_d(t), \tag{10}$$

$\dot{z}_1$ can be obtained as

$$\begin{aligned} \dot{z}_1 &= z_2 + \alpha_1 + f_1\left(x_1(t)\right) + h_1\left(x_1(t-\tau)\right) - \dot{y}_d\left(t\right) \\ &= z_2 - c_1 z_1 - \tilde{W}_{11}^T S_{11}(Z_{11}) - \tilde{W}_{12}^T S_{12}(Z_{12}) + e_1, \end{aligned} \tag{11}$$

where $\tilde{W}_{11} = \hat{W}_{11} - W_{11}^*$, $\tilde{W}_{12} = \hat{W}_{12} - W_{12}^*$. Through out this paper, we shall define $\tilde{(\cdot)} = \hat{(\cdot)} - (\cdot)^*$.

Consider the following Lyapunov function candidate:

$$V_1 = \frac{1}{2} z_1^2 + \frac{1}{2} \tilde{W}_{11}^T \Gamma_{11}^{-1} \tilde{W}_{11} + \frac{1}{2} \tilde{W}_{12}^T \Gamma_{12}^{-1} \tilde{W}_{12}, \tag{12}$$

where $\Gamma_{11} = \Gamma_{11}^T > 0$, $\Gamma_{12} = \Gamma_{12}^T > 0$ are adaptation gain matrices.

The derivative of V_1 is

$$\begin{aligned} \dot{V}_1 &= z_1 \cdot \dot{z}_1 + \tilde{W}_{11}^T \Gamma_{11}^{-1} \dot{\tilde{W}}_{11} + \tilde{W}_{12}^T \Gamma_{12}^{-1} \dot{\tilde{W}}_{12} \\ &= z_1 z_2 - c_1 z_1^2 + z_1 e_1 + \tilde{W}_{11}^T \Gamma_{11}^{-1} \left[\dot{\hat{W}}_{11} - \Gamma_{11} S_{11}(Z_{11}) z_1\right] \\ &\quad + \tilde{W}_{12}^T \Gamma_{12}^{-1} \left[\dot{\hat{W}}_{12} - \Gamma_{12} S_{12}(Z_{12}) z_1\right]. \end{aligned} \tag{13}$$

Consider the following adaptation laws:

$$\begin{aligned} \dot{\hat{W}}_{11} &= \dot{\tilde{W}}_{11} = \Gamma_{11}\left[S_{11}(Z_{11}) z_1 - \sigma_{11} \hat{W}_{11}\right], \\ \dot{\hat{W}}_{12} &= \dot{\tilde{W}}_{12} = \Gamma_{12}\left[S_{12}(Z_{12}) z_1 - \sigma_{12} \hat{W}_{12}\right], \end{aligned} \tag{14}$$

where $\sigma_{11} > 0$, $\sigma_{12} > 0$ are small constants. Formula (14) is so-called σ-modification, introduced to improve the robustness in the presence of the NN approximation error e_1[15], and avoid the weight parameters to drift to very large values.

Let $c_1 = c_{10} + c_{11}$, with c_{10} and $c_{11} > 0$. Then, (13) become

$$\dot{V}_1 = z_1 z_2 - c_{10} z_1^2 - c_{11} z_1^2 + z_1 e_1 - \sigma_{11} \tilde{W}_{11}^T \hat{W}_{11} - \sigma_{12} \tilde{W}_{12}^T \hat{W}_{12}. \tag{15}$$

By completion of squares, we have

$$-\sigma_{11} \tilde{W}_{11}^T \hat{W}_{11} \le -\frac{\sigma_{11} \left\| \tilde{W}_{11} \right\|^2}{2} + \frac{\sigma_{11} \left\| W_{11}^* \right\|^2}{2}, \tag{16}$$

$$-\sigma_{12} \tilde{W}_{12}^T \hat{W}_{12} \le -\frac{\sigma_{12} \left\| \tilde{W}_{12} \right\|^2}{2} + \frac{\sigma_{12} \left\| W_{12}^* \right\|^2}{2}, \tag{17}$$

$$-c_{11} z_1^2 + z_1 e_1 \le -c_{11} z_1^2 + |z_1 e_1| \le \frac{e_1^2}{4c_{11}} \le \frac{\varepsilon_1^{*2}}{4c_{11}}. \tag{18}$$

Substituting (16) (17) (18) into (15), we have the following inequality:

$$\begin{aligned} \dot{V}_1 \le{}& z_1 z_2 - c_{10} z_1^2 - \frac{\sigma_{11} \left\| \tilde{W}_{11} \right\|^2}{2} - \frac{\sigma_{12} \left\| \tilde{W}_{12} \right\|^2}{2} \\ &+ \frac{\sigma_{11} \left\| W_{11}^* \right\|^2}{2} + \frac{\sigma_{12} \left\| W_{12}^* \right\|^2}{2} + \frac{\varepsilon_1^{*2}}{4c_{11}}, \end{aligned} \tag{19}$$

where the coupling term $z_1 z_2$ will be canceled in the next step.

Step i $(2 \le i \le n-1)$: The derivative of $z_i = x_i - \alpha_{i-1}$ is $\dot{z}_i = x_{i+1}(t) + f_i(\bar{x}_i(t)) + h_i(\bar{x}_i(t-\tau)) - \dot{\alpha}_{i-1}$. Similarly, choose the virtual control

$$\alpha_i = -z_{i-1} - c_i z_i - \hat{W}_{i1}^T S_{i1}(Z_{i1}) - \hat{W}_{i2}^T S_{i2}(Z_{i2}) + \dot{\alpha}_{i-1}, \tag{20}$$

where $c_i > 0$, $Z_{i1} = [x_1, x_2, \cdots, x_i]^T \subset R^i$,
$Z_{i2} = [x_1(t-\tau), x_2(t-\tau), \cdots, x_i(t-\tau)]^T \subset R^i$.

Then, we have

$$\begin{aligned} \dot{z}_i &= z_{i+1} + \alpha_i + f_i(\bar{x}_i(t)) + h_i(\bar{x}_i(t-\tau)) - \dot{\alpha}_{i-1} \\ &= z_{i+1} - z_{i-1} - c_i z_i - \tilde{W}_{i1}^T S_{i1}(Z_{i1}) - \tilde{W}_{i2}^T S_{i2}(Z_{i2}) + e_i, \end{aligned} \tag{21}$$

where $z_{i+1} = x_{i+1} - \alpha_i$.

Consider the Lyapunov function candidate

$$V_i = V_{i-1} + \frac{1}{2} z_i^2 + \frac{1}{2} \tilde{W}_{i1}^T \Gamma_{i1}^{-1} \tilde{W}_{i1} + \frac{1}{2} \tilde{W}_{i2}^T \Gamma_{i2}^{-1} \tilde{W}_{i2}. \tag{22}$$

Consider the following adaptation laws:

$$
\begin{aligned}
\dot{\hat{W}}_{i1} &= \dot{\tilde{W}}_{i1} = \Gamma_{i1}\left[S_{i1}(Z_{i1})z_i - \sigma_{i1}\hat{W}_{i1}\right], \\
\dot{\hat{W}}_{i2} &= \dot{\tilde{W}}_{i2} = \Gamma_{i2}\left[S_{i2}(Z_{i2})z_i - \sigma_{i2}\hat{W}_{i2}\right],
\end{aligned}
\tag{23}
$$

where $\sigma_{i1} > 0$, $\sigma_{i2} > 0$ are small constants. Let $c_i = c_{i0} + c_{i1}$, where c_{i0} and $c_{i1} > 0$. By using (19), (21), and (23), and with some completion of squares and straightforward derivation similar to those employed in the former steps, the derivative of V_i becomes

$$
\begin{aligned}
\dot{V}_i < z_i z_{i+1} - \sum_{k=1}^{i} c_{k0} z_k^2 - \sum_{k=1}^{i} \frac{\sigma_{k1}\left\|\tilde{W}_{k1}\right\|^2}{2} - \sum_{k=1}^{i} \frac{\sigma_{k2}\left\|\tilde{W}_{k2}\right\|^2}{2} \\
+ \sum_{k=1}^{i} \frac{\sigma_{k1}\left\|W_{k1}^*\right\|^2}{2} + \sum_{k=1}^{i} \frac{\sigma_{k2}\left\|W_{k2}^*\right\|^2}{2} + \sum_{k=1}^{i} \frac{\varepsilon_k^{*2}}{4c_{k1}}.
\end{aligned}
\tag{24}
$$

Step n: This is the final step. The derivative of $z_n = x_n - \alpha_{n-1}$ is $\dot{z}_n = u + f_n(\bar{x}_n(t)) + h_n(\bar{x}_n(t-\tau)) - \dot{\alpha}_{n-1}$.

Similarly, choosing the practical control law as

$$
u = -z_{n-1} - c_n z_n - \hat{W}_{n1}^T S_{n1}(Z_{n1}) - \hat{W}_{n2}^T S_{n2}(Z_{n2}) + \dot{\alpha}_{n-1}, \tag{25}
$$

where $c_n > 0$, $Z_{n1} = [x_1, x_2, \cdots, x_n]^T \subset R^n$,
$Z_{n2} = [x_1(t-\tau), x_2(t-\tau), \cdots, x_n(t-\tau)]^T \subset R^n$.

We have

$$
\begin{aligned}
\dot{z}_n &= u + f_n(\bar{x}_n(t)) + h_n(\bar{x}_n(t-\tau)) - \dot{\alpha}_{n-1} \\
&= -z_{n-1} - c_n z_n - \tilde{W}_{n1}^T S_{n1}(Z_{n1}) - \tilde{W}_{n2}^T S_{n2}(Z_{n2}) + e_n.
\end{aligned}
\tag{26}
$$

Consider the overall Lyapunov function candidate

$$
V_n = V_{n-1} + \frac{1}{2}z_n^2 + \frac{1}{2}\tilde{W}_{n1}^T \Gamma_{n1}^{-1}\tilde{W}_{n1} + \frac{1}{2}\tilde{W}_{n2}^T \Gamma_{n2}^{-1}\tilde{W}_{n2}. \tag{27}
$$

Consider the following adaptation laws:

$$
\begin{aligned}
\dot{\hat{W}}_{n1} &= \dot{\tilde{W}}_{n1} = \Gamma_{n1}\left[S_{n1}(Z_{n1})z_n - \sigma_{n1}\hat{W}_{n1}\right], \\
\dot{\hat{W}}_{n2} &= \dot{\tilde{W}}_{n2} = \Gamma_{n2}\left[S_{n2}(Z_{n2})z_n - \sigma_{n2}\hat{W}_{n2}\right],
\end{aligned}
\tag{28}
$$

where $\sigma_{n1} > 0$, $\sigma_{n2} > 0$ are small constants. Let $c_n = c_{n0} + c_{n1}$, where c_{n0} and $c_{n1} > 0$. By using (24), (26), and (28), and with some completion of squares and straightforward derivation similar to those employed in the former steps, the derivative of V_n becomes

$$
\dot{V}_n < -\sum_{k=1}^{n} c_{k0} z_k^2 - \sum_{k=1}^{n} \frac{\sigma_{k1}\left\|\tilde{W}_{k1}\right\|^2}{2} - \sum_{k=1}^{n} \frac{\sigma_{k2}\left\|\tilde{W}_{k2}\right\|^2}{2}
$$

$$+\sum_{k=1}^{n}\frac{\sigma_{k1}\left\|W_{k1}^{*}\right\|^{2}}{2}+\sum_{k=1}^{n}\frac{\sigma_{k2}\left\|W_{k2}^{*}\right\|^{2}}{2}+\sum_{k=1}^{n}\frac{\varepsilon_{k}^{*2}}{4c_{k1}}. \tag{29}$$

Let $\delta \triangleq \sum_{k=1}^{n}\frac{\sigma_{k1}\|W_{k1}^{*}\|^{2}}{2}+\sum_{k=1}^{n}\frac{\sigma_{k2}\|W_{k2}^{*}\|^{2}}{2}+\sum_{k=1}^{n}\frac{\varepsilon_{k}^{*2}}{4c_{k1}}$. If we choose c_{k0} such that $c_{k0}>\frac{\gamma}{2}$, $k=1,2,\cdots,n$, where γ is a positive constant, and choose σ_{k1}, σ_{k2}, Γ_{k1} and Γ_{k2} such that $\sigma_{k1}\geq\gamma\lambda_{\max}\left\{\Gamma_{k1}^{-1}\right\}$, $\sigma_{k2}\geq\gamma\lambda_{\max}\left\{\Gamma_{k2}^{-1}\right\}$, $k=1,2,\cdots,n$, then from (29) we have the following inequality:

$$\begin{aligned}\dot{V}_{n}&<-\sum_{k=1}^{n}c_{k0}z_{k}^{2}-\sum_{k=1}^{n}\frac{\sigma_{k1}\left\|\tilde{W}_{k1}\right\|^{2}}{2}-\sum_{k=1}^{n}\frac{\sigma_{k2}\left\|\tilde{W}_{k2}\right\|^{2}}{2}+\delta\\&<-\sum_{k=1}^{n}\frac{\gamma}{2}z_{k}^{2}-\sum_{k=1}^{n}\frac{\gamma\tilde{W}_{k1}^{T}\Gamma_{k1}^{-1}\tilde{W}_{k1}}{2}-\sum_{k=1}^{n}\frac{\gamma\tilde{W}_{k2}^{T}\Gamma_{k2}^{-1}\tilde{W}_{k2}}{2}+\delta\\&=-\gamma V_{n}+\delta.\end{aligned} \tag{30}$$

The following theorem shows the stability and control performance of the closed-loop adaptive system.

Theorem 1. Consider the closed-loop system consisting of the plant (1), the controller (25), and the NN weight updating laws (14) (23) and (28). Assume that there exists sufficiently large compact sets $\Omega_i\in R^i$, $i=1,2,\cdots,n$ such that $Z_{i1}\in\Omega_i$, $Z_{i2}\in\Omega_i$ for all $t\geq 0$. Then, for bounded initial conditions, we have the following:

1) All signals in the closed-loop system remain semiglobally uniformly ultimately bounded;

2) The output tracking error $y(t)-y_d(t)$ converges to a small neighborhood around zero by appropriately choosing design parameters.

Proof. 1) From (30), using the boundedness theorem (e.g., [13]), we have that all z_i, $\hat{W}_{i1}$ and $\hat{W}_{i2}$ are uniformly ultimately bounded. Since $z_1=x_1-y_d$ and y_d are bounded, we have that x_1 is bounded. From $z_i=x_i-\alpha_{i-1}, i=1,2,\cdots,n$, and the definitions of virtual controls (10), (20) we have that $x_i, i=2,3,\cdots,n$ remain bounded. Using (25), we conclude that control is also bounded. Thus, all the signals in the closed-loop system remain bounded.

2) Let $\rho={}^{\delta}\!/_{\gamma}>0$, then (30) satisfies

$$0\leq V_n(t)\leq\rho+(V_n(0)-\rho)\exp(-\gamma t). \tag{31}$$

From (31), we have

$$\sum_{i=1}^{n}\frac{1}{2}z_{k}^{2}<\rho+(V_n(0)-\rho)\exp(-\gamma t)<\rho+V_n(0)\exp(-\gamma t). \tag{32}$$

That is

$$\sum_{i=1}^{n}z_{k}^{2}<2\rho+2V_n(0)\exp(-\gamma t), \tag{33}$$

which implies that given $\mu > \sqrt{2\rho}$, there exists T such that for all $t \geq T$, the tracking error satisfies

$$|z_1(t)| = |x_1(t) - y_d(t)| < \mu, \tag{34}$$

where μ is the size of a small residual set which depends on the NN approximation error e_i and controller parameters c_i, σ_{i1}, σ_{i2}, Γ_{i1} and Γ_{i2}. It is easily seen that the increase in the control gain c_i, adaptive gain Γ_{i1} , Γ_{i2} and NN node number l_j will result in a better tracking performance.

Remark 1. In [6], one neural network is adopted to approximate the unknown smooth function $(f_i(\bar{x}_i) - \dot{\alpha}_{i-1})/g_i(\bar{x}_i)$ in every design step. However, because the derivatives of the virtual control α_{i-1} are include in the NNs, the dimensions of input vectors of the NNs become twice as much as those of corresponding state vectors and these additional inputs must be computed online too. Therefore, the approach is still difficult to implement and apply in practice. In this paper, two NNs are adopted to approximate the unknown smooth functions and unknown time delay functions, respectively, in every design step, but there are no dimensional increments and no additional parameters must be calculated.

Remark 2. Compared with the work in [7], the proposed adaptive neural network controller in this paper can cope with nonlinear time-delay systems.

Remark 3. Compare with reference [11], the method presented in this paper is much simpler to understand, and the system model presented in this paper is more general. The time-delay exists in state variable other than output variable.

Remark 4. Compared with the work in [12], the unknown time-delay functions in this paper are approximated by neural networks. However, in [12], by constructing proper Lyapunov-Krasovskii functionals, the uncertainties of unknown time delays are compensated. So, the requirement on the unknown time delay functions is relaxed in our paper.

4 Simulation

Consider the following strict-feedback system:

$$\begin{aligned} \dot{x}_1(t) &= x_2(t) + 0.5x_1(t), \\ \dot{x}_2(t) &= u(t) + x_1(t) \cdot x_2(t) + \sin\left(x_1(t-\tau) \cdot x_2(t-\tau)\right), \\ y &= x_1, \end{aligned} \tag{35}$$

where $\tau = 5$. The initial condition $[x_1(0), x_2(0)]^T = [0.8, 0]^T$ and the desired reference signal of system is $y_d(t) = \cos(t)$.

The adaptive neural network controller is chosen according to (25) as follows:

$$u = -z_1 - c_2 z_2 - \hat{W}_{21}^T S_{21}(Z_{21}) - \hat{W}_{22}^T S_{22}(Z_{22}) + \dot{\alpha}_1, \tag{36}$$

where $z_1 = x_1 - y_d$, $z_2 = x_2 - \alpha_1$, $\alpha_1 = -c_1 z_1 - \hat{W}_{11}^T S_{11}(Z_{11}) + \dot{y}_d(t)$, $Z_{11} = [x_1]^T$, $Z_{21} = [x_1, x_2]^T$, $Z_{22} = [x_1(t-\tau), x_2(t-\tau)]^T$, and neural network weights $\hat{W}_{11}$, $\hat{W}_{21}$ and $\hat{W}_{22}$ are updated by (14) and (23) correspondingly.

Neural networks $\hat{W}_{11}^T S_{11}(Z_{11})$ contains 13 nodes (i.e., $l_1 = 13$), with centers $\mu_l (l = 1, 2, \cdots, l_1)$ evenly spaced in $[-6, 6]$, and widths $\eta_l = 1 (l = 1, 2, \cdots, l_1)$. Neural network $\hat{W}_{21}^T S_{21}(Z_{21})$ and $\hat{W}_{22}^T S_{22}(Z_{22})$ contains 169 nodes (i.e., $l_2 = 169$), with centers $\mu_l (l = 1, 2, \cdots, l_2)$ evenly spaced in $[-6, 6] \times [-6, 6]$, and widths $\eta_l = 1 (l = 1, 2, \cdots, l_2)$. The design parameters of above controller are $c_1 = 4$, $c_2 = 4$, $\Gamma_{11} = \Gamma_{21} = \Gamma_{22} = diag\{2.0\}$, $\sigma_{11} = \sigma_{21} = \sigma_{22} = 0.2$. The initial weights $\hat{W}_{11} = 0.5$, $\hat{W}_{21} = 0$, $\hat{W}_{22} = 0$. Fig 1 shows the simulation result of applying controller (36) to system (35) for tracking desired signal y_d. From Fig 1, we can see that good tracking performance is obtained.

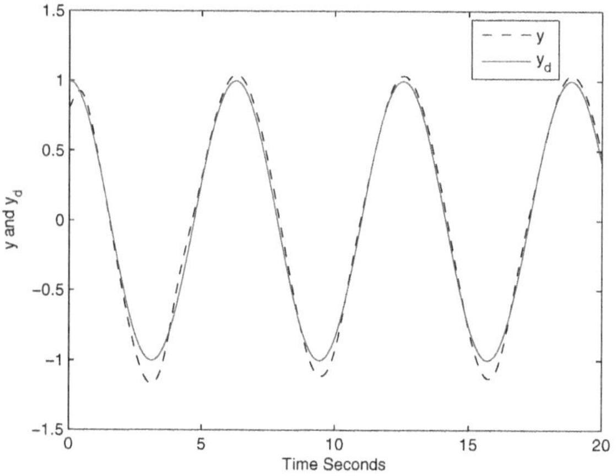

Fig. 1. Output tracking performance

5 Conclusion

In this paper, an adaptive neural network control approach is proposed for a class of strict-feedback nonlinear time-delay systems. The unknown time delay functions are approximated by neural networks, such that the requirement on unknown time delay functions is relaxed. Finally, a numerical simulation is given to show the effectiveness of the approach.

Acknowledgment. This work was supported by the National Natural Science Foundation of China under grant 60574042.

References

1. Ge, S.S., Hang, C.C., Lee, T.H., Zhang, T.: Stable Adaptive Neural Network Control. Norwell, MA: Kluwer (2002)
2. Krstič, M., Kanellakopoulos, I., Kokotovič, P.: Nonlinear and Adaptive Control Design. New York: Wiley (1995)

3. Polycarpou, M.M., Mears, M.J.: Stable Adaptive Tracking of Uncertain Systems Using Nonlinearly Parametrized On-Line Approximators. International Journal of Control **70** (3) (1998) 363-384
4. Zhang, Y., Peng, P.Y., Jiang, Z.P.: Stable Neural Controller Design for Unknown Nonlinear Systems Using Backstepping. IEEE Transactions on Neural Networks **11** (2000) 1347-1359
5. Zhang, T., Ge, S.S., Hang, C.C.: Adaptive Neural Network Control for Strict-feedback Nonlinear Systems Using Backstepping Design. Automatica **36** (2000) 1835-1846
6. Ge, S.S., Wang, C.: Direct Adaptive NN Control of a Class of Nonlinear Systems. IEEE Transactions on Neural Networks **13** (1) (2002) 214-221
7. Li, Y.H., Qiang, S., Zhuang, X.Y., Kaynak, O.: Robust and Adaptive Backstepping Control for Nonlinear Systems Using RBF Neural Networks. IEEE Transactions on Neural Networks **15** (3) (2004) 693-701
8. Cao, J.D., Ho, Daniel W.C.: A General Framework for Global Asymptotic Stability Analysis of Delayed Neural Networks Based on LMI Approach. Chaos, Solitons & Fractals **24** (5) (2005) 1317-1329
9. Xu, S.Y., Lam, J.: A New Approach to Exponential Stability Analysis of Neural Networks with Time-varying Delays. Neural Networks **19** (1) (2006) 76-83
10. Zeng, Z.G., Wang, J., Liao, X.X.: Global Exponential Stability of Neural Networks with Time-Varying Delays. IEEE Transactions on Circuits and Systems-I: Fundamental Theory and Applications **50** (10) (2003) 1353-1358
11. Chen, W.S., Li, J.M.: Adaptive Neural Network Backstepping Control for Nonlinear Time-Delay Systems. Electric Machines and Control **9** (5) (2005) 500-503, 511 (in Chinese)
12. Ge, S.S., Hong, F., Lee, T.H.: Adaptive Neural Network Control of Nonlinear Systems with Unknown Time Delays. IEEE Transactions on Automatic Control **48** (11) (2003) 2004-2010
13. Qu, Z.: Robust Control of Nonlinear Uncertain Systems. New York: Wiley (1998)
14. Haykin, S.: Neural Networks: A Comprehensive Foundation, 2nd ed. Upper Saddle River, NJ: Prentice-Hall (1999)
15. Ioannou, P.A., Sun, J.: Robust Adaptive Control. Englewood Cliffs, NJ: Prentice-Hall (1995)

A Nonlinear ANC System with a SPSA-Based Recurrent Fuzzy Neural Network Controller

Qizhi Zhang[1], Yali Zhou[1], Xiaohe Liu[1], Xiaodong Li[2], and Woonseng Gan[3]

[1] Department of Computer Science and Automation, Beijing Institute of Machinery, P. O. Box 2865, Beijing, 100085, China
zqzbim@yahoo.com.cn
[2] Institute of Acoustics, Academia Sinica, China
[3] School of EEE, Nanyang Technological University, Singapore

Abstract. In this paper, a feedforward active noise control (ANC) system using a recurrent fuzzy neural network (RFNN) controller based on simultaneous perturbation stochastic approximation (SPSA) algorithm is considered. Because RFNN can capture the dynamic behavior of a system through the feedback links, only one input node is needed, and the exact lag of the input variables need not be known in advance. The SPSA-based RFNN control algorithm employed in the ANC system is first derived. Following this, computer simulations are carried out to verify that the SPSA-based RFNN control algorithm is effective for a nonlinear ANC system. Simulation results show that the proposed scheme is able to significantly reduce disturbances without the need to model the secondary-path and has better tracking ability under variable secondary-path. This observation implies that the SPSA-based RFNN controller eliminates the need of the modeling of the secondary-path.

1 Introduction

The active noise control (ANC) using feedforward control techniques has attracted much research attention because it can complement traditional passive techniques and attain better performance on attenuation of low-frequency noises [1]. When the ANC system exhibits nonlinear response characteristics, the most common form of adaptive algorithm/architecture combination is the feedforward neural network (NN) using the gradient descent-based back-propagation (BP) algorithm [2], [3], where the NN would be trained to derive an output signal to cancel the noise. But, in this control method, in order to update the weights of the NN, we need a gradient of the error function, namely, we must know the model of the secondary-path [4] or approximate the model by another NN [2].

However, characteristics of the secondary-path usually vary with respect to temperature or other environments, that is, the secondary-path is time-variant. Therefore, it is difficult to estimate the exact characteristics of the secondary-path accurately. To solve this problem, a model-free (MF) control scheme based on the simultaneous perturbation stochastic approximation (SPSA) algorithm is presented here [5]. This approach is based on the output error of the system to update the weights of the NN

D. Liu et al. (Eds.): ISNN 2007, Part I, LNCS 4491, pp. 176–182, 2007.

without the need to model the secondary-path [6], [7]. The drawback in NN-based ANC system is that the exact lag of the input variables must be known in advance. A RFNN-based ANC system can capture the dynamic behavior of the ANC system through the feedback links, only one input node is needed, and the exact lag of the input variables need not be known in advance [8].

In addition to being able to update the weights of the RFNN without the need to model the secondary-path, the presented algorithm can also give more simple formulae for updating the weights of the RFNN compared to the back-propagation neural network (BPNN) algorithm. Because the SPSA algorithm requires only two objective function measurements regardless of the number of weights being updated, it uses "simultaneous perturbation (SP)" to update all weights of the RFNN simultaneously and can simplify the derivation of the adaptation algorithm of weights greatly, this will be discussed in the following section.

2 Control Algorithm

The block diagram of a feedforward ANC system using the SPSA-based RFNN algorithm and the RFNN controller are shown in Fig.1 and Fig.2, respectively. The primary-path $P(x)$ is from the noise source to the error microphone, and the secondary-path $S(u)$ is from the canceling loudspeaker to the error microphone. The SPSA-based RFNN controller is used to generate an anti-noise signal $u(t)$. Note that this algorithm does not use the estimator of the secondary-path [4]. The RFNN has five layers, and a model with two inputs and a single output is considered here for convenience [8]. The nodes in layer 1 are input nodes that directly transmit input signals to the next layer. Layer 5 is the output layer. The layer 3 performs a fuzzy operation to calculate the firing strength, and "sum" operation is used in this paper.

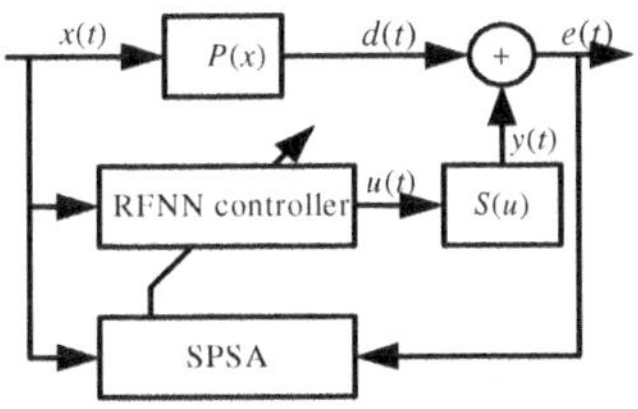

Fig. 1. The block diagram of an ANC system

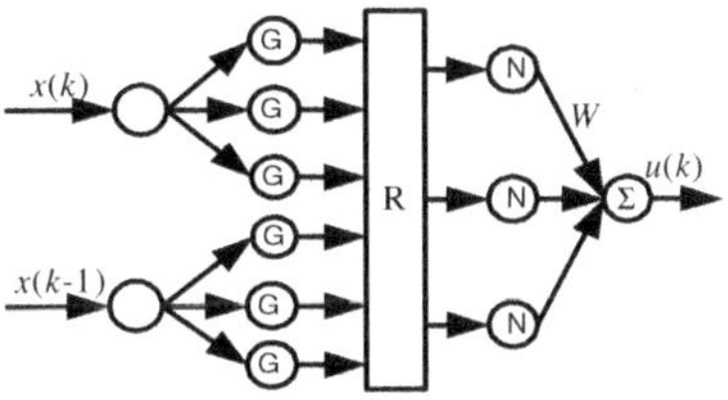

Fig. 2. The RFNN controller

It is common knowledge that if the secondary-path of the ANC system is completely unknown, it is impossible to use usual gradient method as a learning rule to update the controller coefficients [3], [9]. In this case, an estimator of the gradient of the error function is needed. The SPSA which was introduced by J. C. Spall [5] is a well-known gradient approximation approach that relies on measurements of the objective function, not on measurements of the gradient of the objective function.

A SPSA-based NN algorithm is presented in Ref. [7]. The similar steps are used to develop the SPSA-based RFNN algorithm to improve the noise cancellation capability of a nonlinear ANC system.

Step 1: Define the error function

Note that in ANC system, each sampling error signal does not contain enough information as an evaluation function to be optimized. That is, the expectation of the error signal has to be used as the evaluation function. For practicality, the sum of the error signal for a certain interval is used to approximate the expectation of the error signal. Thus, the error function is defined as:

$$J(u(t)) = \frac{1}{2}\sum_{t=1}^{\lambda} e^2(t) = \frac{1}{2}\sum_{t=1}^{\lambda}[y(t)+d(t)]^2. \tag{1}$$

where t is the sampling number in a block interval, and λ is the total sampling number of one block interval.

Step 2: Compute the control signal $y(t)$

The nodes in Layer 2 are "term nodes" (G), and they act as membership functions expressing the input fuzzy linguistic variables. Each node in Layer 3 is called a "rule node" (R) and represents a single fuzzy rule. A fully feedback connection is introduced to give the feed-forward fuzzy NN a temporal processing capability. The nodes in Layer 4 (N) perform the normalization of firing strengths from Layer 3, and the input links are fully connected. In the following descriptions, the symbol $v_i^{(k)}$ denotes the ith input of a node in the kth layer, and the symbol $a^{(k)}$ denotes the output of a node in the kth layer. To provide a clear understanding of an RFNN, the functions of Layer 1 to Layer 5 are defined as follows [8]:

$$a_i^{(1)}(t) = v_i^{(1)}(t), \tag{2}$$

$$a_j^{(2)}(t) = exp\left\{-\frac{(v_i^{(2)}(t)-m_{ij})^2}{\sigma_{ij}^2}\right\}, \tag{3}$$

$$a_i^{(3)}(t) = S_i(t) + f(net_i(t)),\ S_i(t) = \prod_j v_j^{(3)}(t),\ net_i(t) = \sum_j V_{ij} a_j^{(3)}(t-1). \tag{4}$$

$$a_i^{(4)}(t) = \frac{v_i^{(4)}(t)}{\sum_j v_j^{(4)}(t)}, \tag{5}$$

$$u(t) = a^{(5)}(t) = \sum_i v_i^{(5)}(t) W_i\ . \tag{6}$$

According to Eqs. (2)-(6), the output of the RFNN can be represented as

$$u(t)=f(x(t),W). \tag{7}$$

Where $W = (w_1, w_2 \cdots, w_J, v_{11}, v_{12}, \cdots, v_{JJ})^T = (w^1, \cdots, w^n)^T$ is the general weight vector in the output layer and recurrent layer. The superscript n denotes the number of weights to be estimated, $n=JJ+J$, and J is the number of neurons in layer 3. Superscript T is transpose of a vector. The control signal $y(t)$ can be calculated using $y(t)=S(u(t))$.

Step 3: Generation of SP vector
The following perturbation vector Δ_k is generated as independent Bernoulli random variables with outcomes of ±1 that gives small disturbances to all weights [5]

$$\Delta_k = (\Delta_k^1, \cdots, \Delta_k^n)^T . \tag{8}$$

Where the subscript k denotes an iteration.

Step 4: Error function evaluations
Obtain two measurements of the error function $J(\bullet)$ based on the SP: $J(u(W))$ and $J(u(W+c_k\Delta_k))$ with the Δ_k from step 3. Where c_k is a positive scalar and represents a magnitude of the perturbation.

Step 5: Gradient approximation
Generate the SP approximation to the unknown gradient $\frac{\partial J(u(W))}{\partial W}$ as [3], [5], [7]:

$$\Delta W(t) = \frac{J(u(W + c_k\Delta_k)) - J(u(W))}{c_k\Delta_k} . \tag{9}$$

Step 6: Update the weight vector W of the RFNN
Weights of the RFNN are updated in the following manner:

$$W(t+1) = W(t) - a_k\Delta W(t) . \tag{10}$$

Where a_k is a positive learning coefficient.

From Eqs. (9) and (10), it can be seen that the weights of the RFNN controller is updated without the need to model the secondary-path, so this algorithm is called MFRFNN control algorithm. At the same time, the conclusion can be derived that compared to the BPNN algorithm which were reported in Ref. [2], [4], the SPSA-based MFRFNN algorithm has more simple formulae for updating the weights of the RFNN.

3 Simulation Examples

Some simulations are presented to illustrate the noise-canceling performance of the SPSA-based MFRFNN algorithm on a nonlinear ANC system, and at the same time, a comparison between the SPSA-based MFRFNN algorithm and the SPSA-based MFNN algorithm is made. A 300Hz sinusoidal signal is used to generate the primary disturbance signal and also used as the reference signal to the control algorithm. There is only one input node and one output node in RFNN. The input space is partitioned to eight fuzzy sets, and the means and widths of the Gaussian membership functions are selected as [8]:

m=[-0.65,-5/8,-3/8,-1/8,1/8,3/8,5/8,0.65], σ=[-20,0.14,0.14,0.14,0.14,0.14,0.14,20].

The NN used in these simulations is three-layered feedforward network and the number of neurons is set as 15-10-1[7]. The sampling frequency used is 3kHz, and the total sampling number of one block interval λ is set as 30. c_k is set as 0.01, a_k is set as 0.001 for NN and 0.01 for RFNN, respectively. The total simulation duration is 4.5 second.

The model used in this simulation has the following expressions with nonlinear terms:

The primary disturbance *d(t)* is expressed as [4], [7]:

$$d_{t+1} = 0.8x_t + 0.6x_{t-1} - 0.2x_{t-2} - 0.5x_{t-3} - 0.1x_{t-4} \\ + 0.4x_{t-5} - 0.05_{t-6}. \tag{11}$$

The control signal *y(t)* is expressed as [4], [7]:

$$y_{t+1} = 0.9u_t + 0.6u_{t-1}^3 + 0.1u_{t-2}^3 - 0.4u_{t-3}^3 - 0.1u_{t-4}^3 \\ + 0.2u_{t-5}^3 + 0.1u_{t-6}^2 + 0.01u_{t-7}^2 + 0.001u_{t-8}^2. \tag{12}$$

Case 1: A simple static ANC example is first considered to illustrate the SPSA-based MFRFNN algorithm effectiveness by comparison with the result given by the MFNN algorithm, the secondary-path is assumed to be time-invariant. Fig.3 presents the simulation result of the stable canceling errors in the last 100 iterations. It can be found that the MFRENN algorithm yields a lower level of steady-state error than the MFNN algorithm. Fig.4 presents the simulation result of the canceling errors in the frequency domain. The thin solid line shows the power spectrum of active noise canceling error when the ANC system is turned off, and the thick solid line shows the power spectrum of active noise canceling error when the SPSA-based MFRFNN algorithm is used to adapt the coefficients of the controller. From the results shown in Fig.4, it can be clearly seen that the major disturbance frequency are attenuated by approximately 30 dB.

Case 2: Next, we deal with a tracking problem. Using the same settings as in case 1, after the system has entered into steady-state phase, the secondary-path is altered by letting *S*(*u*)= -*S*(*u*). Fig.5 shows the error signal in error microphone versus the number of iterations. When the number of iteration reaches 15,000, the secondary-path is changed. From the result shown in Fig.5, it can be seen that the system has a good tracking ability of the secondary-path. This simulation shows that the SPSA-based RFNN controller can eliminates the need of the modeling of the secondary-path.

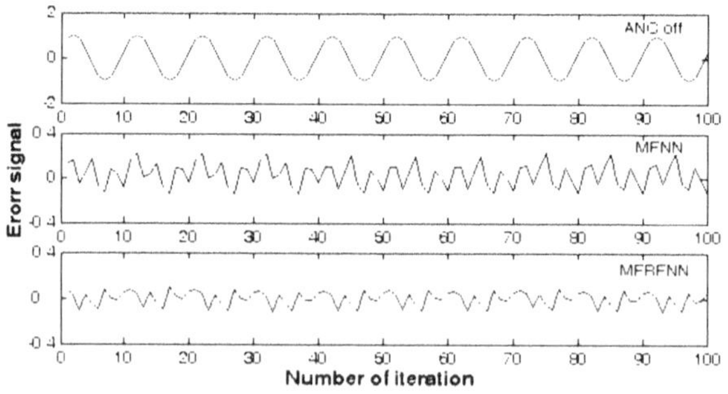

Fig. 3. The stable error signal in the last 100 iterations

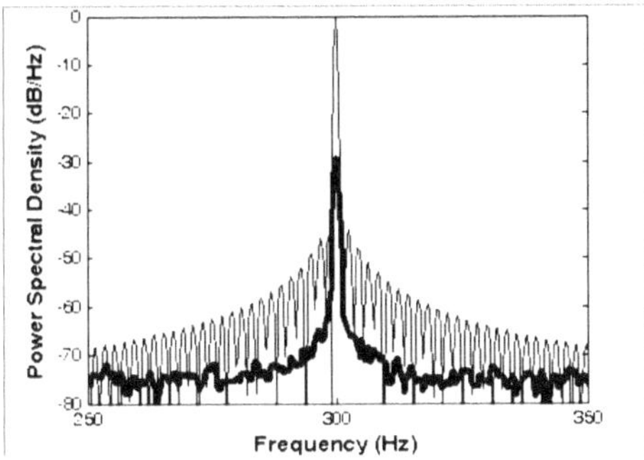

Fig. 4. The error signal spectrum for case 1 (RFNN)

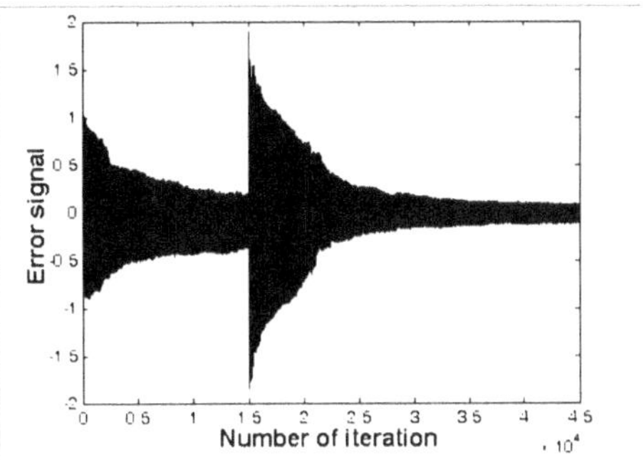

Fig. 5. The error signal versus number of iterations when the secondary-path is changed (RFNN)

4 Conclusions

The RFNN controller based on the SPSA algorithm has been developed for use in a nonlinear ANC system. This approach optimizes error function without using derivative of the error function. Therefore, the presented ANC algorithm does not require any estimation of the secondary-path. The RFNN controller can capture the dynamic behavior of the ANC system through the feedback links, only one input node is needed, and the exact lag of the input variables need not be known in advance. Some simulations were presented to verify that this algorithm is effective. The simulation results indicated that this algorithm was able to significantly reduce disturbances and an output error attenuation of approximately 30dB was achieved.

Acknowledgments

This research is supported by Scientific Research Common Program of Beijing Municipal Commission of Education (KM200511232008, KZD200611232020) and Training Funds for Elitist of Beijing.

References

1. Nelson, P.A., Elliott, S.J.: Active Sound Control. Academic Press, London (1991)
2. Snyder, S.D., Tanaka, N.: Active Control of Vibration using a Neural Network. IEEE Trans On Neural Networks **6** (4) (1995) 819-828
3. Maeda, Y., De Figueiredo, R.J.P.: Learning Rules for Neuro-Controller via Simultaneous Perturbation. IEEE Transactions On Neural Networks **8** (5) (1997) 1119-1130
4. Zhou, Y.L., Zhang, Q.Z., Li, X.D., Gan, W.S.: Analysis and DSP Implementation of an ANC System using a Filtered-Error Neural Network. Journal of Sound and Vibration **285** (1) (2005) 1-25
5. Spall, J.C.: Multivariate Stochastic Approximation using Simultaneous Perturbation Gradient Approximation. IEEE Transactions On Automatic Control **37** (3) (1992) 332-341

6. Maeda, Y., Yoshida, T.: An Active Noise Control without Estimation of Secondary-Path. ACTIVE1999, USA (1999) 985-994
7. Zhou, Y.L., Zhang, Q.Z., Li, X.D., Gan, W.S.: Model-Free Control of a Nonlinear ANC System with a SPSA-based Neural Network Controller. ISNN 2006, LNCS **3972**, 1033 – 1038
8. Zhang, Q.Z., Gan, W.S. and Zhou Y.L.: Adaptive Recurrent Fuzzy Neural Networks for Active Noise Control. Journal of Sound and Vibration **296** (2006) 935-948
9. Spall, J.C., Cristion, J.A.: A Neural Network Controller for Systems with Unmodeled Dynamics with Applications to Wastewater Treatment. IEEE Transactions on Systems. Man. And Cybernetics **27** (3) (1997) 369-375

Neural Control Applied to Time Varying Uncertain Nonlinear Systems

Dingguo Chen[1], Jiaben Yang[2], and Ronald R. Mohler[3]

[1] Siemens Power Transmission and Distribution Inc., 10900 Wayzata Blvd., Minnetonka, Minnesota 55305, USA

[2] Department of Automation, Tsinghua University, Beijing 100084, People's Republic of China

[3] Department of Electrical and Computer Engineering, Oregon State University, Corvallis, OR 97330, USA

Abstract. This paper presents a neural network based control design to handle the stabilization of a class of multiple input nonlinear systems with time varying uncertain parameters while assuming that the range of each individual uncertain parameter is known. The proposed design approach allows incorporation of complex control performance measures and physical control constraints whereas the traditional adaptive control techniques are generally not applicable. The desired system dynamics are analyzed, and a collection of system dynamics data, that represents the desired system behavior and approximately covers the region of stability interest, is generated and used in the construction of the neural controller based on the proposed neural control design. Furthermore, the theoretical aspects of the proposed neural controller are also studied, which provides insightful justification of the proposed neural control design. The simulation study is conducted on a single-machine infinity-bus (SMIB) system with time varying uncertainties on its parameters. The simulation results indicate that the proposed design approach is effective.

1 Introduction

It is noted that the hierarchical neural network structure can be viewed as a kind of generalized parametric control which allows for improved controllability and transient stabilization. In real applications, situations arise where system uncertainties exist. In order to handle these uncertainties, various intelligent schemes have been proposed, to name a few, a control-switching scheme, a multiplicative control scheme, hierarchical neural control [1], [2], [4], [6], [5] (and references therein). In particular, novel techniques were developed in [5], [1], [2], [4], [6], [7] to synthesize hierarchical neural controllers for stabilizing post-fault power systems with unknown load and even with load dynamics as well as generator side transients. From control engineering point of view, these techniques represent innovations that are applicable to real world applications.

It would be ideal if the traditional adaptive control theory can be applied to solve the above mentioned control problems. Despite the tremendous progress in

D. Liu et al. (Eds.): ISNN 2007, Part I, LNCS 4491, pp. 183–192, 2007.

the area of neural network based adaptive control [8], the control designs based on the adaptive control schemes [8] however, have the drawback that the control signal is not constrained within a pre-designated desired physical range. Further, additional control performance criteria, e.g., optimal control performance, are difficult to incorporate within the framework of the traditional adaptive control schemes. It becomes apparent that the popular adaptive control schemes can not be directly applied to solve the practical problems that require that the control signal be bounded by a given number, and a new theory tailored for the application has yet to be worked out.

Along with a parallel effort [9] with an emphasis on the nonlinear systems with unknown, fixed parameters, this paper is focused on neural control of a class of nonlinear systems with unknown, time-varying parameters yet to achieve a near optimal control performance in some sense. These endeavors represent an attempt to narrow the gap of developing applicable near optimal adaptive neural control theory and the real world applications that contain time varying parameters, require constrained control, and yet dictate the control design to meet certain control performance criteria.

This paper is organized as follows: Section 2 describes the class of uncertain nonlinear systems to be studied in this paper. The relevant sub-control problems of the main control problem are studied in the context of time optimal control, and a brief review of the switching-times-variation method (STVM) is conducted in section 3. The design methodology that employs neural networks, and in particular, the so-called hierarchical neural networks is presented in section 4. The design procedures, in particular to deal with the time varying parameters, are supported by a theory developed in section 5. A case study is presented in section 6 to illustrate how the proposed control design can be used to adaptively control the SMIB system. Finally, some conclusions are drawn.

2 Problem Formulation

Studied in this paper is a special class of nonlinear systems that feature parametric uncertainties, confined control inputs, time varying parameters. When the control inputs are not constrained, this type of nonlinear systems have been widely studied in the context of adaptive control. When the control inputs are constrained, adaptive control schemes are not readily applicable. When additional control performance measures are considered, it becomes even more apparent that popular adaptive control schemes are no longer conveniently applicable. From engineering point of view however, the control inputs in most practical applications are usually physically restricted. Motivated by this consideration, we take a different approach, yet to be revealed in later sections, to address the key issues: (a) adaptive control of time-varying, uncertain nonlinear systems and (b) optimal control performance measures for this type of nonlinear systems, which can be characterized by the following state equation of a finite dimensional differential system linear in control and linear in parameters.

$$\dot{x} = a(x) + C(x)p(t) + B(x)u \tag{1}$$

where $x \in G \subseteq R^n$ is the state vector, $p(t) \in \Omega_p \subset R^l$ is the time-varying bounded parameter vector, $u \in R^m$ is the control vector, which is confined to an admissible control set U, $a(x) = [a_1(x)\ a_2(x)\ \cdots\ a_n(x)]^\tau$ is an n-dimensional vector function of x, $C(x)$ is an $n \times l$-dimensional matrix function of x, and $B(x)$ is an $n \times m$-dimensional matrix function of x.

The control objective is to follow a theoretically sound control design methodology to design the controller such that the system is adaptively controlled with respect to parametric uncertainties and yet achieves a desired control performance.

To facilitate the the theoretical derivations, several conventional assumptions are made in the following and applied throughout the paper.

AS1: It is assumed that $a(.)$, $C(.)$ and $B(.)$ have continuous partial derivatives with respect to the state variables on the region of interest.

AS2: Without loss of generality, assume that the admissible control set U is characterized by $U = \{u : |u_i| \leq 1, i = 1, 2, \cdots, m\}$ where u_i is u's ith component.

AS3: It is assumed that the system is controllable.

AS4: The control performance criteria is $J = \int_{t_0}^{t_f} [a_0(x(s)) + b_0^\tau(x(s))u(s)]ds$ where t_0 and t_f are the initial time and the final time, respectively, and $a_0(.)$ and $b_0(.)$ are continuous.

AS5: The target set θ_f is defined as $\theta_f = \{x : \Psi(x(t_f)) = 0\}$ where Ψ_i's $(i = 1, 2, \cdots, q)$ are the components of $\Psi(.)$.

Remark 1: As a step of our approach to address the control design for the system (1), the above same control problem is studied with the only difference that the parameters in Eq. (1) are given. An optimal solution is sought to the following control problem:

The optimal control problem (P_0) consists of the system equation (1) with fixed and known parameter vector p, the initial time t_0, the variable final time t_f, the initial state $x_0 = x(t_0)$, together with the assumptions AS1, AS2, AS3, AS4, AS5 satisfied such that the system state conducts to a pre-specified terminal set θ_f at the final time t_f while the control performance index is minimized.

AS6: There do not exist singular solutions to the optimal control problem (P_0) described in Remark 1 (referenced as the control problem (P_0) later on distinct from the original control problem (P)).

AS7: $\frac{\partial x}{\partial p}$ is bounded on $p \in \Omega_p$ and $x \in \Omega_x$.

Remark 2: For any continuous function $f(x)$ defined on the compact domain $\Omega_x \subset R^n$, there exists a neural network characterized by $NN_f(x)$ such that for any positive number ϵ_f^*, $|f(x) - NN_f(x)| < \epsilon_f^*$.

AS8: Let the well offline-trained neural network be denoted by $NN(x, \Theta_s)$, and the neural network with the ideal weights and biases by $NN(x, \Theta_*)$ where Θ_s and Θ_* designate the parameter vectors comprising weights and biases of the corresponding neural networks. The approximation of $NN_f(x, \Theta_s)$ to $NN_f(x, \Theta_*)$

is measured by $\delta NN_f(x;\Theta_s;\Theta_*) = |NN_f(x,\Theta_s) - NN_f(x,\Theta_*)|$. Assume that $\delta NN_f(x;\Theta_s;\Theta_*)$ is bounded by a pre-designated number $\epsilon^s > 0$.

3 The Main Control Problem (*P*) and Its Sub-Control Problems (P_0)

It is worthwhile to point out that the main difference between the original control problem (P) and the control problem (P_0) is that the parameter vector p is unknown for the former and known for the latter. To address the original control problem (P), the proposed approach is to start with the control problem (P_0). The control problem (P) can be viewed as a family of the control problems (P_0). Each individual control problem (P_0) corresponds to a different known parameter vector p.

It is shown in [9] that there exists a time optimal control solution to the control problem (P_0) with the assumptions AS1 - AS5. The application of maximum principle gives rise to the so-called two-point boundary-value problem (TPBVP) which must be satisfied by an optimal solution. In general, an analytic solution to the TPBVP is extremely difficult, and usually practically impossible to obtain.

Remark 3: As has been pointed out, the original control problem (P) with the unknown parameter vector can be decomposed into a series of control problems (P_0) with distinct known parameter vectors. The decomposition is conducted in conjunction with the tessellation of the bounded convex parameter space. The result of the tessellation is a family of disjointed convex sub-regions. For each individual sub-region, each vertex corresponds to a known parameter vector, which in turn specifies a control problem (P_0). For the original control problem (P) with the unknown parameter located within one of these sub-regions, say $\Omega_{p,k}$, the vertices of $\Omega_{p,k}$ and the desired system control and dynamic behaviors of the corresponding control problems (P_0) are available and can be utilized to construct the controller for the control problem (P). When the diameter of the sub-region $\Omega_{p,k}$ (defined as $D(\Omega_{p,k}) = max||p_i - p_j|| : p_i, p_j \in \Omega_{p,k})$ tends to zero, the unknown parameter vector approaches one of the vertices of $\Omega_{p,k}$. The system behavior of the control problem (P_0) tends to dictate that of the original control problem (P). The detailed control design will be presented in section 4.

4 Near Optimal Neural Control Design

As discussed in [5], the hierarchical neural network based control design was applied to the single-machine infinity-bus (SMIB) power system. In this paper, a class of uncertain nonlinear systems with multiple inputs is considered. The hierarchical neural control design needs to be extended to take into account the nonlinear structure and the multiple control inputs. From the architectural point of view, the hierarchical neural control design involves two parts: the design of lower-level neural controllers and the design of upper-level neural controllers.

From the practical implementation point of view, the hierarchical neural control design involves: (a) tessellation of the bounded, convex parameter space into a collection of disjointed, convex sub-regions; (b) the identification of the switching manifold for each individual control problem (P_0); (c) construction of lower-level neural controllers for all nominal cases; (d) construction of upper-level neural controllers to coordinate the contributions of the control efforts from lower-level neural controllers.

The tessellation of the parameter vector space needs to be conducted so that a collection of convex, disjointed sub-regions are resulted. Generally speaking, the control performance of the neural controller is associated with the granularity of the tessellation. A compromise is sought between the level of control performance and the level of implementation complexity. In addition, the implementation complexity may be reduced based on the qualitative analysis of the dynamics of the system for all involved nominal cases and the assessment of the different effects that each parameter component of the parameter vector has on the system behavior.

For each individual control problem (P_0), the bang-bang control is resulted. Consequently, the switching manifold can be identified after using numerical methods to generate the optimal control and state trajectories that cover the stability region of interest and applying these trajectories. Mathematically, this is equivalent to say $u_i = sgn(S_i(x))$ or $u_i = -sgn(S_i(x))$ where $S(x)$ is the switching function with $S(x) = 0$ identifying the switching manifold; the function $sgn(.)$ is defined as: $sgn(S) = 1$ if $S > 0$; or $Sgn(S) = -1$ if $S < 0$.

The design of the lower-level neural controllers utilizes the off-line generated optimal control and state trajectories to approximate the switching manifolds. The trained neural network tends to produce the outputs closely approximating the optimal control on each side of the switching manifold while it is likely that some mismatch error may occur in some neighborhood contraining the switching manifold. When the desired control takes a positive (or negative) limit, the output of the neural network tends to take a positive (or negative) value.

The design of the upper-level neural controllers also utilizes the off-line generated optimal state trajectories. In addition, it makes use of the outputs of the lower-level neural controllers. When using optimal control and the resulting optimal state trajectories, the output of the corresponding upper-level neural network is 1; the output of rest upper-level neural network varies depending on the distance between the resulted non-optimal state trajectory and the optimal state trajectory.

Each component of the final control vector is the respective sum of the lower-level neural control signals modulated by the corresponding coordinating signals of the upper-level neural networks. For detailed descriptions of hierarchical neural controllers and the hierarchical neural control diagram, the reader is referred to [5].

To address the time varying feature of the uncertain parameter vector $p(t)$, the time optimal control makes more sense as it brings the system sufficiently closer to the system's equilibrium before a significant parameter change may possibly take the system farther away. The time varying property of the uncertain parameter vector $p(t)$ make the neural control behave differently that in the case of uncertain but fixed parameter vector in the sense that once the sub-region is identified to contain the unknown fixed parameter vector, both the lower-level and upper-level neural networks corresponding to the the vertices of the sub-region are activated and will remain active; yet in the case of time varying parameter vector, one sub-region is identified to contain the unknown parameter vector for a sustained period of time yet it is moving to another sub-region afterwards and keeps moving to yet another different sub-region. In other words, different sets of lower-level and upper-level neural networks are activated from time to time so far as time varying parameter vector is concerned. To achieve better stability, a new procedure is introduced: for each sub-region, identify the center point of the sub-region, obtain the optimal control and state trajectories corresponding to this center point. Add a lower-level neural network and a upper-level neural network. The operation of this extra pair of neural networks is as follows:

1. Under normal operation of the hierarchical neural network, the upper-level coordinator has no difficulty in deciding on the control direction (either positive or negative);
2. Under some circumstances where the upper-level coordinator is not so sure of the control direction to take, the extra pair of neural networks introduced above come into the picture and decide on the control direction to follow.

5 Near Optimal Adaptive Neural Controller

This section presents the main theoretical result to support the near optimal adaptive control scheme developed in the last section to deal with the time varying parameters. The main results for the control problem with unknown fixed parameters are presented in [9] and have concluded:

1. The switching manifold for a particular control problem (P_0) can be modeled by a neural network with a sufficiently small error in the L_1 sense; and the lower-level neural controllers can be constructed using the optimal control and state trajectories. For the tessellation of the parameter space, for any sub-region, as long as the unknown parameter vector is close enough to one of the vertices in the sub-region, the hierarchical neural control determined switching-vector and control are close enough to their optimal counterparts.
2. A hierarchical neural network can be utilized to approximate the dynamic behavior of the system represented by the control problem (P).

These results form a base upon which the adaptive control for the systems with time varying parameters is addressed.

To address the near optimal adaptive control for the control problem (P) with the time-varying parameter vector $p(t)$, we present the following result.

Proposition 1. *For the control problem (P) with the assumptions AS1 through AS8 satisfied, suppose Ω is a compact region where with proper control the optimal trajectories starting in the compact region will still remain in it. Let $\hat{p}$ designate the estimate of $p(t)$. Then for any $\epsilon_1 > 0$, there exist $\epsilon_2 > 0$ such that if $||\hat{p} - p(t)|| < \epsilon_2$, then $||x(x_0, \hat{p}, t) - x(x_0, p, t)|| < \epsilon_1$ where $x(x_0, \hat{p}, t)$ is the state trajectory starting from x_0 for the control problem (P_0) with the parameter vector $\hat{p}$ and $x(x_0, p(t), t)$ is the state trajectory starting from x_0 for the control problem (P_0) with the time varying parameter $p(t)$.*

Proof: In the following, it will be shown that a bounded error involved in the identification of the parameters only results in a bounded deviation from the desired trajectory. Here, $\dot{x} = a(x) + C(x)p(x) + B(x)u$ where p is not fixed.

Define the approximation error of $p(t)$ as $e_p = \hat{p} - p$. It follows that $\dot{x} = a(x) + C(x)(\hat{p} + e_p) + B(x)u$.

Note that the optimal control can be obtained for $\dot{x} = a(x) + C(x)\hat{p} + B(x)u$ through the method described before.

With the given initial condition $x(t_0) = x_0$, we have, by integration of the above two equations from t_0 to t, $x_1(t) = x_1(t_0) + \int_{t_0}^{t} [a(x_1(s)) + C(x_1(s))(\hat{p}(x_1(s)) + e_p) + B(x_1(s))u(s)]ds$, and $x_2(t) = x_2(t_0) + \int_{t_0}^{t} [a(x_2(s)) + C(x_2(s))\hat{p}(x_2(s)) + B(x_2(s))u(s)]ds$.

By noting that $x_1(t_0) = x_2(t_0) = x_0$, subtraction of the above two equations yields

$$\begin{aligned} x_1(t) - x_2(t) = \int_{t_0}^{t} \{ & a(x_1(s)) - a(x_2(s)) + C(x_1(s))e_p + C(x_1(s))\hat{p}(x_1(s)) \\ & - C(x_2(s))\hat{p}(x_2(s)) + [B(x_1(s)) - B(x_2(s))]u(s)\}ds \end{aligned} \tag{2}$$

Note that, by Taylor's theorem, $a(x_1(s)) - a(x_2(s)) = a_T(x_1(s) - x_2(s))$, $B_j(x_1(s)) - B_j(x_2(s)) = B_{T,j}(x_1(s) - x_2(s))$,
and $C(x_1(s))\hat{p}(x_1(s)) - C(x_2(s))\hat{p}(x_2(s)) = C_T(x_1(s) - x_2(s))$ where $a_T = \frac{\partial a(x)}{\partial x}|_{x=\eta x_1 + (1-\eta)x_2}$ $(0 < \eta < 1)$,
and $B_{T,j} = \frac{\partial B_j(x)}{\partial x}|_{x=\xi_j x_1 + (1-\xi_j)x_2}$ for $j = 1, 2, \cdots, m$ $(0 < \xi_j < 1)$,
and $C_T = \frac{\partial C(x)\hat{p}(x)}{\partial x}|_{x=\mu x_1 + (1-\mu)x_2}$ $(0 < \mu < 1)$.

Define $\Delta x(t) = x_1(t) - x_2(t)$. Then we have $\Delta x(t) = \int_{t_0}^{t} C(x_1(s))e_p ds + \int_{t_0}^{t} [a_T(x(s))\Delta x(s) + C_T \Delta x(s) + \sum_{j=1}^{m} B_{T,j}(x(s))\Delta x(s)u_j(s)]ds$.

If the appropriate norm of both sides of the above equation is taken and the triangle inequality is applied to it, the following result is obtained: $||\Delta x(t)|| \leq \int_{t_0}^{t} ||C(x_1(s)e_p||ds + || \int_{t_0}^{t} [a_T(x(s))\Delta x(s) + C_T \Delta x(s) + \sum_{j=1}^{m} B_{T,j}(x(s))\Delta x(s)u_j(s)]ds||$.

Note that e_p is uniformly bounded (i.e., $|e_p| < \epsilon_2$), $|u_j(t)| \leq 1$, $||a_T|| = \sup_{x \in \Omega} a_T(x) < \infty$, $||B_{T,j}|| = \sup_{x \in \Omega} B_T(x) < \infty$,
and $||C_T|| = \sup_{x \in \Omega} C_T(x) < \infty$.

It follows that

$$||\Delta x(t)|| \leq ||C||\epsilon_2(t-t_0) + \int_{t_0}^{t} ||[a_T(x(s))\Delta x(s) + C_T \Delta x(s) + \sum_{j=1}^{m} B_{T,j}(x(s))\Delta x(s)u_j(s)]||ds$$

$$\leq ||C||\epsilon_2(t-t_0) + (||a_T|| + ||C_T|| + \sum_{j=1}^{m} ||B_{T,j}||) \int_{t_0}^{t} ||\Delta x(s)||ds \quad (3)$$

Application of *Gronwall-Bellman* Inequality yields

$$||\Delta x(t)|| \leq ||C||\epsilon_2(t-t_0) + \int_{t_0}^{t} (||a_T|| + ||C_T|| + \sum_{j=1}^{m} ||B_{T,j}||)\epsilon_2||C||(s-t_0) \times \exp\{\int_{s}^{t} (||a_T|| + ||C_T|| + \sum_{j=1}^{m} ||B_{T,j}||)d\sigma\}ds$$

$$\leq ||C||\epsilon_2(t-t_0) + \epsilon||C||(||a_T|| + ||C_T|| + \sum_{j=1}^{m} ||B_{T,j}||)\frac{(t-t_0)^2}{2} \times \exp\{(||a_T|| + ||C_T|| + \sum_{j=1}^{m} ||B_{T,j}||)(t-t_0)\} \leq K\epsilon_2 \quad (4)$$

where $K = (t-t_0)||C||[1 + (||a_T|| + ||C_T|| + \sum_{j=1}^{m} ||B_{T,j}||)\frac{(t-t_0)}{2} \exp\{(||a_T|| + ||C_T|| + \sum_{j=1}^{m} ||B_{T,j}||)(t-t_0)\}$, and $K < \infty$, for all $t \in [t_0,\ t_f]$.

Note that K is a monotonous increasing function of t. Let $K_0 > K(t_f)$. Choose $\epsilon_2 = \epsilon_1/K_0$. It follows immediately that $||\Delta x(t)|| < K_0\epsilon_2 = \epsilon_1$.

This completes the proof.

Remark 4: For the original control problem (P) with the unknown time-varying parameter vector located within one of these sub-regions, say $\Omega_{p,k}$, the vertices of $\Omega_{p,k}$ and the desired system control and dynamic behaviors of the corresponding control problems (P_0) are available and can be utilized to construct the controller for the control problem (P). When the diameter of the sub-region $\Omega_{p,k}$ tends to zero, as long as $p(t)$ is correctly classified as within its sub-region $\Omega_{p,k}$ (note that this does not mean that $p(t)$ is identified with zero identification error), then $\hat{p}$ is within the same correct $\Omega_{p,k}$. Consequently, $||p(t) - \hat{p}||$ tends to zero. The above theorem guarantees that the resulted state trajectory for the control problem (P) with fixed $\hat{p}$ is sufficiently close to that of the control problem (P) with the time-varying parameter vector $p(t)$. In a parallel effort, [9] states that with the designed controller the system is controlled to achieve near optimal control performance. The theoretical result presented in this paper further states that for the nonlinear system with time-varying parameters, the system performance is reasonably close to that achievable for a corresponding system with successive estimates $\hat{p}$ of $p(t)$, which is controlled like $\hat{p}$ is fixed as the exact identification of $\hat{p}$ is not needed and only the sub-region which $\hat{p}$ and $p(t)$ are located within is identified. Therefore, the adaptive control of the time-varying nonlinear systems achieves a reasonably satisfactory level of near optimal control performance.

6 Simulation Study

The SMIB system is used for simulation and is briefly described as follows due to page limitations (The reader is referred to [5] for details):

$$\begin{cases} \dot{\delta} = \omega_b(\omega - 1) \\ \dot{\omega} = \frac{1}{M}(P_m - P_c - (D + D_c)(\omega - 1) - \frac{V_t V_\infty}{X_d + (1-s)Xe} \sin\delta) \\ y = \delta \end{cases} \quad (5)$$

where $P_m = 0.3665$, $D = 2.0$, $M = 3.5$, $V_t = 1.0$, $V_\infty = 0.9$, $X_d = 2.0$, $X_e = 0.35$, $s \in [s_{min} = 0.2,\ s_{max} = 0.75]$ with $s = s_e = 0.4$ at the equilibrium; P_c and D_c are unknown with D_c time varying, and y is the system output.

Following the proposed neural control design procedures, all the lower-level and upper-level neural networks are trained properly. When finished with training, they work together to act as a near time optimal neural controller. This adaptive neural controller is examined for a severe short-circuit fault for an unknown time-varying load ($P_l = P_m \times 35\% + D \times r \times \omega$) where r varies with time (for illustration purpose, r takes small step changes starting from 15%;

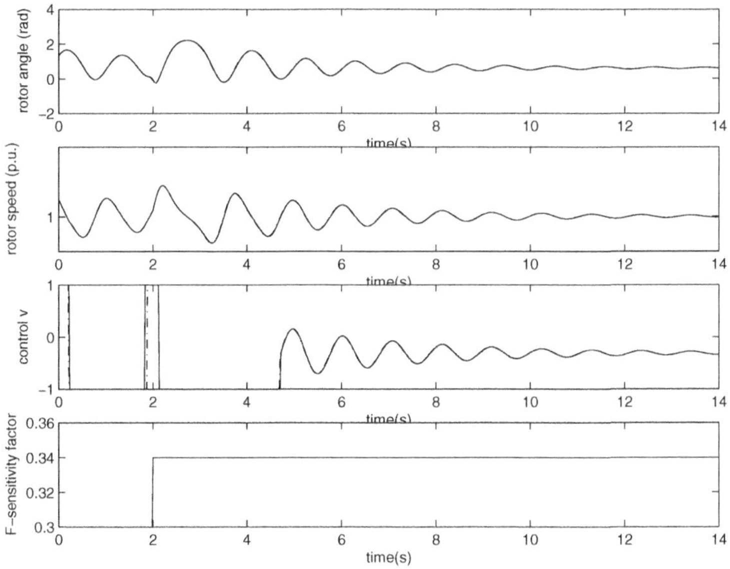

Fig. 1. Performance of the adaptive neural controller for the case of an unknown time varying parameter vector; solid—the resulting trajectories from the neural controller; dashed—the optimal trajectories; F-sensitivity factor represents the frequency-sensitivity factor D_c

and $D_c = D \times r$), which constitutes a non-nominal case from which none of the corresponding optimal control and output data has been used for training the nominal neural controllers and the upper-level neural networks. The resulting output and control trajectories are shown in Fig. 1 along with the off-line calculated optimal trajectories. It is observed that the adaptive neural controller achieves near optimal control performance.

7 Conclusions

Near optimal adaptive control is studied for a class of nonlinear systems with unknown,time-varying parameters and multiple inputs. The design procedures proposed in this paper enables one to simultaneously consider complex control performance measures and physical constraints. The proposed design approach is backed up by a developed theory to particularly deal with the time-varying property of nonlinear systems, which is viewed as an advancement of a parallel effort focused on the control of nonlinear systems with unknown, fixed parameters. The system dynamics are utilized in the process of the proposed neural control design, in particular, the construction of lower-level neural network based controllers and the upper-level neural network based coordinators. The control design is illustrated on a SMIB system with an unknown, time-varying load. The simulation results indicate that the proposed design methodology is effective.

References

1. Chen, D., Mohler, R., Chen, L.: Neural-Network-Based Adaptive Control with Application to Power Systems. Proc. 1999 American Control Conf., San Diego (1999) 3236–3240
2. Chen, D., Mohler, R.: Nonlinear Adaptive Control with Potential FACTS Applications. Proc. 1999 American Control Conf., San Diego (1999) 1077–1081
3. Chen, D., Mohler, R.: The Properties of Latitudinal Neural Networks with Potential Power System Applications. Proc. 1998 American Control Conf., Philadelphia (1998) 980–984
4. Chen, D., Mohler, R., Chen, L.: Synthesis of Neural Controller Applied to Power Systems. IEEE Trans. Circuits and Systems I **47** (2000) 376–388
5. Chen, D.: Nonlinear Neural Control with Power Systems Applications. Ph.D. Dissertation, Oregon State University (1998)
6. Chen, D., Mohler, R., Shahrestani, S., Hill, D.: Neural-Net-Based Nonlinear Control for Prevention of Voltage Collapse. Proc. 38th IEEE Conference on Decision and Control, Phoenix (1999) 2156–2161
7. Chen, D., Mohler, R.: Theoretical Aspects on Synthesis of Hierarchical Neural Controllers for Power Systems. Proc. 2000 American Control Conference, Chicago (2000) 3432–3436
8. Chen, D., Yang, J.: Robust Adaptive Neural Control Applied to a Class of Nonlinear Systems. Proc. 17th IMACS World Congress: Scientific Computation, Applied Mathematics and Simulation, Paris (2005) T5-I-01-0911
9. Chen, D., Yang, J., Mohler, R.: On Near Optimal Neural Control of a Class of Nonlinear Systems with Multiple Inputs, Int. J. Neural Computing and Applications (in press)

Constrained Control of a Class of Uncertain Nonlinear MIMO Systems Using Neural Networks

Dingguo Chen[1] and Jiaben Yang[2]

[1] Siemens Power Transmission and Distribution Inc., 10900 Wayzata Blvd., Minnetonka, Minnesota 55305, USA
[2] Department of Automation, Tsinghua University, Beijing, 100084, People's Republic of China

Abstract. This paper attempts to present a neural inverse control design framework for a class of nonlinear multiple-input multiple-output (MIMO) system with uncertainties. This research effort is motivated by the following considerations: (a) An appropriate reference model that accurately represents the desired system dynamics is usually assumed to exist and to be available, and yet in reality this is not the case often times; (b) In real world applications, there are many cases where controls are constrained within a physically allowable range, which presents another layer of difficulties to directly apply the reference model based inverse control; (c) It is difficult to consider optimal control even for the reference model as in general the analytic solution to the optimal control problem is not available. The simulation study is conducted on a single-machine infinite-bus (SMIB) system to illustrate the proposed design procedure and demonstrates the effectiveness of the proposed control approach.

1 Introduction

Several studies have been conducted to address the control of nonlinear uncertain systems using hierarchical neural networks [2], [13], [11], [15], [14]. In these efforts, the system state information is assumed available and used to construct state-feedback hierarchical neural controllers. There are situations where the system state information is not available. This is the focus of this paper which is devoted to the development of a design that utilizes the system outputs to construct output-feedback hierarchical neural controllers to achieve adaptive control of unknown systems.

In recent years, remarkable progresses have been witnessed in constructing the so-called adaptive controllers for uncertain nonlinear systems with the employment of neural networks. Neural adaptive control has enhanced the traditional feedback control [8]. In particular, when the plant under study is unknown, identification of the plant together with the controller based on neural networks constitutes so-called internal model control scheme [7]. Numerous studies have

D. Liu et al. (Eds.): ISNN 2007, Part I, LNCS 4491, pp. 193–202, 2007.

been conducted on identification and control of nonlinear dynamical systems with uncertainties and disturbances. When the plant is known, a direct inverse control can be applied [5]. Apart from the extensive application of adaptive neural control design in solving real world problems, several studies have been performed to gain better understanding of the theoretical issues involved in the adaptive neural control design and attempt to provide theoretical foundations that are essential for the efficient design of neural networks controllers based on inverse control [6],[9].

The drawbacks of conventional adaptive neural control include (a) a reference model exists and is available for use; (b) additional performance criteria (e.g., minimum time control) are difficult to incorporate; (c) physical constraints are difficult to incorporate (e.g., for many practical systems, the control effort is limited to a physically allowable range). A different approach is taken for the neural inverse control design that utilizes off-line generated optimal trajectories and takes into account different nominal cases that altogether approximately represent desired system behaviors. The scope of this paper is to study a specific class of nonlinear systems 1, as shown in Section 2.

In this paper, a neural inverse control methodology will be adopted to stabilize a class of uncertain systems with multiple inputs and multiple outputs. The neural inverse control is attempted to adjust the amount of control effort based on the system outputs so that the system outputs track respective desired trajectories. To empower the neural controller with the capability of self adjustment, hierarchical neural networks will be constructed (the reader is referred to [13] for a detailed description of hierarchical neural networks).

The hierarchy of the so-called hierarchical neural networks features a two-tier neural network based architecture. The lower-level neural networks correspond to individual nominal cases where the desired control and output trajectories are utilized to construct a neural inverse controller for the corresponding nominal case. The upper-level neural networks coordinates the control efforts contributed by individual nominal neural inverse controllers. The outstanding features of the proposed control design approach include: (a) attempting to achieve desired control performance even with parameter uncertainties; (b) eliminating the need for a popular parameter estimator in many adaptive control designs; (c) physical constraints are respected.

This paper is organized as follows. In Section 2, the control problem is formally presented with several conventional assumptions made. The time optimal control and the Switching-Times Variation method (STVM) are briefly discussed. In Section 3, the neural network based inverse control design is provided. The optimal control and output trajectories are used to establish a mapping from the space of the output and its derivatives of up to a certain order to the space of the input. To show how a hierarchical neural inverse controller can be constructed, a simulation study is presented in Section 4 on a popular single-machine infinite-bus power system. Finally, some conclusions are presented in Section 5.

2 Problem Formulation

The following MIMO system, typifying many practical nonlinear dynamical systems (e.g., power systems [13]), is considered in this paper.

$$\begin{aligned} \dot{x} &= a(x) + C(x)p + B(x)u \\ y &= h(x) \end{aligned} \tag{1}$$

where $x \in G \subseteq R^n$ is the state vector, $p \in \Omega_p \subset R^l$ is a fixed but unknown bounded parameter vector, $u \in R$ is the control variable, $y \in R^{n_y}$ is the output vector, $a(x) = [a_1(x)\, a_2(x)\, \cdots\, a_n(x)]^\tau$ is an n-dimensional vector function of x, $C(x) = \begin{bmatrix} C_{11}(x)\, C_{12}(x) \cdots C_{1l} \\ C_{21}(x)\, C_{22}(x) \cdots C_{2l} \\ \cdots \quad \cdots \quad \cdots\cdots \\ C_{n1}(x)\, C_{n2}(x) \cdots C_{nl}(x) \end{bmatrix}$ is an n-dimensional vector function of x, and $B(x) = \begin{bmatrix} B_{11}(x)\, B_{12}(x) \cdots B_{1m} \\ B_{21}(x)\, B_{22}(x) \cdots B_{2m} \\ \cdots \quad \cdots \quad \cdots\cdots \\ B_{n1}(x)\, B_{n2}(x) \cdots B_{nm} \end{bmatrix}$ is an $n \times m$-dimensional matrix function of x.

To facilitate the analysis of the system (1) and synthesis of a desired controller, a few conventional assumptions are made similarly to those made in [3].

AS1: It is assumed that $a(.)$, $C(.)$ and $B(.)$ have continuous partial derivatives with respect to the state variables on the region of interest. In other words, $a_i(x)$, $C_{is}(x)$, $B_{ik}(x)$, $\frac{\partial a_i(x)}{\partial x_j}$, $\frac{\partial C_{is}(x)}{\partial x_j}$, and $\frac{\partial B_{ik}(x)}{\partial x_j}$ for $i, j = 1, 2, \cdots, n$; $k = 1, 2, \cdots, m$; $s = 1, 2, \cdots, l$ exist and are continuous and bounded on the region of interest.

It should be noted that the above conditions imply that $a(.)$, $C(.)$, and $B(.)$ satisfy the Lipschitz condition which in turn implies that there always exists a unique and continuous solution to the differential equation given an initial condition $x(t_0) = \xi_0$ and a bounded control $u(t)$.

AS2: In practical applications, control effort is usually confined due to the limitation of design or conditions corresponding to physical constraints. Without loss of generality, assume that control vector u is confined to the following admissible control set U:

$$U = \{u : |u_i| \leq 1, i = 1, 2, \cdots, m\} \tag{2}$$

where u_i is u's ith component.

AS3: The system is controllable.

AS4: The control performance criteria is $J = \int_{t_0}^{t_f} [a_0(x(s)) + b_0^\tau(x(s))u(s)]ds$ where t_0 and t_f are the initial time and the final time, respectively, and $a_0(.)$ and $b_0(.)$ are continuous.

AS5: The target set θ_f is defined as $\theta_f = \{x : \Psi(x(t_f)) = 0\}$ where Ψ_i's $(i = 1, 2, \cdots, q)$ are the components of $\Psi(.)$.

AS6: Each component y_i $(1 \leq i \leq y_{n_y})$ of y along with its derivatives up to the $n-1$th order forms a new local coordinates.

An optimal solution is sought to the following control problem:

(P) The optimal control problem consists of the system equation (1) with fixed and known parameter vector p, the initial time t_0, the variable final time t_f, the initial state $x_0 = x(t_0)$, together with the assumptions AS1, AS2, AS3, AS4, AS5 satisfied such that the system state conducts to a pre-specified terminal set θ_f at the final time t_f while the control performance index is minimized.

It is noted that the system of interest has an unknown fixed parameter vector, which is bounded as stated in the assumptions. The space spanned by the bounded parameter vector can be tessellated into non-overlapping convex sub-regions. The tessellation can be conducted down to a granularity as small as desired. The parameter vector at each vertex point of individual sub-regions is apparently known. Therefore, the original optimal control problem consisting of the system equation (1) with unknown fixed parameter vector is decomposed into a number of optimal control problems consisting of the system equation (1) with fixed and known parameter vector.

It can be shown that the existence of the solution to the above optimal control problem is guaranteed. The analytic solution, however, is usually unobtainable due to the nature of the problem and the complexity.

Thanks to the Switching-Times Variation Method (STVM), the approximate, numerical solution is obtainable. Simply speaking, the optimal switching vector is approached iteratively based on a gradient method or one of its numerous variations.

Remark 1: According to the previous studies [13], [15], [14], [16], a state-feedback neural controller can be constructed to achieve near optimal adaptive control. In the context of this paper, it implies that each component of the control vector can be designed as a state-feedback neural controller. With the assumption AS6, a non-singular mapping exists between the system state and any of the system outputs and the corresponding derivatives up to the $n-1$th order. The existence of these mappings is utilized whereas these non-singular mappings (particularly from the system output and its derivatives to the system state) are usually complicated and even intractable and are not of this paper's interest. Their existence implies that each component of the control vector can be constructed as a neural controller with the system output and its derivatives up to the $n-1$th order. This is important since the system state is not measureable in real time and only the system output information is available in the construction of the neural controllers.

3 Hierarchical Neural Inverse Control Design

The hierarchical neural inverse control design comprises several major components as depicted in 1.

The designed hierarchical neural controller has to go through the off-line training using the off-line training data and performance evaluation using the off-line

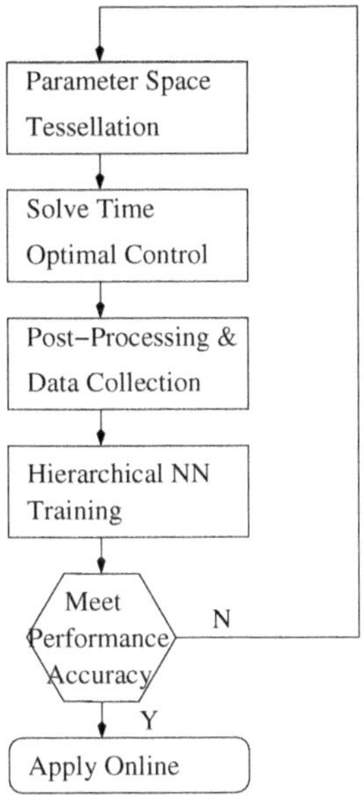

Fig. 1. Hierarchical neural inverse control design procedures

validation data. If the evaluation result is satisfactory, the hierarchical neural controller can be applied on-line.

The four major design steps included in the above diagram reflect several important features of the proposed design methodology. Each design step is associated with a different phase of the design, deals with a different technical problem, and yet works together to assure that the design is technically sound and practically realizable.

Parameter Space Tessellation: The parameter space is divided into non-overlapped convex sub-regions. After tessellation, identify one of the sub-regions and its vertices. The tessellation can be refined for better control performance.

Solve Time Optimal Control: For each fixed parameter vector (i.e., one of vertices identified in the previous design step), employ the STVM to obtain the numerical solution to the corresponding optimal control problem.

Post-Processing & Data Collection: The optimal state trajectories are used to calculate the optimal output trajectories and the trajectories of the output's derivatives of up to the $n-1$th order. All the optimal trajectories are evenly sampled at a sample rate that is sufficient for the control of the studied system. The derived optimal trajectories for the output and its derivatives are sorted in such a manner that they correspond to the optimal control trajectories and the parameter vector.

Hierarchical NN Training: The optimal output (and its derivatives) trajectories are used as the inputs to the lower-level neural networks, and the corresponding optimal control trajectories are used as the outputs. For the upper-level neural network training, the training patterns will be formed to include the system output and its previous values, the system control input and its previous values, and the indicator whether a particular subset of training patterns formed for lower-level neural network training is used for the corresponding upper-level neural network - if so, 1 is assigned, otherwise, 0 is assigned.

4 A Simulation Study

The SMIB system with both frequency-sensitive and frequency-insensitive load components, is used for simulation. This SMIB system is described by the following equation.

$$\begin{cases} \dot{\delta} = \omega_b(\omega - 1) \\ \dot{\omega} = \frac{1}{M}(P_m - P_c - (D + D_c)(\omega - 1) - \frac{V_t V_\infty}{X_d + (1-s)Xe}\sin\delta) \\ y = \delta \end{cases} \tag{3}$$

where
δ - rotor angle (rad);
ω - rotor speed (p.u.);
$\omega_b = 2\pi \times 60$ - synchronous speed as base (rad/sec);
$P_m = 0.3665$ - mechanical power input assumed to be constant (p.u.);
P_c - the frequency-insensitive fixed, unknown component of an unknown load P_l (p.u.);
D_c - the fixed, unknown frequency-sensitivity factor of the unknown load P_l, relating to the frequency-sensitive load component;
$D = 2.0$ - damping factor;
$M = 3.5$ - system inertia referenced to the base power;
$V_t = 1.0$ - terminal bus voltage (p.u.);
$V_\infty = 0.9$ - infinity bus voltage (p.u.);
$X_d = 2.0$ - transient reactance of the generator (p.u.);
$X_e = 0.35$ - transmission reactance (p.u.);
$s \in [s_{min} = 0.2, \ s_{max} = 0.75]$ - series compensation degree ($-sXe$ is the reactance of the TCSC, and often $0 < s < 1$);
y - the system output.
The system is desired to be driven, after a transient period, to its equilibrium $(\delta_e, \ \omega_e)$ by the admissible control $s \in [s_{min}, \ s_{max}]$ and stay in the equilibrium thereafter by the fixed compensation $s_e = 0.4 \in [s_{min}, \ s_{max}]$. Note that the equilibrium differs for different load specification.

The above equation can be transformed to the following equation:

$$\begin{cases} \dot{\delta} = \omega_b\omega \\ \dot{\omega} = \frac{1}{M}(P_m - P_c - (D + D_c)\omega - (V_t V_\infty)(Y_0 + Y_a u)\sin\delta) \\ y = \delta \end{cases} \tag{4}$$

where Y_0 and Y_a are computable constants.

Through additional some algebra, it can be readily shown that the above equation can be converted to the following:

$$\dot{x} = a(x) + b(x)v$$
$$y = x_1 \tag{5}$$

where $x = [x_1 \ \ x_2]^\tau = [\delta \ \ \omega]^\tau$, $a(x) = [\omega_b\omega \ \ c_1 - c_{10} - c_{20}\omega - c_3\sin(\delta_e + \delta)]^\tau$, $b(x) = [0 \ \ c_4\sin(\delta_e + \delta)]^\tau$, $v \in [-1, \ \ 1]$; c_1,c_3,c_4 and δ_e are computable constants, and c_{10} and c_{20} are unknown constants that relate to the load.

Note that the above equation can be rewritten to conform to the form of the system Eq. (1) by decomposing $a(x)$ to a component, relating to c_{10} and c_{20}, and the remainder that is irrelevant to the unknown parameters c_{10} and c_{20}.

4.1 Minimal Time Control

Consider Eq. (4) for minimal time control. The optimal time performance index can be expressed as $J(t_0) = \int_{t_0}^{T} 1dt$. Define the Hamiltonian function as

$$H(x, u, t) = 1 + \lambda^\tau f \tag{6}$$

where $x^\tau = [\delta \ \ \omega]$; $\lambda^\tau = [\lambda_1 \ \ \lambda_2]$; and $f(x, u, t)^\tau = [\omega_b\omega \ \ \frac{1}{M}(P_m - P_c - (D + D_c)\omega - (V_t V_\infty(Y_0 + Y_a u)\sin\delta))]$.

The final-state constraint is $\Psi(x(T), T) = x(T) - x_e = 0$ where $x_e^\tau = [\delta_e \ \ \omega_e]$ is the desired equilibrium point.

The costate equation can be written as

$$\begin{cases} \dot{\lambda_1} = \frac{1}{M} V_t V_\infty (Y_0 + Y_a u)\lambda_2 \cos\delta \\ \dot{\lambda_2} = -\omega_b\lambda_1 + \frac{D + D_c}{M}\lambda_2 \end{cases} \tag{7}$$

Applying the Pontryagin minimum principle [4] yields the time-optimal control as follows:

$$u^* = \begin{cases} u_{max}, & \lambda_2 \sin\delta > 0 \\ u_{min}, & \lambda_2 \sin\delta < 0 \end{cases} \tag{8}$$

Note that the possibility of a singular solution, i.e., $\lambda_2(t)\sin\delta(t) \equiv 0$ for some finite time interval, can be excluded, which has been proved in [13].

The boundary condition be given by $(\Psi_t^T \mu + H)|_T = 0$ which in turn gives

$$\lambda_2(T) = -\frac{M}{P_m - P_c - V_t V_\infty (Y_0 + Y_a u(T))\sin\delta(T)} \tag{9}$$

4.2 The Switching Time Variation Method (STVM)

It is observed that the system described by Eq. (5) is a nonlinear system but linear in control. Since there only exist non-singular optimal controls for this system, the requirements for the STVM's application can then be met. The optimal switching-time vector can be obtained by using a gradient-based method. The convergence of the STVM is guaranteed if there are no singular solutions. Details are available in [1], [3].

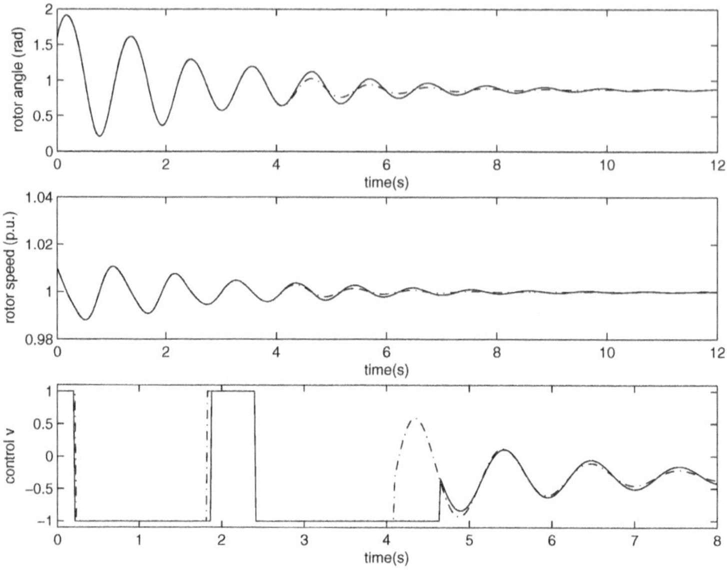

Fig. 2. Performance of the hierarchical neural inverse controller for SMIB with an unknown load after experiencing a short-circuit fault; solid—the resulting trajectories from the neural inverse controller; dashed—the optimal trajectories

4.3 Construction of Hierarchical Neural Inverse Control

Since the range of the uncertain load is known, the nominal cases can be set up by setting P_c and D_c to different sets of values. For instance, P_c may be set as 0, $P_m \times 10\%$, $P_m \times 20\%$, till its maximum possible value, say $P_m \times r_P\%$; and D_c may be set as 0, $D \times 10\%$, $D \times 20\%$, till its maximum possible value, say $D \times r_D\%$. For each different combination of (P_c,D_C), the optimal control and output trajectories can be computed numerically.

Note that the system under study automatically meets the assumption AS6. Therefore, $y(k)$ and $y(k-1)$ will be used as part of the inputs fed to both lower-level and upper-level neural networks. For each nominal case, and form the training patterns as $(y(k-1), y(k); u(k))$ which are employed to training a lower-level nominal neural inverse controller. For the upper-level neural network training, the training patterns are forms as such: $(y(k-1), y(k), u(k); 1)$ if $(y(k-1), y(k); u(k))$ is a training pattern for the corresponding lower-level neural network; $(y(k-1), y(k), u(k); 0)$ if $(y(k-1), y(k); u(k))$ is not a training pattern for the corresponding lower-level neural networks, but is a training pattern for one of the rest lower-level neural networks.

After the completion of the off-line training, the hierarchical neural inverse controller is examined for a severe short-circuit fault for an unknown load ($P_l = P_m \times 35\% + D \times 15\% \times \omega$ or in other words, $P_c = P_m \times 35\%$ and $D_c = D \times 15\%$), which constitutes a non-nominal case from which none of the corresponding optimal control and output data has been used for training the nominal neural inverse controllers and the upper-level neural networks. The resulting output and control trajectories are shown in Fig. 2 along with the off-line calculated optimal trajectories. It is observed that the hierarchical neural inverse controller achieves near optimal control performance.

5 Conclusions

A neural network based inverse control design has been presented in this paper in an aim to control a class of nonlinear, dynamical systems with multiple inputs and multiple outputs that include uncertain components. Compared to the existing techniques, the proposed design respects the reality that control constraints due to physical limitations have to be met, and yet allows for the synthesis of a controller to meet pre-designated control performance criteria, such as stability, minimum time control, adaptive control to handle system parametric uncertainties. The major procedures for construction of an adaptive controller for the nonlinear MIMO's include the following: (a) Identify a number of nominal cases that, combining together, approximately represent the system dynamical behaviors; (b) Determine the optimal control for systems without uncertainties; (c) Utilize effective numerical methods to obtain optimal control and optimal output trajectories; (d) Train the hierarchical neural inverse controller based on the calculated optimal trajectories.

Furthermore, a key assumption (i.e., AS6) was made to justify the use of the system output and its derivatives in the training of hierarchical neural networks. This assumption may be loosened and instead the system's vector relative degree may be used in the establishment of the mappings between the system outputs and system inputs.

The proposed design methodology was applied on a single-machine infinite-bus power system with a load that comprises unknown frequency insensitive component and unknown frequency sensitive component. The simulation results indicate that the control design presented in this paper is effective and can be applied to a class of uncertain nonlinear dynamical systems.

References

1. Mohler, R.R.: Bilinear Control Processes. Academic Press, New York (1973)
2. Zakrzewski, R.R., Mohler, R.R., Kolodziej, W.J.: Hierarchical Intelligent Control with Flexible AC Transmission System Application. IFAC J. Control Engineering Practice **2** (1994) 979–987
3. Moon, S.F.: Optimal Control of Bilinear Systems and Systems Linear in Control. Ph.D. Dissertation, The University of New Mexico (1969)
4. Lee, E.B., Markus, L.: Foundations of Optimal Control Theory. Wiley, New York (1967)
5. Deng, H., Li, H.: A Novel Neural Approximate Inverse Control for Unknown Nonlinear Discrete Dynamical Systems. IEEE Trans. System, Man and Cybernetics Part B **35** (2005) 115–123

6. Cabrera, J., Narendra, K.: Issues in the Application of Neural Networks for Tracking Based on Inverse Control. IEEE Trans. Automatic Contr. **44** (1999) 2007–2027
7. Rivals, I., Personnaz, L.: Nonlinear Internal Model Control Using Neural Networks: Application to Processes with Delay and Design Issues. IEEE Trans. Neural Networks **11** (2000) 80–90
8. Widrow, B., Walach, E.: Adaptive Inverse Control. Prentice-Hall, Englewood Cliffs, New Jersey (1996)
9. Narendra, K., Mukhopadhyay, S.: Adaptive Control Using Neural Networks and Approximate Modes. IEEE Trans. Neural Networks **8** (1997) 475–485
10. Chen, D., Mohler, R.: Load Modelling and Voltage Stability Analysis by Neural Networks. Proc. American Control Conf., Albuquerque (1997)
11. Chen, D., Mohler, R., Chen, L.: Neural-Network-Based Adaptive Control with Application to Power Systems. Proc. American Control Conf., San Diego (1999) 3236–3240
12. Chen, D., Mohler, R.: Nonlinear Adaptive Control with Potential FACTS Applications. Proc. American Control Conf., San Diego (1999) 1077–1081
13. Chen, D.: Nonlinear Neural Control with Power Systems Applications. Ph.D. Dissertation, Oregon State University (1998)
14. Chen, D., Mohler, R., Shahrestani, S., Hill, D.: Neural-Net-Based Nonlinear Control for Prevention of Voltage Collapse. Proc. 38th IEEE Conference on Decision and Control, Phoenix (1999) 2156–2161
15. Chen, D., Mohler, R., Chen, L.: Synthesis of Neural Controller Applied to Power Systems. IEEE Trans. Circuits and Systems I **47** (2000) 376–388
16. Chen, D., Yang, J., Mohler, R.: On Near Optimal Neural Control of a Class of Nonlinear Systems with Multiple Inputs. To appear in Int. J. Neural Computing and Applications.

Sliding Mode Control for Missile Electro-hydraulic Servo System Using Recurrent Fuzzy Neural Network

Huafeng He, Yunfeng Liu, and Xiaogang Yang

Xi'an Research Inst. Of High-tech, Hongqing Town 710025, China
hhf0903@163.com

Abstract. The position tracking control of a missile electro-hydraulic servo system is studied. Since the dynamics of the system are highly nonlinear and have large extent of model uncertainties, such as big changes in parameters and external disturbance, a design method of sliding mode control (SMC) using recurrent fuzzy neural network (RFNN) is proposed. First a SMC system, which is insensitive to uncertainties including parameter variations and external disturbance, is introduced. Then, to overcome the problems with SMC, such as the assumption of known uncertainty bounds and the chattering phenomena in the control signal, an RFNN is introduced in conventional SMC. An RFNN bound observer is utilized to adjust the uncertainty bounds in real time. Simulation results verify the validity of the proposed approach.

1 Introduction

The electro-hydraulic servo system has been frequently used in the position servo system of a missile thanks to their capability of providing large driving forces or torques, rapid response and a continuous operation [1]. However, electro-hydraulic servo system inherently has many uncertainties and highly nonlinear characteristics, which results from the flow-pressure relationship, oil leakage, and etc. Furthermore, the system is subjected to load disturbances [2]. Consequently, the conventional control approaches based on a linearized model near the operating point of interest may not guarantee satisfactory control performance for the system. Since the variable structure control strategy using the sliding mode can offer many good properties, such as insensitivity to parameter variations, external disturbance rejection and fast dynamic response [3], SMC has been studied by many researchers for the control of electro-hydraulic servo system [4-6]. However, SMC may suffer from the main disadvantage associated with the chattering control input due to its discontinuous switching control used to deal with the uncertainties. The most commonly used method for attenuating the chattering control input is the boundary layer method [6]. The control input is smoother than that without using a boundary layer. However, its stability is guaranteed only outside of the boundary layer, and its tracking error is bounded by the width of boundary layer.

Recently, much research has been done on using RFNN to identify and control dynamic systems [7-9]. RFNN is a modified version of recurrent neural network, which use recurrent network to realize fuzzy inference. It is possible to train RFNN

D. Liu et al. (Eds.): ISNN 2007, Part I, LNCS 4491, pp. 203–212, 2007.

using the experience of human operators expressed in term of linguistic rules, and interpret the knowledge acquired from training data in linguistic form. Moreover, with its internal feedback connections, RFNN can temporarily store dynamic information and cope with temporal problems efficiently.

In this paper, a design method of SMC using RFNN is proposed. First a SMC system is introduced. Then, to overcome the problems with SMC, such as the assumption of known uncertainty bounds and the chattering phenomena in the control signal, an RFNN bound observer is utilized to adjust the uncertainty bounds in real time. The simulation results show the advantages of the approach.

2 Problem Statement and Design of Conventional SMC

For a kind of missile electro-hydraulic servo system, which is a typical electro-hydraulic position servo system [1], Fig.1 shows a structure diagram of missile electro-hydraulic servo system.

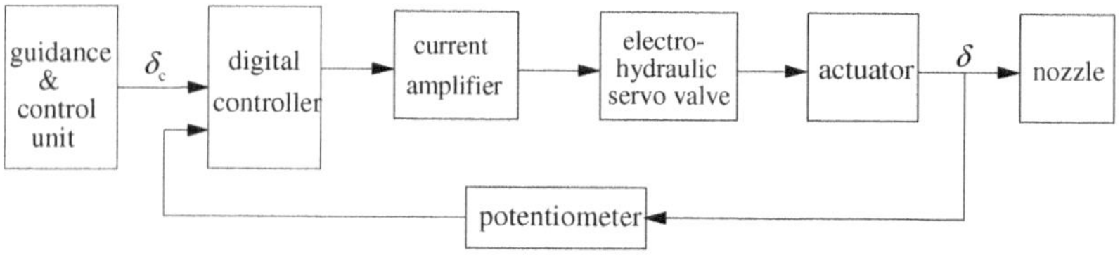

Fig. 1. Structure diagram of Missile electro-hydraulic servo system

The closed loop of control system is composed of a digital controller [10], a current amplifier, an electro-hydraulic servo valve, an actuator, and a potentiometer. The objective of the control is to generate the input current such that the angular position of the nozzle is regulated to the desired position. The piston position of the actuator is controlled as follows: Once the voltage input corresponding to the position input δ_c is transmitted to the digital controller, the input current is generated in proportion to the error between the voltage input and the voltage output from the potentiometer. Then the valve spool position is controlled according to the input current applied to the torque motor of the servo valve. Depending on the spool position and the load conditions of the piston, the rate as well as the direction of the flows supplied to each cylinder chamber is determined. The motion of the piston then is controlled by these flows, and then swing angle δ of the nozzle is achieved. At the same time, the piston is influenced by an external disturbance generated from the nozzle. The whole system dynamics model is given by the following derivation equations [1]

$$\begin{aligned} K_{ui} K_V K_Q u &= ARs\delta + p_L\left(\frac{V_T}{4B}s + K_{ce}\right) \\ ARP_L &= Is^2\delta + ns\delta + K_\delta\delta + M \end{aligned} \tag{1}$$

where K_{ui} —servo amplifier gain, K_V —servo valve gain, P_L —load pressure, K_Q —valve flow gain, A —pressure area in the actuator, R —effective torque arm of the linkage, V_T —effective system oil volume, $K_{ce} = C_e + K_c$ (C_e —leakage coefficient of cylinder, K_c —valve pressure gain), B —oil effective bulk modulus, n —coefficient of viscous friction, I —moment of inertia, M —load torque, K_δ —coefficient of position torque, u —input voltage, δ —swing angle of the nozzle, s —Laplace operator.

Choose system state: $\boldsymbol{X} = [x_1 \quad x_2 \quad x_3]^T = [\delta \quad \dot{\delta} \quad \ddot{\delta}]^T$, then the system state-space equation is

$$\begin{cases} \dot{x}_1 = x_2 \\ \dot{x}_2 = x_3 \\ \dot{x}_3 = f(\boldsymbol{X}) + gu - d \end{cases} \tag{2}$$

where $a_1 = \dfrac{4BK_\delta K_{ce}}{IV_T}$, $a_2 = \dfrac{4B(AR)^2 + 4BK_{ce}n + K_\delta V_T}{IV_T}$, $a_3 = \dfrac{n}{I} + \dfrac{4BK_{ce}}{V_T}$,

$$f(\boldsymbol{X}) = -a_1x_1 - a_2x_2 - a_3x_3, \quad g = \frac{4BAR}{IV_T}K_QK_VK_{ui}, \quad d = \frac{4BK_{ce}}{IV_T}M + \frac{1}{I}\cdot\frac{dM}{dt}$$

The parameters a_1, a_2, a_3, g, d are all uncertainties due to the variations of K_Q, B, C_e, K_{ui} and M. It is assumed that δ_d is the desired angle, and has up to 3rd derivative. All state variables are measurable and bound. The objective is to let the state vector $\boldsymbol{X}$ track $\boldsymbol{X}_d$ under the condition of parameter variations and external disturbances, where $\boldsymbol{X}_d = (\delta_d, \dot{\delta}_d, \ddot{\delta}_d)$. Define the tracking error $e_1 = x_1 - \delta_d$, and the error vector

$$\boldsymbol{e} = [e_1 \quad e_2 \quad e_3]^T = [e_1 \quad \dot{e}_1 \quad \ddot{e}_1]^T \tag{3}$$

In the conventional SMC design, we usually assume

$$a_i = a_{i0} + \Delta a_i, \quad g = g_0 + \Delta g \tag{4}$$

where a_{i0}, g_0 is the nominal parameters of a_i and g, and Δa_i, Δg is the model uncertainty. Let $\alpha_i(t)$, $\beta(t)$ and $r(t)$ are the upper bound function of Δa_i, Δg and d respectively, i.e. $|\Delta a_i| \le \alpha_i(t)$, $|\Delta g| \le \beta(t)$, $|d| \le r(t)$. Take

$$S(\boldsymbol{e}) = c_1e_1 + c_2e_2 + e_3 \tag{5}$$

where c_1, c_2 are constants and $\lambda^2 + c_2\lambda + c_1$ is Hurwitz polynomial. Then the sliding surface is

$$c_1e_1 + c_2e_2 + e_3 = 0 \tag{6}$$

Take the derivative of (**5**), set $\dot{S}(\boldsymbol{e}) = 0$, then the equivalent control can be obtained as

$$u_{equ} = g_0^{-1}[\sum_{i=1}^{3} a_{i0}x_i - c_1e_2 - c_2e_3 + \dddot{\delta}_d] \tag{7}$$

From (**2**) and (**7**), the control law is taken as

$$u = u_{equ} + u_N = u_{equ} + K\,\mathrm{sgn}(S) \tag{8}$$

where $K = -[\sum_{i=1}^{3} \alpha_i(t)|x_i| + \beta(t)|u_{equ}| + r(t) + \eta] \Big/ [g_0 - \beta(t)]$, $\mathrm{sgn}(S) = \begin{cases} 1, S > 0 \\ 0, S = 0 \\ -1, S < 0 \end{cases}$

From the analysis above, we get $S\dot{S} \le -\eta|S| < 0$, where $\eta > 0$ is constant. So under the control law (**8**), the sliding surface exists and is reachable. Since $\lambda^2 + c_2\lambda + c_1$ is Hurwitz polynomial, the sliding surface is stable.

3 Design of SMC Using RFNN

From (**8**) it can be seen that the undesirable control input chattering in the conventional SMC is caused by the discontinuous sign term $\mathrm{sgn}(S)$. The switching control law u_N which guarantees the reach ability and existence of the sliding mode is in proportion to the uncertainty bound including $\alpha_i(t)|x_i|$, $\beta(t)|u_{equ}|$ and $r(t)$. However, the bound of parameter variations and the external disturbance are difficult to know in advance for practical applications. Therefore, usually a conservative control law with large control gain K is selected. However, it will yield unnecessary deviations from the switching surface, causing a large amount of chattering.

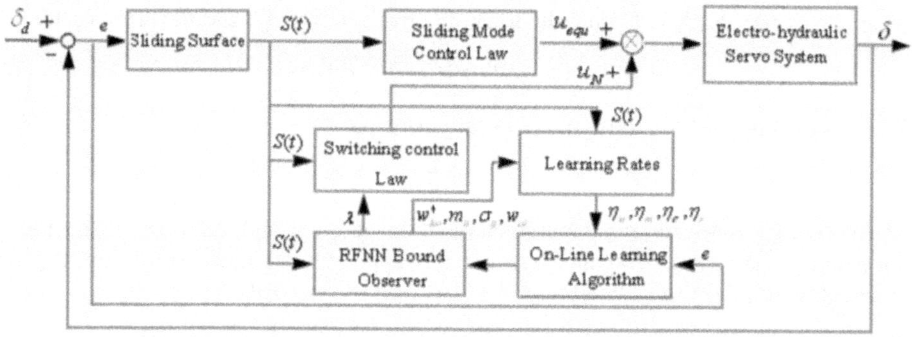

Fig. 2. Missile electro-hydraulic servo system with SMC using RFNN

Therefore, an RFNN is adopted in this study to facilitate adaptive switching control gain adjustment. The control block diagram of the SMC using RFNN is shown in Fig.2. The inputs of the RFNN are S and its derivative $\dot{S}$, and the output of the FNN is the substituted sliding switching control gain λ. The adjustment of λ is stop when the output error between the position command and the actual plant is zero. If the output error $e \to 0$ as $t \to \infty$, it implies S and $\dot{S} \to 0$ as $t \to \infty$. If the uncertainties are absent, once the switching surface is reached initially, a very small value of λ would be sufficient to keep the trajectory on the switching surface, and the amplitude of chattering is small. However, when the uncertainties are present, deviations from the switching surface will require a continuous updating of λ produced by the RFNN to steer the system trajectory quickly back into the switching surface. Though the true value of the lumped uncertainty cannot be obtained by the RFNN, a less conservative control is resulted to achieve minimum control signal according to S and its derivative $\dot{S}$ [11]. Replacing K by λ in (**8**), the control law is taken as

$$u = u_{equ} + \lambda \operatorname{sgn}(S) \tag{9}$$

3.1 Description of the RFNN

A four-layer RFNN, which comprises the input (the i layer), membership (the j layer), rule (the k layer) and output layer (the o layer), is adopted to implement the RFNN bound observer in this paper.

Layer 1: Input layer. For every node i in this layer, the net input and the net output are represented as

$$net_i^1(N) = \prod x_i^1(N) w_{oi} y_o^4(N-1),\ y_i^1(N) = f_i^1(net_i^1(N)) = net_i^1(N)\quad i = 1,2 \tag{10}$$

where $x_1^1 = S(t)$ and $x_2^1 = \dot{S}$. N denotes the number of iterations, w_{oi} is the recurrent weights for the units in the output layer, and y_o^4 is the output of the RFNN.

Layer 2: Membership layer. Each node performs a membership function. The Gaussian function is adopted as the membership function. For the j th node

$$net_j^2(N) = -\frac{(x_i^2 - m_{ij})^2}{(\sigma_{ij})^2},\ y_i^2(N) = f_j^2(net_j^2(N)) = \exp(net_j^2(N))\quad j = 1,\cdots,n \tag{11}$$

where x_i^2 represents i th input vector to the node of layer 2, m_{ij} and σ_{ij} are, respectively, the mean and the standard deviation of the Gaussian function, and n is the total number of the linguistic variables with respect to the input nodes.

Layer 3: Rule layer. Each node k in this layer is denoted by Π, which multiplies the input signals and outputs the product. For the k th rule node

$$net_k^3(N) = \prod_j w_{jk}^3 x_j^3(N),\ y_k^3(N) = f_k^3(net_k^3(N)) = net_k^3(N)\quad k = 1,\cdots,l \tag{12}$$

where x_j^3 represents the j th input to the node of layer 3. w_{jk}^3 is assumed to be unity, and $l = (n/i)^i$ is the number of rules with complete rule connection if each input node has the same linguistic variables.

Layer 4: Output layer. The single node o in this layer is labeled with Σ, which computes the overall output as the summation of all input signals

$$net_o^4(N) = \sum_k w_{ko}^4 x_k^4(N),\ y_o^4(N) = f_o^4(net_o^4(N)) = \left|net_o^4(N)\right|\quad o = 1 \tag{13}$$

where the connecting weight w_{ko}^4 is the output action strength of the o th output associated with the k th rule. x_k^4 represents the k th input to the node of layer 4, $|\cdot|$ is the absolute value, and $y_o^4 = \lambda$.

3.2 Online Learning Algorithm

To describe the online learning algorithm of the RFNN using the supervised gradient descent method, first, the energy function E chosen as

$$E = \frac{1}{2}(\delta - \delta_d)^2 = \frac{1}{2}e^2 \tag{14}$$

Then, the learning algorithm based on backpropagation method is described below.
Layer 4: The error term to be propagated is given by

$$\delta_o^4 = -\frac{\partial E}{\partial net_o^4} = \left[-\frac{\partial E}{\partial e}\frac{\partial e}{\partial \delta}\frac{\partial \delta}{\partial u}\frac{\partial u}{\partial y_o^4}\frac{\partial y_o^4}{\partial net_o^4} \right] \tag{15}$$

and the weight is updated by the amount

$$\Delta w_{ko}^4 = -\eta_w \frac{\partial E}{\partial w_{ko}^4} = \left[-\eta_w \frac{\partial E}{\partial y_o^4}\frac{\partial y_o^4}{\partial net_o^4} \right]\left(\frac{\partial net_o^4}{\partial w_{ko}^4} \right) = \eta_w \delta_o^4 x_k^4 \tag{16}$$

where η_w is the learning-rate parameter of the connecting weights of the RFNN. The weights of the output layer are updated according to the following equation:

$$w_{ko}^4(N+1) = w_{ko}^4(N) + \Delta w_{ko}^4 \tag{17}$$

Layer 3: Since the weights in this layer are unified, only the error term needs to be calculated and propagated:

$$\delta_o^3 = -\frac{\partial E}{\partial net_k^3} = \left[-\frac{\partial E}{\partial y_o^4}\frac{\partial y_o^4}{\partial net_o^4} \right]\left(\frac{\partial net_o^4}{\partial y_k^3}\frac{\partial y_k^3}{\partial net_k^3} \right) = \delta_o^4 w_{ko}^4 \tag{18}$$

Layer 2: The multiplication operation is done in this layer. The error term is computed as follows:

$$\delta_j^2 = -\frac{\partial E}{\partial net_j^2} = \left[-\frac{\partial E}{\partial y_o^4}\frac{\partial y_o^4}{\partial net_o^4}\frac{\partial net_o^4}{\partial y_k^3}\frac{\partial y_k^3}{\partial net_k^3} \right]\left(\frac{\partial net_k^3}{\partial y_j^2}\frac{\partial y_j^2}{\partial net_j^2} \right) = \sum_k \delta_k^3 y_k^3 \tag{19}$$

and the update law of m_{ij} is:

$$\Delta m_{ij} = -\eta_m \frac{\partial E}{\partial m_{ij}} = \left[-\eta_m \frac{\partial E}{\partial y_j^2}\frac{\partial y_j^2}{\partial net_j^2}\frac{\partial net_j^2}{\partial m_{ij}} \right] = \eta_m \delta_j^2 \frac{2(x_i^2 - m_{ij})}{(\sigma_{ij})^2} \tag{20}$$

where η_m is the learning-rate parameter of the mean of the Gaussian functions. The update law of σ_{ij} is:

$$\Delta\sigma_{ij} = -\eta_\sigma \frac{\partial E}{\partial \sigma_{ij}} = \left[-\eta_\sigma \frac{\partial E}{\partial y_j^2} \frac{\partial y_j^2}{\partial net_j^2} \frac{\partial net_j^2}{\partial \sigma_{ij}} \right] = \eta_\sigma \delta_j^2 \frac{2(x_i^2 - m_{ij})^2}{(\sigma_{ij})^3} \tag{21}$$

where η_σ is the learning-rate parameter of the standard deviation of the Gaussian functions. The mean and standard deviation of the hidden layer are updated as follows:

$$m_{ij}(N+1) = m_{ij}(N) + \Delta m_{ij}\ ,\ \sigma_{ij}(N+1) = \sigma_{ij}(N) + \Delta\sigma_{ij} \tag{22}$$

The update law of the weight w_{oi} can be obtained by the following equation:

$$\Delta w_{oi} = -\eta_r \frac{\partial E}{\partial w_{oi}} = \left[-\eta_r \frac{\partial E}{\partial net_j^2} \frac{\partial net_j^2}{\partial y_i^1} \frac{\partial y_i^1}{\partial net_i^1} \frac{\partial net_i^1}{\partial w_{oi}} \right] = \sum_j \eta_r \delta_j^2 \frac{2(m_{ij} - x_i^2(N))}{(\sigma_{ij})^2} x_i^1(N) y_o^4(N-1) \tag{23}$$

where η_r is the learning-rate parameter of the recurrent weights. The recurrent weights are updated as follows:

$$w_{oi}(N+1) = w_{oi}(N) + \Delta w_{oi} \tag{24}$$

The exact calculation of the Jacobian of the plant, $\partial\delta/\partial u$, cannot be determined due to the uncertainties of the plant dynamics. Although the intelligent identifier can be implemented to calculate the Jacobian of the plant, heavy computation effort is required. To overcome this problem and to increase the online learning rate of the parameters of the RFNN, (15) can be rewritten as

$$\delta_o^4 = \left[-\frac{\partial E}{\partial e} \frac{\partial e}{\partial \delta} \frac{\partial \delta}{\partial u} \frac{\partial u}{\partial y_o^4} \frac{\partial y_o^4}{\partial net_o^4} \right] = e\left|\partial\delta/\partial u\right| \operatorname{sgn}(\partial\delta/\partial u) B_{pn}^{-1} \operatorname{sgn}(S) \operatorname{sgn}(net_o^4)$$

$$\equiv e\beta \operatorname{sgn}(\partial\delta/\partial u) \operatorname{sgn}(S) \operatorname{sgn}(net_o^4) \tag{25}$$

where $\beta = \left|\partial\delta/\partial u\right|$ is defined as a positive constant designed by the user. The positive magnitude quantity of $\partial\delta/\partial u$ can be absorbed within β. Therefore, the Jacobian of the plant is only needed to compute the $\operatorname{sgn}(\partial\delta/\partial u)$ term. According to the qualitative knowledge of the dynamic behavior of the plant, δ will increase or decrease as u increases or decreases. In this paper, $\operatorname{sgn}(\partial\theta_r/\partial u) = +1$ is used for simplicity in practical implementation.

3.3 Convergence Analyses

Selection of the values for the learning-rate parameters has a significant effect on the network performance. In order to train the RFNN effectively, η_w, η_m, η_σ and η_r, the four varied learning rates, which guarantee convergence of the tracking error based on the analyses of a discrete-type Lyapunov function. Because of the limited page, the detail of convergence analyses is omitted. Please refer to [12].

4 Simulation Results and Discussion

For a missile electro-hydraulic servo system (**1**), the nominal value [1] of some parameters are assumed as $k_{ui} = 5mA/V$, $K_Q = 12cm^3/(s \cdot mA)$, $A = 10cm^2$, $R = 17cm$. Substituting the values into (**2**), we get $a_{10} = 0$, $a_{20} = 8873.64$, $a_{30} = 37.68$, $g_0 = 179425$, $d = 0.86\dot{M} + 9.73M$, where $M = M_{f0}Sgn\dot{\delta} + M_d$, M_{f0} is frictional torque amplitude, M_d is position torque. Desired trajectory $\delta_d(t) = \sin 2\pi t$. The sampling period $t = 1ms$.

Assume $\Delta a_i = 0.5\sin(2\pi t)a_{i0}$, so $|\Delta a_i| \le \alpha_i(t) = 0.5 \times a_{i0}$, $\Delta g = 0.2\sin(2\pi t)g_0$, so $|\Delta g| \le \beta(t) = 0.2 \times g_0$, $M_{f0} = 3000 + 1000\sin 2\pi t$, $M_d = 500 + 100\sin 2\pi t$.

Choose the poles of the system as described by (6) at $-80, -80$, we can obtain $c_1 = 6400$, $c_2 = 160$. The initial values of system state variables $\boldsymbol{X}(0) = [1 \quad 0 \quad 0]^T$. The means of the Gaussian functions are set at $-1, 0, 1$ for the N, Z, and P neurons and the standard deviations of the Gaussian functions are set at 1. Moreover, the connecting weights between the output and rule layers, and the recurrent weights are initialized with random number $[0,1]$.

We do simulation research and compare results with that of conventional SMC under the same condition of parameter variations and external disturbances. Simulation results are indicated in Fig. 3—Fig. 6. Fig. 3 shows the tracking response of the system. Fig. 4 shows the tracking error. Fig. 5 and Fig. 6 show the control input where the controller is taken as SMC using RFNN or the conventional SMC.

Simulation analysis: From the simulation results, we can conclude that:

1) If the controller is the conventional SMC, the tracking error is small and there are serious high frequency chattering in the control signal due to the sign function in the switching control.

2) If the controller is the SMC using RFNN, chattering phenomena is attenuated, the control input is smooth and the strength of the control signal can also be significantly

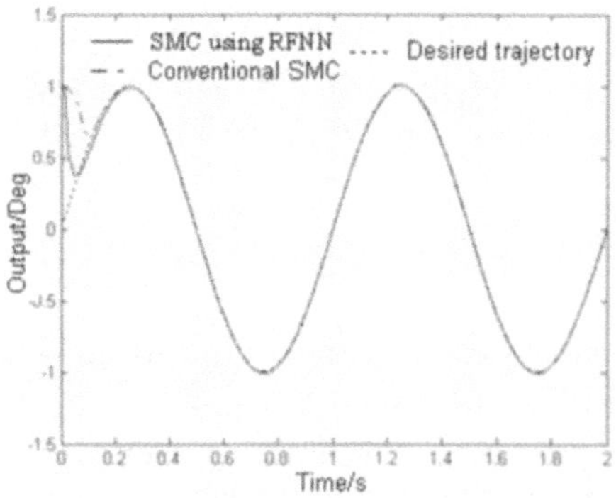

Fig. 3. Tracking response of system

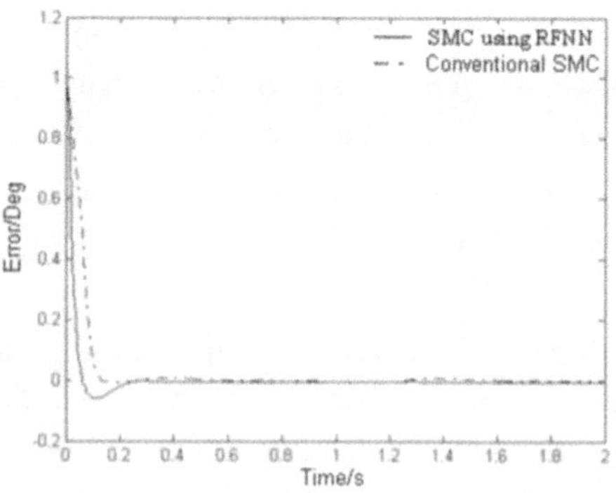

Fig. 4. Tracking error of system

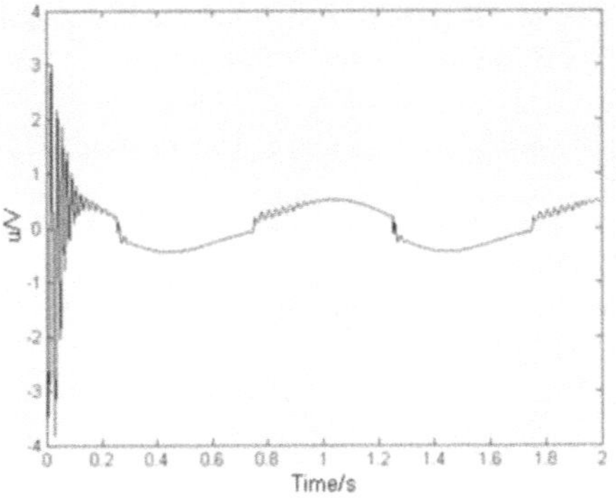

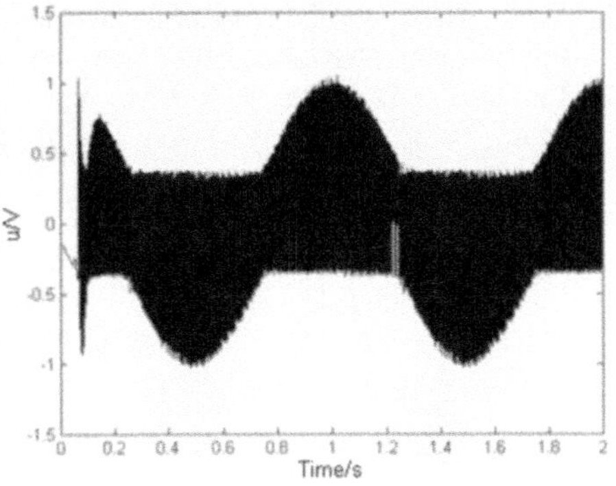

Fig. 5. Control input with SMC using RFNN **Fig. 6.** Control input with conventional SMC

reduced. The transient deviation of tracking error and control input, which are depicted in Fig.4 and Fig.5, respectively, are induced owing to the parameters initialization of the membership functions and connective weights especially under the occurrence of uncertainties. The tracking error is small because the adjusted parameter in the online training of the RFNN can deal with the uncertainty of the system effectively.

5 Conclusions

In this study, a design method of SMC using RFNN is proposed to control the position of missile electro-hydraulic servo system. An RFNN is introduced in conventional SMC to adjust the uncertainty bounds in real time. It delete the assumption of known uncertainty bounds, and the high frequency chattering brought by sliding mode switching control can be effectively minimized, without sacrificing the robustness of sliding mode control. Simulation results indicate that the control approach can cope with uncertainties to obtain an excellent tracking result without the occurrence of chattering control input.

References

1. Zhu, Z.H.: Thrust Vector Control Servo System. Astronautics Press, Beijing (1995)
2. Wang, Z.L.: Control on Modern Electrical and Hydraulic Servo. Beijing University of Aeronautics and Astronautics Press, Beijing (2004)
3. Hung, J.Y., Gao, W.B., Hung, J.C.: Variable Structure Control: A Survey. IEEE Trans. Ind. Electron. **40(2)** (1993) 2-22
4. Mohamed, A.G.: Variable Structure Control for Electro-hydraulic Position Servo System. The 27th Annual Conference of the IEEE Industrial Electronics Society (2001) 2195-2198
5. Liu, Y.F., Dong, D.: Research on Variable Structure Robust Control for Electro-hydraulic Servo System. Journal of Second Artillery Engineering Institute **19(4)** (2005) 12-14
6. Duan, S.L., An, G.C.: Adaptive Sliding Mode Control for Electro-hydraulic Servo Force Control Systems. Chinese Journal of Mechanical Engineering **38(5)** (2002) 109-113

7. Lee, C.H., Teng, C.C.: Identification and Control of Dynamic Systems Using Recurrent Fuzzy Neural Networks. IEEE Trans. on Fuzzy Systems **8** (2000) 349-366
8. Lin, F.J., Lin, C.H., Shen, P.H.: Variable-structure Control for a Linear Synchronous Motor Using a Recurrent Fuzzy Neural Network. IEE Proc. Control Theory Appl. **151(4)** (2004) 395-406
9. Sun, W., Wang, Y.N.: An Adaptive Control for AC Servo System Using Recurrent Fuzzy Neural Network. ICNC 2005, LNCS 3611 (2005) 190-195
10. Liu, Y.F., Dong, D.: 1553B Bus and Its Application in Electro-hydraulic Servo System. Machine Tool & Hydraulics **38(9)** (2004) 106-108
11. Karakasoglu, A., Sundareshan, M.K.: A Recurrent Neural Network-based Adaptive Variable Structure Model Following Control of Robotic Manipulators. Automatica **31(5)** (1995) 1495-1507
12. Wai, R.J.: Total Sliding-mode Controller for PM Synchronous Servo Motor Drive Using Recurrent Fuzzy Neural Network. IEEE Trans. Industrial Electronics **48(5)** (2001) 926-944

Modeling and Control of Molten Carbonate Fuel Cells Based on Feedback Neural Networks

Yudong Tian and Shilie Weng

Power Engineering Department, Shanghai Jiao Tong University, 200030 Shanghai, China
tyd@sjtu.edu.cn

Abstract. The molten carbonate fuel cell (MCFC) is a complex system, and MCFC modeling and control are very difficult in the present MCFC research and development because MCFC has the complicated characteristics such as nonlinearness, uncertainty and time-change. To aim at the problem, the MCFC mechanism is analyzed, and then MCFC modeling based on feedback neural networks is advanced. At last, as a result of applying the model, a new MCFC control strategy is presented in detail so that it gets rid of the limits of the controlled object, which has the imprecision, uncertainty and time-change, to achieve its tractability and robustness. The computer simulation and the experiment indicate that it is reasonable and effective.

1 Introduction

The fuel cell (FC) is a new electric power generation device that directly transforms chemical energy of the fuel and the oxidant into electrical energy though the electrochemical reaction without burning. Its most major characteristic is that it does not pass through the heat engine process, therefore, it jumps from the limit of heat engine cycle, and its energy conversion efficiency is high, moreover its pollution is very small, and the noise is low, so it is considered as the future first-choice electricity generation technology. As a high-effective and clean energy technology, FC has become a rising research field. In all kinds of FC, the molten carbonate fuel cell (MCFC) has many advantages such as low fuel request, high waste gas use value and none precious metal catalyst, therefore it meets with much recognition of the developed countries' government and many scientific research institutes. Now some 1kW~2MW electric power facilities are carrying on the experiment in the region power plant and the distributional generating system all over the world [1].

As an advanced technology, MCFC is listed into '863 Project' in China, and MCFC research & development are independently processed in several institutes. However, MCFC is a complex system, and MCFC modeling is a key problem in the whole of MCFC research & development [2]. In this paper feedback neural networks-based MCFC modeling is presented through analyzing the MCFC mechanism, and then a new MCFC control system is designed as a result of applying the MCFC feedback neural network model.

D. Liu et al. (Eds.): ISNN 2007, Part I, LNCS 4491, pp. 213–221, 2007.

2 MCFC Mechanism Process

A set of MCFC is composed of many MCFC cells which structure is same. Fig. 1 shows the structure of the basic MCFC cell.

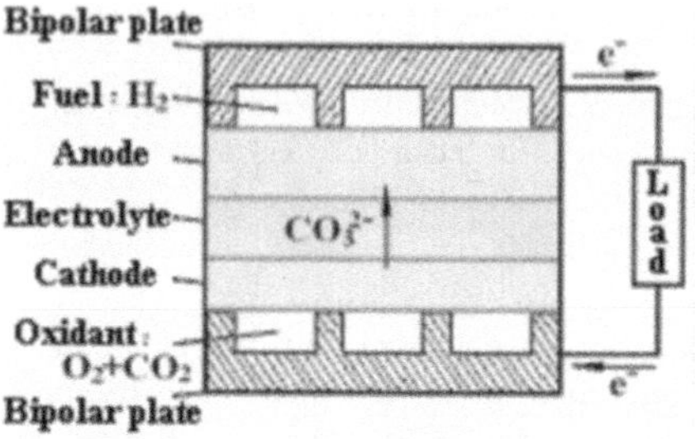

Fig. 1. Schematic diagram of basic MCFC cell structure

In the MCFC cell, its outer layer is the separator plate, its inner layer is the bipolar plate, and its center part is the electrolyte plate where the outer parts are the anode and cathode. When the fuel gases and oxidant gases enter at the gas inlets, they flow in the gas channels of the bipolar plates and filter into the porous anode/cathode, and then react with the aid of the catalyst that is nickel and nickel protoxide in the high temperature. The electrolyte plate obstructs the gases and the electron of the anode, and only permits the carbonic acid ion of the cathode to pass. The electron of the anode can flow across the load of the circuit to the cathode. Thus, MCFC produces direct current electricity.

MCFC electrochemical reactions are described below [3]:

$$\text{Anode reaction: } H_2 + CO_3^{2-} \rightarrow H_2O + CO_2 + 2e^- \tag{1}$$

$$\text{Cathode reaction: } CO_2 + \frac{1}{2}O_2 + 2e^- \rightarrow CO_3^{2-} \tag{2}$$

$$\text{Overall reaction: } H_2 + \frac{1}{2}O_2 + CO_2(c) \rightarrow H_2O + CO_2(a) + E^0 + Q^0 \tag{3}$$

where c and a are respectively the cathode and the anode in the parenthesis. Namely, MCFC anode consumes 1mol hydrogen, at the same time MCFC cathode consumes 0.5mol oxygen and 1mol carbon dioxide, as a result, MCFC can produce 2mol electron, and give out the heat of 246KJ in 0.1MPa, 650□. MCFC standard potential is 1.19V [4].

In conclusion, MCFC can generate electricity continuously as long as the fuel gas and oxidant gas are supplied in the definite operating condition. The main factors that determine the amount of MCFC generating power are the active working area and working status of the electrolyte plate. The amount of electricity that a single MCFC cell produces is very low commonly (A MCFC cell has only a potential of 0.5~0.8V when generating a current density of approximately 0.15~0.25A/cm^2). However, many MCFC cells are connected in series to form a MCFC stack to meet the required power for a specific application. The keys that determine MCFC running normally rest with the internal working status and external operating condition of MCFC stack.

3 Analysis of MCFC Modeling

MCFC is a complex nonlinear system of the multi-input and multi-output that has uncertain factor and random disturbance. There are extremely complicated nonlinear relations among MCFC temperature, pressure, flux, load and so on. Moreover, MCFC temperature and pressure have the characteristics of distributed parameters and the relations of close couplings in the MCFC interior, and the change of MCFC internal component status (especially molten carbonate electrolyte) is dynamic uncertain.

MCFC mechanism modeling is a kind of mathematical modeling that are based on the basic conversation laws, the mass and heat transfer equations and the electrochemical reaction equations, and combined with the internal working principle of MCFC on the definite assumptions. However, it is so difficult as to have to use the methods of numerical analysis and numerical simulation. Unfortunately, its compute is very slowly.

At present, MCFC mathematical models are too complicated to be used to the practice because MCFC is the multi-component, phase changing and multi-dimension flowing mass & heat transfer complex process [5]. Thus, these models only can be taken to the references. However, MCFC control needs a good MCFC model in order to master the MCFC practical working process.

4 Feedback Neural Networks-Based MCFC Modeling

As a parallel computing method, the artificial neural network (ANN) is a good modeling tool. An important function of ANN is that it may highly approach a nonlinear mapping of two different dimensions spaces [6]. Theoretically it was already proved that a tri-layer feed-forward network is capable of approximating any continuous functions by the free precision after training [7].

4.1 MCFC Feedback Neural Network Model

In the multi-layer feed-forward neural networks, the feedback neural network is a kind of ANN that can process the feedback information of the network condition in the interior of the network layer in order to have the dynamic handling ability of the nonlinear. It avoids the information loss phenomenon of the historical network pattern, thus the research becomes more and more important. In recent years, the feedback neural network has obtained the satisfactory effect in the modeling and identification of the nonlinear system.

Here, a feedback neural network with the bias node is created to model MCFC. Its model structure is shown in Fig. 2. A feedback layer is increased before the hidden layer and the corresponding bias node is increased respectively in the input of the hidden layer and the output layer. Thus, the hidden layer can be able to receive the own output signal of one-step time delay through the feedback layer. Moreover, the network study is sped up through the bias node. Therefore, the feedback neural network model with the bias node can save the historical network pattern of the input-output information, and has the memory ability.

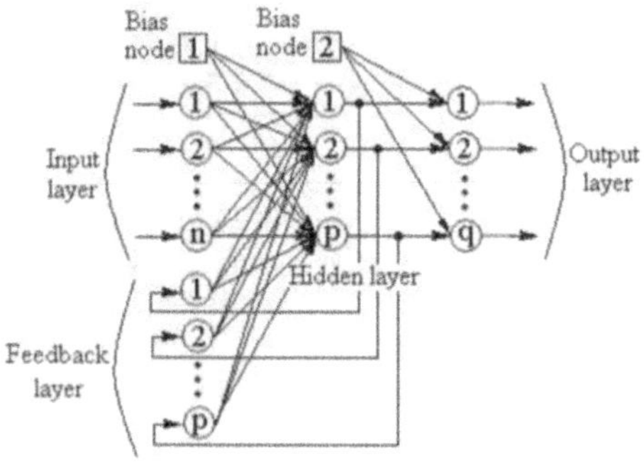

Fig. 2. Structure of feedback neural networks-based MCFC model

4.2 MCFC Feedback Neural Network Study Algorithm

Here, let set up every variable of the feedback neural network model. The input vector of the feedback neural network model is $A_k = (a_1^k, a_2^k, \cdots, a_n^k)$, and the expected output vector of the feedback neural network model is $Y_k = (y_1^k, y_2^k, \cdots, y_n^k)$; the input vector of the hidden layer is $S_k = (s_1^k, s_2^k, \cdots, s_p^k)$, and the output vector of the hidden layer is $B_k = (b_1^k, b_2^k, \cdots, b_p^k)$; the input vector of the output layer is $L_k = (l_1^k, l_2^k, \cdots, l_q^k)$, and the output vector of the output layer is $C_k = (c_1^k, c_2^k, \cdots, c_q^k)$; the connection weight of the input layer to the hidden layer is $\{W_{ij}\}$, $i = 1, 2, \cdots, n$, $j = 1, 2, \cdots, p$; the connection weight of the hidden layer to the output layer is $\{V_{jt}\}$, $j = 1, 2, \cdots, p$, $t = 1, 2, \cdots, q$; the threshold of the hidden layer is $\{\theta_j\}$, $j = 1, 2, \cdots, p$; the threshold of the output layer is $\{\gamma_t\}$, $t = 1, 2, \cdots, q$; the weight coefficients of the input layer to the hidden layer, the feedback layer to the hidden layer and the hidden layer to the output layer, respectively, are WI, WR and WO; the input vector of the bias node is I; and the weight coefficients of the bias node 1 to the hidden layer and the bias node 2 to the output layer are WI_{bias} and WO_{bias} respectively; and the vector number of the above input-output is $k = 1, 2, \cdots, m$.

The Sigmoid function is used for the actuating function of computation node of the network, namely $\sigma(\bullet)$. The study algorithm of the feedback neural network model is as follows:

Firstly, the input vector of the hidden layer $\{S_j\}$ is computed when the study sample A_k is inputted into the input layer of the feedback neural network model, and then the output vector of the hidden layer, namely $\{B_j\}$, is computed through the Sigmoid function. That is:

$$B_j = \sigma(S_j) = \sigma(\sum_{i=1}^{n} W_{ij} \cdot a_j - \theta_j) \tag{4}$$

Secondly, the input vector of the output layer $\{L_t\}$ is computed, and then the output vector of the output layer, namely $\{C_t^k\}$ is computed through the Sigmoid function. That is:

$$C_t^k = \sigma(L_t) = \sigma(\sum_{j=1}^{p} V_{jt} \cdot b_j - \gamma_t) \tag{5}$$

Thirdly, the error of the network node is calculated backward according to the gradient drop law, and then the connection weights of the network are corrected through the accumulated error backward propagation. The network study repeats constantly till the k groups of the pattern training completes. In the end, the overall root-mean-square error of the whole network, namely E, is obtained as follows,

$$E = \sum_{k=1}^{m} \sum_{t=1}^{q} \sqrt{(y_t^k - c_t^k)^2} \tag{6}$$

Finally, the output vector of the feedback neural network with the bias node is

$$\begin{aligned} Y(k) &= \sum_{j=1}^{p} WO_j \cdot \sigma(S_j(k)) + WO_{bias}, \\ S_j(k) &= \sum_{i=1}^{p} WR_{ij} \cdot \sigma(S_j(k-1)) + \sum_{i=1}^{n} WI_{ij} \cdot I_i(k) + WI_{jbias} \end{aligned} \tag{7}$$

4.3 Training of MCFC Feedback Neural Network Model

On the basis of many experiment of 1kW MCFC stack, the change of many variables is so small that these variables are omitted, such as the pressure of the hydrogen entrance and the temperature/humidity of the air entrance. Therefore, the temperature/pressure/flux of the fuel gas entrance and oxidant gas entrance, and the average temperature of the MCFC stack are defined to the input variables of the IRNN model, while the generation voltage/current/resistance of the MCFC stack are defined to the output variables of the feedback neural network model.

7-p-3 structure of the feedback neural network model is created according to the number of the input variables and the output variables of the feedback neural network model, in which 7 neurons of the input layer separately correspond seven input variables and 2 neurons of the output layer separately correspond two output variables. The data sample is the normalized processing data that is composed of 16 groups of MCFC experimental data. In a PC with PIV 1.7MHz CPU, the program based on C language is established, and then large numbers of the simulation are processed. At last, 5 neurons of the hidden layer are determined under synthetic consideration of the performance and the efficiency of the feedback neural network model according to the epoch number and time of the network study, namely p=5.

When the data sample is inputted into the feedback neural network model and the mean-root-square error (MRSE) is set to 1.0×10^{-2} (namely $E\leq1\%$), the system meets the requirement through 100 epochs (i.e. iterations) and the average time of the network study is 0.307 second. The MRSE falling curve of the network training is shown in Fig. 3. It indicates that the output of the MCFC feedback neural network model approaches the target extremely while it is consistent with the analytic solution of the mathematical model greatly.

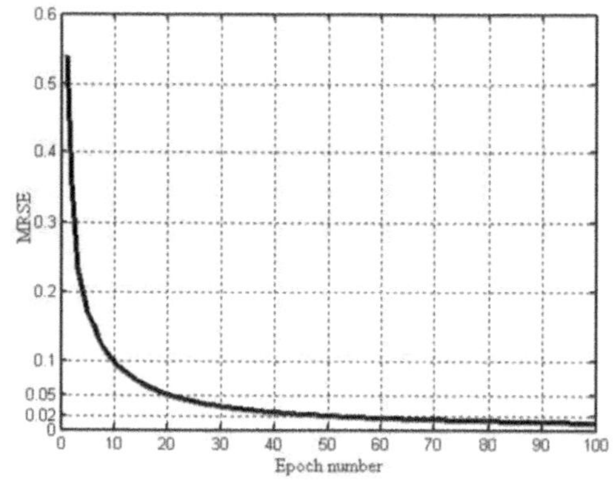

Fig. 3. MRSE falling curve of MCFC feedback neural network model training

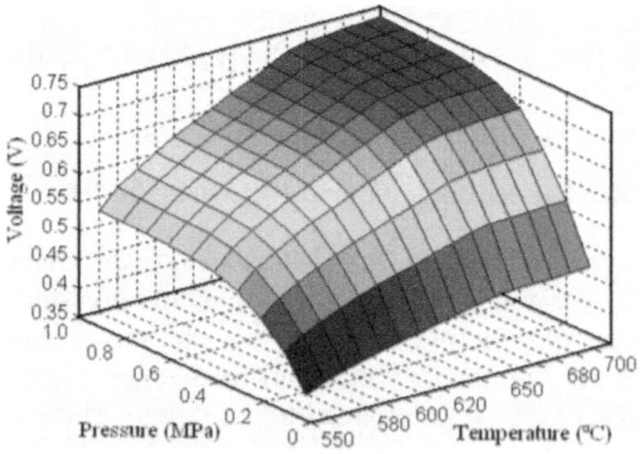

Fig. 4. A simulation result of MCFC feedback neural network model

Fig. 4 shows a simulation result of MCFC feedback neural network model that investigates the average temperature and fuel gas pressure to the voltage.

5 Feedback Neural Networks-Based MCFC Control

MCFC generation system is composed by MCFC stack and many auxiliary facilities. There are some different sensors and adjusting apparatuses of water, heat and gas linked with many different valves to constitute a set of complete MCFC control system. [9].

5.1 Feedback Neural Networks-Based MCFC Control Strategy

MCFC system control is generally preset according to the experimental data and the operation experience. This experience control method both wastes energy and does not achieve the good control quality. For the sake of a study of MCFC control, here presents a new MCFC control strategy based on feedback neural networks. The concrete control architecture is shown in Fig. 5.

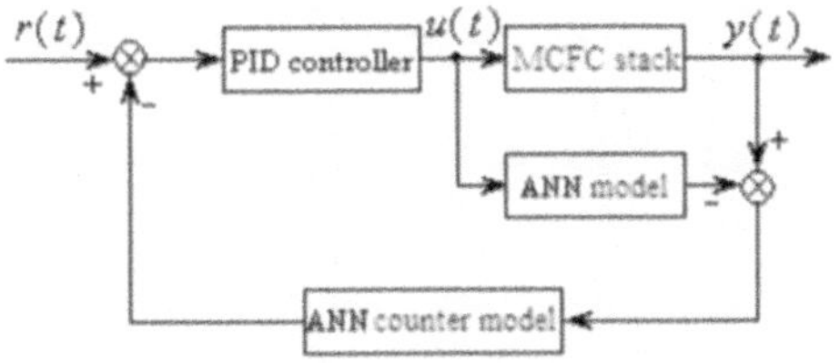

Fig. 5. Architecture of MCFC control strategy based on feedback neural networks

On the basis of the experimental data and the operation experience of 1kW MCFC stack, the MCFC PID feed-forward control is set. According to the actual running condition of the MCFC stack, the MCFC feedback neural network model is used as the contrast model for computing the control error. The error adjusts the input dynamically via applying the feedback of the MCFC feedback neural network counter model. Thus, the control error is decreased and the adaptive robust control is realized.

5.2 Design of MCFC PID-ANN Compound Control

At first, unified the experiment data of the MCFC stack, the controllable operation variables are selected for the variables of the MCFC PID feed-forward control in the variable space (for example, the reactant temperature, flux, the load power, etc.), and the unimportant operation variables are simplified to the system disturbance (for instance, the reactant pressure, the environment variables, etc.). Therefore, the PID controller needs control 7 input variables of the MCFC stack. And the PID parameter can be obtained through the computer identification after the sample data is inputted to the workspace under Matlab/Simulink. For example, the PID parameter of the fuel gas flux to power is 1.428, 0.850 and 0.252 respectively.

Secondly, according to the operation experience of the system, the output of the PID controller not only control the valves of the MCFC system, but also is inputted to the MCFC feedback neural network model to obtain the error of MCFC PID feed-forward control in realtime.

Finally, on the basis of the system stabilization, an MCFC feedback neural network counter model is used in the feedback control of the closed loop system, and the feed-forward control error, which is caused by the system disturbance and the uncertain factor, is decreased, and the output of the MCFC control is continually adjusted. Therefore, the MCFC PID-ANN compound control is more suitable for the MCFC stack after the dynamic readjustment.

5.3 Design of MCFC PID-ANN Compound Control

The simulation of artificial neural networks-based MCFC PID-ANN compound control is programmed on the basis of Matlab/Simulink software. Then, the experimental data of a 1kW MCFC are inputted into the computer from the workspace to carry on the simulation.

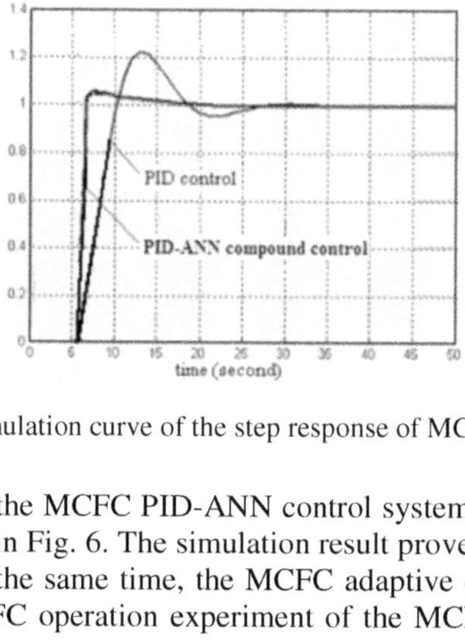

Fig. 6. Simulation curve of the step response of MCFC control

The step response of the MCFC PID-ANN control system that is obtained through the simulation is shown in Fig. 6. The simulation result proves that it is better than the MCFC PID control. At the same time, the MCFC adaptive control plan is simulated on the basis of the MCFC operation experiment of the MCFC experiment platform.

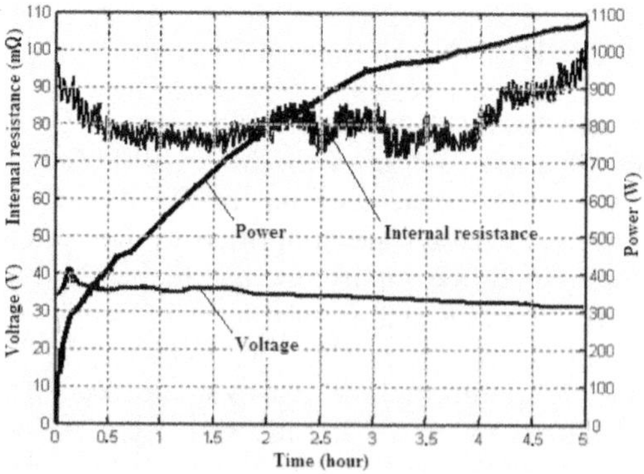

Fig. 7. Experimental curve of MCFC PID-ANN compound control

Its experimental curve is shown in Fig. 7. From the figure, it is shown that the internal resistance and the voltage are very stable with the increasing of the power in the entire process.

6 Conclusion

MCFC modeling and control are the complex problems. On the basis of the research and development of 1kW MCFC stack, feedback neural networks-based MCFC modeling and control are presented. It changes the experience control into the accuracy control, gets rid of the limit that the precise mathematical model is very difficult to build, and solves the problem of MCFC control. The computer simulation and the experiment finally indicate that it is reasonable and effective, and satisfies the requirement of the MCFC system.

References

1. Huijsmans, J.P.P., Kraaij, G.J., Makkus, R.C.: An Analysis of Endurance Issues for MCFC. J. of Power Sources **86** (2000) 117-121
2. Lee, Y.R., Kim, I.G., Chung, G.Y.: Studies on the Initial Behaviors of the Molten Carbonate Fuel Cell. J. of Power Sources **137** (2004) 9-16
3. Shen, C., Cao, G.Y., Zhu, X.J.: Nonlinear Modeling of MCFC Stack Based on RBF Neural Networks Identification. Simulation Modeling Practice & Theory **10** (2002) 109-119
4. Shen, C., Cao, G.Y., Zhu, X.J.: Nonlinear Modeling and Adaptive Fuzzy Control of MCFC Stack. J. of Process Control **12** (2002) 831-839
5. Yi, B.L.: Fuel Cells-theory, Technology and Application. Chinese Chemical Industry Press, Beijing (2003)
6. Hagan, M.T., Demuth, H.B., Beale, M.H.: Neural Network Design. PWS Publishing Company, Boston, MA (1996)
7. Tian, Y.D., Zhu, X.J., Cao, G.Y.: Proton Exchange Membrane Fuel Cells Modeling Based on Artificial Neural Networks. J. of Univ. of S&T Beijing **12(1)** (2005) 72-77
8. Tian, Y.D., Zhu, X.J., Cao, G.Y.: Proton Exchange Membrane Fuel Cells Modeling with IRN Based on Artificial Neural Networks. High Technology Letters **10** (2004) 341-347
9. Tian, Y.D., Weng, S.L., Su, M.: Molten Carbonate Fuel Cells Modeling and Simulation Based on Artificial Neural Networks. Dynamics of Continuous, Discrete and Impulsive Systems **6** (2006) 684-687

An Improved Approach of Adaptive Control for Time-Delay Systems Based on Observer

Lin Chai and Shumin Fei

School of Automation, Southeast University,
Nanjing 210096, P.R. China
chailin_1@163,smfei@seu.edu.cn

Abstract. This paper is concerned with the problem of observer-based stabilization for time-delay systems. Both the state delay and input delay under consideration are assumed to be a constant time-delays, but not known exactly. A new design method is proposed for an observer-based controller with adaptation to the time-delays. The designed controller simultaneously contains both the current state and the past information of systems. The design for adaptation law to delay constants is more concise than the existing conclusions. The controller can be derived by solving a set of linear matrix inequalities (LMIs).

1 Introduction

During the past decade, the Lyapunov-Krasovskii functional, which is based on "descriptor form", has attracted much attention as a powerful tool to deal with some time-delay systems, see for examples [1,4,5], and references therein. Among these various kinds of time-delay systems, some observer-based stabilization methods for delay systems have been proposed in [3, 5-9]. There are two kinds of observer-based controllers for time-delay systems. One is that the design of the observer is memory feedback, while the other is that the observer is memoryless feedback. For the former case, Azuma proposed an observer-based stabilization method for time-delay systems when the constant time-delay is known exactly [7]. An observer design for network time-delay systems is given by Naghshtabrizi [5], where the delay constant needs to be known exactly for the realization of this kind of controller. For the latter case, Wang [8] and Ma [9] presented some observer-based controller design methods for neutral time-delay systems and discrete time-delay singular systems, respectively. These design methods often tend to be more conservative than the former case. In general, it is impossible to know or measure this time-delay exactly. Sugimoto proposed a continuous-time adaptive observer for linear systems with unknown time-delay in [6]. But the input is demanded to be a known scalar function with respect to time and time-delay constant, which results in difficulties of selecting the appropriate function. Jiang proposed some design method of observer-based controllers for linear and nonlinear systems with unknown time-delay [3]. However, the observer controller is memoryless, so the feasibility for solution will to be

D. Liu et al. (Eds.): ISNN 2007, Part I, LNCS 4491, pp. 222–230, 2007.

weak when the time-delay constant influences system greatly. Moreover, there are many parameters needed to be known to realize this kind of controller, so the result isn't practical. To the author's best knowledge, there are few results available about the memory observer-based controller design with adaptation to time-delay for time-delay systems with unknown constant time-delay, where the past information of the systems is used and no additional parameters needed to be known to realize the controller, which motivates the present study.

In this paper, we will discuss the design problem of memory observer-based controller with adaptation to time-delays for time-delay systems, where the time-delay constants both for state and for input are not known exactly but their upper bound and lower bound are available. Unlike the memoryless observer [3], the observer in this paper will simultaneously contain both the current state and the past information of systems. Therefore, it is less conservative than the memoryless observer. Using the "descriptor form"-based Lyapunov-Krasovskii functional, and a linear matrix inequality (LMI) approach, the memory observer-based feedback controller, where the delay parameters of the observer contain not only the estimations of the unknown time-delay constants, but also the difference values, which are between the estimation values of the unknown time-delay and the mean values gained from their upper bounds and lower bounds. There will be three advantages in our method. First, more information on the observer state will be used to implement the controller. Second, the exact values of time-delay will not be known exactly. Third, design for the adaptation law to delay parameters is more concise than the old designs for normal time-delay systems [3,4]. So it is convenient to design a memory observer-based feedback controller which includes the delayed information no matter whether the time-delay constants are available or not.

2 Paper Preparation

Consider the following time-delay systems with input delay:

$$\begin{cases} \dot{x}(t) = Ax(t) + A_1x(t-\tau_1) + Bu(t-\tau_2), \\ \tilde{y}(t) = Cx(t), \\ x(t) = \phi(t), \forall t \in [-\tau, 0], \tau = max\{\tau_1+\tau_2\}+\tau_2, \end{cases} \quad (1)$$

where $x(t) \in \mathbb{R}^n$is the state vector,$u(t) \in \mathbb{R}^{n_1}$is the control input vector, and $\tilde{y}(t) \in \mathbb{R}^{n_2}$ is the measured output vector. A, A_1, B and C are known constant matrices with appropriate dimensions. $\tau_1 > 0$ and $\tau_2 > 0$ are delay constants, which are not known exactly, but the upper bound τ_i^* and the lower bound τ_{i*} are available, i.e. $\tau_{i*} \leq \tau_i \leq \tau_i^* (i = 1, 2)$. $\phi(t) \in C[-\tau, 0]$ is a given continuous vector-valued initial function of system (1). Moreover, there exist positive constants $\bar{\tau}_i, \bar{\tau}_{i1}$, and $\bar{\tau}_{i2}$ such that $0 < \tau_i^* - \bar{\tau}_i \leq \bar{\tau}_{i1}$ and $0 < \bar{\tau}_i - \tau_{i*} \leq \bar{\tau}_{i2}$ hold. Generally, the value $\bar{\tau}_i$ can be chosen as the mean value between τ_i^* and τ_{i*}, i.e. $\bar{\tau}_i = (\tau_i^* + \tau_{i*})/2 (i = 1, 2)$.

As an output feedback controller of system (1), we consider the following memory observer-based output controller:

$$\begin{cases} \dot{\hat{x}}(t) = A\hat{x}(t) + A_1\hat{x}(t-\tau_1) + Bu(t-\tau_2) + L_1[\tilde{y}(t) - \hat{\tilde{y}}(t)] + \sum_{i=1}^{2} L_{i+1}[\tilde{y}(t-\tau_i) \\ \qquad -\hat{\tilde{y}}(t-\tau_i)], \\ u(t) = K\hat{x}(t), \\ \hat{x}(t) = \psi(t), \forall t \in [-\bar{\tau}^*, 0], \bar{\tau}^* = \max_{i=1,2}(\tau_i), \end{cases} \tag{2}$$

By using memory observer, controller (2), which was designed for system (1), is less conservative than memoryless observer-based controllers (e.g. [5]). The shortcoming of controller (2) is that the time delays must be assumed to be known exactly. So controller (2) can't be realized if the delay constants are not available. In general, the time-delay constants can be hardly obtained exactly in the engineering systems. In order to overcome this shortcoming, we can construct the memory observer-based feedback controller as follows:

$$\begin{cases} \dot{\hat{x}}(t) = A\hat{x}(t) + A_1\hat{x}(t - a_1\hat{\tau}_1(t) - (\hat{\tau}_1(t) - h_1)^2) + Bu(t - a_2\hat{\tau}_2(t) - (\hat{\tau}_2(t) - \\ \qquad h_2)^2) + L_1[\tilde{y}(t) - C\hat{x}(t)] + \sum_{i=1}^{2} L_{i+1}[\tilde{y}(t - a_i\hat{\tau}_i(t) - (\hat{\tau}_i(t) - h_i)^2) \\ \qquad -C\hat{x}(t - a_i\hat{\tau}_i(t) - (\hat{\tau}_i(t) - h_i)^2)], \\ u(t) = K\hat{x}(t), \\ \hat{x}(t) = \psi(t), \forall t \in [-\tau^*, 0], \tau^* = \max_{i=1,2}\{\bar{\tau}_i^2 + 2\tau_i^*(\sqrt{\bar{\tau}_i + \bar{\tau}_i^2} - \bar{\tau}_i), \tau_i^*\}, \end{cases} \tag{3}$$

where the constants $a_i, h_i (i = 1, 2)$, and the matrices $K, L_i (i = 1, 2, 3)$ are to be determined. $\hat{\tau}_i(t) (i = 1, 2)$ are the estimation value of the unknown delay constant τ_i satisfying $\dot{\hat{\tau}}_i(t)[2(\hat{\tau}_i(t) - h_i) + a_i] \leq 0 (i = 1, 2), \forall t \geq 0$. The objective of this paper is to stabilize the system (1) by using the controller (3), obtaining the adaptation law for $\hat{\tau}_i(t) (i = 1, 2)$ at the same time.

In order to prove our results, we introduce the following Lemmas.

Lemma 1. Moon-park inequality [2] For any $a \in \mathbb{R}^n, b \in \mathbb{R}^{2n}, N \in \mathbb{R}^{2n \times n}, \mathbb{R} \in \mathbb{R}^{n \times n}, Y \in \mathbb{R}^{n \times 2n}, Y \in \mathbb{R}^{n \times 2n}$, the following holds:

$$-2b^T Na \leq \begin{bmatrix} a \\ b \end{bmatrix}^T \begin{bmatrix} R & Y - N^T \\ Y^T - N & Z \end{bmatrix} \begin{bmatrix} a \\ b \end{bmatrix}, \tag{4a}$$

where

$$\begin{bmatrix} R & Y \\ Y^T & Z \end{bmatrix} \geq 0. \tag{4b}$$

Lemma 2. If

$$h_i = \sqrt{\bar{\tau}_i + \bar{\tau}_i^2}, a_i = 2(\sqrt{\bar{\tau}_i + \bar{\tau}_i^2} - \bar{\tau}_i), \tag{5}$$

there exists $T_i > 0$ such that $\hat{\tau}_i(t) = a_i\hat{\tau}_i + (\hat{\tau}_i - h_i)^2 = \bar{\tau}_i, t \geq T_i$.

Proof. Considering $\hat{\tau}_i(t)$ satisfies$\dot{\hat{\tau}}_i(t)[2(\hat{\tau}_i(t) - h_i) + a_i] \leq 0 (i = 1, 2), \forall t \geq 0$, there exists a time-variant constant $m_i(t) \geq 0$ such that

$$\dot{\hat{\tau}}_i(t) = -[2(\hat{\tau}_i(t) - h_i) + a_i]m_i(t).$$

Apparently, it can be obtained from (6) that there exists T_i such that $\hat{\tau}_i(t) = \frac{1}{2}(2h_i - a_i), t \geq T_i$. Then from (4) yields $\hat{\tau}_i(t) = \frac{1}{2}(2h_i - a_i) = \frac{1}{2}(2\sqrt{\bar{\tau}_i + \bar{\tau}_i^2} - 2(\sqrt{\bar{\tau}_i + \bar{\tau}_i^2} - \bar{\tau}_i)) = \bar{\tau}_i, t \geq T_i$.It can also be obtained that $a_i\hat{\tau}_i(t) + (\hat{\tau}_i(t) - h_i)^2 = a_i\frac{1}{2}(2h_i - a_i) + (\frac{1}{2}(2h_i - a_i) - h_i)^2 = \frac{1}{4}(4a_ih_i - a_i^2) = \bar{\tau}_i, t \geq T_i$.

3 Main Results

If the time-delay constants τ_1 and τ_2 of system (1) are not known exactly, which has been introduced above, our main result on memory observer-based controller with adaptation to delay parameters for system (1) is presented as follows. From (1)-(3), we have

$$\begin{cases} \begin{bmatrix} \dot{\hat{x}}(t) \\ \dot{e}(t) \end{bmatrix} = \begin{bmatrix} A \; L_1C & \\ 0 \;\; A - L_1C \end{bmatrix} \begin{bmatrix} \hat{x}(t) \\ e(t) \end{bmatrix} + \begin{bmatrix} 0 & 0 \\ A_1 & A_1 \end{bmatrix} \begin{bmatrix} \hat{x}(t-\tau_1) \\ e(t-\tau_1) \end{bmatrix} + \begin{bmatrix} 0 & 0 \\ BK & 0 \end{bmatrix} \\ \qquad \begin{bmatrix} \hat{x}(t-\tau_2) \\ e(t-\tau_2) \end{bmatrix} + \begin{bmatrix} A_1 & L_2C \\ -A_1 & -L_2C \end{bmatrix} \begin{bmatrix} \hat{x}(t - a_1\hat{\tau}_1(t) - (\hat{\tau}_1(t) - h_1)^2) \\ e(t - a_1\hat{\tau}_1(t) - (\hat{\tau}_1(t) - h_1)^2) \end{bmatrix} \\ \qquad + \begin{bmatrix} BK & L_3C \\ -BK & -L_3C \end{bmatrix} \begin{bmatrix} \hat{x}(t - a_2\hat{\tau}_2(t) - (\hat{\tau}_2(t) - h_2)^2) \\ e(t - a_1\hat{\tau}_2(t) - (\hat{\tau}_2(t) - h_2)^2) \end{bmatrix}, \\ \begin{bmatrix} \hat{x}(t) \\ e(t) \end{bmatrix} = \begin{bmatrix} \psi(t) \\ \phi(t) - \psi(t) \end{bmatrix}, t \in [-\tau^*, 0]. \end{cases} \tag{6}$$

where $e(t) = x(t) - \hat{x}(t)$is the observer error. Considering the "descriptor form" in [1], we can rewrite (6) as $\begin{bmatrix} \hat{x}(t) \\ e(t) \end{bmatrix} = \tilde{x}(t)$,and $\begin{bmatrix} \dot{\hat{x}}(t) \\ \dot{e}(t) \end{bmatrix} = y(t)$ which yields consequently

$$\begin{cases} \begin{bmatrix} \dot{\hat{x}}(t) \\ \dot{e}(t) \end{bmatrix} = \begin{bmatrix} A + A_1 + BK & (L_1 + L_2 + L_3)C \\ 0 & A + A_1 - (L_1 + L_2 + L_3)C \end{bmatrix} \begin{bmatrix} \hat{x}(t) \\ e(t) \end{bmatrix} \\ \qquad - \sum_{i=1}^{4} \bar{A}_i \int_{t-\tau_i}^{t} y(s)ds, \\ \begin{bmatrix} \hat{x}(t) \\ e(t) \end{bmatrix} = \begin{bmatrix} \psi(t) \\ \phi(t) - \psi(t) \end{bmatrix}, t \in [-\tau^*, 0], \end{cases} \tag{7}$$

where
$\bar{A}_1 = \begin{bmatrix} 0 & 0 \\ A_1 & A_1 \end{bmatrix}, \bar{A}_2 = \begin{bmatrix} 0 & 0 \\ BK & 0 \end{bmatrix}, \bar{A}_3 = \begin{bmatrix} A_1 & L_2C \\ -A_1 & -L_2C \end{bmatrix}, \bar{A}_4 = \begin{bmatrix} BK & L_3C \\ -BK & -L_3C \end{bmatrix}$, $\tau_3 = a_1\hat{\tau}_1(t) + (\hat{\tau}_1(t) - h_1)^2, \tau_4 = a_2\hat{\tau}_2(t) + (\hat{\tau}_2(t) - h_2)^2$. For system (7), consider the following Lyapunov-Krasovskii functional

$$V(x_t) = V_1(x_t) + V_2(x_t) + V_3(x_t) + \frac{l_1}{2}[2(\hat{\tau}_1(t) - h_1) + a_1]^2 + \frac{l_2}{2}[2(\hat{\tau}_1(t) - h_1) + a_1]^2, \tag{8}$$

where $V_1(x_t) = \tilde{x}^T P \tilde{x} = \begin{bmatrix} \tilde{x}^T & y^T \end{bmatrix} E \bar{P}^T \begin{bmatrix} \tilde{x} \\ y \end{bmatrix}, E = \begin{bmatrix} I & 0 \\ 0 & 0 \end{bmatrix}, \bar{P} = \begin{bmatrix} P & P_1 \\ 0 & P_2 \end{bmatrix}, V_3(x_t) = \sum_{i=1}^{4} \int_{t-\tau_i}^{t} \tilde{x}(s)^T S_i \tilde{x}(s)ds, V_2(x_t) = \sum_{i=1}^{4} \int_{-\tau_i}^{0} \int_{t+\theta}^{t} y(s)^T Q_i y(s)dsd\theta, l_i > 0 (i = 1, 2)$

are constants, $h_i > 0, a_i > 0 (i = 1, 2)$ are selected as (5) in lemma2, and $P > 0$, $Q_i > 0, S_i > 0 (i = 1, \cdots, 4)$ stand for positive-definite matrices to be decided. Hence, the derivative of $V_1(x_t)$ alone with system (7) is given by

$$\begin{aligned}\dot{V}_1(x_t) &= 2\tilde{x}^T P y = 2\begin{bmatrix}\tilde{x}^T & y^T\end{bmatrix}\begin{bmatrix}P & P_1\\ 0 & P_2\end{bmatrix}\begin{bmatrix}y\\ 0\end{bmatrix}\\ &= 2\begin{bmatrix}\tilde{x}^T & y^T\end{bmatrix}\begin{bmatrix}P & P_1\\ 0 & P_2\end{bmatrix}\{\begin{bmatrix}y\\ \bar{A}\tilde{x}-y\end{bmatrix} - \sum_{i=1}^{4}\begin{bmatrix}0\\ \bar{A}_i\end{bmatrix}\int_{t-\tau_i}^{t} y(s)ds\,.\\ &= 2\begin{bmatrix}\tilde{x}^T & y^T\end{bmatrix}\begin{bmatrix}P & P_1\\ 0 & P_2\end{bmatrix}\begin{bmatrix}y\\ \bar{A}\tilde{x}-y\end{bmatrix} - \sum_{i=1}^{4}\eta_i\end{aligned} \tag{9}$$

where $\bar{A} = \begin{bmatrix}A + A_1 + BK & (L_1+L_2+L_3)C\\ 0 & A + A_1 - (L_1+L_2+L_3)C\end{bmatrix}$, $\eta_i = 2\begin{bmatrix}\tilde{x}^T & y^T\end{bmatrix}\begin{bmatrix}P & P_1\\ 0 & P_2\end{bmatrix}$ $\begin{bmatrix}0\\ \bar{A}_i\end{bmatrix}\int_{t-\tau_i}^{t} y(s)ds$. By using Lemma 1, η_i can be found as follows:

$$\begin{aligned}\eta_i &\le \int_{t-\tau_i}^{t}\begin{bmatrix}y(s)\\ \bar{x}(s)\end{bmatrix}^T\begin{bmatrix}R_i & T_i - \begin{bmatrix}0 & \bar{A}_i^T\end{bmatrix}^T\bar{P}^T\\ * & Z_i\end{bmatrix}\begin{bmatrix}y(s)\\ \bar{x}(s)\end{bmatrix}ds\\ &= \int_{t-\tau_i}^{t} y(s)^T R_i y(s)ds + 2\int_{t-\tau_i}^{t} y(s)^T(T_i - \begin{bmatrix}0 & \bar{A}_i^T\end{bmatrix})\bar{P}^T\bar{x}(t)ds + \tau_i\bar{x}(t)^T Z_i\bar{x}(t)\,.\\ &\quad (i = 1, \cdots, 4)\end{aligned} \tag{10}$$

where $\bar{x}(t) = \begin{bmatrix}\tilde{x}^T & y^T\end{bmatrix}$, $\begin{bmatrix}R_i & T_i\\ * & Z_i\end{bmatrix} \ge 0$, and $R_i \in \mathbb{R}^{2n\times 2n}, T_i \in \mathbb{R}^{2n\times 4n}, Z_i \in \mathbb{R}^{4n\times 4n}$. Noting that

$$\int_{t-\tau_i}^{t} y(s)ds = \tilde{x}(t) - \tilde{x}(t-\tau_i), \tag{11}$$

then

$$\begin{aligned}\bar{x}(t)^T Z_i\bar{x}(t) &= \begin{bmatrix}\tilde{x}^T & y^T\end{bmatrix}\begin{bmatrix}Z_{i1} & Z_{i2}\\ * & Z_{i3}\end{bmatrix}\begin{bmatrix}\tilde{x}\\ y\end{bmatrix}\\ &= \tilde{x}^T Z_{i1}\tilde{x} + 2\tilde{x}^T Z_{i2}y + y^T Z_{i3}y = \begin{bmatrix}\hat{x}(t)^T & e(t)^T\end{bmatrix}\begin{bmatrix}Z_{i1} & Z_{i2}\\ * & Z_{i3}\end{bmatrix}\begin{bmatrix}\hat{x}(t)\\ e(t)\end{bmatrix} + 2\tilde{x}^T Z_{i2}y\,.\\ &\quad + y^T Z_{i3}y\\ &= \hat{x}(t)^T Z_{i1}\hat{x}(t) + \begin{bmatrix}\hat{x}(t)^T & e(t)^T\end{bmatrix}\begin{bmatrix}0 & Z_{i2}\\ * & Z_{i3}\end{bmatrix}\begin{bmatrix}\hat{x}(t)\\ e(t)\end{bmatrix} + 2\tilde{x}^T Z_{i2}y + y^T Z_{i3}y, i = 3, 4,\end{aligned} \tag{12}$$

And

$$\tau_{i+2}(t) = a_i\hat{\tau}_i(t) + (\hat{\tau}_i(t) - h_i)^2 = \frac{1}{4}\{[2(\hat{\tau}_i(t) - h_i)^2) + a_i]^2 - a_i^2 + 4h_i a_i\}, i = 1, 2, \tag{13}$$

Besides,

$$\frac{d(\int_{-\tau_i}^{0}\int_{t+\theta}^{t} y(s)^T Q_i y(s)dsd\theta)}{dt} = \dot{\tau}(t)\int_{t-\tau_i}^{t} y(s)^T Q_i y(s)ds$$

$$+\int_{-\tau_i}^{0}[y(t)^TQ_iy(t)-y(t+\theta)^TQ_iy(t+\theta)]d\theta,$$
$$i=3,4.$$

As $\dot{\hat{\tau}}_i(t)[2(\hat{\tau}_i(t)-h_i)^2)+a_i]\leq 0(i=1,2)$, and

$$\begin{aligned}\dot{\tau}_{i+2}(t)&=\frac{d\big(a_i\hat{\tau}_i(t)+(\hat{\tau}_i(t)-h_i)^2\big)}{dt}\\&=a_i\dot{\hat{\tau}}_i(t)+2(\hat{\tau}_i(t)-h_i)\dot{\hat{\tau}}_i(t)\\&=\dot{\hat{\tau}}_i(t)[a_i+2(\hat{\tau}_i(t)-h_i)]\leq 0,\qquad i=1,2,\end{aligned}$$

so we have

$$\dot{V}_2(t)\leq\sum_{i=1}^{4}[\tau_iy(t)^TQ_iy(t)-\int_{t-\tau_i}^{t}y(s)^TQ_iy(s)ds].\tag{14}$$

The derivative of $V_3(x_t)$ is

$$\dot{V}_3(x_t)=\sum_{i=1}^{4}[\tilde{x}(t)^TS_i\tilde{x}(t)-\tilde{x}(t-\tau_i)^TS_i\tilde{x}(t-\tau_i)].\tag{15}$$

Let $R_i=Q_i(i=1,\cdots,5)$, according to (8)-(15), the following inequalities are obvious by means of Schur complement lemma

$$\begin{aligned}\dot{V}(x_t)\leq\bar{x}(t)^T\Xi_0\bar{x}(t)+\sum_{i=1}^{2}[2(\hat{\tau}_i(t)-h_i)^2)+a_i]\{l_i\dot{\hat{\tau}}_i(t)+\tfrac{1}{4}[2(\hat{\tau}_i(t)-h_i)^2)+a_i]\\ \hat{x}(t)^TZ_{(i+2)1,1}\hat{x}(t)\},\end{aligned}\tag{16a}$$

$$M_i=\begin{bmatrix}R_i & T_i\\ * & Z_i\end{bmatrix}\geq 0,\ i=1,\cdots,4,\tag{16b}$$

where $\bar{x}(t)^T=\begin{bmatrix}\tilde{x}(t)^T & y(t)^T & \tilde{x}(t-\tau_1)^T & \cdots & \tilde{x}(t-\tau_4)^T\end{bmatrix}$,
$\Xi_0=\begin{bmatrix}\Psi & \bar{P}\begin{bmatrix}0\\ \bar{A}_1\end{bmatrix}-T_1^T & \cdots & \bar{P}\begin{bmatrix}0\\ \bar{A}_4\end{bmatrix}-T_4^T\\ * & -S_1 & \cdots & 0\\ * & * & \ddots & \vdots\\ * & * & * & -S_4\end{bmatrix}$, $\Psi=\bar{P}\begin{bmatrix}0 & I\\ \bar{A}_0 & -I\end{bmatrix}+\begin{bmatrix}0 & I\\ \bar{A}_0 & -I\end{bmatrix}^T\bar{P}^T+$
$\sum_{i=1}^{2}\tau_iZ_i+\sum_{i=3}^{4}\bar{\tau}_i\begin{bmatrix}\bar{Z}_{i1} & Z_{i1}\\ * & Z_{i3}\end{bmatrix}+\begin{bmatrix}\sum_{i=1}^{4}S_i & 0\\ 0 & \sum_{i=1}^{4}\tau_iQ_i\end{bmatrix}+\sum_{i=1}^{4}\begin{bmatrix}T_i\\ 0\end{bmatrix}+\sum_{i=1}^{4}\begin{bmatrix}T_i\\ 0\end{bmatrix}^T$, where $\bar{A}_0=$ $\begin{bmatrix}A & L_1C\\ 0 & A-L_1C\end{bmatrix}$, $Z_i=\begin{bmatrix}Z_{i1} & Z_{i2}\\ * & Z_{i3}\end{bmatrix}$, $Z_{i1}=\begin{bmatrix}Z_{i1,1} & Z_{i1,2}\\ * & Z_{i1,3}\end{bmatrix}$, and $\bar{Z}_{i1}=\begin{bmatrix}0 & Z_{i1,2}\\ * & Z_{i1,3}\end{bmatrix}$ $(i=3,4)$. The following result is then obtained.

Theorem 1. The time-delay system (1) with observer-based controller (3) is asymptotically stabilizable if there exist matrices $T_i\in\mathbb{R}^{2n\times 4n}$, $Z_i\in\mathbb{R}^{4n\times 4n}$,

$P_1 \in \mathbb{R}^{2n\times 2n}$, $P_2 \in \mathbb{R}^{2n\times 2n}$, $K \in \mathbb{R}^{n_1\times n}$, $L_i \in \mathbb{R}^{n\times n_2}$ $(i = 1, 2, 3)$ and positive-definite matrices $P \in \mathbb{R}^{2n\times 2n}$, $S_i \in \mathbb{R}^{2n\times 2n}$, $Q_i \in \mathbb{R}^{2n\times 2n}$ $(i = 1, \cdots, 4)$, such that the linear matrix inequalities (18) hold. Moreover, the adaptive control about delay constants can be obtained from (17).

Proof. Consider the following adaptive control

$$\dot{\hat{\tau}}_i(t) = -\frac{1}{4l_i}[a_i + 2(\hat{\tau}_i(t) - h_i)]\hat{x}(t)^T Z_{(i+2)1,1}\hat{x}(t), \qquad (i = 1, 2), \tag{17}$$

then by using (16), we have $\dot{V}(x_t) \leq \bar{x}(t)^T \Xi_0 \bar{x}(t)$. So if $S \triangleq \Xi_0 < 0$ and $M_i \geq 0\,(i = 1, \cdots, 4)$, then under the action of the control (3), the system (1) is asymptotically stable. The most important work of the observer-based control problem is that how to solve the matrix inequalities $Xi_0 < 0$ and $M_i \geq 0\,(i = 1, \cdots, 4)$. Obviously, there exists $S(\tau_1, \tau_2) \leq S(\tau_1^*, \tau_2^*)$, for $\tau_i \leq \tau_i^*$ $(i = 1, 2)$. So $S(\tau_1^*, \tau_2^*) < 0$ and $M_i \geq 0\,(i = 1, \cdots, 4)$ can guarantee (16) satisfied, which means the time-delay system (1) is asymptotically stabilizable by using feedback controller (3). Let $\Xi \triangleq S(\tau_1^*, \tau_2^*)$. After substituting $h_i = \sqrt{\bar{\tau}_i + \bar{\tau}_i^2}, a_i = 2(\sqrt{\bar{\tau}_i + \bar{\tau}_i^2} - \bar{\tau}_i)$ into $\Xi < 0$ and $M_i \geq 0\,(i = 1, \cdots, 4)$, the following linear matrix inequalities can be obtained.

$$\Xi = \begin{bmatrix} \bar{\Psi} & \bar{P}\begin{bmatrix} 0 \\ \bar{A}_1 \end{bmatrix} - T_1^T & \cdots & \bar{P}\begin{bmatrix} 0 \\ \bar{A}_4 \end{bmatrix} - T_4^T \\ * & -S_1 & \cdots & 0 \\ * & * & \ddots & \vdots \\ * & * & * & -S_4 \end{bmatrix} < 0 \tag{18a}$$

$$M_i = \begin{bmatrix} R_i & T_i \\ * & Z_i \end{bmatrix} \geq 0,\ i = 1, \cdots, 4 \tag{18b}$$

where $\bar{\Psi} = \bar{P}\begin{bmatrix} 0 & I \\ \bar{A}_0 & -I \end{bmatrix} + \begin{bmatrix} 0 & I \\ \bar{A}_0 & -I \end{bmatrix}^T \bar{P}^T + \Phi, \Phi = \sum_{i=1}^{2} \tau_i^* Z_i + \sum_{i=3}^{4} \bar{\tau}_i \begin{bmatrix} \bar{Z}_{i1} & Z_{i1} \\ * & Z_{i3} \end{bmatrix} +$
$\begin{bmatrix} \sum_{i=1}^{4} S_i & 0 \\ 0 & \sum_{i=1}^{4} \tau_i^* Q_i \end{bmatrix} + \sum_{i=1}^{4} \begin{bmatrix} T_i \\ 0 \end{bmatrix} + \sum_{i=1}^{4} \begin{bmatrix} T_i \\ 0 \end{bmatrix}^T.$

Remark 1. For $\tau_{i+2}(t) = a_i\hat{\tau}_i(t) + (\hat{\tau}_i(t) - h_i)^2 = \frac{1}{4}\{[2(\hat{\tau}_i(t) - h_i)^2) + a_i]^2 - a_i^2 + 4h_i a_i\}, i = 1, 2$, as derivative of $\tau_{i+2}(t)$ can be obtained for $\hat{\tau}_i(t) \in [\tau_{i*}, \tau_i^*]$, and we have $\frac{d\tau_{i+2}(\hat{\tau}_i(t))}{d\hat{\tau}_i(t)} = 2\hat{\tau}_i(t) - 2h_i + a_i = 0$, so $\tau_{i+2}(t)$ can achieve extremum when $\hat{\tau}_i(t) = (2h_i - a_i)/2\,(i = 1, 2)$. Furthermore, as $\tau_{i+2}''((2h_i - a_i)/2) = 2 > 0$, so $\tau_{i+2}(t)$ can achieve minimum when $\hat{\tau}_i(t) = (2h_i - a_i)/2 = \bar{\tau}_i$ $(i = 1, 2)$. As a result, the maximum for $\tau_{i+2}(t)$, i.e. $\tau_{i+2}^* = max\{\tau_{i+2}(\hat{\tau}_i(t) = \{\tau_{i*}, \tau_i^*\})\}$ $(i = 1, 2)$. Apparently, if $\bar{\tau}_i$ is selected as the mean value between τ_{i*} and τ_i^*, then $\tau_{i+2}^* = \tau_{i+2}(\hat{\tau}_i(t) = \tau_{i*}) = \tau_{i+2}(\hat{\tau}_i(t) = \tau_i^*)$ $(i = 1, 2)$.

The matrix inequalities in Theorem 1 are BMIs, and there is no efficient numerical method to solve them. Using the similar proof in [5], the theorem can read as the following theorem:

Theorem 2. The controller (3) with parameters K and L asymptotically stabilizes the plant with state space model given by (1) for unknown time-delay constants τ_1 and τ_2, if there exist $2n \times 2n$ matrices $P > 0$, $X_1 > 0$, $X_2 > 0$, Y_1, Y_2, $S_i \geq 0$, $Q_i \geq 0\,(i = 1, \cdots, 4)$, $4n \times 4n$ matrices $Z_i\,(i = 1, \cdots, 4)$, $2n \times 4n$ matrices $T_i\,(i = 1, \cdots, 4)$, $n_1 \times n$ matrix K, $n \times n_2$ matrix L and constant $\alpha > 0$ that satisfy the following matrix inequalities:

$$\begin{bmatrix} U^T X U - N & J^T + \alpha U^T \\ * & Y \end{bmatrix} > 0 \tag{19a}$$

$$\begin{bmatrix} Q_i & T_i \\ * & Z_i \end{bmatrix} \geq 0, i = 1, \cdots, 4, \tag{19b}$$

where $U = \begin{bmatrix} I\,0\,0\,0\,0\,0 \\ 0\,I\,0\,0\,0\,0 \end{bmatrix}$, $N = \begin{bmatrix} \Gamma & -T_1^T & \cdots & -T_4^T \\ * & -S_1 & \cdots & 0 \\ * & * & \ddots & \vdots \\ * & * & * & -S_4 \end{bmatrix}$, $\Gamma = \begin{bmatrix} 0 & P \\ P & 0 \end{bmatrix} + \Phi$,

$J(K, L) = \begin{bmatrix} A_0 & -I & \bar{A}_1 & \cdots & \bar{A}_4 \\ A_0 & -I & \bar{A}_1 & \cdots & \bar{A}_4 \end{bmatrix}$, $X = \begin{bmatrix} X_1 & 0 \\ 0 & X_2 \end{bmatrix}$, $Y = \begin{bmatrix} Y_1 & 0 \\ 0 & Y_2 \end{bmatrix}$, $X = \alpha^2 Y^{-1}$.

Remark 2. Theorem 2 transforms the matrix inequalities (18a) to (19a). Since $\bar{A}_i\,(i = 1, \cdots, 4)$ are linear functions of K and L, (19) are LMIs. However, the fact that $X = \alpha^2 Y^{-1}$ is not convex constraint, makes the whole set of matrix inequalities non-convex. Next we introduce a numerical procedure to solve such a non-convex problem.

Similar to [5], the cone complementarity linearization algorithm introduced in [5] changes the non-convex feasibility problem in Theorem 2 to the following linear minimization problem:

① Choose α;
② Find a feasible point X_0, Y_0 for the set of LMIs (19a) and (19b) and

$$\begin{bmatrix} X & I \\ I & \alpha^{-2} Y \end{bmatrix} \geq 0. \tag{20}$$

Set $X_j = X_{j-1}, Y_j = Y_{j-1}$, and find X_{j+1}, Y_{j+1} that solves the LMI problem

$$\sum_j : \, min\, trace(X_j Y + X Y_j) \, \text{subject to (19),(20).}$$

③ If stopping criterion is satisfied, exit. Otherwise set $j = j + 1$ and go to step 3 if $j < c$ (a preset number) or increase α with a proper amount and go to step 2.

If the minimum is equal to $8n \times \alpha^{-2}$, then (19a) and (19b) with $X = \alpha^2 Y^{-1}$ are satisfied and the controller with parameters K and L can stabilize the plant with adaptive controller (17). Since it is numerically difficult to obtain $trace(X_jY + XY_j) = 8n \times \alpha^{-2}$, (18) can be chosen as the stopping criterion.

Acknowledgments. The research work of Lin Chai and Shumin Fei was supported by the National Natural Science Foundation of China (Grant No. 60574006) and the Foundation of Doctor (Grant No. 20030286013).

References

1. Fridman, E.: New Lyapunov-Krasovskii Functionals for Stability of Linear Retarded and Neutral Type Systems. Systems and Control Letters **43** (2001) 309-319
2. Moon, Y. S., Park, P., Kwon, W., Lee, Y.: Delay-Dependent Robust Stabilization of Uncertain State-Delayed Systems. Int. J. Control **74** (2001) 1447-1455
3. Jiang, X., Xu, W., Han, Q.: Observer-Based Fuzzy Control Design with Adaptation to Delay Parameter for Time-Delay Systems. Fuzzy Sets and Systems **152** (2005) 637-649
4. Chai, L., Fei, S., Xin, Y.: An Approach of Adaptive H^{∞} Control for a Class of Nonlinear Time-delay System with Uncertain Input Delay. Acta Automatic Sinica **32** (2006) 237-245
5. Naghshtabrizi, P., Hespanha, J. P.: Designing an Observer-Based Controller for a Network Control System. Proceedings of the 44th IEEE Conference on Decision and Control, and the European Control Conference 2005, Seville, Spain, December 12-15 (2005) 848-853
6. Michiru, S., Hiromitsu, O., Akira, S.: Continuous-Time Adaptive Observer for Linear System with Unknown Time Delay. Proceedings of the 39th IEEE Conference on Decision and Control Sydney, Australia December (2000) 1104-1109
7. Azuma, T., Sagara, S.: Output Feedback via Control Synthesis for Linear Time-Delay Systems Infinite-dimensional LMI Approach. Proceedings of the 42nd IEEE Conference on Decision and Control, Maui, Hawaii, USA, December (2003) 4206-4231
8. Wang, Z., Lam, J., Burnham, K. J.: Stability Analysis and Observer Design for Neutral Delay Systems. IEEE Transaction on Automatic Control **47** (2002) 478-483
9. Ma, S., Cheng, Z.: Observer Design for Discrete Time-Delay Singular Systems with Unknown Inputs. 2005 American Control Conference, Portland, OR, USA, June 8-10 (2005) 4215-4219.

Vibration Control of Block Forming Machine Based on an Artificial Neural Network

Qingming Wu, Qiang Zhang, Chi Zong, and Gang Cheng

College of Power & Mechanical Engineering, Wuhan University, Wuhan 430072, China
engine2003@126.com

Abstract. A two-stage structure model was developed for the vibration control of an actuator platform and a controller based on a three-layer neural network was applied to realize high performance control for the kickstand disturbance of a block forming machine. This paper presents a survey of the basic theory of the back-propagation(BP) neural network architecture including its architectural design, BP algorithm, the root mean square error (RMSE) and optimal model establishment. The situ-test data of the control system were measured by acceleration transducer and the experimental results indicates that the proposed method was effective.

1 Introduction

In many vibration engineering applications, a lot of vibration problems were solved by experience. How to establish the structure of vibration, mathematic model, controller model and solve the problems were the goals that many researchers pursued. At the same time, many theories and methods were presented and applied. They are linear controllers in most cases. In general, linear controllers have been already used widely and great success has been achieved. However, it had lower efficiency, because it hadn't considered nonlinearity factors. Consequently, nonlinear control techniques were used to solve theses problems in this paper. The artificial neural network (ANN) was the main method of the nonlinear control. It can pass nonlinear value by its neurons. But this method need to predefine some rules that direct ANN propagation its information. For the sake of solving these problems, it is very important to establish the structure of ANN. An optimal topology structure was discussed in this study. It had some advantages: faster self-learning abilities, improved efficiency, preferable nonlinear mapping capacity etc.

The ANN technique has expanded the range of fields which vibration control can be applied, and it has generated many successful demonstrations systems. For example, K.G. Ahn and H.J. Pahk [1] used neural networks on hybrid-type active vibration isolation. X.Q. L [2] reported fault detection and diagnosis based on parameter estimation and neural network. Q. Chen [3] used neural networks on structural fault diagnosis. C.L. Zhang [4] depended on micro-manufacturing platform realized active vibration isolation. Zhang Lei [5] developed CMAC neural network to control vibration. Mahmod M. [6] solved vibration signals by a hybrid analysis method that based on neural networks and pattern recognition techniques and so on.

D. Liu et al. (Eds.): ISNN 2007, Part I, LNCS 4491, pp. 231–240, 2007.

In this study, a block forming machine, which can make incompact material molding and compact is presented. The core part of this machine is vibrant apparatus consisting of an actuator platform, a moulding board, a press module, a kickstand and two rubber layers and so on. The actuator provides simple harmonic vibration and vibration power can compact building block. Moreover, the kickstand must eliminate the vibration from the actuator on a large scale. This part determines the performance and the product quality of the machine. Therefore, it is very important to solve the problem about vibration and vibration isolation.

The rest of this paper is organized as follows. Section II provides the structure and model of a vibration. Section III presents the architecture of an artificial neural network. Section IV establishes the optimal ANN. Section V shows the situ-test experimental results and discussions. Finally, some conclusions are made.

2 Vibration Control System

In block forming machine, the actuator platform is the most important part. Furthermore, we are only interested in the vertical vibration, which determine the effect of the block forming. Vertical vibration is only considered for the sake of making the problem simple. The dynamic model of the vibration control system is presented in Fig. 1.

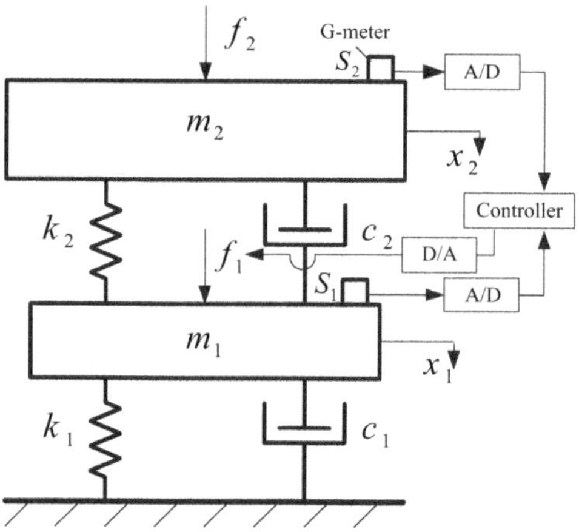

Fig. 1. The model of the vibration system

Where S_1 and S_2 are the acceleration transducer, m_1 the mass of the kickstand, k_1 and c_1 the equivalent spring stiffness and damping coefficient of the first rubber layer between the kickstand and base, m_2 the total mass of the actuator platform and moulding board, k_2 and c_2 the equivalent spring stiffness and damping coefficient of the second rubber layer between the actuator platform and the kickstand, x_1 the displacement of the kickstand vibration, x_2 the displacement of the actuator platform vibration, f_1 the control force produced by the controller, and f_2 the direct vibration force acting on the actuator platform.

The dynamic equation of the vibration control system is as follows:

$$\begin{bmatrix} m_1 & 0 \\ 0 & m_2 \end{bmatrix}\begin{Bmatrix} \ddot{x}_1 \\ \ddot{x}_2 \end{Bmatrix}+\begin{bmatrix} c_1+c_2 & -c_2 \\ -c_2 & c_2 \end{bmatrix}\begin{Bmatrix} \dot{x}_1 \\ \dot{x}_2 \end{Bmatrix}+\begin{bmatrix} k_1+k_2 & -k_2 \\ -k_2 & k_2 \end{bmatrix}\begin{Bmatrix} x_1 \\ x_2 \end{Bmatrix}=\begin{Bmatrix} f_1 \\ f_2 \end{Bmatrix} \tag{1}$$

If the state variable is defined as follows:

$$X^T=\begin{bmatrix} x_1 & x_2 & \dot{x}_1 & \dot{x}_2 \end{bmatrix}$$

The state equation is obtained as follows:

$$\dot{X}=AX+bu \tag{2}$$

where:

$$A=\begin{bmatrix} 0 & 0 & 1 & 0 \\ 0 & 0 & 0 & 1 \\ \dfrac{k_1+k_2}{m_1} & -\dfrac{k_2}{m_1} & \dfrac{c_1+c_2}{m_1} & -\dfrac{c_2}{m_1} \\ -\dfrac{k_2}{m_2} & \dfrac{k_2}{m_2} & -\dfrac{c_2}{m_2} & \dfrac{c_2}{m_2} \end{bmatrix},\ b=\begin{bmatrix} 0 & 0 \\ 0 & 0 \\ \dfrac{1}{m_1} & 0 \\ 0 & \dfrac{1}{m_2} \end{bmatrix},\ u=\begin{bmatrix} f_1 \\ f_2 \end{bmatrix}$$

The data processed by computer are numeric signals and not only are quantitative in numerical value, but also are discrete in time. Therefore the continuous state equation must be transferred into discrete state equation. Discrete state equation is shown as follows:

$$\boldsymbol{x}(k+1)=\boldsymbol{G}(T)\boldsymbol{x}(k)+\boldsymbol{H}(T)\boldsymbol{u}(k) \tag{3}$$

where:

$$G(T)=e^{AT},\ H(T)=\int_0^T e^{AT}dt\cdot b,$$ $\boldsymbol{T}$ is a sampling period.

The vibration of block forming machine is propagated by the kickstand. So as long as it decreases vibration to null based on the vibration signals from the mass m_2, the optimal control force that presses on m_1 can decrease the kickstand vibration and vibration control goal will be achieved.

3 Artificial Neural Network

3.1 The Architecture of BP Network

The foundation of the artificial neural network (ANN) paradigm was laid in the 1950s. ANN is non-logarithmic, non-digital and intensely parallel systems consisting of a number of very simple and highly interconnected processors called neurons, which are analogous to the biological neural cells in the brain. In this paper, a typical feedforward neural network (FNN) topology is introduced. It is comprised of the input layer, one or more hidden layers and the output layer. A topology of a simple FNN is presented in Fig. 2. Each layer includes a certain number of neurons that will transfer signals from one neuron to next.

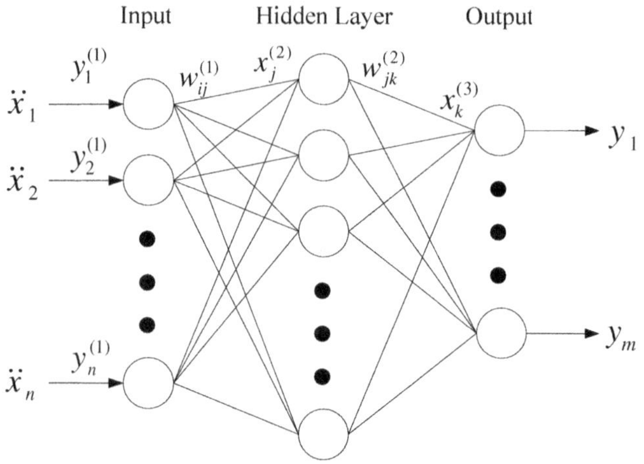

Fig. 2. A topology of BP neural network

Fig. 2 also shows the process of the neuron feedforward. So the output of each neuron is:

$$\begin{cases} y_k^{(3)} = f\left(\sum_{j=0}^{n_1-1} w_{jk}^{(2)} x_j^{(2)} - \theta_k^{(2)}\right), & k = 0,1,\cdots,m-1 \\ x_j^{(2)} = f\left(\sum_{i=0}^{n-1} w_{ij}^{(1)} x_i^{(1)} - \theta_j^{(1)}\right), & j = 0,1,\cdots,n_1-1 \end{cases} \tag{4}$$

Clearly, equation (4) indicates that the neurons have responsibility for mapping n-dimension into m-dimension.

There are a lot of types ANN that have been used in many applications and projects. However, the back propagation network (BP neural network) is known as one of the most popular types. Fig. 3 describes a general model of one BP neuron, where x =input value; w =weight; Σ=summation; θ=bias; f =activation or transformation function and y =output value. BP neurons are similar to other neurons, but the transformation function of BP neuron is non-linear, which is shown in equation (5,6):

$$u_j = \sum_{i=1}^{n} w_i x_i - \theta_j \tag{5}$$

$$y_j = f(u_j) = \frac{1}{1+e^{-x}} \tag{6}$$

where: y_j is the sigmoid function.

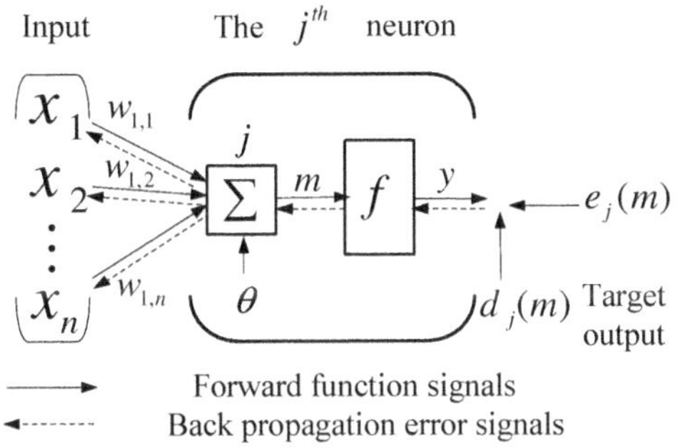

Fig. 3. The model of the jth BP neuron and signal-flow

3.2 The Back-Propagation Algorithm and RMSE

In this study the back-propagation algorithm which uses the parameter update rule is:

$$w_{ij}^{(m)}(n_0+1) = w_{ij}^{(m)}(n_0) + \eta \cdot \delta_{ij}^{(m)} \cdot x_i^{(m)} \tag{7}$$

where: $\delta_{ij}^{(m)}$ is the error in the output of the ith neuron in m layer and n_0 is the index for the iteration.

The update used in this study is modified by the inclusion of an additional momentum term μ which allows previous updates to persist:

$$\Delta w_{ij}^{(1)}(n_0+1) = \eta \cdot \delta_{ij}^{(1)} \cdot x_i^{(1)} + \mu \cdot \Delta w_{ij}^{(1)}(n_0) \tag{8}$$

The training rate of an ANN is sensitive to the learning rate η and momentum coefficient μ. The larger of learning rate is selected, the quicker of the training rate, because large η value causes more changes to weights in the network. However the training phase can cause oscillations when η is selected too large. Therefore, μ is introduced into equation (8). To a certain extent, the momentum coefficient μ has a stabilizing effect and makes curves smoothness.

What's more, the root mean square error (RMSE) is also important to ANN model, the learning rules are based on RMSE, the equation of RMSE is:

$$E = \frac{1}{2}(D-Y)^2 \tag{9}$$

Where: ***D*** and ***E*** are the output of expectation and RMSE respectively. The weight coefficient will be adapted by ***E*** and keep ***Y*** close to ***D***.

Obviously, when the signals flow forward to output layer and the results are compared with target output ***D***. Fig. 3 also denoted the iterative process of one neuron. Two states were chosen by ***E***:

- If ***E*** are less than the predefined threshold then the training will be finished and the result signals will be output,
- Contrarily, if ***E*** are different from the desired response, the error corrections signals will be propagated back to adjust weights and bias levels of each layer in accordance with the error value. Then the signal flow forward again, the process will be repeated until ***E*** achieves the target.

In order to reduce RMSE to minimum it is necessary to have correct network architecture. Therefore the sigmoid equation (6) is used as the transfer function in this study [4], [7], [10], [11] just as the most common transfer function implemented in the literatures. Furthermore, the number of training neurons of the hidden layer is an important factor that will determine the training efficiency and optimization. Generally, excessive neurons of the hidden layer, which are also called over-fitting, can conduce near-zero error on predicting training data, or may lead to a longer training time and slower training speed and result in the process whereby the network learns the training data well but has no ability to meet results for test data. When training set size is too small, the network cannot learn effectively, this will lead to under-fitting and weak generation. In a word, the closest point to the training process and the error of the test data is considered to be the optimal ANN architecture.

4 The Establishment of ANN Optimal Model

There are many scholars and engineers are engaged in researching the number of the hidden layer. For example: Hecht-Nielsen [7] proposed that there exists a three-layer back propagation neural network that can approximate f to within ***E*** mean squared error accuracy. Yu-Jhih and Paul [8] showed there is a three-layer model can solve many questions. Hush [9] developed a one hidden layer of his network and discussed the efficient algorithms for function approximation and so on. Therefore, one hidden layers was preferred in this study. What's more, the acceleration data of actuator and kickstand, for instance, at $time(t), time(t+1), \cdots, time(t+n-1)$ are chosen as the input signals of the neural network controller, Fig. 4 shows it, i.e.

$$\left[y_1^{(1)} \quad y_2^{(1)} \quad \cdots \quad y_n^{(1)}\right]^T = \left[\ddot{x}_1(t) \quad \ddot{x}_1(t+1) \quad \cdots \quad \ddot{x}_1(t+n-1)\right]^T \tag{10}$$

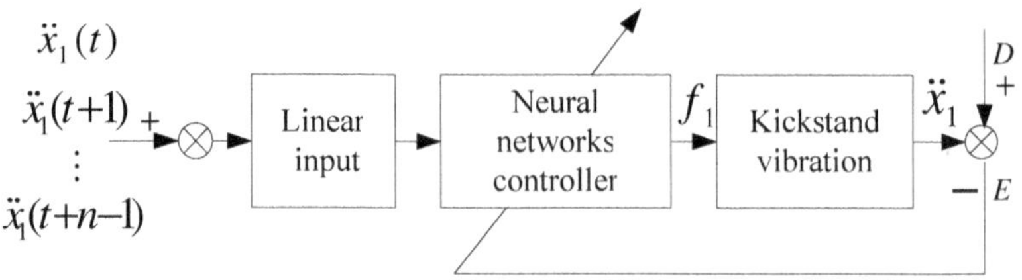

Fig. 4. The structure of the controller

In this study, seven input values and one output value, which is shown in Fig. 4, namely N_i =7, and N_o =1 are presented. Two hundred acceleration data of each neuron are provided.

In order to obtain a good performance of the ANN, it is indispensable to have an optimal ANN model. The empirical calculated number of hidden layer neuron is given in Table 1. Details on the implementation of this system are addressed in [10].

Table 1. The empirical calculated number of neuron of hidden layer(s) (N_i: number of input neuron, N_o :number of output neuron)

The empirical formula	Calculated number of neuron for this study
$\leq 2\times N_i+1$	$\leq$15
$3N_i$	21
$(N_i+N_o)/2$	4
$\frac{2+N_o\times N_i+0.5N_o\times(N_o^2+N_i)-3}{N_i+N_o}$	2
$2N_i/3$	5
$\sqrt{N_i\times N_o}$	3
$2N_i$	14

As can be seen from Table 1, the number of neurons that may be used in the hidden layer varies between 2 and 21. So 2, 4, 14 and 21 are selected as the number of hidden layer. The optimal ANN model will be established:

- Numbers of hidden layers: 1
- Numbers of hidden neurons in each hidden layer: 2,4, 14, 21
- The goal of the training: 0.01
- The epochs of the training: 20000
- Initial μ =0.001.

Table 2. Performance of the neural network models:

Model	Network architecture	RMSE	Results
1	2 hidden neurons	46.8184	Maximum epoch reached.
2	4 hidden neurons	38.5864	Maximum epoch reached
3	14 hidden neurons	7.9022	Maximum μ reached
4	21 hidden neurons	2.0275	Performance goal met

Table 2 shows the optimal models of the ANN. Each ANN model is trained with the training set until it reaches pre-defined training goal. The data are fed to corresponding model, and then the signals flow from input layer to output layer via hidden layer. RMSE is a popular method that can be compared with each ANN model. To evaluate the network architecture, each RMSE is used to be compared with other models in Table 2, the results of ANN architecture models are shown in Fig. 5. The training error vs. training epochs of model 4 is presented in Fig.6.

The number of hidden layers and neurons are important variables that are shown in Table 2. The more the number of hidden layers and neurons, the better the ANN fits the training data. Model 4 has the lower RMSE in these models, higher accuracy and efficiency. In a word, it is the optimal ANN model in this study.

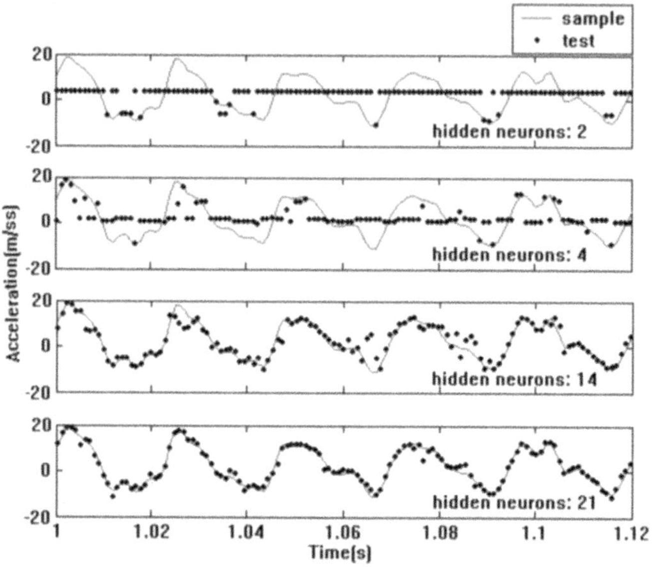

Fig. 5. Performance of each ANN architecture model

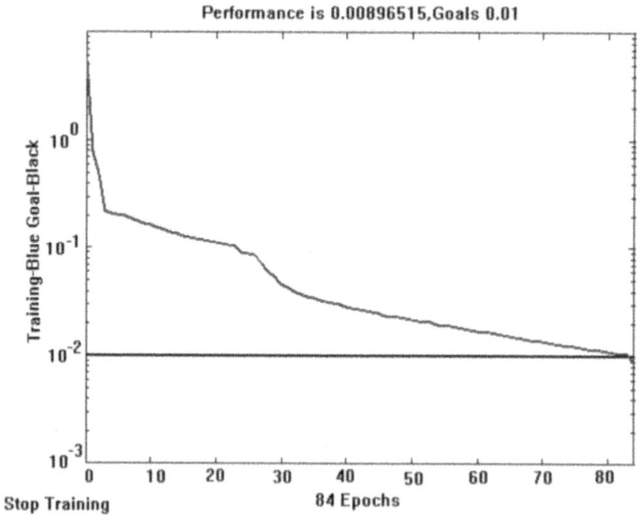

Fig. 6. Training error vs. training epochs of model 4

5 Experimental Results and Discussion

The results of real-time measurements were used to assess the performance of the control method presented in this paper. The sampling frequency is 1024 Hz. Fig. 7 shows the acceleration output response of the kickstand caused by f_1 . As can be seen from Fig. 7, the convergence of vibration signal from 0 to 0.2 is quicker than that from 0.2 to 0.8, but when the amplitude of the following data reached ± 1 , the oscillation become stable. It illuminates that the ANN can guide the controller better in real-time, and shows that the developed vibration control system has good performance against the disturbance that comes from the actuator platform to the floor.

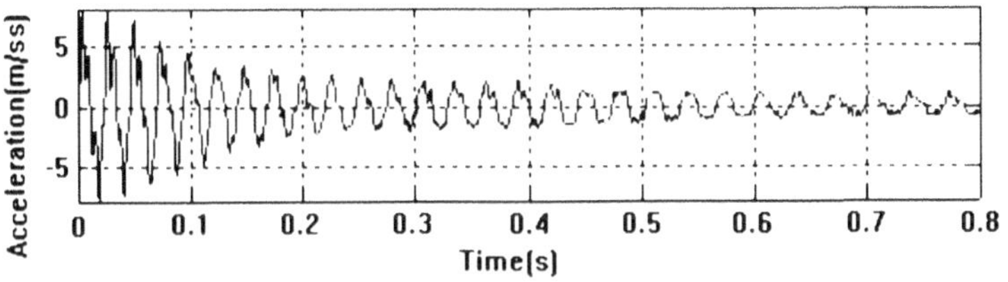

Fig. 7. Real-time measurement of the controller

Resonance frequency located between 22 Hz and 36 Hz in terms of theoretic calculation. The vibration as a function of the excitation frequency was obtained, as shown in Fig. 8. It shows the kickstand's spectrum of with control and without control. In this figure, two peaks can be noticed on no control condition. These are caused by the kickstand and the rubber layer; the maximum values were 4 m/ss at 62 Hz and 11 m/ss at 70 Hz approximately. It's clear that the power of these frequencies is large, and it might damage the vibration system in long running. While one peak can be seen on control condition, and the maximum value was about 3 m/ss at 60 Hz, that is, the vibration of kickstand is weakened when the controller presses downward control force. Obviously, the main frequency decreased by the controller. The vibration can be control better.

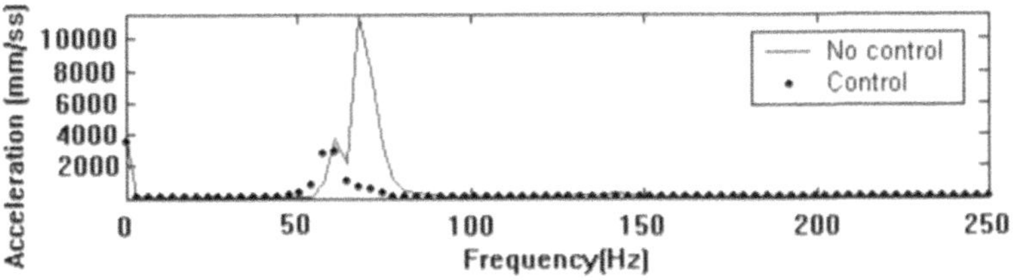

Fig. 8. The kickstand's spectrum of control and no control (n =4200 r/min)

6 Conclusions

In terms of the nonlinearity in vibration control system and the deficiencies of existing linear control methods, a new neuron-control method has been developed. In order to realize high performance in vibration control of a block forming machine, a three-layer BP network is employed, which is adopted as the controller that is presented to update controller weights and momentum coefficient. The optimal model of ANN system demonstrates very satisfactory results in vibration control aspect. The resulting remarks can be drawn hereinafter:

A. This ANN system gives a fairly fast control response for actuator vibration.
B. This ANN control scheme efficiently learns from situ-test data, the result of RMSE is lower than other models and achieves performance goal.
C. This ANN system can suppress the transmissibility from the actuator vibration to kickstand. The results show the effectiveness of the presented control method.
D. The open source code increases the optimal model's flexibility, allowing also the insertion of additional data to enhance the control accuracy and efficiency.

References

1. Ahn, K.G., Pahk, H.J., Jung, M.Y.: a hybrid-type active vibration isolation system using neural networks. Journal of Sound and Vibration **192** (1996) 793-805
2. Liu, X.Q., Zhang, H.Y.: Fault Detection and Diagnosis of Permanent-Magnet DC Motor Based on Parameter Estimation and Neural Network. IEEE Transaction on Industrial Electronics **47** (2000) 1021-1030
3. Chen, Q., Chan, Y.W., Worden, K.: Structural fault diagnosis and isolation using neural networks based on response-only data. Computers and Structures **81** (2003) 2165-2172
4. Zhang, C.L., Mei, D.Q., Chen, Z.C.: Active vibration isolation of a micro-manufacturing platform based on a neural network. Journal of Materials Processing Technology **129** (2002) 634-639
5. Zhang, L., Fu, Y.L., He, L.: A New Active Vibration Isolation Control Method Based on CMAC Neural Network. IEEE International Conference on Industrial Technology (2005) 1280-1282
6. Mahmod, M. Samman: A Hybrid Analysis Method for Vibration Signals Based on Neural Networks and Pattern Recognition Techniques. Journal of Vibration and Acoustics **123** (2001) 122-124
7. Robert Hecht-Nielsen: Theory of the Backpropagation Neural Network. International Joint Conference on Neural Networks **1** (1989) 593-650
8. Wu, Y.J., Chau, P.M., Robert Hecht-Nielsen: A Supervised Learning Neural Network Coprocessor for Soft-Decision Maximum-Likelihood Decoding. IEEE Transactions on Neural Networks **6** (1995) 986-992
9. Hush, D.R., Horne, B.: Efficient Algorithms for Function Approximation with Piecewise Linear Sigmoidal Networks. IEEE Transactions on Neural Networks **9** (1998) 1129-1141
10. Sonmez, H., Gokceoglu, C., Nefeslioglu, H. A., Kayabasi, A.: Estimation of rock modulus: For intact rocks with an artificial neural network and for rock masses with a new empirical equation. International Journal of Rock Mechanics & Mining Sciences **43** (2006) 224-235
11. Karri, V.: Drilling Performance Prediction Using General Regression Neural Networks. IEA/AIE 2000, LNAI 1821. Springer-Verlag, Berlin Heidelberg (2000) 67-73

Global Asymptotical Stability of Internet Congestion Control*

Hong-yong Yang[1,2], Fu-sheng Wang[2], Xun-lin Zhu[1], and Si-ying Zhang[1]

[1] School of Information Science and Engineering, Northeastern University, Shenyang, 110006, China
hyyang@yeah.net
[2] School of Computer Science and Technology, Ludong University, Yantai, 264025, China

Abstract. A class of Internet congestion control algorithms with communication delays is studied. The algorithm is a pieced continuous function that will be switched on the rate of the source. Based on the Lyapunov theorem, the Lyapunov stability of the system is analyzed. By applying Barbalat Lemma, the global asymptotical stability (GAS) of the algorithm is proved, and a more concise criterion is presented.

1 Introduction

Nowadays, in order to ensure the quality of service and the capacity of Internet, the sources of Internet apply TCP congestion control algorithms to avoid network congestion, such as TCP Reno[1] (and its variants), and the link nodes use active queue management (AQM) schemes to improve the serving capacity of Internet, such as DropTail[1], RED[2]. However, the existing congestion control algorithm which are based on "trial-and-error" methods employed on small test beds may be ill-suited for future networks where both communication delay and network capacity can be large, which has been proved by the fact of two times to revise the parameters of the RED algorithm. This has motivated the research on theoretical understanding of TCP congestion control and the search for protocols that scale properly so as to maintain stability in the presence of these variations.

In order to ensure the quality of service, improve the throughout and decrease the queue vibrating of Internet, there are many new Internet congestion control algorithms presented [3,4,5,6]. Based on the views of the network's optimization, Kelly et al [7] have developed a network framework with an interpretation of various congestion control mechanisms. They proposed a prime algorithm for TCP rate control and a dual algorithm for AQM scheme, which generalize the Additive Increase/Multiplicative Decrease (AIMD) congestion avoidance strategy [1] to large-scale networks. The advances in mathematical modelling of Kelly's

* This work is supported by the National Postdoctoral Science Foundation of China (under the grant 20060390972), and the Science Foundation of Office of Education of Shandong Province (under the grant J06G03) and the Postdoctoral Science Foundation of Northeastern University (under the grant 20060110)of China.

D. Liu et al. (Eds.): ISNN 2007, Part I, LNCS 4491, pp. 241–248, 2007.

primal algorithm have stimulated the research on the analysis of the behavior such as stability, robustness and fairness.

Since there exist the communication delays in Internet, the nonlinear delay-differential equation need to be analyzed when studying the dynamics of the algorithm. The convergence of Kelly's primal algorithm has been established in the absence of the communication delays in [7]. The stability of this algorithm with the communication delays has drawn much attention in the past few years. The continuous-time model and the discrete-time model of Kelly's primal algorithm with homogenous communication delays for different TCP connections have been investigated in [8], respectively. For a more general case of networks with heterogeneous round-trip delays, they proposed a conjecture on the local stability of the algorithm, respectively. Recently, their conjectures have received much attention [9,10], where Tian and Yang [10] have studied their conjectures and obtained a more general stability criterion. The new criterion in [10] is stronger than the conjecture, and enlarges the stability region of control gains and admissible communication delays.

In this paper, we study the global asymptotical stability of Kelly's primal algorithm with the communication delay in a single link accessed by a single source. The algorithm model is described as

$$\dot{x}(t) = \begin{cases} f(x), & x > 0; \\ (f(x))^+, & x = 0. \end{cases} \tag{1}$$

where $f(x) = \kappa(w - x(t-D)p(x(t-D)))$, $\kappa > 0$ is the control gain of the system, D is the communication delay, $x(t)$ is the sending rate of the source at time t. The function $p(\cdot)$ is the congestion indication probability (or congestion control rate) back from the link node, which is assumed to be increasing, nonnegative, concave and not identically zero, satisfying $0 \le p(\cdot) \le 1$. $x(t)p(t)$ denotes the marked packets number of the source at time t, w is a desired target value of marked packets received back at source. $(f(x))^+ = max\{f(x), 0\}$. From the description of the system (1), we know that the solution of Eq.(1) should be $x(t) \ge 0$.

We note that the GAS problem of the system (1) has been studied in [11,12], which pointed out that the global asymptotic stability of system (1) can be ensured if the product of the control gain and the delay constant, κD, is upper bounded. However, the upper bound given in [11] is very complicated and might be inconvenient for practical application. In [12], a simpler and more explicit formula of the GAS condition was proposed as $\kappa D < \frac{1}{4}$. In this paper, we consider the problem based on the switched model of the original Kelly's algorithm, and obtain a less conservative GAS criterion.

The rest of this paper is organized as follows. The Lyapunov stability of Kelly's prime algorithm is analyzed by applying Lyapunov-Razumikhin theorem in Section 2. In Section 3, based on the global attractability of the algorithm from Barbalat's Lemma, the criteria of the global asymptotical stability(GAS) is presented. The conclusions are showed in Section 4.

2 Analysis of Lyapunov Stability of the System

In this section, we discuss the Lyapunov stability of the system (1) by applying Lyapunov-Razumikhin theorem. From the description of the system (1), we know that there exists a unique equilibrium point x^* satisfying

$$w = x^* p^*. \tag{2}$$

where $p^* = p(x^*)$ is the congestion rate at the equilibrium point.

From the description of the system (1), all solutions $x \geq 0$. Since $p(x)$ is a increasing, concave function, satisfying $0 \leq p(x) \leq 1,\ p'(x) \geq 0,\ p"(x) \leq 0$. Then

$$xp'(x) \leq \int_0^x p'(s)ds = p(x) - p(0) \leq 1.$$

we obtain

$$p(x) + xp'(x) \leq 2.$$

The following Lemma can be gotten.

Lemma 1. Suppose $f(x) = \kappa(w - xp(x))$. Then $|f'(x)| \leq 2\kappa$.

Now, we define the derivative

$$\dot{x}(t_0) = \dot{x}(t_0^+) = \lim_{t \to t_0^+} \frac{x(t) - x(t_0)}{t - t_0}.$$

Knowing from the system (1), there is $\dot{x}(t) = 0$ or $\dot{x}(t) = f(x(t - D))$, and $f(x^*) = 0$. Suppose $x(t) = x^* + \hat{x}(t)$, when $\dot{x}(t) = f(x(t - D))$, we obtain

$$|\dot{x}(t_0)| = |f(x(t - D)) - f(x^*)| \leq 2\kappa|\hat{x}(t - D)|. \tag{3}$$

where Eq.(3) is correct for $\dot{x}(t) = 0$, too.

Theorem 1. When $\kappa D < \frac{1}{2}$, the system (1) is Lyapunov stable.

proof. Let $x(t) = x^* + \hat{x}(t)$, the system (1) can be transformed into

$$\dot{\hat{x}}(t) = \begin{cases} f(\hat{x}(t - D)), & \hat{x}(t) > -x^*; \\ (f(\hat{x}(t - D)))^+, & \hat{x}(t) = -x^*. \end{cases} \tag{4}$$

where

$$\begin{aligned} f(\hat{x}(t - D)) = & -\kappa\hat{x}(t - D)p(\hat{x}(t - D) + x^*) \\ & -\kappa x^*(p(\hat{x}(t - D) + x^*) - p^*). \end{aligned} \tag{5}$$

It follows that

$$f(\hat{x}(t - D)) = -\kappa\hat{x}(t - D)[p(\hat{x}(t - D)) + x^* p'(\tilde{x})]. \tag{6}$$

where $p(\hat{x}(t - D)) = p(\hat{x}(t - D) + x^*)$, $\tilde{x} = \alpha\hat{x}(t - D) + x^*$, $\alpha \in [0, 1]$.

Supposing the solution $\hat{x}(t) = \hat{x}(t, \varphi)$ of the system (4) with $\varphi \in [-2D, 0]$, and Lyapunov function $V(\hat{x}(t)) = \frac{1}{2}\hat{x}^2(t)$, we know that the function $V(\hat{x}(t))$

is a continuous function in $t \in [-2D, \infty)$. Based on the Lyapunov-Razumikhin theorem [14], let $V(\hat{x}(t+\theta)) \leq V(\hat{x}(t))$, for any $\theta \in [-2D, 0]$, implying

$$|\hat{x}(t+\theta)| \leq |\hat{x}(t)|, \ \forall\ \theta \in [-2D, 0]. \tag{7}$$

In the following, we will compute the derivative of the Lyapunov function along the trajectories of (4). Firstly, when $\hat{x}(t) = -x^*$, we derive from the derivative definition,

$$\dot{V}(t) = \hat{x}(t)\dot{\hat{x}}(t) = -x^*[f(\hat{x}(t-D)]^+ \leq 0. \tag{8}$$

Then, when $\hat{x}(t) > -x^*$, it follows from Eq.(6) that

$$\dot{V}(t) = -\kappa\hat{x}(t)\hat{x}(t-D)[p(\hat{x}(t-D)) + x^* p'(\tilde{x})].$$

Since

$$\hat{x}(t-D) = \hat{x}(t) - \int_{t-D}^{t} \dot{\hat{x}}(s)ds. \tag{9}$$

we have

$$\begin{aligned}\dot{V}(t) = &-\kappa\hat{x}^2(t)[p(\hat{x}(t-D)) + x^* p'(\tilde{x})] \\ &+\kappa\hat{x}(t)[p(\hat{x}(t-D)) + x^* p'(\tilde{x})] \textstyle\int_{t-D}^{t} \dot{\hat{x}}(s)ds.\end{aligned}$$

Substituting inequation (3) and inequation (7), we obtain

$$\begin{aligned}\dot{V}(t) \leq &-\kappa\hat{x}^2(t)[p(\hat{x}(t-D)) + x^* p'(\tilde{x})] \\ &+2\kappa^2 D\hat{x}^2(t)[p(\hat{x}(t-D)) + x^* p'(\tilde{x})]\end{aligned}$$

When $\kappa D < \frac{1}{2}$, we get

$$\begin{aligned}\dot{V}(t) &\leq -\kappa(1-2\kappa D)\hat{x}^2(t)[p(\hat{x}(t-D)) + x^* p'(\tilde{x})] \\ &\leq 0.\end{aligned} \tag{10}$$

Therefore, the system (4) is Lyapunov stable by the Lyapunov-Razumikhin theorem[14]. We finish the Proof of the theorem 1.

3 Analysis of the Global Asymptotical Stability

In Section 2, we have known that the system (4) is Lyapunov stable. If the system (4) is the global attractable, the global asymptotical stability can be derived immediately from the definition of the GAS. In the following, we are going to prove the global attractability of the system (4). Now, let's firstly investigate several important results.

Lemma 2. All solutions $x(t)$ of the system (1) will never go to infinite at a finite time t_0.

Lemma 3. All solutions $x(t)$ of the system (1) does not escape to infinite when $t \to \infty$.

The results of Lemma 2 and Lemma 3 show that when $x(t)$ is increasing, it is going to achieve a maximum at some time. After that time, it will decrease. For convenience of further discussion we denote

$$\begin{aligned} T_0 &= \inf\{t > D : \dot{x}(t) < 0\}, \\ T &= \inf\{t > T_0 : \dot{x}(t) > 0\}, \\ T_1 &= \inf\{t > T : \dot{x}(t) \le 0\} \\ T_2 &= \inf\{t > T_1 : \dot{x}(t) \ge 0\}. \end{aligned}$$

Lemma 4. There exists a positive number M such that for any $t > T$, $x(t)$ satisfies $0 < x(t) \le M$.

Proof. As $x(t)$ is differentiable, and we have excluded the possibility of escaping to infinity for $x(t)$, to prove this lemma it suffices to show all the extreme values of $x(t)$ are greater than zero and upper bounded by M.

Now, let us consider all the staying points of $x(t)$. Suppose $x(t)$ reaches a staying point at $t = t_1$, i.e.,

$$\dot{x}(t_1) = \kappa(w - x(t_1 - D)p(x(t_1 - D))) = 0,$$

then, by the uniqueness of the equilibrium we have

$$x(t_1 - D) = x^*.$$

There are three possibilities for the derivative of $x(t_1 - D)$ at the time $t_1 - D$, namely, $\dot{x}(t_1 - D) > 0$, $\dot{x}(t_1 - D) < 0$, or $\dot{x}(t_1 - D) = 0$. We discuss these three cases below.

(1) $\dot{x}(t_1 - D) > 0$. Without loss of generality we assume that $t_1 - D \in (T, T_1)$. Next we prove that $x(t)$ achieves a maximum at t_1 in this case. Indeed, when $t_1 - D < t < t_1$ and $\max(T, t_1 - 2D) < t - D < t_1 - D$, since $x(t - D) < x(t_1 - D)$, we obtain

$$\dot{x}(t) > \dot{x}(t_1) = 0.$$

When $t_1 < t < t_1 + D$ and $t_1 - D < t - D < \min(t_1, T_1)$, since $x(t - D) > x(t_1 - D)$, we get

$$\dot{x}(t) < \dot{x}(t_1) = 0.$$

Therefore, $x(t)$ achieves a maximum at $t = t_1$.

Now, we prove that $x(t_1)$ is upper bounded by a positive number denoted by M. Since

$$\begin{aligned} x(t_1) &= x(t_1 - D) + \int_{t_1 - D}^{t_1} \kappa(w - x(s - D)p(x(s - D)))ds \\ &= x^* + \kappa w D - \int_{t_1 - D}^{t_1} \kappa x(s - D)p(x(s - D)))ds, \end{aligned}$$

where $\int_{t_1 - D}^{t_1} \kappa x(s - D)p(x(s - D)))ds \ge 0$, we get

$$x(t_1) \le x^* + \kappa w D \doteq M.$$

By carrying out this procedure we can prove that any maximum achieved by $x(t)$ for $t > T$, satisfies

$$x(t) \le M.$$

(2) $\dot{x}(t_1-D)<0$. Without loss of generality we assume that $t_1-D\in(T_1,T_2)$. Using a procedure similar to that used in case (1) we can show that $x(t)$ achieves a minimum at t_1 in this case.

Since

$$\begin{aligned} x(t_1) &= x(t_1-D)+\int_{t_1-D}^{t_1}\kappa(w-x(s-D)p(x(s-D)))ds \\ &= x^*+\kappa wD-\int_{t_1-D}^{t_1}\kappa x(s-D)p(x(s-D)))ds, \end{aligned}$$

where $x(t)\le M$, $p(x)\le 1$, it follows that

$$\begin{aligned} x(t_1) &\ge x^*+\kappa wD-\kappa MD \\ &= x^*(1-\kappa D)+\kappa wD(1-\kappa D). \end{aligned}$$

When $\kappa D<\frac{1}{2}$, we have

$$x(t_1)>0.$$

(3) $\dot{x}(t_1-D)=0$. In this case, by integrating Eq. (1), we get $x(t_1)=x(t_1-D)=x^*\in(0,M]$.

Summarizing the above three cases, we know the extreme points of $x(t)$ are greater than zero and upper bounded by M. So the lemma 4 is proved.

Theorem 2. when $\kappa D<\frac{1}{2}$, the solution of the system (4) is global attractable.

proof. We will split the proof of this theorem into four parts.

1). For all $t>T+D$, the function $\hat{x}^2(t)P(\hat{x}(t-D))$ is uniformly continuous.

Based on the Lemma 3 and Lemma 4, for all $t>T+D$, there are $0<x(t)\le M$. Since $\hat{x}(t)=x(t)-x^*$, we obtain $-x^*\le\hat{x}(t)\le\kappa wD$, i.e. $|\hat{x}(t)|\le M$. Eq (3) implies $|\dot{x}(t)|\le 2\kappa M$. For any positive number $\varepsilon>0$, let $\delta=\varepsilon/(2\kappa M)$, when $|t_1-t_2|<\delta$, we have

$$|x(t_1)-x(t_2)|<\varepsilon.$$

So $x(t)$ is uniformly continuous, i.e. $\hat{x}(t)$ is uniformly continuous.

For all $t\in(T+D,+\infty)$, we derive $x(t)>0$ from the Lemma 4. Since $p'(x)\ge 0$ and $p"(x)\le 0$, it follows that

$$0\le p'(x(t))\le p'(0),\ \forall\, t\in(T+D,+\infty).$$

Because $\hat{x}(t)$ is uniformly continuous, this implies that $\hat{x}(t)p(\hat{x}(t-D))$ is uniformly continuous, for all $t\in(T+D,+\infty)$. Therefore, the function $\hat{x}^2(t)p(\hat{x}(t-D))$ is uniformly continuous since $\hat{x}(t)$ and $\hat{x}(t)p(\hat{x}(t-D))$ are bound and uniformly continuous, for all $t\in(T+D,+\infty)$.

2). $\lim_{t\to+\infty}\hat{x}^2(t)p(\hat{x}(t-D))=0$.

Based on the Lemma 4, $x(t)>0$ for all $t>T+D$ when $\kappa D<\frac{1}{2}$. Since $p(x)\ge 0$, $p'(x)\ge 0$, we derive from Eq. (10)

$$\dot{V}(\hat{x}(t))\le-\kappa(1-2\kappa D)\hat{x}^2(t)p(\hat{x}(t-D)).$$

By integrating, one deduces that for all $t>T+D$

$$V(\hat{x}(t))\le V(\hat{x}(T+D))-\kappa(1-2\kappa D)\int_{T+D}^{t}\hat{x}^2(s)p(\hat{x}(s-D))ds.$$

Since $V(\hat{x}(t))$ is bounded with $|\hat{x}(t)| \leq M$, it follows that

$$\lim_{t\to+\infty} \int_{T+D}^{t} \hat{x}^2(s)p(\hat{x}(s-D))ds < +\infty.$$

When $\hat{x}^2(t)p(\hat{x}(t-D))$ is uniformly continuous for all $t > T + D$, Barbalat's Lemma[13] ensures that

$$\lim_{t\to+\infty} \hat{x}^2(t)p(\hat{x}(t-D)) = 0.$$

3). $\lim_{t\to+\infty} p(x(t)) \neq 0$.
Supposing $\lim_{t\to+\infty} p(x(t)) = 0$, then there exists $T_3 > T + D$ for any $\varepsilon_0 > 0$, when $t > T_3 - D$, we have $P(x(t)) < \varepsilon_0$. i.e. for all $t > T_3$, $P(x(t-D)) - p^* < \varepsilon_0 - p^*$. Since

$$\begin{aligned}\hat{x}(t) &= \hat{x}(T_3) - \kappa \textstyle\int_{T_3}^{t} [\hat{x}(s-D)p(x(s-D)) + x^*(p(x(s-D)) - p^*)]ds \\ &> \hat{x}(T_3) + \kappa(t-T_3)x^*(P^* - \varepsilon_0) - \kappa \textstyle\int_{T_3}^{t} \hat{x}(s-D)p(x(s-D))ds.\end{aligned}$$

where $|\hat{x}(t)| \leq M$, for all $t > T_3$, it follows that

$$\hat{x}(t) > \hat{x}(T_3) + \kappa(t-T_3)(x^*P^* - x^*\varepsilon_0 - M\varepsilon_0).$$

If $\varepsilon_0 < \frac{x*p^*}{x^*+M}$, we yield

$$\lim_{t\to+\infty} \hat{x}(t) = +\infty.$$

This is in contradiction with $|\hat{x}(t)| \leq M$, i.e. $\lim_{t\to+\infty} p(x(t)) \neq 0$.
4). The system (4) is global attractable.
We have known that

$$\lim_{t\to+\infty} \hat{x}^2(t)p(\hat{x}(t-D)) = 0,$$

but

$$\lim_{t\to+\infty} p(\hat{x}(t)) = \lim_{t\to+\infty} p(x(t)) \neq 0.$$

it implies

$$\lim_{t\to+\infty} \hat{x}^2(t) = 0.$$

Since $\hat{x}(t)$ is bounded when $t > T$, it follows that

$$\lim_{t\to+\infty} \hat{x}(t) = 0.$$

Therefore, the system (4) is global attractable. Then, we finish the proof of the Theorem 2.

Now we give the main result of this section

Theorem 3. The system (1) is GAS, if $\kappa D < \frac{1}{2}$.

4 Conclusion

In this paper, we have studied the GAS of Kelly's primal algorithm with communication delay in a single link accessed by a single source. The Lyapunov stability of Kelly's prime algorithm is analyzed by applying Lyapunov-Razumikhin theorem. Based on the global attractability from Barbalat's Lemma, the criteria of the GAS of the algorithm is obtained. at last, an simple ensured upper delay bound guaranteeing the GAS is presented.

References

1. Jacobson, V.: Congestion Avoidance and Control, Proceedings of ACM SIGCOMM'88, Stanford, CA (1988) 314-329
2. Floyd, S., Jacobson, V.: Random Early Detection Gateways for Congestion Avoidance. IEEE/ACM Trans. Networking **1** (1993) 397-413
3. Hollot, V.C., Misra, V., Towsley, D., et al.: Analysis and Design of Controllers for AQM Routers Supporting TCP Flows. IEEE Trans. Automatic Control **47** (2003) 945-959
4. Athuraliya, S., Li, V., Low, S., Yin, Q.: REM: Active Queue Management. IEEE Network **15** (2001) 48-53
5. Gibbens, R., Kelly, F.: Resource Pricing and the Evolution of Congestion Control. Automatica **35** (1999) 1969-1985
6. Kunniyur, S., Srikant, R.: An Adaptive Virtual Queue (AVQ) Algorithm for Active Queue Management. IEEE/ACM Trans. Networking **12** (2004) 286-299
7. Kelly, F.P., Maulloo, A., Tan, D.: Rate Control for Communication Networks: Shadow Prices, Proportional Fairness, and Stability. J. Oper. Res. Soc. **49** (1998) 237-252
8. Johari, R., Tan, D.: End-to-end Congestion Control for the Internet: Delays and Stability. IEEE/ACM Trans. Networking **9** (2001) 818-832
9. Massoulie, L.: Stability of Distributed Congestion Control with Heterogeneous Feedback Delays. IEEE Trans. Automatic Control **47** (2002) 895-902
10. Tian, Y.P., Yang, H.Y.: Stability of the Internet Congestion Control with Diverse Delays. Automatica **40** (2004) 1533-1541
11. Deb, S., Srikant, R.: Global Stability of Congestion Controllers for the Internet. IEEE Trans. Automatic Control **48** (2003) 1055-1060
12. Mazenc, F., Niculescu, S.: Remarks on the Stability of A Class of TCP-like Congestion Control Models. In: Proc. of the 42nd IEEE CDC, Maui, Hawaii USA (2003) 5591-5594
13. Niculescu, S.: Delay Effects on Stability: A Robust Control Approach. Spring-Verlag: Berlin Heidelberg New York (2001)
14. Hale, J.: Theory of Functional Differential Equations, Spring-Verlag, Berlin Heidelberg New York (1977)

Dynamics of Window-Based Network Congestion Control System*

Hong-yong Yang[1,2], Fu-sheng Wang[1], Xun-lin Zhu[2], and Si-ying Zhang[2]

[1] School of Computer Science and Technology, Ludong University,
Yantai, 264025, China
hyyang@yeah.net

[2] School of Information Science and Engineering, Northeastern University,
Shenyang, 110004, China

Abstract. A class of window-based network congestion control system with communication delays is studied. By analyzing the network system with communication delay, a critical value of the window size to ensure the stability of network is obtained, and a critical value of the delay to ensure the system stability is presented. Enlarging the delay across the critical value, we find that the congestion control system exhibits Hopf bifurcation.

1 Introduction

Network congestion control algorithm is important for Internet to improve the capacity and the quality of service (QoS). With the rapid development of the technique of communication networks, especially Internet, it becomes more and more crucial to analyze the dynamics of the network. For ensuring the QoS of Internet, the sources of Internet apply TCP (Transmission Control Protocol) congestion control algorithm, which was presented by Jacobson [1] in last century. Jacobson's Internet congestion avoidance scheme is named as TCP Reno applied in the transmission layer to adjust the rate. In Internet, a source can connect with a lot of link nodes (such as switcher, router et al.), and a link node can be shared by many sources. The packets are sent from the sources to the link nodes. If the packets full the buffers in link nodes, the arriving packet will be discarded, so a congestion marking is raised. When a congestion marking is feedback from the link to the source, the rate of the source will be decreased. TCP Reno uses Additive Increase/Multiplicative Decrease method to adjust the sizes of the sending windows avoiding the congestion.

However, the development of the Internet is so fast that its scale is almost exploded. When the scale and the bandwidth of Internet are small, it is satisfied for the capacity of TCP Reno. With the development of Internet and the increase of the network users, TCP Reno has not adapted to the requirement of

* This work is supported by the National Postdoctoral Science Foundation of China (under the grant 20060390972), and the Science Foundation of Office of Education of Shandong Province (under the grant J06G03) of China.

D. Liu et al. (Eds.): ISNN 2007, Part I, LNCS 4491, pp. 249–256, 2007.

Internet. In order to ensure the QoS of Internet and improve the throughput of network, there are many new TCP congestion control algorithms proposed, such as TCP-friendly rate-based flow control (TFRC)[2], equation-based congestion control[3], binomial congestion algorithm[4], general AIMD congestion control[5] and TCP Vegas[6] so on. Based on the views of the network's optimization, Kelly et al[7] have developed a network framework with an interpretation of various congestion control mechanisms. They proposed a prime algorithm for TCP rate control and a dual algorithm for AQM scheme, which generalize the Additive Increase/Multiplicative Decrease (AIMD) congestion avoidance strategy[1] to large-scale networks. The algorithms associate a utility function with each flow and maximize the aggregate system utility function subject to link capacity constraints.

Since there exist the communication delays in Internet network, we need to analyze the nonlinear delayed system when studying the dynamics of Internet. Johari and Tan[8] investigated the Kelly's prime algorithm[7] with communication delays, and derived some sufficient conditions for the local stability of networks with the same round-trip delays for different TCP connections. For a more general case of networks with heterogeneous round-trip delays, they proposed a conjecture on the local stability of the algorithm. This conjecture has the elegance of being decentralized and locally implemented: each end system needs knowledge only of its own round-trip delay. Recently, this conjecture is studied in [9,10], a weaker result is obtained in [9], a more general criterion is given in [10] and the validity of the conjecture is proved.

In this paper, the stability of a window-based congestion control system is studied, and the dynamics of this system is discussed when the stable conditions are destroyed. The rest of this paper is organized as follows. In the next section we study the stability of window-based congestion control system by computing the critical value of communication delay. The new feature of the algorithm is analyzed when the stable conditions are destroyed in Section 3. Finally, conclusions are showed in Section 4.

2 Stability of Window-Based Congestion Control System

In this section, we discuss a class of window-based network congestion control system with the communication delays. Suppose the set of the sources is $N = \{1, 2, ..., N_0\}$, the rate of the source i is $x_i(t)$. In the congestion control system, the source will increase the sizes of sending windows with aw^n when receiving a acknowledge information, and the size of the windows will decrease bw^m when receiving a congestion information, where $w(t)$ is the sizes of sending windows of the source. The model of the window-based congestion control system is described as the following system, for all $i \in N$,

$$\dot{w}_i(t) = x_i(t)(aw_i^n(t)(1 - q_i(t)) - bw_i^m(t)q_i(t)). \tag{1}$$

where $w_i(t)$ is the sizes of sending windows of the source i at time t, $q_i(t)$ is the congestion indication received by the source i at time t. The constants a, b, m

are larger than zero, and $m > n$. In TCP Reno, there are $a = 1, b = \frac{1}{2}, m = 1, n = -1$.

With the increase of the bandwidth in Internet, the information flow can be approximated to the liquid flow, so we have

$$w_i(t) = x_i(t)T_i.$$

where T_i is the RTT (the round-trip time) of the source i. Then the Eq. (1) is changed, for all $i \in N$,

$$\dot{w}_i(t) = \frac{w_i(t)}{T_i}(aw_i^n(t)(1 - q_i(t)) - bw_i^m(t)q_i(t)). \tag{2}$$

In communication network, the communication delay is the inherent characteristic of the network system. Let T_{i1} denote the forward delay spend by the packets from the source i to the link, and T_{i2} the backward delay of sending the feedback signal from the link to the source i. For all $i \in N$, $T_i = T_{i1} + T_{i2}$. If we consider the effect of communication delay on the system states, Eq (2) can be rewritten as, for all $i \in N$,

$$\dot{w}_i(t) = \frac{w_i(t - T_i)}{T_i}(aw_i^n(t)(1 - q_i(t)) - bw_i^m(t)q_i(t)), \tag{3}$$

where $w_i(t - T_i)$ is the data windows sizes of the source i at time $t - T_i$.

In this paper, we discuss the network with one link shared by a lot of sources, and the model of network is $M/M/1$. The function of the congestion marking (or congestion probability) used as active queue management scheme is

$$p(t) = (\frac{y(t)}{C})^B. \tag{4}$$

where $y(t) = \sum_i x_i(t - T_{i1})$ is the network loadings, C is the capacity of the link, B is the sizes of the buffer. The feedback congestion marking at the sources is

$$q(t) = p(t - T_{i2}) = (\frac{\sum_i x_i(t - T_i)}{C})^B. \tag{5}$$

Assume that system has an isolated equilibrium point w_i^* satisfying

$$\frac{a(w_i^*)^n}{a(w_i^*)^n + b(w_i^*)^m} = q^*, \tag{6}$$

and

$$q^* = (\frac{\sum_i x_i^*}{C})^B. \tag{7}$$

where $x_i^* = w_i^*/T_i$. We linearize the Eq.(3-4) in a neighbour field of the equilibrium point, and obtain

$$\delta\dot{w}_i(t) = -\frac{a(w_i^*)^{n+1}}{T_i q^*}\delta q(t) - \frac{b(m - n)(w_i^*)^m q^*}{T_i}\delta w_i(t), \tag{8}$$

and

$$\delta q(t) = \frac{Bq^*}{y^*}\delta y(t - T_{i2}), \tag{9}$$

where $\delta w_i(t) = w_i(t) - w_i^*$, $\delta q(t) = q(t) - q^*$, $y^* = \sum_i x_i^*$, $\delta y(t - T_{i2}) = \sum_i \delta x_i(t - T_i)$, and $w_i(t - T_i) = x_i(t - T_i)/T_i$. If we suppose each source is fairness in TCP-Like congestion control algorithm, we obtain from Eq.(9)

$$\delta q(t) = \frac{Bq^*}{w_i^*}\delta w_i(t - T_i). \tag{10}$$

Substituting (10) into (8), have

$$\delta \dot{w}_i(t) = -\frac{aB(w_i^*)^n}{T_i}\delta w_i(t - T_i) - \frac{b(m-n)(w_i^*)^m q^*}{T_i}\delta w_i(t), \tag{11}$$

where the dynamic of the time-delay is shown by $\delta w_i(t - T_i)$ in the system. If the effect of the delay is neglected, Eq. (11) becomes

$$\delta \dot{w}_i(t) = -(\frac{aB(w_i^*)^n}{T_i} + \frac{b(m-n)(w_i^*)^m q^*}{T_i})\delta w_i(t), \tag{12}$$

and the system (12) is stable. Next, we discuss the dynamic of the delayed system (11) affected by the delay. The characteristic equation of the linearized equation (11) is

$$(T_i\lambda + \alpha_1)e^{\lambda T_i} + \alpha_2 = 0, \tag{13}$$

where $\alpha_1 = b(m-n)(w_i^*)^m q^*$, $\alpha_2 = aB(w_i^*)^n$. We denote $C(\lambda, T_i) = T_i\lambda + \alpha_1)e^{\lambda T_i} + \alpha_2$.

Let $\lambda = \pm j\omega_0$, where j is the unit of imaginary number, substitute into (13), we have

$$\begin{aligned} \alpha_1 \cos(\omega_0 T_i) - \omega_0 T_i \sin(\omega_0 T_i) + \alpha_2 = 0, \\ \omega_0 T_i \cos(\omega_0 T_i) + \alpha_1 \sin(\omega_0 T_i) = 0. \end{aligned} \tag{14}$$

and obtain

$$\begin{aligned} \omega_0 T_i &= \sqrt{\alpha_2^2 - \alpha_1^2}, \\ \cos(\omega_0 T_i) &= -\tfrac{\alpha_1}{\alpha_2}. \end{aligned} \tag{15}$$

Since $\omega_0 > 0$, we have $\alpha_2 > \alpha_1$ and

$$\frac{\alpha_1}{\alpha_2} = \frac{m-n}{B}(1 - q^*). \tag{16}$$

Since $0 < \frac{\alpha_1}{\alpha_2} < 1$, we obtain $\frac{\pi}{2} < \omega_0 T_i < \pi$ following from Eq. (14), and the critical value of the delay satisfies

$$w_{i0}^* = (\frac{arccos(-\frac{\alpha_1}{\alpha_2})}{bq^*(m-n)\sqrt{(\frac{\alpha_2}{\alpha_1})^2 - 1}})^{\frac{1}{m}}. \tag{17}$$

Applying $w_i = x_i T_i$, we obtain

$$T_{i0} = \frac{1}{x_i^*}(\frac{arccos(-\frac{\alpha_1}{\alpha_2})}{bq^*(m-n)\sqrt{(\frac{\alpha_2}{\alpha_1})^2 - 1}})^{\frac{1}{m}}. \tag{18}$$

Theorem 1. *For all $i \in N$, when $w_i < w_{i0}^*$, the system (3-4) is stable.*

Theorem 2. *For all $i \in N$, when $T_i < T_{i0}$, the system (3-4) is stable.*

Proof. If all roots of the characteristic equation (13) have negative real parts when $T_i < T_{i0}$ for all $i \in N$, the result of Theorem 1 will be obtained. The following Lemma will be used in this proof.

Lemma 1. *[11] For the characteristic equation (13), we define*

$$M(T) = \#\{\lambda : Re(\lambda) \geq 0, C(\lambda, T) = 0\},$$

which denotes the number of the roots with nonnegative real part. Let $0 \leq \bar{T}_1 < \bar{T}_2$. Suppose that for $T \in [\bar{T}_1, \bar{T}_2]$, there are no roots on the imaginary axis. Then $M(\bar{T}_1) = M(\bar{T}_2)$.

Continue to prove the Theorem 1. When $T = 0$, there are no effect of the delay on the system state. Following Eq. (12), we have $C(\lambda, 0) = T\lambda + \alpha_1 + \alpha_2 = 0$. Since $\alpha_1 > 0$ and $\alpha_2 > 0$, the roots of the characteristic equation have negative real parts, that is $M(0) = 0$. When $T_i < T_{i0}$, it is known from the discuss there is no root on the imaginary axis. Following from the Lemma 1, there is $M(T_i) = M(0) = 0$. Therefor, all roots of the characteristic equation (13) have negative real parts when $T_i < T_{i0}$, for all $i \in N$. Then, we finish the proof of Theorem 2.

The next corollary can be obtained following Eq. (17).

Corollary 1. *The system (3-4) is locally asymptotically stable. For all $i \in N$, if*

$$aB(x_i^* T_i)^n < \frac{arccos(-\frac{(m-n)(1-q^*)}{B})}{\sqrt{1 - (\frac{(m-n)(1-q^*)}{B})^2}}. \tag{19}$$

Since q^* is the congestion probability of the system in the equilibrium point, it should be satisfied $q^* \ll 1$. Supposing $q^* < \frac{1}{2}$, then $1 - q^* > \frac{1}{2}$, we have

$$\frac{\alpha_1}{\alpha_2} = \frac{(m-n)(1-Q^*)}{B} > \frac{m-n}{2B}. \tag{20}$$

The following corollary will be get from the Corollary 1.

Corollary 2. *The system (3-4) is locally asymptotically stable. For all $i \in N$, if*

$$aB(x_i^* T_i)^n < arccos(-\frac{m-n}{2B}). \tag{21}$$

Since

$$1 > \frac{m-n}{2B} > 0,$$

have

$$\frac{\pi}{2} < \arccos(-\frac{m-n}{2B}) < \pi.$$

The following corollary will be get from the Corollary 2.

Corollary 3. *The system (3-4) is locally asymptotically stable. For all $i \in N$, if*

$$aB(x_i^* T_i)^n < \frac{\pi}{2}. \tag{22}$$

Note: The result in Corollary 3 is consistent with that in [12,13]. Known following from the analysis, the stable range in Corollary 1 and Corollary 2 is larger than that in [12,13]. So we obtain the stable range in Theorem 2 is larger than that in [12,13].

3 Dynamics of Window-Based Congestion Control System

In this section, we discuss what the dynamics of window-based congestion control system will be shown when the stability is broken. Known from the discuss in last section, the system (3-4) is locally asymptotically stable when $T_i < T_{i0}$, for all $i \in N$. Next, we will study the dynamics of the system when $T_i = T_{i0}$.

Lemma 2. *When $T_i = T_{i0}$, there are a pair of purely imaginary roots $\lambda = \pm j\omega_0$ (where $\omega_0 > 0$) in the characteristic equation, and the pure imaginary root is simple, other roots have strictly negative real parts.*

Proof. Known from the discuss in last section, when $T_i = T_{i0}$ there are a couple of purely imaginary roots in the characteristic equation. We will show that the purely imaginary root $\lambda = \pm j\omega_0$ is simple. Since $C(\lambda, T_i) = (T_i\lambda + \alpha_1)e^{\lambda T_i} + \alpha_2$, derivative

$$\frac{\partial C(\lambda)}{\partial \lambda} = T_{i0}e^{\lambda T_{i0}} + T_{i0}(\lambda T_{i0} + \alpha_1)e^{\lambda T_{i0}}.$$

Let $\lambda = j\omega_0$, we obtain

$$\frac{\partial C(\lambda)}{\partial \lambda}|_{\lambda=j\omega_0} = T_{i0}(jT_{i0}\omega_0 + \alpha_1 + 1)e^{jT_{i0}\omega_0}.$$

So

$$\frac{\partial C(\lambda)}{\partial \lambda}|_{\lambda=j\omega_0} = T_{i0}(\cos(T_{i0}\omega_0) - \alpha_2) + jT_{i0}\sin(T_{i0}\omega_0) \neq 0.$$

Similarly, we have

$$\frac{\partial C(\lambda)}{\partial \lambda}|_{\lambda=-j\omega_0} \neq 0.$$

Therefor, $\lambda = \pm j\omega_0$ is simple root.

Finally, we show that the other roots of Eq. (13) have strictly negative real parts when $T_i = T_{i0}$. Suppose to the contrary that there exist a pair of roots of Eq. (13) $\lambda_{1,2} = \beta \pm j\omega$, where $\beta > 0$. Since the roots are continuous in parameter T_i, for any sufficiently small positive number ϵ, there exists a positive number δ which depends on ϵ, such that $| Re(\lambda_1 - \beta) | < \epsilon$ holds when $T_i \in (T_{i0} - \delta, T_{i0} + \delta)$. Let $\epsilon = \frac{\beta}{2}$, we have $Re(\lambda_1) > \frac{\beta}{2}$ when $T \in (T_{i0} - \delta, T_{i0})$. This contradicts with the conclusion of Theorem 1: For all $i \in N$, when $T_i < T_{i0}$, all roots of the characteristic equation (13) have negative real parts. Thus, we complete the proof of Lemma 2.

Lemma 3. *Suppose* $T_i = T_{i0} + \mu$, *let* $\lambda(\mu) = \beta(\mu) + j\omega(\mu)$ *be the root of Eq. (12) satisfying* $\beta(0) = 0, \omega(0) = \omega_0$. *Then* $\frac{dRe(\lambda)}{d\mu}\mid_{\mu=0} > 0$.

Proof. Comparing Eq.(2) with Eq.(3), we know that the dynamic of the delay is shown in $x_i(t - T_i)$. It is shown by $e^{\lambda T_i}$ in the characteristic equation (13). Since $\lambda(\mu) = \beta(\mu) + i\omega(\mu)$ is the root of Eq. (13), it satisfies

$$(T_i\lambda + \alpha_1)e^{\lambda T_i} + \alpha_2 = 0.$$

Applying the Implicit Function Theorem, we gained

$$\frac{d\lambda}{d\mu} = -\frac{\lambda(T_i\lambda + \alpha_1)}{T_i(T_i\lambda + \alpha_1 + 1)}.$$

Let $\mu = 0$, we have

$$\frac{dRe(\lambda)}{d\mu}\mid_{\mu=0} > 0.$$

Then we finish the proof of Lemma 3.

Based on the conclusions of Lemma 2 and Lemma 3, we obtain the following bifurcation theorem for Eq. (1) by applying the Hopf bifurcation theorem on the delay differential equation [14].

Theorem 3. *Let* $T_i = T_{i0}$, *Eq. (3-4) exhibits the Hopf bifurcation.*

4 Conclusion

In this paper, a class of window-based network congestion control system with communication delays is studied. A critical value of the delay to ensure the stability is presented by analyzing the algorithm with communication delay. The stabilities of the algorithm are presented by comparing the communication delay with the critical value of delay. The new feature of the algorithm is analyzed when the stable conditions are destroyed.

References

1. Jacobson, V.: Congestion Avoidance and Control. Proceedings of ACM SIGCOMM'88, Stanford CA (1988) 314-329
2. Mahdavi, J., Floyd, S.: TCP-friendly Unicast Rate-based Flow Control. http://www.psc.edu/networking /tcp_friendly.html. January (1997)
3. Floyd, S., Handley, M., Padhye, J.: Equation-based Congestion Control for Unicast Applications. In: Proc SIGCOMM Symposium on Communications Architecture and Protocol August (2000) 43-56
4. Bansal, D., Balakrishnan, H.: Binomial Congestion Algorithms. In: IEEE INFOCOM 2001, Anchorage, AK April (2001) 631-640
5. Yang, Y., Lam, S.: General AIMD Congestion Control. Technical Report TR-2000-09, University of Texas at Austin May (2000) http://www.cs.utexas.edu/users/lam/NRL/

6. Brakmo, L., Peterson, L.: TCP Vegas: End to End Congestion Avoidance on a Global Internet. IEEE J. Select. Areas in communications **13** (1995) 1465-1480
7. Kelly, F.P., Maulloo, A., Tan, D.: Rate Control for Communication Networks: Shadow Prices, Proportional Fairness, and Stability. J. Oper. Res. Soc. **49** (1998) 237-252
8. Johari, R., Tan, D.: End-to-end Congestion Control for the Internet: Delays and Stability. IEEE/ACM Trans. Networking **9** (2001) 818-832
9. Massoulie, L.: Stability of Distributed Congestion Control with Heterogeneous Feedback Delays. IEEE Trans. Automatic Control **47** (2002) 895-902
10. Tian, Y., Yang, H.: Stability of the Internet Congestion Control with Diverse Delays. Automatica **40** (2004) 1533-1541
11. Cooke, K., Grossman, Z.: Discrete Delay, Distributed Delay and Stability Switches. J. Math. Anal. Appl. **86** (1982) 592-627
12. Vinnicombe, G.: On the Stability of Networks Operating TCP-like Congestion Control. In Proceedings of IFAC World congress (2002)
13. Kelly, F.: Fairness and Stability of End-to-end Congestion Control. European Journal of Control **9** (2003) 159-176
14. Hale, J.: Theory of functional differential equations. Spring-Verlag, Berlin (1977)

Realization of Neural Network Inverse System with PLC in Variable Frequency Speed-Regulating System

Guohai Liu, Fuliang Wang, Yue Shen, Huawei Zhou, Hongping Jia, and Mei Kang

School of Electrical and Information Engineering, JiangSu University
Zhenjiang 212013, China
ghliu@ujs.edu.cn

Abstract. The variable frequency speed-regulating system which consists of an induction motor and a general inverter, and controlled by PLC is widely used in industrial field. However, for the multivariable, nonlinear and strongly coupled induction motor, the control performance is not good enough to meet the needs of speed-regulating. The mathematic model of the variable frequency speed-regulating system in vector control mode is presented and its reversibility has been proved. By constructing a neural network inverse system and combining it with the variable frequency speed-regulating system, a pseudo-linear system is completed, and then a linear close-loop adjustor is designed to get high performance. Using PLC, a neural network inverse system can be realized in actural system. The results of experiments have shown that the performances of variable frequency speed-regulating system can be improved greatly and the practicability of neural network inverse control was testified.

1 Introduction

In recent years, with power electronic technology, microelectronic technology and modern control theory infiltrating into AC electric driving system, inverters have been widely used in speed-regulating of AC motor. The variable frequency speed-regulating system which consists of an induction motor and a general inverter is used to take the place of DC speed-regulating system. Because of terrible environment and severe disturbance in industrial field, the choice of controller is an important problem. In reference [1][2][3], Neural network inverse control was realized by using industrial control computer and several data acquisition cards. The advantages of industrial control computer are high computation speed, great memory capacity and good compatibility with other software etc. But industrial control computer also has some disadvantages in industrial application such as instability and fallibility and worse communication ability. PLC control system is special designed for industrial environment application, and its stability and reliability are good. PLC control system can be easily integrated into field bus control system with the high ability of communication configuration, so it is wildly used in recent years, and deeply welcomed. Since the system composed of normal inverter and induction motor is a complicated nonlinear system, traditional PID control strategy could not meet the requirement for further control. Therefore, how to enhance control performance of this system is very urgent.

D. Liu et al. (Eds.): ISNN 2007, Part I, LNCS 4491, pp. 257–266, 2007.

The neural network inverse system [4][5] is a novel control method in recent years. The basic idea is that: for a given system, an inverse system of the original system is created by a dynamic neural network, and the combination system of inverse and object is transformed into a kind of decoupling standardized system with linear relationship. Subsequently, a linear close-loop regulator can be designed to achieve high control performance. The advantage of this method is easily to be realized in engineering. The linearization and decoupling control of normal nonlinear system can realize using this method.

Combining the neural network inverse into PLC can easily make up the insufficiency of solving the problems of nonlinear and coupling in PLC control system. This combination can promote the application of neural network inverse into practice to achieve its full economic and social benefits.

In this paper, firstly the neural network inverse system method is introduced, and mathematic model of the variable frequency speed-regulating system in vector control mode is presented. Then a reversible analysis of the system is performed, and the methods and steps are given in constructing NN-inverse system with PLC control system. Finally, the method is verified in experiments, and compared with traditional PI control and NN-inverse control.

2 Neural Network Inverse System Control Method

The basic idea of inverse control method [6] is that: for a given system, an α-th integral inverse system of the original system is created by feedback method, and combining the inverse system with original system, a kind of decoupling standardized system with linear relationship is obtained, which is named as a pseudo linear system as shown in Fig.1. Subsequently, a linear close-loop regulator will be designed to achieve high control performance.

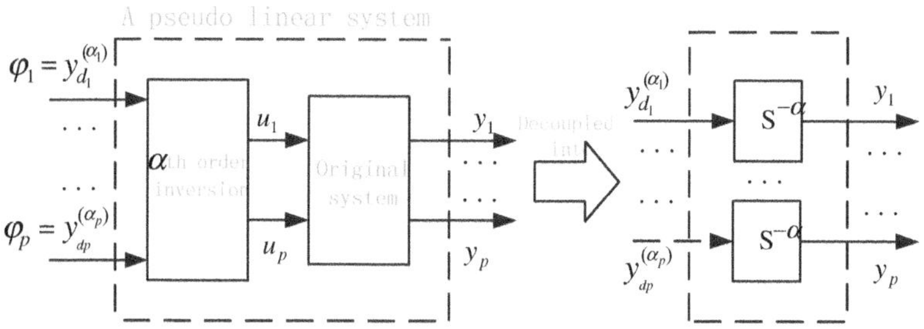

Fig. 1. Linearization and decoupling based on α-th order inversion

Inverse system control method with the features of direct, simple and easy to understand does not like differential geometry method [7], which is discusses the problems in "geometry domain". The main problem is the acquisition of the inverse model in the applications. Since non-linear system is a complex system, and desired strict analytical inverse is very difficult to obtain, even impossible. The engineering application of inverse system control doesn't meet the expectations. As neural network has non-linear approximate ability, especially for nonlinear complexity system, it becomes the powerful tool to solve the problem.

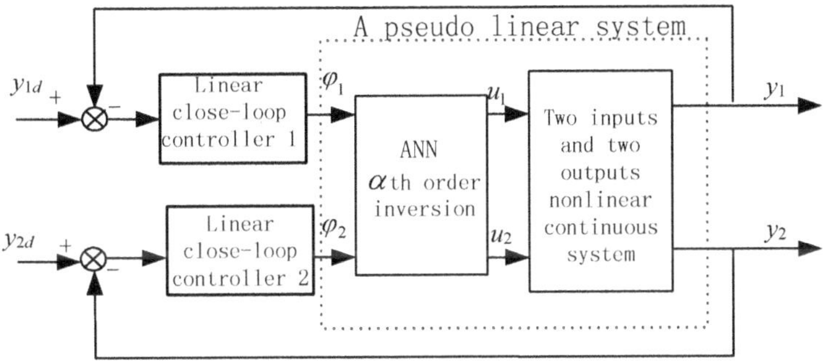

Fig. 2. Compound pseudo linear system control diagram of two inputs and two outputs system

$a-th$ NN inverse system integrated inverse system with non-linear approximate ability of the neural network can avoid the troubles of inverse system method. Then it is possible to apply inverse control method to a complicated non-linear system. $a-th$ NN inverse system method needs less system information such as the relative order of system, and it is easy to obtain the inverse model by neural network training. Cascading the NN inverse system with the original system, a pseudo-linear system is completed (Figure 2 is an example of a two inputs and two outputs system). Subsequently, a linear close-loop regulator will be designed.

3 Mathematic Model of Induction Motor Variable Frequency Speed-Regulating System and Its Reversibility

Induction motor variable frequency speed-regulating system supplied by the inverter of tracking current SPWM can be expressed by 5-th order nonlinear model in d-q two-phase rotating coordinate. The model was simplified as a 3-order nonlinear model. If the delay of inverter is neglected, the model is expressed as follows:

$$\begin{cases} \dot{\omega}_r = \dfrac{n_p^2 L_m}{J L_r}\left(\psi_{rd} i_{sq} - \psi_{rq} i_{sd}\right) - \dfrac{n_p}{J} T_L \\ \dot{\psi}_{rd} = -\dfrac{\psi_{rd}}{T_r} + (\omega_1 - \omega_r)\psi_{rq} + \dfrac{L_m}{T_r} i_{sd} \\ \dot{\psi}_{rq} = -\dfrac{\psi_{rq}}{T_r} - (\omega_1 - \omega_r)\psi_{rd} + \dfrac{L_m}{T_r} i_{sq} \end{cases} \tag{1}$$

where ω_1 denotes synchronous angle frequency, and ω_r is rotate speed. i_{sd}, i_{sq} are stator's current, and ψ_{rd}, ψ_{rq} are rotor's flux linkage in (d,q)axis. n_p is number of poles. L_m is mutual inductance, and L_r is rotor's inductance. J is moment of inertia. T_r is rotor's time constant, and T_L is load torque.

In vector mode, then

$\psi_r = \psi_{rd} \bullet \psi_{rq} = 0$ and $\omega_1 = \omega_r + \frac{L_m}{T_r \psi_r} i_{sq}$

So $i_{sq} = (\omega_1 - \omega_r)\frac{T_r}{L_m}\psi_r$

Substituted it into formula (1), then

$$\begin{cases} \dot{\omega}_r = \frac{n_p}{J}[(\omega_1 - \omega_r)\frac{n_p T_r}{L_r}\psi_r^2 - T_L] \\ \dot{\psi}_r = \frac{-1}{T_r}\psi_r + \frac{L_m}{T_r} i_{sd} \end{cases} \tag{2}$$

Taking reversibility analyses of forum (2), then

$$\dot{\boldsymbol{x}} = f(\boldsymbol{x}, \boldsymbol{u}) = \begin{bmatrix} \frac{n_p}{J}[(u_1 - x_1)\frac{n_p T_r}{L_r} x_2^2 - T_L] \\ -\frac{1}{T_r} x_2 + \frac{L_m}{T_r} u_2 \end{bmatrix} \tag{3}$$

$$\boldsymbol{y} = h(\boldsymbol{x}) = [y_1, y_2]^T = [x_1, x_2]^T = [\omega_r, \psi_r]^T \tag{4}$$

The state variables are chosen as follows

$\boldsymbol{x} = [x_1, x_2]^T = [\omega_r, \psi_r]^T$

Input variables are

$\boldsymbol{u} = [u_1, u_2]^T = [\omega_1, i_{sd}]^T$

Taking the derivative on output in formula(4), then

$$y_1^{(1)} = \frac{n_p}{J}[(u_1 - x_1)\frac{n_p T_r}{L_r} x_2^2 - T_L] \tag{5}$$

$$y_2^{(1)} = -\frac{1}{T_r} x_2 + \frac{L_m}{T_r} u_2 \tag{6}$$

Then the Jacobi matrix is

$$A(\boldsymbol{x},\boldsymbol{u}) = \begin{bmatrix} \frac{\partial y_1^{(1)}}{\partial u_1} & \frac{\partial y_1^{(1)}}{\partial u_2} \\ \frac{\partial y_2^{(1)}}{\partial u_1} & \frac{\partial y_2^{(1)}}{\partial u_2} \end{bmatrix} = \begin{bmatrix} \frac{n_p^2 T_r}{JL_r} x_2^2 & 0 \\ 0 & \frac{L_m}{T_r} \end{bmatrix} \tag{7}$$

$$Det(A(\boldsymbol{x},\boldsymbol{u})) = \frac{n_p^2 L_m}{JL_r} x_2^2 \tag{8}$$

As $\boldsymbol{x} \in \Omega = \{\boldsymbol{x} \in R^2 : x_2^2 \neq 0\}$ so Det(A(x,u)) $\neq$ 0, and system is reversible.

Relative-order of system is $\alpha = (1,1)$, and $\alpha_1 + \alpha_2 = 2 = n$.

When the inverter is running in vector mode, the variability of flux linkage can be neglected (considering the flux linkage to be invariableness and equal to the rating). The original system was simplified as an input and an output system concluded by forum (2).

According to implicit function ontology theorem, inverse system of formula (3) can be expressed as:

$$u = \xi(x, y, \dot{y},) \tag{9}$$

When the inverse system is connected to the original system in series, the pseudo linear compound system can be built as the type of $y = S^{-1}\varphi(S)$.

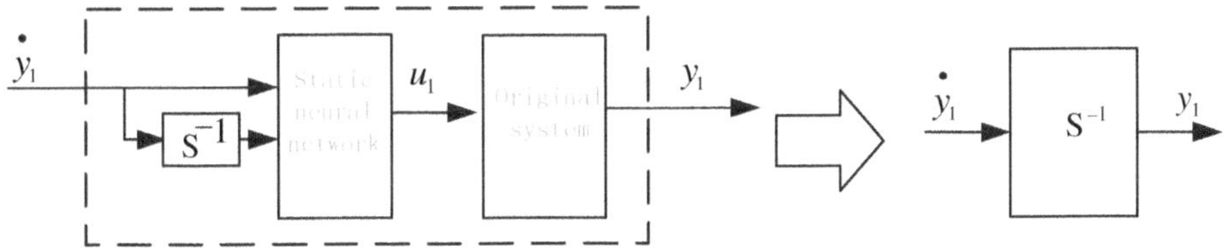

Fig. 3. Compound pseudo linear system of variable frequency speed-regulating system with single input and single output

4 Realization Steps of Neural Network Inverse System

4.1 Acquisition of the Input and Output Training Samples

Training samples are extremely important in the reconstruction of neural network inverse system. It is not only need to obtain the dynamic data of the original system, but also need to obtain the static date. Reference signal should include all the work region of original system, which can be ensure the approximate ability. Firstly the step of actuating signal is given corresponding every 10 HZ form 0HZ to 50HZ, and the responses of open loop are obtain. Secondly a random tangle signal is input, which is a random signal cascading on the step of actuating signal every 10 seconds, and the close loop responses is obtained. Based on these inputs, 1600 groups training samples are gotten.

4.2 The Construction of Neural Network

A static neural network and a dynamic neural network composed of integral is used to construct the inverse system. The structure of static neural network is 2 neurons in input layer, 3 neurons in output layer, and 12 neurons in hidden layer. The excitation function of hidden neuron is monotonic smooth hyperbolic tangent function. The output layer is composed of neuron with linear threshold excitation function. The training datum are the corresponding speed of open-loop, close-loop, first order derivative of these speed, and setting reference speed. After 50 times training, the training error of neural network achieves to 0.001. The weight and threshold of the neural network are saved. The inverse model of original system is obtained.

4.3 System Integration

Cascading neural network inverse system before the original system forms a pseudo linear system, and then a PI adjustor of the speed close-loop is design, as shown in Fig 4.

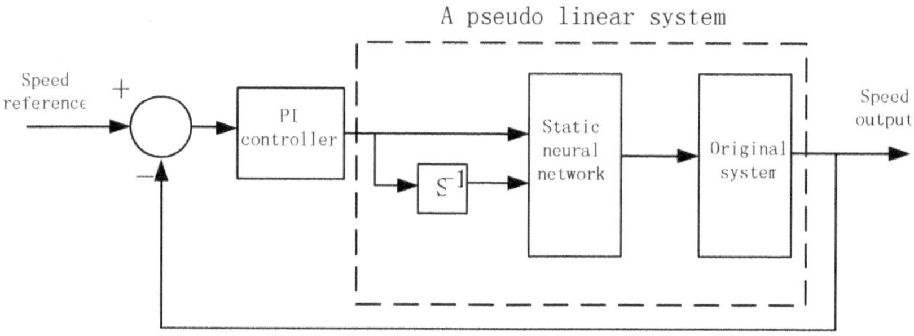

Fig. 4. Variable frequency speed-regulating system of neural network inverse in vector mode

5 Experiments and Results

5.1 Hardware of the System

The hardware of the experiment system is shown in Fig 5. The hardware system includes upper computer installed with Supervisory & Control configuration software WinCC6.0 [8], and S7-300 PLC of SIEMENS, inverter, induction motor and photoelectric coder.

PLC controller chooses S7-315-2DP, which has a PROFIBUS-DP interface and a MPI interface. Speed acquisition module is FM350-1. WinCC is connected with S7-300 by CP5611 using MPI protocol.

The type of inverter is MMV of SIEMENS. It can communicate with SIEMENS PLC by USS protocol. A CB15 [9] module is added on the inverter in this system.

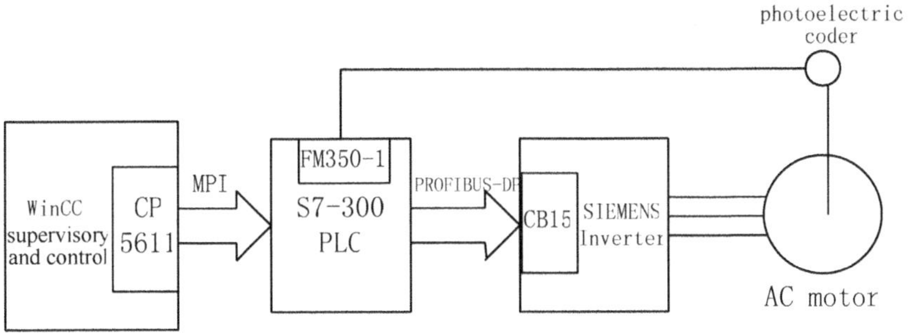

Fig. 5. Diagram of experiment system hardware

5.2 Software Program

5.2.1 Communication Introduction

MPI (MultiPoint Interface) is a simple and inexpensive communication strategy using in slowly and non-large data transforming field. The data transforming between WinCC and PLC is not large, so the MPI protocol is chosen.

The MMV inverter is connected to the PROFIBUS network as a slave station, which is mounted with CB15 PROFIBUS module. PPO1 or PPO3[9] data type can be chosen. It permits to send the control data directly to the inverter addresses, or to use the system function blocks of STEP7V5.2 SFC14/15.

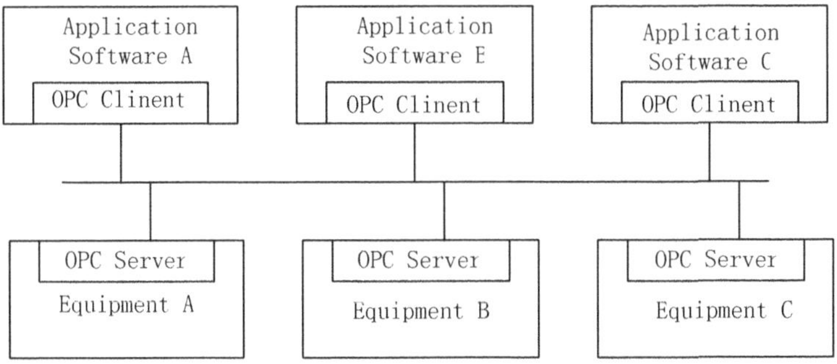

Fig. 6. OPC interface of industry control software and hardware

OPC can efficiently provide data integral and intercommunication. Different type servers and clients can access data sources of each other as shown in Fig.6. Comparing with the traditional mode of software and hardware development, equipment manufacturers only need to develop one driver. This can short the development cycle, save manpower resources, and simplify the structure of the entire control system.

Variety data of the system is needed in the neural network training of Matlab, which can not obtain by reading from PLC or WinCC directly. So OPC technology can be used 1 to obtain the needed data between WinCC and Exce. Setting WinCC as OPC DA server, an OPC client is constructed in Excel by VBA. System real time data is readed and writen to Excel by WinCC, and then the data in Excel is transform to Matlab for offline training to get the inverse system of original system.

5.2.2 Control Program

Used STL to program the communication and data acquisition and control algorithm subroutine in STEP7 V5.2, velocity sample subroutine and storage subroutine are programmed in regularly interrupt OB35, and the interrupt cycle chooses 100ms. In order to minimum the cycle time of OB35 to prevent the run time of OB35 exceeding 100ms and system error, the control procedure and neural network algorithm are programmed in main procedure OB1. Main procedure block diagram is shown in Fig. 7.

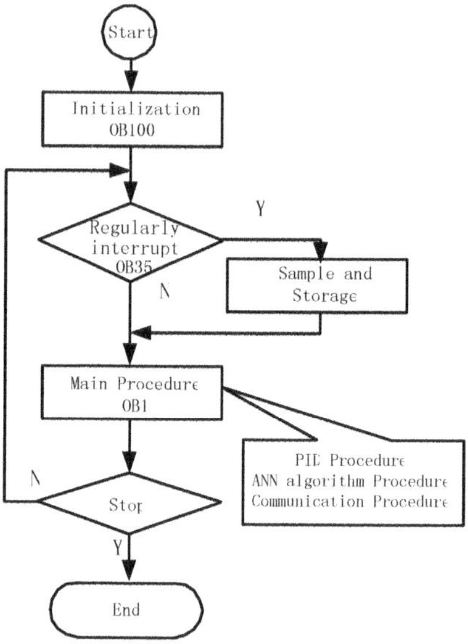

Fig. 7. Diagram of main procedure

In neural network algorithm normalized the training samples is need to speed up the rate of convergence by multiplying a magnification factor in input and output data before the final training.

5.3 Experiment Results

When speed reference is square wave signal with 100 seconds cycle, where the inverter is running in vector mode, the responses of traditional PI control and neural network control are shown in Fig. 9 and 10 respectively. The results show that the tracking performance of neural network control is better than traditional PI control.

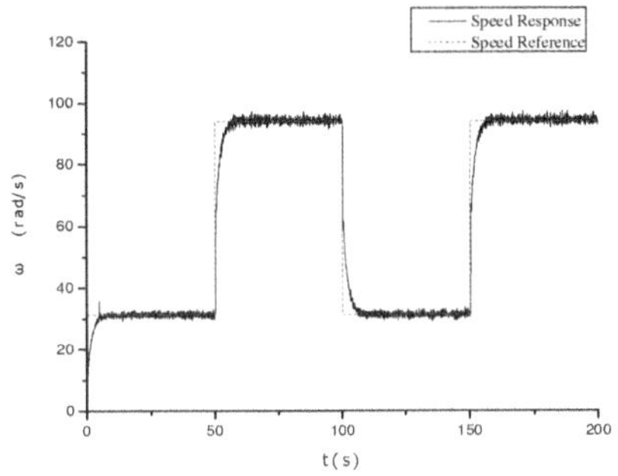

Fig. 9. Response of square wave in PI control

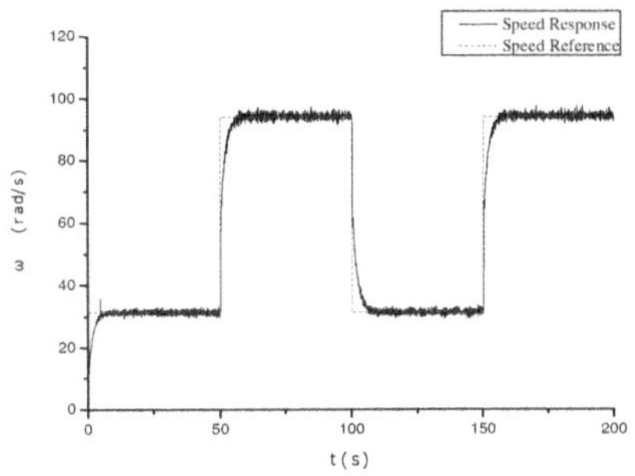

Fig. 10. Response of square wave in neural network inverse control

When speed reference keeps in constant, and the load is reduced to no load at 80 seconds, and increased to full load at 120 seconds, the response curves of speed with traditional PI control and neural network inverse control are shown in Fig. 11 and 12 respectively. It is clearly that the performance of resisting the load disturbing with neural network inverse control is better than the traditional PI control.

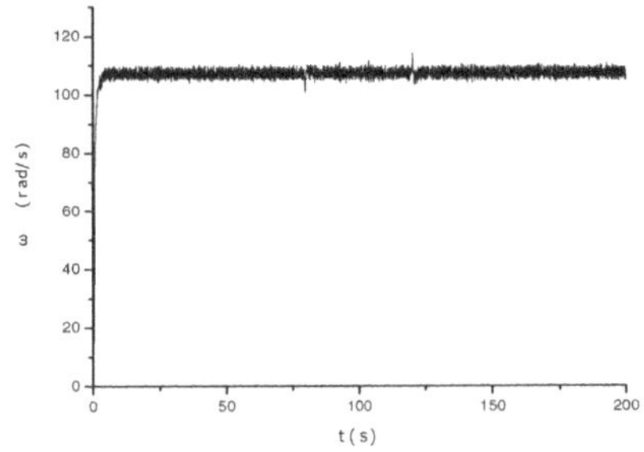

Fig. 11. Speed response in PI control

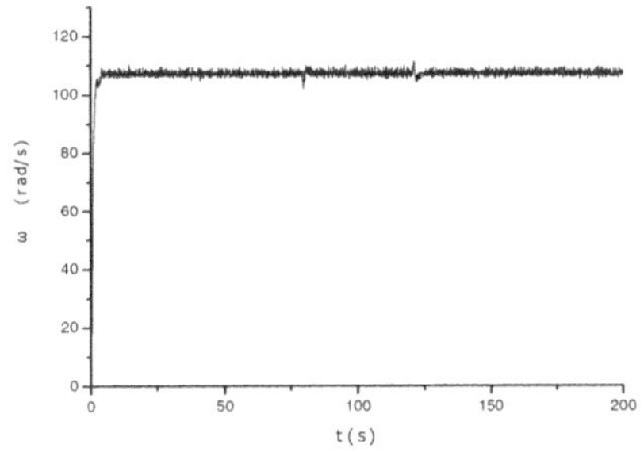

Fig. 12. Speed response in neural network inverse control

6 Conclusion

In order to improve the control performance of PLC Variable Frequency Speed-regulating System, neural network inverse system is used. A mathematic model of variable frequency speed-regulating system was given, and its reversibility was testified. The inverse system and original system is compound to construct the pseudo linear system and linear control method is design to control. With experiment, neural network inverse system with PLC has its effectiveness and its feasibility in industry application.

References

1. Dai, X.Z., Liu, G.H., Zhang, H., Shen, Y.: Neural Network Inverse Control of Variable Frequency Speed-regulating System in V/F Mode. Transactions of China Electrotechnical Society **25(7)** (2005) 109-114
2. Liu, G.H., Dai, X.Z.: Decoupling Control of an Induction Motor Speeding System. Transactions of China Electrotechnical Society **16(5)** (2001) 30-34
3. Zhang, H., Liu, G.H., You, D.T.: The Decoupling Control of AC Variable Frequency Motor System Based on Artificial Neural Network Inverse System Method. Journal of Jiangsu University **23(2)** (2002) 88-91
4. Dai, X.Z., Liu, J.E.: Neural Network αth Order Inverse System Method for the Control of Nonlinear Continuous Systems. IEE Proc. Control Theory and Application **145(6)** (1998) 519-522.
5. Dai, X., He, D., Zhang, X., etal: MIMO System Invertibility and Decoupling Control Strategies Based on ANN a-order Inversion. IEE Proc. Control Theory Appl. **148(2)** (2001) 125-136
6. Li, C.W., Feng, Y.K., Inverse Control of Multivariable Nonlinear System. Tsinghua University Press (1991)
7. Xia, X.H., Gao, W.B.: Nonlinear Control and Decoupling Control. Science Press (1997)
8. A&D Group Siemens Ltd. China. Explain the Profound Things in a Simple Way of SIEMENS S7-300 PLC. Beijing University of Aeronautics and Astronautics (2004)
9. SIEMENS Electrical Driver Ltd. SIEMENS Communication Manual of Standard Driver (2000)

Neural-Network-Based Switching Control for DC Motors System with LFR

Jianhua Wu, Shuying Zhao, Lihong He, Lanfeng Chen, and Xinhe Xu

Northeastern University, School of Information Science and Engineering
Shenyang, Liaoning 110004, China
wujianhua@mail.neu.edu.cn

Abstract. The Loss-Free resistor (LFR) is applied to DC motor speed control system. A compensation Control algorithm based on LFR is proposed. The LFR is realized by means of switching network whose characteristics are nonlinear, thus, a neural network was designed and used. The switching on-off time can be instantaneously calculated by using the Neural-network-based algorithm proposed. The varies of the motor speed is realized by controlling the output of the LFR, the energy loss in the system is reduced compared with using a conventional power amplifier, and the dynamic characteristics of the system are also improved. The validation of the simulation results is proposed as well.

1 Introduction

Loss - free resistor (LFR) is an ideal element, which has a resistive *I* -*V* characteristic and stores the absorbed power instead of converting it into heat [1] [2]. The realization has been achieved by the combination of a time variable transformer (TVT) controlled by a signal processing circuit [3]. This LFR was based on a controlled switched mode converter, which realized the TVT and transferred the absorbed power to output, so that losses are eliminated [1]-[4].

The voltage type Loss -Free resistor (LFR) is composed of a power conditioning system (PCS) and a storage element capacitor. The PCS is a controlled -coupling network realized by means of switches network [5]. By controlling the on-off time of charges and discharges the storage element, the output voltages of the LFR are generated in any form.

The realization of LFR was motivated by the need to replace a conventional resistor. Since the speed control of DC motor may be realized by adjusting a variable resistor which joined in the armature loop of the motor, the LFR is used in this case. The use of the LFR in DC motors system may realize the purposes of regulation and the improvement of dynamic characteristics of the system.

After analyzed characteristics of the voltage type loss - free resistor, the mathematical models of the Loss-Free resistor were set up based on the average value in one-cycle [6]. Since switches network is used in the LFR system, the input-output

D. Liu et al. (Eds.): ISNN 2007, Part I, LNCS 4491, pp. 267–274, 2007.

characteristic of the LFR system is nonlinear. Therefore, a neural network was proposed for getting an inverter model of the nonlinear system and processing the some uncertainty that the switches network has. The neural network makes it possible, in the case of LFR system, to estimate a switching time corresponding to multi- input data with sufficient accuracy.

2 The Mathematical Model of the LFR

The model of the LFR, which was chosen, is a two-port network consisting of an emulated resistance at the input and a power source at the output as shown in Fig. 1 [4]

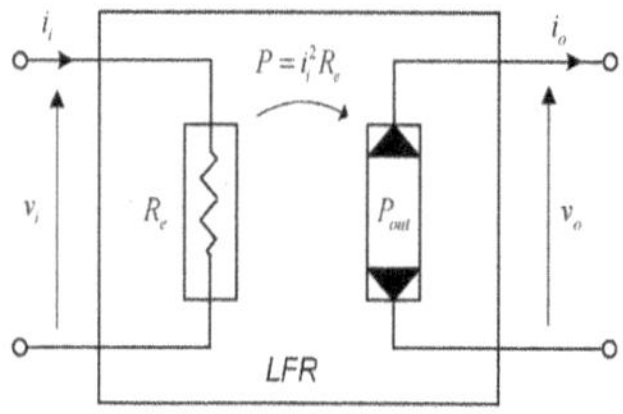

Fig. 1. LFR model as a two-port element

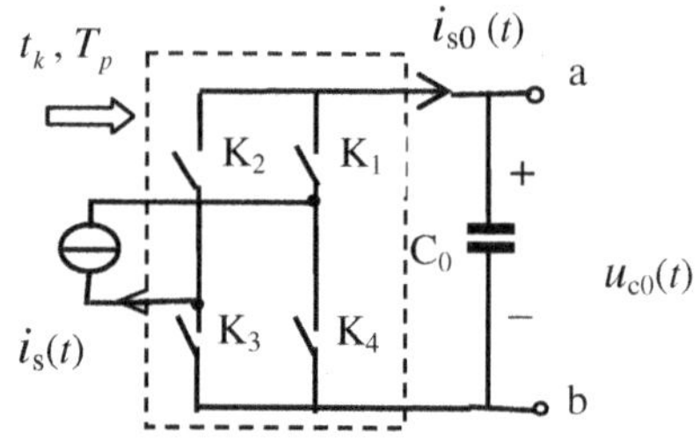

Fig. 2. The principle scheme of the LFR

The electric circuit principle scheme of the voltage type LFR showed in Fig .2, is composed of a capacitor and an ideal switch network, The output voltage (the voltage of the capacitor) of the LFR is controlled by the on-off states of the switch network [6] [7].

Suppose M_i (i = 1, 2) expresses the state of the ideal switch network. M_1 expresses the state of the ideal switch network, when K_1, K_3 are turn on and K_2, K_4 are off; M_2 expresses the state when K_2, K_4 are on and K_1, K_3 are off. T_p expresses the switching period. In the kth switching period, the state of M_1 continues for t_k and the state of M_2 continues for T_P - t_k. T_P is the switch period and T_P is constant.

For the ease of analyzing, we suppose that the ideal current source $i_s(t)$ in Fig.2 is a constant value I_S ; the state of M_1 was first and the state of M_2 is last in every switch period T_P; In this way, after determining the values of I_S, T_P and C_0, the output voltage of the LFR is controlled by changing the on - off continues time t_k ($k=1,2,\cdots,N$) of the ideal switching network. The curve of the $i_{s0}(t)$ in Fig. 2 showed in Fig. 3.

According to the electric circuit theory, the output voltage of the LFR in Fig.2 is

$$u_{C_0}(t) = u_{C_0}(t_0) + \frac{1}{C_0}\int_{t_0}^{t} i_{s0}(t)dt \tag{1}$$

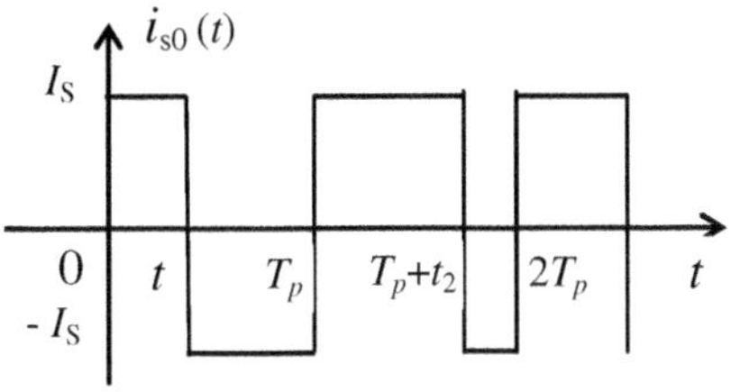

Fig. 3. The curve of the $i_{s0}(t)$

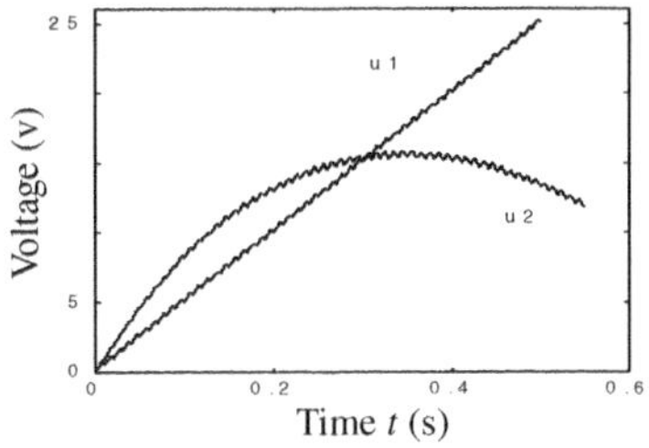

Fig. 4. $u_{c0}(t)$ controlled by t_k

We suppose $I_S / C_0 = A$ and analyze the relation of $u_{c0}(t)$ and t_k ($k = 1, 2, \cdots, N$),

$$0 \le t \le t_1, M = M_1 \qquad u_{c0}(t) = u_{c0}(t_0) + A \cdot t$$

$$t_1 \le t \le T_p, M = M_2 \qquad u_{c0}(t) = u_{c0}(t_0) + A \cdot (2t_1 - t)$$

$$T_p \le t \le T_p + t_2, M = M_1 \qquad u_{c0}(t) = u_{c0}(t_0) + A \cdot (2t_1 - 2T_p + t)$$

$$T_p + t_2 \le t \le 2T_p, M = M_2 \qquad u_{c0}(t) = u_{c0}(t_0) + A \cdot (2t_1 + 2t_2 - t)$$

Time t

……

$$(k-1)T_p \le t \le (k-1)T_p + t_k, M = M_1,$$

$$u_{c0}^{(1)}(t) = u_{c0}(t_0) + A \cdot \left(\sum_{j=1}^{k-1} 2t_j - 2(k-1)T_p + t \right)$$

$$(k-1)T_p + t_k \le t \le kT_p, M = M_2,$$

$$u_{c0}^{(2)}(t) = u_{c0}(t_0) + A \cdot \left(\sum_{j=1}^{k} 2t_j - t \right) \tag{2}$$

The average value of $u_{c0}(t)$ in the kth period can be computed by using following formula,

$$\overline{u}_{c0}(kT_P) = \frac{1}{T_P} \left[\int_{(k-1)T_P}^{(k-1)T_P + t_1} u_{c0}^{(1)}(t)\,dt + \int_{(k-1)T_P + t_k}^{kT_P} u_{c0}^{(2)}(t)\,dt \right] \tag{3}$$

The average value of $u_{c0}(t)$ in the kth period

$$\overline{u}_{c0}(kT_p) = u_{c0}(t_0) + A \cdot \left(\sum_{j=1}^{k} 2t_j - kT_p + \frac{T_p}{2} - \frac{t_k^2}{T_p} \right) \tag{4}$$

$$k = 1, 2, \cdots, N$$

If the switching period T_P is short enough, the sequence $\overline{u}_{c0}(kT_P)$ ($k =$ 1,2, …, N) can approach $u_{c0}(t)$. Suppose $I_s / C_0 = 10^2$, $T_P = 10^{-2}$ s, when $k = 1, 2, \ldots, 50$, $t_k = 0.75\ T_P$, $u_{c0}(t)$ showed as curve u_1 in Fig.4; when $k = 1,2 \ldots 50$, $t_k = T_p e^{-0.01[2(k-1)+1]}$, $u_{c0}(t)$ showed as curve u_2 in Fig.4.

The on-off continues time t_k (k = 1, 2,..., N) of the switch network for every switching period could be got according to the expected $u_{c0}(t)$ (or $\overline{u}_{c0}(kT_P)$) by taking inverse function on formula (4).

3 Application of the Neural Network

Deducing from the formula 4 may get,

$$\overline{u}_{c0}\left((k+1)T_p\right) = u_{c0}(0) + A\cdot\left(\sum_{j=1}^{k+1} 2t_j - (k+1)T_p + \frac{T_p}{2} - \frac{t_{k+1}^2}{T_p}\right)$$

Set $D_k = t_k / T_P$,

$$\overline{u}_{c0}((k+1)T_P) = \overline{u}_{c0}(kT_P) + AT_P(2D_{k+1} - 1 + D_k^2 - D_{k+1}^2) \tag{5}$$

From the formula 5, an inverse function showed in the formula 6,

$$D_{k+1} = f\left(\overline{u}_{c0}((k+1)T_P),\ \overline{u}_{c0}(kT_P),\ D_k\right) \tag{6}$$

The on-off continues time t_{k+1} ($t_{k+1} = D_{k+1}T_P$, k = 1, 2,..., N) of the switching network for every switching period could obtain according to the expected $\overline{u}_{c0}((k+1)T_p)$ by using inverse calculation for the formula (5). The formula (5) has a nonlinear characteristic. If solve directly D_{k+1} from formula (5), some wrongs will occur, For example, It is possible that D_{k+1} is a negative number, or a complex number, the system will can't normally work in this case. Especially, some uncertainty exists in the switching system. Thus a neural network controller was selected and designed, and some practice data is used to solve the nonlinear and uncertainty problems. The neural network digram showed in Fig. 5.

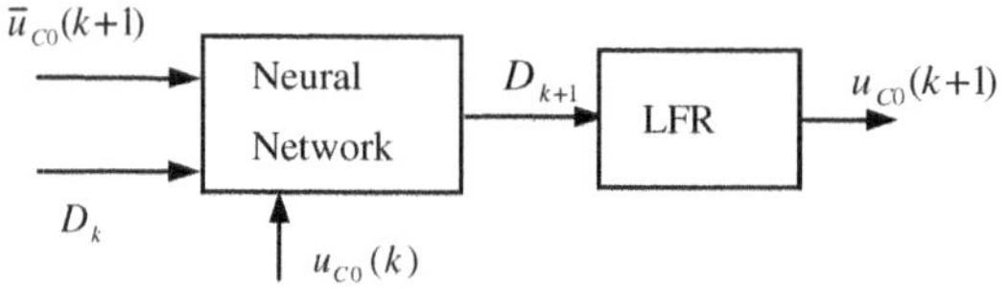

Fig. 5. Neural network controller

$\overline{u}_{c0}(k+1)$ is the expected output of the LFR in k+1th period. $u_{c0}(k+1)$ is output of the LFR in k+1th period. $u_{c0}(k)$ is output of the LFR in k[th] period. D_{k+1} is output of the neural network controller in (k+1)[th] period. D_k is output of the neural network controller in k[th] period.

A training sample collection and a testing sample collection are constructed by synthesis data given from the formula (4). $\overline{u}_{c0}(k+1) - u_{c0}(k)$ and D_k are the input of the neural network designed; D_{k+1} is the output of the neural network designed. 1 divide 10 intervals for $0 \le D_k \le 1$, D_k has 11 values; $\overline{u}_{c0}(k+1) - u_{c0}(k)$ divide 100 intervals for

$0 \le |\bar{u}_{c0}(k+1) - u_{c0}(k)| \le 22.1$, $|\bar{u}_{c0}(k+1) - u_{c0}(k)|$ has 101 values; D_{k+1} is calculated out 101 values according to the 101 values of $|\bar{u}_{c0}(k+1) - u_{c0}(k)|$ for every D_k by using the formula (5). The normalized data for $|\bar{u}_{c0}(k+1) - u_{c0}(k)|$ is used.

The neural network used three layers BP networks. Input layer has two neurons. Output layer has 1 neuron. From many times of tests, 9, a best number of neural neurons in the hidden layer, was decided. Output layer neural neuron use liner function. Hidden layer neurons used sigmoid function improved $f(x)=2/(1+\exp(-x))-1$.

Dynamic adjusting leaning-rate method is used to increase convergence speed. For batch processing training mode, enlighten adjustment can be made according to the variation of total error. The adjustment of leaning-rate can be made in every step of leaning. Target function is also a factor influencing convergence speed and final approaching accuracy. If BP network use residual square summation as target function, it will magnify influence and effect of maximum value, especially when the difference between maximum and minimum values is very large. In this case, it greatly decreases the leaning ability of training examples for the smaller actual output values and the generalization property of testing examples. Based on the characteristics of the system, this paper made both residual square summation and relative error square summation become smallest as target function:

$$J = 1/2\left(\frac{y_i - \hat{y}_i}{y_i}\right)^2 + 1/2(y_i - \hat{y}_i)^2$$

After training 6000 times, in best situation, the mean square error of the results is 0.764%, relative error is 0.41%.

4 DC Motor Speed Control by LFR

The principle scheme of the motor speed control by LFR showed by Fig. 5, where $u_{ref}(t)$ is the reference speed signal, $s(t)$ is the motor speed, $u_o(t)$ is the output of the speed measure generator, $u_{c0}(t)$ is the output voltage of the LFR, $u_a(t)$ is the armature voltage, the magnetic - flux is constant, and the DC voltage source E is according with the speed rating of the DC motor.

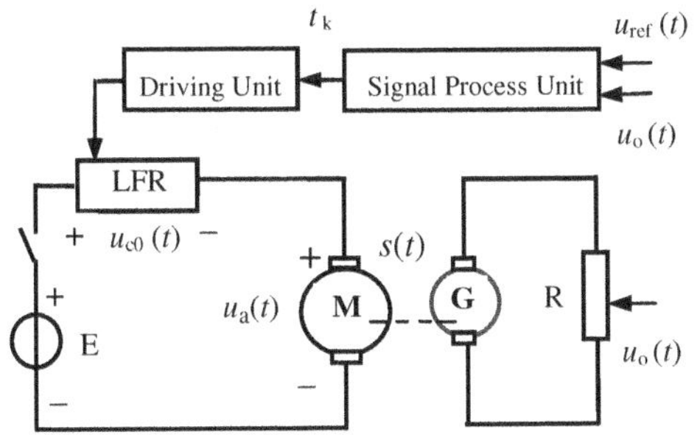

Fig. 6. The Motor Speed Control with LFR

The mathematical model of the system composed of the motor and the generator for measuring speed (no load) is given by

$$\frac{S(s)}{U_a(s)} = \frac{1}{K_e\left(T_m T_e s^2 + T_m s + 1\right)} \tag{7}$$

Where T_m is electromechanical coefficient, T_e is electromagnet coefficient, k_e is anti-potential coefficient. $u_a(t)$ can be given by

$$u_a(t) = E - u_{c0}(t)$$

We may change the armature voltage $u_a(t)$ to control the motor speed by changing the voltage u_{c0} (t) of the LFR basing on the speed reference $u_{ref}(t)$. If the system is supposed to have damped oscillation characteristic, the requirement for dynamic characteristics is to eliminate the overshoot and to decrease the response time.

According to the theory of circuit, when C_0 in Fig. 2 is big enough , the output $u_{c0}(t)$ of the LFR in Fig. 6 can be regard as a ideal controlled - voltage source and output $u_o(t)$ of the system can be considered as the response that is excited by both E and $u_{c0}(t)$. The step response produced alone by $E \cdot 1(t)$ is $s_E(t) \cdot 1(t)$ (section 1 of the $u_o(t)$).

Set that $s_E(t) \cdot 1(t)$ correspond a voltage value $u_E(t) \cdot 1(t)$ (the section 1 of the output of the speed measure generator), $u_E(t) \cdot 1(t)$ is given by curve 1 in Fig. 7 ; The response produced alone by u_{c0} (t) is $s_b(t) \cdot 1(t - t_0)$ (section 2 of the $u_o(t)$). Set that $s_b(t) \cdot 1(t - t_0)$ correspond a voltage value $u_b(t) \cdot 1(t - t_0)$ (the section 2 of the output of the speed measure generator). The sum of the 2 response sections is $u_o(t)$,

$$u_o(t) = u_E(t) \cdot 1(t) + u_b(t) \cdot 1(t - t_0)$$

Here, $u_E(t) \cdot 1(t)$ express that $u_E(t) \cdot 1(t) = 0$, when $t < 0$ and $u_E(t) \cdot 1(t) = u_E(t)$, when $t > 0$. $u_b(t) \cdot 1(t - t_0)$ express that $u_b(t) \cdot 1(t - t_0) = 0$, when $t < t_0$ and $u_b(t) \cdot 1(t - t_0) = u_b(t)$, when $t > t_0$.

On the supposition that the expected output voltage of the system (corresponding a speed given) is u_{ref}, t_0 is the initial time of $u_b(t)$ produced, and $u_b(t)$ is given by

$$u_b(t) \cdot 1(t - t_0) = \left[u_{ref} - u_E(t) \cdot 1(t)\right] \cdot 1(t - t_0) \tag{8}$$

The $u_b(t) \cdot 1(t - t_0)$ is produced by $u_{c0}(t) \cdot 1(t - t_0)$. The output voltage $u_o(t)$ of the system is illustrated by curve 2 in Fig. 7. When $0 < t < t_0$, the system output $u_o(t) = u_E(t)$; when $t > t_0$, $u_o(t) = u_{ref}$. The overshoot is eliminated, and the response time is shorten, because of the compensation produced by $u_{c0}(t)$.

The signal processing unit realized an algorithm. The calculating steps of the algorithm are: firstly, calculating the compensation response $u_b(t) \cdot 1(t - t_0)$ according to the speed signal given u_{ref} by using formula 6, $u_E(t) \cdot 1(t)$ is the step response of the system; secondly, calculating $u_{c0}(t) \cdot 1(t - t_0)$ according to $u_b(t) \cdot 1(t - t_0)$ by using formula 5. This is an inverse calculation. The t_0 is the initial time when $u_E(t)$ is first time to be the expected output $u_{ref.}$; thirdly, calculating the switching time sequence t_k (k =1 , 2 ,..., N) according to the $u_{c0}(t) \cdot 1(t - t_0)$ from formula 4, This is also a inverse calculation. Driving unit will control the on – off time of the switching network on the basis of the switching time sequence t_k.

When the speed reference signal u_{ref} is changed every time, the system begins a dynamics. When it is measured out that the output speed signal $u_o(t)$ reach to $u_{ref}(t)$, the LFR begins to compensate dynamically. The polarity of DC voltage source E in the electric circuit may be changed by adding switches or choosing source E with 3 levels, so, the acceleration, the deceleration and the change the revolving direction of the DC motor can be controlled according to the variation value of the reference speed signal d. We supposed $t_1 < t_2$, and $d = u_{ref}(t_2) - u_{ref}(t_1)$, if $d > 0$, E is positive ; if $d < 0$, E is negative. The processes completed in the signal processing unit and the drive unit.

5 Simulation Test

The simulation system shows in Fig. 7. The voltage source E is 10V, the switching period of the switch network T_p is 0.002 S, $C_0 / I_s = 10^{-4}/3\,(F/A)$, the main parameters of the motor system are T_e = 0.0099 S , T_m = 0.024 S , K_e = 0.42 V·s. The mathematic model of motor-generator system is

$$\frac{U_0(s)}{U_a(s)} = \frac{1}{10^{-4}s^2 + 0.0101s + 1}$$

The reference speed signal u_{ref} is 5V. The curve of the output response $u_o(t)$ with dynamic compensation is shown as curve 2 in Fig. 7. When the armature voltage $u_a(t)$ is $5 \cdot 1(t)$ V, the output $u_o(t)$ without dynamic compensation is shown as curve 3 in Fig .7.

It can be known from Fig 7, the output voltage of generator for measuring speed (according to the motor speed) begins to change when $t = 0$ (seeing curve1) until it reaches to the reference speed (reference signal is 5V) , and stabilize on curve 2, the response speed is faster , and the overshoot isn't exist. Switching time sequence t_k, which realized the control result is shown in Fig 8.

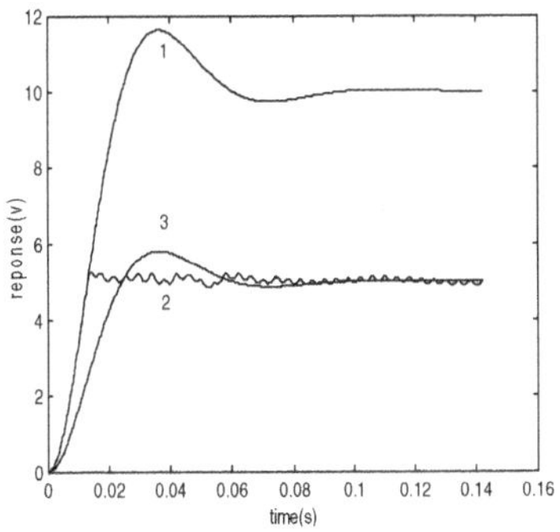

Fig. 7. The system output $u_o(t)$

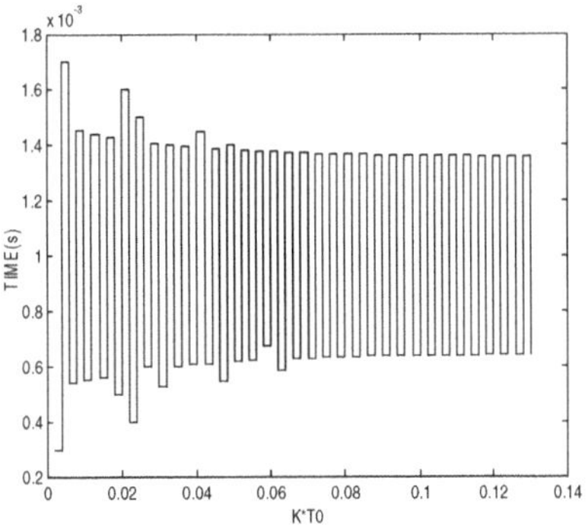

Fig. 8. The switching time sequence t_k

The ripple of the curve 2 in the Fig .7 can be improved by accurately adjusting the switching time sequence t_k or by using power sharing interleaved mode.[8]

6 Conclusions

The application of the LFR in the DC motor speed control system is possible. The principle analysis and simulation results proved the effectiveness of the neural network algorithm proposed. Using the algorithm, the switching on-off time of the switch network in the LFR can be instantaneously calculated according to the compensation voltage. The algorithm may be used in the others switching circuits and systems.

References

1. Singer, S.: Realization of Loss-Free Resistive Elements. IEEE Trans. CAS **37** (1990) 54-58
2. Singer, S.: Smilovitz, Transmission Line-Based Loss-Free Resistor. IEEE Trans. CAS **41** (1994) 120-126
3. Wang, A., Yin, H.: Realization of Source with Internal Loss-free Resistive characteristic. IEEE. Trans. CAS **48** (2001) 830-839
4. Singer, S.: A Pure Realization of Loss-free Resistive Elements. IEEE Trans. CAS **51(8)** (2004) 1639-1647
5. Wu, J., Xu, X., Yin, H.: Character and the Application of Loss-less Resistor. Journal of Northeastern University **22** (2001) 370 –372
6. Smedley, K.M., Zhou, L.W., Qiao, C.M.: Unified Constant-frequency Integration Control of Active Power Filters: Steady-state and Dynamics. IEEE Trans. Power Elec. **16** (2001) 428-436
7. Wu, J., Zhang, S., Song, J., Xu, X.: Algorithm for Computing Switching Time in VAPF. In: Proceedings of the 2004 IEEE Workshop on Computers in Power Electronics COMPEL '04. Illinois, USA (2004) 119-122
8. Zhang, M.T., Jovanovic, M.M., Lee, F.C.Y.: Analysis and Evaluation of Interleaving Techniques in Forward Converters. IEEE Trans. Power Electron. **13** (1998) 690–698

Adaptive Robust Motion Controller with Friction and Ripple Disturbance Compensation Via RBF Networks

Zi-Jiang Yang, Shunshoku Kanae, and Kiyoshi Wada

Department of Electrical and Electronic Systems Engineering,
Graduate School of Information Science and Electrical Engineering,
Kyushu University
744 Motooka, Nishi-ku, Fukuoka, 819-0395 Japan
yoh@ees.kyushu-u.ac.jp

Abstract. In this paper, a practical adaptive robust nonlinear controller is proposed for motion control of an SISO nonlinear mechanical system, where the distrubances due to ripple force and friction are compensated by the RBF networks. Rigorous analysis of transient performance and ultimate bound is given. Numerical examples are included to verify the theoretical results.

1 Introduction

In this paper, a practical adaptive robust nonlinear controller by backstepping design is proposed for motion control of an SISO nonlinear mechanical system, where the distrubances due to ripple force and friction are compensated by the RBF networks. To overcome the main obstacles that prevent the adaptive control techniques from coming into wide use in the industrial side, our attention is focused on the guaranteed transient performance and transparent structure of the control system. The controller is designed in a backstepping manner. At the first step, a PI controller is designed to stabilize the position error. Then at the second step, an adaptive robust nonlinear controller is designed to stabilize the velocity error, where the input-to-state stability (ISS) property is first achieved by the nonlinear damping terms. Then the adaptive laws are adopted to achieve a small ultimate error. The complicated looking adaptive robust nonlinear controller can be explained as hierarchical modifications of the coventional PI position controller with minor-loop. Therefore it is believed that the proposed controller may gain wide acceptance of the engineers. Finally, numerical examples are included to verify the theoretical results.

2 Statement of the Problem

Consider the following SISO nonlinear mechanical system:

$$\begin{aligned} \dot{x}_1 &= x_2 \\ \dot{x}_2 &= F(\boldsymbol{x}) + d(\boldsymbol{x}, t) + M^{-1}u \end{aligned} \tag{1}$$

D. Liu et al. (Eds.): ISNN 2007, Part I, LNCS 4491, pp. 275–284, 2007.

where, $\boldsymbol{x} = [x_1, x_2]^T$, x_1 and x_2 are the position and velocity respectively, u is the control input; M is the mass; $F(\boldsymbol{x})$ is the modelable disturbance function; and $d(\boldsymbol{x}, t)$ is the lumped unmodelable disturbance term.

Consider a motion control problem where the friction and periodic ripple disturbances affect the control performance simultaneously. The nonlinear functions in system model (1) are described as follows [1].

$$
\begin{aligned}
F(\boldsymbol{x}) &= F_f(\boldsymbol{x}) + F_r(\boldsymbol{x}), \qquad d(\boldsymbol{x}, t) = d(x_2, \varepsilon) \\
F_r(\boldsymbol{x}) &= -\frac{2.5\sin(100x_1)}{M} - \frac{3.1\sin(200x_1 + 0.05\pi)}{M} \\
F_f(\boldsymbol{x}) &= -\frac{\sigma_2 x_2 + \left[F_c + (F_s - F_c)e^{-(x_2/\dot{x}_s)^2}\right]\mathrm{sgn}(x_2)}{M} \\
d(x_2, \varepsilon) &= -\frac{\sigma_0 \varepsilon}{M}\left[1 - \frac{|\sigma_1 x_2|}{F_c + (F_s - F_c)e^{-(x_2/\dot{x}_s)^2}}\right], \qquad \varepsilon = z - z_s \\
&|d(x_2, \varepsilon)| \le \Delta_{d1}|x_2| + \Delta_{d2}, \quad {}^{\exists}\Delta_{d1}, {}^{\exists}\Delta_{d2} > 0 \\
\dot{z} = x_2 - \frac{|x_2|}{h(x_2)}z, &\quad z_s = h(x_2)\mathrm{sgn}(x_2), \quad h(x_2) = \frac{F_c + (F_s - F_c)e^{-(x_2/\dot{x}_s)^2}}{\sigma_0}
\end{aligned}
\tag{2}
$$

where $\sigma_0, \sigma_1, \sigma_2, F_c, F_s, \dot{x}_s$ are physical parameters; $F_f(\boldsymbol{x})$ and $d(x_2, \varepsilon)$ are respectively the modelable and unmodelable effects of friction [2]; $F_r(\boldsymbol{x})$ represents the periodic ripple disturbance.

We can model $F(\boldsymbol{x})$ by the following network:

$$
\begin{aligned}
&\widehat{F}(\boldsymbol{x}, \boldsymbol{w}_F) = \boldsymbol{\phi}_F^T(\boldsymbol{x})\boldsymbol{w}_F \\
&\boldsymbol{w}_F = [w_{\sigma_2}, w_{FC}, \boldsymbol{w}_b^T, \boldsymbol{w}_a^T], \quad \boldsymbol{\phi}_F^T(\boldsymbol{x}) = [-x_2, -\mathrm{sgn}(x_2), -\boldsymbol{R}_b^T(x_2)\mathrm{sgn}(x_2), -\boldsymbol{R}_a^T(x_1)] \\
&\boldsymbol{R}_a^T(x_1) = [r(x_1 - p_{1a}), \cdots, r(x_1 - p_{Na})], \quad \boldsymbol{R}_b^T(x_2) = [r(x_2 - p_{1b}), \cdots, r(x_2 - p_{Nb})] \\
&\boldsymbol{w}_a = [w_{1a}, \cdots, w_{Na}]^T, \quad \boldsymbol{w}_b = [w_{1b}, \cdots, w_{Nb}]^T
\end{aligned}
\tag{3}
$$

where $\boldsymbol{R}_a^T(x_1)\boldsymbol{w}_a$ and $\boldsymbol{R}_b^T(x_1)\boldsymbol{w}_b$ are RBF networks. $r(x_1 - p_{na}) = \exp[-(x_1 - p_{na})^2/(2\sigma_a^2)]$ are equidistantly located in $X_1 = \{x_1 \,|\, 0 \le x_1 \le 0.5[\mathrm{m}]\}$, and $r(x_1 - p_{nb}) = \exp[-(x_1 - p_{nb})^2/(2\sigma_b^2)]$ are equidistantly located in $X_2 = \{x_2 \,|\, -0.1[\mathrm{m/s}] \le x_2 \le 0.1[\mathrm{m/s}]\}$. The numbers of the basis functions are chosen as $N_a = 51, N_b = 6$, and $\sigma_a = (\sqrt{2}/\pi)(p_{na} - p_{(n-1)a})$, $\sigma_b = (\sqrt{2}/\pi)(p_{nb} - p_{(n-1)b})$.

Some assumptions are made here.

Assumption 1. *The networks are sufficiently complex such that the approximation errors are sufficiently small on the desired domain of operation Ω_X, i.e., there exists $\boldsymbol{w}_F^*$ satisfying*

$$
\sup_{\boldsymbol{x} \in \Omega_X} |\eta_F(\boldsymbol{x}, \boldsymbol{w}_F^*)| = \left|F(\boldsymbol{x}) - \widehat{F}(\boldsymbol{x}, \boldsymbol{w}_F^*)\right| \le {}^{\exists}\varepsilon_F > 0 \tag{4}
$$

Assumption 2. *The lower and upper bounds of the parameter vectors are known a priori:*

$$
\underline{\boldsymbol{w}}_F \le \boldsymbol{w}_F \le \overline{\boldsymbol{w}}_F, \quad \underline{M^{-1}} \le M^{-1} \le \overline{M^{-1}} \tag{5}
$$

Assumption 3. *There exists a known bounding funvtion* $\overline{d}(\boldsymbol{x},t) > 0$ *such that*

$$\frac{|d(\boldsymbol{x},t)|}{\overline{d}(\boldsymbol{x},t)} \leq {}^{\exists}M_d < \infty \tag{6}$$

3 Controller Design

The controller is designed in a backstepping manner as follows.

Step 1: Define the error signals as

$$z_1 = x_1 - y_r, \qquad z_2 = x_2 - \alpha_1 \tag{7}$$

where α_1 is the virtual input to stabilize z_1.

Then we have subsystem $\mathcal{S}1$ as the following.

$$\mathcal{S}1: \ \dot{z}_1 = \alpha_1 + z_2 - \dot{y}_r \tag{8}$$

The virtual input α_1 is designed based on the common PI control technique.

$$\alpha_1 = -c_{1p}z_1 - c_{1i}\int_0^t z_1 dt + \dot{y}_r \tag{9}$$

where $c_{1p} > 0, \ c_{1i} > 0$.

Step 2: The second subsystem $\mathcal{S}2$ is obtained as

$$\mathcal{S}2: \ \dot{z}_2 = \widehat{F}(\boldsymbol{x},\boldsymbol{w}_F) - \dot{\alpha}_1 + d(\boldsymbol{x},t) + F(\boldsymbol{x}) - \widehat{F}(\boldsymbol{x},\widehat{\boldsymbol{w}_{Ft}}) + \widehat{M^{-1}}_t u + (M^{-1} - \widehat{M^{-1}}_t)u \tag{10}$$

To stabilize the subsystem we design the control input as

$$\begin{gathered}
u = u_l + u_r, \quad u_l = \frac{\alpha_{20}}{\widehat{M^{-1}}_t}, \quad u_r = \frac{-(u_{d1} + u_{d2} + u_{d3})z_2}{\widehat{M^{-1}}_t} \\
\alpha_{20} = -c_2 z_2 + \dot{\alpha}_1 - \widehat{F}(\boldsymbol{x},\widehat{\boldsymbol{w}_{Ft}}) \\
u_{d1} = \kappa_{21}\overline{F}(\boldsymbol{x}), \quad u_{d2} = \kappa_{22}|\alpha_{2d}|, \quad u_{d3} = \kappa_{23}\overline{d}(\boldsymbol{x},t) \\
\alpha_{2d} = |-c_2 z_2 + \dot{\alpha}_1| + \overline{F}(\boldsymbol{x}), \quad \overline{F}(\boldsymbol{x}) = |x_2| + e^{-(x_2/0.1)^2} + 1, \quad \overline{d}(\boldsymbol{x},t) = |x_2| + 1
\end{gathered} \tag{11}$$

where $c_2, \kappa_{21}, \kappa_{22}, \kappa_{23} > 0$; α_{20} is a feedback controller with model compensation; $u_{d1}z_2$C$u_{d2}z_2$ and $u_{d3}z_2$ are nonlinear damping terms [3] to counteract the effects of $\eta(\boldsymbol{x},\widehat{\boldsymbol{w}_{Ft}}) = F(\boldsymbol{x}) - \widehat{F}(\boldsymbol{x},\widehat{\boldsymbol{w}_{Ft}})$, $(M^{-1} - \widehat{M^{-1}}_t)u$ and $d(\boldsymbol{x},t)$ respectively.

Applying u to $\mathcal{S}2$, we have

$$\dot{z}_2 = -c_2 z_2 + \eta_F(\boldsymbol{x},\boldsymbol{w}_F^*) - \boldsymbol{\phi}_F^T(\boldsymbol{x})\widetilde{\boldsymbol{w}_{Ft}} + d(\boldsymbol{x},t) - \widetilde{M^{-1}}_t u_l + M^{-1}u_r \tag{12}$$

where $\widetilde{\boldsymbol{w}_{Ft}} = \widehat{\boldsymbol{w}_{Ft}} - \boldsymbol{w}_F^*, \ \ \widetilde{M^{-1}}_t = \widehat{M^{-1}}_t - M^{-1}$.

To let the adaptive parameters stay in a prescribed range we adopt the following adaptive laws with projection [4]:

$$\dot{\widehat{w_{Fnt}}} = \begin{cases} 0 & \text{for } \widehat{w_{Fnt}} = \underline{w_{F_n}},\ \phi_{Fn}(\boldsymbol{x})z_2 < 0 \\ 0 & \text{for } \widehat{w_{Fnt}} = \overline{w_{Fn}},\ \phi_{Fn}(\boldsymbol{x})z_2 > 0 \\ \gamma_F\ \phi_{Fn}(\boldsymbol{x})z_2 & \text{otherwise} \end{cases} \tag{13}$$

where $n = 1, \cdots, N_a + N_b + 2$, $\gamma_F \geq 0$ and $\phi_{Fn}(\boldsymbol{x})$ is the nth entry of $\boldsymbol{\phi}_F(\boldsymbol{x})$.

$$\dot{\widehat{M^{-1}}}_t = \begin{cases} 0 & \text{for } \widehat{M^{-1}}_t = \underline{M^{-1}},\ u_l z_2 < 0 \\ 0 & \text{for } \widehat{M^{-1}}_t = \overline{M^{-1}},\ u_l z_2 > 0 \\ \gamma_M\ u_l z_2 & \text{otherwise} \end{cases} \tag{14}$$

where $\gamma_M \geq 0$.

In the case of $\gamma_F = \gamma_M = 0$, we have a fixed robust controller.

4 Stability Analysis

Applying the virtual input α_1 to subsystem $\mathcal{S}1$, we have

$$\dot{z}_1 = z_2 - c_{1p}z_1 - c_{1i}\int_0^t z_1 dt \tag{15}$$

Equation (15) can be rewritten into the state-space form:

$$\dot{\boldsymbol{z}}_{1a} = \boldsymbol{A}\ \boldsymbol{z}_{1a} + \boldsymbol{B}\ z_2, \quad \boldsymbol{A} = \begin{bmatrix} 0 & 1 \\ -c_{1i} & -c_{1p} \end{bmatrix}, \quad \boldsymbol{B} = [0\ 1]^T \tag{16}$$

where $\boldsymbol{z}_{1a} = [\int_0^t z_1 dt, z_1]^T$. Then we have:

Lemma 1. *If the virtual input α_1 is applied to subsystem $\mathcal{S}1$, and if z_2 is made uniformly bounded at the next step, then $\mathcal{S}1$ is ISS, i.e., for $^\exists\lambda_0 > 0, ^\exists\rho_0 > 0$,*

$$|\boldsymbol{z}_{1a}(t)| \leq \lambda_0 e^{-\rho_0 t}|\boldsymbol{z}_{1a}(0)| + \frac{\lambda_0}{\rho_0}\left[\sup_{0\leq\tau\leq t} |z_2(\tau)|\right]$$

Since there exists a positive definite symmetric matrix $\boldsymbol{P}$ satisfying $\boldsymbol{A}^T\boldsymbol{P}+\boldsymbol{P}\boldsymbol{A} = -\boldsymbol{Q}$ for any positive definite symmetric matrix $\boldsymbol{Q}$, we have

$$\frac{d}{dt}\left(\frac{\boldsymbol{z}_{1a}^T\boldsymbol{z}_{1a}}{2}\right) = -\frac{1}{2}\boldsymbol{z}_{1a}^T\boldsymbol{Q}\boldsymbol{z}_{1a} + \boldsymbol{P}\boldsymbol{B}z_2 \leq -\frac{\lambda_{Qmin}}{2}|\boldsymbol{z}_{1a}| + |\boldsymbol{P}\boldsymbol{B}||z_2| \tag{17}$$

where λ_{Qmin} is the minimal eigenvalue of $\boldsymbol{Q}$. Then we have:

Lemma 2. *If z_2 is made uniformly ultimately bounded with ultimate bound $\overline{z}_2^u$ at the next step, then Subsystem $\mathcal{S}1$ controlled by α_1 is uniformly ultimately bounded such that*

$$|\boldsymbol{z}_{1a}(t)| \leq C_1\overline{z}_2^u \quad as\ t \geq {}^\exists T_1 > 0, \quad {}^\exists C_1 > 0$$

Next we show that the boundedness and transient performance of z_2 can be achieved by the nonlinear damping terms. From (10), we have

$$\frac{d}{dt}\left(\frac{z_2^2}{2}\right) \le -\frac{c_2}{2}z_2^2 - \left[\frac{c_2}{2} + D_2\right]|z_2|\left[|z_2| - \mu_2\right] \tag{18}$$

$$\mu_2(t) = \frac{|\widehat{\eta}_F(\boldsymbol{x}, \widehat{\boldsymbol{w}_F}_t)| + \left|\dfrac{M^{-1} - \widehat{M^{-1}}_t}{\widehat{M^{-1}}_t}\alpha_{20}\right| + |d(\boldsymbol{x},t)|}{\dfrac{c_2}{2} + D_2} \tag{19}$$
$$D_2 = \frac{M^{-1}}{\widehat{M^{-1}}_t}\left(\kappa_{21}\overline{F}(\boldsymbol{x}) + \kappa_{22}\alpha_{2d} + \kappa_{23}\overline{d}(\boldsymbol{x},t)\right)$$

It is trivial to verify that μ_2 is uniformly bounded since the nonlinear damping terms in the denominator grows at least as the same order as the numerator grows. Then we have

Lemma 3. *Let assumptions 1∼3 hold. Then Subsystem $\mathcal{S}2$ controlled by u is ISS such that*

$$|z_2(t)| \le |z_2(0)|e^{-c_2 t/2} + \sup_{0\le\tau\le t} \mu_2(\tau)$$

Furthermore, to analyze the ultimate bound of $|z_2|$ achieved by the adaptive laws, we define the following Lyapunov function:

$$V_2 = \frac{z_2^2}{2} + \frac{\widetilde{\boldsymbol{w}_F}_t^T\widetilde{\boldsymbol{w}_F}_t}{2\gamma_f} + \frac{\widetilde{M^{-1}}_t\widetilde{M^{-1}}_t}{2\gamma_M} \tag{20}$$

whose derivative satisfies

$$\dot{V}_2 \le -\left[c_2 + D_2\right]|z_2|(|z_2| - \delta_{2t}) \tag{21}$$

$$\delta_{2t} = \frac{|\widehat{\eta}_F(\boldsymbol{x}, \boldsymbol{w}_F^*)| + |d(\boldsymbol{x},t)|}{c_2 + D_2} \le \frac{C_{21}\varepsilon_F}{\kappa_{21}} + + \frac{C_{23}M_d}{\kappa_{23}}, \quad {}^{\exists}C_{21}, {}^{\exists}C_{23} > 0 \tag{22}$$

Then we have

Lemma 4. *Let the conditions and results of lemma 3 hold. If the control input u and adaptive laws are applied to subsystem $\mathcal{S}2$, then we have*

$$|z_2| \le \frac{C_{21}\varepsilon_F}{\kappa_{21}} + \frac{C_{23}M_d}{\kappa_{23}} \quad \textit{as } t \ge {}^{\exists}T_2 > 0$$

Lemmas 1 and 3 implies that the overall error system is a cascade of two ISS subsystems. Then along the same lines of the proof of lemma C.4 in [3], we have the following results:

$$|\boldsymbol{z}(t)| \le \beta_1 e^{-\rho_1 t}|\boldsymbol{z}(0)| + \beta_2\left[\sup_{0\le\tau\le t}\mu_2(\tau)\right] \tag{23}$$

where $\boldsymbol{z}(t) = \left[\boldsymbol{z}_{1a}^T(t), z_2(t)\right]^T$, and

$$\beta_1 = \sqrt{2}\left(\lambda_0^2 + 3\frac{\lambda_0^2}{\rho_0} + 3\frac{\lambda_0}{\rho_0} + 3\right), \quad \rho_1 = \min(\rho_0/2, c_2/4), \quad \beta_2 = \frac{\lambda_0^2}{\rho_0} + \frac{\lambda_0}{\rho_0} + 1 \tag{24}$$

Furthermore, from lemmas 2 and 4, we have the ultimate error bound as

$$|z(t)| \leq C_1 \left(\frac{C_{21}\varepsilon_F}{\kappa_{21}} + \frac{C_{23}M_d}{\kappa_{23}} \right) \quad \text{as } t \geq {}^{\exists}T_m > 0 \tag{25}$$

Finally, the results are summarized as follows.

Theorem 1. *Let assumptions 1∼3 hold. All the internal signals are unifomly bounded and the following results hold:*

1. *The overall error system is ISS such that*

$$|z(t)| \leq \beta_1 e^{-\rho_1 t}|z(0)| + \beta_2 \left[\sup_{0\leq\tau\leq t} \mu_2(\tau) \right]$$

2. *The ultimate bound of $|z(t)|$ can be made sufficiently small such that*

$$|z(t)| \leq C_1 \left(\frac{C_{21}\varepsilon_F}{\kappa_{21}} + \frac{C_{23}M_d}{\kappa_{23}} \right) \quad \text{as } t \geq {}^{\exists}T_m > 0$$

3. *The steady offset (zero frequency) component of z_1 converges to zero.*

5 Comments on the Controller Structure

One of the main obstacles that prevent the adaptive control techniques from coming into wide use in the industrial side is that the controller structure seems much more complicated compared to the conventional PI controller with minor-loop. It is commented here that the complicated looking controller designed in section 3 can be however explained as hierarchical modifications of the conventional PI controller with minor-loop. Therefore, it is belived that the proposed controller may gain wide acceptance of the engineers of various levels.

The conventional PI controller with minor-loop and nominal model compensation by the nominal mass M_0 and nominal disturbance model $F_0(\boldsymbol{x})$ is shown in Fig. 1 (a). The structure is easy to understand and is widely used in the industrial side. To improve trajectory tracking performance, we can add feedforward compoments to each control loop, as shown in Fig. 1 (b). If neither modelling error nor disturbance exists, the system is strictly linearized and the output x_1 will track y_r perfectly. This is however nothing else but the backstepping approach studied by the communities of adaptive control and nonlinear control. Notice if we set $\kappa_{21} = \kappa_{22} = \kappa_{23} = 0$, $\widehat{M^{-1}}_t = M_0^{-1}$ and $\widehat{F}(\boldsymbol{x}, \widehat{\boldsymbol{w}_{Ft}}) = F_0(\boldsymbol{x})$ in controller (11), the control system coincises with Fig. 1 (b). This explanation clarifies close relation between the backstepping approach with the conventional PI controller with minor-loop.

To counteract the modelling errors or disturbances we can adopt the nonlinear damping terms by setting $\kappa_{21}, \kappa_{22}, \kappa_{23} > 0$, $\widehat{M^{-1}}_t = M_0^{-1}$ and $\widehat{F}(\boldsymbol{x}, \widehat{\boldsymbol{w}_{Ft}}) = F_0(\boldsymbol{x})$ in controller (11). Then the control system becomes Fig. 1 (c). In this case, the ISS property is achieved by the nonlinear damping terms.

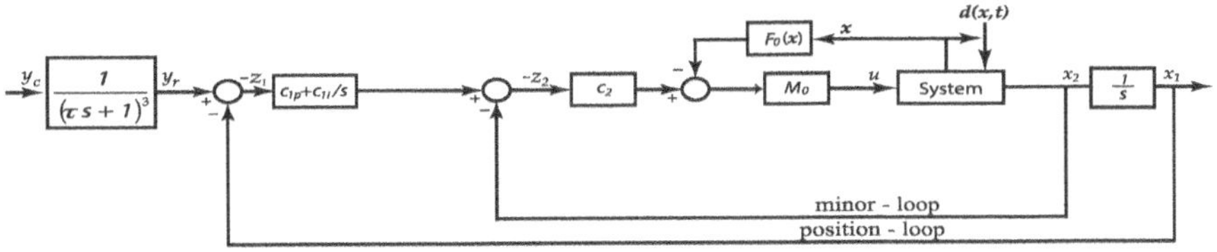

(a) Conventional nonlinear motion controller with minor-loop.

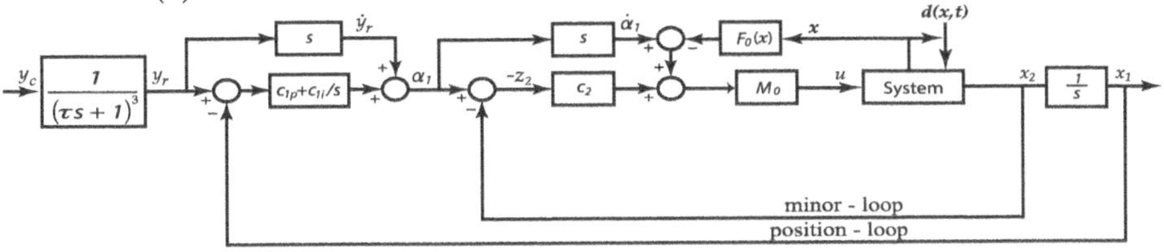

(b) Nominal nonlinear motion controller by backstepping design.

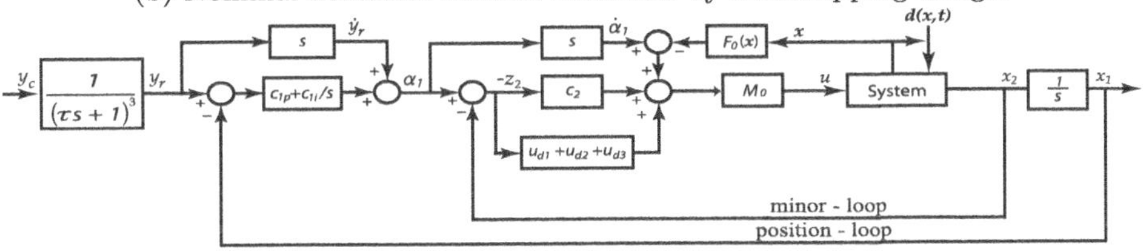

(c) Robust nonlinear motion controller by backstepping design.

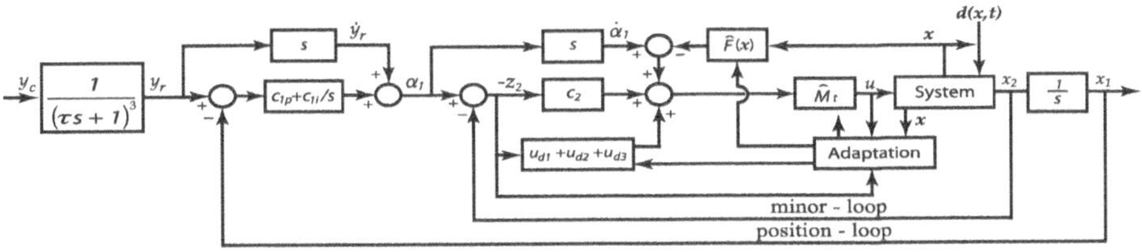

(d) Adaptive robust nonlinear motion controller by backstepping design.

Fig. 1. Evolutionary development of controller structure

To furthermore reduce the control error, we activate the adaptive laws (13) and (14) for the controller. Then the control system becomes Fig. 1 (d).

6 Numerical Studies

The physical parameters in (1)∼(3) are as follows.

$$\begin{array}{c} M=1[\mathrm{kg}],\quad \sigma_0=10^5[\mathrm{N/m}],\quad \sigma_1=\sqrt{10^5}[\mathrm{Ns/m}],\quad \sigma_2=0.8[\mathrm{Ns/m}],\\ F_c=1[\mathrm{N}],\quad F_s=2[\mathrm{N}],\quad \dot{x}_s=0.01[\mathrm{m/s}] \end{array} \tag{26}$$

The nominal value of M is given as $M_0 = 3M$, and all the other parameters' nominal values are zero.

Details of the controller are given as follows.

Bounds of the unknown parameters:

$$0 = \underline{w_{\sigma_2}} \leq w_{\sigma_2} \leq \overline{w_{\sigma_2}} = 20,\ 0 = \underline{w_{FC}} \leq w_{FC} \leq \overline{w_{FC}} = 20,\ 0 = \underline{\boldsymbol{w}_b} \leq \boldsymbol{w}_b \leq \overline{\boldsymbol{w}_b} = 20$$
$$-20 = \underline{\boldsymbol{w}_a} \leq \boldsymbol{w}_a \leq \overline{\boldsymbol{w}_a} = 20,\quad 0.2 = \underline{M^{-1}} \leq M^{-1} \leq \overline{M^{-1}} = 20 \tag{27}$$

Initial values of the unknown parameters:

$$\widehat{M^{-1}}_0 = 1/M_0,\ \widehat{\boldsymbol{w}_F}_0 = [0, \cdots, 0]^T \tag{28}$$

Three controllers are implemented:
(1) Nominal controller:

$$c_{1p}=40,\ c_{1i}=20^2,\ c_2=20,\quad \kappa_{21}=0,\ \kappa_{22}=0,\ \kappa_{23}=0, \gamma_F=0,\ \gamma_G=0 \tag{29}$$

(2) Robust controller:

$$c_{1p} = 40,\ c_{1i} = 20^2,\ c_2 = 20,\quad \kappa_{21} = 5,\ \kappa_{22} = 5,\ \kappa_{23} = 5,\ \gamma_F = 0,\ \gamma_G = 0 \tag{30}$$

(3) Adaptive robust controller:

$$c_{1p} = 40,\ c_{1i} = 20^2,\ c_2 = 20,\ \kappa_{21} = 5,\ \kappa_{22} = 5,\ \kappa_{23} = 5,\ \gamma_F = 5000,\ \gamma_G = 50 \tag{31}$$

The reference trajectory y_r shown in Fig. 2 together with its velocity are obtained by passing a rectangular wave to a low-pass filter $1/(0.1s + 1)^3$. The controllers are implemented at a sampling period of $T = 0.2$[ms]. A uniformly distributed stochastic noise between -10^{-6}[m] and 10^6[m] is added to the measurement of the position x_1. The measurement of velocity x_2 is obtained by pseudo-differentiation $sx_1/(0.0004s + 1)$.

The results are shown in Figs. 2∼4. In each figure, from top to bottom are the error signals z_1 and z_2, and the control input u. It can be seen in Fig. 3 that by the nominal controller, the error signals are quite significant. In Fig. 4, we can find that owing to the nonlinear damping terms employed in the controllers, the error signals are reduced. However, by the fixed robust controller, the error signals are not suppressed enough. Of course, increasing the gains of the nonlinear damping

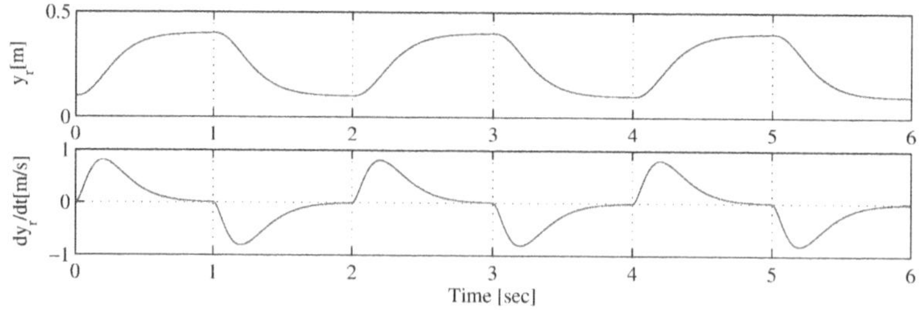

Fig. 2. Reference trajectory and its velocity

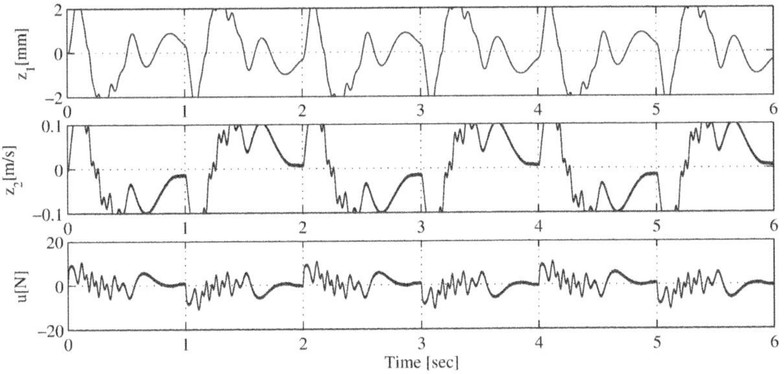

Fig. 3. Results of the nominal controller by backstepping design

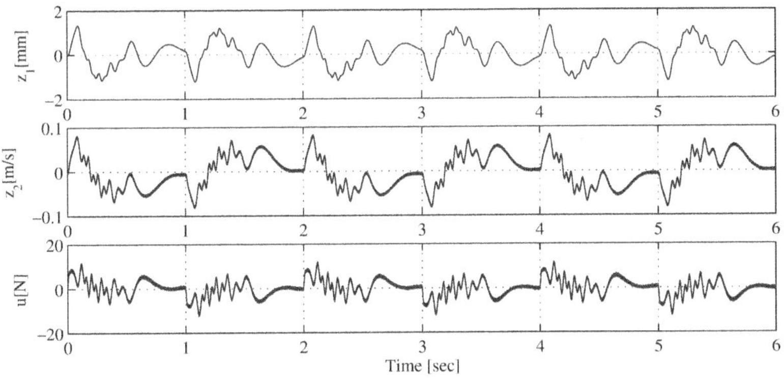

Fig. 4. Results of the robust controller by backstepping design

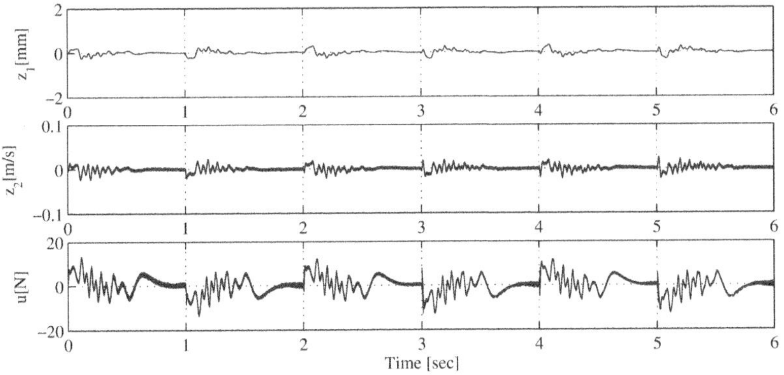

Fig. 5. Results of the adaptive robust controller by backstepping design

terms may lead to smaller error signals. However, the control signal may be quite noisy and even cause saturation of the actuator in the situation of real applications. On the other hand, by the adaptive robust controller, we can find in Fig. 5 that the error signals become much smaller compared to the case of the fixed robust controller. Notice that the results reflect the theoretical results of Theorem 1 quite well.

7 Conclusions

In this paper, a practical adaptive robust nonlinear controller by backstepping design has been proposed for motion control of an SISO nonlinear mechanical system, where the distrubances due to ripple force and friction are compensated by the RBF networks. To overcome the main obstacles that prevent the adaptive control techniques from coming into wide use in the industrial side, our attention is focused on the guaranteed transient performance and transparent structure of the control system. It has been commented that the complicated looking adaptive robust nonlinear controller can be explained as hierarchical modifications of the coventional PI position controller with minor-loop. This strategy is contrastive to those in [1] and [4]. Therefore it is believed that the proposed controller may gain wide acceptance of the engineers. Extensive simulation studies have been carried out to verify the theoretical results.

References

1. Huang, S. N., Tan K. K., Lee T. H.: Adaptive motion control using neural network approximations. Automatica **38** (2002) 227-233
2. Canudas de Wit, C., Olsson, H., Åström, K. J., Lischinsky, P.: A new model for control of systems with friction. IEEE Transactions on Automatic Control **40** (1995) 419-425
3. Krstic, M., Kanellakopoulos, L., Kokotovic, P.: Nonlinear and Adaptive Control Design, John Wiley & Sons, Inc. (1995).
4. Xu, L., Yao, B.: Adaptive robust precision motion control of linear motors with negligible electrical dynamics: theory and experiments. IEEE Transactions on Mechatronics **4** (2001) 444-452

Robust Adaptive Neural Network Control for a Class of Nonlinear Systems with Uncertainties

Hai-Sen Ke and Hong Xu

College of Mechanical and Electrical Engineering,
China Jiliang University, Hangzhou 310018, P.R. China
hske@cjlu.edu.cn

Abstract. In this note, robust adaptive neural network (NN) control scheme is constructed for a class of unknown nonlinear systems with drift terms. The robust adaptive NN control laws are developed using backstepping technique which does not require the unknown parameters to be linear parametrizable and no regression matrices are needed. All the signals in the resulting closed-loop system are proved to be ultimately uniform bounded, and the system states are guaranteed to converge to zero.

Keywords: adaptive control, backstepping, neural network, nonlinear system.

1 Introduction

Over the past several decades, adaptive control theory has evolved as a powerful methodology for designing of nonlinear systems with parametric uncertainty. Perhaps the most important achievement in the design of adaptive controllers for nonlinear system is the development of global adaptive controllers for nonlinear systems in so-called parametric-strict-feedback (PSF) form[1-3]. In the original PSF form, there exists only unknown parameters and the unknown parameters are required to enter the state equations linearly[1,4]. Since then, a considerable amount of the adaptive control research has been devoted to the development of so-called robust adaptive control systems, where the closed-loop stability properties are retained in the presence not only of large parametric uncertainty, but also modeling errors such as additive disturbance and unmodelled dynamics[5-9]. However, the proposed robust adaptive control approaches require a priori knowledge on the system nonlinearities. In order to cope with highly uncertain nonlinear systems, as an alternative, approximation-based adaptive control approaches have been extensively studied in the past decades using Lyapunov stability theory[10-12]. In the previous works, such assumption usually made in adaptive neural network control algorithm is that a bound on the network reconstruction error (also referred to as "approximation error" or "modeling error") is known. In this paper, we present a robust adaptive neural network control strategy to solve the stabilization of a class of nonlinear systems with strong drift nonlinearities.

The remainder of the paper is organized as follows. In Section 2, we describe the class of nonlinear system to be considered and the structure of the linearly parameterized NN used in controller design. A robust adaptive controller design

D. Liu et al. (Eds.): ISNN 2007, Part I, LNCS 4491, pp. 285–291, 2007.

procedure is developed in Section 3, and the stability of the closed-loop system is analyzed also in Section 3. Finally, the note is concluded in Section 4.

2 Preliminaries

Consider the control problem of a class of nonlinear system with strong drift nonlinearities:

$$\begin{aligned}\dot{x}_i &= x_{i+1} + f_i(\bar{\mathbf{x}}_i) + g_i(\bar{\mathbf{x}}_i),\ 1 \le i \le n-1 \\ \dot{x}_n &= u + f_n(\mathbf{x}) + g_n(\mathbf{x})\end{aligned}, \tag{1}$$

where $\mathbf{x} \triangleq [x_1, \cdots, x_n]^{\mathrm{T}} \in \mathbb{R}^n$ is the system state, $\bar{\mathbf{x}}_i \triangleq [x_1, \cdots, x_i]^{\mathrm{T}} \in \mathbb{R}^i$, $u \in \mathbb{R}$ is the control input, the nonlinear functions $f_i(\bar{\mathbf{x}}_i), 1 \le i \le n,$ are known, smooth and satisfy $f_i(0) = 0, \forall \bar{\mathbf{x}}_i \in \mathbb{R}^i$, the unknown nonlinear functions $g_i(\bar{\mathbf{x}}_i), 1 \le i \le n,$ represent the unmodelled dynamics or external disturbances and satisfy $g_i(0) = 0, \forall \bar{\mathbf{x}}_i \in \mathbb{R}^i$.

We consider the control problem of a class of perturbed nonlinear system in the form (1). The control objective is to construct a robust adaptive nonlinear control law of the form

$$\begin{aligned}u &= u(\mathbf{x}, \mu) \\ \dot{\mu} &= v(\mathbf{x}, \mu)\end{aligned}, \tag{2}$$

such that the states $\boldsymbol{x}$ of the system (1) are driven to the equilibrium $\boldsymbol{x}$=0, while keep all other signals in the closed-loop system bounded.

In this paper, for the unknown nonlinear functions $g_i(\bar{\mathbf{x}}_i), 1 \le i \le n,$ we have the following approximation over the compact sets Ω_i

$$g_i(\bar{\mathbf{x}}_i) = w_i^{\mathrm{T}} s_i(\bar{\mathbf{x}}_i) + \varepsilon_i(\bar{\mathbf{x}}_i)\ \ \forall \bar{\mathbf{x}}_i \in \Omega_i \in \mathbb{R}^i, \tag{3}$$

Where $s_i(\bar{\mathbf{x}}_i): \mathbb{R}^i \to \mathbb{R}^{l_i}$ is the known basis function vector, $w_i \in \mathbb{R}^{l_i}$ is the weight vector, $\varepsilon_i(\bar{\mathbf{x}}_i)$ is the approximation error, the NN node number $l_i > 1$.

The optimal weight vector in (3) is an "artificial" quantity required only for analytical purposes. Typically, w_i^* is chosen as the value of w_i that minimizes $|\varepsilon_i(\bar{\mathbf{x}}_i)|$ for all $\bar{\mathbf{x}}_i \in \Omega_i \in \mathbb{R}^i$, i.e.,

$$w_i^* \triangleq \arg\min_{w_i \in \mathbb{R}^{l_i}} \{\sup_{\bar{\mathbf{x}}_i \in \Omega_i} \left| g_i(\bar{\mathbf{x}}_i) - w_i^{\mathrm{T}} s_i(\bar{\mathbf{x}}_i) \right| \}, \tag{4}$$

Assumption1: over a compact region $\Omega_i \in \mathbb{R}^i$

$$|\varepsilon_i(\bar{\mathbf{x}}_i)| \le \varepsilon_i^*,\ \ \forall \bar{\mathbf{x}}_i \in \Omega_i, 1 \le i \le n, \tag{5}$$

Where ε_i^* is an unknown constant.

3 Robust Adaptive NN Controller Design

In this section, we proceed to design a robust adaptive NN controller using backstepping method. The control law u and the adaptive laws are developed based on a change of coordinate

$$\begin{aligned} z_1 &= x_1 \\ z_2 &= x_2 - \alpha_1 \\ &\vdots \\ z_n &= x_n - \alpha_{n-1} \end{aligned} \tag{6}$$

Where the functions $\alpha_i, 1 \le i \le n-1$, are referred to as intermediate control inputs, which will be designed using backstepping approach. At the nth step, the actual control u appears and the design is completed.

Step1: using α_1 as a control to stabilize the z_1-subsystem

$$\dot{z}_1 = \alpha_1 + z_2 + f_1(x_1) + g_1(x_1), \tag{7}$$

Since $g_1(x_1)$ is a smooth function of z_1 and satisfies $g_1(0) = 0$, we can write

$$g_1(x_1) = z_1 \phi_1(z_1) \tag{8}$$

Where $\phi_1(z_1)$ is a smooth function of z_1, which can be approximated by

$$\phi_1(z_1) = w_1^{\mathrm{T}} s_1(z_1) + \varepsilon_1(z_1)$$

Accordingly, system (7) can be further expressed as

$$\dot{z}_1 = \alpha_1 + z_2 + f_1(x_1) + z_1[w_1^{\mathrm{T}} s_1(z_1) + \varepsilon_1(z_1)] \tag{9}$$

Define $w_{a,1} = w_1, d_1 = \varepsilon_1$, let $\tilde{w}_{a,1} = \hat{w}_{a,1} - w_{a,1}$ and $\tilde{d}_1 = \hat{d}_1 - d_1$.

We now design the intermediate α_1 as the following stabilizing function

$$\alpha_1(x_1, \hat{w}_{a,1}, \hat{d}_1) = -k_1 z_1 - f_1(x_1) - \hat{w}_{a,1}^{\mathrm{T}} s_1(z_1) z_1 - \hat{d}_1 z_1, \tag{10}$$

$$\dot{\hat{w}}_{a,1} = \Gamma_1 z_1^2 s_1(z_1) \tag{11}$$

$$\dot{\hat{d}}_1 = \gamma_1 z_1^2 \tag{12}$$

Where $\Gamma = \Gamma^{\mathrm{T}} > 0, \gamma_1 > 0$ are design constants. Clearly, $\alpha_1(0, \hat{w}_{a,1}, \hat{d}_1) = 0$.

The time derivative of

$$V_1 = \frac{1}{2} z_1^2 + \frac{1}{2} \tilde{w}_{a,1}^{\mathrm{T}} \Gamma_1^{-1} \tilde{w}_{a,1} + \frac{1}{2\gamma_1} \tilde{d}_1^2 \tag{13}$$

is given by

$$\dot{V}_1 \le -k_1 z_1^2 + z_1 z_2 - \tilde{w}_{a,1}^{\mathrm{T}} s_1(z_1) z_1^2 - \tilde{d}_1 z_1^2 + \tilde{w}_{a,1}^{\mathrm{T}} \Gamma_1^{-1} \dot{\hat{w}}_{a,1} + \frac{1}{\gamma_1} \tilde{d}_1 \dot{\hat{d}}_1 \le -k_1 z_1^2 + z_1 z_2 \tag{14}$$

Step *i* $(2 \le i < n)$: using the definition of z_i, we have

$$\begin{aligned} \dot{z}_i = \dot{x}_i - \dot{\alpha}_{i-1} = [x_{i+1} + f_i(\bar{\mathbf{x}}_i) + g_i(\bar{\mathbf{x}}_i)] \\ -\sum_{j=1}^{i-1} [\frac{\partial \alpha_{i-1}}{\partial x_j}(x_{j+1} + f_j(\bar{\mathbf{x}}_j) + g_j(\bar{\mathbf{x}}_j)) + \frac{\partial \alpha_{i-1}}{\partial \hat{w}_{a,j}} \dot{\hat{w}}_{a,j} + \frac{\partial \alpha_{i-1}}{\partial \hat{d}_j} \dot{\hat{d}}_j] \end{aligned} \tag{15}$$

Noting the change of coordinate (6), and $g_i(0) = 0$, $\alpha_{j-1}(0, \hat{w}_{a,1} \cdots \hat{w}_{a,j-1}, \hat{d}_1 \cdots \hat{d}_j) = 0$, $2 \le j \le i$, $\forall \bar{\mathbf{x}}_i \in \mathbb{R}^i$, we can write

$$g_i(\bar{\mathbf{x}}_i) = \sum_{j=1}^{i} z_j g_{ij}(\bar{\mathbf{z}}_i) \tag{16}$$

Where $g_{ij}(\bar{\mathbf{z}}_i), 1 \le j \le i$ are smooth continuous functions. We then define

$$\phi_i(\bar{\mathbf{z}}_i) = \frac{i}{4} \max_{1 \le j \le i} \{g_{ij}^2(\bar{\mathbf{z}}_i)\} \tag{17}$$

Using (3), the unknown function can be approximated by

$$\phi_i(\bar{\mathbf{z}}_i) = w_i^{\mathrm{T}} s_i(\bar{\mathbf{z}}_i) + \varepsilon_i(\bar{\mathbf{z}}_i) \tag{18}$$

Define $\bar{f}_i(\bar{\mathbf{x}}_i) = f_i(\bar{\mathbf{x}}_i) - \sum_{j=1}^{i-1} [\frac{\partial \alpha_{i-1}}{\partial x_j}(x_{j+1} + f_j(\bar{\mathbf{x}}_j)) + \frac{\partial \alpha_{i-1}}{\partial \hat{w}_{a,j}} \dot{\hat{w}}_{a,j} + \frac{\partial \alpha_{i-1}}{\partial \hat{d}_j} \dot{\hat{d}}_j]$, $w_{a,i} = [w_1^{\mathrm{T}}, \cdots w_{i-1}^{\mathrm{T}}, w_i^{\mathrm{T}}]^{\mathrm{T}}$, $d_i = \max\{\left|\frac{\partial \alpha_{i-1}}{\partial x_1}\right| \varepsilon_1, \cdots, \left|\frac{\partial \alpha_{i-1}}{\partial x_{i-1}}\right| \varepsilon_{i-1}, \varepsilon_i\}$, $s_{a,i} = [-\frac{\partial \alpha_{i-1}}{\partial x_1} s_1^{\mathrm{T}}, \cdots, -\frac{\partial \alpha_{i-1}}{\partial x_{i-1}} s_{i-1}^{\mathrm{T}}, s_i^{\mathrm{T}}]^{\mathrm{T}}$. Let $\tilde{w}_{a,i} = \hat{w}_{a,i} - w_{a,i}$ and $\tilde{d}_i = \hat{d}_i - d_i$.

We now design the intermediate α_i as the following stabilizing function

$$\alpha_i = -k_i z_i - z_{i-1} - \bar{f}_i(\bar{\mathbf{x}}_i) - \hat{w}_{a,i}^{\mathrm{T}} s_{a,i}(\bar{\mathbf{z}}_i) z_i - \hat{d}_i z_i \tag{19}$$

$$\dot{\hat{w}}_{a,i} = \Gamma_i z_i^2 s_i(\bar{\mathbf{z}}_i) \tag{20}$$

$$\dot{\hat{d}}_i = \gamma_i z_i^2 \tag{21}$$

Where $\Gamma_i = \Gamma_i^{\mathrm{T}} > 0$, $\gamma_i > 0$ are design constants. Clearly, $\bar{\mathbf{z}}_i = 0$ can guarantee $\alpha_i = 0$.

The time derivative of

$$V_i = \frac{1}{2} z_i^2 + \frac{1}{2} \tilde{w}_{a,i}^{\mathrm{T}} \Gamma_i^{-1} \tilde{w}_{a,i} + \frac{1}{2\gamma_i} \tilde{d}_i^2 \tag{22}$$

is given by

$$\dot{V}_i = -k_i z_i^2 - z_{i-1} z_i + z_i z_{i+1} + z_i \sum_{j=1}^{i} z_j g_{ij}(\bar{z}_i) - \hat{w}_{a,i}^{\mathrm{T}} s_i(\bar{z}_i) z_i^2 - \hat{d}_i z_i^2$$

$$+\tilde{w}_{a,i}^{\mathrm{T}} \Gamma_i^{-1} \dot{\hat{w}}_{a,i} + \frac{1}{\gamma_i} \tilde{d}_i \dot{\hat{d}}_i \le \tilde{w}_{a,i}^{\mathrm{T}} \Gamma_i^{-1} \dot{\hat{w}}_{a,i} + \frac{1}{\gamma_i} \tilde{d}_i \dot{\hat{d}}_i$$

$$-k_i z_i^2 - z_{i-1} z_i + z_i z_{i+1} + \sum_{j=1}^{i} z_j^2 + \frac{z_i^2}{4} \sum_{j=1}^{i} g_{ij}^2(\bar{z}_i) - \hat{w}_{a,i}^{\mathrm{T}} s_i(\bar{z}_i) z_i^2 - \hat{d}_i z_i^2 \tag{23}$$

Taking (17), (18), (20) and (21) into account, (23) can be reduced to

$$\dot{V}_i \le -k_i z_i^2 - z_{i-1} z_i + z_i z_{i+1} + \sum_{j=1}^{i} z_j^2 \tag{24}$$

Step *n*: in the final step, the actual control u appears. Employing a similar procedure as before, we can design the control u as the following stabilizing function:

$$u = -k_n z_n - z_{n-1} - \bar{f}_n(\mathbf{x}) - \hat{w}_{a,n}^{\mathrm{T}} s_{a,n}(\mathbf{z}) z_n - \hat{d}_n z_n \tag{25}$$

$$\dot{\hat{w}}_{a,n} = \Gamma_n z_n^2 s_n(\mathbf{z}) \tag{26}$$

$$\dot{\hat{d}}_n = \gamma_n z_n^2 \tag{27}$$

Where $\Gamma_n = \Gamma_n^{\mathrm{T}} > 0,\ \gamma_n > 0$ are design constants.
Consider the Lyapunov function candidate

$$V = \frac{1}{2} z_n^2 + \frac{1}{2} \tilde{w}_{a,n}^{\mathrm{T}} \Gamma_n^{-1} \tilde{w}_{a,n} + \frac{1}{2\gamma_n} \tilde{d}_n^2 + \sum_{i=1}^{n-1} V_i \tag{28}$$

The time derivative of V is given by

$$\dot{V} \le -\sum_{i=1}^{n} k_i z_i^2 + (n-1) z_1^2 + \sum_{j=2}^{n} (n-j+1) z_j^2 \tag{29}$$

Choosing $k_i, 1 \le i \le n$, as

$$k_1 = n,\ k_i = n - i + 2,\ 2 \le i \le n, \tag{30}$$

(29) can be further expressed as

$$\dot{V} \le -\sum_{i=1}^{n} z_i^2 \le 0 \tag{31}$$

From (31), we conclude that V and $z_i(t), 1 \le i \le n$, are all bounded. Furthermore, $z_i(t), 1 \le i \le n$, are square integrable on $[0, \infty)$. It can be shown from the above design procedure that all estimated parameters $\hat{w}_{a,i}, \hat{d}_i, 1 \le i \le n$, and in turn $\alpha_i(t), 1 \le i \le n-1$, the system states $x_i(t), 1 \le i \le n,$ are also bounded on $[0,\infty)$. Therefore, no finite-time escape phenomenon may occur. Since $\dot{z}_i(t), \dot{x}_i(t), 1 \le i \le n,$ are bounded on $[0,\infty)$. Thus, using the Barbalat's lemma, we can conclude that $Lim_{t\to\infty} z_i(t) = 0, 1 \le i \le n$. Noting that the change of coordinate (6) and $\bar{\mathbf{z}}_i = 0$ can guarantee $\alpha_i = 0, 1 \le i \le n-1$, we have that $Lim_{t\to\infty} x_i(t) = 0, 1 \le i \le n$.

We now summarize the result of this paper in the following theorem.

Theorem 1. under assumption1, suppose the robust adaptive stabilizer developed above is applied to system (1), then for bounded initial conditions,

1) There exists a sufficiently large compact set Ω such that $\mathbf{x} \in \Omega \in \mathbb{R}^n$ for all t>0, and all the signals in the closed-loop system remain bounded.
2) The system states eventually converge to zero, i.e., $x(t) \to 0$ as $t \to \infty$.

4 Conclusions

In this paper, a constructive robust adaptive NN control strategy has been presented for a class of nonlinear system with unknown drift term without imposing any restriction on the system order and the growth of the nonlinear drift uncertain. By using NN, our proposed controller is free of the linear-in parameters property of the nonlinear drifts.

Acknowledgment. This work is supported by National Natural Science Foundation of China (No. 60674023).

References

1. Krstic M., Knallakopoulos, I., Kokotovic, P. V.: Nonlinear and adaptive control design. New York: Wiley-interscience, 1995.
2. Ye, X. D.: Adaptive nonlinear output-feedback control with unknown high-frequency gain signs. IEEE Transactions on Automatic Control **46** (2001) 112-115
3. Ye, X. D.: Global adaptive control of nonlinearly parametrized systems. IEEE Transactions on Automatic Control **48** (2003) 169-173
4. Kanellakopoulos, I., Kokotovic, P.V., Morse, A.S.: Systematic design of adaptive controller for feedback linearizable systems. IEEE Transactions on Automatic Control **36** (1991) 1241-1253
5. Zhang, K. J., Feng, C. B., Fei, S. M.: Robust output feedback tracking for a class of uncertain nonlinear systems. Control Theory & Applications **39** (2003) 173-179
6. Qu, Z. H., Jin, Y. F.: Robust control of nonlinear systems in the presence of unknown exogenous dynamics. IEEE Transactions on Automatic Control **48** (2003) 336-343
7. Ge, S. S., Wang, J.: Robust adaptive tracking for time-varying uncertain nonlinear system with unknown control coefficients. IEEE Transactions on Automatic Control **48** (2003) 1463-1469

8. Qu, Z. H.: Global stabilization and convergence of nonlinear systems with uncertain exogenous dynamics. IEEE Transactions on Automatic Control **49** (2004) 1852-1858
9. Ke, H. S., Ye, X. D.: Robust adaptive controller design for a class of nonlinear systems with unknown high frequency gains Journal of Zhejiang University SCIENCE A **7** 315-320
10. Polycarpou, M. M.: Stable adaptive neural control scheme for nonlinear systems. IEEE Transactions on Automatic Control **41** (1996) 447-451
11. Wang, Z. P., Ge, S. S., Lee, T. H.: Robust adaptive neural network control of uncertain nonholonomic systems with strong nonlinear drifts. IEEE Transactions on Systems, Man, and Cybernetics-part b: Cybernetics **34** (2004) 2048-2059
12. Li, Y. H., Qiang, S., Zhuang, X. Y., Kaynak, O.: Robust and adaptive backstepping control for nonlinear systems using RBF neural networks. IEEE Transactions on Neural Network **15** (2004) 693-701

On Neural Network Switched Stabilization of SISO Switched Nonlinear Systems with Actuator Saturation

Fei Long and Wei Wei

The Provincial Key Lab of Playing-Go Strategy and Control System, College of Science, Guizhou University, Guiyang 550025, Guizhou, P.R. China
feilong73@56.com, flong1973@yahoo.com.cn

Abstract. As we know, saturation, deadzone, backlash, and hysteresis are the most common actuator nonlinearities in practical control system applications. Saturation nonlinearity is unavoidable in most actuators. In this paper, we address the Neural Network saturation compensation for a class of switched nonlinear systems with actuator saturation. An actuator saturation compensation switching scheme for switched nonlinear systems with its subsystem in Brunovsky canonical form is presented using Neural Network. The actuator saturation is assumed to be unknown and the saturation compensator is introduced into a feed-forward path. The scheme that leads to switched stability and disturbance rejection is rigorously proved. The tracking performance of switched nonlinear system is guaranteed based on common Lyapunov approach under the designed switching strategy.

1 Introduction

The history of hybrid system research can be traced back at least to the 1950's with the study of engineering systems that contain relays. However, hybrid systems began to attract researcher's attention in the early 1990's, mainly because of the vast development and implementation of digital micro controllers and embedded devices. The last decade has seen considerable research activities in the field of hybrid systems involving researchers from some traditionally distinct fields, such as computer science, control systems engineering, and mathematics [1], [2].

Switched nonlinear system is a hybrid system that comprises a collection of nonlinear subsystems together with a switching rule that specifies the switching among the subsystems. It is well known that different switching strategy would produce different behavior of system and hence lead to different system performances. As a result, how to choose suitable switching law that makes switched system to attain certain performance is indeed an important and well-motivated problem. However, the design of switching strategy is generally very challenging. As a switching strategy is a discontinuous function of time and it may be highly nonlinear, the design problem is very difficult to handle.

As we know, saturation, deadzone, backlash, and hysteresis are the most common actuator nonlinearities in practical control system applications. Saturation nonlinearity is unavoidable in most actuators. Categories of saturation nonlinearities include

D. Liu et al. (Eds.): ISNN 2007, Part I, LNCS 4491, pp. 292–301, 2007.

constraints of the magnitude and the rate of actuator inputs. When an actuator has reached such an input limit, it is said to be "saturated," since efforts to further increase the actuator output would not result in any variation in the output. Due to the non-analytic nature of the actuator nonlinear dynamics and the fact that the exact actuator nonlinear functions are unknown, such systems present a challenge to the controller design and provide an application field for adaptive control, sliding mode control and neural network-based control. Hu and Lin [3] proposed a systematic controller design to compensate the saturation nonlinearity for continuous and discrete time linear systems. Annaswamy et al. [4] developed an adaptive controller to accommodate saturation constraints in the presence of time delays in linear systems. In some recent seminal work, several rigorously derived adaptive control schemes have been presented for actuator nonlinearity compensation. Compensation for non-symmetric deadzone is considered in [5] for linear systems and in [6] for nonlinear systems in Brunovsky form with known nonlinear functions. Universal approximation property and learning capability of Neural Networks have proven to be a powerful tool to control complex dynamical nonlinear systems with parameter uncertainty. Although persistent problems such as approximation of non-smooth functions and offline weights initialization requirement still exist, Neural Networks have been widely used in adaptive and robust adaptive control. In general, Neural Networks are used to estimate the unknown nonlinear dynamics and/or functions and to compensate for their parasitic effects. Unlike the standard adaptive control schemes, Neural Network can cope with a nonlinear system that is linearly un-parameterizable. Recently, many researchers [7]-[16] have used Neural Networks to incorporate the Lyapunov theory in order to ensure the overall system stabilization and disturbance rejection.

However, most of these good results are restricted in non-switched systems. Due to the difference between non-switched systems and switched systems, a stable controller designed in non-switched system may become unstable in switched system via unsuitable switching rule, thus we may run into troubles when we implement these networks controllers in switched system in which the data are typically available only at switching time instants. Therefore, the study for switched system based on neural network is necessary and significant. In some recent works, adaptive Neural Network control schemes have been presented for switched nonlinear systems. Switching stabilization is considered in [17] for switched nonlinear system in Brunovsky form, in [18] for switched nonlinear system in trigonal form, in [19] for switched nonlinear system with time-delay and in [20] for switched nonlinear system with impulsive effect. Messai et al. [21] developed an identification strategy for a class of hybrid dynamic systems using neural networks. This paper proposes a Neural Networks-based scheme for saturation switching control for a class of switched nonlinear systems in the Brunovsky canonical form with actuator saturation.

This paper is organized as follows. Section 2 provides some preliminaries and definitions. Section 3 presents the saturation nonlinearity and converted expressions. Section 4 discusses SISO switched nonlinear systems in the presence of saturation, the design process of outer-loop tracking adaptive switching Neural Network controller and compensator, and the rigorous proof of the tracking performance. Finally, the conclusion is drawn in Section 5.

2 Preliminaries & Definitions

Let $\mathbb{R}$ denote real numbers, $\mathbb{R}^n$ denote the real vector, and $\mathbb{R}^{n\times m}$ denote the real matrices. Suppose that Ω is a compact simply connected subset in $\mathbb{R}^n$.

$\|\cdot\|$ is denoted any suitable vector norm. When it is required to be specific we denote the p-norm by $\|\cdot\|_p$. The supremum norm of vector value function $f(\cdot):\Omega \mapsto \mathbb{R}^n$, over Ω, is defined as $\sup_{x\in\Omega}\|f(x)\|$. Given a matrix $A=[a_{ij}]_{n\times m}$, the Frobenius norm of matrix A is defined by

$$\|A\|_F^2 = tr(A^T A) = \sum_{i=1}^{n}\sum_{j=1}^{m} a_{ij}^2 \tag{2.1}$$

with $tr(A)$ is the trace of the matrix A. Note that the Frobenius norm is compatible with the 2-norm so that $\|Ax\|_2 \le \|A\|_F \|x\|_2$, and the following properties is obvious.

i) $tr(AB) = tr(BA)$.

ii) $tr(A^T BA) \ge 0$, as B is positive–definite matrix.

iii) $\frac{d}{dt}tr(A(x)) = tr(\frac{dA(x)}{dt})$, as $A(x)$ is differentiable matrix-value function.

iv) according to Cauchy inequality, the following inequality is obvious

$$tr(A^T(B-A)) \le \|B\|_F \|A\|_F - \|A\|_F^2 . \tag{2.2}$$

Definition 2.1 (Switched Stabilization): Consider the switched nonlinear system $\dot{x} = g_{\sigma(t)}(x)$, with $x\in\mathbb{R}^n$ is system state, function $\sigma(\cdot):[0,+\infty)\mapsto\{1,2,\cdots,N\}=\underline{\mathbb{N}}$ is the piecewise constant switching signal. The switched system $\dot{x} = g_{\sigma(t)}(x)$ is said to be switched stabilization if there exists a switching rule $\sigma(t)$ such that switched system $\dot{x} = g_{\sigma(t)}(x)$ is asymptotically stable.

Definition 2.2 (Switching sequence): The sequence $\{(t_k, r_k)\}, r_k \in \underline{\mathbb{N}}, k\in\{1,2,\cdots\}$ is said to be switching sequence, if i): $\sigma(t_k^-)\ne\sigma(t_k^+)$; ii): $\sigma(t)=\sigma(t_k^+)=r_k, t\in[t_k, t_{k+1})$. Moreover, the interval $[t_k, t_{k+1})$ is said to be dwell time interval of the r_k -th subsystem.

3 Actuator Saturation

In control engineering, the most commonly used actuators are continuous drive devices, along with some incremental drive actuators such as stepper motors [22]. Saturation nonlinearity with its maximum and minimum operation limits is unavoidable in such devices. This paper investigates actuator saturation that appears in the switched nonlinear system plant and the way of its compensation based on Neural Network, which is shown in Fig. 1.

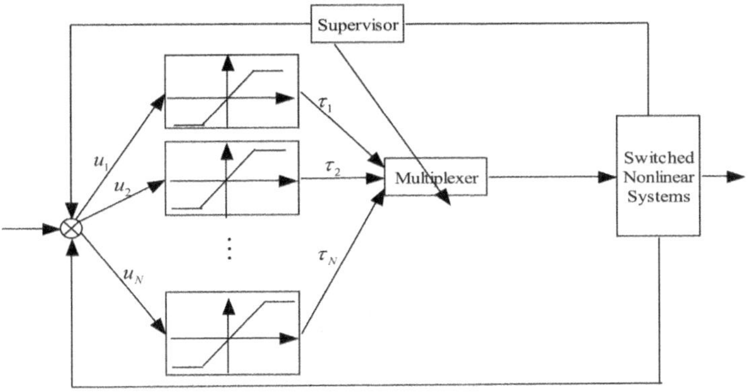

Fig. 1. Switched Nonlinear System with Actuator Saturation

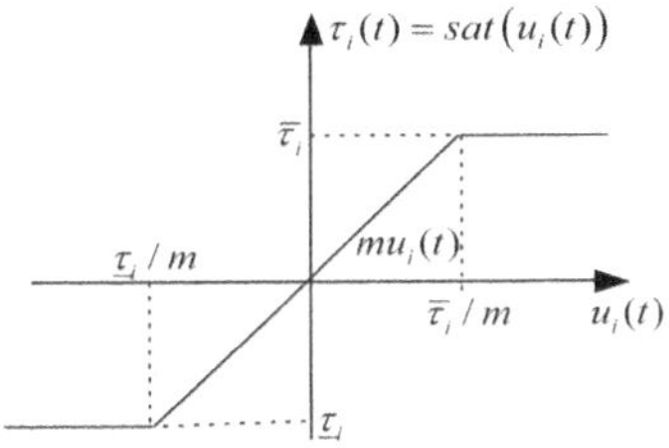

Fig. 2. Symmetric Saturation Nonlinearity

Assuming ideal saturation, shown in Fig. 2, the output of the actuator $\tau_i(t), i \in \underline{\mathbb{N}}$ are given by

$$\tau_i(t) = \begin{cases} \overline{\tau}_i; & u_i(t) \geq \overline{\tau}_i / m \\ mu_i(t); & \underline{\tau}_i / m \leq u_i(t) \leq \overline{\tau}_i / m \\ \underline{\tau}_i; & u_i(t) \leq \underline{\tau}_i / m \end{cases}, \tag{3.1}$$

where $\overline{\tau}_i$ and $\underline{\tau}_i$ is the chosen positive and negative saturation limits, respectively. If the control input $u_i(t), i \in \underline{\mathbb{N}}$ falls outside the range of the actuator, actuator saturation occurs and $u_i(t), i \in \underline{\mathbb{N}}$ can not be fully implemented by the device. The control signal that can not be implemented by the actuator, denoted as $\delta_i(t), i \in \underline{\mathbb{N}}$, is given by

$$\delta_i(t) = \tau_i(t) - u_i(t) = \begin{cases} \overline{\tau}_i - u_i(t); & u_i(t) \geq \overline{\tau}_i / m \\ (m-1)u_i(t); & \underline{\tau}_i / m \leq u_i(t) \leq \overline{\tau}_i / m \\ \underline{\tau}_i - u_i(t); & u_i(t) \leq \underline{\tau}_i / m \end{cases}. \tag{3.2}$$

From (3.2), the nonlinear actuator saturation can be described using $\delta_i(t), i \in \underline{\mathbb{N}}$ [4]. In this note, Neural Network is used to approximate modified saturation nonlinear functions $\delta_i(t), i \in \underline{\mathbb{N}}$.

4 Saturation Compensation for Switched Nonlinear Systems

4.1 Switched Nonlinear Systems Dynamics and Tracking Error Dynamics

Consider the following SISO switched nonlinear system

$$\begin{cases} \dot{x}_j = x_{j+1}, 1 \le j \le n-1 \\ \dot{x}_n = f_\sigma(x) + g_\sigma(x)\tau_\sigma \\ y = x_1 \end{cases}, \tag{4.1}$$

where $x = (x_1, x_2, \cdots, x_n)^T \in \mathbb{R}^n$ denotes system state; function $\sigma(\cdot):[0,+\infty) \mapsto \underline{\mathbb{N}}$ is the piecewise constant switching signal; $f_i(\cdot):\mathbb{R}^n \mapsto \mathbb{R}, i \in \underline{\mathbb{N}}$ are unknown smooth functions that contain the parameter uncertainties; $g_i(\cdot):\mathbb{R}^n \mapsto \mathbb{R}, i \in \underline{\mathbb{N}}$ are some known smooth functions; and there exists a constant $l > 0$ such that $|g_i(x)| \ge l, i \in \underline{\mathbb{N}}$ for every $x \in \mathbb{R}^n$.

Define the tracked reference signal $x_r(t)$ as

$$x_r = \left(y_r, y_r^{(1)}, \cdots, y_r^{(n-1)}\right)^T, \tag{4.2}$$

where the desired trajectory $x_r(t)$ is bounded and continuous; and there exists a known scalar bound $\mu > 0$ such that $\|x_r(t)\| \le \mu$.

Define the state tracking error vector $e \in \mathbb{R}^n$ as

$$e(t) = x(t) - x_r(t), \tag{4.3}$$

where $e = (e_1, e_2, \cdots, e_n)^T$, $e_j = x_j - y_r^{(j-1)}$, $j = 1, 2, \cdots, n.$

Define a filtered tracking error $d \in \mathbb{R}$ as

$$d = \Lambda e, \tag{4.4}$$

where $\Lambda = (\lambda_1, \lambda_2, \cdots, \lambda_{n-1}, 1)$ is appropriately chosen coefficient vector so that $e \to 0$ exponentially as $d \to 0$. Then the filtered tracking error d satisfies the following dynamics equation:

$$\dot{d} = f_\sigma(x) + g_\sigma(x)\tau_\sigma + Y_r, \tag{4.5}$$

where $Y_r = -y_r^{(n)} + \sum_{i=1}^{n-1} \lambda_i e_{i+1}$.

Consider the saturation nonlinearity (3.2). The filtered tracking error dynamics (4.5) can be rewritten as:

$$\dot{d} = f_\sigma(x) + g_\sigma(x)(u_\sigma + \delta_\sigma) + Y_r. \tag{4.6}$$

4.2 Design of Neural Network Saturation Compensator

Many well-known results showed that any sufficiently smooth function can be approximated arbitrary closely on a compact set using a 2-layer Neural Network with appropriate weights [23]. Basis function could be any continuous sigmoidal function. The Neural Network universal approximation property specifies that any continuous function can be approximated arbitrarily well using a linear combination of sigmoidal functions. As a result, there exists 2-layer Neural Network that closely approximates the modified saturation nonlinear functions $\delta_i(x), i\in \underline{N}$ and $f_i(x), i\in \underline{N}$, respectively. The structure of 2-lay Neural Networks is shown in Fig. 3.

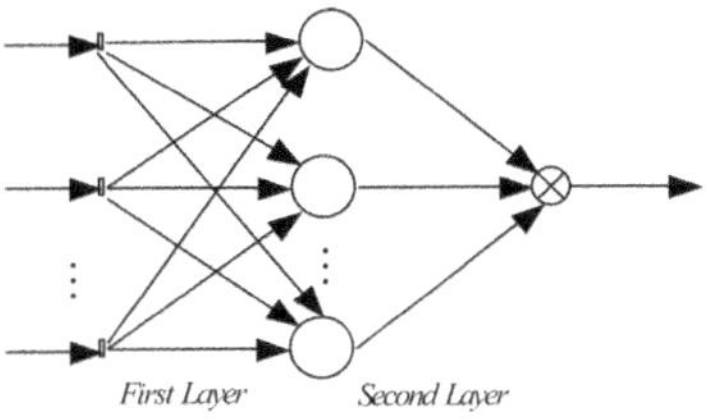

Fig. 3. the structure of 2-lay Neural Networks

Then,

$$\delta_i(x)=\theta_\delta{}^T\varphi_{\delta i}(V_i^T\underline{x})+\varepsilon_{\delta i}=\hat{\delta}_i+\varepsilon_{\delta i}, \tag{4.7}$$

where the input to the Neural Network saturation compensators (Fig. 3) is chosen as $\underline{x}=(x_r,e)^T$, Neural Network weights approximation error is $\tilde{\theta}_\delta=\theta_\delta-\hat{\theta}_\delta$.

$$f_i(x)=\theta_{fi}{}^T\varphi_{fi}(W_i^T x+b_i)+\varepsilon_{fi}=\hat{f}_i(x)+\varepsilon_{fi}. \tag{4.8}$$

For the approximation of $\delta_i(x)$, the first layer weights $V_i, i\in \underline{N}$ are selected randomly and will not be tuned in the note. The second layer weights θ_δ are tunable. The approximation weights θ_δ are ideal target weights, and they are assumed to be bounded so that $\|\theta_\delta\|_F\le\theta_*$ with θ_* a known scalar bound. The network reconstruction errors $\varepsilon_{\delta i}, i\in \underline{N}$ are bounded by ε_δ on a compact set.

For the approximation of $f_i(x), i\in \underline{N}$, we give the following assumption.

Assumption 4.1: The estimate of the unknown functions $f_i(x)\,(i\in \underline{N})$, $\hat{f}_i(x)\,(i\in \underline{N})$, are assumed to be known so that the estimate error $\tilde{f}_i(x)\,(i\in \underline{N})$ satisfy

$$\left\|\tilde{f}_i(x)\right\|\le h_i(x) \tag{4.9}$$

for some known functions $h_i(x)\,(i\in \underline{N})$ [24].

This assumption is reasonable, as in practical systems the bound $h_i(x)\,(i\in \underline{N})$ can be computed knowing the upper bound of variables such as payload masses, frictional effects, and so on [5].

Choose the tracking sub-control law as

$$u_i = g_i^{-1}(x)\left(-\hat{f}_i(x) - Y_r + u_{ci} - Kd\right) - \hat{\delta}_i, i \in \underline{\mathrm{N}}, \tag{4.10}$$

where $\hat{\delta}_i$ is the approximation of modified saturation nonlinear function $\delta_i(x)$. $\hat{f}_i(x)$ is the approximation of function $f_i(x)$. Approximation $\hat{f}_i(x)$ is assumed to be fixed in the note and will not be adapted. Robust term u_{ci} is chosen for the disturbance rejection. The control input u_i is composed of the tracking controller with the saturation compensator, as shown in the Fig. 4.

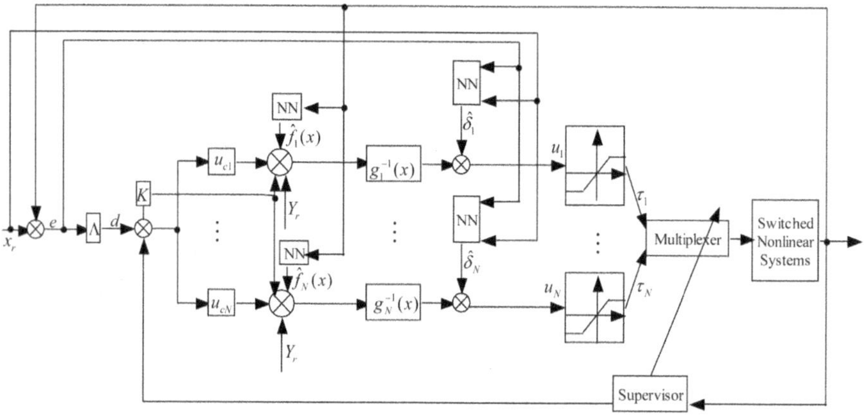

Fig. 4. Switched Nonlinear System and Neural Network Saturation Compensator

4.3 Tracking Performance of Switched Nonlinear System

Substituting sub-controller (4.10) into filtered tracking error dynamics (4.6), overall closed-loop filtered error dynamics is

$$\dot{d} = \tilde{f}_\sigma(x) + g_\sigma(x)\tilde{\theta}_\delta \varphi_{\delta\sigma}(V_\sigma^T \underline{x}) + u_{c\sigma} - Kd + g_\sigma(x)\varepsilon_{\delta\sigma}. \tag{4.11}$$

For the switched dynamic system (4.11), we have the following result.

Theorem 4.1: Consider the switched system (4.11) and suppose that, assumption 4.1 holds. Choose the robust term and Neural Network weights tuning law as follows.

$$u_{ci} = -h_i(x)sign(d), i \in \underline{\mathrm{N}}, \tag{4.12}$$

$$\theta_\delta = \Gamma\varphi_{\delta i}(V_i^T \underline{x})dg_i(x) - k_i\Gamma|d|\hat{\theta}_\delta, i \in \underline{\mathrm{N}}, \tag{4.13}$$

where the symbol "$sign(\cdot)$" denotes a standard sign function and the functions $h_i(x), i \in \underline{\mathrm{N}}$ are the bound on the estimate error $\tilde{f}_i(x), i \in \underline{\mathrm{N}}$; $\Gamma = \Gamma^T > 0$ denotes a constant matrix representing the learning rates of Neural Network; and $k_i, i \in \underline{\mathrm{N}}$ are some small scalar positive design parameters. Let the following condition holds.

$$\bigcup_{i=1}^N \left\{x \in \mathbb{R}^n \middle| \|x\| > \left(k_i\theta_*^2 + 4g_i(x)\varepsilon_\delta\right)\big/\left(4K\left(1+\lambda_1^2+\cdots+\lambda_{n-1}^2\right)\right) + \mu\right\} = \mathbb{R}^n . \tag{4.14}$$

Then, there exists a switching rule (4.15) such that system (4.11) is asymptotically stable.

$$\sigma(t) = \arg\max_{i\in \underline{N}} \{g_i(x)\}. \tag{4.15}$$

Proof: Consider Lyapunov function candidate as

$$V = 2^{-1}d^2 + 2^{-1}tr(\tilde{\theta}_\delta^T \Gamma^{-1} \tilde{\theta}_\delta). \tag{4.16}$$

Let $\{(t_m, r_m) | r_m \in \underline{N}, m \in \{1,2,\cdots\}\}$ be switching sequence that is generated by switching rule (4.15) on the time interval $[t_0, +\infty)$. Then, on the time interval $[t_m, t_{m+1})$, the time derivative of Lyapunov function (4.16) alone with system dynamics (4.11) yields

$$\dot{V} = -Kd^2 + d(\tilde{f}_{r_m}(x) + u_{cr_m}) + dg_{r_m}(x)\varepsilon_{\delta r_m} + dg_{r_m}(x)\tilde{\theta}_\delta \varphi_{\delta r_m}(V_{r_m}\underline{x}) + tr(\tilde{\theta}_\delta^T \Gamma^{-1}\dot{\tilde{\theta}}_\delta) \tag{4.17}$$

By view of the Neural Network tuning law (4.13) and the robust controller (4.12), the equality (4.17) is simplified to

$$\dot{V} = -Kd^2 + d(\tilde{f}_{r_m}(x) + u_{cr_m}) + dg_{r_m}(x)\varepsilon_{\delta r_m} + k_{r_m}|d| tr(\tilde{\theta}_\delta^T \hat{\theta}_\delta), \tag{4.18}$$

$$\dot{V} \le -Kd^2 + |d| g_{r_m}(x)\varepsilon_\delta + k_{r_m}|d| tr\left(\tilde{\theta}_\delta^T\left(\theta_\delta - \tilde{\theta}_\delta\right)\right). \tag{4.19}$$

Applying the inequality (2.2), the inequality (4.19) can be rewritten as

$$\dot{V} \le -Kd^2 + |d| g_{r_m}(x)\varepsilon_\delta + k|d|\left(\|\tilde{\theta}_\delta\|_F \|\theta_\delta\|_F - \|\tilde{\theta}_\delta\|_F^2\right)$$

$$= |d|\left(-Kd + g_{r_m}(x)\varepsilon_\delta + 1/(4k\theta_*^2) - k\left(\|\tilde{\theta}_\delta\|_F - 1/(2\|\theta_\delta\|_F)\right)^2\right)$$

Consequently, for every $t \in [t_m, t_{m+1})$, $\dot{V} < 0$, if the following inequality holds.

$$|d| > \left(k_{r_m}\theta_*^2 + 4g_{r_m}(x)\varepsilon_\delta\right)/(4K). \tag{4.20}$$

By (4.3)-(4.4) and $\|x_r\| \le \mu$, the inequality (4.20) is equivalent to the following inequality.

$$\|x\| > \left(k_{r_m}\theta_*^2 + 4g_{r_m}(x)\varepsilon_\delta\right)/\left(4K\left(1 + \lambda_1^2 + \cdots + \lambda_{n-1}^2\right)\right) + \mu. \tag{4.21}$$

Therefore, according to the condition (4.14) and switching rule (4.15), $\dot{V} < 0$ for every $t \in [t_0, +\infty)$. Hence, this complete the proof. ■

Remark 4.1: The right-hand side of inequality (4.20) can be taken as a practical bound on the tracking error in the sense that will never stray far above it. Note that, by tuning properly the PD gain K and Neural Network learning rates $k_i, i \in \underline{N}$, the energy attenuation domain of every sub-system may be not null and their sum may be covered with the state space of system. PD, PID, or any other standard controller does not posses this property when saturation nonlinearity is present in the switched system. Moreover, it is difficult to guarantee the stability of such a highly nonlinear switched system using only PD controller. Using the Neural Network saturation compensation and Neural Network switching rule, stability of such switched system is rigorously proven and the tracking error can be kept arbitrarily small by tuning the gain K and learning rates $k_i, i \in \underline{\mathbb{N}}$, under the action of designed Neural Network switching rule.

Remark 4.2: It is shown in [25] that for 2-lay Neural Network, termed random variable functional link, the approximation property holds. The weights are initialized at zero. Then the PD switched loop in Fig. 4 holds the switched system stable until the NN begins to learn.

5 Conclusion

For a class of SISO switched nonlinear systems with actuator saturation represented by Brunovsky canonical form, an actuator saturation compensation switching scheme is presented based on Neural Network in this note. The actuator saturation is assumed to be unknown and the saturation compensator is introduced into design of switched stabilizer. The scheme that leads to switched stability and disturbance rejection is rigorously proved. The tracking performance of switched nonlinear system is guaranteed based on common Lyapunov approach under the action of designed switching strategy. The Neural Network switched stabilization for MIMO switched nonlinear system with actuator saturator is our future research interest.

Acknowledgement

The authors would like to thank the anonymous reviewers for their constructive and insightful comments for further improving the quality of this work. This work was partially supported by the National Science Foundation of China under Grant 10661004, the Nomarch Foundation of Guizhou province under Grant No. 2001055 and the Doctor's Startup Foundation of Guizhou University (2007).

References

1. Antsaklis, P. (Ed.): Special Issue on Hybrid Systems. Proceedings of the IEEE **88** (2000)
2. Schaft, A.V., Schumacher, H.: An Introduction to Hybrid Dynamical Systems (Lecture Notes in Control and Information Sciences, Vol. 251). London: Springer-Verlag (2000)
3. Hu, T., Lin, Z.: Control Systems with Actuator Saturation: Analysis and Design. Boston, MA: Birkhauser (2001)
4. Annaswamy, A. M., Evesque, S., Niculescu, S., Dowling, A. P.: Adaptive Control of a Class of Time-Delay Systems in the Presence of Saturation. In Tao, G., Lewis, F. (eds.): Adaptive Control of Non-smooth Dynamic Systems. Springer-Verlag, New York (2001)
5. Selmic, R. R., Lewis, F. L.: Deadzone Compensation in Motion Control Systems Using Neural Networks. IEEE Trans. Autom. Control **45** (4) (2000) 602–613
6. Recker, D. A., Kokotovic, P. V., Rhode, D., Winkelman, J.: Adaptive Nonlinear Control of Systems Containing a Deadzone. In Proc. IEEE Conf. Decision Control (1991) 2111–2115
7. Lewis, F. L., Yesildirek, A., Liu, K.: Multilayer Neural-Net Robot Controller with Guaranteed Tracking Performance. IEEE Trans. Neural Networks **7** (2) (1996) 1–11

8. Polycarpou, M. M.: Stable Adaptive Neural Control Scheme for Nonlinear Systems. IEEE Trans. Autom. Control **41** (3) (1996) 447–451
9. Gao, W., Selmic, Rastko, R.: Neural Network Control of a Class of Nonlinear Systems with Actuator Saturation. IEEE Trans. Neural Networks **17** (1) (2006) 147-156
10. Liu, G. P., et al: Variable Neural Networks for Adaptive Control of Nonlinear Systems. IEEE Trans Systems, man, Cybermetics-Part C **29** (1999) 34-43
11. Patino, H. D., Liu, D.: Neural Network-Based Model Reference Adaptive Control Systems. IEEE Trans Systems, man, Cybermetics-Part B **30** (2001) 198-204
12. Sridhar, S., Hassan, K. K.: Output Feedback Control of Nonlinear Systems Using RBF Neural Networks. IEEE Trans. Neural Networks **11** (2000) 69-79
13. Levin, A. U., Narendra, K. S.: Control of Nonlinear Dynamical Systems Using Neural Networks-Part II: Observability, Identification, and Control. IEEE Trans. Neural Networks **7** (1996) 30-42
14. Lewis, F. L., et al: Multilayer Neural-Net Robot Controller with Guaranteed Tracking Performance. IEEE Trans. Neural Networks **7** (1999) 388-398
15. Polycarpou, M. M.: Stable Adaptive Neural Control Scheme for Nonlinear Systems. IEEE Trans. Automatic Control **41** (1996) 447-450
16. Ge, S. S., et al: Stable Adaptive Neural Network Control. MA: Kluwer, Norwell (2001)
17. Long, F., Fei, S. M.: State Feedback Control for a Class of Switched Nonlinear Systems Based on RBF Neural Networks. In Proc. 23rd Chinese Control Conference **2** (2004) 1611-1614
18. Long, F., Fei, S. M., Fu, Z. M., Zheng, S. Y.: Adaptive Neural Network Control for Switched System with Unknown Nonlinear Part By Using Backstepping Approach: SISO Case. In: Wang, J. et al (eds.): Advance in Neural Networks---ISNN2006. Lecture Notes in Computer Science, Springer-Verlag, Berlin Heidelberg **3972** (2006) 842-848
19. 19 Long, F., Fei, S. M.: Tracking Stabilization for a Class of Switched Nonlinear Systems with Time Delay Based on RBF Neural Network. In Proceedings of 2005 International Conference on Neural Networks & Brain **2** (2005) 930-934
20. Long, F., Fei, S. M.: Tracking Stabilization for a Class of Switched Impulsive Systems Using RBF Neural Networks. Dynamics of Continuous Discrete and Impulsive Systems---Series A: Mathematical Analysis **13** (Suppl., Part 1) (2006) 356-363
21. Messai, M., Zaytoon, J., Riera, B.: Using Neural Networks for the Identification of a Class of Hybrid Dynamic Systems. In Proceeding of IFAC Conference on Analysis and Design of Hybrid Systems (2006) 217-222
22. Astrom, K. J., Wittenmark, B.: Computer-Controlled Systems: Theory and Design (3rd ed.). Englewood Cliffs, NJ: Prentice Hall (1996)
23. Simon, Haykin: Neural Networks: A Comprehensive Foundation (2nd edition). New York: Prentice Hall (1994)
24. Narendra, K. S.: Adaptive Control Using Neural Networks. In Miller, W. T. et al. (eds.): Neural Networks for Control. Cambridge: MIT Press (1991) 115–142
25. Igelnik, B., Pao, Y. H.: Stochastic Choice of Basis Functions in Adaptive Function Approximation and the Functional-Link Net. IEEE Trans. Neural Networks **6** (6) (1995) 1320–1329
26. Narendrk, K. S., Mukhopadhyay, S.: Adaptive Control of Nonlinear Multivariable System Using Neural Network. Neural Network **7** (1994) 737-752
27. Sun, Z., Ge, Shuzhi, S.: Switched Linear Systems: Control and Design. London: Springer (2005)
28. Song, Y., et al: Control of Switched Systems with Actuator Saturation. Journal of Control Theory and Applications **1** (2006) 38-43

Reheat Steam Temperature Composite Control System Based on CMAC Neural Network and Immune PID Controller

Daogang Peng[1,2], Hao Zhang[1,2], and Ping Yang[1]

[1] College of Electric Power and Automation Engineering, Shanghai University of Electric Power, Shanghai 200090, China
[2] CIMS Research Center, Tongji University, Shanghai 200092, China
jypdg@163.com, hzhangk@yahoo.com.cn, yangping1201@126.com

Abstract. Reheat steam circle system is usually used in modern super-high parameters unit of power plant, which has the characteristics of long process channel, large inertia and long time lag, etc. Thus conventional PID control strategy cannot achieve good control performance. Prompted by the feedback regulation mechanism of biology immune response and the virtues of CMAC neural network, a composite control strategy based on CMAC neural network and immune PID controller is presented in this paper, which has the effect of feed-forward control for load changes as the unit load channel signal of reheat steam temperature is transmitted to the CMAC neural network to take charge of load change effects. The input signal of the controlled system are weighted and integrated by the output signals of CMAC neural network and immune PID controller, and then a variable parameter robust controller is constituted to act on the controlled system. Thus, good regulating performance is guaranteed in the initial control stage and also in case of characteristic deviations of the controlled system. Simulation results show that this control strategy is effective, practicable and superior to conventional PID control.

Keywords: CMAC neural network; Immune PID controller; Composite control; Reheat steam temperature system.

1 Introduction

With the increase of steam pressure in modern power plant, reheat steam circle system is usually used in super-high parameters unit in order to improve economy of the unit heat circle and decrease steam humidity of the terminal lamina of the turbine. Generally speaking, the reheat temperature changes largely with the changes of load. For example, the steam temperature at the exit point of the boiler will reduce 28~35℃ when the load of unit reduce 30% if the reheat temperature system hasn't been controlled. So, the reheat steam temperature system must be controlled accurately for the large unit. The task of reheat steam temperature control system is to keep the temperature at the exit point of reheater equaling to the set point. While the process channel of reheat steam temperature system is long and has the characteristics of large inertia, long time

D. Liu et al. (Eds.): ISNN 2007, Part I, LNCS 4491, pp. 302–310, 2007.

lag, etc, conventional PID control cannot achieve good control performance. So, it is necessary to research some other good control strategies.

Cerebellar model articulation controller (CMAC) was put forward by J.S.Albus in 1975. It is a table query adaptive neural network to describe complex nonlinear functions. It can change the contents of the table by its learning algorithms and it has the ability of information classify and storage. Because it is a part learning neural network, its learning rate is rapid and suit for real-time control system.

Biology immune system is characterized by its strong robustness and self-adaptability even when encountering amounts of disturbances and uncertain conditions. Artificial immune system is a new research field as an intelligent information process system. It is a conventional designation of intelligent system and developed by studying and utilizing the principles and mechanisms of biology immune system. Though immune system is very complicated, its self-adaptability of resisting against antigen is clearly. These intelligent behaviors of the biology information system provide various theory reference and technology method for science and engineering field.

Aiming at the characteristics of reheat steam temperature system in power plant, a composite control strategy based on CMAC neural network and immune PID controller is presented in this paper. It has the effect of feed-forward control for load changes as the unit load channel signal of reheat steam temperature is transmitted to the CMAC neural network to take charge of load change effects. Simulation results show that this control strategy can achieve much more satisfied control performance compared with conventional PID control.

2 CMAC Neural Network

2.1 Structure of CMAC Neural Network

The input state of the system is regarded as a pointer by the CMAC neural network and the correlative information is regarded as a group of distributed deposit memory cell. In essence, it is a kind of check table technique that is used to map complicated non-linear functions. The concretely course of action is that the input space is divided into several pieces and each piece is appointed to an actual memory location. The learned information of each piece is distributed and deposited in an adjacent piece place. The number of location is usually smaller than considered most input interspaces, so it is realized as a multi-to-one mapping, that is to say, several pieces are mapped to the same memory address. The basic structure of CMAC neural network is shown in Fig.1.

2.2 Design Steps of CMAC Neural Network

The design steps of CMAC are as follows:

(1) Conception allusive

Conception allusive is to carve up the N dimension input interspaces in the input layer of CMAC and land every input in the overtake cube cell of the N dimension grid group. The middle layer is made up of some estimated zones. There are only a few nonzero outputs for any arbitrary input and the number of nonzero zones is functionality parameter c, which prescribes the size of output area that affected by the CMAC inner network.

(2) Address allusive
Address allusive is to map the input samples to the address of concept memory by adopting the divided remainder method. Then divide one number and get the remainder as the real memory address. That is to say, the c amount cells of concept memory are mapped to c amount addresses of real memory.

(3) Functional calculating of CMAC
Functional calculating of CMAC is to map the input of CMAC to real memory of c amount units and every unit stores the corresponding weights. The input of CMAC is the summation of c amount real memory units.

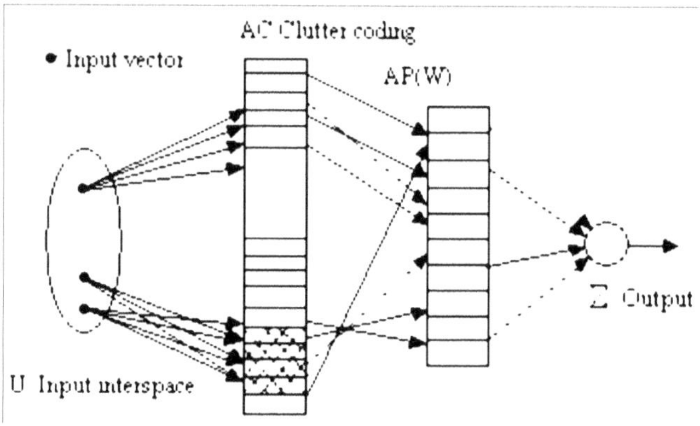

Fig. 1. Structure of CMAC neural network

3 Immune Feedback Mechanism

Immune system is mainly composed of some apparatus, organizes, cells, molecules and related genes, etc, which can protect the antibody from infraction by the pathogen, deleterious object, cancer cell, and so on. The most important cells of immune system are lymphocytes, which have two classes, namely as B cells and T cells. Due to the key role of T cells in immune response, the immune feedback algorithm is mainly based on T cells' feedback regulating principle of biology immune system.

Based on the feedback regulating law of T cells, defining the amount of the antigens at the k generation as follow

$$\varepsilon(k) = \gamma\varepsilon(k-1) - u_{kill}(k-d) \tag{1}$$

where γ is the multiplication factor of antigens, $u_{kill}(k)$ is the amount of killer T cells and d is death time. Defining the output of $T_H(k)$ stimulated by the helper T cells as follow

$$T_H(k) = K_1 \varepsilon(k) \tag{2}$$

where K_1 is a stimulation factor of T_H cells.

Though suppressor T cells forbid the activities of other cells, as they are used in feedback control, supposing the effect of suppressor T cells on B cells as $T_s(k)$ is

$$T_s(k) = K_2 f[\Delta u_{kill}(k)]\varepsilon(k) \tag{3}$$

where K_2 is a suppression factor of T_s cells. And $\Delta u_{kill}(k)$ is defined as

$$\Delta u_{kill}(k) = u_{kill}(k-d) - u_{kill}(k-d-1) \tag{4}$$

In formula (3), $f(\bullet)$ is a nonlinear function. By importing the killer T cells and the reaction at $(k-d)$ generation, $f(\bullet)$ is defined as

$$f(x) = 1.0 - \exp(-x^2/a) \tag{5}$$

where α is a parameter and $a > 0$. For deferent a, the relation of input and output of function $f(\bullet)$ is deferent too. And the curve of $f(\bullet)$ will be more smoothness when a is more bigness. But for all the values of x, it exists $f(x) \in [0,1]$.

The total stimulation received by B cells is as follow

$$S(k) = T_H(k) - T_s(k) \tag{6}$$

where the activity of B cells is gained by the integral of $S(k)$. Supposing the amount of killer T cells is gained by the differential of B cells, then killer T cells $u_{kill}(k)$ is

$$\begin{aligned} u_{kill}(k) &= K_1\varepsilon(k) - K_2 f[\Delta u_{kill}(k)]\varepsilon(k) \\ &= K\{1-\eta_0 f[\Delta u_{kill}(k)]\}\varepsilon(k) \end{aligned} \tag{7}$$

where $K = K_1$ and $\eta_0 = K_2 / K_1$. Formula (7) is the law of immune feedback, and according to it we can know that the parameter K controls the response speed and parameter η_0 controls the stabilization action. Thus, the performance of immune feedback law depends largely on how to select these factors.

4 Reheat Steam Temperature Composite Control Based on CMAC and Immune PID Controller

4.1 Model of Reheat Steam Temperature System

Through some testing data based on the dynamic characteristics of one power plant unit, we can gain the system model by using experimental modeling method for thermal plant. The model system is shown in Fig.2. The transfer function of main control channel is as follow

$$G_m(s) = \frac{-0.55}{(1.05s+1)^6} \qquad \frac{C}{\%}$$

And the transfer function of load channel is

$$G_d(s) = \frac{0.1336}{(3.55s+1)^2} \qquad \frac{C}{MW}$$

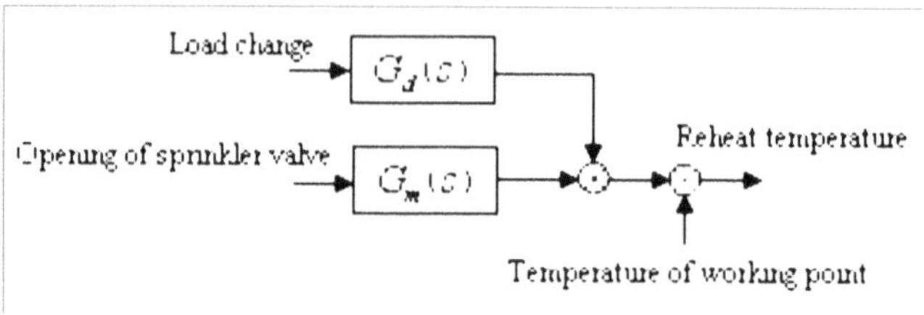

Fig. 2. Model of reheat steam temperature system

4.2 CMAC and Immune PID Composite Control Strategy

The control structure of reheat steam temperature composite control system based on CMAC neural network and immune PID controller is shown in Fig.3. Considering the characteristics of uncertain and time-varying of the reheat steam temperature system, especially at the initial control stage, the CMAC can't be trained well enough to approach the inverse dynamic characteristics of the controlled object, and in order to ensure the stability of the closed-loop control system and obtain good control performance, the system controller is constituted by CMAC neural network and immune PID controller. The CMAC neural network realizes the function of feed-forward control and approaches the inverse model of the controlled object, while the immune PID controller is to ensure the stability of the control system. Then the input signal of the controlled object are integrated by the output signals of the two controllers. And then a variable parameter robust controller is constituted. In addition, the unit load channel signal of reheat steam temperature is transmitted to the CMAC neural network to take charge of load change effects, which has the effect of feed-forward control for load changes.

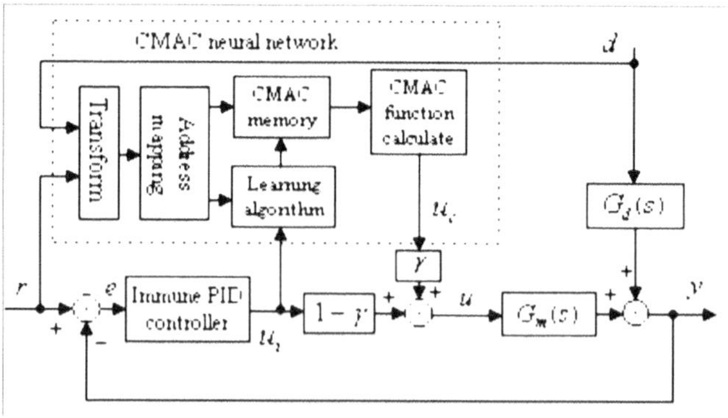

Fig. 3. CMAC and immune PID composite control for reheat steam temperature system

In Fig.3, the output signals of CMAC neural network and immune PID controller are integrated as the input of controlled object. Its express form is as follow

$$u(k) = \gamma u_c(k) + (1-\gamma)u_i(k) \tag{8}$$

Where $u_c(k)$ is the output signal of CMAC neural network. $u_i(k)$ is the output signal of immune PID controller. γ is named robust gene, which reflects the approaching precision of CMAC neural network for inverse system model. In order to change γ self-adaptively with the approaching precision of CMAC, it is defined as

$$\gamma = \exp(-\tau E_m) \tag{9}$$

Where τ is the variable robust coefficient of γ, and $\tau \in (0,1)$. E_m is defined as

$$E_m = \frac{1}{2}[u(k) - u_c(k)]^2 \tag{10}$$

From the formulas (9) and (10) we know that the scope of γ is $\gamma \in [0,1]$, and when $\gamma = 1$, that is to say $u(k) = u_c(k)$, it shows that the approaching precision of CMAC is the best and approached the inverse dynamic characteristics of the controlled object completely.

4.3 Algorithms of Immune PID Controller

Combining biology immune mechanism with conventional PID controller can improve the system control performance. Immune PID controller is a nonlinear controller which is designed by using of biology immune mechanism.

Based on the immune feedback principle, an immune P controller can be gained. That is to say, considering the amount of the antigens $\varepsilon(k)$ as the control error $e(k)$ between the set-point and the output of a control system, and the total stimulation $S(k)$ received by B cells as the control input $u_i(k)$, then the feedback control law is defined as follow

$$u_i(k) = K\{1 - \eta f[\Delta u_i(k)]\}e(k) = K' e(k) \tag{11}$$

where $K' = K\{1-\eta f[\Delta u(k)]\}$ is a proportion coefficient, parameter K is used to control the speed of response and $K = K_1$, suppressor parameter η is used to control the stabilization effect and $\eta = K_2 / K_1$. From formula (11) we can know that the controller based on immune feedback mechanism is a nonlinear P-type adaptive controller, where the proportional coefficient changes with the output of the controller and parameter K is adjusted by its own output.

But the P-type immune adaptive controller isn't suit for controlling a plant whose order is higher than two, and it can't compensate for the yawp or control error caused by nonlinear disturbances. In order to overcome these problems, improve the P-type immune controller as PID-type immune controller defined as

$$u_i(k) = u_i(k-1) + K'[e(k) - e(k-1)] + K_i' e(k) + K_d'[e(k) - 2e(k-1) + e(k-2)] \tag{12}$$

where $K' = K_p\{1 - \eta f[\Delta u_i(k)]\}$ and $K_i' = K' K_i$, $K_d' = K' K_d$ 。

From formula (12) we can know when $0 < \eta f[\Delta u_i(k)] \le 1$, the immune PID controller expresses negative feedback control, while when $1 < \eta f[\Delta u_i(k)]$, it expresses positive feedback control. The upper limit of gene η is used to keep the stabilization of the control system and the immune PID controller equals the conventional PID controller when $\eta = 0$.

Supposing exists parameters K_{p0}, K_{i0} and K_{d0}, which are used to ensure the system stabilization of the conventional PID controller, then, the stabilization conditions of immune PID controller are as follows

$$\begin{cases} 0 < K_p \le K_{p0}, 0 < K_i \le K_{i0}, 0 < K_d \le K_{d0} \\ 0 \le \eta \le \dfrac{1}{\sup f[\Delta u_i(k)]} \end{cases} \tag{13}$$

4.4 Algorithms of CMAC Neural Network

The learning algorithms of CMAC neural network has tutors, that is to say, at the end of every control period, it should calculate the corresponding output signal $u_c(k)$ of CMAC neural network and compare with the input signal $u(k)$ of controlled object. Then, adjusting the weights of CMAC neural network and turning into learning process. The aim in learning is to minimize the error of the input signal of controlled object and the output signal of CMAC neural network. By learning, the whole system input signal is produced by CMAC neural network.

The output signal of CMAC neural network is defined as

$$u_c(k) = \sum_{i=1}^{c} w_i a_i \tag{14}$$

Where a_i is a binary selective vector, c is the functionality parameter of CMAC neural network.

The regulative target of CMAC neural network is

$$E(k) = \frac{1}{2}[u(k) - u_c(k)]^2 \cdot \frac{a_i}{c} \tag{15}$$

By using grads descend methods we can gain the weights adjusting expressions of CMAC neural network as follows

$$\Delta w(k) = \eta \frac{u(k) - u_c(k)}{c} a_i \tag{16}$$

$$w(k) = w(k-1) + \Delta w(k) + \delta(w(k) - w(k-1)) \tag{17}$$

Where η is the learning rate of CMAC neural network and $\eta \in (0,1)$, δ is inertia quantum and $\delta \in (0,1)$.

Set $w = 0$ when control system starts running. At this time, $u_c = 0$ and $u = u_r$. Then the control system action is just immune PID control. By learning of CMAC neural network, the control effect $u_i(k)$ of immune PID becoming to zero gradually, so will the control effect $u_c(k)$ of CMAC neural network approach the whole output signal $u(k)$ of the controller. Then it can realize high precision feed-forward tracking control by CMAC neural network alone.

5 Simulation Study

In order to illustrate the validity of the control strategy proposed in Fig.3 and compare with other methods, this paper does the simulation researches of conventional PID control at the same time.

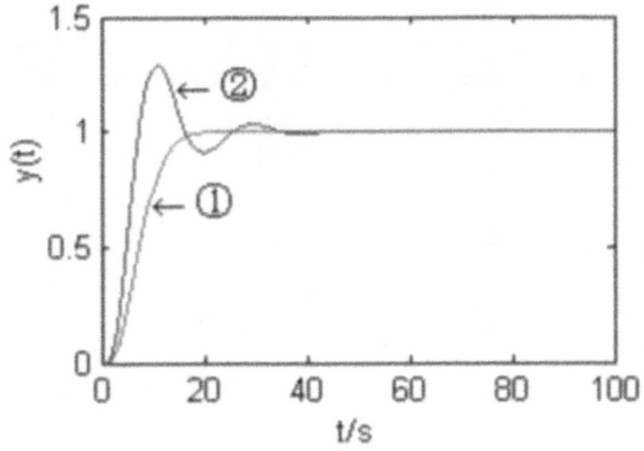

Fig. 4. Response curves of matched model

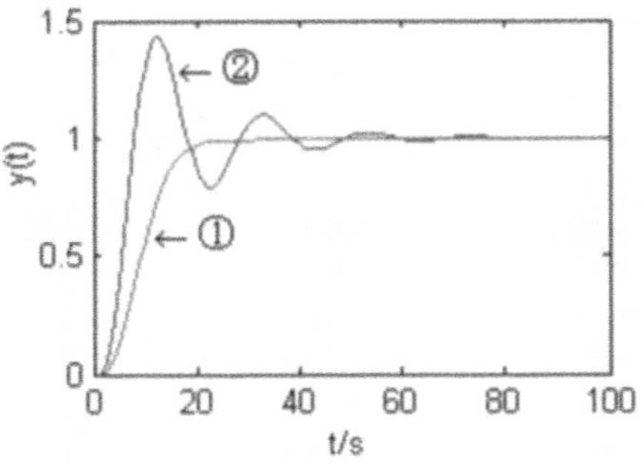

Fig. 5. Response curves of parameters increased model

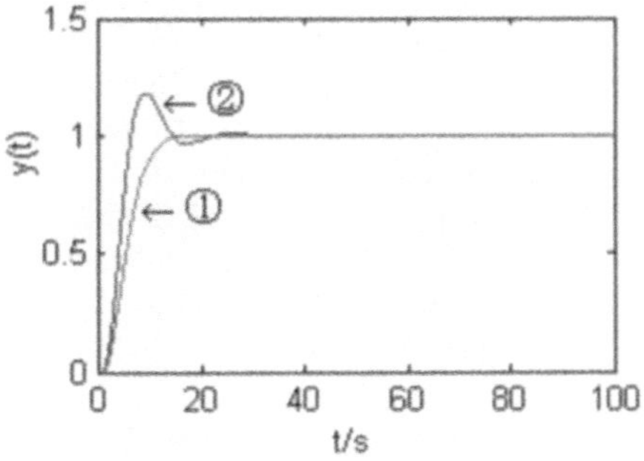

Fig. 6. Response curves of parameters decreased model

The parameters of conventional PID controller are $k_p = 0.65$, $k_i = 0.60$ and $k_d = 0.20$. Fig. 4 shows the response curves of matched model of the reheat steam temperature system. Due to the main factors affect the stability of the control system is the main control channel, this paper just consideration the parameters changing of the main control channel. Fig.5 and Fig.6 are the response curves of the time constant and gain constant of the main control channel of reheat steam temperature system all increased 20% and decreased 20% respectively. In each picture, Curve ① is the response of the strategy put forward in this paper and curve ② is the response curve of conventional PID control. It can be found out from the simulation results that the CMAC neural network and immune PID composite control strategy can achieve better control performance than conventional PID control strategy and has strong robustness and self-adaptability.

6 Conclusions

The simulation results of reheat steam temperature composite control system based on CMAC neural network and immune PID controller in power plant show that this control strategy is effective and practicable. In addition, the unit load channel signal of reheat steam temperature is transmitted to the CMAC neural network to take charge of load change effects, which has the effect of feed-forward control for load changes. Simulation results show that this control strategy can achieve much more satisfied control performance compared with conventional PID control.

Acknowledgments

This work was supported by Shanghai Education Committee Project (No.05LZ06) and supported by Shanghai Leading Academic Discipline Project (No.P1303).

References

[1] Peng, D., Zhang, H., Yang, P. et al.: Reheat steam temperature system based on CMAC neural network robust control. Proceedings of International Conference on Complex Systems and Applications Huhhot, China (2006) 82-85
[2] Takahashi, K., Yamada, T.: Application of an immune feedback mechanism to control systems. JSME Int J, Series C **41** (1998) 184-191
[3] Kim, D.H.: Tuning of a PID controller using immune network model and fuzzy set. IEEE International Symposium on Industrial Electronics (2001) 1656 – 1661
[4] Peng, D., Yang, P., Wang, Z. et al.: Immune PID cascade control of fresh steam temperature control system in fossil-fired power plant. [J].Power Engineering **25** (2005) 234-238
[5] Jiang, Z., Lin, T., Huang, X.: A new self-learning controller based on CMACneural network. [J]. ACTA AUTOMATIC SINICA **26** (2000) 542-546
[6] Yang, P., Peng, D., Yang, Y. et al.: CMAC neural network and PID combined control of water level in power plant boiler drums. [J]. Power Engineering **24** (2004) 805-808

Adaptive Control Using a Grey Box Neural Model: An Experimental Application

Francisco A. Cubillos and Gonzalo Acuña

Facultad de Ingeniería, Universidad de Santiago de Chile,
Casilla 10233, Santiago, Chile
fcubillo@lauca.usach.cl

Abstract. This paper presents the application of a Grey Box Neural Model (GNM) in adaptive-predictive control of the combustion chamber temperature of a pilot-scale vibrating fluidized dryer. The GNM is based upon a phenomenological model of the process and a neural network that estimates uncertain parameters. The GNM was synthesized considering the energy balance and a radial basis function neural network (RBF) trained on-line to estimate heat losses. This predictive model was then incorporated into a predictive control strategy with one step look-ahead. The proposed system shows excellent results with regard to adaptability, predictability and control when subject to setpoint and disturbances changes.

1 Introduction

The nonlinear predictive control problem results in a complex nonlinear constrained dynamic optimization. Although there exist efficient numerical methods and a computational capacity unthinkable a few year ago, a long period of time will still pass until a nonlinear predictive control problem of industrial process can be completely solved, [1]. An approach that seems to be quite promising is to use semi-empirical models to describe the predicted dynamic behavior of the process, adjusting their parameters in an adaptive way to permanently maintain the prediction quality [2, 3].

The use of a neural network as internal model for a nonlinear predictive controller presents its main advantages in the use of simple models and simplified prediction calculations. Meanwhile, this simplicity is associated with limitations in terms of applicability as they can only be used in the operation regions for which they were design (adjusted from data in these regions). Also, the parameters of these models generally have no physical meaning. Hybrid or grey box models, combining fundamental and empirical models, can partially reduce these limitations [4, 5]. One of the most popular commercial software for nonlinear predictive control is based on this philosophy, using neural networks (the empirical part) to represent the dynamic behavior of the process, adjusting the static gains (the fundamental part) in such a way that previous knowledge about the process is satisfied. Looking for a contribution to nonlinear predictive control approximation techniques for systems of reduced dimensions, this work presents the development and practical implementation of a grey box neural model based algorithm.

D. Liu et al. (Eds.): ISNN 2007, Part I, LNCS 4491, pp. 311–318, 2007.

In the developed technique, applied to the combustion system of a vibrating dryer pilot plant, the prediction model combines fundamental aspects, through mass and energy balances, with empirical ones, through neural networks linear in their parameters. Using this last characteristic the algorithm can be adapted to satisfactorily work in unknown operation regions, conferring considerable robustness to the control system [6, 7].

2 Grey Box Modeling

In predictive control, the amount of each kind of knowledge used in any case depends on the available information and the desired complexity for the resulting model. In many cases it is previously known that some fundamental variables balances should be satisfied by the model (mass and energy, in many practical cases). On the other hand it is normal to have no information about such phenomena as kinetics or transfer mechanisms. In a Grey box model these phenomena are described by empirical relations. Because they have a fundamental basis, they can be used to predict process behavior in operation regions not perfectly described by the empirical counterpart. This property can be improved if the empirical portion of the model is adjusted to new regions in an adaptive way [2].

Grey Box neural models, where the empirical contribution is provided by neural networks, have been used in the chemical process area, with different configurations, since 1992 [2, 3, 4, 5]. Figure 1-a, shows the "series scheme" for a Grey Box model as used in this work.

Because neural networks can be considered today as classical mathematical tools, we will only describe some relevant aspects to their use in this work. More details can be obtained elsewhere [8]. There are many different neural networks structures available, and the radial basis function (RBF) networks belong to a category that has the important property of linearity in terms of their parameters, favoring an adaptive adjusting procedure.

RBF networks are structured in three layers of processing elements (neurons) as shown in Figure 1-b. The vector of input variables is distributed by the elements of the input layer to an inner layer. Each processing element of this second layer is associated to a RBF (a Gaussian function, for example), centered in an appropriate point (center) of the input variables space. The distances between the input variables vector and the corresponding center are calculated in these elements, and these values are used as arguments of the RBFs.

$$o_j = \exp\left(-\frac{\|\underline{x} - \underline{c}_j\|}{s_j}\right) \tag{1}$$

The functions values are sent to each processing element of an output layer where they are multiply by weighting factors producing the network output variables.

$$y_i = \sum_{j=1}^{n} w_{i,j} o_j \tag{2}$$

These last linear operations are responsible for an important property in terms of network parameters adjustment. Once the centers and certain weighting factor for the inner layer have been preliminary determined, the network parameters can be calculated in only one step through the solution of a linear quadratic optimization problem [8]. Equations (1) and (2) indicate the relation between output variable y_j and the vector of input variables $\underline{x}$.

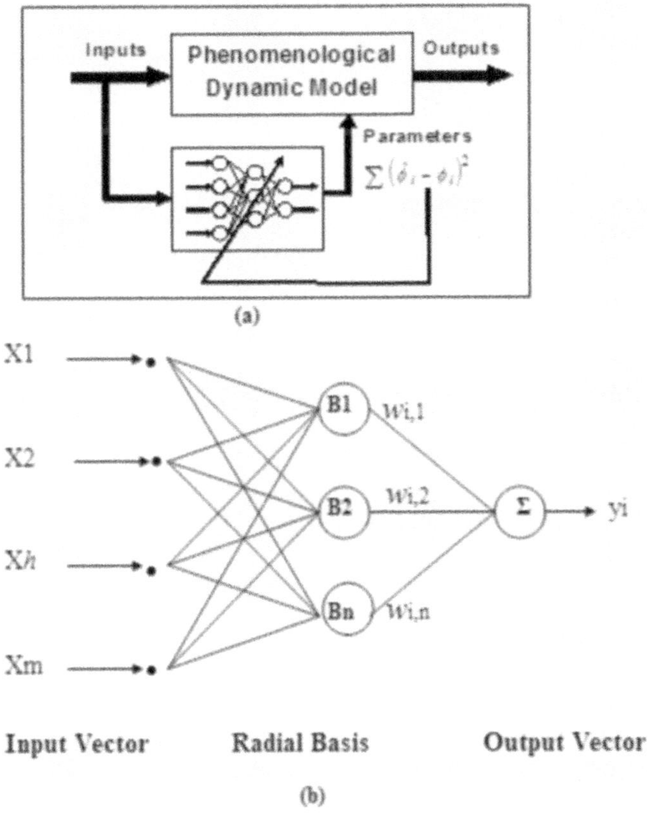

Fig. 1. GNM approach and RBF scheme

3 Experimental System

The system to be controlled is the air heating process of a pilot-scale vibrating fluidized dryer shown in Figure 2, formed by a combustion chamber, a solid material feeder, a drying chamber with a vibrating conveyor, the dried solids discharge, a gas solid cyclone to separate small solid particles from the exhausted gases and a blower to induce the gases flow. The dryer operates through the contact of a mixture of hot dry air and combustion gases with solid particles to be dried, in a cross flow configuration.

The building characteristics of the system, the adopted drying procedure and the environmental conditions introduce significant variations to the drying chamber output temperature (the variable to be controlled), caused by variations of the input air pressure, air looses through the solid feeding system, fed solid particles temperature variations and other unknown perturbations. It also should be noted that this type of combustion and drying systems have nonlinear static and dynamic behaviors [6]. Finally, the system presents large thermal inertia and hysteresis at the control valve, characteristics that represent great challenges for any control system.

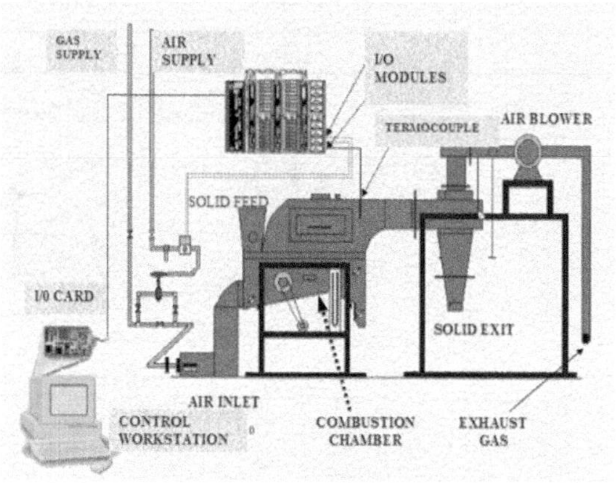

Fig. 2. The experimental combustion chamber

The objective in this work is to control the drying chamber output temperature (T(t); controlled variable) using the current signal to the I/P transducer (C(t); manipulated variable). An OPTO22 Mistic200SX system was used as an interface between the process and a process computer, where data are dynamically interchanged with Matlab´s Simulink via the DDE (Dynamic Data Exchange) protocol of Windows O. S.

4 System Model

For the Grey Box modeling of the heating system it is a priori known that mass and energy balances must be satisfied. The mass balance is assumed to be in a quasi steady state, represented by the following equation, Where w are the mass flows.

$$0 = w_g(t) + w_a(t) - w(t) \tag{3}$$

The energy balance can be described by

$$\rho V c_p \frac{dT(t)}{dt} = w_g(t) c_{p_g} T_{g_i}(t) + w_a(t) c_{p_a} T_{a_i}(t) + w_g(t)\lambda_c - w c_p T(t) + Q(t) \tag{4}$$

with Q, the heat losses, and λ_c, the combustion heat. Putting the input gas flow rate in terms o the manipulated variable through the linear relation,

$$w_g(t) = \alpha C(t) \tag{5}$$

A final expression for the chamber temperature is obtained (6), where the third term on the right hand side is not known and will be represented by a RBF neural network (NN).

$$\frac{dT(t)}{dt} = \frac{1}{\tau(t)}[T_i(t) - T(t)] + \frac{\alpha C(t)\lambda_c}{\rho Vc_p} + NN(t) \tag{6}$$

For computational purposes a discrete version of this model can be obtained using a descent difference approximation for the derivative term, resulting

$$T(k+1) = \left\{ \frac{1}{\tau(k)}[T_i(k) - T(k)] + \frac{\alpha C(k)\lambda_c}{\rho Vc_p} + NN(k) \right\} \Delta t + T(k) \tag{7}$$

The drying chamber output temperature at time $T(k+1)$ can be obtained from Equation (7) given $T, T_{i,} C$ and NN at time k. For training the NN, the unknown neural network output values can be obtained directly solving (7) to obtain NN.

In this work the topology of the neural network, i.e., the input variables and the number of processing elements in the inner layer, was obtained through a rigorous trial and error procedure, performed off-line, using several data sets obtained from different gas valve openings (current signal to I/P transducer). Different RBFs neural networks were trained using the Matlab's Neural Network Toolbox. The best obtained topology has 4 processing elements in the inner layer and uses delayed values of the drying chamber output temperature and current signal to the I/P transducer as the input variables. Fixing the calculated centers for the RBFs, an adaptive procedure was implemented to iteratively calculate new weighting factor using a recursive lest square algorithm whit a constant forgetting factor.

5 Control Algorithm

A general SISO nonlinear predictive controller solves the following nonlinear constrained optimization problem [1],

$$\min_{\Delta u(k)\cdots\Delta u(k+M-1)} \sum_{i=1}^{P} \gamma_i \left[\hat{y}(k+i) + d(k+i) - y^r(k+1)\right]^2 + \sum_{i=1}^{M} \lambda_i \Delta u^2(k+i-1) \tag{8}$$

s. t. Manipulated movements, outputs predictions

Because the objective of this work is to evaluate the use of a simple nonlinear algorithm to control SISO systems, we use the minimal form i.e.: P=M=1. Also, the difference between measured and estimated controlled variable values was not used in our algorithm, as it was assumed that this information is introduced by the adaptive strategy. So, for the heating system, considering constraints on the manipulated variable, the nonlinear control problem takes the following form:

$$\min_{\Delta C(k)} \left\{ \gamma\left[\hat{T}(k+1) - T^{r}(k+1)\right]^2 + \lambda[\Delta C(k)]^2 \right\}$$

$$\text{s.t}\ \hat{T}(k+1) = \left\{ \frac{1}{\tau(k)}[T_i(k) - T(k)] + \frac{\alpha C(k)\lambda_c}{\rho V c_p} + NN(k) \right\} \Delta t + T(k) \qquad (9)$$

$$NN(k) = f[T(k-1), C(k-1)];\ \ 4 \le C(k) \le 20$$

6 Experimental Results

The proposed control algorithm was implemented in the pilot unit using the Matlab Simulink simulation environment. A sampled period $\Delta t = 0.5$ min was found appropriated to balance process dynamics and computational effort. At each sampling time the nonlinear constrained optimization problem described by Equation (9) was solved using a Sequential Quadratic Programming (SQP) routine. Control parameters used during a continuous experimental run where fixed in: λ=2; γ=1. The experiment starts with unknown RBF network weights assumed to be $w_{1,1} = w_{1,2} = w_{1,3} = w_{1,4} = 1$. Figure 3 shows an initialization step in closed loop, showing that, although the system was operating in a closed loop form, the model parameters satisfactorily converged to their "correct" values in a short time interval.

The experimental system was run in a industrial environment, suffering from many unknown (uncontrolled) perturbations. This situation results in general slightly noisy signals and some periods of strong noisy signals. Notwithstanding these real problems, the controlled variable is always close to the desired setpoint.

Figure 4 illustrate the performance to tracking setpoints of the controlled system where the controlled and the manipulated variables behavior are shown from sampling time 1165 to sampling time 1365. It can be observed that, for different operation regions, between 160 °C and 135 °C, the control system satisfactorily maintain the

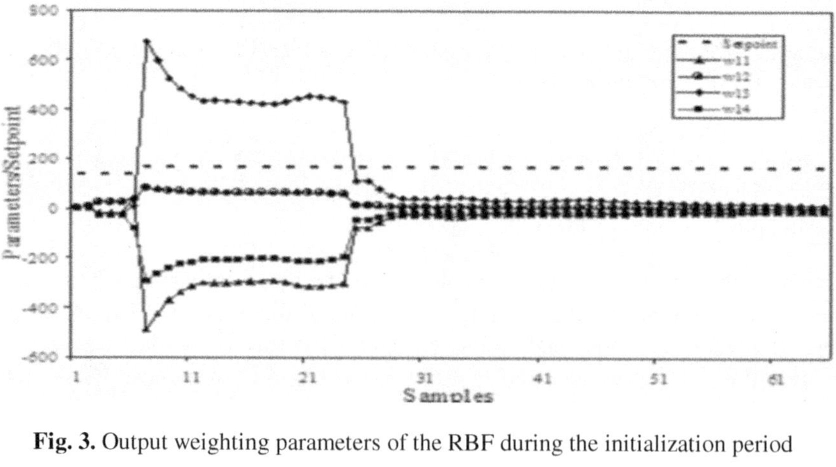

Fig. 3. Output weighting parameters of the RBF during the initialization period

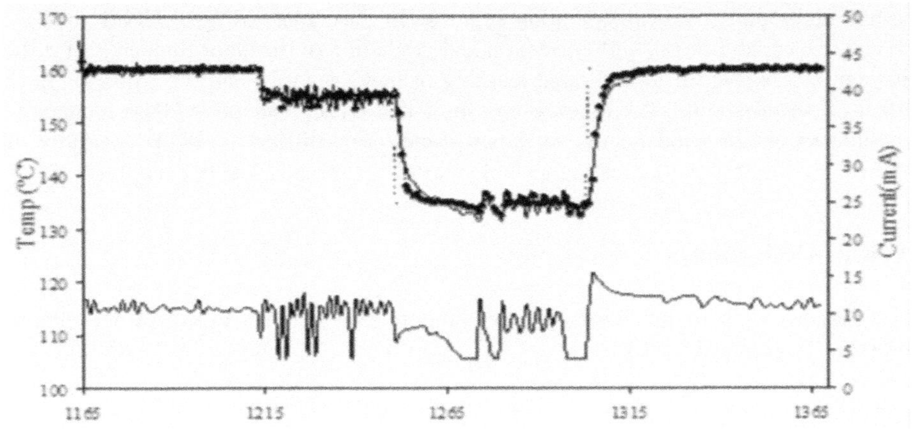

Fig. 4. Controlled system behavior (temperature and I/P output current) during setpoint changes

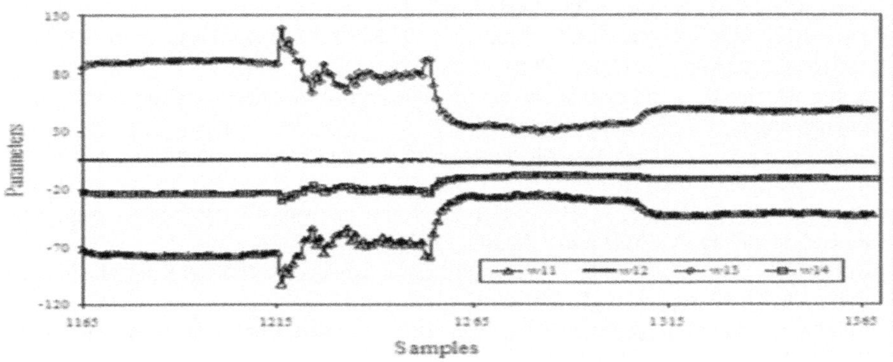

Fig. 5. Output weighting parameters of the RBF during the setpoint changes

controlled variable close to the setpoint. For the same time period depicted in Figure 4, the adaptive characteristic of the algorithm can be appreciated from Figure 5, where the behavior of the model parameters (network weights) is shown.

7 Conclusions

This work presents the development and practical implementation of a adaptive controller based on a Grey box-neural model. Using a neural network linear in its parameters the algorithm could be adapted to satisfactorily work in unknown operation regions, conferring considerable robustness to the control system. The experimental system was the nonlinear combustion system of a semi industrial vibrating dryer, operating in a difficult (noisy) neighborhood.

From the analysis of the obtained results it can be concluded that the proposed formulation is adequate - as an alternative to classical and predictive linear controllers - for controlling simple nonlinear systems, subject to a noisy environment, significant perturbations and different operation regions. In particular, using a GNM improves notably the capabilities of self-correction and prediction of the algorithm, characteristics that can be observed from the rapid learning of new scenarios and the reduction of the effect of perturbations. The main results show that through adaptive GNM models it is possible to obtain good results for small dimensions difficult systems (partially unknown, nonlinear, noisy), creating a viable alternative for classical linear approaches.

Acknowledgments

The authors wish to acknowledge the support provided by FONDECYT (Project 1040208 and Dicyt-Usach).

References

1. Qin, S.J., Badgwell, T.A.: A Survey of Industrial Model Predictive Control Technology. Control Engineering Practice **11** (2003) 733-764
2. Thompson, M.L., Kramer, M.A.: Modeling Chemical Processes Using Prior Knowledge and Neural Networks. AIChE J. **40(8)** (1994) 1328-1340
3. Xiong, Q., Jutan, A.: Grey-box Modelling and Control of Chemical Processes. Chemical Engineering Science **57(6)** (2002) 1027-1039
4. Sohlberg, R.: Hybrid Grey Box Modeling of a Pickling Process. Control Engineering Practice **13(9)** (2005) 1093-1102
5. Psichogios, D.C., Ungar, L. H.: A Hybrid Neural Network-first Principles Approach to Process Modeling. AIChE Journal **38(10)** (1992) 1499-1511
6. Cubillos, F., Lima, E.L.: Adaptive Hybrid Neural Models for Process Control. Computers chem. Engng. **22** Suppl. (1998) s989-s992
7. Cubillos, F., Callejas, H., Lima, E.L., Vega, M.P.: Adaptive Control Using a Hybrid Neural Model: Application to a Polymerization Reactor. Brazilian Journal of Chemical Engineering **18(01)** (2001) 113-120
8. Nahas, E.P., Henson, M.A., Seborg, D.E.: Nonlinear Internal Model Control Strategy for Neural Network Models. Computers chem. Engng **16(12)** (1992) 1039-1057

H_∞ Tracking Control of Descriptor Nonlinear System for Output PDFs of Stochastic Systems Based on B-Spline Neural Networks

Haiqin Sun[1], Huiling Xu[1], and Chenglin Wen[2]

[1] Research Institute of Automation Southeast University, Nanjing, 210096, P.R. China
l.guo@seu.edu.cn
[2] School of Automation, Hangzhou Dianzi University, Hangzhou 310018, P.R. China

Abstract. For stochastic systems with non-Gaussian variables, a descriptor nonlinear system model based on linear B-spline approximation is first established. A new tracking strategy based on H_∞ state feedback control for the descriptor nonlinear system is proposed, with which the probability density functions (PDFs) tracking control problem of the non-Gaussian stochastic systems can be solved. Necessary and sufficient condition for the existence of H_∞ state feedback controller of the problem is presented by linear-matrix-inequality (LMI). Furthermore, simulations on particle distribution control problems are given to demonstrate the efficiency of the proposed approach and encouraging results have been obtained.

1 Introduction

For stochastic systems with non-Gaussian variables, the classical approaches may not be able to cover the requirement of the closed loop control, where only the output mean and covariance are controlled. Recently, probability density function (PDF) control (or stochastic distribution control) methods has been proposed for general stochastic systems with non-Gaussian variables, where the control objective focused on the shape control of output PDF rather than its mean and variance [1]. In order to provide realizable PDF control methods, recently B-spline expansions (see, e.g., [14]) have been introduced for the output PDF modeling and controller design problem in both theoretical studies and practical applications [1-4]. It is shown that linear B-spline NN models result in positive constraints and square root B-spline NN models lead to episode constraints. In [4], stochastic distribution control on descriptor systems has been discussed. However, it is shown that only numeral optimization algorithms were given. Especially, the positive constraint cannot be eliminated by using the descriptor systems.

In this paper, a new design framework is established for the NN-based approaches of PDF tracking theory. Linear B-spline NN models are adopted for the output PDF approximations. Based on the characteristics of the output PDFs, it is shown that the weighting dynamics can be modeled via a descriptor system without any constraints. Consequently, PDF tracking can be transformed into the weighting tracking subject to a descriptor system.

D. Liu et al. (Eds.): ISNN 2007, Part I, LNCS 4491, pp. 319–328, 2007.

Descriptor systems describe a broad class of systems which are not only of theoretical interest but also have great practical significance. Models consist of differential equations and additional algebraic equations. H_∞ control theory and structural analysis for descriptor systems have been paid much attention based on algebraic approaches [7-10]. In the past two decades, there are quite a few studies related to the control of stochastic descriptor systems [5-6]. Moreover, the concerned dynamical models between the control input and the weight corresponding to the output PDF were confined to be linear systems without any uncertainties, while actually modeling errors, uncertainty and non-linearity exist in most of these practical modeling procedures. Consequently, some PDF control strategies with simple structures for non-linear or uncertain dynamical models need to be developed, with which the stability, tracking performance and robustness should be guaranteed.

In the following, if not otherwise stated, matrices are supposed to have compatible dimensions. The identity and zero matrices are denoted by I and 0, respectively, with appropriate dimensions. For a symmetric matrix M, the notations $M > (\geq)0$ are used to denote that is positive definite (positive semi-definite). The case $M < (\leq)0$ for follows similarly. For a vector $v(t)$, it is denoted that $||v(t)||_\infty := sup_{t\geq 0} \sqrt{v(t)^\tau v(t)}$.

The rest of this paper is organized as follows. In Section 2, the model of problem is formulated, which includes two parts. The first step is to use the B-spline expansion technique to model the relationships between the measurable output PDF and the constrained weights. The second one is further to establish the descriptor nonlinear system model for the input and the dynamical weights error. Thus, PDF tracking control problem of the non-Gaussian stochastic systems can be transformed to H_∞ state feedback control problem of the descriptor nonlinear system. Especially, the uncertain dynamics and the exogenous disturbance are considered in the model. With such a model, the corresponding H_∞ static output feedback control schemes to descriptor nonlinear systems are studied in Section 3. In the meantime, the H_∞ static output feedback tracking controller design method is presented based on LMIs. Finally, simulations are given to demonstrate the feasibility of the results in Section 4 and the main results are concluded in Section 5.

2 Problem Formulation and Preliminaries

2.1 B-Spline Expansion and Weight Error Dynamics Model

For some general stochastic systems, the control objective turns to the shape control of the conditional output PDFs, rather than the output mean and variance. In order to simplify the modeling and control methods, B-spline expansions have been adopted to model the measured output PDFs, so that the PDF control problem can be reduced to the classical control problem for dynamical weights error.

Consider Fig. 1 which represents a general stochastic system, where $w(t)$ is the stochastic input, $u(t) \in \mathbb{R}^m$ is the control input. it is supposed that $z(t) \in [a, b]$ is the system output and the probability of output $z(t)$ lying inside $[a, \sigma]$ can be described as

$$P(a \leq z(t) < \sigma, u(t)) = \int_a^\sigma \gamma(y, u(t))dy,$$

where $\gamma(y, u(t))$ is the PDF of the stochastic variable $z(t)$ under control input $u(t)$.

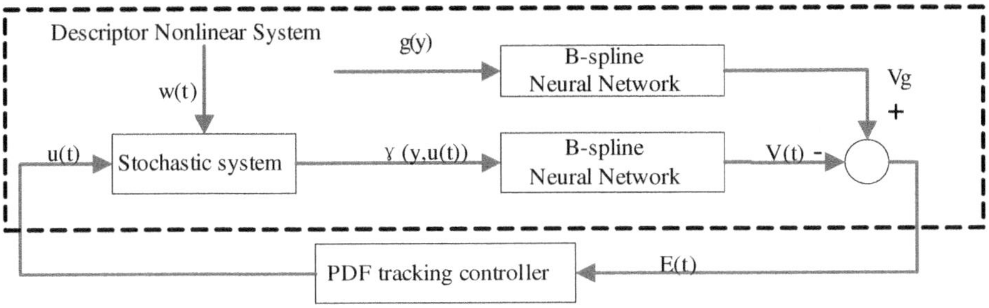

Fig. 1. PDF tracking control for a stochastic system using measured PDFs

To avoid the complex computation involved in partial differential equations and provide crisp control strategies, the linear B-spline approximation has been presented [1], where $\gamma(y, u(t))$ can be represented by

$$\gamma(z, u(t)) = B(y)V(t), \tag{1}$$

where

$$B(y) = [b_1(y) \cdots \; b_{n-1}(y) \; b_n(y)], V(t) = [v_1 \; \cdots \; v_{n-1} \; v_n]^T, \tag{2}$$

$b_i(y)(i = 1, \cdots, n-1, n)$ are pre-specified basis function defined on $y \in [a, b]$, $v_i(u(t) := v_i(t)(i = 1, \cdots, n-1, n)$ are the weights of the such an expansion.

Corresponding to (1), a given desired PDF $g(y)$ can also be expressed by

$$g(y) = B(y)Vg, \tag{3}$$

where Vg is the desired weight vector corresponding to the same group of $b_i(y)(i = 1, \cdots, n-1, n)$. The purpose of the controller design is to find $u(t)$ so that $\gamma(y, u(t))$ can follow $g(y)$. The error between the output PDF and the target one can be formulated as

$$e(y, t) = g(y) - \gamma(y, u(t)) = B(y)E(t), \tag{4}$$

which is a function of both $y \in [a, b]$ and the time instant t, where the weight error vector is defined as $E(t) = Vg - V(t)$, where $E(t) = [e_1(t) \cdots \; e_{n-1}(t) \; e_n(t)]$.

After the basis functions are determined, it is noted that only $n-1$ weight errors are independent due to constraint

$$\int_a^b e(y, t)dy = \int_a^b B(y)E(t)dy = \sum_{i=1}^{n} e_i(t) \int_a^b b_i(y)dy = 0. \tag{5}$$

In this case, equation (5) can be rewritten as

$$e_n(t) = -\sum_{i=1}^{n-1} \frac{b_i}{b_n} e_i(t), \tag{6}$$

where it can be supposed that $\int_a^b b_n(y)dy := b_n \neq 0$.

The next step is establish the dynamic model between the control input and the weight errors, this procedure has been widely used in PDF control. This procedure can be carried out by the corresponding identification processes in [1]. To simply the design algorithm, originally only linear models were considered, where the shape of output PDFs actually cannot be changed [1, 3]. In this paper, we adopt the following descriptor nonlinear model with exogenous disturbances

$$\begin{cases} E\dot{x}(t) = Ax(t) + f(x(t)) + B_1 w(t) + B_2 u(t), \\ z(t) = Cx(t), \end{cases} \tag{7}$$

where $x(t) \in \mathbb{R}^n$ is the state input vector with $x(t) = [x_1 \ \ x_2]^\tau$, $x_1(t) = [e_1 \ \cdots \ e_{n-1}]^\tau$, $x_2(t) = e_n(t)$. $u(t) \in \mathbb{R}^q$, $w(t) \in \mathbb{R}^m$ and $z(t) \in \mathbb{R}$ represent the control,the exogenous input and the controlled output vector, respectively. Constant square matrices E, A, and constant matrices B_1, B_2, C have compatible dimension with

$$A = \begin{bmatrix} A_1 & 0 \\ A_2 & -1 \end{bmatrix}, A \in \mathbb{R}^{n \times n}, A_1 \in \mathbb{R}^{(n-1)\times(n-1)}, A_2 \in \mathbb{R}^{1\times(n-1)};$$

$$E = \begin{bmatrix} I & 0 \\ 0 & 0 \end{bmatrix}, E \in \mathbb{R}^{n \times n}, I \in \mathbb{R}^{(n-1)\times(n-1)}; C = [1 \ \cdots \ 1 \ \ 0], C \in \mathbb{R}^{1 \times n};$$

$$B_1 = \begin{bmatrix} B_{11} \\ 0 \end{bmatrix}, B_1 \in \mathbb{R}^{n \times m}, B_{11} \in \mathbb{R}^{(n-1)\times m};\ B_2 = \begin{bmatrix} B_{21} \\ 0 \end{bmatrix}, B_1 \in \mathbb{R}^{n \times q}, B_{21} \in \mathbb{R}^{(n-1)\times q}.$$

$f(x(t))$ is a nonlinear function satisfying

$$\|f(x(t))\| \leq \|Ux(t)\| \tag{8}$$

with $f(x(t)) = [f_1(x_1(t)) \ \ 0]$, $f(x(t)) \in \mathbb{R}^n$, $f_1(x_1(t)) \in \mathbb{R}^{n-1}$, and for any $x(t)$, where U is a constant matrix with appropriate dimension. It is noted $f(x(t))$ that can also be regarded as a kind of unknown modeling uncertainty.

Based on the continuity theory of functions, it is noted that $e(y,t) \to 0$, if and only if $x(t) \to 0$. As a result, after establishing dynamic models (1) and (7) which combines the output PDFs with the control input through the weight error vector, a new robust tracking performance problem is investigated for the weighting systems with the exogenous disturbances and the model uncertainties. The control objective is to find H_∞ state feedback controllers such that the closed-loop systems is asymptotically stable which can achieve the output PDFs tracking control and the disturbance can be restrained the descriptor nonlinear system (7).

2.2 Descriptor Nonlinear System

Consider the following continuous-time descriptor system with nonlinear perturbations.

$$\begin{cases} E\dot{x}(t) = Ax(t) + f(x(t)) + Bw(t) \\ z(t) = Cx(t) + Dw(t) \end{cases}, \|f(x(t))\| \leq \alpha\|Ux(t)\|, f(0) = 0 \tag{9}$$

where U is a constant matrix with appropriate dimension, α is a positive scalar.

Definition 1. [12] *1) The pair (E, A) is said to be regular if $det(sE - A)$ is not identical zero.*

2) The pair (E, A) is said to be impulse free if $degdet(sE - A) = rank(E)$.

3) System (9) is said to be generalized quadratic stability with degree α there exists a matrix P such that $E^\tau P = P^\tau E$, and

$$\Delta := [Ax + f(x(t))]^\tau Px + x^\tau P^\tau [Ax + f(x(t))] < 0. \tag{10}$$

Lemma 1. [12] *If system (9)with $u(t) = 0$ is generalized quadratically stable with degree α, then*

i) The nominal system of (9)(that is $E\dot{x}(t) = Ax(t),$) is regular and impulse free;

ii) For any given initial condition $x(0)$, the solution $x = x(t)$ of system (9) is globally exponentially stable.

The proof can see [12].

Theorem 1. *Given constants $\gamma > 0$ and $\lambda > 0$. For the system(7), the following statements (1) and (2)are equivalent.*

(1) System (7) is generalized quadratically stable with degree α (α=1)and satisfies $||z(t)||_2 \le \gamma ||w(t)||_2$.

(2) There exists a non-singular constant matrix satisfying

$$E^\tau P = P^\tau E \ge 0, \tag{11}$$

$$\begin{bmatrix} A^\tau P + P^\tau A & P^\tau & P^\tau B_1 & C^\tau & U^\tau \\ P & -\lambda^2 I & 0 & 0 & 0 \\ B_1^\tau P & 0 & -\gamma^2 I & 0 & 0 \\ C & 0 & 0 & -I & 0 \\ U & 0 & 0 & 0 & -\lambda^2 I \end{bmatrix} < 0. \tag{12}$$

Proof. See Appendix.

3 H_∞ State Feedback Controller Design

Now, consider the following state feedback controller

$$u(t) = Kx(t), K \in \mathbb{R}^{q\times n}. \tag{13}$$

Applying this controller to (7) results in the following closed-loop system:

$$\begin{cases} E\dot{x}(t) = A_k x(t) + f(x(t)) + B_1 w(t) \\ z(t) = Cx(t) \end{cases}, A_k = A + B_2 K. \tag{14}$$

Remark 1. Given a constant $\gamma > 0$. For the system (14), the state feedback (13) is said to be an H_∞ controller if system (14) is generalized quadratically stable with degreeα (α=1) and satisfies $||z(t)||_2 \le \gamma ||w(t)||_2$ .The objects of this paper are to find the necessary and sufficient conditions for the existence of an H_∞ state feedback controller in terms of LMIs. Then, we have the following H_∞ control result.

Theorem 2. *Given constants $\gamma > 0$ and $\lambda > 0$. For the system (7), the following statements are equivalent:*

(1)There exists an H_∞ state feedback controller described by (13);

(2) There exist matrix $W \in \mathbb{R}^{q\times n}$ and nonsingular matrix $X \in \mathbb{R}^{n\times n}$ described by

$$X = \begin{bmatrix} X_1 & 0 \\ X_{21} & X_2 \end{bmatrix} \tag{15}$$

which satisfy $X_{21} \in \mathbb{R}^{1\times(n-1)}$, $X_2 \in \mathbb{R} \neq 0$, nonsingular matrix $X_1 \in \mathbb{R}^{(n-1)\times(n-1)}$ and

$$X_1 < 0 \tag{16}$$

$$\begin{bmatrix} (AX + B_2W) + (AX + B_2W)^\tau & I & B_1 & (CX)^\tau & (UX)^\tau \\ I & -\lambda^2 I & 0 & 0 & 0 \\ B_1^\tau & 0 & -\gamma^2 I & 0 & 0 \\ CX & 0 & 0 & -I & 0 \\ UX & 0 & 0 & 0 & -\lambda^2 I \end{bmatrix} < 0 \tag{17}$$

Proof. See Appendix.

Based on this result, the LMI-toolbox in Matlab can be applicable directly and the feasible design steps can be given as follows:

(i) Solve the LMIs (16), (17) to obtain a pair of W and X with the description of(15);

(ii) Construct $K = WX^{-1}$, then we get an H_∞ state feedback control law $u = WX^{-1}x$.

4 Simulation

In the particle distribution control problems, the shape of output PDF usually has two or three peaks[1]. For a stochastic system with non-Gaussian process, it is supposed that the output PDF can be formulated to be (1) with

$$V(t) = [v_1(t) \;\; v_2(t) \;\; v_3(t)]^\tau, V(0) = [1.1283 \;\; 2.26 \;\; 0.385]^\tau,$$

$$b_i = \begin{cases} |\sin 2\pi y|, & y \in [0.5(i-1),\ 0.5i] \\ 0, & y \in [0.5(j-1),\ 0.5j],\ i \neq j \end{cases}.$$

The desired weight vector value is set to be $V_g = [\frac{\pi}{5} \;\; \frac{2\pi}{5} \;\; \frac{3\pi}{5}]^\tau$ corresponding to the desired PDF.

In this context, the dynamical relations with respective to $x(t)$ and $u(t)$ is described by (7) with the selections

$$A = \begin{bmatrix} -3 & 1 & 0 \\ 2 & -6 & 0 \\ -1 & -1 & -1 \end{bmatrix}, B_1 = [0 \;\; 0.2 \;\; 0]^\tau, B_2 = [0.3 \;\; 0 \;\; 0]^\tau, C = [1 \;\; 1 \;\; 0],$$

$$f(x(t)) = \begin{bmatrix} 0.2\sqrt{x_1^2 + x_2^2} sinx_1 \\ 0.2x_2cosx_1 \\ 0 \end{bmatrix}, U = [0.2 \;\; 0.2 \;\; 0]^T.$$

In simulations, for $\gamma = 1.0$ and $\lambda = 0.6$, solving (17) together with (15) and (16) yields

$$K = [-8.3869 \;\; -37.8183 \;\; 3.1002]. \tag{18}$$

Correspondingly, H_∞ state feedback controller can be obtained by using Theorem 2. When the H_∞ state feedback control law is applied, the closed-loop system responses for the dynamical weighting are shown in Fig.2. The control gains are shown

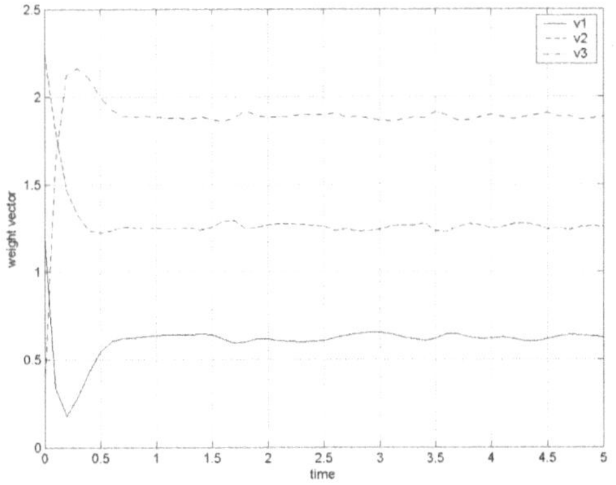

Fig. 2. Response of the weight vector

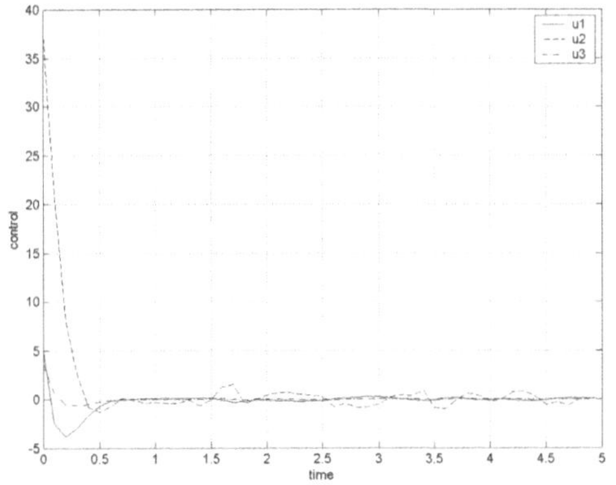

Fig. 3. Control input of the dynamical system

in Fig.3.The practical PDFs for the descriptor uncertain weighting error system and under the proposed robust control strategy is shown in Fig.4. It is demonstrated that the satisfactory tracking performance and robustness are achieved.

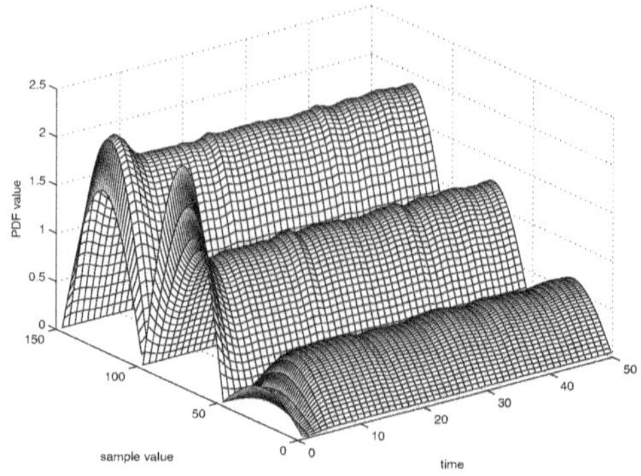

Fig. 4. 3-D-mesh plot of the output function

5 Conclusion

This paper considers the robust tracking problem for the output PDFs of non-Gaussian processes by using H_∞ state feedback controllers. B-spline NN expansions and descriptor nonlinear weighting error systems are applied to formulate the tracking problem. Different from the previous related works, descriptor nonlinear weighting error systems and exogenous disturbances are considered to enhance the robustness, and the constraints of the weighting error vectors are guaranteed by the H_∞ state feedback control law. Feasible controller design procedures are provided to guarantee the closed loop stability and the tracking convergence. Different from the existing results on PDF control, the control strategy proposed in this paper has a simple fixed structure and can guarantee both stability and robustness of the closed loop system. Simulations are provided to show the effectiveness and advantages of the proposed approach.

Acknowledgment

This paper is supported by NSFC (60474050) and NCET program. The authors would like to thank Professor L Guo for his assistance for its first version.

References

1. Wang, H.: Bounded Dynamic Stochastic Systems. Modeling and Control Springer- Verlag (2000)
2. Guo, L., Wang, H.: Applying Constrained Nonlinear Generalized PI Strategy to PDF Tracking Control Through Square Root B-Spline Nodels. Int. J. control **77** (2004) 1481-1492
3. Guo, L., Wang, H.: PID Controller Design for Output PDFs of Stochastic Systems using Linear Matrix Inequalities. IEEE Trans. Systems, Man and Cybernetics **B 35** (2005) 65-71
4. Yue, H., Leprand, A.J.A., Wang, H.: Stochastic Distribution Control of Singular Systems: Output PDF Shaping. ACTA Automatica Sinica **31** (2005) 151-160

5. Dai, L.: Filtering and LQG Problems for Discrete-Time Stochastic Singular Systems. IEEE Trans. Automatic Control **34** (1989) 1105-1108
6. Nikoukhah, R., Campbell, S.L., Delebecque, F.: Kalman Filtering for General Discrete-Time Linear Systems. IEEE Trans. Automatic Control **44** (1999) 1829 -1839
7. Rehm, A., Allgower, F.: H_∞ Control for Descriptor Systems with High Index. Proc. 14th IFAC World Congress, Beijing, China (1999) 31-36
8. Wang, H.S., Yung, C.Y., Chang, F.R.: Bounded Real Lemma and H_∞ Control for Descriptor Systems. IEE Proc. Control Theory App. **145** (1998) 316-322
9. Uezato, E., Ikeda, M.: Strict LMI Conditions for Stability, Robust Stabilization, and H_∞ Control of Descriptor Systems. Proc. 38th Conf. Decis. Contr., Phoenix, USA (1999) 4092-4097
10. Masubuchi, I., Kamitane, Y., Ohara, A., Suda, N.: H_∞ Control for Descriptor Systems: A Matrix Inequalities Approach. Automatica **33** (1997) 669-673
11. Xu, S., Lam, J.: Robust Control and Filtering of Singular Systems. LNCIS **332** (2006) 11-29
12. Lu, G.P., Ho, D.W.C., Yeung, L.F.: Generalized Quadratic Stability for Perturbated Singular Systems. Proc. 42nd IEEE Conf. Decis. Contr., Maui, Hawaii USA (2003)
13. Wang, H.: Model Reference Adaptive Control of the Output Stochastic Distributions for Unknown Linear Stochastic Systems. Int. J. Syst. Sci. **30** (1999) 707-715
14. Brown, M., Harris, C.J.: Neurofuzzy Adaptive Modeling and Control. Englewood Cliffs, NJ: Prentice-Hall (1994)
15. Takaba, K.: Robust H_2 Control of Descriptor System with Time-Varying Unceratinty. Int. J. Control **71** (1998) 559-579

Appendix

Proof of Theorem 1

Considering the Lyapunov function candidate as follows:

$$V(x(t),t) = (Ex(t))^\tau Px(t) + \int_0^t (\lambda^2 \|Ux(\tau)\|^2 - \lambda^2 \|f(x(\tau))\|^2) d\tau$$

$$= x(t)^\tau (E^\tau P) x(t) + \int_0^t (\lambda^2 \|Ux(\tau)\|^2 - \lambda^2 \|f(x(\tau))\|^2) d\tau.$$

So the derivative of V along system (7) with $u(t) = 0$ yields to

$$\dot{V}(x(t),t) = x^\tau A^\tau Px + x^\tau P^\tau Ax + x^\tau \lambda^2 U^\tau Ux + f^\tau Px + x^\tau P^\tau f - \lambda^2 f^\tau f$$

$$= x^\tau (A^\tau P + P^\tau A + \lambda^2 U^\tau U)x + f^\tau Px + x^\tau P^\tau f - \lambda^2 f^\tau f = p_0^\tau \Phi_0 p_0,$$

where

$$p_0 = [x^\tau \ \ f^\tau]^\tau, \Phi_0 = \begin{bmatrix} A^\tau P + P^\tau A + \lambda^2 U^\tau U & P^\tau \\ P & -\lambda^2 I \end{bmatrix},$$

$\Phi_0 < 0$ due to (12) implies that $\dot{V}(x(t),t) = (Ax+f)^\tau Px + x^\tau P^\tau(Ax+f) + \lambda^2(\|Ux\|^2 - \|f\|^2) < 0$. With (8), it can be obtained that $(Ax+f)^\tau Px + x^\tau P^\tau(Ax+f) < 0$. Similar to the proof of Lemma 1, for any given initial condition $x(0)$ and for all tolerable perturbations (8), the solution $x = x(t)$ of system (7) is globally exponentially stable.

The next step is to focus on the condition of disturbance attenuation. To this end, we consider the following auxiliary function (known as the storage function)

$$S(x(t)) = V(x(t), t) + \int_0^t (\|z(\tau)\|^2 - \gamma^2\|w(\tau)\|^2)d\tau,$$

which satisfies that $S(x(t)) = \int_0^t \varepsilon(\tau)d\tau$ with the zero initial condition, where $\varepsilon(\tau) = \dot{V}(x(t), t) + \|z(\tau)\|^2 - \gamma^2\|w(\tau)\|^2$. If $\varepsilon(\tau) \leq 0$, it can be easily obtained that $S(x(t)) \leq 0$, which further leads to $\|z(t)\|_2 \leq \gamma\|w(t)\|_2$ by letting $t \to \infty$.

Similarly to the derivation of $V(x(t), t)$, it can be verified that $\varepsilon(x(t)) = p_1^\tau \Phi_1 p_1$, where $p_1 = [x^\tau \;\; f^\tau \;\; w^\tau]^\tau$ and

$$\Phi_1 = \begin{bmatrix} A^\tau P + P^\tau A + \lambda^2 U^\tau U + C^\tau C & P^\tau & P^\tau B \\ P & -\lambda^2 I & 0 \\ B^\tau P & 0 & -\gamma^2 I \end{bmatrix}.$$

Multiple applications of the Schur complement on $\Phi_1 < 0$, it can be verified that $\Phi_1 < 0$ holds if and only if (12)holds with which $\varepsilon(\tau) \leq 0$ can be guaranteed. On the other hand, it is shown that $\varepsilon(\tau) \leq 0$ implies $\|z(t)\|_2 \leq \gamma\|w(t)\|_2$. Q.E.D.

Proof of Theorem 2

According to theorem1, There exists a non-singular constant matrix satisfying (11) and

$$\begin{bmatrix} (A + B_2K)^\tau P + P^\tau(A + B_2K) & P^\tau & P^\tau B_1 & C^\tau & U^\tau \\ P & -\lambda^2 I & 0 & 0 & 0 \\ B_1^\tau P & 0 & -\gamma^2 I & 0 & 0 \\ C & 0 & 0 & -I & 0 \\ U & 0 & 0 & 0 & -\lambda^2 I \end{bmatrix} < 0 \qquad (19)$$

Denote $P^{-1} = X$. By pre-multiplying P^{-1^τ} and post-multiplying P^{-1} to (11), it can be obtained that

$$EX = (EX)^\tau \geq 0 \qquad (20).$$

(20) becomes

$$\begin{bmatrix} I & 0 \\ 0 & 0 \end{bmatrix}\begin{bmatrix} X_1 & X_{12} \\ X_{21} & X_2 \end{bmatrix} = \begin{bmatrix} X_1 & X_{12} \\ X_{21} & X_2 \end{bmatrix}^\tau \begin{bmatrix} I & 0 \\ 0 & 0 \end{bmatrix} \geq 0, \qquad (21)$$

where

$$X := \begin{bmatrix} X_1 & X_{12} \\ X_{21} & X_2 \end{bmatrix}.$$

(21) implies $X_1 = X_1^\tau$, $X_1 > 0$, $X_2 \neq 0$ and $X_{12} = 0$.

Denote $X = P^{-1}$ and $W = KX$. By pre-multiplying $diag\{p^{-\tau} \;\; I \;\; I \;\; I \;\; I\}$ and post-multiplying $diag\{p^{-1} \;\; I \;\; I \;\; I \;\; I\}$ to (19) and substituting the closed loop system matrices into the result. Q.E.D.

Steady-State Modeling and Control of Molecular Weight Distributions in a Styrene Polymerization Process Based on B-Spline Neural Networks

Jinfang Zhang and Hong Yue

Department of Automation North China Electric Power University Beijing, Beijing 102206 P.R. China
Manchester Interdisciplinary Biocentre, The University of Manchester, 131 Princess street, Manchester M1 7ND UK
jfzhang@ncepu.edu.cn,
h.yue@manchester.ac.uk
http://www.springer.com/lncs

Abstract. The B-spline neural networks are used to model probability density function (PDF) with least square algorithm, the controllers are designed accordingly. Both the modeling and control methods are tested with molecular weight distribution (MWD) through simulation.

1 Introduction

The bounded stochastic control theory was put forward in 1998, and many works have been developed based on B-spline models, ARMAX models, and other neural network models. The system in study ranged form ordinary stochastic systems to the generalized stochastic systems, from the deterministic systems to the stochastic systems, from the discrete control sequence to structural controller. A series of control theories and approaches including the minimum entropy control and other integrated research structure has been established [1,2,3,4,5].

Stochastic distribution control has a lot of typical objects in chemical, papermaking, combustion and food manufacturing processes. For all these processes, the output probability density function (PDF) of certain product qualities is subjected to specified requirements. The control target can't be reached by the common control strategies. MWD control in chemical processes is very important as the quality of the products is mainly decided by the MWD of the polymer. It is a typical PDF control problem. Some open-loop control of MWD for pilot scale process has been proposed. As the on-line measurement of MWD is still difficult [6,7,8,9,10], the information of MWD is mainly obtained by the mathematical model. In this paper, the new developed stochastic control theory is used to control the MWD of the polymer in a styrene polymerization. The B-spline neural network is used to model the MWD.

D. Liu et al. (Eds.): ISNN 2007, Part I, LNCS 4491, pp. 329–338, 2007.

2 Mechanical Model of MWD in the Styrene Polymerization

The polymerization takes place in a continuous stirred-tank reactor (CSTR), as shown in figure 1. The monomer and the initiator are sent into the reactor with a controlled ratio C, the temperature of the reactor is kept constant, the monomer is styrene and the initiator is azobisisobutyronitrile.

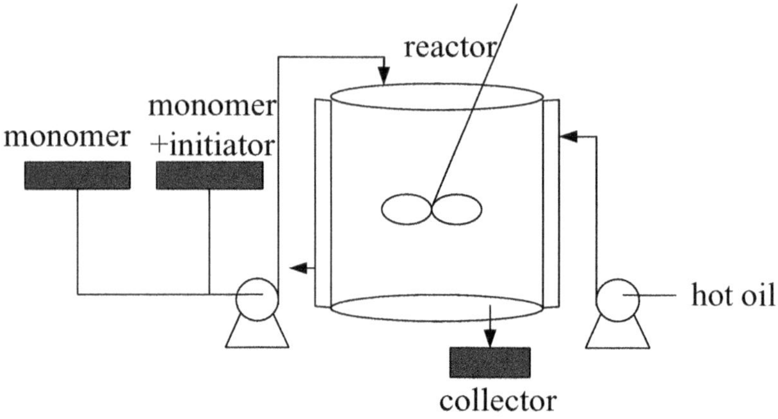

Fig. 1. The sketch of the styrene polymerization

In this system, the total input volume flow rate F and the ratio C can be changed by the controller. The ratio C is defined as $F_m/(F_i + F_m)$ and $F = F_i + F_m$, where F_m and F_i are the flow rates of monomer and initiator. C is the main control input of the system.

For the static system, the concentration of the initiator I, the active free radical R, the monomer M and the polymer P can be expressed as follows:

$$I = \frac{I_0}{1 + K_d\theta}, \tag{1}$$

$$R = \frac{-\frac{1}{\theta} + \sqrt{\frac{1}{\theta^2} + 8K_tK_iI}}{2K_t}, \tag{2}$$

$$M = \frac{M_0}{1 + (K_p + K_{trm})R\theta}, \tag{3}$$

$$P = \theta \cdot (K_{trm}MR + \frac{K_t}{2}R^2), \tag{4}$$

where I_0 and M_0 are the initial concentration of the initiator and the initial concentration of the monomer respectively, V is the volume of the reactor, $\theta = V/F$ is the average stay time of the reactants in the reactor. K_d, K_t, K_p and K_{trm} are rate constants of chain initiation, chain termination, chain increase and chain transfer reactions, R is the ideal gas constant.

Considering the mechanical analysis and the above model, the following concentration of the polymer with chain length j can be obtained

$$P_j = \begin{cases} \frac{\theta}{P}\left(\alpha^{-1} K_{trm} M R_1 + K_t R_1^2\right), & j = 2; \\ \frac{\theta}{P}\left(\alpha^{-(j-1)} K_{trm} M R_1 + \frac{j-1}{2}\alpha^{-(j-2)} K_t R_1^2\right), & j \geq 3, \end{cases} \quad (5)$$

where

$$R_1 = \frac{2K_i I + K_{trm} MR}{K_p M\alpha}, \quad (6)$$

$$\alpha = 1 + \frac{K_{trm}}{K_p} + \frac{K_t R}{K_p M} + \frac{1}{K_p M\theta}. \quad (7)$$

For each group of the reaction and operation conditions, the chain length j changes from 2 to a large number, the concentration of P_j will construct the distribution curve of the number-average MWD. With the definition of the PDF, the number-average MWD can be expressed as follows through normalization

$$\sum_{j=2}^{\infty} P_j = 1. \quad (8)$$

3 Static Modeling of PDF with B-Spline Neural Network

With linear B-spline neural networks, the output PDF of a stochastic system can be expressed as a linear combination of some pre-specified basis functions. Once the basis functions are specified, the shape of the output PDF can be realized by controlling the weights of the neural network.

In this paper, the modeling of PDF for both single input single output (SISO) systems and multi inputs single output (MISO) systems are discussed.

The definition of the linear B-spline functions can be expressed as [11]

$$B_{i,0}(y) = \begin{cases} 1, & y_i < y < y_{i+1}; \\ 0, & otherwise, \end{cases} \quad (9)$$

$$B_{i,q}(y) = \frac{y - y_i}{y_{i+q} - y_i} B_{i,q-1}(y) + \frac{y_{i+q+1} - y}{y_{i+q+1} - y_{i+1}} B_{i+1,q-1}(y), \quad (10)$$

where $B(y)$ stands for the B-spline function, the subscript i is the ith B-spline basis function, q ($q \geq 1$) stands for the order of the B-spline basis function, y_i ($i = 1, 2, $) is the knot that divides the definition domain.

3.1 Static Modeling of PDF with B-Spline Neural Network for SISO System

In this part, the model of PDF is set up with B-spline neural networks through recursive least square algorithm.

The Weights of B-spline Neural Network and the Static PDF. To represent the model, $v \in [a, b]$ notes as the bounded stochastic variable whose distribution can be expressed by its PDF $\gamma(y, u)$, where $u \in R^1$ is the control input. It is obviously that the shape of PDF is controlled by u. Supposed that $\gamma(y, u)$ is a continuous and bounded function, with the general rule of function approximation, the following linear B-spline neural networks can be used to approximate $\gamma(y, u)$ [1]

$$\gamma(y, u) = \sum_{i=1}^{n} \omega_i(u) B_i(y) + e_0, \tag{11}$$

where $\omega_i(u)(i = 1, 2, \cdots, n)$ is the weight, $B_i(y) \geq 0$ is the pre-specified basis function which is defined in $[a, b]$, e_0 is the approximation error. The error can often be neglected in the discussion of the closed-loop systems. As the integral of PDF over the definition domain is 1, the linear B-spline neural work approximation has only $(n-1)$ independent weights.

The above approximation can reach any precision. It should be noted that equation (11) is an instant or static PDF expression.

Denote

$$L(y) = \frac{B_n(y)}{\int_a^b B_n(y)dy}, \tag{12}$$

$$C_i(y) = B_i(y) - \frac{B_n(y)}{\int_a^b B_n(y)dy} \int_a^b B_i(y)dy, \; i = 1, 2, \cdots, (n-1), \tag{13}$$

then the approximation of B-spline neural networks for a static PDF can be expressed in a compact form [1]

$$\gamma(y, u) = C(y)V(u) + L(y), \tag{14}$$

where $C(y) = [C_1, C_2, \cdots, C_{n-1}]$ is the vector corresponding to the $(n-1)$ B-spline basis functions, $V(u) = [\omega_1(u), \omega_2(u), \cdots, \omega_{n-1}(u)]^T$ is the vector of the $(n-1)$ independent weights.

The Control Input of the Process and the Weights of B-spline Neural Network. For the static system, the relationship between the control input u and the weights of the B-spline neural network can also be expressed with B-spline neural networks

$$\omega_i(u) = \sum_{k=1}^{m} v_{ik} \varphi_k(u), \;\; i = 1, 2, \cdots, n-1. \tag{15}$$

Similarly to the B-spline neural network approximation to PDF in previous, $\varphi_k(k = 1, 2, \cdots, m)$ is the pre-specified B-spline neural network defined in the domain of the control input, v_{ik} is the corresponding weight.

Parameter Identification of the Static Model. Incorporating (15) into (11)and neglecting e_0, the static model of PDF can be obtained

$$\gamma(y,u) - L(y) = \sum_{i=1}^{n-1}\sum_{k=1}^{m} v_{ik}\varphi_k(u)B_i(y) = \theta^T\phi(u,y), \tag{16}$$

where $\phi(u,y)$ is the product of $B_i(y)$ and $\varphi_k(u)$, θ is the vector to be identified whose dimension is $m \times (n-1)$.

For equation (16), the standard recursive least square algorithm can be used to train the weights of the B-spline neural networks, so as to set up the model of PDF.

3.2 Modeling of PDF in MISO Static System

As the control inputs are more than one in many cases, the modeling with B-spline neural networks through least square algorithm for PDF in MISO systems is developed in this part.

Description of the Model. The static PDF model of an MISO system based on B-spline neural networks can be expressed as

$$\gamma(y,\mathbf{u}_k) = \sum_{i=1}^{n} \omega_i(\mathbf{u}_k)B_i(y), \tag{17}$$

$$\omega_i(\mathbf{u}_k) = D(\mathbf{u}_k)\varphi_i, \tag{18}$$

where $B_i(y) \in \mathbb{R}^{1\times 1}(i = 1, 2, \cdots, n)$ is the B-spline basis function defined on the independent variable y and $y \in [a, b]$. $B(y) = [B_1(y), B_2(y), \cdots, B_n(y)]^T \in \mathbb{R}^{n\times 1}$ is the vector of B-spline basis functions, n stands for the number of the B-spline functions defined on y. $\omega(\mathbf{u}_k) = [\omega_1(\mathbf{u}_k), \omega_2(\mathbf{u}_k), \cdots, \omega_n(\mathbf{u}_k)] \in \mathbb{R}^{1\times n}$ is the vector of weights. $D(\mathbf{u}_k)$ are the B-spline functions defined on control input $\mathbf{u}_k$, $\mathbf{u}_k = [u_{1k}, u_{2k}, \cdots, u_{lk}]^T$. For each element of $\mathbf{u}_k$, the number of B-spline functions is denoted by m_i, i.e. $D(u_{ik}) = [D(u_{ik}, 1), D(u_{ik}, 2), \cdots, D(u_{ik}, m_i)] \in \mathbb{R}^{1\times m_i}$, $(i = 1, 2, \cdots, l)$, and $u_{1k} \in [c_1, d_1]$,$\cdots$, $u_{lk} \in [c_l, d_l]$. $m = m_1 + m_2 + \cdots + m_l$ is the total number of the B-spline functions defined on $\mathbf{u}_k$, $D(\mathbf{u}_k) = [D(u_{1k}), D(u_2k), \cdots, D(u_{lk})] \in \mathbb{R}^{1\times m}$, $\varphi_i \in \mathbb{R}^{m\times 1}$.

Denote

$$\begin{aligned} \mathit{\Gamma}(y) &= [\gamma(y,\mathbf{u}_1), \gamma(y,\mathbf{u}_2), \cdots, \gamma(y,\mathbf{u}_{nn})]^T \in \mathbb{R}^{nn\times 1}, \\ D &= [D(\mathbf{u}_1), D(\mathbf{u}_2), \cdots, D(\mathbf{u}_{nn})]^T \in \mathbb{R}^{nn\times m}, \\ W &= [w(\mathbf{u}_1)^T, w(\mathbf{u}_2)^T, \cdots, w(\mathbf{u}_{nn})^T]^T \in \mathbb{R}^{nn\times n}, \\ \mathit{\Phi} &= [\varphi_1, \varphi_2, \cdots, \varphi_n] \in \mathbb{R}^{m\times n}, \end{aligned}$$

where nn is the number of the sampling data for modeling, and equation (17) and (18) can be rewritten in the following form

$$\mathit{\Gamma}(y) = WB(y), \tag{19}$$

$$W = D\mathit{\Phi}. \tag{20}$$

Steps for PDF Model Parameters Identification of MISO System. For the stochastic system described by equation (17)-(20), the least square algorithm is adopted to identify the model parameters. The steps of the identification are

Step1: Sample the data $\gamma(y, \mathbf{u}_k)$ and $\mathbf{u}_k (k = 1, 2, \cdots, nn)$ of the system to form $\Gamma(y)$;

Step2: Define the B-spline functions $B(y)$ and D in the form of equations (9) and (10);

Step3: With the sampling data and the specified B-spline functions defined on y, the least square algorithm is chosen to decide the weights vector W in equation (19)

$$W = \Gamma(y) B(y)^T (B(y) B(y)^T)^{-1}; \tag{21}$$

Step4: With the weights vector W and the B-spline functions defined on $\mathbf{u}$, the least square algorithm can be used again to identify Φ in equation (20)

$$\Phi = (D^T D)^{-1} D^T W. \tag{22}$$

With the steps mentioned above, the weights of the PDF model based on B-spline neural networks for MISO stochastic system can be identified, so the model can be set up accordingly.

Based on the models for SISO and MISO systems, the controllers can be designed to shape the output PDF of the systems.

4 Controllers Design of PDF Shaping for Static Stochastic Systems

When the model is set up, the controller can be designed to control the shape of PDF. The aim is to chosen a crisp control input to make the output PDF of the system to follow a given PDF. The controllers design for both SISO system and the MISO system are derived separately in following.

4.1 Controller Design of PDF for SISO System

To make the output PDF $\gamma(y, u_k)$ to follow the shape of a given PDF $g(y)$ as close as possible, a performance function is chosen as follows:

$$J(u_k) = \int_a^b (\gamma(y, u_k) - g(y))^2 dy, \tag{23}$$

where k is the time instant, u_k can be calculated with the gradient method:

$$u_{k+1} = u_k - \lambda \left. \frac{\partial J}{\partial u} \right|_{u=u_k}, \tag{24}$$

where λ is the pre-specified factor. Denote

$$\Sigma = \int_a^b (C(y)^T C(y)) dy, \tag{25}$$

$$\eta = \int_a^b (g(y) - L(y))C(y)dy \in \mathbb{R}^{1\times(n-1)}, \tag{26}$$

then the control input u_k is

$$u_{k+1} = u_k + 2\lambda(V(u)^T \Sigma - \eta) \left. \frac{\partial V(u)}{\partial u} \right|_{u=u_k}, \tag{27}$$

where

$$\frac{\partial V(u)}{\partial u} = [\frac{\partial w_1(u)}{\partial u}, \frac{\partial w_2(u)}{\partial u}, \dots, \frac{\partial w_{n-1}(u)}{\partial u}]^T, \tag{28}$$

$$\frac{\partial w_i(u)}{\partial u} = \sum_{k=1}^{m} v_{ik}(\frac{n}{u_{k+n} - u_k} B_{k,n-1}(u) - \frac{n}{u_{k+n+1} - u_{k+1}} B_{k+1,n-1}(u)). \tag{29}$$

As the B-spline neural networks are pre-specified, the control input can be obtained to shape PDF.

4.2 Controller Design of PDF for MISO System

For the PDF Control of a MISO system, the following performance function is chosen:

$$J = \int_a^b (\gamma(y, \mathbf{u}_k) - g(y))^2 dy + \mathbf{u}_k^T R \mathbf{u}_k, \tag{30}$$

where the first term makes the output PDF of the system to trace the given shape $g(y)$, and the second term is a constraint to the control input $\mathbf{u}$. Submit equation (19) and (20) into (30), equation (30) can be rewritten as

$$J = \int_a^b (D(\mathbf{u}_k)\Phi B(y) - g(y))^2 dy + \mathbf{u}_k^T R \mathbf{u}_k. \tag{31}$$

To get the optimal control law, the performance function should be made minimum, that is $\frac{\partial J}{\partial \mathbf{u}_k} = 0$, then

$$2\frac{\partial D(\mathbf{u}_k)}{\partial \mathbf{u}_k}\Phi(\int_a^b (B(y)B(y)^T)dy\Phi^T D(\mathbf{u}_k)^T - \int_a^b (B(y)g(y))dy) + 2R\mathbf{u}_k = 0. \tag{32}$$

As $D(\mathbf{u}_k)$ is often a nonlinear function of $\mathbf{u}_k$, it is difficult to calculate the derivatives, therefore, the gradient method is used to get the control input of the system

$$\begin{aligned}\mathbf{u}_{k+1} = \mathbf{u}_k &- 2\mu \left. \frac{\partial D(\mathbf{u})}{\partial \mathbf{u}} \right|_{\mathbf{u}=\mathbf{u}_k} \Phi \int_a^b \left(B(y)B(y)^T\right) dy \Phi^T D(\mathbf{u}_k)^T \\ &+ 2\mu \left. \frac{\partial D(\mathbf{u})}{\partial \mathbf{u}} \right|_{\mathbf{u}=\mathbf{u}_k} \Phi \int_a^b \left(B(y)g(y)^T\right) dy - 2\mu R \mathbf{u}_k,\end{aligned} \tag{33}$$

where μ is the pre-specified factor and

$$\frac{\partial D(\mathbf{u})}{\partial \mathbf{u}} = \begin{bmatrix} \frac{\partial D(\mathbf{u})}{\partial u_1} \\ \frac{\partial D(\mathbf{u})}{\partial u_2} \\ \vdots \\ \frac{\partial D(\mathbf{u})}{\partial u_l} \end{bmatrix} = \begin{bmatrix} \frac{\partial D(u_1)}{\partial u_1} & 0 & \cdots & 0 \\ 0 & \frac{\partial D(u_2)}{\partial u_2} & 0 & 0 \\ \vdots & \vdots & \ddots & \vdots \\ 0 & \cdots & 0 & \frac{\partial D(u_l)}{\partial u_l} \end{bmatrix}, \tag{34}$$

$$\frac{\partial D(u_i)}{\partial u_i} = \begin{bmatrix} \frac{\partial D(u_i,1)}{\partial u_i} & \frac{\partial D(u_i,2)}{\partial u_i} & \cdots & \frac{\partial D(u_i,m_i)}{\partial u_i} \end{bmatrix} \; i = 1, 2, \cdots, l. \tag{35}$$

5 Simulation Study of Static PDF Modeling and Control Methods in Styrene Polymerization

The above modeling and control methods are used to study the modeling and control of MWD in the styrene polymerization process shown in figure 1. Both SISO and MISO systems are discussed.

5.1 MWD Modeling and Controller Design for SISO Polymerization

For the SISO process, the control input is C, the output is the MWD of the polymer obtained from the static mechanical model: equation (1),(2),(3),(4),(5).

The definition domain of the control input is $C \in [0.2, 0.8]$, the chain length y of polymer varies from 2 to 1000, the B-spline functions for control input C and chain length y take the form in equations (9) and (10). The orders and numbers of B-spline functions for y and C are 2, 40 and 1, 60 respectively. The performance function takes the form of equation (23). The modeling and control simulation results are shown in figure 2.

The first plot in figure 2 is the training data for modeling, the second plot is the modeling result. From these plots, it can be seen that the modeling precision is satisfactory. The third plot is the control input, the fourth one in figure 2 is the output MWD during the control process. From these two plots, it can be seen that the control input converges to a constant and the output MWD reaches a stable shape.

5.2 MWD Modeling and Controller Design for Two Inputs Single Output Polymerization Process

For the MISO process study, the inputs are C and F, the output is still the MWD of the polymer. $C \in [0.2, 0.8]$ and $F \in [2.856, 57.12] ml/min$. The numbers of B-spline basis functions and orders for chain length y and control input C and F are 40, 2, 45, 1 and 60, 1 respectively.

For the two input single output process, the simulation result of the modeling and control study for the MWD is shown in figure 3. The first plot in figure 3 shows the training data for modeling and the second one is the B-spline modeling result, the third plot shows the control inputs, and the fourth one is the output MWD during the control process. From figure 3, it can be seen that the modeling and control algorithm can set up a satisfied model and obtain a satisfied control for the shape of the output MWD.

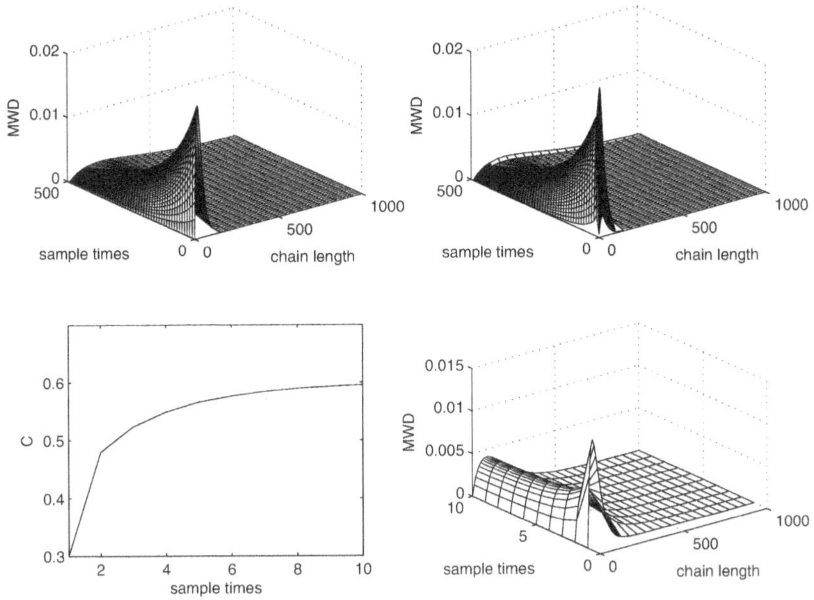

Fig. 2. SISO simulation: modeling data, modeling result, control input, control output

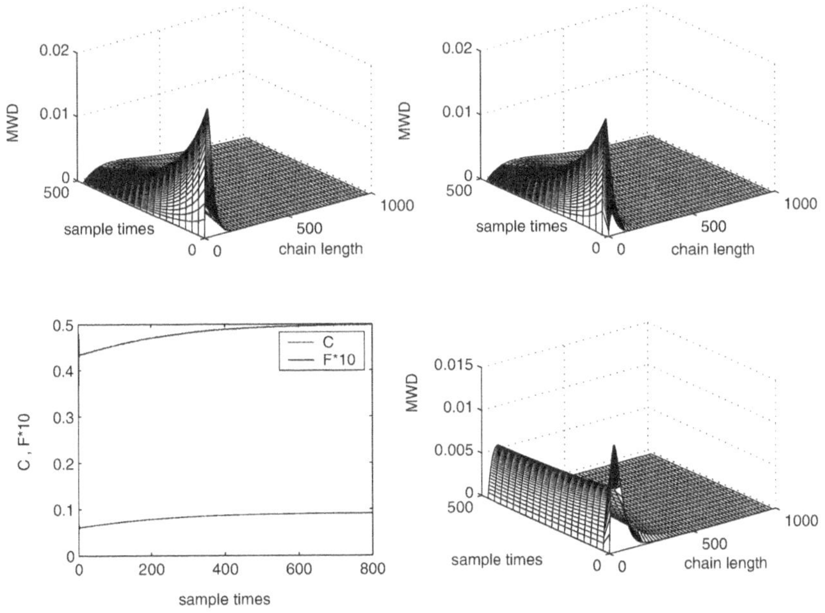

Fig. 3. MISO simulation: modeling data, modeling result, control input, control output

6 Conclusions

In this paper, the PDF models of both SISO and MISO static systems are set up with B-spline neural networks through least square algorithm. The modeling and control methods are applied to a simulation system of MWD control in a styrene polymerization process. Both SISO and MISO processes are discussed and satisfactory results are obtained. The modeling and control algorithms of static PDF base on B-spline neural networks provide the methods for closed-loop control of output distribution control problems. The dynamic PDF modeling and control algorithms should be studied and other advanced control algorithm should be used to the PDF control.

Acknowledgements. This work is supported by the Doctor Degree Fund of the North China Electric Power University and National Natural Science Foundation of China under grant (No. 60674051). These are gratefully acknowledged.

References

1. Wang, H.: Bounded Dynamic Stochastic Distributions Modeling and Control. London: Springer-Verlag Ltd (2000)
2. Yue, H., Wang, H.: Recent Developments in Stochastic Distribution Control: A Review. Journal of Measurement and Control **36** (2003) 209-215
3. Guo, L., Wang, H.: Pseudo-PID Tracking Control for a Class of Output PDFs of General Non-Gaussian Stochastic Systems. Proceedings of the 2003 American Control Conference, Denver, Colorado, USA (2003) 362-367
4. Wang, H.: Control of the Output Probability Density Functions For a Class of Nonlinear Stochastic Systems. Proceedings of the IFAC Workshop on Algorithms and Architectures for Real-time Control, Cancun, Mexico (1998) 95-99
5. Wang, H., Zhang, J.H.: Control of the Output Stochastic Distributions Via Lyapunov Function Analysis. Proceedings of IEEE International Conference on Control Applications CACSD, Glasgow (2002) 927-931
6. Clarke-Pringle, T.L., MacGregor, J.F.: Optimization of Molecular-Weight Distribution using Batch-to-Batch Adjustments. Industrial and Engineering Chemistry Research **37** (1998) 3660-3669
7. Crowley, T.J., Choi, K.Y.: Calculation of Molecular Weight Distribution from Molecular Weight Moments in Free Radical Polymerization. Industrial and Engineering Chemistry Research **36** (1997) 1419-1423
8. Soares, J.B., Kim, J.D., Rempel, G.L.: Analysis and Control of the Molecular Weight and Chemical Composition Distributions of Polyolefins Made with Metallocene and Ziegler-Natta Catalysts. Industrial and Engineering Chemistry Research **36** (1997) 1144-1150
9. Vicente, M., BenAmor, S., Gugliotta, L.M., Leiza, J.R., Asua, J.M.: Control of Molecular Weight Ddistribution in Emulsion Polymerization using On-Line Reaction Calorimetry. Industrial and Engineering Chemistry Research **40** (2001) 218-227
10. Vicente, M., Sayer, C., Leiza, J.R., Arzamendi, G., Lima, E.L., Pinto, J.C., Asua, J.M.: Dynamic Optimization of Non-Linear Emulsion Co-Polymerization Systems Open-Loop Control of Composition and Molecular Weight Distribution. J. Chemical Engineering **85** (2002) 339-349
11. Chen, Z.X., Li, S.G.: Data Approximation and the Numerical Solution to the Common Differential Equations. Xi'an Jiaotong University Publishing Company (2000)

A Neural Network Model Based MPC of Engine AFR with Single-Dimensional Optimization

Yu-Jia Zhai and Ding-Li Yu

Control Systems Research Group, School of Engineering Liverpool John Moores University, Byrom Street, Liverpool L3 3AF, UK
D.Yu@ljmu.ac.uk
http://www.ljmu.ac.uk/ENG/72971.htm

Abstract. This paper presents a model predictive control (MPC) based on a neural network (NN) model for air/fuel ration (AFR) control of automotive engines. The novelty of the paper is that the severe nonlinearity of the engine dynamics are modelled by a NN to a high precision, and adaptation of the NN model can cope with system uncertainty and time varying effects. A single dimensional optimization algorithm is used in the paper to speed up the optimization so that it can be implemented to the engine fast dynamics. Simulations on a widely used mean value engine model (MVEM) demonstrate effectiveness of the developed method.

1 Introduction

Many of the current production fuel injection controllers utilize feed-forward control based on a mass airflow sensor located upstream of the throttle plus a proportional integral (PI) type feedback control. The feed-forward control with look-up tables requires a laborious process of calibration and tuning. Furthermore, it is difficult to apply this method since it needs the output magnitude information that is not available in the A/F ratio control[1]. A variety of researches have been conducted during past decade on advanced control strategies on AFR. Onder and Geering[2] made an LQR regulator to improve the air-fuel ratio control. It obtained fairly good AFR when throttle angle ranging from 4° to 8° , but is impractical due to heavy computations resulting from the high order of linearized model.

A nonlinear MPC control scheme for air-fuel ratio based on a RBF model is developed in this paper. The RBF network is on-line adapted to model engine parameter uncertainties and severe nonlinear dynamics in different operating regions. Based on the multiple-step-ahead prediction of the air fuel ratio, an optimal control is obtained to maintain the stoichiometric value when throttle angle changes. A single dimensional optimization algorithm, Secant method, is used to reduce the optimization time, so that the developed method can be implemented to the fast dynamics of automotive engines. Satisfactory AFR control results are obtained by using the developed MPC scheme, as demonstrated on the MVEM[3] .

D. Liu et al. (Eds.): ISNN 2007, Part I, LNCS 4491, pp. 339–348, 2007.

2 Engine Dynamics

The engine dynamics concerned with air/fuel ratio control include air intake manifold, fuel injection, crankshaft speed, and exhaust oxygen measurement. A schematic diagram of the engine dynamics is shown in Fig.1.

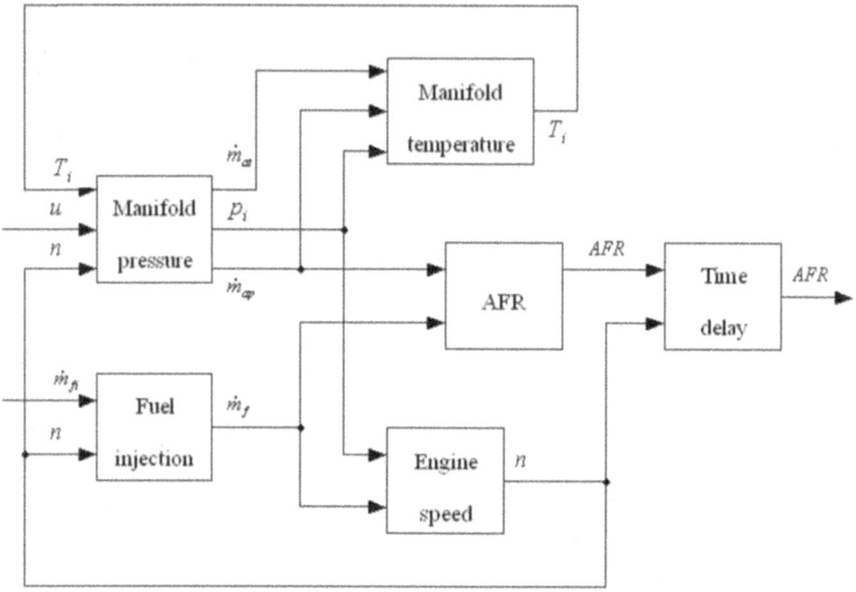

Fig. 1. Schematic diagram of engine dynamics

The system has one input, the injected fuel mass flow rate $\dot{m}_{fi}$ and one output, air/fuel ratio AFR. Besides, the system is subjected to a significant disturbance, the throttle angle u. Due to the space limitation, the dynamics of each of the four sub-systems, a number of differential and algebraic equations, are not included. The interested reader can refer to [4].

The manifold filling dynamics can be described by manifold pressure and temperature dynamics,

$$\dot{p}_i = \frac{\kappa R}{V_i}(-\dot{m}_{ap}T_i + \dot{m}_{at}T_a + \dot{m}_{EGR}T_{EGR}) \tag{1}$$

$$\dot{T}_i = \frac{RT_i}{p_iV_i}[-\dot{m}_{ap}(\kappa-1)T_i + \dot{m}_{at}(\kappa T_a - T_i) + \dot{m}_{EGR}(\kappa T_{EGR} - T_i)] \tag{2}$$

The crankshaft speed dynamics can be described as

$$\dot{n} = -\frac{1}{In}(P_f(p_i,n) + P_p(p_i,n) + P_b(n)) + \frac{1}{In}H_u\eta_i(p_i,n,\lambda)\dot{m}_f(t-\Delta\tau_d) \tag{3}$$

Both the friction power P_f and the pumping power P_p are related with the manifold pressure p_i and the crankshaft speed n. The fuel injection dynamics are

$$\ddot{m}_{ff} = \frac{1}{\tau_f}(-\dot{m}_{ff} + X_f \dot{m}_{fi}) \tag{4}$$

$$\dot{m}_{fv} = (1 - X_f)\dot{m}_{fi} \tag{5}$$

$$\dot{m}_f = \dot{m}_{fv} + \dot{m}_{ff} \tag{6}$$

where the model is based on keeping track of the fuel mass flow. The parameters in the model are the time constant for fuel evaporation τ_f , and the proportion χ_f of the fuel which is deposited on the intake manifold, $\dot{m}_{ff}$, or close to the intake valves, $\dot{m}_{fv}$.

3 Adaptive Neural Network Model

The advantage of using adaptive neural network is that it can track the time-varying properties of the process to provide efficient information to the controller, under circumstances where the process parameters change. Radial basis function networks (RBFN) with Gaussian transfer function are chosen in this application as it has been shown to map a nonlinear function arbitrarily well, and possess the best approximation property[5].

3.1 Data Collection

A set of random amplitude signal (RAS) combining short pulse width (transient state) and long pulse width (steady state) was designed for throttle angle and fuel injection, therefore the RBFN model after trained would produce adequate transient and steady state performance. Throttle angle was bounded between 20° and 40° and the range of fuel injection is from 0.0014 to 0.0079 kg/s, the sample time is set to be 0.1 s. The excitation signal is shown in Fig.2 partially, consisting of two parts. The length of square waves is set 0.3 s in the first part and 1.5 s in the second part. A set of 3000 data samples of AFR obtained was divided into two groups. The first 1500 data samples were used for training RBFN model and the rest would be remained for model validation.

3.2 Engine Modelling

Given the expanded engine model as shown Fig.1, the RBFN engine model has 6 inputs and one output as shown in Fig.3, where orders and delays are determined through experiments. The centers c and the width σ in hidden layer nodes of the RBFN were determined using K-means algorithm and ρ-nearest neighborhood heuristic respectively. RLS algorithm was used for training the neural network and the corresponding parameters were set as follows, $\mu = 0.99$, $w(0) = 2.22^{16} \times U_{n_h \times 2}$ and $P(0) = 1 \times 10^4 \times I_{n_h \times n_h}$, where I is the identity matrix and U stands for a matrix whose components are ones.

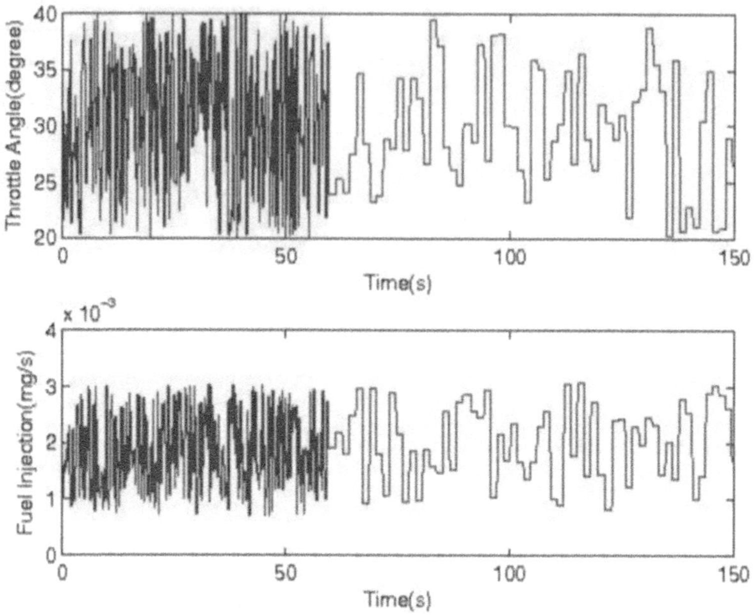

Fig. 2. Training Data with Mixed Pulse Width

After training with the training data set and test with the test data, the modelling error of the AFR in the normalized value with the mean absolute error, MAE = 0.0265.

4 MPC of Air Fuel Ratio

4.1 Control System Structure

The idea of model predictive control with neural network has been introduced in details by Draeger[6]. The strategy is shown in Fig.4. The obtained adaptive RBF neural network is used to predict the engine output for N_2 steps ahead. The nonlinear optimizer minimizes the errors between the set point and the engine output by using the cost function,

$$J(k) = \sum_{i=k+N_1}^{k+N_2} [msp(i) - \hat{y}(i)]^2 + \xi \sum_{i=k}^{k+N_u} [\dot{m}_{fi}(i) - \dot{m}_{fi}(i-1)]^2 \quad (7)$$

Here, N_1 and N_2 define the prediction horizon. ξ is a control weighting factor which penalizes excessive movement of the control input, the fuel injection $\dot{m}_{fi}$. N_u is the control horizon. Then the remaining main problem of MPC is to solve the nonlinear optimization problem, i.e. in each sample period, calculate a series of optimal $\dot{m}_{fi}(k)$, $\dot{m}_{fi}(k+1)$, $\cdots$, $\dot{m}_{fi}(k+N_2-1)$, from which the neural network model generates outputs to minimize $J(k)$ in Equation (7). Finally the

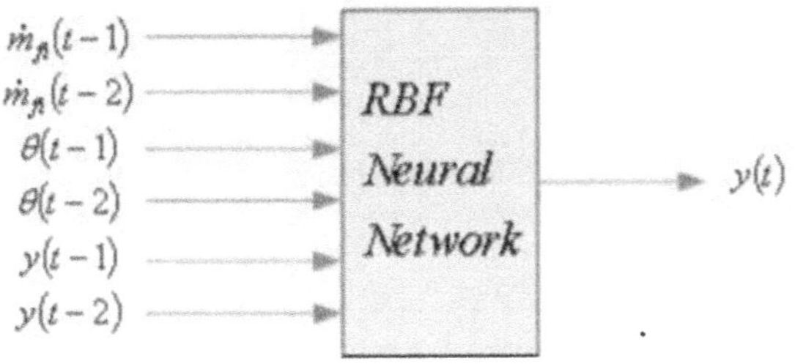

Fig. 3. RBFN Structure

first control variable $\dot{m}_{fi}$ is used to control the process and this procedure is repeated in the next sample period.

4.2 Single-Dimensional Optimization Approach

As second-order RBFN structure was chosen to achieve the minimum prediction error in engine modelling, the optimization problem involved in the paper is multi-dimensional and constrained. That is, we are going to find the future input $\dot{m}_{fi}(k)$, $\dot{m}_{fi}(k+1)$, $\cdots$, $\dot{m}_{fi}(k+N_2-1)$ that can minimize $J(k)$ such that the predicted outputs $\hat{y}(k)$, $\hat{y}(k+1)$, $\cdots$, $\hat{y}(k+N_2)$ coincides with the modified set-point input $mspi(k)$, $mspi(k+1)$, $\cdots$,$mspi(k+N_2)$ here the fuel injection rate is bounded within the region from 0.0014 to 0.0079 kg/s. Sequential Quadratic Programming (SQP) can be used to acquire the accurate solution, which is perhaps one of the best methods of optimization, would be shown in next section. However, the multi-dimensional optimization always requires heavy computation, especially when constraints exist.

Practical applications often place emphasis on computation speed on the premise that all the performance requirements are met. Therefore, we chose the simplest structure in this paper and assumed that the input $\dot{m}_{fi}$ will remain constant over the prediction horizon: $\dot{m}_{fi}(k)=\dot{m}_{fi}(k+1)=\cdots=\dot{m}_{fi}(k+N_2-1)$, in this case there is only one parameter that we are going to find. The optimization problem to be solved is reduced as one-dimensional. Secant method is chosen to find the solution of this nonlinear programming (NLP) problem and our experiments show that it is more efficient and reliable in this application if compared with the other interpolation methods.

Secant Method. The general nonlinear programming problem could be defined as,

$$\min_{x \in R^n} J(x) \tag{8}$$

subject to

$$c_{eq} = 0 \tag{9}$$

$$c_{in} \leq 0 \tag{10}$$

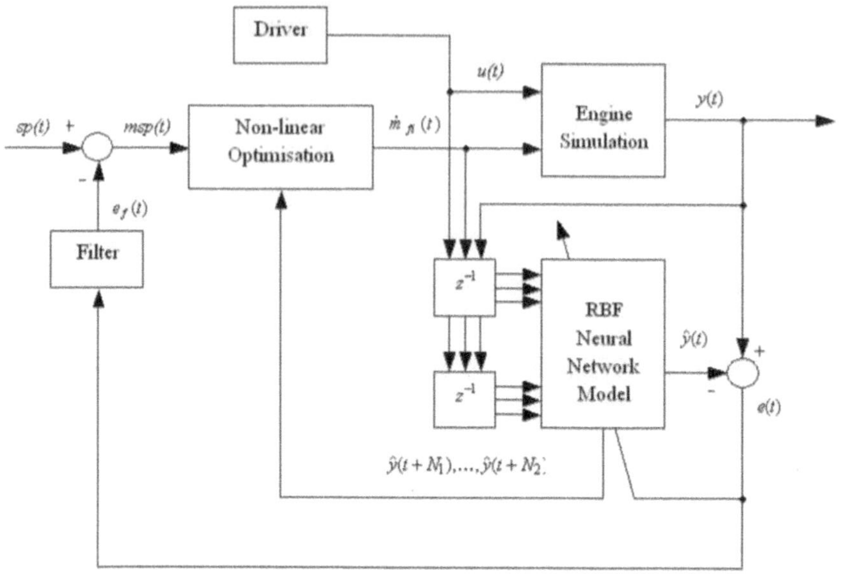

Fig. 4. Configuration of Model Predictive Control on AFR

where $J : R^n \rightarrow R$ is the objective function, $c_{eq} : R^n \rightarrow R^m$ and $c_{in} : R^n \rightarrow R^p$ are constraint functions. All of these functions are smooth. Only inequality constraint applied in our case, as fuel injection rate is bounded within a region.

The Secant Method is to find the improved design vector X_{i+1} from the current design vector X_i using the formula

$$X_{i+1} = X_i + \xi_i^* S_i \tag{11}$$

where S_i is the know search direction and ξ_i^* is the optimal step length found by solving the one-dimensional minimization problem as

$$\xi_i^* = \min_{\xi_i}[J(X_i + \xi_i^* S_i)] \tag{12}$$

Here the objective function J is to be evaluated at any trial step length t_0 as

$$J(t_0) = J(X_i + t_0 S_i) \tag{13}$$

Similarly, the derivative of the function J with respect to ξ corresponding to the trial step length t_0 is to be found as

$$\left.\frac{dJ}{d\xi}\right|_{\xi=t_0} = S_i^T \Delta J\big|_{\xi=t_0} \tag{14}$$

The necessary condition for $J(\xi)$ to have a minimum of ξ^* is that $J'(\xi^*) = 0$. The secant method seeks to find the root of this equation[7] . The equation is given with the form as follows,

$$J'(\xi) = J'(\xi_i) + s(\xi - \xi_i) = 0 \tag{15}$$

where s is the slope of the line connecting the two points $(A, J'(A))$ and $(B, J'(B))$, where A and B denote two different approximations to the correct solution ξ^*. The slope s can be expressed as

$$s = \frac{J'(B) - J'(A)}{B - A} \tag{16}$$

Equation (15) approximates the function $J'(\xi^*)$ between A and B as a linear equation and the solution of Equation (15) gives the new approximation to the root of $J'(\xi^*)$ as

$$\xi_{i+1} = \xi_i - \frac{J'(\xi_i)}{s} = A - \frac{J'(A)(B - A)}{J'(B) - J'(A)} \tag{17}$$

The iteration process given in Equation (16) is illustrated in Fig.5.

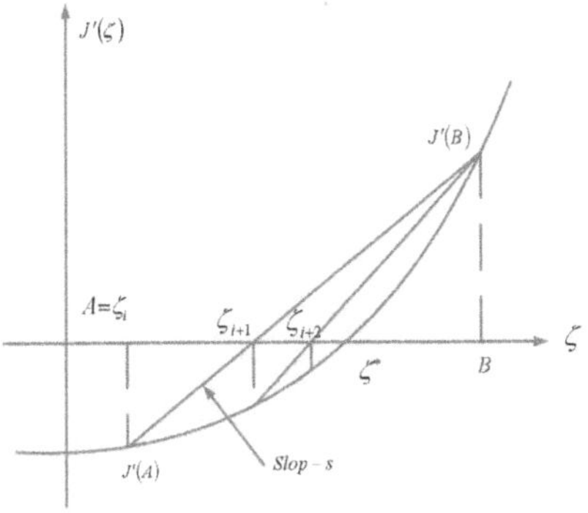

Fig. 5. Iterative process of Secant method

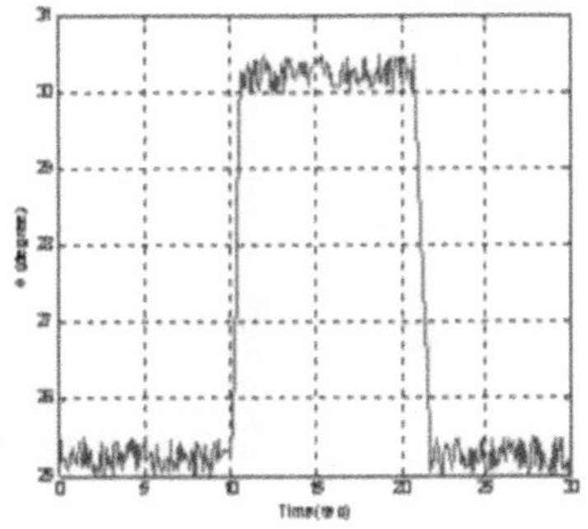

Fig. 6. Throttle angle pattern

Simulation Result Using Secant Method. In the simulation, the set-point of the system is set to be the constant stoichiometric value 14.7. The throttle angle u is set as disturbance, a change from 25° to 30° with 0.5% uncertainty as shown in Fig.6. This is to evaluate the tracking performance and the robustness to throttle angle change of the designed system.

The AFR is to be controlled between the ±1% bounds of the stoichiometric value (14.7). Choosing the sampling time to be $0.1s$. The parameters of nonlinear optimization were chosen as $N_1 = 1$, $N_2 = 6$, $\xi = 1$, $N_u = 0$, then the MPC of SI engines can be considered as a sub-problem of NLP problems:

$$\min_{x \in R^n} f(\dot{m}_{fi}) \tag{18}$$

subject to

$$\dot{m}^l_{fi} \leq \dot{m}_{fi} \leq \dot{m}^u_{fi} \tag{19}$$

where $f : R^n \rightarrow R$, $\dot{m}^l_{fi}$ and $\dot{m}^u_{fi}$ represent the lower bound and the upper bound of the control variable $\dot{m}_{fi}$.

The system output under the developed MPC is displayed in Fig.7, together with the associated manipulated variable displayed in Fig.8. The mean absolute error (MAE) of the AFR tracking is 0.4464. One can see that the air-to-fuel ratio is regulated within a neighborhood of stoichiometric. This performance is much better than that of PI controller[8] that is widely used in automotive industry.

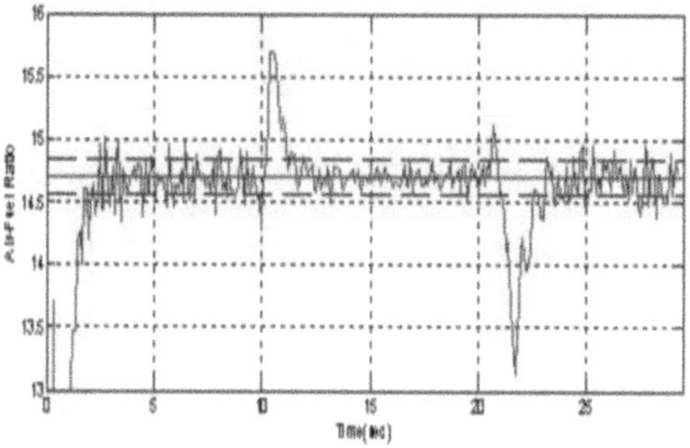

Fig. 7. MPC on AFR using Secant method

The time cost in optimization in each sample period is shown in Fig.9. The mean time cost in one sample period is 0.0277 seconds. Since the whole simulation was running in **Matlab** environment, we feel that the further reduction on time cost of optimization could be achieved if optimization algorithm is realized by **C** code in real application. The multi-dimensional approach for MPC

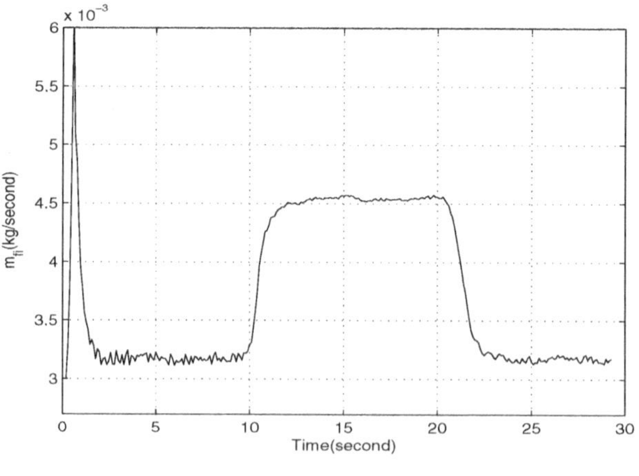

Fig. 8. Fuel injection using Secant method

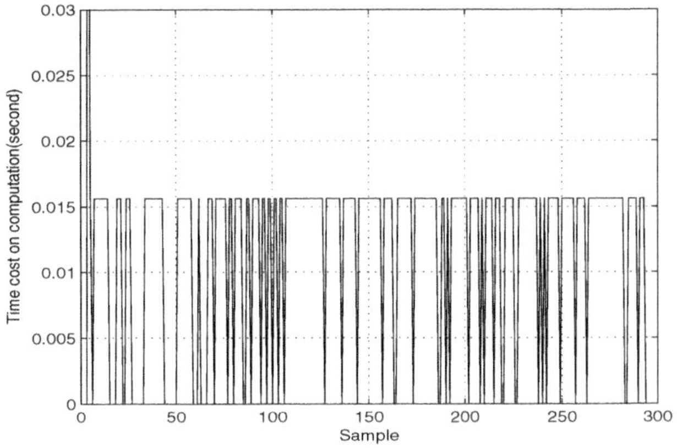

Fig. 9. Time cost on optimization using Secant method

was implemented using Reduced Hessian Method and is compared with Secant Method, in terms of the control performance and time consumptions on optimization. The simulation results show that Reduced Hessian Method has the similar tracking performance of Secant Method, however, its time consumption in optimization is much more than that of previous method. In our experiment, the mean time cost in one sample period using this method is 0.0473 s that is nearly twice as many as that used by Secant Method.

5 Conclusions

In this paper, adaptive RBF model based MPC is applied to AFR control of automotive engines. The simulation results validated that the developed method can control the AFR to track the set-point value under disturbance of changing throttle angle. To meet the requirement for fast optimization in engine control, a one-dimensional optimization method, Secant Method, is implemented in the MPC and is compared with the multi-dimensional method, Reduced Hessian Method. Simulations show a much shorter optimization time using Secant Method and the achieved tracking control with similar performance to that in Reduced Hessian Method.

References

1. Mooncheol, W., Seibum, B.C., Hedrick, J.K.: Air-to-Fuel Ratio Control of Spark Ignition Engines Using Gaussian Network Sliding Control. IEEE Transactions on Control Systems Technology **6(5)** (1998) 678-687
2. Onder, C.H., Geering, H.P: Model-based Multivariable Speed and Air-to-Fuel Ratio Control of an SI Engine. SAE Technical Paper No. 930859 (1993)

3. Hendricks, E., Engler, D., Fam, M.: A Generic Mean Value Engine Model for Spark Ignition Engines. 41st Simulation Conference, SIMS, DTU, Lyngby, Denmark (2000) 18-19
4. Vinsonneau J.A.F., Shields D.N., King P.J., Burnham K.J.: Polynomial and Neural Network sSpark Ignition Engine Intake Manifold Modeling. Proc. 16th Int. Conf. on Systems Engineering, ICSE **2** (2003) 718-723
5. Girosi, F., Poggio, T.: Network and the Best Approximation Property. Biological Cybernetics **63** (1999) 169-176
6. Draeger, A., Engell, S., Ranke, H.: Model Predictive Control Using Neural Networks. IEEE Control Systems Magazine **15** (1995) 61-66
7. Singiresu, S.R.: Engineering Optimization. John Wiley Sons, Inc. (1996) 100-123
8. Wang, S.W., Yu, D.L., Gomm, J.B., Page, G.F., Douglas, S.S.: Adaptive Neural Network Model Based Predictive Control for Air-fuel Ratio of SI Engine. Engineering Application of Artificial Intelligence **19** (2006) 189-200

Approximate Dynamic Programming for Ship Course Control

Xuerui Bai, Jianqiang Yi, and Dongbin Zhao

Key Lab of Complex Systems and Intelligence Science,
Institute of Automation, Chinese Academy of Sciences
95 Zhongguancun East Road, Haidian District,
Beijing 100080, P.R. China
baixuerui05@yahoo.com.cn

Abstract. Dynamic programming (DP) is a useful tool for solving many control problems, but for its complexity in computation, traditional DP control algorithms are not satisfactory in fact. So we must look for a new method which not only has the advantages of DP but also is easier in computation. In this paper, approximate dynamic programming (ADP) based controller system has been used to solve a ship heading angle keeping problem. The ADP controller comprises successive adaptations of two neural networks, namely action network and critic network which approximates the Bellman equations associated with DP. The Simulation results show that the ship keeps the desired heading satisfactorily.

1 Introduction

There have been lots of conventional ship autopilots which use proportional integral and derivative (PID) control algorithms to keep a ship on a fixed heading angle, since the first autopilot was implemented by Sperry.

In 1970s [1] [2], adaptive autopilots were designed and their control parameters were adjusted automatically in accordance with conditions and therefore were able to serve well under different circumstances. However, they have a big disadvantage which is their complexity in computation.

Since 1980s, intelligent autopilots have begun to attract wide attention. Three intelligent theories, Genetic Algorithm, Fuzzy Logic Control [3] and Neural Network play important parts in many applications and result in more sophisticated and reliable control systems [4].

Modern sea going ships all have their own characteristics, and they are more and more complex in their structure. It is impossible to obtain all of the models when we design the controllers. Approximate Dynamic Programming (ADP), also called adaptive critic designs (ACD), is suitable for learning in noisy, nonlinear, and nonstationary environments [5]. We present a new kind of ADP controller in this paper. The major advantage is that we don't need to know the exact mathematic model of the ship and the control system is trained on line.

D. Liu et al. (Eds.): ISNN 2007, Part I, LNCS 4491, pp. 349–357, 2007.

This paper is organized as follows: Section 2 presents two dynamic models of ship steering used in this study, and show the figure of steering the ship. Section 3 presents the ADP controller designed in this study and gives an overview of the theory based on ADHDP used in this paper. In section 4, numerical simulations are made to demonstrate the control performance of ADHDP controller. Conclusions are summarized in section 5.

2 Ship Steering Control

We use two ship models to make a comparison of the simulation results. In Fig. 1, θ_d is the desired heading angle, and θ is the current heading angle. The first ship model is introduced [6]:

$$\begin{cases} \theta(t+1) = \theta(t) + \Delta\dot{\theta}(t) \\ \dot{\theta}(t+1) = \dot{\theta}(t) + \Delta[K\delta(t) - \dot{\theta}(t)]/T \end{cases} \tag{1}$$

where T is the time constant of convergence to desired turning rate, Δ is the sampling interval, and K is the gain of the rudder. The control signal is the rudder angle $\delta(t)$ of the ship. The rudder angle $\delta(t)$ is constrained to $-35^\circ \sim 35^\circ$.

The second ship model is nonlinear Nomoto equation [7]:

$$T\ddot{\theta}(t) + \alpha\dot{\theta}(t) + \beta\dot{\theta}^3(t) = K\delta(t) \tag{2}$$

where T is the time constant of convergence to desired turning rate, K is the gain of the rudder, α and β are called the parameters of Norbbin. The control signal is the rudder angle of the ship.

3 The Design of ADHDP Controller

More and more people adopt ADP to solve different kinds of nonlinear problems. ADP is defined as a scheme that approximates dynamic programming in the general case, i.e., approximate optimal control over time in noisy, nonlinear environment [5]. Generally speaking, there are three design families: Heuristic dynamic programming (HDP), dual heuristic programming (DHP), and globalized dual heuristic dynamic programming (GDHP). The action dependent (AD) versions of the above architectures are also often used nowadays. AD means that the action value is an additional input to the critic network [8] [9]. Our proposed method in this paper is ADHDP controller.

Approximate dynamic programming stems from the idea of dynamic programming. Dynamic programming was proposed by Bellman to solve the problems of dynamic systems in the 1950's. Dynamic programming is a robust tool for solving simple and small problems. However, it is well known that the computation costs of dynamic

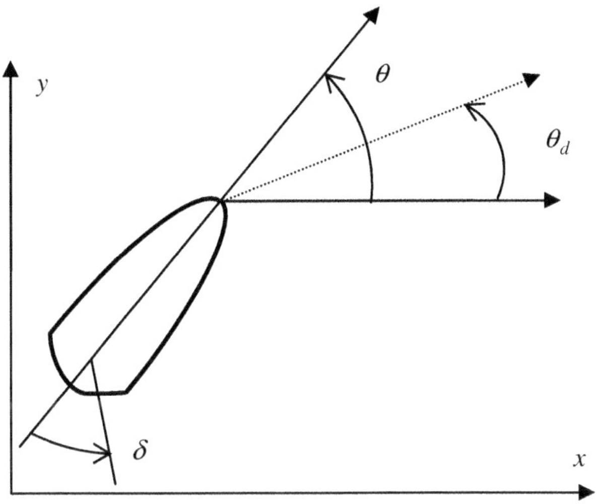

Fig. 1. Ship motion in Earth-fixed coordinate frame

programming are very high for many important problems. Then we just need to find approximate solutions to dynamic programming. Estimating the cost function in dynamic programming is the key step in finding approximate solutions to dynamic programming. Then the optimal control signal can be obtained by minimizing the cost function. As we all know, a three layers neural network can approximate any nonlinear functions with any desired precision, so the artificial neural networks are often used to represent the cost function in dynamic programming.

A typical structure of ACD has two components, critic network and action network, as is shown in Fig. 2. In this figure, the action network outputs the control signal, and the critic network outputs an estimate of cost function.

Fig.3 is a schematic diagram of our proposed ADHDP controller scheme. When the heading angle $\theta(t)$ is out of the boundary $-90^{\circ} \sim 90^{\circ}$, we set the reinforcement signal *r(t)* to -1; otherwise, we set *r(t)* to $-(\theta(t)-\theta_d)^2/90^2$. The weights/parameters of the action network and the critic network are initialized randomly. The controller output the control signal based on the parameters in the action network. When a system state is observed, an improved control signal will lead to a more balanced equation of the principle of optimality. Association between states and control output in the action network will reinforce this series of system operations. Otherwise, the weights in the action network will be tuned in order to adjust the control value and to make the equation of the principle of optimality more balanced.

The critic network outputs *J*, which approximates the discounted total reward-to-go. To be more quantitative, it approximates *R(t)* at time t given by

$$R(t) = r(t+1) + \alpha r(t+2) + \cdots \tag{3}$$

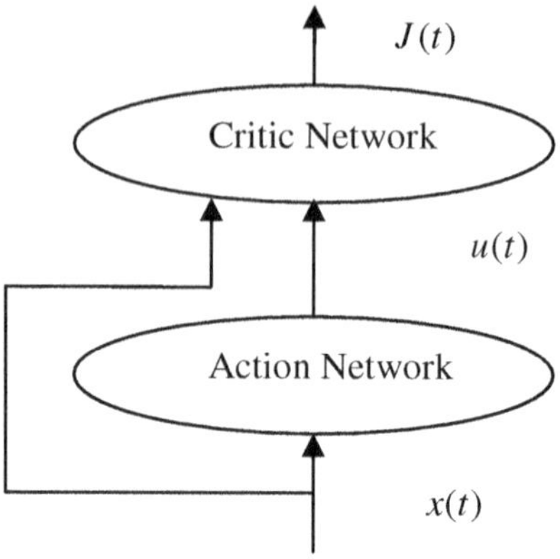

Fig. 2. The two modules in a typical adaptive critic design

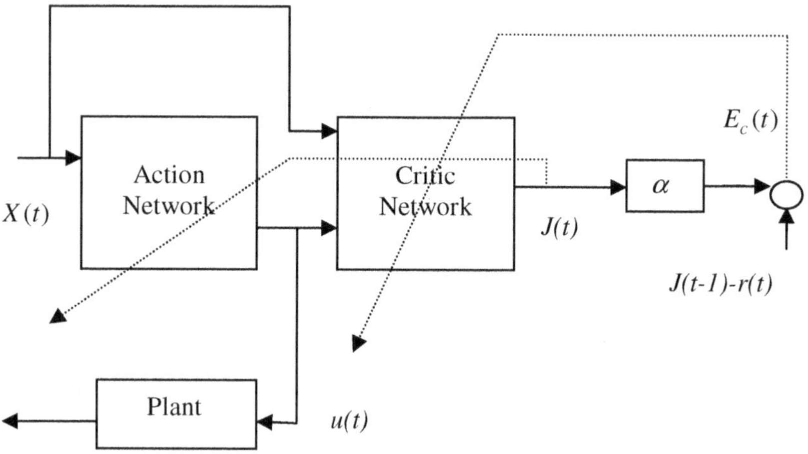

Fig. 3. Schematic diagram for implementations of ADHDP

where *R(t)* is the future accumulative reward-to-go value at time t, α is a discount factor for the infinite-horizon problem($0<\alpha<1$). The discount factor α is chosen as 0.95 in this paper.

A The Critic Network

We use critic network to provide *J(t)* as an approximate of *R(t)*. Our aim is minimizing the following error measure over time [8],

$$e_c(t) = \alpha J(t) - [J(t-1) - r(t)] \tag{4}$$

$$E_c(t) = \frac{1}{2} e_c^{\,2}(t) \tag{5}$$

The weight update rule for the network is gradient descent rule, which is given by the following equations,

$$w_c(t+1) = w_c(t) + \Delta w_c(t) \tag{6}$$

$$\Delta w_c(t) = l_c(t)\left[-\frac{\partial E_c(t)}{\partial w_c(t)}\right] \tag{7}$$

$$\frac{\partial E_c(t)}{\partial w_c(t)} = \left[-\frac{\partial E_c(t)}{\partial J(t)}\frac{\partial J(t)}{\partial w_c(t)}\right] \tag{8}$$

where $l_c(t) > 0$ is the learning rate of the critic network at time t, and w_c is the weight vector in the critic network.

B The Action Network

The weights updating equations in the action network are as follows [8].

$$e_a(t) = J(t) \tag{9}$$

$$E_a(t) = \frac{1}{2}e_a^2(t) \tag{10}$$

$$w_a(t+1) = w_a(t) + \Delta w_a(t) \tag{11}$$

$$\Delta w_a(t) = l_a(t)\left[-\frac{\partial E_a(t)}{\partial w_a(t)}\right] \tag{12}$$

$$\frac{\partial E_a(t)}{\partial w_a(t)} = \left[\frac{\partial E_a(t)}{\partial J(t)}\frac{\partial J(t)}{\partial u(t)}\frac{\partial u(t)}{\partial w_a(t)}\right] \tag{13}$$

where $l_a(t) > 0$ is the learning rate of the action network at time t, and w_a is the weight vector in the action network.

4 Simulation Results

We implement the ADHDP controller to the two different ship dynamic models to make a comparison. The first model is linear, but the other one is nonlinear. The two networks of the controller are both implemented using multilayer feedforward neural networks. The action network has two inputs= (θ $\dot{\theta}$), and the critic network has three inputs= (θ $\dot{\theta}$ u). Hidden layers of two networks both have 6 neurons. The critic network outputs J, and the action network outputs u.

4.1 Simulation Results for the First Ship Model

The parameters of the first ship model are given as

$$T = 5,\ \Delta = 0.02,\ K=0.211,$$

Fig. 4 shows the result for the ship course angle regulation. The initial angle is 30° and the desired angle is 0°. It takes 6 seconds to change the heading angle from 30° to 0° with the allowed error 0.01°.

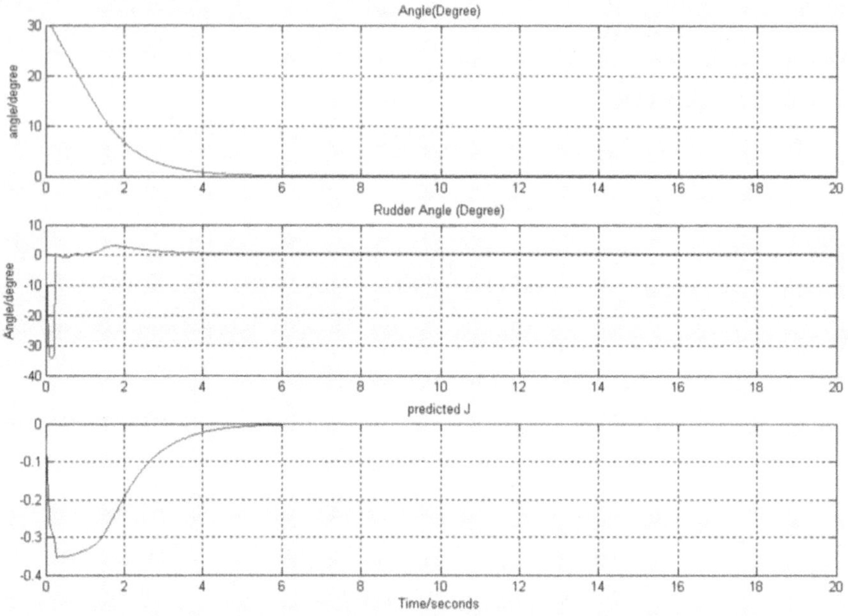

Fig. 4. Result for the first ship model: heading angle, rudder angle and predicted J

4.2 Simulation Results for Nomoto Ship Model

The parameters of the Nomoto model are given as

$$T = 5, K = 0.211, \alpha = 2.2386, \beta = 1998.4$$

Fig. 5 shows the results for the ship course angle regulation. The initial angle is 30° and the desired angle is 0°. It takes 16 seconds to change the heading angle from 30° to 0° with the allowed error 0.01°.

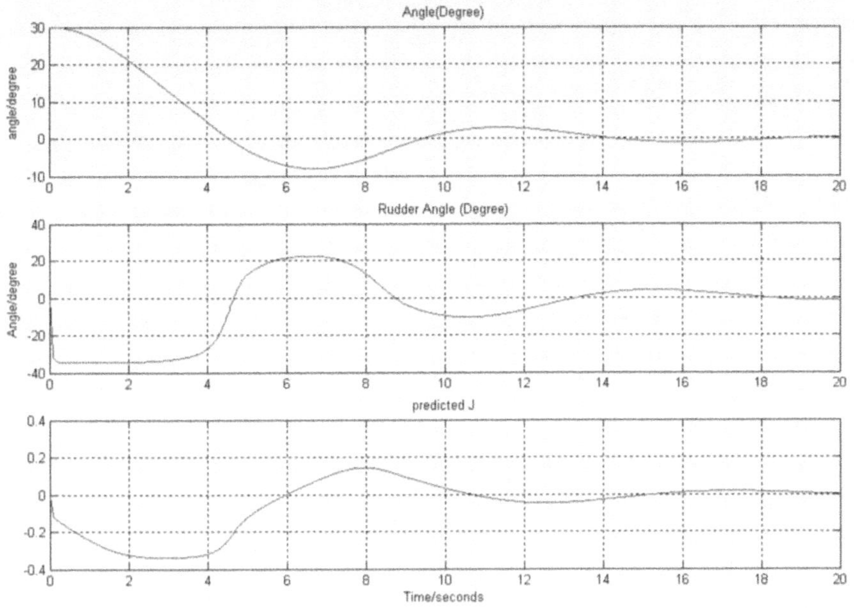

Fig. 5. Result for the second ship model: heading angle, rudder angle and predicted J

4.3 The Generalization of ADHDP Controller

In order to test the generalization of the ADHDP controller, we set the initial angles to 40° and -30° separately which are never used in the training of the controllers. Then

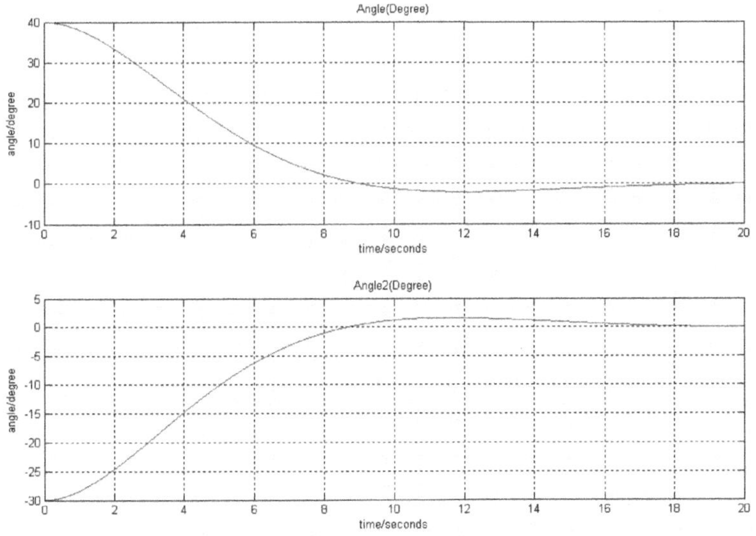

Fig. 6. Generalization test for the first model

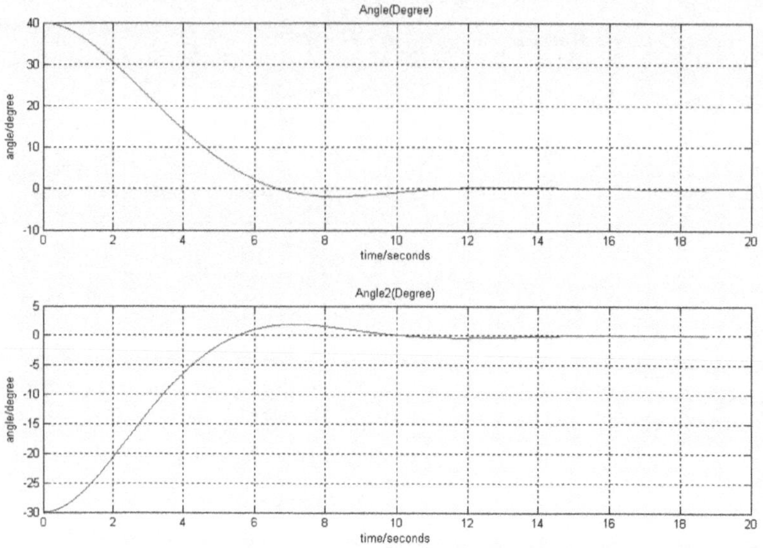

Fig. 7. Generalization test for the second model

we use the controllers trained in 4.1 and 4.2 to control the initial heading angles to the desired angle 0° separately. Figure 6 and Figure 7 show the simulation results.

4.4 Analysis to the ADHDP Controller

In our study a run consists of a maximum of 1000 consecutive trials. A trial consists of 6000 steps, and every step is 0.02 seconds. It is considered successful if the heading angle in last trial (trial number less than 1000) of the run becomes 0° with the allowed error 0.01°. Otherwise, the run is considered unsuccessful. As we can see from the figures above and the table below, when the dynamic model of the ship is more complex, it takes more time for the ADHDP controller to tune its network weights. On the other hand, we also find that the ADHDP controller still has good performance, although the model of the ship is more complex.

Table 1. The summary of simulaions

	The first model	The second model
Percentage of successful runs	100%	100%
Average number of trials to success	7	10

5 Conclusion

A new ship course angle controller has been presented in this article. The ADHDP controller has its obvious advantages over traditional controllers. It doesn't need the model of the controlled object, and it is trained on-line. We also find that it has good performance in both in both a simple and a complex ship models However, we find that the ship may not change its heading angle towards the right direction at the training beginning. The phenomenon exists because the weights matrixes of the action and critic networks are initialized randomly. Then future directions of investigation will be oriented to mix the ADHDP with fuzzy control, and to make the ship change its heading angle towards the right direction as soon as possible.

Acknowledgement

This work was partly supported by the NSFC Projects under Grant No. 60621001, 60575047 and 60475030, the National 973 Project No. 2006CB705500, the Outstanding Overseas Chinese Scholars Fund of Chinese Academy of Sciences (No. 2005-1-11), and the International Cooperative Project on Intelligence and Security Informatics by Chinese Academy of Sciences, China.

References

1. Aseltine, J.A., Mancini, A.R., Sarture, C.W.: A Survey of Adaptive Control Systems. IEEE Transactions on Automatic Control (1958) 102-108
2. Arie, T., Itoh, M., Senoh, A., Takahashi, N., Fujii, S., Mizuno, N.: An Adaptive Steering System for a Ship. Control Systems Magazine **6** (1986)
3. Yi, J.Q., Yubazaki, N., Hirota K.: Trajectory Tracking Control of Unconstrained Object by the SIRMs Dynamically Connected Fuzzy Inference Model. Journal of Advanced Computational Intelligence **4** (2000) 302-312
4. Witt, N.A.J., Miller, K.M.: A Neural Network Autopilot for Ship Control. Proceedings of Maritime Communications and Control Conference (1993) 13-19
5. Liu, D.: Action-Dependent Adaptive Critic Designs. Proceedings of International Joint Conference on Neural Networks **2** (2001) 15-19
6. Liu, D.: Adaptive Critic Designs for Self-Learning Ship Steering Control. Proceedings of the 1999 IEEE International Symposium on Intelligent Control/Intelligent Systems and Semiotics (1999) 46-51
7. Cheng, J., Yi., J.Q., Zhao, D.B.: A New Fuzzy Autopilot for Way-point Tracking Control of Ships. Proceedings of 2006 IEEE International Conference on Fuzzy Systems (2006) 16-21
8. Si, J., Wang, Y.T.: On-line Learning Control by Association and Reinforcement. IEEE Transactions on Neural Network **12** (2001)
9. Prokhorov, D.D., Wunsch, D.C.: Adaptive critic designs. IEEE Transactions on Neural Networks **8** (1997) 997-1007

Traffic Signal Timing with Neural Dynamic Optimization

Jing Xu[1], Wen-Sheng Yu[1], Jian-Qiang Yi[1], and Zhi-Shou Tu[2]

[1] The Key Laboratory of Complex Systems and Intelligence Science, Institute of Automation, Chinese Academy of Sciences, Beijing 100080, China
{jing.xu, jianqiang.yi, wensheng.yu}@ia.ac.cn
[2] Experiment and Practice Center, Chongqing Technology and Business University, Chongqing 400067, China

Abstract. With the discrete traffic model of an oversaturated intersection, the technique of neural dynamic optimization is used to approximate the optimal signal timing strategy which can lead the minimal delay time while considering the whole congestion period. Our approach can provide an approximation of the optimal timing split in each cycle, as well as the most reasonable number of cycles for specific oversaturated traffic inflows. Specifically, for the two-phase case, we are interested to find that the optimal timing strategy is a bang-bang like control instead of a strict bang-bang one as proposed in relative literature. Moreover, our approach is evaluated with a general four-phase case, and its optimal strategy appears also to be a bang-bang like control, which may illuminate the traffic signal timing in practice.

1 Introduction

With the rapid societal development, the number of vehicles and the need for mobility increase more quickly than the road capacity could hold, which results in congestion, and consequent excess delays, reduced safety and increased environmental pollution. This phenomenon occurs typically and always periodically at an urban intersection in daily rush hours, when traffic flow exceeds intersection capacity causing queuing of vehicles that can not be eliminated in one signal cycle.

For an oversaturated intersection, nearly all conventional signal theories, such as those developed by Webster [1], May [2] and Allsop [3], tend to provide an equal time-sharing signal timing strategy. Obviously, this strategy can not efficiently handle oversaturated traffic because it provides no timing optimization at all. Those commonly used softwares, such as SOAP [4] and TRANSYT [5], can not adequately handle heavy traffic either. While addressing the limitations of conventional signal timing strategies, researchers have made great endeavor in searching for timing optimization. Typical results include the work of Cronje [6], and the knowledge-based system SCII developed by Elahi et al. [7]. However, these optimal methods only plan for the next single cycle after the executing one, not concurrently for the whole congestion period.

D. Liu et al. (Eds.): ISNN 2007, Part I, LNCS 4491, pp. 358–367, 2007.

While considering the entire congestion period, Michalopoulos and Stephanopolos proposed an efficient two-stage timing method, termed bang-bang control [8][9]. Their method uses the continuous traffic model and attempts to find an optimal switch-over point during the oversaturated period to interchange the timing of the approaches. However, it is not proper to use the continuous model for the general signal timing because the switch-over point does not necessarily occur at the end of a cycle, neither does the termination of the oversaturated period occur only at the end of the final cycle. Thereby, Chang and Lin proposed an optimal signal timing plan based on a discrete model [10][11]. In their work, the feedforward optimal control (FOC) method was used to find the optimal signal timing. With the objective of minimizing the total delay, they also obtained a bang-bang control for the two-phase case.[1] However, for general three- and four-phase cases, this method seems too complicated. On the other hand, the method of FOC finds the sequence of optimal control values with a specified initial state instead of the feedback solution for all possible initial states. Then, its solution can be very sensitive to disturbances and model uncertainties.

Though traditional dynamic programming (DP) can yield an optimal feedback controller, the real implementation of DP is always difficult due to the high computation and storage complexity of DP for high-order nonlinear systems, which is known as the curse of dimensionality [12]. Thereby, we adopt the technique of neural dynamic optimization (NDO) for solving the optimal signal timing problem. This approach can provide an approximation of the optimal feedback solution whose existences DP justifies [13][14][15]. Another reason of choosing NDO is that this approach can be readily extended to multi-phase timing plans.

In the following section, we describe our approach for the two-phase signal timing. And four-phase signal timing in common use is introduced later. In these two works, the discrete traffic model is used, which overcomes limitations with the continuous one.

2 Two-Phase Signal Timing with NDO

2.1 Statement of the Control Problem

For a four-leg intersection with two-phase signal control as shown in Fig. 1, during oversaturation, the queuing and dispersing situation is as indicated in Fig. 2. Without loss of generality, the cumulative demand on all the approaches is assumed herein to be a linear asymptotic function of time.

According to Fig. 2, the dynamics of the oversaturated intersection can be represented by the following discrete equations,

$$l_1(k+1) = l_1(k) + q_1(k-1)g_2(k-1) + [q_1(k) - s_1] \cdot [c - g_2(k)], \tag{1}$$

$$l_2(k+1) = l_2(k) + q_2(k-1)g_2(k-1) + [q_2(k) - s_2] \cdot [c - g_2(k)], \tag{2}$$

$$l_3(k+1) = l_3(k) + q_3(k)c - s_3 g_2(k), \tag{3}$$

$$l_4(k+1) = l_4(k) + q_4(k)c - s_4 g_2(k), \tag{4}$$

[1] In fact, the optimal control should be a bang-bang like control, not a strict bang-bang one. The reason will be introduced later.

Fig. 1. Four-leg intersection with two-phase signal control

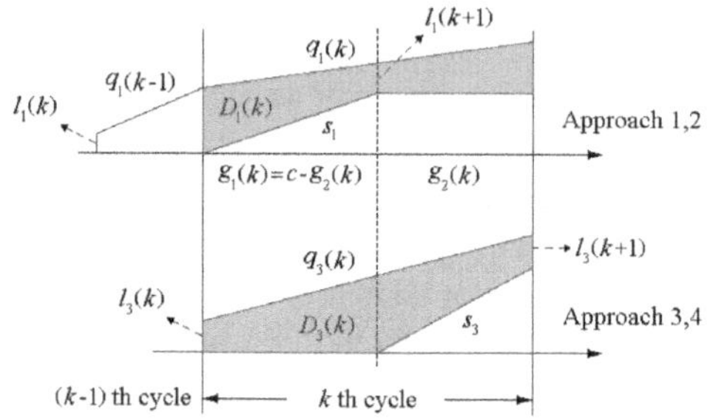

Fig. 2. Queue and delay of the four-leg intersection with two-phase control

where $l_i(k)$, $q_i(k)$, and s_i are the queue length at the beginning of the kth cycle, the input flow rate in the kth cycle, the saturated flow rate of approach i, respectively, for $i = 1, 2, 3, 4$. The parameter of c is the cycle length which is a constant and also the minimal control step, and $g_2(k)$ is the effective green time for the second phase in the kth cycle. From Fig. 2, the delay of each movement can geometrically be calculated as

$$\begin{aligned}
D_1(k) &= l_1(k)c + q_1(k-1)g_2(k-1)c + \frac{1}{2}q_1(k)c^2 - \frac{1}{2}s_1c^2 + \frac{1}{2}s_1g_2^2(k),\\
D_2(k) &= l_2(k)c + q_2(k-1)g_2(k-1)c + \frac{1}{2}q_2(k)c^2 - \frac{1}{2}s_2c^2 + \frac{1}{2}s_2g_2^2(k),\\
D_3(k) &= l_3(k)c + \frac{1}{2}q_3(k)c^2 - \frac{1}{2}s_3g_2^2(k),\\
D_4(k) &= l_4(k)c + \frac{1}{2}q_4(k)c^2 - \frac{1}{2}s_4g_2^2(k).
\end{aligned}$$

For convenience, we define the state vector $x(k) \triangleq [l_1(k), l_2(k), l_3(k), l_4(k)]^T$, and the control $u(k) \triangleq g_2(k)$. Then equations (1)-(4) can be rewritten in the vector form as

$$x(k+1) = f[x(k), u(k), u(k-1)], \tag{5}$$

where $f(\cdot)$ is the corresponding vector function, which can also be denoted as $f(k)$.

If any queue becomes negative, it implies the end of the oversaturated period, and the active controller is just switched off. At this time, all queues are expected to be equivalent and minimal. For this purpose, while considering to minimize the total delay time, we can define the following system performance,

$$J = \Phi(N) + \sum_{k=1}^{N-1} D(k), \tag{6}$$

where $D(k) = D[x(k), u(k), u(k-1)] = a_1D_1(k) + a_2D_2(k) + a_3D_3(k) + a_4D_4(k)$, and $\Phi(N) = \frac{1}{2}x(N)diag\{b_1, b_2, b_3, b_4\}x(N)^T$, where a_i and b_i are the number of lanes and queue penalty of approach i, respectively, for $i = 1, 2, 3, 4$.

The traffic input flow rates may be predicted ones, used for the controller's adaptation to upcoming congestions. They can also be measured ones, saved for the controller developing for an periodically congested intersection such as that in daily rush hours. A specific instance (extracted from Michalopoulos and Stephanopolos) is as shown in Fig. 3.

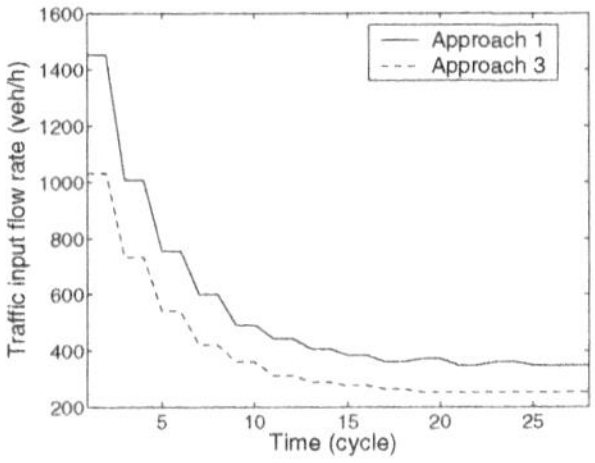

Fig. 3. Sequences of traffic input flow rates extracted from Michalopoulos and Stephanopolos

And now, for given traffic inflows $\{q_i(k), k = 1, 2, \ldots\}$, our problem is to find the proper cycle number N for the oversaturated condition, as well as the optimal timing split in each cycle with any initial system state.

2.2 Controller Developing with Neural Dynamic Optimization

Since the problem formulated above is not a regular optimal control, it is not easy to get the solution using conventional optimal control theory. However, we can provide an accurate approximation of the optimal solution. The technique used herein is the neural dynamic optimization. The basic idea of our approach is to build a neural network which learns to minimize the system performance (6) after each controlled process for oversaturation. Fig. 4 shows the configuration of the control system.

The neural controller can be structured by the multilayer feedforward network. For present problem, specifically, the network is chosen with 5 external inputs, which consists of four queue lengths and the time information (scaled properly before feeding), 10 neurons (recommended) in the hidden layer, and one neurons in the output layer. The activation functions of all neurons can be sigmoidal nonlinearities such as $logsig(\cdot)$ which have built-in saturation limits between 0 and 1. Thus, the control value turns to be

$$u(k) = t_{\min} + (t_{\max} - t_{\min}) \cdot h[x(k), W, k], \tag{7}$$

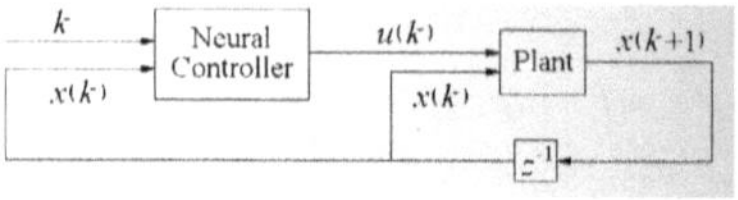

Fig. 4. Configuration of the control system

where t_{min}, t_{max} are specified minimal and maximal green times respectively, $h[x(k), W, k]$ denotes the nonlinear mapping of the neural network, and W is the corresponding weight vector. For convenience, the right side of (7) is denoted as $N_c(k)$.

For given traffic inflows and any initial state, the neural controller should learn to minimize the system performance (6) with the system equation (5) and the control (7). In order to get the updating rule, we adjoin the system and control equations to the system performance with Lagrange multiplier sequences $\{\delta_x(k)\}$ and $\{\delta_u(k)\}$, respectively. Then, the calculus of variations is exploited. Detailed procedure can be referred in [14]. Here, we only provide the trimmed results as follows,

the adjoint system equations:

$$\delta_x(k) = \frac{\partial f(k)^T}{\partial x(k)}\delta_x(k+1) + \frac{\partial D(k)}{\partial x(k)} + \frac{\partial N_c(k)}{\partial x(k)}\delta_u(k), \quad (8)$$

$$\delta_u(k) = \frac{\partial f(k)^T}{\partial u(k)}\delta_x(k+1) + \frac{\partial f(k+1)^T}{\partial u(k)}\delta_x(k+2) + \frac{\partial D(k)}{\partial u(k)} + \frac{\partial D(k+1)}{\partial u(k)}, \quad (9)$$

with $\delta_x(N) = \partial\Phi(N)/\partial x(N)$, $\delta_x(N+1) = 0$, and $D(N) = 0$ for $k = N-1, N-2, \ldots, 1$,

the optimality condition:

$$E\left\{\sum_{k=1}^{N-1} \frac{\partial N_c(k)}{\partial W}\delta_u(k)\right\} = 0. \quad (10)$$

The expectation operator is employed in (10) because we optimize the system performance for any initial state x_0 which subjects to a certain probability distribution $P(x_0)$. For this optimality condition, we may solve it numerically using the stochastic steepest descent method. Specifically, the weights of the neural network are initialized randomly. Then, the following three steps are repeated until the weights remain stable.

***Step* 1)** Forward sweep
The initial state x_0 is picked according to the specified probability distribution which is often recommended to be uniform. Equations here consist of the system equation (5) and control law (7). Accordingly, we can compute the state $x(k)$ and the control $u(k)$ until the oversaturation ends at the Nth cycle.

***Step* 2)** Backward sweep
Equations here are the same as the adjoint system equations (8) and (9). We can compute $\delta_x(k)$ and $\delta_u(k)$ backward in time because the sequences of $x(k)$ and $u(k)$ have been obtained from the forward sweep.

***Step* 3)** Update of the weight vector

$$\Delta W = -\eta \sum_{k=1}^{N-1} \frac{\partial N_c(k)}{\partial W} \delta_u(k), \tag{11}$$

$$W \leftarrow W + \Delta W, \tag{12}$$

where $\eta > 0$ is the learning rate.

At the initial moment, the neural controller is naive because its weights are chosen randomly, which makes the oversaturation end too early while leading great discrepancy between queues of the two approaches. However, the neural controller can learn to reduce this discrepancy, which will bring the oversaturated period with certain additional cycles. At the same time, the neural controller tries to meet the optimality condition.

In fact, besides the initial state, the traffic inflows can also be specified with different sequences during the learning period. Assuming they are stochastic processes subjecting to certain probability distributions, we can apply the stochastic steepest descent algorithm to approximate the optimal controller in the average sense.

2.3 Evaluation with a Case

The case, proposed by Michalopoulos and Stephanopolos, is applied herein. For an intersection of two one-way streets with a two-phase signal control, relative parameters are specified as follows, c = 150 s, s_1 = 1400 pcu/h, s_2 = 1000 pcu/h, $t_{min} = 0.35c$, $t_{max} = 0.6c$, $a_1 = a_2 = 2$, $a_3 = a_4 = 1$, and $b_1 = b_2 = b_3 = b_4 = 10$. The input traffic flows of approach 1 and 3 are as shown in Fig. 3. For simplicity, we assume the other two input traffic flows are the same as those in their opposite directions. We pick a learning rate $\eta = 10^{-6}$ by experiment. The total number of iterations is 5×10^5. Initial queues are chosen uniformly over the range of $[0, 20]$.

Evaluation results are reported in Fig. 5 and 6. With initial zero queues, we conduct four control processes using the initial naive neural controller, the trained NDO controller, the bang-bang control, and the bang-bang like control, respectively. We note that the bang-bang like control is tentatively determined, and it is used as an optimal benchmark. It can be seen that the oversaturated period ends at the 14th cycle with the initial neural controller while leading a great discrepancy between the final queues, which is not a desired feature. However, the NDO controller can handle this problem well at the cost of prolonging the oversaturated period with three cycles. This fact justifies the capability of our approach in searching the proper number of cycles. For the bang-bang control, the NDO controller, and the bang-bang like control, the total delay times are 815,780 s, 811,650 s, and 807,790 s, respectively. Obviously, the optimal strategy for the discrete model is not the same as the bang-bang control in the continuous case. The reason may lie in the importing of the discrete mechanism. Instead, a bang-bang like control can be the optimal strategy for the discrete case. We believe that our NDO controller can approximate this optimal strategy more accurately with further learning operations.

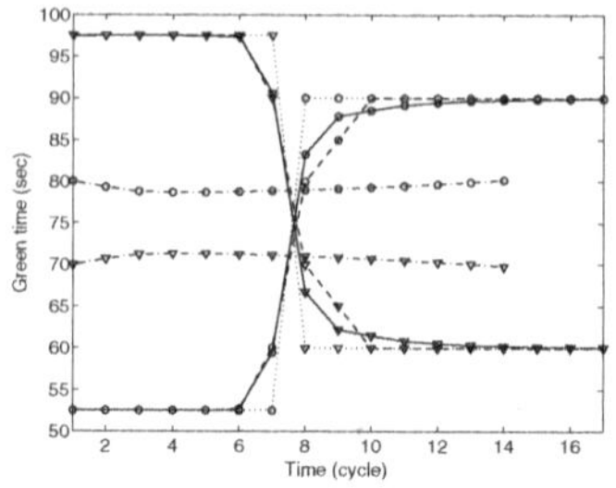

Fig. 5. Effective green times allocated for two phases during the oversaturated period. Lines with triangle markers correspond to phase 1, and lines with circle markers to phase 2. The dash-dot lines denote the signal timing with the initial neural controller. The solid lines are produced by the trained neural controller. The dotted lines denote the bang-bang control proposed by Chang and Lin. And the dashed lines are the bang-bang like control obtained tentatively.

Finally, we indicate that the reason for Chang's suboptimal result (a bang-bang control) may be that the effect of the control in the preceding cycle was just neglected in their works.

3 Four-Phase Signal Timing with NDO

In this section, we apply our approach for the four-phase signal timing in common use. Fig. 7 illustrates the four-phase signal with left-turn protection and Fig. 8 shows the queue and dispersion situation.

The discrete traffic model can be obtained by simply extending the model for the two-phase case. And the control performance is also defined similarly as the one in (6). For the implementation of NDO for the four-phase case, the following two key problems need to be considered. One is about the determination of traffic input flow sequences. For present case, these sequences are assumed to be stochastic processes. Specifically, for through traffic, the sequence is picked uniformly around the mean one, $\{0.6q_1(k), k = 1, 2, \ldots\}$, and for left-turn traffic, it is $\{0.6q_3(k), k = 1, 2, \ldots\}$, where $q_1(k)$ and $q_3(k)$ are as shown in Fig. 3. The other problem is about the setting of controls. Due to the constraint of the minimal green time for each phase and the constraint of all phase times summed to be the cycle length, it is not easy to determine the controls using conventional optimal methods. In our work, we exploit the built-in nature of neurons for limiting their outputs, and provide the following formulation for the controls,

$$u_i(k) = g_i(k) = t_{\min i} + (c - \sum_{j=1}^{4} t_{\min j}) \frac{h_i[x(k), W, k]}{\sum_{j=1}^{4} h_j[x(k), W, k]}, \quad i = 1, 2, 3, 4, \quad (13)$$

where $t_{\min i}$ is the minimal green time of phase i, and $h_i[x(k), W, k]$ is the ith output of the neural controller which is limited between 0 and 1. We can see that the control constraints can be readily satisfied with this formulation.

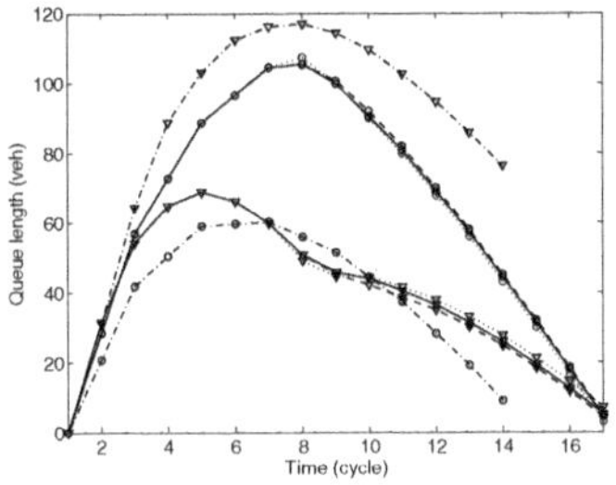

Fig. 6. Queue lengthes correspond to different signal timing strategies. Lines with triangle markers correspond to approach 1, and lines with circle markers to approach 3. The dash-dot lines denote queue lengthes controlled by the initial neural controller. The solid lines are those using the trained neural controller. The dotted lines correspond to the bang-bang control. And the dashed lines are those under the bang-bang like control.

Fig. 7. A four-phase signal with left-turn protection

Relative parameters are specified as follows, $c = 180$ s, and $t_{\min i} = 30$ s. The saturated flow rate is 1400 pcu/h for through traffic, and 1000 pcu/h for left-turn traffic. There are two lanes for through traffic, and one for left-turns. We pick a learning rate $\eta = 10^{-7}$ by experiment. The total number of iterations is 5×10^6. Initial queues are chosen uniformly over the range of $[0, 20]$.

For the oversaturated traffic inflows with four-phase signals, the comparative final queues can hardly be desired using the equal time-sharing control. However, with our NDO controller, this oversaturation can be handled much well. A typical signal timing plan is illustrated in Fig. 9, and the corresponding queues in Fig. 10. Interesting results can be found in Fig. 9, where we can see that the green time of phase 3 is not the minimal, but a little higher value when phase 1 has the maximal

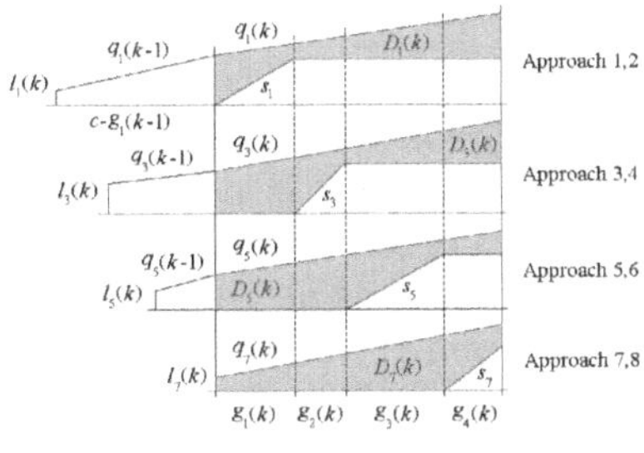

Fig. 8. Queue and delay of a four-phase signal with left-turn protection

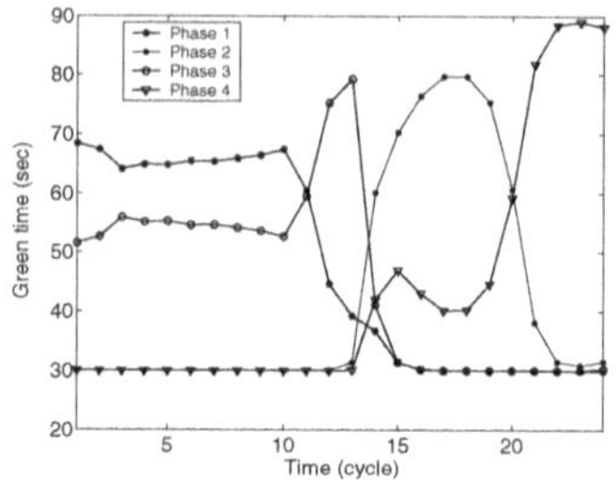

Fig. 9. Effective green times allocated for four phases during the oversaturated period

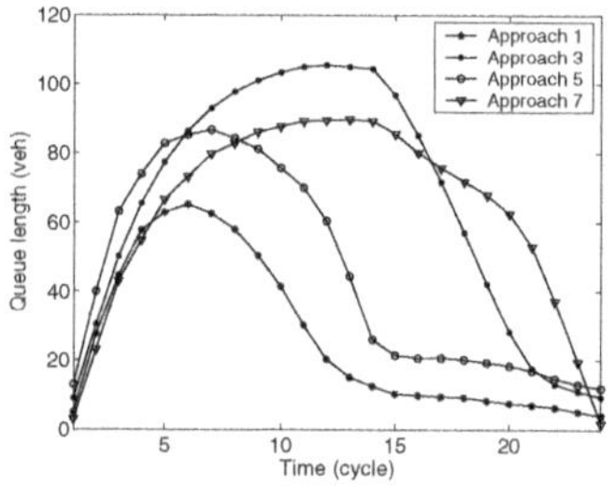

Fig. 10. Queue lengthes under the signal timing plan shown in Fig. 9

green allocation. Similar phenomena exists between phase 2 and 4. Moreover, though all phases have the same minimal value, they do not appear with the same maximal value. Since the maximal, intermediate and minimal control values arise alternately, the four-phase signal timing can also be thought approximately as a bang-bang like control.

4 Conclusions

Based on the technique of NDO, we have proposed approximately optimal solutions for signal timing at oversaturated urban intersections. Through simulation studies with two-phase and four-phase cases, our approach has been demonstrated with many desirable features such as optimality, feedback, and numerical properties.

Acknowledgments. This work was partly supported by the NSFC Projects under Grant No. 60621001, China. And it was also supported by the National Natural Science Foundation of China (No. 60572056, 60621001 and 60334020), the Outstanding Overseas Chinese Scholars Fund of Chinese Academy of Sciences(No. 2005-1-10), and the Open Program of United Laboratory of Intelligence Science and Technology, Institute of Automation, Chinese Academy of Sciences University of Science and Technology of China.

References

1. Webster, F.V.: Traffic Signal Settings. Road Res. Tech. Paper **39** (1958) Great Britain Road Res. Lab. London
2. May, A.D., Jr.: Traffic Flow Theory-The Traffic Engineer's Challenge. Proc. Inst. Traf. Eng. (1965) 290-303
3. Allsop, R.E.: Delay at a Fixed Time Traffic Signal I: Theoretical Analysis. Transp. Sci. **6(3)** (1972) 260-285
4. SOAP 84: User's Manual. (1985) Fed. High. Admi.
5. TRANSYT-7F: User's Manual. Release **5.0**. (1987) Fed. High. Admi.
6. Cronje, W.B.: Optimization Model for Isolated Signalized Traffic Intersections. Transp. Res. Rec. **905** (1983) 80-83
7. Elahi, S.M., Radwan, A. E., Goul, K. M.: Knowledge-based System for Adaptive Traffic Signal Control. Transp. Res. Rec. **1324** (1991) 115-122
8. Michalopoulos, P.G., Stephanopolos, G.: Oversaturated Signal System with Queue Length Constraints-I. Transp. Res. **11** (1977) 413-421
9. Michalopoulos, P.G., Stephanopolos, G.: Optimal Control of Oversaturated Intersections Theoretical and Practical Considerations. Traf. Eng. and Ctrl. (1978) 216-221
10. Chang, T.H., Lin, J.T.: Optimal Signal Timing for an Oversaturated Intersection. Transp. Res. Part B **34** (2000) 471-491
11. Chang, T.H., Sun, G.Y.: Modeling and Optimization of an Oversaturated Signalized Network. Transp. Res. Part B **38** (2004) 687-707
12. Bellman, R.E.: Dynamic Programming. Princeton Univ. Press, Princeton, NJ (1957)
13. Seong, C., Widrow, B.: Neural Dynamic Optimization for Control Systems-Part I: Background. IEEE Trans. Syst. Man Cybern. B **31** (2001) 482-489
14. Seong, C., Widrow, B.: Neural Dynamic Optimization for Control Systems-Part II: Theory. IEEE Trans. Syst. Man Cybern. B **31** (2001) 490-501
15. Seong, C., Widrow, B.: Neural Dynamic Optimization for Control Systems-Part III: Applications. IEEE Trans. Syst. Man Cybern. B **31** (2001) 502-513

Multiple Approximate Dynamic Programming Controllers for Congestion Control

Yanping Xiang, Jianqiang Yi, and Dongbin Zhao

Key Lab of Complex Systems and Intelligence Science,
Institute of Automation, Chinese Academy of Sciences
95 Zhongguancun East Road, Haidian District,
Beijing 100080, China
{yanping.xiang, jianqiang.yi, dongbin.zhao}@ia.ac.cn

Abstract. A communication network is a highly complex nonlinear dynamical system. To avoid congestion collapse and keep network utilization high, many congestion control methods have been proposed. In this paper, a new framework, using Adaptive Critic Designs (ACD) based on Approximate Dynamic Programming (ADP) theory, is presented for network congestion control. At the present time, almost all ACD controllers are designed for centralized control system. In the new frame, the whole network is considered as a multiple non-cooperative ACDs control system, wherein, each source controller is governed by an ACD. This frame provides a new approach to solve the congestion control problem of the networks.

1 Introduction

A communication network may experience periods where the traffic load offered to it exceeds the available transmission capacity; during such periods the network is said to be congested. To avoid congestion collapse and keep high network utilization, users response to the congestion and adapt their transfer rates. Congestion control is a distributed algorithm to share network resources among competing users [7]. How the available bandwidth within the network should be shared among these competing users is the key issue concerned by researchers.

By using an adaptive critic control method [3] based on Approximate Dynamic Programming (ADP) [4] theory, a new framework, including multiple non-cooperative ADP controllers, is presented for network congestion control in this paper. Adaptive critic control is an advanced control technology developed for nonlinear dynamical systems in recent years. It can easily be applied to nonlinear systems with or without constraints on the control and state variables. This technology provides a new approach to solve the congestion control problem of the networks which are highly complex nonlinear dynamical system.

The organization of this paper is as follows. In section 2 we give a common network flow control model. In section 3, approximate dynamic programming (ADP) based on neural networks is described. In section 4, we propose a source update law based on ACD. In section 5, conclusions are given.

D. Liu et al. (Eds.): ISNN 2007, Part I, LNCS 4491, pp. 368–373, 2007.

2 Network Flow Control Model and Optimization Problem

Network flows are commonly modeled as the interconnection of information sources and communication links through the routing matrices as shown in Fig.1 [1, 2, 5, 6].

R is the forward routing matrix and R^T is the return routing matrix. For an arbitrary source *r*, a rate x_r is allocated and $x \in R^N$ is the source rate vector. For an arbitrary link *l*, the aggregate link rate is y_l and $y \in R^L$ is the vector of aggregate link rate. Link *l* has a fixed capacity c_l. Based on its congestion degree and queue size, a link price p_l is computed. The link price information is then sent back to sources which utilize this link with aggregate source prices. We can get the following relationship [2]:

$$y = Rx \qquad q = R^T p \tag{1}$$

where $p \in R^L$ is the link price vector and $q \in R^N$ is the vector of the aggregate price.

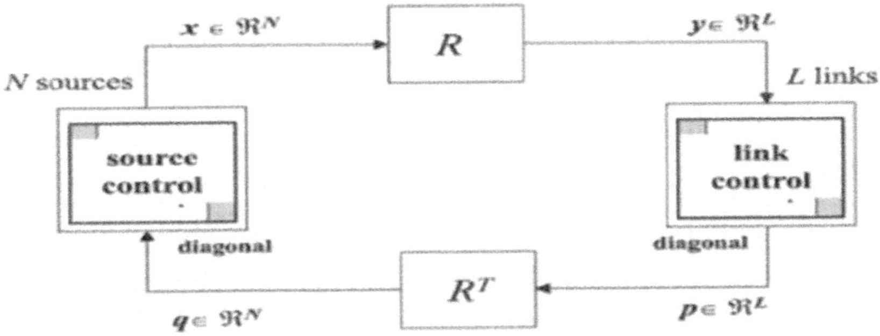

Fig. 1. Network flow control model [2]

To get decentralized source and link control law, commonly, the network flow control problem is decomposed into a static optimization problem and a dynamic stabilization problem [1],[5]. The static optimization problem computes the optimal equilibrium condition by maximizing the sum of the source utility functions $U_r(x_r)$, and satisfying the capacity constraints in the links; that is,

SYSTEM (U, R, C)

$$\max_{x \geq o} \sum_{r=1}^{N} U_r(x_r) \tag{2}$$

subject to $\quad Rx \leq C \quad$ over $\quad x \geq 0$

where $C \in R^L$ is a vector of link capacities and $U_r(x_r)$ is an increasing, strictly concave and continuous differentiable function of x_r [1]. As shown in [2], a unique equilibrium exists. The utility function determines the equilibrium, and consequently the steady state fairness and utilization. The dynamic problem has been explored in several papers, including [1], [6], [8], [9], which present source and link control laws and provide stability proofs.

3 ADP Controller

In this section, adaptive critic control based on approximate dynamic programming (ADP) [4] is described. Consider a nonlinear system of the form

$$\frac{d\eta(t)}{dt} = f(\eta(t), u(t), t) \tag{3}$$

We want to find a stabilizing controller $u = k^0(\eta)$, which minimizes the cost function $J(\eta(0), u) = \int_0^\infty l(\eta(t), u(t))dt$ where $l(\eta(t), u(t))$ is utility function. The Dynamic Programming solution to this problem is obtained via the solution of the Hamilton-Jacobi-Bellman equation

$$-\frac{\partial J^*}{\partial t} = \min_{u(t)}[l(\eta(t), u(t)) + (\frac{\partial J^*}{\partial \eta})^T f(\eta, u, t)] \qquad x \in R^n \tag{4}$$

Here, $J^* = \min J(\eta(0), u(t))$ is the optimal cost function for the problem.

In the above, if f and J are known, the solution for k^0 is a simple optimization problem. However, the solution of the Hamilton-Jacobi-Bellman equation is often computationally untenable as a result of the well-known "curse of dimensionality"[10]. The adaptive critic control employs an iterative solution for the problem, to approximate the cost function and learn J^* and k^0 in real-time. The main idea is to approximate dynamic programming solutions by using a function approximation structure such as neural networks to approximate the cost function.

Typically, an Adaptive Critic Design (ACD) consists of three modules-Critic, Model, and Action. Literature [11] proposes a model-free action-dependent ACD version. This design includes the control action signal as input to the critic network. In this paper, we will use this ACD version as the source rate controller of the network.

4 Multiple ADP Controllers for Congestion Control

At the present time, almost all ACD controllers are designed for centralized control system. ACD controllers have been applied in the form of one controller-one plant or one controller-multiple plants. This section will present a multiple ACDs control system and apply it to the network congestion control, wherein, each source controller is governed by an ACD.

4.1 Decomposition of the Optimization Problem

Since congestion control is a distributed algorithm, the problem *SYSTEM (U, R, C)* can be decomposed into two sub problems [1]. Assume each user corresponds to a source of the network. If user r is charged with a price λ_r, then the optimization problem for user r is as follows.

$$USER_r(U_r;\lambda_r)$$
$$\max[U_r(x_r)-\lambda_r x_r] \quad \text{over} \quad x_r \geq 0. \tag{5}$$

If the network receives revenue λ_r per unit flow from user r, then the revenue optimization problem for the network is as follows.

$$NETWORK(R,C;\lambda)$$
$$\max \sum_r \lambda_r x_r \log x_r \tag{6}$$
$$\text{subject to} \quad Rx \leq c \quad \text{over} \quad x \geq 0$$

It is shown [12] that there always exist vectors λ and x solving $USER_r(U_r;\lambda_r)$ and $NETWORK(R,C;\lambda)$; further, the vector x is then the unique solution to *SYSTEM (U, R, C)*. So the control laws can be designed separately to solve these two problems respectively.

4.2 Source Control Law Based on ACD

A source update law based on ACD is designed below. For an arbitrary user r, the transmission rate is generated by the action network of an ACD, while the action network is trained with the objective of minimizing the critic networks outputs J_r. To match the objective of ACD with the objective of $USER_r(U_r;\lambda_r)$, the utility function $l(x_r,\lambda_r)$ is set as (7).

$$l(x_r,\lambda_r) = -(U_r(x_r)-\lambda_r x_r) \tag{7}$$

So the cost function J_r can be obtained as follow.

$$J_r = \int_0^\infty l(x_r,\lambda_r)dt = \int_0^\infty -(U_r(x_r)-\lambda_r x_r)dt \tag{8}$$

When an ACD minimizes J_r, $(U_r(x_r)-\lambda_r x_r)$ is maximized at the same time.

The link price update law is given by a static function shown as (9)

$$p = h(y) \tag{9}$$

where $h(y)\in R^L$, with lth component $h_l(y)$, is a penalty function that enforces the link capacity constraint, $y_l \leq c_l$. We assume that each penalty function is monotonically nondecrease, such as the following used in [1]:

$$h_l(y) = \frac{(y_l - c_l + \varepsilon)^+}{\varepsilon^2} \tag{10}$$

A flow control law (see Fig.2) based on charging scheme to the optimization problem mentioned above is also proposed in [1]. This control consists of a first order

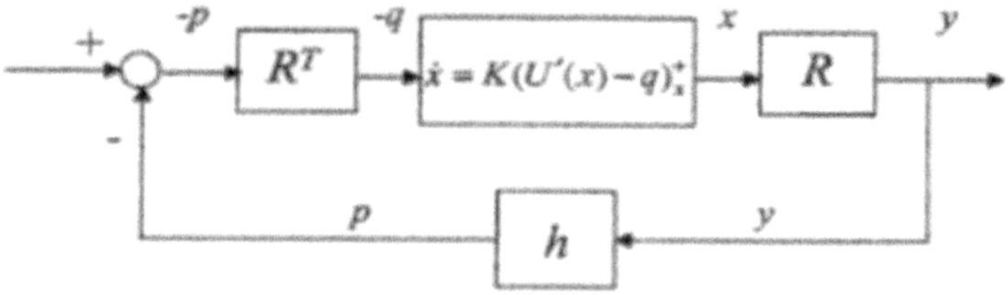

Fig. 2. A flow control law based on [1]

source update law, and a static penalty function for the link to keep the aggregate rate below its capacity. We replace its first order source update law by ACDs and keep the link controller without modification.

Since the users are non-cooperative, each user is controlled by an ACD. The whole system is shown in Fig 3. In this frame, the link controller can be replaced by ACDs too and this is our future work.

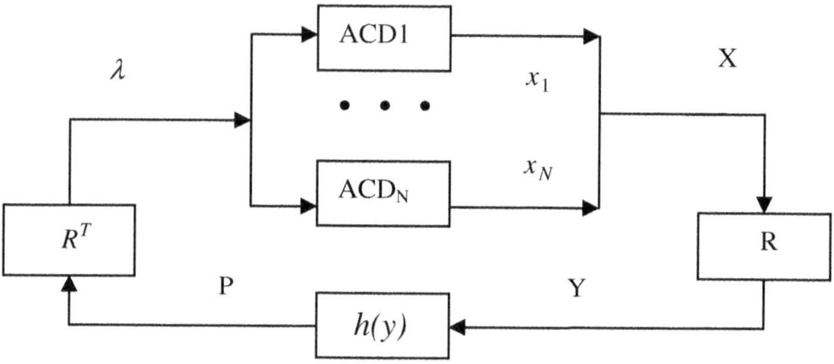

Fig. 3. ACD based flow control

Most rate control laws subject to the common constraint that the routing matrix R is fixed and the network is a deterministic system. However, in practical networks, numbers of users and links vary randomly and topology of the network nodes is time variable. The whole network system may have multiple equilibrium points. It is very difficult to follow the networks' variation. ACDs can be applied to plants with completely unknown dynamics and are not limited by the constraints mentioned above. Neural networks embedded in ACDs are trained on-line and adaptively follow the changes of the network system. For each ACD rate controller, the main task is to optimize its own performances through minimizing the cost function *J*. In fact, the links controller can also be replaced by ACDs. Thus the whole system is constructed as a pure multiple noncooperative ACDs system.

5 Conclusion

By using an adaptive critic control method based on Approximate Dynamic Programming (ADP) theory, a new framework, including multiple non-cooperative ADP controllers, is presented for network congestion control in this paper. This frame provides a new approach to solve the congestion control problem of the networks which is a highly complex nonlinear dynamical system. However, only preliminary analysis and designs is given. Our work is still on early phase, more detailed analysis and designs remain to be done in the future.

Acknowledgement

This work was partly supported by the Outstanding Overseas Chinese Scholars Fund of Chinese Academy of Sciences (No. 2005-1-11), the NSFC Projects under Grant No. 60334020, 60440420130, and 60621001, and the National 973 Project No. 2006CB705500, China.

References

1. Kelly, F., Maulloo, A., Tan, D.: Rate Control in Communication Networks: Shadow Prices, Proportional Fairness and Stability. J. Oper. Res. Soc. **49** (1998) 237-252
2. Wen, J., Arcak, M.: A Unifying Passivity Framework for Network Flow Control. IEEE Transactions on Automatic Control **49(2)** (2004) 162-174
3. Balakrishnan, S., Biega, V.: Adaptive-critic-based Neural Networks for Aircraft Optimal Control. Journal of Guidance, Control, and Dynamics **19** (1996) 893-898
4. Werbos, P.: Approximate Dynamic Programming for Real-time Control and Neural Modeling. Handbook of Intelligent Control: Neural, Fuzzy, and Adaptive Approaches, Chapter 13 (1992)
5. Steven, L., Lapsley, D.: Optimization Flow Control-I: Basic Algorithm and Convergence. IEEE/ACM Transaction on Networking **7(7)** (1999) 861-874
6. Kunniyur, S., Srikant, R.: End-to-end Congestion Control: Utility Functions Random Losses and ecn Marks. IEEE/ACM Transactions on Networking (2003) 689-702
7. Cheng, J., David, X., Steven, L.: Fast TCP: Motivation, Architecture, Algorithms, Performance. IEEE INFOCOM **4** (2004) 2490-2501
8. Fan, X., Arcak, M., Wen, J.: Robustness of Network Flow Control against Disturbances and Time-delay. System and Control Letters **53(1)** (2004)13-29
9. Paganini, F.: A Global Stability Result in Network Flow Control. Systems Control Lett. **46** (2002) 165-172
10. Bellman, R.: Dynamic Programming. Princeton University Press, Princeton, NJ (1957)
11. Liu, D., Xiong, X., Zhang, Y.: Action-dependent Adaptive Critic Designs. Proceedings IJCNN 01 **2** (2001) 990-995
12. Kelly, F.: Charging and Rate Control for Elastic Traffic. European Transactions on Telecommunications **8** (1997) 33-37

Application of ADP to Intersection Signal Control

Tao Li, Dongbin Zhao, and Jianqiang Yi

Laboratory of Complex Systems and Intelligence Science,
Institute of Automation, Chinese Academy of Sciences
95 Zhongguancun East Road, Haidian District, Beijing 100080, China
xpnslt@eyou.com

Abstract. This paper discusses a new application of adaptive dynamic programming (ADP). Meanwhile, traffic control as an important factor in social development is a valuable research topic. Considering with advancement of ADP and importance of traffic control, this paper present a new signal control in a single intersection. Simulation results show that the proposed signal control is valid.

1 Introduction

With the economic development and the growing citification, the urban traffic flow increases quickly. Traffic congestion has become a worldwide problem today. The negative impacts include the loosing of leisure time, increase of fuel consumption, air pollution, etc.

In the short term, the most effective measure to reduce the traffic congestion might be a better traffic control. Traffic signal control which retimes and coordinates existing signals has been proven to bring about substantial reductions in traffic delay, and considerable energy savings. As a complex system with randomness, the traffic system can be taken as an uncertain system. So it is very difficult to give this system an accurate model.

With the emergence of the car, the urban traffic control technology has been, and is still being developed. In 1930, the first signal controller was developed in USA. The first artery control system in the world was applied in Salt Lake City, and it was widely used later in UK, Japan, etc. The term Intelligent Traffic Control (ITC) has been paid more and more attention with recent development of intelligent control theory. ITC is adopted to address the latest generation of traffic control methods to meet the demand for a more efficient and effective traffic network management. In all of these methods, fuzzy control and neural network control are dominant [1, 2].

Despite all achievements, a number of problems need to be solved respectively. For example, fuzzy control is so mainly dependent on expert experience that design and optimization are difficult. Normal neural network considers few of unitary optimization.

Dynamic programming has been applied in different field of engineering, economics, and so on for many years. Werbos proposed a family of approximate dynamic designs [3] as a new optimization technique combining concepts of reinforcement learning and

D. Liu et al. (Eds.): ISNN 2007, Part I, LNCS 4491, pp. 374–379, 2007.

dynamic programming (DP). The adaptive critic designs supply optimal control law for a system by successively adapting two ANNs. These two neural networks are used to approximate the Hamilton-Jacobi-Bellman (HJB) equation associated with optimal control theory.

The critic network is trained to approximate the cost-to-go or strategic utility function. During the adaptation, the networks do not need any "information" of an optimal trajectory, only the desired cost needs to be known. An optimal control policy for the entire range of initial conditions can be obtained by ADP. Since any differentiable structure is suitable as a building block, we do not require exclusively neural network implementations in applications. A series of applications and/or experiments applying various ADP has been approved [4-5].

This paper focuses on discussing the application of ADP in signal control of single intersection. In this section, background in traffic control and adaptive dynamic programming (ADP) is introduced. In section II, application of ADP for isolated intersection traffic signal control is discussed. Action-dependent ADP is constructed for signal control. Finally, simulations are presented in section III.

2 Intersection Model

As a basic part of traffic control, signal control of single intersection has been paid many attentions to. In general, the aim of signal control is to minimize average delay time or number of stops for vehicles passing through the junction.

There are four phases in a typical four-direction intersection. In general the right-turning movement has special passing rule in china. Thus there are four phases movement left to be controlled (straight-going in south-north, left-turning in south-north, straight-going in east-west, left-turning in east-west). There are only two directions in one phase in which cars can pass through at the same time. These four phase states are shown in Fig. 1.

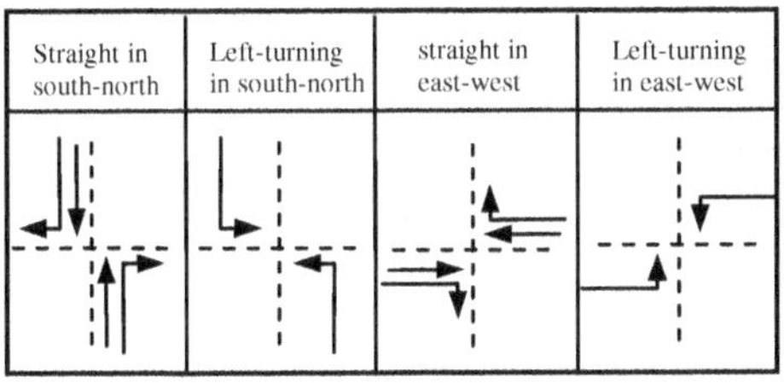

Fig. 1. Four movements phases

For collecting traffic data, each lane is equipped with two inductive loops (Upstream loop and Stop-line loop). The upstream loop is used to measure the number of vehicles entering the intersection: $I_{north}(t)$, $I_{west}(t)$, $I_{south}(t)$, $I_{east}(t)$; the stop-line loop is used to measure the number of vehicles leaving the intersection: $O_{north}(t)$, $O_{west}(t)$, $O_{south}(t)$, $O_{east}(t)$.

At the time t the waiting queue length of each lane of the intersection are defined as: $Q_{ij}(t)$ $i \in$ (up, down), $j \in$ (No.1, No.2, No.3, No.4). We define the distance between upstream loop and stop-line loop as D, the sum of average vehicle length and head-distance as L. Then the maximum detectable queue by the system for each lane is given by:

$$Q_{\text{limit}} = D / L \ . \tag{1}$$

The parameter of the waiting queue length should be constricted by

$$0 \leq Q_{ij} \leq Q_{\text{limit}} \ . \tag{2}$$

The queue length $Q_{ij}(t)$ in every phase j is time-varying, and can be achieved by

$$Q_{ij}(t+\Delta t) = \max\left\{\min\left\{I_{ij}\left(t,t+\Delta t\right)+Q_{ij}(t)-O_{ij}\left(t,t+\Delta t\right),Q_{\text{limit}}\right\},0\right\} \ , \tag{3}$$

where $I_{ij}(t+\Delta t)$ denotes the number of cars which get in i lane during time $\left(t,t+\Delta t\right)$; $Q_{ij}(t)$ denotes the number of cars which are in i lane at time t; $O_{ij}(t+\Delta t)$ denotes the number of cars which get out i lane during time $\left(t,t+\Delta t\right)$.

It is logical that there are only two directions in which cars are allowed to move at the same time. Queue length in two directions in which cars can pass through the intersection at the same time are defined as $Q_{\text{up},j}(t)$ and $Q_{\text{down},j}(t)$ $j \in$ (No.1, No.2, No.3, No.4). The maximum between $Q_{\text{up},j}(t)$ and $Q_{\text{down},j}(t)$ is the main factor to be considered in signal control. Thus the maximum is chosen as input variables:

$$Q_{\max,j}(t) = \max\left\{Q_{\text{up},j}(t), Q_{\text{down},j}(t)\right\}, \quad j \in \text{(No.1, No.2, No.3, No.4)} . \tag{4}$$

3 Signal Controller Design

In this section, the main aim is to discuss signal control of the isolated intersection. A model-free action-dependent ADP (ADHDP) is applied to deal with signal control. Action network and critic network in ADP are described simply in the following sections:

3.1 Action Neural Network

The action neural network is a three-layer feedforward neural network with four inputs, a single hidden layer with five neurons, and a single output neuron. Gradient descent algorithm is implied to train the action network. The inputs are the maximum of queue length defined in equation (4). The input items are looped in order No.1, No.2, No.3, No.4 and No.1 to make the green phase as first neuron. The green time $u(t)$ for passing phase is defined as:

$$u^*(t) = \begin{cases} \text{minGreen}, & u(t) < \text{minGreen}, \\ u(t), & \text{minGreen} \le u(t) \le \text{maxGreen}, \\ \text{maxGreen}, & u(t) > \text{maxGreen}. \end{cases} \tag{5}$$

Delay time can be gotten as following.

This paper assumes that only one car can arrived in one second. The number of arrived cars in a certain second is defined by

$$q_n = \begin{cases} 1, & \textit{one car} \text{ arrived}, \\ 0, & \text{other}. \end{cases} \tag{6}$$

The entering number of cars during $(t, t+\Delta t)$ is defined by

$$I(t, t+\Delta t) = \sum_{i=1}^{\Delta t} q_i . \tag{7}$$

Length of queue after $n(s)$ red light is defined by

$$Q_n = Q_G + \sum_{i=1}^{n} q_i , \tag{8}$$

where Q_G is waiting length after last green time.

Total delay time during red time is defined by

$$D_R = \sum_{j=1}^{n} (Q_G + \sum_{i=1}^{j} q_i) . \tag{9}$$

The number of cars which go out during $(t, t+\Delta t)$ is defined by.

$$O(t, t+\Delta t) = s \times \Delta t . \tag{10}$$

Length of queue after $n(s)$ green light is defined by

$$S_n = \max\{[Q_R + \sum_{i=1}^{n} q_i - s \times n], 0\} , \tag{11}$$

where Q_R is waiting length after last red time and s is the saturated flow.

Total delay time during green light is defined by

$$D_G = \sum_{j=1}^{n} S_j . \tag{12}$$

Average delay time is defined by

$$T_{delay} = (D_R + D_G) \Big/ \sum_{n=1}^{R+G} q_n . \tag{13}$$

The control objective is to reduce delay time T_{delay}, the length of the time a vehicle waiting before a red light.

3.2 Critic Neural Network

The utility function is defined by

$$U(t) = T_{delay}(t) \ . \tag{14}$$

The critic network in this case will be trained by minimizing the following error measure over time,

$$\begin{aligned} \|E_c\| &= \tfrac{1}{2}\sum_t E_c^2(t) \\ &= \tfrac{1}{2}\sum_t [J(t) - U(t) - \gamma J(t+1)]^2 \ . \end{aligned} \tag{15}$$

When $E_c(t) = 0$ for all t, (15) implies that

$$\begin{aligned} J(t) &= U(t) + \gamma J(t+1) \\ &= U(t) + \gamma[U(t+1) + \gamma J(t+2)] \\ &= \cdots \\ &= \sum_{k=0}^{N} \gamma^k U(t+k) \ . \end{aligned} \tag{16}$$

The structure of the critic neural work is chosen as a three-layer feedforward network with five inputs, a single hidden layer with seven neurons, and a single output neuron. The inputs to the critic are the queue length in every phase and output of the action network.

By minimizing the output J , the action network is adjusted to get the optimum result.

4 Simulation

For simulating presented control algorithm, this paper applied binomial distribution to describe arrival rate. Min green time is 15s, max green time is 50s, losing time is 4s in a cycle, and saturated velocity is 3600 veh/h . Discount factor γ is chosen as: $\gamma = 0.95$.

Table 1. Simulation result

Enter flow in Phases (veh/h)	No.1	360	450	550	650	750	750	750
	No.2	360	360	360	360	360	450	500
	No.3	360	360	360	360	360	360	360
	No.4	360	360	360	360	360	360	360
Average delay time (s/veh)	Proposed	9.3	9.4	9.5	11.3	13.1	13.2	13.4
	Fixed 30s	20.2	20.4	21.0	21.3	22.9	23.1	23.6
	Fixed 40s	27.3	27.7	28.2	29.1	29.5	29.7	31.2

From simulation result, it can be concluded that signal control applied ADHDP has achieved a shorter average delay time than control with fixed timing plan.

5 Conclusion

In this paper, an ADP approach for the signal control of the isolated intersection has been introduced and discussed. From the simulation result, we can conclude that the average time delay is smaller in application of ADHDP than control with fixed timing. Five state variables are taken into account for the signal control in a single intersection at the same time in this case.

Acknowledgment

This work was partly supported by the NSFC Projects under Grant No. 60621001, the National 973 Project No. 2006CB705500, the Outstanding Overseas Chinese Scholars Fund of Chinese Academy of Sciences (No. 2005-1-11), and the International Cooperative Project on Intelligence and Security Informatics by Chinese Academy of Sciences, China.

References

1. Dipti, S., Min, C.C., Ruey, L.C.: Neural Networks for Real-Time Traffic Signal Control. IEEE Transactions on Intrlligent Transportation Systems **7** (2006) 261-272
2. Zu, Y.Y., Xi, Y.H., Hong, F.L., Chang, C.X.: Multi-phase Traffic Signal Control for Isolated Intersections Based on Genetic Fuzzy Logic. Proceedings of the 6th World Congress on Intelligent Control and Automation, Dalian, China (2006) 3391-3395
3. Werbos, P.: Approximate Dynamic Programming for Real-time Control and Neural Modeling. Handbook of Intelligent Control, White and Sofge, Eds. New York, Van Nostrand Reinhold (1992) 493-525
4. Padhi, R., Unnikrishnan, N., Balakrishnan, S.N.: Optimal Control Synthesis of a Class of Nonlinear Systems Using Single Network Adaptive Critics. Proceeding of the 2004 American Control Conference, Boston, Massachusetts **2** (2004) 1592-1597
5. Olesia, K., De, R.L., Hossein, J.: Neural Network Modeling and Adaptive Critic Control of Automotive Fuel-Injection Systems. Proceedings of the 2004 IEEE International Symposium on Intelligent Control, Taipei, Taiwan (2004) 368-373

The Application of Adaptive Critic Design in the Nosiheptide Fermentation

Dapeng Zhang[1], Aiguo Wu[1], Fuli Wang[2], and Zhiling Lin[2]

[1] School of Electrical Engineer and Automation, Tianjin University
300072, Tianjin, China
zdp1995@163.com
[2] College of Information Science and Engineering, Northeast University
110004, Shenyang, China

Abstract. An adaptive critic design is used in the nosiheptide fermentation process to solve the intractable optimization problem. The utility function is defined as the increment of biomass concentration at the adjacent intervals. The state variables are chosen as the biomass concentration, the substrate concentration, the dissolved oxygen concentration and the inhibition concentration. The decision variables are chosen as the temperature, the stirring speed, the airflow and the tank pressure. The adaptive critic method determines optimal control laws for a system by successively adapting the critic networks and the action network. The simulation shows at the same initial conditions this technique can make the fermentation shorten 6 hours.

1 Introduction

Nosiheptide, a novel type of sulfur-containing peptide antibiotics, is a perfect no-attenuating feed additive which can promote animal growth and have no residual in animal body [1]. The nosiheptide fermentation is a very sensitive biochemical process and the same materials cause the different outputs in the different condition.

The optimal decision variable of fermentation is conventionally obtained by the experiments of shake flask [2,3]. There are many limits from the shake flask to the industrial production. So the optimization of industrial scale cannot be assured only by the experiments of shake flask. Recently the system theory and model based control techniques have been used to the optimization of biotechnological process [4-7]. But the accuracy of model cannot meet the conventional optimized methods' needs because of the complexity of fermentation. The adaptive critic designs (ACDs) technique, which was proposed by Werbos, is a new optimization technique to handle the nonlinear optimal control problem using ANNs [8]. The ACDs can be used to maximize or minimize any utility function of a system over time in a noisy non-stationary environment [9,10]. So it is determined as a tool for the optimization of the nosiheptide fermentation process.

D. Liu et al. (Eds.): ISNN 2007, Part I, LNCS 4491, pp. 380–386, 2007.

2 ACDs Technique

The ACDs technique handles the classical optimal control problem by combining concepts of reinforcement learning and approximate dynamic programming (ADP). The ACDs technique determines optimal control laws for a system by successively adapting two neural networks, namely, an action neural network (which dispenses the control signal) and a critic neural network (which "learns" the desired performance index for some function associated with the index). The critic network can guide the action network towards the optimal solution at each successive adaptation. Detailed derivations of ACDs technique can be found in Prokhorov [11] and Liu [12,13].

The ACDs is classed as the heuristic dynamic programming (HDP), the dual heuristic programming (DHP) and the globalized dual heuristic programming (GDHP). Because the DHP can effectively make use of the information of the change rate of goal, it can arrive the approximate accuracy of the GDHP but a little complexity. So the DHP is determined as the application of the fermentation.

3 Application of DHP in Nosiheptide Fermentation Process

3.1 Determining of the Utility Function

In the process of nosiheptide fermentation the goal is to gain the higher biomass concentration. So the utility function is defined as the increment of biomass concentration at the adjacent intervals according to formula (1)

$$U = c_x(t+1) - c_x(t), \tag{1}$$

where c_x is the biomass concentration.

3.2 Determining of State Variables and Decision Variables

Many studies have done in the process of nosiheptide fermentation [14-16]. Accordingly the biomass concentration c_x, the substrate concentration c_s, the dissolved oxygen concentration c_o and the inhibition concentration c_i are determined as the state variables of nosiheptide fermentation. And the temperature *T*, the stirring speed *n*, the airflow *Q* and the tank pressure *P* are determined as the decision variables of nosiheptide fermentation.

3.3 Prediction of State Variables Values

The model of nosiheptide fermentation has been built according to formula (2) to formula (6) [15].

$$\frac{dc_x}{dt} = \dot{A} e^{-\dot{E}_a / R(273+T)} \left(1 - \frac{c_x}{x_{max}}\right) \times \frac{c_s}{k_{sc} c_x + c_s} \times \frac{c_o}{k_{so} + c_o} \times (1 - b c_i) c_x - \ddot{A} e^{-\dot{E}_a / R(273+T)} \left(1 - \frac{c_o}{k_d + c_o}\right) c_x, \tag{2}$$

$$\frac{dc_s}{dt} = -(m_s c_x + \frac{1}{Y_{gs}} \frac{dc_x}{dt} + \frac{1}{Y_{ps}} \frac{dc_p}{dt}), \tag{3}$$

$$\frac{dc_i}{dt} = a \frac{dc_x}{dt}, \quad 4)$$

$$\frac{dc_O}{dt} = K n^{0.27} Q^{0.2488} [\varsigma_L (1+2.5c_x)]^{0.7} [\frac{p}{(-0.00063 \times T^2 + 0.14 \times T + 4.2) \times 10^{K c_s}} - c_o] - q_{O^2} c_x, \tag{5}$$

$$\frac{dc_p}{dt} = \beta c_x (0.0042 \times T^3 - 0.3771 \times T^2 + 11.3415 \times T - 112.3671). \tag{6}$$

This model can simulate the nosiheptide fermentation in the range of allowed errors. So it is used to predict the values of state variables at the next time t+1.

3.4 Application of DHP Method

3.4.1 Establishment of the Critic Network

A three layers BP network is determined as the critic network. The input of this network is four state variables and four decision variables. The output is four costate variables. The number of nodes of the hidden layer is obtained by the method of trials and errors. The structure of critic network is in figure 1.

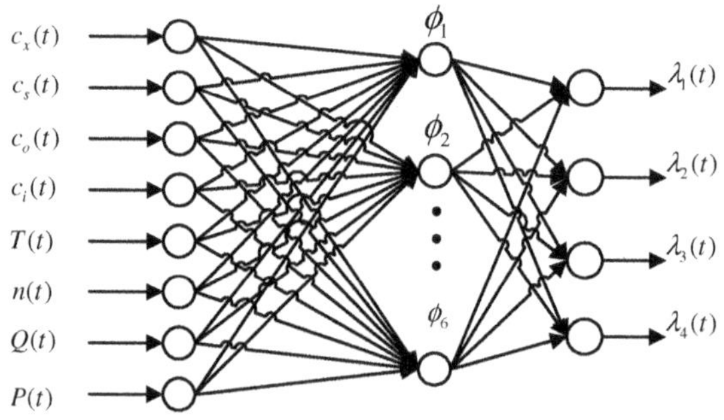

Fig. 1. The structure of critic neural network

3.4.2 Establishment of the Action Network

The action network is also made up of the three layers BP network. The input of this network is four states variables. The output is four decision variables. The number of nodes of the hidden layer is obtained by the method of trials and errors. The structure of action network is in figure 2.

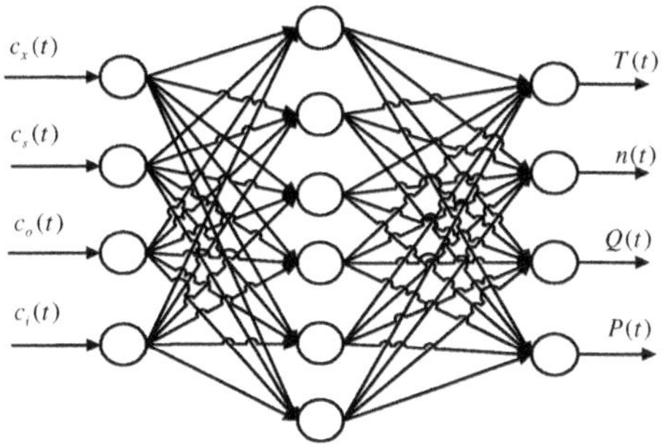

Fig. 2. The structure of action neural network

3.4.3 Training of the Critic Network

The training procedure of critic network is as follows.

① A group of $X(t)$ is randomly yielded from the feasible zone. To every $X(t)$

a) get the decision variable $A(t)$ from the action network;

b) get $X(t+1)$ from the process model;

c) input $X(t+1)$ to the critic network and get $\lambda(t+1)$;

d) get $\lambda^*(t)$ from the costate equation, i.e.

$$\lambda^*(t) = \frac{\partial U(t)}{\partial X(t)} + [\frac{\partial X(t+1)}{\partial X(t)}]^T \lambda(t+1).$$

② training the critic network with the couple data $[X(t), \lambda^*(t)]$.

3.4.4 Training of the Action Network

The action network is trained to reach the goal as formula (7) under the unchanged weight of critic network.

$$\min J = \frac{\partial U(t)}{\partial A(t)} + \frac{\partial J(t+1)}{\partial A(t)} + \mu\varphi, \tag{7}$$

where $\varphi = \sum_{i=1}^{m} x_i^2 \phi(x_i)$, in which $\phi(x) = \begin{cases} 0, & x \in D, \\ 1, & x \notin D, \end{cases}$ and D is the given ranges:

$$100r/\min \le n \le 400r/\min,\ 24m^3/h \le Q \le 3.2m^3/h,\ 27^\circ C \le T \le 32^\circ C,\ 15Pa < P < 50Pa.$$

The weight of action network is updated by formula (8)

$$\Delta W_A = -\alpha[\frac{\partial U(t)}{\partial A(t)} + \frac{\partial J(t+1)}{\partial A(t)}]\frac{\partial A(t)}{\partial W_A} = -\alpha[\frac{\partial U(t)}{\partial A(t)} + \frac{\partial J(t+1)}{\partial R(t+1)}\frac{\partial R(t+1)}{\partial R(t)}\frac{\partial R(t)}{\partial A(t)}]\frac{\partial A(t)}{\partial W_A}. \quad (8)$$

3.4.5 Training of DHP

The integrity training of DHP includes the two processed of the critic network training and the action network training[11]. Firstly the pre-determined decision variables and the corresponding state variables are together used to train the critic network. Suppose the training converges after N_c steps. Then keep the weight of critic network unchanged and train the action network. Suppose the action network converges after N_A steps running. The output of the action network is the optimal decision value $A^*(t)$ at time t. Put the new state variables and the decision variables into the critic network and repeat the above two processes. Then get the optimal decision variable of the next time, $A^*(t+1)$. This process repeats again and again. Finally the optimal decision variable sequence can be obtained.

4 Simulations

The initial values of states variables: the biomass concentration 0.06 g/ml, the substrate concentration 5 mg/ml, the dissolved oxygen concentration 100%, the inhibition concentration 0 mg/ml. The initial values of decision variables: the temperature 305 °K, the stirring speed 200 r/min, the airflow 2.6 m^3/h, the tank pressure 35 Pa. The optimized variable values with the method of DHP are in table 1.

Table 1. Optimized variable values

	State variables				Decision variables			
t	c_x	c_s	c_o	c_i	T	n	Q	P
0	0.06	5	1	0	305	200	2.6	35
6	0.1059	4.9366	0.8327	0.0023	303.2962	183.7731	2.9550	30.3559
12	0.1702	4.8429	0.6554	0.0055	303.9626	104.2358	2.4026	31.2913
18	0.2518	4.7100	0.4749	0.0097	302.3237	219.6443	2.7965	30.0917
24	0.3357	4.5524	0.3404	0.0139	300.0000	400.0000	2.4000	48.7621
30	0.4212	4.3653	1.2147	0.0181	300.4929	106.3731	2.7856	37.9796
36	0.4917	4.1696	0.9989	0.0217	303.7730	100.0274	2.4009	42.9249
42	0.5343	3.9779	0.5006	0.0239	303.9986	245.8901	2.8005	45.6298
48	0.5567	3.6890	0.3929	0.0245				

It is seen from the table 2 that the biomass concentration can reach 0.5567 g/ml after 48 hours. If the DHP was not accepted, the final biomass concentration can also reach 0.5546 g/ml. But it at least spends 54 hours to reach. The cost time with the DHP method is 6 hours shorter than that without this method.

5 Conclusions

The fermentation is a complex biochemical process influenced by many factors. It is difficult for the conventional optimization methods. The adaptive critic design is proposed to apply to the process of nosiheptide fermentation. The simulation shows at the same initial conditions the process of nosiheptide fermentation can be 11 percent shorter with the ACD optimization than that without its optimization.

References

1. Zhou, P., Jiang, Y.F., Li, X., *et al.*: Studies on Submerged Culture Production of Nosiheptide-a Feed Additive. Industrial Microbiology **20** (1990) 7-11
2. Feng, W.X., Li, X.H., Shao, Y.Y., *et al*: Optimum Control in the Production of FDP by Fermentation. Journal of East China University of Science and Technology **26** (1997) 688-691
3. Ni, Y.F., Zhu, J.P.: Optimum Control in Sisomicm Fermentation. Biotechnology **10** (2000) 25-28
4. Lee, J.: Control of Fed-Batch Fermentations. Biotechnology Advances **17** (1999) 29-48
5. Berber, A., Cenk, P., Mustafa, T.: Optimization of Feeding Profile for Baker's Yeast Production by Dynamic Programming. Proceedings of the American Control Conference. Philadelphia, Pennsylvania (1998) 811-81
6. Dhir, S., Morrow, K.J., Rhinehart, R.R.: Dynamic Optimization of Hybridoma Growth in a Fed-Batch Bioreactor. Biotechnology and Bioengineering **67** (2000) 197-205
7. Wang, B., Wang, S.A.: Temperature Control Model in Fermentation Process. Journal of Xi'an Jiaotong University **38** (2004) 737-740
8. Prokhorov, D.V., Santiago, R.A., Wunsch, D.C.: Adaptive Critic Designs: a Case Study for Neurocontrol. Neurral Networks **8** (1995) 1367-1372
9. Venayagamoorthy, G.K., Harley, D.G., Wunsch, D.C.: Comparison of Heuristic Dynamic Programming and Dual Heuristic Programming Adaptive Critics for Neurocontrol of a Turbogenerator. IEEE Transactions on Neural Networks **13** (2002) 764-773
10. Park, J.W., Harley, R.G., Venayagamoorthy, G.K.: New Internal Optimal Neurocontrol for a Series FACTS Device in a Power Transmission Line. Neural Networks **16** (2003) 881-890
11. Prokhorov, D.V., Wunsch, D.C.: Adaptive Critic Designs. IEEE Transactions on Neural Networks **26** (1997) 997-1007
12. Liu, D.R., Xiong, X.X., Zhang, Y.: Action-Dependent Adaptive Critic Designs. International Joint Conference on Neural Networks **2** (2001) 990 – 995
13. Liu, D.R.: Neural Network-Based Adaptive Critic Designs for Self-Learning Control. Proceedings of the 9th International Conference on Neural Information Processing (2002)
14. Zhang, D.P., Wang, F.L., He, J.Y., *et al.*: Mathematical Modeling of Batch Fermentation Process for Nosiheptide. Acta Simulata Systematica Sinica **18** (2006) 2311-2313

15. Zhang, D.P, Wang, F.L., He, J.Y., *et al.*: Modeling and Simulation of Influence of Temperature on Batch Fermentation Process of Nosiheptide. Journal of Northeastern University(Natural Science) **27** (2006) 363-366
16. Sang, H.F., Wang, F.L., He, D.K., *et al.*: Hybrid Modeling of Fermentation Process Based on Least Square Support Vector Machines. Chinese Journal of Scientific Instrument **27** (2006) 629-633

On-Line Learning Control for Discrete Nonlinear Systems Via an Improved ADDHP Method

Huaguang Zhang[1,2], Qinglai Wei[1], and Derong Liu[3]

[1] School of Information Science and Engineering, Northeastern University
Shenyang, Liaoning, 110004, People's Republic of China
hgzhang@ieee.org
[2] Key Laboratory of Process Industry Automation, Ministry of Education, People's Republic of China
qinglai_wei@163.com
[3] Department of Electrical and Computer Engineering University of Illinois at Chicago 60607-7053 Chicago, USA

Abstract. This paper mainly discusses a generic scheme for on-line adaptive critic design for nonlinear system based on neural dynamic programming (NDP), more exactly, an improved action-depended dual heuristic dynamic programming (ADDHP) method. The principal merit of the proposed method is to avoid the model neural network which predicts the state of next time step, and only use current and previous states in the method, as makes the algorithm more suitable for real-time or on-line application for process control. In this paper, convergence proof of the method will also be given to guarantee the control to reach the optimal. At last, simulation result verifies the performance.

1 Introduction

The Adaptive Critic Design (ACD) structures developed from a combination of dynamic programming and back-propagation with a long and solid history of works [1]-[9]. Dynamic programming is a very useful tool in solving the nonlinear MIMO control cases most of which can be formulated as cost minimization or maximization problem.

Suppose that a discrete-time nonlinear system is described by

$$x(t+1) = a[x(t)] + b[x(t)]u(t), \tag{1}$$

where state vector $x \in R^n$ and control vector $u \in R^m$. Define the performance index or cost function associated the system in the form

$$P[x(t)] = \sum_{k=t}^{\infty} \gamma^{k-t} U[x(k), u(k)], \tag{2}$$

where $U[x(k), u(k)]$ is called utility function and $0 < \gamma \leq 1$ is discount factor. The objective is to choose a sequence of the control vector $u(k)$, $k = t, t+1, ...$ in order to make the performance index function P to be minimized (without

D. Liu et al. (Eds.): ISNN 2007, Part I, LNCS 4491, pp. 387–396, 2007.

loss of generality, we suppose here the minimized P is the optimal performance index).

Based on Bellman's optimal principle: An optimal (control) policy has the property that no matter what previous decisions have been, the remaining decisions must constitute an optimal policy with regard to the state resulting from those previous decisions, we can get the optimal performance index function formulated by the following:

$$P^*(t) = \arg\min_{u(t)} \{U[x(t), u(t)] + \gamma P^*(t+1)\}, \tag{3}$$

where $P^*(t)$ is a function of $x(t)$, and for convenience we take the form for all the following performance index functions. The above equation indicates that we should search all the possible states $x(t+1)$, and then find a most optimal control sequence $u^*(t+1), u^*(t+2), ...$, to apply to the system. Unfortunately, the time backward process required for running dynamic programming makes the computation and storage burden seriously heavy, especially for high order nonlinear systems i.e. the problem commonly known as "curse of dimensionality". Over the years, progress had been made to overcome this problem by building a system called critic to approximate the cost function of the dynamic programming in the form of ACDs [2]. The basic structures of ACD proposed in literature were Heuristic Dynamic Programming (HDP), Dual Heuristic Programming (DHP) and Globalized Dual Heuristic Programming (GDHP), and their action-dependent (AD) versions, that means Action Dependent Heuristic Dynamic Programming (ADHDP), Action Dependent Dual Heuristic Programming (ADDHP) and Action Dependent Globalized Dual Heuristic Programming (ADGDHP). A typical ACD consists [2-4] of three neural network modules called action (the decision making module), critic (the evaluation module) and model (the prediction module). In the action dependent versions action is directly connected to the critic without using models.

For HDP, $P(t)$ is the direct output of the critic network, while for DHP, the output of the critic network is the derivative of $P(t)$ which is $P(t)/x(t)$. For GDHP, both $P(t)$ and the derivative of $P(t)$ are outputs of the network. Theoretically, GDHP will obtain the best characters. But GDHP needs to compute the second derivative $\partial^2 P(t)/\partial[x(t)]\partial\omega(t)$ every time step and the calculating burden will be seriously heavy. The AD version [5, 6] is just use the outputs of the action network as the partial inputs of the critic network, and correspondingly become ADHDP, ADDHP and so on.

However in ACDs, the model is not easy to obtain, especially when the complication of the nonlinear system is increasing. Some methods have been introduced to control nonlinear system without the model network, most of which are grounded on ADHDP [7, 8], and achieved some good results. However, since the output of ADHDP is $P(t)$, the sensitivity of system for time t is worse than ADDHP whose output is $P(t)/x(t)$ which is directly built on time.But unlike ADHDP, ADDHP still needs a model network in the propagating [2]. A new method based on ADDHP will be introduced in this paper, and the main difference against DHP and ADDHP [9] is that we avoid the model network which predicts the state of next time $t+1$ at time t. We just need to know the state t and the previous state at time $t-1$ to calculate $P(t)/x(t)$. Many lectures dealing with control problems using ACDs mentioned previously have just focused on algorithm improving and simulation results, while little algorithm analysis is discussed. In this paper, the adaptive critic method will be examined by convergence proof to enhance theoretic completeness. And finally, one application example is given to verify the new method proposed in this paper.

2 General Framework of the Method

In this paper, suppose the ultimate optimal performance index function is formulated by

$$\begin{aligned} V(t) &= U[x(t+1), u(t+1)] + \gamma U[x(t+2), u(t+2)] + \gamma^2 U[x(t+3), u(t+3)] + \cdots \\ &= \sum_{k=t+1}^{\infty} \gamma^{k-t-1} U[x(k), u(k)], \end{aligned} \tag{4}$$

where $0 < \gamma \leq 1$ is a discount factor, $U[x(k), u(k)]$ is a utility function or local cost, which is defined by user for the specific application. From (4), we can get

$$V(t) = \frac{1}{\gamma}(V(t-1) - U[x(t), u(t)]). \tag{5}$$

From (5) we can see that if we get $U[x(t), u(t)]$ at time t, and $V(t-1)$ at time $t-1$, the ultimate optimal cost will be obtained.

Our destination is to build the neural networks which defined the performance index function J to approximate the optimal function V. If the error between J and V is zero, then we announce the approximation is accurate. The structure of the neural network expressed as fig.1.

2.1 Action Network Training

The main objective of action network (action NN) is to generate a sequence control signal $u(t), u(t+1), u(t+2) \cdots$ to make the performance index be optimal. When the approximation of neural network is accurate, J function can be formulated as

$$J(t) = \sum_{k=t+1}^{\infty} \gamma^{k-t-1} U[x(k), u(k)], \tag{6}$$

and so, we can obtain the following iterative HJB function

$$J(t) = \frac{1}{\gamma}(J(t-1) - U[x(t), u(t)]). \tag{7}$$

For the action NN, the weights are updated by the following train rules

$$\omega_a(t+1) = \omega_a(t) + \Delta\omega_a(t), \tag{8}$$

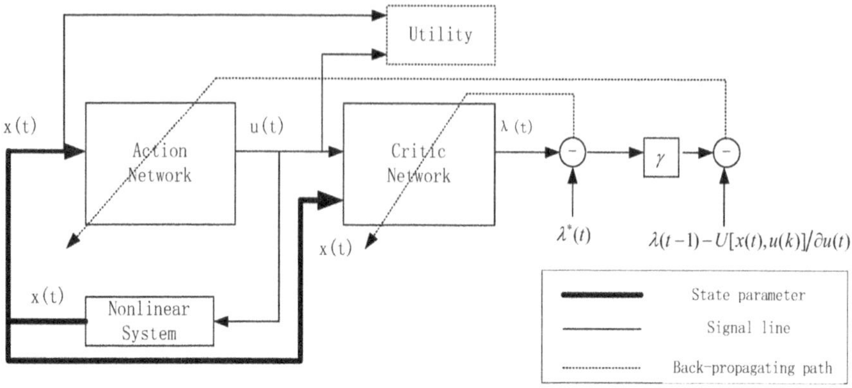

Fig. 1. Schematic diagram of main components of the method process

where ω_a denotes the weight of the action network.

$$\Delta\omega_a(t) = -\beta_1 \frac{\partial J(t)}{\partial \omega_a(t)} = -\beta_1 \sum_{k=1}^{m} \frac{\partial J(t)}{\partial u_k(t)} \cdot \frac{\partial u_k(t)}{\partial \omega_a(t)}, \tag{9}$$

where $0 < \beta_1 \leq 1$ is a given learning rate for the action network. And m is the dimension of the control signal $u(t)$. Expanding $\partial J(t)/\partial u_k(t)$ in the form

$$\frac{\partial J(t)}{\partial u_k(t)} = \frac{\partial(\frac{1}{\gamma}(J(t-1) - U[x(t), u(t)]))}{\partial u_k(t)} = \frac{1}{\gamma}(\frac{\partial J(t-1)}{\partial u_k(t)} - \frac{\partial U[x(t), u(t)]}{\partial u_k(t)}), \tag{10}$$

$$\frac{\partial(J(t-1))}{\partial u_k(t)} = \sum_{i=1}^{n} \frac{d(J(t-1))}{dx_i(t-1)} \cdot \frac{\partial x_i(t-1)}{\partial u_k(t)} = \sum_{i=1}^{n} \lambda_i(t-1) \frac{\partial x_i(t-1)}{\partial u_k(t)}, \tag{11}$$

where n is the dimension of the state vector. Here we define

$$\partial J(t)/\partial x_i(t) = \lambda_i(t), \tag{12}$$

where $\lambda_i(t)$ is approximated by the critic network in response to the state $x_i(t)$ at time t. Other terms above can be computed through the propagation path.

2.2 Critic Network Training

For the critic network (critic NN), whose output is $\lambda(t)$, the weight update needs an error between the real output and the "desired output" of the critic NN. So a value must be calculated as the "desired output", defined as λ^*. And the control vector $u(t)$ is also partial inputs for the critic NN. Using (10), we can write

$$\lambda_i^*(t) = \frac{\partial J(t)}{\partial x_i(t)} = \frac{\partial(\frac{1}{\gamma}(J(t-1) - U[x(t), u(t)]))}{\partial x_i(t)} = \frac{1}{\gamma}(\frac{\partial J(t-1)}{\partial x_i(t)} - \frac{\partial U[x(t), u(t)]}{\partial x_i(t)}). \tag{13}$$

Expanding (13) we get

$$\lambda_i^*(t) = \frac{1}{\gamma}\{\sum_{k=1}^{n}(\lambda_k(t-1)\cdot\frac{\partial x_k(t-1)}{\partial x_i(t)}) + \sum_{k=1}^{n}(\sum_{j=1}^{m}(\lambda_k(t-1)\cdot\frac{\partial x_k(t-1)}{\partial u_j(t)}\cdot\frac{\partial u_j(t)}{\partial x_i(t)}))$$
$$= -\frac{\partial U[x(t),u(t)]}{\partial x_i(t)} - \sum_{j=1}^{m}(\frac{\partial U[x(t),u(t)]}{\partial u_j(t)}\cdot\frac{\partial u_j(t)}{\partial x_i(t)})\}. \tag{14}$$

The partial derivative team $\partial u_j(t)/\partial x_i(t)$ is computed by back-propagation through the action NN and $\lambda_k(t-1)$ can be obtained at time $t-1$. Then training critic network, the error team is calculated in the form

$$e_c(t) = \lambda^*(t) - \lambda(t), \tag{15}$$

$$E(t) = \frac{1}{2}e_c^2(t). \tag{16}$$

The critic network weights updating rule can be expressed as

$$\omega_c(t+1) = \omega_c(t) + \Delta\omega_c(t), \tag{17}$$

where ω_c denotes the weight of the critic NN

$$\Delta\omega_c(t) = -\beta_2\frac{\partial E(t)}{\partial\omega_c(t)} = -\beta_2\sum_{i=1}^{n}\frac{\partial E_i(t)}{\partial\omega_c(t)}, \tag{18}$$

where $0 < \beta_2 \leq 1$ is also a given learning rate and can be computed like action NN. From (18), we can see that for obtaining $\lambda_i(t)$ at time t, we just need to conserve all the $\lambda_k(t-1)$, $k = 1, 2, ..., n$ instead of computing or searching all the $\lambda_k(t+1)$ at next time step, and the model NN which predicts the state for the next time step is omitted. For the control of nonlinear system, especially for the on-line control, the proposed method in this paper will emerge superior performance.

Remark. Differing from HDP and ADHDP based methods, the derivative of performance index function $\lambda(t)$ and $\lambda^*(t)$ in this paper is a vector instead of a scalar quantity. The dimension of performance index function equals to the dimension of the state vector, and so the structure of the critic NN should be a MIMO neural network, unlike the HDP and ADHDP which output is just a scalar, as the critic NN is just structured of MISO neural network.

3 The Analysis of Convergence About the Method

In the above section, we expatiate how the neural networks are trained using the improved ADDHP method. As the objective of the method is to use the performance function $J(t)$ to approximate the ultimate optimal function, the convergence of $J(t)$ in the method is most important performance. In this section, the analysis of convergence will be discussed.

Theorem. For the discrete-time nonlinear system (1), when the utility function is defined as the quadratic form

$$U[x(t),u(t)] = q[x(t)] + u^T(t)r[x(t)]u(t), \tag{19}$$

where $q[x(t)] \geq 0$ is a matrix function and $q[x(t)] = 0$ if and only if $x(t) = 0$, $r[x(t)]$ is a positive defined matrix function for all the $t \geq 0$, then performance function $J(t)$ is convergent.

Proof. From (9) and (19), the performance index function can be expressed as

$$J(t) = \sum_{k=t+1}^{\infty} \gamma^{k-t-1} U[x(k),u(k)] = \sum_{k=t+1}^{\infty} \gamma^{k-t-1}(q[x(t)] + u^T(t)r[x(t)]u(t)). \tag{20}$$

The difference form of the HJB function can be expressed as

$$DJ(t) = \gamma J(t+1) - J(t) = -U[x(t+1),u(t+1)], \tag{21}$$

$$\begin{aligned} DJ(t) &= \frac{DJ(t)}{Dx(t+1)} \cdot f[x(t+1),u(t+1)] = -U[x(t+1),u(t+1)] \\ &= -q[x(t+1)] - u^T(t+1)r[x(t+1)]u(t+1), \end{aligned} \tag{22}$$

where

$$f[x(t+1),u(t+1)] = Dx(t+1) = \gamma x(t+2) - x(t+1), \tag{23}$$

$$\begin{aligned} f[x(t+1),u(t+1)] &= \gamma a[x(t+1)] + \gamma b[x(t+1)]u(t+1) - x(t+1) \\ &= \bar{a}[x(t+1)] + \gamma b[x(t+1)]u(t+1), \end{aligned} \tag{24}$$

where $\bar{a}[x(t+1)] = \gamma a[x(t+1)] - x(t+1)$. So(22) becomes

$$\begin{aligned} &\frac{DJ(t)}{Dx(t+1)} \cdot \{\bar{a}[x(t+1)] + \gamma b[x(t+1)]u(t+1)\} \\ &= -q[x(t+1)] - u^T(t+1)r[x(t+1)]u(t+1). \end{aligned} \tag{25}$$

Differentiating the equation (25) by $u(t+1)$, yields

$$\frac{DJ(t)}{Dx(t+1)} b[x(t+1)] = -\frac{2}{\gamma} u^T(t+1)r[x(t+1)]. \tag{26}$$

Then $u(t+1)$ can be expressed as fowling formulation

$$u(t+1) = k[x(t+1)] = -\frac{\gamma}{2} r^{-1}[x(t+1)]b^T[x(t+1)][\frac{DJ(t)}{Dx(t+1)}]^T. \tag{27}$$

Substitute (27) to (24), then let

$$\begin{aligned} F[x(t+1)] &= f(x(t+1),k[x(t+1)]) \\ &= \bar{a}[x(t+1)] - \frac{\gamma}{2} b[x(t+1)]r^{-1}[x(t+1)]b^T[x(t+1)][\frac{DJ(t)}{Dx(t+1)}]^T. \end{aligned} \tag{28}$$

As such, (22) can be expressed as

$$\begin{aligned} DJ(t) &= \frac{DJ(t)}{Dx(t+1)} \cdot f(x(t+1), k[x(t+1)]) = \frac{DJ(t)}{Dx(t+1)} \cdot F[x(t+1)] \\ &= -U[x(t+1), u(t+1)]. \end{aligned} \tag{29}$$

Let $x = (x_1, x_2, ..., x_j, ...)$ and every x_j has the same dimension n, defining

$$x_{j+1}(t) = x_j(t+1) = a[x_j(t)] + b[x_j(t)]u_j(t). \tag{30}$$

For $j = 1, 2, ...$ and $i = 1, 2, ...n$, we can write

$$DJ_j(t) = \sum_{i=1}^{n} \frac{DJ_j(t)}{Dx_i(t+1)} \cdot F_j[x_i(t+1)], \tag{31}$$

where

$$\begin{aligned} &F_j[x_i(t+1)] = f(x_i(t+1), k_j[x_i(t+1)]) = \\ &a[x_i(t+1)] - \tfrac{\gamma}{2} b[x_i(t+1)] r^{-1}[x_i(t+1)] b^T[x_i(t+1)] [\frac{DJ_j(t)}{Dx_i(t+1)}]^T, \end{aligned} \tag{32}$$

and we define

$$\lambda_{ji}(t) = DJ_j(t)/Dx_i(t+1). \tag{33}$$

So (31) becomes

$$\begin{aligned} DJ_j(t) = &\sum_{i=1}^{n} \{\lambda_{ji}(t+1)\bar{a}[x_i(t+1)] - \\ &\frac{\gamma}{2}\lambda_{ji}(t+1) b[x_i(t+1)] r^{-1}[x_i(t+1)] b^T[x_i(t+1)] \lambda_{ji}^T(t+1)\}. \end{aligned} \tag{34}$$

From (32) we may substituting the following deducing equation

$$\begin{aligned} F_{j+1}[x_i(t)] &= f(x_i(t), k_{j+1}[x_i(t)]) \\ &= \bar{a}[x_i(t)] - \tfrac{\gamma}{2} b[x_i(t)] r^{-1}[x_i(t)] b^T[x_i(t)] [\tfrac{DJ_{j+1}(t)}{Dx_i(t)}]^T, \end{aligned} \tag{35}$$

into (29), then we have

$$-\frac{1}{\gamma}U(x_i(t), k_{j+1}[x_i(t)]) = \sum_{i=1}^{n} \{\lambda_{ji}(t)\bar{a}[x_i(t)] - \frac{\gamma}{2}\lambda_{ji} b[x_i(t)] r^{-1}[x_i(t)] b^T[x_i(t)] \lambda_{(j+1)i}^T(t)\}. \tag{36}$$

where $\lambda_{(j+1)i}(t) = dJ_{j+1}(t)/dx_i(t)$. As we can see (34) can be replaced by the same term $\lambda_{ji}(t)\bar{a}[x_i(t)]$ in (36), then we have

$$\begin{aligned} DJ_j(t) = &\sum_{i=1}^{n} \{\frac{\gamma}{2}\lambda_{ji}(t) b[x_i(t)] r^{-1}[x_i(t)] b^T[x_i(t)] \lambda_{(j+1)i}^T(t) \\ &- \frac{\gamma}{2}\lambda_{ji}(t) b[x_i(t)] r^{-1}[x_i(t)] b^T[x_i(t)] \lambda_{ji}^T(t)\} - \frac{1}{\gamma}U(x_i(t), k_{j+1}[x_i(t)]). \end{aligned} \tag{37}$$

The expression of $U(x_i(t), k_{j+1}[x_i(t)])$ can be derived from (22) and (27), we have

$$U(x_i(t), k_{j+1}[x_i(t)]) = \sum_{i=1}^{n} \{q[x_i(t)] + \frac{\gamma^2}{4}\lambda_{(j+1)i}(t)b[x_i(t)]r^{-1}[x_i(t)]b^T[x_i(t)]\lambda_{(j+1)i}^T(t)\}. \tag{38}$$

Take (38) into (37), then obtain

$$\begin{aligned} DJ_j(t) &= \sum_{i=1}^{n} \{\frac{\gamma}{2}\lambda_{ji}(t)b[x_i(t)]r^{-1}[x_i(t)]b^T[x_i(t)]\lambda_{(j+1)i}^T(t) \\ &= -\frac{\gamma}{2}\lambda_{ji}(t)b[x_i(t)]r^{-1}[x_i(t)]b^T[x_i(t)]\lambda_{ji}^T(t) \\ &- \frac{\gamma}{4}\lambda_{(j+1)i}(t)b(x_i(t))r^{-1}(x_i(t))b^T(x_i(t))\lambda_{(j+1)i}^T(t) - \frac{1}{\gamma}q[x_i(t)]\}. \end{aligned} \tag{39}$$

Finally, clear up (40) and we can obtain

$$\begin{aligned} DJ_j(t) &= \sum_{i=1}^{n} \{-\frac{1}{\gamma}q[x_i(t)] - \frac{\gamma}{4}\lambda_{ji}(t)b[x_i(t)]r^{-1}[x_i(t)]b^T[x_i(t)]\lambda_{ji}^T(t) \\ &= -\frac{\gamma}{4}[\lambda_{(j+1)i}(t) - \lambda_{ji}(t)]b[x_i(t)]r^{-1}[x_i(t)]b^T[x_i(t)][\lambda_{(j+1)i}(t) - \lambda_{ji}(t)]^T\}. \end{aligned} \tag{40}$$

As such, $DJ_j(t) < 0$ for all the $x(t) \neq 0$, is a negative defined matrix function, while $J_j(t) > 0$, and then proves $J_j(t)$ is a convergence performance index function (end proof).

4 Example Application

In this section, an example is used to illustrate the effectiveness of the proposed method for discrete nonlinear system. Suppose the nonlinear plant is formulated as

$$\begin{aligned} x_1(t+1) &= 0.1(x_1(t) + x_2(t)) + 0.05x_2^2(t) + 0.1x_1(t)u(t), \\ x_2(t+1) &= 0.2x_1(t) - 0.15x_2(t) + 0.3x_2(t)u(t). \end{aligned} \tag{41}$$

In neural network implementation, two training plans can be adopted. One is training one network while the other's weight is unchanged. The other plan is that both the action and critic NN are trained simultaneously. In this paper, we adopt the latter. The structure of both action and critic NN are three layers and the hidden layer are composed of 8 neurons. The input of action NN contains 2 neurons while the input of the critic NN is 3, including the analog action signal $u(t)$ from the action network, and the output layer of action NN has one neuron, while critic NN are 2. Set initial state to be random state during [0, 1], and the same set for all the weights of both action and critic network. Then the training strategy can be formulated as following:

(1) Apply $x(t)$ to action NN, and obtain $u(t)$.
(2) Apply $u(t)$ and $x(t)$ to critic NN, then obtain $\lambda(t)$.

(3) Calculate desired $\lambda^*(t)$ for critic NN and obtain the error.
(4) Calculate weights changes for critic NN.
(5) Calculate weights changes for action NN.
(6) Increment t and go to (1).

In the training propagating path, for the hidden layers of both action and critic network, we adopt standard log-sigmoid transformation function. The simulation results are shown in the following figure.

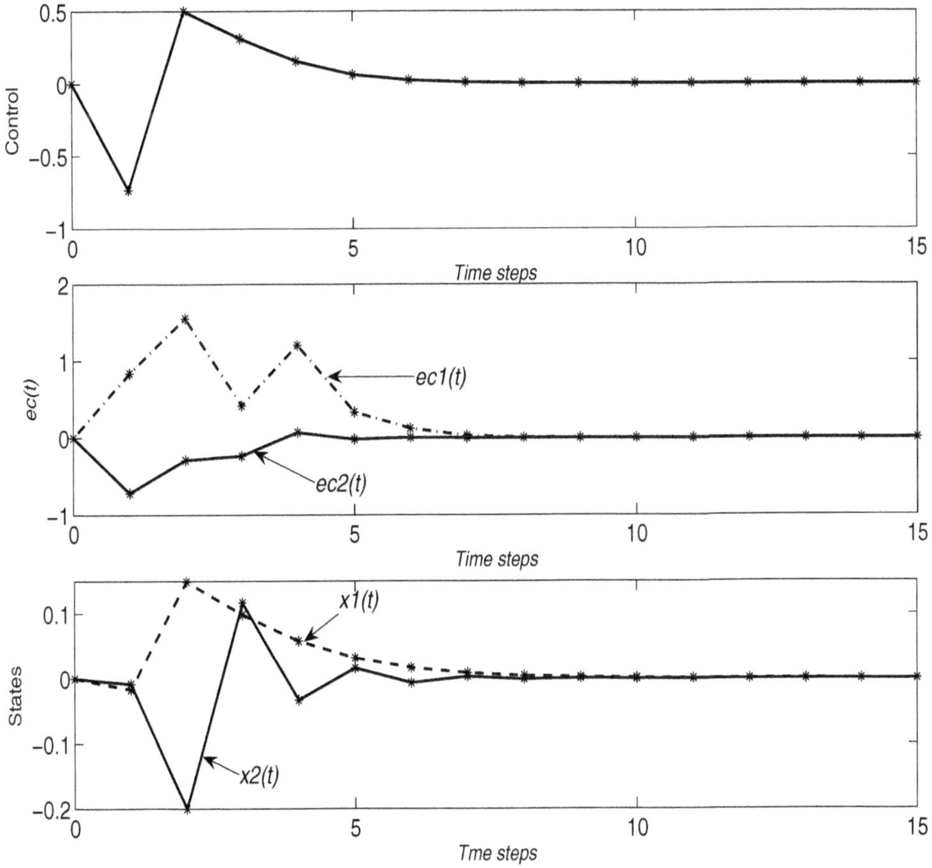

Fig. 2. Typical trajectories of control, critic error $e_c(t)$ and states respectively

5 Conclusions

This paper focuses on providing a systematic treatment of an improved DHP method, from architecture to the algorithm, convergence proof, and finally given an example simulation result. From the theory to the example application, our design presented in this paper has several principal advanced aspects from the existing results. First, the proposed configuration is simpler than the traditional DHP or ADDHP. The improved method avoid predicting the state of next time step, instead in this method, we use present and previous time step to calculate the derivative of the performance function. And so the model network is omitted. Second, this method is directly built on the derivative of the performance index which can behave more sensitivity than HDP or ADHDP. Furthermore, a proof is given out to guarantee the convergence in theory, and simulation result also proves the correctness and effectiveness. Above all, the new on-line adaptive critic design for discrete nonlinear systems proposed in this paper has more advanced practice value.

Acknowledgements

This work was supported in part by the National Nature Science Foundation of China (60325311, 60534010, 60572070), the Funds for Creative Research Groups of China (Grant No. 60521003) and the Program for Changjiang Scholars and Innovative Research Team in University (Grant No. IRT0421).

References

1. Seong, C.Y., Bermard, W.: Neural Dynamic Optimization for Control Systems-Part I: Background. IEEE Transactions on Systems, Man, and Cybernetics-Part B: Cybernetics **31** (4) (2001) 482-489
2. Danil, P., Don, W.: Adaptive Critic Designs. IEEE Transactions on Neural Networks **8** (5) (1997) 997-1007
3. John, J.M., Chadwick, J.C., George, G.L., Richard, S.: Adaptive Dynamic Programming. IEEE Transactions on Systems, Man, and Cybernetics-Part C: Applications and Reviews **32** (2) (2002) 140-153
4. Zhang, H.G., Luo, Y.H., Liu, D.R.: A New Fuzzy Identification Method Based on Adaptive Critic Designs. Lecture notes in Computer Science **3971** (2006) 804-809
5. Liu, D.R., Xiong, X.X., Zhang, Y.: Action-Dependent Adaptive Critic Designs. IEEE Neural Networks Proceedings (2001) 990-995
6. Liu, D.R., Zhang, H.G.: A Neural Dynamic Programming Approach for Learning Control of Failure Avoidance Problems. International Journal of Intelligence Control and Systems **10** (1) (2005) 21-32
7. Liu, D.R., Zhang. Y., Zhang, H.G.: A Self-learning Call Admission Control Scheme for CDMA Cellular Networks. IEEE Transactions on Neural Network **16** (5) (2006) 804-809
8. Jennie, S., Wang, Y.T.: On-Line Learning Control by Association and Reinforcement. IEEE Transactions on Neural Networks **12** (2) (2001) 264-276
9. George, G. L., Christian, P.: Training Strategies for Critic and Action Neural Networks in Dual Heuristic Programming Method. IEEE Neural Networks International Conference **2** (1997) 712-717

Reinforcement Learning Reward Functions for Unsupervised Learning

Colin Fyfe and Pei Ling Lai

Southern Taiwan Institute of Technology,
Tainan, Taiwan
pei_ling_lai@hotmail.com.

Abstract. We extend a reinforcement learning algorithm, REINFORCE [13] which has previously been used to cluster data [10]. By using base Gaussian learners, we extend the method so that it can perform a variety of unsupervised learning tasks such as principal component analysis, exploratory projection pursuit and canonical correlation analysis.

1 Introduction

There has been a great deal of recent interest in exploratory data analysis mainly because we are automatically acquiring so much data from which we are extracting little information. Such data is typically high-dimensional and high volume, both of which features cause substantial problems. We consider the reinforcement learning paradigm interesting as a potential tool for exploratory data analysis: the exploitation-exploration trade-off is exactly what is required in such situations: it precisely matches what a human would do to explore a new data set - investigate, look for patterns *of any identifiable type* and follow partial patterns till they become as clear as possible.

Many of such methods try to project the data to lower dimensional manifolds; an alternative is to find clusters in a data set. In this paper, we investigate the REINFORCE [13] algorithm which has previously been applied to (unsupervised) clustering of data [10]. By changing the base learner from a Bernoulli unit to a Gaussian unit, we show how unsupervised projection methods may be implemented. We chose these methods since we have experience implementing them with artificial neural networks e.g. for PCA see [3], for EPP see [4], for CCA see [8]. We do not report results on real data sets in this paper since the main thrust of the paper is exposition of new methods which is most clearly seen when we have artificial data on which the correct answer is evident.

2 Immediate Reward Reinforcement Learning

[13] investigated a particular form of reinforcement learning in which reward for an action is immediate which is somewhat different from mainstream reinforcement learning [12,7]. Williams [13] considered a stochastic learning unit in which

D. Liu et al. (Eds.): ISNN 2007, Part I, LNCS 4491, pp. 397–402, 2007.

the probability of any specific output was a parameterized function of its input, $\mathbf{x}$. For the ith unit, this gives

$$P(y_i = \zeta|\mathbf{w}_i, \mathbf{x}) = f(\mathbf{w}_i, \mathbf{x}) \tag{1}$$

where, for example,

$$f(\mathbf{w}_i, \mathbf{x}) = \frac{1}{1 + \exp(- \parallel \mathbf{w}_i - \mathbf{x} \parallel^2)} \tag{2}$$

Williams [13] considers the learning rule

$$\Delta w_{ij} = \alpha_{ij}(r_{i,\zeta} - b_{ij})\frac{\partial \ln P(y_i = \zeta|\mathbf{w}_i, \mathbf{x})}{\partial w_{ij}} \tag{3}$$

where α_{ij} is the learning rate, $r_{i,\zeta}$ is the reward for the unit outputting ζ and b_{ij} is a reinforcement baseline which in the following we will take as the reinforcement comparison, $b_{ij} = \bar{r} = \frac{1}{K}\sum r_{i,\zeta}$ where K is the number of times this unit has output ζ. ([13], Theorem 1) shows that the above learning rule causes weight changes which maximizes the expected reward.

[13] gave the example of a Bernoulli unit in which $P(y_i = 1) = p_i$ and so $P(y_i = 0) = 1 - p_i$. Therefore

$$\frac{\partial \ln P(y_i)}{\partial p_i} = \begin{cases} -\frac{1}{1-p_i} & \text{if } y_i = 0 \\ \frac{1}{p_i} & \text{if } y_i = 1 \end{cases} = \frac{y_i - p_i}{p_i(1 - p_i)} \tag{4}$$

[10] applies the Bernoulli model to clustering with

$$p_i = 2(1 - f(\mathbf{w}_i, \mathbf{x})) = 2(1 - \frac{1}{1 + \exp(- \parallel \mathbf{w}_i - \mathbf{x} \parallel^2)}) \tag{5}$$

The environment identifies the p_{i^*} which is maximum over all output units and y_{i^*} is then drawn from this distribution. Rewards are given such that

$$r_i = \begin{cases} 1 \ \text{ if } i = i^* \text{ and } y_i = 1 \\ -1 \text{ if } i = i^* \text{ and } y_i = 0 \\ 0 \qquad\quad \text{if } i \neq i^* \end{cases} \tag{6}$$

This is used in the update rule

$$\Delta w_{ij} = \alpha r_i(y_i - p_i)(x_j - w_{ij}) \tag{7}$$

$$= \alpha|y_i - p_i|(x_j - w_{ij}) \text{ for } i = i^* \tag{8}$$

which is shown to perform clustering of the data set.

3 Principal Component Analysis

Principal component analysis (PCA)[11] is a standard technique for multivariate data analysis; indeed, it is often the first technique which a data analyst will use on a new data set. It finds the linear projection of a data set which contains maximal variance; alternatively it may be described as the linear projection which minimizes the mean squared error between the original data points and their projections. Of course we require to make this a constrained optimization (otherwise the weights could just increase without bound, thereby increasing the variance) and so we constrain $WW^T = I$, where W is the weight matrix of the projection and I is the identity matrix. We will show how to use the REINFORCE algorithm to perform various forms of linear multivariate analysis; we begin with PCA and examine only the first principal component - the extension to further principal components is obvious.

One advantage that the REINFORCE representation has is that it is very simple to change the base learner. One type of learner which is appropriate is the Gaussian. Thus we are implicitly saying that $\mathbf{w} \sim N(\mathbf{m}, \beta^2 I)$, the Gaussian distribution with mean $\mathbf{m}$ and variance β^2. The learner now has two parameters to learn, its mean and variance. The learning rules can be derived [13] as

$$\Delta \mathbf{m} = \alpha_m (r - \overline{r}) \frac{\mathbf{w} - \mathbf{m}}{\beta^2} \tag{9}$$

$$\Delta \beta = \alpha_\beta (r - \overline{r}) \frac{(\mathbf{w} - \mathbf{m})^2 - \beta^2}{\beta^3} \tag{10}$$

To perform PCA, we wish to maximize variance and so we can take the sample weights, $\mathbf{w}$, drawn from $N(\mathbf{m}, \beta^2 I)$ and multiply by the data sample at each instant in time. Since we wish to maximize $E(\mathbf{w}^T(\mathbf{x} - \overline{\mathbf{x}})(\mathbf{x} - \overline{\mathbf{x}})^T \mathbf{w})$ where $\mathbf{x}$ is the current sample data, we can use $r = (\mathbf{w}^T \mathbf{x})^2$ for centered data. This is certainly effective with artificial data however we are concerned that this may be unstable with a real data set and so in the experiments we report on the reward function, $r = \frac{1}{1+\exp(-\gamma(\mathbf{w}^T\mathbf{x})^2)}$ which is still greatest when $\mathbf{w}^T\mathbf{x}$ is largest but is bounded.

To illustrate this reward function, we create an artificial 5 dimensional data set of 1000 samples such that each element of $\mathbf{x}$ is drawn independently from a Gaussian distribution with $x_i \sim N(0, i)$ i.e. greatest variance is in x_5 and least in x_1. We show final estimates of $\mathbf{m}$ and β^2 in Table 1. We see that the first principal component, that with greatest variance, is clearly identified.

We may also perform a Minor Component Analysis (MCA) by using a reward function of the form $r = \exp(-(\mathbf{w}^T\mathbf{x})^2)$. Using the same type of data as before, we get results such as shown in the bottom line of Table 1. Again the principal

Table 1. Top: The weights and residual standard deviation from the artificial data experiment with the PCA reward function. Bottom: the weights and residual standard deviation when using the MCA reward function.

mean	weights				std. dev.
0.0545619	-0.0123173	-0.0469857	0.0383933	**0.996589**	4.34423
0.997412	-0.0693583	0.0176544	-0.00459576	0.0051755	0.305068

component with smallest variance is clearly identified. Note that, for both PCA and MCA, we specifically perform a normalization so that $\| \mathbf{w} \|_2 = 1$.

In the above, we have constrained the learner to be an isotropic Gaussian distribution. Refinements would be to use $\mathbf{w} \sim N(\mathbf{m}, D)$ where D is a diagonal matrix or even $\mathbf{w} \sim N(\mathbf{m}, \Sigma)$ where Σ is a full covariance matrix.

4 Exploratory Projection Pursuit

Exploratory Projection Pursuit (EPP) [2,1,6] is another linear projection technique which, instead of maximizing variance, maximizes some index which defines how "interesting" the projection is: most projections of a high dimensional data set tend to be Gaussian and somewhat uninteresting, therefore measures of interestingness usually are quantified as to how different the distribution is from a Gaussian distribution. The simplest such functions would be the third moment (measuring skewness) and the fourth moment (measuring kurtosis). Generally the data is first of all sphered (or whitened) to remove any differences in the first or second moments of the data stream.

Table 2. The weights and residual standard deviation from two runs of the artificial data experiment. Top row used $r = \frac{1}{1+\exp(-\gamma(\mathbf{w}^T\mathbf{x})^4)}$. Bottom used $r = \gamma(\mathbf{w}^T\mathbf{x})^4$.

mean	weights				variance
0.995115	-0.0367719	0.0132427	0.0613801	0.0667128	1.93178
-0.961274	0.136552	0.201865	-0.128263	0.0102482	1.81945

To illustrate this method, we create 1000 samples of a 5 dimensional data set in which 4 elements of each vector are drawn iid from N(0,1) while the fifth is given positive kurtosis (Gaussian has kurtosis=0) but still mean=0 and variance=1. Table 2 shows the outcome of two simulations in both of which the kurtotic distribution, corresponding to the first element of the vector, was >3. We see the kurtotic distribution is easily identified in both cases but the second is much less accurate than the first. The former simulation used $r = \frac{1}{1+\exp(-\gamma(\mathbf{w}^T\mathbf{x})^4)}$ while the latter used $r = \gamma(\mathbf{w}^T\mathbf{x})^4$ as the reward function. This accords with our previous experience of EPP [4], that stability and accuracy of convergence are very simulation-dependent and multiple runs of any experiment particularly with different indices of interestingness may lead to very different results even though the indices may be attempting to measure the same statistic.

5 Canonical Correlation Analysis

Canonical Correlation Analysis is used when we have two data sets which we believe have some underlying correlation. Consider two sets of input data, $\mathbf{x}_1 \in X_1$ and $\mathbf{x}_2 \in X_2$. Then in classical CCA, we attempt to find the linear combination of the variables which gives us maximum correlation between the combinations. Let $y_1 = \mathbf{w}_1^T \mathbf{x}_1$ and $y_2 = \mathbf{w}_2^T \mathbf{x}_2$. Then, for the first canonical correlation, we find those values of $\mathbf{w}_1$ and $\mathbf{w}_2$ which maximizes $E(y_1 y_2)$ under the constraint that $E(y_1^2) = E(y_2^2) = 1$. Therefore, for each of our two data streams, we draw samples $\mathbf{w}_i$ from $N(\mathbf{m}_i, \beta_i^2), i = 1, 2$. We update the parameters using

$$\Delta \mathbf{m}_i = \alpha_{m_i}(r_i - \overline{r_i}) \frac{\mathbf{w}_i - \mathbf{m}_i}{\beta_i^2} \tag{11}$$

$$\Delta \beta_i = \alpha_{\beta_i}(r_i - \overline{r_i}) \frac{(\mathbf{w}_i - \mathbf{m}_i)^2 - \beta_i^2}{\beta_i^3} \tag{12}$$

where $r_i = \exp(-\gamma \parallel \mathbf{w}_1^T \mathbf{x}_1 - \mathbf{w}_2^T \mathbf{x}_2 \parallel^2)$, i.e. the closer the projections of the two data points are to each other, the greater the reward,. We note that in this case $r_1 = r_2$ i.e. the rewards to each Gaussian learner are equal (though the actual updates are dependent on the parameters of the individual learner). Clearly we could insert an additional parameter with $\gamma_1 \neq \gamma_2$.

Note the effects of the learning rules on the Gaussian parameters. If a value $\mathbf{w}_i$ is chosen which leads to a better reward than has typically been achieved in the past, the change in the mean is towards $\mathbf{w}_i$; if the reward is less than the previous average, change is away from $\mathbf{w}_i$. Also, if the reward is greater than the previous average, the variance will decrease if $\parallel \mathbf{w}_i - \mathbf{m}_i \parallel^2 < \beta_i^2$ i.e. narrowing the search while it will increase if $\parallel \mathbf{w}_i - \mathbf{m}_i \parallel^2 > \beta_i^2$, thus widening the search volume. We have found that it is best to update the variances more slowly than the means and again we normalise after each weight update.

Table 3. The weights found for the artificial data and (last column) the estimated standard deviation in the weight vector

	weights	std. dev.
$\mathbf{m}_1$	**0.997125** 0.0427865 0.0543842 0.0308709	0.38558
$\mathbf{m}_2$	**0.995058** 0.0432388 0.089392	0.372447

We illustrate this method on a simulation with artificial data: we create two sets of data $\mathbf{x}_{1,i}, \mathbf{x}_{2,i}, i = 1, ..., 1000$ i.e. 1000 samples of $\mathbf{x}_1$ and 1000 samples of $\mathbf{x}_2$ where $\mathbf{x}_1$ is 4 dimensional data with each element randomly drawn from a uniform distribution in [0,1] and $\mathbf{x}_2$ is 3 dimensional also with each element randomly drawn from a uniform distribution in [0,1]. We create a correlation between the two data streams by drawing a further random sample and adding it to the first element of each vector for each $i \in \{1, ..., 1000\}$ and dividing the result by $\sqrt{2}$ to normalize the variances. Results are shown in Table 3: we see that the first element in each data set is very clearly identified.

6 Conclusion

In this paper, we first reviewed the use which [10] had made of the REINFORCE algorithm to perform clustering. By adopting a different type of underlying model, a Gaussian unit rather than a Bernoulli unit, and using a variety of appropriate reward functions, we were able to have the REINFORCE algorithm find optimal linear projections in terms of

- maximizing the variance of the projections which is equivalent to minimizing the mean square error between the data and its projections.
- maximizing some index of interestingness in the data so as to present a user with a projection in which he/she can search for structure by eye.
- maximizing the correlation between two twinned data streams simultaneously.

Given the early success of the method, we are encouraged to investigate more mappings with the method. We will investigate both linear manifold methods such as Independent Component Analysis [9] as well as nonlinear manifolds [5]. We will also investigate stability with respect to the Gaussian learner and further investigate combinations of unsupervised with reinforcement learning.

References

1. Friedman, J.H.: Exploratory Projection Pursuit. Journal of the American Statistical Association **82** (397)(1987) 249–266
2. Friedman, J.H., Tukey, J.W.: A Projection Pursuit Algorithm for Exploratory Data Analysis. IEEE Transactions on Computers, c-**23** (9) Sept. (1974) 881–889
3. Fyfe, C.: Introducing Asymmetry into Interneuron Learning. Neural Computation **7** (6) (1995) 1167–1181
4. Fyfe, C.: A Comparative Study of Two Neural Methods of Exploratory Projection Pursuit. Neural Networks **10** (2) (1997) 257–262
5. Fyfe, C.: Two topographic maps for data visualization. Data Mining and Knowledge Discovery, 2006.
6. Jones, M.C., Sibson, R.: What Is Projection Pursuit. Journal of The Royal Statistical Society (1987) 1–37
7. Kaelbling, L.P., Littman, M.L., Moore, A.W.: Reinforcement Learning: A survey. Journal of Artificial Intelligence Research **4** (1996) 237–285
8. Lai, P.L., Fyfe, C.: A Neural Network Implementation of Canonical Correlation Analysis. Neural Networks **12** (10) Dec. (1999) 1391–1397
9. Lai, P.L., Fyfe, C.: Kernel and Nonlinear Canonical Correlation Analysis. International Journal of Neural Systems **10** (5) (2001) 365–377
10. Likas, A.: A Reinforcement Learning Approach to On-Line Clustering. Neural Computation 2000
11. Mardia, K.V., Kent, J.T., Bibby, J.M.: Multivariate Analysis. Academic Press, 1979.
12. Sutton, R.S., Barto, A.G.: Reinforcement Learning: an Introduction. MIT Press, 1998.
13. Williams, R.: Simple Statistical Gradient-Following Algorithms for Connectionist Reinforcement Learning. Machine Learning **8** (1992) 229–256

A Hierarchical Learning System Incorporating with Supervised, Unsupervised and Reinforcement Learning

Jinglu Hu[1], Takafumi Sasakawa[1], Kotaro Hirasawa[1], and Huiru Zheng[2]

[1] Waseda University, Kitakyushu, Fukuoka, Japan
[2] University of Ulster, N. Ireland, UK
jinglu@waseda.jp, sasa@moegi.waseda.jp,
hirasawa@waseda.jp, h.zheng@ulster.ac.uk

Abstract. According to Hebb's *Cell assembly theory*, the brain has the capability of function localization. On the other hand, it is suggested that the brain has three different learning paradigms: supervised, unsupervised and reinforcement learning. Inspired by the above knowledge of brain, we present a hierarchical learning system consisting of three parts: supervised learning (SL) part, unsupervised learning (UL) part and reinforcement learning (RL) part. The SL part is a main part learning input-output mapping; the UL part realizes the function localization of learning system by controlling firing strength of neurons in SL part based on input patterns; the RL part optimizes system performance by adjusting parameters in UL part. Simulation results confirm the effectiveness of the proposed hierarchical learning system.

1 Introduction

In the book of *Organization of Behavior* [1], D.O. Hebb proposed two radically new theories about how the brain worked. The first idea is later known as *Hebbian learning*; the second one is known as *Cell assemblies*. Figure 1(a) shows an image of cell assemblies. Neurons would form many groups thanks to Hebbian learning. Specific groups of neurons would be activated corresponding to specific sensory information the brain receives [2]. Additionally, the formed groups would have neuron overlapped between the other groups. That is, neurons would be mutually connected "functionally" rather than "structurally", and the connections would vary appropriately according to the sensory information. If we consider the formed loops as modules, the brain may be seen to consist of many overlapping modules.

On the other hand, it is recently suggested that the three parts of brain: cerebellum, cerebral cortex and basal ganglia are specialized, respectively, in supervised, unsupervised and reinforcement learning, see Fig.1(b) [3]. Our brain is a highly complicated structure and has many capabilities which are not entirely clear. The motivation of this research is intended to introduce a brain-like neural network that has capabilities of function localization as well as learning, by

D. Liu et al. (Eds.): ISNN 2007, Part I, LNCS 4491, pp. 403–412, 2007.

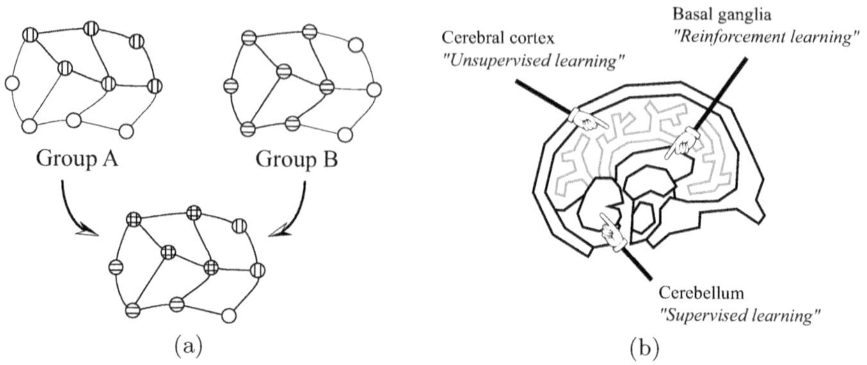

Fig. 1. The knowledge of brain: (a) Cell assemblies; (b) Three learning paradigms

developing a hierarchical learning system incorporating supervised, unsupervised and reinforcement learning.

The proposed hierarchical learning system consists of three parts: supervised learning (SL) part, unsupervised learning (UL) part and reinforcement learning (RL) part. The SL part is a main part learning input-output mapping; Structurally, it is the same as an ordinary 3-layer feedforward neural network, but each neuron in its hidden layer contains a signal from the UL part, controlling its firing strength. The UL part is a competitive learning network whose outputs associate with the hidden neurons in the SL part one by one; It divides the input space to subspaces by unsupervised learning and control the firing strength of neurons in SL part according to input patterns; In this way, the learning system realizes function localization. The RL part is a reinforcement learning algorithm that adjusts parameters of UL part to optimize the whole system performance automatically. The brain-like learning system constructed in this way not only has capability of function localization, but also can optimize its performance automatically. Simulation results confirm the effectiveness of the proposed learning system and show that it has superior performance to an ordinary neural network.

2 Structure of the Learning System

The hierarchical learning system is a function localization neural network (FLNN) inspired from Hebb's *Cell assemblies* theory [1,4]. Figure 2 shows the structure of learning system; It consists of three parts: SL part, UL part and RL part.

2.1 The SL Part

The SL part is a main part of the learning system. Structurally, it is the same as an ordinary 3-layer feedforward neural network; but each of its hidden neurons contains a signal from the UL part, controlling its firing strength.

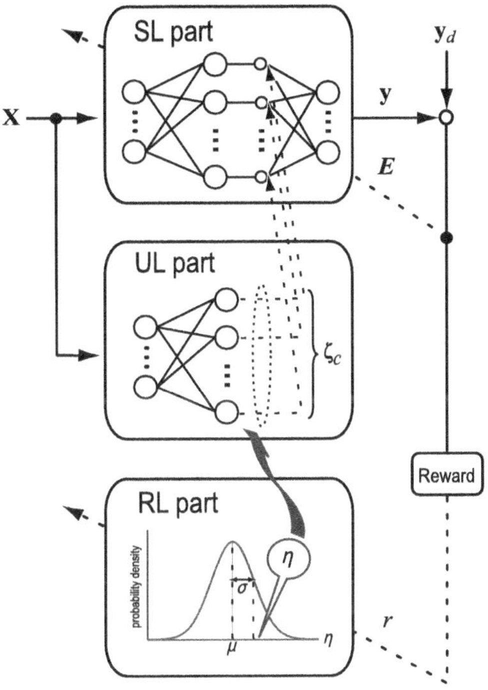

Fig. 2. Structure of the hierarchical learning system

Let us denote the system input vector by $\mathbf{x} \in \mathcal{R}^n$ where $\mathbf{x} = [x_1, \cdots, x_n]^T$, the output vector by $\mathbf{y} \in \mathcal{R}^m$ where $\mathbf{y} = [y_1, \cdots, y_m]^T$. The input-output mapping of the learning system is defined by

$$y_k = f_o\left(\sum_{j=1}^{l} w_{kj}^{(2)} O_j + \theta_k^{(2)}\right) \tag{1}$$

$$O_j = \zeta_{cj} \cdot f_h\left(\sum_{i=1}^{n} w_{ji}^{(1)} x_i + \theta_j^{(1)}\right) \tag{2}$$

where $f_o(\cdot)$ and $f_h(\cdot)$ are the node functions of output and hidden layers respectively, $w_{ji}^{(1)}$'s and $w_{kj}^{(2)}$'s are the weights of input layer and output layer respectively, $\theta_j^{(1)}$'s and $\theta_k^{(2)}$'s are the biases of hidden node and output node respectively, O_j's are the outputs of hidden node, l is the number of hidden nodes, and $0 \leq \zeta_{cj} \leq 1$ is the signal controlling firing strength of hidden node j. The firing signal vector $\boldsymbol{\zeta_c} = [\zeta_{c1}, \zeta_{c2}, \ldots, \zeta_{cl}]^T \in \mathcal{R}^l$ is output vector of the UL part. It can be seen from (2), with with properly defined firing signal ζ_{cj}, the SL part can be seen as a neural network consists of many overlapping modules; When $\zeta_{cj} = 1$ for all $j = 1, \cdots, l$, the SL part is exactly the same as an ordinary feedforward neural network.

2.2 The UL Part

The UL part is a competitive network. It divides the input space into subspaces and controls firing strength of each hidden neuron in the SL part according to the subspaces.

The competitive network has one layer of input neuron and one layer of output neuron. It has same input vector as that of SL part, and its output neuron has same number as the hidden neuron of the SL part. The inputs and outputs are connected by weight vectors called reference vectors $\mathbf{m}_j = [\mu_{j1}, \mu_{j2}, \ldots, \mu_{jn}]$ $(j = 1, \ldots, l)$, where l is the total number of hidden neuron of the SL part.

For a given input vector, UL part first determines the best-matching neuron c associated with reference vector $\mathbf{m}_c$, by

$$c = \arg\min_j \{\|\mathbf{x} - \mathbf{m}_j\|\}. \tag{3}$$

Then based on the best-matching neuron c, UL part calculates its outputs in a similar, but simplified way as a self-organizing mapping (SOM) [5]. That is, the output of UL part is given by

$$\zeta_{cj} = \exp\left(-\frac{2\eta^2}{(l-1)^2} f_d\left(r_j - r_c\right)^2\right) \tag{4}$$

where $0 < \zeta_{cj} \leq 1$ is the output of the jth neuron, $\eta \geq 0$ is a parameter that determines the shape of Gaussian function, $f_d(\cdot)$ is a function that calculates the distance between two neurons (e.g. Euclidean distance, link distance, etc.), and r_j, r_c are the positions of neuron j and neuron c in the array of output units, respectively. $f_d = 1$ for the best-matching neuron, and the value decreases as for a unit far away from the best-matching unit.

The outputs of UL part ζ_{cj} are then used to control firing strength of hidden neurons of SL part according to input patterns. This realizes the capability of function localization of the learning system. The parameter η in (4) is an important one determining the shape of control signal ζ_{cj}. It has been found that an optimal value of η exists [6]. The learning in the RL part will search for such an optimal value so as to optimize performance of the learning system.

2.3 The RL Part

The RL part is a learning scheme that performs a simplified reinforcement learning as to find an optimal value of η such that the whole system has the best performance. The simplified reinforcement learning used is called *evaluative feedback* [7]. The policy used to determine parameter η is defined as a probability function by

$$\eta(k) = \mathcal{N}(\mu(k), \sigma(k)^2) \tag{5}$$

where $\mu(k)$ is the mean denoting the most suitable value of parameter η at kth step, and $\sigma(k)$ is the variance denoting the confidence.

During the reinforcement learning, $\mu(k)$ and $\sigma(k)$ are updated in such a way that it has higher probability to obtain a value of η close to its optimal value.

3 Learning Strategies of the System

The learning of the system consists of three stages. In the first stage, an unsupervised learning is carried out in the UL part to extract structural information in the input space; In the second stage, supervised learning and reinforcement learning are carried out in the SL part and the RL part, respectively, in order to improve system performance by optimizing the value of parameter η; Finally, in the third stage, based on an optimized value of η, the system performs a supervised learning in the SL part to realize an input-output mapping.

3.1 Learning in the UL Part

The learning in the UL part performs an unsupervised learning such as Winner-Take-All (WTA) type competitive learning or SOM learning, referred to [5] for details.

3.2 Learning in the SL Part

The learning in the SL part performs a supervised learning, realized by back-propagation (BP) algorithm similar to an ordinary neural network. It can be formulated as a nonlinear optimization problem, defined by

$$\mathbf{\Theta} = \arg\min_{\mathbf{\Theta}}\{E\}, \qquad \mathbf{\Theta} \in \mathcal{W} \tag{6}$$

where $\mathbf{\Theta} = \{w_{ji}^{(1)},\ w_{kj}^{(2)},\ \theta_j^{(1)},\ \theta_k^{(2)},\ i = 1, \cdots, n,\ j = 1, \cdots, l,\ k = 1, \cdots, m\}$ is parameter vector. $\mathcal{W}$ denotes a compact region of parameter space, and E is a cost function defined by

$$E = \sum_{d \in \mathcal{D}} \|\mathbf{y}_d(d) - \mathbf{y}(d)\|^2, \tag{7}$$

where $\mathcal{D}$ is a set of training data, $\mathbf{y}_d(d)$ and $\mathbf{y}(d)$ are teacher signal and the output of SL part for the element $d \in \mathcal{D}$.

3.3 Learning in RF Part

The leaning in the RL part performs a reinforcement learning based on a reward defined by

$$r(k) = \frac{1}{E(k)} \tag{8}$$

where $r(k)$ is the reward received at step k, $E(k)$ is a learning error at step k, calculated by (6). The reinforcement learning is described by the following steps.

1) Determine $\eta(k)$ according to the policy (5);
2) Perform learning in SL part with a firing signal calculated basing on $\eta(k)$;
3) Calculate the reward $r(k)$ by (8) based on the learning error of SL part;

4) Update the policy by renewing $\mu(k)$ and $\sigma(k)$ based on the reward $r(k)$;
5) Repeat from Step 1).

In Step 4), the policy is updated based on a method called *Reinforcement comparison* [7]. It introduces a reference level called *reference reward* $\bar{r}(k)$. If the received reward $r(k) > \bar{r}(k)$, then update the policy by

$$\mu(t+1) = \mu(t) + \left(1 - e^{-(r(t)-\bar{r}(t))^2}\right) \cdot (\eta(t) - \mu(t)) \tag{9}$$

$$\sigma(t+1) = \begin{cases} \gamma_{inc} \cdot \sigma(t) \text{ if } \sigma(t) \leq \Delta\sigma(t) \\ \gamma_{dec} \cdot \sigma(t) \text{ if } \sigma(t) > \Delta\sigma(t) \end{cases} \tag{10}$$

$$\Delta\sigma(t) = |\eta(t) - \mu(t+1)| \tag{11}$$

where γ_{inc} and γ_{dec} are ratios for increasing and decreasing the variance σ, respectively. And the reference reward $\bar{r}(k)$ is updated by

$$\bar{r}(t+1) = \bar{r}(t) + \alpha_r(r(t) - \bar{r}(t)) \tag{12}$$

where α_r $(0 < \alpha_r \leq 1)$ is a step-size parameter.

4 Numerical Simulations

4.1 Task and Simulation Conditions

The learning system is applied to a two-nested-spiral problem. The task is to separate two nested spirals. The training sets consist of 152 associations formed by assigning the 76 points belonging to each of the nested spirals to two classes.

Table 1. Conditions for the three kinds of learning

• UL part	
Learning algorithm	Competitive learning (WTA)
Learning steps	100,000
• SL part	
Learning algorithm	Levenberg-Marquardt[9]
Learning steps	100
• RL part	
Learning steps	5000
Initial value of μ	$\mu(0) \in (0, 10)$
Initial value of σ	1
Initial value of $\bar{r}$	1
γ_{inc}	1.04
γ_{dec}	0.99
α_r	0.1

Table 2. Neighborhood distances of nodes

Node No.	1	2	3	4	5	6	7	8	9
1	0	1	2	1	2	3	2	3	4
2	1	0	1	2	1	2	3	2	3
3	2	1	0	3	2	1	4	3	2
4	1	2	3	0	1	2	1	2	3
5	2	1	2	1	0	1	2	1	2
6	3	2	1	2	1	0	3	2	1
7	2	3	4	1	2	3	0	1	2
8	3	2	3	2	1	2	1	0	1
9	4	3	2	3	2	1	2	1	0

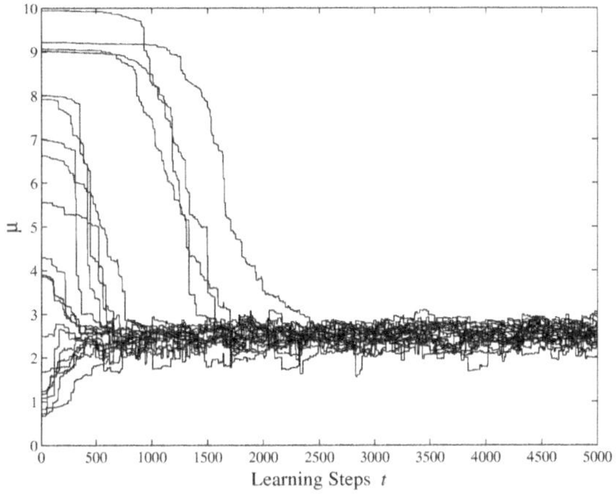

Fig. 3. Learning curves of μ

The problem has been extensively used as a benchmark for evaluation of neural network training [8].

The SL part used in simulations is a layered network of $N_{2\times9\times1}$ with 2 input neurons, 9 hidden neurons and 1 output neuron. The number of hidden neuron is chosen in such a way that an ordinary neural network of $N_{2\times9\times1}$ are not able to solve the problem, while the proposed learning system with proper value of η can solve the problem successfully.

Conditions used for the learning of three parts are shown in Tab.1. The algorithms modified from MATLAB neural network toolbox [10] are used for the learning in SL part and UL part. And neighborhood distances of nodes shown in Tab.2 are used to calculate the output of UL part.

4.2 Simulation Results

Figure 3 and 4 shows learning curves of μ and σ in the reinforcement learning of RL part. 20 trials are carried out with random initial values. From Fig.3 and 4,

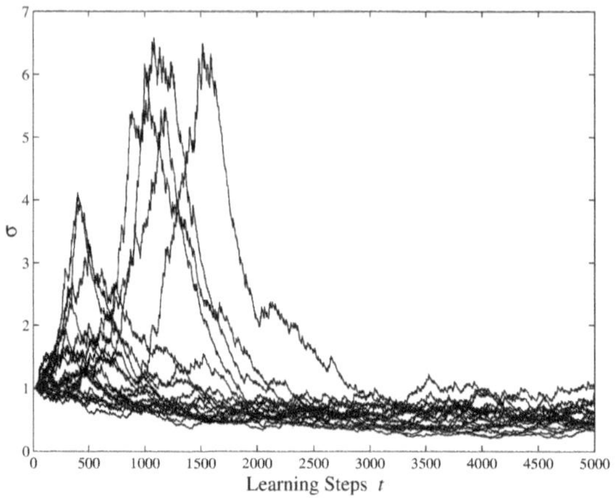

Fig. 4. Learning curves of σ

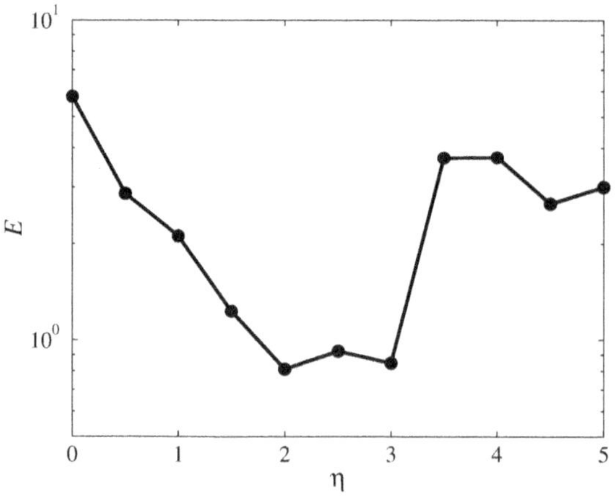

Fig. 5. Learning error for $\eta = 0, 0.5, \ldots, 5$ (averaged on the best 3 trials)

we can see that the value of μ converges to $2 < \mu < 3$ and σ converges to small value. It follows from Eq.(5) that the optimal value of η should be $2 < \eta < 3$.

To confirm the result, we carry out a set of simulations by fixing $\eta = 0$, 0.5, ..., 5, respectively, instead of using the RL part. Since the BP learning in SL part is easy to get stuck in local minimum, we did 50 trials and averaged the learning error of the best 3 trials as the result for each case. Figure 5 shows the results. From Fig.5, we can see that $2 < \eta < 3$ gives smaller learning error. This shows that the reinforcement learning in RL part is able to optimize the performance of learning system.

When $\eta = 0$ the proposed learning system reduces to an ordinary neural network. Figure 6 shows histogram of learning errors for the cases of $\eta = 0$ and

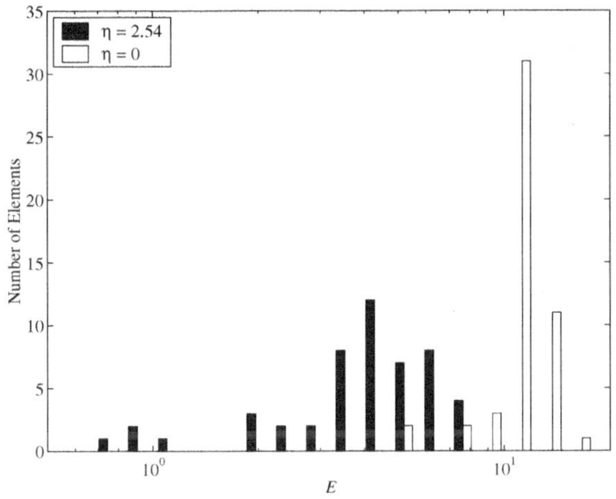

Fig. 6. Histogram of learning errors of the cases $\eta = 0$ and $\eta = 2.54$

$\eta = 2.54$. We can see that the proposed learning system with proper value of η has better representation ability than an ordinary neural network.

5 Conclusions

Inspired by Hebb's *Cell assembly theory* that brain has capability of function and the suggestion that brain has three different learning paradigms, we present a brain-like neural network that has the capabilities of function localization as well as learning. The proposed learning system has three parts: SL part, UL part and RF part and combines three kinds of learning: supervised learning, unsupervised learning and reinforcement learning. The SL part is similar to a 3-layer neural network, but its neurons in hidden layer are controlled by signals from UL part. The UL part is a competitive network extracts information of input patterns and control the SL part to realize function localization. The RF part is a reinforcement learning algorithm that optimizes the performance of learning system. It has been shown through numerical simulations that the proposed brain-like learning system optimizes its performance automatically, and has superior performance to an ordinary neural network.

References

1. Hebb, D.: The Organization of Behavior–A Neuropsychological Theory. John Wiley & Son, New York (1949)
2. Sawaguchi, T.: Brain Structure of Intelligence and Evolution. Kaimeisha, Tokyo (1989)
3. Doya, K.: What are the Computations of the Cerebellum, the Basal Ganglia, and the Cerebral Cortex. Neural Networks **12** (1999) 961-974

4. Sasakawa, T., Hu, J., Hirasawa, K.: Self-organized Function Localization Neural Network. Proc. of International Joint Conference on Neural Networks (IJCNN'04), Budapest (2004)
5. Kohonen, T.: Self-Organizing Maps (3ed.). Springer, Heidelberg (2000)
6. Sasakawa, T., Hu, J., Hirasawa, K.: Performance Optimization of Function Localization Neural Network by Using Reinforcement Learning. Proc. of International Joint Conference on Neural Networks (IJCNN'05), Montreal (2005) 1314-1319
7. Sutton, R., Barto, A.: Reinforcement Learning: An Introduction. MIT Press, Cambridge (1998)
8. Solla, S., Fleisher, M.: Generalization in Feedforwad Neural Networks. Proc. the IEEE International Joint Conference on Neural Networks, Seattle (1991) 77-82
9. Hagan, M., Menhaj, M.: Training Feedforward Networks with the Marqurdt Algorithm. IEEE Trans. Neural Networks **5** (1994) 989-993
10. Demuth, H., Beale, M.: Neural Network Toolbox: for use with MATLAB. The MATH WORKS Inc. (2000)

A Hierarchical Self-organizing Associative Memory for Machine Learning

Janusz A. Starzyk[1], Haibo He[2], and Yue Li[3]

[1] School of Electrical Engineering and Computer Science
Ohio University, OH 45701 USA
starzyk@bobcat.ent.ohiou.edu
[2] Department of Electrical and Computer Engineering
Stevens Institute of Technology, NJ 07030 USA
Haibo.He@stevens.edu
[3] O_2 Micro Inc., Santa Clara, CA 95054 USA
yue.lily.lee@gmail.com

Abstract. This paper proposes novel hierarchical self-organizing associative memory architecture for machine learning. This memory architecture is characterized with sparse and local interconnections, self-organizing processing elements (PE), and probabilistic synaptic transmission. Each PE in the network dynamically estimates its output value from the observed input data distribution and remembers the statistical correlations between its inputs. Both feed forward and feedback signal propagation is used to transfer signals and make associations. Feed forward processing is used to discover relationships in the input patterns, while feedback processing is used to make associations and predict missing signal values. Classification and image recovery applications are used to demonstrate the effectiveness of the proposed memory for both hetero-associative and auto-associative learning.

1 Introduction

Associative memory is of critical importance for machine learning, information representation, signal processing and a wide range of applications. Therefore, it has attracted extensive research in engineering and science. There are two types of associative memories: hetero-associative (HA) memory makes associations between paired patterns, such as words and pictures, while auto-associative (AA) memory associates a pattern with itself, recalling stored patterns from fractional parts of the pattern as in image recovery.

Both types of memories have attracted a significant attention in recent literature. For instance, among HA studies, J. Y. Chang and C. W. Cho proposed adaptive local training rules for second-order asymmetric bidirectional associative memory (BAM) in [1]. Simulation results of this BAM on color graphics adapter (CGA) fonts illustrate the effectiveness of this memory. Salih et al. proposed a new approach for bidirectional associative memories (BAM) using feedback neural networks [2]. The perceptron training algorithm was used to solve a set of linear inequalities for the BAM neural

D. Liu et al. (Eds.): ISNN 2007, Part I, LNCS 4491, pp. 413–423, 2007.

network design. In [3], Wang presented a multi-associative neural network (MANN) and showed its application to learning and retrieving complex spatio-temporal sequences. Simulation results show that this system is characterized by fast and accurate learning, and has the ability to store and retrieve a large number of complex sequences of nonorthogonal spatial patterns. Hopfield's paper [4] is a classic reference for auto-associative studies. Since that paper, many research results have been reported. For instance, Vogel presented an algorithm for auto-associative memory in sparsely connected networks [5]. The resulting networks have large information storage capacities relative to the number of synapses per neuron. Vogel et al. derived a lower bound on the storage capacities of two-layer projective networks (P-nets) with binary Hebbian synapses [6]. Recently, Wang et al. proposed an enhanced fuzzy morphological auto-associative memory based on the empirical kernel map [7].

In this paper, we developed a probability based associative memory algorithm and memory architecture that is capable of both hetero-associative (HA) and auto-associative (AA) memory. This paper is organized as follows. In section 2, a new probability based associative learning algorithm is proposed. In section 3, we discuss the network architecture and its associative mechanism. In section 4, classification and image recovery applications are used to illustrate the HA and AA applications of the proposed memory structure. Finally, conclusions are given in section 5.

2 Associative Learning Algorithm

The proposed memory architecture consists of a multilayer array of the processing elements (PE). Its organization follows a general self-organizing learning array concept presented in [8]. Fig. 1 gives the interface model of an individual PE, which consists of two inputs (I_1 and I_2) and one output (O). Each PE stores observed probabilities P_{00}, P_{01}, P_{10} and P_{11}, corresponding to four different combinations of inputs I_1 and I_2 ($\{I_1 I_2\} = \{00\}, \{01\}, \{10\}, \{11\}$), respectively.

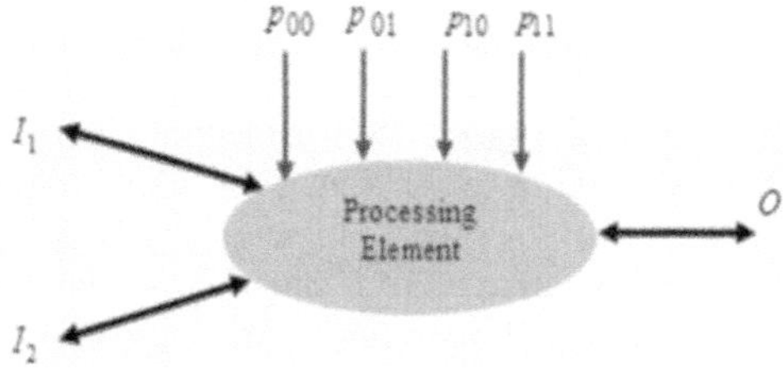

Fig. 1. Individual PE interface model

Fig.2 gives an example of possible distribution of the observed input data points (scaled to the range [0 1]). Probabilities are estimated from $p_{00} = n_{00} / n_{tot}$, $p_{01} = n_{01} / n_{tot}$, $p_{10} = n_{10} / n_{tot}$ and $p_{11} = n_{11} / n_{tot}$, where n_{00}, n_{01}, n_{10} and n_{11} is the number of data points located in $I_1 < 0.5 \& I_2 < 0.5$, $I_1 < 0.5 \& I_2 > 0.5$,

$I_1 > 0.5 \& I_2 < 0.5$ and $I_1 > 0.5 \& I_2 > 0.5$, respectively. n_{tot} is the total number of data points defined as $n_{tot} = n_{00} + n_{01} + n_{10} + n_{11}$.

Based on the observed probability distribution p_{00}, p_{01}, p_{10} and p_{11} of an individual PE, each PE decides its output function value F by specifying in its truth table as shown in Table 1.

The output function values f_{00}, f_{01}, f_{10} and f_{11} are decided as follows:

(1) The input, (I_1, I_2), that is associated with the largest probability, $p_{ij}, (i, j = 0,1)$, is assigned a corresponding output function F value of 0.

(2) If the largest probability is less than 0.5, then the input (I_1, I_2), that is associated with smallest probability is also assigned a corresponding F value of 0;

(3) If the sum of the largest and smallest probabilities is less than 0.5, then the input, (I_1, I_2), that is associated with the second-smallest probability $p_{ij}, (i, j = 0,1)$ is also assigned a corresponding F value of 0;

(4) All input combinations not assigned corresponding F value of 0 by the above rules are assigned a corresponding F value of 1.

Table 1. Self-determination of function value F

Probability	p_{00}	p_{01}	p_{10}	p_{11}
I_1	0	0	1	1
I_2	0	1	0	1
Function value	f_{00}	f_{01}	f_{10}	f_{11}

The probability that the neuron is active is smaller than 0.5. This type of assignment is motivated by the sparse activity of biological neurons [9]. In addition to biological motivation, lower activities are preferable for efficient power consumption. Probabilities p_{ij} can be efficiently estimated in real time hardware using dynamic probability estimator [10]. Table 2 shows two examples of this self-determination of the function value F.

Table 2. Two examples of setting F value

p_{00}	p_{01}	p_{10}	p_{11}		F			
0.4	0.2	0.3	0.1	⟶	0	1	1	0
0.4	0.05	0.3	0.25	⟶	0	0	1	0

During training, each PE counts its input data points in n_{00}, n_{01}, n_{10} and n_{11} and estimates their corresponding probabilities p_{00}, p_{01}, p_{10} and p_{11}. The objective of the training stage for each PE is to discover the potential relationship between its

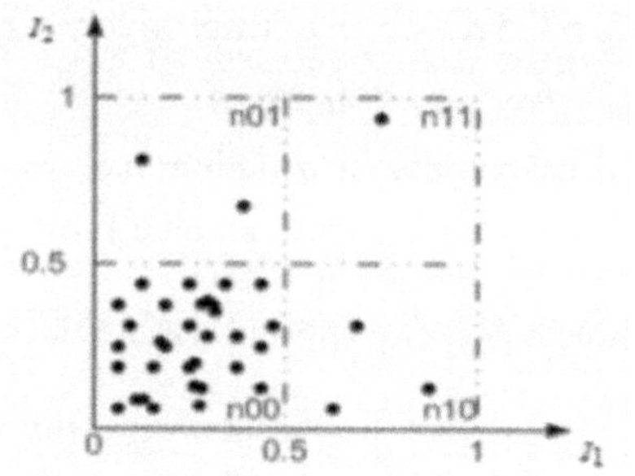

Fig. 2. An example of input space distribution of PE

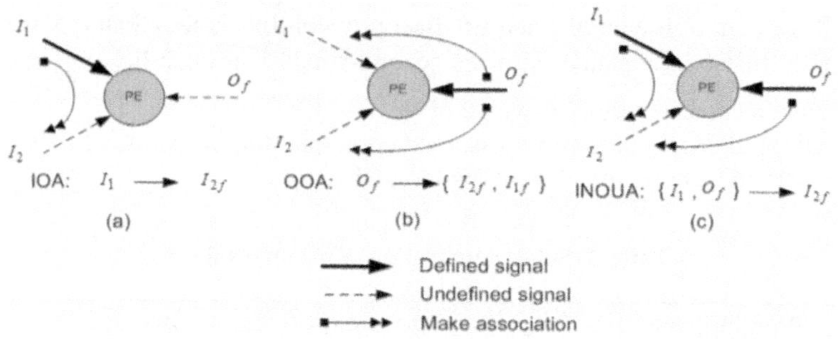

Fig. 3. Three types of associations of processing element

inputs. This relationship is remembered as the corresponding probabilities and is used to make associations during the testing stage.

Considering the example in Fig. 2, this particular PE finds that most of its input data points are distributed in the lower-left corner ($I_1 < 0.5 \,\&\, I_2 < 0.5$). Therefore, if this PE only knows one of the input signal is $I_1 < 0.5$, it will associatively predict that the other signal most likely should also be $I_2 < 0.5$. The developed algorithm allows all the PEs in the network to make such associations between different input signals.

Fig. 3 illustrates the three types of associations used in the proposed memory model. The undefined signal means its value is equal to 0.5, in such way, 0 and 1 represents the strongest signal. There are three types of associations used in the testing stage to infer the undefined signal value.

(1) Input only association (IOA). If, in the testing stage, one input is defined while the other input and the received output feedback signal O_f from other PEs are undefined (for instance, if $I_1 = 0$, $I_2 = 0.5$ and $O_f = 0.5$ as in Fig. 3(a)), this PE will determine I_2 through association with I_1, driving I_2 towards logic 0.

(2) Output only association (OOA). If both inputs, I_1 and I_2, are undefined, a defined feedback signal, O_f, will determine both inputs (Fig 3(b)). For instance, if

$O_f = 0$, based on PE function F= {0, 1, 1, 1}, then this PE will set both inputs, I_{1f} and I_{2f} to 0. (Here we use I_{1f} and I_{2f} to denote the feedback signals of inputs 1 and 2 to distinguish them from the corresponding feed forward signals). On the other hand, if F sets the received output feedback signal to $O_f = 1$, the input feedback values, I_{1f} and I_{2f}, are intermediate and their values will be estimated according to data distribution probabilities.

(3) Input–output association (INOUA). If one input and the output feedback signal, O_f, are defined and the other input is undefined, the PE will set the other input signal according to its observed probabilities, as shown in Fig. 3(c).

This probability based associative learning algorithm can be described as follows:

Case 1: Given the semi-logic values of both inputs $V(I_1)$ and $V(I_2)$, decide the output value $V(O)$

Assume one PE received input values $V(I_1) = m$ and $V(I_2) = n$, then

$$V(O) = \frac{p(I_1 = 1, I_2 = 1, F = 1)}{p(I_1 = 1, I_2 = 1)} \bullet V_{11} + \frac{p(I_1 = 0, I_2 = 1, F = 1)}{p(I_1 = 0, I_2 = 1)} \bullet V_{01} + \frac{p(I_1 = 1, I_2 = 0, F = 1)}{p(I_1 = 1, I_2 = 0)} \bullet V_{10} + \frac{p(I_1 = 0, I_2 = 0, F = 1)}{p(I_1 = 0, I_2 = 0)} \bullet V_{00}, \quad (1)$$

where V_{11}, V_{01}, V_{10} and V_{00} are defined as

$$V_{11} = mn, \quad V_{01} = (1-m)n, \qquad V_{10} = m(1-n), \quad V_{00} = (1-m)(1-n), \quad (2)$$

and $p(I_1 = 1, I_2 = 1, F = 1)$, $p(I_1 = 1, I_2 = 1)$ etc. are joint probabilities. Case 1 is used when a signal is propagated forward.

Case 2: Given the values of one input, $(V(I_1)\ or\ V(I_2))$, and an undefined output $V(O)$, decide the value of the other input.

Case 2 corresponds to input-only-association (IOA) when a signal is propagated backwards, as shown in Fig. 3(a). We can use a given $V(I_1)$ to decide an unknown $V(I_2)$ as follows:

$$V(I_2) = \frac{p(I_1 = 1, I_2 = 1)}{p(I_1 = 1)} \bullet V(I_1) + \frac{p(I_1 = 0, I_2 = 1)}{p(I_1 = 0)} \bullet (1 - V(I_1)), \quad (3)$$

where $p(I_1 = 1) = p_{10} + p_{11}$, $p(I_1 = 0) = p_{00} + p_{01}$. In the case in which $V(I_2)$ is given and determines $V(I_1)$, I_1 and I_2 are switched in equation (3).

Case 3: Given the value of the output $V(O)$, decide the value of both inputs $V(I_1)$ and $V(I_2)$.

$$V(I_1) = \frac{p(F = 1, I_1 = 1)}{p(F = 1)} \bullet V(O) + \frac{p(F = 0, I_1 = 1)}{p(F = 0)} \bullet (1 - V(O)),$$

$$V(I_2) = \frac{p(F=1, I_2=1)}{p(F=1)} \bullet V(O) + \frac{p(F=0, I_2=1)}{p(F=0)} \bullet (1 - V(O)) \ . \tag{4}$$

Case 3 corresponds to output-only-association (OOA) when a signal is propagated backwards as shown in Fig. 3(b). $p(F=1)$ and $p(F=0)$ are determined by the output of each PE.

Case 4: Given the values of one input, ($V(I_1)$ or $V(I_2)$, and the output, $V(O)$, decide the other input value, $V(I_2)$ or $V(I_1)$;

Case 4 corresponds to the input-output-association (INOUA) when a signal is propagated backwards (Fig. 3(c)). For example, we can use a given $V(I_1)$ and $V(O)$ to decide $V(I_2)$ as follows:

$$\begin{aligned} V(I_2) = & \frac{p(I_1=1, F=1, I_2=1)}{p(I_1=1, F=1)} \bullet \hat{V}_{11} + \frac{p(I_1=0, F=1, I_2=1)}{p(I_1=0, F=1)} \bullet \hat{V}_{01} \\ & + \frac{p(I_1=1, F=0, I_2=1)}{p(I_1=1, F=0)} \bullet \hat{V}_{10} + \frac{p(I_1=0, F=0, I_2=1)}{p(I_1=0, F=0)} \bullet \hat{V}_{00} \end{aligned} \tag{5}$$

where $\hat{V}_{11}, \hat{V}_{01}, \hat{V}_{10}$ and $\hat{V}_{00}$ are determined in the following way:

$$\hat{V}_{11} = \begin{cases} V(I_1) \bullet V(O) & \begin{cases} X & X & 0 & 1 \\ X & X & 1 & 0 \end{cases} \\ 0 & X \ X \ 0 \ 0 \\ V(I_1) & X \ X \ 1 \ 1 \end{cases} \qquad \hat{V}_{10} = \begin{cases} V(I_1) \bullet (1 - V(O)) & \begin{cases} X & X & 0 & 1 \\ X & X & 1 & 0 \end{cases} \\ V(I_1) & X \ X \ 0 \ 0 \\ 0 & X \ X \ 1 \ 1 \end{cases}$$

$$\hat{V}_{01} = \begin{cases} (1 - V(I_1)) \bullet V(O) & \begin{cases} 0 & 1 & X & X \\ 1 & 0 & X & X \end{cases} \\ 0 & 0 \ 0 \ X \ X \\ (1 - V(I_1)) & 1 \ 1 \ X \ X \end{cases} \qquad \hat{V}_{00} = \begin{cases} (1 - V(I_1)) \bullet (1 - V(O)) & \begin{cases} 0 & 1 & X & X \\ 1 & 0 & X & X \end{cases} \\ (1 - V(I_1)) & 0 \ 0 \ X \ X \\ 0 & 1 \ 1 \ X \ X \end{cases} \tag{6}$$

The conditions in equation (6) refer to the function value of F for each particular PE, where " X " is a do not care, which means its value can be either '0' or '1'. For example, if one PE received $V(I_1) = m$ and $V(O) = t$, and the function value of this PE is F= {0 1 1 1}, we will get the following results:

$$\hat{V}_{11} = m; \quad \hat{V}_{10} = 0; \quad \hat{V}_{01} = (1-m) * t; \quad \hat{V}_{00} = (1-m) * (1-t)$$

When $V(I_2)$ and $V(O)$ are given one only needs to switch I_1 and I_2 in equations (5) and (6) to decide $V(I_1)$.

3 Memory Network Architecture

The overall memory network is a hierarchical structure of sparsely connected self-organizing PE's. Each layer of this hierarchical structure contains a number of PEs connected to the primary inputs or to the outputs of other processing elements from lower layers of hierarchy. For n-dimensional input, the network should have at least n/2 PEs in each layer. The required number of layers depends on the problem complexity and may be determined through simulation. In practice, the number of layers grows logarithmically with the size of the input vector.

Each PE in the array can self-organize by dynamically adapting its function in response to the input data. The hierarchical connections are suitable for hardware implementation, time control, and correlate well to complexity of object representation. The further away a PE is from the sensory input, the more abstract and invariant is the representation of objects or their features captured by the PE. Each PE is more likely to connect to other PEs within a short Euclidean distance. This organization is observed in biological memory where neurons tend to have mostly local connections.

(1) Feed forward operation
Fig 5 shows a feed forward network structure for the proposed memory architecture. For simplification, we only illustrate 4 layers with 6 PEs per layer and 6 input signals. The bold lines from PE 1 to PE 11 and from PE18 to PE21 are two examples of the distant connections.

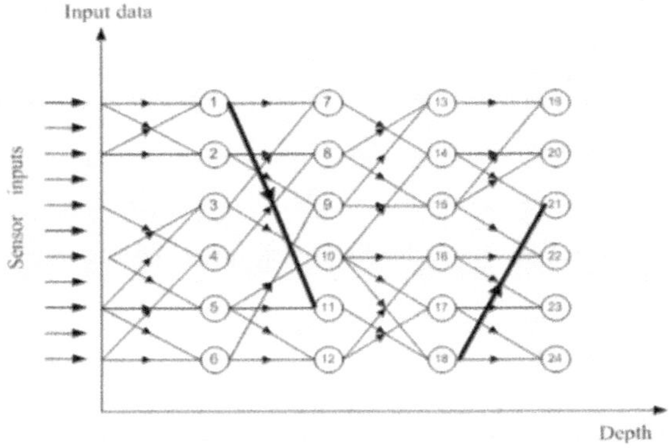

Fig. 4. An example of feed forward operation network

During training, all external input data are presented to the network. Each PE counts activities on its inputs to estimate the corresponding probabilities, $p_{ij}, (i, j = 0,1)$, and decide its output function as in case 1 of Section 2. This probability information will be used to make associations in the feedback operation.

(2) Feedback operation
Feedback operation is essential for the network to make correct associations and to recover the missing parts (undefined signals) of the input data. Fig. 5 shows a feedback structure. Assume that signals 1, 2 and 3 are undefined as would be the case in a classification application where all the class ID code inputs are undefined and only the feature input values are available, and in the image recovery application, part of the image could be blocked or undefined. In both cases, the network will use the associations mechanism as discussed in Section 2 to determine these undefined signal values.

In Fig. 5, the shaded PEs are associative and will use associations to recover the undefined values. For instance, PE4 received one defined signal and one undefined signal. In this situation, PE4 will use the IOA to associatively recover this undefined signal based on the information it learned in the training stage. Some associations will also happen in a deeper layer. Considering PE22, it will use IOA to associatively recover the other input signal I_{2f}. This feedback signal will back propagate to PE15 (it will become the O_f for PE15). Therefore, based on the OOA, PE15 will associatively recover both input signals of PE15. In this way, these feedback signals will further back propagate to other hierarchical layers in the network. Therefore, the missing information in the sensor input will be recovered.

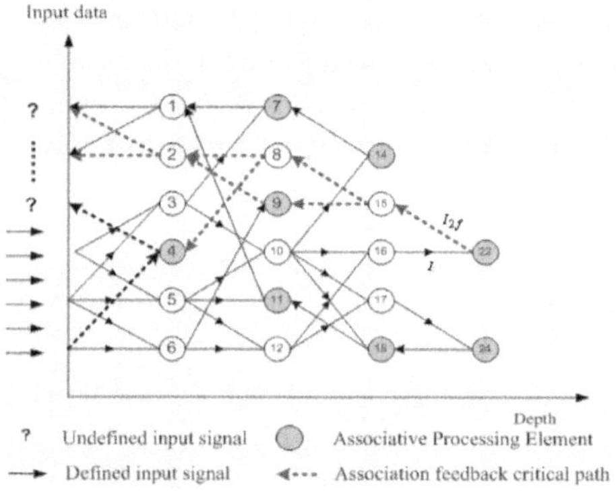

Fig. 5. Example of feedback structure in testing stage

4 Simulation Results

The proposed probability based self-organizing associative memory is capable of both hetero and auto-associations. In this section, the Iris database and an image recovery problem are used to illustrate the HA and AA applications.

(1) Hetero-associative memory: Iris database classification
The Iris database [11] developed by R. A. Fisher was used to test the classification performance of the proposed associative memory. We used an N-bits sliding bar coding mechanism to code the input data. Assume that the maximum and minimum values to be coded are $V_{\max}$ and $V_{\min}$ respectively. We set $N - L = V_{\max} - V_{\min}$, where L is the length of the sliding bar. Assume that the value of the scaled feature to be coded is V. In the coded input we set bits numbered from $(V - V_{\min}) + 1$ to $(V - V_{\min}) + L$ to 1s, while the remaining bits were set to 0. The class ID is coded in a similar way using M bit code redundancy. Since there are 3 classes in this database, we use M*3 bits to code the class ID, maximizing their Hamming distance. This was achieved by filling the M bits from position $(C_i - 1) * M$ to $C_i * M$ with 1's, while filling the remaining $M * 2$ bits with 0's. Here $C_i = 1,2$ and 3 for this 3 classes Iris database.

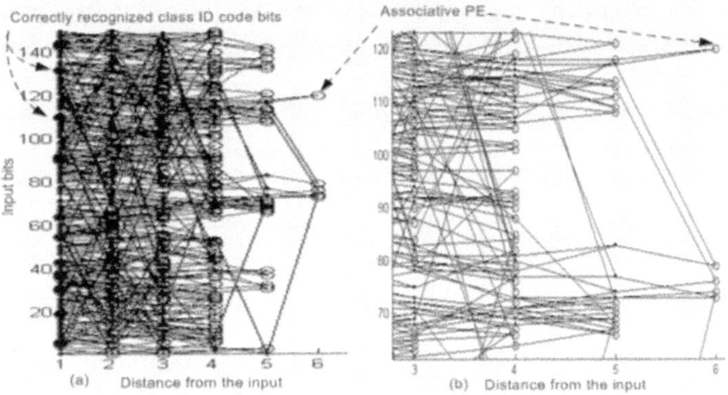

Fig. 6. Associative PEs and their inter connection structure

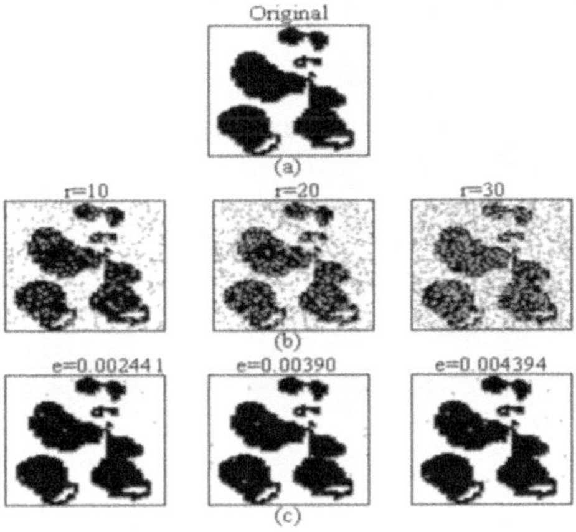

Fig. 7. (a) The original image; (b) Blocked image with r% of undefined values (r= 10, 20 and 30 respectively); (c) Recovered image and the recovery error

Since there are only 150 instances in the Iris database, the ten-fold cross validation method was used to handle this small sample dataset. Our memory network achieved an overall of 96% correct classification accuracy. Fig. 6(a) shows the associative PEs and their connection structure, and Fig. 6 (b) shows associative PE firing activity for part of the network. The Y-axis represents the input bits, and the X-axis represents the distance from the input (association depth). The associative PEs are represented by circles and their backward propagation paths are marked. The large dots at the input layer represent correctly recognized class ID code bits. It may be seen that only 6 layers are needed for the network to learn the associations in the Iris database.

(2) Auto-associative memory: image recovery

An image recovery problem was used to test the effectiveness of the proposed memory for auto-associative applications. We used the proposed memory to associate parts of images, and then recall the images from fractional parts of the images. This is necessary for applications where only partial images are available without specifying class identities. Our model can learn features of the training data using unsupervised learning, self-determine the feedback depth, and make correct associations to recover the original images.

We used a 64 x 64 binary panda image [12] to illustrate the auto-associative application of the proposed memory architecture. The panda image is represented by a vector $p_i = (x_1 \quad x_2 \quad \ldots \quad x_n), n = 4096$, with x_i =1 for a black pixel and x_i =0 for a white pixel. In testing, $r\%$ percentage ($r = 10{,}20$ and 30) of the panda image was randomly blocked. The original panda image and samples of its blocked image are shown in Figs. 7(a) and (b), respectively. Fig. 7(c) shows images recovered through our associative memory. We evaluate image recovery performance by computing the ratios of the number of incorrectly recovered pixels (both erroneous pixels and pixels remaining undefined after recovery) over the total number of pixels. As we can see from Fig. 7, the recovery error bits of our associative memory is range from 0.2% ~ 0.4%.

5 Conclusion

In this paper, we proposed a hierarchical associative memory architecture for machine learning that uses probability based associations. Through the associative learning algorithm, each processing element in the network learns the statistical data distribution, and uses such information for input data association and prediction. Simulation results on both classification and image recovery applications show the effectiveness of the proposed method.

References

1. Chang, J. Y., Cho, C. W.: Second-order Asymmetric BAM Design with a Maximal Basin of Attraction. IEEE Trans. on System, Man, and Cybernetics, Part A: Systems and Humans **33** (2003) 421-428
2. Salih, I., Smith, S. H., Liu, D.: Synthesis Approach for Bidirectional Associative Memories Based on the Perceptron Training Algorithm. Neurocomputing **35** (2000) 137-148
3. Wang, L.: Multi-associative Neural Networks and Their Applications to Learning and Retrieving Complex Spatio-temporal Sequences. IEEE Trans. on System, Man, and Cybernetics, part B-Cybernetics **29** (1999) 73-82
4. Hopfield, J. J.: Neural Networks and Physical Systems with Emergent Collective Computational Abilities. in Proc. Nat. Acad. Sci. USA **79** (1982) 2554-2558
5. Vogel, D.: Auto-associative Memory Produced by Disinhibition in a Sparsely Connected Network. Neural Networks **5** (11) (1998) 897-908
6. Vogel, D., Boos, W.: Sparsely Connected, Hebbian Networks with Strikingly Large Storage Capacities. Neural Networks **4** (10) (1997) 671-682
7. Wang, M., Chen, S.: Enhanced EMAM Based on Empirical Kernel Map. IEEE Trans. on Neural Network **16** (2005) 557-563
8. Starzyk, J. A., Zhu, Z., Liu, T.-H.: Self-Organizing Learning Array. IEEE Trans. on Neural Networks **16** (2) (2005) 355-363
9. Triesch, J.: Synergies between Intrinsic and Synaptic Plasticity in Individual Model Neurons. Neural Information Processing System (NIPS) **17** (2004)

10. Starzyk, J. A., Wang, F.: Dynamic Probability Estimator for Machine Learning. IEEE Trans. on Neural Networks **15** (2) (2004) 298-308
11. Fisher, R. A.: The Use of Multiple Measurements in Taxonomic Problem. Ann. Eugenics 7 (2) (1936) 179-188
12. Djuric, P. M., Huang, Y., Ghirmai, E.: Perfect Sampling: a Review and Applications to Signal Processing. IEEE Trans. on Signal Processing **50** (2) (2002) 345 – 356

Enclosing Machine Learning for Class Description

Xunkai Wei[1], Johan Löfberg[2], Yue Feng[1], Yinghong Li[1], and Yufei Li[1]

[1] School of Engineering, Air Force Engineering University,
Shanxi Province, Xian 710038, China
xunkai.wei@gmail.com, fengy1228@163.com,
yinghong_li@126.com, horizon_lyf@hotmail.com
[2] Automatic Control Laboratory, ETHZ,
CH-8092 Zürich, Switzerland
loefberg@control.ee.ethz.ch

Abstract. A novel machine learning paradigm, i.e. enclosing machine learning based on regular geometric shapes was proposed. It adopted regular minimum volume enclosing and bounding geometric shapes (sphere, ellipsoid, and box) or their unions and so on to obtain one class description model and thus imitate the human "Cognizing" process. A point detection and assignment algorithm based on the one class description model was presented to imitate the human "Recognizing" process. To illustrate the concept and algorithm, a minimum volume enclosing ellipsoid (MVEE) strategy for enclosing machine learning was investigated in detail. A regularized minimum volume enclosing ellipsoid problem and dual form were presented due to probable existence of zero eigenvalues in regular MVEE problem. To solve the high dimensional one class description problem, the MVEE in kernel defined feature space was presented. A corresponding dual form and kernelized Mahalanobis distance formula was presented. We investigated the performance of the enclosing learning machine via benchmark datasets and compared with support vector machines (SVM).

1 Introduction

Cognitive processing is the instinct learning ability of human being. Human always transfers the feature information to the brain through perception, and the brain then extract the feature information and remember this for the given objects. According to the cognitive science theory, the human brain can be imitated but can not be completely reproduced. Currently, artificial intelligence is an important direction for function imitation of the human brain.

Neural-computing and neural networks (NN) families, based on the neuron working mechanism, have made great progress in various aspects. Recently, statistical learning and support vector machines (SVM) draw extensive attention and show attractive and excellent performances in various areas compared with NN, which imply that artificial intelligence can also be made via advanced statistical computing theory.

It should be noted that as for both NN and SVM, the function imitation of human cognitive process for pattern classification can be explained as follows [1]. Given the training pairs (sample features, class indicator), we can train a NN or a SVM learning

D. Liu et al. (Eds.): ISNN 2007, Part I, LNCS 4491, pp. 424–433, 2007.

machine. The training process of these learning machines actually imitates the learning ability of human. For clarity, we call this process "cognizing". Then, the trained NN or SVM can be used for testing an unknown sample and determine the class it belongs to. The testing process of an unknown sample actually imitates the recognizing process of human being. We call this process "recognizing".

From a mathematical point of view, both these two learning machines are based on the hyperplane adjustment, and obtain the optimum or sub-optimum hyperplane combinations after the training process. As for NN, each neuron acts as a hyperplane in the feature space. The feature space is divided into many partitions according to the selected training principle. Each feature space partition is linked with a corresponding class, which accomplishes the "cognizing" process. Given an unknown sample, it only detects the partition where the sample locates in and then assigns the indicator of this sample, which accomplishes the "recognizing" process. Like NN, SVM is based on the optimum hyperplane. Unlike NN, standard SVM determines the hyperplane via solving a convex optimization problem. They have the same "cognizing" and "recognizing" process except different solving strategies.

If a complete unknown and novel sample comes, both SVM and NN will not recognize it correctly and prefer to assign it to the closest indicator in the learned classes [2-3]. This is generally a wrong classification, and here comes the topic which this paper concerns with. The root cause of this phenomenon is the learning principle, which is based on feature space partition. This kind of learning principle may amplify each class's region especially when the samples are small due to incompleteness. This makes it impossible to automatically detect the novel samples. Here comes the problem: how to make it clever enough to automatically identify the novel samples and cut down the misclassification errors.

The rest of this paper is organized as follows. Section 2 gives the basic concepts of enclosing learning machine and reviews some related works. Section 3 describes the proposed one class description algorithm based on MVEE cognitive learner, and shows how this can be used to build cognitive learner in kernel defined feature space. Experimental results are presented in Section 4, and Section 5 gives some conclusions.

2 Enclosing Machine Learning

2.1 Basic Concepts

Humans generally cognize things of one kind and recognize completely unknown things of a novel kind easily. So the answer is why not make the learning machine "cognize" or "recognize" things like human being. In other words, the learning machine should "cognize" the training samples of the same class. Each class is cognized or described by a cognitive learner. It uses some kind of model to describe each class instead of using feature space partition so as to imitate the "cognizing" process. The bounding and closing boundary of each cognitive learner scatters in the feature space. For an unknown sample, the cognitive class recognizer then detects whether the unknown sample is located inside a cognitive learner's boundary to imitate the "recognizing" process. If the sample is completely new (i.e., none of the trained cognitive learner contains the sample), it can be again described by a new cognitive learner and the new obtained learner can be added to the feature space without changing others. We call this feedback process active self-learning. This concludes the basic concepts of enclosing machine learning [4]. Now, we can investigate the definition of the cognitive learner. The cognitive learner should have at least the following features:

A. regular and convenient to calculate,
B. bounding and closing with the minimum volume,
C. convex bodies to guarantee optimality,
D. fault tolerant to guarantee generalization performance.

Basic geometric shapes are probably are the best choices since they are convex bodies and operations such as intersection, union or complement of the basic geometric shapes can be implemented easily. Hence, we propose to use basic geometric shapes such as spheres, cubes or ellipsoids. The cognitive learner can then use these geometric shapes to enclose all the given samples with minimum volume in the feature space. This is the most important reason why we call this learning paradigm enclosing machine learning. Here we give the definition of the cognitive learner and recognizer.

Definition 1. A cognitive learner is defined as the bounding boundary of a minimum volume set enclosing all the given samples. The cognitive learner can be either a sphere or an ellipsoid or their combinations. For illustration, we only investigate ellipsoids here.

Definition 2. A cognitive recognizer is defined as the point detection and assignment algorithm.

2.2 Related Works

The minimum volume set has achieved broad attention by many scholars, and there is a large body of interesting prior work. But to the knowledge of us, we are the first to apply minimum volume sets to cognitive process modeling for class cognitive description. Here we will shortly review the most influential related works.

L.Vandenberghe and S.Boyd [5] give a stable interior-point method for determinant maximization with linear matrix inequality constraints problem, which is closely related to minimum volume enclosing ellipsoid problem. D.M.J. Tax and R.P.W. Duin [6] propose a convex QP based minimum volume enclosing sphere method (called Support Vector Data Description, SVDD) that works both in Euclid space and kernel feature space. Kaspar Fischer [7] et al develops a simple combinatorial algorithm for computing the smallest enclosing ball of a set of points in high dimensional Euclidean space. Piyush Kumar et al [8] propose a $(1+\varepsilon)$-approximation to the minimum volume enclosing sphere using second-order cone programming technique and core sets. Peng Sun and Robert M.Freund [9] propose a combined interior point and active set method for solving exact minimum volume enclosing ellipsoid problem. Rina Panigrahy [10] proposed a greedy iteration algorithm for minimum enclosing polytope in high dimensions. This work generalizes the minimum volume enclosing sets to arbitrary shape. P. Kumar and E. A. Yildirim [11] presented a

$(1+\varepsilon)$-approximation to the minimum-volume enclosing ellipsoid, in which the size of core sets depends on only the dimension d and ε, but not on the number of points.

In this paper, we only study the minimum volume enclosing ellipsoid problem for class description, but the idea is quite straightforward for other cases such as minimum volume enclosing sphere, bounding box and so on.

3 MVEE Cognitive Learner for One Class Description

3.1 Preliminaries

Our concern is with covering m given points $x_i \in \Re^k$, $X := [x_1 \; x_2 \; \cdots \; x_n]$ with an ellipsoid of minimum volume [5][9]. To avoid trivialities, we also make the following assumption for the remainder of this paper, to guarantee that any ellipsoid containing $X := [x_1 \; x_2 \; \cdots \; x_n]$ has positive volume:

Assumption 1. The affine hull of $X := [x_1 \; x_2 \; \cdots \; x_n]$ spans $\Re^k$.

Definition 3. For $c \in \Re^k$ and $E \in S_{++}^{k\times k}$, we define the ellipsoid

$$\varepsilon(E,c) := \left\{ x \in \Re^k \mid (x-c)^T E (x-c) \le 1 \right\}.$$

where $E \in S_{++}^{k\times k}$ determines the shape and directions of the ellipsoid. The length of the axes is given by $\left[\sqrt{\lambda_1}, \sqrt{\lambda_2}, \cdots, \sqrt{\lambda_k}\right]$, where $[\lambda_1, \lambda_2, \cdots, \lambda_k]$ are the corresponding eigenvalues of the matrix E.

Definition 4. For $x_i \in \Re^k$, $X := [x_1 \; x_2 \; \cdots \; x_n]$, A MVEE cognitive learner is defined as the boundary of all the possible enclosing ellipsoids with the minimum volume.

Under Assumption 1, a natural formulation of a minimum volume ellipsoid enclosing can be formulated as following convex minimization problem,

$$\begin{aligned} \min_{M} \quad & -\ln \det M, \\ s.t. \quad & (Mx_i - z)^T (Mx_i - z) \le 1, \forall i = 1,2,\cdots,n, \\ & M \succ 0, \end{aligned} \tag{1}$$

where $M = \sqrt{E}, z = c\sqrt{E}$, square root of X is defined as : $\sqrt{X} = V^T \sqrt{D_{[d_{ii}]}} V$, V is eigenvectors, $D_{[d_{ii}]}$ is an element-wise square root of eigenvalues.

Definition 5. Decompose $\tilde{M} = \left(\sum_{i=1}^{n} \alpha_i \tilde{x}_i \tilde{x}_i^T \right)^{-1} = \begin{pmatrix} s & v^T \\ v & F \end{pmatrix}$, $\tilde{x}_i = \begin{pmatrix} 1 \\ x_i \end{pmatrix}$, where $v = -F\tilde{z}$, $F \in \Re^{k\times k}$, $v \in \Re^k$, s is a constant. Denote $\delta = 1 - s + \tilde{z}^T F \tilde{z}$, the linear transformation $f : \varepsilon(\tilde{M}, 0) \to \varepsilon(M, z)$ is defined as $\begin{cases} z = \tilde{z} \\ M = \delta^{-1} F \end{cases}$. The ellipsoid $\varepsilon(E, c)$ can be computed from $\begin{cases} E = (\delta^{-1}F)^T (\delta^{-1}F) \\ c = -F^{-1} v(E)^{-\frac{1}{2}} \end{cases}$.

Lemma 1. Minimization of the volume of the ellipsoid $\varepsilon(M, z)$ in $\Re^k$ space is equivalent to minimization of the volume of the ellipsoid $\varepsilon(\tilde{M}, 0)$ in augmented $\Re^{k+1}$ space centered at the origin using linear transformation f.

Proof. The proof is straightforward. Since there exists a linear transformation f, we can conclude for a given ellipsoid $\varepsilon(M, z)$ in $\Re^k$ space that there always exists an accordingly augmented $\varepsilon(\tilde{M}, 0)$ in $\Re^{k+1}$ space and vice versa.

According to Lemma 1, (1) can be rewritten as:

$$\begin{aligned} \min_{\tilde{M}} \quad & -\ln \det \tilde{M}, \\ s.t. \quad & \tilde{x}_i^T \tilde{M} \tilde{x}_i \le 1, \forall i = 1, 2, \cdots, n, \\ & \tilde{M} \succ 0. \end{aligned} \tag{2}$$

The dual form is:

$$\begin{aligned} \max_{\alpha_i} \quad & \ln \det \sum_{i=1}^{n} \alpha_i \tilde{x}_i \tilde{x}_i^T, \\ s.t. \quad & \sum_{i=1}^{n} \alpha_i = k + 1, \\ & 0 \le \alpha_i \le 1, \forall i = 1, 2, \cdots, n. \end{aligned} \tag{3}$$

According to KKT, we get $\tilde{M}^{-1} = \sum_{i=1}^{n} \alpha_i \tilde{x}_i \tilde{x}_i^T$. Define $A : A_{ii} = a_i = \sqrt{\alpha_i} \ge 0$, then $\sum_{i=1}^{n} \alpha_i \tilde{x}_i \tilde{x}_i^T = X^T A^2 X$, $(AX)^T AX = X^T A^2 X$, $AX\,(AX)^T = AXX^T A$.

Using singular value decomposition, we can get $X^T A^2 X = P\Lambda P^T$, $AXX^T A = AKA = Q\Lambda Q^T$, $P = X^T AQ\Lambda^{-\frac{1}{2}}$.

Using eigenspectrum analysis, we can infer following important lemma.

Lemma 2. For $\tilde{M}^{-1} = (AX)^T (AX) = X^T A^2 X$, $(AX)(AX)^T = AXX^T A$, following identities are concluded [12-13]:

$$\begin{aligned} \ln\det\left(AXX^T A\right) &= \sum_{i:\lambda_i \neq 0} \ln(\lambda_i) + \left(k+1-\#\{\lambda_i \neq 0\}\right)\ln(0), \\ \ln\det\left(X^T A^2 X\right) &= \sum_{i:\lambda_i \neq 0} \ln(\lambda_i) + \left(n-\#\{\lambda_i \neq 0\}\right)\ln(0), \end{aligned} \tag{4}$$

where λ_i is nonzero eigenvalue, # denotes their total number.

3.2 Regularized MVEE Cognitive Learner

According to lemma 2, it is obvious that there are probable existences of zero eigenvalues. It is therefore recommended to add a regularized item μI in the $\ln\det(\bullet)$ objective function. According to lemma 2, we can easily conclude following identities:

$$\begin{aligned} \ln\det\left(X^T A^2 X + \mu I\right) &= \sum_{i:\lambda_i \neq 0} \ln(\lambda_i + \mu) + \left(k+1-\#\{\lambda_i \neq 0\}\right)\ln(\mu), \\ \ln\det\left(AXX^T A + \mu I\right) &= \sum_{i:\lambda_i \neq 0} \ln(\lambda_i + \mu) + \left(n-\#\{\lambda_i \neq 0\}\right)\ln(\mu). \end{aligned} \tag{5}$$

To realize this regularized operation, we can add the item $\mu\,\mathrm{trace}(\tilde{M})$ due to $\mathrm{trace}(\tilde{M}) = \sum_{i=1}^{k+1} \frac{1}{\lambda_i}$, where λ_i is eigenvalue of M^{-1}. Then the primal regularized MVEE can be written as:

$$\begin{aligned} \min_{\tilde{M},\xi_i} \quad & -\ln\det\tilde{M} + \mu\,\mathrm{trace}(\tilde{M}) + \frac{1}{n}\sum_{i=1}^{n}\xi_i + \nu\rho, \\ s.t. \quad & \tilde{x}_i^T \tilde{M} \tilde{x}_i \le \rho + \xi_i, \forall i = 1, 2, \cdots, n, \\ & \tilde{M} \succ 0, \xi_i \ge 0, \rho \ge 0, \forall i = 1, 2, \cdots, n, \end{aligned} \tag{6}$$

where $\nu \geq 0$ is now a user specified parameter that equals the fraction of sample points outside the optimized ellipsoid. ρ is a variable that controls the volume according to ν, ξ_i is a slack variable that adjusts for misclassified samples.

By introducing dual variables $\alpha_i, \beta_i, \gamma \geq 0$, the Lagrange dual form is:

$$\begin{aligned} &\max_{\alpha_i} \quad \ln\det\left(\sum_{i=1}^{n} \alpha_i \tilde{x}_i \tilde{x}_i^T + \mu I\right), \\ &s.t. \quad \sum_{i=1}^{n} \alpha_i = \nu, \\ &\qquad 0 \leq \alpha_i \leq \frac{1}{n}, \forall i = 1, 2, \cdots n, \end{aligned} \tag{7}$$

where ρ^*, ξ_i^* can be determined using following KKT conditions:

$$\begin{cases} \alpha_i^* \left(\rho^* + \xi_i^* - \tilde{x}_i^T U^* U^{*T} \tilde{x}_i\right) = 0, \\ \beta_i^* \xi_i^* = 0, \left(\frac{1}{n} - \alpha_i^*\right) \xi_i^* = 0, \\ \gamma^* \rho^* = 0. \end{cases} \tag{8}$$

Thus, for a given sample $\tilde{x}_i$, we only need to check whether it is located inside the MVEE as mentioned before. If it satisfies $\tilde{x}_i^T \tilde{M} \tilde{x}_i \leq 1$, then the sample is inside the MVEE. Otherwise the sample is outside it, which forms the basic idea of the recognizing algorithm.

3.3 Kernel Regularized MVEE Cognitive Learner

The matrices $X^T A^2 X$ and $AXX^T A$ have the same nonzero eigenvalues. According to (5), we have following identity:

$$\ln\det\left(AXX^T A + \mu I\right) = \ln\det\left(X^T A^2 X + \mu I\right) + \left(n - (k+1)\right)\ln(\mu). \tag{9}$$

As for the inner product XX^T, we can always find a kernel K which satisfies Mercer condition to replace it, i.e. $XX^T = K$. Equation (9) can be rewritten as:

$$\ln\det(AKA + \mu I) = \ln\det(X^T AAX + \mu I) + (n - k - 1)\ln(\mu). \tag{10}$$

Hence, we can optimize $\ln\det(AKA + \mu I)$ instead of $\ln\det(X^T AAX + \mu I)$. The corresponding kernel regularized MVEE can be written as:

$$\max_{\alpha_i} \quad \ln\det\left(AKA+\mu I\right),$$
$$s.t. \quad \sum_{i=1}^{n}\alpha_i = \nu, \tag{11}$$
$$0 \le \alpha_i \le \frac{1}{n}, \forall i = 1,2,\cdots n.$$

To connect (11) with the dual variable α, we can define $G \in R^{n\times n}$ by $K = GG^T$. We then have $AKA = AGG^T A$, $G^T A^2 G = \sum_{i=1}^{n}\alpha_i g_i g_i^T$. According to (9), we obtain the final dual kernel regularized MVEE:

$$\max_{\alpha_i} \quad \ln\det\left(\sum_{i=1}^{n}\alpha_i g_i g_i^T + \mu I\right),$$
$$s.t. \quad \sum_{i=1}^{n}\alpha_i = \nu, \tag{12}$$
$$0 \le \alpha_i \le \frac{1}{n}, \forall i = 1,2,\cdots n.$$

The equation (12) is convex, and can be solved via state of the arts of convex programming solvers such as SeDuMi [14] and YALMIP [15]. From the solution, we obtain the kernel regularized MVEE cognitive learner. Now we should consider the kernelized recognizing algorithm. For a given ellipsoid $\varepsilon(\tilde{M},0)$, the Mahalannobis distance is defined as $d\left(\tilde{x},\tilde{M}\right) = \tilde{x}^T\tilde{M}\tilde{x}, \tilde{x}\in R^{k+1}$. Yet, $d\left(\tilde{x},\tilde{M}\right)$ can not be directly expressed in kernel form. However, by noting that $X^T A^2 X = P\Lambda P^T$, $P = X^T AQ\Lambda^{-\frac{1}{2}}$, the Woodbury formula shows that the Mahalanobis distance can be rewritten as:

$$d\left(\tilde{x},\tilde{M}\right) = \frac{1}{\mu}k\left(\tilde{x},\tilde{x}\right) - \frac{1}{\mu}\mathbf{k}^T AQ\left(\Lambda+\mu I\right)^{-1} Q^T A\mathbf{k}, \tag{13}$$

where $\mathbf{k} = \left(k\left(\tilde{x}_1,x\right),k\left(\tilde{x}_2,x\right),\cdots,k\left(\tilde{x}_n,x\right)\right)^T$, Q,Λ can be determined via $AKA = Q\Lambda Q^T$.

4 Experiments

This section investigates the enclosing learning machine on a ball bearing dataset for novelty detection [16]. One class SVM (OCSVM) is adopted for performance comparison. We use LIBSVM [17] for implementation of OCSVM. The MVEE are programmed in MATLAB via YALMIP. Both linear and RBF kernel for the two methods are investigated. The optimum parameters of OCSVM and MVEE are determined via Cross-Validation. The dataset consisted of 5 categories: Normal data from new ball bearings and 4 types of abnormalities, i.e. Fault 1 (outer race completely broken), Fault 2 (broken cage with one loose element), Fault 3 (damaged cage with four loose elements) and Fault 4 (a badly worn ball bearing with no evident damage). Each instance consisted of 2048 samples of acceleration. After preprocessing with a discrete Fast Fourier Transform, each such instance had 32 attributes.

Table 1. Ball bearing dataset novelty detection success rate

Learner	σ	ν	μ	Normal	Fault 1	Fault 2	Fault 3	Fault 4
OCSVM	—	0.001	—	89.7	99.7	94.9	54.9	52.1
OCSVM	320	0.001	—	90.3	100	95.2	68.3	85.1
MVEE	—	0.001	0.03	92.5	100	97.1	90.8	97.2
MVEE	320	0.001	0.03	90.1	100	100	97.5	98.1

5 Conclusions

We proposed a novel machine learning paradigm based on minimum volume enclosing shapes called enclosing machine learning, and illustrated the concept and algorithm using minimum volume enclosing ellipsoid. We have developed MVEE class description algorithm for cognitive process modeling, and validated the algorithm via benchmark dataset. The results prove the proposed MVEE enclosing learning machine is comparable or even better than SVMs in the dataset studied.

Acknowledgements

This paper is jointly supported by NSFC and CAAC under Grant #60672179, and also supported by the Doctorate Foundation of the Engineering College, Air Force Engineering University of China under Grant #BC0501.

References

1. Li, Y.H., Wei, X.K, Liu, J.X.: Engineering Applications of Support Vector Machines. 1st edn. Weapon Industry Press, Beijing China (2004)
2. Li, Y.H., Wei X.K.: Fusion Development of Support Vector Machines and Neural Networks. Journal of Air Force Engineering University **4** (2005) 70-73
3. Wei, X.K., Li, Y.H, Feng, Y.: Comparative Study of Extreme Learning Machine and Support Vector Machines. Advances in Neural Networks-ISNN 2006 (Lecture Notes in Computer Science), Springer-Verlag, Berlin Heidelberg New York. **3971** (2006) 1089-1095
4. Wei, X.K., Li, Y.H.: Enclosing Machine Learning: Concepts and Algorithms. Technique Report AFEC-2006-1, Air Force Engineering University, Xi'an, China (2006)

5. Vandenberghe, L., Boyd, S., Wu, S.P.: Determinant Maximization with Linear Matrix Inequality Constraints. SIAM Journal on Matrix Analysis and Applications **2** (1998) 499-533
6. Tax, D.M.J., Duin, R.P.W.: Support Vector Domain Description. Pattern Recognition Letters **20** (1999) 1191-1199
7. Fischer, K., Gartner, B., Kutz, M.: Fast Smallest Enclosing Ball Computation in High Dimensions. Lecture Notes in Computer Science, Algorithms - ESA 2003, Springer-Verlag, Berlin Heidelberg New York **2832** (2003) 630-641
8. Kumar, P., Mitchell, J. S. B., Yıldırım, E. A.: Approximate Minimum Enclosing Balls in High Dimensions Using Core-sets. The ACM Journal of Experimental Algorithmics **8** (1) (2003) 1-29
9. Sun, P., Freund, R.M.: Computation of Minimum Volume Covering Ellipsoids. Operations Research **5** (2004) 690-706
10. Panigrahy, R.: Minimum Enclosing Polytope in High Dimensions. Preprints, available at http://arxiv.org/abs/cs.CG/0407020 (2004)
11. Kumar, P., Yıldırım, E.A.: Minimum Volume Enclosing Ellipsoids and Core-sets. Journal of Optimization Theory and Applications **126** (1) (2005) 1-21
12. Shawe-Taylor, J., Williams, C., Cristianini, N., Kandola, J. S.: On the Eigenspectrum of the Gram Matrix and Its Relationship to the Operator Eigenspectrum. In: N. CesaBianchi et al (eds.): Proceedings of the 13th International Conference on Algorithmic Learning Theory (ALT2002). Lecture Notes in Artificial Intelligence, Vol. 2533. Springer-Verlag, Berlin Heidelberg New York (2002) 23-40
13. Dolia, A.N., Page, S.F., White, N.M., Harris, C.J.: D-optimality for Minimum Volume Ellipsoid with Outliers. In: Proceedings of the Seventh International Conference on Signal/Image Processing and Pattern Recognition (UkrOBRAZ2004) (2004)73-76
14. Sturm, J.F.: Using SeDuMi 1.02, A MATLAB Toolbox for Optimization over Symmetric Cones. Optimization Methods and Software **11&12** (1999) 625-653
15. Löfberg, J.: YALMIP: A Toolbox for Modeling and Optimization in MATLAB. http://control.ee.ethz.ch/ ~joloef (2006)
16. Structural Integrity and Damage Assessment Network. http://www.sidanet.org
17. Chang, C.C., Lin, C.J.: LIBSVM: a Library for Support Vector Machines. http://www.csie.ntu.edu.tw/ ~cjlin/libsvm (2001)

An Extremely Simple Reinforcement Learning Rule for Neural Networks

Xiaolong Ma

Stony Brook University, Stony Brook, NY 11794-3800, USA
xma@grad.physics.sunysb.edu

Abstract. In this paper we derive a simple reinforcement learning rule based on a more general form of REINFORCE formulation. We test our new rule on both classification and reinforcement problems. The results have shown that although this simple learning rule has a high probability of being stuck in local optimum for the case of classification tasks, it is able to solve some global reinforcement problems (e.g. the cart-pole balancing problem) directly in the continuous space.

1 Introduction

In contrast to supervised training methods such as error back-propagation [1], for global reinforcement learning [2], a "learning agent" is provided with a global evaluative feedback r ("reward"), rather than with examples of correct answers. Under these conditions, some randomness is generally needed in order to explore the space of all possible "policies" (i.e. the set of agent's internal parameters $\boldsymbol{\theta}$). This paper concerns reinforcement learning in neural networks consisting of neural cells whose output signals y are sent to inputs of other cells through synapses with certain weights w_{ij}:

$$x_i = \sum_j w_{ij} y_j \ . \tag{1}$$

In this case, the policy is just the input-output (i.e., the state-action) mapping, which is determined by the set of synaptic weights.

Williams [3] has derived a general class of "REINFORCE" (REward Increment = Nonnegative Factor × Offset Reinforcement × Characteristic Eligibility) learning algorithms for neural networks with random cell outputs. In our previous work [4], [5] we have shown that the REINFORCE approach can be extended to networks with randomness coming from cell outputs, cell inputs or synaptic weights; and very simple and hardware friendly learning rules can be derived from the more general form of REINFORCE. (This goal is motivated by our group's work on CMOL CrossNets, a specific nanoelectronic implementation of neural networks [6].)

This paper focus on the simplest one of the new learning rules, namely Rule B, and its application to the reinforcement learning tasks. In Sec. 2 we will briefly

D. Liu et al. (Eds.): ISNN 2007, Part I, LNCS 4491, pp. 434–440, 2007.

discuss the derivation of rule B. In Sec. 3 and Sec. 4 we apply rule B to some popular classification and reinforcement benchmark learning tasks. The results are summarized in Sec. 5.

2 Derivation of Rule B

In our recent work [4] we have shown that for any stochastic system with a set of random signals $\mathbf{v} = \{v_1, v_2, ...\}$, and a probability function $p(\mathbf{v}, \boldsymbol{\theta})$ controlled by a set of deterministic internal parameters $\boldsymbol{\theta}$, the following general stochastic learning rule [7]

$$\Delta\theta_k = \eta_k r e_k \ , \tag{2a}$$
$$e_k = \nabla_{\theta_k} \ln[p(\mathbf{v}, \boldsymbol{\theta})] \ , \tag{2b}$$

performs a statistical ascent on the average reward:

$$\Delta E\{r|\boldsymbol{\theta}\} \approx \boldsymbol{\nabla}_{\boldsymbol{\theta}} E\{r|\boldsymbol{\theta}\} \cdot E\{\Delta\boldsymbol{\theta}|\boldsymbol{\theta}\} \geq 0 \ . \tag{3}$$

In Eqs. (2), $\eta_k > 0$ is the learning rate; the reward is a function of signals $r = r(\mathbf{v})$; e_k is called "characteristic eligibility"; $\mathbf{v}$ can be any random variable, such as y_i, x_i, or even w_{ij}; and $\boldsymbol{\theta}$ can be any deterministic parameter, for example, in the case of random weights it can be average weight $\langle w_{ij} \rangle$ instead of w_{ij}. This makes Eqs. (2) applicable to more general situations.

In the case of neural networks, with their local relation between signals, Eq. (2b) leads to simple learning rules because the partial derivative with respect to a particular weight only depends on local signals (for details, see Ref. [4]). For example, in networks with the "Bernoulli-logistic" stochastic cells (and deterministic synapses), Eqs. (2) (with addition of a small anti-Hebbian term to avoid local maxima) leads to the the famous [1], [2]

$$\textbf{Rule A}_{\textbf{r-p}}\textbf{:} \quad \Delta w_{ij} = \eta \left[r(y_i - \langle y_i \rangle) y_j + \lambda(1-r)(-y_i - \langle y_i \rangle) y_j \right] \ , \tag{4}$$

where λ is a small positive number.

Now let us apply Eqs. (2) to random weights w_{ij} with mean values μ_{ij}. At a fixed network input, we may consider r as a function of the synaptic weight set, $r = r(\mathbf{w})$. From this standpoint, in Eqs. (2) we can replace $\mathbf{v}$ with $\mathbf{w}$, and $\boldsymbol{\theta}$ with $\boldsymbol{\mu}$. Let us assume that the synaptic weights have the Gaussian distribution with variance σ_{ij}^2:

$$p_{ij}(w_{ij}) = \frac{1}{\sqrt{2\pi}\sigma_{ij}} \exp\left[-\frac{(w_{ij} - \mu_{ij})^2}{2\sigma_{ij}^2} \right] \ , \tag{5}$$

then

$$e_{ij} = \frac{\partial \ln p_{ij}}{\partial \mu_{ij}} = \frac{w_{ij} - \mu_{ij}}{\sigma_{ij}^2} \ . \tag{6}$$

With $\eta_{ij} = \eta\sigma_{ij}^2$, we obtain the following simple rule:

$$\textbf{Rule B:} \quad \Delta\mu_{ij} = \eta r(w_{ij} - \mu_{ij}) \; . \tag{7}$$

This is perhaps the simplest learning rule suggested for artificial neural networks so far. Each weight change involves no information other than its own perturbation and the global reward signal.

Rule B can also be applied to binary weights with Bernoulli distribution. For example if the weights w_{ij} can be either 0 or 1, with the probability

$$p(w_{ij}, \mu_{ij}) = \begin{cases} \mu_{ij}, & \text{if } w_{ij} = 1; \\ 1 - \mu_{ij}, & \text{if } w_{ij} = 0, \end{cases} \tag{8}$$

then

$$\frac{\partial \ln p}{\partial \mu_{ij}} = \begin{cases} 1/\mu_{ij}, & \text{if } w_{ij} = 1; \\ -1/(1 - \mu_{ij}), & \text{if } w_{ij} = 0. \end{cases} \tag{9}$$

Therefore

$$e_{ij} = \frac{w_{ij} - \mu_{ij}}{\mu_{ij}(1 - \mu_{ij})} = \frac{w_{ij} - \mu_{ij}}{\sigma_{ij}^2} \; , \tag{10}$$

and the learning rule is the same as Eq. (7).

3 Applying Rule B to the Parity Problem

We tested our learning rule on the parity function problem. Figure 1 shows the learning dynamics of a Multi-Layer Perceptron (MLP) [1] trained by Rule B. In this simple experiment, the inputs and output were binary (-1 and +1),[1] and the single output should tell whether the number of +1s in the inputs is even or odd. The neural cells were deterministic, with the following activation functions:

$$y_i = \tanh(h_i) = \tanh\left(Gx_i/\sqrt{N_{m-1}}\right) \; . \tag{11}$$

where $G = 0.4$ throughout the paper and N_{m-1} is the number of cells in the previous layer (suppose cell i is in layer m). The reward signal was simply $r = +1$ for the correct sign of the output signal, and $r = -1$ for the wrong sign. The network performance was measured by the sliding average reward defined as

$$r_a(t) = 0.99 r_a(t-1) + 0.01 r(t) \; , \tag{12}$$

where $r(t)$ is r averaged for all patterns (16 of them in this case) at the tth epoch. At the iteration of each pattern, the weights were drawn independently from Gaussian distributions with a global variance σ (but different mean values). The following "fluctuation quenching" procedure (strength controlled by positive parameter α) was used to help stabilize output at the end of training:

$$\sigma(t) = \sigma(0)[1 - r_a(t)]^\alpha \; . \tag{13}$$

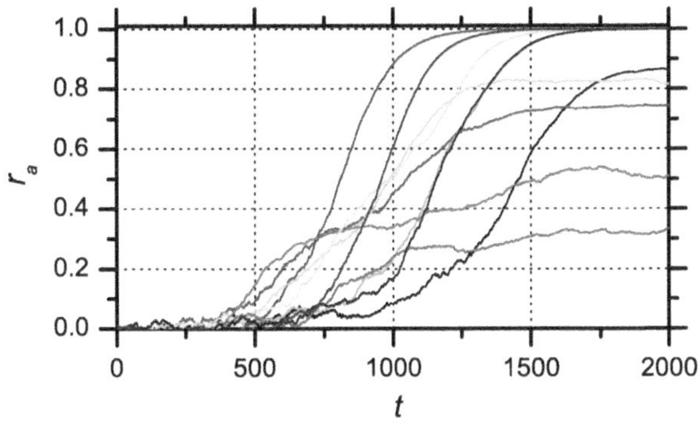

Fig. 1. The process of training a fully connected MLP (4-10-1) with random synaptic weights to implement the 4-input parity function. Plots show the sliding average reward as a function of training epoch number for 10 independent simulations runs. The parameters are $\eta = 0.02$, $\sigma(0) = 10$ and $\alpha = 1$.

Figure 1 shows that Rule B generally works, but it suffers from occasional trapping in local maxima of the effective multidimensional potential profile. This can be a severe problem when Rule B is applied to the classification problems. Moreover, unlike other REINFORCE learning rules, Rule B can not be augmented with an "anti-trapping" term to tackle this problem (see Ref. [5] for details). Fortunately, in the next section we will see that when applied to reinforcement problems, Rule B usually does not get stuck in local maxima.

4 Applying Rule B to the Cart-Pole Balancing Problem

In the cart-pole balancing task [8] the system tries to balance a pole hinged to a cart moving freely on a track (Fig. 2) by applying a horizontal force to the cart. A failure occurs when either the pole incline angle magnitude θ exceeds 12 degrees, or the cart hits one of the walls ($x = \pm 2.4$ m). In our experiments, a reward of $r = -1$ is issued upon failure and $r = 0.1$ otherwise.

To solve this delayed-reward problem (for which the reward is not simply a function of current action), the usual actor-critic method [2] was used. The actor is a 4-30-1 MLP which takes the state vector of the cart-pole system $\{x(t), \dot{x}(t), \theta(t), \dot{\theta}(t)\}$ as input and produces a single output $a(t)$ as the action. This network has been trained by either B or $A_{r\text{-}i}$ (which is just $A_{r\text{-}p}$ with $\lambda = 0$)[2] with the Temporal Difference (TD) error [2]

$$\delta(t) = r(t) + \gamma V(t+1) - V(t) \; , \tag{14}$$

[1] Because of such symmetric data representation, a certain number of bias cells with constant output (+1) had to be added to the input and hidden layer. These biases are not included into the cell count in this paper.

[2] For this task, anti-trapping terms are not necessary.

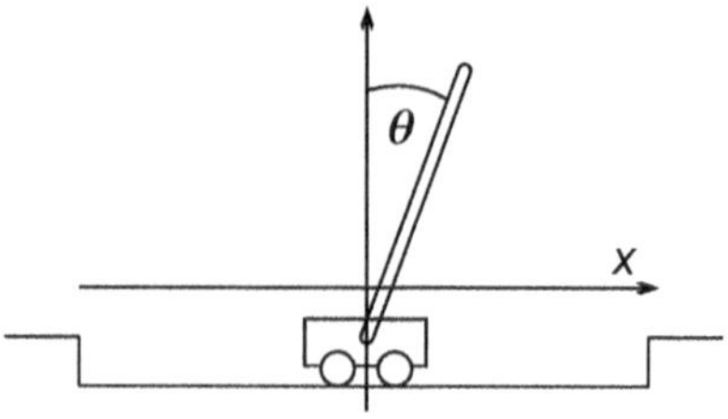

Fig. 2. The cart-pole balancing problem. The force applied to the cart is $a(t)F_{\max}$, where $-1 \leq a(t) \leq 1$ is a function of time. In our particular example $F_{\max} = 10$ N, the masses of the cart and pole are 1.0 kg and 0.1 kg, respectively, and the length of the pole is 1 m. The dynamics of the system is simulated with a time step of 0.02 s which is small in comparison with the dynamics time scales (which are of the order of 1 s).

playing the role of the instant reward signal. In Eq. (14), $r(t)$ is the real reward at time t, $V(t)$ is the value function and γ is the discount factor. For example, in the case of TD(λ), Rule B takes the form

$$\Delta w_{ij}(t) = \eta_a \delta(t) e_{ij}(t) , \tag{15a}$$
$$e_{ij}(t) = \gamma \lambda_{\mathrm{TD}} e_{ij}(t-1) + [w_{ij}(t) - \mu_{ij}(t)] . \tag{15b}$$

One more option here is to use an additional adaptation of individual fluctuation intensities instead of a global quenching Eq. (13). Indeed, by identifying the set of standard deviations σ_{ij} with $\boldsymbol{\theta}$ in Eqs. (2) and letting $\eta_{ij} = \eta_\sigma \sigma_{ij}^2$, one arrives at the following[3] (see Ref. [3] for a similar rule derived for Gaussian random cells):

$$\textbf{Rule } \sigma\textbf{:} \quad \Delta\sigma_{ij} = \eta_\sigma r[(w_{ij} - \mu_{ij})^2 - \sigma_{ij}^2]/\sigma_{ij} . \tag{16}$$

The critic is a 5-30-1 MLP which takes the state-action vector $\{x(t), \dot{x}(t), \theta(t), \dot{\theta}(t), a(t)\}$ as input and produces a single output $V(t)$ as a value function estimate. The critic has been trained by error backpropagation with TD error. In the case of TD(λ),

$$\Delta \mathbf{w}(t) = \eta_c \delta(t) \mathbf{e}(t) , \tag{17a}$$
$$\mathbf{e}(t) = \gamma \lambda_{\mathrm{TD}} \mathbf{e}(t-1) + \boldsymbol{\nabla}_{\mathbf{w}} V(t) . \tag{17b}$$

All the somatic cells in the critic network have the tanh activation function Eq. (11), except for the output cell which is linear: $V = y = 0.1h$.

In Fig. 3, we show simulation results using Rule Bσ (a combo of Rule B and Rule σ) and $A_{\text{r-i}}$. As we can see, although rule $A_{\text{r-i}}$ leads to faster training, the much simpler Rule B is also able to fully solve this problem (i.e., to learn how to balance the pole without failure indefinitely) eventually.

[3] This rule and Eq. (13) both help obtain faster training, but they are not necessary. The activation function naturally quenches the flucturation at the cell output level at the end of training.

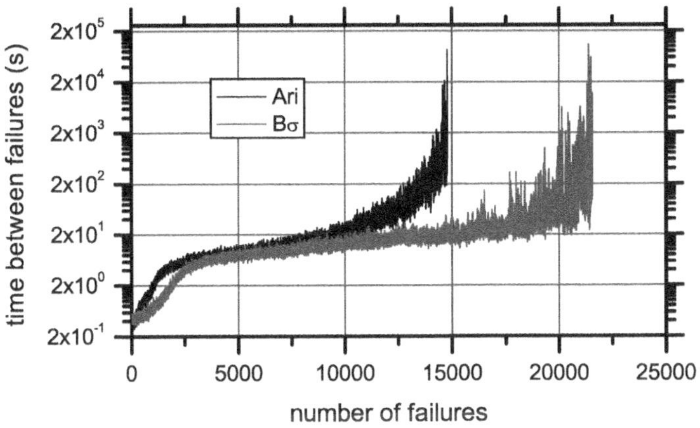

Fig. 3. Training dynamics for the cart-pole balancing task. All results were averaged over 20 independent experiments. After each failure, the system is restored to its initial condition ($x = \dot{x} = \theta = \dot{\theta} = 0$), and the experiment is continued. Parameters used in the training are: for Rule B: $\eta_a = 0.006$; for Rule σ: $\sigma_{ij} = 10$ initially and $\eta_\sigma = 0.00012$; for Rule $A_{\text{r-i}}$: $\eta_a = 0.02$; for Backprop: $\eta_c = 10$; for TD(λ): $\gamma = 0.95$, $\lambda_{\text{TD}} = 0.6$.

In comparison with the usual reinforcement learning using RBF network [2] or CMAC [9], the learning is slow, at least for this particular problem. However, unlike those methods, our method learns directly in the continuous space – no discretization whatsoever is involved.

5 Discussion

In this paper, we have shown that an extremely simple reinforcement learning rule, Rule B, can be used to solve reinforcement control problems. When applied to the parity problem, this gradient following learning rule can easily get stuck in local optima. But our application of Rule B to the cart-pole balancing task was very successful (and quite insensitive to parameter changes), probably due to the random nature of the reinforcement tasks.

Rule B does not assume any knowledge of the structure of the network, therefore it is applicable to any learning model with an arbitrary set of internal parameters (not limited to neural networks). In our simulation of the cart-pole balancing task, the continuous space vector was directly fed to the networks without any preprocessing. We believe this makes our method applicable to a broader range of problems.

Acknowledgments. Valuable discussions with Paul Adams, Dan Hammerstrom, Jung Hoon Lee and Konstantin K. Likharev are gratefully acknowledged. This work was supported in part by AFOSR, NSF, and MACRO via FENA Center.

References

1. Hertz, J., Palmer, R.G., Krogh, A.S.: Introduction to the theory of neural computation. Addison-Wesley Pub. Co., Redwood City, CA (1991)
2. Sutton, R.S., Barto, A.G.: Reinforcement learning : An introduction. MIT Press, Cambridge, MA (1998)
3. Williams, R.J.: Simple statistical gradient-following algorithms for connectionist reinforcement learning. Machine Learning **8** (1992) 229–256
4. Ma, X., Likharev, K.K.: Global reinforcement learning in neural networks with stochastic synapses. In: Proc. of WCCI/IJCNN'06. (2006) 47–53
5. Ma, X., Likharev, K.K.: Global reinforcement learning in neural networks. to be published in IEEE Tran. on Neural Networks (2007)
6. Türel, Ö., Lee, J.H., Ma, X., Likharev, K.K.: Neuromorphic architectures for nanoelectronic circuits. Int. J. Circ. Theory App. **32** (2004) 277–302
7. Baxter, J., Bartlett, P.L.: Infinite-horizon policy-gradient estimation. Journal of Artificial Intelligence Research **15** (2001) 319–350
8. Barto, A.G., Sutton, R.S., Anderson, C.W.: Neuronlike adaptive elements that can solve difficult learning control problems. IEEE Trans. Syst., Man, Cybern. **SMC-13** (1983) 834–846
9. Albus, J.S.: A new approach to manipulator conrol: the cerebellar model articulation controller (CMAC). Trans. of ASME Journal of Dynamic Systems, Measurements, and Control **97** (1975) 220–227

Online Dynamic Value System for Machine Learning

Haibo He[1] and Janusz A. Starzyk[2]

[1] Deptartment of Electrical and Computer Engineering, Stevens Institute of Technology, Hoboken, NJ 07030, USA
hhe@stevens.edu
[2] School of Electrical Engineering and Computer Science, Ohio University, Athens, OH 45701, USA
starzyk@bobcat.ent.ohiou.edu

Abstract. A novel online dynamic value system for machine learning is proposed in this paper. The proposed system has a dual network structure: data processing network (DPN) and information evaluation network (IEN). The DPN is responsible for numerical data processing, including input space transformation and online dynamic data fitting. The IEN evaluates results provided by DPN. A dynamic three-curve fitting (TCF) scheme provides statistical bounds to the curve fitting according to data distribution. The system uses a shift register communication channel. Application of the proposed value system to the financial analysis (bank prime loan rate prediction) is used to illustrate the effectiveness of the proposed system.

1 Introduction

Online value system is useful for machine learning. For instance, in reinforcement learning (RL) a machine learns values of its state/action pairs [1] to direct its actions towards a goal. By analyzing sensory inputs from the external environment, an intelligent system (agent) should evaluate the information received according to its value system, and act to maximize the expected reward. An agent learns from active interaction with its environment, and while acting on the environment, it accumulates knowledge through experience.

A typical reinforcement learning system includes the external environment, a policy, and a value function that describes expected reward. R. S. Sutton argued that in this system the value function is of critical importance as all RL algorithms estimate the state-action values [1]. Although it is important to estimate the value accurately and dynamically, it is difficult to do so in practical learning environment for numerous reasons:

– Limited availability of information;
– Information ambiguity and redundancy;
– High dimensionality of the data set;
– Time variability of the information;

D. Liu et al. (Eds.): ISNN 2007, Part I, LNCS 4491, pp. 441–448, 2007.

Due to the importance of the value systems, many research results have been recently reported in the literature. For instance, paper [2] proposed an artificial neural network value system incorporating multiple regression analysis. This system combined the analysis results of three neural networks, named back-propagation, probabilistic network and self-organizing feature map. Paper [3] proposed a fuzzy-based navigation system for two mobile robots using distributed value function reinforcement learning. This approach enables multiple robots to learn a value function, which is an estimation of future rewards for all robots. In this way, cooperations between two robots are maintained and each robot learns to execute the actions that are good for the team.

In this paper, we propose a novel online feedforward neural network value system capable of estimating the value of multi-dimensional data sets. Similar to learning array presented in [4], this network has a multilayer, regular array structure of processing elements (PE) with local interconnections that can be determined by PE self-organization scheme.

The motivation of this research is to provide a mechanism for the intelligent machines to be able to dynamically estimate the value function in reinforcement learning (specify "good" from "bad"), therefore guiding the machines to adjust its actions to achieve the goal. The "value" in this paper can be a numerical expression of a fundamental principle or desired objective function value in a practical application problem. A user can define his own value for each application. For example, in financial analysis, the value could be a numerical index that reflects the intrinsic value of the analyzed company for an investment decision or a numerical measure of its financial performance. This approach differs from classical backpropagation neural network approach in which a function value is given as a desired output and is used to adjust interconnection weights in the backpropagation process [5].

The main contribution of our research is the proposed dynamic value system and its implementation architecture. This value system is a scheme, not a specific algorithm; therefore, it can be used in different ways, such as selection of input space transform functions, selection of different basis functions or different voting schemes.

2 Online Curve Fitting Principles

Online dynamic curve fitting is the core module of the proposed value system. It contains a network of processing elements (PE) that approximate the incoming data values. In this section, we first show how PEs implement online dynamic curve fitting. We then discuss the proposed three curve fitting (TCF) scheme to fit the statistically distributed incoming data.

2.1 Online Dynamical Curve Fitting

Consider a dynamic adjustment of the fit function described by a linear combination of basis functions $\varphi_i, i = 1, 2, ...q$, where q is the number of basis functions.

This number can be adjusted according to the required accuracy and the data noise level. Our objective is to dynamically fit values of the received data samples. We assume that each PE has two inputs describing its subspace data points with coordinates x and y. Each PE dynamically adjusts its fit function to minimize the least square error in approximated values of all the training data x and y as follows:

$$Y = a_1 \times \varphi_1 + a_2 \times \varphi_2 + ... + a_q \times \varphi_q. \tag{1}$$

We can determine the coefficients $a_1, a_2, ...a_q$ by pseudo inverse of the matrix composed of the basis function values at the input data. To do this dynamically we need to accumulate function values and their combinations for different input samples. Thus the unknown coefficients a_i in equation (1) can be solved as follows:

$$Y = \begin{bmatrix} \varphi_1 \ \varphi_2 \ ... \ \varphi_q \end{bmatrix} = \begin{bmatrix} a_1 \\ a_2 \\ ... \\ a_q \end{bmatrix} = \Phi \times A, \tag{2}$$

then we have

$$\begin{bmatrix} a_1 \\ a_2 \\ ... \\ a_q \end{bmatrix} = \left(\Phi^T \Phi\right)^{-1} \Phi^T Y = \begin{bmatrix} \sum_{i=1}^{n} \Phi_{1i}\Phi_{1i} & \sum_{i=1}^{n} \Phi_{1i}\Phi_{2i} & ... & \sum_{i=1}^{n} \Phi_{1i}\Phi_{qi} \\ \sum_{i=1}^{n} \Phi_{1i}\Phi_{2i} & \sum_{i=1}^{n} \Phi_{2i}\Phi_{2i} & ... & \sum_{i=1}^{n} \Phi_{2i}\Phi_{qi} \\ ... & & & \\ \sum_{i=1}^{n} \Phi_{1i}\Phi_{qi} & \sum_{i=1}^{n} \Phi_{2i}\Phi_{qi} & ... & \sum_{i=1}^{n} \Phi_{qi}\Phi_{qi} \end{bmatrix}^{-1} * \begin{bmatrix} \sum_{i=1}^{n} \Phi_{1i}Y_i \\ \sum_{i=1}^{n} \Phi_{2i}Y_i \\ ... \\ \sum_{i=1}^{n} \Phi_{qi}Y_i \end{bmatrix}, \tag{3}$$

where n is the number of data points. For online implementation, this requires storage of $q \times (q+1)/2 + q$ values of different correlations in equation (4)

$$\begin{cases} \sum_{i=1}^{n} \Phi_{ki}\Phi_{mi}, \\ \sum_{i=1}^{n} \Phi_{ki}Y_i, \end{cases} \tag{4}$$

where $k, m = 1, 2...q$. As new samples arrive, these s values are updated, and equation (3) is solved for new coefficients $a_1, a_2, ...a_q$. In general, for q basis functions we may need to invert $q \times q$ matrix $(\Phi^T \Phi)$ to update the coefficients of the approximating equation.

2.2 Three-Curve Fitting and the Voting Scheme

For noisy data, the single curve fitting technique presented in section 2.1 has its limitations. Fig. 1(a) gives a general idea of such a single curve fit by an individual

PE. As we can see, the fitted curve does not reflect the statistical distribution of the input data values in areas A and B, which will cause poor value fitting in these areas. We could compute a standard deviation of the approximated data from the curve fit, but this would only give a uniform measure of statistical errors that does not reflect the different quality of approximation in different regions of the input space. In order to overcome this limitation, a three-curve-fitting (TCF) scheme is proposed. Fig. 1(b) illustrates how the TCF method, fits the same sample data values by using three curves:

Neutral Curve: that fits to all the data samples in the input space (same as the curve in Fig. 1(a))

Upper Curve: that fits only to the data points that are above the neutral curve.

Lower Curve: that fits only to the data points that are below the neutral curve.

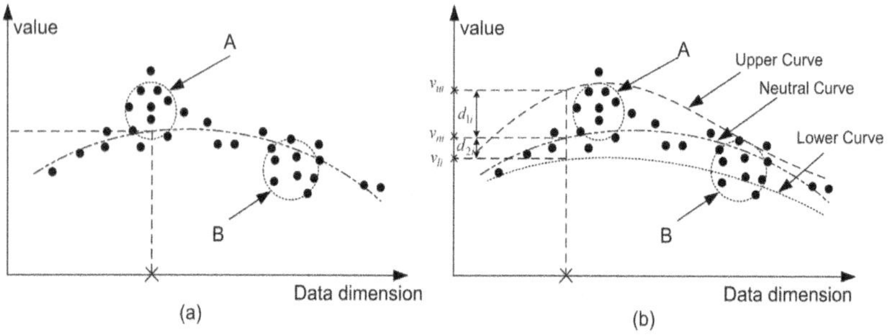

Fig. 1. (a) Single curve fitting; (b) Three curve fitting (TCF) scheme

As we can see from Fig. 1(b), the *neutral curve* provides a rough estimation of the fitted value, while the *upper* and the *lower* curves provide localized statistical distribution information. In a dynamic implementation, when a new data sample is received, we first modify the *neutral curve.* Then we calculate the fitted value of the *neutral curve* v_{ni}. If v_{ni} is smaller than the true value of this new sample, then we continue to modify the coefficients of the *upper curve* and keep the *lower curve* unchanged; otherwise, we modify the *lower curve* and keep the *upper curve* unchanged. Based on these *upper* and *lower* curves, we can locally characterize a statistical deviation of the approximated data from the value estimated by the *neutral curve.* As illustrated in Fig. 1(b), v_{ui}, v_{ni} and v_{li} are the values estimated by the *upper curve*, *neutral curve* and *lower curve*, respectively. The standard deviation of the estimated value is defined in the following way:

$$d_{1i} = |v_{ni} - v_{ui}|, d_{2i} = |v_{ni} - v_{li}|, d_i = (d_{1i} + d_{2i})/2, \tag{5}$$

d_i reflects how accurate the estimated value v_{ni} is compared to its true value. Small values of d_i mean that v_{ni} is obtained with higher confidence and should carry higher weight in the voting scheme at the information evaluation network.

However, when d_i is large, it means that v_{ni} is not so accurate and should contribute less to the final result. Therefore, the weights for each PE are calculated as $w_i = 1/d_i$. For a value system with k processing elements, the voting mechanisms used in the IEN network is implemented through:

$$v_{vote} = \sum_{i=1}^{k}(v_{ni}w_i)/\sum_{i=1}^{k}(w_i). \tag{6}$$

The average weight of all inputs processed by a PE can be used as a measure of quality of the local fit to the function approximated by this PE. This in turn, can be used by the PE to select a subset of inputs from all inputs connected to this PE, and to perform the dynamic function approximation in the subspace based on the selected inputs. Each PE selects its inputs such that its average weight is maximized. It can do it locally, independent on the state and the interconnection scheme of other PEs. This results in topological self-organization similar to the one presented in [4].

3 Value System Architecture

Fig. 2 shows the architecture of the proposed value system with dual network structure composed of DPN and IEN. DPN contains multiple layers of data processing elements (DPE). Each DPE selects its inputs, conducts the three-curve fitting as discussed in Section 2, and outputs their fitted values v_{ni}, v_{ui} and v_{li} to be processed by voting processing element (VPE) in IEN. The VPE establishes the final value using equations (5)-(6).

This architecture channels the information in a way similar to a hybrid shift-register structure. Each DPE has a set of inputs, that are pseudo randomly connected to local input channels. At the first clock cycle, input data is available at the first layer channel, and the first layer DPEs read this data as their inputs. DPEs output the transformed data (using their local transformation functions) into the same locations at the input channel. They also output their estimated values v_{ni} and their corresponding weights w_i to the VPE in IEN network. IEN network is composed of the sequence of VPE elements terminated with an element that computes the final value. The VPE combines values and weights received from a single layer of DPEs according to the following equations and passes them to the next layer of VPEs at the next clock cycle.

$$\hat{v}_l = \sum(w_i v_i)_l + \hat{v}_{l-1}, \tag{7}$$

$$\hat{w}_l = \sum(w_i)_l + \hat{w}_{l-1}, \tag{8}$$

here the subscript "l" means the values obtained from the layer l. Therefore, $\hat{v}_l$ and $\hat{w}_l$ represent the combined value and weight information for layer l. Final values are estimated by computing the ratio of the last layer $\hat{v}_l$ and $\hat{w}_l$.

At the next clock cycle, the transformed data (the output data of the DPE in the first layer) is shifted as the input data to the DPEs in the second layer, while another set of input data samples are sent to the first layer channel. The VPEs in the second layer combine the results obtained in the second layer with that passed from the previous layer. Other layers process their corresponding information concurrently, implementing a hybrid pipeline structure for function vale estimation. In this way, all processing elements in the system are active during all clock cycles making this architecture suitable for the dynamic online processing.

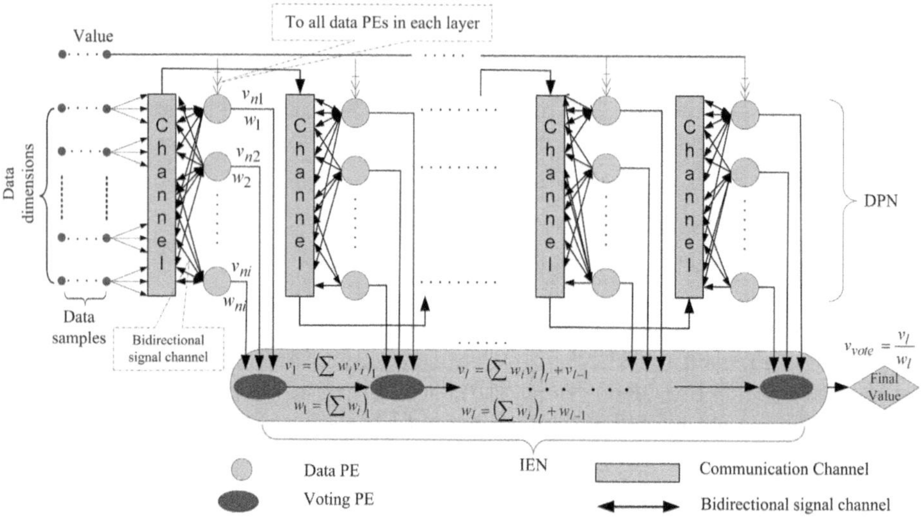

Fig. 2. Dynamic value system architecture

4 Simulation Results

We illustrate the application of the proposed value system to the financial data analysis - bank prime loan rate prediction. Financial data prediction is difficult due to the inherent noise, non-stationary, and non-linear characteristics of such data sets. The neural network based approach is a powerful tool financial data analysis and many research results have been reported recently. For instance, in [6], three neural network based learning mechanisms, including standard backpropagation (SBP), scaled conjugate gradient (SCG) and backpropagation with Bayesian regularization (BPR) were used to predict the foreign currency exchange rates. In [7], foreign exchange rate prediction was analyzed using recurrent neural networks and grammatical inference. In [8], an iterative evolutionary learning algorithm using fuzzy logic, neural networks, and genetic algorithm was proposed for the financial data prediction, and prediction results were compared with those obtained by classical fuzzy neural networks as in [9]. In [10], J. Yao

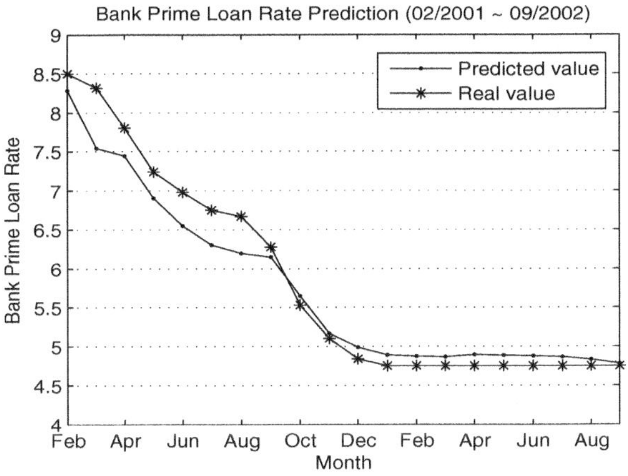

Fig. 3. Bank prime loan rate prediction by value system from February 2001 to September 2002

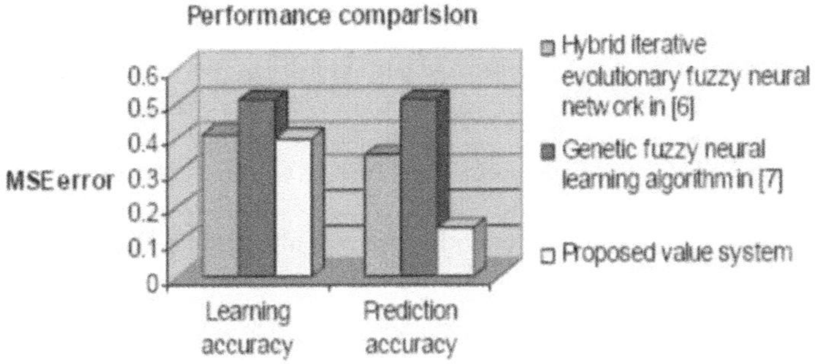

Fig. 4. Performance comparison of the proposed value system with those of [8] and [9]

and C. Tan presented empirical evidence that a neural network model is capable of foreign exchange rates prediction, and they also discussed the network architecture, model parameters, and performance evaluation methods.

In this paper, we have used the dataset from Financial Forecast Center (www.forecasts.org) and compared our prediction results with those of [8] [9]. The feature vector has four dimensions (monthly bank prime loan rate, discount rate, federal funds rate and ten-year treasury constant maturity rate) and the prediction value is the next month's bank prime loan rate. We use the data set from January 1995 to December 2000 for training, and February 2001 to September 2002 for testing. Fig. 4 shows the testing performance of the bank prime loan rate.

Fig. 4 shows the mean square error (MSE) comparison of the proposed value system with the best results of hybrid evolutionary fuzzy neural network and genetic fuzzy neural learning algorithm (both of them with 300 training iterations) as presented in [8] and [9], respectively. As we can see from Fig. 3 and Fig. 4, the proposed value system can effectively learn and predict the signal values.

5 Conclusion

A novel online value system for machine learning is proposed. The proposed system combines a data processing network and information evaluation network. A dynamic three-curve fitting scheme is proposed to improve the fitting quality based on the statistical distribution of the data samples. In addition, a hardware-oriented system level architecture with hybrid shift-register channel structure is also presented. Simulation results on the financial data prediction illustrate the effectiveness of the proposed value system. Motivated by the results presented in this paper, we believe that this approach may benefit research of value based reinforcement learning schemes.

References

1. Sutton, R.S., Barto, A.G.: Reinforcement Learning: An Introduction. MIT Press, Cambridge, MA (1998)
2. Panayiotou, P.A., Pattichis, C., Jenkins , D., Plimmer, F.: A Modular Artificial Neural Network Valuation System. The 10th Mediterranean Electrotechnical Conference (MELECON) (2000) 457-460
3. Babyey, S., Momtahan, O., Meybodi, M.R.: Multi Mobile Robot Navigation using Distributed Value Function Reinforcement Learning. Proceedings of IEEE International Conference on Robotics and Automation (ICRA '03) (2003) 14-19
4. Starzyk J. A., Zhu Z., Liu T.-H.: Self-Organizing Learning Array. IEEE Trans. on Neural Networks **16** (2) (2005) 355-363
5. Hecht-Nielsen, R.: Theory of the Backpropagation Neural Network Int. Joint Conf. Neural Networks (IJCNN) **1** (1989) 593-605
6. Kamruzzaman, J.: ANN-based Forecasting of Foreign Currency Exchange Rates. Neural Information Processing **3** (2) (2004) 49-58
7. Giles, C., Lawrence, S., Tsoi, A.: Noisy Time Series Prediction using Recurrent Neural Networks and Grammatical Inference. Machine Learning **44** (2001) 161-183
8. Yu, L., Zhang, Y: Evolutionary Fuzzy Neural Networks for Hybrid Financial Prediction. IEEE Trans. on Systems, Man, and Cybernetics, part C (2005) 1-5
9. Zhang, Y., Kandel, A.: Compensatory Genetic Fuzzy Neural Networks and Their Applications. Series in Machine Perception and Artificial Intelligence, Singapore: World Scientific (1998)
10. Yao, J., Tan, C. L.: A Case Study on using Neural Networks to Perform Technical Forecasting of Forex. Neurocomputing **34** (2000) 79-98

Extensions of Manifold Learning Algorithms in Kernel Feature Space

Yaoliang Yu, Peng Guan, and Liming Zhang

Dept. E.E, Fudan University, Shanghai 200433, China
{052021037,052021025,lmzhang}@fudan.edu.cn

Abstract. Manifold learning algorithms have been proven to be capable of discovering some nonlinear structures. However, it is hard for them to extend to test set directly. In this paper, a simple yet effective extension algorithm called PIE is proposed. Unlike LPP, which is linear in nature, our method is nonlinear. Besides, our method will never suffer from the singularity problem while LPP and KLPP will. Experimental results of data visualization and classification validate the effectiveness of our proposed method.

1 Introduction

It has been proven that manifold learning algorithms: ISOMAP [1], LLE [2], LE [3] are capable of discovering some nonlinear structures; in the meantime, they also share the same computational effectiveness as classical subspace methods. However, manifold learning algorithms yield maps that are *only* defined on the training set, so they can not extend to test set straightforwardly. This disadvantage heavily limits their application in pattern recognition.

LPP [4], as a linearization counterpart of LE [3], is proposed to tackle the extension problem. LPP is well defined in the *whole* data space, so it can effectively extend to test set. However, when the sample dimension is high, LPP suffers from the *singularity* problem (also known as *small sample size* problem). Besides, LPP is linear in nature.

Kernel method [6] is another popular nonlinear dimensionality reduction technique. [5] proposed a variant of LPP based on kernels. However, [5] maps samples to a very high (possibly infinite) dimensional feature space, making it always suffer from the *singularity* problem.

In this paper, a new extension algorithm called PIE is proposed. We first map samples to the feature space to make sure they are linearly independent. With the help of manifold learning algorithms such as LE, we can obtain the dimensionality reduction results of the training set. Due to the linearly independent characteristic of samples in feature space, we can easily construct a linear transformation through the *pseudo inverse* of sample matrix. Note that our proposed method employs the manifold learning algorithms in the *kernel feature space*.

The remainder of this paper is arranged as follows: In section II, we briefly review LE, LPP and KLPP. A new extension algorithm is presented in section III. Two experiments will be given to verify the effectiveness of our method in Section IV. Finally, conclusions will be drawn in section V.

D. Liu et al. (Eds.): ISNN 2007, Part I, LNCS 4491, pp. 449–454, 2007.

2 Overviews of LE, LPP and KLPP

Let the training set $X = [x_1, x_2, \cdots, x_n] \in \mathbb{R}^{m\times n}$, $x_i \in \mathbb{R}^m$, $i = 1, 2, \cdots n$, be composed of n m-dimensional samples which possibly reside on a manifold. Our objective is to reduce samples to d-dimensional space. Typically, we have $d << m$.

LE [3] tries to keep the relative distance between nearby samples in the high dimensional space when mapping them to the reduced low dimensional space. It first constructs an adjacency weighted graph $S \in \mathbb{R}^{n\times n}$. The weights S_{ij} for neighboring samples are simply put 1, while for faraway samples, the weights are always 0. Then it constructs the Laplacian matrix $L = D - S$, where D is a diagonal matrix and its diagonal elements are column (or row, since S is symmetric) sums of matrix S. LE turns out to be the following generalized eigen-problem:

$$Ly = \lambda \cdot Dy\,. \tag{1}$$

Let $y_0, y_1, \cdots, y_d$ be the solutions of (1), ordered according to their eigenvalues: $0 = \lambda_0 \le \lambda_1 \le \cdots \le \lambda_d$ .We drop the trivial eigenvector y_0 corresponding to zero eigenvalue and use the next d eigenvectors for reducing the training samples to d-dimensional space: $Y_{LE} = [y_1, y_2, \cdots, y_d]^T \in \mathbb{R}^{d\times n}$. Note that S and L are usually very sparse, and the diagonal matrix D is always positive definite. We can simply solve equation (1) by eigen-decomposition of $D^{-1} \cdot L$. It is worthwhile to point out that, it is sufficient to implement LE when the distances between samples are known. In section 3 we will see that this characteristic becomes particularly important in our algorithm, where we work in a very high dimensional feature space. When a new test sample enters, the new adjacency weighted graph S must be re-calculated. Therefore, LE has to pool the training data and test data together to re-run the whole procedure. So it is very ineffective for LE to extend to test set.

As a linearization counterpart of LE, LPP [4] makes a linear transformation constraint $y = X^T w$, which was inserted in (1), and we get a similar generalized eigen-problem:

$$XLX^T w = \lambda XDX^T w\,. \tag{2}$$

When w_{LPP} is computed by equation (2), it can directly extend to the test set. Nevertheless, we should point out that, as $X \in \mathbb{R}^{m\times n}$ is involved in LPP, it concludes as a $m \times m$ generalized eigen-decomposition, while LE concludes as a $n \times n$ generalized eigen-decomposition. When $m >> n$, XDX^T will be singular, and LPP will suffer from the singularity problem. [4] utilized PCA to project the sample set to a subspace so that the resulting matrix XDX^T is nonsingular, however, it will increase computational cost. The objective of PCA is to keep the global structure in terms of variance as much as possible, while the objective of LPP is to preserve the local structure in image space, so it still remains unclear whether this kind of cascade system degrades the performance. Besides, in LE, matrix S, D and L are very sparse, but in LPP, matrix XLX^T and XDX^T are no longer sparse. This makes LPP more expensive when solving equation (2).

[5] proposed a variant of LPP, called KLPP. It maps samples to the kernel feature space, and then employs LPP in the feature space. However, due to the high dimensionality (possibly infinite) of the feature space, KLPP always suffers from the singularity problem. [5] adopted the similar strategy as in [4], so it can not avoid the problems of LPP mentioned above.

3 Proposed Method

Given the training sample matrix $X = [x_1, \cdots, x_n] \in \mathbb{R}^{m \times n}$, assume that we have obtained the corresponding dimensionality reduction results $Y_{LE} = [y_1, \cdots, y_n] \in \mathbb{R}^{d \times n}$ by LE (or the other manifold learning algorithms). Is it possible to find a simple linear transformation $W \in R^{m \times d}$ satisfying $Y_{LE} \approx W^T \cdot X$? This problem can be converted to the following least-square problem:

$$\min_{W} \ \varepsilon = \| Y_{LE} - W^T \cdot X \|^2 . \tag{3}$$

Obviously, $W^T = YX^\dagger$ is one solution (the solution may not be unique). $X^\dagger$ denotes the pseudo inverse of matrix X [7]. More specifically, we have the following theorem [7]:

Theorem 1: If X has full column rank, there exists a group of optimal solutions:

$$W_{opt}{}^T = Y_{LE} \cdot (X^T EX)^{-1} X^T E \tag{4}$$

$E \in \mathbb{R}^{m \times m}$ satisfying $rank(X^T EX) = rank(X) = n$, n is the number of samples. Among those optimal solutions, there exists one has the minimal F-norm (Let E be an identity matrix):

$$W^T = Y_{LE} X^\dagger = Y_{LE} \cdot (X^T X)^{-1} X^T . \tag{5}$$

□

When X has full column rank, it is easy to see that (4) is the optimal solution of (3). Actually, by (4), we have $\varepsilon = 0$, which indicates the linear transformation (4) has the same dimensionality reduction results as LE in training set.

Theorem 1 implies us maybe we can just simply use (4) as an effective linear transformation for dimensionality reduction if X has full column rank. However, there exists an arbitrary matrix E in (4) to be determined. In this paper, we adopt the criterion which favors (5) that has the minimal F-norm among (4), and this is highly related to the regularization term in SVM [6].

However, if samples are not linearly independent, which means X is not of full column rank, for example, when $n >> m$. Then linear transformations (5) can not guarantee $\varepsilon = 0$ any more, which means (5) is not equivalent to LE even in the training set.

Fortunately, although samples $x_1, \cdots, x_n$ may not be linearly independent, they are at least different to each other ($i \neq j$, $x_i \neq x_j$). So we can map samples to the feature space by some nonlinear mappings, to make sure samples in the feature space are linearly independent. This is always possible if we choose some positive definite kernels [8], for example, Gaussian kernels.

For solving (5), we need to obtain Y_{LE} in training set, however, it is not easy to implement LE in the high dimensional feature space. We adopt the kernel trick to avoid computing the nonlinear mapping $\varphi(x)$ explicitly. According to Mercer's theorem [8], it is possible to use some appropriate kernel functions to obtain the inner products in the kernel feature space: $k(x_i, x_j) = \varphi(x_i)^T \varphi(x_j)$. We can easily convert inner products to Euclidean distances between samples, for example, $\| \varphi(x_i) - \varphi(x_j) \|^2 = \varphi(x_i)^T \varphi(x_i) + \varphi(x_j)^T \varphi(x_j) - 2 \cdot \varphi(x_i)^T \varphi(x_j)$, making it sufficient for us to perform LE in kernel feature space to obtain Y_{LE}. Gaussian kernels $\exp(-\gamma \| x_i - x_j \|^2)$ and polynomial kernels $(x_i^T x_j)^p$ are two widely used kernels, which will also be adopted in the next experiments.

We conclude our proposed method, which we call Pseudo Inverse Extension (PIE), as follows:

Step 1: Choosing some appropriate kernels to map samples to the kernel feature space to make sure they are linearly independent in that space.

Step 2: Employing LE in the *kernel feature space* to obtain the low dimensional coordinates Y_{LE} of training set.

Step 3: Although in the feature space we can not get the explicit representation of (5), we can still map test set to the reduced low dimensional space:

$$Y_{test} = W_{PIE}^T \cdot \varphi(X_{test}) = Y_{LE} \varphi(X_{train})^{\dagger} \cdot \varphi(X_{test}) = Y_{LE} \cdot (K_{train})^{-1} \cdot K_{test}, \tag{6}$$

$K_{train} = \varphi(X_{train})^T \cdot \varphi(X_{train})$, $K_{train} = \varphi(X_{train})^T \cdot \varphi(X_{test})$ can be easily computed through kernel functions.

4 Experimental Results

We give two experiments to demonstrate that PIE is capable of extending LE to the test set: one for visualization and the other for pattern classification.

4.1 Visualization Experiment

We choose the classical S-Curve in manifold learning area to compare our proposed method with LPP and KLPP. In this experiment, we adopt the Gaussian kernel for PIE and KLPP. We randomly sample 4000 points on S-Curve: 2000 points for training and the other 2000 points for testing.

Note that it is impossible for LE to extend to test set directly, in this experiment we pool the training set and test set together to re-run LE. However, this strategy costs too much time to apply in practice. From Fig. 1 we can clearly see that, neither of LPP and KLPP could unfold the manifold faithfully both in training set and test set. As expected, PIE gives the same result as LE in the training set. In the test set, PIE also gives a satisfying result, while keeping the whole extension procedure simple yet effective.

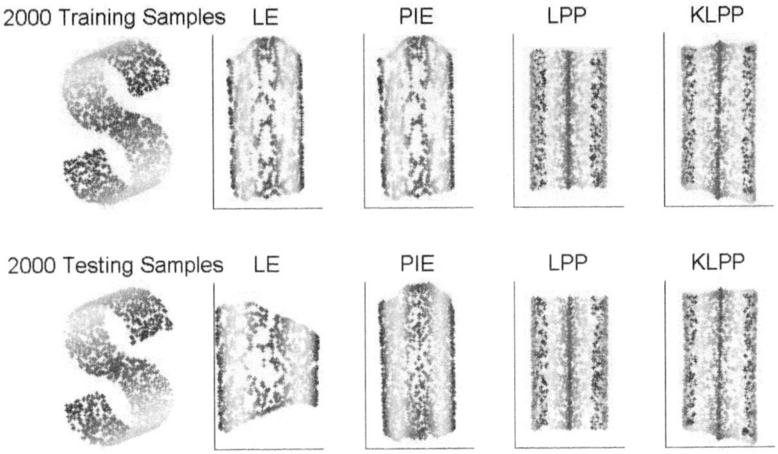

Fig. 1. Experimental Results on S-Curve
(First Row: Results on Training Set Second Row: Results on Test Set)

4.2 Classification Experiment

This experiment is performed on the well-known ORL face database. There are 400 112 x 92 gray images of 40 individuals. We randomly select 5 images per person as training set, while keep the rest images as test set. We first utilize LPP, KLPP and PIE to reduce the dimensionality, and then in the reduced space we adopt two classifiers: nearest neighbor classifier and nearest centroid classifier, to classify which class the test sample belongs to. In both classifiers, Euclidean distance is adopted. Note that, both LPP and KLPP will suffer from the singularity problem in this experiment. We repeat the experiment for 5 times and average the results. In this experiment, we adopt the polynomial kernel for PIE and KLPP.

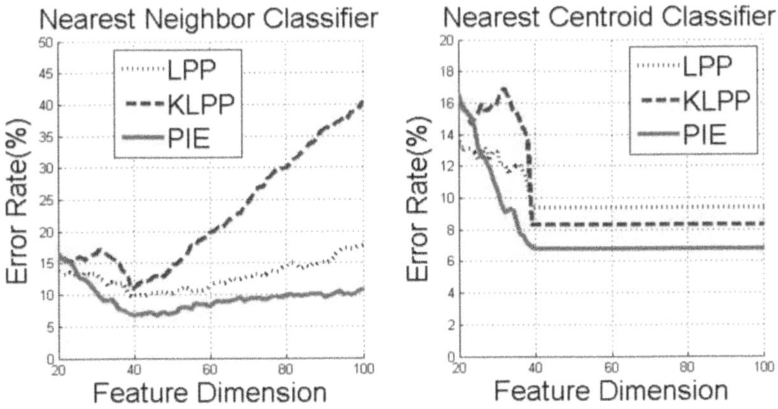

Fig. 2. Classification Error Rate on ORL Face Database

Table 1. Lowest Classification Error Rate on ORL Face Database

	LPP	KLPP	PIE
Nearest Neighbor Classifier	9.8%	11%	6.8%
Nearest Centroid Classifier	9.4%	8.3%	6.8%

Fig. 2 shows the average error rates versus feature dimensions, while Table.1 lists the lowest error rates. It is clear to see that in both classifiers, our method exhibits a significant improvement over LPP and KLPP. It is interesting to note that in the nearest neighbor classifier, KLPP is even inferior to LPP.

5 Conclusion

In this paper, we proposed a new algorithm to extend the manifold learning algorithms to test set. We summarize the main advantages of our proposed method as follows:

1. In the kernel feature space where samples are linearly independent, PIE is guaranteed to be equivalent to LE in the training set, in the meantime, PIE can simply yet effectively extend to test set.
2. PIE is nonlinear by the explicit nonlinear mapping, so it is possible for PIE to discover some nonlinear structures hidden in data distribution. However, LPP as a linearization counterpart of LE is linear in nature.
3. Compared with LPP and KLPP, PIE will never suffer from the singularity problem. Besides, PIE has a lower computational cost than KLPP.

References

1. Tenenbaum, J.B., Silva, V.de, Langford, J.C.: A Global Geometric Framework for Nonlinear Dimensionality Reduction. Science **290** (2000) 2319-2323
2. Roweis, S.T., Saul, L.K.: Nonlinear Dimensionality Reduction by Locally Linear Embedding. Science **290** (2000) 2323-2326
3. Belkin, M., Niyogi, P.: Laplacian Eigenmap and Spectral Techniques for Embedding and Clustering. Advances in Neural Information Processing Systems **14** (2001) 585-591
4. He, X., Yan, S., Hu, Y., Niyogi, P., Zhang, H.: Face Recognition Using Laplacianfaces. IEEE Trans. Pattern Analysis and Machine Intelligence **27** (2005) 328-340
5. Feng, G., Hu, D. Zhang, D., Zhou, Z.: An Alternative Formulation of Kernel LPP with Application to Image Recognition. Neurocomputing **69** (2006) 1733-1738
6. Müller, K.R, Mika, S., Rätsch, G., Tsuda, K., Schölkopf, B.: An Introduction to Kernel-based Learning Algorithms. IEEE Trans. Neural Networks **12** (2001) 181-201
7. Golub, G.H., Van Loan, C.F.: Matrix Computations. 3rd Ed., Johns Hopkins Univ. Press (1996)
8. Simon Haykin.: Neural Networks - A Comprehensive Foundation. 2nd Ed., Prentice-Hall (1999)

A Kernel-Based Reinforcement Learning Approach to Dynamic Behavior Modeling of Intrusion Detection*

Xin Xu[1] and Yirong Luo[2]

[1] Institute of Automation, National University of Defense Technology, 410073, Changsha, P.R. China
xuxin_mail@263.net
[2] Network Center, Hunan Agriculture University, 410080, Changsha, P.R. China

Abstract. As an important active defense technique for computer networks, intrusion detection has received lots of attention in recent years. However, the performance of current intrusion detection systems (IDSs) is far from being satisfactory due to the increasing number of complex sequential attacks. Aiming at the above problem, in this paper, a novel kernel-based reinforcement learning method for sequential behavior modeling in host-based IDSs is proposed. Based on Markov process modeling of host-based intrusion detection using sequences of system calls, the performance optimization of IDSs is transformed to a sequential prediction problem using evaluative reward signals. By using the kernel-based learning prediction algorithm, i.e., the kernel least-squares temporal-difference (kernel LS-TD) algorithm, which implements LS-TD learning in a kernel-induced feature space, the nonlinear modeling and prediction problem for sequential behaviors in IDSs is efficiently solved. Experiments on system call data from the University of New Mexico illustrate that the proposed kernel-based RL approach can achieve better detection accuracy than previous sequential behavior modeling methods including Hidden Markov Models (HMMs) and linear TD algorithms.

1 Introduction

With the fast development of the Internet and related applications, security problems in computer networks have become more and more critical to our information society since the number of computer attacks has increased a lot by exploiting vulnerabilities in network protocols and operating systems, which has led to great losses of many e-commerce companies as well as banks and governments. In order to defend computer networks and information systems from cyber attacks as well as computer viruses, intrusion detection systems (IDSs) [1] have been considered to be one of the most promising techniques for active defense because based on early detection of attack behaviors, various response techniques can be employed to stop and trace attacks. Therefore, lots of advances in intrusion detection techniques have been made in the last decade [2-5].

* Corresponding author.

D. Liu et al. (Eds.): ISNN 2007, Part I, LNCS 4491, pp. 455–464, 2007.

Generally speaking, there are three types of detection strategies for IDSs, which are misuse detection, anomaly detection and hybrid modeling approaches. In misuse detection, only abnormal behaviors are modeled by their known signatures. On the contrary, only the profiles of normal behaviors are described in anomaly detection and attacks are detected as deviations from normal profiles. Although misuse detection and anomaly detection have been widely studied in the literature, both of them have some difficulties to be solved. For misuse detection, it is very hard to detect deformed or new attacks, which have different signatures from misuse detection models. On the other hand, although anomaly detection systems can detect new attacks, there are usually lots of false alarms due to incomplete modeling of normal behaviors. Hybrid modeling approaches are very promising to overcome the difficulties of misuse detection and anomaly detection but there are still many problems to be solved since it is also very challenging to realize hybrid modeling of complex sequential behaviors.

Aiming at the above problems, adaptive intrusion detection using machine learning and data mining methods has been an important research topic not only in computer security but also in the neural network community. Earlier works on adaptive intrusion detection include misuse detection based on neural networks [2] and fuzzy logic models [3]. To detect new attacks, several adaptive anomaly detection methods were proposed based on statistics [4], or clustering techniques [5]. There also have been several efforts in designing anomaly detection algorithms using neural networks [6]. However, since multi-stage attacks consist of sequences of temporally related observation features, it will be very difficult for previous supervised or unsupervised learning methods to give accurate predictions for such kind of attacks. Therefore, dynamic behavior modeling approaches for intrusion detection are very important to improve the performance of IDSs in complex environments. In [7] and [8], it was pointed out that using a series of consecutive events during a given period could produce better performance than just using a single event at a given time. The Markov chain model studied in [8] was based on this idea and it was applied as an anomaly-detection method. Nevertheless, since it was an anomaly detection approach, the Markov-chain technique is not robust to noise in the training data (the mixture level of normal activities and intrusive activities) and it only produced desirable performance at a low noise level.

To overcome the weakness of existing supervised and unsupervised learning approaches to adaptive intrusion detection, especially the challenges of dynamic modeling for sequential behaviors, in this paper, a kernel-based reinforcement learning method of sequential behavior modeling is proposed for host-based IDSs using sequences of system calls. The kernel-based reinforcement learning method for intrusion detection makes use of least-squares temporal difference (LS-TD) [9] prediction in a kernel-induced feature space to realize nonlinear sequential behavior modeling [10], which is an important extension of our previous work on linear TD learning prediction for intrusion detection [11]. Experimental results on system call data from University of New Mexico (UNM) demonstrate that the kernel LS-TD method can achieve better prediction accuracy than previous methods.

This paper is organized as follows. In Section 2, a brief introduction on Markov modeling of IDSs is given. In Section 3, the kernel-based RL algorithm for sequential behavior modeling in adaptive intrusion detection is presented. In Section 4, experimental results on host-based intrusion detection are provided to illustrate the effectiveness of the proposed method. Some conclusions and remarks on future work are given in Section 5.

2 A Markov Reward Model for Host-Based IDSs

According to the data sources used for detecting computer attacks, existing IDSs can be divided into two classes, i.e., host-based IDSs and network-based IDSs. Network-based IDSs use various features of traffic flow as observation data and host-based IDSs usually monitor system calls or shell command data in different operating systems. However, the intrinsic problem of IDSs is to determine efficient detection strategies to differentiate attack behaviors from normal behaviors. In this paper, we will mainly focus on sequential modeling and prediction techniques of complex multi-stage attacks. Although the proposed model and algorithm can be applied both in network-based IDSs and host-based IDSs, to facilitate description, we only restrict our discussions on host-based IDSs using sequences of system calls. For other intrusion detection problems of sequential behavior modeling, similar methods can be used by selecting different features and state representations.

For host-based intrusion detection, the observation data are usually obtained by monitoring execution trajectories of processes or user commands in a host computer [12]. As discussed in previous works [7][11-12], each trace is defined as the sequence of system calls issued by a single process from the beginning of its execution to the end. If m successive system calls $(o_{t-m+1}, o_{t-m+2},\ldots,o_t)$ are selected as a state at time step t, and a sliding window with length 1 is defined, the traces of system calls can be transformed to corresponding state transition sequences. The length m of the sequence for a single state can be selected using some criteria based on information theory, which has been studied in [12].

After the definition of states in system call traces, a Markov reward model M for host-based IDSs can be defined as follows. In general, M can be denoted as a tuple $\{S, R, P\}$, where S is the state space, R is the reward function, P is the state transition probability. Let $\{x_t \mid t=0,1,2,\ldots;\ x_t \in S\}$ denote a trajectory generated by M. For each state transition from x_t to x_{t+1}, a scalar reward r_t is defined. The state transition probabilities satisfy the following Markov property:

$$P\{x_{t+1} \mid x_t, x_{t-1}, \ldots, x_1, x_0\} = P\{x_{t+1} \mid x_t\}. \tag{1}$$

By selecting the states appropriately, it is easy to make the state transitions of system call traces satisfy the above Markov property. In addition, the state transition probabilities are assumed to be unknown since only observation data are available for host-based IDSs. Then, the reward function of the Markov reward model plays an important role for dynamic behavior modeling in intrusion detection problems. As described in the following Fig.1, in a Markov reward model for intrusion detection based on system calls, each state is defined as a short sequence of successive system calls and after each state transition, a scalar reward r_t is given to indicate whether there is a possibility to be normal or attack behaviors. The design of the reward function can make use of available *a priori* information so that the anomaly probability of a whole state trajectory can be estimated based on the accumulated rewards for state transitions. However, in many real situations, due to the sequential properties of system call data and the vague distinctions between normal traces and abnormal traces, it is usually not appropriate or even impossible to indicate whether an intermediate state to be normal or abnormal. Therefore, it is more reasonable to develop dynamic behavior modeling approaches which not only incorporate the temporal properties of state transitions but also need little *a priori* knowledge for class labeling. One simple way is to provide evaluative signals to a whole state transition trajectory, i.e., only a whole state trajectory is indicated to be normal or abnormal while the intermediate states are not definitely labeled. For example, in the following Fig.1, zero rewards can be assigned to every intermediate states and only the reward at the terminal state r_T is given as:

$$r_T = \begin{cases} -1, & \textit{for normal trace}, \\ 1, & \textit{for abnormal trace}. \end{cases} \tag{2}$$

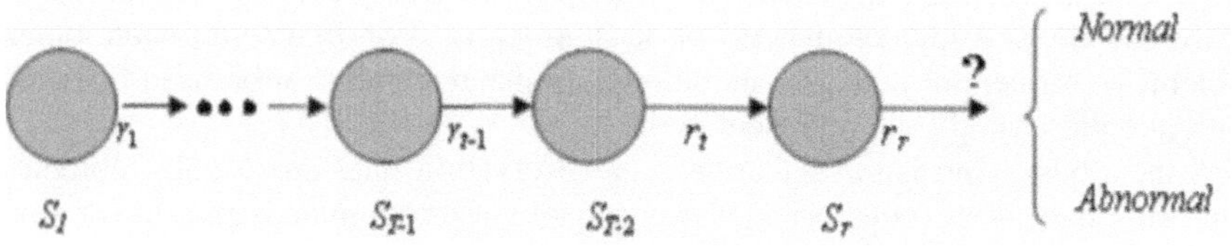

Fig. 1. A Markov reward model for intrusion detection

As studied and verified in [9], dynamic behavior models for sequential pattern prediction are superior to static models when description and modeling of temporal relationships among state transitions are necessary, which is true for host-based IDSs using system calls. In our previous work [11], by introducing the above Markov reward model to the IDS problem, the intrusion detection problem can be transformed to the learning prediction of state anomaly probabilities, which can be approximated by the accumulated rewards received after a state and computed by the state value functions defined as follows:

$$V(s) = E[\sum_{t=0}^{T} r_t | s_0 = s]. \tag{3}$$

Based on the Markov property, the value function satisfies the following equation:

$$V(s_t) = R(s_t, s_{t+1}) + \gamma E[V(s_{t+1})], \tag{4}$$

where $R(s_t, s_{t+1})$ is the expected reward received after the state transition from s_t to s_{t+1}.

According to the above Markov model, the learning prediction for anomaly probabilities of system call traces can be formally described as follows: Given observation data from state transitions of the Markov reward model $\{(x_i, r_i, x_{i+1}), i=1,2,\ldots t\}$, the goal of learning prediction is to compute the value functions without knowing the state transition probabilities as *a priori*. Since learning prediction methods for value function estimation have been widely studied in reinforcement learning, the learning prediction algorithms, e.g., temporal-difference (TD) methods, developed in the RL literature [13] can be applied to the IDS problem. In the following Section 3, we will present a novel kernel-based LS-TD learning approach for sequential behavior modeling of host-based IDSs.

3 Kernel-Based LS-TD Learning for Intrusion Detection

After introducing the above Markov reward model, the intrusion detection problem using system call traces can be solved by a class of reinforcement learning algorithms called temporal-difference (TD) learning, which was originally proposed by R. Sutton [14]. The aim of TD learning is to predict the state value functions of a Markov reward process by updating the value function estimations based on the differences between temporally successive predictions rather than using errors between the real values and the predicted ones. And it has been verified that TD learning is more efficient than supervised learning in multi-step prediction problems [14].

Until now, TD learning algorithms with linear function approximators have been widely studied in the literature [9]. In one of our previous works [11], a linear TD learning algorithm was applied to host-based intrusion detection using sequences of system calls and very promising results have been obtained. Nevertheless, the approximation ability of linear function approximators is limited and the performance of linear TD learning is greatly influenced by the selection of linear basis functions. In the following, we will use a sparse kernel-based LS-TD(λ) algorithm for value function prediction in host-based IDSs. The sparse kernel-based LS-TD algorithm was recently developed by the author(s) [10] and it was demonstrated that by realizing least-squares TD learning in a kernel-induced high-dimensional feature space, nonlinear value function estimation can be implicitly implemented by a linear form of computation with high approximation accuracy. Therefore, by making use of the kernel-based LS-TD learning algorithm, the predictions of anomaly probabilities for intrusion detection will have higher precision and it will be more beneficial to realize high-performance IDSs based on dynamic behavior modeling.

In the kernel-based LS-TD learning method proposed in [10], the same solution to the following LS-TD problem was considered:

$$E_0[\vec{z}_t(\phi^T(x_t) - \gamma\phi^T(x_{t+1}))]W^* - E_0[\vec{z}_t r_t] = 0, \tag{5}$$

where the corresponding value functions are estimated by

$$V(x_t) = \phi^T(x_t)W^*. \tag{6}$$

Using the average value of observations as the estimation of expectation $E_0[\cdot]$, equation (5) can be expressed as follows:

$$\sum_{i=1}^{N}[\vec{z}(s_i)(\phi^T(s_i) - \gamma\phi^T(s_{i+1})]W = \sum_{i=1}^{N}\vec{z}(s_i)r_i. \tag{7}$$

Based on the idea of kernel methods, a high-dimensional nonlinear feature mapping can be constructed by selecting a Mercer kernel function $k(x_1, x_2)$ in a reproducing kernel Hilbert space (RKHS). In the following, the nonlinear feature mapping based on the kernel function $k(.,.)$ is also denoted by $\phi(s)$ and according to the Mercer Theorem [15], the inner product of two feature vectors is computed by

$$k(x_i, x_j) = \phi^T(x_i)\phi(x_j). \tag{8}$$

Due to the properties of RKHS [15], the weight vector W can be represented by the weighted sum of the state feature vectors:

$$W = \Phi_N \alpha = \sum_{t=1}^{N} \phi(s(x_t))\alpha_i, \tag{9}$$

where $x_i (i = 1,2,..., N)$ are the observed states, N is the total number of states and $\alpha = [\alpha_1, \alpha_2, ..., \alpha_N]^T$ are the corresponding coefficients, and the matrix notation of the feature vectors is denoted as

$$\Phi_N = (\phi(s_1), \phi(s_2), ..., \phi(s_N)). \tag{10}$$

For a state sequence $x_i (i = 1,2,..., N)$, let the corresponding kernel matrix K be denoted as $K=(k_{ij})_{N\times N}$, where $k_{ij}=k(x_i, x_j)$.

By substituting (8), (9) and (10) into (7), and multiplying the two sides of (7) with Φ_N^T we can get

$$\bar{Z}_N H_N K\alpha = \bar{Z}_N R_N, \tag{11}$$

where Z_N, H_N and R_N are defined as

$$\bar{Z}_N = [\bar{z}_1, \bar{z}_2, ..., \bar{z}_{N-1}] = [k_{1N}, k_{2N} + \gamma\lambda\bar{z}_1, ..., k_{(N-1)N} + \gamma\lambda\bar{z}_{N-2}], \tag{12}$$

$$R_N = [r_1, r_2, ..., r_{N-1}]^T, \tag{13}$$

$$H_N = \begin{bmatrix} 1 & \beta_1\gamma & & & \\ & 1 & \beta_2\gamma & & \\ & & \cdots & & \\ & & & 1 & \beta_{N-1}\gamma \end{bmatrix}_{(N-1)\times N}. \tag{14}$$

In (14), the values of β_i (i=1,2,...,N-1) are determined by the following rule: when state x_{i-1} is not an absorbing state, β_i is equal to -1, otherwise, β_i is set to zero.

As discussed in [10], by using the techniques of generalized inverse matrix in [16], the kernel-based LS-TD solution to (11) is as follows:

$$\alpha = (H_N K)^+ \bar{Z}_N^+ \bar{Z}_N R_N, \tag{15}$$

where $(.)^+$ denotes the generalized inverse of a matrix.

One problem remained for the above kernel-based LS-TD learning algorithm is that the dimension of the kernel-based LS-TD solution is equal to the number of state transition samples, which will cause huge computational costs when the number of observation data is large. To make the above algorithm be practical, one key problem is to decrease the dimension of kernel matrix K as well as the dimensional of α . The problem has been studied in our previous work [10] by employing an approximately linear dependence (ALD) analysis method [17] for the sparsification of kernel matrix K.

The main idea of ALD-based sparcification is to represent the feature vectors of the original data samples by an approximately linearly independent subset of feature vectors, which is to compute the following optimization problem

$$\delta_t = \min_{\vec{a}} \left\| \sum_j \vec{a}_j \phi(x_j) - \phi(x_t) \right\|^2 . \tag{16}$$

During the sparsification procedure, a data dictionary is incrementally constructed and every new data sample x_t is tested by compute the solution δ_t of (16). Only if δ_t is greater than a predefined threshold, the tested data sample x_t will be added to the dictionary. For detailed discussion of the sparsification process, please refer to [17] and [10]. After the sparsification procedure, a data dictionary D_N with reduced number of feature vectors will be obtained and the approximated state value function can be represented as:

$$\tilde{V}(x) = \sum_{j=1}^{n(D_N)} \hat{\alpha}_j \hat{k}(x_j, x), \tag{17}$$

where $n(D_N)$ is the size of the dictionary.

When the above learning and sparcification process is completed, a value function model of the IDS problem can be obtained. And the accumulated anomaly probability of a state sequence $S_n=\{x_1, x_2, \ldots x_n\}$ can be computed as

$$P(s) = \frac{1}{n} \sum_{i=1}^{n} \tilde{V}(x_i). \tag{18}$$

By selecting an appropriate threshold μ, the detection output of the adaptive IDS can be simply determined as follows:

If $P(S_n) > \mu$, then raise alarms.

4 Experimental Results

In this section, to compare the performance between the kernel LS-TD approach with the previous linear LS-TD [11] and the HMM-based approach [7], experiments on host-based intrusion detection using system calls were conducted. In the experiments, the data set of system call traces generated from the *Sendmail* program [18] was used.

This data set is publicly available at the website of University of New Mexico [18]. The system call traces were divided into two parts. One part is for model training and threshold determination and the other part is for performance evaluation. The normal trace numbers for training and testing are 13 and 67, respectively. The numbers of attack traces used for training and testing are 5 and 7. The total number of system calls in the data set is 223733. During the threshold determination process, the same traces were used as the training process. The testing data are different from those in model training and their sizes are usually larger than the training data. In the testing stage, two criterions for performance evaluations were used, which include the detection rate *Dr* and the false alarm or false positive rate *Fp*, and they are computed as follows:

$$Dr = \frac{n_d}{n_a}, \tag{19}$$

$$F_p = \frac{N_a}{N}, \tag{20}$$

where n_d is the number of abnormal traces that have been correctly identified by the detection model and n_a is the total number of abnormal traces, N_a is the number of normal states that have been incorrectly identified as anomaly by the detection model, and N is the total number of normal states.

Table 1. Performance comparisons between different methods

	Kernel LS-TD		LS-TD[11]		HMM [7]	
	Dr	*Fp*	*Dr*	*Fp*	*Dr*	*Fp*
sendmail	1.00	0.00016	1.00	0.0029	0.615	0.05*
					0.846	0.10*
					0.923	0.20*

* The false alarm rates were only computed for trace numbers, not for single state.

In the learning prediction experiments for intrusion detection, the kernel LS-TD algorithm and previous linear TD(λ) algorithms, i.e., LS-TD(λ), are all implemented for the learning prediction task. In the kernel-based LS-TD algorithm, a radius basis function (RBF) kernel is selected and its width parameter is set to 0.8 in all the experiments. A threshold parameter δ=0.001 is selected for the sparsification procedure of the kernel-based LS-TD learning algorithm. The LS-TD(λ) algorithm uses a linear function approximator, which is a polynomial function of the observation states and has a dimension of 24.

The experimental results are shown in Table 1. It can be seen from the results that both of the two RL methods, i.e., the kernel LS-TD and linear LS-TD, have 100% detection rates and the kernel-based LS-TD approach has better performance in false alarm rates than the linear LS-TD method. The main reason is due to the learning prediction accuracy of kernel-based LS-TD for value function estimation. It is also illustrated that the two TD learning prediction methods have much better performance than the previous HMM-based method. Therefore, the applications of kernel-based reinforcement learning methods, which are based on the Markov reward model, will be very promising to realize dynamic behavior modeling and prediction for complex multi-stage attacks so that the performance of IDSs can be efficiently optimized.

5 Conclusion

Due to the increasing amounts of complex multi-stage attacks, dynamic behavior modeling has been an important and challenging problem for intrusion detection. Although several sequential prediction methods have been proposed for dynamic behavior modeling in IDSs, the performance of existing methods still needs to be improved in order to detect novel attacks with high accuracy and low false alarm rates. In this paper, a kernel-based reinforcement learning approach, which makes use of the kernel LS-TD learning prediction algorithm, is applied to dynamic behavior modeling for intrusion detection. As analyzed in [10], kernel methods in reinforcement learning, especially in temporal difference learning, are very beneficial to improve the generalization ability of RL algorithms in large and nonlinear spaces. Thus, by employing the kernel LS-TD algorithm for sequential behavior prediction, better modeling and prediction accuracy can be realized in host-based intrusion detection using sequences of system calls. Experimental results demonstrated that the proposed kernel LS-TD method not only has better detection accuracy than previous HMMs but also has lower false alarms than linear LS-TD studied in [11]. The application of the proposed method to network-based intrusion detection is our future research work.

Acknowledgement

This work was supported by the National Natural Science Foundation of China Under Grant 60303012, National Fundamental Research 973 Program Under Grant 2005CB321801.

References

1. Denning, D.: An Intrusion-Detection Model. IEEE Transactions on Software Engineering **13** (2) (1987) 222-232
2. Ryan, J., Lin, M.J., Miikkulainen, R.: Intrusion Detection with Neural Networks. Proceedings of AAAI-97 Workshop on AI Approaches to Fraud Detection and Risk Management, AAAI Press (1997) 72-77
3. Luo, J., Bridges, S.M.: Mining Fuzzy Association Rules and Fuzzy Frequency Episodes for Intrusion Detection. International Journal of Intelligent Systems (2000) 687-703
4. Barbara, D., Wu, N., Jajodia, S.: Detecting Novel Network Intrusions Using Bayes Estimators. First SIAM Conference on Data Mining, Chicago, IL (2001).
5. Shah, H., Undercoffer, J., Joshi, A.: Fuzzy Clustering for Intrusion Detection. In: Proceedings of the 12th IEEE International Conference on Fuzzy Systems (2003) 1274-1278

6. Ghosh, A.K., Schwartzbard, A.: A Study in Using Neural Networks for Anomaly and Misuse Detection. Proceedings of the 8th USENIX Security Symposium, Washington D C:ASME Press (1999) 23-26
7. Yeung, D.Y., Ding, Y.X.: Host-Based Intrusion Detection Using Dynamic and Static Behavioral Models. Pattern Recognition **36** (2003) 229 – 243
8. Ye, N., Zhang, Y., Borror, C.M.: Robustness of the Markov-Chain Model for Cyber-Attack Detection. IEEE Transactions on Reliability **53** (1) (2004) 116-123
9. Boyan, J.A.: Technical Update: Least-Squares Temporal Difference Learning. Machine Learning **49** (2002) 233-246
10. Xu, X.: A Sparse Kernel-Based Least-Squares Temporal Difference Algorithm for Reinforcement Learning. In: Proceedings of International Conference on Intelligent Computing. 2006, Lecture Notes in Computer Science, LNCS 4221 (2006) 47-56
11. Xu, X.: A Reinforcement Learning Approach for Host-Based Intrusion Detection Using Sequences of System Calls. In: Proceedings of International Conference on Intelligent Computing. 2005, Lecture Notes in Computer Science, LNCS 3644 (2005) 995 –1003
12. Hofmeyr, S., et al.: Intrusion Detection Using Sequences of Systems Call. Journal of Computer Security **6** (1998) 151-180
13. Kaelbling, L.P., Littman, M.L., Moore, A.W.: Reinforcement Learning: A Survey. Journal of Artificial Intelligence Research **4** (1996) 237-285
14. Sutton, R.: Learning to Predict by the Method of Temporal Differences. Machine Learning **3** (1) (1988) 9-44
15. Schölkopf, B., Smola, A.: Learning with Kernels. Cambridge, MA: MIT Press (2002)
16. Nashed, M.Z., ed.: Generalized Inverses and Applications. Academic Press, New York (1976)
17. Engel, Y., Mannor, S., Meir, R.: The Kernel Recursive Least-Squares Algorithm. IEEE Transactions on Signal Processing **52** (8) (2004) 2275-2285
18. http://www.cs.unm.edu /˜immsec/data-sets.html

Long-Term Electricity Demand Forecasting Using Relevance Vector Learning Mechanism

Zhi-gang Du[1,2], Lin Niu[1], and Jian-guo Zhao[1]

[1] School of Electrical Engineering, Shandong University, Jinan 250061, China
du_zhigang@163.com
[2] State Grid Corporation of China, Beijing 100031, China

Abstract. In electric power system, long term peak load forecasting plays an important role in terms of policy planning and budget allocation. The planning of power system expansion project starts with the forecasting of anticipated load requirement. Accurate forecasting method can be helpful in developing power supply strategy and development plan, especially for developing countries where the demand is increased with dynamic and high growth rate. This paper proposes a peak load forecasting model using relevance vector machine (RVM), which is based on a probabilistic Bayesian learning framework with an appropriate prior that results in a sparse representation. The most compelling feature of the RVM is, while capable of generalization performance comparable to an equivalent support vector machine (SVM), that it typically utilizes dramatically fewer kernel functions. The proposed method has been tested on a practical power system, and the result indicates the effectiveness of such forecasting model.

1 Introduction

Forecasting of the power system load expected at a certain period in the future is indispensable because generating plant capacity must be available to balance exactly any network load whenever it occurs. Long-term peak load forecasting plays an important role in the context of generation, transmission and distribution network planning, and future recurred investment cost in a power system. Thus, every electric utility would be able to have an idea about the amount of required power in order to prepare for the maximum electric load demand ahead of time. However, because power system long-term load forecasting was an uncertain, nonlinear, dynamic and complicated system, it was difficult to describe such a nonlinear characteristics of this system by traditional methods, so the load forecasting could not be accurately forecasted.

Some forecasting methods based on macro-analysis, in which the total system demand is forecast using historical load data together with socio-economic forecasts, have been so far investigated for the long-term load forecasting [1,2,3,4]. In [1], a methodology of mathematical modeling for global forecasting based on regression analysis was presented. In [2], an extended logistic model with varying asymptotic

D. Liu et al. (Eds.): ISNN 2007, Part I, LNCS 4491, pp. 465–472, 2007.

upper bound for long-range peak demand forecasting was described. However, in this reference a large error associated with some years peak demand have been reported. In [3], the authors introduce their future demand with upper and lower bandwidth up to year 20erformance. However, the SVM has a number of the significant and practical limitations. In the SVM, the predictions are not probabilistic and the kernel function $K(x, x_i)$ must satisfy Mercer's condition. That is, it must be a positive definite continuou00. In their model, the effect of weather on these bandwidths was not considered. In [4], traditional forecasting methods based on average historical growth rates or historical relations between electricity consumption and key economic and demographic variables do not perform well. The average forecast error of these methods was reported to be within 2.12~20.18%.

From the above-mentioned references, we can conclude that the electric loads depend on a number of complex factors that have non-linear characteristics, and good results may not be obtained using traditional methods. A better method of forecasting would be one that could find non-linear relations between load and various economic and other factors and is adaptable to changes. In [5], the state-of-the-art SVM has been used to load forecasting with good p s symmetric function. It is also necessary to estimate the error/margin tradeoff parameter C. The number of the found support vector is sensitive to given error bound ε.

In this paper, we propose a new long-term peak load forecasting model, which performs system optimization and generalization simultaneously using relevance vector machine (RVM) with a probabilistic Bayesian learning framework that does not suffer from above disadvantages. This paper takes full advantages of RVM to solve the problem, such as probabilistic predictions, automatic estimation of 'nuisance' parameters, and the facility to utilize fewer arbitrary basis functions (*e.g.* non-'Mercer' kernels). The proposed model was tested on Shandong Province power system in China, it was shown that the forecasting model could generalize well and provide accurate forecasting results at low computational cost.

2 Factors Affecting Electricity Demand

As mentioned earlier, the peak load demand was affected by the weather conditions and changes of economic factors. With a careful investigation on the selection of related parameters for long-term load forecasting, the following factors were thought to influence the electric power demand [6]. These factors are later used as inputs to the forecasting model.

- Economic factors: Gross National Product (GNP), Gross Domestic Product (GDP), Population (Pop), Index of Industrial Production (IIP), Coal Price (CP), Electricity Price (EP), Number of households (NH).
- Weather conditions: Summer Degree Days (SDD), Cool Degree Days (CDD).

In this paper, an important index called *contribution factor* [6] is presented to determine the level of influences of selected inputs on output. The contribution factor is

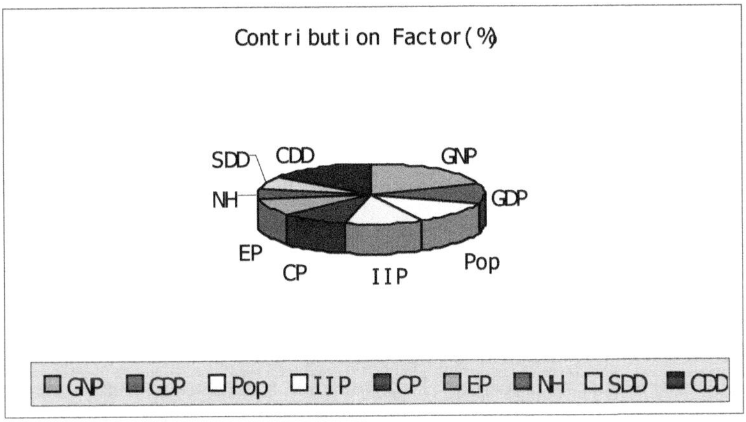

Fig. 1. Contribution factor of the selected inputs

the sum of the absolute values of the weights leading from the particular variable. This function produces a number for each input variable called a "contribution factor", that is, a rough measure of the importance of that variable in predicting the model's output relative to the other input variables in the same model. The higher the number, the more the variable is contributing to the prediction. The above input variables have been tested for their contribution to the peak load forecasting using RVM, as shown in Figure.1.

3 Long-Term Peak Load Forecasting Model Based on RVM

The RVM has an exploited probabilistic Bayesian learning framework. It acquires relevance vectors and weights by maximizing a marginal likelihood. The structure of the RVM is described by the sum of product of weights and kernel functions. A kernel function means a set of basis function projecting the input data into a high dimensional feature space.

Given a data set of input-target pairs $\{\mathbf{x}_n, t_n\}_{n=1}^{N}$ of size N, where $\mathbf{x}_n = (x_{n-(d-1)\tau}, x_{n-(d-2)\tau}, \cdots, x_n) \in R^d$ (d is the embedding dimension and τ is the delay) and $t_n \in R$, and assuming that the targets are independent and contaminated with mean-zero Gaussian noise ε_n with variance σ^2:

$$t_n = \mathrm{y}(\mathbf{x}_n; \mathbf{w}) + \varepsilon_n \tag{1}$$

The RVM with a bias term can be represented as follows [7], [8]:

$$\mathrm{y}(\mathbf{x}; \mathbf{w}) = \sum_{i=1}^{N} w_i K(\mathbf{x}, \mathbf{x}_i) + w_0 = \mathbf{\Phi w} \tag{2}$$

where N is the length of the data, weight vector $\mathbf{w}=[w_0,\cdots,w_N]^{\mathrm{T}}$ and $\mathbf{\Phi}$ is the $N\times(N+1)$ design matrix with $\mathbf{\Phi}=[\phi(\mathbf{x}_1),\phi(\mathbf{x}_2),\cdots,\phi(\mathbf{x}_N)]^{\mathrm{T}}$, wherein $\phi(\mathbf{x}_n)=[1,K(\mathbf{x}_n,\mathbf{x}_1),K(\mathbf{x}_n,\mathbf{x}_2),\cdots,K(\mathbf{x}_n,\mathbf{x}_N)]^{\mathrm{T}}$ and $K(\mathbf{x},\mathbf{x}_i)$ is a kernel function.

Due to the assumption of independence of the t_n, the likelihood of the measured training data set is written as:

$$p(\mathbf{t}|\mathbf{w},\sigma^2)=(2\pi\sigma^2)^{-\frac{N}{2}}\exp\left\{-\frac{1}{2\sigma^2}\left\|\mathbf{t}\text{-}\mathbf{\Phi}\mathbf{w}\right\|^2\right\} \tag{3}$$

where target vector $\mathbf{t}=[t_1,\cdots,t_N]^{\mathrm{T}}$. Maximizing likelihood estimation of $\mathbf{w}$ and σ^2 from (3) leads to severe over-fitting. To avoid this, a zero-mean Gaussian prior distribution over $\mathbf{w}$ with variance $\boldsymbol{\alpha}^{-1}$ is added as:

$$p(\mathbf{w}|\boldsymbol{\alpha})=\prod_{i=0}^{N}N(w_i|0,\alpha_i^{-1})=\prod_{i=0}^{N}\sqrt{\frac{\alpha_i}{2\pi}}\exp(-\frac{\alpha_i}{2}w_i^2) \tag{4}$$

where hyperparameter $\boldsymbol{\alpha}=[\alpha_0,\alpha_1,\cdots,\alpha_N]^{\mathrm{T}}$. An individual hyperparameter associates independently with every weight.

The posterior distribution over the weight form Bayes rule is thus given by:

$$\begin{aligned}p(\mathbf{w}|\mathbf{t},\boldsymbol{\alpha},\sigma^2)&=\frac{\text{Likelihood}\times\text{Prior}}{\text{Normalizing factor}}=\frac{p(\mathbf{t}|\mathbf{w},\sigma^2)p(\mathbf{w}|\boldsymbol{\alpha})}{p(\mathbf{t}|\boldsymbol{\alpha},\sigma^2)}\\&=(2\pi)^{-\frac{N+1}{2}}\left|\mathbf{\Sigma}\right|^{-\frac{1}{2}}\exp\left\{-\frac{1}{2}(\mathbf{w}-\boldsymbol{\mu})^{\mathrm{T}}\mathbf{\Sigma}^{-1}(\mathbf{w}-\boldsymbol{\mu})\right\}\end{aligned} \tag{5}$$

where the posterior mean $\boldsymbol{\mu}$ and covariance $\mathbf{\Sigma}$ are as follows:

$$\boldsymbol{\mu}=\sigma^{-2}\mathbf{\Sigma}\mathbf{\Phi}^{\mathrm{T}}\mathbf{t} \tag{6}$$

$$\mathbf{\Sigma}=(\sigma^{-2}\mathbf{\Phi}^{\mathrm{T}}\mathbf{\Phi}+\mathbf{A})^{-1} \tag{7}$$

with $\mathbf{A}=\mathrm{diag}(\alpha_0,\alpha_1,\cdots,\alpha_N)$.

The likelihood distribution over the training targets (3) can be marginalized with respect to the weights to obtain the marginal likelihood, which is also a Gaussian distribution:

$$p(\mathbf{t}|\boldsymbol{\alpha},\sigma^2)=\int p(\mathbf{t}|\mathbf{w},\sigma^2)p(\mathbf{w}|\boldsymbol{\alpha})d\mathbf{w}=(2\pi)^{-\frac{N}{2}}\left|\mathbf{C}\right|^{-\frac{1}{2}}\exp(-\frac{1}{2}\mathbf{t}^{\mathrm{T}}\mathbf{C}^{-1}\mathbf{t}) \tag{8}$$

with covariance $\mathbf{C}=\sigma^2\mathbf{I}+\mathbf{\Phi}\mathbf{A}^{-1}\mathbf{\Phi}^{\mathrm{T}}$.

Values of $\boldsymbol{\alpha}$ and σ^2 that maximize the marginal likelihood cannot be obtained in closed form, and an iterative re-estimation method is required. The following approach gives:

$$\alpha_i^{new} = \frac{\gamma_i}{\mu_i^2} \tag{9}$$

$$(\sigma^2)^{new} = \frac{\|\mathbf{t}\text{-}\boldsymbol{\Sigma}\boldsymbol{\mu}\|^2}{N - \Sigma_i \gamma_i} \tag{10}$$

where μ_i is the i th posterior mean weight (6) and the quantities $\gamma_i \equiv 1 - \alpha_i \Sigma_{ii}$ with the i th diagonal element Σ_{ii} of the posterior weight covariance (7).

In practice, since many of the hyperparameter α_i tend to infinity during the iterative re-estimation, the posterior distribution (5) of the corresponding weight w_i becomes highly peak at zero. In this optimization process, the vector from the training set that associates with the remaining nonzero weights w_i is called RVM.

At convergence of the hyperparameter estimation procedure, we make predictions based on the posterior distribution over the weights, conditioned on the maximizing values $\boldsymbol{\alpha}_{\mathrm{MP}}$ and σ^2_{MP}. We can then compute the predictive distribution for a new datum $\mathbf{x}_*$:

$$p(t_* \mid \mathbf{t}, \boldsymbol{\alpha}_{\mathrm{MP}}, \sigma^2_{\mathrm{Mp}}) = \int p(t_* | \mathbf{w}, \sigma^2_{\mathrm{Mp}})\, p(\mathbf{w} | \mathbf{t}, \boldsymbol{\alpha}_{\mathrm{MP}}, \sigma^2_{\mathrm{Mp}})\, d\mathbf{w} \tag{11}$$

Since both terms in the integrand are Gaussian, the predictive distribution is readily computed, giving:

$$p(t_* \mid \mathbf{t}, \boldsymbol{\alpha}_{\mathrm{MP}}, \sigma^2_{\mathrm{MP}}) = N(t_* \mid \mathrm{y}_*, \sigma^2_*) \tag{12}$$

with

$$\mathrm{y}_* = \boldsymbol{\mu}^T \phi(\mathbf{x}_*) \tag{13}$$

$$\sigma^2_* = \sigma^2_{\mathrm{MP}} + \phi(\mathbf{x}_*)^{\mathrm{T}} \Sigma \phi(\mathbf{x}_*) \tag{14}$$

So the predictive mean is intuitively $\mathrm{y}(\mathbf{x}_*; \boldsymbol{\mu})$, or the basis functions weighted by the posterior mean weights, many of which will typically be zero. The predictive variance ('error-bars') comprises the sum of two variance components: the estimated noise on the data and that due to the uncertainty in the prediction of the weights.

From the above characteristics of RVM, we perceive that it is possible to obtain a better solution for prediction type problem with a large number of inputs. In long-term load forecasting, it is usually used a huge quantity of data which is suitable for RVM architecture. Thus, RVM is proposed for the long-term load forecasting.

4 Simulation Analysis

In this section, three typical simulations based on the above-mentioned forecasting model have been carried out and some of the comparisons were held between RVM and state-of-the-art SVM.

4.1 Data Preparation

In this research, the output was the maximum demand for power, which should be produced by Shandong Electric Power Corporation, and the number of inputs was equal to the number of 'factors' mentioned above. Furthermore, in order to increase the numbers of sample sets, the monthly data have been taken as inputs for forecasting the maximum electric load, which was extrapolated out of yearly data for RVM to predict the loads in the future.

4.2 Simulation Results

Here the simulation was held to check the validity of the forecasting model by taking Shandong Electric Power Corporation (SEPCO) load data for learning from the year 1990 to 2005.

4.2.1 Validation Evaluation of Forecasting Model Based on RVM and SVM

We applied actual monthly data from 1990 to 2000 for both RVM and SVM, and compared their forecasting performance for following 5 years (2001-2005) in Fig.2. The trained RVM used 6 vectors, compared to 29 for the SVM. The root-mean-square (RMS) deviation from the true value for the RVM was 0.0245, while for the SVM it was 0.0291. Note that for the latter model, it was necessary to tune the parameter C and ε, in this simulation using 5-fold cross-validation. For the RVM, the analogues of these parameters (α and σ^2) are automatically estimated by the learning procedure.

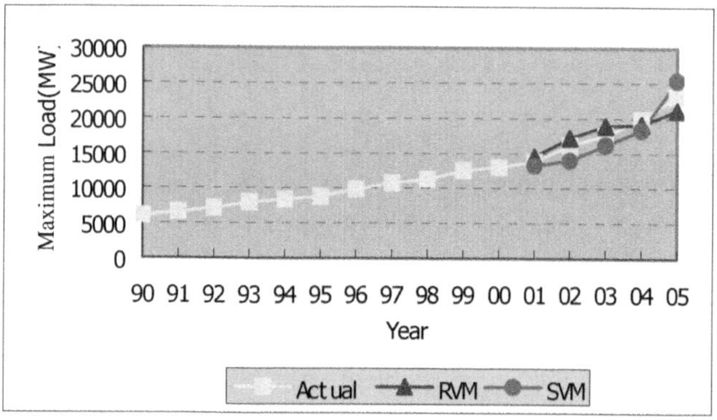

Fig. 2. Forecasting the peak loads of 2001-2005(training data: 1990-2000)

4.2.2 Long-Term Load Forecasting Result from the Year 2006 to 2010

The year 2006 to 2010 has been taken as target years to predict the loads in Fig.3.

In this stage, the error calculation of RVM was shown in Table1. Here validation as well as training years was taken from the year 1990 to 2005 and target years

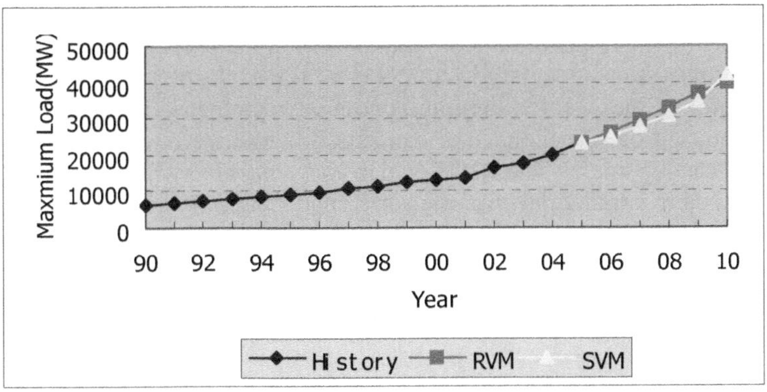

Fig. 3. Forecasting the peak loads of 2006-2010(training data: 1990-2005)

(average test) were taken from the year 2006 to 2010. It showed the training error was 1.05% and the target error was calculated for those years were 2.634%, whereas from SVM we obtained the average error was 6.35%.

Table 1. Out error measures of RVM

	Root Mean Squared Error	Mean Absolute Squared Error	Mean Absolute Error (%)
Training (1990-2005)	0.024	0.113	1.05
Average Test (2006-2010)	0.628	0.869	2.634

4.2.3 Super Long-Term Load Forecasting Result in the Planning Year

It should be noted that based on the forecasting model obtained form RVM, the loads of every 5 years, 2010,2015,and 2020 were predicted (Table2). In other words, the intermediate years were not predicted. We determine that the loads are increasing with mean annual growth rate (MAGR) of about 7.97% up to the year 2020. The result is according with the economy forecasting of Shandong Province.

Table 2. Electricity demand forecasting of SEPCO in the planning year

	2005	MAGR	2010	MAGR	2015	MAGR	2020
Maxmium load(MW)	23234	11.5%	40000	7.0%	56000	5.4%	73000

5 Conclusion

In this paper, we have introduced a new approach to long-term electricity demand forecasting using the relevance vector learning mechanism based on a probabilistic Bayesian learning framework. Our main concern is to find the best structure of the forecasting model for modeling nonlinear dynamic systems with measurement error.

The number of rules and the parameter values of membership functions can be found as optimizing the marginal likelihood of the RVM in the proposed model. Because the RVM is not necessary to satisfy Mercer's condition, selection of kernel function beyond the limit of the positive definite continuous symmetric function of SVM. The relaxed condition of kernel function can satisfy various types of membership functions in forecasting model. The RVM, which was compared with support vector learning mechanism in simulations, had the small model capacity and described good generalization. Simulation results showed the effectiveness of the proposed method for modeling of nonlinear dynamic systems with noise. Hopelly, the process of this study will be useful for the power companies.

References

1. Vlahovic, V. M., Vujosevic, I. M.: Long-term Forecasting: a Critical Review of Direct-trend Extrapolation Methods. Electrical Power Energy Systems **9**(1) (1987) 2-8
2. Barakat, E.H., Al-Rashed, S. A.: Long Range Peak Demand Forecasting under Conditions of High Growth. Power Systems **7**(4) (1992) 1483-1486
3. Gen, M. R.: Electric Supply and Demand in the United States: Next 10 years. IEEE Power Eng Rev (1992) 8-13
4. Leung, P.S., Miklius, W.: Accuracy of Electric Power Consumption Forecasts Generated by Alternative Methods: the Case of Hawaii. Energy Source **16** (1994) 289-299
5. Chen, B. J.: Load Forecasting using Supporting Vector Machines: a Sstudy on Eunite Competition 2001. Power Systems **19**(4) (2004) 1821-1829
6. Al-Alawi, S. M.: Principles of Electricity Demand Forecasting Parting I Methodologies. IEEE Power Engineering Journal **7** (1996) 1-6
7. Tipping, M. E.: Sparse Bayesian Learning and the Relevance Vector Machine. Mach Learning **1** (2001) 211-244
8. Muller, K. R.: An Introduction to Kernel-based Learning Algorithm. Neural Network **12**(2) (2001) 181-201

An IP and GEP Based Dynamic Decision Model for Stock Market Forecasting

Yuehui Chen, Qiang Wu, and Feng Chen

School of Information Science and Engineering
University of Jinan, Jinan 250022, P.R. China
yhchen@ujn.edu.cn,
qwu_master@163.com

Abstract. The forecasting models for stock market index using computational intelligence such as Artificial Neural networks(ANNs) and Genetic programming(GP), especially hybrid Immune Programming (IP) Algorithm and Gene Expression Programming(GEP) have achieved favorable results. However, these studies, have assumed a static environment. This study investigates the development of a new dynamic decision forecasting model. Application results prove the higher precision and generalization capacity of the predicting model obtained by the new method than static models.

1 Introduction

Time series forecasting[1] is an integral part of everyday life. The analysis of time series may include many statistical methods that aim to understand such data by constructing a model. Such as exponential smoothing methods, autoregressive integrated moving average (ARIMA) methods, generalized autoregressive conditionally heteroskedastic (GARCH) methods. The objective of modeling problem is to find a suitable mathematical model that can roughly explain the behavior of dynamic system. The system can be seen as in Eq.1

$$\mathbf{x(t) = F[x(t-1), x(t-2), ..., x(t-p)]}. \tag{1}$$

The function $F(\cdot)$ and constant p are the "center of the storm". Several studies have a full insight of dynamic system to describe $F(\cdot)$. Evolutionary computation models[2][3] have been used in the past, mainly for chaotic, nonlinear and empirical time series. Recently several researchers have used a hybrid algorithm such as Gene Expression Programming (GEP)[5][6] and Immune Programming(IP) Algorithm[7]. However, unfortunately, one does not have a good method to decide the size of p.

The aim of this study is to investigate the development of a new adaptive model that is specifically tailored for forecasting time series. The proposed model is based on Gene Expression Programming (GEP)and Immune Programming (IP) with additional features that seek to capture better accuracy. The rest of this paper is organized as follows: First, we cite the forecasting model of GEP

D. Liu et al. (Eds.): ISNN 2007, Part I, LNCS 4491, pp. 473–479, 2007.

and IP. Next, describe the dynamic decision model based on GEP and IP. Then the results of some computational experiments are reported. Finally, we present our conclusions.

2 A Dynamic Decision Model of GEP and IP

To forecast stock prices, a dynamic decision model of GEP and IP is developed in this study. In the proposed model, IP is used to find out the optimal structure of GEP. With each slide of the forecast window, the model adjusts it dynamically. There are two phases in this section:

(1) Optimization of GEP; (2) Obtaining the forecast sequence using dynamic decision model. These phases are described in detail below.

2.1 The Forecasting Model of GEP and IP

Gene expression programming(GEP), is a population-based evolutionary algorithm developed by Ferreira(2001)[6] and it is a direct descendent of genetic programming(Koza,1992). GEP genes are composed of a head and a tail[6]. The head contains symbols that represent both functions(elements from the function set F) and terminals (elements from the terminal set T), whereas the tail contains only terminals. For each problem, the length of head h is chosen, here the length of tail t is a function of h and the number of arguments of the function with the most arguments n, and is evaluated by the equation

$$\mathbf{t} = \mathbf{h}(\mathbf{n} - \mathbf{1}) + \mathbf{1}. \tag{2}$$

Consider a gene composed of {Q, *, /, -, +, a, b}. In this case $n = 2, h = 10$ then $t = 11$, the length of the gene is 10+11=21. One such gene is shown below (the tail is shown in bold): {+Q-/b*aaQb**aabaabbaaab**}.

Immune programming(IP), is a novel paradigm combining the program-like representation of solutions to problems with the principles and theories of the immune system. It is briefly described as follows:

1. **Initialization.** An initial repertoire (population), AB, of n antibodies, $Ab_i, i = 1, 2, ..., n$, is generated. The generation counter is set to $G = 1$;

2. **Evaluation.** An antigen, Ag, representing the problem to be solved, is presented. Ag is compared to all antibodies $Ab_i \in AB$ and their affinity, f_i, with respect to the antigen is determined;

3. **Replacement.** With a certain probability, P_r, a new antibody is generated and placed into the new repertoire. This way, low affinity antibodies are implicitly replaced. The parameter P_r is the probability of replacement;

4. **Cloning.** If a new antibody has not been generated, an antibody, is drawn from the current repertoire with a probability directly proportional to its antigenic affinity. With a probability, P_c, this antibody is cloned and placed in the new repertoire. The parameter P_c is termed probability of cloning;

5. **Hypermutation.** If the high-affinity antibody selected in the previous step has not been cloned, it is submitted for hypermutation with a probability inversely proportional to its antigenic affinity. If the antibody is selected for hypermutation, each component of its attribute string is mutated with probability of mutation P_m;

6. **Iteration-repertoire.** Steps 3-5 are repeated until a new repertoire AB' of size n is constructed;

7. **Iteration-algorithm.** The generation counter is increased, $G = G + 1$, and the new repertoire is submitted to step 2, evaluation. The process continues iteratively until a stopping criteria is met.

GEP structure is optimized by IP Algorithm. For static model, analysis window starts at the beginning of available historical data, then data window slides to include the next time series observation. Several generations are run with the new data value and then window slides again. This process is repeated until all available data have been analyzed up to and including all historical data. In our experiments the constant windows size is $p = 10$.

2.2 The Dynamic Decision Model

As expounded in section 1, designating the correct size of analysis window is critical to the success for any forecasting model. Automatic discovery of the windowsize is indispensable when forecasting concern is not well understood. With each slide of window, the model adjusts its windowsize dynamically. This is accomplished in following way.

1. Select two initial windowsizes, one of size n and one of size $n + i$ or $n - i$, where n and i are positive integers.

2. Run dynamic generations at the beginning of time series data with windowsizes n and $n + i$, use the best solution for each of these two independent runs to predict the future data points.

3. Select another two windowsizes based on which windowsize had better accuracy. If the smaller of the two windowsizes (size n) predicted more accurately, then choose current windowsizes, one of size n and one of size $n+i$; If the larger of the two windowsizes (size $n + i$) predicted more accurately, then choose new windowsizes $n + i$ and $n + 2i$.

4. Slide the analysis window to include the next value. Using the two selected windowsizes to run another two dynamic generations, predict future data, and measure their prediction accuracy.

5. Repeat the previous steps until the analysis window reaches the end of historical data.

Thus, at each slide of the analysis window, prediction accuracy is used to determine the direction in which to adjust the windowsize. Suppose the following time series is to be analyzed and forecasted.

{22, 33, 30, 27, 24, 20, 21, 20, 23, 26, 29, 30, 28, 29, 30, 31}

The model starts by selecting two initial windowsizes. Then, two separate dynamic generations are run at the beginning of data, each with its own size. After each dynamic generation, the best solution is used to predict the future data and accuracy of this prediction is measured. Fig.1 illustrates this step.

In the initial step, if *win2's* prediction accuracy is better, two new windowsizes for *win1* and *win2* are selected with sizes of 3 and 4, respectively. Then the analysis window slides to include the next time series value, two new dynamic generations are run, and the best solutions for each are used to predict future data. As shown in Fig.2, *win1* and *win2* now include the next time series value, 27, and *pred* has shifted one value to the right(above); if *win1's* prediction accuracy is better, *win1* and *win2* with the current windowsizes just slide to the next value 27(below).

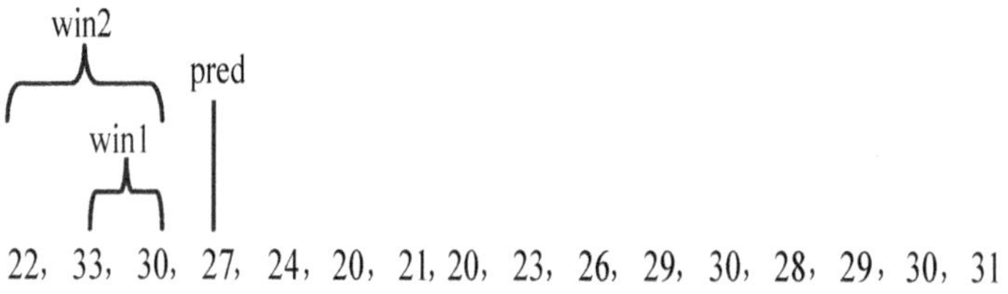

Fig. 1. Initial steps, *win1* and *win2* represent data analysis windows of size 2 and 3, respectively, and *pred* represents the future data predicted

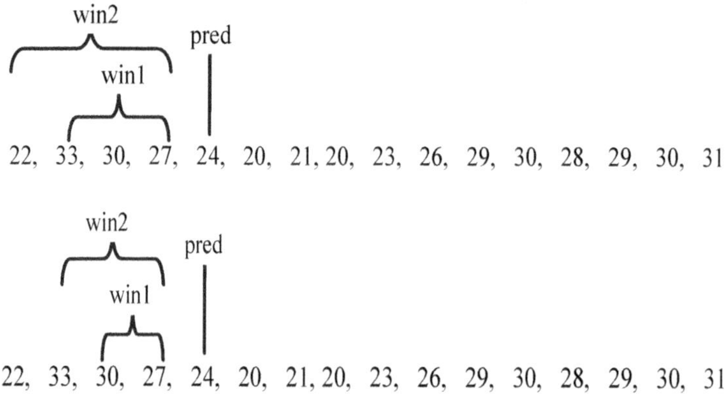

Fig. 2. Data analysis windows slide to new value

These processes of selecting two new windowsizes, sliding the analysis window, running two new dynamic generations, and predicting future data are repeated until the analysis window reaches the end of time series data.

3 Experiments and Results

To test the efficacy of this proposed method we have used stock prices in IT sector: the daily stock price of Apple Computer Inc., IBM Corp. and Dell Inc.[8].

As shown above, we don't need all of stock data like previous study but just use the close price from the daily stock market. The forecast variable here is also the close price. In this case, set of functions $F = \{\sin(x), \cos(x), +, -, \times, \div\}$ and set of terminals $T = \{x_0, x_1, ..., x_n\}$.

The dynamic decision model requires that a number of parameters be specified before a run. Some of these are general GEP and IP parameters and Some are special parameters only used by the dynamic decision model. Table 1 gives the parameter values.

Table 1. Parameter setting

Parameter	Apple Inc.	IBM Corp.	Dell Inc.
Population size	60	60	50
Window slide increment	1	1	1
Start windowsize	2	2	2
Probability of replacement	0.25	0.25	0.2
Probability of cloning	0.2	0.2	0.2
Probability of mutation	0.15	0.15	0.1
Min tree depth	4	4	4
Max tree depth	8	8	8
Number Of function	6	6	6

The performance of the method is measured in terms of Mean Absolute Percentage Error (MAPE)

$$MAPE = \frac{1}{N}\sum_{i=1}^{N}\left(\frac{|y_i - p_i|}{y_i}\right) \times 100\% \tag{3}$$

N: total number of test data sequences;
y_i: actual stock price on day i;
p_i: forecast stock price on day i.

Comparing the dynamic decision model with static model to predict the three stock indices, the results are shown in Fig.3. To show the validity of this algorithm, feed forward neural networks predicting model is also compared. Obviously the accuracy of the proposed model is better than other models.

For a set of runs, forecasting performance is measured by calculating MAPE value over all runs. Tables 2 lists the observed results, respectively comparing with static GEP and IP model.

Table 2. The performance improvement of the dynamic decision model

Experiment	Dynamic Test	Static Test
App Inc	1.843144	2.703284
IBM Corp	0.650613	0.701826
Dell Inc	0.847629	1.229218

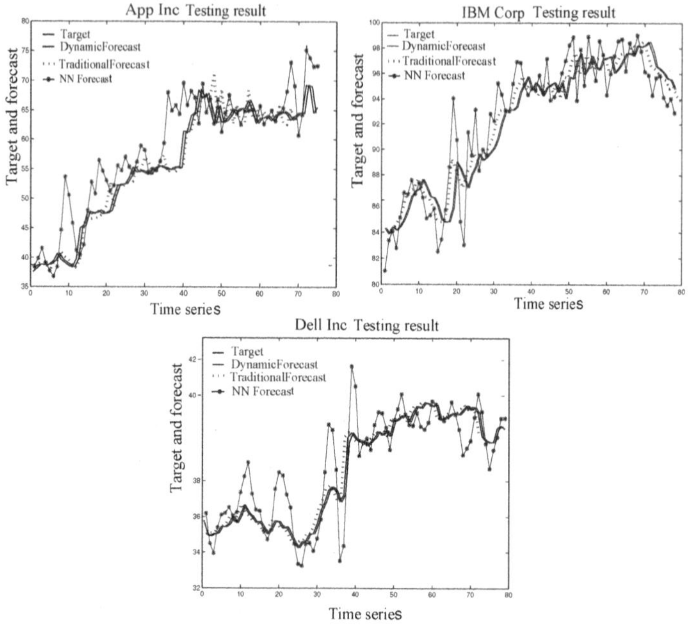

Fig. 3. Forecasting accuracy comparison between different methods

4 Conclusion

In this study a dynamic decision model is developed and tested for prediction accuracy on stock market indices. Results show that this model outperforms static models for all experiments. These findings affirm the potential as an adaptive, non-linear model for real-world forecasting applications and suggest further investigations. The dynamic decision model presents an attractive forecasting alternative.

Acknowledgment

This research was supported by the NSFC under grant No. 60573065 and the Key Subject Research Foundation of Shandong Province.

References

1. Noakes, D.J. and McLeod, A.I.: Forecasting monthly riverflow time series, International Journal of Forecasting **1** (1985) 179-190.
2. Back, T.: Evolutionary algorithms in theory and practice: evolution strategies, Evolutionary Programming, and Genetic Algorithms, Oxford University Press, 1996.
3. Back, B., Laitinen, T. and Sere, K.: Neural networks and genetic algorithms for bankruptcy predictions, Expert Systems with Applications **11** (1996) 407-413.

4. Kaboudan, M.: A measure of time series predictability using genetic programming applied to stock returns, Journal of Forecasting **18** (1999) 345-357.
5. Ferreira, C.: Gene expression programming: a new adaptive algorithm for solving problems, Complex Systems **13**(2) (2001) 87-129.
6. Ferreira, C.: Gene expression programming: mathematical modeling by an artificial intelligence, Angra do Heroismo, Portugal, 2002.
7. Petr, M., Adriel, L. and Marek R.: Immune programming, Information Sciences **176** (2006) 972-1002.
8. Md. Rafiul, H., Baikunth, N. and Michael, K.: A fusion model of HMM: ANN and GA for stock market forecasting, Expert Systems with Applications, 2006.

Application of Neural Network on Rolling Force Self-learning for Tandem Cold Rolling Mills

Jingming Yang, Haijun Che, Fuping Dou, and Shuhui Liu

Institute of Electrical Engineering, Yanshan University, Qinhuangdao, Hebei 066004, China

Abstract. All the factors that influence the rolling force are analyzed, and the neural network model which uses the back propagation (BP) learning algorithm for the calculation of rolling force is created. The initial network's weights corresponding to the input material grades are taught by the traditional theoretical model, and saved in the database. In order to increase the prediction accuracy of rolling force, we use the measured rolling force data to teach the neural network after several coils of the same input material are rolled down.

1 Introduction

The calculation of rolling schedule plays an important role in the process of tandem rolling. Only when it is connected with the basic automation can the high-quality steel rolling be achieved. The basic item of it is the calculation of rolling force. Before rolling, the rolling force only can be pre-calculated from the parameters such as the input and output thicknesses at each stand. The theoretical calculation methods of rolling force had been well researched already, and there are several famous formulas for the rolling force calculation in tandem cold rolling such as the Целиков, А.И. formula, Bland-Ford formula and STONE formula. The widely used Bland-Ford formula is:

$$P = B_m l_c^{'} Q_p K_T K \times 1000 \tag{1}$$

Where P is the rolling force (kN); B_m is the average width of the rolled piece (mm); $l_c^{'}$ is the contact arc length with the working roll bruise considered (mm); Q_p is the stress state factor; K_T is the tension affecting factor; K is the deformation resistance (MPa).

However, there have not been any formulas that can describe the real rolling force exactly due to the complexity of the rolling process. The prime reasons can be generalized as follows. First, there are too many factors affecting the rolling force, such as friction factor, the work-hardness of material, the forward and backward tension and so on. In fact, the changes of lubricate condition, which is connected with emulsion and equipments, cannot but infect the friction factor at rolling. Second, the basic material strength is quite dispersive, so we cannot get the real rolling force if it is

D. Liu et al. (Eds.): ISNN 2007, Part I, LNCS 4491, pp. 480–486, 2007.

calculated by a fixed numeric value. Last, there is seldom scientific reason to consider the yield stress minus the mean tensile stress as the infection of the strip's forward and backward tension towards rolling force. Consider of all the limitations, Hill presents a regression formula $Q_p = 1.08 + 1.79\mu\varepsilon\sqrt{1-\varepsilon}\sqrt{\frac{R'}{h}} - 1.02\varepsilon$ of stress state coefficient combined with the theoretical calculation results. Where μ is the friction factor; R' is the roller radius considering bruise(mm); ε is the relative reduction; h is the output thickness(mm). If the Hill formula is practical, we expect the regression coefficient can be properly modified under the certain rolling force of the mill.

Now that the rolling force can be expressed by the regression formula, it might as well be expressed by the high-regressed neural network[1-2]. We take the factors that infect the rolling force as the input node of the neural network, and teach the network with the measured rolling force. Then modify the weights of the neural network with the deviations.

2 The Neural Network Model for Rolling Force

The feed forward neural network can approach to arbitrary function with an arbitrary accuracy when its hidden layers and hidden nodes are not restricted. The typical feed-forward neural networks are the BP neural network and RBP neural network. The neural network rolling force model calls for high ability of normalization while the RBP neural network cannot satisfy this requirement; therefore we choose the BP neural network to create the rolling force model. The input nodes are consisted of: material original thickness H ; output thickness and input thickness at each rolling mill h and h_0 ; the forward and backward tension T_f and T_b ; the basic material strength K ; strip width B ; and working roll radius R . H, h and h_0 are used to describe the character of work-hardening, because when the same h and h_0 are in different rolling passes, the rolling force behaves a significant difference. We can show the accumulating deforming degree by the warp between H and h, h_0 . The friction factor is not considered as an input node due to the emulsion used in the rolling mills does not change frequently. As a result, the neural network itself can find the affecting regular patterns of the friction factor towards rolling force. In other words, the friction factor is included in the network. Although the working roll radius changes a little, it infects the rolling force a lot so is considered as an input node as well.

The neural network has one hidden layer, and to decide the numbers of the hidden nodes both the model accuracy and calculation complexity are considered. Test shows that when the hidden nodes increased to 16 from 15, the increase of model accuracy is not marked. So we have the hidden nodes as 15. The structure of the neural network is shown below.

For this neural network the transfer function for the first layer is $\log sig(x)$ and the transfer function for the second layer is linear $purelin(x)$ [3].

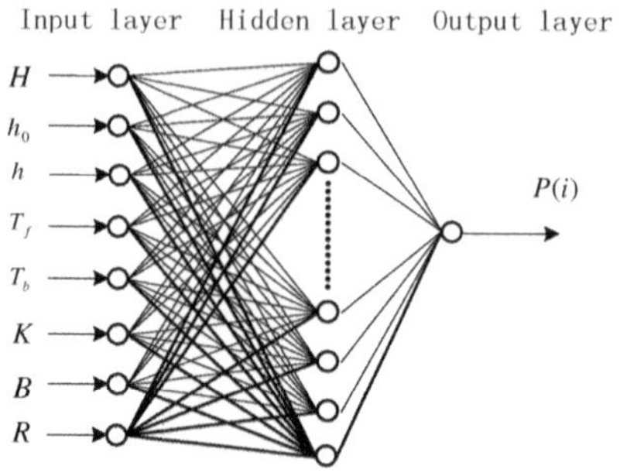

Fig. 1. The structure of the BP network for rolling force

3 The Training of Neural Network

Although the principle of the feedforward neural network BP algorithm is simple and easy to be realized, the features of slower convergence and astringing to the local optimum easily make it not perfect in practice. Levenberg-Marquardt is the compromise of Newton's method and the steepest descent. It is used to minimize the functions of other nonlinear functions' quadratic sum. We use object's second derivative in the process of iteration so that the iteration can get superlinear. Therefore the training speed and efficiency of the network can be improved potently. The rationale is:

Given $E(x)$ is the sum of square functions:

$$E(X) = F^T(X)F(X) \tag{2}$$

Levenberg-Marquardt algorithm iteration function:

$$X_{k+1} = X_k + \Delta X_k \tag{3}$$

$$\Delta X_k = -\left[J^T(X_k)J(X_k) + \mu_k I\right]^{-1} J^T(X_k)F(X_k) \tag{4}$$

Where $J(X_k)$ is the Jacobian matrix of $E(X)$ at X_k; I is the identity matrix; and μ_k is a coefficient. This algorithm has the very useful feature that as μ_k is increased it approaches the steepest descent algorithm with small learning rate, while as μ_k is decreased to zero the algorithm becomes Gauss-Newton.

In actual practice, we choose the neural network as 8-15-1（ $R - s^1 - s^2$. R is the number of the input node and S^m means that there are S neurons at the mth layer）.Capability target function is the sum of square errors:

$$E(X) = \sum_{q=1}^{N} (t_q - y_q)^2 = F^T F \tag{5}$$

Where N is the sample number; y_q and t_q are the network's output and target according to each sample.

The error vector is: $F^T = [f_1, f_2, \cdots f_N] = [e_1, e_2, \cdots e_N]$; and the parameter vector is: $X^T = [x_1, x_2 \cdots x_n] = [w^1_{1,1}, w^1_{1,2}, \cdots w^1_{1,8}, w^1_{2,1}, \cdots w^1_{2,8}, \cdots w^1_{15,8}, b^1_1, \cdots b^1_{15}, w^2_{1,1}, \cdots w^2_{1,15}, b^2_1]$

Where $w^m_{i,j}$ is the weight between the mth neuron i and the *(m-1)*th neuron j ; and b^m_i is the bias of the neuron i in layer m.

Given $m = 1,2$ and $n = s^1(R+1) + s^2(s^1+1) = 15\times(8+1) + 1\times(15+1) = 151$.

Therefore the Jacobian matrix for the training network is:

$$J = \begin{bmatrix} \frac{\partial e_1}{\partial w^1_{1,1}} & \cdots & \frac{\partial e_1}{\partial w^1_{1,8}} & \cdots & \frac{\partial e_1}{\partial w^1_{15,8}} & \frac{\partial e_1}{\partial b^1_1} & \cdots & \frac{\partial e_1}{\partial b^1_{15}} & \frac{\partial e_1}{\partial w^2_{1,1}} & \cdots & \frac{\partial e_1}{\partial w^2_{1,15}} & \frac{\partial e_1}{\partial b^2_1} \\ \frac{\partial e_2}{\partial w^1_{1,1}} & \cdots & \frac{\partial e_2}{\partial w^1_{1,8}} & \cdots & \frac{\partial e_2}{\partial w^1_{15,8}} & \frac{\partial e_2}{\partial b^1_1} & \cdots & \frac{\partial e_2}{\partial b^1_{15}} & \frac{\partial e_2}{\partial w^2_{1,1}} & \cdots & \frac{\partial e_2}{\partial w^2_{1,15}} & \frac{\partial e_2}{\partial b^2_1} \\ \vdots & \vdots & \vdots & \vdots & \vdots & \vdots & \vdots & \vdots & \vdots & \vdots & \vdots & \vdots \\ \frac{\partial e_N}{\partial w^1_{1,1}} & \cdots & \frac{\partial e_N}{\partial w^1_{1,8}} & \cdots & \frac{\partial e_N}{\partial w^1_{15,8}} & \frac{\partial e_N}{\partial b^1_1} & \cdots & \frac{\partial e_N}{\partial b^1_{15}} & \frac{\partial e_N}{\partial w^2_{1,1}} & \cdots & \frac{\partial e_N}{\partial w^2_{1,15}} & \frac{\partial e_N}{\partial b^2_1} \end{bmatrix} \quad (6)$$

For the weight x_l : $[J]_{q,l} = \frac{\partial e_q}{\partial w^m_{i,j}} = \frac{\partial e_q}{\partial n^m_{i,q}} \times \frac{\partial n^m_{i,q}}{\partial w^m_{i,j}} = \tilde{s}^m_{i,q} \times a^{m-1}_{j,q}$; and for the bias x_l : $[J]_{q,l} = \frac{\partial e_q}{\partial b^m_i} = \frac{\partial e_q}{\partial n^m_{i,q}} \times \frac{\partial n^m_{i,q}}{\partial b^m_i} = \tilde{s}^m_{i,q} \times \frac{\partial n^m_{i,q}}{\partial b^m_i} = \tilde{s}^m_{i,q}$.

Where define $\tilde{s}^m_{i,q} \equiv \frac{\partial e_q}{\partial n^m_{i,q}}$ as the sensitivity; $n^m_{i,q}$ is the input of the i th neuron of the mth layer; and $a^{m-1}_{j,q}$ is the output of the j th neuron of the *(m-1)*th layer.

The steps of LM algorithm can be summarized as follows:

1. Initialize each training parameters, including $\mu = 0.1$; $\theta = 2$; and network's initial weights X_k randomly;
2. Consider that Levenberg-Marquardt algorithm is a batch processing training algorithm, all the data in the sample should be calculated to get the error e of each sample and target function value E ;
3. If $E \le E_{\min}$, we have got to the target, stop. Else switch to 4.;
4. Calculate the Jacobian matrix $J(X)$ and solve ΔX_k ;
5. Reset the weights $X_{k+1} = X_k + \Delta X_k$ and recalculate the target function value E' . If $E' < E$, then $\mu = \mu / \theta$ and switch to 2. Else, give $\mu = \mu \times \theta$, solve ΔX_k , reset the weight $X_{k+1} = X_k + \Delta X_k$ and switch to 2.

4 Neural Network Real-World Application and Self-learning Result Analyses Optimized Schedules

During the training of the neural network, the training samples need some data pretreatments such as dimension regulation and the input and output parameters normalization of the neural network. At the meantime, the selection of samples possesses vital significance in whether the training of the neural network would success.

Take the 1450 five-stand tandem cold rolling mills in an steel plant for example, the material specification is: 08AL, 2.25mm×900mm(H×B); finished product thickness: 0.5mm; the working roll diameter is 520 mm.

There are 11 measure points at the main rolling line of the 1450 tandem cold rolling mill, which are at the front of the first stand, among each stand, at the top of each stand and at the back of the last stand. A group of 17 dimension measured data is collected from the 11 measure points every 5ms, including 6 thicknesses, 5 rolling force values and 6 tension values. If there are malformed or paradoxical data in the measured data, they should be repealed. Average all the measured data every 1s and save the result as one group of the samples. The sampling time sustains to the end of the coil. The data that sampled incomplete 1s is not included. Then train the neural network with samples made up of all the measured data, and correct the network's parameters real time finally. Two third of the sample data are used as the network training and one third as testing for every type of steel.

The predicted rolling forces calculated by the traditional model Bland-Ford formula and neural network model of the same measured data are compared in figure 2.

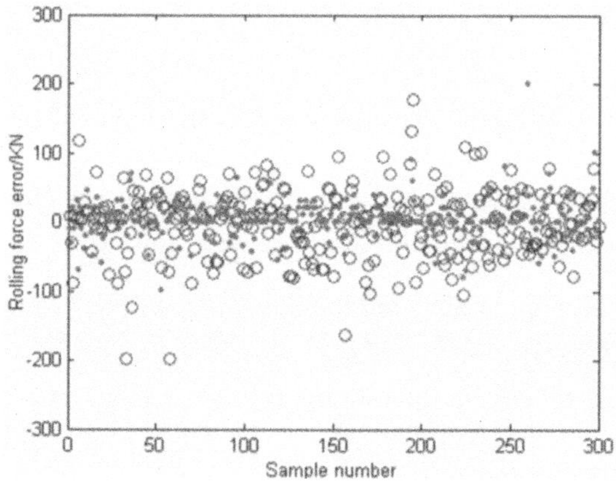

Fig. 2. The error comparison between Bland-Ford and NN model

Where the curve of '•' is the error curve between the neural network's predicted value and measured value; while the curve of '◦' is the error curve between the Bland-Ford model's predicted value and measured value. Obviously, the former error is far smaller than the latter.

The weights of each layer in the neural network will be changed after trained by the measured value off-line. Consequently, the precision of network's prediction will be higher and the predicted value will approach to the measured value.

5 Establish Mode Database (Self-learning Data)

The main function of the neural network model database are to save and refresh the result gained from the online primitive data training network and to provide each type of strip with the online neural network's parameters[4-5]. The parameters are saved in the database by the record number in the model database.

5.1 The Algorithm of the Data Record Number

Based on the recorded data from a steel plant these years, the strips are divided into different grades as follows: the basic intension K (MP), 270~420 are divided into five grades; incoming thickness H (mm),1.70~3.5 are divided into six grades; finished strip thickness h (mm), 0.18~1.68 are divided into twelve grades and strip's width B (mm), 880~1380 are divided into four grades. So there are 1440 records in self-learning files, in other words, there are 1440 neural networks. Suppose that 5, 6,12,4 are the maximum subscripts of a four-dimensional array, so we can create a four-dimensional array $S(5,6,12,4)$. Apparently, there are $5\times6\times12\times4=1440$ elements in the array. The sequential arrange is: $S(1,1,1,1)$; ...... $S(1,1,1,4)$; $S(1,1,2,1)$; ...... $S(1,1,2,4)$; ...... $S(5,6,12,4)$. As a result, the record number of an arbitrary element in the array is:

$$\text{Record number}=(i-1)\times6\times12\times4+(j-1)\times12\times4+(k-1)\times4+(l-1)+1 \quad (7)$$

Here we can solve the record number for any type of steel based on function (7).

5.2 Self-learning Data File

The neural network weights are saved in the self-learning file. The total number is 1440. Table 1 shows a portion of it.

Besides the neural network's weights which are mentioned above, there are records of other elements such as ID, the record number of each type of steel; maximum value and minimum value of input thickness; output thickness; forward tensile stress; backward tensile stress; material basic intensity; strip's width and rolling force.

We use Levenberg- Marquardt algorithm to train the neural network off-line for every type of steel. Then we save the calculated neural network's parameters into the data base. When using it online, we call the parameters saved in the data base according to the record number that represents the type of the steel and acquire the predicted value of the rolling force. For the new type of steel, we use the rolling force value acquired from the rolling force theoretical model as a target to train the neural network at first. Then modify the network's parameters after we got the real rolling data, and save the new parameters into the database at last.

Table 1. The parameter of self-learning

ID	$w^1_{1,1}$	$w^1_{1,2}$	…	b^2_1	h_{0min}	h_{0max}	h_{max}	h_{min}	T_{fmax}	T_{fmin}	…
1	-0.6617	3.76057		0.68075	1.89	0.22	1.15	0.18	16.32	3.97	
2	-0.1215	0.00961	…	2.69165	1.89	0.22	1.15	0.20	16.21	3.97	…
3	1.38044	2.20887		-1.4787	1.89	0.3	1.21	0.20	16.38	3.97	
4	1.64365	1.79041		-2.9102	1.89	0.28	1.21	0.18	16.67	3.83	
5	0.2307	-1.4251		1.5326	1.89	0.25	1.18	0.24	16.16	3.83	
6	1.98021	-0.7124		-3.0625	1.87	0.3	1.12	0.24	16.23	3.87	
7	-0.6322	1.67286	…	-4.444	1.87	0.3	1.15	0.24	16.32	3.93	…
8	1.30313	1.74435		4.63212	1.87	0.32	1.15	0.24	16.47	4.07	
⋮			⋮					⋮			

6 Conclusion

In this paper, we trained the neural network with the Levenberg-Marquardt algorithm and established the neural network rolling force model for each type of steel. The analyses of the neural network's emulator show that it possesses an excellent traceability. We also designed the self-learning system which can increase the precision of the rolling force setting value. In general, the neural network model is better than Bland-Ford model.

References

1. Lee, D.M., Choi, S.G.: Application of On-line Adaptable Neural Network for Rolling Force Set-up of A Plate Mill. Engineering Applications of Artificial Intelligence. **17**(5) (2004) 557-565
2. Larkiola, J., Myllykoski, P., Korhonen, A. S., Cser, L.: The Role of Neural Networks in the Optimization of Rolling Processes. Journal of Materials Processing Technology. **80-81**(1998) 16-23
3. Yang, J., Che, H., Xu, Y.et al.: Application of Adaptable Neural Networks for Rolling Force Set-UP in Optimization of Rolling Schedule. Advances in Neural Networks-ISNN 2006[C].2006,864-869
4. Wang, L., Frayman, Y.: A Dynamically Generated Fuzzy Neural Network and Its Application to Torsional Vibration Control of Tandem Cold Rolling Mill Spindles. Engineering Applications of Artificial Intelligence. **15** (2002) 541-550
5. Wang, D.D., Tieu, A.K., de Boer, F.G., Ma, B., Yuen, W.Y.D.: Toward a Heuristic Optimum Design of Rolling Schedules for Tandem Cold Rolling Mills. Engineering Applications of Artificial Intelligence **13** (2000) 397-406

Recurrent Fuzzy CMAC for Nonlinear System Modeling

Floriberto Ortiz[1], Wen Yu[1], Marco Moreno-Armendariz[2], and Xiaoou Li[3]

[1] Departamento de Control Automtico, CINVESTAV-IPN
A.P. 14-740, Av.IPN 2508, Mxico D.F., 07360, Mxico
yuw@ctrl.cinvestav.mx
[2] Centro de Investigacin en Computacin-IPN
Unidad Profesional "Adolfo Lpez Mateos", Mxico, D. F. C. P. 07738, Mxico
[3] Departamento de Computacin, CINVESTAV-IPN
A.P. 14-740, Av.IPN 2508, Mxico D.F., 07360, Mxico

Abstract. Normal fuzzy CMAC neural network performs well because of its fast learning speed and local generalization capability for approximating nonlinear functions. However, it requires huge memory and the dimension increases exponentially with the number of inputs. In this paper, we use recurrent technique to overcome these problems and propose a new CMAC neural network, named recurrent fuzzy CMAC (RFCMAC). Since the structure of RFCMAC is more complex, normal training methods are difficult to be applied. A new simple algorithm with a time-varying learning rate is proposed to assure the learning algorithm is stable.

1 Introduction

The Cerebellar Model Articulation Controller (CMAC) presented by Albus [1] is an auto-associative memory feedforward neural network, which is a simplified mode of the cerebellar based on the neurophysiological theory. A very important property of CMAC is that it has faster convergence speed than MLP neural networks. Many practical applications have been presented in recent literatures [9].

Since the data in CMAC is quantized and the knowledge information cannot be presented, fuzzy CMAC(FCMAC) was proposed in [2] where fuzzy set (fuzzy label) is used as the input clusters instead of crisp set. Compared to normal CMAC which associates with numeric values, FCMAC can model a problem using linguistic variables based a set of If-Then fuzzy rules. Also FCMAC is more robust, highly intuitive and easily comprehended [9].

A major drawback of FCMAC is that its application domain is limited to static problems due to its feedforward networks structure. Recurrent techniques incorporate feedback, they have powerful representation capability and can overcome disadvantages of feedforward networks [4]. Recurrent CMAC network naturally involves dynamic elements in the form of feedback connections. Its architecture is a modified model of the conventional CMAC network to attain a small number of memory, it includes delay units in CMAC. There are several types of recurrent

D. Liu et al. (Eds.): ISNN 2007, Part I, LNCS 4491, pp. 487–495, 2007.

structures, for examples, external feedback [18], internal recurrence [11], the recurrent loop is in the premise part [8], or the recurrent loop is in the consequence part [7]. In this paper, we apply recurrent method on CMAC and propose a new CMAC, named recurrent fuzzy CMAC (RFCMAC).

It is well known that normal identification algorithms (for example, gradient decent and least square) are stable for ideal conditions. In the presence of unmodeled dynamics, these adaptive procedures can go to instability easily. The lack of robustness of the parameter identification was demonstrated in [3] and became a hot issue in 1980s, when some robust modification techniques for adaptive identification was suggested [5]. Some robust modifications must be applied to assure stability with respect to uncertainties. Projection operator is an effective tool to guarantee fuzzy modeling bounded [14]. It was also used by many fuzzy-neural systems [10]. Another general approach is to use robust adaptive techniques [5] in fuzzy neural modeling. For example, applied a switch $\sigma-$modification to prevent parameters drift. By using passivity theory, we successfully proved that for continuous-time recurrent neural networks, gradient descent algorithms without robust modification were stable and robust to any bounded uncertainties [16], and for continuous-time identification they were also robustly stable [17]. Nevertheless, do recurrent fuzzy CMAC (RFCMAC) has the similar characteristics?.

In this paper backpropagation-like approach is applied to nonlinear system modeling via RFCMAC, where feedback is in the fuzzification layer of RFCMAC. The gradient decent learning is used. Time-varying learning rate is obtained by input-to-state stability (ISS) approach to update the parameters of the membership functions, this learning law can assure stability in the training process.

2 Preliminaries

The main concern of this section is to understand some concepts of ISS. Consider the following discrete-time state-space nonlinear system

$$\begin{array}{l} x(k+1) = f\left[x\left(k\right), u\left(k\right)\right] \\ y(k) = x\left(k+1\right) \end{array} \tag{1}$$

where $u\left(k\right) \in \Re^{m}$ is the input vector, $x\left(k\right) \in \Re^{n}$ is a state vector, and $y\left(k\right) \in \Re^{l}$ is the output vector. f is general nonlinear smooth function $f \in C^{\infty}$. Let us recall the following definitions.

Definition 1. *(a) If a function $\gamma(s)$ is continuous and strictly increasing with $\gamma(0) = 0$, $\gamma(s)$ is called a $\mathcal{K}$-function (b) For a function $\beta\left(s,t\right)$, $\beta\left(s,\cdot\right)$ is $\mathcal{K}$-function, $\beta\left(\cdot,t\right)$ is decreasing and $\lim_{t\to\infty} \beta\left(\cdot,t\right) = 0$, $\beta\left(s,t\right)$ is called a $\mathcal{KL}$ -function. (c) If a function $\alpha(s)$ is $\mathcal{K}$-function and $\lim_{s\to\infty} \alpha\left(s\right) = \infty$, $\alpha(s)$ is called a $\mathcal{K}_{\infty}$-function.*

Definition 2. *(a) A system (1) is said to be input-to-state stable if there is a $\mathcal{K}$-function $\gamma(\cdot)$ and $\mathcal{KL}$ -function $\beta\left(\cdot\right)$, such that, for each $u \in L_{\infty}$, i.e.,* $\sup\left\{\left\|u(k)\right\|\right\} < \infty$, *and each initial state $x^{0} \in R^{n}$, it holds that*

$$\left\|x\left(k, x^0, u\left(k\right)\right)\right\| \leq \beta\left(\left\|x^0\right\|, k\right) + \gamma\left(\left\|u\left(k\right)\right\|\right)$$

(b) A smooth function $V\Re^n \rightarrow \Re \geq 0$ *is called a ISS-Lyapunov function for system (1) if there is* $\mathcal{K}_\infty$*-functions* $\alpha_1(\cdot)$, $\alpha_2(\cdot)$ *and* $\alpha_3(\cdot)$, *and* $\mathcal{K}$*-function* $\alpha_4(\cdot)$ *such that for any* $s \in \Re^n$, *each* $x\left(k\right) \in \Re^n, u\left(k\right) \in \Re^m$

$$\begin{array}{l} \alpha_1(s) \leq V\left(s\right) \leq \alpha_2(s) \\ V_{k+1} - V_k \leq -\alpha_3(\left\|x\left(k\right)\right\|) + \alpha_4(\left\|u\left(k\right)\right\|) \end{array}$$

These definitions imply that for the nonlinear system (1), the following are equivalent: a) it is ISS; b) it is robustly stable; c) it admits a smooth ISS-Lyapunov function.

3 Recurrent Fuzzy CMAC Neural Networks

In order to identify the nonlinear system (1), we use a recurrent fuzzy CMAC (RFCMAC). This network can be presented in Fig. 1.

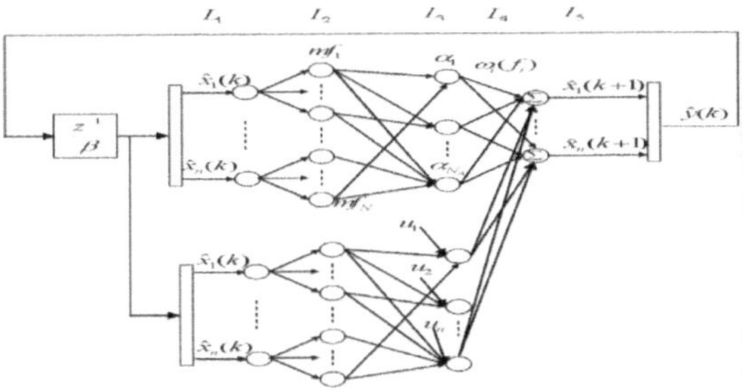

Fig. 1. Recurrent fuzzy CMAC neural networks

This network can be divided into five layers Input Layer (L_1), Fuzzified Layer (L_2), Fuzzy Association Layer (L_3), Fuzzy Post-association Layer (L_4) and Output Layer (L_5) and β is a scale constant, $\beta > 0$. The input Layer transfers input $x = (x_1, x_2, \ldots, x_n)^T$ to the next layer $mf_i = x_i$, $i = 1, ...n$, n is the number of input variables. Each node at Fuzzified Layer corresponds to a linguistics variable which are expressed by membership functions $\mu_{A_j^i}$, there are m quantizations (membership functions) for each input. The number of the nodes in this layer is n^m. Fuzzified Layer accomplishes the fuzzification of input variables. And it corresponds to both sensor layer of CMAC and fuzzifier of fuzzy logic controller. Fuzzy Association Layer connects fuzzified layer and accomplishes the matching of precondition of fuzzy logic rule. Each node at this layer completes fuzzy implication operation (flo) to obtain firing strength $\alpha_j = \pi\left(x\right) flo\{mf_i\left(x_1\right), ..., mf_n\left(x_n\right)\}$. If we use product rule for flo,

$$\alpha_k = \prod_{i=1}^{n} \lambda_k\left(\mu_{A_j^i}\right)$$

where k is association times, $k = 1 \cdots l$, l is association number, λ is the selection vector of association memory which is defined as

$$\lambda_k \left(\mu_{A_j^i} \right) = \mu_{A_{j,k}^i} = [0, 0 \cdots 1, 0 \cdots] \begin{bmatrix} \mu_{A_1^i}, \\ \vdots \\ \mu_{A_m^i} \end{bmatrix}$$

where $i = 1 \cdots n$. Fuzzy post-association layer will calculate the normalization of firing strength and prepare for fuzzy inference,

$$\bar{\alpha}_k = \alpha_k / \sum_{k=1}^{l} \alpha_k = \left(\prod_{i=1}^{N_i} \mu_{A_j^i}, \right) / \left(\sum_{j=1}^{N_A} \prod_{i=1}^{N_i} \mu_{A_j^i}, \right)$$

In the output layer, Takagi fuzzy inference will be used, that is, consequence of each fuzzy rule is defined as a function of input variables (with control input)

$$\begin{array}{l} R^j \text{ IF } \widehat{x}_1 \text{ is } A_j^1 \cdots \text{ and } \widehat{x}_n \text{ is } A_j^n \text{ THEN } \beta \widehat{x}_1 (k+1) \text{ is } f_1 (x_1, x_2, ..., x_n) \\ \quad \text{IF } \widehat{x}_1 \text{ is } A_j^1 \cdots \text{ and } \widehat{x}_n \text{ is } A_j^n \text{ THEN } \beta \widehat{x}_1 (k+1) \text{ is } f_2 (x_1, x_2, ..., x_n) u \end{array}$$

The output of the recurrent fuzzy CMAC can be expressed in a vector notation as

$$\begin{array}{l} \beta \widehat{x} (k+1) = \sum_{i=1}^{l} w_{1,i} \varphi_{1,i} [x(k)] + \sum_{i=1}^{l} w_{2,i} \varphi_{2,i} [x(k)] u(k) \\ \text{or } \beta \widehat{x} (k+1) = W_1^T \varphi_1 [x(k)] + W_2^T \varphi_2 [x(k)] U(k) \\ \widehat{y}(k) = \widehat{x} (k+1) \end{array} \qquad (2)$$

where w_k plays the role of connective weight, W_j $(j = 1, 2)$ is adjustable weight values, $\varphi_j(x)$ is base function defined as

$$\varphi_k = \frac{\prod_{i=1}^{n} \lambda_k \left(\mu_{A_j^i}, \right)}{\sum_{k=1}^{l} \prod_{i=1}^{n} \lambda_k \left(\mu_{A_j^i}, \right)}$$

We use l $(k = 1 \cdots l)$ times to perform association from an input vector $X = [x_1, \cdots, x_n] \in \Re^n$ to an output linguistic y. Each input variable x_i $(i = 1 \ldots n)$ has m quantizations.

4 System Identification Via RFCMAC with Stable Learning

We assume the base function φ_k of CMAC is known, only the weights need to be updated for system identification. We will design a stable learning algorithm such that the output $\widehat{y}(k)$ of recurrent fuzzy CMAC neural networks (2) can follow the output $y(k)$ of nonlinear plant (1). Let us define identification error vector $e(k)$ as

$$e(k) = \widehat{y}(k) - y(k)$$

According to function approximation theories of fuzzy logic and neural networks [14], the identified nonlinear process (1) can be represented as

$$\begin{array}{l}\beta x(k+1) = Ax(k) + W_1^{*T}\varphi_1[x(k)] + W_2^{*T}\varphi_2[x(k)]U(k) + \nu(k) \\ y(k) = x(k+1)\end{array} \tag{3}$$

Where W_1^* and W_2^* are unknown weights which can minimize the unmodeled dynamic $\nu(k)$. The identification error can be represented by (2) and (3),

$$\beta e_p(k+1) = Ae_p(k) + \widetilde{W}_1(k)\varphi_1[x(k)] + \widetilde{W}_2^T\varphi_2[x(k)]U(k) - \nu(k) \tag{4}$$

where $\widetilde{W}_1(k) = W_1(k) - W_1^*$, $\widetilde{W}_2(k) = W_2(k) - W_2^*$. In this paper we are only interested in open-loop identification, we assume that the plant (1) is bounded-input and bounded-output (BIBO) stable, *i.e.*, $y(k)$ and $u(k)$ in (1) are bounded. By the bound of the base function φ_k, $\nu(k)$ in (3) is bounded. The following theorem gives a stable gradient descent algorithm for fuzzy neural modeling.

Theorem 1. *If the Recurrent Fuzzy CMAC neural network (2) is used to identify nonlinear plant (1) and the eigenvalues of A is selected as* $-1 < \lambda(A) < 0$, *the following gradient updating law without robust modification can make the identification error* $e(k)$ *bounded (stable in an* L_∞ *sense)*

$$\begin{array}{l}W_1(k+1) = W_1(k) - \eta(k)\varphi_1[x(k)]e^T(k) \\ W_2(k+1) = W_2(k) - \eta(k)\varphi_2[x(k)]U(k)e^T(k)\end{array} \tag{5}$$

where $\eta(k)$ *satisfies*

$$\eta(k) = \begin{cases} \dfrac{\eta}{1 + \|\varphi_1\|^2 + \|\varphi_2 U\|^2} & \text{if } \beta\|e(k+1)\| \geq \|e(k)\| \\ 0 & \text{if } \beta\|e(k+1)\| < \|e(k)\| \end{cases}$$

$0 < \eta \leq 1$.

Proof. Select Lyapunov function as

$$V(k) = \left\|\widetilde{W}_1(k)\right\|^2 + \left\|\widetilde{W}_2(k)\right\|^2$$

where $\left\|\widetilde{W}_1(k)\right\|^2 = \sum_{i=1}^n \widetilde{w}_1(k)^2 = tr\left\{\widetilde{W}_1^T(k)\widetilde{W}_1(k)\right\}$. From the updating law (5)

$$\widetilde{W}_1(k+1) = \widetilde{W}_1(k) - \eta(k)\varphi_1[x(k)]e^T(k)$$

So

$$\begin{array}{l}\Delta V(k) = V(k+1) - V(k) = \left\|\widetilde{W}_1(k) - \eta(k)\varphi_1 e(k)^T\right\|^2 - \left\|\widetilde{W}_1(k)\right\|^2 \\ + \left\|\widetilde{W}_2(k) - \eta(k)\varphi_2 U(k)e^T(k)\right\|^2 - \left\|\widetilde{W}_2(k)\right\|^2 \\ = \eta^2(k)\|e(k)\|^2\|\varphi_1\|^2 - 2\eta(k)\left\|\varphi_1\widetilde{W}_1(k)e^T(k)\right\| \\ + \eta^2(k)\|e(k)\|^2\|\varphi_2 U(k)\|^2 - 2\eta(k)\left\|\varphi_2 U(k)\widetilde{W}_2(k)e^T(k)\right\|\end{array}$$

There exist a constant $\beta > 0$, such that

If $\|\beta e(k+1)\| \geq \|e(k)\|$, using (4) and $\eta(k) \geq 0$,

$$\begin{array}{l} -2\eta(k)\left\|\varphi_1 \widetilde{W}_1(k) e^T(k)\right\| - 2\eta(k)\left\|\varphi_2 U(k) \widetilde{W}_2(k) e^T(k)\right\| \\ \leq -2\eta(k)\left\|e^T(k)\right\| \|\beta e(k+1) - Ae(k) - \nu(k)\| \\ = -2\eta(k)\left\| \begin{array}{c} e^T(k)\beta e(k+1) \\ -e^T(k) Ae(k) - e^T(k)\nu(k) \end{array} \right\| \\ \leq -2\eta(k)\left\|e^T(k)\beta e(k+1)\right\| + 2\eta(k) e^T(k) Ae(k) + 2\eta(k)\left\|e^T(k)\nu(k)\right\| \\ \leq -2\eta(k)\|e(k)\|^2 + 2\eta(k)\lambda_{\max}(A)\|e(k)\|^2 + \eta(k)\|e(k)\|^2 + \eta(k)\|\nu(k)\|^2 \end{array}$$

Since $0 < \eta \leq 1$

$$\begin{array}{l} \Delta V(k) \leq \eta^2(k)\|e(k)\|^2\|\varphi_1\|^2 + \eta^2(k)\|e(k)\|^2\|\varphi_2 U(k)\|^2 \\ -\eta(k)\|e(k)\|^2 + 2\eta(k)\lambda_{\max}(A)\|e(k)\|^2 + \eta(k)\|\nu(k)\|^2 \\ = -\eta(k)\left[\begin{array}{c} (1 - 2\lambda_{\max}(A)) \\ -\eta \dfrac{\|\varphi_1\|^2 + \|\varphi_2 U(k)\|^2}{1 + \|\varphi_1\|^2 + \|\varphi_2 U(k)\|^2} \end{array} \right] e^2(k) \\ +\eta_k^2 \nu^2(k) \leq -\pi e^2(k) + \eta \nu^2(k) \end{array} \tag{6}$$

where $\pi = \dfrac{\eta}{1+\kappa}\left[1 - 2\lambda_{\max}(A) - \dfrac{\kappa}{1+\kappa}\right]$, $\kappa = \max_k\left(\|\varphi_1\|^2 + \|\varphi_2 U(k)\|^2\right)$.
Since $-1 < \lambda(A) < 0$, $\pi > 0$

$$n \min\left(\widetilde{w}_i^2\right) \leq V_k \leq n \max\left(\widetilde{w}_i^2\right)$$

where $n \times \min\left(\widetilde{w}_i^2\right)$ and $n \times \max\left(\widetilde{w}_i^2\right)$ are $\mathcal{K}_\infty$-functions, and $\pi e^2(k)$ is an $\mathcal{K}_\infty$-function, $\eta \nu^2(k)$ is a $\mathcal{K}$-function, so V_k admits the smooth ISS-Lyapunov function as in *Definition 2*. From *Theorem* 1, the dynamic of the identification error is input-to-state stable. The "INPUT" is corresponded to the second term of the last line in (6), *i.e.*, the modeling error $\nu(k)$, the "STATE" is corresponded to the first term of the last line in (6), *i.e.*, the identification error $e(k)$. Because the "INPUT" $\nu(k)$ is bounded and the dynamic is ISS, the "STATE" $e(k)$ is bounded.

If $\beta\|e(k+1)\| < \|e(k)\|$, $\Delta V(k) = 0$. $V(k)$ is constant, $W_1(k)$ is constant. Since $\|e(k+1)\| < \frac{1}{\beta}\|e(k)\|$, $\frac{1}{\beta} < 1$, $e(k)$ is bounded.

Remark 1. The condition "$\eta(k) = 0$ if $\beta\|e(k+1)\| < \|e(k)\|$" is dead-zone. If β is selected big enough, the dead-zone becomes small.

5 Simulations

We will use the nonlinear system which proposed [12] and [13] to illustrate the training algorithm for recurrent fuzzy CMAC. The identified nonlinear plant is

$$\begin{array}{l} x_1(k+1) = x_2(k) \\ x_2(k+1) = x_3(k) \\ x_3(k+1) = \frac{x_1(k)x_2(k)x_3(k)u(k)[x_3(k)-1]+u(k)}{1+x_2(k)^2+x_3(k)^2} \\ y(k) = [x_1(k), x_2(k), x_3(k)]^T \end{array}$$

The input signal is selected the same as [12][13](training)

$$u(k) = \begin{cases} 0.8\sin(\frac{2\pi}{25}k) + 0.2\sin(\frac{2\pi}{10}k) & k \le 200 \\ \sin(\frac{2\pi}{25}k) & k > 200 \end{cases} \quad (7)$$

We use the following recurrent fuzzy CMAC neural networks to identify it, see Fig.1. The quantization $m = 10$. The association times $l = 10$. We use 1000 data to train the model, the training input is used as in (7), then we use another 1000 data to test the model ($u(k) = \frac{1}{2}\cos(\frac{t}{35}) + \frac{1}{2}\sin(\frac{t}{10})$. The identification results are shown in Fig. 2 and Fig. 3.

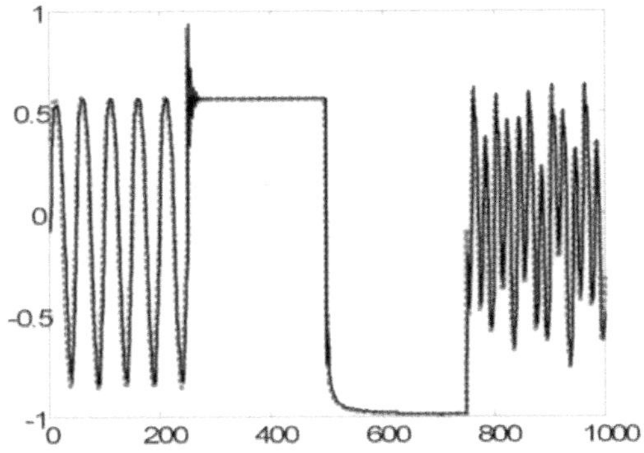

Fig. 2. RFCMAC training

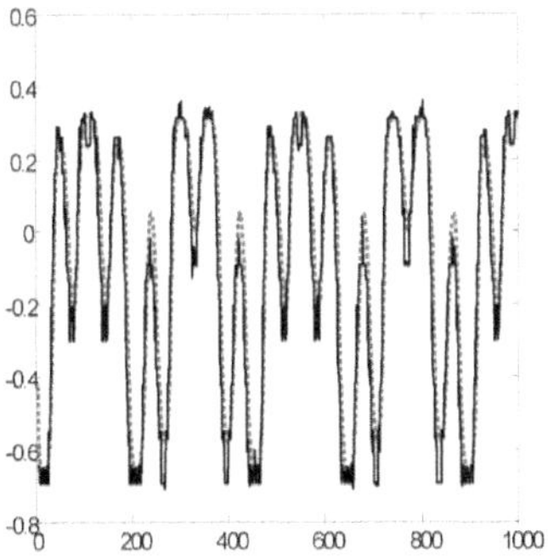

Fig. 3. RFCMAC testing

Now we compare our algorithm with normal fuzzy CMAC neural networks [2]. The training rule is (5). The identification results are shown in Fig.44 and Fig. 5. We can see that compared to normal fuzzy CMAC neural networks, recurrent fuzzy CMAC neural networks can model nonlinear system with more acurrancy. By the training algorithm proposed in this paper, the convergence speed is faster than the normal one.

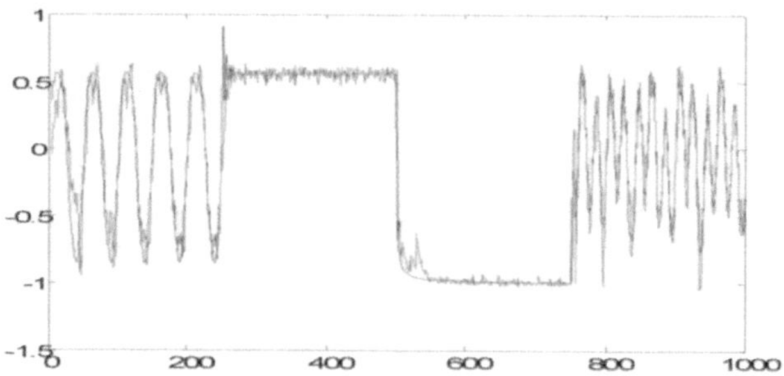

Fig. 4. Normal FCMAC training

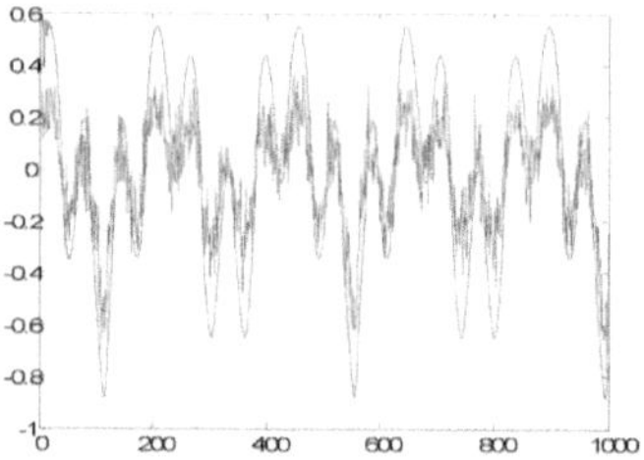

Fig. 5. Normal FCMAC testing

6 Conclusion

In this paper we propose a new CMAC structure for system identification and a simple training algorithm for recurrent fuzzy CMAC. The new stable algorithms with time-varying learning rates are stable. Further works will be done on structure training and adaptive control. Also an FPGA real-time implementation will be tested.

Acknowledgment

Dr. M.A.Moreno-Armendariz would like to thank to SIP-IPN for grant project 20062090.

References

1. Albus, J.S.: A New Approach to Manipulator Control The Cerebellar Model Articulation Controller (CMAC). Journal of Dynamic Systems, Measurement, and Control, Transactions of ASME (1975) 220-227
2. Chiang, C.-T., Lin, C.-S.: CMAC with General Basis Functions. Neural Networks **9** (1996) 1199–1211

3. Egardt, B.: Stability of Adaptive Controllers. Lecture Notes in Control and Information Sciences **20** (1979)
4. Haykin, S.: Neural Networks- A Comprehensive Foundation. Macmillan College Publ. Co. (1994)
5. Ioannou, P.A., Sun, J.: Robust Adaptive Control. Prentice-Hall, Inc, Upper Saddle River NJ (1996)
6. Jiang, Z.P., Wang, Y.: Input-to-State Stability for Discrete-Time Nonlinear Systems. Automatica **37** (2001) 857-869
7. Juang, C.F.: A TSK-type Recurrent Fuzzy Networks for Dynamic Systems Processing by Neural Network and Genetic Algorithms. IEEE Trans. Fuzzy Syst **10** (2002) 155-170
8. Lee, C.H., Teng,C.C.: Identification and Control of Dynamic System Using Recurrent Fuzzy Neural Networks. IEEE Trans. Fuzzy Syst. **8** (2000) 349-366
9. Kim, Y. H., Lewis, F. L.: Optimal Design of CMAC Neural-Network Controller for Robot Manipulators. IEEE Transactions On Systems, Man, And Cybernetics—Part C Applications And Reviews **30** (2000), 123-142
10. Leu, Y.G., Lee, T.T., Wang, W.Y.: Observer-based Adaptive Fuzzy-neural Control for Unknown Nonlinear Dynamical Systems. IEEE Trans. Syst., Man, Cybern. B **29** (1999) 583-591
11. Mastorocostas, P.A., Theocharis, J.B.: A Recurrent Fuzzy-neural Model for Dynamic System Identification, IEEE Trans. Syst., Man, Cybern. B **32** (2002) 176-190
12. Narendra, K.S., Parthasarathy, K.: Identification and Control of Dynamical Systems Using Neural Networks, IEEE Trans. Neural Networks **1** (1990) 4-27
13. Sastry, P. S., Santharam, G., Unnikrishnan, K. P.: Memory Neural Networks for Identification and Control of Dynamic Systems. IEEE Trans. Neural Networks **5** (1994) 306-319
14. Wang, L.X.: Adaptive Fuzzy Systems and Control. Englewood Cliffs NJ Prentice-Hall (1994)
15. Wang,W.Y., Leu , Y.G., Hsu,C.C.: Robust Adaptive Fuzzy-neural Control of Nonlinear Dynamical Systems Using Generalized Projection Updated Law and Variable Structure Controller. IEEE Trans. Syst., Man, Cybern. B **31** (2001) 140-147
16. Yu, W., Li, X.: Some Stability Properties of Dynamic Neural Networks. IEEE Trans. Circuits and Systems, Part I **48** (2001) 256-259
17. Yu, W., Li, X.: Some New Results on System Identification with Dynamic Neural Networks. IEEE Trans. Neural Networks **12** (2001) 412-417
18. Zhang, J., Morris, A.J.: Recurrent Neuro-fuzzy Networks for Nonlinear Process Modeling, IEEE Trans. Neural Networks **10** (1999) 313-326

A Fast Fuzzy Neural Modelling Method for Nonlinear Dynamic Systems*

Barbara Pizzileo, Kang Li, and George W. Irwin

School of Electronics, Electrical Engineering & Computer Science
Queen's University Belfast, Belfast BT9 5AH, UK
bpizzileo01@qub.ac.uk, {k.li, g.irwin}@ee.qub.ac.uk

Abstract. The identification of nonlinear dynamic systems using fuzzy neural networks is studied. A fast recursive algorithm (FRA) is proposed to select both the fuzzy regressor terms and associated parameters. In comparison with the popular orthogonal least squares (OLS) method, FRA can achieve the fuzzy neural modelling with high accuracy and less computational effort.

1 Introduction

Fuzzy neural networks represent a large class of neural networks that combine the advantages of associative memory networks (e.g. B-splines, radial basis functions and support vector machines) with improved transparency, a critical issue for nonlinear modelling using conventional neural networks. For associative neural networks, the advantage is that the linear parameters can be trained online with good convergence and stability properties. However, they produce essentially black box models with poor interpretability. By contrast, for FNNs, the basis functions are associated with some linguistic rules, and thus every numerical result can admit a linguistic interpretation [1].

One of the main obstacles in the application of fuzzy neural networks is the 'curse of dimensionality' problem. In fuzzy-neural modelling of nonlinear dynamic systems, it is usual that an excessive number of fuzzy regressor terms have to be considered initially. From these a useful fuzzy model is then generated based on the parsimonious principle, of selecting the smallest possible network which explains the data. Given some fuzzy model construction criterion, this can be achieved by an exhaustive search of all possible combinations of regressor terms using the least-squares methods. To reduce the computational complexity involved in this process, efficient suboptimal search algorithms have been proposed, among which the orthogonal least-squares method is perhaps the most popular [2]-[7]. OLS was first applied to nonlinear dynamic system identification [2][3] and has now been widely used in many other areas [4]-[7], including fuzzy neural networks. For example, after slight modification, the orthogonal least squares (OLS) method has been applied to select both the input variables and the rules [1][11].

* This work was jointly supported by the European Social Fund, the UK Engineering and Physical Sciences Research Council (EPSRC Grant GR/S85191/01).

D. Liu et al. (Eds.): ISNN 2007, Part I, LNCS 4491, pp. 496–504, 2007.

In general, OLS approaches are derived from an orthogonal (or ***QR***) decomposition of the regression matrix [2]-[8]. The elegance of the OLS approach lies in the fact that the net decrease in the cost function can be explicitly formulated as each new regressor term is selected for inclusion in the fuzzy model, with the parameters then obtained using backward substitution [2]. In this way the computational burden is significantly reduced. A number of fast orthogonal least-squares algorithms have been proposed to further improve in certain cases the efficiency [6]-[7]. In this paper, a fast recursive algorithm (FRA) [10][18] is used to build fuzzy neural networks for modelling nonlinear dynamic systems, with improved accuracy and numerical efficiency. Unlike orthogonal least squares the FRA solves the least-squares problem recursively over the model order without requiring matrix decomposition and transformation.

In literature, the main objective to reduce the model complexity in FNNs is the rules' selection. The paper shows that the FRA selects the fuzzy regressors with less effort than the OLS and with the same model complexity achieved selecting the fuzzy rules.

The paper is organised as follows. The problem statement is given first in section 2, while section 3 presents the fast recursive algorithm and its application to nonlinear modelling. A numerical example is provided in section 4 to illustrate the effectiveness of the approach. Section 5 gives the conclusion.

2 Problem Statement

In fuzzy neural networks, for a given set of m inputs and N samples, each input variable $x_i(t)$, $t=1,...,N$ is classified by k_i , $i=1,..,m$ fuzzy sets, denoted as $A_i(j_i)$, $j_i=1,...,k_i$ [14]. For every input value $x_i(t)$, its membership degree in $A_i(j_i)$ is denoted as $0 \le \mu_i^{A_i(j_i)}(t) \le 1$

$$\sum_{j_i=1}^{k_i} \mu_i^{A_i(j_i)}(t) = 1 \tag{1}$$

The construction of a fuzzy neural network mainly involves the following three steps:

Fuzzification. Each variable is classified into a certain number of fuzzy sets which involves choosing the number of the fuzzy sets and selecting the shapes of the membership functions. There are a number of different methods for handling the first issue, such as the fuzzy C-means algorithm [15], an iterative method which although computationally efficient is very sensitive in the choice of the initial iterative matrix. The membership functions are commonly chosen as the B-spline [12] for convenience. Expert knowledge can also usefully be employed [16].

Rule evaluation. There are two main types of fuzzy neural models [17]:

Mandami: IF $x_1(t)$ is $A_1(j_1)$ AND/OR $x_2(t)$ is $A_2(j_2)$ AND/OR... AND/OR $x_m(t)$ is $A_m(j_m)$, THEN y(t) is $B(j)$ ($\forall t=1,...,N$; $j_i=1,..., k_i$; $i=1,..,m$; $j=1,..., k$; k: number of the output fuzzy sets and $B(j)$ is the fuzzy set for the output).

Takagi-Sugeno: IF $x_1(t)$ is $A_1(j_1)$ AND/OR $x_2(t)$ is $A_2(j_2)$ AND/OR... AND/OR $x_m(t)$ is $A_m(j_m)$ THEN $y(t)=f_z\ (x_1, x_2, ..., x_m)$, ($\forall t=1,...,N$; $j_i=1,..., k_i$; $i=1,..,m$, where $f_z\ (\bullet)$ is some linear or nonlinear output function).

Aggregation of rules and defuzzification. The next step is to aggregate all the rules and to defuzzify the final value to produce a crisp quantity [15]. (The Takagi-Sugeno gives a crisp value directly).

Such defuzzification is achieved by the centroid technique that mathematically produces the centre of gravity. Thus,

$$y(t)=\frac{\sum_{r=1}^{NR} W_t^r y^r(t)}{\sum_{r=1}^{NR} W_t^r}=\sum_{r=1}^{NR} \Phi_t^r y^r(t) \tag{2}$$

where $y(t)$ is the crisp output for the tth sample; N_R is the total number of rules; $\Phi_t^r = W_t^r \Big/ \sum_{r=1}^{NR} W_t^r$ is the fuzzy basis function [13] associated with the rth rule ($\forall r=1,...,N_R$), where:

$$W_t^r=\prod_{i=1}^{m} \mu_i^{A_i^r}(t) \tag{3}$$

In (3) A_i^r is the fuzzy set of the ith variable associated with the rth rule and must be one of $A_i(j_i)$, $j_i=1,..., k_i$; $y^r(t)$ is the output associated with the rth rule and the tth sample whose expression depends on the particular choice of model structure. For example, an ARX model will have $\forall r=1,...,N_R$

$$y^r(t)=\sum_{i=1}^{m} g_i^r x_i(t) \tag{4}$$

where g_i^r is the consequence parameter associated with the ith input and the rth rule [14]; $x_i(t)$ is the ith input in the tth sample.

From (2) and (4) it follows that $\forall t=1,...,N$ the crisp output can be expressed in the extended form:

$$y(t)=W_t^1\left(g_1^1 x_1(t)+...+g_m^1 x_m(t)\right)+...+W_t^{NR}\left(g_1^{NR} x_1(t)+...+g_m^{NR} x_m(t)\right) \tag{5}$$

There is no denominator in (5) as $\sum_{r=1}^{NR} W_t^r = 1$.

The total number of unknown parameters g_i^r is $n=mN_R$ where $N_R=\prod_{i=1}^{m} k_i$, N_R being the total number of rules. So for instance, given $m=6$ inputs and $k_i=4$, $i=1,...,6$, the total number of parameters associated with the fuzzy model is 24576. This has to be estimated, usually by some least-squares method. Due to the very large number of parameters involved, one of the main objectives in this paper is to reduce the number of fuzzy model regressor terms in (5). Although this selection could be achieved using Orthogonal Least Squares, the new Fast Recursive Algorithm to be introduced in the next section allows these parameters to be computed with much less effort [10].

3 A Fast Recursive Algorithm

3.1 Problem Formulation and OLS

Consider a nonlinear discrete-time dynamic system [2][6]:

$$y(t) = f(y(t-1),\ldots, y(t-n_y), u(t-1),\ldots, u(t-n_u)) = f(\mathbf{x}(t)) \tag{6}$$

where $u(t)$ and $y(t)$ are system input and output variables at time instant t, n_u and n_y are the corresponding maximal lags, $\mathbf{x}(t) = [y(t-1),\ldots, y(t-n_y), u(t-1),\ldots,u(t-n_u)]^T$ is model 'input' vector, and $f(\bullet)$ is some unknown nonlinear function.

Suppose a linear-in-the-parameters model is used to represent system (6) such that

$$y(t) = \textstyle\sum_{i=1}^{n} \theta_i \varphi_i(\mathbf{x}(t)) + \varepsilon(t), \quad t = 1,\ldots,N \tag{7}$$

where $\varphi_i(\bullet), i = 1,\ldots,n$ are all candidate model terms, and $\varepsilon(t)$ is the model residual sequence. Here n can initially be a significantly large value and thus it is desirable to find a smaller number of terms n_1 [2]-[9].

If N data samples $\{\boldsymbol{x}(t), y(t)\}_{t=1}^{N}$ are used for model identification, (7) can then be formulated as

$$\boldsymbol{y} = \boldsymbol{\Phi\Theta} + \boldsymbol{\Xi} \tag{8}$$

where $\boldsymbol{\Phi} = [\boldsymbol{\varphi}_1,\ldots,\boldsymbol{\varphi}_n]$, $\boldsymbol{\varphi}_i = [\varphi_i(\boldsymbol{x}(1)), \ldots, \varphi_i(\boldsymbol{x}(N))]^T$, $i = 1,\ldots,n$, $\boldsymbol{\Phi} \in \Re^{N\times n}$, $\boldsymbol{y}^T = [y(1),\cdots, y(N)] \in \Re^N$, $\boldsymbol{\Xi}^T = [\varepsilon(1), \varepsilon(2),\ldots, \varepsilon(N)] \in \Re^N$, $\boldsymbol{\Theta} = [\theta_1, \theta_2,\ldots,\theta_n]^T \in \Re^n$.

If the modelling cost function E is defined as

$$E = \textstyle\sum_{t=1}^{N} (y(t) - \sum_{i=1}^{n} \theta_i \varphi_i(\boldsymbol{x}(t)))^2 \tag{9}$$

This can be reformulated as

$$E = (\boldsymbol{\Phi\Theta} - \boldsymbol{y})^T (\boldsymbol{\Phi\Theta} - \boldsymbol{y}) \tag{10}$$

If $\boldsymbol{\Phi}$ is of full column rank, the least-squares estimate of $\boldsymbol{\Theta}$ that minimizes this cost function is then given by [8]

$$\hat{\boldsymbol{\Theta}} = \arg \min_{\boldsymbol{\theta}} \|\boldsymbol{y} - \boldsymbol{\Phi\Theta}\|_2 = (\boldsymbol{\Phi}^T \boldsymbol{\Phi})^{-1} \boldsymbol{\Phi}^T \boldsymbol{y} \tag{11}$$

where $\|\bullet\|_2$ denotes the Euclidean norm and $\boldsymbol{\Phi}^T \boldsymbol{\Phi}$ is sometimes called the information matrix. The associated minimal cost function is

$$E(\hat{\boldsymbol{\Theta}}) = \boldsymbol{y}^T \boldsymbol{y} - \hat{\boldsymbol{\Theta}}^T \boldsymbol{\Phi}^T \boldsymbol{y} \tag{12}$$

Amongst the numerical methods available for computing $\hat{\boldsymbol{\Theta}}$ and $E(\hat{\boldsymbol{\Theta}})$, matrix decomposition methods have been widely used [7]. In particular, a ***QR*** decomposition of $\boldsymbol{\Phi}$ leads to the well-known orthogonal least-squares (OLS) method [2] for modelling and identification of nonlinear dynamic systems.

In conventional OLS [2]-[4], an orthogonal transformation is applied to (7) to produce

$$f(t) = \sum_{i=1}^{n} g_i \psi_i(\boldsymbol{x}(t)) + \varepsilon(t) \tag{13}$$

The estimated parameters in (13) are then computed as

$$\hat{\boldsymbol{G}} = [\hat{g}_1 \cdots \hat{g}_n]^T = \left[\boldsymbol{\Psi}^T \boldsymbol{\Psi}\right]^{-1} \boldsymbol{\Psi}^T \boldsymbol{Y} \tag{14}$$

where

$$\boldsymbol{\Psi} = \boldsymbol{\Phi} \boldsymbol{A}^{-1} = [\psi_1 \quad \cdots \quad \psi_n] \tag{15}$$

is an orthogonal regression matrix, and

$$\boldsymbol{A} = \begin{bmatrix} 1 & \alpha_{12} & \cdots & \alpha_{1n} \\ 0 & 1 & \cdots & \alpha_{2n} \\ \vdots & \vdots & \vdots & \vdots \\ 0 & 0 & 0 & 1 \end{bmatrix} \tag{16}$$

is a unit upper triangular matrix. The parameter estimates in (11) can be recovered by

$$\hat{\boldsymbol{\Theta}} = \boldsymbol{A}^{-1} \hat{\boldsymbol{G}} \tag{17}$$

and the related cost function is computed as

$$E(\hat{\boldsymbol{\Theta}}) = \boldsymbol{y}^T \boldsymbol{y} - \sum_{i=1}^{n} ((\boldsymbol{y}^T \psi_i)^2 / (\psi_i^T \psi_i)) \tag{18}$$

Note that according to (18), the net contribution of an orthogonal term ψ_i to the cost function can be explicitly computed as $\delta E_i = -(\boldsymbol{y}^T \psi_i)^2 / (\psi_i^T \psi_i)$ without explicitly solving the least-squares problem. OLS is therefore a computationally efficient subset selection method for nonlinear system modelling.

3.2 The Fast Recursive Algorithm

As shown in the above subsection, in order to select the model terms and to identify the model parameters efficiently, the net contribution δE_k of each term chosen for the model needs to be computed explicitly [2]-[3]. In OLS this is done using an orthogonal transformation of the regression matrix $\boldsymbol{\Phi}$. It will now be shown that this contribution can be computed by solving the least-squares problem recursively.

The development of the complete method can be found in [10]. In the following, the algorithm will only be briefly outlined.

Define:

$$\begin{cases} a_{k,i} \triangleq (\boldsymbol{\varphi}_k^{(k\text{-}1)})^{\mathrm{T}} \boldsymbol{\varphi}_i^{(k\text{-}1)}, & a_{1,i} \triangleq \boldsymbol{\varphi}_1^{\mathrm{T}} \boldsymbol{\varphi}_i \\ a_{1,y} \triangleq (\boldsymbol{\varphi}_1^{(0)})^{\mathrm{T}} \mathbf{y} = \boldsymbol{\varphi}_1^{\mathrm{T}} \mathbf{y}, & a_{k,y} \triangleq (\boldsymbol{\varphi}_k^{(k\text{-}1)})^{\mathrm{T}} \mathbf{y} \\ i = k, \cdots, n; \; k = 1, 2, \cdots, n & \end{cases} \tag{19}$$

The net contribution of a selected model term φ_{k+1}, $k=0,1,\ldots,n-1$ to the cost function can be explicitly expressed as

$$\delta E_{k+1} = -\frac{\left(\mathbf{y}^{\mathrm{T}}\varphi_{k+1} - \sum_{j=1}^{k}(a_{j,y}a_{j,k+1}/a_{j,j})\right)^2}{(\varphi_{k+1})^{\mathrm{T}}\varphi_{k+1} - \sum_{j=1}^{k}(a_{j,k+1}^2/a_{j,j})} \tag{20}$$

The effective formula for the estimation of parameters is

$$\hat{\theta}_j = \left(a_{j,y} - \sum_{i=j+1}^{k}\hat{\theta}_i a_{j,i}\right)/a_{j,j},\ j=k,k-1,\cdots,1 \tag{21}$$

Equation (21) describes the fast algorithm for computing the required model parameter estimates.

The computational efficiency and numerical stability have been detailed in [10]. It has been shown that for $N >> n_1$ and for $n >> n_1$ in the OLS method the computational effort is of the order nNn_1^2, while for the FRA it is of the order nNn_1. The FRA therefore can reduce the computation by a factor of n_1 times compared to OLS.

The FRA has been initially applied to nonlinear system identification. In this paper, it will be shown that, after modification, FRA can be effectively applied to FNN modelling. To achieve this, (5) is rewritten using the notation in (8). If for notational convenience the quantity $W^r x_i$ is defined as the fuzzy model terms, then these n vectors will constitute the new candidate fuzzy model regressor terms as in (7), i.e. $\varphi_i = W^r x_i,\ i=1,\ldots,n$. The new regression matrix $\boldsymbol{\Phi}$ can then be expressed as follows

$$\begin{cases} \Phi = \begin{pmatrix} W_1^1 x_1(1) & \ldots & W_1^1 x_m(1) & W_1^{NR} x_1(1) & \ldots & W_1^{NR} x_m(1) \\ \ldots & \ldots & \ldots & \ldots & \ldots & \ldots & \ldots \\ W_N^1 x_1(N) & \ldots & W_N^1 x_m(N) & W_N^{NR} x_1(N) & \ldots & W_N^{NR} x_m(N) \end{pmatrix} \\ \Theta^T = \begin{bmatrix} g_1^1 & \ldots & g_m^1 & \ldots & g_1^{NR} & \ldots & g_m^{NR} \end{bmatrix} \end{cases} \tag{22}$$

4 Numerical Example

Here the *Membrane* function shown in Fig 1 was approximated using a fuzzy neural network, the objective being the selection of the fuzzy model terms as defined in (22). The function inputs are x_1 and x_2 which lie in the range [0 1]. The membership functions are 1-D piecewise quadratic B-splines [12], which were generated using the recursive Cox-De Boor algorithm [12]:

$$\left[\mu_i^z(t)\right]_d = \frac{x_i(t)-\tau_z}{\tau_{z+m}-\tau_z}\cdot\left[\mu_i^z(t)\right]_{d-1} + \frac{\tau_{z+m+1}-x_i(t)}{\tau_{z+m+1}-\tau_{z+1}}\cdot\left[\mu_i^{z+1}(t)\right]_{d-1} \tag{23}$$

$$\left[\mu_i^z(t)\right]_0 = \begin{cases} 1, & \tau_z \le x_i(t) \le \tau_{z+1} \\ 0, & otherwise \end{cases}$$

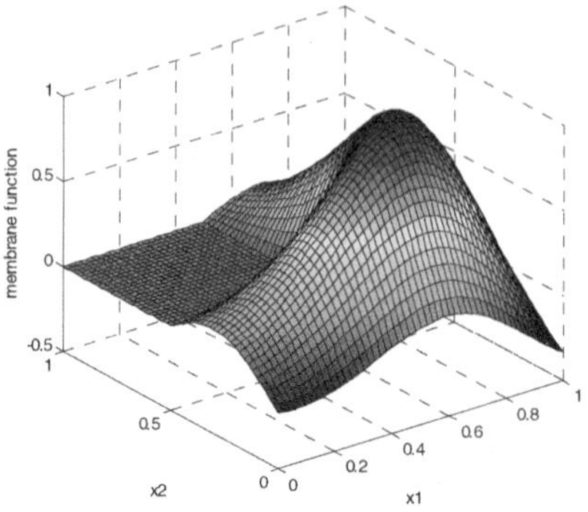

Fig. 1. Target function

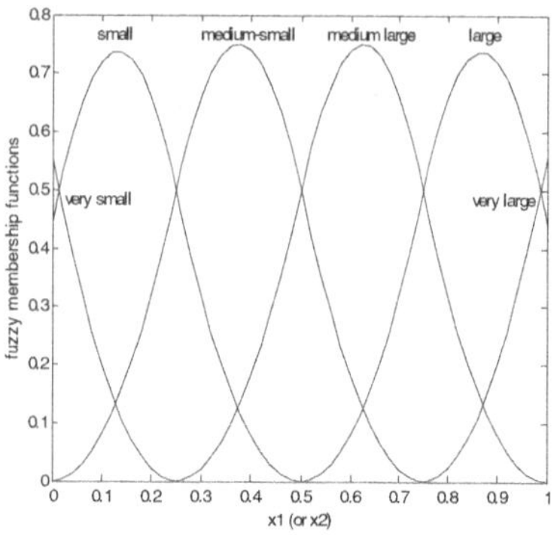

Fig. 2. 1-D piecewise quadratic B-spline fuzzy membership functions

where d is the degree of the B-spline, τ is a knot vector defined to be of dimension equal to $k_i + d + 1$, $\lfloor \mu_i^z(t) \rfloor_d$ is the B-spline with degree d of the input value $x_i(t)$ in the iteration step z. Choosing $k_i = 6 \ (i = 1,2)$, $N_R = k_1 \times k_2 = 6 \times 6 = 36$, $d = 2$ (being the B-spline quadratic) and $\tau = [-0.2, 0, 0.2, 0.4, 0.6, 0.8, 1, 1.2, 1.4]$, the above method gave the membership functions shown in Fig 2. Table 1 summarises the cardinality of rules associated with their linguistic interpretations. Using the FRA and choosing the stop

Table 1. Association of rules r to fuzzy sets

r		X_1					
		Very small	*Small*	*Medium-small*	*Medium-large*	*Large*	*Very large*
X_2	***Very small***	1	7	13	19	25	31
	Small	2	8	14	20	26	32
	Medium-small	3	9	15	21	27	33
	Medium-large	4	10	16	22	28	34
	Large	5	11	17	23	29	35
	Very large	6	12	18	24	30	36

Table 2. MSE in the selection of regressors

Selected regressors	$W^{21}x_1$	$W^{14}x_2$	$W^{28}x_2$	$W^{20}x_1$	$W^{27}x_1$	$W^{34}x_2$	$W^{23}x_1$	$W^{9}x_1$
[MSE]	0.046983	0.028632	0.018252	0.013684	0.010755	0.0076301	0.0060025	0.0050675
Selected regressors	$W^{25}x_1$	$W^{22}x_1$	$W^{16}x_1$	$W^{35}x_2$	$W^{33}x_1$	$W^{15}x_1$	$W^{16}x_2$	$W^{10}x_1$
$[MSE]_j$	0.0042134	0.0035378	0.0029761	0.0027244	0.0024922	0.0021804	0.0019271	0.0016712
Selected regressors	$W^{1}x_1$	$W^{2}x_2$	$W^{19}x_1$	$W^{31}x_1$	$W^{17}x_1$	$W^{3}x_1$	$W^{18}x_2$	$W^{1}x_2$
$[MSE]_j$	0.0014903	0.001249	0.0010839	0.00093415	0.00078773	0.00070816	0.00067721	0.00065102
Selected regressors	$W^{2}x_1$	$W^{32}x_1$	$W^{7}x_1$	$W^{13}x_2$	$W^{4}x_1$	$W^{26}x_1$	$W^{20}x_2$	$W^{29}x_2$
$[MSE]_j$	0.0006234	0.0005974	0.00057175	0.00051519	0.00048893	0.0004649	0.00043883	0.00042232
Selected regressors	$W^{28}x_1$	$W^{35}x_1$	$W^{19}x_2$	$W^{8}x_2$	$W^{3}x_2$	$W^{8}x_1$	$W^{7}x_2$	$W^{21}x_2$
$[MSE]_j$	0.00039868	0.00038711	0.00037986	0.0003735	0.00036845	0.00036415	0.00034822	0.00033935

criteria as $[MSE]_j - [MSE]_{j+1} \leq 1\times10^{-5}$ (MSE is the mean-squared-error), the selection of the fuzzy regressors ended at the 40th term, where the $MSE = 3.34\times10^{-4}$, as shown in Table 2 and Fig 3. The final fuzzy model output is shown in Fig 4.

With the following PC's specifications:
CPU: Intel(R) Pentium(R) 4 CPU 3.20GHz
Main board: Intel Corporation; model D975XBX2
Speed: 512 MB; max bandwidth 266MHz

the simulation of the OLS method, applied in the selection of the fuzzy model terms as defined in (22), has required 5.8 sec whilst the FRA required just 0.7 sec. Moreover the same simulation example was applied in reference [1] for the selection of rules and produces very similar results ($MSE = 3.45\times10^{-4}$, the number of rules selected was 20 and hence the number of parameters was $N_R m = 40$).

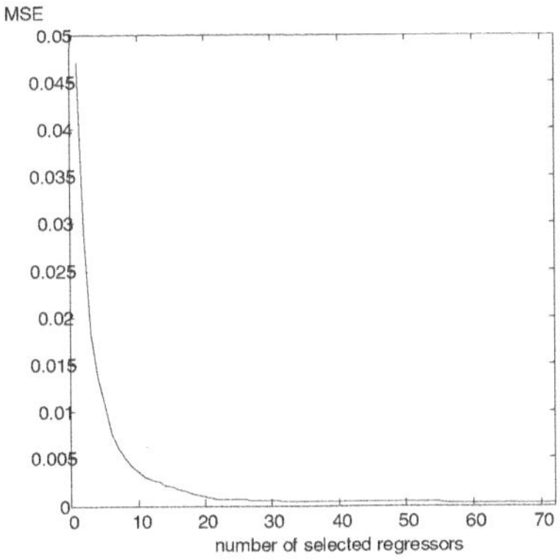

Fig. 3. MSE in the selection of regressors

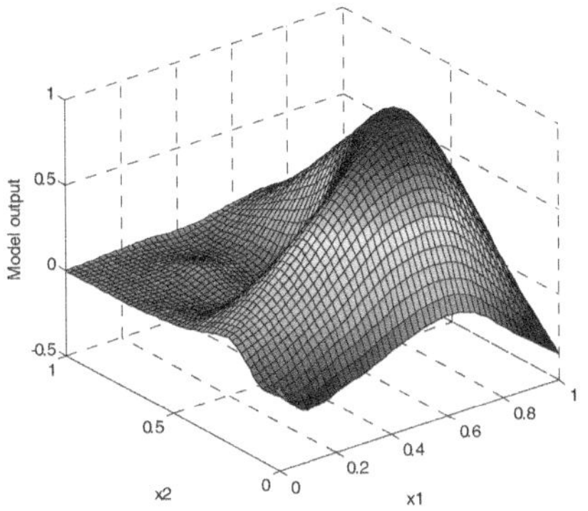

Fig. 4. Fuzzy model output

5 Conclusion

In this paper, fuzzy neural modelling of nonlinear dynamic systems has been studied. A fast recursive algorithm (FRA), initially proposed for nonlinear system identification using linear-in-the-parameters models [10], has been modified and extended for fuzzy neural modelling in selecting significant fuzzy model terms. Unlike orthogonal least squares the FRA solves the least-squares problem recursively over the model order, selecting both the fuzzy model terms and identifying the model parameters with reduced computational complexity. A simulation example shows that the selection of fuzzy model terms can lead to the same equivalent model complexity and performance achieved by selecting the fuzzy rules, but with reduction in computation compared with the OLS.

References

[1] Hong, X., Harris, C.J., Chen, S.: Robust Neurofuzzy Rule Base Knowledge Extraction and Estimation Using Subspace Decomposition Combined with Regularization and D-Optimality. IEEE Trans. Systems, Man and Cybernetics Part B **34** (2004) 598-608
[2] Chen, S., Billings, S.A., Luo, W.: Orthogonal Least Squares Methods and Their Application to Nonlinear System Identification. Int. J. Contr. **50** (1989) 1873-1896
[3] Chen, S., Cowan, C.F.N., Grant, P.M.: Orthogonal Least Squares Learning Algorithm for Radial Basis Function Networks. IEEE Trans. Neural Networks **2** (1991) 302 –309
[4] Drioli, C., Rocchesso, D.: Orthogonal Least Squares Algorithm for the Approximation of A Map and Its Derivatives with A RBF Network. Signal Processing **83** (2003) 283 – 296
[5] Chen, S., Wigger, J.: Fast Orthogonal Least Squares Algorithm for Efficient Subset Model Selection. IEEE Trans. Signal Processing **43** (1995) 1713-1715
[6] Zhu, Q.M., Billings, S.A.: Fast Orthogonal Identification of Nonlinear Stochastic Models and Radial Basis Function Neural Networks. Int. J. Contr. **64** (1996) 871–886
[7] Mao, K.Z.: Fast Orthogonal Forward Selection Algorithm for Feature Subset Selection. IEEE Trans. Neural networks **13** (2002) 1218-1224
[8] Lawson, L., Hanson, R.J.: Solving Least Squares Problem. Englewood Cliffs, NJ: Prentice-Hall (1974)
[9] Ljung, L.: System Identification: Theory for the User. Englewood Cliffs, N.J.: Prentice Hall (1987)
[10] Li, K., Peng, J., Irwin, G.: A Fast Nonlinear Model Identification Method. IEEE Trans. Automatic Control **50**(8) (2005) 1211-1216
[11] Hong, X., Harris, C.J.: A Neurofuzzy Network Knowledge Extraction and Extended Gram-Schmidt Algorithm for Model Subspace Decomposition. IEEE Trans. Fuzzy Syst.**11** (2003) 528-541
[12] Wang, C.H., Wang, W.Y., Lee, T.T., Tseng, P.S.: Fuzzy B-spline Membership Function (BMF) and Its Applications in Fuzzy-neural Control. IEEE Trans. Systems, Man and Cybernetics **25** (1995) 841-851
[13] Kim, H.M., Mendel, J.M.: Fuzzy Basis Functions: Comparison with Other Basis Functions. IEEE Trans. Fuzzy Syst. **3**(2) (1995) 158-168
[14] Sàez, D.: Takagi-Sugeno Fuzzy Model Structure Selection Based on New Sensitivity Analysis. Proc. IEEE International Conference on Fuzzy Systems (2005) 501-506
[15] Ross, T.J.: Fuzzy Logic with Engineering Applications. John Wiley & Son Ltd (2004)
[16] Makrehchi, M., Basir, O., Kamel, M.: Generation of Fuzzy Membership Function Using Information Theory Measures and Genetic Algorithm. 10th International Fuzzy Systems Association World Congress **2715** (2003) 603-610
[17] Yu, W., Li, X.: Fuzzy Identification Using Fuzzy Neural Networks with Stable Learning Algorithm. IEEE Trans. Fuzzy Syst. **12**(3) (2004) 411-420
[18] Li, K., Peng, J., Bai, E.W.: A Two-stage Algorithm for Identification of Nonlinear Dynamic Systems. Automatica **42**(7) (2006) 1189-1197

On-Line T-S Fuzzy Model Identification with Growing and Pruning Rules

Longtao Liao and Shaoyuan Li

Department of Automation, Shanghai Jiao Tong University, 200240 Shanghai, China
syli@sjtu.edu.cn

Abstract. This paper focuses on seeking an appropriate number of rules for a T-S inference system. A growing and pruning strategy in neural network is employed, which relates one fuzzy rule's contribution to the modeling accuracy by a statistic criterion, such that fuzzy rules is added/removed, whereas all the parameters can learn using EKF, both absolutely on-line and with small computation. A simulation for nonlinear system identification illustrates the good performance.

1 Introduction

T-S fuzzy inference system, which is firstly explored by Tanaka and Sugeno[1], combines the reasoning capabilities of fuzzy logic to capture uncertainty in the system and the general linear model to obtain a concise mathematical relationship between the input and output of the system. With good approximation[2] ability and analytic convenience [3], it is particularly advantageous in complex system identification. Although we are sophisticated in the parameter identification of T-S fuzzy system by introducing some algorithms of neural networks to learn the parameters [4], in the structure identification, the context of the approximation accuracy and the number of fuzzy rules has not been theoretically revealed. Several approaches are utilized to find an appropriate number of rules for T-S fuzzy systems, such as grid-partitioning [5] of input space, clustering [6], self-organizing [7], *etc.* They need a priori knowledge and training samples to find the number offline. Therefore the real-time adaptability of the rule number is lost. Recently, a novel algorithm, called Generalized Growing and Pruning (GGAP) was proposed to adjust the neuron's number of a RBF neural network on-line [8], and then been applied in Mamdani-type fuzzy inference system to deal with the number of fuzzy rules [9]. Both perform well and suggest it applicable for T-S fuzzy systems. Therefore, in this paper, an on-line learning algorithm, combining GGAP and EKF, will be proposed to adjust both the number of fuzzy rules and all the parameters, such that the structure and parameter identification of T-S fuzzy model can be fulfilled on-line. It constructs a Growing And Pruning T-S fuzzy system (GGAP-TS), which outperforms in adaptability and simplicity. A simulation illustrates it good behavior.

D. Liu et al. (Eds.): ISNN 2007, Part I, LNCS 4491, pp. 505–511, 2007.

2 T-S Fuzzy Model Identification

T-S fuzzy model is generally of the following form:

$$\begin{aligned} &\textit{Model rule } k: \text{ IF } x_1 \text{ is } A_1^k \text{ and} \ldots \text{and } x_l \text{ is } A_l^k \\ &\qquad\qquad\qquad \text{THEN } y_{jk} = p_{j0}^k + p_{j1}^k x_1 + \ldots + p_{jl}^k x_l, j = 1, 2, \ldots, m \end{aligned} \tag{1}$$

where $k = 1, 2, \ldots, K$, K is the number of rules and $A_i^k (i = 1, 2, \ldots, l)$ is the fuzzy set corresponding to the known premise variables $x_1, x_2, \ldots, x_l$, which can be denoted by vector $\mathbf{x} = (x_1, x_2, \ldots, x_l)^{\mathrm{T}} \in \mathbf{X} \subseteq \mathbf{R}^l$. y_{jk} is the jth output of T-S model from kth fuzzy rule, $p_{ji}^k (i = 0, 1, \ldots, l) \in R$.

With Gaussian membership function, singleton fuzzifier, product inference and centroid defuzzifier, the output of T-S model is derived as

$$\mathbf{y} = \sum_{k=1}^{K} \bar{R}_k(\mathbf{x}) \bar{P}_k^{\mathrm{T}} \bar{\mathbf{x}} = \frac{\sum_{k=1}^{K} R_k(\mathbf{x}) \bar{P}_k^{\mathrm{T}} \bar{\mathbf{x}}}{\sum_{k=1}^{K} R_k(\mathbf{x})} = \frac{\sum_{k=1}^{K} \exp\left(-\frac{\|\mathbf{x}-\mu_k\|^2}{\sigma_k^2}\right) \bar{P}_k^{\mathrm{T}} \bar{\mathbf{x}}}{\sum_{k=1}^{K} \exp\left(-\frac{\|\mathbf{x}-\mu_k\|^2}{\sigma_k^2}\right)} \tag{2}$$

where $\mathbf{y} = (y_1, y_2, \ldots, y_m)^{\mathrm{T}}$, R_k is the firing strength, $\bar{R}_k$ is the normalized R_k. μ_k and σ_k are the kth rule's center and width. $\|\cdot\|$ represents the Euclidean distance of two vectors, $\bar{\mathbf{x}}=(1, \mathbf{x}^{\mathrm{T}})^{\mathrm{T}}$, $\bar{P}_k = (\mathbf{p}_1^k, \ldots, \mathbf{p}_m^k)$ and $\mathbf{p}_j^k = \left(p_{j0}^k, p_{j1}^k, \ldots, p_{jl}^k\right)^{\mathrm{T}}$.

Let $(\mathbf{x}_n, \mathbf{y}_n^*)$ which is the input and output of the actual MIMO (Multiple-Input Multiple-Output) process at epoch, be the teaching signal for T-S fuzzy system, then the T-S model can be identified to approximate the actual system.

3 GGAP-TS Fuzzy System

3.1 *Influence* of One Fuzzy Rule

According to (2), for the output $\mathbf{y}$ and at epoch, kth rule contributes:

$$\mathbf{y}_n^k = \frac{R_k(\mathbf{x}_n) \bar{P}_k^{\mathrm{T}} \bar{\mathbf{x}}_n}{\sum_{k=1}^{K} R_k(\mathbf{x}_n)} \tag{3}$$

calculate all the N samples received so far, (3) can be rewritten as

$$\mathbf{y}^k = \bar{P}_k^{\mathrm{T}} \frac{\sum_{n=1}^{N} R_k(\mathbf{x}_n) \bar{\mathbf{x}}_n}{\sum_{k=1}^{K} \sum_{n=1}^{N} R_k(\mathbf{x}_n)} \tag{4}$$

dividing both the numerator and denominator in (4) by N, and when $N \to \infty$, above equation becomes:

$$\mathbf{y}_{\mathrm{inf}}^k = \lim_{N\to\infty} \mathbf{y}^k = \lim_{N\to\infty} \bar{P}_k^{\mathrm{T}} \frac{\sum_{n=1}^{N} R_k(\mathbf{x}_n) \bar{\mathbf{x}}_n / N}{\sum_{k=1}^{K} \sum_{n=1}^{N} R_k(\mathbf{x}_n) / N} \tag{5}$$

assume that $\mathbf{x}_n$ satisfies sampling density function $\varphi(\mathbf{x})$, where $\mathbf{x} \in \mathbf{X}$. We divide $\mathbf{X}$ into M small spaces $\Delta_t (t = 1, \ldots, M)$ and let $S(\Delta_t)$ denotes its size. Thus for the samples in Δ_t , $N \cdot \varphi(\mathbf{x}_t) \cdot S(\Delta_t)$ and $\mathbf{x}_t$ can represent their amount and value, while M goes extremely large, which implies Δ_t becomes extremely small [8]. We have

$$\mathbf{y}_{\inf}^k \approx \lim_{M \to \infty} \bar{P}_k^{\mathrm{T}} \frac{\sum_{t=1}^{M} R_k(\mathbf{x}_t)\, \bar{\mathbf{x}}_t \cdot N \cdot \varphi(\mathbf{x}_t) \cdot S(\Delta_t)/N}{\sum_{k=1}^{K} \sum_{t=1}^{M} R_k(\mathbf{x}_t) \cdot N \cdot \varphi(\mathbf{x}_t) \cdot S(\Delta_t)/N} \tag{6}$$

which suggests the contribution of k rule over all the output can be constructed in integral by

$$\mathbf{y}_{\inf}^k = \bar{P}_k^{\mathrm{T}} \frac{\int_{\mathbf{X}} \bar{\mathbf{x}} \exp\left(-\frac{\|\mathbf{x}-\mu_k\|^2}{\sigma_k^2}\right) \varphi(\mathbf{x})\, d\mathbf{x}}{\sum_{k=1}^{K} \int_{\mathbf{X}} \exp\left(-\frac{\|\mathbf{x}-\mu_k\|^2}{\sigma_k^2}\right) \varphi(\mathbf{x})\, d\mathbf{x}} \tag{7}$$

where $\mathbf{y}_{\inf}^k$ is a m-dimensional vector and statistically describes the contribution. For convenience and following application, we quantify it with a *q-norm* [8], i.e.

$$\|\mathbf{y}\|_q = \sqrt[q]{(\sum_{j=1}^{m} |y_j|^q)/m}, \mathbf{y} = (y_1, y_2, \ldots, y_m)^{\mathrm{T}} \tag{8}$$

Thus the *influence* of fuzzy rule can be defined as

$$E(k) = \left\|\mathbf{y}_{\inf}^k\right\|_q \tag{9}$$

which represents the average contribution of kth fuzzy rule over all the output. When relating it with modeling accuracy, its value reflects whether it is significant enough to influence the output's error. Thus adding/removing rules can be based on that.

Without loss of generality, σ_k is much less than the size of $\mathbf{X}$, and assume that $\mathbf{x}_n$ satisfies the uniform distribution, where $\varphi(\mathbf{x}) = (S(\mathbf{x}))^{-1}$. For MISO (Multiple-Input Single-Output) system, $\bar{P}_k = \mathbf{p}_1^k$ and $\|\mathbf{y}\|_q = |y|$, so (9) can be approximated by

$$E(k) \approx \left| \frac{(\mathbf{p}_1^k)^{\mathrm{T}} (\sqrt{\pi}\sigma_k)^l \bar{\mu}_k / S(\mathbf{x})}{\sum_{k=1}^{K} \left(\int_{-\infty}^{+\infty} \exp\left(-\frac{x^2}{\sigma_k^2}\right) dx\right)^l / S(\mathbf{x})} \right| = \frac{\sigma_k^l}{\sum_{k=1}^{K} \sigma_k^l} \left| (\mathbf{p}_1^k)^{\mathrm{T}} \bar{\mu}_k \right| \tag{10}$$

where $\bar{\mu}_k = (1, \mu_k^{\mathrm{T}})^{\mathrm{T}}$.

3.2 GGAP-TS Algorithm

According to (10) and combine a heuristic criterion for distance. After nth teaching signal $(\mathbf{x}_n, \mathbf{y}_n^*)$ is received, we add a new rule in addition to K rules, if

$$\|\mu_{K+1} - \mu_{nr}\| > \varepsilon_n \text{and} \frac{\sigma_{K+1}^l \left| (\mathbf{p}_1^{K+1})^{\mathrm{T}} \bar{\mu}_{K+1} \right|}{\sum_{k=1}^{K} \sigma_k^l + \sigma_{K+1}^l} > e_c \tag{11}$$

where μ_{nr} is the nearest center to $\mathbf{x}_n$, required modeling accuracy e_c is set previously, and distance threshold ε_n is given by

$$\varepsilon_n = \max\left\{\varepsilon_{\max} \times \gamma^n, \varepsilon_{\min}\right\} \tag{12}$$

In (12), $\varepsilon_{\max}$, $\varepsilon_{\min}$, and the decay factor γ decide the spatial layout of the rules. It implies that the first criterion in (11) will guarantee a new rule is sufficiently far from present rules spatially. Then the second criterion in (11) ensures the newly added rule will decrease modeling error greater than e_c, and its parameter is obtained:

$$\mathbf{p}_1^{K+1} = (y_n^* - y_n, 0, \ldots, 0)^{\mathrm{T}}_{(l+1)\times 1}, \mu_{K+1} = \mathbf{x}_n, \sigma_{K+1} = \kappa \left\|\mathbf{x}_n - \mu_{nr}\right\| \tag{13}$$

where the overlap factor κ needs to be selected appropriately.

If (11) is not satisfied, a competitive EKF is utilized only to adjust the parameters of the fuzzy rules that approaches $\mathbf{x}_n$ most [8], as the gradients of other rules almost remain zero. Let the parameter vector

$$\theta_n = ((\mathbf{p}_1^1)^{\mathrm{T}}, \mu_1^{\mathrm{T}}, \sigma_1, \ldots, (\mathbf{p}_1^{nr})^{\mathrm{T}}, \mu_{nr}^{\mathrm{T}}, \sigma_{nr}, \ldots, (\mathbf{p}_1^K)^{\mathrm{T}}, \mu_K^{\mathrm{T}}, \sigma_K)^{\mathrm{T}} \tag{14}$$

and its gradient is given by

$$\mathbf{B}_n \approx (0, \ldots, 0, (\dot{\mathbf{p}}_1^{nr})^{\mathrm{T}}, \dot{\mu}_{nr}^{\mathrm{T}}, \dot{\sigma}_{nr}, 0, \ldots, 0)^{\mathrm{T}} \tag{15}$$

where in GGAP-TS model:

$$\begin{cases} \dot{\mathbf{p}}_1^{nr} = \frac{\partial \mathbf{y}_n}{\partial \mathbf{p}_1^{nr}} = \bar{R}_{nr}\bar{\mathbf{x}}_n \\ \dot{\mu}_{nr} = \frac{\partial \mathbf{y}_n}{\partial \mu_{nr}} = \frac{\partial \mathbf{y}_n}{\partial R_{nr}}\frac{\partial R_{nr}}{\partial \mu_{nr}} = ((\mathbf{p}_1^{nr})^{\mathrm{T}}\bar{\mathbf{x}}_n - \mathbf{y}_n)\frac{2\bar{R}_{nr}(\mathbf{x}_n - \mu_{nr})}{\sigma_{nr}^2} \\ \dot{\sigma}_{nr} = \frac{\partial \mathbf{y}_n}{\partial \sigma_{nr}} = \frac{\partial \mathbf{y}_n}{\partial R_{nr}}\frac{\partial R_{nr}}{\partial \sigma_{nr}} = ((\mathbf{p}_1^{nr})^{\mathrm{T}}\bar{\mathbf{x}}_n - \mathbf{y}_n)\frac{2\bar{R}_{nr}\|\mathbf{x}_n - \mu_{nr}\|^2}{\sigma_{nr}^3} \end{cases} \tag{16}$$

and the learning of θ_n can be iteratively given by

$$\begin{array}{l} \mathbf{K}_n = \mathbf{P}_{n-1}\mathbf{B}_n\left(\mathbf{R}_n + \mathbf{B}_n^{\mathrm{T}}\mathbf{P}_{n-1}\mathbf{B}_n\right)^{-1} \\ \theta_n = \theta_{n-1} + \mathbf{K}_n e_n \\ \mathbf{P}_n = \left(\mathbf{I}_{Z\times Z} - \mathbf{K}_n\mathbf{B}_n^{\mathrm{T}}\right)\mathbf{P}_{n-1} + q\mathbf{I}_{Z\times Z} \end{array} \tag{17}$$

where q is the step along EKF's gradient, Z and Z_1 is the dimension of θ_n and θ_n^{nr} respectively, and p_0 represents the uncertainty and set to 1. whereas a rule is pruned, $\mathbf{P}_n$ will be shrink the size of $\mathbf{I}_{Z_1\times Z_1}$.

EKF learning for the nearest rule may cause its parameters to shift, which leads the rule's *influence* less than the modeling error, thus it can be pruned by

$$\frac{\sigma_{nr}^l\left|(\mathbf{p}_1^{nr})^{\mathrm{T}}\bar{\mu}_{nr}\right|}{\sum_{k=1}^K \sigma_k^l} \le e_c \tag{18}$$

This GGAP-TS algorithm is executed on-line whenever current sample is received. It is obvious that it needs small scale of data and matrix calculation, such that the computation reduces a lot and thus it fits well for real-time application.

4 Simulation and Performance Comparison

In this section, an identification problem of CSTR (Continuous Stirred-Tank Reactor) system, which has strong nonlinearity, is presented. Two indices, MAE and RMSE, are used to verify the performance of our GGAP-TS algorithm and compare with two other T-S model: (1) FCM-TS: off-line clustering with Fuzzy C-mean, on-line parameter learning with EKF [10]; (2) SAFIS: A Mamdani fuzzy system with growing and pruning rules.

In CSTR, an irreversible exothermic reaction can be described by

$$\begin{cases} \dot{C}_a = \frac{q}{V}(C_{af} - C_a) - a_0 C_a e^{-\frac{E}{RT_a}} \\ \dot{T}_a = \frac{q}{V}(T_f - T_a) + a_1 C_a e^{-\frac{E}{RT_a}} + a_3 q_c (1 - e^{-\frac{a_2}{q_c}})(T_{cf} - T_a) \end{cases} \tag{19}$$

The objective is to control the effluent concentration C_a by manipulating the coolant flow rate q_c . The sampling time set $0.1s$, 1000 and 200 samples are used for training and verifying. The input for T-S model is $(C_a(n-1), C_a(n-2), q_c(n-1))^{\mathrm{T}}$, where the initial conditions $C_{a0} = 0.1mol/L, T_{a0} = 440.0K, q_{c0} = 100ml/\min$, q_c generated from $[90, 110]$ uniformly randomly, and parameters in 19 according to [11]. The learning parameters $\varepsilon_{\max} = 0.1, \varepsilon_{\min} = 0.01, \gamma = 0.999$, $\kappa = 0.87$, $q = 0.0002$ both for GGAP-TS and SAFIS, and FCM-TS are classified to 20 clusters, required accuracy $e_c = 0.001$.

Table 1. Performance comparison (normalized to $[0, 1]$)

Algorithm	Rule's	Training error		(P1)		(P2)		CPU
	Number	RMSE	MAE	RMSE	MAE	RMSE	MAE	time
GGAP-TS	16	0.0575	0.0397	0.0445	0.0351	0.0640	0.0379	4.001
FCM-TS	20	0.0530	0.0331	0.0396	0.0298	0.1193	0.0436	103.640
SAFIS	17	0.0655	0.0456	0.0516	0.0419	0.0716	0.0427	2.018

Furthermore, we divide the verification into two phases, P1 and P2, where the feed temperature T_f rises by $10K$ at the end of P1 ($110s$), in order to test the adaptability.

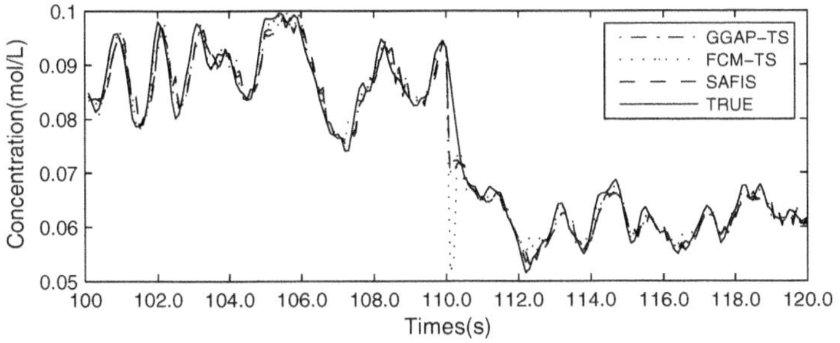

Fig. 1. Result comparison of three algorithms

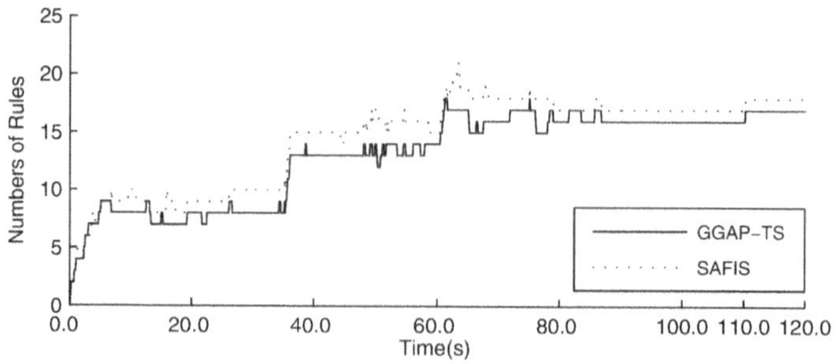

Fig. 2. Number of Rules

According to Table 1, most of the indices of FCM-TS are minimum, but it gets a long off-line clustering and abruptly performs bad when condition altered at P2. The other two algorithms show good adaptability at this point, where GGAP-TS achieves better indices. Figure 1 indicates that FCM-TS performs bad at 110s where condition altered, however, due to growing and pruning, GGAP-TS and SAFIS adapt well both by adding a new rule, which shows on Figure 2.

5 Conclusion

In this paper, a Generalized Growing And Pruning T-S fuzzy system, GGAP-TS, is developed based on a novel strategy for adding/removing neurons in a RBF network. With the idea of *influence*, which quantitatively defines single rule's contribution over all the output of T-S model, adding/removing rules is fulfilled on-line to satisfy the required modeling error. Besides, due to the competitive EKF, real-time parameter learning is carried out with small data and computation. Simulation results indicate the GGAP-TS algorithm has good accuracy and on-line adaptability on a nonlinear system identification problem.

Acknowledgments. This research is supported by the National Natural Science Foundation of China(Grant: 60474051) and by the Specialized Research Fund for the Doctoral Program of Higher Education of China(Grant: 20060248001), and partly by the program for New Century Excellent Talents in University of China(NCET).

References

1. Takagi, T., Sugeno, M.: Fuzzy Identification of Systems and Its Applications to Modeling and Control. IEEE Trans. on Syst., Man and Cybern **15** (1985) 116-132.
2. Hao Y.: Sufficient Conditions on Uniform Approximation of Multivariate Functions by General Takagi-Sugeno Fuzzy Systems with Linear Rule Consequent. IEEE Transactions on Systems, Man, and Cybernetics Part A:Systems and Humans **28** (1998) 515-520

3. Wang, H., Tanaka, K., Griffin, M.: Parallel Distributed Compensation of Nonlinear Systems by Takagi-Sugeno Fuzzy Model. Proc IEEE Int Conf on Fuzzy Systems, Yokohama, Jpn (1995) 531-538
4. Jang J.S. Roger: ANFIS: Adaptive-Network-Based Fuzzy Inference System. IEEE Trans on Syst, Man and Cybern **23** (1993) 665-685
5. Wang L.X., Mendel, J. M.: Generating Fuzzy Rules by Learning from Examples. IEEE Transactions on Systems, Man and Cybernetics **22** (1992) 1414-1427
6. Setnes, M.: Supervised Fuzzy Clustering for Rule Extraction. IEEE Transactions on Fuzzy Systems **8** (2000) 416-424
7. Nie, J.H.: Constructing Fuzzy Model by Self-Organizing Counterpropagation Network. IEEE Transactions on Systems, Man and Cybernetics **25** (1995) 963-970
8. Huang G.B., Saratchandran, P., Sundararajan, N.: A Generalized Growing and Pruning RBF (GGAP-RBF) Neural Network for Function Approximation. IEEE Trans on Neural Networks **16** (2005) 57-67
9. Rong H.J., Sundararajan, N., Guangbin Huang: Sequential Adaptive Fuzzy Inference System (SAFIS) for Nonlinear System Identification and Prediction. Fuzzy Sets and Systems **157** (2006) 1260-1275
10. Karayiannis, N.B., Weiqun Mi: Growing Radial Basis Neural Networks: Merging Supervised and Unsupervised Learning with Network Growth Techniques. IEEE Trans on Neural Networks **8** (1997) 1492-1506
11. Ge, S.S., Hang, C.C., Zhang, T.: Nonlinear Adaptive Control Using Neural Networks and Its Application to CSTR Systems. Journal of Process Control **9** (1999) 313-323

Improvement Techniques for the EM-Based Neural Network Approach in RF Components Modeling

Liu Tao[1], Zhang Wenjun[1], Ma Jun[2], and Yu Zhiping[1]

[1] Institute of Microelectronics, Tsinghua University
zwj@tsinghua.edu.cn
[2] Dept. of Computer Science, Xi'an Univ. of the Finance and Economics,
Xi'an 710061, China

Abstract. Electromagnetic (EM)–based neural network (NN) approaches have recently gained recognition as unconventional and useful methods for radio frequency (RF) components modeling. In this paper, several improvement techniques including a new data preprocessing technique and an improved training algorithm are presented. Comprehensive cases are compared in this paper. The experimental results indicate that with these techniques, the modified model has better performance.

1 Introduction

The lack of fast and accurate models for passive components is one of the bottlenecks in the design of Radio-Frequency (RF) circuits. Conventional empirical models do not accurately account for parasitic and coupling effects[1]. Detailed numerical techniques suffer from large computation time required to solve algebraic and differential equations, therefore are not practical for interactive CAD [3]. Thus, neural network approach, with attractive performance on both speed and accuracy [4], is drawing intense attention in RF components modeling.

Many scholars have applied the EM-based NN method in the modeling of RF components. However, NN predication often exhibits fierce vibrations in high frequency domain. Besides, the training algorithms frequently utilized in the modeling procedures are linear or super-linear algorithm, which are comparatively slow. In this paper, the vibration of model output which results from the wide frequency range is eliminated using a data preprocessing technique. Meanwhile, some second-order gradient algorithms are employed in NN training. Training speed is substantially improved.

2 Neural Network Modeling Approach

An excellent presentation of neural network technique is presented in [4]. The most frequently used neural network structure in RF components modeling is multilayer perceptron (MLP). As shown in Fig.1, MLP typically consists of an input layer, one or more hidden layers, and an output layer.

D. Liu et al. (Eds.): ISNN 2007, Part I, LNCS 4491, pp. 512–518, 2007.

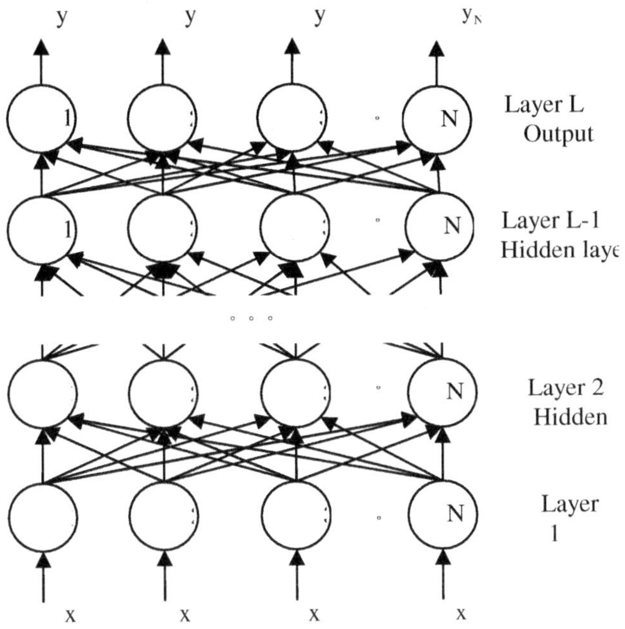

Fig. 1. MLP neural network structure

Suppose the total number of layers is L, the number of neurons in lth layer is NL. Let wijl represent the weight that connects the jth neuron of l-1th layer and the ith neuron of lth layer, bil stands for the bias of the ith neuron of lth layer and Zil denotes the output of the ith neuron of lth layer. The computation formula for the input layer is given by

$$z_i^l = x_i \quad i = 1, 2, ..., N_1. \tag{1}$$

The output of each hidden neuron is expressed as

$$z_i^l = f(\sum_{j=0}^{N_{L-1}} w_{ij}^l z_j^{l-1} + b_i^l) \qquad i = 1, 2, ..., N_l, \quad l = 2, 3, ..., L-1, \tag{2}$$

where f is the transfer function. One of the typical transfer functions is hyperbolic tangent sigmoid function given by

$$f(x) = \text{th}\, x = \frac{e^x - e^{-x}}{e^x + e^{-x}}. \tag{3}$$

The formula for the output layer is,

$$y_i = \sum_{j=0}^{N_{L-1}} w_{ij}^L z_j^{L-1} + b_i^L \quad i = 1, 2, ..., N_L. \tag{4}$$

Thus, an analytical expression can be established between input parameters (x) and output parameters (y) of the neural network. The unfixed variables in the expression are weights and bias. The universal approximation theorem[5] states that there always exists a three-layered MLP neural network that can approximate any arbitrary nonlinear continuous multidimensional function to any desired accuracy.

3 Data Preparation

To demonstrate the effectiveness of the improvement techniques described in later parts, on-chip square spiral inductors are used as test examples. In RF circuit design, the electrical characteristics of passive components are generally represented by S-parameters for the ease of measurement. Therefore, S-parameters are often used as model outputs. Since S12 and S21 are equal for passive two-port reciprocal networks such as spiral inductors, only S12 is included in the output parameters. Meanwhile, the model inputs often include design parameters such as geometrical parameters and operation frequency.

Fig. 2 is the physical layout of the spiral inductor used in this paper. The design parameters used as model input parameters are: line width (W), spacing (S), inner radius (R), and number of turns (N). The operation frequency (F) is also an input parameter.

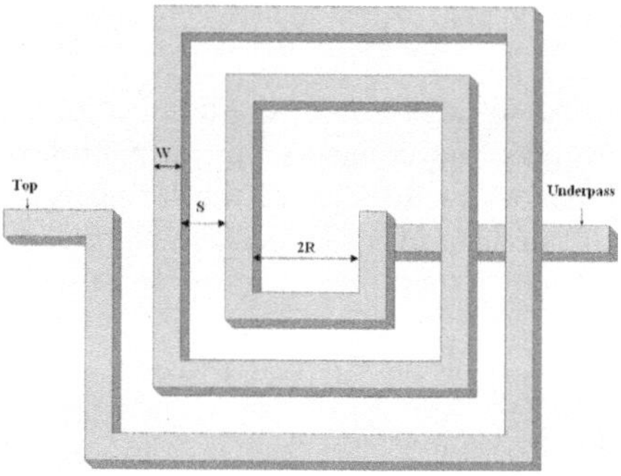

Fig. 2. Physical layout of square spiral inductor with 2.5 turns, where W, S, and R stand for width, spacing, inner radius respectively

Table 1. Training data and test data

	Training Data			Test Data		
Parameters	Min	Max	Step	Min	Max	Step
Width(μm)	5	35	10	10	30	10
Spacing(μm)	1	7	2	2	6	2
Inner Radius(μm)	20	140	40	40	120	40
Number of Turns	1.5	7.5	2	2.5	6.5	2
Frequency (GHZ)	0.5	20	0.5	0.5	20	0.5

The training and validation data are gathered using the scheme shown in Table 1. The S-parameters are simulated using the EM simulation tools - ADS Momentum from Agilent [6].

To prevent the activation values from becoming too large and the occurrence of neuron saturation during training, both the inputs and outputs data are transformed to [$\tilde{x}_{\min}$, $\tilde{x}_{\max}$] by means of linear scaling-[4]:

$$\tilde{x} = \tilde{x}_{\min} + \frac{x - x_{\min}}{x_{\max} - x_{\min}} (\tilde{x}_{\max} - \tilde{x}_{\min}), \tag{5}$$

where $x_{\min}$ $x_{\max}$ are the minimum and maximum values of original data respectively, x is a generic element in original data. In this paper, $\tilde{x}_{\min}$ =-1, $\tilde{x}_{\max}$ =1.

4 Data Preprocessing Technique

The S-parameters are often represented by magnitude and phase parts in most previous modeling works [2-4]. Unfortunately, this data format will produce peaks of neural network computed output. To illustrate the details of this drawback, a typical spiral inductor (R=50μm, N=5, W=12μm, S=7μm) is used as a test case in our experiment. A neural network model is established for this inductor. The input parameter is simply the operation frequency and the output parameters are S-parameters at corresponding frequencies.

At first, the S-parameters are represented by magnitude and phase parts. The data are obtained over a frequency range of 1 to 20GHz at 0.5GHz intervals. Among them, the S-parameters at frequency points from 1 to 20GZH at 2GHz intervals sever as training data while others sever as test data. All data are transformed to [-1, 1]. The size of the hidden layer is determined experimentally as 7. After training, the NN model is tested. Results are given in Fig. 3. Because S22 is similar to S11, only S11 is given.

It's shown in Fig.3 that owing to the sharp variation on phase S12, some conspicuous peaks occur on every output parameter near the frequency at which sharp variation occurs, in this case, about 9GHZ. However, the magnitude of S11, the magnitude of S12, and the phase of S11 are smooth. Besides, the NN computed MagS22 even exhibits negative values. Obviously, sharp variation in one parameter brings negative influences to all output parameters. Thus, traditional data format is not suitable for wide frequency range modeling.

A data preprocessing technique is utilized in this paper to solve the problem. The technique simply replaces the magnitude and phase parts with real and imaginary parts. To demonstrate the merit of the technique, the same neural network is trained with new data, i.e. scaled S-parameters represented by real and imaginary parts. Fig. 4 illustrates the results.

It can be seen from Fig. 4 that equipped with the new data preprocessing technique, every output parameter becomes very smooth. Meanwhile, the S-parameters computed by neural network shows excellent agreement with desired target and the peaks in traditional approach are eliminated.

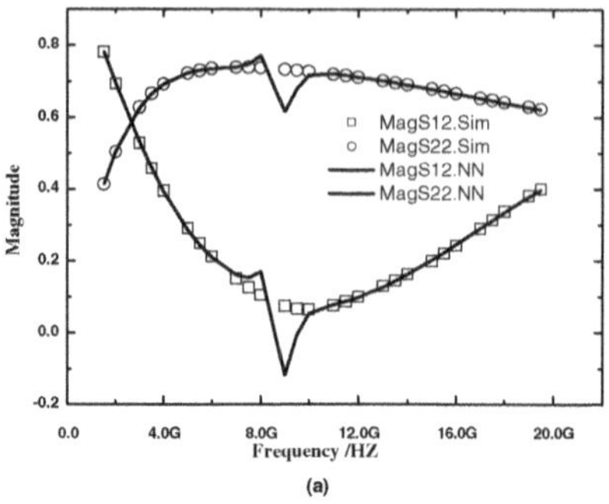

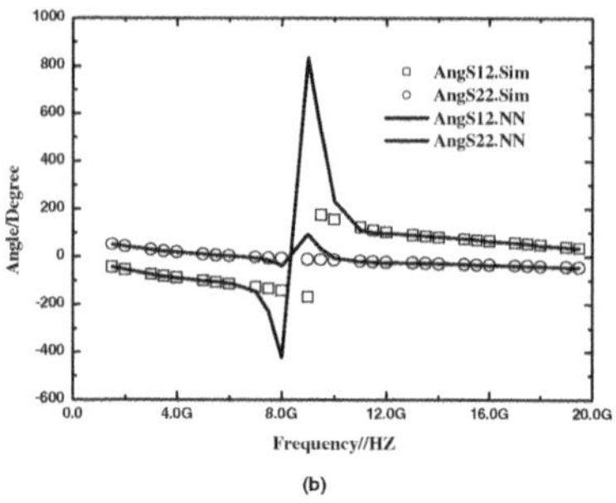

Fig. 3. Comparison of simulated and NN computed S-parameters (a) MagS12 and Mag S22 (b) AngS12 and Ang S22

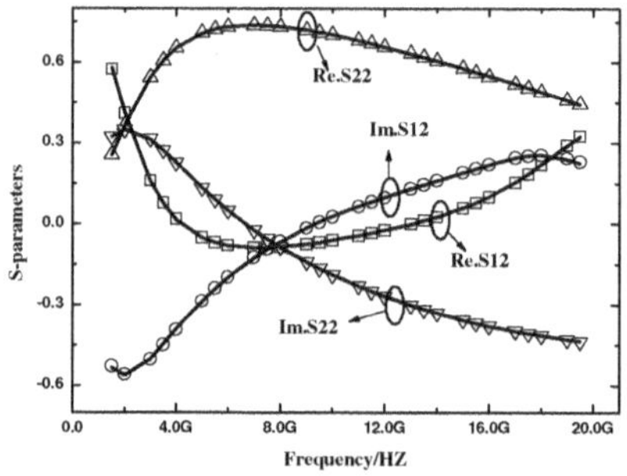

Fig. 4. Comparison of simulated (symbol) and NN computed S-parameters (line)

5 Training Algorithms Improvement

In previous works, the most frequently used training algorithms are either back-propagation (BP) algorithm[2,3] or BP-momentum (BPm) algorithm[4]. Although adding the momentum term helps the network from being trapped in a local minimum, their convergence speeds are still very slow. In order to improve the training speed, numerous other algorithms have been proposed. However, neural network are prone to exhibit over-learning only using too fast optimization methods such as Levenberg-Marquardt (LM) algorithm. When neural network is over-learning, it can fit training data very well, yet without generalizing satisfactorily.

In this paper, a validation procedure is embedded in order to incorporate fast training algorithms. In this technique, the data are divided into three parts, training data, validation data, and test data. Validation data may be the same as test data when data are limited. During the training process, the mean square error (MSE) for the validated data is monitored to avoid over-fitting of the training data by performing simultaneous training and testing of NN. Once the error on the validated data set begins to increase or stay flat, the training process is terminated.

The validation data is the same as test data shown in Table 1. Due to the relatively large number of training and validation data, 35 hidden layer neurons are used. The BPm (momentum factor is 0.9) algorithm, the CG algorithm, and the LM algorithm are utilized to train the same neural network respectively with the same initial settings. The convergence details are given in Fig. 5, where Tr and Te strand for training and test error respectively.

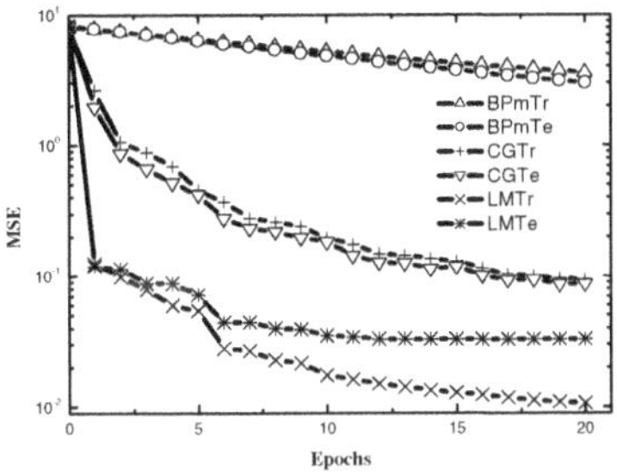

Fig. 5. Comparison of different training algorithms with validation procedure

As can be seen in Fig. 5, after 20 iterations, the BPm algorithm achieves training MSE of ~100, and the CG algorithm achieves the order of 10-1. The LM algorithm has the best performance and converges to ~10-2 both on training and test MSE.

6 Conclusions

Some improvement techniques for EM-based neural network modeling approaches including the data preprocessing technique and the fast training algorithm are proposed in this paper. Various comparisons demonstrate the superiorities of these techniques. This excellent predication ability of EM-based NN modeling approach combined with these improvements indicates that NN is advantageous for RF components modeling.

References

1. Goldfarb, M., Platzker, A.: The Effects of Electromagnetic Coupling on MMIC Design. International Journal of Microwave and Millimeter-Wave Computer-Aided Engineering **1(1)** (1991) 38-47
2. Creech, G.L., Paul, B.J., Lesniak, C.D., etal.: Artificial Neural Networks for Accurate High Frequency CAD Applications. IEEE International Symposium on Circuits and Systems (1996) 317-320
3. Creech, G.L., Paul, B.J., Lesniak, C.D., etal.: Artificial Neural Networks for Fast and Accurate EM-CAD of Microwave Circuits. IEEE Transactions On Microwave Theory And Techniques **45(5)** (1997) 794-802

4. Zhang, Q., Gupta, K.C., Devabhaktuni, V.K.: Artificial Neural Networks for RF and Microwave Design - from Theory to Practice. IEEE Transactions on Microwave Theory and Techniques **51(4)** (2003) 1339-1350
5. Hornik, K., Stinchcombe, M., White, H.: Multilayer Feedforward Networks Are Universal Approximators. Neural Networks **2** (1989) 359-366
6. Momentum Agilent EESOF EDA, Agilent Corporation, Palo Alto, CA, 2004A

A Novel Associative Memory System Based Modeling and Prediction of TCP Network Traffic

Jun-Song Wang[1,2], Zhi-Wei Gao[1], and Ning-Shou Xu[3]

[1] Department of Automation, Tianjin University, Tianjin 300270 China
wjsong2004@126.com
[2] Department of Automation, Tianjin University of Technology and Education
Tianjin 300222 China
[3] Department of Automation, Beijing Polytechnic University, Beijing 100086, China

Abstract. This paper proposes a novel high-order associative memory system (AMS) based on the Newton's forward interpolation (NFI), The Interpolation Polynomials and training algorithms for the new AMS scheme are derived. The proposed novel AMS is capable of implementing error-free approximation to complex nonlinear functions of arbitrary order. A method Based on NFI-AMS is designed to model and predict network traffic dynamics, which is capable of modeling the complex nonlinear behavior of a traffic time series and capturing the properties of network traffic. The simulation results showed that the proposed scheme is feasible and efficient. Furthermore, the NFI-AMS based traffic prediction can be used in more fields for network design, management and control.

1 Introduction

Associative memory system (AMS) is a perceptron-like double-layer feedforward neural network (NN) which is now widely used in constructing mathematical mappings of real-time systems with complicated physical characteristics. It stores information in the associative cells within the partitioned input space and recovers the global mapping from a portion of its own. Only a small portion of information is affected during an individual learning run, thus, the learning interference is far less than that of the globally generalized networks such as multilayer perceptrons (MLPs).

As one type of AMSs, the cerebellar model articulation controller (CMAC) proposed by Albus in 1975 [2]. Nonetheless, the CMAC suffers from large memory request and relatively low learning precision. Its hash coding and addressing mechanism for reducing memory request may cause data-collision problem, hence destroying the previously trained weights stored in the memory. Therefore, in the past ten years, research works had been done to reduce the memory request [3], to avoid the data-collision [4], to improve the approximation capability, to speed up the learning convergence rate [5]. Moreover, the original motivation and rationale for using hash-coding in CMAC were questioned due to certain undesirable loss in approximation ability [6].

TCP is a reliable, connection-oriented, Transport Control Protocol providing a reliable network managing mechanism, which is considered as the mainly used way to

D. Liu et al. (Eds.): ISNN 2007, Part I, LNCS 4491, pp. 519–527, 2007.

manage the network traffic [11]. However, with constantly growing of TCP applications, The traditional TCP is faced with more and more challenge, and the accuracy of the network traffic prediction will directly impact the stableness of the Internet. So the accurate model of TCP network traffic is required for predicting the future traffic to manage the network and design novel TCP. In fact, the behavior of TCP network traffic is so complex, nonlinear and time-variant that it is impossible to be precisely modeled by normal linear or nonlinear equations.

In this paper, we propose a novel high-order AMS based on the newton's forward interpolation (NFI) for implementing an error-free approximation to a polynomial function with an arbitrary order. We obtain a higher learning capability of the NFI - AMS by using the interpolation algorithm developed. The advantages it offers over conventional CMAC neural network are: high-precision of learning, much smaller memory requirement without the data-collision problem, much less computational effort for training and faster convergence rates than that attainable with multi-layer BP neural networks. Based on NFI-AMS, a method is proposed to model and predict TCP network traffic, which is more effective and accurate thanks to the NFI-AMS's adaptive learning, good approximating ability and generalization features.

2 Newton's Forward Interpolation Based Associative Memory System (NFI-AMS)

2.1 Newton's Forward Interpolation Polynomials

To approximate a μ -th order N -variables dimensional polynomial, the Newton's forward interpolation polynomial is given by[1]

$$
\begin{aligned}
&\phi_\mu(s_1, s_2, \cdots s_N) \\
&= \Delta^{(0,0,\cdots,0)} f(s_{1_0}, s_{2_0} \cdots, s_{N_0}) + (s_2 - s_{2_0})\Delta^{(0,1,\cdots,0)} f(s_{1_0}, s_{2_0} \cdots, s_{N_0} \\
&+\cdots+(s_N - s_{N_0})\Delta^{(0,0,\cdots,1)} f(s_{1_0}, s_{2_0} \cdots, s_{N_0}) + \cdots + \sum_{l_1+l_2+\cdots+l_N=0}^{\mu} (\prod_{j=0}^{l_1-1} t_{1_j} \prod_{j=0}^{l_2-1} t_{2_j} \cdots \prod_{j=0}^{l_N-1} t_{N_j})
\end{aligned} \tag{1}
$$

$$
\Delta^{(l_1,l_2,\cdots,l_N)} f(s_{1_0}, s_{2_0} \cdots, s_{N_0}) = \sum_{l_1+l_2+\cdots+l_N=0}^{\mu} c_{l_1 l_2 \cdots l_N} \Delta^{(l_1,l_2,\cdots,l_N)} f(s_{1_0}, s_{2_0} \cdots, s_{N_0}) \tag{2}
$$

where $c_{l_1,l_2,\cdots l_N} = \prod_{j=0}^{l_1-1} t_{1_j} \prod_{j=0}^{l_2-1} t_{2_j} \cdots \prod_{j=0}^{l_N-1} t_{N_j}$, $t_{i_j} = s_i - s_{i_j}, i = 1,2\cdots, N$ and $\Delta^{(l_1,l_2,\cdots,l_N)} f(s_1, s_2, \cdots, s_N)$ is the $(l_1 + l_2 + \cdots + l_N)$ -*th* order discrete difference of $f(s_1, s_2, \cdots, s_N)$.

2.2 Interpolation Algorithm

The weights number strictly depended on the coefficient of a N -variable polynomial function in the form of

$$f(s)=f(s_1,s_2,\cdots,s_N)=a_0+\sum_{j=1}^{N}a_js_j+\sum_{j=1}^{N}\sum_{j_2=j_1}^{N}a_{j_1}a_{j_2}s_{j_1}s_{j_2}+\cdots$$
$$+\sum_{j=1}^{N}\sum_{j_2=j_1}^{N}\cdots\sum_{j_\mu=j_\mu-1}^{N}a_{j_1}a_{j_2}\cdots a_{j_\mu}s_{j_1}s_{j_2}\cdots s_{j_\mu} \tag{3}$$

where $s=[s_1,s_2,\cdots,s_N]$. Thus, the total number of its coefficients is

$$n_\mu=1+C_N^1+C_{N+1}^2+\cdots+C_{N+\mu-1}^{\mu}+=C_{N+\mu}^{\mu} \tag{4}$$

So n_μ pieces of independent information is enough to approximate μ -th order polynomial function in a specified N -dimensional input space[7].

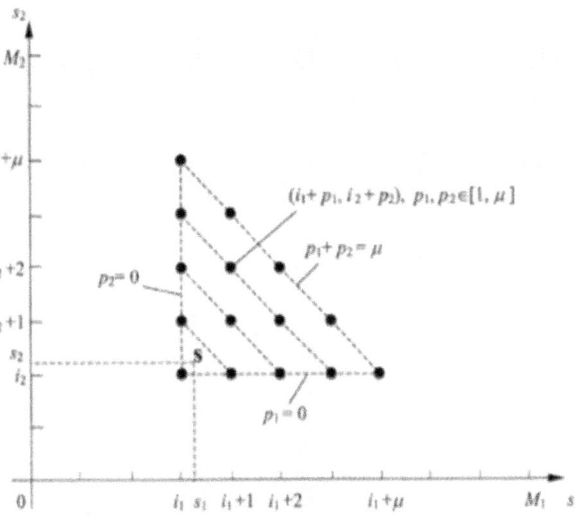

Fig. 1. Distribution of the active cells of the two-dimensional cases

In order to reduce the computational effort, it is necessary to develop an implicit expression, in which all the discrete differences (instead of the function values themselves) of $f(s_1,s_2,\cdots,s_N)$ are directly used in the interpolation algorithm[9].

A N -variable high-order polynomial function can be approximated by

$$f(s_1,s_2,\cdots,s_N)=\phi_\mu(s_1,s_2,\cdots,s_N)+R_\mu(s_1,s_2,\cdots,s_N)$$

$$\approx \phi_\mu(s_1, s_2, \cdots, s_N)$$
$$= \sum_{l_1+l_2+\cdots+l_N=0}^{\mu} c_{l_1,l_2\cdots l_N} \Delta^{(l_1,l_2,\cdots,l_N)} f(s_1, s_2, \cdots, s_N) \quad (5)$$

It can be easily proved that the total number of its coefficients is

$$n_\mu = 1 + C_N^1 + C_{N+1}^2 + \cdots + C_{N+\mu-1}^\mu + = C_{N+\mu}^\mu \quad (6)$$

The distribution of the active cells of the two-dimensional cases is illustrated as Fig.1.

2.3 Training Algorithm

When applying NFI-AMS to learn a mapping in the form of

$$s = (s_{1_0}, s_{2_0}, \cdots, s_{N_o}) \Rightarrow f(s) = f(s_1, s_2, \cdots, s_N) = p \quad (7)$$

a sufficient amount of sample data-pairs is required to properly determine the weights to be stored in the memory cells. Two training algorithms can be used according to the concrete practical cases[1].

1) Regular Training: If the input data exactly locate on the nodes of lattices formed by the memory cells, then the sampled function values, , may be directly stored in the corresponding memory cells addressed by $(a_{i_1}, a_{i_2}, \cdots, a_{i_N})$ as their respective weights

$$w(a_{i_1}, a_{i_2}, \cdots, a_{i_N}) = f_{i_1,i_2,\cdots,i_N}, \quad (8)$$

where $i_j = 0, 1, \cdots, M;\ j = 0, 1, \cdots, N$.

2) *Irregular training:* In the irregular training (which is suitable for most practical applications), the actual output of the NFI-AMS should be firstly calculated as follows

$$\hat{p} = \sum_{p_1+p_2+\cdots+p_N=0}^{\mu} c_{p_1p_2\cdots p_N} w^{(N)}(\alpha_{p_1p_2\cdots p_N}) \quad (9)$$

where $w^{(N)}(\alpha_{p_1p_2\cdots p_N})$ is the original weights stored in the memory cell addressed by $\alpha_{p_1p_2\cdots p_N}$ in the last training run. Furthermore, all the weights of the conceptual receptive field are updated according to the error between the teaching signal ρ and actual output $\hat{\rho}$. $e = p - \hat{p}$ as follows

$$w^{(N+1)}(\alpha_{p_1p_2\cdots p_N}) = w^{(N)}(\alpha_{p_1p_2\cdots p_N}) + \frac{c^2_{p_1p_2\cdots p_N}}{C_e} e \quad (10)$$

where

$$C_e = \sum_{l_1+l_2+\cdots+l_N=0}^{\mu} c^2_{l_1 l_2 \cdots l_N} \tag{11}$$

And $p_1, p_2, \cdots, p_N$ satisfy

$$(p_1 \geq 0) \cap (p_2 \geq 0) \cap \cdots \cap (p_N \geq 0) \cap (p_1 + p_2 + \cdots + p_N \leq \mu) \tag{12}$$

2.4 Content-Addressing Mechanism

The content-addressing mechanism is implemented such that the address of the cell storing the function value, $f(i_1, i_2, \cdots, i_N)$, directly depends on the contents of the information carried by the input vector $[i_1, i_2, \cdots, i_N]$, $i_j = 0,1,\cdots,M_j$, $j = 1,2,\cdots,N$ in the following simple way.

$$\alpha_{i_1, i_2, \cdots, i_N} = \sum_{j=2}^{N} i_j \prod_{j=1}^{j-1} (M_j + 1) + i_1 + 1 \tag{13}$$

Where $i_j = 0,1,\cdots,M_j$, $j = 1,2,\cdots,N$, and M_j is the segmentation number of the variation range of the j-th component s_j.

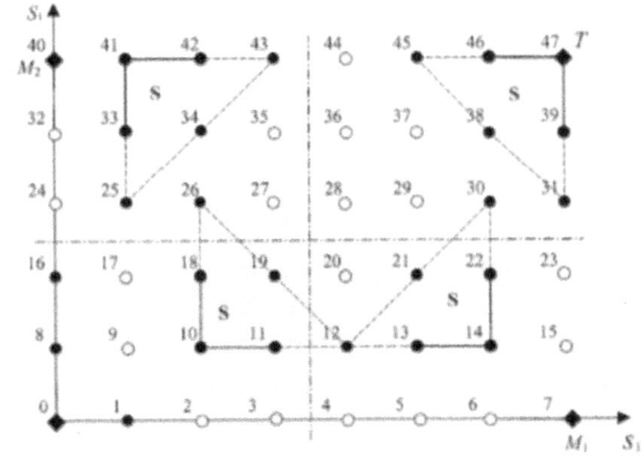

Fig. 2. Input space and memory cell of the two-dimensional cases

In a practical use, if the active cells in a receptive field are restricted to a hyper lower triangular sub-domain, the following problem will be encountered with the content-addressing method: Once the input vector enters into the right upper marginal region:

$$D'_{i_1,i_2,\cdots,i_N} = (M_1 - s_1 \le \mu) \cap (M_2 - s_2 \le \mu) \cap \cdots \cap (M_N - s_N \le \mu) \quad (14)$$

Then the information stored in the corresponding memory cells will no longer be enough for recalling the desired output[10].

To overcome this limit, a novel strategy for addressing the active cells in the receptive field is chosen as follows: The input space is uniformly divided into *2N* parts, and thus each part must share one comer of the input space. Within each part, the right angle of a hyper triangular sub-domain should always be pointed to the comer of input space, and thus the corresponding active cells can be determined with easy. The Input space and memory cell of NFI-AMS with two dimension input is shown as Fig.2.

3 Modeling and Prediction of TCP Network Traffic Based on FNI-AMS

As discussed above, the proposed NFI-AMS has a strong nonlinear function approximating ability, in this section, it is employed to model the TCP network traffic property, as is well known, which is nonlinear and time variant. Generally speaking, the TCP network traffic dynamics in discrete form can expressed as[12] [13]

$$y(k) = f[y(k-1), y(k-2), y(k-3), y(k-4), y(k-5)] \quad (15)$$

Where $y(k), y(k-1), y(k-2), y(k-3), y(k-4)$ and $y(k-5)$ is respectively the traffic value for the sampling instants $k, k-1, k-2, k-3, k-4$ and $k-5$. In fact, it is impossible to obtain the accurate form of function $f()$, so the NFI-AMS is used to approximate $f()$ as follows:

$$\hat{y}(k) = \hat{f}[y(k-1), y(k-2), y(k-3), y(k-4), y(k-5)] \quad (16)$$

Where $\hat{f}()$ is the NFI-AMS based traffic model, $\hat{y}(k)$ represents an estimate of the traffic of *k-th* sampling instant, so the input of the NFI-AMS is $y(k-1), y(k-2)$, $y(k-3), y(k-4), y(k-5)$, and the output $\hat{y}(k)$ respectively.

After the NFI-AMS based traffic model is established, which can be employed to predict the future network traffic. Each predict cycle consists of a training period and a predicting period. At the beginning of the $(k+1)$*-th* instant predict cycle, a training step is executed firstly, the observed traffic value of the TCP network during the previous cycle is input to the NFI-AMS, and the training data pairs for the NFI-AMS is chosen as follows:

$$(y(k-1), y(k-2), y(k-3), y(k-4), y(k-5)) \Leftrightarrow y(k) \quad (17)$$

The error between the predicted traffic $\hat{y}(k)$ and the actual traffic is used for adjusting the weights stored in the NFI-AMS cells. After the training period, the predicting step is executed. The previous traffic values $y(k), y(k-1), y(k-2), y(k-3), y(k-4)$ are used as input to NFI-AMS and predict the next instant traffic value $\hat{y}(k+1)$ $y(k+1)$ as follows.

$$(y(k), y(k-1), y(k-2), y(k-3), y(k-4)) \Rightarrow \hat{y}(k+1) \tag{18}$$

4 Simulations and Conclusions

A novel Associative memory system is presented via newton's forward interpolation, and the error-free approximation results can be obtained for multi-variable polynomial functions with arbitrarily given order. Based on the developed NFI-AMS, we have studied a method to model and predict network traffic, which is capable of modeling the complex nonlinear behavior of a time series and capturing the properties of network traffic. Numerical simulations are conducted to study the performance of NFI-AMS based TCP network traffic predict scheme. The simulation results of modeling and predicting error of the TCP network traffic based on NFI-AMS are shown as Fig.3 and Fig.4 respectively. The simulation results showed that the proposed scheme is feasible and efficient. Furthermore, the NFI-AMS based traffic prediction can be used in more fields for network design, management and control, which make the network can provide better quality of service and decrease the requirement of resources with lower network delay and higher resource utilization.

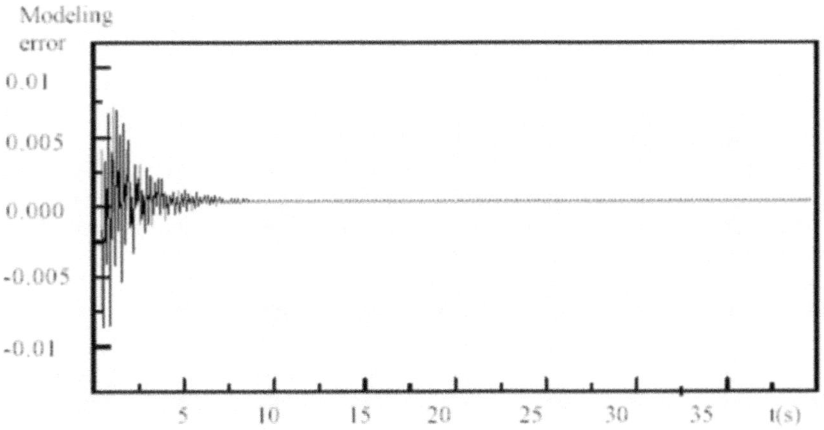

Fig. 3. Modeling error of the TCP network traffic

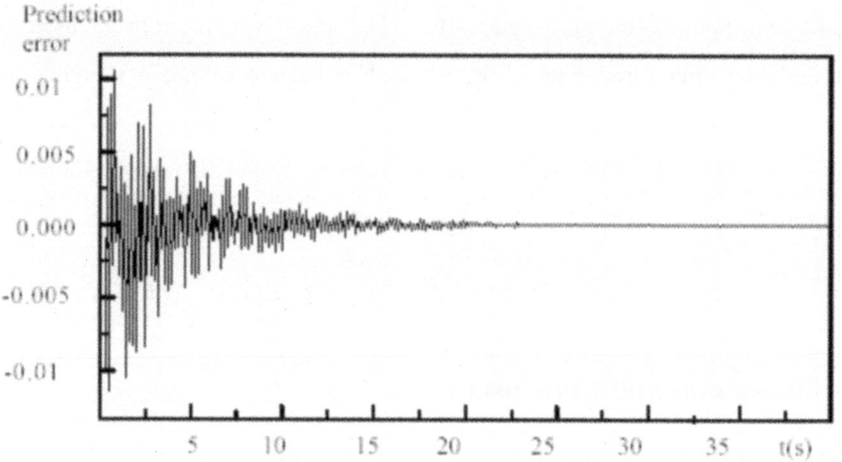

Fig. 4. Predicting error of the TCP network traffic

References

1. Wang, J.S.: Associative Memory Systems-based Robot Intelligent Control System. BeiJing Poly-technique University Master Paper (1998)
2. Albus, J.S.: A New Approach to Manipulator Control: The Cerebella Model Articulation Controller (CMAC). Trans. ASME, Series G, J. Dynamics Systems, Measurement, Control **97** (1975) 220–227
3. Xu, N.S., Wu, Z.L., Jia, R.X., Zhang, H.: A New Content-addressing Mechanism of CMAC-type Associative Memory Systems for Reducing the Required Memory Size. In: Proc. 13th IFACWorld Congr., San Francisco, CA (1996) 357–362
4. Thompson, D.E., Sunggyu, K.: Neighborhood Sequential and Random Training Techniques for CMAC. IEEE Trans. Neural Networks **6** (1995) 196–202
5. Gonzalez-Serrano, F.J., Figueiras-Vidal, A.R., Artes-Rodriguez, A.: Generalizing CMAC Architecture and Training. IEEE Trans. Neural Networks **9** (1998) 1509–1514
6. Wang, Z.Q., Schiano, J.L., Ginsberg, M.: Hash-coding in CMAC Neural Networks. In: Proc. IEEE Int. Conf. Neural Networks **3** (1996) 1698–1703
7. Xu, N.S., Bai, Y.F., Zhang, L.: A Novel High-order Associative Memory System via Discrete Taylor Series. IEEE Trans. Neural Networks **14** (2003) 734–747
8. Shu, Y.T., Wang, L., Zang, L.F.: Internet Traffic Modeling and Prediction Using FARIMA Model. Chinese Journal of Computers **24(1)** (2001) 46–54
9. Xu, N.S., Bai, Y.F., Lin, Q.: Model Parameter Estimation Based on Associative Memory System. J. Beijing Polytech. Univ. **22(4)** (1996) 134–143
10. Xu, N.S., Wang, J.S., Feng, W.N.: Associative Memory-based Robotic Manipulator Intelligent Control System. In: Proc. 19th Chinese Control Conf., Hong Kong (2000) 558–563.
11. Chen, H.L., Liu, Z.X., Chen, Z.Q., Yuan, Z.Z.: Estimating TCP Throughput: A Neural Network Approach. In: Proc 6th World Congress on Control and Automation, Dalian (2006) 2850–2854

12. 12.Christos Douligerisa, Brajesh Kumar Singhb: Analysis of neural-network-based congestion control algorithms for ATM networks. Engineering Applications of Artificial Intelligence, 12 (1999) 453-470.
13. 13.Hyun C. Cho, M. Sami Fadali, Hyunjeong Lee: Neural Network Control for TCP Network Congestion. 2005 American Control Conference, 2005. Portland, OR, USA, 3480-3485.

A Hybrid Knowledge-Based Neural-Fuzzy Network Model with Application to Alloy Property Prediction

Min-You Chen, Quandi Wang, and Yongming Yang

Key Lab of High Voltage Eng. and Electric New Tech., Education Ministry,
School of Electrical Engineering, Chongqing University,
Chongqing 400044, China
{mchencqu@126.com,wangquandi@yahoo.com.cn,yangyym@cqu.edu.cn}

Abstract. This paper presents a hybrid modeling method which incorporates knowledge-based components elicited from human expertise into underlying data-driven neural-fuzzy network models. Two different methods in which both measured data and a priori knowledge are incorporated into the model building process are discussed. Based on the combination of fuzzy logic and neural networks, a simple and effective knowledge-based neural-fuzzy network model has been developed and applied to the impact toughness prediction of alloy steels. Simulation results show that the model performance can be improved by incorporating expert knowledge into existing neural-fuzzy models.

1 Introduction

Neural-fuzzy networks have been widely used in a variety of engineering areas such as process control, pattern recognition and classification, system identification, image processing and materials property prediction [1], [2], [3], [4]. This is mainly due to the rapid development of intelligent modeling techniques based on fuzzy logic, neural networks and evolutionary algorithms. However, many existing neural-fuzzy models concentrate on prediction accuracy, paying less attention to interpretability and reliability of the obtained models. In engineering practice, users usually require the model not only predict the system's output accurately but also provide useful physical description of the system that generated the data. Such description can be elicited and possibly combined with the knowledge of domain experts, helping not only to understand the system but also to validate the model acquired from data. It is desirable to establish a hybrid model which combines accuracy, simplicity and transparency in a unified framework. This paper presents a hybrid neural-fuzzy model which incorporates knowledge-based components elicited from human expertise into underlying data-driven neural-fuzzy network models. In the following sections, a general approach of constructing neural fuzzy model is described. Then two different methods in which both measured data and a priori knowledge are incorporated into the model building process are discussed. The effectiveness of the developed models is verified in the application of mechanical properties prediction of alloy steels based on the collected industrial data.

D. Liu et al. (Eds.): ISNN 2007, Part I, LNCS 4491, pp. 528–535, 2007.

2 Generating Neural-Fuzzy Models

A neural-fuzzy model can be viewed as a neural network-based fuzzy logic system whose rules are automatically generated and optimized through network training. Compared to pure neural network models, neural-fuzzy models possess some distinctive advantages, such as the capacity of taking linguistic information from human experts and combining it with numerical data, and the ability of approximating complicated non-linear functions with simpler models.

Fuzzy radial basis function network (RBFN) is one of the most commonly used neural-fuzzy networks due to its ability to integrate the logical processing of information with mathematical properties of general function approximation [5]. Consider a collection of n data points $P(\boldsymbol{x}, y)$ in a m+1 dimensional space that combines both input and output dimensions. The input-output data pair of a multi-input and single-output system can be represented as

$$\boldsymbol{P}_k=(\boldsymbol{x}_{1k}, \boldsymbol{x}_{2k}, \ldots, \boldsymbol{x}_{mk}, y_k),\ P_k \in R^{m+1},\ k=1,2,\ldots,n.$$

According to the neural-fuzzy modeling paradigm previously proposed in [4], a data-driven fuzzy RBFN model can be generated through following steps: 1) generating an initial RBFN model from data using self-organizing network; 2) determining the number of hidden neurons (fuzzy rules) via fuzzy clustering; and 3) optimizing the network parameters through back-propagation learning; 4) simplifying the obtained model by similarity analysis to increase the model interpretability. The acquired fuzzy RBFN model can be presented as:

$$y=\sum_{i=1}^{p} v_i g_i(x);\quad i=1, 2, \ldots p;\ j=1, 2, \ldots m.$$

Where the radial basis functions are defined as

$$g_i(x)=\frac{\exp(-\| x\text{-}c_i \|^2 / \sigma_i^2)}{\sum_{i=1}^{p}\exp(-\| x\text{-}c_i \|^2 / \sigma_i^2)}$$

where $\boldsymbol{c}_i$ is the center of the ith radial unit (the ith cluster center) obtained from fuzzy clustering, $\boldsymbol{\sigma}_i$ is the unit width which determines over what distance in the input space the unit will have a significant influence. It is easy to see that the above model is functionally and structurally equivalent to the fuzzy logic system with center of average defuzzification, product-inference-rule and singleton fuzzification, consisting of a collection of p fuzzy rules in the form of:

If x_1 is A_{i1} **and** x_2 is A_{i2} ...**and** x_m is A_{im} **then** $y_i = v_i(\boldsymbol{x})$

where $\boldsymbol{x} = (x_1, x_2, \ldots, x_m) \in U$ and $y \in V$ are linguistic variables, A_{ij} are fuzzy sets of the universes of discourse $U_i \in R$, and $v_i(\boldsymbol{x})$ is a function of input variables.

To enhance the model reliability, a measure of prediction confidence for RBF networks proposed in [6] is introduced to calculate the confidence interval of the model output. It is seen that the above neural fuzzy model possesses the learning ability of neural network and semantic meaning of fuzzy rule-based system.

3 Incorporating Knowledge into Neural-Fuzzy Models

In the modeling of engineering processes, there are two kinds of information available. One is numerical information from measurements and the other is linguistic information from human experts. The aforementioned neural-fuzzy model is designed for data-driven models and cannot directly deal with fuzzy information. To enable the model to utilize expert knowledge presented by fuzzy if-then rules, an information processing mechanism must be established.

The use of linguistic qualitative terms in the rules can be regarded as a kind of information quantization. Generally, there are two different ways to incorporate knowledge into neural-fuzzy models, as shown in Fig.1. The first one is to encode expert knowledge in the form of If-Then rules into input-output fuzzy data, and then to use both numerical and fuzzy data to train the neural-fuzzy model, as shown in Fig.1(a). In cases where the data obtained from the system are incomplete but some expert knowledge regarding to the relationship between system input and output is available, this method can be used to incorporate linguistic knowledge into data-driven neural fuzzy models. Fuzzy sets can be defined by a collection of α-cut sets according to the resolution identity theorem. Linguistic information can be represented by α-cut sets of fuzzy numbers. Expert knowledge represented in the form of If-Then rules can be converted to fuzzy clusters in the input and output spaces. The neural-fuzzy model can be trained using both numerical data and fuzzy data which complement each other. To illustrate the effectiveness of the knowledge incorporation approach, a non-linear function approximation example is presented as follows.

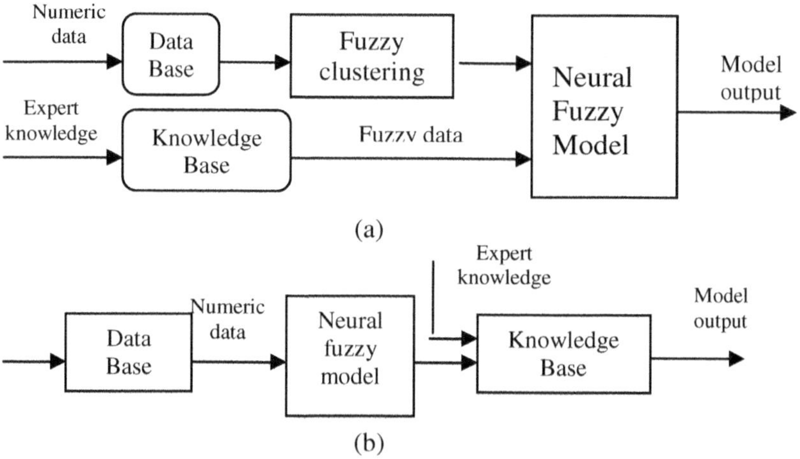

Fig. 1. Two approaches to the knowledge incorporation into neural-fuzzy models

A non-linear function is given as: $y=0.2+0.8e^{-x}+0.4\sin(2\pi(1-x^2))$

A neural fuzzy model will be generated to approximate the non-linear function in the range of $0 \leq x \leq 1$. Firstly, 100 training data were chosen randomly from the range of $0.3 \leq x \leq 1$, and 100 evenly distributed data in the range of $0 \leq x \leq 1$ and their corresponding function values were selected as testing data. Using the proposed neural fuzzy model generation approach, a FRBFN model with 4 hidden neurons (fuzzy rules) was generated, and represented in fuzzy rules as depicted in Fig. 2(a). The model testing result with RMSE=0.124 is shown in Fig. 3(a). It is clear that the model performance in the region of [0, 0.3] is very poor because of missing training data in this local region. However, if we know something about the input-output relation in the specific region, even though it is qualitative knowledge such as, *if x is A then y is B*, that will be useful to model training. In this example, we use 90 data randomly selected in the range of [0.3, 1] and one linguistic rule: *if x is small then y is large*, to train the FRBFN model. The fuzzy terms *small* and *large* are denoted as the fuzzy number *zero* and *one*, which are represented as membership functions $\mu_A(x)$ and $\mu_B(y)$, defined by α–cut sets:

$$\mu_A(x) = \bigcup_{0<\alpha\leq 1}[\alpha \cdot \mu_{A_\alpha}(x)] \quad \text{and} \quad \mu_B(y) = \bigcup_{0<\alpha\leq 1}[\alpha \cdot \mu_{B_\alpha}(y)]$$

Based on the fuzzy data and numeric data, a 4-rule neural fuzzy model was built as shown in Fig. 2(b). It is easy to see that the acquired FRBFN model correctly reflects the input-output relation in a very interpretable way. The model testing result with RMSE=0,0155 is shown in Fig. 3(b). Obviously, knowledge incorporation improved the model accuracy significantly.

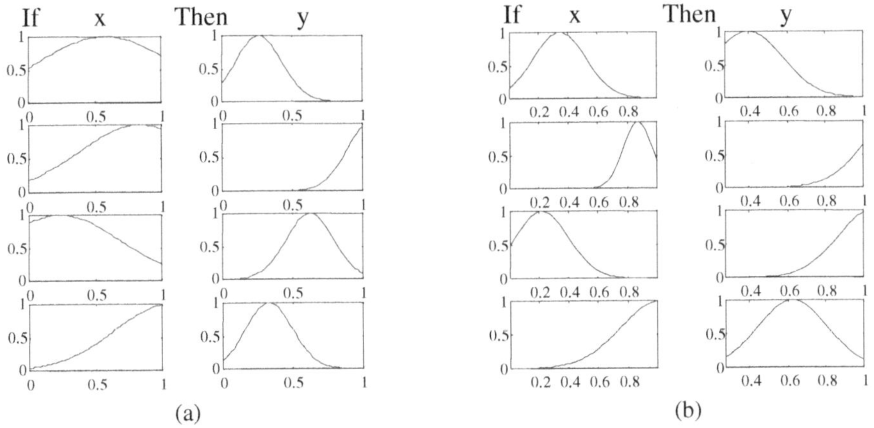

Fig. 2. Self-generated fuzzy models. (a) Modeling from 100 data, (b) Modeling from 90 data and one fuzzy rule.

On the other hand, in many cases the knowledge that links the system input and output is not available or not sufficient to generate fuzzy relations between system input and output. However, it is still possible to use expert knowledge to improve the model performance if some knowledge about model output assessment and

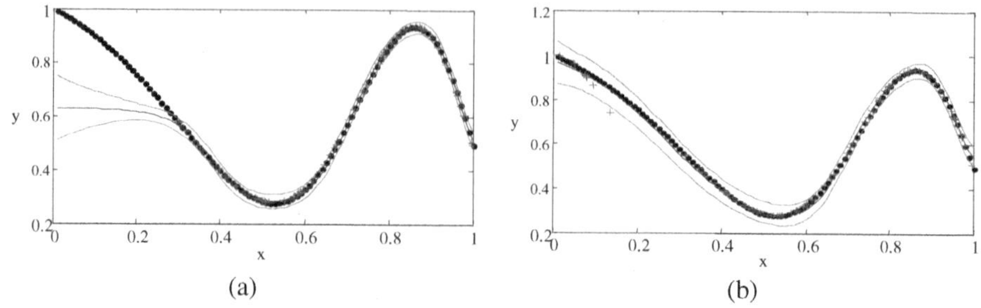

Fig. 3. Neural-fuzzy model testing results. Cross: training data, Solid line: model output, Dotted line: function values, Dashed and dotted line: confidence bounds.

adjustment rules is available from domain experts. In such cases, the fuzzy rule-base is generated from expert knowledge. Firstly, the neural-fuzzy model is trained using numerical data, and then the obtained model output is assessed and adjusted by the established fuzzy rule-base, as shown in Fig.1 (b). This method is useful for knowledge-based model modification, and will be demonstrated in the next section.

4 Impact Toughness Prediction of Alloy Steels

In material engineering, there is much interest in characterizing the fracture behavior of structural steels since, at low temperatures, such materials undergo a ductile to brittle transition. This can be determined by the Charpy V-notch impact test, which is widely used for material selection and toughness evaluation. The absorbed energy versus temperature curve for the test is often used to characterize the ductile-brittle transition in steels [7]. However, the value of Charpy energy only allows a rather qualitative description of toughness because of its complex and subtle connection with material composition and microstructure. The prediction and assessment of toughness properties of hot rolled steel products is of great interest [8][9][10].

The proposed hybrid neural fuzzy model has been used to impact toughness prediction for lowalloy steels. 408 experimental data including 22 types of steels provided were used to develop the prediction models. The data set contains chemical compositions, processing parameters and Charpy energy C_v(J) tested at different temperatures (between -120 and 60 °C). 60% of the data were used for model training and 40% of the data were used as testing data. Steel compositions C, Si, Mn, S, Ni, Nb, V, processing variables RHT (Reheating Temperature), FRT (Finish Rolling Tenperature) and Charpy test temperature were selected as the model inputs.

Based on the data-driven neural-fuzzy modeling approach mentioned in previous sections, a 6-rule fuzzy model was developed to predict Charpy impact energy. The predicted result at -50°C with Root-Mean-Square-Error RMSE=22.5(J) is shown in Fig. 4. The resultant mean transition curve of Charpy impact energy versus test temperature is displayed in Fig. 5. This curve was generated on the basis of fixing all input variables at their mean values while the test temperature was varied from –120 to 60°C. It can be seen that the model prediction is quite satisfactory for intrinsically scattered Charpy test data. Since the developed neural fuzzy model can generate the full brittle to ductile transition curves for specific steels, we can apply the same model to predict Impact Transition Temperature (ITT) at a pre-defined impact energy level without model re-training. ITT_p in Table 1 shows the predicted ITT of AM8 steels at 60J energy level. It is seen that the predicted ITT based on the Charpy energy prediction model are scattered and non-conservative (i.e. generally, ITT prediction is below the actual values). To improve the model performance, we incorporated a small knowledge-base into the obtained fuzzy model, which consists of three knowledge-based fuzzy rules:

R1: If RHT is Low and FRT is Low Then increase ITT by about X%
R2: If RHT is Medium and FRT is Medium Then increase ITT by about Y%
R3: If RHT is High and FRT is High Then ITT increase ITT by about Z%

The membership functions of terms Low, Medium and High are defined by expert knowledge. The values of X, Y and Z in the consequent part of the rules were initialized by prior knowledge and then optimized via a simulated annealing algorithm.

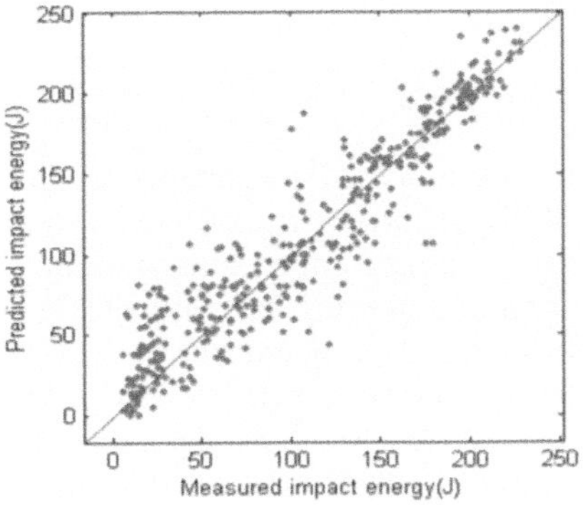

Fig. 4. Impact energy prediction for TMCR steels

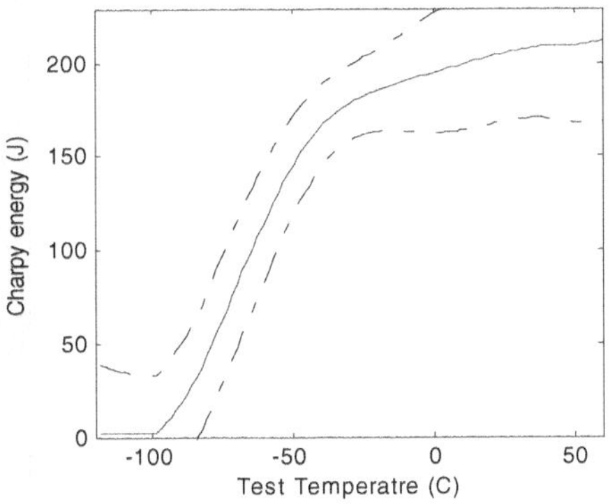

Fig. 5. Impact transition curve with confidence interval generated by the neural fuzzy model

The knowledge-base was incorporated into a data-based fuzzy model in the second mode shown in Fig. 1(b). After knowledge-based model modification, the RMSE of the ITT predictions was reduced from 8.5°C to 6.2°C. In Table 1, the ITT predictions with knowledge incorporation are represented by ITT_K, The model predictions without knowledge incorporation are denoted by ITT_P, It can be seen that compared to the ITT_P, the modified ITT predictions ITT_K are more accurate and reasonable. Based on the developed fuzzy model, we can predict impact energy and transition temperature effectively. Table 1 shows the predicted impact energies of different steels at a temperature of -50°C with corresponding 95% confidence intervals CI, and the transition temperature at 60J energy level. It is seen that the model predicted energy values are mostly all between the two measured Charpy test energy values, and the prediction of ITT(60J) is also quite encouraging.

Table 1. Charpy impact properties prediction for different steels

Steel Type	Charpy Energy (-50°C)			ITT (60J)		
	Measured	Predicted	CI	ITT	ITT_K	ITT_P
A8M101 (V-Ti)	204, 178	185	33	-70	-64	-75
A8M98(V)	154, 148	152	30	-65	-66	-71
A8M90 (Nb-Ni)	125, 62	84	44	-50	-55	-64
A8M94 (Nb-V-Ni)	80, 54	83	28	-45	-50	-53
A8M95 (Nb-Ni)	98, 63	81	28	-50	-48	-56
A8M96(V-Ni)	176, 152	171	27	-80	-84	-89
A8M100 (0.045%Nb-V)	97, 51	75	38	-45	-55	-60
A8M102(Nb-V)	170, 135	143	47	-55	-65	-74
A8M105(Nb-V-Ti)	192, 176	190	35	-80	-80	-89
A8M104 (0.2%Mo-Nb-V)	104, 85	84	20	-45	-50-	-54
A8M92(Low C Mn-Nb-V)	185, 184	182	23	-70	-70	-82
A8M93(Low C Nb-V)	206, 197	200	21	-70	-75	-83
A8M99 (LowC 0.045%Nb-V)	199, 194	196	37	-85	-78	-89
A8M97(Nb-Ni-Ti)	186, 173	168	34	-90	-87	-101
A8M91(Cu-Nb-V)	157, 138	136	35	-75	-70	-74

5 Conclusion

In the field of intelligent modeling, it is a current trend to establish a unified predictive and descriptive model with the capability of processing different types of information, including numeric data and symbolic or fuzzy information from human priori knowledge and empirical heuristics. In order to develop an interpretable and applicable prediction model, a hybrid knowledge-based neural-fuzzy modeling approach is presented. In the proposed modeling framework, expert knowledge in the form of If-Then rules can be incorporated into data-driven RBFN models in different way. Simulation experiments show that the developed FRBFN model has satisfactory prediction accuracy and good interpretation property. The model performance can be improved by knowledge incorporation. The proposed modeling approach has been successfully applied to alloy toughness prediction. Experimental results show that the developed impact toughness prediction model not only predicts the impact properties of alloy steels, but also provides a useful description of the link between composition-process conditions and Charpy toughness. Further improvement in the model reliability by incorporating a confidence bound to the model prediction would be beneficial.

References

1. Juang, C., Lin, C.: An On-line Self-Constructing Neural Fuzzy Inference Network and Its Applications. *IEEE Trans. On Fuzzy Systems* **6** (1998) 12-32
2. Wu, S., Er, M.: Dynamic Fuzzy Neural Networks—A Novel Approach to Function Approximation. IEEE Transactions on Systems, Man and Cybernetics, Part B **30** (2000) 358-364

3. Ahtiwash, O., Abdulmuin, M., Siraj, S.: A Neural-fuzzy Logic Approach for Modeling and Control of Nonlinear Systems. IEEE International Symposium on Intelligent Control. OCT 27-30, Canada (2002) 270-275
4. Chen, M., Linkens, D.: A Systematic Neurofuzzy Modelling Framework with Application to Material Property Prediction. IEEE Transactions on Systems, Man and Cybernetics, Part B **31** (2001) 781-790
5. Jin, Y., Sendhoff, B.: Extracting Interpretable Fuzzy Rules from RBF Networks. Neural Process. Lett. **17** (2003) 149–164
6. Leonard, A., Kramer, M., Ungar, L.: A Neural Network Architecture That Computes Its Own Reliability. Computers Chem. Eng. **16** (1992) 819-835
7. Needleman, A., Tvergaard, V.: Numerical Modeling of the Ductile-brittle Transition. International-Journal-of-Fracture **101** (2000) 73-97
8. Moskovic, R., Windle, P., Smith, A.: Modeling Charpy Impact Energy Property Changes Using A Bayesian Method. Metallurgical and Materials Transactions A **28** (1997) 1181-.1193
9. Todinov, M., Novovic. M., Bowen, P., Knott, J.: Modelling the Impact Energy in The Ductile/Brittle Transition Region of C-Mn Multi-run Welds. Materials-Science-&-Engineering-A **287** (2000) 116-124
10. Chen, M., Linkens, D., Howarth, D., Beynon, J.: Fuzzy Model-based Charpy Impact Toughness Assessment for Ship Steels. ISIJ International **44** (2004) 1108-1113

A Novel Multiple Improved PID Neural Network Ensemble Model for pH Value in Wet FGD

Shen Yongjun, Gu Xingsheng, and Bao Qiong

[1]Research Institute of Automation, East China University of Science and Technology, Shanghai, 200237, China
gufanbaihe@sohu.com, xsgu@ecust.edu.cn, baihegufan@sohu.com

Abstract. In the limestone/gypsum wet flue gas desulphurization (FGD) technology, the change of slurry pH value in absorber is a nonlinear and time-variation process with a large number of uncertainties, so it's difficult to acquire satisfying mathematical model. In this paper, a novel multiple improved PIDNN ensemble model is proposed to establish the model of slurry pH value. In this model, the concepts of variable integral and partial differential are introduced in the design of hidden-layer of PIDNN, and the concept of output feedback is utilized to improve the ability of PIDNN for dynamic modeling, then multiple improved PIDNN are dynamic combined to get the system output. The results of simulation with field data of wet FGD indicate the validity of this modeling approach.

1 Introduction

The emission of sulfur dioxide (SO_2) is known to have detrimental effects on human health and the environment such as the formation of acid rain. In recent years, there are several methods for removing SO_2 from the flue gas in coal-fired power plants. Of these processes, wet scrubbers, especially the limestone/gypsum wet flue gas desulphurization (FGD), is by far the most popular technique [1, 2]. In this technology, the control of slurry pH value in absorber is one of the key factors to affect desulphurization rate and the quality of gypsum [3], so it is very important to establish the pH model. But in wet FGD, the complexity, nonlinearity and multivariable nature of slurry pH value makes it difficult or even impossible to develop its mathematical model. Therefore, alternative approaches, such as neural modeling, become an important possibility for this process.

Although neural networks for dynamic system modeling have been investigated widely in recent years [4, 5], it is important to notice that there are still many problems that need to be solved, such as the choice of net structure, generalization ability of networks and so on.

PID Neural Network (PIDNN), proposed by Shu Huailin in 1997[6], is a new type of dynamic feed-forward net structure which integrates conventional PID control strategy into neural networks. In this paper, the concepts of variable integral and partial differential are introduced in the design of hidden-layer of PIDNN to enhance net performance, and the concept of output feedback is

D. Liu et al. (Eds.): ISNN 2007, Part I, LNCS 4491, pp. 536–545, 2007.

utilized to improve the ability of PIDNN for dynamic modeling. Furthermore, different from the single model approach which can't represent the characteristics of complex system integrally, a multi-model approach based on improved PIDNN is proposed to increase the net generalization ability, and a dynamic ensemble method is used to obtain the system output. The results of simulation with field data of wet FGD indicate the validity of this modeling approach.

The rest of the paper is organized as follows: Section 2 presents the improvement of PIDNN, and Section 3 depicts the technology of wet FGD. Section 4 specifies the system modeling based on multi-model approach, and the simulation results are introduced in Section 5. Finally, the conclusions are given in Section 6.

2 Improvement of PIDNN

PIDNN is a new type of dynamic feed-forward network with the fixed structure of 2-3-1, and the output functions of the neurons in its hidden layer are different from each other, which are proportional (P) function, integral (I) function and differential (D) function so that they are named as P-neuron, I-neuron and D-neuron respectively, and its structure is shown in Fig.1. [7, 8]

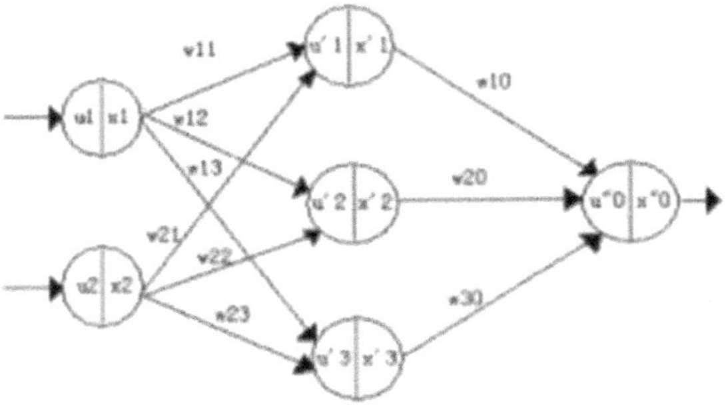

Fig. 1. Structure of PIDNN

The perfect combination of conventional PID control strategy and feed-forward networks makes PIDNN perform a perfect function in system identification and process control [9]. But the limitations of conventional PID control, such as the condition of integral saturation and high-frequency interference, restrict the development of PIDNN to some extent. In this paper, the concepts of variable integral and partial differential are introduced in the design of hidden-layer of PIDNN to overcome above problems and enhance net performance, and the functions are as follows:

P'-neuron:

$$x'_1(k) = \begin{cases} 1, & u'_1(k) > 1, \\ u'_1(k), & -1 \le u'_1(k) \le 1, \\ -1, & u'_1(k) < -1, \end{cases} \tag{1}$$

I'-neuron:

$$x_2'(k) = \begin{cases} 1, & x_2'(k) > 1, \\ x_2'(k-1) + f(u_2'(k))u_2'(k), & -1 \le x_2'(k) \le 1, \\ -1, & x_2'(k) < -1, \end{cases} \quad (2)$$

where $f(u_2'(k)) = \begin{cases} 1, & |u_2'(k)| \le B \\ \frac{A-|u_2'(k)|+B}{A}, & B < |u_2'(k)| \le A+B \\ 0, & |u_2'(k)| > A+B \end{cases}$, and A and B are constants.

D'-neuron:

$$x_3'(k) = \begin{cases} 1, & x_3'(k) > 1, \\ (1-\alpha)(u_3'(k) - u_3'(k-1)) + \alpha x_3'(k-1), & -1 \le x_3'(k) \le 1, \\ -1, & x_3'(k) < -1, \end{cases} \quad (3)$$

where $0 \le \alpha \le 1$.

3 Technology of Wet FGD

In recent years, FGD technology, especially the wet limestone/gypsum process, is advocated to equip in coal-fired power plants to remove SO2 from the flue gas. The wet FGD pilot plant is shown in Fig.2, and the main components are a natural gas burner, a falling film column and a holding tank [3]. SO2, produced by the gas burner, is removed from the flue gas by absorption into limestone/gypsum slurry in the falling film column, and the main solid product gypsum (CaSO4·2H2O) is produced by the blow of forced oxidation air [10].

Chemical reactions between SO2 and the limestone/gypsum slurry ensure an effective absorption process, which can be disassembled into four steps. They are absorption of SO2, oxidation of HSO_3^-, dissolution of limestone and crystallization of gypsum [11]. Because of the complicated chemical reactions, the slurry pH value becomes one of the key factors to obtain high desulphurization rate and high-quality of gypsum. Most wet FGD projects operating experience [3,11,12] indicates that under the stability of other parameters such as inlet flue gas concentration of SO2, reactor temperature and residual limestone in the gypsum,, rising slurry pH value can to some extent improve desulphurization efficiency, but prolonged operation with high slurry pH value will lead to a decline in gypsum quality. On the other hand, the absorption of SO_2 will be restrained with the low slurry pH value, and if its value is below 4.0, almost no absorption will happen, moreover, the equipment corrosion will increase. Therefore the measurement and control of the slurry pH value is of great significance.

4 System Modeling

To well control the slurry pH value in wet FGD, the system model is very important to be established. System modeling based on neural network is one of the

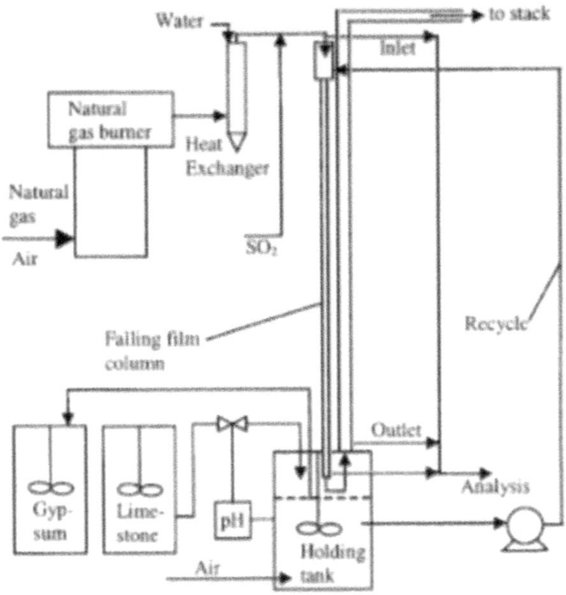

Fig. 2. Schematic illustration of the wet FGD pilot plant

most significant applications of neural network in the domain of process control. Through lots of investigations and researches, we consider that the success in obtaining a reliable and robust network depends strongly on the improvement of net generalization ability, as well as the choice of net structure and process variables. In this paper, a multi-model approach is proposed, which is shown in Fig.3, and each module is composed of improved PIDNN introduced above, and output feedback is utilized to improve the ability of PIDNN for dynamic modeling. Its structure is shown in Fig.4. Besides, a dynamic ensemble method is used to obtain the system output by adding a gating network, whose structure is shown in Fig.5. It's a two-layer feed-forward network, and its inputs are whole process variables and its outputs are dynamic weights of each module. The algorithm is as follows:

$$y = \sum_{i=1}^{k} g_i y_i, \tag{4}$$

where $y_i$is the output of each module, and g_i is the ith output of gating network, and y is the system output.

Besides, g_i must satisfy:

$$\begin{aligned} &0 \le g_i \le 1, \;\; 1 \le i \le k, \\ &\sum_{i=1}^{k} g_i = 1. \end{aligned} \tag{5}$$

So a middle variable ξ_i is introduced, and the ith output of gating network is as follows:

$$g_i = \frac{e^{\xi_i}}{\sum_k e^{\xi_k}}. \tag{6}$$

In Eq. (6),

$$\xi_i = v_i^T x, \tag{7}$$

where $\boldsymbol{v}_i$ is the weight coefficients of gating network.

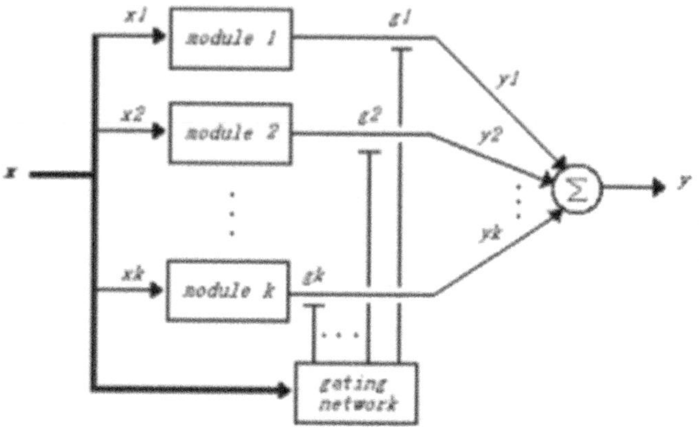

Fig. 3. Structure of multi-model approach

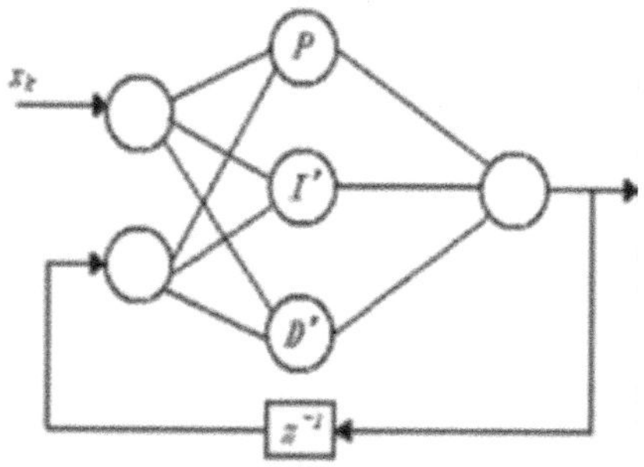

Fig. 4. Improved PIDNN with output feedback

5 Simulation Research

Considering the dissolution of $CaCO_3$ and all three phase balance reactions of gas, liquid and solid, we acknowledge that there are four process variables which have main effect on the change of slurry pH value, and they are gross of flue gas, concentration of SO_2, flux of limestone slurry and concentration of limestone slurry. So we utilize four improved PIDNN to establish the pH model. 150 groups of field data which is sampled with each one minute's interval are used to model

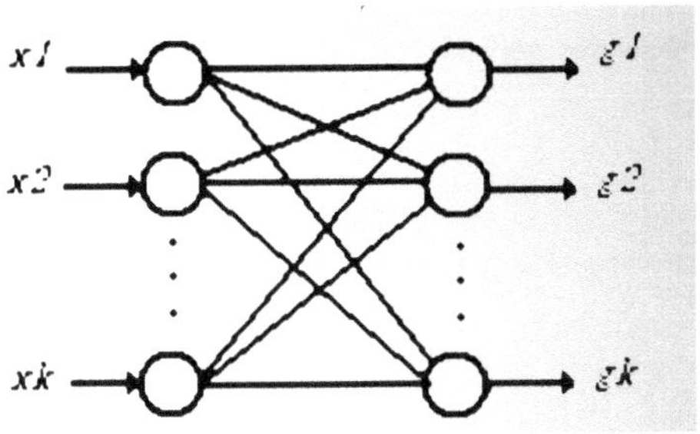

Fig. 5. Structure of gating network

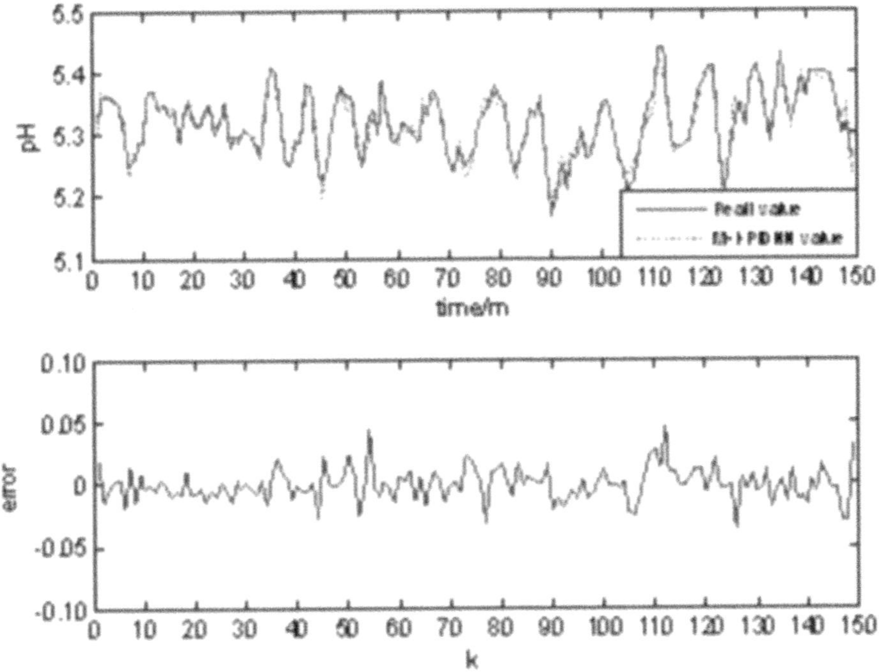

Fig. 6. Simulation result of system modeling based on M-I-PIDNN

the slurry pH value in wet FGD based on above multi-model approach, and other 90 groups of field data as the out-of-sample data set (i.e., testing set), are used to evaluate the good or bad performance of predictions.

The values of the parameters are selected as follows:

1)$\alpha = 0.2$, A=0.4, B=0.6.
2)$\eta(1) = 0.08$, $\gamma = 0.8$.
3) Initial values of PIDNN weights are:
wlj(0)=+1, w2j(0) =-1, where j=1, 2, 3, and wio(0)= 0.1, where i=1, 2, 3.
4) Initial values of gating network weights are zeros.

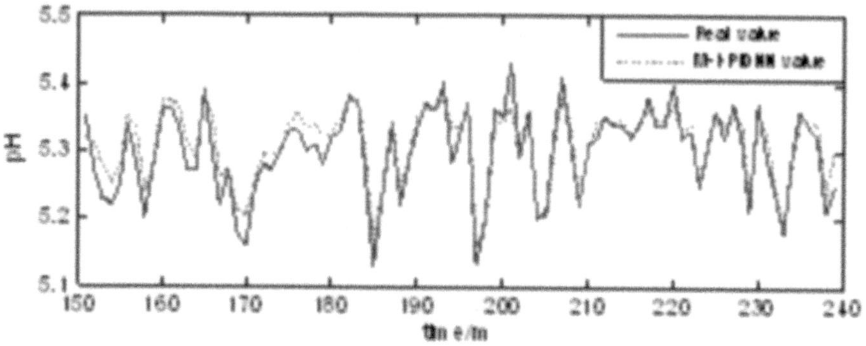

Fig. 7. Simulation result of model prediction based on M-I-PIDNN

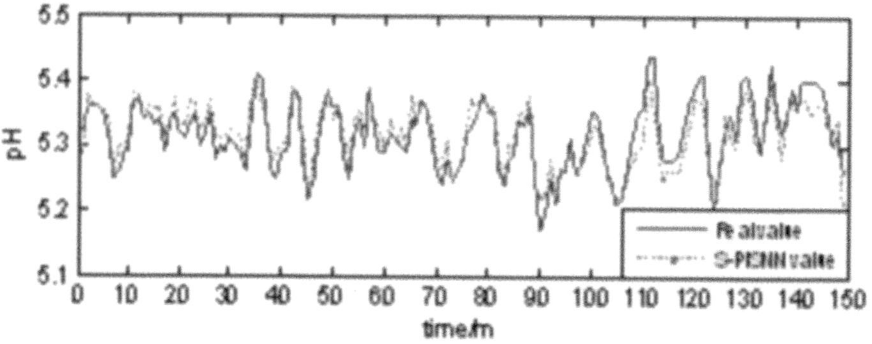

Fig. 8. Simulation result of system modeling based on S-PIDNN

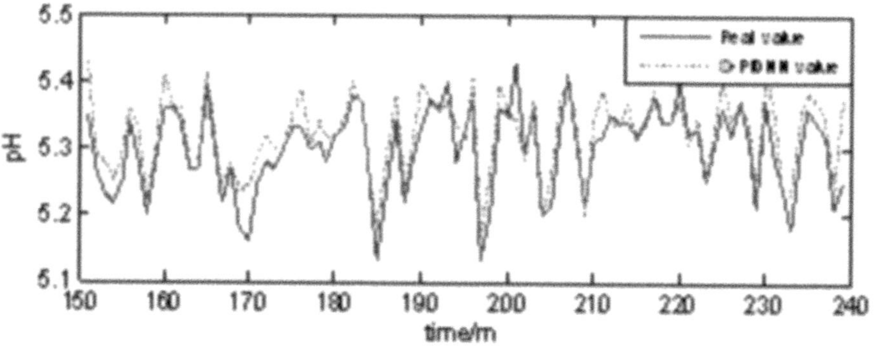

Fig. 9. Simulation result of model prediction based on S-PIDNN

Simulation result based on above parameters is depicted in Figs.6, and the test result is shown in Fig.7.

As can be seen from the Figures 6 and 7, the modeling approach based on Multiple Improved PIDNN (M-I-PIDNN) is well approximate to the real change

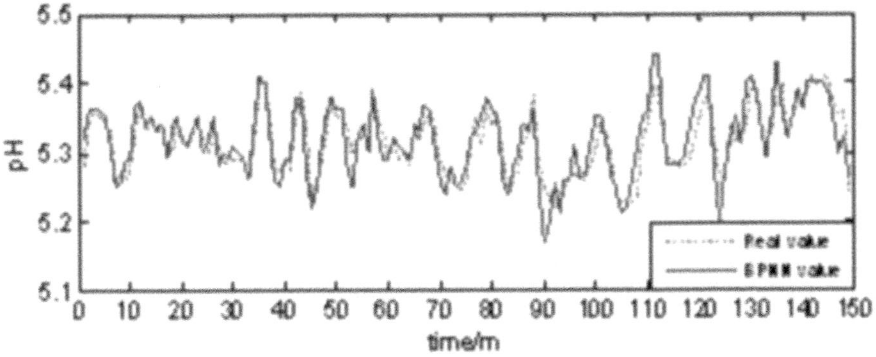

Fig. 10. Simulation result of system modeling based on BPNN

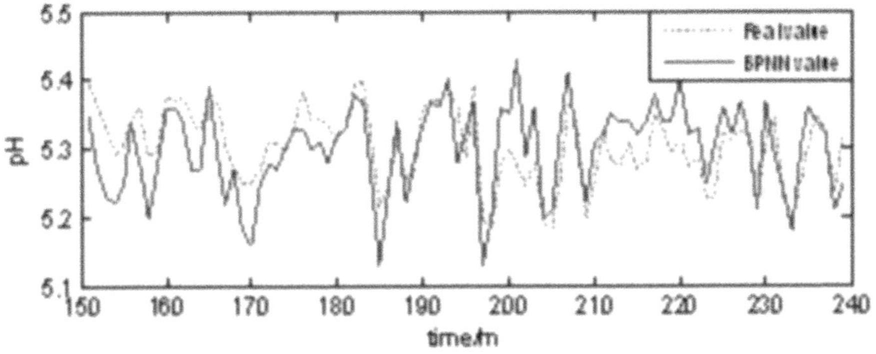

Fig. 11. Simulation result of model prediction based on BPNN

of the slurry pH value and the result of model prediction indicates the validity of this approach.

Besides, other approaches such as Single PIDNN (S-PIDNN) and BPNN are utilized to establish and predict the pH model, where the input of S-PIDNN is the main variable of slurry pH value, viz. flux of limestone slurry. The structure of BPNN is $4\times20\times1$, and its inputs are identical with the inputs of M-I-PIDNN. The simulation results are shown in Figures 8 to 11.

Compared the modeling approach based on M-I-PIDNN with the one based on S-PIDNN and BPNN from Figures 6 to 11, we can see that the modeling method based on M-I-PIDNN can describe the change process of pH value more integrally, and identify nonlinear dynamic system with large time delay process more effectively. To further describe the superiority of M-I-PIDNN, the corresponding results are reported in Tables 1.

where *T-MSE* means the mean square error of system training, and *T-MaxE* means the maximum error in system training, and *S-MSE* means the mean square error of system testing.

Table 1. Comparison among several methods of network modeling

	BPNN	*S -PIDNN*	*M-I-PIDNN*
Training sample	150	150	150
Testing sample	90	90	90
Iteration number	2000	200	200
T-MSE	0.0184	0.0176	0.0083
T-MaxE	0.0808	0.0780	0.0462
S-MSE	0.0440	0.0260	0.0157

From Table 1, we can conclude that in all methods the multiple improved PIDNN ensemble model performs the best with the least MSE of prediction, which means the highest net generalization ability, though its iteration time is a little more than the one of S-PIDNN method.

6 Conclusion

In this study, we propose a novel multiple improved PIDNN ensemble model for pH value in wet FGD. The experimental results with the field data of wet FGD reported in this paper demonstrate the effectiveness of the proposed ensemble approach, implying that the proposed nonlinear ensemble model can be used as a feasible approach to a great number of industrial nonlinear processes like pH value.

References

1. Soud, H.N.: Developments in FGD. IEA Coal Research, CCC/29, London, UK (2000)
2. Zheng, Y.J., Kiil, S., Johnson, J.E.: Experimental Investigation of a Pilot-Scale Jet Bubbling Reactor for Wet Flue Gas Desulphurization. J. Chemical Engineering Science **58** (2003) 4695-4703
3. Frandsen, J.B.W., Kiil, S., Johnson, J.E.: Optimization of a Wet FGD Pilot Plant using Fine Limestone and Organic Acids. J. Chemical Engineering Science **56** (2001) 3275-3287
4. Hunt, K.J., Sbarbaro, D., Zbikowski, R., Gawthrop, P.J.: Neural Networks for Control Systems – A Survey. Automatica **28** (1992) 1083-1112
5. Yu, D.L., Gomm, J.B.: Enhanced Neural Network Modeling for a Real Multivariable Chemical Process. J. Neural Computing and Applications **10** (2002) 289-299
6. Shu, H.L., Pi, Y.G.: The Analysis of PID Neurons and PID Neural Networks. Proceedings of '98 Chinese Control Conference **9** (1998) 607-613
7. Shu, H.L., Pi, Y.G.: PID Neural Networks for Time-Delay Systems. J. Computers and Chemical Engineering **24** (2000) 859-862
8. Guo, Q.G., Shu, H.L.: The Research and Simulation on a New Type of Neural Network Structure PID Controller. J. Electrical Drive Automation **21** (1999) 29-32
9. Shen, Y.J., Gu, X.S.: Identification and Control of Nonlinear System based on PID Neural Networks. J. East China University of Science and Technology **32** (2006) 860-863

10. Nygaarda, H.G., Kiil, S., Johnson, J.E., et al: Full-Scale Measurements of SO2 Gas Phase Concentrations and Slurry Compositions in a Wet Flue Gas Desulphurization Spray Absorber. J. Fuel **83** (2004) 1151-1164
11. Warek, J., et al: Optimum Values of Process Parameters of the Wet Limestone Flue Gas Desulphurization System. J. Chemical Engineering Technology **25** (2002) 427-432
12. Ylen, J.P: Measuring Modeling and Controlling the ph Value and the Dynamic Chemical State. PhD: Helsinki University of Technology (2001)

Acoustic Modeling Using Continuous Density Hidden Markov Models in the Mercer Kernel Feature Space

R. Anitha and C. Chandra Sekhar

Department of Computer Science and Engineering,
Indian Institute of Technology Madras, Chennai, India
{anitha,chandra}@cs.iitm.ernet.in

Abstract. In this paper, we propose an approach for acoustic modeling using Hidden Markov Models (HMMs) in the Mercer kernel feature space. Acoustic modeling of subword units of speech involves classification of varying length sequences of speech parametric vectors belonging to confusable classes. Nonlinear transformation of the space of parametric vectors into a higher dimensional space using Mercer kernels is expected to lead to better separability of confusable classes. We study the performance of continuous density HMMs trained using the varying length sequences of feature vectors obtained from the kernel based transformation of parametric vectors. Effectiveness of the proposed approach to acoustic modeling is demonstrated for recognition of spoken letters in E-set of English alphabet, and for recognition of consonant-vowel type subword units in continuous speech of three Indian languages.

1 Introduction

Speech recognition involves conversion of the input speech signal to a sequence of symbols corresponding to subword units of speech of a particular language. The confusability among sound units makes the speech recognition a complex task. The speech segment of a subword unit is represented by a sequence of parametric vectors extracted from the speech signal. The durations of two different utterances of a given word, uttered by the same speaker in different context can be different. As the durations of speech segments of subword units vary, the lengths of the sequences of parametric vectors also vary. Development of discriminative training based models for classification of varying length patterns is important for acoustic modeling.

Complex pattern classification tasks typically involve nonlinearly separable patterns. According to Cover's theorem on the separability of patterns, an input space made up of nonlinearly separable patterns may be transformed into a feature space where the patterns are more easily separable, provided the transformation is nonlinear and the dimensionality of the feature space is high enough [1]. The innerproduct kernels or Mercer kernels can be used for nonlinear transformation from the input space to a high-dimensional feature space [2]. Because

D. Liu et al. (Eds.): ISNN 2007, Part I, LNCS 4491, pp. 546–552, 2007.

of the easier separability of classes in the feature space, the nondiscriminative training based models such as HMMs can be used to solve complex pattern classification tasks involving varying length patterns. The goal of this paper is to construct HMMs in kernel feature space for recognition of varying length patterns extracted from the data of confusable classes in the input space. Results of our studies demonstrate the potential of our proposed approach for acoustic modeling of highly confusable sounds such as spoken letters in E-set of English alphabet, and Consonant-Vowel (CV) type subword units in continuous speech of three Indian languages.

In section 2, we briefly explain the hidden Markov models for speech recognition task in the input space. Section 3 explains the acoustic modeling using HMMs in the kernel feature space. We address the issues involved in construction of discrete and continuous density HMMs in the feature space of an explicit mapping kernel. In section 4, we present our studies using the proposed approach for E-set recognition and CV segment recognition.

2 Hidden Markov Models in the Input Space

Hidden Markov models (HMMs) have been extensively used for speech recognition tasks [3]. The HMMs have been used for acoustic modeling at subword unit level in vocabulary independent continuous speech recognition systems. An HMM for a subword unit is trained using the observation sequences corresponding to the sequences of speech parametric vectors extracted from the speech signal data of multiple examples of the unit. The HMM for a unit is trained to maximize the likelihood of the model generating the observation sequences of that unit. During recognition, the observation sequence of a test pattern is given as input to the HMM of each unit, to compute the probability of the test sequence being generated by that model. Then the class of the model with the highest probability is assigned to the test pattern. An HMM is characterized by the number of states in the model, the state-transition probability distribution, the observation symbol probability distribution for each state, and the initial state probability distribution.

3 Hidden Markov Models in the Kernel Feature Space

Recently, kernel methods have been considered for pattern analysis [4]. The kernel based methods for pattern classification involve nonlinear mapping from the input space to a high dimensional feature space using innerproduct kernels. Often the patterns are nonlinearly separable in the input space. Nonlinear mapping using the innerproduct kernels is expected to lead to linear separability of patterns in the kernel feature space.

The innerproduct kernel function $\mathcal{K}(\mathbf{x}_i, \mathbf{x}_j)$ for two vectors $\mathbf{x}_i$ and $\mathbf{x}_j$ in the d-dimensional input space is defined as follows:

$$\mathcal{K}(\mathbf{x}_i, \mathbf{x}_j) = \Phi(\mathbf{x}_i)^t \Phi(\mathbf{x}_j). \tag{1}$$

Here $\Phi(\mathbf{x}_i)$ and $\Phi(\mathbf{x}_j)$ are the vectors in the D-dimensional kernel feature space induced by the nonlinear mapping kernel, due to vectors $\mathbf{x}_i$ and $\mathbf{x}_j$, respectively. Depending on whether the feature space representation is explicit or not, the innerproduct kernels can be either explicit mapping kernels or implicit mapping kernels. In this paper, we explore the possibility of constructing HMMs in the feature space of explicit mapping kernels.

For explicit mapping kernels such as the polynomial kernel function, the feature space representation is explicitly known. The polynomial kernel is defined by:

$$\mathcal{K}(\mathbf{x}_i, \mathbf{x}_j) = (a.\mathbf{x}_i{}^t\mathbf{x}_j + c)^g, \qquad (2)$$

where g is the degree of the polynomial kernel, and a and c are constants. The vector $\Phi(\mathbf{x})$ in the feature space of the polynomial kernel corresponding to the input space vector $\mathbf{x}$ includes the monomials upto order g of elements in $\mathbf{x}$. For a d-dimensional input space vector, the dimension D of the vector in the feature space of polynomial kernel of degree g is given by:

$$D = \frac{(d+g)!}{d!g!}. \qquad (3)$$

3.1 Discrete HMMs (DHMMs) in Polynomial Kernel Feature Space

Construction of a DHMM involves estimation of initial state probabilities, state transition probabilities, and discrete observation symbol probabilities for each state in the DHMM. The parameters of the DHMM for a class are estimated using the codebook index sequences derived from the sequences of observation data vectors extracted from the training examples of the class.

As the feature space representation for a polynomial kernel is explicitly known, the methods used for clustering and vector quantization in the input space can be used for clustering and vector quantization in the feature space of explicit mapping kernels. The K-means clustering method can be used for forming clusters in the kernel feature space. For performing vector quantization in kernel feature space, a vector $\mathbf{x}$ is assigned the index of the cluster whose center, $\mathbf{m}^{\Phi}$, has the highest similarity to $\Phi(\mathbf{x})$. Let $\{\mathbf{x}_1, \mathbf{x}_2, \ldots, \mathbf{x}_j, \ldots, \mathbf{x}_T\}$ be a sequence of T observation vectors in the input space. The corresponding sequence of vectors in the kernel feature space is given by $\{\Phi(\mathbf{x}_1), \Phi(\mathbf{x}_2), \ldots, \Phi(\mathbf{x}_j), \ldots, \Phi(\mathbf{x}_T)\}$. This sequence of kernel feature space vectors is represented by a sequence of codebook indices, $\{i_1^{\Phi}, i_2^{\Phi}, \ldots, i_j^{\Phi}, \ldots, i_T^{\Phi}\}$ derived by vector quantization in the kernel feature space. Such sequences of codebook indices are used to build the DHMMs in the polynomial kernel feature space [5].

3.2 Continuous Density HMMs (CDHMMs) in Polynomial Kernel Feature Space

It is important to note that the performance of DHMMs in the kernel feature space is not expected to be as good as the performance of the CDHMMs in the input space. This is mainly due to the significant loss of information incurred in discretization of continuous signal representations using vector quantization in construction of the DHMMs. Performance of the CDHMMs in the kernel feature space is expected to be better than that of the CDHMMs in the input space.

Construction of a CDHMM in the input space involves estimation of initial state probabilities, state transition probabilities, and continuous observation probability density functions for each state in the CDHMM. For construction of CDHMMs in the feature space of an explicit mapping kernel, the methods used for construction of the CDHMMs in the input space can be used. However, as the dimension of the feature space is high, construction of CDHMMs in the feature space would have a significantly higher computational complexity and would need larger training data sets for proper estimation of the parameters of the HMMs. In the next section, we present our studies on acoustic modeling using the HMMs in the feature space of explicit mapping kernels.

4 Studies on Acoustic Modeling

In this section, we study the performance of HMMs using the proposed approach for various speech recognition tasks. The tasks include recognition of isolated utterances of the E-set of English alphabet and recognition of CV segments from a continuous speech corpus of broadcast news in three Indian languages, namely, Tamil, Telugu and Hindi.

In our studies, the short-time analysis of the speech signal is performed for each utterance of the speech signal using a frame size of 25 milliseconds with a frame shift of 10 milliseconds. Each frame is represented by a 39-dimensional feature vector consisting of 12 Mel-Frequency Cepstral Coefficients (MFCC), energy, their first order time derivatives (delta coefficients) and their second order time derivatives (acceleration coefficients). In our studies, we use the polynomial kernel of degree 2. The 820 dimensional feature vector includes the monomials of order 0, 1 and 2 derived from the 39-dimensional input space data vector. For vector quantization, a codebook of size 64 is constructed. A 5-state, left-to-right HMM is constructed for each class. The CDHMMs in the input space use two mixtures per state. The CDHMMs in the polynomial kernel feature space use one mixture per state.

4.1 Recognition of E-Set Data

We study the performance of HMM based systems for recognition of highly confusable subset of spoken letters in English alphabet, namely, E-set. The E-set includes the following 9 letters: {B,C,D,E,G,P,T,V,Z}. The Oregon Graduate Institute (OGI) spoken letter database [6] is used in the study on recognition of E-set. The training data set consists of 240 utterances from 120 speakers for each letter, and the test data set consists of 60 utterances from 30 speakers for each letter.

The performance of different approaches to build HMMs is given in Table 1. It is seen that the performance of DHMMs in the kernel feature space is better than that of DHMMs in the input space (by about 2.5%). The performance DHMMs in the kernel feature space is inferior (by 15%) to that of the performance of the CDHMMs in the input space. However, the performance of the CDHMMs in the kernel feature space is better than that of the CDHMMs in the input space (by about 10%).

Table 1. Performance (in %) of different models in E-set recognition

Classification model	Classification accuracy (in%)
DHMMs in the input space	65.19
DHMMs in the polynomial kernel feature space	67.96
CDHMMs in the input space	82.96
CDHMMs in the polynomial kernel feature space	93.24

The confusion matrix for E-set recognition using the CDHMMs in input space and the CDHMMs in the polynomial kernel feature space is given in Table 2. Out of 540 test examples, 92 examples are misclassified (17.04% misclassification) using CDHMMs in the input space. The sound unit /B/ is mostly misclassified as /D/ or /E/ or /V/. The sound unit /P/ is mostly misclassified as /V/ or /B/. The sound unit /C/ is mostly confused with /G/ or /Z/. It is seen that the confusable sound units are better classified with an increase in classification accuracy by more than 10% by the CDHMMs in the kernel feature space.

Table 2. Confusion matrix (classification and misclassification rates in %) for CDHMMs in the input space and CDHMMs in the polynomial kernel space for E-set recognition

CDHMMs in the input space										CDHMMs in kernel feature space									
Class	B	C	D	E	G	P	T	V	Z	Class	B	C	D	E	G	P	T	V	Z
B	**77**	0	10	6	0	0	2	5	0	B	**90**	0	6	4	0	0	0	0	0
C	0	**92**	0	0	5	0	0	0	3	C	0	**96**	0	0	2	0	0	0	2
D	2	0	**75**	8	5	0	6	4	0	D	0	0	**90**	5	2	0	3	0	0
E	0	0	4	**93**	0	0	0	3	0	E	0	0	0	**98**	0	0	0	2	0
G	0	5	2	0	**83**	0	6	0	4	G	0	3	0	0	**93**	0	2	0	2
P	4	2	0	2	0	**80**	2	10	0	P	0	0	0	2	0	**92**	0	6	0
T	0	2	0	4	10	4	**80**	0	0	T	0	0	0	2	6	0	**92**	0	0
V	8	0	0	5	0	0	2	**82**	3	V	5	0	0	3	0	0	0	**92**	0
Z	0	6	0	4	0	0	2	3	**85**	Z	0	3	0	2	0	0	0	0	**95**

4.2 Recognition of CV Segments in Indian Languages

Next, we study the performance of different approaches on recognition of CV segments from a continuous speech corpus of broadcast news in three Indian languages, namely, Tamil, Telugu and Hindi. This study involves recognition of a large number of subword units of speech with high similarity among several units. In these studies, we consider the CV classes with atleast 50 examples in the training data set that results in 123, 86 and 103 CV classes for Tamil, Telugu and Hindi respectively. The Tamil data set consists of 43,541 CV segments for training and 10,293 CV segments for testing. The Telugu data set consists of 35,448 CV segments for training and 12,799 CV segments for testing. The Hindi data set consists of 20,236 CV segments for training and 4,137 CV segments for testing.

Classification performance for different classification models in recognition of CV segments for three languages is given in Table 3. It is seen that the CDHMMs in the polynomial kernel feature space give a significantly better performance (by about 13%, 14% and 16% for Tamil, Telugu and Hindi respectively) compared to the CDHMMs in the input space. An analysis of the performance has shown that the CDHMMs in the kernel feature space give a significantly higher classification accuracy for confusable CV units. This result demonstrates the effectiveness of CDHMMs in the kernel feature space for better discrimination of confusable units.

Table 3. Performance (in %) of different models in recognition of CV segments

Classification model	Classification accuracy (in %)		
	Tamil	Telugu	Hindi
DHMMs in the input space	50.55	46.48	40.09
DHMMs in the polynomial kernel feature space	52.73	49.13	41.04
CDHMMs in the input space	64.24	60.55	52.93
CDHMMs in the polynomial kernel feature space	77.36	74.62	68.79

5 Summary and Conclusions

In this paper, we have proposed an approach for acoustic modeling using HMMs in the feature space of explicit mapping kernels. Results of our studies show that CDHMMs in polynomial kernel feature space perform better than the CDHMMs in the input space. A further step is to explore building CDHMMs in the feature space of implicit mapping kernels such as the Gaussian kernels. Construction of CDHMMs in the implicit kernel feature space will involve developing techniques for probability density estimation in the feature space using innerproduct kernel operations [7].

References

1. Haykin, S.: Neural Networks: A Comprehensive Foundation. Prentice-Hall International, New Jersey (1999)
2. Schölkopf, B., Mika, S., Burges, C., Knirsch, P., Miiller, K. R., Ratsch, G., Smola, A.: Input Space Vs. Feature Space in Kernel-based Methods. IEEE Transactions on Neural Networks **10** (1999) 1000-1017
3. Rabiner, L. R., Juang, B. H.: Fundamentals of Speech Recognition. Prentice-Hall (1993)

4. Shawe-Taylor,J., Cristianini, N.: Kernel Methods for Pattern Analysis. Cambridge University Press (2004)
5. Srikrishna Satish, D., Chandra Sekhar, C.: Discrete Hidden Markov Models in Kernel Feature Space for Speech Recognition. In: Proceedings of the International Conference on Systemics, Cybernetics and Informatics (2004) 653-658
6. ISOLET Corpus, Release 1.1. Center for Spoken Language Understanding, Oregon Graduate Institute (2000)
7. Vapnik, V. N., Mukherjee, S.: Support Vector Method for Multivariate Density Estimation. In: Advances in Neural Information Processing Systems (1999) 659-665

TS-Neural-Network-Based Maintenance Decision Model for Diesel Engine

Ying-kui Gu and Zhen-Yu Yang

School of Mechanical and Electronical Engineering,
Jiangxi University of Science and Technology,
Ganzhou 341000, China
guyingkui@163.com

Abstract. To decrease the influence of fuzzy and uncertain factors on the maintenance decision process of diesel engine, a fuzzy-neural-network-based maintenance decision model for diesel engine is presented in this paper. It can make the maintenance of diesel engine follow the prevention policy and take the technology and economy into account at the same time. In the presented model, the fuzzy logic and neural network is integrated based on the state detection technology of diesel engine. The maintenance decision process of diesel engine is analyzed in detail firstly. Then, the fuzzy neural network model of maintenance decision is established, including an entire network and two module sub-networks, where the improved T-S model is used to simply the structure of neural networks. Finally, an example is given to verify the effective feasibility of the proposed method. By training the network, the deterioration degree of the diesel engine and its parts can be obtained to make the right maintenance decision.

Keywords: Fuzzy neural network, T-S model, maintenance decision, diesel engine.

1 Introduction

Because the diesel engine has many advantages, such as high thermal efficiency, energy conservation and good energy use efficiency, it already becomes the main power of automobile, agricultural machinery, construction machinery, ship, internal combustion locomotion, drilling machinery, power generation, and so on. However, the technology state of the diesel engine would become bad and its performance would decrease in the using process because of the weary, fatigue, deformation or the damage. Maintenance is one of the most important steps in the equipment management. It is also an important guarantee to prolong the life of the equipment and to prevent the accident [1]. Therefore, modern enterprises pay more attention to the high efficiency and low consumption of the equipment than any time before, and regard the equipment management as the important part of the business management. Moreover, the equipment management mainly is the equipment maintenance management,

D. Liu et al. (Eds.): ISNN 2007, Part I, LNCS 4491, pp. 553–561, 2007.

and modern enterprises need more advanced maintenance management patterns to realize the scientific management.

In order to make the diesel engine maintenance work under the prevention policy, it is necessary to make each kind of maintenance decision correctly at the right moment. During the maintenance decision process, one important task is to identify the specific time for maintenance and repair. Nowadays, point to the diesel engine maintenance decision, many maintenance management modes appeared, such as TNPM (Total normalized productive maintenance), SMS (Selective maintenance system), PMS (Preventive maintenance system), and so on [2]. However, maintenance decision is a time-consuming process, and there exist many fuzzy and uncertain factors which have important influence on the decision-making.

In recent years, the advances in computing intelligence, such as fuzzy set theory, neural network, genetic algorithms, rough set theory, and so on, provide a stronger tool for decreasing the influence of these fuzzy and uncertain factors in maintenance decision modeling. Compared with the other non-linear modeling methods, the advantages of fuzzy modeling are that the structure of the model and the physical meaning of the parameters are easily to understand. The reason for incorporating neural network into fuzzy logic system is that neural network has the characteristics of self-learning capacity, fault tolerance and model-free. Many new maintenance decision models have been presented in recent years, such as fuzzy-logic-based model, neural network model, and rough-set-based model, and so on [3-6].

In this paper, we present a fuzzy neural network decision model for diesel engine using fuzzy logic and improved T-S neural network. The structures of this paper are as follows. The maintenance decision theory for the diesel engine is analyzed in detail in section 2. A fuzzy-neural-network-based maintenance decision model is developed in section 3. An example is given in section 4 to verify the feasibly of the model. Conclusions are given in section 5.

2 Maintenance Decision Process of Diesel Engine

The equipment deterioration degree indicates the deviation degree from good state to the fault state. The value of the deterioration degree varies from 0 to 1. If the value of the deterioration degree is 0, it shows that the equipment state is good. If the value of deterioration degree is 1, it shows that the fault appears, and the equipment state is worse.

The maintenance decision theory of diesel engine is as follows. According to the oil monitor results and performance parameter monitor data, we can decide whether the main engine needs to be serviced or not, or whether the examination cycle needs to be prolonged or not. When it approach the legal examination cycle, if parameter monitoring and oil analysis results indicate that the technical state of the main engine is normal, the examination cycle should be prolonged. Otherwise, the engine and its parts should be serviced. We can decide whether the engine or its parts needs to maintenance based on the monitor results.

The maintenance decision process generally includes the following steps.

(1) Determination the state monitoring performance parameters and standards of maintenance object.

(2) Calculating the deterioration degree of performance parameters of maintenance object based on the technical parameter monitor value.

(3) Carrying on the maintenance decision.

When the deterioration degree of the diesel engine and its modules is obtained, the maintenance decision can be carried on, the decision steps are as follows:

(1) Let b be the deterioration degree of entire machine and b_i ($i > 1$) be the deterioration degree of its each module.

(2) Set threshold value. The threshold value can be regarded as the criteria to evaluate the technical state of maintenance object.

(3) Let the threshold value be 0.6. If b <0.6, it indicates that the technical state of the diesel engine is good, and does not need to maintenance. If b >0.8, it indicates that the technical state of the diesel engine is bad, and needs to maintenance. If 0.8> b >0.6, it indicates that whether the diesel engine needs to maintenance or not depends on the state of the module. If b_{imax}>0.8 and 0.6< b_{iover}<0.8 (where b_{imax} is the maximal value of the deterioration degree of each module and b_{iover} is the average value of the deterioration degree), it indicates that the state of most parts of the engine is worse, and needs to maintenance. Otherwise, it only needs to strengthen the monitoring of the engine. When the main engine needs to maintenance, which parts need to maintenance is decided by the technical state of the engine, the parts which need to maintenance are that their deterioration degree are bigger than 0.8, i.e. b_i>0.8.

3 The Fuzzy Neural Network Model of the Maintenance Decision for Diesel Engine

Fuzzy rule-based models and especially Takagi-Sugeno (TS) fuzzy models have gained significant impetus due to their flexibility and computational efficiency. They have a quasi-linear nature and use the idea of approximation of a nonlinear system by a collection of fuzzily mixed local linear models. The TS fuzzy model is attractive because of its ability to approximate nonlinear dynamics, multiple operating modes and significant parameter and structure variations.

The T-S fuzzy model approximates a nonlinear system with a combination of several linear systems. The overall T-S fuzzy model is formed by fuzzy partitioning of the input space. The premise of a fuzzy implication indicates a fuzzy subspace of the input space and each consequent expresses a local input-output relation in the subspace corresponding to the premise part. The set of fuzzy implications shown in the Takagi-Sugeno (T-S) model can express a highly nonlinear functional relation in spite of a small number of fuzzy implication rules [7-10].

The Takagi-Sugeno fuzzy network with M inputs and one output is shown in Figure1. The fuzzy controller has M+ 1 linguistic variable: M input ones and one output variable. Linguistic variables x_i are A_{1i}, A_{2i}, …, A_{ni}.

The number of linguistic rules is $p = n^M$.

The output from Takagi-Sugeno controller from Figure 1 is:

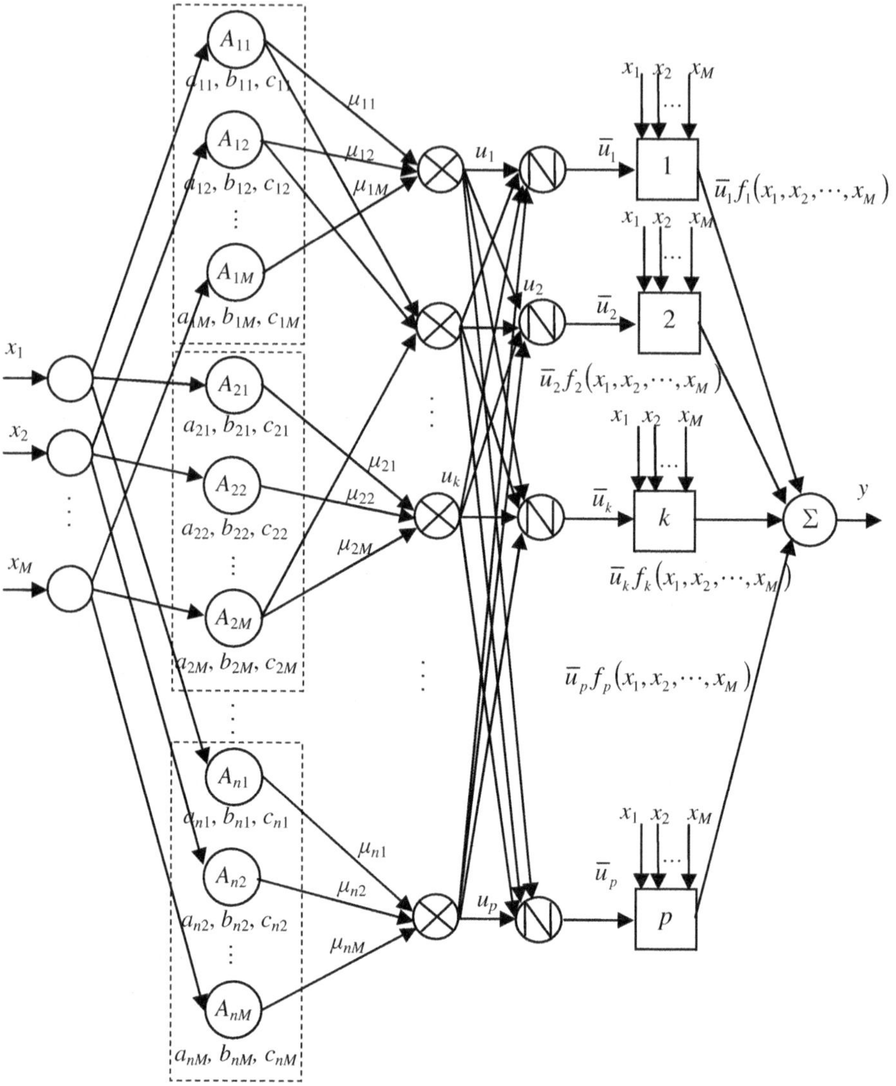

Fig. 1. The structure of TS fuzzy network

$$y = \sum_{i=1}^{p} \bar{u}_i f_i \ . \tag{1}$$

where

$$\bar{u}_i = \frac{u_i}{\sum_{i=1}^{p} u_i} \ . \tag{2}$$

If the membership functions are taken in the Gaussian form then:

$$\mu_{ij} = \frac{1}{1 + \left[\left(\frac{x_j - a_{ij}}{ci_j} \right)^2 \right]^{b_{ij}}}, \ i = 1,2,\cdots,n, \ j = 1,2,\cdots,M . \tag{3}$$

Consequences functions of the fuzzy rules are of the form:

$$fi = \sum_{j=1}^{M} p_{ij} x_j + c_i . \tag{4}$$

Substituting (2) into (1) the network output is obtained as:

$$y = \frac{1}{\sum_{i=1}^{p} u_i} \sum_{i=1}^{p} u_i f_i . \tag{5}$$

or, substituting (4) into (5), the output of the Takagi-Sugeno network is:

$$y = \frac{1}{\sum_{i=1}^{p} u_i} \sum_{i=1}^{p} u_i \left(\sum_{i=1}^{p} p_{ij} x_j + c_i \right) . \tag{6}$$

Because the technical state monitor parameters of the diesel engine mainly are the performance parameters of the crankshaft-bearing module and piston-cylinder module, and these performance parameters exist some overlapping, an improved fuzzy T-S neural network is proposed as shown in Figure.2. The improved fuzzy T-S network have some advantages, such as simple structure, quick convergence, high precision, good study ability and fault-tolerant ability, and so on. By using this fuzzy neural network, the maintenance decision of diesel engine could be realized based on the state monitoring data.

The maintenance decision model of the diesel engine includes two sub-networks and one combination network. The inputs of sub-network are the deterioration degree of the correlation performance parameters, and the output is the technical state deterioration degree of the module. The input of the combination network is the output of each sub-network, the output of the combination network is the technical state deterioration degree of the main engine. The structure of each fuzzy neural network is similar. The differences of the networks lie in the number of the input nodes. Take the sub-network of cylinder-piston module as an example to illustrate the fuzzy neural network which has four layers.

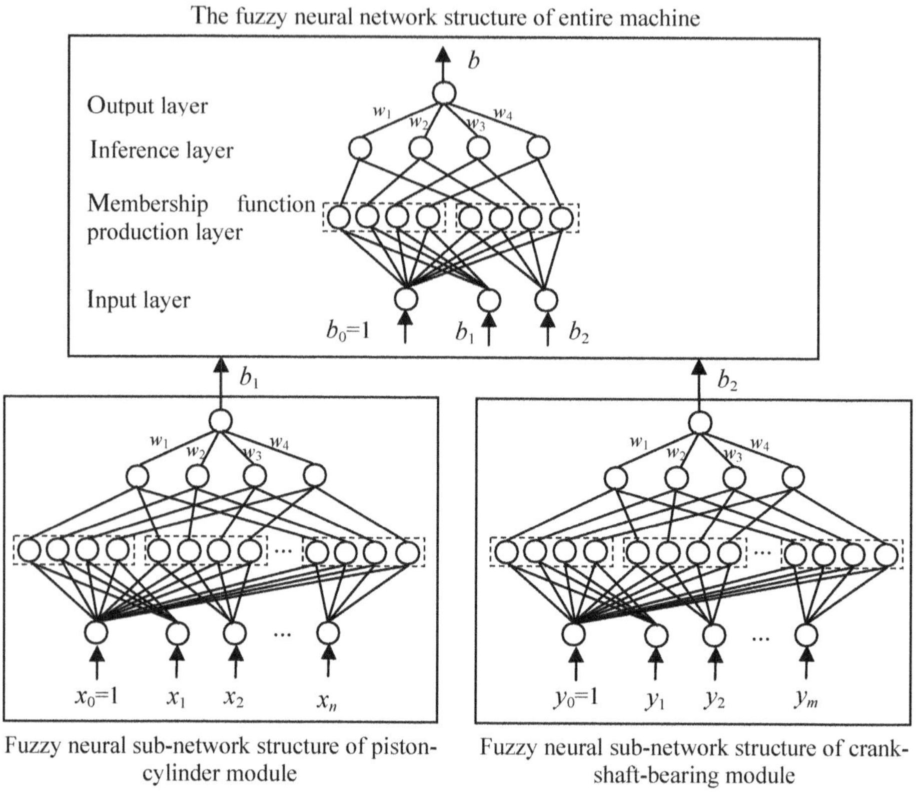

Fig. 2. Fuzzy-neural-network-based maintenance decision T-S model for diesel engine

The first layer is the input layer. The input is the deterioration degree of each performance parameter, and the number of input nodes is n. The output is the deterioration degree of this module.

The second layer is the membership function production layer. The input is the output of the first layer, and the number of input nodes is $n\times4$, where 4 indicates that there are 4 technical state, i.e. good, better, general, bad.

The third layer is the reasoning layer.

The fourth layer is the output layer. The output of this layer is the technical state deterioration degree of the maintenance object.

The input and output of piston-cylinder module fuzzy neural sub-network are as shown in Table 1.

By training the fuzzy neural network, the technical state deterioration degree of maintenance object can be obtained, and the maintenance decision can be carried on based on these deterioration degrees. The study rule of fuzzy neural network adopts the error function adjustment method.

Table 1. The input-output of piston-cylinder module fuzzy neural sub-network structure

Layer	The number of the nodes	Input	Output
The first layer(input layer)	n	$I_i^{(1)} = x_i$, $i = 1,2,\cdots,n$	$O_i^{(1)} = x_i$, $i = 1,2,\cdots,n$
The second layer(membership function production laye)	n×4	$I_{ij}^{(2)} = O_i^{(1)}$ $i = 1,2,\cdots,n$, $j = 1,2,3,4$	$O_{ij}^{(2)} = \exp\left(-\left(\frac{x_i - m_{ij}}{\delta_{ij}}\right)^2\right)$ $i = 1,2,\cdots,n$, $j = 1,2,3,4$
The third layer(reasoning layer)	4	$I_{ij}^{(3)} = O_{ij}^{(2)}$ $i = 1,2,\cdots,n$, $j = 1,2,3,4$	$O_j^{(3)} = \prod_{i=1}^{n} \mu_{ij}(x_i)$ $i = 1,2,\cdots,n$, $j = 1,2,3,4$
The fourth layer(output layer))	1	$I_{ij}^{(4)} = O_j^{(3)}$	$O^{(4)} = \sum_{j=1}^{4} w_j I_j^{(4)}$, $j = 1,2,3,4$

4 Case Study

Take the maintenance decision of YC6100 as an example to illustrate the established fuzzy-neural-network model. According to the Maintenance record, there are 9 performance parameters. i.e. the input of the subnet of cylinder-piston module are the deterioration degree of the pressure of the piston coollant (x_1), the deterioration degree of the pressure of the cylinder liner coollant (x_2), the deterioration degree of the temperature of the air cylinder coollant outlet (x_3), the deterioration degree of the piston

Table 2. The samples of the piston- cylinder module

	The piston- cylinder module performance parameter deterioration degree						Deterioration degree of the module
	x_1	x_2	x_3	x_4	x_5	x_6	b_1
1	0.3901	0.4100	0.3721	0.5625	0.6225	0.4523	0.5242
2	0.3912	0.4112	0.3741	0.5678	0.6220	0.4534	0.5251
3	0.3922	0.4201	0.3745	0.5672	0.6232	0.4546	0.3976
4	0.3923	0.4212	0.3735	0.5671	0.6251	0.4537	0.4650
5	0.3914	0.4118	0.3746	0.5680	0.6245	0.4542	0.5321
6	0.3925	0.4125	0.3748	0.5721	0.6222	0.4551	0.6352
7	0.3913	0.4120	0.3751	0.5625	0.6232	0.4518	0.4321
8	0.3932	0.4211	0.3745	0.5679	0.6235	0.4557	0.6365
9	0.3927	0.4217	0.3727	0.5642	0.6231	0.4561	0.6221
10	0.3915	0.4116	0.3742	0.5655	0.6242	0.4523	0.5276

Table 3. The samples of the crankshaft-bearing module and the entire engine

	The crankshaft-bearing module					Deterioration degree of the entire engine
	The technical parameter deterioration degree of the crankshaft-bearing module				Deterioration degree of the module	
	y_1	y_2	y_3	y_4	b_2	b
1	0.4815	0.2112	0.1525	0.2320	0.4850	0.5127
2	0.4618	0.2132	0.1521	0.2321	0.5265	0.5362
3	0.4725	0.2120	0.1519	0.2319	0.5306	0.5410
4	0.4667	0.2125	0.1517	0.2317	0.5405	0.5312
5	0.4669	0.2118	0.1524	0.2325	0.4998	0.5601
6	0.4660	0.2123	0.1510	0.2323	0.5150	0.6215
7	0.4721	0.2120	0.1527	0.2317	0.5272	0.6327
8	0.4716	0.2119	0.1522	0.2316	0.4925	0.5106
9	0.4732	0.2117	0.1515	0.2332	0.5532	0.6356
10	0.4756	0.2118	0.1513	0.2345	0.5545	0.6678

Table 4. The training results of the entire engine fuzzy neural network

	The training results of the piston- cylinder module		The training results of the crankshaft-bearing module		The training results of the entire engine	
	b_1	error_1	b_2	error_2	b	error
1	0.5256	9.982E-07	0.4863	9.962E-07	0.5140	9.947E-07
2	0.5264	9.983E-07	0.5279	9.981E-07	0.5376	9.954E-07
3	0.3990	9.972E-07	0.5321	9.962E-07	0.5303	9.987E-07
4	0.4666	9.964E-07	0.5418	9.975E-07	0.5328	9.946E-07
5	0.5336	9.982E-07	0.5012	9.987E-07	0.5615	9.957E-07
6	0.6366	9.987E-07	0.5163	9.968E-07	0.6201	9.962E-07
7	0.4334	9.984E-07	0.5287	9.953E-07	0.6341	9.949E-07
8	0.6341	9.943E-07	0.4939	9.943E-07	0.5119	9.980E-07
9	0.6234	9.958E-07	0.5546	9.959E-07	0.6341	9.958E-07

coollant outlet temperature (x_4), the deterioration degree of the discharge temperature (x_5), the deterioration degree of the main engine power (x_6); The output of this sub-network is the deterioration degree of the cylinder-piston module (b_1). The input of the crankshaft-bearing module subnet are the deterioration degree of the pressure of the main bearing the lubricating oil (y_1), the deterioration degree of the pressure of the crosshead bearing the lubricating oil (y_2), the deterioration degree of the lubricating oil cooler inlet temperature (y_3), the deterioration degree of the temperature difference of the lubricating oil import and export (y_4); The outputs is the deterioration degree of crankshaft bearing module (b_2). The input of the combination network is the output of the two sub-network, i.e. (b_1), (b_2). The output is the deterioration degree of the main engine, i. e. b. The sample sets are shown in Table 2 and Table 3.

The network precision of training parameter is 10^{-6} and the study rate is 0.001. We can take one sample as examination sample, such as sample No.10, and the other nine samples can be as training samples. The training results including expectation output, network error and system error are listed in Table 4.

From Table 4 we can see, the expectation output and the actual output of each sample are very close. The absolute error is smaller than 0.01. Moreover, the system average error is in 10^{-6} magnitude.

5 Conclusions

From the simulation result we can see, the combination fuzzy neural network model not only could reflect the fuzzy characteristic and the logic behavior of the main engine system structure, but also could assist the decision-maker to carry on the maintenance decision. The training results are very accurate. Therefore the combination fuzzy neural network model could offer theory base for the maintenance decision for the diesel engine.

Acknowledgments. This research was supported by China Postdoctoral Science Foudation under the contract number 20060391029.

References

1. Yan, L., Zhu, X.H., Yan, Z.J., Xu, J.J.: The Status and Prospect of Transportation Machine Equipment Repairing in 21th Century. Journal of Traffic and Transportation Engineering **1** (2001) 47-51
2. Hu, Y.P., Yan, L., Zhu, X.H.: Application of Neuro-net in Maintenace Decision for Ship Diesel Engine. Journal of Traffic and Transportation Engineering **1** (2001) 69-73
3. Yue, Y.F., Mao, J.Q.: An Adaptive Modeling Method Based on T-S Fuzzy Models. Control and Decision **17** (2002) 155-158
4. Sun, Z.Q., Xu, H.B.: Neuro-Fuzzy Network Based on T-S Fuzzy Models. Journal of Tsinghua University **37** (1997) 76-80
5. Xu, C.M., Er, L.J., Hu, H.J.: TSK-FNN and Its Constrained Optimization Algorithm. Journal of Beijing University of Aeronautics and Astronautics **31** (2005) 595-600
6. Gómez-Skarmeta, A.F, Delgado, M., Vila, M.A.: About the Use of Fuzzy Clustering Techniques for Fuzzy Model Identification. Fuzzy Sets and Systems **106** (1999) 179-188
7. Takagi, T., Sugeno, M.: Fuzzy Identification of Systems and Its Application to Modeling and Control. IEEE Transactions on Systems, Man, and Cybernetics **51** (1985) 116-132
8. Zhu, H.X., Shen, J., Li, Y.G.: A Novel Dynamic Clustering Algorithm and Its Application in Fuzzy Modeling for Thermal Processes. Proceedings of the CSEE **25** (2005) 34-39
9. Jang, C.H.: A TSK-Type Recurrent Fuzzy Network for Dynamic Systems Processing by Neural Network and Genetic Algorithms. IEEE Transaction on Fuzzy Systems **10** (2002) 155-169
10. Tor, A. J.: On the Interpretation and Identification of Dynamic Takagi–Sugeno Fuzzy Models. IEEE Transactions on Fuzzy Systems **8** (2000) 297-213

Delay Modelling at Unsignalized Highway Nodes with Radial Basis Function Neural Networks

Hilmi Berk Celikoglu[1] and Mauro Dell'Orco[2]

[1] Technical University of Istanbul, Faculty of Civil Engineering, Department of Civil Engineering, Division of Transportation, Ayazaga Campus, 34469, Maslak, Istanbul, Turkey
celikoglu@itu.edu.tr

[2] Technical University of Bari, Department of Highways and Transportation, Via Orabona 4, 70125, Bari, Italy
dellorco@poliba.it

Abstract. In vehicular traffic modelling, the effect of link capacity on travel times is generally specified through a delay function. In this paper, the Radial Basis Function Neural Network (RBFNN) method, integrated into a dynamic network loading process, is utilized to model delays at a highway node. The results of the model structure have then been compared to evaluate the relative performance of the integrated neural network method.

1 Introduction

In most of the traffic assignment models, the effect of link capacity on travel times is specified through a delay function. This kind of functions usually consists of the product of the free-flow time multiplied by a normalized congestion function, whose argument is the ratio flow-volume/capacity. Research in which delay scheme at unsignalized junctions is explicitly analyzed includes Bureau of Public Roads (BPR) functions, delay models, assignment models, and queuing models. However, none of these incorporated any soft computing method. In the literature, there exist no studies utilizing neural network methods for delay modelling scheme in a network loading frame. So, the motivation for this study has been on the integration of a NN based delay model to a dynamic network loading (DNL) process. In this study a dynamic node model (DNM) is set up to compute, within a DNL framework, the time-varying traffic flows conflicting at a node. A delay modelling component is added downstream the DNM, thus obtaining an integrated model structure.

The following section describes the integrated structure of the proposed model summarizing the link, the node rules and the delay models components. The third section gives the simulation results and the comparisons. Finally the last section concludes the paper with findings, discussion and the possible future extensions.

2 Model Structure

The presented DNM has two components: a link model, set up with a linear travel time function, and a node model with a set of rules considering constraints such as

D. Liu et al. (Eds.): ISNN 2007, Part I, LNCS 4491, pp. 562–571, 2007.

conservation and non-negativity of flows. The link model has been set up with the point-queuing (PQ) assumption, and therefore the delays are calculated in presence of possible vertical queues as a result of over-saturation and flows conflicts. At the beginning, the inputs for the link model are the flows entering simultaneously the merging links (MLs), inflows. The flows exiting from these links, outflows, are computed with respect to the link and flow characteristics. Then, taking these outflows as inflows to the node, the possible conflicting flows are processed according to the predefined characteristics of the diverging links (DLs), and the exiting flows from the node are computed. Afterwards, we calculate the delays resulting from flows conflicts and capacity constraints through both a conical delay function and an RBFNN. Finally, a comparison of the performances of the integrated model structure is carried out.

2.1 Components of Node Model

The travel time function based link modelling component and the delay modelling component are summarized in this section.

2.1.1 Link Model Component

In the travel time formulation, the travel time needed to traverse a link starting at time t is introduced as $\tau(t)$. Since the traffic flow is continuous and is resembled as a fluid, the user is treated just as a particle of this fluid. The flow propagation through a link can be described by the relationships between control variables such as travel time $\tau(t)$, inflow $u(t)$, outflow $w(t)$ or the number of vehicles on the link at each point in time $x(t)$; therefore, the travel time $\tau(t)$ for traffic entering a link at time t can be expressed as given in Equation 1 [1].

$$\tau(t) = T(u(t),\ x(t),\ w(t))\ . \tag{1}$$

Considering the "instantaneous flow-dependent running time" and the "instantaneous queuing delay" as non-negative, increasing functions of the variables x, u and x, w respectively, the general form of a travel time function can be written as given in Equation 2, where α is the free-flow travel time [1,2].

$$\tau(t) = \alpha + f(u(t)) + g(x(t)) + h(w(t))\,. \tag{2}$$

Using the hydrodynamic analogy to represent the traffic, the model should respect the conservation principle derived from continuity equations considering the control variables as given in Equation 3.

$$x(t) = \int_0^t (u(\omega) - w(\omega))d\omega\,. \tag{3}$$

A vehicle entering a link at time t exits at time $t+\tau(t)$. If the FIFO rule and conservation hold, the number of vehicles entering the link by time t must be equal to the number of vehicles that have left by time $t+\tau(t)$ as given in Equation 4.

$$\int_{-\infty}^{t} u(\omega)d\omega = \int_{-\infty}^{t+\tau(t)} w(\omega)d\omega\,. \tag{4}$$

Differentiating and rearranging Equation 4 with respect to t give Equation 5, which calls for the respect of the FIFO rule in the travel time formulation [3].

$$w(t+\tau(t)) = \frac{u(t)}{1+\frac{d\tau(t)}{dt}}. \tag{5}$$

In this study, we consider a travel time function, derived from Equation 2, solely dependent on the number of vehicles on the link as given in Equation 6.

$$\tau(t) = \alpha + \beta(x(t)), \quad \alpha, \beta > 0. \tag{6}$$

This linear form of the travel time function given in Equation 7 is widely used in most of the DNL studies based on travel time formulation [3-7].

2.1.2 Node Rules Component

The notations and the variables used in the proposed DNM are as follows:

- FW_k : set of links diverging from a node k;
- BW_k : set of links merging to a node k;
- i : generic link included in the set of MLs to node k ($i \in BW_k$);
- r : generic link included in the set of DLs from node k ($r \in FW_k$);
- $w^{ik}(t)$: flow exiting from link i and entering node k at time t ($i \in BW_k$);
- $u^{kr}(t)$: total flow entering link r, and exiting from node k at time t ($r \in FW_k$);
- C^r : capacity of link r ($r \in FW_k$);
- $NK^r(t)$: total number of vehicles stored in the PQ at the entrance of link r at time t ($r \in FW_k$);
- $G^r(t)$: delay occurring due to capacity constraint on link r at time t ($r \in FW_k$);
- ξ^{ir} : partial flow splitting rate from an ML i to a DL r ($i \in BW_k$, $r \in FW_k$).

For the node rules component, another set of constraints has been defined considering the flow conservation principle, the capacities of DLs, and the splitting rates at a node. In the structure of the integrated model, the link model component has been set up with the PQ assumption. Therefore, the delays are calculated in the presence of these possible vertical queues instead of physical ones.

The first constraint expresses the non-negativity of the partial flow exiting from the link i, and entering the link r at time t:

$$u^{ikr}(t) \geq 0, \quad \forall i \in BW_k \quad \text{and} \quad \forall r \in FW_k. \tag{7}$$

Another constraint, derived from the capacity of DL, requires that the total flow entering the link r at time t should not be greater than the capacity of the link r:

$$u^{kr}(t) = \sum_{i \in BW_k} w^{ikr}(t) \leq C^r, \quad \forall i \in BW_k \quad \text{and} \quad \forall r \in FW_k. \tag{8}$$

According to the flow conservation principle, the total flow entering all DLs at time t should not be greater than the amount of total flows exiting from all the MLs at time t:

$$\sum_{r \in FW_k} u^{kr}(t) \leq \sum_{i \in BW_k} w^{ik}(t), \quad \forall i \in BW_k \text{ and } \forall r \in FW_k. \tag{9}$$

The total flow entering all DLs, respecting the capacity constraint and the above mentioned delay assumption can be determined by the following relationship:

$$\sum_{r \in FW_k} u^{kr}(t) = \begin{cases} \sum_{i \in BW_k} w^{ik}(t), & \text{if } \sum_{i \in BW_k} w^{ik}(t) \leq \sum_{r \in FW_k} C^r, \\ \sum_{r \in FW_k} C^r, & \text{if } \sum_{i \in BW_k} w^{ik}(t) > \sum_{r \in FW_k} C^r, \end{cases} \quad \forall i \in BW_k \text{ and } \forall r \in FW_k. \tag{10}$$

When the whole node modelling process is considered 'till the time horizon T, the inequality constraint given with Equation 9 turns out to an equality constraint, given in Equation 11.

$$\sum_t^T \sum_{r \in FW_k} u^{kr}(t) = \sum_t^T \sum_{i \in BW_k} w^{ik}(t), \quad t \in [0, T]. \tag{11}$$

A possible PQ at the entrance of a DL can be determined assuming that there is no initial queue. Then, for t=0, $\forall r \in FW_k$, and $t \in [0, T]$, the Equation 12 gives the number of vehicles stored in such a buffer area:

$$NK^r(t) = \begin{cases} 0, & \text{if } (t=0) \vee \left(u^{kr}(t) + \frac{NK^r(t-\Delta t)}{\Delta t} \leq C^r \right), \\ \left(u^{kr}(t) - C^r\right) \cdot \Delta t + NK^r(t-\Delta t), & \text{if } u^{kr}(t) + \frac{NK^r(t-\Delta t)}{\Delta t} > C^r. \end{cases} \tag{12}$$

In the proposed model, the partial splitting rates from MLs to diverging ones are assumed to be known upon current information on link flows. The partial flow splitting rate from an ML i to a DL r is calculated as given in Equation 13, and should satisfy the inequality given in Equation 14 due to delaying phenomena.

$$\xi^{ir} = \frac{u^{ikr}(t)}{w^{ik}(t)}, \quad \xi^{ir} \geq 0. \tag{13}$$

$$\sum_{r \in FW_k} \xi^{ir} \leq 1. \tag{14}$$

2.2 Delay Modelling Component

In the past, delays due to capacity constraints have been represented with varying types of proposed and applied flow-rate delay functions [8]. These functions are usually expressed as the product of the free-flow time multiplied by a normalized congestion function $f(x)$:

$$delay(x) = t_{free-flow} \cdot f(x). \tag{15}$$

where the argument of the delay function is the ratio volume/capacity. BPR functions [9], conventionally used to represent delays have many drawbacks as explained in details by Spiess [10]. Therefore, to overcome the drawbacks inherent these conventional functions, we selected a conical delay function (CDF) with desirable properties. Note that, depending on the vertical queuing process, in this case the traffic volume is considered as inflow $u^{kr}(t)$, requiring to enter a DL. Moreover, since the free-flow time of a link is constant during the modelling horizon, the performances of congestion functions are evaluated to model delays.

2.2.1 Conical Delay Function

Since the free-flow travelling time on links is constant, we can skip it in the congestion function. In this study, we used a CDF having the desired properties of a delay function, whose form is given by Equation 16 [10]:

$$f\left(\frac{u^{kr}(t)}{C^r}\right) = 2 + \sqrt{a^2\left(1-\frac{u^{kr}(t)}{C^r}\right)^2 + b^2} - a\left(1-\frac{u^{kr}(t)}{C^r}\right) - b \tag{16}$$

where $b=(2a\text{-}1)/(2a\text{-}2)$, and parameter a is any number larger than 1 that defines how sudden the congestion effects change when the capacity is reached.

2.2.2 Radial Basis Function Neural Networks

The radial basis function (RBF) method, as Artificial Neural Network (ANN), is made of three layers: a layer of input neurodes (neural nodes, rather than biological neurons) feeding the feature vectors into the network; a hidden layer of RBF neurodes, calculating the outcome of the basis functions; and a layer of output neurodes, calculating a linear combination of the basis functions (see Fig. 1, in which the structure of an RBFNN is shown).

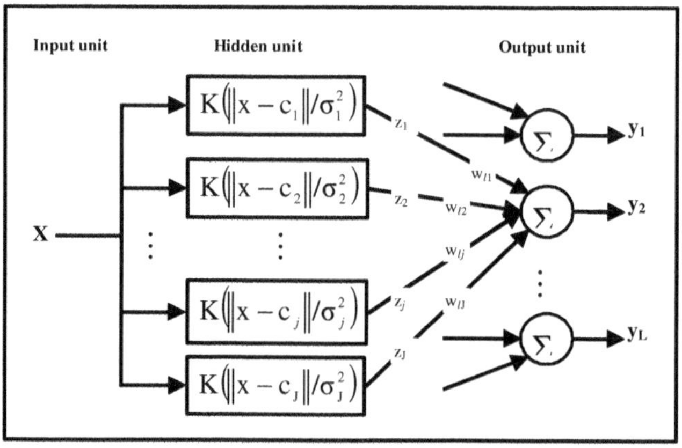

Fig. 1. Structure of a radial basis function (RBF) neural network

RBF networks are generally used for function approximation, pattern recognition, and time series prediction problems. Such networks have the universal approximation property [11], arise naturally as regularized solutions of ill-posed problems [12] and are dealt well in the theory of interpolation [13].

Their simple structure enables learning in stages, gives a reduction in the training time, and this has led to the application of such networks to many practical problems. The adjustable parameters of such networks are the centres (the location of basis functions), the width of the receptive fields (the spread), the shape of the receptive field and the linear output weights.

An RBF network is a feed-forward network [14] with a single layer of hidden units that are fully connected to the linear output units. Equation 17 shows the output units (Ψ_j) form a linear combination of the basis (or kernel) functions computed by the hidden layer nodes.

$$\Psi_j(x) = K\left(\frac{\left\|x - c_j\right\|}{\sigma_j^2}\right). \tag{17}$$

Each hidden unit output Ψ_j is obtained by calculating the closeness of the input to a n-dimensional parameter vector $\mathbf{c}_j$ associated with the jth hidden unit. K is a positive radially symmetric function (kernel) with a unique maximum at its centre $\mathbf{c}_j$, dropping off rapidly to zero away from the centre. Activations of such hidden units decrease monotonically with the distance from a central point or prototype (local) and are identical for inputs that lie at a fixed radial distance from the centre.

Assume that a function f: $R^n \rightarrow R^1$ is to be approximated with an RBF network, whose structure is given below. Let $\mathbf{x} \in R^n$ be the input vector, $\Psi(\mathbf{x}, \mathbf{c}_j, \sigma_j)$ be the jth function with centre $\mathbf{c}_j \in R^n$, and width σ_j, $\mathbf{w}=(w_1, w_2, \ldots, w_M) \in R^M$ be the vector of linear output weights and M be the number of basis function used. We concatenate the M centres $\mathbf{c}_j \in R^n$, and the widths σ_j to get $\mathbf{c}=(c_1, c_2, \ldots, c_M) \in R^{nM}$ and $\boldsymbol{\sigma}=(\sigma_1, \sigma_2, \ldots, \sigma_M) \in R^M$, respectively. The output of the network for $\mathbf{x} \in R^n$ and $\boldsymbol{\sigma} \in R^M$ is shown in Equation 18.

$$F(x, c, \sigma, w) = \sum_{j=1}^{M} w_j \cdot \Psi(x, c_j, \sigma_j). \tag{18}$$

Now, let $(\mathbf{x}_i, y_i)$: i=1, 2, …, N be a set of training pairs and $\mathbf{y}=(y_1, y_2, \ldots, y_N)^T$ the desired output vector, in which the superscript ^{T} denotes the vector/matrix transpose. For each $\mathbf{c} \in R^{nM}$, $\mathbf{w} \in R^M$, $\boldsymbol{\sigma} \in R^M$ and for arbitrary weights λ_i, i=1, 2, …, N, set Equation 19. Note that λ_i are nonnegative numbers, chosen to emphasize certain domains of the input space.

$$E(c, \sigma, w) = \frac{1}{2}\sum_{i=1}^{N}\left[\lambda_i\left(y_i - F(x_i, c, \sigma, w)\right)\right]^2. \tag{19}$$

3 Application to Case Study

RBFNN delay modelling process consists of two steps, respectively the training and testing step. The prediction problem is transformed into a minimum norm problem: the search for one RBFNNs for each DL, that minimizes the Euclidean distance, $\|P(\mathbf{UI}_k) - NN(\mathbf{UI}_k)\|$, where $\mathbf{UI}_k$ and $\mathbf{Y}_k$ are the actual values vector series (obtained by the first run of the model with CDF component), $\mathbf{UI}=[u^{11}(t), u^{21}(t), u^{31}(t), u^{41}(t), \xi^{11}, \xi^{21}, \xi^{31}, \xi^{41}, C^{1}]$ is the input vector of NN for first DL (note that, the delay occurring as a result of flow conflicts are considered through the bias), and $\mathbf{Y}_{k+1}=P(\mathbf{UI}_k)$.

With RBFNN method, the solution to the minimum norm problem involves a number of steps. The first one is the choice of the model inputs, and the second step is the attainment of parameters that minimize the norm given above. In order to obtain an approximation to the CDF function, the input variables are selected on purpose to accurately represent the argument of the CDF. Each of the merging flows $u^{ikr}(t)$, the partial splitting rates ξ^{ir} and the DL capacities C^{r} is selected as an input node. So, the input layer of all NN configurations consists of 9 nodes and the unique output layer node represents the congestion function value for a DL.

Since the success of an NN approximator depends heavily on the availability of a good subset of training data, data partitioning for the NN approximator is carried out considering explicitly the error term computations in all available partitions. The iterative structure of the training process needs a threshold value to stop learning; performance criteria for varying RBFNN configurations require convergence to some selected error term targets. One SSE value is targeted for RBFNN training processes. During the training stage the first 80 out of 146 values were analyzed, the last 66 were then used to examine the performance of the testing phase. The optimum number of training pairs has been selected considering the minima existing after the plot of MSE terms obtained by scaled training pairs. Following the training period, the networks are applied to the testing data and RBFNN performance is evaluated with the selected statistical criteria. The training vectors formed the initial centres of the Gaussian RBFs. The initial process of the training procedure was the determination of the hidden layer besides the number of nodes in the input layer, providing best training results. The target for SSE, to be reached at the end of the simulations, was set equal to 0.0001. Because the second step is largely a trial-and-error process, and runs involving RBFNNs with hidden layer node were more than 18, any sizeable improvement in prediction accuracy was not observed. The selected number of units (the number of RBFs) for the single hidden layer was 18. The optimum spread parameter (the width of the RBFs') has been selected as 0.16, after the trials with the selected hidden layer node number. In the training process, 36 iterations were found sufficient with respect to the minimum SSE term (0.0003453).

4 Simulation Results and Comparisons

To represent the deviation of predictions from the data generated by the CDF, predictions and previously generated values have been used to calculate the root mean squared percent error (RMSPE), the root mean squared error (RMSE), and the coefficient of determination (R^2), shown in Table 1.

Table 1. Statistical evaluation criteria for RBFNN integrated model

	RMSPE (percent)	R^2	RMSE (congestion value)
DL 1	1.426	0.980	3.013
DL 2	3.712	0.946	0.455
DL 3	5.872	0.978	0.518

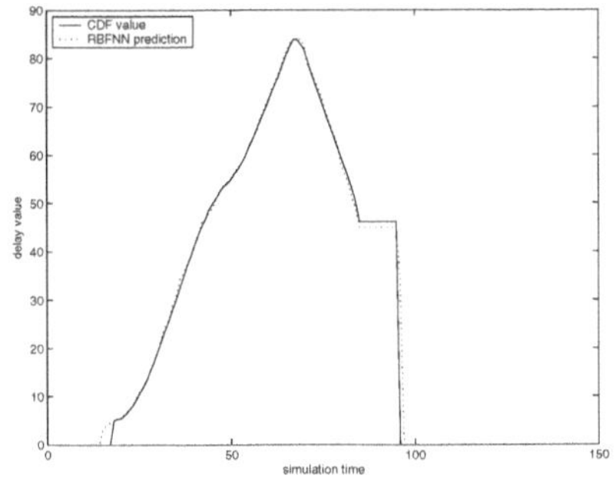

Fig. 2. RBFNN prediction of delay function with original CDF on DL 1

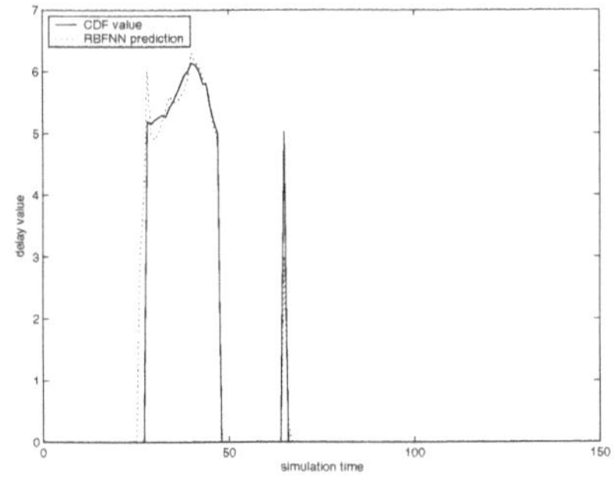

Fig. 3. RBFNN prediction of delay function with original CDF on DL 2

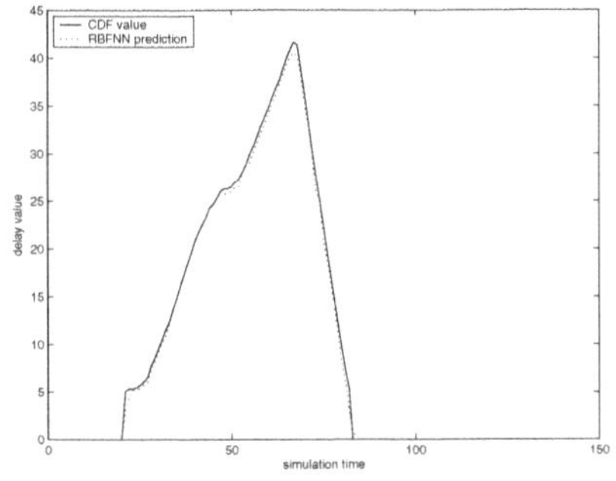

Fig. 4. RBFNN prediction of delay function with original CDF on DL 3

To clearly show the deviations of results from observations, the RBFNN predicted congestion function is plotted with the corresponding CDF for each DL in figures 2, 3 and 4.

In both periods including the minima and the maxima of the field data, the configured RBFNN provided pretty close estimates. The results point out that the function approximation by RBFNN is closer to the original one. Efficiency of NNs can be attributed to the capability of NNs to capture the nonlinear dynamics and generalize the structure of the whole data set. In the congestion function representing delays, the nonlinear relationship of traffic flow variables and physical link characteristics with each other is modelled more appropriately with utilizing nonlinear transfer functions in the nodes of the hidden layer of NN configuration.

5 Conclusions

The results highlight the fact that, the methodologically different NN estimating method is able to provide accurate delay computation. Among this performed simulation, it is seen that approximating with radial basis functions leads to significantly considerable predictions. This is due to RBF's flexibility to adapt to nonlinear traffic flow relationships. After the non-linearity is worked in the neurodes of hidden layer, the linear filtering is applied by the summing up nodes in the output layer.

NNs have a distributed processing structure in which each individual processing unit or the weighted connection between two units is responsible for one small part of the input-output mapping system. Therefore, each component has no more than a marginal influence with respect to the complete solution. As a result, the neural mechanism will still function and generate reasonable mappings where the CDF has some constraints to fit the desired mathematical properties.

References

1. Ran, B., Boyce, D.E., LeBlanc L.J.: A new class of instantaneous dynamic user-optimal traffic assignment models. Oper. res. **41** (1993) 192-202
2. Boyce, D.E., Ran, B., LeBlanc L.J.: Solving an instantaneous dynamic user optimal route choice model. Transp. sci. **29** (1995) 128-142
3. Astarita, V.: A continuous time link model for dynamic network loading based on travel time function. In: Lesort, J.B. (ed.): Proc. 13th Internat. Sympos. Theory Traffic Flow. Elsevier, Exeter (1996) 79-102
4. Ran, B., Rouphail, N.M., Tarko, A., Boyce, D.E.: Toward a class of link travel time functions for dynamic assignment models on signalized networks. Transp. res., Part B: methodol **31**(1997) 277-290
5. Wu, J.H., Chen, Y., Florian, M.: The continuous dynamic network loading problem: A mathematical formulation and solution method. Transp. res., Part B: methodol. **32** (1998) 173-187
6. Xu, Y.W., Wu, J.H., Florian, M., Marcotte, P., Zhu, D.L.: Advances in the continuous dynamic network loading problem. Transp. sci. **33** (1999) 341-353
7. Adamo, V., Astarita, V., Florian, M., Mahut, M., Wu, J.H.: Modelling the spill-back of congestion in link based dynamic network loading models: A simulation model with application. In: Ceder, A. (ed.): Proc. 14th Internat. Sympos. Transportation Traffic Theory. Elsevier, Amsterdam (1999) 555-573
8. Branston, D.: Link capacity functions: a review. Transp. Res. **10** (1976) 223-236
9. Bureau of Public Roads 1964. Traffic Assignment Manual. U.S. Department of Commerce, Urban Planning Division, Washington D.C.

10. Spiess, H.: Conical volume-delay functions. Transp. sci. **24** (1999) 153-158
11. Park, I., Sandberg, I.W.: Universal approximation using radial basis function networks. Neural comput. **3** (1991) 246-257
12. Poggio, T., Girosi, F.: Networks for approximation and learning. In: Proc. IEEE. **78** (1990) 1481-1497
13. Powell, M.J.D.: Radial basis function for multivariate interpolation. In: Mason, J.C., Cox, M.G. (eds.): A review algorithms for the approximation of functions and data. Clarendon, Oxford U.K. (1987)
14. Poggio, T., Girosi, F.: A theory of networks for approximation and learning. MIT Al Memo, no. 1140, MIT Press, Cambridge (1989)

Spectral Correspondence Using the TPS Deformation Model

Jun Tang[1], Nian Wang[1], Dong Liang[1], Yi-Zheng Fan[1,2], and Zhao-Hong Jia[1]

[1] Key Lab of Intelligent Computing & Signal Processing Ministry of Education, Anhui University, Hefei 230039, China
[2] Department of Mathematics, Anhui University, Hefei 230039, China
{tangjun,wn_xlb,dliang,fanyz,jiazhaohong}@ahu.edu.cn

Abstract. This paper presents a novel algorithm for point correspondences using spectral graph analysis. Firstly, the correspondence probabilities are computed by using the modes of proximity matrix and the method of doubly stochastic matrix. Secondly, the TPS deformation model is introduced into the field of spectral correspondence to estimate the transformation parameters between two matched point-sets. The accuracy of correspondences is improved by bringing one point-set closer to the other in each iteration with transformation parameters estimated from the current correspondences. Experiments on both real-world and synthetic data show that our method possesses comparatively high accuracy.

1 Introduction

Point pattern matching is often encountered in computer vision, image analysis and pattern recognition. In recent years, there has been a boost of interest for the application of graph spectra theory in this field. Scott and Longuet-Higgins [1] are among the first to use spectral theory for correspondence analysis. They have proposed a method to recover correspondence by singular value decomposition (SVD) of an appropriate correspondence strength matrix, which can cope with point-sets of different size but is sensitive to the degree of rotation. In order to overcome the above problem, Shapiro and Brady [2] have presented a method of the intra-image point proximity matrix which commences by computing a point-proximity matrix based on Gaussian function of the distance between points in the same image. A modal representation is constructed by finding the eigenvalues and eigenvectors of the proximity matrix. Correspondences are located by comparing the ordered eigenvectors of proximity matrices for different images. Provided that the point-sets are of the same size, the correspondences delivered by the Shapiro-Brady method are relatively robust to random point jitter, affine rotations and scaling. In order to render the spectral method robust, Carcassoni and Hancock [3] have embedded the Shapiro-Brady method within the framework of the EM algorithm. The global structural properties of the point pattern are represented as the eigenvalues and eigenvectors of the point proximity matrix. The influence of the contamination and drop-out in the

D. Liu et al. (Eds.): ISNN 2007, Part I, LNCS 4491, pp. 572–581, 2007.

point pattern is discounted via the EM algorithm, so the accuracy of correspondences is increased. Carcassoni and Hancock [4] have adopted a hierarchical approach to the correspondence problem, which is based on the observation that the modes of the proximity matrix can be viewed as pairwise clusters. They aim to characterize the potential groupings in an implicit or probabilistic way and to exploit their arrangement to provide constraints on the pattern of correspondences.

In [3], affine transformation is introduced to explicitly align the matched point-sets. But in many circumstances, especially in the case of non-rigid deformation, affine transformation is insufficient to describe the spatial transformation between two matched point-sets. On the other hand, although the combination of the EM algorithm can improve the accuracy of spectral correspondence to some extent, the exclusion principle (keeping one-to-one match) incorporated in the Shapiro-Brady method has been discarded.

Aiming at the two problems mentioned above, we propose a new algorithm for correspondence based on graph spectrum. Here, we adopt the strategy of iterated point matching and spatial transformation estimations. Correspondence probabilities are computed from the modes of the proximity matrix, and the method of doubly stochastic matrix is used to normalize both the rows and columns, then correspondence is judged from its corresponding row and column to enforce one-to-one match. In particular, we use a somewhat different method for the problem of recovering transformational geometry. The TPS (thin plate spline) deformation model is exploited to bring two matched point-sets closer. In each iteration, the accuracy of spectral correspondence is improved when the point-sets become closer.

2 The Probabilities of Spectral Correspondence

Given two matched point-sets $\boldsymbol{I},\boldsymbol{J}$ of the same size, Shapiro and Brady [2] define the proximity matrix $\boldsymbol{H}^I=\left[h_{pq}^I\right]$ for the point-set $\boldsymbol{I}$ containing m points $\boldsymbol{I}_p(p=1,2,\cdots,m)$ as:

$$h_{pq}^I=\begin{cases}H_D(p,q) & if\ p\neq q\ ,\\ 0 & if\ p=q\ .\end{cases}\quad p,q=1,2,\cdots m\ , \tag{1}$$

where $H_D(p,q)$ is the weighting function between two points $\boldsymbol{I}_p$ and $\boldsymbol{I}_q$. Carcassoni and Hancock [3] have pointed out that the increasing weighting function $H_D(p,q)=[1+\frac{1}{\sigma}\|\boldsymbol{I}_p-\boldsymbol{I}_q\|]^{-1}$ performs best for positional jitter, so we use this style of weighting function through this paper. And $\boldsymbol{H}^I$ can be decomposed as:

$$\boldsymbol{H}^I=\boldsymbol{U}\boldsymbol{\Delta}_I\boldsymbol{U}^T\ , \tag{2}$$

where $\boldsymbol{\Delta}_I=\mathrm{diag}\{\lambda_1,\cdots,\lambda_m\}$ ($\lambda_1\geq\cdots\geq\lambda_{m-1}>\lambda_m=0$) is a diagonal matrix whose diagonal entries are eigenvalues of $\boldsymbol{H}^I$, and $\boldsymbol{U}=(\boldsymbol{u}_1,\cdots\boldsymbol{u}_m)$ is an orthogonal matrix whose column $\boldsymbol{u}_i$ is an eigenvector of $\boldsymbol{H}^I$ corresponding to the eigenvalue λ_i for each $i = 1, 2, \cdots, m$. Similarly, we have:

$$\boldsymbol{H}^J = \boldsymbol{V}\boldsymbol{\Delta}_J\boldsymbol{V}^T , \tag{3}$$

where $\boldsymbol{\Delta}_J = \text{diag}\{\gamma_1, \cdots, \gamma_m\}$ ($\gamma_1 \geq \cdots \geq \gamma_{m-1} > \gamma_m = 0$), $\boldsymbol{V} = (\boldsymbol{v}_1, \ldots, \boldsymbol{v}_m)$. If the point-sets are of different size (for example, point-set $\boldsymbol{I}$ has m points and $\boldsymbol{J}$ has n points with $m < n$), the Shapiro-Brady method only handles the former m rows and columns and truncates the last $n - m$ rows and columns of $\boldsymbol{V}$ to keep the consistency of dimensionality. We can use the ith row vector $\boldsymbol{u}^{(i)}$ of $\boldsymbol{U}$ (and respectively, the ith row vector $\boldsymbol{v}^{(i)}$ of $\boldsymbol{V}$) to represent the mode of the ith point of point-set $\boldsymbol{I}$ (and respectively, the ith point of point-set $\boldsymbol{J}$). Let p_{ij} denote the correspondence probability between the points $\boldsymbol{I}_i \in \boldsymbol{I}$ and $\boldsymbol{J}_j \in \boldsymbol{J}$, then all the correspondence probabilities between point-sets $\boldsymbol{I}$ and $\boldsymbol{J}$ can be organized as a matrix $\boldsymbol{P}$ with dimension $m \times m$. Carcassoni and Hancock [3] have given an approach to computing p_{ij} from the modes of the proximity matrix as follows:

$$p_{ij} = \frac{\sum_{l=1}^{m} \exp[-\alpha \| \boldsymbol{u}^{(i)}(l) - \boldsymbol{v}^{(j)}(l) \|^2]}{\sum_{j' \in J} \sum_{l=1}^{m} \exp[-\alpha \| \boldsymbol{u}^{(i)}(l) - \boldsymbol{v}^{(j')}(l) \|^2]} . \tag{4}$$

The weakness of this method is that only one-way normalization constraint is enforced and correspondences are selected on the basis of maximum probability in their corresponding rows, which may lead to many-to-one matches. But in many applications one-to-one match is desired.

Here, in order to enforce one-to-one match, we propose a different method via doubly stochastic matrix to compute the spectral correspondence probabilities. The doubly stochastic matrix is a square matrix, where the sum of each row and column is one. Sinkhorn [5] has noted that the iterative process of alternated row and column normalization can convert a matrix with positive elements to the form of doubly stochastic matrix. We compute the element p_{ij} of the probability matrix $\boldsymbol{P}$ as follows:

$$p_{ij} = \exp(-\beta \| \boldsymbol{u}^{(i)} - \boldsymbol{v}^{(j)} \|^2) , \tag{5}$$

where β is the smoothing coefficient. We set $\beta = 10$ in the following experiments. Then we perform alternated row and column normalization on matrix $\boldsymbol{P}$. Generally, a few rounds of normalization are enough to bring matrix $\boldsymbol{P}$ close to a doubly stochastic matrix. With the doubly stochastic matrix $\boldsymbol{P}$ to hand, we decide the correspondence from both the row and column directions, which keeps the spirit of the exclusion principle in the earlier spectral method: If p_{ij} is the greatest element both in its corresponding row and column, we conclude that the *i*th point of $\boldsymbol{I}$ matches the *j*th point of $\boldsymbol{J}$; otherwise, there is no correspondence.

3 The TPS Deformation Model

The TPS deformation model is widely used for flexible coordinate transformations. Bookstein [6] has found it is of high effectiveness for modeling changes with a physical explanation and given a closed-form representation. It has been successfully applied to non-rigid shape matching in [7], [8], [9]. Let z_i denote the target function value at the corresponding location $\boldsymbol{I}_i = (x_i, y_i)$, with $i = 1, 2, \cdots m$. And two TPS models are applied to the 2D coordinate transformation. Suppose point $\boldsymbol{I}_i = (x_i, y_i)$ is matched to point $\boldsymbol{J}_i = (s_i, t_i)$, we set z_i equal to s_i and t_i respectively in turn to obtain one continuous transformation for each coordinate. The TPS interpolant $f(x, y)$ minimizes the bending energy

$$I_f = \iint_{R^2} (\frac{\partial^2 f}{\partial x^2})^2 + 2(\frac{\partial^2 f}{\partial x \partial y})^2 + (\frac{\partial^2 f}{\partial y^2})^2 dxdy \tag{6}$$

and has the closed-form solution:

$$f(x, y) = a_1 + a_x x + a_y y + \sum_{i=1}^{m} w_i U(\| (x_i, y_i) - (x, y) \|) \ , \tag{7}$$

where $U(r)$ is the kernel function with the form of $U(r) = r^2 \log r^2$. The TPS coefficients $\boldsymbol{w}$ and $\boldsymbol{a}$ are the solutions of the linear equation

$$\begin{bmatrix} \boldsymbol{K} & \boldsymbol{P} \\ \boldsymbol{P}^T & \boldsymbol{0} \end{bmatrix} \begin{bmatrix} \boldsymbol{w} \\ \boldsymbol{a} \end{bmatrix} = \begin{bmatrix} \boldsymbol{z} \\ \boldsymbol{0} \end{bmatrix} , \tag{8}$$

where $\boldsymbol{K}_{ij} = U(\| (x_i, y_i) - (x_j, y_j) \|)$, the *i*th row of $\boldsymbol{P}$ is $(1, x_i, y_i)$, $\boldsymbol{w}$ and z are column vectors formed from $\boldsymbol{w}_i$ and z_i respectively, and $\boldsymbol{a}$ is the column vector with elements a_1, a_x, a_y .

If there are errors in the matching results, regularization is used to relax the exact interpolation requirement, which is achieved by minimizing the bending energy as follows:

$$H[f] = \sum_{i=1}^{m} [z_i - f(x_i, y_i)]^2 + \lambda E_f \ . \tag{9}$$

The regularization parameter λ , a positive coefficient, controls the amount of smoothing. The TPS coefficients in the regularized form can be solved by substituting the matrix $\boldsymbol{K} + \lambda \boldsymbol{E}$ for $\boldsymbol{K}$, where $\boldsymbol{E}$ is the $m \times m$ identity matrix [10], [11]. We set $\lambda = 1$ through this paper.

4 Spectral Correspondence Using the TPS Deformation Model

In this section, we detail the steps of our algorithm.

1) Set iteration number to one.
2) Construct proximity matrixes $\boldsymbol{H}^I, \boldsymbol{H}^J$ on point-sets $\boldsymbol{I}$ and $\boldsymbol{J}$ respectively.
3) Perform singular value decomposition on $\boldsymbol{H}^I, \boldsymbol{H}^J$ respectively, and get the modes $\boldsymbol{U}, \boldsymbol{V}$ of point-sets $\boldsymbol{I}$ and $\boldsymbol{J}$.
4) Use Eq.(5) to initialize the matching probability matrix $\boldsymbol{P}$, and convert it to a doubly stochastic matrix via alternated row and column normalizations.
5) According to the magnitude of p_{ij} in its corresponding row and column, decide the correspondences between two point-sets.
6) Use Eq.(8) to estimate the TPS deformation model between the correspondences obtained in step 5.
7) Transform point-set $\boldsymbol{I}$ to $\boldsymbol{I}^D$ using the transformation parameters acquired in step 6.
8) Set the transformed point-set $\boldsymbol{I}^D$ as $\boldsymbol{I}$.
9) Increase the iteration number by one. If the iteration number is less than $N_{\max}$ ($N_{\max} = 10$), go to step 2.

In summary, our spectral method jointly solves the correspondence and the geometric transformation between two matched point-sets. The TPS deformation model is used to estimate the transformation parameters between the current correspondences obtained from spectral analysis. The transformation parameters are applied to bring one point-set closer to the other. The closer two matched point-sets are, the higher accuracy of correspondences can be acquired by means of spectral analysis.

The computational complexity of our algorithm depends heavily on the implementation of SVD and TPS deformation model. Suppose both point-sets contain m points, SVD takes $O(m^3)$ time. And the computational cost of TPS deformation model is in the order of $O(m^3)$. Therefore, our algorithm can be computed in $O(m^3)$ per iteration. With our un-optimized Matlab implementation, matching two point-sets (each with 100 points) takes about 5.1 seconds on a P4 2.4GHZ 512M RAM PC.

5 Performance Analysis

Firstly, we compare the results using Eq.(4) and our method for computing correspondence probabilities. We use only the SVD method of Shapiro and Brady to find correspondences. Here we investigate the effect of positional jitter. The experiments are conducted with random point-sets. We subject the positions of points to Gaussian measurement error with zero mean and controlled standard deviation. The positional jitter is generated by displacing the points from their original positions by Gaussian measurement errors. The parameter of the noise process is the standard deviation of the point position error. The standard deviation is recorded as a fraction of the average closest point distance. Fig. 1 shows the fraction of correct correspondences as a function of the standard deviation of the added Gaussian position errors. We can see that our probabilistic method gives slightly better results, which demonstrate the effectiveness of our approach.

Next, we turn our attention to the performance of several different spectral methods. The first of these is the Shapiro-Brady method. The second is the EM method of Carcassoni and Hancock [3]. The last is our spectral method using the TPS deformation model. Fig. 2 shows the effect of positional jitter on the fraction of correct correspondences. As the noise standard deviation increases, all the three methods degrade, whereas our iterative approach offers significant improvement over the other two spectral methods. And our method outperforms the Carcassoni method by about 6%.

Finally we investigate the effect of controlled affine skew of the point-sets. Fig. 3 shows the fraction of correct correspondences as a function of the skew angle in degree. It is worth noting that affine skew has little influence on our approach. However, once the skew angle is greater than 40 degrees, the performance of the other two spectral methods degrades rapidly.

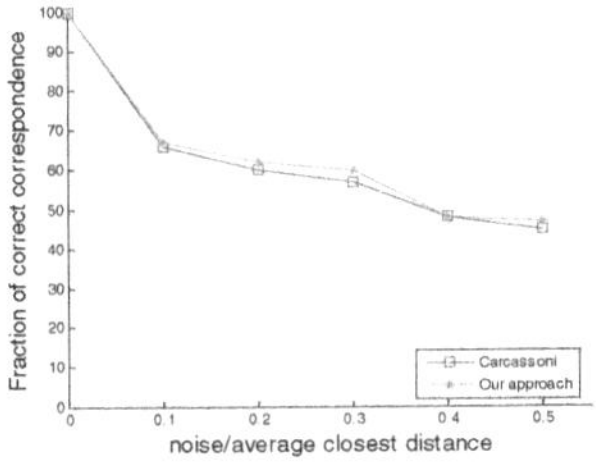

Fig. 1. Effect of matching probabilities on correspondence error

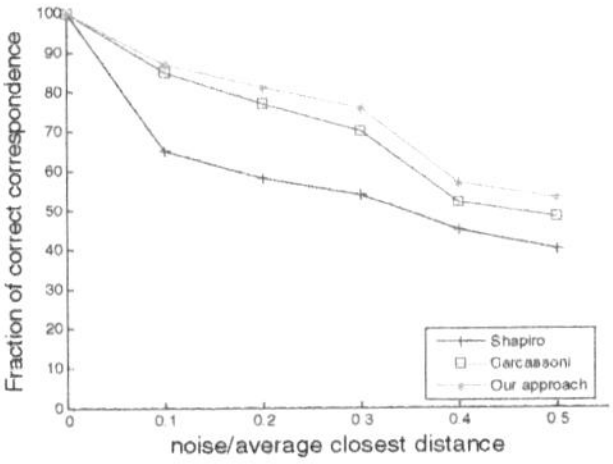

Fig. 2. Effect of positional jitter on correspondence error

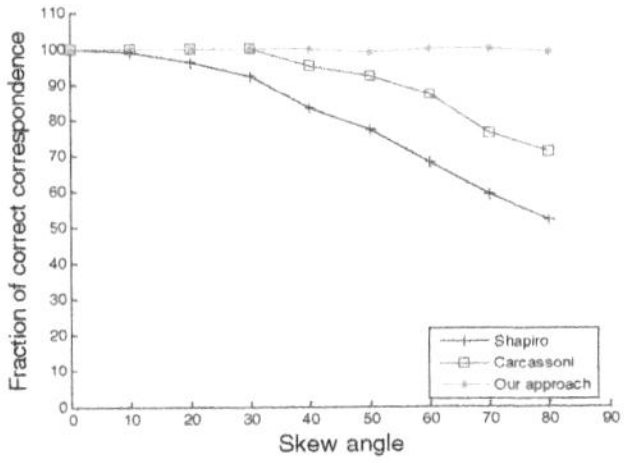

Fig. 3. Effect of skew angle on correspondence error

6 Experiments

Firstly, we test our algorithm on six successive frames of the CMU/VASC house sequence, and compare our results with those of the Shapiro-Brady method and the Carcassoni algorithm [3]. 30 corner points are detected in each image respectively. The images used in our experiments correspond to different viewing directions. And the first frame is tested against the remaining five. Fig. 4 shows the correspondences when we match the first frame to the fourth and the sixth frame respectively. The experimental results are summarized in Table 1.

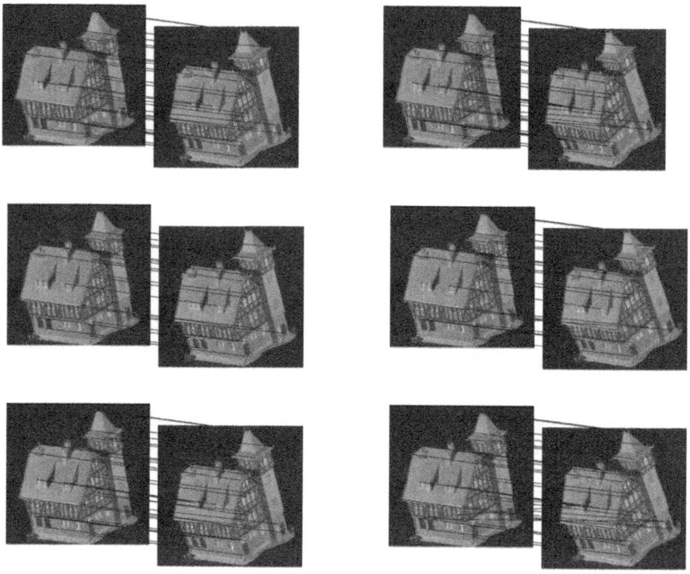

Fig. 4. Point matching results on the CMU/VASC house sequence. Top row: our approach. Middle row: the Shapiro-Brady method. Bottom row: the Carcassoni method.

Table 1. Summary of the experimental results on the CMU/VASC house sequence

Image	1	2	3	4	5	6
Correctly matched(Our approach)	-	30	30	28	30	30
%matched(Our approach)	-	100%	100%	93%	100%	100%
Correctly matched(Shapiro-Brady)	-	24	27	20	20	22
%matched(Shapiro-Brady)	-	80%	90%	67%	67%	73%
Correctly matched(Carcassoni)	-	29	28	27	29	26
%matched(Carcassoni)	-	97%	93%	90%	97%	87%

From these results, it is clear that our method performs best in all the experiments. Although the Carcassoni method offers significant improvement over the Shapiro-Brady method, we can observe that there are many-to-one matches. On the contrary, our method keeps one-to-one match neatly. The main conclusion to note from these experiments is that our method not only improves the correspondence accuracy but also has the advantage of achieving rigorous one-to-one match under rigid transformation.

Secondly, we test our algorithm on several synthetic data (available in http://www.ece.umd.edu/~zhengyf/PointMatching.htm) made by Dr. Haili Chui and Prof. Anand Rangarajan. Here we aim to measure the performance of our algorithm under non-rigid deformation. The chosen data-sets are save_chinese_def_1_1.mat, save_chinese_def_3_1.mat and save_chinese_def_5_1.mat. The final matching results are illustrated in Fig. 5 and the numbers of correct matches are summarized in Table 2.

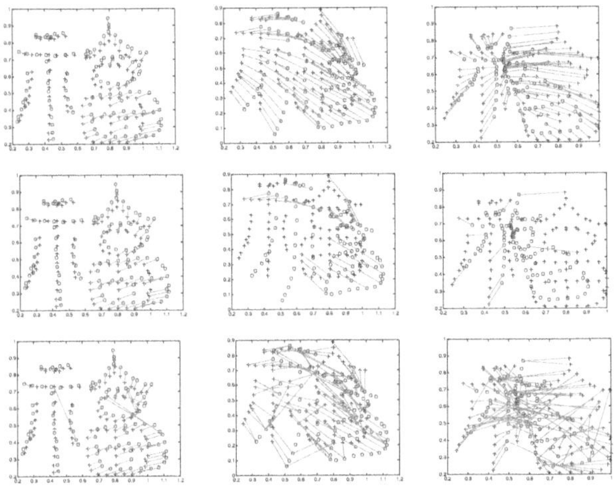

Fig. 5. Point matching results on the Chui-Rangarajan's synthetic data. Top row: our approach. Middle row: the Shapiro-Brady method. Bottom row: the Carcassoni method.

The Shapiro-Brady method fails abruptly when the non-rigid deformation becomes larger. As the non-rigid deformation increases, the performance of the Carcassoni method degrades. Especially in the last data-set, the Carcassoni method breaks down. Meanwhile, in the case of the Carcassoni method, many-to-one matches have occurred in all the data-sets. However, our spectral method, using the TPS deformation model, possesses high accuracy in the overall experiments. Moreover, there is no many-to-one match in the case of our approach. The experimental results verify that our approach outperforms the two alternatives considerably under non-rigid deformation.

Table 2. Summary of the experimental results on the Chui-Rangarajan's synthetic data

Data-set	1_1.mat	3_1.mat	5_1.mat
Correctly matched(Our approach)	105	105	104
%matched(Our approach)	100%	100%	99%
Correctly matched(Shapiro-Brady)	78	25	9
%matched(Shapiro-Brady)	74%	24%	9%
Correctly matched(Carcassoni)	97	47	11
%matched(Carcassoni)	92%	45%	10%

Fig. 6 shows the fraction of correct correspondences as a function of iteration number for the experiment on the synthetic data save_chinese_def_5_1.mat. Our method only takes about 6 iterations to converge. Meanwhile, there is significant improvement in each iteration.

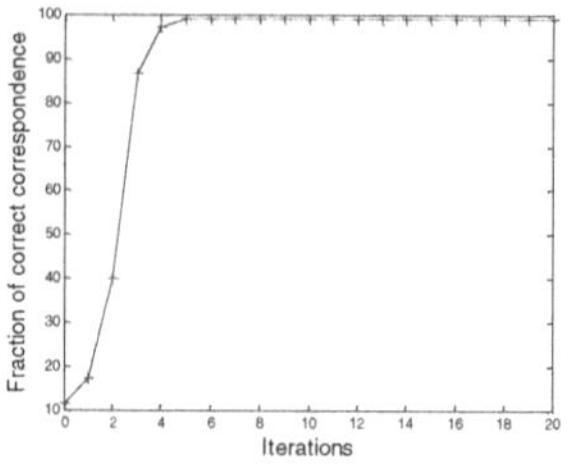

Fig. 6. Convergence rate

7 Conclusion

Our main contributions in this paper are twofold: Firstly, we describe a new approach to computing the correspondence probabilities using the modes of proximity matrix and the method of doubly stochastic matrix. Secondly, we introduce the TPS deformation model into the field of spectral correspondence, which makes the spectral method more robust. Finally, our theoretical results are supported by experiments on both real-world and synthetic data.

Acknowledgements. The authors gratefully acknowledge the financial support of National Science Foundation of China (Grant No.10601001), Anhui Provincial Nature Science Foundation (Grant No.05046012), Natural Science Foundation of Anhui Provincial Education Department (Grant No.2006KJ030B), Foundation for University Young Teachers of Anhui Provincial Education Department (Grant No.2006jq1034) and Innovative Research Team of 211 Project in Anhui University.

References

1. Scott, G.L., Longuet-Higgins, H. C.: An Algorithm for Associating the Features of Two Images. Proc. Roy. Soc. London Ser. B **244** (1991) 21-26
2. Shapiro, L.S., Brady, J.M.: Feature-based Correspondence: an Eigenvector Approach. Image and Vision Computing **10** (5) (1992) 283-288
3. Carcassoni, M., Hancock, E.R.: Spectral Correspondence for Point Pattern Matching. Pattern Recognition **36** (11) (2003) 193-204
4. Carcassoni, M., Hancock, E.R.: Correspondence Matching with Modal Clusters. IEEE Pattern Analysis and Machine Intelligence **25** (12) (2003) 1609-1615
5. Sinkhorn, R.: A Relationship between Arbitrary Positive Matrices and Doubly Stochastic Matrices. The Annals of Mathematical Statistics **35** (2) (1964) 876-879
6. Bookstein, F.L.: Principal Warps: Thin-plate Splines and the Decomposition of Deformation. IEEE Pattern Analysis and Machine Intelligence **11** (6) (1989) 567-585

7. Zheng, Y., Doermann, D.: Robust Point Matching for Nonrigid Shapes by Preserving Local Neighborhood Structures. IEEE Pattern Analysis and Machine Intelligence **28** (4) (2006) 643-649
8. Chui, H., Rangarajan, A.: A New Point Matching Algorithm for Non-rigid Registration. Computer Vision and Image Understanding **89** (2) (2003) 114-141
9. Belongie, S., Malik, J., Puzicha, J.: Shape Matching and Object Recognition Using Shape Contexts. IEEE Pattern Analysis and Machine Intelligence **24** (4) (2002) 509-522
10. Wabha, G.: Spline Models for Observational Data. Soc. Industrial and Applied Math. (1990)
11. Girosi, F., Jones, M., Poggio, T.: Regularization Theory and Neural Networks Architectures. Neural Computation **2** (7) (1995) 219-269

Dynamic Behavioral Models for Wideband Wireless Transmitters Stimulated by Complex Signals Using Neural Networks

Taijun Liu[1], Yan Ye[1], Slim Boumaiza[2], and Fadhel M. Ghannouchi[2]

[1] College of Information Science and Engineering, Ningbo University
Ningbo, Zhejiang 315211, China
{liutaijun, yeyan}@nbu.edu.cn
[2] Department of Electrical and Computer Engineering, University of Calgary
Calgary, Alberta, T2N 1N4, Canada
{sboumaiz, fghannou}@ucalgary.ca
http://www.iradio.ucalgary.ca/

Abstract. In this paper, a time-delay structure is included in the neural network architecture to emulate the memory effects of wideband wireless transmitters. A simplified analysis approach is proposed to illustrate that the Real-Valued Time-Delay Neural Network (RVTDNN) is one of the most promising neural networks for modeling a complex dynamic nonlinear system. Then the RVTDNN is utilized to build the complex signal dynamic behavioral model of a wideband transmitter. Finally, a behavioral model with three-layer RVTDNN is employed in an experimental system to demonstrate the effectiveness of RVTDNNs in mimicking the dynamic behaviors of a wideband wireless transmitter.

1 Introduction

In modern wireless communication systems, the base-band signals transmitted in a wideband wireless transmitter are complex signals. A complex signal behavioral model of a wideband transmitter, which is extracted from a practical transmitter with the modulated signals, is playing an important role for simulating the communication systems, correcting the system transmission errors and suppressing the out-of-band emission caused by the system nonlinearities. For the wideband signals such as Multi-Carrier CDMA and OFDM, the transmitter will exhibit strong memory effects so that the output of the transmitter will not only depend on the current input signal but also the previous signals. Therefore, the complex signal behavioral model should have memory mechanism to reflect the dynamic characteristics of the wideband wireless transmitter. Moreover, in order to improve the power efficiency of the transmitter, the power amplifier in the transmitter last stage will work near to its saturated area. Then the transfer function of the transmitter will be strong nonlinear. Consequently, modeling the wideband transmitter becomes a dynamic nonlinear system modeling problem.

D. Liu et al. (Eds.): ISNN 2007, Part I, LNCS 4491, pp. 582–591, 2007.

Recently, there are many researchers who are applying the neural networks to model the dynamic behaviors of high power amplifiers. For instance, a radial-basis function neural network (RBFNN) in [1] has been used for characterizing the dynamic nonlinear property of an RF power amplifier for third generation (3G) wireless communication systems. A fully recurrent neural network (RNN) with Gamma tapped-delay lines was presented in [2] for predicting spectral regrowth of 3G handset power amplifiers with memory effects. Moreover, a Real-Valued Time-Delay Neural Network (RVTDNN) based model was proposed in [3] to build dynamic behavioral models of 3G power amplifiers.

In this paper, some mathematical analysis will be made to the three kinds of simplified neural network models for processing complex signals so as to demonstrate the advantages of RVTDNN. Then a three-layer RVTDNN will be selected to model a practical wideband transmitter stimulated by a two-carrier 3GPP WCDMA FDD signal. Finally, the extracted RVTDNN model will be validated experimentally to illustrate the effectiveness of this neural network model in simulating the dynamic behaviors of the wideband wireless transmitters.

2 TDNN Models for Complex Signal Processing

In time domain, the memory effects of a transmitter will result in that the equivalent base-band output of the transmitter will not only depend on the current base-band input of the transmitter, but also the previous inputs of the transmitter. Then the problem converts from a one-dimension static problem to a multi-dimension dynamic problem. To physically implement this conversion, a tapped delay line (TDL) is often used. Therefore, the TDL is combined into the neural network to form a time-delay neural network (TDNN) so as to make a general neural network to be able to model the memory effects of the transmitter.

There are different TDNN structures that have been implemented to build the complex signal models of power amplifiers or transmitters. The first conventional neural network approach is that two TDNNs are used to model the transfer characteristics of in-phase (I, real part) and quadrature (Q, imaginary part) components of the complex signal separately. Another conventional neural network approach is that the transfer characteristics of the complex signals in the device concerned are modeled by a complex-valued TDNN (CVTDNN). With this complex value based neural network model, a cumbersome complex training algorithm such as the complex backpropagation training algorithm [4] has to be employed to extract the complex model parameters (complex weights). Therefore, a real-valued TDNN was proposed and implemented successfully to construct a complex model of a 3G power amplifier [3]. In fact, this real value based TDNN is fully compatible with the complex-valued TDNN. For the convenience of comparing the differences among these structures, one-layer neural networks with the pure linear activation functions shown in Fig.1 are selected here as examples to illustrate the benefits of the RVTDNN model.

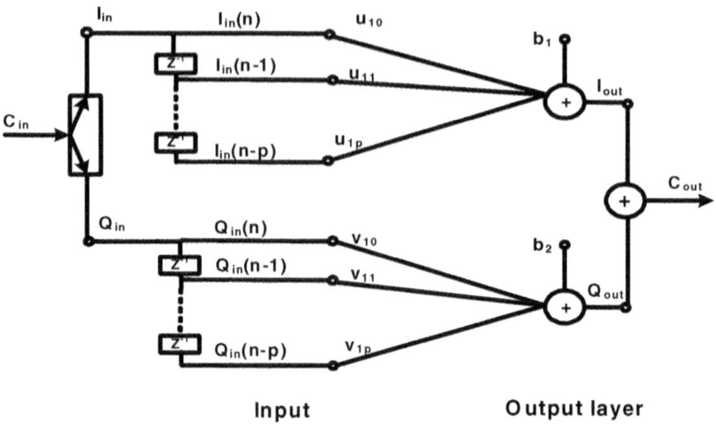

(a) The conventional two-TDNN model

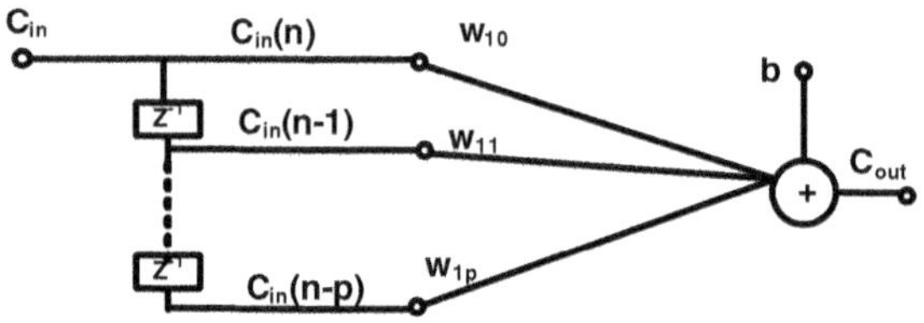

(b) The conventional complex-valued TDNN model

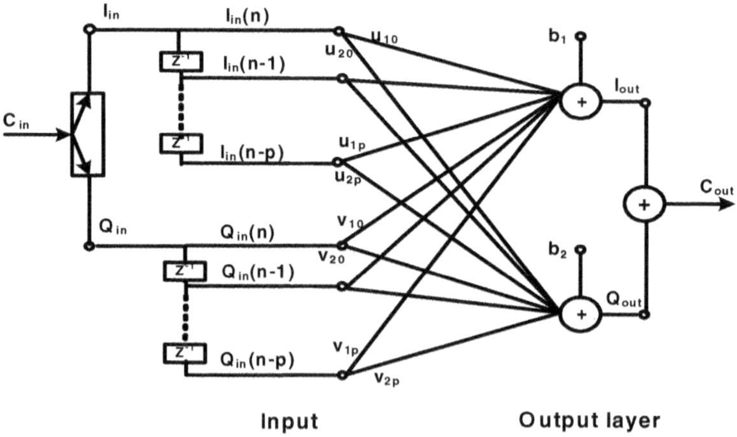

(c) The real-valued TDNN model

Fig. 1. TDNN models for complex signal processing

Since there are memory effects for the wideband transmitters, the base-band complex signal output C_{out} at instant n are a function of past values of the base-band complex signal input C_{in} as follows:

$$C_{out}(n) = f\left[C_{in}(n), C_{in}(n-1), ..., C_{in}(n-p)\right] \tag{1}$$

Suppose

$$C_{in}(n) = I_{in}(n) + jQ_{in}(n) \tag{2}$$

$$C_{out}(n) = I_{out}(n) + jQ_{out}(n) \tag{3}$$

where $I_{in}(n)$, $Q_{in}(n)$, $I_{out}(n)$, $Q_{out}(n)$ are in-phase and quadrature components of the inputs and the outputs.

For the conventional two-TDNN model as shown in Fig.1 (a), we have

$$I_{out}(n) = \sum_{i=0}^{p} u_{1i} I_{in}(n-i) + b_1 \tag{4}$$

$$Q_{out}(n) = \sum_{i=0}^{p} v_{2i} Q_{in}(n-i) + b_2 \tag{5}$$

For the complex-valued TDNN model as shown in Fig.1 (b), the output of the network is as follows:

$$C_{out}(n) = \sum_{i=0}^{p} w_i C_{in}(n-i) + b \tag{6}$$

Suppose $w_i = u_i + jv_i$ and $b = b_1 + jb_2$, then substitute equation (2) and (3) into equation (6) we can obtain the quadrature expression of the network output as:

$$I_{out}(n) = \sum_{i=0}^{p} u_i I_{in}(n-i) + \sum_{i=0}^{p} (-v_i) Q_{in}(n-i) + b_1 \tag{7}$$

$$Q_{out}(n) = \sum_{i=0}^{p} v_i I_{in}(n-i) + \sum_{i=0}^{p} u_i Q_{in}(n-i) + b_2 \tag{8}$$

For the real-valued TDNN model as shown in Fig.1 (c), the quadrature output expression is as follows:

$$I_{out}(n) = \sum_{i=0}^{p} u_{1i} I_{in}(n-i) + \sum_{i=0}^{p} v_{1i} Q_{in}(n-i) + b_1 \tag{9}$$

$$Q_{out}(n) = \sum_{i=0}^{p} u_{2i} I_{in}(n-i) + \sum_{i=0}^{p} v_{2i} Q_{in}(n-i) + b_2 \tag{10}$$

Let $v_{1i} = -u_{2i}$ and $v_{2i} = u_{1i}$, then Eq.(9) and (10) becomes

$$I_{out}(n) = \sum_{i=0}^{p} u_{1i} I_{in}(n-i) + \sum_{i=0}^{p} v_{1i} Q_{in}(n-i) + b_1 \tag{11}$$

$$Q_{out}(n) = \sum_{i=0}^{p} (-v_{1i}) I_{in}(n-i) + \sum_{i=0}^{p} u_{1i} Q_{in}(n-i) + b_2 \tag{12}$$

Comparing equation (7) and (8) with equation (11) and (12), it is clear that equation (7) and (8) are only a special case of equation (9) and (10). Equation (9) and (10) are general expressions of the TDNN model for complex signal processing. This also means that the real-valued TDNN model has more freedom than the complex-valued TDNN, which is only a special case of the RVTDNN model. Moreover, comparing equation (4) and (5) with equation (9) and (10), we can easily find that equation (4) and (5) are also a special case of equation (9) and (10). This implies that the two-TDNN model is a special case of the RVTDNN model too. In fact, there are quadrature coupling components in the expressions of both RVTDNN model and CVTDNN model except for those of the two-TDNN model. This characteristic suggests that both RVTDNN and CVTDNN model are more suitable for solving the complex signal modeling problems than the two-TDNN model. Since RVTDNN model is a fully real-valued neural network, many mature training algorithms can be applied to extract the model parameters. Consequently, RVTDNN model is more powerful and suitable for processing complex signal than other two kinds of TDNN model.

3 RVTDNN Behavior Models of Wideband Wireless Transmitters

As shown in Fig.2, a three-layer RVTDNN is implemented to build a base-band behavioral model of a wideband transmitter in a real operation condition. The complex input signal C_{in} is divided into two branches with its quadrature components: I_{in} and Q_{in}. The complex output C_{out} is composed of the quadrature output components I_{out} and Q_{out}. The two TDLs are used to account for the memory effects of the transmitter. The memory depth is determined by the length of the TDL taps.

Based on the RVTDNN shown in Fig. 2, the baseband output components I_{out} and Q_{out} can be written as follow:

$$I_{out}(n) = \sum_{k=1}^{N} w_{1k}^{3} f_2(net_k^2(n)) + b_1^3 \tag{13}$$

$$Q_{out}(n) = \sum_{k=1}^{N} w_{2k}^{3} f_2(net_k^2(n)) + b_2^3 \tag{14}$$

where

$$net_k^2(n) = \sum_{j=1}^{M} w_{kj}^2 f_1(net_j^1(n)) + b_k^2 \tag{15}$$

$$net_j^1(n) = \sum_{i=0}^{p} u_{ji}^1 I_{in}(n-i) + \sum_{i=0}^{p} v_{ji}^1 Q_{in}(n-i) + b_j^1 \tag{16}$$

The activation functions are:

$$f_1(x) = f_2(x) = \tanh(x) = \frac{(1-e^{-2x})}{(1+e^{-2x})} \tag{17}$$

Where p stands for the memory depth of the transmitter and is determined through an optimization process. In other word, mean-squared-errors (MSEs) of the RVTDNN for the validation sequence are calculated and compared for different values of p while keeping other parameters of the RVTDNN fixed. The value of p that produces a minimum MSE is taken as the optimal memory depth. When p=0, the RVTDNN is degenerated to a real-valued multi-Layer perceptron (RVMLP), and consequently changed to a memoryless model. Moreover, the number of layers of the RVTDNN and the neuron numbers in the two hidden layers are also optimized with the similar method.

4 Experimental Validation with 3GPP-WCDMA FDD Signals

The prototype of the wideband wireless transmitter used in the experimental validation process consists of an Electronic Signal Generator (ESG: Agilent E4438C) and a LDMOS PA from Freescale Semiconductor Inc. The PA has 51dB gain and 49dBm saturated power, and is suitable for 3G wireless base station transmitters operating at the band of 2110–2170 MHz. The test signal has two neighboring 3GPP WCDMA FDD carriers (carrier spacing 5MHz), each carrier transmitting a signal according to 3GPP test model 3, 32 code channels [5]. This test signal will be synthesized in Agilent Advanced Design System (ADS) with the corresponding ADS libraries and is saved as two txt files corresponding to the quadrature I and Q components of the test signal in PC. These data will be taken as the base-band input I and Q components of the transmitter during the model extraction procedure. After that, the generated test signal is up-loaded to the ESG where the signal is modulated onto an RF carrier. Finally, the RF output of the ESG is fed to the PA. The equivalent base-band output I and Q components of the transmitter at the output of the PA can be obtained by a base-band data acquisition system, which contains a spectrum analyzer (PSA, Agilent E4446A) and a Vector Signal Analyzer (VSA, Agilent 89611A), and saved in PC with the VSA data acquisition software.

A three-layer RVTDNN, as shown in Fig.2, contains two neurons with a pure linear activation function in the output layer. The RVTDNN model, having six neurons in the two hidden layers and three taps in the two input TDLs, is found to be appropriate for the LDMOS PA driven by 3GPP signals through the optimization procedure as mentioned in Section 3. Trading off the training speed and the model

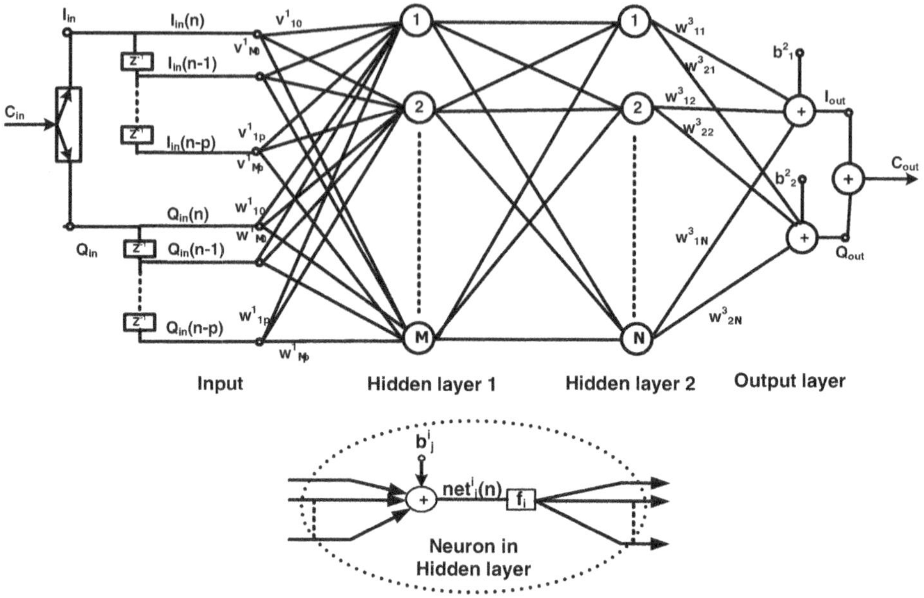

Fig. 2. A three-layer RVTDNN behavioral model

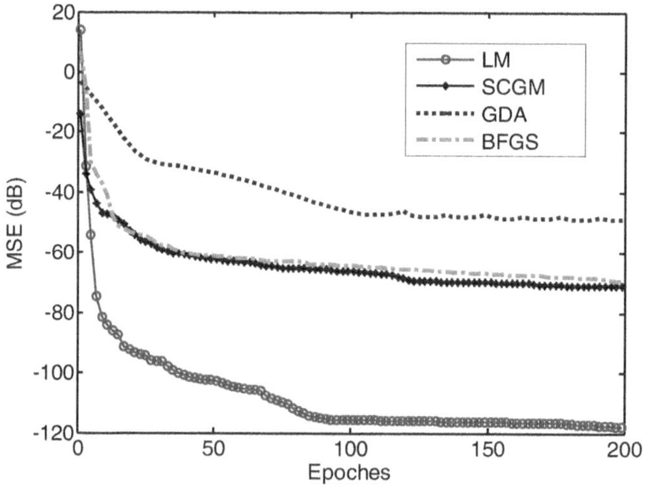

Fig. 3. Convergence curves with different training algorithms

accuracy, we select 2560 data as the training sequence from the recorded baseband data. The Scaled Conjugate Gradient Method (SCGM) [6], Gradient Descent with daptive learning rate backpropagation (GDA), BFGS quasi-Newton backpropagation [7] and Levenberg-Marquardt (LM) algorithm [8] are selected to train the RVDNN to obtain the model parameters. The convergence curves for these algorithms are shown

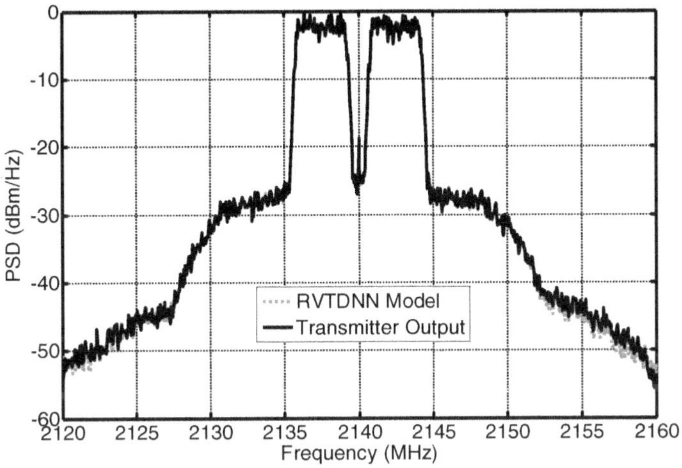

(a) Original size spectrum

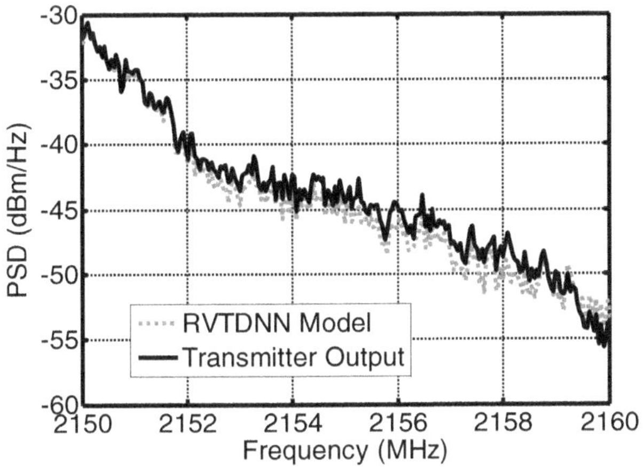

(b) Spectrum zoomed-in

Fig. 4. Spectrum comparison of the test results

in Fig.3. One can clearly observe that LM algorithm permits the fastest convergence and lower Mean-Square-Error (MSE). Therefore, the LM algorithm is selected in this work for training the RVTDNN. The training program is developed in Matlab (Mathworks Inc.).

After training RVTDNN, its weight and bias values are determined and the RVTDNN becomes the transmitter behavioral model for the 3GPP signal. Then this RVTDNN model is implemented in ADS and the synthesized two-carrier 3GPP signal s applied to this model. The output of the model will be up-loaded to the ESG and the spectrum of the RF output of the ESG will be compared with the spectrum of the transmitter prototype to validate the accuracy of the RVTDNN behavioral model.

The power spectral density (PSD) comparison between the RVTDNN behavioral model and the practical transmitter is shown in Fig.4. It can be observed that the RVTDNN model can precisely predict the output spectrum of the transmitter. This demonstrates that the RVTDNN behavioral model can effectively simulate the dynamic behaviors of a wideband wireless transmitter.

5 Conclusion

In this paper, a time-delay structure is included in the neural network architectures to emulate the memory effects of wideband wireless transmitters. Through the mathematical analysis for the three kinds of simplified TDNN models for processing complex signals, it can be concluded that the RVTDNN is the most suitable model architecture among them for modeling a complex dynamic nonlinear system such as a wideband wireless transmitter. A three-layer RVTDNN is trained with a Matlab program and implemented in ADS to demonstrate the accuracy of the model for mimicking the baseband behaviors of a wideband transmitter prototype, which makes up of Agilent ESG signal generator and an 90-watt LDMOS power amplifier, with a two-carrier 3GPP WCDMA FDD signal. The validation results illustrate that the RVTDNN model can precisely predict the dynamic behaviors of the wideband wireless transmitter.

Acknowledgement

Authors would like to thank Alberta's Informatics Circle of Research Excellence (iCORE), Natural Sciences and Engineering Research Council of Canada (NSERC), Communications Research Centre Canada (CRC), TRLabs, and National Natural Science Foundation of China (NSFC, 60671037) for their financial support. The authors also want to acknowledge Agilent Technologies for software donation.

References

1. Isaksson, M., Wisell, D., Ronnow, D.: Wide-band Dynamic Modeling of Power Amplifiers Using Radial-basis Function Neural Networks. IEEE Trans. Microwave Theory Tech. **53(11)** (2005) 3422-428
2. Luongvinh, D., Kwon, Y.: A Fully Recurrent Neural Network-based Model for Predicting Spectral Regrowth of 3G Handset Power Amplifiers with Memory Effects. IEEE Microwave and Wireless Components Letters **16(11)** (2006) 621-623
3. Liu, T., Boumaiza, S., Ghannouchi, F.M.: Dynamic Behavioral Modeling of 3G Power Amplifiers Using Real-valued Time-delay Neural Networks. IEEE Trans. Microwave Theory Tech. **52(3)** (2004) 1025-1033
4. Leung, H., Haykin, S.: The Complex Backpropagation Algorithm. IEEE Trans. Signal Processing **39(9)** (1991) 2101-2104
5. 3GPP specifications: TS 25.104 v4.2.0, TS 25.141 v4.2.0, 2002

6. Moller, M.F.: A Scaled Conjugate Gradient Algorithm for Fast Supervise Learning. Neural Networks **6(4)** (1993) 525-533
7. Gill, P.E., Murray, W.: Wright, M.H.: Practical Optimization. Academic Press, New York (1981)
8. Marquardt, D.W.: An Algorithm for Least-squares Estimation of Nonlinear Parameters. Journal of the Society for Industrial and Applied Mathematics **11(2)** (1963) 431-441

An Occupancy Grids Building Method with Sonar Sensors Based on Improved Neural Network Model

Hongshan Yu, Yaonan Wang, and Jinzhu Peng

College of Electrical and Information Engineering, Hunan University,
Changsha Hunan, P.R. China, 410082
yuhongshancn@hotmail.com,yaonan@hnu.cn,pengjinzhu1980@126.com
http://www.springer.com/lncs

Abstract. This paper presents an improved neural network model interpretating sonar readings to build occupancy grids of mobile robot. The proposed model interprets sensor readings in the context of their space neighbors and relevant successive history readings simultaneously. Consequently the presented method can greatly weaken the effects by multiple reflections or specular reflection. The output of the neural network is the probability vector of three possible status(empty, occupancy, uncertainty) for the cell. As for sensor readings integration, three probabilities of cell's status are updated by the Bayesian update formula respectively, and the final status of cell is defined by Max-Min principle.Experiments performed in lab environment has shown occupancy map built by proposed approach is more consistent, accurate and robust than traditional method while it still could be conducted in real time.

1 Introduction

Recent research has produced two fundamental paradigms for mobile robot mapping: metric maps and topological maps. In metric maps, the positions of some objects, mainly the obstacles that the robot can encounter are stored in a common reference frame. On the contrary, topological maps store definitions of places the robot can reach along with information about their relative positions [1-2].

Due to its efficiency, one of the most popular and successful methods is occupancy grid introduced by Moravec and Elfes[3-4]. Occupancy grid divides space into regular grid of cells in 2D representation, and estimates the probability of any cell being occupied based on sensor readings.Occupancy grids could be built based on laser range-finders, stereo vision, and sonar sensors[5]. Laser range-finders have high angular resolution, but they are more expensive than most other spatial sensors. The stereo vision depends on the lighting, smoke, mist, etc, and very sensitive to errors, as the process of the collapsing data from 3D to 2D encourages errors. Sonar sensors are commonly used due to operation simplicity, robustness, and low price. However sonar readings are prone to several measuring errors due to various phenomena (e.g., multiple reflections, wide radiation cone, low angular resolution).

D. Liu et al. (Eds.): ISNN 2007, Part I, LNCS 4491, pp. 592–601, 2007.

Many efforts are directed to process uncertainty information obtained from sonar reading. Moravec and Elfes presented probabilistic Bayesian updating approach using Gaussian noise distribution with large variance to account for the gross errors entailed by multiple reflections[3-4]. However it has lots of undesirable features: Firstly, modeling multiple reflections as Gaussian distributed is not realistic, since typically it gives highly-correlated readings from nearby positions; Further, using Gaussian distribution implies an averaging model.

To correct those problems, Konlige.K presented MURIEL method which analyzes how refined sensor models and independence assumptions for occupancy grid building[6]. Thrun.S used feed-forward neural network to create robust local occupancy grids modeling the space surrounding the robot[7].Following the Thrun.S's idea, Arleo.A introduced the neural network sonar model to variable-resolution mapping[8]. K.M. Chow turned probability distribution function by fuzzy rules formed from the information obtained from the environment at each sonar data scan[9]. Different with popular inverse models, Thrun.S present a forward model to interpret the sonar data for occupancy grids building[10].

As for integrating the sensor reading to build global map, techniques are commonly divided into three classes, probabilistic theory, D-S evidence theory, and fuzzy settheory[11]. The Bayesian method rules the greatest part of the work related to the probabilistic sensor fusion in building OGs. This attraction stems from the recursive and incremental property of the Bayesian updating rule.

The paper is organized as follows. Section 2 describes the proposed map building architecture. Section 3 introduces the improved neural network model for sonar readings interpretation. Different with those presented methods in paper[7-8], the proposed model takes the space distribution and time series of sonar readings into consideration simultaneously. Section 3 discusses the selection of training dataset and training method for the neural network. Experimental results are shown in section 4.Conclusions are reported in section 5.

2 The Proposed Map Building Architecture

Fig.1 shows the map building architecture. While mobile robot explores the environment, it collects and stores the three consecutive sonar reading series with different time and robot global pose. Local map could be constructed based on those sensors reading. Updating process of a given cell (x, y) in local map starts with the Sensors Selector module that chooses three sensors with orientations closest to the orientation of the cell. As a result, nine sensor reading are selected as neural network inputs. The output of neural network (Fig.2) is probability vector $[prob_{emp}(x, y), prob_{occ}(x, y), prob_{unc}(x, y)]$ representing the probabilities for three possible status of cell (x, y) respectively. Though the transform form local to global coordinate reference frame, the cell (x, y) is projected to the global cell (i, j).Then according to equation (1), the probability for three status of cell (i, j) is updated by the Bayesian update formula respectively .

$$
\begin{aligned}
prob\left(occ_{i,j}|s\right) &= 1 - \left(1 + \frac{prob(occ_{i,j})}{1-prob(occ_{i,j})} \prod_{t=1}^{T} \frac{prob(occ_{i,j})|s^t}{1-prob(occ_{i,j}|s^t)} \frac{1-prob(occ_{i,j})}{prob(occ_{i,j})}\right)^{-1} \\
prob\left(emp_{i,j}|s\right) &= 1 - \left(1 + \frac{prob(emp_{i,j})}{1-prob(emp_{i,j})} \prod_{t=1}^{T} \frac{prob(emp_{i,j})|s^t}{1-prob(emp_{i,j}|s^t)} \frac{1-prob(emp_{i,j})}{prob(emp_{i,j})}\right)^{-1} \\
prob\left(unc_{i,j}|s\right) &= 1 - \left(1 + \frac{prob(unc_{i,j})}{1-prob(unc_{i,j})} \prod_{t=1}^{T} \frac{prob(unc_{i,j})|s^t}{1-prob(unc_{i,j}|s^t)} \frac{1-prob(unc_{i,j})}{prob(unc_{i,j})}\right)^{-1}
\end{aligned} \tag{1}
$$

where $s = s^{(1)}, \cdots, s^{(T)}$ is sensor readings for time span T, $prob(occ_{i,j})$, $prob(emp_{i,j})$, $prob(unc_{i,j})$ denotes the prior probability for three possible status of cell(i, j).

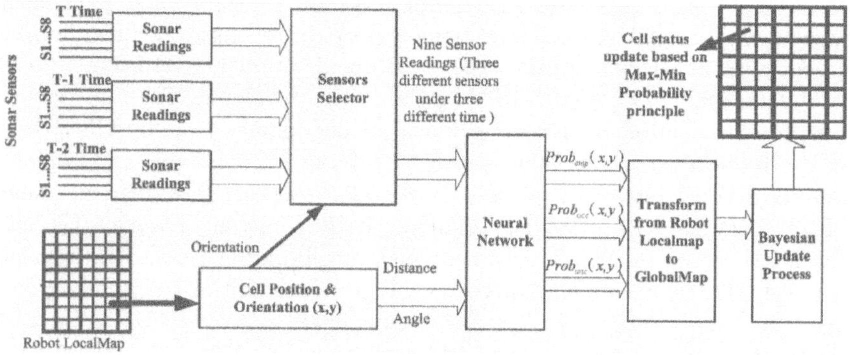

Fig. 1. The map building architecture

Finally, the final status of cell(i, j) is defined by Max-Min principle as equation(2):

$$
S(c_{i,j}) = \begin{cases} occupied \quad ifmax(prob(occ_{i,j}), prob(emp_{i,j}), prob(unc_{i,j}))isprob(occ_{i,j}) \\ empty \quad ifmax(prob(occ_{i,j}), prob(emp_{i,j}), prob(unc_{i,j}))isprob(emp_{i,j}) \\ uncertain \; ifmaxprob(occ_{i,j}), prob(emp_{i,j}), prob(unc_{i,j})isprob(unc_{i,j}) \end{cases} \tag{2}
$$

3 Improved Neural Network Model for Sonar Sensors Interpretation

To build metric maps, sensor reading must be translated into occupancy values for each grid cell. As sonar sensor has many problems, such as multiple reflections, specular reflection, wide radiation cone, low angular resolution and so on. Consequently defining a high accuracy mathematical model to interpret sonar reading is impossible.

As multi-layer neural network can be trained to approximate any distribution, training an artificial neural network to map sonar measurements to occupancy values is feasible[7-8]. It must be noticed that the input for neural network introduced in those paper are current sensor readings. As a result, the neural network can solve the interpretation very well if the current reading is valid. However in

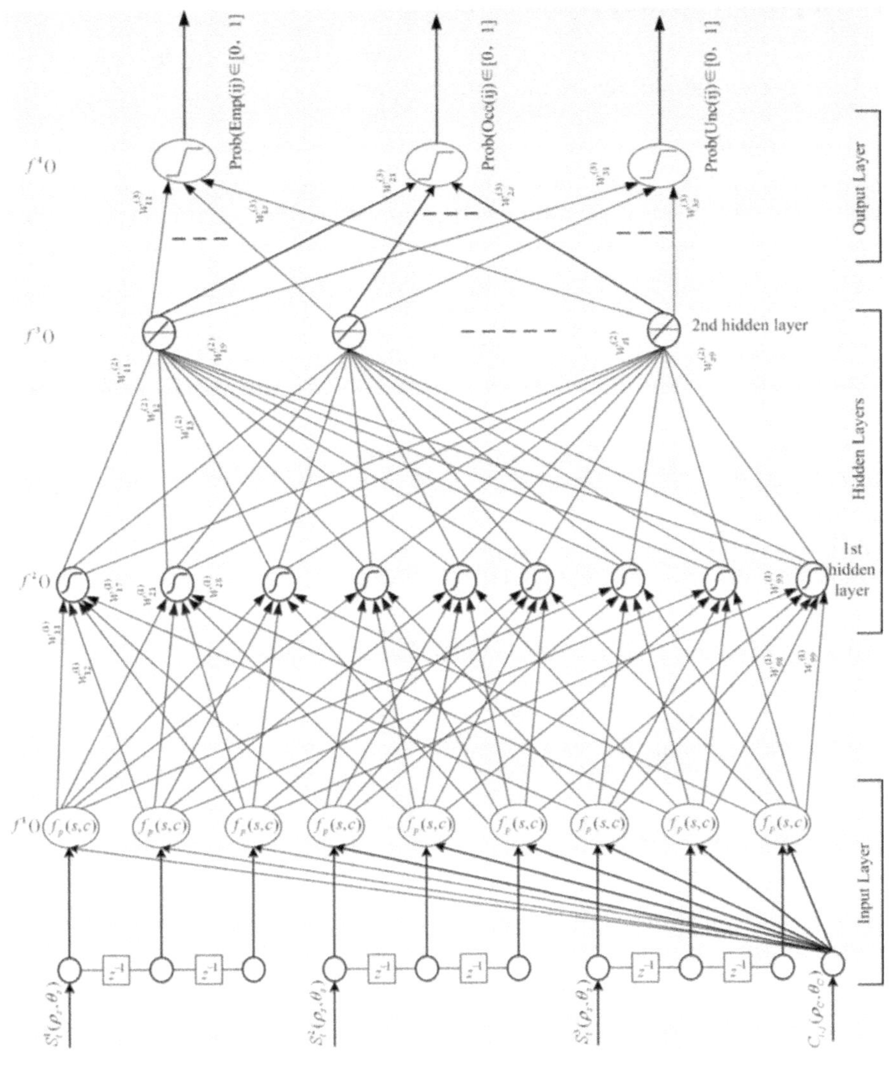

Fig. 2. The structure of improved neural network for sonar sensors interpretation

office-like environments, a lot of sonar reading are suffered from multiple reflections or specular reflection. To overcome those problems, this paper proposed an improved neural network model which takes the space relevant and time series of sonar readings into consideration simultaneously. So if current readings are produced by multiple reflections or specular reflection, this neural network model could depend on relevant history reading to obtain the correct occupancy values. Consequently the presented method in this paper can greatly weaken the effect by multiple reflections and specular reflection.

3.1 Improved Neural Network Design

As shown in Fig.2, the proposed neural network has four layers consisting of input layer, two hidden layers, and output layer.

(1) Input layer
As explained in above sections, input vector has ten elements for the neural network. For given cell(i, j), nine occupancy values can be computed based on 9 sensor reading applied the sonar space distribution function$f_p(s, c)$, and the results are provided as input vector of 1st hidden layer.

For a given cell(i, j), the input layer of the neural networks consists of:

1) The observation $S = (S^1, S^2, S^3)$ of the three sensors oriented in the direction of cell(i, j), and every sonar sensor has three different consistent readings $S^i = (S^i_{t-2}, S^i_{t-1}, S^i_t)$. Every sensor reading has the form $S^i_t(\rho_s, \theta_s)$,where ρ_s, θ_sdenotes the distance and angle sensor reading.

2) $C_{i,j}(\rho_m, \theta_m)$, the distance and angle of the center of the cell (i, j) with respect to the mobile robot coordinate system, as illustrated in Fig.5 for pioneer 2-DXE mobile robot used in the experiments.

Then input layer can be expressed as following:

$$a^1 = f^1(n^1) = f^1(W^1 p) \tag{3}$$

where n^1 is the 1st layer input vector,a^1 is output vector, p denotes the neural network input vector$p = [S^1_t, S^1_{t-1}, S^1_{t-2}, S^2_t, S^2_{t-1}, S^2_{t-2}, S^3_t, S^3_{t-1}, S^3_{t-2}, C_{i,j}]^T$. W^1 is known weight matrix of 1st layer input:

$$\forall w_{i,j} \in W^1 (i = 1, \cdots, 9, j = 1, \cdots, 10), w_{i,j} = \begin{cases} 1, \ i = j \\ 1, \ j = 10 \\ 0, \ else \end{cases} \tag{4}$$

In 1st layer, transfer function is selected as sonar space distribution[6,9], which has the following form:

$$f(s,c) = \begin{cases} 1 - exp\left(\frac{-(\theta_m - \theta_s)^2}{2\sigma_\theta^2}\right), & if \ 0 \le \rho_m < a \ and \ exp\left(\frac{-(\theta_m - \theta_s)^2}{2\sigma_\theta^2}\right) > 0.5 \\ 1 - \left(1 - exp\left(\frac{-(\rho_m - \rho_s)^2}{2\sigma_r^2}\right)\right) exp\left(\frac{-(\theta_m - \theta_s)^2}{2\sigma_\theta^2}\right), & if \ a \le \rho_m < b \ and \\ \quad \left(1 - exp\left(\frac{-(\rho_m - \rho_s)^2}{2\sigma_r^2}\right) exp\left(\frac{-(\theta_m - \theta_s)^2}{2\sigma_\theta^2}\right)\right) > 0.5 & \\ exp\left(\frac{-(\rho_m - \rho_s)^2}{2\sigma_r^2}\right) exp\left(\frac{-(\theta_m - \theta_s)^2}{2\sigma_\theta^2}\right), & if \ b \le \rho_m < c \ and \\ \quad exp\left(\frac{-(\rho_m - \rho_s)^2}{2\sigma_r^2}\right) exp\left(\frac{-(\theta_m - \theta_s)^2}{2\sigma_\theta^2}\right) & \\ 0.5, & else \end{cases} \tag{5}$$

Where σ_r is the variance of the sonar measurement, σ_θ corresponds to the variance of the angular probability. According to experiments by trail and error, we got the following selection for pioneer robot,$\sigma_\theta = 12^o$,$\sigma_r = 0.01 + 015\rho_s$,$a = 0.6\rho_s$,$b = \rho_s - \sigma_r$,$c = \rho_s + \rho_r$. A plot of $f(s, c)$ corresponding to sensor measurement of 1m and 2m is shown in Fig.3.

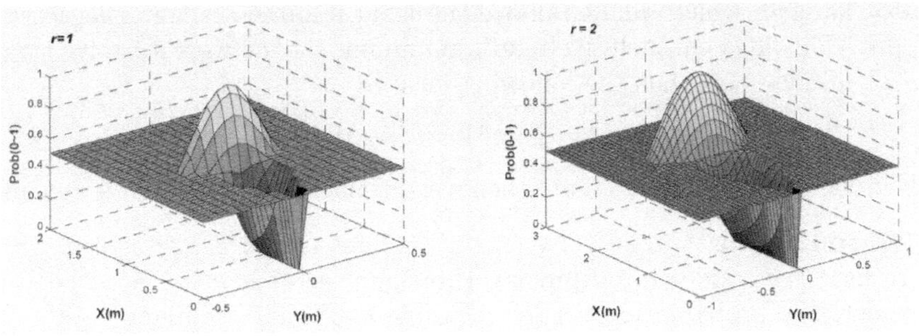

Fig. 3. The occupancy probabilities $f(s, c)$ profile in two-dimensional case

(2)The second layer-1st hidden layer
From 1st layer, hidden layer get nine probability values about cell(i, j) from nine sensor readings as input. Though the relationship analysis of nine sensor readings, it could be found that five of them are most relevant. For any sonar reading, it has close relation with other two sonar reading in the same time, and other two reading of the same sensor in different time. Based on this fact, this paper establishes this special hidden layer. In this layer, output node num is the same with input node num. The layer can be represented in the following form:

$$a^2 = f^2(n^2) = \frac{1}{1+e^{-n^2}} \tag{6}$$

where $n^2 = W^2a^1$ denotes the input of 2^{nd}layer, a^2 is output of 2nd layer.

As shown in Fig.2, weight matrix of 2^{nd}layerW^2 can be defined in equation (7) based on the relationship of sonar measurements

$$W^2 = \begin{bmatrix} w^{(2)}_{1,1} & w^{(2)}_{1,2} & w^{(2)}_{1,3} & w^{(2)}_{1,4} & 0 & 0 & w^{(2)}_{1,7} & 0 & 0 \\ w^{(2)}_{2,1} & w^{(2)}_{2,2} & w^{(2)}_{2,3} & 0 & w^{(2)}_{2,5} & 0 & 0 & w^{(2)}_{2,8} & 0 \\ w^{(2)}_{3,1} & w^{(2)}_{3,2} & w^{(2)}_{3,3} & 0 & 0 & w^{(2)}_{3,6} & 0 & 0 & w^{(2)}_{3,9} \\ w^{(2)}_{4,1} & 0 & 0 & w^{(2)}_{4,4} & w^{(2)}_{4,5} & w^{(2)}_{4,6} & w^{(2)}_{4,7} & 0 & 0 \\ 0 & w^{(2)}_{5,2} & 0 & w^{(2)}_{5,4} & w^{(2)}_{5,5} & w^{(2)}_{5,6} & 0 & w^{(2)}_{5,8} & 0 \\ 0 & 0 & w^{(2)}_{6,3} & w^{(2)}_{6,4} & w^{(2)}_{6,5} & w^{(2)}_{6,6} & 0 & 0 & w^{(2)}_{6,9} \\ w^{(2)}_{7,1} & 0 & 0 & w^{(2)}_{7,4} & 0 & 0 & w^{(2)}_{7,7} & w^{(2)}_{7,8} & w^{(2)}_{7,9} \\ 0 & w^{(2)}_{8,2} & 0 & 0 & w^{(2)}_{8,5} & 0 & w^{(2)}_{8,7} & w^{(2)}_{8,8} & w^{(2)}_{8,9} \\ 0 & 0 & w^{(2)}_{9,3} & 0 & 0 & w^{(2)}_{9,6} & w^{(2)}_{9,7} & w^{(2)}_{9,8} & w^{(2)}_{9,9} \end{bmatrix} \tag{7}$$

where $w^n_{i,j}$ is the weigh of the ith node of n layer to the jth node of $n-1$ layer.

(3) The third layer- 2nd hidden layer
Initially the network has just one hidden layer. As learning process performs, this new layer is added when the current network cannot reduce the error E any further. By modifying the network architecture, the shape of the weight space

is also changed, which might remove the local minimum where the network is trapped. This idea can help to build a minimum-size network to solve the task. The 2nd hidden layer has the following form:

$$a^3 = f^3(n^3) = n^3 = W^3 a^2 \tag{8}$$

If the node num of this layer is n ,then weight matrix W^3 with $n \times 9$ dimensions.

(4) Output layer
There has three nodes in output layer, output vector is $a^4 = [a_1^4, a_2^4, a_3^4]$ which represents the probabilities of three possible status for occupancy grids, where a_1^4 is the probability for empty,a_2^4 is the probability for occupancy,a_1^4 is the probability for empty. According to Fig.2, the output layer has the following form:

$$a^4 = f^4(n^4) = f^4(W^4 a^3) = \begin{cases} 0, & n^4 < 0 \\ n^4, & 0 \le n^4 \le 1 \\ 1, & n^4 > 1 \end{cases} \tag{9}$$

3.2 Neural Network Training

Training dataset are in the form as following equation:

$$< S_t^1, S_{t-1}^1, S_{t-2}^1, S_t^2, S_{t-1}^2, S_{t-2}^2, S_t^3, S_{t-1}^3, S_{t-2}^3, \theta_{i,j}, \rho_{i,j}, o_{i,j} > \tag{10}$$

where $o_{i,j} = [o_{occ}, o_{unc}, o_{emp}]$ is the desired output, its possible value are $[1,0,0]$, $[0,1,0]$, $[0,0,1]$. They are generated by randomly placing and orienting the robot in a known environment (i.e., an environment where the position of obstacles is known). For each position and orientation of the robot, the elements (i,j) of the local grid are randomly sampled. The target output $o_{i,j}$ is the true occupancy state of (i,j) computed by considering the intersection between the cell (i,j) and the known obstacles.

Fig.4a is the part of the environment and the robot within it. Fig.4b is ideal status of map. Fig.4(c-e)is probability distribution according to sonar space model based on three consistent sonar reading in the environment.

It must be noted that speed of the mobile robot should be slow enough during the collection training data, which is to assure three sensor measurement to be consistent and the three area to be overlapped each other.

The neural network is trained off-line by Levenberg-Marquardt method[12]. As for series of training dataset:$\{p_1, t_1\}, \{p_2, t_2\}, \cdots, \{p_q, t_q\}$, p_q is input vector, t_q is corresponding desired output. The input dataset corresponds an actual output dataset by neural network processing, that is $\{p_1, o_1\}, \{p_2, o_2\}, \cdots, \{p_q, o_q\}$, o_q is actual network output. As for any training data, desired output and actual output is different. Therefore, error is available :

$$e(q) = t_q - o_q = \{t_1 - o_1, t_2 - o_2, t_3 - o_3\}^T \tag{11}$$

In this paper, error function is defined as equation (12).

$$\begin{aligned} E &= \textstyle\sum_{q=1}^{Q}(t_q - o_q)^T = \sum_{q=1}^{Q} e_q^T e_q = \sum_{q=1}^{Q}\sum_{j=1}^{SM}(e_{jq}^2) = \sum_{i=1}^{N} v_i^2 \\ &= \textstyle\sum_{q=1}^{Q}[(t_1^q - o_1^q)^2 + (t_2^q - o_2^q)^2 + (t_3^q - o_1^q)^3] \end{aligned} \tag{12}$$

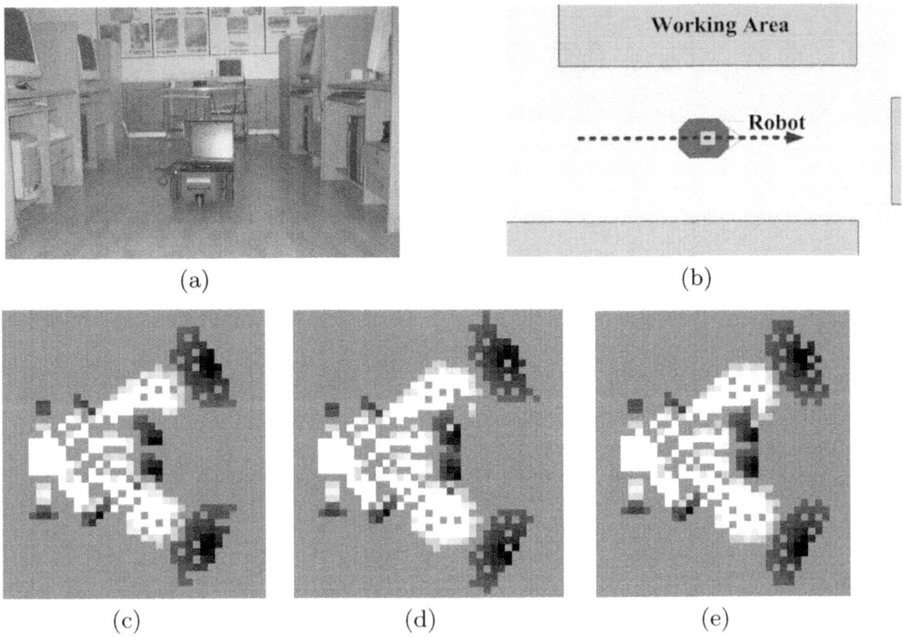

Fig. 4. Parts of enviroment and sonar measurement for training neural network

In order to improve the training speed while keeping good convergence, training process is divided into two phases in this paper. 1) In the first phase, smaller training data set Q_1 is selected. When the neural network converges to accuracy range reaching to $\triangle\delta_1$, the training process is stopped and turns to next step. 2) In the second phase, full-size scale training data set Q_2 is used to train the network until reaching the expected accuracy $\triangle\delta_2$.

4 Experiments Results

In this section, we present experimental results obtained with pioneer 2-DXE mobile robot in lab environment. As shown in Fig.5, this robot is equipped with 8 sonar sensors. Fig.6 shows the image of lab environment, and its occupancy status is depicted in Fig.7. During the mobile robot exploration the environment, safety and avoiding obstacles programme is embedded. As the robot only equipped with the front sonar ring, so we defined the safe and reachable area which is no obstacle in 1m front and 0.5m in both side.

In the experiment, occupancy grid resolution is defined as $10cm * 10cm$.Based on trained neural network, the mobile robot projects the sonar readings into occupancy grids while exploring the environment. The final occupancy grids about lab environment is shown as Fig.8, where the dark area is obstacle, red area is uncertain and white area is empty.

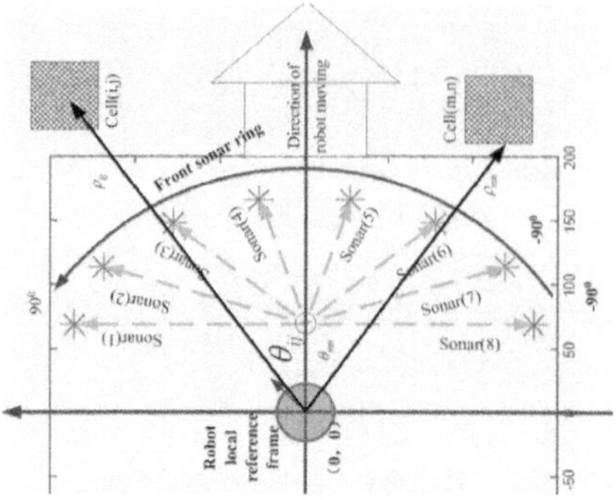

Fig. 5. The sonar ring configuration of Pioneer2-DXE mobile robot

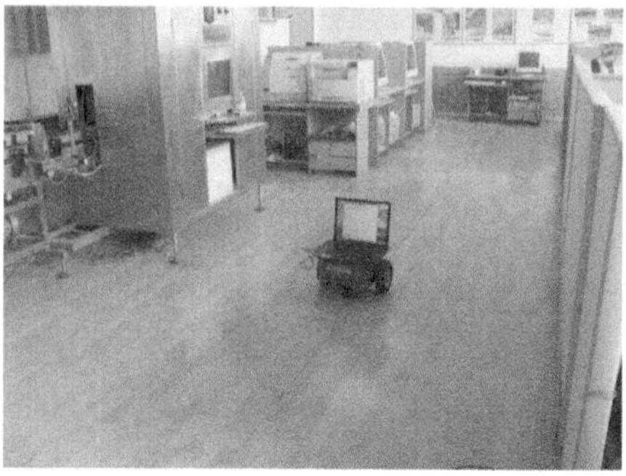

Fig. 6. Real Lab environment

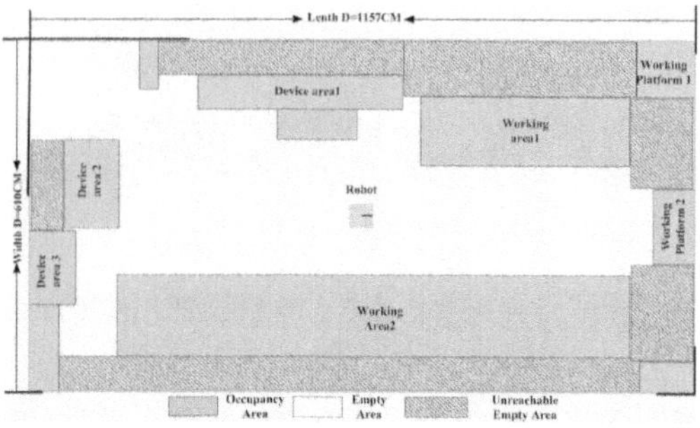

Fig. 7. Actual occupancy status of lab environment

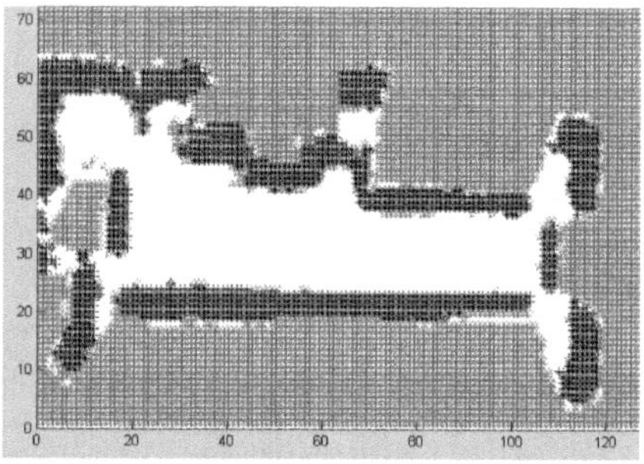

Fig. 8. Occupancy grids built by the proposed approach

From Fig.6-8, we can see that the proposed algorithm is able to model the environment. The produced metric map is more consistent, accurate and robust than traditional method. Moreover it has advantage on adaptive to environments. The shortcoming of the presented method is that enough time is needed to train the neural network. However, once the neural networks are trained, it can easily be adapted to new circumstances. The computation time of the method is longer than traditional method, but it still could be used in real time.

5 Conclusion

This paper presents an improved neural network for sonar readings interpretation to build occupancy grids of mobile robot. The proposed model interprets sensor readings in the context of their space neighbors and relevant successive history readings simultaneously while previous neural network sonar models only consider current readings. So if current readings are produced by multiple reflections or specular reflection, this neural network model could depend on relevant history readings to obtain the correct occupancy values. Moreover, using the trained neural network based on the proposed approach can easily be adapted to new circumstances since neural network is trained based on examples. Even though time is short, the neural network could quickly be retrained to accommodate this new situation.

Acknowledgments. This work is supported by the National Natural Science Foundation of China (60375001) and Hunan Provinxcial Science Foundation of China(06JJ50121).

References

1. Meyer, J.A. and Filliat, D.: Map-based navigation in mobile robots: a review of map-learning and path-planning strategies, Cognitive systems Research **4** (2003) 283–317.
2. Thrun, S.: Robotic mapping: a survey, (G. Lakemeyer and B. Nebel (eds.)) Exploring Artificial Intelligence in the New Millenium, Morgan Kaufmann, San Francisco (2002) 1–35.
3. Elfes, A.: Sonar based real-world mapping and navigation: cognitive systems research, IEEE Transactions on Robotics and Automation **RA-3**(3) (1987) 249–265.
4. Moravec, H.P. and Elfes, A.: High resolution maps form wide angle sonar, Proc. Of the IEEE Conference on Robotic and Automation, IEEE Press, Washtington, DC (1985) 116–121.
5. Noykov, Sv. and Roumenin, Ch.: Occupancy grids building by sonar and mobile robot, Robotics and autonomous systems (2006) (doi:10.1016/j.robot.2006.06.004).
6. Kurt, K.: Improved occupancy grids for map building, Autonomous Robots **4** (1987) 351–367.
7. Thrun, S.: Learning metric-topological maps for indoor mobile robot navigation, Artifical Intelligence **99**(1) (1998) 21–71.
8. Angelo, A., Jose, del R.M. and Dario, F.: Efficient learning of variable-resolution cognitive maps for autonomous indoor navigation, IEEE Transactions on Robotics and Automation **15**(6) (1999) 990–1000.
9. CHOW, K.M., RAD, A.B. and IP, Y.L.: Enhancement of probabilistic grid-based map for mobile robot applications, Journal of Intelligent and Robotic Systems **34** (2002) 155–174.
10. Thrun, S.: Learning occupancy grids with forward sensor models, Autonomous Systems **15** (2003) 111–127.
11. Miguel, R. and Axel, P.: A comparison of three uncertainty calculi for building sonar-based occupancy grids, Robotics and Autonomous Systems **35** (2001) 201–209.
12. Martin, T.H., Howard, B.D. and Mark H.B.: Neural network design, 1st edn. PWS Publishing Company, 1996.

Adaptive Network-Based Fuzzy Inference Model of Plasma Enhanced Chemical Vapor Deposition Process

Byungwhan Kim[1] and Seongjin Choi[2]

[1] Department of Electronic Engineering, Sejong University,
Seoul, Korea
kbwhan@sejong.ac.kr
[2] Department of Electronics and Information Engineering, Korea University,
Yeongi, Korea
choisj@korea.ac.kr

Abstract. In this study, a prediction model of plasma enhanced chemical deposition (PECVD) data was constructed by using an adaptive network-based fuzzy inference system (ANFIS). The PECVD process was characterized by means of a Box Wilson statistical experiment. The film characteristics modeled are deposition rate and stored charge. The prediction performance of ANFIS models was evaluated as a function of training factors, including the step-size, type of membership functions, and normalization factor of inputs-output pairs. The effects of each training factor were sequentially optimized. The root mean square errors of optimized deposition rate and charge models were 11.94 Å/min and $1.37\times10^{12}/cm^2$, respectively. Compared to statistical regression models, ANFIS models yielded an improvement of more than 20%. This indicates that ANFIS can effectively capture nonlinear plasma dynamics.

1 Introduction

In manufacturing integrated circuits, plasmas play a crucial role in etching and depositing thin films. Prediction models of plasma processes are in demand for characterization, diagnosis, and control of plasma equipment. First principle models are subject to many simplifying assumptions due to the lack of understanding of physical and chemical processes. As alternative methods, intelligent techniques such as neural network and fuzzy logic have been applied to model plasma data [1-9]. In the context of fuzzy logic applications, adaptive network fuzzy inference system (ANFIS) and a neuro-fuzzy system called FALCON were applied to plasma etch [8] and deposition data [7], respectively.

In this study, the ANFIS was applied to plasma enhanced chemical vapor deposition (PECVD) process. The PECVD process was characterized by a statistical experiment. It should be noted that FALCON was applied to the same data examined in this study [7]. However, this study is differentiated from that study in that ANFIS was applied to model with different film characteristics, including the deposition rate and stored charge. The prediction performance of ANFIS model was optimized as a function of training factors such as step-size, type of membership functions, and

D. Liu et al. (Eds.): ISNN 2007, Part I, LNCS 4491, pp. 602–608, 2007.

normalization factor of inputs-output pairs. The optimized models were also compared to statistical regression models.

2 Experimental Data

The SiN films were deposited by using a Plasma-Therm 700 series batch reactor operating at 13.56 MHz. The distance between the electrodes was 2.29 cm and the electrode diameter was 11 inches. The PECVD process was characterized by the face-centered central composite circumscribed experimental design, which consisted of 2^{6-1} fractional experiment and 12 axial points [10]. The resulting 33 experiments, including one experiment corresponding to one center point, were used to train the ANFIS. Prediction performance of the trained ANFIS was tested with additional 12 experiments not pertaining to the training data. The experimental parameters and their operational ranges in the experimental design are included in Table 1. Four-inch, float zone p-type silicon wafers of (100) orientation with a resistivity of 2.0 Ω-cm were used as the substrate. During the deposition, SiH_4 was diluted to 2% in nitrogen.

Table 1. Experimental parameters and ranges

Parameters	Range	Unit
Substrate Temperature	200-400	^oC
Pressure	0.6-1.2	Torr
RF Power	20-40	Watts
NH_3 Flow	1-1.4	Sccm
SiH_4 Flow	180-260	Sccm
N_2 Flow	0-1000	Sccm

Using a Metricon 2010 Prism Coupler, the deposition rate was measured. The Metricon 2010 can measure the thickness with the resolution and accuracy of ±0.3% and ±(0.5% + 50Å), respectively. The stored charge was estimated by means of typical C-V measurement. The inputs to ANFIS models were the substrate temperature, pressure, RF power, NH_3 flow, SiH_4 flow, and N_2 flow as shown in Table 1. The output of ANFIS model was the deposition rate or the stored charge. The chosen inputs and output from each experimental data constituted inputs-output pairs to train and evaluate ANFIS models.

3 Results

3.1 Training Factors

Figure. 1 shows the schematic of ANFIS. Each layer has its functional role to train ANFIS model. Each node in layer 1 refers to the membership function assigned to the input parameter with a linguistic level. Each node in layer 2 represents the firing

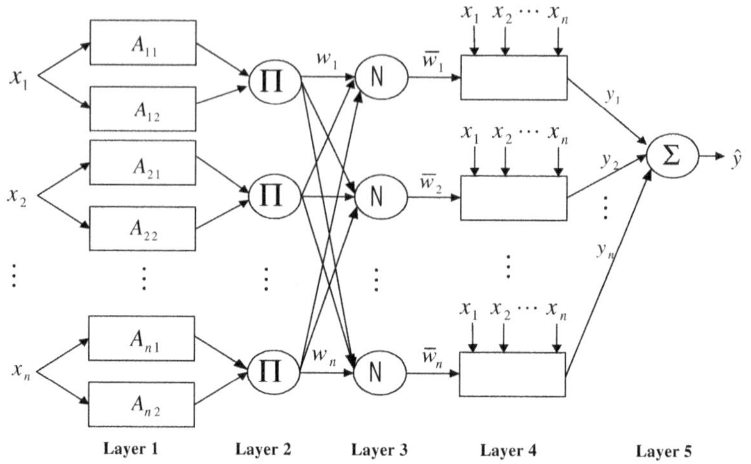

Fig. 1. Schematic of ANFIS

strength of a rule with a multiplication of the incoming signals, and each node in layer 3 calculates the ratio of the firing strength of the *i*th rule to the summation of the firing strength of each rule. The node in layer 4 represents a product of firing strength ratio and the output of the *i*th rule of ANFIS. The node in layer 5 produces a prediction by computing the overall output with the summation of all incoming signals. The operating principles were detailed in [11]. In ANFIS, a generic parameter denoted as α is formed by the union of the premise and consequent parameters. The update formula for α is described as

$$\Delta\alpha = -\eta \frac{\partial E}{\partial \alpha}, \tag{1}$$

where η is a learning rate and is adjusted according to:

$$\eta = \frac{\delta}{\sqrt{\sum_{\alpha}\left(\frac{\partial E}{\partial \alpha}\right)^2}}, \tag{2}$$

where δ is the *step-size*, the length of each gradient transition. During the training phase, the α is updated at each training epoch in a hybrid fashion. More specifically, the consequent parameters of α are updated first using a least square algorithm and the premise parameters are then adjusted by backpropagating the errors. In this study, the effect of step-size is optimized by experimentally from 0.2 to 2.0 with an increment of 0.1. This was followed by optimizing the effect of membership functions, including gaussian, generalized bell, and sigmoid functions.

Both training and evaluation data consisted of inputs-output pairs from the experimental data and were normalized by the maximum value of each variable. Also, the number of training epoch and the step-size for training ANFIS models were set 100

Table 2. Initial training conditions for training ANFIS models

Training Factor	Initial Training Factor
Membership Function	Gaussian function
Step-size	1.0
Normalization Factor	max(variable)

and 1.0, respectively, as initial training conditions. The initial training conditions for ANFIS models are summarized in Table 2.

3.2 Optimization

The effect of step-size on ANFIS models is shown in Fig. 2. In Fig. 2, the circles and triangles represent the prediction accuracy of deposition rate and stored charge, respectively. The prediction accuracy was quantified by the root mean square error (RMSE). As shown in Fig. 2, the RMSE varies considerably with the variations in the step-size. For a certain band of step-sizes, the prediction errors show a band of the minimum values compared to other chosen step-sizes. We chose the step-size from a lower boundary value of a certain band of step-sizes to yield the minimum prediction error in RMSE. The selected step-sizes for the deposition rate and stored charge are 0.5 and 0.8, respectively. The corresponding RMSEs are 14.26 Å/min and $1.63\times10^{12}/cm^2$, respectively.

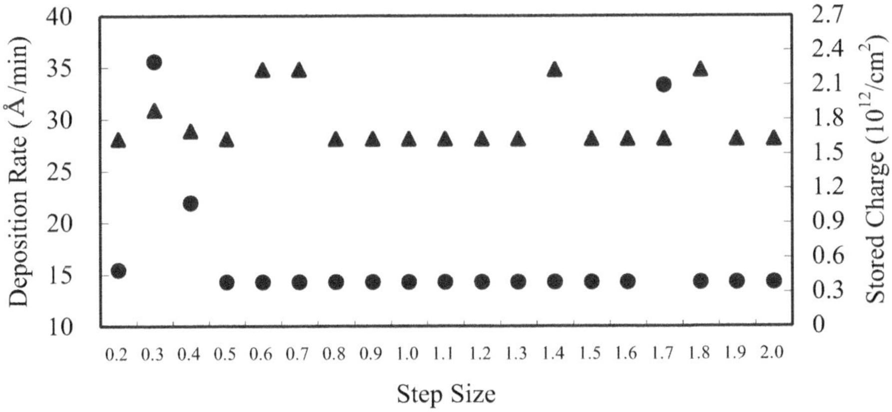

Fig. 2. Prediction performance of ANFIS models

Next, the type of membership functions for training ANFIS models was changed to compare the prediction performances with chosen step-sizes. Table 3 shows the prediction results according to the types of membership functions as Gaussian, generalized bell, and closed sigmoidal functions. In Table 3, the prediction performance for the deposition rate model was improved from 14.26 to 13.70 Å/min with an

Table 3. RMSEs as a function of membership functions at specified step-sizes

Step-size	Membership Function	Deposition Rate (Å/min)	Stored Charge ($10^{12}/cm^2$)
0.5	Gaussian function	14.26	
	Generalized bell function	**13.70**	
	Sigmoidal function	47.68	
0.8	Gaussian function		**1.63**
	Generalized bell function		1.63
	Sigmoidal function		1.64

adaptation of the generalized bell function as membership function. For the stored charge model, Gaussian function provides comparable or better prediction performances compared to the other two functions. Therefore, the generalized bell and Gaussian functions were adopted as the membership functions for the deposition rate and stored charge models, respectively.

Lastly, in order to perceive effect of normalization factor in training ANFIS models, the values of inputs-output pairs were normalized by product of scalar k and maximum values of the variables under the chosen training conditions. The value of k was changed from 1 to 1/4, 1/3, 1/2, 2, 3, and 4, to provide the trained ANFIS model with several prediction performances. The prediction performances of the trained ANFIS models according to variations in normalization factor are summarized in Table 4. As shown in Table 4, the deposition rate and stored charge models achieve the smallest RMSEs of 11.94 Å/min and $1.37\times10^{12}/cm^2$, respectively. Compared to the RMSEs determined earlier, these are much smaller. This indicates that the normalization factor is the most significant factor influencing ANFIF model performance. It seems that the normalization factor of 2×max(variable) could provide a more proper training data distribution with a limited number of inputs-output pairs to construct a better ANFIS model for the deposition rate. Similarly, the normalization factor of 1/3×max(variable) could provide a better training data distribution to construct the ANFIS model for the stored charge. In consequence, variations in the distribution of training inputs-output pairs due to the variations in normalization factor contributed to achieve improved ANFIS prediction models.

3.3 Comparisons with Statistical Regression Models

The film characteristics, used here for the development of the ANFIS prediction models, were once employed in constructing other type of neural network model [9]. In that study, for the purpose of comparison of prediction performances, four statistical regression models (SRMs) were built and the established smallest RMSEs of the SRMs are shown in Table 5. As shown in Table 5, for either film characteristic, ANFIS models yield an improvement of more than 20% over SRMs. The comparison demonstrates that ANFIS can effectively learn nonlinear plasma dynamics.

Table 4. RMSEs of ANFIS models with different normalization factors at the predetermined step-sizes and membership functions

Step-size	Membership Function	Normalization Factor	Deposition Rate (Å/min)	Stored-Charge ($10^{12}/cm^2$)
0.5	**General-ized bell function**	1/4 x max(variable)	51.27	
		1/3 x max(variable)	18.80	
		1/2 x max(variable)	34.15	
		max(variable)	13.70	
		2 x max(variable)	**11.94**	
		3 x max(variable)	14.00	
		4 x max(variable)	13.91	
0.8	**Gaussian function**	1/4 x max(variable)		1.63
		1/3 x max(variable)		**1.37**
		1/2 x max(variable)		1.86
		max(variable)		1.63
		2 x max(variable)		1.63
		3 x max(variable)		1.63
		4 x max(variable)		1.63

Table 5. Prediction performance comparison of ANFIS models and SRMs with RMSE

Film Characteristics	ANFIS	SRM	Improvement (%)
Deposition Rate (Å/min)	11.94	16.50	26
Stored Charge ($10^{12}/cm^2$)	1.37	1.78	23

4 Conclusions

ANFIS was used to construct prediction models of plasma process data, particularly plasma-enhanced chemical vapor deposition data. For a systematic modeling, the deposition process was characterized by using a statistical experiment. The effects of three training factors on ANFIS models were evaluated experimentally. This revealed that the normalization factor was determined as the most influencing factor on ANFIS prediction. Comparisons with statistical regression models also revealed that the ANFIS models are much more accurate in predicting parameter effects on plasma dynamics. The constructed ANFIS models can be used to interpret parameter effects on film characteristics.

Acknowledgements

This work was supported by the Seoul Research and Business Development Program (Grant No.10583), and partly by the MIC (Ministry of Information and

Communication), Korea, under the ITRC (Information Technology Research Center) support program supervised by the IITA (Institute of Information Technology Advancement) (IITA-2006-C109006030030).

References

1. Kim, B., Kwon, K.H., Kwon, S.K., Park, J.M., Yoo, S.W., Park, K.S., You, I.K., Kim, B.W.: Modeling Etch and Uniformity of Oxide via Etching in a CHF_3/CF_4 Plasma Using Neural Network. Thin Solid Films **426** (2003) 8-15
2. Kim, B., Kim, S.: GA-optimized Backpropagation Neural Network with Multi-parameterized Gradients and Applications to Predicting Plasma Etch Data. Chemometr. Intell. Lab. Syst. **79** (2005) 123-128
3. Kim, B., Kim, S.: Plasma Diagnosis by Recognizing in-situ Data Using a Modular Backpropagation Network. Chemomemetr. Intell. Lab. Syst. **65** (2) (2003) 231-240
4. Kim, B., Kim, S., Lee, B.T.: Modeling SiC Surface Roughness Using Neural Network and Atomic Force Microscopy. J. Vac. Sci. Technolo. B **22** (5) (2004) 2467-2472
5. Kim, B., Kim, K.: Prediction of Profile Surface Roughness in CHF3/CF4 Plasma Using Neural Network. Applied Surface Science **222** (1-4) (2004) 17-22
6. Kim, B., Lee, B. T.: Prediction of SiC Etching in a NF3/CH4 Plasma Using Neural Network. Journal of Vaccum Science and Technology A **22** (6) (2004) 2517-2522
7. Geisler, J. P., Lee, C. S. George, May, G. S.: Neurofuzzy Modeling of Chemical Vapor Deposition Processes. IEEE Trans. Semicond. Manufact **13** (2000) 46-59
8. Kim, B., Park, J. H.: Qualitative Fuzzy Logic Model of Plasma Etch Process. IEEE Trans. Plasma Science **30** (2) (2002) 673-678
9. Kim, B., Lee, D., Han, S. S.: Prediction of Plasma Enhanced Deposition Process Using GA-optimized GRNN. LNCS (2006)
10. Montgomery, D. C.: Design and Analysis of Experiments. John Wiley & Sons, Singapore (1991)
11. Jang, J. R.: ANFIS: Adaptive-network-based Fuzzy Inference System. IEEE Trans. Syst. Man. Cybern. **23** (3) (1993) 665-685

Hybrid Intelligent Modeling Approach for the Ball Mill Grinding Process

Ming Tie[1], Jing Bi[2], and Yushun Fan[1]

[1] National CIMS Engineering Research Center, Tsinghua University, Beijing, 100084, China
tming7611@sina.com
[2] Software College, Northeastern University, Shenyang 110004, China
bijing39@sina.com

Abstract. Modeling for the ball mill grinding process is still an imperative but difficult problem for the optimal control of mineral processing industry. Due to the integrated complexities of grinding process (strong nonlinearity, unknown mechanisms, multivariable, time varying parameters, etc.), a hybrid intelligent dynamic model is presented in this paper, which includes a phenomenological ball mill grinding model with a neurofuzzy network to describe the selection function of different operating conditions, a populace balance based sump model, a phenomenological hydrocyclone model with some reasoning rules for its parameters correction and a radius basis function network (RBFN) for fine particle size error compensation. With the production data from a ball mill grinding circuit of an iron concentration plant, the experiments and simulation results show the proposed hybrid intelligent modeling approach effective.

1 Introduction

As the most important and most energy-intensive production process in mineral processing industry, the ball mill grinding process is to reduce the size of the mineral to facilitate its concentration in the next stage. To improve the economic recovery of the valuable minerals as well as energy efficiency, an appropriate model is highly desirable for optimum operation [1].

There have been various attempts to develop equations or associations for grinding process, such as the ball mill models [2] [3] based on the dynamic population balance concept, the perfect mixing assumption with the grinding theory, and the hydrocyclone models [4] based on the classification theory or experiences. Nonetheless, the above physically-based models can not give accurate descriptions of the grinding process because of the overlooked variation of the selection function in different operation conditions and the low precision of those hydrocyclone models. Due to the complex and poorly understood underlying mechanisms, modeling grinding process with data-driven techniques draws more attention. But modeling the whole grinding process in black-box structure [5] is often unappreciative in industrial application, as the model is not embedded with physical knowledge and not adapted to different operating conditions.

D. Liu et al. (Eds.): ISNN 2007, Part I, LNCS 4491, pp. 609–617, 2007.

The present work proposes a hybrid intelligent model for grinding process with the combination of the phenomenological models, the knowledge from experiences and the intelligent modeling techniques. The presented paper is organized as follows. A brief description of the grinding circuit with grinding process is given in Section 2. Then in Section 3, the dynamic hybrid intelligent model for the grinding process is developed detailedly on the sequence of the modeling strategy, the phenomenological models, the radius basis function network (RBFN) for error compensation and parameters reasoning rules, the neurofuzzy network for selection function.

2 Description of the Ball Mill Grinding Process

The ball mill grinding process we studied is shown in Fig.1, and its operation is as follows. The fresh ore feed into a ball mill for grinding with the water fed to control the mill slurry concentration. Then the discharge slurry from the mill pours into a sump. The pump takes out the slurry to feed a hydrocyclone with the sump water fed to control the concentration of the cyclone feed slurry. Through the classification of the hydro-cyclone, the overflow fine slurry is the product, and the recycle coarse slurry return to the ball mill.

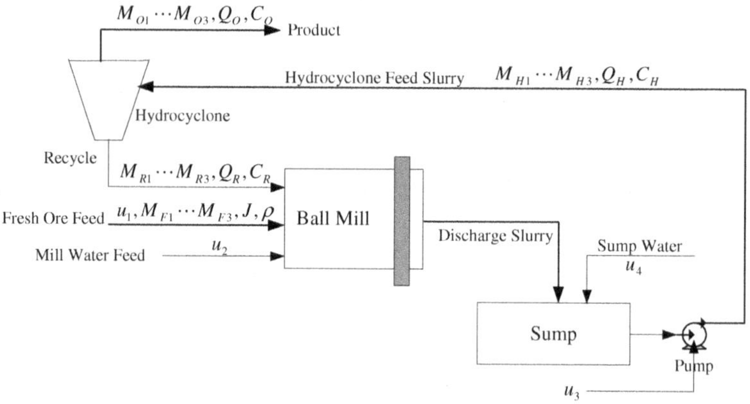

Fig. 1. Schematic diagram of the ball mill grinding process

As shown in fig. 1, the outputs of the grinding process are Q_O, C_O and $M_{O1}...M_{O3}$, which represent the overflow flow rate, concentration and particle size (solid particle size distribution) respectively. The inputs are $u_1...u_4$, which represent the fresh ore feed rate, the mill water feed rate, the pump rate, the sump water feed rate, respectively. The boundary conditions are J, $M_{F1}...M_{F3}$, ρ, which represent the fresh ore rigidity, the particle size and the density respectively. The definition of particle size fraction is as table1.

Table 1. The particle size fraction definition

Particle size fraction	M_1	M_2	M_3
Particle diameter fraction (micron)	150~300	75~150	35~75

Q_H, C_H and $M_{H1}...M_{H3}$ are the hydrocyclone slurry feed rate, concentration and particle size respectively. Q_R, C_R and $M_{R1}...M_{R3}$ are hydrocyclone slurry recycle rate, concentration and particle size respectively.

3 The Hybrid Intelligent Model for the Ball Mill Grinding Process

The hybrid intelligent modeling strategy for the ball mill grinding process
To achieve higher precision than those of the phenomenological models [2-4] and the neural networks [5], the hybrid intelligent modeling strategy for grinding process is presented as shown in fig. 2. The ultimate result is the sum of the outputs of the phenomenological model and the compensation RBFN. ϕ_1 represents the selection function whose real value can be achieved from assay of the sump slurry. Due to the quick response time of classification [6], both the hydrocyclone phenomenological model and its compensation RBFN are static one. ϕ_1 varies with different operating conditions and it can not be described with accurate mathematical models, thus a neurofuzzy model is developed for its capability of combination with phenomenological models.

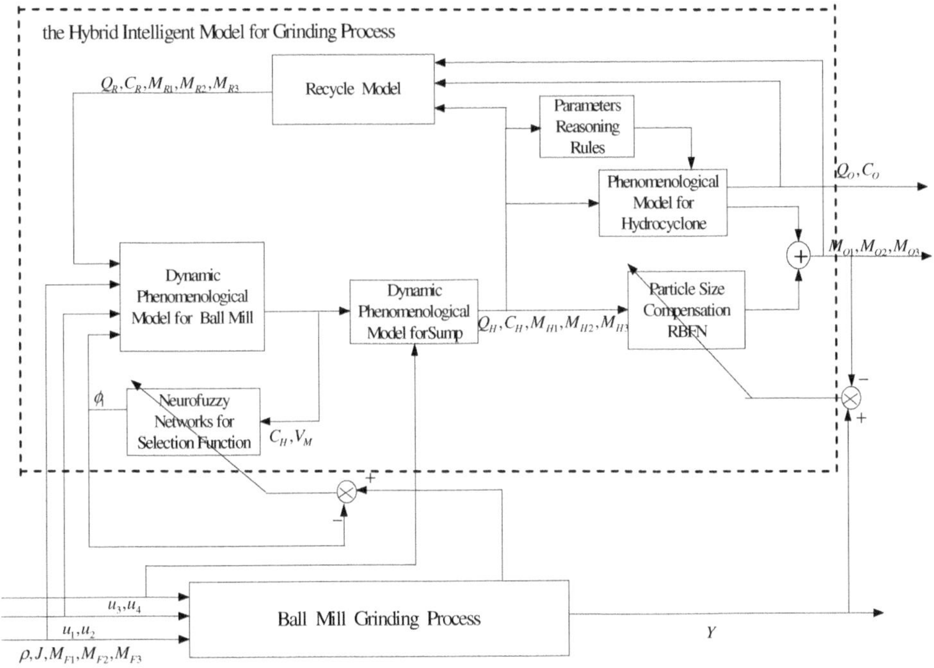

Fig. 2. The hybrid intelligent modeling strategy diagram for the ball mill grinding process

The phenomenological models for the ball mill grinding process
The phenomenological models include the dynamic ball mill model, the dynamic sump model, the recycle model and the static hydrocyclone model.

Due to the classification models [2] [4], the hydrocyclone model is described as

$$Q_O = Q_H C_H (1 - M_{H1}E_1 - M_{H2}E_2 - M_{H3}E_3) + \alpha_1 Q_H (1 - C_H) + \alpha_2, \quad (1)$$

$$Q_O C_O = Q_H C_H - Q_H C_H (M_{H1}E_1 + M_{H2}E_2 + M_{H3}E_3), \quad (2)$$

$$Q_O C_O M_{O,i} = Q_H C_H M_{H,i}(1 - E_i), \qquad (i = 1,2,3), \quad (3)$$

$$E_i = 1 - \exp[-0.693(\frac{\bar{d}_i}{d_{50}})^{a_5}] + (a_3 + a_4 \frac{Q_O - Q_O C_O}{Q_H - Q_H C_H})\exp[-0.693(\frac{\bar{d}_i}{d_{50}})^{a_5}], \quad (4)$$

$$\log(d_{50}) = a_6 + a_7 C_H Q_H + a_8 Q_H, \quad (5)$$

where d_{50} is the cut size, $\bar{d}_i$ is the average size, E_i is the separation efficiency, $\alpha_1 \ldots \alpha_8$ are the coefficients.

The inputs of the hydrocyclone model, its slurry feed rate, concentration, particle size are the same with u_3 and the outputs of sump model. And the sump dynamic model can be described as the following populace balance equations,

$$\dot{V}_S = Q_D + u_4 - u_3, \quad (6)$$

$$\frac{dV_S C_S}{dt} = Q_D C_D - u_3 C_H, \quad (7)$$

$$\frac{dV_S C_S M_{S,i}}{dt} = Q_D C_D M_{D,i} - u_3 C_H M_{H,i}, \qquad (i = 1,2,3), \quad (8)$$

where V_S, C_S and $M_{S,i}$ are the sump slurry volume, concentration and particle size respectively, Q_D, C_D and $M_{D,i}$ are the mill slurry discharge rate, concentration and particle size of the ball mill respectively. The sump can behave as a perfect fixer with slurry suspended [2], thus Eq.(9) and Eq.(10) can be derived

$$\dot{C}_H = \frac{1}{V_S}(Q_D C_D - u_4 C_H - Q_D C_H), \quad (9)$$

$$\dot{M}_{H,i} = \frac{Q_D C_D}{V_S C_H}(M_{D,i} - M_{H,i}), \qquad (i = 1,2,3). \quad (10)$$

As parts of the inputs of the sump model, Q_D, C_D and $M_{D,i}$ can be achieved from the dynamic ball mill model.

Based on the grinding theory, the populace equations and the mixer concept [3] [7] [8], together with our analysis of the dynamics of this ball mill grinding process, the ball mill model can be derived as

$$\dot{V}_M = \frac{u_1}{\rho} + u_2 + Q_R - Q_D, \tag{11}$$

$$\dot{C}_D = \frac{1}{V_M}(u_1 + Q_R C_R - Q_R C_D - u_2 C_D - u_1 C_D), \tag{12}$$

$$\dot{M}_{D1} = \frac{Q_R C_R + u_1}{V_M C_D}\left(\frac{Q_R C_R M_{R1} + u_1 M_{F1}}{Q_R + u_1 + u_2} - M_{D1}\right) - \phi_1 M_{D1}, \tag{13}$$

$$\dot{M}_{D2} = \frac{Q_R C_R + u_1}{V_M C_D}\left(\frac{Q_R C_R M_{R2} + u_1 M_{F2}}{Q_R + u_1 + u_2} - M_{D2}\right) - \xi_1 \phi_1 M_{D2} + b\phi_1 M_{D1}, \tag{14}$$

$$M_{D1}+M_{D2}+M_{D3}=1, \tag{15}$$

where V_M represents the volume of the slurry in the mill, ξ_1 is a constant coefficient, b represents the breakage function. For a given type of ore, b can be described by [1]

$$b = (\xi_2 + \xi_3 C_D)\left(\frac{\bar{d}_3}{\bar{d}_2}\right)^{\xi_4} + (1 - \xi_2 - \xi_3 C_D)\left(\frac{\bar{d}_3}{\bar{d}_2}\right)^{\xi_5}, \tag{16}$$

where $\xi_2 \ldots \xi_5$ are constant coefficients.

As a 1-order model [9], Q_D can be described as

$$T_Q \dot{Q}_D = \frac{u_1}{\rho} + u_2 + Q_R - Q_D, \tag{17}$$

where T_Q can be simplified as a time constant.

Q_R, C_R and $M_{R1} \ldots M_{R3}$ can be derived from the recycle model, which is described as

$$Q_R = Q_H - Q_O, \tag{18}$$

$$Q_R C_R = Q_H C_H - Q_O C_O, \tag{19}$$

$$Q_R C_R M_{R,i} = Q_H C_H M_{H,i} - Q_O C_O M_{O,i}. \tag{20}$$

The particle size compensation RBFN and the parameters reasoning system for the hydrocyclone phenomenological model

In the hydrocyclone phenomenological model Eq.(1)~Eq.(5), the parameters α_1 and α_2 vary with Q_H and cyclone structure parameters; α_3 and α_4 vary with Q_H, C_H and cyclone structure parameters; α_5 varies with fresh ore property; $\alpha_6 \ldots \alpha_8$ vary with Q_H, C_H, $M_{H,i}$ and cyclone structure parameters. To optimal operations, the ore type and structure parameters are constant, thus the parameters reasoning rules are described with the expert experiences as

Rule i:

$$\text{If } Q_H \in [265, 300] \text{ and } C_H \in [30\%, 40\%] \text{ and } M_{H1} \in [45\%, 70\%], \tag{21}$$
$$\text{then } [\alpha_1 \ldots \alpha_8] = [0.67, -0.0044, 1.2, 0.0025, -0.45, 2.3, 0.9, 0.75].$$

As the classification theory is too complex and the expert experiences are not accurate enough, the above reasoning rules can not satisfy the precision demand, especially for the particle size $M_{O,i}$. Therefore, the RBFN for error compensation of the fine particle size computation is presented due to the advantages of RBFN [11].

According to the classification theory and the analysis by experts [6] [10], the fine particle size error ΔM_{O3} can decided mainly by Q_H, C_H , Q_O and $M_{H,i}$. Thus the error compensation RBFN can be described as

$$\Delta M_{O3} = w_0 + \sum_{j=1}^{n} w_j \exp[-\frac{1}{(2\sigma_j)^2}(X - c_j)^T (X - c_j)], \tag{22}$$
$$X = [Q_H, C_H, M_{H1}, M_{H2}, Q_O]^T,$$

where n is the num of nodes of hidden layer, σ_j is the distance of centers, c_j is called RBF center, w_j is the tap weight, w_0 is the threshold.

The centers of hidden layer are determined by fuzzy c-means clustering approach [12], and w_j can be identified with the least square recursive method.

The neurofuzzy network for selection function

Due to the analysis of relationship between the selection function and the net mill power [3] [13], ϕ_1 can be described as

$$\log\phi_1 = k_1 \log P - k_2 \log(C_D V_M) + \log J + k_3, \tag{23}$$

where P represents the net mill power draft, $k_1 \ldots k_3$ are parameters.

According to the experience equations for the net mill power draft [14] [15], P can be described by

$$P = f_1(C_D, V_M, \rho, V_B, \rho_B, L, D), \tag{24}$$

where V_B represents the ball volume in mill, ρ_B is the ball density, L and D are the length and the width of the mill respectively, f_1 is an unknown nonlinear function. ρ, ρ_B , L, D can be considered as constants in this equation, and V_B can be described as

$$\frac{dV_B}{dt} = -k_4 V_B - k_5 (V_B)^{0.66} + k_6, \tag{25}$$

where $k_4 \ldots k_6$ are parameters related to physics properties of the slurry.

According to the Eq.(23) and Eq.(24) together with grinding energy draft theory, the inputs and the output of the neurofuzzy network model can be determined as

$$X=[x_1, x_2, x_3]^T=[log(V_B), log(C_D), log(V_M)]^T, \quad y=log(\Phi_1). \tag{26}$$

Based on experiences and grinding theory, we determined 2 fuzzy sets for x_1, 4 fuzzy sets for x_2 and 3 fuzzy sets for x_3.

The neurofuzzy network has a 5-layed structure, which is described as follows.

Layer 1, compute the matching degree to a fuzzy condition with Gauss-shaped node function.

Layer 2, compute the firing strength of each rule as the minimum value of its inputs.

Layer 3, compute the normalized matching degree for each rule as

$$v_i = \mu_i \Big/ \sum_{j=1}^{24} \mu_j \quad (i=1,2, \ldots 24), \tag{27}$$

where μ_i is an output of the layer 2.

Layer 4, compute the conclusion inferred by each fuzzy rule

$$z_i = v_i(\lambda_{i,0} + \lambda_{i,1}x_1 + \lambda_{i,2}x_2 + \lambda_{i,3}x_3) \quad (i=1,2, \ldots 24). \tag{28}$$

Layer 5, combine the conclusions of all fuzzy rules and obtain the network output.

The fuzzy c-means clustering approach [12] is applied to identify the clustering centroids of the network. The parameters of the network are achieved with a hybrid learning algorithm [16] which combines a recursive SVD-based least square method and the gradient descent method.

4 Results and Discussions

The data used for experiments comes from the ball mill grinding process of an iron concentration plant. The ball mill is 3.5 meter in length, 3.2 meter in diameter, and its percent critical velocity is 0.45.

450 groups of data are used for hybrid intelligent modeling of the grinding process, and the following results are achieved. The parameters $\xi_1 \ldots \xi_5$ are 0.57, 0.1, 0.25, 0.12, 1.1 respectively, and $k_4 \ldots k_6$ are 0.0014, 0.00065, 0.002 respectively. There are 5 rules in the parameter reasoning system of the hydrocyclone model. The error compensation RBFN has 13 hidden layer nodes while the neurofuzzy network for selection function has 24 fuzzy rules.

Another 70 groups of data, which is sampled with the variation of fresh ore feed and water feed, are used for the simulation with the proposed hybrid intelligent model. In Fig. 3, the contrast curves between the outputs of the proposed hybrid intelligent model and the real value are presented. From the contrast curves shown in Fig. 3, it's easy to see that the simulation result of the proposed model gives a correct dynamic characteristic description of the real outputs of the ball mill grinding process.

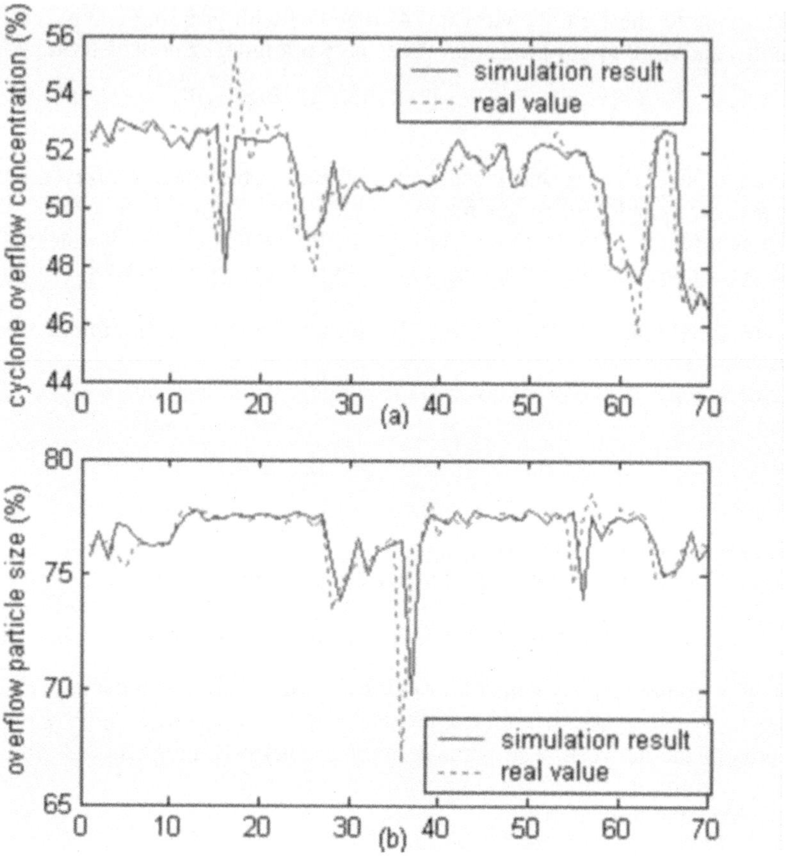

Fig. 3. Contrast curves of the real value with simulation results from the proposed model (a) overflow concentration simulation; (b) overflow fine particle size simulation

5 Conclusions

This work proposes a hybrid intelligent model to solve the difficult problem of modeling grinding process accurately. Due to the integrated complexities in grinding process, the presented model combines the phenomenological models, expert experiences, the NN technique and the fuzzy logic. The simulation and experiments show its high precision and adaptation for different operating conditions. The proposed hybrid intelligent modeling approach is important not only for the dynamics research and the optimization research of grinding process, but also for grinding simulation system construction.

Acknowledgements

This work is supported by National Basic Research Development Program of China (2006CB705400) and National High Technology Research and Development Program of China (2004AA412030).

References

1. Radhakrishnan, V.: Model Based Supervisor Control of a Ball Mill Grinding Circuit. Journal of Process Control **9** (1999) 195-211
2. Rajamani, K., Herbst, J.: Grinding Circuit Modeling and Dynamic Simulation. Chemical Engineering Science **46** (1991) 861-870
3. Morrell, S., Man, Y.: Using Modeling and Simulation for the Design of Full Sscale Ball Mill Circuits. Minerals Engineering **10** (1997) 1311-1327
4. Brayshaw, M.: Numerical Model for the Inviscid Flow of a Fluid in a Hydrocyclone to Demonstrate the Effects of Changes in the Vorticity Function of the Flow Field on Particle Classification. International Journal of Mineral Processing **29** (1990) 51-75
5. Kishalay, M., Mahesh, G.: Modeling of an Industrial Wet Grinding Operation Using Data-driven Techniques. Computers & Chemical Engineering **30** (2006) 508-520
6. Plitt, L.:Cyclone Modeling, a Review of Present Technology. CIM Bulletin **80** (1987) 39-50
7. Tie, M., Yue, H., Chai, T.Y.: A Hybrid Intelligent Soft-sensor Model for Dynamic Particle Size Estimation in Grinding Circuits. Lecture Notes in Computer Science **3498** (2005) 871-876
8. Hoyer, D.: Batch Grinding Simulation, Population Balance Models and Self-similar Size Distributions. Minerals Engineering **8** (1995) 1275-1284
9. Dubé, Y, Lanther, R.: Computer Aided Dynamic Analysis and Control Design for Grinding Circuits. CIM Bulletin **80** (1987) 65-70
10. Du, Y., et al.: Neural Net-based Soft-sensor for Dynamic Particle Size Estimation in Grinding Circuits. International Journal of Mineral Processing **52** (1997) 121-135
11. James, S., Legge, R.: Comparative Study of Black-box and Hybrid Estimation Methods in Fed-batch Fermentation. Journal of Process Control **12** (2002) 113-121
12. Teppola, P., Mujumen, S., Minkkinen, P.: Adaptive Fuzzy C-means Clustering in Process Monitoring. Chemometrics and Intelligent Systems **45** (1999) 23-38
13. Rajiamani, K., Herbst, J.: Simultaneous Estimation of Selection and Breakage Functions from Batch and Continuous Grinding Data. Transactions of the Instituttion of Mining & Metallurgy **93** (1984) 74-85
14. Tangsathitkulchai, C.: Effects of Slurry Concentration and Powder Filling on the Net Mill Power of a Laboratory Ball Mill. Powder Technology **137** (2003) 131-138
15. Tito, V., Karina, C., Gonzalo, A., et al.: Neural Grey Box Model for Power Estimation in Semi-autogenous Mill. International Symposium on Neural Networks, Chongqing (2005)
16. Lee, S., Ouyang, C.: A Neurofuzzy System Modeling with Self-constructing Rule Generation and Hybrid SVD-based Learning. IEEE Transactions on Fuzzy Systems **11** (2003) 341-353

Nonlinear Systems Modeling Using LS-SVM with SMO-Based Pruning Methods

Changyin Sun[1,2], Jinya Song[1], Guofang Lv[1], and Hua Liang[1]

[1] College of Electrical Engineering,
Hohai University, Nanjing 210098, P.R. China
[2] School of Automation,
Southeast University, Nanjing 210096, P.R. China
cysun@seu.edu.cn

Abstract. This paper firstly provides a short introduction to least square support vector machine (LS-SVM), then provides sequential minimal optimization (SMO) based on Pruning Algorithms for LS-SVM, and uses LS-SVM to model nonlinear systems. Simulation experiments are performed and indicated that the proposed method provides satisfactory performance with excellent accuracy and generalization property and achieves superior performance to the conventional method based on common LS-SVM and neural networks.

1 Introduction

Many, but not the most, physical systems exhibit some degree of nonlinearity. Nonlinear systems are simply those systems whose input-output relationship does not possess the property of superposition. In contrast to linear systems, the output of a nonlinear system in response to a weighted sum of several signals is not the weighted sum of the responses to each of those signals. Here we mainly want to study control systems [1]. Obtaining an accurate model of a complex, nonlinear, dynamic system is the basic step towards the creation of high performance controllers. In system modeling field, researchers are very enthusiastic about the potential of neural networks especially regarding the multilayer perceptron (MLP)[2][3][4]. However, their performance is not always satisfactory. Some inherent drawbacks, e.g., the multiple local minima problem, the choice of the number of hidden units and the danger of over fitting, etc., would make it difficult to put the MLP into some practice. In order to overcome those hard problems, major breakthroughs are obtained at this point, such as support vector machine (SVM), developed within the area of statistical learning theory and structural risk minimization. SVM has many advantages, such as nonexistence of curse of dimensionality, possessing good generalization performance and so on. As an interesting variant of the standard support vector machines, least squares support vector machines (LS-SVM) have been proposed by Suykens and Vandewalle [5][6] for solving pattern recognition and nonlinear function estimation problems. Standard SVM formulation is modified in the sense of ridge regression. LS-SVM taking equality instead of inequality constraints of SVM in the problem

D. Liu et al. (Eds.): ISNN 2007, Part I, LNCS 4491, pp. 618–625, 2007.

formulation. As a result one solves a linear system instead of a QP problem, so LS-SVM is easy to training. Accuracy is very important to the model problem, and different training method lead to different training accuracy, so we must find the methods which produce high accuracy. The pruning method is based on Keerthi's SMO formulation, which has been successfully applied to find non-sparse LS-SVM solutions. The sparseness is imposed by subsequently omitting data that introduce the smallest training errors and retraining the remaining data. Iterative retraining requires more intensive computations than training a single non-sparse LS-SVM. This paper we use the methods proposed in [7]. Both the computational costs and regression accuracy are solved. The effectiveness of the proposed method is demonstrated by numerical experiments. This paper is organized into five sections. Section 2 introduces LS-SVM used for regression. The Modeling and Control Framework with LS-SVM are described in Section 3. The proposed pruning algorithm is described in Section 4. In Section 5, we present the experimental results. Finally, conclusions are drawn in Section 6.

2 LS-SVM Regression

In the following, we briefly introduce LS-SVM regression. Consider a given training set of N data points $\{x_i, y_i\}_{i=1}^{N}$, with input data $x_i \in R$ and output $y_i \in R$. In feature space LS-SVM models take the form:

$$y(x) = w^T \psi(x) + b \tag{1}$$

where the nonlinear mapping $\psi(\cdot)$ maps the input data into a higher dimensional feature space. Note that the dimensional of w is not specified (it can be infinite dimensional). In LS-SVM for function estimation the following optimization problem is formulated

$$\min \frac{1}{2} w^T w + \frac{C}{2} \sum_{K=1}^{N} e_i^2 \tag{2}$$

subject to the equality constrains

$$y_i(w \cdot \varphi(x_i + b) = 1 - e_i, i = 1, \cdots, N \tag{3}$$

or an equivalent constraint used in [7]:

$$y_i = w \cdot \varphi(x_i) + b + e_i \tag{4}$$

where C is a regularization factor and e_i is the difference between the desired output and the actual output. For simplicity, we consider the problem without a bias term, as did in [8]. The Lagrangian for problem (2) is

$$R(w, e_i; \alpha_i) = \frac{1}{2} w^T w + \frac{1}{2} C \sum_i e_i^2 + \sum_i \alpha_i [y_i - w \cdot \varphi(x_i) - e_i] \tag{5}$$

where α_i are Lagrangian multiplier.

The nonlinear regression function (the output of LS-SVM) can be formulated by:

$$y(x) = \sum_{i=1}^{l} \alpha_i k(x, x_i) + b \tag{6}$$

$k(x_i, x_j)$ is a symmetric function which satisfies Mercer conditions. Some useful kernels are as following:
1) Polynomial kernel:

$$k(x, x_i) = [(x \cdot x_i) + 1]^q \tag{7}$$

2) RBF kernel:

$$k(x, x_i) = exp(-^{\|x - x_i\|^2/\sigma^2}) \tag{8}$$

3) Sigmoid kernel:

$$k(x, x_i) = \tanh(v(x \cdot x_i + c)) \tag{9}$$

Formula ($7 \sim 9$), parameter q, R, c are all real constant. In actual application, usually we must choice appropriate kernel function as well as the corresponding parameter according to the certain condition. The choice of the kernel function has several possibilities. In this work, the radial basis function (RBF) is used as the kernel function of the LS-SVM because RBF kernels tend to give good performance under general smoothness assumptions.

3 Modeling and Control Framework with LS-SVM

In this section we will use the LS-SVM regression algorithm, SMO-Based pruning Methods for LS-SVM are described previously to: (1) model a nonlinear dynamical system (design of the modeling LS-SVM block), (2) generate the control input to the nonlinear plant (design of the control block). Here we use a typical modeling and control framework [9], as sketched in the Fig.1. Suppose the actual plant is known. According to the data sets of the model, we design the LS-SVM block to model the plant's dynamical behavior. Where, r is the reference input of the plant, u is the control signal to the plant, e is the control error, t is the actual output of a plant, $\hat{t}$ is the output of the sparse LS-SVM, $\hat{e}$ is the error between the output of the plant and the output of the LS-SVM, it

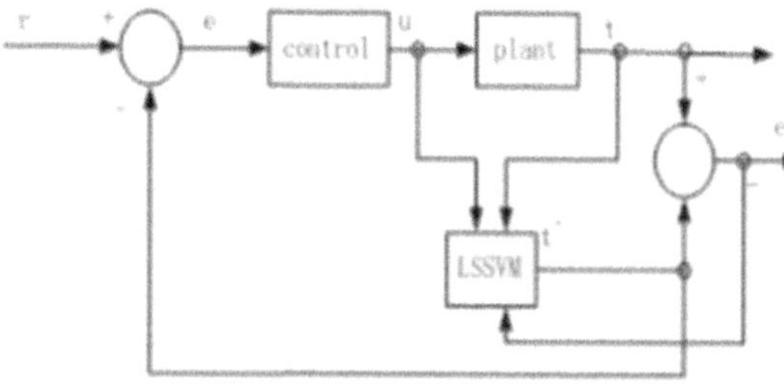

Fig. 1. A modeling-control framework with LS-SVM

is the identification error, and depends on the approximation capability of the LS-SVM itself.

The problem is to derive a self-tuning controller that minimizes a quadratic cost function based on a very general class of nonlinear models that can include nonlinear functions of old inputs, old outputs as well as the products of these functions and any power of most recent inputs. It is assumed that the nonlinear process can be described adequately by discrete time models which is linear in parameters and allows for time delays. Such a general model is known as NARMAX and can be represented in the following general form: Here, we use the NARMAX model [9], The relationship is:

$$\begin{cases} t(k+1) = f(x(k)) \\ x(k) = [t(k), \cdots, t(k-n+1), u(k), \cdots, u(k-m+1)]^T \end{cases} \tag{10}$$

$x \in R^{n+m}, n, m \in N$ at the time $k+1$,the sparse LS-SVM gives an estimate of $t(k+1)$, called $\hat{t}(k+1)$:

$$\hat{t}(k+1) = \hat{f}(x(k), w)) \tag{11}$$

To design the sparse LS-SVM block (namely solving the parameter vector), firstly we must build a training set, as follow:

$$(x_1, t_1) = [(t(n-1), \cdots, t(0), u(n-1), \cdots, u(n-m)), t(n)]$$

$$(x_2, t_2) = [(t(n), \cdots, t(1), u(n-1), \cdots, u(n-m+1)), t(n+1)]$$

$$(x_3, t_3) = [(t(n+1), \cdots, t(2), u(n+1), \cdots, u(n-m+2)), t(n+2)]$$

$$\vdots$$

$$(x_n, t_n) = [(t(3n-2), \cdots, t(n-1), u(2n-2), \cdots, u(2n-m-1)), t(2n-1)]$$

and so on. So we can get the training data set, preparing for the next section.

4 SMO-Based Pruning Algorithms for LS-SVM

The sparseness is very important for LS-SVM regression. The sparseness is imposed by subsequently omitting data that introduce the smallest training errors and retraining the remaining data. In the following, the SMO-Based pruning Algorithms are given in [7]. From (5), the Karush-Kuhn-Tucker (KKT) conditions for optimality are:

$$\begin{cases} \dfrac{\partial R}{\partial w} = 0 \quad w = \sum\limits_i \alpha_i \varphi(x_i) \\ \dfrac{\partial R}{\partial e_i} = 0 \quad \alpha_i = Ce_i \\ \dfrac{\partial R}{\partial \alpha_i} = 0 \quad y_i - w \cdot \varphi(x_i) - e_i = 0 \end{cases} \tag{12}$$

By substituting the KKT conditions (10) into the Lagrangian (5), the dual problem is to maximize the flowing objective function:

$$\max(L(\alpha)) = -\frac{1}{2}\sum_j\sum_i \alpha_i\alpha_j Q(x_i, x_j) + \sum_i \alpha_i y_i \tag{13}$$

where $Q(x_i, x_j) = K(x_i, x_j) + \sigma_{ij}/C$,and if $i = j, \sigma_{ij} = 1$;otherwise $\sigma_{ij} = 0$.

The SMO algorithm works by optimizing only one α_i at a time keeping the others fixed,i.e., α is adjusted by a step tt as follows:

$$\alpha_i^{new} = \alpha_i + tt; \alpha_j^{new} = \alpha_j, \forall j \neq i \tag{14}$$

Define

$$f_i = -\frac{\partial L}{\partial \alpha_i} = -y_i + \sum_{j=1}^{N} \alpha_j Q(x_i, x_j) \tag{15}$$

the tt can be suggested as in [7],

$$tt = \frac{-f_i}{Q(x_i, x_i)} \tag{16}$$

The criterion for determination of pruning points is a crucial factor in pruning process. In this section, we detail a new criterion that is directly based on the dual objective function and easy to compute in SMO formulation. To derive the proper criterion for pruning, the dual objective function (13) is rewritten using the definition of f_i .

$$L(\alpha) = \frac{1}{2}\sum_i \alpha_i(y_i - f_i) \tag{17}$$

Along with the idea of SMO, we consider that the removal of a sample k does not directly affect the support values of other samples, but it introduces the update of all f_i , which leads to a difference in the objective function, it is suggested in [7] , as :

$$d(L) = \frac{1}{2}\alpha_k^2 Q(x_k, x_k) - \alpha_k F_k \tag{18}$$

A summary of this training algorithm are as follows:

Step 1) Train the initial non-sparse LS-SVM using the SMO formulation as described in 4, using the training data set got in section 3.

Step 2) Repeat the following inner loop by time tt:

·remove a sample from the training set using criterion (18);

·update $f_i, \forall i \neq k$,of the remaining samples in the training set using $f_i^{'} = f_i - \alpha_k Q(x_i, x_k)$, where k is the omitted data point.

Step 3) Retrain the LS-SVM using the SMO formulation based on the support values α and the updated f of the remaining data set, where $\alpha^{'} = (\alpha_1, \cdots, \alpha_{k-1}, \alpha_{k+1}, \cdots, \alpha_N)$ and$f^{'} = (f_1^{'}, \cdots, f_{k-1}^{'}, f_{k+1}^{'}, \cdots, f_N^{'})$.

Step 4) Repeat Step 3) and Step 4) until the defined termination condition is satisfied.

Step 5) Get the LSSVM model.

5 Numerical Results

In this section we will take a concrete example to illustrate proposed control method. The plant under consideration is a spring-mass-damper system with a hardening spring:

$$\ddot{y}(t) + \dot{y}(t) + y(t) + y^3(t) = u(t)$$

We begin by generating 100 data dot $\{u(k), t(k)\}$ in the Fig.2. The data set is then split into two portions: one for training and one for testing, as follow: We

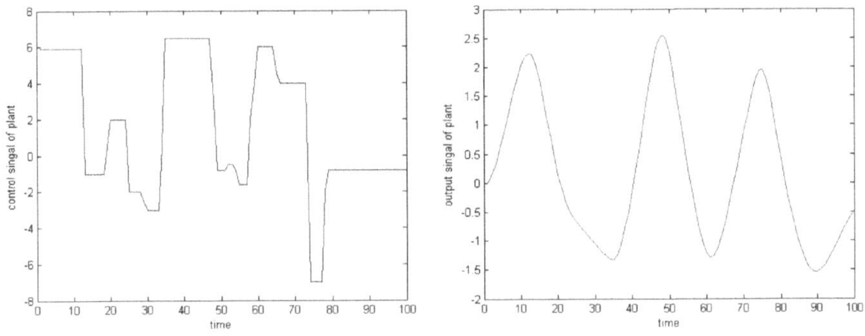

Fig. 2. Training and testing data set

got the train sets using two past outputs and two past controls as regression, namely [9]:

$$x(k) = [y(k), y(k-1), u(k), u(k-1)]^T$$

$$t(k) = y(k+1)$$

Getting those data sets$\{x(k), t(k)\}$, then we can train and test LS-SVM.

We use the SMO-Based on pruning methods to train the LS-SVM. The parameter are :

$\sigma = 3, C = 150$. In order to show the superiority of this method, we use the LS-SVM and RBF neural network to model the above plant. The following table illustrates training and testing error of the proposed method, LS-SVM and NN.

Table 1. The simulation error of Proposed method, LSSVM and RBF NN

Training	Testing	Proposed method		SVM		RBF NN	
LS-SVM	sets	Training error	Testing error	Training error	Testing error	Training error	Testing error
20	20	0.0053	0.0856	0.0406	0.1494	6.5143e-015	0.3305
30	30	0.0034	0.0794	0.0391	0.1050	5.7547e-014	0.3880
40	40	0.0029	0.0681	0.0478	0.0882	4.7945e-014	0.1094
50	50	0.0033	0.0618	0.0470	0.0698	9.0175e-014	0.1635
60	40	0.0030	0.0618	0.0440	0.0654	1.6850e-013	0.1021
70	30	0.0028	0.0668	0.0419	0.0975	3.0721e-013	0.1454

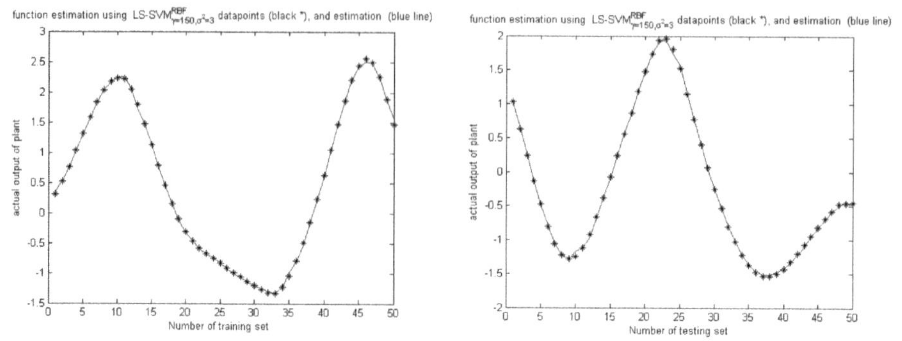

Fig. 3. LS-SVM training and testing results

As shown in Table 1, we can see the performance of the proposed method is superior to the performance of common LS-SVM and RBF NN Methods. Both the accuracy and generalization of training and testing error are best.

Fig.3 illustrate LS-SVM's training samples and testing samples fitting quality in the case of 50 training samples, 50 testing samples, we can see that trained LS-SVM block can exactly model the plant's dynamical behavior. From the obtained simulation results, we deduce that the method based on proposed LS-SVM can model a nonlinear unknown system efficiently.

6 Conclusion

In this paper we introduce the use of SMO-Based pruning methods for least square support vector machines for solving nonlinear systems' modeling problems. An introduction to LS-SVM is given at first, then gives its training algorithm, and uses it to build a modeling framework to control a nonlinear system, the numerical experiment has shown the efficiency of the LS-SVM based modeling method.

Acknowledgement

The authors would like to thank the reviewers for their helpful comments and constructive suggestions, which have been very improving the presentation of this paper. This work was supported by the Natural Science Foundation of Jiangsu province, China under Grant BK2006564 and China Postdoctoral Science Foundation under Grant 20060400274.

References

1. Slotine, L.J.E., Li, W.: Applied Nonlinear Control. Prentice-Hall. Englewood Cliffs. NJ. 1991
2. Lu, S., Basar, T.: Robust Nonlinear System Identification Using Neural Network Models. IEEE Trans on Neural Networks **9** (1998) 407-429

3. Narendra, K.S., Parthasarathy, K.: Identification and Control of Dynamical Systems Using Neural Networks. IEEE Trans on Neural Networks **1** (1990) 4-26
4. Grin, R., Cembrano, G., Torras, C.: Nonlinear System Identification Using Additive Dynamic Neural Networks–Two On-Line Approaches. IEEE Trans on Circuits and Systems-I **47** (2000) 150-165
5. Suykens, J.A.K., Vandewalle, J.: Least Squares Support Vector Machine Classifiers. Neural Processing Letter **9**(3) (1999) 293-300
6. Gritianini, N., Shaw e-Taylor, J.: An Introduction to Support Vector Machines. Cambridge University Press, 2000
7. Zeng, X.Y., Chen, X.W.: SMO-Based Pruning Methods for Sparse Least Squares Support Vector Machines. IEEE transactions on Neural Networks **16** (2005) 1541-1546
8. Radhakrishnan, T.K., Sundaram, S., Chidambaram, M.: Non-Linear Control of Continuous Bioreactors. Springer-Berlin/Heidelberg **20** (2005) 173-178
9. Zhang, H.R., Wang, X.D.: Nonlinear Systems Modeling and Control Using Support Vector Machine Technique. Springer-Berlin/Heidelberg **3967** (2006) 660-669

Pattern-Oriented Agent-Based Modeling for Financial Market Simulation

Chi Xu and Zheru Chi

Department of Electronics and Information Engineering,
The Hong Kong Polytechnic University,
Hung Hom, Kowloon, Hong Kong
chi.xu@polyu.edu.hk, enzheru@inet.polyu.edu.hk

Abstract. The paper presents a pattern-oriented agent-based model to simulate the dynamics of a stock market. The model generates satisfactory market macro-level trend and volatility while the agents obey simple rules but follow the behaviors of the neighbors closely. Both the market and the agents are made to evolve in an environment where Darwin's natural selection rules apply.

1 Introduction

Although a stock exchange market is a complex system, it obeys the rules a common market supplies to all buyers and sellers. The stock market attracts a lot of investors, and some of them deal with the market profitably from intensive analysis and research into the information with which they make successful judgment towards the state and trend of the market. The intensive research for finding the right stock to buy and the right time to buy it is not fruitless. Financial practitioners use different trading and forecasting strategies, say different agents, such as short horizons investors tend to use extrapolative chartists' trading rules, and long horizons investors tend to use mean reverting fundamentalists' trading rules. Generally the approaches to describe such a complex market system include the efforts on analysis of previous market data and charts, and some intelligent computational finance solutions to construct agent-based models to simulate the market behaviors. The developments in the latter area focus on two aspects. The economic dynamics approaches use heterogeneous economical price-explanation models to simulate the market and generate artificial time series as outputs. The alternative approach, the econometric model, describes the market prices by fitting agent's behavior to simulate the real-world economic relationships [1].

The financial market is a complex system, which cannot be represented by a simple mathematical or statistical model. To provide a better overview and more accurate prediction of the financial market, much effort has been put into research and development in intelligent computational finance. Intelligent computational finance employs a bottom-up approach, i.e., it uses heterogeneous or non-rational agent-based models to describe the traders who represent different opinions among market participants, and the outcome from a well constructed market simulation system reflects well tremendous trading volume in real markets. One of the benefits of using

D. Liu et al. (Eds.): ISNN 2007, Part I, LNCS 4491, pp. 626–631, 2007.

heterogeneous or non-rational agents is that the constructed models can give better explanations of asset price movements for the empirical observations. In the heterogeneous world, the agents can form the expectations equilibrium, which is some degree of consistency between expectations and realizations [5]. Another benefit is that the evolutionary approach can play important role in construction of the models that artificial intelligence makes the models act and learn independently to adapt to changing circumstances in behavioral economics, which has many degrees of freedom, hence to clearly forecast the price or trend of the stocks in a market. The individuals often do not behave rationally, and this causes bubbles of price in the markets. This behavioral agent-based computational approach has changed the way of thinking about the financial markets, and has become an important research area in economics known as economic dynamics. The development of economic dynamics contributes to the analysis of the complex economics and finance systems like stock market. Computational tools and numerical simulation analysis can be applied with heterogeneous agent model, hence to adapt nonlinear dynamics, chaos and complex systems for analysis [9].

In the real market, it is almost impossible to identify a pacemaker among all the trading agents, although the wealthiest stock dealers play an important role in the market. Although LeBaron [7] tried to explain the market performances by distinguishing multiple short- and long-memory investors as different pattern agents in SF-ASM, he did not probe in depth about how the different agents could switch their pattern during the simulation. Our project tries to implement the model using a scheme called pattern-oriented modeling from the study of ecology [4], in which evolution is not a process designed to produce any particular species but sort of rearrangement approaching the optimal or stronger total structures in species [3]. This scheme attempts to analyze the movement of the market price and trading volume as results of a pattern-amplifying machine. The agents have simple rules in their minds, to make the profit as large as possible. The key feature of pattern-oriented modeling is that a single agent notices simply the behaviors of his neighborhoods for a large collectivity of emergence [6], because small shifts in an agent's behavior can quickly escalate into larger movements of pushing up or drawing down the market price and generating trading volumes when the patterns are fed back to the community. Such scheme should be able to improve the performance of agent-based modeling by emphasizing on analyzing and validating the applicability of models to real problems. The more the model constructed approaches the real market, the more accurate predictive output can be obtained from the simulation. The pattern-oriented approaches should deal with the time series properties better and easier.

2 Construction of a Market

The agent-based simulation platform, StarLogo from MIT Media Lab, is applied in the research. Thanks to its powerful ability in modeling the behavior of decentralized systems, we can construct the market with non-pacemaker and non-rational trading agents.

2.1 Agents and Trading Rules

The goal of an agent is simple: to buy and sell stocks for making money. He knows not much about the market situation but the price of the stock and how agents around him are dealing with their stocks. Hence, the movement of agents in the market is a random walk on StarLogo canvas for a possible stock trading opportunity, almost as same as an ant is looking for food. The agents are designed to carry out the trading transaction according to the simple rule of "buy low and sell high", and the objective is to make them wealthier. For better simulation of pattern-oriented feature of mass attraction by making money among the agents, the agents are also designed to be able to sense the chemical that nearby agents issue as buying or selling stocks and they are always making efforts to approach the scent from stocks.

During the setting up process of the market, each agent is assigned randomly the buying power. When the buying power is less than zero, the agent is bankrupt and driven out of the market. An agent performs purchase activity if the stock he meets has a price lower than his power. The agent sells the stock when he finds out that the stock price is twice as much as his buying power or his buying power closes to zero.

The wealth $w_{i,t}$ of the agent i at time t for selling a stock can be expressed as:

$$w_{i,t} = E\sum_{s=1}^{\infty}\{(w_{i,t-s} + p_{j,t}) - \beta \log c_{i,t}\}$$

in which, $p_{j,t}$ represents the price of the stock j at time t, $c_{i,t}$ is the consumption, performing as logarithmic utility for agent's optimal choice and occupying a constant proportion of wealth [2], at time t that the agent i needs to maintain his life. The time rate of discount β can be set to $(1+0.27)^{1/12}$, which corresponds to an effective monthly rate of 0.02.

During the trade, the agents buying power decreases when a purchase is made, and increases when a selling is made. When an agent holds the stock, he still needs to pay off the loss from his living consumptions.

2.2 Market Information

According to the principle of macro economics, the stocks should maintain price trends over time if the company behind a stock is in a constant condition, so that it is possible for an agent to outperform the market by carefully selecting entry and exit points for equity investments.

In addition, the market should not be a zero-sum game place, where one participant's gains result only from another participant's equivalent losses, so some stocks have been put into the market by being assigned random values, say stock prices. After the commencement of the simulation, volume of the stock being traded should be in a dynamic balance state, in which equal amount of stocks are sold and bought simultaneously. The stock price hikes when an agent made the purchase, and the cash carried by the agent descends accordingly.

The necessary information that an agent needs for the stock trade can be expressed as $I(p_{j,t}; chemical; scent)$, so the stock price and chemical information are the guidance for agent's decision to buy, sell, or hold a stock.

The aggregate demand for stocks can be given by a demand function:

$$D(p_t) = \sum_{i=0}^{N} \frac{I(p_t; chemical; scent)\beta w_{i,t}}{p_t}$$

in which, N is the number of agent whose buying power is greater than a stock price.

The stock price hikes by adding a constant to its original value when an agent makes a purchase of the stock, and price falls by subtracting a constant accordingly.

2.3 Evolution of Agent and Market

The evolution conforms simply to Darwin's theory of survival of the fittest. If an agent's holding cash value becomes zero, he is driven out of the market. On the contrary, if agent's wealth breaks the top limit value, he becomes a super agent who can purchase stock as twice as a normal agent. The more super agents appear in the market, the faster stock price changes.

3 Simulation Results

At present stage, only two variables are measured for validation of the model. One is the average of stock price, and the other is the average of the amount that agent holds the cash.

The moving trends for both stock price and cash amount hike with a gentle slope, but when the agents sell the stocks eventually at almost the same moment, the stock price falls very steeply.

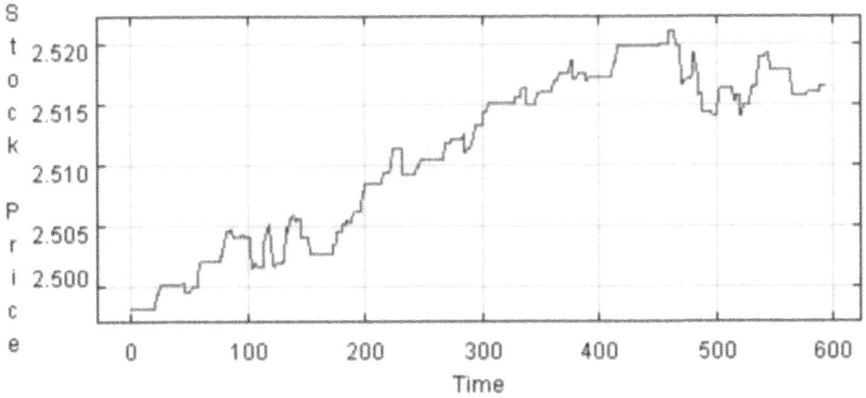

Fig. 1. Time series of average stock price volatility

The cash held in agents' hands has a steep rise corresponding to the moment that stock price descends, because agents sells out the stocks at a higher price, which means a good cross correlation exists between these two variables.

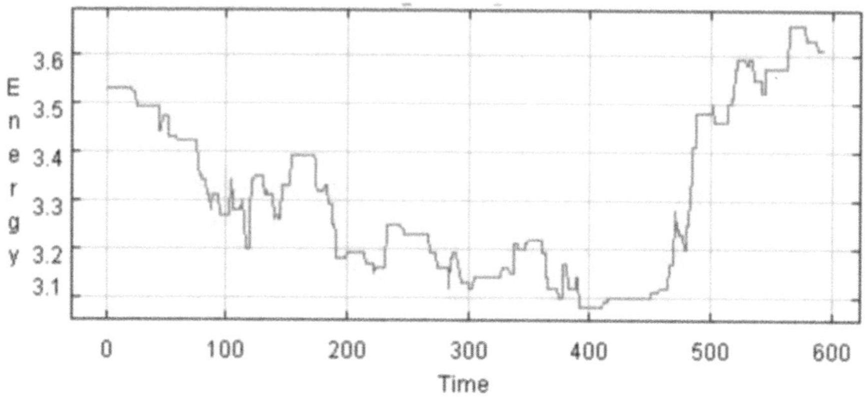

Fig. 2. Time series of agent's average buy power

4 Discussion and Future Research

Our research is the first step into pattern-oriented approach to explore complex financial market system. The outcome from the experiments shows that the model behaves several macro level phenomena in real market, especially the ideal moving trend of market, and cross correlation between the market price and the cash volume in buyers' hands.

At this point, the micro level agent behavior is almost as simple as an ant who is seeking for food. If the agents do more and more complicated consideration before any decision making, the interaction between the macro level stock market might carry out different dynamics. In the mean time, our present market is lack of strong impact from social or political elements, which might bring complete different outcome from the market dynamics.

The analytical research in the area of stock market is for purpose of making forecasting. During the evolution of the stock market, it is necessary to use existing data to train and testing market model, and the pattern-orientation should be strengthen, so agents are able to learn by their neighborhood faster.

References

1. Zimmermann, H.G., Neuneier, R., Grothmann, R.: Multiagent Modeling of Multiple FX-Markets by Neural Networks. IEEE Transactions on Neural Networks **12** (2001) 735-743
2. Mullainathan, S.: A Memory Based Model of Bounded Rationality, Massachusetts Institute of Technology Technical Report, Cambridge, MA (1998)
3. Dennett D.: Darwin's Dangerous Idea, Simon & Schuster 48-60 New York, 1995
4. Grimm, V., et al.: Pattern-Oriented Modeling of Agent-Based Complex Systems: Lessons from Ecology, Science, **310** (2005) 987-991
5. Hommes C.: Heterogeneous Agent Models in Economics and Finance, Handbook of Computational Economics, North-Holland **2** (2005)

6. Johnson, S.: Emergence, the connected lives of ants, brains, cities, and software, Scribner, New York (2001)
7. LeBaron, B.: Empirical Regularities From Interacting Long- and Short-Memory Investors in an Agent-Based Stock Market, IEEE Transactions on Evolutionary Computation **5** (2001) 442-455
8. LeBaron, B.: Building the Santa Fe Artificial Stock Market (2002)
9. LeBaron, B.: Agent-based Computational Finance. The Handbook of Computational Economics II (2005)

Non-flat Function Estimation Using Orthogonal Least Squares Regression with Multi-scale Wavelet Kernel

Meng Zhang[1], Lihua Fu[2], Tingting He[1], and Gaofeng Wang[3]

[1] Department of Computer Science,
Central China Normal University, 430079 Wuhan, P.R. China
[2] School of Mathematics and Physics,
Chinese University of Geosciences, 430074 Wuhan, P.R. China
fulihua9270@yahoo.com.cn
[3] CJ Huang Information Technology Research Institute,
Wuhan University, 430072 Wuhan, P.R. China

Abstract. Estimating the non-flat function which comprises both the steep variations and the smooth variations is a hard problem. The existing kernel methods with a single common variance for all the regressors can not achieve satisfying results. In this paper, a novel multi-scale model is constructed to tackle the problem by orthogonal least squares regression (OLSR) with wavelet kernel. The scheme tunes the dilation and translation of each wavelet kernel regressor by incrementally minimizing the training mean square error using a guided random search algorithm. In order to prevent the possible over-fitting, a practical method to select termination threshold is used. The experimental results show that, for non-flat function estimation problem, OLSR outperforms traditional methods in terms of precision and sparseness. And OLSR with wavelet kernel has a faster convergence rate as compared to that with conventional Gaussian kernel.

1 Introduction

In science and engineering areas, there are a lot of concerns about the problem of estimating non-flat functions which comprise both the steep variations and smooth variations. It is unsuitable to use the conventional kernel methods, such as support vector regression (SVR) [1], least squares support vector machines (LS-SVM) [2], linear programming (LP) [3] and so on. Those methods adopt a single common variance for all kernel regressors and estimate both the steep and smooth variations using an unchanged scale. Recently, a revised version of SVR, namely multi-scale support vector regression (MSSVR) [4, 5], is proposed by combining several feature spaces rather than a single feature space in standard SVR. The constructed multi-feature space is induced by a set of kernels with different scales. MSSVR outperforms traditional methods in terms of precision and sparseness, which will also be illuminated in our experiments.

Kernel basis pursuit (KBP) algorithm [6] is another possible solution which enables us to build a l_1-regularized multiple-kernel estimator for regression. However, KBP is prone to over-fit the noisy data. We will compare its performance with our new algorithm.

D. Liu et al. (Eds.): ISNN 2007, Part I, LNCS 4491, pp. 632–641, 2007.

Orthogonal least squares regression (OLSR) is an efficient learning procedure for constructing sparse regression models [7-9]. A key feature of OLSR is its ability to select candidate model regressors with different scales and centres, which allows the produced model to fit different parts of original function with different scales. Some global searching algorithms, such as the genetic algorithm, adaptive simulated annealing and repeating weighted boosting search (RWBS), can be used to determine the parameters of regressor [9-11].

When applying OLSR, many researchers usually regard Gaussian function as the first choice for kernel function, for its good generalized ability. But estimating a non-flat function requires the kernel function holds good local property to describe the local character of original function.

Wavelet techniques have shown promise for non-flat function estimation [12, 13]. Since the local property of wavelet makes efficient the estimation of the function having local characters, it is valuable for us to study the combination of wavelet and OLSR.

In this paper, a multi-scale model with wavelet kernel is constructed by use of OLSR. OLSR algorithm used here tunes the dilation parameter and translation parameter of individual wavelet regressors by incrementally minimizing the training mean square error (MSE) using RWBS.

In modeling noisy dataset, OLSR can fit a non-flat function by any precision which is prone to cause over-fitting. So when the user should stop selecting regressors is also a problem. By virtue of cross validation, an algorithm to select termination threshold is presented in order to prevent possible over-fitting.

The simulations are performed on the non-flat function estimation problem of both artificial dataset and real dataset. The experimental results show that

1 The OLSR model outperforms traditional ones by precious and sparseness.
2 OLSR with wavelet kernel has much faster convergence than that with Gaussian kernel.

2 Theory

Consider the problem of fitting the N pairs of training data $\{\mathbf{x}(l), y(l)\}_{l=1}^{N}$ with the regression model

$$y(l) = \hat{y}(l) + e(l) = \sum_{i=1}^{M} w_i \phi_i(l) + e(l), \quad l = 1, 2, \cdots, N \ , \tag{1}$$

where $\hat{y}(l)$ denotes the "approximated" model output, w_i's the model weights, $e(l)$ the modeling error at $\mathbf{x}(l)$ and $\phi_i(l) = k(\mathbf{c}(i), \mathbf{x}(l))$ are the regressors generated from a given kernel function $k(\cdot,\cdot)$ with centre vector $\mathbf{c}(i)$. If we choose $k(\cdot,\cdot)$ as a Gaussian kernel and $\mathbf{c}(i) = \mathbf{x}(i)$, then model (1) describes a RBF network with each data as a RBF centre and a fix RBF width. We are to find the best model $\sum_{i=1}^{M} w_i \phi_i(l)$ to describe the mapping $f(\mathbf{x})$ between the input $\mathbf{x}(l)$ and the output $y(l)$.

Let $\mathbf{\Phi}_i = [\phi_i(1), \cdots, \phi_i(N)]^T = [k(\mathbf{c}(i), \mathbf{x}(1)), \cdots, k(\mathbf{c}(i), \mathbf{x}(N))]^T$, $i = 1, 2, \cdots, M$, and then the regression matrix $\mathbf{\Phi} = [\mathbf{\Phi}_1, \cdots, \mathbf{\Phi}_M]$, weight vector $\mathbf{w} = [w_1, \cdots, w_M]^T$, output vector $\mathbf{y} = [y(1), \cdots, y(N)]^T$, and error vector $\mathbf{e} = [e(1), \cdots, e(N)]^T$.

Then the regression model (1) can be presented as following matrix form

$$\mathbf{y} = \mathbf{\Phi w} + \mathbf{e} . \tag{2}$$

The goal of modeling data is to find the best linear combination of the column of $\mathbf{\Phi}$ (i.e. the best value for $\mathbf{w}$) to explain $\mathbf{y}$ according to some criteria. The popular criteria is to minimize the sum of squared errors $E = \mathbf{e}^T \mathbf{e}$.

By OLSR algorithm, the solution is searched in a transformed orthogonal space. In more detail, let an orthogonal decomposition of the regression matrix $\mathbf{\Phi}$ be $\mathbf{\Phi} = \mathbf{HA}$, where $\mathbf{A}$ is an upper triangular matrix with the unit diagonal element and $\mathbf{H} = [\mathbf{H}_1, \mathbf{H}_2, \cdots, \mathbf{H}_M]$ with the orthogonal columns that satisfy $\mathbf{H}_i^T \mathbf{H}_j = 0$ if $i \neq j$. The regression model (2) can alternatively be expressed as

$$\mathbf{y} = \mathbf{H\theta} + \mathbf{e} , \tag{3}$$

where the new weight vector $\mathbf{\theta} = [\theta_1, \cdots, \theta_M]^T$ satisfies the triangular system $\mathbf{\theta} = \mathbf{Aw}$. Although the problem is converted to find the best solution in the linear space spanned by the column of $\mathbf{H}$ (i.e. the best value for $\mathbf{\theta}$), the resulting model remains equivalent to the solution of (2), which is still an element in the original space. For the orthogonal regression model (3), the training MSE can be expressed as

$$J = \mathbf{e}^T \mathbf{e} / N = \mathbf{y}^T \mathbf{y} / N - \sum_{i=1}^{M} \mathbf{H}_i^T \mathbf{H}_i \theta_i^2 / N . \tag{4}$$

Thus the training MSE for the k -term subset model can be expressed as $J_k = J_{k-1} - \mathbf{H}_i^T \mathbf{H}_i \theta_i^2 / N$ with $J_0 = \mathbf{y}^T \mathbf{y} / N$.

At the k th stage of regression, the k th regressor is determined by maximizing the error reduction criterion $E_k = \mathbf{H}_k^T \mathbf{H}_k \theta_k^2 / N$ with respect to the kernel centre $\mathbf{c}_k$ and its scale parameter d_k. The selection procedure is determined at k th step if $J_k < \xi$ is satisfied. A practical method to select a proper tolerance ξ is also presented in this paper (Subsection 3.2). Generally, Gaussian kernel is often the first choice of kernel because of its excellent generalized ability. Since the local property of wavelet makes efficient the estimation of the function having local characters, this paper will also study the OLSR with wavelet kernel, and compare it with the case with Gaussian kernel.

Wavelet transform turns to be a useful tool in time series analysis and signal processing for its excellent localization property [12, 13]. The idea behind the wavelet analysis is to express or approximate a signal or function by a family of functions generated by dilations and translations of a function $h(x)$ called mother wavelet:

$$h_{c,d}(x) = | d |^{-1/2} h(\frac{x - c}{d}) , \tag{5}$$

where $x, d, c \in R$, d is a dilation factor, and c is a translation parameter or centre parameter. A multidimensional wavelet function can be written as the product of 1-d wavelet function $h(\mathbf{x}) = \prod_{i=1}^{N} h(x_i)$ with $\{\mathbf{x} = (x_1, \cdots, x_N) \in R^N\}$. In this paper, we use the same mother wavelet as in [14], that is $h(x) = \cos(1.75x)\exp(-x^2/2)$.

3 Algorithm

Some guided random search methods can be used to determine the parameters of the k th wavelet regressor, that is d_k and $\mathbf{c}_k$, such as the genetic algorithm and adaptive simulated annealing. RWBS is recently proposed global searching algorithms [11]. It is extremely simple and easy to implement, involving a minimum programming effort. So, we perform this optimization by RWBS.

3.1 The k th Wavelet Regressor Selection

Let the vector $\mathbf{u}_k$ contain both centre parameters and translation parameters of k th wavelet regressor, that is $\mathbf{u}_k = [d_k, \mathbf{c}_k]^T$. Given the data $\{\mathbf{x}(l), y(l)\}_{l=1}^{N}$, and randomly selecting Ps parameter vectors $\{\mathbf{u}_i \mid i = 1, \cdots, Ps\}$, the basic weighted boosting search algorithm is summarized as Figure 1.

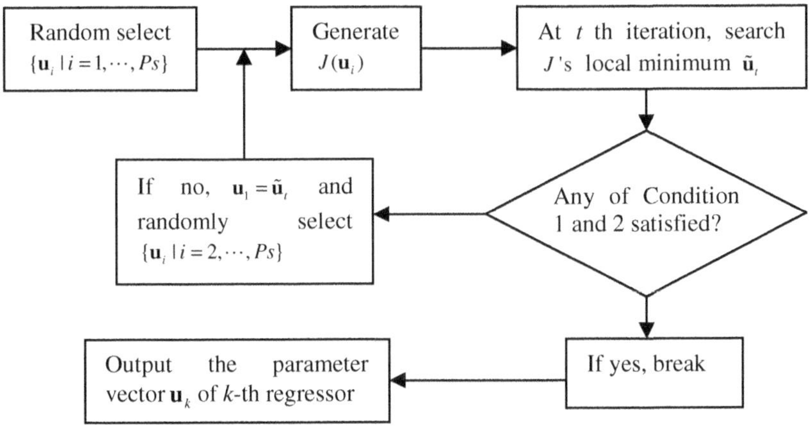

Fig. 1. The scheme of basic weighted boosting search algorithm

In Figure 1, Condition 1 means that the local minimums obtained at two continuous steps is close enough, that is $\|\tilde{\mathbf{u}}_t - \tilde{\mathbf{u}}_{t+1}\| < \varsigma$. Condition 2 means that the iteration number reaches the threshold Nb. The method of searching local minimum of J can refer to [11].

The cost function $J(\mathbf{u}_i)$ is generated according to the following steps:

Step 1 for $1 \le i \le Ps$, generate $\mathbf{\Phi}_i$ from $\mathbf{u}_i$, the candidates for the *k*-th model column,
Step 2 Orthogonalise $\mathbf{\Phi}_i$:

$$\alpha_j^i = \mathbf{H}_j^T \mathbf{\Phi}_i / (\mathbf{H}_j^T \mathbf{H}_j),\quad 1 \le j < k\, ,\, \tilde{\mathbf{H}}_i = \mathbf{\Phi}_i - \sum_{j=1}^{k-1} \alpha_j^i \mathbf{H}_j\, ,$$

where $\{\mathbf{H}_j \mid j = 1, \cdots, k-1\}$ denote the already-selected regressors of equation (3) while $\{\tilde{\mathbf{H}}_i \mid i = 1, \cdots, Ps\}$ mean the candidates for the *k*th regressor.

Step 3 Generate $J(\mathbf{u}_i)$

$$\gamma_i = (\tilde{\mathbf{H}}_i)^T \tilde{\mathbf{H}}_i\, , \theta_i = (\tilde{\mathbf{H}}_i)^T \mathbf{y} / \gamma_i\, , \text{ and}$$
$$J(\mathbf{u}_i) = J_{k-1} - \gamma_i (\theta_i)^2 / N$$

with J_{k-1} refers to the training MSE for the $k-1$-term subset model.

The above basic weighted boosting search algorithm performs a guided random search and solution obtained may depend on the initial choice of the population. To derive a robust algorithm that ensures a stable and global solution, RWBS algorithm is used by applying the basic weighted boosting search for NG times.

Using RWBS, one can obtain the best dilation and translation factors of the *k*th wavelet regressor.

Remark 1. To guarantee a global optimal solution as well as to achieve a fast convergence, the algorithmic parameters, NG, Nb, Ps and ς, need to be set carefully. The appropriate values of these parameters depend on the dimension of $\mathbf{u}$ and how hard the objective functions to be optimized. In this paper, in order to assure a global optimal solution, the thresholds NG, Nb and the size of generation size Ps are assigned to a little larger than needed.

In theory, this procedure can generate a model, by any precision, approximating the original mapping $f(\mathbf{x})$ between input $\mathbf{x}(l)$ and output $y(l)$. It will cause over-fitting in noisy setting. So it is necessary to preset a threshold ξ and if the condition $J_k < \xi$ is satisfied, we can stop the regressor selecting procedure before the model is fitted into the noise.

The procedure to generate the whole regression model can be described as:

```
For n=1:N
            Repeated Basic weighted boosting search
If   J_n > J_{n-1}  or If  J_n ≤ ξ
Break
End if
End for
```

Here, the largest iteration number N can be designed as the size of the training set. Usually the procedure will be ended at n-th when any of the two termination conditions satisfied, that is $J_n > J_{n-1}$ and $J_n \le \xi$ with $n << N$.

3.2 Threshold Selection

In theory, one can simply assign the threshold ξ is equal to the noise variance σ^2_{noise}. Unfortunately, the noise variance σ^2_{noise} usually remains unknown to us. One can select a proper ξ by cross validation.

First, divide the whole dataset into two equal-size groups which are training dataset and test dataset respectively. Given a candidate set for ξ, say $\{2^0, 2^{-1}, \cdots, 2^{-10}\}$. And train the model using every ξ in the candidate set, then evaluate the model using the test error obtained by test dataset. From the trend of test error for different values for ξ, one can get a proper value for ξ readily.

As a byproduct, we are able to estimate the noise level through the selected threshold.

4 Simulations

To demonstrate the practicability and the good performance of OLSR with wavelet kernel for non-flat function estimations, we applied to some simulated data and real data. Both Gaussian kernels and wavelet kernel were used in our experiment. In all experiments following, we assign $Ps = 10$ $NG = 50$, $Nb = 9$ and $\varsigma = 0.02$

4.1 Example 1: Oscillating Function

The first experiment was performed with the function used in [11].

$$f(u) = \begin{cases} -[\sin[6(u+5)]/2(u+5)]+4 & -8 \le u \le -2, \\ -4\exp[-5(u+1)^2 + 4.16] & -2 < u \le 0, \\ -[\sin[6(u-4)]/2(u-4)]+4 & 0 < u \le 8. \end{cases}$$

We generated the training dataset and test dataset of equal size $N = 160$ by an additive noise process $y_i = f(u_i) + n_i$, where the inputs u_i were uniformly sampled from the domain [-8, 8] and the noise $n_i \sim N(0, 0.2^2)$.

As the method described in Subsection 3.2, Figure 2 shows the procedure to select the threshold ξ. In Figure 2 (left), n label indicates the threshold ξ assigned as 2^{-n}. The results are obtained by both OLSR with Gaussian kernel (GOLSR) and OLSR with wavelet kernel (WOLSR). Both of the two lines indicate that the threshold should be assigned as 2^{-5}. A rough estimation for noise level also can be given as $\delta_{\text{noise}} \approx 2^{-5/2} = 0.18$ (actually $\delta_{\text{noise}} = 0.2$).

Figure 2 (right) shows the convergence of the GOLSR and WOLSR. WOLSR reaches the given threshold at 31 step while GOLSR at 38 step. After a further observation of Figure 2 (right), one can find that WOLSR is very close to the threshold at 18th step, while GOLSR at 27th step approximately. That is, for a given precision, WOLSR is capable to obtain a much sparser model than GOLSR, which means more efficient computation and better generalized ability of WOLSR.

The estimations of the non-flat function $f(x)$ are shown in Figure 3. The left shows the performance of GOLSR and the right for WOLSR. Input noisy data and kernel centre are shown by 'star' and 'circle' respectively. The original non-flat function $f(x)$ and the regression model are denoted by solid line and dashed line respectively. Due to the good local property of wavelet kernel, One can find that WOLSR fit the original function $f(x)$ better than GOLSR in the domains where the original function oscillates severely, such as $\{x \mid x \in (0,2) \cup (4,8)\}$ in this example.

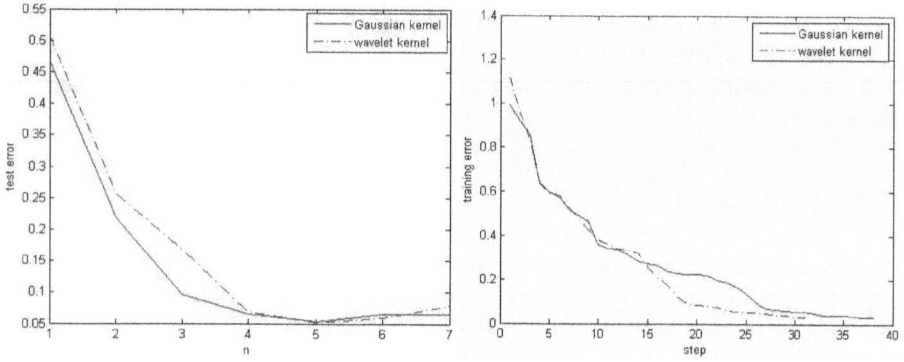

Fig. 2. Left shows the test error of GOLSR and WOLSR when assigning $\xi = 2^{-n}$; Right shows the training error of GOLSR and WOLSR at each step

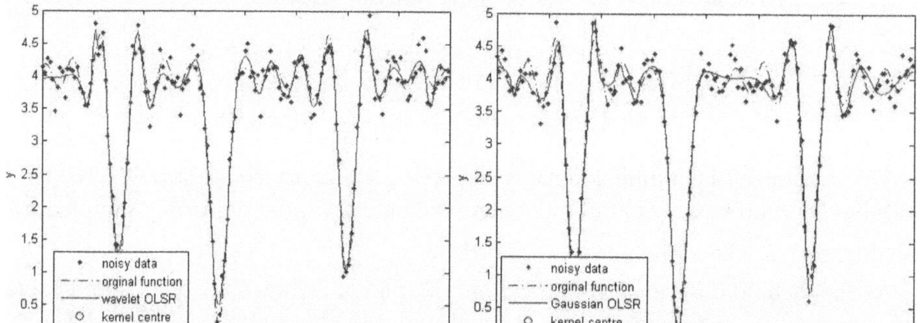

Fig. 3. Estimating the non-flat function: the left is for WOLSR and the right for GOLSR

4.2 Example 2: Mixture Function

A simple mixture of two Gaussian distributions in [4]

$$f(x) = \{4\exp[(x-2)^2/0.18] + \exp[(x-7)^2/2.88]\}/(\sqrt{2\pi} \times 1.2)$$

was studied. We generated training set and test set with equally size $N = 100$ by an additive noise process $y_i = f(x_i) + n_i$, where the input x_i were uniformly sampled from the domain [0, 10] and the noise $n_i \sim N(0, 0.05^2)$.

According to the procedure described in subsection 3.2, we assigned the threshold $\xi = 2^{-9}$ (which indicates that the noise level was roughly $\delta_{\text{noise}} \approx 2^{-9/2} = 0.044$ with the true value $\delta_{\text{noise}} = 0.05$). Figure 4 shows the performances of GOLSR and WOLSR. We repeated the experiment for 30 times. Table 1 shows the comparison with traditional algorithms as published in [4], which indicates that the OLSR algorithm produces the sparsest model.

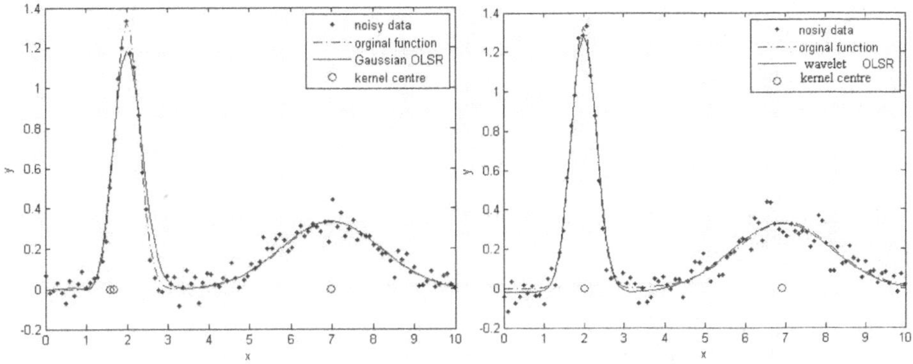

Fig. 4. Estimating the non-flat function: the left is for WOLSR and the right for GOLSR

Table 1. The averaged experimental results for experiment 2. The first 7 results were quoted from [4].

	Model Size	RMSE
SVR	63.9	0.0274 ± 0.0041
LPR	22.3	0.0268 ± 0.0037
LS-SVM	100.0	0.0250 ± 0.0034
KBP	14.7	0.0177 ± 0.0032
MS-SVR(E)	15.8	0.0202 ± 0.0035
MS-SVR(Q)	13.4	0.0171 ± 0.0031
MS-SVR(H)	10.9	0.0169 ± 0.0031
Gaussian OLSR	3.4	0.0370 ± 0.0052
Wavelet OLSR	3.1	0.0364 ± 0.0062

4.3 Example 3: Motorcycle Data

In this simulation, the motorcycle data [4] was used to evaluate the performance GOLSR and WOLSR. It consisted of a sequence of accelerometer readings through time following a simulated motorcycle crash during an experiment to deter mine the efficiency of crash helmets. There were a total of 133 samples in this dataset. As in [4], we preprocessed the dataset $x_i' = x_i / 6; \quad y_i' = y_i / 100$. As the threshold selection procedure, we selected $\xi = 2^{-4}$ which indicates that the noise level can be roughly estimated as $\delta_{\text{noise}} \approx 2^{-4/2} = 0.25$ compared with the result $\delta_{\text{noise}} = 0.2230$ published in [4]

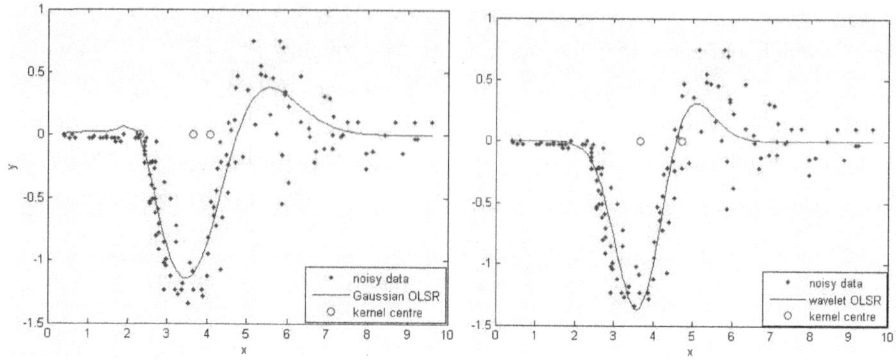

Fig. 5. Estimating the non-flat function: the left is for WOLSR and the right for GOLSR

Table 2. The averaged experimental results for experiment 3. The first 7 results were quoted from [4].

	Model Size	RMSE
SVR	49.3	0.2334
LPR	12.0	0.2343
LS-SVM	100.0	0.2330
KBP	9.0	0.2324
MS-SVR(E)	8.0	0.2322
MS-SVR(Q)	7.5	0.2329
MS-SVR(H)	9.0	0.2329
Gaussian OLSR	3.4	0.2258
Wavelet OLSR	3.1	0.2298

value. Figure 4 shows the performance of GOLSR and WOLSR. Table 2 shows the comparison with some traditional algorithms published as [4].

5 Conclusions

In order to estimate the non-flat function, this paper proposes a novel multi-scale model by orthogonal least squares regression (OLSR) with wavelet kernel. Unlike most of the other kernel method, the wavelet kernel's centres are not restricted to the training input data points and each wavelet kernel has an individually adjusted translations. Due to the flexibility of OLSR and good local prosperity of wavelet kernel, this new approach outperforms some traditional methods in our simulation experiments. And OLSR with wavelet kernel has a much faster convergence than that with traditional Gaussian kernel.

Acknowledgements

This work was supported by the National Natural Science Foundation of China No.60442005, 60673040 and SRF for OYT, CUG-Wuhan under grant CUGQNL0520.

References

1. Smola, A.: Regression Estimation with Support Vector Learning Machines. Master's Thesis, Technische University Műnchen, (1996) Available at (http:// www.kernel-machines.org)
2. Suykens, J.A.K., Vandewalle J.: Least Squares Support Vector Machine Classifiers. Neural Process. Lett. **9** (1999) 293–300
3. Smola, A., Schőlkopf, B., Rätsch, G.: Linear Programs for Automatic Accuracy Control in Regression. Proceedings of the Ninth International Conference on Artificial Neural Networks, London (1999) 575–580
4. Zheng, D., Wang, J., Zhao Y.: Non-flat Function Estimation with A Multi-scale Support Vector Regression. Neurocomputing, in press
5. Zheng, D., Wang, J., Zhao Y.: Training Sparse MS-SVR with an Expectation-Maximization Algorithm. Neurocomputing **69** (2006) 1659-1664
6. Guigue, V., Rakotomamonjy, A., Canu, S.: Kernel Basis Pursuit. Proceedings of the 16th European Conference on Machine Learning, Porto (2005)
7. Chen, S., Billings, S.A., Luo, W.: Orthogonal Least Squares Methods and Their Application to Non-linear System Identification. Int. J. Control **50** (1989) 1873–1896
8. Chen, S., Cowan, C.F.N., Grant, P.M.: Orthogonal Least Squares Learning Algorithm for Radial Basis Function Networks. IEEE Trans. Neural Networks **2** (1991) 302–309
9. Chen, S. Wang, X.X. Brown, D.J.: Orthogonal Least Squares Regression with Tunable Kernels. Electronics Letters **41** (8) (2005)
10. Chen, S.Y. Wu and B.L. Luk,: Combined Genetic Algorithm Optimization and Regularized Orthogonal Least Squares Learning for Radial Basis Function Networks. IEEE Trans. Neural Networks **10** (5) (1999) 1239-1243
11. Chen, S., Wang, X.X. Harris, C.J.: Experiments with Repeating Weighted Boosting Search for Optimization in Signal Processing Applications. IEEE Trans. Syst. Man Cybern. B, Cybern. **35** (4) (2005) 682-693
12. Mallat, S.: A Wavelet Tour of Signal Processing. Academic Press, (1999)
13. Daubechies, I. : Ten Lectures on Wavelets. CBMS, 61, SIAM, Philadelphia, (1992)
14. Zhang, L., Zhou, W., Jiao, L.: Wavelet Support Vector Machine. IEEE Trans. on System, Man and Cybernetics-Part B: Cybernetics **34** (2004) 34–39

Tension Identification of Multi-motor Synchronous System Based on Artificial Neural Network*

Guohai Liu, Jianbing Wu, Yue Shen, Hongping Jia, and Huawei Zhou

School of Electrical and Information Engineering
Jiangsu University, Zhenjiang 212013, Jiangsu Province, China
{ghliu, wjb714}@ujs.edu.cn

Abstract. Sensorless tension control of multi-motor synchronous system with closed tension loop is required in many fields. How to identify the knowledge of instantaneous magnitude of tension is key. In this paper the tension identification is managed on the base of stator currents and its previous values with neural network. According to the fundamental state equations of multi-motor system for tension control, the novel method of tension identification using neural network is presented .A multi-layer feed-forward neural network (MFNN) is trained by Back Propagation Levenberger-Marquardt's method. Simulation and experiment results show that the system with tension identification via a neural network has better performance, and it can be used in many application fields.

1 Introduction

The processed objects are rolled in continuous production line such as producing steel line, printing and dyeing or papermaking. To ensure product quality, the tension acting on the processed objects must be constant. Because the thickness of the processed objects will be uneven if the tension is too large or the processed objects will be piled up if the tension is too small.

Dancer rolls or pressure sensors [1-3] are used to measure the tension in the usual tension control system. Mechanical potentiometers detect the position of dancer rolls and compare with the given position. The errors are sent to position controller to keep the tension constant. Pressure sensors are installed under the detected roll to measure the tension of the processed objects. The tension signal is sent to tension controller to obtain constant tension. But the installation of dancer rolls and pressure sensors results in complex equipment, more cost and difficult maintenance. At the same time, the mechanical error influences on the precision of detection and the performance of control to cause the decline of system's reliability. More over, dancer rolls and pressure sensors sometimes can't be installed account for speed and material. So the method of identification tension is necessary.

* Project supported by China ministry fund of education(20050299009) and Jiangsu Nature Science Foundation No.BK2003049.

D. Liu et al. (Eds.): ISNN 2007, Part I, LNCS 4491, pp. 642–651, 2007.

Previous literature [4] about tension identification is based on motor stator current and rotor speed by a tension observer. In this paper a method to identify tension by artificial neural network based on motor stator currents only is presented. Simulation and experiment results show that the system with tension identification via a neural network has better performance.

2 Building the Tension Model of Multi-motor Synchronous System

A two-motor system is the smallest one of multi-motor synchronous systems. So a two-motor system is studied in this paper to illustrate the method of tension identification.

2.1 Mathematic Model of Induction Motor

The fundamental state equations of induction motors and inverter controlled by current in two-phase rotating dq coordinates are as follows [5].

$$\begin{cases} \dfrac{d\omega_r}{dt} = \dfrac{n_p{}^2 L_m}{JL_r}(\psi_{rd} i_{sq} - \psi_{rq} i_{sd}) - \dfrac{n_p}{J} T_L \\ \dfrac{d\psi_{rd}}{dt} = -\dfrac{1}{T_r}\psi_{rd} + (\omega_1 - \omega_r)\psi_{rq} + \dfrac{L_m}{T_r} i_{sd} \\ \dfrac{d\psi_{rq}}{dt} = -\dfrac{1}{T_r}\psi_{rq} - (\omega_1 - \omega_r)\psi_{rd} + \dfrac{L_m}{T_r} i_{sq} \\ \dfrac{di_{sd}}{dt} = -\dfrac{1}{T_s} i_{sd} + \omega_1 i_{sq} + \dfrac{1}{T_s} i_{sd}^* \\ \dfrac{di_{sq}}{dt} = -\dfrac{1}{T_s} i_{sq} - \omega_1 i_{sd} + \dfrac{1}{T_s} i_{sq}^* \end{cases} \tag{1}$$

Where ω_r is the angular velocity of rotor. n_p is the number of pole-pairs of induction motor. L_m is the mutual inductance between stator winding and rotor winding in dq coordinates system. L_r is the self-inductance of rotor winding in dq coordinates system. J is the rotary inertia of induction motor. ψ_{rd}, ψ_{rq} is the rotor flux linkage of axes d and q, respectively. i_{sd}, i_{sq} is the stator current of axes d and q, respectively. T_L is the load torque, and T_r is the time constant of rotor. ω_1 is the synchronous speed. T_s is the delayed time constant of inverter. i_{sd}^*, i_{sq}^* is the given stator current of axes d and q, respectively .

$i_{sd} \approx i_{sd}^*, i_{sq} \approx i_{sq}^*$ when the delay of inverter is ignored ($T_s \approx 0$) and then the state equations become

$$\begin{cases} \dfrac{d\omega_r}{dt} = \dfrac{n_p{}^2 L_m}{JL_r}(\psi_{rd} i_{sq} - \psi_{rq} i_{sd}) - \dfrac{n_p}{J} T_L \\ \dfrac{d\psi_{rd}}{dt} = -\dfrac{1}{T_r}\psi_{rd} + (\omega_1 - \omega_r)\psi_{rq} + \dfrac{L_m}{T_r} i_{sd} \\ \dfrac{d\psi_{rq}}{dt} = -\dfrac{1}{T_r}\psi_{rq} - (\omega_1 - \omega_r)\psi_{rd} + \dfrac{L_m}{T_r} i_{sq} \end{cases} \tag{2}$$

According to the rotor flux orientation scheme ($\psi_r = \psi_{rd}$, $\psi_{rq} = 0$), the state equations are as follows.

$$\begin{cases} \dfrac{d\omega_r}{dt} = \dfrac{n_p{}^2 L_m}{JL_r}\psi_r i_{sq} - \dfrac{n_p}{J} T_L \\ \dfrac{d\psi_r}{dt} = -\dfrac{1}{T_r}\psi_r + \dfrac{L_m}{T_r} i_{sd} \\ \omega_1 = \omega_r + \dfrac{L_m}{T_r \psi_r} i_{sq} \end{cases} \tag{3}$$

The third formula of the equations set (3) is solved. $i_{sq} = (\omega_1 - \omega_r)T_r\psi_r / L_m$ is substituted to the first formula of the equations set (3).

$$\frac{d\omega_r}{dt} = \frac{n_p{}^2 L_m}{JL_r}\psi_r[(\omega_1 - \omega_r)T_r\psi_r / L_m] - \frac{n_p}{J} T_L \tag{4}$$

2.2 Model of Tension

Fig. 1 shows the tension model of two-motor system.

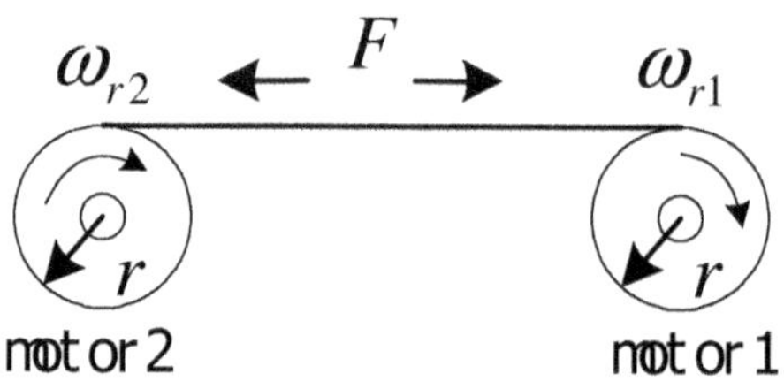

Fig. 1. Tension model of two-motor system

The motion equation of motor 1 is

$$\dot{\omega}_{r1} = \frac{n_p{}^2 L_m}{JL_r}\psi_1[(\omega_{11} - \omega_{r1})\frac{L_r\psi_1}{L_m R_r}] - n_p\frac{T_{L1}}{J} - \frac{n_p r F}{J} \tag{5}$$

where ω_{r1}, ψ_1, ω_{11}, T_{L1} is the speed, rotor flux linkage, synchronous rotating velocity and load torque of motor 1, respectively. r is the radius of roller, and F is the tension.

The motion equation of motor 2 is

$$\dot{\omega}_{r2} = \frac{n_p^{\,2} L_m}{J L_r}\psi_2[(\omega_{12}-\omega_{r2})\frac{L_r\psi_2}{L_m R_r}] - n_p\frac{T_{L2}}{J} + \frac{n_p rF}{J} \tag{6}$$

where ω_{r2}, ψ_2, ω_{12}, T_{L2} is the speed, rotor flux linkage, synchronous rotating velocity and load torque of motor 2, respectively.

The mathematic model of tension is

$$\dot{F} = K_c(rk\omega_{r1} - rk\omega_{r2})\frac{1}{T_c} - F\frac{1}{T_c} \tag{7}$$

where K_c, k, T_c is the transport coefficient, velocity ratio and the time constant of tension variation.

The second formula of the equations (3), formula (5), formula (6) and formula (7) constitute the mathematic model of tension when induction motor is in the rotating $d-q$ coordinates system according to the rotor flux orientation.

$$\begin{cases}
\dot{\psi}_1 = -\frac{R_r}{L_r}\psi_1 + L_m\frac{R_r}{L_r}i_{sd1} \\
\dot{\omega}_{r1} = n_p^{\,2}\frac{L_m}{JL_r}\psi_1[(\omega_{11}-\omega_{r1})\frac{L_r\psi_1}{L_m R_r}] - n_p\frac{T_{L1}}{J} - \frac{n_p rF}{J} \\
\dot{\theta}_1 = \omega_{11} \\
\dot{\psi}_2 = -\frac{R_r}{L_r}\psi_2 + L_m\frac{R_r}{L_r}i_{sd2} \\
\dot{\omega}_{r2} = n^2_{\ p}\frac{L_m}{JL_r}\psi_2[(\omega_{12}-\omega_{r2})\ \frac{L_r\psi_2}{L_m R_r}] - n\frac{T_{L2}}{J} + \frac{n_p rF}{J} \\
\dot{\theta}_2 = \omega_{12} \\
\dot{F} = K_c(rk\omega_{r1} - rk\omega_{r2})\frac{1}{T_c} - F\frac{1}{T_c}
\end{cases} \tag{8}$$

where i_{sd1}, i_{sd2} is the stator current in axes d of motor 1 and motor 2, respectively. θ_1, θ_2 is the electrical rotary angular position of motor 1 and motor 2 , respectively.

According to the mathematic model of tension (formula 7), there is a nonlinear mapping relationship between the tension and the rotary speed of motor 1 and motor 2.

That is to say

$$F = f(\omega_{r1}, \omega_{r2}) \tag{9}$$

In the synchronous rotary coordinates system, the stator current of axes $\alpha\beta$ in two-phase stationary coordinates system is obtained through the transformation formula from two-phase rotary coordinates system to two-phase stationary coordinates system ($C_{2r/2s}$) [6]:

$$\begin{bmatrix} i_{sd} \\ i_{sq} \end{bmatrix} = \begin{bmatrix} \cos\theta & \sin\theta \\ -\sin\theta & \cos\theta \end{bmatrix} \begin{bmatrix} i_{s\alpha} \\ i_{s\beta} \end{bmatrix} \tag{10}$$

where θ is the rotor angular position. For a certain θ, $C_{2r/2s}$ is a constant matrix. So there is a nonlinear mapping relationship between the motor speed and stator current of axes $\alpha\beta$ in two-phase stationary coordinates system (including its previous value) [7-8].

$$\omega_r = g[i_{s\alpha}(k), i_{s\alpha}(k-1), i_{s\alpha}(k-2), i_{s\beta}(k), i_{s\beta}(k-1), i_{s\beta}(k-2)] \tag{11}$$

where $i_{s\alpha}(k), i_{s\alpha}(k-1), i_{s\alpha}(k-2), i_{s\beta}(k), i_{s\beta}(k-1), i_{s\beta}(k-2)$ is respectively the stator current in $\alpha\beta$ axes and its value at the time before one or two intervals, when the motor is in two-phase stationary coordinates system.

According to formula (9) and (11), there is a nonlinear mapping relationship between the tension and the stator current of axes $\alpha\beta$ in two-phase stationary coordinates system (including its previous value) of motor 1 and motor 2.

$$\begin{aligned} F = h[&i_{s\alpha1}(k), i_{s\alpha1}(k-1), i_{s\alpha1}(k-2), i_{s\beta1}(k), i_{s\beta1}(k-1), i_{s\beta1}(k-2), i_{s\alpha2}(k), i_{s\alpha2}(k-1), \\ &i_{s\alpha2}(k-2), i_{s\beta2}(k), i_{s\beta2}(k-1), i_{s\beta2}(k-2)] \end{aligned} \tag{12}$$

where subscript 1 and 2 represent motor 1 and motor 2.

A multilayer feed-forward neural network (MFNN) is the most common and effective network. It can approach random linear or nonlinear function at any precision

A MFNN can substitute the nonlinear relationship between the tension and the stator current of axes $\alpha\beta$ in two-phase stationary coordinates system (including its previous value) of motor 1 and motor 2. A MFNN information is transferred in positive direction. The input information is transferred from input layer to output layer through hidden layer. The neuron status of every layer only influences the neuron status of the next layer. There is no any connection between the same layer neurons.

3 Learning and Training of a MFNN and the Model of Tension Identification

3.1 Training a MFNN with the Stator Current

According to the equations set (8), S-function is built in Matlab 6.5. The SIMULINK module of tension model is produced when the multi-motor synchronous system is in the rotary dq coordinates system oriented by the rotor flux. This module is used to gain the information for computer simulation.

According to the formula (12), the learning and training model to identify the tension via a MFNN is shown as figure 2. The input information is the stator currents and its previous values of motor 1 and motor 2.

The neural network's input samples for learning are the stator currents and its previous values of $\alpha\beta$ axes of two motors in two-phase stationary coordinates system.

The neural network is trained by the back propagation (BP) algorithm that is supervised by tutors. The neural network's output sample is the expected output of neural network that is the output of tension control model $F(k)$. The course of neural network learning is to modify the neural network's weight and deviation by the error

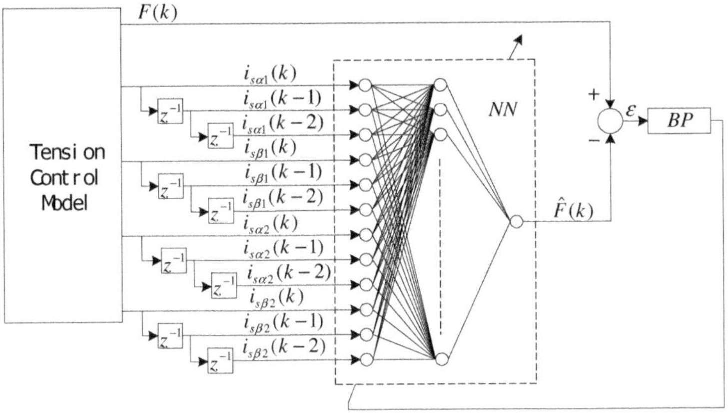

Fig. 2. Neural network training model of tension identification

ε between the neural network's output vector $\hat{F}(k)$ and the target vector $F(k)$. It tries its best to make the neural network's output closest to the expected target.

The variation of neural network's weight and deviation are calculated continuously in the direction of error function's descending slope. The variation of weight and deviation are proportional to the influence on the network error every time. The error variation of output layer is calculated if the expected output of output layer is not obtained. The error variation is back propagated by neural network along with the original connection. The weight of every layer neuron is modified till the neural network output reaches the expected value.

In the control system's toolbox of Matlab 6.5, calculation such as the output of hidden layer and output layer, error function and the variation of weight and deviation have been composed in function form. It is convenient to call the function to obtain the calculation results. Matrix form is adopted during the course of training to make training simple and fast.

At the beginning of MFNN training, the weight and deviation of every layer is initialized with a random number. The neural network weighted input vector, output vector, error vector and the sum of error square are calculated. The training stops when the sum of error-square is smaller than the expected one or the training reaches the designed step numbers. Otherwise, the error variation is calculated in output layer and the weight and deviation is adjusted using the learning rule of back propagation. This course repeats till the error is smaller than the expected one. The number of repeat times is the number of steps.

During the course of computer simulation, the motor parameter is as follows: P_e =1.1KW, R_s =6.1 Ω, R_r =5.6 Ω, L_s =0.573H, L_r =0.58H, L_m =0.55H. The rated speed is 1400r/min, and the sampling period is T=1ms. The simulation time is from 0 second to 1 second.

The sample data for learning is 1000 sets. The trained input and output sample datum are normalized to accelerate the convergence of neural network training. This neural network consists of two hidden layers (including 20 and 7 neurons, respectively). All of the hidden layer neurons adopt hyperbolic tangential Sigmoid function.

This neural network has only an output vector. Input and output layer neurons adopt pure linear transformation function. Trainlm(Levenberg-Marquardt rule) is used to train the BP neural network. The error between the neural network output $\hat{F}(k)$ and the expected value $F(k)$ (the output of tension control model) together with the error variation acts as the back propagated signal to adjust the weight and deviation of neural network in order to continuously reduce the error. The sum of error-square is 2.8×10^{-7} when the neural network is trained 1000 times. It is shown that the training process is convergent in the course of computer simulation.

Thus a BP neural network module is produced to realize the nonlinear mapping relationship between tension and stator current in $\alpha\beta$ axes (including its previous value) of motor 1 and 2 when the multi-motor synchronous system is in the two-phase stationary coordinates system.

3.2 Tension Identification Model Using a Neural Network

A neural network module has been trained to identify the tension. The identification model is shown as figure 3.

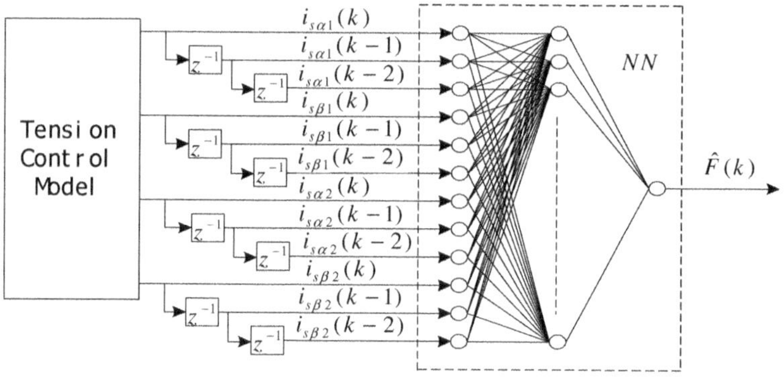

Fig. 3. Neural network model of tension identification

The input channel of the trained neural network module is feeded with the stator current in $\alpha\beta$ axes (including its previous value) of motor 1 and 2, when the multi-motor synchronous system is in the two-phase stationary coordinates system.

The neural network output is produced by computer simulation to identify the tension of multi-motor synchronous system through a neural network.

The neural network's output must be multiplicated a corresponding coefficient to eliminate the influence of normalization. Because input and output sample datum are normalized to accelerate training convergence before training.

Figure 4 shows the result of computer simulation. The neural network's output tracks the output of tension control model closely. The identification error is less than 0.3%.

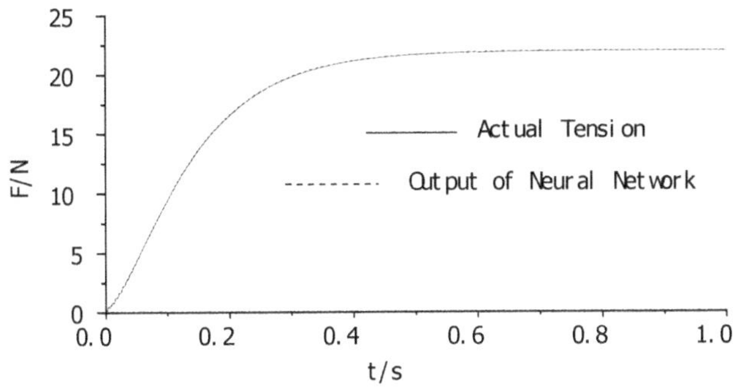

Fig. 4. Tension identification via neural network

4 Experiment Results

The motor parameters in actual experiment are the same as those used in computer simulation.

Stator winding adopts Y-connection without zero line. So $i_A + i_B + i_C = 0$. Then the transformation of two-phase stationary coordinates system to three-phase stationary coordinates system is as follows.

$$\begin{bmatrix} i_{s\alpha} \\ i_{s\beta} \end{bmatrix} = \begin{bmatrix} \sqrt{\frac{3}{2}} & 0 \\ \frac{1}{\sqrt{2}} & \sqrt{2} \end{bmatrix} \begin{bmatrix} i_A \\ i_B \end{bmatrix} \tag{13}$$

Formula (13) shows that there is only a constant matrix between the stator current in $\alpha\beta$ axes of two-phase stationary coordinates system and the stator current in AB axes of three-phase stationary coordinates system. So the stator current in AB axes of three-phase stationary coordinates system can act as direct experimental data of neural network's input.

In the experiment, Hall current sensors detect the stator currents of the two-motors synchronous system in AB axes of three-phase stationary coordinates system.

The oscilloscope records the stator current in AB axes from motor starting to motor stabilization.

The stator currents in AB axes from 0 second to 10 seconds are obtained, and used as sample datum.

These datum act as sample input of neural network to produce a neural network module by training.

A pressure sensor is used to detect the strip tension every 0.1 second. The tension from 0 second to 10 seconds is obtained when the two-motor synchronous system begins from start till stabilization. These data act as the target tension sample of neural network's output.

Interpolation method is used to ensure that the dimension of target output is the same as the one of neural network's input. The training course to produce a neural network module is the same as the above one in simulation.

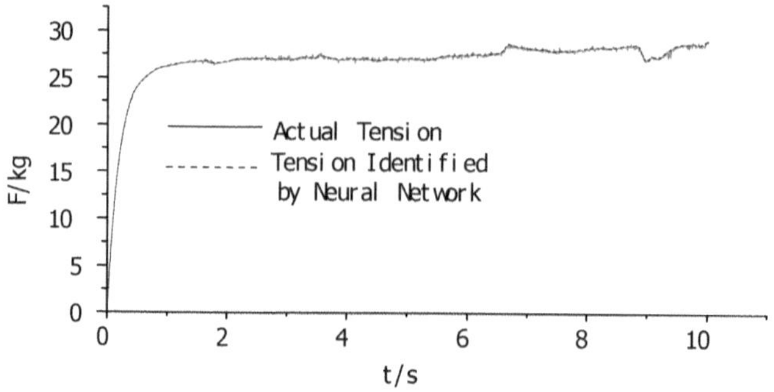

Fig. 5. Experiment of tension tracking via neural network

Figure 5 shows the actual output and neural network's outputof tension wave in two-motor synchronous system. The tension identified by neural network tracks the actual tension closely. The identification error is less than 3%.

5 Conclusion

The fundamental state equations of a multi-motor system controlled by current inverter and oriented by rotor flux are obtained in rotating coordinates system. According to the state equations, a method of tension identification using a neural network is presented. The identification model is built up based on the stator current in $\alpha\beta$ axes and its previous value. Even thouth when a multi-motor synchronous system is in three-phase stationary coordinates system, this method can be used. The proposed method is practically effective from the results of simulation and experiment. The tension identified by a neural network tracks the actual tension closely and the system with neural network identification model has better dynamic performance.

References

1. Thiffault, C., Sicard, P., Bouscayrol, A.: Desensitization to Voltage Sags of a Rewinder by Using an Active Dancer Roll for Tension Control. In: IEEE International Electric Machines and Drives Conference. 15-18 May 2005, Hilton Palacio Del Rio San Antonio, TX, USA (2005) 466-473
2. Li, J.H., Shi, Y.Y.: Design of the Tension Control System for a Cloth Winding Machine. Mechanical Science and Technology (2005) 1127-1129
3. Ebler, N.A., Arnason, R., Michaelis, G., D'Sa, N.: Tension Control: Dancer Rolls or Load Cells. IEEE Transactions on Industry Applications **29 (4)** (1993) 727-739
4. Song, S.H, Sul, S.K.: A New Tension Controller for Continuous Strip Processing Line. IEEE Transactions on Industry Applications **36 (2)** (2000) 633 –639
5. Chen, J.: The Mathematic Model of an Induction Motor and Its Speed Regulating System. National Defence Industry Press, Beijing (1989)

6. Chen, B.S.: Automatic Control System of Motor Drives. Mechanical Industry Press (1992)
7. Orlowska-Kowalska, T., Kowalski, C.T.: Neural Network Application for Flux and Speed Estimation in the Sensorless Induction Motor Drive. ISIE '97., Proceedings of the IEEE International Symposium on Industrial Electronics **3** (1997) 1253-1258
8. Wu, J.B., Liu, G.H.: Rotor Flux and Speed Estimation of Induction Motor Using BP Neural Network. Power Electronics **4** (2002) 27-30

Operon Prediction Using Neural Network Based on Multiple Information of Log-Likelihoods

Wei Du, Yan Wang, Shuqin Wang, Xiumei Wang, Fangxun Sun, Chen Zhang, Chunguang Zhou, Chengquan Hu, and Yanchun Liang

College of Computer Science and Technology, Jilin University, Key Laboratory of Symbol Computation and Knowledge Engineering of Ministry of Education, Changchun 130012, China
ycliang@jlu.edu.cn

Abstract. Operon represents a basic organizational unit in microbial genomes. Operon prediction is an important step to study genic transcriptional and regulatory mechanism in microbial genomes. This paper predicted operons in the Escherichia coli K12 genome using neural network based on four types of genomic log-likelihood data. First this method estimated the log-likelihood values for intergenic distances, COG gene functions, conserved gene pairs and phylogenetic profiles, and then used these information by a generalized regression neural network to discriminate pairs of genes within operons (WO pairs) or transcription unit borders (TUB pairs). We test the method on E. coli K12 and find that it can obtain average sensitivity, specificity and accuracy at 85.9%, 89.2% and 87.9% respectively, which indicates that the proposed method has a powerful capability for operon prediction.

1 Introduction

The concept of operon first appeared in the theory about protein regulatory mechanism proposed by F. Jacob and J. Monod. An operon represents a basic transcriptional unit of genes in the complex biological processes of microbial genomes [1]. Therefore, predicting operons is one of the most fundamental but an important research in microbial genomes [2].

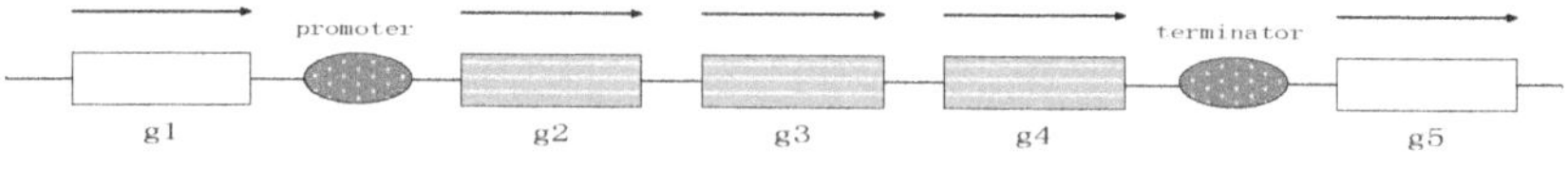

Fig. 1. The structure of operon: g2, g3, g4 composing an operon

In general, an operon is a cluster of one or more tandem genes delimited by a promoter and a terminator, and its structure is shown in Figure 1. It usually has the same properties mostly [3]: two adjacent pairs in a same operon tend to have shorter intergenic distances, a higher probability being in the same COG functional categories, easier to be conserved and more similar phylogenetic profiles.

D. Liu et al. (Eds.): ISNN 2007, Part I, LNCS 4491, pp. 652–657, 2007.

In this paper, we propose a method for operon prediction (for short, OPNN) which uses four types log-likelihood values of genomic information by a Generalized Regression Neural Network. They are intergenic distances, COG gene functions, conserved gene pairs and phylogenetic profiles. This method needn't use experimental data so that it can be well applied to new genomes which are not fully annotated.

2 Methodologies

In OPNN, two adjacent genes which belong to the same operons are pairs of genes within operons (WO pairs); to the contrary, if they belong to different operons, they are transcription unit borders (TUB pairs) [2]. The WO pairs tend to have shorter intergenic distances, higher probability being in the same COG functional categories, easier to be conserved and more similar phylogenetic profiles than TUB pairs.

2.1 Data Preparation

All of 304 microbial genomes and their annotated data are downloaded from NCBI database (ftp://ftp.ncbi.nih.gov/genomes) (as of April 10, 2006) which are used for the conserved gene pairs and phylogenetic profiles calculation.

E. coli K12 (GenBank accession number NC 000913) is the target genome. Its operons data are obtained from RegulonDB database [4]. In the corresponding version of the database, E. coli K12 contains 770 operons that are experimentally validated including 902 WO pairs and 697 TUB pairs.

2.2 Log-Likelihoods

For a typical prediction problem with many different types of data, the relations between them must be identified. In this paper, we estimate probability measurements by Log-Likelihoods [2]. The formula is then defined as follows:

$$LL(WO \mid d(g_a, g_b)) = \log \frac{P(d(g_a, g_b) \mid WO)}{P(d(g_a, g_b) \mid TUB)}, \quad (1)$$

where P (d (ga, gb) |WO) and P (d (ga, gb) |TUB) are respectively the anterior probability that a specific relation d (ga, gb) could be observed in WO genes pairs or TUB genes pairs, and ga and gb are the property values observed for two adjacent genes. LL (WO|d (ga, gb)) is the logarithm likelihood score, which expresses the probability of an adjacent genes pair belonging to the same operons. Adjacent genes pairs with the higher log-likelihood scores are the more likely to be WO pairs [2].

2.3 Intergenic Distances

The property of intergenic distances is frequently used in operon predictions [5]. Researches show that if two adjacent genes on the same strand belong to the same operon, then their intergenic distance tends to be shorter [3]. We first estimate the anterior probability of different intergenic distances in WO pairs and TUB pairs, and then calculate their log-likelihood scores using Eq. (1) described above.

In Figure 2, we can see that the intergenic distances of most WO pairs are between -10bp and 20bp, while the distances between TUB pairs are more than 100bp. In Figure 3, the log-likelihood scores between -10bp and 20bp are comparatively high, so we can assume that, although not strictly true, two adjacent genes with the intergenic distance between -10bp and 20bp are more likely to form a WO pair.

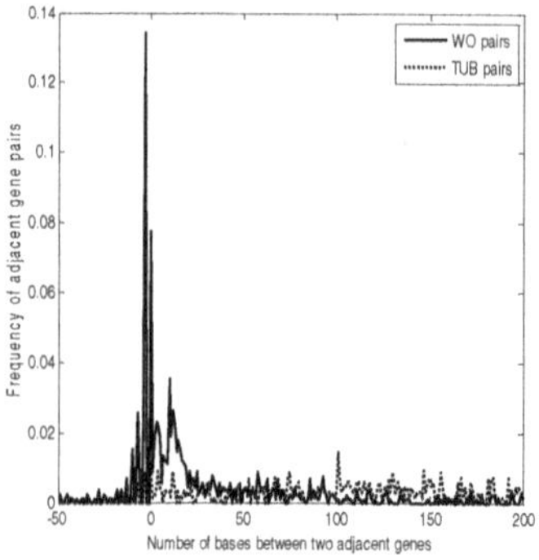

Fig. 2. Frequency distribution of different intergentic distances

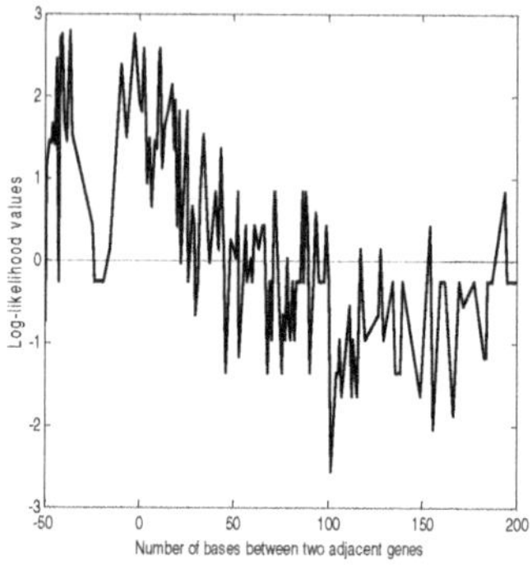

Fig. 3. Log-likelihoods for intergenetic distances

2.4 COG Functions

The COGs (clusters of orthologous groups) comprise a framework for the analysis of evolutionary and functional relationships among homologous genes from multiple genomes [6].

We estimate COG function category of each genes using our own program COGOVER. The program calculates COG function by comparing the genes of the COG database from NCBI (ftp://ftp.ncbi.nih.gov/pub/COG/COG). COGOVER bases on BLASTp and carries out by Perl. Its results are consistent with COGnitor program (http://www.ncbi.nlm.nih.gov/COG/), but our procedure supports batch process.

There are three major levels in COG function category [6]. In the OPNN, only the first and the second level are used. After estimating the anterior probability of different COG function categories in WO pairs and TUB pairs, their log-likelihood scores are computed by using Eq. (1) described above.

2.5 Conserved Gene Pairs

Operons prediction using conserved gene pairs first appeared in [7]. The procedure of calculating the conserved gene pairs is as follows. Firstly, compare each adjacent gene pairs to the 304 whole-genomes [7], then calculate statistically the number of conserved gene pairs for them. The number is between 0-304. Secondly, estimate the anterior probability of different numbers of conserved gene pairs in WO pairs and TUB pairs. Finally, calculate the log-likelihood scores using Eq. (1).

2.6 Phylogenetic Profile

The phylogenetic profile of a gene is a binary string, each bit of which represents the presence or absence of the gene in the comparative genome [8]. The phylogenetic profiles, which reflect similarity of homology gene, can provide certain information indicating genes function categories and metabolism pathways.

The key issue of using phylogenetic profiles is how to measure the co-relation between two profiles. In OPNN, Hamming distance was introduced [8]. The Hamming distance simply counts the number of different bits between two profile strings. The equation is defined as follows:

$$d_H = \sum_{1}^{n} P_i, \tag{2}$$

where $P_i = 0$ if the phylogenetic profiles of two adjacent genes in the ith position is the same, otherwise $P_i = 1$.

After calculating the distances, anterior probability of different distances in WO pairs and TUB pairs are estimated respectively. Finally, their log-likelihood scores are evaluated using Eq. (1).

2.7 Fusing Multiple Types of Data

In many typical prediction problems in bioinformatics, various disparate types of property information are used. The key is how to integrate them in an efficient way to improve the prediction performance. We choose a simple but powerful tool, Generalized Regression Neural Network (GRNN) to integrate all the types of data used in the prediction process.

GRNN is a kind of change form of RBF Neural Network, which was proposed by Specht D. F. in 1991[9].GRNN has faster training speed and stronger ability for nonlinear mapping, therefore it was used frequently. We use GRNN to integrate each type of data. In the network, there are four nodes in the input layer, which respectively are the log-likelihood values of four types of genomic data. There is only one node in the output layer of this network, which is a probability value indicating whether an adjacent genes pair is WO pair or not.

There is only one parameter that needs to be adjusted in GRNN, which is called smoothing factor b_1, which can improve the sensitivity of the network. In the experiment, the most effective spread value is 1.0 when the prediction is best.

3 Experimental Results

We predict operons in the Escherichia coli K12 genome by GRNN based on the log-likelihood values of the four types of genomic data. 50% of E. coli K12 genes are selected for training, and the other 50% is used to predicted operons. The average sensitivity, specificity, and accuracy of the prediction are 85.9%, 89.2% and 87.9%, respectively. The results of JPOP [2], OFS [10] and those from the proposed OPNN algorithm are shown in Table 1. It can be seen that the average sensitivity, specificity and accuracy of GRNN are higher than JPOP and OFS.

Table 1. The result of three prediction method

Method of prediction	Sensitivity	Specificity	Accuracy
JPOP	82.4%	85.1%	83.8%
OFS	80.1%	85.4%	82.1%
OPNN	85.9%	89.2%	87.9%

The Receiver Operating Curve(ROC) is shown in Figure 4. We can see that the overall results of OPNN are better than OFS. As shown in Figure 5, the ROC below shows the performances of predicting operons with all types of data or with only one data, which indicate that integrating multiple types of data can significantly improve the prediction result.

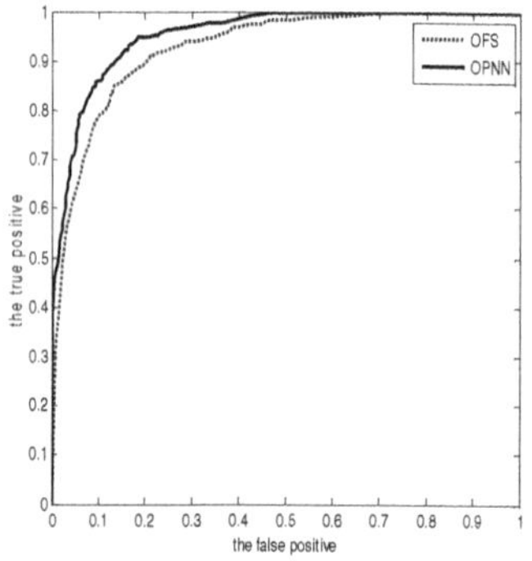

Fig. 4. The ROC of OFS and OPNN

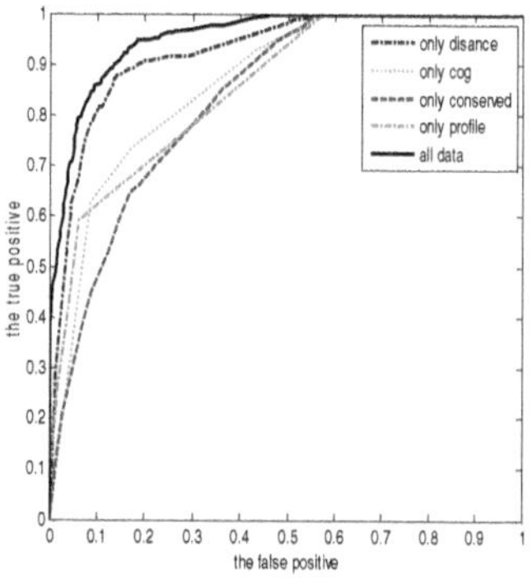

Fig. 5. The ROC of all data and only any one

4 Discussions and Conclusions

Operon provides highly useful information for characterizing or constructing the regulatory network in microbial genome. Therefore, the operon prediction for microbial genomes has been considered as one of the most fundamental and challenging bio-informatics problems.

The OPNN, proposed in this paper, predicts operons in E. coli K12 genome by Neural Network based on four types of genomic data with their log-likelihood values. After estimating log-likelihood values distribution for intergenic distances, COG gene functions, conserved gene pairs and phylogenetic profiles respectively, their results are inputted into GRNN, which integrates the four types of data to predict the operon structure of the genome.

The OPNN is used to obtain the average sensitivity, specificity and accuracy, which are 85.9%、89.2% and 87.9% respectively. The experimental results show that the prediction based on multiple information is better than single information. The genomic information obtained from the calculation is independent to the experiment which makes the OPNN more flexible to be applied to new sequence genomes.

GPSR Compliance

The European Union's (EU) General Product Safety Regulation (GPSR) is a set of rules that requires consumer products to be safe and our obligations to ensure this.

If you have any concerns about our products, you can contact us on ProductSafety@springernature.com

In case Publisher is established outside the EU, the EU authorized representative is:

Springer Nature Customer Service Center GmbH
Europaplatz 3
69115 Heidelberg, Germany

Batch number: 10761593

Printed by Printforce, the Netherlands

Lecture Notes in Computer Science 4491

Commenced Publication in 1973
Founding and Former Series Editors:
Gerhard Goos, Juris Hartmanis, and Jan van Leeuwen

Derong Liu Shumin Fei
Zeng-Guang Hou Huaguang Zhang
Changyin Sun (Eds.)

Advances in Neural Networks – ISNN 2007

4th International Symposium
on Neural Networks, ISNN 2007
Nanjing, China, June 3-7, 2007
Proceedings, Part I

Springer

Volume Editors

Derong Liu
University of Illinois at Chicago, IL 60607-7053, USA
E-mail: dliu@ece.uic.edu

Shumin Fei
Southeast University, School of Automation, Nanjing 210096, China
E-mail: smfei@seu.edu.cn

Zeng-Guang Hou
Chinese Adacemy of Sciences, Institute of Automation, Beijing 100080, China
E-mail: hou@compsys.ia.ac.cn

Huaguang Zhang
Northeastern University, Shenyang 110004, China
E-mail: zhanghuaguang@ise.neu.edu.cn

Changyin Sun
Hohai University, School of Electrical Engineering, Nanjing 210098, China
E-mail: cysun@hhu.edu.cn

Library of Congress Control Number: 2007926816

CR Subject Classification (1998): F.1, F.2, D.1, G.2, I.2, C.2, I.4-5, J.1-4

LNCS Sublibrary: SL 1 – Theoretical Computer Science and General Issues

ISSN 0302-9743
ISBN-10 3-540-72382-X Springer Berlin Heidelberg New York
ISBN-13 978-3-540-72382-0 Springer Berlin Heidelberg New York

Springer is a part of Springer Science+Business Media

springer.com

Printed in Germany

Typesetting: Camera-ready by author, data conversion by Scientific Publishing Services, Chennai, India
Printed on acid-free paper SPIN: 12060757 06/3180 5 4 3 2 1 0

Preface

ISNN 2007 – the Fourth International Symposium on Neural Networks—was held in Nanjing, China, as a sequel of ISNN 2004/ISNN 2005/ISNN 2006. ISNN has now become a well-established conference series on neural networks in the region and around the world, with growing popularity and increasing quality. Nanjing is an old capital of China, a modern metropolis with a 2470-year history and rich cultural heritage. All participants of ISNN 2007 had a technically rewarding experience as well as memorable experiences in this great city.

A neural network is an information processing structure inspired by biological nervous systems, such as the brain. It consists of a large number of highly interconnected processing elements, called neurons. It has the capability of learning from example. The field of neural networks has evolved rapidly in recent years. It has become a fusion of a number of research areas in engineering, computer science, mathematics, artificial intelligence, operations research, systems theory, biology, and neuroscience. Neural networks have been widely applied for control, optimization, pattern recognition, image processing, signal processing, etc.

ISNN 2007 aimed to provide a high-level international forum for scientists, engineers, and educators to present the state of the art of neural network research and applications in diverse fields. The symposium featured plenary lectures given by worldwide renowned scholars, regular sessions with broad coverage, and some special sessions focusing on popular topics.

The symposium received a total of 1975 submissions from 55 countries and regions across all six continents. The symposium proceedings consists of 454 papers among which 262 were accepted as long papers and 192 were accepted as short papers. We would like to express our sincere gratitude to all reviewers of ISNN 2007 for the time and effort they generously gave to the symposium. We are very grateful to the National Natural Science Foundation of China, K. C. Wong Education Foundation of Hong Kong, the Southeast University of China, the Chinese University of Hong Kong, and the University of Illinois at Chicago for their financial support. We would also like to thank the publisher, Springer, for cooperation in publishing the proceedings in the prestigious series of *Lecture Notes in Computer Science*.

Derong Liu
Shumin Fei
Zeng-Guang Hou
Huaguang Zhang
Changyin Sun

ISNN 2007 Organization

General Chair

Derong Liu, University of Illinois at Chicago, USA, and Yanshan University, China

General Co-chair

Marios M. Polycarpou, University of Cyprus

Organization Chair

Shumin Fei, Southeast University, China

Advisory Committee Chairs

Shun-Ichi Amari, RIKEN Brain Science Institute, Japan
Chunbo Feng, Southeast University, China
Zhenya He, Southeast University, China

Advisory Committee Members

Hojjat Adeli, Ohio State University, USA
Moonis Ali, Texas State University-San Marcos, USA
Zheng Bao, Xidian University, China
Tamer Basar, University of Illinois at Urbana-Champaign, USA
Tianyou Chai, Northeastern University, China
Guoliang Chen, University of Science and Technology of China, China
Ruwei Dai, Chinese Academy of Sciences, China
Dominique M. Durand, Case Western Reserve University, USA
Russ Eberhart, Indiana University Purdue University Indianapolis, USA
David Fogel, Natural Selection, Inc., USA
Walter J. Freeman, University of California-Berkeley, USA
Toshio Fukuda, Nagoya University, Japan
Kunihiko Fukushima, Kansai University, Japan
Tom Heskes, University of Nijmegen, The Netherlands
Okyay Kaynak, Bogazici University, Turkey
Frank L. Lewis, University of Texas at Arlington, USA
Deyi Li, National Natural Science Foundation of China, China
Yanda Li, Tsinghua University, China
Ruqian Lu, Chinese Academy of Sciences, China

John MacIntyre, University of Sunderland, UK
Robert J. Marks II, Baylor University, USA
Anthony N. Michel, University of Notre Dame, USA
Evangelia Micheli-Tzanakou, Rutgers University, USA
Erkki Oja, Helsinki University of Technology, Finland
Nikhil R. Pal, Indian Statistical Institute, India
Vincenzo Piuri, University of Milan, Italy
Jennie Si, Arizona State University, USA
Youxian Sun, Zhejiang University, China
Yuan Yan Tang, Hong Kong Baptist University, China
Tzyh Jong Tarn, Washington University, USA
Fei-Yue Wang, Chinese Academy of Sciences, China
Lipo Wang, Nanyang Technological University, Singapore
Shoujue Wang, Chinese Academy of Sciences
Paul J. Werbos, National Science Foundation, USA
Bernie Widrow, Stanford University, USA
Gregory A. Worrell, Mayo Clinic, USA
Hongxin Wu, Chinese Academy of Space Technology, China
Youlun Xiong, Huazhong University of Science and Technology, China
Lei Xu, Chinese University of Hong Kong, China
Shuzi Yang, Huazhong University of Science and Technology, China
Xin Yao, University of Birmingham, UK
Bo Zhang, Tsinghua University, China
Siying Zhang, Qingdao University, China
Nanning Zheng, Xi'an Jiaotong University, China
Jacek M. Zurada, University of Louisville, USA

Steering Committee Chair

Jun Wang, Chinese University of Hong Kong, China

Steering Committee Co-chair

Zongben Xu, Xi'an Jiaotong University, China

Steering Committee Members

Tianping Chen, Fudan University, China
Andrzej Cichocki, Brain Science Institute, Japan
Wlodzislaw Duch, Nicholaus Copernicus University, Poland
Chengan Guo, Dalian University of Technology, China
Anthony Kuh, University of Hawaii, USA
Xiaofeng Liao, Chongqing University, China
Xiaoxin Liao, Huazhong University of Science and Technology, China
Bao-Liang Lu, Shanghai Jiaotong University, China

Chenghong Wang, National Natural Science Foundation of China, China
Leszek Rutkowski, Technical University of Czestochowa, Poland
Zengqi Sun, Tsinghua University, China
Donald C. Wunsch II, University of Missouri-Rolla, USA
Gary G. Yen, Oklahoma State University, Stillwater, USA
Zhang Yi, University of Electronic Science and Technology, China
Hujun Yin, University of Manchester, UK
Liming Zhang, Fudan University, China
Chunguang Zhou, Jilin University, China

Program Chairs

Zeng-Guang Hou, Chinese Academy of Sciences, China
Huaguang Zhang, Northeastern University, China

Special Sessions Chairs

Lei Guo, Beihang University, China
Wen Yu, CINVESTAV-IPN, Mexico

Finance Chair

Xinping Guan, Yanshan University, China

Publicity Chair

Changyin Sun, Hohai University, China

Publicity Co-chairs

Zongli Lin, University of Virginia, USA
Weixing Zheng, University of Western Sydney, Australia

Publications Chair

Jinde Cao, Southeast University, China

Registration Chairs

Hua Liang, Hohai University, China
Bhaskhar DasGupta, University of Illinois at Chicago, USA

Local Arrangements Chairs

Enrong Wang, Nanjing Normal University, China
Shengyuan Xu, Nanjing University of Science and Technology, China
Junyong Zhai, Southeast University, China

Electronic Review Chair

Xiaofeng Liao, Chongqing University, China

Symposium Secretariats

Ting Huang, University of Illinois at Chicago, USA
Jinya Song, Hohai University, China

ISNN 2007 International Program Committee

Shigeo Abe, Kobe University, Japan
Ajith Abraham, Chung Ang University, Korea
Khurshid Ahmad, University of Surrey, UK
Angelo Alessandri, University of Genoa, Italy
Sabri Arik, Istanbul University, Turkey
K. Vijayan Asari, Old Dominion University, USA
Amit Bhaya, Federal University of Rio de Janeiro, Brazil
Abdesselam Bouzerdoum, University of Wollongong, Australia
Martin Brown, University of Manchester, UK
Ivo Bukovsky, Czech Technical University, Czech Republic
Jinde Cao, Southeast University, China
Matthew Casey, Surrey University, UK
Luonan Chen, Osaka-Sandai University, Japan
Songcan Chen, Nanjing University of Aeronautics and Astronautics, China
Xiao-Hu Chen, Nanjing Institute of Technology, China
Xinkai Chen, Shibaura Institute of Technology, Japan
Yuehui Chen, Jinan University, Shandong, China
Xiaochun Cheng, University of Reading, UK
Zheru Chi, Hong Kong Polytechnic University, China
Sungzoon Cho, Seoul National University, Korea
Seungjin Choi, Pohang University of Science and Technology, Korea
Tommy W. S. Chow, City University of Hong Kong, China
Emilio Corchado, University of Burgos, Spain
Jose Alfredo F. Costa, Federal University, UFRN, Brazil
Mingcong Deng, Okayama University, Japan
Shuxue Ding, University of Aizu, Japan
Meng Joo Er, Nanyang Technological University, Singapore
Deniz Erdogmus, Oregon Health & Science University, USA

Shuxian Lun, Bohai University, China
Fa-Long Luo, Anyka, Inc., USA
Jinwen Ma, Peking University, China
Xiangping Meng, Changchun Institute of Technology, China
Kevin L. Moore, Colorado School of Mines, USA
Ikuko Nishikawa, Ritsumeikan University, Japan
Stanislaw Osowski, Warsaw University of Technology, Poland
Seiichi Ozawa, Kobe University, Japan
Hector D. Patino, Universidad Nacional de San Juan, Argentina
Yi Shen, Huazhong University of Science and Technology, China
Daming Shi, Nanyang Technological University, Singapore
Yang Shi, University of Saskatchewan, Canada
Michael Small, Hong Kong Polytechnic University
Ashu MG Solo, Maverick Technologies America Inc., USA
Stefano Squartini, Universita Politecnica delle Marche, Italy
Ponnuthurai Nagaratnam Suganthan, Nanyang Technological University, Singapore
Fuchun Sun, Tsinghua University, China
Johan A. K. Suykens, Katholieke Universiteit Leuven, Belgium
Norikazu Takahashi, Kyushu University, Japan
Ying Tan, Peking University, China
Yonghong Tan, Guilin University of Electronic Technology, China
Peter Tino, Birmingham University, UK
Christos Tjortjis, University of Manchester, UK
Antonios Tsourdos, Cranfield University, UK
Marc van Hulle, Katholieke Universiteit Leuven, Belgium
Dan Ventura, Brigham Young University, USA
Michel Verleysen, Universite Catholique de Louvain, Belgium
Bing Wang, University of Hull, UK
Dan Wang, Dalian Maritime University, China
Pei-Fang Wang, SPAWAR Systems Center-San Diego, USA
Zhiliang Wang, Northeastern University, China
Si Wu, University of Sussex, UK
Wei Wu, Dalian University of Technology, China
Shunren Xia, Zhejiang University, China
Yousheng Xia, University of Waterloo, Canada
Cheng Xiang, National University of Singapore, Singapore
Daoyi Xu, Sichuan University, China
Xiaosong Yang, Huazhong University of Science and Technology, China
Yingjie Yang, De Montfort University, UK
Zi-Jiang Yang, Kyushu University, Japan
Mao Ye, University of Electronic Science and Technology of China, China
Jianqiang Yi, Chinese Academy of Sciences, China
Dingli Yu, Liverpool John Moores University, UK
Zhigang Zeng, Wuhan University of Technology, China

Guisheng Zhai, Osaka Perfecture University, Japan
Jie Zhang, University of Newcastle, UK
Liming Zhang, Fudan University, China
Liqing Zhang, Shanghai Jiaotong University, China
Nian Zhang, South Dakota School of Mines & Technology, USA
Qingfu Zhang, University of Essex, UK
Yanqing Zhang, Georgia State University, USA
Yifeng Zhang, Hefei Institute of Electrical Engineering, China
Yong Zhang, Jinan University, China
Dongbin Zhao, Chinese Academy of Sciences, China
Hongyong Zhao, Nanjiang University of Aeronautics and Astronautics, China
Haibin Zhu, Nipissing University, Canada

Table of Contents – Part I

Neural Fuzzy Control

Neural Networks for Control Applications

Adaptive Dynamic Programming and Reinforcement Learning

Neural Networks for Nonlinear Systems Modeling

Robotics

Stability Analysis of Neural Networks

Learning and Approximation

RST-Based RBF Neural Network Modeling for Nonlinear System*

Tengfei Zhang[1], Jianmei Xiao[1], Xihuai Wang[1], and Fumin Ma[2]

[1] Department of Electrical and Automation, Shanghai Maritime University, Shanghai 200135, China
[2] CIMS Research Center, Tongji University, Shanghai 200092 China
wxh@shmtu.edu.cn

Abstract. Due to its structural simplicity and good properties, radial basis function (RBF) neural network has increasingly been used in many areas for the solution of difficult real-world problems, especially the nonlinear system dynamic modeling. However, the major problem toward using RBF network is the appropriate selection of radial basis function parameters. The basis function parameters are in general the centers and the widths. Our attention in this paper is focused on the configuring the optimal set of parameters to make the networks small and efficient based on rough set theory (RST), which is a valid mathematical tool to perform data reduction. RST is first applied to extract the underlying rules from data. The condition components of the rules are then mapped into network centers. For improve performance, the width parameter of each hidden neuron is initialized individual. The valid of this algorithm is illustrated by an example on the modeling of a ship synchronous generator.

1 Introduction

Radial basis function (RBF) neural network has been successfully applied to many practical problems with the good properties such as simple network structure, strong nonlinear approximation ability, rapid convergence speed and global convergence property, and so on [1-3]. While in RBF networks an optimal set of parameters would be required to make the networks small and efficient. The basis function parameters are in general the centers and the widths.

Placement of centers is said to have a significant effect on the performance of the network [2]. Though both supervised and unsupervised methods are used to determine the location of centers, unsupervised methods are preferred in practical applications due to faster training time. Meanwhile, the width parameter is usually taken to be a constant that depend on the input data. However, as we all known that randomly selected centers cannot guarantee to well cover the training data and fixed width parameter cannot preventing local data features from fading away. In addition, a

* This work was supported by Science Project of Shanghai Education (04FA02, 05FZ06) and Shanghai Leading Academic Discipline Project (T0602).

D. Liu et al. (Eds.): ISNN 2007, Part I, LNCS 4491, pp. 658–666, 2007.

network configured by such parameters may be has a small training error with a poor generalization capability, which so-called over fitting. The situation is possibly caused by the following reasons [3]. For one thing, the training data may usually have irrelevant or superfluous knowledge; in other words, some of the input variables may be unnecessary. Secondly, some incompatible input-output relations can be found in the training data. Otherwise, certain input factors may not be significant and should be deleted in a certain center vector; namely, the width parameter should not be the same for all center vectors. The mentioned above tend to mislead local learning and result in poor property.

The contribution of this article is the introduction of a rough set approach to configuring the RBF network parameters. In this algorithm, rough set theory (RST) is first applied to data analysis, including attribute reduction, incompatible data elimination and decision rules extraction. The centers are then determined by the decision rules directly. Meanwhile, the width parameter vector of each hidden neuron is initialized individual.

The rest of paper is organized as follows. The fundamental of RBF network is briefly given in Section 2. The detail of configuring function parameters is presented in Section 3. Simulation result in Section 4 show that the approach presented in this paper leads to a RBF network with a perfect performance.

2 Radial Basis Function Neural Networks

A standard radial basis function network consists of three layers of neurons as depicted in Fig.1.

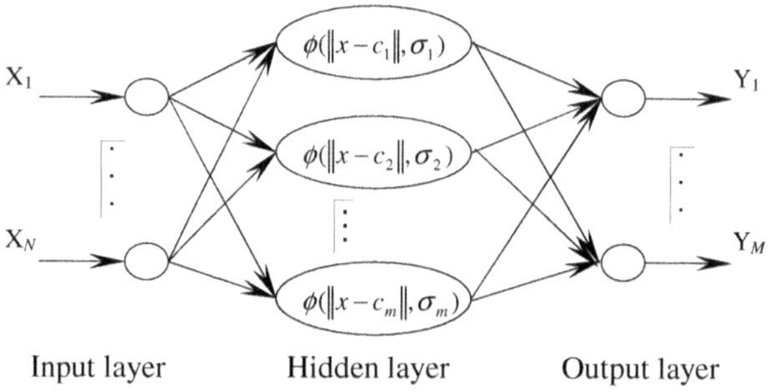

Fig. 1. Architecture of a standard RBF network

It is easy to see that RBF neural network is a three-architecture with no feedback, which has only one hidden layer, can approach any non-linear functions precisely. The input layer units and the output units are determined by the practical problems and only the number of hidden layer units can be variable. A typical RBF neural network has H processing nodes in the hidden layer and M summing nodes in the output layer. The input sample is an N-dimensional vector.

The detailed description of the network structure can be found in many inferences and can be considered as a multidimensional interpolation technique implementing general mappings $f: R^N \rightarrow R$ with a linear equation of the form according to:

$$f(x) = w_0 + \sum_{i=1}^{H} w_i \varphi_i(x) \quad (1)$$

The linearity is with respect to the weights of the output layer, w_i, not the input variable x. The basis function, $\varphi(\cdot)$, is the basis function of the hidden unit in the network.

The performance of RBF neural network is decided by the hidden layer, which consists of H hidden neurons (radial basis units) with radial activation functions, and each neuron only responds to an input that is close to its center. Different basis functions $\varphi(\cdot)$ can be adopted; a typical choice for this function is the Gaussian function, which is given by the following equation:

$$\varphi(x) = \exp(-\| x - c \|^2 / 2r^2) \quad (2)$$

where $c = [c_1, c_2, \cdots, c_n]^T$ is the center vectors and $r = [r_1, r_2, \cdots, r_n]^T$ is the width parameter, which scale each dimension; $\|*\|$denotes the Euclidean norm

The output layer simply performs a linear combination of the (nonlinearly) transformed inputs and thus the weights w_i can be obtained by using lots of algorithms, such as the standard LMS algorithm or its momentum version, the well known gradient descent method, orthogonal least squares learning, and so on.

3 RBF Neural Networks Configuring Based on RST

The training for neural network is to make it provide the required output when the certain input is given, through the adjustment of the parameter of the network. So, the network can reach our intention if we can configure it directly. Rough set theory put forward by Z. Pawlak in the early 1980s has proved its usefulness as a new mathematical tool for data analysis [4,5]. In this paper, we utilize RST to construct RBF network.

A. To establish the information table

Knowledge representation in rough set theory is done via information system. An information system can be characterized as a decision table, where the information is stored in a table. Each row in the table represents an individual record. Each column represents some an attribute of the records or a field.

A simple information table is shown in Table.1.

Table 1. A Simple Information Table

U/A	x_1	x_2	x_3	y
1	0.2	0.5	-0.4	2
2	0.1	0.3	0.8	-1
3	1	0.6	-2	0.9

B. To be discrete for Continuous Attributes

It is a pity that rough set theory can't deal with continuous attributes. When the value set of any attribute in a decision table is continuous valued or real numbers, then it is likely that there will be very few objects that will have the same value of the corresponding attribute. In such a situation the number of equivalence classes based on that attribute will be large and there will be very few elements in each of such equivalence class. This leads to the generation of a large numbers of antecedents in the classification rule, thereby making rough set theoretic classifiers inefficient. Therefore, to be discrete is necessary for continuous attributes.

A rough set theory based software *RSES* [6], is a toolkit for analysis of information table data. With use of *Discretize/Generate cuts* from data table context menu we may generate decompositions of attribute value sets. With these descriptions, further discretization of continuous attributes or grouping (quantization) of nominal attributes referred to as cuts we may perform.

C. To use RST to get suitable centers

It is noted that the most important consideration in configuring an RBF network is the determination of the number and centers of the hidden units. For the existing centers chosen methods, most of them are in essence still analytical method based on clustering, which make center vectors of RBF lie in the important area of input space. Usually, it needs abundant, accurate data to set up a good network model, and these data should cover all main situations.

It is well known that an information system or a decision system may usually have irrelevant or superfluous knowledge (attributes), which is inconvenient for us to get concise and meaningful decision. Therefore, the reduction of attributes is demanded greatly. In rough set theory, the reduction is defined as a minimal set of attributes that enables the same classification of elements of the universe as the whole set of attributes.

Thus, we can use the RST to get the key attributes. Here, a reduction strategy based on the positive region is adopted [7]. After this operation, we get n factors as the input of the networks.

Each row in decision table can regard as a decision rule. If the condition attributes of any two rules are equal whereas the decision attributes are not, we say that the two rules are inconsistent. When this happens both rules should be removed from the training data because they will mislead the local leaning. If two or more rules represent the same class, all but one of the rules should be eliminated. This reduces computational time.

Using the advantage of rough set theory in data processing, we can extract the connotative rules from the training data, via relative attribute reduction and rule extraction algorithms, and each rule is compact formally. Such every rule represents a certain class of the batch of data. Because the decision rules set is minimum and has covered all relations in initial data, the condition of rules is exactly the ideal center vector space in RBF that we should seeking for based on clustering method, and each vector is a representative point in initial data. So, it is reasonable to regard each rule's condition as a hidden neuron center vector in RBF neural networks.

D. To tune some very closer centers

Large number of rules will occur for different reasons, for example, if the modeling system is very complex with strong non-linear characteristic, or the continue attributes are discretized with more demarcation points, and so on. That is to say, the RBF centers will be excessive, which mapping into more hidden neurons. For further simplify the neural networks, the very closer centers should be clustering as a new center if not affecting the performance of networks in evidence. The criterion of tuning closer centers varies with the actual problems.

E. To configuring the widths of each hidden unit

Each hidden neuron may have a different width and for some application adjusting individual width can often improve performance. That is to say, the width parameter should not be the same for all center vectors. In this paper, we configure the width parameter of each hidden neuron individual.

For better explanation, denote $c_i=[c_{i1}, c_{i2}, \ldots c_{in}]^T$ as center and $r_i=[r_{i1}, r_{i2}, \cdots, r_{in}]^T$ width of the ith-hidden neuron. Then the jth element of width vector can be initialized as: $r_{ij} = | c_{kj} - c_{lj} | / 2$, where c_{kj}, c_{lj} are the nearest values of c_{ij}, which are the centers of kth and lth hidden neurons.

The basic network structure is configured when confirmed the centers and widths of the hidden layer in RBF neural network according to the above-mentioned method. Then we can adjust the network parameters and the train the weight matrix w by gradient descent method.

The objective is to minimize the squared norm of the residuals:

$$E = \|Y - wz\|^2 \tag{3}$$

where Y is the $M\times K$ matrix of training targets, $z=(z_1, z_2, \ldots, z_k)$ is a $H\times K$ matrix, and w is the $M\times H$ weight matrix.

Then the parameters and the weights w (initialized randomly) are updated by:

$$\begin{aligned} c(n+1) &= c(n) - \eta \frac{\partial E^n}{\partial c} \\ r(n+1) &= r(n) - \eta \frac{\partial E^n}{\partial r} \\ w(n+1) &= w(n) - \eta \frac{\partial E^n}{\partial w} \end{aligned} \tag{4}$$

where η is the learning rate, which is small enough and decreases gradually.

4 Simulation Results

In this section, to show the key features of a RBF neural networks that use the new configuring method, the system with a 3.125 MVA, 2400V ship synchronous generator, driven by a diesel engine is under consideration. The dynamic characteristic of ship power system depends on the diesel engine generator. According to working principles of diesel generator, we select excitation voltage (v_f) and diesel engine mechanical torque (P_m) as the input parameters; and the output parameters are terminal voltage (vt), current (i) and frequency (f) of the ship synchronous generator. We selected some different running states of ship power system, such as part of rated load increasing or decreasing suddenly, two-phase and three-phase ground fault at generator terminal. The parameters of input and output were measured as sampling data, and used for neural network training and testing. There are 10000 dada points sampled, where 5000 samples data for training and 3000 samples data testing. It is important that proper selection of input variables for building the dynamic network model to meet the highly non-linear characteristics of ship synchronous generator. Initially, the following notation is used for creating the input variables candidate set:

$$X(t) = [v_f(t-1), v_f(t-2), v_f(t-3), P_m(t-1), P_m(t-2), y(t-1), \\ y(t-2), \cdots, y(t-n_y)]^T . \quad (5)$$

$$y(t) = f[X(t)]$$

where $y(t\text{-}1)$, $y(t\text{-}2)$,…, $y(t\text{-}n_y)$ are the actual terminal output values when the time t-1, t-2, t-n_y. While $y(t)$ is the output of model at the moment t. The system dynamic modeling structure based on RST and RBF neural network is shown in Fig.2.

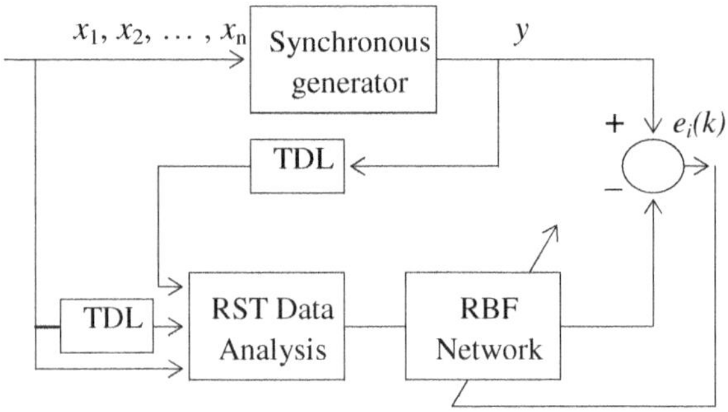

Fig. 2. The dynamic modeling structure based on RST and RBF neural network

RST is used first to determine the significant lag n_y and then to select the most appropriate regressors which make up the RBF network.

As an example, terminal voltage parameter of generator is selected to demonstrate the precision of model presented in this paper. At the beginning, the maximum lag was set to be 7. Thus, the establishing information table consists of 12 condition attributes. After analysis by RST, the 8 input nodes [$v_f(t\text{-}1)$, $v_f(t\text{-}2)$, $v_f(t\text{-}3)$, $P_m(t\text{-}1)$, $P_m(t\text{-}2)$, $v_t(t\text{-}1)$, $v_t(t\text{-}2)$,$v_t(t\text{-}3)$] were selected as the best subset. The RBF network was then initialized from the extracted 746 rules by RST and then tuned as 159 centers to configure the hidden units. The simulation results under different running states are shown in Fig.3 ((a), (b)), Fig.4 ((a), (b)) respectively (black real line denote the actual system output, blue line RST based RBF network output). In order to verify this method being superior, conventional RBF neural network model was presented suing all the training data without pre-treatment. Fig.5 show the relative errors by both two methods. Here, the relative error defined as:

$$RE = | y_i - \hat{y}_i | / y_i \quad (6)$$

where y_i is actual voltage values, and $\hat{y}_i$ is the test value.

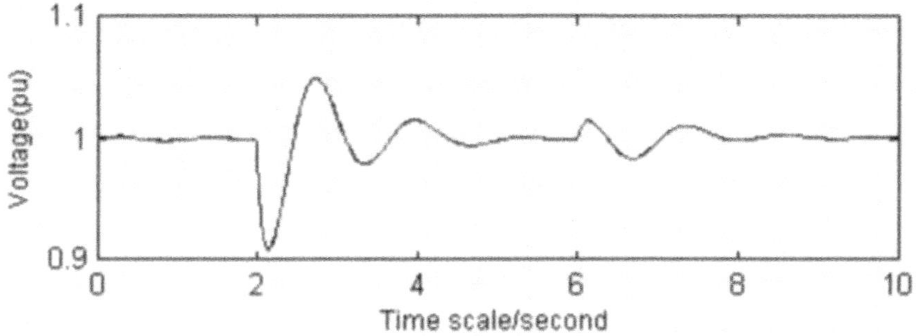

(a). Voltage output of RST based RBF model

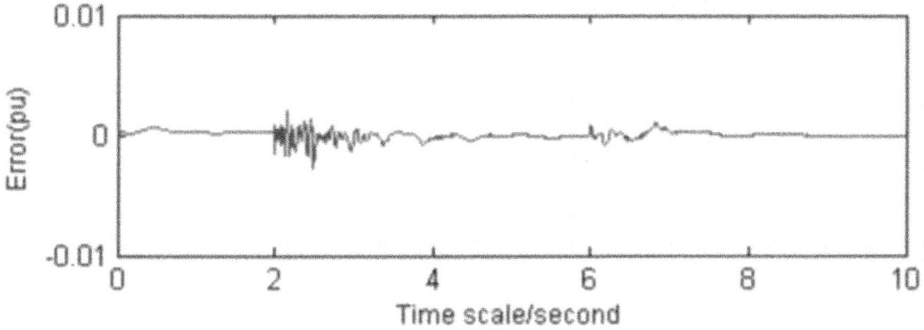

(b). Model voltage output error

Fig. 3. 100 percent of rated load suddenly applied and 50 percent decreased at t=2s, 6s

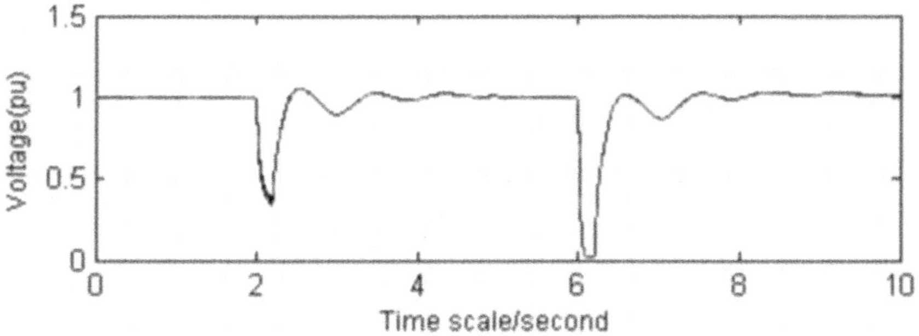

(a). Voltage output of RST based RBF model

Fig. 4. Two-phase and three-phase grounded fault at t=2s, 6s respectively

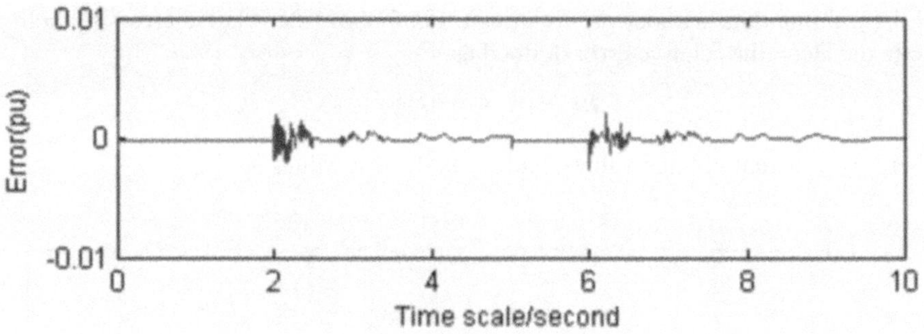

(b). Model voltage output error

Fig. 4. (*Continued*)

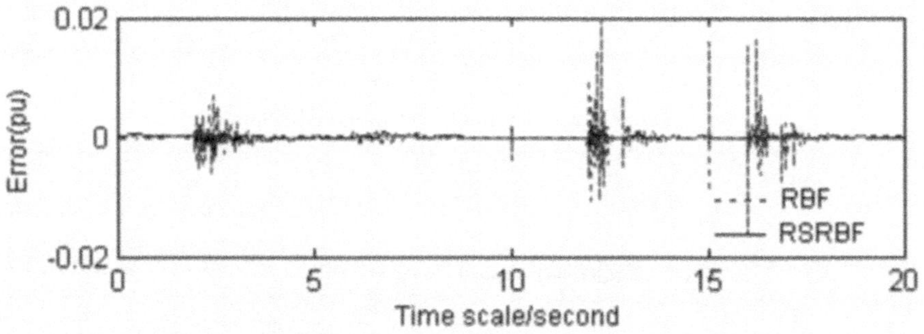

Fig. 5. Errors between model outputs with actual terminal voltage values

From the simulation result all above, we can see that the RST based RBF neural network model could achieve the dynamic characteristic of ship synchronous generator with sufficient accuracy and better generalization.

5 Conclusions

A novel rough set theory based RBF neural network configuring approach applied to nonlinear dynamic modeling is studied in this paper. The simulation results show that the method succeeds in combining the RBF neural network study and analysis ability with RST for building an excellent small dynamic model mapping the complex non-linear relationships.

References

1. Kenne, G., Ahmed-Ali, T. et al.: Nonlinear systems parameters estimation using radial basis function network. Control Engineering Practice **14** (2006) 819–832
2. Sarimveis, H., Doganis, P., Alexandridis, A.: A classification technique based on radial basis function neural networks. Advances in Engineering Software **37** (2006) 218–221

3. Guo, J., Peter B.: Selecting Input Factors for Clusters of Gaussian Radial Basis Function Networks to Improve Market Clearing Price Prediction. IEEE Transactions on Power Systems **18** (2003) 665-671
4. Pawlak, Z.: Some Issues on Rough Sets. In: Peters, J.F. et al. (Eds.): Transactions on Rough Sets I. Lecture Notes in Computer Science, Vol. 3100. Springer-Verlag, Berlin Heidelberg New York (2004) 1-58
5. Pawlak, Z.: Rough Sets and Intelligent Data Analysis. Information Sciences **147** (2002) 1-12
6. Bazan, J.G., Szczuka, M.: The Rough Set Exploration System. In: Peters, J.F. et al. (Eds.): Transactions on Rough Sets III, Lecture Notes in Computer Science, Vol. .3400, Springer-Verlag, Berlin Heidelberg New York (2005) 37-56
7. Xiao, J., Zhang, T.: New Rough Set Approach to Knowledge Reduction in Decision Table. In: Proceedings of the Third International Conference on Machine Learning and Cybernetics (2004) 2208-2211
8. Wang, X., Zhang, T., Xiao, J.: Ship Synchronous Generator Modeling Based on RST and RBF Neural Networks. In: Wang, J. et al. (Eds.): Advances in Neural Networks. Lecture Notes in Computer Science, Vol. 3972, Springer-Verlag, Berlin Heidelberg New York (2006) 1363-1369

A New Method for Accelerometer Dynamic Compensation Based on CMAC

Mingli Ding, Qingdong Zhou, and Kai Song

Dept. of Automatic Test and Control, Harbin Institute of Technology,
150001 Harbin, China
dingml@hit.edu.cn

Abstract. To acquire a satisfied accelerometer dynamic compensation effect, the accelerometer model should be with high precision using the traditional method of zero-pole assignment (ZPS). But in the accelerometer output, the noises, the drift error and the disturbances of the system which make the low precision of the built accelerometer model, can not be avoidable. In this paper, an accelerometer dynamic compensation method based on CMAC neural network is proposed. In this method, a dynamic compensation model can be set up according to the measurement data of dynamic response of the accelerometer without knowing its dynamic model. The dynamic compensation model parameters are trained by CMAC neural network. To a kind of micro-silicon piezoresistance accelerometer, the simulation results show that the proposed new dynamic compensation method has the advantages of fast training process, high precision and easy realization of the dynamic compensation device.

1 Introduction

With the development of micro-electro-mechanical system (MEMS), the accelerometer is developing into small structure, low power consumption, high reliability and large measuring range. Now the accelerometers are widely used in the application fields of navigation, transportation, robotics and other correlative respects. People want the accelerometer not only has a stable static characteristic, but also an excellent dynamic characteristic. But in fact, due to the limitation of the structure of the accelerometer itself, the accelerometer has small frequency range. For the navigation application, if the dynamic characteristic can not fulfill the requirement of the frequency range of the input signal, the dynamic error of the system arisen from the accelerometer will greatly drop down the system navigation precision. The dynamic error has become an obstacle to the accelerometer's development obviously [1]. So it is necessary to improve the dynamic characteristic of the accelerometer.

The traditional method of zero-pole assignment (ZPS) Method should be designed based on a sensor model with a high precision [2, 3]. But due to many error sources, a high precision accelerometer model is always difficult to be obtained. In this paper, a new dynamic compensation method for accelerometer using the cerebella model articulation controller (CMAC) is proposed Among the many neural networks, we choose CMAC neural network for its good generalization ability, fast convergence

D. Liu et al. (Eds.): ISNN 2007, Part I, LNCS 4491, pp. 667–675, 2007.

rate and being suitable for on-line applications [4-9]. The method has advantages of fast learning speed and can not only extend the measured frequency range, but also improve the damping ratio. A simulation case is also investigated to verify the feasibility of the method.

2 Pole-Zero Assignment Method

The accelerometer can be considered as a typical two-order system from the view of mechanical structure. The accelerometer can be simplified as a single-degree-of-freedom system, which includes an inertial mass, an elastic beam and a damper as shown in Fig. 1. When the sensor sense a external acceleration, the inertial mass of the sensor may drive the elastic beam to produce a corresponding deformation. And then the sensing part on the elastic beam can convert the deformation or the displacement into an electrical output, which is can be measured.

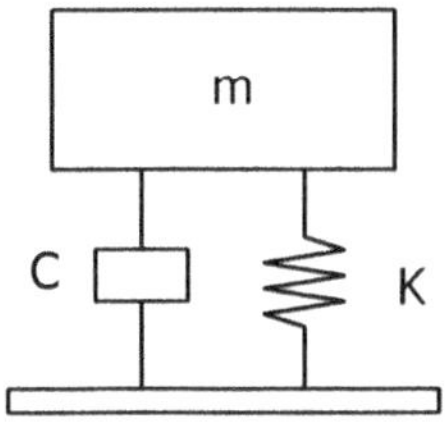

Fig. 1. Equivalent model of accelerometer

When the sensor senses the acceleration, the differential equation of the relationship between the input and output is

$$m\frac{d^2y}{dt^2}+c\frac{dy}{dt}+ky(t)=m\ddot{x}\,, \tag{1}$$

where m is the inertial mass; $y(t)$ is the displacement of the measured mass; c is the damping coefficient of the system; k is the elasticity coefficient of the system; $\ddot{x}$ is the acceleration.

The transform function of the system can be get using the Laplace transformation to Eq. (1)

$$\frac{Y(s)}{X(s)}=\frac{m}{k}\cdot\frac{1}{\frac{1}{\omega_0^2}s^2+\frac{2\xi}{\omega_0}s+1}\,, \tag{2}$$

where $\omega_0=\sqrt{\frac{k}{m}}$ is the natural resonant frequency of the system; $\xi=\frac{c}{2\sqrt{mk}}$ is the damping ratio of the system.

Using Eq. (2), we can get the amplitude-frequency response of the system

$$A(\omega_d) = \frac{m}{k} \cdot \frac{1}{\sqrt{\left[1 - \left(\frac{\omega_d}{\omega_0}\right)^2\right]^2 + 4\xi^2 \left(\frac{\omega_d}{\omega_0}\right)^2}}, \tag{2}$$

where ω_d is the angular frequency of the input signal.

From the above equations, it is obvious that the amplitude of the output signal is changing with the frequency changing of the input signal. That is to say the output equation is a kind of frequency function. To an accelerometer, its natural resonant frequency determines the range of the working frequency. The larger the natural resonant frequency is, the wider the range of the working frequency is.

But now, in many application areas, the accelerometer used in the engineering survey have the disadvantages of low natural resonant frequency and the nonideal damping ratio, which are seriously affect the application range of the accelerometer. So, in order to ensure the measure precision and the real-time working of the accelerometer, we usually let the damping ratio ξ be around 0.707 when using it. But ξ=0.707 is a theoretic equation to some extent. Obviously, it is necessary to improve the dynamic characteristic to enlarge the application areas of the accelerometer. Generally, there are two ways to arrive at the purpose. One is to change the structure of accelerometer. But this way is limited by the level of the technique and the corresponding material. It is difficult to turn it into reality now. The other way is to design the dynamic characteristic compensator. With the rapid developing of the micro processing system, the compensator realization is feasible.

From Eq. (2), we find that if we change the pole and zero point of the accelerometer, the dynamic characteristic of the accelerometer can be improved. So we replace the pole of the sensor model and hold the zero point, and build a compensator to improve the dynamic characteristic. This is the idea of the zero-pole assignment (ZPS) method. ZPS method can fulfill the requirements only by changing the parameters a little. Also it can not increase the order of the sensor model, which ensures the real-time working of the system.

The compensating part can be expressed as

$$H_n(s) = \frac{(s^2 + 2\xi\omega_0 s + \omega_0^2)\omega_n^2}{(s^2 + 2\xi_n\omega_n s + \omega_n^2)\omega_0^2} = \frac{C_0 s^2 + C_1 s + C_2}{s^2 + B_1 s + B_2}, \tag{4}$$

where $B_1 = 2\xi_n\omega_n$; $B_2 = \omega_n^2$; $C_0 = \frac{\omega_n^2}{\omega_0^2}$; $C_1 = \frac{2\xi\omega_n^2}{\omega_0}$; $C_2 = \omega_n^2$.

We select a kind of micro-silicon piezoresistance accelerometer (3022 style) whose ω_0=42 Hz and $\xi = 0.25$ get from the datasheet, to discuss the method. If we let $\xi_n = 0.707$ and $\omega_n = 10$ kHz, we are easy to get the compensation part using Eq. (4). The accelerometer dynamic response curve after compensating using ZPS method is shown in Fig. 2.

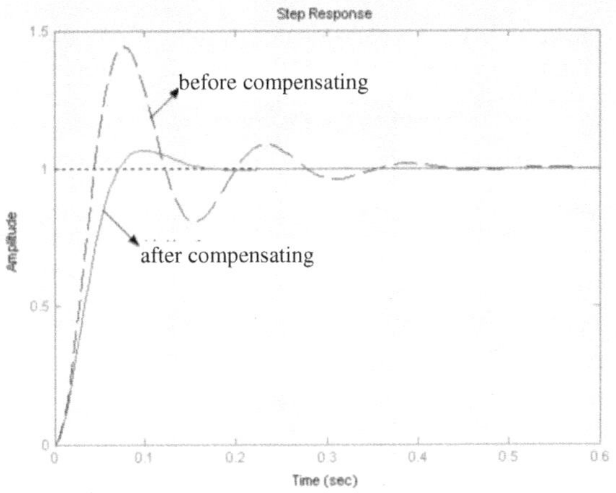

Fig. 2. Dynamic response of the accelerometer when the input signal is step signal

From Fig. 2, before compensating the maximal overshoot of the dynamic response is $\sigma = 49\%$, after compensating $\sigma = 1\%$. Before compensating, the surge times of the dynamic response is 5(*N*=5), while after compensating, the times is 2(*N*=2). So we can conclude that the zero-pole assignment method can compensate the dynamic characteristic of the accelerometer effectively. But we should not ignore a fact in this simulation: the sensor model is exactly known before using the compensating method. That is to say the method depends on a high precision sensor mathematic model. However, in most of the application areas, we can not get the exact mathematic model of the accelerometer easily. So the method is not practical.

3 Compensation Method Base on CMAC Neural Network

CMAC neural networks were originally developed by Albus [4, 5] for the on-line control of robotic manipulator. In general, a CMAC neural network works like a distributive table look-up method. The memory storage space of a direct table look-up method increases exponentially with the input data and resolutions and its usefulness is limited to very simple cases. In a CMAC neural network, its output information is stored in a relatively small region of the network, with similar inputs mapping to similar regions, while dissimilar inputs map to completely different regions of the network. The output of a CMAC neural network is then formed from a linear combination of nonlinearly transformed inputs. CMAC can overcome the shortcoming of low learning speed of the most forward neural network and is very suitable for the fast response system. So in this paper, CMAC neural network is applied to design the compensation part of the accelerometer.

The structure of CMAC is shown in Fig. 3. The main parts are network input, conceptual mapping, physical mapping and network output.

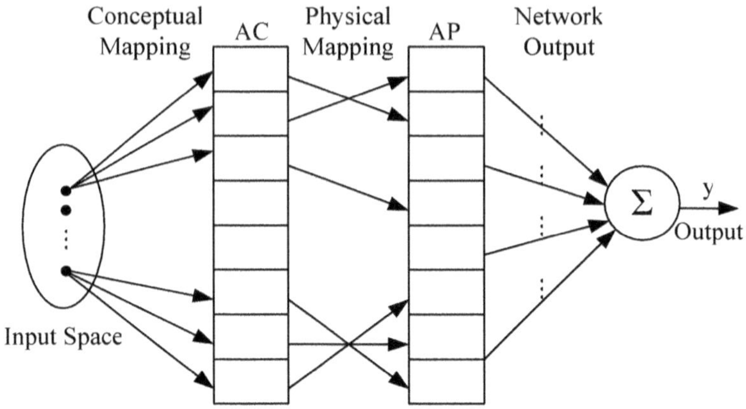

Fig. 3. Structure of CMAC neural network

Let U be an N-dimensional real value input vector

$$\boldsymbol{U} = [u_1, u_2, \cdots, u_N]^T . \tag{5}$$

Assume that the number of simultaneously excited receptive fields for each input is C . A classic CMAC neural network computation can be described as follows.

Step 1) Quantization. For example, normalize the input vector by dividing each component u_i by an appropriate quantization parameter Δ_i

$$\begin{aligned} \boldsymbol{U}' &= \begin{bmatrix} u_1', & u_2', & \cdots, & u_N' \end{bmatrix}^T \\ &= \left[\text{int}\left(\frac{u_1}{\Delta_1}\right), \ \text{int}\left(\frac{u_2}{\Delta_2}\right), \ \cdots, \ \text{int}\left(\frac{u_N}{\Delta_N}\right) \right], \end{aligned} \tag{6}$$

where int() is integer function that only takes the integer part of a real number.
Step 2) Association. Form the vector addresses A_i of the C receptive fields that contain the input point $\boldsymbol{U}'$

$$\begin{aligned} A_i &= \begin{bmatrix} u_1' - (u_1' - i)\%C, & u_2' - (u_2' - i)\%C, & \cdots, & u_N' - (u_N' - i)\%C \end{bmatrix} \\ &= \begin{bmatrix} a_{i1}, & a_{i2}, & \cdots, & a_{iN} \end{bmatrix}, \end{aligned} \tag{7}$$

where $X\%C$ is the modulo function that takes the remainder part of X/C . The above equation is only valid for positive $u_j' - i$. However, if $u_j' - i$ is negative, a similar expression can be easily formulated.
Step 3) Hashing. Since the total number of receptive fields in a space of dimension N can be quite large, the receptive field addresses A_i are typically considered as virtual

rather than physical address. Hashing is to form the scalar physical addresses $A_i^{'}$ of the actual adjustable weights to be used in the output computation

$$A_i^{'} = h(a_{i1}, a_{i2}, \cdots, a_{iN}), \quad (8)$$

where $h()$ represents any pseudorandom hashing function which operates on the components a_{ij} of the virtual addresses of the receptive fields, producing uniformly distributed scalar addresses in the physical weight memory of smaller size.
Step 4) Output. The CMAC scalar output is the average of the addressed weights

$$y(\boldsymbol{U}) = \sum_{i=1}^{C} \frac{W[A_i^{'}]}{C}, \quad (9)$$

Note that a vector CMAC output can be produced by simply considering the weight memory locations to contain vector rather than scalar values and by performing a vector rather than scalar average in the above equation.
Step 5) Learning. CMAC learning is typically based on observed training data pairs $\boldsymbol{U}$ and $y_d(\boldsymbol{U})$, where $y_d(\boldsymbol{U})$ is the desired network output in response to the vector input $\boldsymbol{U}$. The memory learning adjustment ΔW is given by

$$\Delta W = \alpha * \left[y_d(\boldsymbol{U}) - y(\boldsymbol{U}) \right], \quad (10)$$

where α is a learning factor and the same value ΔW is added to each of the C memory locations $W\left[A_i^{'}\right]$ accessed in the computation of $y(\boldsymbol{U})$.

We use the accelerometer model in part 1 to discuss the method. When $\omega_n = 42$ and $\xi = 0.25$, the transfer function of the accelerometer is

$$H(s) = \frac{1764}{s^2 + 21s + 1764}. \quad (11)$$

We let $\omega_n = 42$ and $\xi = 0.65$, the transfer function of the referenced model is

$$H_0(s) = \frac{1764}{s^2 + 54.6s + 1764}. \quad (12)$$

The block diagram of the compensation parts, which includes CMAC neural network, is shown in Fig. 4. to design the compensation parts. $u(t)$ is a step input signal, $e(t)$ is the step response of the system, $e_0(t)$ is the step response of the reference model, $y(t)$ is the system output.

In each end of the calculating period, first we can get an output $y(t)$ from CMAC. And then, we compare $y(t)$ with $e_0(t)$ to get the difference value between them. Next, we modify the weight value using the difference value between $y(t)$ and $e_0(t)$. Finally, we use the learning process to make the difference minimal so we can get an ideal output $y(t)$.

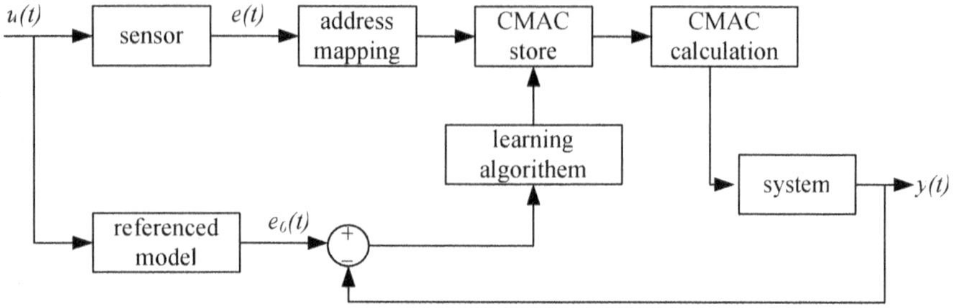

Fig. 4. Diagram of the compensation part

4 Simulation and Results

In the simulation, the initial weigh value is 0, the receptive fields number C is 5, the learning factor $\alpha = 0.05$, the learning iteration number is 200, the simple interval is 1ms. Fig. 5 and Fig. 6 are the simulation results.

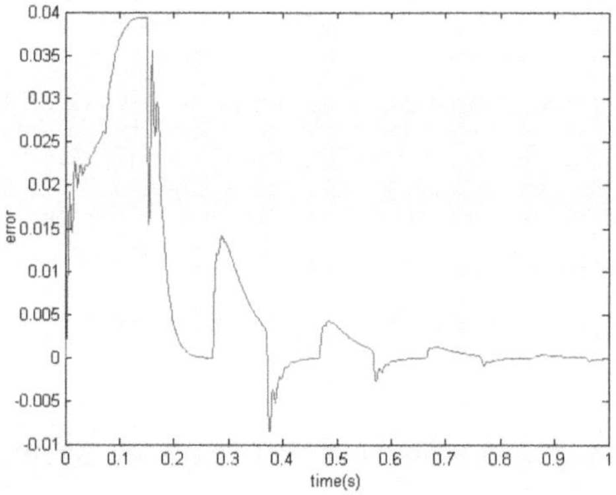

Fig. 5. Training process of CMAC neural network

Fig. 5 shows the training process of CMAC neural network to the system. From the curve, we can find that the difference value between $y(t)$ and $e_0(t)$ drops to zero quickly in 1 second. The curve shows the fast convergence rate of the CMAC neural network.

Fig. 6 shows the dynamic response of the accelerometer before using the CMAC neural network and after using the CMAC neural network. From the curves, it is obvious that, before compensating the maximal overshoot of the dynamic response is $\sigma = 49\%$, after compensating $\sigma = 0.5\%$. Before compensating, the surge times of the dynamic response is 5(N=5), while after compensating, the times is 2(N=2). Comparing with the zero-pole assignment method, the method based CMAC neural network also can reach the equivalent compensating precision and don't need the exact

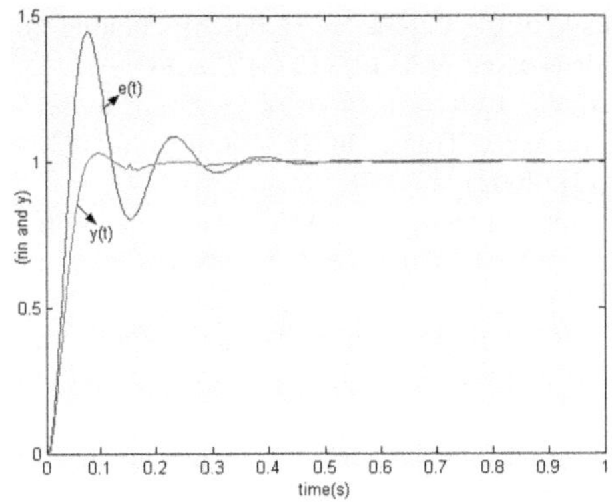

Fig. 6. Dynamic response of the accelerometer

mathematic model of the system. So we can conclude that the method based CMAC neural work is feasible for the sensor dynamic compensation. From the simulation results of compensating the dynamic error, we can observe that CMAC has a good convergence rate and the scheme is operating satisfactorily.

5 Conclusions

In the applications, the accelerometer should be selected carefully according to the frequency range of the input signal. Once the dynamic characteristic of the accelerometer can not meet the system requirements, the compensation to the dynamic error is necessary.

We can get a satisfied result using the ZPS compensation method. But the method needs a high precision sensor model, which is very difficult to be acquired in most of the application areas.

In this paper, a dynamic compensation method for the accelerometer based on CMAC neural network is proposed. The CMAC is more suitable for real time implementation, since it does not contain time consuming sigmoid activation functions. It can overcome the shortcomings of the ZPS method and also compensate the dynamic error effectively.

References

1. Meydan, T.: Recent Trends in Linear and Angular Accelerometers. Sensors and Actuators A **59** (1997) 43-50
2. Lim, C.M., Elangovan, S.: Pole Assignment of SISO System Using Dynamic Compensation with Prespecified Poles. IEEE Trans. on Circuit and System **31** (11) (1984) 990-991
3. Cai, H., Zhou, Z., Li, Y., Zhang, W.: A Study on Software Compensation Method of Accelerometer' Dynamic Characteristic. Chinese Journal of Scientific Instrument **19** (3) (1998) 263-267
4. Albus, J.S.: A New Approach to Manipulator Control: The Cerebellarmodel Articulation Controller (CMAC). Trans. ASME, J. Dyn. Syst.Meas. Control **63** (3) (1975) 220-227

5. Albus, J.S.: Data Storage in the Cerebellar Model Articulation Controller (CMAC). Trans. ASME, J. Dyn. Syst.Meas. Control **63** (3) (1975) 228-233
6. Lin, J., Song, S.: Modeling Gait Transitions of Quadrupeds and Their Generalization with CMAC Neural Network. IEEE Trans. On System, man, and Cybernetics-Part C: Applications and Reviews **32** (3) (2002) 177-189
7. Kim, Y., Lewis, F.L.: Optimal Design of CMAC Neural-Network Controller for Robot Manipulators. IEEE Trans. on System, man, and Cybernetics-Part C: Applications and Reviews **30** (1) (2000) 22-31
8. Christophe, S., Olivier, B.: Robustness of the Dynamic Walk of a Biped Robot Subjected to Disturbing External Forces by Using CMAC Neural Network. Robotics and Autonomous Systems **51** (2005) 81-99
9. Erkan, M.: A Rotor Position Estimator for Switched Relutance Motros Using CMAC. Energy Conversion & Management **44** (2003) 1229-1245

Modelling of Dynamic Systems Using Generalized RBF Neural Networks Based on Kalman Filter Mehtod

Jun Li and You-Peng Zhang

School of Automation and Electrical Engineering,
Lanzhou Jiaotong University, Lanzhou 730070, China
lijun691201@yahoo.com.cn

Abstract. A novel multi-input, multi-output generalized radial basis function (RBF) neural networks for nonlinear system modelling is presented in the paper, which uses extend Kalman filter to sequentially update both the output weights and the centers of the network. Simultaneously, such RBF models employ radial basis functions whose form is determined by admissible exponential generator functions. To test the validity of the proposed method, this paper demonstrates that generalized RBF neural networks with the extended Kalman filter can be used effectively for the identification and modelling of nonlinear dynamical systems. Simulation results reveal that the new generalized RBF networks guarantee faster learning and very satisfactory function approximation capability in modeling nonlinear dynamic systems.

1 Introduction

Nonlinear system identification and modeling via neural networks enables finding non-linear models of reality where the outputs of neural networks and the plant are matched [1], [2]. A Radial basis function (RBF) neural network is usually trained to perform a mapping from an q-dimensional input space to an l-dimensional output space. The performance of a RBF network depends on the number and position of the radial basis functions, their shape, and the method used for learning the input-output mapping. RBF neural networks have obtain successful results in system identification and control because they provide good approximations when applied to highly nonlinear and complex systems and it has been used as an alternative to conventional neural networks [3], [4]. Furthermore, it has improved training characteristics in comparison with feedforward neural networks due to their localized nature and the fact that they are linear in weighs.

For classify problems, supervised learning algorithms based on Gradient decent for training reformulated RBF networks has proven to be much more effective than conventional Gauss RBF networks. This approach results in a broad variety of admissible RBF models whose form is determined by admissible generator functions [5]. Kalman filters are attractive theoretically and have been used extensively to train neural net-works or fuzzy systems due to its optimality properties. To improve the performance and accelerate speed of RBF neural networks for nonlinear system identification and modelling, a new multi-input, multi-output RBF neural networks based on extend Kalman filter method is proposed in this paper, which is an extension of the work presented in [5], [6].

D. Liu et al. (Eds.): ISNN 2007, Part I, LNCS 4491, pp. 676–684, 2007.

The proposed RBF networks with Gaussian radial basis functions are generated by admissible exponential generator functions. Hence, it is also suitable for gradient decent learning.

To illustrate the performance of the generalized RBF networks with extend Kalman filter algorithm, several examples are provided, which include identification of a multi-input multi-output (MIMO) nonlinear dynamic systems as well as modelling the behaviour of a hydraulic robot arm.

2 New Generalized RBF Neural Networks for Function Approximation

2.1 Generalized RBF Neural Networks for Function Approximation

Consider the $\mathbb{R}^q \rightarrow \mathbb{R}^l$ mapping implemented by the following form:

$$\hat{y}_i = f\left(w_{i,0} + \sum_{j=1}^{c} w_{i,j} g(\|\boldsymbol{x} - \boldsymbol{v}_j\|^2)\right), 1 \leq i \leq l . \quad (1)$$

where $f(\cdot)$ is a non-decreasing, continuous and differentiable everywhere function. The model (1) describes an RBF neural network with q inputs, c radial basis functions, and l output units if $g(x^2) = \phi(x)$, and $\{ \phi(\cdot) \}$ are radial basis functions. Let $f(\cdot)$ be a line function of the form $f(x) = x$, in such a case, the response of the RBF neural networks to the input vector $\boldsymbol{x}_k$ is $\hat{y}_{i,k} = \sum_{j=1}^{c} w_{i,j} h_{j,k}$, $1 \leq i \leq l$, $h_{0,k} = 1$ and $h_{j,k} = g(\|\boldsymbol{x}_k - \boldsymbol{v}_j\|^2)$, $1 \leq j \leq c$, $1 \leq k \leq N$, where $w_{i,j}$ is the weight that connects the ith output unit with the jth radial basis function, and $h_{j,k}$ represents the response of the radial basis function centered at the prototype $\boldsymbol{v}_j$ to the input vector $\boldsymbol{x}_k$.

New generalized RBF neural networks are developed to facilitate the training of RBF models by learning algorithms based on gradient decents [4], [5]. This is done by including the centers of the radial basis functions in the adjustable model parameters and searching for radial basis functions that improve the effectiveness of gradient decent learning. The search for any admissible radial basis functions can be simplified by considered basis functions of the form $\phi(x) = g(x^2)$, with $g(\cdot)$ defined in terms of generator function $g_0(\cdot)$ as $g(x) = (g_0(x))^{\frac{1}{1-r}}$, $r \neq 1$.

For admissible exponential generator function $g_0(x) = \exp(\beta x), \beta > 0$ and $r > 1$, then we have

$$g(x) = (\exp(\beta x))^{\frac{1}{1-r}} . \quad (2)$$

If $\sigma^2 = (r-1)/\beta$, the corresponding radial basis function $\phi(x) = g(x^2)$ can be obtained from (2) as follows

$$\phi(x) = \exp\left(-\frac{x^2}{\sigma^2}\right). \quad (3)$$

which corresponds to Gaussian radial basis function of the form. The reformulated RBF networks generated by exponential generator functions can be trained by gradi-ent descent and perform considerably better than conventional RBF networks [5].

2.2 Training Generalized RBF Neural Networks Based on Extend Kalman Filter

New generalized RBF neural networks can be trained to map $\boldsymbol{x}_k \in \mathbb{R}^q$ into $\boldsymbol{y}_k = [\, y_{1,k} \;\; y_{2,k} \;\; \cdots \;\; y_{l,k} \,]^T \in \mathbb{R}^l$, where the vector pairs $(\boldsymbol{x}_k, \boldsymbol{y}_k), k = 1, \ldots, N$, form the training set. If $\boldsymbol{x}_k$ is the input to an generalized RBF neural networks, its actual output is $\hat{\boldsymbol{y}}_k = [\, \hat{y}_{1,k} \;\; \hat{y}_{2,k} \;\; \cdots \;\; \hat{y}_{l,k} \,]^T$, where $\hat{y}_{i,k} = \boldsymbol{w}_i^T \boldsymbol{h}_k$, $\boldsymbol{w}_i = [\, w_{i,0} \;\; w_{i,1} \;\; \cdots \;\; w_{i,c} \,]^T$, $\boldsymbol{h}_k = [\, h_{0,k} \;\; h_{1,k} \;\; \cdots \;\; h_{c,k} \,]^T$, $h_{0,k} = 1$, $h_{j,k} = g\left(\|\boldsymbol{x}_k - \boldsymbol{v}_j\|^2\right)$, $1 \le j \le c$. We can view the optimization of the weight vector and the prototypes as a weighted least-squares minimization problem, where the error vector is the difference between the RBF outputs and the desired values for those outputs.

In order to apply extend Kalman filtering algorithms, we let the elements of the weight vector $\boldsymbol{w}_i$ and the elements of the prototypes $\boldsymbol{v}_j$ constitute the state of a nonlinear system, and let the output of the generalized RBF network constitute the output of the nonlinear system. The state of the nonlinear system can then be represented as

$$\theta = [\, \boldsymbol{w}_1{}^T \;\; \cdots \;\; \boldsymbol{w}_l{}^T \;\; \boldsymbol{v}_1{}^T \;\; \cdots \;\; \boldsymbol{v}_c{}^T \,]^T . \tag{4}$$

The vector θ thus consists of all $(l(c + 1) + qc)$ of the RBF parameters. The nonlinear system model to which the Kalman filter can be applied is

$$\begin{cases} \theta_{n+1} = \theta_n + \omega_n \\ \boldsymbol{y}_n = h(\theta_n) + \zeta_n, \end{cases} \tag{5}$$

where $h(\theta_n)$ is the RBF networks nonlinear mapping between its parameters and its output, ω_n and ζ_n are artificially added noise processes. Hence, It can be shown that the desired parameter estimate $\hat{\theta}_n$ can be obtained by the recursion extend Kalman filtering algorithms as follows [6], [7]

$$\begin{cases} \hat{\theta}_n = \hat{\theta}_{n-1} + \boldsymbol{K}_n \left[\boldsymbol{y}_n - h(\hat{\theta}_{n-1})\right] \\ \boldsymbol{K}_n = \boldsymbol{P}_n \boldsymbol{H}_n \left(\boldsymbol{R} + \boldsymbol{H}_n{}^T \boldsymbol{P}_n \boldsymbol{H}_n\right)^{-1} \\ \boldsymbol{P}_{n+1} = \boldsymbol{P}_n - \boldsymbol{K}_n \boldsymbol{H}_n{}^T \boldsymbol{P}_n + \boldsymbol{Q}, \end{cases} \tag{6}$$

where $\boldsymbol{K}_n$ is the Kalman gain matrix, $\boldsymbol{Q}$ and $\boldsymbol{R}$ matrices are diagonal covariance matrices that provides a mechanism by which the effects of artificial processes noise are included in the Kalman recursion. $h(\hat{\theta}_n)$ is the actual output of the RBF networks. $\boldsymbol{H}_n$ is the partial derivative of the RBF output with respect to the RBF network parameters at the nth iteration of the recursion equation (6), it is defined as

$$\boldsymbol{H}_n = \begin{bmatrix} \boldsymbol{H}_w \\ \boldsymbol{H}_v \end{bmatrix}, \tag{7}$$

where $\boldsymbol{H}_w$ and $\boldsymbol{H}_v$ are given as follows [6]

$$\boldsymbol{H}_w = \begin{bmatrix} \boldsymbol{H} & 0 & \cdots & 0 \\ 0 & \boldsymbol{H} & \cdots & 0 \\ \vdots & \vdots & \vdots & \vdots \\ 0 & \cdots & 0 & \boldsymbol{H} \end{bmatrix}, \tag{8}$$

$$\boldsymbol{H}_v = \begin{bmatrix} -2w_{1,1}g'_{1,1}(\boldsymbol{x}_1 - \boldsymbol{v}_1) & \cdots & -2w_{1,1}g'_{N,1}(\boldsymbol{x}_N - \boldsymbol{v}_1) & \cdots \\ \vdots & & \vdots & \vdots \\ -2w_{1,c}g'_{1,c}(\boldsymbol{x}_1 - \boldsymbol{v}_c) & \cdots & -2w_{1,c}g'_{N,c}(\boldsymbol{x}_N - \boldsymbol{v}_c) & \cdots \end{bmatrix} \tag{9}$$

$$\begin{matrix} -2w_{l,1}g'_{1,1}(\boldsymbol{x}_1 - \boldsymbol{v}_1) & \cdots & -2w_{l,1}g'_{N,1}(\boldsymbol{x}_N - \boldsymbol{v}_1) \\ \vdots & \vdots & \vdots \\ -2w_{l,c}g'_{1,c}(\boldsymbol{x}_1 - \boldsymbol{v}_c) & \cdots & -2w_{l,c}g'_{N,c}(\boldsymbol{x}_N - \boldsymbol{v}_c) \end{bmatrix}.$$

Note that $\boldsymbol{H} = [\boldsymbol{h}_1 \quad \boldsymbol{h}_2 \quad \cdots \quad \boldsymbol{h}_n]$, it is the $(c+1) \times N$ matrix, and the first-order derivatives $g'_{i,j} = g'\left(\|\boldsymbol{x}_k - \boldsymbol{v}_j\|^2\right)$. $\boldsymbol{H}_w$ in (8) is an $l(c+1) \times lN$ matrix, $\boldsymbol{H}_v$ in (9) is an $qc \times lN$ matrix.

For generalized RBF neural networks with exponential generator function, then $g_0(x) = \exp(\beta x)$, in this case, the first-order derivatives $g'(x) = \frac{1}{1-r}(g(x))^r g'_0(x)$, and we have

$$\begin{aligned} g'_{i,j} &= g'\left(\|\boldsymbol{x}_k - \boldsymbol{v}_j\|^2\right) \\ &= \frac{\beta}{1-r} h_{j,i} \end{aligned} . \tag{10}$$

The Kalman filter parameters of (6) are initialized with $P_0 = 40I$, $Q = 40I$, and $R = 40I$, where I is the identity matrix of appropriate dimensions.

Now that we have have the H_n matrix, we can apply the recursion of (6) using extend Kalman filter training algorithms to determine the weight vectors and the prototypes vectors. The algorithms are initialized with prototype vectors randomly selected from the input data, and with the weight vectors set to 0. Furthermore, the computational expense of the Kalman filter is on the order of $lN \times [l(c+1) + qc]^2$.

3 Application to Modelling of Nonlinear Dynamic Systems

The following dynamic modeling examples illustrate the effectiveness of the proposed generalized RBF networks with extend Kalman filter. The generalized RBF network is trained using the hidden layer function of (3) with the exponential generator function of (2) with r=3 and β=0.5, as well as normalized version of the input data. The accuracy of model is assessed using root mean square error (RMSE).

3.1 Example1: The Identification of the Nonlinear Mulitivariable Plant with Two Inputs and Two Outputs

In the example, the difference equation describing the plant was assumed to be of the following form [8]:

$$\begin{bmatrix} y_{p1}(k+1) \\ y_{p2}(k+1) \end{bmatrix} = \begin{bmatrix} \dfrac{0.8y_{p1}^3(k) + u_1(k)^2 u_2(k)}{2 + y_{p2}^2(k)} \\ \dfrac{y_{p1}(k) - y_{p1}(k)y_{p2}(k) + (u_1(k) - 0.5)(u_2(k) + 0.8)}{1 + y_{p2}^2(k)} \end{bmatrix} + \begin{bmatrix} u_1(k) \\ u_2(k) \end{bmatrix}. \tag{11}$$

A series-parallel identification model based on generalized RBF network is described by the equation

$$\begin{bmatrix} \hat{y}_{p1}(k+1) \\ \\ \hat{y}_{p2}(k+1) \end{bmatrix} = N\left[y_{p1}(k), y_{p2}(k), u_1(k), u_2(k)\right] . \quad (12)$$

The generalized RBF networks model tested in the example consists of four inputs, two outputs units, and c=35 radial basis functions. 1600 simulated data points are generated from the plant model (11). The first 1000 data points are obtained by assuming random inputs $u_1(k)$ and $u_2(k)$ uniformly distributed in the interval [-1, 1], and the last 600 data points are obtained by a sinusoid vector input $[\sin(2\pi k/250), \cos(2\pi k/250)]^T$. The performance of the generalized RBF networks is tested using the remaining 600 data points. After it was trained, the sum of square error(SSE) versus epoch is shown in Fig. 1. It shows that Kalman filter training only needs to iterate 16 epochs for convergence.

The final responses of the plant and the identification model in test region are shown in Fig. 2. The RMSE in training region for y_{p1} and y_{p2} is 0.0048, 0.0401, respectively. Fig. 2 (a) and Fig. 2 (b) show that the performance of the generalized RBF model is very good.

For the purpose of comparison, the generalized RBF network is also trained by gradient descent learning and the learning rate is set to 0.001. Note that the convergence criterion is identical for different training methods, that is, training is terminated when the SSE decreased by less than 0.1%. In the case, 377 epochs are required to identify the system and the RMSE in training region for for and is 0.0179, 0.0510, respectively.

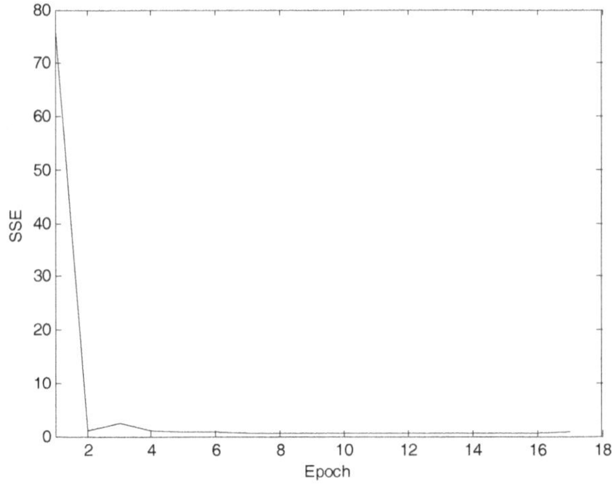

Fig. 1. SSE versus the number of the training epoch in example 1

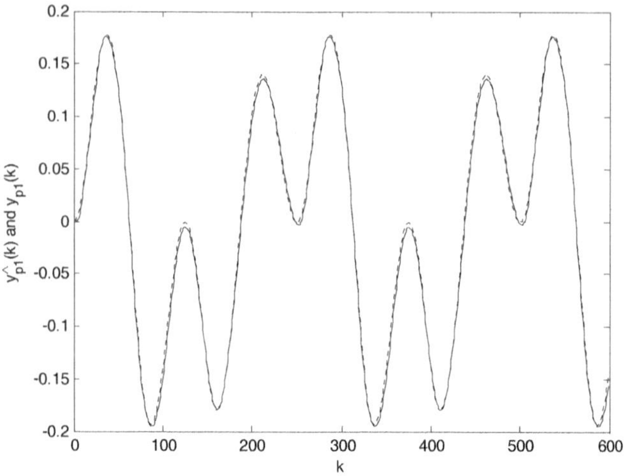

(*a*)

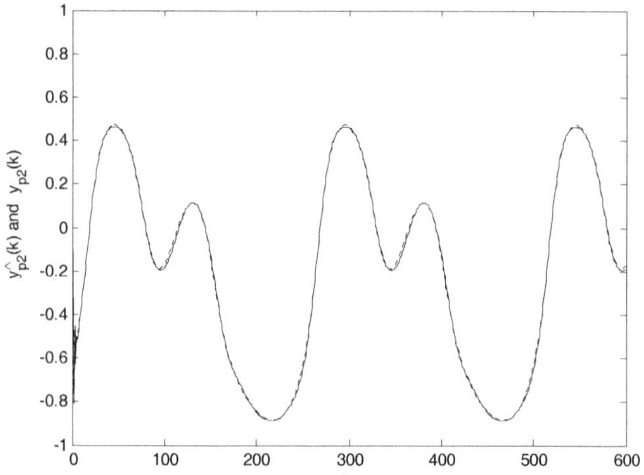

(*b*)

Fig. 2. MIMO system identification:(a) outputs of the identified model [$\hat{y}_{p1}$, dashed line] and the system model [y_{p1}, solid line]; (b) outputs of the identified model [$\hat{y}_{p2}$, dashed line]and the system model [y_{p2}, solid line]

3.2 Example 2: Modelling a Hydraulic Robot Actuator

In the example, the position of a robot arm is controlled by a hydraulic actuator. The oil pressure in the actuator is controlled by the size of the valve opening through which the oil flows into the actuator. The position of a robot arm is then a function of the oil pressure. We apply generalized RBF networks with extend Kalmn filter to modeling behaviour of a hydraulic robot arm. Our results will be compared with neural network NARX and wavelet network NARX models in [9]. Measured values of the valve size u

and the oil pressure y are input and output signals, respectively. For the purpose of comparison, we used the following series-parallel identification model based on generalized RBF network, since this is also used in [9].

$$\hat{y}(k) = N\left[y(k-3), y(k-2), y(k-1), u(k-2), u(k-1)\right] . \tag{13}$$

The generalized RBF networks model tested in the example consists of five inputs, one outputs units, and c=35 radial basis functions. The entire data set consists of 1024 samples data pairs. We also used half the data set for training, and half as validation data, again following the procedure of Sjoberg et al [9].

After generalized RBF networks with Kalman filter was trained, the sum of square error(SSE) versus training epoch is shown in Fig. 3. It shows that Kalman filter training only needs to iterate 17 epochs for convergence.

The output of the true oil pressure and the identification model on the training data is shown in Fig. 4 (a). The output of the true oil pressure and the identification model on the testing data is shown in Fig. 4 (b). It can be seen from Fig. 4 (a) and Fig. 4 (b) that the performance of the generalized RBF model is very satisfactory. The RMSE of this prediction on the test set is 0.0326, which is lower than both the wavelet network RMSE (0.579), and the prediction made by a one-hidden-layer sigmoid neural network with ten hidden units(0.467) in [9], although the RMSE can be further reduced to 0.328 on the data set in [9].

The same generalized RBF network is also trained by gradient descent approach, and the learning rate is set to 0.001. In the case, 478 epochs are required to modelling the system and the RMSE in test region is 0.1765.

The employed RBF network with Kalman filter algorithm improves greatly on previously results, and the estimate accuracy one order of magnitude increases over the neural networks.

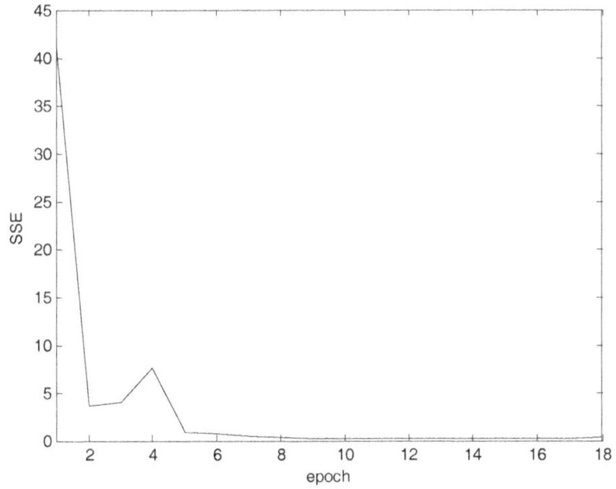

Fig. 3. SSE versus the number of the training epoch in example 2

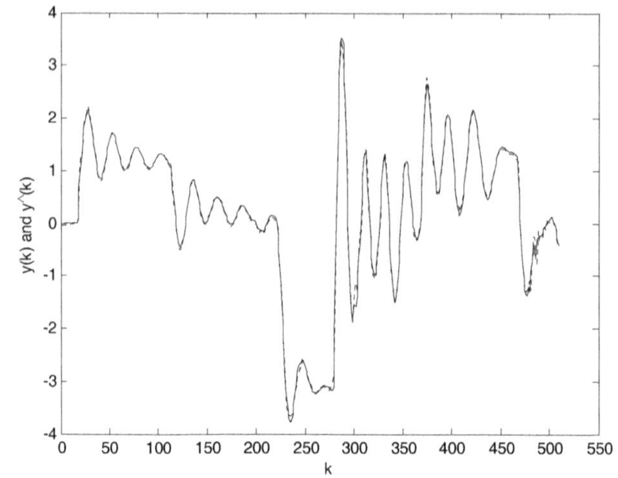

(a)

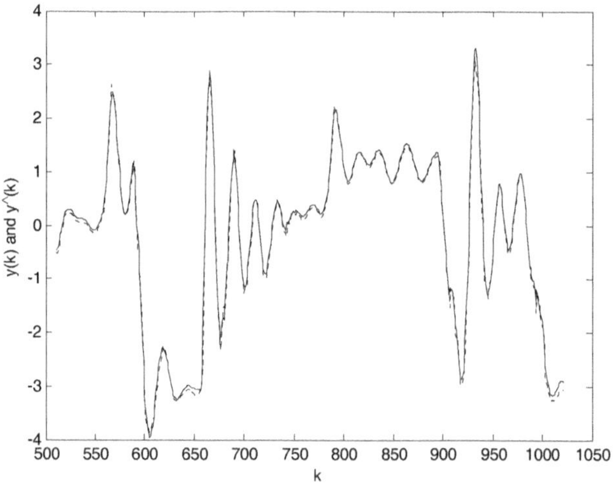

(b)

Fig. 4. Robert arm data and generalized RBF network model: (a) outputs of the identified model [$\hat{y}$, dashed line] and the oil pressure [y, solid line] on training data; (b) outputs of the identified model [$\hat{y}$, dashed line] and the oil pressure[y, solid line] on validation data

4 Conclusion

The paper develops a novel generalized RBF network using extend Kalman filtering approach to dynamical system modeling and identification, in which the hidden layer RBF is constructed by exponential generator functions. Compared with conventional Gaussian RBF neural networks and feedforward neural networks, the proposed gen-eralized RBF network with exponential form of generator function improves computational cost and shows better performance with higher approximation accuracy. The simulation results demonstrate that the performances of identification and modeling scheme based on proposed RBF network are considerably satisfactory. Hence, it is an attractive new method for multivariable nonlinear system modeling.

Acknowledgments. This research is in part supported by the National Science Foundation of Gansu Province under grant no.3ZS042-B25-026 and supported by 'Qinglan' project of Lanzhou Jiaotong University.

References

1. Pham, D.T., Liu, X.: Neural Network for Identification, Prediction and Control. Springer-Verlag, Berlin Heidelberg (1995)
2. Narendra, K., Parthasarathy, K.: Identification and Control of Dynamical Systems Using Neural Networks **1(1)** (1990) 4-27
3. Haykin, S.: Neural Networks: A Comprehensive Foundation. 2rd edn. Prentice-Hall, Upper saddle River, NJ (1999)
4. Karayiannis, N.B., Mi, G.W.: Growing Radial Basis Neural Networks: Merging Supervised and Unsupervised Learning with Network Growth Techniques. IEEE Trans. Neural Networks **8(6)** (1997) 1492-1506
5. Karayiannis, N.B.: Reformulated Radial Basis Function Neural Networks Trained by Gradient Descent. IEEE Trans. Neural Networks **10(3)** (1999) 657-671
6. Simon, D.: Training Radial Basis Neural Networks with the Extended Kalman Filter. Neurocomputing **48** (2002) 455-475
7. Dennis, W.R., Steven Rogers, K., Matthew, K. et al.: Comparative Analysis of Backpropagation and the Extended Kalman Filter for Training Multilayer Perceptrons. IEEE Trans. Pattern Analysis and Machine Intelligence **14(6)** (1992) 686-691
8. Li, H.X., Philip Chen, C.L., Huang, H.P.: Fuzzy Neural Intelligent Systems: Mathe-matical Foundation and the Applications in Engineering. CRC Press, Boca Raton, FL (2001)
9. Sjoberg, J., Zhang, Q., Ljung, L. et al.: Nonlinear Black-box Modeling in System Identification: A Unified Overview. Automatica **31(12)** (1995) 1691-1724

Recognition of ECoG in BCI Systems Based on a Chaotic Neural Model*

Ruifen Hu[1], Guang Li[2], Meng Hu[3], Jun Fu[1], and Walter J. Freeman[4]

[1] Biomedical Engineering Department, Zhejiang University, Hangzhou 310027, China
[2] National Laboratory of Industrial Control Technology, Zhejiang University, Hangzhou 310027, China,
[3] Department of Physics, Zhejiang University, Hangzhou 310027, China
[4] Department of Molecular & Cell Biology, University of California at Berkeley Berkeley CA 94720-3206 USA
guangli@zju.edu.cn

Abstract. For the practical use of brain-computer interface systems, one of the most significant problems is the generalizing ability of the classifiers, since the states of both people and instruments are altering as time goes on. In this paper, a novel chaotic neural network termed KIII model, is introduced to classify single-trial ECoG during motor imagery, acquired in two different sessions. Then, by comparing with other three traditional classifiers, KIII model shows a greater ability to generalize, which demonstrates that KIII model is an effective tool for brain-computer interfaces systems.

Keywords: brain-computer interface, chaotic neural model, KIII model.

1 Introduction

Nowadays, brain-computer interface (BCI) system has become one of the hottest topics in academic fields, since its prospective applications in various domains, like facilitating disabled people [1]. In present, BCI systems generally can be classified into two groups: invasive and non-invasive. Invasive systems usually use electrocorticogram (ECoG), which is recorded directly from the surface of cortex, to extract pattern information, while the analyzed data of the non-invasive systems are electroencephalogram (EEG), electrical signals from scalp. By training subjects to do specific motion or motor imagery according to specific cues and recording their ECoG or EEG simultaneously, BCI systems can obtain specific feature information corresponding to the states of subjects, and then realize the communication between human beings and machines

* This work is supported in part by National Natural Science Foundation of China Grant #60421002 and the National Basic Research Program of China (973 Program) Grant #2004CB720302. The authors would like to thank Wolfgang Rosenstiel, Niels Birbaumer, Bernhard Schölkopf and Christian Elger for providing the dataset.

D. Liu et al. (Eds.): ISNN 2007, Part I, LNCS 4491, pp. 685–693, 2007.

without physical contacts. Although non-invasive systems impose less hurt on subjects, the advantage of higher signal-to-noise ratio and better spatial resolution [2] makes ECoG still attractive.

However, the robustness of BCI systems is still a problem. For example, because of the state change of both subjects and recording systems [3], a well-trained system may fail in test datasets recorded several days later. Studies show that KIII model has a good performance in noise-resistance and error-tolerance [4], and therefore, this paper employs it for BCI systems.

2 KIII Model

KIII model is a neurodynamics model of olfactory system developed by Freeman and his colleagues over the last 30 years [5]. It consists of three key parts: the topology, the mathematical expression and the learning and classification algorithms.

2.1 The Topology

KIII model is established based on physiological structure of mammalian olfactory system. In accordance with the anatomic architecture, KIII network is a multi-layer neural network model, which is composed of several K0, KI, and KII units [6]. The topology of KIII model is as Fig.1:

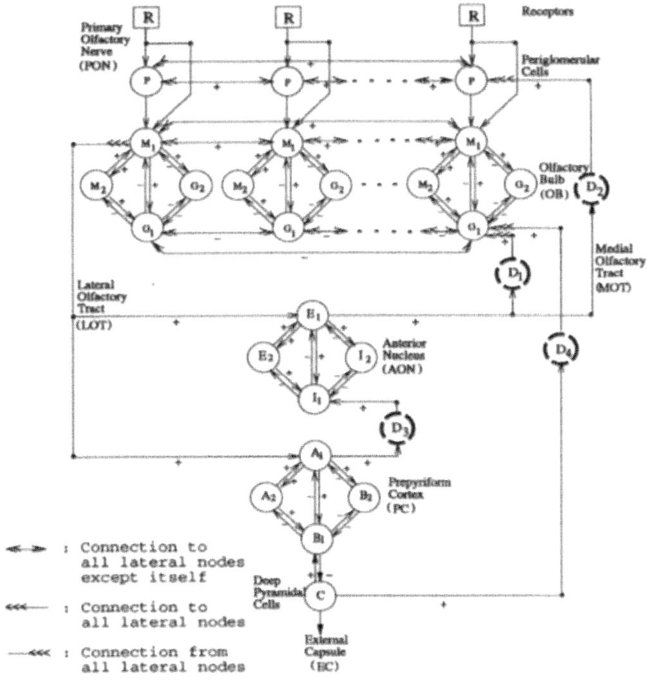

Fig. 1. The topology of KIII model [6]

2.2 Mathematical Expression

As in Fig.1, the basic cells of KIII model are several nodes, each of which represents a population of neurons. All the nodes are modeled by a second order differential equation (ODE) as Eq. (1):

$$\frac{1}{a \cdot b}[x_i''(t)+(a+b)x_i'(t)+a \cdot b \cdot x_i(t)] = \sum_{j \neq i}^{N} [W_{ij} \cdot Q(x_j(t), q_j)] + I_i(t) \quad (1)$$

$$Q(x_j(t), q_j) = \begin{cases} q_j(1-e^{-(e^{x_j(t)}-1)/q_j}) & x_j(t) > x_0 \\ -1 & x_j(t) < x_0 \end{cases} \quad (2)$$

$$x_0 = \ln(1 - q_j \ln(1+1/q_j))$$

where x_i (t) represents the state variable of the *ith* node, x_j (t) represents the state variable of the *jth* node, which is connected to the *ith*, while W_{ij} is the connection weight from *j* to *i*. I_i(t) is an external input signal to the *ith* channel, a and b are two rate constants, determined by physiological experiments, and Q (-) is a static sigmoid function derived from the Hodgkin-Huxley equation; see Eq. (2), where q_j is an adjustable parameter to fit the I/O transformation to the experimental data. All the parameters of the model in this paper are come from [5].

In addition, one of the most attractive features of KIII model is the introduction of additive noises. A low level Gaussian noise is imported respectively into the receptors (R) and anterior olfactory nucleus (AON) to enhance the robustness of the model; the model can hold a stable output under possible perturbations of initial conditions and parameter values. The above step introduces an alternative to the deterministic chaos, which is named "Stochastic Chaos" by Freeman [4].

2.3 Learning and Classification Algorithm

KIII model, with the above topology and mathematical model, has the ability of learning and remembering patterns. The operation of the KIII dynamic memory can be described as follow [4]: the system holds a high-dimensional state of spatially coherent basal activity when stimulus is absence, and once the external stimulus is coming, transfers into a local memory basin of a lower-dimension attractor. The duration of the localized basin is consistent with that of the stimulus, after which the system returns back into the basic state. Each of the lower-dimension attractor is corresponding to a specific pattern.

In the learning process of the algorithm used [7], merely the activity states of the mitral nodes (M) are analyzed and only the weights between M nodes are regulated. That is because, in accord with physiological processes, the affected synapses are located not at the R synapses on M nodes, but at the synapses of M nodes on other nodes in the olfactory bulb (OB) [4]. There are two basic rules for training: Hebbian reinforcement is used for establishing the memory basins of certain patterns, while the habituation is employed to reduce the impact of environment noises or other irrelative network inputs. The two rules modify the network weights as in Eq. (3):

$$\begin{array}{llr} \text{If} \quad SD_{ai} > (1+K)\,SD_a^{\ m} \quad \text{and} \quad SD_{aj} > (1+K)\,SD_a^{\ m} & & \\ \text{Then} \quad \omega\,(mml)_{ij} = hebb \cdot \omega\,(mml)_{ij} & \text{(Hebb rule)} & (3) \\ \text{Otherwise} \quad \omega\,(mml)_{ij} = hab^{TT} \cdot \omega\,(mml)_{ij} & \text{(Hab rule),} & \end{array}$$

where SD_{ai} is the standard deviation (SD) of the output of the *ith* M node (M_i), representing the activity measure of M_i, and $SD_a^{\ m}$ represents the mean activity measure over the whole OB layer. K is a bias coefficient to avoid the saturation of the weight space, $\omega\ (mml)_{ij}$ is the lateral connection between two M nodes, and TT is the duration of stimulus. *hebb* and *hab* are two constants used to strength or reduce the weights.

At the end of learning, the connection weights are fixed to perform pattern classification. During training, the activity vector for each trial is calculated, and the mean vector through each class is defined as the corresponding cluster center. When in test, the Euclidean distances from the activity vector of each test trial to those cluster centers are estimated, and the minimum distance determines the classification.

3 Methods

3.1 Data Acquisition and Preprocessing

The dataset used in this paper was provided by the organizers of the BCI data competition III. All its data were from a subject with an 8*8 ECoG platinum electrode grid which was placed on the contra lateral (right) motor cortex. All the recordings were performed with a sampling rate of 1000Hz, amplified and stored as microvolt values. Every trial consisted of either an imagined tongue or an imagined finger movement and was recorded for 3 seconds duration. Furthermore, visually evoked potentials have been avoided. The detail on the dataset can be found in [3].

The specialty of this dataset is that training data (278 trials) and test data (100 trials) were recorded on two different days with about 1 week in between, though from the same subject and with the same tasks. The state changes of both the subject and the recording system raise a challenge.

In order to decrease the computation load and reduce artifacts' effects, original data are preprocessed. Firstly, the original data are decimated from 3000 sampling points to 300. Then, common average reference is employed to re-reference the re-sampled data. Besides of reducing artifacts' effects, it can also enhance differences between two patterns.

3.2 Feature Exaction

Besides the Bereitschaftspotential, two other phenomena in the EEG are characteristic for self-paced movement; one is the event-related desynchronization (ERD) and the other is the event-related synchronization (ERS) [8]. On the other hand, some studies demonstrate that recordings from subdural electrodes indicate behaviors analogous to EEG data [9]. Therefore, the electrophysiological features (ERD/ERS) can also be used in ECoG classification. Since recent studies show that when performing motor imagination, mu (8-12Hz) and beta (18-26Hz) rhythms are found to reveal ERD/ERS over sensorimotor cortex [10], the relative pattern information is extracted with a band pass (8~30Hz) Chebysheve filter [11].

In order to simplify the expression, in the following paper, task of imaged finger movement and task of imaged tongue movement are represented as class -1 and class 1 respectively.

3.2.1 Common Spatial Subspace Decomposition

In this paper, a novel method Common Spatial Subspace Decomposition (CSSD), proposed by Wang etc. [12] is utilized to extract most distinctive components between data of two cognitive tasks. The principle of CSSD is to extract source components of signals by estimating their spatial filters:

$$S_1 = F_1 \cdot X_1, \qquad S_{-1} = F_{-1} \cdot X_{-1}, \tag{4}$$

where S_i is the source component of class i, F_i is the spatial filter of class i, and X_i is the data from class i (i=-1, 1). The detail of the method can be found in [12]. By using this algorithm, spatial factors can be selected from their corresponding spatial filters, and then target signal components can be extracted by applying the resulting spatial factors to the ECoG data matrix. Moreover, studies proved that in the early period of pre-movement, the distribution of the ERD responses is spatially limited [9]. Hence, CSSD will be an appropriate way to analyze the ERD components.

In this paper, four spatial factors, F_{1_1}, F_{1_-1}, F_{2_1}, F_{2_-1} (F_{1_1} and F_{2_1} are the first and the fourth column of F_1 respectively; F_{1_-1} and F_{2_-1} are the last and the last fourth column of F_{-1} respectively), are obtained by CSSD. The projections of signals on specific spatial factor (the target signal components) are extracted by multiplying the corresponding spatial factor with signals (for each trial, the signal is a 64*300 array, where 64 is the number of ECoG channels and 300 is the length of the time series.). For classification, in this study, variance analysis and autoregressive (AR) spectral analysis are applied on those target signal components, extracting a four-dimension feature vector for each trial.

3.2.2 Variance Analysis

Variance analysis means to calculate the variance (square of SD) of the target signal components, which are time series. By employing this method, two dimensions of feature vectors are defined as Eq. (5) and the feature values of training data are shown in Fig. 2:

$$f_i = \frac{\mathrm{var}(F_{i\_1}X)}{\mathrm{var}(F_{i\_1}X) + \mathrm{var}(F_{i\_-1}X)} \qquad i = 1, 2\,, \tag{5}$$

3.2.3 Autoregressive Prediction Model Analysis

Autoregressive (AR) prediction model is a parametric spectral analysis method, estimating the power spectral density (PSD) of time series. By employing AR model (tool in MATLAB6.5), the other two dimensions of feature vectors are obtained. Average PSD of through each task in training data is illustrated in Fig. 3, showing that great difference between two tasks exists from about 8Hz to 11Hz.

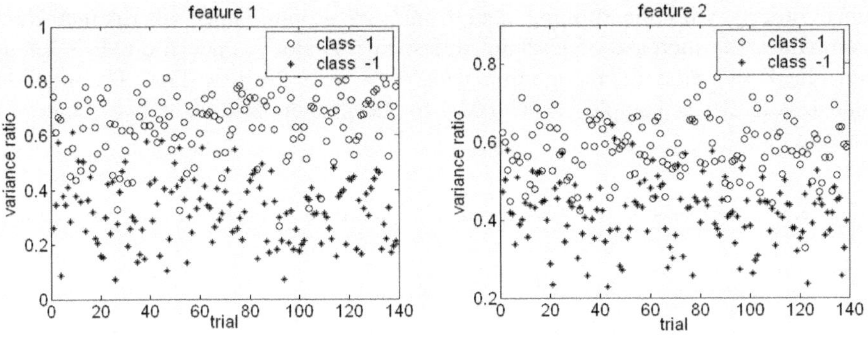

(a) (b)

Fig. 2. (a) shows the calculating results of Eq. 5 for each trial when i=1, (b) shows the calculating results of Eq. 5 for each trial when i=2; class 1 is expressed with 'o', and class -1 is expressed with '*'

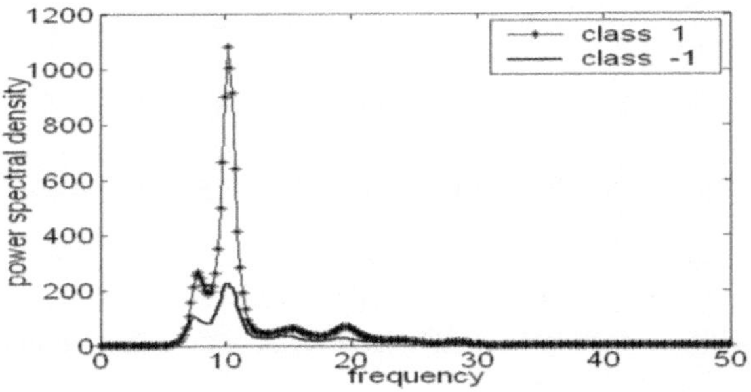

Fig. 3. Average PSD through Trials of two tasks

Procedure of calculating the other two dimension of feature vectors:

(1) Filter data with spatial filters F_{1_1} and F_{1_-1} (the same with those mentioned above), generating two series TS_1 and TS_2.

(2) Calculate the spectral power of TS_1 and TS_2 in band range of 9.7-10.7Hz, P_1 and P_2, respectively.

(3) Then, calculate our one feature as follow:

$$f_3 = \frac{P_1}{P_1 + P_2}, \qquad (6)$$

(4) The other feature is extracted by almost the same way, with only a distinct band range (10.7-50Hz).

3.3 Experiment Results

KIII model is trained and tested following the rules mentioned before. As our feature space is four-dimension, a 4-channel model is used. During learning, KIII model is trained orderly with the training data from two groups; since no distinct change accompanies the increase of learning times, in this study, once training is enough. Then check the model's recognition capability with the test data. The Euclidean distances of all test samples (100 trials) to the cluster centers of two patterns are shown in Fig. 4:

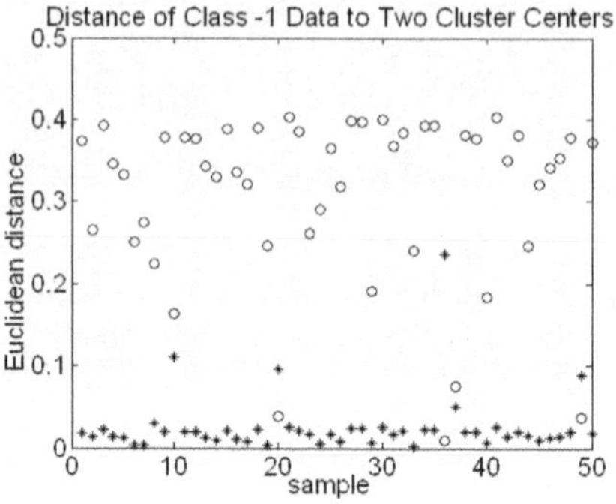

(a)

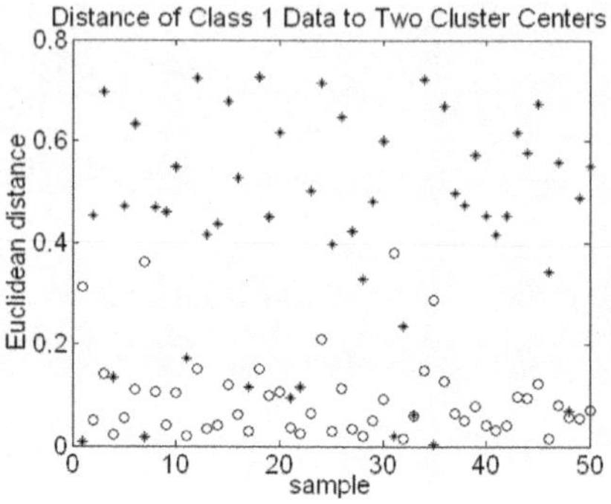

(b)

Fig. 4. (a) shows the Euclidean distance of trials from class -1 to the cluster centers of two patterns, (b) shows the Euclidean distance of trials from class 1 to the cluster centers of two patterns; distance to class 1 center is expressed with 'o', and to class -1 is expressed with '*'

In order to analyze the performance of KIII model, we compare it with other three methods [13]: Fisher Discriminate Analysis (FDA), k-nearest neighbor classifier and Back-Propagation Neural Network (BP).

Since training data cluster well in feature space, FDA, a simple discriminator is tried firstly. FDA is a kind of linear discriminate classifier proposed by Fisher. The principle of this algorithm is to project multi-dimension data features onto one dimension space, seeking for the optimal space for classifying. The optimization procedure is based on training data.

Then as the distribution of training data in feature space is not clear exactly, and therefore, k-nearest neighbor classifier is taken into account. The basic conception of k-nearest neighbor classifier is to analyze the k (in this paper, the optimal k is 36) nearest points of certain sample in the feature space. The sample will be categorized into the group to which most k-nearest points belong.

Finally, in order to make a comparison between KIII model and traditional neural network, BP is checked. BP is a sort of artificial neural network (ANN), in which the output error is propagated back and employed to modify the weights of the network. In our experiments, the final optimized architecture of BP network is 4-2-1; more hidden units will introduce overfitting.

Our experiments are all operated and optimized on MATLAB6.5. See the results in Table 1:

Table 1. Comparison between KIII model and other algorithms

Algorithm	Training Data	Test Data
KIII model	86.3%	92%
FDA	91.0%	87%
k-nearest neighbor classifier	91.0%	89%
BP	92.6±0.5%	87.5±0.7%

As we can see, KIII model do the best performance in test data.

The experimental results indicate that KIII model maintains its recognition capacity when differences between training and test data are distinct. Since one of the challenges for BCI systems is how to guarantee their generalization, the above results are hopeful.

Looking closer into the algorithms demonstrates that the distinction between machine learning principle of KIII model and that of other conventional classifiers makes KIII model outperform. As we know, FDA and BP optimize themselves with Empirical Risk Minimization (ERM), which imports a problem of overfitting. And although k-nearest neighbor classifier does not work based on ERM, it also implements classification merely lying on the statistic distribution of training data. Fortunately, KIII model should not be confined.

With a different data modeling way, KIII model extracts global chaotic attractors corresponding to different patterns from training data by simulating the action of biological neural systems. Its optimizing process is based on the active state of the neural nodes, without any supervisor. And moreover, additive noises as mentioned above enhance KIII model's ability of noise-resistance and error-tolerance. Hence, KIII model holds an advantage to deal with the generalization problem in BCI systems.

For neural network, the training time, which is relative to the computation complexity, is one of the most significant parameters to judge the recognition capability of certain network. In this study, compared with BP, KIII model shows another advantage, needing less training times; see Table 2. More experiments show that the classification accuracy of BP will be stable and get its best recognition accuracy until the training times increases to ten, and the result is 87.5±0.7%.

Table 2. Classification Accuracy under Different Training Times

Training time	KIII model	BP
1	92%	64.5±17.5%
2	92%	83.9±6.9%
3	92%	77.5±16.7%

4 Conclusions

In this paper, firstly CSSD is employed to derive source components of our ECoG data, and then four-dimension features are extracted. For classification, a neural network, named KIII model, is used, and compared its recognition performance with other three algorithms. The experiment results indicate that KIII model is competent in ECoG pattern classification, despite the unavoidable differences in both subjects' state and recording conditions between training data and test data. Furthermore, KIII model needs much less training times than BP; as in this paper, only once is enough. Thereby, as we can see, KIII model is a promising tool for BCI systems.

References

1. Wolpaw, J.R., Birbaumer, N., McFarland, D.J., Pfurtscheller, G., Vaughan, T.M.: Brain-computer interfaces for communication and control: Clinical Neurophysiology **113** (2002) 767-791
2. Graimann, B., Huggins, J.E., Levine, S.P., Pfurtscheller G.: Detection of ERP and ERD/ERS patterns in single ECoG channels: Proceedings of the 1st International IEEE EMBS Conference on Neural Engineering (2003)
3. Lal, T., Hinterberger, T., Widman, G., Schröder, M., Hill, J., Rosenstiel, W., Elger, C., Schölkopf, B., Birbaumer, N.: Methods Towards Invasive Human Brain Computer Interfaces: Advances in Neural Information Processing Systems (2004)
4. Kozma, R., Freeman, W.J.: Chaotic Resonance-Methods and Application for Robust Classification of Noisy and Variable Patterns: International Journal of Bifurcation and Chaos **11** (2001) 1607-1629
5. Chang, H.J., Freeman, W.J.: Biologically modeled noise stabilizing neurodynamics for pattern recognition: International Journal of Bifurcation and Chaos 8 (1998) 321-345
6. Chang, H.J., Freeman, W.J., Burke, B.C.: Optimization of olfactory model in software to give $1/f$ power spectra reveals numerical instabilities in solutions governed by aperiodic (chaotic) attractors: Neural Networks **11** (1998) 449-466
7. Li, G., Lou, Z., Wang. L., Li X., Freeman, W.J.: Application of Chaotic Neural Model Based on Olfactory System on Pattern Recognitions: Lecture Notes in Computer Science **3610** (2005) 378-381
8. Pfurtscheller, G., Zalaudek, K., Neuper C.: Event-related beta synchronization after wrist, finger and thumb movement: Electroencephalography and clinical Neurophysiology **109** (1998) 154-160
9. Toro, C., Deuschl, G., Robert, T., Sato, S., Kufta, C., Hallett, M.: Event-related desynchronization and movement-related cortical potentials on the ECoG and EEG:Electroencephalography and clinical Neurophysiology **93** (1994) 380-389
10. Pfurscheller, G., Lopes da Silva, F.H.: Event-related EEG/MEG synchronization and desynchronization: basic principles: Clinical Neurophysiology **110** (1999) 1842-1857
11. Li, Y., Gao, X., Liu, H., Gao, S.: Classification of Single-Trial Electroencephalogram During Finger Movement: IEEE Transactions on Biomedical Engineering **51** (2004)
12. Wang, Y., Berg, P., Scherg, M.: Common spatial subspace decomposition applied to analysis of brain responses under multiple task conditions: a simulation study: Clinical Neurophysiology **110** (1999) 604-614
13. Bian, Z., Zhang, X. etc.: Pattern Recognition: second edition, Tsinghua press, Peking, China, (2000)

Plan on Obstacle-Avoiding Path for Mobile Robots Based on Artificial Immune Algorithm

Yen-Nien Wang[1], Tsai-Sheng Lee[1], and Teng-Fa Tsao[2]

[1] Department of Electronic Engineering, Lunghwa University of Science and Technology, Taoyuan, Taiwan 33306, R.O.C.
ynwang@mail.lhu.edu.tw G94232119@ms.lhu.edu.tw
[2] Department of Electrical Engineering, Nan Kai Institute of Technology, Nantou, Taiwan 54210, R.O.C.
tdf@nkc.edu.tw

Abstract. This paper aims to plan the obstacle-avoiding path for mobile robots based on the Artificial Immune Algorithm (AIA) developed from the immune principle; AIA has a strong parallel processing, learning and memorizing ability. This study will design and control a mobile robot within a limited special scale. Through a research method based on the AIA, this study will find out the optimum obstacle-avoiding path. The main purpose of this study is to make it possible for the mobile robot to reach the target object safely and successfully fulfill its task through optimal path and with minimal rotation angle and best learning efficiency. In the end, through the research method proposed and the experimental results, it will become obvious that the application of the AIA after improvement in the obstacle-avoiding path planning for mobile robots is really effective.

1 Introduction

In the recent years, with the rapid development in the research of basic medicine, the functional mechanism of the immune system becomes more and more clear. The numerous good characteristics of the immune system, especially its information processing ability, are being studied with great emphasis. Some scholars have designed many kinds of immune algorithms and models based on the mechanism of the immune system. And the research achievements involve many aspects, including control, optimized learning and troubleshooting. These aspects have become the forefront of the current researchers [1]; However, some algorithms and models are still in the research stage, and there are still many research methods to be improved. Planning of the optimal path has always been the target pursued by many researchers, and its application in mobile robot is one of the most important research topics! Hence, this paper starts from this topic and gradually realizes the planning on obstacle-avoiding path for mobile robots by applying the AIA.

From many studies concerning similar topics, a few typical papers are introduced below. In 1995, A. Ishiguro et al. [2] proposed to build a behavior controller for the artificial immune network, so as to determine the obstacle-avoiding mode of independent mobile robots. This paper generated a dynamic model for the artificial

D. Liu et al. (Eds.): ISNN 2007, Part I, LNCS 4491, pp. 694–703, 2007.

immune system based on the idiotypic network hypothesis of biological immune system developed by Jerne [3] [4]. Later, Ishiguro et al. [5] [6] made it possible for the robots to have different plans and fulfill different assignments based on this dynamic model of artificial immune system. In addition, the paper [7] applied the dynamic model of artificial immune system in the navigation control of the robots and really achieved good results in actual operation. Apart from planning of unsuitability improvements for the mobile robot and the sensing area, this paper will also add expected direction selection design and adaptation learning design. The expected direction selection design can quickly determine the movement direction and reduce collisions in a dynamic environment so that the mobile robot can fulfill its assignments more effectively. The adaptation learning design is to strengthen the adaptability of the robot to all kinds of environment, so that it can adapt to the changes of environment and generate effective reaction. Also, with the adaptation learning design, the robot will have an effective learning ability to meet the demand of the memorizing characteristics of the immune system and strengthened learning effect of the mobile robot. In the end, the results will be compared to the path planning literatures based on the genetic algorithm proposed by Tu [8].

From the above research papers, it is obvious that the application of AIA in the behavior control of mobile robots is a new and unique research aspect.

2 Introduction of Artificial Immune Algorithm

AIA is a unique mechanism of calculation model developed from the concept of biological immune system and applied in calculations in actual engineering. This paper generally considers all algorithms developed based on the immunology principle and applied to engineering as AIA.

2.1 Biological Immune System

Biological immune system is a highly evolved, complicated adaptability system in the body of advanced spinal animals, which can identify and resist antigenic foreign bodies like bacteria and viruses and maintain the stability of the in vivo environment. Independent of any center, it has a distributed processing ability. It can take intelligent actions locally, and through network communication, global reaction can be generated.

2.2 Jerne's Idiotypic Network Hypothesis

According to the sketch map of the idiotypic network hypothesis developed by Jerne, as shown in Fig. 1, an immune network is formed through interaction of stimulation and suppression between the B cells, antibody, antigen epitope, idiotope and antibody paratope. Each antibody does not exist independently in the biological body, but is bound with other antibodies. In the figure, the relation between the antigen and antibody is similar to the relation between key and lock, and the relation between one antibody and the other is also similar to the relation between antigen and antibody.

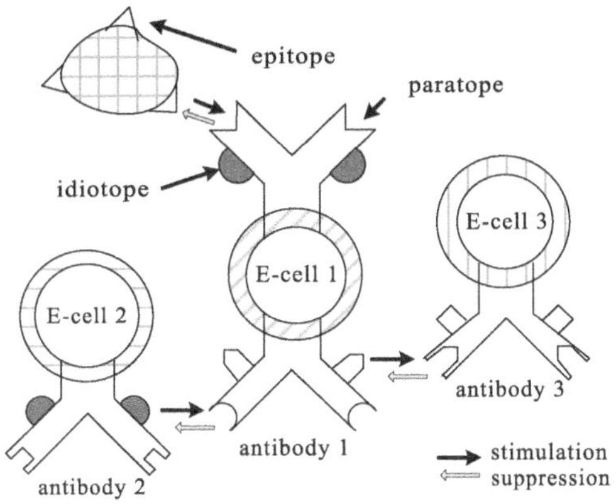

Fig. 1. Sketch map of Jerne's idiotypic network hypothesis

2.3 Mathematical Models of Artificial Immune Network

The characteristics of memorizing immune reaction and learning of the AIA can adjust the concentration of antibody to adapt to various antigens. Firstly, the affinity between antibody j and antibody i is defined as m_{ji} and the affinity between the detected antigen and antibody i is defined as m_i. The definitions are as follows:

$$m_{ji} = \sum_{k}^{L} I_j(k) \oplus \overline{P_i(k)}, \quad (1)$$

$$m_i = \sum_{k}^{L} E(k) \oplus \overline{P_i(k)}. \quad (2)$$

Here, $I_j(k)$ represents the kth binary value of the idiotope of antibody j, $P_i(k)$ represents the kth binary value of the antibody paratope of antibody i, $E(k)$ represents the kth binary value of the antigen epitope. The lengths of the binary antigen epitope, idiotope and antibody paratope are all L. The sign $\oplus$ represents XOR logic operator. More cases of true XOR with the corresponding digit means stronger reaction. With different affinity, the antibody stimulation and suppression of the different activated antibodies within the artificial immune network will also change.

Next, the dynamic model of the artificial immune network generated based on the idiotypic network proposed by Jerne is applied to the concentration of the ith antibody, which is represented by a_i:

$$\frac{dA_i(t)}{dt} = \left(\frac{\sum_{j}^{N} m_{ji} a_j(t)}{N} - \frac{\sum_{k}^{N} m_{ik} a_k(t)}{N} + m_i - k_i \right) \times a_i(t). \quad (3)$$

In this formula, $i, j, k = 0, \cdots, N-1$, N is the total number of antibody types in the network, the first and second items on the right of represent stimulation and suppression between the antibodies respectively, the third item represents the intensity of stimulation by the antigen on the antibody, and the fourth item represents dissipation factor.

$$a_i\left(t+\Delta t\right) = \frac{1}{1+\exp\left(0.5 - A_i\left(t+\Delta t\right)\right)}. \tag{4}$$

Here, Eq. (4) is a squashing function. This function can ensure that the concentration value will not diverge.

In the end, the concentration values of obstacle a_i^o and target object a_i^g are calculated, the overall antibody concentration value a_i^x is calculated through combination in Eq. (5):

$$a_i^x = \left(1 - r_i\right) \cdot a_i^o + r_i \cdot a_i^g. \tag{5}$$

Here, r_i is proportional distribution combination of the antibody i obstacle concentration and target object concentration. It can be set as a fixed constant.

3 Control Design of Artificial Immune Algorithm

Apart from some improved designs for the problems encountered by the mobile robot based on the research method by Ishiguro, expected direction selection design and adaptability learning design are added in this study for the purpose of strengthening the ability of the mobile robot to effectively detect and judge the obstacles and environment so as to determine the advance direction of the mobile robot, as well as for the purpose of strengthening the learning ability of the robot.

3.1 Improved Mobile Robot Planning

The originally designed eight detection zones are changed in size, and are divided into two sections – near zone and far zone. The main purpose of this is to make up the insufficient sensitivity in the detection zone behind the mobile robot, because after adjustment of the detection zone sizes, the sensitivity is lowered. Division of far and near zones can help strengthen the sensitivity. In the advancement direction, the mobile robot has eight omnibearing movements, so that the mobile robot can have more effective obstacle-avoiding reactions towards the obstacles, as shown in Fig. 2. When the mobile robot detection zone is in the angle of 45° .

3.2 Expected Direction Selection Design

To strengthen the environment judgment ability of the mobile robot, in the experiment method, the obstacle restriction method (ORM) by Minguez [9] is adopted for the expected direction selection design for this experiment. Here, it is considered as the

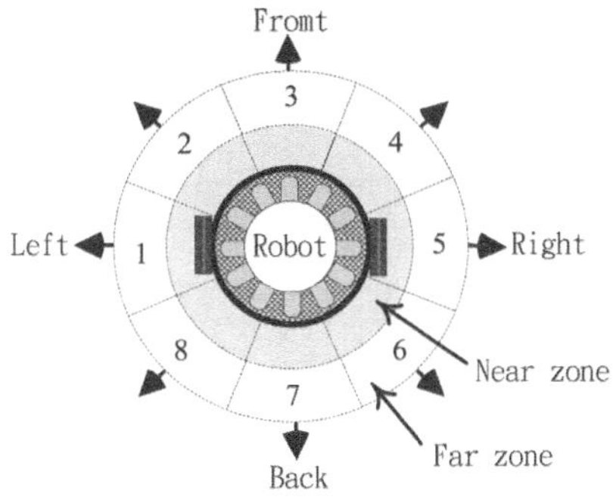

Fig. 2. Mobile robot detection zone and advance direction

expected movement behavior of the robot. With this design, when an obstacle enters the near zone, in emergency, an immediate avoiding action can be made through the behavior reaction of the expected direction, instead of making a decision on the mobile robot movement direction after complicated calculation by the AIA.

In this paper, the expected direction selection design is explained in three steps:

Step 1: As for the approaching principle of the robot toward the target object, a target zone S_{goal} is planed through Eq. (6), as shown in Fig. 3(a).

$$S_{goal} = \theta_{goal}. \tag{6}$$

Step 2: Eq. (7) only judges a single obstacle or multiple neighboring obstacles, defined as danger zone S_1, within the detection zone, as shown in Fig. 3(b). In addition, through Eq. (8), the S_2 zone is planned. In Eq. (8), n is the number of detection zones planned, as shown in Fig. 3(c). In the end, S_1 and S_2 are combined to form the restricted behavior zone S_{obst} of the mobile robot, represented by Eq. (9), as shown in Fig. 3(d).

$$S_1 = \{\theta_{obst1} \cup \theta_{obst2} \cup \cdots \cup \theta_{obst8}\}, \tag{7}$$

$$S_2 = \begin{cases} r_R = \left[\min(\theta_{S1}), \min(\theta_{S1}) - \dfrac{2\pi}{n}\right], \\ r_L = \left[\max(\theta_{S1}), \max(\theta_{S1}) + \dfrac{2\pi}{n}\right], \end{cases} \tag{8}$$

$$S_{obst} = S_1 \cup S_2. \tag{9}$$

Step 3: The target zone S_{goal} and the restricted behavior zone S_{obst} obtained from **Step 1** and **Step 2** will form the behavior decision zone S_{md} of the robot with strengthened learning ability. The formation of the decision zone is determined by Eq.(10).

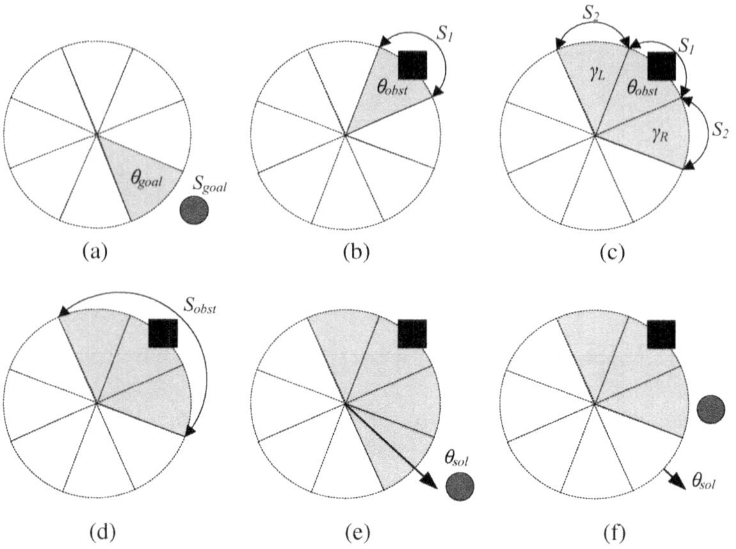

Fig. 3. Sketch drawings of the expected direction decision mechanism

$$S_{md} = S_{obst} \cap S_{goal}. \tag{10}$$

Here, Eq. (10) can have the following two possibilities:

(1) $S_{md} = \varnothing$, means the target zone does not overlap with the restricted behavior zone, then the expected decision direction θ_{sol} of the robot with strengthened learning ability is the same as the target direction, represented by Eq.(11), as shown in Fig. 3(e).

$$\theta_{sol} = \theta_{goal}. \tag{11}$$

(2) $S_{md} \neq \varnothing$, means the target zone and the restricted behavior zone interfere with each other, hence the expected decision direction of the mobile robot with strengthened learning ability is calculated by Eq.(12), as shown in Fig. 3(f).

$$\theta_{sol} = \begin{cases} \min(\theta_{Sobst}) - \dfrac{\pi}{n}, if \left|\theta_{goal} - \min(\theta_{Sobst})\right| < \left|\theta_{goal} - \max(\theta_{Sobst})\right|, \\ \max(\theta_{Sobst}) + \dfrac{\pi}{n}, if \left|\theta_{goal} - \min(\theta_{Sobst})\right| \geq \left|\theta_{goal} - \max(\theta_{Sobst})\right|. \end{cases} \tag{12}$$

The adoption of the expected direction selection mechanism helps to improve the insufficient detection sensitivity behind the mobile robot, and meanwhile reduces the complicated calculations by AIA, so as to take immediate avoiding actions to decide the movement direction of the mobile robot.

3.3 Adaptability Learning Design

The additional adaptability learning design is to strengthen the learning ability of the mobile robot. This study constructs the mobile robot learning model and achieves the strengthened learning effect based on the artificial immune strengthened learning mechanism proposed by Luh [10].

In the design of strengthened learning ability, if the actual movement direction of the robot is the same as the expected movement direction, an awarding will be granted. The awarding behavior is represented by Eq. (13), indicating increase of affinity m_{ji} between one antibody and the other antibody. If the actual movement direction of the robot is not the same as the expected movement direction, a punishment will be imposed. The punishment behavior is represented by Eq. (14), indicating decrease of affinity m_{ik} between one antibody and the other antibody.

$$m_{ji} = m_{ji} \cdot (1 + learning\_rate), \tag{13}$$

$$m_{ik} = \frac{m_{ik}}{(1 + learning\_rate)}. \tag{14}$$

Here, when there is no obstacle within the mobile robot detection zone, its learning rate will be calculated by Eq. (15) and Eq. (16).

For the obstacle:

$$learning\_rate = 1. \tag{15}$$

For the target object:

$$learning\_rate = \frac{d_{goal} + d_g}{\alpha \cdot d_{goal} + \beta}. \tag{16}$$

If there are obstacles within the mobile robot detection zone, the learning rate will be calculated by Eq.(17) and Eq.(18).

For the obstacle:

$$learning\_rate = \frac{d_{goal} - \alpha \cdot d_{obst}}{\alpha \cdot d_{goal} + \beta}. \tag{17}$$

For the target object:

$$learning\_rate = \frac{d_{goal} + d_g}{\alpha \cdot d_{goal} + 2 \cdot d_{obst} + \beta}. \tag{18}$$

Here, d_{goal} represents the distance between the robot i direction and the target object, d_{obst} represents the distance between the mobile robot i direction and obstacle, d_g represents the distance from the central point of the mobile robot to the target object, and α , β are fixed parameters.

4 Simulation Results and Analysis

This simulation experiment mainly aims to compare and analyze the differences between the artificial immune research method proposed by Ishiguro, the improved artificial immune research method, and the genetic algorithm proposed by Tu [8], and present the experiment results. Mobile robot needs to avoid eight static obstacles and three dynamic obstacles to reach the target object and fulfill its task.

4.1 Experiment 1: Artificial Immune Algorithm Simulation Results Developed by Ishiguro and Simulation Results of Genetic Algorithm Proposed by Tu

Fig. 4(a) is the simulation result of the eighth generation number of the mobile robot, and 62 steps will be needed before reaching the target object. Fig. 4(b) is the simulation drawing of the genetic algorithm proposed by Tu. This is the simulation result of the eighth generation number of the mobile robot, and 57 steps will be needed before reaching the target object.

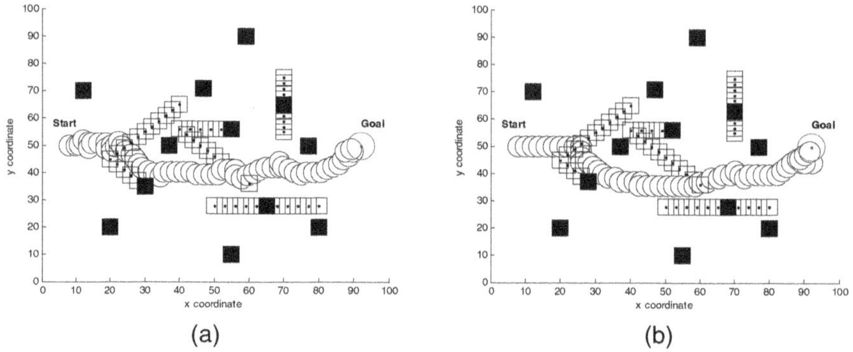

Fig. 4. Research technique simulation chart (a) proposed by Ishiguro, (b) proposed by Tu

4.2 Experiment 2: Simulation Results of the Artificial Immune Algorithm of Strengthened Learning

Fig. 5 shows the simulation results of the AIA of strengthened learning proposed by this study. This is the simulation result of the eighth generation number of the mobile robot, and 50 steps will be needed before reaching the target object.

4.3 Discussion of the Simulation Results

The results of three research methods are shown in Table 1. The results of simulation experiment of this study shows that the artificial immune robot with strengthened learning ability can efficiently reach the destination in a complicated environment. And the expectation mechanism and adaptation mechanism designed are both effective. The results show that the artificial immune robot with strengthened learning ability can quickly adapt to the environment, and after learning, it can also avoid obstacles to approach the target object via the shortest path.

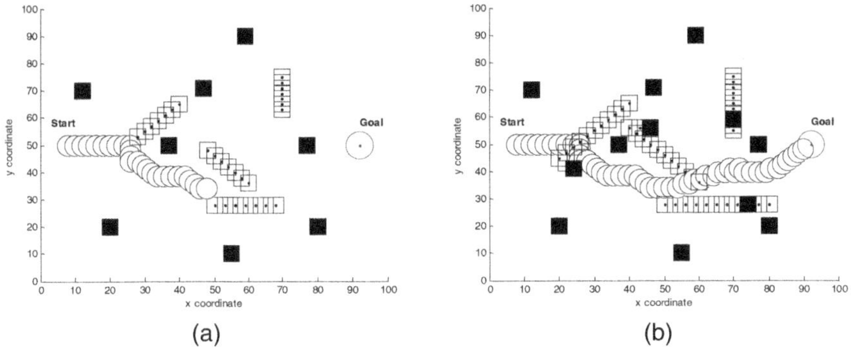

Fig. 5. MATLAB Simulation Breakdown Drawings ((a) -> (b)) Based on the Research Method Proposed by This Study

Table 1. Comparison of the path learning results of the robot

Robot training value and generation number	2	3	8	20	30	50
Ishiguro's artificial immune learning results	77	62	62	59	60	60
Tu's genetic literature learning results	66	68	57	53	51	51
Learning results of the artificial immune of strengthened learning	**58**	**55**	**50**	**50**	**50**	**50**

5 Conclusions

Algorithms based on the biological system are a major topic in current researches and applications of computer intelligence. This study applies the new AIA design in mobile robot. The simulation experiments have proved that this algorithm is really effective in identifying diversity and learning. Future researches will be directed to more complicated environments and assignments of tasks for the mobile robot to fulfill.

Acknowledgment

Support for this research by the National Science Council of the Republic of China under Grant No. NSC 95-2221-E-262-013 is gratefully acknowledged.

References

1. Dipankar Dasgupta, Ed.: Artificial Immune Systems and Their Applications. Springer (1998) 3-21
2. Ishiguro, A., Watanabe, R., Uchikawa, Y.: An Immunological Approach to Dynamic Behavior Control for Autonomous Mobile Robots. Proceedings of the IEEE/RSJ International Conference on Intelligent Robots and Systems **1** (1995) 495-500

3. Jerne, N. K.: The Immune System. Scientific American **229** (1) (1973) 52-60
4. Jerne, N. K.: Idiotypic Networks and Other Preconceived Ideas. Immunological Rev. **79** (1984) 5-24
5. Ishiguro, A., Kondo, T., Watanabe, Y., Uchikawa, Y.: Dynamic Behavior Arbitration of Autonomous Mobile Robots using Immune Networks. Proceedings of the IEEE International Conference on Evolutionary Computation **2** (1995) 722-727
6. Ishiguro, A., Watanabe, Y., Kondo, T., Uchikawa, Y.: Decentralized Consensus-making Mechanisms based on Immune System-application to a Behavior Arbitration of an Autonomous Mobile Robot. Proceedings of the IEEE International Conference on Evolutionary Computation (1996) 82-87
7. Vargas, P.A., de Castro, L.N., Michelan, R., Von Zuben, F.J.: Implementation of an Immuno-genetic Network on a Real Khepera II robot. Proceedings of the Congress on Evolutionary Computation **1** (2003) 420-426
8. Tu, Jianping, Yang, S. X.: Genetic Algorithm based Path Planning for a Mobile Robot. Proceedings of the 2003 IEEE International Conference on Robotics And Automation **1** (2003) 1221- 1226
9. Minguez, J.; The Obstacle-restriction Method for Robot Obstacle Avoidance in Difficult Environments. Proceedings of the IEEE/RSJ International Conference on Intelligent Robots and Systems (2005) 2284- 2290
10. Luh, G. C., Cheng, W. C.: Behavior-Based Intelligent Mobile Robot Using Immunized Reinforcement Adaptive Learning Mechanism. Advanced Engineering Informatics **16** (2002) 85-98

Obstacle Avoidance Path Planning for Mobile Robot Based on Ant-Q Reinforcement Learning Algorithm

Ngo Anh Vien, Nguyen Hoang Viet, SeungGwan Lee, and TaeChoong Chung*

Artificial Intelligence Lab, Department of Computer Engineering,
School of Electronics and Information, Kyunghee University
1-Seocheon, Giheung, Yongin, Gyeonggi, 446-701, South Korea
{vienna,vietict,leesg,tcchung}@khu.ac.kr

Abstract. Path planning is an important task in mobile robot control. When the robot must move rapidly from any arbitrary start positions to any target positions in environment, a proper path must avoid both static obstacles and moving obstacles of arbitrary shape. In this paper, an obstacle avoidance path planning approach for mobile robots is proposed by using Ant-Q algorithm. Ant-Q is an algorithm in the family of ant colony based methods that are distributed algorithms for combinatorial optimization problems based on the metaphor of ant colonies. In the simulation, we experimentally investigate the sensitivity of the Ant-Q algorithm to its three methods of delayed reinforcement updating and we compare it with the results obtained by other heuristic approaches based on genetic algorithm or traditional ant colony system. At last, we will show very good results obtained by applying Ant-Q to bigger problem: Ant-Q find very good path at higher convergence rate.

1 Introduction

Mobile robots are expected as the attractive tool for operation in a wide variety of application domains such as manufacturing, space exploration, deep ocean and nuclear plants, medical surgery and the assistance of movement for the elder and handicapped people. Developing such complex autonomous robots requires research in areas of control, automated reasoning, and perception. Current and future research efforts in this area will continue to strive for increased robustness and flexibility, better reliability, and greater autonomy. The path planning problem arises in attempts to develop more autonomous robotic systems. This capability is necessary for autonomous robots since it is essential for a robot to accomplish tasks by moving in the real world. This requires the ability to plan a path. Path planning problem is an optimization problem with the objective of computing an appropriate path between two specific locations. A feasible path is a path which does not collide with static and/or dynamic obstacles in the environment, regardless of different existing constraints that can be applied

* The corresponding author.

D. Liu et al. (Eds.): ISNN 2007, Part I, LNCS 4491, pp. 704–713, 2007.

during path planning. There are many studies on robot-path planning using various approaches, such as the grid-based A*(A-star) algorithm [1], [2], road maps (Voronoi diagrams and visibility graphs) [3], [4], cell decomposition [5], [6], [7], and artificial potential field [8], [9], [10], [13].

The path can be generated using an internal representation of the environment under consideration. There are several different types of representation. Among them, *composite space map* [11] method is the most popular way to represent a given environment. In this method, the environment is discretized into a grid of rectangular cells (or voxels) and each cell is marked as an obstacle or a non-obstacle one based on the actual position of the obstacles. By such representation, the path can be defined as a consecutive sequence of cells which begins at the start cell and ends at the destination cell. Using this kind of representation method, there are many different algorithms developed such as: Graph search methods [12], A* algorithm, artificial potential fields, ant colony [14], [15] and genetic algorithm [16], [17], [18] are some of these approaches.

In this paper, we use Ant-Q algorithm [19], [20] to optimize the mobile robot's path. Ant-Q is a new family of algorithms inspired by both the Q-learning and by the observation of ant colonies behavior. We already know that many researchers already applied Q-learning algorithm in the robot's obstacle avoidance problem. So Ant-Q based obstacle avoidance path planning algorithm is proposed. Using this algorithm, we must have a good cost function of the moving path and an effective updating Ant-Q value method. The purpose of this cost function is to evaluate and choose only the collision and shortest path for mobile robots. And the purpose of the updating method is to make the algorithm find only optimum solution and converge faster. Results obtained have shown that Ant-Q is very effective in finding very good, often optimum rather than ACS, especially with bigger problem instances.

The rest of this paper is organized as follow. In Section 2, we introduce Ant-Q algorithm. In Section 3, Ant-Q based obstacle avoidance path planning algorithm is presented. Sections 4 reports simulations and experimental results respectively. Concluding remarks follow in Section 5.

2 Ant-Q Algorithm

Ant-Q learning method [19], [20] that is proposed by Colorni, Dorigo and Maniezzo is an extension of Ant System (AS) [21], [22]. It is reinforcement learning reinterpreting in view of Q-learning. In Ant-Q, an agent(k) situated in node(r) moves to node(s) using the following rule, called pseudo-random proportional action choice rule (or state transition rule) as Eq.(1):

$$s = \begin{cases} \arg\max\limits_{u \in J_k(r)} \{[AQ(r,u)]^{\delta}.[HE(r,u)]^{\beta}\}, & if\ q \leq q_0(exploitation), \\ S, & otherwise. \end{cases} \quad (1)$$

AQ(r, u) is Ant-Q value, a positive real value associated to the edge(r, u), It is counterpart of Q-learning Q-values, and is intended to indicate how useful it is to move to node (u) when in node (r). $AQ(r, u)$ is changed at run time. $HE(r, u)$ is a heuristic value associated to edge(r, u) which allows a heuristic evaluation of which moves are better (in the TSP [19], the inverse of the distance).

Let k be an ant making a tour. $J_k(r)$ is a set of nodes to be visited from the current node(r). δ and β are parameters which weigh the relative importance of the learned AQ-values and the heuristic values. q is a value chosen randomly with uniform probability in [0,1], $q_0 (0 \leq q_0 \leq 1)$ is a parameter, and S is a random variable selected according to the distribution given by Eq.(2) which gives the probability with which an ant in node(r) choose the node(s) to move to

$$p_k(r,s) = \begin{cases} \frac{[AQ(r,s)]^{\delta}.[HE(r,s)]^{\beta}}{\sum\limits_{u \in J_k(r)} [AQ(r,u)]^{\delta}.[HE(r,u)]^{\beta}}, & if\ s \in J_k(r), \\ 0, & otherwise. \end{cases} \tag{2}$$

The goal of Ant-Q is to learn AQ-values to find better solution as stochastic. AQ-values are updated by the following Eq.(3)

$$AQ(r,s) \leftarrow (1-\alpha)AQ(r,s) + \alpha.(\Delta AQ(r,s) + \gamma. \underset{z \in J_k(s)}{Max}\ AQ(s,z)). \tag{3}$$

The update term is computed of a reinforcement term and of the discounted evaluation of the next state. α $(0 < \alpha < 1)$ is the pheromone decay parameter and γ is discount rate. Also, $\Delta(AQ)$ is reinforcement value, local reinforcement is always zero. While global reinforcement, which is given after all the agents have finished their tour, is computed by the following Eq.(4)

$$\Delta AQ(r,s) = \begin{cases} \frac{W}{L_{k_{ib}}}, & if (r,s) \in path\ done\ by\ the\ best\ agent\ k_{ib}, \\ 0, & otherwise, \end{cases} \tag{4}$$

where L_{kib} is the length of the tour done by the best agent which did the shortest tour in the current iteration, and W is a parameter.

3 Ant-Q Based Path Planning Algorithm

Fig. 1 shows the representation of environment already discretized into a grid. The robot is assigned a task that moves from point X to Y. The direct distance between X and Y is m cells, the width of the grid is n cells. So the possible paths lie in this rectangular area. The static obstacles' coordinates in environment are known. The rectangle is divided as follow: The x-axis is divided equally into m parts by the m vertical lines $L_1, L_2, ..., L_m$. The y-axis is also divided the same into n parts. So we have a grid $m \times n$. Hence, a found path by the kth ant in the colony is like this:

$\mathrm{Path_k} = \{\mathrm{X_1(x_1, y_{k1}), X_2(x_2, y_{k2}), ..., X_m(x_m, y_{km})}\}(\mathrm{k_i} = 1, 2, ..., \mathrm{n})$,
where $X_i(x_i, y_{ki})$ is point i^{th} of the path on the vertical line L_i. The beginning point is $X = X_1(x_1, y_{k1})$ and destination point is $Y = Y_m(x_m, y_{km})$. So the task of each ant is to find respective point consecutively on the vertical lines from X to Y.

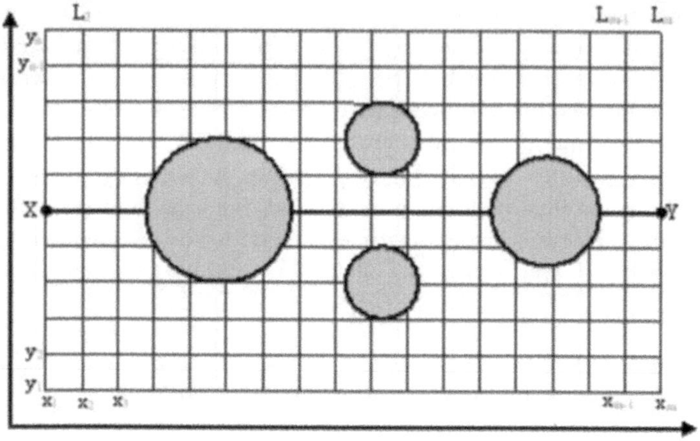

Fig. 1. An example of an initial discretized environment

3.1 The Proposed Cost Function of the Path

In this section we will define the cost function for the path. The uppermost purpose of path planning is to search the collision-free and shortest path for mobile robots. In this paper, we proposed the following cost function

$$J_k = Len_k + f_{ob}, \tag{5}$$

where the first term is the length of the tour. It makes the robot move to the target with the shortest path. The second term forces the robot to follow a collision-free path. Len_k= the length of the tour from X to Y. The tour's length is counted by summing Euclidean distances between nodes of the path. The Euclidean distance from node $X_i(x_i, y_{ki})$ to next node $X_{i+1}(x_{i+1}, y_{k(i+1)})$ is: $\sqrt{1^2 + (y_{k(i+1)} - y_{ki})^2}$. Consequently, we have formula of Len_k

$$\begin{aligned} Len_k = \sum_{i=1}^{m-1} |X_i X_{i+1}| &= \sum_{i=1}^{m-1} \sqrt{(x_{i+1} - x_i)^2 + (y_{k(i+1)} - y_{ki})^2} \\ &= \sum_{i=1}^{m-1} \sqrt{1 + (y_{k(i+1)} - y_{ki})^2}. \end{aligned} \tag{6}$$

We assume here that the edge of each cell is 1-unit. The term f_{ob} helps the robot avoid obstacles. There are q static obstacles each of which is represented by a circle with a centre point (x_j, y_j) and a radius r_j. Supposed that the obstacles' coordinates in environment are known

$$f_{ob} = \begin{cases} 0, & \textit{no obstacle collision}, \\ \frac{K}{\sum_{i=1}^{m} d(X_i, obstacle)}, & \textit{otherwise}, \end{cases}$$

where :

$$d(X_i, obstacle) = \begin{cases} 0, & X_i \; is \; safe \; node, \\ \sqrt{(x_i - x_j)^2 + (y_{ki} - y_j)^2}, & X_i \; in \; obstacle \, (x_j, y_j)'s \; area. \end{cases} \tag{7}$$

Where K is an obstacle avoidance coefficient, the bigger it is, the safer the path (in the simulation, K=20.000). Here, if node X_i in an obstacle (x_j, y_j)'s area then $d(X_i, obstacle)$ is the distance of the node X_i to the center of that obstacle. If the path is collision-free, the distance summation in Eq.(7) is zero.

3.2 Action Choice Rule

In this section we will realize the node choosing rule for each ant. Each ant starts from initial point and finds the collision-free shortest path to the destination. At time t, the k_{th} ant at the point (x_i, y_{ki}) on the vertical line L_i must choose the next point $y_{k(i+1)}$ on the vertical line L_{i+1} to move to. To solve this problem, we will use pseudo-random proportional action choice rule of Ant-Q algorithm. So each ant will move to next $y_{k(i+1)}$ Busing the following rule:

$$y_{k(i+1)} = \begin{cases} \arg\max_{u \in J_k(y_{ki})} \{[AQ(y_{ki}, u)]^{\delta}.[HE(y_{ki}, u)]^{\beta}\}, & if\ q \leq q_0(exploitation), \\ Y, & otherwise. \end{cases} \tag{8}$$

$AQ(y_{ki}, u)$ is Ant-Q value, a positive real value associated with the edge(y_{ki}, u). It is intended to indicate how useful to move from point(y_{ki}) to point(u) when the ant is situated at point(y_{ki}). $HE(y_{ki}, u)$ is a heuristic value associated with the edge (y_{ki}, u), in the path planning problem, it is the inverse of the Euclidean distance between two points. k is the ant making the path, $J_k(y_{ki})$ are set of points to be able to be visited from the current point(y_{ki}).

δ, β, q, q_0 are parameters already mentioned previously in the section 2. Y is a random point selected according to the distribution given by Eq.(9) which gives the probability with which an ant in point(y_{ki}) choose the point$(y_{k(i+1)})$ to move to

$$p(y_{ki}, y_{k(i+1)}) = \begin{cases} \frac{[AQ(y_{ki}, y_{k(i+1)})]^{\delta}.[HE(y_{ki}, y_{k(i+1)})]^{\beta}}{\sum_{y_u \in J_k(y_{ki})} [AQ(y_{ki}, y_u)]^{\delta}.[HE(y_{ki}, y_u)]^{\beta}}, & if\ y_{k(i+1)} \in J_k(y_{ki}), \\ 0, & otherwise. \end{cases} \tag{9}$$

3.3 Updating Rule

AQ-values are updated by the following Eq.(10)

$$\begin{aligned} AQ(y_{ki}, y_{k(i+1)}) \leftarrow & (1-\alpha)AQ(y_{ki}, y_{k(i+1)}) + \\ & \alpha.(\Delta AQ(y_{ki}, y_{k(i+1)}) + \gamma. \underset{z \in J_k(y_{k(i+1)})}{Max} AQ(y_{k(i+1)}, z)), \end{aligned} \tag{10}$$

$\alpha(0 < \alpha < 1)$ is the pheromone decay parameter and γ is discount rate. $MaxAQ(y_{k(i+1)}, z)$ is the evaluation of the next state. ΔAQ is reinforcement value, it can be local (immediate) or global (delayed).

We already know that reinforcement learning algorithm has recently been receiving increased attention as a method for robot learning with little or no a priori knowledge and higher capability of reactive and adaptive behavior. The robot senses the current state of the environment and selects an action. Based on the state and the action, the environment makes a transition to a new state and a reward is passed to the robot. So in this section, we want to survey that when Q-learning reward is applied to ant colony system, how the updating rule of Ant-Q value is. Here we define three methods of updating rule for path planning problem.

Local updating. It works like Q-learning algorithm. Each ant receives reinforcement after each moving action and Ant-Q is updated immediately when each move is made. The immediate reinforcement is computed as follows

$$\begin{aligned} &\Delta AQ(y_{ki}, y_{k(i+1)}) = \frac{W}{d(X_i, X_{i+1}) + d_{ob}}, \\ &where \\ &d_{ob} = \begin{cases} 0, & edge\ (X_i, X_{i+1})\ in\ safe\ area, \\ \frac{K}{d(X_i, obstacle)}, & otherwise, \end{cases} \end{aligned} \tag{11}$$

$d(X_i, X_{i+1})$ is Euclidean distance between two points $X_i(x_i, y_{ki})$ and $X_{i+1}(x_{i+1}, y_{k(i+1)})$. $d(X_i, obstacle)$, K is the same as in Eq.(7). Eq.(11) means that ΔAQ becomes larger when the move distance is shorter and this move makes no collision, so this edge has more possibility of being chosen.

Global updating. After all the agents have finished their tour, Ant-Q value will be updated. The reinforcement is computed by the formula:

$$\Delta AQ(y_{ki}, y_{k(i+1)}) = \begin{cases} \frac{W}{J_{k_{ib}}}, \ if(y_{ki}, y_{k(i+1)}) \in path\ done\ by\ the\ best\ agent\ k_{ib}, \\ 0, \qquad otherwise, \end{cases} \tag{12}$$

where k_{ib} is the ant who made the best tour in the current iteration of the trial, and J_{kib} is the cost value of the best path done. W in Eqs.(11), (12) is a parameter, set to 10.

Mixture updating. Using both updating methods. Local updating is used at each move while global updating is used at the end of each tour.

Table 1. Comparisons of updating rule methods. It was run for 500 iterations (with set 4 and 5, we run 2000 iterations and 4000 respectively) and the results are averaged over 10 trials (see section 4 about sets of simulation).

	Mixture		Local updating		Global updating	
	Mean	Best	Mean	Best	Mean	Best
Set 1	24.75	23.95	24.17	23.32	23.63	**23.14**
Set 2	38.86	**36.85**	39.78	38.59	38.13	37.80
Set 3	55.29	53.47	56.37	54.26	52.41	**51.33**
Set 4	88.25	84.05	87.33	86.17	84.25	**83.39**
Set 5	155.10	146.69	169.87	160.33	140.43	**136.12**

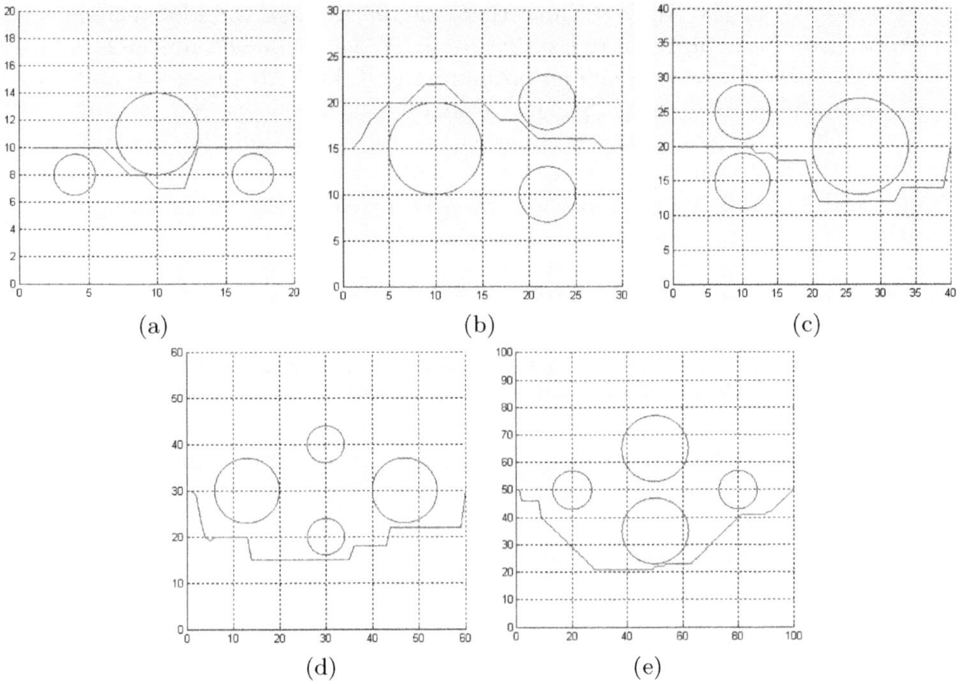

Fig. 2. Problem set and optimum solutions. Numbers in parenthesis are iterations which return above optimum solutions. (a) Set 1: grid $20 \times 20(500)$; (b) Set 2: grid $30 \times 30(500)$; (c) Set 3: grid $40 \times 40(500)$; (d) Set 4: grid $60 \times 60(2000)$; (e) Set 5: grid $100 \times 100(4000)$.

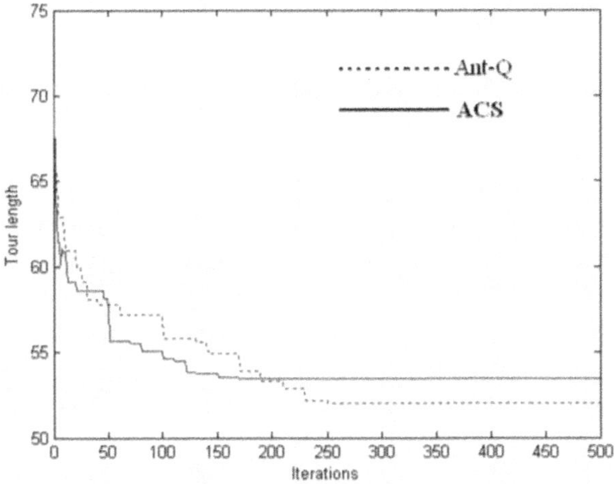

Fig. 3. Convergence speed of each method Ant-Q, ACS. Mean length of best tour, averaged over 10 trials. Using problem set 3: (40×40).

The results in the Table 1 show that, on the average the global updating method gives the optimum and stable result. When the environment is small, the three methods give very similar optimum result, but only the global updating method is stable on all trials (because of small deviation between mean and best value). When the environment becomes larger, the global updating method gives not only the best result but stable also. Moreover, it was slightly faster in finding solutions by global updating method. So we decided to use the global updating method in the simulation section of the paper.

4 Performance Evaluation

Algorithm performance was evaluated repeating each trial 10 times. We report mean and best values. The mean performance is computed by taking the best result obtained in each of 10 trials and computing the mean. The best performance is given by the best result in the 10 trials. In the experiments we reported in this paper, the value of parameter was set to: $\delta = 3, \beta = 3, q_0 = 0.9, AQ_0 = 1/(average\_length\_of\_edges * m), \alpha = 0.1, \delta = 0.3, W = 10$. In the simulation, we used 5 sets of environment with grids: set 1: 20×20; set 2: 30×30; set 3: 40×40; set 4: 60×60; set 5: 100×100. 5 sets of environment including obstacles are shown in Fig. 2. In each figure, there is the optimum path found by Ant-Q algorithm. Fig. 3 shows the convergence speed in case of 500 iterations and using problem set 3. The convergence speed of ACS is faster in beginning. However, we can see that Ant-Q convergences faster to the optimum solution. In Table 2, we compare Ant-Q with other approaches using 5 sets of problem. We compared Ant-Q with Ant colony system (ACS), genetic algorithm (GA). In Table 2 we report the best path length, the average path length over 5 trials and the number of iterations. We can see clearly that Ant-Q always outperformed ACS and GA, especially with large problems.

Table 2. Comparison on average result obtained on different sets of problem. GA: Genetic algorithm, ACS: Ant Colony System. The results are averaged over 5 trials.

	GA		ACS		Ant-Q	
	Mean	Best	Mean	Best	Mean	Best
Set 1	26.93	25.20 (8.000)	23.80	**23.14**	23.63	**23.14** (500)
Set 2	42.10	41.43 (13.000)	40.89	38.03	38.13	**37.80** (500)
Set 3	59.60	55.89 (18.000)	58.85	56.51	52.41	**51.33** (2000)
Set 4	96.04	93.61 (25.000)	94.07	91.99	84.25	**83.39** (2000)
Set 5	162.55	157.53 (50.000)	155.90	149.76	133.43	**130.12** (4000)

5 Conclusion and Future Work

In this paper, we proposed an obstacle avoidance path planning algorithm for mobile robot by using Ant-Q. Simulations gave the results, especially in large problems, which have shown that Ant-Q is very effective in finding collision-free path with very good quality. Our further work will concentrate more on the application of Ant-Q in path planning problem, with different kinds of representation of the environment. In particular, we plan to extend the application of Ant-Q in dynamically changing environment. In dynamically changing environment, there are both static and dynamic obstacles. Moreover, we will improve the path's smoothness. At first glance, we thought that mixture updating method would give the best result but the simulation gave differently. So that we will study why mixture updating method is worse than global updating method as shown in the simulation in the next research.

References

1. Hart, P.E., Nilsson, N.J., Raphael, B.: A Formal Basis for the Heuristic Determination of Minimum Cost Paths. IEEE Trans. Syst. Sci. Cybern. **4** (1968) 100-107
2. Warren, C.W.: Fast Path Planning using Modified A* Method. Proc. IEEE Int. Conf. Robotics and Automation, Atlanta, GA (1993) 662-667
3. Latombe, J.C.: Robot Motion Planning. Boston, MA: Kluwer (1991)
4. Takahashi, O., Schilling, R.J.: Motion Planning in A Plane using Generalized Voronoi Diagrams. IEEE Trans. Robot. Autom. **11** (1989) 143-150
5. Hou, E., Zheng, D.: Mobile Robot Path Planning based on Hierarching Hexagonal Decomposition and Artificial Potential Fields. J. Robot. Syst. **11** (1994) 605-614
6. Schwartz, J.T., Sharir, M.: On the Piano Movers' Problem: I. The Case If a Two-Dimensional Rigid Polygonal Body Moving Amidst Polygonal Barriers. IEEE Trans. Robot. Autom. **36** (1983) 345-398
7. Leven, D., Sharir, M.: An Efficient and Simple Motion Planning Algorithms for a Ladder Moving in Two-Dimensional Space Amidst Polygonal Barriers. Proc. 1st ACM Symp. Computational Geometry, Nice, France (1997) 1208-1213
8. Khatib, O.: Real-Time Obstacle Avoidance for Manipulators and Mobile Robots. Int. J. Rob. Res. **5** (1986) 90-98
9. Chuang, J., Ahuja, N.: An Analytically Tractable Potential Field Model of Free Space and Its Application in Obstacle Avoidance. IEEE Trans. Syst., Man, Cybern. **B 28** (1998) 729-736
10. Valavanis, K.P., Hebert,T., Kolluru, R., Tsourveloudis, N.: Mobile Robot Navigation in 2-D Dynamic Environments using an Electrostatic Potential Field. IEEE Trans. Syst., Man, Cybern. **A 30** (2000) 187-196
11. Bandi, S., Talmann, D.: Space Discretization for Efficient Human Navigation. Computer Graphic Forums **17** (1998) 195-206
12. Russell, S., Norvig, P.: Artificial Intelligence: A Modern Approach. USR, New Jersey: Prentice Hall (1995)
13. Borenstein, J., Koren, Y.: The Vector Field Histogram-Fast Obstacle Avoidance for Mobile Robots. IEEE Trans. Robotics and Automation **7** (1991) 278-288
14. Dorigo, M., Bonabeau, E., Theraulaz, G.: Ant Algorithms and Stigmergy. Future Generation Computer Systems **16** (2000) 851-871

15. Wen, Y, Dengwu, M., Hongda, F.: Path Panning for Space Robot based on the Self-Adaptive Ant Colony Algorithm. IEEE, 1st International Symposium on Systems and Control in Aerospace and Astronautics (2006)
16. Hocaoglu, C., Sanderson, A.C.: Planning Multiple Paths with Evolutionary Speciation. IEEE Trans. Evolutionary Computation **5** (2001) 169-191
17. Gemeinder, M., Gerke, M.: GA-based Path Planning for Mobile Robot Systems Employing an Active Search Algorithm. Applied Soft Computing **3** (2003) 149-158
18. Tian, L., Collins, C.: An Effective Robot Trajectory Planning Method using a Genetic Algorithm. Mechatronics **14** (2004) 455-470
19. Gambardella, L.M., Dorigo, M.: Ant-Q: A Reinforcement Learning Approach to the Traveling Salesman Problem. In: Prieditis, A., Russell, S.(eds.): Proceedings of ML-95, Twelfth International Conference on Machine Learning. Morgan Kaufmann (1995) 252-260
20. Dorigo, M., Gambardella, L.M.: A Study of Some Properties of Ant-Q. Lecture Notes in Computer Science (1996) 656-665
21. Colorni, A., Dorigo, M., Maniezzo, V.: An Investigation of Some Properties of an Ant Algorithm. Proceedings of the Parallel Problem Solving from Nature Conference (1992) 509-520
22. Colorni, A., Dorigo, M., Maniezzo, V.: Distributed Optimization by Ant Colonies. Proceedings of the First European Conference of Artificial Life (1991) 134-144

Monocular Vision Based Obstacle Detection for Robot Navigation in Unstructured Environment

Yehu Shen, Xin Du, and Jilin Liu

Department of Information Science and Electronic Engineering, Zhejiang University, Hangzhou, 310027 Zhejiang, P.R. China
{paulsyh, duxin, liujl}@zju.edu.cn

Abstract. This paper proposes an algorithm to detect the obstacles in outdoor unstructured environment with monocular vision. It makes use of motion cues in the video streams. Firstly, optical flow at feature points is calculated. Then rotation of the camera and FOE(focal of expansion) are evaluated separately. A non-linear optimization method is adopted to refine the rotation and FOE. Finally, we get inverse TTC(time to contact) with rotation and FOE and detect the obstacles in the scene. The algorithm doesn't need any assumption that the ground is flat or partially flat as the conventional methods. So it is suitable for outdoor unstructured environment. Qualitative and quantitative experiment results show that our algorithm works well on different kinds of terrains.

1 Introduction

The development of mobile robot which can move autonomously in different environments is one of the important fields for robotic researchers in recent years. In order to move safely, obstacle detection is a crucial part. Researchers have proposed several methods to detect obstacles.

Some algorithms recover the range information of the environment. With 3D information, obstacles can be found by the height of the object. In [1][2], LADAR is used to generate the range map. Though the accuracy is high, it is quite power consuming and expensive. So it is mainly used on the earth. For the power restricted field such as planetary navigation, stereo vision is widely used[3][4].

Other algorithms just extract 2D information from the images without any 3D reconstruction. Early work[5] uses color based pixel classification. Later, many algorithms calculate optical flow with single camera or several cameras with little common field of view. Unfortunately, until now, most approaches restrict their usages to indoor environments. Enkelmann[6] computes a reference flow related to the motion of the planar ground and compares it with flow from image streams. In [7], a method that combines central flow divergence and peripheral flow is proposed. The paper claims that the robot wanders around the room for about 20 minutes. Recently, Low et al[8] propose a method to get the TTC of the obstacle by optical flow. They compensate for the rotation of the robot with the help of the gyroscope.

D. Liu et al. (Eds.): ISNN 2007, Part I, LNCS 4491, pp. 714–722, 2007.

Very few researches were done on monocular vision based obstacle detection in outdoor unstructured environments. The algorithm in [9] doesn't restrict the motion of the robot. But the experiment is elementary and whether it could be used in unstructured environment is still a question. In [10], the authors also don't restrict the motion. But it needs at least 5 points with known 3D position which are hard to get in practical usage. So it only demonstrates the indoor results. Demonceaux et al[11] detect obstacles in the roads with frame motion. They assume that the roads were relatively flat so it can't be used in unstructured environment.

In this paper, we detect obstacles with monocular vision. The only assumption is that there are more than two distant points in the images. The distance not less than 10 meters is sufficient. This is reasonable because in unstructured environments especially those on lunar, there must be some far away points in the scene. Fig. 1 sketches the overview of our system. There are two features in our algorithm:

(1) We need only one camera and we don't need to recover the full 3D information.
(2) We don't need any terrain model and make no restriction on the robot motion.

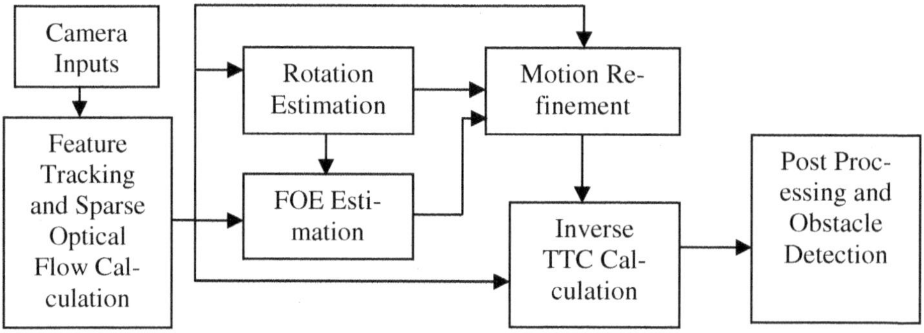

Fig. 1. System overview

2 Preliminary

We assume $\mathbf{P} = [X, Y, Z]^T$ to be a 3D scene point in the camera reference frame. The camera moves with a translational velocity $\mathbf{T} = [T_x, T_y, T_z]^T$ and a rotational velocity $\boldsymbol{\omega} = [\omega_x, \omega_y, \omega_z]^T$. The motion of this point is $\mathbf{V} = -\mathbf{T} - \boldsymbol{\omega} \times \mathbf{P}$.

We apply the pinhole model to the camera and the projection of the point $\mathbf{P}$ onto 2D image plane with homogeneous coordinates can be written as:

$$\mathbf{p} = f\,\mathbf{P}/Z \tag{1}$$

Where f is the focal length. Projecting the motion of the point $\mathbf{P}$ results in the optical flow of $\mathbf{p}$. Differentiating both sides of Eq. (1) can obtain:

$$v_x = (T_z x - T_x f)/Z - \omega_y f + \omega_z y + \omega_x xy/f - \omega_y x^2/f$$
$$v_y = (T_z y - T_y f)/Z + \omega_x f - \omega_z x - \omega_y xy/f + \omega_x y^2/f \quad (2)$$

In the above equation, $\mathbf{v} = [v_x, v_y]^T$ is the optical flow of the point $\mathbf{p}$.

From Eq. (2) we find that Z cannot be measured since it is always divided by the translational velocity. It is called the velocity scaling property[12].

We define $k_{xz} = T_x/T_z$ and $k_{yz} = T_y/T_z$ so that $[k_{xz}, k_{yz}]^T$ is the FOE.

TTC was first introduced by Lee[13]. Given the distance s between the robot and point, the speed of the robot u, TTC is defined as: $TTC = s/u$. In Eq. (2), Z/T_z has the similar meaning as that of TTC, so we call T_z/Z inverse TTC.

3 Obstacle Detection with Monocular Vision

3.1 Feature Tracking and Sparse Optical Flow Calculation

In order to generate optical flow with high quality and accelerate the computational speed, we apply the sparse optical flow calculation to the image streams. We use the KLT feature tracker[14] for this purpose. It tracks the features between two consecutive frames. Fig. 2 shows an example of two consecutive frames.

Fig. 2. Example of two consecutive frames

In the original version of the KLT tracker, the features are chosen according to their goodness for tracking. It doesn't take into account the position of the feature. As a result, sometimes there are too few features on the obstacles. We do some modifications to the KLT tracker. First we choose ROI (regions of interest) in which the obstacles will be detected. Then we force a preset number of feature points, for example, 50% of the total feature points, to generate from these regions in the feature choosing step of KLT tracker. This simple mechanism greatly increases the possibility to detect the obstacles in the images. After tracking, optical flow can simply be generated from the relative positions between the matched points. In all of our experiments, we

Fig. 3. The optical flow generated by KLT tracker and the region with distant points

choose 2500 feature points for tracking in every frame. Fig. 3 is the optical flow of the first frame in Fig. 2. The features are sub-sampled for clarity.

3.2 Rotation Estimation

If we find some feature points that Z is large, Eq. (2) can be approximated as:

$$\begin{aligned} v_x &\approx -\omega_y f + \omega_z y + \omega_x xy/f - \omega_y x^2/f \\ v_y &\approx \omega_x f - \omega_z x - \omega_y xy/f + \omega_x y^2/f \end{aligned} \tag{3}$$

In our experiment, the focal length of the camera is about 1000 pixels. The pixel coordinates x and y are no more than 400. The translation of the camera between two consecutive frames $T \sim 10^{-2} m$. If $Z \sim 10^1 m$, we have $(T_z x - T_x f)/Z < 1 pixel$ and $(T_z y - T_y f)/Z < 1 pixel$. So we can find that it's quite safe for the approximation in Eq. (3).

From the above discussion, we know that the key problem is to find some distant feature points. In unstructured environments, it is not very difficult to find distant feature points that $Z > 10m$. In this paper, we choose the feature points in the up-left and up-right corners. As is shown in Fig. 3, we choose distant feature points inside two green rectangles. This scheme works well in all our experiments.

Assume we have N distant feature points in the first frame. We rearrange Eq. (3) and write them in the matrix form:

$$\mathbf{A}\boldsymbol{\omega} = \mathbf{v}_o$$

$$\mathbf{A} = \begin{bmatrix} x_1 y_1/f & -f - x_1^2/f & y_1 \\ f + y_1^2/f & -x_1 y_1/f & -x_1 \\ \vdots & \vdots & \vdots \\ x_N y_N/f & -f - x_N^2/f & y_N \\ f + y_N^2/f & -x_N y_N/f & -x_N \end{bmatrix}, \boldsymbol{\omega} = [\omega_x, \omega_y, \omega_z]^T, \mathbf{v}_o = [v_{1x}, v_{1y}, \cdots v_{Nx}, v_{Ny}]^T \tag{4}$$

We have 3 unknowns and every feature point can provide 2 independent equations, so we can determine these unknowns if $N \geq 2$. Since KLT tracker will unavoidably give false matches and the distant points may not be very ideal. If we simply use least square method to solve Eq. (4), the results will be greatly affected by the outliers. To cope with these problems, we solve Eq. (4) with RANSAC[15].

3.3 FOE Estimation

First we compensate for the rotation effect in the optical flow at feature points except those distant ones:

$$\begin{aligned} v'_x &= v_x - (-\omega_y f + \omega_z y + \omega_x xy/f - \omega_y x^2/f) = (T_z x - T_x f)/Z \\ v'_y &= v_y - (\omega_x f - \omega_z x - \omega_y xy/f + \omega_x y^2/f) = (T_z y - T_y f)/Z \end{aligned} \tag{5}$$

After compensation, the resulting optical flow is mainly caused by the translation. We divide the second equation in Eq. (5) with the first one and rearrange it:

$$v'_y f k_{xz} - v'_x f k_{yz} = v'_y x - v'_x y \tag{6}$$

Assume there are M non-distant tracked feature points in the first frame. We rewrite Eq. (6) and stack them in the matrix form similarly:

$$\mathbf{BK} = \mathbf{C}$$

$$\mathbf{B} = \begin{bmatrix} v'_{1y} f & -v'_{1x} f \\ \vdots & \vdots \\ v'_{My} f & -v'_{Mx} f \end{bmatrix}, \mathbf{K} = [k_{xz}, k_{yz}]^T, \mathbf{C} = [v'_{1y} x_1 - v'_{1x} y_1, \cdots, v'_{My} x_1 - v'_{Mx} y_1]^T \tag{7}$$

There are 2 unknowns and every feature point can provide 1 equation. If $M \geq 2$, we can determine the FOE. Since we track 2500 feature points in each frame, it can be solved efficiently with Iterative Reweighted Least Squares technique[16].

3.4 Motion Refinement

In the above sections, the rotation and FOE are calculated by linear methods. The evaluation of rotation depends on the distant feature points. The calculation of FOE uses the rotational components. They are evaluated separately. So if there are some errors in the rotational components, they will propagate to the FOE. In order to diminish this effect and refine the results, we need to optimize them simultaneously.

Start from Eq. (2), we move the rotational components to the left side of the equations and divide the second equation with the first one. We define:

$$F_i(\boldsymbol{\omega}, \mathbf{K}) = \frac{v_{iy} - (f + y_i^2/f)\omega_x + x_i y_i / f \omega_y + x_i \omega_z}{v_{ix} - x_i y_i / f \omega_x + (f + x_i^2/f)\omega_y - y_i \omega_z} - \frac{y_i - f k_{yz}}{x_i - f k_{xz}} \quad (1 \leq i \leq M + N)$$

We minimize the following function to get the true value of $\boldsymbol{\omega}$ and $\mathbf{K}$:

$$\sum_{i=1}^{M+N} \left\| F_i(\boldsymbol{\omega}, \mathbf{K}) \right\|^2 \tag{8}$$

In this paper, Levenberg-Marquardt algorithm is used to minimize Eq. (8). The results of $\boldsymbol{\omega}$ and $\mathbf{K}$ evaluated from the previous sections are set as the initial value.

3.5 Inverse TTC Calculation

Again, we start from Eq. (2) and rewrite them as the following:

$$\begin{aligned} v_x &= T_z/Z\,(x - k_{xz} f) - \omega_y f + \omega_z y + \omega_x xy/f - \omega_y x^2/f \\ v_y &= T_z/Z\,(y - k_{yz} f) + \omega_x f - \omega_z x - \omega_y xy/f + \omega_x y^2/f \end{aligned} \tag{9}$$

The only unknown is the inverse TTC. As every feature point can provide 2 independent equations, the inverse TTC can be calculated without difficulty in Eq. (9).

The difference of inverse TTC is caused by the difference of Z. So from the inverse TTC, we can discriminate the closer objects from the far ones. Fig. 4 shows the resulting inverse TTC. The length of the arrow is proportional to inverse TTC. The features are also sub-sampled for clarity. From figure we conclude that in most part of the image the result is correct. The arrows on the obstacle are longer than those on the building and trees. They are also longer than the lawn behind the obstacle.

Fig. 4. The inverse TTC of two consecutive frames in Fig. 2

3.6 Post Processing and Obstacle Detection

The obstacle is defined as the object above the ground. Because the unavoidable noise in the inverse TTC, it is not wise to classify every feature point as obstacle or not directly. Instead, we use the following scheme.

Step 1. Divide the image into square blocks. We determine whether the block belongs to obstacle or not. In our experiment, the size is 20×20. We also determine the ROI in which to detect the obstacles. ROI is not fixed during the detection. For example, if the optical flow in the current image is moving up, the ROI will also move up in the next frame. It guarantees to keep in trace with the moving of the obstacle.

Step 2. For every block with feature points, calculate the average inverse TTC. Assume there are N_b blocks with feature points and average inverse TTC for the ith block is $ITTC_i \quad (1 \le i \le N_b)$. We calculate $ITTC_{max} = \max ITTC_i \quad (1 \le i \le N_b)$, $ITTC_{min} = \min ITTC_i \quad (1 \le i \le N_b)$, $NITTC_i = (ITTC_i - ITTC_{min})/(ITTC_{max} - ITTC_{min}) \quad (1 \le i \le N_b)$. For the block in the ROI which satisfies: $NITTC_i > thresh$, this block is the suspect obstacle region. The larger the $thresh$ is, the nearer the obstacle we want to detect. In our experiment, we choose $thresh = 0.5$.

Step 3. For every block in the ROI that belongs to the suspect obstacle region, if the block is above the half of the ROI, it belongs to the obstacle region. Otherwise, if there are any blocks in the same column that is above the half of the ROI belong to obstacle region, it will be in the obstacle region. Fig. 5 demonstrates the region that belongs to obstacle, this region is colored red.

Step 4. In Fig. 5, most of the region on the obstacle is colored red. There are also some missed parts on the obstacle. This is because in those regions there are no feature points or the wrong inverse TTC calculated. In order to give a more pleasant way to describe the obstacle, we use rectangle to approximate the position of the obstacle. The rectangle is just the smallest bounding box of the obstacle region. The result can be found in the first image of the first row of Fig. 7.

Fig. 5. Obstacle region labeled in red

Fig. 6. Pioneer 3 robot with camcorder

4 Experiment Results

We test our algorithm on three different videos. The videos are all captured by single camcorder SONY HDR-HC1E. The resolution of the video is 720×576. The camera is equipped on the Pioneer 3 robot as shown in Fig. 6. The algorithm is conducted offline on a 2.8GHz Pentuim 4 PC. Fig. 7 demonstrates the obstacle detection results. The obstacles are labeled in green rectangles.

The first two videos are captured on an uneven lawn. The motion of Pioneer 3 robot is a complex combination of large size translation and rotation. In the first row of Fig. 7, the building and trees in the background are about tens of meters away. The effectiveness of our algorithm shows that our assumption of the existence of distant points is not difficult to meet. In the second row, the building is standing against the sky. It is difficult for the conventional color based pixel classification to tell whether it's an obstacle or not. But with our method, it can be classified as distant object successfully. The third video is captured on a playground. Though the ground is mainly flat, there are many small stones on it. The motion of the robot is full of small vibrations. The results convinced us that our algorithm is also suitable for this kind of terrain.

In order to provide some quantitative intuition, we evaluate the rotational and FOE components since they're crucial to obstacle detection. As we can't get the ground truth in experiment videos, we compare the results calculated by our algorithm with the method proposed in [17] since the authors claimed it to be accurate and robust and the method is totally different from ours. We evaluate the results on 30 frame pairs. Since the two algorithms are both written in Matlab, it's also fair to compare the computational times of the two algorithms. The result is shown in table 1.

From the table we find that the differences are quite small so that the FOE and rotational component calculated by two algorithms agree very well. On the other hand, our method is much faster than the method in [17] so ours is more efficient.

Fig. 7. Obstacle detection results

Table 1. Quantitative evaluation between two algorithms

Average difference of FOE component	Average difference of rotational component	Average computational time of our algorithm	Average computational time of algorithm in [17]
3.75%	6.04%	5.3(s)	157.1(s)

5 Conclusion

This paper proposes an algorithm to detect obstacles in outdoor unstructured environment with a single camera. We don't assume that the ground is flat or partially flat. All we need is more than two distant feature points in the image. The experiments show that this assumption is reasonable. The algorithm doesn't need full 3D reconstruction as the Structure from Motion does so our method is simpler. In the future, we plan to integrate our algorithm into the robot navigation system.

Acknowledgments. This work is supported by Grant No. 60534070 and No. 60502006 from the National Science Foundation of China and also by Grant No. 2005C14008 from Zhejiang Province.

References

1. Urmson, C., Anhalt, J., Clark, M. et al.: High Speed Navigation of Unrehearsed Terrain: Red Team Technology for Grand Challenge 2004. CMU-RI-TR-04-37 (2004)
2. Dickinson, S., Davis, L., DeMenthon, D., and Veatch, P.: Algorithms for Road Navigations. In Masaki, I.(eds.): Vision-based Vehicle Guidance. (1990) 83–110
3. Maurette, M.: Mars Rover Autonomous Navigation. Autonomous Robots. 14 (2003) 199-208
4. Goldberg, S., Maimone, M., Matthies, L.: Stereo Vision and Rover Navigation Software for Planetary Exploration. IEEE Aerospace Conference Proceedings (2002) 2025-2036
5. Gremban, K. et al.: Vits—a Vision System for Autonomous Land Vehicle Navigation. IEEE Trans. Pattern Anal. Machine Intell. **10** (1988) 342–361
6. Enkelmann, W.: Obstacle Detection by Evaluation of Optical Flow Fields from Image Sequences. Image and Vision Computing **9** (1991) 160-168
7. Coombs, D. et al.: Real-time Obstacle Avoidance Using Central Flow Divergence and Peripheral Flow. Proc. Int. Conf. on Computer Vision (1995) 276-283
8. Low, T., Wyeth, G.: Obstacle Detection Using Optical Flow. Proc. Australasian Conference on Robotics and Automation (2005)
9. Nelson, R., Aloimonos, J.: Obstacle Avoidance Using Flow Field Divergence. IEEE Trans. Pattern Anal. Machine Intell. **11** (1989) 1102–1106
10. Rodrigo, R., Samarabandu, J.: Monocular Vision for Robot Navigation. Proc. IEEE International Conference on Mechatronics and Automation (2005) 707-712
11. Demonceaux, C., Potelle, A., Akkouche, D.: Obstacle Detection in a Road Scene Based on Motion Analysis. IEEE Trans. Vehicular Technology **53** (2004) 1649-1656
12. Dev, A., Krose, B., Groen, F.: Navigation of a Mobile Robot on the Temporal Development of the Optic Flow. Proc. Int. Conf. on Intelligent Robots and Systems (1997) 558-563
13. Lee, D., Young, D.: Brain Mechanisms and Spatial Vision. Springer-Verlag (1985)
14. Shi, J., Tomasi, C.: Good Features to Track. Proc. IEEE Conf. Computer Vision and Pattern Recognition (1994) 593-600
15. Fischler, M., Bolles, R.: Random Sample Consensus: A Paradigm for Model Fitting with Applications to Image Analysis and Automated Cartography. Commun. Assoc. Comput. Mach. **24** (1981) 391-395
16. Beaton, A., Turkey, J.: The Fitting of Power Series, Meaning Polynomials, Illustrated on Band-spectroscopic Data. Technometrics **16** (1974) 147-185
17. Domke, J., Aloimonos, Y.: A Probabilistic Framework for Correspondence and Egomotion. Proc. ICCV Workshop on Dynamical Vision (2005)

Attention Selection with Self-supervised Competition Neural Network and Its Applications in Robot *

Chenlei Guo and Liming Zhang

Electronic Engineering Department, Fudan University Shanghai, China 200433
chenlei.guo@gmail.com, lmzhang@fudan.edu.cn

Abstract. This paper proposes a novel attention selection system with competition neural network supervised by visual memory. As compared with others, this system can not only attend some salient regions randomly according to sensory information but also mainly focus on some learned objects by the visual memory. So it can be applied in robot self-localization or object tracking. The weights of neural networks can be adapted in real time to environment change.

1 Introduction

It is well known that human or animal vision system uses attention mechanism to select important information from environment. Thus they can efficiently process huge information every day. How to use the attention mechanism to computer vision is a research hotspot on both biological and engineering areas.

Psychologists put forward some attention theories and early concept models [1] [2], which integrated multi-features from bottom-up and subjective intent from top-down together. The computational model was proposed by Koch & Ullman[3]. After that, Itti et al[4] developed a NVT system that is basis of many current attention system. But they did not mention top-down part more. Other attention systems which incorporate top-down information were proposed in [5][6][7][8][9][10], but these top-down methods only use some cues by human, so they could not be implemented in real world. Ouerhani et al proposed an approach to objects tracking and visual landmarks selection based on visual attention [11-12]. However, they only used visual attention to reduce the input information and engineering means were still applied to solve object tracking or landmark selection.

This paper presents a novel attention selection system with self-supervised competition neural network called SSCNN. A visual memory with long-term and short-term is implemented to remember important message and to learn by selecting the focus of attention (FOA), which is the novelty of our model. We also introduce an eyeball movement prediction mechanism to estimate the movement of eyeball of human beings. Some tries of our system are applied to robot self-localization and to eyeballs' gaze of an object.

* No.60571052, 30370392 supported by NSF; No.045115020 supported by Shanghai Science and Technology Committee.

D. Liu et al. (Eds.): ISNN 2007, Part I, LNCS 4491, pp. 723–732, 2007.

The rest of this paper is organized as follows. In section 2, we show the architecture of whole system. Section 3 introduces our self-supervised competition neural network and long-term and short-term memory. Section 4 explains eyeball movement prediction mechanism. Experimental results and conclusion are given finally.

2 System Architecture

Fig. 1 shows the architecture of our attention selection system, which is divided into 3 parts: sensory mapping, cognitive mapping and motor mapping.

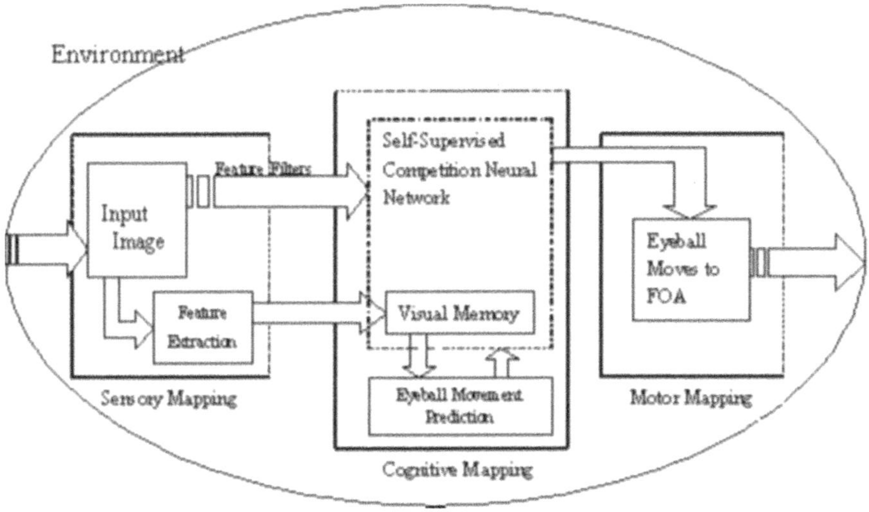

Fig. 1. The architecture of our attention selection system

In sensory mapping module, there are two sub-modules. Some feature maps such as motion, color, intensity and local orientation is got from input image by feature filters, which directly connect to competition neural network (SSCNN) for the computation of current FOA. Feature extraction module is to extract a feature vector by a feature extraction model such as SIFT[13] on $l \times l$ child image which is located at the focus area of attention, where l is much less than the height or width of input image, which is related to the size of input image. In general, the child image is the center of input image by eyeball's moving. The feature vector will be inputted into visual memory to learn or to search.

Cognitive mapping module includes self-supervised computation neural network and eyeball movement prediction map that will be introduced in section 3-4. Under the supervision of visual memory, the prediction of eyeball movement is calculated, which is regarded as key information to be inputted into SSCNN. SSCNN integrates all the feature information both from sensory mapping and cognitive mapping to select the FOA by competition. Then it will update its weights according to the rule of Hebbian learning.

In motor mapping module, our camera gets coordinates of winner neuron from SSCNN and move the centre of camera to FOA, which responds to the outside world.

3 Self-supervised Competition Neural Network

Fig. 2 shows the structure of SSCNN under the supervision of visual memory. Suppose five feature maps: motion, color, intensity, orientation and eye movement prediction are acquired. Every pixel of all feature maps is considered as a neuron which is connected to the corresponding neuron in CFM map shown in Fig.2. Those neurons in the same feature map share the same weights $w_1 \sim w_5$. Let neuron i on the CFM map get inputs both from feature maps and its surrounding neurons in a $N \times N$

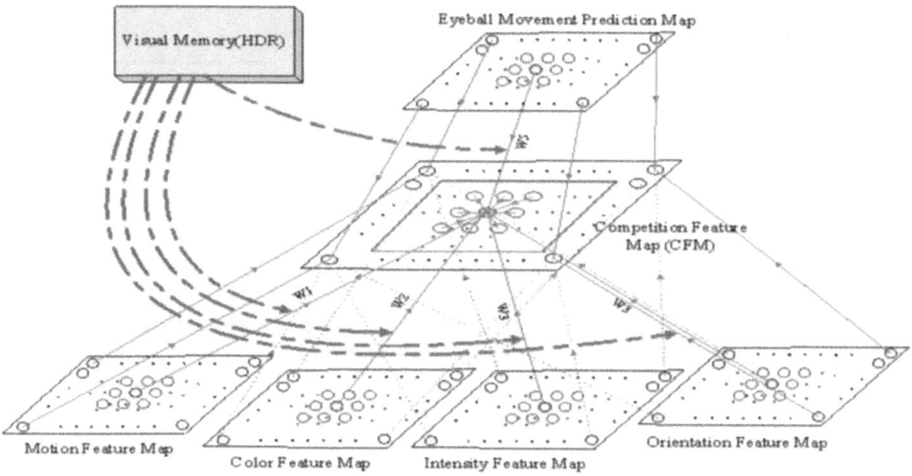

Fig. 2. The structure of self-supervised competition neural network

neighborhood N_i, which are modulated by connection weights S_{ik}. The output of neuron i is calculated by the following equation:

$$\tilde{y}_i(t) = \sum_{j=1}^{5} w_j u_{ji} + \sum_{k \in N_i} s_{ik} (\sum_{j=1}^{5} w_j u_{jk}) = \sum_{j=1}^{5} w_j (u_{ji} + \sum_{k \in N_i} s_{ik} u_{jk}) = \sum_{j=1}^{5} \tilde{D}_j(t), \quad (1)$$

where u_{ji} is the output of the ith neuron in the jth feature map, and $j = 1 \sim 5$ represent the five feature maps in Fig 2. S_{ik} is connection weights between neuron i and the neurons within neighborhood N_i, which has Gaussian shape. $\tilde{D}_j, j = 1,2,3,4,5$ is absolute contribution of each feature map. Then the FOA can be calculated by neuron's competition on CFM. The attention focus is obtained by the winner neuron in position (x_{win}, y_{win}) and at time t on CFM map by

$$y(x_{win}, y_{win}, t) = \arg\max_i (\sum_{i \in N_i} \tilde{y}_i(t)) . \quad (2)$$

Normalize $\tilde{D}_j, j = 1,2,3,4,5$ as relative contribution of each feature map:

$$D_j = \tilde{D}_j / \sum_{j=1}^{5} \tilde{D}_j, \ j = 1,2,3,4,5. \quad (3)$$

Visual memory consists of two kinds of memory: long-term and short-term memory shown in Fig. 3. A Hierachical Discriminant Regression(HDR) tree[15] is applied here to mimic visual memory of human beings. Some predefined examples which are learned by HDR algorithm[15] are considered as long-term memory, and those samples around FOA which are learned in real time by Incremental Hierachical Discriminant Regression (IHDR) algorithm[16] are regarded as short-term memory.

In training stage, the child images with $l \times l$ size are selected by sensory mapping, which belong to interesting objects. Feature vectors $FVec(n)$ are extracted by a feature extraction model, where $n = 1...L$ and L is the total number of child images. These feature vectors are clustered into m classes just like m normal nodes in Fig. 4. Each class represents a specified interesting object. If a normal node contains feature vectors that only belong to one interesting object, such node is called a leaf node.

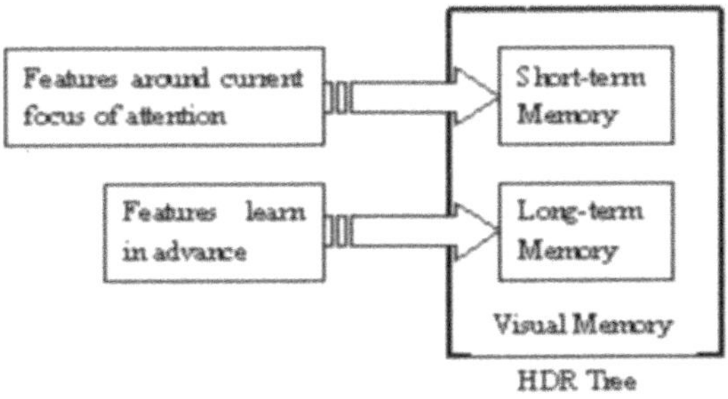

Fig. 3. Two parts of visual memory: short-term memory and long-term memory

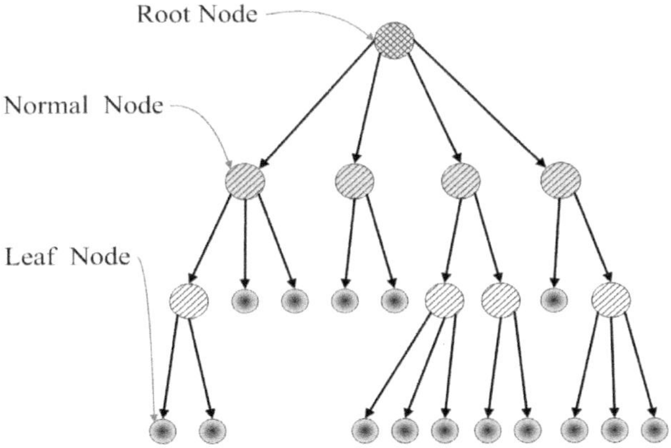

Fig. 4. Structure of a HDR tree. The circle with cross lines is a root node. Circles with oblique lines are normal nodes. Normal nodes contain statistical values of their samples. Circles with a dot in the middle are a leaf node, which contains samples within one class.

Otherwise, the node is continually split until it becomes a leaf node. Fig. 4 shows the structure of a HDR tree (IHDR tree). Every node is represented by the mean and variance of the feature vectors it contains. In testing period, search the tree by the comparison of testing feature vector and node's statistical values to decide if the child image needs to be focused on.

Brief introduction of HDR trees as top-down visual memory is shown in our previous work [14]. IHDR is an incremental algorithm which is used to update HDR tree. IHDR algorithm has such merits of online learning and fast search that leads visual memory to be able to adjust the weights between feature maps and CFM in time.

Suppose the winner neuron on CFM is in position of (x_{win}, y_{win}) at time *t*. A feature vector *FVec*(*t*) is extracted by a feature extraction model (eg. SIFT) from a region (child image) around the winner. *FVec*(*t*) is sent into visual memory tree (HDR) to test. If *FVec*(*t*) matches the object in HDR, the connection weights between CFM and the feature maps with large contributions will be strengthened. On the contrary, their connection weights will be depressed, which satisfies the famous Hebbian rule. The steps to update the weights are shown as follows:

1. Initialization: $w_j = random(\cdot)$, $j = 1,2,3,4,5$ where $\sum_{i=1}^{5} w_i = 1$.
2. The weights are adjusted according to the following equation

$$\tilde{w}_j = w_j \cdot (1 + \alpha_j D_j) \ , \quad w_j = \tilde{w}_j / \sum_{j=1}^{5} \tilde{w}_j, \qquad j = 1,2,3,4,5 \,. \tag{4}$$

Here α_j, $j = 1,2,3,4,5$ are coefficients of each feature map.

3. Define the label of interesting objects in visual memory as $IntObj = \{\phi_i\}$. The decision result of HDR is Φ .If Φ belongs to $IntObj$ $\alpha_j > 0$, otherwise $\alpha_j < 0$.

4. If *FVec*(*t*) is similar to feature vectors of learned objects in HDR tree, but does not match it, HDR tree will be updated by IHDR algorithm so that visual memory can adapt to the change of both objects and environment.

4 Eyeball Movement Prediction Via Visual Memory

The eyeball movement prediction is applied to estimate the movement of eyeball based on visual memory. Note that eyeball moving always makes the attention area in the centre of the test image. Let (x_c, y_c) be the centre of input image and (x_{t-1}, y_{t-1}) be the position of winner neuron on CFM at time (*t*-1). Thus instinct prediction of eyeball can be calculated as:

$$(\hat{x}_t, \hat{y}_t) = (x_c, y_c) + \beta \cdot ((x_{t-1}, y_{t-1}) - (x_c, y_c)) \,, \tag{5}$$

where β is adaptation coefficient, typically $0.5 \le \beta \le 1.5$. β helps to adjust the accuracy of prediction smoothly. The predicted child image $R(\hat{x}_t, \hat{y}_t, t)$ is shown as

$$R(\hat{x}_t, \hat{y}_t, t) = \{x, y / |x - \hat{x}_t|, |y - \hat{y}_t| \le l\} \,. \tag{6}$$

A feature vector $\hat{F}(t)$ is calculated from $R(\hat{x}_t, \hat{y}_t, t)$ by a feature extraction module in Fig.1 and sent to HDR tree. HDR will decide whether $\hat{F}(t)$ matches those objects in visual memory. If so, return $(\hat{x}_t, \hat{y}_t)$ as the result. Otherwise, search a larger area around (x_c, y_c) to find the most similar child image whose centre is considered as new $(\hat{x}_t, \hat{y}_t)$. Thus, the output of neurons on eyeball movement prediction map whose coordinates are (x, y) can be calculated below:

$$u_{eyeball}(x, y, t) = \exp\left[-\frac{(x-\hat{x}_t)^2 + (y-\hat{y}_t)^2}{2\sigma^2}\right]. \tag{7}$$

5 Experimental Results

Three video sequences are applied to test the effects of our system in robot vision. Experiment 1 shows the effect of long-term memory. Experiment 2 gives a target tracking application when object's color changes. A robot self-localization application gives finally.

Experiment 1: An image sequence with size of 640×480 at 10f/S taken from Lab is considered as test samples. Three interesting objects (flower, box and cup) were learned by HDR in advance as long-term memory. The test results of 855 pictures are shown in Fig. 5&6. When there are no interesting objects in scene for long-term memory, the robot will randomly select the attention areas (eg. telephone and power plug) according to salient area in SSCNN shown in Fig 6. At about 42th frame, when the learned cup appeared, the robot's eyeball shifted its attention on cup, then flower, cup and box so on, shown in Fig. 5&6. All the learned objects disappeared from 360th frame to 400th frame and our system selected objects randomly (eg. papers and books) and tried to find interesting objects. The flower appeared again in 400th frame and our system shifted its attention to the flower immediately. Fig. 7 shows the online change of the connection weights between CFM and five feature maps. It can see that different feature map plays various roles for different frames.

In this experiment the size of child image is $l\times l, l=32$ and neighbor area in SSCNN is $N\times N, N=11$.

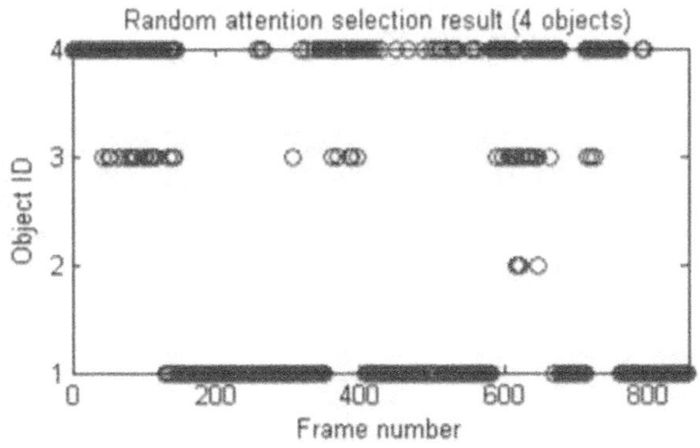

Fig. 5 Output of our attention selection system in HDR tree. Object ID 1~4 represents flower, bottle, cup and environment respectively.

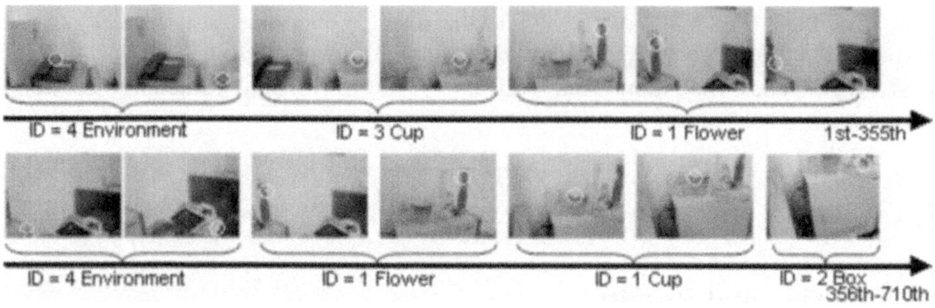

Fig. 6. Process of random attention selection. Frames range from 1st to 710th. The red dotted yellow circle represents current FOA.

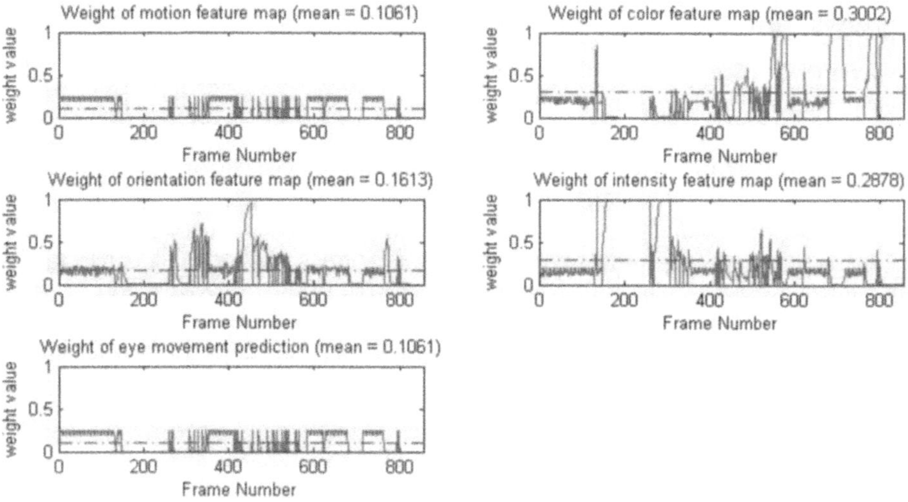

Fig. 7. Online change of connection weights. Different feature map plays various roles (eg. Intensity feature map plays the most important role from 200th frame to 300th frame, however, its influence decreases for the change of environment).

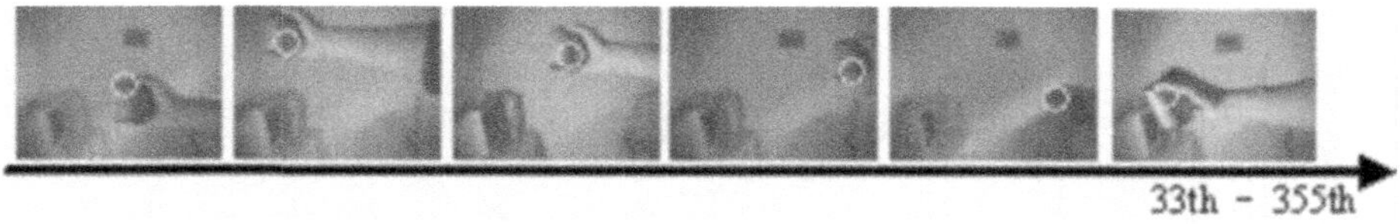

Fig. 8. Tracking results in an environment with light change. The yellow circle represents the focus of attention on target.

Experiment 2: A video sequence (526 frames) in size of 320×240 with a red cup moving is captured in an environment with light luminance changes. In Fig 8 the cup's color changes severely, but the proposed system still can track the cup

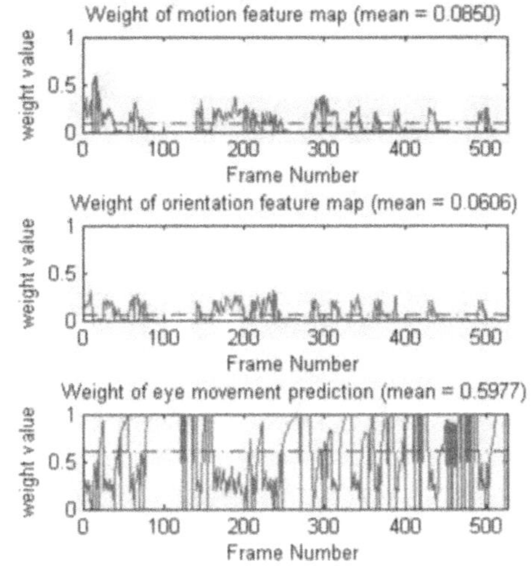

Fig. 9. Online change of weights. Red line shows the instantaneous value of weight and blue dot line is the mean weight value.

successfully in correct rate 89.35% because of short-term memory. Fig.9 shows the online change of weights. The eyeball movement prediction map gives more contribution than other features, for its mean weight is the highest. Here $l = 16, N = 11$.

Experiment 3: Robot self-localization is a difficult task. Five scenes are learned by HDR beforehand. The robot itself can extract key areas automatically for each scene. Then, when robot walks to these scenes, it can recognize them. Fig. 10 shows the recognition result in 2703 frames (size 640×480 at 10f/S). As compared with ideal result, its correct rate is 90.15% .Fig. 12 shows the test images. Here $l = 64, N = 11$.

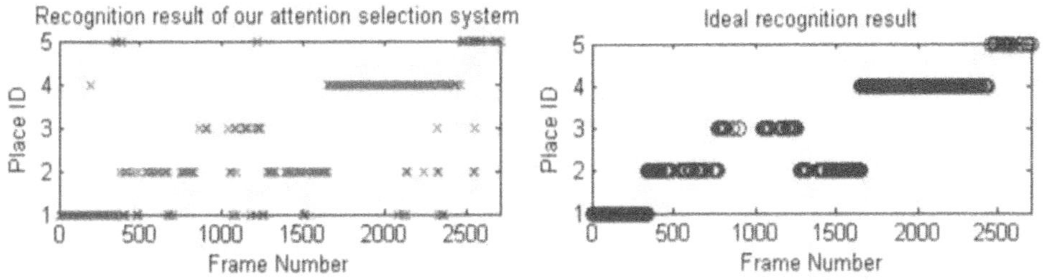

Fig. 10. The recognition result of our system and the ideal recognition result. ID1~5 represents our laboratory, corridor, elevator, outside of physical building and teacher office.

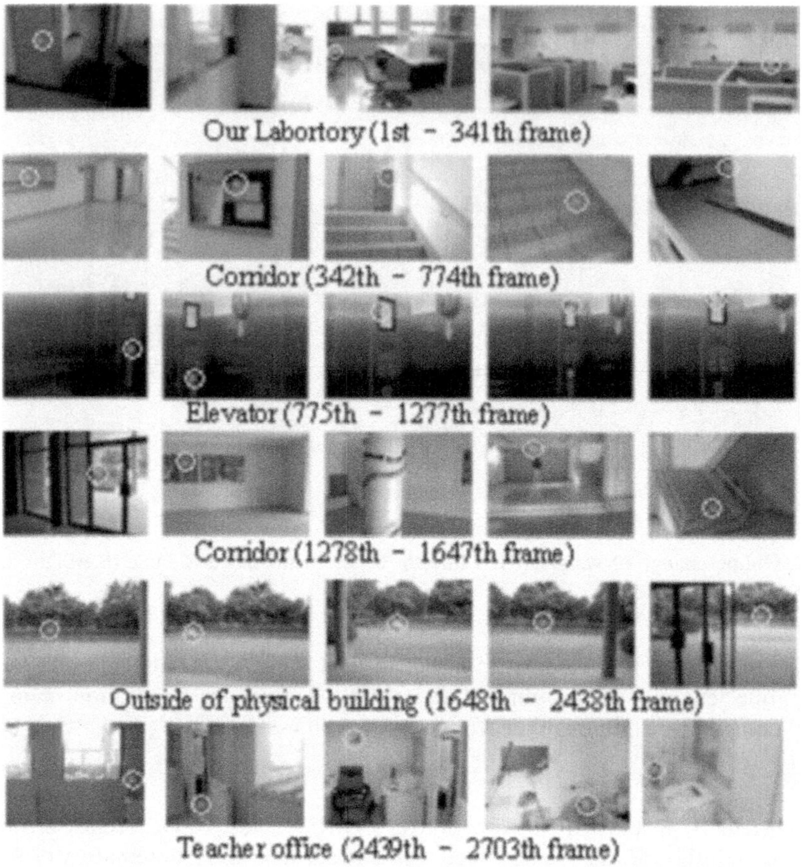

Fig. 12. Robot self-localization for five scenes. The red dotted yellow circle is the current FOA.

6 Conclusions

In this paper, an attention selection system with self-supervised competition neural network is proposed. It can integrate information from both sensory mapping and cognitive mapping to calculate current FOA by competition. The learning rule to update the connection weights is given. Experimental results show that both visual memory and eyeball movement prediction play important roles in the process of attention selection. Finally, our attention selection system is applied to perform object tracking and self-localization task.

References

1. Treisman, A., Gelade, G.: A Feature-integration Theory of Attention. Cognit. Psychol. **12** (1980) 97–136
2. Wolfe, J.M.: Guided Search 2.0: A Revised Model of Visual Search. Psychonomic Bulletin & Review **1** (2) (1994) 202-238

3. Koch, C., Ullman, S.: Shifts in Selective Visual Attention: Towards the Underlying Neural Circuitry. Human Neurobiology **4** (4) (1985) 219-227
4. Itti, L., Koch, C., Niebur, E.: A Model of saliency-based visual attention for rapid scene analysis. IEEE Transactions on PAMI **20** (11) (1998) 1254-1259
5. Breazeal, C., Scassellati, B.: A Context-dependent Attention System for A Social Robot. in Proc. Int. Joint Conf. Artificial Intelligence (1999) 1146--1151
6. Navalpakkam, V., Itti, L.: Modeling The Influence of Task On Attention. Visual Search 45 (2) (2005) 205-231
7. Hamker, F.H.: The Emergence of Attention By Population-based Inference and Its Role in Distributed Processing and Cognitive Control of Vision. Journal of Computer Vision and Image Understanding. Special Issue on Attention and Performance **100** (1-2) (2005) 64-106
8. Simone Frintrop.: VOCUS: A Visual Attention System for Object Detection and Goal-Diected Search. PhD thesis. University of Bonn, Germany (2005)
9. Dong, L., Ban, S.W., Lee, I., Lee, M.: Incremental Knowledge Representation Model based on Visual Selective Attention. Neural Information Processing – Letters and Reviews. **10** (4-6) (2006) 115-124
10. Sun, Y., Fisher, R.: Object-based Visual Attention for Computer Vision. Artificial Intelligence **146** (1) (2003) 77-123
11. Ouerhani, N., Hügli, H.: A model of dynamic visual attention for object tracking in natural image sequences. In 4th International Conference on Artificial and Natural Neural Network (IWANN). Springer Verlag, Lecture Notes in Computer Science **2686** (2003) 702-709
12. Ouerhani, N., Hügli, H., Gruener, G., Codourey, A.: A Visual Attention-based Approach for Automatic Landmark Selection and Recognition WAPCV 2004, LNCS **3368** (2005) 183-195
13. David, G. Lowe.: Distinctive Image Features from Scale-Invariant Keypoints. International Journal of Computer Vision **60** (2) (2004) 91-110
14. Guo, C.L., Zhang, L.M.: An Attention Selection System based on Neural Network and Its Application in Tracking Objects. Lecture Notes in Computer Science **3972** (2006) 404-410
15. Hwang, W.S., Weng, J.Y.: Hierachical Discriminant Regression. IEEE Trans. PAMI **22** (11) (2000)
16. Weng, J.Y., Zhang, Y.L., Hwang, W.S.: Incremental Hierarchical Discriminant Regression for Online Image Classification. Proceedings of Sixth International Conference on Document Analysis and Recognition (2001) 476 – 480

Kinematic Analysis, Obstacle Avoidance and Self-localization for a Mobile Robot

Hongbo Wang, Xingbin Tian, and Zhen Huang

Robotics Institute, Yanshan University,
Qinhuangdao, Hebei Province 066004, P.R. China
wzfamily@hotmail.com, xbtian@ysu.edu.cn, huangz@ysu.edu.cn

Abstract. This paper presents a novel omni-directional mobile robot with obstacle avoidance and self-location function. Since a special transmission mechanism is designed, we introduce the mechanism design of the mobile robot and analyze its kinematics. A fuzzy-neural control algorithm is presented to realize the obstacle avoidance of the mobile robot. The obstacle avoidance system based on FKCN and the control algorithm for obstacle avoidance are described. A self-localization technique that uses the ceiling light for the landmark is proposed. Using this self-location function, the mobile robot could locate itself in a world coordinate system.

1 Introduction

A self-steering mobile robot must have two important basic functions: obstacle avoidance and autonomous navigation.

When a mobile robot navigates itself automatically, it will unavoidably encounter stationary and moving obstacles. Obstacle avoidance is one of the fundamental requirements for the automatic navigation of mobile robots.

The visual sensors provide the richer source of useful information about the surroundings. However, a problem of visual sensors is that they are slow in computing data and are expensive in cost. In comparison with the visual sensors, the ultrasonic sensors are accomplished in small cost. Although ultrasonic range measurement instruments suffer from some fundamental drawbacks which limit their usefulness in mapping or in any other task requiring high accuracy in a domestic environment, many researchers used ultrasonic sensors for obstacle avoidance [1]-[3]. Borenstein and Koren [4] summarized relevant obstacle avoidance methods using ultrasonic sensors into edge detection, certainty grids and the potential field method. Also, they proposed a vector field histogram, a virtual force field and a histogramic in-motion mapping method. In our robot system, the ultrasonic sensor is used to detect the obstacle and a fuzzy-neural control algorithm is presented to realize the obstacle avoidance.

To navigate effectively in a fairly complex workspace, an intelligent autonomous mobile robot system needs to be able to determine the state of its world and its own location with respect to its immediate surroundings. Such self-location is critical for the reliable performance of an autonomous system. Many

D. Liu et al. (Eds.): ISNN 2007, Part I, LNCS 4491, pp. 733–742, 2007.

methods are used to realize self-location for mobile robots. These methods can be roughly categorized as relative position measurements (Odometry, Inertial navigation) and absolute position measurements (Magnetic compasses, Active beacons, Global positioning systems, Landmark navigation, Model matching) [5]. Among the above methods, landmarks probably provide the best information for the location of mobile robots. Several researchers have approached the problem of self-location in mobile robots by employing landmarks [6]-[8]. The key idea of self-location is to use special marks that include a wealth of geometric information under perspective projection so that the camera location can be easily computed from the image of the guide marks.

Researchers have chosen ceiling lights as landmarks since they can be easily detected due to the high contrast between the light and the ceiling surface. Also, the ceiling lights do not require to be specially installed [9], [10]. Therefore, we choose ceiling lights as landmarks to realize the self-localization. With this self-location function, the mobile robot developed here is capable of moving along any path in an environment where ceiling lights have been installed.

This paper presents a novel omni-directional mobile robot and describes the kinematics, obstacle avoidance and self-localization of the mobile robot.

2 Mechanism and Kinematics of Mobile Robot

2.1 Mechanism

The mechanism of a mobile robot with 8 wheels is shown in Fig.1 and Fig.2. The 8 wheels are divided into 4 groups that have same transmission mechanism as shown in Fig. 3. The one wheel of each group is driving wheel and another is free wheel. Since two belts driven by two motors make the four driving wheels move in synchronous way, the motion of the robot platform is a plane translation. To obtain the motion of the mobile robot, we need only to analyze the motion of one wheel.

The transmission mechanism between the driving belt and one driving wheel is a gear train with two degrees of freedom. The coordinate system of the gear train is shown in Fig. 4. In the following kinematics analysis, the velocity and

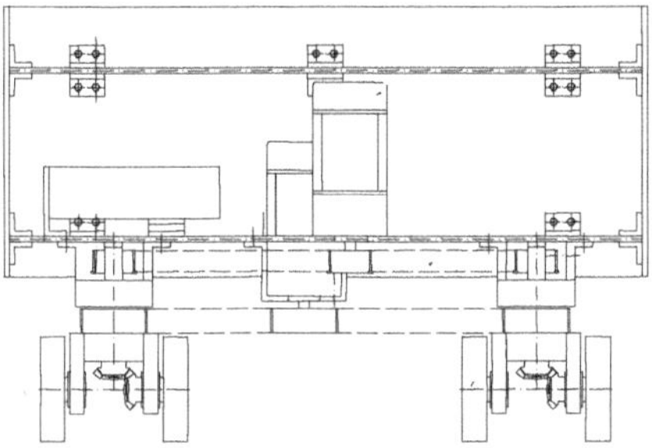

Fig. 1. A omni-directional mobile robot

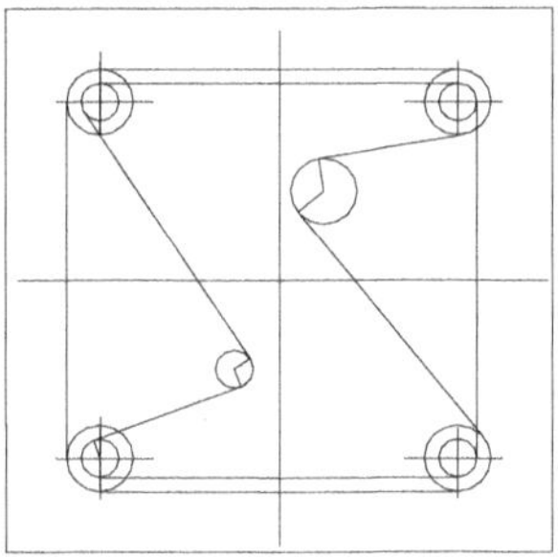

Fig. 2. Driving system of mobile robot

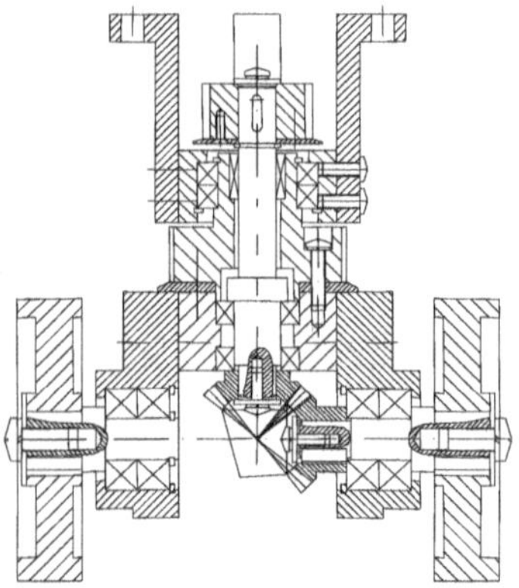

Fig. 3. Transmission mechanism

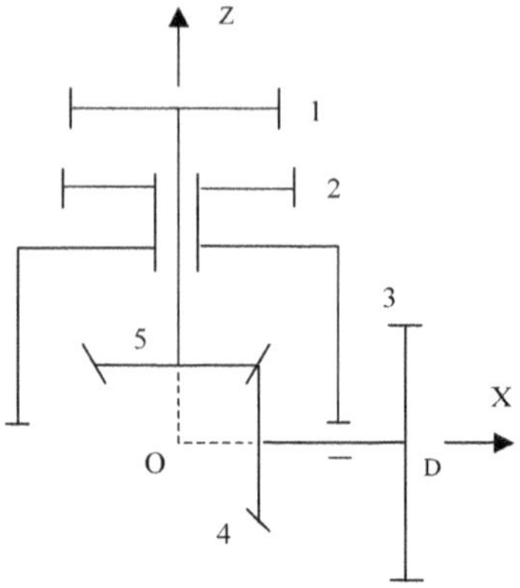

Fig. 4. Diagram of gear train

angular velocity are regarded as vectors. The forward direction of the velocity and angular velocity is identical with the direction of the coordinate axis.

2.2 Kinematics

Linear Motion. When the angular velocity ω_2 of the pulley 2 is zero, and the angular velocity of the pulley 1 is ω_1, the angular velocity of the driving wheel 3 can be obtained as follows

$$\omega_3 = -\omega_1. \tag{1}$$

The negative sign in the above equation indicates that the direction of angular velocity ω_3 and the direction of x-axis are reverse. The velocity V_O of the point O that is the intersection point of the x-axis and the z-axis can be expressed as

$$V_O = V_D = -\omega_3 R_W = \omega_1 R_W, \tag{2}$$

where, V_D and R_W are the center point velocity and the radium of the driving wheel 3. The direction of the velocity is along the y-axis that can be determined by right hand rule.

Since the motion of four driving wheels is identical, the mobile robot moves along a straight line (y-axis).

Circular Motion. When the angular velocity ω_1 of the pulley 1 is zero and the angular velocity of the pulley 2 is ω_2, the driving wheel 3 has a rotation around the x-axis with angular velocity $\omega_3 = \omega_2$ due to the engagement of the bevel gears 4 and 5. The velocity of central point of the wheel 3 can be obtained as follows

$$V_D = -\omega_3 R_W = -\omega_2 R_W. \tag{3}$$

In this case, there is a point C that locates in the x-axis of the driving wheel 3 and whose absolute velocity is zero. Therefore, the velocity V_D can also be expressed as

$$V_D = \omega_2 L_{CD}, \tag{4}$$

where L_{CD} is the distance between points C and D, and can be obtained as follows

$$L_{CD} = V_D/\omega_2 = -R_W. \tag{5}$$

The negative sign in the above equation indicates that the point C locates at the forward direction of x-axis (the right of wheel 3). The x coordinate of the point C is

$$X_C = L_{OD} - L_{CD} = L_{OD} + R_W. \tag{6}$$

It should be noted that the point C is also the center of rotation of the total gear train. The total gear train rotates around the point C with the angular velocity ω_2. The velocity of the total gear train can be written as

$$V_O = \omega_2(0 - X_C) = -\omega_2(L_{OD} + R_W). \tag{7}$$

When the angular velocities ω_1 and ω_2 are not equal to zero, the angular velocity of the driving wheel 3 is $\omega_3 = \omega_2 - \omega_1$. The velocity of center point of the driving wheel 3 can be obtained as follows

$$V_D = -\omega_3 R_W = (\omega_1 - \omega_2)R_W. \tag{8}$$

The distance between the points C and D, and the x coordinate of the point C can be obtained as follows

$$L_{CD} = V_D/\omega_2 = (\omega_1 - \omega_2)R_W/\omega_2, \tag{9}$$

$$X_C = L_{OD} - L_{CD} = L_{OD} - (\omega_1 - \omega_2)R_W/\omega_2. \tag{10}$$

The velocity of the total gear train can be written as

$$V_O = \omega_2(0 - X_C) = -\omega_2(L_{OD} - (\omega_1 - \omega_2)R_W/\omega_2). \tag{11}$$

Since the motion of the mobile robot is a plane translation, the velocity of the total gear train is the velocity of the mobile robot.

3 Obstacle Avoidance

3.1 Arrangement of Ultrasonic Sensors

To avoid obstacle, 12 ultrasonic sensors (BTE054: US Sensor 2) are installed in the mobile robot. The operating frequency of the sensor is 40kHz. The maximum effective examination distances under short distance pattern and long distance pattern are 1500mm and 3000mm respectively. In this paper, the measurement mode of the short distance pattern is used. The 12 ultrasonic sensors are divided into 8 groups to arrange at the four corners and four sides of mobile robot with square platform ($400 \times 400mm$) as shown in Fig. 5. The groups A, C, E, G at four sides include 2 sensors, and the groups B, D, F, H at four corners include 1 sensor respectively. In order to reduce the computing data, only 8 sensors (5 groups) are used to avoid obstacle during navigation. When the navigation direction of the mobile robot is between OB and OD, the groups A, B, C, D and E are used. When the robot navigates in the direction between OH and OB, the groups G, H, A, B and C are used. For the direction between OF and OH (OD and OF), the groups E, F, G, H and A (C, D, E, F and G) are used.

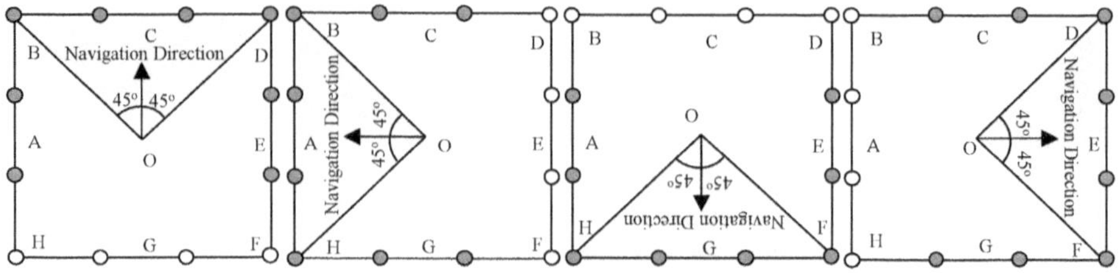

Fig. 5. Arrangement of ultrasonic sensors installed in mobile robot

3.2 Obstacle Avoidance System Based on FKCN

In order to enable the mobile robot to avoid the obstacle in the navigational path with rapid reaction capacity, the better mapping relation between the sensor data input and the control output must be established. Since this mapping relation is extremely complex and nonlinear, it is very inconvenient to solve this problem using the general control method. However, the artificial neural network has the astonishing ability to deal with nonlinear problem. Using the merit of artificial neural network, we successfully establish the mapping relations. The control system structure of obstacle avoidance is shown in Fig. 6. The left side of the control system structure is FKCN (fuzzy Kohonen cluster network) and the right side is the calculation of speed output. Since we choose five groups of ultrasonic sensors as the input of obstacle avoidance system to avoid obstacle, the input vector can be expressed as follows

$$\mathbf{S}_i = (i_1, i_2, i_3, i_4, i_5)^T. \tag{12}$$

3.3 The Mapping Relations Between the Sensor Input and the Control Output

As shown in Fig. 6, the hidden layer is to use for the comparison of the input pattern $\mathbf{S}$ and the prototype pattern $\mathbf{W}$. When the input pattern $\mathbf{S}_i$ and the

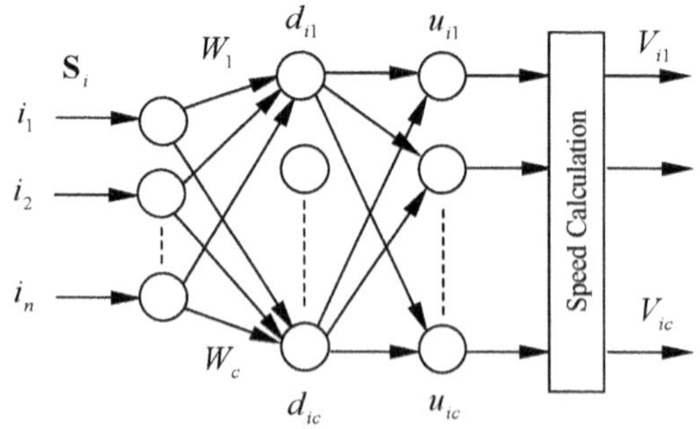

Fig. 6. The diagram of the obstacle avoidance system based on FKCN

prototype pattern $\mathbf{W}_j$ are completely consistent, the output d_{ij} of j-th node in hidden layer is zero. The output of the hidden layer can be expressed in the following equation

$$d_{ij} = |\mathbf{S}_i - \mathbf{W}_j|^2 = (\mathbf{S}_i - \mathbf{W}_j)^T(\mathbf{S}_i - \mathbf{W}_j), \tag{13}$$

where $\mathbf{W}_j$ is j-th prototype pattern.

The output value u_{ij} in output layer is determined based on d_{ij}, that is, if some input pattern $\mathbf{S}_i$ is different from the prototype pattern $\mathbf{W}_j$, the similarity between them is expressed using the output value $u_{ij}(0 \leq u_{ij} \leq 1)$. The output value u_{ij} can be obtained as follows

$$u_{ij} = \left(\sum_{l=1}^{c} \frac{d_{ij}}{d_{il}}\right)^{-1}. \tag{14}$$

When the input pattern and the prototype pattern are completely consistent, we can obtain the following equation

$$u_{ij} = 1, u_{ik} = 0, \tag{15}$$

where $k \neq j$ and $1 \leq k \leq c$. From the above two equations, we know that the bigger the output value is, the higher the similarity between the input pattern and the prototype pattern is. Since u_{ij} indicates the membership degree of input pattern $\mathbf{S}_i$ to prototype pattern $\mathbf{W}_j$. Each kind of prototype pattern $\mathbf{W}_j$ is corresponding to a fuzzy control rule, and each fuzzy control rule is corresponding to a speed vector. The control output $\mathbf{V}_i$ can be determined in the following equation

$$\mathbf{V}_i = \sum_{j=1}^{c} \mathbf{V}_j u_{ij}, \tag{16}$$

where $\mathbf{V}_j$ is the velocity of mobile robot.

4 Self-localization

4.1 Position Vector of Ceiling Light

To calculate the position vector of ceiling light, the two images whose image planes are parallel to ceiling plane are used. The camera coordinate system and the image plane coordinate system are shown in Fig. 7. The z-axes of two coordinate systems are coincided with the optical axis of camera. In the first image plane coordinate system $o_u^1 - x_u^1 y_u^1 z_u^1$, the coordinates of the two ends $Q_i(i = 1, 2)$ of the light are expressed as $Q_i(x_i^1, y_i^1, z_i^1)$, and the coordinates of the projection points P_i of light ends Q_i are expressed as $P_i^1(x_{ui}^1, y_{ui}^1, z_{ui}^1)$. Since the three points Q_i, P_i^1, and o_c^1 are in the same line, the following line equations can be obtained

$$z_i^1 = ax_i^1 + b, 0 = ax_{ui}^1 + b, -f = b, z_i^1 = cy_i^1 + d, 0 = cy_{ui}^1 + d, -f = d, \tag{17}$$

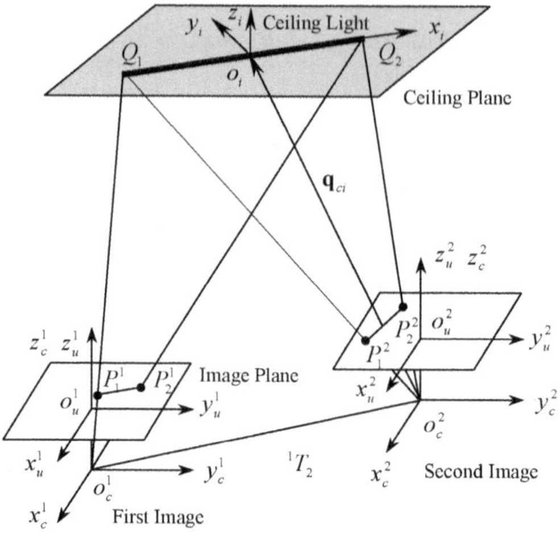

Fig. 7. Coordinate systems of ceiling light and camera

where a, b, c, d are constant and f is the focal distance of the camera. From the above equation, the equation of the line that is through the three points Q_i,P_i^1 and o_c^1 can be obtained as follows

$$z_i^1 = f\frac{x_i^1}{x_{ui}^1} - f, z_i^1 = f\frac{y_i^1}{y_{ui}^1} - f. \tag{18}$$

In the second image plane coordinate system $o_u^2 - x_u^2 y_u^2 z_u^2$, using similar method, the equation of the line that is through the three points Q_i, P_i^2 and o_c^2 can be written as follows

$$z_i^2 = f\frac{x_i^2}{x_{ui}^2} - f, z_i^2 = f\frac{y_i^2}{y_{ui}^2} - f. \tag{19}$$

The relation between $Q_i(x_i^1, y_i^1, z_i^1)$ and $Q_i(x_i^2, y_i^2, z_i^2)$ can be written as

$$(x_i^1, y_i^1, z_i^1, 1)^T = {}^1T_2(x_i^2, y_i^2, z_i^2, 1)^T, \tag{20}$$

where ${}^1T_2$ is the homogenous transformation matrix of the second image plane coordinate system relative to the first image plane coordinate system. Since the motion of the mobile robot is a plane translation, the rotation angle between two image plane coordinate systems is zero. Therefore, ${}^1T_2$ can be expressed in the following equation

$${}^1T_2 = \mathbf{I}(d_x, d_y, 0, 1)^T, \tag{21}$$

where $\mathbf{I}$ is a unit transformation matrix, d_x and d_y are the translation distances of the second image plane coordinate system relative to the first image plane coordinate system. The translation distances d_x and d_y can be calculated from the information of the encoders installed on two driver wheels.
From the equations (20) and (21), the following equation can be obtained

$$x_i^1 = x_i^2 + d_x, y_i^1 = y_i^2 + d_y, z_i^1 = z_i^2. \tag{22}$$

From the equations (18), (19) and (22), the following equation can be derived

$$x_i^2(x_{ui}^1 - x_{ui}^2) = x_{ui}^2 d_x, y_i^2(y_{ui}^1 - y_{ui}^2) = y_{ui}^2 d_y. \tag{23}$$

From the above equation, the x, y coordinates of the two ends $Q_i(i = 1, 2)$ of the light can be obtained as follows

$$x_i^2 = \frac{x_{ui}^2 d_x}{x_{ui}^1 - x_{ui}^2}, y_i^2 = \frac{y_{ui}^2 d_y}{y_{ui}^1 - y_{ui}^2}. \tag{24}$$

From the equations (19) and (24), the z coordinate of the two ends $Q_i(i = 1, 2)$ of the light can be obtained as follows

$$z_i^2 = f\frac{-x_{ui}^1 + x_{ui}^2 + d_x}{x_{ui}^1 - x_{ui}^2}. \tag{25}$$

In the camera coordinate system, the position vector $\mathbf{q}_i$ of the two ends $Q_i(i = 1, 2)$ of the light can be expressed as follows

$$\mathbf{q}_i = (x_{ci}^2, y_{ci}^2, z_{ci}^2)^T = (x_i^2, y_i^2, z_i^2 + f)^T. \tag{26}$$

4.2 Self-location

Once $\mathbf{q}_i$ is known, the light coordinate system can be determined. The origin o of the light coordinate system is at the middle point of points Q_1 and Q_2, that is, $\mathbf{q}_c = (\mathbf{q}_1 + \mathbf{q}_2)/2$. The x-axis is along the direction from Q_1 to Q_2. The unit vector of the z-axis of the light coordinate system is parallel to the z-axis of the camera coordinate system. The transformation matrix of the light coordinate system relative to camera coordinate system can be expressed as follows

$$\mathbf{A} = [\mathbf{e}_x, \mathbf{e}_y, \mathbf{e}_z], \tag{27}$$

where

$$\mathbf{e}_x = \frac{\mathbf{q}_2 - \mathbf{q}_1}{|\mathbf{q}_2 - \mathbf{q}_1|}, \mathbf{e}_z = (0, 0, 1)^T, \mathbf{e}_y = \mathbf{e}_z \times \mathbf{e}_x. \tag{28}$$

The position of the mobile robot in the light coordinate system can be determined as follows

$$\mathbf{r}_c = -\mathbf{A}^T \mathbf{q}_c. \tag{29}$$

When one landmark is in sight of the camera, we assume the landmark is j-th ceiling light. Since the position vector of the j-th ceiling light in the world coordinate system is known, the position vector of the mobile robot in a two-dimensional world coordinate system $o - xy$ can be obtained as follows

$$\mathbf{r}_{oc} = \mathbf{r}_{oj} + \mathbf{r}_{jc}, \tag{30}$$

where $\mathbf{r}$ is a two-dimensional vector with x and y coordinates.

5 Navigation Experiments

We did a navigation test in a room where 9 ceiling lights were installed as shown in Fig. 8. The navigational path is a circle whose center is the center of the light 5 and radium is 2.5m. The mobile robot was able to navigate with a maximum position error of 100mm at the goal node as shown in Fig. 9. If we laid an obstacle in the way, the mobile robot could avoid the obstacle and went to the goal position.

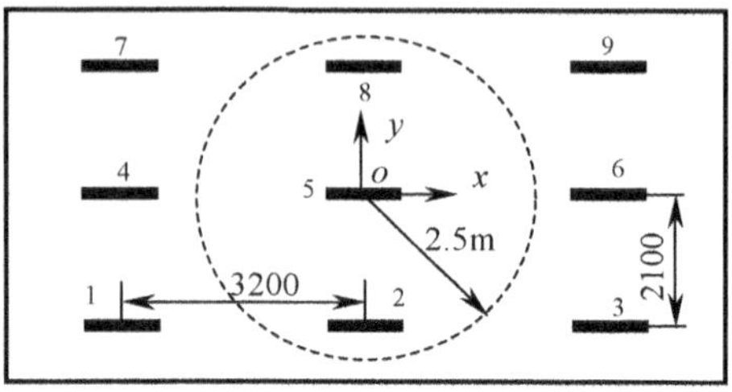

Fig. 8. Environment of navigation experiment

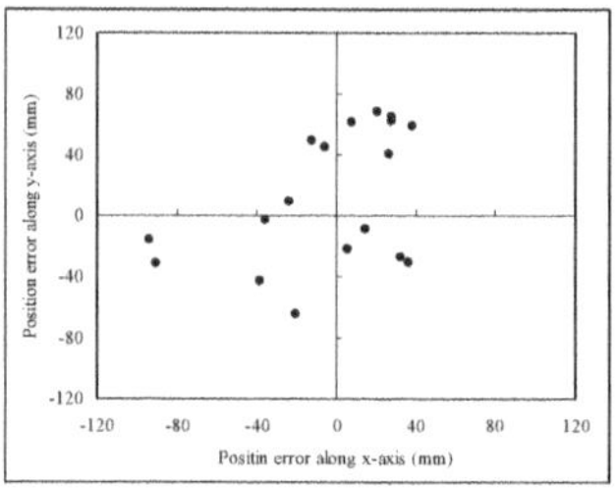

Fig. 9. Experiment result

6 Conclusion

This paper presents a novel omni-directional mobile robot with obstacle avoidance and self-localization. Since the special transmission mechanism is designed for the mobile robot, the motion of the platform of the mobile robot is a plane translation. The mechanism design and the kinematics of the mobile robot are described. In order to detect the obstacle in the way, the ultrasonic sensor is used. A fuzzy-neural control algorithm is used for obstacle avoidance of the mobile robot. The obstacle avoidance system based on FKCN is described and the mapping relation between the sensor input and the control output is established. To enable the mobile robot to locate itself in a world coordinate system, a self-localization technique is proposed. The ceiling light is chosen to function as landmark and the self-localization of the mobile robot is realized using different two images during motion. A navigational experiment indicates the effectiveness of the self-localization method.

References

1. Borgolte, U., Hoyer, H., Buhler, C., Heck, H., Hoelper, R.: Architectural Concepts of a Semiautonomous Wheelchair. Journal of Intelligent and Robotic Systems **22** (1998) 233-253
2. Li, H., yang, S. X.: Ultrasonic Sensor Based Fuzzy Obstacle Avoidance Behaviors. In Proc. of IEEE Intl. Conf. on Systems, Man and Cybernetics. Yasmine Hammamet, Tunisia **2** 644-649

3. Shoval, S., Borenstein, J.: Using Coded Signals to Benefit from Ultrasonic Sensor Crosstalk in Mobile Robot Obstacle Avoidance. Proceedings of the 2001 IEEE International Conference on Robotics and Automation. Seoul, Korea (2001) 2879-2884
4. Borenstein, J., Koren, Y.: The Vector Field Histogram - Fast Obstacle Avoidance for Mobile Robots. IEEE Trans. Robotics and Automation **7** (1991) 278-288
5. Borenstein, J., Everett, H. R., Feng, L., Wahe, D.: Mobile Robot Positioning: Sensors and Techniques. J. Robotic Systems **14** (1997) 231-249
6. Se, S., Lowe, D., Little, J.: Mobile Robot Localization and Mapping with Uncertainty Using Scale-invariant Visual Landmark. International Journal of Robot Research **21** (2002) 735-758
7. Armingol, J. M., Escalera, A. de la, Moreno, L., Salichs, M. A.: Mobile Robot Localization Using a Non-linear Evolutionary Filter. Journal of Advanced Robotics **16** (2002) 629-652
8. Bais, A., Sablatnig, R.: Landmark Based Global Self-localization of Mobile Soccer Robots. In Proceedings of the 7th Asian Conference on Computer Vision. Hyderabad, India, Springer Lecture Notes in Computer Science 3852, **2** (2006) 842-85
9. Dulimarta, H. S., Jain, A. K.: Mobile Robot Localization in Indoor Environment. Pattern Recognition **30** (1997) 99-111
10. Wang, H. B., Ishimatsu, T.: Vision-based Navigation for an Electric Mobile Robot Using Ceiling Light Landmark. Journal of Intelligent and Robotic Systems **41** (2004) 283-314

Mobile Robot Self-localization Based on Feature Extraction of Laser Scanner Using Self-organizing Feature Mapping*

Jinxia Yu[1,2], Zixing Cai[2], and Zhuohua Duan[2,3]

[1] College of Computer Science & Technology, Henan Polytechnic University, Jiaozuo 454003, Henan, China
melissa2002@163.com
[2] College of Information Science & Engineering, Central South University, Changsha 410083, Hunan, China
[3] Department of Computer Science, Shaoguan University Shaoguan 512003, Guangdong, China

Abstract. This paper investigates the use of SOM to process the signal of a 2D laser scanner encountered in feature extraction (corner) and mobile robot self-localization in indoor environments. It presents the method of combining SOM with occupancy grid matching to improve the self-localization performance at the lower computational cost. Experimental results demonstrate that this method can reliably extract the feature of corner point and can effectively improve the self-localization performance of mobile robot.

1 Introduction

Mobile robot self-localization involves matching the current sensory information against prior knowledge of the operating environment in which a crucial issue is how to extract distinct features such as corner points from the indoor environment with all kinds of sensor and to match these features with the environment map to position robot[1]. Compared with stereo vision and sonar, laser measurement system can provide more largely accurate range information for meeting the resolution and velocity requirements, so it has widely been applied to the field of mobile robot self-localization. In recent years, self-organizing feature mapping (SOM)[2] attracts many researchers' attention as an efficient tool to extract the features of input data. SOM is also introduced into the research on mobile robot localization such as Janet[3] and Gerecke[4]. The research of Duckett & Nehmzow[5] discovers that SOM has a better localization performance compared with nearest neighbor classifier and a lower computational cost with occupancy grid matching.

Considering the analyses above, this paper tries to combine SOM with occupancy grid matching to improve the self-localization performance at the lower computational cost. With mobile robot MORCS-1 as experimental platform, experimental

* This work is supported by the National Natural Science Foundation of China (No. 60234030).

D. Liu et al. (Eds.): ISNN 2007, Part I, LNCS 4491, pp. 743–748, 2007.

results demonstrate this method can reliably extract the feature of corner point and can effectively improve the self-localization performance of mobile robot.

2 Laser Sensing

LMS291 made by Sick Corporation is a laser scanner that can provide 2D scanning range (180º). It is based on a time-of-fight (TOF) measurement principle as depicted in figure 1(a). In our research, LMS291 would get 361 measurement data throughout the 180° scanning field with 0.5° resolution every scan. With 500K baud rate, the transmission delay time is 13ms and the scan time of LMS291 is 26.67ms. Thus we use 40ms to undergo the scan of LMS291and deal with the data.

LMS291 is mounted on a high precision rotating table with horizontal and pitch rotation to sense the environment. After coordinates transformation of ranging data from the laser scanning plane to the robot reference plane and then to the world coordinate system, the environment map is built. The operating environment of mobile robot is described by 2D Cartesian grids and a 2D array is taken to storage the environment map in this paper. The knowledge about occupancy condition of a given cell at time t is stored as probability of empty or occupied state, given all the prior sensor observations. Since the size of grids directly has an effect on the resolution of feature extraction, each grid cell corresponding to $10 \times 10 cm^2$ in real environment is adopted considering that LMS291 has high measurement resolution and quick response time. The mapping of corner point by LMS291 is shown in figure1(b).

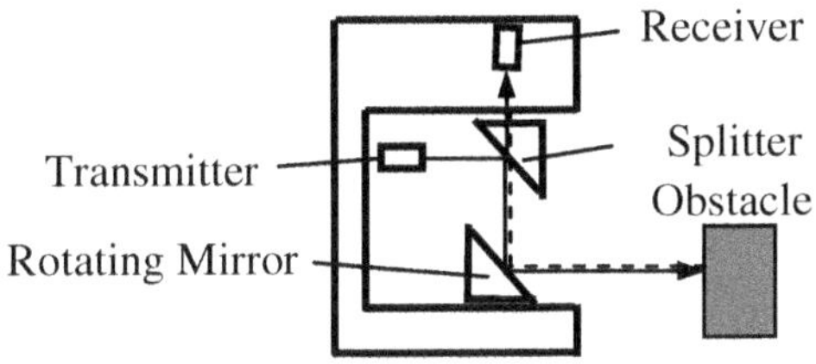

(a) Measurement principle

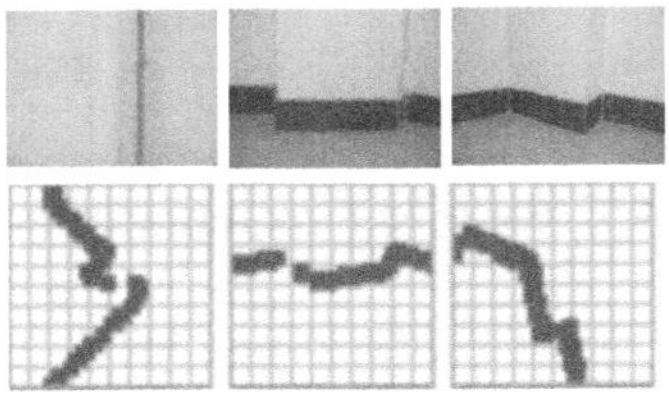

(b) Corner point by LMS291

Fig. 1. Laser scanner LMS291

3 Self-localization with SOM

3.1 Sample Data Collection and Network Training Method

As shown in figure 2, different data sets were collected for the feature of corner point at different locations (r, θ) of mobile robot with 10 different orientation from 0° to 180° in 15° increments and with 5 different distance from 1m to 6m in 1m increments. So, there are 50×*Num* sets to be used for training if the number of different corner point is *Num*. Considering its measurement principle, laser scanner is able to extract 12 sets with 30 data at 15° interval by one scan. In fact, the orientation change of sensing feature in the view of laser scanner reflects that of mobile robot. Therefore, we adopt a 2D scanning data (ρ_n, ϕ_n) $(n = 1*k, \cdots, 30*k)$ $k = 1, \cdots, 12$ as the input of network, where ρ_n is the distance from the origin to the n-th obstacle, ϕ_n is the angle of the n-th laser beam relative to the home orientation. Aimed at the practical environment, 5 typical corners are selected and 6×5 neuron nodes are used as the output of network. It is necessary for a reliable neural network not only to have a good classification to the ideal input but also to have enough accuracy to the noisy input. Therefore, a robust training strategy should be designed to reduce the effect of disturbance factors in sample data set. In order to have the tolerant capacity for the input vector, the best way to train SOM is to use both the ideal and noisy signal to do it, which assure the stability of the network when it distinguished the ideal input.

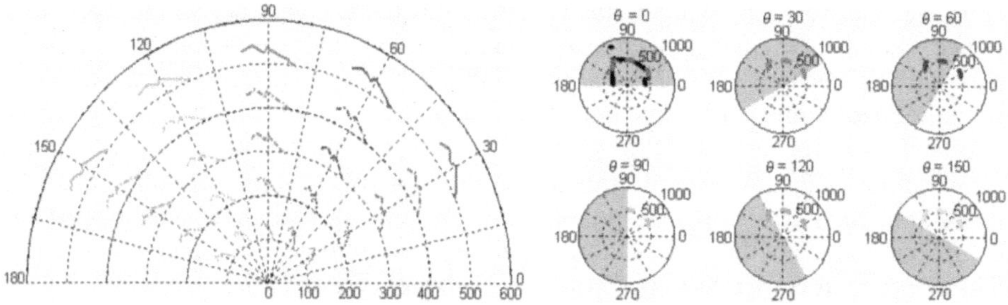

(a) Measurement in different distance (b) Measurement in different orientation

Fig. 2. Collection of sample data set

3.2 Feature Extraction Based on SOM

Assumed the input vector is $X_k = (x_1^k, x_2^k, \cdots x_N^k),\ k = 1,2,\cdots 250$, the neuron vector in competitive layer is $Y_j\,(j = 1,2,\cdots 25)$ and the connection weight vector between the input layer and the $j-th$ competitive layer is $W_j = (w_{j1}, w_{j2}, \cdots w_{ji}, \cdots, w_{jN})$ where $i = 1,2,\cdots,30$ and $j = 1,2,\cdots 25$. At each iteration, SOM algorithm finds a winning neuron node g by minimizing the following formula.

$$Y_g = \min[Y_j],\ j = 1,2,\cdots, M \tag{1}$$

where, $Y_j = [\sum_{i-1}^{N} (\tilde{x}_i^k - \tilde{w}_{ji})^2]^{1/2}$.

Then, the weight values between the neighboring area $N_g(t)$ of the wining neuron g and the input neuron are adjusted according to the learning rate $\eta(t)$ by formula 2.

$$\begin{cases} \tilde{w}_{ji}(t+1) = \tilde{w}_{ji}(t) + \eta(t)(\tilde{x}_i^k(t) - \tilde{w}_{ji}(t)) & j \in N_g(t) \\ \tilde{w}_{ji}(t+1) = \tilde{w}_{ji}(t) & j \notin N_g(t) \end{cases} \tag{2}$$

The learning is carried out for 5000 epochs on the different training sets: the first is the standard input without the dead reckoning error, then is that with the error and last is also the standard input. Estimation of the distance and orientation of mobile robot is shown in figure 3 in which the large error curve (red) is the input with the error, the smaller one (blue) is the standard input at first and the middle one (green) is secondly.

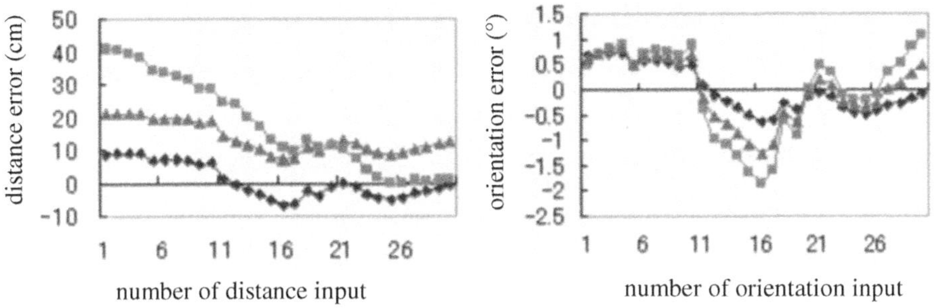

(a) Distance estimation (b) Orientation estimation

Fig. 3. Feature estimation based on SOM

3.3 Self-localization Method

Before the trained SOM can be used in the feature extraction of corner point, its output node must be associated with specific locations in the operating environment. Once the most likely corner point has been found by SOM at the current sample time of laser scanner, we can use it as invariant landmark to compute the relative displacement of mobile robot between the current and previous location.

The realization method in practice is that mobile robot firstly uses laser scanner LMS291 to scan the operating environment with the rotation of the rotating table and to create a local map of occupancy grids. The feature of corner points is extracted by SOM without any localization errors so that we have an initial position of corner point to compare against. Then, mobile robot whose velocity is selectable from 20cm/s to 40cm/s explores the environment at some strategy and detects the feature of corner points at different positions. Because it exists the difference for the feature of same corner points represented by occupancy grids at different time due to localization errors in robot's motion, these corner points under the influence of uncertain disturbance are matched into the environment map with the initial corner points to search the counterpoint. The robot relative displacement at current time associated on the previous position, given the information of dead-reckoning information and the feature of corresponding corners, is estimated. The errors between the estimated and the real values of robot relative displacement are computed to improve the self-localization precision.

4 Experiments and Analysis

To validate the effectiveness of this method, we uses mobile robot MORCS-1 as the test platform to implement the following experiment in a large indoor environment. Figure 4(a) shows the self-localization experiment of mobile robot MORCS-1 from the beginning point A to the ending point B, which is randomly run in different scene at 5 times to compare the self-localization performance with different method. There

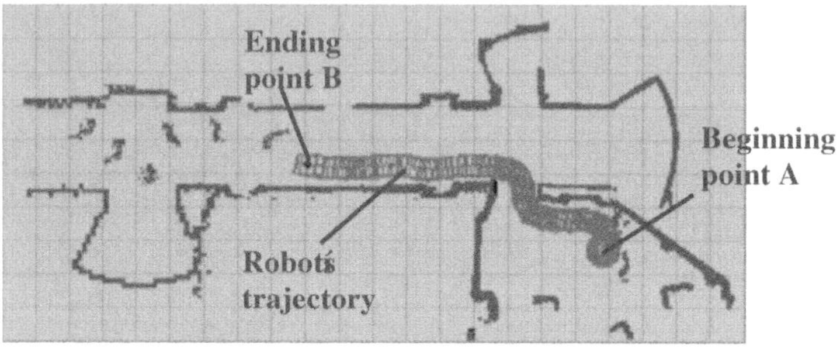

(a) Experiment in indoor environment

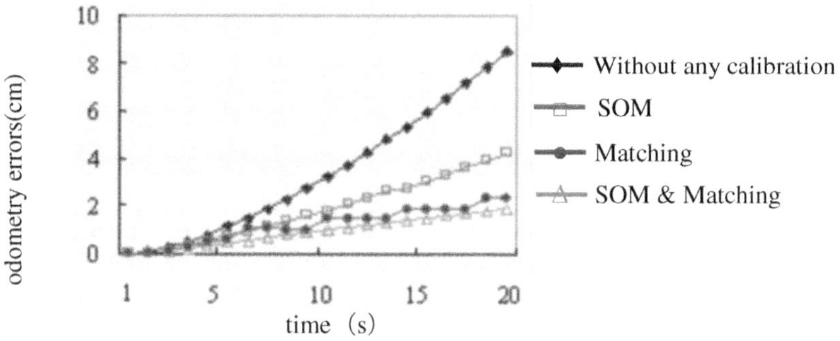

(b) Odometry errors by different methods

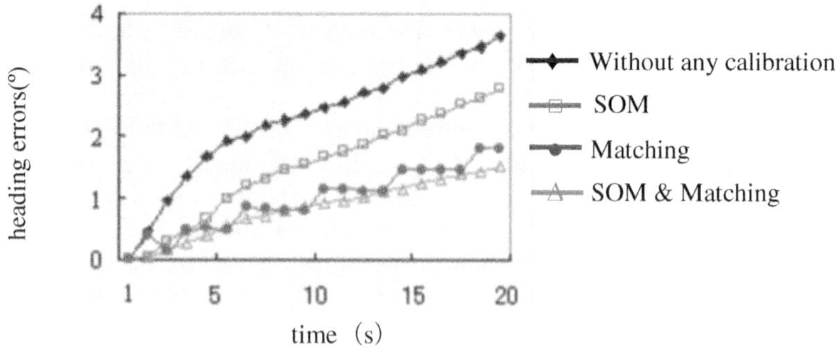

(c) Heading errors by different methods

Fig. 4. Self-localization experiments with MORCS-1

are three methods to be adopted to reduce the localization error: SOM, grid map matching, and SOM with grid map matching. Considering the need of computation time, grid map matching is implemented every several sampling intervals, whereas SOM is also done at each sampling intervals. Figure 4(b) and (c) show the error curve of odometry and heading of mobile robot, respectively. From it, we can see the localization error without any calibration is rather large, and the performance of grid map matching is better than that of SOM although the computational cost of grid map matching is expensive because it needs use the total map information to compare. The method based on SOM combined with grid map matching is better than other two methods whether in the localization performance or in computational cost. Experimental results discovered the effectiveness of this method.

5 Conclusions

Aimed at the self-localization problem of mobile robot navigation, it present a method by integrating grid map matching into the feature of corner point based on SOM to reduce robot's localization errors and improve its precision. In order to verify the effectiveness of this method, the experiment with mobile robot MORCS-1 is implemented in indoor environment. Experimental result discovered that this method could build more accurate environmental map, reduce sensors' noise and the uncertainty from the ambient environment. It also demonstrated that this method can reliably extract the feature of corner point and can effectively improve the self-localization performance of mobile robot.

References

1. Cai, Z.X., He, H.G., Chen, H.: Some Issues for Mobile Robot Navigation under Unknown Environments (in Chinese). Control and Decision **17(4)** (2002) 385-391
2. Kohonen, T.: Self-organized Formation of Topologically Correct Feature Maps. Biological Cybernetics, **43(1)** (1982) 59-69
3. Janet, J.A., Gutierre, R., Chase T.A., et al.: Autonomous Mobile Robot Global Self-Localization Using Kohonen and Region-Feature Neural Networks. Journal of Robotic Systems, **14(4)** (1997) 263-282
4. Gerecke, U., Sharkey, N.: Quick and Dirty Localization for a Lost Robot. In: Proceedings of the 1999 IEEE Int. Symp. on Computational Intelligence in Robotics and Automation(CIRA-99), Piscataway, NJ, (1999) 262-267
5. Duckett, T., Nehmzow, U.: Performance Comparison of Landmark Recognition Systems for Navigating Mobile Robots. In: Proceedings of the 17th National Conf. on Artificial Intelligence(AAAI'2000), Austin, TX, (2000) 826-831

Generalized Dynamic Fuzzy Neural Network-Based Tracking Control of Robot Manipulators

Qiguang Zhu[1], Hongrui Wang[2], and Jinzhuang Xiao[2]

[1] Institute of Information Science and Engineering, Yanshan University, Qinhuangdao 066004, China
zhu7880@ysu.edu.cn
[2] Institute of Electronic and Information Engineering, Hebei University, Baoding 071002, China

Abstract. A robust adaptive control based on generalized dynamic fuzzy neural network (GD-FNN) is presented for robot manipulators. Fuzzy control rules can be generated or deleted automatically according to their significance to the control system, and no predefined fuzzy rules are required. Being use of radial basis function neural network (RBFNN) the learning speed is very fast. The asymptotic stability of the control system is established using Lyapunov theorem. Simulations are given for a two-link robot in the end of paper, and validated the control arithmetic.

1 Introduction

Generally, robot manipulators used as industrial automatic elements are known as system with high nonlinearities that are often unknown and time-varying. The conventional feedback controllers such as PID controllers are commonly used in the field of industries because their control architectures are very simple and easy to implement. But when these conventional feedback controllers are directly applied to nonlinear systems, they suffer from the poor performance and low robustness due to the nonlinearities and the external disturbances [1], [2], [3].

During the past decade, much research effort has been put into the design of intelligent controllers using fuzzy logic. But most adaptive fuzzy controllers have difficulties in determining suitable fuzzy control rules and membership functions [4]. Recently, hybrid control laws containing neural networks have attracted more and more attention. Neural networks were used to adjust and optimize parameters of fuzzy controllers through online of offline. However these control methods all require predefined and fixed fuzzy rules or neural network structure. Therefore there is a need for fuzzy neural controllers with structure adaptation capability [5], [6].

In this paper, the GD-FNN algorithm is used as the structure adaptation mechanism for the adaptive fuzzy neural controller (AFNC). The AFNC is able to learn unmodeled disturbance of the robot manipulator online. The error convergence of the trained fuzzy neural network is found to be fast.

D. Liu et al. (Eds.): ISNN 2007, Part I, LNCS 4491, pp. 749–756, 2007.

2 Robot Manipulator Dynamics

In this paper, the robot manipulator is modeled as a set of n rigid bodies connected in series with one end fixed to the ground and the other end free. The dynamic equations of robot manipulator motion are a set of highly nonlinearly coupled differential equations. Using the Lagrange-Euler formulation, the dynamic equation of a n-joint robot arm can be expressed as

$$D(q)\ddot{q}+C(q,\dot{q})\dot{q}+G(q)+F_f(\dot{q})=\tau-\tau_e \tag{1}$$

Where the vector q is the $n\times 1$ joint angle, the vector $\dot{q}$ is the $n\times 1$ joint angular velocity, and the vector $\ddot{q}$ is the $n\times 1$ joint angular acceleration, $D(q)$ is the $n\times n$ symmetric positive definite inertia matrix, $C(q,\dot{q})$ is the $n\times 1$ vector of Coriolis and centrifugal matrix, $G(q)$ is the $n\times 1$ gravitational torques, $F_f(\dot{q})$ is the $n\times 1$ vector of dynamic and static friction forces, τ is the $n\times 1$ vector of joint torques supplied by the actuators, τ_e is the $n\times 1$ vector compensating the unmodeled dynamics and external disturbances.

3 Adaptive Fuzzy Neural Control System

3.1 Fuzzy Neural Control Structure

The structure of the robust AFNC system including a PD controller and a FNN controller, as showed in Fig.1. The FNN controller is connected in parallel with the PD controller to generate a compensated control signal.

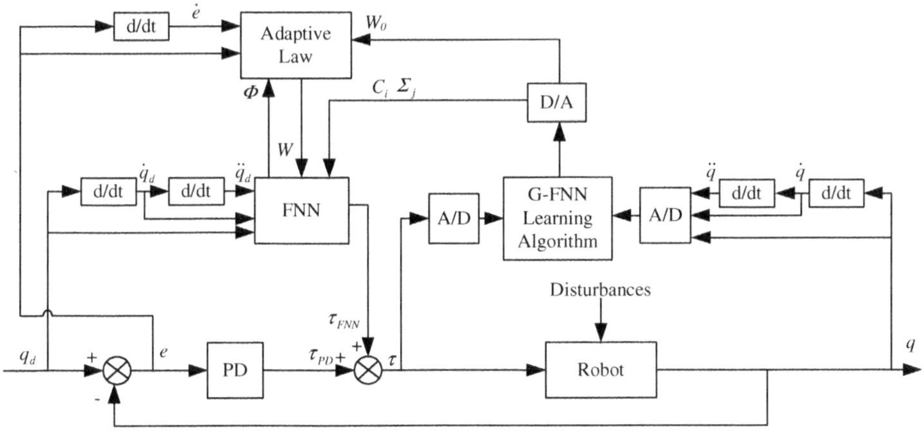

Fig. 1. Robust adaptive control based on GD-FNN control structure

The control law is given by

$$\tau = \tau_{FNN} + \tau_{PD} \tag{2}$$

where τ_{FNN} is the output torque of the AFNC, and τ_{PD} is the torque produced by the PD controller. The tracking error vector is defined as

$$E = [q_d - q \quad \dot{q}_d - \dot{q}]^T = [e \quad \dot{e}]^T \tag{3}$$

where q_d is the desired joint angel, and e is the tracking error. If the parameters of the robot dynamics are all known, the control torque can be designed as

$$\tau^* = D(q)\ddot{q}_d + C(q,\dot{q})\dot{q} + G(q) + F_f(\dot{q}) + \tau_e + D(q)KE \tag{4}$$

where $K = [k_2 \quad k_1]$, and k_2, k_1 are positive real numbers. Substituting (4) into (1) yields

$$\ddot{e} + k_1\dot{e} + k_2 e = 0 \tag{5}$$

If we choose the proper K, the tracking error will converge to zero. However, external disturbances and unmodeled dynamics are unknown in practice. So we cannot implement the perfect control law. We proposed the FNN controller's output torque is τ^*_{FNN}, and the perfect control law is executed by

$$\tau^* = \tau^*_{FNN} + D(q)KE \tag{6}$$

3.2 Fuzzy Neural Network Structure

The FNN is constructed based on Radial Basis Function, and it is equivalent to T-S model-based fuzzy system. The inputs of the FNN are the joint positions, velocities and acceleration, and the outputs are joint torques, as showed in Fig.2.

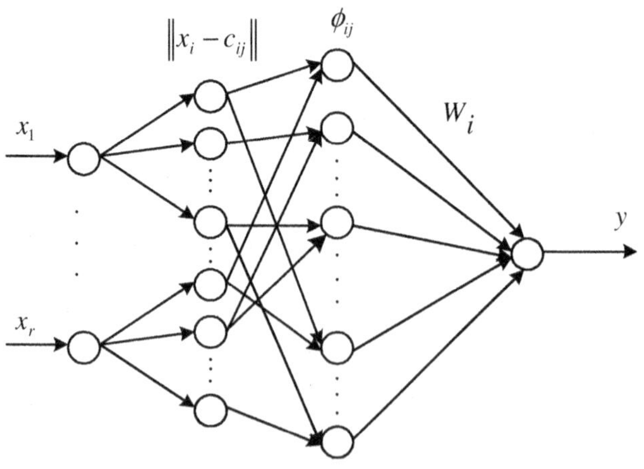

Fig. 2. Fuzzy RBF neural network model structure

a) Layer 1 defines input variables $X = [x_1, x_2 \cdots x_r]^T$
b) Layer 2 represents the membership function associated with the input variables. The membership function is given by a Gaussian function as

$$F_{ij}(x_i) = \exp(-\|x_i - C_{ij}\| / \sigma_{ij}^{\ 2}) \qquad \imath = 1, \cdots r; j = 1, \cdots u \tag{7}$$

where i is the number of input variables, u is the number of membership functions, C_{ij} is the center vector and σ_{ij} is the width vector.

c) Layer 3 represents the IF part of the fuzzy rules. The number of RBF units indicates the number of fuzzy rules. The outputs are given by

$$\phi_j = \prod_{i=1}^{r} F_{ij} \tag{8}$$

d) Layer 4 completes the THEN part of the fuzzy rules.

$$y_k = \sum_{j=1}^{u} \phi_j w_{jk}, k = 1,2,\ldots, s \tag{9}$$

where $w_{jk} = k_0^{jk} + k_1^{jk} x_1 + \ldots + k_r^{jk} x_r$,and s is the number of output variables. It can conclude that

$$\tau_{FNN} = Y = W^T \Phi \tag{10}$$

where

$$W = \begin{bmatrix} K_0^{11} & \cdots & K_0^{u1} & \cdots & K_r^{11} & \cdots & K_r^{u1} \\ \cdots & \cdots & \cdots & \cdots & \cdots & \cdots & \cdots \\ K_0^{1n} & \cdots & K_0^{un} & \cdots & K_r^{1n} & \cdots & K_r^{un} \end{bmatrix}^T$$

$$\Phi = [\phi_1 \quad \cdots \quad \phi_u \quad \cdots\cdots \quad \phi_1 \mathrm{x}_r \quad \cdots \quad \phi_u \mathrm{x}_r]^T$$

The unique feature of the GD-FNN learning algorithm is that the system starts with no hidden units. Fuzzy rules and RBF units can be recruited and deleted dynamically according to their Error Reduction Ratio to the system performance. The weight matrix is calculated based on the Linear Least Square method.

3.3 Stability Analysis of Fuzzy Neural Control System

From (1)-(4), we obtain the error tracking equation as follows:

$$\dot{E} = A^* E + B(\tau^* - \tau_{FNN} - \tau_{PD}) \tag{12}$$

where $A^* = -\begin{bmatrix} 0 & -I \\ k_2 & k_1 \end{bmatrix}, B = \begin{bmatrix} 0 \\ D(q)^{-1} \end{bmatrix}$. Substituting (6) and (11) into (12), we have

$$\begin{aligned}\dot{E} &= AE + B\left[\tau^*_{FNN} - \tau_{FNN} + D(q)KE - K_{PD}E\right] \\ &= AE + B\left[\left(W^*\right)^T \Phi^* - W^T \Phi\right]\end{aligned} \tag{13}$$

where $A = -\begin{bmatrix} 0 & -I \\ D(q)^{-1}K_P & D(q)^{-1}K_V \end{bmatrix}$ W^* and Φ^* are the optimal value of the weight and the regression, $K_{PD} = [K_P \quad K_V]$ is the PD controller gain. We assume that the regression value we obtained from the GD-FNN learning algorithm is the optimal one, $\Phi^* = \Phi$. So (13) can be written as

$$\dot{E} = AE + B\left[\left(W^* - W\right)^T \Phi\right] \tag{14}$$

The adaptive law of W is designed as

$$\dot{W}_i = k\Phi E^T PB_i \quad i = 1,2,\ldots n \tag{15}$$

Where k is a positive constant and P is a symmetric positive definite matrix that satisfies the following relationship:

$$PA + A^T P = -Q \tag{16}$$

Here, Q is a symmetric positive definite matrix. In order to guarantee stability of the control system, the outputs of the FNN must be bounded which requires the weights W to be bounded. Define the constraint set Γ for W as

$$\Gamma = \left\{\|W_i\| \le \|W_0\|\right\} \quad i = 1,2,\ldots n \tag{17}$$

According to the projection algorithm [7], the adaptation law (15) can be modified as

$$\dot{W}_i = \begin{cases} k\Phi E^T PB_i \\ if(\|W_i\| < \|W_0\|) \\ or(\|W_i\| = \|W_0\| and W_i^T \Phi E^T PB_i \le 0) \quad (a) \\ k\left(I - \dfrac{W_i W_i^T}{\|W_i\|^2}\right)\Phi E^T PB_i \\ if(\|W_i\| = \|W_0\| and W_i^T \Phi E^T PB_i > 0) \quad (b) \end{cases} \tag{18}$$

Theorem. Concerning the robot manipulator system represented by (1). If the robust control law of (2) and the adaptive law of (18) are applied, asymptotic stability is guaranteed.

Proof. We consider the Lyapunov function candidate based on (14),

$$V(t) = 0.5k^{-1}tr\left[\left(W^* - W\right)^T\left(W^* - W\right)\right] + 0.5E^T PE \tag{19}$$

Taking the derivative of (19) and using (14) and (16), we have

$$\begin{aligned}\dot{V}(t) &= 0.5\dot{E}^T PE + 0.5E^T P\dot{E} - k^{-1}tr\left[(W^* - W)^T \dot{W}\right] \\ &= -0.5E^T QE + E^T PB(W^* - W)^T \Phi - k^{-1}tr\left[(W^* - W)^T \dot{W}\right]\end{aligned} \tag{20}$$

Under condition 1 of (18), (20) becomes

$$\begin{aligned}\dot{V}(t) &= -0.5E^T QE + E^T PB(W^* - W)^T \Phi - tr\left[(W^* - W)^T \Phi E^T PB\right] \\ &= -0.5E^T QE + E^T PB(W^* - W)^T \Phi - tr\left[E^T PB(W^* - W)^T \Phi\right] \\ &= -0.5E^T QE \le 0\end{aligned}$$

Under condition 2 of (18), (20) becomes

$$\dot{V}(t) = -0.5E^T QE + E^T PB(W^* - W)^T \Phi - k^{-1}\sum_{i=1}^{n}\left[(W^* - W)^T k\left(I - \frac{W_i W_i^T}{\|W_i\|^2}\right)\Phi E^T PB_i\right]$$

$$= -0.5E^T QE - \sum_{i=1}^{n}\left[\left(I - \frac{W_i W_i^T}{\|W_i\|^2}\right)W_i^T \Phi E^T PB_i\right] = -0.5E^T QE \le 0$$

If and only if $E = 0 \bullet V(t) = 0$. Therefore, global stability is guaranteed by the Lyapunov theorem.

4 Simulation Results

Simulations were carried out to verify that the proposed AFNC could compensate for unmodeled disturbances. The manipulator used for the simulation study is a typical two degrees-of-freedom robot. The dynamic equation of the manipulator and the parameters were taken from [8], [9]

$$D(q)\ddot{q} + C(q,\dot{q})\dot{q} + G(q) + F_f(\dot{q}) = \tau - \tau_e$$

where

$$D(q) = \begin{bmatrix} 3.8 + 2\cos q_2 & 0.9 + \cos q_2 \\ 0.9 + \cos q_2 & 0.9 \end{bmatrix} \quad C(q) = \begin{bmatrix} -\dot{q}_2 \sin q_2 & -(\dot{q}_1 + \dot{q}_2)\sin q_2 \\ \dot{q}_1 \sin q_2 & 0 \end{bmatrix}$$

$$G(q) = 0,\ F_f(\dot{q}) = diag[2sig(\dot{q}), 2sig(\dot{q})]\ K_p = diag[25,25],\ K_v = diag[7,7]$$

$$q_d = [0.5\cos t + 0.2\sin 3t, -0.2\sin t - 0.5\cos t]^T$$

and q_d is the desired trajectory. The simulation result was show as follow:

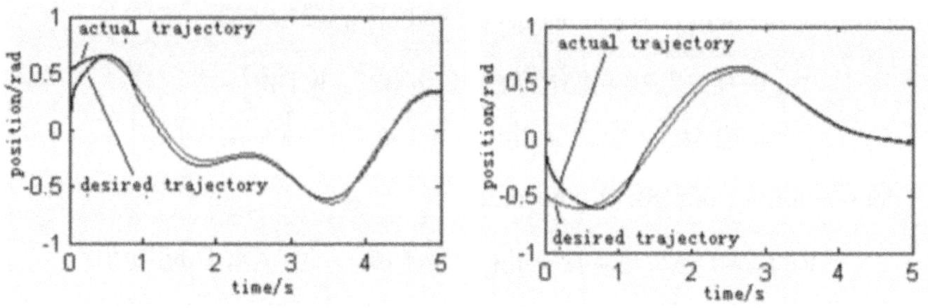

(a) (b)

Fig. 3. The tracking of using PD control arithmetic Tracking of q_1, (b) Tracking of q_2

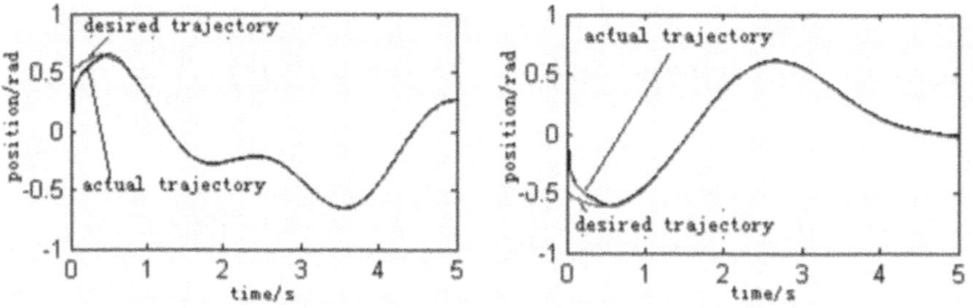

(a) (b)

Fig. 4. The tracking of using this paper's control arithmetic(a) Tracking of q_1, (b) Tracking of q_2

5 Conclusions

In this paper, an adaptive fuzzy neural control scheme using generalized dynamic fuzzy neural network was proposed and its adaptive capability to handle modeling errors and external disturbances was demonstrated. The errors convergence rate with AFNC was found to be fast. Asymptotic stability of the control system is established using the Lyapunov approach. Computer simulation studies of a two-link robot manipulator verify the adaptation and tracking performance of this model-free adaptive control system.

References

1. Hu, s.: Model-following Robust Adaptive Control Based on Neural Networks. Acta Automatica Sinica. **5** (2000) 623-629
2. Hunt, K.J.: Extending The Functional of Equivalence of Radial Basis Function Networks and Fuzzy Inference System. IEEE J. Trans. On Neural Networks **3** (1996) 776-781
3. Chen, C.: Intelligent Process Control Using Neural Fuzzy Techniques. Journal of Process Control **9** (1999) 493-503
4. Lee, M.: An Adaptive Control Method for Robot Manpipulators Using Radial Basis Function Networks. ISIE ,Pusan, Korea. (2001)

5. Menhaj, M.B.: A Novel Neuro-based Model Reference Adaptive Control for A Two Link Robot Arm. IEEE Trans. On Neural Networks **1** (2002) 47-52
6. Chien, C.: A Neural Network Based Learning Controller for Robot Manipulators. Proceeding of the 39th IEEE Conference on Decision and Control Sydney. Australia. (2000)
7. Wu, S., Er, M.J.: A Fast Approach for Automatic Generation of Fuzzy Rules by Generalized Dynamic Fuzzy Neural Networks. IEEE Trans. On Fuzzy System **4** (2001) 578-594
8. Ying, D.: Robust Adaptive Control Strategies for Robot Manipulators with Uncertainties. Xi'an: Xi'an Jiao Tong University. (1998)
9. Meng, Y.: MATLAB 5.X APPLP AND SKILL. Beijing: Science Publishing Company. (1999) 124-133

A 3-PRS Parallel Manipulator Control Based on Neural Network

Qingsong Xu and Yangmin Li

Department of Electromechanical Engineering, Faculty of Science and Technology, University of Macau, Av. Padre Tomás Pereira, Taipa, Macao SAR, P.R. China
{ya47401,ymli}@umac.mo
http://www.sftw.umac.mo/~yangmin/

Abstract. Due to the time-consuming calculation for the forward kinematics of a 3-PRS (prismatic-revolute-spherical) parallel manipulator, neither the kinematic nor dynamic control algorithm can be implemented on real time. To deal with such problem, the forward kinematics is solved by means of artificial neural network (NN) approach in this paper. Based on the trained NN, the kinematic control of the manipulator is carried out by resorting to an ordinary control algorithm. Simulation results illustrate that the NN can approximate the forward kinematics perfectly, which leads to ideal control results of the parallel manipulator.

1 Introduction

A parallel manipulator (PM) typically consists of a moving platform and a fixed based, which are connected by more than one limbs or legs actuating in parallel. Generally, PMs can provide some attractive merits over their serial counterparts in terms of high speed, high rigidity, and high accuracy, etc., which enable them become challenging alternatives for wide applications such as flight simulators, machine tools, ultra-precision instruments, medical devices, and so on [15]. Recently, limited-DOF (degree of freedom) PMs with less than six DOF are drawing attentions of more and more researchers because these limited-DOF PMs own several other additional advantages in terms of total cost reduction in manufacture and operations. As a result, a lot of limited-DOF PMs have been proposed and investigated for various applications [13, 2, 14, 16, 12].

The 3-PRS architecture parallel mechanism possesses three spatial DOF with two rotations about the x and y axes, and one translational motion along the z axis. Thus, this tripod-based PM has potential applications where the orientation and reachable distance in z direction are more important than the translations in x and y directions. In previous works concerning this type of PM, the forward displacement kinematics problem is solved in [11] by using the Sylvester dialytic elimination method, and the compliance analysis of the PM is carried out in [9]. After that, the forward kinematics problem is dealt with by the authors via the Newton iterative algorithm in [10], where the inverse dynamic model is established as well. In addition, the manipulator workspace is generated by adopting the NN approach in [8] successfully.

D. Liu et al. (Eds.): ISNN 2007, Part I, LNCS 4491, pp. 757–766, 2007.

However, up to now, there are no efforts made towards neither kinematic nor dynamic control of such type of PM in the literatures yet. The main reason is that both the forward displacement kinematics (FDK) and inverse dynamics of the manipulator are too complicated to be calculated online in a control procedure, which prohibit further applications of the promising 3-PRS PM. In addition, the FDK has multiple solutions. Recently, NNs are used to solve the FDK problems of PMs and attract many researchers due to their considerable ability to approximate nonlinear maps or functions, and their generalization capacities and structures which make them robust and fault-tolerant in algorithms. And NN has been successfully applied to compute the FDK problems of some kinds of PMs, e.g., [5,6,7].

In this paper, to solve the real time control problem of a 3-PRS PM, the FDK is solved by resorting to the artificial neural network (NN) approach. It is shown that, once properly trained, the NN can approximate the FDK quickly and can be adopted in real-time control of the manipulator.

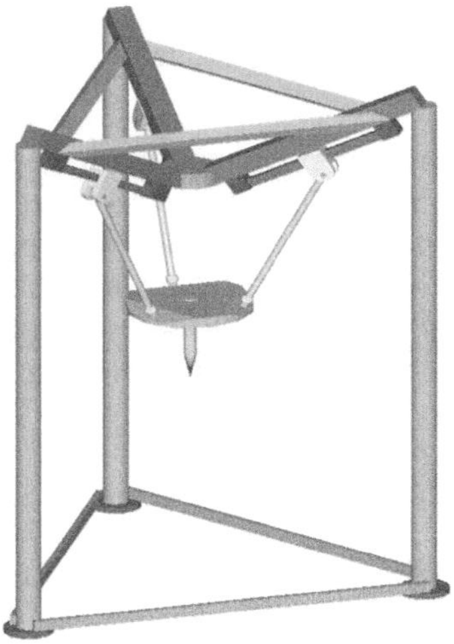

Fig. 1. CAD model of a 3-PRS parallel manipulator

2 Displacement Kinematics Analysis of the Manipulator

The CAD model of a 3-PRS PM is shown in Fig. 1, which is composed of a moving platform and a fixed base, which are connected by three supporting limbs with identical kinematic structure. Each limb consists of a prismatic (P) joint, a revolute (R) joint, and a spherical (S) joint in sequence, where the P joint is actuated by a linear actuator.

As depicted in Fig. 2, a fixed reference frame $O\{x,\ y,\ z\}$ is attached at the centered point O of the fixed base platform $\Delta A_1 A_2 A_3$. Similarly, a moving frame $P\{u,\ v,\ w\}$ is attached at point P which is the centered point of the moving

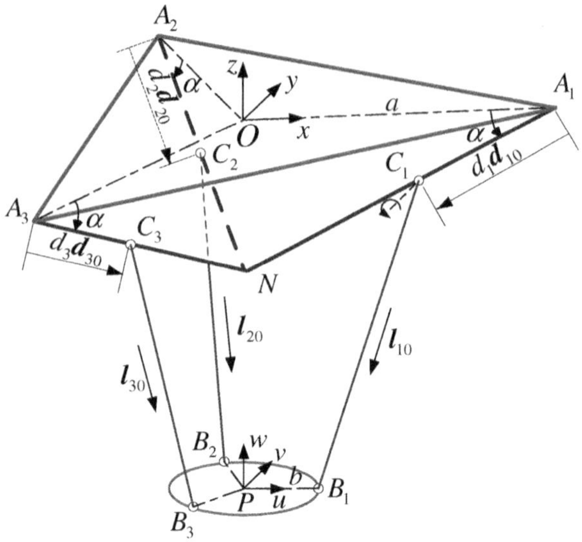

Fig. 2. Schematic diagram of a 3-PRS PM

platform $\Delta B_1B_2B_3$. Without lose of generality, let the x axis point along vector $\overrightarrow{OA_1}$, and the u axis along $\overrightarrow{PB_1}$. Both $\Delta A_1A_2A_3$ and $\Delta B_1B_2B_3$ are designed to be equilateral triangle to yield a symmetrical workspace of the manipulator.

Let $d = [d_1 \ d_2 \ d_3]^T$ be the vector of the three actuated joint variables, and $X_a = [p_x \ p_y \ p_z \ \theta_1 \ \theta_2 \ \theta_3]^T$ be the vector of constrained and unconstrained variables, which describes the pose (position and orientation) of the moving platform with θ_1, θ_2, and θ_3 representing the Z-Y-Z Euler angles. The transformation from the moving frame to fixed frame can be described by a position vector $p = [p_x \ p_y \ p_z]^T$, and a 3×3 rotation matrix ${}^OR_P$, which can be expressed by

$$ {}^OR_P = R_z(\theta_1)\, R_{y'}(\theta_2)\, R_{z''}(\theta_3)\,. \tag{1}$$

Meanwhile, assign u, v, and w be three unit vectors directing along the u, v, and w axes of the moving frame P. Then the rotation matrix can be expressed in terms of the direction cosines of u, v, and w as follows:

$$ {}^OR_P = \begin{bmatrix} u_x \ v_x \ w_x \\ u_y \ v_y \ w_y \\ u_z \ v_z \ w_z \end{bmatrix}. \tag{2}$$

The position vector q_i pointing from O to the i-th S joint B_i can be expressed by

$$ q_i = p + b_i, \tag{3}$$

where $b_i = {}^OR_P\, {}^Pb_i$ and $q_i = [q_{ix} \ q_{iy} \ q_{iz}]^T$, for $i = 1$, 2, and 3.

Considering the mechanical constraints imposed by the R joint, the S joint B_i can only move in a plane defined by the i-th linear actuator and i-th link C_iB_i. Therefore the following three equations hold

$$q_{1y} = 0, \tag{4a}$$
$$q_{2y} = -\sqrt{3}\, q_{2x}, \tag{4b}$$
$$q_{3y} = \sqrt{3}\, q_{3x}. \tag{4c}$$

Substituting the components of q_i from (3) into (4), yields

$$p_y + b u_y = 0, \tag{5a}$$
$$v_x = u_y, \tag{5b}$$
$$p_x = \frac{b}{2}(u_x - v_y), \tag{5c}$$

which impose three constraints on the motion of the moving platform.

2.1 Inverse Displacement Kinematics (IDK) Modeling

The inverse displacement kinematics (IDK) problem solves the actuated variables from a given position and orientation of the output platform.

With reference to Fig. 2, we can obtain that

$$L_i - d_i d_{i0} = l\, l_{i0}, \tag{6}$$

where $L_i = q_i - a_i$ denotes a vector pointing from point A_i to B_i, with q_i expressed by (3).

Then, solving (6) allows the generation of the IDK solutions:

$$d_i = L_i^T d_{i0} \pm \sqrt{(L_i^T d_{i0})^2 - L_i^T L_i + l^2}\ . \tag{7}$$

It is seen that there exist two solutions for each actuator, thus there are total of eight possible solutions for a given platform position and orientation. To enhance the stiffness of the manipulator, only the negative square roots are selected to yield a unique set of solutions where the three legs are inclined inward from top to bottom.

2.2 Forward Displacement Kinematics (FDK) Modeling

Given a set of actuated inputs, the position and orientation of the output platform is solved by the forward displacement kinematics (FDK).

Traditionally, the FDK problem is solved by considering that the geometric distance between two S joints B_i and B_j is equal to a constant, i.e., $\|\overrightarrow{B_i B_j}\| = \sqrt{3}\, b$, i.e.,

$$[q_i - q_j]^T [q_i - q_j] - 3b^2 = 0, \tag{8}$$

where $i \neq j$ and $i, j = 1, 2$, and 3.

Equation (8) represents three nonlinear equations, which can be solved by resorting to Sylvester dialytic elimination method [11] or numerical method such as Newton iterative algorithm [10]. However, these approaches are time-consuming and can not be applied to real-time control of a 3-PRS parallel manipulator.

3 FDK Calculation Via NN Approach

In this section, the NN is applied to calculate the complicated FDK problems of the 3-PRS PM.

3.1 NN Structure and Setup

It has been shown that multi-layer feedforward network with only one layer of hidden units is sufficient to approximate any function provided the activation functions of the hidden units are non-linear [4]. Multi-layer feedforward network with back propagation (BP) learning is used here to solve FDK problem of a 3-PRS PM, and the Levenberg-Marquardt (L-M) training algorithm is adopted to speed up the convergence of the BP learning algorithm, since it has been shown that L-M algorithm is the fastest method for training moderate-sized (up to several hundred weights) feedforward neural networks [3]. The structure of the feedforward neural network is shown in Fig. 3.

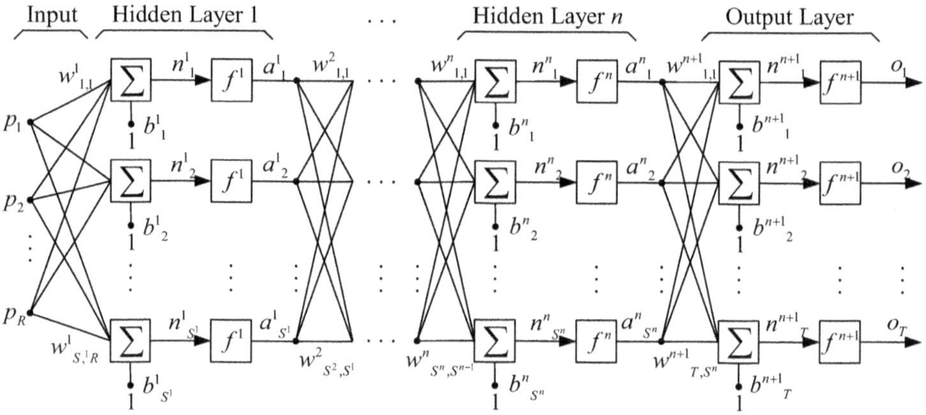

Fig. 3. Multi-layer feedforward NN structure

In view of the constraint conditions (5), one can obtain the relationships between the constraint and unconstraint variables for the a 3-PRS PM:

$$p_x = -\frac{b}{2}(1 - c\theta_2)c(2\theta_1), \tag{9a}$$

$$p_y = \frac{b}{2}(1 - c\theta_2)s(2\theta_1), \tag{9b}$$

$$\theta_3 = -\theta_1, \tag{9c}$$

where c stands for cosine and s stands for sine functions.

The input layer in NN has three nodes which take joint space variables $d = [d_1 \;\; d_2 \;\; d_3]^T$ as the input, and the output layer also has three nodes to give the independent position and orientation variables $X = [p_z \;\; \theta_1 \;\; \theta_2]^T$ of the moving platform.

The number of hidden layers, number of neurons in each layer, transfer function used in each layer, training algorithm, and the performance function are designed parameters for a feedforward neural network. However, there is no basis at all to determine which configuration is the most efficient one for any particular problem. Basically, the numbers of weights and the training time will increase with more neurons in hidden layers. And a large number of hidden units leads to a small error on the training set but not necessarily leads to a small error on the test set. Simulation results in [5] have shown that the performance of the NN with two hidden layers is slightly better. Hence, we use two hidden layers with the neurons of 9 and 11, respectively. That is a moderate-sized network, and the L-M training algorithm is more suitable for training it. The transfer functions used in two hidden layers are all hyperbolic tangent sigmoid transfer function, and the output layer uses linear transfer function. We use mean square error (MSE), which is the average squared error between the network outputs and the target outputs, as the performance function. The weights and biases are randomly initialized. A random generator generates the moving platform vector X within the workspace of the manipulator. The IDK solution described in (7) then gives the actuator variable vector d for any valid X. The data pair is then used to train the network, during which the weights and biases of the network are iteratively adjusted to minimize the network performance function through the errors back propagating process (the back propagation method).

Before the training of the network, the inputs and targets are scaled so that they fall within a specified range, which can make the NN training more efficient. Here, the mean and standard deviations of the training set are normalized by normalizing the inputs and targets so that they will have zero mean and unity standard deviation. After the network is trained, the outputs are then converted into the same units that are used for the original targets.

Table 1. Architectural parameters of a 3-PRS PM

Parameter	Value
a	400 mm
b	200 mm
l	550 mm
α	30 °

3.2 NN Training and Results

Considering a 3-PRS PM with the architectural parameters described in Table 1, a total of 1000 pairs of data uniformly distributed within the manipulator workspace are selected to train the 3-layer feedforward network with the MSE performance of 10^{-5}. The network is created and trained in Matlab environment using the Neural Network Toolbox, and the training curve is plotted in Fig. 4.

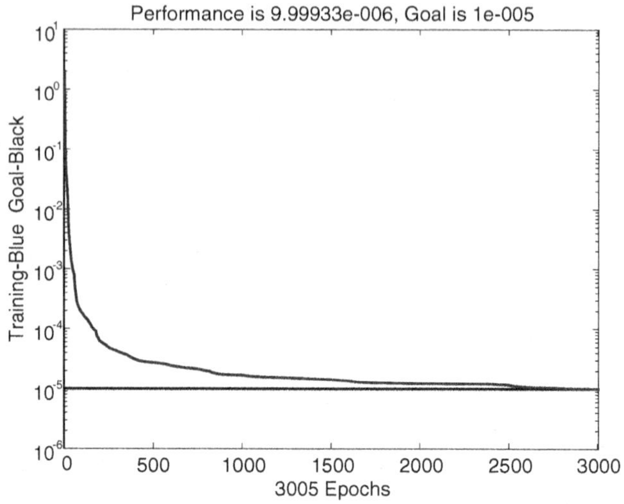

Fig. 4. Training curve of the NN

To illustrate the efficiency of the neural network for solving FDK of the manipulator, the trained NN is employed to the kinematic control of the 3-PRS PM in the following section.

4 Control Algorithm and Results

Controlling a PM to track trajectories in task space is required for a lot of applications such as welding, spray painting, or laser cutting, etc. Various types of control methods can be employed into the control of a PM. Here, for the sake of demonstrating the efficiency of the NN method, we just adopt an ordinary kinematic control algorithm for the tracking control of a 3-PRS PM. The control block diagram is represented in Fig. 5, with reference to which the following equation can be derived.

$$\dot{X} = \dot{X}_d + Ke, \tag{10}$$

that is equivalent to

$$\dot{e} + Ke = 0, \tag{11}$$

where $e = X_d - X$ represents the error vector. It has been shown that if K is a positive definite matrix, the system is asymptotically stable [1].

The kinematic control with the aforementioned control algorithm is performed such that the moving platform can track a trajectory given by the following equations:

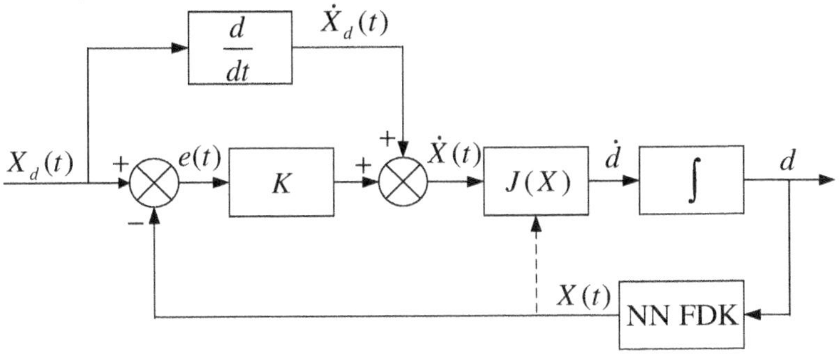

Fig. 5. Block diagram of kinematic control for a 3-PRS PM

$$p_z = -470 + 10\cos(\pi t/10), \tag{12a}$$

$$\theta_1 = \frac{\pi}{2}\sin(\pi t/20), \tag{12b}$$

$$\theta_2 = -0.6\sin(\pi t/20). \tag{12c}$$

The simulation of kinematic control is implemented with Matlab and Simulink software. In the simulation, the matrix gain K is taken as `diag{100, 500, 500}`. Figure 6 illustrates the simulation results of the tracking in task space. It can be observed that the tracking errors are convergent to steady state values after 0.3 second. In addition, it is seen that the steady state errors for p_z, θ_1, and θ_2

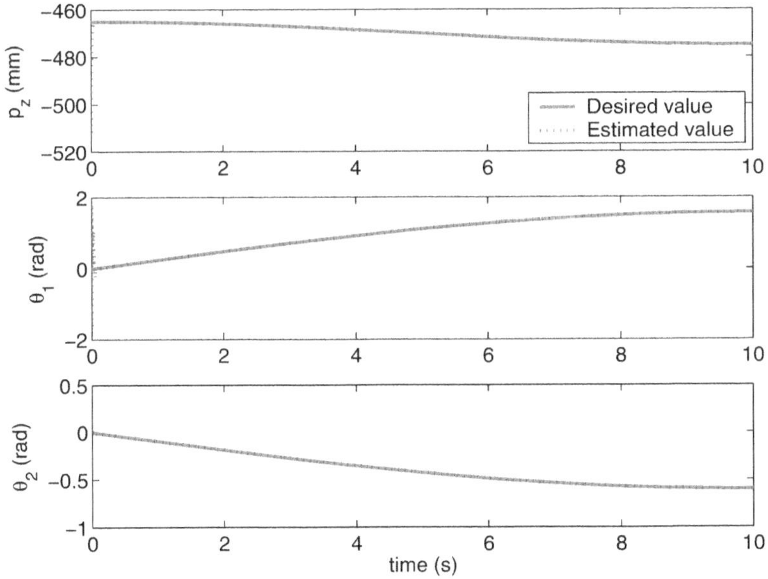

Fig. 6. Tracking results for a 3-PRS PM

are below 0.1 mm, 0.01 rad, and 0.001 rad, respectively, which mean relatively high accuracy.

The results imply that the system is stable enough, and the moving platform can track the desired trajectory well, which also validates the efficiency of NN in solving the FDK of a 3-PRS PM.

5 Conclusions

In this paper, the complicated forward displacement kinematics problem of a 3-PRS spatial parallel manipulator is solved by adopting the neural network method. The trained neural network provides a fast and reliable solution to the FDK and makes the real-time control possible for this type of parallel manipulator. Consequently, it is employed to the kinematic control of the manipulator. The simulation results demonstrate both the well performance of the used control algorithm and the efficiency of NN approach in solving the FDK problem. The results presented in this paper provide a sound approach for real-time control of a 3-PRS parallel manipulator. Moreover, the method can also be applied to other types of parallel manipulator with complicated forward kinematics problem as well. Our further works will concentrate on NN control of the parallel manipulator with the consideration of dynamic characteristics.

Acknowledgments. The authors appreciate the fund support from the research committee of University of Macau under grant no.: RG068/05-06S/LYM/FST and Macao Science and Technology Development Fund under grant no.: 069/2005/A.

References

1. Sciavicco, L., Siciliano, B.: Modeling and Control of Robot Manipulators. McGraw-Hill Book Company,New York (1996)
2. Gosselin, C., Angeles, J.: The Optimum Kinematic Design of a Spherical Three-Degree-of-Freedom Parallel Manipulator. ASME J. Mech. Transm. Autom. Des. **111** (1989) 202-207
3. Hagan, M.T., Menhaj, M.B.: Training Feedforward Networks with the Marquardt Algorithm. IEEE Trans. Neural Networks **5** (1994) 989-993
4. Hornik, K.M., Stinchcombe, M., White, H.: Multilayer Feedforward Networks are Universal Approximators. Neural Networks **2** (1989) 359-366
5. Yee, C.S., Lim, K.B.: Forward Kinematics Solution of Stewart Platform using Neural Network. Neurocomputing **16** (1997) 333-349
6. Lee, H.S., Han, M.C.: The Estimation for Forward Kinematic Solution of Stewart Platform using the Neural Network. Proc. of IEEE/RSJ Int. Conf. on Intelligent Robots and Systems, Kyongju, Korea (1999) 501-506
7. Li, T., Li, Q., Payendeh, S.: NN-based Solution of Forward Kinematics of 3-DOF Parallel Spherical Manipulator. Proc. of IEEE/RSJ Int. Conf. on Intelligent Robots and Systems, Edmonton, Canada (2005) 827-832

8. Cheng, X., Huang, Y.M., Fan, Z.M., Su, J.H.: Workspace Generation of the 3-PRS Parallel Robot based on the NN. Proc. of 1st Int. Conf. on Machine Learning and Cybernetics, Beijing (2002) 2087-2089
9. Xi, F., Zhang, D., Mechefske, C.M., Lang, S.Y.T.: Global Kinetostatic Modelling of Tripod-based Parallel Kinematic Machine. Mech. Mach. Theory **39** (2004) 357-377
10. Li, Y., Xu, Q.: Kinematics and Inverse Dynamics Analysis for a General 3-PRS Spatial Parallel Mechanism. Robotica **23** (2005) 219-229
11. Tsai, M.S., Shiau, T.N., Tsai, Y.J., Chang, T.H.: Direct Kinematic Analysis of a 3-PRS Parallel Mechanism. Mech. Mach. Theory **38** (2003) 71-83
12. Li, Y., Xu, Q.: Kinematic Analysis and Design of a New 3-DOF Translational Parallel Manipulator. ASME J. Mech. Des. **128** (2006) 729-737
13. Clavel, R.: DELTA, A Fast Robot with Parallel Geometry. Proc. of 18th Int. Symp. Industrial Robots, Lausanne, Switzerland (1988) 91-100
14. Tsai, L.W., Walsh, G.C., Stamper, R.E.: Kinematics of a Novel Three DOF Translational Platform. Proc. of IEEE Int. Conf. on Robotics and Automation, Minneapolis, Minnesota (1996) 3446-3451
15. Merlet, J.P.: Parallel Robots. Kluwer Academic Publishers, London (2000)
16. Li, Y., Xu, Q.: A Novel Design and Analysis of a 2-DOF Compliant Parallel Micromanipulator for Nanomanipulation. IEEE Trans. Automation Science and Engineering **3** (2006) 248-254

Neural Network Based Kinematic Control of the Hyper-Redundant Snake-Like Manipulator

Jinguo Liu[1,3], Yuechao Wang[1], Bin Li[1], and Shugen Ma[1,2]

[1] Robotics Laboratory of Chinese Academy of Sciences, Shenyang Institute of Automation, Shenyang, China
{liujinguo,ycwang,libin,sgma}@sia.cn
[2] Center for Promotion of the COE Program, Ritsumeikan University, Shiga-ken, Japan
shugen@fc.ritsumei.ac.jp
[3] Graduate School of Chinese Academy of Sciences, Bejing, China

Abstract. In a sinusoid like curve configuration, the snake-like manipulator (also called snake arm) has a wide range of potential applications for its redundancy to overcome conventional industrial robot's limitation when carrying out a complex task. It can perform many kinds of locomotion like the nature snake or the animal's tentacle to avoid obstacles, follow designated trajectories, and grasp objects. Effectively control of the snake-like manipulator is difficult for its redundancy. In this study, we propose an approach based on BP neural network to kinematic control the hyper-redundant snake-like manipulator. This approach, inspired by the Serpenoid curve and the concertina motion principle of the nature snake, is completely capable of solving the control problem of a planar snake-like manipulator with any number of links following any desired direction and trajectory. With shape transformation and base rotation, the manipulator's configuration changes accordingly and moves actively to perform the designated tasks. By using BP neural networks in modeling the inverse kinematics, this approach has such superiorities as few control parameters and high precision. Simulations have demonstrated that this control technique for the snake-like manipulator is available and effective.

1 Introduction

The bio-inspired snake-like robots have attracted a great deal of attention in recent years [1]-[9]. They can be mainly classified into two categories: the snake-like mobile robot and the snake-like manipulator with fixed base. The snake-like manipulator, also called snake arm, have been potentially applied in many kinds of missions such as welding, painting, assembling, space exploration, grasping, and deep sea exploring [1,3,8,9]. Given a desired workspace trajectory, the task to find out the corresponding joint space trajectory is a complex problem since the snake-like manipulators usually have more than the least necessary degrees of freedom (DOF) and their inverse kinematics' problems have multiple or infinite number of solutions. Moreover, the inverse kinematics equations are usually nonlinear and are difficult to find closed-form solutions. In the initial research, the Jacobian pseudo inverse approach had been widely applied [10]. Whereas in these techniques there are a lot of matrix computations since

D. Liu et al. (Eds.): ISNN 2007, Part I, LNCS 4491, pp. 767–775, 2007.

the complexity increases greatly with the redundancy increasing. Recently, neural networks (NN) have been successfully applied in intelligent control, whose unique capacities and structures make them robust and fault-tolerant in algorithms, which also are able to solve the problem difficult to handle before, for instance, the highly nonlinear problems. The properties of NN make them so promising that they are widely applied to robotic control problems [11]-[16].

Since the hyper-redundant manipulators are analogous in shape and operation to snakes, elephant trunks and tentacles, many shape constrained control techniques have been proposed to turn the hyper-redundant manipulator into non-redundant manipulators according to the task's need and then solve the subsystems' kinematics and dynamics. In [3], Chirikjian proposed the backbone curve to control the hyper-redundant manipulator. Kobyashi also adopted the shape control technique to control the continuous manipulator [17]. Ma efficiently controlled the hyper-redundant manipulator by Serpenoid curve [18]. In this paper, we have proposed an approach based on BP neural networks and the principle of snake's concertina locomotion to control the hyper-redundant manipulators. The rest of this paper is organized as follows. We first discuss the kinematics of the snake-like manipulator in Section 2. Next we model the inverse kinematics of the manipulator based on BP neural network in Section 3. In section 4, several simulations have been provided to validate this control technique. Finally, we conclude this paper in the Section 5.

2 Kinematics of the Snake-Like Manipulator

The nature snake can adopt manifold gaits such as serpentine motion, concertina motion, side-winding motion, rectilinear motion, pushing motion, jumping motion and grasping under different environments [1]. During these gaits, the concertina motion has the best precision in object achieving mission [6]. In this paper, we concentrate on application of the concertina motion in snake-like manipulator's control.

2.1 Kinematics of Snake-Like Robot in Concertina Motion

Based on experiment and observation over the nature snake, Prof. Hirose proposed Serpenoid curve (as shown in Fig.1 (a)) which changes like a sine wave along the central axis to describe the winding motion [1]. Having these characteristics of catholicity, ground adaptability and high efficiency, concertina motion (as shown in Fig.1 (b)) is another basic gait of snake. A demonstration of the snake-like robot in concertina motion is given in Fig.1(c). In this description, the muscle processes the constringency and extension regularly like sine wave. It matches snake's usual motion gait. The concertina motion of Serpenoid curve can be given by the curvature function

$$\rho(s_p) = -\frac{2K_n\pi\alpha}{L}\sin(\frac{2K_n\pi s_p}{L}) \tag{1}$$

where K_n gives number of the S-shape, α is a periodically changeable winding angle, L is the whole length of the robot body, and s_p is the body length along the body curve.

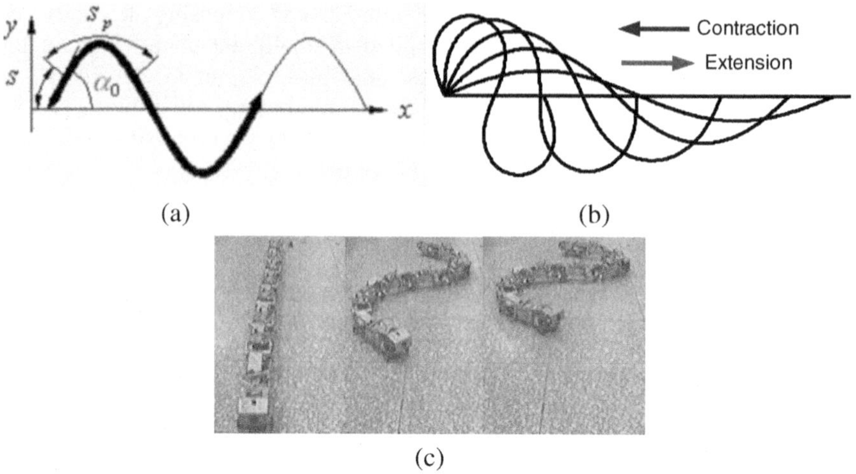

Fig. 1. (a) Serpenoid curve of the snake-like robot. (b) Conceptual demonstration of the concertina motion. (c) The snake-like robot in concertina motion.

Relative rotation angle of the joints which indicates the locomotion range can be got by integral calculus of the curvature

$$\theta_i(s) = \int_{s+s_{pi-1}+\frac{1}{2}l}^{s+s_{pi}+\frac{1}{2}l} \rho(u)du = -2\alpha \sin(\frac{K_n\pi}{n})\sin(\frac{2K_n\pi}{L}s + \frac{2K_n\pi}{n}i) \tag{2}$$

where s is the ideal displacement along the tail, l is the single unit length, and i is the unit number respectively. While α changes, the robot's configuration varies correspondingly. And locomotion velocity can be altered by α 's change during a locomotion period. To get the relationship more directly, we convert (1) into 2D-coordinate as follows description

$$x(s,\alpha) = \int_0^s \cos(\alpha\cos(\frac{2K_n\pi u}{L}))du \tag{3}$$

$$y(s,\alpha) = \int_0^s \sin(\alpha\cos(\frac{2K_n\pi u}{L}))du \tag{4}$$

It can be seen from (3) and (4) that the end of the curve is decided by s , K_n and α , while the shape or the amplitude of the curve is mainly decided by α .

2.2 Kinematics of Snake-Like Manipulator in Concertina Motion

As shown in Fig.2 (a), a snake-like manipulator has a lot of links with the same length. In the concertina motion as show in Fig2. (b), the end-effector of the robot is kept in the horizontal line. With the base rotation, the snake-like manipulator has enough flexibility in the planar space. It is composed of a serial chain of links where the internal variables $\theta_i (i = 1 \sim n)$ are the relative angles between adjacent links given by (2). The absolute link angles are given by

$$\psi_i = \sum_{j=1}^{i} -2\alpha \sin(\frac{K_n \pi}{n}) \sin(\frac{2K_n \pi}{n} j) + \theta_{rb} \quad (i = 1 \sim n) \tag{5}$$

where θ_{rb} is the base rotation angle.

The end effector's position is given by

$$\begin{pmatrix} x_n \\ y_n \end{pmatrix} = \left(\sum_{i=1}^{n} l \cos \psi_i , \sum_{i=1}^{n} l \sin \psi_i \right)^T \tag{6}$$

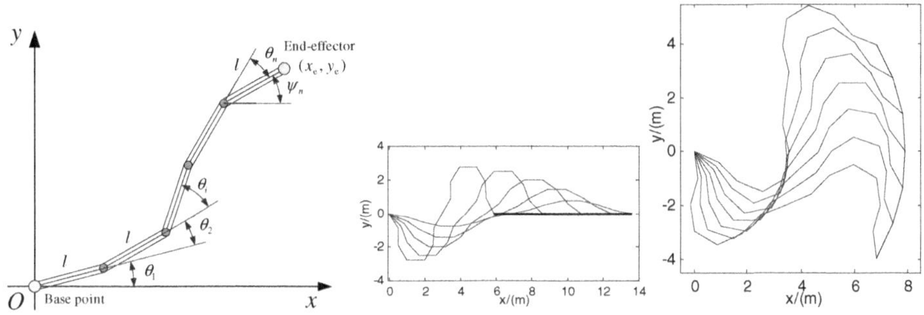

(a) (b) (c)

Fig. 2. (a) Representation of the snake-like manipulator. (b) The snake-like manipulator in concertina motion. (c) The snake-like manipulator in base rotation.

To sum up, the end-effector's position and orientation are decided by three parameters: the extension coefficient α, the base rotation angle θ_{rb} and the number of links *n*. To get a stable and safe kinematics performance, α and θ_{rb} are planed in the same manner by

$$\alpha(t) = \alpha_{ei} + (\alpha_{ef} - \alpha_{ei}) \times (1 + \sin((t - t_{ei})\pi / (t_{ef} - t_{ei}) - \pi / 2)) / 2 \tag{7}$$

$$\theta_{rb}(t) = \theta_{ei} + (\theta_{ef} - \theta_{ei}) \times (1 + \sin((t - t_{ei})\pi / (t_{ef} - t_{ei}) - \pi / 2)) / 2 \tag{8}$$

where α_{ei} is initial α, α_{ef} is final α, θ_{ei} is initial θ_{rb}, θ_{ef} is final θ_{rb}, t_{ei} is initial time, t_{ef} is final time during extension respectively. In (7), it aims to guarantee that $\dot{\alpha}(t)$ is zero at the initial and the final time of an extension/contraction stage. And in (8), it aims to guarantee that $\dot{\theta}_{rb}(t)$ is zero at the initial and the final time of a rotation stage [15].

When the snake-like manipulator in extension/contraction and rotation, the velocities of the manipulator joints are given by

$$\dot{\varphi}_1 = \frac{d\theta_1}{d\alpha} \frac{d\alpha}{dt} + \frac{d\theta_{rb}}{dt} \tag{9}$$

$$\dot{\varphi}_i = \frac{d\theta_i}{d\alpha}\frac{d\alpha}{dt} \quad (i = 2 \sim n) \tag{10}$$

where $d\alpha / dt$ and $d\theta_{rb} / dt$ come from planning the extension/ contraction and rotation. The angle accelerations come from differential of equation (9) and (10).

3 BP Neural Network Based Inverse Kinematics Solution

To control the hyper-redundant snake-like manipulator's end effector to realize the tasks such as point to point and path following, we have to calculate all the joints' relative angles, that is, to solve the inverse problem basically. According to the shape control technique presented above, the inverse kinematics of can be given by

$$(\alpha, \theta_{rb}) = f(x_n, y_n, K_n, n) \tag{11}$$

where *f*(.) denotes the inverse kinematics function of the manipulator. It is apparent that the equation (11) has indeterminate solutions. This inverse kinematics problem is difficult because it is nonlinear and cannot be solved directly.

Nowadays, artificial neural networks are studied and applied in various areas such as computer science, cognitive science, engineering, economics, medicine, etc. The back propagation (BP) algorithm is on of the most popular choices in engineering applications. As a neural network model of extensive application, BP algorithm is attracting more and more attention, especially in the field of approximation of nonlinear function and pattern recognition. The application of BP algorithm can be simply divided into three stages: modeling stage, training stage and calculating stages. The modeling stage is to use the BP neural network to model the relactionship between the input and the output acoording to the experience which usually are original data. The

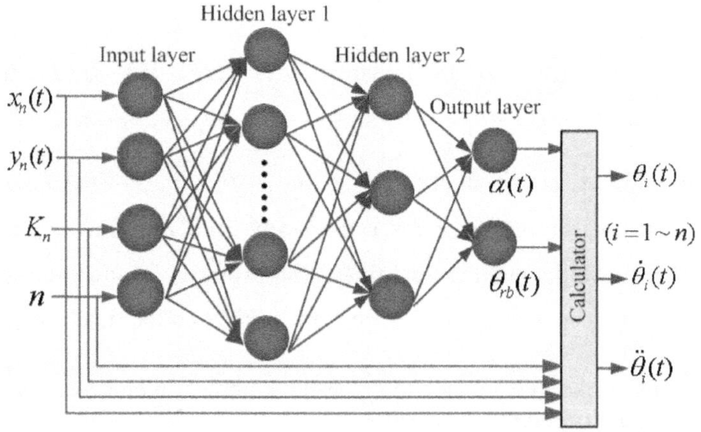

Fig. 3. Architecture of the BP neural networks based inverse kinematics controller

training stage is composed of forward direction calculation and reverse error correction, while calculation stage only includes the forward direction calculation process in the learning stage. Although training stage of BP algorithm needs many samples and consumes a long time, the efficiency of problem resolving is satisfactory if learning is successfully completed. In this paper, we construct a multiple BP neural networks kinematics controller as shown in Fig3.

The principles of BP neural networks have been detailedly introduced in [19]. The architecture of a BP network refers to the way it decodes information, that is the direction of information during recall. In a BP neural network the nodes are organized in the input layer, the hidden layer, and the output layer. The input layer acts as an input data holder that distributes the input to the first hidden layer. The outputs from the first hidden layer then become the inputs to the second layer and so on. The last layer acts as the network output layer. The BP networks are typical multi-layered perceptron feed forward neural networks with one or more hidden layers. Cybenko [20] and Funahashi [21] have proved that the multi-layered perceptron network is a general function approximator and that one hidden layer networks will always be sufficient to approximate any continuous function up to certain accuracy. According to [22] and [23], the functions of the perceptron in the four-layer BP neural networks can be given as follows. The output of the j-th neuron of the k-th hidden layer is given by

$$v_j^k(t) = F\left(\sum_{i=1}^{n_{k-1}} w_{ij}^k v_i^{k-1}(t) + b_j^k\right) \quad (1 \le j \le n_k) \tag{12}$$

And if the m-th layer is the output layer then the output of the u-th neuron of the output layer is given by

$$\hat{y}_u(t) = \sum_{i=1}^{n_{m-1}} w_{ij}^m v_i^{m-1}(t) \quad (1 \le u \le n_o) \tag{13}$$

where n_k , n_o w's, b's and $F(.)$ are the number of neurons in k-th layer, number of neurons in output layer, weights, thresholds and an activation function respectively [22].

The activation function F (.) is selected to be

$$F(v(t)) = \frac{1}{1+e^{-v(t)}} \tag{14}$$

The weights w_i and threshold b_j are unknown and should be selected to minimize the prediction errors defined as

$$\varepsilon(t) = y(t) - \hat{y}(t) \tag{15}$$

where y (t) is the actual output and $\hat{y}(t)$ is the network output [23].

For an n-link snake-like manipulator to carry out a designated mission, we use the forward kinematics data to train the neural networks BP off-line, and then use the trained BP neural networks model to control the correspondence manipulator on-line. As shown in Fig.3, the BP neural networks based inverse kinematics controller includes a calculator which synthesizes the equations from (5) to (10) following the output of neural networks.

4 Simulations

As mentioned previously, hyper-redundant manipulator has several potential advantages over non-redundant manipulator. The extra degrees of the freedom can be used to achieve some special goals. In this paper, we mainly discussed the end-effector of the manipulator has position requirement. For any n-link snake-like manipulator, we use $K_n = 1:0.1:2$, $\alpha = -1{:}0.01{:}1$ and $\theta_{rb} = 0 \sim 2\pi$ to train the neural network offline. Simulations of different operations with trajectory following of the end-effector are given in Fig.4. After 10000-epoch training of the neural networks model, the position errors of the end-effectors in Fig.4 range from 0.006m to 0.013m in x-axis, and from 0.015m to 0.024m in y-axis. It can be seen that the manipulator shows

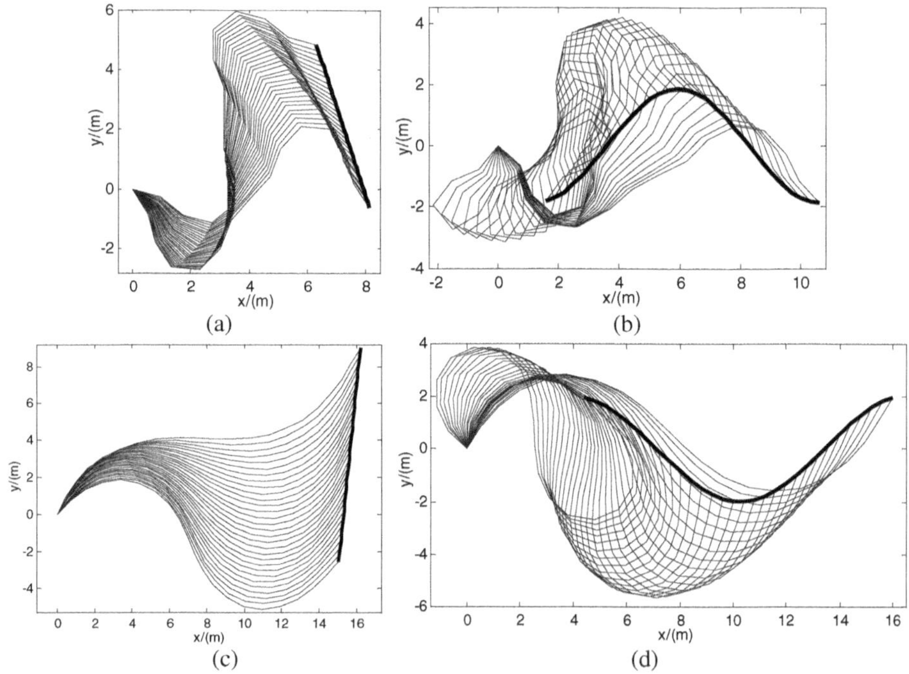

Fig. 4. Simulations of snake-like manipulators in trajectory following. (a) A 14-link snake-like manipulator carries out a trajectory following operation along an inclined line; (b) A 14-link snake-like manipulator carries out a trajectory following operation along a sine curve; (c) A 20-link snake-like manipulator carries out a trajectory following operation along an inclined line; (d) A 20-link snake-like manipulator carries out a trajectory following operation along a sine curve.

excellent flexibility of its end-effector. The BP neural network controller, simple as it is, has stable and high precision solution in the inverse kinematics solution. With shape transformation and base rotation, the manipulator's end-effector moves actively to follow the desired path. It is also available for the snake-like mobile robot with part of its body fixed to the ground and the rest controlled by this approach.

5 Conclusions and Future Works

Based on BP neural networks and shape control technique, the approach presented in this paper can kinematic control the end-effector's position of the hyper-redundant snake-like manipulator with few control parameters and high precision. This approach has been demonstrated by several simulations and it has the potential application in more complicated environment. It can be used in two cases. It can be efficiently used to obtain the inverse kinematics solution of the planar snake-like manipulator with any number of links following any desired path or trajectory. Moreover, it is also available for the snake-like mobile robot when it carries out the nearby object achieving mission with part of its body fixed to the ground and the rest controlled by this approach. The approach is proposed to kinematic control the snake-like manipulator. When the manipulator joint increases, joint torque will increase quickly. How to dynamic control and apply it are desired in the future study.

Acknowledgment

This research is supported in part by China National High-Technology 863 Program No.2001AA422360, the Chinese Academy of Sciences Advanced Manufacturing Technology R&D Base Fund No.F050103, and the GUCAS-BHPB Billiton Scholarship. The authors would like to thank the anonymous reviewers for their kind and insightful comments.

References

1. Hirose, S.: Biologically Inspired Robot—Snake-like Locomotors and Manipulators, Oxford University Press (1993)
2. Burdick, J., Radford, J., Chirikjian, G. S.: A Side-Winding Locomotion Gait for Hyper-Redundant Robots. Advanced Robotics **9** (1995) 195-216
3. Chirikjian, G. S., Burdick, J.W.: The Kinematics of Hyper-Redundant Robot Locomotion. IEEE Transactions on Robotics and Automation **11** (1995) 781-793
4. Dowling, K.: Limbless Locomotion: Learning to Crawl. In: Proc. of IEEE International Conference on Robotics and Automation, Detroit, MI (1999) 3001-3006
5. Ma, S.: Analysis of Creeping Locomotion of a Snake-Like Robot. Advanced Robotics **15** (2001) 205-224.
6. Liu, J., Wang, Y., Li, B., et al.: Path Planning of a Snake-Like Robot Based on Serpenoid Curve and Genetic Algorithms. In: Proc. of the 5th World Congress on Intelligent Control and Automation, Hangzhou (2004) 4860-4864

7. Collection of Snake-like Robots. http://www.ais.fraunhofer.de/~worst/snake-collection.htm. October 2006
8. Buckingham, R.: Snake-Arm Robots for Flexible Delivery. Insight **44** (2002) 150-152
9. Snake-Arm Robots. http://www.ocrobotics.com/snakearms/index.html. October, 2006.
10. 10.Baker, D. R., Wampler, C. W.: On the Inverse Kinematics of Redundant Manipulators. International Journal of Robotics Research **7** (1988) 3-21
11. 11.Liu J.,Wang, Y., Ma, S., et al.: RBF neural network based shape control of hyper-redundant manipulator with constrained end-effector. J. Wang et al. (Eds.): ISNN 2006, LNCS **3972** (2006) 1146-1152
12. 12.Zhang, Y., Wang, J., Xu, Y.: A Dual Neural Network for Bi-criteria Kinematic Control Redundant Manipulators. IEEE Transactions on Robotics and Automation **18** (2002) 923-931
13. Xia, Y., Wang, J., Fok, L. M.: Grasping Force Optimization of Multi-Fingered Robotic Hands Using a Recurrent Neural Network. IEEE Transactions on Robotics and Automation **20** (2004) 549-554
14. Yang, Simon X., Meng, M.: Neural Network Approaches to Dynamic Collision-Free Trajectory Generation. IEEE Transactions on Systems, Man, and Cybernetics, Part B **31** (2001) 302-318
15. Liu, J., Wang, Y., Ma, S., Li, B.: Shape Control of Hyper-Redundant Modularized Manipulator Using Variable Structure Regular Polygon. In: Proc. of IEEE/RSJ International Conference on Intelligent Robots and Systems, Sendai (2004) 3924-3929
16. Liu, Y., Li, Y.: Sliding Mode Adaptive Neural-network Control for Nonholonomic Mobile Modular Manipulators. Journal of Intelligent & Robotic Systems **44** (2005) 203-224
17. 17.Kobyashi, H., Ohtake, S.: Shape Control of Hyper Redundant Manipulator. In: Proc. of IEEE International Conference on Robotics and Automation, Nagoya (1995) 2803-2808
18. 18 Ma, S., Kobayashi, I., Hirose, S., Yokoshima K.: Control of a Multijoint Manipulator: Moray Arm, IEEE/ASME Trans. on Mechatronics **7** (2002) 304-317
19. Rumelhart, D. E., McClelland, J. L.: Parallel Distributed Processing: Explorations in the Microstructure of Cognition, I & II, MIT Press, Cambridge, MA (1986)
20. 20.Funahashi, K.: On the Approximate Realisation of Continuous Mappings by Neural Networks . Neural Networks **2** (1989) 183-192
21. 21.Cybenko, G.: Approximations by Superposition of a Sigmoidal Function. Mathematics of Control, Signal and Systems **2** (1989) 303-314
22. 22.Esugasini, S., Mashor, MY., Mat-Isa, NA., Othman, NH.: Performance Comparison for MLP Networks Using Various Back Propagation Algorithms for Breast Cancer Diagnosis. KES 2005, LNAI **3682** (2005) 123-130
23. 23.Mashor, MY.: Hybrid Multilayered Perceptron Networks. International Journal of System Science **31** (2000) 771-785

Neural Network Based Algorithm for Multi-Constrained Shortest Path Problem

Jiyang Dong[1,2], Junying Zhang[2], and Zhong Chen[1]

[1] Department of Physics, Fujian Engineering Research Center for Solid-State Lighting, Xiamen University, Xiamen 361005, P.R. China
[2] National Key Laboratory for Radar Signal Processing, Xidian University, Xi'an 710071, P.R. China
jydong@xmu.edu.cn

Abstract. Multi-Constrained Shortest Path (MCSP) selection is a fundamental problem in communication networks. Since the MCSP problem is NP-hard, there have been many efforts to develop efficient approximation algorithms and heuristics. In this paper, a new algorithm is proposed based on vectorial Autowave-Competed Neural Network which has the characteristics of parallelism and simplicity. A nonlinear cost function is defined to measure the autowaves (*i.e.*, paths). The *M*-paths limited scheme, which allows no more than *M* autowaves can survive each time in each neuron, is adopted to reduce the computational and space complexity. And the proportional selection scheme is also adopted so that the discarded autowaves can revive with certain probability with respect to their cost functions. Those treatments ensure in theory that the proposed algorithm can find an approximate optimal path subject to multiple constraints with arbitrary accuracy in polynomial-time. Comparing experiment results showed the efficiency of the proposed algorithm.

1 Introduction

Providing Quality-of-Service (QoS) guarantees in packet networks gives rise to several challenging issues. One of them is how to determine a feasible path that satisfies a set of constraints while maintaining high utilization of network resources. The latter objective implies the need to impose an additional optimality requirement on the feasibility problem. This can be done through a primary cost function according to which the selected feasible path is optimization [1,2]. In general, multi-constrained path selection, with or without optimization, is an NP-complete problem that cannot be exactly solved in polynomial time [3]. Heuristics and approximation algorithms with polynomial- and pseudo-polynomial-time complexities are often used to deal with this problem. One common heuristic approach is to find the *k*-shortest paths with respect to a cost function defined based on the link weights and the given constraints, hoping that one of these paths is feasible [4]. Increasing *k* improves the performance of this approach, but the computational cost becomes excessive that cannot be used for online network operation. Another approach is to exploit the dependencies among the constraints, and to solve the path selection problem assuming specific scheduling schemes at network routers [5]. Specifically in QoS

D. Liu et al. (Eds.): ISNN 2007, Part I, LNCS 4491, pp. 776–785, 2007.

routing, if weighted fair queuing service discipline is being used and the constraints are bandwidth, queuing delay, delay-jitter, and loss, then the problem can be reduced to standard shortest path problem by expressing all the constraints in terms of bandwidth. As we can see, this approach only deals with special cases of the problem.

Jaffe proposed a new solution called multi-label routing for the MCSP problem [6]. It is simple, easy to implement. And most importantly, it can find the approximate optimal path with multiple constraints. Unfortunately, the computational complexity is also exponentially increased with the network scale for the reasons of too many labels to be reserved and too many vectors to be compared. Some modifications are suggested for the multi-label method to reduce the computational and space complexity [7], *e.g.*, limiting the total labels of each node, and the algorithmic loops are reduced to about 1/3 of the original multi-label algorithm. However, a large number of labels are needed for the improved multi-label routing algorithm to ensure the approximate optimal paths can be found. Therefore, the computational and space complexity are not reduced essentially.

In this paper, the Autowave-Competed Neural Network (ACNN) [8,9] is vectorized and applied successfully to the MCSP problem. The *M*-paths limited scheme is adopted to reduce the computational and space complexity. All the autowaves propagating to a neuron have to compete with the paths reserved on the neuron's threshold, and only *M* autowaves (*i.e.*, paths) can survive the competition. The winners will be reserved for the next competition by replacing the old paths on the neuron's threshold, while the losers will be discarded. At the same time, the winners will propagate forward to its adjacent neuron. A nonlinear cost function is defined to measure the paths. However, differing from the traditional MCSP algorithms which focus on finding the paths with minimum cost function, the proposed algorithm uses the cost function to build on a proportional path selection scheme. The discarded paths can revive with certain probability. So the optimal path can also be found in theory by the proposed algorithm even with small *M*. Furthermore, the vectorial ACNN based algorithm is parallel in architecture and in running mode.

The rest of paper is organized as follows. Section 2 gives the definition of MCSP problems. Section 3 introduces the scalar ACNN for traditional shortest path problem (no constraint). The vectorial ACNN, the nonlinear cost function and the new algorithm of MCSP is defined in section 4. Simulation results are presented in section 5 and some concluding remarks are drawn in section 6.

2 Problem Definition

Consider a network that is represented by a directed graph $G = (V,E)$, where V is the set of nodes which represent switches, routers, and hosts and E is the set of edges which represent communication links. Each edge $(i,j) \in E$ is associated with a primary cost parameter $c(i,j)$ and K QoS weights, $\omega_k(i,j),\ k = 1,2,\cdots,K$, all parameters are non-negative. Given K constraints, $C_k,\ k = 1,2,\cdots,K$. The MCSP problem is to find a path P from a source node s to the destination node d such that [1]:

$$\text{(i)}\ \omega_k(P) = \sum_{(i,j)\in P} w_k(i,j) \le C_k \qquad for\ k = 1,2,......,K$$

$$\text{(ii)}\ c(P) = \sum_{(i,j)\in P} c(i,j) \text{ is minimized over all feasible paths} \tag{1}$$

satisfying (i)

For the simplicity of expression and computation, the cost parameter can be regarded as one of the constraints on the link, *e.g.*, let $\omega_0(i,j) = c(i,j)$. Then the MCSP problem is to find a path satisfied all K+1 constraints and with the minimum cost from node s to node d. To solve the MCSP problem, one can find firstly all the paths satisfied the constraints, then chooses the one with minimum cost from those paths [6]. This paper also treats the cost parameter as a constraint of the link.

Each link in the network is associated with multiple parameters which can be roughly classified into additive and non-additive [10]. For the additive parameters (e.g., delay, jitter, administrative weight), the cost of an end-to-end path is given by the sum of the individual link values along that path. In contrast, the cost of a path with respect to a non-additive parameter, such as bandwidth, is determined by the value of that constraint at the bottleneck link. It is known that constraints associated with non-additive parameters can be easily deal with a preprocessing step by pruning all links that do not satisfy these constraints [5]. Hence, in this paper we will mainly focus on additive QoS parameters and assume that the optimality of a path is evaluated based on an additive parameter.

3 ACNN for Shortest Path Problem

The Shortest Path (SP) problem is well-documented. Many practical algorithms have been developed for shortest path problem. Recently, we proposed a neural network model called autowave-competed neural network (ACNN) for the SP problem [8,9]. The ACNN neuron is consisted of three parts, *i.e.*, the minimum selector, the autowave generator and the threshold updater, see Fig.1.

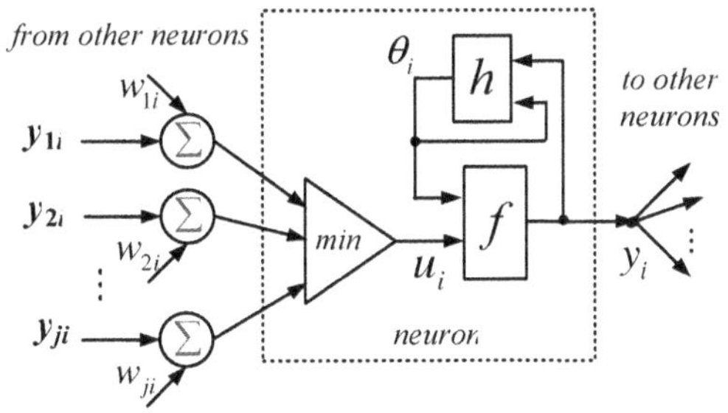

Fig. 1. Neuron model of ACNN

The ACNN neuron can be described with the following equations,

$$Z_i(t) = \{ j \mid w_{ji} \neq \infty \quad and \quad y_j(t-1) > 0 \} \tag{2}$$

$$u_i(t) = \begin{cases} 0 & Z_i(t) = \phi \\ \min\limits_{j \in Z_i(t)} (y_j(t-1) + w_{ji}) & otherwise \end{cases} \tag{3}$$

$$y_i(t) = f[u_i(t), \theta_i(t-1)] = \begin{cases} u_i(t) & u_i(t) < \theta_i(t-1) \\ 0 & otherwise \end{cases} \tag{4}$$

$$\theta_i(t) = h[y_i(t), \theta_i(t-1)] = \begin{cases} \theta_i(t-1) & y_i(t) = 0 \\ y_i(t) & otherwise \end{cases} \tag{5}$$

where i is the index of neuron, t is the time (or says the iterations). $u_i(t)$, $\theta_i(t)$ and $y_i(t)$ are the internal activity, the threshold and the output of neuron i at time t respectively. w_{ij} is the connection weight from neuron i to neuron j. $z_i(t)$ is the set of neurons which fired at time t and is reachable to neuron i.

When applied to the SP problem, an ACNN isomorphic to the weighted graph G should be constructed, *i.e.*, each node of G is corresponding to a unique neuron of the network, and w_{ij} is associated with the weight of the edge (i, j) in G, see Fig. 2. All neurons are initialized with infinite threshold and zero-internal-activity. Fire the source neuron to run the network, and the firing would inspire some autowaves

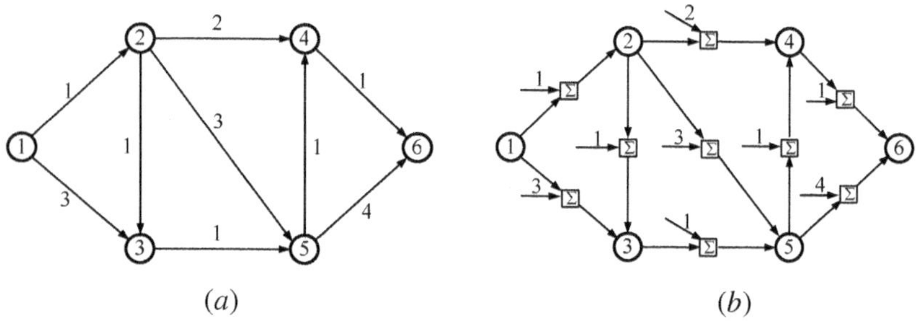

Fig. 2. ACNN topology for SP problem. (*a*) A weighted digraph. The circles with numbers inner are the vertexes, and the numbers on edges are the costs associated with the corresponding edges. (*b*) The ACNN model for the SP problem of the graph shown in (*a*). The circles with numbers inner are the neurons, and the squares with "Σ" inner are the summators on the corresponding links.

propagating through the whole network. When passed through the neuron i, the traveling distance of an autowave would be recorded on the threshold $\theta_i(t)$ if it is the shortest. All neurons decrease their thresholds progressively until the network stops.

When the network stops, the threshold θ_i is corresponding to the distance of the shortest path from the source neuron to the neuron i.

The ACNN based shortest path algorithm is parallel, non-parameter, and flexible, which can easily be modified to suit for the other problems concerned with shortest path [9]. Furthermore, it is suitable for large-scale network. Reader can refer to the reference [8] for more details about ACNN.

4 Vectorial ACNN for MCSP Problem

4.1 Vectorial ACNN

In order to solve the MCSP problem, a vectorial ACNN must be established, in which the connection weight, the threshold, the internal activity and the output of the neurons should be vectored.

(1) The vector-formed connection weight from neuron i to the reachable neuron j should be $\vec{\mathbf{w}}_{ij} = [\omega_1(i,j), \omega_2(i,j), \cdots, \omega_K(i,j)]$, where $\omega_k(i,j), k=1,\cdots,K$ is the kth QoS parameters on the link (i, j).

(2) A path $(s \to \cdots \to i \to j \to \cdots)$ can be written as a vector-formed $\mathbf{P} = (\vec{\mathbf{p}}, \vec{\mathbf{d}})$, where $\vec{\mathbf{d}} = (d_1, d_2, \cdots, d_K)$ and $\vec{\mathbf{p}} = (s, \cdots, i, j, \cdots)$ are the total weights and the node sequence of the path respectively, and $d_k = \sum_{(i,j)\in \mathbf{P}} \omega_k(i,j)$, $k = 1, \cdots, K$.

(3) If an autowave with traveling path $\mathbf{P} = (s \to \cdots \to i)$ reaches neuron i, the internal activity of neuron i would be $\vec{\mathbf{u}}_i^m = \mathbf{P} = (\vec{\mathbf{p}}, \vec{\mathbf{d}})$ if $\vec{\mathbf{d}}$ satisfies the QoS constraints, where m is a temporal label for the autowave.

(4) When passed through neuron i, the autowave whose traveling path is $\mathbf{P}$ may be recorded on the threshold $\vec{\boldsymbol{\theta}}_i^m = \mathbf{P} = (\vec{\mathbf{p}}, \vec{\mathbf{d}})$, where m is the index for path $\mathbf{P}$.

(5) Neuron i outputs simultaneously all paths in its threshold to its neighborhood. *i.e.*, $\vec{\mathbf{y}}_i = \{\mathbf{P} = (\vec{\mathbf{p}}, \vec{\mathbf{d}}) \mid \mathbf{P} \in \vec{\boldsymbol{\theta}}_i\}$.

Furthermore, a cost function must be defined for a path $\mathbf{P}$ so that we can make a comparison between two different paths. It is documented that the performance of an MCSP algorithm is closely concerned with the cost function. A variety of cost functions have been proposed and discussed. For example, in [6] the author proposed for a two-constraint problem a cost function $f(\mathbf{P}) = \alpha\omega_1(\mathbf{P}) + \beta\omega_2(\mathbf{P})$, then one can search fast for a feasible path using Dijkstra's shortest path algorithm by minimizing the cost function $f(\mathbf{P})$. However, no performance guarantee for path $\mathbf{P}$ is given with respect to individual constraints. In [11] the authors proposed an algorithm that dynamically adjusts the values of α and β within a logarithmic number of calls to Dijkstra's shortest path algorithm. However the problem is still unsolved. Some researchers [1] have recently proposed a nonlinear cost function whose minimization provides a continuous spectrum of solutions. In this paper, the cost function of a path $\mathbf{P}$ is defined as:

$$f(\mathbf{P})=\begin{cases}\sum_{k=1}^{K}\left(\frac{d_k}{C_k}\right)^{\lambda_k} & \mathbf{P}\ \textit{is feasible}\\ \infty & \textit{otherwise}\end{cases} \tag{6}$$

Where $C_1, C_2, \cdots, C_K$ are the K QoS constraints, $\lambda_k \geq 1$ is the significant coefficient of the kth QoS parameter.

4.2 Vectorial ACNN Based Algorithm

There are two improvements in vectorial ACNN based algorithm as to the original multi-label algorithm.

One is to limit the total labels of each node. In the original multi-label algorithm, a path can be discarded only when its constraints are all worse than others or don't satisfy the QoS constraints. So a large number of paths would be reserved on each node, which results in an exponential complexity of the algorithm. However, it is proved that the exponential complexity can be reduced to a polynomial one if the total labels of each node are limited to be no more than *M* (constant). In our algorithm, a maximal *M* thresholds on each neuron is limited.

Another is to select the paths for reserving. Because of the randomicity of the QoS parameters on each link, no cost function can provide an exactly scalar evaluation for a multi-parameter path. A path $\mathbf{P}(s,i)=(s\rightarrow\cdots\rightarrow i)$ with good fitness from the source node s to a middle node i may become feasible when outspread to the destination node d and *vice versa*. So the optimal path might not be found if we merely reserve paths according to their fitness. The proportional selection scheme is a simple and effective way to solve this kind of problem, which is well-documented in the heuristic algorithms, such as genetic algorithms [12]. In the proposed algorithm, a proportional selection scheme is used to prevent the optimal solution from being discarded.

The proposed algorithm can be described as following:

Step 1: Initialization. Let $\vec{\mathbf{y}}_i(0)=\phi$, $\vec{\boldsymbol{\theta}}_i(0)=\phi$, $\vec{\mathbf{u}}_i(0)=\phi$, $\forall i\in V$.

Step 2: Network starting. Let $\vec{\mathbf{y}}_s(1)=\{(s,0,\cdots,0)\}$, then fire the source neuron s and run the network.

Step 3: Internal activity calculating. Calculate the internal activity according to the following equation,

$$\vec{\mathbf{u}}_i(t)=\{((\vec{\mathbf{p}},i),\vec{\mathbf{d}}+\vec{\mathbf{w}}_{ji})\mid(\vec{\mathbf{p}},\vec{\mathbf{d}})\in\vec{\mathbf{y}}_j(t-1),\ \vec{\mathbf{w}}_{ji}\neq 0,\forall j\in V\}.$$

Noted that $\vec{\mathbf{u}}_i(t)$ is a path set which collects the arriving autowaves at time t.

Step 4: Thresholds updating.

(1) Combine the two path set $\vec{\mathbf{u}}_i(t)$ and $\vec{\boldsymbol{\theta}}_i(t-1)$ into a new set $\vec{\mathbf{A}}_i(t)$.

(2) Remove from $\vec{\mathbf{A}}_i(t)$ the paths whose distance components $d_1, d_2, \cdots, d_K$ are all worse than others. And remove the unfeasible paths from $\vec{\mathbf{A}}_i(t)$.

(3) Evaluate the paths in the new path set $\vec{\mathbf{A}}_i(t)$ according to the Eq.(6).

(4) Update the threshold $\vec{\mathbf{\theta}}_i(t)$ with M paths (if possible) chosen from $\vec{\mathbf{A}}_i(t)$ with the probabilities in inverse proportion to their fitness.

Step 5: Autowaves generating. The neurons generate M autowaves according to the paths in their threshold (or less than M if there are not enough paths in their threshold), and propagate those autowaves to their neighborhoods, *i.e.*,

$$\vec{\mathbf{y}}_i(t) = \{\mathbf{P} = (\vec{\mathbf{p}}, \vec{\mathbf{d}}) \mid \forall \mathbf{P} \in \vec{\mathbf{\theta}}_i(t)\}.$$

Step 6: Network stop condition. Let $t = t + 1$. If the optimal path from the source neuron s to the destination neuron d is obtained or $t > Maxtime$, stop the network, otherwise, go to step 3.

The algorithm mentioned-above is called vectorial ACNN. One can have an intuitionistic interpretation of the vectorial ACNN as follows:

Once produced from the source neuron (see Step 2), the autowaves reproduce themselves and propagate from one neuron to another along with the links (see Step 3). When gathering in a neuron, the autowaves compete with each other (see Step 4). The M winners occupy the neuron's threshold waiting for the next competition (see Step 4), and their duplicates will propagate to the next neurons for another competition (see Step 5). This process is recurring again and again until the approximate optimal path is found (see Step 6).

The computation of this algorithm focuses on the threshold updating step (*i.e.*, Step 4). Assuming the average number of adjacent nodes to be m, and the node number of the network to be n, the algorithm needs to calculate and compare $n \cdot (m+1) \cdot M$ paths at Step 4 in each algorithmic loop. So the computational complexity of the algorithm is $O(n^2(m+1)M)$ to find a feasible path, which is a polynomial complexity.

5 Computer Simulation

Fig. 3 shows a network with 10 nodes. Let node 3 be the source, and node 10 be the destination. The problem is to find a shortest (cheapest) path from node 3 to node 10, whose cost does not exceed 80 and whose constraint does not exceed 60, *i.e.*, the QoS request is R=(source=3,destination=10,cost$\leq$80,constraint$\leq$60).

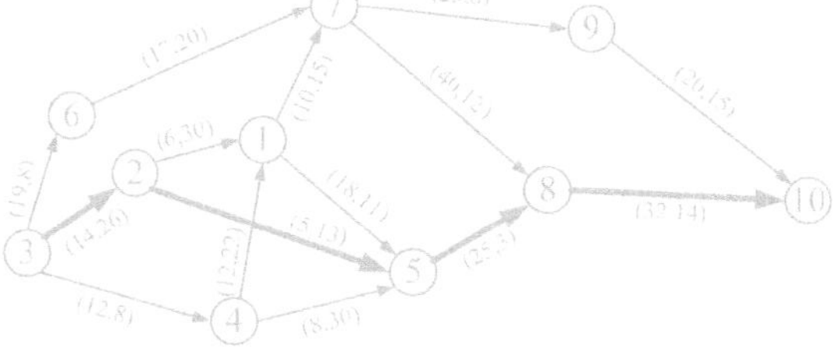

Fig. 3. A 10-node digraph. The two numbers on links are the cost and the constraint respectively, both of which are additive.

The vectorial ACNN based algorithm is used to solve the problem. Where the maximal number of paths reserved on each neuron is K=2, and the significant coefficients of the two weights are $r_1 = 2$ and $r_2 = 1$ respectively (see Eq.(6)). Tables 1 and 2 show the result path sets of the neuron's thresholds at first five iterations. Where, a path is denoted as [(*s*, *node* 1, *node* 2,…), (*cost*, *constraint*)], *i.e.*, the number in first bracket are the nodes sequence of the path, and the two numbers in second bracket are the cost and the constraint of the path respectively.

Table 1. Path sets on the threshold of different neurons (1~5) at first five iterations (*t*)

t	Node 1	Node 2	Node 3	Node 4	Node 5
1	ϕ	ϕ	[(3),(0,0)]	ϕ	ϕ
2	ϕ	[(3,2),(14,26)]	[(3),(0,0)]	[(3,4),(12,8)]	ϕ
3	[(3,4,1),(24,30)] [(3,2,1),(20,56)]	[(3,2),(14,26)]	[(3),(0,0)]	[(3,4),(12,8)]	[(3,2,5),(19,39)] [(3,4,5),(20,38)]
4	[(3,4,1),(24,30)] [(3,2,1),(20,56)]	[(3,2),(14,26)]	[(3),(0,0)]	[(3,4),(12,8)]	[(3,2,5),(19,39)] [(3,4,5),(20,38)]
5	[(3,4,1),(24,30)] [(3,2,1),(20,56)]	[(3,2),(14,26)]	[(3),(0,0)]	[(3,4),(12,8)]	[(3,2,5),(19,39)] [(3,4,5),(20,38)]

Table 2. Path sets on the threshold of different neurons (6~10) at first five iterations (*t*)

t	Node 6	Node 7	Node 8	Node 9	Node 10
1	ϕ	ϕ	ϕ	ϕ	ϕ
2	[(3,6),(19,8)]	ϕ	ϕ	ϕ	ϕ
3	[(3,6),(19,8)]	[(3,6,7),(36,28)]	ϕ	ϕ	ϕ
4	[(3,6),(19,8)]	[(3,6,7),(36,28)] [(3,4,1,7),(34,45)]	[(3,4,5,8),(45,41)] [(3,2,5,8),(44,42)]	[(3,6,7,9),(51,36)]	ϕ
5	[(3,6),(19,8)]	[(3,6,7),(36,28)] [(3,4,1,7),(34,45)]	[(3,4,5,8),(45,41)] [(3,2,5,8),(44,42)]	[(3,6,7,9),(51,36)]	**[(3,2,5,8,10),(76,56)]** [(3,4,5,8,10),(77,55)]

Tables 1 and 2 show that the optimal solution (the underline path on the threshold of neuron 10 at fifth iteration) was found after five iterations. This result path is [(3,2,5,8,10),(76,56)], see the thick path on Fig. 3, which is better than the result path found by Liu's algorithm [7], *i.e.*, the path [(3,4,5,8,10),(77,55)].

Furthermore, comparative simulation experiments of Jaffe's algorithm [6], Liu's algorithm [7] and our algorithm (the vectorial ACNN based algorithm) are done in different scale networks with different constraint number. Table 3 shows the iteration number of one constraint problem in the networks with 10, 100 and 200 nodes. While Table 4 shows the results of two constraint problem in different scale networks. The cost and additive constraints on the links of those networks are all produced randomly by computer, which subject to Gaussian distribution ranging from 1 to 100 with mean 50 and stand deviation 20. The adjacent nodes number is also random ranging from 1 to 5 with mean 3. M is set to be 2, 10 and 20 for 10-node networks, 100-node networks and 200-node networks respectively. For the sake of simplicity, the significant coefficients of the constraints λ_k are all set to be 1. The parameters of Liu's algorithm are set to be the same to document [7].

Tables 3 and 4 show that the iteration number is dramatically reduced using the vectorial ACNN based algorithm. With the increasing of networks scale, the iteration number increases slower than that of the Liu's algorithm. The results are all better than that of Liu's algorithm.

Table 3. Iteration number of one constraints problem in 3 different scale networks

method	10 nodes	100 nodes	200 nodes
Jaffe's algorithm	9	167	532
Liu's algorithm	5	89	204
Our algorithm	5	76	159

Table 4. Iteration number of two constraints problem in 3-different scale networks

method	10 nodes	100 nodes	200 nodes
Jaffe's algorithm	39	716	6113
Liu's algorithm	25	328	1527
Our algorithm	21	262	967

6 Conclusions

Multi-constraint quality-of-service (QoS) routing, which is in fact an MCSP problem, will become increasingly important as the Internet evolves to support real-time services. In this paper, a vectorial ACNN based algorithm is proposed for MCSP problem, which can find the approximate optimal path in polynomial time. Performance of the proposed algorithm is improved greatly comparing to the original multi-label routing algorithm (*i.e.*, Jaffe's algorithm). Firstly, the proposed algorithm updates all neurons' threshold synchronously, *i.e.*, it is a parallel algorithm. Secondly, only the maximal M paths are reserved on the threshold of each neuron, which reduces greatly the memory space and the computation of the algorithm needed. Thirdly, although the M-paths limited scheme may discard the optimal paths sometimes because of the randomicity of the constraints and the rigidness of the cost function, the proportional selection scheme would afford the optimal autowaves (*i.e.*, paths) enough opportunity to revive or to survive the search process. Comparing experiments are given in the paper, and the results show the efficiency of the proposed algorithm. In conclusion, the proposed algorithm has the characteristics of parallelism, efficiency and lower computational complexity.

Acknowledgment

This work was supported by the National Natural Science Foundation of China (Grant No. 60574039) and the "863" Project of National Ministry of Science and Technology (Grant No. 2006AA03A175).

References

1. Korkmaz, T., Krunz, M.: Multi-constrained optimal path selection. The 20th Annual Joint Conference of the IEEE Computer and Communications Societies **2** (2001) 834-843.
2. Xu, D., Chen, Y., Xiong, Y., Qiao, C.: On the Complexity of and Algorithms for Finding the Shortest Path With a Disjoint Counterpart, IEEE/ACM Trans. Networking.**14** (2006) 147-158
3. Wang, Z., Croweroft, J.: Quality-of-service routing for supporting multimedia applications. IEEE J. Select. Area. Commun.**14** (1996) 1219-1234
4. Jia, Z., Varaiya,P.: Heuristic Methods for Delay Constrained Least Cost Routing Using k-Shortest-Path, IEEE Trans. AC.**17** (2006) 707-712
5. Dumitrescu, I., Boland, N.: Improved Preprocessing, Labeling and Scaling Algorithms for the Weight-Constrained Shortest Path Problem, Networks **42** (2003) 135-153
6. Jaffe, J.M.: Algorithm for finding paths with multiple constraints, Networks **14** (1984) 95-116
7. Liu, J., Niu, Z., Zheng, J.: An improved routing algorithm subject to multiple constraints for ATM networks. ACTA ELECTRONICA SINICA **27** (1999) 4-8 (In Chinese)
8. Dong, J., Wang, W., Zhang, J.: Accumulative competition neural network for shortest path tree computation. International Conference on Machine Learning and Cybernetics,Vol.III, Xi'an China (2003) 1157-1161
9. Dong, J., Zhang, J.: Accumulating Competition Neural Networks based Multiple Constrained Routing Algorithm. Control and Decision **19** (2004) 751-755
10. Wang, Z.: On the complexity of quality of service routing. Information Processing Letters **69** (1999) 111-114
11. Korkmaz, T., Krunz, M., Tragoudas, S.: An Efficient Algorithm for Finding a Path Subject to Two Additive Constraints. Proceedings of the ACM SIGMENTRICS **1** (2000) 318-327
12. Gelenbe, E., Liu, P., Laine, J.: Genetic algorithms for route discovery. IEEE Trans. on SMC--Part B **99** (2006) 1247 – 1254

Neuro-Adaptive Formation Control of Multi-Mobile Vehicles: Virtual Leader Based Path Planning and Tracking

Z. Sun[1], M.J. Zhang[2], X.H. Liao[1], W.C. Cai[2], and Y.D. Song[1,2]

[1] Center for Cooperative Systems, National Institute of Aerospace, 100 Exploration Way, Hampton
{zhaosun,cindyliao,song}@nianet.org
[2] Electrical Engineering Department, North Carolina A&T State University Greensboro,NC,27411, USA
cheungmingjin@hotmail.com,caiwch@ncat.edu

Abstract. This paper presents a neuro intelligent virtual leader based approach for close formation of a group of mobile vehicles. Neural Network-based trajectory planning is incorporated into the leading vehicle so that an optimal reference path is generated automatically by the virtual leader, which guides the whole team vehicles to the area of interest as precisely as possible. The steering control scheme is derived based on the structural properties of the vehicle dynamics. Simulation on multiple vehicles formation is conducted as a verification of the effectiveness of the proposed method.

1 Introduction

Many systems in nature exhibit stable formation behaviors such as swarms, schools and flocks [1]-[2]. In these systems, without colliding with neighbors, individuals approach to the target in close formations. Inspired by this phenomena, formation control for multiple mobile vehicles emerged.

Typical formation control methods include "Behavior-Based Strategy" [3]-[5], "Multi-Agent System Strategy" [6]-[7], "Virtual Structure Strategy" [8]-[9], and "Leader-Follower Strategy" [10]-[14], among which the "Leader-Follower Strategy" is of particular interest to us because it is characterized by simplicity with no need for global knowledge and computation. However, most existing "leader-follower" formation control methods are either model based, thus lack robustness to unknown environment and disturbances, or suffer from the singularity problem. Moreover, actuation dynamics are usually ignored.

In this work, we investigate a virtual leader based neuro-adaptive path planning and tracking scheme for multiple mobile vehicles. Generally, virtual leader is adopted in multi-agent system such as a school, a swarm or a flock to coordinate and distribute multiple autonomous mobile robots and/ or manipulate the group [19]-[21]. The main idea behind this approach is to set a virtual leader in the forefront of the group in the running direction so that the desired separation distance and orientation could be specified without involving singularity. Furthermore, by using Neural Network-based

D. Liu et al. (Eds.): ISNN 2007, Part I, LNCS 4491, pp. 786–795, 2007.

optimal control, the virtual leader is able to autonomously generate a reference path for the formation to reach the area of interest quickly and precisely. Also, the proposed formation control algorithm is able to generate the control current directly for the driving motor of the vehicle to steer the vehicle toward the desired position precisely. The singularity issue inherent in traditional leader-follower method is removed with the virtual leader concept and the orthogonal transformation [15]-[16]. The effectiveness of the proposed strategy is confirmed by theoretical analysis and computer simulation.

2 Virtual Leader Based Neuro-Adaptive Path Planning

Consider the formation as shown in Fig. 1 where the virtual leader's trajectory is automatically generated via a neural network based optimal path planning strategy as described below. The kinematics model of the vehicle is

$$\dot{x}_{VL} = u_{VL} \cos\theta_{VL}; \ \ \dot{y}_{VL} = u_{VL} \sin\theta_{VL}; \ \ \omega_{VL} = \dot{\theta}_{VL} \tag{1}$$

where (x_{VL}, y_{VL}) is the global position, u_{VL} is the constant linear velocity, and ω_{VL} is the angular velocity of the virtual leader. The dynamics of the vehicle which contains a dc motor, a dc amplifier and a gear transmission system is given by:

$$\omega_{VL} = (0.45 \times 102.6) \times \omega_{DT} /(s^2 + 9.21s + 102.6) \tag{2}$$

where ω_{DT} is the desired turning rate of the vehicle [24].

This virtual leader is desired to autonomously obtain a sequence of optimal control actions in order to reach the target point as fast as possible. In this paper, the virtual leader's linear speed u_{VL} is taken as constant. It is assumed that the virtual leader is under the constrain of pure rolling, nonslipping and free space without obstacles. Thus, minimum-time control is equivalent to shortest-path control. The state of the virtual leader for optimal control is given by $\Re = (x_{VL}, y_{VL}, \theta_{VL}, \omega_{VL})$ and the control signal is ω_{DT}.

Given a discrete-time nonlinear time varying system:

$$\Re(k+1) = F[\Re(k), \omega_{DT}(k), k] \tag{3}$$

And, it is desired to minimize for a cost index

$$J[\Re(i), i] = \sum_{k=i}^{\infty} \gamma^{k-i} U[\Re(k), \omega_{DT}(k), k] \tag{4}$$

where U is the utility function which indicates the performance of the overall system and γ is the discount factor ($0 < \gamma \le 1$). In this study, U is chosen as:

$$U(k) = \frac{1}{2}[(x_{VL}(k) - x_{VL}{}^{T})^2 + (y_{VL}(k) - y_{VL}{}^{T})^2] \tag{5}$$

where $(x_{VL}{}^{T}, y_{VL}{}^{T})$ is the "*target*" point of the virtual leader.

The objective here is to choose the control sequence $\omega_{DT}(k), k = i, i+1, \ldots$ so that the J function is minimized. The cost in this case accumulates indefinitely and this kind of problem is referred to as infinite horizon problems in dynamic programming [25]. Apparently,

$$J^*[\Re(k),k] = \min_{u(k)} J[\Re(k),k] = \min_{u(k)}(U([\Re(k),\omega_{DT}(k),k) + \gamma J^*[\Re(k+1),k+1])$$

The proposed neural network-based adaptive path planning consists of three modules: Critic, Model, and Action. The three networks are all implemented using multilayer-feedforward neural networks. Each neural network has six inputs: $x_{VL}^T, y_{VL}^T, x_{VL}, y_{VL}, \theta_{VL}, \omega_{VL}$.

Firstly, the critic neural network is used to approximate the J function by minimizing the following error measure over time:

$$\|E_m\| = \sum_k E_m(k) = \frac{1}{2}\sum_k [J(k) - U(k) - \gamma J(k+1)]^2 \tag{6}$$

$$\text{with} \quad J(k) = J[\Re(k),k,W_c] \tag{7}$$

where W_c is the weight vector of the critic network updated by gradient descendent method as follows:

$$W_{c,j}^{(p+1)} = W_{c,j}^{(p)} - \eta_c \frac{\partial E_m(k)}{\partial W_{c,j}^{(p)}} = W_{c,j}^{(p)} - \eta_c [J(k) - U(k) - \gamma J(k+1)] \frac{\partial J(k)}{\partial W_{c,j}^{(p)}} \tag{8}$$

where η_c is the learning rate of the critic network, $W_{c,j}$ is the jth component of W_c. The training samples for the critic network are obtained over a trajectory from $k = 0$ until the target is reached. The training process will be repeated until no more weight update is needed.

The model network in an adaptive critic design predicts $\Re(k+1)$ given $\Re(k)$ and $\omega_{DT}(k)$; it is needed for the computation of

$$J(k+1) = J(\Re(k+1), k+1, W_c^{(p-1)}) \tag{9}$$

In (8) for the weight update. The model network learns the mapping given in equation (3); it is trained in parallel with the critic and action networks [26].

After the critic network's training is finished, the action network's training starts with the objective of minimizing J(k+1). The action network generates an action signal

$$\omega_{DT}(k) = u[\Re(k),k,W_A] \tag{10}$$

whose training follows a similar procedure to the one for the critic network's training. The training process will be repeated until no more weight update is needed while keeping the critic network's weights fixed. The weight update is given as:

$$W_{A,i}^{(p+1)} = W_{A,i}^{(p)} - \eta_a \frac{\partial J(k+1)}{\partial W_{A,i}^{(p)}} = W_{A,i}^{(p)} - \eta_a \sum_{j=1}^{n} \frac{\partial x_j(k+1)}{\partial u_k(k)} \sum_{k=1}^{m} \frac{\partial x_j(k+1)}{\partial u_k(k)} \frac{\partial u_k(k)}{\partial W_{A,i}^{(p)}} \tag{11}$$

where η_a is the learning rate of the action network and W_a is the weight vector.

It can be seen that information is propagated backward through the critic network to the model network and then to the action network, as if three networks formed one large feed forward network, by which the reference trajectory was given by a set of points $(x^d_{VL}(k), y^d_{VL}(k))$, together with the reference orientation angle $\theta_{VL}(k)$ and angle velocity $\omega_{VL}(k)$ series of the virtual leader $(k = 0,1,2....)$.

3 Virtual Leader Based Formation Control

The virtual leader able to generate desired path as described above is introduced into the robot team, as conceptually shown in Fig. 1, where the real leader was titled as Follower 0, and its desired relative distance and orientation with respect to the virtual leader are $\rho_0{}^d = \rho_0$ and $\phi_0{}^d = 0$. The desired position for each vehicle is represented by desired relative distance $\rho_i{}^d$ and relative orientation angle $\phi_i{}^d$ with respect to the virtual leader, where

$$\rho_i{}^d = (\rho_{i0}{}^d)^2 + (\rho_0{}^d)^2 + 2\rho_{i0}{}^d \rho_0{}^d \cos\phi_{i0}$$

$$\phi_i{}^d = \phi_{i0} - \cos^{-1}\left(\frac{(\rho_{i0}{}^d)^2 + (\rho_i{}^d)^2 - (\rho_0{}^d)^2}{2\rho_{i0}{}^d \rho_i{}^d}\right)$$

Apparently, $\phi_i{}^d \in (-\pi/2, \pi/2)$; and thus avoid the problem of singularity.

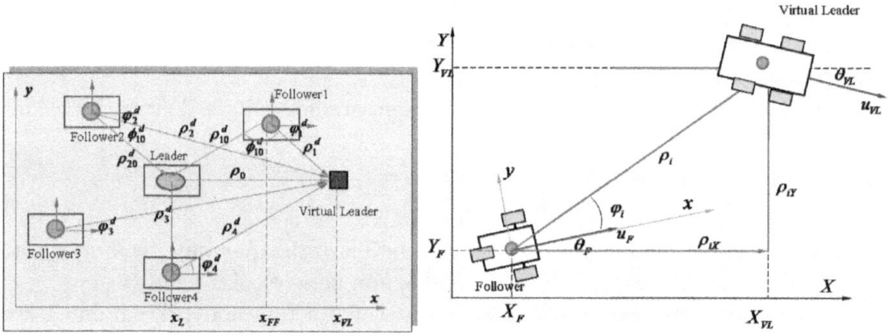

Fig. 1. Virtual Leader-follower Configuration

Now we begin to derive the kinematics and dynamic equations of system. Fig. 1 illustrates the formation geometry which is determined by the relative position between the leader and the follower, where *X*-*Y* is the ground coordinates; *x*-*y* is the Cartesian coordinates fixed on the leader's body; (x_{VL}, y_{VL}) is the global position of the virtual leader's driving axes center and (x_F, y_F) is that of the followers' (including the real leader); u_{VL} and u_F are Virtual leader's and follower's linear velocities; θ_{VL} and θ_F are their orientation angles respectively; ρ and φ are the follower's relative distance and orientation angle with respect to the virtual leader. With the reference trajectory generated by the intelligent virtual leader, what we need to do is to resolve the trajectory-tracking problem which is to control the relative position and angle of the followers to the desired value, i.e., $\rho_i \to \rho_i^d, \varphi_i \to \varphi_i^d$.

Projecting the relative distance ρ_i along *x*, *y* direction and we have ρ_{ix} and ρ_{iy} as follows:

$$\rho_{ix} = (X_{VL} - X_F)\cos\theta_F + (Y_{VL} - Y_F)\sin\theta_F \,;\, \rho_{iy} = -(X_{VL} - X_F)\sin\theta_F + (Y_{VL} - Y_F)\cos\theta_F \quad (12)$$

And, the follower's desired position along *x* and *y* directions with respect to the virtual leader are:

$$\rho_{ix}^{\ d} = \rho_i^d \cos\varphi_i^d \quad , \rho_{iy}^{\ d} = \rho_i^d \sin\varphi_i^d \quad (13)$$

Referring to the coordinate system as shown in the right of Fig. 1, we define a state variable to represent the difference of the orientation angles between the virtual leader and the follower as:

$$e_\theta = \theta_F - \theta_{VL} \quad (14)$$

It is noted that in this paper, we use the Cartesian coordinates fixed on the follower's body. Thus, the formation control objective is to steer the follower to maintain separation distance in longitudinal and lateral directions, i.e., $\rho_{ix} \to \rho_{ix}^{\ d}, \rho_{iy} \to \rho_{iy}^{\ d}$, and the relative bearing angle e_θ should be stable at least.

From equation (12)-(14), we can derive the system's kinematics model described as below:

$$\dot{\rho}_{ix} = \rho_{iy}\omega_F + u_{VL}\cos\theta_e - u_F \,,\ \dot{\rho}_{iy} = -\rho_{ix}\omega_F - u_{VL}\sin\theta_e \quad (15)$$

Introduce relative position error variables e_x and e_y as following:

$$e_x = \rho_{ix} - \rho_{ix}^{\ d}, \qquad e_y = \rho_{iy} - \rho_{iy}^{\ d} \quad (16)$$

Substitute that into equations (15) and (16), we derive the error dynamics of the system model as following:

$$\begin{bmatrix}\dot{e}_x & \dot{e}_y\end{bmatrix}^T = R\begin{bmatrix}u_F & \omega_F\end{bmatrix}^T + A \quad (17)$$

With: $$R = \begin{bmatrix}-1 & \rho_{iy} \\ 0 & -\rho_{ix}\end{bmatrix}, A = \begin{bmatrix}u_{VL}\cos\theta_e \\ -u_{VL}\sin\theta_e\end{bmatrix} - \frac{d}{dt}\begin{bmatrix}\rho_i^d \cos\varphi_i^d \\ \rho_i^d \sin\varphi_i^d\end{bmatrix}$$

Note that the inverse of *R* is not defined at $\rho_{ix} = 0$ (this physically corresponds to the situation that the relative distance in *x*-axis from the follower toward the Leader becomes zero, which could happen anytime), direct stabilization of e_x and e_y is undesirable. Following the same idea as in [15]-[16], we carry out an orthogonal coordinate transformation

$$E = \begin{bmatrix}E_x & E_y\end{bmatrix}^T = B(\theta_F)\begin{bmatrix}e_x & e_y\end{bmatrix}^T \quad (18)$$

Where $B(\theta_F)$ is defined as:

$$B(\theta_F)=\begin{bmatrix}\sin\theta_F & \cos\theta_F\\ \cos\theta_F & -\sin\theta_F\end{bmatrix} \qquad (B(\theta_F)^T B(\theta_F)=\begin{bmatrix}1&0\\0&1\end{bmatrix}) \tag{19}$$

Consequently, we have

$$\|E\|^2=\left\|\begin{matrix}e_x\\ e_y\end{matrix}\right\|^2 \tag{20}$$

Therefore the problem of stabilizing e_x and e_y boils down to stabilizing E_x and E_y. Our controller design now is focused on the transformed error dynamics. From equation (18)-(19), we have

$$\dot{E}=BA+C\left[u_F \quad \omega_F\right]^T \tag{21}$$

Where

$$C=\begin{bmatrix}-\sin\theta_F & -\rho_i^d\cos(\varphi_i^d+\theta_F)\\ -\cos\theta_F & \rho_i^d\sin(\varphi_i^d+\theta_F)\end{bmatrix} \qquad (|C|=\rho_i^d\cos\varphi_i^d\neq 0) \tag{22}$$

From our previous work on single vehicle [16], we have derived the dynamics :

$$\begin{bmatrix}\dot{u}_F\\ \dot{\omega}_F\end{bmatrix}=K_t\begin{bmatrix}\dfrac{R}{\Theta_u} & 0\\ 0 & \dfrac{Rd}{\Theta_\omega}\end{bmatrix}\begin{bmatrix}1 & 1\\ 1 & -1\end{bmatrix}u+F+\delta \tag{23}$$

with

$$F=\begin{bmatrix}\dfrac{mbR^2}{\Theta_u}\omega_F{}^2 & \dfrac{-2bmR^2}{\Theta_\omega}u_F\omega_F\end{bmatrix}^T;\ \Theta_u=mR^2+2I_e\ ;\Theta_\omega=I_ed^2+2R^2(I_c+mb^2)$$

$$\delta=\begin{bmatrix}\{mR^2\omega_Fv^s+R^2[F_{ex}+\eta_{ex}]-B_e[2u_F+u_{Fl}{}^s+u_{Fr}{}^s]-I_e[\dot{u}_{Fl}{}^s+\dot{u}_{Fr}{}^s]\}/\Theta_u\\ \{-B_ed^2\omega_F-B_ed[u_{Fr}{}^s-u_{Fl}{}^s]-I_ed[\dot{u}_{Fr}{}^s-\dot{u}_{Fl}{}^s]-2mbR^2\dot{v}^s+2R^2[(b-e)\eta_{ey}\\ +(a+b)F_{cy}+\tau_e]\}/\Theta_\omega\end{bmatrix}$$

Take the derivative of equation (9), we got

$$\ddot{E}=\frac{d}{dt}(BA)+\dot{C}\left[u_F \quad \omega_F\right]^T+C\left[\dot{u}_F \quad \dot{\omega}_F\right]^T \tag{24}$$

Substitute (10) into (13), we derived the following equation:

$$\ddot{E}=D(.)\left[i_r \quad i_l\right]^T+Q(.)+L(.) \tag{25}$$

With

$$D(.) = CM\begin{bmatrix} 1 & 1 \\ 1 & -1 \end{bmatrix}; C = \begin{bmatrix} -\sin\theta_F & -\rho_i^d \cos(\varphi_i^d + \theta_F) \\ -\cos\theta_F & \rho_i^d \sin(\varphi_i^d + \theta_F) \end{bmatrix};$$

$$M = K_t \begin{bmatrix} \frac{R}{\Theta_u} & 0 \\ 0 & \frac{Rd}{\Theta_\omega} \end{bmatrix} = \begin{bmatrix} m_x & 0 \\ 0 & m_y \end{bmatrix}$$

$$Q(.) = \begin{bmatrix} \dot{u}_{VL}\sin\theta_{VL} + \omega_{VL}u_{VL}\cos\theta_{VL} \\ \dot{u}_{VL}\cos\theta_{VL} - \omega_{VL}u_{VL}\sin\theta_{VL} \end{bmatrix} + \dot{C}\begin{bmatrix} u_F \\ \omega_F \end{bmatrix} - \frac{d}{dt}\left[B\frac{d}{dt}\begin{bmatrix} \rho_i^d \cos\varphi_i^d \\ \rho_i^d \sin\varphi_i^d \end{bmatrix}\right]$$

$$L(.) = CF + C\delta$$

Apparently, the real control variable for the formation control is $[i_r \quad i_l]^T$, which are the current through the armature of the motors. We assume $\rho_i^d, \varphi_i^d, \omega_{VL}$ and $u_{VL}e^{i\theta}$ all have sufficiently smooth derivatives. The control task is to design control law for $u = [i_r \quad i_l]^T$ in order to make $e_x, e_y \to 0$ and e_θ stable.

As analyzed in [16], $Q(.)$ and $D(.)$ contains uncertain parameters (such as K_t, Θ_u, Θ_ω etc.), they should not be used directly in the control design. Here we introduce a virtual current vector to substitute the real one via the linear transformation:

$$i' = \begin{bmatrix} i'_1 \\ i'_2 \end{bmatrix} = \begin{bmatrix} 1 & 1 \\ 1 & -1 \end{bmatrix}\begin{bmatrix} i_r \\ i_l \end{bmatrix} = T\begin{bmatrix} i_r \\ i_l \end{bmatrix} \tag{26}$$

which leads to the transferred system model

$$\ddot{E} = CMi' + Q(.) + L(.) \tag{27}$$

It can be seen that by introducing the virtual current vector as the control variable, known matrix T is combined into the control signal and thus make the gain matrix to be the product of an invertible matrix J and a positive diagonal matrix M. Such treatment allows a new adaptive tracking control scheme to be developed for the vehicle based on [17], which will be addressed in next section.

4 Adaptive Control Algorithm and Stability Analysis

Now we are in the position to present the control scheme:
Define

$$s = \dot{E} + \beta E \tag{28}$$

Then (16) can be expressed in terms of s as follows,

$$\dot{s} = CMi' + P(.) + L(.) \tag{29}$$

where :

$$P(.) = Q(.) + \beta(BA + C[u_F \quad \omega_F]^T)$$

Following the idea as reported in [17], the adaptive control scheme for the vehicle is derived as

$$i' = C'^{T}[-ks - P(.) - u_c],\ k > 0 \tag{30}$$

where

$$u_c = \frac{\hat{a}^T \varphi s}{\|s\|} \quad \dot{\hat{a}} = \|s\| \varphi (\hat{a} \in R_{6\times 1}) \tag{31}$$

$$\varphi = \left[\left\| \begin{bmatrix} \omega_F^{\ 2} \\ u_F \omega_F \end{bmatrix} \right\| \quad \|\omega_F\| \left\| \begin{bmatrix} u_F \\ \omega_F \end{bmatrix} \right\| \quad \|\omega_F\| \quad \|\omega_F\| \quad \|\dot{e}\| \quad \left\| \begin{bmatrix} \ddot{x}_{VL}^{\ d} \\ \ddot{y}_{VL}^{\ d} \end{bmatrix} \right\| \right]^T \tag{32}$$

We can conclude that e (i.e., e_x and e_y) converges to zero as time increases, see [16] for detail proof of the stability. The real control signal (currents of the motors for the left and right wheels) can be obtained from (26) and (30).

5 Simulation Results

We simulate with the proposed control scheme to guide five mobile vehicles to a target positions with a " Δ -shape" formation. The systems parameters are partly shown as: $m = 1301$kg $I_z = 1627$kg-m $K_t = 1$Nm/A.

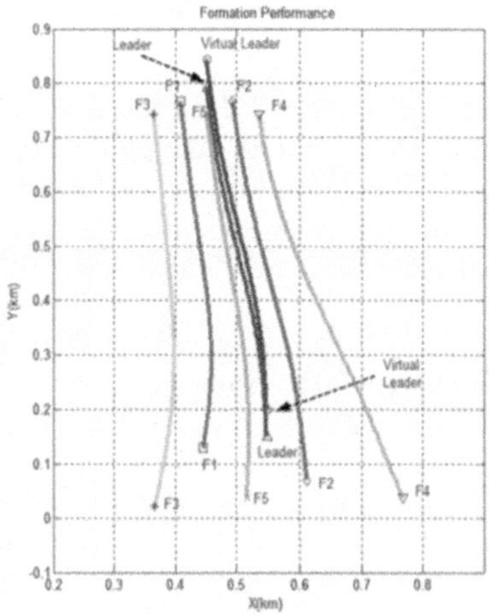

Fig. 2. Formation Performance

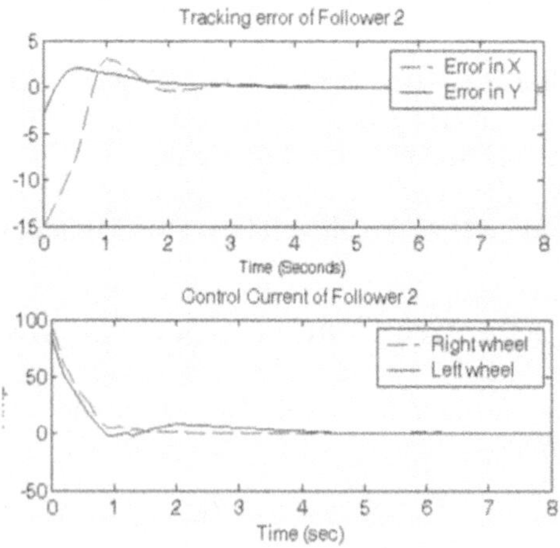

Fig. 3. Control Error and Control Signal

The control and parameters are chosen as: $\beta = 2; k = 10; \sigma = 0.2; g = 1; \varepsilon = 0.001$ Initially, the six vehicles (including a real-leader) are located at different places, as shown in Fig. 2. The separation distances and relative orientation angular of the vehicle to the virtual leader are: $\rho_0 = 50(m), \phi_0^d = 0(\text{deg})$; $\rho_5 = 60(m), \phi_5^d = 0(\text{deg})$,

$$\rho_1^d = \rho_2^d = 60(m), \quad \varphi_1^d = \varphi_3^d = 30(\text{deg}) \; \rho_3^d = \rho_4^d = 120(m), \quad \varphi_2^d = \varphi_4^d = 150(\text{deg})$$

Under the control of the proposed algorithm, the six vehicles (one leader, five followers) quickly achieve Δ-shape formation from their original positions and reach the target following the virtual leader as recorded in Fig. 2 The tracking error and the control signal for Follower 2 are depicted in Fig. 3, which shows good tracking precision and smooth control action.

6 Concluding Remarks

A neuro virtual leader based approach for close formation of a group of mobile vehicles is investigated in this paper. A neural network-based trajectory planning is proposed to generate reference path for the virtual leader automatically to guide the whole team vehicles to the area of interest as precisely as possible. The proposed method avoids the singularity issue inherent in formation. The overall control scheme demands little system dynamic information. Simulation on six mobile vehicles formation demonstrates that the proposed method is effective and feasible. We are currently developing a real-time testbed (see [22] for more detail) for experiment verification of the developed algorithm.

References

[1] Mizuno, Y., Kato, S., Mutoh, A., Itoh, H.: A Behavioral Model Based on Meme and Qualia for Multi-agent Social Behavior. Proc. 19th Int. Conf. Advanced Information Networking and Applications 2 (2005) 181 – 184
[2] Wang, Y.X.: Sociological Models of Software Engineering. Proc. Canadian Conf. Electrical and Computer Engineering. (2005) 1819 – 1822
[3] Arkin, R.C.: Behavior-Based Robotics. Cambridge, MA: MIT Press, 1998
[4] Balch, T., Dellaert, F., Feldman, A., Guillory, A. et.al.: How Multirobot Systems Research Will Accelerate Our Understanding of Social Animal Behavior. Proceedings of IEEE. 94(7) (2006) 1445 – 1463
[5] E.Monteiro, S., Bicho, E.: A Dynamical Systems Approach to Behavior Based Formation Control. Proc. IEEE Int. Conf. Robotics and Automation, Washington, D.C. 3 (2002) 2606–2611
[6] Egerstedt, M., Hong, X.M.: Formation Constrained Multi-agent Control. IEEE Transaction of Robotics and Automation 17 (2001) 947–951
[7] Jongusuk, J., Mita, T.: Tracking Control of Multiple Mobile Robots. Proc. IEEE Int. Conf. Robotics and Automation, Seoul, Korea 3 (2001) 2885–2890
[8] Young, B.J., Beard, R.W., Kelsey, J.M.: A Control Scheme for Improving Multi-vehicle Formation Maneuvers. Proc. American Control Conf. 2 (2001) 704–709

[9] Esposito, J.M., Kumar, V.: Closed Loop Motion Plans for Mobile Robots. Proc. IEEE Int. Conf. Robotics and Automation, San Francisco, CA 3 (2000) 2777–2782
[10] Desay, J.P., Ostrowski, J.P., Kumar, V.: Modeling and Control of Formations of Nonholonomic Mobile Robots. IEEE Trans. Robot. and Automat. 17 (2001) 905–908
[11] Desay, J.P., Kumar, V., Ostrowski, P.: Control of Change in Formation for a Team of Mobile Robots. Proc. IEEE Int. Conf. Robot. and Automat. Detroit, MI 2 (1999), 1556–1561
[12] Fierro, R., Das, A.K., Kumar, V. et.al.: A Vision-based Formation Control Framework. IEEE Trans. Robot. and Automat. 18 (2002) 813–825
[13] Lemay, M., Michaud, F., Letourneau, D., Valin, J.M.: Autonomous Initialization of Robot Formations. Proc. IEEE Conf. Robot. and Automat. 3 (2004) 3018 – 3023
[14] Shao, J.Y., Xie, G.M., Yu, J.Z., Wang, L.: Leader-Follower Formation Control of Multiple Mobile Robots. Proc. IEEE, Int. Symposium on Intelligent Control, Limassol, Cyprus (2005) 803-813
[15] Song, Y.D., Li, Y., Liao, X.H.: Orthogonal Transformation Based Robust Adaptive Close Formation Control of Multi-UAVs. Proc. American Control Conf., Portland , Oregon 5 (2005) 2983 – 2988
[16] Sun, Z., Cai, W.C., Liao, X.H., Dong, T., Song, Y.D.: Adaptive Path Control of Unmanned Ground Vehicles. Proc. 38th Southeastern Symposium on System Theory , Cookeville, TN (2006) 507-511
[17] Song, Y.D.: Neuro-Adaptive Control with Application to Robotic Systems. J. Robotic Systems 14 (6) (1997) 433-447
[18] Slotine, J.J., Li, W.: Applied Nonlinear Control, Prentice-Hall, Inc, 1991
[19] Jadbabaie, A., Lin, J., Morse, A.S.: Coordination of Groups of Mobile Autonomous Agents Using Nearest Neighbor Rules. IEEE Trans. Autom. Control 48(6) (2003) 988-1001
[20] Lenonard, N.E., Fiorelli, E.: Virtual Leaders, Artificial Potentials and Coordinated Control of Groups. Proc. 40th IEEE Conf. Decision and Control, Orlando, FL (2001) 2968-2973
[21] Tanner, H., Jadbabaie, A., Pappas, G. J.: Stable Flocking of Mobile Agentsrt I: Fixed Topology. Proc. Conf. Decision and Control, Maui, HI (2003) 2010-2015
[22] Cai, W.C., Weng, L.G., Zhang, R. et.al.: Development of Real-time Control Test-bed for Unmanned Mobile Vehicles. Proc. 32 Int.. Conf. IEEE Industrial Electronics Paris, FRANCE, 2006
[23] Koh, K.C., Beom, H.R., Kim, J.S., Cho, H.S.: A Neural Network-Based Navigation System for Mbile Robots. KACC 94, 2709-2714
[24] Patifo, H.D., Carelli, R.: Neural Network-Based Optimal Control for autonomous Mobile Vehicle Navigation.Pro. of the 2004 IEEE International Symposium on Intelligent Control ,Taipei, Taiwan (2004) 391-396
[25] Liu, D.R.: Neural Network-Based Adaptive Critic Designs for Self-Learning Control. Proc. of the 9th Int. Conf. on Neural Information Processing. 3(18-22) (2002) 1252 – 1256
[26] Seaks, R.E., Cox, C.J., Mathia, K., Maren, A.J.: Asymptotic Dynamic Programming: Preliminary Concepts and Results. Proc. of Int. Conf. of Neural Networks, Houston, Tx, USA (1997) 2273-2278

A Multi-stage Competitive Neural Networks Approach for Motion Trajectory Pattern Learning*

Hejin Yuan[1], Yanning Zhang[1], Tao Zhou[1,2], Fang'an Deng[2], Xiuxiu Li[1], and Huiling Lu[3]

[1] School of Computer Science, Northwestern Polytechnical University
Xi'an 710072, China
[2] Department of Maths, Shanxi University of Technology
Hanzhong, Shanxi 723000, China
[3] Department of Computer, Shanxi University of Technology
Hanzhong , Shanxi 723000, China

Abstract. This paper puts forward a multi-stages competitive neural networks approach for motion trajectory pattern analysis and learning. In this method, the rival penalized competitive learning method, which could well overcome the competitive networks' problems of the selection of output neurons number and weight initialization, is used to discover the distribution of the flow vectors according to the trajectories' time orders. The experiments on different sites with CCD and infrared cameras demonstrate that our method is valid for motion trajectory pattern learning and can be used for anomaly detection in outdoor scenes.

1 Introduction

The increasing demand for security by society leads to a growing need for surveillance activities in many sensitive and public environments. Intelligent visual surveillance systems are just generated for this purpose. Different from the traditional video surveillance system, visual surveillance system can automatically perceive the varies of the environments, detect, recognize and track moving objects from the image sequences, and even more to understand and describe their behaviors, then provides useful clues to the operators in advance if some emergences or abnormal behaviors are taking place or will occur. The typical configuration of processing modules for visual surveillance task includes moving object detection, recognition, tracking, behavior analysis, and anomaly detection. Among them, a significant amount of work has been done on the low-level processing steps, such as moving object detection, tracking and recognition, and many valid solutions have been proposed [1~3]. Behavior understanding is a very important part of visual surveillance system and its

* This work was supported by the National Natural Science Foundation of China (NSFC) under Grants 60472072, the Specialized Research Foundation for the Doctoral Program of Higher Education under Grant 20040699034 , the Aeronautical Science Foundation of China under Grant 04I50370 and the Natural Science Foundation of Shan'xi Province.

D. Liu et al. (Eds.): ISNN 2007, Part I, LNCS 4491, pp. 796–803, 2007.

function is to analyze and recognize human or vehicle motion patterns and then produce some high-level semantic description of actions and interactions. Motion trajectory is a very useful feature for object behaviors analysis. This paper puts forward a multi-stages neural network approach for motion trajectory pattern learning and then gives a correspondent anomaly detection method. The experiments on different sites with CCD and infrared cameras demonstrate our method is effective and efficiency for trajectory pattern learning and anomaly detection.

The rest of this paper is organized as follows: section 2 provides a brief review for some previous efforts have been made to handle this problem; section 3 presents our trajectory pattern learning approach with multi-stage competitive neural networks, and the specific details of the correspondent anomaly detection method is also included in this section; section 4 shows the experiment results and analysis of our method on different sites with CCD and infrared cameras; finally, we draw a conclusion in section 5.

2 Related Works

Many unsupervised learning methods have been proposed for motion trajectory analysis. These methods needn't any information about the scene and needn't predefine the object's behavior manners. The following paragraph provides a brief introduction and summary about them. Johnson [4] proposed a two-layer competitive neural networks model, which is connected by leaky neurons. In this approach, the first network is used to model the distribution of flow vectors, and the second to model the trajectory distribution. The main problems of this method are its slow training speed and information distortion caused by the leaky neurons. Sumpter [5] introduced a feedback mechanism to the second competitive network of Johnson. In this way, it could predict the object's behavior more efficiently. However, the number of the input and output neurons are all same as the number of flow vectors, so it can not be quickly trained for its too complex structure and the anomaly detection method is also not given in their paper. Owens [6] applied the Kohonen Self-Organizing Feature Mapping to reflect the flow vectors distribution. This method allows novelty detection to be applied on a point-by-point basis in real time. Its network structure is much simpler than Johnson. Whereas the time orders information among the flow vectors is completely ignored. So it can not give any prediction information about the behavior. For these drawbacks, Hu [7] provided a hierarchical Self-Organizing Neural Network model to learn the trajectory distribution patterns. In their approach, some lines are formed through linking the side neurons and each line is an internal net which is correspondent to a class of trajectory pattern. Fu [8] gave a hierarchical clustering framework to classify vehicle motion trajectories in real traffic video based on their pair-wise similarities. Then spectral clustering is used to group trajectories with similar spatial patterns. Dominant paths and lane can be distinguished as a result of two-layer hierarchical clustering. Unlike above methods, which use the single flow vector as the processing unit, Tan [9] offered a fuzzy self-organizing neural networks and batch learning mechanism to discover the trajectory distribution. In their model, the network uses the whole trajectories as input and has a much simpler structure. Each input vector corresponds to a complete trajectory and the weight vectors represent their distribution pattern. However, it needs to preprocess the training trajectories with same length. Differ from flow vector coding scheme, Khalid et al [10] considered the trajectory as two time series and modeled them with the leading Fourier coefficients obtained by Discrete Fourier Transform. Trajectory clustering is then carried out in the Fourier coefficient space to discover patterns of similar object motions. It is robust to the tracking noise since the global features of motion trajectories are represented by Fourier approximations.

Though the unsupervised methods could discover the trajectory distribution pattern automatically, they deeply depend on the training data. And the largest problem with these model-free approaches to novelty detection is that when the training data is not enough, then the novel but actually acceptable behavior will be classified as suspicious. This implies a requirement to update the neural online with newly detected normal trajectory when they occur.

3 Multi-stage Competitive Neural Networks Approach

3.1 The Model of Multi-stage Neural Networks

As well known, the trajectory patterns are not only dependent on the distribution of the flow vectors, but also their time orders. Unlike linking the side neurons to construct internal net in [7], this paper proposed a multi-stage competitive neural network, as shown in Fig.1, to model the distribution of the motion trajectory, i.e. the ith network is only used to learn the distribution of the ith flow vectors of the trajectory in the training set. The number of the neurons in each network can be different. It should be determined by the distribution of the trajectories. This multi-stage model can not only discover the distribution pattern of the flow vectors, but also reflect their time orders automatically. The training speed is very quickly since each neural network only needs to cluster the flow vectors with the same sequence number. Moreover, the newly observed flow vector can be quickly determined whether it is abnormal or not with this pipe-line structure rather than checking the whole trajectory again each time. So it could well satisfy the requirements of real-time applications.

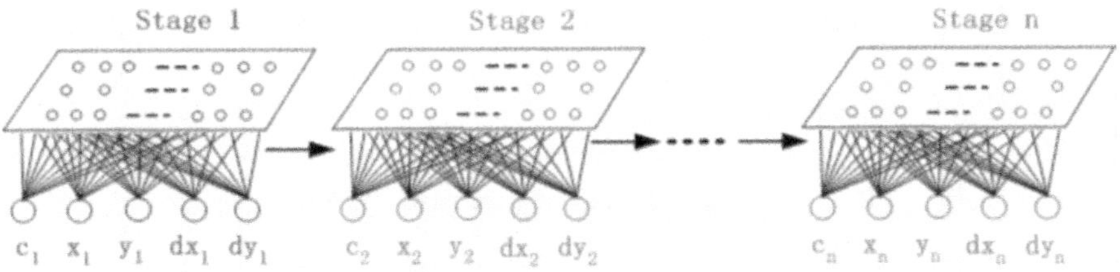

Fig. 1. Multi-stage competitive neural networks model in this paper

3.2 Trajectory Coding

It is important to represent the motion trajectory in a reasonable manner. Our training data are composed of the features of trajectory and the features of moving objects. The feature vector contains the position, velocity in the image plane and the class information of the moving object.

Suppose the ith centroid of the object is (x_i, y_i), then the trajectory can be represented as $T = \{(x_1, y_1), (x_2, y_2), \cdots, (x_n, y_n)\}$ .Besides the position, velocities are also important features for the description of trajectory pattern. They can be simply denoted as $dx = x_{i+1} - x_i, dy = y_{i+1} - y_i$ since the trajectory is resampled at fixed time interval. We know different objects, such as pedestrian and vehicle, have different behavior manners. So, the class information should be considered when extracting the distribution of the trajectory. Then the flow vector can be denoted as $f = (c, x, y, dx, dy)$, here c is the class label. Thus for any motion trajectory, we can use a flow vector sequence $Q = [f_1, f_2, \cdots, f_n]$ to represent it. In order to reflect the greater difference between the flow vectors of objects with different class labels, we adopt the following formula to measure their distance:

$$d(f_i, f_j) = \beta\sqrt{(x_i - x_j)^2 + (y_i - y_j)^2 + (dx_i - dx_j)^2 + (dy_i - dy_j)^2} \tag{1}$$

Here, if $c_i = c_j$ then $\beta = 1.0$ else $\beta = 1.5$.

3.3 The Rival Penalized Competitive Learning Method for Neural Networks

Comparing to K-means and other clustering methods, competitive learning neural networks has the advantages of robustness and the on-line learning ability. But how to select an appropriate number of output neurons and avoid the influence of weight initialization are two difficult problems. For these problems, Xu had put forward an effective method named rival penalized competitive learning (RPCL) algorithm [11]. Its basic idea is that for each input, not only the weights of the winner unit are modified to adapt to the input, but also the weights of its rival are delearned by a smaller learning rate.

The primary steps of RPCL algorithm are as follows:

1) Selecting a relative great neuron number k and initialize their weights;

2) Choosing an input sample x randomly from the training set, and calculating the following formula for $i = 1, 2, ..., k$:

$$u_i = \begin{cases} 1 & \text{if } i = c \text{ that } \gamma_c \|x - w_c\|^2 = \min_j \gamma_j \|x - w_j\|^2 \\ -1 & \text{if } i = r \text{ that } \gamma_r \|x - w_r\|^2 = \min_{j, j \neq c} \gamma_j \|x - w_j\|^2 \\ 0 & \text{otherwise} \end{cases} \tag{2}$$

Where $\gamma_j = n_j / \sum_{i=1}^{K} n_i$ and n_i is the cumulative number of the occurrence of $u_i = 1$.

The import of parameter γ_j overcomes the "dead node" problem, by which the influence of the neuron weight initialization is eliminated.

3) Adjusting the weight of the competitive neuron according to the following formula:

$$\Delta w_i = \begin{cases} a_c(x-w_i) & \text{if } u_i = 1 \\ -a_r(x-w_i) & \text{if } u_i = -1 \\ 0 & \text{otherwise} \end{cases} \quad (3)$$

$$w_i = w_{i-1} + \Delta w_i \quad (4)$$

Where $0 \le a_r, a_c \le 1$ are the learning rates for the winner and rival unit respectively and it should be hold that $a_r << a_c$ in practice.

4) Running step 2) and 3) repeatedly until reach the maximal iteration or the weights of the neurons don't change obviously.

When using RPCL algorithm to train the neural network, a relative larger k is given at the beginning. With the development of the training, the redundant neurons will be repelled far from the training data. After the algorithm ends, re-labeling the training data and deleting the neurons which only have only very small number of training data corresponding to them. Then the remainder neurons will be the final result.

3.4 Anomaly Detection

Firstly, we resample the testing trajectory at the same time and space interval as the training trajectories. As traffic patterns are complex in real traffic video, the first flow vector of the test trajectory may not start from the position represented by the neurons of the first stage neural work. So, we establish point correspondence between the current trajectory and the multi-stage neural network simply by aligning. The algorithm for flow vector aligning can be described as following:

Algorihtm of flow vector aligning

Input:
$NS = \{W_1, W_2, \cdots, W_n\}$:the weight sets of the neurons in each stage neural network, here $W_i = (w_{i,1}, w_{i,2}, \cdots, w_{i,m_i})$ is the weights of the ith network
n :the number of the neural networks stages
m_i :the number of the neurons in the ith network
f_1 :the first flow vector of the test trajectory
Output:
t_s :the most correspondent neural network to f_1
t_n :the most correspondent neuron in the t_s network to f_1
```
Procedure Initial_flow_vector_aligning
Begin
    For i=1:n
        For j=1:m_i
```

```
            d_{i,j} = dis(f_1, w_{i,j}) ;
            // calculate the distance between f_1 and w_{i,j}
        end
        d_i = min_j d_{i,j} ;
    end
    t_s = arg min_i d_i ;
    t_n = arg min_j d_{t_s,j} ;
End.
```

Suppose the best correspondent network to f_1 is the t_s neural network. Then for each flow vector $f_i = (c_i, x_i, y_i, dx_i, dy_i)$ $(i = 1, 2, \cdots, k)$ of testing trajectory, finding the nearest neuron to it in the $t_s + i - 1$ neural network, and calculating the distance d between them. Checking the condition $\left|d - u_{ic'}^{*}\right| \leq 2\sigma_{ic'}^{*}$, if it is satisfied, then the flow vector is normal, else abnormal. Here $u_{ic'}^{*}$ is the average value of the distance between flow vectors f_i in the training trajectories and its correspondent neuron in the $t_s + i - 1$ network, and $\sigma_{ic'}^{*}$ is the standard variance. Counting the number m of the flow vectors being classified as abnormal and checking whether the condition $m / k > \delta$ is satisfied, i.e. when the abnormal flow vector in the test trajectory exceeds certain percentage, the test trajectory is classified as abnormal. For real-time application, the system needs to monitor trajectories as they are generated rather than waiting until a complete path is created. Obviously, our method can meet this requirement excellently.

4 Experiments and Analysis

We implemented the algorithm with Visual C++6.0 on the framework of DirectX9.0 and Windows XP platform. In the experiments, we tested different sites with CCD and infrared cameras respectively. As shown in Fig.2 (a),(c),(e),(g), the blue curves in the images are the training trajectories, and the white blocks shows the trajectory patter extracted by our method. From these figures, we can obviously find the primary trajectory patterns are ideally extracted in our method. Fig. 2 (b),(d),(f) and (h) give the testing trajectories and the novelty detection results. In these figures, the yellow curves are the trajectories detected as abnormal and the blue are normal. The green snippets represent the flow vectors which are classified as normal in the abnormal trajectories. From these, we can find the novel trajectories such as entering the park, passing across the intersection and promenades away from the road are all detected accurately. Because of the constrictions of experiments condition, the image sequences we test only include few moving objects and the trajectory pattern are also relatively simple. We still can draw the conclusion that despite of the complexity of outdoor scenes and tracking noises, our method also can learn the patterns hidden behind the training trajectories and detect the novel behavior satisfactorily.

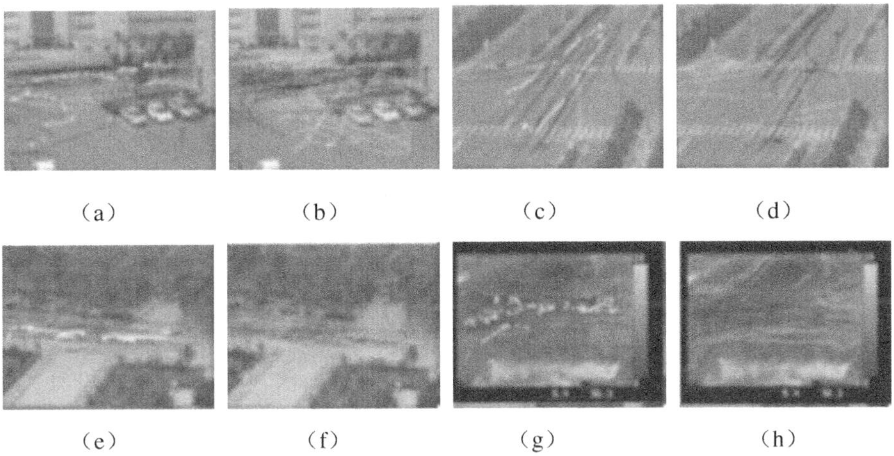

Fig. 2. The training and testing results of the proposed algorithm in the natural scenes with CCD and Infrared Cameras

5 Conclusions

In this paper, a new multi-stage competitive neural networks approach is proposed to learn the motion trajectory pattern and then a method based on flow vector aligning is given to detect the abnormal trajectory and the abnormal flow vectors in it. This approach has a very simple structure and can be successfully applied to real traffic video system. It not only can study the distribution of the flow vectors, but also can reflects the connotative time orders among the flow vectors of the trajectory. Comparing to the existing techniques, our approach has such advantages: 1) it overcomes the competitive networks' problems of output neuron number selection and reasonable weight initialization by using the rival penalized competitive learning method; 2) it can process the incoming data as the observed object is moving without the need to wait for the full track being acquired; 3) this method can not only detect the abnormal trajectories , but also can point out the points which are classified as abnormity in the abnormal trajectories; 4) it needn't to preprocess the training trajectories with same length. However, it is a pity in our approach, that we haven't expressed the trajectory pattern explicitly. So we can not provide the information for the activity prediction for the moment. Our future effort will involve this problem.

References

1. Collins, R., et al. A system for video surveillance and monitoring: VSAM final report. Carnegie Mellon University, Technical Report: CMU-RI-TR-00-12, 2000.
2. Haritaoglu, I., Harwood, D., Davis, L. W4: Real-time surveillance of people and their activities, IEEE Trans. on Pattern Analysis and Machine Intelligence 2000 **22** (2000) 809-830

3. Foresti, G.L.: Object recognition and tracking for remote video surveillance. IEEE Trans on Circuits and Systems for Video Technology **9** (1999) 1045-1062
4. Johnson, N., Hogg, D.: Learning the distribution of object trajectories for event recognition. Image and Vision Computing **14** (1996) 609-615
5. Sumpter, N., Bulpitt, A.: Learning spatio-temporal patterns for predicting object behavior. Image and Vision Computing **18** (2000) 697-704
6. Ownes, J., Hunter, A.: Application of the self-organizing map to trajectory classification. In: Proceedings of IEEE Workshop on Visual Surveillance (2000) 77-83
7. Hu,W.M., Dan Xie, D., Tieniu Tan, T.N.: A hierarchical self-organizing approach for learning the patterns of motion trajectories. IEEE Trans on Neural Networks **15**(2004) 135-144.
8. Zhouyu Fu, Z.Y., Hu,W.M., Tieniu Tan, T.N.: Similarity based vehicle trajectory clustering and anomaly detection. IEEE Conference on Image Processing (2005) 602-605
9. Hu,W.M., Dan Xie, D., Tieniu Tan, T.N. et al.: Learning activity patterns using fuzzy self-organizing neural works. IEEE Trans on Systems, Man and Cybernetics-Part B:Cybernetics. **34** (2004) 1618-1626
10. Khalid, S., Nafterl, A.: Classifying spatiotemporal object trajectories using unsupervised learning of basis function coefficients. VSSN (2005) 45-51
11. Xu, L., Krzyzak, A., Erkki, Oja.: Rival Penalized Competitive Learning for Clustering Analysis, RBF Net and Curve Detection. IEEE Trans on Neural Networks **4** (1993) 636-649

Neural Network-Based Robust Tracking Control for Nonholonomic Mobile Robot

Jinzhu Peng, Yaonan Wang, and Hongshan Yu

College of Electrical and Information Engineering, Hunan University, Changsha Hunan, P.R. China, 410082
pengjinzhu1980@126.com, yaonan@hnu.cn, yuhongshancn@hotmail.com

Abstract. A robust tracking controller with bound estimation based on neural network is proposed to deal with the unknown factors of nonholonomic mobile robot, such as model uncertainties and external disturbances. The neural network is to approximate the uncertainties terms and the interconnection weights of the neural network can be tuned online. And the robust controller is designed to compensate for the approximation error. Moreover, an adaptive estimation algorithm is employed to estimate the bound of the approximation error. The stability of the proposed controller is proven by Lyapunov function. The proposed neural network-based robust tracking controller can overcome the uncertainties and the disturbances. The simulation results demonstrate that the proposed method has good robustness.

1 Introduction

The tracking control of nonholonomic mobile robot has been a topic of research during recent years. The characteristic of the nonholonomic system is that the constraints, which are imposed on the motion, are not integratable, i.e., the constraints cannot be written as time derivatives of some functions of the generalized co-ordinates. It is a typical nonholonomic mechanical system with high nonlinearity and its control is very difficult. It is also a typical nonlinear uncertain system with both the parametric uncertainty in the dynamic model of the robot including motor dynamics and disturbances from the external environment or unmodelled dynamics.

For the tracking control problem of the mobile robot, lots of control methods have been applied. J. M. Yang and J. H. Kim [1] proposed a robust tracking controller for nonholonomic wheeled mobile robots using sliding mode technique. Y. Kanayama et al. [2] developed smooth static time invariant state feedback for a velocity-controlled mobile robot with nonholonomic constraint. In [3-8], the backstepping technique was used to design the adaptive and robust controller for the nonholonomic system. M.S. Kim et al. [9] applied a robust adaptive dynamic controller for a nonholonomic mobile robot with modeling uncertainty and disturbances.

In recent years, intelligent systems, such as fuzzy logic [10] and neural network [5, 11-13], have been applied to approximate the models or to deal with the

D. Liu et al. (Eds.): ISNN 2007, Part I, LNCS 4491, pp. 804–812, 2007.

disturbances and dynamic uncertainties of dynamic systems [14, 15]. F. M. Raimondi, M. Melluso [10] developed a new theoretical control method based on the dynamic behavior of a wheeled vehicle, where a mechanism of fuzzy inference for designing a robust control system was present. In [5], a robust motion controller based neural network and backstepping technique is proposed for a two-DOF low-quality mobile robot. In [11-13], the neural network controllers in the proposed control structure were to deal with unmodeled bounded disturbances and unstructured unmodeled dynamics in the vehicle.

In this paper, we proposed a neural network-based robust tracking controller for a mobile robot with nonholonomic constrains. The proposed controller can guarantee robustness to parametric and dynamics uncertainties and also rejects any bounded, immeasurable disturbances entering the system. The stability is proved by the Lyapunov theory.

2 Dynamic Model of a Nonholonomic Mobile Robot

2.1 Preliminary Definitions

A mobile robot is shown in Fig.1, which contains two driven wheels mounted on the same axis and a castor. It is a typical example of a nonholonomic mechanical system.

An inertial Cartesian frame $\{O, X, Y\}$ linked to the world and $\{C, X_C, Y_C\}$ linked to the mobile platform are used here. It is assumed that the center of mass of the mobile robot is local in C. The pose of the mobile robot is completely specified by:

$$\boldsymbol{q} = [x, y, \theta]^T \tag{1}$$

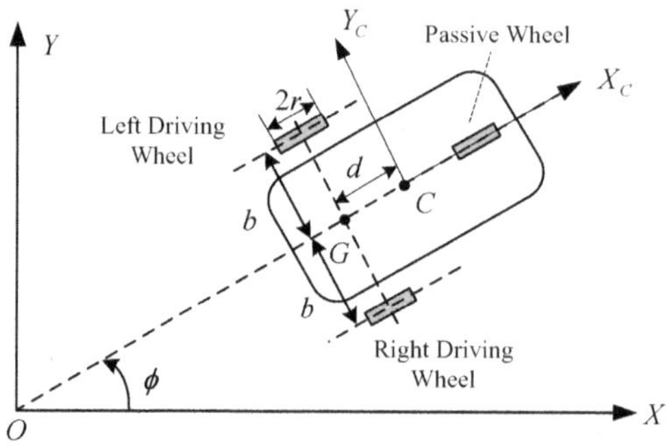

Fig. 1. A nonholonomic mobile robot

The nonholonomic constraint states that the mobile robot satisfies the conditions of pure rolling and non-slipping, i.e., the mobile robot can only move in the direction normal to the axis of the driving wheels:

$$\dot{y}\cos\theta - \dot{x}\sin\theta - d\dot{\theta} = 0 \tag{2}$$

2.2 Dynamic Model of a Nonholonomic Mobile Robot

Consider a nonholonomic mobile robot system with n generalized coordinate q and subject to m constrains can be described by [12]:

$$\boldsymbol{M}(\boldsymbol{q})\ddot{\boldsymbol{q}} + \boldsymbol{C}(\boldsymbol{q}, \dot{\boldsymbol{q}})\dot{\boldsymbol{q}} + \boldsymbol{F}(\boldsymbol{q}, \dot{\boldsymbol{q}}) + \tau_d = \boldsymbol{B}(\boldsymbol{q})\tau - \boldsymbol{A}^T(\boldsymbol{q})\lambda \tag{3}$$

$$\boldsymbol{A}(\boldsymbol{q})\dot{\boldsymbol{q}} = \mathbf{0} \tag{4}$$

where $\boldsymbol{M}(\boldsymbol{q}) \in \Re^{n\times n}$ is a symmetric, positive definite inertia matrix, $\boldsymbol{C}(\boldsymbol{q}, \dot{\boldsymbol{q}}) \in \Re^{n\times n}$ is the centripetal and coriolis matrix, $\boldsymbol{F}(\boldsymbol{q}, \dot{\boldsymbol{q}}) \in \Re^{n}$ denotes the surface friction and the gravitational vector, $\tau_d \in \Re^n$ denotes bounded unknown disturbances including unstructured unmodeled dynamics, $\boldsymbol{B}(\boldsymbol{q}) \in \Re^{n\times r}$ is the input transformation matrix, $\tau \in \Re^r$ is the input vector, $\boldsymbol{A}^T(\boldsymbol{q}) \in \Re^{m\times n}$ is the matrix associated with the constrains, $\lambda \in \Re^m$ is the vector of constrain forces.

Let $\boldsymbol{S}(\boldsymbol{q}) = [s_1(q), \cdots, s_{n-m}(q)]$ be a set of smooth and linearly independent vector fields in $\emptyset(A)$, the null space of $\boldsymbol{A}$, i.e.,

$$\boldsymbol{S}^T(\boldsymbol{q})\boldsymbol{A}^T(\boldsymbol{q}) = 0 \tag{5}$$

It is possible to find a velocity vector $\boldsymbol{v}(t) \in \Re^{n-m}$, such that

$$\dot{\boldsymbol{q}} = \boldsymbol{S}(\boldsymbol{q})\boldsymbol{v}(t) \tag{6}$$

Multiplying both sides by $\boldsymbol{S}^T(\boldsymbol{q})$ and using (5), we have

$$\boldsymbol{S}^T\boldsymbol{M}\boldsymbol{S}\dot{\boldsymbol{v}} + \boldsymbol{S}^T(\boldsymbol{M}\dot{\boldsymbol{S}} + \boldsymbol{C}\boldsymbol{S})\boldsymbol{v} + \boldsymbol{S}^T\boldsymbol{F} + \boldsymbol{S}^T\tau_d = \boldsymbol{S}^T\boldsymbol{B}\tau \tag{7}$$

$$\bar{\boldsymbol{M}}\dot{\boldsymbol{v}} + \bar{\boldsymbol{C}}\boldsymbol{v} + \bar{\boldsymbol{F}} + \bar{\tau}_d = \bar{\tau} \tag{8}$$

where $\boldsymbol{v} = [v, \omega]^T$, v is the velocity of mobile robot, ω is the angle velocity, $\bar{\boldsymbol{M}} = \boldsymbol{S}^T\boldsymbol{M}\boldsymbol{S}$, $\bar{\boldsymbol{C}} = \boldsymbol{S}^T(\boldsymbol{M}\boldsymbol{S}+\boldsymbol{C}\boldsymbol{S})$, $\bar{\boldsymbol{F}} = \boldsymbol{S}^T\boldsymbol{F}$, $\bar{\tau}_d = \boldsymbol{S}^T\tau_d$, $\bar{\tau} = \boldsymbol{S}^T\boldsymbol{B}\tau$.

Property 1. $\bar{\boldsymbol{M}}$ is a symmetric positive definite matrix.

Property 2.

$$\bar{M}_{\min} \le \|\bar{\boldsymbol{M}}(\boldsymbol{q})\| \le \bar{M}_{\max}, \|\bar{\boldsymbol{C}}(\boldsymbol{q}, \dot{\boldsymbol{q}})\| \le \bar{C}_b\|\dot{\boldsymbol{q}}\| \tag{9}$$

where $\bar{M}_{\min}$, $\bar{M}_{\max}$, $\bar{C}_b$ are some positive constants that assumed to be unknown. and $\|\cdot\|$ denotes Euclid norm.

Property 3. The matrix $\left(\dot{\bar{\boldsymbol{M}}}(\boldsymbol{q}) - 2\bar{\boldsymbol{C}}(\boldsymbol{q}, \dot{\boldsymbol{q}})\right)$ is skew-symmetric.

Assumption 1. The friction and gravity are bounded by $\|\bar{\boldsymbol{F}}(\boldsymbol{q}, \dot{\boldsymbol{q}})\| \le \xi_0 + \xi_1\|\dot{\boldsymbol{q}}\|$, where ξ_0 and ξ_1 are some positive constants.

Assumption 2. Disturbance is bounded by $\|\bar{\tau}_d\| \le \bar{\tau}_D$, where $\bar{\tau}_D$ is a positive constant.

For a two-wheeled mobile robot, the kinematic model can be given as [2]:

$$\begin{bmatrix} \dot{x} \\ \dot{y} \\ \dot{\theta} \end{bmatrix} = \begin{bmatrix} \cos\theta & -d\sin\theta \\ \sin\theta & d\cos\theta \\ 0 & 1 \end{bmatrix} \cdot \begin{bmatrix} v \\ \omega \end{bmatrix} \tag{10}$$

In order to simplify the problem formulation, it is assumed that $d = 0$. The alternative formulations can be readily deduced when $d \neq 0$[5].

Suppose the mobile robot is required to follow a reference trajectory, with position and velocity are

$$\begin{cases} \boldsymbol{q}_r = [x_r, y_r, \theta_r]^T \\ \boldsymbol{v}_r = [v_r, \omega_r]^T \end{cases} \tag{11}$$

Then the tracking error expressed with respect to a frame fixed on the mobile robot are given as [2]

$$\boldsymbol{e}_q = \begin{bmatrix} e_1 \\ e_2 \\ e_3 \end{bmatrix} = \begin{bmatrix} \cos\theta & \sin\theta & 0 \\ -\sin\theta & \cos\theta & 0 \\ 0 & 0 & 1 \end{bmatrix} \cdot \begin{bmatrix} x_r - x \\ y_r - y \\ \theta_r - \theta \end{bmatrix} \tag{12}$$

The Lyapunov candidate is chosen as

$$L_1 = \frac{1}{2}\left({e_1}^2 + {e_2}^2\right) + \frac{(1 - \cos e_3)}{k_2} \tag{13}$$

Differentiating L_1, then we obtain

$$\begin{aligned} \dot{L}_1 &= e_1\dot{e}_1 + e_2\dot{e}_2 + \frac{\sin e_3}{k_2}\dot{e}_3 \\ &= e_1(-v + v_r\cos e_3) + \frac{\sin e_3}{k_2}(\omega_r - \omega + k_2 e_2 v_r) \end{aligned} \tag{14}$$

The velocity control law $\boldsymbol{v}_d$ achieves stable tracking of the mobile robot for the kinematic model (10) as:

$$\boldsymbol{v}_d = \begin{bmatrix} k_1 e_1 + v_r\cos e_3 \\ \omega_r + k_2 e_2 v_r + k_3 \sin e_3 \end{bmatrix} \tag{15}$$

where $k_1 > 0, k_2 > 0, k_3 > 0$ are the controller gains.

Then, the equation (14) can be rewritten as

$$\dot{L}_1 = -k_1{e_1}^2 - \frac{k_3}{k_2}\sin{e_3}^2 \leq 0 \tag{16}$$

The velocity control law (15) may achieve theoretical stability with respect to a reference trajectory. In practice, however, the velocity $\boldsymbol{v}_d$ cannot be generated directly by the motors. Instead, the motor provide a control torque to the wheels, which will result in an actual velocity $\boldsymbol{v}$. So it is necessary to design the torque for the robot system.

3 Neural Network-Based Robust Control with Bound Estimation

Dynamics of mobile robotic are highly nonlinear and may contain uncertain elements. Many efforts have been made in developing control schemes to achieve the precise tracking control of mobile robot [8]. In order to control the mobile robot effectively, a neural network-based robust controller with bound estimation is proposed in this paper.

First, we define the velocity tacking error as

$$\boldsymbol{e} = \boldsymbol{v} - \boldsymbol{v}_d \tag{17}$$

Then, define a filtered tracking error as

$$\boldsymbol{r} = \boldsymbol{k}_4 \boldsymbol{e} \tag{18}$$

where $\boldsymbol{k}_4$ is a positive coefficient vector. The time derivative of the filtered tracking error can be written as

$$\dot{\boldsymbol{r}} = -\bar{\boldsymbol{M}}^{-1}(\bar{\boldsymbol{C}}\boldsymbol{v} + \bar{\boldsymbol{F}} + \tau_d) + \bar{\boldsymbol{M}}^{-1}\tau - \dot{\boldsymbol{v}}_d \tag{19}$$

In general, the inertia matrix is known while uncertainties in the centripetal and coriolis matrix are sometimes difficult to compute. So, the first term of the equation (19), denoted by f, is an unknown smooth function.

$$f = \boldsymbol{W}^T \sigma(\boldsymbol{V}^T x) + \varepsilon \tag{20}$$

where the neural network approximation error ε is assumed to be bounded by $\|\varepsilon\| \leq \Delta.\sigma(\cdot)$ is a continuous sigmoid activation function. The first layer weights $\boldsymbol{V}$ are selected randomly and will not be tuned while the second layer weights $\boldsymbol{W}$ are tunable. The ideal neural network weights in vectors $\boldsymbol{W}$ that are needed to best approximate the given function f are difficult to determine. All one needs to know for control purposes is that, for a specified value of E some ideal approximating neural network weights exist. Then, an estimate value of f can be given by

$$\hat{f} = \hat{\boldsymbol{W}}^T \sigma(\boldsymbol{V}^T x) \tag{21}$$

where $\hat{\boldsymbol{W}}$ is the estimated value of $\boldsymbol{W}$.

Choose the tracking control law as

$$\tau = \bar{\boldsymbol{M}}\left\{-\hat{\boldsymbol{W}}^T \sigma(\boldsymbol{V}^T x) - \varphi\right\} \tag{22}$$

where φ is the robust controller, and the first term is the neural network controller.

Then, equation (19) can be rewritten as

$$\dot{\boldsymbol{r}} = \tilde{\boldsymbol{W}}^T \sigma(\boldsymbol{V}^T x) - \varphi + \varepsilon_d \tag{23}$$

where $\tilde{W} = W - \hat{W}$ is the estimation error, $\varepsilon_d = \varepsilon - \bar{M}\tau_d$ is the uncertain term of the approximation error and the disturbances. According to (21) and Assumption 2, we can know that the uncertain term is bounded, that is,

$$\|\varepsilon_d\| = \|\varepsilon\| + \|\bar{M}\|\|\tau_d\| \leq \Delta + \bar{M}_{\max}\tau_D = E \tag{24}$$

Theorem: Given the system (8), choose the velocity control law (15), the tracking control law (23), and the adaptation law of the neural network as

$$\dot{\hat{W}} = -\dot{\tilde{W}} = \Gamma r \sigma(V^{\mathrm{T}}\mathrm{x}) \tag{25}$$

where $\Gamma > 0$ is the learning rate of the neural network.

In (23), the robust controller is designed as

$$\varphi = -\hat{E}\mathbf{sgn}(r) \tag{26}$$

where $\hat{E}$ is the estimated value of E, $\mathbf{sgn}(\cdot)$ is a standard sign function. And the bound estimation law is chosen as

$$\dot{\hat{E}} = -\dot{\tilde{E}} = \eta r\mathbf{sgn}(r) \tag{27}$$

where $\tilde{E} = E - \hat{E}$ is the estimation error, η is a positive constant.

Then, the closed-loop system (8) and (23) is asymptotically stable, the filtered error r, the neural network weights error $\tilde{W}$ and the bounded estimation error $\tilde{E}$ are all bounded.

Proof: Choose Lyapunov function candidate as

$$L = L_1 + \frac{1}{2}r^2 + \frac{1}{2}\tilde{W}^T\Gamma^{-1}\tilde{W} + \frac{1}{2}\tilde{E}^{\mathrm{T}}\eta^{-1}\tilde{E} \tag{28}$$

Differentiating yields

$$\dot{L} = \dot{L}_1 + r\dot{r} + \tilde{W}^T\Gamma^{-1}\dot{\tilde{W}} + \tilde{E}^{\mathrm{T}}\eta^{-1}\dot{\tilde{E}} \tag{29}$$

Substituting (16), (24), and (25)-(28) into (30), we can obtain:

$$\begin{aligned}
\dot{L} &= \dot{L}_1 + r(-\tilde{W}^T\sigma(V^Tx) - \varphi + \varepsilon_d) + \tilde{W}\Gamma^{-1}\dot{\tilde{W}} - \tilde{E}^{\mathrm{T}} r\mathbf{sgn}(r) \\
&= \dot{L}_1 + r(\varepsilon_d - \hat{E}\mathbf{sgn}(r)) - \tilde{E}^T r\mathbf{sgn}(r) - \tilde{W}^T(r\sigma(V^Tx) + \Gamma^{-1}\dot{\tilde{W}}) \\
&\leq r\varepsilon_d - \tilde{E}^T r\mathbf{sgn}(r) - (E - \hat{E})r\mathbf{sgn}(r) \\
&\leq -\|r\|(E - \|\varepsilon_d\|) = -\alpha\|r\| \leq 0
\end{aligned} \tag{30}$$

where $\alpha = E - \|\varepsilon_d\| > 0$ is a small positive constant. Since $\dot{L} \leq 0$, it can be inferred that the filtered error r, the neural network weights error $\tilde{W}$ and the bounded estimation error $\tilde{E}$ are all bounded. Let function $\Xi(\mathrm{t}) = -\dot{L} = \alpha\|r\|$, and integrate function $\Xi(\mathrm{t})$ with respect to time[14, 15]

$$\int_0^t \Xi(\tau)\mathrm{d}\tau \leq L(r(0), \tilde{W}, \tilde{E}(0)) - L(r(\mathrm{t}), \tilde{W}, \tilde{E}(\mathrm{t})) \tag{31}$$

Because $L(\boldsymbol{r}(0), \tilde{\boldsymbol{W}}, \tilde{E}(0))$ is bounded, and $L(\boldsymbol{r}(t), \tilde{\boldsymbol{W}}, \tilde{E}(t))$ is nonincreasing and bounded, the following result is obtained

$$\lim_{t\to\infty} \int_0^t \Xi(\tau)\mathrm{d}\tau \leq 0 \tag{32}$$

In addition, $\dot{\Xi}(t)$ is bounded, by Barbalat's Lemma, it can be show that $\lim_{t\to\infty} \Xi(\tau) \leq 0$. That is, $\boldsymbol{r}(t) \to 0$ as $t \to 0$. As a result, the closed-loop system (8) and (23) is asymptotically stable.

4 Simulation Results

In order to verify the validity of the proposed controller, a nonholonomic mobile robot is used for illustration in this paper, as shown in Fig. 1. The dynamical equations of the mobile robot can be expressed in (1) where [12]

$$\boldsymbol{M}(\boldsymbol{q}) = \begin{bmatrix} m & 0 & md\sin\theta \\ 0 & m & -md\cos\theta \\ md\sin\theta & -md\cos\theta & \mathtt{I} \end{bmatrix}, \boldsymbol{C}(\boldsymbol{q}, \dot{\boldsymbol{q}}) = \begin{bmatrix} 0 \; 0 \; md\dot{\theta}\cos\theta \\ 0 \; 0 \; md\dot{\theta}\sin\theta \\ 0 \; 0 \quad\; 0 \end{bmatrix},$$

$$\boldsymbol{B}(\boldsymbol{q}) = \frac{1}{r}\begin{bmatrix} \cos\theta \; \sin\theta \\ \sin\theta \; \sin\theta \\ b \quad b \end{bmatrix}, \tau = \begin{bmatrix} \tau_r \\ \tau_l \end{bmatrix}, \boldsymbol{A}^T(\boldsymbol{q}) = \begin{bmatrix} -\sin\theta \\ \cos\theta \\ -d \end{bmatrix}, \lambda = -m(\dot{x}_c\cos\theta + \dot{y}_c\sin\theta)\dot{\theta}$$

where $m = 10\mathtt{kg}, I = 5\mathtt{kg}\cdot\mathtt{m}^2, b = 0.25\mathtt{m}, r = 0.05\mathtt{m}$, and $v_r = 0.5\mathtt{m/s}$. The external disturbance $|\tau_{di}| \leq 3.0$ is a random noise with the magnitude bounded.

The initial values of neural network weights $\boldsymbol{W}$ are selected randomly in [-1, 1], and the estimations are $\hat{E}(0) = [0, 0]^T$, The controller gains are $k_1 = 10, k_2 = 5, k_3 = 4$ and $\boldsymbol{k}_4 = \mathtt{diag}\{10, 10\}$.

Defining a straight line, starting from $\boldsymbol{q}_r(0) = [x_r(0), y_r(0), \theta_r(0)]^T = [0, 1, 45°]^T$, The mobile robot, however, is initially at $\boldsymbol{q}(0) = [x(0), y(0), \theta(0)]^T = [1, 0, 0°]^T$. where $\theta(0) = 0°$ indicates that the robot is heading toward positive direction of x.

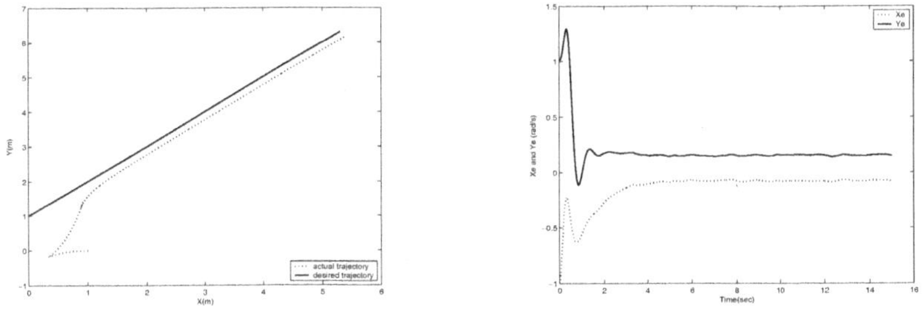

(a) Trajectory in the (x, y) plane

(b) Position errors

Fig. 2. Results by computed torque controller

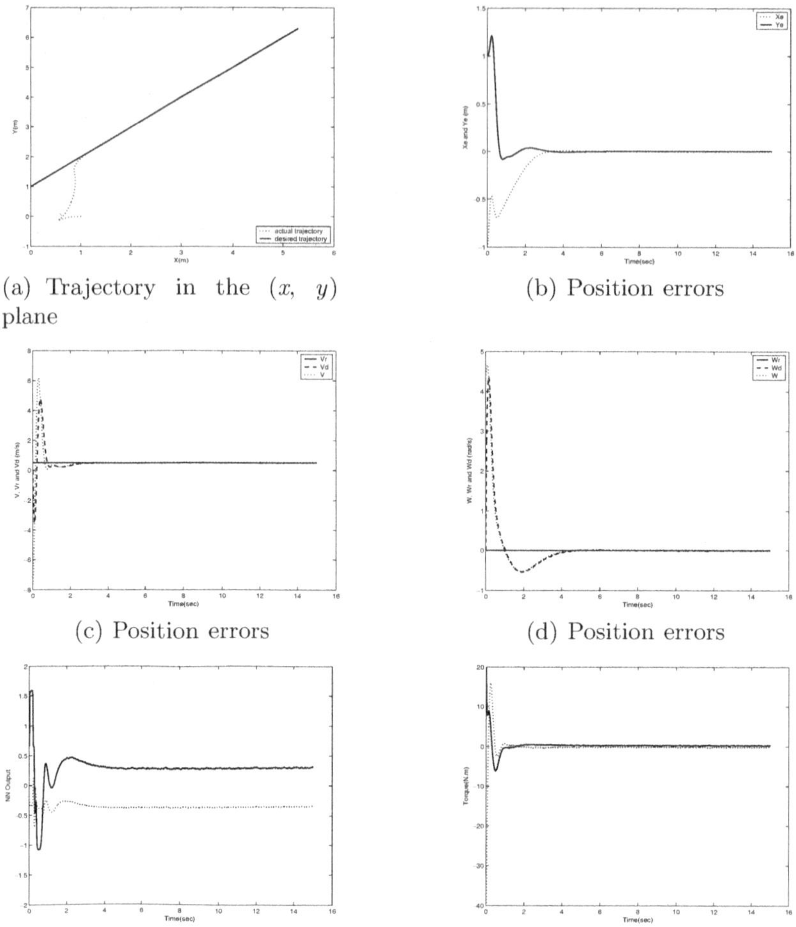

(a) Trajectory in the $(x,\ y)$ plane

(b) Position errors

(c) Position errors

(d) Position errors

(e) Position errors

(f) Position errors

Fig. 3. Results by the proposed method

Fig. 2 shows the simulation results for tracking a straight line using computed torque method. Since there are the uncertainties and disturbance, the mobile robot cannot track the trajectory and exhibit a steady state error. Under the same conditions, Fig. 3 shows the results for tracking a straight line using the proposed method. As it can be seen from the figure, the mobile robot can reach the line quickly and continues to track it.

5 Conclusions

Using the robust and neural network methods, a robust tracking controller with bounded estimation based on neural network is proposed for a nonholonomic mobile robot. This controller can guarantee robustness to parametric and dynamics uncertainties and also rejects any bounded, immeasurable disturbances entering the system. The stability is proven using the Lyapunov method. The velocity error, the neural network weights error and the bounded estimation error are all bounded. Finally, some simulation examples are utilized to illustrate the control performance.

Acknowledgments. The authors would like to acknowledge the support of the National Natural Science Foundation of China (No. 60375001).

References

1. Yang, J.M. and Kim, J.H.: Sliding mode control for trajectory of nonholonomic wheeled mobile robots, IEEE Trans. on Robotics and Automation **15**(3) (1999) 578–587.
2. Kanayama, Y., Kimura, Y., Miyazaki, F. and Noguchi, T.: A stable tracking control method for an autonomous mobile robot, Proc. IEEE Int. Conf Robot. Automar (1990) 384–389.
3. Jiang, Z.P. and Nijmeijer, H.: Tracking control of mobile robots: a case study in backstepping, Automatica **33**(7) (1997) 1393–1399.
4. Fierro, R. and Lewis, F.L.: Control of a nonholonomic mobile robot: backstepping kinematics into dynamics, Journal of Robotic Systems **14**(3) (1997) 149–163.
5. Zhang, Q., Shippen, J. and Jones, B.: Robust backstepping and neural network control of a low-quality nonholonomic mobile robot, Int. J. of machine Tools and manufacture **39** (1999) 1117–1134.
6. Lee, T.C., Song, K.T., Lee, C.H. and Teng, C.C.: Tracking control of mobile robots using saturation feedback controller, Proc. of IEEE International Conf. on Robotics and Automation (1999) 2639–2644.
7. Kim, M.S., Shin, J.H., Hong, S.G. et al.: Designing a robust adaptive dynamic controller for nonholonomic mobile robots under modeling uncertainty and disturbances, Mechatronics **13** (2003) 507–519.
8. Sarker, N., Yun, X. and Kumar, V.: Control of mechanical system with rolling constraints: application to dynamic control of mobile robots, Int. J. Robot. Res. **13**(1) (1994) 55–69.
9. Yildirim, S.: Adaptive robust neural controller for robots, Robotics and Autonomous Systems **46** (2004) 175–184.
10. Raimondi, F.M. and Melluso, M.: A new fuzzy robust dynamic controller for autonomous vehicles with nonholonomic constraints, Robotics and Autonomous Systems **52** (2005) 115–131.
11. Lin, S. and Goldenberg, A.A.: Neural-network control of mobile manipulators, IEEE Trans. Neural Networks **12**(5) (2001) 1121–1133.
12. Fierro, R. and Lewis, F.L.: Control of a nonholonomic mobile robot using neural networks, IEEE Trans. Neural Networks **9**(4) (1998) 589–600.
13. Oh, C., Kim, M.S., Lee, J.Y. et al.: Control of mobile robots using RBF network, Proc. IEEE/RSJ Conf. on Intelligent Robots and Systems (2003) 3528–3533.
14. Lin, C.M. and Hsu, C.F.: Neural-network-based adaptive control for induction servomotor drive system, IEEE Trans. on Industrial Electronics **49**(1) (2002) 115–123.
15. Wai, R.J.: Tracking control based on neural network strategy for robot manipulator, Neurocomputing **51** (2003) 425–445.

Enhance Computational Efficiency of Neural Network Predictive Control Using PSO with Controllable Random Exploration Velocity

Xin Chen and Yangmin Li

Department of Electromechanical Engineering, Faculty of Science and Technology, University of Macau, Av. Padre Tomás Pereira S.J., Taipa, Macao SAR, P.R.China
{ya27407, ymli}@umac.mo

Abstract. NNPC has been used widely to control nonlinear systems. However traditional gradient decent algorithm (GDA) needs a large computational cost, so that NNPC is not acceptable for systems with rapid dynamics. To apply NNPC in fast control of mobile robots, the paper proposes an improved optimization technique, particle swarm optimization with controllable random exploration velocity (PSO-CREV), to replace of GDA in NNPC. Therefore for one cycle of control, PSO-CREV needs less iterations than GDA, and less population size than conventional PSO. Hence the computational cost of NNPC is reduced by using PSO-CREV, so that NNPC using PSO-CREV is more feasible for the control of rapid processes. As an example, a test of trajectory tracking using mobile robots is chosen to compare performance of PSO-CREV with other algorithms to show its advantages, especially on the aspect of computational time.

1 Introduction

For many industrial processes which usually exhibit multivariable and nonlinear dynamical behavior, neural network predictive control (NNPC) is widely used to generate control signals. NNPC gradient decent algorithm (GDA) is viewed as the basic iterative method to derive control signals, where a well-known Jacobian matrix should be calculated. If prediction horizon is chosen relatively large, the computational cost induced by the complex Jacobian becomes so large that it is difficult to employ NNPC for nonlinear systems with rapid dynamics.

To speed up the computation of NNPC, some improvements have been developed. For example, a recursive form for computation of elements of Jacobian matrix is proposed [1]. At the same time, some improvements about NN structure have also been developed to reduce complexity of NN structure [2]. From the view of optimization, the process of computation of future control signals is also the process to optimize certain performance criterions. Therefore some evolutionary computation technologies, such as GA, can be employed to find out proper control signals. However due to relative slow convergence of GA, using it to obtain control signals is not able to make NNPC be feasible in rapid dynamic processes yet. Instead of GA, particle swarm optimization (PSO) [3,4] looks like a good alternative

D. Liu et al. (Eds.): ISNN 2007, Part I, LNCS 4491, pp. 813–823, 2007.

because of its fast convergence, so that the complex computation of Jacobian can be avoided.

The computational cost at each iteration of PSO is determined by the population size. To make PSO alternate GDA in NNPC, we need to improve PSO exploration ability in order that a swarm with relative small population size can find the optimal solution. For this purpose, a PSO with controllable random exploration velocity (PSO-CREV) is proposed in this paper.

2 Overview on Neural Network Predictive Control (NNPC)

2.1 Conventional Algorithm of NNPC

Given an unknown nonlinear system, its model can be expressed as

$$y(t+1) = f(q) = f(y(t-1), y(t-2), \cdots, y(t-n_y), u(t-d-1), \cdots, u(t-d-n_u)). \tag{1}$$

A multi-layer perceptron neural network (MLP) is designed as neural network model (NNM) to approximate the plant, i.e.,

$$\hat{y}(t+1) = \hat{f}(q) = W[tanh(Vq + b_1)] + b_2,$$

where W, V, b_1, and b_2 are the weights and biases matrices.

As a result, the well trained NNM can be chosen as alternative to the real plant. If let T denote a prespecified predictive range, then the reference point in future is denoted by $R_t = [r(t+1), r(t+2), \cdots, r(t+T)]^T$. And $\hat{Y}(t) = [\hat{y}(t+1), \hat{y}(t+2), \cdots, \hat{y}(t+T)]^T$ denotes the predictive output of NNM. The error vector of predictive control is denoted by $E_t = [e(t+1), e(t+2), \cdots, e(t+T)]^T$, where $e(t+i) = r(t+i) - \hat{y}(t+i)$. The purpose of NNPC is to derive a control vector denoted by $U_t = [u(t-d+1), u(t-d+2), \cdots, u(t-d+T)]^T$,such that the following objective function is minimized.

$$J_C = \frac{1}{2}[E_t^T E_t]. \tag{2}$$

The conventional algorithm to derive control vector is based on gradient decent rule, which is in the form of

$$U_t(k+1) = U_t(k) - \eta \frac{\partial J_C}{\partial U_t}(k) \tag{3}$$

where

$$\frac{\partial J_C}{\partial U_t}(k) = \begin{bmatrix} \frac{\partial J_C}{\partial u(t+1)} \\ \vdots \\ \frac{\partial J_C}{\partial u(t+T)} \end{bmatrix} = \begin{bmatrix} \frac{\partial \hat{y}(t+1)}{\partial u(t+1)} & 0 & \cdots & 0 \\ \frac{\partial \hat{y}(t+2)}{\partial u(t+1)} & \frac{\partial \hat{y}(t+2)}{\partial u(t+2)} & \cdots & 0 \\ \vdots & \vdots & \ddots & \vdots \\ \frac{\partial \hat{y}(t+T)}{\partial u(t+1)} & \frac{\partial \hat{y}(t+T)}{\partial u(t+2)} & \cdots & \frac{\partial \hat{y}(t+T)}{\partial u(t+T)} \end{bmatrix}^T \cdot$$

This matrix is called the Jacobian matrix, which implies that NNPC requires a huge computational effort to generate control signals using the iterative algorithm in (3).

2.2 Problems of NNPC in Real-Time Applications

The primary shortage of NNPC results from its computational cost. Summarily the two important reasons resulting in huge computation cost are listed as follows:

- To avoid premature convergence, the fixed learning rate η should not be selected too large, so that the convergence of gradient decent rule is relative slow;
- Due to Jacobian matrix, if it is required to derive predictive control signals in future, the dimension of Jacobian becomes too large to be computed rapidly.

3 NNPC Based on Particle Swarm Optimization with Controllable Random Exploration Velocity (PSO-CREV)

To apply NNPC in plants with rapid dynamics, the iterative algorithm should be speeded up to overcome the both shortages mentioned in Section 2.2. PSO algorithm looks like a good alterative to gradient decent rule to avoid computing Jacobian and reduce iterations referred in (3).

Roughly speaking, if the control vector U_t is viewed as a solution vector in a certain solution space, a group of particles represent potential solutions of U_t. When they converge to a position in solution space, then it is the optimal control signal found by PSO. Since there is no gradient decent algorithm involved, the computation of Jacobian can be avoided.

In conventional PSO, the exploration of particles is determined by the cognitive and the social components. To improve exploration ability, the only method for traditional PSO is to increase the number of particles. But for application involving rapid dynamics, increasing population size means increasing computational burden and reducing the feasibility of PSO-NNPC! Hence instead of adding particles, improving exploration ability of particles seems a direct way to improve accuracy of PSO-NNPC, while the computational cost is kept acceptable. For this purpose, an improved PSO algorithm, named PSO with controllable random exploration velocity (PSO-CREV) is proposed.

3.1 Definition of PSO with Controllable Random Exploration Velocity

A *PSO with Controllable Random Exploration Velocity* (PSO-CREV) is described as follows. Let $\mathcal{F}(n)$ be a sequence of sub-σ-algebra of $\mathcal{F}$ such that $\mathcal{F}(n) \subset \mathcal{F}(n+1)$, for all n. For a swarm including M particles, the position of particle i is defined as $X_i = [\, x_{i1} \quad x_{i2} \quad \cdots \quad x_{iD} \,]^T$, where D represents the dimension of swarm space. The updating principle for individual particle is defined as

$$\begin{aligned} v_{id}(n+1) &= \varepsilon(n)\left[v_{id}(n) + c_1 r_{1id}(n)(P_{id}^d(n) - X_{id}(n)) \right. \\ &\left. \quad + c_2 r_{2id}(n)(P_{id}^g(n) - X_{id}(n)) + \xi_{id}(n)\right] \\ X_{id}(n+1) &= \alpha X_{id}(n) + v_{id}(n+1) \\ &\quad + \tfrac{1-\alpha}{\phi_{id}(n)}(c_1 r_{1id}(n) P_{id}^d(n) + c_2 r_{2id}(n) P_{id}^g(n)), \end{aligned} \tag{4}$$

where $d = 1, \cdots, D$, c_1 and c_2 are positive constants; $r_{1id}(n)$ and $r_{2id}(n)$ are $\mathcal{F}(n)$-measurable random variables; $P_i^d(n)$ represents the best position that particle i has found so far, which is of the form $P_i^d(n) = \arg\min_{k \le n} F(X_i(k))$, where $F(\cdot)$ represents a fitness function to be decreased; $P_i^g(n)$ represents the best position found by particle i's neighborhood, which is of the form $P_i^g(n) = \arg\min_{j \in \Pi_i} F(X_j(n))$; $\phi_i(n) = \phi_{1i}(n) + \phi_{2i}(n)$, where $\phi_{1i}(n) = c_1 r_{1i}(n)$, $\phi_{2i}(n) = c_2 r_{2i}(n)$.

Suppose the following assumptions hold:

(1) $\xi_i(n)$ is a bounded random variable with continuous uniform distribution. It has a constant expectation denoted by $\Xi_i = \mathbf{E}\xi_i(n)$;
(2) $\varepsilon(n) \to 0$ with n increasing, and $\sum_{n=1}^{\infty} \varepsilon(n) = \infty$;
(3) $0 < \alpha < 1$;
(4) $r_{1id}(n)$ and $r_{2id}(n)$ are independent random variables satisfying continuous uniform distribution in $[0, 1]$, or $r_{1id} \sim U(0,1)$ and $r_{2id} \sim U(0,1)$. And denote $\Phi_{1i} = \mathbf{E}\phi_{1i}(n)$ and $\Phi_{2i} = \mathbf{E}\phi_{2i}(n)$ respectively.

Then swarm must converge with probability one.

In addition, let $P^* = \inf_{\lambda \in (\mathbf{R}^D)} F(\lambda)$ represent the global optimal position in solution space. Then swarm must converge to P^* if $\lim_{n\to\infty} P_i^d(n) \to P^*$ and $\lim_{n\to\infty} P_i^g(n) \to P^*$.

3.2 Major Characters of PSO-CREV

The random velocity $\xi(n)$ is the key improvement of PSO-CREV. Without the additional stochastic behavior, or $\xi(n) = 0$, PSO-CREV behaves much like the traditional PSO with relatively fast convergence. Since the global optimal solution is unknown, the way to make particles converge to P^* is improving opportunities to approach (explore) the vicinity of P^*. Hence a nonzero $\xi(n)$ is very useful to drive particles into unknown solution space, so that the exploration ability of individual particle is enhanced significantly. Therefore there are fewer particles in a swarm than conventional PSO to save computational time.

From the assumption 2) of the definition of PSO-CREV, it is noted that the only constraint of $\xi(n)$ is that it is bounded. Hence to control convergence rate of PSO-CREV (to speed up convergence indeed), a time-varying $\xi(n) = w(n)\bar{\xi}(n)$

is proposed, where $\bar{\xi}(n)$ represents a stochastic velocity with zero expectant and constant value range, $w(n)$ represents a time-varying positive coefficient, which is defined as

$$w(n) = \begin{cases} 1, & n < \frac{3}{4}N_b; \\ \lambda_1 w(n-1), & n \geq \frac{1}{4}N_b, n < \frac{3}{4}N_b; \\ \lambda_2 w(n-1), & n \geq \frac{3}{4}N_b, \end{cases} \tag{5}$$

where N_b represents the total number of iterations, and λ_1 and λ_2 are positive constants less than 1. For example, if the total iterations for one time optimization is set to 200, we chose $\lambda_1 = 0.99$ and $\lambda_2 = 0.8$. Hence during iteration 1 to 50, the intensity of $\xi(n)$ is strong, so that particles have more opportunities to reach unknown solution space. And after that, the bound of $\xi(n)$ decreases iteration by iteration. Especially in the last quarter of iterations, $\xi(n)$ has trivial effect on the convergence of PSO-CREV finally. In one word, such a time-varying bound of random search velocity makes PSO-CREV meet the both requirements of strong exploration ability and fast convergence.

4 The Algorithm of NNPC Using PSO-CREV

Comparing the introduction of NNPC in Section 2 with PSO-CREV in Section 3, we know that the control vector U_t at time t should be chosen as solution vector in PSO-CREV, which is denoted by X in the definition of PSO-CREV. And the fitness function that should be optimized by PSO-CREV is of the form in (2). Consequently the algorithm of NNPC using PSO-CREV is summarized as follows.

1) Select T.
2) Initialize PSO-CREV, where the dimension of solution space equals the predictive horizon T multiplying with the dimension of control signal.
3) At the time t, initialize particles' positions. Each particle represents a potential predictive control signals, i.e. $X_i = [u(t-d+1), \cdots, u(t-d+T)]$.
4) Use PSO-CREV to optimize predictive control signals. To evaluate fitness, the input of NNM consists of three parts, the potential control signals, the real output of plant, and the predictive output of NNM.
5) When PSO-CREV converges, apply $u(t-d+1)$ found by PSO-CREV to close the control loop.
6) Return to 3).

5 Application on Mobile Robot Control

5.1 NNPC for Mobile Robot

In this section, a simulation is proposed where NNPC using PSO-CREV algorithm is applied to control a mobile robot with nonholonomic constraints. There

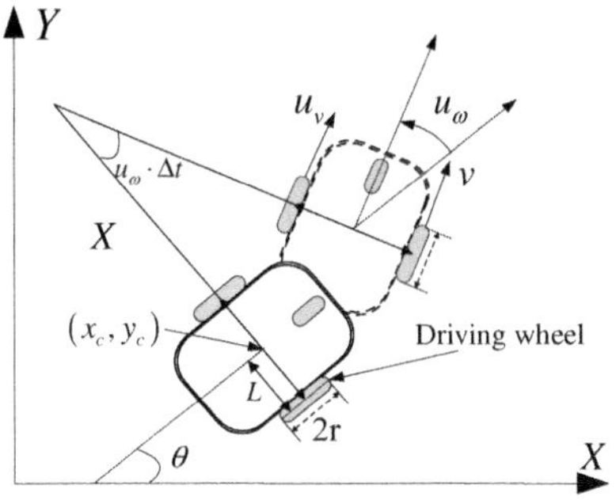

Fig. 1. Kinematics of a mobile robot

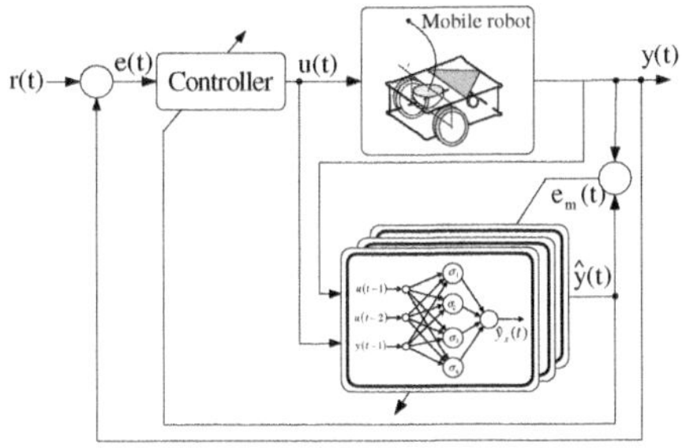

Fig. 2. The diagram of NNPC using PSO-CREV for mobile robot

is a car-like mobile robot shown in Fig. 1, whose kinematic model of the mobile robot is expressed as

$$\begin{bmatrix} x_c(t+1) \\ y_c(t+1) \\ \theta(t+1) \end{bmatrix} = \begin{bmatrix} x_c(t) \\ y_c(t) \\ \theta(t) \end{bmatrix} + \begin{bmatrix} u_v \cos(\theta + \frac{u_\omega}{2}) \\ u_v \sin(\theta + \frac{u_\omega}{2}) \\ u_\omega \end{bmatrix} \quad (6)$$

where u_v denotes the translational velocity, u_ω denotes the rotational velocity.

To improve the accuracy of the NNM, we choose three NNs to model three coordinates respectively. Assume that it is known that the time delay of the system, d, equals 0, but the order of mobile robot is unknown. Then the NNM can be expressed as

$$\hat{y}_j(t) = \hat{f}(q) = \hat{f}(y(t-1), u(t), u(t-1)),$$

where $y(t) = [y_x(t), y_y(t), y_\theta(t)]^T$, the subscript j represents the subscript of three general coordinates x, y, and θ respectively. Therefore the dimension of input to NNM equals 7 (two control signals with dimension two and one output history with dimension three). And the structure of single NNM is chosen as $7-5-1$ with biases in hidden and output layers.

When NNM is trained to approximate a plant, there are two kinds of feedback methods which send different signals to NN for training. The first method is called the parallel connection for system identification, where the output of NNM is feedback to training NN, and the second method is called the series-parallel connection for system identification, where the real output of the plant is sent to NNM for training [5]. Since the series-parallel configuration is simple and easy to be realized, in this paper NNM is trained by series-parallel way. Hence the structure diagram of the mobile robot control is shown in Fig. 2.

5.2 Test Setup and Results

In the test, a mobile robot is required to follow two kinds of trajectories, a circle one and a sine one. In the test of cycle trajectory, the linear velocity and steering velocity driving a robot must be constant under desired circumstances. Hence the test referring circle trajectory can test that whether the outputs of NNPCs using different learning algorithms are stable or not. Obviously given a learning algorithm, if control signals generated by NNPC using this learning algorithm are stable (in other words the control signals are less fluctuant), it means that this learning algorithm is with high ability to find the desired control signals. On the other hand, the test referring sine trajectory will test the performance of NNPCs when the desired control signals are time-varying. The parameters involved in the the two tests are shown as follows:

- **Desired cycle trajectory**: The desired cycle trajectory results from the movement of a virtual mobile robot with the constant velocities, $[\,0.1 \quad \pi/10\,]$ in interval $[\,0s \quad 20s\,]$.
- **Desired sine trajectory**: The desired sine trajectory results from the movement of a virtual mobile robot with the following velocities in interval $[\,0s \quad 21s\,]$.

$$u_v(t) = 0.2; \qquad u_\omega(t) = -2.5^2 \sin(5x^d(t-1)) \cdot u_v \cos(\theta^d(t-1))^3,$$

where $p^d(t) = [\,x^d(t) \quad y^d(t) \quad \theta^d(t)\,]^T$ represents the coordinate of the virtual robot at time t.

And it is assumed that the sample interval is $0.1s$. That means NNPC outputs control signals for every $0.1s$. In both tests, the predictive horizon is set to 2, i.e., $T = 2$.

Besides PSO-CREV, three other algorithms, including GDA, PSO with linear decreasing inertia weight (PSO-LDIW) [6], and GA, are selected as comparisons, so that we can compare them in two aspects – the accuracy of NNPCs and computational cost. The configurations of these algorithms are chosen as:

- PSO-CREV: The updating principle is of the form (4) with a decreasing $\xi(n)$ whose $w(n)$ is in the form of (5). All parameters are chosen as: $c_1 = c_2 = 3.5$, $\alpha = 0.95$, $a = 3$, $b = 0.35$, $\lambda_1 = 0.99$, and $\lambda_2 = 0.8$. The *lbest* version [7] of neighborhood is used in which each particle exchanges information with four other particles. Due to high exploration ability of PSO-CREV, there are only 5 particles enclosed in PSO-CREV that is even less than the dimension of input of NNM (the dimension of each NN is 7). At each sample time,
- PSO-LDIW: A conventional PSO with time-varying inertia weights is chosen, whose parameters are chosen as: $c_1 = c_2 = 2$. The inertial weight is set to change from 0.9 to 0.4 over the iterations. The *lbest* PSO is used to realize social component. In order to obtain acceptable results, two population sizes, including 10 and 20 particles, are used in the conventional PSO paradigm.
- GA: A standard GA algorithm with selection, crossover, and mutation operations is employed to optimize control signals. The crossover probability is set to 0.8, while the mutation probability is chosen as 0.1.
- Conventional GDA: The details of iterative algorithm of GDA can be found in [1]. Here η is set to 0.03.

Each optimization process of PSO-CREV or PSO-LDIW includes only 200 iterations of fitness evaluation at each sample time, while the maximal iterations of GDA and GA are set to 2000 and 3000 iterations respectively. Since in fact GA converges so slowly that there is seldom opportunity to evolve acceptable control signal, in the results the performance of GA learning will be ignored. All tests will be run for 20 times, and the runs with the best performance with respect to each algorithm are selected as the results.

1) Performances of NNPCs Using Different Learning Algorithms
The results of both tests are shown in Figs. 3 and 4, where the traces resulted from NNPCs based on all four algorithms (here two PSO-LDIWs with two population size are employed, and the results of GA are ignored) are shown in figure (a) with respect to different desired trajectories, the tracking errors over sample times are shown in figure (b) (Here the tracking error is defined as $E = \|p^d - p\|$, where $p^d = [\, x^d \quad y^d \quad \theta^d \,]^T$, $p = [\, x_c \quad y_c \quad \theta \,]^T$), finally the control signals generated by NNPCs over time are shown in Fig. 3 (c).

From the traces and relative errors over time, it is observed that there is not too much difference between the performance of all NNPCs using four learning algorithms. All traces of mobile robots are closed enough to the desired trajectories. And the curves of relative errors are somewhat similar. Comparing all curves, it is observed that the shapes of all curves have similar trends. Especially in the test of cycle trajectory, the tracking errors of all algorithms increase slightly within the end period. Hence such increasing errors must be induced by the approximation error of NNM. Hence only from the traces it seems that PSO-CREV learning algorithm is not superior to other algorithms.

However from Fig. 3 (c), it is obviously observed that comparing with other algorithms, the curve of control signals generated by NNPC using PSO-CREV looks more smooth, or with less fluctuation. That means at every sample time, PSO-CREV converges to the best solution very closely. Therefore the control signals are almost the same at every sample time. As a comparison, although the total iterations for NNPC using PSO-LDIW is set to 200, that is the same as PSO-CREV, the curves with respect to NNPCs using PSO-LDIW also show some sudden changes of control signals. Hence from the Fig. 3 (c) it is concluded that PSO-CREV is more efficient to find the best solution within only 200 iterations.

Similarly from the Fig. 4 (c) showing control signals, the output of NNPC using PSO-CREV looks more smooth than other algorithms, especially near the peaks of the rotational velocities v_ω. Hence it is verified again that if the number of iterations is set to only 200, PSO-CREV is more effective to find optimal control signals within such a short period.

It should be noted if population size of PSO-LDIW is set to 5, NNPC using PSO-LDIW always prematurely converge within 200 iterations. Hence NNPC using PSO-LDIW with small population size (for example 5 particles) can hardly generate proper control signals to realize trajectory tracking. Comparing with PSO-LDIW, PSO-CREV undoubtedly has higher exploration ability, which

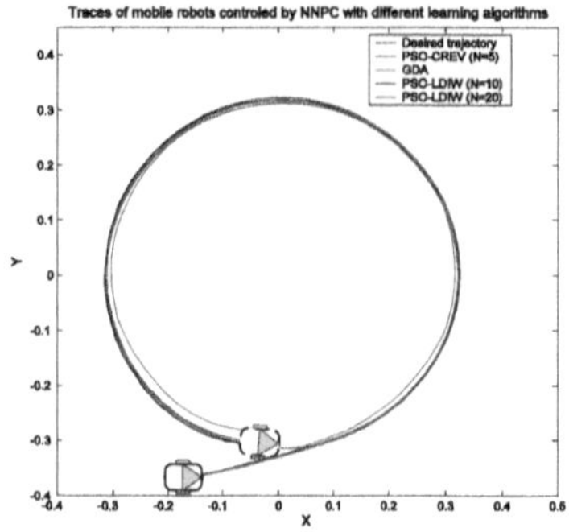

(a) The trace of robot tracking cycle trajectory.

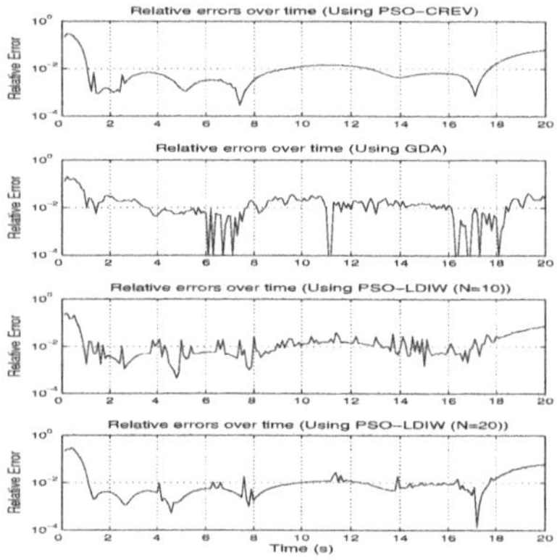

(b) The relative errors over time.

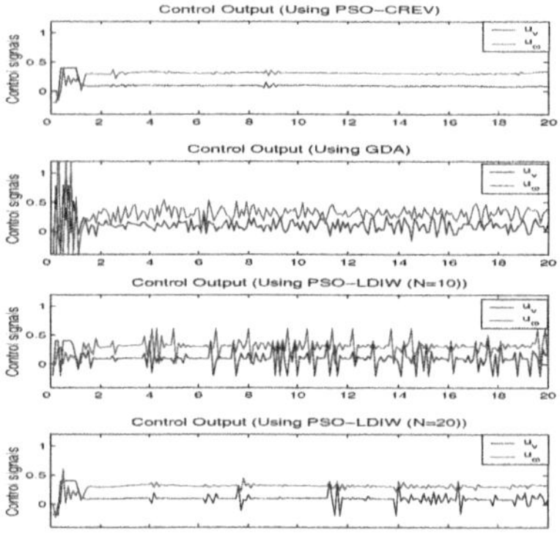

(c) The control signals over time.

Fig. 3. The simulation results of the test referring cycle trajectory

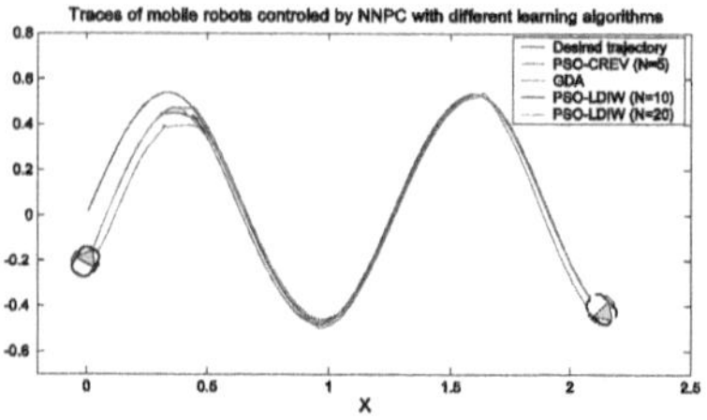

(a) The trace of robot tracking sine trajectory.

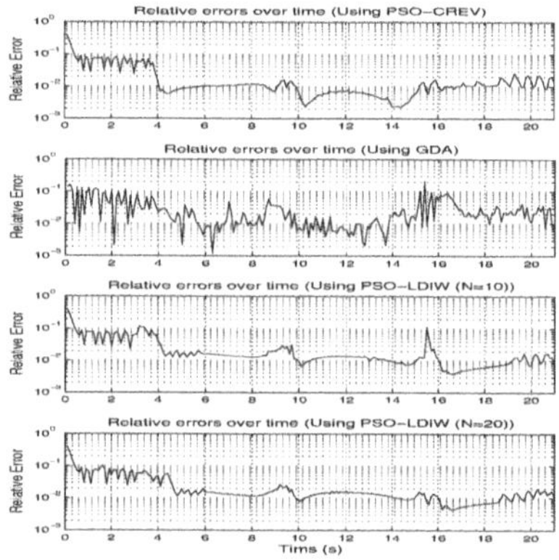

(b) The relative errors over time.

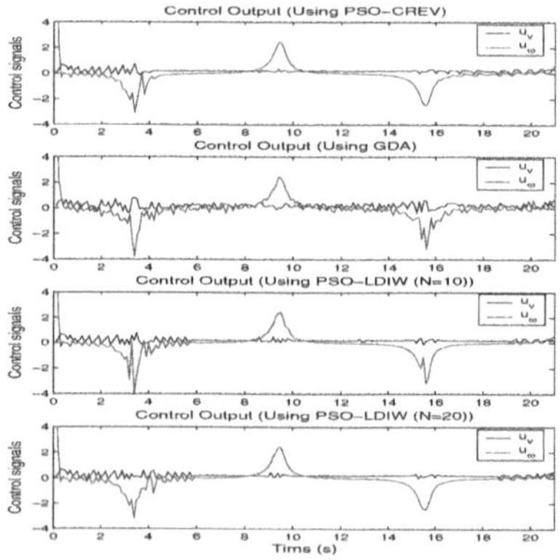

(c) The control signals over time.

Fig. 4. The simulation results of the test referring sine trajectory

Table 1. Comparison of computational cost of NNPCs using different algorithms

Algorithm	Number of fitness evaluation	Relative computation time
PSO-CREV (N=5)	$5 \times 200 = 1000$	1
GDA	2000	10.2654
PSO-LDIW (N=10)	$10 \times 200 = 2000$	1.7161
PSO-LDIW (N=20)	$20 \times 200 = 4000$	4.4596
GA	$30 \times 3000 = 90000$	13.3430

makes PSO-CREV perform much better than traditional PSO with the same population size.

2) Comparison of Computational Cost

Now let's consider the computational cost of all NNPCs. Since at each sample time, all algorithms need to evaluate fitness, i.e. the predictive tracking errors, it is easy to estimate the numbers of computing outputs of NNM for every optimization iteration. If let computational time of PSO-CREV algorithm be one unit, the numbers of fitness evaluation and computational times for one optimization process with respect to all algorithms are listed in Table. 1, which results from the average of all 20 runs in the test referring sine trajectory. Obviously computational time of PSO-CREV is far less than other algorithms, especially only ten percentage of the time consumed by GDA. Hence we can conclude that using PSO-CREV algorithm can enhance optimization efficiency of NN training on line, so that it is more feasible to be employed in NNPC than traditional NNPC based on gradient decent algorithm in rapid dynamics.

6 Conclusions

This paper proposes a useful optimization method, PSO-CREV, to enhance computational efficiency of NNPC, so that NNPC can be used in the systems with rapid dynamics. The tests show that PSO-CREV converges very quickly, while the computational time consumed by it is far less than conventional PSO and GDA. Hence this PSO-CREV enhances efficiency of NNPC very much. And NNPC based on PSO-CREV becomes feasible for rapid dynamic processes.

References

1. Noriega, J.R., Wang, H.: A Direct Adaptive Neural-Network Control for Unknown Nonlinear Systems and Its Application. IEEE Tran. Neural Networks **9** (1998) 27-34
2. Yoo, S.J., Choi, Y.H., Park, J.B.: Generalized Predictive Control Based on Self-Recurrent Wavelet Neural Network for Stable Path Tracking of Mobile Robots: Adaptive Learning Rates Approach. IEEE Trans. Circuits and Systems **53** (2006) 1381-1394
3. Eberhart, R.C., Kennedy, J.: A New Optimizer Using Particle Swarm Theory. Proceedings of the 6th International Symposium on Micro Machine and Human, Science, Nagoya, Japan (1995) 39-43

4. Kennedy, J., Eberhart, R.C.: Particle Swarm Optimization. Proceedings of IEEE International Conference on Neural Network, Perth, Australia (1995) 1942-1948
5. Plett, G.L.: Adaptive Inverse Control of Linear and Nonlinear Systems Using Dynamic Neural Networks. IEEE Trans. Neural Networks **14** (2003) 360-376
6. Shi, Y., Eberhart, R.C.: Parameter Selection in Particle Swarm Optimization. Proceedings of the 7th Annual Conference on Evolutionary Programming, New York (1998) 591-600
7. Shi, Y., Eberhart, R.: An Empirical Study of Particle Swarm Optimization. Proceedings of IEEE Congress on Evolutionary Computation, Washington, DC (1999) 1945-1949

Ultrasonic Sensor Based Fuzzy-Neural Control Algorithm of Obstacle Avoidance for Mobile Robot

Hongbo Wang, Chaochao Chen, and Zhen Huang

Robotics Institute, Yanshan University,
Qinhuangdao, Hebei Province 066004, P.R. China
wzfamily@hotmail.com, 086403w@gs.kochi-tech.ac.jp, huangz@ysu.edu.cn

Abstract. This paper presents a novel fuzzy-neural control algorithm to realize obstacle avoidance of a mobile robot. A heuristic fuzzy-neural network is developed based on heuristic fuzzy rules and the Kohonen clustering network. By applying the off-line and unsupervised training method to this network, the pattern mapping relation between ultrasonic sensory input and velocity command is established. This paper describes mechanical design of the mobile robot, the arrangement of ultrasonic sensors, the obstacle avoidance system based on FKCN, classification of obstacle, the control algorithm for obstacle avoidance and training data library. In order to verify the effectiveness of this algorithm, we give the results of simulation in a computer virtual environment.

Keywords: Fuzzy-neural control, network, obstacle avoidance, ultrasonic sensor.

1 Introduction

When a mobile robot navigates itself automatically, it will unavoidably encounter stationary and moving obstacles. Obstacle avoidance is one of the fundamental requirements for the automatic navigation of mobile robots. Several studies have been done on the obstacle avoidance problem for mobile robots.

The visual sensors provide the richer source of useful information about the surroundings. However, a problem of visual sensors is that they are slow in computing data and are expensive in cost [1]. In comparison with the visual sensors, the ultrasonic sensors are accomplished in small cost. Although ultrasonic range measurement instruments suffer from some fundamental drawbacks which limit their usefulness in mapping or in any other task requiring high accuracy in a domestic environment, many researchers used ultrasonic sensors for obstacle avoidance [2-4]. Borenstein and Koren [5] summarized relevant obstacle avoidance methods using ultrasonic sensors into edge detection, certainty grids and the potential field method. Also, they proposed a vector field histogram, a virtual force field and a histogramic in-motion mapping method.

D. Liu et al. (Eds.): ISNN 2007, Part I, LNCS 4491, pp. 824–833, 2007.

In recent years, researchers used the artificial neural networks and the fuzzy logic control to realize the obstacle avoidance, and achieved better results [6-7]. The two methods have some same properties that make it an ideal tool for handling these complex tasks in uncertain changing environment. Artificial neural networks have the ability to self-study and fault tolerant characteristics in parallel computation that leads to fast data processing [8]. Fuzzy logic is the way the human brain works, and we can mimic this in machines so they will perform somewhat like humans [9]. However, neural network has some disadvantages such as the local minimum points, the slow astringency [10], and fuzzy systems have a drawback when applied to many applications, their rules are often very difficult or even impossible to be determined [11].

This paper presents a novel pattern recognition algorithm that use fuzzy-neural control technique to realize the obstacle avoidance of a mobile robot in real-time. This method mainly applies fuzzy Kohonen cluster network (FKCN) to replace the general fuzzy controller, thus many complex useless fuzzy controls rules become unnecessary.

2 Mechanical Design and Arrangement of Ultrasonic Sensors

2.1 Mechanical Design

The mechanism of a mobile robot with 8 wheels is shown in Fig.1 and Fig.2. The 8 wheels are divided into 4 groups that have same transmission mechanism. The one wheel of each group is driving wheel and another is free wheel. Since two belts driven by two motors make the four driving wheels move in synchronous way, the motion of the robot platform is a plane translation.

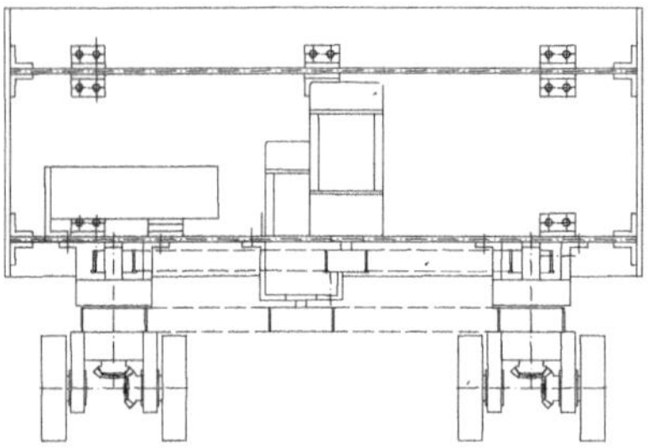

Fig. 1. A omni-directional mobile robot

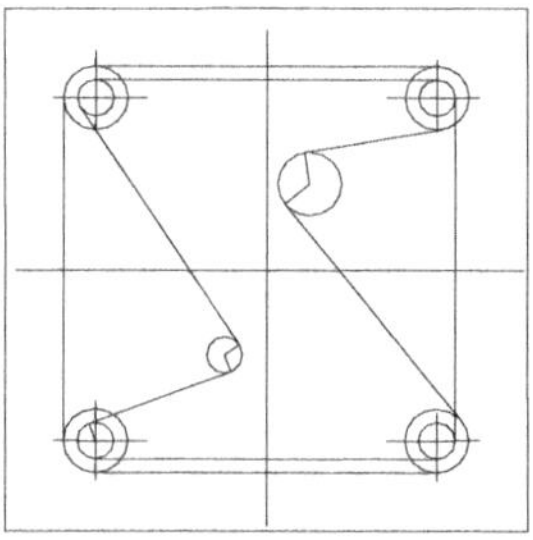

Fig. 2. Driving system of mobile robot

2.2 Arrangement of Ultrasonic Sensors

To avoid obstacle, 12 ultrasonic sensors (BTE054: US Sensor 2) are installed in the mobile robot. The operating frequency of the sensor is 40kHz. The maximum effective examination distances under short distance pattern and long distance pattern are 1500mm and 3000mm respectively. In this paper, the measurement mode of the short distance pattern is used. The 12 ultrasonic sensors are divided into 8 groups to arrange at the four corners and four sides of mobile robot with square platform ($400 \times 400mm$) as shown in Fig. 3. The groups A, C, E, G at four sides include 2 sensors, and the groups B, D, F, H at four corners include 1 sensor respectively. In order to reduce the computing data, only 8 sensors (5 groups) are used to avoid obstacle during navigation. When the navigation direction of the mobile robot is between OB and OD, the groups A, B, C, D and E are used. When the robot navigates in the direction between OH and OB, the groups G, H, A, B and C are used. For the direction between OF and OH (OD and OF), the groups E, F, G, H and A (C, D, E, F and G) are used.

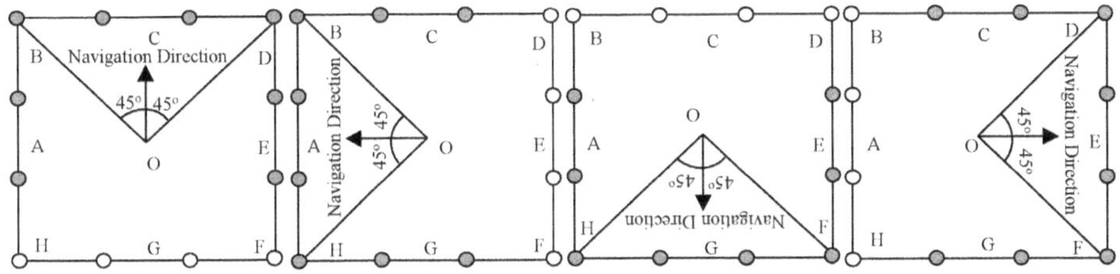

Fig. 3. Arrangement of ultrasonic sensors installed in mobile robot

3 Obstacle Avoidance System and Classification of Obstacle

3.1 Obstacle Avoidance System Based on FKCN

In order to enable the mobile robot to avoid the obstacle in the navigational path with rapid reaction capacity, the better mapping relation between the sensor data input and the control output must be established. Since this mapping relation is extremely complex and nonlinear, it is very inconvenient to solve this problem using the general control method. However, the artificial neural network has the

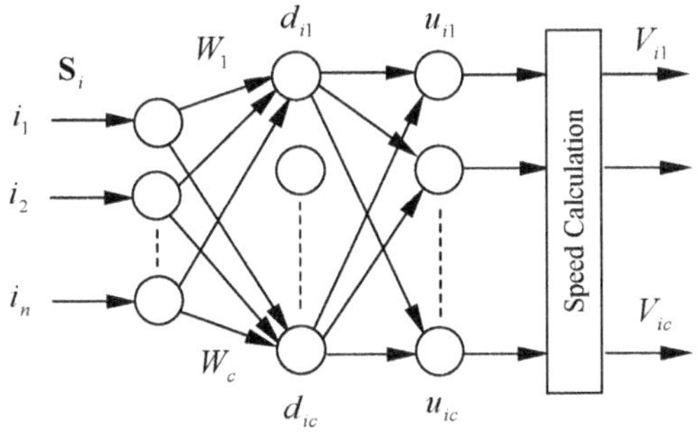

Fig. 4. The diagram of the obstacle avoidance system based on FKCN

astonishing ability to deal with nonlinear problem. Using the merit of artificial neural network, we successfully establish the mapping relations. The control system structure of obstacle avoidance is shown in Fig. 4. The left side of the control system structure is FKCN and the right side is the calculation of speed output. Since we choose five groups of ultrasonic sensors as the input of obstacle avoidance system to avoid obstacle, the input vector can be expressed as follows

$$\mathbf{S}_i = (i_1, i_2, i_3, i_4, i_5)^T. \tag{1}$$

3.2 Classification of Obstacle

Obstacle of Normal Mode. In navigation environment of a mobile robot, the distribution of obstacle is complex. If the robot has detected an obstacle and can find the navigation path in navigational direction, the obstacle is classified as obstacle of normal mode. From the arrangement and group of ultrasonic sensors, the obstacle of normal mode in navigation environment that is sensed from ultrasonic sensors installed in mobile robot can be classified as 8 kinds of categories as shown in Fig. 5.

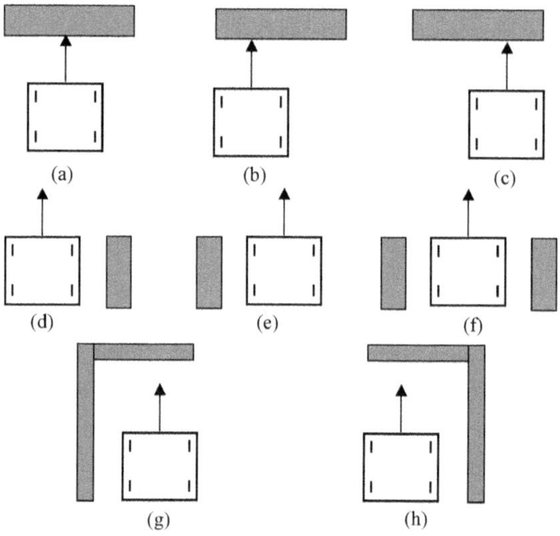

Fig. 5. Obstacles of normal mode

In Fig.5 (a), (b) and (c), since the obstacle exists in the navigational direction of mobile robot and the width of obstacle is less than the measurement width of ultrasonic sensors, the mobile robot can find the navigation path in the left or right of the obstacle. If the mobile robot detects an obstacle in the left or right or two sides, it can determine the navigation path in the front as shown in Fig. 5 (d), (e) and (f). In the case as shown in Fig. 5 (g) and (h), the mobile robot can find a navigation path in the right or left.

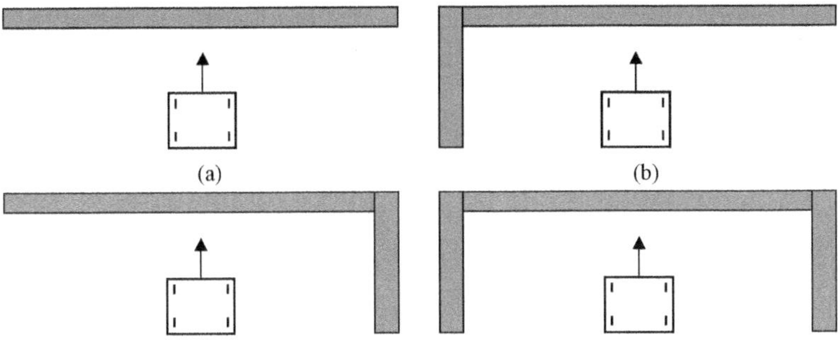

Fig. 6. Obstacles of danger mode

Obstacle of Danger Mode. If the mobile robot detects an obstacle and cannot find a path in navigational direction, this kind of obstacle is classified as obstacle of danger mode as shown in Fig. 6. For the obstacle of danger mode, some special control scheme has to be given. Fig. 6(a) shows that the obstacle in the front of mobile robot is larger than the measurement range of ultrasonic sensors and the robot cannot determine in which side the obstacle can be avoided. In this case, we specify the mobile robot to turn to the right. When the obstacle in the front is larger than the measurement range of ultrasonic sensors and an obstacle in the left is detected as shown in Fig. 6(b), the mobile robot turns to the right to avoid the obstacle. A similar processing is done for the case in Fig. 6(c). If the mobile robot detects simultaneously the obstacle in the front, left and right as shown in Fig. 6(d), the mobile robot will go back first and then turn to the right.

When the obstacle avoidance in danger mode is terminated, the obstacle avoidance in normal mode will be immediately carried out.

4 Control Algorithm for Obstacle Avoidance

4.1 The Determination of Weight Vector

The FKCN structure has the function of pattern recognition and includes three layers: input layer, hidden layer and output layer [12]. In this network, all prototype patterns reflect on weight vector $\mathbf{W}_j (1 \leq j \leq c)$ in hidden layer. The weight vector $\mathbf{W}_j$ and the number c of prototype pattern are determined using Kohonen's self organizing feature map algorithm [13].

The initial value of weight vector in hidden layer is given by operator from the real track of obstacle avoidance. The initial value of weight vector $\mathbf{W}_j$ is an n-dimensional vector and its number is c.

$$\mathbf{W}_j = (w_{1j}, w_{2j}...w_{nj})^T, n = 5, j = 1, 2, ..., c. \quad (2)$$

The FKCN input value is

$$\mathbf{S} = (i_1, i_2 ... i_n)^T, n = 5. \tag{3}$$

In order to update the weight value $\mathbf{W}_j$ according to the Kohonen's self organizing feature map algorithm, the "winner" node can be obtained by the following expression

$$|\mathbf{W}_x - \mathbf{S}|^2 = min|\mathbf{W}_j - \mathbf{S}|^2, j = 1, 2, ..., c, \tag{4}$$

where x is the "winner" neuron.

In this paper, the one-dimensional Kohonen's self organizing feature map algorithm is used. Since the number of weight vector $\mathbf{W}_j$ in hidden layer is equal to the number of the neuron in hidden layer, once the "winner" neuron is obtained, the new weight value can be obtained according to the following equation

$$\mathbf{W}_j(t+1) = \mathbf{W}_j(t) + \eta(t)(\mathbf{W}_x - \mathbf{W}_j(t)), j \in A_{ix}(t), t = 1, 2, ..., mc, \tag{5}$$

where $\mathbf{W}_j(t)$ is j-th weight vector at time t, $A_{ix}(t)$ is a neighborhood function that is a decreasing discrete time function and defines the size of neighboring region around the "winner" neuron, $\eta(t)$ is the learning rate at time t, mc is the number of circles.

4.2 The Number of Prototype Pattern

After above calculation, the renewed weight value $\mathbf{W}_j$ is obtained. If the Euclidean distance between two weight values in hidden layer is very small, it means two nodes in the hidden layer give the similar result associated with the same input. In order to guarantee the convergence of this network and reduce the computing time during obstacle avoidance, the following algorithm is presented.

(1) To calculate the Euclidean distance between j-th neuron and $j+k$-th neuron in hidden layer using the following equation

$$D_{j,j+k} = |\mathbf{W}_j - \mathbf{W}_{j+k}|, j = 1, k = 1, 2, ..., c - j. \tag{6}$$

(2) If $D_{j,j+k}$ is smaller than the specified threshold, the two neurons in hidden layer and the corresponding speed output vectors $\mathbf{V}_j$ can be combined as follows:
(a) Two new values are assigned to the weight vector $\mathbf{W}_j$ and speed vector $\mathbf{V}_j$ respectively

$$\mathbf{W}_j = \frac{\mathbf{W}_j + \mathbf{W}_{j+k}}{2}, \mathbf{V}_j = \frac{\mathbf{V}_j + \mathbf{V}_{j+k}}{2}, \tag{7}$$

where $\mathbf{V}_j$ is the velocity vector of the mobile robot.
(b) The nodes in hidden layer are reduced as $c = c - 1$.
(c) If $j + k$ is smaller than c, the weight vector $\mathbf{W}_j$ and the speed vector $\mathbf{V}_j$ are updated using the following equation

$$\mathbf{W}_n = \mathbf{W}_{n+1}, \mathbf{V}_n = \mathbf{V}_{n+1}, \tag{8}$$

where $n = j + k, ..., c, j = j + 1$.
(3) Repeat the step (1) and step (2) until $j = c$.

4.3 The Mapping Relations Between the Sensor Input and the Control Output

As shown in Fig. 4, the hidden layer is to use for the comparison of the input pattern $\mathbf{S}$ and the prototype pattern $\mathbf{W}$. When the input pattern $\mathbf{S}_i$ and the prototype pattern $\mathbf{W}_j$ are completely consistent, the output d_{ij} of j-th node in hidden layer is zero. The output of the hidden layer is expressed in the following equation

$$d_{ij} = |\mathbf{S}_i - \mathbf{W}_j|^2 = (\mathbf{S}_i - \mathbf{W}_j)^T(\mathbf{S}_i - \mathbf{W}_j), \tag{9}$$

where $\mathbf{W}_j$ is j-th prototype pattern.

The output value u_{ij} in output layer is determined based on d_{ij}, that is, if some input pattern $\mathbf{S}_i$ is different from the prototype pattern $\mathbf{W}_j$, the similarity between them is expressed using the output value $u_{ij}(0 \le u_{ij} \le 1)$. The output value u_{ij} can be obtained as follows

$$u_{ij} = \left(\sum_{l=1}^{c} \frac{d_{ij}}{d_{il}} \right)^{-1}. \tag{10}$$

When the input pattern and the prototype pattern are completely consistent, we can obtain the following equation

$$u_{ij} = 1, u_{ik} = 0, \tag{11}$$

where $k \neq j$ and $1 \le k \le c$. From the above equations, we know that the bigger the output value is, the higher the similarity between the input pattern and the prototype pattern is. Since u_{ij} indicates the membership degree of input pattern $\mathbf{S}_i$ to prototype pattern $\mathbf{W}_j$. Each kind of prototype pattern $\mathbf{W}_j$ is corresponding to a fuzzy control rule, and each fuzzy control rule is corresponding to a speed vector. The control output $\mathbf{V}_i$ can be determined in the following equation

$$\mathbf{V}_i = \sum_{j=1}^{c} \mathbf{V}_j u_{ij}. \tag{12}$$

5 Training Data Library and Simulation

5.1 Training Data Library

To obtain the weight value of pattern clustering network, the training data library has to be established so that the network is trained. Although the data of training sample library has no rule, the classified training data enables training process to be more distinct and effective. Therefore, the measurement distances for the above 8 kinds of obstacle categories under normal mode are applied as the training data. The detailed training sample library is shown in Table. 1.

Table 1. Training data library

Class a					Class b					Class c					Class d				
Lfet	Left corner	Front	Right corner	Right	Lfet	Left corner	Front	Right corner	Right	Lfet	Left corner	Front	Right corner	Right	Lfet	Left corner	Front	Right corner	Right
85	100	40	122	87	90	111	40	52	89	97	50	40	111	98	75	102	85	85	89
90	120	40	135	88	99	123	40	85	99	99	62	40	120	99	56	111	78	100	95
94	110	40	140	95	123	145	40	75	105	93	75	42	145	124	42	145	56	120	102
86	140	38	109	123	145	105	40	88	145	120	65	41	150	150	32	123	62	145	124
100	135	38	124	126	150	109	42	90	122	110	82	38	123	111	56	150	88	123	135
120	145	37	145	95	88	108	41	95	145	145	78	39	122	124	56	98	97	109	148
130	150	45	120	124	96	111	42	85	123	124	80	40	145	135	32	85	105	125	150
98	120	43	100	150	143	150	40	77	95	95	95	41	120	94	21	122	111	135	100
94	130	43	128	97	111	128	41	45	96	93	98	40	140	98	22	100	78	134	78
150	124	41	109	120	89	135	38	56	135	124	77	40	123	140	15	95	58	134	111
120	134	40	105	140	90	125	39	77	144	111	56	40	111	135	45	84	66	97	124
Class e					Class f					Class g					Class h				
Lfet	Left corner	Front	Right corner	Right	Lfet	Left corner	Front	Right corner	Right	Lfet	Left corner	Front	Right corner	Right	Lfet	Left corner	Front	Right corner	Right
75	102	120	85	89	78	110	123	88	80	80	95	40	100	89	99	98	40	96	78
56	111	135	100	95	56	100	119	95	75	75	85	41	124	90	98	99	41	85	70
42	145	90	120	102	80	105	80	102	65	62	65	42	135	100	120	124	40	78	68
32	123	120	145	124	48	135	110	135	62	50	45	39	147	125	142	120	40	65	69
56	150	88	123	135	50	125	80	145	68	42	82	39	98	136	135	111	40	88	56
56	105	97	109	148	55	111	100	122	61	55	78	40	99	124	111	132	39	90	66
32	110	105	125	150	42	109	110	134	55	62	88	40	135	145	125	145	39	58	45
61	122	111	135	100	62	105	140	125	68	35	68	41	120	99	99	135	41	68	67
33	100	120	134	78	53	110	105	110	78	55	90	40	133	98	135	95	42	59	39
15	95	90	134	111	30	99	100	120	62	75	96	40	95	133	102	89	40	67	78
62	84	106	97	124	68	105	105	88	35	45	99	40	99	111	109	111	40	68	65

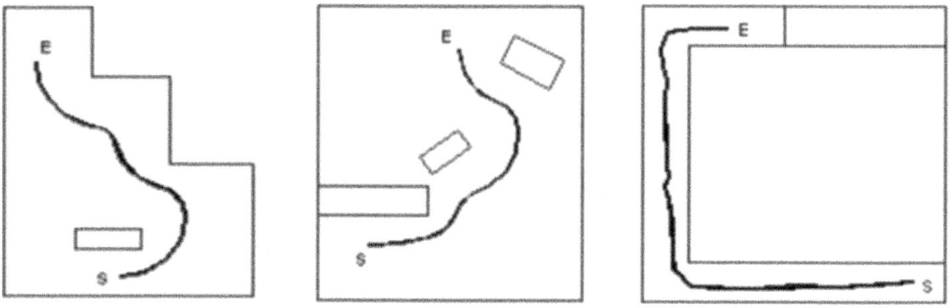

Fig. 7. Simulation results of obstacle avoidance

5.2 Simulation

To reduce the measurement error of the ultrasonic sensor and simplify the control algorithm, the distance input of the every ultrasonic sensor is divided into 5 grades using the following equation

$$L = \begin{cases} 1 & 0 < d \leq 15 \\ 2 & 15 < d \leq 30 \\ 3 & 30 < d \leq 60 \\ 4 & 60 < d \leq 100 \\ 5 & 100 < d \leq 150 \end{cases}, \qquad (13)$$

where L and d are the grade value and the distance measured from ultrasonic sensor. Similarly, the weight vector $\mathbf{W}_j$ obtained using the Kohonen's self organizing feature map algorithm is also divided into 5 grades.

In order to verify the feasibility of the method presented in the paper, a simulation of obstacle avoidance was done on the PC. The simulation results are shown in Fig. 7. In this figure, S and E are the start point and end point, the continuous line is the navigational track of the mobile robot. Fig. 7 indicates that the fuzzy-neural control algorithm proposed in the paper can efficiently realize the obstacle avoidance of a mobile robot.

6 Conclusions

This paper presents a novel pattern recognition algorithm to use for autonomous obstacle avoidance of a mobile robot. Through the combination of heuristic fuzzy rules and the Kohonen clustering network, a heuristic fuzzy neural network is developed. This method also builds up pattern mapping relation between ultrasonic sensor input and velocity command by applying the off-line and unsupervised training method to this network. From the track given by operator, the algorithm can automatically adjust the weight vector and the number of prototype pattern. Since the adjustment ability of parameter, the fuzzy-neural network can produce less prototype pattern and corresponding control rule so that the computing time is reduced and the reality of obstacle avoidance is enhanced. Compared with conventional obstacle avoidance methods, this pattern recognition algorithm developed in this paper enables a mobile robot to have much faster response to unexpected events during navigation. The simulation results indicate the effectiveness of the control algorithm.

References

1. Wang, H. B., Ishimatsu, T.: Vision-based Navigation for an Electric Wheelchair Using Ceiling Light Landmark. Journal of Intelligent and Robotic Systems **41** (2004) 283-314
2. Borgolte, U., Hoyer, H., Buhler, C., Heck, H., Hoelper, R.: Architectural Concepts of a Semiautonomous Wheelchair. Journal of Intelligent and Robotic Systems **22** (1998) 233-253
3. Tahboub, K. A., Asada, H. H.: A Semiautonomous Control Architecture Applied to Robotic Wheelchairs. In Proceedings of the 1999 IEEE/RSJ International Conference on Intelligent Robots and Systems. Kyongju, Korea (1999) 906-911,
4. Shoval, S., Borenstein, J.: Using Coded Signals to Benefit from Ultrasonic Sensor Crosstalk in Mobile Robot Obstacle Avoidance. In Proceedings of the 2001 IEEE International Conference on Robotics and Automation. Seoul, Korea (2001) 2879-2884
5. Borenstein, J., Koren, Y.: The Vector Field Histogram - Fast Obstacle Avoidance for Mobile Robots. IEEE Transactions on Robotics and Automation **7** (1991) 278-288
6. Benreguieg, M., Hoppenot, P., Maaref, H., Colle, E., Barret, C.: Fuzzy Navigation Strategy: Application to Two Distinct Autonomous Mobile Robots. Robotica **15** (1997) 609-615
7. Bezdek, J. C.: Fuzzy models for pattern recognition. IEEE Press, New York 1992

8. Papadourakis, M. G., Tsagatakis, G.: Applications of Neural Network to Robotics. In Proceedings of 2nd International Workshop on Embedded Systems, Internet Programming and Industrial IT. Kiel, Germany (2003) 20-22
9. Sowell, T.: Fuzzy Logic for Just Plain Folks. Tesco Pub. (1997)
10. Wu, H., Cong, Y., Jiang, G. Y., Wang, H. Y.: Study on Short-Term Prediction Methods of Traffic Flow on Expressway based on Artificial Neural Network. In Proceedings of The seventh IASTED International Conference on Signal and Image Processing. Honolulu, Hawaii, USA (2005) 479-088
11. Mills, D. J., Harris, C. J.: Neurofuzzy Modelling and Control of a Six Degree of Freedom AUV. Technical Report, University of Southampton (1995)
12. Huntsberger, T., Ajjimarangsee, P.: Parallel Self-organizing Feature Maps for Unsupervised Pattern Recognition. International Journal of General Systems **16** (1990) 357-372
13. Kohonen, T.: Self-Organizing Maps. Springer, London (1997)

Appearance-Based Map Learning for Mobile Robot by Using Generalized Regression Neural Network*

Ke Wang, Wei Wang, and Yan Zhuang

Research Center of Information and Control, Dalian University of Technology,
116024 Dalian, China
qingshang00@163.com

Abstract. Regression analysis between features of high-dimension is receiving attention in environmental learning of mobile robot. In this paper, we propose a novel framework, namely General regression neural network (GRNN), for approximating the functional relationship between high-dimensional map features and robot's states. We firstly adopt PCA to preprocess images taken from omnidirenctional vision. The method extracts map features optimally and reduces the correlated features while keeping the minimum reconstruction error. Then, the robot states and corresponding features of the training panoramic snapshots are used to train the given neural network. This enables robot to memorize the environmental features as well as to predict available scene given its location information. Experimental results are shown finally.

1 Introduction

Appearance-based map learning, in its simplest form, involves taking reference snapshots at according reference locations and storing them into a visual database along with positioning information. With this technique, when the robot moves in the environment, it searches this database to find out which memorized images is more similar to the current view, and then infer its position in the environment [1,2,3]

We are quite aware of the side effect in appearance-based method if the reference locations are scattered. In this case, further localization algorithm is not likely to generate accurate pose estimate without adequate training samples. Therefore, we need to form a general smooth mapping which not only represents the functional relationship between robot poses and its observations, but also spans the intermediate information between references. One solution is to construct a hypersurface by nonparametric estimate [4], yet another promising approach is to use a generalized regression neural network. Through neural network training, the feature patterns and corresponding robot poses are stored in the neural network [5,6]. In this case, the intermediate information between tow adjacent reference locations can be inferred. From such a scheme, the robot has the prior information about environment and the prediction ability to some extend.

* The author greatly appreciate for the support by National Natural Science Foundation of China (Grant No. 60605023) and the Specialized Research Fund for the Doctoral Program of Higher Education of China (Grant No. 20060141006).

D. Liu et al. (Eds.): ISNN 2007, Part I, LNCS 4491, pp. 834–842, 2007.

This paper is organized as follows: Section 2 details how to construct eigenspace (global map) by using PCA in an efficient fashion for our platform shown in Fig. 1. Then we describe how to design a GRNN-based neural predictor for nonlinear regression problem with uncertainty propagation scheme. The rest of the paper shows the experimental results conducted on our robot.

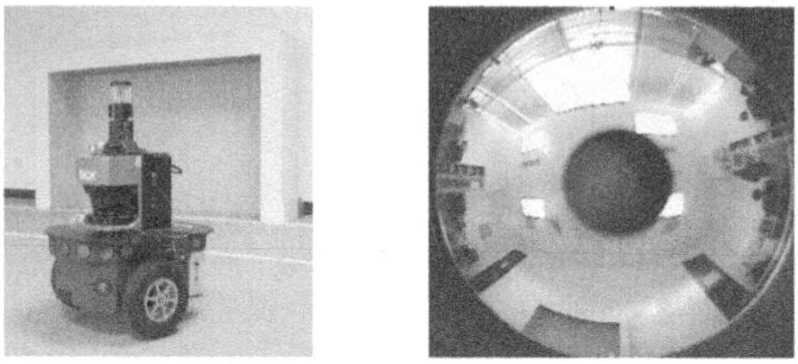

Fig. 1. Pioneer 3-DX robot with omnidirectional camera(*left*), and a panoramic snapshot taken by omnidirectional camera(*right*)

2 Framework of Appearance-Based Map Learning

Fig. 2 demonstrates the learning framework for mobile robot. In this scheme, the robot is made to capture the environmental information from certain reference locations. It simulates the rotation versions (phantasms) of original reference images by shifting a certain angle and, and then stores all the information in robot memory together with its according states. However, it is necessary that a large number of reference images be reduced adequately. In this paper, PCA is used to extract features by projecting high-dimensional data into a low-dimensional global map(Eigenspace)[7]. A second step,

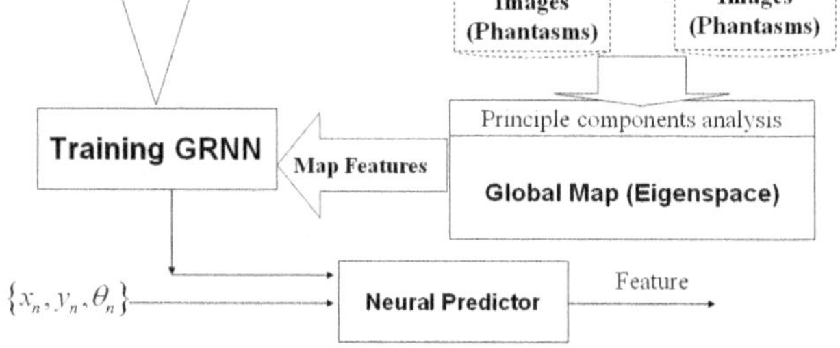

Fig. 2. Scheme of appearance-based map learning for mobile robot by using GRNN

during the learning process, is to specify the regression relation between the stored robot prior states and the corresponding dependent features in the global map. This complicated MIMO nonlinear mapping is approximated by generalized regression neural networks. Through networks training, the patterns are memorized, and a smooth hyper-surface is provided. In this case, the matured GRNN serves as Neural Predictor which makes the robot have the primary intelligence to predict environmental information according to its states.

3 Eigenspace Constructions

When the images are collected (See Fig. 3 for example) by the robot, we calculate the eigenspace which describes the acquired data. We describe a method which is based on recognizing a panoramic view using a model built from 2D panoramic images acquired in the learning phase. The model is constructed by using an approximation of the image set, which is represented by a low-dimensional set of principal components and eigenvectors that span the low-dimensional eigenspace.

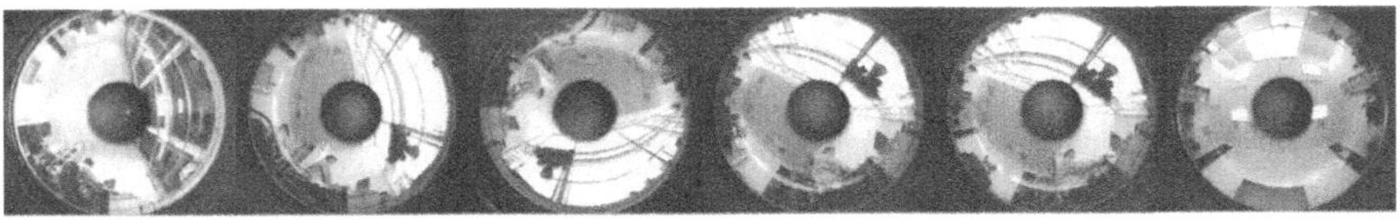

Fig. 3. Images captured taken from six different locations

3.1 Dimension Reduction of Data Through PCA

Suppose we have obtained M images at corresponding reference locations $\{x_i, y_i\}$. Each image vector of dimension n needs to be rotated N times respectively and subtracted the mean, in order to formulate the normalized training set I_g with the form

$$I_g = [I^1, I^2, \cdots, I^M] \in R^{n\times(M\cdot N)} \quad . \tag{1}$$

Thus, the covariance matrix $\mathbb{C}$ can be formulated as

$$\mathbb{C} = \frac{1}{M \cdot N} \cdot I_g \cdot (I_g)^T \quad . \tag{2}$$

Traditional PCA computes the subspace of covariance matrix $\mathbb{C}$ through SVD decomposition [8]. which, however, is time consuming. Notice that $M \cdot N \ll n$, then an very efficient way for calculating the eigenvectors $v_i' = 1, 2, \cdots, MN$, can be formed by analyzing the properties of inner product matrix

$$\Gamma_g = I_g \cdot (I_g)^T \quad . \tag{3}$$

This enables us to directly choose the complex-valued assembled orthogonal eigenvectors $v_j, j = 1, 2, \cdots, N$ from weighted j column of block Fourier Matrix $F \in \Re^{N \cdot N}$.

And Leonardis [1] presented a detailed solution for solving the eigen-problem of such matrix.

For finding principle components, we keep K largest eigenvalues which specify a certain percentage of dominant energy in eigenvalue spectrum of training set. In this manner, Estimation of eigenvectors u_i of $\mathbb{C}$ can be formed as

$$u_i = \frac{1}{\sqrt{N\lambda'_i}} I' v'_i \tag{4}$$

where λ'_i is an eigenvalue of Γ_g. Notice that, we can generate the real-valued eigenvectors from the real and the imaginary parts of v'_i due to the properties of real symmetrical matrix Γ_g.

3.2 Generation of Global Map

Once the eigenvectors of matrix Γ_g and covariance matrix of training images I_g are sequentially determined, we can form the global map features

$$A_g = (U_g)^T \cdot I_g \tag{5}$$

where coefficient matrix $A_g \in \Re^{K\times(M\cdot N)}$. Therefore, our global map for training images, is represented as the set of projective coefficients in K-dimensional subspace spanned by eigenvectors U_g. Fig. 4 shows the first three principle coefficients which represent the training images in Fig.3. As one can see, coefficients for a single location, representing one panoramic image that is rotated N times , form a regular closed loop.

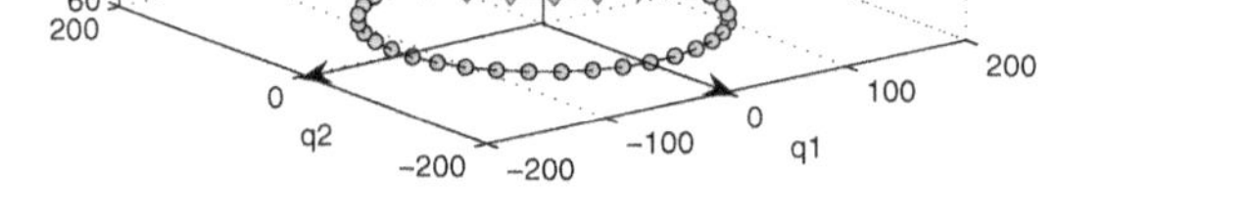

Fig. 4. First three principle components representing the rotation versions of images

4 Nonlinear Mapping by Using GRNN

The Global map represents partial information of the working environment because the training images are taken only from finite reference locations. Without intermediate information between each reference location, robot has to attach the novel observation to the nearest map elements for localization. Although this might lead to a passable accurate performance with massive training samples, we would prefer the neural network to approximate the regression relationship between robot pose and the corresponding coefficient.

Given a regression problem with n training samples including robot poses X and map features $y \in A_g$, the nonlinear regression of y given X is defined by

$$E(y/X) = \hat{y}(X) = \frac{\int_{-\infty}^{\infty} y f(X, y) dy}{\int_{-\infty}^{\infty} f(X, y) dy} \tag{6}$$

where $f(X, y)$ represents the joint PDF of X and y. By using Parzen nonparametric estimator[9], we obtain an equivalent formula from the training set with multivariate Gaussian kernel Ψ

$$\hat{A}(X) = \frac{\sum_{i=1}^{n} A_i \Psi_i(-\frac{1}{2\sigma^2}(X - X_i)^T \Lambda_X^{-1}(X - X_i))}{\sum_{i=1}^{n} \Psi_i(-\frac{1}{2\sigma^2}(X - X_i)^T \Lambda_X^{-1}(X - X_i))} \tag{7}$$

where σ is smooth parameter, Λ_X stands for the covariance.

The resulting regression equation can be implemented in a parallel MIMO generalized regression neural network(as shown in Fig. 5) which is a memory-based feed-forward network often used for function approximation without precision physical mathematic model. With increasing number of training samples, the GRNN asymptotically converges to the optimal regression surface [10], which spans the information

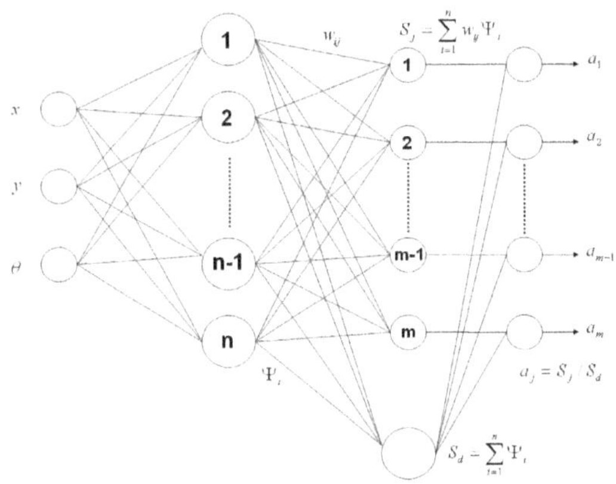

Fig. 5. GRNN for nonlinear mapping the robot pose to the map feature

coming from reference locations and the association information at intervals between each reference locations.

The network consists of four layers, input, pattern, summation and output layer. The pattern layer receives the input vector $X = \{x, y, \theta\}$ coming from robot states and generates output Ψ_i. Neurons of summation layer compute the weighted sum of Ψ_i for numerator S_i and simple arithmetic sum denominator S_d. The output layer then performs a division between S_i and S_d to obtain system output A. The networks have the ability to compute the gradient of the regression surface directly without the need of numerical approximation, if Gaussian Kernel is provided, the gradient of equation yields

$$\nabla A(X)\,|_x = \frac{1}{\sigma^2}\left[\frac{\sum\limits_{i=1}^{n} A_i(X-X_i)^T\Psi_i - \sum\limits_{i=1}^{n} A_i\Psi_i \sum\limits_{i=1}^{n}(X-X_i)^T\Psi_i}{\sum\limits_{i=1}^{n}\Psi_i}\right]\Lambda_x^{-1}\,. \quad (8)$$

It is very convenient for calculating scene prediction error through above equation whenever a robot state is provided. And for manipulation of error propagation, we can use the gradient of the network to propagate the pose error through network

$$Q = (\nabla A)P(\nabla A)^T + R \quad (9)$$

where Q means prediction covariance, P stands for state covariance of robot and R is additive white noise.

5 Experimental Results

We setup our experiments in office as shown in Fig. 6, the laser scanner provide a geometric primitive descriptions of the environment. The robot went through the given reference positions marked as circles, sampled images of environment from 25 different reference locations and recorded corresponding poses. We then simulate the rotation versions for each image of dimension 76×360 every 30 degree. This generates $n = 300$ samples. The samplesare then processed by the proposed PCA to extract the map features in subspace of dimension 47×300. After this, all the available data are normalized

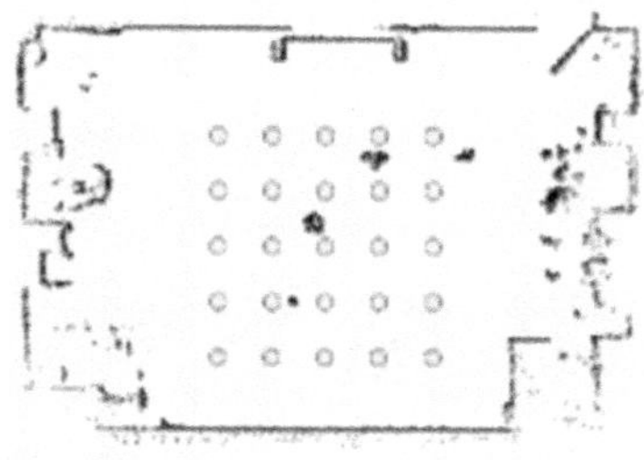

Fig. 6. Experimental setup in our lab

accordingly. Through this procedure, the majority of correlation features in original images are reduced, which further guarantee the regression performance of GRNN.

5.1 Training Network

The features and corresponding robot poses are used to train GRNN. The only parameter for training is kernel width σ, we here use leave-one-out cross validation approach

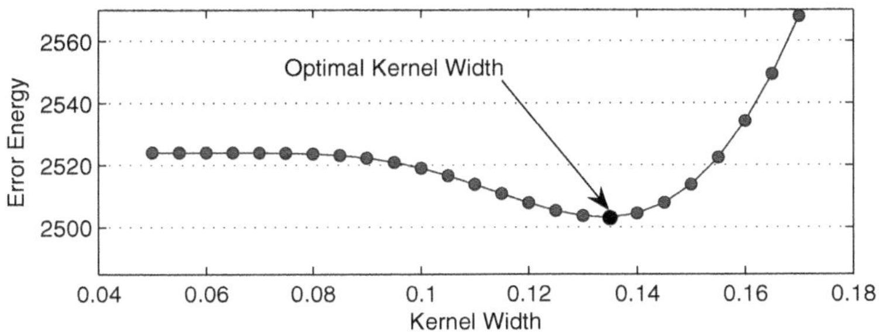

Fig. 7. Leave-one-out cross validation for finding *optimal kernel width*

[11] to prevent overfitting of the training data. The training algorithms that evaluate error energy function will require $n \times (n-1)$ times calculations of kernel function. As shown in Fig. 7, network training starts from an initial sigma, and sample the error function in equidistant steps for finding the optimal value.

The proposed learning method makes the network approximate the desired map feature elements given the corresponding robot pose. Fig. 8 shows the accurate interpolation performance of GRNN for the elements of one map feature.

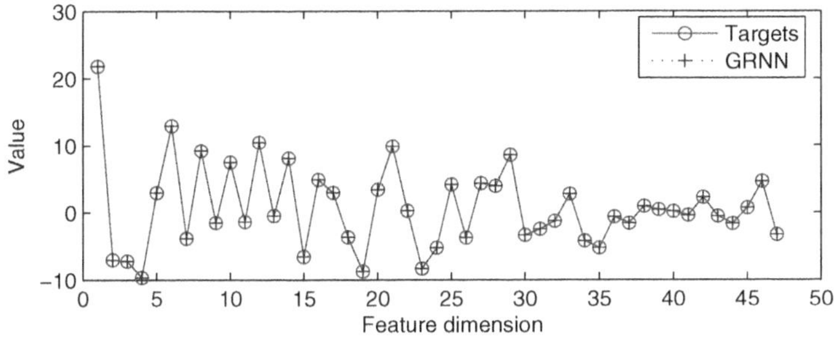

Fig. 8. Approximation result for one map feature using GRNN. Feature elements (*red circles*) and output (*blue crosses*) of GRNN

5.2 Testing Network Performance

To further confirm the network ability, we have conducted two procedures, the first one is to generate 300 states as testing samples from 25 locations previously mentioned, each is rotated a 15 degree from corresponding training state. Because the clusters of the training samples and their corresponding testing data are generated from the same locations, the network prediction should be situated in the middle of two adjacent map points of the same closed loop. As shown in Fig. 9, we take 72 test data belong to 6 different locations for example. The outputs (blue square makers) of GRNN locate in the centers of two adjacent training feature points (red circle makers), which satisfies our proposed regression problem.

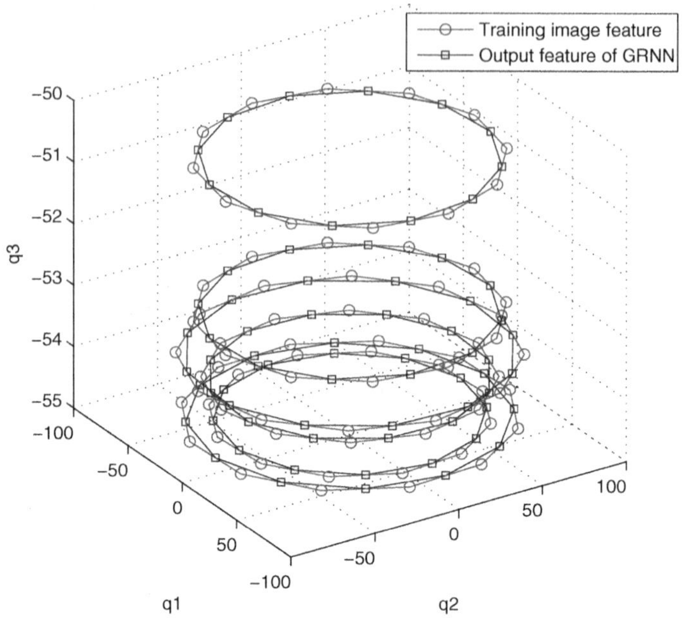

Fig. 9. Testing generalization performance of GRNN. *Red circles* stands for the training map features in eigenspace while *the blue square makers* represent output of neural networks given new inputs

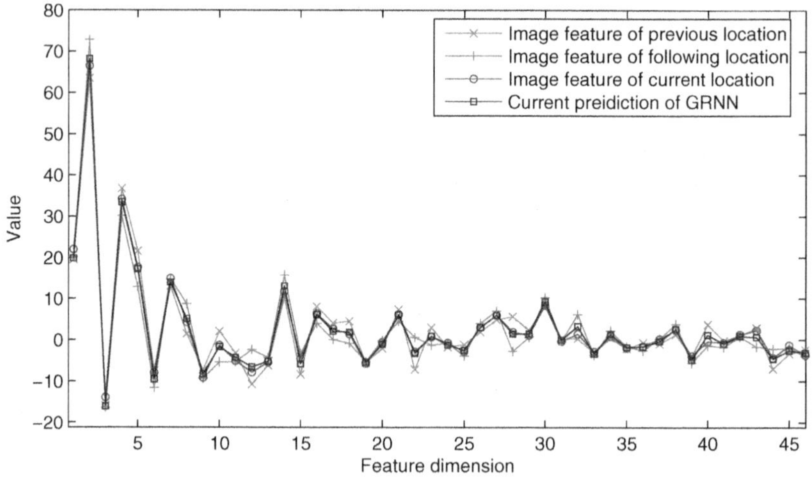

Fig. 10. Interpolation performance of GRNN when robot situates in the middle of two adjacent reference locations

Another procedure is to examine the interpolation performance of the network when we make the robot move along a straight trajectory without changing its' heading. That is to say, the network inputs are different from reference locations in parameters $\{x, y\}$ while the θ component keeps the same. In Fig. 10, we illustrate the interpolation performance of the network when robot is situated in the middle of some two adjacent reference locations.

6 Conclusions

In this paper, we proposed a method based on neural network, generalized regression neural network, to approximate the MIMO nonlinear mapping in map learning for mobile robot. To guarantee the regression performance, we primarily use principle components analysis to reduce the dimension of original training data. The resulting features are used to train the network with according robot poses. Therefore, the network not only stores the memorized training pattern, but has the generalization ability to predict the scene which represents interval information between reference locations.

Further research includes the investigation of the real-time learning ability of neural network. Also, the methods dealing with occlusions and reduction of computation of PCA are to be explored.

References

1. Matjaž, J., Aleš, L.: Robust Localization Using an Omnidirectional Appearance-based Subspace Model of Environment. Robotics and Autonomous Systems **45** (2003) 51-72
2. Emanuele, M., Takeshi, M., Hiroshi, I.: Image-based Memory for Robot Navigation Using Properties of Omnidirectional Images. Robotics and Autonomous Systems **47** (2004) 251-267
3. Nayar, S.K., Nene, S.A., Murase, H.: Subspace Methods for Robotvision. IEEE Trans. Robotics and Automation **12** (1996) 750-758
4. Ben, K., Nikos, V., Roland, B., Motomura, Y.: A Probabilistic Model for Appearance-based Robot Localization. Image and Vision Computing **19** (2001) 381-391
5. Dirk, T., Andreas, S.: A Modified General Regression Neural Network (MGRNN) with New, Efficient Training Algorithms as a Robust 'black box'-tool for Data Analysis. Neural Networks **14** (2001) 1023-1034
6. Donald, F.S.: A General Regression Neural Network. IEEE Trans. Neural Networks **2** (1991) 568-576
7. Yang, J., Zhang, D., Frangi, A.F., Yang J.Y.: Two-dimensional PCA: a New Approach to Appearance-based Face Representation and Recognition. IEEE Trans. on Pattern Analysis and Machine Intelligence **26** (2004) 131-137
8. Richard, L., Burden, J., Douglas, F.: Numerical Analysis. 7th edn. Higher Education Press Beijing (2001)
9. Sergios, T., Konstantinos, K.: Pattern Recognition. 3rd edn. China Machine Press Beijing (2006)
10. Feng, Z.P., Chu, F.L., Song, X.G.: Application of General Regression Neural Network to Vibration Trend Prediction of Rotating Machinery. Lecture Notes in Computer Science. Springer-Verlag, Berlin Heidelberg New York **3174** (2004) 767-772
11. Simon, H.: Neural Networks: A Comprehensive Foundation. 2nd edn. Tsinghua University Press Beijing (2003)

Design of Quadruped Robot Based Neural Network

Lei Sun[1,2], Max Q.-H. Meng[1,3], Wanming Chen[1,2], Huawei Liang[1], and Tao Mei[1]

[1] Center for Biomimetic Sensing and Control Research
Institute of Intelligent Machines, Chinese Academy of Sciences, Hefei 230031
[2] University of Science and Technology of China, Hefei 230021
[3] The Chinese University of Hong Kong, Hong Kong
sunustc@126.com

Abstract. The paper proposed a method for a quadruped robot control system based Central Pattern Generator (CPG) and fuzzy neural networks (FNN). The common approach for the control of a quadruped robot includes two methods mainly. One is the CPG that is based the bionics, the other is the dynamic control that is based the model of quadruped robot. The control result of CPG is decided by the gait data of the quadruped and the parameters of the CPG are choosing manually. Modeling a quadruped robot is difficult because it is a high nonlinear system. This paper presents a much simpler method for the control of a quadruped robot. A simple CPG is adopted for a timing oscillator; it generates the motion periodic pattern of legs. The FNN is used to control the joint motion in order to get a desired stable trajectory motion.

1 Introduction

Walking machine has the advantage over the wheeled machine in that it used but the isolated point to support the trunk not the continuous terrain that is needed by wheeled machine. It can get steady walking on uneven terrain and avoid the obstacle by keeping clear of it, and can get omnidirectional motion by keeping the terrain intact. It can climb stair, stride brook, trudge swamp. Walking machine can choose the effective stand point on the badly terrain. So it can find stable standpoint even on the rugged mountain.

All of the advantages make the walking robot become a important study branch in the robot studying field. Because the quadruped robot has more carrying capacity and good stability than the biped robot, and has the more simple structure than the 6-legged robot and 8-legged robot. So the quadruped robot arouses extensive attention. The common approach for the control of a quadruped robot includes two methods mainly. One is the CPG that is based the bionics, the other is the dynamic control that is based the model of quadruped robot.

Several works on legged robots have used biological principles as a source for solutions to common problems with biomechanical systems [1]. CPG is the usual method. CPG is a local oscillation network consisting of interneurons, it can generate steady phase-lock signal sequence, and inspire pertinent part of the body to put up the rhythmic movement. There exist manifold output modalities for CPG nerve network, corresponding to manifold locomotion modalities of animal's limbs or other parts respectively, moreover, can transfer from one to another reciprocally[2].

D. Liu et al. (Eds.): ISNN 2007, Part I, LNCS 4491, pp. 843–851, 2007.

Kimura designed a 12-DOF quadruped robot Patrush and a 16-DOF quadruped robot Tekken [3-5]. Kimura adopt neural oscillator to build the CPG. And the quadruped robots can walk, trot, pace and gallop. But each gait need a CPG network and this make the control system become very complicated. Collins J J used four nonlinear oscillators to make up hardware CPG, with it the quadruped robot can walk, trot and gallop [6], Billard A constituted the CPG with eight oscillators [7]. Each oscillator controls a joint and implemented walk, trot and gallop on the AIBO quadruped robot platform. Zhang designed a quadruped robot and used the CPG to control it and implemented four typed gaits on flat ground [2]. Carlos Queiroz studied the static gait of a quadruped robot [8]. He presented a method of analysis of stability and an algorithm of searching for static gaits.

Takashi simplified the quadruped robot as a reaction-wheel type inverted pendulum [9], and the trunk is regarded as the reaction-wheel, and the supporting legs as the inverted pendulum. But the reaction-wheel type inverted pendulum model is too simple to describe the quadruped system in that the quadruped robot is a complicated motion system. Costas Tzafestas adopt the adaptive impedance method to solve the model error and uncertain parameters of the quadruped system [10].The fuzzy control method is used to control the motion of a quadruped robot by Duane W, David E.Orin and Luther R.Palmer[11],[12],[13].

All of the methods mentioned above have problems. CPG need many parameters to work. And the parameters have to be decided by manually. Dynamic modelling of a quadruped robot is difficult because the quadruped system is nonlinear, time varying, high coupling system. So many researchers studied the system on a simplified model. This brings many uncertainty and error.

This paper proposed a much simpler method to develop the quadruped robot. A simple CPG is used for a timing oscillator to generate the motion timing of every leg. Then the FNN control the joints to get a stable motion. The paper is organized as follows. In Section II, the quadruped robot system is introduced. The design of CPG and FNN is presented in Section III and Section IV. The Section V is simulation and experiment results. Conclusion and future plan of research will be discussed in Section VI.

2 Design of Quadruped Robot System

The quadruped robot TIM-1 has been designed for the study of home robotic pets. It must look like a pet dog, and it has the nurse ability and can carry a payload.

So we designed a quadruped robot showed in Fig.1, the quadruped robot is 505mm height, 570mm length and 250mm width, the gross weight of body is 10Kg. The weigh of a leg is 2Kg. So the total weight is 18Kg. Twelve DOF (Degree of Freedom) distribute in four legs averagely. Eight DOF are set in four hip joints and four knee joints; four passive DOF are set in ankle joints to accommodate badly terrain using the spring mechanism fixed on the bottom of foot.

The hip joints of the four legs are driven by a 15 Watt Maxon 236655 DC servo motor, reducing the speed directly by planetary gear reducer box (reduction

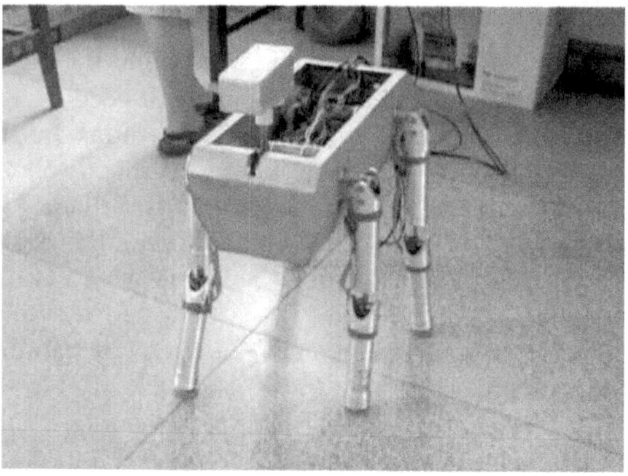

Fig. 1. The robot TIM-1

ration:50:1). To measure the turn angle and angular velocity of motor, the photoelectric encoder (Maxon 110514) is used. The four knee joints are driven by a 20 Watt Maxon 236669 DC servo motor, reducing the speed by the planetary gear reducer box (reduction ration:100:1), and the photoelectric encoders are identical to those of the hip joints. The inclinometer is to measure the deflection angle between the robot's stance and ground. To get the home position of the joint the limit switch is set. Every foot of the robot is equipped by a touch switch in order to make sure the robot's foot has contacted the ground.

In order to make the control system robust, we design a master-slave mode control technology.The control system is consisted of upper master controller and lower controller, the PC/104 computer stack acts as the upper controller. Four motion control card with DSP are the lower controller respectively control the three joins of a leg. The communication between upper controller and lower controller is implemented by the CAN bus.

The robot's each leg is controlled by a control card with a DSP core. It receives the command from upper controller, and then controls the three joint motors to implement a wanted trajectory curve. It includes a 16-bits Fixed-Point DSP TMS320C2407A with 64K flash memory, 64K RAM, DAC, ADC, I/O and encoder interface, etc. To get a high speed and reliable communication, the CAN bus is adopted. The PC/104 computer stack contains a Pentium-compatible processor board running the Linux operating system, a CAN bus interface card PCI-5110, a PCMCIA board with wireless Ethernet card for teleportation, a power supply board.

3 Design of the Simple CPG

Biologist think that the rhythm motion of animals, such as locomotion, breathing and chewing, are controlled by CPG which is located in spinal cord of vertebrate. CPG generates a periodic rhythmic pattern of nerve activity that activates motor neurons, and generates rhythm movements of animals [14].

A number of methods have been studied to simulate the function of CPG since CPG has been realized by people, such as the nonlinear differential equation, artificial neural network, and so on. Now using the Continuous Time Recurrent Neural Networks (CTRNN) as a sequence rhythmic and dynamical behaviour of CPG is the most familiar choice. But if there are many joins to be controlled, the CTRNN will become more complicated. So this paper proposed a new way to implement the function of CPG.

We chose the Amari-Hopfield model as the neural oscillator. The model consists of an excitatory neuron and an inhibitory neuron. The model is represented by the following non-linear differential equations:

$$\begin{cases} \tau\dot{u} = -u + Af_{\mu}(u) - Cf_{\mu}(u) + S_{u}(t) \\ \tau\dot{v} = -v + Bf_{\mu}(u) - Df_{\mu}(v) + S_{v}(t) \end{cases} \quad (1)$$

where u and v express the activities of the excitatory neurons and the inhibitory neurons, respectively. $S_{u}(t)$ and $S_{v}(t)$ are the external inputs. A, B, C, D can alter the dynamic character of the model. $f(u)$ is the transfer function, and

$$f(u) = \frac{1+\tanh(\mu u)}{2} \quad (2)$$

Depending on the parameters A through D and the external inputs, the model can generates the periodic pattern automatically [15].

4 Fuzzy Neural Network

Cerebellar model articulation controller (CMAC) is a kind of self-associative memory network, which is firstly proposed by Albus [16]. CMAC is capable of nonlinear approximating and mapping inputs from multidimensional space to a lower finite region. As a local network, CMAC has the advantage of fast learning. But there are other problems to be overcome: (1) CMAC divides the input space into several blocks. The relation between the input state and the blocks is binary relation. So the relation isn't differential, and then it can't be modulated on line and the study ability is poor. (2) It suffers from the disadvantage of over-large dimensionality if the number of receptive field functions growth [17].

4.1 Structure of FCMAC

In order to improve the generalization of CMAC and solve the problems mentioned above, FCMAC is proposed [18]. Uncertainty with the mechanics of the robot TIM-1 and the friction between joints make the system hard to control. Self study ability of FCMAC can improve the robust performance of the system. The structure of FCMAC is showed in Fig.2. It consists of five layers: input layer, fuzzified layer, fuzzy association layer, fuzzy post association layer and output layer.

The input vector $X = (x_1, x_2, \cdots x_m)$ is transmitted to the fuzzified layer. Each node in fuzzified layer implies a lingual variable, fuzzed the input vector, and the

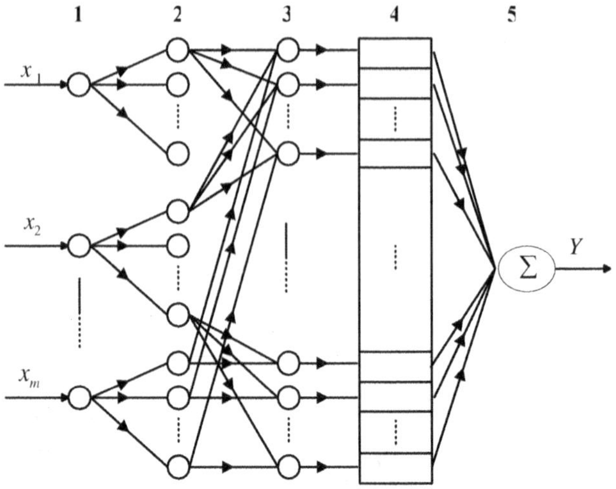

Fig. 2. Structure of FCMAC

membership function is Gaussian function. Fuzzy association layer construct the fuzzy logic rule through the AND operation. Fuzzy Post association Layer calculates the normalization of firing strength and prepare for fuzzy inference. Output layer is:

$$y = \sum_{i=1}^{m} a_i w_i \tag{3}$$

where w is the weights vector of output layer of FCMAC, and a_i is the association cell vector corresponding to input X .

4.2 Learn Algorithm of FCMAC

The learning of FCMAC includes the weights of FCMAC, centers and width of Gaussian membership function. Suppose y_d is the desired output, y is the real output of FCMAC, the error performance function is:

$$J(t) = (y_d(t) - y(t))^2 / 2 \tag{4}$$

The centers and width of Gaussian membership function are decided off line. In order to get a high speed, so only the weights of output layers are modulated on line. The learn algorithm is shown as follows:

$$\Delta w = \frac{\partial E}{\partial w} = \frac{\partial E}{\partial y} \cdot \frac{\partial y}{\partial w} = (y_d(t) - y(t)) \cdot a \tag{5}$$

$$w_i = w_{i-1} + \eta \cdot \Delta w \tag{6}$$

5 Simulation and Experiment

5.1 Simulation of FCMAC Control Leg

The lower controller receives the command from upper controller through CAN bus. The command indicates this leg is in swing phase or in supporting phase, and the swing period and supporting period is fixed. The trajectory of hip joint and knee joint is sinusoidal curve. Then the FCMAC control the joint motors to realize the motion curve. In walk and trot gait, the yaw joint in hip's function is little for no turn action happened. In order to simplify the control system, here we take attention to the hip pitch joint and knee joint. And the simulation model is showed in Fig.3.

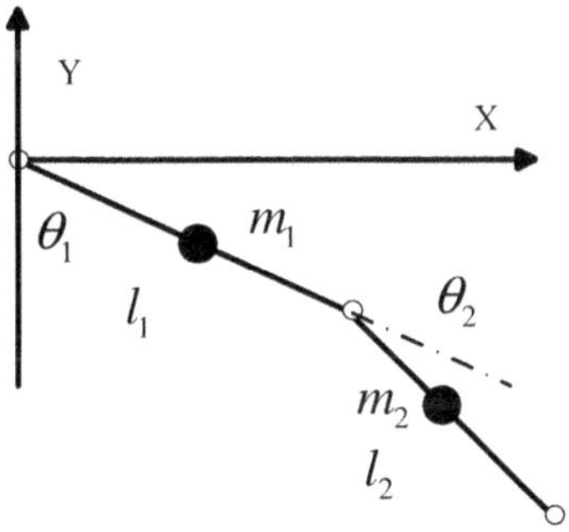

Fig. 3. Simulation model of a leg

m_1 is mass of thigh, l_1 is the length of thigh. θ_1 is the angle between the thigh and vertical line. m_2 is mass of crus , l_2 is the length of crus. θ_2 is the angle between the crus and extend line of thigh. The dynamics equation of the model is build as follows based on Lagrange function.

$$D(\theta)\cdot\ddot{\theta}+H(\theta,\dot{\theta})\cdot\dot{\theta}+G(\theta)=T_\theta \tag{7}$$

where $D(\theta)$ is the 2×2 definite and symmetric inertia matrix, $H(\theta,\dot{\theta})$ is the 2×2 matrix related to centrifugal and coriolis terms, $G(\theta)$ is the 2×1 matrix of gravity terms, θ, $\dot{\theta}$, $\ddot{\theta}$, T_θ are the 2×1 vectors of coordinates, velocities, accelerations and torques, respectively. We use the same method as that Wang used [19]. Fig.4 gives the system based on FCMAC.

The simulation is implemented in Matlab 6.5. The parameters of the quadruped robot are as follows:

$l_1=l_2=0.2m$, $m_1=m_2=1Kg$

The initial value: $\theta_1(0)=0, \theta_2(0)=1, \dot{\theta}_1(0)=0, \dot{\theta}_2(0)=0$

The desired trajectory is $\theta_1^d=\sin(2\pi t), \theta_2^d=\cos(2\pi t)$, the sampling time is 1ms. Fig.12 shows the hip joint and knee join motion curve.

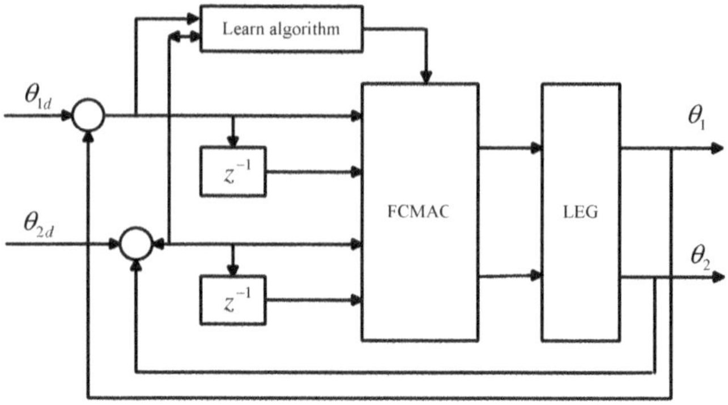

Fig. 4. The leg control system based FCMAC

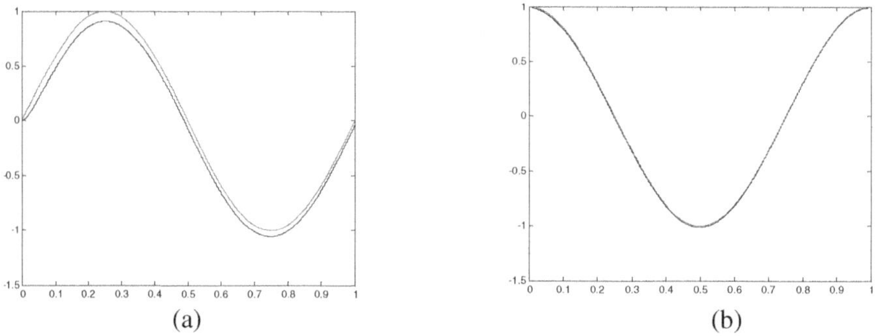

Fig. 5. (a) is the hip join motion curve; (b) is the knee motion curve

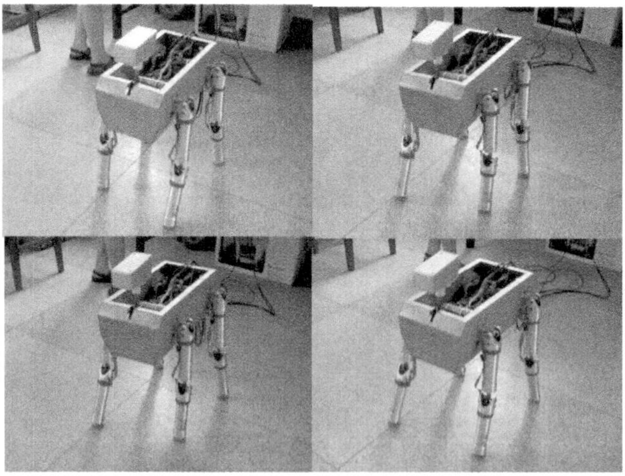

Fig. 6. Walking on flat terrain

5.2 Experiments

We do some experiments to test robot's motion performance. The robot can walk on flat terrain at velocity about 0.1m/s with walk gait. Fig.6 shows that the quadruped robot is walking on flat terrain with walk gait.

6 Conclusion

Generating movement gait is an important thing in legged robot studying. This paper presents a much simpler method for the control of a quadruped robot. A simple CPG is adopted for a timing oscillator; it generates the motion periodic pattern of legs. The FNN is used to control the joint motion in order to get a desired stable trajectory motion. Simulation and experiments show that the method is valid. But there are many things to do in future. (1) How to implement the autonomous movement in an unknown environment. (2) Using few sensors to locate itself in outdoor. (3) Based on CPG, control the robot accurately.

Acknowledgement

This work is partially supported by National Science Fund of China under Grant 50275141 and 60475027.

References

1. Jose, C.: Gait Synthesis and Modulation for Quadruped Robot Locomotion Using a Simple Feed-forward Network. ICAISC (2006) 731-739
2. Cheng, Z.F., Zhang, X.L.: The CPG-based Bionic Quadruped System. In: IEEE International Conference on Systems, Man and Cybernetics (2003) 1828-1833.
3. Kimura, H., Fukuoka, Y.: Biologically Inspired Dynamic Walking of a Quadruped Robot on Irregular Terrain-adaptation at Spinal Cord and Brain Stem. Adaptive Motion of Animals and Machines [C], Montreal Canada (2000)
4. Kimura, H., Fukuoka, Y.: Adaptive Dynamic Walking of a Quadruped Robot on Irregular Terrain by Using Neural System Model. Mechatronics Systems, IFAC, Darmstadt (2000) 641-646.
5. Kimura, H., Fukuoka, Y.: Adaptive Dynamic Walking of the Quadruped on Irregular Terrainautonomous Adaptation Using Neural System Model. IEEE Robotics and Automation, ICRA (2000) 436-443
6. Collins, J.J., Richmond, S. A.: Hard-wired Central Pattern Generators for Quadrupedal Locomotion. Biologically Cybernetics **71(5)** (1994) 375-385
7. Billard, A., Ijspeert, A.J.: Biologically Inspired Neural Controllers for Motor Control in a Quadruped Robot. Neural Network, IJCNN (2000) 637-641
8. Carlos, Q., Nuno, G., Paulo, M.: A Study on Static Gaits for a Four Leg Robot. www.deec.uc.pt/~nunogon/PUBS/goncalves_n_control2000.pdf
9. Takashi, E., Akira, A.: Attitude Control of a Quadruped Robot during Twolegs Supporting. In: Proceeding of 5th Intenational Conference on Advanced Robotics (1991) 711-716
10. Elcomyn, F.: Neural Basis of Rhythmic Behavior in Animals. Science **210** (1998) 492-498

11. Marhefka, S.D.W., Orin, D.E., Waldron, K.J.: Intelligent Control of Quadruped Gallops. IEEE/ASME Transactions of Mechatronics **8(4)** (2003) 446-456
12. Palmer, L.R., Orin, D.E. etal: Intelligent Control of an Experimental Articulated Leg for a Galloping Machine. In: Proceedings of the 2003 IEEE International Conference on Robotics & Automation (2003) 3821-3827
13. Mu, X.P., Wu, Q.: Dynamic Modeling and Sliding Mode Control of a Five-link Biped during the Double Support Phase. In: Proceeding of the 2004 American Control Conference (2004) 2069-2014
14. Kazuki, N., Yoshihito, A.: An Analog CMOS Central Pattern Generator for Interlimb Coordination in Quadruped Locomotion. IEEE Transactions on Neural Networks **14(5)** (2003) 1356-1365
15. Hiroshi, K., Seiichi, A.: Realization of Dynamic Walking and Running of the Quadruped Using Neural Oscillator. Autonomous Robots 7 (1999) 247-258
16. Albus, J.S.: A New Approach to Manipulator Control: The Cerebellar Model Articulation Controller (CMAC). Trans of the ASME Dyn. Sys. Meas **97** (1975) 220-227
17. Du, W.L., Qian, F.: 4-CBA Soft Sensor Based on Fuzzy CMAC Neural Networks. Chinese J. Chem. Eng **13(3)** (2005) 437-440
18. Deng, Z. D, Sun, Z. Q, Zhang, Z. X.: A Fuzzy CMAC Neural Network. ACTA Automatic Sinica **21** (1995) 288-294
19. Wang, Y.N.: Intelligent Robot Control Project. Science Press (In Chinese) (2004)

A Rough Set and Fuzzy Neural Petri Net Based Method for Dynamic Knowledge Extraction, Representation and Inference in Cooperative Multiple Robot System*

Hua Xu[1], Yuan Wang[1,2], and Peifa Jia[1,2]

[1] State Key Laboratory of Intelligent Technology and Systems, Tsinghua University, Beijing 100084, China
[2] Department of Computer Science and Technology, Tsinghua University, Beijing 100084, China
{xuhua,yuanwang05,dcsjpf}@mail.tsinghua.edu.cn

Abstract. In cooperative multiple robot systems (CMRS), dynamic knowledge representation and inference is the key in scheduling robots to fulfill the cooperation requirements. The first goal of this work is to use rough set based rules generation method to extract dynamic knowledge of our CMRS. Kang's rough set based rules generation method is used to get fuzzy dynamic knowledge from practical decision data. The second goal of this work is to use Fuzzy Neural Petri nets (FNPN) to represent and infer the dynamic knowledge on the base of dynamic knowledge extraction with self-learning ability. In particular, we investigate a new way to extract, represent and infer dynamic knowledge with self-learning ability in CMRS. Finally, the effectiveness of the dynamic knowledge extraction, representation and inference procedure are demonstrated.

1 Introduction

In cooperative multiple robot systems (CMRS), dynamic knowledge extraction, representation and inference is the key in dispatching robots for cooperation requirements. The scheduling is conducted by the decision making modules distributed in all robots. According to the environment information, system states, task aims and other decision factors, the decision dispatches robots effectively. However, in different situations, the knowledge about the environment and tasks is various. At the instance of decision making, dynamic production rules generation is required for extracting the dynamic knowledge from dynamic environments. Then the extracted dynamic knowledge is used in the decision making module and the system schedule is conducted. On the other hand, since the knowledge of dynamic environments may be incomplete or uncertain, the knowledge may be fuzzy and the corresponding inference may be a kind of fuzzy reasoning. What's more, the

* This work is jointly supported by the National Nature Science Foundation (Grant No: 60405011, 60575057) and the China Postdoctoral Foundation for China Postdoctoral Science Fund (Grant No: 20040350078).

D. Liu et al. (Eds.): ISNN 2007, Part I, LNCS 4491, pp. 852–862, 2007.

uncertain knowledge may also need to be adjusted according to the dynamic changes of the environment. Therefore, it needs a new method to extract, represent and reason dynamic knowledge in the decision making of CMRS. At the same time, the method also requires self-learning ability.

As an important modeling and computational paradigm, Petri nets (PN) [1, 2] have been widely used. In order to deal with uncertain information or knowledge, Fuzzy Petri Nets (FPNs) [3-18] have been introduced, which can be used to represent Horn clauses or Non-Horn clauses [4, 5] and represent and execute the fuzzy rules [14-16]. To improve the FPN adjusting (or learning) ability [13, 17], a generalized fuzzy Petri net (GFPN) [17] is proposed. To solve the learning problem, an adaptive fuzzy Petri net (AFPN) [9] is proposed by adjusting weights the same as those in a neural network (NN). But the weight adjustment is off-line. To overcome this weakness, neuron concept is introduced into FPN [17] on the base of the proposed fuzzy object-oriented Petri net (FOPN) [25], where neurons are regarded as a special type of components or objects, and Fuzzy neural Petri net (FNPN) is proposed.

Although FNPN has both the dynamic ability and the learning ability, it requires the dynamic systems to be described with weighted fuzzy production rules (WFPR). In actual dynamic systems, dynamic knowledge represents the facts, not the production rules. That is to say, there exists a rules generation problem. At the same time, in order to improve the system performance, the optimal and minimum production rules set is needed in actual systems. These dynamic systems mentioned above may include dynamic decision making systems and on-line expert system. For example, the decision making subsystem (or module) in our CMRS is a typical one.

The rough set (RS) theory [19] has been proved to be a useful mathematical tool for the analysis of vague description of objects. To some extent, RS is superior to other knowledge extraction methods [19], because no other antecedent knowledge is required in advance except the knowledge (or facts) to be processed. Wu, etc al [24] have used the RS based algorithm to extract multi-knowledge from a decision system or a training set. But the extracted knowledge is accurate without any vague features. In order to analyze vague information and get optimal fuzzy production rules, Kang, etc al [20], propose a rules generation method on the base of fuzzy set (FS) and RS.

This paper proposes a dynamic knowledge extraction, representation and inference method on the base of RS and FNPN, which aims to model the decision making sub-system of our CMRS. In our method, Kang's rules generation method [20] is firstly used to generate fuzzy production rules from the practical decision data. Then, on the base of the fuzzy production rules and FOPN [25], FNPN [9] is used to model the decision making subsystem of our cooperative multiple robot system.

This paper is organized as the following. In section 2, the RS based fuzzy rules extraction method (RFREM) and FNPN are illustrated. Section 3 proposes the decision making subsystem in our CMRS. Then in section 4, the rules generation of the decision making system is discussed. Section 5 proposes the FNPN model of the decision making sub-system. Finally, the conclusion and future work are summarized in section 6.

2 Preliminary Notions of RFREM and FNPN

2.1 RFREM

Definition 1: The fuzzy equivalence class ***R***(k) on the universal ***U*** is defined as ***R***(k)=U/ind*(***R***), where ind(***R***) is the equivalence relation set on the universal ***U***.

Definition 2: The fuzzy lower approximation **R$_*$(X)** is defined as:

$R_*(X)=\{(R_k(x),\mu_{Rk}(x)) \mid x\in U, R_k(x)\subseteq X, 1\le k \le|R(X)|\}$,

where $\mu_{Rk}(x)$ is the member function.

Definition 3: The fuzzy upper approximation **R*(X)** is defined as:

$R^*(X)=\{(R_k(x),\mu_{Rk}(x)) \mid x\in U, R_k(x)\cap X\neq\varnothing, 1\le k \le|R(X)|\}$,

where $\mu_{Rk}(x)$ is the member function.

On the base of FS and RS theory, the Kang's fuzzy rules generation method [20] is just like the following:

1) Suppose the discussed object set is U^i (i=1 to m), and then the conditional property set is A_j (j=1 to n). The decision property set is C.
2) Decide the member function and language variable (the number of fuzzy intervals) of every conditional property.
3) Depict the trend figure of the member function for every language variable of every conditional property.
4) for i = 1 to m
   for j = 1 to n

Transform the conditional property values to the following form:

$$\frac{f_{j1}^{i}}{R_{j1}} + \frac{f_{j2}^{i}}{R_{j2}} + \cdots + \frac{f_{jl}^{i}}{R_{jl}}, \tag{1}$$

where R_{jk} is the kth fuzzy interval of the conditional property A_j. f_{jk}^{i} is the member function value of the object U^i on the fuzzy interval R_{jk}^{i}. $l=|A_j|$ is the number of language variable of the conditional property A_j.

5) Transform the original data table to the decision table with the member function values. And calculate the fuzzy equivalence matrix.
6) for i = 1 to m
   for j = 1 to n
   Set $f_{jk}^{i}=\max(f_{j1}^{i},\ldots, f_{jl}^{i})$, where $1 \le k \le l$.
   $R'_j= R_{jk}$

Then the new decision table can be got.

7) Simplify the new decision table using the RS reduction algorithm, and the reduction rules can be got.
8) The uncertain rule table can be got, which is made up of the incompatible decision and the decision conflicted with the incompatible decision.
9) List the equivalence class of conditional properties in the uncertain rule table. According to its fuzzy equivalence class, the fuzzy equivalence class A′(k) can be got.
10) Calculate the $A'^*(k)$.
11) Suppose

$$D_p = A'^*(C) \cap (U/ind(C)), \tag{2}$$

where $1 \le p \le |A'^*(k)|$.

The selection function of D_p^d is defined as the following:

$$F(D_p^d) = \sum_{k=1}^{|D_p^d|} \frac{\min(f_j^{i*})}{\sum_{j=1}^{n} f_j^{i*}} \times \frac{D_p^d}{|U/ind^*(C)|}, \tag{3}$$

where $1 \le d \le |D_p^d|$. f_j^{i*} is the fuzzy interval value of A′(k). $|U / ind^*(C)|$ is the number of decision equivalence class of D_p^{dk}.

12) Select the maximum $A'^*(C)$ in $F(D_p)$ as the final decision, where $1 \le p \le |A'^*(C)|$.
13) Combine the results in step 7) with those in step 12), and the final decision rules can be got.

2.2 FNPN

A FNPN is an 11-tuple:

$$FNPN=(P,T,D,A,M_0,K_p,K_t, th, f, W, \beta)$$

Where $P=\{p_1,p_2,\ldots,p_n\}$ denotes a finite set of places, $T=\{t_1,t_2,\ldots,t_n\}$ denotes a finite set of transitions, $D=\{d_1, d_2,\ldots, d_n\}$ denotes a set of propositions, $P \cap T \cap D = \varnothing$ and $P \cup T \neq \varnothing$. $A=\{P \times T\} \cup \{T \times P\}$ denotes a set of arcs. M_0: $p \rightarrow \{0,1,2,3,\ldots\}$ denotes the initial marking. K_p is the state set of hidden layer and output layer. K_t is the mapping from T to rule sets. *th*: $T \rightarrow [0,1]$ denotes the mapping from transitions to thresholds. f:$T \rightarrow [0,1]$ denotes the mapping from transitions to confidence level value. W:$P \rightarrow [0,1]$ is the mapping from propositions represented by places to truth values. It represents the supporting level of every place representing the corresponding proposition condition for transition firing. β:$P \rightarrow D$ is the mapping from places to propositions.

In FNPN, the truth value of the place p_i, $p_i \in P$ is denoted by $W(p_i)$, $W(p_i)=\omega_i$ and $\omega_i \in [0,1]$. If $W(p_i)= \omega_i$ and $\beta(p_i)=d_i$. This configuration states that the confidence level of the proposition d_i is ω_i. A transition t_i with only one input place is enabled, if for the input place $p_j \in I(t_i)$, $\omega_i \geqslant th(t_i)= \lambda_i$ where λ_i is the threshold value. If the transition t_i fires, the truth value of its output place is $\omega_i \bullet \mu_i$.

For the self-learning of neurons in FNPN, BP based learning algorithm [26] can be used to adjust the corresponding parameters. Suppose the FNPN model to be studied is n-layered with b ending places p_j, where j=1,…,b. And r learning samples are used to train the FNPN model. The performance evaluation function is defined as the following:

$$E = \frac{\sum_{i=1}^{r} \sum_{j=1}^{b} (M_i(p_j) - M_i'(p_j))^2}{2} \tag{4}$$

Where $M_i(p_j)$ and $M_i'(p_j)$ represent the actual marking value and the expected one of the ending place p_j respectively.

Suppose $t_i^{(n)}$ is one transition on the nth layer $t_i^{(n)} \in T_n$. The weights of the corresponding input arcs are $\omega_{i1}^{(n)}, \omega_{i2}^{(n)}, \ldots, \omega_{im}^{(n)}$. The weights can be adjusted as the following:

$$\frac{dE}{d\omega_{ix}^{(n)}} = \delta^{(n)} \times \frac{d(M^{(n)}(p_j))}{dw_{ix}^{(n)}} \quad x=1,2,\ldots,m-1 \tag{5}$$

$$\delta^{(n)} = \frac{dE}{d(M^{(n)}(p_j))} \tag{6}$$

The adjusting algorithm of the weight parameters of the transition of $t_i^{(q)}$ can be got as the following:

$$\omega_{ix}^{(q)}(k+1) = \omega_{ix}^{(q)}(k) - \eta dE/d\omega_{ix}^{(q)} \tag{7}$$

Where $x=1,\ldots,m-1$, $q=n,\ldots,1$ and $\sum \omega_{ix}^{(q)}=1$. In the above equations, η is the learning rate.

The adjusting algorithms of the threshold $\lambda_i^{(n)}$ and the truth $\mu_i^{(n)}$ value owns similar forms of the weight $\omega_{ix}^{(n)}$.

3 The Decision Making in Our CMRS

In CMRS, decision making is one of the important parts. However, this kind of robot systems always works in dynamic environments. The dynamic knowledge can not be summarized accurately beforehand. Usually it is fuzzy, incomplete or uncertain. What's more, decision making rules in dynamic environments stem from practical data in dynamic states. In order to generate the decision making rules from practices, the FS and RS based rules generation method [20] is used to get decision rules.

Our CMRS is a kind of distributed one. In every robot sub-system, there is a kind of a two-level system illustrated in Fig.1. The lower level is the knowledge extraction layer. The higher level is the control decision making layer. According to the robot working states, velocity, movement direction and aim distance, decisions of every robot are made respectively in the case that a new task needs to be performed.

In our CMRS, the system control decisions are made according to the working states mentioned above. In order to get the decision knowledge—the fuzzy decision rules of our CMRS, a set of typical practical decision data are chosen. Then Kang's RS based rules generation method [20] is used to get the CMRS fuzzy decision rules set in the knowledge extraction, which is an optimal and minimum rules set. After the dynamic knowledge of the CMRS control decision has been extracted from the typical practical data, FNPN is used to describe and model dynamic knowledge representation and inference procedure in the control decision level. In the control, the corresponding CMRS model can be adjusted, because FNPN owns self-learning ability. Then in the actual CMRS control, the FNPN control model has just depicted a cooperation control model.

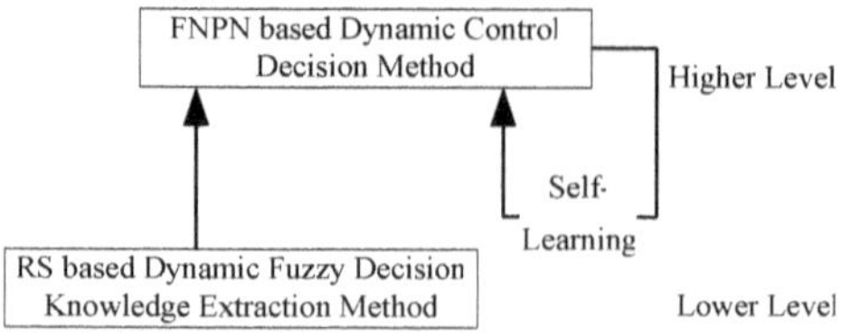

Fig. 1. The Decision Control Architecture of Our CMRS

4 Rules Generation in Our CMRS

In the decision making sub-system of CMRS, the form of dynamic knowledge represents as production rules. In order to generate decision making rules, a set of practical decision data have been got, which is shown in Table.1. In Table.1, Y means that the object will be scheduled to perform the task. W means that it will be regarded as an alternate. N means that it will not be scheduled. Then Kang's rules generation method [20] reviewed beforehand can be used.

Table 1. Data for Objects

Object	VE (m/s)	DA (Degree)	AD (m)	PE
1	0.5	30	1	Y
2	2	110	3	N
3	1	100	2	Y
4	2	80	2	N
5	0.5	45	3	N
6	1	90	2	W

1) From the object data, it is clear that velocity (VE), direction angle (DA) and aim distance (AD) are the condition properties set $\boldsymbol{A}$, while perform (PE) is the decision property set $\boldsymbol{C}$.
2) The number of selected fuzzy interval of VE, DA and AD is 3. They are represented as L (low-level), M (middle-level) and H (high-level).
3) The member function for VE, DA and AD is the same, which is the triangle function depicted as in Fig.2.

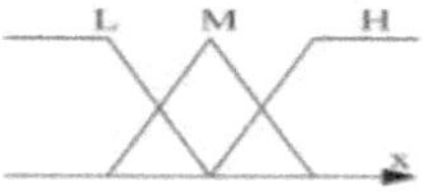

Fig. 2. Member Function

4) According to the member function, the properties can be transformed into the following form in Table.2.

Table 2. Data Transformed From Member Functions

Object	VE	DA	AD	PE
1	0.5/L+0.5/M	0.67/L+0.33/M	0.5/L+0.5/M	Y
2	1/H	0.67/M+0.33/H	0.5/M+0.5/H	N
3	1/M	0.78/M+0.22/H	1/M	Y
4	1/H	0.11/L+0.89/M	1/M	N
5	0.5/L+0.5/M	0.5/L+0.5/M	0.5/M+0.5/H	N
6	1/M	1/M	1/M	W

5) The fuzzy equivalence class matrix can be got from Table.2. It is listed in Table.3.

Table 3. The Fuzzy Equivalence Class Matrix

	(L,L)	(L,M)	(L,H)	(M,L)	(M,M)	(M,H)	(H,L)	(H,M)	(H,H)
L	(1)		(5)	(1)		(5)			
M	(1)		(5)	(1)	(1,3)	(5)		(3)	
H		(4)			(2,4)	(2)		(3)	(2)

6) The decision table (Table.4) can be got from Table.2 according to the formula reviewed beforehand.

Table 4. Data Transformed From Member Functions

Object	VE	DA	AD	PE
1	L(or M)	L(or M)	L(or M)	Y
2	H	M(or H)	M(or H)	N
3	M	M(or H)	M	Y
4	H	L(or M)	M	N
5	L(or M)	L(or M)	M(or H)	N
6	M	M	M	W

7) Simplify the Table.4 and the certain rules can be got as
Rule 1: IF (VE=H) and (DA=M) and (AD=M) THEN N,
Rule 2: IF (VE=H) and (DA=M) and (AD=H) THEN N,
Rule 3: IF (VE=L) and (DA=L) and (AD=L) TEHN Y,
Rule 4: IF (VE=L) and (DA=L) and (AD=M) TEHN Y,
Rule 5: IF (VE=M) and (DA=L) and (AD=L) TEHN Y,
Rule 6: IF (VE=M) and (DA=L) and (AD=M) TEHN Y.

8) According to the results of step 7), the uncertain decision table (see Table.5) can be got as

Table 4. Uncertain Decision Table

Object	VE	DA	AD	PE
1	M	M	M	Y
3	M	M	M	Y
6	M	M	M	W

9) The property equivalence class: K={(M,M,M)},
The fuzzy equivalence class: A'(K)={(1,3,6)}
10) U / ind(C)={(1,3), (2,4,5),(6)},
$A^*(C)$= {(1,3,6)}.
11) From the formula reviewed beforehand, the following results can be got.
D_l={(1,3),(6)} ; $F(D_l^1)$=0.18; $F(D_l^2)$=0.17.

So the uncertain rules can be got as:
Rule 7: IF (VE=M) and (DA=M) and (AD=M) THEN Y (CF=0.18),
Rule 8: IF (VE=M) and (DA=M) and (AD=M) THEN W (CF=0.17).

12) From the results of Step 7) and 11), the rule set of our decision system can be got as:

Rule 1: IF (VE=H) and (DA=M) and (AD=M) THEN N (CF=1),
Rule 2: IF (VE=H) and (DA=M) and (AD=H) THEN N (CF=1),
Rule 3: IF (VE=L) and (DA=L) and (AD=L) TEHN Y (CF=1),
Rule 4: IF (VE=L) and (DA=L) and (AD=M) TEHN Y (CF=1),
Rule 5: IF (VE=M) and (DA=L) and (AD=L) TEHN Y (CF=1),
Rule 6: IF (VE=M) and (DA=L) and (AD=M) TEHN Y (CF=1),
Rule 7: IF (VE=M) and (DA=M) and (AD=M) THEN Y (CF=0.18),
Rule 8: IF (VE=M) and (DA=M) and (AD=M) THEN W (CF=0.17).

5 The FNPN Model of the Decision Making System

In this section, FNPN is used to model the decision making sub-system in our CMRS. Based on the FNPN translation principle, the decision making sub-system can be mapped into the FNPN model as shown in Fig.3.

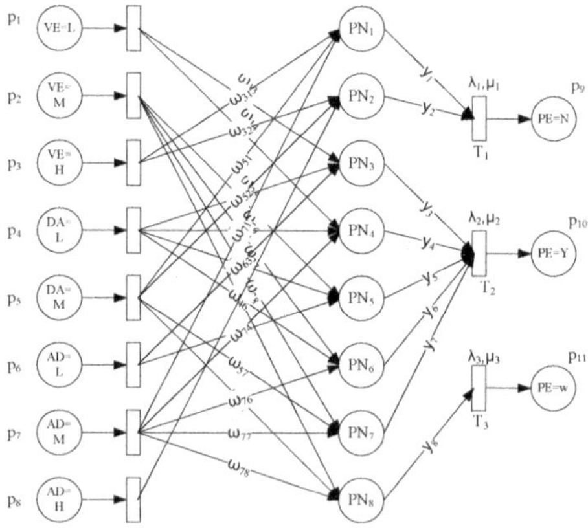

Fig. 3. The FNPN Model of the Decision Making Sub-system in Our CMRS

According to the requirements of the decision making subsystems, the thresholds can be got easily from the member function of VE, DA and AD.

For the weights in FNPN model, the initial weights can be set to an arbitrary value firstly, and then the learning method of FNPN is used to train the model and the converged weights can be got. The actual parameters of the FNPN model are set randomly, which also fulfill the weight requirements.

The FNPN model is trained on the base of 100 groups of testing data, where b=1000, η=0.03. After the model has been trained, the fuzzy reasoning of CMRS decision is conducted on the base of the trained FNPN. The actual and expected reasoning results are listed in Table.6.

Table 5. The Actual Output and the Expected Output of The FNPN Model

No.	p_9		p_{10}		p_{11}	
	Actual Output	Expected Output	Actual Output	Expected Output	Actual Output	Expected Output
1	1.0000	1	0.0000	0	0.0000	0
2	1.0000	1	0.0000	0	0.0000	0
3	0.0000	0	1.0000	1	-0.0000	0
4	0.0000	0	1.0000	1	0.0000	0
5	-0.0000	0	1.0000	1	0.0000	0
6	0.0000	0	1.0000	1	-0.0000	0
7	0.0000	0	0.5000	0.5	0.5000	0.5
8	0.0000	0	0.5000	0.5	0.5000	0.5

From Table.6, it is clear that the FNPN model can complete the fuzzy reasoning effectively. From the actual application of the proposed method in this paper, compared with traditional methods such as RS, FS and NN, the RS and FNPN based method can not only extract fuzzy dynamic knowledge, but also adjust the decision procedure through learning by itself. So the proposed method manifests powerful advantages in making decision of CMRS.

6 Conclusions and Future Work

Dynamic knowledge and uncertain rules are usual phenomena in dynamic systems, especially in cooperative multiple robot systems. Therefore, dynamic knowledge extraction, representation and inference are the critical problems to be solved in the modeling and analysis of dynamic systems. This paper presents a new combined method on the base of Kang's rules generation method [20] and Xu's FOPN [25] based FNPN. The dynamic fuzzy production rules of dynamic systems can be got according to the Kang's rules generation method. Then based on the rules extracted from the practical data, FNPN with self-learning ability is used to model the dynamic systems formally. The validity of this method has been demonstrated by using it in modeling the decision making sub-system in our CMRS. Compared with traditional method, the method proposed in this paper synchronously manifests the following advantages in solving decision making problems: dynamic knowledge extraction and self-learning ability.

The improvement of the FNPN modeling ability will be studied in the future. Via the improvement, it is to realize the aim of modeling complex dynamic systems and overcoming complexity phenomena. For the future work, we hope that Object-oriented concept can be completely introduced into the FNPN.

References

[1] Murata, T.: Petri Nets: Properties, Analysis and Applications. Proceedings of IEEE **77** (1989) 541-580
[2] Peterson, J. L.: *Petri Net Theory and the Modeling of Systems*. Prentice-Hall, New Jersey (1991)
[3] Chen, S., Ke, J., Chang, J.: Knowledge Representation using Fuzzy Petri Nets, IEEE Trans. Knowledge Data Engineering **2** (1990) 311–319
[4] Jeffrey, J., Lobo, J., & Murata, T.: A High-level Petri Net for Goal Directed Semantics of Horn Clause logic. IEEE Transactions on Knowledge Data Engineering, **8** (1996) 241–259
[5] Chaudhury, A., Marinescu, D. C., Whinston, A.: Net-based Computational Models of Knowledge-processing Systems. IEEE Expert **5** (1993) 79–86
[6] Bugarn, A. J., Barro, S.: Fuzzy Reasoning Supported by Petri Nets. IEEE Trans. Fuzzy Systems **2** (1994) 135–150
[7] Cao, T., Sanderson, A. C.: Representation and Analysis of Uncertainty using Fuzzy Petri Nets. Journal of Intelligent Fuzzy Systems **3** (1995) 3–19
[8] Chen, S., Ke, J., Chang, J.: Knowledge Representation using Fuzzy Petri Nets. IEEE Tran. Knowledge Data Engineering **2** (1990) 311–319
[9] Looney, C. G.: Fuzzy Reasoning in Information, decision and control systems. Kluwer, Dordrecht (1994)
[10] Scarpelli, H., Gomide, F.: Fuzzy Reasoning and Fuzzy Petri Nets in Manufacturing Systems Modeling. Journal of Intelligent Fuzzy Systems **1** (1993) 225–241
[11] Scarpelli, H., Gomide, F., Yager, R. R.: A Reasoning Algorithm for High-level Fuzzy Petri Nets. IEEE Trans. Fuzzy Systems **4** (1996) 282–293
[12] Yeung, D. S., Tsang, E. C. C.: Fuzzy Knowledge Representation and Reasoning using Petri Nets. Expert System Application **7** (1994) 281–290
[13] Yeung, D. S., Tsang, E. C. C.: A Multilevel Weighted Fuzzy Reasoning Algorithm for Expert Systems. IEEE Trans. SMC—Part A: Systems Humans **28** (1998) 149–158
[14] Garg, M. L., Ahson, S. I., Gupta, P. V.: A Fuzzy Petri Net for Knowledge Representation and Reasoning. Information Processing Letters **39** (1991) 165–171
[15] Konar, A., Mandal, A. K.: Uncertainty Management in Expert Systems using Fuzzy Petri Nets. IEEE Trans. Knowledge Data Engineering **8** (1996) 96–105
[16] Scarpelli, H., Gomide, F.: A High-level Fuzzy Net Approach for Ddiscovering Potential Inconsistencies in Fuzzy Knowledge Bases. Fuzzy Sets and Systems **64** (1994) 175–193
[17] Pedrycz, W., Gomide, F.: A Generalized Fuzzy Petri Net Model. IEEE Tran. Fuzzy Systems **2** (1994) 295–301
[18] Li, X., Lara-Rosano, F.: Adaptive Fuzzy Petri Nets for Dynamic Knowledge Representation and Inference. Expert Systems with Applications **19** (2000) 235–241
[19] Pawlak, Z.: Rough Sets: Theoretical Aspects of Reasoning About Data. Kluwer Academic Publisher , Boston (1991)
[20] Kang, S. W, Wang, Y. M., Cai. Z. F.: An Approach to Generating Rules Based on Rough and FSs Theories. Journal of Xiamen University (Natural Science) **41** (2002) 173-176
[21] Zadeh, L. A.: Fuzzy sets. Information and Contro*l* **8** (1965)338-353
[22] Bonarini, A.: An Introduction to Learning Fuzzy Classifier Systems. LCS '99, LNAI, **1813**(2000) 83-104

[23] Gu, X.P., Tso, S. K.: Applying Rough-set Concept to Neural-network-based Transient-stability Classification of Power Systems. In: Proceedings of the 5th International Conference on Advances in Power System Control, Operation and Management, Hong Kong(2000) 400-404

[24] Wu, Q. X., Bell, D.: Multi-knowledge Extraction and Application, G. Wang et al. (Eds.): RSFDGrC 2003, LNAI, **2639** (2003) 274–278

[25] Xu, H., Jia, P., Fuzzy Timed Object-Oriented Petri Net, Artificial Intelligence Applications and Innovations (Proceedings of AIAI2005), Springer(2005):148-160

[26] Gallant, S., Neural Network Learning and Expert Systems, Cambridge, Mass. : MIT Press, (1993)

Hybrid Force and Position Control of Robotic Manipulators Using Passivity Backstepping Neural Networks

Shu-Huan Wen and Bing-yi Mao

Yanshan University, Qinhuangdao 066004, China
wenshuhuan@sohu.com

Abstract. This paper presents a method of force/position control by using the backstepping and passivity strict-feedback neural networks technique; passivity monitor can evaluate stability of a system based on the concept of passivity. The parameters estimation for the design is made by the neural networks technology, using the decouple method and matrix transforming technology, decomposing the robot system as the position subsystem and the force subsystem, then the control law of these subsystems are designed respectively. The results obtained are satisfactory by using hybrid force and position control, the error is negligible and the global stability of the system can also be obtained.

1 Introduction

In recent years, working environment for robot become more and more complex, and the requirements for a robot system control are made more and more imperceptible. In the same time, a robot system that never becomes unstable under a certain ambit is required. In such cases, many control strategies are given out. Reference [1] presents a position control method for robot systems, and gains the satisfactory system performance; reference [2] attains the asymptotic stability of robot systems by using fuzzy unidirectional force control; references [3], [4] guarantee asymptotic stability of robot systems based on observer technique. All the methods did not involve the problem of nonlinear force control of robotic systems by the passivity backstepping strict-feedback. A major advantage of this method is its flexibility to build the control by avoiding the cancellation of useful non-linearities.

Based on above, this paper presents a hybrid force/position control method, which considers not only the contact force between the workspace and the robot, but also the trajectory tracking of the robot. By decoupling the robot system, this paper gains the two tasks space, the force workspace and the position workspace, and then designs the control law for each space respectively. The parameters estimation for the design is made by the neural networks technique, and the concept of passivity makes the stability for each recursive step.

D. Liu et al. (Eds.): ISNN 2007, Part I, LNCS 4491, pp. 863–870, 2007.

2 Problem Description

Considering a kind of nonlinear systems

$$\dot{x} = f(x) + g(x)u, \quad y = y(x) \tag{1}$$

where $x \in R^n$ is the state, $u \in R^m$ is the control input, y is the system output. System (1) is passive, and there is a storage function W satisfying the following terms.

$$\dot{W} \le y^T u \tag{2}$$

$$y = \frac{\partial W}{\partial x} g(x) \cdot \tag{3}$$

Choosing proper input can attain the stability of system. The following section will depict a kind of robot systems, and the system dynamics equation can be written as

$$M(q)\ddot{q} + C(q,\dot{q})\dot{q} + G(q) = \tau + J^T f, \tag{4}$$

where $q, \dot{q}, \ddot{q} \in R^n$ denote the vector of joint position, the velocity, and acceleration respectively. $M(q) \in R^{n \times n}$ is the inertia matrix (shortening for M), which is symmetric and positive definite; $C(q,\dot{q}) \in R^{n \times n}$ is the centripetal-Coriolis matrix (shortening for C); $G(q) \in R^n$ is the gravity vector (shortening for G); τ and f are the joint input torque and the input vector of force controller respectively.

Considering the equation of the actuator and the relation between the joint input torque and the actuators current: $\tau = k_r I$, the dynamic has a new form.

$$\begin{aligned} &M(q)\ddot{q} + C(q,\dot{q})\dot{q} + G(q) = K_r I + J^T f \\ &L\dot{I} + RI + K_m \dot{q} = v \end{aligned} \tag{5}$$

where L, R, K_m and $K_r \in R^{n \times n}$ are positive definite diagonal matrices, which represent the actuator inductance, the actuator resistance, the constant coefficient of the actuator and the conversion coefficient between the current and the torques. $I \in R^n$ is the vector armature current, and $v \in R^n$ is the vector armature voltage.

The dynamic system in Cartesian space can be expressed as

$$\begin{aligned} &M_x(q)\ddot{x} + C_x(q,\dot{q})\dot{x} + G_x(q) = F_x + f \\ &F_x = J^{-T} K_r I, \quad M_x(q) = J^{-T}(q) M(q) J^{-1}(q) \\ &C_x(q) = J^{-T}(q) C(q,\dot{q}) J^{-1}(q) + M_x(q) J(q) \dot{J}^{-1}(q) \\ &G_x(q) = J^{-T}(q) G(q), \quad \dot{q} = J^{-1}\dot{x} \end{aligned} \tag{6}$$

Choose a transformation matrix R_0, which is made up of two sub-matrices of the same rows. Considering that the force is constraint perpendicular to the contact surface, so we choose the matrix R_0 as follows: $R_0 = \begin{bmatrix} R_1 & 0 \\ 0 & R_2 \end{bmatrix}$, the relation between the decoupling states and original states can be written as:

$$\dot{x} = R_0 \dot{x}_c, \quad \ddot{x} = R_0 \ddot{x}_c + \dot{R}_0 \dot{x}_c \tag{7}$$

Then the equation (8) can be obtained from (7) and (6):

$$M_x(q)(R_0\ddot{x}_c + \dot{R}_0\dot{x}_c) + C_x(q,\dot{q})R_0\dot{x}_c + G_x(q) = F_x + f \tag{8}$$

Multiply (8) on both sides by R_0^T, then

$$M_{cx}\ddot{x}_c + C_{cx}\dot{x}_c + G_{cx} = F_{cx} + f_c \tag{9}$$

where

$$F_{ci} = R_0^T F_i, \quad M_{cx} = R_0^T M_x R_0$$

$$C_{cx} = R_0^T C_x R_0 + R_0^T M_x \dot{R}_0, \quad G_{cx} = R_0^T G_x$$

Defining $J_c^{-T} = R_0^T J^{-T}$, the active force in the task space can be written as:

$$F_{cx} = R_0^T F_x = R_0^T J^{-T}\tau = J_c^{-T}\tau = J_c^{-T} K_r I \tag{10}$$

Thus equation (9) can be expressed as:

$$M_{cx}\ddot{x}_c + C_{cx}\dot{x}_c + G_{cx} = J_c^{-T} K_r I + f_c \tag{11}$$

According to the motion of the robot end-effecter in the constrained direction is negligible compared with the motion in the unconstrained direction. In other words, the direction of the motion is vertical to the constrained. So the velocity vector $\dot{x}_c = \begin{bmatrix} \dot{x}_p \\ 0 \end{bmatrix}$, and then system (11) can be decoupled as the following two sub-equations by using orthogonal principle, the position control movement in the unconstrained direction, the force control movement in the constrained direction. The first one is for the position control and the other is for the force control:

$$\begin{aligned} M_{cx11}\ddot{x}_p + C_{cx11}\dot{x}_p + G_{cx1} &= J_{c1}^{-T} K_r I \\ M_{cx21}\ddot{x}_p + C_{cx21}\dot{x}_p + G_{cx2} &= J_{c2}^{-T} K_r I + f_{ci} \end{aligned} \tag{12}$$

where f_{ci} is the interaction force between the tool and the surface in force-controlled direction M_{cx11}, M_{cx21}, C_{cx11}, C_{cx21}, G_{cx1}, G_{cx2}, J_{c1}, J_{c2} are the sub-matrix take into account decoupling respectively. Considering $\dot{q} = J^{-1}R_0\dot{x}_c = J_c\dot{x}_c$, then the second equation of the equation (5) can be written as:

$$J_c^T L\dot{I} + J_c^T RI + J_c^T K_m J_c \dot{x}_c = J_c^T v \tag{13}$$

Supposing $v_c = J_c^T v$, $I = J_c I_c$ and $\dot{I} = J_c \dot{I}_c + \dot{J}_c I_c$, equation (13) can be expressed as the following:

$$J_c^T L J_c \dot{I}_c + J_c^T (L\dot{J}_c + RJ_c) I_c + J_c^T K_m J_c \dot{x}_c = v_c \tag{14}$$

$$J_c\dot{x}_c = \begin{bmatrix} J_{c1}\dot{x}_p \\ 0 \end{bmatrix},\quad L_c = J_c^T L J_c,\quad R_c = J_c^T\left(L\dot{J}_c + RJ_c\right),\quad K_{mc} = J_c^T K_m J_c$$

We can obtain the finally decoupling force and position control systems respectively. The position control subsystem

$$\begin{aligned} M_{cx11}\ddot{x}_p + C_{cx11}\dot{x}_p + G_{cx1} &= K_{rc1}I_{c1} \\ L_{c1}\dot{I}_{c1} + R_{c1}I_{c1} + K_{mc1}\dot{x}_p &= v_{c1} \end{aligned} \tag{15}$$

The force control subsystem

$$\begin{aligned} M_{cx21}\ddot{x}_p + C_{cx21}\dot{x}_p + G_{cx2} &= K_{rc2}I_{c2} + f_{ci} \\ L_{c2}\dot{I}_{c2} + R_{c2}I_{c2} + K_{mc2}\dot{x}_p &= v_{c2} = v_{c21} + v_{c22} \end{aligned} \tag{16}$$

where v_{c21} is used to compensate the dynamic coupling forces and gravity, v_{c22} is used for force tracking.

3 Neural Network Backstepping Control Law

This section will give out the neural network backstepping control law by direct adaptive method. Choosing some appropriate parameters, the dynamics (15) can be written as

$$K_{rc1}\varepsilon_{c31} = \Phi_{cx1}\theta \tag{17}$$

where Φ_{cx1} is the regresses matrix of known function, θ is the system parameters. The position control subsystem (15) can be expressed as:

$$\begin{cases} \dot{\varepsilon}_{c1} = \varepsilon_{c2} \\ \dot{\varepsilon}_{c2} = -M_{cx11}^{-1}\left(C_{cx11}\varepsilon_{c2} + G_{cx1}\right) + M_{cx11}^{-1}K_{rc1}\varepsilon_{c31} \\ \dot{\varepsilon}_{c31} = -L_{c1}^{-1}\left(R_{c1}\varepsilon_{c31} + K_{mc1}\varepsilon_{c2}\right) + L_{c1}^{-1}v_{c1} \end{cases} \tag{18}$$

where $\varepsilon_{c1}, \varepsilon_{c2}, \varepsilon_{c31}$ and v_{c1} represent the position vector (x_p), speed vector$\left(\dot{x}_p\right)$, current vector$\left(I_c\right)$and state vectors and the voltages respectively. Because the systems (18) satisfy the requirements of the strict-feedback, we can use backstepping control to design the control law.

Step 1: considering the first equation of system (18)

$$\dot{\varepsilon}_{c1} = \varepsilon_{c2} \tag{19}$$

Equation (19) can also be written as:

$$\begin{aligned} \dot{\varepsilon}_{c1} &= f\left(\varepsilon_{c1}\right) + g\left(\varepsilon_{c1}\right)\varepsilon_{c2} \\ f\left(\varepsilon_{c1}\right) &= 0,\ g\left(\varepsilon_{c1}\right) = I \end{aligned} \tag{20}$$

Now we can regard ε_{c2} as the dummy input u_1, and use it to stabilize the subsystem (20), then we can give out the following control equation:

$$\dot{\varepsilon}_{c1} = u_1 = -K_0\left(\varepsilon_{c1} - \varepsilon_{c1d}\right) + \dot{\varepsilon}_{c1d} \tag{21}$$

where K_0 is a diagonal positive definite matrix. Choosing the storage function W_0 as: $W_0 = \frac{1}{2}(\varepsilon_{c1} - \varepsilon_{c1d})^T(\varepsilon_{c1} - \varepsilon_{c1d})$ and using equation (3), we can attain the output y_0:

$$y_0 = \varepsilon_{c1} - \varepsilon_{c1d} \tag{22}$$

Supposing $u_1 = \alpha_0 + \beta_0$, $\alpha_0 = -K_0(\varepsilon_{c1} - \varepsilon_{c1d})$, $\beta_0 = \dot{\varepsilon}_{c1d}$, $e = \varepsilon_{c1} - \varepsilon_{c1d}$, the following section we will prove that the dummy input u_1 stabilizes the subsystem (20).

The error equation of subsystem (20) can express as: $\dot{e} + K_0 e = 0$. Choosing a suitable matrix K_0, subsystem (20) will be asymptotic.

Step 2: now considering the following subsystem of equation (18):

$$\begin{cases} \dot{\varepsilon}_{c1} = \varepsilon_{c2} \\ \dot{\varepsilon}_{c2} = -M_{cx11}^{-1}(C_{cx11}\varepsilon_{c2} + G_{cx1}) + M_{cx11}^{-1}K_{rc1}\varepsilon_{c31} \end{cases} \tag{23}$$

In the same way, we regard ε_{c31} as the dummy control input, and using it to stabilizes the subsystem (23), now we choose the output and the storage function as

$$W_1 = \frac{1}{2}(\varepsilon_{c1} - \varepsilon_{c1d})^T(\varepsilon_{c1} - \varepsilon_{c1d}) + \frac{1}{2}y_1^T y_1$$

$$y_1 = \varepsilon_{c2} - \varepsilon_{c2d} - \alpha_0$$

Then we can gain the following results:

$$\alpha_1 = M_{cx11}^{-1}(C_{cx11}\varepsilon_{c2} + G_{cx1}) - y_0 + \dot{u}_1$$
$$\beta_1 = -K_1 y_1,\ u_2 = \left(M_{cx11}^{-1}K_{rc1}\right)^{-1}(\alpha_1 + \beta_1) \tag{24}$$

We can also prove that the control input u_2 stabilizes the subsystem (23).

Step 3: in this step we will consider the whole system (18), and based on the same reasoning, we attain the finally results:

$$y_2 = \varepsilon_{c3} - \varepsilon_{c3d} - \alpha_1,\ \alpha_2 = L_{c1}^{-1}(R_{c1}\varepsilon_{c31} + K_{mc1}\varepsilon_{c2}) - y_1 + \dot{u}_2$$
$$\beta_2 = -K_2 y_2$$
$$\dot{u}_2 = \frac{\partial u_2}{\partial \varepsilon_{c1}}\dot{\varepsilon}_{c1} + \frac{\partial u_2}{\partial \varepsilon_{c2}}\dot{\varepsilon}_{c2} + \frac{\partial u_2}{\partial \varepsilon_{c1d}}\dot{\varepsilon}_{c1d} + \frac{\partial u_2}{\partial \varepsilon_{c2d}}\dot{\varepsilon}_{c2d} \tag{25}$$
$$v_{c1} = L_{c1}(\alpha_2 + \beta_2)$$

where v_{c1} is the position control input. For the force control system (16), combined with equation (17) we can gain the following forms:

$$M_{cx21}\dot{\varepsilon}_{c2} + C_{cx21}\varepsilon_{c2} + G_{cx2} = K_{rc2}\varepsilon_{c32} + f_{ci}$$
$$L_{c2}\dot{\varepsilon}_{c32} + R_{c2}\varepsilon_{c32} + K_{mc2}\varepsilon_{c2} = v_{c21} + v_{c22} \tag{26}$$

Considering that the input of the force controller is produced from the real actuator current $f_i = k_{rc2}\varepsilon_i$, we can attain the following result

$$\varepsilon_{c32} = K_{rc2}^{-1}(M_{cx21}\dot{\varepsilon}_{c2} + C_{cx21}\varepsilon_{c2} + G_{cx2} - K_{rc2}\varepsilon_i) \tag{27}$$

From equation (26) and (27), we gain the force control inputs:

$$v_{c21} = L_{c2}\dot{\varepsilon}_{c32} + R_{c2}K_{rc2}^{-1}(M_{cx21}\dot{\varepsilon}_{c2} + C_{cx21}\varepsilon_{c2} + G_{cx2}) + K_{mc2}\varepsilon_{c2} \tag{28}$$

$$v_{c22} = R_{c2}[\varepsilon_{c32d} - k_p(\varepsilon_i - \varepsilon_{c32d}) - k_v(\dot{\varepsilon}_i - \dot{\varepsilon}_{c32d})] \tag{29}$$

$$v_{c2} = L_{c2}\dot{\varepsilon}_{c32} + R_{c2}K_{rc2}^{-1}(M_{cx21}\dot{\varepsilon}_{c2} + C_{cx21}\varepsilon_{c2} + G_{cx2} - K_{rc2}\varepsilon_i) + K_{mc2}\varepsilon_{c2} \tag{30}$$

Supposing

$$M_{cx11} = \hat{M}_{cx11} + \Delta M_{cx11}$$
$$C_{cx11} = \hat{C}_{cx11} + \Delta C_{cx11}$$
$$G_{cx1} = \hat{G}_{cx1} + \Delta G_{cx1}$$

where $\hat{M}_{cx11}$, $\hat{C}_{cx11}$, $\hat{G}_{cx1}$ represent the estimated matrices, and ΔM_{cx11}, ΔC_{cx11}, ΔG_{cx1} represent the error matrices respectively, then

$$\dot{\varepsilon}_{c2} = -\hat{M}_{cx11}^{-1}(\Delta M_{cx11}\dot{\varepsilon}_{c2} + \Delta C_{cx11}\varepsilon_{c2} + \Delta G_{cx1}) - \hat{M}_{cx11}^{-1}(\hat{C}_{cx11}\varepsilon_{c2} + \hat{G}_{cx1} - K_{rc1}u_2) \tag{31}$$

$$\begin{cases} \dot{\varepsilon}_{c1} - \dot{\varepsilon}_{c1d} = \varepsilon_{c2} - \varepsilon_{c2d} \\ \dot{\varepsilon}_{c2} - \dot{\varepsilon}_{c2d} = -M_{cx11}^{-1}(\Delta M_{cx11}\dot{\varepsilon}_{c2} + \Delta C_{cx11}\varepsilon_{c2} + G_{cx1}) - (\varepsilon_{c1} - \varepsilon_{c1d}) - \\ \quad k_0(\varepsilon_{c2} - \varepsilon_{c2d}) - K_1(\varepsilon_{c2} - \varepsilon_{c2d} - K_0(\varepsilon_{c1} - \varepsilon_{c1d})) \end{cases} \tag{32}$$

Rewritten the position system (32) as the error form

$$\dot{e}_p = Ae_p + W\Delta\theta$$

$$e_p = \begin{bmatrix} \varepsilon_{c1} - \varepsilon_{c1d} \\ \varepsilon_{c2} - \varepsilon_{c2d} \end{bmatrix} A = \begin{bmatrix} 0 & I \\ -I - K_1K_0 & -K_0 - K_1 \end{bmatrix} W = \begin{bmatrix} 0 \\ -\hat{M}_{cx11}^{-1}\Phi_{cx1}\Delta\theta \end{bmatrix} \tag{33}$$

In the same way, we can gain the force error system and the adaptive control inputs:

$$\dot{e}_p = Ae_p + W\Delta\theta \tag{34}$$

$$\dot{e}_f = -\frac{k_p + 1}{k_v}e_f - \frac{K_{rc2}^{-1}}{k_v}\Phi_{cx2}\Delta\theta \tag{35}$$

$$\hat{v}_{c2} = L_{c2}\dot{\varepsilon}_{c32d} + R_{c2}K_{rc2}^{-1}(M_{cx21}\dot{\varepsilon}_{c2} + C_{cx21}\varepsilon_{c2} + G_{cx2}) + K_{mc2}\varepsilon_{c2} + v_{c22} \tag{36}$$

$$\hat{v}_{c21} = L_{c2}\dot{\varepsilon}_{c32d} + R_{c2}K_{rc2}^{-1}(M_{cx21}\dot{\varepsilon}_{c2} + C_{cx21}\varepsilon_{c2} + G_{cx2}) + K_{mc2}\varepsilon_{c2} \tag{37}$$

$$\hat{v}_{c22} = R_{c2}[\varepsilon_{c32d} - k_p(-\varepsilon_i - \varepsilon_{c32d}) - k_v(-\dot{\varepsilon}_i - \dot{\varepsilon}_{c32d})] \tag{38}$$

Now putting (33) and (35) in matrix form as the following

$$\begin{bmatrix} \dot{e}_p \\ \dot{e}_f \end{bmatrix} = \begin{bmatrix} A & 0 \\ 0 & -\dfrac{k_p+1}{k_v} \end{bmatrix} \begin{bmatrix} e_p \\ e_f \end{bmatrix} + \begin{bmatrix} W \\ -\dfrac{K_{rc2}\Phi_{cx2}}{k_v} \end{bmatrix} \Delta\theta \tag{39}$$

In the following, a RBF neural network is used to learn the system parameters θ adaptively. $\theta(k) = W^T \phi_i(p) + \varepsilon(k)$, $\|\varepsilon(k)\| \le \varepsilon_M$, where $\varepsilon(k)$ is the neural network functional reconstruction error; ε_M is a specified value, W is the weight matrix, $\phi(p)$ is a Gaussian function, $\phi_i(p) = \exp(-\|p - C_i\|/\sigma_i^2)$ (i=1,2...m), where C_i is the center of basis function, σ_i is the radius of basic function.

Then the functional estimation $\hat{\theta}(k)$ can be written as $\hat{\theta}(k) = \hat{W}^T \phi_i(p)$, where $\hat{W}$ is an estimation of the ideal neural network weights that are provided by some on-line weight tuning algorithms.

Then we can attain the finally adaptive control law as:

$$\hat{v}_{c1} = L_{c1}[L_{c1}^{-1}(R_{c1}\varepsilon_{c31} + K_{mc1}\hat{\varepsilon}_{c2}) - y_1 + \frac{\partial \hat{u}_2}{\partial \varepsilon_{c1}}\dot{\varepsilon}_{c1} + \frac{\partial \hat{u}_2}{\partial \hat{\varepsilon}_{c2}}\dot{\hat{\varepsilon}}_{c2} + \frac{\partial \hat{u}_2}{\partial \varepsilon_{c1d}}\dot{\varepsilon}_{c1d} + \frac{\partial \hat{u}_2}{\partial \varepsilon_{c2d}}\dot{\varepsilon}_{c2d} - K_2 y_2] \tag{40}$$

$$\hat{v}_{c21} = L_{c2}\dot{\varepsilon}_{c32} + R_{c2}K_{rc2}^{-1}(M_{cx21}\dot{\varepsilon}_{c21} + C_{cx21}\hat{\varepsilon}_{c21} + G_{cx2}) + K_{mc2}\hat{\varepsilon}_{c2} \tag{41}$$

$$\hat{v}_{c22} = R_{c2}[\varepsilon_{c32d} - k_p(-\varepsilon_i - \varepsilon_{c32d}) - k_v(-\dot{\varepsilon}_i - \dot{\varepsilon}_{c32d})] \tag{42}$$

$$\hat{v}_{c21} = L_{c2}\dot{\varepsilon}_{c32d} + R_{c2}K_{rc2}^{-1}(M_{cx21}\dot{\varepsilon}_{c2} + C_{cx21}\varepsilon_{c2} + G_{cx2}) + K_{mc2}\varepsilon_{c2} \tag{43}$$

4 Simulation Experiment

Comprehensive simulation studies have been carried out using a two-link robot manipulator to identify this approach. The dynamic equation and the parameters of the manipulator adopt those in literature [6].We choose $F_d = diag[2,2]$, $F_s = diag[1,1]$. The

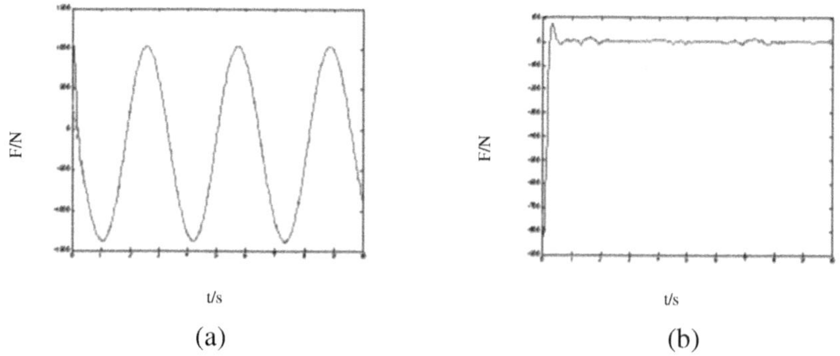

Fig. 1. The force tracking error: (a) Tracking of variable force, (b) Error curve of tracking of variable force

structure of RBF neural network is 4-3-1, that is to say, the number of input nodes is 4, which is [$x, \dot{x}, e_f, \dot{e}_f$]; the number of concealed nodes is 3, the number of output nodes corresponding to $h(s)$ is 1, $L=[3.1\ 0;0\ 3.1]$, $R=[1 0;0 1]$, $K_r=[2.2\ 0;0\ 2.2]$, $K_m=[2\ 0;0\ 2]$. Thus the simulation figure 1 are gained as following:

5 Conclusion

In this paper, a backstepping and passivity strict-feedback force/position control scheme based on neural network is proposed. It imposes desired stability properties by fixing an output given, the storage, input and stabilizing functions with each recursive step of the system. A RBF neural network is used to learn the parameters of the system. The proposed scheme is feasible for the robot force/position control.

References

1. Nganga-Kouya, D.: Backstepping Passivity Adaptive Position Control for Robotic Manipulators, Proceedings of the American Control Conference, AK, May (2002) 4607-4611
2. Xie, M.J. Nonlinear H_∞ State Feedback Control of Robot Manipulators, Robot, **23** (2) (2001) 161-165
3. Kanaoka, K., Yoshikawa, T.: Passivity Monitor and Software Limiter Which Guarantee Asymptotic Stability of Robot Control Systems, International Conference on Robotics & Automation, Taiwan, Sep (2003) 4366-4373
4. Slotine, J.J.E.: Adaptive Strategies in Constrained Manipulation, IEEE Conference on Robotics and Automation, 595-601
5. Qin, H S.: Passivity, Stability and Optimality, Control Theory and Application, **11** (4) Aug (1994) 421-427
6. Meng, Y. Application and Skill of MATLAB5.X. Science Publishing Company (1999) Istenbul: 124-133

New Global Asymptotic Stability Criterion for Uncertain Neural Networks with Time-Varying and Distributed Delays

Jiqing Qiu, Jinhui Zhang, Zhifeng Gao, and Hongjiu Yang

College of Sciences, Hebei University of Science and Technology, Shijiazhuang, Hebei 050018, P.R. China
qiujiqing@263.net

Abstract. This paper investigates the problem of global asymptotic-stability for a class of uncertain neural networks with time-varying and distributed delays. The uncertainties we considered in this paper are norm-bounded, and possibly time-varying. By Lyapunov-Krasovskii functional approach and S-procedure, a new stability criteria for the asymptotic stability of the system is derived in terms of linear matrix inequalities (LMIs). Two simulation examples are given to demonstrate the effectiveness of the developed techniques.

1 Introduction

Neural networks have been investigated extensively in recent years since its successful applications in pattern recognition, image processing, optimization problems, and so on. Moreover, time delays as a source of instability and bad performance always appear in many neural networks, such as Hopfield neural networks, cellular neural networks and bi-directional associative memory networks. On the other hand, uncertainties are unavoidable due to modeling errors, measurement errors, linearization approximations, and so on. Therefore, it is important to perform robust stability analysis for uncertain delayed neural networks, and a large number of conditions have been reported to guarantee the robust stability, see for example, [1-4] and the references therein.

To the best of our knowledge, there are two kinds of delays usually studied by many researchers, that is, discrete delays and distributed delays, and some neural networks model possesses both of them. It should be noticed that most of existing stability condition for delayed neural networks are concerned with the discrete delays case only. For the distributed delays case, there have been some initial studies on the stability analysis problem for various neural networks, and some sufficient conditions have been derived [5-7]. As pointed in [7], the stability problem for neural networks with discrete and distributed delays has not been fully investigated yet, and very few results are available for the global asymptotic stability of neural networks with discrete and distributed delays. In the work of [7], the global asymptotic stability problem for a class of neural networks with discrete and distributed time-delays are studied, but the discrete

D. Liu et al. (Eds.): ISNN 2007, Part I, LNCS 4491, pp. 871–878, 2007.

delay is a constant delay, and no uncertainties exist in the system matrices. So the problem of global asymptotic stability for uncertain neural networks with both discrete and distributed time-delays still remains open, but challenging.

In this paper, we tackle the problem of global asymptotic stability for uncertain neural networks with time-varying and distributed delays. Based on Lyapunov-Krasovskii functional approach and S-procedure, we propose a new stability criteria in terms of linear matrix inequalities which can be solved numerically using the Matlab LMI control toolbox. Two numerical examples are given to illustrate the feasibility and effectiveness of the proposed technique.

Notations: The notation in this paper is quite standard. The superscripts "⊤" stands for the transpose of a matrix; $\mathbb{R}^n$ and $\mathbb{R}^{n\times n}$ denote an n-dimensional Euclidean space and the set of all $n \times n$ real matrices, respectively; the notation $X > Y$ ($X \geq Y$) means that the matrix $X - Y$ is positive definite ($X - Y$ is semi-positive definite, respectively); I is the identity matrix of appropriate dimension; $diag\{\cdots\}$ denote the block diagonal matrix; $|\cdot|$ is the Euclidean vector norm, and the symmetric terms in a symmetric matrix are denoted by $\star$.

2 Problem Formulation

In this section, we consider the following uncertain neural networks with time-varying and distributed delays:

$$\begin{aligned}\dot{u}(t) &= -(A + \varDelta A(t))u(t) + (W_0 + \varDelta W_0(t))g(u(t)) \\ &\quad + (W_1 + \varDelta W_1(t))g(u(t - \tau(t))) + (W_2 + \varDelta W_2(t)) \int_{t-h}^{t} g(u(s))ds + I,\end{aligned} \tag{1}$$

where $u(t) = [u_1(t),\ u_2(t),\ \cdots,\ u_n(t)]^\top \in \mathbb{R}^n$ is the neural state vector, $A = diag\{a_1, a_2, \cdots, a_n\}$ is a diagonal matrix, where $a_i > 0$, $i = 1, \cdots, n$. $W_0 \in \mathbb{R}^{n\times n}$, $W_1 \in \mathbb{R}^{n\times n}$ and $W_2 \in \mathbb{R}^{n\times n}$ are the interconnection weight matrices, $\varDelta A(t)$, $\varDelta W_0(t)$, $\varDelta W_1(t)$, $\varDelta W_2(t)$ are parametric uncertainties, and $\tau(t)$ is the time-varying delay satisfies $0 \leq \tau(t) \leq \bar{\tau}$, $\dot{\tau}(t) \leq d < 1$, where $\bar{\tau}$ and d are constants, $h \geq 0$ is a distributed delay. I is a constant external input vector, and $g(u) = [g_1(u_1),\ g_2(u_2),\ \cdots,\ g_n(u_n)]^\top \in \mathbb{R}^n$ is the neuron activation function.

The time-varying uncertainties $\varDelta A(t)$, $\varDelta W_0(t)$, $\varDelta W_1(t)$, $\varDelta W_2(t)$ are defined by

$$[\varDelta A(t),\ \varDelta W_0(t),\ \varDelta W_1(t),\ \varDelta W_2(t)] = EF(t)\left[G_1,\ G_2,\ G_3,\ G_4\right], \tag{2}$$

where E, G_i, $i = 1, 2, 3, 4$, are known constants real matrices with appropriate dimensions, and $F(t)$ is the unknown time-varying matrix satisfying

$$F^\top(t)F(t) \leq I. \tag{3}$$

Remark 1. The parameter uncertainty structure as in (2)-(3) has been widely exploited in the problems of robust control and robust filtering of uncertain systems, see e.g., [8, 10]. Many practical systems possess parameter uncertainties which can be either exactly modeled or overbounded by (3).

For the neuron activation functions $g_i(u_i)$, in most of the existing results, it is assumed to be continuous, differentiable, monotonically increasing and bounded. However, in many practical systems, such as electronic circuits, the function $g_i(u_i)$ may be neither monotonically increasing nor continuously differentiable. Therefore, throughout this paper, $g_i(u_i)$ is assumed to be bounded and satisfying

$$0 \leq \frac{g_i(x) - g_i(y)}{x - y} \leq k_i, \quad \forall x, y \in \mathbb{R}, \quad x \neq y, \quad i = 1, 2, ..., n, \tag{4}$$

where k_i are constants for $i = 1, 2, \cdots, n$.

It is well known that from the above assumption and Brouwer's fixed point theorem, for a given constant I, system (1) has at least one equilibrium point $u^* = [u_1^*,\ u_2^*,\ \cdots,\ u_n^*]^\top \in \mathbb{R}^n$.

For the sake of convenience in the following discussion, we shift the equilibrium point u^* to the origin by the transformation $x = u - u^*$ and after some calculations, we can rewrite system (1) into the following:

$$\begin{aligned}\dot{x}(t) &= -(A + \Delta A(t))x(t) + (W_0 + \Delta W_0(t))f(x(t)) \\ &\quad +(W_1 + \Delta W_1(t))f(x(t - \tau(t))) + (W_2 + \Delta W_2(t)) \int_{t-h}^{t} f(x(s))ds, \end{aligned} \tag{5}$$

where $x = [x_1,\ x_2,\ \cdots,\ x_n]^\top \in \mathbb{R}^n$, and $f(x) = [f_1(x_1),\ f_2(x_2),\ \cdots,\ f_n(x_n)]^\top \in \mathbb{R}^n$. From (1), we can see that the transformed neuron activation functions are $f_i(x_i) = g_i(x_i + u_i^*) - g(u_i^*)$, $i = 1, 2, \cdots, n$, and the following conditions are satisfied

$$0 \leq \frac{f_i(x_i)}{x_i} \leq k_i, i = 1, 2, \cdots, n, \tag{6}$$

where k_i are known constants.

For the sake of presentation simplicity, we denote:

$$A(t) = A + \Delta A(t), \quad W_i(t) = W_i + \Delta W_i(t) \quad (i = 0, 1, 2).$$

Remark 2. It can be easily seen that the equilibrium u^* of system (1) is globally asymptotically stable if and only if the trivial solution of (5) is globally asymptotically stable.

Before ending this section, we recall the following lemma which will be used in next section.

Lemma 1. *[7] For any constant matrix $M \in \mathbb{R}^{n\times n}$, $M = M^\top > 0$, scalar $\gamma > 0$, vector function $\omega : [0, \gamma] \longrightarrow \mathbb{R}^n$ such that the integrations are well defined, the following inequality holds:*

$$\left[\int_0^\gamma \omega(s)ds\right]^\top M \left[\int_0^\gamma \omega(s)ds\right] \leq \gamma \int_0^\gamma \omega^\top(s) M \omega(s)ds.$$

3 Main Results

In this section, we will perform global robust stability analysis for uncertain neural networks (5), based on Lyapunov-Krasovskii stability theorem, we have the following result.

Theorem 1. *The uncertain neural networks (5) is robustly asymptotically stable, if there exist symmetric positive definite matrices* P, R, S *and real matrices* P_1, P_2 *and diagonal matrices* $D = diag\{d_1, d_2, \cdots, d_n\} \geq 0$ *and* $M = diag\{m_1, m_2, \cdots, m_n\} \geq 0$ *and scalars* $\varepsilon_i > 0$, $i = 1, 2, 3, 4$ *such that the following LMI holds:*

$$\Sigma = \begin{bmatrix} (1,1) & (1,2) & (1,3) & P_1^\top W_1 & P_1^\top W_2 & -P_1^\top E & P_1^\top E & P_1^\top E & P_1^\top E \\ \star & -P_2 - P_2^\top & P_2^\top W_0 + D & P_2^\top W_1 & P_2^\top W_2 & -P_2^\top E & P_2^\top E & P_2^\top E & P_2^\top E \\ \star & \star & (3,3) & 0 & 0 & 0 & 0 & 0 & 0 \\ \star & \star & \star & (4,4) & 0 & 0 & 0 & 0 & 0 \\ \star & \star & \star & \star & (5,5) & 0 & 0 & 0 & 0 \\ \star & \star & \star & \star & \star & -\varepsilon_1 I & 0 & 0 & 0 \\ \star & \star & \star & \star & \star & \star & -\varepsilon_2 I & 0 & 0 \\ \star & \star & \star & \star & \star & \star & \star & -\varepsilon_3 I & 0 \\ \star & \star & \star & \star & \star & \star & \star & \star & -\varepsilon_4 I \end{bmatrix} < 0 \quad (7)$$

where

$$\begin{aligned} &(1,1) = -P_1^\top A - AP_1 + \varepsilon_1 G_1^\top G_1, \quad (1,3) = P_1^\top W_0 + KM \\ &(1,2) = P - P_1^\top - AP_2, \quad (3,3) = R + hS - 2M + \varepsilon_2 G_2^\top G_2 \\ &(4,4) = -(1-d)R + \varepsilon_3 G_3^\top G_3, \quad (5,5) = -h^{-1}S + \varepsilon_4 G_4^\top G_4 \\ &K = diag\{k_1, k_2, \cdots, k_n\}. \end{aligned}$$

Proof. First of all, we define the following positive define Lyapunov-Krasovskii functional,

$$\begin{aligned} V(x(t)) = {} & x^\top(t)Px(t) + 2\sum_{j=1}^{n} d_j \int_0^{x_j} f_j(s)ds + \int_{t-\tau(t)}^{t} f^\top(x(s))Rf(x(s))ds \\ & + \int_{t-h}^{t} (s - t + h) f^\top(x(s))Sf(x(s))ds. \end{aligned}$$

Taking the derivative of $V(x(t))$ along the trajectory of system (5), then we have

$$\begin{aligned} \dot{V}(x(t)) \leq {} & 2x^\top(t)P\dot{x}(t) + 2f^\top(x(t))D\dot{x}(t) + f^\top(x(t))Rf(x(t)) \\ & -(1-d)f^\top(x(t-\tau(t)))Rf(x(t-\tau(t))) + hf^\top(x(t))Sf(x(t)) \\ & - \int_{t-h}^{t} f^\top(x(s))Sf(x(s))ds. \end{aligned}$$

From Lemma 1, we have that

$$-\int_{t-h}^{t} f^\top(x(s))Sf(x(s))ds \le -\frac{1}{h}\left[\int_{t-h}^{t} f(x(s))ds\right]^\top S\int_{t-h}^{t} f(x(s))ds. \quad (8)$$

It is obvious that

$$\begin{aligned}
0 &= 2[x^\top(t)P_1^\top + \dot{x}^\top(t)P_2^\top][-\dot{x}(t) - A(t)x(t) + W_0(t)f(x(t)) \\
&\quad + W_1(t)f(x(t-\tau(t))) + W_2(t)\int_{t-h}^{t} f(x(s))ds] \\
&= -2x^\top(t)P_1^\top\dot{x}(t) - 2x^\top(t)P_1^\top A(t)x(t) + 2x^\top(t)P_1^\top W_0(t)f(x(t)) \\
&\quad + 2x^\top(t)P_1^\top W_1(t)f(x(t-\tau(t))) + 2x^\top(t)P_1^\top W_2(t)\int_{t-h}^{t} f(x(s))ds \\
&\quad - 2\dot{x}^\top(t)P_2^\top\dot{x}(t) - 2\dot{x}^\top(t)P_2^\top A(t)x(t) + 2\dot{x}^\top(t)P_2^\top W_0(t)f(x(t)) \\
&\quad + 2\dot{x}^\top(t)P_2^\top W_1(t)f(x(t-\tau(t))) + 2\dot{x}^\top(t)P_2^\top W_2(t)\int_{t-h}^{t} f(x(s))ds. \quad (9)
\end{aligned}$$

Then add up (8) and (9) to $\dot{V}(x(t))$, we have that

$$\dot{V}(x(t)) \le \zeta^\top \Xi\zeta,$$

where

$$\Xi = \begin{bmatrix} (1,1) & (1,2) & P_1^\top W_0(t) & P_1^\top W_1(t) & P_1^\top W_2(t) \\ \star & -P_2 - P_2^\top & P_2^\top W_0(t) + D & P_2^\top W_1(t) & P_2^\top W_2(t) \\ \star & \star & R + hS & 0 & 0 \\ \star & \star & \star & -(1-d)R & 0 \\ \star & \star & \star & \star & -h^{-1}S \end{bmatrix},$$

$$(1,1) = -P_1^\top A(t) - A^\top(t)P_1, \quad (1,2) = P - P_1^\top - A^\top(t)P_2,$$

$$\zeta = \left[x^\top(t), \dot{x}^\top(t), f^\top(x(t)), f^\top(x(t-\tau(t))), \left(\int_{t-h}^{t} f(x(s))ds\right)^\top\right]^\top.$$

From (6), we can obtain the following inequalities easily,

$$f_i(x_i(t))[f_i(x_i(t)) - k_i x_i(t)] \le 0, \quad i = 1, 2, ..., n. \quad (10)$$

Now, based on S-procedure, we have that

$$\dot{V}(x(t)) - 2\sum_{i=1}^{n} m_i f_i(x_i(t))[f_i(x_i(t)) - k_i x_i(t)] \le \zeta^\top \Sigma^* \zeta, \quad (11)$$

where

$$\Sigma^* = \begin{bmatrix} (1,1) & (1,2) & P_1^\top W_0(t) + KM & P_1^\top W_1(t) & P_1^\top W_2(t) \\ \star & -P_2 - P_2^\top & P_2^\top W_0(t) + D & P_2^\top W_1(t) & P_2^\top W_2(t) \\ \star & \star & R + hS - 2M & 0 & 0 \\ \star & \star & \star & -(1-d)R & 0 \\ \star & \star & \star & \star & -h^{-1}S \end{bmatrix}, \quad (12)$$

$$(1,1) = -P_1^\top A(t) - A^\top(t)P_1, \quad (1,2) = P - P_1^\top - A^\top(t)P_2.$$

Notice that, if we multiply the left and the right hand sides of (7) by $\eta^\top$ and η, respectively, we can get $\dot{V}(x(t)) \le \zeta^\top \Sigma^* \zeta < 0$, where

$$\eta = \Big[x^\top(t), \dot{x}^\top(t), f^\top(x(t)), f^\top(x(t-\tau(t))), \left(\int_{t-h}^{t} f(x(s))ds\right)^\top, \\ x^\top(t)G_1^\top F^\top(t), f^\top(x(t))G_2^\top F^\top(t), f^\top(x(t-\tau(t)))G_3^\top F^\top(t), \\ \left(\int_{t-h}^{t} f(x(s))ds\right)^\top G_4^\top F^\top(t)\Big]^\top.$$

Therefore, based on Lyapunov stability theorem, the uncertain neural works (5) is robustly asymptotically stable. □

From the proof of Theorem 1, we can see that if there are no uncertainties in system (5), then we have the following corollary:

Corollary 1. *The neural networks (5) with $\Delta A(t) = 0$, $\Delta W_0(t) = 0$, $\Delta W_1(t) = 0$, $\Delta W_2(t) = 0$ is asymptotically stable, if there exist symmetric positive definite matrices P, R, S and real matrices P_1, P_2 and diagonal matrices $D = diag\{d_1, d_2, \cdots, d_n\} \ge 0$ and $M = diag\{m_1, m_2, \cdots, m_n\} \ge 0$ such that the following LMI holds:*

$$\begin{bmatrix} (1,1) & (1,2) & P_1^\top W_0 + KM & P_1^\top W_1 & P_1^\top W_2 \\ \star & -P_2 - P_2^\top & P_2^\top W_0 + D & P_2^\top W_1 & P_2^\top W_2 \\ \star & \star & R + hS - 2M & 0 & 0 \\ \star & \star & \star & -(1-d)R & 0 \\ \star & \star & \star & \star & -h^{-1}S \end{bmatrix} < 0, \qquad (13)$$

$$(1,1) = -P_1^\top A - AP_1, \quad (1,2) = P - P_1^\top - AP_2,$$
$$K = diag\{k_1, k_2, \cdots, k_n\}.$$

Remark 3. The neural networks with constant delay, multiple time delays or polytopic uncertainties can also be studied similarly.

4 Numerical Examples

In this section, a examples is introduced to illustrate the effectiveness of our results.

Example 1. Consider the following neural networks with time-varying and distributed delays

$$\dot{x}(t) = -Ax(t) + W_0 f(x(t)) + W_1 f(x(t-\tau(t))) + W_2 \int_{t-h}^{t} f(x(s))ds, \quad (14)$$

with

$$A = \begin{bmatrix} 3.1 & 0 & 0 \\ 0 & 3.4 & 0 \\ 0 & 0 & 2.5 \end{bmatrix}, \quad W_0 = \begin{bmatrix} 0.9 & -1.5 & 0.1 \\ -1.2 & 1 & 0.3 \\ 0.2 & 0.3 & 0.1 \end{bmatrix}, \quad W_1 = \begin{bmatrix} 0.7 & 0.6 & 0.2 \\ 0.5 & 0.7 & 0.1 \\ 0.2 & 0.1 & 0.5 \end{bmatrix},$$

$$W_2 = \begin{bmatrix} 0.3 & 0.2 & 0.2 \\ 0.1 & 0.2 & 0.1 \\ 0.1 & 0.1 & 0.5 \end{bmatrix}, \quad K = \begin{bmatrix} 0.2 & 0 & 0 \\ 0 & 0.2 & 0 \\ 0 & 0 & 0.2 \end{bmatrix}, \quad h = 1.2.$$

the neuron activation function is $f(x) = [|x+1| - |x-1|]/2$.

Using Corollary 1, we conclude that the neural networks (14) is asymptotically stable, the solutions of LMI (13) is given as follows:

$$P = \begin{bmatrix} 74.4699 & 23.8605 & -5.7717 \\ 23.8605 & 86.7348 & -8.0547 \\ -5.7717 & -8.0547 & 103.9226 \end{bmatrix}, \quad R = \begin{bmatrix} 91.9921 & 21.5162 & 1.5299 \\ 21.5162 & 91.0592 & -0.7279 \\ 1.5299 & -0.7279 & 98.8928 \end{bmatrix},$$

$$S = \begin{bmatrix} 88.8681 & 17.7964 & -1.5791 \\ 17.7964 & 86.7115 & -3.6930 \\ -1.5791 & -3.6930 & 107.6655 \end{bmatrix}, \quad P_1 = \begin{bmatrix} 14.8256 & 3.3368 & 1.2579 \\ -0.2617 & 13.8819 & 0.9944 \\ -1.5521 & -1.7726 & 20.8157 \end{bmatrix},$$

$$P_2 = \begin{bmatrix} 25.1119 & 6.3721 & -0.3314 \\ 5.0460 & 26.5116 & -0.8410 \\ -1.5272 & -2.0785 & 40.5331 \end{bmatrix}, E = 173.8826I, D = 28.4880I.$$

The dynamical behavior is shown in the following figure 1. The simulation results imply that the neural network (14) is indeed asymptotically stable.

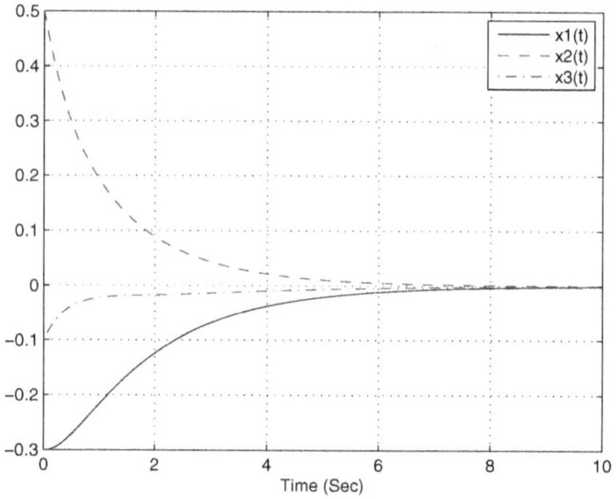

Fig. 1. The dynamical behavior of the neural network (14)

Example 2. Consider a second-order neural networks with time delays and the parameters are as follows [9]:

$$A = \begin{bmatrix} 1 & 0 \\ 0 & 1 \end{bmatrix}, \ W_0 = \begin{bmatrix} 0 & 0.5 \\ -0.5 & -0.5 \end{bmatrix}, \ W_1 = \begin{bmatrix} 0.4 & 0.5 \\ 1 & 0 \end{bmatrix}, \ W_2 = \begin{bmatrix} 0 & 0 \\ 0 & 0 \end{bmatrix}, \ K = \begin{bmatrix} 1 & 0 \\ 0 & 1 \end{bmatrix},$$

the neuron activation function is $f(x) = [|x+1| - |x-1|]/2$. Using Corollary 1, the solutions of LMI (13) is given as follows:

$$S = \begin{bmatrix} 3.0223 & 0 \\ 0 & 3.0223 \end{bmatrix}, P_1 = \begin{bmatrix} 3.3835 & -1.2177 \\ 1.0980 & 1.9486 \end{bmatrix}, P = \begin{bmatrix} 2.1516 & -0.2832 \\ -0.2832 & 2.4316 \end{bmatrix},$$
$$E = \begin{bmatrix} 2.9931 & 0 \\ 0 & 2.9931 \end{bmatrix}, P_2 = \begin{bmatrix} 1.9608 & -1.4226 \\ 1.0307 & 1.3587 \end{bmatrix}, R = \begin{bmatrix} 3.2541 & 0.0984 \\ 0.0984 & 3.0011 \end{bmatrix},$$
$$D = \begin{bmatrix} 1.6509 & 0 \\ 0 & 1.6509 \end{bmatrix},$$

which implies that the neural networks in this example is asymptotically stable, as shown in [9], several of the previous criteria fail in this example.

5 Conclusions

In this paper, we investigate the robust stability problem for uncertain neural networks with time-varying and distributed delays. Based on the Lyapunov-Krasovskii functional approach, a sufficient condition for the asymptotic stability of the uncertain neural networks is developed in terms of linear matrix inequalities (LMIs). The efficiency of our method is demonstrated by two numerical examples.

References

1. Zhang, H., Li, C., Liao, X.: A Note on the Robust Stability of Neural Networks with Time Delay. Chaos, Solitons & Fractals **25** (2005) 357-360
2. Xu, S., Lam, J.: A New Approach to Exponential Stability Analysis of Neural Networks with Time-Varying Delays. Neural Networks **19** (2006) 76-83
3. Singh, V.: Global Robust Stability of Delayed Neural Networks: an LMI Approach. IEEE Transactions on Circuits and Systems **52** (2005) 33-36
4. Zhang, H., Liao, X.: LMI-Based Robust Stability Analysis of Neural Networks with Time-Varying Delay. Neurocomputing **67** (2005) 306-312
5. Liang, J., Cao, J.: Global Asymptotic Stability of Bi-Directional Associative Memory Networks with Distributed Delays. Applied Mathematics and Computation **152** (2004) 415-424
6. Zhao, H.: Global Asymptotic Stability of Hopfield Neural Network Involving Distributed Delays. Neural Networks **17** (2004) 47-53
7. Wang, Z., Liu, Y., Liu, X.: On Global Asymptotic Stability of Neural Networks with Discrete and Distributed Delays. Physics Letters A **345** (2005) 299-308
8. Lam, J., Gao, H., Wang, C.: H_∞ Model Reduction of Linear Systems with Distributed Delay. IEE Proceedings-Control Theory And Applications **152** (2005) 662-674
9. Cao, J., Ho, Daniel W.C.: A General Framework for Global Asymptotic Stability Analysis of Delayed Neural Networks Based on LMI Approach. Chaos, Solitons & Fractals **24** (2005) 1317-1329
10. Gao, H., Lam, J., Xie, L., Wang, C.: New Approach to Mixed H_2/H_∞ Filtering for Polytopic Discrete-Time Systems. IEEE Transactions on Signal Processing **53** (2005) 3183-3192

Equilibrium Points and Stability Analysis of a Class of Neural Networks*

Xiaoping Xue

Department of Mathematics, Harbin Institute of Technology, Harbin 150001, China
xiaopingxue@263.net

Abstract. This paper discusses a mathematical model of network, which is more general than the cellular neural networks(CNNs). In this study, we discuss some dynamical properties of this type of network, such as the distribution of equilibrium points and the influence of external input on stability. Moreover, we give some criterions, which ensure the complete stability of this network.

1 Introduction

We are concerned with the following mathematical model:

$$\dot{x}(t) = -x(t) + Af(x(t)) + b, \tag{1}$$

where $A = [a_{ij}]_{n\times n}$ is a real $n \times n$ matrix, $x(t) = (x_1(t), x_2(t), \cdots, x_n(t))^T \in R^n$ is a stable variable, $f(x(t)) = (f_1(x_1(t)), f_2(x_2(t)), \cdots, f_n(x_n(t)))^T$ is a neuron activation function, $b = (b_1, b_2, \cdots, b_n)^T \in R^n$ is an external input. According to the difference of the activation functions, the system (1) can stand for not only a Hopfield network, but also a cellular neural netwrok [2]. In cellular neural networks, the activation functions $f_i(x_i(t)) = \frac{1}{2}[|x_i(t)+1| - |x_i(t)-1|]$ $(i = 1, 2, \cdots, n)$ are all piecewise-linear, meaning that the activation functions of neuron are the same. In recent years, there are many papers about the various dynamical properties analysis of cellular neural networks [1, 3, 4, 5, 7]. It is well known that this type of network appears in a wide variety of applicaitons [2, 7]. The cellular neural networks are achieved by the connected components, so it is very necessary to concern the nonlinear property of the connected components. The purpose of this paper is to study the dynamical properties of a general dynamical neural network, considering the nonlinear property of activation function within the non-saturated region.

2 Preliminaries

In this paper, $f_i(x_i) : R \to R$, the activation function of every neuron, satisfies the following assumptions:

* This work was supported by the national natural science fund of grant(10571035).

D. Liu et al. (Eds.): ISNN 2007, Part I, LNCS 4491, pp. 879–889, 2007.

(1) $f_i(x_i)$ is continuous and differentiable on $[-r_i, r_i]$, $f_i^{'}(x_i) > 0(x_i \in [-r_i, r_i])$, $f_i(0) = 0$, and $f_i(x_i)$ is symmetric with respect to zero point.
(2) $f_i(x_i) \equiv f(r_i) = \alpha_i > 0 \quad (x_i > r_i)$. $f_i(x_i) \equiv -f(r_i) = f(-r_i) = -\alpha_i < 0 \ (x_i < -r_i)$.
(3) $m_i = \min\left\{f_i^{'}(x_i)\middle|x_i \in [-r_i, r_i]\right\} > 0$, $M_i = \max\left\{f_i^{'}(x_i)\middle|x_i \in [-r_i, r_i]\right\} \leqslant 1$.
$x^* = (x_1^*, x_2^*, \cdots, x_n^*)^T$ is said to be an equilibrium point of (1) if $-x^* + Af(x^*) + b = 0$.

Let E_N denote the set consisted of all the equilibrium points of the dynamical system (1).

Definition 1. *Dynamical system (1) is said to be completely stable if, for each initial value x_0, the solution $x(t, x_0)$ of the system (1) satisfies the following condition:*

$$\lim_{t\to\infty} \text{dist}(x(t, x_0), E_N) = 0,$$

where dist$(\cdot, \cdot)$ *is the distance from $x(t, x_0)$ to E_N .*

Property 1. $E_N \neq \phi$.

Proof. It is direct to get this property from Brouwer fixed-point theorem.

Property 2. For each initial value x_0, the solution $x(t, x_0)$ of the system (1) is bounded.

Proof. It is easy to get this property from the method variation of parameter of differential equation.

3 The Criterions of Complete Stability

Let us denote $k_i = \max\left\{\left|a_{ii}f_i^{'}(x_i) - 1\right| : x_i \in [-r_i, r_i]\right\}$, and assume $k_i > 0 \ \ (i = 1, 2, \cdots, n)$.

Theorem 1. *If $|b_i| \geqslant r_i k_i + \sum\limits_{j\neq i} |a_{ij}|\alpha_i \quad (i = 1, 2, \cdots, n)$, the system (1) is completely stable.*

Proof. **Step 1.** If $x^* \in E_N$, then we will prove that $|x^*| \geqslant r_i(i = 1, 2, \cdots, n)$. In fact, if there exists i_0 such that $\left|x_{i_0}^*\right| < r_{i_0}$, then according to the mean value theorem, we can get that $f_{i_0}(x_{i_0}^*) = f_{i_0}^{'}(\xi_{i_0})x_{i_0}^*$ and since $x_{i_0}^* = a_{i_0 i_0} f(x_{i_0}^*) + \sum_{j\neq i_0} a_{i_0 j} f_j(x_j^*) + b_{i_0}$, we obtain

$$\left|x_{i_0}^*\right| = \frac{\left|\sum\limits_{j\neq i_0} a_{i_0 j} f_j(x_j^*) + b_{i_0}\right|}{\left|a_{i_0 i_0} f_{i_0}^{'}(\xi_{i_0}) - 1\right|} \geqslant \frac{|b_{i_0}| - \sum\limits_{j\neq i_0} |a_{i_0 j}|\alpha_j}{k_{i_0}} \geqslant r_{i_0},$$

which is contradictive to the assumption.

Step 2. If $x(t) = (x_1(t), x_2(t), \cdots, x_n(t))^T$ is an arbitrary solution of the system (1), then we will prove that, for every element $x_i(t)$, either $\lim_{t\to+\infty} x_i(t)$ exists, or there exists $T > 0$, such that when $t \geqslant T$, $|x_i(t)| \geqslant r_i$.

(a) If there exists $\underline{t}$ such that, when $t \geqslant \underline{t}$, $|x_i(t)| < r_i$, then $|\dot{x}_i(t)| \neq 0$. If it is not in this case, there will exist $\bar{t} > \underline{t}$ such that $\dot{x}_i(\bar{t}) = 0$. According to the similar proof to Step 1, we get $|x_i(\bar{t})| \geqslant r_i$, which is contradictive to the assumption. Because of the continuity of $\dot{x}_i(t)$ and intermediate value theorem we obtain that the sign of $\dot{x}_i(t)$ is fixed, which means that $x_i(t)$ is monotonous. From Property 2, that $x(t)$ is bounded, we obtain that $\lim\limits_{t\to+\infty} x_i(t)$ exists.

(b) If there exists $\underline{t} < T < t_1$ such that $|x_i(\underline{t})| < r_i$, $|x_i(t_1)| < r_i$, and $|x_i(T)| \geqslant r_i$. Then, from the proof of step (a), we get $|\dot{x}_i(\underline{t})| \neq 0$, $|\dot{x}_i(t_1)| \neq 0$, and the signs of $\dot{x}_i(\underline{t})$ and $\dot{x}_i(t_1)$ are opposite. We denote

$$P = (p_1, p_2, \cdots, p_n)^T \triangleq (x_1(\underline{t}), x_2(\underline{t}), \cdots, x_n(\underline{t}))^T,$$
$$Q = (q_1, q_2, \cdots, q_n)^T \triangleq (x_1(t_1), x_2(t_1), \cdots, x_n(t_1))^T,$$

and construct a function

$$F(\lambda) = -[\lambda p_i + (1-\lambda)q_i] + \sum_{j=1}^{n} a_{ij} f_j(\lambda p_j + (1-\lambda)q_j) + b_i,$$

then, $F(0) = \dot{x}_i(t_1)$, $F(1) = \dot{x}_i(\underline{t})$ and thus, the signs of $F(0)$ and $F(1)$ are oppositive. Hence, there exists $0 < \lambda_0 < 1$ such that $F(\lambda_0) = 0$. We denote

$$\overline{P} = (\lambda_0 p_1 + (1-\lambda_0)q_1, \lambda_0 p_2 + (1-\lambda_0)q_2, \cdots, \lambda_0 p_n + (1-\lambda_0)q_n)^T$$
$$\triangleq (\overline{p_1}, \overline{p_2}, \cdots, \overline{p_n}).$$

then

$$0 = F(\lambda_0) = -\overline{p_i} + \sum_{j=1}^{n} a_{ij} f_j(\overline{p_j}) + b_i. \tag{2}$$

According to (2), we obtain $\overline{|p_i|} \geqslant r_i$, which is contradictive to $|\overline{p_i}| \leqslant \lambda_0 |p_i| + (1-\lambda_0)|q_i| < r_i$.

From (a) and (b), we know that the claim of Step 2 is true.

Step 3. According to the proof of Step 2, let $x(t) = (x_1(t), x_2(t), \cdots, x_n(t))^T$ be an arbitrary solution of the system (1). Without loss of generality, we can suppose $\lim\limits_{t\to+\infty} x_i(t) = x_i^*$ $(1 \leqslant i \leqslant m)$ and $|x_i(t)| \geqslant r_i$ $(t \geqslant T, m+1 \leqslant i \leqslant n)$, then we will prove that, for each $m+1 \leqslant i \leqslant n$, $\lim\limits_{t\to+\infty} x_i(t) = x_i^*$ is still right.

We denote $\beta_i = f(x_i^*)$ $(1 \leqslant i \leqslant m)$. For any $\varepsilon > 0$, take $T_0 \geqslant T$ such that

$$|f_i(x_i(t)) - \beta_i| < \varepsilon,\ t \geqslant T_0,\ i = 1, 2, \cdots, m.$$

Let $y_i(t) = x_i(t + T_0)$, then

$$|f_i(y_i(t)) - \beta_i| < \varepsilon \text{ for } t \in [0, \infty),\ i = 1, 2, \cdots, m.$$

From the method of variation of parameter, we get

$$y_i(t) = e^{-t} y_i(0) + \sum_{j=1}^{m} a_{ij} \int_0^t e^{-(t-s)} f_j(y_j(s))ds + (1 - e^{-t}) \left(\sum_{j=m+1}^{n} a_{ij}\alpha_j + b_i \right)$$

with $m + 1 \leqslant i \leqslant n$.

Set $\hat{y}_i(t) = e^{-t} y_i(0) + (1 - e^{-t}) \left(\sum_{j=1}^{m} a_{ij}\beta_j + \sum_{j=m+1}^{n} a_{ij}\alpha_j + b_i \right)$, then

$$|y_i(t) - \hat{y}_i(t)| = \left| \sum_{j=1}^{m} a_{ij} \int_0^t e^{-(t-s)} \left[f_j(y_j(s)) - \beta_j \right] ds \right|$$
$$\leqslant \sum_{j=1}^{m} |a_{ij}| (1 - e^{-t})\varepsilon < \sum_{j=1}^{m} |a_{ij}|\varepsilon.$$

We also have that $\lim_{t\to+\infty} \hat{y}_i(t) = \sum_{j=1}^{m} a_{ij}\beta_j + \sum_{j=m+1}^{n} a_{ij}\alpha_j + b_i$. Therefore, $\lim_{t\to+\infty} y_i(t) = \lim_{t\to+\infty} \hat{y}_i(t)$.

Step 4. According to Step 3, we can get that every solution $x(t)$ of the system (1) satisfies $\lim_{t\to+\infty} x(t) = x^*$ and then $\lim_{t\to+\infty} \dot{x}(t)$ exists. Using Property 2, that $x(t)$ is bounded, we obtain $\lim_{t\to+\infty} \dot{x}(t) = 0$. which means that x^* is an equilibrium point of the system (1), thus, $\lim_{t\to+\infty} \operatorname{dist}(x(t), E_N) = 0$.

Remark 1. When the system (1) represents cellular neural network, the conditions of Theorem 1 become $k_i = |a_{ii} - 1| > 0$ and $|b_i| \geqslant k_i + \sum_{j\neq i} |a_{ij}|$. Meanwhile, the system (1) is complete stable.

The authors of [5] realized the influence from external input to cellular neural network, but they only showed this result by several examples. However, this paper is the first one that gives the qualitative result.

Lemma 1. *Let $h(t) : [0, +\infty) \to [0, +\infty)$ be bounded and satisfy Lipschitz condition. If $\int_0^{+\infty} h(t)dt < +\infty$, then $\lim_{t\to+\infty} h(t) = 0$.*

Proof. We only prove that $\overline{\lim_{t\to\infty}} h(t) = 0$.

If it is not in this case, denote $\overline{\lim_{t\to\infty}} h(t) = \alpha > 0$ and take $T_1 > 0$ such that $h(T_1) > \dfrac{\alpha}{2}$.

Because $h(t)$ satisfies Lipschitz condition, we get

$$h(T_1) - h(t) \leqslant |h(t) - h(T_1)| \leqslant L|t - T_1|,$$

and then $h(t) > \frac{\alpha}{4}$ with $t \in \left[T_1, T_1 + \frac{\alpha}{4L}\right]$. So we can construct a serious of intervals $I_i = \left[T_i, T_i + \frac{\alpha}{4L}\right]$ $(T_i \to +\infty, i = 1, 2, \cdots)$, which are mutually disjoint and satisfy $h(t) > \frac{\alpha}{4}$ with $t \in I_i$. Therefore,

$$\int_0^{+\infty} h(t)dt \geqslant \sum_{i=1}^{+\infty} \int_{I_i} h(t)dt = \sum_{i=1}^{+\infty} \frac{\alpha}{4}\frac{\alpha}{4L} = \sum_{i=1}^{+\infty} \frac{\alpha^2}{16L} = +\infty,$$

which is contradictive to the assumption.

Theorem 2. *The system (1) satisfies one of the following two conditions:*

(A) A is invertible, and we denote H as the inverse matrix of A, then $H + H^T < 0$ *(negative definite).*
(B) A is invertible, and we denote H as the inverse matrix of A, then the minimal eigenvalue of $H + H^T$ *is bigger than* 1.

then the system (1) is completely stable.

Proof. Firstly, let us consider the case that $b = 0$.
If the system (1) satisfies condition (A), then we construct a functional

$$V(x) = x^T Hx - \sum_{i=1}^{n} \int_0^{x_i} f_i(s)ds.$$

Let us calculate the derivative of $V(x)$ respect to time, along with the trajectory of the system (1). Note that

$$\begin{aligned}
\frac{dV(t)}{dt} &= x^T H\dot{x} + x^T H^T \dot{x} - f^T(x)\dot{x} \\
&= x^T H\dot{x} + (Hx - f(x))^T \dot{x} \\
&= x^T H(-x + Af(x)) + [-H(-x + Af(x))]^T \dot{x} \\
&= -x^T Hx + x^T f(x) - \dot{x}^T H^T \dot{x} \\
&= -x^T \left(\frac{H + H^T}{2}\right) x + x^T f(x) - \dot{x}^T \left(\frac{H + H^T}{2}\right) \dot{x}.
\end{aligned}$$

Since $H + H^T < 0$, *and* $x^T f(x) \geqslant 0$, we get $\frac{dV(t)}{dt} \geqslant \beta\|\dot{x}(t)\|^2$, where β is the minimal eigenvalue of $-\left(\frac{H + H^T}{2}\right)$. Because $V(t)$ is bounded and $V(t) - V(0) \geqslant \beta \int_0^t \|\dot{x}(s)\|^2 ds$. We get $\|\dot{x}(t)\| \in L^2[0, +\infty)$.

Noting that $\|\dot{x}(t)\|$ is bounded, we see that $x(t)$ satisfies Lipschitz condition and thus $\|\dot{x}\|^2$ satisfies Lipschitz condition. Therefore, from Lemma 1, we get $\lim_{t\to+\infty} \|\dot{x}\| = 0$ and then, $\lim_{t\to+\infty} [-x(t) + Af(x(t))] = 0$. According to the fact that $x(t)$ is bounded, we can also get $\lim_{t\to+\infty} \mathrm{dist}(x(t), E_N) = 0$.

If $b \neq 0$, we take an equilibrium point x^* of the system (1) arbitrarily, and let $y(t) = x(t) - x^*$, then $y(t)$ satisfies

$$\dot{y} = -y + Ag(y), \tag{3}$$

where $g(y) = (f_1(y_1 + x_1^*) - f_1(x_1^*), f_2(y_2 + x_2^*) - f_2(x_2^*), \cdots, f_n(y_n + x_n^*) - f_n(x_n^*))^T$. Let E_{N_1} denote the set consisted of the equilibrium points of the system (3). Because $g(y)^T \cdot y \geqslant 0$, according to the proof above, we obtain

$$\lim_{t\to+\infty} \mathrm{dist}(y(t), E_{N_1}) = 0.$$

Noting that $x^* + E_{N_1} = \{x^* + y^* | y^* \in E_{N_1}\} \subset E_N$, we see

$$\lim_{t\to+\infty} \mathrm{dist}(x(t), E_N) = 0.$$

If the system (1) satisfies the condition (B), then we construct a function $(b = 0)$

$$V(x) = -x^T Hx + \sum_{i=1}^{n} \int_0^{x_i} f_i(s)ds.$$

Let us calculate the derivative of V along with the trajectory of (1).

$$\begin{aligned}\frac{dV(t)}{dt} &= -x^T H\dot{x} - x^T H^T \dot{x} + f^T(x)\dot{x} \\ &= -x^T H(-x + Af(x)) + [H(-x + Af(x))]^T \dot{x} \\ &= x^T Hx - x^T f(x) + \dot{x}^T H^T \dot{x}.\end{aligned}$$

According to the assumptions of f, the activation function, we get $x^T f(x) \leqslant x^T x$. Thus

$$\frac{dV}{dt} \geq x^T \left(\frac{H + H^T}{2} - I\right) x + \dot{x}^T \left(\frac{H + H^T}{2}\right) \dot{x} \geqslant \|\dot{x}(t)\|^2.$$

The remainder proof is similar to the case of (A).

Remark 2. The original definition of complete stability [7] is to say that any trajectory of the system tends to some equilibrium point relatively. The meaning of complete stability in this paper is weaker, only demanding any trajectory of the system tends to the set of the equilibrium points. In truth, according to the proof of Theorem 1, it is the complete stability in the sense of original. There are some examples in [5], which indicate that the no-trivial periodic trajectory of the system in E_N exists. So there are some differences between the two definitions.

4 The Distribution of Equilibrium Points

In this section, we will study the distribution of equilibrium points. Let

$$D_{ts} = \left\{x = (x_1, x_2, \cdots, x_n)^T \mid |x_i| \geqslant r_i, \ \ i = 1, 2, \cdots, n\right\},$$
$$D_{nts} = \left\{x = (x_1, x_2, \cdots, x_n)^T \mid |x_i| < r_i, \ \ i = 1, 2, \cdots, n\right\},$$

and we call D_{ts} and D_{nts} saturation region and non-saturated region of the system (1) respectively. And $D_{pts} = IR^n \backslash (D_{ts} \cup D_{nts})$ is called partial saturation region.

Now let us recall some concepts about matrix. A matrix $A = (a_{ij})_{n\times n}$ is called an M-matrix if $|A| \neq 0$, $A^{-1} \geqslant 0$, and $a_{ij} \leqslant 0$ for all $i \neq j$; A matrix $A = (a_{ij})_{n\times n}$ is said to be diagonally dominant if $a_{ii} \geqslant \sum\limits_{j\neq i} |a_{ij}|$ $(i = 1, 2, \cdots, n)$. The properties of M-matrix are introduced in [6]. Let us denote $W = (w_{ij})_{n\times n}$, where

$$w_{ij} = \begin{cases} (a_{ii}m_i - 1)r_i, & i = j \\ -|a_{ij}|r_j, & i \neq j \end{cases}.$$

Theorem 3. *If W is an M-matrix, then there is one equilibrium point of the system (1) in D_{nts} at most.*

Proof. Let x^* and y^* be two equilibrium points in D_{nts}, then $x^* = Af(x^*) + b$, $y^* = Af(y^*) + b$, that is, $x^* - y^* = A\left[f(x^*) - f(y^*)\right]$. Because W is an M-matrix, there is a positive diagonal matrix $D = \mathrm{diag}(d_1, d_2, \cdots, d_n)$ such that

$$d_i(a_{ii}m_i - 1)r_i > \sum_{j\neq i} |a_{ij}|r_j d_j \;\; (i = 1, 2, \cdots, n).$$

According to mean value theorem, we get

$$f_i(x_i^*) - f_i(y_i^*) = f_i^{'}(\xi_i^*)(x_i^* - y_i^*) \triangleq \beta_i(x_i^* - y_i^*) \; (i = 1, 2, \cdots, n).$$

Let $z_i = x_i^* - y_i^*$, $Z = (z_1, z_2, \cdots, z_n)^T$, $\beta = \mathrm{diag}(\beta_1, \beta_2, \cdots, \beta_n)$, then $Z = A\beta Z$.

Suppose $\dfrac{|z_{i_0}|}{d_{i_0}r_{i_0}} = \max\left\{\dfrac{|z_i|}{d_i r_i} \Big| \, 1 \leqslant i \leqslant n\right\} > 0$, then

$$z_{i_0} = a_{i_0i_0}\beta_{i_0}z_{i_0} + \sum_{j\neq i_0} a_{i_0j}\beta_j z_j,$$

i.e.

$$(a_{i_0i_0}\beta_{i_0} - 1)|z_{i_0}| = \left|\sum_{j\neq i_0} a_{i_0j}\beta_j z_j\right|.$$

Thus

$$\begin{aligned} 0 &= (a_{i_0i_0}\beta_{i_0} - 1)|z_{i_0}| - \left|\sum_{j\neq i_0} a_{i_0j}\beta_j z_j\right| \\ &\geqslant (a_{i_0i_0}\beta_{i_0} - 1)d_{i_0}r_{i_0}\left(\frac{|z_{i_0}|}{d_{i_0}r_{i_0}}\right) - \sum_{j\neq i_0} |a_{i_0j}|d_j r_j \left(\frac{|z_{i_0}|}{d_{i_0}r_{i_0}}\right) \\ &\geqslant \left[(a_{i_0i_0}m_{i_0} - 1)d_{i_0}r_{i_0} - \sum_{j\neq i_0} |a_{i_0j}|d_j r_j\right]\left(\frac{|z_{i0}|}{d_{i_0}r_{i_0}}\right) > 0, \end{aligned}$$

which is contradictive to itself. Therefore, $x^* = y^*$.

Property 3. If there is an equilibrium point x^* in D_{nts} under the assumption of Theorem 3, then x^* is unstable.

Proof. Suppose that at some time the solution $x(t) = (x_1(t), x_2(t), \cdots, x_n(t))^T$ of the system satisfies that $x(t) \in D_{nts}$. Then we construct a function

$$V(t) = \max_{1 \leqslant i \leqslant n} \frac{|x_i(t) - x_i^*|}{d_i r_i},$$

and there is a time point t, satisfying

$$V(t) = \frac{|x_{i_0}(t) - x_{i_0}^*|}{d_{i_0} r_{i_0}}.$$

Now let us calculate the derivative of V along with the solution trajectory at this time point t

$$\begin{aligned}
\frac{dV(t)}{dt} &= \frac{\dot{x}_{i_0}}{d_{i_0} r_{i_0}} \operatorname{sgn}(x_{i_0} - x_{i_0}^*) \\
&= \frac{1}{d_{i_0} r_{i_0}} \left[-x_{i_0} + x_{i_0}^* + \sum_{j=1}^{n} a_{ij}(f(x_j) - f(x_j^*)) \right] \operatorname{sgn}(x_{i_0} - x_{i_0}^*) \\
&\geqslant \left[-|x_{i_0} - x_{i_0}^*| + a_{i_0 i_0} \beta_{i_0} |x_{i_0} - x_i^*| - \sum_{j \neq i_0} |a_{i_0 j}| |x_j - x_j^*| \right] \frac{1}{d_{i_0} r_{i_0}} \\
&= \left[(a_{i_0 i_0} \beta_{i_0} - 1) d_{i_0} r_{i_0} - \sum_{j \neq i_0} |a_{ij}| d_j r_j \left(\frac{\dfrac{|x_j - x_j^*|}{d_j r_j}}{\dfrac{|x_{i_0} - x_{i_0}^*|}{d_{i_0} r_{i_0}}} \right) \right] \frac{|x_{i_0} - x_{i_0}^*|}{d_{i_0}^2 r_{i_0}^2} \\
&\geqslant \left[(a_{i_0 i_0} m_{i_0} - 1) d_{i_0} r_{i_0} - \sum_{j \neq i_0} |a_{ij}| d_j r_j \right] \frac{|x_{i_0} - x_{i_0}^*|}{d_{i_0}^2 r_{i_0}^2} > 0.
\end{aligned}$$

Therefore, x^* is an unstable equilibrium point.

Theorem 4. *If W is an M-matrix, then there is one equilibrium point of the system (1) in D_{ts} at least.*

Proof. Because W is an M-matrix, there is a positive diagonal matrix $D = \operatorname{diag}(d_1, d_2, \cdots, d_n)$ such that $d_i r_i (a_{ii} m_i - 1) > \sum_{j \neq i} |a_{ij}| d_j r_j$. Noting that $\frac{\alpha_i}{r_i} = \frac{f(r_i)}{r_i} \leqslant 1$, we get $r_i d_i (a_{ii} m_i - 1) > \sum_{j \neq i} |a_{ij}| d_i \alpha_j$ and

$$\alpha_i a_{ii} - r_i = f(r_i) a_{ii} - r_i = f'(\xi_i) a_{ii} r_i - r_i \geqslant (a_{ii} m_i - 1) r_i,$$

thus, $(\alpha_i a_{ii} - r_i)d_i > \sum_{j\neq i} |a_{ij}| d_i \alpha_j$. Let $\widetilde{W} = (\widetilde{w_{ij}})$, where

$$\widetilde{w}_{ij} = \begin{cases} \alpha_i a_{ii} - r_i, & i = j \\ -|a_{ij}|\alpha_j, & i \neq j \end{cases},$$

then we can see that $\widetilde{W}$ is an M-matrix. From [7], we know there is a permutation $\pi:(1,2,\cdots,n)\rightarrow(\pi(1),\pi(2),\cdots,\pi(n))$ such that $\widetilde{w}_{\pi(i)\pi(i)} > \sum_{j=i+1}^{n} \left|\widetilde{w}_{\pi(i)\pi(j)}\right|$. Without loss of generality, we can suppose that $\widetilde{W}$ satisfies $\widetilde{w}_{ii} > \sum_{j=i+1}^{n} |\widetilde{w}_{ij}|$. Let $x^* = (x_1^*, x_2^*, \cdots, x_n^*)^T$ satisfy

$$x_i^* = \begin{cases} \alpha_i, & \text{if } \sum_{j=1}^{i-1} a_{ij}x_j^* + b_i \geqslant 0 \\ -\alpha_i, & \text{other cases} \end{cases} \qquad (i = 1, 2, \cdots, n).$$

If $x_i^* = \alpha_i$, then

$$\sum_{j=1}^{n} a_{ij}x_j^* + b_i = \sum_{j=1}^{i-1} a_{ij}x_j^* + a_{ii}\alpha_i + \sum_{j=i+1}^{n} a_{ij}x_j^* + b_i \geqslant a_{ii}\alpha_i - \sum_{j=i+1}^{n} |a_{ij}|\alpha_j > r_i.$$

If $x_i^* = -\alpha_i$, then

$$\begin{aligned} \sum_{j=1}^{n} a_{ij}x_j^* + b_i &= \sum_{j=1}^{i-1} a_{ij}x_j^* - a_{ii}\alpha_i + \sum_{j=i+1}^{n} a_{ij}x_j^* + b_i \\ &\leqslant -a_{ii}\alpha_i + \sum_{j=i+1}^{n} |a_{ij}|\alpha_j \\ &< -r_i. \end{aligned}$$

Then, we can see that

$$f_i\left(\sum_{j=1}^{n} a_{ij}x_j^* + b_i\right) = x_i^*.$$

Let $y_i^* = \sum_{j=1}^{n} a_{ij}x_j^* + b_i$, then

$$y_i^* = \sum_{j=1}^{n} a_{ij}y_j^* + b_i \ \ (i = 1, 2, \cdots, n).$$

Therefore, $y^* = (y_1^*, y_2^*, \cdots, y_n^*)^T$ is an equilibrium point of the system (1).

Theorem 5. *If W is diagonally dominant and $b = 0$, then there are 2^n equilibrium points of the system (1) in D_{ts}.*

Proof. In D_{ts}, we take every $I_i = \pm\alpha_i$, then the number of $(I_1, I_2, \cdots, I_n)^T$ is 2^n. For each fixed $(I_1, I_2, \cdots, I_n)$, we denote $x_i^* = \sum_{j=1}^n a_{ij} I_j$, then $x^* = (x_1^*, x_2^*, \cdots, x_n^*)^T$ is an equilibrium point of the system (1).
If $I_i > 0$, then

$$\sum_{j=1}^n a_{ij} I_j = a_{ii}\alpha_i + \sum_{j\neq i} a_{ij} I_j = a_{ii} f'(\xi_i) r_i + \sum_{j\neq i} a_{ij} I_j \geqslant a_{ii} m_i r_i - \sum_{j\neq i} |a_{ij}|\alpha_j \geqslant r_i.$$

If $I_i < 0$, then

$$\sum_{j=1}^n a_{ij} I_j = -a_{ii}\alpha_i + \sum_{j\neq i} a_{ij} I_j \leqslant -a_{ii} m_i r_i + \sum_{j\neq i} |a_{ij}|\alpha_j \leqslant -r_i.$$

Then, we get $f(x_i^*) = I_i (i = 1, 2, \cdots, n)$, which means that x^* is an equilibrium point of the system (1). Then we will prove that

$$AI \neq AI' \ \text{ with } I = (I_1, I_2, \cdots, I_n)^T \neq I' = (I_1', I_2', \cdots, I_n')^T.$$

If $I_{i_0} \neq I_{i_0}'$, then let $I_{i_0} = \alpha_{i_0}, I_{i_0}' = -\alpha_{i_0}$, and since $\sum_{j=1}^n a_{i_0 j} I_j - \sum_{j=1}^n a_{i_0 j} I_j' = 0$, we obtain

$$a_{i_0 i_0}(I_{i_0} - I_{i_0}') = \sum_{j\neq i_0} a_{i_0 j}(I_j' - I_j),$$
$$2a_{i_0 i_0}\beta_{i_0} r_{i_0} = \sum_{j\neq i_0} a_{i_0 j}(I_j' - I_j).$$

Thus, $2a_{i_0 i_0} m_{i_0} r_{i_0} \leqslant 2\sum_{j\neq i_0} |a_{i_0 j}|\alpha_j \leqslant 2\sum_{j\neq i_0} |a_{i_0 j}| r_j$, which is contradictive to diagonally dominant matrix.

5 Concluding Remarks

In this paper, we analyze a model of network, which is more general than the cellular neural networks, and get the distribution of equilibrium points by M-matrix and some other conditions. Firstly, complete stability of the networks is proved when external input satisfies some conditions. Besides, the conditions of complete stability, which have no relationship with external input, are given. We can find that the conditions given in this paper are very simple, so it is easy to be applied to engineering.

References

1. Arik, S., Tavsanoglu, V.: Equilibrium Analysis of Non-Symmetric CNNs. Int. J. Circuit Theory Appl. **24** (1996) 269-274
2. Chua, L.O., Yang, L.: Cellular Neural Networks: Theory and Application. IEEE Trans. Circuits Syst. **35** (1988) 1257-1290

3. Givalleri, P., Gilli, M.: On the Dynamic Bechavious of Two-Cell Cellular Neural Networks. Int. J. Circuit Theory Appl. **21** (1993) 251-271
4. Liao, X.: Mathematical Theory of CNNs. Science in China (Series A) **24** (1994) 1037-1046
5. Li, X., Huang, L., Wu, J.: External Inputs, Stable Equilibria and Complete Stability of CNNs. Int. J. Circuit Theory Appl. **31** (2003) 133-138
6. Siljak, D.D.: Large-Scale Dynamical Systems. Stability and Structure, North-Holland (1978)
7. Takahashi, N., Chua, L.O.: On the Complete Stability of Nonsymetric Cellular Neural Networks. IEEE Trans. Circuits Syst. **45** (1998) 754-758

Global Exponential Stability of Fuzzy Cohen-Grossberg Neural Networks with Variable Delays

Jiye Zhang[1], Keyue Zhang[2], and Dianbo Ren[1]

[1] Traction Power State Key Laboratory, Southwest Jiaotong University, Chengdu 610031, China
[2] Southwest Jiaotong University, Emei 614202, China
jyzhang@home.swjtu.edu.cn

Abstract. In this paper, we extend the Cohen–Grossberg neural networks from classical to fuzzy sets, and propose the fuzzy Cohen–Grossberg neural networks (FCGNN). The global exponential stability of FCGNN with time-varying delays is studied. Without assuming the boundedness and differentiability of the activation functions, based on the properties of M-matrix, by constructing vector Liapunov functions and applying differential inequalities, the sufficient conditions ensuring existence, uniqueness, and global exponential stability of the equilibrium point of fuzzy Cohen–Grossberg neural networks with variable delays are obtained.

1 Introduction

In the vehicle design and system analysis, there are many optimization problems. Since Cohen and Grossberg proposed a class of neural networks in 1983 [1], this model have attracted the attention of the scientific community due to their promising potential for tasks of classification, associative memory, and parallel computation and their ability to solve difficult optimization problems. In applications to parallel computation and signal processing involving solution of optimization problems, it is required that the neural network should have a unique equilibrium point that is globally asymptotically stable. Thus, the qualitative analysis of dynamic behaviors is a prerequisite step for the practical design and application of neural networks [2-15]. The stability of Cohen–Grossberg neural networks with delays has been investigated in [6-10]. Yang extended the cellular neural networks(CNNs) from classical to fuzzy sets, and proposed the fuzzy cellular neural networks (FCNNs), and applied it to the image processing [11,12]. Some conditions ensuring the global exponential stability of FCNNs with variable time delays were given in [13-15].

In the paper, we extend the Cohen–Grossberg neural networks from classical to fuzzy sets, and propose the fuzzy Cohen–Grossberg neural networks (FCGNN), which contains variable delays. By constructing proper nonlinear integro-differential inequalities involving both variable delays, applying the idea of vector Liapunov method, we obtain the sufficient conditions of global exponential stability of FCGNN.

D. Liu et al. (Eds.): ISNN 2007, Part I, LNCS 4491, pp. 890–896, 2007.

2 Notation and Preliminaries

For convenience, we introduce some notations. x^{T} and A^{T} denote the transpose of a vector x and a matrix A, where $x \in R^n$ and $A \in R^{n\times n}$. $[A]^s$ is defined as $[A]^s = [A^{\mathrm{T}} + A]/2$. $|x|$ denotes the absolute-value vector given by $|x| = (|x_1|, |x_2|, \cdots |x_n|)^{\mathrm{T}}$ and $|A|$ denotes the absolute-value matrix given by $|A| = (|a_{ij}|)_{n\times n}$. $\|x\|$ denotes the vector norm defined by $\|x\| = (x_1^2 + \cdots + x_n^2)^{1/2}$ and $\|A\|$ denotes the matrix norm defined by $\|A\| = (\max\{\lambda : \lambda$ is an eigenvalue of $A^{\mathrm{T}}A\})^{1/2}$. $\wedge$ and $\vee$ denote the fuzzy AND and fuzzy OR operation, respectively.

The dynamical behavior of FCGNNs with variable time delays can be described by the following nonlinear differential equations:

$$\dot{x}_i = \theta_i(x)[-c_i(x_i(t)) + \sum_{j=1}^{n} a_{ij} f_j(x_j(t)) + \bigwedge_{j=1}^{n} \alpha_{ij}(f_j(x_j(t-\tau_{ij}(t))))$$

$$+ \bigvee_{j=1}^{n} \beta_{ij}(f_j(x_j(t-\tau_{ij}(t))) + J_i], \quad (i = 1,2,\cdots n). \tag{1}$$

where x_i is the state of neuron i, $i = 1,2,\cdots,n$, and n is the number of neurons; J_i denotes bias of the ith neuron, respectively; $\theta_i(x)$ is an amplification function; f_i is the activation function of the ith neuron; a_{ij} are elements of feedback template; α_{ij} and β_{ij} are elements of fuzzy feedback MIN template, fuzzy feedback MAX template, respectively; $\tau_{ij}(t)$ denote the variable time delays. Assume that delays $\tau_{ij}(t)$ are bounded, continuous with $\tau_{ij}(t) \in [0,\tau]$ for all $t \ge 0$, where τ is a constant, $i, j = 1,2,\cdots,n$.

The initial conditions associated with equation (1) are of the form

$$x_i(s) = \phi_i(s), \; s \in [-\tau, 0], \; \phi_i \in C([-\tau,0], R), \; i = 1,2,\cdots,n,$$

where it is assumed that $\phi_i \in C([-\tau,0], R)$, $i = 1,2,\cdots,n$. Let $A = (a_{ij})_{n\times n}$, $\alpha = (\alpha_{ij})_{n\times n}$, $\beta = (\beta_{ij})_{n\times n}$, $J = (J_1, J_2, \cdots, J_n)^{\mathrm{T}}$, $f(x) = (f_1(x_1), f_2(x_2), \cdots, f_n(x_n))^{\mathrm{T}}$.

Assumption 1. For each $i \in \{1,2,...,n\}$, $f_i : R \to R$ is globally Lipschitz with Lipschitz constants $L_i > 0$, i.e. $|f_i(x) - f_i(y)| \le L_i |x - y|$ for all x, y. Let $L = \mathrm{diag}(L_1, L_2, \cdots, L_n) > 0$.

Assumption 2. For each $i \in \{1,2,\cdots,n\}$, $c_i : R \to R$ is strictly monotone increasing, i.e., there exists constant $d_i > 0$ such that, $[c_i(x) - c_i(y)]/(x - y) \ge d_i$ for all x, y $(x \ne y)$. Let $D = \mathrm{diag}(d_1, d_2, \cdots, d_n)$.

Assumption 3. For each $i \in \{1,2,\cdots,n\}$, $\theta_i : R^n \to R$ is a continuous function and satisfies $0 < \sigma_i \le \theta_i$, where σ_i is a constant, i=1,2,…,n.

Note. In papers [6-8], the boundedness of function θ_i was assumed. However, in this paper, the Assumption 3 is only needed. It is obvious that the function θ_i satisfied Assumption 3 maybe an unbounded one.

Definition 1. The equilibrium point x^* of (1) is said to be globally exponentially stable, if there exist constant $\lambda > 0$ and $M > 0$ such that $|x_i(t) - x_i^*|$ $\le M \| \phi - x^* \| e^{-\lambda t}$ for all $t \ge 0$, where $\| \phi - x^* \| = \max\limits_{1 \le i \le n} \{ \sup\limits_{s \in [-\tau, 0]} |\phi_i(s) - x_i^*| \}$.

Lemma 1. [3]. If $H(x) \in C^0$ is injective on R^n, and $\| H(x) \| \to \infty$ as $\| x \| \to \infty$, then $H(x)$ is a homeomorphism of R^n.

Lemma 2. [11]. Suppose x and y are two states of system (1), then

$$|\bigwedge_{j=1}^{n} \alpha_{ij} f_j(x_j) - \bigwedge_{j=1}^{n} \alpha_{ij} f_j(y_j)| \le \sum_{j=1}^{n} |\alpha_{ij}| \, | f_j(x_j) - f_j(y_j)|,$$

$$|\bigvee_{j=1}^{n} \beta_{ij} f_j(x_j) - \bigvee_{j=1}^{n} \beta_{ij} f_j(y_j)| \le \sum_{j=1}^{n} |\beta_{ij}| \, | f_j(x_j) - f_j(y_j)|.$$

3 Existence and Uniqueness of the Equilibrium Point

In the section, we study the existence and uniqueness of the equilibrium point of (1). We firstly study the nonlinear map associated with (1) as follows:

$$H_i(x_i) = -c_i(x_i) + \sum_{j=1}^{n} a_{ij} f_j(x_j) + \bigwedge_{j=1}^{n} \alpha_{ij} f_j(x_j) + \bigvee_{j=1}^{n} \beta_{ij} f_j(x_j) + J_i, \quad i = 1,2,\cdots n. \tag{2}$$

Let $H(x) = (H_1(x_1), H_2(x_2), ..., H_n(x_n))^{\mathrm{T}}$. It is known that the solutions of $H(x) = 0$ are equilibriums in (1). If map $H(x)$ is a homeomorphism on R^n, then there exists a unique point x^* such that $H(x^*) = 0$, i.e., systems (1) have a unique equilibrium x^*. Based on the Lemma 1, we get the conditions of the existence of the equilibrium for system (1) as follows.

Theorem 1. If Assumption 1-3 are satisfied, and $D - (|A| + |\alpha| + |\beta|)L$ is an M-matrix, then for each *J*, system (1) has a unique equilibrium point.

Proof. In order to prove that systems (1) have a unique equilibrium point x^*, it is only need to prove that $H(x)$ is a homeomorphism on R^n. In the following, we shall prove that map $H(x)$ is a homeomorphism in two steps.

Step 1. We prove that $H(x)$ is an injective on R^n. For purposes of contradiction, suppose that there exist $x, y \in R^n$ with $x \ne y$, such that $H(x) = H(y)$, i.e,

$$c_i(x_i)-c_i(y_i)=\sum_{j=1}^{n}a_{ij}[f_j(x_j)-f_j(y_j)]+\bigwedge_{j=1}^{n}\alpha_{ij}f_j(x_j)-\bigwedge_{j=1}^{n}\alpha_{ij}f_j(y_j)$$
$$+\bigvee_{j=1}^{n}\beta_{ij}f_j(x_j)-\bigvee_{j=1}^{n}\beta_{ij}f_j(y_j),\ i=1,2,\cdots n.$$

From Lemma 2, and Assumption 1-3, we have

$$[D-(|A|+|\alpha|+|\beta|)L]|x-y|\le 0. \tag{3}$$

Because of $D-(|A|+|\alpha|+|\beta|)L$ being an M-matrix, we know that all elements of $(D-(|A|+|\alpha|+|\beta|)L)^{-1}$ are nonnegative [14]. Therefore $|x-y|\le 0$, i.e., $x=y$. From the supposition $x\neq y$, thus this is a contradiction. So map $H(x)$ is injective.

Step 2. We prove that $\|H(x)\|\to\infty$ as $\|x\|\to\infty$.

Let $\overline{H}(x)=H(x)-H(0)$. From (2), we get

$$\overline{H}_i(x_i)=-[c_i(x_i)-c_i(0)]+\sum_{j=1}^{n}a_{ij}[f_j(x_j)-f_j(0)]+\bigwedge_{j=1}^{n}\alpha_{ij}f_j(x_j)-\bigwedge_{j=1}^{n}\alpha_{ij}f_j(0)$$
$$+\bigvee_{j=1}^{n}\beta_{ij}f_j(x_j)-\bigvee_{j=1}^{n}\beta_{ij}f_j(0),\ (i=1,2,\cdots n). \tag{4}$$

Since $D-(|A|+|\alpha|+|\beta|)L$ is an M-matrix, there exists a diagonal matrix $T=\text{diag}\{T_1,T_2,\cdots,T_n\}>0$, such that

$$[T(-D+(|A|+|\alpha|+|\beta|)L)]^s\le-\varepsilon E_n<0, \tag{5}$$

where $\varepsilon>0$ and E_n is the identity matrix [14]. From equation (4) and Lemma 3, we get

$$[Tx]^{\mathrm{T}}\overline{H}(x)=\sum_{i=1}^{n}x_iT_i\{-[c_i(x_i)-c_i(0)]+\sum_{j=1}^{n}a_{ij}[f_j(x_j)-f_j(0)]$$
$$+\bigwedge_{j=1}^{n}\alpha_{ij}f_j(x_j)-\bigwedge_{j=1}^{n}\alpha_{ij}f_j(0)+\bigvee_{j=1}^{n}\beta_{ij}f_j(x_j)-\bigvee_{j=1}^{n}\beta_{ij}f_j(0)\}$$
$$\le|x|^{\mathrm{T}}[T(-D+(|A|+|\alpha|+|\beta|)L)]^s|x|\le-\varepsilon\|x\|^2. \tag{6}$$

Using Schwarz inequality, and from (6), we get $\varepsilon\|x\|^2\le\|T\|\|x\|\|\overline{H}(x)\|$, so $\|\overline{H}(x)\|\ge\varepsilon\|x\|/\|T\|$. Therefore, $\|\overline{H}(x)\|\to+\infty$, i.e., $\|H(x)\|\to+\infty$ as $\|x\|\to+\infty$.

Based on Lemma 1, from steps 1 and 2, we know $\overline{H}(x)$ is a homeomorphism and for every input *J*, map $H(x)$ is a homeomorphism on R^n. So system (1) has a unique equilibrium point. The proof is completed.

4 Global Exponential Stability of the Equilibrium Point

Theorem 2. If Assumptions 1-3 are satisfied and $D-(|A|+|\alpha|+|\beta|)L$ is an M-matrix, then for each J, system (1) has a unique equilibrium point, which is globally exponentially stable.

Proof. Since $D-(|A|+|\alpha|+|\beta|)L$ is an M-matrix, from Theorem 1, system (1) has a unique equilibrium x^*. Let $y(t)=x(t)-x^*$, we have

$$\dot{y}_i(t)=\overline{\theta}_i(y(t))[-\overline{c}_i(y_i)+\sum_{j=1}^{n} a_{ij}\overline{f}_j(y_j(t)) + \bigwedge_{j=1}^{n} \alpha_{ij} f_j(y_j(t-\tau_{ij}(t))+x_j^*) - \bigwedge_{j=1}^{n} \alpha_{ij} f_j(x_j^*)$$
$$+\bigvee_{j=1}^{n} \beta_{ij} f_j(y_j(t-\tau_{ij}(t))+x_j^*) - \bigvee_{j=1}^{n} \beta_{ij} f_j(x_j^*)], \quad (i=1,2,\cdots,n). \tag{7}$$

Here $\overline{\theta}_i(y)=\theta_i(y+x^*)$, $\overline{c}_i(y_i)=c_i(y_i+x_i^*)-c_i(x_i^*)$, $\overline{f}_j(y_j)=f_j(y_j+x_j^*)-f_j(x_j^*)$. The initial conditions of equation (7) are $\Psi(s)=\phi(s)-x^*$, $s\in[-\tau,0]$. Systems (7) have a unique equilibrium at $y=0$.

Let

$$V_i(t)=e^{\lambda t}|y_i(t)|, \tag{8}$$

where λ is a constant to be given. Calculating the upper right derivative of $V_i(t)$ along the solutions of (7), we have

$$D^+V_i(t)=e^{\lambda t}\,\mathrm{sgn}(y_i(t))[\dot{y}_i(t)+\lambda y_i(t)] \le e^{\lambda t}\{\overline{\theta}_i(y(t))[-|\overline{c}_i(y_i(t))|+\sum_{j=1}^{n}|a_{ij}||\overline{f}_j(y_j(t))|$$
$$+\sum_{j=1}^{n}(|\alpha_{ij}|+|\beta_{ij}|)\,|\overline{f}_j(y_j(t-\tau_{ij}(t))|] +\lambda|y_i(t)|\},\ (i=1,2,\cdots,n).$$

From Assumption 3, we know that $0<\sigma_i\le\theta_i(y(t)+x^*)$, so $\theta_i(y(t)+x^*)/\sigma_i\ge 1$. Thus, from Assumption 1 and Lemma 2, we get

$$D^+V_i(t)\le\theta_i\{(-d_i+\lambda/\sigma)V_i(t)+\sum_{j=1}^{n}L_j[|a_{ij}|V_j(t)+e^{\lambda\tau}(|\alpha_{ij}|+|\beta_{ij}|)\sup_{t-\tau\le s\le t}V_j(s)]\}, \tag{9}$$

where τ is a fixed number. Due to $D-(|A|+|\alpha|+|\beta|)L$ is an M-matrix, from the properties of M-matrix [14], there exist positive constant numbers $\xi_i, i=1,2,\cdots n,$ and $\lambda>0$ satisfying

$$-\xi_i(d_i-\lambda/\sigma)+\sum_{j=1}^{n}\xi_j[|a_{ij}|+e^{\lambda\tau}(|\alpha_{ij}|+|\beta_{ij}|)]L_j<0 \quad (i=1,2,\cdots n). \tag{10}$$

Define the curve $\gamma=\{z(l): z_i=\xi_i l, l>0, i=1,2,\cdots,n\}$ and the set $\Omega(z)=\{u: 0\le u\le z, z\in\gamma\}$. Let $\xi_M=\max_{i=1,\ldots,N}\xi_i$, $\xi_m=\min_{i=1,\ldots,N}\xi_i$, taking $l_0=(1+\delta)e^{\lambda\tau}\|\Psi\|/\xi_m$, $\delta>0$ be a constant. Defining set $O=\{V: V=e^{\lambda\tau}\|\Psi_1(s)\|,|\cdots,\|\Psi_n(s)\|)^{\mathrm{T}}, -\tau\le s\le 0\}$. So, $O\subset\Omega(z_0(l_0))$, namely

$$V_i(s)\le e^{\lambda\tau}\|\Psi_i(s)\|<\xi_i l_0,\ -\tau\le s\le 0, \quad (i=1,2,\cdots n). \tag{11}$$

In the following, we shall prove

$$V_i(t)<\xi_i l_0, \quad t>0, \quad (i=1,2,\cdots n). \tag{12}$$

If (12) is not true, then from (11), there exist $t_1 > 0$ and some index i such that

$$V_i(t_1) = \xi_i l_0,\ D^+(V_i(t_1)) \geq 0,\ V_j(t) \leq \xi_j l_0,\ j = 1,2,\cdots n,\ t \in [-\tau, t_1]. \quad (13)$$

However, from (9), and (10), we get

$$D^+(V_i(t_1)) \leq \overline{\theta}_i\{-\xi_i(d_i - \lambda/\sigma) + \sum_{j=1}^{n}\xi_j[|a_{ij}| + e^{\lambda\tau}(|\alpha_{ij}| + |\beta_{ij}|)]L_j\}l_0 < 0.$$

This is a contradiction. So $V_i(t) < \xi_i l_0$, for $t > 0$ ($i = 1,2,\cdots n$). Furthermore, from (8), and (12), we obtain

$$|y_i(t)| \leq \xi_i l_0 e^{-\lambda t} \leq (1+\sigma)e^{\lambda\tau}\xi_M/\xi_m \|\Psi\| e^{-\lambda t} \leq M\|\Psi\| e^{-\lambda t},\ t \geq 0\ (i = 1,2,\cdots n).$$

where $M = (1+\sigma)e^{\lambda\tau}\xi_M/\xi_m$. So $|x_i(t) - x_i^*| \leq M\|x_i(t) - x_i^*\| e^{-\lambda t}$, and the equilibrium point of (1) is globally exponentially stable. The proof is completed.

5 Conclusions

In this paper, we analyze the existence, uniqueness, and global exponential stability of the equilibrium point of FCGNN with variable delays. Applying the idea of vector Liapunov function method, by analyzing proper nonlinear integro-differential inequalities involving both variable delays, we obtain sufficient conditions for global exponential stability. The results obtained are basic to construct a novel procedure to deal with the optimization problems in vehicle design and system analysis.

Acknowledgments

This work is supported by Natural Science Foundation of China (No. 50525518), Natural Science Foundation of China (No. 50521503), and National Program for New Century Excellent Talents in University (No. NCET-04-0889)

References

1. Cohen, M.A. and Grossberg, S.: Absolute Stability and Global Pattern Formation and Parallel Memory Storage by Competitive Neural Networks. IEEE Trans. Syst., Man, Cybern. **13** (1983) 815-826.
2. Arik, S.: An Improved Global Stability Result for Delayed Cellular Neural Networks. IEEE Trans. Circuits and Systems **I 49** (2002) 1211-1214.
3. Forti, M. and Tesi, A.: New Conditions for Global Stability of Neural Networks with Application to Linear and Quadratic Programming Problems. IEEE Trans. Circuits and Systems **I 42** (1995) 354-366.
4. Zhang, J.: Globally Exponential Stability of Neural Networks with Variable Delays. IEEE Trans. Circuits and Systtems **I 50** (2003) 288-291.
5. Xu, D., Zhao, H. and Zhu, H.: Global Dynamics of Hopfield Neural Networks Involving Variable Delays. Computers and Mathematics with Applications **42** (2001) 39-45.

6. Wang, L. and Zou, X.: Exponential Stability of Cohen-Grossberg Neural Networks. Neural Networks **15** (2002) 415-422.
7. Chen, T. and Rong, L.: Robust Global Exponential Stability of Cohen- Grossberg Neural Networks with Time-Delays. IEEE Trans. Neural Networks **15** (2004) 203-206.
8. Xiong, W. and Cao, J.: Absolutely Exponential Stability of Cohen-Grossberg Neural Networks with Unbounded Delays. Neurocomputing **68** (2005) 1-12.
9. Song, Q. and Cao, J.: Stability Analysis of Cohen–Grossberg Neural Network with Both Time-Varying and Continuously Distributed Delays, Journal of Computational and Applied Mathematics **197** (2006) 188-203.
10. Zhang, J., Suda, Y. and Komine, H.: Global Exponential Stability of Cohen-Grossberg Neural Networks with Variable Delays. Phys. Lett. A **338** (2005) 44-50.
11. Yang, T. and Yang, L.B.: Exponential Stability of Fuzzy Cellular Neural Networks with Constant and Time-Varying Delays. IEEE Trans. Circuits and Systems **I 43** (1996) 880-883.
12. Yang, T. and Yang, L.B.: Fuzzy Cellular Neural Networks: A New Paradigm for Image Processing. Int. J. Circ. Theor. Appl. **25** (1997) 469-481.
13. Liu, Y. and Tang, W.: Exponential Stability of Fuzzy Cellular Neural Networks with Constant and Time-Varying Delays. Phys. Lett. A **323** (2004) 224-233.
14. Zhang, J., Ren, D. and Zhang, W.: Global Exponential Stability of Fuzzy Cellular Neural Networks with Variable Delays. Lecture Notes in Computer Science **3971** (2006) 236-242.
15. Yuan, K., Cao, J. and Deng, J.: Exponentially Stability and Periodic Solutions of Fuzzy Cellular Neural Networks with Time-Varying Delays. Neurocomputing **69** (2006) 1619-1627.

Some New Stability Conditions of Delayed Neural Networks with Saturation Activation Functions

Wudai Liao[1], Dongyun Wang[1], Jianguo Xu[1], and Xiaoxin Liao[2]

[1] Zhongyuan University of Technology, Zhengzhou Henan 450007, China
{wdliao,wdy}@zzti.edu.cn
[2] Huazhong University of Science and Technology, Wuhan Hubei 430074, China
xiaoxin-liao@hotmail.com

Abstract. Locally and globally asymptotical stability on equilibria of delayed neural networks with saturation activation functions are studied by the Razumikhin-type theorems, which are the main approaches to study the stability of functional differential equations, and some new stability conditions are obtained, which are constructed by the networks' parameters. In the case of local stability conditions, the attracted fields of equilibria are also estimated. All results obtained in this paper need only to compute the eigenvalues of some matrices or to verify some inequalities to be holden.

1 Introduction

The stability problem of delayed neural networks with saturation activation functions has been studied by some scholars [1,2,3,4,5,6,7,8,9,10] and the results obtained in these papers mainly use the Lyapunov direct method and the Razumikhin-type theorems. We will work on this field and obtain some new stability conditions by using the saturation characteristic and the matrix analysis method.

We are going to research our work by two steps. First, by constructing an appropriate Lyapunov function and using the Razumikhin-type theorem, global stability conditions are examined; and in the next, we study the local stability conditions by rewriting the neural network's mathematical equations to be locally linear differential equations. In this case, we also estimate the attracted fields of the equilibria of the neural networks.

Consider the delayed neural networks with saturation activation functions as following

$$\dot{x}(t) = -Bx(t) + Af(x(t-\tau)) + I, \tag{1}$$

where $x = (x_1, x_2, \cdots, x_n)^{\mathrm{T}} \in \mathbb{R}^n$ is the state vector of the neural networks, $x(t-\tau) = (x_1(t-\tau_1), x_2(t-\tau_2), \cdots, x_n(t-\tau_n))^{\mathrm{T}}$, $\tau_i \geq 0$ is the time delay of the neuron i and $0 \leq \tau_i \leq \tau$, $i = 1, 2, \cdots, n$. $f(\cdot)$ is the vector of the output functions of the neurons, $f(x) = (f_1(x_1), f_2(x_2), \cdots, f_n(x_n))^{\mathrm{T}}$, $f_i(\cdot)$ has the saturation form

$$f_i(u) = \frac{1}{2}(|u+1| - |u-1|), \quad u \in \mathbb{R}. \tag{2}$$

D. Liu et al. (Eds.): ISNN 2007, Part I, LNCS 4491, pp. 897–903, 2007.

$B = \text{diag}(b_1, b_2, \cdots, b_n)$ denotes a diagonal matrix, $b_i > 0, i = 1, 2, \cdots, n$, $A = (a_{is})_{n\times n}$ is the weight matrix between neurons, I is the bias vector of the neurons in the delayed neural networks.

It is easily to see that the saturation activation function vector f satisfies the following Lipschitz condition:

$$\|f(x) - f(y)\| \le \|x - y\|, \quad x, y \in \mathbb{R}^n. \tag{3}$$

In this paper, for a vector $x \in \mathbb{R}^n$ and a matrix $A \in \mathbb{R}^{n\times n}$, we define the norms as following respectively

$$\|x\| = x^T x, \quad \|A\| = \sqrt{\lambda_{\max}(A^T A)}.$$

2 Definition and Lemma

In this section, we give the stability definition of delayed differential equations and the Razumikhin-type theorem [11] which plays a very important role in this paper.

Denote $C = C([-\tau, 0]; \mathbb{R}^n)$ to be the space of all continuous functions ϕ from $[-\tau, 0]$ to $\mathbb{R}^n$ with a norm $\|\phi\| = \sup_{-\tau\le\theta\le 0} \|\phi(\theta)\|$, $x_t \in C$ is defined by $x_t(\theta) = x(t+\theta), -\tau \le \theta \le 0$. Suppose $f : \mathbb{R}^n \times \mathbb{R}^n \times \mathbb{R}_+ \to \mathbb{R}^n$ is continuous and consider the following delayed differential equations

$$\frac{\mathrm{d}x(t)}{\mathrm{d}t} = f(x(t), x(t-\tau), t), x_{t_0} = \xi \in C, \tag{4}$$

where $x(t-\tau) = (x_1(t-\tau_1), \cdots, x_n(t-\tau_n))^T$, $\tau_i \ge 0$ is the delay of the state $x_i, i = 1, 2, \cdots, n$, and $\tau = \max_{1\le i\le n}\{\tau_i\}$. Denote $x(t) := x(t; t_0, \xi)$.

For a continuous function $V : \mathbb{R}^n \times \mathbb{R} \to \mathbb{R}$, we define a differential operator related to Equation (4) as following

$$LV(x, y, t) = V_t(x, t) + V_x(x, t) f(x, y, t), x, y \in \mathbb{R}^n.$$

Definition 1. *Suppose $f(0, 0, t) = 0$ for all $t \in \mathbb{R}$. The solution $x = 0$ of Equation (4) is said to be stable if for any $t_0 \in \mathbb{R}_+, \epsilon > 0$, there is a $\delta = \delta(\epsilon, t_0)$ such that $\|\xi\| < \delta$ implies $\|x(t)\| < \epsilon$. The solution $x = 0$ of Equation (4) is said to be asymptotically stable if it is stable and there is a $b_0 = b(t_0) > 0$ such that $\|\xi\| < b_0$ implies $\|x(t)\| \to 0$ as $t \to \infty$.*

Lemma 1 (Razumikhin). *Suppose $u, v, w : \mathbb{R}_+ \to \mathbb{R}_+$ are continuous, non-decreasing functions, $u(s), v(s), w(s)$ positive for $s > 0$, $u(0) = v(0) = 0$. If there is a continuous function $V : \mathbb{R}^n \times \mathbb{R} \to \mathbb{R}$ such that*

$$u(\|x\|) \le V(x, t) \le v(\|x\|), \quad t \in \mathbb{R}, x \in \mathbb{R}^n,$$

and

$$LV(x, y, t) \le -w(\|x\|), \text{ if } V(y, t-\tau) < qV(x, t), \ q > 1,$$

then the solution $x = 0$ of Equation (4) is asymptotically stable. If $u(s) \to \infty$ as $s \to \infty$, then the solution $x = 0$ of Equation (4) is also a global attractor.

3 Main Results

In this section, we will set up some sufficient algebraic criteria ensuring the equilibrium of System (1) to be asymptotically delay-independent stability. This section is classified to two parts: local stability and global stability of equilibria of Equation (1).

3.1 Local Stability Conditions

We will first rewrite Equation (1) to be linear form in the neighborhood of an equilibrium of System (1). And then, based on it, we can get some new local stability conditions.

According to decomposing $\mathbb{R} = (-\infty, \infty)$ into three intervals $(-\infty, -1)$, $[-1, 1]$ and $(1, \infty)$, the n-dimension Euclidean space $\mathbb{R}^n$ can be divided into 3^n sub-regions $V_k, k = 1, 2, \cdots, 3^n$.

Suppose that $x^* = (x_1^*, x_2^*, \cdots, x_n^*)^{\mathrm{T}}$ is an arbitrary equilibrium of System (1), which is an interior point in some sub-region V_{k_0}, $N(x^*) \subset V_{k_0}$ is the greatest neighborhood of the point x^*. Take the transform $z = x - x^*$ and the function

$$\phi(u) = \begin{cases} 1, & |u| < 1 \\ 0, & |u| > 1. \end{cases}$$

By using the characteristic of the output functions (see formula (2)), for any $x \in N(x^*)$, we have

$$f_i(x_i) - f_i(x_i^*) = \begin{cases} x_i - x_i^*, & |x_i^*| < 1 \\ 0, & |x_i^*| > 1 \end{cases} = \phi(x_i^*)(x_i - x_i^*), \quad i = 1, 2, \cdots, n.$$

Furthermore, they can be rewritten as the vector form

$$f(x(t-\tau)) - f(x^*) = \Phi(x^*)\big(x(t-\tau) - x^*\big),$$

where the matrix $\Phi(x^*) = \mathrm{diag}(\phi(x_1^*), \phi(x_2^*), \cdots, \phi(x_n^*))$ is a diagonal matrix, the elements of which are either 0 or 1.

Thus, in order to discuss the stability of the equilibrium x^* of System (1), we need only to study the same property of the trivial equilibrium $z = 0$ of the system

$$\dot{z}(t) = -Bz(t) + A^* z(t-\tau), \tag{5}$$

where $A^* = A\Phi(x^*) = \big(a_{is}\phi(x_s^*)\big)_{n\times n}$ related to the equilibrium x^*.

Denote $\lambda := \lambda_{\min}((B - A^*) + (B - A^*)^T)$, that is, λ is the smallest eigenvalue of the matrix $(B - A^*) + (B - A^*)^T$.

Theorem 1. *If $\lambda > 2\|A^*\| + \|A^* + A^{*^T}\|$, then the equilibrium x^* of System (1) is locally asymptotical stability, and the attracted domain is $N(x^*)$.*

Proof. For the Lyapunov function $V(z) = z^T z, z \in \mathbb{R}^n$, the differential operator related to the equation (5) is as following:

$$LV(z, y) = 2z^T(-Bz + A^*y), \ \ z, y \in \mathbb{R}^n.$$

We have the following estimation

$$\begin{aligned} LV(z, y) &= -z((B - A^*) + (B - A^*)^T)z^T + 2z^T A^* y - z^T(A^* + A^{*^T})z \\ &\leq -\lambda\|z\|^2 + 2\|A^*\| \cdot \|z\| \cdot \|y\| + \|A^* + A^{*^T}\| \cdot \|z\|^2. \end{aligned}$$

Choose a real number $q > 1$ such that $\lambda > 2q\|A^*\| + \|A^* + A^{*^T}\|$. If $V(y) < q^2V(z)$, that is, $\|y\| < q\|z\|$, then, we have

$$LV(z, y) < -(\lambda - 2q\|A^*\| - \|A^* + A^{*^T}\|)\|z\|^2 := -w(\|z\|),$$

here $w(s) = (\lambda - 2q\|A^*\| - \|A^* + A^{*^T}\|)s$ is positive for $s > 0$.

According to Lemma 1, we see that the equilibrium $z = 0$ of System (5), equivalently, x^* of System (1) is asymptotically stable. The proof is complete.

Example 1. Let $x^* = (x_1^*, x_2^*, \cdots, x_n^*)^T$ be an equilibrium of System (1) with all $|x_i^*| > 1$, then, the equilibrium is asymptotically stable.

In this case, in fact, $\Phi(x^*) = 0$ and $A^* = A\Phi(x^*) = 0$, $\lambda = \lambda_{\min}(2B) = 2\min\{b_i\} > 0 = 2\|A^*\| + \|A^* + A^{*^T}\|$, this is the required condition of Theorem 1, so, the equilibrium x^* is locally asymptotical stability.

Theorem 2. *Select an optional diagonal $S = diag(s_1, s_2, \cdots, s_n)$, construct a matrix*

$$T = \begin{pmatrix} -2B & A^* \\ A^{*^T} & -S \end{pmatrix}.$$

Denote $-\lambda = \lambda_{\max}(T)$, the biggest eigenvalue of the matrix T. If we can choose the diagonal matrix S such that

$$2\lambda > \max_{1 \leq i \leq n} \{s_i\},$$

then, the equilibrium x^* of System (1) is locally asymptotical stability, the attracted domain is $N(x^*)$.

Proof. For Lyapunov function $V(z) = \|z\|^2$, the differential operator along the solutions of System (5) has the following estimation:

$$\begin{aligned} LV(x, y) &= -2z^T Bz + 2z^T A^* y \\ &= (z^T, y^T) \begin{pmatrix} -2B & A^* \\ A^{*^T} & -S \end{pmatrix} \begin{pmatrix} z \\ y \end{pmatrix} + y^T Sy \\ &\leq -\lambda(\|z\|^2 + \|y\|^2) + y^T Sy \\ &= -\lambda\|z\|^2 + \sum_{i=1}^{n} (s_i - \lambda)y_i^2. \end{aligned}$$

Obviously, $s_i - \lambda \geq 0$ for all i, so

$$LV(x,y) \leq -\lambda\|z\|^2 + \max_{1\leq i\leq n}\{s_i - \lambda\}\|y\|^2.$$

By using the condition $2\lambda > \max_{1\leq i\leq n}\{s_i\}$ in this theorem, we have $\lambda > \max_{1\leq i\leq n}\{s_i - \lambda\}$, we can choose a real number $q > 1$, such that $\lambda > q\max_{1\leq i\leq n}$ $\{s_i - \lambda\}$. For this q, if $V(y) < qV(z)$, then

$$LV(x,y) \leq -[\lambda - q\max_{1\leq i\leq n}\{s_i - \lambda\}]\|z\|^2 := -w(\|z\|),$$

$w(s)$ is positive for $s > 0$. According to Lemma 1, we see that the equilibrium $z = 0$ of System (5), equivalently, x^* of System (1) is asymptotically stable. The proof is complete.

3.2 Global Stability Conditions

Assume that x^* is the unique equilibrium of System (1). Let $z = x - x^*$, the equation of System (1) is rewritten the following form:

$$\dot{z}(t) = -Bz(t) + A\big(f(z(t-\tau) + x^*) - f(x^*)\big). \tag{6}$$

In order to study the globally asymptotical stability of the equilibrium x^* of System (1), we need only to examine the same property on $z = 0$ of System (6).

Theorem 3. *Select a diagonal $R = diag(r_1, r_2, \cdots, r_n)$, construct a matrix*

$$H = \begin{pmatrix} -2B & A \\ A^T & -R \end{pmatrix}.$$

Denote $-\lambda = \lambda_{\max}(H)$, the biggest eigenvalue of the matrix H. If we can choose the diagonal matrix R such that

$$\lambda > \max_{1\leq i\leq n}\{r_i/2\},$$

then, the equilibrium $z = 0$ is globally asymptotical stability.

Proof. For the Lyapunov function $V(z) = \|z\|^2$, its differential operator along the solutions of System (6) has the following estimation:

$$\begin{aligned}
LV(z,y) &= 2z^T\big(-Bz + A(f(y+x^*) - f(x^*))\big) \\
&= -2z^TBz + z^TA(f(y+x^*) - f(x^*)) + (f(y+x^*) - f(x^*))^TA^Tz \\
&= \big(z^T, (f(y+x^*) - f(x^*))^T\big)\begin{pmatrix} -2B & A \\ A^T & -R \end{pmatrix}\begin{pmatrix} z \\ f(y+x^*) - f(x^*) \end{pmatrix} \\
&+ (f(y+x^*) - f(x^*))^TR(f(y+x^*) - f(x^*)) \\
&\leq -\lambda(\|z\|^2 + \|f(y+x^*) - f(x^*)\|^2) + \sum_{i=1}^n r_i|f_i(y_i + x_i^*) - f_i(x_i^*)|^2 \\
&= -\lambda\|z\|^2 + \sum_{i=1}^n (r_i - \lambda)|f_i(y_i + x_i^*) - f_i(x_i^*)|^2.
\end{aligned}$$

From the structure of the matrix H, it is easy to see that $r_i - \lambda \geq 0$, $i = 1, 2, \cdots, n$. By the assumption of (3), we have

$$\begin{aligned} LV(z, y) &\leq -\lambda\|z\|^2 + \sum_{i=1}^{n}(r_i - \lambda)|y_i|^2 \\ &\leq -\lambda\|z\|^2 + \max_{1\leq i\leq n}\{r_i - \lambda\}\|y\|^2. \end{aligned}$$

By the assumption $\lambda > \max\{r_i/2\}$, we have $\lambda > \max_{1\leq i\leq n}\{r_i - \lambda\}$, and this implies that there exits a real number $q > 1$ such that $\lambda > q\max_{1\leq i\leq n}\{r_i - \lambda\}$. For this $q > 1$, if $V(y) < qV(z)$, that is, $\|y\|^2 < q\|z\|^2$, then

$$LV(z, y) \leq -(\lambda - q\max_{1\leq i\leq n}\{r_i - \lambda\})\|z\|^2 := -w(\|z\|).$$

Here, $w(s)$ is positive for $s > 0$. According to Lemma 1, the conclusion of the theorem is true. The proof is complete.

Remark 1. From the structure of the matrix H and the theorem, it is easy to deduce that $\max_{1\leq i\leq n}\{r_i/2\} < \lambda \leq \min\{r_i, 2b_i\}$, $i = 1, 2, \cdots, n$. Hence, the optional matrix R in Theorem 3 should necessarily satisfy the conditions:

$$\max_{1\leq i\leq n}\{r_i\} < 2\min_{1\leq i\leq n}\{r_i\}, \text{ and } \max_{1\leq i\leq n}\{r_i\} < 4\min_{1\leq i\leq n}\{b_i\}.$$

Example 2. Consider the following 2-neuron neural networks:

$$\begin{aligned} \dot{x_1} &= -x_1 + \frac{1}{2}f(x_1(t - \tau_1)) + \frac{1}{3}f(x_2(t - \tau_2)) \\ \dot{x_2} &= -x_2 + \frac{1}{3}f(x_1(t - \tau_1)) + \frac{1}{2}f(x_2(t - \tau_2)). \end{aligned}$$

The activation function f is saturation linear form (see (2)).The unique equilibrium is $x_1^* = x_2^* = 0$. Now, we choose $R = \operatorname{diag}(r, r)$, r satisfies the necessary condition $0 < r < 4$. Here, we select $r = 1$ and construct the matrix:

$$H = \begin{pmatrix} -2 & 0 & 1/2 & 1/3 \\ 0 & -2 & 1/3 & 1/2 \\ 1/2 & 1/3 & -1 & 0 \\ 1/3 & 1/2 & 0 & -1 \end{pmatrix}.$$

The eigenvalues of the matrix H are $-2.4718, -2.0270, -0.9730, -0.5282$. $-\lambda = \lambda_{\max}(H) = -0.5282$, that is, $\lambda = 0.5282 > 0.5 = r/2$. By using Theorem 3, the equilibrium $(0, 0)$ is globally asymptotical stability.

Acknowledgment

This work was supported in part by the National Natural Science Foundation of China (60474001, 10572156) and the Natural Science Foundation of Henan province of China (0611054500).

References

1. Liao, W., Xu, Y., Liao, X.: Exponential Stability of Delayed Stochastic Cellular Neural Networks. Lecture Notes in Computer Scicence **3971** (2006) 224-229
2. Liao, W., Liao, X.: Stability Analysis of Cellular Neural Networks. Control Theroy and Applications **20** (2003) 89-92
3. Liao, W., Wang, Z., Liao, X.: Almost Sure Exponential Stability on Interval Stochastic Neural Networks with Time-Varying Delays. Lecture Notes in Computer Science **3971** (2006) 159-164
4. Liao, X.: Mathematical Theory of Cellular Neural Networks 1. China Science **24** (1994) 902-910
5. Cao, J., Zhou, D.: Stability Analysis of Delayed Celluar Neural Networks. Neural Networks **11** (1998) 1601-1605
6. Liao, T., Wang, F.: Global Stability for Cellular Neural Networks with Time Delay. IEEE Trans. Neural Networks **11** (2000) 1481-1484
7. Zeng, Z. G., Wang, J.: Complete Stability of Cellular Neural Networks with Time-Varying Delays. IEEE Trans. Circuits and Systems **I 53** (2006) 944-955
8. Zeng, Z. G., Wang, J.: Multiperiodicity and Exponential Attractivity Evoked by Periodic External Inputs in Delayed Cellular Neural Networks. Neural Computation **18** (2006) 848-870
9. Zeng, Z. G., Wang, J., Liao, X. X.: Stability Analysis of Delayed Cellular Neural Networks Described using Cloning Templates. IEEE Trans. Circuits and Systems **I 51** (2004) 2313-2324
10. Shen, Y., Jiang, M. H., Liao, X. X.: Global Exponential Stability of Cohen-Grossberg Neural Networks with Time-Varying Delays and Continuously Distributed Delays. Lecture Notes in Computer Science **3496** (2005) 156-161
11. Hale, J.: Theory of Functional Differential Equations. Springer-Verlag New York (1977)

Finite-Time Boundedness Analysis of Uncertain Neural Networks with Time Delay: An LMI Approach

Yanjun Shen[1], Lin Zhu[2], and Qi Guo[3]

[1,2] The Institute of Nonlinear Complex System,
China Three Gorges University, YiChang, 443002, China
shenyj@ctgu.edu.cn, rigss84@yahoo.com.cn
[3] School of Economic & Management, Three Gorges University,
YiChang, 443002, China
guoqi@ctgu.edu.cn

Abstract. This paper considers the problem of finite-time boundedness (FTB) of the general delayed neural networks with norm-bounded parametric uncertainties. The concept of FTB for time delay system is extended first. Then, based on the Lyapunov function and linear matrix inequality (LMI) technique, some delay-dependent criteria are derived to guarantee FTB. The conditions can be reduced to a feasibility problem involving linear matric inequalities (LMIs). Finally, two examples are given to demonstrate the validity of the proposed methodology.

1 Introduction

In recent years, artificial neural networks have been widely studied due to their extensive applications in pattern recognition, image processing, association memories, optimal computation and other areas. Time delays are unavoidably encountered in implementation of artificial networks. As is well known, time delays may degrade system performance and induce oscillation in a network, causing instability. So, it is very important to study time delays effects on stability and convergent dynamics of neural networks. It has received considerable attention in the past decades [1-8]. In many practical applications, some systems may be unstable, in this case, the main concern is the behavior of the system over a fixed finite time interval, it could be required that the trajectories of the controlled system do not exceed given bounds. In order to deal with this problem, Peter Dorato [9] presented the concept of finite-time stability (FTS). After that, Amato [10]-[13] extended the definition of FTS to the definition of finite-time boundedness (FTB), which takes into external constant disturbances.

In this paper, we further extend the results of FTB to the general delayed neural networks with norm-bounded parametric uncertainties. Some sufficient conditions are presented to ensured the delayed neural networks is FTB. The conditions can be reduced to a feasibility problem involving LMIs[14]. Finally, two examples are given to demonstrate the validity of the proposed methodology.

D. Liu et al. (Eds.): ISNN 2007, Part I, LNCS 4491, pp. 904–909, 2007.

The following notations will be used through out the papers: R denotes the set of real numbers, R^n denotes the $n-$dimensional Euclidean space, and $R^{n\times m}$ denotes the set of all $n \times m$ real matrices. λ_{max} and λ_{min} denotes the maximum and minimum eigenvalue of a matrix. The superscript X^T denotes the transpose of X. I denotes the identity matrix with appropriate dimension. The notation $\|x\|$ denotes a vector norm defined by $\|x\| = (\sum_{i=1}^{n} x_i^2)^{\frac{1}{2}}$.

2 Problem Statement

Consider the following time-delay neural networks with norm-bounded parametric uncertainties which is described by a nonlinear delay differential equation of the form:

$$\dot{u}(t) = -Au(t) + (W + \Delta W)f(u(t)) + (W_1 + \Delta W_1)f(u(t - d(t))) + J,$$

$$u(t) = \varphi(t), t \in [-\overline{d}, 0], \tag{1}$$

where $u(t) = [u_1(t), u_2(t), \ldots, u_n(t)]^T$ is the state vector associated with n neurons, the diagonal matrix $A = diag(a_1, a_2, \ldots, a_n)$. W and W_1 are interconnection weight matrices. $\varphi(t)$ denotes the initial condition. J is a constant input vector. ΔW, ΔW_1 are parametric uncertainties, and $f(u(t)) = [f_1(u(t)), f_2 (u(t)), \ldots, f_n(u(t))]^T$ denotes the neuron activation. $d(t)$ denotes the time-varying delay. The scalar $\overline{d} > 0$ denotes the time delay. As in many papers, we make the following assumption on the activation function:

Assumption 1. The activation function $f(u)$ is bounded and global Lipschitz with Lipschitz constant $\sigma_i \in R^+$, that is

$$|f_i(x) - f_i(y)| \leq \sigma_i |x - y|, i = 1, 2, \ldots, n.(\forall x, y \in R). \tag{2}$$

The uncertainties $\Delta W, \Delta W_1$ are defined by

$$\Delta W = HFE, \Delta W_1 = H_1 F_1 E_1, \tag{3}$$

where H, H_1, E, E_1 are known constant matrices of appropriate dimensions, and F, F_1 are unknown matrices representing the parameter uncertainties, which satisfy

$$F^T F \leq I, F_1^T F_1 \leq I. \tag{4}$$

Then by Ou Ou [4], it can be seen that there exist an equilibrium for (1). Let u^* is the equilibrium point of (1). Letting $x(t) = u(t) - u^*$, then it transforms model (1) to the following:

$$\dot{x}(t) = -Ax(t) + (W + \Delta W)g(x(t)) + (W_1 + \Delta W_1)g(x(t - d(t))),$$

$$x(t) = \phi(t), t \in [-\overline{d}, 0], \tag{5}$$

where $\phi(t)$ denotes the initial condition. $g_i(x_i) = f_i(x_i + u_i^*) - f_i(u_i^*)$, $i = 1, 2, \ldots, n$. $x(t) = [x_1(t), x_2(t), \ldots, x_n(t)]^T$, $g(x(t)) = [g_1(x_1(t)), g_2(x_2(t)), \ldots, g_n(x_n(t))]^T$.

Note that $g_i(0) = 0$, and g_i also satisfies a sector condition in the form of

$$|g_i(x_i)| \leq \sigma_i |x_i|. \tag{6}$$

The problem to be addressed in this paper is to develop some sufficient conditions which guarantee that the state of time-delay neural networks with norm-bounded parametric uncertainties is finite time boundedness (FTB).

Definition 1. *System (5) is said to be finite time boundedness (FTB) with respect to* (c_1, c_2, T), *if*

$$\sup_{t \in [-d, 0]} \|\phi^T(t)\|^2 \leq c_1^2 \Rightarrow \|x^T(t)\|^2 \leq c_2^2, \forall t \in [0, T]. \tag{7}$$

3 Main Result

We will give the main results in this section.

Theorem 1. *System (5) with time-varying delay* $d(t)$, *which is a differentiable function satisfies for all* $t \geq 0$, $0 \leq d(t) \leq \overline{d}$, $\dot{d}(t) \leq \mu < 1$, *is FTB with respect to* (c_1, c_2, T), *if there exist two symmetric positive define matrices* P, Q, *two diagonal matrices* $Y > 0$, $Q_1 > 0$, *and a positive scalar* α *such that the following conditions hold*

$$\begin{bmatrix} (1,1) & PH_1 & PH & PW & PW_1 \\ H_1^T P & -Q_1 & 0 & 0 & 0 \\ H^T P & 0 & -Q_1 & 0 & 0 \\ W^T P & 0 & 0 & Q - Y & 0 \\ W_1^T P & 0 & 0 & 0 & -(1-\mu)Q \end{bmatrix} < 0, \tag{8}$$

and

$$\frac{e^{\alpha T} c_1^2 [\lambda_{max}(P) + \overline{d}\lambda_{max}(Q)\lambda_{max}(\Sigma^T \Sigma)]}{\lambda_{min}(P)} < c_2^2, \tag{9}$$

where $(1,1) = -A^T P - PA + \Sigma^T Y \Sigma + \Sigma^T E^T Q_1 E \Sigma + \Sigma^T E_1^T Q_1 E_1 \Sigma - \alpha P$, $\Sigma = diag(\sigma_1, \sigma_2, \ldots, \sigma_n)$.

Proof. Let $V(x(t)) = x^T(t)Px(t) + \int_{t-d(t)}^{t} g^T(x(s))Qg(x(s))ds$. Then, the time derivative of $V(x(t))$ along the solution of (5) gives

$$\begin{aligned} \dot{V} &= \dot{x^T}(t)Px(t) + x^T(t)P\dot{x}(t) + g^T(x(t))Qg(x(t)) - (1-\dot{d}(t))g^T(x(t-d(t)))Q \\ &\quad \times g(x(t-d(t))) \\ &\leq -x^T(t)(A^T P + PA)x(t) + 2x^T(t)P(W + HFE)g(x(t)) + g^T(x(t))Qg(x(t)) \\ &\quad -(1-\mu)g^T(x(t-d(t)))Qg(x(t-d(t))) - g^T(x(t))Yg(x(t)) + 2x^T(t)P(W_1 \\ &\quad +H_1F_1E_1)g(x(t-d(t))) + g^T(x(t))Yg(x(t)). \end{aligned} \tag{10}$$

Noting that $Y > 0$ is a diagonal matrix and using (6), we can obtain

$$g^T(x(t))Yg(x(t)) \leq x^T(t)\Sigma^T Y \Sigma x(t). \tag{11}$$

Then, we have the following inequalities:

$$2x^T(t)PHFEg(x(t)) \leq x^T(t)PHQ_1^{-1}H^TPx(t) + x^T(t)\Sigma^T E^T Q_1 E\Sigma x(t), \tag{12}$$

$$2x^T(t)PH_1F_1E_1g(x(t)) \leq x^T(t)PH_1Q_1^{-1}H_1^TPx(t) + x^T(t)\Sigma^T E_1^T Q_1 E_1\Sigma x(t). \tag{13}$$

Taking (11)-(13) into (10), we can get

$$\begin{aligned}\dot{V} \leq & -x^T(t)(A^TP + PA)x(t) + 2x^T(t)PWg(x(t)) + 2x^T(t)PW_1g(x(t-d(t))) \\ & +g^T(x(t))Qg(x(t)) - g^T(x(t))Yg(x(t)) - (1-\mu)g^T(x(t-d(t)))Qg(x(t \\ & -d(t))) + x^T(t)\Sigma^T Y\Sigma x(t) + x^T(t)PHQ_1^{-1}H^TPx(t) + x^T(t)\Sigma^TE^TQ_1E \\ & \times\Sigma x(t) + x^T(t)PH_1Q_1^{-1}H_1^TPx(t) + x^T(t)\Sigma^TE_1^TQ_1E_1\Sigma x(t).\end{aligned} \tag{14}$$

By Schur complement, we can obtain that (14) is equivalent to the following matrix inequality:

$$\dot{V} \leq \xi^T \begin{bmatrix} (1,1)+\alpha P & PH_1 & PH & PW & PW_1 \\ H_1^TP & -Q_1 & 0 & 0 & 0 \\ H^TP & 0 & -Q_1 & 0 & 0 \\ W^TP & 0 & 0 & Q-Y & 0 \\ W_1^TP & 0 & 0 & 0 & -(1-\mu)Q \end{bmatrix} \xi, \tag{15}$$

where $\xi = [x^T(t), g^T(x(t), g^T(x(t-d(t)))]^T$.

Condition (8) implies that

$$\dot{V} < \alpha x^T(t)Px(t) \leq \alpha[x^T(t)Px(t) + \int_{t-d(t)}^t g^T(x(s))Qg(x(s))ds] = \alpha V, \tag{16}$$

multiplying (16) by $e^{-\alpha t}$, we can obtain

$$\frac{d}{dt}(e^{-\alpha t}V) < 0. \tag{17}$$

Integrating (17) from 0 to t, with $t \in [0, T]$, we have $e^{-\alpha t}V(x(t)) < V(x(0))$. Then

$$\begin{aligned}V(x(t)) < e^{\alpha t}V(x(0)) &= e^{\alpha t}[x^T(0)Px(0) + \int_{-d(t)}^0 g^T(x(s))Qg(x(s))ds] \\ &\leq e^{\alpha T}[\lambda_{max}(P)x^T(0)x(0) + \lambda_{max}(Q)\lambda_{max}(\Sigma^T\Sigma)\int_{-d(t)}^0 x^T(s)x(s)ds] \\ &\leq e^{\alpha T}c_1^2[\lambda_{max}(P) + \bar{d}\lambda_{max}(Q)\lambda_{max}(\Sigma^T\Sigma)].\end{aligned} \tag{18}$$

Noting that

$$x^T(t)Px(t) \leq V(x(t)) \Longrightarrow \lambda_{min}(P)x^T(t)x(t) \leq V(x(t)). \tag{19}$$

Putting together (18) and (19), we have

$$\|x^T(t)\|^2 < \frac{e^{\alpha T}c_1^2[\lambda_{max}(P) + \bar{d}\lambda_{max}(Q)\lambda_{max}(\Sigma^T\Sigma)]}{\lambda_{min}(P)}.$$

Condition (9) implies, for all $t \in [0, T]$, $\|x^T(t)\|^2 < c_2^2$. Therefore, the proof follows. □

Notes and Comments. If the conditions in Theorem 1 are satisfied for $\alpha = 0$, then the neural networks system is global exponential stable via the Lyapunov theory [5].

It is easy to check that condition (9) is guaranteed by imposing the conditions

$$\lambda_1 I < P < \lambda_2 I, \lambda_3 I < Q < \lambda_4 I,$$

$$-e^{-\alpha T} c_2^2 \lambda_1 + c_1^2 \lambda_2 + d c_1^2 \lambda_4 \lambda_{max}(\Sigma^T \Sigma) < 0.$$

From a computational point of view, it is important to note that, once we have fixed a value for α, the feasibility of conditions stated in Theorem 1 can be turned into LMIs based feasibility problem.

4 Illustrative Examples

Example 1. Consider a delayed neural networks in (5) with parameters as

$$A = \begin{bmatrix} 1 & 0 \\ 0 & 1.5 \end{bmatrix}, W = \begin{bmatrix} 0.5 & 0.9 \\ -0.2 & 0.5 \end{bmatrix}, W1 = \begin{bmatrix} 0.9 & 0.1 \\ -0.1 & 0.1 \end{bmatrix}, H = \begin{bmatrix} 0.2 & 0.2 \\ 0.1 & -0.3 \end{bmatrix},$$

$$H1 = \begin{bmatrix} 0.4 & 0.3 \\ 0.3 & 0.4 \end{bmatrix}, E = \begin{bmatrix} 0.2 & 0.5 \\ 0.1 & -0.3 \end{bmatrix}, E1 = \begin{bmatrix} 0.4 & 0.3 \\ 0.3 & 0.4 \end{bmatrix}, \Sigma = \begin{bmatrix} 0.4 & 0 \\ 0 & 0.9 \end{bmatrix},$$

and $\bar{d} = 1, c_1 = 1, \mu = 0.25$.

When $\alpha = 0$, we can get the neural networks system is global exponential stable, and we can also get the minimum boundedness of $c_2 = 2$. In this case, system (5) is also FTB with respect to (c_1, c_2, T) for a maximum $T_{max} = 0.9s$, obtain for $\alpha = 1.1$, $c_2 = 2$.

Example 2. Consider a delayed neural networks in (5) with parameters as

$$A = \begin{bmatrix} 0.9 & 0 \\ 0 & 1.0 \end{bmatrix}, W = \begin{bmatrix} 0.5 & 0.9 \\ -0.2 & 0.5 \end{bmatrix}, W1 = \begin{bmatrix} 0.9 & 0.1 \\ -0.1 & 0.1 \end{bmatrix}, H = \begin{bmatrix} 0.2 & 0.2 \\ 0.1 & -0.3 \end{bmatrix},$$

$$H1 = \begin{bmatrix} 0.4 & 0.3 \\ 0.3 & 0.4 \end{bmatrix}, E = \begin{bmatrix} 0.2 & 0.5 \\ 0.1 & -0.3 \end{bmatrix}, E1 = \begin{bmatrix} 0.4 & 0.3 \\ 0.3 & 0.4 \end{bmatrix}, \Sigma = \begin{bmatrix} 0.4 & 0 \\ 0 & 0.9 \end{bmatrix},$$

and $\bar{d} = 1, c_1 = 1, \mu = 0.25$.

In this example, when $\alpha = 0$, we will find condition (8) is infeasible. So we can not guarantee whether system (5) is global exponential stable or not. But if we fix $\alpha = 1.3$, $c_2 = \sqrt{6}$, then system (5) is FTB with respect to (c_1, c_2, T) for a maximum $T_{max} = 1.0s$. Or if we fix $\alpha = 1.3$, $T = 1.0s$, we can get the minimum $c_2 = 2.4$.

From this example, we can know that, although we can not guarantee whether system (5) is global exponential stable or not, if we choose appropriate α and T, we can make system (5) is bounded over a fixed finite time interval.

5 Conclusion

This paper has studied the problems of finite-time boundedness of the general delayed neural networks with norm-bounded parametric uncertainties. Based on LMI technique, some sufficient conditions are derived. Two examples have been provided to illustrates the proposed methodology.

Acknowledgments

This work was supported by the Science Foundation of Education Commission of Hubei Province D200613002, the Doctoral Pre-research Foundation of the Three Gorges University.

References

1. Liao, X., Chen, G., Sanchez, E.: LMI-Based Approach for Asymptotically Stability Analysis of Delayed Neural Networks. IEEE Trans Circ Syst I **49** (2002) 1033–1039
2. Liao, X., Chen, G., Sanchez, E.: Delay-Dependent Exponential Stability Analysis of Delayed Neural Network:An LMI Apprroach. Neural Networks **15** (2002) 855–866
3. Singh, V.: A Generalized LMI-Based Approach to the Global Asymptotic Stability of Delayed Cellular Neural Networks. IEEE Trans Neural networks **15** (2004) 223–225
4. Ou, ou.: Global Robust Exponential Stability of Delayed Neural Networks: An LMI Approach. Chaos, Soliton and Fractals. (2006)
5. Xu, S., James, L.: A New Approach to Exponential Stability Analysis of Neural Networks with Time-Varying Delays. Neural Networks **19** (2006) 76–83
6. Sun, C., Feng, C.: Exponential Periodicity of Continuous-Time and Discrete-Time Neural Networks with Delays. Neural Processing Letters **19** (2) (2004) 131-146
7. Sun, C., Feng, C.: Exponential Periodicity and Stability of Delayed Neural Networks. Mathematics and Computers in Simulation **66** (6) (2004) 469-47
8. Sun, C., Zhang, K., Fei, S., Feng, C.: On Exponential Stability of Delayed Neural Networks with a General Class of Activation Functions. Physics Letters A **298** (2/3) (2002) 122-132
9. Dorato, P.: Short Time Stability in Linear Time-Varying System. Proc. IRE International Convention Record Part 4 (1961) 83-87
10. Amato, F., Ariola, M., Dorato, P.: Finite-Time Control of Linear Systems Subject to Parameteric Uncertainties and Disturbances. Automatica **37** (2001) 1459-1463
11. Amato, F., Ariola, M., Abdallah, C.T., Dorato, P.: Dynamic Output Feedback FiniteTimeControl of LTI SystemsSubject to Parametric Uncertainties and Disturbances. Proc. European control Conference Karlsruhe, CA (1999) 1176-1180
12. Amato, F., Ariola, M., Cosentino, C.: Finite-Time Control of Linear Time-Varying Systems via Output Feedback. American Control Conference June 8-10,Portland. OR. USA (2005)
13. Amato, F., Ariola, M., Dorate, P: Finite-Time Stabilization via Dynamic Output Feedback. Automatica **42** (2006) 337-342
14. Boyd, S., El Ghaoui, L., Feron, E., Balakrishnan, V.: Linear Matrix Inequalities in System and Control Theory. Philadelhhia: SIAM (1994)

Global Asymptotic Stability of Cellular Neutral Networks With Variable Coefficients and Time-Varying Delays

Yonggui Kao[1,3], Cunchen Gao[2], and Lijing Zhang[1]

[1] College of Information Science and Engineering, Ocean University of China, China
ygkao2006@yahoo.com.cn
[2] Department of Mathematics, Ocean University of China, Qingdao 266071, China
[3] Department of Mathematics, ZaoZhuang University , ZaoZhuang 277160, China

Abstract. In this paper, we study the global asymptotic stability properties of cellular neural networks with variable coefficients and time varying delays. We present sufficient conditions for the global asymptotic stability of the neural networks . The proposed conditions, which are applicable to all continuous nonmonotonic neuron activation functions and do not require the interconnection matrices to be symmetric, establish the relationships between network parameters of the neural systems and the delay parameters. Some examples show that our results are new and improve the previous results derived in the literature.

1 Introduction

Cellular neural networks with delays (DCNNs) first introduced in [1] have found many important applications in motion-related areas such as classification of patterns, processing of moving images and recognition of moving objects, psychophysics, speech, perception, robotics, adaptive pattern recognition, and image processing. Recently, many researchers have studied the equilibria and stability properties of neural networks and presented various criteria for the uniqueness and global asymptotic stability of the equilibrium point of different classes of neural networks with or without time delays [1]-[15].

We will present new sufficient conditions for the global asymptotic stability of neural networks with variable coefficients and time varying delays. We consider a neural network model whose dynamical behavior is assumed to be governed by the following set of ordinary differential equations:

$$\dot{x}_i(t) = -c_i(t)x_i(t) + \sum_{i=1}^{n} a_{ij}(t)f_j(x_j(t)) + \sum_{i=1}^{n} b_{ij}(t)g_j(x_j(t-\tau_j(t))) + u_i(t), \tag{1}$$

where $i = 1, 2, ..., n$. n denotes the number of the neurons, $x_i(t)$ denotes the state of the neuron i at time t , $f_i(\cdot)$ and $g_i(\cdot)$ denote some bounded nonlinear output functions (also called activation functions), $a_{ij}(t)$ and $b_{ij}(t)$ denote the strengths of connectivity between neurons j and i at time t and $t-\tau_j(t)$, respectively; $\tau_j(t)$

D. Liu et al. (Eds.): ISNN 2007, Part I, LNCS 4491, pp. 910–919, 2007.

represents the time delays, u_i is the external constant input to the neuron i, $c_i(t)$ is the charging rate for the neuron i.

Accompanying the neural system (1) is an initial condition of the form

$$x(\theta) = \varphi(\theta)\ ,\ -\tau \leq \theta \leq 0\ ,\ \tau = \max_{1\leq j\leq n} \tau_j(t)$$

where $\varphi(.)$ is continuous on $[-\tau, 0]$. If $x(t) = (x_1(t), x_2(t), \ldots, x_n(t))^T$ is the vector solution of system (1), then $x(t) = x(t, \varphi)$ for $t > 0$ and $x(\theta))$ for $t \in [-\tau, 0]$.

System (1) can be written in the vector-matrix form as follows:

$$\dot{x}_i(t) = -C(t)x(t) + A(t)f(x(t)) + B(t)g(x(t - \tau(t))) + u(t), \tag{2}$$

where $f(x(t)) = (f_1(x_1(t)), f_2(x_2(t)), \ldots, f_n(x_n(t)))^T$, $A(t) = (a_{ij}(t))_{n\times n}$, $B(t) = (b_{ij}(t))_{n\times n}$, $C(t) = diag(c_1(t), x(t) = (x_1(t), x_2(t), \ldots, x_n(t))^T, c_2(t), \ldots, c_n(t))$, $g(x(t-\tau)) = (g_1(x_1(t-\tau_1(t))), g_2(x_2(t-\tau_2(t))), \ldots, g_n(x_n(t-\tau_n(t))))^T$ and $u(t) = (u_1(t), u_2(t), \ldots, u_n(t))^T$.

In most of the results derived in the literature, the activation functions have been assumed to be continuously differentiable, monotonic increasing and bounded [1]-[3]. However, as pointed out in [4], in some applications of neural networks, one may require to use the class of nondecreasing activation functions. In [5], neural networks with bounded and Lipschitzian activation functions have been considered. In a most recent paper [6], the authors have considered neural networks with globally Lipschitz activation functions without requiring them to be bounded, nondecreasing and differentiable. Such an assumption allows more activation functions to be employed in neural networks. From a mathematical point of view, the functions $f_i(\cdot)$ and $g_i(\cdot)$are said to be globally Lipschitz if there exist positive constants k_i and ℓ_i such that

$$|f_i(\xi_1) - f_i(\xi_2)| \leq k_i|\xi_1 - \xi_2| \quad |g_i(\xi_1) - g_i(\xi_2)| \leq \ell_i|\xi_1 - \xi_2| \tag{3}$$

$\forall \xi_1, \xi_2 \in R$ and $\xi_1 \neq \xi_2, i = 1, 2, \cdots, n$.

The aim of this paper is to derive new sufficient conditions for the uniqueness and global asymptotic stability of the neural network model defined by (1) with respect to the bounded activation functions satisfying the conditions given by (3), we assume activation functions are only bounded at zero point, but not differentiable and nondecreasing.

In order to derive the stability conditions for neural system (1) and make a precise comparison between our results and the previous results, We choose $\underline{c}_i, \overline{a}_{ij}, \overline{b}_{ij}$ and $\overline{u}_i$,and let $\inf_{t\in R} c_i(t) = \underline{c}_i, \sup_{t\in R} a_{ij}(t) = \overline{a}_{ij}, \sup_{t\in R} b_{ij}(t)| = \overline{b}_{ij}, \sup_{t\in R} |u_i(t)| = \overline{u}_i$. Let $K = diag(k_1, k_2, \cdots, k_n), L = diag(\ell_1, \ell_2, \cdots, \ell_n)$. $\overline{A} = (\overline{a}_{ij})_{n\times n}, \overline{B} = (\overline{b}_{ij})_{n\times n}, \underline{C} = diag(\underline{c}_1, \underline{c}_2, \ldots, \underline{c}_n), U = (\overline{u}_1, \overline{u}_2, \cdots, \overline{u}_n), y = (y_1, y_2, \cdots, y_n) \in R^n$. The three commonly used vector norms are

$$\|y\|_1 = \sum_{i=1}^{n} |y_i|\ ,\ \|y\|_2 = (\sum_{i=1}^{n} y_i^2)^{\frac{1}{2}}\ ,\ \|y\|_\infty = \max_{1\leq i\leq n} |y_i|$$

For any matrix $V = (v_{ij})_{n\times n}$, the following norms are defined

$$\|V\|_2 = \sqrt{\lambda_M(V^TV)} \qquad \|V\|_1 = \max_{1\le i\le n} |v_{ij}|$$

where $\lambda_M(V^TV)$ denotes the maximum eigenvalue of the matrix V^TV.

Now, let us defined the following mapping associated with (2):

$$H(x,t) = -C(t)x + A(t)f(x) + B(t)g(x) + u(t) \tag{4}$$

which can be written in the scaler form as follows:

$$h_i(x,t) = -c_i(t)x_i + \sum_{i=1}^{n} a_{ij}(t)f_j(x_j(t)) + \sum_{i=1}^{n} b_{ij}(t)g_j(x_j) + u_i(t), \tag{5}$$

where $x(t) = (x_1(t), x_2(t), \ldots, x_n(t))^T, A(t) = (a_{ij}(t))_{n\times n}, B(t) = (b_{ij}(t))_{n\times n}, C(t) = diag(c_1(t), c_2(t), \ldots, c_n(t)), f(x) = (f_1(x), f_2(x), \ldots, f_n(x)^T, g(x) = (g_1(x_1), g_2(x_2), \ldots, g_n(x_n))^T$ and $H(x,t) = (h_1(x,t), h_2(x,t), \ldots, h_n(x,t))^T$. When x is a globally asymptotically stable solution of system (2), it is clearly necessary that the solutions of $H(x,t) = 0$ be unique. It is known that if $H(x,t) : R^{n+1} \longrightarrow R^{n+1}$ is homeomorphism, then the solutions of $H(x,t) = 0$ are unique. In this context, we find the following results useful for establishing the existence and uniqueness of the equilibrium point for neural networks.

Definition 1. [4] *A mapping $H(x,t) : R^{n+1} \to R^{n+1}$ is homeomorphism of R^{n+1} onto itself if $H(x,t) \in C^0$, $H(x,t)$ is one-to-one and the inverse mapping $H^{-1}(x,t) \in C^0$.*

Lemma 1. [4] *If $H(x,t) \in C^0$ satisfies the following conditions:*
i) $H(x,t) \ne H(y,t)$ for all $x \ne y, t \in R$;
ii) $\|H(x,t)\| \to \infty$ as $\|x\| \to \infty$, for any $t \in R$;
then $H(x,t)$is homeomorphism of R^{n+1}.
It is now possible to state the following lemma.

Lemma 2. [10] *For the neural system (2), there exists a unique equilibrium point for every input vector u if $H(x,t)$ given by (4) is homeomorphism of R^{n+1}.*

2 Existence and Uniqueness Analysis

In this section, we obtain delay independent sufficient conditions under which the neural network model (2) has a unique equilibrium point.

Theorem 1. *Let the assumptions given by (3) hold. The neural network (2) has a unique equilibrium point for every input u if there exists positive diagonal matrices $D = diag(d_1, d_2, \ldots, d_n)$,and $P = diag(p_1, p_2, \ldots, p_n)$,such that*

$$\gamma_i = 2\underline{c}_i - d_i - p_i - k_i^2\|D^{-\frac{1}{2}}\overline{A}\|_2^2 - \iota_i^2\|P^{-\frac{1}{2}}\overline{B}\|_2^2 > 0 \;\; \forall i$$

Proof: For the map $H(x,t) = -C(t)x + A(t)f(x) + B(t)g(x) + u(t)$, we have

$$H(x,t) - H(y,t) = -C(t)(x-y) + A(t)(f(x) - f(y)) + B(t)(g(x) - g(y)) \quad (6)$$

If we multiply both sides of (6) by $2(x-y)^T$, and then add and subtract the term $(x-y)^T(D+P)(x-y)$, we get

$$\begin{aligned}&2(x-y)^T(H(x,t)-H(y,t)) = -2(x-y)^T C(t)(x-y) + 2(x-y)^T A(t)(f(x)-f(y))\\&+ 2(x-y)^T B(t)(g(x)-g(y)) + (x-y)^T(D+P)(x-y)\\&- (x-y)^T D(x-y) - (x-y)^T P(x-y)\end{aligned}$$

We note that the following inequalities hold:

$$\begin{aligned}&-(x-y)^T D(x-y) + 2(x-y)^T A(t)(f(x)-f(y))\\&\leq (f(x)-f(y))^T \overline{A}^T D^{-\frac{1}{2}} D^{-\frac{1}{2}} \overline{A}(f(x)-f(y))\\&- (x-y)^T P(x-y) + 2(x-y)^T B(t)(g(x)-g(y))\\&\leq (g(x)-g(y))^T \overline{B}^T P^{-\frac{1}{2}} P^{-\frac{1}{2}} \overline{B}(g(x)-g(y))\end{aligned}$$

Hence, we can write

$$\begin{aligned}&2(x-y)^T(H(x,t)-H(y,t)) \leq -\sum_{i=1}^{n}(2\underline{c}_i - d_i - p_i)(x_i-y_i)^2 + (f(x)-f(y))^T \overline{A}^T D^{-\frac{1}{2}}\\&D^{-\frac{1}{2}}\overline{A}(f(x)-f(y)) + (g(x)-g(y))^T \overline{B}^T P^{-\frac{1}{2}} P^{-\frac{1}{2}} \overline{B}(g(x)-g(y))\\&\leq -\sum_{i=1}^{n}(2\underline{c}_i - d_i - p_i)(x_i-y_i)^2 + \|D^{-\frac{1}{2}}\overline{A}\|_2^2 \sum_{i=1}^{n}(f_i(x_i) - f_i(y_i))^2 + \|P^{-\frac{1}{2}}\overline{B}\|_2^2\\&\sum_{i=1}^{n}(g_i(x_i) - g_i(y_i))^2\\&\leq -\sum_{i=1}^{n}(2\underline{c}_i - d_i - p_i)(x_i-y_i)^2 + \|D^{-\frac{1}{2}}\overline{A}\|_2^2 \sum_{i=1}^{n} k_i^2(x_i-y_i)^2 + \|P^{-\frac{1}{2}}\overline{B}\|_2^2\\&\sum_{i=1}^{n} \ell_i^2(x_i-y_i)^2\\&= -\sum_{i=1}^{n}(2\underline{c}_i - d_i - p_i - k_i^2\|D^{-\frac{1}{2}}\overline{A}\|_2^2 - \ell_i^2\|P^{-\frac{1}{2}}\overline{B}\|_2^2)(x_i-y_i)^2\\&= -\sum_{i=1}^{n}\gamma_i(x_i-y_i)^2 \leq -\gamma_m(x-y)^T(x-y)\end{aligned}$$

where $\gamma_m = \min \gamma_i$. From this previous inequality, we have

$$\gamma_m\|x-y\|_2^2 \leq 2\|x-y\|_\infty\|H(x,t) - H(y,t)\|_1 \leq 2\|x-y\|_2\|H(x,t) - H(y,t)\|_1$$

resulting in

$$\gamma_m\|x-y\|_2 \leq 2\|H(x,t) - H(y,t)\|_1 \quad (7)$$

where from Lemma 1, $x \neq y$ implies that $H(x,t) \neq H(y,t)$. It will now been shown that $\|H(x,t)\| \to \infty$, $\|x\| \to \infty$ for any $t \in R$. For $y = 0$, (7) takes the form

$$\gamma_m\|x\|_2 \leq 2\|H(x,t) - H(0,t)\|_1 \leq 2\|H(x,t)\|_1 + 2\|H(0,t)\|_1$$

implying that $2\|H(x,t)\|_1 \geq \gamma_m\|x\|_2 - 2\|H(0,t)\|_1$. Since the variable coefficients are bounded and activation functions are bounded at zero point , $\|H(0,t)\|_1$ is finite, it follows that $\|H(x,t)\| \to \infty$ as $\|x\| \to \infty$ for any $t \in R$. From lemma 2, This completes the proof.

3 Global Stability Analysis

In this section, we will derive some sufficient conditions for the global asymptotic stability of the equilibrium point for neural system (1). To this end, we first simplify system (1) by using the transformation$z_i(\cdot) = x_i(\cdot) - x_i^*, i = 1, 2, \ldots, n$,which puts system (1) into the form

$$\dot{z}_i(t) = -c_i(t)z_i(t) + \sum_{j=1}^{n} a_{ij}(t)\phi_j(z_j(t)) + \sum_{j=1}^{n} b_{ij}(t)\psi_j(z_j(t-\tau_j(t))), i = 1, 2, \ldots, n \tag{8}$$

where

$$\phi_i(z_i(t) = f_i(z_i(t) + x_i^*) - f_i(x_i^*) \quad i = 1, 2, \ldots, n,$$
$$\psi_i(z_i(t-\tau_j(t))) = g_i(z_i(t-\tau_j(t)) + x_i^*) - g_i(x_i^*) \quad i = 1, 2, \ldots, n.$$

It can easily be verified that the functions$\phi_i$and$\psi_i$satisfy the following conditions:

$$|\phi_i(z_i)| \leq k_i|z_i| \quad and \quad \phi_i(0) = 0, \quad i = 1, 2, \ldots, n$$

$$|\psi_i(z_i)| \leq \ell_i|z_i| \quad and \quad \psi_i(0) = 0, \quad i = 1, 2, \ldots, n$$

It is ,therefore, sufficient to consider the stability of the origin of the transformed system(8)in order to consider the stability of $x^*$of the original system(1). Neural network(8)can be expressed in the form

$$\dot{z}(t) = -C(t)z(t) + A(t)\Phi(z(t)) + B(t)\psi(z(t-\tau(t))) \tag{9}$$

where

$$\Psi(z(t-\tau(t))) = (\psi_1(z_1(t-\tau_1(t))), \psi_2(z_2(t-\tau_2(t))), \ldots, \psi_1(z_n(t-\tau_n(t))),$$

$$z(t) = (z_1(t), z_2(t), \ldots, z_n(t))^T, \Phi(z(t)) = (\phi_1(z_1(t)), \phi_2(z_2(t))), \ldots, \phi_n(z_n(t)).$$

We are now in a position to prove the following theorems.

Theorem 2. *Let $\dot{\tau}_j(t) \leq \mu < 1$. Under the assumptions (3), the origin of neural system (8)is global asymptotically stable if there exists positive diagonal matrices $D = diag(d_{1,2}, \ldots, d_n)$,and $P = (p_1, p_2, \ldots, p_n)$, such that, for i=1,2,...,n,*

$$\gamma_i^* = 2\underline{c}_i - d_i - p_i - k_i^2\|D^{-\frac{1}{2}}\overline{A}\|_2^2 - \frac{1}{1-\mu}\ell_i^2\|P^{-\frac{1}{2}}\overline{B}\|_2^2 > 0$$

Proof: The condition $0 \leq \mu < 1$ implies that

$$2\underline{c}_i - d_i - p_i - k_i^2\|D^{-\frac{1}{2}}\overline{A}\|_2^2 - \ell_i^2\|P^{-\frac{1}{2}}\overline{B}\|_2^2$$

$$\geq 2\underline{c}_i - d_i - p_i - k_i^2 \|D^{-\frac{1}{2}}\overline{A}\|_2^2 - \frac{1}{1-\mu}\ell_i^2 \|P^{-\frac{1}{2}}\overline{B}\|_2^2 > 0.$$

Therefore, the uniqueness of the equilibrium point directly follows from the result of Theorem 1. In order to prove the global asymptotic stability of the equilibrium point, we will use the following positive-definite Lyapunov functional:

$$V(z(t)) = z(t)^T z(t) + \frac{\|P^{-\frac{1}{2}}\overline{B}\|_2^2}{1-\mu} \sum_{j=1}^{n} \int_{t-\tau_j(t)}^{t} \psi_j^2(z_j(\xi))d\xi.$$

Taking the time derivative of $V(z(t))$ along the trajectories of the system (8), and then adding and subtracting the term $z(t)^T(D+P)z(t)$ results in

$$\begin{aligned}
&\dot{V}(z(t)) = -2z(t)^T C(t)z(t) + 2z(t)^T A(t)\Phi(z(t)) + 2z(t)^T B(t)\Psi(z(t-\tau(t))) \\
&+ \tfrac{1}{1-\mu}\|P^{-\frac{1}{2}}\overline{B}\|_2^2 \Psi^T(z(t))\Psi(z(t)) + z(t)^T(D+P)z(t) \\
&- \|P^{-\frac{1}{2}}\overline{B}\|_2^2 \sum_{j=1}^{n} \tfrac{1-\dot{\tau}_j(t)}{1-\mu}\psi_j^2(z_j(t-\tau_j)) - z(t)^T Dz(t) - z(t)^T Pz(t) \\
&\leq -2z(t)^T \underline{C}z(t) + 2z(t)^T \overline{A}\Phi(z(t)) + 2z(t)^T \overline{B}\Psi(z(t-\tau)) \\
&+ \tfrac{1}{1-\mu}\|P^{-\frac{1}{2}}\overline{B}\|_2^2 \Psi^T(z(t))\Psi(z(t)) + z(t)^T(D+P)z(t) \\
&- \|P^{-\frac{1}{2}}\overline{B}\|_2^2 \Psi^T(z(t-\tau))\Psi(z(t-\tau)) - z(t)^T Dz(t) - z(t)^T Pz(t).
\end{aligned}$$

we have

$$\begin{aligned}
&-z(t)^T Dz(t) + 2z(t)^T A(t)\Phi(z(t)) \leq \|D^{-\frac{1}{2}}\overline{A}\|_2^2 \Psi^T(z(t))\Psi(z(t)) \\
&- z(t)^T Pz(t) + 2z(t)^T B(t)\Psi(z(t-\tau)) \leq \|P^{-\frac{1}{2}}\overline{B}\|_2^2 \Psi^T(z(t-\tau))\Psi(z(t-\tau)).
\end{aligned}$$

Hence, $\dot{V}(z(t))$ becomes

$$\begin{aligned}
&\dot{V}(z(t)) \leq -z(t)^T(2\underline{C} - D - P)z(t) + \|D^{-\frac{1}{2}}\overline{A}\|_2^2 \Psi^T(z(t))\Psi(z(t)) \\
&+ \tfrac{1}{1-\mu}\|P^{-\frac{1}{2}}\overline{B}\|_2^2 \Psi^T(z(t))\Psi(z(t)) \\
&\leq -z(t)^T(2\underline{C} - D - P)z(t) + \|D^{-\frac{1}{2}}\overline{A}\|_2^2 z^T(t)K^2 z(t) \\
&+ \tfrac{1}{1-\mu}\|P^{-\frac{1}{2}}\overline{B}\|_2^2 z^T(t)L^2 z(t) = -\sum_{i=1}^{n} \gamma_i^* z_i^2(t).
\end{aligned}$$

which guarantees the negative definiteness of $\dot{V}(z(t))$ for all $z(t) \neq 0$.Now, consider the case where $z(t) = 0$,(implying that $\Phi(z(t)) = \Psi(z(t)) = 0$). In this case, $\dot{V}(z(t))$ is the form

$$\dot{V}(z(t)) = -\|P^{-\frac{1}{2}}\overline{B}\|_2^2 \sum_{j=1}^{n} \tfrac{1-\dot{\tau}_j(t)}{1-\mu}\psi_j^2(z_j(t-\tau_j)) \leq -\|P^{-\frac{1}{2}}\overline{B}\|_2^2 \sum_{j=1}^{n} \psi_j^2(z_j(t-\tau_j)).$$

It is obvious that $\dot{V}(z(t)) < 0$ for all $\Psi(z(t-\tau_j)) \neq 0$. Therefore, $\dot{V}(z(t)) = 0$ if and only if $z(t) = \Phi(z(t)) = \Psi(z(t)) = 0$. Hence, it immediately follows that the origin of system(8), or equivalently, the equilibrium point of system (1) is globally asymptotically stable.

Theorem 3. *Assume that* $\gamma_j(t) = \tau$ *is a positive constant. Under the assumptions (3), the origin of neural system (9) is a unique equilibrium point and it is globally asymptotically stable if there exists positive diagonal matrices* $D = diag(d_1, d_2, \ldots, d_n)$*,and* $P = diag(p_1, p_2, \ldots, p_n)$*, such that*

$$\gamma_i = 2\underline{c}_i - d_i - p_i - k_i^2\|D^{-\frac{1}{2}}\overline{A}\|_2^2 - \ell_i^2\|P^{-\frac{1}{2}}\overline{B}\|_2^2 > 0 \quad \forall i.$$

Proof: The proof of Theorem 3 follows from the fact that $\mu = 0$ when $\tau_j(t) = \tau$ is a constant.

4 Comparisons And Examples

We will compare our results with the previous results derived in the literature.

Theorem 4 [6]. *Let* $c_i(t) = c_i, a_{ij}(t) = a_{ij}, b_{ij}(t) = b_{ij}, \dot{\tau}_j(t) \leq \mu < 1$*. Under the assumptions (3), the origin of neural system (8) is a unique equilibrium point and it is globally asymptotically stable if there exists positive constant* $r_i > 0$ *such that*

$$r_i c_i - \sum_{j=1}^{n} r_j|a_{ij}|k_j - \sum_{j=1}^{n} r_j|b_{ij}|\ell_j > 0, \ i = 1, 2, \ldots, n.$$

It is known that the previous result holds if and only if $(C - |A|K - (1/(1-\mu))|B|L)$ *is a nonsingular M-matrix,(a matrix with positive diagonal elements and nonnegative off-diagonal elements is called a nonsingular M-matrix if the matrix have all positive real-parted eigenvalues*[16].

Theorem 5 [6]. *Assume that* $c_i(t) = c_i, a_{ij}(t) = a_{ij}, b_{ij}(t) = b_{ij}$ *are constants,* $\tau_j(t) = \tau$ *is a positive constant. Under the assumptions (3),the origin of neural system (8) is a unique equilibrium point and it is globally asymptotically stable if* $(C - |A|K - |B|L)$*is a nonsingular M-matrix.*

Theorem 6 [7]. *Assume that* $c_i(t) = c_i, a_{ij}(t) = a_{ij}, b_{ij}(t) = b_{ij}$ *are constants,* $\tau_j(t) = \tau$ *is a positive constant. Under the assumptions (3),the origin of neural system (8) is a unique equilibrium point and it is globally asymptotically stable if* $(C - \hat{A}K - |B|L)$*is a nonsingular M-matrix, where* $\hat{A}$*is the comparison matrix of A, which is defined as* $\hat{a_{ii}} = a_{ii}$ *and* $\hat{a_{ij}} = -|a_{ij}|$[16].

Theorem 7 [8]. *Assume that* $c_i(t) = c_i, a_{ij}(t) = a_{ij}, b_{ij}(t) = b_{ij}$ *are constants,* $\tau_j(t) = \tau$ *is a positive constant. Under the assumptions (3),the origin of neural system (8) is a unique equilibrium point and it is globally asymptotically stable if there exists constants* $p_m > 0 (m = 1, 2, \ldots, L_1)$*,*$q_k > 0 (k = 1, 2, \ldots, L_2)$*,* $\alpha_{ij}, \alpha_{ij}^*, \beta_{ij}, \beta_{ij}^*, \xi_i j, \xi_{ij}^*, \eta_i j, \eta_{ij}^* \in R (i, j = 1, 2, \ldots, n)$ *such that*

$$\sum_{j=1}^{n}(\sum_{m=1}^{L_1} p_m|a_i j|^{\frac{r\alpha_i j}{p_m}} k_j^{r\xi_{ij}} + k_i^{r\xi_{ji}^*}|a_j i|^{r\alpha_j i^*} + \sum_{m=1}^{L_2} q_m|b_i j|^{\frac{r\beta_i j}{q_m}} \ell_j^{r\eta_{ij}} + \ell^{r\eta_{ji}^*}|b_j i|^{r\beta_j i^*})$$

$< rc_i, \forall i$, *where* $\alpha_{ij}, \alpha^*_{ij}, \beta_{ij}, \beta^*_{ij}, \xi_{i}j, \xi^*_{ij}, \eta_{i}j, \eta^*_{ij} \in R(i,j = 1,2,\ldots,n)$ *are constants with* $L_1\alpha_{i}j + \alpha_{i}j^*1, L_2\beta_{i}j + \beta_{i}j^*1, L_1\xi_{i}j + \xi_{i}j^*1, L_2\eta_{i}j + \eta_{i}j^*1$,*and* $r = \sum_{m=1}^{L_1} p_m + 1 = \sum_{m=1}^{L_2} q_m + 1.$

Now, consider the following examples.

Example 1: Assume that the network parameters of neural network (1) are given as follows:

$$A = B = \begin{bmatrix} p & p & p & p \\ p & -p & p & -p \\ -p & -p & p & p \\ -p & p & p & -p \end{bmatrix} \qquad C = K = L = \begin{bmatrix} 1&0&0&0 \\ 0&1&0&0 \\ 0&0&1&0 \\ 0&0&0&1 \end{bmatrix} \qquad \mu = \tfrac{1}{2}$$

where $p > 0$ is a real number. When applying the result of Theorem 4 to this example, the stability condition is obtained as $p < \frac{1}{12}$. Let $D = P = \frac{1}{2}I$. The conditions obtained in Theorem 2 can be expressed as follows:

$$\gamma_i^* = 2\underline{c}_i - d_i - p_i - \|D^{-\frac{1}{2}}\overline{A}\|_2^2 - 2\|P^{-\frac{1}{2}}\overline{B}\|_2^2 > 0.$$

For the network parameters of this example, one can obtain

$$\|D^{-\frac{1}{2}}\overline{A}\|_2^2 = \|P^{-\frac{1}{2}}\overline{B}\|_2^2 = 8p^2.$$

We now have

$$\gamma_1^* = \gamma_2^* = \gamma_3^* = \gamma_4^* = 1 - 24p^2 > 0$$

yielding $p < \frac{1}{2\sqrt{6}}$. Hence, for this example, Theorem 2 imposes less restrictive constraint conditions on the network parameters than the constraint conditions imposed by Theorem 4. Now, assume that $\tau_j(t) = \tau$ is a positive constant. In this case, Theorem 5 requires that $p < \frac{1}{8}$ and Theorem 5 requires that $p < \frac{1}{6}$.

If we choose $D = P = \frac{1}{2}I$. Then, for the same network parameters, the conditions obtained in Theorem 3 can be expressed as follows:

$$\gamma_i^* = 1 - \|D^{-\frac{1}{2}}A\|_2^2 - 2\|P^{-\frac{1}{2}}B\|_2^2 > 0, i = 1,2,3,4$$

which are calculated as:

$$\gamma_1^* = \gamma_2^* = \gamma_3^* = \gamma_4^* = 1 - 16p^2 > 0,$$

from which the stability condition is obtained as : $p < \frac{1}{4}$. Hence, for the case of constant time delays, our results are weaker than those obtained in [6] and [7] for the neural network possessing the previous network parameters.

Example 2: Now consider the example where the network parameters of the neural network(1) are given as follows:

$$A = B = \begin{bmatrix} 1 & 1 & 1 & 1 \\ 1 & -1 & 1 & -1 \\ -1 & -1 & 1 & 1 \\ -1 & 1 & 1 & -1 \end{bmatrix} \quad K = L = \begin{bmatrix} 1 & 0 & 0 & 0 \\ 0 & 1 & 0 & 0 \\ 0 & 0 & 1 & 0 \\ 0 & 0 & 0 & 1 \end{bmatrix} \quad \mu = 0 \quad C = \begin{bmatrix} c & 0 & 0 & 0 \\ 0 & c & 0 & 0 \\ 0 & 0 & c & 0 \\ 0 & 0 & 0 & c \end{bmatrix}$$

where c is a positive constant. If we let $D = P = I$, then we obtain

$$\|D^{-\frac{1}{2}}\overline{A}\|_2^2 = \|P^{-\frac{1}{2}}\overline{B}\|_2^2 = 4.$$

Hence, Theorem 3 implies that

$$\gamma_1^* = \gamma_2^* = \gamma_3^* = \gamma_4^* = 2c - 10 > 0$$

from which the stability condition is obtained as $c > 5$. For the previous network parameters, the conditions of Theorem 7 are stated as

$$\sum_{j=1}^{4}(\sum_{m=1}^{L_1} p_m + 1 + \sum_{m=1}^{L_2} q_m + 1) < rc$$

Since

$$r = \sum_{m=1}^{L_1} p_m + 1 = \sum_{m=1}^{L_2} q_m + 1$$

it follows that $\sum_{j=1}^{4} 2r < rc$ from which one would obtain the stability condition as $c > 8$. Hence, if $5 < c < 8$, then Theorem 3 is applicable to this example,whereas the conditions of Theorem 7 are not satisfied.

Example 3: Assume that the network parameters of neural network (1) are given as follows:

$$A = B = \begin{bmatrix} p+|sint| & p+|sint| & p+|sint| & p+|sint| \\ p+|sint| & -p-|cost| & p+|sint| & -p-|cost| \\ -p-|cost| & -p-|cost| & p+|sint| & p+|sint| \\ -p-|cost| & p+|sint| & p+|sint| & -p-|cost| \end{bmatrix}$$

$$C = \begin{bmatrix} 1+|sint| & 0 & 0 & 0 \\ 0 & 1+|cost| & 0 & 0 \\ 0 & 0 & 1+|sint| & 0 \\ 0 & 0 & 0 & 1+|cost| \end{bmatrix} \quad K = L = \begin{bmatrix} 1 & 0 & 0 & 0 \\ 0 & 1 & 0 & 0 \\ 0 & 0 & 1 & 0 \\ 0 & 0 & 0 & 1 \end{bmatrix} \quad \mu = \frac{1}{2}$$

where $p > 0$ is a real number. Let$D = P = \frac{1}{2}I$. The conditions obtained in Theorem 2 can be expressed as follows:

$$\gamma_i^* = 2\underline{c}_i - d_i - p_i - \|D^{-\frac{1}{2}}\overline{A}\|_2^2 - 2\|P^{-\frac{1}{2}}\overline{B}\|_2^2 > 0.$$

so we can see our results are applicable, but others can't deal with this example. All the three examples show our results are new and improve the previous results.

5 Conclusion

The main contribution of this paper ensures the existence, uniqueness and global asymptotic stability of neural networks with time-varying delays. The results do not require the activation functions to be continuously differentiable or monotone increasing. Some examples have been given to prove that our results are new and efficient.

Acknowledgment. The work are supported by the national nature science foundation of CHINA under grant 60674020 and the nature science foundation of SHANDONG under grant Z2006G11.

References

1. Roska, T., Boros, T., Thiran P., Chua, L.O.: Detecting Simple Motion Using Celluar Neural Networks. Proc. IEEE int. workshop on Celluar Neural Networks and Their Applications (1990) 127–138
2. Arik, S.: Global Asymptotic Stability of A Class of Synsmical Neural Networks. IEEE Trans. Circuits Syst I **4** (2000) 568–571
3. Arik, S., Tavsanoglu, V.: Equilibrium Analysis of Delayed CNNs. IEEE Trans. Circuits Syst I **2** (1998) 168–171
4. Forti, M., Tesi, A.: New Conditions for Global Stability of Neural Networks with Applicatons to Linear and Quadractic Programming Problems. IEEE Trans. Circuits Syst I **7** (1995) 354–365
5. Gao, J.: Exponential Stability and Periodic Oscillatory Solution in BAM Networks with Delays. IEEE Tran. Neural Networks **2** (2002) 457–463
6. Gao, J., Wang, J.: Global Asymptotic Stability of a General Class of Recurrent Neural Networks with Time-Varying Delays. IEEE Trans. Circuits Syst I **1** (2003) 34–44
7. Lu, H., Chung, F.L., He, Z.: Some Sufficient Conditions for Global Exponential Stability of Delayed Neural Networks. Neural Networks (2004) 437–544
8. Zhang, Q., Ma, R., Wang, C., Xu, J.: On the Global Stability of Delayed Neural Networks. IEEE Trans. Autom. Control **5** (2003) 794–797
9. Chen, T.: Global Conbergence of Delayed Dynamical Systems. IEEE Trans. Neural Networks **6** (2001) 1532–1536
10. Ensari, T., Arik, S.: Global Stability Analysis of Neural Networks with Multiple Time Varying Delays. IEEE Trans. Autom. Control **11** (2005) 1781–1785
11. Zeng, Z., Wang, J., Liao, X.: Global Exponential Stability of A General Class of Recurrent Neural Networks with Unbounded Time-Varying Delays. IEEE Transactions on Circuits and Systems-Part II Express Briefs **52 (3)** (2005) 168-173
12. Wang, L.S., Xu, D.Y.: Stability Analysis of Hopfield Neural Networks with Time Delay. Applied Mathematics and Mechanics **23** (2002) 250–252
13. Hu, S., Wang, J.: Global Robust Stability of A Class of Discrete-Time Interval Neural Networks. IEEE Transactions on Circuits and Systems–Part I: Regular Papers **53** (2006) 129–138
14. Zeng, Z., Wang, J.: Complete Stability of Cellular Neural Networks with Time-Varying Delays. IEEE Transactions on Circuits and Systems-Part I: Regular Papers **53** (2006) 944–955
15. Liao, X., Wang, J., Zeng, Z.: Global Asymptotic Stability and Global Exponential Stability of Delayed Cellular Neural Networks. IEEE Transactions on Circuits and Systems-Part II Express Briefs **52 (7)** (2005) 403-409
16. Horn, R.A., Johnson, C.R.: Topic in Matrix Analysis. Cambridge, U. K.: Cambridge Univ. Press (1991)

Exponential Stability of Discrete-Time Cohen-Grossberg Neural Networks with Delays

Changyin Sun, Liang Ju, Hua Liang, and Shoulin Wang

College of Electric Engineering, Hohai University, Nanjing 210098
cysun@hhu.edu.cn

Abstract. Discrete-time Cohen-Grossberg neural networks(CGNNs) are studied in this paper. Several sufficient conditions are obtained to ensure the global exponential stability of the discrete-time systems of CGNNs with delays based on Lyapunov methods. The obtained results have not assume the symmetry of the connection matrix, and monotonicity, boundness of the activation functions.

1 Introduction

In 1983, Cohen and Grossberg presented a kind of neural system, which is now called Cohen-Grossberg neural networks (CGNNs) [1]. It can be described by the following system

$$\frac{dx_i}{dt} = -a_i(x_i)[b_i(x_i) - \sum_{j=1}^{m} c_{ij} S_j(x_j)] \tag{1}$$

where $i \in \{1, 2, \cdots, m\}$, x_i denotes the state variable associated with the ith neuron, a_i represents an amplification function, b_i is an appropriately behaved function.The n×n connection matrix $T = (t_{ij})$ tell show the neurons are connected in the network, and the activation function S_i shows how the jth neuron reacts to the input. In [2]-[3], some sufficient conditions have been obtained for globally asymptotic stability of delayed Cohen-Grossberg neural networks. In [4], the following system of delay differential equations:

$$\frac{dx_i}{dt} = -a_i(x_i)[b_i(x_i) - \sum_{j=1}^{m} d_{ij} S_j(x_j(t - \tau_{ij})) + J_i] \tag{2}$$

where $i \in \{1, 2, \cdots, M\}$, was considered. In [5], the author consider the exponential stability of discrete-time Cohen-Grossberg neural networks. In implementing the continuous-time network (2) for computer simulation, experimental or computational purposes, it is common to discretize the continuous-time networks. So consider the discrete-time system is necessary. As we know, the CGNNs are very general, and include the well-known Hopfield neural networks, cellur neural networks. Based on the above discussion, we go to consider the existence and

D. Liu et al. (Eds.): ISNN 2007, Part I, LNCS 4491, pp. 920–925, 2007.

exponential stability of solutions of the discrete general cohen-grossberg neural networks with delays, which are described as follows:

$$\frac{dx_i}{dt} = -a_i(x_i)[b_i(x_i) - \sum_{j=1}^{m} c_{ij}S_j(x_j) - \sum_{j=1}^{m} d_{ij}S_j(x_j(t-\tau_{ij})) - \sum_{j=1}^{m} e_{ij}S_j(\int_1^{\infty} K_{ij}(s)x_j(t-s)ds) + J_i] \tag{3}$$

$i \in \{1, 2, \cdots, M\}$.

With a similar discretizing method of [6], we obtain the discrete-time analogue of the continuous-time networks (4) given

$$x_i(n+1) = x_i(n) - a_i(x_i(n))[b_i(x_i(n)) - \sum_{j=1}^{m} c_{ij}S_j(x_j(n)) - \sum_{j=1}^{m} d_{ij}S_j(x_j(n-k_{ij})) - \sum_{j=1}^{m} e_{ij}S_j(\sum_{s=1}^{\infty} K_{ij}(s)x_j(n-s)) + J_i] \tag{4}$$

where $x_i(l) = \varphi_i(l)$, $l \in (-\infty, 0]_z, k_{ij} \in Z^+, i \in \{1, 2, \cdots, m\}$.

(H1): $0 < \underline{a_i}(u) \leq a_i \leq \overline{a_i}, i \in \{1, 2, \cdots, m\}$.

(H2): $b_i(\bullet)$ is Lipschitz continuous with Lipschit constant l_i and $[b_i(u) - b_i(v)](u-v) \geq \gamma_i(u-v)^2, i \in \{1, 2, \cdots, m\}$.

(H3): S_i is Lipschitz continuous with Lipschit constant L_i, $i \in \{1, 2, \cdots, m\}$.

(H4): $\sum_{p=1}^{\infty} K_{ij}(p) = 1$.

The rest of this paper is organized as follows: in Section 2we give some notations and provide several sufficient conditions to ensure the existence and the global exponential stability of the unique equilibrium point for network (4); one example is given to illustrate the effectiveness of our results in Section 3; in Section 4 the conclusion are given.

2 Main Results

Lemma 1. *[7]: If $H(x) \in C^0$ and satisfies the following conditions.*

(1) $H(x)$ is infective on R^n.

(2)

$$\lim_{\|x\| \to \infty} \|H(x)\| = \infty \tag{5}$$

then $H(x)$ is a homeomorphism of R^n.

Theorem 1. *Under the assumptions $(H1) - (H3)$ and $L_i \sum_{j=1}^{m} \overline{a_j}(|c|_{ji} + |d|_{ji} + |e|_{ji}) < \overline{a_i}\gamma_i(i \in \{1, 2, M\})$ the neural network (4) has an unique equilibrium point.*

Proof. Clearly, the equilibrium point x^* of system (4) satisfy the following equation:

$$b_i(x^*) - \sum_{j=1}^{m} c_{ij} S_j(x^*) - \sum_{j=1}^{m} d_{ij} S_j(x^*) - \sum_{j=1}^{m} e_{ij} S_j(x^*) + J_i. \quad (6)$$

$i \in \{1, 2, \cdots, m\}$. let

$$H(x) = b_i(x) - \sum_{j=1}^{m} c_{ij} S_j(x) - \sum_{j=1}^{m} d_{ij} S_j(x) - \sum_{j=1}^{m} e_{ij} S_j(x) + J_i. \quad (7)$$

$i \in \{1, 2, \cdots, m\}$. Note that ,to prove that system (4) has an equilibrium point, by Lemma 1, we only need to show that H(x) is a homeomorphism.

First, we prove that H is an injective on R^n. In fact, if there exist $x \neq y \in R^n$such that $H(x) = H(y)$, then

$$\sum_{j=1}^{m} \overline{a_i} \gamma_i |x_i - y_i| \leq \sum_{j=1}^{m} \overline{a_i} |b_i(x_i) - b_i(y_i)| = U \quad (8)$$

$$\begin{aligned} U &\leq \sum_{j=1}^{m} \overline{a_i} sign(b_i(x_i) - b_i(y_i)) b_i(x_i) - b_i(y_i) \\ &\leq \sum_{j=1}^{m} \overline{a_i} sign(b_i(x_i) - b_i(y_i)) [\sum_{j=1}^{m} (c_{ij} + d_{ij} + e_{ij})(S_j(x_j)) - (S_j(x_j))] \\ &\leq \sum_{i=1}^{m} \overline{a_i} [\sum_{j=1}^{m} L_j (|c_{ij}| + |d_{ij}| + |e_{ij}|) |x_j - y_j|] \\ &\leq \sum_{i=1}^{m} [\sum_{j=1}^{m} \overline{a_i} (|c_{ji}| + |d_{ji}| + |e_{ji}|)] L_i |x_i - y_i|. \end{aligned} \quad (9)$$

Note that the condition $L_i \sum_{j=1}^{m} \overline{a_j}(|c|_{ji} + |d|_{ji} + |e|_{ji}) < \overline{a_i}\gamma_i$, So we have a contradiction. there, we have $x = y$, This implies that map H is an injective on R^n. Second, we prove that $\lim_{\|x\| \to \infty} \|H(x)\| = \infty$.

To show that $\lim_{\|x\| \to \infty} \|H(x)\| = \infty$, it suffices to show that

$$\lim_{\|x\| \to \infty} \left\| \widetilde{H}(x) \right\| = \infty$$

where are

$$\begin{aligned} \sum_{i=1}^{m} \overline{a_i} sign(x_i) \widetilde{H}(x) &\leq \sum_{i=1}^{m} -\overline{a_i} [|x_i| + \sum_{j=1}^{m} (|c|_{ij} + |d|_{ij} + |e|_{ij}) L_j x_j] \\ &\leq \sum_{i=1}^{m} \{-\overline{a_i}\gamma_i + L_i [\sum_{j=1}^{m} -\overline{a_j} (|c|_{ji} + |d|_{ji} + |e|_{ji})]\} |x_i| \\ &\leq -\varepsilon \sum_{i=1}^{m} |x_i|. \end{aligned} \quad (10)$$

$$\varepsilon = \min_{1\leq i\leq n}[-\overline{a_i}\gamma_i + L_i[\sum_{j=1}^{m} -\overline{a_j}(|c|_{ji} + |d|_{ji} + |e|_{ji})] > 0 \tag{11}$$

Therefore, we have

$$\sum_{i=1}^{m} |x_i| \leq \frac{1}{\varepsilon} \max_{1\leq i\leq m} \{\overline{a_i}\} \sum_{i=1}^{m} |\widetilde{H}(x)| \tag{12}$$

so

$$\|x\| \leq \frac{1}{\varepsilon} \max_{1\leq i\leq m} \{\overline{a_i}\} \|\widetilde{H}(x)\| \tag{13}$$

It implies that

$$\lim_{\|x\|\rightarrow\infty} \|H(x)\| = \infty \tag{14}$$

From lemma 1, we know that for every input J, map H is a homeomorphism. Thus the neural network (4) has an unique equilibrium point.

The proof of Theorem 1 is complete.

Let x^* be a solution of system (4), and denote $u_i(n) = x_i(n) - x_i^*$, then the system(4) becames

$$\begin{aligned} u_i(n+1) = u_i(n) - \alpha_i(u_i(n))\{\beta_i(u_i(n)) \\ - \sum_{j=1}^{m} c_{ij} g_j(u_j(n)) - \sum_{j=1}^{m} d_{ij}[S_j(x_j(n-k_{ij})) - S_j(x_j^*)] \\ - \sum_{j=1}^{m} e_{ij}[S_j(\sum_{s=1}^{\infty} K_{ij}(s)x_j(n-s)) - S_j(x_j^*)]\}. \end{aligned} \tag{15}$$

where $\alpha_i(u_i(n)) = a_i(u_i(n)+x_j^*), \beta_i(u_i(n)) = b_i(u_i(n)+x_j^*) - b_i(x_j^*), g_i(u_i(n)) = S_i(u_i(n)+x_j^*) - S_i(x_j^*)$ $i \in \{1,2,\cdots,m\}$.

Under the conditions of Theorem 2, we will easily know that system (4) has an equilibrium point. then using Lyapunov methods, the Theorem 2 can be obtained easily.

Theorem 2. *Under the assumptions* $(H1)-(H3)$, $\overline{a_i}l_i < 1$ *and* $L_i\sum_{j=1}^{m}\overline{a_j}(|c|_{ji} + |d|_{ji} + |e|_{ji}) < \underline{a_i}\gamma_i(i \in \{1,2,M\})$ *hold, the neural network (4) is unique and globally Exponential stable for every input J,*

$$\sum_{i=1}^{m} |x_i(n) - x_i^*| \leq v(\frac{1}{\xi})^n \sum_{i=1}^{m} (\sup_{l\in[-k,0]_z} |\phi_i(l) - x_i^*|), i \in \{1,2,\cdots,m\}. \tag{16}$$

where v, ξ *are constants, and* $v > 1, \xi > 1$.

Remark 1. Contrast to [5], the results in this paper don't need $a_i(x)$ to be Lipschitz continuous and the activation functions $S_i(x)$ to be bounded, so the results obtained are new.

3 An Illustrative Example

$$\begin{pmatrix} U_1(n+1) \\ U_2(n+1) \end{pmatrix} = \begin{pmatrix} U_1(n) \\ U_2(n) \end{pmatrix} - \begin{bmatrix} \frac{5+\sin u_1(n)}{20} & 0 \\ 0 & \frac{5+\cos u_2(n)}{20} \end{bmatrix}$$
$$* \left\{ \begin{bmatrix} 0.25 & 0 \\ 0 & 0.25 \end{bmatrix} \begin{pmatrix} U_1(n) \\ U_2(n) \end{pmatrix} - \begin{bmatrix} -0.12 & -0.12 \\ -0.12 & -0.12 \end{bmatrix} \begin{pmatrix} S_1(u_1(n-1)) \\ S_2(u_2(n-1)) \end{pmatrix} \right\}$$

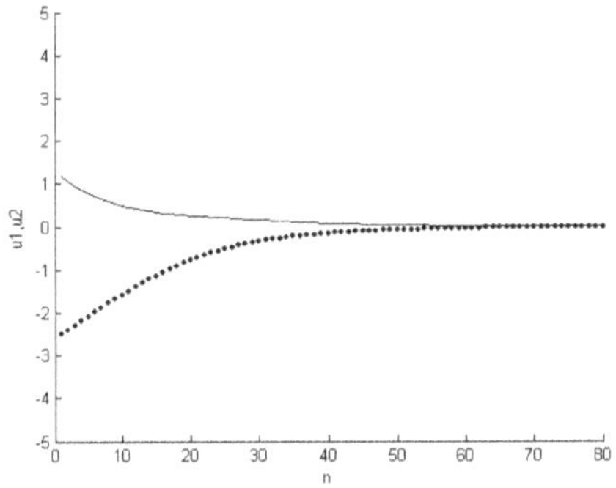

Fig. 1. Numeric simulation

where $S_1(u), S_2(u)$ satisfy $H(3)$, one can easily check that the above system satisfies all the conditions of Theorem 1. Thus, (4) has a unique solution and all solutions of (4) globally exponentially converge to it.

4 Conclusions

Several sufficient conditions have been obtained to ensure the global exponential stability of the equilibrium point of discrete-time CGNNS. The derived criteria do not require the differentiability, boundness and monotonicity of the activation functions. In addition, an example is given to show the effectiveness of the obtained result.

Acknowledgement

The authors would like to thank the reviewers for their helpful comments and constructive suggestions, which have been very improving the presentation of this paper. This work was supported by the Natural Science Foundation of Jiangsu province, China under Grant BK2006564 and China Postdoctoral Science Foundation under Grant 20060400274.

References

1. Cohen, M., Grossberg, S.: Absolute Stability and Global Pattern Formation and Parallel Memory Storage by Competitive Neural Networks. IEEE Trans. Syst Man Cybernet **13** (1983) 815-826
2. Ye, H., Michel, A.N., Wang, K.: Qualitative Analysis of Cohen-Grossberg Neural Networks with Multiple Delays. Phys. Rev. E **51** (1995) 2611-2618
3. Wang, L., Zou, X.: Harmless Delays in Cohen-Grossberg Neural Networks. Physica D **170** (2002) 162-173
4. Wang, L., Zou, X.: Exponential Stability of Cohen-Grossberg Neural Networks. Neural networks **15** (2002) 415-422
5. Xiong, W.J., Cao, J.D.: Exponential Stability ofDiscrete-Time Cohen-Grossberg Neural Networks. Neurocomputing **64** (2005) 433-446
6. Mohamad, S., Naim, A.: Discrete-Time Analogues of Integro-Differential Equations Modelling Bidirectional Neural Networks. Journal of Computational and Applied Mathematics **138** (2002) 1-20
7. Forti, M. AND Tesi, A.: New Conditions for Global Stability of Neural Networks with Application to Linear and Quadratic Programming Problems. IEEE Trans. Circuits and Syst.-I **42** (1995) 354-366

The Tracking Speed of Continuous Attractors

Si Wu[1], Kosuke Hamaguchi[2], and Shun-ichi Amari[2]

[1] Department of Informatics, University of Sussex, UK
[2] Amari Research Unit, RIKEN Brain Science Institute, Japan

Abstract. Continuous attractor is a promising model for describing the encoding of continuous stimuli in neural systems. In a continuous attractor, the stationary states of the neural system form a continuous parameter space, on which the system is neutrally stable. This property enables the neutral system to track time-varying stimulus smoothly. In this study we investigate the tracking speed of continuous attractors. In order to analyze the dynamics of a large-size network, which is otherwise extremely complicated, we develop a strategy to reduce its dimensionality by utilizing the fact that a continuous attractor can eliminate the input components perpendicular to the attractor space very quickly. We therefore project the network dynamics onto the tangent of the attractor space, and simplify it to be a one-dimension Ornstein-Uhlenbeck process. With this approximation we elucidate that the reaction time of a continuous attractor increases logarithmically with the size of the stimulus change. This finding may have important implication on the mental rotation behavior.

1 Introduction

External stimuli are encoded in neural activity patterns in the brain. The brain can reliably retrieve the stored information even when external inputs are incomplete or noisy, achieving the so-called associative memory or invariant object recognition. Mathematically, this can be described as attractor computation, that is, the network dynamics enables the neural system to reach the same stationary state once external inputs fall into its basin of attraction. In the conventional models for attractor computation, such as the Hopfield model [1], it is often assumed that the stationary states of the neural system are discretely distributed in the state space, which are called discrete attractors.

Recently, the progress in both experimental and theoretical studies have suggested that there may exist another form of attractor, called continuous attractors, in biological systems [2,3,4,5,6,7,8,9,10,11,12,13,14,15]. This type of attractor is appealing for encoding continuous stimuli, such as the orientation, the moving direction and the spatial location of objects, or the continuous features that underlying the categorization of objects. In a continuous attractor, the stationary states of the neural system are properly aligned in the state space according to the stimulus values they represent. They form a continuous parameter space, on which the neural system is neutrally stable. Fig.1 illustrates the

D. Liu et al. (Eds.): ISNN 2007, Part I, LNCS 4491, pp. 926–934, 2007.

typical structure difference between a continuous and a discrete attractors. We see that in a discrete point attractor, the system is only stable at the bottom of the bowl, whereas, in a continuous line attractor, the system is neutrally stable on the one-dimensional valley.

Neutral stability is the key that distinguishes a continuous attractor from a discrete one. Intuitively think, neutral stability implies that the system state can be easily moved along the attractor space under the driving of external inputs. This property enables the neural system to track time-varying stimuli in real-time, a capacity which is crucial for the brain to carry out many important computational tasks, such as motion control and spatial navigation. Although this property has been widely pointed out in the literature (see, e.g., [2,15]), a careful investigation about the speed of continuous attractors tracking moving stimulus is still lacking. The goal of this work is to fill in this gap. By tracking speed, we mean the reaction time for the neural system to catch up the change of external inputs.

When analyzing the behavior of a large-size network, the main challenge is to handle the high dimensionality of the system dynamics. Here, by utilizing the specific nature of continuous attractor, we develop a strategy to reduce its dimensionality. It utilizes the facts that neural population dynamics is extremely fast and that continuous attractor can clean those noise components perpendicular to the attractor space very quickly. Therefore, we can project the network dynamics onto the tangent of the attractor space, and simplify it to be a one-dimensional Ornstein-Uhlenbeck process. Simulation shows that this method works efficiently.

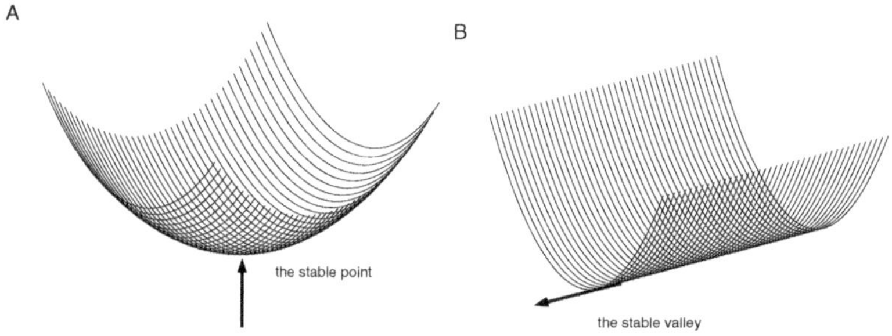

Fig. 1. An illustration of the structural difference between discrete and continuous attractors. (a) An example of a discrete point attractor. The system is only stable at the bottom of the bowl. (b) An example of line attractor, the one-dimensional version of continuous attractor. The stationary states of the system form a one-dimensional valley. Along the valley the system is neutrally stable.

2 The Model

Although diverse models exist for continuous attractors in the literature, they all share two common features, namely, 1) the network should have properly balanced excitatory and inhibitory interactions, so that it can hold persistent activities after external inputs are removed; and 2) the neuronal interactions should be translationally invariant, so that the network can have a continuous family of stationary states. In this study, we will consider a simple firing-rate-based model for continuous attractor. The advantage of this model is that: 1) it allows us to compute the network dynamics analytically; and 2) its main conclusions can be extended to general cases, since they only depend on the common features of continuous attractor.

Consider a one-dimensional continuous stimulus x is encoded by an ensemble of neurons. The neuronal preferred stimulus is denoted as c. We assume $c \in (-\infty, \infty)$ for convenience. The neurons are clustered according to their preferred stimuli, mimicking a column structure. The clusters are uniformly distributed in the parameter space c with the density ρ. We denote γ_c to be the firing rate of the cluster c, and U_c the population-averaged input. The interaction between the two clusters c and c' is written as $J_{c,c'}$.

The dynamics of the network, in the unit of a cluster, is given by

$$\tau \frac{dU_c}{dt} = -U_c + \rho \int_{c'} J_{c,c'} \gamma_{c'} dc' + I_c^{ext}, \tag{1}$$

$$\gamma_c = \frac{U_c^2}{1 + k\rho \int_{c'} U_{c'}^2 dc'}, \tag{2}$$

where k is a small positive constant and I_c^{ext} the external input. The parameter τ is the time constant for the population dynamics, which is in the order of 1ms.

The recurrent interaction is set to be

$$J_{c,c'} = \frac{J}{\sqrt{2\pi} a} e^{-(c-c')^2/2a^2}, \tag{3}$$

where J is a constant that controls the magnitude of the recurrent interactions. $J_{c,c'}$ is the decay function of the difference between the preferred stimuli of the clusters, $(c-c')$. Here, $J_{c,c'}$ has only the excitatory components. The contribution of inhibition is achieved indirectly through the divisive normalization in eq.(2).

When $I_c^{ext} = 0$, the stationary states of the network, referred to as $\bar{U}_c$ and $\bar{\gamma}_c$, satisfy the following conditions,

$$\bar{U}_c = \rho \int_{c'} J_{c,c'} \bar{\gamma}_{c'} dc', \tag{4}$$

$$\bar{\gamma}_c = \frac{\bar{U}_c^2}{1 + k\rho \int_{c'} \bar{U}_{c'}^2 dc'}. \tag{5}$$

It is straightforward to check that the network holds a continuous family of stationary states[13,15],

$$\bar{U}_c(z) = \frac{A\rho J}{\sqrt{2}} e^{-(c-z)^2/4a^2}, \tag{6}$$

$$\bar{\gamma}_c(z) = A e^{-(c-z)^2/2a^2}, \tag{7}$$

where $A = (1 - \sqrt{1 - 8\sqrt{2\pi}ak/(J^2\rho)})/(2\sqrt{2\pi}ak\rho)$, and $z \in (-\infty, \infty)$ is a free parameter. These states are of a Gaussian bell-shape (see Fig.2A), and can be retained after removing external inputs, if $0 < k < J^2\rho/(8\sqrt{2\pi}a)$ (note k controls the amount of inhibition). The parameter z is the peak position of the bump, which indicates the network representation of the external stimulus.

The stimulus information is conveyed to the neural system through the external input I_c^{ext}. Without loss of generality, we choose I_c^{ext} to be of the following form

$$I_c^{ext} = \alpha \bar{U}_c(x) + \sigma \xi_c(t), \tag{8}$$

where both α and σ are small positive constants, and $\xi_c(t)$ is Gaussian white noise with zero mean and unit variance. The first term, $\alpha\bar{U}_c(x)$, represents the stimulus signal, whose contribution is to drive the system to the location of the stimulus x. The second term, $\sigma\xi_c(t)$, represents the input noise with σ the noise strength. For simplicity, we assume ξ_c and $\xi_{c'}$, for $c \neq c'$, are independent to each other.

3 The Dynamics of Continuous Attractor

In general it is difficult to solve the dynamics of a large-size fully connected network. Here, by utilizing the specific features of a continuous attractor, we develop a strategy to assess its dynamics approximately.

To proceed, let us first check how neutral stability shapes the dynamics of a continuous attractor. Consider the network state to be initially at a position z. An input variation induces small fluctuations on the network state and the stationary inputs, which are denoted as $\delta\gamma_c(z)$ and $\delta U_c(z)$ for the cluster c, respectively. Then, according to the stability conditions in eqs.(4) and (5), we have

$$\begin{aligned}
\delta\gamma_c(z) &= \int_{c'} \frac{\partial \bar{\gamma}_c(z)}{\partial \bar{U}_{c'}(z)} \delta U_{c'}(z) dc', \\
&= \int_{c',c''} \frac{\partial \bar{\gamma}_c(z)}{\partial \bar{U}_{c'}(z)} \rho J_{c',c''} \delta\gamma_{c''}(z) dc' dc'', \\
&= \int_{c'} F_{c,c'}(z) \delta\gamma_{c'}(z) dc',
\end{aligned} \tag{9}$$

where the matrix $\mathbf{F}(z)$ is calculated to be

$$\begin{aligned}
F_{c,c'}(z) &= \rho \int_{c''} \frac{\partial \bar{\gamma}_c(z)}{\partial \bar{U}_{c''}(z)} J_{c',c''} dc'' \\
&= \frac{A\rho^2 J^2}{B\sqrt{\pi}a} e^{-(c-z)^2/4a^2} e^{-(c-c')^2/2a^2} \\
&\quad - \frac{kA^3\rho^5 J^4}{\sqrt{3}B^2} e^{-(c-z)^2/2a^2} e^{-(c'-z)^2/6a^2},
\end{aligned} \tag{10}$$

with $B = 1 + A^2 J^2 \sqrt{2\pi}ak\rho^3/2$.

Neutral stability implies that if the change of the network state is along the attractor space (i.e., only the peak position is moved whereas the bump shape is unchanged), then the network is stable at the new position; Otherwise, the system will return to its original shape. Intuitively stated, a continuous attractor will only clean those inputs components perpendicular to the attractor space. Mathematically, this means that the matrix $\mathbf{F}(z)$ has one eigenvector whose eigenvalue is one and all other eigenvalues are smaller than one. The eigenvector, belonging to unit eigenvalue, referred to as $\mathbf{e}^r(z)$, is along the tangent of the attractor space and is dependent of the position z, whose component is given by

$$\begin{aligned} e_c^r(z) &\sim \bar{\gamma}_c'(z), \\ &= D_r(c-z)e^{-(c-z)^2/2a^2}, \end{aligned} \tag{11}$$

where D_r is a constant (the exact value of D_r is not important here). It is straightforward to check that $\mathbf{e}^r$ is indeed the right eigenvector of $\mathbf{F}$ with eigenvalue unit (i.e., $\int_c F_{c,c'} e_{c'}^r dc' = e_c^r$).

The vector $\mathbf{e}^r(z)$ specifies the direction in the state space along which the network state is neutrally stable. In the input space I_c, the corresponding direction, referred to as $\mathbf{e}^I(z)$, is given by

$$\begin{aligned} e_c^I(z) &\sim \bar{U}_c'(z), \\ &= D_I(c-z)e^{-(c-z)^2/4a^2}, \end{aligned} \tag{12}$$

where D_I is a constant. Similarly, it can be checked that, $\mathbf{e}^I$ is the right eigenvector of the matrix, $G_{c,c'} = \int_{c''} J_{c,c''}(\partial\bar{\gamma}_{c'}/\partial\bar{U}_{c''})dc''$, with eigenvalue unit.

We note that the neural population dynamics, in the unit of a cluster, is extremely fast, which is in the order of τ ($1 \sim 2$ms), much smaller than the membrane time constant of single neurons ($10 \sim 20$ms). Combining this with the special stability of continuous attractor, it means that the network dynamics can clean those inputs components perpendicular to $\mathbf{e}^I(z)$ very quickly. Thus, if time is sufficiently long (e.g., much larger than τ), we can reasonably assume the network dynamics is mainly driven by the projection of external inputs on the direction $\mathbf{e}^I(z)$, and ignore the contribution of other components. Behaviorly, this implies that the network bump has only its position shifted, whereas its shape is unchanged. By this approximation, we reduce the dimensionality of the network dynamics from the original value of infinity (since $c \in (-\infty, \infty)$) to unity.

Now we come to analyze the dynamics of the continuous attractor under the driving of varying external inputs. Without loss of generality, we consider the following scenario: the bump position is initially at $z(0)$; then the stimulus value is abruptly changed to $x = 0$. Under the driving of the stimulus signal, the bump will move to the new position at $z = 0$. The reaction time is measured by the time needed for finishing this tracking process. For the simplicity of analysis, we assume $z(0)$ is small compared with the tuning width a (this can be extended in practice).

Assuming the peak position is at z at time t, we project both sides of eq.(1) on the direction $\mathbf{e}^I(z)$, and obtain

$$\begin{aligned}\text{Left-hand side} &= \tau \int_c \frac{dU_c}{dt} e_c^I(z) dc, \\ &= [\frac{\tau A J \rho}{2\sqrt{2}a^2} \int_c (c-z) e^{-(c-z)^2/4a^2} e_c^I dc] \frac{dz}{dt}, \\ &= [\frac{\tau A J \rho \sqrt{\pi} D_I a}{2}] \frac{dz}{dt}, \end{aligned} \quad (13)$$

and

$$\begin{aligned}\text{Right-hand side} &= -\int_c (U_c - \sum_{c'} J_{c,c'} \gamma_{c'}) e_c^I(z) dc + \alpha \int_c \bar{U}_c(0) e_c^I(z) dc \\ &\quad + \sigma \int_c \xi_c(t) e_c^I(z) dc, \\ &= -\frac{A J a \rho \sqrt{\pi} D_I}{2} \alpha z + \sigma (2\pi)^{1/4} a^{3/2} D_I \epsilon(t). \end{aligned} \quad (14)$$

To obtain the above results, we have used the approximations: $U_c \approx \bar{U}_c(z)$, $\gamma_c \approx \bar{\gamma}_c(z)$, and $\bar{U}_c(0) \approx \bar{U}_c(z) - zAJ\rho/(2\sqrt{2}a^2 D_I) e_c^I(z)$, for $|z| \ll a$. The second term in eq.(14) is the projection of the input noise on $\mathbf{e}^I(z)$, where $\epsilon(t)$ the Gaussian white noise of zero mean and unit variance.

Combining the above results, we get,

$$\tau \frac{dz}{dt} = -\alpha z + \beta \epsilon(t), \quad (15)$$

where β is a positive number and β^2 is given by

$$\begin{aligned}\beta^2 &= \frac{\sigma^2}{\int_c [\bar{U}_c'(z)]^2 dc}, \\ &= \frac{4\sqrt{2} a \sigma^2}{A^2 J^2 \rho^2 \sqrt{\pi}}. \end{aligned} \quad (16)$$

Eq.(16) is the one-dimensional Ornstein-Uhlenbeck process. The meaning of this equation is straightforward: when the bump is not at the stimulus position, the stimulus signal generates a force, $-\alpha z$, which pulls the bump to the stimulus position ($z = 0$). The noise effect, $\beta\epsilon(t)$, on the other hand, tends to shift the bump position randomly.

From eq.(15), we see that when z is approaching zero, the driving force $-\alpha z$ becomes smaller and smaller, which implies that when there is no noise, it takes $t \to \infty$ for the bump reaches the stimulus position. However, this should not be a problem in practice, since the neural system has not to wait for the bump to be exactly at $z = 0$ in order to make judgement. We therefore assume after the absolute value $|z|$ is below a threshold θ, a predefined small positive number, the tracking process is finished. Thus, the reaction time of the continuous attractor is given by the first passage time for $|z|$ reaching the threshold θ. Following the standard procedure for solving the Ornstein-Uhlenbeck process [16], we get the mean of the reaction time T to be

$$\langle T \rangle = \frac{\tau}{\alpha}\sqrt{\pi}\int_{d_1}^{d_2} e^{u^2}[1+erf(u)]du, \tag{17}$$

where $d_1 = -z(0)\sqrt{\alpha\tau}/\beta$ and $d_2 = -\theta\sqrt{\alpha\tau}/\beta$.

To see this relationship more clearly, we consider the noise is sufficiently small and can be ignored, then the above equation can be written as

$$\langle T \rangle = \frac{\tau}{\alpha}\ln\frac{|z(0)|}{\theta}. \tag{18}$$

This equation reveals that the reaction time of continuous attractors increases logarithmically with the size of abrupt stimulus change (here $|x - z(0)| = |z(0)|$ quantifies the change size). This result is confirmed by the simulation result (see Fig.2B).

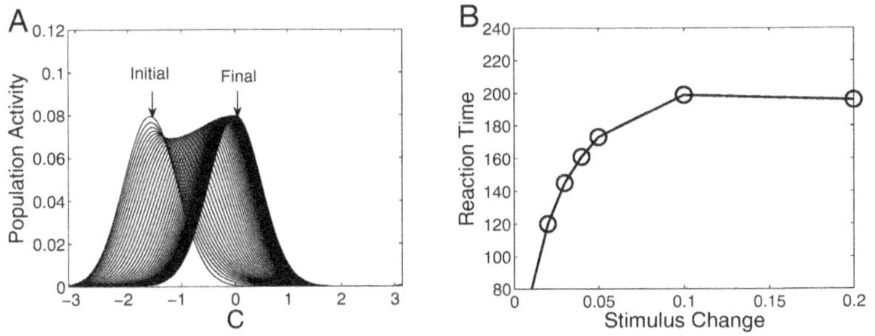

Fig. 2. (A) Illustrating the smooth tracking process. The stationary state of the network has the Gaussian bell-shape. The stimulus value is abruptly changed from -0.5π to 0. (B) The reaction time vs. the abrupt stimulus change. In the simulation, we consider 101 neural clusters uniformly and periodically distributed in the range $(-\pi, \pi]$. The parameters are: $\tau = 1, k = 10, J = 50, a = 1, \alpha = 0.1, \sigma^2 = 0.1$ and $\theta = 0.001$.

4 Conclusions and Discussions

The present study investigates the dynamics of continuous attractors under the driving of external inputs. It has two main contributions. Firstly, we develop a strategy to analytically calculate the dynamics of continuous attractors by utilizing the fact that continuous attractors only retain the input components along the attractor space. Therefore, we can project the network dynamics onto the tangent of the attractor space and simplify it to be a one-dimensional Ornstein-Uhlenbeck process. We expect this strategy can be used for analyzing general attractor network dynamics. Secondly, we show that the reaction time of continuous attractors increases logarithmically with the abrupt stimulus change. This is an important finding. This property is associated with the specific nature of continuous attractors, i.e., neutral stability, and hence can serve as an important clue for experimental data to check whether continuous attractors are really applied in neural systems. Indeed, we have found an encouraging supporting evidence in a special type of mental rotation, called backward alignment [17]. In this experiment, human subjects are instructed to judge whether the presented rotated letter is the one shown previously. The data tends to support that the reaction time of the subjects increases logarithmically with the angle of the letter being rotated (however, since this experiment was not designed for checking this property, the authors did not measure the relationship between the reaction time and the rotation angle carefully. It is not clear how accurately the data fits the logarithm function). We plan to carry out psychophysical experiments to further check this property.

Acknowledgement

We acknowledge Royal Society for supporting K. H. visiting Sussex when this work was conducted.

References

1. Hopfield, J. J.: Neurons with Graded Responses Have Collective Computational Properties Like those of Two-State Neurons. Proc. Natl. Acad. Sci. USA **81** (1984) 3088-3092
2. Amari, S.: Dynamics of Pattern Formation in Lateral-Inhibition Type Neural Fields. Biological Cybernetics **27** (1977) 77-87
3. Georgopoulos, A. P., Kalaska, J. F., Caminiti, R., Massey, J. T.: On the Relations between the Direction of Two-Dimensional Arm Movements and Cell Discharge in Primate Motor Cortex. J. Neurosci. **2** (1982) 1527-1537
4. Maunsell, J. H. R., Van Essen, D. C.: Functional Properties of Neurons in Middle Temporal Visual Area of the Macaque Monkey. I. Selectivity for Stimulus Direction, Speed, and Orientation. J. Neurophysiology **49** (1983) 1127-1147
5. Funahashi, S., Bruce, C., Goldman-Rakic, P.: Mnemonic Coding of Visual Space in the Monkey's Dorsolateral Prefrontal Cotex. J. Neurophysiology **61** (1989) 331-349
6. Wilson, M. A., McNaughton, B. L.: Dynamics of Hippocampal Ensemble Code for Space. Science **261** (1993) 1055-1058
7. Zhang, K. C.: Representation of Spatial Orientation By the Intrinsic Dynamics of the Head-Direction Cell Ensemble: A Theory. J. Neuroscience **16** (1996) 2112-2126
8. Seung, H. S.: How the Brain Keeps the Eyes Still. Proc. Acad. Sci. USA **93** (1996) 13339-13344
9. Ermentrout, B.: Neural networks as Spatial-Temporal Pattern-Forming Systems. Reports on progress in physics **61** (1998) 353-430
10. Taube, J. S.: Head Direction Cells and the Neurophysiological Basis for A Sense of Direction. Prog. Neurobiol. **55** (1998) 225-256
11. Deneve, S., Latham, P. E., Pouget, A.: Reading Population Codes: A Neural Implementation of Ideal Observers. Nature Neuroscience **2** (1999) 740-745
12. Wang, X. J.: Synaptic Reverberation Underlying Mnemonic Persistrent Activitity. Trends in Neuroscience **24** (2001) 455-463

13. Wu, S., Amari, S., Nakahara, H.: Population Coding and Decoding in A Neural Field: A Computational Study. Neural Computation **14** (2002) 999-1026
14. Trappenberg, T.: Continuous Attractor Neural Networks. Recent Developments in biologically inspired computing. Leandro Nunes de Castro and Fernando J. Von Zuben eds. 2003
15. Wu, S., Amari, S.: Computing with Continuous Attractors: Stability and On-Line Aspects. Neural Computation **17** (2005) 2215-2239
16. Tuckwell, H.: Introduction to Theoretical Neurobiology. Cambridge University Press, Cambridge. 1988
17. Koriat, A., Norman, J.: Establishing Global and Local Correspondence Between Successive Stimuli: The Holistic Nature of Backward Alignment. J. of Experimental Psychology **15** (1989) 480-494

Novel Global Asymptotic Stability Conditions for Hopfield Neural Networks with Time Delays

Ming Gao, Baotong Cui, and Li Sheng

Research Center of Control Science and Engineering, Southern Yangtze University, 1800 Lihu Rd., Wuxi, Jiangsu 214122, P.R. China
btcui@vip.sohu.com

Abstract. In this paper, the global asymptotic stability of Hopfield neural networks with time delays is investigated. Some novel sufficient conditions are presented for the global stability of a given delayed Hopfield neural networks by constructing Lyapunov functional and using some well-known inequalities. A linear matrix inequality (LMI) approach is developed to establish sufficient conditions for the given neural networks. An illustrative example is provided to demonstrate the effectiveness of our theoretical results.

1 Introduction

Hopfield neural networks which were first proposed in [1] have been used in various applications such as designing associative memories and solving optimization problems. Since time delay exists in many fields of our society, neural network models with time delay have received important attention in recent years. Particularly, a large number of studies have been devoted to global asymptotic stability or global exponential stability of neural networks with delays [2,3,4,5,6]. In general, the delay type can be constant, time-varying, or distributed, and the stability criteria can be delay-dependent or delay-independent.

In this paper, we study the global asymptotic stability problem for the Hopfield neural networks with time delays. By utilizing Lyapunov functional methods and some well-known inequalities, the linear matrix inequality (LMI) approach is developed to establish sufficient conditions for the given delayed neural networks to be globally asymptotically stable. LMI-based techniques have been successfully used to solve various stability problems for neural networks with time delays [7,8]. The main advantage of the LMI-based approaches is that the LMI stability conditions can be checked numerically very efficiently by resorting to recently developed standard algorithms such as interior-point methods [9].

Notations. Throughout this paper, N^T and N^{-1} denote respectively the transpose and the inverse of any square matrix N. The notation $N > 0$ $(N < 0)$ means that N is a positive-(negative-) definite matrix. I is used to denote the identity matrix, and diag[$\cdot$] denotes a diagonal matrix. R^n and $R^{n\times m}$ denote the n-dimensional Euclidean space and the set of all $n\times m$ real matrices, respectively.

D. Liu et al. (Eds.): ISNN 2007, Part I, LNCS 4491, pp. 935–940, 2007.

2 Model Descriptions and Preliminaries

The DHNN model to be investigated presently can be described as follows:

$$\dot{x}(t) = -Cx(t) + As(x(t-\tau)) + u \,, \tag{1}$$

where $x(t) = [x_1(t), x_2(t), \ldots, x_n(t)]^T$ is the state vector associated with the neurons; $C = \mathrm{diag}(c_i)$ is a positive diagonal matrix; $s_j(\cdot)$ denotes the neuron activation function, and $s(x(t-\tau)) = [s_1(x_1(t-\tau)), s_2(x_2(t-\tau)), \ldots, s_n(x_n(t-\tau))]^T \in R^n$; $A = (a_{ij})_{n\times n}$ represents the connection weights; $u = [u_1(t), u_2(t), \ldots, u_n(t)]^T$ is a constant external input vector and $\tau > 0$ represents the delay parameter.

Throughout this paper, the activation functions are assumed to satisfy the following conditions:

(H_1) $s_j(\cdot)$ is a bounded function for any $j = 1, 2, \ldots, n$;

(H_2) There exist constant numbers $L_j > 0$ such that $|s_j(\xi_1) - s_j(\xi_2)| \leqslant L_j|\xi_1 - \xi_2|$, $j = 1, 2, \ldots, n$, for all ξ_1, $\xi_2 \in R$. Let $L = \mathrm{diag}(L_j)$, $j = 1, 2, \ldots, n$.

Lemma 1 [10]. Assume the function $s_j(\cdot)$ satisfies the hypotheses (H_1) and (H_2) above, then there exists an equilibrium point for system (1).

According to Lemma 1, system (1) has one equilibrium point x^* at least. The transformation $y(t) = x(t) - x^*$ put (1) into the following form:

$$\dot{y}(t) = -Cy(t) + Af(y(t-\tau)) \,, \tag{2}$$

where $y(t) = [y_1(t), y_2(t), \ldots, y_n(t)]^T$ is the state vector of the transformed system, and $f(y(t)) = [f_1(y_1(t)), f_2(y_2(t)), \ldots, f_n(y_n(t))]^T$, with $f_j(y_j(t)) = s_j(y_j(t) + x_j^*) - s_j(x_j^*)$, $j = 1, 2, \ldots, n$. Since each function $s_j(\cdot)$ satisfies the hypotheses (H_1) and (H_2), each $f_j(\cdot)$ satisfies

$$f_j^2(y_j) \leqslant L_j^2 y_j^2, \quad f_j(0) = 0, \quad \forall\, y_j \in R, \quad j = 1, 2, \ldots, n \,. \tag{3}$$

Lemma 2 [11]. For any vectors a, $b \in R^n$, the inequality

$$\pm 2a^T b \leqslant a^T X a + b^T X^{-1} b$$

holds, in which X is any matrix with $X > 0$.

Lemma 3 [12]. Consider an operator $D(\cdot) : \mathcal{C}_{n,\tau} \to R^n$ with $D(x_t) = x(t) + B\int_{t-\tau}^{t} x(s)ds$, where $x(t) \in R^n$ and $B \in R^{n\times n}$. For a given scalar ρ, where $0 < \rho < 1$, if a positive definite symmetric matrix Γ exists, such that

$$\begin{bmatrix} -\rho\Gamma & \tau B^T \Gamma \\ \tau \Gamma B & -\Gamma \end{bmatrix} < 0$$

holds, then the operator $D(x_t)$ is stable.

Lemma 4 [13]. For any positive definite matrix $M > 0$, scalar $\gamma > 0$, vector function $\omega : [0, \gamma] \to R^n$ such that the integrations concerned are well defined, the following inequality holds:

$$\left(\int_0^\gamma \omega(s)ds\right)^T M \left(\int_0^\gamma \omega(s)ds\right) \leqslant \gamma \left(\int_0^\gamma \omega^T(s)M\omega(s)ds\right) .$$

3 Main Results

Theorem 1. Suppose the hypotheses (H_1) and (H_2) hold. The equilibrium point x^* of (1) is globally asymptotically stable if there exist a positive scalar ε, a positive definite matrix P, positive diagonal matrices Q_1, Q_2, Q_3, and $M = \text{diag}(m_1, m_2, \ldots, m_n)$ such that the following LMIs hold:

$$\Omega = \begin{bmatrix} \Sigma_1 & -MC & 2PC \\ -CM & MAQ_2A^TM + 3Q_2^{-1} - \varepsilon I & 0 \\ 2CP & 0 & \Sigma_2 \end{bmatrix} < 0 , \tag{4}$$

$$\begin{bmatrix} -P & \tau(-C)P \\ \tau P(-C) & -P \end{bmatrix} < 0 , \tag{5}$$

where

$$\Sigma_1 = -2(PC + CP) + 2Q_1^{-1} + PAQ_2A^TP + PCQ_1CP + \tau C^TQ_3C + \varepsilon L^TL ,$$
$$\Sigma_2 = PAQ_2A^TP + PCQ_1CP - \tau^{-1}Q_3 .$$

Proof. Define the operator

$$D(y_t) = y(t) - \int_{t-\tau}^t Cy(s)ds ,$$

where

$$y_t = y(t+s) , \ s \in [-\tau, 0] .$$

From the definition of $D(y_t)$, we have

$$\dot{D}(y_t) = -2Cy(t) + Af(y(t-\tau)) + Cy(t-\tau) . \tag{6}$$

Now, we choose the following Lyapunov functional

$$V(y(t)) = D^TPD + 2\int_{t-\tau}^t y^T(s)Q_1^{-1}y(s)ds + 2\sum_{i=1}^n \int_0^{y_i(t)} m_if_i(s)ds$$
$$+ 3\int_{t-\tau}^t f^T(y(s))Q_2^{-1}f(y(s))ds + \int_{-\tau}^0 \int_{t+s}^t y^T(\eta)C^TQ_3Cy(\eta)d\eta ds .$$

By using Lemma 2, the time derivative of $V(y(t))$ along the trajectories of Eqs. (2) and (6) takes the form

$$\begin{aligned}\dot{V}(y(t)) \leqslant\, & y^T(t)\left[-2(PC+CP)+2Q_1^{-1}+PAQ_2A^TP+PCQ_1CP+\tau C^TQ_3C\right.\\ & \left.+\varepsilon L^TL\right]y(t)+f^T(y(t))\left(MAQ_2A^TM+3Q_2^{-1}-\varepsilon I\right)f(y(t))\\ & +\left(\int_{t-\tau}^{t}Cy(s)ds\right)^T\left(PAQ_2A^TP+PCQ_1CP\right)\left(\int_{t-\tau}^{t}Cy(s)ds\right)\\ & +4\left(\int_{t-\tau}^{t}Cy(s)ds\right)^T PCy(t)-2f^T(y(t))MCy(t)\\ & -\int_{t-\tau}^{t}y^T(s)C^TQ_3Cy(s)ds\,. \end{aligned} \tag{7}$$

In view of Lemma 4, we obtain

$$\int_{t-\tau}^{t}y^T(s)C^TQ_3Cy(s)ds \geqslant \tau^{-1}\left(\int_{t-\tau}^{t}Cy(s)ds\right)^T Q_3\left(\int_{t-\tau}^{t}Cy(s)ds\right)\,. \tag{8}$$

Substituting (8) into (7), we have

$$\dot{V}(y(t)) \leqslant \eta^T(t)\Omega\eta(t),$$

where

$$\eta(t)=\left[y^T(t),f^T(y(t)),\left(\int_{t-\tau}^{t}Cy(s)\,\mathrm{d}s\right)^T\right]^T\,. \tag{9}$$

If (5) holds, we can prove that the positive scalar ρ which is less than one exists such that

$$\begin{bmatrix}-\rho P & \tau(-C)P\\ \tau P(-C) & -P\end{bmatrix}<0\,.$$

According to Lemma 3, the operator $D(y_t)$ is stable. In light of the Theorem 9.8.1 in [14], if LMIs (4) and (5) hold, then the solution of (2) is asymptotically stable. That is to say, the equilibrium point x^* of system (1) is asymptotically stable. This completes the proof of Theorem 1.

The conditions in Theorem 1 possess some adjustable parameters. To make the conditions in Theorem 1 more testable, we choose $Q_1=Q_2=Q_3=I$ and $M=I$, then we have

Corollary 1. Suppose the hypotheses (H_1) and (H_2) hold. The equilibrium point x^* of (1) is globally asymptotically stable if there exist a positive scalar ε, a positive definite matrix P, such that the following LMIs hold:

$$\Omega=\begin{bmatrix}\Sigma_1^* & -C & 2PC\\ -C & AA^T+3I-\varepsilon I & 0\\ 2CP & 0 & \Sigma_2^*\end{bmatrix}<0\,, \tag{10}$$

$$\begin{bmatrix} -P & \tau(-C)P \\ \tau P(-C) & -P \end{bmatrix} < 0 \,, \tag{11}$$

where

$$\Sigma_1^* = -2(PC + CP) + 2I + PAA^T P + PCCP + \tau C^T C + \varepsilon L^T L \,,$$
$$\Sigma_2^* = PAA^T P + PCCP - \tau^{-1} I \,.$$

4 Example

Example 1. Consider a simple two-neuron DHNN, and the network parameters are given as follows

$$C = \begin{pmatrix} 1 & 0 \\ 0 & 0.9 \end{pmatrix} , \quad A = \begin{pmatrix} 1.16 & 1.14 \\ 0.18 & 1.43 \end{pmatrix} , \quad u = \begin{pmatrix} 1.5 \\ 2.5 \end{pmatrix} .$$

The activation function is described by $s_j(x) = (e^x - e^{-x})/(e^x + e^{-x})$, then we let $L = I$, and the initial values $\phi_1(s) = 0.8$, $\phi_2(s) = 0.2$, $s \in [-\tau, 0]$. In Corollary 1 above, let $\tau = 0.5$. By utilizing the MATLAB LMI toolbox, we can get the following solutions which satisfies LMIs (10) and (11):

$$P = \begin{pmatrix} 1.4347 & -0.6734 \\ -0.6734 & 1.7188 \end{pmatrix} > 0 \,, \quad \varepsilon = 3.6189 \,.$$

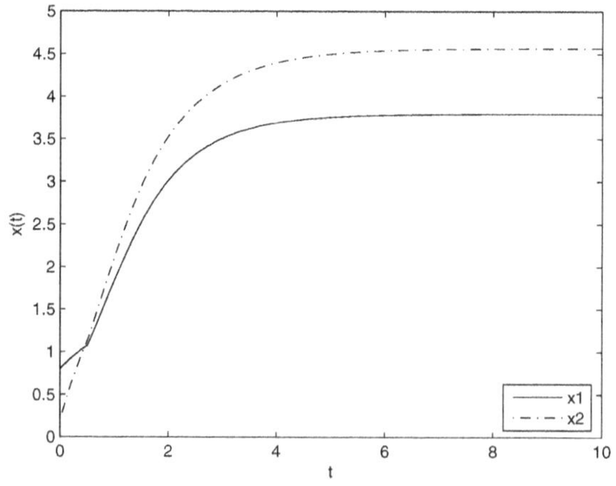

Fig. 1. Numeric simulation for the asymptotic stability of the given system

Therefore, the given system has a unique asymptotically stable equilibrium point $x^* = [3.7983, 4.5653]^T$. The numerical simulation is illustrated in Fig.1.

5 Conclusions

In this paper, new conditions for global asymptotic stability of delayed Hopfield neural networks are obtained based on Lyapunov functional and linear matrix inequality technique. More general results related the case of multiple time-varying delays would be the extension of the present results to more general cases.

Acknowledgments

This work was supported by the National Natural Science Foundation of China (No. 60674026) and the Science Foundation of Southern Yangtze University.

References

1. Hopfield, J.J.: Neural Networks and Physical Systems with Emergent Collective Computational Abilities. In: Proceedings of the National Academy of Sciences, USA **79** (1982) 2554-2558
2. Arik, S.: Global Asymptotic Stability of a Larger Class of Neural Networks with Constant Time Delay. Phys. Lett. A **311** (2003) 504-511
3. Park, J.H.: A Novel Criterion for Global Asymptotic Stability of BAM Neural Networks with Time Delays. Chaos, Solitons & Fractals **29** (2006) 446-453
4. Lou, X.Y., Cui, B.T.: On the Global Robust Asymptotic Stability of BAM Neural Networks with Time-Varying Delays. Neurocomputing **70** (2006) 273-279
5. Lou, X.Y., Cui, B.T.: Absolute Exponential Stability Analysis of Delayed Bi-Directional Associative Memory Neural Networks. Chaos, Solitons & Fractals **31** (2007) 695-701
6. Zhang, Q., Wei, X.P., Xu, J.: Global Exponential Convergence Analysis of Delayed Neural Networks with Time-Varying Delays. Phys. Lett. A **318** (2003) 537-544
7. Lou, X.Y., Cui, B.T.: New LMI Conditions for Delay-Dependent Asymptotic Stability of Delayed Hopfield Neural Networks. Neurocomputing **69** (2006) 2374-2378
8. Liao, X.F., Li, C.D.: An LMI Approach to Asymptotical Stability of Multi-Delayed Neural Networks. Phys. D **200** (1-2) (2005) 139-155
9. Boyd, S., Ghaoui, E.I.L., Feron, E., Balakrishnan, V.: Linear Matrix Inequalities in System and Control Theory. SIAM Studies in Applied Mathematics, SIAM, Philadelphia, PA (1994)
10. Cao, J.D., Zhou, D.M.: Stability Analysis of Delayed Cellular Neural Networks. Neural Networks **11** (1998) 1601-1605
11. Liao, X.F., Chen, G., Sanchez, E.N.: LMI-Based Approach for Asymptotically Stability Analysis of Delayed Neural Networks. IEEE Trans. Circuits Systems-I **49** (2002) 1033-1039
12. Yue, D., Won, S.: Delay-Dependent Robust Stability of Stochastic Systems with Time Delay and Nonlinear Uncertainties. Electron Lett. **37** (2001) 992-993
13. Gu, K.: An Integral Inequality in the Stability Problem of Time-Delay Systems. In: Proc IEEE CDC. Australia (2000) 2805-2810
14. Hale, J., Verduyn Lunel, S.M.: Introduction to Functional Differential Equations. New York: Springer-Verlag (1993)

Periodic Solution of Cohen-Grossberg Neural Networks with Variable Coefficients*

Hongjun Xiang[1,2] and Jinde Cao[1]

[1]Department of Mathematics, Southeast University, Nanjing 210096, China
[2]Department of Mathematics, Xiangnan University, Chenzhou, 423000, China
hunxhjxhj67@126.com, jdcao@seu.edu.cn

Abstract. In this paper, the periodic solution for a class of Cohen-Grossberg neural networks with variable coefficients is discussed. By using inequality analysis technique and matrix theory, some new sufficient conditions are obtained to ensure the existence, uniqueness, global attractivity and exponential stability of the periodic solution. An example is given to show the effectiveness of the obtained results.

1 Introduction

Cohen-Grossberg neural network [1], which was described as follows

$$\frac{dx_i}{dt} = -a_i(x_i)\left[b_i(x_i) - \sum_{j=1}^{m} t_{ij}s_j(x_j) + I_i\right], \quad i = 1, 2, \cdots, m, \tag{1}$$

has aroused a tremendous surge of investigation for decades due to its promising potential for applications in classification, parallel computation, associative memory and optimization. Many scholars researched the existence or uniqueness of equilibrium points, and the qualitative properties of stability for C-G neural networks with delays (see [2-4,6-13]). However, experimental and theoretical studies (see [14-15]) have indicated that a mammalian brain may be exploiting dynamical attractors for its storage associative memories rather than static attractors as it has been supposed in most investigations of artificial neural networks. Recently, based on Lyapunov method, J.Cao et al [16,17,22] studied stability and periodicity of the delayed interval cellular neural network(CNNs). In [19], the authors investigated the exponential convergence and the exponential stability of the periodic solution for a general class of non-autonomous competitive-cooperative neural networks via the decomposition approach. In [5], the authors discussed the existence and the global stability of periodic solution for dynamical systems with periodic interconnections, inputs and self-inhibition.

* This work was jointly supported by the National Natural Science Foundation of China under Grant No. 60574043, the National Science Foundation of Hunan Provincial Education Department (06C792), and the Foundation of Professor Project of Xiangnan University.

D. Liu et al. (Eds.): ISNN 2007, Part I, LNCS 4491, pp. 941–951, 2007.

To the best of our knowledge, there are few results of periodic oscillation for Cohen-Grossberg neural networks with delays. In this paper, Our main aim is to obtain some sufficient conditions for the existence, uniqueness and globally exponential stability of periodic oscillatory solution of the following system:

$$\dot{x_i}(t)=-a_i(x_i(t))\Big[b_i(x_i(t))-\sum_{j=1}^{n}a_{ij}(t)f_j(x_j(t))-\sum_{j=1}^{n}b_{ij}(t)g_j(x_j(t-\tau_{ij}(t)))-I_i(t)\Big], \tag{2}$$

where $i = 1, 2, \cdots, n$; $n \geq 2$ is the number of neurons in the networks, $x_i(t)$ denotes the neuron state vector; $a_i(\cdot)$ is an amplification function, $b_i(\cdot)$ denotes an self-signal function; $A = (a_{ij}(t))_{n\times n}$ and $B = (b_{ij}(t))_{n\times n}$ respectively denote the normal and the delayed connection weight matrix; $f_i, g_i : R \to R$ denote the normal and the delayed activation functions and satisfy $f_i(0) = g_i(0) = 0$; $\tau_{ij}(t) \geq 0$ is the time-varying delay caused during the switching and transmission processes; and $I_i(t)$ is the external input. $\dot{x_i}(t)$, $a_{ij}(t), b_{ij}(t)$, $\tau_{ij}(t)$ and $I_i(t)$ are continuously periodic functions in $t \in [t_0, \infty)$ with a common period $\omega > 0$. Clearly, system (2) includes many previous models as its special cases (see [1-4, 6-13, 18-22]). We do not use topological degree theory, fixed point theorem, and so on. Instead, we will apply a general and very concise approach. By this method, we obtain some sufficient conditions ensuring the existence, uniqueness, global attractivity and globally exponential stability of the periodic solution of system (2).

The rest of this paper is organized in the following way: In Section 2, some notations, definitions and preliminaries are introduced. In Section 3, some novel sufficient conditions are obtained ensuring the global exponential stability of the periodic oscillatory solution. In Section 4, an example is given to illustrate the effectiveness of the obtained results. Finally, we give the conclusion in Section 5.

2 Notations, Definitions and Preliminaries

Let $\tau = \max\{\tau_{ij}(t),\ i, j = 1, 2, \cdots, n,\ t \in [t_0, \infty)\}$, $\mathbb{C} = C([-\tau,\ 0]; R^n)$ denotes the set of all continuous mapping from $[-\tau,\ 0]$ to R^n. For any $\phi \in \mathbb{C}$, we define $\|\phi\| = \max\limits_{1\leq i\leq n}\{\|\phi_i\|\}$, which $\|\phi_i\| = \sup\limits_{-\tau\leq s\leq 0}\{|\phi_i(s)|\}$, then $\mathbb{C}$ be the Banach space of continuous functions which map $[-\tau,\ 0]$ into R^n with the topology of uniform convergence. For matrix $A = (a_{ij}(t))_{n\times n}$, let $\rho(A)$ denote the spectral radius of A. A matrix or a vector $A \geq 0$ means that all the elements of A are greater than or equal to zero, similarly define $A > 0$. For $x(t) = (x_1(t), x_2(t), \cdots, x_n(t))^T$ define $|x(t)| = \sum\limits_{i=1}^{n}|x_i(t)|$, $\|x(t)\| = \max\limits_{1\leq i\leq n}\{\|x_i(t)\|\}$, $[x(t)]^+ = (\|x_1(t)\|, \|x_2(t)\|, \cdots, \|x_n(t)\|)^T$, where $\|x_i(t)\| = \sup\limits_{-\tau\leq s\leq 0}\{|x_i(t+s)|\}$, $i = 1, 2, \cdots, n$.

We give some usual assumptions as follows:

$(A1)$: The functions $a_i(\cdot)(i = 1, 2, \cdots, n)$ are continuously bounded, i.e., there exist positive constants a_i^- and a_i^+ such that

$$0 < a_i^- \leq a_i(x) \leq a_i^+, \ \forall x \in R, \ i = 1, 2, \cdots, n.$$

$(A2)$: Each $b_i(\cdot)$ is continuous and there exists a constant λ_i such that

$$\frac{b_i(x) - b_i(y)}{x - y} \geq \lambda_i > 0, \ \forall x, y \in R, \ x \neq y, \ i = 1, 2, \cdots, n.$$

$(A3)$: For activation function $f_i(x)$ and $g_i(y)$, there exist $\delta_i > 0$ and $\sigma_i > 0$ such that

$$\delta_i = \sup_{x \neq y} \Big| \frac{f_i(x) - f_i(y)}{x - y} \Big|, \ \sigma_i = \sup_{x \neq y} \Big| \frac{g_i(x) - g_i(y)}{x - y} \Big|, \ i = 1, 2, \cdots, n.$$

In addition, we shall use the following notations:

$$a_{ij} = \sup_{t \geq t_0} |a_{ij}(t)|, \ b_{ij} = \sup_{t \geq t_0} |b_{ij}(t)|, \ I_i = \sup_{t \geq t_0} |I_i(t)|.$$

Definition 1. Let $R^+ = [0, \infty)$. Suppose that $\mathbb{C}$ is a Banach space and that $u : R \times \mathbb{C} \times R^+ \to \mathbb{C}$ is a given mapping. Define $U(\xi, t) : \mathbb{C} \to \mathbb{C}, \ \forall \xi \in R, t \in R^+$ by $U(\xi, t)x = u(\xi, x, t)$. A process on $\mathbb{C}$ is a mapping $u : R \times \mathbb{C} \times R^+ \to \mathbb{C}$ satisfying the following properties: (i) u is continuous; (ii) $U(\xi, 0) = E$ is the identity; (iii) $U(\xi + s, t)U(\xi, s) = U(\xi, s + t)$.

A process u is said to be an $\omega-$periodic process if there is an $\omega > 0$ such that $U(\xi + \omega, t) = U(\xi, t), \forall \xi \in R$ and $t \in R^+$.

Definition 2. A continuous map $T : \mathbb{C} \to \mathbb{C}$ is said to be point dissipative if there exists a bounded set $\mathbb{C}_0 \subset \mathbb{C}$ such that $\mathbb{C}_0$ attracts each point of $\mathbb{C}$.

Lemma 1.(Hale and Verduyn Lunel[23]) If an $\omega-$periodic retardant functional differential equation f (RFDE(f)) generates an $\omega-$periodic process u on $\mathbb{C}$, $U(\xi, t)$ is a bounded map for each ξ, t and is point dissipative, then there exists a compact, connected, global attractor. Also, there is an $\omega-$periodic solution of the RFDE(f).

Remark 1. For a given $s \in [0, \infty)$ and a continuous function $x : [-\tau, \infty) \to R^n$, we define $x_s : [-\tau, 0] \to R^n$ by $x_s(\theta) = x(s + \theta)$ for $\theta \in [-\tau, 0]$. One can solve system (2) by method of steps to obtain a unique mapping $x : [t_0 - \tau, \infty) \to R^n$. Suppose that $f : R \times \mathbb{C} \to R^n$ is completely continuous and let $x(t_0, \varphi)$ denote a solution of the RFDE(f) : $\dot{x}(t) = f(t, x_t)$ through (t_0, φ) and assume x is uniquely defined for $t \geq t_0 - \tau$. If $u(t_0, \varphi, t) = x_{t_0 + t}(t_0, \varphi)$ for $(t_0, \varphi, t) \in R \times \mathbb{C} \times R^+$, then u is a process on $\mathbb{C}$. If there is an $\omega > 0$ such that $f(t_0 + \omega, \varphi) = f(t_0, \varphi)$ for all $(t_0, \varphi) \in R \times \mathbb{C}$, then the process generated by the RFDE(f) is an $\omega-$periodic process.

Remark 2. For a RFDE(f), Lemma 1 provides the existence of a periodic solution under the weak assumption of point dissipative. Under some fit assumptions, we can view system (2) as a dissipative system and apply Lemma 1 to the system. It is clearly to show that system (2) can generate an $\omega-$periodic process u on $\mathbb{C}$.

Lemma 2.(La Salle[24]) If $\rho(A) < 1$ for $A \geq 0$, then $(E - A)^{-1} \geq 0$, where E denotes the identity matrix of size n.

Lemma 3.([25]) A denotes a matrix of size n and α is a vector. If $\alpha \leq A\alpha$, we have $\rho(A) \geq 1$.

Definition 3. For model (2), an $\omega-$periodic solution $x^*(t) = (x_1^*(t), x_2^*(t), \cdots, x_n^*(t))^T$ is said to be globally exponentially stable if there exist constants $\lambda > 0$ and $M > 0$ such that

$$|x(t) - x^*(t)| \leq M\|\phi - x^*\|e^{-\lambda(t-t_0)},\ t \geq t_0,$$

for the solution $x(t)$ of system (2) with any initial value $\phi \in \mathbb{C}$. Moreover, λ is called to be globally exponentially convergent rate.

3 Main Results

Theorem 1. Suppose that the system (2) satisfy $(A1) - (A3)$ and the following condition:

$(A4)$: if $\rho(P) < 1$, where $P = (p_{ij})_{n\times n}$ and $p_{ij} = (a_{ij}\delta_j + b_{ij}\sigma_j)\frac{a_i^+}{\lambda_i a_i^-}$;
then the set $\Omega = \{\phi \in \mathbb{C};\ [\phi]^+ \leq Q = (E - P)^{-1}M\}$ is a positively invariant set of system (2), where vector $M = (M_1, M_2, \cdots, M_n)^T$, and $M_i = \frac{a_i^+ I_i}{\lambda_i a_i^-}$, E denotes an identity matrix of size n. Moreover, the set Ω is a global attractivity set of system (2).

Proof. By system (2) and $(A1) - (A3)$, we have

$$\begin{aligned}|x_i(t)| \leq \exp[-\lambda_i a_i^-(t-t_0)]|x_i(t_0)| + a_i^+ \textstyle\int_{t_0}^t \exp[-\lambda_i a_i^-(t-s)]\times \\ \{\sum_{j=1}^n [|a_{ij}(s)|\delta_j + |b_{ij}(s)|\sigma_j]\|x_i(s)\| + |I_i(s)|\}ds.\end{aligned} \tag{3}$$

In view of $\rho(P) < 1$ and Lemma 2, one have $(E - P)^{-1} \geq 0$ and set $Q = (E - P)^{-1}M$. We will prove that: if $[\phi]^+ \leq Q$, then

$$[x(t)]^+ \leq Q,\ t \geq t_0. \tag{4}$$

It suffices to show that $\forall \eta > 1$, if $[\phi]^+ < \eta Q$, then

$$[x(t)]^+ < \eta Q,\ t \geq t_0. \tag{5}$$

If not, then there exists a $i \in \{1, 2, \cdots, n\}$ and $t_1 \geq t_0$ such that

$$|x_i(t_1)| = \eta Q_i, \ |x_i(t)| \leq \eta Q_i, \ t \in [t_0 - \tau, t_1) \tag{6}$$

and

$$[x(t)]^+ \leq \eta Q, \ t \in [t_0, t_1]. \tag{7}$$

Where Q_i is the ith component of vector Q. Noticing that $Q = (E - P)^{-1}M$, i.e. $PQ + M = Q$ or $\sum_{j=1}^{n} p_{ij}Q_j + M_i = Q_i$, then by (6) and (3), we have

$$\begin{aligned}\eta Q_i = |x_i(t_1)| \leq \exp[-\lambda_i a_i^-(t_1 - t_0)]\Big[\eta Q_i - (\sum_{j=1}^{n} p_{ij}\eta Q_j + M_i)\Big] + \\ \Big(\sum_{j=1}^{n} p_{ij}\eta Q_j + M_i\Big).\end{aligned} \tag{8}$$

One imply that $\eta Q_i < \eta Q_i$ from (8), which is a contradiction. Therefore, (5) holds. Let $\eta \to 1$, then one can prove that (4) holds. The proof of the first part is completed.

In the following, we will discuss the global attractivity of the invariant set Ω.

From the above arguments, for any given initial value $\phi \in \mathbb{C}$, there exists some constant $\eta > 1$ such that if $[\phi]^+ < \eta Q$, then

$$[x(t)]^+ < \eta Q, \ t \geq t_0. \tag{9}$$

If not, there exists a nonnegative vector $\alpha = (\alpha_1, \alpha_2, \cdots, \alpha_n)^T$ such that

$$\lim_{t\to\infty} \sup[x(t)]^+ = \alpha + Q. \tag{10}$$

By the definition of upper limit of $x(t)$ and (9), for any sufficiently small value $\varepsilon > 0$, there exists a $t_2 \geq t_0$ such that

$$|x_i(t - \tau_{ij}(t))| < Q_i + (1 + \varepsilon)\alpha_i, \ t \geq t_2, \ i, j = 1, 2, \cdots, n. \tag{11}$$

According to the continuity of function $\exp(\cdot)$, set $T > -\frac{\ln \varepsilon}{\lambda_i a_i^-} > 0$, then

$$\exp[-\int_{t-T}^{t} \lambda_i a_i^- ds] < \varepsilon, \ or \ \exp(-T\lambda_i a_i^-) < \varepsilon, \tag{12}$$

for all $t > t_2 + T$. Together with (2) and (10)-(12), we have

$$|x_i(t)| \leq \exp[-\lambda_i a_i^-(t - t_0)]\|\phi_i\| + \sum_{j=1}^{n} p_{ij}Q_j + \sum_{j=1}^{n} p_{ij}(1 + \varepsilon)\alpha_j + M_i.$$

Therefore, by (10) and the properties of upper limit of $x(t)$, we can imply that there exists a sequence t_k such that $t_k \geq t_2 + T$ and

$$\lim_{t_k\to\infty} |x_i(t_k)| = \alpha_i + Q_i.$$

Then, we get

$$\alpha_i + Q_i \leq \sum_{j=1}^{n} p_{ij} Q_j + \sum_{j=1}^{n} p_{ij}\alpha_j + M_i,\ t_k \to \infty,\ \varepsilon \to 0,$$

which concludes that $\alpha_i \leq \sum_{j=1}^{n} p_{ij}\alpha_j,\ i = 1, 2, \cdots, n$. i.e. $\alpha \leq P\alpha$. By Lemma 3, we have $\rho(P) \geq 1$, which is a contradiction. Thus, $\alpha \equiv 0$, This completes the proof of this Theorem.

Theorem 2. Assume that the assumptions of Theorem 1 hold, then there exists a global periodic attractor of system (2). Moreover, it is $\omega-$periodic and belongs to the positively invariant set Ω.

Proof. It is clearly to see that system (2) can generate an $\omega-$periodic process by Lemma 1 and Theorem 1. Moreover, together with the assumptions of Theorem 1, we can draw that there exists $\omega-$periodic solution denoted by $x^*(t)$.

Let $x(t)$ be an arbitrary solution of model (2) and use the transformation $z(t) = x(t) - x^*(t)$, the neural network model (2) can be rewritten as:

$$\dot{z}_i(t) = -\gamma_i(z_i(t))\left[\beta_i(z_i(t)) - \sum_{j=1}^{n} a_{ij}(t)F_j(z_j(t)) - \sum_{j=1}^{n} b_{ij}(t)G_j(z_j(t-\tau_{ij}(t)))\right], \tag{13}$$

where $\gamma_i(z_i(t)) = a_i(z_i(t) + x_i^*(t))$, $\beta_i(z_i(t)) = b_i(z_i(t) + x_i^*(t)) - b_i(x_i^*(t))$, $F_j(z_j(t)) = f_j(z_j(t) + x_j^*(t)) - f_j(x_j^*(t))$, $G_j(z_j(t-\tau_{ij}(t))) = g_j(z_j(t-\tau_{ij}(t)) + x_j^*(t)) - g_j(x_j^*(t))$, $i = 1, 2, \cdots, n$.

One can easily see that system (13) satisfies all conditions of Theorem 1 with $M = \{0\}$. Thus, $\lim_{t\to\infty} z_i = 0,\ i = 1, 2, \cdots, n$ which derives that the periodic solution $x^*(t)$ is globally attracting. So, this is completed the proof of this theorem.

Theorem 3. Under the assumptions of Theorem 1, the $\omega-$periodic solution $x^*(t)$ of system (2) is globally exponentially stable if the following condition holds:

$$\lambda_i a_i^- > a_i^+ \sum_{j=1}^{n} \left[a_{ij}\delta_j + b_{ij}\sigma_j\right],\ i = 1, 2, \cdots, n. \tag{14}$$

Proof. From Theorem 2, we only prove that $(0, 0, \cdots, 0)^T$ of the model (13) is globally exponentially stable. Considering the function $h_i(\mu)$ given by

$$h_i(\mu) = \lambda_i a_i^- - \mu - a_i^+ \sum_{j=1}^{n} \left[a_{ij}\delta_j + b_{ij}\sigma_j e^{\mu\tau}\right],\ i = 1, 2, \cdots, n.$$

Associating with (14),we note that

$$h_i(0) = \lambda_i a_i^- - a_i^+ \sum_{j=1}^{n} \left[a_{ij}\delta_j + b_{ij}\sigma_j\right] > 0$$

and $h_i(\mu)$ is continuous, $h_i(\mu) \to -\infty$ as $\mu \to +\infty$. Thus, there exists a $\mu_i > 0$ such that $h_i(\mu_i) = 0$. Without loss of generality, set $\mu_i = min\{\mu > 0 | h_i(\mu) = 0\}$, so $h_i(\mu) > 0$ when $\mu \in (0,\ \mu_i)$. Now let $\bar{\mu} = min\{\mu_i,\ i = 1, 2, \cdots, n\}$, when $\mu_0 \in [0, \bar{\mu})$, we have $h_i(\mu_0) > 0,\ i = 1, 2, \cdots, n$. i.e.

$$\lambda_i a_i^- - \mu_0 - a_i^+ \sum_{j=1}^{n} \Big[a_{ij}\delta_j + b_{ij}\sigma_j e^{\tau\mu_0} \Big] > 0,\ i = 1, 2, \cdots, n. \tag{15}$$

Let $z(t) = (z_1(t), z_2(t), \cdots, z_n(t))^T$ be a solution of system (13) with any initial value $\phi \in \mathbb{C}$, then we have

$$\begin{aligned} |z_i(t)| \le \exp\Big[-\lambda_i a_i^-(t - t_0) \Big] |z_i(t_0)| + a_i^+ \textstyle\int_{t_0}^{t} \exp\Big[-\lambda_i a_i^-(t - s) \Big] \times \\ \Big\{ \textstyle\sum_{j=1}^{n} \big[a_{ij}\delta_j |z_j(s)| + b_{ij}\sigma_j |z_j(s - \tau_{ij}(s))| \big] \Big\} ds, \end{aligned} \tag{16}$$

for $t \ge t_0$ and $i = 1, 2, \cdots, n$. Let

$$y_i(t) = \begin{cases} e^{\mu_0(t - t_0)} |z_i(t)| & t > t_0, \\ |z_i(t)| & t_0 - \tau \le t \le t_0. \end{cases} \tag{17}$$

Then from (16) and (17), we can obtain

$$\begin{aligned} y_i(t) \le \exp\big[-(\lambda_i a_i^- - \mu_0)(t - t_0) \big] y_i(t_0) + a_i^+ \textstyle\int_{t_0}^{t} \exp[-(\lambda_i a_i^- - \mu_0)(t - s)] \times \\ \Big\{ \textstyle\sum_{j=1}^{n} \big[a_{ij}\delta_j + b_{ij}\sigma_j e^{\tau\mu_0} \big] \|y_j(s)\| \Big\} ds, \end{aligned} \tag{18}$$

for $t \ge t_0,\ i = 1, 2, \cdots, n$.

For the initial value $\phi \in \mathbb{C}$, there must exist $L > 0$ and $r \in \{1, 2, \cdots, n\}$ such that $\|\phi_r(0)\| = L$ and $\|\phi_i(0)\| \le L$ for $i = 1, 2, \cdots, n$.

We will show that for any sufficiently small constant $\varepsilon > 0$,

$$y_i(t) < L + \varepsilon,\ t \ge t_0,\ i = 1, 2, \cdots, n. \tag{19}$$

Using the method of contrary, then there must exist some $t_1 \ge t_0$ and $k \in \{1, 2, \cdots, n\}$ such that

$$y_k(t_1) = L + \varepsilon,\ y_i(t) \le L + \varepsilon,\ t \in [t_0,\ t_1),\ i = 1, 2, \cdots, n. \tag{20}$$

Therefore, by (18) and (15), we have

$$\begin{aligned} L + \varepsilon = y_k(t_1) \le (L + \varepsilon) \exp\Big[-(\lambda_i a_i^- - \mu_0)(t_1 - t_0) \Big] + (L + \varepsilon) \frac{a_i^+}{\lambda_i a_i^- - \mu_0} \times \\ \sum_{j=1}^{n} \Big[a_{ij}\delta_j + b_{ij}\sigma_j e^{\tau\mu_0} \Big] \Big[1 - \exp\big[-(\lambda_i a_i^- - \mu_0)(t_1 - t_0) \big] \Big] < L + \varepsilon, \end{aligned}$$

which is a contradiction. Thus (19) holds. Let $\varepsilon \to 0$, we have

$$y_i(t) \le L,\ t \ge t_0,\ i = 1, 2, \cdots, n.$$

This implies that there exists $\Gamma > 1$ such that

$$|y(t)| \leq \Gamma \|\phi(0)\|, \; t \geq t_0.$$

From (17), there exist some constant $\zeta \geq 1$ such that

$$|z(t)| \leq \zeta \|\phi(0)\| e^{\mu_0 (t_0 - t)}, \; t \geq t_0,$$

which implies the solution $(0, 0, \cdots, 0)^T$ of the model (13) is globally exponentially stable. The proof of Theorem 3 is completed.

Clearly, the systems of references [16,17,19,22] are just a special case of our model (2). Therefore, our results can be applied to a broad range of neural networks. When $a_i(x_i(t)) \equiv 1, \; b_i(x_i(t)) = c_i(t)x_i(t), \; i = 1, 2, \cdots, n$, system (2) reduces to the following system:

$$\dot{x}_i(t) = -c_i(t)x_i(t) + \sum_{j=1}^{n} a_{ij}(t) f_j(x_j(t)) + \sum_{j=1}^{n} b_{ij}(t) f_j(x_j(t - \tau_{ij}(t))) + I_i(t), \quad (21)$$

where $c_i(t) > 0, \forall t \geq t_0, \; i = 1, 2, \cdots, n$. In [22], system (21) was considered and some sufficient conditions were derived guaranteeing the existence and exponential stability of the periodic solution. By Theorem 1 and Theorem 2, one can easily derive the following conclusions:

Corollary 1. Assume that $(A3)$ hold. Moreover,

$(H1) : c_i(t) \geq c_i > 0, \; \forall t \in R, \; i = 1, 2, \cdots, n.$

$(H2)$: if $\rho(P) < 1$, where $P = (p_{ij})_{n \times n}$ and $p_{ij} = \frac{a_{ij}\delta_j + b_{ij}\sigma_j}{c_i}$; set $M = (M_1, M_2, \cdots, M_n)^T$, and $M_i = \frac{I_i}{c_i}$;

then the set $\Omega = \{\phi \in C; \; [\phi]^+ \leq Q = (E - P)^{-1} M\}$ is a positively invariant set of system (21), which is a globally attracting set. Moreover, it is $\omega-$periodic and belongs to the positively invariant set Ω.

Corollary 2. Under the assumptions of Corollary 1, the $\omega-$periodic solution $x^*(t)$ of system (21) is globally exponentially stable if the following condition holds:

$$\sum_{j=1}^{n} \left[a_{ij}\delta_j + b_{ij}\sigma_j \right] < c_i, \; i = 1, 2, \cdots, n.$$

4 Simulation Example

Example. Consider the following Cohen-Grossberg neural networks with variable coefficients:

$$\frac{dx(t)}{dt} = -\begin{pmatrix} 3-cos(x_1(t)) & 0 \\ 0 & 2-sin(x_2(t)) \end{pmatrix} \Bigg[\begin{pmatrix} 8 & 0 \\ 0 & 8 \end{pmatrix} \begin{pmatrix} x_1(t) \\ x_2(t) \end{pmatrix} - \begin{pmatrix} sin(t) & -cos(t) \\ 0 & \frac{1}{2}cos(t) \end{pmatrix} \begin{pmatrix} tanh(x_1(t)) \\ tanh(x_2(t)) \end{pmatrix} - \begin{pmatrix} cos(t) & -\frac{1}{2}sin(t) \\ \frac{1}{2}cos(t) & sin(t) \end{pmatrix} \times \begin{pmatrix} tanh(x_1(t-5)) \\ tanh((x_2(t-3)) \end{pmatrix} - \begin{pmatrix} 2sin(t) \\ cos(t) \end{pmatrix} \Bigg]. \quad (22)$$

By simple computation, one can get $\rho(P) = 0.7963 < 1$. By Theorem 1 and Theorem 2, we know that the set $\Omega = \{\phi \in \mathbb{C};\ [\phi]^+ \leq Q = (E-P)^{-1}M = (1.6842,\ 1.5789)^T\}$ is a positively invariant set of system (22), which is a global attracting set. In addition, one can easily obtain

$$a_1^+ \sum_{j=1}^{2} \left[a_{1j}\delta_j + b_{1j}\sigma_j\right] = 14 < \lambda_1 a_1^- = 16,\ a_2^+ \sum_{j=1}^{2} \left[a_{2j}\delta_j + b_{2j}\sigma_j\right] = 6 < \lambda_2 a_2^- = 8.$$

According to Theorem 3, the $2\pi-$periodic solution $x^*(t)$ of system (22) is globally exponentially stable.

For numerical simulation, the following five cases are given: case 1 with the initial state $[\phi_1, \phi_2]^T = [-1, 2]^T$ for $t \in [-5, 0]$; case 2 with the initial state $[\phi_1, \phi_2]^T = [1, 1]^T$ for $t \in [-5, 0]$; case 3 with the initial state $[\phi_1, \phi_2]^T = [-0.5, -0.5]^T$ for $t \in [-5, 0]$; case 4 with the initial state $[\phi_1, \phi_2]^T = [0.6, -0.4]^T$ for $t \in [-5, 0]$; case 5 with the initial state $[\phi_1, \phi_2]^T = [-0.3, 0.3]^T$ for $t \in [-5, 0]$. Fig.1 depicts the time responses of state variables of $x_1(t)$ and $x_2(t)$ with step $h = 0.01$, and Fig. 2 depicts the phase responses of system (22). It confirms that the proposed conditions in our results are effective for model (22).

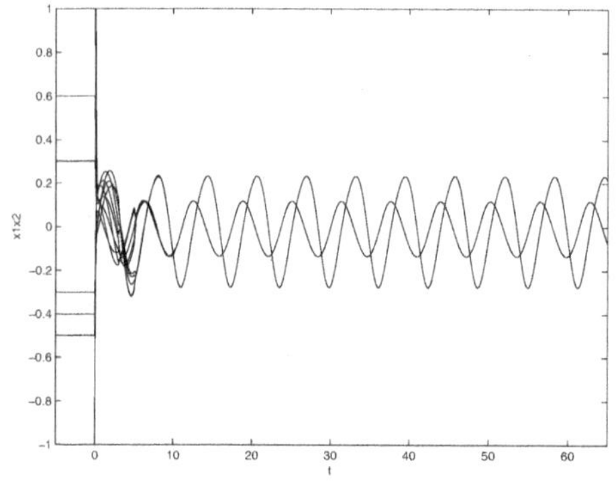

Fig. 1. Transient response of state $x_1(t)$, $x_2(t)$

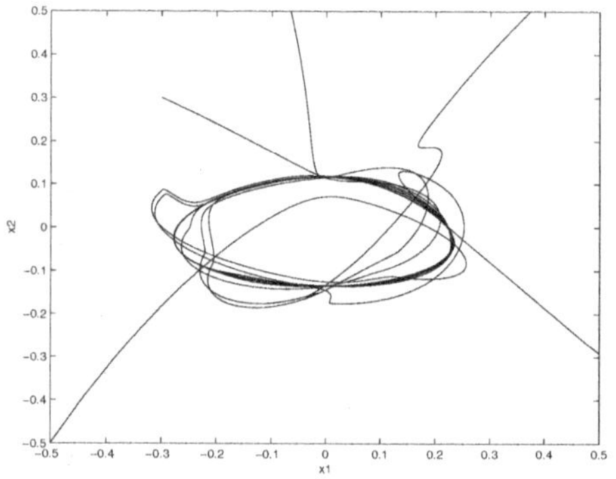

Fig. 2. Phase response of system (22)

5 Conclusion

In this paper, some novel sufficient conditions are derived ensuring the existence, uniqueness, global atractivity and globally exponential stability of the periodic solution for Cohen-Grossberg neural networks with variable coefficients. The results are not only less restrictive conditions than previously known criteria but also extend and improve some results in existent literatures. In addition, the method in this paper is general and very concise.

References

1. Cohen, M.A., Grossberg, S.: Absolute Stability and Global Pattern Formation and Partial Memory Storage by Compeitive Neural Networks. IEEE Trans. Systems, Man and Cybernetics **13** (1983) 815-826.
2. Cao, J., Li, X.: Stability in Delayed Cohen-Grossberg Neural Networks: LMI Optimization Approach. Physica D **212** (2005) 54-65
3. Xiong, W., Cao, J.: Exponential Stability of Discrete-time Cohen-Grossberg Neural Networks. Neurocomputing **64** (2005) 433-446
4. Cao, J., Liang, J.: Boundedness and Stability for Cohen-Grossberg Neural Network with Time-varying Delays. J. Math. Anal. Appl. **296** (2004) 665-685
5. Lu, W., Chen, T.P.: On Periodic Dynamical Systems. Chinese Annals of Mathematics Series B **25**(4) (2004) 455-462
6. Xiong, W., Cao, J.: Absolutely Exponential Stability of Cohen-Grossberg Neural Networks with Unbounded Delays. Neurocomputing **68** (2005) 1-12
7. Liu, Z.: Global Stability Analysis of Cohen-Grossberg Neural Networks Involving Multiple Delays. Journal of Xiangnan University **26** (2005) 1-8
8. Chen, T., Rong, L.: Delay-independent Stability Analysis of Cohen-Grossberg Neural Networks. Phys. Lett. A **317** (2003) 436-449
9. Wang, L., Zou, X.F.: Exponential Stability of Cohen-Grossberg Neural Networks. Neural Networks **15** (2002) 415-422
10. Wang, L., Zou, X.F.: Harmless Delays in Cohen-Grossberg Neural Networks. Physica D **170**(2) (2002) 162-173
11. Ye, H., Michel, A.N., Wang, K.: Qualitative Analysis of Cohen-Grossberg Neural Networks with Multiple Delays. Phys. Rev. E **51** (1995) 2611-2618
12. Lu, W.L., Chen, T.P.: New Conditions on Global Stability of Cohen-Grossberg Neural Networks. Neural Computation **15**(5) (2003) 1173-1189
13. Liao, X.F., Li, C.G., Wong, K.W.: Criteria for Exponential Stability of Cohen-Grossberg Neural Networks. Neural Networks **17** (2004) 1401-1414
14. Freeman, W.J.: The Physiology of Perception. Sci. Am. (1991)
15. Skarda, C.A., Freman, W.J.: How Brains Make Chaos in Order to Make Sense of the World. Brain Behav. Sci. **10** (1987) 161-195
16. Cao, J., Chen, T.: Global Exponentially Robust Stability and Periodicity of Delayed Neural Networks. Chaos, Solitons & Fractals **22** (2004) 957-963
17. Huang, H., Ho, D.W.C., Cao, J.: Analysis of Global Exponential Stability and Periodic Solutions of Neural Networks with Time-varying Delays. Neural Networks **18** (2005) 161-170
18. Cao, J., Li, Q., Wan, S.: Periodic Solutions of The Higher-dimensional Non-autonomous Systems. Applied Mathematics and Computation **130** (2002) 369-382
19. Yuan, K., Cao, J.: Periodic Oscillatory Solutions in Delayed Competitive-cooperative Neural Networks : A Decomposition Approach. Chaos, Solitons & Fractals **27** (2006) 223-231
20. Cao, J., Jiang, Q.: An Analysis of Periodic Solutions of Bi-directional Associative Memory Networks with Time-varying Delays. Phys. Lett. A **330** (2004) 203-213

21. Cao, J., Wang, L.: Periodic Oscillotory Solution of Bi-directional Associative Memory Networks with Delays. Phys. Rev. E **61** (2000) 1825-1828
22. Guo, S., Huang, L.: Periodic Oscillotory for a Class of Neural Networks with Variable Coefficients. Nonlinear Anal. **6** (2005) 545-561
23. Hale, J., Verduyn Lunel, S.M.: Introduction to Functional Differential Equations. Springer, New York (1993)
24. Lasalle, J.P.: The Stability of Dynamical System. SIAM, Philadelphia, PA (1976)
25. Horn, R.A., Johnson, C.R.: Matrix Analysis. Cambridge University Press, London (1990)

Existence and Stability of Periodic Solution of Non-autonomous Neural Networks with Delay

Minghui Jiang[1], Xiaohong Wang[1], and Yi Shen[2]

[1] Institute of Nonlinear Complex Systems, China Three Gorges University, Yichang, Hubei, 443000, China
jmh1239@hotmail.com
[2] Department of Control Science and Engineering, Huazhong University of Science and Technology, Wuhan, Hubei, 430074, China
lhfu@hust.edu.cn

Abstract. The paper investigates the existence and global stability of periodic solution of non-autonomous neural networks with delay. Then the existence and uniqueness of periodic solutions of the neural networks are discussed in the paper. Moreover, criterion on stability of periodic solutions of the neural networks is obtained by using matrix function inequality, and algorithm for the criterion on the neural networks is provided. Result in the paper generalizes and improves the result in the existing references. In the end, an illustrate example is given to verify our results.

1 Introduction

In recent years, recurrent neural networks are widely investigated in [1-3], because of their immense potentials of application perspective. The Hopfield neural network is typical representative recurrent neural networks among others, and has been successfully applied to signal processing, especially in image processing, and to solve nonlinear algebraic and transcendental equations ([4,5]). Therefore, the stability analysis of Hopfield neural networks is important from both theoretical and applied points of view([1-11]). To our knowledge, few authors have considered global stability of periodic oscillatory solutions for the non-autonomous neural networks with delays[12-14]. Therefore, the stability analysis of non-autonomous neural networks with delays is important from both theoretical and applied points of view. It is well known that the research of neural networks with delays involves not only the stability analysis of equilibrium points but also that of periodic solution[13-14]. In particular, global stability of periodic solution of non-autonomous neural networks with delays is important since the global stability of equilibrium points can be considered as a special case of periodic solution with zero period[7-8]. Hence, the stability analysis of periodic solutions is more general than that of equilibrium points. In this paper, based on the matrix function inequality, sufficient condition on the global exponential stability of non-autonomous neural networks with delays is proposed. Furthermore, stability of non-autonomous neural networks with delays can be viewed

D. Liu et al. (Eds.): ISNN 2007, Part I, LNCS 4491, pp. 952–957, 2007.

as robust stability of autonomous neural networks in a manner. In the end, an example is given to demonstrate the feasibility of our main result.

2 Preliminaries

The model on non-autonomous neural networks with delays to be considered here is described by the following differential equation

$$\dot{x}_i(t) = -d_i(t)x_i(t) + \sum_{j=1}^{n} a_{ij}(t)f_j(x_j(t)) + \sum_{j=1}^{n} b_{ij}(t)f_j(x_j(t-\tau(t))) + I_i(t), \quad i = 1, 2, \ldots, n, \tag{1}$$

where $\dot{x}_i(t)$ denote the derivative of $x_i(t)$. We call the variables $x_i(t), i = 1, \cdots, n$ the state variables. $d_i(t) > 0$ denote the passive decay rates; $I_i(t)$ are external inputs; $a_{ij}(t)$ and $b_{ij}(t)$ are connection weights of the network; the delays $\tau(t)$ responding to the finite speed of axonal signal transmission, which is nonnegative. We assume that $d_i(t), a_{ij}(t), b_{ij}(t), I_i(t)$ are continuous periodic function with period T. In addition, the following conditions are satisfied.

(**H1**) There exist positive constants $M_j, j = 1, \cdots, n$ such that for any $\theta, \vartheta \in R$

$$0 \le \frac{f_j(\theta) - f_j(\vartheta)}{\theta - \vartheta} \le M_j, \ \theta \neq \vartheta,$$

(**H2**) There are constants d_i^M, d_i^m, a_{ij}^M, a_{ij}^m, b_{ji}^M, b_{ji}^m, τ such that continuous functions $d_i(t)$, $a_{ij}(t), b_{ij}(t), \tau(t)$ satisfy

$$d_i^M \ge d_i(t) \ge d_i^m > 0, \ 0 \le \tau(t) \le \tau, \ \text{and}\, \tau'(t) \le 0;$$

$$a_{ij}^M \ge a_{ij}(t) \ge a_{ij}^m \ \text{and} \ b_{ji}^M \ge b_{ji}(t) \ge b_{ji}^m.$$

(**H3**) Continuous functions $f_j(\cdot)$ are bounded on R, in other other, there are constants f_j^m, f_j^M such that the following inequalities hold

$$f_j^m \le f_j(\cdot) \le f_j^M.$$

The initial conditions associated with non-autonomous neural networks (1) are of the form

$$x_i(s) = \phi_i(s), s \in [-\tau, 0], \, i = 1, \cdots, n.$$

where $\phi_i(s)$ are continuous T-periodic functions.

Define $x(t) = [x_1(t), \cdots, x_n(t)]^T \in R^n$. For any solution $x(t)$ with initial conditions $x_i(s) = \phi_i(s), s \in [-\tau, 0]$ and periodic solution $x^*(t)$, define $\|(\phi(t))^T - (x^*(t))^T\| = \sum_{i=1}^{n} \max_{-\tau \le t \le 0} |\phi_i(t) - x_i^*(t)|$.

For simplicity, the neural networks (1) can be rewritten as the following vector form:

$$\dot{x}(t) = -D(t)x(t) + A(t)f(x(t)) + B(t)f(x(t-\tau(t))) + I(t), \tag{2}$$

where $D(t) = \text{diag}(d_1(t), \cdots, d_n(t))$, $A(t) = (a_{ij}(t))_{n\times n}$, $B(t) = (b_{ji}(t))_{n\times n}$, $f(y) = (f_1(y_1), \cdots, f_n(y_n))^T$, $I(t) = (I_1(t), \cdots, I_n(t))$.

3 Existence and Stability

In this section, in order to investigate the stability of periodic solution of the neural network (1), we firstly discuss the existence, uniqueness and boundedness of solutions of the neural networks (1), we have the following result.

Theorem 1. *Assume that* (**H1**)*-*(**H3**) *hold, then there is an unique solution through* (t_0, ϕ), *and all solutions of the neural networks (1) are bounded for all* $t \geq t_0$.
The proof is easy and omitted here.

Now we shall consider the following existence of the periodic solution of the neural networks (1).

Theorem 2. *The neural networks (1) has at least one* $T-$*periodic solution if the neural networks (1) satisfies Assumption* (**H1**)*-*(**H3**).
The proof is similar to that of Theorem 1 in [8] and omitted here.

To show the uniqueness of the periodic solution, we use the transformation $u_i(t) = x_i(t) - x_i^*(t)$, $i = 1, \cdots, n$ to shift the periodic solution $(x_1^*(t), \cdots, x_n^*(t))$ of the neural networks (1) to the origin, and get the following form of the neural networks (1):

$$\dot{u}_i(t) = -d_i(t)u_i(t) + \sum_{j=1}^{n} a_{ij}(t)s_j(u_j(t - \tau(t))) + \sum_{j=1}^{n} b_{ij}(t)s_j(u_j(t - \tau(t))), \quad i = 1, 2, \ldots, n. \tag{3}$$

where $s_j(u_j(t)) = f_j(u_j(t) + x_j^*(t)) - f_j(x_j^*(t)), j = 1, 2, \ldots, n.$

It is obvious that the functions $s_j(\cdot)$ satisfy the same assumptions (**H1**) and (**H3**). The neural networks (11) can be rewritten as the following vector form:

$$\dot{u}(t) = -D(t)u(t) + A(t)S(u(t)) + B(t)S(u(t - \tau(t))). \tag{4}$$

where $u(t) = (u_1(t), \cdots, u_n(t))^T$, $D(t) = \text{diag}(d_1(t), \cdots, d_n(t))$, $A(t) = (b_{ij}(t))_{n \times n}$, $B(t) = (b_{ji}(t))_{n \times n}, S(z) = (s_1(u_1), \cdots, s_n(u_n))^T$.

We next prove that the $T-$periodic solution $(x_1^*(t), \cdots, x_n^*(t))$ of the neural networks (1) satisfying some conditions is unique and stable in the following Theorem 3.

Theorem 3. *Assume that Assumptions* (**H1**), (**H2**) *and* (**H3**) *hold, the neural networks (1) has an unique* $T-$*periodic solution and it is globally exponential stability if there exist positive definite matrices* P, $E \in R^{n \times n}$ *and a positive definite diagonal matrix* $Q = diag(q_1, \cdots, q_n)$ *such that the following matrix function inequality holds*

$$\Omega(t) = \begin{pmatrix} PD(t) + D^T(t)P & -PA(t) & -PB(t) \\ -A^T(t)P & 2QD(t)M^{-1} - E - QA(t) - A^T(t)Q & -QB(t) \\ -B^T(t)P & -B^T(t)Q & E \end{pmatrix} > 0, \tag{5}$$

where $M = diag(M_1, \ldots, M_n)$.

Proof. From (5), we have

$$\begin{pmatrix} 2QD(t)M^{-1} - QA(t) - A^T(t)Q - E & -QB(t) \\ -B^T(t)Q & E \end{pmatrix} > 0. \tag{6}$$

Using the Schur Complement and (6), we get

$$2QD(t)M^{-1} - E - QA(t) - A^T(t)Q - QB(t)E^{-1}B^T(t)Q > 0. \tag{7}$$

For any $x \in R^n$, we have

$$\begin{aligned}&(E^{\frac{1}{2}}x - E^{-\frac{1}{2}}B^T(t)Qx)^T(E^{\frac{1}{2}}x - E^{-\frac{1}{2}}B^T(t)Qx)\\ &= x^TEx - x^TB^T(t)Qx - x^TQB(t)x + x^TQB(t)E^{-1}B^T(t)Qx \geq 0.\end{aligned}$$

That is to say

$$E + QB(t)E^{-1}B^T(t)Q - B^T(t)Q - QB(t) \geq 0. \tag{8}$$

Applying (15) to (16), we obtain

$$2QD(t)M^{-1} - Q(A(t) + B(t)) - (A(t) + B(t))^TQ > 0. \tag{9}$$

Take

$$J(x) = -D(t)x + (A(t) + B(t))f(x), \tag{10}$$

then by reduction to absurdity we can prove that J is injective in R^n.

It is easy to show

$$\lim_{|x| \longrightarrow \infty} |J(x)| = \infty. \tag{11}$$

In fact, $\|D(t)\|$, $\|A(t)\|$, $\|B(t)\|$ and $\|f(\cdot)\|$ are bounded by the conditions (**H1-H3**), then (11) holds. Therefore, it follows that $J(x)$ is a homeomorphism from R^n to itself. Therefore, the equation $J(x) = 0$ has an unique solution; i.e., the origin of the neural network (4) is an unique equilibrium. This complete the proof.

The rest of the proof is similar to the second part of the proof of Theorem 2 in [7] and omitted here.

Remark 1. If period T is zero and $d_i^M = d_i^m$, $a_{ij}^M = a_{ij}^m$, and $b_{ji}^M = b_{ji}^m$, then the Theorem 3 in this paper reduces to the Theorem 2 in [7]. Therefore, the Theorem 3 in this paper generalize and improve the results in [7].

4 Algorithm for Criterions

In this section, we give an algorithm on the matrix function inequality (5) to verify it. Applying the continuous properties of matrix function and linear matrix inequality (LMI) technique, we can derive the following Algorithm **A** on the matrix function inequality (5).

Algorithm **A:**

Step 1: Let initial time $t_0 = 0$, maximum iterative number $N = N_0$ (for example $N_0 = 200$), then go to next.

Step 2: If there are one feasible solution P_0, Q_0, E_0 about the matrix function inequality (5) by LMI toolbox in Matlab while take $t = t_0$, then go to step 3, or the matrix function inequality (5) don't hold, stop.

Step 3: Set $P = P_0$, $Q = Q_0$, $E = E_0$ in the matrix inequalities (5), there must exit $\delta > 0$ such that the determinant of the kth leading principal minor of the matrix continuous function on left side of the matrix inequalities (5) are positive in $[t_0, t_0 + \delta]$. If $\delta \geq T$, then the matrix inequalities (5) hold, stop; If $\delta < T$ and $i < N_0$, then $t_0 = t_0 + \delta$ and go to step 2; If $i \geq N$, then fail and stop.

Remark 2. The above algorithm provides the tool to verify the global stability criterions on the T-periodic solution of the neural networks (1). In the next section, we demonstrate the effective of the above algorithm.

5 Illustrative Example

Example 1. Consider the following neural networks:

$$\begin{cases} \dot{x}_1(t) = -(2 + 0.3\sin(t))x_1(t) + (1 + 0.1\sin(t))/3\tanh(y_2) + \frac{1}{2}\tanh(y_1 - 1) \\ \qquad + \sin(t), \\ \dot{x}_2(t) = -1.5x_2(t) + (1 - 0.1\cos(t))/2\tanh(y_1) + \frac{1}{3}\tanh(y_2 - 1) - \cos(t). \end{cases}$$

It is obvious that the above neural network satisfies Assumption (**H1**), (**H2**) and (**H3**). Applying the Algorithm **A**, in the first, we can get one feasible solution by computation:

$$P_0 = \begin{pmatrix} 8.7124 & 0 \\ 0 & 11.6232 \end{pmatrix},\ Q_0 = \begin{pmatrix} 15.5534 & 0 \\ 0 & 20.1932 \end{pmatrix},\ E_0 = \begin{pmatrix} 31.0314 & -4.7571 \\ -4.7571 & 30.4867 \end{pmatrix}$$

in the matrix function inequalities (5) for $t_0 = 0$.

Take $P = P_0$, $Q = Q_0$, $E = E_0$ in the inequalities (5), we obtain

$$\Omega(t) = \begin{pmatrix} 8 + 1.2\sin(t) & 0 & 0 & 2/3 + 0.67\sin(t) & 1.00 & 0 \\ 0 & 6.00 & 1 - 0.10\cos(t) & 0 & 0 & 0.67 \\ 0 & 1 - 0.10\cos(t) & 11 + 1.8\sin(t) & -3.50 + 0.15\cos(t) & 1.50 & 0 \\ 2/3 + 0.67\sin(t) & 0 & -3.5 + 0.15\cos(t) & 8.00 & 0 & 1.00 \\ 1.00 & 0 & 1.50 & 0 & 31.03 & -4.75 \\ 0 & 0.67 & 0 & 1.00 & -4.75 & 30.48 \end{pmatrix}.$$

It is obvious that the $\Omega(t)$ is positive for $t \in [0, 2\pi]$. Therefore, the matrix function inequality (5) hold, then the neural network in Example 1 exponentially converges to the unique 2π-periodic solution by Theorem 3. But, the results in [9,10,12] are very difficult to judge the stability of periodic solution of the neural networks in Example 1.

Acknowledgments

The work was supported by Natural Science Foundation of China Three Gorges University(No.604114), Natural Science Foundation of Hubei (Nos.2004ABA055, D200613002) and National Natural Science Foundation of China (No.60574025).

References

1. Cao, J., Wang, J.: Global Asymptotic Stability of Recurrent Neural Networks with Lipschitz-continuous Activation Functions and Time-Varying Delays. IEEE Trans. Circuits Syst I **50** (2003) 34-44
2. Liao, X., Chen, G., Sanchez, E.: Delay-Dependent Exponential Stability Analysis of Delayed Neural Networks: an LMI Approach. Neural Networks **15** (2002) 855-866
3. Zeng, Z., Wang, J., Liao X.: Global Exponential Stability of a General Class of Recurrent Neural Networks with Time-Varying Delays. IEEE Trans. Circuits Syst **50** (2003) 1353-1359
4. Chua, L.O., Yang, L.: Cellular Neural Networks: Theory. IEEE Trans. Circuits Syst **35** (1988) 1257-1272
5. Hopfiels, J.J.: Neurons with Graded Response Have Collective Computational Properties like Those of Two-State Neurons. Porc.Natl Acad.Sci.USA **81** (1984) 3088-3092
6. Zeng, Z., Wang, J.: Global Exponential Stability of Recurrent Neural Networks with Time-Varying Delays in the Presence of Strong External Stimuli. Neural Networks **19** (2006) 1528-1537
7. Zeng, Z., Wang, J.: Improved Conditions for Global Exponential Stability of Recurrent Neural Networks with Time-Varying Delays. IEEE Transactions on Neural Networks **17** (2006) 1141-1151
8. Jiang, M., Shen, Y., Liao, X.: Global Stability of Periodic Solution for Bidirectional Associative Memory Neural Networks with Varying-time Delay. Applied Mathematics and Computation **182** (2006) 509-520
9. Jiang, M., Shen, Y., Liu, M.: Global Exponential Stability of Non-autonomous Neural Networks with Variable Delay. Lecture Notes in Computer Science **3496** (2005) 108-113
10. Sun, C., Feng, C.: Exponential Periodicity and Stability of Delayed Neural Networks. Mathematics and Computers in Simulation **66** (2004) 469-478
11. Liao, T., Wang, F.: Global Stability for Cellular Neural Networks with Time Delay. IEEE Trans. Neural Networks **11** (2000) 1481-1484
12. Jiang, H., Li, Z., Teng, Z.: Boundedness and Stability for Nonautonomous Cellular Neural Networks with Time Delay. Physics Letter A **306** (2003) 313-325
13. Guo, S., Huang, L., Dai, B., Zhang, Z.: Global Existence of Periodic Solutions of BAM Neural Networks with Variable Coefficients. Physics Letter A **317** (2003) 97-106
14. Xiang, H., Yan, K., Dai, B., Wang, B.: Existence and Global Exponential of Periodic Solutions for Delayed High-order Hopfield-Type Neural Networks. Physics Letter A **352** (2006) 341-349
15. Rouche, N., Mawhin, J.: Ordinary Differential Equations: Stability and Periodic Solutions. Pitman, Boston (1980)

Stability Analysis of Generalized Nonautonomous Cellular Neural Networks with Time-Varying Delays

Xiaobing Nie[1], Jinde Cao[1], and Min Xiao[1,2]

[1]Department of Mathematics, Southeast University, Nanjing 210096, China
[2]Department of Mathematics, Nanjing Xiaozhuang College, Nanjing 210017, China
{xbnie,jdcao}@seu.edu.cn

Abstract. In this paper, a class of generalized nonautonomous cellular neural networks with time-varying delays are studied. By means of Lyapunov functional method, improved Young inequality $a^m b^n \leq ma\,t^{-n} + nb\,t^m$ $(0 \leq m \leq 1, m+n=1, t>0)$ and the homeomorphism theory, several sufficient conditions are given guaranteeing the existence, uniqueness and global exponential stability of the equilibrium point. The proposed results generalize and improve previous works. An illustrative example is also given to demonstrate the effectiveness of the proposed results.

1 Introduction

The dynamics of autonomous cellular neural networks have been extensively studied in the past decades, and many important results for checking global asymptotic/exponential stability of equilibrium have been presented, see, for example, [1-7] and references cited therein. However, to the best of our knowledge, few studies have considered dynamics of nonautonomous cellular neural networks with delays [8-10].

In this paper, we consider a class of generalized nonautonomous neural networks with time-varying delays which described by the following functional differential equations

$$\begin{aligned} x_i'(t) = &-c_i(t)h_i(x_i(t)) + \sum_{j=1}^{n} a_{ij}(t)f_j(x_j(t)) \\ &+ \sum_{j=1}^{n} b_{ij}(t)g_j(x_j(t-\tau_{ij}(t))) + I_i(t), \quad i=1,2,\ldots,n. \end{aligned} \tag{1}$$

where $\tau_{ij}(t)$ is time delay and $0 \leqslant \tau_{ij}(t) \leqslant \tau$.

Through this paper, we will assume that the real value functions $c_i(t), a_{ij}(t)$, $b_{ij}(t), I_i(t)$ are bound continuous functions. Further, in order to convenient description we introduce the following assumptions.

(H_1) $h_i(0)=0$ and there exists constants m_i, M_i such that

$$0 < m_i \leq \frac{h_i(x)-h_i(y)}{x-y} \leq M_i, \quad (x \neq y).$$

D. Liu et al. (Eds.): ISNN 2007, Part I, LNCS 4491, pp. 958–967, 2007.

(H_2) There are constants $k_i > 0, l_i > 0$ such that

$$|f_i(x_1) - f_i(x_2)| \leq k_i|x_1 - x_2|, \quad |g_i(x_1) - g_i(x_2)| \leq l_i|x_1 - x_2|,$$

for all $x_1, x_2 \in \mathbb{R}$, and $i = 1, 2, \cdots, n$.
(H_3) $\tau_{ij}(t)$ is differentiable and $0 \leq \tau_{ij}'(t) \leq \tau_{ij}^* < 1$, where $\tau_{ij}^* = \sup_{t \in \mathbb{R}} \tau_{ij}'(t)$.
(H_3') $\tau_{ij}(t)$ is differentiable and $\tau_{ij}'(t) \leq 0$.

In [1, 8-10], by utilizing Young inequality and Lyapunov method, the authors established some sufficient conditions for global exponential stability of the special cases of system (1). However, the existence of the equilibrium point wasn't discussed. Our main purpose in this paper is to establish a series of new criteria on the existence, uniqueness and global exponential stability of the equilibrium point for system (1) by means of Lyapunov functional method, the elementary inequality $a^m b^n \leq mat^{-n} + nbt^m$ $(0 \leq m \leq 1, m + n = 1, t > 0)$ and the homeomorphism theory. Different from [7], [10], we will not require functions $h_i(u)$ in system (1) to be differential and thus the condition (H_1) is less restrictive. In addition, these criteria posses infinitely adjustable real parameters, which are of highly important significance in the designs and applications of networks. The criteria obtained in this paper generalize and improve previous works.

2 Preliminaries

The initial conditions associated with system (1) are of the form

$$x_i(t) = \phi_i(t), \quad -\tau \leq t \leq 0,$$

in which $\phi_i(t)$ are continuous functions, $i = 1, 2, \cdots, n$.

Let $C = C([-\tau, 0], R^n)$ be the Banach space of continuous functions which maps $[-\tau, 0]$ into R^n with the topology of uniform convergence. for any $\varphi \in C$, we define

$$\|\varphi\| = \sup_{-\tau \leq \theta \leq 0} [\sum_{i=1}^{n} |\varphi_i(\theta)|^p]^{1/p},$$

where $p \geq 1$ is constant.

Define $x_t = x(t + \theta), \quad \theta \in [-\tau, 0], t \geq 0$.

Definition 1. A vector $x^* = (x_1^*, x_2^*, \cdots, x_n^*)^T$ is said to be an equilibrium point of system (1) if the following equation

$$c_i(t)h_i(x_i^*) = \sum_{j=1}^{n} a_{ij}(t)f_j(x_j^*) + \sum_{j=1}^{n} b_{ij}(t)g_j(x_j^*) + I_i(t),$$

holds for all $t \geq 0$ and $i = 1, 2, \cdots, n$.

Definition 2[11]. A map $H : \mathbb{R}^n \to \mathbb{R}^n$ is a homeomorphism of $\mathbb{R}^n$ onto itself, if $H \in C^0, H$ is one-to-one, H is onto and the inverse map $H^{-1} \in C^0$.

Lemma 1[11]. If $H(x) \in C^0$ and satisfied the following conditions:
i) $H(x)$ is infective on $\mathbb{R}^n$;
ii) $\lim\limits_{\|x\| \to +\infty} \|H(x)\| = +\infty$, then $H(x)$ is a homeomorphism of $\mathbb{R}^n$.

Lemma 2[12]. Assume that $a > 0, b > 0, t > 0, 0 \leq m \leq 1, m + n = 1$, then the following inequality

$$a^m b^n \leq m a t^{-n} + n b t^m,$$

holds.

3 Existence, Uniqueness and Global Exponential Stability of the Equilibrium Point

Theorem 1. Under Assumptions $(\mathrm{H}_1) - (\mathrm{H}_3)$, the network model (1) has a unique equilibrium point x^* which is globally exponentially stable if there constants α_{ij}, α^*_{ij}, β_{ij}, β^*_{ij}, ξ_{ij}, ξ^*_{ij}, η_{ij}, η^*_{ij} $\in \mathbb{R}$, $w_i > 0, \lambda > 0, \mu > 0$, and $\sigma > 0, p \geq 1,\ i, j = 1, 2, \cdots, n$ such that

$$-c_i(t)\frac{m_i^p}{M_i^{p-1}} + \sum_{j=1}^{n}[\tfrac{p-1}{p}\lambda^{-\frac{1}{p}}|a_{ij}(t)|^{p\alpha_{ij}}k_j^{p\beta_{ij}} + \tfrac{1}{p}\lambda^{\frac{p-1}{p}}\tfrac{w_j}{w_i}(\tfrac{M_j}{M_i})^{p-1}|a_{ji}(t)|^{p\alpha^*_{ji}}k_i^{p\beta^*_{ji}}] + \sum_{j=1}^{n}[\tfrac{p-1}{p}\mu^{-\frac{1}{p}}(b^+_{ij})^{p\xi_{ij}}l_j^{p\eta_{ij}} + \tfrac{1}{p}\mu^{\frac{p-1}{p}}\tfrac{1}{1-\tau'_{ji}(\psi^{-1}_{ji}(t))}\tfrac{w_j}{w_i}(\tfrac{M_j}{M_i})^{p-1}(b^+_{ji})^{p\xi^*_{ji}}l_i^{p\eta^*_{ji}}] < -\sigma, \tag{2}$$

holds for all $t \geq 0$ and $i = 1, 2, \cdots, n$, where $\psi^{-1}_{ij}(t)$ denotes the inverse function of $\psi_{ij}(t) = t - \tau_{ij}(t)$, $b^+_{ij} = \sup\limits_{t\in\mathbb{R}} |b_{ij}(t)|$, $(p-1)\alpha_{ij} + \alpha^*_{ij} = 1$, $(p-1)\beta_{ij} + \beta^*_{ij} = 1$, $(p-1)\xi_{ij} + \xi^*_{ij} = 1$, $(p-1)\eta_{ij} + \eta^*_{ij} = 1$.

proof. Let $H_t(x) = (H_{t1}(x), H_{t2}(x), \cdots, H_{tn}(x))^T$, where

$$H_{ti}(x) = -c_i(t)h_i(x_i) + \sum_{j=1}^{n} a_{ij}(t)f_j(x_j) + \sum_{j=1}^{n} b_{ij}(t)g_j(x_j) + I_i(t), \quad i = 1, 2, \cdots, n.$$

Note that, to prove that system (1) has a unique equilibrium point, by Lemma 1, we only need to show that $H_t(x)$ with respect to x is a homeomorphism for all $t \geq 0$.

First, we prove that $H_t(x)$ is an injective map on R^n for all $t \geq 0$. In fact, if there exists $x \neq y \in R^n$ and $t \geq 0$ such that $H_t(x) = H_t(y)$, then

$$\begin{aligned}
&\sum_{i=1}^{n} w_i c_i(t) m_i^p |x_i - y_i|^p \le \sum_{i=1}^{n} w_i c_i(t) |h_{ti}(x_i) - h_{ti}(y_i)|^p \\
&= \sum_{i=1}^{n} w_i |h_{ti}(x_i) - h_{ti}(y_i)|^{p-1} \mathrm{sign}(h_{ti}(x_i) - h_{ti}(y_i)) [\sum_{j=1}^{n} a_{ij}(t)(f_j(x_j) - f_j(y_j)) \\
&\quad + \sum_{j=1}^{n} b_{ij}(t)(g_j(x_j) - g_j(y_j))] \\
&\le \sum_{i=1}^{n} w_i M_i^{p-1} |x_i - y_i|^{p-1} [\sum_{j=1}^{n} |a_{ij}(t)| k_j |x_j - y_j| + \sum_{j=1}^{n} |b_{ij}(t)| l_j |x_j - y_j|] \\
&\le \sum_{i=1}^{n} w_i M_i^{p-1} [\sum_{j=1}^{n} |a_{ij}(t)| k_j |x_j - y_j| |x_i - y_i|^{p-1} + \sum_{j=1}^{n} b_{ij}^{+} l_j |x_j - y_j| |x_i - y_i|^{p-1}] \\
&= \sum_{i=1}^{n} w_i M_i^{p-1} \{ \sum_{j=1}^{n} [(|a_{ij}(t)|^{p\alpha_{ij}} k_j^{p\beta_{ij}} |x_i - y_i|^p)^{\frac{p-1}{p}}] \times [(|a_{ij}(t)|^{p\alpha_{ij}^*} k_j^{p\beta_{ij}^*} |x_j - y_j|^p)^{\frac{1}{p}}] \\
&\quad + \sum_{j=1}^{n} [((b_{ij}^{+})^{p\xi_{ij}} l_j^{p\eta_{ij}} |x_i - y_i|^p)^{\frac{p-1}{p}}] \times [((b_{ij}^{+})^{p\xi_{ij}^*} l_j^{p\eta_{ij}^*} |x_j - y_j|^p)^{\frac{1}{p}}] \}.
\end{aligned}$$

By using the improved Young inequality in Lemma 2, we have

$$\begin{aligned}
&\sum_{i=1}^{n} w_i c_i(t) m_i^p |x_i - y_i|^p \\
&\le \sum_{i=1}^{n} w_i M_i^{p-1} \{ \sum_{j=1}^{n} [\tfrac{p-1}{p} \lambda^{-\frac{1}{p}} |a_{ij}(t)|^{p\alpha_{ij}} k_j^{p\beta_{ij}} |x_i - y_i|^p + \tfrac{1}{p} \lambda^{\frac{p-1}{p}} |a_{ij}(t)|^{p\alpha_{ij}^*} k_j^{p\beta_{ij}^*} \\
&\quad |x_j - y_j|^p] + \sum_{j=1}^{n} [\tfrac{p-1}{p} \mu^{-\frac{1}{p}} (b_{ij}^{+})^{p\xi_{ij}} l_j^{p\eta_{ij}} |x_i - y_i|^p + \tfrac{1}{p} \mu^{\frac{p-1}{p}} (b_{ij}^{+})^{p\xi_{ij}^*} l_j^{p\eta_{ij}^*} |x_j - y_j|^p] \} \\
&= \sum_{i=1}^{n} \{ \sum_{j=1}^{n} [\tfrac{p-1}{p} \lambda^{-\frac{1}{p}} |a_{ij}(t)|^{p\alpha_{ij}} k_j^{p\beta_{ij}} + \tfrac{1}{p} \lambda^{\frac{p-1}{p}} \tfrac{w_j}{w_i} (\tfrac{M_j}{M_i})^{p-1} |a_{ji}(t)|^{p\alpha_{ji}^*} k_i^{p\beta_{ji}^*}] \\
&\quad + \sum_{j=1}^{n} [\tfrac{p-1}{p} \mu^{-\frac{1}{p}} (b_{ij}^{+})^{p\xi_{ij}} l_j^{p\eta_{ij}} + \tfrac{1}{p} \mu^{\frac{p-1}{p}} \tfrac{w_j}{w_i} (\tfrac{M_j}{M_i})^{p-1} (b_{ji}^{+})^{p\xi_{ji}^*} l_i^{p\eta_{ji}^*}] \} w_i M_i^{p-1} |x_i - y_i|^p.
\end{aligned}$$

Note that $\frac{1}{1-\tau_{ji}'(\psi_{ji}^{-1}(t))} \ge 1$ and thus the following condition

$$\begin{aligned}
&\sum_{j=1}^{n} [\tfrac{p-1}{p} \lambda^{-\frac{1}{p}} |a_{ij}(t)|^{p\alpha_{ij}} k_j^{p\beta_{ij}} + \tfrac{w_j}{p w_i} \lambda^{\frac{p-1}{p}} (\tfrac{M_j}{M_i})^{p-1} |a_{ji}(t)|^{p\alpha_{ji}^*} k_i^{p\beta_{ji}^*}] + \\
&\sum_{j=1}^{n} [\tfrac{p-1}{p} \mu^{-\frac{1}{p}} (b_{ij}^{+})^{p\xi_{ij}} l_j^{p\eta_{ij}} + \tfrac{w_j}{p w_i} \mu^{\frac{p-1}{p}} (\tfrac{M_j}{M_i})^{p-1} (b_{ji}^{+})^{p\xi_{ji}^*} l_i^{p\eta_{ji}^*}] < c_i(t) \frac{m_i^p}{M_i^{p-1}},
\end{aligned}$$

holds for all $i = 1, 2, \cdots, n$, we have

$$0 < \sum_{i=1}^{n} w_i c_i(t) m_i^p |x_i - y_i|^p < \sum_{i=1}^{n} w_i M_i^{p-1} c_i(t) \frac{m_i^p}{M_i^{p-1}} |x_i - y_i|^p = \sum_{i=1}^{n} w_i c_i(t) m_i^p |x_i - y_i|^p.$$

a contradiction. This implies that map $H_t(x)$ is an injection on R^n for all $t \ge 0$.

Second, we prove that $\lim_{\|x\| \to +\infty} \|H_t(x)\| = +\infty$.

To show that $\|H_t(x)\| \to +\infty (\|x\| \to +\infty)$, it suffices to show that $\|\widetilde{H}_t(x)\| \to +\infty (\|x\| \to +\infty)$, where

$$\begin{aligned}
&\widetilde{H}_t(x) = (\widetilde{H}_{t1}(x), \widetilde{H}_{t2}(x), \cdots, \widetilde{H}_{tn}(x))^T, \\
&\widetilde{H}_{ti}(x) = -c_i(t) h_i(x_i) + \sum_{j=1}^{n} a_{ij}(t)(f_j(x_j) - f_j(0)) + \sum_{j=1}^{n} b_{ij}(t)(g_j(x_j) - g_j(0)).
\end{aligned}$$

In fact

$$\begin{aligned}
&\sum_{i=1}^{n} w_i m_i^{p-1}|x_i|^{p-1}\mathrm{sign}(x_i)\widetilde{H}_{ti}(x)\\
\le& -\sum_{i=1}^{n} w_i c_i(t) m_i^{p-1}|x_i|^{p-1}\mathrm{sign}(x_i)h_i(x_i) + \sum_{i=1}^{n} w_i m_i^{p-1}|x_i|^{p-1}(\sum_{j=1}^{n}|a_{ij}(t)|\\
&|f_j(x_j)-f_j(0)| + \sum_{j=1}^{n}|b_{ij}(t)||g_j(x_j)-g_j(0)|)\\
\le& -\sum_{i=1}^{n} w_i c_i(t) m_i^{p}|x_i|^{p} + \sum_{i=1}^{n} w_i M_i^{p-1}[\sum_{j=1}^{n}|a_{ij}(t)|k_j|x_j||x_i|^{p-1}\\
&+\sum_{j=1}^{n}|b_{ij}(t)|l_j|x_j||x_i|^{p-1}]\\
\le& -\sum_{i=1}^{n} w_i c_i(t) m_i^{p}|x_i|^{p} + \sum_{i=1}^{n} w_i M_i^{p-1}\{\sum_{j=1}^{n}[(|a_{ij}(t)|^{p\alpha_{ij}}k_j^{p\beta_{ij}}|x_i|^p)^{\frac{p-1}{p}}]\times[(|a_{ij}(t)|^{p\alpha_{ij}^*}\\
&k_j^{p\beta_{ij}^*}|x_j|^p)^{\frac{1}{p}}] + \sum_{j=1}^{n}[((b_{ij}^+)^{p\xi_{ij}}l_j^{p\eta_{ij}}|x_i|^p)^{\frac{p-1}{p}}]\times[((b_{ij}^+)^{p\xi_{ij}^*}l_j^{p\eta_{ij}^*}|x_j|^p)^{\frac{1}{p}}]\}\\
\le& -\sum_{i=1}^{n} w_i c_i(t) m_i^{p}|x_i|^{p} + \sum_{i=1}^{n} w_i M_i^{p-1}\{\sum_{j=1}^{n}[\tfrac{p-1}{p}\lambda^{-\frac{1}{p}}|a_{ij}(t)|^{p\alpha_{ij}}k_j^{p\beta_{ij}}|x_i|^p + \tfrac{1}{p}\lambda^{\frac{p-1}{p}}\\
&|a_{ij}(t)|^{p\alpha_{ij}^*}k_j^{p\beta_{ij}^*}|x_j|^p] + \sum_{j=1}^{n}[\tfrac{p-1}{p}\mu^{-\frac{1}{p}}(b_{ij}^+)^{p\xi_{ij}}l_j^{p\eta_{ij}}|x_i|^p + \tfrac{1}{p}\mu^{\frac{p-1}{p}}(b_{ij}^+)^{p\xi_{ij}^*}l_j^{p\eta_{ij}^*}|x_j|^p]\}\\
=& \sum_{i=1}^{n}\{-c_i(t)\frac{m_i^p}{M_i^{p-1}} + \sum_{j=1}^{n}[\tfrac{p-1}{p}\lambda^{-\frac{1}{p}}|a_{ij}(t)|^{p\alpha_{ij}}k_j^{p\beta_{ij}} + \tfrac{1}{p}\lambda^{\frac{p-1}{p}}\tfrac{w_j}{w_i}(\tfrac{M_j}{M_i})^{p-1}|a_{ji}(t)|^{p\alpha_{ji}^*}\\
&k_i^{p\beta_{ji}^*}] + \sum_{j=1}^{n}[\tfrac{p-1}{p}\mu^{-\frac{1}{p}}(b_{ij}^+)^{p\xi_{ij}}l_j^{p\eta_{ij}} + \tfrac{1}{p}\mu^{\frac{p-1}{p}}\tfrac{w_j}{w_i}(\tfrac{M_j}{M_i})^{p-1}(b_{ji}^+)^{p\xi_{ji}^*}l_i^{p\eta_{ji}^*}]\}w_i M_i^{p-1}|x_i|^p\\
\le& -\sigma\min_{1\le i\le n}\{w_i M_i^{p-1}\}\sum_{i=1}^{n}|x_i|^p.
\end{aligned}$$

Therefore, we have

$$\begin{aligned}
&\sigma\min_{1\le i\le n}\{w_i M_i^{p-1}\}\sum_{i=1}^{n}|x_i|^p \le -\sum_{i=1}^{n} w_i m_i^{p-1}|x_i|^{p-1}\mathrm{sign}(x_i)\widetilde{H}_{ti}(x)\\
&\le \max_{1\le i\le n}\{w_i m_i^{p-1}\}\sum_{i=1}^{n}|x_i|^{p-1}|\widetilde{H}_{ti}(x)|.
\end{aligned}$$

By using *Hölder* inequality, we get

$$\sum_{i=1}^{n}|x_i|^p \le \frac{\max_{1\le i\le n}\{w_i m_i^{p-1}\}}{\sigma\min_{1\le i\le n}\{w_i M_i^{p-1}\}}(\sum_{i=1}^{n}|x_i|^p|)^{\frac{p-1}{p}}(\sum_{i=1}^{n}|\widetilde{H}_{ti}(x)|^p)^{\frac{1}{p}}.$$

So, we have

$$\|x\| \le \frac{\max_{1\le i\le n}\{w_i m_i^{p-1}\}}{\sigma\min_{1\le i\le n}\{w_i M_i^{p-1}\}}\|\widetilde{H}_t(x)\|.$$

It implies that $\lim_{\|x\|\to+\infty}\|\widetilde{H}_t(x)\| = +\infty$. Thus, we have $\lim_{\|x\|\to+\infty}\|H_t(x)\| = +\infty$. Form Lemma 1, we know that map $H_t(x)$ with respect to x is a homeomorphism on R^n for all $t \ge 0$. Thus, (1) has a unique equilibrium point.

In the following, we prove that the unique equilibrium point is globally exponentially stable.

Since condition (2) holds, we can choose a small $\varepsilon > 0$, such that

$$\frac{1}{p}\varepsilon(\frac{m_i}{M_i})^{p-1} - c_i(t)\frac{m_i^p}{M_i^{p-1}} + \sum_{j=1}^{n}[\frac{p-1}{p}\lambda^{-\frac{1}{p}}|a_{ij}(t)|^{p\alpha_{ij}}k_j^{p\beta_{ij}} + \frac{1}{p}\lambda^{\frac{p-1}{p}}\frac{w_j}{w_i}(\frac{M_j}{M_i})^{p-1}$$
$$|a_{ji}(t)|^{p\alpha_{ji}^*}k_i^{p\beta_{ji}^*}] + \sum_{j=1}^{n}[\frac{p-1}{p}\mu^{-\frac{1}{p}}(b_{ij}^+)^{p\xi_{ij}}l_j^{p\eta_{ij}} + \frac{1}{p}\mu^{\frac{p-1}{p}}\frac{1}{1-\tau_{ji}'(\psi_{ji}^{-1}(t))}$$
$$\frac{w_j}{w_i}(\frac{M_j}{M_i})^{p-1}(b_{ji}^+)^{p\xi_{ji}^*}l_i^{p\eta_{ji}^*}e^{\varepsilon\tau}] < -\sigma,$$

holds for all $t \geq 0$ and $i = 1, 2, \cdots, n$.

Let $x^* = (x_1^*, x_2^*, \cdots, x_n^*)^T$ be an equilibrium point of the network model (1), one can derive form (1) that the derivations $y_i(t) = x_i(t) - x_i^*$ $(i = 1, 2, \cdots, n)$ satisfy

$$\begin{aligned} y_i'(t) = &-c_i(t)[h_i(x_i^* + y_i(t)) - h_i(x_i^*)] + \sum_{j=1}^{n} a_{ij}(t)[f_j(x_j^* + y_j(t)) - f_j(x_j^*)] \\ &+ \sum_{j=1}^{n} b_{ij}(t)[g_j(x_j^* + y_j(t - \tau_{ij}(t))) - g_j(x_j^*)]. \end{aligned} \quad (3)$$

By equation (3), we have

$$D^+|y_i(t)| \leq -m_i c_i(t)|y_i(t)| + \sum_{j=1}^{n}|a_{ij}(t)|k_j|y_j(t)| + \sum_{j=1}^{n}|b_{ij}(t)|l_j|y_j(t - \tau_{ij}(t))|.$$

Now we consider the following Lyapunov functional:

$$V(t) = \frac{1}{p}\sum_{i=1}^{n} w_i m_i^{p-1}[|y_i(t)|^p e^{\varepsilon t} + \mu^{\frac{p-1}{p}}\sum_{j=1}^{n}(b_{ij}^+)^{p\xi_{ij}^*}l_j^{p\eta_{ij}^*}\int_{t-\tau_{ij}(t)}^{t}\frac{|y_j(s)|^p e^{\varepsilon(s+\tau)}}{1 - \tau_{ij}'(\psi_{ij}^{-1}(s))}ds].$$

Calculate the upper right Dini-derivative D^+V of V along the solution of (3), we have

$$\begin{aligned} &D^+V(t) \\ \leq &\frac{1}{p}\sum_{i=1}^{n} w_i m_i^{p-1}\{pe^{\varepsilon t}|y_i(t)|^{p-1}[-m_i c_i(t)|y_i(t)| + \sum_{j=1}^{n}|a_{ij}(t)|k_j|y_j(t)| \\ &+ \sum_{j=1}^{n}|b_{ij}(t)|l_j|y_j(t - \tau_{ij}(t))|] + \varepsilon e^{\varepsilon t}|y_i(t)|^p + \mu^{\frac{p-1}{p}}\sum_{j=1}^{n}(b_{ij}^+)^{p\xi_{ij}^*}l_j^{p\eta_{ij}^*} \\ &\frac{|y_j(t)|^p}{1-\tau_{ij}'(\psi_{ij}^{-1}(t))}e^{\varepsilon t}e^{\varepsilon\tau} - \mu^{\frac{p-1}{p}}\sum_{j=1}^{n}(b_{ij}^+)^{p\xi_{ij}^*}l_j^{p\eta_{ij}^*}|y_j(t - \tau_{ij}(t))|^p e^{\varepsilon t}\} \end{aligned}$$

$$\leq e^{\varepsilon t}\sum_{i=1}^{n} w_i m_i^{p-1}[(\tfrac{1}{p}\varepsilon - m_i c_i(t))|y_i(t)|^p + \tfrac{p-1}{p}\lambda^{-\frac{1}{p}}\sum_{j=1}^{n}|a_{ij}(t)|^{p\alpha_{ij}}k_j^{p\beta_{ij}}|y_i(t)|^p$$
$$+\tfrac{1}{p}\lambda^{\frac{p-1}{p}}\sum_{j=1}^{n}|a_{ij}(t)|^{p\alpha^*_{ij}}k_j^{p\beta^*_{ij}}|y_j(t)|^p + \tfrac{p-1}{p}\mu^{-\frac{1}{p}}\sum_{j=1}^{n}(b_{ij}^{+})^{p\xi_{ij}}l_j^{p\eta_{ij}}|y_i(t)|^p$$
$$+\tfrac{1}{p}\mu^{\frac{p-1}{p}}\sum_{j=1}^{n}(b_{ij}^{+})^{p\xi^*_{ij}}l_j^{p\eta^*_{ij}}|y_j(t-\tau_{ij}(t))|^p + \tfrac{1}{p}\mu^{\frac{p-1}{p}}\sum_{j=1}^{n}(b_{ij}^{+})^{p\xi^*_{ij}}l_j^{p\eta^*_{ij}}\tfrac{|y_j(t)|^p}{1-\tau'_{ij}(\psi_{ij}^{-1}(t))}$$
$$e^{\varepsilon\tau} - \tfrac{1}{p}\mu^{\frac{p-1}{p}}\sum_{j=1}^{n}(b_{ij}^{+})^{p\xi^*_{ij}}l_j^{p\eta^*_{ij}}|y_j(t-\tau_{ij}(t))|^p]$$
$$\leq e^{\varepsilon t}\sum_{i=1}^{n} w_i[(\tfrac{1}{p}\varepsilon m_i^{p-1} - m_i^p c_i(t))|y_i(t)|^p + \tfrac{p-1}{p}\lambda^{-\frac{1}{p}}M_i^{p-1}\sum_{j=1}^{n}|a_{ij}(t)|^{p\alpha_{ij}}k_j^{p\beta_{ij}}$$
$$|y_i(t)|^p + \tfrac{1}{p}\lambda^{\frac{p-1}{p}}M_i^{p-1}\sum_{j=1}^{n}|a_{ij}(t)|^{p\alpha^*_{ij}}k_j^{p\beta^*_{ij}}|y_j(t)|^p + \tfrac{p-1}{p}\mu^{-\frac{1}{p}}M_i^{p-1}$$
$$\sum_{j=1}^{n}(b_{ij}^{+})^{p\xi_{ij}}l_j^{p\eta_{ij}}|y_i(t)|^p + \tfrac{1}{p}\mu^{\frac{p-1}{p}}M_i^{p-1}\sum_{j=1}^{n}(b_{ij}^{+})^{p\xi^*_{ij}}l_j^{p\eta^*_{ij}}\tfrac{|y_j(t)|^p}{1-\tau'_{ij}(\psi_{ij}^{-1}(t))}e^{\varepsilon\tau}]$$
$$= e^{\varepsilon t}\sum_{i=1}^{n}[\tfrac{1}{p}\varepsilon(\tfrac{m_i}{M_i})^{p-1} - c_i(t)\tfrac{m_i^p}{M_i^{p-1}} + \tfrac{p-1}{p}\lambda^{-\frac{1}{p}}\sum_{j=1}^{n}|a_{ij}(t)|^{p\alpha_{ij}}k_j^{p\beta_{ij}}$$
$$+\tfrac{1}{p}\lambda^{\frac{p-1}{p}}\sum_{j=1}^{n}\tfrac{w_j}{w_i}(\tfrac{M_j}{M_i})^{p-1}|a_{ji}(t)|^{p\alpha^*_{ji}}k_i^{p\beta^*_{ji}}] + \tfrac{p-1}{p}\mu^{-\frac{1}{p}}\sum_{j=1}^{n}(b_{ij}^{+})^{p\xi_{ij}}l_j^{p\eta_{ij}}$$
$$+\tfrac{1}{p}\mu^{\frac{p-1}{p}}\sum_{j=1}^{n}\tfrac{1}{1-\tau'_{ji}(\psi_{ji}^{-1}(t))}\tfrac{w_j}{w_i}(\tfrac{M_j}{M_i})^{p-1}(b_{ji}^{+})^{p\xi^*_{ji}}l_i^{p\eta^*_{ji}}e^{\varepsilon\tau}]w_i M_i^{p-1}|y_i(t)|^p \leq 0.$$

So

$$V(t) \leq V(0), \quad t \geq 0.$$

Since

$$\tfrac{1}{p}e^{\varepsilon t}\min_{1\leq i\leq n}\{w_i m_i^{p-1}\}\sum_{i=1}^{n}|y_i|^p \leq \tfrac{1}{p}e^{\varepsilon t}\sum_{i=1}^{n}w_i m_i^{p-1}|y_i|^p \leq V(t), \quad t\geq 0.$$
$$V(0) = \tfrac{1}{p}\sum_{i=1}^{n}w_i m_i^{p-1}[|y_i(0)|^p + \mu^{\frac{p-1}{p}}\sum_{j=1}^{n}(b_{ij}^{+})^{p\xi^*_{ij}}l_j^{p\eta^*_{ij}}\int_{-\tau_{ij}(0)}^{0}\tfrac{|y_j(s)|^p}{1-\tau'_{ij}(\psi_{ij}^{-1}(s))}e^{\varepsilon(s+\tau)}ds]$$
$$\leq \tfrac{1}{p}\max_{1\leq i\leq n}\{w_i m_i^{p-1}\}[1 + \mu^{\frac{p-1}{p}}\tau e^{\varepsilon\tau}\sum_{i=1}^{n}\max_{1\leq j\leq n}(b_{ij}^{+})^{p\xi^*_{ij}}l_j^{p\eta^*_{ij}}h_{ij}]\|\phi - x^*\|^p,$$

where $h_{ij} = \max_{s\in[-\tau,0]}\frac{1}{1-\tau'_{ij}(\psi_{ij}^{-1}(s))}$.
Then, we easily get

$$\sum_{i=1}^{n}|x_i(t) - x_i^*|^p \leq M\|\phi - x^*\|^p e^{-\varepsilon t},$$

for all $t \geq 0$, where $M \geq 1$ is a constant. It follows that

$$\|x_t - x^*\| \leq M^{1/p}\|\phi - x^*\|e^{-\varepsilon t/p},$$

for all $t \geq 0$, where $M \geq 1$ is a constant. This implies that the equilibrium point $x^* = (x_1^*, x_2^*, \cdots, x_n^*)^T$ is globally exponentially stable.

Let $p = 1$ in the condition (2), then we have

$$-c_i(t)m_i + \sum_{j=1}^{n} \frac{w_j}{w_i}|a_{ji}(t)|\, k_i + \sum_{j=1}^{n} \frac{w_j}{w_i} \frac{b_{ji}^{+} l_i}{1 - \tau_{ji}^{'}(\psi_{ji}^{-1}(t))} < -\sigma. \tag{4}$$

Theorem 2. Under Assumptions (H_1), (H_2) and ($H_3^{'}$), the network model (1) has a unique equilibrium point x^* which is globally exponentially stable if there constants α_{ij}, α_{ij}^*, β_{ij}, β_{ij}^*, ξ_{ij}, ξ_{ij}^*, η_{ij}, $\eta_{ij}^* \in R$, $w_i > 0, \lambda > 0, \mu > 0$, and $\sigma > 0$, $p \geq 1$, $i, j = 1, 2, \cdots, n$ such that

$$-c_i(t)\frac{m_i^p}{M_i^{p-1}} + \sum_{j=1}^{n}[\tfrac{p-1}{p}\lambda^{-\frac{1}{p}}|a_{ij}(t)|^{p\alpha_{ij}}k_j^{p\beta_{ij}} + \tfrac{1}{p}\lambda^{\frac{p-1}{p}}\tfrac{w_j}{w_i}(\tfrac{M_j}{M_i})^{p-1}|a_{ji}(t)|^{p\alpha_{ji}^*}$$
$$k_i^{p\beta_{ji}^*}] + \sum_{j=1}^{n}[\tfrac{p-1}{p}\mu^{-\frac{1}{p}}(b_{ij}^{+})^{p\xi_{ij}}l_j^{p\eta_{ij}} + \tfrac{1}{p}\mu^{\frac{p-1}{p}}\tfrac{w_j}{w_i}(\tfrac{M_j}{M_i})^{p-1}(b_{ji}^{+})^{p\xi_{ji}^*}l_i^{p\eta_{ji}^*}] < -\sigma, \tag{5}$$

holds for all $t \geq 0$ and $i = 1, 2, \cdots, n$, where $b_{ij}^{+} = \sup_{t\in\mathbb{R}}|b_{ij}(t)|$, $(p-1)\alpha_{ij} + \alpha_{ij}^* = 1$, $(p-1)\beta_{ij} + \beta_{ij}^* = 1$, $(p-1)\xi_{ij} + \xi_{ij}^* = 1$, $(p-1)\eta_{ij} + \eta_{ij}^* = 1$.

Proof. We only need to consider the following Lyapunov function:

$$V(t) = \frac{1}{p}\sum_{i=1}^{n} w_i m_i^{p-1}[|y_i(t)|^p e^{\varepsilon t} + \mu^{\frac{p-1}{p}}\sum_{j=1}^{n}(b_{ij}^{+})^{p\xi_{ij}^*}l_j^{p\eta_{ij}^*}\int_{t-\tau_{ij}(t)}^{t}|y_j(s)|^p e^{\varepsilon(s+\tau)}ds].$$

The remaining details of the proof is similar to those in theorem 1 and here are omitted.

Remark 1. In [7] and [10], under the assumption $m_i = \inf_{u\in R} h_i^{'}(u) > 0$, for some special cases of system (1), the exponential stability are studied. Different form [7] and [10], our condition presented here does not require that $h_i(u)$ is differential.

In the following, we consider the autonomous neural networks which are investigated in [1-4]:

$$x_i^{'}(t) = -c_i x_i(t) + \sum_{j=1}^{n} a_{ij} f_j(x_j(t)) + \sum_{j=1}^{n} b_{ij} f_j(x_j(t-\tau_{ij})) + I_i, \qquad i = 1, 2, \ldots, n. \tag{6}$$

System (6) is a special case of system (1). Noting that $m_i = M_i = 1, k_i = l_i, \tau_{ij}^{'}(t) = 0 \quad (i, j = 1, 2, \ldots, n.)$ and applying Theorem 1 above, we can easily obtain the following corollary.

Corollary 1. Under Assumptions (H_2) , the network model (6) has a unique equilibrium point x^* which is globally exponentially stable if there constants α_{ij}, α_{ij}^*, β_{ij}, β_{ij}^*, ξ_{ij}, ξ_{ij}^*, $\eta_{ij}, \eta_{ij}^* \in \mathbb{R}$, $w_i > 0$, $\lambda > 0$, $\mu > 0$, and $p \geq 1$, $i, j = 1, 2, \cdots, n$ such that

$$-c_i + \sum_{j=1}^{n}[\tfrac{p-1}{p}\lambda^{-\frac{1}{p}}|a_{ij}|^{p\alpha_{ij}}k_j^{p\beta_{ij}} + \tfrac{1}{p}\lambda^{\frac{p-1}{p}}\tfrac{w_j}{w_i}|a_{ji}|^{p\alpha_{ji}^*}k_i^{p\beta_{ji}^*}] + \sum_{j=1}^{n}[\tfrac{p-1}{p}\mu^{-\frac{1}{p}}$$
$$|b_{ij}|^{p\xi_{ij}}k_j^{p\eta_{ij}} + \tfrac{1}{p}\mu^{\frac{p-1}{p}}\tfrac{w_j}{w_i}|b_{ji}|^{p\xi_{ji}^*}k_i^{p\eta_{ji}^*}] < 0, \quad i = 1, 2, \cdots, n, \tag{7}$$

where $(p-1)\alpha_{ij} + \alpha_{ij}^* = 1$, $(p-1)\beta_{ij} + \beta_{ij}^* = 1$, $(p-1)\xi_{ij} + \xi_{ij}^* = 1$, $(p-1)\eta_{ij} + \eta_{ij}^* = 1$.

Remark 2. In condition (7), if take $\lambda = \mu = 1$, $\alpha_{ij} = \frac{p-r_{ij}^*}{p(p-1)}$, $\alpha_{ij}^* = \frac{r_{ij}^*}{p}$, $\beta_{ij} = \frac{p-q_{ij}^*}{p(p-1)}$, $\beta_{ij}^* = \frac{q_{ij}^*}{p}$, $\xi_{ij} = \frac{p-r_{ij}^*}{p(p-1)}$, $\xi_{ij}^* = \frac{r_{ij}^*}{p}$, $\eta_{ij} = \frac{p-q_{ij}^*}{p(p-1)}$, $\eta_{ij}^* = \frac{q_{ij}^*}{p}$, then the result in [2] is derived, this implies that the result in Ref.[2] is actually a special case of Corollary 1 here.

4 An Illustrative Example

We consider the following nonautonomous 2-dimensional cellular neural networks with delays.

$$\begin{aligned} x_1'(t) &= -c_1(t)h_1(x_1(t)) + a_{11}(t)f_1(x_1(t)) + a_{12}(t)f_2(x_2(t)) \\ &\quad + b_{11}(t)f_1(x_1(t-\tau_{11}(t))) + b_{12}(t)f_2(x_2(t-\tau_{12}(t))), \\ x_2'(t) &= -c_2(t)h_2(x_2(t)) + a_{21}(t)f_1(x_1(t)) + a_{22}(t)f_2(x_2(t)) \\ &\quad + b_{21}(t)f_1(x_1(t-\tau_{21}(t))) + b_{22}(t)f_2(x_2(t-\tau_{22}(t))), \end{aligned} \tag{8}$$

where $h_i(x) \equiv h(x) = 2x - |\sin x|$ $(i = 1, 2)$. Thus, we have $m_i = 1$, $M_i = 3$ $(i = 1, 2)$.

Example. For system (8), take $c_1(t) = 8 + \cos t$, $c_2(t) = 8 + \sin t$, $a_{11}(t) = 1 + \sin t$, $a_{12}(t) = 1 + \cos t$, $a_{21}(t) = 1 + \cos t$, $a_{22}(t) = 1 + \sin t$, $b_{11}(t) = \sin t$, $b_{12}(t) = \cos t$, $b_{21}(t) = \cos t$, $b_{22}(t) = \sin t$, $f_i(x) \equiv f(x) = \frac{1}{2}(|x+1| - |x-1|)$, $\tau_{ij}(t) \equiv \tau(t) = 1 - \frac{1}{2}e^{-t}$ $(i, j = 1, 2)$, then we have $k_i = l_i = 1$, $b_{ij}^+ = 1$, $\tau_{ij}'(t) = \frac{1}{2}e^{-t} > 0$, $\tau_{ij}^* = \frac{1}{2} < 1$ $(i, j = 1, 2)$. Furthermore, we choose $w_1 = w_2 = 1$ in (4), then we can easily check that

$$-c_1(t)m_1 + |a_{11}|k_1 + |a_{21}|k_1 + \frac{l_1 b_{11}^+}{1-\tau_{11}'(\psi_{11}^{-1}(t))} + \frac{l_1 b_{21}^+}{1-\tau_{21}'(\psi_{21}^{-1}(t))} \leq -2 + \sin t \leq -1,$$
$$-c_2(t)m_2 + |a_{12}|k_2 + |a_{22}|k_2 + \frac{l_2 b_{12}^+}{1-\tau_{12}'(\psi_{12}^{-1}(t))} + \frac{l_2 b_{22}^+}{1-\tau_{22}'(\psi_{22}^{-1}(t))} \leq -2 + \cos t \leq -1.$$

Note that $(0, 0)$ is an equilibrium point of system (8), so it follows form Theorem 1 that the equilibrium point $(0, 0)$ is unique and globally exponentially stable.

Remark 3. Since $h(x)$ is not differential at the point $x = 0$, the results in [7, 10] can't be applied to the above example.

5 Conclusion

Several new sufficient conditions have been derived for ascertaining global exponential stability of a class of generalized nonautonomous cellular neural networks with time-varying delays by constructing suitable Lyapunov function and applying an elementary inequality with the homeomorphism theory. The obtained criteria are shown to be generalized and improved upon existing ones.

Acknowledgement

This work was jointly supported by the National Natural Science Foundation of China under Grant 60574043, the Natural Science Foundation of Jiangsu Province of China under Grant BK2006093.

References

1. Cao, J.: New Results Concerning Exponential Stability and Periodic Solutions of Delayed Cellular Neural Networks. Physics Letters A **307** (2-3) (2003) 136-147
2. Cao, J., Chen, T.: Globally Exponentially Robust Stability and Periodicity of Delayed Neural Networks. Chaos, Solitons & Fractals **22** (2004) 957-963
3. Cao, J., Wang, J.: Global Exponential Stability and Periodicity of Recurrent Neural Networks with Time Delays. IEEE Trans. Circuits Syst. I **52** (5) (2005) 920-931
4. Zhao, H., Cao, J.: New Conditions for Global Exponential Stability of Cellular Neural Networks with Delays. Neural Networks **18** (2005) 1332-1340
5. Huang, C., Huang, L., Yuan, Z.: Global Stability Analysis of A Class of Delayed Cellular Neural Networks. Mathematics and Computers in Simulation **70** (3) (2005) 133-148
6. Zhang, Q., Wei, X., Xu, J.: On Global Exponential Stability of Delayed Cellular Neural Networks with Time-Varying Delays. Applied Mathematics and Computation **162** (2005) 679-686
7. Sun, C., Feng, C.: Exponential Periodicity and Stability of Delayed Neural Networks. Mathematics and Computers in Simulation **66** (2004) 469-478
8. Zhang, Q., Wei, X., Xu, J.: Gloabl Exponential Stability for Nonautonomous Cellular Neural Networks with Delays. Physics Letter A **351** (3) (2006) 153-160
9. Liang, J., Cao, J.: Boundeness and Stability for Recurrent Neural Networks with Variable Coefficients and Time-Varying Delays. Physics Letters A **318** (2003) 53-64
10. Jiang, H., Teng, Z.: Some New Results for Recurrent Neural Networks with Varying-Time Coefficients and Delays. Physics Letters A **338** (2005) 446-460
11. Forti, M., Tesi, A.: New Conditions for Global Stability of Neural Networks with Application to Linear and Quadratic Programming Problems. IEEE Trans. Circuits Systems I Fund. Theory Appl **42** (1995) 354-366
12. Kuang, J.: Applied Inequalities. Shandong Science and Technology Press (2004) (in Chinese)

LMI-Based Approach for Global Asymptotic Stability Analysis of Discrete-Time Cohen-Grossberg Neural Networks

Sida Lin[1], Meiqin Liu[2,*], Yanhui Shi[3],
Jianhai Zhang[2], Yaoyao Zhang[2], and Gangfeng Yan[2]

[1] Office of Zhejiang Provincial Natural Science Foundation, Hangzhou 310007, China
linsida@zjinfo.gov.cn
[2] College of Electrical Engineering, Zhejiang University, Hangzhou 310027, China
liumeiqin@zju.edu.cn
[3] Shijiazhuang Railway Institute, Shijiazhuang 050043, China
shiyanhui408@163.com

Abstract. The global asymptotic stability of discrete-time Cohen–Grossberg neural networks (CGNNs) with or without time delays is studied in this paper. The CGNNs are transformed into discrete-time interval systems, and several sufficient conditions of asymptotic stability for these interval systems are derived by constructing some suitable Lyapunov functionals. The obtained conditions are given in the form of linear matrix inequalities that can be checked numerically and very efficiently by resorting to the MATLAB LMI Control Toolbox.

1 Introduction

Cohen–Grossberg neural networks (CGNNs) were first introduced by Cohen–Grossberg [1] in 1983. The class of networks have been the subject of extensive investigation because of their many important applications, such as pattern recognition, associative memory and combinatorial optimization, etc. Such applications heavily depend on the dynamical behaviors. Thus, the analysis of the dynamical behaviors such as stability is a necessary step for practical design of neural networks. Recently, many scientific and technical workers have been joining the study fields with great interest, and various interesting results for CGNNs with delays or without delays have been reported [2-7]. In general, the continuous-time CGNN model is described by the set of ordinary differential equations [1]:

$$\dot{x}_i(t) = -a_i(x_i(t))\left[b_i(x_i(t)) - \sum_{j=1}^{n} c_{ij} f_j(x_j(t)) + J_i \right], \; i = 1,2,\ldots,n, \tag{1}$$

where $x_i(t)$ is the state variable of the ith neuron, $a_i(\cdot)$ represents an amplification function, $b_i(\cdot)$ is the behaved function, $(c_{ij})_{n\times n}$ denotes the connection matrix in which

[*] Corresponding author.

D. Liu et al. (Eds.): ISNN 2007, Part I, LNCS 4491, pp. 968–976, 2007.

c_{ij} represents the connection strength from neurons i to j, $f_j(\cdot)$ is called an activation functions indicating how the jth neuron responses to its input, and J_i denotes the ith component of an external input source introduced from outside the network to the cell i. When the delays are introduced in CGNN (1), we obtain the following system [6]:

$$\dot{x}_i(t) = -a_i(x_i(t))\left[b_i(x_i(t)) - \sum_{j=1}^{n} c_{ij} f_j(x_j(t-\eta_{ij}(t))) + J_i\right],\ i=1,2,\ldots,n, \tag{2}$$

where $\eta_{ij}(t)$ is the transmission delay. Xiong *et al* [2] formulated the following discrete-time versions of the system (1) and (2):

$$x_i(k+1) = x_i(k) - a_i(x_i(k))\left[b_i(x_i(k)) - \sum_{j=1}^{n} c_{ij} f_j(x_j(k)) + J_i\right],\ i=1,2,\ldots,n, \tag{3}$$

and

$$x_i(k+1) = x_i(k) - a_i(x_i(k))\left[b_i(x_i(k)) - \sum_{j=1}^{n} c_{ij} f_j(x_j(k-\eta_{ij}(k))) + J_i\right],\quad i=1,2,\ldots,n, \tag{4}$$

where $\eta_{ij}(k)$ represent the time delay and is a positive integer with $\eta_{ij}(k)\leq h$. The initial conditions associated with Eq. (3) are of the form

$$x_i(k) = x_i(0),\ i=1,2,\ldots,n, \tag{5}$$

and the initial conditions associated with Eq. (4) are of the form

$$x_i(k) = \varpi_i(k),\ \forall k\in[-h,0],\ i=1,2,\ldots,n, \tag{6}$$

where $\varpi_i(k)$ is the given discrete-time function on $[-h, 0]$.

For system (3) and (4), we make the following assumptions:

Assumption A. Suppose that $a_i(\cdot)$, $b_i(\cdot)$ and $f_j(\cdot)$ are Lipschitz continuous, furthermore, $0<\underline{a}_i \leq a_i(x_i(k)) \leq \bar{a}_i$, $0<\gamma_i \leq b_i(x_i(k))/x_i(k) < +\infty$, and $0 \leq f_j(x_j(k))/x_j(k) \leq \sigma_j$, $i=1,2,\ldots,n, j=1,2,\ldots,n$.

It is worth noting that although Ref. [2] has investigated the global exponential stability of CGNN (3) and (4), the asymptotic stability conditions are also required in wide engineering fields. In this paper, our main purpose is to derive some criteria of global asymptotic stability for the discrete-time CGNN (3) and (4) by a new method. We first transform the CGNNs into the interval systems, and analyze their stability based on the linear matrix inequality (LMI) approaches [8]. The global asymptotic stability of the discrete-time CGNNs is judged by solving some LMIs by using MATLAB LMI Control Toolbox [9].

2 Main Results

Throughout this paper, $\boldsymbol{I}$ denotes identity matrix of appropriate order, $*$ denotes the symmetric parts. If $\boldsymbol{\Lambda}$ is a diagonal positive (or semi-positive) definite matrix, $\Lambda^{1/2}$ denotes a diagonal positive (or semi-positive) definite matrix of which the diagonal element is square root of $\boldsymbol{\Lambda}$'s. The notations $\boldsymbol{X}>\boldsymbol{Y}$ and $\boldsymbol{X}\geq\boldsymbol{Y}$, respectively, where $\boldsymbol{X}$ and $\boldsymbol{Y}$ are matrices of same dimensions, mean that the matrix $\boldsymbol{X}-\boldsymbol{Y}$ is positive definite and semi-positive definite, respectively.

From Theorem 2.1 in [2], system (3) (or system (4)) always has an equilibrium point under *Assumption A*. Suppose $\boldsymbol{x}^*=(x^*_1(k), x^*_2(k), \ldots, x^*_n(k))^{\mathrm{T}}$ be an equilibrium point of system (3) (or (4)), let $y_i(k)=x_i(k)-x_i^*$ ($i=1,2,\ldots,n$), then we can rewrite Eqs. (3) and (4) into

$$y_i(k+1)=y_i(k)-\alpha_i(y_i(k))\left[\beta_i(y_i(k))-\sum_{j=1}^{n}c_{ij}g_j(y_j(k))\right], \tag{7}$$

and

$$y_i(k+1)=y_i(k)-\alpha_i(y_i(k))\left[\beta_i(y_i(k))-\sum_{j=1}^{n}c_{ij}g_j(y_j(k-\eta_{ij}(k)))\right], \tag{8}$$

where $\alpha_i(y_i(k))=a_i(y_i(k)+x_i^*)$, $\beta_i(y_i(k))=b_i(y_i(k)+x_i^*)-b_i(x_i^*)$, $g_j(y_j(k))=f_j(y_j(k)+x_j^*)-f_j(x_j^*)$. If Assumption A is satisfied, $\alpha_i(\cdot)$, $\beta_i(\cdot)$ and $g_j(\cdot)$ are Lipschitz continuous, furthermore, $0<\underline{\alpha}_i\leq\alpha_i(y_i(k))\leq\bar{\alpha}_i$, $0<\underline{\gamma}_i\leq\beta_i(y_i(k))/y_i(k)\leq\bar{\gamma}_i$, and $0<\underline{\sigma}_j\leq g_j(y_j(k))/y_j(k)\leq\bar{\sigma}_j$. Let $\theta_i(y_i(k))=\alpha_i(y_i(k))\beta_i(y_i(k))/y_i(k)$, then Eqs.(7) and (8) can be rewritten as

$$y_i(k+1)=(1-\theta_i(y_i(k))y_i(k)+\alpha_i(y_i(k))\sum_{j=1}^{n}c_{ij}g_j(y_j(k)), \tag{9}$$

and

$$y_i(k+1)=(1-\theta_i(y_i(k))y_i(k)+\alpha_i(y_i(k))\sum_{j=1}^{n}c_{ij}g_j(y_j(k-\eta_{ij}(k))), \tag{10}$$

where $0<\underline{\alpha}_i\underline{\gamma}_i\leq\theta_i(y_i(k))\leq\bar{\alpha}_i\bar{\gamma}_i$. Noting $\boldsymbol{y}(k)=(y_1(k), y_2(k), \ldots, y_n(k))^{\mathrm{T}}$, $\boldsymbol{g}(\cdot)=(g_1(\cdot), g_2(\cdot), \ldots, g_n(\cdot))^{\mathrm{T}}$, $A(k)=\mathrm{diag}(1-\theta_1(y_1(k)), 1-\theta_2(y_2(k)), \ldots, 1-\theta_n(y_n(k)))$, $\boldsymbol{B}(k)=\mathrm{diag}(\alpha_1(y_1(k)), \alpha_2(y_2(k)), \ldots, \alpha_n(y_n(k)))\times(c_{ij})_{n\times n}$, $\boldsymbol{\eta}(k)=(\eta_{ij}(k))_{n\times n}$, then system (9) and (10) can be respectively written as

$$\boldsymbol{y}(k+1)=\boldsymbol{A}(k)\boldsymbol{y}(k)+\boldsymbol{B}(k)\boldsymbol{g}(\boldsymbol{y}(k)), \tag{11}$$

and

$$\boldsymbol{y}(k+1)=\boldsymbol{A}(k)\boldsymbol{y}(k)+\boldsymbol{B}(k)\boldsymbol{g}(\boldsymbol{y}(k-\eta(k))). \tag{12}$$

Since the time-varying state matrices $\boldsymbol{A}(k)$ and $\boldsymbol{B}(k)$ in (11) (or (12)) satisfy the following constraints: $1-\bar{\alpha}_i\bar{\gamma}_i=\underline{a}_{ii}\le a_{ii}(k)\le\bar{a}_{ii}=1-\underline{\alpha}_i\underline{\gamma}_i$, $\underline{\alpha}_i c_{ij}=\underline{b}_{ij}\le b_{ij}(k)\le\bar{b}_{ij}=\bar{\alpha}_i c_{ij}$, i=1,2,…,n, j=1,2,…,n, system (11) (or (12)) is a interval system without (or with) time delays. For convenience, let

$$A^0=\frac{1}{2}(\underline{A}+\bar{A}),\ B^0=\frac{1}{2}(\underline{B}+\bar{B}),$$

$$A^*=\frac{1}{2}(\bar{A}-\underline{A})=(a_{ij}^*)_{n\times n},\ B^*=\frac{1}{2}(\bar{B}-\underline{B})=(b_{ij}^*)_{n\times n},$$

where $\underline{A}=diag(\underline{a}_{11},\underline{a}_{22},\cdots,\underline{a}_{nn})$, $\bar{A}=diag(\bar{a}_{11},\bar{a}_{22},\cdots,\bar{a}_{nn})$, $\underline{B}=(\underline{b}_{ij})_{n\times n}$, $\bar{B}=(\bar{b}_{ij})_{n\times n}$.

Note that each element of matrices A^* and B^* is nonnegative. So we can define

$$E_1=[\sqrt{a_{11}^*}\boldsymbol{I}_1,\cdots,\sqrt{a_{1n}^*}\boldsymbol{I}_1,\cdots,\sqrt{a_{n1}^*}\boldsymbol{I}_n,\cdots,\sqrt{a_{nn}^*}\boldsymbol{I}_n]_{n\times n^2},$$
$$F_1=[\sqrt{a_{11}^*}\boldsymbol{I}_1,\cdots,\sqrt{a_{1n}^*}\boldsymbol{I}_n,\cdots,\sqrt{a_{n1}^*}\boldsymbol{I}_1,\cdots,\sqrt{a_{nn}^*}\boldsymbol{I}_n]^{\mathrm{T}}_{n^2\times n},$$
$$E_2=[\sqrt{b_{11}^*}\boldsymbol{I}_1,\cdots,\sqrt{b_{1n}^*}\boldsymbol{I}_1,\cdots,\sqrt{b_{n1}^*}\boldsymbol{I}_n,\cdots,\sqrt{b_{nn}^*}\boldsymbol{I}_n]_{n\times n^2},$$
$$F_2=[\sqrt{b_{11}^*}\boldsymbol{I}_1,\cdots,\sqrt{b_{1n}^*}\boldsymbol{I}_n,\cdots,\sqrt{b_{n1}^*}\boldsymbol{I}_1,\cdots,\sqrt{b_{nn}^*}\boldsymbol{I}_n]^{\mathrm{T}}_{n^2\times n},$$

where $\boldsymbol{I}_i$ denotes the ith column vector of the identity matrix.

From Lemma 1 in [10], system (11) and (12) are respectively equivalent to the following system:

$$\begin{cases}\boldsymbol{y}(k+1)=(\boldsymbol{A}^0+E_1\Sigma_1F_1)\boldsymbol{y}(k)+(B^0+E_2\Sigma_2F_2)\boldsymbol{g}(\xi(k)),\\ \xi(k)=y(k),\end{cases} \tag{13}$$

and

$$\begin{cases}\boldsymbol{y}(k+1)=(\boldsymbol{A}^0+E_1\Sigma_1F_1)\boldsymbol{y}(k)+(B^0+E_2\Sigma_2F_2)\boldsymbol{g}(\xi(k)),\\ \xi(k)=\boldsymbol{y}(k-\eta(k)),\end{cases} \tag{14}$$

where Σ_i (i=1, 2) are diagonal matrices of appropriate dimension, and absolute values of their diagonal elements are not larger than 1.

In this paper, we assume that the delays in system (12) (or (14)) are constant, i.e., $\eta(\cdot)=h>0$. We will first analyze the stability of system (12) (i.e. system (14)). Before stating the main results, we first need the following lemma.

Lemma 1 [11]. Let $\boldsymbol{D}$ and $\boldsymbol{E}$ be real matrices of appropriate dimensions. Then, for any scalar $\delta>0$,

$$\boldsymbol{DE}+\boldsymbol{E}^{\mathrm{T}}\boldsymbol{D}^{\mathrm{T}}\le\delta\boldsymbol{DD}^{\mathrm{T}}+\delta^{-1}\boldsymbol{E}^{\mathrm{T}}\boldsymbol{E}. \tag{15}$$

Theorem 1. If there exist symmetric positive definite matrices $\boldsymbol{P}$ and $\boldsymbol{\Gamma}$, and diagonal semi-positive definite matrices $\boldsymbol{\Lambda}$ and $\boldsymbol{T}$, and positive scalars δ that satisfy

$$\boldsymbol{G}=\begin{bmatrix}\boldsymbol{G}_{11} & \boldsymbol{G}_{12}\\ * & \boldsymbol{G}_{22}\end{bmatrix}<0, \tag{16}$$

the equilibrium point of system (14) (i.e. system (12)) is globally asymptotically stable. The submatrices of $\boldsymbol{G}$ are

$$G_{11}=\begin{bmatrix}-\boldsymbol{P} & \boldsymbol{P}\boldsymbol{A}^0 & 0 & \boldsymbol{P}\boldsymbol{B}^0\\ * & -\boldsymbol{P}+\boldsymbol{\Gamma}+\delta(F_1^TF_1) & 0 & 0\\ * & * & -\boldsymbol{\Gamma}-2TUV & \Lambda+T(U+V)\\ * & * & * & -2T+\delta(F_2^TF_2)\end{bmatrix},$$

$$G_{12}=\mathrm{diag}(P(E_1E_1^T+E_2E_2^T)^{\frac{1}{2}},0,0,0),$$
$$G_{22}=\mathrm{diag}(-\delta I,-\delta I,-\delta I,-\delta I),$$

where $V=\mathrm{diag}(\underline{\sigma}_1,\underline{\sigma}_2,\cdots,\underline{\sigma}_n)$, $\boldsymbol{U}=\mathrm{diag}(\bar{\sigma}_1,\bar{\sigma}_2,\cdots,\bar{\sigma}_n)$.

Proof. For system (14), we choose the following positive definite Lyapunov functional:

$$V(y(k),\xi(k))=y^T(k)Py(k)+\sum_{i=k-h}^{k-1}y^T(i)\boldsymbol{\Gamma}y(i)+2\sum_{i=1}^{n}\lambda_i\sum_{j=0}^{k-1}g_i(\xi_i(j))\xi_i(j),$$

where $\boldsymbol{P}>0$, $\boldsymbol{\Gamma}>0$, $\lambda_i\geq 0$. Thus, $\forall y(k)\neq 0$, $\forall\xi(k)\neq 0$, $V(y(k),\xi(k))>0$, and $V(y(k),\xi(k))=0$ iff $y(k)=0$, $\xi(k)=0$. The difference of $V(y(k),\xi(k))$ along the solution to (14) is

$$\begin{aligned}\Delta V(y(k),\xi(k))&=V(y(k+1),\xi(k+1))-V(y(k),\xi(k))=y^{\mathrm{T}}(k+1)Py(k+1)\\ &-y^{\mathrm{T}}(k)Py(k)+y^{\mathrm{T}}(k)\boldsymbol{\Gamma}y(k)-y^{\mathrm{T}}(k-h)\boldsymbol{\Gamma}y(k-h)+2\sum_{i=1}^{n}\lambda_ig_i(\xi_i(k))\xi_i(k)\\ &=\left[(\boldsymbol{A}^0+E_1\Sigma_1F_1)\boldsymbol{y}(k)+(B^0+E_2\Sigma_2F_2)\boldsymbol{g}(\boldsymbol{\xi}(k))\right]^{\mathrm{T}}\times P\\ &\times\left[(\boldsymbol{A}^0+E_1\Sigma_1F_1)\boldsymbol{y}(k)+(B^0+E_2\Sigma_2F_2)\boldsymbol{g}(\boldsymbol{\xi}(k))\right]-y^{\mathrm{T}}(k)Py(k)+y^{\mathrm{T}}(k)\boldsymbol{\Gamma}y(k)\\ &-y^{\mathrm{T}}(k-h)\boldsymbol{\Gamma}y(k-h)+2\sum_{i=1}^{n}\lambda_ig_i(\xi_i(k))y_i(k-h)=\begin{bmatrix}y(k)\\ y(k-h)\\ g(\xi(k))\end{bmatrix}^{\mathrm{T}}\cdot T_0\cdot\begin{bmatrix}y(k)\\ y(k-h)\\ g(\xi(k))\end{bmatrix},\end{aligned}$$

where

$$T_0 = \begin{bmatrix} \begin{pmatrix} (\boldsymbol{A}^0 + E_1\Sigma_1 F_1)^{\mathrm{T}} \boldsymbol{P} (\boldsymbol{A}^0 + E_1\Sigma_1 F_1) \\ -\boldsymbol{P} + \boldsymbol{\Gamma} \end{pmatrix} & 0 & (\boldsymbol{A}^0 + E_1\Sigma_1 F_1)^{\mathrm{T}} \boldsymbol{P} (B^0 + E_2\Sigma_2 F_2) \\ 0 & -\boldsymbol{\Gamma} & \boldsymbol{\Lambda} \\ (B^0 + E_2\Sigma_2 F_2)^{\mathrm{T}} \boldsymbol{P} (\boldsymbol{A}^0 + E_1\Sigma_1 F_1) & \boldsymbol{\Lambda} & (B^0 + E_2\Sigma_2 F_2)^{\mathrm{T}} \boldsymbol{P} (B^0 + E_2\Sigma_2 F_2) \end{bmatrix},$$

$\boldsymbol{\Lambda} = \mathrm{diag}(\lambda_1, \lambda_2, \cdots, \lambda_n)$ and $\boldsymbol{\Lambda} \geq 0$.

The sector conditions, $0 < \underline{\sigma}_j \leq g_j(\xi_j(k))/\xi_j(k) \leq \bar{\sigma}_j$, $j=1,2,\ldots,n$, can be rewritten as follows

$$[g_j(\xi_j(k)) - \underline{\sigma}_j \xi_j(k)] \cdot [g_j(\xi_j(k)) - \bar{\sigma}_j \xi_j(k)] \leq 0, \quad j=1,2,\ldots, n, \tag{17}$$

which is equivalent to

$$2g_j^2(\xi_j(k)) - 2(\underline{\sigma}_j + \bar{\sigma}_j) y_j(k-h) g_j(\xi_j(k)) + 2\underline{\sigma}_j \bar{\sigma}_j y_j^2(k-h) \leq 0, \; j=1,2, \ldots, n, \tag{18}$$

Rewrite (18) in the matrix form

$$\begin{bmatrix} y(k) \\ y(k-h) \\ g(\xi(k)) \end{bmatrix}^{\mathrm{T}} \cdot T_j \cdot \begin{bmatrix} y(k) \\ y(k-h) \\ g(\xi(k)) \end{bmatrix} \leq 0,$$

where $T_j = \begin{bmatrix} 0 & 0 & 0 \\ 0 & 2U_j V_j & -(U_j + V_j) \\ 0 & -(U_j + V_j) & 2I \end{bmatrix}$, $U_j = diag(0, \cdots, \bar{\sigma}_j, \cdots, 0)$, and $V_j = diag(0, \cdots, \underline{\sigma}_j, \cdots, 0)$.

By the S-procedure [8], if there exist $\tau_j \geq 0$ ($j=1,2,\ldots,n$), such that the following inequality holds

$$\boldsymbol{T}_0 - \sum_{j=1}^{n} \tau_j \boldsymbol{T}_j$$

$$= \begin{bmatrix} (\boldsymbol{A}^0 + E_1\Sigma_1 F_1)^{\mathrm{T}} \boldsymbol{P} (\boldsymbol{A}^0 + E_1\Sigma_1 F_1) - \boldsymbol{P} + \boldsymbol{\Gamma} & 0 & (\boldsymbol{A}^0 + E_1\Sigma_1 F_1)^{\mathrm{T}} \boldsymbol{P} (B^0 + E_2\Sigma_2 F_2) \\ 0 & -\boldsymbol{\Gamma} & \boldsymbol{\Lambda} \\ (B^0 + E_2\Sigma_2 F_2)^{\mathrm{T}} \boldsymbol{P} (\boldsymbol{A}^0 + E_1\Sigma_1 F_1) & \boldsymbol{\Lambda} & (B^0 + E_2\Sigma_2 F_2)^{\mathrm{T}} \boldsymbol{P} (B^0 + E_2\Sigma_2 F_2) \end{bmatrix}$$

$$- \begin{bmatrix} 0 & 0 & 0 \\ 0 & 2TUV & T(U+V) \\ 0 & T(U+V) & 2T \end{bmatrix}$$

$$
=\begin{bmatrix}
\begin{pmatrix}(\boldsymbol{A}^0+E_1\Sigma_1F_1)^{\mathrm T}\boldsymbol{P}(\boldsymbol{A}^0+E_1\Sigma_1F_1)\\ -\boldsymbol{P}+\boldsymbol{\Gamma}\end{pmatrix} & 0 & (\boldsymbol{A}^0+E_1\Sigma_1F_1)^{\mathrm T}\boldsymbol{P}(B^0+E_2\Sigma_2F_2)\\
0 & \begin{pmatrix}-\boldsymbol{\Gamma}\\ -2TUV\end{pmatrix} & \Lambda+T(U+V)\\
(B^0+E_2\Sigma_2F_2)^{\mathrm T}\boldsymbol{P}(\boldsymbol{A}^0+E_1\Sigma_1F_1) & \begin{pmatrix}\Lambda\\ +T(U+V)\end{pmatrix} & (B^0+E_2\Sigma_2F_2)^{\mathrm T}\boldsymbol{P}(B^0+E_2\Sigma_2F_2)-2T
\end{bmatrix}
$$

$$<0, \tag{19}$$

where $\boldsymbol{T}=\mathrm{diag}(\tau_1,\tau_2,\cdots,\tau_n)$ and $\boldsymbol{T}\geq 0$, then $T_0<0$, that is, $\forall y(k)\neq 0$, $\forall \xi(k)\neq 0$, $\Delta V(y(k),\ \xi(k))<0$ and $\Delta V(\boldsymbol{x}_k,\ \xi_k)=0$ iff $\boldsymbol{y}(k)=0$, $\xi(k)=0$. Therefore, we can conclude that the equilibrium point of system (14) (i.e. system (12)) is globally asymptotically stable. Using the Schur complements [8], the inequality (19) is equivalent to

$$
\begin{bmatrix}
-\boldsymbol{P} & \boldsymbol{P}(\boldsymbol{A}^0+E_1\Sigma_1F_1) & 0 & \boldsymbol{P}(B^0+E_2\Sigma_2F_2)\\
* & -\boldsymbol{P}+\boldsymbol{\Gamma} & 0 & 0\\
* & * & -\boldsymbol{\Gamma}-2TUV & \Lambda+T(U+V)\\
* & * & * & -2T
\end{bmatrix}<0. \tag{20}
$$

The inequality (20) can be rewritten as

$$
\begin{bmatrix}
-\boldsymbol{P} & \boldsymbol{P}\boldsymbol{A}^0 & 0 & \boldsymbol{P}B^0\\
* & -\boldsymbol{P}+\boldsymbol{\Gamma} & 0 & 0\\
* & * & -\boldsymbol{\Gamma}-2TUV & \Lambda+T(U+V)\\
* & * & * & -2T
\end{bmatrix}+M^{\mathrm T}+M<0, \tag{21}
$$

where $M=\begin{bmatrix}\boldsymbol{P}E_1 & \boldsymbol{P}E_2\\ 0 & 0\\ 0 & 0\\ 0 & 0\end{bmatrix}\cdot\begin{bmatrix}0 & \Sigma_1F_1 & 0 & 0\\ 0 & 0 & 0 & \Sigma_2F_2\end{bmatrix}$. Using Lemmal 1,

$$M^{\mathrm T}+M\leq \delta^{-1}diag(P(E_1E_1^T+E_2E_2^T)P,0,0,0)+\delta diag(0,F_1^TF_1,0,F_2^TF_2).$$

Therefore, if

$$\begin{bmatrix} -\boldsymbol{P} & \boldsymbol{P}\boldsymbol{A}^0 & 0 & \boldsymbol{P}\boldsymbol{B}^0 \\ * & -\boldsymbol{P}+\boldsymbol{\Gamma} & 0 & 0 \\ * & * & -\Gamma-2TUV & \Lambda+T(U+V) \\ * & * & * & -2T \end{bmatrix} + \delta^{-1}diag(P(E_1E_1^T+E_2E_2^T)P,0,0,0)$$

$$+\delta diag(0,F_1^TF_1,0,F_2^TF_2) = G_{11} + diag(P(E_1E_1^T+E_2E_2^T)^{\frac{1}{2}},0,0,0)$$

$$\times diag(\delta^{-1}I,\delta^{-1}I,\delta^{-1}I,\delta^{-1}I)\times diag((E_1E_1^T+E_2E_2^T)^{\frac{1}{2}}P,0,0,0)<0 \tag{22}$$

holds, the inequality (21) is true. By applying the Schur complements [8], the inequality (22) can be expressed as (16). This completes the proof.

For system (13) (i.e. system(11)), we have the following stability theorem.

Theorem 2. If there exist symmetric positive definite matrices $\boldsymbol{P}$, and diagonal semi-positive definite matrices Λ and T, and positive scalars δ that satisfy

$$\begin{bmatrix} -P & PA^0 & PB^0 & P(E_1E_1^T+E_2E_2^T)^{\frac{1}{2}} & 0 & 0 \\ * & \begin{pmatrix} -P-2TUV \\ +\delta(F_1^TF_1) \end{pmatrix} & \Lambda+\boldsymbol{T}(\boldsymbol{U}+\boldsymbol{V}) & 0 & 0 & 0 \\ * & * & -2T+\delta(F_2^TF_2) & 0 & 0 & 0 \\ * & * & * & -\delta I & 0 & 0 \\ * & * & * & * & -\delta I & 0 \\ * & * & * & * & * & -\delta I \end{bmatrix} < 0, \tag{23}$$

where $V=\mathrm{diag}(\underline{\sigma}_1,\underline{\sigma}_2,\cdots,\underline{\sigma}_n)$, $\boldsymbol{U}=\mathrm{diag}(\bar{\sigma}_1,\bar{\sigma}_2,\cdots,\bar{\sigma}_n)$, the equilibrium point of system (13) (i.e. system (11)) is globally asymptotically stable.

The proof of Theorem 2 follows the same idea as that for Theorem 1, and is omitted here. For system (13), we adopt the following Lyapunov function:

$$V(y(k),\xi(k)) = y^T(k)Py(k)+2\sum_{i=1}^{n}\lambda_i\sum_{j=0}^{k-1}g_i(\xi_i(j))\xi_i(j).$$

3 Conclusion

We propose a novel method to analyze the global asymptotic stability of discrete-time CGNNs with or without time delays. We converting the CGNNs into the interval systems and analyze the stability of interval systems based on the LMI approaches. Our stability conditions are less conservative and easily checked by solving LMIs. Moreover, our results are applicable to both symmetric and asymmetric interconnection matrices and can be conveniently used to design some neural networks with global asymptotic stability property.

Acknowledgment. This work was supported in part by the National Natural Science Foundation of China under Grant 60504024, in part by the Research Project of Zhejiang Provincial Education Department under Grant 20050905, in part by the Zhejiang Provincial Natural Science Foundation of China under Grant Y106010, and in part by the Specialized Research Fund for the Doctoral Program of Higher Education (SRFDP), China under Grant 20060335022.

References

1. Cohen, M.A., Grossberg, S.: Absolute Stability of Global Pattern Formation and Parallel Memory Storage by Competitive Neural Network. IEEE Trans. Systems, Man, and Cybern. **13** (1983) 815-826
2. Xiong, W., Cao, J.: Global Exponential Stability of Discrete-time Cohen–Grossberg Neural Networks. Neurocomputing **64** (2005) 433-446
3. Wang, W., Cao, J.: LMI-based Criteria for Globally Robust Stability of Delayed Cohen-Grossberg Neural Networks. IEE Proc.-Control Theory Appl. **153** (2006) 397-402
4. Guo, S., Huang, L.: Stability Analysis of Cohen-Grossberg Neural Networks. IEEE Trans. Neural Networks **17** (2006) 106-117
5. Chen, Y.: Global Asymptotic Stability of Delayed Cohen-Grossberg Neural Network. IEEE Trans. Circuits and Systems **I 53** (2006) 351-357
6. Cao, J., Liang, J.: Boundedness and Stability for Cohen- Grossberg Neural Networks with Time-varying Delays. Journal of Mathematical Analysis and Applications **296** (2004) 665-685
7. Zeng, Z. G., Wang, J.: Improved Conditions for Global Exponential Stability of Recurrent Neural Networks with Time-varying Delays. IEEE Trans. Neural Networks **17** (2006) 623-635
8. Boyd, S. P., Ghaoui, L. E., Feron, E., Balakrishnan, V.: Linear Matrix Inequalities in System and Control Theory. SIAM, Philadelphia (1994)
9. Gahinet, P., Nemirovski, A., Laub, A. J., Chilali, M.: LMI Control Toolbox- for Use with Matlab. The MATH Works, Inc., Natick, MA (1995)
10. Li, C. D., Liao, X. F., Zhang, R.: Global Asymptotical Stability of Multi-delayed Interval Neural Networks: an LMI Approach. Phys. Lett. A **328** (2004) 452-462
11. Khargonekar, P. P., Petersen, I. R., Zhou, K.: Robust Stabilization of Uncertain Linear Systems: Quadratic Stability and H_∞ Control Theory. IEEE Trans. Automatic Control **35** (1990) 356-361

Novel LMI Criteria for Stability of Neural Networks with Distributed Delays

Qiankun Song[1] and Jianting Zhou[2]

[1] Department of Mathematics, Chongqing Jiaotong University, Chongqing 400074, China
qiankunsong@163.com
[2] College of Civil Engineering and Architecture, Chongqing Jiaotong University, Chongqing 400074, China

Abstract. In this paper, the global asymptotic and exponential stability are investigated for a class of neural networks with distributed time-varying delays. By using appropriate Lyapunov-Krasovskii functional and linear matrix inequality (LMI) technique, two delay-dependent sufficient conditions in LMIs form are obtained to guarantee the global asymptotic and exponential stability of the addressed neural networks. The proposed stability criteria do not require the monotonicity of the activation functions and the differentiability of the distributed time-varying delays, which means that the results generalize and further improve those in the earlier publications. An example is given to show the effectiveness of the obtained condition.

1 Introduction

Time delays inevitably exist in neural networks due to various reasons. For example, time delays can be caused by the finite switching speed of amplifier circuits in neural networks [1] or deliberately introduced to achieve tasks of dealing with motion-related problems such as moving image processing [2]. The existence of time delay may lead to some complex dynamic behaviors such as oscillation, divergence, chaos, instability or other poor performance of the neural networks [3]. Therefore, stability analysis for neural networks with delays has been an attractive subject of research in the past few years. Various sufficient conditions, either delay-dependent or delay-independent, have been proposed for stability of neural networks with constant and time-varying delays, for example, see [1]-[10] and references therein.

On the other hand, as pointed out in [11]-[14], neural networks usually have a spatial extent due to the presence of a multitude of parallel pathways with a variety of axon sizes and lengths, and hence there is a distribution of propagation delays over a period of time. In [11], a neural circuit was designed with distributed delays, which solves a general problem of recognized patterns in time-dependent signal. In [12]-[18], the global stability, convergency and boundedness were investigated for neural networks with distributed delays, some sufficient conditions were given to ensure the stability, convergency and boundedness. However, the

D. Liu et al. (Eds.): ISNN 2007, Part I, LNCS 4491, pp. 977–985, 2007.

given criteria were based upon certain diagonal dominance or M-matrix conditions on weight matrices of the networks, which only depend on absolute values of the weights and ignore the signs of the weights, and hence are somewhat conservative.

Motivated by the above discussions, the objective of this paper is to study the asymptotic and exponential stability of neural networks with distributed time-varying delays by employing a new Lyapunov-Krasovskii functional. The obtained sufficient conditions are expressed in terms of LMI, which can be checked numerically using the effective LMI toolbox in MATLAB. An example is given to show the effectiveness of the proposed criteria.

2 Problem Formulation and Preliminaries

In this paper, we consider the following model

$$\frac{dx(t)}{dt} = -Cx(t) + Af(x(t)) + B\int_{t-\sigma(t)}^{t} f(x(s))ds + J, \quad t \geq 0, \tag{1}$$

where n corresponds to the number of neurons; $x(t) = (x_1(t), \cdots, x_n(t))^T \in R^n$ is the state vector of the network at time t; $f(x(t)) = (f_1(x_1(t)), \cdots, f_n(x_n(t)))^T$ denotes the neuron activation at time t; $C = \mathrm{diag}(c_1, \cdots, c_n) > 0$ is a positive diagonal matrix, $A = (a_{ij})_{n\times n}$ and $B = (b_{ij})_{n\times n}$ represent the connection weight matrix and the distributively delayed connection weight matrix, respectively; $J = (J_1, \cdots, J_n)^T \in R^n$ is a constant external input vector. $\sigma(t)$ denotes the distributed time-varying delay, and is assumed to satisfy $0 \leq \sigma(t) \leq \sigma$, where σ is constant.

The initial condition associated with model (1) is

$$x_i(s) = \varphi_i(s), \quad i = 1, 2, \cdots n, \tag{2}$$

where $\varphi_i(s)$ is bounded and continuous on $[-\sigma, 0]$.

Throughout this paper, we make the following assumptions:

(H1) The activation functions are bounded.
(H2) There exists a positive diagonal matrix $F = \mathrm{diag}(F_1, F_2, \cdots, F_n)$ such that

$$|f_j(v_1) - f_j(v_2)| \leq F_j|v_1 - v_2|$$

for all v_1, $v_2 \in R$, $j = 1, 2, \cdots, n$.

Since activation functions are bounded, by employing the Brouwer's fixed point theorem, one can easily prove that there exists an equilibrium point for model (1). In the sequel we shall analyze the global asymptotic and exponential stability of the equilibrium point, which in turn implies the uniqueness of the equilibrium point.

To simplify the stability analysis of model (1), we let x^* be the equilibrium point of model (1), and shift the intended equilibrium point x^* to the origin by letting $y = x - x^*$, and then model (1) can be transformed into:

$$\frac{dy(t)}{dt} = -Cy(t) + Ag(y(t)) + B\int_{t-\sigma(t)}^{t} g(y(s))ds, \quad t \geq 0, \tag{3}$$

where $g_j(y_j(t)) = f_j(x_j(t) + x_j^*) - f_j(x_j^*)$. It follows from assumption (**H2**) that

$$|g_j(y_j(t))| \leq F_j|y_j(t)|, \quad j = 1, 2, \cdots, n.$$

Thus, for any positive diagonal matrix Λ, we can obtain that

$$g^T(y(t))\Lambda g(y(t)) \leq y^T(t)F\Lambda Fy(t). \tag{4}$$

To obtain our main results, the following lemmas are necessary.

Lemma 1. *([7]) Let $a, b \in R^n$, P be a positive definite matrix, then*

$$2a^Tb \leq a^TP^{-1}a + b^TPb.$$

Lemma 2. *([6]) For any constant matrix $W \in R^{m\times m}$, $W^T = W > 0$, scalar $h > 0$, vector function $\omega : [0, h] \to R^m$ such that the integrations concerned are well defined, then*

$$\Big(\int_0^h \omega(s)ds\Big)^T W\Big(\int_0^h \omega(s)ds\Big) \leq h\int_0^h \omega^T(s)W\omega(s)ds.$$

Lemma 3. *([9]) Given constant matrices P, Q and R, where $P^T = P$, $Q^T = Q$, then*

$$\begin{pmatrix} P & R \\ R^T & -Q \end{pmatrix} < 0$$

is equivalent to the following conditions

$$Q > 0, \quad P + RQ^{-1}R^T < 0.$$

3 Main Results

Theorem 1. *Under assumptions (**H1**) and (**H2**), the equilibrium point 0 of model (3) is globally asymptotically stable if there exist a symmetric positive definite matrix P, four positive diagonal matrices Y_i $(i = 1, 2, 3, 4)$, and two matrices Q_1, Q_2 such that the following LMI holds:*

$$\Omega = \begin{pmatrix} \Omega_1 & \Omega_2 & Q_1^TA & Q_1^TB & 0 & 0 \\ \Omega_2^T & -Q_2^T - Q_2 & 0 & 0 & Q_2^TA & Q_2^TB \\ A^TQ_1 & 0 & -Y_1 & 0 & 0 & 0 \\ B^TQ_1 & 0 & 0 & -Y_2 & 0 & 0 \\ 0 & A^TQ_2 & 0 & 0 & -Y_3 & 0 \\ 0 & B^TQ_2 & 0 & 0 & 0 & -Y_4 \end{pmatrix} < 0, \tag{5}$$

where $\Omega_1 = -Q_1^TC - CQ_1 + FY_1F + FY_3F + \sigma^2F(Y_2 + Y_4)F$, $\Omega_2 = P - Q_1^T - CQ_2$.

Proof. Consider the following Lyapunov-Krasovskii functional for model (3) as

$$V(t) = V_1(t) + V_2(t),$$

where

$$V_1(t) = y^T(t)Py(t), \tag{6}$$

$$V_2(t) = \sigma \int_{t-\sigma}^{t} \int_{\xi}^{t} g^T(y(s))(Y_2 + Y_4)g(y(s))dsd\xi, \tag{7}$$

Evaluating the time derivative of $V_1(t)$ along the trajectories of model (3), we obtain

$$\begin{aligned}
\frac{dV_1(t)}{dt} &= 2y^T(t)P\dot{y}(t) \\
&= 2y^T(t)P\dot{y}(t) + 2\Big(y^T(t)Q_1^T + \dot{y}^T(t)Q_2^T\Big)\Big(- \dot{y}(t) - Cy(t) \\
&\quad +Ag(y(t)) + B\int_{t-\sigma(t)}^{t} g(y(s))ds\Big) \\
&= 2y^T(t)P\dot{y}(t) - 2y^T(t)Q_1^T\dot{y}(t) - 2y^T(t)Q_1^TCy(t) + 2y^T(t)Q_1^TAg(y(t)) \\
&\quad +2y^T(t)Q_1^TB\int_{t-\sigma(t)}^{t} g(y(s))ds - 2\dot{y}^T(t)Q_2^T\dot{y}(t) - 2\dot{y}^T(t)Q_2^TCy(t) \\
&\quad +2\dot{y}^T(t)Q_2^TAg(y(t)) + 2\dot{y}^T(t)Q_2^TB\int_{t-\sigma(t)}^{t} g(y(s))ds \\
&\le 2y^T(t)(P - Q_1^T - CQ_2)\dot{y}(t) - 2y^T(t)Q_1^TCy(t) + g^T(y(t))Y_1g(y(t)) \\
&\quad +y^T(t)Q_1^TAY_1^{-1}A^TQ_1y(t) + y^T(t)Q_1^TBY_2^{-1}B^TQ_1y(t) \\
&\quad +\Big(\int_{t-\sigma(t)}^{t} g(y(s))ds\Big)^TY_2\Big(\int_{t-\sigma(t)}^{t} g(y(s))ds\Big) - 2\dot{y}^T(t)Q_2^T\dot{y}(t) \\
&\quad +\dot{y}^T(t)Q_2^TAY_3^{-1}A^TQ_2\dot{y}(t) + \dot{y}^T(t)Q_2^TBY_4^{-1}B^TQ_2\dot{y}(t) \\
&\quad +g^T(y(t))Y_3g(y(t))) + \Big(\int_{t-\sigma(t)}^{t} g(y(s))ds\Big)^TY_4\Big(\int_{t-\sigma(t)}^{t} g(y(s))ds\Big) \\
&\le y^T(t)\Big(- 2Q_1^TC + Q_1^TAY_1^{-1}A^TQ_1 + FY_1F + Q_1^TBY_2^{-1}B^TQ_1 \\
&\quad +FY_3F\Big)y(t) + y^T(t)\Big(2P - 2Q_1^T - 2CQ_2\Big)\dot{y}(t) \\
&\quad +\dot{y}^T(t)\Big(- 2Q_2^T + Q_2^TAY_3^{-1}A^TQ_2 + Q_2^TBY_4^{-1}B^TQ_2\Big)\dot{y}(t) \\
&\quad +\sigma(t)\int_{t-\sigma(t)}^{t} g^T(y(s))(Y_2 + Y_4)g(y(s))ds.
\end{aligned} \tag{8}$$

In deriving the above inequalities, we have made use of Lemma 1, inequality (4) and Lemma 2.

Calculating the time derivatives of $V_2(t)$, we have

$$\frac{dV_2(t)}{dt} = \sigma^2y^T(t)F(Y_2 + Y_4)Fy(t)$$

$$-\sigma \int_{t-\sigma}^{t} g^T(y(s))(Y_2 + Y_4)g(y(s))ds. \tag{9}$$

It follows from inequalities (8), (9) and $\sigma(t) \leq \sigma$ that

$$\begin{aligned}\frac{dV(t)}{dt} &\leq y^T(t)\Big(- 2Q_1^T C + Q_1^T A Y_1^{-1} A^T Q_1 + F Y_1 F + Q_1^T B Y_2^{-1} B^T Q_1 \\ &\quad + F Y_3 F + \sigma^2 F(Y_2 + Y_4)F\Big)y(t) + y^T(t)\Big(2P - 2Q_1^T - 2CQ_2\Big)\dot{y}(t) \\ &\quad + \dot{y}^T(t)\Big(- 2Q_2^T + Q_2^T A Y_3^{-1} A^T Q_2 + Q_2^T B Y_4^{-1} B^T Q_2\Big)\dot{y}(t) \\ &= (y^T(t), \dot{y}^T(t))\Omega^*(y^T(t), \dot{y}^T(t))^T,\end{aligned}$$

where

$$\Omega^* = \begin{pmatrix} \Omega_1^* & P - Q_1^T - CQ_2 \\ P - Q_1 - Q_2^T C & \Omega_2^* \end{pmatrix}$$

with

$$\begin{aligned}\Omega_1^* &= -Q_1^T C - CQ_1 + Q_1^T A Y_1^{-1} A^T Q_1 + F Y_1 F + Q_1^T B Y_2^{-1} B^T Q_1 \\ &\quad + F Y_3 F + \sigma^2 F(Y_2 + Y_4)F, \\ \Omega_2^* &= -Q_2^T - Q_2 + Q_2^T A Y_3^{-1} A^T Q_2 + Q_2^T B Y_4^{-1} B^T Q_2.\end{aligned}$$

It is easy to verify the equivalence of $\Omega < 0$ and $\Omega^* < 0$ by using Lemma 3. Thus, from condition (5), we get

$$\frac{dV(t)}{dt} < 0$$

for all $y(t) \neq 0$, which implies that the origin of model (3) is globally asymptotically stable. The proof is completed.

Next, we are now in a position to discuss the exponential stability of model (3) as follows.

Theorem 2. *Under the conditions of Theorem 1, model (3) is globally exponentially stable and the exponential convergence rate index ε can be estimated from the inequality*

$$\Pi < 0, \tag{10}$$

where

$$\Pi = \begin{pmatrix} \Pi_1 & P - Q_1^T - (C - \varepsilon I)Q_2 \\ P - Q_1 - Q_2^T(C - \varepsilon I) & \Pi_2 \end{pmatrix},$$

with

$$\begin{aligned}\Pi_1 &= -(C - \varepsilon I)Q_1 - Q_1^T(C - \varepsilon I) + Q_1^T A Y_1^{-1} A^T Q_1 + F Y_1 F \\ &\quad + Q_1^T B Y_2^{-1} B^T Q_1 + F Y_3 F + \sigma^2 e^{2\sigma\varepsilon} F(Y_2 + Y_4)F, \\ \Pi_2 &= -Q_2 - Q_2^T + Q_2^T A Y_3^{-1} A^T Q_2 + Q_2^T B Y_4^{-1} B^T Q_2.\end{aligned}$$

Proof. From $\Omega < 0$, we have $\Omega^* < 0$. Thus, we can choose a sufficiently small constant $\varepsilon > 0$ such that $\Pi < 0$.

Letting $z(t) = e^{\varepsilon t}y(t)$, then model (3) can be transformed into model:

$$\frac{dz(t)}{dt} = -(C-\varepsilon I)z(t) + e^{\varepsilon t}Ag(e^{-\varepsilon t}z(t)) + e^{\varepsilon t}B\int_{t-\sigma(t)}^{t} g(e^{-\varepsilon s}z(s))ds \quad (11)$$

for $t \geq 0$. Consider the following Lyapunov-Krasovskii functional candidate for model (11) as

$$V(t) = V_1(t) + V_2(t),$$

where

$$V_1(t) = z^T(t)Pz(t), \quad (12)$$

$$V_2(t) = \sigma e^{2\sigma\varepsilon}\int_{t-\sigma}^{t}\int_{\xi}^{t} z^T(s)F(Y_2+Y_4)Fz(s)dsd\xi. \quad (13)$$

Along the trajectories of model (11), we can obtain the time derivative of $V_1(t)$ as follows

$$\begin{aligned}
\frac{dV_1(t)}{dt} &= 2z^T(t)P\dot{z}(t) \\
&= 2z^T(t)P\dot{z}(t) + 2\Big(z^T(t)Q_1^T + \dot{z}^T(t)Q_2^T\Big)\Big(-\dot{z}(t) - (C-\varepsilon I)z(t) \\
&\quad + e^{\varepsilon t}Ag(e^{-\varepsilon t}z(t)) + e^{\varepsilon t}B\int_{t-\sigma(t)}^{t} g(e^{-\varepsilon s}z(s))ds\Big) \\
&\leq 2z^T(t)(P - Q_1^T - (C-\varepsilon I)Q_2)\dot{z}(t) - 2z^T(t)Q_1^T(C-\varepsilon I)z(t) \\
&\quad + z^T(t)Q_1^TAY_1^{-1}A^TQ_1z(t) + e^{2\varepsilon t}g^T(e^{-\varepsilon t}z(t))Y_1g(e^{-\varepsilon t}z(t)) \\
&\quad + z^T(t)Q_1^TBY_2^{-1}B^TQ_1z(t) \\
&\quad + e^{2\varepsilon t}\Big(\int_{t-\sigma(t)}^{t} g(e^{-\varepsilon s}z(s))ds\Big)^TY_2\Big(\int_{t-\sigma(t)}^{t} g(e^{-\varepsilon s}z(s))ds\Big) \\
&\quad - 2\dot{z}^T(t)Q_2^T\dot{z}(t) + \dot{z}^T(t)Q_2^TAY_3^{-1}A^TQ_2\dot{z}(t) \\
&\quad + e^{2\varepsilon t}g^T(e^{-\varepsilon t}z(t))Y_3g(e^{-\varepsilon t}z(t)) + \dot{z}^T(t)Q_2^TBY_4^{-1}B^TQ_2\dot{z}(t) \\
&\quad + e^{2\varepsilon t}\Big(\int_{t-\sigma(t)}^{t} g(e^{-\varepsilon s}z(s))ds\Big)^TY_4\Big(\int_{t-\sigma(t)}^{t} g(e^{-\varepsilon s}z(s))ds\Big) \\
&\leq z^T(t)\Big(-2Q_1^T(C-\varepsilon I) + Q_1^TAY_1^{-1}A^TQ_1 + FY_1F + Q_1^TBY_2^{-1}B^TQ_1 \\
&\quad + FY_3F\Big)z(t) + z^T(t)\Big(2P - 2Q_1^T - 2(C-\varepsilon I)Q_2\Big)\dot{z}(t) \\
&\quad + \dot{z}^T(t)\Big(-2Q_2^T + Q_2^TAY_3^{-1}A^TQ_2 + Q_2^TBY_4^{-1}B^TQ_2\Big)\dot{z}(t) \\
&\quad + \sigma e^{2\sigma\varepsilon}\int_{t-\sigma(t)}^{t} z^T(s)F(Y_2+Y_4)Fz(s)ds. \qquad (14)
\end{aligned}$$

In deriving the above inequalities, we have made use of Lemma 1, inequality (4) and Lemma 2.

Calculating the time derivatives of $V_2(t)$, we have

$$\begin{aligned}\frac{dV_2(t)}{dt} &= \sigma e^{2\sigma\varepsilon}\Big(\int_{t-\sigma}^{t} z^T(t)F(Y_2+Y_4)Fz(t)d\xi \\ &\quad - \int_{t-\sigma}^{t} z^T(s)F(Y_2+Y_4)Fz(s)ds\Big) \\ &\leq \sigma^2 e^{2\sigma\varepsilon} z^T(t)F(Y_2+Y_4)Fz(t) \\ &\quad -\sigma e^{2\sigma\varepsilon}\int_{t-\sigma(t)}^{t} z^T(s)F(Y_2+Y_4)Fz(s)ds. \end{aligned} \tag{15}$$

It follows from inequalities (14) and (15) that

$$\begin{aligned}\frac{dV(t)}{dt} &\leq z^T(t)\Big(-2Q_1^T(C-\varepsilon I)+Q_1^TAY_1^{-1}A^TQ_1+FY_1F+Q_1^TBY_2^{-1}B^TQ_1 \\ &\quad +FY_3F+\sigma^2e^{2\sigma\varepsilon}F(Y_2+Y_4)\Big)z(t) \\ &\quad +z^T(t)\Big(2P-2Q_1^T-2(C-\varepsilon I)Q_2\Big)\dot{z}(t) \\ &\quad +\dot{z}^T(t)\Big(-2Q_2^T+Q_2^TAY_3^{-1}A^TQ_2+Q_2^TBY_4^{-1}B^TQ_2\Big)\dot{z}(t) \\ &= (z^T(t),\dot{z}^T(t))\Pi(z^T(t),\dot{z}^T(t))^T,\end{aligned}$$

which indicates from $\Pi<0$ that

$$\frac{dV(t)}{dt}\leq 0,\quad t\geq 0.$$

Hence,

$$V(t)\leq V(0),\quad t\geq 0.$$

It is easy to compute

$$V(t)\geq \lambda_{\min}(P)\|z(t)\|^2,\quad t\geq 0,$$

and

$$V(0)\leq (\lambda_{\max}(P)+\sigma^3e^{2\sigma\varepsilon}\lambda_{\max}(FY_2F+FY_4F))\sup_{-\sigma\leq s\leq 0}\|z(s)\|^2.$$

Take $M=\left(\frac{\lambda_{\max}(P)+\sigma^3e^{2\sigma\varepsilon}\lambda_{\max}(FY_2F+FY_4F)}{\lambda_{\min}(P)}\right)^{1/2}$, then

$$\|z(t)\|\leq M\sup_{-\sigma\leq s\leq 0}\|z(s)\|,\quad t\geq 0.$$

It follows from $z(t)=e^{\varepsilon t}y(t)$ that

$$\|y(t)\|\leq Me^{\varepsilon t}\sup_{-\sigma\leq s\leq 0}\|y(s)\|,\quad t\geq 0.$$

Therefore, model (3) is globally exponentially stable and the exponential convergence rate index ε can be estimated from (10). The proof is completed.

4 An Example

Consider a two-neuron neural network (3), where

$$C=\begin{pmatrix}0.7 & 0\\ 0 & 0.6\end{pmatrix}, A=\begin{pmatrix}1 & -1.4\\ -1.3 & 1\end{pmatrix}, B=\begin{pmatrix}1 & 0.3\\ 0.5 & 0.2\end{pmatrix},$$

$$f_1(x)=f_2(x)=0.1(|x+1|-|x-1|), \quad \sigma(t)=0.1|\cos t|.$$

Obviously, Assumptions (**H1**) and (**H2**) are satisfied with $F=\text{diag}\{0.2, 0.2\}$ and $\sigma=0.1$. By the Matlab LMI Control Toolbox, we find a solution to the LMI in (5) as follows:

$$P=\begin{pmatrix}19.0432 & 1.2367\\ 1.2367 & 24.4531\end{pmatrix}, Q_1=\begin{pmatrix}10.6745 & 1.1123\\ 2.3425 & 13.6792\end{pmatrix}, Q_2=\begin{pmatrix}4.4539 & 3.5641\\ 1.7741 & 7.3547\end{pmatrix},$$

$$Y_1=\begin{pmatrix}135.7836 & 0\\ 0 & 167.0942\end{pmatrix}, \quad Y_2=\begin{pmatrix}69.3874 & 0\\ 0 & 78.4538\end{pmatrix},$$

$$Y_3=\begin{pmatrix}28.0325 & 0\\ 0 & 34.1137\end{pmatrix}, \quad Y_4=\begin{pmatrix}34.1574 & 0\\ 0 & 23.8179\end{pmatrix}.$$

From Theorem 1 and Theorem 2, we know that model (3) is globally asymptotically and exponentially stable. Moreover, from (10), we can also get that the exponential convergence index $\varepsilon=0.08729$.

5 Conclusions

In this paper, the global asymptotic and exponential stability have been investigated for a class of neural networks with distributed time-varying delays. Two delay-dependent sufficient conditions in LMI form have been obtained for global asymptotic and exponential stability of such systems by using appropriate Lyapunov-Krasovskii functional and linear matrix inequality technique. The proposed results do not require the differentiability of the distributed time-varying delays. An example has been provided to demonstrate the effectiveness of the obtained results.

Acknowledgments

The authors would like to thank the editor and the reviewers for their detailed comments and valuable suggestions which have led to a much improved paper. This work was jointly supported by the National Natural Science Foundation of China under Grant 50608072, and in part by the Department of Education of Zhejiang Province under Grant 20060315.

References

1. Lu, H., Chung, F.L., He, Z.: Some Sufficient Conditions for Global Exponential Stability of Hopfield Neural Networks. Neural Networks **17** (2004) 537-544
2. Arik, S.: An Analysis of Exponential Stability of Delayed Neural Networks with Time Varying Delays. Neural Networks **17** (2004) 1027-1031
3. Cao, J., Li, X.: Stability in Delayed Cohen-Grossberg Neural Networks: LMI Optimization Approach. Physica D **212** (2005) 54-65
4. Cao, J., Wang, J.: Global Asymptotic and Robust Stability of Recurrent Neural Networks with Time Delays. IEEE Trans. Circuits and Systems **I 52** (2005) 417-426
5. Li, C., Liao, X., Zhang, R.: A Global Exponential Robust Stability Criterion for Interval Delayed Neural Networks with Variable Delays. Neurocomputing **69** (2006) 803-809
6. Park, J. H.: A Novel Criterion for Global Asymptotic Stability of BAM Neural Networks with Time Delays. Chaos, Solitons and Fractals **29** (2006) 446-453
7. Xu, S., Lam, J., Ho, D.W.C.: A New LMI Condition for Delay-dependent Asymptotic Stability of Delayed Hopfield Neural Networks. IEEE Transactions on Circuits and Systems **II 53** (2006) 230-234
8. Zeng, Z., Wang J.: Improved Conditions for Global Exponential Stability of Recurrent Neural Networks with Time-varying Delays. IEEE Transactions on Neural Networks **17** (2006) 623-635
9. Yang, H., Chu, T., Zhang, C.: Exponential Stability of Neural Networks with Variable Delays via LMI Approach. Chaos, Solitons and Fractals **30** (2006) 133-139
10. Singh, S.: Simplified LMI Condition for Global Asymptotic Stability of Delayed Neural Networks. Chaos, Solitons and Fractals **29** (2006) 470-473
11. De, V. B., Principe, J.C.: The Gamma Model-A New Neural Model for Temporal Processing. Neural Networks **5** (1992) 565-576
12. Zhao, H.: Global Asymptotic Stability of Hopfield Neural Network Involving Distributed Delays. Neural Networks **17** (2004) 47-53
13. Liao, X., Liu, Q., Zhang, W.: Delay-dependent Asymptotic Stability for Neural Networks with Distributed Delays. Nonlinear Analysis: Real World Applications **7** (2006) 1178-1192
14. Liu, B., Huang, L.: Global Exponential Stability of BAM Neural Networks with Recent-history Distributed Delays and Impulses. Neurocomputing **69** (2006) 2090-2096
15. Huang, T.: Exponential Stability of Fuzzy Cellular Neural Networks with Distributed Delay. Physics Letters A **351** (2006) 48-52
16. Jiang, H., Teng, Z.: Boundedness and Global Stability for Nonautonomous Recurrent Neural Networks with Distributed Delays. Chaos, Solitons and Fractals **30** (2006) 83-93
17. Liang, J., Cao, J.: Global Output Convergence of Recurrent Neural Networks with Distributed Delays. Nonlinear Analysis: Real World Applications **8** (2007) 187-197
18. Cao, J., Yuan, K., Li, H.: Global Asymptotical Stability of Recurrent Neural Networks with Multiple Discrete Delays and Distributed Delays. IEEE Transactions on Neural Networks **17** (2006) 1646-1651

Asymptotic Convergence Properties of Entropy Regularized Likelihood Learning on Finite Mixtures with Automatic Model Selection

Zhiwu Lu, Xiaoqing Lu, and Zhiyuan Ye

Institute of Computer Science and Technology, Peking University,
Beijing 100871, China
zhiwu.lu@yahoo.com.cn

Abstract. In finite mixture modelling, it is crucial to select the number of components for a data set. We have proposed an entropy regularized likelihood (ERL) learning principle for the finite mixtures to solve this model selection problem under regularization theory. In this paper, we further give an asymptotic analysis of the ERL learning, and find that the global minimization of the ERL function in a simulated annealing way (i.e., the regularization factor is gradually reduced to zero) leads to automatic model selection on the finite mixtures with a good parameter estimation. As compared with the EM algorithm, the ERL learning can go across the local minima of the negative likelihood and keep robust with respect to initialization. The simulation experiments then prove our theoretic analysis.

1 Introduction

As a typical statistical model, the finite mixture model has been widely used in a variety of practical applications. Although there have been several unsupervised learning approaches to finite mixture modelling, such as the k-means algorithm [1] and the EM algorithm [2], the number k of components in the data set is usually assumed to be pre-known. Unfortunately, in many instances, this key information is not available, and then we have to select the number of components in the finite mixtures before or during parameter estimation. The traditional approach to solve this kind of model selection problem is to choose the optimal number k^* of components with some statistical criteria such as the Akaike's information criterion [3] and the Bayesian inference criterion [4]. However, since we need to repeat the entire parameter estimation at a number of different values of k, the process of evaluating these criteria incurs a large computational cost.

Some more efficient approaches have also been developed to make automatic model selection on the finite mixtures using a mechanism that an appropriate number of components can be automatically selected during parameter learning, with the mixing proportions of the extra mixture components being reduced to zero. From the Bayesian Ying-Yang (BYY) harmony learning theory [5], a gradient BYY harmony learning algorithm [6] was proposed for Gaussian mixture as

D. Liu et al. (Eds.): ISNN 2007, Part I, LNCS 4491, pp. 986–993, 2007.

a special case of the finite mixture model, which can detect the number of Gaussians (i.e., model scale) automatically during parameter estimation. However, the number of components in the finite mixtures can not be detected accurately by this BYY harmony learning algorithm when the overlap among the actual components becomes larger.

Under the regularization theory [7], we have proposed an entropy regularized likelihood (ERL) learning principle for the finite mixtures in [8,9,10] to solve the above problems. In this paper, we further give an asymptotic analysis of the ERL learning, and find that the global minimization of the ERL function in a simulated annealing way (i.e., the regularization factor is gradually reduced to zero) leads to automatic model selection on the finite mixtures with a good parameter estimation even when the finite mixture model has a certain degree of overlap among the actual components. As compared with the EM algorithm, the ERL learning can go across the local minima of the negative likelihood and keep robust with respect to initialization. The simulation experiments then prove our theoretic analysis.

2 Entropy Regularized Likelihood Learning

We consider the following finite mixture model:

$$p(x|\Theta_k) = \sum_{l=1}^{k} \alpha_l p(x|\theta_l), \sum_{l=1}^{k} \alpha_l = 1,\ \alpha_l \geq 0, \tag{1}$$

where $p(x|\theta_l)(l = 1, ..., k)$ are densities from the same parametric family, and k is the number of components.

Given a sample data set $S = \{x_t\}_{t=1}^{N}$ generated from a finite mixture model with k^* true components and setting $k \geq k^*$, the negative log-likelihood function on the finite mixture model $p(x|\Theta_k)$ is given by

$$L(\Theta_k) = -\frac{1}{N} \sum_{t=1}^{N} \ln(\sum_{l=1}^{k} (p(x_t|\theta_l)\alpha_l)). \tag{2}$$

The well-known EM algorithm is just an implementation of minimizing $L(\Theta_k)$.

With the posterior probability that x_t arises from the l-th component in the finite mixture model

$$P(l|x_t) = p(x_t|\theta_l)\alpha_l / \sum_{j=1}^{k} p(x_t|\theta_j)\alpha_j, \tag{3}$$

we have the following discrete Shannon entropy of these posterior probabilities for the sample x_t

$$E(x_t|\Theta_k) = -\sum_{l=1}^{k} P(l|x_t) \ln P(l|x_t), \tag{4}$$

which is globally minimized at $P(l_0|x_t) = 1, P(l|x_t) = 0(l \neq l_0)$, i.e., the sample x_t is totally classified into or subject to the l_0-th component.

We now consider the average or mean entropy over the sample set S:

$$E(\Theta_k) = -\frac{1}{N}\sum_{t=1}^{N} E(x_t|\Theta_k) = -\frac{1}{N}\sum_{t=1}^{N}\sum_{l=1}^{k} P(l|x_t)\ln P(l|x_t), \tag{5}$$

and use it to regularize the log likelihood function by

$$H(\Theta_k) = L(\Theta_k) + \gamma E(\Theta_k), \tag{6}$$

where $\gamma > 0$ is the regularization factor. That is, $E(\Theta_k)$ is a regularization term to reduce the model complexity such that the finite mixture model can be made as simple as possible by minimizing $H(\Theta_k)$.

Typically, we can obtain the well-known Gaussian mixture model by setting $p(x|\theta_l)$ as a Gaussian probability density function (pdf), that is,

$$p(x|\theta_l) = \frac{1}{(2\pi)^{n/2}|\Sigma_l|^{1/2}} \exp\{-(1/2)(x - m_l)^T \Sigma_l^{-1}(x - m_l)\}, \tag{7}$$

where $x \in R^n$, and $\theta_l = (m_l, \Sigma_l)$ are the mean vectors and covariance matrices of the Gaussian distributions.

In light of the above ERL learning principle, the minimization of $H(\Theta_k)$ should be able to make automatic model selection on the finite mixtures since it requires the least complexity of model structure. Actually, the property of automatic model selection was demonstrated well in [8,9,10] via the gradient and iterative ERL learning algorithms in the case of Gaussian mixture modelling. Moreover, the regularization factor γ can be reduced gradually during iteration in a simulated annealing way, i.e., $\gamma = c\gamma$ $(0 < c < 1)$, and we find that the ERL learning can still make automatic model selection on the finite mixtures with a good parameter estimation.

3 Asymptotic Analysis of ERL Learning

In this section, we try to give an asymptotic analysis of the ERL learning with the regularization factor γ reduced to zero gradually, and then prove the promising property of automatic model selection when the finite mixture model has a certain degree of overlap among the actual components.

To avoid the effect of the randomness in the sample data set, we have to consider the ERL learning asymptotically, i.e., we let $N \to \infty$. The object function $H(\Theta_k)$ of the ERL learning estimated on the sample set S is rewrote as $H_N(\Theta_k)$. Likewise, the estimated functions $L(\Theta_k)$ and $E(\Theta_k)$ are also rewrote as $L_N(\Theta_k)$ and $E_N(\Theta_k)$, respectively. According to probability theory, we have

$$H(\Theta_k) = \lim_{N\to\infty} H_N(\Theta_k) = \lim_{N\to\infty}(L_N(\Theta_k) + \gamma E_N(\Theta_k)) = L(\Theta_k) + \gamma E(\Theta_k), \tag{8}$$

with $L(\Theta_k)$ and $E(\Theta_k)$ now updated as

$$L(\Theta_k) = \lim_{N\to\infty} L_N(\Theta_k) = -\int p(x|\Theta_{k^*}^*) \ln p(x|\Theta_k)dx, \tag{9}$$

$$E(\Theta_k) = \lim_{N\to\infty} E_N(\Theta_k) = \int E(x|\Theta_k)p(x|\Theta_{k^*}^*)dx, \tag{10}$$

where $\Theta_{k^*}^* = \{\alpha_l^*, \theta_l^*\}_{l=1}^{k^*}$ denotes the set of the true parameters in the finite mixtures which the sample data come from. Specifically, k^* is the number of the actual components and $\{\alpha_l^*, \theta_l^*\}$ is the set of true parameters of the l-th component for the actual finite mixture pdf. Here, we always assume that these actual components in the finite mixtures are different.

Furthermore, we consider the case that the finite mixture model $p(x|\Theta_{k^*}^*)$ has a certain degree of overlap among the actual components. According to the information theory, $E(x|\Theta_{k^*}^*)$ is high when the belonging component of x is obscure, i.e., the overlap among these actual components is large; otherwise, $E(x|\Theta_{k^*}^*)$ is low when the belonging component of x is clear, i.e., the overlap among the actual components is small. Hence, the average entropy $E(\Theta_{k^*}^*)$ can be used to measure the overlap of the finite mixtures. In this paper, we suppose that the overlap of the true finite mixtures $p(x|\Theta_{k^*}^*)$ should not be too high, i.e., the average entropy $E(\Theta_{k^*}^*)$ should be bounded as $|E(\Theta_{k^*}^*)| < M$.

Just as the Gaussian mixture model, all the components in the finite mixtures we consider are supposed to have the same functional form. Based on this constraint, we then assume the finite mixtures are discriminant. That is, in the cases that all the components are different, $p(x|\Theta_k) = p(x|\Theta'_{k'})$ if and only if $\Theta_k \supseteq \Theta'_{k'}$ with $k \geq k'$ and the mixing proportions of the other $k - k'$ extra components in Θ_k being zero (i.e., these components have no contribution to the finite mixture pdf). We now investigate the asymptotic convergence properties of the ERL learning for the finite mixtures and have the following theorem.

Theorem 1. *Suppose that the finite mixtures $p(x|\Theta_k)$ are discriminant, and the overlap of the true finite mixtures $p(x|\Theta_{k^*}^*)$ is bounded, i.e., $|E(\Theta_{k^*}^*)| < M$. If $\Theta_{k^h}^h(\gamma) = arg \min_{\Theta_k} H(\Theta_k)$, we have $\Theta_{k^h}^h(\gamma) \supseteq \Theta_{k^*}^*$ with $k^h \geq k^*$ and the mixing proportions of the other $k^h - k^*$ components in $\Theta_{k^h}^h(\gamma)$ being zero, when $\gamma \to 0$.*

Proof. Since $\Theta_{k^h}^h(\gamma) = arg \min_{\Theta_k} H(\Theta_k)$, we have $H(\Theta_{k^h}^h(\gamma)) \leq H(\Theta_{k^*}^*)$, that is,

$$L(\Theta_{k^h}^h(\gamma)) - L(\Theta_{k^*}^*) \leq \gamma[E(\Theta_{k^*}^*) - E(\Theta_{k^h}^h(\gamma))]. \tag{11}$$

Under the information theory, $E(\Theta_{k^*}^*) \geq 0$ and $E(\Theta_{k^h}^h(\gamma)) \geq 0$. Hence, with $|E(\Theta_{k^*}^*)| < M$, it follows that $0 \leq E(\Theta_{k^*}^*) < M$. According to (11), we have

$$L(\Theta_{k^h}^h(\gamma)) - L(\Theta_{k^*}^*) \leq \gamma E(\Theta_{k^*}^*) < \gamma M. \tag{12}$$

Let $D_{KL}(p(x|\Theta_{k^*}^*), p(x|\Theta_{k^h}^h(\gamma))) = \int p(x|\Theta_{k^*}^*) \ln \frac{p(x|\Theta_{k^*}^*)}{p(x|\Theta_{k^h}^h(\gamma))}dx$, where $D_{KL}(\cdot,\cdot)$ is the Kullback-Leibler distance between two probability densities and it always keeps $D_{KL}(\cdot,\cdot) \geq 0$. Since $D_{KL}(p(x|\Theta^*_{k^*}), p(x|\Theta^h_{k^h}(\gamma))) = L(\Theta^h_{k^h}(\gamma)) - L(\Theta^*_{k^*})$, according to (12), we have

$$0 \leq D_{KL}(p(x|\Theta^*_{k^*}), p(x|\Theta^h_{k^h}(\gamma))) < \gamma M. \tag{13}$$

When $\gamma \to 0$, $D_{KL}(p(x|\Theta^*_{k^*}), p(x|\Theta^h_{k^h}(\gamma))) = 0$, i.e., $p(x|\Theta^*_{k^*}) = p(x|\Theta^h_{k^h}(\gamma))$ under the information theory. Based on the discrimination of the finite mixtures, we then have $\Theta^h_{k^h}(\gamma) \supseteq \Theta^*_{k^*}$ with $k^h \geq k^*$ and the mixing proportions of the other $k^h - k^*$ components in $\Theta^h_{k^h}(\gamma)$ being zero, when $\gamma \to 0$. □

According to Theorem 1, we have actually proved that the global minimization of the ERL function in a simulated annealing way leads to the property of automatic model selection on the finite mixtures if we let $k > k^*$ and cancel the components with negligible mixing proportions. That is, if the model scale is actually defined by the number of positive mixing proportions in a finite mixture model, it will be equal to k^* via globally minimizing the ERL function. Thus, the true model scale can be correctly detected through the global minimization of the ERL function.

From the above proof, we can also find that though the asymptotic negative log-function $L(\Theta_k)$ is globally minimized at $\Theta^*_{k^*}$, the global minimum of the asymptotic ERL function $H(\Theta_k)$ may has some deviation from $\Theta^*_{k^*}$ when $\gamma > 0$, since $E(\Theta_k)$ may be globally minimized at some point nearby $\Theta^*_{k^*}$. However, according to (13), this deviation is dominated by the regularization factor γ. That is, as $\gamma \to 0$, the minimum ERL estimates tend to the true parameters in the actual finite mixtures. Specially, when the components are well-separated in the finite mixtures, each posterior probability $p(l|x)$ $(l = 1, ..., k^*)$ at a sample x is either 1 or 0. Hence, $E(x|\Theta^*_{k^*}) = 0$ for all $x \in R^n$, i.e., the asymptotic entropy $E(\Theta^*_{k^*}) = 0$. In this case, the asymptotic ERL function $H(\Theta_k)$ is just globally minimized at $\Theta^*_{k^*}$ even when $\gamma > 0$ and we no longer need to reduce the regularization factor to zero gradually.

Though we originally introduce entropy regularization into the maximum likelihood estimation (by EM algorithm) for automatic model selection on the finite mixtures, it can also be observed that the minimization of the ERL function $H(\Theta_k)$ is robust with respect to initialization and the drawbacks of EM algorithm may be avoided. That is, when local minima of the negative likelihood $L(\Theta_k)$ arise during minimizing the ERL function, the average entropy $E(\Theta_k)$ may still be large and we can then go across these local minima by minimum $H(\Theta_k)$.

4 Simulation Results

In order to show the convergence properties of the iterative ERL learning algorithm, several simulation experiments are carried out on four different sets of sample data just generated from Gaussian mixtures. Moreover, we also make a

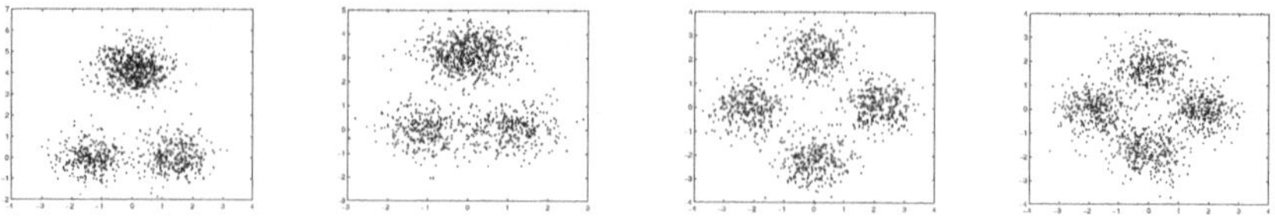

(a) (b) (c) (d)

Fig. 1. Four sets of sample data used in the experiments

comparison with the EM algorithm, especially in the cases that the Gaussian mixture has a certain degree of overlap.

The four sets of sample data for the simulation experiments are drawn from a mixture of four or three bivariate Gaussian densities (i.e., n = 2). As shown in Fig.1, each sample data set is generated at different degree of overlap among the clusters(i.e., Gaussians) in the Gaussian mixture by controlling the mean vectors and covariance matrices of the Gaussian distributions, and with equal or unequal mixing proportions of the clusters in the mixture by controlling the number of samples from each Gaussian density.

We implement the iterative ERL learning algorithm always with $k \geq k^*$ (e.g., $k = 8$) and $\gamma \in [0.3, 0.8]$. The regularization factor γ is reduced to zero gradually by $\gamma = 0.98\gamma$ during iteration. Moreover, the other parameters are initialized randomly within certain intervals. In all the experiments, the iterative ERL learning algorithm is stopped if $|\Delta H| < 10^{-6}$. The EM algorithm has the same initialization except that we must set $k = k^*$. During the iterative ERL learning process, all of the samples are continuously classified into some Gaussians, which can cause other Gaussians to have few samples. Hence, the the mixing proportions of some Gaussians may be reduced to a small value(e.g., below 0.001) after certain iterations, and then these Gaussians are discarded and not shown in the following figures.

The experiments on each data set are repeated 20 times with random initializations for the two learning algorithms. Their average deviation errors between the estimated parameters and the true parameters on the four data sets are listed in Table 1, and the results of one trial by the iterative ERL learning algorithm on each data set are shown in Fig. 2. Actually, the iterative ERL learning algorithm with $k = 8$ can almost always successfully detect the correct number of Gaussians during the 20 trials.

Table 1. The average deviation errors between the estimated parameters by the two algorithms and the true parameters $\Theta^*_{k^*}$ on the four data sets from Fig.1 ($\Theta^h_{k^h}$ estimated by the ERL learning and $\Theta^l_{k^l}$ estimated by the EM algorithm)

Data Set	$\|\|\Theta^h_{k^h} - \Theta^*_{k^*}\|\|^2$	$\|\|\Theta^l_{k^l} - \Theta^*_{k^*}\|\|^2$	$\|\|\Theta^h_{k^h} - \Theta^l_{k^l}\|\|^2$
(a)	0.0138	0.0137	0.0000002
(b)	0.0180	0.0181	0.0000009
(c)	0.0198	0.0198	0.0000004
(d)	0.0332	13.9472	13.4819

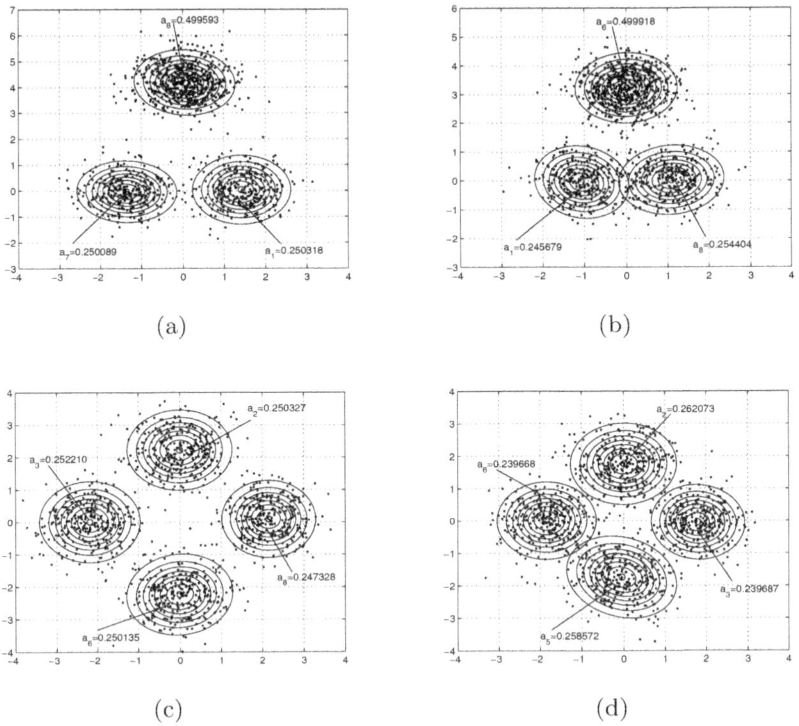

Fig. 2. The experiment results of automatic detection of the number of Gaussians on the four sample sets from Fig.1 by the iterative ERL learning algorithm

Here, the deviation error is used to evaluate the performance of a learning algorithm, which is just the mean square error between the converged parameters and the true parameters. From Table 1, we conclude that the iterative ERL learning algorithm can always converge to the true parameters with a lower deviation error even in the case that there is a certain high degree of overlap among the actual Gaussians. Moreover, we can find that the difference between the parameters estimated by the iterative ERL learning algorithm and the ones estimated by the EM algorithm is very small if the EM algorithm converges to the true parameters correctly, which just proves that the minimum ERL estimation is actually the maximum likelihood estimation when the regularization factor is reduced to zero.

Note that the EM algorithm can not converge to the true parameters on the data set from Fig. 1(d), i.e., some local minima of the negative likelihood have been obtained, while the ERL learning can go across these local minima and make automatic model selection on the Gaussian mixture with a good parameter estimation. Hence, the drawbacks of the standard EM algorithm may be avoided and the ERL learning can be robust with respect to initialization.

5 Conclusions

We have investigated the automatic model selection and parameter estimation on the finite mixtures through implementing some kind of entropy regularized likelihood (ERL) learning principle. We further give an asymptotic analysis of the ERL learning and find that the global minimization of the ERL function in a simulated annealing way leads to automatic model selection on the finite mixtures with a good parameter estimation. As compared with the EM algorithm, the ERL learning can go across the local minima of the negative likelihood and keep robust with respect to initialization. The simulation experiments then prove our theoretic analysis.

References

1. Devijver, P.A., Kittter J.: Pattern Recognition: A Statistical Approach. Prentice Hall, Englewood Cliffs, N. J. (1982)
2. Render, R.A., Walker, H.F.: Mixture Densities, Maximum Likelihood and the EM Algorithm. SIAM Review **26** (2) (1984) 195–239
3. Akaike, H.: A New Look at the Statistical Model Identification. IEEE Transactions on Automatic Control **19** (6) (1974) 716–723
4. Schwarz, G.: Estimating the Dimension of a Model. The Annals of Statistics **6** (2) (1978) 461–464
5. Xu, L.: BYY Harmony Learning, Structural RPCL, and Topological Self-Organizing on Mixture Modes. Neural Networks **15** (8–9) (2002) 1231–1237
6. Ma, J., Wang, T., Xu, L.: A Gradient BYY Harmony Learning Rule on Gaussian Mixture with Automated Model Selection. Neurocomputing **56** (2004) 481–487
7. Dennis, D.C., Finbarr, O.S.: Asymptotic Analysis of Penalized Likelihood and Related Estimators. The Annals of Statistics **18** (6) (1990) 1676–1695
8. Lu, Z.: Entropy Regularized Likelihood Learning on Gaussian Mixture: Two Gradient Implementations for Automatic Model Selection. Neural Processing Letters **25** (1) (2007) 17–30
9. Lu, Z.: An Iterative Algorithm for Entropy Regularized Likelihood Learning on Gaussian Mixture with Automatic Model Selection. Neurocomputing **69** (13-15) (2006) 1674–1677
10. Lu, Z.: A Regularized Minimum Cross-Entropy Algorithm on Mixtures of Experts for Time Series Prediction and Curve Detection. Pattern Recognition Letters **27** (9) (2006) 947–955

Existence and Stability of Periodic Solutions for Cohen-Grossberg Neural Networks with Less Restrictive Amplification

Haibin Li and Tianping Chen

Key Laboratory of Nonlinear Science of Chinese Ministry of Education, Institute of Mathematics, Fudan University, Shanghai, 200433, P.R. China
happyli.happyli@gmail.com,
tchen@fudan.edu.cn

Abstract. The existence and global asymptotic stability of a large class of Cohen-Grossberg neural networks is discussed in this paper. Previous papers always assume that the amplification function has positive lower and upper bounds, which excludes a large class of functions. In our paper, it is only needed that the amplification function is positive. Also, the model discussed is general, the method used is direct and the conditions needed are weak.

1 Introduction

It is well known that Cohen-Grossberg neural networks, proposed by Cohen and Grossberg in [1,2], have been extensively studied both in theory and applications. In recent years, there has been increasing interest in the study of periodic solutions of Cohen-Grossberg neural networks([3-15]). In this article, we investigate the following generalized Cohen-Grossberg neural networks

$$\frac{dx_i(t)}{dt} = -a_i(x_i(t)) \left[b_i(x_i(t)) - \sum_{j=1}^{n} c_{ij}(t) g_j(x_j(t)) \right. \\ \left. - \sum_{j=1}^{n} \int_0^{\infty} f_j(x_j(t-s)) d_s K_{ij}(t,s) + I_i(t) \right], \ i = 1, 2, \cdots, n \quad (1)$$

where $x_i(t)$ denotes the state variable of the ith neuron at time t, $a_i(x)$ represent the amplification functions. And for $i, j = 1, 2, \cdots, n$ and $t \geqslant 0$, $d_s K_{ij}(t,s)$ are Lebesgue-Stieljies measures satisfying $d_s K_{ij}(t+\omega, s) = d_s K_{ij}(t,s)$, and there exist measures dK_{ij} satisfying $|d_s K_{ij}(t,s)| \leqslant |dK_{ij}(s)|$ for any t, $\int_0^{\infty} |dK_{ij}(s)| < \infty$, and $\int_0^{\infty} s|dK_{ij}(s)| < \infty$. $I_i(t)$ and $c_{ij}(t)$ are all periodic with period $\omega > 0$.

To the best of our knowledge, when discussing periodic sulotions, previous papers ([3-15]) always make the assumption that $0 < \underline{\alpha}_i \leqslant a_i(x) \leqslant \bar{\alpha}_i$. This assumption excludes a large class of functions, and is actually not necessary. In our paper, we only assume that $a_i(x) > 0$, which is more sensible and realistic.

D. Liu et al. (Eds.): ISNN 2007, Part I, LNCS 4491, pp. 994–1000, 2007.

This assumption allows our model to represent a larger class of Cohen-Grossberg neural networks than those in previous papers.

In this paper, we do not use complicated theories which are commonly used in previous papers, such as Mawhin's continuation theory and Lyapunov method. By developing ideas in [16-18], we use a direct method to discuss the problem. The resulted conditions needed for existence and global asymptotic stability are weak, and does not depend on the amplification function or the external inputs.

2 Main Results

Definition 1. *$\{\xi,\infty\}$–norm: $\|x(t)\|_{\{\xi,\infty\}} = \max_{i=1,2,\cdots,n} |\xi_i^{-1}x_i(t)|$, where $\xi_i > 0$, $i = 1,2,\cdots,n$.*

Assumption 1. $g_i \in Lip(G_i)$, $f_i \in Lip(F_i)$, $i = 1,2,\cdots,n$, where $G_i > 0$, $F_i > 0$. $Lip(G)$ denotes Lipschitz functions with Lipschitz constant $G > 0$.

Assumption 2. $a_i(x) > 0$, $i = 1,2,\cdots,n$ are continuous functions. $b_i(t)$ is continuous and $\frac{b_i(u)-b_i(v)}{u-v} \geqslant \gamma_i > 0$, $i = 1,2,\cdots,n$.

Since $I_i(t)$ and $c_{ij}(t)$ are period functions, they are all bounded. Denote

$$c_{ij}^* = \sup_{t\in\mathbb{R}} |c_{ij}(t)| < \infty, \ I_i^* = \sup_{t\in\mathbb{R}} |I_i(t)| < \infty$$

Lemma 1. *If there exist constants $\xi_i > 0$, $i = 1,2,\cdots,n$, such that for all $i = 1,2,\cdots,n$ and $0 \leqslant t < \omega$,*

$$\gamma_i\xi_i - \sum_{j=1}^{n} |c_{ij}(t)|G_j\xi_j - \sum_{j=1}^{n} F_j\xi_j \int_0^\infty |d_sK_{ij}(t,s)| > 0, \tag{2}$$

then any solution $x(t)$ of system (1) is bounded.

Proof. Since $c_{ij}(t)$ are continuous and periodic with period ω, $d_sK_{ij}(t,s)$ are ω-periodic with respect to t, then there (2) implies that exists a constant $\eta > 0$, s.t., $\eta = \min_i \min_{0\leqslant t<\omega} \left\{\gamma_i\xi_i - \sum_{j=1}^n |c_{ij}(t)|G_j\xi_j - \sum_{j=1}^n F_j\xi_j \int_0^\infty |d_sK_{ij}(t,s)|\right\}$. So, we have

$$-\gamma_i\xi_i + \sum_{j=1}^{n} |c_{ij}(t)|G_j\xi_j + \sum_{j=1}^{n} F_j\xi_j \int_0^\infty |d_sK_{ij}(t,s)| \leqslant -\eta < 0.$$

Let $M(t) = \max_{s\leqslant t} \|x(s)\|_{\{\xi,\infty\}}$. Clearly, $M(t)$ is non-decreasing, and $\|x(t)\|_{\{\xi,\infty\}} \leqslant M(t)$. Denote $H = \max_i \left\{|b_i(0)| + |I_i^*| + \sum_{j=1}^n c_{ij}^*|g_j(0)| + \sum_{j=1}^n |f_j(0)| \int_0^\infty |d_sK_{ij}(t,s)|\right\}$.

Now we prove that $M(t) \leqslant \max\{M(0), H/\eta\}$.

For any $t_0 \geqslant 0$, consider the following two cases:

1) If $\|x(t_0)\|_{\{\xi,\infty\}} < M(t_0)$, then there exists $\delta > 0$, such that in $(t_0, t_0+\delta)$, $\|x(t)\|_{\{\xi,\infty\}} < M(t_0)$, and hence $M(t) = M(t_0)$.

2) If $\|x(t_0)\|_{\{\xi,\infty\}} = M(t_0)$, then let i_0 be an index such that $\xi_{i_0}^{-1}|x_{i_0}(t_0)| = \|x(t_0)\|_{\{\xi,\infty\}}$. Note that from Assumption 1 and 2, we have

$$|g_i(s)| \leqslant G_i|s| + |g_i(0)|,\ |f_i(s)| \leqslant F_i|s| + |f_i(0)|,\ i = 1, 2, \cdots, n,\ s \in \mathbb{R},$$

and

$$sign(s)b_i(s) \geqslant \gamma_i|s| + sign(s)b_i(0),\ \ s \in \mathbb{R},$$

then we have

$$\begin{aligned}
\{\frac{d}{dt}|x_{i_0}(t)|\}_{t=t_0} &= a_{i_0}(x_{i_0}(t_0))sign(x_{i_0}(t_0))\Bigg[-b_{i_0}(x_{i_0}(t_0)) + \sum_{j=1}^{n} c_{i_0j}(t_0)g_j(x_j(t_0)) \\
&\quad + \sum_{j=1}^{n}\int_0^\infty f_j(x_j(t_0-s))d_sK_{i_0j}(t_0,s) + I_{i_0}(t_0)\Bigg] \\
&\leqslant a_{i_0}(x_{i_0}(t_0))\Bigg[-\gamma_{i_0}\xi_{i_0}|x_{i_0}(t_0)|\xi_{i_0}^{-1} + \sum_{j=1}^{n}|c_{i_0j}(t_0)|G_j\xi_j|x_j(t_0)|\xi_j^{-1} \\
&\quad + \sum_{j=1}^{n}F_j\xi_j\int_0^\infty |x_j(t_0-s)|\xi_j^{-1}|d_sK_{i_0j}(t_0,s)| + |b_{i_0}(0)| + |I_{i_0}(t_0)| \\
&\quad + \sum_{j=1}^{n}|c_{i_0j}(t)||g_j(0)| + \sum_{j=1}^{n}|f_j(0)|\int_0^\infty |d_sK_{i_0j}(t_0,s)|\Bigg] \\
&\leqslant a_{i_0}(x_{i_0}(t_0))\Bigg\{\Bigg[-\gamma_{i_0}\xi_{i_0} + \sum_{j=1}^{n}|c_{i_0j}(t_0)|G_j\xi_j \\
&\quad + \sum_{j=1}^{n}F_j\xi_j\int_0^\infty |d_sK_{i_0j}(t_0,s)|\Bigg]\|x(t_0)\|_{\{\xi,\infty\}} + H\Bigg\} \\
&\leqslant a_{i_0}(x_{i_0}(t_0))(-\eta M(t_0) + H).
\end{aligned}$$

If $M(t_0) \geqslant H/\eta$, then there exists a small interval $(t_0, t_0+\delta_1)$, in which $M(t)$ is non-increasing. Otherwise, if $M(t_0) < H/\eta$, then there exists a small interval $(t_0, t_0+\delta_2)$, in which we have $\|x(t)\|_{\{\xi,\infty\}} < H/\eta$. So, there exists a small interval $(t_0, t_0+\delta)$, in which $M(t) \leqslant \max\{M(t_0), H/\eta\}$.

So, in either case, we have that $M(t) \leqslant \max\{M(t_0), H/\eta\}$ holds in a small interval $(t_0, t_0+\delta)$. It can be seen from the above discussion that if $M(0) > H/\eta$, then for any $t > 0$, $M(t) = M(0)$; if $M(0) \leqslant H/\eta$, then for any $t > 0$, $M(t) \leqslant H/\eta$. So, for any $t > 0$, we have $M(t) \leqslant \max\{M(0), H/\eta\}$, and hence $x(t)$ is bounded. Lemma is proved.

Remark 1. *The proof of lemma 3 can be found in Theorem 2.3, chapter 6 of [19].*

Theorem 1. *If there exist constants $\xi_i > 0,\ \zeta_i > 0,\ i = 1, 2, \cdots, n$, such that for all $i = 1, 2, \cdots, n$ and $0 \leqslant t < \omega$,*

$$\xi_i\gamma_i - \sum_{j=1}^{n} |c_{ij}(t)|G_j\xi_j - \sum_{j=1}^{n} F_j\xi_j \int_0^{\infty} |dK_{ij}(s)| > 0, \tag{3}$$

$$\zeta_i\gamma_i - \sum_{j=1}^{n} |c_{ji}(t)|\zeta_j G_i - \sum_{j=1}^{n} \zeta_j F_i \int_0^{\infty} |dK_{ji}(s)| > 0. \tag{4}$$

Then system (1) has a periodic solution with period ω, which is globally asymptotically stable.

Proof. It's easy to see that the condition (3) is more restrictive than the condition (2). By Lemma 1, any solution of system (1) is bounded.

Let $\lambda = \min_i \inf_t \left\{\zeta_i\gamma_i - \sum_{j=1}^{n} |c_{ji}(t)|\zeta_j G_i - \sum_{j=1}^{n} \zeta_j F_i \int_0^{\infty} |dK_{ji}(s)|\right\}$, we have $\lambda > 0$.

For a specific solution $x(t)$ of system (1), let $u_i(t) = x_i(t+\omega) - x_i(t)$, and $v_i(t) = \int_{x_i(t)}^{x_i(t+\omega)} \frac{1}{a_i(\rho)} d\rho$, $i = 1, 2, \cdots, n$. Note that $a_i(\cdot)$ is continuous, $a_i(x) > 0$, and x_i is bounded, thus $\int_{x_i(t)}^{x_i(t+\omega)} \frac{1}{a_i(\rho)} d\rho$ exists. By mean-value theorem for integral, $v_i(t) = \frac{1}{a_i(\xi)}(x_i(t+\omega) - x_i(t)) = \frac{1}{a_i(\xi)} u_i(t)$, where $\xi \in [\min\{x_i(t), x_i(t+\omega)\}, \max\{x_i(t), x_i(t+\omega)\}]$, then we have $sign(v_i(t)) = sign(u_i(t))$.

Since $c_{ij}(t)$, $I_i(t)$, and $d_s K_{ij}(t,s)$ are all ω-periodic with respect to t, direct calculation gives

$$\begin{aligned}
\frac{d|v_i(t)|}{dt} &= sign(v_i(t)) \left[\frac{1}{a_i(x_i(t+\omega))}\{\frac{dx_i(s)}{ds}\}_{s=t+\omega} - \frac{1}{a_i(x_i(t))}\{\frac{dx_i(s)}{ds}\}_{s=t}\right] \\
&= sign(u_i(t)) \Bigg[-(b_i(x_i(t+\omega)) - b_i(x_i(t))) + \sum_{j=1}^{n} c_{ij}(t)(g_j(x_j(t+\omega)) \\
&\quad - g_j(x_j(t))) + \sum_{j=1}^{n} \int_0^{\infty} (f_j(x_j(t+\omega-s)) - f_j(x_j(t-s))) d_s K_{ij}(t,s)\Bigg] \\
&\leqslant -\gamma_i|u_i(t)| + \sum_{j=1}^{n} |c_{ij}(t)|G_j|u_j(t)| + \sum_{j=1}^{n} \int_0^{\infty} F_j|u_j(t-s)||dK_{ij}(s)|.
\end{aligned}$$

Define $L(t) = \sum_{i=1}^{n} \zeta_i|v_i(t)| + \sum_{i,j=1}^{n} \zeta_i F_j \int_0^{\infty} \int_{t-s}^{t} |u_j(\rho)| d\rho |dK_{ij}(s)|$.

Differentiate $L(t)$ along the trajectory $x(t)$ of system (1) and we have

$$\begin{aligned}\frac{dL(t)}{dt} &= \sum_{i=1}^{n}\zeta_i\frac{d}{dt}|v_i(t)| + \sum_{i,j=1}^{n}\zeta_i F_j\left[\int_0^\infty |u_j(t)||dK_{ij}(s)| - \int_0^\infty |u_j(t-s)||dK_{ij}(s)|\right]\\ &\leqslant \sum_{i=1}^{n}\zeta_i\left[-\gamma_i|u_i(t)| + \sum_{j=1}^{n}|c_{ij}(t)|G_j|u_j(t)| + \sum_{j=1}^{n}\int_0^\infty F_j|u_j(t-s)||dK_{ij}(s)|\right]\\ &\quad + \sum_{i,j=1}^{n}\zeta_i F_j\left[\int_0^\infty |u_j(t)||dK_{ij}(s)| - \int_0^\infty |u_j(t-s)||dK_{ij}(s)|\right]\\ &= \sum_{i=1}^{n}\left[-\zeta_i\gamma_i + \sum_{j=1}^{n}|c_{ji}(t)|\zeta_j G_i + \sum_{j=1}^{n}\zeta_j F_i\int_0^\infty |dK_{ji}(s)|\right]|u_i(t)|\\ &\leqslant -\lambda\sum_{i=1}^{n}|u_i(t)| = -\lambda\|u(t)\|_1. \end{aligned} \tag{5}$$

Since $L(t)\geqslant 0$, Integrate both sides of (5) form 0 to ∞ and we have

$$\int_0^\infty \sum_{i=1}^{n}|u_i(t)|dt \leqslant \frac{1}{\lambda}L(0) < +\infty, \tag{6}$$

From the definition of $u_i(t)$, (6) is

$$\sum_{n=1}^{\infty}\int_0^\omega \|x(t+n\omega) - x(t+(n-1)\omega)\|_1 dt < +\infty.$$

By Cauchy criterion, we have that $x(t+n\omega)$ converges in $L^1[0,\omega]$ as $n\to\infty$. Since $x(t)$ is bounded, we have that $a_i(x_i(t))$, $i=1,2,\cdots,n$ are also bounded, and seen from the form of system (1), we have $x(t)$ is uniformly continuous. Then the sequence $\{x(t+n\omega)\}$ is uniformly bounded and equicontinuous. Thus, by Arzéla-Ascoli theorem, there exists a subsequence $\{x(t+n_k\omega)\}$ converging on any compact set of $\mathbb{R}$. Denote this limit by $x^*(t)$. We have that $x^*(t)$ is also the limit of $\{x(t+n\omega)\}$ in $L^1[0,\omega]$, i.e.,

$$\lim_{n\to\infty}\int_0^\omega \|x(t+n\omega) - x^*(t)\|_1 dt = 0.$$

Then, we have that $x(t+n\omega)\to x^*(t)$ uniformly on $[0,\omega]$. Similarly, $x(t+n\omega)\to x^*(t)$ uniformly on any compact set of $\mathbb{R}$.

Now we will prove that $x^*(t)$ is periodic solution with period ω of system (1). Since

$$x^*(t+\omega) = \lim_{n\to\infty}x(t+(n+1)\omega) = \lim_{n\to\infty}x(t+n\omega) = x^*(t),$$

we have that $x^*(t)$ is periodic with period ω. Then, replace $x(t)$ with $x(t+n_k\omega)$ in system (1), let $k\to\infty$, and we have that $x^*(t)$ is a solution of system (1).

Let $t = t_1 + n\omega$, where $0\le t_1<\omega$. Then

$$\|x(t) - x^*(t)\|_1 = \|x(t_1+n\omega) - x^*(t_1)\|_1.$$

And the uniform convergence of $\{x(t + n\omega)\}$ on $[0, \omega]$ leads to

$$\lim_{t\to\infty} \|x(t) - x^*(t)\|_1 = 0. \tag{7}$$

Finally, we prove that any solution of system (1) converges to $x^*(t)$. Suppose $y(t)$ is another solution of system (1). Redefine $u_i(t) = y_i(t) - x_i(t)$, $v_i(t) = \int_{x_i(t)}^{y_i(t)} \frac{1}{a_i(\rho)} d\rho$, $i = 1, 2, \cdots, n$. Using the same method above, it is easy to prove that

$$\lim_{t\to\infty} \|y(t) - x(t)\|_1 = 0.$$

In conjunction with (7), we conclude that

$$\lim_{t\to\infty} \|y(t) - x^*(t)\|_1 = 0.$$

Theorem is proved.

3 Conclusions

In this paper, we investigated the solution of periodic Cohen-Grossberg neural networks with unbounded amplifications. By using a direct method, sufficient conditions for the existence and global asymptotic stability of periodic solution is derived. This criterion does not depend on the amplification functions or the external inputs, and is rather weak.

References

1. Cohen, M. A., Grossberg, S., Absolute Stability and Global Pattern Formation and Parallel Memory Storage by Competitive Neural Networks. IEEE Trans. Syst. Man Cybern. B. 13 (1983) 815-821
2. Grossberg, S., Nonlinear Neural Networks, Principles, Mechanisms, and Architectures. Neural Networks 1 (1988) 17-61
3. Li, Y., Existence and stability of periodic solutions for Cohen-Grossberg neural networks with multiple delays, Chaos, Solitons and Fractals 20 (2004) 459-466
4. Huang, C., Huang, L., Dynamics of a class of Cohen-Grossberg neural networks with time-varying delays, Nonlinear Analysis: Real World Applications, ARTICLE IN PRESS, Available on line at www.sciencedirect.com
5. Liu, B., Huang, L., Existence and exponential stability of periodic solutions for a class of Cohen-Grossberg neural networks with time-varying delays, Chaos, Solitons and Fractals, ARTICLE IN PRESS, Available on line at www.sciencedirect.com
6. Yuan, Z., Hu, D., Huang, L., Guohua D., Existence and global exponential stability of periodic solution for Cohen-Grossberg neural networks with delays, Nonlinear Analysis: Real World Applications 7 (2006) 572-590
7. Sun, J., Wan, L., Global exponential stability and periodic solutions of Cohen-Grossberg neural networks with continuously distributed delays, Physica D 208 (2005) 1-20
8. Chen, Z., Ruan, J., Global stability analysis of impulsive Cohen-Grossberg neural networks with delay, Physics Letters A 345 (2005) 101-111

9. Zhao, H., Wang, L., Hopf bifurcation in Cohen-Grossberg neural network with distributed delays, Nonlinear Analysis: Real World Applications, ARTICLE IN PRESS, Available on line at www.sciencedirect.com
10. Cao, J., Li, X., Stability in delayed Cohen-Grossberg neural networks: LMI optimization approach, Physica D 212 (2005) 54-65
11. Wang, L., Stability of Cohen-Grossberg neural networks with distributed delays, Applied Mathematics and Computation 160 (2005) 93-110
12. Yuan Z., Yuan, L., Huang L., Dynamics of periodic Cohen-Grossberg neural networks with varying delays, Neurocomputing, ARTICLE IN PRESS, Available on line at www.sciencedirect.com
13. Long, F., Wang, Y., Zhou, S., Existence and exponential stability of periodic solutions for a class of Cohen-Grossberg neural networks with bounded and unbounded delays, Nonlinear Analysis: Real World Applications, ARTICLE IN PRESS, Available on line at www.sciencedirect.com
14. Wu, C., Ruan, J., Lin, W., On the existence and stability of the periodic solution in the Cohen-Grossberg neural network with time delay and high-order terms, Applied Mathematics and Computation 177 (2006) 194-210
15. Chen, A., Cao, J., Periodic bi-directional Cohen-Grossberg neural networks with distributed delays, Nonlinear Analysis, ARTICLE IN PRESS, Available on line at www.sciencedirect.com
16. Lu, W, Chen, T., On periodic dynamical systems, Chin. Ann. Math. 25B: 4 (2004) 455-462
17. Lu, W., Chen, T., Global exponential stability of almost periodic solution for a large class of delayed dynamical systems 48 (8) (2005) 1015-1026
18. Lin, W., Chen, T., Dynamics of periodic Lotka-Volerra systems with general time delays, to appear in Dynamics of Continuous, Discrete and Implusive Systems
19. Berman, A., Plemmons, R.J., Nonnegative Matrices in the Mathematical Sciences, Academic Press, New York, 1979.

Global Exponential Convergence of Time-Varying Delayed Neural Networks with High Gain

Lei Zhang and Zhang Yi

Computational Intelligence Laboratory, School of Computer Science and Engineering, University of Electronic Science and Technology of China, Chengdu 610054, P. R. China
{leilazhang, zhangyi}@uestc.edu.cn
http://cilab.uestc.edu.cn

Abstract. This paper studies a general class of neural networks with time-varying delays and the neuron activations belong to the set of discontinuous monotone increasing functions. The discontinuities in the activations are an ideal model of the situation where the gain of the neuron amplifiers is very high. Because the delay in combination with high-gain nonlinearities is a particularly harmful source of potential instability, in the paper, conditions which ensure the global convergence of the neural network are derived.

1 Introduction

So far many fundament results have been established on global stability and global exponential stability of the equilibrium point for a class of neural networks with Lipschitz continuous neuron activations (see, e.g., [1]-[5] and references therein). However, recent work has demonstrated the interest in studying global stability of the equilibrium point for neural networks with discontinuous neuron activations which described by differential equations with a discontinuous right-hand side. This is an ideal model for the case where the gain of the neuron amplifiers is very high (see, e.g., [6,7]). As shown by Hopfield neural network, under the standard assumption of high-gain amplifiers, the sigmoid neural activations closely approach a discontinuous hard-comparator function and favor binary output formation. When dealing with dynamical systems possessing high-slope nonlinear elements it is often advantageous to model them with a system of differential equations with discontinuous right-hand side, rather than studying the case where the slope is high but of finite value. The main advantage of analyzing the ideal discontinuous case is that such analysis is usually able to give a clear picture of the salient features of motion, such as the presence of sliding models, i.e., the possibility that trajectories be confined for some time intervals on discontinuity surfaces.

In addition, as we all know, delays are important parameters of neural networks. They also affect the dynamical properties of networks. The presence of

D. Liu et al. (Eds.): ISNN 2007, Part I, LNCS 4491, pp. 1001–1007, 2007.

switching delay in a high-gain neuron amplifier is a particularly harmful source of potential instability. In fact, it is clear that a constant delay is only a special case of neural networks. In most situations, delays are variable. The problem of exponential stability for neural networks with distributed delays is studied in some literatures. However, those methods and results are devoted to the case when the neuron activations are continuous.

On the base of the discussion above, in this paper, we introduce a class of neural networks with time-varying delays in the neuron interconnections, and neuron activations modelled by a class of discontinuous monotone increasing functions. The model differs from those considered in some quoted papers on global stability of delayed neural networks, where smooth Lipschitz continuous functions are employed. It is also more general than the discontinuous neural network model in [7], where the delays were assumed to be constants.

The structure of this paper is outlined as follows. Section 2 discusses the neural network model studied in the paper with some preliminaries. Then, the main results on global exponential convergence of network are given in Section 3. Section 4 illustrates examples with simulation. Finally, conclusion is provided in Section 5.

2 Preliminaries

A class of neural networks considered in this paper described by the system of differential equations

$$\dot{x}_i(t) = -d_i x_i(t) + \sum_{j=1}^{n} \Big[a_{ij} g_j\big(x_j(t)\big) + b_{ij} g_j\big(x_j(t-\tau_{ij}(t))\big) \Big] + I_i \tag{1}$$

for all $t \geq 0$ and $i = 1, 2, \cdots, n$, where each $x_i(t)$ is the state of neuron i, denote $x(t) = (x_1(t), \cdots, x_n(t))^T$. $d_i > 0 (i = 1, 2, \cdots, n)$ are the neuron self-inhibitions. a_{ij} and $b_{ij} (i, j = 1, 2, \cdots, n)$ are connection weights which are all constants. Throughout this paper, we assume that the time-varying delays $\tau_{ij}(t) (i, j = 1, 2, \cdots, n)$ are also continuous differentiable functions with $0 \leq \tau_{ij}(t) \leq \tau$, where τ is a constant. The initial conditions for (1) is defined as $x(s) = \phi(s), s \in [-\tau, 0]$. The diagonal mapping $g(x) = (g_1(x_1), g_2(x_2), \cdots, g_n(x_n))^T$ has components $g_i(x_i)$ that model the nonlinear input-output activations of the neurons. In the paper, we assume that $g(x)$ belongs to the following class of discontinuous functions.

Definition 1. *We say that $g \in \mathcal{G}_\mathcal{D}$ if and only if, for $i = 1, 2, \cdots, n$, g_i satisfies the following assumptions.*

a) g_i is piecewise continuous, i.e., g_i is continuous in $\mathbb{R}$ except a countable set of points of discontinuity, ρ_k, where there exist finite right and left limits, $g_i^+(\rho_k)$ and $g_i^-(\rho_k)$, respectively, with $g_i^+(\rho_k) > g_i^-(\rho_k)$; moreover, g_i has a finite number of discontinuous on any compact interval of $\mathbb{R}$.

b) g_i is bounded.

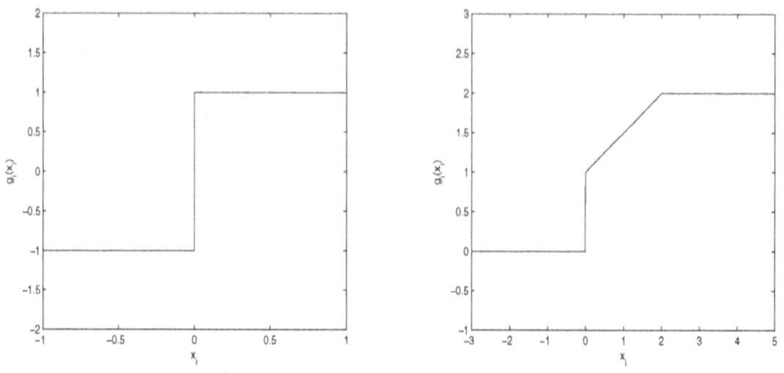

Fig. 1. Examples of discontinuous functions in the class $\mathcal{G}_{\mathcal{D}}$: (a) hard comparator (signum) function; (b) piecewise linear monotonically nondecreasing function

c) g_i is nondecreasing, i.e., for any ρ_a and ρ_b such that $\rho_a > \rho_b$ and g_i is continuous at ρ_a and ρ_b, it results $g_i^+(\rho_a) \geq g_i^-(\rho_b)$.

The class of discontinuous functions $\mathcal{G}_{\mathcal{D}}$ includes a number of neuron activations of interest for the applications. Fig. 1 shows some kinds of them.

Since for $g \in \mathcal{G}_{\mathcal{D}}$ the right-hand side of Eqn. (1) is a discontinuous function of the state x, it is needed to explain what is meant by a solution of a Cauchy problem associated to Eqn. (1). We adopt a possible definition from the theory of differential equations with discontinuous right-hand side as introduced by Filippov [8]. This theory has become a standard mathematical tool in a number of engineering applications. We note that if g satisfies Definition 1, then any $g_i (i = 1, 2, \cdots, n)$, possesses only isolated jump discontinuities where g_i is not necessarily defined. Hence for all $x \in \mathbb{R}^n$, we have $K[g_i(x_i)] = [g_i^-(x_i), g_i^+(x_i)] (i = 1, 2, \cdots, n)$, where $K[E]$ denotes the closure of the convex hull of set $E \subset \mathbb{R}^n$, see [6]. Then we introduce the following definition of solution of Eqn. (1) and an output associated to that solution.

Definition 2. *A function $x : [-\tau, T) \to \mathbb{R}^n, T \in (0, +\infty]$, is a solution of (1) on $[-\tau, T)$ if:*

1) x is continuous on $[-\tau, T)$ and absolutely continuous on $[0, T)$;

2) there exists a measurable function $\gamma = (\gamma_1, \gamma_2, \cdots, \gamma_n)^T : [-\tau, T) \to \mathbb{R}^n$, such that for almost all (a.a.) $t \in [-\tau, T)$, $\gamma(t) \in K[g(x(t))]$ and for a.a. $t \in [0, T)$ it holds that

$$\dot{x}_i(t) = -d_i x_i(t) + \sum_{j=1}^{n} \Big[a_{ij}\gamma_j(t) + b_{ij}\gamma_j(t - \tau_{ij}(t)) \Big] + I_i. \tag{2}$$

Any function γ as in (2) is called an output associated to the solution x. Clearly, γ actually represents the vector of neural network output.

Then it is straightforward that the existence of the solutions of Eqn. (1) since g is bounded on $\mathbb{R}^n$.

3 Global Exponential Convergence

In this section, we first address the existence of the equilibrium point and the corresponding output equilibrium point of neural network (1).

Define a set-valued map

$$\mathcal{F}(x) = D^{-1}\big[(A+B)K[g(x)] + I\big] : \mathbb{R}^n \to \mathbb{R}^n,$$

which is an upper semi-continuous map with nonempty compact convex values. From Kakutani's fixed point theorem, it is straightforward that there exists at least one fixed point of $\mathcal{F}$, i.e., a point $\xi \in \mathbb{R}^n$ such that $\xi \in \mathcal{F}(\xi)$, which also represents an equilibrium point of (1). Then from Definite 2, it is subsequence that the existence of output equilibrium point η.

Next, we establish some results on global convergence of network (1) using an approach based on the concept of Lyapunov function.

Theorem 1. *The network (1) will have a unique equilibrium point ξ which is globally exponentially stable, if for all $t \geq 0$, it holds that for all $t \geq 0$, and $\Omega = (\omega_{ij})_{n\times n}$ is a* M*-matrix where*

$$\omega_{ij} = \begin{cases} -a_{ii} - \frac{|b_{ii}|}{1-l_{ii}}, & \mathit{if}\ i = j, \\ -|a_{ij}| - \frac{|b_{ij}|}{1-l_{ij}}, & \mathit{if}\ i \neq j. \end{cases}$$

and $\dot{\tau}_{ij}(t) \leq l_{ij} < 1 (i,j = 1,2,\cdots,n)$.

Proof. Since the matrix $(\omega_{ij})_{n\times n}$ is a M-matrix, there exists a positive constant vector $\beta = (\beta_1, \beta_2, \cdots, \beta_n)^T \in \mathbb{R}^n$, such that $\beta^T\Omega > 0$. For all $\epsilon \in [0, \epsilon_M)$, where

$$\epsilon_M = \frac{1}{\tau} \min_{1\leq j\leq n} \ln \frac{-\beta_j a_{jj} - \sum_{i=1}^n \beta_i\big[|a_{ij}|(1-\delta_{ij})\big]}{\sum_{i=1}^n \frac{|b_{ij}|}{1-l_{ij}}},$$

fix $\epsilon > 0$ such that $\epsilon \in \big(0, \min\{\epsilon_M, d_1, d_2, \cdots, d_n\}\big)$, let us consider the matrix $\Omega_\epsilon = (\omega_{ij}^\epsilon)_{(n\times n)}$, where

$$\omega_{ij}^\epsilon = \begin{cases} -a_{ii} - \frac{|b_{ii}|e^{\epsilon\tau}}{1-l_{ii}}, & \text{if } i = j, \\ -|a_{ij}| - \frac{|b_{ij}|e^{\epsilon\tau}}{1-l_{ij}}, & \text{if } i \neq j. \end{cases}$$

Being $\Omega_0 = \Omega$, by a continuity argument we have $\beta^T\Omega_\epsilon > 0$.

Then consider a candidate Lyapunov functional using the ϵ presented above

$$V[x,\gamma](t) = \sum_{i=1}^n \beta_i \left[|x_i(t)|e^{\epsilon t} + e^{\epsilon\tau}\sum_{j=1}^n \frac{|b_{ij}|}{1-l_{ij}} \int_{t-\tau_{ij}(t)}^t |\gamma_j(s)|e^{\epsilon s} ds\right].$$

For $t \geq 0$, we have

$$
\begin{aligned}
& D^+V[x-\xi,\gamma-\eta](t) \\
& \leq -e^{\epsilon t}\sum_{i=1}^{n}\beta_i(d_i-\epsilon)|x_i(t)-\xi_i| \\
& \quad +e^{\epsilon t}\sum_{i=1}^{n}\beta_i\left[a_{ii}|\gamma_i(t)-\eta_i|+\sum_{j=1}^{n}\left(|a_{ij}|(1-\delta_{ij})+\frac{e^{\epsilon\tau}|b_{ij}|}{1-l_{ij}}\right)|\gamma_j(t)-\eta_j|\right] \\
& = -e^{\epsilon t}\sum_{i=1}^{n}\beta_i(d_i-\epsilon)|x_i(t)-\xi_i| - e^{\epsilon t}\beta^T\Omega_\epsilon\big(|\gamma_1(t)-\eta|,|\gamma_2(t)-\eta|,\cdots, \\
& \quad |\gamma_n(t)-\eta|\big)^T \\
& \leq 0.
\end{aligned}
$$

It follows that

$$
\begin{aligned}
\|x(t)-\xi\|_\beta \triangleq \sum_{i=1}^{n}\beta_i|x_i(t)-\xi_i| & \leq e^{-\epsilon t}V[x-\xi,\gamma-\eta](t) \\
& \leq e^{-\epsilon t}V[x-\xi,\gamma-\eta](0).
\end{aligned}
$$

It implies the global exponential stability of the equilibrium point of Eqn. (1). The proof is completed.

4 Illustrative Example

In this section, an example will be given to further illustrate the results above.

Example 1. Consider the seconder-order neural network defined as follows:

$$
\begin{aligned}
\dot{x}_1(t) &= -x_1(t) - \frac{5}{3}g_1(x_1(t)) + \frac{1}{3}g_1(x_1(t-\tau)), \\
\dot{x}_2(t) &= -x_2(t) - \frac{2}{3}g_2(x_2(t)) + \frac{1}{4}g_1(x_1(t-\tau)),
\end{aligned}
$$

where $g_1(\theta) = g_2(\theta) = \mathrm{sign}(\theta)$ which is a discontinuous function. For the reason of simplicity, $\tau = 10$ is a constant.

It is seen that $\xi = 0$ is an equilibrium point, with corresponding output equilibrium point $\eta = 0$. Since the assumptions of Theorem 1 are satisfied, the state x converges to the equilibrium point ξ. Fig. 2 shows the time-domain behavior of the state variables x_1 and x_2 with the initial conditions $[\phi_1(t),\phi_2(t)]^T = [5\cos 10t, -5\cos 10t]^T$ for all $t \in [-10, 0)$.

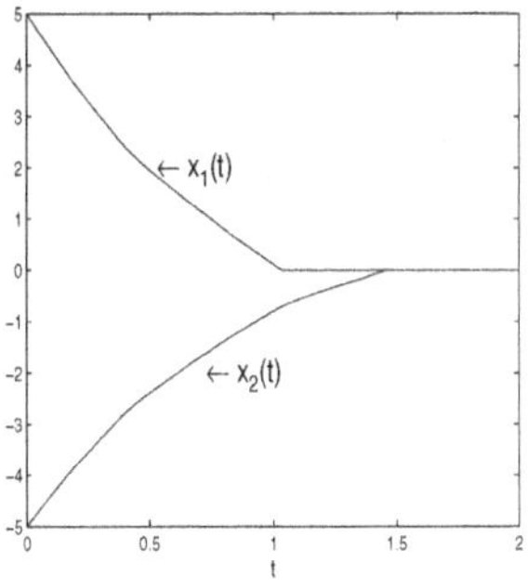

Fig. 2. Behavior of the state $x_1(t)$ and $x_2(t)$ for the neural network in Example 1

5 Conclusions

This paper has introduced a general class of neural networks model with time-varying delays in the neuron amplifiers response a neuron activations possessing jump discontinuities. The discontinuities of the neuron activations are a model for neuron amplifiers with very high gain. Easily testable conditions have been established that ensure global exponential stability of the state with a known convergence rate.

Acknowledgments

This work was supported by National Science Foundation of China under Grant 60471055 and Specialized Research Fund for the Doctoral Program of Higher Education under Grant 20040614017.

References

1. Yi, Z., Heng, P. A., Vadakkepat, P.: Absolute Periodic and Absolute Stability of Delayed Neural Networks. IEEE Trans. Circuits and Systems **I 49** (2002) 256-261
2. Yi, Z., Heng, P. A., Leung, K. S.: Convergence Analysis of Celluar Neural Networks with Unbounded Delay. IEEE Trans. Circuits and Systems **I 48** (2001) 680-687
3. Yi, Z., Heng, P. A., Fu, A. W. C.: Estimate of Exponential Convergence Rate and Exponential Stability of Neural Networks with Unbounded Delay. IEEE Trans. Neural Networks **10** (1999) 1487-1493
4. Liao, X., Wang, J.: Algebraic Crieria for Global Exponential Stability of Celluar Neural Networks with Multiple Time Delays: IEEE Trans. Circuits and Systems **I 50** (2003) 268-285
5. Qi, H., Qi, L.: Deriving Sufficient Conditions for Global Asymptotic Stability of Delayed Neural Networks via Nonsmooth Analysis. IEEE Trans. Neural Networks **15** (2004) 99-109
6. Forti, M., Nistri, P.: Global Convergence of Neural Networks with Discontinuous Neural Activation. IEEE Trans. Circuits and Systems **I 50** (2003) 1421-1435

7. Forti, M., Nistri, P., Papini, D.: Global Exponential Stability and Global Convergence in Finite Time of Delayed Neural Networks with Infinite Gain. IEEE Trans. Neural Networks **16** (2005) 1449-1463
8. Filippov, A. F.: Differential Equations with Discontinuous Right-Hand Side. Mathematics and its Applications (Soviet Series) Boston (1988).

Global Asymptotic Stability of Cohen-Grossberg Neural Networks with Mixed Time-Varying Delays

Haijun Jiang and Xuehui Mei

College of Mathematics and System Sciences, Xinjiang University,
Urumqi 830046, China
jianghai@xju.edu.cn

Abstract. In this paper, we study the Cohen-Grossberg neural networks with mixed time-varying delays. By applying the Lyapunov functional method and combining with the inequality $3abc \leq a^3+b^3+c^3$ $(a, b, c > 0)$ technique, a series of new and useful criteria on the existence of equilibrium point and its global asymptotical stability are established. The results obtained in this paper extend and generalize the corresponding results existing in previous literature.

1 Introduction

In recently years, the dynamical characteristic such as stability and periodicity of Hopfield networks, cellular neural networks, bidirectional associative memory neural networks and Cohen-Grossberg neural networks play an important role in the pattern recognition, associative memory, and combinatorial optimization (see [1-5]).

In reality, due to the finite switching speeds of neurons and amplifiers, time delays inevitably exist in biological and artificial neural networks and thus should be incorporated into the model. In [6], the authors observed that time delays could induce instability, causing sustained oscillations which may be harmful to a system by experiment and numerical analysis. In this paper, we consider the following Cohen-Grossberg neural networks with mixed time-varying delays

$$\begin{aligned}\frac{dx_i(t)}{dt} &= -e_i(x_i(t))\Big[h_i(x_i(t)) - \sum_{j=1}^{n} a_{ij} f_j(x_j(t)) - \sum_{j=1}^{n} b_{ij} f_j(x_j(t-\tau_{ij}(t))) \\ &\quad - \sum_{j=1}^{n} d_{ij} \int_{-\infty}^{t} K_{ij}(t-s) g_j(x_j(s))ds + I_i\Big], \quad i = 1, 2, \cdots, n.\end{aligned} \tag{1}$$

where n denotes to the number of units in a neural networks; $x_i(t)$ is the state of the ith unit at time t; $f_j(x_j(t))$ is the output of the jth unit at time t; a_{ij}, b_{ij}, d_{ij} are constants, I_i is constant input, $\tau_{ij}(t)$ is nonnegative continuous function and $e_i(\cdot) > 0$.

D. Liu et al. (Eds.): ISNN 2007, Part I, LNCS 4491, pp. 1008–1014, 2007.

Let $\tau = \sup\{\tau_{ij}(t) : t \in [0,\ +\infty),\ i,j = 1,2,...,n\}$, in this paper we always assume that all solutions of system (1) satisfy the following initial conditions

$$x_i(\theta) = \phi_i(\theta), \quad \theta \in [-\tau,\ 0],\ i = 1,2,...,n, \tag{2}$$

where $\phi_i :\ [-\tau,\ 0] \to \mathbb{R}$ are continuous functions. It is well known that by the fundamental theory of functional differential equations [7], system (1) has a unique solution $x(t) = (x_1(t), x_2(t), ..., x_n(t))$ satisfying the initial condition.

The main purpose of this paper is to study the dynamic behavior of the general Cohen-Grossberg neural networks with mixed delays. By applying the Lyapunov functional method and inequality $3abc \le a^3+b^3+c^3$ techniques, we will establish a series of new and useful criteria on the existence of equilibrium point and its global asymptotic stability for system (1). We will see that the results obtained in this paper will extend and generalize the corresponding results existing in [8-10].

The rest of this paper is organized as follows. In Section 2, we will give a description for system (1). In Section 3, we will establish new conditions ensuring the global asymptotic stability of system (1). Section 4 given a example to illustrate the effectiveness of the result given in this paper. In Section 5, we will give some concluding remarks of the results.

2 Preliminaries

We assume that each of the relations between the output of the cell f_i and g_i $(i = 1, 2, \cdots, n)$ and the state of the cell possess the following properties.

(H_1) $h_i(u) : \mathbb{R} = (-\infty,\ +\infty) \to \mathbb{R}$ $(i = 1, 2, \cdots, n)$ is differentiable function, $\gamma_i = \inf_{u\in\mathbb{R}} h_i'(u) > 0$ and $h_i(0) = 0$, where $h_i'(u)$ represent the derivative of $h_i(u)$.

(H_2) $\tau_{ij}(t)$ $(i = 1, 2, \cdots, n)$ are nonnegative, bounded and continuous differentiable function defined on $\mathbb{R}_+ = [0,\ +\infty)$ and $\inf_{t\in\mathbb{R}_+}\{1 - \dot{\tau}_{ij}(t)\} > 0$.

(H_3) Functions $k_{ij}(s)$ $(i, j = 1, 2, \cdots, n)$ satisfy $\int_0^\infty k_{ij}(s)ds = 1$.

(H_4) f_i and g_i are bounded on $\mathbb{R}$;

(H_5) There are positive constants k_i and p_i $(i = 1, 2, \cdots, n)$ such that

$$|f_i(u) - f_i(u^*)| \le k_i|u - u^*| \quad \text{and} \quad |g_i(u) - g_i(u^*)| \le p_i|u - u^*|$$

for all $u,\ u^* \in \mathbb{R}$ and $i = 1, 2, \cdots, n$.

For system (1), making use of the approach in Wang and Zou [8], it is easy to obtain the following lemma.

Lemma 1. *If (H_1)-(H_5) hold, then for every input I_i, there exists an equilibrium point $x^* = (x_1^*, x_2^*, \cdots, x_n^*)$ for system (1).*

In order to simplify our proofs, we shall shift the equilibrium point x^* of system (1) to the origin. Using the following transformation

$$y_i(t) = x_i(t) - x_i^*, \quad y_j(t - \tau_{ij}(t)) = x_j(t - \tau_{ij}(t)) - x_j^*$$

the system (1) can be transformed into the following form

$$\frac{dy_i(t)}{dt} = -E_i(y_i(t))\Big[H_i(y_i(t)) - \sum_{j=1}^{n} a_{ij}\Phi_j(y_j(t)) - \sum_{j=1}^{n} b_{ij}\Phi_j(y_j(t-\tau_{ij}(t))) - \sum_{j=1}^{n} d_{ij}\int_{-\infty}^{t} K_{ij}(t-s)\Psi_j(y_j(s))ds\Big], \qquad (3)$$

in which $i = 1, 2, \cdots, n$, $E_i(y_i(t)) = e_i(y_i(t) + x_i^*)$ $H_i(y_i(t)) = h_i(y_i(t) + x^*) - h_i(x_i^*)$, $\Phi_j(y_j(t)) = f_j(y_j(t) + x_j^*) - f_j(x_j^*)$, $\Psi(y_j(t)) = g_j(y_j(t) + x_j^*) - g_j(x_j^*)$ with $\Phi(0) = 0$ and $\Psi(0) = 0$.

Lemma 2. *All solutions of the system (1) remain bounded on* $\mathbb{R}_+$.

The proof of Lemma 2 is easy, here, we omit it.

3 Main Results

Theorem 1. *Suppose that* (H_1)*-*(H_5) *hold, if the system parameters* a_{ij}, b_{ij}, d_{ij} $(i, j = 1, 2, \cdots, n)$ *satisfy the following condition*

$$\gamma_i - \frac{2}{3}\sum_{j=1}^{n}|a_{ij}| - \frac{1}{3}\sum_{j=1}^{n}|a_{ji}|k_i^3 - \frac{2}{3}\sum_{j=1}^{n}|b_{ij}| - \frac{2}{3}\sum_{j=1}^{n}|d_{ij}| - \frac{1}{3}\sum_{j=1}^{n}\frac{|b_{ji}|}{1-\dot{\tau}_{ji}(\psi_{ji}^{-1}(t))}k_i^3 - \frac{1}{3}\sum_{j=1}^{n}|d_{ji}|p_i^3 > \sigma,$$

where $\sigma > 0$ *is a constant. Then the origin of system (3) is globally asymptotically stable.*

Proof. We construct the Lyapunov functional as follows

$$V(t) = \sum_{i=1}^{n}\{\int_0^{y_i(t)}\frac{s^2}{E_i(s)}ds + \frac{1}{3}\sum_{j=1}^{n}\int_{t-\tau_{ij}(t)}^{t}\frac{|b_{ij}|}{1-\dot{\tau}_{ij}(\psi_{ij}^{-1}(s))}|\Phi_j(y_j(s))|^3 ds\} + \sum_{i=1}^{n}\sum_{j=1}^{n}\frac{1}{3}|d_{ij}|\int_0^{\infty}K_{ij}(s)(\int_{t-s}^{t}|\Psi_j(y_j(\xi))|^3 d\xi)ds,$$

where $\psi_{ij}^{-1}(t)$ is the inverse function of $\psi_{ij}(t) = t - \tau_{ij}(t)$. Calculating the derivative $V(t)$ along the solution of Eq. (3) by using the inequality $a^2b \leq \frac{1}{3}(2a^3 + b^3)$ which is derived from the inequality $3abc \leq a^3 + b^3 + c^3$ $(a, b, c \geq 0)$ as $a = c$, we get

$$D^+V(t) = \sum_{i=1}^{n}|y_i(t)|^2\{-H_i(y_i(t)) + \sum_{j=1}^{n}a_{ij}\Phi_j(y_j(t)) + \sum_{j=1}^{n}b_{ij}\Phi_j(y_j(t-\tau_{ij}(t))) + \sum_{j=1}^{n}d_{ij}\int_{-\infty}^{t}K_{ij}(t-s)\Psi_j(y_j(s))ds\Big]\} + \frac{1}{3}\sum_{i=1}^{n}\sum_{j=1}^{n}\frac{|b_{ij}|}{1-\dot{\tau}_{ij}(\psi_{ij}^{-1}(t))}|\Phi_j(y_j(t))|^3$$

$$
\begin{aligned}
&-\frac{1}{3}\sum_{i=1}^{n}\sum_{j=1}^{n}|b_{ij}||\Phi_j(y_j(t-\tau_{ij}(t)))|^3\\
&+\sum_{i=1}^{n}\sum_{j=1}^{n}\frac{1}{3}|d_{ij}|\int_0^{\infty}K_{ij}(s)|\Psi_j(y_j(t))|^3d)ds\\
&-\sum_{i=1}^{n}\sum_{j=1}^{n}\frac{1}{3}|d_{ij}|\int_0^{\infty}K_{ij}(s)|\Psi_j(y_j(t-s))|^3ds\\
\le&\sum_{i=1}^{n}\{-\gamma_i|y_i(t)|^3+\sum_{j=1}^{n}|a_{ij}||y_i(t)|^2|\Phi_j(y_j(t))|\\
&+\sum_{j=1}^{n}|b_{ij}||y_i(t)|^2|\Phi_j(y_j(t-\tau_{ij}(t)))|\\
&+\sum_{j=1}^{n}|d_{ij}|\int_{-\infty}^{t}K_{ij}(t-s)|y_i(t)|^2|\Psi_j(y_j(s))|ds\Big]\}\\
&+\frac{1}{3}\sum_{i=1}^{n}\sum_{j=1}^{n}\frac{|b_{ij}|}{1-\dot{\tau}_{ij}(\psi_{ij}^{-1}(t))}|\Phi_j(y_j(t))|^3\\
&-\frac{1}{3}\sum_{i=1}^{n}\sum_{j=1}^{n}|b_{ij}||\Phi_j(y_j(t-\tau_{ij}(t)))|^3\\
&+\sum_{i=1}^{n}\sum_{j=1}^{n}\frac{1}{3}|d_{ij}|\int_0^{\infty}K_{ij}(s)|\Psi_j(y_j(t))|^3d)ds\\
&-\sum_{i=1}^{n}\sum_{j=1}^{n}\frac{1}{3}|d_{ij}|\int_0^{\infty}K_{ij}(s)|\Psi_j(y_j(t-s))|^3ds\\
\le&\sum_{i=1}^{n}\{-\gamma_i|y_i(t)|^3+\sum_{j=1}^{n}|a_{ij}|\frac{2}{3}|y_i(t)|^3+\sum_{j=1}^{n}|a_{ij}|\frac{1}{3}k_j^3|y_j(t)|^3\\
&+\sum_{j=1}^{n}|b_{ij}|\frac{2}{3}|y_i(t)|^3+\frac{1}{3}\sum_{j=1}^{n}|b_{ij}||\Phi_j(y_j(t-\tau_{ij}(t)))|^3\\
&+\frac{1}{3}\sum_{j=1}^{n}|d_{ij}|\int_{-\infty}^{t}K_{ij}(t-s)|\Psi_j(y_j(s))|^3ds\\
&+\frac{2}{3}\sum_{j=1}^{n}|d_{ij}|\int_{-\infty}^{t}K_{ij}(t-s)|y_i(t)|^3ds\}\\
&+\frac{1}{3}\sum_{i=1}^{n}\sum_{j=1}^{n}\frac{|b_{ij}|}{1-\dot{\tau}_{ij}(\psi_{ij}^{-1}(t))}|\Phi_j(y_j(t))|^3\\
&-\frac{1}{3}\sum_{i=1}^{n}\sum_{j=1}^{n}|b_{ij}||\Phi_j(y_j(t-\tau_{ij}(t)))|^3\\
&+\sum_{i=1}^{n}\sum_{j=1}^{n}\frac{1}{3}|d_{ij}|\int_0^{\infty}K_{ij}(s)|\Psi_j(y_j(t))|^3d)ds\\
&-\sum_{i=1}^{n}\sum_{j=1}^{n}\frac{1}{3}|d_{ij}|\int_0^{\infty}K_{ij}(s)|\Psi_j(y_j(t-s))|^3ds
\end{aligned}
$$

$$= \sum_{i=1}^{n}\{-\gamma_i + \frac{2}{3}\sum_{j=1}^{n}|a_{ij}| + \frac{1}{3}\sum_{j=1}^{n}|a_{ji}|k_i^3 + \frac{2}{3}\sum_{j=1}^{n}|b_{ij}| + \frac{2}{3}\sum_{j=1}^{n}|d_{ij}| + \frac{1}{3}\sum_{j=1}^{n}\frac{|b_{ji}|}{1-\dot{\tau}_{ji}(\psi_{ji}^{-1}(t))}k_i^3 + \frac{1}{3}\sum_{j=1}^{n}|d_{ji}|p_i^3\}|y_i(t)|^3$$
$$\leq -\sigma\sum_{i=1}^{n}|y_i(t)|^3 < 0,$$

for all $t \geq 0$. From this, we obtain

$$V(t) + \sigma\int_0^t \sum_{i=1}^{n}|y_i(s)|^3 ds \leq V(0),$$

for all $t \geq 0$. Consequently,

$$\sum_{i=1}^{n}\int_0^{\infty}|y_i(s)|^3 ds < \infty. \tag{4}$$

The boundedness of $f_i(\cdot)$ and $g_i(\cdot)$ implies that each $y_i(t)$ is bounded on R_+. From directly system (3) we obtain that $\frac{dy_i(t)}{dt}$ also is bounded on $\mathbb{R}_+$. Hence, $|y_i(t)|^3$ $(i = 1, 2, \cdots, n)$ are uniformly continuous on $\mathbb{R}_+$. Therefore, by (4) we obtain $\lim_{t\to\infty} y_i(t) = 0$ $(i = 1, 2, \cdots, n)$. This shows that the equilibrium point of system (1) is globally asymptotically stable. This completes the proof.

Remark 1. Compare the result obtained in this paper with ones given in [9-10], we can find that the result obtained in this paper improve and extend the corresponding result given in [9-10], and different from the result given in [11-12].

Remark 2. If $E_i(y_i(t)) \equiv 1$, then system (3) reduce to the following recurrent neural networks with mixed time-varying delays

$$\frac{dy_i(t)}{dt} = -H_i(y_i(t)) + \sum_{j=1}^{n} a_{ij}\Phi_j(y_j(t)) + \sum_{j=1}^{n} b_{ij}\Phi_j(y_j(t-\tau_{ij}(t))) + \sum_{j=1}^{n} d_{ij}\int_{-\infty}^{t} K_{ij}(t-s)\Psi_j(y_j(s))ds, \quad i = 1, 2, \cdots, n. \tag{5}$$

For system (5), we have the following theorem.

Theorem 2. *Under the conditions of Theorem 1, the equilibrium point of system (5) is globally asymptotically stable.*

Proof. The details of the proof of Theorem 2 are similar to those of Theorem 1. In order to obtain this result we only need consider the following Lyapunov functional $V_1(t)$ defined by

$$V_1(t) = \sum_{i=1}^{n}\{\frac{1}{3}|y_i(t)|^3 + \frac{1}{3}\sum_{j=1}^{n}\int_{t-\tau_{ij}(t)}^{t}\frac{|b_{ij}|}{1-\dot{\tau}_{ij}(\psi_{ij}^{-1}(s))}|\Phi_j(y_j(s))|^3 ds\} + \sum_{i=1}^{n}\sum_{j=1}^{n}\frac{1}{3}|d_{ij}|\int_0^{\infty}K_{ij}(s)(\int_{t-s}^{t}|\Psi_j(y_j(\xi))|^3 d\xi)ds,$$

where $\psi_{ij}^{-1}(t)$ is the inverse function of $\psi_{ij}(t) = t - \tau_{ij}(t)$. This completes the proof.

4 Simulation Result

Example 1. Consider the following 2-dimensional Cohen-Grossberg neural networks with delays

$$\begin{aligned}\frac{dx_1(t)}{dt} &= -(8+\sin x_1(t))[10x_1(t) - \sum_{j=1}^{2} a_{1j}f_j(x_j(t)) - \sum_{j=1}^{2} b_{1j}f_j(x_j(t-\tau_{1j}(t))) \\ &\quad -\sum_{j=1}^{2} d_{1j}\int_{-\infty}^{t} K_{1j}(t-s)g_j(x_j(s))ds + 2], \\ \frac{dx_2(t)}{dt} &= -(5+\cos x_2(t))[11x_2(t) - \sum_{j=1}^{2} a_{2j}f_j(x_j(t)) - \sum_{j=1}^{2} b_{2j}f_j(x_j(t-\tau_{1j}(t))) \\ &\quad -\sum_{j=1}^{2} d_{2j}\int_{-\infty}^{t} K_{2j}(t-s)g_j(x_j(s))ds + 3],\end{aligned} \tag{6}$$

where $f_j(x) = g_j(x) = \frac{1}{2}(|x+1| - |x-1|)$ $(j = 1, 2)$. Obviously, $f_j(x)$ and $g_j(x)$ are bounded and satisfies the Lipschitz condition with Lipschitz constant $k_i = p_i = 1$ $(i = 1, 2)$.

In (6), we take $a_{11} = a_{22} = 1$, $a_{12} = a_{21} = 1$, $b_{11} = b_{22} = 1$, $b_{12} = b_{21} = 2$, $d_{11} = d_{22} = 1$, $d_{12} = d_{21} = 2$, $\tau_{ij}(t) = 1 + \frac{1}{2}\sin t$ and $k_{ij}(s) = e^{-s}$ $(i, j = 1, 2)$ satisfy $\int_0^\infty k_{ij}(s)ds = 1$. Further, choose $\sigma = \frac{1}{3}$, then we have $\gamma_1 = 10$, $\gamma_2 = 11$, and

$$\begin{aligned}&\gamma_1 - \frac{2}{3}\sum_{j=1}^{2}|a_{1j}| - \frac{1}{3}\sum_{j=1}^{2}|a_{j1}| - \frac{2}{3}\sum_{j=1}^{2}|b_{1j}| \\ &-\frac{2}{3}\sum_{j=1}^{2}|d_{1j}| - \frac{1}{3}\sum_{j=1}^{2}\frac{|b_{j1}|}{1-\dot{\tau}_{j1}(\psi_{j1}^{-1}(t))} - \frac{1}{3}\sum_{j=1}^{2}|d_{j1}| > 1 > \sigma \\ &\gamma_2 - \frac{2}{3}\sum_{j=1}^{2}|a_{2j}| - \frac{1}{3}\sum_{j=1}^{2}|a_{j2}| - \frac{2}{3}\sum_{j=1}^{2}|b_{2j}| \\ &-\frac{2}{3}\sum_{j=1}^{2}|d_{2j}| - \frac{1}{3}\sum_{j=1}^{2}\frac{|b_{j2}|}{1-\dot{\tau}_{j2}(\psi_{j2}^{-1}(t))} - \frac{1}{3}\sum_{j=1}^{2}|d_{j2}| > 2 > \sigma\end{aligned}$$

Therefore, by Theorem 1, systems (6) is globally asymptotically stable.

5 Conclusion

In this paper, we have investigated the global asymptotical stability for Cohen-Grossberg neural networks with mixed time delays. Using Lyapunov functional method and inequality technique, we gave a sufficient criterion ensuring the global asymptotical stability of the equilibrium point. The obtained result improves and extend several earlier publications and is useful in applications of manufacturing high quality neural networks.

Acknowledgement

This work was supported by the National Natural Science Foundation of P.R. China (10361004), the Major Project of The Ministry of Education of P.R. China and the Funded by Scientific Research Program of the Higher Education Institution of Xinjiang (XJEDU2004I12 and XJEDU2006I05), and the doctoral Foundation of Xinjiang University under Grants 070171.

References

1. Hopfield, J. J.: Neurons with Graded Response Have Collective Computational Properties Like Those of Two-Stage Neurons. Proc. Nat. Acad. Sci.-Biol. **81** (1984) 3088-3092
2. Chua, L.O., Yang, L.: Cellular Neural Networks: Theory. IEEE Trans. Circuits and Systems **35** (1988) 1257-1272
3. Gopalsamy, K., He, X. Z.: Delay-Independent Stability in Bidrectional Associative Memories Networks. IEEE Trans. Neurak networks **5** (1994) 998-1002
4. Cao, J., Wang, J.: Global Asymptotic Stability of a General Class of Recurrent Neural Networks with Time-Varying Delays. IEEE Trans. Circuits and Systems **I 50** (2003) 34-44
5. Cohen, M. A., Grossberg, S.: Absolute Stability and Global Pattern Formation and Parallel Memory Storage by Competitive Neural Networks. IEEE Trans. Syst. Man Cybern. **13** (1983) 815-826
6. Marcus, C., Westervelt, R.: Stability of Analog Neural Networks with Delay. Phys. Rev. A **39** (1989) 347-359
7. Hale, J.: Theory of Functional Differential Equations. Springer New York 1977
8. Wang, L., Zou, X.: Exponential Stability of Cohen-Grossberg Neural Networks. Neural Networks **15** (2002) 415-422
9. Wang, L.: Stability of Cohen-Grossberg Neural Networks with Distributed Delays. Appl. Math. Computa. **160** (2005) 93-110
10. Hwang, C., Cheng, C., Li, T.: Globally Exponential Stability of Generalized Cohen-Grossberg Neural Networks with Delays. Phys. Lett. A **319** (2003) 157-166
11. Liao, X., Li, C., Wong, K.: Critria for Exponential Stability of Cohen-Grossberg Neural Networks. Neural Networks **17** (2004) 1401-1414
12. Yuan, K., Cao, J.: An Analysis of Global Asymptotic Stability of Delayed Cohen-Grossberg Neural Networks via Nonsmooth Analysis. IEEE Trans. Circuits and Systems **I 52** (2005) 1854-1861

Differences in Input Space Stability Between Using the Inverted Output of Amplifier and Negative Conductance for Inhibitory Synapse

Min-Jae Kang[1], Ho-Chan Kim[1], Wang-Cheol Song[2], Junghoon Lee[3], Hee-Sang Ko[4], and Jacek M. Zurada[5]

[1] Faculty of Electrical and Electronic Engineering, Cheju National University, Jeju, 690-756, Korea
{minjk, hckim}@cheju.ac.kr
[2] Department of Computer Engineering, Cheju National University, Jeju, 690-756, Korea
philo@cheju.ac.kr
[3] Department of Computer Science, Cheju National University, Jeju, 690-756, Korea
jhlee@cheju.ac.kr
[4] Dept. of Electrical and Computer Engineering, The University of British Columbia, Vancouver, BC V6T 1Z4, Cananda
Hee-sang@ece.ubc.ca
[5] Department of Electrical Engineering, University of Louisville, Louisville, KY, 40292, U.S.A
j.zurada@ieee.org

Abstract. In this paper, the difference between using the inverted neuron output and negative resistor for expressing inhibitory synapse is studied. We analyzed that the total conductance seen at the neuron input is different in these two methods. And this total conductance has been proved to effect on the system stability in this paper. Also, we proposed the method how to stabilize the input space and improve the system's performance by adjusting the input conductance between neuron input and ground. Pspice is used for circuit level simulation.

1 Introduction

The synapse is a contact organ which connects the axon of neuron to the neighboring neurons' dendrites. There are two kinds of synapses, one is an excitatory one which causes a neuron firing, the other is an inhibitory one which hinders the firing of neuron [1-3].

We often see the negative weights in optimization problem using neural networks. These negative weights can be regarded as inhibitory synapses. When designing Hopfield neural networks, the synapse (w_{ij}) which connects the j-th neuron output to the i-the neuron input can be expressed by the resistance ($R_{ij} = 1/w_{ij}$). Therefore, negative resistor is needed for expressing an inhibitory synapse.

D. Liu et al. (Eds.): ISNN 2007, Part I, LNCS 4491, pp. 1015–1024, 2007.

However, the commonly used method for expressing an inhibitory synapse is that positive resistor is connected to the inverted output of neuron. To provide for both excitatory and inhibitory synaptic connections between neurons while using conventional electrical components, each amplifier used for neuron has two outputs: a normal output (+) and an inverted output (-) of the same magnitude but opposite in sign. If the synapse is excitatory ($w_{ij} > 0$), this resister ($R_{ij} = 1/w_{ij}$) is connected to the normal output (+) of the j-th amplifier. For an inhibitory synapse ($w_{ij} < 0$), it is connected to the inverted output (-) of the j-th amplifier. Thus, the normal and inverted outputs for each neuron allow for the construction of both excitatory and inhibitory connections through the use of normal (positive valued) resistors [1-3].

From the viewpoint of the current flowing through synapse, this method is same as using negative resistor for an inhibitory synapse. However, the total conductance seen at the neuron input is different. In this paper, this total conductance has been proved to effect on the system stability. It has been shown that the input space of the system is stable only for this total conductance being positive. Also, it has been proved that the input conductance g_i plays a major role in the input space stability and system's performance.

2 The Two Models for Inhibitory Synapse: Using Inverting Output and Negative Conductance

Fig. 1 shows the input node of the i-th neuron where the neural network model is implemented using electrical component. C_i represents the nonzero input capacitance of the i-th neuron. Similarly, g_i represents the input conductance between the i-th neuron input and ground. These components have been known partially to define the time constant of the neuron and to provide for the integrative analog summation of the synaptic input currents from other neurons in the networks. Conductance w_{ij} connects the output of the j-th neuron to the input of i-th neuron. Current i_i is the bias coming

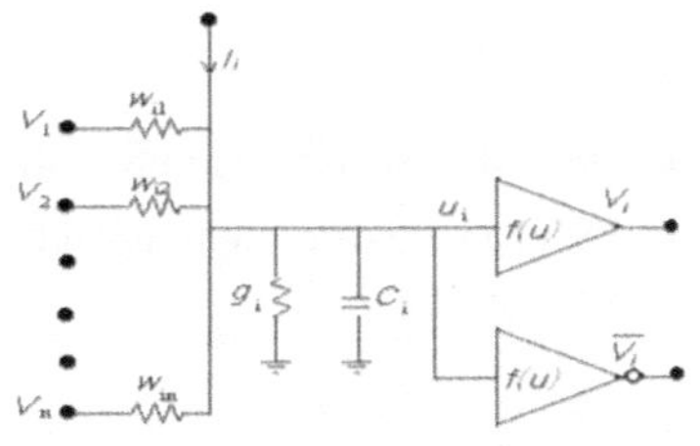

Fig. 1. Input node of the i-th neuron in the electrical component neural network model

current into the i-th neuron input. Also, each neuron maps its input voltage u_i into the output voltage V_i through the activation function $f(u_i)$. V_i is the neuron's output and $\bar{V_i}$ is the neuron's inverting output [3].

Fig. 2 shows the Norton equivalent circuits of the neural network at the i-th input node. Let us assume w_{ik} an inhibitory synapse, and two methods can be used for representing this synapse. One method is as shown in Fig. 2a, using negative conductance w_{ik} and the k-th neuron's normal output. The other one is as shown in Fig. 2b, using normal conductance w_{ik} and the k-th neuron's inverted output. For two circuits to be same, Norton's equivalent currents and conductance at the i-th input node should be same. As shown in Fig. 2a, the Norton's equivalent current and conductance of the first method are as follows

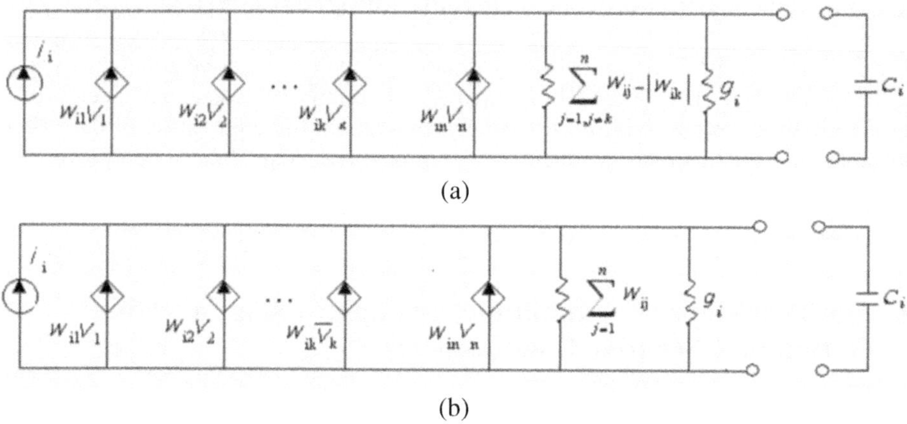

Fig. 2. Equivalent circuits seen from input node i where w_{ik} is inhibitory synapse: (a) Using negative conductance, (b) Using the k-th inverting output neuron

$$I_{na} = \sum_{j=1, j\neq k}^{n} w_{ij}v_j - |w_{ik}|v_k + i_i, \tag{1}$$

$$G_{na} = \sum_{j=1, j\neq k}^{n} w_{ij} - |w_{ik}| + g_i. \tag{2}$$

In equation (1) and (2), the reason of using absolute value of w_{ik} is because the w_{ik} has negative value. And the Norton's equivalent currents and conductance of the second method are follows as shown in Fig. 2b

$$I_{nb} = \sum_{j=1, j\neq k}^{n} w_{ij}v_j + w_{ik}\overline{v_k} + i_i, \tag{3}$$

$$G_{nb} = \sum_{j=1}^{n} w_{ij} + g_i. \tag{4}$$

By comparing equation (1) and (3), we can notice that the Norton's equivalent current is same for both methods. The second terms in the right hand sides of equations (1) and (3) are the currents flowing through inhibitory synapse. The direction and magnitude of these currents are same even though they look different. However, as known in equation (2) and (4), the Norton's equivalent conductance of two methods is different.

3 Stability Analysis of Input Space

Our objective in this section is to analyze how the Norton's equivalent conductance affects the stability of the neuron input space. The different methods for representing inhibitory synapse could have different stability in neuron input space because of the different Norton's equivalent conductance. The Norton's equivalent circuit in Fig. 2 can be more simplified as shown in Fig. 3. The Norton's equivalent current I_{ni} denotes the total current in the left hand sides of equations (1) and (3). Also the Norton's equivalent conductance G_{ni} denotes the total conductance in the left hand sides of equation (2) and (4). As seen in equation (2) and (4), the Norton's equivalent conductance G_{ni} is the sum of input conductance g_i and synapses which are connected to the i-th neuron input. Therefore, the Norton's equivalent conductance can be any sign of value, because there exists inhibitory synapses and normal synapses. Here, we are going to analyze the input space stability in the three cases ($G_{ni} > 0$, $G_{ni} = 0$ and $G_{ni} < 0$).

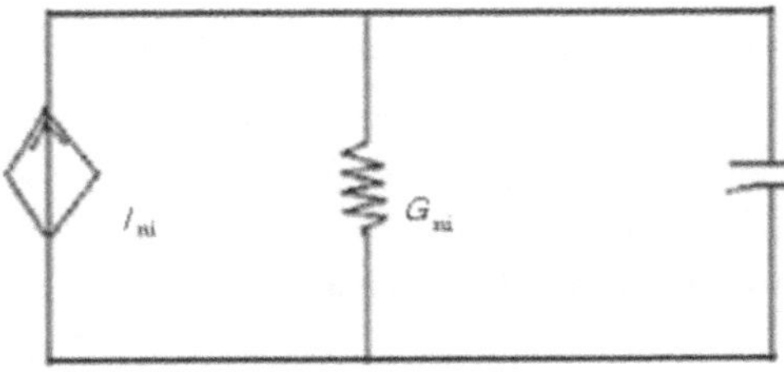

Fig. 3. The Norton's equivalent circuit seen from input node i

Propositions. In the continuous Hopfield type neural network, the following statements about the stability of input space are true:

(1) If the equivalent conductance G_{ni} is great than zero, then the input space is exponentially stable.
(2) If the equivalent conductance G_{ni} is equal to zero, then the input space is linearly unstable.
(3) If the equivalent conductance G_{ni} is less than zero, then the input space is exponentially unstable.

Proof
Proof of (1). Using KCL (Kirchoff Current Law) at the input of the neuron, the following equation can be obtained [3]

$$C_i \frac{du_i}{dt} = -G_{ni} u_i + I_{ni} . \tag{5}$$

Because that Hopfield neural network has a Lyapunov energy function, the output of system becomes asymptotically stabilized in time decreasing the energy function. This means that the output is converged to the certain value and stabilized after some time. Also, the flowing current through the synapses is stabilized and becomes constant because the synapse flowing current is the product of synapse's weight and neuron output. Therefore, the equivalent current I_{ni} can be regarded as constant after neuron output is stabilized. The equivalent current I_{ni} is the sum of bias current i_i and the synapse flowing current as seen in equation (1) and (3). The solution of equation (5) can be obtained by taking Laplace transform,

$$C_i(sU(s)-u_i(0)) = -G_{ni}U_i(s)+\frac{I_{ni}}{s}. \tag{6}$$

Equation (6) can be rearranged as

$$U_i(s) = (u_i(0)-\frac{I_{ni}}{G_{ni}})\frac{1}{s+\frac{G_{ni}}{C_i}}+\frac{I_{ni}}{G_{ni}}\frac{1}{s}. \tag{7}$$

The inverse Laplace transform of equation (7) gives

$$u_i(t) = (u_i(0)-\frac{I_{ni}}{G_{ni}})e^{-\frac{G_{ni}}{C_i}t}+\frac{I_{ni}}{G_{ni}}. \tag{8}$$

If the total conductance G_{ni} is greater than zero, the first term of equation (8) is disappearing exponentially and finally $u_i(t)$ becomes as

$$\lim_{t\to\infty} u_i(t) = \frac{I_{ni}}{G_{ni}}. \tag{9}$$

Therefore, If the total conductance G_{ni} is great than zero, then the input space is exponentially stable.

Proof of (2). If $G_{ni}=0$, the following equation is obtained by using equation (6)

$$U(s) = u_i(0)\frac{1}{s}+\frac{I_{ni}}{C_i}\frac{1}{s^2}. \tag{12}$$

The inverse Laplace transform of equation (12) gives

$$u_i(t) = u_i(0)+\frac{I_{ni}}{C_i}t. \tag{13}$$

Same as before, by regarding I_{ni} as a constant after the output space being stabilized, the second term in equation (13) continues to increase or decrease linearly in time, so the input space is linearly unstable.

Proof of (3). If the total conductance is less than zero, the first term of equation (8) is increasing exponentially. Therefore, if the total conductance is less than zero, then the input space is exponentially unstable.

Case study of (1)
The 2-bit A/D converter is selected for a case study [2]. The connecting weights and the bias currents for 2bit A/D converter are as follow, where x is analog input.

$$w = \begin{bmatrix} 0 & -2 \\ -2 & 0 \end{bmatrix}, \quad i = \begin{bmatrix} x - \frac{1}{2} \\ 2x - 2 \end{bmatrix}. \tag{10}$$

Pspice is used to simulate transient analysis. Fig. 4 shows the schematic diagram for $C_1 = C_2 = 1[\mu F]$, $x = 1.6[V]$ and $R_1(1/g_1) = R_2(1/g_2) = 0.4[\Omega]$. In this schematic diagram, negative resistor R_{12} and R_{21} are used for the inhibitory synapses for $w_{12} = w_{21} = -2$. These resistors values are the reciprocal values of w_{12}, w_{21}, g_1 and g_2. The g_1 and g_2 's values are assigned to 2.5℧ for the total conductance G_{ni} being positive. The bias currents $i_1 = 1.1[A]$ and $i_2 = 1.2[A]$ are obtained by using equation (10) for $x = 1.6[V]$.

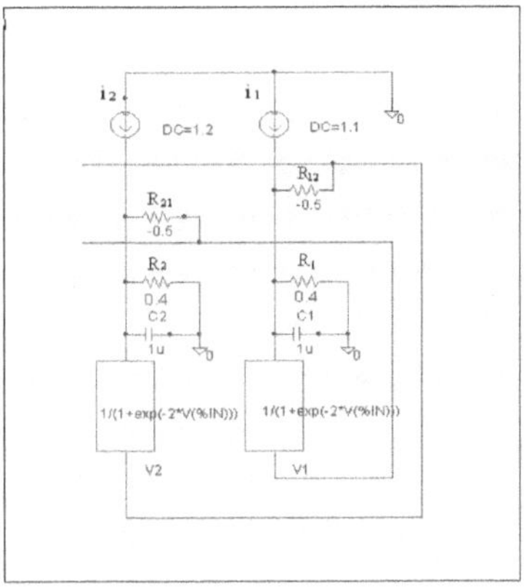

Fig. 4. Pspice schematic diagram for 2bit A/D converter in case $x = 1.6[V]$, $C_1 = C_2 = 1[\mu F]$ and $g_1 = g_2 = 2.5[℧]$

The neural network is expected to converge to digital value 2 $[V]$ for analog input $x = 1.6[V]$, therefore output V_1 and V_2 are expected to converge to 0 and 1 respectively. In Pspice simulation, V_1 and V_2 converge to $0.029[V]$ and $0.989[V]$ respectively shown in Fig. 5b, which can be assumed digital value 0 and 1. Using equation (9), the stabilized inputs for this case are obtained as below

$$\lim_{t\to\infty}\begin{bmatrix}u_1(t)\\u_2(t)\end{bmatrix}=\begin{bmatrix}\dfrac{i_1+w_{12}\times V_2}{G_1}\\\dfrac{i_2+w_{21}\times V_1}{G_2}\end{bmatrix}=\begin{bmatrix}\dfrac{1.1-2\times 0.989}{0.5}\\\dfrac{1.2-2\times 0.029}{0.5}\end{bmatrix}=\begin{bmatrix}-1.756[V]\\2.284\ [V]\end{bmatrix}. \tag{11}$$

As shown in Fig. 5a, the inputs u_1 and u_2 are stabilized exponentially and converge to -1.7584$[V]$ and 2.2839$[V]$ respectively, which are almost same as calculated analytically in equation (11).

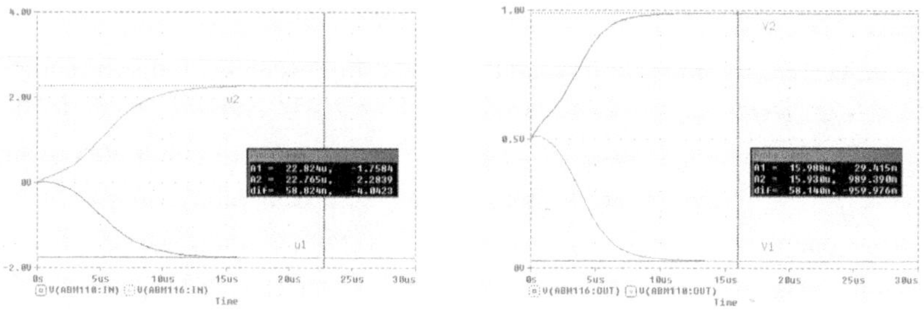

Fig. 5. Transient simulation for 2bit A/D converter in case $x=1.6[V]$, $C_1=C_2=1[\mu F]$ and $g_1=g_2=2.5[\mho]$: (a) input, (b) output

Case study of (2)

All other condition are same as the previous case study except g_1 and g_2. To make the total conductance G_{ni} being zero, g_1 and g_2 's values are assigned to $2[\mho]$. Fig. 6 shows the transient simulations of input and output spaces for $g_1=g_2=2[\mho]$. As seen in output space simulation, the output can be regarded being stabilized after $6[\mu s]$. Therefore, after $6[\mu s]$, equation (13) for this case can be expressed as

$$\begin{bmatrix}u_1(t)\\u_2(t)\end{bmatrix}=\begin{bmatrix}u_1(6\times 10^{-6})\\u_2(6\times 10^{-6})\end{bmatrix}+\begin{bmatrix}-0.9\times 10^{6}(t-6\times 10^{-6})\\1.2\times 10^{6}(t-6\times 10^{-6})\end{bmatrix}\quad \text{for}\quad t\geq 6\times 10^{-6}[s] \tag{14}$$

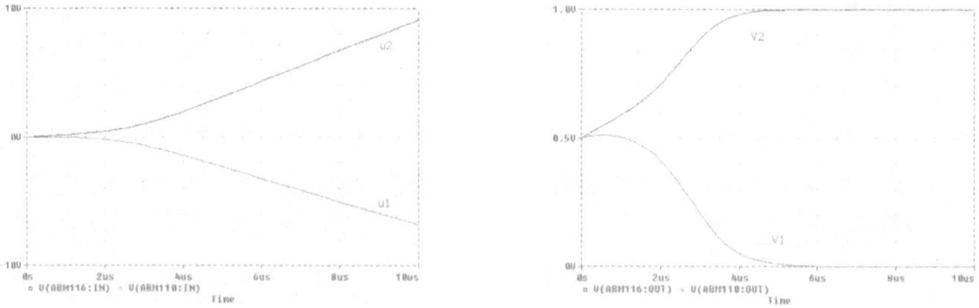

Fig. 6. Transient simulation for 2bit A/D converter in case $x=1.6[V]$, $C_1=C_2=1[\mu F]$ and $g_1=g_2=2[\mho]$: (a) input, (b) output

As seen in Fig. 6a, u_1 and u_2 are changing linearly and diverging. Hopfield neural network has been known as always stable system because it has the Lyapunov type energy function. However, this stability holds only in output space not in input space. Therefore, parasitic input conductance g_i has to be carefully selected to ensure the total conductance G_{ni} being positive.

Case study of (3)
The g_1 and g_2 's values are assigned to $0.5[\mho]$ to make the total conductance being negative ($G_{n1} = G_{n2} = -1.5[\mho]$). The output is stabilized ($V_1 = 0,\ V_2 = 1$) after $2.5[\mu s]$ as shown in Fig. 7b. Therefore, the total currents I_{n1} and I_{n2} become -0.9[A] and 1.2[A] respectively after $2.5[\mu s]$. By using equation (8), the input for the negative total conductance is as follows

$$\begin{bmatrix} u_1(t) \\ u_2(t) \end{bmatrix} = \begin{bmatrix} (u_1(2.5\times10^{-6}) - \frac{0.9}{1.5})\exp^{1.5\times10^6(t-2.5\times10^{-6})} \\ (u_2(2.5\times10^{-6}) + \frac{1.2}{1.5})\exp^{1.5\times10^6(t-2.5\times10^{-6})} \end{bmatrix} + \begin{bmatrix} -\frac{0.9}{1.5} \\ \frac{1.2}{1.5} \end{bmatrix}, \quad (15)$$

$$\text{for } t \geq 2.5\times10^{-6}[s].$$

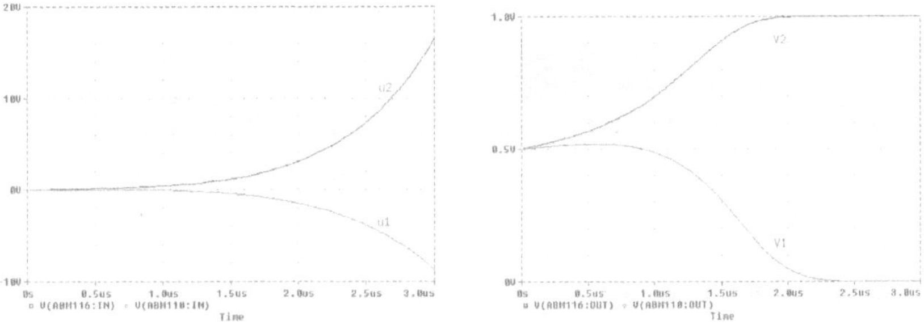

Fig. 7. Transient simulation for 2bit A/D converter in case $x = 1.6[V]$, $C_1 = C_2 = 1[\mu F]$ and $g_1 = g_2 = 0.5[\mho]$: (a) input, (b) output

4 The Effect of Total Conductance on System's Performance

Hopfield (1984) has introduced the Lyapunov like energy function E(v) which is defined by [1]

$$E(v) = -\frac{1}{2}\sum_{i=1}^{n}\sum_{j=1, j\neq i}^{n} w_{ij}V_iV_j - \sum_{i=1}^{n} i_iV_i + \sum_{i=1}^{n} G_i \int f^{-1}(z)dz. \quad (16)$$

The first two terms in equation (16) are used for mapping to the objective function which is to be minimized. The third term exists in the continuous type Hopfield network satisfying energy function as Lyapunov function [4]. Therefore, it is recommendable to reduce this term as small as possible for forcing this energy function matched to the objective function [7-10]. As seen equation (16), we can notice that the third term of E(v) can be ignored by making the total conductance zero. As discussed in the above section, the input space is unstable when the total conductance is zero. Therefore, it is recommendable to make the total conductance small as possible but positive.

As mentioned earlier, the total conductance is the sum of the input conductance (g_i) and synapses (w_{ij}). Because the synapse' values are determined according to application systems and can not be changed, we have to adjust the input conductance (g_i) in order for the total conductance (G_{ni}) being positive.

Case study

Except g_1 and g_2, all other conditions are same as in the Fig. 5. Two different values of the input conductance are tested; one value is a little bigger ($g_1 = g_2 = 2.8[\mho]$) than in Fig. 5 and the other is a little smaller ($g_1 = g_2 = 2.1[\mho]$). Then the total conductance (G_{ni}) becomes 0.8 $[\mho]$ and 0.1 $[\mho]$ respectively because of $w_{12} = w_{21} = -2[\mho]$. Fig. 8 and 9 show the energy map and transient result for these two cases. As seen in

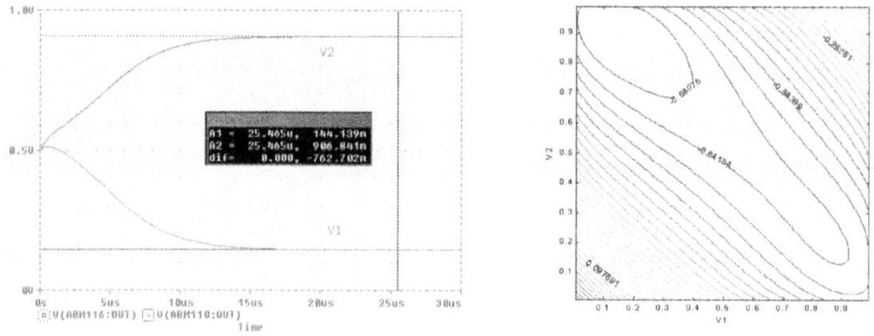

Fig. 8. 2bit A/D converter in case $x = 1.6[V]$ and $g_1 = g_2 = 2.8[\mho]$: (a) Transient simulation of output, (b) The energy map

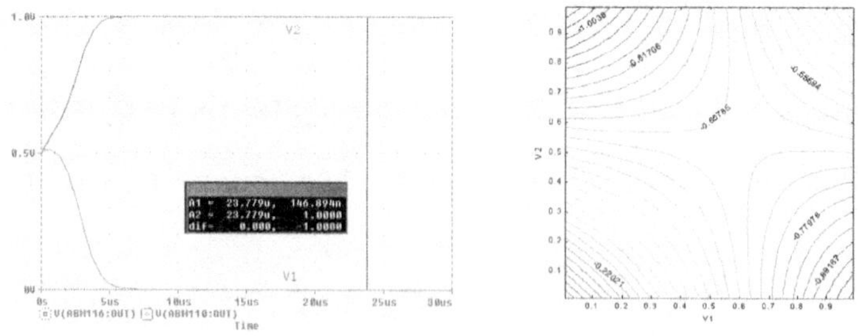

Fig. 9. 2bit A/D converter in case $x = 1.6[V]$, $g_1 = g_2 = 2.1[\mho]$, and α=2: (a) Transient simulation of output, (b)The energy map

Fig. 8, the output converges to $V_1 \approx 0.14$ and $V_2 \approx 9.06$ which are inferior result than in Fig. 5. However, the output in Fig. 9 converges to $V_1 \approx 0.0001$ and $V_2 \approx 1.0$ which can be regarded almost as digital value. As expected, the better result is achieved when the smaller total conductance is used as in Fig. 9. Also we can notice that the minima of energy function in Fig. 9 is located more closely at the correct answer compared as in Fig. 8.

6 Conclusions

Two methods can be used to represent an inhibitory synapse. One is to use negative resistor and the other is the inverted output of neuron. These two methods are same in the viewpoint of the current flowing through synapse. However, the total conductance seen at the neuron input is different. This total conductance has been proved to effect on the system stability. If the total conductance $G_{ni} > 0$, the input space converges exponentially. In case of $G_{ni} = 0$, the input space is linearly unstable. In case of $G_{ni} < 0$, the input space is exponentially unstable. Therefore, the input conduce g_i has to be adjusted in order that the total conductance is slightly greater than zero for the input space stability and the system's performance.

Acknowledgement

The part of researchers participating in this study are supported by the grant from "the 2nd phase BK21 project".

References

1. Hopfield, J.J., Tank, D.W.: Neural Computation of Decisions in Optimization Problems. Biolog. Cybern. **52** (1985) 141-152
2. Hopfield, J.J., Tank, D.W.: Computing with Neural Circuits: A Model. Science **233** (1986) 625-633
3. Zurada, J.M.: Introduction to Artificial Neural Systems. West, St. Paul MN (1992)
4. Peng, M., Gupta, N.K., Armitage, A.F.: An Investigation into the Improvement of Local Minima of the Hopfield Network. Neural Networks **9** (1996) 1241-1253
5. Golub, G.H., Van Loan, C.F.: Matrix Computations. John Hopkins Univ. Press (1996)
6. Chen, T., Amari, S.: Stability Asymmetric Hopfield Networks. IEEE Trans. Neural Networks **12** (2001) 159-163
7. Xue, B., Wang, J.: A Recurrent Neural Networks for Nonlinear Optimization with a Continuously Differentiable Objective Function with Bound Constraints. IEEE Trans. Neural Networks **11** (2000) 1251-1262
8. Xia, Y.S., Wang, J.: On the Stability of Globally Projected Dynamical Dystems. J. Optimizat. Theory Applicat. **106** (2000) 129-160
9. Arik, S.: An Analysis of Global Asymtotic Stability for Cellur Delayed Neural Networks. IEEE Trans. Neural Networks **13** (2002) 1239-1342
10. Gao, X.B.: A Novel Neural Network for Nonlinear Convex Programming. IEEE Trans. Neural Networks **11** (2004) 613-621

Global Asymptotical Stability for Neural Networks with Multiple Time-Varying Delays

Jianlong Qiu[1,2], Jinde Cao[1], and Zunshui Cheng[1]

[1] Department of Mathematics, Southeast University, Nanjing 210096, China
[2] Department of Mathematics, Linyi Normal University, Linyi 276005, China
{qjl9916, jdcao}@seu.edu.cn, qjl9916@gmail.com

Abstract. In this paper, the global uniform asymptotical stability is studied for neural networks with multiple time-varying delays by constructing appropriate Lyapunov-Krasovskii functional and using the linear matrix inequality (LMI) approach. The restriction on the derivative of the time-varying delay function $\tau_{ij}(t)$ to be less than unit is removed by using slack matrix method. A numerical example is provided to demonstrate the effectiveness and applicability of the proposed criteria.

1 Introduction

It is well known that the dynamics of delayed recurrent neural networks such as cellular neural networks (CNNs) and Hopfield neural networks (HNNs) have been deeply investigated, in recent years, many researchers have studied the global asymptotical stability and global exponential stability analysis for delayed neural networks. A great number of results on this topic have been reported in the literature; see, [1]-[11] and the references therein. In [2], the authors considered the existence and uniqueness of the equilibrium point and its global asymptotic stability of a general class of recurrent neural networks with time-varying delays as $\tau_i(t), i = 1, 2, \cdots, n$ and Lipschitz continuous activation functions. In [5], [7], [10] and [11], the authors investigated the stability of a class of neural networks with time-varying or multiple time-varying delays. Motivated by the above discussions, the aim of the present paper is to consider a class of neural networks with multiple time-varying delays as $\tau_{ij}(t), i, j = 1, 2, \cdots, n$. By using matrix method, we give a novel criteria which don't require the derivative of the time-varying delay function $\tau_{ij}(t)$ to be less than unit.

In this paper, we consider the following cellular neural network model with multiple time-varying delays

$$\dot{u}_i(t) = -a_i u_i(t) + \sum_{j=1}^{n} w_{ij}^0 g_j(u_j(t)) + \sum_{j=1}^{n} w_{ij} g_j(u_j(t - \tau_{ij}(t))) + I_i, \qquad (1)$$

where $u_i(t, x)$ denotes the state of the ith neuron at time t; g_j are activation functions of the jth neuron; the scalar $a_i > 0$ is the rate with which the ith unit will reset its potential to the resting state in isolation when disconnected

D. Liu et al. (Eds.): ISNN 2007, Part I, LNCS 4491, pp. 1025–1033, 2007.

from the network and external inputs at time t; $w_{ij}^0, w_{ij}, i, j = 1, 2, \cdots, n$ are known constants denoting the strength of the ith neurons on the jth neurons; I_i denote the ith component of an external input source; $\tau_{ij}(t) \geq 0$ denotes the transmission delay,, $j = 1, 2, \cdots, n$.

To proceed conveniently, let $W_0 = (w_{ij}^0)_{n\times n}$. Let $W_k = (w_{ij}^k)_{n\times n}$, whose kth row is composed by the kth row of matrix $W = (w_{ij})_{n\times n}$ and the other rows are zeros, $k = 1, 2, \cdots, n$. let $u(t) = (u_1(t), u_2(t), \cdots, u_n(t))^T$, $\tau_i(t) = (\tau_{i1}(t), \tau_{i2}(t), \cdots, \tau_{in}(t))^T$, $A = diag(a_1, a_2, \cdots, a_n)$, $I = (I_1, I_2, \cdots, I_n)$, $g(u(t)) = (g_1(u_1(t)), g_2(u_2(t)), \cdots, g_n(u_n(t)))^T$, $g(u(t-\tau_i(t))) = (g_1(u_1(t-\tau_{i1}(t))), g_2(u_2(t-\tau_{i2}(t))), \cdots, g_n(u_n(t-\tau_{in}(t))))^T$, $i, j = 1, 2, \cdots, n$. So, the system (1) can be rewritten in a matrix format as follows

$$\dot{u}(t) = -Au(t) + W_0 g(u(t)) + \sum_{i=1}^{n} W_i g(u(t-\tau_i(t))) + I. \tag{2}$$

Notation. Throughout this paper, vector $X \in \mathbb{R}^n$(n-dimensional Euclidean space), its norm is defined as $\|X\| = \sqrt{X^T X}$. The notation $A \geq B$ (respectively, $A \geq B$) means that the matrix $A - B$ is symmetric positive definite (respectively, positive semi-definite), where A and B are matrices of the same dimensions. An asterisk $*$ represents the elements below the main diagonal of a symmetric matrix.

2 Preliminaries

Throughout the paper we need the following assumption and lemmas.

(H_1) The neurons activation functions g_j are continuous, and there exist constants $k_j > 0$ such that

$$0 \leq \frac{g_j(\xi_1) - g_j(\xi_2)}{\xi_1 - \xi_2} \leq k_j$$

for any $\xi_1, \xi_2 \in \mathbb{R}, \xi_1 \neq \xi_2, j = 1, 2, \cdots, n$.

Note that the assumption H_1 guarantee there has at least one equilibrium point for neural network (1).

(H_2) The multiple time-varying delay $\tau_{ij}(t)$ are continuous function, and there exist constants $\tau_i(t) = \max_{1\leq j\leq n} \tau_{ij}(t)$, $h_i = \sup_{t\in[0,+\infty)} \tau_i(t)$ and $h = \max_{1\leq i\leq n} h_i$ such that

$$0 \leq \tau_{ij}(t) \leq h.$$

Definition. A vector $U^* = (u_1^*, u_2^*, \cdots, u_n^*)^T \in \mathbb{R}^n$ is said to be an equilibrium point of system (1) if it satisfies

$$-a_i u_i^* + \sum_{j=1}^{n} w_{ij}^0 g_j(u_j^*) + \sum_{j=1}^{n} w_{ij} g_j(u_j^*) + I_i = 0, i = 1, 2, \cdots, n.$$

For convenience, we will make the following transformation to system (1)

$$x(t) = u(t) - u^*$$

where $x(t) = (u_1(t) - u_1^*, u_2(t) - u_2^*, \cdots, u_n(t) - u_n^*)^T$. Under this transformation, neural network (1) can be rewritten as follows

$$\dot{x}(t) = -Ax(t) + W_0 f(x(t)) + \sum_{i=1}^{n} W_i f(x(t - \tau_i(t))), \tag{3}$$

where $f(x(t)) = (f_1(x_1(t)), f_2(x_2(t)), \cdots, f_n(x_1 n(t)))^T$, $f_j(x_j(t)) = g_j(x_j(t) + u_j^*) - g_j(u_j^*)$, $f(x(t - \tau_i(t))) = (f_1(x_1(t - \tau_{i1}(t))), f_2(x_2(t - \tau_{i2}(t))), \cdots, f_n(x_n(t - \tau_{in}(t))))^T$, $f_j(x_j(t - \tau_{ij}(t))) = g_j(x_j(t - \tau_{ij}(t)) + u_j^*) - g_j(u_j^*)$, $j = 1, 2, \cdots, n$. Obviously $f_j(0) = 0$, and it is easy to verify that .

$$0 \le \frac{f_j(x_j)}{x_j} \le k_j, \forall x_j \ne 0, j = 1, 2, \cdots, n \tag{4}$$

Based on (4), we easily know that for any scalars $s_{0j} \ge 0, s_{ij} \ge 0, i = 1, 2, \cdots, n,$

$$2\sum_{j=1}^{n} s_{0j} f_j(x_j(t))[k_j x_j(t) - f_j(x_j(t))] \ge 0, \tag{5}$$

$$2\sum_{j=1}^{n} s_{ij} f_j(x_j(t - \tau_{ij}(t)))[k_j x_j(t - \tau_{ij}(t)) - f_j(x_j(t - \tau_{ij}(t)))] \ge 0, \tag{6}$$

which can be rewritten in a matrix format as

$$2f^T(x(t))S_0 K x(t) - 2f^T(x(t))S_0 f(x(t)) \ge 0, \tag{7}$$

$$2f^T(x(t - \tau_i(t)))S_i K x(t - \tau_i(t)) - 2f^T(x(t - \tau_i(t)))S_i f(x(t - \tau_i(t))) \ge 0, \tag{8}$$

where $S_0 = diag(s_{01}, s_{02}, \cdots, s_{0n}) \ge 0; S_i = diag(s_{i1}, s_{i2}, \cdots, s_{in}) \ge 0, i = 1, 2, \cdots, n.$

By system (3), we have the fact that for any matrices $M_l, N_l, l = 0, 1, 2, \cdots, n$ and Q with appropriate dimension

$$\begin{aligned} 2\big[x^T(t)M_0 + &\sum_{i=1}^{n} x(t - \tau_i(t))M_i + f^T(x(t))N_0 \\ &+ \sum_{i=1}^{n} f^T(x(t - \tau_i(t)))N_i + \dot{x}^T(t)Q\big] \\ &\times \big[\dot{x}(t) + Ax(t) - W_0 f(x(t)) - \sum_{i=1}^{n} W_i f(x(t - \tau_i(t)))\big] = 0, \end{aligned} \tag{9}$$

Lemma 1. [12] The equilibrium of the system is globally uniformly asymptotically stable if there exists a C^1 function $V : \mathbb{R}^+ \times \mathbb{R}^n \to \mathbb{R}$ such that (i) V is a positive definite, decrescent and radially unbounded, and (ii) $-\dot{V}$ is positive definite.

3 Main Results

Theorem. Under the assumption $(H_1), (H_2)$, the equilibrium point U^* of system (1) is globally uniformly asymptotically stable if there exist matrices $P > 0$, diagonal matrices $D > 0, S_l \geq 0 (l = 0, 1, \cdots, n)$ and appropriate dimensional matrices $M_l, N_l (l = 0, 1, \cdots, n)$,$X_{ks}^{(i)} (i = 1, \cdots, n; k, s = 1, 2, 3)$ such that the following hold

$$\Xi = \begin{pmatrix} \Pi_0 & \Pi_1 & \Pi_2 & \cdots & \Pi_n & \Pi_{(n+1)} & \Sigma_1 & \Sigma_2 & \cdots & \Sigma_n & \Sigma_{(n+1)} \\ * & \Phi_{11} & 0 & \cdots & 0 & \Gamma_1 & \Omega_{11} & \Omega_{12} & \cdots & \Omega_{1n} & H_1 \\ * & * & \Phi_{22} & \cdots & 0 & \Gamma_2 & \Omega_{21} & \Omega_{22} & \cdots & \Omega_{2n} & H_2 \\ \vdots & \vdots & \vdots & \ddots & \vdots & \vdots & \vdots & \vdots & \ddots & \vdots & \vdots \\ * & * & * & \cdots & \Phi_{nn} & \Gamma_n & \Omega_{n1} & \Omega_{n2} & \cdots & \Omega_{nn} & H_n \\ * & * & * & \cdots & * & \Gamma_{(n+1)} & \Theta_1 & \Theta_2 & \cdots & \Theta_n & \Theta_{(n+1)} \\ * & * & * & \cdots & * & * & \Psi_{11} & \Psi_{12} & \cdots & \Psi_{1n} & \Lambda_1 \\ * & * & * & \cdots & * & * & * & \Psi_{22} & \cdots & \Psi_{2n} & \Lambda_2 \\ \vdots & \vdots & \vdots & \ddots & \vdots & \vdots & \vdots & \vdots & \ddots & \vdots & \vdots \\ * & * & * & \cdots & * & * & * & * & \cdots & \Psi_{nn} & \Lambda_n \\ * & * & * & \cdots & * & * & * & * & \cdots & * & \Upsilon \end{pmatrix} < 0, \quad (10)$$

$$X^{(i)} = \begin{pmatrix} X_{11}^{(i)} & X_{12}^{(i)} & X_{13}^{(i)} \\ * & X_{22}^{(i)} & X_{23}^{(i)} \\ * & * & X_{33}^{(i)} \end{pmatrix} > 0, \quad (11)$$

where

$$\begin{aligned} \Pi_0 &= M_0 A + A M_0^T + \sum_{i=1}^{n} (h_i X_{11}^{(i)} + 2X_{13}^{(i)}); \\ \Pi_i &= M_i A + h_i X_{12}^{(i)} + X_{23}^{(i)} - X_{13}^{(i)T}, i = 1, 2, \cdots, n; \\ \Pi_{(n+1)} &= A N_0^T + K S_0 - M_0 W_0; \\ \Sigma_i &= A N_i - M_0 W_i, i = 1, 2, \cdots, n; \Sigma_{(n+1)} = P + M_0 + QA; \\ \Phi_{ii} &= h_i X_{22}^{(i)} - X_{23}^{(i)} - X_{23}^{(i)T}, i = 1, 2, \cdots, n; \\ \Gamma_i &= -M_i W_0, i = 1, 2, \cdots, n; \quad \Gamma_{(n+1)} = -N_0 W_0 - W_0^T N_0^T - 2S_0; \end{aligned}$$

$$H_i = M_i, i = 1, 2, \cdots, n;$$
$$\Theta_i = -W_0^T N_i^T - N_0 W_i, i = 1, 2, \cdots, n; \quad \Theta_{(n+1)} = D + N_0 - QW_0;$$
$$\Lambda_i = N_i - QW_i, i = 1, 2, \cdots, n; \quad \Upsilon = \sum_{i=1}^{n} h_i X_{33}^{(i)} + Q + Q^T;$$
$$\Omega_{ii} = KS_i - M_i W_i; \quad \Omega_{ij} = -M_i W_j, i \neq j, i, j = 1, 2, \cdots, n;$$
$$\Psi_{ii} = -2S_i - N_i W_i - W_i^T N_i^T; \quad \Psi_{ij} = -N_i W_j, i \neq j, i, j = 1, 2, \cdots, n.$$

Proof. Consider Lyapunov-krasovskii functional as

$$V(x(t)) = x^T(t)Px(t) + 2\sum_{i=1}^{n} d_i \int_0^{x_i(t)} f_i(s)ds + \sum_{i=1}^{n} \int_{t-h_i}^{t} \int_{\sigma}^{t} \dot{x}^T(s) X_{33}^{(i)} \dot{x}(s) ds d\sigma$$
$$+ \sum_{i=1}^{n} \int_0^t \int_{\sigma-\tau_i(\sigma)}^{\sigma} \Psi_i^T(\sigma, s) X^{(i)} \Psi_i(\sigma, s) ds d\sigma, \qquad (12)$$

where $\Psi_i(\sigma, s) = [x^T(\sigma), x^T(\sigma - \tau_i(\sigma)), \dot{x}^T(s)]$.

By (7),(8) and (9), the time-derivative of $V(x(t))$ along the solution of system (3) is given

$$\begin{aligned}
\dot{V}(x(t)) = {} & 2x^T(t)P\dot{x}(t) + 2f^T(x(t))D\dot{x}(t) \\
& + \sum_{i=1}^{n} h_i \dot{x}^T(t) X_{33}^{(i)} \dot{x}(t) - \sum_{i=1}^{n} \int_{t-h_i}^{t} \dot{x}^T(s) X_{33}^{(i)} \dot{x}(s) ds \\
& + \sum_{i=1}^{n} \Bigg[\tau_i(t) \begin{pmatrix} x(t) \\ x(t-\tau_i(t)) \end{pmatrix}^T \begin{pmatrix} X_{11}^{(i)} & X_{12}^{(i)} \\ * & X_{22}^{(i)} \end{pmatrix} \begin{pmatrix} x(t) \\ x(t-\tau_i(t)) \end{pmatrix} \\
& + 2x^T(t) X_{13}^{(i)} x(t) - 2x^T(t) X_{13}^{(i)} x(t-\tau_i(t)) + 2x^T(t-\tau_i(t)) X_{23}^{(i)} x(t) \\
& - 2x^T(t-\tau_i(t)) X_{23}^{(i)} x(t-\tau_i(t)) + \int_{t-\tau_i(t)}^{t} \dot{x}^T(s) X_{33}^{(i)} \dot{x}(s) ds \Bigg] \\
\leq {} & 2x^T(t)P\dot{x}(t) + 2f^T(x(t))D\dot{x}(t) \\
& + \sum_{i=1}^{n} h_i \dot{x}^T(t) X_{33}^{(i)} \dot{x}(t) - \sum_{i=1}^{n} \int_{t-h_i}^{t} \dot{x}^T(s) X_{33}^{(i)} \dot{x}(s) ds \\
& + \sum_{i=1}^{n} \Bigg[\tau_i(t) \begin{pmatrix} x(t) \\ x(t-\tau_i(t)) \end{pmatrix}^T \begin{pmatrix} X_{11}^{(i)} & X_{12}^{(i)} \\ * & X_{22}^{(i)} \end{pmatrix} \begin{pmatrix} x(t) \\ x(t-\tau_i(t)) \end{pmatrix} \\
& + 2x^T(t) X_{13}^{(i)} x(t) - 2x^T(t) X_{13}^{(i)} x(t-\tau_i(t)) + 2x^T(t-\tau_i(t)) X_{23}^{(i)} x(t) \\
& - 2x^T(t-\tau_i(t)) X_{23}^{(i)} x(t-\tau_i(t)) \Bigg] + \sum_{i=1}^{n} \int_{t-\tau_i(t)}^{t} \dot{x}^T(s) X_{33}^{(i)} \dot{x}(s) ds \\
\leq {} & 2x^T(t)P\dot{x}(t) + 2f^T(x(t))D\dot{x}(t) \\
& + \sum_{i=1}^{n} h_i \dot{x}^T(t) X_{33}^{(i)} \dot{x}(t) + \sum_{i=1}^{n} h_i x^T(t) X_{11}^{(i)} x(t) + 2\sum_{i=1}^{n} h_i x^T(t-\tau_i(t)) X_{12}^{(i)} x(t)
\end{aligned}$$

$$
\begin{aligned}
&+\sum_{i=1}^{n} h_i x^T(t-\tau_i(t))X_{22}^{(i)}x(t-\tau_i(t)) + 2\sum_{i=1}^{n} x^T(t)X_{13}^{(i)}x(t) \\
&-2\sum_{i=1}^{n} x^T(t)X_{13}^{(i)}x(t-\tau_i(t)) + 2\sum_{i=1}^{n} x^T(t-\tau_i(t))X_{23}^{(i)}x(t) \\
&-2\sum_{i=1}^{n} x^T(t-\tau_i(t))X_{23}^{(i)}x(t-\tau_i(t)) + 2x^T(t)M_0\dot{x}(t) + 2\sum_{i=1}^{n} x^T(t-\tau_i(t))M_i\dot{x}(t) \\
&+2f^T(x(t))N_0\dot{x}(t) + 2\sum_{i=1}^{n} f^T(x(t-\tau_i(t)))N_i\dot{x}(t) + 2\dot{x}^T(t)Q\dot{x}(t) \\
&+2\sum_{i=1}^{n} x^T(t-\tau_i(t))M_iAx(t) + 2x^T(t)M_0Ax(t) + 2f^T(x(t))N_0Ax(t) \\
&+2\sum_{i=1}^{n} f^T(x(t-\tau_i(t)))N_iAx(t) + 2\dot{x}^T(t)QAx(t) - 2x^T(t)M_0W_0f(x(t)) \\
&-2\sum_{i=1}^{n} x^T(t-\tau_i(t))M_iW_0f(x(t)) - 2f^T(x(t))N_0W_0f(x(t)) \\
&-2\sum_{i=1}^{n} f^T(x(t-\tau_i(t)))N_iW_0f(x(t)) - 2\dot{x}^T(t)QW_0f(x(t)) \\
&-2\sum_{i=1}^{n} x^T(t)M_0W_if(x(t-\tau_i(t))) - 2\sum_{i=1}^{n}\sum_{j=1}^{n} x^T(t-\tau_i(t))M_iW_jf(x(t-\tau_i(t))) \\
&-2\sum_{i=1}^{n} f^T(x(t))N_0W_if(x(t-\tau_i(t))) - 2\sum_{i=1}^{n}\sum_{j=1}^{n} f^T(x(t-\tau_i(t)))N_iW_jf(x(t-\tau_i(t))) \\
&-2\sum_{i=1}^{n} \dot{x}^T(t)QW_if(x(t-\tau_i(t))) + 2f^T(x(t))S_0Kx(t) - 2f^T(x(t))S_0f(x(t)) \\
&+2\sum_{i=1}^{n} f^T(x(t-\tau_i(t)))S_iKx(t-\tau_i(t)) - 2\sum_{i=1}^{n} f^T(x(t-\tau_i(t)))S_if(x(t-\tau_i(t))), \\
&= x^T(t)[2M_0A + \sum_{i=1}^{n} h_iX_{11}^{(i)} + 2\sum_{i=1}^{n} X_{13}^{(i)}]x(t) \\
&+\sum_{i=1}^{n} x^T(t)[-2X_{13}^{(i)} + 2h_i(X_{12}^{(i)})^T + 2(X_{23}^{(i)})^T + 2AM_i^T]x(t-\tau_i(t)) \\
&+x^T(t)[-2M_0W_0 + 2AN_0^T + 2KS_0]f(x(t)) - 2\sum_{i=1}^{n} x^T(t)M_0W_if(x(t-\tau_i(t))) \\
&+x^T(t)[2M_0 + 2P + 2AQ^T]\dot{x}x(t) + \sum_{i=1}^{n} x^T(t-\tau_i(t))[h_iX_{22}^{(i)} - 2X_{23}^{(i)}]x(t-\tau_i(t)) \\
&-2\sum_{i=1}^{n} x^T(t-\tau_i(t))M_iW_0f(x(t)) - 2\sum_{i=1}^{n}\sum_{j=1}^{n} x^T(t-\tau_i(t))M_iW_jf(x(t-\tau_j(t)))
\end{aligned}
$$

$$+2\sum_{i=1}^{n} x^T(t-\tau_i(t))M_i\dot{x}(t) + 2\sum_{i=1}^{n} f^T(x(t-\tau_i(t)))S_iKx(t-\tau_i(t))$$
$$-2f^T(x(t))[N_0W_0 + S_0]f(x(t)) - 2\sum_{i=1}^{n} f^T(x(t))N_0W_if(x(t-\tau_i(t)))$$
$$+2f^T(x(t))[N_0 + D]\dot{x}(t) + 2\sum_{i=1}^{n} f^T(x(t-\tau_i(t)))N_i\dot{x}(t) + \dot{x}^T(t)[\sum_{i=1}^{n} h_iX_{33}^{(i)} + 2Q]\dot{x}(t)$$
$$+2\sum_{i=1}^{n} f^T(x(t-\tau_i(t)))N_iAx(t) - 2\sum_{i=1}^{n} f^T(x(t-\tau_i(t)))N_iW_0f(x(t))$$
$$-2\sum_{i=1}^{n}\sum_{j=1}^{n} f^T(x(t-\tau_i(t)))N_iW_jf(x(t-\tau_j(t))) - 2\sum_{i=1}^{n} f^T(x(t-\tau_i(t)))S_if(x(t-\tau_i(t)))$$
$$-2\dot{x}^T(t)QW_0f(x(t)) - 2\sum_{i=1}^{n} \dot{x}^T(t)QW_if(x(t-\tau_i(t))) = \eta^T(t)\Xi\eta(t), \quad (13)$$

where

$$\eta^T(t) = (x^T(t), x^T(t-\tau_1(t)), \cdots, x^T(t-\tau_n(t)), f^T(x(t)),$$
$$f^T(x(t-\tau_1(t))), \cdots, f^T(x(t-\tau_n(t))), \dot{x}^T(t)).$$

By the condition (10), we have $\dot{V}(t) < 0$ for any $\eta(t) \neq 0$. From Lemma 1, we follow that the equilibrium of system (1) is globally uniformly asymptotically stable. □

4 Simulation Example

Example. Consider a two-state cellular neural networks defined by (1) with parameters

$$A = \begin{pmatrix} 4.2 & 0 \\ 0 & 3.6 \end{pmatrix}, W_0 = \begin{pmatrix} -3 & 0.8 \\ 0.01 & -1.8 \end{pmatrix}, W = \begin{pmatrix} -1.8 & 0.1 \\ -0.2 & -0.5 \end{pmatrix},$$

and the time-varying delayed as

$$\tau_{11}(t) = 1 + 0.2\sin(10t), \tau_{12} = 0, \tau_{21} = 0, \tau_{22}(t) = 1 + 0.2\sin(15t),$$

It is obvious that the derivative of the time-varying delay of $\tau_{11}(t), \tau_{22}(t)$ do not always less than unit. By Theorem , we can conclude that this delayed neural network is globally asymptotically stable. Using the Matlab LMI Control Toolbox, we can also see there exist feasible solution of condition (10). The numerical simulations is shown in Fig. We can see that the solution with external input $(0.01, 0.02)$ converge to the unique equilibrium. This validate our results.

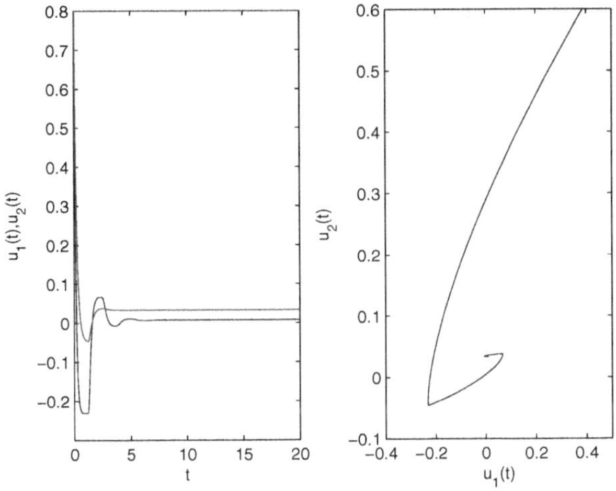

Fig. 1. Time responses of state $u_1(t)$, $u_2(t)$ and phase portrait in Example

5 Conclusion

In this paper, we have investigated the global uniform asymptotical stability for neural networks with multiple time-varying delays. Using Lyapunov functional method and the LMI approach, we gave a sufficient criterion ensuring the global asymptotical stability of the equilibrium point. And by using slack matrix method, the obtained result improves and extend several earlier publications and is useful in applications of manufacturing high quality neural networks.

Acknowledgement

This work was supported by the National Natural Science Foundation of China under Grant 60574043, and the Natural Science Foundation of Jiangsu Province of China under Grant BK2006093.

References

1. Cao, J., Zhou, D.: Stability Analysis of Delayed Cellular Neural Networks. Neural Networks **11** (1998) 1601-1605
2. Cao, J., Wang, J.: Global Asymptotic Stability of a General Class of Recurrent Neural Networks with Time-varying Delays. IEEE Trans. Circuits Syst. **I 50(1)** (2003) 34-44
3. Cao, J., Wang, J.: Global Exponential Stability and Periodicity of Recurrent Neural Networks with Time Delays. IEEE Trans. Circuits Syst. **I 52(5)** (2005) 925-931
4. Cao, J., Ho, D.W.C.: A General Framework for Global Asymptotic Stability Analysis of Delayed Neural Networks Based on LMI Approach. Chaos, Solitons and Fractals **24(5)** (2005) 1317-1329

5. He, Y., Wang, Q., Wu, M.: LMI-based Stability Critera for Neural Networks with Multiple Time-varing Delays. Phys. D. **212** (2005) 126-136
6. Liang, J., Cao, J.: A Based-on LMI Stability Criterion for Delayed Recurrent Neural Networks. Chaos, Solitons and Fractals **28** (2006) 154-160
7. Liao, X. and, Li, C.: An LMI Approach to Asymptotic Stability of Multiple-delayed Neural Networks. Phys. D. **200** (2005) 139-155
8. Xu, S., Lam, J., Ho, D.W.C., Zou, Y.: Delay-dependent Exponential Stability for a Class of Neural Networks with Time Delays. IEEE Trans. Circuits Syst. **II 52(6)** (2005) 349-353
9. Yang, H., Chu, T., Zhang, C.: Expotential Stability of Neural Networks with Variable Delays via LMI Approach. Chaos, Solitons and Fractals **24(5)** (2005) 1317-1329
10. Zhang, H., Wang, Z., Liu, D.: Expotential Stability Analysis of Delayed Neural Networks with Multiple Time Delays. Lecture Notes in Computer Science **3519** (2005) 142-148
11. Zhang, Q., Wei, X., Xu, J.: On Globally Expotential Stability of Delayed Neural Networks with Time-varying Delays. Appl. Math. Comput. **162** (2005) 679-686
12. Vidyasagar M.(ed.): Nonlinear System Analysis. Englewood Cliffs, NJ: Prentice Hall (1993)

Positive Solutions of General Delayed Competitive or Cooperative Lotka-Volterra Systems

Wenlian Lu[1,2] and Tianping Chen[1]

[1] Key Laboratory of Nonlinear Science of Chinese Ministry of Education, Institute of Mathematics, Fudan University, Shanghai, 200433, P.R. China
021018041@fudan.edu.cn wenlian@mis.mpg.de Wenlian.Lu@mis.mpg.de
[2] Max Planck Institute for Mathematics in the Sciences, Leipzig, German
tchen@fudan.edu.cn

Abstract. In this paper, we investigate dynamical behavior of a general class of competitive or cooperative Lotka-Volterra systems with delays. Positive solutions and global stability of nonnegative equilibrium are discussed. Sufficient condition independent of delays guaranteeing existence of globally stable equilibrium is given. A Simulation verifying theoretical results is given, too.

1 Introduction

Modelling and analysis of the dynamics of biological populations by means of differential equations is one of the primary concerns in population growth problem. A well-known and extensively studied class of models in population dynamics is the competitive or cooperative Lotka-Volterra system (see [1]), which describes the interaction among various species. This model can be realized by analog integrated circuits (see [2]) and the analysis of the model presents a technique for a class of substitution models, such as Gompertz, Bass, Non-Symmetrical Responding Logistic (NSRL) and Sharif-Kabir substitution models (see [3]). Recently, a lot of papers have studied the dynamical behavior of this Lotka-Volterra system (see [4,5,7,8,9,10,11,12,13]). In [4], the author only considered the case of the system without delays. In [5,6,7,8,11], the authors focused on the dynamical behavior of low-dimensional systems, such as two or three dimensional cases. Some of them considered the case of positive coefficients which implies only competition occurs. In most papers, authors considered the global stability of the positive equilibrium, which means all species exist at certain quantities after a long period of competition or cooperation. However, in real world, there do exist cases that both competition and cooperation occur, the number of some species in a biological system is very large. Instead, some others vanish in the end. Therefore, we should study dynamical behavior of general class of Lotka-Volterra systems of high dimensional systems with delays, even infinite delays, which includes both competition and cooperation, and has non-positive equilibrium. In this paper, we address a general class of Lotka-Volterra systems.

D. Liu et al. (Eds.): ISNN 2007, Part I, LNCS 4491, pp. 1034–1044, 2007.

Without discussing the permanence or persistence of the system, we present a sufficient condition independent of delays for global convergence directly.

Lotka-Volterra system with delays can be described by the following system of differential equations:

$$\frac{dx_i(t)}{dt} = x_i(t)\left[b_i - \sum_{j=1}^{n} a_{ij}x_j(t) - \sum_{j=1} b_{ij}x_j(t-\tau_{ij})\right], \tag{1}$$

where $x_i(t)$ represents number of species i at time t, b_i represents its birth-rate or death-rate, depending on $b_i > 0$ or $b_i < 0$, $i = 1, \cdots, n$.

If all $a_{ij} > 0$, $b_{ij} > 0$, $i, j = 1, \cdots, n$, the system (1) is the Competitive Lotka-Volterra system with time delays. Instead, if $a_{ii} > 0$, $a_{ij} < 0$, $b_{ij} < 0$, for $i, j = 1, \cdots, n$, $i \neq j$, then the system (1) is the Cooperative Lotka-Volterra system with time delays.

Because $x_i(t)$ represents the number of the i-th species, which is positive. Therefore, we should discuss positive solution of the system (1). Moreover, if $x_i(t) \to 0$, then the i-th species will be distinct in the competition process.

In most papers, for the system (1), the authors assumed that the system of linear equations

$$b_i - \sum_{j=1}^{n} a_{ij}x_j - \sum_{j=1}^{n} b_{ij}x_j = 0 \tag{2}$$

has a positive solution $x_i^* > 0$, $i = 1, \cdots, n$, and discussed its asymptotic stability, which implies the permanence of all populations in the system. In this case, the system (1) is a special case of the delayed Cohen-Grossberg neural networks discussed in [14] and others. However, in real competition process, some species might be extinct. Therefore, we must discuss its positive solutions and nonnegative equilibrium.

In this paper, we discuss dynamical behavior of positive solution of the following general competitive cooperative Lotka-Volterra system:

$$\frac{dx_i(t)}{dt} = a_i(x_i(t))\left[b_i - \sum_{j=1}^{n} a_{ij}x_j(t) - \sum_{j=1}^{n}\int_0^\infty x_j(t-\tau_{ij}-s)dk_{ij}(s)\right], \tag{3}$$

where $dk_{ij}(s)$, for $i, j = 1, \cdots, n$, are Lebesgue-Stieljies measures and $a_i(\rho)$ are amplifier functions, in particular, $a_i(\rho) = \rho$.

The initial condition is $\phi(t) = [\phi_1(t), \cdots, \phi_n(t)]^T$, where

$$x_i(s) = \phi_i(s) \quad for \quad s \in (-\infty, 0], \quad (1.6)$$

and every $\phi_i \in C(-\infty, 0]$, $i = 1, \cdots, n$, is positive and bounded.

If $dk_{ij}(0) = b_{ij}$, $dk_{ij}(s) = 0$ for $s \neq 0$, and $a_i(\rho) = \rho$, for $i = 1, \cdots, n$, then system (3) reduces to (1). If $\tau_{ij} = 0$ and $dk_{ij}(s) = k_{ij}(s)ds$. Then, we obtain Lotka-Volterra system with distributed delays

$$\frac{dx_i(t)}{dt} = a_i(x_i(t))\left[b_i - \sum_{j=1}^{n} a_{ij}x_j(t) - \sum_{j=1}^{n}\int_0^\infty x_j(t-s)k_{ij}(s)ds\right]. \tag{4}$$

By taking different $dk_{ij}(s)$, we can derive various models. Here, we omit the details.

2 Preliminaries

In this section, we make some assumptions, which will be used throughout the paper.

$H1:$

$$\int_0^\infty s|dk_{ij}(s)| < +\infty, \quad i,j = 1,\cdots,n;$$

$H2:$ *$a_i(\rho)$ is continuous and for $\rho \geq 0$, $a_i(\rho) > 0$ for $\rho > 0$, $a_i(0) = 0$, and*

$$\int_0^\epsilon \frac{d\rho}{a_i(\rho)} = +\infty, \quad i = 1,\cdots,n$$

where ϵ is an arbitrary positive numbers.

We also need following definitions.

Definition 1. *We say a matrix $C = (c_{ij}) > 0$ if and only if $c_{ij} > 0$ for all i,j. In the similar way, we can define $C \geq 0$, $C < 0$, $C \leq 0$ respectively. A vector $x \in R^n$ is said $x \geq 0$, if and only if all $x_i \geq 0$ for $i = 1,\cdots,n$.*

Definition 2. $x(t) = x(t,\phi) = [x_1(t),\cdots,x_n(t)]^T$ *is said to be a positive solution of system(3) with initial condition $\phi(\theta) \in C([-\infty,0],R^n)$ if and only if $x(t) > 0$, where $x(t) > 0$ means every $x_i(t) > 0$, $-\infty < t < \infty$, $i = 1,\cdots,n$, and $x(\theta) = \phi(\theta)$, for $\theta \in (-\infty,0]$.*

Definition 3. *Let $\zeta_i > 0$, for $i = 1,\cdots,n$, we define norms for $x \in R^n$ by*

$$\|x\|_{\{\zeta,1\}} = \sum_{i=1} \zeta_i|x_i|,$$
$$\|x\|_{\{\zeta,\infty\}} = \max_{i=1,\cdots,n} \zeta_i^{-1}|x_i|.$$

Definition 4. *We say that an equilibrium x^* is globally asymptotically stable for system (3), if and only if for any* **positive** *solution $x(t)$ of system (3), we have*

$$\lim_{t\to\infty} x(t) = x^*.$$

Moreover, we say (3) is globally exponentially asymptotically stable, if and only if for each positive solution $x(t)$, there exist $M > 0$, $\epsilon > 0$ and $T > 0$ such that

$$\|x(t) - x^*\| \leq Me^{-\epsilon t} \quad for\ t > T.$$

3 Main Results

In this section, we give main results of this paper by proving several theorems concerning with existence of nonnegative equilibrium and its stability.

Theorem 1. *If assumptions H1, H2 are satisfied, and there exist constants* $\zeta_j > 0$, $j = 1, \cdots, n$, *such that*

$$\zeta_j a_{jj} - \sum_{i=1, i\neq j}^{n} \zeta_i |a_{ij}| - \sum_{i=1}^{n} \zeta_i \int_0^\infty |dk_{ij}(s)| > 0, \tag{5}$$

Or equivalently, there exist constants $\xi_i > 0$, $i = 1, \cdots, n$, *such that*

$$\xi_i a_{ii} - \sum_{j=1, j\neq i}^{n} \xi_j |a_{ij}| - \sum_{j=1}^{n} \xi_j \int_0^\infty |dk_{ij}(s)| > 0. \tag{6}$$

Then system (3) has a unique nonnegative equilibrium $x^* \in R^n$, *and for any positive solution* $x(t)$ *of system (3)*

$$\lim_{t\to\infty} x(t) = x^*.$$

Moreover, if $x^* > 0$. *Then* $x(t)$ *converges to* x^* *exponentially*

Before presenting the main theorems, we prove two lemmas.

Lemma 1. *If Hypothesis H1 and H2 are satisfied and initial condition is positive. Then the solution of the system (3) is positive.*

Proof: Suppose that for some index i, we have $x_i(T) = 0$ and $x_i(t) > 0$, for $0 < t < T$. Then

$$\int_0^T \left[b_i - \sum_{j=1}^{n} a_{ij} x_j(s) - \sum_{j=1}^{n} \int_0^\infty x_j(s - \tau_{ij} - \theta) dk_{ij}(\theta) \right] ds$$
$$= \int_0^T \frac{\dot{x}_i(s) ds}{a_i(x_i(s))} = - \int_{x_i(T)}^{x_i(0)} \frac{d\rho}{a_i(\rho)} = -\infty,$$

which contradicts that $x_i(s)$ is continuous and bounded in $[0, T]$. Therefore, $x_i(t) > 0$, for $i = 1, \cdots, n$. Lemma is proved.

For the simplicity, in the sequel, when we say that $x(t)$ is a positive solution of the system (3), it is always assumed that the initial condition is positive.

Lemma 2. *Under assumptions of Theorem 1. Any positive trajectory* $x(t)$ *of system (3) is bounded, i.e., there exists* $N > 0$ *such that* $0 < x_i(t) \le N$, *for all* $t > 0$ *and* $i = 1, \cdots, n$.

Proof: From Lemma 1, $x_i(t) > 0$ for all $t > 0$ and $i = 1, \cdots, n$.

Denote

$$\alpha = \min_i \left[\xi_i a_{ii} - \sum_{j=1,j\neq i} \xi_j |a_{ij}| - \sum_{j=1}^{n} \xi_j \int_0^\infty |dk_{ij}(s)|\right],$$

$$M = \frac{\max_i |b_i|}{\alpha}, \qquad M(t) = \sup_{s\in(-\infty,0]} \|x(t+s)\|_{\{\xi,\infty\}}.$$

It is easy to see that $\|x(t)\|_{\{\xi,\infty\}} \leq M(t)$.

We claim that if $M(\bar{t}) > M$, $M(t)$ is non-increasing at point $\bar{t}$.

In fact, there are two cases.

Case 1. $\|x(\bar{t})\|_{\{\xi^{-1},\infty\}} < M(\bar{t})$. In this case, there exists a small interval $[\bar{t},\bar{t}+\delta]$ such that $\|x(t)\|_{\{\xi^{-1},\infty\}} < M(\bar{t})$, which implies $M(t) = M(\bar{t})$ for all $t \in [\bar{t},\bar{t}+\delta]$.

Case 2. $\|x(\bar{t})\|_{\{\xi,\infty\}} = M(\bar{t})$. In this case, let i_t be the index such that $\|x(t)\|_{\{\xi^{-1},\infty\}} = \xi_{i_t}^{-1}|x_{i_t}|$. Then we have

$$\left\{\frac{d|x_{i_{\bar{t}}}(s)|}{ds}\right\}_{s=\bar{t}} = sign(x_{i_{\bar{t}}})a_{i_{\bar{t}}}(x_{i_{\bar{t}}})\Big[b_{i_{\bar{t}}} - a_{i_{\bar{t}}i_{\bar{t}}}x_{i_{\bar{t}}}(\bar{t})$$

$$- \sum_{j=1,j\neq i_{\bar{t}}} a_{i_{\bar{t}}j}x_j(\bar{t}) - \sum_{j=1}^{n}\int_0^\infty x_j(\bar{t}-\tau_{ij}-s)dk_{i_{\bar{t}}j}(s)\Big]$$

$$\leq a_{i_{\bar{t}}}(x_{i_{\bar{t}}}(t))\Big\{\Big[-(\xi_{i_{\bar{t}}}a_{i_{\bar{t}}i_{\bar{t}}} - \sum_{j=1,j\neq i_{\bar{t}}} \xi_j|a_{i_{\bar{t}}j}| - \sum_{j=1}^{n}\xi_j\int_0^\infty |dk_{i_{\bar{t}}j}(s)|)\Big]M(\bar{t}) + |b_i|\Big\}$$

$$\leq a_{i_{\bar{t}}}(x_{i_{\bar{t}}}(\bar{t}))\Big[-\alpha M + |b_i|\Big] < 0,$$

which implies that there is a small δ, such that for every $t \in [\bar{t},\bar{t}+\delta_0]$, $\|x(t)\|_{\{\xi^{-1},\infty\}} \leq \|x(\bar{t})\|_{\{\xi^{-1},\infty\}} = M(\bar{t})$, which implies $M(t) = M(\bar{t})$ for all $t \in [\bar{t},\bar{t}+\delta]$.

In summary, $M(t) \leq \max\{M, M(0)\}$. As a direct result, $x(t)$ is bounded. Lemma 2 is proved.

Now, we are to prove Theorem 1.

Proof of Theorem 1: By Lemma 2, every positive trajectory $x(t)$ of the system (3) is bounded.

First, we assume that every initial function $\phi_i(t)$, $i = 1,\cdots,n$ is differentiable. Denote $y_i(t) = \frac{\dot{x}_i(t)}{a_i(x_i(t))}$. Then, we have

$$\frac{dy_i(t)}{dt} = -\sum_{j=1}^{n} a_{ij}\dot{x}_j(t) - \sum_{j=1}^{n}\int_0^\infty \dot{x}_j(t-\tau_{ij}-s)dk_{ij}(s). \tag{7}$$

Let

$$\beta = \min_i \left[\zeta_j a_{jj} - \sum_{i=1,i\neq j}^{n} \zeta_i |a_{ij}| - \sum_{i=1}^{n} \zeta_i \int_0^\infty |dk_{ij}(s)|\right].$$

Then, by (5), we have $\beta > 0$. Define function

$$L(t) = \sum_{i=1}^{n} \zeta_i |y_i(t)| + \sum_{i,j=1}^{n} \zeta_i \int_0^\infty |dk_{ij}(s)| \int_{t-\tau_{ij}-s}^{t} |\dot{x}_j(\theta)| d\theta.$$

Differentiating $L(t)$, we have

$$\frac{dL(t)}{dt} = \sum_{i=1}^{n} \zeta_i sign(y_i(t)) \Big[-a_{ii}\dot{x}_i(t) - \sum_{j=1, j\neq i}^{n} a_{ij}\dot{x}_j(t) - \sum_{j=1}^{n} \int_0^\infty \dot{x}_j(t-\tau_{ij}-s) dk_{ij}(s) \Big]$$
$$+ \sum_{i,j=1}^{n} \zeta_i \Big[|\dot{x}_j(t)| \int_0^\infty |dk_{ij}(s)| - \int_0^\infty |\dot{x}_j(t-\tau_{ij}-s)||dk_{ij}(s)| \Big]$$
$$\leq -\sum_{j=1}^{n} \Big[\zeta_j a_{jj} - \sum_{i=1, i\neq j}^{n} \zeta_i |a_{ij}| - \sum_{j=1}^{n} \zeta_i \int_0^\infty |dk_{ij}(s)| \Big] |\dot{x}_j(t)| \leq -\beta \sum_{j=1}^{n} |\dot{x}_j(t)|, \tag{8}$$

which implies

$$\int_0^{+\infty} \sum_{i=1}^{n} |\dot{x}_i(s)| ds < \frac{1}{\beta} L(0) < +\infty. \tag{9}$$

Therefore, for any $\epsilon > 0$, there is $T > 0$, such that for any $t_1 > T$ and $t_2 > T$,

$$\sum_{i=1}^{n} |x_i(t_2) - x_i(t_1)| \leq \int_{t_1}^{t_2} \sum_{i=1}^{n} |\dot{x}_i(s)| ds < \epsilon.$$

By Cauchy convergence criterion, there must exist $x^* \in R^n$ such that

$$\lim_{t\to\infty} x(t) = x^*. \tag{10}$$

Now, we lift the constraints that every initial function ϕ_i is differentiable.

Let $x^{\{1\}}(t)$ and $x^{\{2\}}(t)$ are two solutions of system the (3), $z(t) = x^{\{1\}}(t) - x^{\{2\}}(t)$. Then

$$\frac{d}{dt} \int_{x^{\{2\}}(t)}^{x^{\{1\}}(t)} \frac{1}{a_i(\rho)} d\rho = -\sum_{j=1}^{n} a_{ij} z_j - \sum_{j=1}^{n} \int_0^\infty z_j(t-s) dk_{ij}(s), \quad i = 1, \cdots, n \tag{11}$$

Define another Lyapunov function

$$\bar{L}(t) = \sum_{i=1}^{n} \zeta_i \Big| \int_{x^{\{2\}}(t)}^{x^{\{1\}}(t)} \frac{1}{a_i(\rho)} d\rho \Big| + \sum_{i,j=1}^{n} \zeta_i \int_0^\infty |dk_{ij}(s)| \int_{t-s}^{t} |z_j(\theta)| d\theta, \tag{12}$$

By similar argument, we can prove

$$\int_0^{+\infty} \sum_{i=1}^{n} \Big| \int_{x^{\{2\}}(s)}^{x^{\{1\}}(s)} \frac{1}{a_i(\rho)} d\rho \Big| ds < \frac{1}{\alpha} \bar{L}(0) < +\infty, \tag{13}$$

which implies.

$$\lim_{s\to\infty}\sum_{i=1}^{n}\left|\int_{x^{\{2\}}(s)}^{x^{\{1\}}(s)}\frac{1}{a_i(\rho)}d\rho\right|ds=0. \tag{14}$$

Because both $x^{\{1\}}(s)$ and $x^{\{2\}}(s)$ are bounded. Therefore,

$$\lim_{s\to\infty}\sum_{i=1}^{n}\left|x^{\{2\}}(s)-x^{\{1\}}(s)\right|=0. \tag{15}$$

It means that every solution of (3) converges to x^*.

Let x^* be the equilibrium point defined above. $x^*(t)$ be any solution of the system the (3), $z(t)=x^*(t)-x^*$. Then

$$\frac{d}{dt}\int_{x_i^*}^{x_i^*(t)}\frac{1}{a_i(\rho)}d\rho=-\sum_{j=1}^{n}a_{ij}z_j(t)-\sum_{j=1}^{n}\int_0^{\infty}z_j(t-s)dk_{ij}(s).$$

Define another function

$$\bar{L}(t)=\sum_{i=1}^{n}\zeta_i\left|\int_{x_i^*}^{x_i^*(t)}\frac{1}{a_i(\rho)}d\rho\right|+\sum_{i,j=1}^{n}\zeta_i\int_0^{\infty}|dk_{ij}(s)|\int_{t-\tau_{ij}-s}^{t}|z_j(\theta)|d\theta. \tag{16}$$

By similar argument, we have

$$\frac{d\bar{L}(t)}{dt}\le-\beta\sum_{j=1}^{n}|z_j(t)|, \tag{17}$$

and

$$\int_0^{+\infty}\sum_{i=1}^{n}|x^*(s)-x^*|ds<\frac{1}{\beta}\bar{L}(0)<+\infty. \tag{18}$$

Because $x^*(s)$ is bounded, which implies that all $x_i(s)$ are uniformly continuous. Therefore,

$$\lim_{s\to\infty}\sum_{i=1}^{n}|x^*(s)-x^*|=0. \tag{19}$$

It means that every solution of (3) converges to x^*.

If $x^*>0$. By previous proof, we have $\lim_{t\to\infty}x(t)=x^*$. Thus, there exists $T>0$ such that $a_i(x_i)(t)>\frac{1}{2}a_i(x^*)>0$ for $t>T$, $i=1,\cdots,n$. Then, exponential stability of the equilibrium can be derived directly from the result given in [14]. The theorem is proved.

As a special case, for the system (1), we have

Corollary 1. *If there exist constants $\zeta_i>0$, $i=1,\cdots,n$, such that*

$$a_{ii}\zeta_i-\sum_{j=1,j\ne i}^{n}\zeta_j|a_{ji}|-\sum_{j=1}^{n}\zeta_j|b_{ji}|>0. \tag{20}$$

Then system (1) has a unique nonnegative equilibrium $x^* \in R^n$, and for any positive solution $x(t)$ of (1), we have

$$\lim_{t\to\infty} x(t) = x^*.$$

Remark 1. The systems discussed in this paper are different from Cohen-Grossberg neural networks discussed in literature, where $\bar{a}_i > a_i(\rho) > \underline{a}_i > 0$ is assumed. In this case, under the condition (5), the linear equations

$$b_i - \sum_{j=1}^{n} a_{ij}x_j - \sum_{j=1}^{n} \int_0^{\infty} dk_{ij}(s)x_j = 0, \tag{21}$$

has a solution, which might not be non-negative but exponentially stable. However, here, $\bar{a}_i > a_i(\rho) > \underline{a}_i > 0$ does not hold any more. We still prove that under condition (5), the system (3) has a nonnegative equilibrium x^*, and every positive solution converges to x^*.

Define $H = (H_{ij})$ for the system (3), where

$$H_{ij} = \begin{cases} a_{ii} - \int_0^{\infty} |dk_{ii}(s)|, & i = j \\ -|a_{ij}| - \int_0^{\infty} |dk_{ij}(s)|, & i \neq j \end{cases} \tag{22}$$

and $\bar{H} = (\bar{H}_{ij})$ for the system (1), where

$$\bar{H}_{ij} = \begin{cases} a_{ii} - |b_{ii}|, & i = j \\ -|a_{ij}| - |b_{ij}|, & i \neq j \end{cases} \tag{23}$$

In terms of M-matrix, we have

Theorem 2. *Suppose that $H1$, $H2$ are satisfied. If H is an M matrix. Then for all b_i, $i = 1, \cdots, n$, the Lotka-Volterra system (3) has a unique nonnegative equilibrium x^* such that any positive solution $x(t)$ converges to x^*.*

Theorem 3. *Suppose that $H1$, $H2$ are satisfied. If $\bar{H}$ is an M matrix.. Then for all b_i, $i = 1, \cdots, n$, the Lotka-Volterra system (1) has a unique nonnegative equilibrium x^*, such that any positive solution $x(t)$ converges to x^*.*

Instead, for the cooperative systems, we have following stronger results

Theorem 4. *Suppose that $a_{ii} > 0$, $a_{ij} < 0$, $i \neq j$, the measure $dk_{ij}(s) < 0$, $i, j = 1, \cdots, n$, and $H1$, $H2$ are satisfied. Then for all b_i, $i = 1, \cdots, n$, the Cooperative Lotka-Volterra system (3) has a unique nonnegative equilibrium x^* such that any positive solution $x(t)$ converges to x^*, if and only if G is an M matrix, where $G = (G_{ij})_{i,j=1}^{n}$, and $G_{ij} = a_{ij} + \int_0^{\infty} dk_{ij}(s)$.*

Theorem 5. *Suppose that $a_{ii} > 0$, $a_{ij} < 0$, $i \neq j$, $b_{ij} < 0$, $i, j = 1, \cdots, n$. Then for all b_i, $i = 1, \cdots, n$, the Cooperative Lotka-Volterra system (1) has a unique nonnegative equilibrium x^*, such that any positive solution $x(t)$ converges to x^*, if and only if $\bar{G}$ is an M matrix, where $\bar{G} = (\bar{G}_{ij})_{i,j=1}^{n}$, and $\bar{G}_{ij} = a_{ij} + b_{ij}$.*

4 Numerical Example

In this section, we present an example to verify theoretical results given in previous sections. we consider the following competitive and cooperative system with delays:

$$\dot{x}_i(t) = x_i(t)\left[b_i - \sum_{j=1}^{5} a_{ij}x_j(t) - \sum_{j=1}^{5} b_{ij}x_j(t-\tau_{ij})\right], \tag{24}$$

where $i = 1,2,3,4,5$, $b = [b_1, \cdots, b_5]^T = [3,6,0,-5,1]^T$,

$$\tau = \begin{bmatrix} 3\,2\,3\,4\,4 \\ 2\,4\,2\,2\,3 \\ 4\,4\,1\,4\,2 \\ 4\,3\,2\,3\,3 \\ 3\,4\,3\,2\,3 \end{bmatrix}, \quad A = \begin{bmatrix} 3 & 1 & 1 & 0 & 2 \\ -1 & 10 & 0 & 2 & 2 \\ 0 & 2 & 12 & 3 & 0 \\ 1 & 1 & -3 & 14 & 2 \\ 0 & 3 & -4 & 2 & 24 \end{bmatrix},$$

and

$$B = \begin{bmatrix} -1 & 1 & 1 & 1 & 1 \\ -1 & 0 & -1 & 0 & 1 \\ -0.3 & 0 & -2 & 2 & 1 \\ 0 & 1 & -1 & -2 & 0 \\ 1 & 0 & -1 & 1 & -2 \end{bmatrix}.$$

We choose the initial values as follows:

$$x_1(\theta) = \sin^2(\theta) + 1; \quad x_2(\theta) = -\theta + 0.5;$$
$$x_3(\theta) = (1-\theta)^2; \quad x_4(\theta) = |\log(-10\theta)|;$$
$$x_5(\theta) = e^{-\frac{\theta}{10}} - 0.5. \quad for\ \theta \in [-4, 0]$$

Direct calculation gives

$$\bar{H} = \begin{bmatrix} 2 & -2 & -2 & -1 & -3 \\ -2 & 10 & -1 & -2 & -3 \\ -0.3 & -2 & 10 & -5 & -1 \\ -1 & -2 & -4 & 12 & -2 \\ -1 & -3 & -5 & -3 & 22 \end{bmatrix}.$$

Its eigenvalues are $(0.0685, 23.0648, 5.8486, 11.2715, 15.7467)$, all are positive. Therefore, $\bar{H}$ is an M-matrix. By Theorem 3, the system (24) is globally asymptotically stable.

By Matlab Optimization Toolbox, we obtain the equilibrium of system (24) is $\bar{x} = [1.4563, 0, 0.0437, 0, 0]^T$. Let

$$\sigma(t) = \sqrt{\sum_{i=1}^{5}(x_i(t) - \bar{x}_i)^2},$$

Fig.1 indicates that $\sigma(t)$ converges to zero as time tends to infinity, which verifies

$$\lim_{t\to\infty} x(t) = \bar{x}.$$

Instead, the solution of the linear equations in the square bracket of (24) is $x^* = [1.4556, -0.2355, 0.2896, -0.4247, 0.1351]^T$. It is clear that $x^* \neq \bar{x}$.

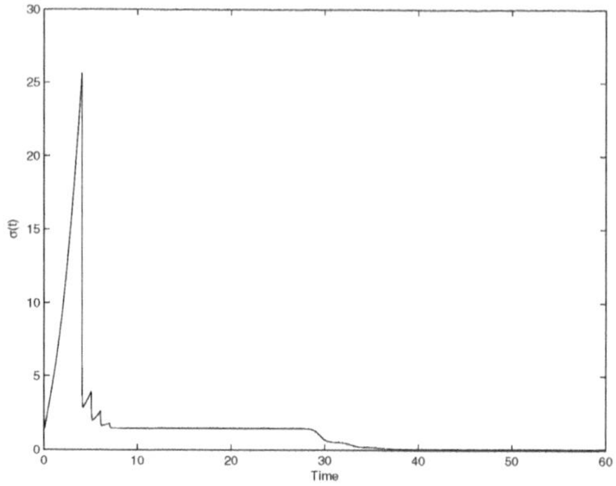

Fig. 1. Change $\sigma(t)$ through Time

5 Conclusion

In this paper, we investigate dynamical behavior of a general class competitive or cooperative Lotka-Volterra systems with delays. Sufficient conditions independent of delays guaranteeing existence of globally stable equilibrium is given.

References

1. Lotka, A. J.,: Elements of Physical Biology. Repubilic as Elements of Mathematics biology. Dover Publications, New York, (1956)
2. Asai, T., Ohtani M., and Yonezu, H.,: Analog Integrated Circuits for The Lotka-Volterra Competition Neural Networks. IEEE Trans. Neural Networks 10 (5) (1999) 1222-1231
3. Morris, S. A., and Pratt, D.,: Analysis of the Lotka-Volterra Competion Equations as A Technological Substitution Models. Tech. Fore. Social Change 70 (2003) 103-133
4. Goplasamy, K.,: Global Asymptotic Stability in Volterra's Population Systems. J. Math. Biol. 19 (1964) 157-168
5. Zhen, J., and Ma, Z.,: Stability for A Competition Lotka-Volterra System with Delays. Nonlinear Analysis 51 (2002) 1131-1142
6. Tineo, A.,: On The Convexity of The Carrying Simplex of Planar Lotka-Volterra Competitive Systems. Appl. Math. Comp. 123 (2001) 93-108

7. Tang, X. H., and Zou, X.: 3/2-type Criteria for Global Attractivity of Lotka-Volterra Competition System without Instantaneous Negative Feedbacks. J. diff. Equat. 186 (2002) 420-439
8. Saito, Y.,: The Necessary and Sufficient Condition for Global Stability of Lotka-Volterra Cooperative or Competition System With Delays. J. Math. Annal. Appl. 268 (2002) 109-124
9. Pao, C. V.,: Global Asymptotic Stability of Lotka-Volterra Competition Systems with Diffusion and Time Delays. Nonlinear Anal.: Real World Appl. 5 (2004) 91-124
10. Hou, Z.,: Global Attractive for a Retarded Competitive Lotka-Volterra Systems. Nonlinear Anal. 47 (2001) 4037-4048
11. Ma, W., and Takeuchi, Y.: Stability Snalysis on a Predator-Prey System with Distributed Delays. J. Comp. Appl. Math. 88 (1998) 79-94
12. Liao, X. X.,: Robust Interval Stability, Persistence, and Partial Stability on Lotka-Volterra Systems with Time-Delay. Appl. Math. Comp. 73 (1996) 103-115
13. Takeuchi, Y.,: Global Dynamical Properties of Lotka-Volterra Systems. World Scientific, Singapore, Rive Edge NJ (1996)
14. Chen, Tianping, Rong, Libin,: Robust Global Exponential Stability of Cohen-Grossberg Neural Networks with Time-Delays. IEEE Transactions on Neural Networks 15 (1) (2004) 203-206

An Improvement of Park-Chung-Cho's Stream Authentication Scheme by Using Information Dispersal Algorithm

Seok-Lae Lee[1], Yongsu Park[2,*], and Joo-Seok Song[3]

[1] Korea Information Security Agency
78 Garak-Dong, Songpa-Gu, Seoul, 138-803, Korea
sllee@kisa.or.kr
[2] College of Information and Communications, Hanyang University
17 Haengdang-dong Seongdong-gu, Seoul 133-791, Korea
yongsu@hanyang.ac.kr
[3] Yonsei University
134 Shinchon-Dong, Seodaemoon-Gu, Seoul, 120-749, Korea
jssong@emerald.yonsei.ac.kr

Abstract. We present an efficient stream authentication scheme that improves the verification probability of Park-Chung-Cho's scheme [1] by using Rabin's Information Dispersal Algorithm. It is shown that under the same communication overhead the verification probability of the proposed scheme is higher than those of SAIDA as well as Park-Chung-Cho's scheme, and that the execution time of the proposed scheme is smaller than that of SAIDA.

1 Introduction

Today, data streaming services such as stock quotes and real-time news are widely used over the Internet. To enable widespread commercial stream services, it is crucial to ensure data integrity and source authentication, e.g., a listener may feel the need to be assured that news streams have not been altered and were made by the original broadcast station.

There are three issues to consider for authenticating live streams. First, for a receiver, the authentication scheme must assure a high verification probability, the ratio of verifiable packets to received packets, even against high packet loss. Second, for a sender, the computational cost should be small to support fast packet rates. Third, communication overhead should be small as well. In this regard, SAIDA (Signature Amortization using Information Dispersal Algorithm) [2] is claimed to be the best algorithm in terms of the verification probability.

In this paper, we present an efficient stream authentication scheme that is based on Park-Chung-Cho's (hereafter "PCC") scheme [1]. In PCC scheme, the loss of signature packet is the main cause of the drastic decrease in the verification probability. To overcome this problem, we process the content of a signature

* Corresponding author.

D. Liu et al. (Eds.): ISNN 2007, Part I, LNCS 4491, pp. 1045–1053, 2007.

packet by introducing some amount of redundancy and splitting the result into pieces, which are then transmitted. The receiver can reconstruct the content of a signature packet if the number of received pieces is larger than the threshold.

The advantages of the proposed scheme are as follows. First, simulation results showed that the verification probability of the proposed scheme was higher than those of SAIDA as well as PCC scheme. Second, unlike the previous schemes [1,3,5,6], the proposed scheme does not have signature packets that require reliable transmission for high verification probability. Third, experimental results showed that although the execution time of the proposed scheme was longer than that of PCC scheme, it was shorter than that of SAIDA.

2 The Proposed Scheme

In Section 2.1, we describe PCC scheme, and in Section 2.2 we enhance PCC scheme by using IDA. We use the following notations: S and R denote a sender and a receiver, respectively. $h(x)$ denotes a one-way hash function. $SIG(M)$ stands for the digital signature of data M signed by a sender S. $C||D$ denotes the concatenation of data C and D, and $|E|$ denotes the byte size of data E. If an integer f is divisible by an integer e, we write $e|f$.

2.1 Park-Chung-Cho's (PCC) Scheme

For the first n stream chunks among $M_0, M_1, \cdots, M_{tn-1}$ ($n > 0$, $t > 0$), S computes authentication information (n_s authentication trees) and generates n packets. Each packet is sent to R as soon as it is generated, and then S transmits a signature packet. S repeats this procedure for the next n chunks (we call each n chunks as a group). When R receives some packets and the signature packet for a certain group, R attempts to verify them. It is assumed that S's packet buffer size is not less than 2^b ($\geq n/n_s$). In this scheme, $n_s|n$ and $2^l|(n/n_s)$ should be satisfied. The selection of n and n_s that satisfy these conditions is described in detail in [1].

Authentication tree. For each group, S constructs n_s authentication trees [7]. Each authentication tree A_m ($0 \leq m < n_s$) has n/n_s leaves. A_m is not a full binary tree: the root node of A_m has $n/(2^{l-1}n_s)$ children, each of which is the root of a full binary subtree T_k ($nm/(2^{l-1}n_s) \leq k < n(m+1)/(2^{l-1}n_s)$) that has 2^{l-1} leaves (see Figure 1). In an authentication tree, a hash value is associated with each node. Let $e_{c\sim c}$ and $E_{c\sim c}$ denote a leaf and its value ($= h(M_i)$), respectively. Similarly, let $e_{c\sim c'}$ and $E_{c\sim c'}$ ($c < c'$) represent an internal node and its value that is equal to $h(E_{c_1\sim c'_1}||E_{c_2\sim c'_2}||\cdots||E_{c_j\sim c'_j})$ ($c = c_1 \leq c'_1 < c_2 \leq c'_2 < \cdots < c_j \leq c'_j = c'$), where $E_{c_1\sim c'_1}, E_{c_2\sim c'_2}, \cdots, E_{c_j\sim c'_j}$ are values associated with the children of $e_{c\sim c'}$, i.e., $E_{0\sim 1} = h(E_{0\sim 0}||E_{1\sim 1})$ in Figure 1. $Siblings(e, e')$ is defined as the set of values associated with all siblings of each node in the path from e to e', i.e., $Siblings(e_{0\sim 0}, e_{0\sim 1}) = \{E_{1\sim 1}\}$ in Figure 1. A pair of subtrees is defined as $Pair_j = (T_{2j}, T_{2j+1})$ ($nm/(2^l n_s) \leq$

$j < n(m+1)/(2^l n_s)$), i.e., $Pair_0 = (T_0, T_1), Pair_1 = (T_2, T_3)$ in Figure 1. The value associated with the root node of A_m is the hash of the concatenation of the values associated with the roots of all T_k. To provide high verification probability against bursty loss patterns over the Internet, the ith node among 2^b leaf nodes is associated with $f(i)$ $(= (i \bmod 2^l)(2^{b-l}) + \lfloor i/2^l \rfloor)$th chunk among 2^b chunks, i.e., $e_{0\sim0}, e_{1\sim1}, e_{2\sim2}, \cdots, e_{(2^b-1)\sim(2^b-1)}$ are associated with $M_{f(0)}, M_{f(1)}, M_{f(2)}, \cdots, M_{f(2^b-1)}$, respectively.

Example 1. *Figure 1 shows one of the four authentication trees for the first group when* $l = 2, n = 32, n_s = 4$, *and* $2^b = 8$. *It has four subtrees* (T_0, T_1, T_2, *and* T_3). $E_{0\sim7} = h(E_{0\sim1}||E_{2\sim3}||E_{4\sim5}||E_{6\sim7})$.

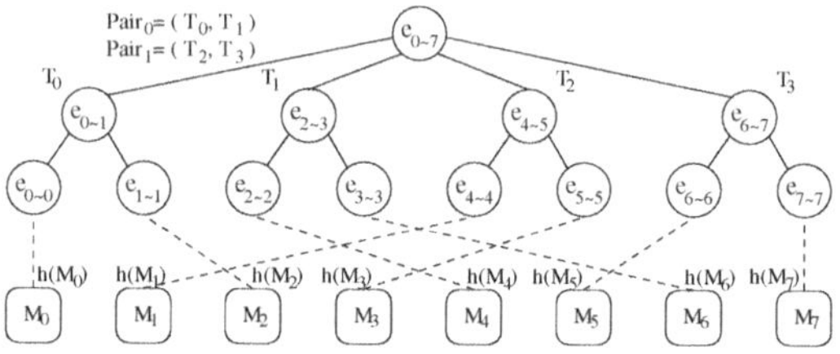

Fig. 1. Example of the authentication tree

Packet generation. For a group $G_0 = \{M_i | (0 \le i < n)\}$, S builds n_s authentication trees as follows. For every n/n_s chunks, S calculates the hash values associated with all internal and leaf nodes of an authentication tree A_m, and then calculates the hash value associated with the root of A_m. S generates and sends each packet $P_i = (M_i, Siblings(e_{c\sim c}$, the root of $T_a) \cup \{$the value associated with the root of $T_b\})$. The number of hash values in P_i is l. After constructing all A_m, S sends the signature packet of G_0. Then, S repeats the above process with $G_1, G_2, \cdots, G_{t-1}$.

Example 2. *When* $l = 2, n = 32, n_s = 4$, *and* $2^b = 8$, *for the first 32 chunks S builds four trees including the tree described in Example 1, and then generates 32 packets and a signature packet.* $h(M_2)$ *is associated with* $e_{1\sim1}$ *of* T_0. T_0 *is paired with* T_1. $P_2 = (M_2, \{E_{0\sim0}, E_{2\sim3}\})$. *The signature packet is* $(E_{0\sim7}||E_{8\sim15}||E_{16\sim23}$ $||E_{24\sim31}, SIG(E_{0\sim7}||E_{8\sim15}||E_{16\sim23}||E_{24\sim31}))$.

Packet verification. If R does not receive the signature packet of a group, then R cannot verify all the received packets for the group. Otherwise, R attempts to verify the received packets for the group as follows. First, R verifies the digital signature in the signature packet. If R fails to verify the signature, all the packets of the group will not be verifiable. Otherwise, verification is done in the unit of n/n_s chunks of each A_m. This scheme can either verify all n/n_s chunks or unable to verify them at all. For each received packet $P_i = (M_i, Siblings(e_{c\sim c}$, the root of $T_a) \cup \{$the value associated with the root of $T_b\})$, R can obtain two values associated with the roots of T_a and T_b (the latter is in P_i and the former is computed using $Siblings()$ and $E'_{c\sim c} = h(M_i)$). If R obtains the values associated with the roots of all subtrees of A_m using the received packets, R can compute the value of the root of A_m and verify any packets corresponding to the leaves of A_m.

The value associated with the root of A_m can be computed even if R obtains only $n/(2^l n_s)$ packets and these packets contain a chunk ($= M_i$) from every pair of subtrees. However, R will fail to compute it if all the chunks of a pair are lost. To avoid this worst case scenario, the mapping between the packets and the leaf nodes of A_m by $f(i)$ ($= (i \bmod 2^l)(2^{b-l}) + \lfloor i/2^l \rfloor$) is used.

Example 3. *In Example 2, assume that R receives only P_2 among P_0, P_2, P_4, and P_6. $P_2 = (M_2, \{E_{0\sim0}, E_{2\sim3}\})$. R obtains $E'_{0\sim1}$ (from $E_{0\sim0}$ and $E'_{1\sim1} = h(M_2)$) and $E_{2\sim3}$. If P_1, P_3, P_5, and P_7 are lost during the transmission, R can obtain neither $E_{4\sim5}$ nor $E_{6\sim7}$ and would be unable to verify P_2. However, if R receives P_5, R will be able to obtain $E_{4\sim5}$ and $E'_{6\sim7} = h(E'_{6\sim6}||E_{7\sim7})$ using $\{E_{7\sim7}, E_{4\sim5}\}$ in P_5. Therefore, R can compute $E'_{0\sim7} = h(E'_{0\sim1}||E_{2\sim3}||E_{4\sim5}||E'_{6\sim7})$. If $E'_{0\sim7}$ is identical to $E_{0\sim7}$ in the signature packet, P_2 and P_5 are successfully verified.*

2.2 The Proposed Scheme

In most of the stream authentication schemes including PCC scheme [1,3,5,6], the loss of signature packet is the main cause of the drastic decrease in the verification probability. To address this problem, the authors of [1,3] suggested that a sender repeatedly transmits the same signature packet after some amount of time delay. However, this method has the following problems: communication overhead increases due to the repeated transmission of the signature packet, and significant verification delay occurs if the first signature packet is lost. To overcome these problems, in the proposed scheme, the content of a signature packet is encoded with Rabin's Information Dispersal Algorithm (IDA) [4].

IDA consists of the following two procedures. $Dispersal(F, m', n)$ splits the data F with some amount of redundancy resulting in n pieces F_i ($0 \le i < n$), where $m' \le n$ and $|F_i|$ is $|F|/m'$. Reconstruction of F is possible with any combination of m' pieces by calling $Recovery(\{F_{i_j}|(0 \le j < m'), (0 \le i < n)\}, m', n)$. Thus, IDA is resistant to the loss of $(n - m')$ pieces.

Packet generation. We set S's packet buffer size, 2^b, as n. For a group $G_0 = \{M_i|(0 \le i < n)\}$, S builds n_s authentication trees by PCC scheme. Then, S generates the signature packet $SIG = (E_{0\sim n/n_s-1}||E_{n/n_s\sim 2n/n_s-1}||\cdots|| E_{(n_s-1)n/n_s\sim n-1}, SIG(E_{0\sim n/n_s-1}||E_{n/n_s\sim 2n/n_s-1}||\cdots||E_{(n_s-1)n/n_s\sim n-1}))$. The signature packet is dispersed to n pieces (SIG_i ($0 \le i < n$)) by calling $Dispersal$ (SIG, m', n). Then, S generates and sends each packet $P_i = (M_i, Siblings(e_{c\sim c}$, the root of $T_a) \cup \{$the value associated with the root of $T_b\}, SIG_i)$. The sender repeats the above process with $G_1, G_2, \cdots, G_{t-1}$.

Example 4. *When $l = 2, n = 32, n_s = 4, m' = 16$, and $2^b = n$, the packet generation process is as follows. If we use 128-bit MD5 as the hash function and 1024-bit RSA as the signature algorithm, $|h()| = 16$ and $|SIG()| = 128$. For a group G_0,*

S builds four trees including the tree A_0 described in Example 1, and then generates 32 packets and the signature packet $SIG = (E_{0\sim7}||E_{8\sim15}||E_{16\sim23}||E_{24\sim31}, SIG(E_{0\sim7}||E_{8\sim15}||E_{16\sim23}\,||E_{24\sim31}))$ ($|SIG| = 192bytes$). Finally, the signature packet is dispersed to 32 pieces (SIG_i $(0 \leq i < 32)$) by calling $Dispersal(SIG, 16, 32)$. The size of each SIG_i is 12bytes ($= 192/16$). Figure 2 describes an example of generation of the packets associated with the authentication tree A_0. In this figure, each packet P_i contains two hashes and one SIG_i, e.g., $P_8 = (M_8, \{E_{0\sim0}, E_{2\sim3}\}, SIG_8)$.

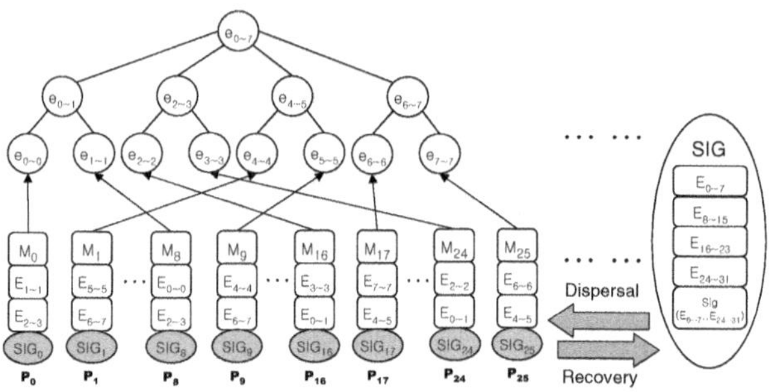

Fig. 2. Example of packet generation

Packet verification. If the number of received packets for a certain group G_r $(0 \leq r < t)$ is less than m', R is unable to verify the received packets because it becomes impossible to reconstruct the signature packet of G_r. Otherwise, the signature packet is reconstructed by calling $Recovery(\{SIG_{i_j}|(0 \leq j < m'), (0 \leq i < n)\}, m', n)$, and the remaining verification process is the same as that of PCC scheme.

Example 5. *In Example 4, assume that R receives packets more than 16 among 32 packets for the group G_0 and only P_8 among $P_0, P_8, P_{16},$ and P_{24}. First, R reconstructs the signature packet with any combination of 16 pieces included in the received packets by calling $Recovery(\{SIG_{i_j}|(0 \leq j < 16), (0 \leq i < 32)\}, 16, 32)$. Using $P_8 = (M_8, \{E_{0\sim0}, E_{2\sim3}\}, SIG_8)$, R obtains $E'_{0\sim1}$ (from $E_{0\sim0}$ and $E'_{1\sim1} = h(M_8)$) and $E_{2\sim3}$. If $P_1, P_9, P_{17},$ and P_{25} are lost during the transmission, R can obtain neither $E_{4\sim5}$ nor $E_{6\sim7}$ and would be unable to verify P_8. However, if R receives P_{17}, R will be able to obtain $E_{4\sim5}$ and $E'_{6\sim7} = h(E'_{6\sim6}||E_{7\sim7})$ using $\{E_{7\sim7}, E_{4\sim5}\}$ in P_{17}. Therefore, R can compute $E'_{0\sim7} = h(E'_{0\sim1}||E_{2\sim3}||E_{4\sim5}||E'_{6\sim7})$. If $E'_{0\sim7}$ is identical to $E_{0\sim7}$ in the signature packet, P_8 and P_{17} are successfully verified.*

3 Performance Analysis

In Section 3.1, we mathematically analyze the verification probability of our scheme. We compare the simulation results on the verification probability of

PCC scheme, SAIDA, and our scheme in Section 3.2. In Section 3.3, we analyze the computation cost of PCC scheme, SAIDA, and our scheme. As in [2], we use 128-bit MD5 as the hash function and 1024-bit RSA as the signature algorithm.

3.1 Mathematical Analysis on the Verification Probability

In this section, we analyze the verification probability of our scheme, mathematically, under the assumption that each packet loss occurs independently with the probability of $(1-p)$. Note that the minimal number of pieces that is required to recover the signature packet of a group is m'.

Theorem 1. *Under the above packet loss model, the verification probability of our scheme, P_{ePCC}^{ind}, is as follows:*

$$P_{ePCC}^{ind} > (1-(1-p)^{2^l})^{n/(2^l n_s)-1}\{1-\sum_{i=0}^{m'-n/(2^l n_s)-1}\binom{n-n/n_s}{i}p^i(1-p)^{n-n/n_s-i}\}.$$

Proof. The receiver R can successfully verify a received packet P_k if the following two conditions are satisfied. First, as in PCC scheme, R must receive at least one packet from every subtree pairs of the authentication tree A_m that contains the leaf node associated with the received packet P_k. Second, R must receive at least m' packets among n packets associated with the group G_j that includes the received packet P_k. Let E_1 and E_2 denote the events that correspond to the first and second conditions, respectively. We define the sample space $S = \{a_0 a_1 \cdots a_{n-1} | a_i = 0 \text{ or } 1\ (0 \le i < n)\}$, where each element represents the receiving status of n packets associated with G_j. Then, $E_1, E_2 \subset S$. The probability of the occurrence of E_1, $P(E_1)$, is $(1-(1-p)^{2^l})^{n/(2^l n_s)-1}$ by Theorem 2 of [1]. The occurrence of E_1 means that R receives at least one packet every 2^l packets among n/n_s packets associated with A_m, so under the occurrence of E_1 the number of received packets is at least $n/(2^l n_s)$. Consider the new event E_2' ($\subset E_2$) that R receives at least $m' - n/(2^l n_s)$ packets among $n - n/n_s$ packets associated with all the remaining authentication trees except for A_m in G_j. All the elements included in $E_1 \cap E_2'$ satisfy the second condition as well as the first condition because the number of received packets among n packets associated with G_j is at least m'. Hence, $(E_1 \cap E_2') \subset (E_1 \cap E_2)$. Note that E_1 and E_2' are statistically independent events. Thus, $P_{ePCC}^{ind} = P(E_1 \cap E_2) > P(E_1 \cap E_2') = P(E_1)P(E_2')$. $P(E_2')$ is $1-\sum_{i=0}^{m'-n/(2^l n_s)-1}\binom{n-n/n_s}{i}p^i(1-p)^{n-n/n_s-i}$ by equation 2.5-3 of [8]. Therefore,

$$P(E_1)P(E_2') = (1-(1-p)^{2^l})^{n/(2^l n_s)-1}\{1-\sum_{i=0}^{m'-n/(2^l n_s)-1}\binom{n-n/n_s}{i}p^i(1-p)^{n-n/n_s-i}\}.$$

3.2 Simulation Results on the Verification Probability

In this section, we analyze the simulation results on the verification probability of each scheme. The simulation of PCC scheme is carried out for the following two cases: in the first case, the signature packet for a certain group is transmitted only once, and in the second case the signature packet is transmitted twice with a delay of 128 packets. To simulate the general pattern of the packet transmission over the Internet, we adopt 2-state Markov model for generating packet loss [1,2,3]. As in [2], we use the following parameters: group size $n = 128$; packet loss probability= 20%; average length of burst loss= 8 packets.

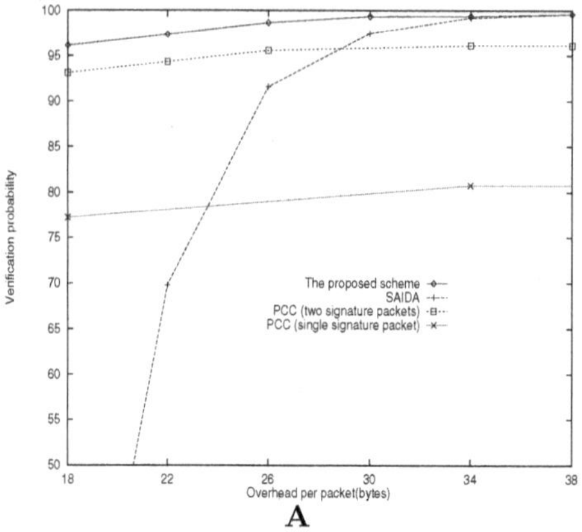

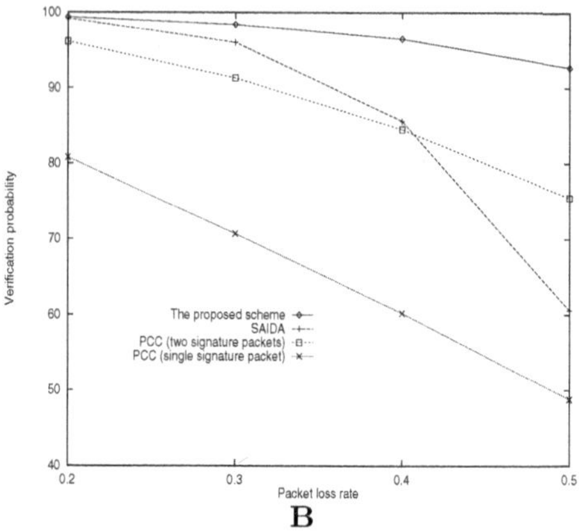

Fig. 3. Verification probability

The graph A on Figure 3 shows the verification probability of each scheme when the communication overhead is from 18 to 38 bytes. For all communication overheads, the verification probability of the proposed scheme is higher than those of any other schemes. The graph B on Figure 3 shows the verification probability of each scheme when the communication overhead is 34 bytes and the packet loss rate is from 20% to 50%. In this graph, curves for SAIDA and PCC scheme drop steeply to unacceptable levels as the loss rate is increased. In contrast, the curve for the proposed scheme drops much more moderately, maintaining a verification probability of over 92% even though the packet loss rate is 50%.

3.3 Comparison of the Computation Cost

In this section, we analyze the computation cost on a sender and receiver of PCC, SAIDA, and our scheme. Let C_{hash}, $C_{Dispersal(|F|,m',n)}$, $C_{Recovery(|F|,m',n)}$, C_{sign}, and C_{ver} denote the computation cost of the hash function, $Dispersal(|F|, m', n)$, $Recovery(|F|, m', n)$, signature generation and verification, respectively. r denotes the number of received packets, and we assume that $r \geq m'$. Table 1 shows the comparison results of the computation cost on a sender and receiver.

We implemented each scheme and measured the execution time for a single group. As in [2], we set the group size n to be 128. Table 2 shows the comparison results of the elapsed time on a sender and receiver when the communication overhead was from 22 to 38 bytes. Experimental environments are as follows: CPU, RAM, OS, crypto-library and compiler are Pentium 4 2.4GHz, 512 MBytes, Linux version 2.4.20, Crypto^{++} 4.2, gcc version 2.96, respectively. We used 128-bit MD5 as the hash function and 1024-bit RSA as the signature algorithm [2]. The elapsed time on a sender and receiver of our scheme was about two and one-and-a-half times as long as that of PCC scheme, respectively. This is due to the fact that in our scheme IDA, which is not involved in PCC scheme, is used. The elapsed time on a sender and receiver of our scheme was about 28% and 10% smaller than that of SAIDA, respectively. This is presumed to be due to the fact that the hash operation is much faster than $Dispersal()$ and $Recovery()$ which rely on finite field operations.

Table 1. Comparison of the computation cost for a single group

Scheme	Sender	Receiver
SAIDA	$(n+1)C_{hash}+$ $C_{Dispersal(n\|h()\|+\|SIG()\|,m',n)}+C_{sign}$	$(r+1)C_{hash}+$ $C_{Recovery(n\|h()\|+\|SIG()\|,m',n)}+C_{ver}$
PCC	$(2n-\frac{n}{2^{l-1}}+n_s+1)C_{hash}+C_{sign}$	$(lr+n_s+1)C_{hash}+C_{ver}$
Proposed scheme	$(2n-\frac{n}{2^{l-1}}+n_s+1)C_{hash}+$ $C_{Dispersal(n_s\|h()\|+\|SIG()\|,m',n)}+C_{sign}$	$(lr+n_s+1)C_{hash}+$ $C_{Recovery(n_s\|h()\|+\|SIG()\|,m',n)}+C_{ver}$

Table 2. Comparison of the elapsed time for a single group

Overhead per packet (bytes)	Scheme	Sender (ms)	Receiver (ms)
22	SAIDA	62.29	3.64
	PCC	23.33	1.76
	Proposed scheme	50.32	2.88
26	SAIDA	68.45	3.46
	PCC	23.44	1.78
	Proposed scheme	53.88	3.04
34	SAIDA	77.68	2.95
	PCC	23.72	1.81
	Proposed scheme	53.11	2.63
38	SAIDA	80.33	2.85
	PCC	23.93	1.86
	Proposed scheme	49.48	2.98

4 Conclusion

We presented an efficient stream authentication scheme that improved the verification probability of Park-Chung-Cho's (PCC) scheme [1] by using Rabin's Information Dispersal Algorithm. We provided a mathematical analysis on the verification probability of the proposed scheme. The advantages of the proposed scheme are as follows:

- Compared with PCC scheme and SAIDA, simulation results showed that under the same communication overhead the proposed scheme has the highest verification probability. When the overhead per packet was 34 bytes and the packet loss rate was 50%, the verification probability of the proposed scheme was 92%, whereas those of PCC scheme and SAIDA were 75% and 60%, respectively.

- Unlike the previous schemes [1,3,5,6], the proposed scheme does not have signature packets that require reliable transmission for high verification probability.
- Experimental results showed that the elapsed time of the proposed scheme is smaller than that of SAIDA. The sender and receiver programs of the proposed scheme were about 28% and 10% faster than those of SAIDA, respectively.

Acknowledgement

This work was supported by the Korea Research Foundation Grant funded by the Korean Government (MOEHRD) (KRF-2006-331-D00556). Also, this work was supported by the Korea Research Foundation Grant funded by the Korean Government (MOEHRD) (KRF-2005-042-D00294). Also, this work was supported by the Korea Information Security Agency (KCAC 06-01).

References

1. Park, Y., Chung, T., Cho, Y.: An efficient Stream Authentication Scheme Using Tree Chaining. Information Processing Letters **86** (1) (2003) 1-8
2. Park, J., Chong, E., Siegel, H.: Efficient Multicast Packet Authentication Using Signature Amortization. Proceedings of IEEE Security and Privacy Symposium (2002) 227-240
3. Perrig, A., Canetti, R., Song, D., Tygar, J.: Efficient Authentication and Signing of Multicast Streams over Lossy Channels. Proceedings of IEEE Security and Privacy Symposium (2000)
4. Michael, O.R.: Efficient Dispersal of Information for Security, Load Balancing and Fault Tolerance. Journal of the Association for Computing Machinery **36** (2) (1989) 335-348
5. Philippe, G., Nagendra, M.: Authenticating Streamed Data in the Presence of Random Packet Loss. NDSS'01 (2001) 13-22
6. Sara, M., Jessica, S.: Graph-Based Authentication of Digital Streams. Proceedings of IEEE Security and Privacy Symposium (2001) 232-246
7. Merkle, R.C.: A Certified Digital Signature. CRYPTO'89 (1989)
8. Peebles, P.Z.: Probability, Random Variables and Random Signal Principles. McGraw-Hill International Edition (2001)

Dynamics of Continuous-Time Neural Networks and Their Discrete-Time Analogues with Distributed Delays

Lingyao Wu[1], Liang Ju[2], and Lei Guo[1]

[1] Research Institute of Automation, Southeast University, Nanjing 210096, P.R. China
[2] College of Electrical Engineering, Hohai University, Nanjing 210098, P.R. China
boguanban@seu.edu.cn

Abstract. Discrete-time analogues of continuous-time neural networks with continuously distributed delays and periodic inputs are introduced. The discrete-time analogues are considered to be numerical discretizations of the continuous-time networks and we study their dynamical characteristics. By employing Halanay-type inequality, we obtain easily verifiable sufficient conditions ensuring that every solutions of the discrete-time analogue converge exponentially to the unique periodic solutions. It is shown that the discrete-time analogues preserve the periodicity of the continuous-time networks.

1 Introduction

There have been active investigations recently into the dynamics and applications of neural networks with delays because the dynamical properties of equilibrium points of neural systems play an important role in some practical problems [1-4], such as optimization solvers, associative memories, image compression, speed detection of moving objects, processing of moving images, and pattern classification. It is well known that an equilibrium point can be viewed as a special periodic solution of continuous-time neural networks with arbitrary period [5-7]. In this sense the analysis of periodic solutions of neural systems may be considered to be more general sense than that of equilibrium points. Among the most previous results of the dynamical analysis of continuous-time neural systems, a frequent assumption is that the activation functions are differentiable and bounded, such as the usual sigmoid-type functions in conventional neural networks. But, in some application, the differentiability and boundedness of the activation function are not satisfied. Since the boundedness and differentiability assumption is not always practical, it is necessary and important to investigate the dynamical properties of the continuous-time neural systems in both theory and applications without assuming the smoothness and boundedness of the activation functions [5-7]. Moreover, validity of the mathematical theorem on neural

D. Liu et al. (Eds.): ISNN 2007, Part I, LNCS 4491, pp. 1054–1060, 2007.

systems without assuming the boundedness and differentiability of the activation functions, will offer better foundation in practice than that with bounded activation functions.

On the other hand, in implementing the continuous-time neural system for simulation or computational purposes, it is essential to formulate a discrete-time system which is an analogue of the continuous-time system. Mohamad and Gopalsamy [8] have formulated the discrete-time analogues of continuous-time neural systems and addressed their stability properties. Sun [5] has extended the discrete-time analogues with constant inputs to the discrete-time analogues with periodic inputs, and further investigated their periodicity properties.

The results on dynamical analysis of continuous-time neural systems are mostly assumed that the time delays are discrete. However, neural networks usually have spatial extent due to the presence of a multitude of parallel pathways with a variety of axon sizes and lengths. Thus, there will be a distribution of propagation delays. This means that the signal propagation is no longer instantaneous and is better represented by a model with continuously distributed delays. In this paper, our focus is on periodicity of continuous-time neural networks with continuously distributed delays. By using the Halanay-type inequality, we will derive some easily checkable conditions to guarantee exponential periodicity of continuous-time neural networks with continuously distributed delays. Discrete-time analogues of integro-differential equations modeling neural networks with periodic i nputs are introduced, and we will also study the exponential periodicity of the discrete-time analogues.

The rest of the paper is organized as follows.The neural network model and preliminaries will be given in Section II. In Section III, periodicity of continuous-time and discrete-time neural networks will be presented and discussed. One example is given to illustrate the effectiveness of our results in Section IV. Finally, Section V is the conclusion.

2 Preliminaries

Consider the model of continuous-time neural networks described by the following integro-differential equations of the form

$$\begin{aligned} \dot{x_i}(t) &= -d_i x_i(t) + \sum_{j=1}^{n} a_{ij} g_j(x_j(t)) \\ &\quad + \sum_{j=1}^{n} b_{ij} g_j\Big(\int_0^{\infty} K_{ij}(s) x_j(t-s)ds\Big) + I_i(t), \end{aligned} \tag{1}$$

for $i \in \{1, 2, \ldots, n\}$, $x_i(t)$ is the state vector of the $i_t h$ unit at time t, a_{ij}, b_{ij} are constants. For simplicity, let D be an constant $n \times n$ diagonal matrix with diagonal elements $d_i > 0, i = 1, 2, \ldots, n$, $A = (a_{ij})$ and $B = (b_{ij})$ are $n \times n$ constant interconnection matrices, $g_j(x_j(t))$ denotes the nonlinear activation function of the j-th unit at time t, I(t) is an input periodic vector function with period ω. Suppose for each $i \in \{1, 2, \ldots, n\}$, g_i is globally Lipschitz continuous (GLC) with Lipschitz constant L_i. Some researchers have studied the pure-delay model (with A=0) [9]. However, neural network model with periodic inputs in which instantaneous signaling as well as delayed signaling both occur (with $A \neq 0$, $B \neq 0$) has not been investigated yet. $K_{ij}(\cdot)$, $i, j = 1, 2, \ldots, n$ are the delay kernels which are assumed to satisfy the following conditions simultaneously:

(1)K_{ij} are bounded and continuous on $[0, +\infty]$ and $\int_0^\infty K_{ij}(s)ds = 1$;
(2)There exists a positive number μ such that $\int_0^\infty K_{ij}(s)e^{\mu}ds < \infty$.

We refer [8] to propose a semi-discretization technique in formulating a discrete-time analogue of the continuous-time system (1). We can easily obtain the following the discrete-time analogue of the integro-differential system (1)

$$x_i(m+1) = x_i(m)e^{-dh} + \theta_i(h)\{\sum_{j=1}^{n} a_{ij}g_j(x_j(m)) + \sum_{j=1}^{n} b_{ij}g_j(\sum_{j=1}^{\infty} K_{ij}(p)x_j(m-p)) + I_i(m)\}, \tag{2}$$

where$I_i(m) = I_i(m + \overline{\omega})$,$\overline{\omega}$ is a positive integer.

$$\theta_i(h) = \frac{1 - e_{-d_i h}}{d_i}, \quad i = 1, 2, \ldots, n.$$

(1)$K_{ij}(p)$ are bounded for $p = 1, 2, \ldots$, and $\sum_{p=1}^{\infty} K_{ij}(p)dp = 1$;
(2)There exists a positive number $v > 1$ such that $\sum_{p=1}^{\infty} K_{ij}(p)v^p ds < \infty$.

For the proof of the main results in this paper, the following Halanay-type inequality is needed. Halanay-type inequality is rarely used in the literature on stability of dissipative dynamical systems in spite of their possible generalizations and applications (see [9] and its references therein for detail). This inequality is narrated from [8] as follows:

Lemma 1. *Let $v(t) > 0$,for $t \in R, \tau \in [0, +\infty)$ and $t_0 \in R$. assume*

$$\frac{dv(t)}{dt} \leq -av(t) + b(\sup_{s \in [t-\tau, \tau]} v(s)), \quad t > t_0,$$

if $a > b > 0$, then there exist cnstants $\gamma > 0$ and $k > 0$ such that

$$v(t) \leq ke^{-\gamma}(t - t_0).$$

3 Main Result

In this section, we will use norms of vector x in R^n, which is defined as $\|x\| = max_{1 \leq i \leq n}|x_i|$. Let C be the Banach space of all continuous function from $(-\infty, 0]$ to R^n with the topology of uniform convergence. Given any ϕ and φ, let $x(t, \phi)$ and $x(t, \varphi)$ be the solutions of (1) starting from ϕ and φ respectively.

Theorem 1. *Suppose $g \in GLC$. If there exist positive constants λ_j such that the following conditions*

$$\lambda_i d_i - \sum_{j=1}^{n} \lambda_j a_{ij} L_j - \sum_{j=1}^{n} \lambda_j b_{ij} L_j > 0 \quad i, j = 1, 2, \ldots, n,$$

hold, then for every periodic input $I(t)$, the delayed neural system (1) is exponentially periodic.

Proof. It follows from (1) that

$$x_i(\dot{t}, \phi) - x_i(\dot{t}, \varphi) = -d_i(x_i(t, \phi) - x_i(t, \varphi)) + \sum_{j=1}^{n} a_{ij}[g_j(x_j(t, \phi)) - g_j(x_j(t, \varphi))]$$
$$+ \sum_{j=1}^{n} b_{ij} g_j(\int_0^{\infty} K_{ij}(s)[x_j(t-s, \phi) - x_j(t-s, \varphi)]ds, \qquad (3)$$

for $t > 0, i = 1, 2, \ldots, n$.

We consider function $G(\cdot)$ defined by

$$G_i(\varepsilon_i) = \lambda_i(d_i - \varepsilon_i) - \sum_{j=1}^{n} \lambda_j |a_{ij}| L_j - \sum_{j=1}^{n} \lambda_j |b_{ij}| L_j \int_0^{\infty} K_{ij}(s) e^{\varepsilon_i s},$$

for $\varepsilon \in [0, \mu), i = 1, 2, \ldots, n$.

We have that $G_i(0) > 0, i = 1, 2, \ldots, n$. and hence by the continuity of $G_i(\cdot)$ on $[0, \mu)$, there exists a constant ε_i such that

$$G_i(\varepsilon) = \lambda_i(d_i - \varepsilon) - \sum_{j=1}^{n} \lambda_j |a_{ij}| L_j - \sum_{j=1}^{n} \lambda_j |b_{ij}| L_j \int_0^{\infty} K_{ij}(s) e^{\varepsilon s} > 0.$$

We then consider functions $z_i(\cdot)$ defines by

$$z_i(t) = \frac{1}{\lambda_i} e^{\varepsilon t} |x_i(t, \phi) - x_i(t, \varphi)|, \qquad t \in (-\infty, +\infty), \quad i = 1, 2, \ldots, n. \qquad (4)$$

Define the upper right derivative of $z_i(t)$ by

$$D^+ z_i(t) = \limsup_{h \to 0^+} \frac{z_i(t+h) - z_i(t)}{h}.$$

By using (3) and (4) we derive that

$$D^+ z_i(t) \leq -(d_i - \varepsilon) z_i(t) + \frac{1}{\lambda_i} \sum_{j=1}^{n} |a_{ij}| L_j \lambda_j z_i(t)$$
$$+ \frac{1}{\lambda_i} \sum_{j=1}^{n} |b_{ij}| L_j \lambda_j (\int_0^{\infty} K_{ij}(s) L_j \lambda_j [x_j(t-s, \phi) - x_j(t-s, \varphi)]ds$$

$$\leq -(d_i - \varepsilon)z_i(t) + \frac{1}{\lambda_i}\sum_{j=1}^{n}|a_{ij}|L_j\lambda_j z_i(t)$$

$$+\frac{1}{\lambda_i}\sum_{j=1}^{n}|b_{ij}|L_j\lambda_j(\int_0^{\infty} K_{ij}(s)L_j\lambda_j e^{\varepsilon s}ds(\sup_{s\in(-\infty,t]} z_i(s)), \tag{5}$$

where $t > 0, i = 1, 2, \ldots, n$.

We note that the system (5) denotes one of the generalizations of Halanay-type inequality.

Let $\delta > 1$ denote an arbitrary real number and let

$$K = \max_{1\leq i\leq n}\sup_{s\in 9-\infty,0]}\frac{1}{\lambda_i}|\phi_i(s) - \psi_i(s)| > 0. \tag{6}$$

It follows from (4) and (6) that $z_i(t) < \delta K$, for $i = 1, 2, \ldots, n$ and $t \in (-\infty, 0]$. We claim

$$z_i(t) < \delta K,$$

for $i = 1, 2, \ldots, n$, and $t > 0$.

Suppose this is not the case. Let there be an k and a first time $t_1 > 0$ such that

$z_i(t) < \delta K$ for $i \neq k, i = 1, 2, \ldots, n$ and $t \in (-\infty, t_i]$.

$$z_k(t) < \delta K, \quad z_k(t_1) = \delta K, \quad D^+ z_k(t_1) \geq 0. \tag{7}$$

By using (5) and (7) we can easily obtain

$$D^+ z_k(t_1) \leq -\frac{1}{\lambda_k}\{\lambda_i(d_i - \varepsilon) - \sum_{j=1}^{n}\lambda_j|a_{ij}|L_j - \sum_{j=1}^{n}\lambda_j|b_{ij}|L_j\int_0^{\infty} K_{ij}(s)e^{\varepsilon s}\}\delta K$$
$$< 0. \tag{8}$$

So which is a contradiction. Hence the claim is true. Since $\delta > 1$, we then use (4) to obtain

$$|x_i(t,\phi) - x_i(t,\varphi)| \leq \frac{\lambda_{max}}{\lambda_{min}}e^{\varepsilon t}\max_{1\leq i\leq n}\{\sup_{s\in(-\infty,t]}|\phi_i(s) - \varphi_i(s)|\}, \tag{9}$$

where

$$\lambda_{min} = \min_{1\leq i\leq n}\lambda_i, \lambda_{max} = \max_{1\leq i\leq n}\lambda_i.$$

We can choose a positive integer m such that

$$\frac{\lambda_{max}}{\lambda_{min}}e^{\varepsilon m\omega} \leq \frac{1}{4}.$$

Define a Poincaré mapping $H : C \to C$ by $H(\phi) = x_\omega(\phi)$. Then from (8), we can derive that

$$\|H^m(\phi) - H^m(\varphi)\| \leq \frac{1}{4}\|\phi - \varphi\|,$$

where $H^m(\phi) = H_{m\omega}(\phi)$. This implies that H^m is a contraction mapping. Therefore, there exists a unique fixed pointϕ^* such that $H^m\phi^* = \phi^*$. So $H^m(H\phi^*) = H(H^m\phi^*) = H\phi^*$. This shows that $H^m\phi^*$ is also a fixed point of H^m, hence, $H\phi^* = \phi^*$, that is $x_\omega(\phi^*) = \phi^*$. Let $x(t, \phi^*)$ be the solution of (1) through $(0, \phi)$. By using $I(t+\omega) = I(t)$ for $t \geq 0$, $x(t+\omega, \phi^*)$ is also a solution of (1). Note that $x_{t+\omega}(\phi^*) = x_t(x_\omega(\phi^*)) = x_t(\phi^*)$ for $t \geq 0$, then $x(t+\omega, \phi^*) = x(t, \phi^*)$ for $t \geq 0$. This shows that $x(t, \phi^*)$ is a periodic solution of (1) with period ω. From (9), it is easy to see that all other solutions of (1) converge to this periodic solution exponentially as $t \to \infty$.

Remark 1. When $I = (I_1, I_2, \ldots, I_n)$ is a constant vector, then a unique periodic solution becomes a periodic solution with any positive constant as its period. So, the periodic solution reduced to a constant solution, that is, an equilibrium point. Furthermore, all other solutions globally exponentially converge to this equilibrium point. Obviously, the results obtained when let $\lambda_i = 1, i = 1, 2, \ldots, n$, are consistent with the exponential stability results that were recently reported in [9].

Similarly to the proof of theorem 1 ,we can proof exponential periodicity of the discrete-time neural networks:

Theorem 2. *Suppose $g \in GLC$. If there exist positive constants λ_j such that the following conditions*

$$\lambda_i d_i - \sum_{j=1}^{n} \lambda_j a_{ij} L_j - \sum_{j=1}^{n} \lambda_j b_{ij} L_j > 0, \quad i, j = 1, 2, \ldots, n,$$

hold, then for every periodic input $I(m)$, the delayed neural system (2) is exponentially periodic.

4 Illustrative Example

Consider the dynamical systems (1) or (2) where $d_1 = 10, d_2 = 3, b_{11} = b_{22} = a_{11} = a_{22} = 1, b_{21} = a_{12} = 1, b_{12} = a_{21} = 2, L_1 = 0.25, L_2 = 0.5$. There exist

$$\lambda_j > 0, j = 1, 2. \quad \frac{3}{19} \leq \frac{\lambda_1}{\lambda_2} \leq \frac{8}{3}.$$

The criteria in Theorem 1 or Theorem 2 are effective.

5 Conclusion

In this paper, we formulated a discrete-time analogue of continuous-time neural networks with continuously distributed delays and periodic input. Exponential periodicity of continuous-time and discrete-time neural networks has been investigated using Halanay-type inequality. The easily checked conditions ensuring the exponential periodicity of neural system are obtained. It is shown that the discrete-time analogue preserves the dynamical characteristics of the continuous-time neural systems. In addition, the results are applicable to neural networks with both symmetric and nonsymmetric interconnection matrices.

References

1. Bouzerman, A., Pattison, T.: Neural Network for Quadratic Optimization with Bound Constraints. IEEE Trans. Neural Networks **4** (1993) 293-303
2. Forti, M.: On Global Asymptotic Stability of a Class of Nonlinear Systems Arising in Neural Network Theory. Journal of differential equations **113** (1994) 246-264
3. Cao, J.: On Exponential Stability and Periodic Solutions of Delayed Celluler Neural Networks. Science in China **30** (2000) 541-549
4. Liao, X., Wang, J.: Algebraic Criteria for Global Exponential Stability of Cellular Neural Networks with Multiple Time Delays. IEEE Trans. Circuits Systems **I 50** (2003) 268-275
5. Sun, C., Feng, C.: Exponential Periodicity of Continuous-Time and Discrete-Time Neural Networks with Delays. Neural Processing Letters **19** (2004) 131-146
6. Sun, C., Feng, C.: On Robust Exponential Periodicity of Interval Neural Networks with Delays. Neural Processing Letters **20** (2004) 32-37
7. Sun, C., Feng, C.: Exponential Periodicity and Stability of Delayed Neural Networks. Mathematics and Computers in Simulation **66** (2004) 469-478
8. Mohamad, S., Gopalsamy, K.: Dynamics of a Class of Discrete-Time Neural Networks and Their Continuous-Time Counterparts. Mathematics and Computers in simulation **53** (2000) 1-39
9. Mohamad, S., Gopalsamy, K.: Exponential Stability of Contiuous-Time and Discrete-Time Cellular Neural Networks with Delays. Applied Mathematics and Computation **135** (2003) 17-38

Dynamic Analysis of a Novel Artificial Neural Oscillator

Daibing Zhang, Dewen Hu, Lincheng Shen, and Haibin Xie

College of Mechatronic Engineering and Automation,
National University of Defense Technology, Changsha, P.R. China
swimsword@163.com

Abstract. This paper proposes a novel artificial neural oscillator consisted of two neurons with excellent control properties. The mutual connections between the neurons are just linear functions and determine the oscillation angular frequency. And each neuron has a nonlinear self-feedback connection to hold up oscillation amplitude. The dynamics of the neural oscillator was modelled with nonlinear coupling functions. And the stability, amplitude, angular frequency of the oscillator are determined independently by three parameters of the functions. Since it has simple structure and favorable control advantages, it can be used in bionic robot's locomotion control system. The first application is an artificial central pattern generator (CPG) controller for bionic robot's joint. The second is a bionic neural network for fish-robot's locomotion control.

1 Introduction

Oscillation is universal behavior and almost exists in all biological neural systems, such as control system of respiration, heartbeat, run, walk, swimming. Lots of experimental results indicated that neural oscillation plays a major role in locomotion, sensory and memory [1]. The biological tissues which produce oscillation are called "neural oscillator" or "neuronal oscillator". They often consist of many neurons which have complicated nonlinear connections, and cannot be modeled with accurate mathematic functions. An important thought is use a few artificial neurons to construct a bionic neural oscillator which has similar properties with the biologic oscillator. It can be used not only to analysis the principles of biologic oscillator, but also to design some bionic neural control system for bionic robots [2,3,4].

The pioneer neural oscillator was proposed by Wilson and Cowan in 1972 [5]. It consists of one excitatory neuron and one inhibitory neuron. Since the oscillatory status have no direct mapping with the dynamic functions parameters, the modulation of oscillatory status parameters depends on difficult manual tests [6]. The second neural oscillator consists of four neurons was proposed in 1987 by Matsuoka [7]. The added two neurons are used to simulate the fatigue property of biologic neurons. Matsuoka oscillator has been widely adopted in locomotion control systems of bionic robots. Its oscillatory parameters are also modulated

D. Liu et al. (Eds.): ISNN 2007, Part I, LNCS 4491, pp. 1061–1068, 2007.

hardly for its complicated nonlinear dynamics. So our purpose in this paper is design a novel artificial neural oscillator which consists of just two neurons and has excellent control properties.

2 Design of Novel Artificial Neural Oscillator

The novel artificial neural oscillator was designed as shown in Fig.1. It consists of one excitatory neuron u and one inhibitory neuron v. The mutual connections are linear and continuous to simplify the dynamics of oscillation. The self-feedback connections are special designed to form and sustain of steady oscillation.

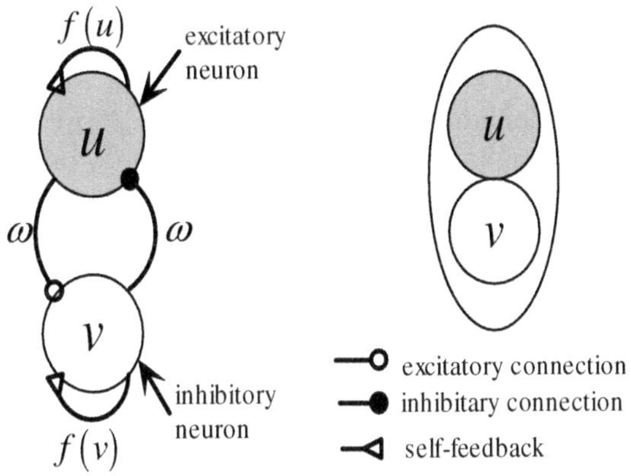

Fig. 1. Model of the novel artificial neural oscillator

The dynamics of the oscillator is described as following equations:

$$\begin{cases} \dot{u} = -\omega v + f(u), \\ \dot{v} = \omega u + f(v), \end{cases} \tag{1}$$

where u denotes the activity of excitatory neuron, v denotes the activity of inhibitory neuron, ω is the strength of mutual connections, and function f represents the self-feedback property. If the self-feedback function is a linear function, the oscillator becomes to a typical two dimensional linear system and cannot produces oscillation. Once the self-feedback doesn't exist, the oscillator becomes to the original sine-cosine oscillatory model. The status parameters such as amplitude, frequency and phase are absolutely determined by the initial conditions. It has an integral effect of the external disturbing noise and cannot forms steady oscillation. The nonlinear function we chose is described as following equation:

$$f(x) = kr[-\frac{x}{r} + \frac{4}{\pi}\tan^{-1}(\frac{x}{r})], \tag{2}$$

where r denotes the oscillatory amplitude control parameter, k denotes the convergence speed to limit cycles. The nonlinear function output is shown in Fig.2(a).

Since the output of nonlinear function is connect with each neuron, it has a obvious effect to keep the oscillatory amplitude $A = |x|$ around the oscillatory control parameter r. It is described as following equations:

$$\begin{cases} \dot{A} < 0 \ , |x| > r, \\ \dot{A} = 0 \ , |x| = r \ or \ x = 0, \\ \dot{A} > 0 \ , 0 < |x| < r, \end{cases} \tag{3}$$

The complete dynamic model of the oscillator is described as follows:

$$\begin{cases} \dot{u} = -\omega v + kr[-\frac{u}{r} + \frac{4}{\pi}\tan^{-1}(\frac{u}{r})], \\ \dot{v} = \omega u + kr[-\frac{v}{r} + \frac{4}{\pi}\tan^{-1}(\frac{v}{r})]. \end{cases} \tag{4}$$

The curves of oscillator in phase plane is shown in Fig.2(b), where $k = 10$, $r = 1$, $\omega = 2\pi$. The default restrictions of the oscillator are $\omega \geq 0$, $k \geq 0$, $r > 0$. The oscillatory angular frequency is mainly determined by ω. All points of the phase plane except the original point convergence to the limit cycles in anticlockwise direction. Even a small disturbing input will induce the oscillator leave the origin point and fall down to the limit cycles, so the original point is also an unsteady balance point.

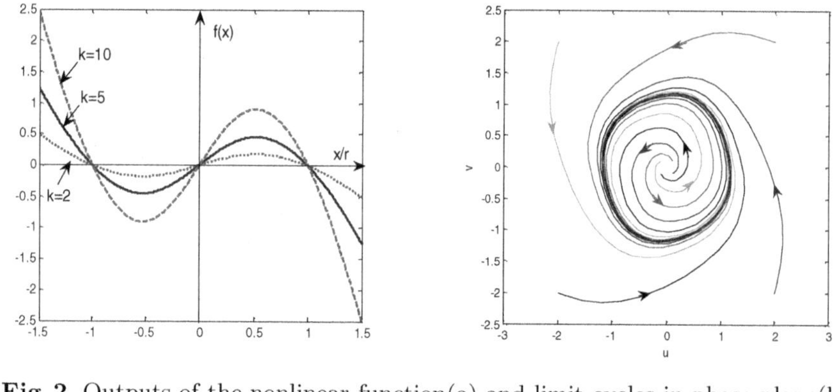

Fig. 2. Outputs of the nonlinear function(a) and limit cycles in phase plane(b)

3 Dynamic Analysis of the Oscillator

3.1 Stability Analysis of the Oscillation

To convenient for analysis the stability of oscillation, we convert the dynamic functions into polar coordinates as following equations:

$$\begin{cases} \dot{\rho} = k[-\rho + \frac{4r}{\pi}[tan^{-1}\frac{\rho\sin\theta}{r}\sin\theta + \tan^{-1}\frac{\rho\cos\theta}{r}\cos\theta]], \\ \dot{\theta} = \omega + \frac{4kr}{\pi\rho}[\tan^{-1}(\frac{\rho\sin\theta}{r})\cos\theta - \tan^{-1}(\frac{\rho\cos\theta}{r})\sin\theta], \end{cases} \tag{5}$$

where ρ is polar radius, θ is the polar angle. Assumed $D(\rho,\theta)|\rho \leq 3r)$ is a ring area in phase plane, any points on its boundary curve $C(\rho,\theta)|\rho = 3r)$ has a pole radius derivative which described as follows:

$$\dot{\rho}_C = kr[-3 + \frac{4}{\pi}[\tan^{-1}(3\sin\theta) + \tan^{-1}(3\cos\theta)\cos\theta]] < kr(-3 + 2\sqrt{2}) < 0. \quad (6)$$

It means that all curves pass through the boundary from the outside to inside, and there is no steady balance point inside the boundary. From the second Bendixon theorem, we proved that steady limit cycles in the ring area is certainly exist. Since there is one sole unsteady balance point "origin point", its must be the sole limit cycles in the whole phase plane.

3.2 Amplitude Analysis of Oscillation

The oscillatory amplitude indicates the oscillator's oscillatory energy intensity. It cannot be deduced from the dynamic functions or limit cycles in phase plane. Since the shape of limit cycles is abnormity and close to circle, the integral of polar radius derivative along a certain circle which inside of the limit cycles is positive, and which outside of the limit cycles is negative. The integral of polar radius derivative along different radius circles on the phase plane is shown in Fig.3(a). We supposed the average oscillatory amplitude is $\bar{R} = ar$, and the integral of polar radius derivative along the circle $C_\rho = \bar{R}$is zero. It is described as following equation:

$$\oint_{C(\rho,\theta)|\rho=ar} \dot{\rho}d\theta = 0. \quad (7)$$

The solution of former equation is obtained by computational mathematic method and described as follows:

$$a \approx 1.18, \bar{R} \approx 1.18r, r \approx 0.8475\bar{R}. \quad (8)$$

It is obvious that the oscillatory amplitude can be modulated as any value by change the parameter r. And it accords with the phenomenon that oscillatory amplitude almost holds up geometric proportion with amplitude control parameter which we observed in simulation of Matlab/Simulink. So we can rewrite the dynamics of the oscillator as following functions:

$$\begin{cases} \dot{u} = -\omega v + 0.8475kr[-\frac{u}{0.8475r} + \frac{4}{\pi}\tan^{-1}(\frac{u}{0.8475r})], \\ \dot{v} = \omega u + 0.8475kr[-\frac{v}{0.8475r} + \frac{4}{\pi}\tan^{-1}(\frac{v}{0.8475r})]. \end{cases} \quad (9)$$

Then the actual oscillatory amplitude is equal to the amplitude control parameter r. So it has a clearer physical meaning than function (4).

3.3 Frequency Analysis of Oscillation

It is observed from function (5) that the real-time oscillatory frequency is composed of constant item of ω and period item of phase angle θ. The oscillatory

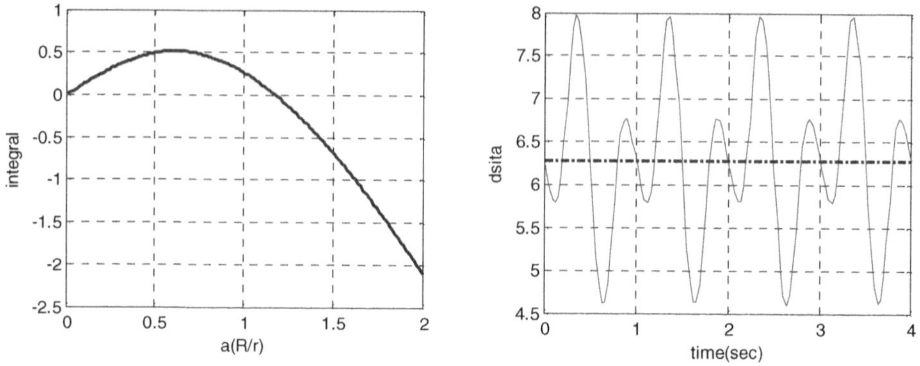

Fig. 3. Integral of radius derivative(a) and change of oscillatory frequency(b)

frequency change is shown in Fig.3(b), where $k = 2$, $\omega = 2\pi$ and $r = 1$. The oscillatory frequency periodically changes with frequency which twice of oscillatory frequency. An appropriate restriction of convergence speed parameter k is needed to limit the change of oscillatory frequency in allowable range.

4 Application of the Novel Neural Oscillator

4.1 Artificial Central Pattern Generator Controller for Bionic Robot's Joint

The first application of the oscillator was to be the core module of an artificial central pattern generator (CPG) for bionic robot's joint which in shown in Fig.4 [8]. On the left side of the figure, there are three control parameters include of expect swing amplitude, swing angular frequency and trigger signal of startup or stop. In the middle of the figure, there is an oscillator which analyzed in this paper to generate rhythmic pattern signals. On the right side, there is a dynamic model of bionic robot's joint which connects with an artificial neural estimator. The frequency modulation properties of the CPG control were investigated by several computational simulations. It is shown that the CPG controller can modulate oscillation angular frequency to hold up swing amplitude with changing parameters in dynamic model.

4.2 Bionic Neural Network for Fish-Robot's Locomotion

The second application is to construct a bionic neural network for fish-robot's locomotion. Vertebrates which include most fishes control the rhythmic locomotion such as swimming, walking, running and flying by neural network called by "central pattern generators" (CPGs) [9, 10]. The evidence for CPGs governing

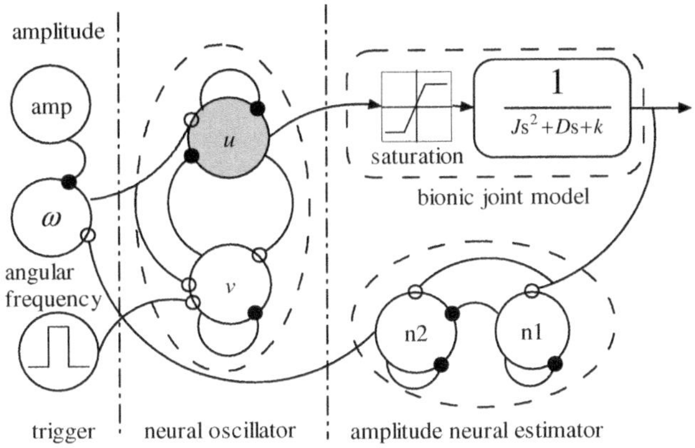

Fig. 4. CPG controller based on novel oscillator

locomotion were reviewed by Lyons [11], and the neural circuitry of lamprey has been worked out in experiments [12, 13]. The CPGs are often modelled as a chain of coupled nonlinear neural oscillators [14], and the outputs are shaped by sensory and neuro-modulatory inputs to allow the animal to adapt its locomotion to changing needs [15]. We have proposed a bionic neural network which is shown in Fig.5(a) [16]. The forward connections transfer the propulsive wave from tail to head to make fish-robot swimming backward, and the swimming status parameters are controlled by the high level controller directly.

We have also built a fish-robot which has an undulatory fin propulsor and shown in Fig.5(b). The bionic neural network control method presents better performances than conventional planning method in both startup or stop process and steady swimming process.

The bionic network presents many similar behaviors as CPGs of fish. For example, the startup process of the bionic neural network and fish-robot are shown in Fig.6.

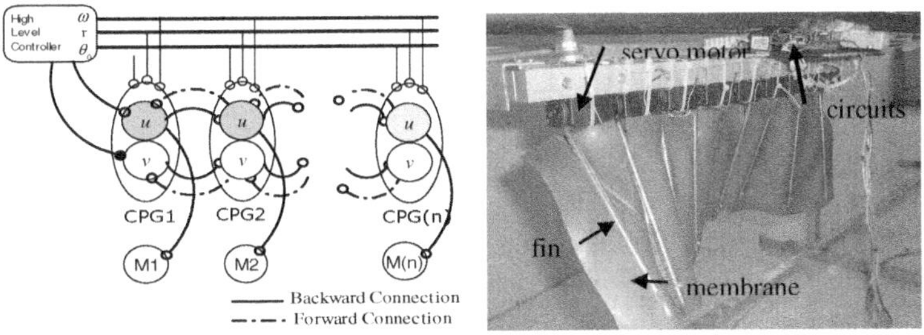

Fig. 5. A bionic neural network for fish-robot(a) and unduatory fin bionic propulsor(b)

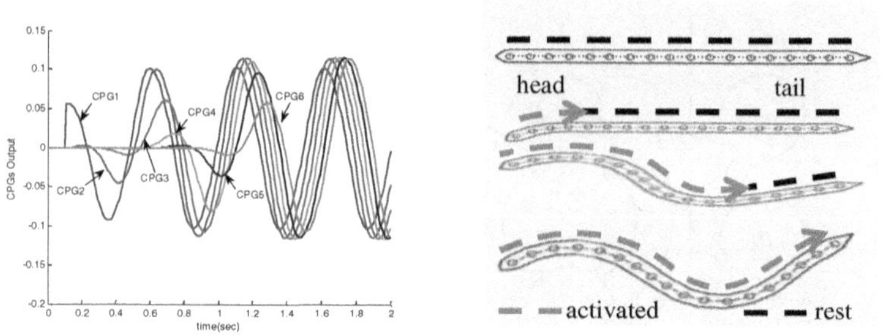

Fig. 6. Startup process of the bionic neural network(a) and the undulatory fin(b)

5 Conclusions

We present a novel artificial neural oscillator in this paper. It consists of two neurons with simple mutual connections and self-feedback connections. It has steady limit cycles in the phase plane, and the oscillatory stability, frequency, amplitude were analyzed in detail. And each of the oscillatory status parameters can be modulated individually. Despite of the essential differences with biological neural oscillator, the novel neural oscillator has more advantages in design of artificial neural control system for bionic robots. The successful applications include CPG controller in bionic robot's joint and bionic neural network for fish-robot's locomotion. In the future we will take the sensory feedback into account and use an automatic modulating circuit to optimize the locomotion.

References

1. Ermentrout, G., Chow, C.C.: Modeling Neural Oscillations. Physiology and Behavior **77** (2002) 629-633
2. Kazuki, N., Tesuya, A., Yoshihito, A.: Design of an Artificial Central Pattern Generator with Feedback Controller. Intelligent Automation and Soft Computing **10** (2004) 185-192
3. Kimura, H., Akiyama, S., Sakurama, K.: Realization of Dynamic Walking and Running of the Quadruped Using Neural Oscillator. Autonomous Robots **7** (1999) 247-258
4. Endo, G., et al.: An Empirical Exploration of a Neural Oscillator for Biped Loco-motion Control. Proceedings of the 2004 IEEE International Conference on Robotics and Automation (2004) 3036-3042
5. Wilson, H.R., Cowan, J.D.: Excitatory and Inhibitory Interactions in Localized Populations of Model Neurons. Biophys. **12** (1972) 1-24
6. Ueta, T., Chen, G.: On Synchronization and Control of Coupled Wilson Cowan Neural Oscillators. Int. J. Bifur. Chaos. **13** (2003) 163-175
7. Matsuoka, K.: Mechanisms of Frequency and Pattern Control in the Neural Rhythm Generators. Biolog. Cybern. **56** (1987) 345-353
8. Zhang, D.B., et al.: Design of a Central Pattern Generator for Bionic-robot Joint with Angular Frequency Modulation. IEEE International Conference on Robotics and Biomimetics, Kunming (2006)

9. Feldman, J.L., Grillner, S.: Control of Vertebrate Respiration and Locomotion: A Brief Account. The Physiologist **26** (1983) 310-316
10. Marder, E., Bucher, D.: Central Pattern Generators and the Control of Rhythmic Movements. Current Biology **11** (2001) 986-996
11. MacKay-Lyons, M.: Central Pattern Generation of Locomotion: A Review of the Evidence. Physical Therapy **82** (2002) 69-83
12. Grillner, S.: Neural Networks for Vertebrate Locomotion. Scientific American (1996) 64-69
13. Zhaoping, L., Lewis A., Scarpetta, S.: Mathematical Analysis and Simulations of the Neural Circuit for Locomotion in Lampreys. Physical Review Letters **92** (2004) 198106(1-4)
14. Ijspeert, A.J., Crespi, A., Cabelguen, J.M.: Simulation and Robotics Studies of Salamander Locomotion. Neuroinformatics **3** (2005) 171-196
15. Cohen, A.H., Lewis, M.A.: Sensorimotor Integration in Lampreys and Robot I: CPG Principles. Physiological Reviews **76** (1996)
16. Zhang, D.B., et al.: A Bionic Neural Network for Fish-Robot Locomotion. Journal of bionic engineering **4** (2006) 187-194

Ensembling Extreme Learning Machines

Huawei Chen[1], Huahong Chen[2], Xiaoling Nian[1], and Peipei Liu[1]

[1] School of Information Science & Technology, Southwest Jiaotong University, P.O. Box 406, Chengdu, Sichuan 610031, China
chw.mail@163.com
[2] School of Economics & Management, Southwest Jiaotong University, Chengdu, Sichuan 610031, China
huahongchen@126.com

Abstract. Extreme learning machine (ELM) is a novel learning algorithm much faster than the traditional gradient-based learning algorithms for single-hidden-layer feedforward neural networks (SLFNs). Neural network ensemble is a learning paradigm where several neural networks are jointly used to solve a problem. In our work, we investigated the performance of ELMs ensemble on regression problems. A simple ensembling approach Product Index based Excluding ensemble(PIEx) was proposed to ensemble accurate and diverse member networks. The experimental results show that the ensemble can effectively improve the performance compared with the generalization ability of single ELM and PIEx outperforms Bagging and Simple Averaging. The results also show ELM training can generate diverse neural networks even though using the same training set.

1 Introduction

In past years neural network ensembles have been used increasingly to improve neural networks' generalization ability [1-4]. Ensemble algorithms firstly train multiple individual neural networks and then combine their outputs, so ensemble means that a group of neural networks jointly solve a problem. Since neural network ensembles offer a number of advantages over a single neural network system, neural networks ensemble has been a hot topic and effective method in application fields.

Extreme learning machine (ELM) is a new learning scheme for single hidden layer feedforward neural networks (SLFNs) [5-9]. In ELM, the output weights (connecting the hidden layer to the output layer) are directly calculated by using Moore–Penrose generalized inverse, while the input weights (connecting the input layer to the hidden layer) and hidden biases can be randomly chosen. It's interesting and surprising that ELM can learn very fast as well as achieve good generalization performance.

In this paper, ELM learning is used to train individual networks and then the ELMs are grouped to constitute the ensemble. The rest of this paper is organized as follows: Section II briefly introduces ELM. Section III overviews the neural network ensemble and the bias-variance trade-off. Section IV proposes a simple ensemble principle. Section V presents the experimental results on some regression problems and some discussions. Finally, Section VI concludes with a summary of the paper and a few remarks.

D. Liu et al. (Eds.): ISNN 2007, Part I, LNCS 4491, pp. 1069–1076, 2007.

2 Extreme Learning Machine

Extreme learning machine (ELM) was proposed by Huang et al [5-7]. Suppose learning N arbitrary distinct samples (x_i, t_i), where $x_i = [x_{i1}, x_{i2}, ...x_{in}]^T \in R^n$, and $t_i = [t_{i1}, t_{i2}, ...t_{im}]^T \in R^m$, standard SLFNs with $\tilde{N}$ hidden neurons and activation function g(x) are mathematically modeled as a linear system

$$\sum_{i=1}^{\tilde{N}} \beta_i g(w_i \bullet x_j + b_i) = o_j, \quad j = 1, ..., N, \tag{1}$$

where $w_i = [w_{i1}, w_{i2}, ...w_{in}]^T$ is the weight vector linking the ith hidden neuron and the input neurons, $\beta_i = [\beta_{i1}, \beta_{i2}, ...\beta_{im}]^T$ is the weight vector linking the ith hidden neuron and output neurons, and b_i is the threshold of the ith hidden neuron. $w_i * x_j$ denotes the inner product of w_i and x_j.

If the SLFN with N hidden neurons with activation function g(x) can approximate N distinct samples(x_i, t_i) with zero error means that

$$H\beta = T, \tag{2}$$

H is the hidden layer output matrix of the neural network. So for fixed arbitrary input weights w_i and the hidden layer biases b_i, to train a SLFN is simply equivalent to finding a least-squares solution $\hat{\beta}$ of the linear system $H\beta = T$. $\hat{\beta} = H^{+}T$ is the best weights, where H^+ is the Moore-Penrose(MP) generalized inverse. As pointed out by Huang, ELM using such MP inverse method to obtain good generalization performance with dramatically fast learning speed. Unlike traditional approaches, such as standard BP algorithm and its variants, ELM avoids difficulties in tuning control parameters (learning rate, learning epochs, etc) and sticking to local minima.

3 Neural Network Ensemble

Since there is no rigorous theoretical framework for neural computing now, whether a neural network based application will be successful or not is almost fully determined by user's experience or trials. Sometimes building an appropriate neural network for real work is more like handcraft than science. However ensemble approach can lessens this trouble.

Hansen and Salamon showed that the generalization ability of a neural network system can be significantly improved through ensembling neural networks [1]. Afterward many works have been done in investigating why and how neural network ensemble works well.

Suppose $x \in R^m$ is randomly sampled according to a distribution p(x). The expected output of x is d(x). The actual output of the i-th individual neural network is f_i (x). Then the output of the neural network ensemble is:

$$f(x) = \frac{1}{N}\sum_{i=1}^{N} f_i(x). \tag{3}$$

The generalization error $E_i(x)$ of the i-th individual network on input x and the generalization error E(x) of the ensemble on input x are respectively:

$$E_i(x) = (f_i(x) - d(x))^2, \tag{4}$$

$$E(x) = (f(x) - d(x))^2. \tag{5}$$

The diversity term or ambiguity term $A_i(x)$ between the i-th individual network on input x and the ensemble on input x is:

$$A_i(x) = (f_i(x) - f(x))^2. \tag{6}$$

So average generalization ability $\overline{E}$ of all individual network over all input data and the average ambiguity $\overline{A}$ of all individual network over all input data are shown below:

$$\overline{E} = \frac{1}{N}\sum_{i=1}^{N} \int dx p(x) E_i(x), \tag{7}$$

$$\overline{A} = \frac{1}{N}\sum_{i=1}^{N} \int dx p(x) A_i(x). \tag{8}$$

After those definitions above, the famous equation was derived by Krogh and Vedels[2].

$$E = \overline{E} - \overline{A}. \tag{9}$$

The equation demonstrates that the generalization ability of the ensemble E is determined by the average generalization ability $\overline{E}$ and the average ambiguity $\overline{A}$ of the individual neural networks that constitutes the ensemble.

According the theory of the ensemble, generating and selecting accurate and diverse members of a neural-network ensemble is most important to improve generalization ability. Many algorithms were proposed to train the individual and ensemble the members. Bagging based on bootstrap sampling is a common technique to create diverse learners in machine learning. Negatively correlation (NC) learning was proposed to simultaneously train the member networks in an ensemble [3]. A genetic algorithm based approach named GASEN (Genetic Algorithm based Selective ENsemble) is proposed by Zhou [4], which employs genetic algorithm to select an optimum set of individuals.

Although neural networks ensemble is a good idea for application, training many neural networks requires long time and computational cost. Due to good attributes of ELM learning, it is suitable to use for training the individual networks. But ELM learning only trains the networks once, negatively correlation learning can't be applied on the ELMs. As analyzed by Zhou, selecting some individuals is always better than selecting all individuals in the ensemble [4]. How to ensemble the ELM individuals or select appropriate members from trained ELM candidates is the important method to control the performance of the ensemble. To be noted, good accuracy obtained from the individual member is necessary to better the performance of the ensemble in practice.

4 A Simple Selecting Principle for Ensembling

Although GASEN has preferable performance in generating ensembles with strong generalization ability, it needs added gene coding and genetic algorithm operation. As mentioned above, generalization ability of the ensemble depends on average generalization ability and the average ambiguity of the individual neural networks that constitutes the ensemble.

Considering that the neural networks are working in finite discrete data set in real applications, the alternative version of equation (9) is presented here

$$\begin{aligned}\overline{E} &= \frac{1}{N}\sum_{i=1}^{N}(E_i - A_i)\\ &= \frac{1}{N}\sum_{i=1}^{N}\Big(\frac{1}{L}\sum_{n=1}^{L}(f_i(n)-d(n))^2 - \frac{1}{L}\Big(\sum_{n=1}^{L}(f_i(n)-f(n))^2\Big)\Big).\end{aligned} \tag{10}$$

where L is the size of the data set used here. $f_i(n)$ is n-th output of the individual network f_i, $d(n)$ is the n-th expected output; $f(n)$ represents n-th output of the ensemble system.

Since the first term E_i is constant for each network after the individual networks have been trained, the performance of the ensemble is determined by the second term. Greater the diversity, the better generalization ability the ensemble will obtain. For an individual, small training error and big diversity can help the ensemble better the performance. Based on this understanding, we present a simple principle for how to choose the individual as the member of the ensemble -- an individual with great training error and small diversity should be excluded from the ensemble.

Pearson correlation coefficients are used here to measure the diversity between outputs of the individual f_i and outputs of the ensemble system f

$$\rho_i = \frac{\sum_{n=1}^{L}(f_i(n)-\overline{f_i})(f(n)-\overline{f})}{\sqrt{\sum_{n=1}^{L}(f_i(n)-\overline{f_i})^2\sum_{n=1}^{L}(f(n)-\overline{f})^2}}, \tag{11}$$

where $\bar{f_i}$ is the mean value of all output samples from f_i, $\bar{f}$ is the mean value of all output samples from ensemble system.

After the measurement of the diversity, a product index is defined as below

$$P_i = mse_i * \rho_i, \tag{12}$$

where mse_i is the mean square error of the individual f_i, ρ_i measures the diversity between the individual f_i and the ensemble f. So greater mse_i and greater ρ_i lead to greater P_i.

Therefore a guideline for ensemble is described here, if the P_i of the individual f_i is greater than the mean value of all P_i, the individual f_i will be excluded from the ensemble members. Because P_i is the product value and this method actually excludes some individuals from the ensemble, so this method is called Product Index based Excluding ensemble (PIEx).

5 Experiments for Ensembling ELMs

In this section, the performance of the Bagging, Simple Averaging, PIEx used to ensemble ELMs are tested on some regression problems. Firstly, two different methods to generate individual ELM networks are used in the experiments, one is bootstrap sampling from the training set to form the individual training sets (Bagging), another way is using the same whole training set for each individual training. Two methods are also used to combine the member into the ensemble, one is averaging all individuals, the other is Product Index based Excluding ensemble (PIEx). Simple Averaging here refers to averaging all the individuals.

5.1 Data Set

Four regression problems have been used to compare the performance the ensemble approaches as mentioned above. The data sets function and the constraints on the variables are shown in Table 1, where U[x,y] means a uniform distribution over the interval determined by x and y.

Table 1. Data sets

Data set	function	variable
Friedman #1	$y = 10\sin(\pi x_1 x_2) + 20(x_3 - 0.5)^2 + 10x_4 + 5x_5 + \varepsilon$	$x_i \sim U[0,1]$
SinC	$y = \frac{\sin(x)}{x} + \varepsilon$	$x_i \sim U[-10,10]$
Multi	$y = 0.79 + 1.27x_1x_2 + 1.56x_1x_4 + 3.42x_2x_5 + 2.06x_3x_4x_5 + \varepsilon$	$x_i \sim U[0,1]$
Gabor	$y = \frac{\pi}{2}\exp[-2(x_1^2 + x_2^2)]\cos[2\pi(x_1 + x_2)] + \varepsilon$	$x_i \sim U[0,1]$

Note that in our experiments ε is a normal noise term that has been added to the functions.

Table 2 shows the size of training set and testing set as well as number of neurons used in ELM for each data set. Due to ELM learns the samples only once, no validate sets used here to avoid over-fitting or over-learning.

Table 2. Size of Data set and design of ELM

Data set	Size		no. of neurons used in ELM
	Training set(plus noise)	Testing set(noise-free)	
Friedman #1	5000	1000	20
SinC	5000	5000	10
Multi	4000	1000	15
Gabor	3000	1000	15

5.2 Results

Twenty ELMs are trained whatever method used to ensemble. The mean generalization results over 20 trials are tabulated in Table 3. Individual means the mean MSEs of the individual networks trained, which represents the generality ability of the individual networks. To compare the relative performance of these approaches clearly, all the relative performances have been normalized according to the individual errors. The comparison results are shown in Fig.1.

Table 3. Generalization performance of the ensemble approaches over the testing sets

Data set	Mean Square Error(MSE)				
	Bagging	No Bootstrap sampling + Simple Averaging	PIEx		Individual
			Bootstrap sampling	No Bootstrap sampling	
Friedman #1	2.6831	2.6703	2.2139	2.1495	3.7218
SinC	0.0178	0.0174	0.0128	0.0124	0.0260
Multi	0.1798	0.1727	0.1449	0.1393	0.2460
Gabor	0.1660	0.1655	0.1496	0.1481	0.1841

Fig. 1 and Table 3 show that all the ensemble approaches are consistently better than the individual neural network in tests.

To be noted, although Bagging is considered as the effective ensemble technique, Fig. 1 and Table 3 show that there is no evident difference on the generalization ability between using bootstrap sampling and not using bootstrap sampling in ELM training. On the contrary no bootstrap sampling outperforms Bagging slightly. These results give us the assumption that initial weights can play an important role in creating diversity among individual neural networks although they were thought as least effective way. Traditionally neural networks are trained by gradient-descent algorithms, so

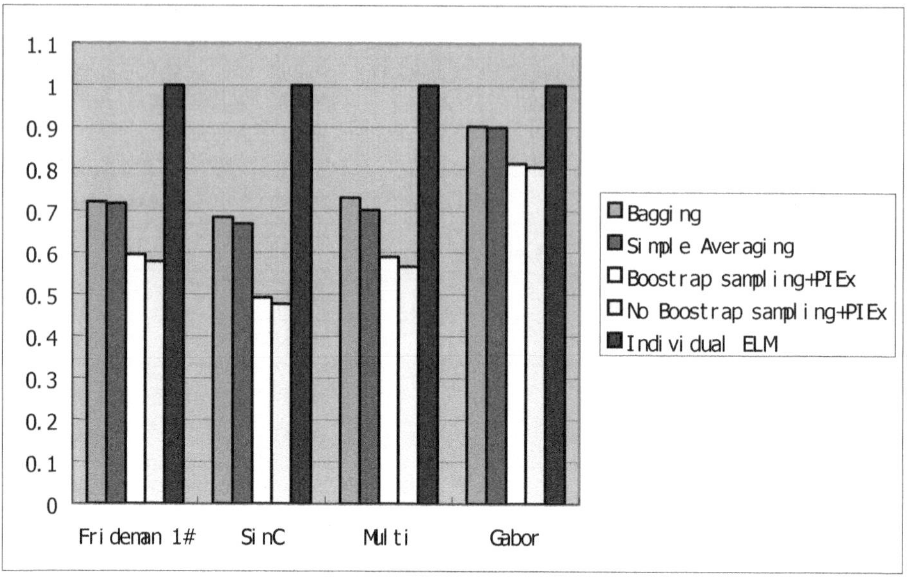

Fig. 1. Comparison of the relative error of the ensemble approaches tested on the data sets

they maybe converge to the same local minima or near to the same local minima despite different weights initialization in neural networks. So Bagging is effective because it changes the training set for each neural networks to obtain the diversity by avoiding the same local minima. While using ELM, this benefit didn't occur. ELM training computes only once, the output weights are determined by hidden layer output matrix that will be changed if input weights vary, therefore the diversity is easily generated by ELMs even though the same training set is used for each ELM.

Fig. 1 and Table 3 also show that PIEx is superior to both Bagging and Simple Averaging in all tests, which supports the conclusion that our ensembling approach is a simple but good method for selecting accurate and diverse members. According to PIEx, the half of member candidates will be included into the ensemble statistically.

6 Conclusion

This paper investigated ensembling ELMs to get better performance than single ELM. A simple ensembling approach PIEx was proposed to ensemble accurate and diverse member networks. The experimental results show that Bagging, Simple Averaging and PIEx improve the performance over all tests compared with the generalization ability of single ELM. It is noticeable that ELM training can generate effective diversity even though using the same training set. This benefit adds the good to use ELM learning to train the candidates. PIEx outperforms Bagging and Simple Averaging, which supports its simplicity and effectiveness.

Although ensembling ELMs has many benefits, it's obvious that the number of hidden nodes in ELM affects the learning capability and generalization ability of the ELM neural network. How to determine the appropriate number of hidden nodes still need more study. The data sets in this paper are restricted within the regression problems, further works will be done on classification tasks.

Acknowledgements

We would like to thank Dr. Huang for providing free ELM source codes on the website and relevant papers.

References

1. Hansen, L.K., Salamon, P.: Neural Network Ensembles. IEEE Trans. Pattern Analysis and Machine Intelligence **12** (1990) 993-1001
2. Krogh, A., Vedelsby, J.: Neural Network Ensembles, Cross Validation, and Active Learning. In: Advances in Neural Information Processing Systems **7** (1995) 231-238
3. Liu, Y., Yao, X.: Ensemble Learning via Negative Correlation. Neural Networks **12** (1999) 1399-1404
4. Zhou, Z. H., Wu, J., Tang, W.: Ensembling Neural Networks: Many Could be Better Than All. Artificial Intelligence **137** (2002) 239-263
5. Huang, G.B., Zhu, Q.Y., Siew, C.K.: Extreme Learning Machine: A New Learning Scheme of Feedforward Neural Networks. In: Proceedings 2004 International Joint Conference on Neural Networks **2** (2004) 985-990
6. Huang, G.B., Zhu, Q.Y., Siew, C.K.: Saratchandran P., Sundararajan N.: Can Threshold Networks Be Trained Directly? IEEE Trans. Circuits and Systems **II 53** (2006) 187-191
7. Huang, G.B., Zhu, Q.Y., Siew, C.K.: Extreme Learning Machine: Theory and Applications. Neurocomputing **70** (2006) 489-501
8. Huang, G.B., Chen, L., Siew, C.K.: Universal Approximation using Incremental Constructive Feedforward Networks with Random Hidden Nodes. IEEE Trans. Neural Networks **17** (2006) 879-892
9. Huang, G.B., Chen, L., Siew, C.K.: Real-Time Learning Capability of Neural Networks. IEEE Trans. Neural Networks **17** (2006) 863-878

A Robust Online Sequential Extreme Learning Machine

Minh-Tuan T. Hoang, Hieu T. Huynh, Nguyen H. Vo, and Yonggwan Won*

Department of Computer Engineering, Chonnam National University
300 Yongbong-Dong, Buk-Gu, Kwangju 500-757 Korea
{minhtuanht, hthieu, vohanguyen}@gabriel.chonnam.ac.kr,
ykwon@chonnam.ac.kr

Abstract. Online-sequential extreme learning machine (OS-ELM) shows a good solution to online learning using extreme learning machine approach for single-hidden-layer feedforward network. However, the algorithm tends to be data-dependent, i.e. the bias values need to be adjusted depending on each particular problem. In this paper, we propose an enhancement to OS-ELM, which is referred to as robust OS-ELM (ROS-ELM). ROS-ELM has a systematic method to select the bias that allows the bias to be selected following the input weights. Hence, the proposed algorithm works well for every benchmark dataset. ROS-ELM has all the pros of OS-ELM, i.e. the capable of learning one-by-one, chunk-by-chunk with fixed or varying chunk size. Moreover, the performance of the algorithm is higher than OS-ELM and it produces a better generalization performance with benchmark datasets.

1 Introduction

Conventional feedforward neural networks have been extensively studied and have been used in many applications [1], [2], [3]. On the other hand, one main weak-point of commonly used feedforward neural network is the slow training speed. It is still far slower than required and causes a bottle-neck in using neural network in real applications. The recently proposed algorithm by Huang et al. [4], [5], called extreme learning machine (ELM) with a very fast learning speed and high accuracy, has opened a new stage in development and applications of neural networks. ELM has been extensively tested with various benchmark dataset. The results have proved the dominant of ELM, as compared with other gradient-descent-based learning algorithms, not only in generalization performance at higher learning speed but also in the high accuracy.

Like most of conventional feedforward neural networks, ELM is only suitable for batch training in which the applications' training data sets are all available. In cases where the training data sets are large or the boundary decision is complicated, the ELM algorithm requires much high load in computing the pseudo-inverse of the weight matrix between the hidden layer and the output layer of the neural network.

* To whom all correspondences should be sent.

D. Liu et al. (Eds.): ISNN 2007, Part I, LNCS 4491, pp. 1077–1086, 2007.

For further study of the ELM algorithm, Liang et al. [6] has proposed a sequential learning algorithm for single-hidden layer feedforward neural networks which is referred to as online sequential extreme learning machine (OS-ELM). OS-ELM proves to be good in case of online learning with different sizes of new training data sets, i.e. training data can be one-by-one or chunk-by-chunk (with fixed or varying size.) And it discards the data for which the training has already been done.

In OS-ELM, it has most of the strong-points of ELM, and also has some drawbacks. We leave out the lower training speed as compared to ELM because this is obvious in online learning. As in Liang et al [6], the bias values need to be carefully adjusted so that the matrix multiplication between the transpose of the hidden layer output matrix and the hidden layer output matrix not to be singular. This causes the OS-ELM not to be a good generalization algorithm.

In this paper, we propose a method for selection of the bias values in accordance with the weights. The enhanced algorithm, referred to as robust online sequential extreme learning machine (ROS-ELM) has been extensively tested with various benchmark dataset and it produces not only good generalization but also higher testing results with less variance as compared to OS-ELM.

The paper is organized as follows. Section 2 is a brief review of ELM algorithm. Section 3 is the detailed description of the ROS-ELM. Then, the performance evaluation is described in section 4. Finally, section 5 is the conclusion.

2 Review of ELM

2.1 Single-Hidden Layer Feedforward Neural Network (SLFN)

Assume we have N distinct samples $(\mathbf{x}_i, \mathbf{t}_i)$, $\mathbf{x}_i = [x_{i1}, x_{i2}, ..., x_{in}]^T \in \mathbb{R}^N$ and $\mathbf{t}_i = [t_{i1}, t_{i2}, ..., t_{im}]^T$, where $\mathbf{x}_i$ is an input vector and $\mathbf{t}_i$ is the corresponding desired output vector. An output value of standard single-layer feedforward neural networks (SLFN) with $\tilde{N}$ hidden units and activation function g(x) can be represented by

$$f_{\tilde{N}}(\mathbf{x}_j) = \sum_{i=1}^{\tilde{N}} \boldsymbol{\beta}_i g(\mathbf{w}_i \mathbf{x}_j + b_i) = \mathbf{o}_j \ . \quad (1)$$

where β_i, $\mathbf{w}_i$ and b_i are the hidden-to-output weights, the input-to-hidden weights and bias values, respectively. The activation functions are sigmoids, and the output neurons are chosen linear in this paper.

The ultimate purpose of SLFNs is to find out the values of β, $\mathbf{w}_i$ and b_i such that $\sum_{j=1}^{\tilde{N}} \|\mathbf{o}_j - \mathbf{t}_j\| = 0$, or

$$\sum_{i=1}^{\tilde{N}} \boldsymbol{\beta}_i g(\mathbf{w}_i \mathbf{x}_j + b_i) = \mathbf{t}_j \ . \quad (2)$$

The compact form of the above $\tilde{N}$ equations is

$$\mathbf{H}.\boldsymbol{\beta} = \mathbf{T} \ . \quad (3)$$

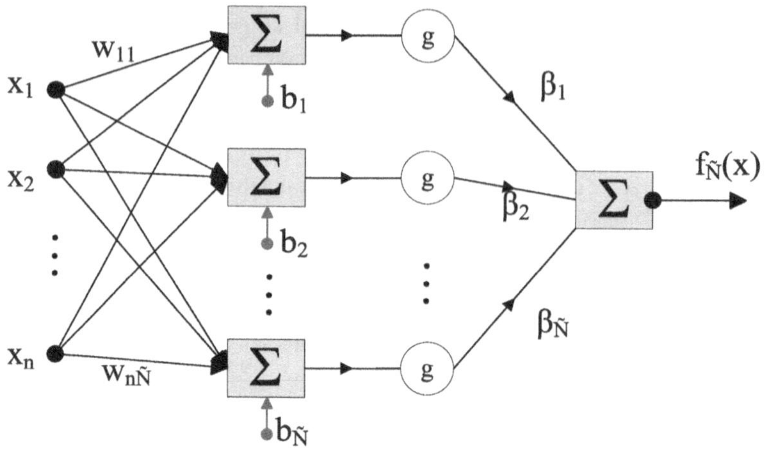

Fig. 1. A sample SLFN with n-Ñ-m network structure

where

$$\underset{N\times\tilde{N}}{\mathbf{H}} = \begin{bmatrix} g\left(\mathbf{w}_1.\mathbf{x}_1 + b_1\right) & \cdots & g\left(\mathbf{w}_{\tilde{N}}.\mathbf{x}_1 + b_{\tilde{N}}\right) \\ \vdots & \cdots & \vdots \\ g\left(\mathbf{w}_1.\mathbf{x}_N + b_1\right) & \cdots & g\left(\mathbf{w}_{\tilde{N}}.\mathbf{x}_N + b_{\tilde{N}}\right) \end{bmatrix} . \tag{4}$$

$$\underset{\tilde{N}\times m}{\boldsymbol{\beta}} = \begin{bmatrix} \boldsymbol{\beta}_1^T \\ \vdots \\ \boldsymbol{\beta}_{\tilde{N}}^T \end{bmatrix} \text{ and } \underset{N\times m}{\mathbf{T}} = \begin{bmatrix} \mathbf{t}_1^T \\ \vdots \\ \mathbf{t}_N^T \end{bmatrix} . \tag{5}$$

As named in Huang et al. [4], $\mathbf{H}$ is called the hidden layer output matrix of the neural network. The i-th column of $\mathbf{H}$ is the output vector of the hidden layer with respect to an input sample $\mathbf{x}_i$.

2.2 Extreme Learning Machine (ELM)

In this part, we skip the rigorous proof for ELM by Huang et al. [4]. The main idea of ELM is that ones do not need to adjust the hidden node parameters $\mathbf{w}_i$ and b_i, i.e. input weights and biases need not to be tuned during training and can be assigned with random values. Under this assumption, $\mathbf{H}$ is completely defined. When we have the number of the input samples N is equal to or larger than the number of hidden units Ñ, the SLFNs can be completely learned by estimating the output weights as

$$\hat{\boldsymbol{\beta}} = \mathbf{H}^{\dagger}.\mathbf{T} . \tag{6}$$

where $\mathbf{H}^{\dagger}$ is the Moore-Penrose generalized inverse [7] of the hidden layer output matrix $\mathbf{H}$.

It should be noted that in order to compute **H**, the system need to have a complete set of training data. ELM is thus a batch learning method, in a natural way of understanding. Universal approximation capability of ELM has also been analyzed in [8] and that SLFNs with randomly generated additive or RBF nodes with a widespread of piecewise continuous activation functions can universally approximate any continuous target function on any compact subspace of $\mathbb{R}^n$. Besides, in the implementations of ELM, the activation functions for additive nodes can be any bounded nonconstant piecewise continuous functions and the activation functions for RBF nodes can be any integrable piecewise continuous functions.

3 Robust Online Sequential Extreme Learning Machine (ROS-ELM)

ELM algorithm requires the whole training set to be available during learning phase. In real applications, the training data may not available at once, i.e. it may arrive one-by-one or chunk-by-chunk. This induces to a new method for sequential learning, first proposed by Liang et al. [6]. In this section, a complete review of the OS-ELM algorithm is described with its pros and cons. The idea of enhancement is motivated by Ferrari and Stengel [1], and we propose a robust-online-sequential extreme learning machine (ROS-ELM).

3.1 Online Sequential Extreme Learning Machine (OS-ELM)

The estimated output weight matrix $\hat{\boldsymbol{\beta}}$ given by 6 is a least-square solution of 3. If the number of input samples is equal to or larger than the number of hidden units ($\mathrm{N} \geq \tilde{\mathrm{N}}$), and $rank(\mathbf{H}) = \tilde{\mathrm{N}}$, the estimated matrix $\mathbf{H}^\dagger$ is given by

$$\mathbf{H}^\dagger = \left(\mathbf{H}^T\mathbf{H}\right)^{-1}\mathbf{H}^T \ . \tag{7}$$

which is called the *left pseudo-inverse* of **H** from the fact that $\mathbf{H}^T\mathbf{H} = \mathbf{I}_{\tilde{N}}$.

Substitute 7 into 6, then the equation for estimating $\hat{\boldsymbol{\beta}}$ is

$$\hat{\boldsymbol{\beta}} = \left(\mathbf{H}^T\mathbf{H}\right)^{-1}\mathbf{H}^T\mathbf{T} \ . \tag{8}$$

From this *least-square solution* to Equ. 3, the sequential learning method of the least-square solution of 8 is derived.

Step 1 (*Boosting phase*). Given a chunk of initial training set $\aleph_0 = \{(\mathbf{x}_i, \mathbf{t}_i)\}_{i=1}^{N_0}$, $\mathrm{N_0} \times \tilde{\mathrm{N}}$.

a) Assign the input weight and bias randomly within the range [-1, 1].
b) Calculate the initial hidden layer output matrix $\mathbf{H}_0$.

$$\mathbf{H}_0 = \begin{bmatrix} g\left(\mathbf{w}_1.\mathbf{x}_1 + b_1\right) & \cdots & g\left(\mathbf{w}_{\tilde{N}}.\mathbf{x}_1 + b_{\tilde{N}}\right) \\ \vdots & \cdots & \vdots \\ g\left(\mathbf{w}_1.\mathbf{x}_{N_0} + b_1\right) & \cdots & g\left(\mathbf{w}_{\tilde{N}}.\mathbf{x}_{N_0} + b_{\tilde{N}}\right) \end{bmatrix} .$$

c) Estimate the initial output weight $\boldsymbol{\beta}^{(0)} = \mathbf{P}_0\mathbf{H}_0^T\mathbf{T}_0$, where $\mathbf{P}_0 = \left(\mathbf{H}_0^T\mathbf{T}_0\right)^{-1}$ and $\mathbf{T}_0 = \left[\mathbf{t}_1, \ldots, \mathbf{t}_{N_0}\right]^T$
d) Set the index for data chunk k to zero ($k = 0$).

Step 2 (*Sequential learning phase*). For each further $(k+1)$-th chunk of new observations

$$\aleph_{k+1} = \{(\mathbf{x}_i, \mathbf{t}_i)\}_{i=(\sum_{j=1}^{k} N_j)+1}^{\sum_{j=1}^{k+1} N_j}.$$

where $N_{(k+1)}$ denotes the number of samples in the $(k+1)$-th chunk.
a) Calculate the partial hidden layer output matrix $\mathbf{H}_{k+1}$ for the $(k+1)$-th chunk, as shown below

$$\underset{N_{k+1}\times\tilde{N}}{\mathbf{H}_{k+1}} = \begin{bmatrix} g\left(\mathbf{w}_1.\mathbf{x}_{(1)} + b_1\right) & \cdots & g\left(\mathbf{w}_{\tilde{N}}.\mathbf{x}_{(1)} + b_1\right) \\ \vdots & \cdots & \vdots \\ g\left(\mathbf{w}_1.\mathbf{x}_{(N_{k+1})} + b_1\right) & \cdots & g\left(\mathbf{w}_{\tilde{N}}.\mathbf{x}_{(N_{k+1})} + b_1\right) \end{bmatrix}.$$

b) Calculate the output weight matrix $\boldsymbol{\beta}^{(k+1)}$.

$$\mathbf{P}_{k+1} = \mathbf{P}_k - \mathbf{P}_k\mathbf{H}_{k+1}^T\left(\mathbf{I} + \mathbf{H}_{k+1}\mathbf{P}_k\mathbf{H}_{k+1}^T\right)^{-1}\mathbf{H}_{k+1}\mathbf{P}_k$$
$$\boldsymbol{\beta}^{(k+1)} = \boldsymbol{\beta}^{(k)} + \mathbf{P}_{k+1}\mathbf{H}_{k+1}^T\left(\mathbf{T}_{k+1} - \mathbf{H}_{k+1}\boldsymbol{\beta}^{(k)}\right).$$

c) Set $k = k + 1$. Go to step 2)

Although, this algorithm seems perfect in theory, it has some difficulties when applied to real world applications. In some real applications, $\mathbf{H}^T\mathbf{H}$ tends to be either a singular or ill-conditioned matrix. In order to evaluate the performance of OS-ELM, Liang et al. [6] have to take care of the bias values, such as the bias values are chosen in the range [0.2, 4.2] for Satellite image, California housing, and within [3, 11] for Image segment, and within [20, 60] for DNA. The generalization characteristic of the OS-ELM algorithm is somehow reduced because of this data-dependent selection.

3.2 Robust Online Sequential Extreme Learning Machine (ROS-ELM)

The weak point of OS-ELM is that the matrix multiplication $\mathbf{H}^T\mathbf{H}$ tends to be ill-conditioned or singular when randomly selecting the input weights and bias values. Hence, before solving that problem, we solve the small problem defined as: "If the m-by-n matrix $\mathbf{H}$ is full rank, then $\mathbf{H}^T\mathbf{H}$ is invertible" with some supplement knowledge.

Definition 1. The rank of an m-by-n matrix $\mathbf{A}$ is the maximum number of linearly independent columns (or rows) of $\mathbf{A}$, denoted by $rank(\mathbf{A})$.

Statement 1. The rank of a matrix $\mathbf{A}$ equals to the rank of its transpose, $\mathbf{A}^T$.
Statement 2. If the square matrix $\mathbf{B}$ is full rank, then $\mathbf{B}$ is invertible.

From statement 1, if $\mathbf{H}$ is full rank, then $\mathbf{H}^T\mathbf{H}$ is full rank and we also have $rank(\mathbf{H}^T\mathbf{H}) = \tilde{N}$. Because $\mathbf{H}^T\mathbf{H}$ is a square matrix of size $\tilde{N}$-by-$\tilde{N}$; then, based on statement 2, $\mathbf{H}^T\mathbf{H}$ is invertible. If we have an invertible matrix $\mathbf{H}^T\mathbf{H}$, then the drawback of OS-ELM can be overcome. Hence, what we need is to find a solution to make sure $\mathbf{H}$ is full rank, i.e. input weights and bias values should not randomly selected, but in a control ways that assure $\mathbf{H}$ is full rank. A strategy for producing a well-conditioned matrix $\mathbf{H}$ as referred in [2] consists of generating the input weights and bias according to:

$$w_{ij} = c.r_{ij} \quad . \tag{9}$$

where r_{ij} is a random variable of normal distribution ($\mu = 0$, $\sigma = 1$); c is a user-defined scalar that can be adjusted to obtain the input-to-hidden weights that do not saturate the sigmoid functions.

Values of r_{ij} can be obtained from a random number generator using single seed value. The factor c is used to scale the distribution of input-to-hidden weights. The sigmoid functions come close to being saturated for inputs whose absolute values is greater than 5. Then, when the input values from the training set are normalized, the factor c should be selected in order for contributing to a smooth approximating function and producing a nonsingular $\mathbf{H}$. The input bias b_i is computed to center each sigmoid at one of the training pair $\{\mathbf{x}_i, \mathbf{t}_i\}$.

$$\mathbf{b} = -diag\left(\mathbf{X}\mathbf{W}^T\right) \quad . \tag{10}$$

The (diag) operator extracts the diagonal of its argument.

With this enhancement, the proposed algorithm, referred to as robust online sequential extreme learning machine (ROS-ELM), can overcome a drawback of OS-ELM. The proposed ROS-ELM is considered a modification of the OS-ELM in the initialization stage for weights and biases. ROS-ELM narrows down the value space of biases in such a way that guarantees the ROS-ELM not only working well with all test cases but also giving a higher accuracy and less variance as compared to the OS-ELM algorithm.

Remark 1. The matrix multiplication in the initialization phase of ROS-ELM is slower than the randomly selection of weights in the one of OS-ELM. Even though, comparison the whole training time between ROS-ELM and OS-ELM, they are almost the same.

4 Performance Evaluation of ROS-ELM

In order to compare the performance of ROS-ELM with OS-ELM, different trials have been performed with various benchmark datasets [9] which is described in

Table 1. Specifications of benchmark data sets with number of hidden nodes for testing

Dataset	Attribute	Classes	Training data	Testing data	Nodes
Auto-MPG	7	-	320	78	25
Abalone	8	-	3,000	1,177	25
Image Segment	19	7	1,500	810	180
Satellite Image	36	6	4,435	2,000	400

Table 1. The experiments are performed with two regression applications (auto-MPG, abalone) and two classification problems (image segment, satellite image).

In our experiments, all the features (input and output) of regression applications are normalized into the range [0, 1] while the input attributes of application problems are normalized into the range [-1, 1]. The weights and biases values are initialized as described in Sect. 3 of ROS-ELM; the bias values do not need to have specific ranges for a certain data. For this reason, the generalization characteristic of the ROS-ELM is better than OS-ELM. The number of hidden nodes is selected for each test case similar to the one in Liang et al. [6]. And also, fifty trials have been conducted for each test case and the average result is computed.

4.1 Regression Problems

Two regression problems, auto-MPG and abalone, were performed to compare the performance between ROS-ELM and OS-ELM. The auto-MPG problem is to predict the fuel consumption (miles/gallon) of different car models based on 7 attributes. The abalone problem is to estimate the age of abalone from the physical measurement based on 8 attributes. The data sets for training and testing were randomly selected for each trial as described in the following paragraph.

To demonstrate the performance of OS-ELM and ROS-ELM in one-by-one learning, we used 300 training samples for boosting phase and the other 20 samples for sequential learning (Auto-MPG case), and 2800 training samples for boosting phase and 200 samples for sequential learning (abalone case). For chunk-by-chunk learning, we tested with fixed chunk size of 20-by-20 and varied chunk size of 20-by-30. In which, 40 samples were used for 20-by-20 learning and 50 samples for 20-by-30 learning (Auto-MPG) and 200 samples were used for both 20-by-20 and 20-by-30 learning (abalone cases).

As observed from Table 2, the root-mean-square error (RMSE) obtained by ROS-ELM is lower than OS-ELM in the every test case. Liang et al. [6] showed that OS-ELM obtained the lowest RMSE as compared with other methods. Thus, based on this comparison results, we can say that ROS-ELM has the lowest RMSE at this time. Moreover, one does not need to adjust the bias values for different application problems anymore. This is very important because it makes ROS-ELM overcome a major weak point of OS-ELM.

Table 2. Performance comparison of OS-ELM and ROS-ELM on Regression Applications

Datasets	Learning mode	Algorithms	RMSE	
			Training	Testing
Auto-MPG	1-by-1	OS-ELM	0.082955	0.091689
		ROS-ELM	0.082419	0.090321
	20-by-20	OS-ELM	0.083818	0.089498
		ROS-ELM	0.083442	0.083744
	20-by-30	OS-ELM	0.082900	0.090005
		ROS-ELM	0.082050	0.089840
Abalone	1-by-1	OS-ELM	0.075512	0.076820
		ROS-ELM	0.075619	0.076102
	20-by-20	OS-ELM	0.075317	0.077300
		ROS-ELM	0.075817	0.075977
	20-by-30	OS-ELM	0.075245	0.077549
		ROS-ELM	0.075547	0.076814

4.2 Classification Problems

For comparison in classification, two benchmark problems of image segment and satellite image were considered for experiments. Image segmentation dataset is the set of regions in outdoor images. The aim is to recognize each region as one of seven categories based on 18 attributes. Training and testing data sets were randomly selected from the database. Satellite image dataset is the set of 3x3 (tiny) subareas of satellite images. The aim is to classify the region into six categories. The training and test sets were fixed according to [9], but the order of training sets was randomly selected for each trial.

In classification domain, we also used the same scheme for testing as in regression problems. Table 3 shows that, as comparison between OS-ELM and

Table 3. Performance comparison of OS-ELM and ROS-ELM on Classification Applications

Datasets	Learning mode	Algorithms	Accuracy	
			Training	Testing
Image Segment	1-by-1	OS-ELM	96.8160	94.3852
		ROS-ELM	97.2320	94.8519
	20-by-20	OS-ELM	96.8147	94.2198
		ROS-ELM	97.3067	94.9852
	20-by-30	OS-ELM	96.7440	94.2519
		ROS-ELM	96.9613	94.9111
Satellite Image	1-by-1	OS-ELM	91.9324	88.9170
		ROS-ELM	92.7806	89.8520
	20-by-20	OS-ELM	91.9436	88.9040
		ROS-ELM	92.7251	89.7690
	20-by-30	OS-ELM	91.9729	88.7860
		ROS-ELM	92.6011	89.6550

ROS-ELM, the accuracy in testing for ROS-ELM is higher than OS-ELM in all test cases. Because OS-ELM and ROS-ELM used the same mechanism for updating, computation time is not concerned in this case. In summary, the comparison results prove that ROS-ELM is an efficient enhancement to OS-ELM not only in generalization characteristic but also in performance. ROS-ELM makes the neural networks based on ELM approach to be more stable for solving various problems.

5 Conclusion

In this paper, an important enhancement to OS-ELM is proposed, which is referred to as ROS-ELM. It overcomes a drawback of OS-ELM, which is to select different bias values for different application data sets. The experimental results indicate that ROS-ELM not only has better generalization as compared with OS-ELM but also increases the performance in both regression and classification applications. This proves again that extreme learning machine approach is an effective evolution in artificial neural networks, particularly in single-hidden-layer feedforward neural networks.

Acknowledgement

This work was supported by grant No. RTI-04-03-03 from the Regional Technology Innovation Program of the Ministry of Commerce, Industry and Energy (MOCIE) of Korea.

References

[1] Ferrari, S., Stengel, R.F.: Smooth Function Approximation using Neural Networks. IEEE Trans. Neural Networks **16** (2005) 24-38
[2] Ferrari, S.: Algebraic and Adaptive Learning in Neural Control Systems. PhD thesis, Princeton University (2002)
[3] Huang, G.B., Chen, Y.Q., Babri, H.A.: Classification Ability of Single Hidden Layer Feedforward Neural Networks. IEEE Trans. Neural Networks **11** (2000) 799-801
[4] Huang, G.B., Zhu, Q.Y., Siew, C.K.: Extreme Learning Machine: Theory and Applications. Neurocomputing **70** (2006) 489-501
[5] Huang, G.B., Zhu, Q.Y., Siew, C.K.: Extreme Learning Machine: A New Learning Scheme of Feedforward Neural Networks. In: International Joint Conference on Neural Networks (IJCNN'2004), Budapest, Hungary (2004)
[6] Liang, N.Y., Huang, G.B., Saratchandran, P., Sundararajan, N.: A Fast and Accurate On-Line Sequential Learning Algorithm for Feedforward Networks. IEEE Trans. Neural Networks **17** (2006) 1411-1423
[7] Golub, G.H., Loan, C.F.V. In: Matrix Computation. 3rd edn. The Johns Hopkins University Press (1996) 257-258

[8] Huang, G.B., Chen, L., Siew, C.K.: Universal Approximation using Incremental Constructive Feedforward Networks with Random Hidden Nodes. IEEE Trans. Neural Networks **17** (2006) 879-892
[9] Blake, C., Merz, C.: UCI Repository of Machine Learning Databases, Online Available: http://www.ics.uci.edu/ mlearn/mlrepository.html (1998)

An Improved On-Line Sequential Learning Algorithm for Extreme Learning Machine

Bin Li[1], Jingming Wang[1], Yibin Li[2], and Yong Song[2]

[1] College of Mathematical and Physical Sciences, Shandong Institute of Light Industry, Jinan, 250353, China
ribbenlee@hotmail.com, jingmingwang98@163.com
[2] Center for Robotics, Shandong University, Jinan 250061, China
{liyb, songyong}@sdu.edu.cn

Abstract. This paper presents an efficient online sequential learning algorithm for extreme learning machine, which can learn data one by one. In this algorithm, the parameters of hidden nodes (the input weights and biases of additive nodes or the centers and impact factors of RBF nodes) are randomly selected and the output weights are analytically determined based on the sequentially arriving data. In the online sequence, the algorithm updates the output-layer weights with a Givens QR decomposition based on the orthogonalized least square algorithm. Simulations on benchmark problems demonstrate that the algorithm produces much better generalization performance than another online sequential extreme learning machine algorithm, or sometimes it has good performance than primitive extreme learning machine algorithm.

1 Introduction

In the past two decades, single hidden layer feedforward neural networks (SLFNs) have been studied thoroughly by many researchers and got many applications in the pattern classification and function approximation area. But, it may take much time to train this neural networks which is the bottleneck constraining application. Another disadvantage of this neural networks is its generalization performance, since as we all know, it is useless for a neural networks with bad generalization performance. Recently, Huang, et al [1] have proposed a new batch learning algorithm called extreme learning machine (ELM) for SLFNs which randomly chooses the input weights and the hidden nodes' biases with additive nodes or centers and impact widths with RBF kernels, and then analytically determines instead of adjusting the output weights of SLFNs. In theory, it has been shown [2] that SLFNs' input weights and hidden neurons' biases need not be adjusted during training and one may simply randomly assign values to them. The ELM has been improved by many people and got many achievements for function approximation and pattern classification area with extreme leaning speed and good generalization performance [3, 4, 5]. But the algorithms belong to the batch learning algorithms which prohibited them further application. In some online industrial applications, sequential learning algorithms may be preferred over batch learning algorithms as they do not require retraining

D. Liu et al. (Eds.): ISNN 2007, Part I, LNCS 4491, pp. 1087–1093, 2007.

whenever a new data is received. In paper [6], an online sequential extreme learning machine based on recursive least-squares (RLS) algorithm was presented, which is called OS-ELM(RLS) . In this paper, an improved online sequential extreme learning machine algorithm called OS-ELM (OLS) for OS-ELM(RLS) is introduced, which carries out sequential modification based on a Givens QR decomposition of the orthogonalized least square algorithm [7, 8, 9], this algorithm can learn the training data one-by-one and discard the training data as long as the training procedure for those data is completed. Simulations for two regression benchmark problems and two chaotic time series prediction benchmark problems demonstrate that the proposed algorithm produces better generalization performance than OS-ELM(RLS) and ELM algorithms.

This paper is organized as follows. Section 2 gives a brief review of the primitive batch ELM. Section 3 presents the derivation of the OS-ELM(OLS) in certain details. The comparison on four benchmark problems between the proposed algorithm and OS-ELM(RLS), ELM is conducted in Section 4. Discussions and conclusions are summarized in Section 5.

2 The Batch Extreme Learning Machine

In this section, we review the primitive mode ELM proposed by Huang, et al [3, 4, 5].

The output of a standard SLFNs with $\tilde{N}$ hidden nodes is

$$f_{\tilde{N}}(\mathbf{x})=\sum_{i=1}^{\tilde{N}}\beta_i G(\mathbf{w}_i,b_i,\mathbf{x}),\quad \mathbf{x}\in R^n,\mathbf{w}_i\in R^n,\beta_i\in R^m\,, \tag{1}$$

where $G(\mathbf{w}_i,b_i,\mathbf{x})$ is the output of the i th hidden node corresponding to the input $\mathbf{x}$, and $\beta_i=[\beta_{i1},\beta_{i2},\cdots,\beta_{im}]^T$ is the weight vector connecting the i th hidden node and the output nodes.

For additive hidden nodes with activation function $g(x):R\rightarrow R$, the output of the i th hidden node is

$$G(\mathbf{w}_i,b_i,\mathbf{x})=g(\mathbf{w}_i\cdot\mathbf{x}+b_i),\quad b_i\in R\,, \tag{2}$$

where $\mathbf{w}_i=[w_{i1},w_{i2},\cdots,w_{in}]^T$ is the weight vector connecting the i th hidden node and the input nodes, and b_i is the threshold of the i th hidden node.

For radial basis function (RBF) hidden nodes with activation function $g(x):R\rightarrow R$, the output of the i th hidden node is

$$G(\mathbf{w}_i,b_i,\mathbf{x})=g(b_i\|\mathbf{x}-\mathbf{w}_i\|),\quad b_i\in R^+\,, \tag{3}$$

where $\mathbf{w}_i$ and b_i are the center and impact factor of the i the RBF hidden node, R^+ is the set of all positive reals.

For N arbitrary samples $(\mathbf{x}_i,\mathbf{t}_i)$,where $\mathbf{x}_i=[x_{i1},x_{i2},\cdots,x_{in}]\in R^n$, $\mathbf{t}_i=[t_{i1},t_{i2},\cdots,t_{im}]\in R^m$, standard SLFNs with $\tilde{N}$ hidden nodes and activation function $G(\mathbf{w}_i,b_i,\mathbf{x})$ can approximate these N samples with zero error means that there exist β_i, $\mathbf{w}_i$ and b_i such that

$$\sum_{i=1}^{\tilde{N}} \beta_i G(\mathbf{w}_i, b_i, \mathbf{x}) = \mathbf{t}_j \quad j = 1, \cdots, N . \tag{4}$$

The above N equation can be written compactly as:

$$\mathbf{H}\beta = \mathbf{T} \tag{5}$$

where $\mathbf{H}(\mathbf{w}_1, \cdots, \mathbf{w}_{\tilde{N}}, b_1, \cdots, b_{\tilde{N}}, \mathbf{x}_1, \cdots, \mathbf{x}_N) = \begin{bmatrix} G(\mathbf{w}_1, b_1, \mathbf{x}_1) \cdots G(\mathbf{w}_{\tilde{N}}, b_{\tilde{N}}, \mathbf{x}_1) \\ \vdots \quad \cdots \quad \vdots \\ G(\mathbf{w}_1, b_1, \mathbf{x}_N) \cdots G(\mathbf{w}_{\tilde{N}}, b_{\tilde{N}}, \mathbf{x}_N) \end{bmatrix}_{N \times \tilde{N}}$,

$$\beta = \begin{bmatrix} \beta_1^T \\ \vdots \\ \beta_{\tilde{N}}^T \end{bmatrix}_{\tilde{N} \times m} \text{ and } \mathbf{T} = \begin{bmatrix} \mathbf{t}_1^T \\ \vdots \\ \mathbf{t}_N^T \end{bmatrix}_{N \times m} .$$

$\mathbf{H}$ is called the hidden layer output matrix of the neural networks; the i th column of $\mathbf{H}$ is the i th hidden node's output with respect to inputs $\mathbf{x}_1, \mathbf{x}_2, \cdots, \mathbf{x}_N$.

The smallest norm of the least squares solution for $\mathbf{H}\beta = \mathbf{T}$ is

$$\hat{\beta} = \mathbf{H}^+ \mathbf{T} . \tag{6}$$

where $\mathbf{H}^+$ is the Moore-Penrose generalized inverse of the hidden layer output matrix $\mathbf{H}$. Thus, the batch ELM learning algorithm can be summarized as follows:

Given a training set $\{(\mathbf{x}_i, \mathbf{t}_i) | \mathbf{x}_i \in R^n, \mathbf{t}_i \in R^m, i = 1, \cdots, N\}$, activation function $g(x)$, and hidden node number $\tilde{N}$.

step 1 Randomly assign the values for parameters $\mathbf{w}_i$ and b_i of the hidden nodes, $i = 1, \cdots, \tilde{N}$.

step 2 Calculate the hidden layer output matrix $\mathbf{H}$.

step 3 Calculate the output weight β: $\beta = \mathbf{H}^+ \mathbf{T}$, where $\mathbf{T} = [\mathbf{t}_1, \cdots, \mathbf{t}_N]^T$.

In theory this algorithm works for any infinitely differential activation function $g(x)$. Such activation functions include the sigmoidal functions as well as the radial basis, sine, cosine, exponential, and other nonregular functions [1]. The universal approximation capability of ELM also has been rigorously proved in an incremental method by Huang, et al [2].

3 Proposed OS-ELM(OLS) Algorithm

In the context of parameter adaptation, it is essential to view the equation (1) as a special case of linear regression model

$$\mathbf{t}_j = \sum_{i=1}^{\tilde{N}} \beta_i G(\mathbf{w}_i, b_i, \mathbf{x}) + e_j = \beta \varphi_j + e_j , \tag{7}$$

where t_j is the desired output, and $\beta = [\beta_1, \cdots, \beta_{\tilde{N}}]$ is a vector of the output weights, $\varphi_j = \left[G(\mathbf{w}_1, b_1, \mathbf{x}), \cdots, G(\mathbf{w}_{\tilde{N}}, b_{\tilde{N}}, \mathbf{x})\right]^T$ is known as the regression vector, and e_j is an error signal which is assumed to be zero mean and uncorrelated with the regression vector.

Instead of RLS in OS-ELM [6], we use a Givens QR decomposition based OLS algorithm to update the output weights of the neural networks [7, 8, 9]. The least square algorithm is motivated by finding the weight vector β such that the weighted sum of squared output errors up to N th presentation

$$V(N,\beta) = \sum_{j=1}^{N} \lambda^{N-j} e_j^2 = \sum_{j=1}^{N} \lambda^{N-j} \left[t_j - \beta\varphi_j\right]^2 \quad (8)$$

is minimized, where $0 < \lambda < 1$ is a forgetting factor.

By introducing $T(N) = [t(1), t(2), \cdots, t(N)]^T$, $E(N) = [e(1), e(2), \cdots, e(N)]^T$,

$$T(N) = \beta\Phi(N) + E(N), \quad (9)$$

and the cost function (8) can be reformulated as

$$V(N,\beta) = [T(N) - \beta\Phi(N)]^T \Lambda(N)[T(N) - \beta\Phi(N)], \quad (10)$$

where $\Lambda(N) = \text{diag}[\lambda^{N-1}, \lambda^{N-2}, \cdots,] = \text{diag}[\lambda\Lambda(N-1)]$. Minimizing over β results in a least-squares solution $\beta(N)$ satisfying

$$\Phi(N)^T \Lambda(N)\Phi(N)\beta(N) = \Phi(N)^T \Lambda(N)T(N). \quad (11)$$

Denote by $\Lambda^{1/2}(N)$ the Cholesky factor of $\Lambda(N)$. Then equation (11) can be rewritten as

$$(\Lambda^{1/2}(N)\Phi(N))^T \Lambda^{1/2}(N)\Phi(N)\beta(N) = (\Lambda^{1/2}(N)\Phi(N))^T \Lambda^{1/2}(N)T(N). \quad (12)$$

Using QR decomposition

$$\Lambda^{1/2}(N)\Phi(N) = Q(N)R(N), \quad (13)$$

where $Q(N)$ is a $N \times \tilde{N}$ orthogonal matrix satisfying $Q(N)^T Q(N) = I$ and $R(N)$ is an $\tilde{N} \times \tilde{N}$ upper triangular matrix, from (13) we get $R(N)\beta(N) = p(N)$,or

$$\beta(N) = R(N)^{-1} p(N), \quad (14)$$

where

$$p(N) = Q(N)^T \Lambda^{1/2}(N)T(N). \quad (15)$$

We rearrange (13) and (15) in the following form:

$$\Lambda^{1/2}(N)\left[\Phi(N), T(N)\right] = Q(N)\left[R(N), p(N)\right]. \quad (16)$$

from which we readily understand the following weight updating procedure.

The process of updating the output weights of the neural networks with Givens QR decomposition is summarized as:

At each instant k time for the sample into the neural networks, update $\lambda(k)=\lambda_0\lambda(k-1)+1-\lambda_0$, where $0<\lambda(0)<1$ and $0<\lambda_0<1$. Create the left-hand matrix of the following equation, and then transform it with Givens rotation into the right-hand matrix in upper triangular form. Finally, compute $\beta(k)$ by (14).

$$\begin{bmatrix} \lambda(k)^{1/2} R(k-1) & \lambda(k)^{1/2} p(k-1) \\ \varphi^T(k) & t(k) \end{bmatrix} \xrightarrow{\text{Givens Rotation}} \begin{bmatrix} R(k) & p(k) \\ 0 & \Delta \end{bmatrix}. \quad (17)$$

Based on the Givens QR decomposition of orthogonalized least square algorithm, the OS-ELM(OLS) algorithm is derived as the following:

Given an activation function $g(x)$ (which may sigmoid function or RBF) and hidden nodes $\tilde{N}$.

step 1 Boosting Phase:

Given a small initial training set $N_0:\{(\mathbf{x}_i,\mathbf{t}_i)|\mathbf{x}_i\in R^n,\mathbf{t}_i\in R^m,i=1,\cdots,N_0,N_0\geq\tilde{N}\}$ to boost the learning algorithm first through the following boosting procedure:

(1) Assign arbitrary input weight $\mathbf{w}_{,i}$ and bias b_i (for additive hidden nodes) or center or center $\mathbf{w}_i$ and impact width b_i (for RBF hidden nodes), $i=1,\cdots,\tilde{N}$.

(2) Calculate the initial hidden layer output matrix $\mathbf{H}_0=[h_1,\cdots,h_{N_0}]^T$ where $h_i=[G(\mathbf{w}_1,b_1,\mathbf{x}_i),\cdots,G(\mathbf{w}_{\tilde{N}},b_{\tilde{N}},\mathbf{x}_i)]^T$, $i=1,\cdots,N_0$.

(3) Estimate the initial output weight $\beta(0)=(\mathbf{H}_0^T\mathbf{H}_0)^{-1}\mathbf{H}_0^T T_0$, where $T_0=[\mathbf{t}_1,\cdots,\mathbf{t}_{N_0}]^T$.

(4) Set $k=0$.

step 2 Sequential Learning Phase:

For each further coming observation $(\mathbf{x}_i,\mathbf{t}_i)$, where, $\mathbf{x}_i\in R^n,\mathbf{t}_i\in R^m,i=N_0+1, N_0+2,\cdots$, do

(1) Calculate the hidden layer output vector $h_{k+1}=[G(\mathbf{w}_1,b_1,\mathbf{x}_i),\cdots,G(\mathbf{w}_{\tilde{N}},b_{\tilde{N}},\mathbf{x}_i)]^T$.

(2) Calculate latest output weight $\beta(k)=R(k)^{-1}p(k)$ by the Givens QR decomposition based OLS algorithm.

(3) Set $k=k+1$, Go to step 2.

Just like OS-ELM(RLS) algorithm, the OS-ELM(OLS) algorithm also consists of two main phases. Boosting phase is to train the SLFNs using the primitive ELM method with some batch of training data in the initialization stage and these boosting training data will be discarded as soon as boosting phase is completed. The required batch of training data are very small, which can be equal to the number of hidden neurons. After boosting phase, the OS-ELM(OLS) will then learn the train data one-by-one and all the training data will be discarded once the learning procedure on these data is completed.

4 Simulations on Benchmark Problems

In this section, the performance of OS-ELM(OLS) algorithm is evaluated with that of OS-ELM(RLS), ELM algorithms with additive hidden nodes(which chooses sigmoid hidden nodes) and radial basis function hidden nodes. The simulation benchmark problems described in table 1 includes two regression applications (Sine, Abalone) and two chaotic time series predication benchmark applications (Mackey-Glass, Box and Jenkins).

Table 1. Specification of benchmark datasets

Dataset	Attributes	Training Data	Testing Data
Sine	1	5000	5000
Abalone	8	3000	1177
Mackey-Glass	4	4000	500
Box and Jenkins	10	200	90

Table 2. Performance comparison of OS-ELM(OLS) algorithm and OS-ELM(RLS),ELM algorithms on benchmark problems

Datasets	Activation Functions	Algorithms	RMSE		No. of Neurons
			Training	Testing	
Sine	Sigmoid	OS-ELM(OLS)	0.1263	0.0483	30
		OS-ELM(RLS)	0.1342	0.0720	30
		ELM	0.1158	0.0081	30
	RBF	OS-ELM(OLS)	0.1163	0.0130	30
		OS-ELM(RLS)	0.1487	0.0934	30
		ELM	0.1586	0.1083	30
Abalone	Sigmoid	OS-ELM(OLS)	0.0798	0.0858	25
		OS-ELM(RLS)	0.0746	0.0860	25
		ELM	0.0742	0.0836	25
	RBF	OS-ELM(OLS)	0.0748	0.0794	25
		OS-ELM(RLS)	0.0741	0.0819	25
		ELM	0.0894	0.0964	25
Mackey -Glass	Sigmoid	OS-ELM(OLS)	2.3129	0.0718	30
		OS-ELM(RLS)	2.1364	2.1335	30
		ELM	0.0150	0.0762	30
	RBF	OS-ELM(OLS)	1.6915	0.0961	30
		OS-ELM(RLS)	0.1591	0.1521	30
		ELM	0.3961	0.3934	30
Box and Jenkins	Sigmoid	OS-ELM(OLS)	0.0168	0.0138	25
		OS-ELM(RLS)	0.0164	0.0141	25
		ELM	0.0164	0.0154	25
	RBF	OS-ELM(OLS)	0.0227	0.0211	25
		OS-ELM(RLS)	0.0261	0.0222	25
		ELM	0.2660	0.2308	25

Table 2 summarizes the results for benchmark problems in terms of the training RMSE, testing RMSE and the hidden nodes for each algorithm. As observed from Table 2, comparing with other algorithms, the OS-ELM(OLS) algorithm produces much better generalization performance than OS-ELM(RLS) algorithm or sometimes has good performance than ELM algorithm, especially in the chaotic time series time predication problems.

5 Conclusions

In this paper, an efficient improved online sequential learning algorithm for OS-ELM (RLS) algorithm is proposed with much better generalization performance, especially in the chaotic time series predication applications. The algorithm can further expand the applications of the ELM algorithm with a good generalization performance. But in the algorithm, the OLS algorithm computational complexity is about $O[N^2]$, more than the RLS algorithm computational complexity [10]. Therefore, the further work is to find a fast Givens QR decomposition of the orthogonalized least square algorithm for reduce the output weights' computational complexity.

References

1. Huang, G.B., Zhu, Q.Y., Siew, C.K.: Extreme learning machine: A New Learning Scheme of Feedforward Neural Networks. Proceedings of International Joint Conference on Neural Networks, Budapest, Hungary (2004) 25-29
2. Huang, G.B., Chen, L., Siew, C.K.: Universal Approximation Using Incremental Networks With Random Hidden Computation Nodes. IEEE Trans. Neural Networks **17** (4) (2006) 879-892
3. Huang, G.B., Siew, C.K.: Extreme Learning Machine With Randomly Assigned RBF Kernels. International Journal of Information Technology **11** (1) (2005) 16-24
4. Zhu, Q.Y., Qin, A.K., Suganthan, P. N., Huang, G.B.: Evolutionary Extreme Learning Mchine. Pattern Recognition **38** (2005) 1759-1763
5. Li, M.B., Huang, G.B., Saratchandran, P., Sundararajan ,N.: Fully Complex Extreme Learning Machine.Neurocomputing **68** (2005) 306-314
6. Liang , N.Y., Huang, G.B., Saratchandran, P., Sundararajan, N.: A Fast and Accurate Online Sequential Learning Algorithm for Feedforward Networks. IEEE Trans. Neural Networks **17** (6) (2006)1411-1423
7. Rosipal, R., Koska, M., Farkas, I.: Prediction of Chaotic Time-Series with a Resource Allocating RBF Network. Neural Processing Letters **7** (1998)185-197
8. Li, B.: An Improvement of the RAN Learning Algorithm. Pattern Recognition and Artificial Intelligence. (2006) 220-226
9. Lai, X.P., Li, B.: An Efficient Learning Algorithm Generating Small RBF Neural Networks. Neural Network World. (2005) 525-533
10. Paulo S.-R.Diniz.: Adaptive Filtering: Algorithms and Practical Implementation. 2rd edn. Kluwer Academic Publishers(2002)

Intelligence Through Interaction: Towards a Unified Theory for Learning

Ah-Hwee Tan[1], Gail A. Carpenter[2], and Stephen Grossberg[2]

[1] Intelligent Systems Centre and School of Computer Engineering
Nanyang Technological University, Nanyang Avenue, Singapore 639798
asahtan@ntu.edu.sg
[2] Center for Adaptive Systems and Department of Cognitive and Neural Systems
Boston University, 677 Beacon Street, Boston, MA 02215, USA
gail,steve@cns.bu.edu

Abstract. Machine learning, a cornerstone of intelligent systems, has typically been studied in the context of specific tasks, including clustering (unsupervised learning), classification (supervised learning), and control (reinforcement learning). This paper presents a learning architecture within which a universal adaptation mechanism unifies a rich set of traditionally distinct learning paradigms, including learning by matching, learning by association, learning by instruction, and learning by reinforcement. In accordance with the notion of embodied intelligence, such a learning theory provides a computational account of how an autonomous agent may acquire the knowledge of its environment in a real-time, incremental, and continuous manner. Through a case study on a minefield navigation domain, we illustrate the efficacy of the proposed model, the learning paradigms encompassed, and the various types of knowledge learned.

1 Introduction

Machine learning, a cornerstone of intelligent system research, has typically been studied in the context of specific tasks, including clustering (unsupervised learning), classification (supervised learning), and control (reinforcement learning). In reality, an autonomous system acquires intelligence through its interaction with the environment. This is in keeping with the view in modern cognitive science that cognition is a process deeply rooted in the body's interaction with the world [1]. Embodied cognition is also akin to the intensive study on reinforcement learning [15] in which an autonomous agent learns to adjust its behaviour according to evaluative feedback received from the environment.

Over the past decades, a family of neural architectures known as Adaptive Resonance Theory (ART) [3,5,8,9], has been steadily developed. With well-founded computational principles, ART has been applied successfully to many pattern analysis, recognition, and prediction applications [6,12]. These successful applications are of particular interest because the basic ART principles have been derived from an analysis of human and animal perceptual and cognitive

D. Liu et al. (Eds.): ISNN 2007, Part I, LNCS 4491, pp. 1094–1103, 2007.

information processing, and have led to behavioral and neurobiological predictions that have received significant experimental support during the last decade; see Grossberg 2003 and Raizada & Grossberg 2003 for reviews. In this paper, we show that Adaptive Resonance Theory lays the foundation of a unified model that encompasses a myriad of learning paradigms, traditionally viewed as distinct. The proposed model is a natural extension of the original ART models from a single pattern field to multiple pattern channels. Whereas the original ART models [2] perform unsupervised learning of recognition nodes in response to incoming input patterns, the proposed neural architecture, known as fusion ART (fusion Adaptive Resonance Theory), learns multi-channel mappings simultaneously across multi-modal pattern channels in an online and incremental manner.

To illustrate the unified model, this paper presents a case study based on a minefield navigation task, which involves an autonomous vehicle (AV) learning to navigate through obstacles to reach a stationary target (goal) within a specified number of steps. The experimental results show that fusion ART is capable of performing a myriad of learning tasks and is able to produce a fast and stable learning performance.

The rest of the paper is organized as follows. Section 2 provides a summary of the fusion ART architecture and the associated system dynamics. Sections 3, 4, 5 and 6 show how fusion ART can be used for various types of learning tasks. Section 7 illustrates the fusion ART functionalities and performance based on the minefield navigation task. The final section concludes and highlights possible future directions.

2 Fusion ART

Fusion ART employs a multi-channel architecture (Figure. 1), comprising a category field F_2 connected to a fixed number of (K) pattern channels or input fields through bidirectional conditionable pathways. The model unifies a number of network designs, most notably Adaptive Resonance Theory (ART) [3,5], Adaptive Resonance Associative Map (ARAM) [16] and Fusion Architecture for Learning, COgnition, and Navigation (FALCON) [20], developed over the past decades for a wide range of functions and applications. The generic network dynamics of fusion ART, based on fuzzy ART operations [4], is summarized as follows.

Input vectors: Let $\mathbf{I}^{ck} = (I_1^{ck}, I_2^{ck}, \ldots, I_n^{ck})$ denote the input vector, where $I_i^{ck} \in [0,1]$ indicates the input i to channel ck. With complement coding, the input vector $\mathbf{I}^{ck}$ is augmented with a complement vector $\bar{\mathbf{I}}^{ck}$ such that $\bar{I}_i^{ck} = 1 - I_i^{ck}$. **Activity vectors:** Let $\mathbf{x}^{ck}$ denote the F_1^{ck} activity vector for $k = 1, \ldots, K$. Let $\mathbf{y}$ denote the F_2 activity vector. **Weight vectors:** Let $\mathbf{w}_j^{ck}$ denote the weight vector associated with the jth node in F_2 for learning the input patterns in F_1^{ck} for $k = 1, \ldots, K$. Initially, F_2 contains only one *uncommitted* node and its weight vectors contain all 1's. **Parameters:** The fusion ART's dynamics is determined by choice parameters $\alpha^{ck} > 0$, learning rate parameters $\beta^{ck} \in [0,1]$, contribution parameters $\gamma^{ck} \in [0,1]$ and vigilance parameters $\rho^{ck} \in [0,1]$ for $k = 1, \ldots, K$.

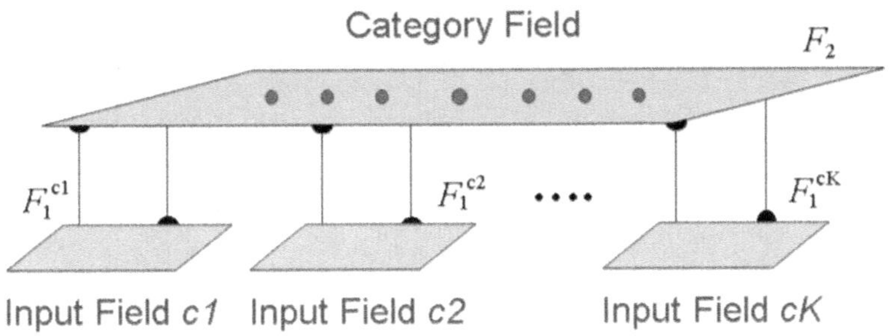

Fig. 1. The fusion ART architecture

As a natural extension of ART, fusion ART responds to incoming patterns in a continuous manner. It is important to note that at any point in time, fusion ART does not require input to be present in all the pattern channels. For those channels not receiving input, the input vectors are initialized to all 1s. The fusion ART pattern processing cycle comprises five key stages, namely code activation, code competition, activity readout, template matching, and template learning, as described below. **Code activation:** Given the activity vectors $\mathbf{I}^{c1}, \ldots, \mathbf{I}^{cK}$, for each F_2 node j, the choice function T_j is computed as follows:

$$T_j = \sum_{k=1}^{K} \gamma^{ck} \frac{|\mathbf{I}^{ck} \wedge \mathbf{w}_j^{ck}|}{\alpha^{ck} + |\mathbf{w}_j^{ck}|}, \tag{1}$$

where the fuzzy AND operation $\wedge$ is defined by $(\mathbf{p} \wedge \mathbf{q})_i \equiv min(p_i, q_i)$, and the norm $|.|$ is defined by $|\mathbf{p}| \equiv \sum_i p_i$ for vectors $\mathbf{p}$ and $\mathbf{q}$. **Code competition:** A code competition process follows under which the F_2 node with the highest choice function value is identified. The winner is indexed at J where

$$T_J = \max\{T_j : \text{for all } F_2 \text{ node } j\}. \tag{2}$$

When a category choice is made at node J, $y_J = 1$; and $y_j = 0$ for all $j \neq J$. This indicates a winner-take-all strategy. **Activity readout:** The chosen F_2 node J performs a readout of its weight vectors to the input fields F_1^{ck} such that

$$\mathbf{x}^{ck} = \mathbf{I}^{ck} \wedge \mathbf{w}_J^{ck}. \tag{3}$$

Template matching: Before the activity readout is stabilized and node J can be used for learning, a template matching process checks that the weight templates of node J are sufficiently close to their respective input patterns. Specifically, resonance occurs if for each channel k, the *match function* m_J^{ck} of the chosen node J meets its vigilance criterion:

$$m_J^{ck} = \frac{|\mathbf{I}^{ck} \wedge \mathbf{w}_J^{ck}|}{|\mathbf{I}^{ck}|} \geq \rho^{ck}. \tag{4}$$

If any of the vigilance constraints is violated, mismatch reset occurs in which the value of the choice function T_J is set to 0 for the duration of the input presentation. Using a *match tracking* process, at the beginning of each input presentation, the vigilance parameter ρ^{ck} in each channel ck equals a baseline vigilance $\bar{\rho}^{ck}$. When a mismatch reset occurs, the ρ^{ck} of all pattern channels are increased simultaneously until one of them is slightly larger than its corresponding match function m_J^{ck}, causing a reset. The search process then selects another F_2 node J under the revised vigilance criterion until a resonance is achieved. **Template learning:** Once a resonance occurs, for each channel ck, the weight vector $\mathbf{w}_J^{ck}$ is modified by the following learning rule:

$$\mathbf{w}_J^{ck(\text{new})} = (1 - \beta^{ck})\mathbf{w}_J^{ck(\text{old})} + \beta^{ck}(\mathbf{I}^{ck} \wedge \mathbf{w}_J^{ck(\text{old})}). \tag{5}$$

When an uncommitted node is selected for learning, it becomes *committed* and a new uncommitted node is added to the F_2 field. Fusion ART thus expands its network architecture dynamically in response to the input patterns.

The network dynamics described above can be used to support a myriad of learning operations. We show how fusion ART can be used for a variety of traditionally distinct learning tasks in the subsequent sections.

3 Learning by Similarity Matching

With a single pattern channel, the fusion ART architecture reduces to the original ART model. Using a selected vigilance value ρ, an ART model learns a set of recognition nodes in response to an incoming stream of input patterns in a continuous manner. Each recognition node in the F_2 field learns to encode a template pattern representing the key characteristics of a set of patterns. ART has been widely used in the context of unsupervised learning for discovering pattern groupings. Please refer to the selected ART literatures [3,5,8,9] for a review of ART's functionalities, interpretations, and applications.

4 Learning by Association

By synchronizing pattern coding across multiple pattern channels, fusion ART learns to encode associative mappings across distinct pattern spaces. A specific instance of fusion ART with two pattern channels is known as Adaptive Resonance Associative Map (ARAM), that learns multi-dimensional supervised mappings from one pattern space to another pattern space [16]. An ARAM system consists of an input field F_1^a, an output field F_1^b, and a category field F_2. Given a set of feature vectors presented at F_1^a with their corresponding class vectors presented at F_1^b, ARAM learns a predictive model (encoded by the recognition nodes in F_2) that associates combinations of key features to their respective classes.

Fuzzy ARAM, based on fuzzy ART operations, has been successfully applied to numerous machine learning tasks, including personal profiling [19], document classification [11], personalized content management [18], and DNA gene expression analysis [22]. In many benchmark experiments, ARAM has demonstrated predictive performance superior to those of many state-of-the-art machine learning systems, including C4.5, Backpropagation Neural Network, K Nearest Neighbour, and Support Vector Machines.

5 Learning by Instruction

During learning, fusion ART formulates recognition categories of input patterns across multiple channels. The knowledge that fusion ART discovers during learning, is compatible with symbolic rule-based representation. Specifically, the recognition categories learned by the F_2 category nodes are compatible with a class of IF-THEN rules that maps a set of input attributes (antecedents) in one pattern channel to a disjoint set of output attributes (consequents) in another channel. Due to this compatibility, at any point of the incremental learning process, instructions in the form of IF-THEN rules can be readily translated into the recognition categories of a fusion ART system. The rules are conjunctive in the sense that the attributes in the IF clause and in the THEN clause have an *AND* relationship. Augmenting a fusion ART network with domain knowledge through explicit instructions serves to improve learning efficiency and predictive accuracy.

The fusion ART rule insertion strategy is similar to that used in Cascade ARTMAP, a generalization of ARTMAP that performs domain knowledge insertion, refinement, and extraction [17]. For direct knowledge insertion, the IF and THEN clauses of each instruction (rule) is translated into a pair of vectors **A** and **B** respectively. The vector pairs derived are then used as training patterns for inserting into a fusion ART network. During rule insertion, the vigilance parameters are set to 1s to ensure that each distinct rule is encoded by one category node.

6 Learning by Reinforcement

Reinforcement learning [15] is a paradigm wherein an autonomous system learns to adjust its behaviour based on reinforcement signals received from the environment. An instance of fusion ART, known as FALCON (Fusion Architecture for Learning, COgnition, and Navigation), learns mappings simultaneously across multi-modal input patterns, involving states, actions, and rewards, in an online and incremental manner. Compared with other ART-based reinforcement learning systems, FALCON presents a truly integrated solution in the sense that there is no implementation of a separate reinforcement learning module or Q-value table. Using competitive coding as the underlying principle of computation, the network dynamics encompasses a myriad of learning paradigms, including unsupervised learning, supervised learning, as well as reinforcement learning.

FALCON employs a three-channel architecture, comprising a category field F_2 and three pattern fields, namely a sensory field F_1^{c1} for representing current states, a motor field F_1^{c2} for representing actions, and a feedback field F_1^{c3} for representing reward values. A class of FALCON networks, known as TD-FALCON [21,23], incorporates Temporal Difference (TD) methods to estimate and learn value function $Q(s, a)$, that indicates the goodness to take a certain action a in a given state s.

The general sense-act-learn algorithm for TD-FALCON is summarized in Table 1. Given the current state s, the FALCON network is used to predict the value of performing each available action a in the action set $\mathcal{A}$ based on the corresponding state vector **S** and action vector **A**. The value functions are then processed by an action selection strategy (also known as policy) to select an action. Upon receiving a feedback (if any) from the environment after performing the action, a TD formula is used to compute a new estimate of the Q-value for performing the chosen action in the current state. The new Q-value is then used as the teaching signal (represented as reward vector **R**) for FALCON to learn the association of the current state and the chosen action to the estimated value.

Table 1. The TD−FALCON algorithm

1. Initialize the FALCON network.
2. Given the current state s, for each available action a in the action set $\mathcal{A}$, predict the value of performing the action Q(s,a) by presenting the corresponding state and action vectors **S** and **A** to FALCON.
3. Based on the value functions computed, select an action a from $\mathcal{A}$ following an action selection policy.
4. Perform the action a, observe the next state s', and receive a reward r (if any).
5. Estimate the value function $Q(s, a)$ following a temporal difference formula given by $\Delta Q(s, a) = \alpha TD_{err}$.
6. Present the corresponding state, action, and reward (Q-value) vectors, namely **S**, **A**, and **R**, to FALCON for learning.
7. Update the current state by s=s'.
8. Repeat from Step 2 until s is a terminal state.

7 Case Study: Minefield Navigation

The minefield simulation task studied here is similar to the underwater navigation and mine avoidance domain developed by Naval Research Lab (NRL) [7,14]. The objective is to navigate through a minefield to a randomly selected target position in a specified time frame without hitting a mine. In each trial, the autonomous vehicle (AV) starts from a random position in the field, and repeats the cycles of sense, act, and learn. A trial ends when the system reaches the target (success), hits a mine (failure), or exceeds 30 sense-act-learn cycles (out of time). The target and the mines remain stationary during the trial.

Minefield navigation and mine avoidance is a non-trivial task. As the configuration of the minefield is generated randomly and changes over trials, the

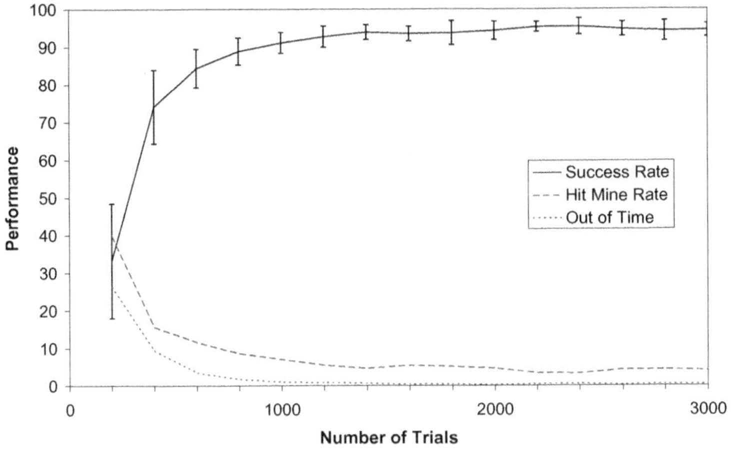

(a) TD-FALCON

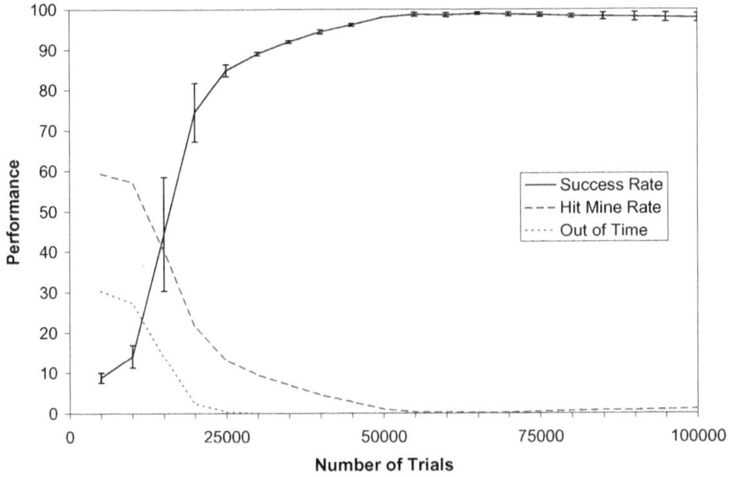

(b) BP-Q Learner

Fig. 2. The task completion performance of TD-FALCON compared with the BP-Q Learner operating with delayed rewards

system needs to learn strategies that can be carried over across experiments. For sensing, the AV has a coarse 180 degree forward view based on five sonar sensors. In each direction i, the sonar signal is measured by $s_i = \frac{1}{d_i}$, where d_i is the distance to an obstacle in the i direction. Other sensory inputs include the current and target bearings. In each step, the system chooses one of the five possible actions, namely move left, move diagonally left, move straight ahead, move diagonally right, and move right.

In this domain, we conduct experiments using TD-FALCON, a three-channel fusion ART model, with both immediate and delayed evaluative feedback. For both reward schemes, at the end of a trial, a reward of 1 is given when the AV reaches the target. A reward of 0 is given when the AV hits a mine. For the immediate reward scheme, a reward is estimated at each step of the trial by computing a utility function $utility = \frac{1}{1+rd}$, where rd is the remaining distance between the AV and the target position.

7.1 Performance Comparison

We compare the performance of TD-FALCON with an alternative reinforcement learning system (hereafter referred to as the BP-Q Learner), in terms of success rate, hit-mine rate, and out-of-time rate in a 16 by 16 minefield containing 10 mines. The BP-Q Learner uses the standard Q-learning rule and a gradient descent based multi-layer feedforward neural network as the function approximator. For illustration purpose, we only show the performance of the two systems operating with delayed rewards (Figure. 2). We can see that the BP-Q Learner generally takes a very large number of (more than $40,000$) trials to reach 90% success rates. In contrast, TD-FALCON consistently achieves the same level of performance within the first 1000 trials. In other words, TD-FALCON learns at least an order of magnitude faster than the BP-Q learner.

Considering network complexity, the BP-Q Learner has the advantage of a highly compact network architecture. When trained properly, a BP network consisting of 36 hidden nodes can produce performance equivalent to that of a TD-FALCON model with say 200 category nodes. In terms of the speed of adaptation, however, TD-FALCON is clearly a faster learner by consistently mastering the task in a much smaller number of trials.

7.2 Knowledge Interpretation

To illustrate the variety of the knowledge learned by TD-FALCON, Table 2 shows a sample set of the knowledge encoded by its recognition nodes. Through learning by similarity matching, TD-FALCON identifies key situations in its environment that are of significance to its mission. Two such typical situations are shown in the first row of the table. Through learning by association (or directly as instructions), TD-FALCON learns the association between typical situations and their corresponding desired actions. Two such association rules are shown in the second row. Finally, through the reinforcement signals given by the environment, TD-FALCON learns the value of performing a specific action in a given situation. The third row shows two extreme cases, one indicating a high payoff for taking an action in a situation and the other giving a severe penalty for taking the same action in a slightly different situation.

Table 2. Sample knowledge learned by FALCON in the minefield navigation domain. $\wedge$ is used here to indicate *AND* operator.

Type of Learning	Knowledge Learned
Similarity matching	FrontSonar=1.0 $\wedge$ Target=Front FrontSonar$\leq$0.5 $\wedge$ Target=Front
Association or Instruction	**IF** FrontSonar$\leq$0.5 $\wedge$ Target=Front **THEN** Move=Front **IF** FrontSonar=1.0 $\wedge$ DRightSonar$\leq$0.5 $\wedge$ Target=Front **THEN** Move=DRight
Reinforcement	**IF** FrontSonar$\leq$0.5 $\wedge$ Target=Front **THEN** Move=Front (Q=1.0) **IF** FrontSonar=1.0 $\wedge$ Target=Front **THEN** Move=Front (Q=0.0)

8 Conclusion

This paper has outlined a generalized neural architecture, known as fusion Adaptive Resonance Theory (fusion ART), that learns multi-dimensional mappings simultaneously across multi-modal pattern channels, in an online and incremental manner. Such a learning architecture enables an autonomous agent to acquire its intelligence in a real-time dynamic environment. Using Adaptive Resonance Theory (ART) as an universal coding mechanism, the proposed model unifies a myriad of traditionally distinct learning paradigms, including unsupervised learning, supervised learning, rule-based knowledge integration, and reinforcement learning. In fact, ART-style learning and matching mechanism seems to be operative in many levels of the cerebral cortex of the brain, especially in the vision system [10]. The proposed framework may thus serve as a foundation model for developing high level cognitive information processing capabilities, including awareness, reasoning, explaining, and surprise handling.

References

1. Anderson, M.L.: Embodied Cognition: A Field Guide. Artificial Intelligence **149** (2003) 91-130
2. Carpenter, G. A., Grossberg, S.: A Massively Parallel Architecture for a Self- organizing Neural Pattern Recognition Machine. Computer Vision, Graphics, and Image Processing **37** (1987) 54-115
3. Carpenter, G. A., Grossberg, S., editors: Pattern Recognition by Self-Organizing Neural Networks. Cambridge, MA: MIT Press (1991)
4. Carpenter, G. A., Grossberg, S., Rosen, D. B.: Fuzzy ART: Fast Stable Learning and Categorization of Analog Patterns by an Adaptive Resonance System. Neural Networks **4** (1991) 759-771
5. Carpenter, G.A., Grossberg, S.: Adaptive Rresonance Theory. In M.A. Ar- bib, editor, The Handbook of Brain Theory and Neural Networks, 87-90. Cambridge, MA: MIT Press (2003)
6. Duda, R. O., Hart, P.E., Stock, D.G., editors. Pattern Classification (2nd edi- tion). New York: John Wiley, Section 10.11.2, (2001)

7. Gordan, D., Subramanian, D.: A Cognitive Model of Learning to Navigate. In Proceedings, Nineteenth Annual Conference of the Cognitive Science Society (1997) 271-276
8. Grossberg, S.: Adaptive Pattern Recognition and Universal Recoding, I: Parallel Development and Coding of Neural Feature Detectors. Biological Cybernetics **23** (1976) 121-134
9. Grossberg, S.: Adaptive Pattern Recognition and Universal Recoding, II: Feedback, Expectation, Olfaction, and Illusion. Biological Cybernetics **23** (1976) 187-202
10. Grossberg, S.: How Does the Cerebral Cortex Work? Development, Learning, Attention, and 3d Vision by Laminar Circuits of Visual cortex. Behavioral and Cognitive Neuroscience Reviews **2** (2003) 47-76
11. He, J., Tan, A-H., Tan, C-L: On Machine Learning Methods for Chinese Documents Classification. Applied Intelligence: Special Issue on Text and Web Mining **18** (3) (2003) 311-322
12. Levine, D. S., editor: Introduction to Neural and Cognitive Modeling. New Jersey: Lawrence Erlbaum Associates, Chapter 6 (2000)
13. Raizada, R., Grossberg, S.: Towards a Theory of the Laminar Architecture of Cerebral Cortes: Computational Clues from the Visual System. Cerebral Cortex **13** (2003) 200-213
14. Sun, R., Merrill, E., Peterson, T.: From Implicit Skills to Explicit Knowledge: A Bottom-up Model of Skill Learning. Cognitive Science **25** (2) (2001) 203-244
15. Sutton, R.S., Barto, A.G.: Reinforcement Learning: An Introduction. Cambridge, MA: MIT Press (1998)
16. Tan, A. H.: Adaptive Resonance Associative Map. Neural Networks **8** (3) (1995) 437-446
17. Tan, A. H.: Cascade ARTMAP: Integrating Neural Computation and Symboli Knowledge Processing. IEEE Transactions on Neural Networks **8** (2) (1997) 237-250
18. Tan, A-H., Ong, H-L., Pan, H., Ng, J., Li, Q-X.: Towards Personalized Web Intelligence. Knowledge and Information Systems **6** (5) 595-616 (2004)
19. Tan, A-H., Soon, H.-S.: Predictive Adaptive Resonance Theory and Knowledge Discovery in Database. In Proceedings, Fourth Pacific-Asia Conference on Knowledge Discovery and Data Mining (PAKDD'00), Kyoto, pages (2000) 173-176
20. Tan, A.H.: FALCON: A Fusion Architecture for Learning, Cognition, and Navigation. In Proceedings, International Joint Conference on Neural Networks (2004) 3297-3302
21. Tan, A.H.: Self-organizing Neural Architecture for Reinforcement Learning. In Proceedings, J. Wang et al. (Eds.): International Symposium on Neural Networks (ISNN'06), Chengdu, China, LNCS **3971** (2006) 470-475
22. Tan, A.H., Pan, H.: Predictive Neural Networks for Gene Expression Data Analysis. Neural Networks **18** (3) (2005) 297-306
23. Tan, A.H., Lu, N., Xiao, D.: Integrating Temporal Difference Methods and Self-organizing Neural Networks for Reinforcement Learning with Delayed Evaluative feedback. IEEE Transactions on Neural Networks, to appear

An Improved Multiple-Instance Learning Algorithm

Fengqing Han[1,2], Dacheng Wang[1], and Xiaofeng Liao[2]

[1] Chongqing Jiaotong University, Chongqing 400074, China
hanfengqing@cqit.edu.cn
[2] Chongqing University, Chongqing 400044, China

Abstract. Multiple-instance learning (MIL) is a variation on supervised learning, where the task is to learn a concept given positive and negative bags of instances. In this paper a novel algorithm has been introduced for multiple-instance learning. This method was inspired by both diverse density (DD) and its expectation maximization version (EM-DD). It converts MIL problem to a single-instance setting. This improved method has better accuracy and time complexity than DD and EM-DD. We apply it to drug activity prediction and image retrieval. The experiments show it has competitive accuracy values compared with other previous approaches.

1 Introduction

Multiple-instance learning [1] is a way to model ambiguity in semi-supervised learning setting, where each training example is a bag of instances and the labels are assigned on the bags instead of on the instances. In the binary case, a bag is labeled positive if at least one instance in that bag is positive, and the bag is labeled negative if all the instances in it are negative. There are no labels on the individual instances. The goal of MIL is to classify unseen bags or instances based on the labeled bags as the training data. Standard supervised learning can be viewed as a special case of MI learning where each bag holds a single instance.

After being introduced by Dietterich et al., MIL has become an active research area and a number of MIL algorithms have been proposed, such as learning axis-parallel concepts [1], diverse density [2], EM-DD [3], extended Citation kNN [4], etc. Meanwhile, the nature of MIL makes it applicable to several applications, ranging from drug activity prediction to text or multimedia information retrieval [1, 2, 5-7].

Diverse Density (DD) was proposed as a general framework for solving multi-instance learning problems. The main idea of DD approach is to find a concept point in the feature space that are close to at least one instance from every positive bag and meanwhile far away from instances in negative bags. The optimal concept point is defined as the one with the maximum diversity density, which is a measure of how many different positive bags have instances near the point, and how far the negative instances are away from that point.

The difficulty of MIL comes from the ambiguity of not knowing which instance is the most likely one. In [3], the knowledge of which instance determines the label of the

D. Liu et al. (Eds.): ISNN 2007, Part I, LNCS 4491, pp. 1104–1109, 2007.

bag is modeled using a set of hidden variables, which are estimated using the Expectation Maximization style approach. It results in an algorithm called EM-DD, which combines this EM-style approach with the DD algorithm.

Inspired by both the Diverse Density and EM-DD, we present a new MIL learning algorithm, which is called improved DD (I-DD) algorithm. This paper is organized as follows. Section 2 introduces Diverse Density and EM-DD algorithm. Section 3 formulates the I-DD algorithm. Section 4 gives the experimental results. Finally, some conclusions are drawn in Section 5.

2 Diverse Density and Its EM Version

A probabilistic derivation of Diverse Density is given below. We denote positive bags as B_i^+, and the j-th instance in that bag as B_{ij}^+. Suppose each instance can be represented by a feature vector (or a point in the feature space), and we use B_{ijk}^+ to denote the value of the k-th feature of instance B_{ij}^+. Likewise, B_i^- denotes a negative bag and B_{ij}^- is the j-th instance in that bag. The true concept is a single point h defined by maximizing the diverse density defined as $DD(h) = P(h \mid B_1^+, \dots, B_n^+, B_1^-, \dots, B_m^-)$ over the feature space. Using Bayes rule and assuming uniform prior over the concept location, this is equivalent to maximizing the following likelihood:

$$\arg\max_h P(B_1^+, \dots, B_n^+, B_1^-, \dots, B_m^- \mid h) \ . \tag{1}$$

By making additional assumption that the bags are conditionally independent given the concept point h, this decomposes to:

$$\arg\max_h \prod_i P(B_i^+ \mid h) \prod_i P(B_i^- \mid h) \ . \tag{2}$$

Using Bayes rule once more with the uniform prior assumption, this is equivalent to:

$$\arg\max_h \prod_i P(h \mid B_i^+) \prod_i P(h \mid B_i^-) \ , \tag{3}$$

which gives a general definition of the Diverse Density. Given the fact that boolean label (say, 1 and 0) of a bag is the result of "logical-OR" of the labels of its instances, $P(h \mid B_i)$ is instantiated using the noise-or model:

$$P(h \mid B_i^+) = 1 - \prod_j (1 - P(h \mid B_{ij}^+)) \text{ and } P(h \mid B_i^-) = \prod_j (1 - P(h \mid B_{ij}^-)) \ . \tag{4}$$

With Eq.(4), model (3) is equivalent to:

$$\arg\max_h \prod_i (1 - \prod_j (1 - P(h \mid B_{ij}^+))) \prod_{i,j} (1 - P(h \mid B_{ij}^-)) \ . \tag{5}$$

Finally, $P(h \mid B_{ij}^{+})$ (or $P(h \mid B_{ij}^{-})$) is estimated (though not necessarily) by a Gaussian-like distribution

$$P(h \mid B_{ij}^{+}) = \exp(-\sum_{k} w_k (B_{ijk}^{+} - h_k)^2) \ , \tag{6}$$

where w_k is a non-negative scaling factor that reflects the degree of relevance of different features.

Without close-form solution to the above maximization problem, gradient ascent method is used to search the feature space for the concept point with (local) maximum DD. Usually the search is repeated using the instances from every positive bag as the starting points.

By removing the noise-or part in the original DD algorithm, EM-DD turns a MI problem into a single-instance one. The model (5) is converted to:

$$\arg\max_{h} \prod_{i} (1 - \min_{j}(1 - P(h \mid B_{ij}^{+}))) \prod_{i} \min_{j}(1 - P(h \mid B_{ij}^{-})) \ . \tag{7}$$

EM-DD starts with an initial guess of the concept point h (which can be obtained using original DD algorithm), and then repeatedly performs the following two steps: in E-step, the current hypothesis of concept h is used to pick the most likely instance from each bag given a generative model; in M-step, the a new concept h' is estimated by maximizing a transformed DD defined on the instances selected in the E-step using the gradient search. Then, the old concept h is replaced by the new concept h' and the two steps are repeated until the algorithm converges.

3 I-DD Model and Algorithm

We now describe our model and algorithm, then compare it with the original DD and EM-DD algorithms.

With Eq. (6), model (5) is converted to:

$$\max_{h,w} \prod_{i} \max_{j} \exp(-\sum_{k} w_k (B_{ijk}^{+} - h_k)^2) \prod_{i,j} (1 - \exp(-\sum_{k} w_k (B_{ijk}^{-} - h_k)^2)) \ . \tag{8}$$

Let

$$f(h, w, s_1, \ldots, s_n) = \prod_{i} \exp(-\sum_{k} w_k (s_{ik} - h_k)^2) \prod_{i,j} (1 - \exp(-\sum_{k} w_k (B_{ijk}^{-} - h_k)^2)) \ . \tag{9}$$

Denote $B_{ij*}^{+} = \arg\min_{j} \sum_{k} w_k (B_{ij*k}^{+} - h_k)^2$, then model (8) is equivalent to:

$$\max_{h,w} F(h, w) \ , \tag{10}$$

where $F(h, w) = f(h, w, B_{1j*}^{+}, \ldots, B_{nj*}^{+})$.

In some aspects, the I-DD algorithm is similar to the EM-DD. I-DD starts with some initial guess of a target point h obtained in the standard way by trying points from positive bags, then repeatedly performs two steps to search for the maximum likelihood hypothesis. In the first step, the a new concept ($h^{(t+1)}, w^{(t+1)}$) is estimated by the gradient ascent algorithm such that $F(h^{(t+1)}, w^{(t+1)}) > F(h^{(t)}, w^{(t)})$. In the second step, the current hypothesis of concept ($h^{(t+1)}, w^{(t+1)}$) is used to pick the most likely instance from each positive bag. The two steps are repeated until the algorithm converges. Pseudo code for this algorithm is followed.

Choose a random initial scaling vector $w^{(0)}$ and a random instance B_{ij}^{+} as $h^{(0)}$;

$t = 0$; $F_1 = F(h^{(0)}, w^{(0)})$;

Repeat

$$\lambda_t = \frac{\nabla F(h^{(t)}, w^{(t)})^T \nabla F(h^{(t)}, w^{(t)})}{\nabla F(h^{(t)}, w^{(t)})^T \nabla^2 F(h^{(t)}, w^{(t)}) \nabla F(h^{(t)}, w^{(t)})} ;$$

$$(h^{(t+1)}, w^{(t+1)}) = (h^{(t)}, w^{(t)}) + \lambda_t \nabla F(h^{(t)}, w^{(t)})^T ;$$

$t = t{+}{+}$;

For each positive bag B_i^{+} ,

$$B_{ij}^{(t)+} = \arg\min_j \sum_k w_k^{(t)} (B_{ijk}^{+} - h_k^{(t)})^2 ;$$

$F_0 = F_1$; $F_1 = F(h^{(t)}, w^{(t)})$;

Until ($F_0 \geq F_1$ or $\left\| \nabla F(h^{(t-1)}, w^{(t-1)}) \right\| < \varepsilon$)

Return F_1 ;

We now briefly provide intuition as to why I-DD improves computation time of the EM-DD algorithm. In EM-DD, the current hypothesis of concept h is used to pick the most likely instance from each bag and two-step gradient ascent search is employed to find a new h' that maximizes DD(h). Once this maximization step is completed, the proposed target h is resetted to h' and return to the first step until the algorithm converges. But in our algorithm, the current hypothesis h is used to pick one instance from each positive bag, not from each bag. In the other hand we do not maximize DD(h), but use the gradient ascent algorithm to find a new h' such that $F(h', w') > F(h, w)$. These improvement results in better time complexity than EM-DD.

We now provide the proof of convergence. Note $F(h,w) \le 1$, and at each iteration we have $F(h^{(t+1)}, w^{(t+1)}) > F(h^{(t)}, w^{(t)})$. It means $\exists F^*, s.t. \lim_{k\to\infty} F(h^{(t)}, w^{(t)}) = F^*$. Since there are finite instances in each bag, the algorithm will terminate after a finite number of iterations.

Note in the I-DD it has some difficulties and calculation error for computing $\nabla^2 F$. So we can use the following model instead of (8) in the algorithm.

$$\min_{h,w,B_{ij}^+} \sum_i \sum_k w_k (B_{ijk}^+ - h_k)^2 - \sum_{i,j} \sum_k w_k (B_{ijk}^- - h_k)^2 , \tag{11}$$

4 Experimental Results

In this section, we have performed experiments on various data sets to evaluate the proposed method and compare it to other methods for MIL. For both drug activity prediction and image retrieval problem, this method achieves competitive accuracy values. And in these experiments I-DD algorithm runs faster than DD and EM-DD.

4.1 Experiments with Drug Activity Prediction Problem

The Musk data sets provided by Dietterich et al.[1] are the benchmark data sets used in all previous approaches. These data sets (Musk1 and Musk2) consist of descriptions of molecules using multiple low-energy conformations. Each conformation is represented by a point with 166 features vector. Musk1 contains on average approximately 6 conformations per bag, while Musk2 has on average more than 60 instances in each bag. The averaged results of 20 runs are summarized in Table1. For both Musk1 and Musk2 data sets our algorithm achieves competitive accuracy values.

Table 1. Accurary comparison on the Musk data sets

	IAPR	Citation-kNN	DD	EM-DD	mi-SVM	I-DD
Musk1	92.4%	91.3%	88.9%	89.8%	87.4%	90.8%
Musk2	89.2%	86.3%	84%	85.9%	83.6%	86.4%

Table 2. Performance comparison on image retrieval

	mi-SVM	Citation-kNN	EM-DD	I-DD
Elephant	82.4%	81.3%	80.8%	81.5%
Fox	59.6%	58.7%	57.1%	57.3%
Tiger	80.6%	78.3%	78.9%	80.7%

4.2 Experiments with Image Retrieval

We conduct the experiments on the images from Corel Image Database, which is popular in the image retrieval area. Specifically, we perform three retrieval tasks, in order to classify images of "elephant", "fox", and "tiger". For each tasks, we use a test data collection of 200 images, with 100 positive and 100 negative images. All the images are segmented using the Blobworld system [8] into a set of regions and for each region, a 320-d feature vector is extracted to represent its color, texture, and shape characteristics. After the segmentation process, the number of instances (regions) for each retrieval tasks is 1390, 1319, and 1219 respectively. We use mi-SVM, Citation-kNN, EM-DD, and compare their performance to our algorithm. Table 2 shows the comparison.

5 Conclusions

In this paper a novel algorithm I-DD has been introduced for MIL. This method was inspired by both DD and EM-DD. Compared with other previous approaches, it has competitive accuracy values. It has better accuracy and time complexity than DD and EM-DD. In the future work we will apply it in some new applications and improve it performance.

Acknowledgement

This work was supported by the Natural Science Foundations of China (50578168) and the ChonqQing Science Technology Project (KJ060416).

References

1. Dietterich, T.G., Lathrop, R.H., Lozano-Perez, T.: Solving the Multiple-Instance Problem with Axis-Parallel Rectangles. Artificial Intelligence Journal **89** (1-2) (1997) 31–71
2. Maron, O., Lozano-Pérez, T.: A Framework for Multiple-instance Learning. In: Jordan MI, Kearns MJ, Solla SA, (eds.): Advances in Neural Information Processing Systems 10. MIT Press, Cambridge (1998) 570–576
3. Zhang, Q., Goldman, S.A.: EM-DD: An Improved Multiple-Instance Learning Technique. In: Dietterich TG, Becker S, Ghahramani Z, (eds.): Advances in Neural Information Processing Systems 14. MIT Press, Cambridge (2001) 1073–1080
4. Wang, J., Zucker, J.-D.: Solving the Multiple-Instance Problem: A Lazy Learning Approach. In: Langley P, (ed.): Proceedings of the 17th International Conference on Machine Learning, San Francisco: Morgan Kaufmann Publishers (2000) 1119–1125
5. Yang, C., Lozano-Pérez, T.: Image Database Retrieval with Multiple-Instance Learning Techniques. In: Proceeding of the 16th International Conference on Data Engineering (2000) 233–243
6. Andrews, S., Tsochantaridis, I., Hofmann, T.: Support Vector Machines for Multiple-Instance Learning. In: Becker S, Thrun S, and Obermayer K, (eds.): Advances in Neural Information Processing Systems 15. MIT Press, Cambridge (2002) 561–568
7. Zhang, Q., Goldman, S.A., Yu, W., Fritts, J.E.: Content-Based Image Retrieval Using Multiple-Instance Learning. In: Proceeding of the 19th International Conference on Machine Learning (2002) 682–689
8. Carson, C., Thomas, M., Belongie, S. et al.: A System for Region-Based Image Indexing and Retrieval. In: Proceeding 3rd International Conference on Visual Information and Information Systems. Lecture Notes in Computer Science, Springer-Verlag, Berlin Heidelberg New York **1614** (1999) 509–516

Uniform Approximation Capabilities of Sum-of-Product and Sigma-Pi-Sigma Neural Networks

Jinling Long, Wei Wu*, and Dong Nan

Dept. Appl. Math., Dalian University of Technology, Dalian 116023, P.R. China
wuweiw@dlut.edu.cn

Abstract. Investigated in this paper are the uniform approximation capabilities of sum-of-product (**SOPNN**) and sigma-pi-sigma (**SPSNN**) neural networks. It is proved that the set of functions that are generated by an **SOPNN** with its activation function in $C(\mathbb{R})$ is dense in $C(\mathbb{K})$ for any compact $\mathbb{K} \in \mathbb{R}^N$, if and only if the activation function is not a polynomial. It is also shown that if the activation function of an **SPSNN** is in $C(\mathbb{R})$, then the functions generated by the **SPSNN** are dense in $C(\mathbb{K})$ if and only if the activation function is not a constant.

1 Introduction

There have been many methods for multivariate function approximation: polynomials, Fourier series, tensor products, wavelets, radial basis functions, ridge functions, etc. In this respect, a current trend is to use artificial neural networks to compute superpositions and linear combinations of simple univariate functions. One of the most important problems for neural networks is their approximation capability. This problem is related to the question that whether, or under what conditions, multivariate functions can be represented or approximated by superpositions of univariate functions. There have been many papers related to this topic: [1,2,7,8,9,11] for feedforward neural networks (**FNN**), [3,4,10,13,16,17] for radial basis function neural networks (**RBFNN**), and [5,15] for Sigma-Pi neural networks. **SOPNN** and **SPSNN** are introduced respectively in [14] and [12], and the aim of this paper is to show their uniform approximation capability.

SOPNN can approximate nonlinear mappings in a similar manner as the **FNN** and **RBF**. The output of **SOPNN** has the form $\sum_{m=1}^{M} \prod_{n=1}^{N} f_{mn}(x_n)$, where x_n's are the inputs, N is the number of inputs, and M is the number of the product terms. The function $f_{mn}(x_n)$ is supposed to have the form $\sum_k \omega_{mnk} B_{nk}(x_n)$, where $B_{nk}(\cdot)$ is a univariate basis function and ω_{mnk}'s are the weights. If $B_{nk}(\cdot)$ is a Gaussian function, the new neural network degenerates to a Gaussian function network. The learning algorithm and the novel performance

* Corresponding author, supported by the National Natural Science Foundation of China (10471017).

D. Liu et al. (Eds.): ISNN 2007, Part I, LNCS 4491, pp. 1110–1116, 2007.

in function approximation, prediction, classification and learning control are presented in [14]. For convenience of hardware implementation, artificial neural networks usually require the nonlinear basis functions (the activation functions) to have a similar structure. So we restrict the form of $B_{nk}(x_n)$ to $g(a_{nk}x_n+\theta_{nk})$ for a given univariate function $g \in C(\mathbb{R})$, and concentrate our attention to the set of functions in the form $\sum_{m=1}^{M}\prod_{n=1}^{N}\sum_{k=1}^{K_n} c_{mnk}g(a_{nk}x_n+\theta_{nk})$, where M and K_n are in $\mathbb{N}$ (natural numbers); c_{mnk}, a_{nk}, $\theta_{nk} \in \mathbb{R}$; $x = (x_1, \cdots, x_N) \in \mathbb{R}^N$. We will prove that this set of functions is dense in $C(\mathbb{K})$ if and only if g is not a polynomial.

We point out that our proof for the approximation capability of **SOPNN** uses some corresponding results obtained in [11] for **FNN**. This fact reveals to some extent the relationship between the two kinds of neural networks.

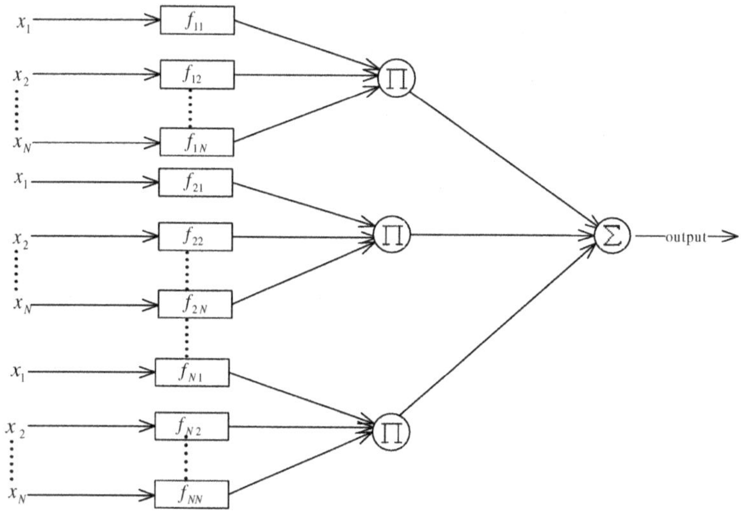

Fig. 1. Structure of sum-of-product neural network, $f_{mn}(x_n) = \sum_k \omega_{mnk}B_{nk}(x_n)$

The organization of this paper is as follows. The main result for the approximation capability of **SOPNN** is presented and proved in Section 2. In Section 3, the main result in Section 2 is generalized to deal with a sigma-pi-sigma neural network (**SPSNN**). Section 4 is devoted to a summary of results.

2 Approximation Capability of SOPNN

Lemma 1. **([6]: Weierstrass Approximation Theorem)** *Let $\mathbb{K}$ be a compact set in $\mathbb{R}^N$. Then, the polynomials in N variables form a dense set in $C(\mathbb{K})$.*

Lemma 2. ([11]) *Let $f(t) \in C(\mathbb{R})$. The set of functions $\{\sum_{k=1}^{K} c_k f(\lambda_k \cdot x + \theta_k)\}$ is dense in $C(\mathbb{K})$ for any compact set $\mathbb{K}$ in $\mathbb{R}^N$, if and only if $f(t)$ is not a polynomial on $\mathbb{R}$, where c_k, $\theta_k \in \mathbb{R}$; x, $\lambda_k \in \mathbb{R}^N$; and $\lambda_k \cdot x$ denotes the inner product of λ_k and x.*

In the following, we show that for a continuous function to be qualified as an activation function in **SOPNN**, the necessary and sufficient condition is that it is not a polynomial. The next theorem is our main result on the approximation capacity of **SOPNN**.

Theorem 1. *Let $g(t) \in C(\mathbb{R})$. The family of the functions*

$$\left\{ \sum_{m=1}^{M} \prod_{n=1}^{N} \sum_{k=1}^{K_n} c_{mnk} g(a_{nk} x_n + \theta_{nk}) \;\middle|\; M, K_n \in \mathbb{N}; c_{mnk}, a_{nk}, \theta_{nk}, x_n \in \mathbb{R} \right\} \tag{1}$$

is dense in $C(\mathbb{K})$ for any compact set $\mathbb{K}$ in $\mathbb{R}^N$, if and only if $g(t)$ is not a polynomial on $\mathbb{R}$.

Proof. Sufficiency.
Since $\mathbb{K}$ is a compact set in $\mathbb{R}^N$, there exists a hypercube $\mathbb{H} = [a_1, b_1] \times \cdots \times [a_N, b_N]$, such that $\mathbb{K} \subset \mathbb{H}$.

Any $f(x) \in C(\mathbb{K})$ can be approximated by a multivariate polynomial thanks to Lemma 1, i.e., for any $0 < \varepsilon < 1$, there exists a multivariate polynomial $P(x) = \sum_{|i|=0}^{I} \alpha_i x^i$, such that

$$|f(x) - P(x)| < \frac{\varepsilon}{2}, \; \forall x \in \mathbb{K}, \tag{2}$$

where $x^i = x_1^{i_1} x_2^{i_2} \cdots x_N^{i_N}$, $i = (i_1, i_2, \cdots, i_N)$ is a multi-index, and $|i| = i_1 + i_2 + \cdots + i_N$.

Suppose $P(x)$ has $\widetilde{m}$ terms, $A = \max_{0 \le |i| \le I} \{|\alpha_i|, 1\}$, $B = \max_{1 \le j,\, k \le N} \{|a_j|, |b_k|, 1\}$, and $L = \max_{\substack{1 \le n \le N \\ 1 \le i_n \le I}} \{|x_n^{i_n}|, a_n \le x_n \le b_n]\}$. We note that each component $x_n^{i_n}$ is a continuous function on $[a_n, b_n]$. Noting that g is not a polynomial and using Lemma 2, we can approximate $x_n^{i_n}$ by an **FNN** with g as its activation function in the following fashion: There exist $N_{i_n} \in \mathbb{N}$ and $c_k^{i_n}, \lambda_k^{i_n}, \theta_k^{i_n} \in \mathbb{R}$, such that

$$\left| x_n^{i_n} - \sum_{k=1}^{N_{i_n}} c_k^{i_n} g\left(\lambda_k^{i_n} x_n + \theta_k^{i_n}\right) \right| \le \frac{\varepsilon}{2^N A \widetilde{m} (L+1)^N B^I}, \forall x_n \in [a_n, b_n]. \tag{3}$$

Write $Q_n^{i_n}(x_n) = \sum_{k=1}^{N_{i_n}} c_k^{i_n} g\left(\lambda_k^{i_n} x_n + \theta_k^{i_n}\right)$. By Equation (3) we have that

$$\left|Q_n^{i_n}(x_n)\right| \le L + 1, \forall x_n \in [a_n, b_n], \tag{4}$$

and that

$$\left|x_n^{i_n} - Q_n^{i_n}(x_n)\right| \le \frac{\varepsilon}{2^N A \widetilde{m} (L+1)^N B^I}, \forall x = (x_1, x_2, \cdots, x_N) \in \mathbb{H}. \tag{5}$$

More generally, let us use an induction argument to establish the estimate

$$\left|x_1^{i_1}\cdots x_q^{i_q} - Q_1^{i_1}(x_1)\cdots Q_q^{i_q}(x_q)\right| \leq \frac{\varepsilon}{2^{(N+1-q)}A\widetilde{m}(L+1)^{(N+1-q)}}, \forall x \in \mathbb{H}. \quad (6)$$

Note that Equation (6) is already valid for $q = 1$ due to (5). In the following, we assume that Equation (6) is valid for q with $1 \leq q \leq N-1$, and we try to show that Equation (6) is also valid for $q+1$. To this end, we use the triangle inequality to get

$$\begin{aligned}
&\left|x_1^{i_1}x_2^{i_2}\cdots x_{q+1}^{i_{q+1}} - Q_1^{i_1}(x_1)Q_2^{i_2}(x_2)\cdots Q_{q+1}^{i_{q+1}}(x_{q+1})\right| \\
&\leq \left|x_1^{i_1}\cdots x_q^{i_q}\right|\left|x_{q+1}^{i_{q+1}} - Q_{q+1}^{i_{q+1}}(x_{q+1})\right| \\
&+ \left|x_1^{i_1}\cdots x_q^{i_q} - Q_1^{i_1}(x_1)\cdots Q_q^{i_q}(x_q)\right|\left|Q_{q+1}^{i_{q+1}}(x_{q+1})\right|.
\end{aligned}$$

It follows from Equations (3, 4, 6) and the above inequality that

$$\begin{aligned}
&\left|x_1^{i_1}x_2^{i_2}\cdots x_{q+1}^{i_{q+1}} - Q_1^{i_1}(x_1)Q_2^{i_2}(x_2)\cdots Q_{q+1}^{i_{q+1}}(x_{q+1})\right| \\
&\leq \frac{\varepsilon}{2^{(N+1-(q+1))}A\widetilde{m}(L+1)^{(N+1-(q+1))}}.
\end{aligned} \quad (7)$$

Here we have made the convention that $\prod_{t=q+2}^{N}(b_t - a_t) = 1$ when $q = N-1$. So we have proved by induction that Equation (6) is valid for all $q = 1, 2, \cdots, N$.

In particular, by Equation (6) with $q = N$, we have

$$\left|x_1^{i_1}x_2^{i_2}\cdots x_N^{i_N} - Q_1^{i_1}(x_1)Q_2^{i_2}(x_2)\cdots Q_N^{i_N}(x_N)\right| \leq \frac{\varepsilon}{2A\widetilde{m}}, \forall x \in \mathbb{H}. \quad (8)$$

Set $Q(x) = \sum_{|i|=0}^{I} \alpha_i Q_1^{i_1}(x_1)Q_2^{i_2}(x_2)\cdots Q_N^{i_N}(x_N)$, then Equation (8) implies

$$\begin{aligned}
|P(x) - Q(x)| &\leq \sum_{|i|=0}^{I} |\alpha_i|\left|x_1^{i_1}x_2^{i_2}\cdots x_N^{i_N} - Q_1^{i_1}(x_1)Q_2^{i_2}(x_2)\cdots Q_N^{i_N}(x_N)\right| \\
&\leq \frac{\varepsilon}{2}, \forall x \in \mathbb{H}.
\end{aligned} \quad (9)$$

A combination of Equations (2) and (9) implies

$$|f(x) - Q(x)| \leq |f(x) - P(x)| + |P(x) - Q(x)| < \frac{\varepsilon}{2} + \frac{\varepsilon}{2} = \varepsilon, \forall x \in \mathbb{K}. \quad (10)$$

Obviously $Q(x) \in \left\{\sum_{m=1}^{M}\prod_{n=1}^{N}\sum_{k=1}^{K_n} c_{mnk}g(a_{nk}x_n + \theta_{nk})\right\}$, and the estimate (10) shows that $C(\mathbb{K}) \subseteq \left\{\sum_{m=1}^{M}\prod_{n=1}^{N}\sum_{k=1}^{K_n} c_{mnk}g(a_{nk}x_n + \theta_{nk})\right\}$. Also note the apparent inclusion relation $\left\{\sum_{m=1}^{M}\prod_{n=1}^{N}\sum_{k=1}^{K_n} c_{mnk}g(a_{nk}x_n + \theta_{nk})\right\} \subseteq C(\mathbb{K})$. Then we see that $\left\{\sum_{m=1}^{M}\prod_{n=1}^{N}\sum_{k=1}^{K_n} c_{mnk}g(a_{nk}x_n + \theta_{nk})\right\} = C(\mathbb{K})$.

Necessity.

Now let us assume $g(t)$ is a univariate polynomial with degree l. Then, all the functions in the form of (1) are polynomials with degrees at most l^N, which of course are not dense in $C(\mathbb{K})$. □

Remark 1. Theorem 1 describes an approximate representation of multivariate functions by univariate functions. It provides an answer to the question that what kinds of univariate functions are qualified to approximate multivariate functions. In our proof, $Q(x)$ as the approximant to a function in $C(\mathbb{K})$ is entirely constructed in terms of univariate functions $Q_n^{i_n}(x_n)$. The role of $Q_n^{i_n}(x_n)$ in the approximant $Q(x)$ is very much like the role of $x_n^{i_n}$ in the polynomial $P(x)$ as an approximant to a general nonlinear function.

3 Approximation Capability of SPSNN

A sigma-pi-sigma neural network (**SPSNN**) is proposed in [12]. Its output is

$$\sum_{m=1}^{M}\prod_{j=1}^{J_m}\sum_{n=1}^{N}\sum_{k=1}^{K_n}\omega_{mjnk}B_{jnk}(x_n), \tag{11}$$

where $x=(x_1,\cdots,x_N)\in\mathbb{R}^N$ is the input, ω_{mjnk}'s are the weights, $B_{jnk}(\cdot)$'s are univariate basis functions, and M, J_m and $K_n\in\mathbb{N}$. Similarly as before, we concentrate our attention to the family of functions of the form

$$\sum_{m=1}^{M}\prod_{j=1}^{J_m}\sum_{n=1}^{N}\sum_{k=1}^{K_n}c_{mjnk}g(a_{jnk}x_n+\theta_{jnk}),$$

where $c_{mjnk}, a_{jnk}, \theta_{jnk}\in\mathbb{R}$.

In this section, we show that for a continuous function to be qualified as an activation function in **SPSNN**, the necessary and sufficient condition is that it is not a constant. The next theorem is the other main result on the approximation capacity of **SPSNN**.

Theorem 2. *Suppose $g(t)\in C(\mathbb{R})$. The family of functions of $x=(x_1,\cdots,x_N)$*

$$\left\{\sum_{m=1}^{M}\prod_{j=1}^{J_m}\sum_{n=1}^{N}\sum_{k=1}^{K_n}c_{mjnk}g(a_{jnk}x_n+\theta_{jnk})\;\middle|\;M,K_n,J_m\in\mathbb{N};c_{mjnk},a_{jnk},\theta_{jnk}\right\} \tag{12}$$

is dense in $C(\mathbb{K})$ for any compact set $\mathbb{K}$ in $\mathbb{R}^N$, if and only if $g(t)$ is not a constant function.

Proof. Sufficiency.

We first consider the case that $g(t)$ is not a polynomial. Note that the family of functions of form (1) is a subset of that of form (12), thus the latter must be dense in $C(\mathbb{K})$ for any compact set $\mathbb{K}$ in $\mathbb{R}^N$ according to Theorem 1.

Next, we turn to the other case that $g(t)$ is a polynomial but not a constant function on $\mathbb{R}$. We claim that $g(t)$ can generate the monomial t by translations, stretching and sums. To see this, let us assume that $g(t) = a_l t^l + a_{l-1} t^{l-1} + \cdots + a_0$ with $l \geq 1$ and $a_l \neq 0$. Then, $g(t+1) = a_l(t+1)^l + a_{l-1}(t+1)^{l-1} + \cdots + a_0$, and we have that $h_1(t) \equiv g(t+1) - g(t) = l a_l t^{l-1} + p_{(l-2)}(t)$, where $p_{(l-2)}(t)$ denotes a polynomial with degree equal to or less than $l-2$. We proceed to observe that $h_2(t) \equiv h_1(t+1) - h_1(t) = l(l-1) a_l t^{(l-2)} + p_{(l-3)}(t)$. And we can repeat this procedure to obtain $h_{l-1} = l(l-1)\cdots 2a_l t + p_0(t) = l(l-1)\cdots 2a_l t + b_0$, where b_0 is a constant. Let $b_1 = l(l-1)\cdots 2a_l$, then $h(t) \equiv \frac{1}{b_1}(h_{(l-1)}(t) - b_0) = t$. This confirms the claim. Next, we notice that the functions of form (12) are all polynomials and they constitute an algebra. On the other hand, it follows from the above claim that all the monomials $x_1, x_2, \cdots, x_N$ are members of the family (12). Thus, the family (12) must contains all the multivariate polynomials. Therefore, by Lemmas 1, the family (12) is also dense in $C(\mathbb{K})$ for any compact set $\mathbb{K}$ in $\mathbb{R}^N$ when $g(t)$ is a nonconstant polynomial.

Necessity.

Otherwise, if $g(t)$ is a constant function on $\mathbb{R}$, then the functions of form (12) are all constant functions on $\mathbb{R}^N$, which of course are not dense in $C(\mathbb{K})$. □

Remark 2. Theorem 2 gives the other approximate representation of multivariate functions by univariate functions. It provides an answer to the question that what kinds of univariate functions are qualified to approximate multivariate functions by the **SPSNN**. We can see that any nonconstant continuous function can be used as an activation function in the **SPSNN**.

4 Conclusion

The approximations of multivariate functions by sum-of-product and sigma-pi-sigma neural networks with a univariate activation function are investigated. This paper solves the problem of whether a function is qualified as an activation function in the two kinds of new structure neural networks. We have proved that all the functions generated by the **SOPNN** with its activation function in $C(\mathbb{R})$ are dense in $C(\mathbb{K})$, if and only if the activation function is not a polynomial. We also show that if the activation function of the **SPSNN** is in $C(\mathbb{R})$, then the functions generated by the **SPSNN** are dense in $C(\mathbb{K})$ if and only if the activation function is not a constant.

References

1. Attali, J.G., Pagès, G.: Approximations of Functions by a Multilayer Perceptron: a New approach. Neural Networks **10** (1997) 1069-1081
2. Chen, T.P., Chen, H., Liu, R.: Approximation Capability in by Multilayer Feedforward Networks and Related Problems. IEEE Transactions on Neural Networks **6** (1) (1995) 25-30

3. Chen, T.P., Chen, H.: Approximation Capability to Functions of Several Variables, Nonlinear Functionals and Operators by Radial Basis Function Neural Networks. IEEE Transactions on Neural Networks **6** (4) (1995) 904-910
4. Chen, T.P., Chen, H.: Universal Approximation Capability of RBF Neural Networks with Arbitrary Activation Functions and Its Application to Dynamical Systems. Circuits, Systems and Signal Processing **15** (5) (1996) 671-683
5. Chen, T.P., Wu, X.W.: Characteristics of Activation Function in Sigma-Pi Neural Networks. Journal of Fudan University **36** (6) (1997) 639-644
6. Cheney, W., Light, W.: A Course in Approximation Theory. Beijing: China Machine Press (2003)
7. Chui, C.K., Li, X.: Approximation by Ridge Functions and Neural Networks with One Hidden Layer. Journal of Approximation Theory **70** (1992) 131-141
8. Cybenko, G.: Approximation by Superpositions of Sigmoidal Functions. Mathematics of Control, Signals, and Systems **2** (1989) 303-314
9. Hornik, K.: Approximation Capabilities of Mutilayer Feedforward Networks. Neural Networks **4** (1991) 251-257
10. Jiang, C.H.: The Approximate Problem on Neural Network. Annual of Math (in Chinese) **19**A (1998) 295-300
11. Leshno, M., Lin, Y.V., Pinkus, A., Schocen, S.: Multilayer Feedforward Networks with a Non-Polynomial Activation Function Can Approximate Any Function. Neural Networks **6** (1993) 861-867
12. Li, C.K.: A Sigma-Pi-Sigma Neural Network (SPSNN). Neural Processing Letters **17** (2003) 1-19
13. Liao, Y., Fang, S., Nuttle, H.L.W.: Relaxed Conditions for Radial-Basis Function Networks to Be Universal Approximators. Neural Networks **16** (2003) 1019-1028
14. Lin, C.S., Li, C.K.: A Sum-of-Product Neural Network(SOPNN). Neurocomputing **30** (2000) 273-291
15. Luo, Y.H., Shen, S.Y.: L^p Approximation of Sigma-Pi Neural Networks. IEEE Transactions on Neural Networks **11** (6) (2000) 1485-1489
16. Park, J., Sandberg, I.W.: Universal Approximation Using Radial- Basis-Function Networks. Neural Computation **3** (1991) 246-257
17. Park, J., Sandberg, I.W.: Approximation and Radial-Basis-Function Networks. Neural Computation **5** (1993) 305-316

Regularization for Regression Models Based on the K-Functional with Besov Norm

Imhoi Koo and Rhee Man Kil

Division of Applied Mathematics
Korea Advanced Institute of Science and Technology
373-1 Guseong-dong, Yuseong-gu, Daejeon 305-701, Korea
haihaba@amath.kaist.ac.kr, rmkil@kaist.ac.kr

Abstract. This paper presents a new method of regularization in regression problems using a Besov norm (or semi-norm) acting as a regularization operator. This norm is more general smoothness measure to approximation spaces compared to other norms such as Sobolev and RKHS norms which are usually used in the conventional regularization methods. In our work, we also suggest a new candidate of the regularization parameter, that is, the trade-off between the data fit and the smoothness of the estimation function. Through the simulation for function approximation, we have shown that the suggested regularization method is effective and the estimated values of regularization parameters are close to the optimal values associated with the minimum expected risks.

1 Introduction

The task of learning from data is minimizing the expected risk (or generalization error) of regression model under the constraints of the absence of a priori model of data generation and of the limited size of data. For this learning problem, the estimation function (or regression model) with enough number of parameters can be established and trained to minimize the square errors between the target values and function estimates. However, the reducing the square errors for training samples does not guarantee to minimize the expected risk which can be decomposed by the bias and variance terms of regression models. If we increase the number of parameters, the bias term is decreased but the variance term is increased or vice versa. If the number of parameters is so small that the performance is not optimal due to large bias term, it is called the under-fitting of regression models. If the number of parameters is too large so that the performance is not optimal due to large variance term, it is called the over-fitting of regression models. So there is a reasonable trade-off between the under-fitting and over-fitting of regression models. To cope with this problem, one of the methods is reducing the error measure composed of train error, that is, the distance between the target response and the fitted value of the regression model, and the regularization term which biases the solution toward functions with a priori desirable characteristics, such as the smoothness of estimation functions. The most popular regularization term is weight decay method[1] or weight elimination method[2] in which the squared norm of the weights (or parameters)

D. Liu et al. (Eds.): ISNN 2007, Part I, LNCS 4491, pp. 1117–1126, 2007.

of the estimation function is used as a regularization term, but it is an ad hoc technique that controls weight values rather than the fit to the data directly. Furthermore, the weight decay does not necessarily mean the smoothness of estimation functions. In this context, the regularization methods[3,4] based on the smoothness of the estimation function were suggested. These methods consider the smoothness measure in the Sobolev space. In this paper, we consider more general form of smoothness measure for the derivation of new regularization forms using the K-functional which is the measure of the minimum distance between two functions using the defined norm. For this derivation, we consider the Besov norms since they are more general than other norms such as RKHS and Sobolev norms which are usually used in the conventional regularization methods. As a result, through the simulation for function approximation, we have shown that the suggested regularization method provides better performance than other regularization methods from the view point of the expected risk. For this simulation, we also provides the method of estimating the regularization parameter, that is, the trade-off between the data fit and the smoothness of the estimation function, using the density of input samples.

2 Regularization for Regression Models

Consider the regression problem of estimating a continuous function f in $C(X, \mathbb{R})$ where X is the compact subset of Euclidean space $\mathbb{R}^m$ ($m \geqslant 1$) and $C(X, \mathbb{R})$ is a class of continuous functions. The observed output y for $\mathbf{x} \in X$ can be represented by

$$y = f(\mathbf{x}) + \epsilon \tag{1}$$

where $f(\mathbf{x})$ is the target function and ϵ is a random noise with mean zero and variance σ_ϵ^2. Here, the input and output samples $(\mathbf{x}_i, y_i)$, $i = 1, \cdots, N$ are randomly generated according to the distribution $P(\mathbf{x})$, $\mathbf{x} \in X$:

$$y_i = f(\mathbf{x}_i) + \epsilon_i, \quad \mathbf{x}_i \in X \tag{2}$$

where ϵ_i, $i = 1, \cdots, N$ are independent and identically distributed random variables with mean zero and variance σ_ϵ^2. For these samples, our goal is to construct an estimation network (or function) $f_n(\mathbf{x})$ in the function space F_n which minimizes the expected risk

$$R(f_n) = \int_{X \times \mathbb{R}} L(y, f_n(\mathbf{x})) dP(\mathbf{x}, y) \tag{3}$$

with respect to the number of parameters n, where $L(y, f_n(\mathbf{x}))$ is a given loss functional, usually the square loss function $L(y, f_n(\mathbf{x})) = (y - f_n(\mathbf{x}))^2$ for regression problems. In general, we can construct an estimation function as

$$f_n(\mathbf{x}) = \sum_{k=1}^{n} w_k \phi_k(\mathbf{x}) \tag{4}$$

where w_k and ϕ_k represent the kth weight value and kernel function respectively.

To minimize the expected risk (3), we have to identify the distribution $P(\mathbf{x}, y)$ but it is usually unknown. Rather, the parameters minimizing the empirical risk

$$R_{emp}(f_n) = \frac{1}{N}\sum_{i=1}^{N} L(y_i, f_n(\mathbf{x}_i)) \tag{5}$$

are obtained for the given samples. If we increase the number of parameters, the empirical risk of (5) is decreased so that the bias of an estimation function is decreased but the variance of an estimation function is increased and vice versa. Therefore, there is a reasonable trade-off between the bias and variance terms to minimize the expected risk composed of these two terms.

To solve this problem, on can use the regularization theory[5], that is, we determined the solution f that minimizes the penalized empirical risk functional

$$\frac{1}{N}\sum_{i=1}^{N} L(y_i, f_n(\mathbf{x}_i)) + \lambda \|f_n\|^2_{\mathcal{H}_K} \tag{6}$$

where λ represents the regulation parameter and $\|\cdot\|_{\mathcal{H}_K}$ represents a norm of the reproducing kernel Hilbert space (RKHS) $\mathcal{H}_K$ with the positive semi-definite, symmetric kernel K. Here, the regularization parameter λ works as a trade-off between the empirical risk and the smoothness of estimation function f_n. Wahba[3] suggested the general cross validation (GCV) for a ridge regression problem to find the optimal penalty parameter λ and also suggested the leave-one-out method to modify the cross validation. These methods consider the smoothness measure in the Sobolev space. In this paper, we consider more general form of smoothness measure for the derivation of new regularization forms using the K-functional which is the measure of the minimum distance between two functions using the defined norm.

3 K-Functionals and Smoothness Measures

Let G and H be normed linear spaces, with H continuously embedded in G, that is, $H \subset G$ and $\|\cdot\|_G \leqslant C\|\cdot\|_H$ where C is a positive constant. Then, for any $t > 0$, we can define the K-functional for $g \in G$:

$$K(g,t) := K(g,t,G,H) := \inf_{h\in H}\{\|g-h\|_G + t\|h\|_H\} \tag{7}$$

where $\|\cdot\|_G$ is the norm on G. Here, $\|\cdot\|_H$ can be replaced by a semi-norm $|\cdot|_H$ on H, that is,

$$K(g,t,G,H) := \inf_{h\in H}\{\|g-h\|_G + t|h|_H\} \tag{8}$$

where a semi-norm $|\cdot|_H$. Indeed, the inequality $K(g,t) < \epsilon$ for some $t > 0$ implies that g can be approximated with error $\|g-h\|_G < \epsilon$ by an element $h \in H$, whose norm is not too large, that is, $\|h\|_H < \epsilon t^{-1}$ or $|h|_H < \epsilon t^{-1}$.

The original form of K-functional defined in (7) or (8) can not be directly applied to regression problems since we don't exactly know the target function

$f \in F$. From this point of view, if spaces F and F_n are based on p-norm, we can consider the following modified K-functional using the empirical risk:

$$\widetilde{K}(f,t,F,F_n)^p = \inf_{f_n \in F_n} \left\{ \frac{1}{N}\sum_{i=1}^{N} |f(x_i) - f_n(x_i)|^p + t^2 |f_n|^p_{F_n} \right\}. \quad (9)$$

Here, let us consider an example of K-functional using the reproducing kernel Hilbert space (RKHS):

Example Let $\mathcal{H}_K$ be a RKHS with reproducing kernel K and the kernel function K be continuous on $\Omega \times \Omega$, symmetric real-valued function such that the integral operator $L_K : L_2(\Omega) \longrightarrow L_2(\Omega)$,

$$(L_K f)(x) = \int_\Omega K(x,y) f(y) d\mu(y)$$

is positive semi-definite. Then, the series with respect to orthogonal eigenvectors ϕ_k with $\|\phi_k\| = 1$ and eigenvalues λ_k of L_K converges uniformly $K(x,y)$, that is,

$$K(x,y) = \sum_{k=1}^{\infty} \lambda_k \phi_k(x) \phi_k(y).$$

This property is called Mercer's theorem. Let us consider a function $g \in \mathcal{H}_K$ such that

$$g(x) = \sum_{k=1}^{\infty} a_k \phi_k(x). \quad (10)$$

Then, we have the norm of $g(x)$ as

$$\|g\|^2_{\mathcal{H}_K} = \sum_{k=1}^{\infty} \frac{a_k^2}{\lambda_k} \quad (11)$$

and the K-functional for $f \in L_2$ becomes

$$\begin{aligned} K(f,t,L_2,\mathcal{H}_K) &= \inf_{g \in \mathcal{H}_K} \{ \|f-g\|_{L_2} + t\,\|g\|_{\mathcal{H}_K} \} \\ &= \inf_{g \in \mathcal{H}_K} \left\{ \|f-g\|_{L_2} + t \left(\sum_{k=1}^{\infty} \frac{a_k^2}{\lambda_k} \right)^{1/2} \right\}. \end{aligned} \quad (12)$$

Let us introduce the smoothness spaces based on the modulus of smoothness. For positive $h \in \mathbb{R}$ and for any positive integer r, the r-th difference operator with step h is defined by

$$\Delta^r_\delta(f,x) := \sum_{k=0}^{r} (-1)^{r-k} \binom{r}{k} f(x + k\delta). \quad (13)$$

If $x + kh$ is not in a domain $\Omega \subset \mathbb{R}^d$ for any $k = 0, \cdots, r$, then $\Delta_h^r(f, x)$ becomes zero. Furthermore, if $f \in L_p(\Omega)$, $0 < p \leqslant \infty$, the r-th order modulus of smoothness of f in $L_p(\Omega)$ is defined by

$$\omega_r(f, h)_p := \sup_{|\delta| \leqslant h} \|\Delta_\delta^r(f, \cdot)\|_{L_p(\Omega)} .$$

Then, we can use $\omega_r(f, h)_p$ to measure the smoothness of function f. Let us also define the Besov space as the class of all functions f satisfying that the following semi-norm described by $\omega_r(f, h)_p$, is finite:

$$|f|_{B_q^\alpha(L_p(\Omega))} := \begin{cases} \left[\int_0^\infty h^{-\alpha}(\omega_r(f,h)_p)^q \frac{dh}{h}\right]^{1/q} & \text{if } 0 < q < \infty \\ \sup\limits_{h>0} \frac{\omega_r(f,h)_p}{h^\alpha} & \text{if } q = \infty \end{cases} ,$$

where $\alpha > 0$, $0 < p \leqslant \infty$, $0 < q \leqslant \infty$. Here, we take $r := \lceil \alpha \rceil + 1$, that is, the smallest integer larger than α. It is easy to check that $|f|_{B_q^\alpha}$ is the semi-norm on $B_q^\alpha(L_p(\Omega))$. Therefore, we can describe the K-functional of f with respect to Besov semi-norm:

$$K(f, t, L_p, B_q^\alpha(L_p)) := \inf_{f_n \in B_q^\alpha} \left\{ \|f - f_n\|_{L_p} + t|f_n|_{B_q^\alpha} \right\} . \tag{14}$$

Here, the modified K-functional with $p > 0$ and Besov semi-norm using the empirical risk is

$$\widetilde{K}(f, t)^p = \inf_{f_n \in B_q^\alpha} \left\{ \frac{1}{N} \sum_{i=1}^N |f(x_i) - f_n(x_i)|^p + t^p |f_n|_{B_q^\alpha}^p \right\} . \tag{15}$$

Since the modulus of smoothness is based on the L_p norm, the Besov norm is independent on the discontinuity of f_n. Here, we take $\alpha = 1 (r = 2)$, $p = 2$, and $q = \infty$ for the easy calculation of Besov semi-norms and for the loss functions using square errors. Thus, we have the following empirical K-functional:

$$\widetilde{K}(f, t)^2 = \inf_{f_n \in B_\infty^1} \left\{ \frac{1}{N} \sum_{i=1}^N |f(x_i) - f_n(x_i)|^2 + t^2 \left(\sup_{h>0} \frac{\omega_2(f_n, h)_2}{h} \right)^2 \right\} . \tag{16}$$

In the above equation, it is important to determine the constant t referred to as the regularization parameter which is usually determined experimentally such as finding the parameter value using the cross-validation technique in other regularization methods. First, let us consider the dependency of the modulus of continuity $\omega_2(f_n, h)_2$ in the mean square error of estimation function f_n. If the estimation function f_n becomes complex, $\omega_2(f_n, h)_2$ increases and the variance of f_n increases. Since the mean square error of f_n is decomposed by the variance of f_n and the square of bias, this results in increasing the mean square error of f_n. Thus, in equation (16), the value of t can be associated with the value of h. In practice, we can give some bound of h at which $\omega_2(f_n, h)_2/h$ has the maximum value. Since, in the experiments, the value of $\omega_2(f_n, h)_2$ should be determined by the samples and h is a distance defined in the input space X, the half of the distance between two adjacent input samples can be a good candidate for h. From this point of view, we consider to use the *density* of input samples as the candidate value for t. Actually, the density gives an important role in the approximation theory on the L_∞ spaces and model selection problems[8]. In practice, the density of one dimensional input space can be described by the mean of the distances of between two adjacent input samples. For instance, for N samples, the estimate $\widehat{t}$ for the parameter t can be determined by

$$\widehat{t} = \frac{B-A}{2(N+1)} \tag{17}$$

assuming the uniform distribution of input samples on $[A, B]$. As a result, the value $\widehat{t}$ is dependent on the length of a input domain and the number of samples. In the case that the density $\widehat{t}$ is small, that is, the small length of input domain or large number of input samples, the empirical error is sufficiently closed to the expected error by the weak law of large numbers or vice versa.

4 Besov Norm Based Regularization for Spline Regression Models

We consider estimation functions using the splines (on the circle) for the approximation of target functions, that is, estimation functions have the following form with trigonometric polynomials:

$$f_n(x) = a_0 + \sum_{k=1}^{n}(a_k \cos kx + b_k \sin kx). \tag{18}$$

For the above estimation functions, we consider the empirical K-functional of (16) and propose the following theorem:

Theorem 1. *Let us define* $B^1_\infty(\mathbb{T}_n, \hat{t}) = \left\{ f_n \in \mathbb{T}_n : \sup_{0<t<\hat{t}} \frac{\omega_2(f_n,t)_2}{t} < \infty \right\}$ *and* f_n *be given by (18). Then, for some* $t > 0$*, the empirical K-functional* $\widetilde{K}(f, \hat{t})^2$ *satisfies the following inequality:*

$$\widetilde{K}(k, \hat{t})^2 \leqslant \tag{19}$$

$$\inf_{f_n \in B^1_\infty} \left\{ \frac{1}{N}\sum_{i=1}^{N}(f(x_i) - g(x_i))^2 + \pi \sum_{k=1}^{n_0}(a_k^2 + b_k^2)k^4\hat{t}^4 + \frac{16}{\pi}(n_0+1)^2\hat{t}^2 \sum_{k=n_0+1}^{n}(a_k^2 + b_k^2) \right\}.$$

where $n_0 = \lfloor \pi/\hat{t} \rfloor$.

Proof. First, let us calculate $\left\|\Delta_h^2 f_n(\cdot)\right\|_2$ for $h(>0)$. Then,

$$\left\|\Delta_h^2 f_n(\cdot)\right\|_2 = \left[\int_{-\pi+h}^{\pi-h} (f_n(x+h) - 2f_n(x) + f_n(x-h))^2 \, dx\right]^{1/2}.$$

For the component of f_n, we calculate $\left\|\Delta_h^2 \cos kx\right\|_2^2$ as

$$\left\|\Delta_h^2 \cos kx\right\|_2^2 = \left[\int_{-\pi+h}^{\pi-h} \left(\cos k(x+h) - 2\cos kx + \cos k(x-h)\right)^2 dx\right]^{1/2}.$$

Here, the bound of $\left\|\Delta_h^2 \cos kx\right\|_2^2$ is given by

$$\left\|\Delta_h^2 \cos kx\right\|_2^2 \leqslant \int_{-\pi}^{\pi} \left(\Delta_h^2 \cos kx\right)^2 dx \leqslant 16\pi \sin^4 \frac{kh}{2}.$$

Thus, the 2nd order modulus of smoothness for f_n is determined by

$$\omega_2(f_n, t)_2 = \sup_{|h|<t} \left\|\Delta_h^2 f(\cdot)\right\|_2 \leqslant \sup_{|h|<t} 4\sqrt{\pi}\sqrt{\sum_{k=1}^{n} (a_k^2 + b_k^2) 16\pi \sin^4 \frac{kh}{2}}.$$

There exists a positive integer n_0 such that $\frac{\pi}{t} - 1 < n_0 \leqslant \frac{\pi}{t}$. This implies that

$$\omega_2(f_n, t)_2 \leqslant 4\sqrt{\pi}\left[\sum_{k=1}^{n_0} (a_k^2 + b_k^2)\sin^4 \frac{kt}{2} + \sum_{k=n_0+1}^{n} (a_k^2 + b_k^2)\right]^{1/2}.$$

Thus,

$$\begin{aligned}\frac{\omega_2(f,t)_2}{t} &\leqslant 4\sqrt{\pi}\left[\frac{1}{t^2}\sum_{k=1}^{n_0} (a_k^2 + b_k^2)\sin^4 \frac{kt}{2} + \frac{1}{t^2}\sum_{k=n_0+1}^{n} (a_k^2 + b_k^2)\right]^{1/2} \\ &\leqslant 4\sqrt{\pi}\left[\sum_{k=1}^{n_0} (a_k^2 + b_k^2)\frac{k^4 t^2}{16} + \frac{1}{t^2}\sum_{k=n_0+1}^{n} (a_k^2 + b_k^2)\right]^{1/2}.\end{aligned}$$

Therefore, for a positive integer n_0 with $\frac{\pi}{t} - 1 < n_0 \leqslant \frac{\pi}{t}$, the Besov semi-norm for f_n with $\alpha = 1, r = 2, p = 2, q = \infty$ satisfies the following inequality:

$$\sup_{0<t<\hat{t}} \frac{\omega_2(f,t)_2}{t} \leqslant 4\sqrt{\pi}\left\{\sum_{k=1}^{n_0} (a_k^2 + b_k^2)\frac{k^4 \hat{t}^2}{16} + \left(\frac{n_0+1}{\pi}\right)^2 \sum_{k=n_0+1}^{n} (a_k^2 + b_k^2)\right\}^{1/2}.$$

This implies that the empirical K-functional $\widetilde{K}(f, \hat{t})^2$ satisfies the following inequality:

$$\widetilde{K}(k, \hat{t})^2 \leqslant$$

$$\inf_{f_n \in B_\infty^1}\left\{\frac{1}{N}\sum_{i=1}^{N} (f(x_i) - g(x_i))^2 + \pi\sum_{k=1}^{n_0} (a_k^2 + b_k^2) k^4 \hat{t}^4 + \frac{16}{\pi}(n_0+1)^2 \hat{t}^2 \sum_{k=n_0+1}^{n} (a_k^2 + b_k^2)\right\}.$$

□

This theorem states that the regularization term is dependent upon the value of n_0 which is related to the number of samples. In the case that the number of basis functions n is less than n_0, the regularization term is similar to Wahba's result [3] using the RKHS. In the case of $n > n_0$, the suggested regularization term is similar to Wahba's result when $k \leqslant n_0$ while the terms not dependent upon k are added to the regularization term when $k > n_0$.

5 Simulation for Function Approximation

We performed the simulation for function approximation using the estimation function with trigonometric polynomials to check the validity of the suggested theorem. For the benchmark data, we choose target functions from Donoho-Johnstone Benchmark data [7], they are blocks, bumps, heavysine, and doppler as illustrated in Figure 1. For each target function, we made samples according

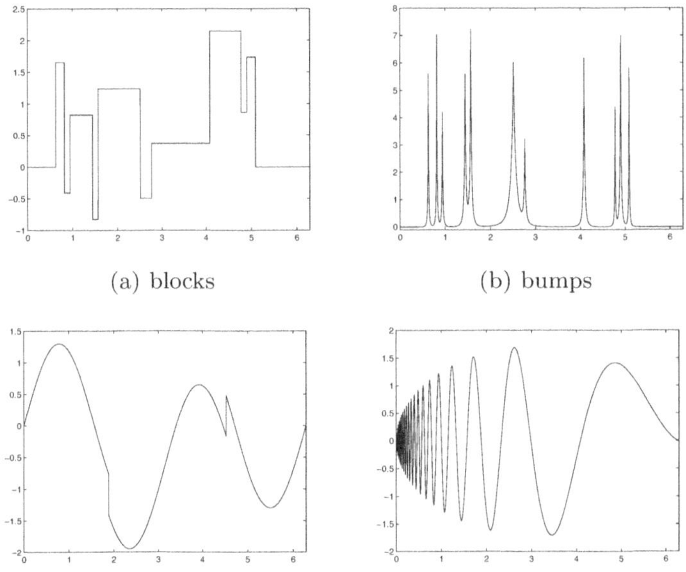

(a) blocks (b) bumps

(c) heavysine (d) doppler

Fig. 1. Donoho-Johnstone Benchmark data

to (2). In this simulation, the input samples were independently generated by a uniform distribution on the range of $[0, 2\pi]$. The noise terms were also independently generated according to the normal distribution with a mean zero and a standard deviation σ_ϵ, and added to the output samples. Here, for each target function, we generated ten sets of training samples of size 100 with $\sigma_\epsilon = 0.2$. To evaluate the performance of regression models, we also generated another set of test samples of size 2000. As the estimation function, we used the spline regression model of (18) with $n = 25$. For the regularization of this estimation model, we used the weight decay method in which the regularization term was given by

$$a_0^2 + \sum_{k=1}^{25} \left(a_k^2 + b_k^2 \right),$$

the Sobolev-norm-based method [3], and the suggested Besov-norm-based method of (19). The simulation results for approximating target functions using the spline regression model with the weight decay as the regularization term and the K-functional method using the Sobolev or Besov semi-norm dependent upon the regularization parameter t were illustrated in Figures 2. In the suggested method, we also considered to measure the risk ratio r_R to represent the effectiveness of $\hat{t}$ which was estimated by the density parameter of (17):

$$r_R = \log_{10} \frac{MSE(f_n)|_{t=\hat{t}}}{\min_{t \in T} MSE(f_n)|_t} \tag{20}$$

where $MSE(f_n)|_t$ represents the mean square error of f_n with the regularization parameter $t \in T = [10^{-5}, 10^0]$. The above risk ratios for each target function were plotted as box-plots as illustrated in Figure 3. These results showed that the K-functional method using the Sobolev or Besov semi-norm achieved better performance than the weight decay as the regularization term when we compared the minimum mean square errors (MMSEs), and the estimated values of regularization parameters using the density parameter $\widehat{t}$ for the Besov semi-norm, were close to the optimal regularization parameter values at which the MMSEs were achieved.

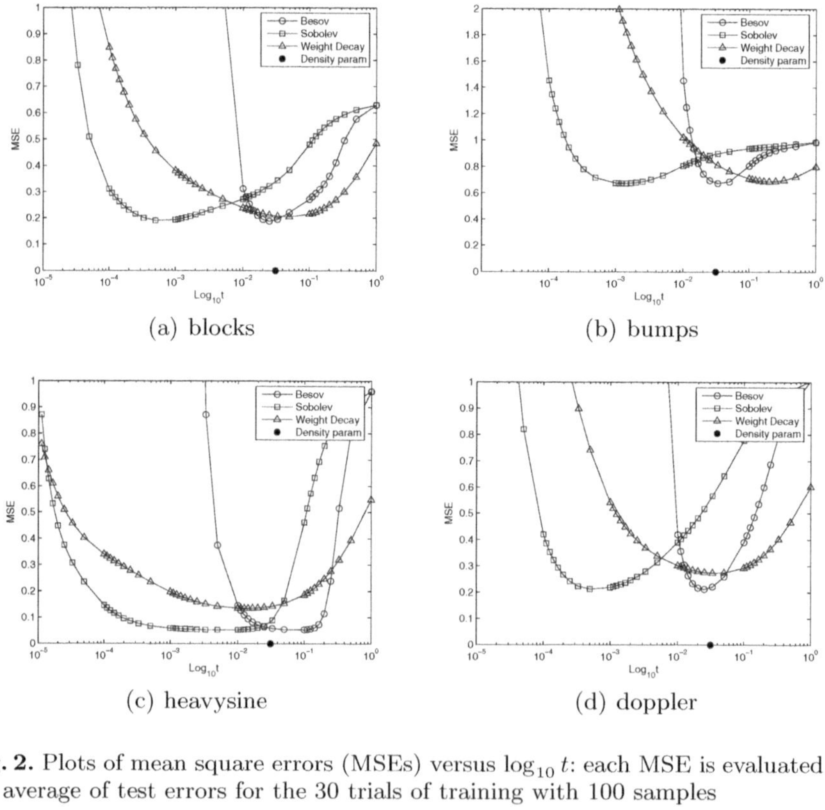

(a) blocks (b) bumps

(c) heavysine (d) doppler

Fig. 2. Plots of mean square errors (MSEs) versus $\log_{10} t$: each MSE is evaluated from the average of test errors for the 30 trials of training with 100 samples

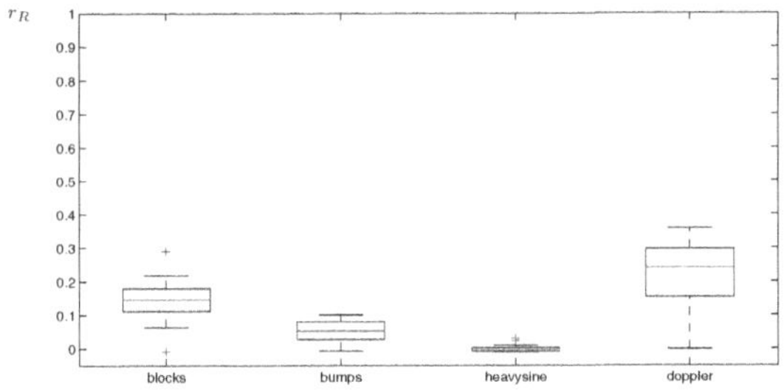

Fig. 3. Risk ratios for Donoho-Johnstone benchmark data

6 Conclusion

In this paper, we presented a new method of regularization in regression problems using a K-functional in Besov spaces. Through the simulation for function approximation, we have shown that the suggested regularization method is effective and the estimated values of regularization parameters are close to the optimal values associated with MMSEs. This approach can be applied to other regression models such as the estimation functions with sigmoid or radial basis functions by computing the appropriate Besov-norms in the estimation function spaces.

References

1. Hinton, G.E.: Connectionist Learning Procedures. Artificial Intelligence **40** (1989) 185–234
2. Weigend, A.S., Rumelhart, D.E., Huberman, B.A.: Generalization by Weight-Elimination with Application to Forecasting. Advances in Neural Information Processing Systems **3** (1991) 875–882
3. Wahba, G.: Splines Models for Observational Date. Series in Applied Mathematics **59** SIAM Philadelphia (1990)
4. Moody, J.E., Rögnvaldsson, T.: Smoothing Regularizers for Projective Basis Function Networks. Advances in Neural Information Processing Systems **9** (1997) 585–591
5. Tikhonov, A.N., Arsenin, V.Y.: Solutions of Ill-Posed Problems. W.H. Winston Washington D.C. (1977)
6. DeVore, R.A.: Nonlinear Approximation. Acta Numerica, Cambridge University Press (1998) 51–150
7. Donoho, D.L., Johnstone, I.M.: Adapting to Unknown Smoothness via Wavelet Shrinkage. Journal of the American Statistical Association **90 (432)** (1995) 1200–1224
8. Koo, I., Kil, R.M.: Nonlinear Model Selection Based on the Modulus of Continuity. International Joint Conference on Neural Networks (2006) 3552–3559

Neuro-electrophysiological Argument on Energy Coding

Rubin Wang and Zhikang Zhang

Institute for Brain Information Processing and Cognitive Neurodynamics, School of Information Science and Engineering, East China University of Science and Technology, Meilong Road 130, Shanghai 200237, P.R. China
rbwang@163.com

Abstract. According to analysis of both neuro-electrophysiological experimental data and the biophysical properties of neurons, in early research paper we proposed a new biophysical model that reflects the property of energy coding in neuronal activity. On the based of the above research work, in this paper the proposed biophysical model can reproduce the membrane potentials and the depolarizing membrane current by means of neuro-electrophysiological experimental data. Combination with our previous research results, the proposed biophysical model is demonstrated again to be more effective compared with known biophysical models of neurons.

1 Introduction

Due to the limitations in current biophysical models of neural coding, research into the mechanisms of neural information processing remain very difficult [1-5]. Because of these limitations, currently, the principles of neural information processing underlying cognitive processes within the brain are not completely understood [3, 6-9]. William B Levy and Robert A. Baxter studied the relationship between neural information and energy consumption, and gave a description of the average energy consumption required for a given level of neural network activity according to Shannon's principle. Here the role of energy efficiency was detected in the process of neural coding [10, 11]. Recently, Simon B. Laughlin and Terrence J. Sejnowski have posited that networks of neurons increase efficiency by distributing signals sparsely in space and time [12]. Although it was already recognized that sparse coding improves energy efficiency. However, the functional relationship between information coding and energy consumption for neurons is not known. Does the energy-efficient cortical neuron select signals from synapses that are most informative? This question draws energy efficiency into one of the most active and important areas of neuroscience: synaptic plasticity [12]. The research in this paper is relevant to this kind of question, and we have demonstrated that energy coding is a new biophysical mechanism in neural information processing [3].

Due to the proposed biophysical model to be different from known neuronal models, in this paper we will give further demonstration for validity of the proposed model and

D. Liu et al. (Eds.): ISNN 2007, Part I, LNCS 4491, pp. 1127–1134, 2007.

bring a new perspective upon global brain information processing. We will demonstrate that this perspective allows us a great comprehension of the role of information coding in neural networks.

2 The Biophysical Model

Analysis of both neuro-electrophysiological experimental data and the biophysical properties of neurons reveal the following physical model that reflects the essential electronic properties of neuronal activity as in figure 1.

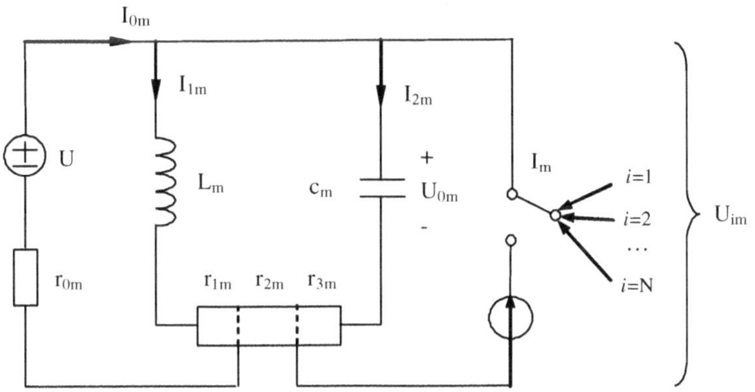

Fig. 1. Physical model of *m*th neuron under case of coupling

This physical model describes the interaction between a single neuron and all other neurons connected to the single neuron. Interaction and mutual coupling among neurons is achieved through the total electrical current formed by the input of N neurons to the m th neuron to generate the sub-threshold current level. The m th neuron in the coupling relationship under the state of firing action potential does not react to external stimulation, hence, stimulation's the electric current I_m becomes a fixed constant (i.e. constant-current source). Similar to the model in reference [10, 11], for the convergence of positive and negative ions inside and outside of the cellular membrane we use C_m to denote this membranous capacitance. For the voltage formed by charges of positive and negative ions at the cellular membrane we use U_{0m} to express this corresponding potential difference. In the resting membrane state the intensity of the magnetic field produced by the motion of the ionic charges upon neuronal activity is very weak, therefore, it can be neglected. However, the magnetic field formed by the violent motion of ionic charges during action potential is much stronger than during the resting state, and can not be ignored. This is because the influx of sodium ions and efflux of potassium ions achieve very violent level. Motion charges formed in this case have to produce the self-induction phenomenon. Accordingly, we use the inductance L_m to denote the intensity of the magnetic field is not only important but also reasonable in a physical sense. And the physical phenomenon of membrane current depending on membrane potential has been confirmed by many experiments [6, 8]. Therefore taking into account the effect of inductance during the neuronal action potential follows naturally. The site of the inductance is designed to parallel the membranous capacitance as in figure 1. Although this is a hypothesis, the computational results obtained from the biophysical model in this paper show that introduction of inductance into the model agrees well with the experimental results of neuro-electrophysiology.

The neuronal action potential requires energy for activity, and voltage source U denotes the total energy supplied by both the sodium-potassium pump, through the production of charge separation and thermal noise energy generated by water molecules having undergone ATP hydrolysis [13]. In addition, that a neuron can maintain its resting membrane potential shows that there exists a current source E of energy within the cytosome. The electric resistance r_{0m} models the loss of energy, and the electric resistance $r_{1m} + r_{2m} + r_{3m}$ as in figure 1, is equivalent to the electrical resistance in [14]. Neurophysiologists pay great attention to these sites in neuronal activities because the highest energetic demand in the brain is centralized to sites of synaptic input [15, 16]. Therefore the internal energy source E for the m th neuron and the site of total temporal-spatial input from N neurons to the m th neuron is designed to different points in the physical model as in figure 1, i.e. the site of internal energy generation and the site of total synaptic current have associated resistances r_{1m} and r_{3m}, respectively, and the electric resistance between r_{1m} and r_{3m} is denoted by r_{2m}. The physical quantities can be observed as the membrane potential U_{im} and the membrane electric current I_{0m} in the physical model, as in figure 1. The work of investigators at Yale University indicates that most of the energy used in the brain is for the propagation of the action potential and for the restoring postsynaptic ion fluxes after the receptors have been stimulated by neurotransmitters [17]. The symbol I_m denotes the total synaptic current formed after temporal-spatial integration of numerous synaptic inputs on the m th neuron — this interaction among neurons in the cerebral cortex is orderable and obeys a self-organizing rule [18]. Hence, the stimulation induced free motion of electric currents does not take arbitrary values. This is because the dynamic mechanism of the ionic channel can greatly restrict the form of the electric current I_m [19]. According to this point of view, the numerical computational results given in the later section of the paper certified that the concrete form of energy consumption for neurons is just the Hamiltonian energy function presented below.

3 Circuit Equations

We obtain the following equations in figure 1

$$U_{im} = C_m r_{3m} \dot{U}_{0m} + U_{0m} \tag{1}$$

$$I_{2m} = C_m \dot{U}_{0m} \tag{2}$$

$$U_{im} = L_m \dot{I}_{0m} + r_{1m} I_{1m} + r_{2m}(I_m - I_{2m}) \tag{3}$$

$$I_{0m} = I_{1m} + I_{2m} - I_m \tag{4}$$

$$U = r_{0m} I_{0m} + r_{1m} I_{1m} + L_m \dot{I}_{1m} \tag{5}$$

Combination equation (1-5) yields the model equations:

$$L_m \dot{I}_{1m} + r_{1m} I_{1m} = C_m (r_{2m} + r_{3m}) \dot{U}_{0m} + U_{0m} - r_{2m} I_m \tag{6}$$

$$I_m = \frac{1}{r_{0m}} (C_m r_{0m} \dot{U}_{0m} - U + (r_{0m} + r_{1m}) I_{1m} + L_m \dot{I}_{1m}) \tag{7}$$

At sub-threshold state the intensity of the magnetic field produced by the motion of the ionic charges upon neuronal activity is very weak, therefore, it can be ignored, i.e.

$$L_m = 0 \tag{8}$$

The electric current I_{1m} is eliminated from (6, 7) we obtain

$$C_m R_2 \dot{U}_{0m} + (r_{0m} + r_{1m}) U_{0m} = R_1 I_m + r_{1m} U \tag{9}$$

Where

$$R_1 = r_{0m} r_{1m} + r_{0m} r_{2m} + r_{1m} r_{2m} \tag{10}$$

$$R_2 = r_{0m} r_{1m} + r_{0m} r_{2m} + r_{0m} r_{3m} + r_{1m} r_{2m} + r_{1m} r_{3m} \tag{11}$$

The electric current $\dot{I}_{1m}$ is eliminated from (6, 7) we obtain

$$I_{1m} = \frac{1}{R_1} (C_m r_{0m} r_{3m} \dot{U}_{0m} + r_{0m} U_{0m} + r_{2m} U) \tag{12}$$

At supra-threshold level, we set

$$I_m = i_0 \tag{13}$$

The m th neuron in the coupling relationship under the state of firing action potential does not react to external stimulation, hence, stimulation's the electric current I_m becomes a fixed constant i_0 (i.e. constant-current source).

4 The Numerical Analysis of the Biophysical Model and Comparison with Neuro-electrophysiological Results

(1) The membrane potential at sub-threshold stimulation
The experimental condition is the following

$$I_m = i_o \tag{14}$$

One obtains the following result from (9)

$$U_{0m} = U_{0m}(\infty) + (U_{0m}(0) - U_{0m}(\infty))e^{-\frac{t}{\tau}} \tag{15}$$

$$U_{im} = U_{0m}(\infty) + (1 - \frac{C_m r_{3m}}{\tau})(U_{0m}(0) - U_{0m}(\infty))e^{-\frac{t}{\tau}} \tag{16}$$

where

$$\tau = \frac{C_m R_2}{r_{0m} + r_{1m}} \tag{17}$$

$$U_{0m}(\infty) = \frac{R_1 i_0 + r_{1m} U}{r_{0m} + r_{1m}} \tag{18}$$

Using equations (14, 16), one obtains the following numerical results

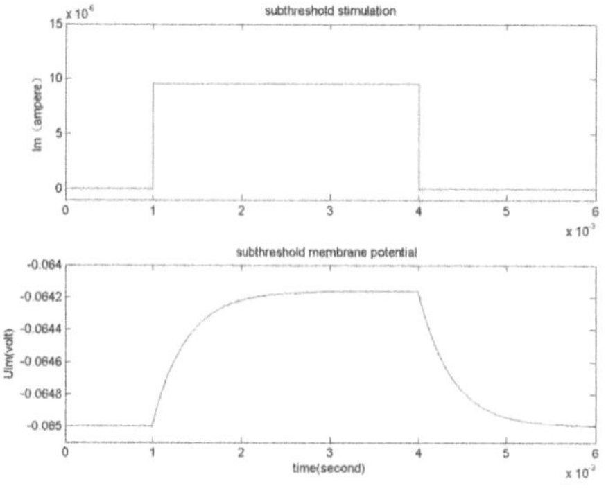

Fig. 2. Reproduce of the subthreshold membrane potential

$$i_0 = 0.954\times10^{-5} A, r_{0m} = 0.004\Omega; r_{1m} = 5\Omega; r_{2m} = 88\Omega;$$
$$r_{3m} = 2.2\Omega; U_{0m}(0) = -65\times10^{-3} V$$

The above result proved that membrane potential at sub-threshold stimulation agrees astonishingly with result given in figure 7.3 in reference [14]. This biophysical model can reproduce all kind of membrane potentials given in figure 3 in reference [14] as long as parameters are changed. Therefore the biophysical model can be used to describe the basic characteristic of neuron's electric activity under condition of various different stimulations. In order to demonstrate the validity of the biophysical model, an example is given again in the below.

(2) The action current at supra-threshold level

The experimental condition is the following

$$U_{im} = Au(t) + U_{0m}(0) \tag{19}$$

Where $u(t)$ is a step function.

Substituting U_{0m} into (2) one obtains

$$I_{2m} = C_m \dot{U}_{0m} = \frac{A}{r_{3m}} e^{-\frac{t}{C_m r_{3m}}} \tag{20}$$

Using above conditions, one obtains solution from equation (6) as follows

$$I_{1m} = I_{1m}(0) e^{-\frac{r_{1m}t}{L_m}} + \frac{A + U_{0m}(0) - r_{2m}i_0}{r_{1m}}(1 - e^{-\frac{r_{1m}t}{L_m}}) + \frac{r_{2m}A}{L_m r_{3m}} \frac{e^{-\frac{t}{C_m r_{3m}}} - e^{-\frac{r_{1m}t}{L_m}}}{\frac{r_{1m}}{L_m} - \frac{1}{C_m r_{3m}}} \tag{21}$$

Inserting (20, 21) into (4) yields the action current

$$I_{0m} = \frac{A - (r_{1m} + r_{2m})i_0 + U_{0m}(0)}{r_{1m}} + \frac{A}{r_{3m}}(1 - \frac{\frac{r_{2m}}{L_m}}{\frac{1}{C_m r_{3m}} - \frac{r_{1m}}{L_m}})e^{-\frac{t}{C_m r_{3m}}} - (\frac{A + U_{0m}(0) - r_{2m}i_0}{r_{1m}} - \frac{A r_{2m}}{L_m r_{3m}(\frac{1}{C_m r_{3m}} - \frac{r_{1m}}{L_m})} - I_{1m}(0))e^{-\frac{r_{1m}t}{L_m}} \tag{22}$$

Using equations (19, 22), one obtains the following numerical results

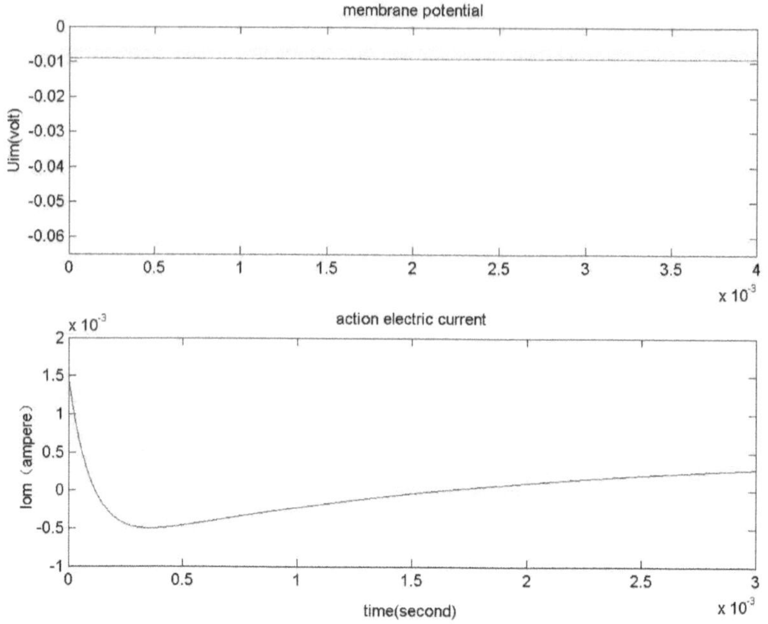

Fig. 3. Reproduce of the depolarizing membrane electric current at supra-threshold stimulation

$$L_m = 10.9\times10^{-3}H, C_m = 8\times10^{-6}F, r_{1m} = 6.5\Omega; r_{2m} = 52.8524\Omega;$$
$$r_{3m} = 12.5167\Omega; i_0 = -2.0639\times10^{-4}A, U_{0m}(0) = -69\times10^{-3}V, I_{1m}(0) = -3.5mA$$

The above result demonstrates that the biophysical model given in Figure 1 can reproduce a depolarizing membrane current. This membrane current is completely in accordance with figure 6.3 in reference [14].

5 Concluding Remarks

In order to prove the validity of the biophysical model proposed in previous paper [3-5, 20], we further performed numerical analysis on the basis of the model. We demonstrated that the proposed model can reproduce the membrane potentials of neuron and the depolarizing membrane current of neuron by means of neuro-electrophysiological experimental data. These results show that we will bring a new perspective upon global brain information processing. We believe that this perspective allows us a great comprehension of the role of information coding in neural networks. In subsequent work many quantitative neural models and analytic results will be given by means of principle of energy coding. For example, using the principle of energetic superposition, we have obtained an evolution of the energy coding principle by observing neuronal ensembles as we varied the intensity of external stimulation continuously, which results in subsets of neurons firing action potentials at supra-threshold and others simultaneously perform activities at sub-threshold level in neural ensembles.

Acknowledgment

This work was supported by the National Natural Science Foundation of China (NSFC). (30270339, 10672057).

References

1. Quiroga, R.Q., Reddy, L., Kreiman, G.., Koch, C., Fried, I.: Invariant Visual Representation by Single Neurons in the Human Brain. Nature **435** (2005) 1102-1107
2. Stein, R.B., Gossen, E.R., Jones, K.E.: Neuronal Variability: Noise or Part of the Signal? Nat. Rev. Neurosci **6** (2005) 389-397
3. Wang, R.B., Zhang, Z.K.: Appl. Phys. Lett. **89** (2006) 123903
4. Jiao, X.F., Wang, R.B.: Appl. Phys. Lett. **87** (2005) 083901
5. Jiao, X.F., Wang, R.B.: Synchronization in Neuronal Population with the Variable Coupling Strength in the Presence of External Stimulus. Appl. Phys. Lett. **88** (2006) 203901
6. Arbib, M.A.: The Handbook of Brain Theory and Neural Networks. The MIT Press, Cambridge, Massachusetts. London, England (2002)
7. Wilson, R.A., Keil, F.C.: The MIT Encyclopedia of the Cognitive Sciences. The MIT Press, Cambridge, Massachusetts. London, England (1999)
8. Freeman, W.J.: Neurodynamics. Springer-Verlag, Berlin (2000)
9. Crotty, P., Levy, W.B.: Energy-efficient Interspike Interval Codes. Neurocomputing **65-66** (2005) 371–378
10. Levy, W.B., Baxter, R.A.: Energy Efficient Neural Codes. Neural Comput. **8** (1996) 531-543
11. Levy, W.B., Baxter, R.A.: Energy-efficient Neuronal Computational via Quantal Synaptic Failures. J. Neurosci. **22** (2002) 4746-4755
12. Laughlin, S.B., Sejnowski, T.J.: Communication in Neuronal Networks. Science **301** (2003) 1870-1874
13. Wang R.B, Hayashi, H., et al:. An Exploration of Dynamics on Moving Mechanism of the Growth Cone. Molecules **8** (2003) 127-138
14. Nicholls, J.G., Martin, A.R., Wallace, B.G.: From Neuron to Brain. 3rd edn. Sunderland: Sinauer (2000)
15. Schwartz, W.J., Smith, C.B., Davidsen, L., Savaki, H.E., Sokoloff, L., Mata, M., Fink, D. J., Gainer, H.: Metabolic Mapping of Functional Activity in the Hypothalamo-neurohypophyseal System of the Rat. Science **205** (1979) 723–725
16. Mata, M., Fink, D.J., Gainer, H., Smith, C.B., Davidsen, L., Savaki, H., Schwartz, W.J., Sokoloff, L.: Activity-dependent Energy Metabolism in Rat Posterior Pituitary Primarily Reflects Sodium Pump Activity. J. Neurochem. **34** (1980) 213-215
17. Raichle, M.E., Gusnard, D.A.: Appraising the Brain's Energy Budget. Proc. Natl. Acad. Sci. PNAS. USA. **99** (2002) 10237-10239
18. Haken, H.: Principles of Brain Functioning. Springer, Berlin (1996)
19. Koch, C., Segev, I.: Methods in Neuronal Modeling. The MIT Press, Cambridge, Massachusetts (1998)
20. Wang, R.B., Zhang, Z.K.: On Energy Principle of Couple Neuron Activities. Acta Biophysica Sinica **21** (2005) 436-442.

A Cognitive Model of Concept Learning with a Flexible Internal Representation System

Toshihiko Matsuka and Yasuaki Sakamoto

Center for Decision Technologies
Wesley J. Howe School of Technology Management
Stevens Institute of Technology, Hoboken, NJ 07030, USA
{tmatsuka, ysakamot}@stevens.edu

Abstract. In the human mind, high-order knowledge is categorically organized, yet the nature of its internal representation system is not well understood. While it has been traditionally considered that there is a single innate representation system in our mind, recent studies suggest that the representational system is a dynamic, capable of adjusting a representation scheme to meet situational characteristics. In the present paper, we introduce a new cognitive modeling framework accounting for the flexibility in representing high-order category knowledge. Our modeling framework flexibly learns to adjust its internal knowledge representation scheme using a meta-heuristic optimization method. It also accounts for the multi-objective and the multi-notion natures of human learning, both of which are indicated as very important but often overlooked characteristics of human cognition.

1 Introduction

High-order human cognition involves processing abstract and categorically represented knowledge. Instead of describing many features of an animal, "dog", (e.g. a hairy animal with pointed teeth, etc), for example, we use a word "dog" to categorically represent the entity in communication. Although some loss of absolute accuracy is often inevitable, the use of categorical knowledge is essential in the high-order cognitive processes which would otherwise result in information over-flow due to limited capacity in our cognitive system [1]. By compressing the vast amount of available information, a cognitive process called *categorization* allows us to process, understand, and communicate complex thoughts and ideas by efficiently utilizing salient and relevant information while ignoring other types.

Because of its importance as a fundamental process in high-order abstract human cognition, categorization has been extensively investigated (e.g. [1], [2], [3], [4]). However, despite this effort, our understanding of the cognitive process is rather limited. In particular, there has been a long heated debate about how categorical knowledge is internally represented in our mind. Traditionally, it has been considered that there is a single innate representation system. There are three main theories on internal representation, namely, *Rules, Prototypes*, and *Exemplars*.

However, a new theoretical stance on human internal representation system has recently emerged. In stead of considering our internal representation mechanism as a

D. Liu et al. (Eds.): ISNN 2007, Part I, LNCS 4491, pp. 1135–1143, 2007.

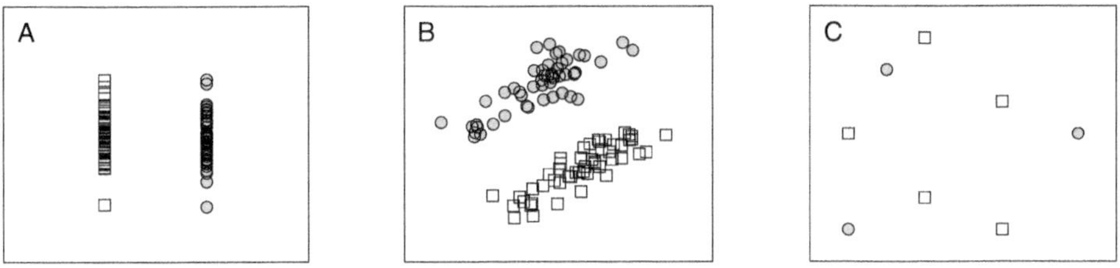

Fig. 1. Hypothetical category structures that would result in different internal representation scheme. Rule or prototype-like representations seem most appropriate for category structure A, prototype-like representation for B, and exemplar-like representation for C.

static system, it considers the representation mechanism as a dynamic system that is capable of altering a representation scheme that meets category structures (e.g. [2], [4]) or situational characteristics (e.g. [5] [6]). For example, we would utilize simple rules or prototypes when the categorical structure is very simple (e.g. category structure depicted in Fig.1A) while exemplar-like representation would be utilized for a complex categorical structures (Fig. 1C) (see [2] for review). Likewise, domain experts would be prone to have an exemplar-like representation scheme in order to be able to differentiate various exemplars.

The emergence of this new stance may not be surprising, because there are inconsistent evaluations of the traditional theories (i.e., rule, prototype, and exemplar), in that a particular theory accounts empirical phenomena better in some situations while others have better accounts in others situations.

Customarily, computational cognitive modeling studies have been used as a means to evaluate theories on human cognition. However, computational models or modeling frameworks accounting for a flexible internal representation system have not been established; and thus, quantitative evaluation of the new theory has been difficult to accomplish. The purpose of the present work is to introduce a new framework for descriptive models of human category learning that integrate a flexible internal representation system, situationally adapting its representation scheme.

1.1 Background

Despite a lack consensus on internal representation, there is a widely accepted standard (except, of course, how knowledge is represented) in how humans integrate information in order to categorize an input stimulus – humans utilize similarities between the input stimulus and internal (memorized) reference points (i.e., rules, prototypes, or exemplars) as evidence to probabilistically assign the input stimulus to an appropriate category. A majority of recent models of category learning can be considered as variants of a radial basis function network (RBF) where the main sources of variation are the type of basis units (e.g. prototypes or exemplars) and sometime selective attention operations.

In the following section, we briefly review categorization process with different types of internal representation of categorical knowledge.

Exemplar Models. Exemplar theory of categorization assumes that humans utilized previously-seen memorized exemplars as reference points (i.e., basis units) in categorization process [3]. The associations between those memorized exemplars and category memberships are to be learned through supervised learning. One important cognitive process in exemplar models (and prototype models) is selective attention that translates physical or logical distances between input stimuli and memorized exemplars into psychological similarities between them [3].

Exemplar models have been the most successful models in replicating observed psychological and cognitive phenomena. However, there are some limitations. The most prominent criticism of exemplar models has been that it assumes that human would utilize many if not all exemplars that individuals encounters (e.g. this model would try to memorize all exemplars in Fig. 1A & 1B). Thus, its modeled knowledge complexity is usually highest. More important, the models have not been successful in replicating empirical data where a knowledge abstraction process is apparently involved [2] [7]. Similarly, the models assume that knowledge complexity almost always increases as an individual develops, and the only way to achieve knowledge abstraction is the selective attention process to ignore irrelevant dimensions (vs. ignoring "irrelevant" exemplars in prototype & rule representation). The abstraction by selective inattention usually results in smaller degrees of abstraction than the abstraction by selective ignorance of useless (e.g. similar) exemplars. Although its implied cognitive process describing cognitive behavior in simple laboratory experiments seems plausible, it demands very high, if not unrealistic, levels of memory capacity and retrieval capability in order to describe human cognition in general.

Prototype Models. The modeled cognitive processes in prototype models are almost identical to those of exemplar models, except reference point that the models utilize. Instead of using many exemplars, prototype models assume that humans utilize very small numbers (usually as small as the number of categories) of prototypical entities. There are two prevalent definitions of a prototype: a) a prototype represents the central tendency of a category and b) a prototype is the most representative exemplar of a category [1].

Although prototype models are appealing for their compact and abstract representation of knowledge, simulation studies with these models are usually less successful than those with exemplar models. A main limitation of prototype models is that their simplistic representation system often cannot account for learning of categories with complex decision boundaries (Fig. 1C). Similarly, it has been argued that a traditional prototype model cannot learn linearly non-separable categories. In other words, the cognitive processes modeled in traditional prototype models have been too simple than real human cognition. However, a recent development in prototype modeling with a complex yet realistic selective attention mechanism showed that a prototype model can account for several empirical phenomena that have previously been considered to be impossible to be replicated (e.g. XOR categorization) by prototype models. [8].

Rule Models. Rule models are the oldest models of categorization. These models assume that there are necessarily and sufficient feature defining each category and learning involves identification of these features [9]. The models assume that category membership has an all-or-none property (i.e., either inside or outside of a category) and thus every member of a category shares the identical similarity and typicality measures (vs. graded similarity in Exemplar and Prototype models). Hence, its input stimuli are most likely either discrete nominal values (vs. numerical values) or preprocessed numerical values (e.g. "bigger than X).

Empirical data on category learning with simple category structures are often consistent with predictions by rule models. However, several studies with complex structures showed that these models could not account for many other empirical findings. One limitation is that many categories do not have the all-or-none property in human mind. Rather, humans find some members of a category more "typical" than others, indicating that categories often have a property of graded similarity and typicality.

2 A New Modeling Framework

Our new model is called SUPERSET, because its knowledge architecture is built on the basis of a superset of exemplar, prototype, and rule models of category learning. SUPERSET's flexible internal representation mechanism is capable of dynamically adjusting its knowledge representation on the basis of situational characteristics, including the complexity of category structure. Learning of category and adjustment of representation scheme are achieved by a multi-objective stochastic optimization method. SUPERSET assumes that humans posses more than one notion of a particular type knowledge and learning involves combinations and modifications of various notions.

2.1 Categorization Algorithm

In SUPERSET, psychological similarity or distance, $s_j(x)$, between an input stimulus, x, and reference point R_j (i.e., rules, prototypes, or exemplars), are defined by the Mahalanobis distance (in quadratic form) between them, allowing for sensitivity to correlations among features dimensions. Thus,

$$s_j^n(x) = \sum_{i=1}^{I} \frac{(R_{ji}^n - x_i)^2}{1 + \exp\left(-D_{ji}^n\right)} + \sum_{i=1}^{I-1} \sum_{m=i+1}^{I} 2C_{jim}^n (R_{ji}^n - x_i)(R_{jm}^n - x_m), \quad (1)$$

where D_j and C_j are R_j's dimensional and correlational selective attention, respectively. Superscript n is an index for different notions. Subscripts i and m indicate feature dimensions, and I is the number of feature dimensions. Note that it is assumed that $C_{jim} = C_{jmi}$, $C_{jim}^2 \leq (a_{jii} \cdot a_{jmm})$, where $a_{jii} = \left(1 + \exp\left(-D_{ji}^n\right)\right)^{-1}$. The correlational attention weights can be a negative value, where its signum indicates direction of attention field while its magnitude indicates the strength of attention. Psychological distance measures activate reference units using the following function

$$h_j^n(x) = \exp\left(-\beta \cdot s_j^n(x)\right), \quad (2)$$

where β controls overall sensitivity. Reference units activations are then fed forward to category output nodes, or

$$O_k^n(x) = \sum\nolimits_j w_{kj}^n h_j^n(x), \quad (3)$$

where w_{kj} is an association weights between R_j and category node *k*. The output activations will be used to obtain response probability by the following function: $P(k) = \exp(\phi O_k)/\sum_l \exp(\phi O_l)$, where ϕ scales decisiveness of response. In short, SUPERSET assumes that humans utilize psychological similarity between input object and reference points, psychologically scaled by correlation sensitive local selective attention processes, as evidence for categorizing the input instance into the most probable category.

2.2 Interpretations of Categorization Process in SUPERSET

SUPERSET can result in acquisitions of knowledge based on either rules, prototypes, exemplars representation, or any combinations of them.

Exemplar Models: A traditional exemplar model of categorization would utilize only the first term of Eq. 1, with $D_j = D_l, \forall j\&l$, and *R* being memorized exemplars. Thus, if SUPERSET learns to pay *no* attention to feature correlations and have an identical dimensional attention for all exemplars, then it would behave like a traditional exemplar model.

Prototype Models: In order for SUPERSET to behave like a traditional prototype model, it needs to learn to 1) acquire and utilize a prototype (e.g. the most "representative" exemplar or the averaged exemplar) for each category; 2) pay NO attention to feature correlations; and 3) have an identical dimensional attention for all prototypes. For SUPERSET to behave like a recently developed prototype model, that successfully replicated many important psychological phenomena [8], it needs only learn to acquire and utilize a prototype for each category without any restrictions on attentional weights

Rule Models: For SUPERSET to behave like a traditional rule model, it needs to learn to: 1) identify an exemplar that is consistent with a category membership rule on diagnostic feature dimension(s); and 2) pay *no* attention to non-diagnostic feature dimensions and correlations. In addition, input features needs to be preprocessed to be in discrete values in order to behave like a traditional rule model.

Hybrid Models: SUPERSET with any coefficient configurations, other than ones described above, can be considered as a hybrid model. SUPERSET can acquire rule-plus-exception, rule-plus-prototype, and/or prototype-plus-exception like representation.

The following sections describe how SUPERSET would achieve context-sensitive learning, resulting in the acquisition of different types of knowledge representation.

2.3 Learning Via Evolutionary Algorithm

SUPERSET utilizes the Evolution Strategy (ES) method for its learning processes. SUPERSET, as in a typical ES application, assumes three key processes in learning: *crossover*, *mutation*, and (survivor) *selection*. In the *crossover* process, the randomly selected notions of categorical knowledge form a pair and exchange gene information, creating a new pair of notions. In human cognition, the crossover process can be interpreted as conceptual combination in which new notions are created based on merging ideas from existing useful notions (e.g., creative discovery). In the *mutation* process, each model coefficient is randomly altered. A mutation can be considered as a modification of a notion by randomly creating a new hypothesis. In the *selection* process, a certain number of notions are deterministically selected on the basis of their fitness in relation to the environment for survival. Those selected notions will be kept in SUPERSET's memory trace (i.e., population space), while non-selected notions become obsolete or are forgotten.

Unlike previous modeling approaches to category learning research, which modify a single notion (i.e., a single set of coefficients), SUPERSET maintains, modifies, and combines a set of notions. The idea of having a population of notions (as opposed to having an individual notion) is important because it allows not only the selection and concept combination in learning, but also the creation of diverse notions, making learning more robust. Thus, unlike previous models, SUPERSET assumes that humans have the potential to maintain a range of notions and are able to apply a notion most suitable for a particular set of situational characteristics.

Although SUPERSET always has multiple notions in its knowledge space, SUPERSET opts for and applies a single notion with the highest predicted utility (e.g., accuracy, score, etc.) to make one response at a time (e.g., categorize an input instance). The functions for estimating the utility for each notion is described in a later section.

Hypotheses Combinations. In SUPERSET, randomly selected pairs of notions exchange information to combine knowledge. For the sake of simplicity, we use the following notation $\{\mathbf{w}^n, \mathbf{D}^n, \mathbf{C}^n\} \in \boldsymbol{\theta}^n$. SUPERSET utilizes discrete recombination of coefficients and intermediary recombination of the coefficient for self-adaptation. Thus, parent notions $\boldsymbol{\theta}^{p1}$ and $\boldsymbol{\theta}^{p2}$ would produce a child notion $\boldsymbol{\theta}^c$, where $\theta_i^c = \theta_i^{p1}$ if UNI $\leq$ 0.5 or θ_i^{p2} otherwise, where UNI is a random number drawn from the Uniform distribution. For self-adapting strategy, $\sigma_i^c = 0.5 \cdot (\sigma_i^{p1} + \sigma_i^{p2})$. This combination process continues until the number of children notions produced reaches the memory capacity of SUPERSET.

Hypotheses Modifications. After the recombination process, SUPERSET randomly modifies its notions, using a self-adapting strategy. Thus,

$$\sigma_{\theta_l}^n(t+1) = \sigma_{\theta_l}^n(t) \cdot \exp(N(0,\gamma)) \tag{4}$$

$$\theta_l^n(t+1) = \theta_l^n(t) + N(0, \sigma_{\theta_l}^n(t+1)) \tag{5}$$

where t indicates time, l indicates coefficients, γ defines search width (via σ's), and $N(0,\sigma)$ is a random number drawn from the Normal distribution with the corresponding parameters.

Selection of Surviving Hypotheses. After creating new sets of notions, SUPERSET selects a limited number of notions to be maintained in its memory. In SUPERSET, the survivor selection is done deterministically, selecting best notions on the basis of estimated utility of concepts or knowledge. The function defining utility of knowledge is described in the next section.

2.4 Knowledge Utility Estimation

The utility of each notion or a set of coefficients determines the selection process in SUPERSET, which occurs twice. During categorization, SUPERSET selects a single notion with the highest predicted utility to make a categorization response (referred to as concept utility for response or UR hereafter). During learning, SUPERSET selects best fit notions to update its knowledge (utility for learning or UL hereafter). In both selection processes, the notion utility is subjectively and contextually defined, and a general function is given as: $U(\boldsymbol{\theta}^n) = \Upsilon(E(\boldsymbol{\theta}^n), Q_1(\boldsymbol{\theta}^n), ..., Q_L(\boldsymbol{\theta}^n))$ where Υ is a function that takes concept inaccuracy (i.e., E) and L contextual factors (i.e., Q) and returns an estimated notion utility value (Note that SUPERSET's learning is framed as a minimization problem).

In SUPERSET, the predicted (in)accuracy of a notion during categorization is estimated based on a retrospective verification function [10], which assumes that humans estimate the accuracies of the notions by applying the current notions to previously encountered instances with a memory decay mechanism. Thus,

$$E(\boldsymbol{\theta}^n) = \sum_{g=1}^{G} \left[\left(\frac{\sum\limits_{\forall i | x^{(i)} = x^{(g)}} (\tau^{(i)} + 1)^{-\delta}}{\sum\limits_{g} \sum\limits_{\forall i | x^{(i)} = x^{(g)}} (\tau^{(i)} + 1)^{-\delta}} \right) \sum_{k}^{K} \left[d_k^{(g)} - O_k^n \left(x^{(g)} \right) \right]^2 \right], \tag{6}$$

where g indicates particular training exemplars, G is the number of unique training exemplars, the last term is the sum of squared error with d being the desired output, and the middle term within a parenthesis is the (training) exemplar retention function defining the strength of the retaining training exemplar $x^{(g)}$. Memory decay parameter, δ, in the exemplar retention function controls speed of memory decay, and τ indicates how many instances were presented since $x^{(g)}$ appeared, with the current training being represented with "0." Thus, $\tau = 1$ indicates $x^{(g)}$ appeared one instance before the current trial. The denominator in the exemplar retaining function normalizes retention strengths, and thus it controls the relative effect of training exemplar, $x^{(g)}$, in evaluating the accuracy of knowledge or concept. $E(\boldsymbol{\theta})$ is strongly influenced by more recently encountered training exemplars in early training trials, but it evenly accounts for various exemplars in later training trials, simultaneously accounting for the Power Law of Forgetting and the Power Law of Learning [11]

There are many functions or sets of functions appropriately defined for describing a variety of contextual factors, including motivation. However, the rudimentary set of objective functions for SUPERSET consists of two elements: concept accuracy and concept simplicity. Thus,

$$U(\boldsymbol{\theta}^n) = E(\boldsymbol{\theta}^n) + \lambda_w S_w(\mathbf{w}^n) + \lambda_D S_D(\mathbf{D}^n) + \lambda_C S_C(\mathbf{C}^n), \tag{7}$$

where S_w, S_D and S_C define knowledge complexity for association weight, dimensional attention, and correlation attention, respectively. In particular,

$$S_w(\mathbf{w}^n) = \sum\nolimits_k \sum\nolimits_j (w_{kj}^n)^2, \tag{8}$$

$$S_D(\mathbf{D}^n) = \sum_c \sum_i \frac{(\bar{a}_{ci}^n)^2 / \sum_{l=1}^{I} (\bar{a}_{cl}^n)^2}{1 + (\bar{a}_{ci}^n)^2 / \sum_{l=1}^{I} (\bar{a}_{cl}^n)^2}, \quad (9)$$

$$\bar{a}_{ci}^n = \frac{1}{N_c} \sum_{j \in c} \frac{1}{1 + \exp(-D_{ji}^n)}, \quad (10)$$

where subscript c indicates clusters. Clusters are to be identified using both dimensional and correlational attention weights associated with *active* reference points that have significant associations with at least one category node (e.g. $\sum_k (w_{kj}^n)^2 > \zeta$, where ζ is a threshold). Similarly,

$$S_C(\mathbf{C}^n) = \sum_c \sum_{i=1}^{I-1} \sum_{m=i+1}^{I} \left(\frac{1}{N_c} \sum_{j \in c} C_{jim}^n \right)^2, \quad (11)$$

where the term inside the parentheses is a simple average correlational attention for dimensions i and m in attention configuration cluster c. Eq. 9 is minimized when there is only one attention configuration cluster and when only one feature dimension is attended. Eq. 10, on the other hand, is minimized when SUPERSET pays no attention to feature correlations.

3 Conclusion

Traditionally, it has been considered that there is a single innate representation system in our mind. However, on the basis of the previous empirical and simulation studies (e.g. [5] [6]), we view the representational system as a dynamic mechanism, capable of selecting a representation scheme that meets situational characteristics, including complexities of category structure. For example, we would create and apply simple rules when a categorical structure is very simple, while we would recall previously seen exemplars and use them as reference for categorization when a categorical structure is complex.

Human learning often involves exploration first and then exploitation. In the exploration stage, we explore to obtain a coarse picture of the concept space. Because of initial uncertainty in category structure, it is possible for individuals to explore the concept space more thoroughly than it is required for "correct" categorization or conceptualization. People then exploit acquired concepts and iteratively adjust them in order to have stable and correct concepts. In the case of "possessing more-than-necessarily complex concepts" (i.e., too thorough exploration), people would simplify them to have more abstract concepts with similar accuracies. For example, in order to derive a simple set of rules, it may be necessary to remember and recognize some exemplars first, then learn to ignore irrelevant information (e.g., non-diagnostic dimension and/or conceptually identical exemplars). This is exactly what was shown in some recent empirical studies (e.g. [12]).

This phenomenon casts serious doubt on the descriptive validity of existing theories of internal representation. Therefore, we developed a cognitive modeling framework that is capable of flexibly adjusting its internal representation scheme that meets situational demands including the complexity of category or knowledge structures. Our modeling framework also accounts for the multi-objective and multi-notion nature of human learning, both of which are indicated as very important but often overlooked characteristics of human cognition (e.g. [6]).

References

1. Estes, W.: Classification and Cognition. New York: Oxford (1996)
2. Minda, J.P., Smith, J.D.: Prototypes in Category Learning: The Effects of Category Size, Category Structure, and Stimulus Complexity. Journal of Experimental Psychology: Learning, Memory, and Cognition **27** (2001) 775-799
3. Kruschke, J.E.: ALCOVE: An Exemplar-Based Connectionist Model of Category Learning. Psychological Review **99** (1992) 22-44
4. Love, B.C., Medin, D.L., Gureckis, T.M.: SUSTAIN: A Network Model of Human Category Learning. Psychological Review **111** (2004) 309-332
5. Barsalou, L.W.: Ad-hoc categories. Memory & Cognition **11** (1983) 211-217
6. Matsuka, T., Sakamoto, Y., Nickerson, J.V., Chouchourelou, A.: A Cognitive Model of Multi-Objective Multi-Concept Formation. In S. Kollias et al., Artificial Neural Networks, ICANN06, LNCS **4131** (2006) 563 -572
7. Feldman, J.: The Simplicity Principle in Human Concept Learning. Current Directions in Psychological Science **12** (2003) 227-232
8. Matsuka, T.: A Model of Category Learning with Attention Augmented Simplistic Prototype Representation. In Advances in Neural Networks, ISNN 2006 LNCS **3971** (2006) 34-40
9. Medin, D.L., Ross, B.H., Markman, A.B.: Cognitive Psychology (4th Ed). Hoboken, NJ: Wiley (2005)
10. Matsuka, T., Chouchourelou, A.: A Model of Human Category Learning with Dynamic Multi-Objective Hypotheses Testing with Retrospective Verification. In Proceedings of the Annual International Joint Conference on Neural Networks (2006) 3648-3656
11. Anderson, J.R., Bothell, D., Lebiere, C.,Matessa, M.: An Integrated Theory of List Memory. Journal of Memory and Language **38** (1998) 341-380
12. Matsuka, T., Corter, J.E.: Process Tracing of Attention Allocation in Category Learning. Under review

Statistical Neurodynamics for Sequence Processing Neural Networks with Finite Dilution

Pan Zhang and Yong Chen*

Institute of Theoretical Physics, Lanzhou University, 730000 Lanzhou, China
july.lzu@gmail.com, ychen@lzu.edu.cn

Abstract. We extend the statistical neurodynamics to study transient dynamics of sequence processing neural networks with finite dilution, and the theoretical results are supported by extensive numerical simulations. It is found that the order parameter equations are completely equivalent to those of the Generating Functional Method, which means that crosstalk noise follows normal distribution even in the case of failure in retrieval process. In order to verify the gaussian assumption of crosstalk noise, we numerically obtain the cumulants of crosstalk noise, and third- and fourth-order cumulants are found to be indeed zero even in non-retrieval case.

1 Introduction

Models of attractor neural networks for processing sequences of patterns, as a realization of a temporal association, have been of great interest over some time [1,2,3,4,5]. On contrary to Hopfield model [6,7], the asymmetry of the interaction matrix in this model leads to violation of detailed balance, ruling out an equilibrium statistical mechanics analysis. Usually, Generating Functional Method and Statistical Neurodynamics are two ways to study this model. Generating Functional Method [2,13,14,15] is also known as e.g. path integral method, and it allows the exact solution of the dynamics and to search all relevant physical order parameters at any time step via the derivatives of a generating functional. Statistical Neurodynamics (so called Signal-To-Noise Analysis) [3,8,9,12], starts from splitting of the local field into a signal part originating from the pattern to be retrieved and a noise part arising from the other patterns, and uses simple assumption that crosstalk noise is normally distributed with zero mean and evaluated variance. In Hopfield model, this treatment is proved to be just an approximation and the Gaussian form of crosstalk noise only holds when pattern recall occurs [10,11,17]. But Kawamuran et al. suggest that Statistical Neurodynamics is exact in fully connected sequence processing model [3]. In this paper we show that treatment of Statistical Neurodynamics is also exact correct in the case of diluted connection.

In the nature of brain, fully-connected structure of network is biological unrealistic. Therefore, as a relaxation of conventional, unrealistic condition that

* Corresponding author.

D. Liu et al. (Eds.): ISNN 2007, Part I, LNCS 4491, pp. 1144–1152, 2007.

every neuron is connected to every other, diluted networks have received lots of attention [18,19,20,21,22]. There are several kinds of dilution, with different average connections cN, where c is connection probability. If $cN = O(\ln n)$, it is called extreme dilution, and in this case, network has a local Caley-Tree structure and almost all pairs of neurons have entirely different sets of ancestors. Thus, the correlations in noise may be neglected. The reported first that the extremely diluted asymmetric network can be calculated exactly is Derrida et al [19]. Extremely diluted symmetric Hopfield network was also studied using replica-symmetric calculation. When $cN = O(1)$, network only have finite connection regardless of its scale, and the equilibrium properties has been calculated using replica-symmetric approximation [23]. In this work, we focus on the case that $cN = O(N)$, and we will show that our theoretical result also suitable for extremely diluted network $cN \ll N$. Recently, Theumann [4] studied this type of dilution using Generating Functional Method, and we will discuss the relationship between our result and those obtained by Theumann.

This paper was organized as follows. In Section 2 we recall the sequence processing neural networks with finite dilution. In Section 3 we use statistical neurodynamics to obtain the order parameter equations which describe dynamics of network, and compare our theoretical result with numerical simulations. Section 4 discuss the relationship between our results and those obtained by Generating Functional Method. Section 5 contains conclusion and discussion.

2 Model Definition

Let us consider a sequence processing model that consists of N spins of neurons, when $N \to \infty$. The state of spins takes $S_i(t) = \pm 1$ and updates the state synchronously with following probability [24]:

$$\mathbf{Prob}[S_i(t+1)\,|h_i(t)] = \frac{1}{2}[1 + S_i(t+1)\tanh \beta h_i(t)], \tag{1}$$

where $\beta = 1/T$ is inverse temperature and local field is given by

$$h_i(t) = \sum_{j=1}^{N} J_{ij} S_j(t). \tag{2}$$

When we use $F(\cdot)$ to express transfer function, the parallel dynamics is expressed by

$$S_i(t+1) = F(h_i(t)). \tag{3}$$

We store $P = \alpha N$ random patterns $\xi^\mu = (\xi_1^\mu, ..., \xi_N^\mu)$ in network, where α is the loading ratio. Interaction matrix J_{ij} is chosen to retrieve the patterns as $\xi^1 \to \xi^2 \to ...\xi^P \to \xi^1$ sequentially. For instance, it is given by [5]

$$J_{ij} = \frac{c_{ij}}{cN} \sum_{\mu=1}^{P} \xi_i^{\mu+1} \xi_j^\mu, \tag{4}$$

with $\xi^{(P+1)} = \xi^1$. c_{ij} takes its value as 0 or 1 probabilities

$$\mathbf{Prob}\left(c_{ij}=1\right) = 1 - \mathbf{Prob}\left(c_{ij}=0\right) = c \tag{5}$$

with $(0 \le c \le 1)$ and $c_{ij} = c_{ji}$ for symmetric dilution.

Let us consider the case to retrieve ξ^q. We define $m^q(t)$ as the overlap parameter between the network state $\mathbf{S}(t)$ and the condensed pattern ξ^q

$$m^q(t) = \frac{1}{N}\sum_{i=1}^{N} \xi_i^q S_i(t). \tag{6}$$

3 Statistical Neurodynamic for Diluted Networks

We start from Somplinsky's idea that in the limit $N \to \infty$ the synaptic matrix in (4) can be written as a fully connected model with synaptic noise [22]

$$J_{ij}^{\mathtt{eff}} = \frac{1}{N}\sum_{\mu=0}^{\alpha N} \xi_i^{\mu+1}\xi_j^{\mu} + \eta_{ij}, \tag{7}$$

where η_{ij} is a complex random variable following Gaussian distribution with mean 0 and variance $\alpha(1-c)/c$. Theumann also proved this relationship by taking expansion of averaging $e^{-i\hat{\mathbf{h}}(t)\cdot\mathbf{JS}(t)}$ over disorder, where $\hat{\mathbf{h}}$ is auxiliary local field (see section 4).

Then local field is decribed by

$$h_i(t) = \xi_i^{q+1} m^q(t) + Z_i(t). \tag{8}$$

Here $Z_i(t)$ denotes the crosstalk noise term from uncondensed patterns,

$$Z_i(t) = \frac{1}{N}\sum_{\mu\neq q}^{\alpha N}\sum_{j\neq i}^{N} \xi_i^{\mu+1}\xi_j^{\mu} S_j(t) + \sum_{j=1}^{N} S_j(t)\,\eta_{ij}. \tag{9}$$

We assume that the noise term $Z_i(t)$ is normally distributed with mean 0 and variance $\sigma_i^2(Z_i(t))$. For Hopfield model, the assumption is shown to be valid within statistical errors by Monte Carlo simulations as long as the memory retrieval is successful [10]. And for sequence processing neural networks, the assumptions is shown to be hold even in the non-retrieval case [3]. We also keep this assumption in our model with presence of c_{ij}, which is verified by numerical simulations of cumulants of $Z_i(t)$.

To obtain the variance $\sigma^2(Z_i(t+1))$ in a recursive form, we express the noise term of the transfer function $F(\cdot)$,

$$\begin{aligned} Z_i(t+1) &= \frac{1}{N}\sum_{\mu\neq q}^{\alpha N}\sum_{j\neq i}^{N} \xi_i^{\mu+1}\xi_j^{\mu} F(h_j(t)) + \sum_{j=1}^{N} F(h_j(t))\,\eta_{ij} \\ &= Y_i(t+1) + \sum_{j=1}^{N} F(h_j(t))\,\eta_{ij}, \end{aligned} \tag{10}$$

where mean of $Y_i(t+1)$ is zero, and

$$h_j(t) = \xi_j^q m^{q-1}(t) + \frac{1}{N}\sum_{k\neq j}^{N}\sum_{\nu\neq q-1}^{\alpha N}\xi_j^{\nu+1}\xi_k^{\nu}S_k(t-1) + \sum_j^N S_k(t-1)\eta_{jk}. \quad (11)$$

Since $h_j(t)$ depends strongly on ξ_j^μ, and the term $\frac{1}{N}\sum_{k\neq j}^{N}\xi_j^\mu\xi_k^{\mu-1}S_k(t)$ can be rationally considered as stochastic order $O(1/\sqrt{N})$, one can expand the transfer function $F(\cdot)$ up to the first order, then obtain

$$Y_i(t+1) = \frac{1}{N}\sum_{j\neq i}^{N}\sum_{\mu\neq q}^{\alpha N}\xi_i^{\mu+1}\xi_j^{\mu}F\left(\hat{h}_j(t)\right) + U(t+1)Y_i(t), \quad (12)$$

where

$$\hat{h}_j(t) = \xi_j^q m^{q-1}(t) + \frac{1}{N}\sum_{k\neq j}^{N}\sum_{\nu\neq q-1,\mu}^{\alpha N}\xi_j^{\nu+1}\xi_k^{\nu}S_k(t-1) + \sum_j^N S_k(t-1)\eta_{jk}, \quad (13)$$

and

$$U(t+1) = \frac{1}{N}\sum_{j\neq i}^{N}F'\left(\hat{h}_j(t)\right). \quad (14)$$

Since crosstalk noise is assumed to be zero mean, one can calculate variance of crosstalk noise directly by

$$\sigma^2(Z_i(t+1)) = \sigma^2(Y_i(t+1)) + \alpha(1-c)/c. \quad (15)$$

Firstly, we have to determine $\sigma^2(Y_i(t+1))$,

$$\begin{aligned}\sigma^2(Y_i(t+1)) = &\frac{1}{N^2}E\left[\sum_j\sum_\mu\left(\xi_i^{\mu+1}\right)^2\left(\xi_j^\mu\right)^2\left(F\left(\hat{h}_j(t)\right)\right)^2\right]\\ &+U^2(t+1)\sigma^2(Y_i(t))\\ &+E\left[2\frac{U(t+1)}{N}\sum_{j,k}\sum_{\mu,\nu}\xi_i^{\mu+1}\xi_i^{\nu+1}\xi_j^{\mu}\hat{S}_j(t+1)\xi_k^{\nu-1}S_k(t)\right],\end{aligned} \quad (16)$$

where we use $\hat{S}_j(t+1) = F\left(\hat{h}_j(t)\right)$. According to the literature [3,12], when expand $S_k(t)$ up to $t=0$, the last term in Eq. (16) becomes

$$\frac{2}{N}E\left[\sum_{n=1}^{t+1}\left(\prod_{\tau=1}^{n}U(t+1-\tau)\sum_j\hat{S}_j(t+1)\hat{S}_j(t+1-n)\right)\sum_\mu\left(\xi_i^{\mu+1}\right)^2\xi_j^\mu\xi_j^{\mu-n}\right]. \quad (17)$$

Here all terms in Eq. (17) are independent of each other except for $n = \alpha N, 2\alpha N, 3\alpha N$.... So when $N \to \infty$, the dependence vanishes, and the last term in Eq. (16) is ignored. The variance of $Y_i\,(t+1)$ term becomes

$$\sigma_i^2\left(Y_i\left(t+1\right)\right) = \alpha + U^2\left(t+1\right)\sigma_i^2\left(Y_i\left(t\right)\right). \tag{18}$$

When variance of crosstalk noise is determined using Eqs. (15, 18), all order parameters can be expressed by the following closed equations

$$m\left(t+1\right) = \int Dz \left\langle \xi F\left(\xi m\left(t\right) + \sigma\left(Z_i\left(t\right)\right) z\right)\right\rangle_\xi , \tag{19}$$

$$U\left(t+1\right) = \frac{1}{\sigma\left(Z_i\left(t\right)\right)} \int Dzz \left\langle F\left(\xi m\left(t\right) + \sigma\left(Z_i\left(t\right)\right) z\right)\right\rangle_\xi , \tag{20}$$

$$\sigma^2\left(Z_i\left(t\right)\right) = \frac{\alpha}{c} + U^2\left(t\right)\left(\sigma^2\left(Z_i\left(t-1\right)\right) - \alpha\left(1-c\right)/c\right), \tag{21}$$

where $\langle\cdot\rangle_\xi$ stands for the average over the retrieval pattern ξ, and $Dz = \frac{1}{\sqrt{2\pi}} exp\left(-z^2/2\right) dz$.

In the case that the temperature is absolute zero, $F(\cdot) = sgn(\cdot)$, we get following expressions of order parameters

$$m\left(t+1\right) = \mathtt{erf}\left(\frac{m\left(t\right)}{\sigma\left(Z_i\left(t\right)\right)}\right), \tag{22}$$

$$U\left(t+1\right) = \frac{1}{\sigma\left(Z_i\left(t\right)\right)}\sqrt{\frac{2}{\pi}} \exp\left(-\frac{m^2}{2\sigma^2\left(Z_i\left(t\right)\right)}\right), \tag{23}$$

where

$$\mathtt{erf}\left(u\right) = \sqrt{2/\pi} \int_0^u \exp\left(-x^2/2\right) dx. \tag{24}$$

This finishes statistical neurodynamics treatment of Sequence Processing Neural Networks with finite synaptic dilution. The above equations form a recursive scheme to calculate the dynamical properties of the systems with an arbitrary time step. The time evolution of overlaps obtained both in theory and numerical simulation are plotted in Fig. 1. Initial overlaps range from 0.1 to 1.0. when initial overlap is smaller than 0.5, network will fail to retrieval, and overlap will finally vanishes. Fig. 2 shows the basin of attraction both in theory and in numerical simulations.

4 Generating Functional Method and Statistical Neurodynamics

The idea of generating functional method is to concentrate on the moment generating function $Z\left[\psi\right]$, which fully captures the statistics of paths,

$$Z\left[\psi\right] = \sum_{\sigma(0),...,\sigma(t)} P\left[\sigma\left(0\right),...,\sigma\left(t\right)\right] e^{-i\sum_{s<t}\mathbf{S}(s)\cdot\psi(s)}. \tag{25}$$

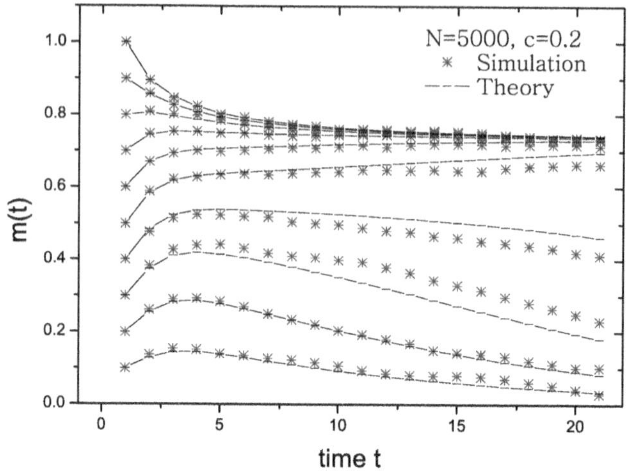

Fig. 1. Temporal evolution of overlap $m(t)$, initial overlap ranging from 1.0 to 0.1 (up to down). The parameters of networks are $N = 5000$, $c = 0.2$ and $\alpha/c = 0.38$.

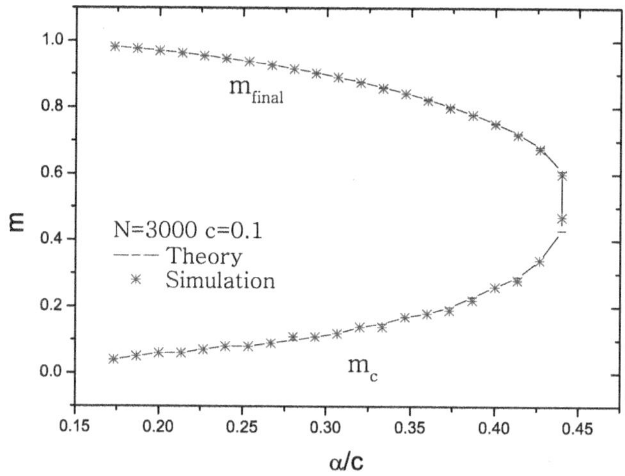

Fig. 2. Basin of attraction obtained by theory and numerical simulations. The parameters of networks are $N = 5000$ and $c = 0.1$.

The generating function $Z[\psi]$ involves the overlap parameter $m(t)$. The response functions $G(t, t')$ and the correlation functions $C(t, t')$ are

$$m(t) = i \lim_{\psi \to 0} \frac{1}{N} \sum_{i=1}^{N} \xi_i^t \frac{\partial Z[\psi]}{\partial \psi_i(t)} = \frac{1}{N} \sum_{i=1}^{N} \xi_i^t \langle S_i(t) \rangle, \tag{26}$$

$$G(t, t') = i \lim_{\psi \to 0} \frac{1}{N} \sum_{i=1}^{N} \frac{\partial^2 Z[\psi]}{\partial \psi_i(t) \partial \theta_i(t')} = \frac{1}{N} \sum_{i=1}^{N} \frac{\partial \langle S_i(t) \rangle}{\partial \theta_i(t')}, \tag{27}$$

$$C\left(t,t'\right) = -\lim_{\psi\to 0}\frac{1}{N}\sum_{i=1}^{N}\frac{\partial^2 Z\left[\psi\right]}{\partial\psi_i\left(t\right)\partial\psi_i\left(t'\right)} = \frac{1}{N}\sum_{i=1}^{N}\langle S_i\left(t\right)S_i\left(t'\right)\rangle. \tag{28}$$

Düring et al. first discussed the sequence processing model using Generating Functional Method and obtained dynamical equations in the form of a multiple Gaussian integral [2], which is too complex to calculate. Then Kawamura et al. simplified those equations and obtained a tractable description of dynamical equations with single Gaussian integral [3]. Theumann discussed the case of finite dilution using Generating Functional Method [4], with idea that

$$\langle e^{-i\hat{\mathbf{h}}(s)\cdot\mathbf{J}^{\mathbf{eff}}\mathbf{S}(t)}\rangle_{c_{ij}} = e^{-i\hat{\mathbf{h}}(t)\cdot\mathbf{J}\mathbf{S}(t)}e^{-\Delta^2\hat{\mathbf{h}}^2(t)/2}, \tag{29}$$

where $\Delta^2 = \alpha\left(1-c\right)/c$ presents variance of independent Gaussian random variables, which is exactly the idea of Sompolinsky that $J_{ij}^{eff} = J_{ij} + \eta_{ij}$ [22]. With this formula, Theumann obtained the temporal evolution of order parameters using the same scheme to that in [2,3],

$$m\left(t\right) = \int Dz\left\langle tanh\beta\left[m\left(t-1\right)+\theta\left(t-1\right)+z\sqrt{\alpha D\left(t-1,t-1\right)}\right]\right\rangle_{\xi}, \tag{30}$$

$$G\left(t,t-1\right) = \beta\left\{1-\int Dz\left\langle tanh^2\beta\left[m\left(t-1\right)+\theta\left(t-1\right)\right.\right.\right.$$
$$\left.\left.\left.+z\sqrt{\alpha D\left(t-1,t-1\right)}\right]\right\rangle_{\xi}\right\}, \tag{31}$$

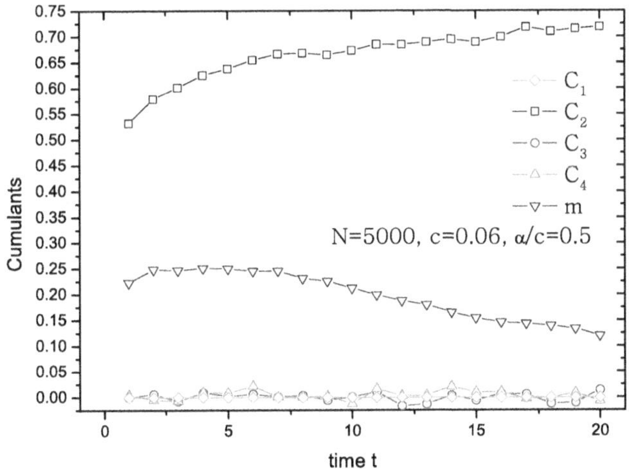

Fig. 3. Temporal evolution of cumulants $C_1\left(t\right), C_2\left(t\right), C_3\left(t\right), C_4\left(t\right)$, and overlap $m\left(t\right)$. The initial overlap is $m\left(0\right) = 0.2$. The parameters of networks are $N = 5000$, $c = 0.06$, and $\alpha/c = 0.5$.

and $C(t,t) = 1$, where $Dz = 1/\sqrt{2\pi}\exp\left(-z^2/2\right)$. Covariance matrix of crosstalk noise is given by

$$D(t,t) = R(t,t) + (1-c)/c, \tag{32}$$

$$R(t,t) = 1 + G^2(t,t-1)\,R(t-1,t-1)\,. \tag{33}$$

From Eq. (19) to Eq. (30), Note that $\sigma^2(Z_i(t))$ corresponds to $\alpha D(t,t')$, and $U(t)$ corresponds to $G(t,t-1)$. It is easy to find that Statistical Neurodynamics and Generating Functional Method present the same order parameter equations of temporal evolution. It means that the Gaussian form of crosstalk noise holds and Statistical Neurodynamics can give the exact solution, comparing with Hopfield model that the crosstalk noise is normally distributed only in retrieval case [10]. To verify the distribution of crosstalk noise, the first, second, third, and fourth cumulants $c_1(t), c_2(t), c_3(t), c_4(t)$ are evaluated numerically, and the third and fourth cumulants are found to be zero even when network fails in retrieval (see Fig. 3).

5 Conclusion and Discussion

In this paper, Statistical Neurodynamics is extended to study the retrieval dynamics of sequence processing neural networks with random synaptic dilution. Our theoretical results are exactly consistent with numerical simulations. The order parameter equations are obtained which is complete equivalent to those obtained by Generating Functional Method. We also present the first, second, third and forth-order cumulants of crosstalk noise to verify the Gaussian distribution of noise.

Finally, note that, in fully connected network or extremely diluted network, one can also obtain the order parameter equations from our Eqs. (19-21). For $c = 1$, $\sigma^2(Z_i(t)) = \sigma^2(Y_i(t))$ and equation (21) corresponds to the order parameter equations in fully connected networks [3]. For the limit $cN \ll N$, $\alpha \ll \alpha/c$, then the variance of crosstalk noise is always α/c, that is exactly the result obtained in extremely diluted network where the local Cayley-Tree structure is held and all correlations in noise are neglected [19].

Acknowledgment

The work reported in this paper was supported in part by the National Natural Science Foundation of China with Grant No. 10305005 and the Special Fund for Doctor Programs at Lanzhou University.

References

1. Sompolinsky, H., Kanter, I.: Temporal Association in Aymmetric Neural Networks. Phys. Rev. Lett. **57** (1986) 2861-2864
2. Düring, A., Coolen, A. C. C., Sherrington, D.: Phase Diagram and Storage Capacity of Sequence Processing Neural Networks. J. Phys. A: Math. Gen. **31** (1998) 8607-8621

3. Kawamura, M., Okada, M.: Transient Dynamics for Sequence Processing Neural Networks . J. Phys. A: Math. Gen. **35** (2002) 253-266
4. Theumann, W. K.: Mean-field Dynamics of Sequence Processing Neural Networks with Finite Connectivity. Physica A **328** (2003) 1-12
5. Yong, C., Hai, W. Y., Qing, Y. K.: The Attractors in Sequence Processing Neural Networks. Int. J. Modern Phys. C **11** (2000) 33-39
6. Hopfield, J. J.: Neural Networks and Physical Systems with Emergent Collective Computational Abilities. Proc. Nat. Acad. Sci. **79** (1982) 2554-2558
7. Amit, D. J., Gutfreund, H., Sompolinsky H.: Spin-glass Models of Neural Networks. Phys. Rev. A. **32** (1985) 1007-1018
8. Amari, S.: Statistical Neurodynamics of Associative Memory. Proc. IEEE Conference on Neural Networks. **1** (1988) 633-640
9. Okada, M.: A Hierarchy of Macrodynamical Equations for Associative Memory. Neural Networks **8** (1995) 833-838
10. Nishimori, H., Ozeki, T.: Retrieval Dynamics of Associative Memory of the Hopfield Type. J. Phys. A: Math. Gen. **26** (1993) 859-871
11. Ozeki, T., Nishimori, H.: Noise Distributions in Retrieval Dynamics of the Hopfield Model. J. Phys. A: Math. Gen. **27** (1994) 7061-7068
12. Kitano, K., Aoyagi, T.: Retrieval Dynamics of Neural Networks for Sparsely Coded Sequential Patterns. J. Phys. A: Math. Gen. **31** (1998) L613-L620
13. Gardner, E., Derrida, B., Mottishaw, P.: Zero Temperature Parallel Dynamics for Infinite Range Spin Glasses and Neural Networks. J. Physique **48** (1987) 741-755
14. Sommers, H. J.: Path-integral Approach to Ising Spin-glass Dynamics. Phys. Rev. Lett. **58** (1987) 1268-1271
15. Gomi, S., Yonezawa F.: A New Perturbation Theory for the Dynamics of the Little-Hopfield Model. J. Phys. A: Math. Gen. **28** (1995) 4761-4775
16. Koyama, H., Fujie, N., Seyama, H.: Results From the Gardner-Derrida-Mottishaw Theory of Associative Memory. Neural Networks **12** (1999) 247-257
17. Coolen, A.C.C.: Statistical Mechanics of Recurrent Neural Networks II. Dynamics. cond-mat/0006011
18. Watkin, T. L. H., Sherrington, D.: The Parallel Dynamics of a Dilute Symmetric Hebb-rule Network. J. Phys A: Math. Gen. **24** (1991) 5427-5433
19. Derrida, B., Gardner, E., Zippelius, A.: An Exactly Solvable Asymmetric Neural Network Model. Europhys. Lett. **4** (1987) 167-173
20. Patrick, A. E., Zagrebnov, V. A.: Parallel Dynamics for an Extremely Diluted Neural Network. J. Phys. A: Math. Gen. **23** (1990) L1323-L1329
21. Castillo, I. P., Skantzos, N. S.: The Little-Hopfield Model on a Random Graph. cond-mat/0307499
22. Sompolinsky, H.: Neural Networks with Nonlinear Synapses and Static Noise. Phys. Rev. A. **34** (1986) 2571-2574
23. Wemmenhove, B., Coolen, A. C. C.: Finite Connectivity Attractor Neural Networks. J. Phys A: Math. Gen. **36** (2003) 9617-9633
24. Chen, Y., Wang, Y. H., Yang, K. Q.: Macroscopic Dynamics in Separable Neural Networks. Phys. Rev. E **63** (2001) 041901-4

A Novel Elliptical Basis Function Neural Networks Model Based on a Hybrid Learning Algorithm*

Ji-Xiang Du[1, 2, 3], Guo-Jun Zhang[3], and Zeng-Fu Wang[2]

[1] Department of Computer Science and Technology, Huaqiao University, China
[2] Department of Automation, University of Science and Technology of China
[3] Intelligent Computing Lab, Hefei Institute of Intelligent Machines, Chinese Academy of Sciences, P.O.Box 1130, Hefei, Anhui 230031, China
du_jx@iim.ac.cn

Abstract. In this paper, a novel elliptical basis function neural networks model (EBFNN) based on a hybrid learning algorithm (HLA) is proposed. Firstly, a geometry analytic algorithm is applied to construct the hyper-ellipsoid units of hidden layer of the EBFNN, i.e., initial the structure of the EBFNN. Then, the hybrid learning algorithm (HLA) is further applied to adjust the centers and the shape parameters. The experimental results demonstrated the proposed hybrid learning algorithm for the EBFNN model is feasible and efficient, and the EBFNN is not only parsimonious but also has better generalization performance than the RBFNN.

1 Introduction

The radial basis function neural network (RBFNN) is a special type of neural network model with several distinctive features. Since firstly proposed, the RBFNN has attracted a high degree of attention and interest in research communities. And currently it has successfully been exploited in many applications. One of the main applications for the RBFNN model is pattern recognition or classification [1]. Usually, when used as pattern classifier, the outputs of the RBFNN represent the posterior probabilities of the training data by a weighted sum of Gaussian basis functions with diagonal covariance matrices., which control the spread of the kernel function (or referred to as transfer function) for the corresponding RBF unit. As a result, the RBF units can perform hyper-spherical division on the input samples. Usually, high recognition accuracy can be achieved when the sample sets are independent. If this case can not be satisfied, more basis functions will be required so that the input data in the region covered by each basis function can still be considered to be independent. In fact, it

* This work was supported by the Postdoctoral Science Foundation of China (NO.20060390180), Youth Technological Talent Innovative Proiect of Fujian Province (NO.2006F3086) and Scientific Research Foundation of Huaqiao University (NO.06BS217).

D. Liu et al. (Eds.): ISNN 2007, Part I, LNCS 4491, pp. 1153–1161, 2007.

would be beneficial if the full covariance matrices could be incorporated into the RBFNN structure so that complex distributions could be well represented without the need for using a large number of basis functions. As a result, the RBF units are in hyper-ellipsoidal shapes, and can enhance the approximation capability of conventional RBFNN models. Thus, the elliptical basis function neural networks (EBFNN) can be considered as an extension of the RBFNN for performing pattern classification or function approximation. This paper, therefore, will introduce a novel EBFNN model with the hyper-ellipsoidal units in an attempt to obtain the better classification capability with respect to the conventional RBFNN.

So far, there have been several works about the studies of how to use covariance matrices of elliptic shapes for the neural networks. For example, Literature [2] proposed a hyper-ellipsoidal clustering algorithm, where the mean vectors and covariance matrices are determined by minimizing the regularized Mahalanobis distance. Literature [3, 4] estimated the parameters of the EBFNN using the expectation-maximization (EM) algorithm and applied it to classification of remote-sensing images and speaker verification, respectively, in which the shape parameters were determined by a heuristic method. In this paper, instead of using EM algorithm to estimate the parameters of the EBFNN, a geometry analytic algorithm [5] was firstly applied to construct the units of hidden layer of the EBFNN. Finally, the hybrid learning algorithm (HLA) is further applied to adjust the centers and the shape parameters. In another words, a new hybrid learning algorithm for the EBFNN is proposed.

This paper is organized as follows. Section 2 introduces how to use the geometry analytic algorithm to initialize the structure of the EBFNN. Section 3 will describe how to further adjust the EBFNN structure by the hybrid learning algorithm. The experimental results are presented in Section 4, and Section 5 concludes the paper and gives related conclusions.

2 Constructing the Hyper-ellipsoid Units of EBFNN

Generally, it would be more reasonable and beneficial if hyper-ellipsoidal units could be adopted to the RBFNN. Consequently, EBFNN can be considered as an extension of the RBFNN. The output of an EBF network can be defined as:

$$y_j(\mathbf{x}) = \sum_{i=1}^{l} w_{ji} h_i(\mathbf{x}), \quad j = 1,2,\cdots,m, \tag{1}$$

$$h_i(\mathbf{x}) = \exp\left\{-\frac{D_i(\mathbf{x})}{\alpha_i^2}\right\}, \quad i = 1,2,\cdots,l, \tag{2}$$

where $\mathbf{x}$ is the input vector, α_i is the shape parameter controlling the spread of the ith basis function. $D_i(\mathbf{x})$ is the distance between the input vector and the ith center of the hyper-ellipsoid unit. Since the unit of the EBFNN is a hyper-ellipsoid basis one, the definition for hyper-ellipsoid can be firstly given as follows:

Definition: A hyper-ellipsoid that is represented by a linear operator $\mathbf{L}$ contains a point $\mathbf{x} = (x_1, x_2, \cdots, x_d)^T$ on its surface if and only if $|\mathbf{L}(\mathbf{x}-\boldsymbol{\mu})| = 1$.

where the matrix $\mathbf{L}$ is non-singular and real valued and $\boldsymbol{\mu}$ is the center of the hyper-ellipsoid.

In other words, a hyper-ellipsoid represented by a certain non-singular matrix $\mathbf{L}$ is a pre-image of the unit hyper-sphere for a linear transformation of the space determined by $\mathbf{L}$. This representation allows the same ellipsoid to be represented by multiple linear operators coinciding up to any rotation of the space. Then, equation (2) can be written as

$$h_i(\mathbf{x}) = \exp\left\{-\frac{\left|\mathbf{L}_i(\mathbf{x}-\boldsymbol{\mu}_i)\right|^2}{\alpha_i^2}\right\}, \quad i = 1, 2, \cdots, l. \tag{3}$$

Suppose that a negative sample $\mathbf{v}$ resides on the surface of the unit hyper-sphere I, and a positive sample $\mathbf{w}$ lies outside it (as shown in Fig 1(a)). The enlargement of the hyper-sphere must be in the direction of the vector e, which is orthogonal to $\mathbf{v}$ and lies in the two dimensional plane defined by v, w and the center μ. The dilation coefficient must be chosen so that w resides on the surface of the resulting hyper-ellipsoid:

$$k = \frac{|w_e|}{|\tilde{w}_e|} = \sqrt{\frac{|w|^2 - b^2}{1-b^2}}, \quad b = |w_v| = \frac{v^T w}{|v|}. \tag{4}$$

And the dilation itself is

$$D = I + \left(\sqrt{\frac{|w|^2 - b^2}{1-b^2}} - 1\right)\frac{ee^T}{|e|^2}, \quad e = v - \frac{|v|^2}{v^T w} w. \tag{5}$$

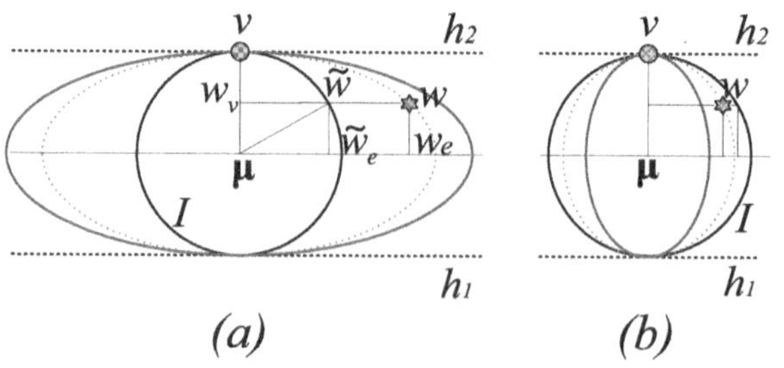

Fig. 1. Illustration of the geometry analytic algorithm (a) enlargement; (b) contraction

In order to modify an arbitrary hyper-ellipsoid L, one must combine L with an operation $C = D^{-1}$, and to replace v and w by $\hat{v} = Lv$ and $\hat{w} = Lw$ respectively, the final formula is:

$$L' = \left[I + \left(\sqrt{\frac{1-b^2}{|w|^2 - b^2}} - 1 \right) \frac{ee^T}{|e|^2} \right] L,$$
$$\hat{v} = Lv, \quad \hat{w} = Lw, \quad b = \frac{\hat{v}^T \hat{w}}{|\hat{v}|}, \quad e = \hat{v} - \frac{|\hat{v}|^2}{\hat{v}^T \hat{w}} \hat{w}. \tag{6}$$

The same computation can be applied without modifications to contraction (as shown in Fig 1(b)). In Fig. 1, the resulting hyper-ellipsoid is figured by a dot line. Moreover, to give the points close to the surface of the hyper-ellipsoid the chance to be assigned to a possibly better hyper-ellipsoid, the equation (7) is modified again as follows

$$L' = \left[I + \left((1+\varepsilon) \sqrt{\frac{1-b^2}{|w|^2 - b^2}} - 1 \right) \frac{ee^T}{|e|^2} \right] L, \tag{7}$$

$$\varepsilon \subset \begin{cases} [0,1] & , \quad \text{Contraction} \\ (-1,0] & , \quad \text{Expansion} \end{cases}. \tag{8}$$

The resulting hyper-ellipsoid modified by equation (7) is figured by a solid line in Fig. 1. This modification method can be always applied to contract a hyper-ellipsoid. However, enlarging a hyper-ellipsoid need certain conditions, which is also a criterion for the creation of a new hyper-ellipsoid:

(1) w must reside between hyper-planes $h1$ and $h2$, that is $|\hat{w}_v| < |\hat{v}|$, which is equivalent to $|\hat{v}^T \hat{w}_v| < |\hat{v}|^2$;

(2) $1 < k \le w_k$, that is to avoid to create the needle-shaped hyper-ellipsoid.

Consequently, the steps of the algorithm can be summarized as follows:

Procedure make_initialization

Input: a set of training samples $S = \{s_1, s_2, \cdots, s_n\}$; parameter value of ε and w_k.

Output: the initial structure of the EBFNN

Begin

for each input training sample $s_i \in S$

Set m_bEnlargement = FALSE;

for each currently existing hyper-ellipsoid E

if S_i is covered by E, i.e., $S_i \in E$, and they don't belong to the same class

Contract E according to equation (7);

else if $S_i \notin E$, and they belong to the same class, and satisfy the conditions

Enlarge E according to equation (7);

Set m_bEnlargement = TRUE;

if m_bEnlargement == FALSE

Construct a new hyper-ellipsoid, which is a hyper-sphere and $\mu = s_i$;

end

3 Further Adjusting the Hyper-ellipsoid Units Parameters with Hybrid Learning Algorithm

The hybrid learning algorithm (HLA) [6], combining the gradient paradigm and the linear least square (LLS) paradigm, can be further used to adjust the initial centers and the shape parameters. Generally, effective initialization of the centers and the shape parameters is required. This algorithm includes two passes. In the forward pass, we supply input data and functional signals to calculate the hidden output $\mathbf{H}$. Then, the weight $\mathbf{W}$ is modified by the LLS method as follows:

$$\mathbf{W}^* = \mathbf{D}\left(\mathbf{H}^T\mathbf{H}\right)^{-1}\mathbf{H}^T, \tag{9}$$

where $\mathbf{D}$ is the target matrix consisting of 1's and 0's. After identifying the weight, the functional signals continue going forward until the error measure is calculated. In the backward pass, the errors propagate from the output end towards the input end. Keeping the weight fixed, the centers and shape parameters of the EBF neurons are modified as follows:

$$E^t = \frac{1}{2}\sum_{k=1}^{m}\left(d_k^t - y_k^t\right)^2, \qquad t = 1,2,\cdots,n, \tag{10}$$

$$\begin{aligned}\Delta\mu^t(i,j) &= -\xi\frac{\partial E^t}{\partial\mu^t(i,j)}\\ &= -\xi\frac{\partial E^t}{\partial y_k^t}\frac{\partial y_k^t}{\partial h_k^t}\frac{\partial h_k^t}{\partial\mu^t(i,j)} \qquad i=1,2,\cdots,d, \quad j=1,2,\cdots,l\\ &= 2\xi\sum_{k=1}^{m}\left(d_k^t - y_k^t\right)\cdot w^t(k,j)\cdot\exp\left\{-\frac{\left|\mathbf{L}_j^t\left(\mathbf{x}^t-\boldsymbol{\mu}_j^t\right)\right|^2}{\left(\alpha_j^2\right)^\gamma}\right\}\cdot\frac{x(i,t)-\mu(i,j)}{\left(\alpha_j^2\right)^\gamma}\cdot\left(\sum_{s=1}^{d}L_j^2(i,s)\right)^t\end{aligned} \tag{11}$$

$$\begin{aligned}\Delta\sigma_j^t &= -\xi\frac{\partial E^t}{\partial\sigma_j^t}\\ &= -\xi\frac{\partial E^t}{\partial y_k^t}\frac{\partial y_k^t}{\partial h_j^t}\frac{\partial h_j^t}{\partial\sigma_j^t},\quad j=1,2,\cdots,l\\ &= 2\xi\sum_{k=1}^{m}\left(d_k^t-y_k^t\right)\cdot w^t(k,j)\cdot\exp\left\{-\frac{\left|\mathbf{L}_j^t\left(\mathbf{x}^t-\boldsymbol{\mu}_j^t\right)\right|^2}{\left(\alpha_j^2\right)^t}\right\}\cdot\frac{\left|\mathbf{L}_j^t\left(\mathbf{x}^t-\boldsymbol{\mu}_j^t\right)\right|^2}{\left(\alpha_j^3\right)^t}\end{aligned} \tag{12}$$

$$\begin{aligned}\Delta L_j^t(i,s) &= -\xi\frac{\partial E^t}{\partial L_j^t(i,s)}\\ &= -\xi\frac{\partial E^t}{\partial y_k^t}\frac{\partial y_k^t}{\partial h_k^t}\frac{\partial h_k^t}{\partial L_j^t(i,s)}\qquad i,s=1,2,\cdots,d,\quad j=1,2,\cdots,l\\ &= -2\xi\sum_{k=1}^{m}\left(d_k^t-y_k^t\right)\cdot w^t(k,j)\cdot\exp\left\{-\frac{\left|\mathbf{L}_j^t\left(\mathbf{x}^t-\boldsymbol{\mu}_j^t\right)\right|^2}{\left(\alpha_j^2\right)^t}\right\}\cdot\frac{L_j(i,s)}{\left(\alpha_j^2\right)^t}\cdot\sum_{j=1}^{d}\left[x(i,t)-\mu(i,j)\right]^2\end{aligned} \tag{13}$$

where E^t is the error function for the t th training pattern; d_k^t is the k th desired output for the t th training pattern; y_k^t is the k th actual output for the t th training pattern; $\Delta\mu^t(i,j)$ is the centre error rate of the i th input variable of the j th HE unit for the t th training pattern; $\Delta\sigma_j^t$ is the shape parameter error rate of the j th HE unit for the t th training pattern; $\Delta L_j^t(i,s)$ is the hyper-ellipsoid error rate of the j th HE unit for the t th training pattern; ξ the learning rate.

4 Experimental Results and Discussions

In order to test our optimization algorithm for the EBFNN, the telling-two-spirals-apart (TTSA) problem and iris classification problem are used.

Firstly, we generated 70 groups of different training samples sets in which the number of samples is changed from 40 to 1400. For each of the training samples sets, the numbers of hyper-ellipsoid units initially constructed by the geometry analytic algorithm, the number of the hyper-sphere constructed by the moving median center hyper-sphere covering (MMCHS) algorithm [7] were recorded. All the results were showed in Fig 3. It can be seen that the number of initially constructed hyper-ellipsoid units is at most about 36 as the centers for the EBFNN. However, there are about 44 hyper-spheres constructed by the MMCHS as the center of the RBFNN. These three methods were also tested by the iris problem and the results were plotted in Fig. 4. Obviously, for the same training samples set, the RBFNN need more centers than the EBFNN to cover the regions of samples space. The constructed hyper-ellipsoids of the telling-two-spirals-apart (TTSA) problem were also plotted in Fig. 5.

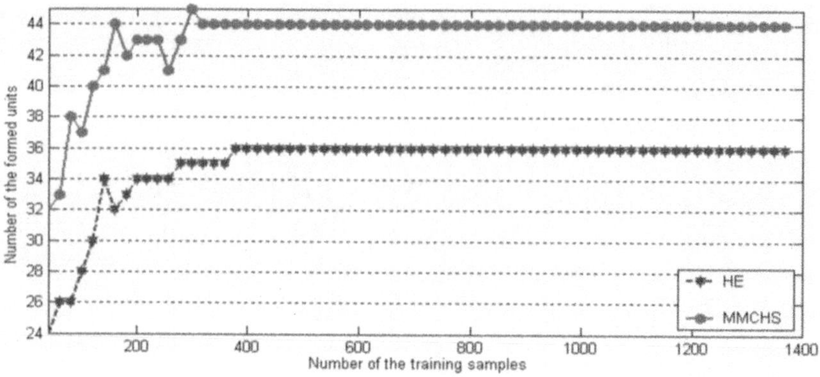

Fig. 2. Number of the formed units for the telling-two-spirals-apart problem

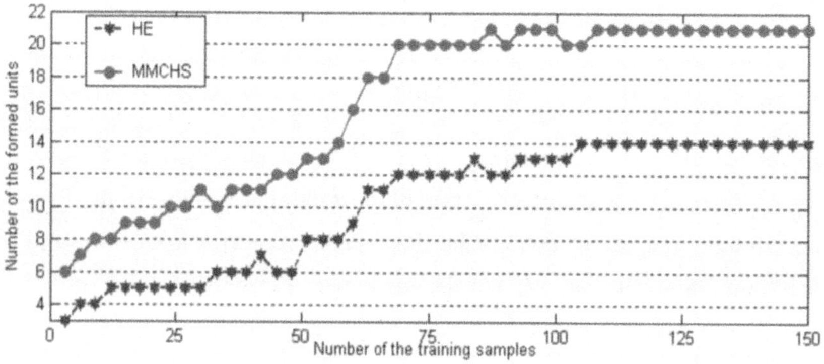

Fig. 3. Number of the formed units for the iris classification problem

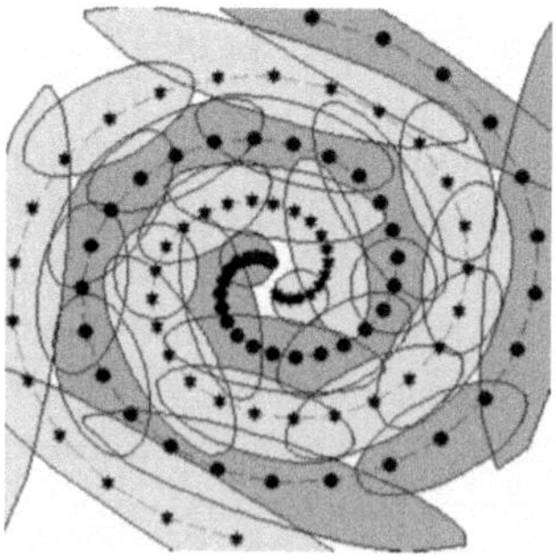

Fig. 4. The constructed hyper-ellipsoids for the TTSA problem

Next, a set of 200 samples for the telling-two-spirals-apart problem was produced. Then we take the parameters of the hyper-ellipsoid units of EBFNN constructed by the method described in section 2 as the initial values of the centers for the HLA. The shape parameters of all hyper-ellipsoid units are set 0.2 as the initial values for the HLA. After further training the parameters of the EBFNN using the HLA, the corresponding mean square errors were further decreased. In addition, the testing results of Gaussian white noisy training samples with zero means but different variances using various methods were shown in Table 1. From Table 1, it can be seen that the EBFNN learned by the HLA after initial optimization by the hybrid PSO has the most reduced structure and the better generalization capability compared with the RBFNN [8].

Table 1. Classification performance comparison for the telling-two-spirals-apart problem

Classifier	Method	Rooted variances of the noises				
		0.010	0.025	0.050	0.075	0.100
EBFNN	HE + HLA	99.0	89.0	75.5	55.5	54.0
RBFNN	MMCHS	97.0	82.5	70.0	51.5	50.5
	MMCHS+HLA	98.5	88.5	74.0	54.0	53.0

5 Conclusions

A two-step learning and optimization scheme for the elliptical basis function neural networks (EBFNN) is proposed. The initial units of hidden layer of the EBFNN were constructed by a geometry analytic algorithm, and then the hybrid learning algorithm was applied to further adjust the centers and the shape parameters of the EBFNN simultaneously. As a result, the experimental results showed that a more efficient and parsimonious structure of EBFNN than the RBFNN with better generalization capability can be designed by our proposed hybrid learning algorithm.

References

1. Oyang, Y.J., Hwang, S.C., Ou, Y.Y., Chen, C.Y., Chen, Z.W.: Data Classification with Radial Basis Function Networks Based on a Novel Kernel Density Estimation Algorithm. IEEE Trans. Neural Networks **16** (2005) 225-236
2. Mao, J., Jain, A.K.: A Self-Organizing Network for Hyperellipsoidal Clustering (HEC). IEEE Trans. Neural Networks **7** (1996) 16-29
3. Luo, J.C., Chen, Q.X., Zheng, J., Leung, Y., Ma, J.H.: An Elliptical Basis Function Network for Classification of Remote-Sensing Images. Proceedings of IEEE International Geoscience and Remote Sensing Symposium '03 (IGARSS03) **6** (2003) 3489-3494
4. Mak, M.W., Li, C.K.: Elliptical Basis Function Networks and Radial Basis Function Networks for Speaker Verification: A Comparative Study. Proceedings of International Joint Conference on Neural Networks **5** (1999) 3034-3039
5. Kositsky, M., Ullman, S.: Learning Class Regions by the Union of Ellipsoids. Proceedings of the 13th International Conference on Pattern Recognition, IEEE Computer Society Press (1996) 750-757

6. Er, M.J., Wu, S.Q., Lu, J.W., Toh, H.L.: Face Recognition with Radial Basis Function (RBF) Neural Networks IEEE Trans. Neural Networks **13** (2002) 697-710
7. Zhang, G.J., Wang, X.F., Huang, D.S., Chi, Z.R., Cheung, Y.M., Du, J.X., Wan, Y.Y.: A hypersphere Method for Plant Leaves Classification. Proceedings of The 2004 International Symposium on Intelligent Multimedia, Video & Speech Processing (ISIMP 2004), Hong Kong, China (2004) 165-168
8. Huang, D.S.: Systematic Theory of Neural Networks for Pattern Recognition. Publishing House of Electronic Industry of China, Beijing (1996)
9. Huang, D.S.: Radial Basis Probabilistic Neural Networks: Model and Application. International Journal of Pattern Recognition and Artificial Intelligence **13** (1999) 1083-1101

A Multi-Instance Learning Algorithm Based on Normalized Radial Basis Function Network

Yu-Mei Chai and Zhi-Wu Yang

School of Information Engineering, Zhengzhou University,
Zhengzhou 450052, Henan, China
ielmwang@zzu.edu.cn,
raynxien@yahoo.com.cn

Abstract. Multiple-Instance Learning is increasingly becoming one of the most promiscuous research areas in machine learning. In this paper, a new algorithm named NRBF-MI is proposed for Multi-Instance Learning based on normalized radial basis function network. This algorithm defined Compact Neighborhood of bags on which a new method is designed for training the network structure of NRBF-MI. The behavior of kernel function radius and its influence is analyzed. Furthermore a new kernel function is also defined for dealing with the labeled bags. Experimental results show that the NRBF-MI is a high efficient algorithm for Multi-Instance Learning.

1 Introduction

Multiple-Instance Learning (MIL) model was firstly formalized by Dietterich et al. [1] from their researches of predicting the musk molecule's activity in 1990s. In MIL model, the training data set comprises of multitude labeled bags (corresponds to the musk molecules) which are sets of unlabeled instances (corresponds to musk molecules). A bag will be positively labeled if and only if there exists at least one positive instance in it, otherwise it will be negatively labeled. The goal of learning is to train a classifier which gives unseen bags correct label. Furthermore, only bag's label is known while the labels of instances in bags remain unknown and the number of positive instances in positive bags varies vastly. These characters bring greater ambiguity to MIL than traditional supervised learning. Hence Maron et al. [2] described MIL as the fourth learning framework besides supervised learning, unsupervised learning and reinforcement learning.

The theoretical researches were focused on the PAC-learnability of MIL. Long et al. [3] proved that APR [1] learning is PAC learnable if the instances in the bags are drawn independently from product distribution and presented a theoretical algorithm. Auer et al. [4] further proved that APR learning is NP-hard if the instances in the bags are not independent. Blum et al. [5] gave a reduction from APR learning under the MIL framework to PAC-learning with one or two-sided random classification noise.

Designing and extending traditional algorithms for MIL also attracted the interests of machine learning community. Maron et al. [6] presented Diverse Density (DD) algorithm and Zhang et al. [7] improved the DD algorithm using the popular EM approach and named their algorithm EM-DD, yet these two algorithms need large

D. Liu et al. (Eds.): ISNN 2007, Part I, LNCS 4491, pp. 1162–1172, 2007.

number of probability computation. Wang et al. [8] successfully extended the traditional k-NN algorithm for MIL with Housdorff metric and presented algorithms Citation-kNN and Bayesian-kNN. As lazy learning algorithms, they need to search the whole training data space for classifying unseen bags. Zhou et al. [9] trained their MIL neural network BP-MIP and achieved higher classification efficiency, yet the classification accuracy of BP-MIP is not as good as those of other MIL algorithms. Therefore, a new MIL algorithm based on Normalized Radial Basis Function network named NRBF-MI is proposed in this paper.

The rest of this paper is organized as follows. In section 2, details of algorithm NRBF-MI is presented. In section 3, experimental results on popular MIL datasets are reported. Finally, a conclusion is given.

2 MIL Algorithm NRBF-MI

Points are the input of traditional Normalized Radial Basis Function (NRBF) network. Yet the NRBF-MI has to deal with bags (sets of unlabeled instances) in MIL. Therefore necessary modifications are to be implemented in order that NRBF network can be used to solve MIL problem.

2.1 Architecture of NRBF-MI

Let $TD =< B_i, l_i >$ denote the training data set in MIL, where B_i ($1 \le i \le N$) is the ith bag composed of m_i instances denoted by $B_{i,j}$, i.e. $B_i = \{B_{i,1}, B_{i,2}, \ldots, B_{i,m_i}\}$, $1 \le j \le m_i$, the label $l_i = 1$ if B_i is positive, otherwise $l_i = 0$.

The MIL classifier constructed based on NRBF network is named NRBF-MI. Its architecture is shown in figure 1. Instead of single vector for traditional NRBF network, the inputs of NRBF-MI are bags.

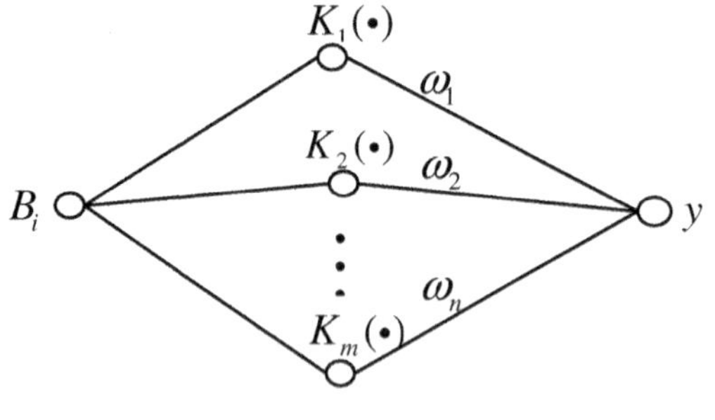

Fig. 1. Architecture of NRBF-MI

Similar to the popular training procedure for NRBF network, a two-stage training procedure is employed to train the first layer and the output layer of NRBF-MI separately.

2.2 First Layer Training

Housdorff distance is introduced into NRBF-MI for measuring the distances between bags.

Given sets A and B and $D(\bullet)$ is a distance measurement function, then the maximal Housdorff distance [10] between A and B is defined as follows:

$$\max HD(A,B) = \max(\max_{p\in A} D(p,B), \max_{p\in B} D(p,A)); \quad (1)$$

And the minimal Housdorff distance [8] is:

$$\min HD(A,B) = \min(\min_{p\in A} D(p,B), \min_{p\in B} D(p,A)). \quad (2)$$

The performance of RBF classifier is greatly influenced by the distribution of its centers [11]. Well selected centers will efficiently improve the classification accuracy; simplify network structure and training process. Although clustering is popular method for training NRBF networks, it is obviously that local maxima of clustering will degrade the performance of NRBF-MI. Besides, kneading positive and negative bags into one cluster will increase the cost of training the weights of NRBF-MI. Hence, a new method for constructing the first layer of NRBF-MI is presented.

As an example, we analyzed the distribution of data in Musk1 [1] based on Housdorff distance and listed the result in table 1. Observing the labels of the nearest neighbors of each bag in Musk1, it can be found that every bag owns a number of neighbors sharing with it the same label and no single neighbor labeled adversely with it appears closer than these same labeled neighbors. The numbers of such neighbors of each bag is given in the column BCN and the Ids of bags in column NO. For example, the positive bag No. 46 has 19 positive neighbors appear closer to it than any negative neighbor, and 16 such negative neighbors for the negative bag No. 61.

Table 1. Distribution of members of each bag's compact neighborhood in Musk1

NO	BCN	NO	BCN	NO	BCN	NO	BCN	NO	BCN	NO	BCN	NO	BCN	NO	BCN	NO	BCN	NO	BCN
0	10	10	18	20	7	30	3	40	13	50	1	60	2	70	1	80	0	90	6
1	10	11	0	21	0	31	3	41	12	51	7	61	16	71	0	81	4	91	0
2	4	12	4	22	7	32	3	42	2	52	0	62	2	72	2	82	4		
3	9	13	0	23	4	33	9	43	18	53	3	63	2	73	4	83	12		
4	4	14	0	24	4	34	12	44	13	54	3	64	1	74	5	84	4		
5	6	15	12	25	10	35	10	45	3	55	2	65	0	75	2	85	9		
6	11	16	5	26	3	36	3	46	19	56	4	66	3	76	1	86	1		
7	12	17	5	27	5	37	4	47	7	57	0	67	5	77	1	87	0		
8	0	18	5	28	3	38	4	48	5	58	3	68	7	78	1	88	10		
9	3	19	0	29	4	39	5	49	1	59	0	69	1	79	1	89	6		

Two definitions are generalized from the analysis above:

Definition 1 Compact Neighborhood: let $\mathbb{R}(B)$ be a set of the nearest neighbors of bag B, if $\forall B' \in \mathbb{R}(B)$ (B' is different from B), satisfying $l_{B'} = l_B$,then $\mathbb{R}(B)$ is a Compact Neighborhood of B ; $\mathbb{R}(B)$'s radius R equals to the distance between B the center and its farthest neighbor in it.

Definition 2 maximal Compact Neighborhood: let $\mathbb{R}(B)$ be a Compact Neighborhood of bag B and R is its radius, for $\forall R' > R$, always $\exists B'$ satisfying $D(B', B) < R'$ and $l_{B'} \neq l_B$, then $\mathbb{R}(B)$ is the maximal Compact Neighborhood of bag B denoted by $\max \mathbb{R}(B_i)$.

Based on the Compact Neighborhood of bags, a new algorithm named Compact Neighborhood Training (CNTR) is designed for training the first layer of NRBF-MI.

Algorithm CNTR:
Input: $< B_i : \{B_{i,1}, B_{i,2}, ..., B_{i,m_i}\}, l_i >, B_i \in TD$;
Output: the first layer of NRBF-MI network;
Gathering all $B_i \in TD$ into set $\{C\}$;
Computing $\max \mathbb{R}(B_i)$ of each bag B_i in $\{C\}$ and its R_{B_i} ;
Do {
Selecting B_i from $\{C\}$ whose $\max \mathbb{R}(B_i)$ is the largest;
Constructing a kernel function using B_i and its $\max \mathbb{R}(B_i)$;
Excluding each member of $\max \mathbb{R}(B_i)$ from $\{C\}$;
} While $\{C\}$ not empty

Algorithm CNTR always choose the maximal Compact Neighborhoods which owns bigger radius and construct kernel functions for each of them.

2.3 Output Layer Optimization

The equation (3) is the formal output of NRBF-MI: where m is the number of its hidden units, ω_j is the weight from the *j*th hidden unit to the output unit, bag B_i is the network input, $K_j(\bullet)$ is the *j*th kernel function with its center is C_j.

$$y(B_i) = \frac{\sum_{j=1}^{m} \omega_j K_j(B_i, C_j)}{\sum_{j=1}^{m} K_j(B_i, C_j)} \tag{3}$$

The Gaussian kernel function is defined in equation (4):

$$K_j(\bullet) = \exp(-HD^2(B_i, C_j)/2\sigma_j^2) \quad 1 \le i \le N \tag{4}$$

In equation (4), minimal Housdorff distance can also be used to measure the distance between bag and kernel function centers.

Substitute equation (4) into equation (3) then the computation model of NRBF-MI is:

$$y(B_i) = \frac{\sum_{j=1}^{m} \omega_j \exp(-HD^2(B_i, C_j)/2\sigma_j^2)}{\sum_{j=1}^{m} \exp(-HD^2(B_i, C_j)/2\sigma_j^2)} \tag{5}$$

The weight vector of NRBF-MI is optimized through minimizing the sum of square error. Here the error function is:

$$E = \frac{1}{2}\sum_{i=1}^{N}(t(B_i) - y(B_i))^2,\ t(B_i) \text{ is target output of bag } B_i \tag{6}$$

Substituting equation (5) into equation (6):

$$E = \frac{1}{2}\sum_{i=1}^{N}[t(B_i) - \sum_{j=1}^{m}\omega_j \exp(-HD^2(B_i, C_j)/2\sigma_j^2) \Big/ \sum_{j=1}^{m}\exp(-HD^2(B_i, C_j)/2\sigma_j^2)]^2 \tag{7}$$

Differentiating with respect to ω_j, the training rule of the weight vector of NRBF-MI is derived from equation (7), where μ is learning rate:

$$\Delta\omega_j = \sum_{i=1}^{N}[\mu(t(B_i) - y(B_i))\exp(-HD^2(B_i, C_j)/2\sigma_j^2) \Big/ \sum_{j=1}^{m}\exp(-HD^2(B_i, C_j)/2\sigma_j^2)] \tag{8}$$

Using stochastic gradient descent training, this rule is rewritten as follows:

$$\Delta\omega_j = [\mu(t(B_i) - y(B_i))\exp(-HD^2(B_i, C_j)/2\sigma_j^2) \Big/ \sum_{j=1}^{m} K_j(B_i) \tag{9}$$

Intuitively smaller radius makes the Gaussian kernel function more localized, but the cost of training will therefore be increased vastly. This can be explained by observing limitation:

$\lim_{\omega\to\infty, \sigma\to 0} \omega \bullet K(\bullet)$, where ω and σ are two real-valued numbers;

It is obviously: $1/\sigma \to \infty$, when $\sigma \to 0$;

$x \sim \omega,\ x \sim 1/\sigma$, when $x \to \infty$;

Without lost generation, rewriting the former limitation: $\lim_{x\to\infty} x \bullet e^{-\lambda/x}$

Using L'Hospital rule: $\lim_{x\to\infty} x \bullet e^{-\lambda/x} = 0$.

This indicates that the contribution of kernel function decays more quickly than its radius decreases, and even the increment of the weight is trivial compared with this tendency.

Considering three representative distribution of Gaussian function $N(0,1)$:

$$P(-\sigma \le \xi \le \sigma) \approx 0.688 \tag{10}$$

$$P(-2\sigma \le \xi \le 2\sigma) \approx 0.955 \tag{11}$$

$$P(-3\sigma \le \xi \le 3\sigma) \approx 0.997 \tag{12}$$

From (10): if $\sigma = R_{Bi}$, then $P(|\xi| > R_{Bi}) = 1 - P(|\xi| \le R_{Bi}) \approx 0.312$;

This probability value indicates that Gaussian kernel function loses its localization because it contributes too much to the points outside its own range when $\sigma = R_{Bi}$;

From (11) and (12): if $\sigma = R_{Bi}/3$, then $P(|\xi| > R_{Bi}) \approx 0.003$, $P(R_{Bi} \ge |\xi| > 2R_{Bi}/3) \approx 0.022$;

These results indicate that kernel function is localized well when $\sigma = R_{Bi}/3$, but its contribution is trivial to the points near the border within its range;

From (11): if $\sigma = R_{Bi}/2$, then $P(|\xi| > R_{Bi}) \approx 0.045$, $P(R_{Bi} \ge |\xi| > R_{Bi}/2) \approx 0.267$;

These values indicate that the kernel function contributes neither too large nor too small.

Therefore the kernel function of NRBF-MI is supposed to contribute best for classification when its radius $\sigma_i \approx R_{Bi}/2$.

3 Experiments Analysis

The NRBF-MI network is trained and tested on MUSK [1] dataset and 4 artificial datasets generalized by Dooly et al. [12]. Experiments are done to testify the analysis in section 2.3 about the relation between the kernel radius and the performance of NRBF-MI. And the NRBF-MI network's performance based on minimal Housdorff distance is also compared with it based on maximal Housdorff distance. All the experiments and tests apply leave-one –out method.

3.1 Experimental Results on MUSK Dataset

MUSK data set is generalized from musk molecules. It comprises of two data sets Musk1 and Musk2. Details are listed in Table 2. There are 47 positive bags (musk molecules) and 45 negative bags (non- musk molecules) in Musk1 and 39 positive bags and 63 negative bags in Musk2.The number of instances in the bags varies between 2 and 40 in Musk1. Although the total number of bags in Musk2 is only 10 bags more than that in Musk1, the instance number of bags in Musk2 varies from 1 to 1044 and the total number of instances is 6598 compared with only 476 instances in Musk1.

Table 2. Distribution of instances in MUSK data set

Data Set	Musk 1	Musk2
Number of bags in Data Set	92	102
Number of positive bags in Data Set	47	39
Number of negative bags in Data Set	45	63
Min number of instances in a bag	2	1
Max number of instances in a bag	40	1044
Number of Instances	476	6598

In contrast with 62 hidden units on Musk1 and 68 hidden units on Musk2 are generated for NRBF-MI using the standard RBF training method, the algorithm CNTR presented in this paper generates 23 hidden units on Musk1 and 32 hidden units on Musk2 respectively. And the classification accuracies of NRBF-MI are at the same level. These results are obtained based on minimal Housdorff distance. As it was reported, BP-MIP used 80 hidden units.

Experiments are designed for testifying the relation between NRBF-MI's performance and the radius of its kernel functions. This testing procedure is controlled by a scalar k i.e. let $\sigma = k \bullet R_{Bi}$. The value of k is initialized to be 1.0 and gradually decreased to be 0.20. The experimental results on Musk1 and Musk2 based on minimal Housdorff distance are shown in figure 2. The figure shows that the best classification accuracy of NRBF-MI is achieved when $\sigma_i \approx 0.45 \bullet R_{Bi}$, and then it degrades with the radius gradually deviates from $0.45 R_{Bi}$. Finally the classification accuracy is only about 60% when $\sigma_i = 1.0 R_{Bi}$ or $\sigma_i = 0.20 R_{Bi}$.

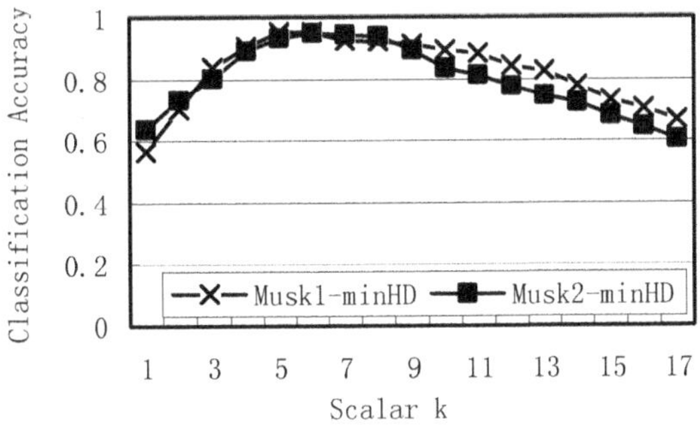

Fig. 2. Relation curves of accuracy and kernel radius in Musk1 and Musk2

Curves representing the experimental results when $\sigma_i = 0.45R_{Bi}$ and $\sigma_i = 0.60R_{Bi}$ are shown in figure 3 and figure 4 respectively where minHD represents the curves obtained based on minimal Housdorff distance and maxHD curves based on maximal Housdorff distance..

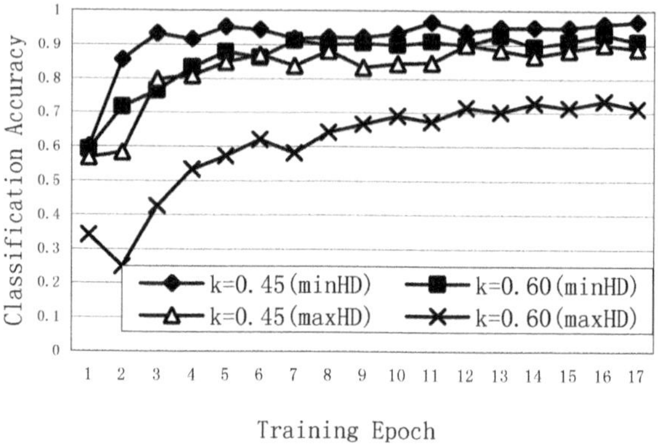

Fig. 3. The curves of relation between the accuracy and training epochs of NRBF-MI on Musk1

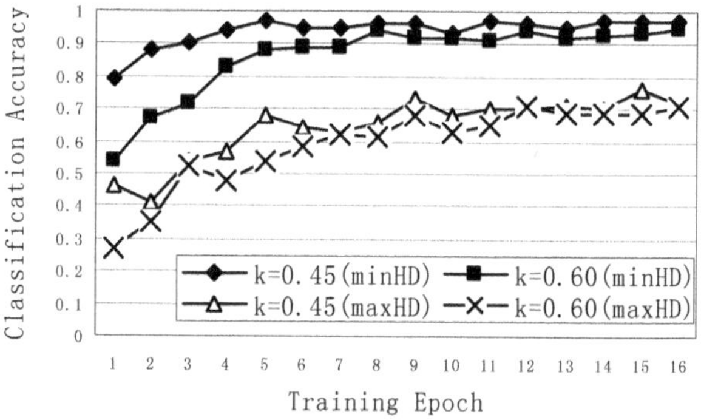

Fig. 4. The curves of relation between the accuracy and training epochs of NRBF-MI on Musk2

From figure 3 and figure 4, it can be observed that the kernel radius also have obviously influent of the convergent speed of the weight vector of the NRBF-MI network. When $\sigma_i = 0.60R_{B_i}$, the weight vectors converge much slower then it does when $\sigma_i = 0.45R_{Bi}$. And the weight vector hardly converges when $\sigma = 1.00R_{B_i}$. These results support the analysis about the relation of the NRBF-MI's performance and its kernel function radius.

The performance of NRBF-MI based on minimal Housdorff distance outperforms that based on maximal Housdorff distance. This phenomenon can be found both on Musk1 and Musk2. As an example, we analyze the curves obtained when the scalar $k = 0.45$ shown in figure 3. The weights of NRBF-MI network are optimized after about 100 epochs of training on Musk1 based on minimal Housdorff distance compared with more than 150 epochs are needed for the optimization of the weights based on maximal Housdorff distance. Meanwhile the classification accuracy based on minimal Housdorff distance is higher than those based on maximal Housdorff distance. The reason of this phenomenon is that a bag is positively labeled if and only if there is at least one positive instance in it. And this character can be correctly represented by minimal Housdorff distance. It was reported that BP-MIP reaches its best accuracy after about 850 epochs of training on Musk1 and Musk2.

Table 3. Comparison of the classification accuracy on MUSK dataset

Algorithm	*Musk1* %correct	Algorithm	*Musk2* %correct
EM-DD[7]	96.8	EM-DD[7]	96.0
NRBF-MI	**95.3**	**NRBF-MI**	**94.8**
iterated-discrim APR [1]	92.4	iterated-discrim APR [1]	89.2
Citation-*k*NN [8]	92.4	Citation-*k*NN [8]	86.3
GFS elim-kde APR [1]	91.3	Diverse Density [6]	82.5
GFS elim-count APR [1]	90.2	Bayesian-*k*NN [8]	82.4
Bayesian-*k*NN [8]	90.2	BP-MIP [9]	80.4
Diverse Density [6]	88.9	GFS elim-kde APR [1]	80.4
BP-MIP[9]	88.0	GFS elim-count APR [1]	75.5

In table 3, the performance of NRBF-MI is compared with those reported in literatures. It can be found that NRBF-MI is among the top-ranked MIL learning algorithms. The classification accuracy of NRBF-MI on Musk1 and Musk2 are 95.3% and 94.8% respectively, which are significantly better than that of BP-MIP, Citation-kNN and Bayesian-kNN, especially on Musk2. Furthermore NRBF-MI performs steadily on both Musk1 and Musk2 as algorithm EM-DD does.

3.2 Experimental Results on Artificial Datasets

NRBF-MI network is also trained and tested on 4 artificial datasets i.e. LJ-160.166.1-s, LJ-160.166.1, LJ-80.166.1-s, and LJ-80.166.1. The name with suffix –s indicates that these datasets are created to mimic the MUSK dataset by using no label near 1/2. Only real-valued labels are used in these datasets. The NRBF-MI classifies a bag by rounding a real-valued label to 0 or 1. Experimental results are listed in table 3. In column Loss listed the square loss and classification error rate in column Err (%).

Table 4. Comparison of the classification error and square loss on artificial datasets

Dataset	NRBF-MI		BP-MIP		Diverse-Density		Citation-KNN	
	Err(%)	Loss	Err%	Loss	Err(%)	Loss	Err(%)	Loss
LJ-160.166.1-s	3.2	0.0283	18.5	0.0731	0.0	0.0052	0.0	0.0022
LJ-160.166.1	4.6	0.041	16.3	0.0398	23.9	0.0852	4.3	0.0014
LJ-80.166.1-s	3.9	0.0573	18.5	0.0752	53.3	N/A	0.0	0.0025
LJ-80.166.1	5.3	0.0414	18.5	0.0487	N/A	0.1116	8.6	0.0109

It can be found that the NRBF-MI performs competitively on these 4 artificial datasets. Firstly, the CNTR generated for NRBF-MI network 19, 21 17 and 16 hidden units on these 4 artificial datasets respectively compared with the number of hidden units generated by traditional training method are 49, 54, 52 and 47 respectively. Secondly, the classification error rate of NRBF-MI on these 4 artificial datasets outperforms BP-MIP, while the square loss of it is comparable with that of algorithm BP-MIP, Diverse-Density and Citation-kNN. Thirdly, the optimization of the weights of NRBF-MI need no more than 150 epochs of training on these 4 datasets, furthermore experiments showed that the network's classification accuracy achieves highest when the radius $\sigma_i = 0.45 R_{Bi}$ and weight vector converge faster under this condition, these results are consistent with the analysis of the relation of the performance of NRBF-MI and its kernel radius. Last, the performance of NRBF-MI based on minimal distance is much better than that based on maximal Housdorff distance, and it is worth noting that although no feature selection techniques were applied, NRBF-MI works well on MUSK dataset and the 4 artificial datasets. This indicates that NRBF-MI is robust in MIL model.

4 Conclusion

In this paper, a new MIL algorithm named NRBF-MI is proposed based on Normalized Radial Basis Function network. Through definition of Compact Neighborhood of bags, a new method named CNTR is designed for training the structure of NRBF-MI. The relation between the performance of NRBF-MI and the radius of its kernel functions is also analyzed and $\sigma_i \approx R_{Bi}/2$ is suggested in order that NRBF-MI's kernel functions contribute best for its performance. Finally, based on Housdorff distance, a new kernel function that can deal with the bags in MIL is defined for NRBF-MI network. Experiments show that NRBF-MI is an efficient algorithm for MIL.

References

[1] Dietterich, T.G., Lathrop, R.H., Pérez, T.L.: Solving the Multiple-instance Problem with Axis-parallel Rectangles. Artificial Intelligence **89**(1-2) (1997) 31-71

[2] Maron, O.: Learning from Ambiguity, PhD Dissertation. Department of Electrical Engineering and Computer Science, MIT (1998)

[3] Long, P.M., Tan, L.: PAC Learning Axis-aligned Rectangles with Respect to Product Distributions from Multiple-instance Examples. Machine Learning **30**(1) (1998) 7-21

[4] Auer, P.: On Learning from Multi-instance Examples: Empirical Evaluation of a Theoretical Approach. in Proceedings of the 14th International Conference on Machine Learning, Nashville, TN (1997) 21-29

[5] Blum, A., Kalai, A.: A Note on Learning from Multiple-instance Examples. Machine Learning **30**(1) (1998) 23-29

[6] Maron, O., Pérez, T.L.: A Framework for Multiple-instance Learning. in: Advances in Neural Information Processing Systems 10, Jordan, M.I., Kearns, M.J., Solla, S.A., Eds. Cambridge, MA: MIT Press (1998) 570-576

[7] Zhang, Q., Goldman, S.A.: EM-DD: An Improved Multiple Instance Learning Technique. In Neural Information Processing Systems **14** (2001)

[8] Wang, J., Zucker, J.D.: Solving the Multiple-instance Problem: A Lazy Learning Approach. in Proceedings of the 17th International Conference on Machine Learning, San Francisco, CA (2000) 1119-1125

[9] Zhou, Z.H., Zhang, M.L.: Neural Networks for Multi-instance Learning. Technical Report, AI Lab, Computer Science & Technology Department, Nanjing University. China, Aug. (2002)

[10] Edgar, G.A.: Measure, Topology, and Fractal Geometry (3rd print). Springer-Verlag (1995)

[11] Kim, N., Byun, H.G., Kwon, K.H.: Learning Behaviors of Stochastic Gradient Radial Basis Function Network Algorithms for Odor Sensing System. ETRI Journal **28**(1) (2006) 59-66

[12] Dooly, D.R., Zhang, Q., Amar, R.A.: Multiple-Instance Learning of Real-Valued Data. in Journal of Machine Learning Research 3 (2002)

Neural Networks Training with Optimal Bounded Ellipsoid Algorithm

Jose de Jesus Rubio and Wen Yu

Departamento de Control Automatico, CINVESTAV-IPN
A.P. 14-740, Av.IPN 2508, Mexico D.F., 07360, Mexico
yuw@ctrl.cinvestav.mx

Abstract. Compared to normal learning algorithms, for example backpropagation, the optimal bounded ellipsoid (OBE) algorithm has some better properties, such as faster convergence, since it has a similar structure as Kalman filter. OBE has some advantages over Kalman filter training, the noise is not required to be Guassian. In this paper OBE algorithm is applied traing the weights of recurrent neural networks for nonlinear system identification. Both hidden layers and output layers can be updated. From a dynamic systems point of view, such training can be useful for all neural network applications requiring real-time updating of the weights. A simple simulation gives the effectiveness of the suggested algorithm.

1 Introduction

Recent results show that neural network technique seems to be very effective to identify a broad category of complex nonlinear systems when complete model information cannot be obtained. Neural networks can be classified as feedforward and recurrent ones [4]. Feedforward networks, for example Multilayer Perceptrons (MLP), are implemented for the approximation of nonlinear functions in the right hand side of dynamic model equations. The main drawback of these neural networks is that the weights' updating do not utilize information on the local data structure and the function approximation is sensitive to the training data [6]. Since recurrent networks incorporate feedback, they have powerful representation capability and can successfully overcome disadvantages of feedforward networks [4]. Even though backpropagation (BP) has been widely used as a practical training method for neural networks, the limitations are that it may converge very slowly, there exists local minima problem and, the training process is sensitive to measurement noise. The stability of modified backpropagation algorithm is proved in [11].

Gradient-like learning laws are relatively slow. In order to solve this problem, many descendent methods in the identification and filter theory have been proposed to estimate the weights of neural networks. For example the extended

D. Liu et al. (Eds.): ISNN 2007, Part I, LNCS 4491, pp. 1173–1182, 2007.

Kalman filter is applied to train neural networks in [5], [6], [8] and [10], they can give solutions of least-square problems. Most of them use static neural networks, sometimes the output layer must be linear and the hidden layer weights are chosen at randomly [1]. A faster convergence with the extended Kalman filter is reached, because it has fewer interactions [5]. However, the computational complexity in each interaction is increased, it require of large amount of memory. Decoupling technique is used to decrease computational burden [7], the decoupled Kalman filter with diagonal matrix P is similar to gradient algorithm [4], but the learning rate is a time-varying matrix. A big drawback of the Kalman filter training is when we analyze the algorithm, we have to require the uncertainty of neural modeling satisfies Gaussian process.

The optimal bounded ellipsoid (OBE) algorithm require the uncertainty of neural modeling is bounded. And it has a similar structure as Kalman filter [9]. In [2] unsupervised and supervised learning laws in the form of ellipsoids are used to find and tune the fuzzy function rules. In [3] ellipsoid type of activation function is proposed for feedforward neural networks. To the best of our knowledge, neural networks training with optimal bounded ellipsoid algorithm has not yet been established in the literature.

In this paper a modified optimal bounded ellipsoid algorithm is proposed such that it can be used for training the weights of a recurrent neural network for nonlinear system identification. Both hidden layers and output layers can be updated. From a dynamic systems point of view, such training can be useful for all neural network applications requiring real-time updating of the weights. A simple simulation gives the effectiveness of the suggested algorithm.

2 Recurrent Neural Networks

Consider following unknown discrete-time nonlinear system

$$x(k+1) = f\left[x(k), u(k)\right], \tag{1}$$

where $u(k) \in \Re^m$ is the input vector, $|u(k)|^2 \leq \overline{u}$, $x(k) \in \Re^n$ is a state vector, $u(k)$ and $x(k)$ are known. f is an unknown general nonlinear smooth function $f \in C^\infty$. We use the following state-space recurrent neural network to identify the nonlinear plant (1)

$$\widehat{x}(k+1) = A\widehat{x}(k) + V_{1,k}\sigma\left[W_{1,k}x(k)\right] + V_{2,k}\phi\left[W_{2,k}x(k)\right]u(k), \tag{2}$$

where $\widehat{x}(k) \in \mathbb{R}^n$ represents the internal state of the neural network. The matrix $A \in \mathbb{R}^{n\times n}$ is a stable matrix. The weights in output layer are $V_{1,k}$, $V_{2,k} \in \mathbb{R}^{n\times m}$, the weights in hidden layer are $W_{1,k}$, $W_{2,k} \in \mathbb{R}^{m\times n}$, σ is $m-$dimension vector function $\sigma = [\sigma_1 \cdots \sigma_m]^T$, $\phi(\cdot)$ is $\mathbb{R}^{m\times m}$ diagonal matrix.

$$\sigma\left[W_{1,k}x(k)\right]=[\sigma_1\left(\sum_{j=1}^{n}w_{1,1,j}x_j\right),\sigma_2\left(\sum_{j=1}^{n}w_{1,2,j}x_j\right),\cdots,\sigma_m\left(\sum_{j=1}^{n}w_{1,m,j}x_j\right)]^T$$
$$\phi\left[W_{2,k}x(k)\right]u(k)=[\phi_1\left(\sum_{j=1}^{n}w_{2,1,j}x_j\right)u_1,$$
$$\phi_2\left(\sum_{j=1}^{n}w_{2,2,j}x_j\right)u_2,\cdots\phi_m\left(\sum_{j=1}^{n}w_{2,m,j}x_j\right)u_m]^T, \tag{3}$$

where σ_i and ϕ_i are Sigmoid functions. According to the Stone-Weierstrass theorem and density properties of recurrent neural networks [4], the unknown nonlinear system (1) can be written in the following form

$$x(k+1)=Ax(k)+V_{1,k}\sigma\left[W_{1,k}x(k)\right]+V_{2,k}\phi\left[W_{2,k}x(k)\right]u(k)-\eta(k), \tag{4}$$

where $\eta(k)=-f\left[x(k),u(k)\right]+Ax(k)+V_{1,k}\sigma\left[W_{1,k}x(k)\right]+V_{2,k}\phi\left[W_{2,k}x(k)\right]u(k)$ is modeling error with respect to the weights $V_{1,k}$, $V_{2,k}$, $W_{2,k}$ and $W_{2,k}$, they are time-varying weights which will be updated by identification error. By [4] we know that the term $\eta(k)$ can be made arbitrarily small by simply selecting appropriate the number of neurons in the hidden layer (in this paper, it is m). In the case of two independent variables, a smooth function f has the following Taylor series expansion

$$f=\sum_{k=0}^{l-1}\frac{1}{k!}\left[\left(x_1-x_1^0\right)\frac{\partial}{\partial x_1}+\left(x_2-x_2^0\right)\frac{\partial}{\partial x_2}\right]_0^k f+\varepsilon, \tag{5}$$

where ε is the remainder of the Taylor formula. If we let x_1 and x_2 correspond $W_{1,k}x\left(k\right)$ and $V_{1,k}$, x_1^0,x_2^0 correspond $W_1^0x\left(k\right)$ and V_1^0, and define $\widetilde{W}_{1,k}=W_{1,k}-W_1^0$, $\widetilde{V}_{1,k}=V_{1,k}-V_1^0$, then we have

$$V_{1,k}\sigma\left[W_{1,k}x(k)\right]=V_1^0\sigma\left[W_1^0x\left(k\right)\right]+\Theta_{1,k}B_{1,k}+\varepsilon_1, \tag{6}$$

where V_1^0, V_2^0, W_1^0 and W_2^0 are set of known initial constant weights, $B_{1,k}=\left[\sigma,\sigma' V_{1,k}^T x\right]^T\in\mathbb{R}^{2m\times 1}$, $\Theta_{1,k}=\left[V_{1,k},W_{1,k}^T\right]^T\in\mathbb{R}^{n\times 2m}$, σ' is the derivative of nonlinear activation function $\sigma\left(\cdot\right)$ with respect to $W_{1,k}x(k)$, the the definition of (3), $\sigma'\in\mathbb{R}^{m\times m}$. Similar

$$V_{2,k}\phi\left[W_{2,k}x(k)\right]u(k)=V_2^0\phi\left[W_2^0x\left(k\right)\right]u(k)+\Theta_{2,k}B_{2,k}+\varepsilon_2, \tag{7}$$

where $B_{2,k}=\left[\phi u,\phi'\mathrm{diag}(u)V_{2,k}^T x\left(k\right)\right]^T$, $\Theta_{2,k}=\left[V_{2,k},W_{2,k}^T\right]^T$. We define the modelling error

$$\zeta(k)=\varepsilon_1+\varepsilon_2-\eta(k),$$

substituting (6) and (7) into (4) we have

$$y(k)=B_k^T\Theta_k+\zeta(k), \tag{8}$$

where $\Theta_k = \begin{bmatrix} \Theta_{1,k} \\ \Theta_{2,k} \end{bmatrix} = \left[V_{1,k}, W_{1,k}^T, V_{2,k}, W_{2,k}^T\right]^T$,

$$B_k = \begin{bmatrix} B_{1,k} \\ B_{2,k} \end{bmatrix} = \left[\sigma, \sigma' V_{1,k}^T x, \phi u, \phi' \mathrm{diag}(u) V_{2,k}^T x\,(k)\right]^T,$$

the output $y(k)$ is

$$y(k) = x(k+1) - Ax(k) - V_1^0 \sigma\left[W_1^0 x\,(k)\right] - V_2^0 \phi\left[W_2^0 x\,(k)\right] u(k).$$

3 Training with Optimal Bounded Ellipsoid Algorithm

Now we use Optimal bounding ellipsoid algorithm (OBE) to train the recurrent neural networks (2) such that the identification error $\epsilon_i\,(k)$ between the plant (1) and the neural networks (2), *i.e.*,

$$\epsilon_i\,(k) = x_i(k) - \widehat{x}_i(k)$$

is bounded. We rewrite (8) in state-space with single output

$$y_i\,(k) = B_k^T \theta_i\,(k) + \zeta_{i,k}, \tag{9}$$

where $i = 1, \cdots, n$, $\theta_i(k) \in \mathbb{R}^{4m\times 1}$, $\Theta_k = [\theta_1(k), \cdots, \theta_n(k)]$, $y(k) = [y_1(k), \cdots,$ $y_n(k)]$, $\zeta = [\zeta_{1,k}, \cdots, \zeta_{n,k}]^T$.

$$y_i(k) = x_i(k+1) - a_i x_i(k) + V_1^0 \sigma\left[W_1^0 x_i\,(k)\right] + V_2^0 \phi\left[W_2^0 x_i\,(k)\right] u(k).$$

A1. It is assumed here that $\zeta_{i,k}$ belong to ellipsoidal sets $\forall k$ according to [9]

$$S_k = \left\{\zeta_{i,k} \in \mathbb{R} : \zeta_{i,k}^T \frac{1}{\gamma} \zeta_{i,k} \leq 1\right\}. \tag{10}$$

A2. Assume that the initial parameter is inside the ellipsoid $E\left(\widehat{\theta}_i\,(1), P_1\right)$ given by

$$E_1 = E\left(\widehat{\theta}_i\,(1), P_1\right) = \left\{\theta_i \in \mathbb{R}^{4m\times 1} : \widetilde{\theta}_i^T(1) P_1^{-1} \widetilde{\theta}_i(1) \leq 1\right\},$$

where $P_1 > 0$ and $P_1 = P_1^T \in \mathbb{R}^{4m\times 4m}$, $\widetilde{\theta}_i(1) = \theta_i - \widehat{\theta}_i(1)$, θ_i is the unknown true parameter to be identified, the center is $\widehat{\theta}_i\,(1)$ of ellipsoid, the orientation of the ellipsoid is given by eigenvectors $(u_1, \ldots, u_{4m})$ of P_1, and the axes are given by eigenvalues $(a_1, \ldots, a_{4m})$ of P_1 as $\frac{1}{\sqrt{a_i}}$, see Fig.1.

Parameters estimation via OBE algorithm typically proceeds by alternatively using recursive method. At time k, the time update equation is used to form the feasible set for the predicted parameter. This is done by a vector summing of ellipsoids: the bounding ellipsoid for the parameter estimate at time k and the ellipsoid bounding 1. The observation equation is then used to update the

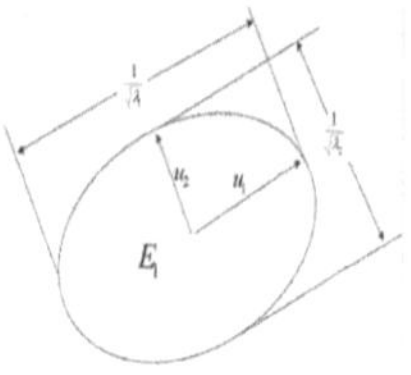

Fig. 1. Initial ellipsoid

Fig. 2. OBE algorithm

predicted parameter estimate by an ellipsoidal intersection: the ellipsoid form the previous step and the ellipsoid obtained by using the bound of 1. In general, the ellipsoidal summing and intersection operations do not yield ellipsoids and thus have to be bounded in some sense.

Substituting (9) into (10) gives:

$$S_k = \{\theta_i \in \mathbb{R}^{4m\times 1} : \left[B_k^T\theta_i - y_i(k)\right]^T \frac{1}{\gamma}\left[B_k^T\theta_i - y_i(k)\right] \leq 1\},$$

without loss of generality we consider $\gamma = 1$, that is, a circle with radius 1 [9], then

$$S_k = \left\{\theta_i \in \mathbb{R}^{4m\times 1} : \left\|y_i(k) - B_k^T\theta_i\right\|^2 \leq 1\right\}. \tag{11}$$

We derive a recursive observation update algorithm.

An ellipsoid that contains $E_k \cap S_k$ is given by [9]:

$$E_{k+1} = \{\theta_i \in \mathbb{R}^{4m\times 1} : (1-\lambda)\,\widetilde{\theta}_i^T(k)P_k^{-1}\widetilde{\theta}_i(k) + \lambda\left\|y_i(k) - B_k^T\theta_i\right\|^2 \leq 1\}, \tag{12}$$

where $E_k = \left\{\theta_i \in \mathbb{R}^{4m\times 1} : \widetilde{\theta}_i^T(k)P_k^{-1}\widetilde{\theta}_i(k) \leq 1\right\}$, $\widetilde{\theta}_i(k) = \theta_i - \widehat{\theta}_i(k)$ and λ is a real number in $(0,1)$. Denote:

$$e_i(k) = y_i(k) - \widehat{y}_i(k), \tag{13}$$

where $\widehat{y}_i(k) = B_k^T\widehat{\theta}_i(k)$.

Theorem 1. *Consider equations (12) to (13), the following modified bounding ellipsoid algorithm*

$$\begin{array}{l} P_{k+1} = \frac{1}{1-\lambda}\left[P_k - \lambda P_k B_k \frac{1}{Q_k} B_k^T P_k\right] \\ \widehat{\theta}_i(k+1) = (1-\lambda)\,P_{k+1}P_k^{-1}\widehat{\theta}_i(k) + \lambda P_{k+1}B_k y_i(k), \end{array} \tag{14}$$

where $Q_k = (1-\lambda) + \lambda B_k^T P_k B_k$, P_k *is a diagonal positive definite matrix,* $\lambda \in (0,1)$ *make the following recursive ellipsoid equation is true*

$$E_{k+1} = \left\{\theta_i \in \mathbb{R}^{4m\times 1} : \widetilde{\theta}_i^T(k+1)P_{k+1}^{-1}\widetilde{\theta}_i(k+1) \leq z_{k+1}\right\}, \tag{15}$$

where $\widetilde{\theta}_i(k+1) = \theta_i - \widehat{\theta}_i(k+1)$, $z_{k+1} = 1 - \lambda(1-\lambda)\,P_k^{-1}P_{k+1}e_i^2(k)$ *and* $e_i(k)$ *is given in (13).*

Proof. We we apply the matrix inversion lemma to first equation of (14) to get P_{k+1}^{-1} *as:*

$$P_{k+1}^{-1} = (1-\lambda)\left[P_k - P_k\lambda B_k\left[B_k^T P_k \lambda B_k + (1-\lambda)\right]^{-1} B_k^T P_k\right]^{-1},$$

gives:

$$P_{k+1}^{-1} = (1-\lambda)\,P_k^{-1} + \lambda B_k B_k^T. \qquad (16)$$

Using (16) and the $\widehat{\theta}_i\,(k+1)$ *of (14) the set (12) can be rewritten as:*

$$\begin{array}{l}
\lambda\left[y_i\,(k) - B_k^T\theta_i\right]^T\left[y_i\,(k) - B_k^T\theta_i\right] + (1-\lambda)\left[\theta_i - \widehat{\theta}_i(k)\right]^T P_k^{-1}\left[\theta_i - \widehat{\theta}_i(k)\right] \leq 1 \\
\theta_i^T\left[\lambda B_k B_k^T + (1-\lambda)\,P_k^{-1}\right]\theta_i - 2\theta_i^T P_{k+1}^{-1} P_{k+1}\left[\lambda B_k y_i\,(k) + (1-\lambda)\,P_k^{-1}\widehat{\theta}_i(k)\right] \\
\leq 1 - \lambda y_i^2\,(k) - (1-\lambda)\,\widehat{\theta}_i^T(k) P_k^{-1}\widehat{\theta}_i(k) \\
\theta_i^T P_{k+1}^{-1}\theta_i - 2\theta_i^T P_{k+1}^{-1}\widehat{\theta}_i\,(k+1) \leq 1 - \lambda y_i^2\,(k) - (1-\lambda)\,\widehat{\theta}_i^T(k) P_k^{-1}\widehat{\theta}_i(k).
\end{array}$$

Adding $\widehat{\theta}_i^T(k+1)P_{k+1}^{-1}\widehat{\theta}_i\,(k+1)$, *both sides gives (15) where:*

$$z_{k+1} = 1 - \lambda y_i^2\,(k) - (1-\lambda)\,\widehat{\theta}_i^T(k)P_k^{-1}\widehat{\theta}_i(k) + \widehat{\theta}_i^T(k+1)P_{k+1}^{-1}\widehat{\theta}_i\,(k+1),$$

but using $\widehat{\theta}_i\,(k+1)$ *of (14):*

$$\begin{array}{l}
\widehat{\theta}_i^T(k+1)P_{k+1}^{-1}\widehat{\theta}_i\,(k+1) = \left[\lambda B_k y_i\,(k) + (1-\lambda)\,P_{k-1}^{-1}\widehat{\theta}_i(k)\right]^T \\
P_{k+1}P_{k+1}^{-1}P_{k+1}\left[\lambda B_k y_i\,(k) + (1-\lambda)\,P_k^{-1}\widehat{\theta}_i(k)\right] \\
= \lambda^2 B_k^T P_{k+1} B_k y_i^2\,(k) + 2\lambda\,(1-\lambda)\,\widehat{\theta}_i^T(k) B_k P_k^{-1} P_{k+1} y_i\,(k) \\
+ (1-\lambda)^2\,\widehat{\theta}_i^T(k) P_k^{-1} P_{k+1} P_k^{-1}\widehat{\theta}_i(k)
\end{array},$$

on other side:

$$\begin{array}{l}
-\lambda y_i^2\,(k) = -\lambda P_{k+1}^{-1} P_{k+1} y_i^2\,(k) \\
= -\lambda\left[(1-\lambda)\,P_k^{-1} + \lambda B_k B_k^T\right] P_{k+1} y_i^2\,(k) \\
= -\lambda\,(1-\lambda)\,P_k^{-1} P_{k+1} y_i^2\,(k) - \lambda^2 B_k^T P_{k+1} B_k y_i^2\,(k)
\end{array}$$

and

$$\begin{array}{l}
-(1-\lambda)\,\widehat{\theta}_i^T(k)P_k^{-1}\widehat{\theta}_i(k) \\
= -(1-\lambda)\,\widehat{\theta}_i^T(k)P_k^{-1}P_{k+1}^{-1}P_{k+1}\widehat{\theta}_i(k) \\
= -(1-\lambda)\,\widehat{\theta}_i^T(k)P_k^{-1}\left[(1-\lambda)\,P_k^{-1} + \lambda B_k B_k^T\right] P_{k+1}\widehat{\theta}_i(k) \\
= -(1-\lambda)^2\,\widehat{\theta}_i^T(k)P_k^{-1}P_{k+1}P_k^{-1}\widehat{\theta}_i(k) - \lambda\,(1-\lambda)\,\widehat{\theta}_i^T(k)B_k P_k^{-1}P_{k+1}B_k^T\widehat{\theta}_i(k)
\end{array}$$

then we have:

$$\begin{array}{r}
z_{k+1} = 1 - \lambda\,(1-\lambda)\,P_k^{-1}P_{k+1}y_i^2\,(k) + 2\lambda\,(1-\lambda)\,\widehat{\theta}_i^T(k-1)B_k P_k^{-1}P_{k+1}y_i\,(k) \\
-\lambda\,(1-\lambda)\,\widehat{\theta}_i^T(k)B_k P_k^{-1}P_{k+1}B_k^T\widehat{\theta}_i(k)
\end{array}$$

and it is z_{k+1} *given in (15).*

Remark 1. From (15) $z_{k+1} \leq 1$, the fusion of E_k and S_k, whose the intersection contain the true parameter θ_i, is E_{k+1} for the value of $\lambda \in (0, 1)$ that minimizes its volume [7], see Fig. 2.

Remark 2. The error $e_i(k)$ of OBE algorithm is not the same as the identification error $\epsilon_i(k) = x_i(k) - \widehat{x}_i(k)$, but they are minimized at the same time. From (2), (4), (9) and (13), we have:

$$\epsilon_i(k+1) = a_i \epsilon_i(k) + e_i(k). \tag{17}$$

By the relation $\epsilon_i(2) = a_i \epsilon_i(1) + e_i(1)$, $\epsilon_i(3) = a_i \epsilon_i(2) + e_i(2) = a_i^2 \epsilon_i(1) + a_i e_i(1) + e_i(2)$, $\epsilon_i(k+1) = a_i^k \epsilon_i(1) + \sum_{j=1}^{k} a_i^{k-j} e_i(j)$. Because $|a_i| < 1$

$$|\epsilon_i(k+1)| \leq |\epsilon_i(1)| + \sum_{j=1}^{k} |e_i(j)|.$$

Since $\epsilon_i(1)$ is a constant, the minimization of the OBE error $e_i(j)$ means the upper bound of the identification error $\epsilon_i(k+1)$ is minimized.

Remark 3. The observer (14) is for each subsystem. This method can decrease computational burden when we estimate the weights of the recurrent neural network, see [7]. By (6) and (7) we know the data matrix B_k depends on the parameters $V_{1,k}^T$ and $V_{2,k}^T$, this will not effect parameter updating algorithm (14), because the unknown parameter $\widehat{\theta}_i(k+1)$ is calculated by the known parameters $\widehat{\theta}_i(k)$ and data B_k. For each element of Θ_k^T and B_k in (14), we have $\widehat{\theta}_i(k+1) = (1-\lambda) P_{k+1} P_k^{-1} \widehat{\theta}_i(k) + \lambda P_{k+1} B_k y_i(k)$

$$\begin{array}{l} V_{1,k+1} = (1-\lambda) P_{k+1} P_k^{-1} V_{1,k} + \lambda P_{k+1} \sigma \left[W_{1,k} x(k) \right] y^T(k), \\ W_{1,k+1} = (1-\lambda) P_{k+1} P_k^{-1} W_{1,k} + \lambda P_{k+1} \sigma' \left[W_{1,k} x(k) \right] V_{1,k}^T x(k) y^T(k). \end{array} \tag{18}$$

It is much more complex to the backpropagation, also the learning rate is not positive constant, in this case we have the element P_{k+1} which makes the OBE algorithm more suitable. That is main reason why OBE algorithm for training has a faster convergence speed.

Remark 4. The extended Kalman filter training algorithm [8] is similar in structure to the OBE algorithm, the extended Kalman filter algorithm is given as:

$$\begin{array}{l} \widehat{\theta}_i(k+1) = \widehat{\theta}_i(k) - P_k B_k \left(R_2 + B_k^T P_k B_k \right)^{-1} e_i(k), \\ P_{k+1} = R_1 + \left[P_k - P_k B_k \left(R_2 + B_k^T P_k B_k \right)^{-1} B_k^T P_k \right]. \end{array}$$

where $e_i(k)$ is as (13), R_1 can be chosen as αI, where α is small and positive. In fact, if we do not have a change during the interval of time, R_1 tends to zero, it becomes the least square algorithm. The plant is not expected to change at all during the time of interest, the covariance of 'process noise' $\zeta_{i,k}$ can be assumed as $E\left\{\zeta_{i,k}\zeta_{i,k}^T\right\} = R_2$ in order to have the OBE algorithm we have $R_1 = 0$, $R_2 = (1-\lambda)$ and we have to multiply

$$\left[P_k - P_k B_k \left(R_2 + B_k^T P_k B_k\right)^{-1} B_k^T P_k\right]$$

for $\frac{1}{(1-\lambda)}$ in the second equation. In the first equation, from (14) we have

$$\widehat{\theta}_i(k+1) = P_{k+1}P_k^{-1}\widehat{\theta}_i(k) + \lambda P_{k+1}\left[B_k y_i(k) - P_k^{-1}\widehat{\theta}_i(k)\right],$$

which is a little similar to the extended Kalman filter but it is more suitable because of the parameter $P_{k+1}P_k^{-1}$ that changes $\widehat{\theta}_i(k)$ in the OBE algorithm.

The following steps show how to train the weights of recurrent neural networks with the OBE algorithm:

1. Construct a recurrent neural networks model (2) to identify an unknown nonlinear system (1). The matrix A is selected such that it is stable.
2. Rewrite the neural network in linear form

$$\begin{aligned} y(k) &= B_k^T \Theta_k + \zeta(k), \\ \Theta_k &= \left[V_{1,k}, W_{1,k}^T, V_{2,k}, W_{2,k}^T\right]^T = \left[\theta_1(k), \cdots \theta_n(k)\right], \\ B_k &= \left[\sigma, \sigma' V_{1,k}^T x, \phi u, \phi' \operatorname{diag}(u) V_{2,k}^T x(k)\right]^T. \end{aligned}$$

3. Train the weights as

$$\begin{aligned} \widehat{\theta}_i(k+1) &= (1-\lambda) P_{k+1}P_k^{-1}\widehat{\theta}_i(k) + \lambda P_{k+1}B_k y_i(k), \\ B_k &= \left[\sigma, \sigma'\widehat{V}_{1,k}^T x, \phi u, \phi' \operatorname{diag}(u)\widehat{V}_{2,k}^T x(k)\right]^T, \quad \lambda \in (0,1). \end{aligned}$$

4. P_k is changed as OBE algorithm:

$$P_{k+1} = \frac{1}{1-\lambda}\left[P_k - \lambda P_k B_k \frac{1}{Q_k} B_k^T P_k\right],$$

where $Q_k = (1-\lambda) + \lambda B_k^T P_k B_k$.

With initial conditions for the weight $\widehat{\theta}_i(1)$ and $P_1 > 0$, we can start the system identification with the recurrent neural networks.

4 Simulations

We will use the nonlinear system which proposed [1] to illustrate the behavior of the optimal bounded ellipsoid training algorithm proposed in this paper

$$x_1(k+1) = \frac{x_1(k)x_2(k)x_3(k)}{1+x_1(k)^2+x_2(k)^2+x_3(k)^2} + 2u(k),$$

where $x_2(k+1) = x_1(k)$, $x_3(k+1) = x_2(k)$. The unknown nonlinear system has the standard form (1), we use the recurrent neural network (series-parallel model) given in (2) to identify it, where $\widehat{x}(k) \in \mathbb{R}^4$, $A \in \mathbb{R}^{4\times 4}$ is a stable diagonal matrix which is specified as $A = diag(0.1)$. In this paper, in order to exam the effectiveness of the OBE algorithm for training, we use 1 node in the hidden layer. The weights in output layer are $V_{1,k} \in \mathbb{R}^{4\times 1}$, the weights in hidden layer are $W_{1,k} \in \mathbb{R}^{1\times 4}$, $\sigma = [\sigma_1]^T$, $\phi(\cdot)$ is an element. The elements of the initial weights $W_{1,0}$ and $V_{1,0}$ are chosen in random number between $(0,1)$. The input is $u(t) = 0.03\sin(3\pi kTs) + 0.01\sin(4\pi kTs) + 0.06\sin(\pi kTs)$. We select $P = diag(10) \in \mathbb{R}^{4\times 4}$, $\lambda = 1\times 10^4$. The identification results for $x_2(k)$ are shown in Fig.3. We can see that OBE algorithm has good behavior. Now we use the backpropagation algorithm [4] with learning rate 0.02, and also recursive least square method [1] to train the neural network given in (2). We define the mean squared error for finite time as $J(N) = \frac{1}{2N}\sum_{k=1}^{N} e^2(k)$. The identification results for $J(N)$ are shown in Fig.4. We find that OBE training has best convergence property.

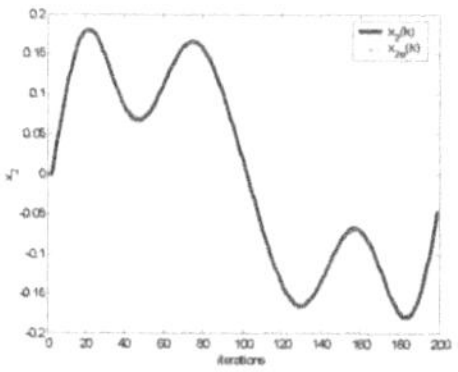

Fig. 3. OBE algorithm for $x_2(k)$

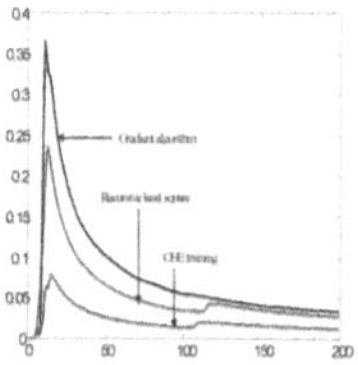

Fig. 4. Errors comparison

5 Conclusions

In this paper a novel training method for neural identification is proposed. We give a modified OBE algorithm for recurrent neural networks training. Both hidden layers and output layers of the state-space recurrent neural networks can be updated. The future work will be on the stability analysis of the algorithm.

References

1. Chowdhury, F.N.: A New Approach to Real-Time Training of Dynamic Neural Networks. Int. J. Adaptive Control and Signal Processing **3** (2003) 509-521
2. Dickerson, J.A., Kosko, B.: Fuzzy Function Approximation with Ellipsoidal Rules. IEEE Trans. Systems, Man and Cybernetics **26** (1996) 542-560
3. Kayuri, N.S., Vienkatasubramanian, V.: Representing Bounded Fault Classes using Neural Networks with Ellipsoidal Activation Functions. Computers & Chemical Engineering **17** (1993) 139-163
4. Kosmatopoulos, E.B., Polycarpou, M.M., Christodoulou, M.A., Ioannou, P.A.: High-Order Neural Network Structures for Identification of Dynamical Systems. IEEE Trans. Neural Networks **6** (1995) 422-431

5. Liguni,Y., Sakai, H., Tokumaru, H.: A Real-Time Learning Algorithm for a Multilayered Neural Network Based on the Extended Kalman Filter. IEEE Trans. Signal Processing **40** (1992) 959-966
6. Parlos, A.G., Menon, S.K., Atiya, A.F.: An Algorithm Approach to Adaptive State Filtering Using Recurrent Neural Network. IEEE Trans. Neural Networks **12** (2001) 1411-1432
7. Puskorius, G.V., Feldkamp, L.A.: Neurocontrol of Nonlinear Dynamical Systems with Kalman Filter Trained Recurrent Networks. IEEE Trans. Neural Networks **5** (1994) 279-297
8. Rubio, J., Yu, W.: Nonlinear System Identification with Recurrent Neural Networks and Dead-Zone Kalman Filter Algorithm. Neurocomputing, in press
9. Schueppe, F.C.: Uncertain Dynamic Systems. Englewood cliffs, Prentice-Hall (1973)
10. Singhal, S., Wu, L.: Training Multilayer Perceptrons with the Extended Kalman Algorithm. Advances in Neural inform. Processing Syst. **I** (1989) 133-140
11. Yu, W.: Nonlinear System Identification using Discrete-Time Recurrent Neural Networks with Stable Learning Algorithms. Information Sciences **158** (2002) 131-147

Efficient Training of RBF Networks Via the BYY Automated Model Selection Learning Algorithms

Kai Huang, Le Wang, and Jinwen Ma*

Department of Information Science, School of Mathematical Sciences
And LMAM, Peking University, Beijing, 100871, China
jwma@math.pku.edu.cn

Abstract. Radial basis function (RBF) networks of Gaussian activation functions have been widely used in many applications due to its simplicity, robustness, good approximation and generalization ability, etc.. However, the training of such a RBF network is still a rather difficult task in the general case and the main crucial problem is how to select the number and locations of the hidden units appropriately. In this paper, we utilize a new kind of Bayesian Ying-Yang (BYY) automated model selection (AMS) learning algorithm to select the appropriate number and initial locations of the hidden units or Gaussians automatically for an input data set. It is demonstrated well by the experiments that this BYY-AMS training method is quite efficient and considerably outperforms the typical existing training methods on the training of RBF networks for both clustering analysis and nonlinear time series prediction.

1 Introduction

The radial basis function (RBF) networks [1]-[3] are a typical class of forward neural networks widely used in the fields of pattern recognition and signal processing. Clearly, it was developed from the approximation theory of radial basis functions for multivariate interpolation. That is, the RBFs are embedded in a two-layer neural network such that each hidden unit implements a radial basis function and the output units implement a weighted sum of hidden unit outputs. With such a structure, a RBF network can approximate any continuous function at a certain degree as long as the number of hidden units are large enough. Moreover, the structure of the input data can be appropriately matched via the selection of receptive fields of the radial basis functions with the hidden units, which leads to a good generalization of the RBF network. Therefore, the RBF network has been widely applied to various fields of pattern recognition and signal processing such as speech recognition, clustering analysis, time series prediction, industrial control etc., and the most commonly used radial basis functions in the RBF networks are Gaussian activation functions.

However, the training of a RBF network of Gaussian activation functions is still a rather difficult task. In fact, the main crucial problem is how to select the number and locations of the hidden units appropriately for a practical problem. In the previous

* The corresponding author.

D. Liu et al. (Eds.): ISNN 2007, Part I, LNCS 4491, pp. 1183–1192, 2007.

approaches, the number of hidden units was just the number of sample data and the locations of the hidden units were those sample data. However, it was proved that such a training or selection is expensive in terms of memory requirement. Moreover, the exact fit to the training set might cause a bad generalization. In order to overcome these problems, many training methods were proposed for training RBF networks and most of them utilized a test-and-growing or evolutionary approach to selecting the number of hidden units on the practical applications (e.g., [4]-[6]). Actually, a good selection of hidden units should appropriately match the structure of the input data associated with the practical problem. If the input data set consists of k clusters, the RBF network should select k hidden units with their locations being the centers of the k clusters, respectively. However, the selection of number of clusters for an input data set is also a difficult problem [7].

With development of competitive learning, the rival penalized competitive learning (RPCL) algorithm was proposed to determine the number of clusters or Gaussians in a sample data automatically [8]-[9]. Therefore, it has provided a new tool for the training of the RBF network (e.g., [10]-[11]). On the other hand, the scatter-based clustering (SBC) method [12] and the least biased fuzzy clustering method [13] were also proposed to determine the best number of hidden units in a RBF network.

Recently, based on the Bayesian Ying-Yang (BYY) harmony learning theory [14]-[16], a new kind of automated model selection (AMS) learning mechanism has been established for the Gaussian mixture modeling [17]-[19]. Actually, this kind of BYY-AMS learning rules can determine the number of Gaussians automatically during the parameter learning, which can be utilized to select the number of Gaussians as the hidden units in the RBF network.

In this paper, we utilize the BYY-AMS adaptive gradient learning algorithm [19] to select the appropriate number and initial locations of the Gaussians automatically on an input data set for the train of the RBF network. It is demonstrated by the experiments that this new training method is quite efficient and considerably outperforms some typical existing methods on the training of a RBF network for both clustering analysis and nonlinear time series prediction.

2 BYY-AMS Adaptive Gradient Learning Algorithm

We begin to introduce the adaptive gradient learning algorithm of automated model selection on the Gaussian mixture model proposed in [19] in the light of the BYY harmony learning theory. A BYY system describes each observation $x \in X \subset R^d$ and its corresponding inner representation $y \in Y \subset R^m$ via the two types of Bayesian decomposition of the joint density $p(x, y) = p(x)p(y \mid x)$ and $q(x, y) = q(x \mid y)q(y)$, being called Yang machine and Ying machine, respectively. For the Gaussian mixture modeling, y is only limited to be an integer variable, i.e., $y \in Y = \{1, 2, \cdots, K\} \subset R$ with m = 1. Given a data set $D_x = \{x_t\}_{t=1}^{N}$, the task of learning on a BYY system consists of specifying all the aspects of

$p(y \mid x)$, $p(x)$, $q(x \mid y)$, $q(y)$ via a harmony learning principle implemented by maximizing the functional:

$$H(p \| q) = \int p(y \mid x) p(x) \ln[q(x \mid y) q(y)] dx dy - \ln z_q, \tag{1}$$

where z_q is a regularization term.

If both $p(y \mid x)$ and $q(x \mid y)$ are parametric, i.e., from a family of probability densities with some parameter θ, the BYY system is called to have a Bi-directional Architecture (or BI-Architecture for short). For the Gaussian mixture modeling, we use the following specific BI-architecture of the BYY system. $q(y = j) = \alpha_j$ with $\alpha_j \geq 0$ and $\sum_{j=1}^{K} \alpha_j = 1$. Also, we ignore the regularization term z_q (i.e., set $z_q = 1$) and let $p(x)$ be the empirical density $p(x) = \frac{1}{N} \sum_{t=1}^{N} \delta(x - x_t)$. Moreover, the BI-architecture is constructed with the following parametric form:

$$p(y = j \mid x) = p(j \mid x) = \frac{\alpha_j q(x \mid \theta_j)}{q(x, \Theta_K)}, \quad q(x, \Theta_K) = \sum_{j=1}^{K} \alpha_j q(x \mid \theta_j), \tag{2}$$

where $q(x \mid \theta_j) = q(x \mid y = j)$ with θ_j consisting of all its parameters and $\Theta_K = \{\alpha_j, \theta_j\}_{j=1}^{K}$ is the set of parameters for the finite mixture model.

Substituting these component densities into Eq. (1), we have

$$H(p \| q) = J(\Theta_K) = \frac{1}{N} \sum_{t=1}^{N} \sum_{j=1}^{K} \frac{\alpha_j q(x_t \mid \theta_j)}{\sum_{i=1}^{K} \alpha_j q(x_t \mid \theta_j)} \ln[\alpha_j q(x_t \mid \theta_j)]. \tag{3}$$

That is, $H(p \| q)$ becomes a harmony function $J(\Theta_K)$ on the parameters Θ_K. Furthermore, we let $q(x \mid \theta_j)$ be a Gaussian density given by

$$q(x \mid \theta_j) = q(x \mid m_j, \Sigma_j) = \frac{1}{(2\pi)^{\frac{n}{2}} |\Sigma_j|^{\frac{1}{2}}} e^{\frac{1}{2}(x - m_j)^T \Sigma_j^{-1} (x - m_j)}, \tag{4}$$

where m_j is the mean vector and Σ_j is the covariance matrix which is assumed positive definite.

According to the harmony function given in Eq.(3), we can construct an adaptive gradient algorithm or rule to search for a maximum of $J(\Theta_K)$ as an estimate of the parameters Θ_K with the sample data set D_x. For convenience of derivation, we let

$$\alpha_j = \frac{e^{\beta_j}}{\sum_{i=1}^{K} e^{\beta_j}}, \quad \Sigma_j = B_j B_j^T, \quad j = 1, 2, \cdots, K,$$

where $-\infty < \beta_1, \cdots, \beta_K < +\infty$ and B_j is a nonsingular square matrix. Via these transformations, the parameters in $J(\Theta_K)$ turn into $\{\beta_j, m_j, B_j\}_{j=1}^{K}$.

Denoting $U_j(x)=\alpha_j q(x \mid m_j,\Sigma_j)$ for $j=1,\cdots,K$, $J(\Theta_K)$ has the following simple expression:

$$J(\Theta_K)=\frac{1}{N}\sum_{t=1}^{N} J_t(\Theta_K),\ J_t(\Theta_K)=\sum_{j=1}^{K}\frac{U_j(x_t)}{\sum_{i=1}^{K} U_i(x_t)}\ln U_j(x_t). \tag{5}$$

Furthermore, we have the derivatives of $J(\Theta_K)$ with respect to β_j, m_j and B_j, respectively, as follows.

$$\frac{\partial J_t(\Theta_K)}{\partial \beta_j}=\frac{1}{q(x_t \mid \Theta_k)}\sum_{i=1}^{K}\lambda_i(t)(\delta_{ij}-\alpha_j)U_i(x_t), \tag{6}$$

$$\frac{\partial J_t(\Theta_K)}{\partial m_j}=p(j \mid x_t)\lambda_j(t)\Sigma_j^{-1}(x_t-m_j), \tag{7}$$

$$vec[\frac{\partial J_t(\Theta_K)}{\partial B_j}]=\frac{\partial (B_j B_j^T)}{\partial B_j} vec[\frac{\partial J_t(\Theta_k)}{\partial \Sigma_j}], \tag{8}$$

where δ_{ij} is the Kronecker function, $vec[A]$ denotes the vector obtained by stacking the column vectors of the matrix A, and

$$\lambda_i(t)=1-\sum_{l=1}^{K}(p(l \mid x_t)-\delta_{il})\ln[\alpha_l q(x_t \mid m_l,\Sigma_l)], \tag{9}$$

$$\frac{\partial J_t(\Theta_K)}{\partial \Sigma_j}=\frac{1}{2}p(j \mid x_t)\lambda_j(t)[\Sigma_j^{-1}(x_t-m_j)(x_t-m_j)^T\Sigma_j^{-1}-\Sigma_j^{-1}], \tag{10}$$

$$\frac{\partial (BB^T)}{\partial B}=I_{d\times d}\otimes B_{d\times d}^T+E_{d^2\times d^2}\bullet B_{d\times d}^T\otimes I_{d\times d},$$

where $\otimes$ denotes the Kronecker product (or tensor product), and

$$E_{d^2\times d^2}=\frac{\partial B^T}{\partial B}=(\Gamma_{ij})_{d^2\times d^2}=\begin{pmatrix}\Gamma_{11} & \cdots & \Gamma_{1d}\\ \vdots & \ddots & \vdots\\ \Gamma_{d1} & \cdots & \Gamma_{dd}\end{pmatrix}_{d^2\times d^2},$$

where Γ_{ij} is an d × d matrix whose $(j,i)^{th}$ element is just 1, with all the other elements being zero. With the above expression of $\frac{\partial (BB^T)}{\partial B}$, we have

$$vec[\frac{\partial J(\Theta_K)}{\partial B_j}] = \frac{1}{2} p(j \mid x_t) \lambda_j(t) (I_{d\times d} \otimes B_{d\times d}^T + E_{d^2\times d^2} \bullet B_{d\times d}^T \otimes I_{d\times d})$$

$$\times vec[\Sigma_j^{-1}(x_t - m_j)(x_t - m_j)^T \Sigma_j^{-1} - \Sigma_j^{-1}]. \tag{11}$$

Based on the above preparations, we have the adaptive gradient learning algorithm as follows.

$$\Delta\beta_j = \frac{\eta}{q(x_t \mid \Theta_k)} \sum_{i=1}^{K} \lambda_i(t)(\delta_{ij} - \alpha_j) U_i(x_t), \tag{12}$$

$$\Delta m_j = \eta p(j \mid x_t) \lambda_j(t) \Sigma_j^{-1}(x_t - m_j), \tag{13}$$

$$\Delta vec B_j = \frac{\eta}{2} p(j \mid x_t) \lambda_j(t) (I_{d\times d} \otimes B_{d\times d}^T + E_{d^2\times d^2} \bullet B_{d\times d}^T \otimes I_{d\times d})$$

$$\times vec[\Sigma_j^{-1}(x_t - m_j)(x_t - m_j)^T \Sigma_j^{-1} - \Sigma_j^{-1}], \tag{14}$$

where η denotes the learning rate that starts from a reasonable initial value and then reduces to zero with the iteration number n in such a way that $0 \le \eta(n) \le 1$, and

$$\lim_{n\to\infty} \eta(n) = 0 \ , \quad \sum_{n=1}^{\infty} \eta(n) = \infty , \tag{15}$$

i.e. , in the way used in the conventional stochastic approximation procedure [20]. The typical example of the learning rate satisfying Eq.(15) is $\eta(n) = \eta_0 / n$, where η_0 is a positive constant, which will be used in the following experiments.

This kind of BYY harmony learning can make model selection automatically on the Gaussian mixture model by forcing the mixing proportions of the extra Gaussians to be reduced to zero during the parameters learning as long as K is set to be larger than the number of actual Gaussians in the sample data. Actually, it had shown by the experiments in [19] that this BYY-AMS adaptive gradient learning algorithm can make model selection efficiently with a good estimate for the true parameters of the Gaussian mixture generating the sample data D_x .

3 Training of the RBF Network

We now consider the training of the RBF network via the above BYY-AMS adaptive gradient learning algorithm. In fact, the RBF network is just a two-layer forward neural network and its outputs are given by

$$y_j(x) = \sum_{j=1}^{n} w_{ij} \phi_i(x), \tag{16}$$

where n is the number of hidden units or RBF's, w_{ij} is the connection weight from the ith hidden unit to the jth output unit. $\phi_i(x)$ is just a Gaussian radial basis function (RBF) as the activation function corresponding to the output of the ith hidden unit given by

$$\phi_j(x) = \phi_j(\| x - m_j \|) = \exp(-\frac{\| x - m_j \|^2}{2\sigma_j^2}), \tag{17}$$

where m_j, σ_j are the center and scale of the Gaussian RBF $\phi_j(x)$, respectively.

Without loss of generality, we just consider the case of the RBF network with one single output unit. In this special case, the output function of the RBF network takes a simple form as follows.

$$y(x) = \sum_{j=1}^{n} \lambda_j \phi_j(x) = \sum_{j=1}^{n} \lambda_j \exp(-\frac{\| x - m_j \|^2}{2\sigma_j^2}), \tag{18}$$

where λ_j is the connection weight from the jth hidden unit or RBF to the output unit. Thus, the parameters of the RBF network are just $\lambda_j, m_j, \sigma_j (> 0)$ for $j = 1, 2, \cdots, n$. Moreover, the mean square error of the RBF network on a sample set $D_{(x,y)} = \{(x_i, y_i) : i = 1, 2, \cdots, N\}$ can be given as follows.

$$\begin{aligned} E &= \frac{1}{2}\sum_{i=1}^{N}[y_i - f(x_i)]^2 = \frac{1}{2}\sum_{i=1}^{N}[y_i - \sum_{j=1}^{n}\lambda_j \phi_j(x_i)]^2 \\ &= \frac{1}{2}\sum_{i=1}^{N}[y_i - \sum_{j=1}^{n}\lambda_j \exp(-\frac{\left\| x_i - m_j \right\|^2}{2\sigma_j^2})]^2 . \end{aligned} \tag{19}$$

According to the least mean square error principle, we have the following learning rules on the parameters of the single-output-unit RBF network as follows:

$$\begin{cases} \Delta\lambda_j = \eta_\lambda \sum_{i=1}^{N}[y_i - \sum_{l=1}^{n}\lambda_l \phi_l(x_i)]\phi_j(x_i); \\ \Delta m_j = \eta_m \sum_{i=1}^{N}[y_i - \sum_{l=1}^{n}\lambda_l \phi_l(x_i)]\phi_j(x_i)(x_i - m_j)\lambda_j / \sigma_j^2; \\ \Delta\sigma_j = \eta_\sigma \sum_{i=1}^{N}[y_i - \sum_{l=1}^{n}\lambda_l \phi_l(x_i)]\phi_j(x_i)(x_i - m_j)^T(x_i - m_j)\lambda_j / \sigma_j^3, \end{cases} \tag{20}$$

where η_λ, η_m, η_σ are the learning rates for the updates of the parameters λ_j, m_j, σ_j, respectively, which are assumed to be invariant with the index j.

Usually, these learning rates are selected as some small positive constants by experience.

However, the LMS learning algorithm given by Eq.(20) has a major disadvantage that it is very sensitive to the selection of n and the initial values of the other parameters. Fortunately, the BYY-AMS adaptive gradient learning algorithm can be utilized to solve this sensitiveness problem. Actually, based on the input data set, the BYY-AMS adaptive gradient learning algorithm can determine an appropriate number of Gaussians, i.e., Gaussian RBF's, for the network. That is, we let $n = K^*$, where K^* is the number of actual Gaussians in the input sample data obtained by the the BYY-AMS adaptive gradient learning algorithm. Moreover, the final values of the mixing proportions and mean vectors can serve the initial values of the weights and centers of the n RBF's, respectively. And the initial value of σ_j can be set by

$$\sigma_j = \frac{1}{N_j} \sum_{x_t \in C_j} (x_t - m_j)^T (x_t - m_j), \tag{21}$$

where C_j is the set of the input sample set x_t with the maximum posteriori probability $p(j \mid x_t)$, N_j is the number of elements in C_j and m_j is the final value of the mean vector obtained by the BYY-AMS adaptive gradient learning algorithm. Augmented with the BYY-AMS adaptive gradient learning algorithm in this way, the LMS learning algorithm becomes very efficient on the training of the RBF network, which will be demonstrated by the experiments in the next section. For clarity, we refer to this compound training method just as the BYY-AMS training method for the RBF network.

4 Experiment Results

In this section, two kinds of experiments are carried out to demonstrate the efficiency of the BYY-AMS adaptive gradient learning algorithm on the training of a RBF network. Moreover, we compare the BYY-AMS training method with some other typical existing training methods.

4.1 On the Noisy XOR Problem

The noisy XOR problem [11] is a typical non-linear classification problem and we use a RBF network to learn it. The sample data are shown in Fig. 1 such that the sample points around the centers (1,0) and (-1,0) are in the first class and their outputs should be 1, while the sample points around the centers (0,1) and (0,-1) are in the second class and their outputs should be 0. For this problem, we generated 800 sample points totally, and 200 sample points per each center. We took 100 points per each center, and total 400 points to form the training set, and let the other 400 points be the test set. The actual outputs of the RBF network obtained by the BYY-AMS training method on the 400 test samples are given in Fig. 2.

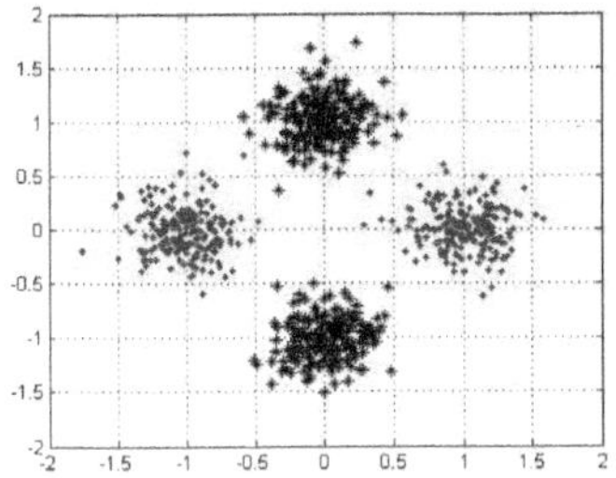

Fig. 1. The sample points of the noisy XOR problem

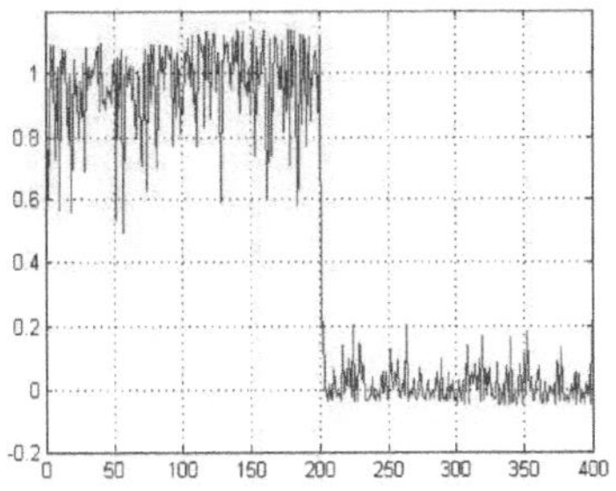

Fig. 2. The outputs of the trained RBF network on the test sample points

When we let the output of RBF network on an input point be processed via a threshold function such that if the output value of the network is over 0.5, the classification output is considered to be 1, otherwise, the classification output is considered to be 0, the correct classification rate of the trained RBF network on the test sample points can reach to 99.75%. However, in the same situation, the correct classification rates of the trained RBF networks with the RPCL and SBC methods can only reach to 97.5% and 97.75%, respectively.

4.2 On the Mackey-Glass Time Series Prediction

We further trained the RBF network with the help of the BYY-AMS adaptive gradient learning algorithm for time series prediction. As shown in Fig. 3, a piece of the Mackey-Glass time series was generated via the following iterative equation:

$$x(t+1)=(1-b)x(t)+\frac{ax(t-\tau)}{1+x(t-\tau)^{10}}, \tag{22}$$

where $a=0.2, b=0.1, \tau=17$. Particularly, 1000 sample data were generated to form pieces of time series as $\{x(t-18), x(t-12), x(t-6), x(t), x(t+6)\}, 118 \le t \le 1117$, where the first four data were considered as an input data of the RBF network, while the last one was considered as the prediction result of the RBF network. Mathematically, the mapping relation behind the Mackey-Glass time series can be given as $y_i = f(x_i)$, where $x_i = [x(t-18), x(t-12), x(t-6), x(t)]^T$, $y_i = x(t+6)$, and $i = 1,\cdots,N$. In our experiment, we divided these 1000 sample data into two sets: the training and test sets with the preceding and remaining 500 sample data, respectively. The mean square error (MSE) was used to measure the prediction accuracy.

We implemented the BYY-AMS training method to train the RBF network for the prediction of this time series and the prediction result on the test data is given in Fig. 4, with the prediction mean square error 0.0033, which may be the lowest prediction error on the the Mackey-Glass time series. For comparison, we also implemented the least biased fuzzy clustering (LBFC) method to train the RBF network on the same data set and obtained the prediction result with the prediction mean square error as 0.2328, which is much greater than that of the RBF network via the BYY-AMS training method.

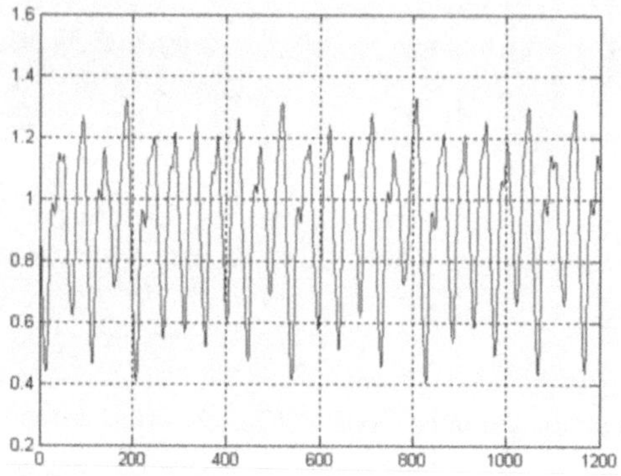

Fig. 3. The sketch of the piece of the Mackey-Glass time series

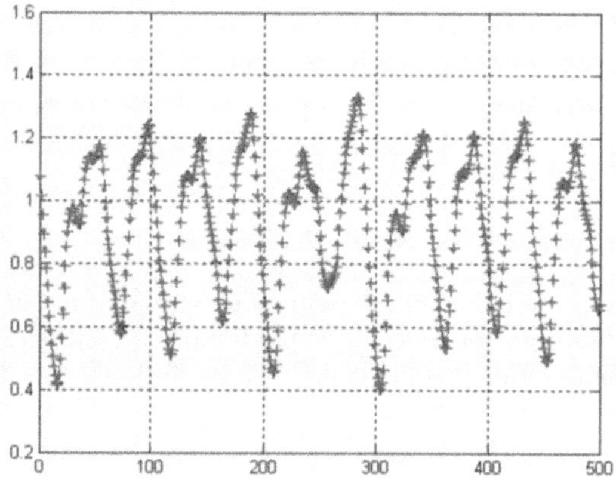

Fig. 4. The prediction result with the BYY-AMS adaptive gradient learning algorithm, where + represents the sample datum, while·represents the prediction datum

5 Conclusions

We have investigated the training of a RBF network with the help of a new kind of automated model selection learning algorithm based on the BYY harmony learning theory. Since this BYY-AMS learning algorithm can detect the structure of the input sample data, it can make the RBF network be more appropriate to a practical problem and improve the approximation and generalization or prediction abilities. The experimental results show that this BYY-AMS training method is very efficient and considerably outperforms the typical existing training methods on the training of a RBF network for clustering analysis and nonlinear time series prediction.

Acknowledgements

This work was supported by the Natural Science Foundation of China for Project 60471054.

References

1. Broomhead, D.S., Lowe, D.: Multivariable Functional Interpolation and Adaptive Networks. Complex System **2** (1988) 521-355
2. Moody, J., Darken, C.: Fast Learning in Networks of Locally Tuned Processing Units. Neural Computation **1** (1989) 281-294
3. Poggio, T., Girosi, F.: Regularization Algorithms for Learning that are Equivalent to Multiplayer Networks. Science **247** (1990) 978-982
4. Platt, J.: A Resource-allocating Network for Function Interpolation. Neural Computation **3** (1991) 213-225
5. Esposito, A., Marinaro, M., Oricchio, D., Scarpetta, S.: Approximation of Continuous and Discontinuous Mapping by a Growing Neural RBF-based Algorithm. Neural Networks **13** (2000) 651-665
6. Sanchez, A.V.D.: Searching for a Solution to the Automatic RBF Network Design Problem. Neurocomputing **42** (2002) 147-170
7. Hartigan, J. A.: Distribution Problems in Clustering. In J. Van Ryzin Editor, Classification and clustering, Academic Press, New York (1977) 45-72
8. Xu, L., Krzyzak, A., Oja, E.: Rival Penalized Competitive Learning for Clustering Analysis, Rbf Net and Curve Detection. IEEE Trans. on Neural Networks **4** (1993) 636–649
9. Ma, J., Wang, T.: A Cost-functionApproach to Rival Penalized Competitive Learning (RPCL). IEEE Transactions on Systems, Man and Cybernetics, Part B: Cybernetics, **36** (4) (2006) 722-737
10. Krzyzak, A., Linder, T., Lugosi, G.: Nonparametric Estimation and Classification Using Radial Basis Function Nets and Empirical Risk Minimization. IEEE Trans. on Neural Networks, 7(1996): 475-487
11. Bors A. G. and Pitas I.: Median radial basis function neural network. IEEE Transactions on Neural Networks **7** (1996) 1351-1364
12. Sohn, I., Ansari, N.: Configure Rbf Neural Networks. Electronics Letters **34** (1998) 684- 685
13. Beni, G., Liu, X.: A Least Biased Fuzzy Clustering Method. IEEE Transactions On Pattern Analysis and Machine Intelligence **16** (1994) 954-960
14. Xu, L.: Ying-Yang Machine: A Bayesian Kullback Scheme for Unified Learning and New Results on Vector Quantization. Proceedings of International Conference on Neural Information Processing (ICONIP95) **2** 977-988
15. Xu, L.: A Unified Learning Scheme: Bayesian-Kullback Ying-Yang Machine. Advances in Neural Information Processing Systems 8 (1996) 444-450
16. Xu, L.: Best Harmony, Unified RPCL and Automated ModelSelection for Unsupervised and SupervisedLearning on Gaussian mixtures,Three-layer Nets and Me-rbf-svm Model. International Journal of Neural System **11** (2001) 43-69
17. Ma, J., Wang, T., Xu, L.: A Gradient BYY Harmony Learning Rule on Gaussian Mixture with Automated Model Selection. Neurocomputing **56** (2004) 481-487
18. Ma, J., Gao, B., Wang, Y., Cheng, Q.: Conjugate and Natural Gradient Rules for BYY Harmony Learning on Gaussian Mixture with Automated Model Selection. International Journal of Pattern Recognition and Artificial Intelligence **19** (2005) 701-713
19. Ma, J., Wang, L.: BYY Harmony Learning on Finite Mixture: Adaptive Gradient Implementation and a Floating RPCL Mechanism. Neural Processing Letters **24** (2006) 19-40
20. Robbins, H., Monro, S.: A Stochastic Approximation Method. Annals of Mathematical Statistics **22** (1951) 400-407

Unsupervised Image Categorization Using Constrained Entropy-Regularized Likelihood Learning with Pairwise Constraints

Zhiwu Lu, Xiaoqing Lu, and Zhiyuan Ye

Institute of Computer Science and Technology, Peking University,
Beijing 100871, China
zhiwu.lu@yahoo.com.cn

Abstract. We usually identify the categories in image databases using some clustering algorithms based on the visual features extracted from images. Due to the well-known gap between the semantic features (e.g., categories) and the visual features, the results of unsupervised image categorization may be quite disappointing. Of course, it can be improved by adding some extra semantic information. Pairwise constraints between some images are easy to provide, even when we have little prior knowledge about the image categories in a database. A semi-supervised learning algorithm is then proposed for unsupervised image categorization based on Gaussian mixture model through incorporating such semantic information into the entropy-regularized likelihood (ERL) learning, which can automatically detect the number of image categories in the database. The experiments further show that this algorithm can lead to some promising results when applied to image categorization.

1 Introduction

Unsupervised image categorization plays an important role in browsing an image database for further image retrieval and query, i.e., we can find the image query of interest by providing first the best overview of the database. To identify "natural" categories in a collection of images, we resort to some clustering algorithms [1] which rely exclusively on similarity measures using the visual features extracted from images. Due to the well-known gap [2] between the semantic features (e.g., categories) and the visual features, the results of image categorization based on clustering may be quite disappointing.

With some supervision provided by the user during image categorization, we can expect to obtain more adequate results. Supervision may consist in class labels for a few data items (not necessarily from all the classes) or in pairwise constraints specifying whether two items should be in a same category or rather in different categories. Such pairwise constraints [3] are indeed much easier to provide than class labels, when the user has little prior knowledge about the image categories in a database. In the case of image collections, pairwise constraints can either be directly provided by users or obtained from the keyword annotations that are usually few and only available for some categories. For example,

D. Liu et al. (Eds.): ISNN 2007, Part I, LNCS 4491, pp. 1193–1200, 2007.

must-link constraints can be defined between images that share many keywords and cannot-link constraints between annotated images that have no keyword in common. The clustering approaches that take into account such simple semantic information during the clustering process are called semi-supervised clustering [4,5]. However, these algorithms tend to be heuristic without employing explicit models or must be provided with the number of clusters firstly.

In this paper, a semi-supervised learning algorithm is then proposed for unsupervised image categorization based on Gaussian mixture model [6] through incorporating such semantic information into the entropy-regularized likelihood (ERL) learning [7,8], which can automatically detect the number of image categories in the database. Actually, the property of automatic model selection was demonstrated well in [9,10] via the iterative ERL learning algorithm in the case of Gaussian mixture modeling. The experiments further show that the constrained ERL Learning algorithm can lead to some promising results when applied to image categorization.

2 Constrained ERL Learning

The well-known Gaussian mixture model is taken into account for unsupervised image categorization:

$$p(x|\Theta) = \sum_{l=1}^{k} \alpha_l p(x \mid \theta_l), \ \sum_{l=1}^{k} \alpha_l = 1, \ \alpha_l \geq 0, \tag{1}$$

$$p(x|\theta_l) = (2\pi)^{-n/2} |\Sigma_l|^{-1/2} \exp\{-\frac{1}{2}(x - m_l)^T \Sigma_l^{-1} (x - m_l)\}, \tag{2}$$

where $x \in R^n$, and k is the number of Gaussians in the mixture. The parameter set Θ consists of the mixing proportions α_l, the mean vectors m_l, and the covariance matrices Σ_l which are assumed positive definite.

We further consider incorporating pairwise constraints into the above Gaussian mixture model. It is important to note that there is a basic difference between must-link and cannot-link constraints: while must-link constraints are transitive (i.e. a group of must-link constraints can be merged using a transitive closure), cannot-link constraints are not transitive. As it turns out, it is much more difficult to incorporate cannot-link constraints into Gaussian mixture modeling, and we require some heavy duty inference machinery such as Markov networks which can incur a large computational cost. Moreover, it has been shown in [3] that most improvement of the EM algorithm can be attributed to the must-link constraints, and in most cases adding the cannot-link constraints contributes a small improvement over results obtained when using only must-link constraints. Hence, we just take advantage of must-link constraints in this paper and the cannot-link constraints can be used in the future work.

Given a sample set $S = \{x_t\}_{t=1}^N$ drawn from $p(x|\Theta)$, we have $S = \bigcup_{i=1}^M X_i$, $M \leq N$, where X_i denotes a subset of samples x_t from the same unknown Gaussian source and may be obtained by applying the transitive closure to the set of must-link constraints (unconstrained samples appear as X_i of size one). We assume that all $X_i (i = 1, ..., M)$ are sampled i.i.d., with respect to the mixing proportion of their corresponding source (samples within each X_i are also sampled i.i.d.). Hence, the negative log-likelihood function on the mixture model $p(x|\Theta)$ is given by

$$L(\Theta) = -\frac{1}{M}\sum_{i=1}^{M} \ln(\sum_{l=1}^{k} \alpha_l \prod_{x_t \in X_i} p(x_t|\theta_l)). \tag{3}$$

The constrained EM algorithm [3] is just an implemented of minimizing $L(\Theta)$.

With the posterior probability that X_i arises from the l-th Gaussian

$$P(l|X_i) = \alpha_l \prod_{x_t \in X_i} p(x_t|\theta_l) / \sum_{j=1}^{k} \alpha_j \prod_{x_t \in X_i} p(x_t|\theta_j), \tag{4}$$

we have the discrete Shannon entropy of these posterior probabilities for X_i

$$E(X_i) = -\sum_{l=1}^{k} P(l|X_i) \ln P(l|X_i), \tag{5}$$

which is globally minimized at $P(l_0|X_i) = 1, P(l|X_i) = 0 (l \neq l_0)$, that is, X_i is totally classified into the l_0-th Gaussian.

We now consider the average entropy over the sample set S:

$$E(\Theta) = \frac{1}{M}\sum_{i=1}^{M} E(X_i) = -\frac{1}{M}\sum_{i=1}^{M}\sum_{l=1}^{k} P(l|X_i) \ln P(l|X_i), \tag{6}$$

and use it to regularize the log-likelihood function by

$$H(\Theta) = L(\Theta) + \gamma E(\Theta), \tag{7}$$

where $\gamma > 0$ is the regularization factor. That is, $E(\Theta)$ is a regularization term to reduce the model complexity such that the Gaussian mixture can be made as simple as possible by minimizing $H(\Theta)$.

In order to solve the minimization problem of $H(\Theta)$ without constraint conditions, we utilize the following substitution: $\alpha_l = exp(\beta_l) / \sum_{j=1}^{k} exp(\beta_j)$, where $-\infty < \beta_l < +\infty$. Using the general methods for matrix derivatives, we are led to the following series of equations:

$$U(l|X_i) = P(l|X_i)(1 + \gamma \sum_{j=1}^{k} (\delta_{jl} - P(j|X_i)) \ln(\alpha_j \prod_{x_t \in X_i} p(x_t|\theta_j))), \tag{8}$$

$$\frac{\partial H(\Theta)}{\partial \beta_l} = \frac{-1}{M}\sum_{i=1}^{M}\sum_{j=1}^{k} U(j|X_i)(\delta_{jl} - \alpha_l) = 0, \tag{9}$$

$$\frac{\partial H(\Theta)}{\partial m_l} = \frac{-1}{M}\sum_{i=1}^{M} U(l|X_i) \sum_{x_t \in X_i} \Sigma_l^{-1}(x_t - m_l) = 0, \tag{10}$$

$$\frac{\partial H(\Theta)}{\partial \Sigma_l} = \frac{-1}{2M}\sum_{i=1}^{M} U(l|X_i) \sum_{x_t \in X_i} \Sigma_l^{-1}[(x_t - m_l)(x_t - m_l)^T - \Sigma_l]\Sigma_l^{-1} = 0, \tag{11}$$

where δ_{jl} is the Kronecker function. Then, the solution of those equations can be given explicitly as follows:

$$\hat{\alpha}_l = \frac{1}{\sum_{i=1}^{M}\sum_{j=1}^{k} U(j|X_i)} \sum_{i=1}^{M} U(l|X_i), \tag{12}$$

$$\hat{m}_l = \frac{1}{\sum_{i=1}^{M} U(l|X_i)|X_i|} \sum_{i=1}^{M} U(l|X_i) \sum_{x_t \in X_i} x_t, \tag{13}$$

$$\hat{\Sigma}_l = \frac{1}{\sum_{i=1}^{M} U(l|X_i)|X_i|} \sum_{i=1}^{M} U(l|X_i) \sum_{x_t \in X_i} (x_t - m_l)(x_t - m_l)^T, \tag{14}$$

where $|X_i|$ denotes the number of samples in X_i.

These explicit expressions give us an iterative algorithm for minimum $H(\Theta)$: for each iteration, we first update P and U according to (4) and (8), respectively, and then update Θ with newly estimated U according to (12)–(14). Hence, this iterative algorithm seems very similar to the EM algorithm on Gaussian mixture. Once the algorithm has converged to a reasonable solution Θ^*, all the samples can then be divided into k clusters(or classes) by

$$C[l] = \{x_t : x_t \in X_i,\ P(l|X_i) = \max_{j=1,...,k} P(j|X_i),\ i = 1, ..., M\}. \tag{15}$$

Due to the regularization mechanism introduced in the iteration process, some clusters may be forced to have no samples and then the desired k^*, that is, the number of true Gaussians, can be selected automatically.

Though we originally introduce entropy regularization into the maximum likelihood estimation (implemented by EM algorithm) for automatic model selection on the Gaussian mixture, it can also be observed that the minimization of the ERL function $H(\Theta)$ is robust with respect to initialization and the drawbacks of EM algorithm may be avoided. That is, when local minima of the negative likelihood $L(\Theta)$ arise during minimizing the ERL function, the average entropy $E(\Theta)$ may still keep large and we can then go across these local minima by minimum $H(\Theta)$. For example, the EM algorithm may not escape one type of local minima when two (or more) components in the Gaussian mixture have similar parameters and then share the same data. However, the ERL learning can promote the competition among these components by minimum $E(\Theta_k)$ as shown in [7], and then only one of them will "win" and the other will be discarded. That is, all of the data are continuously classified into some components in the Gaussian mixture during the ERL learning process, which can cause other components to have few data.

3 Experimental Results

We further apply the constrained ERL learning algorithm to the categorization of a ground truth image database used in [11] to give the summary of it for further browsing, and also make a comparison with the constrained EM algorithm [3] and the unconstrained ERL learning algorithm [9]. One issue of image categorization is the unknown number of natural categories in the database. To test the presented algorithm on this issue, we select three classes ($k^* = 3$) from the image database with each class having 48 images. Some samples of the image database are given in Fig. 1, and each row shows the samples from a different class. In the following experiments, some random pairs of data samples are selected to provide pairwise constraints.

Fig. 1. The image database (each row shows the samples from a different class)

The image features we used in the experiments are the Gabor textures, the Hough histogram (i.e., the shape feature), and a classical color histogram obtained in HSV (hue, saturation, value) color space. Note that the use of Gabor wavelet features for texture analysis been shown to be able to provide the best pattern retrieval accuracy in [12]. The dimension of the joint feature vector is originally above 500 and then reduced by about fifty times using linear principal component analysis.

The ERL learning (constrained or unconstrained) is always implemented with $k \geq k^*$ and $\gamma \in [0.2, 0.5]$, while the other parameters are initialized randomly within certain intervals. In the following experiments, we always set k a relatively larger value (e.g., $k = 6$), and select γ in the empirical range which is obtained by lots of experiment trials. Moreover, the ERL learning is stopped when $|H(\hat{\Theta}) - H(\Theta)| < 10^{-6}$. The constrained EM algorithm has the same initialization except that we must set $k = k^*$. During the iterative ERL learning process, all of the samples are continuously classified into some clusters, which can cause other clusters to have few samples. Hence, the the mixing proportions of some clusters may be reduced to a small value(i.e., below 0.001) after certain iterations, and then these clusters are discarded.

For a statistical observation, we repeat the constrained ERL learning algorithm 20 times always with $k = 6$ and $\gamma = 0.3$. When we randomly select 28 pairwise constraints, the results of one trial by the constrained ERL learning algorithm are given by Fig. 2. We can find that the constrained ERL learning algorithm can detect the three classes in the image database correctly. Actually, it can be observed that the three classes can almost always be detected successfully by the constrained ERL learning algorithm during the 20 trials with different numbers of pairwise constraints.

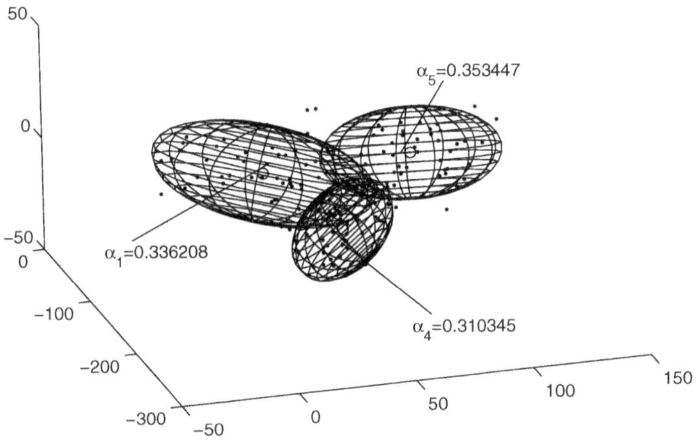

Fig. 2. The experimental results of automatic detection on the number of image classes by the constrained ERL learning algorithm (only in the first 3-dimensional view)

To make a comparison, we further present the average classification accuracies by the three learning algorithms in Fig. 3 with the number of pairwise constraints gradually increased. Note that the classification accuracy is taken into account to evaluate the clustering result, since we exactly know which cluster each sample should be divided into in the data set. That is, the class labels of all the samples are available when the clustering task is considered as classification according to (15). We can find that both the constrained ERL learning algorithm and constrained EM algorithm can benefit from the pairwise constraints. However, the ERL learning algorithm (constrained or unconstrained) performs much better since we can go across these local minima during image database categorization and the three classes can also be detected correctly.

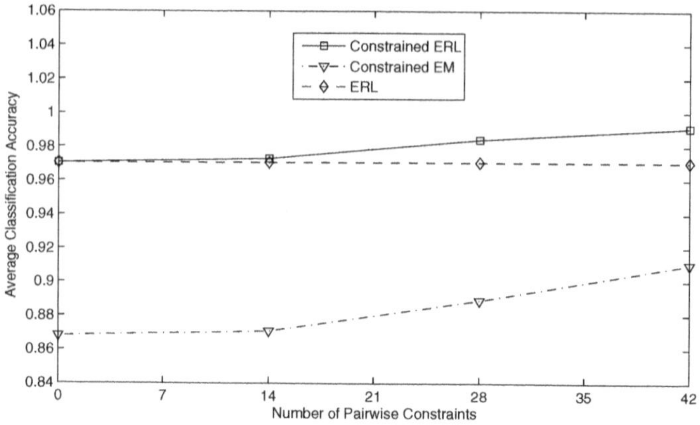

Fig. 3. The average classification accuracies by the three learning algorithms on the image database with the number of pairwise constraints gradually increased

4 Conclusions

We have proposed a semi-supervised learning algorithm for unsupervised image categorization based on Gaussian mixture model through incorporating some semantic information (i.e., pairwise constraints) into the ERL learning, which can automatically detect the number of image categories in the database. The experiments further show that this constrained ERL learning algorithm can lead to some promising results when applied to image categorization.

References

1. Render, R.A., Walker, H.F.: Mixture Densities, Maximum Likelihood and the EM Algorithm. SIAM Review **26** (2) (1984) 195–239
2. Liu, Y., Zhang, D., Lu, G., Ma, W.Y.: A Survey of Content-Based Image Rretrieval with High-Level Semantics. Pattern Recognition **40** (1) (2007) 262–282
3. Shental, N., Bar-Hillel, A., Hertz, T., Weinshall, D.: Computing Gaussian Mixture Models with EM Using Equivalence Constraints. Advances in Neural Information Processing Systems **16** (2004)
4. Wagstaff, K., Cardie, C., Rogers, S., Schroedl, S.: Constrained K-means Clustering with Background Knowledge. In: Proceedings of the 18th International Conference on Machine Learning (2001) 577–584
5. Grira, N., Crucianu, M., Boujemaa, N.: Semi-supervised Fuzzy Clustering with Pairwise-Constrained Competitive Agglomeration. In: Proceedings of the IEEE International Conference on Fuzzy Systems (2005) 867–872
6. Dattatreya, G.R.: Gaussian Mixture Parameter Estimation with Known Means and Unknown Class-Dependent Variances. Pattern Recognition **35**(7) (2002) 1611–1616
7. Lu, Z.: Entropy Regularized Likelihood Learning on Gaussian Mixture: Two Gradient Implementations for Automatic Model Selection. Neural Processing Letters **25**(1) (2007) 17–30

8. Lu, Z., Ma, J.: A Gradient Entropy Regularized Likelihood Learning Algorithm on Gaussian Mixture with Automatic Model Selection. Lecture Notes in Computer Science **3971** (2006) 464–469
9. Lu, Z.: An Iterative Algorithm for Entropy Regularized Likelihood Learning on Gaussian Mixture with Automatic Model Selection. Neurocomputing **69**(13-15) (2006) 1674–1677
10. Lu, Z.: Unsupervised Image Segmentation Using an Iterative Entropy Regularized Likelihood Learning Algorithm. Lecture Notes in Computer Science **3972** (2006) 492–497
11. Li, Y., Shapiro, L.G., Bilmes, J.A.: A Generative/Discriminative Learning Algorithm for Image Classification. In: Proceedings of the Tenth IEEE International Conference on Computer Vision **2** (2005) 1605–1612
12. Manjunath, B.S., Ma, W.Y.: Texture Features for Browsing and Retrieval of Image Data. IEEE Transactions on Pattern Analysis and Machine Intelligence **18**(8) (1996) 837–842

Mistaken Driven and Unconditional Learning of NTC

Taeho Jo[1] and Malrey Lee[2,*]

[1] Advanced Graduate Education Center of Jeonbuk for Electronics and Information Technology-BK21
[2] The Research Center of Industrial Technology, School of Electronics & Information Engineering, ChonBuk National University, 664-14, 1Ga, DeokJin-Dong, JeonJu, ChonBuk, 561-756, South Korea
Fax: 82-63-270-2394
tjo018@naver.com1, mrlee@chonbuk.ac.kr

Abstract. This paper attempts to evaluate machine learning based approaches to text categorization including NTC without decomposing it into binary classification problems, and presents another learning scheme of NTC. In previous research on text categorization, state of the art approaches have been evaluated in text categorization, decomposing it into binary classification problems. With such decomposition, it becomes complicated and expensive to implement text categorization systems, using machine learning algorithms. Another learning scheme of NTC mentioned in this paper is unconditional learning where weights of words stored in its learning layer are updated whenever each training example is presented, while its previous learning scheme is mistake driven learning, where weights of words are updated only when a training example is misclassified. This research will find advantages and disadvantages of both learning schemes by comparing them with each other

1 Introduction

Text categorization refers to the process of assigning one or some of predefined categories to unseen documents. We can consider two environments where machine learning algorithms are applied to text categorization as text classifiers. In one environment, text categorization is decomposed into binary classification problems as many as predefined categories. In the environment, a text classifier is allocated to each category, and it answers, 'yes' or 'no', to whether an unseen document belongs to its corresponding category, or not. In the other environment, machine learning algorithms are applied directly to text categorization, without decomposing it into binary classification tasks. Only single text classifier is given and it generates one of predefined categories as its answer. If there is sufficiently robust applicable directly to text categorization itself, an implementation of text categorization systems becomes very simple.

In classification and nonlinear regression, neural networks have been applied successfully [1]. Among neural network models, back propagation is used most popularly

* Corresponding author.

D. Liu et al. (Eds.): ISNN 2007, Part I, LNCS 4491, pp. 1201–1210, 2007.

as a supervised learning algorithm. Therefore, in 1995, Wiener initially applied back propagation to text categorization and compared it with KNN (K Nearest Neighbor) on the standard test bed, Reuter 21578 [9]. In 1996, Larkey and Croft applied a simple neural network, Perceptron, to text categorization, with its combination with other statistical linear models [5]. In 2000, Wermter applied a recurrent network including context layer to document routing which is a real task of text categorization [8]. In 2002, Ruiz and Srinivasan proposed a hierarchical organization of back propagations to hierarchical text categorization and compared it with a flat organization of them [6].

Neural networks involved in the above research use numerical vectors as their input vectors and weights vectors. Therefore, it is required to represent documents into numerical vectors, in order to apply the neural networks to text categorization. Such representation of documents leads to two main problems: huge dimensionality and sparse distribution. Although Joachims solved one of two problems by proposing SVM (Support Vector Machine) as an approach to text categorization in 1998 [4], it is applicable only to binary classification problems and not tolerant to sparse distribution. Although state of the art feature selection methods were already proposed as solution to the problem, huge dimensionality [7], they have limits in reducing dimensions and are not solutions to the problem, sparse distribution.

In order to address the two problems completely, it was proposed that documents should be represented into string vectors, instead of numerical vectors [2]. A string vector refers to a finite ordered set of terms which are given mono-gram, bi-gram, or n-grams. A concept of string vectors was introduced, when Jo proposed NTC for text categorization in 2000 [2]. In 2004, Jo validated the performance of NTC by comparing it with main traditional machine learning approaches, such as NB (Naïve Bayes), KNN, and back propagation on the test bed, Reuter 21578 [3]. However, he evaluated these approaches to text categorization by decomposing it into binary classification problems, like other literatures, [10] and [7].

This research provides two points concerning the application of NTC to text categorization. The first point is an alternative learning scheme of NTC to previous one described in [2] and [3]. In the previous learning scheme of NTC, weights of words are updated only when a training example is misclassified, and Perceptron adopts this learning scheme. In the proposed learning scheme, weights of words are updated whenever each training example is presented, and back propagation adopts it. This research will find merits and demerits of the two learning schemes by comparing them with each other on two different test beds.

The second point is the evaluation of machine learning based approaches to text categorization including NTC without decomposing it into binary classification problems. Since SVM is applicable only to binary classification problems, it will be excluded in the evaluation. Note that the cost for decomposing text categorization into binary classification problems is not ignorable. Text classifiers must be set up as many as predefined categories and training examples consisting of positive examples and negative examples should be built and allocated to each text classifier, category by category. Furthermore, training examples are learned and unseen documents are classified sequentially along a series of text classifiers, unless they are implemented in parallel or distributed versions. Without the decomposition of text categorization, only single text classifier which provides one of predefined categories is needed.

This paper consists of five sections. Section 2 describes the process of encoding documents into numerical vectors and string vectors. Section 3 describes the architecture and the learning algorithm of NTC. Section 4 presents results of evaluating NB, KNN, BP, and two learning schemes of NTC, on two different test beds. Section 5 mentions the significance of this research and further research for improving current one as the conclusion. The preparation of manuscripts which are to be reproduced by photo-offset requires special care. Papers submitted in a technically unsuitable form will be returned for retyping, or canceled if the volume cannot otherwise be finished on time.

2 Document Encodings

This section describes string vectors as representations of documents for applying the proposed version of SVM to text categorization. A string vector refers to a finite ordered set of words. Each of such words may be mono-gram, bi-gram, or n-gram. A string vector is defined as an ordered set of words with the fixed size, independent of the length of the given document. The string vector representing the document, d_i is denoted by $d_i^s = [w_{i1} \ \ w_{i2} \ \ \dots \ \ w_{in}]$, where n is the dimension of the string vector, d_i^s. From the given document, a bag of words is generated by indexing the document, as an intermediate representation for a string vector. Figure 1 illustrates the process of mapping a bag of words into a string vector. The dimension of a string vector is determined and the properties of words, such as the word with its highest frequency in the document, a random word in the first sentence, the word with its highest weight, or the word with its highest frequency in the first paragraph, are defined with the features of that vector. For simplicity and convenience of implementing the automatic process of encoding documents into string vectors, we defined the properties of string vectors as the highest frequent word, the second highest frequent word, the third highest frequent word, the fourth highest frequency word, and so on, in the document. In general, the dimension of string vectors is smaller than the size of bags of words. To each property given as a feature, its corresponding word is arranged so as to build a string vector.

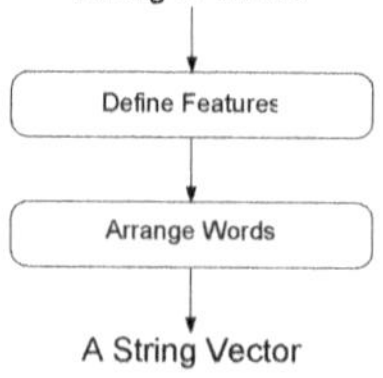

Fig. 1. The process of mapping a bag of words into a string vector

3 NTC (Neural Text Categorizer)

This section describes the architecture and the learning algorithm of NTC. Jo proposed this approach initially in 2000, as a new approach to text categorization, in order to solve the two main problems induced by representing documents into numerical vectors [2]. The approach was validated by comparing it with three main traditional approaches to text categorization, NB, SVM, and BP on the standard test bed, Reuter21578, in 2004 [3]. Its validation will be presented in the next subsection as one of the preliminary experiments of this dissertation proposal.

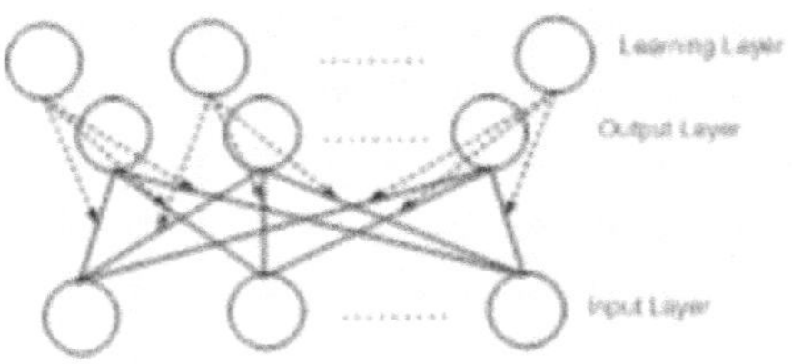

Fig. 2. The Architecture of NTC

Figure 2 illustrates the architecture of NTC consisting of three layers: the input layer, the learning layer, and the output layer. The role of the input layer is to receive the input vector given as a string vector; each node corresponds to each of its elements. The learning layer determines the weights between the input layer and the output layer, whenever an input vector is given. The output layer generates the degree of membership of the input vector encoding a document to each category. The conditions for designing the architecture of NTC are given as follows.

- The number of input nodes should be identical to the dimension of the string vectors encoding documents
- The number of learning nodes should be identical to the number of given categories.
- The number of output nodes should be identical to the number of given categories.

Table 1 defines the nodes in these three layers involved in the architecture of NTC. Each input node corresponds to an element of the input vector given as a string vector denoted by $d_k^s = [w_{k1}, w_{k2}, ..., w_{kn}]$, encoding a document, d_i . Each learning node corresponds to a category and is defined as an unordered set of words and their weights. Given the string vector, d_k^s, the weight between the input node, i_j and the output node, o_r, $weight_r(w_{kj})$, is computed by looking it up in the unordered set expressing the learning node, l_r, as expressed in equation (4). This weight indicates the degree of the membership of the word, w_{kj} in the string vector, d_k^s, in the category, c_r .

$$weight_r(w_{kj}) = \begin{cases} weight(w_{rm}), \text{ if } ((w_{rm}, weight(w_{rm})) \in \mathbf{l}_r) \\ \wedge (w_{kj} = w_{rm}) \\ \qquad 0, \text{ otherwise} \end{cases} \tag{1}$$

The value of each output node, o_r is computed using the equation presented in table 1. Therefore, the classified category of the input vector, $\hat{c}_k$ corresponds to the output node with its highest value. Therefore, the process of computing the values of the given output nodes refers to the classification of string vectors encoding documents with the current weights.

Table 1. The Definition of the Input Nodes and the Output Nodes in NTC

	The Notation of Nodes	The Value of Nodes
Input layer	$\mathbf{i} = \{i_1, i_2, ..., i_{\lvert\mathbf{i}\rvert}\}$	$i_j = w_{kj}$
Learning layer	$\mathbf{l} = \{\mathbf{l}_1, \mathbf{l}_2, ..., \mathbf{l}_{\lvert\mathbf{o}\rvert}\}$	$\mathbf{l}_r = \{(w_{r1}, weight(w_{r1})), ..., (w_{r\lvert\mathbf{l}_r\rvert}, weight(w_{r\lvert\mathbf{l}_r\rvert}))\}$
Output layer	$\mathbf{o} = \{o_i, o_2, ..., o_{\lvert\mathbf{o}\rvert}\}$	$o_r = \sum_{j=1}^{\lvert\mathbf{i}\rvert} weight_r(w_{kj})$

In this approach, learning is the process of updating weights connecting the input layer with the output layer to each word, in order to minimize the number of misclassifications. At first, the weights are initialized with the number of the documents including the word within the category. The learning rate, denoted by η, is given as a parameter of this model, like in BP. The value of each output node is computed using the equation of table 9 with the current weights. The target category of the input vector, d_k^s, is denoted by t_k. In mistaken driven learning, The weights are updated when the target category, t_k is not consistent with the classified category, $\hat{c}_k$. The output node corresponding to the target category is denoted by o_t, and the output node corresponding to the classified category is denoted by $o_{\hat{c}}$. The rule for updating weights is expressed by equation (5),

$$\text{if } o_t \neq o_{\hat{c}} \begin{cases} weight_t(w_{kj}) \leftarrow weight_t(w_{kj}) + \eta weight_t(w_{kj}) \\ weight_{\hat{c}}(w_{kj}) \leftarrow weight_{\hat{c}}(w_{kj}) - \eta weight_{\hat{c}}(w_{kj}) \end{cases} \tag{2}$$

In unconditional learning, weights of words are updated by equation (6).

$$\begin{cases} weight_t(w_{kj}) \leftarrow weight_t(w_{kj}) + \eta weight_t(w_{kj}) \\ weight_{\bar{t}}(w_{kj}) \leftarrow weight_{\bar{t}}(w_{kj}) - \eta weight_{\bar{t}}(w_{kj}) \end{cases} \tag{3}$$

where $\bar{t}$ is one of other categories than the target category. This learning scheme reinforces the sum of weights of words corresponding to target category, and inhibits sum of weights corresponding to non-target categories.

4 Experimental Results

This section presents results of comparing the two learning schemes of NTC with main traditional approaches, NB, KNN, and back propagation and comparing the two learning schemes with each other by varying dimensions of string vectors. Two test beds, NesPage.com and 20NewsGroup, will be used for these evaluations. Since each text classifier provides one of predefined categories as its answer, instead of 'yes' or 'no' to its corresponding category, accuracy is used as evaluation measure, instead of micro-averaged or macro-averaged F1.

In these experiments, documents are represented into string vectors for using two version of NTC or numerical vectors for using the others. Dimensions of numerical vectors and string vectors representing documents are set as 500 and 50 or 10, respectively. For encoding documents into numerical vectors, most frequent 500 words from a given training set for each problem are selected as their features. The values of the features of numerical vectors are binary ones indicating absence or presence of words in a given document. For encoding documents into string vectors, most frequent 50 or 10 words are selected from a given document as values of its corresponding string vector. Here, features of string vectors are the most frequent word, the second most frequent word, the third most frequent word, and so on.

The parameters of the involved approaches involved in these experiments are set by tuning them with a validation set, which is built separately by selecting some documents randomly from training documents. Table 2 shows the definition of the parameters which is obtained through this tuning. With the parameters defined in table 2, the involved approaches to text categorization will be applied to all of two test beds.

Table 2. Parameters of the Involved Approaches

Approaches to Text Categorization	Definition of Parameters
KNN	#nearest number = 3
Back Propagation	Hidden Layer: 10 hidden nodes Learning rate: 0.3 #Iteration of Training: 1000
NTC	Learning rate: 0.3 #Iteration of Training: 10

The first set of this experiment pursues the evaluation of the approaches on the test bed, Newspage.com. This test bed consists of 1,200 news articles in the format of plain texts built by copying and pasting news articles in the web site, www.newspage.com. Table 3 shows the predefined categories, the number of documents of each category, and the partition of the test bed into training set and test set. As shown in table 3, the ratio of training set to test set is set as 7:3. Here, this test bed is called Newspage.com, based on the web site, given as its source.

Table 3. Training Set and Test Set of Newspage.com

Category Name	Training Set	Test Set	#Document
Business	280	120	400
Health	140	60	200
Law	70	30	100
Internet	210	90	300
Sports	140	60	200
Total	840	360	1200

Figure 3 presents the results of evaluating NB, KNN, BP (Back Propagation), and two versions of NTC on the test bed, NewsPage.com. The y-axis in figure 3 indicates accuracy which is the rate of correctly classified test documents to all of test ones. In figure 3, 'NTC-mis' means NTC in conjunction with mistake driven learning and 'NTC-abs' means NTC in conjunction with unconditional learning. As illustrated in figure 3, both versions of NTC have better performance than NB, KNN, and BP in the text categorization on this test bed. The results show that NTC in conjunction with mistake driven learning is better than that in conjunction with unconditional learning.

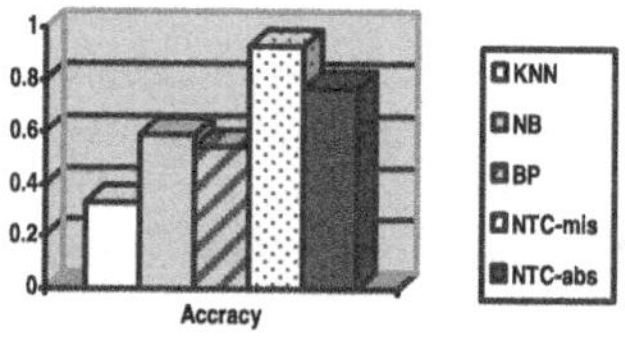

Fig. 3. The Results of comparing the involved Approaches on NewsPage.com

The second set of this preliminary experiment is for the evaluation of the five approaches on the test bed, called '20NewsGroups'. This test bed is obtained by downloading it from the web site, http://kdd.ics.uci.edu/databases/20newsgroups/20newsgroups.html. This test bed consists of 20 categories and 20,000 documents; each category contains 1,000 documents, identically. This test bed is partitioned into training set and test set with the ratio, 7:3; there are 700 training documents and 300 test documents within each category. Hence, 20,000 documents are partitioned into 14,000 training documents and 6000 test documents.

Figure 4 presents the results of evaluating the approaches to text categorization in the second test bed, 20NewsGroups. The results of this experiment set are different from those of the previous one in two points. The first point is that both versions of NTC are better than KNN and NB, and comparable to BP. The second point is that the unconditional learning of NTC is slightly better than mistake driven learning.

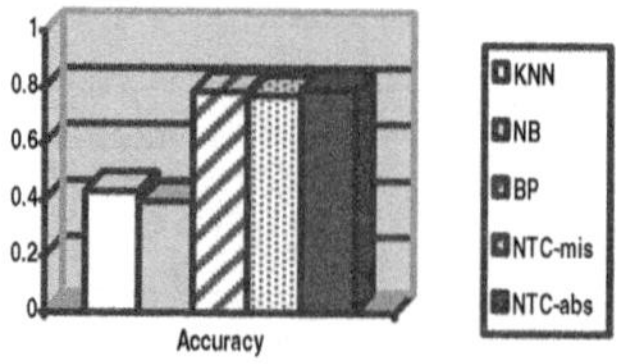

Fig. 4. The Results of comparing the involved Approaches on NewsPage.com

Figure 5 presents the performance of the two learning schemes of NTC depending on dimensions of string vectors on the first test bed, NewsPage.com. In the line graph illustrated in figure 5, the x-axis indicates the dimension of string vectors and the y-axis indicates accuracy of the two learning schemes. In figure 5, the solid line shows the trend of mistake driven learning and the dashed line shows that of unconditional learning. As illustrated in figure 5, mistake driven learning is far better on the first test bed than unconditional learning. As the dimensions of string vectors increases, mistake driven learning gets better, while unconditional learning gets worse. Therefore, in the experiment set corresponding to figure 3, NTC with mistaken driven learning uses 50 dimensional string vectors, while that with unconditional learning uses 10 dimensional string vectors.

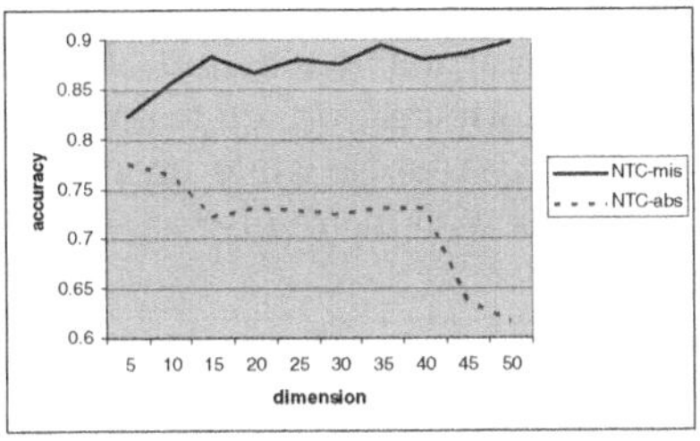

Fig. 5. Trends of two learning schemes of NTC on NewsPage.com

Figure 5 shows an advantage of mistake driven learning and a disadvantage of unconditional learning. When tables of weights and words are built in the learning layer, some words included in these tables may be spanning over categories. In this test bed, there are many overlapping words. The advantage of mistake driven learning is that it optimizes these weights of common words spanning over more than two categories stably. Because of to its ability, mistake driven learning scheme works better than unconditional learning within NTC on the first test bed. However, unconditional learning makes weights of such overlapped words very unstable. As the dimension of string vectors increases, weights of more words are unstable. So unconditional learning is worse as the dimension of string vectors increases.

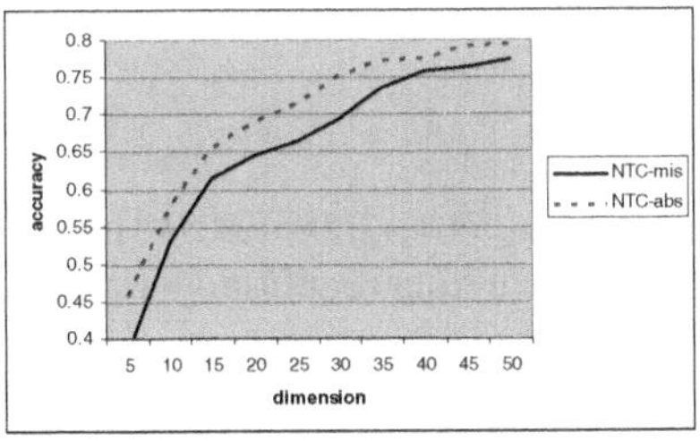

Fig. 6. Trends of two learning schemes of NTC on 20NewsGroups

Figure 6 presents the performance of the two learning schemes of NTC depending on dimensions of string vectors on the second test bed, 20NewsGroups. The comparison of the two learning schemes of NTC depending on input dimensions show the opposite results to those on the previous test bed; unconditional learning is better than mistake driven learning, spanning over all of dimensions of string vectors, as illustrated in figure 6. As the dimension of string vectors increases, accuracy of both learning schemes of NTC increases monotonically. Therefore, both version of NTC use 50 dimensional string vectors in the experiment set corresponding to figure 4.

Figure 6 shows an advantage of unconditional learning and a disadvantage of mistake driven learning. In this test bed, there are fewer overlapping words compared with the first test bed. In this situation, the advantage of unconditional learning is that it reinforces discriminations of words in categories, since it updates weights of words, whenever each training example is presented. If weights of words are updated only when a training example is misclassified, there are only slight difference between its target category and the others with respect to sum of weights; incorrect categories may have more chance to have their maximum sum of weights, when an unseen example with its noise is classified. Mistake driven learning have better ability to optimize weights of overlapping words stably, while unconditional learning have better ability to classify unseen examples with their noise.

5 Conclusion

This research contributes to apply NTC to text categorization in two points. The first point is that it applies another learning scheme to NTC and compares the two learning schemes with each other by varying its input dimension. This comparison of two learning schemes provides a merit and demerit of the two learning schemes. The second point is the evaluation of machine learning based approaches including NTC in text categorization without decomposing it into binary classification problems. By finding an approach feasible to text categorization in the environment, this research provides a way to implement simple and real time text categorization systems.

Although the experiments of this paper validated the performance of NTC, we need to consider its demerit. NTC learns training examples and classifies unseen examples by matching elements of string vectors with words stored in its learning layer, not semantically but lexically. NTC treats two semantically similar words, 'car' and 'automobile', as completely different words. This treatment may lead to misclassification. We need to consider not only lexical matching but also semantic relation to assign weights to words as a further research.

Acknowledgements

This work was supported by grand R01-2006-000-10147-0 from the Basic Research program of the Korea Science of Engineering Foundation

References

1. Hayking, S.: Neural Networks: Comprehensive Foundation Macmillan College Publishing Company (1994)
2. Jo, T.: Neural Text Categorizer: A New Model of Neural Networks for Text Categorization. The Proceedings of the 7th International Conference on Neural Information Processing (2000) 280-285

3. Jo, T.: Machine Learning Based Approaches to Text Categorization with Resampling Methods. The Proceedings of the 8th World Multi-Conference on Systemics (2004) 93-98
4. Joachims, T.: A Statistical Learning Model of Text Classification for Support Vector Machines. The Proceedings of the 24th annual international ACM SIGIR (1998) 128-136
5. Larkey, L.S., Croft W.B.: Combining Classifiers in Text Categorization. The Proceedings of the 19th annual international ACM SIGIR (1996) 289-297
6. Ruiz, M.E., Srinivasan, P: Hierarchical Text Categorization Using Neural Networks. Information Retrieval **5** (2002) 89-118
7. Sebastiani, F.: Machine Learning in Automated Text Categorization. ACM Computing Survey **34** (2002) 1-47
8. Wermter, S.: Neural Network Agents for Learning Semantic Text Classification. Information Retrieval **3** (2000) 87-103
9. Wiener, E.D.: A Neural Network Approach to Topic Spotting in Text. The Thesis of Master of University of Colorado (1995)
10. Yang, Y.: An Evaluation Statistical Approaches to Text Categorization. Information Retrieval **1** (1999) 69-90

Investigation on Sparse Kernel Density Estimator Via Harmony Data Smoothing Learning

Xuelei Hu and Yingyu Yang

School of Computer Science and Technology,
Nanjing University of Science and Technology, Nanjing, China
{xlhu, yangjy}@mail.njust.edu.cn

Abstract. In this paper we apply harmony data smoothing learning on a weighted kernel density model to obtain a sparse density estimator. We empirically compare this method with the least squares cross-validation (LSCV) method for the classical kernel density estimator. The most remarkable result of our study is that the harmony data smoothing learning method outperforms LSCV method in most cases and the support vectors selected by harmony data smoothing learning method are located in the regions of local highest density of the sample.

1 Introduction

The kernel density estimator (KDE) [1-3], also called Parzen window density estimator [4], can be regarded as a most popular non-parametric density estimator. A key problem in kernel density estimate is how to choose the kernel parameters, also called bandwidth. A standard method of automatic bandwidth selection is the least squares cross-validation (LSCV) method [5, 6] which aims to minimizing the mean integrated square error. However, a disadvantage of the kernel density estimator is that it requires large amounts of computational time and space because all training sample points are retained.

Recently, a sparse density estimation method has been developed in [7, 8], named harmony data smoothing learning for density estimation. It employs the weighting coefficients to the kernel density estimator and applies the harmony data smoothing learning principle [9] to estimate the weighting coefficients together with other unknown parameters. An adaptive algorithm was given in [7] and we will present an algorithm for batch case in this paper. Some weighting coefficients will become zero after learning. Thus a sparse representation of the density estimate in terms of only a subset of training data, called support vectors (SVs), can be yield.

We further conduct a comparative simulation study of these two methods with Gaussian kernel. We consider various different populations and make the mean squared error and the mean absolute error between the estimate and the original density function on testing data as metrics. The most remarkable result of our study is that the harmony data smoothing learning methods outperforms

D. Liu et al. (Eds.): ISNN 2007, Part I, LNCS 4491, pp. 1211–1220, 2007.

the least squares cross-validation (LSCV) method in most cases. We also observe that the support vectors selected by harmony data smoothing learning method lie approximately on the local centers of data points.

The remainder of this paper is organized as follows. In Section 2, we introduce conventional KDE methods. and then we provide a sparse kernel density estimator by harmony data smoothing learning in Section 3. Comparative experiments of kernel density estimation are given in Section 4. Finally, a conclusion is made in Section 5.

2 Classical Kernel Density Estimator

The kernel density estimator [1, 4] has been the most popular non-parametric density estimator. The general formula for the kernel density estimator is

$$p(\mathbf{x}) = \frac{1}{n}\sum_{i=1}^{n} K_{\mathbf{\Sigma}}(\mathbf{x}, \mathbf{x}_i). \tag{1}$$

The fundamental problem in kernel density estimator is the choice of $\mathbf{\Sigma}$ which is called bandwidth matrix. Several methods for selecting bandwidth have been proposed, and some reviews and simulation studies also have been provided [2, 10, 11]. Among those methods, the least squares cross-validation (LSCV) method, proposed in [5, 6], is probably the most popular and best studied one. It seeks to estimate $\mathbf{\Sigma}$ by minimizing the function

$$LSCV(\mathbf{\Sigma}) = \int p(\mathbf{x})^2 dx - \frac{2}{n}\sum_{t=1}^{n} p_{-t}(\mathbf{x}_t), \tag{2}$$

where $p_{-t}(\mathbf{x})$ is the density estimate based on the sample with $\mathbf{x}_t$ deleted, often called the "leave-one-out" density estimator [2]. $LSCV(\Sigma)$ is an unbias estimate of $MISE\{\hat{p}(\mathbf{x})\} - \int p(\mathbf{x})^2 dx$ where MISE denotes the mean integrated square error. The asymptotic optimality of the bandwidth chosen by LSCV method with respect to the mean integrated square error has been established under fairly general conditions [12, 13]. Considering the asymptotic optimality and popularity properties of this method we use it to select the bandwidth in our experiments. In the case of gaussian kernel the objective function Eq. 2 can be expressed as follows:

$$LSCV(\mathbf{\Sigma}) = \frac{1}{(2\sqrt{\pi})^d n|\mathbf{\Sigma}|^{\frac{1}{2}}} + \frac{1}{(2\sqrt{\pi})^d n^2|\mathbf{\Sigma}|^{\frac{1}{2}}} \sum_{i=1}^{n}\sum_{j\neq i}(e^{-\frac{1}{4}\mathbf{\Delta}_{ij}} - 2\times 2^{\frac{d}{2}} e^{-\frac{1}{2}\mathbf{\Delta}_{ij}}), \tag{3}$$

where $\mathbf{\Delta}_{ij} = (\mathbf{x}_i - \mathbf{x}_j)^T\mathbf{\Sigma}^{-1}(\mathbf{x}_i - \mathbf{x}_j)$.

An advantage of the kernel density estimator is that it is a most flexible density estimator with only bandwidth matrix $\mathbf{\Sigma}$ unknown. However, since the kernel density estimator in terms of all the observations it is very computationally expensive and its structure is too complex. Moreover, it is not known how many data points and which data point should be collected in order to provide a better representation.

3 Sparse Kernel Density Estimator Via Harmony Data Smoothing Learning

We consider the following extension of the classical kernel density estimator

$$p(\mathbf{x}) = \sum_{i=1}^{n} \alpha_i K_{\mathbf{\Sigma}}(\mathbf{x}, \mathbf{x}_i), \tag{4}$$

where α_i satisfies

$$\sum_{i=1}^{n} \alpha_i = 1, \alpha_i \geq 0, i = 1, ..., n. \tag{5}$$

Unless otherwise stated the kernel $K_{\mathbf{\Sigma}}(\mathbf{x}, \mathbf{u})$ is the gaussian kernel

$$K_{\mathbf{\Sigma}}(\mathbf{x}, \mathbf{u}) = (2\pi)^{-\frac{d}{2}} |\mathbf{\Sigma}|^{-\frac{1}{2}} e^{-\frac{1}{2}(\mathbf{x}-\mathbf{u})^T \mathbf{\Sigma}^{-1}(\mathbf{x}-\mathbf{u})}. \tag{6}$$

In practice, considering the computational cost we usually use the spherical gaussian kernel with $\mathbf{\Sigma} = \sigma^2 \mathbf{I}_d$ for multivariate density estimation.

According to [7, 8], the estimator of the probability density function can be denoted by expression Eq. 4 which employs the concept of weighting comparing with the kernel density estimator. We seek to estimate the unknown parameters $\theta = \{\alpha_i, \mathbf{\Sigma}\}$ by using the harmony learning principle with data smoothing regularization technique [9]. The implementation of data smoothing learning for density estimation can be described as follows: maximize

$$H(\theta, h) = L(\theta) + 0.5h^2 \Pi(\theta) + 0.5d \ln(2\pi h^2) - \ln J(h), \tag{7}$$

where

$$L(\theta) = \frac{1}{n} \sum_{t=1}^{n} \sum_{i=1}^{n} P_t(i) \ln[\alpha_i K_{\mathbf{\Sigma}}(\mathbf{x}_t, \mathbf{x}_i)], \tag{8}$$

$$\Pi(\theta) = \frac{1}{n} \sum_{t=1}^{n} \sum_{i=1}^{n} P_t(i) tr[\frac{\partial^2 \ln(\alpha_i K_{\mathbf{\Sigma}}(\mathbf{x}, \mathbf{x}_i))}{\partial \mathbf{x} \partial \mathbf{x}^T} |_{\mathbf{x}=\mathbf{x}_t}], \tag{9}$$

$$J(h) = \sum_{t=1}^{n} \sum_{r=1}^{n} e^{-0.5 \frac{\|\mathbf{x}_t - \mathbf{x}_r\|^2}{h^2}}, \tag{10}$$

subject to

$$\sum_{i=1}^{n} \alpha_i = 1, \alpha_i \geq 0, i = 1, ..., n. \tag{11}$$

One choice of the term $P_t(i)$ is the Bayesian posteriori probability

$$P_t(i) = \frac{\alpha_i K_{\mathbf{\Sigma}}(\mathbf{x}_t, \mathbf{x}_i)}{\sum_{j=1}^{n} \alpha_j K_{\mathbf{\Sigma}}(\mathbf{x}_t, \mathbf{x}_j)}. \tag{12}$$

We note that the first term of the cost function $H(\theta, h)$ is equivalent to the likelihood and the other terms are regularization terms. This regularization technique avoid over-fitting via smoothing the likelihood in the near-neighbor of sample points. Unlike many other regularization techniques, no prior regularization parameters in this cost function. All the unknown parameters including the smooth parameter h can be estimated by maximizing the cost function. An adaptive algorithm was proposed [7] to solve this optimization problem.

Here we present an algorithm to solve this optimization problem for batch case. If $K(\mathbf{x}, \mathbf{x}_i)$ is a gaussian kernel then

$$\frac{\partial^2 \ln(\alpha_i K_{\mathbf{\Sigma}}(\mathbf{x}, \mathbf{x}_i))}{\partial \mathbf{x} \partial \mathbf{x}^T} = -\mathbf{\Sigma}^{-1}. \tag{13}$$

So we have $\Pi(\theta) = -\Sigma^{-1}$. Note that $P_t(i)$ is calculated before the step of updating parameters in the algorithm and remains constant respect to the parameters when updating parameters. First, we discuss the updating rules of coefficients α_i. Similar with traditional maximum likelihood learning based on gaussian mixture model, we obtain the solution of updating α_i as follows

$$\alpha_i = \frac{1}{n} \sum_{t=1}^{n} P_t(i), \tag{14}$$

where $P_t(i)$ is calculated by Eq. 12. Second, we discuss parameters Σ. The solution that follows from $\frac{\partial H(\theta,h)}{\partial \mathbf{\Sigma}} = 0$ is

$$\Sigma = \frac{1}{n} \sum_{t=1}^{n} \sum_{i=1}^{n} P_t(i)(\mathbf{x}_t - \mathbf{x}_i)(\mathbf{x}_t - \mathbf{x}_i)^T + h^2 \mathbf{I}_d. \tag{15}$$

Third, we discuss how to update smoothing parameters h. To ensure h be positive we introduce variable $\tilde{h}$ with $h = e^{\tilde{h}}$. We update h via updating $\tilde{h}$ as follows

$$h^{new} = e^{\tilde{h}^{new}}, \tilde{h}^{new} = \tilde{h}^{old} + \eta_0 \frac{\partial H(\theta, h)}{\partial \tilde{h}}, \tag{16}$$

where η_0 is a step length constant and

$$\frac{\partial H(\theta, h)}{\partial \tilde{h}} = d - h^{old2} \mathbf{\Sigma}^{-1} - \sum_{t=1}^{n} \sum_{r=1}^{n} \gamma_{t,r} \|\mathbf{x}_t - \mathbf{x}_r\|^2, \tag{17}$$

with

$$\gamma_{t,r} = \frac{e^{-0.5 \frac{\|\mathbf{x}_t - \mathbf{x}_r\|^2}{h^{old2}}}}{\sum_{i=1}^{n} \sum_{j=1}^{n} e^{-0.5 \frac{\|\mathbf{x}_i - \mathbf{x}_j\|^2}{h^{old2}}}}. \tag{18}$$

We describe this harmony data smoothing learning algorithm for density estimation in batch case as follows:

step 1 Initialize all the parameters.
step 2 Calculate $P_t(i)$ with $t = 1, ..., n$ and $i = 1, ..., n$ according to Eq. 12.

step 3 Update $\alpha_i, i = 1, ..., N$ and $\mathbf{\Sigma}$ according to the updating rules described above.
step 4 Update smoothing parameter h using the updating rules described above.
step 5 Go to **step 2** or stop if stop condition satisfied.

We usually initialize coefficients $\alpha_i = 1/n$. The initial value of $\mathbf{\Sigma}$ and h^2 should be larger than the smallest square distant between sample points. One choice of the stop condition is reaching a given iteration number. Another is the cost function $H(\theta, h)$ being stationary. There are also some other conditions can be made.

After learning, some α_i are automatically pushed to zero, and then the contribution of corresponding sample $\mathbf{x}_i$ is discarded. We can get a set $\mathcal{V} = \{i : \alpha_i \neq 0 \text{ or } \alpha_i > \varepsilon\}$ with ε being a pre-specified small positive number [7]. Consequently we obtain a sparse representation of the density estimate denoted by

$$\hat{p}(\mathbf{x}) = \sum_{i \in \mathcal{V}} \alpha_i K_{\mathbf{\Sigma}}(\mathbf{x}, \mathbf{x}_i), \tag{19}$$

which is in terms of only a subset of sample points $\{\mathbf{x}_i : i \in \mathcal{V}\}$. We also observe that α_i corresponding to those data points near the local centers take bigger value and others take smaller or even zero value.

A major advantage of this method is that it can automatically decide how many data points and which data point should be collected during parameter learning to provide a sparse estimator. A disadvantage is that this nonlinear optimization problem suffers some difficulties in practice due to the inherent initialization dependent variability of the solutions.

4 Experiments on Kernel Density Estimation

4.1 Description of the Study

We compare by simulation the two density estimation methods: harmony data smoothing learning for the weighted kernel density estimator denoted by HDS, and the kernel density estimator (KDE) with the bandwidth selected by LSCV method. We consider six different univariate populations and two different multivariate populations:

1. a standard norm $G(x|0,1)$ [10], where $G(\mathbf{x}|\mu, \mathbf{\Sigma})$ denotes a multivariate normal (Gaussian) with mean μ and covariance matrix $\mathbf{\Sigma}$ in this paper,
2. a beta distribution with parameters 2 and 2 $\beta(2,2)$,
3. a Student-t distribution with 5 degrees of freedom t_5 [10],
4. a skewed normal mixture $\frac{3}{4}G(x|0,1) + \frac{1}{4}G(x|1.5, 1/3)$ [14],
5. a normal mixture $\frac{1}{2}G(x|0,1) + \frac{1}{2}G(x|6,2)$ [10],
6. a symmetric normal mixture $\frac{9}{10}G(x|0,1) + \frac{1}{10}G(x|0,4)$ [10],
7. a 2-dimensional normal mixture
$\frac{1}{3}G(\mathbf{x}|(-1,-1), 0.1\mathbf{I}_2) + \frac{1}{3}G(\mathbf{x}|(0,0), 0.1\mathbf{I}_2) + \frac{1}{3}G(\mathbf{x}|(1,-1), 0.1\mathbf{I}_2)$,
8. a 5-dimensional normal mixture
$\frac{1}{2}G(\mathbf{x}|(0,0,0,0,0), 0.1\mathbf{I}_5) + \frac{1}{2}G(\mathbf{x}|(1,1,1,1,1), 0.05\mathbf{I}_5)$.

The first six univariate densities include symmetric and asymmetric, thin-tailed and heavy-tailed, unimodel and bimodel densities as shown in Fig. 1. For multivariate density estimation, we limit our study to the case of the spherical gaussian kernel for computational consideration. A training sample of 100 data points for first seven densities, and 200 data points for the last density are randomly generated from each density distribution. A test sample of 1000 data points for first seven densities and 2000 data points for the last density is then drawn from each distribution.

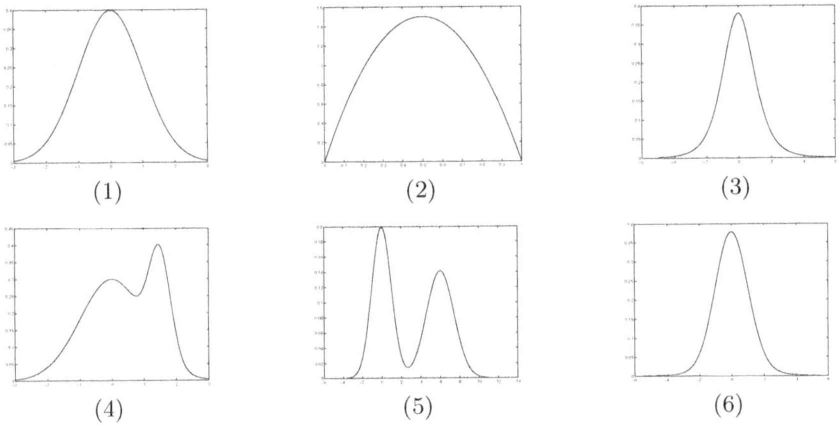

Fig. 1. The univariate densities in our collection

To compare the performance of the two density estimation methods, we calculate the mean squared error (MSE) and mean absolute error (MAE) between the estimate and the original density function on testing data as metrics. Tab. 1 illustrates the experimental results. Fig. 2 shows plots of two density estimates on 2-dimensional normal mixture data. Tab. 2 lists the number of support vectors selected by harmony data smoothing learning methods for each density.

4.2 Discussion and Comments

Next, we summarize the experimental results. First we consider the performance according to the metrics MSE and MAE (see Tab. 1). The harmony data smoothing density estimator shows a better behavior for the both metrics MSE and MAE expect for the data drawn from density 4 and density 8. For density 4, the harmony data smoothing learning has a slightly better performance for the metric MSE, while the kernel density estimator has a slightly better performance for the metric MAE. For density 8 (5-dimensional normal mixture), the performance of the kernel density estimator is better than the harmony data smoothing learning. We observe that among 200 training data points drawn from density 8

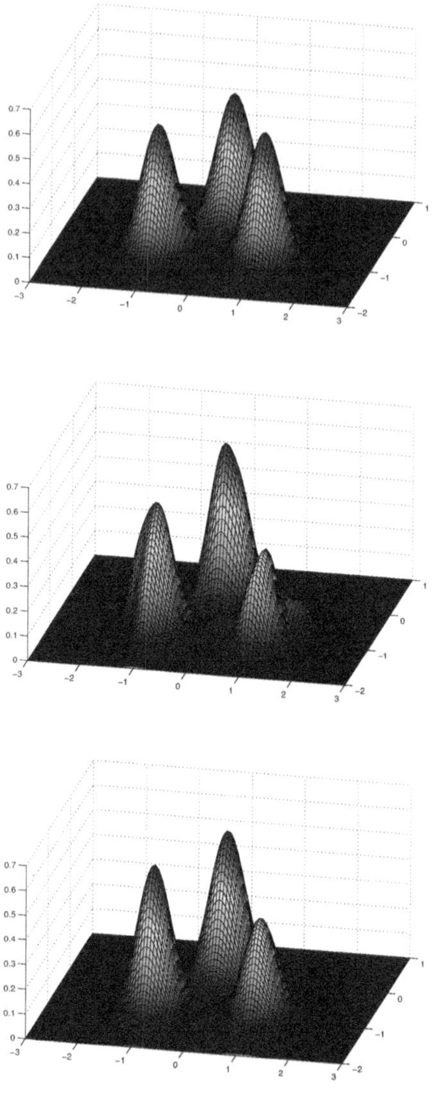

Fig. 2. Density estimation results on 2-dimensional normal mixture data. The true probability density function (top), estimate by the kernel density estimator (middle), estimate by the harmony data smoothing learning (bottom) are represented respectively.

only 13 data points are selected as support vectors. We could point out a possible reason that the number of support vectors selected by the harmony data smoothing learning method on high dimensional data is too small to accurately describe the high dimensional density function.

Table 1. The MSE and MAE of two methods

density index	MSE		MAE	
	KDE	HDS	KDE	HDS
1	0.001567	0.000695	0.034100	0.023166
2	0.022645	0.019877	0.137490	0.118577
3	0.001395	0.001264	0.031630	0.029738
4	0.002452	0.002340	0.034332	0.036834
5	0.000593	0.000391	0.020153	0.016340
6	0.003393	0.001416	0.051172	0.032707
7	0.007187	0.003481	0.062469	0.045381
8	0.050665	0.074793	0.136262	0.167303

Table 2. The number of support vectors selected by harmony data smoothing learning method

density index	1	2	3	4	5	6	7	8
number of SVs	42	28	39	41	42	35	12	13

Second, we consider the computational complexity. In the case of training, for kernel density estimator, most cross-validation bandwidth selector including LSCV method involve the evaluation of the estimator for m candidate bandwidths $\mathbf{\Sigma}_1, \mathbf{\Sigma}_2, ..., \mathbf{\Sigma}_m$. This leads to $O(mn^2)$ calculations of the kernel. In multivariate case, it is too computational expansive to estimate a full bandwidth matrix since the value of k will be very large. Thus usually we assume bandwidth matrix is spherical in practice. For harmony data smoothing algorithm, one iteration involves $O(n^2)$ calculations of kernel. Suppose after r iterations the algorithm converges to a local maximum, then it leads to $O(rn^2)$ calculations of kernel. In practice, it is hard to say which one has less training computational complexity because they depend on different cases. In the case of testing, for the kernel density estimator, the prediction on a new data point takes $O(n)$ calculations of the kernel. While for harmony data smoothing density estimator, the prediction only takes $O(k)$ calculations of the kernel where k is the number of support vectors. Since $k < n$ or even $k << n$, the prediction time of harmony data smoothing density estimator is shorter than the kernel density estimator.

Finally, we make a note of sparsity of harmony data smoothing learning for density estimation. From Tab. 2, the harmony data smoothing method shows a good sparse property with only a small subset of training data retained. We illustrate the weighting coefficients value of training data drawn from density 5 and the positions of support vectors on data drawn from density 7 (2-dimensional normal mixture) in Fig. 3. It is interesting to note that the selected support vectors (non-zero weighting) occur in the regions of local highest density of the sample, and lie approximately on the local centers of the training data. The corresponding weights to those data points near the local centers take bigger value and others take smaller or even zero value.

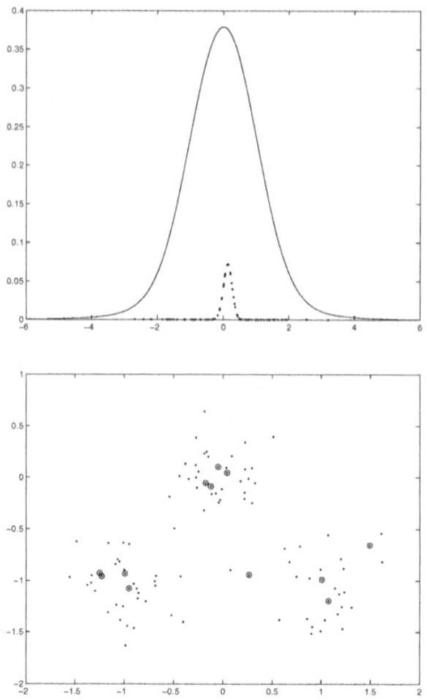

Fig. 3. Sparse properties of harmony data smoothing learning method on data drawn from density 6 (top) and density 7 (bottom). The top figure illustrates the value of weighting coefficients (dot) of the training sample drawn from density 6 and the original density function of density 6 (solid line). The bottom figure shows the training data points (dot) and the support vectors (circle) of density 7.

5 Conclusion

We investigate the harmony data smoothing learning on a weighted kernel density estimator which can automatically obtain a sparse kernel density estimator. We show a comparative simulation study of this method and the kernel density estimator with bandwidth selected by least squares cross-validation method. We could say that the harmony data smoothing learning method showed better performance in the one and two dimensional cases in our experiments. The harmony data smoothing learning method provides a principled way of choosing a subset of training data points to obtain a sparse estimator. Thus the computational cost on prediction is smaller than the kernel density estimator. It was shown in experiments that the selected support vectors (non-zero weighting) by harmony data smoothing learning method occur approximately in the regions of local highest density of the sample.

Acknowledgement

The work described in this paper was supported by grants from the National Natural Science Foundation of China (No. 60632050, No. 60472060, No. 60503026).

The first author would like to express thanks to Prof. Lei Xu at Department of Computer Science and Engineering, the Chinese Univ. of Hong Kong for insightful guidance and valuable feedback on the work of this paper.

References

1. Rosenblatt, M.: Remarks on Some Nonparametric Estimates of a Density Function. Annals Mathematical Statistics **27** (1956) 832-837
2. Scott, D.W.: Multivariate Density Estimation. Wiley, New York (1992)
3. Silverman, B.W.: Density Estimation for Statistics and Data Analysis. Chapman and Hall, London (1986)
4. Parzen, E.: On Estimation of a Probability Density Function and Mode. Annals Mathematical Statistics **33** (1962) 1065-1076
5. Rudemo, M.: Empirical Choice of Histograms and Kernal Density Estimates. Scandinavian Journal of Statistics **9** (1982) 65-78
6. Bowman, A.: An Aternative Method of Cross-validation for the Smooth18 ing of Density Estimates. Biometrika **71** (1984) 353-360
7. Xu, L.: BYY Harmony Learning, Structural RPCL, and Topological Self-organizing on Mixture Models. Neural Networks **15** (2002) 1125-1151
8. Xu, L.: Best Harmony, Unified RPCL and Automated Model Selection for Unsupervised and Supervised Learning on Gaussian Mixtures, Three-layer Nets and Me-rbf-svm Models. International Journal of Neural Systems **11** (2001) 43-69
9. Xu, L.: BYY Harmony Learning, Independent State space, and Generalized APT Financial Analyses. IEEE Tansactions on Neural Networks **12** (2001) 822-849
10. Cao, R., Cuevas, A., Gonzales-Manteiga, W.: A Comparative Study of Several Smoothing Methods in Density Estimation. Computational Statistics and Data Analysis **17** (1994) 153-176
11. B. Park, J.M.: Comparison of Datadriven Bandwidth Selectors. Journal of the American Statistical Association **85** (1990) 66-72
12. Hall, P.: Large-sample Optimality of Least-squares Cross-validation in Density Esti- mation. Annals of Statistics **11** (1983) 1156-1174
13. Stone, C.: An Asymptotically Optimal Window Selection Rule for Kernel Density Estimates. Annals of Statistics **12** (1984) 1285-1297
14. Marron, J., Wand, M.: Exact Mean Integrated Squared Error. Annals of Statistics **20** (1992) 712-736

Analogy-Based Learning How to Construct an Object Model

JeMin Bae

Department of Computer Education, Kwandong University
gemini@kd.ac.kr

Abstract. Code reuse in software reuse has several limitations such as difficulties of understanding and retrieval of the reuse code written by other developers. To overcome these problems, it should be possible to reuse the analysis/design information than source code itself. In this paper, I present analogical matching techniques for the reuse of object models and design patterns. We have suggested the design patterns as reusable components and the representation techniques to store them. The contents of the paper are as follows. 1) Analogical matching functions to retrieve analogous design patterns from reusable libraries. 2) The representation of reusable components to be stored in the library in order to support the analogical matching.

1 Introduction

The reuse of software can contribute to improve the productivity by reusing previous development experiences to the development processes of new software [1, 2, 9]. Up to this point, efforts for the reuse of software have been focused on reuse at the code level, being either reuse of the algorithm library or the reuse of the class library [4, 5, 6]. In order to overcome the limitations which comes with reuse of code [2, 3], continued effort is being exerted for the reuse of the data material created within the analysis/design phase previous to the coding phase.

Recently, attempts for software development with object-oriented methodologies applied have increased in accordance to activation of object-oriented methodologies [1, 9].

In order to support the reuse of object models within the requirement specification level, this paper proposes a reusable environment for object models through analogical matching by introducing the analogy technique[4, 7, 8] being researched in knowledge engineering. The analogy technique is different from the similarity measurement technique being used for the reuse of existing codes.

Although similarity refers to a characteristic or intrinsic commonness which exists between two objects, analogy refers to the consistent structure or confronting nature between two objects which exists from all point of views [9].

In order to support the reuse of object models, this paper proposes an environment which reuses object models and patterns based on analogies by introducing the analogical matching technique. Section 2 will describe the techniques necessary for reuse and section 3 will define the library structure and representation technique for

D. Liu et al. (Eds.): ISNN 2007, Part I, LNCS 4491, pp. 1221–1226, 2007.

model reuse and also define the textual matching technique for patterns as well as the semantic matching techniques based on analogical matching techniques. Section 4 will describe and evaluate the implementation of matching agents which searches for suitable models from the constructed reusable library in a given situation. Section 5 will provide a conclusion.

2 Library Structure

For reusing the object model, the library is essential. Therefore, we define the library structure for storing the object model and related information. In this paper, the library contains template and case for a specific problem. Template means abstract model for domain. Case means the practical solution of one application based on template.

2.1 Object Modeling as Reusable Parts

To be flexible, we didn't choose particular modeling methodologies for our object modeling. Every modeling methodology has little differences but all of them are based on object-oriented concepts. As modeling methodology proposed, we define object model can represent the interface and behavior. Interface means what object should do with what kinds of information. Interface contains a set of sort signatures and a set of function signatures.

Object = (ObjectID, ObjectName, set-of-Attribute, Set-of-Operation, ObjectType)
ObjectType = resource | process | history | logic | role | interaction
Attribute = (ObjectID, AttributeName, AttributeType, AttributeStructure)
AttributeType = identifier | descriptive | referential
AttributeStructure = simple | composite
Operation = (ObjectID, OperationName, ResultType, Set-of-Argument)
Argument = (OperationName, ArgumentName, ArgumentType,
ArgumentStructure)
ArgumentStructure = simple | composite | object

Behavior shows what states, events, and transitions are necessary for. Behavior consists of four components: pre-state, post-state, event, activities.

State = (ObjectID, set-of-StateDomain)
StateDomain = AbstactSpace | Predicate
Invariant = (ObjectID, AttributeName, Relation, StateDomain)
Behavior = (ObjectID, OperationName, Pre-State, Post-State,
set-of-ModifiesAttribute)
Pre-State = StateDomain
Post-State = StateDomain
ModifiesAttribute = Attribute

2.2 Domain Knowledge

Since analogy doesn't mean same appearance, syntactic differences can't mean mismatching. So, we need additional semantic information about general knowledge and application-specific knowledge. We manage two kinds semantic information: is-a lattice and part-of lattice which is corresponding to generalization/specialization and aggregation respectively.

1) Is-a lattice

Is-a lattice shows an abstraction level among the objects or attributes or functions. We can apply inheritance semantic to is-a lattice. It means that subtype and supertype is substitutable. With is-a lattice, we can extend and restrict our query. Extending with supertype results in broadening a matching space and restricting with subtype results in narrowing a matching. For analogical matching we define two kinds of is-a lattice. First is is-a lattice for category. if is-a(A, TYPE) can be defined, we can say it is possible to apply concept in TYPE to concept is A. Second one is for type like is-a(stack, linearList) which means something typed stack can be regarded as typed linearList or its subtype.

2) Part-of lattice

Part-of lattice present in which concepts are composite of concepts of other part among the attributes or objects. In a case of object, it is related with relationship. We can flatten the model with part-of lattice.

3 Analogical Learning of Object Modeling

Analogical learning is a significant part of human problem solving and learning methodologies.

3.1 Defining Analogical Matching

We can classify model match for the four cases where component are either object or pattern and are modeled by interface or behavior. Given a query and component model, Component C is ($C_{interface}$, $C_{Behavior}$, $C_{Framework}$) and Query Q($Q_{interface}$, $Q_{Behavior}$, $Q_{Framework}$), we define a generic match function AMatch.

Definition 1. Analogical Matching Function (AMatch)

$$AMatch(Q,C) = amatch(Q_{interface}, C_{interface}) \vee amatch(Q_{Behavior}, C_{Behavior}) \vee amatch(Q_{Framework}, C_{Framework})$$

Candidate components can be retrieved if the interfaces of components are analogous with that of query or the behavioral characteristics of components are analogous with those of query. Finally, Candidate patterns can be retrieved, if query were pattern and the features of it are analogous with those of patterns stored in library.

1) Object interface matching

Interfaces of two objects are analogous if they have same characteristics as well as they have same interface that exactly match. Same characteristic means when one component do something and the other component do similar things regardless of name. For example, query has resource-type attribute what is also attribute of component even its name and real data type was different such as book or tape. We can match component and query with the following matching functions.

Definition 2. Analogical interface matching by Type Equality $T1 \equiv_e T2$

1) T1 and T2 has same name and type

2) is-a(T_1,T_2) is exist or supertype T is defined and is-a(T_1 ,T) and is-a(T_2 ,T) are defined in design library

3) if both of T1(e_1,e_2,...en), T2(m_1,m_2,...,ms) are composite type, $\exists e_i$,

$\exists m_j . e_i \equiv e\ m_j \quad e_i \equiv_e m_j \quad 1 \leq i,\ j \leq n$ or s

4) if one of T1 and T2 is composite type, $\forall a \in$ part-of(T_1), $\exists a \in T_2$

Definition 3. Analogical interface matching by renaming Renaming $R(E \equiv_r E')$, iff $R(E) \equiv_r E'$

1) if we can define $R(E) \equiv_r E'$ or is-a (E, E') is defined already in the library

2) if E(e_1,e_2,...e_n), E'(m_1,m_2,...,m_s) are composite type, $\exists e_i$, $\exists m_j$. is-a(e_i, m_j) is defined in the library $1 \leq i,\ j \leq n$ or s

Definition 4. Analogical interface matching amatch($Q_{interface}$, $C_{interface}$)

1) ObjectKinds(Q) $\equiv_e$ ObjectKinds(C) $\vee$ ObjectKinds(Q) $\equiv_r$ ObjectKinds(C))

2) (($\forall$attribute$_q$ $\in$ attribute-Set(Q), $\exists$attributec $\in$ attribute-Set(C).
attribute$_q$ $\equiv_e$ attribute$_c$ $\vee$ attribute$_q$ $\equiv_r$ attribute$_c$)
$\wedge$ ($\forall$ operationName$_q$ $\in$ operation-Set(Q), $\exists$ operationName$_c$ $\in$ operation-Set(C), operationName$_q$ $\equiv_e$ operationName$_c$ $\vee$operation$_q$ $\equiv_r$ operation$_c$)) $\wedge$ ($\forall$operation$_q$ $\in$ operation-Set(Q), $\exists$operation$_c$ $\in$ operation-Set(C), operation$_q$ $\equiv_o$ operation$_c$))

2) Object Behavior matching

Object behavioral equivalence and behavioral conformance can decide behavior matching.

Definition 5. Analogical behavior matching by behavioral equivalence amatch (Qbehavior, Cbehavior)

1) if pre-State($Q_{operation}$) $\equiv_e$ pre-State($C_{operation}$) or pre-State($Q_{operation}$) $\equiv_q$ pre-State($C_{operation}$)
then post-State($Q_{operation}$) $\equiv_e$ post-State($C_{operation}$) or post-State($Q_{operation}$) $\equiv_q$ post-State($C_{operation}$)

Definition 6. Analogical behavior matching by behavioral conformance amatch (Qbehavior, Cbehavior)

1) pre-State(Q) $\Rightarrow$ pre-State(C)

2) (pre-State(Q) $\Rightarrow$ post-State(Q)) $\Leftrightarrow$ post-State(C)

4 Evaluation

In order to verify the efficiency of the reuse techniques proposed in this paper, evaluation was conducted through widely used evaluation standards of recall and precision.

Recall is the evaluation standard which tests the capability to remove unrelated components from the search and is defined as a ratio for components which are suitable with the actual query among the searched components. Precision is the evaluation regarding the search capability for related components. It is defined as a ratio of the searched components from all of the components which satisfy the query.

(Table 3) presents the results for full matching and partial matching regarding both the object unit search and pattern unit search.

Table 3. Evaluation Results in accordance to Recall, Precision

(a) Evaluation Results for Recall

	Object Search
Full Search	56.3
Partial Search	75.8

(b) Evaluation Results for Precision

	Object Search
Full Search	78.6
Partial Search	66.3

The evaluation standard dependent upon recall and precision satisfied average standards [7] upon application to the system constructed in this paper. In terms of precision and search scale, it was found that cases for object searches were better than cases for pattern searches.

Models and pattern searches based on matching techniques carry the following advantages when compared to the facet technique or decimal classification used in existing search systems. In the case of the Facet technique or the decimal classification technique, the library developer writes the hierarchy between models by first possessing a classification standard of models and deciding the terms included between Facet and Facet. However, when examining examples proposed for Facets regarding actual software, such as Function, Object, Media, Language, Environment, and others as proposed by Diaz, they represent either the possibilities for operation of a function or subjects for application. In other words, it is extremely difficult to have a set of already defined terms or facets. Also, it is also difficult to define the hierarchy between models. In other words, defining the conceptual or implementation hierarchical structure between models is difficult to do. From this viewpoint, it is difficult to apply the facet technique or the hierarchical classification technique, which defines a fixed frame, during the search or reuse process for a model in which the intentions of the developer are expressed. In comparison to this, after saving a well-established model or previously verified model into the library, the matching technique allows the user who is writing a model within a new domain to conduct searches by matching with similar models already saved in the library, which can be used to serve as a foundation.

5 Conclusion

Many managers want to apply object-oriented development into their real projects. But, the novices to object-oriented as well as modeling, feel that object modeling is too difficult. They can't be sure that their models are going well. So far, we think analogy-based object learning is helpful for the beginners of object modeling and developers without any experiences in given area. It means previous experiences and model can be provided during their modeling when they construct their partial model. In our context, we apply reuse concept to our analogical learning. Partial model can be query and analogical matching function determine which stored model is the best matched with a query. There are two analogical reuse varied on granularity of part: object itself and framework consisting of objects with relationship. For object, interface matching and behavior matching function were defined. We constructed model base and developed analogical agent supporting learning, this makes people to have their pattern for object modeling and provides a better way to understand the application.

References

1. Breu, R.: Algebraic specification techniques in object oriented programming environment, Spring-Verlag, 1991.
2. Fowler, M.: Analysis patterns: reusable object models, Addison-Wesley, 1997.
3. Gamma, E. et. al.: Design patterns: elements of reusable object-oriented software, Addison-Wesley, 1995.
4. Gentner, G.: The mechanisms of analogical learning, Readings in Knowledge Acquisition and Learning, Morgan Kaufmann Publishers, (1993) 673-694.
5. Maiden, N.A.. McDougall: Analogical specification reuse during requirements analysis, Ph.D. thesis, City University, London, July 1992.
6. Alan, R.W.: Systemic software reuse through analogical reasoning, Ph.D. thesis, University of Illinois at Urbana-Champaign, 1995.
7. Feiks, F. and Hemer, D.: Specification matching of object-oriented components, First International Conference on Software Engineering and Formal Methods (2003) 182.
8. Hemer, D.: Specification matching of state-based modular components, 10th Asia-Pacific Software Engineering Conference (2003) 446.
9. Hemer, D. and Lindsay, P.: Specification-based retrieval strategies for module reuse, 13th Australian Software Engineering Conference (2001) 235.

Informative Gene Set Selection Via Distance Sensitive Rival Penalized Competitive Learning and Redundancy Analysis

Liangliang Wang and Jinwen Ma*

Department of Information Science, School of Mathematical Sciences
and LMAM, Peking University, Beijing, 100871, China
jwma@math.pku.edu.cn

Abstract. This paper presents an informative gene set selection approach to tumor diagnosis based on the Distance Sensitive Rival Penalized Competitive Learning (DSRPCL) algorithm and redundancy analysis. Since the DSRPCL algorithm can allocate an appropriate number of clusters for an input dataset automatically, we can utilize it to classify the genes (expressed by the gene expression levels of all the samples) into certain basic clusters. Then, we apply the post-filtering algorithm to each basic gene cluster to get the typical and independent informative genes. In this way we can obtain a compact set of informative genes. To test the effectiveness of the selected informative gene set, we utilize the support vector machine (SVM) to construct a tumor diagnosis system based on the express profiles of its genes. It is shown by the experiments that the proposed method can achieve a higher diagnosis accuracy with a smaller number of informative genes and less computational complexity in comparison with the previous ones.

1 Introduction

Microarray data or gene expression profiles have been widely used in many applications, especially on tumor diagnosis (e.g., [1], [2]). Given a set of samples labelled "tumorous" or "normal", the task of tumor diagnosis is to build a binary classifier as a diagnosis system to predict the unlabelled samples. Mathematically, a microarray dataset of N genes and d samples can be represented by a matrix $(x_i^\mu)_{N\times d}$, where the element x_i^μ represents the expression level of the μ-th gene at the i-th sample. Generally, there are thousands of genes in a microarray chip and so high dimensional data would cause a series of problems, such as high computing complexity, low prediction accuracy and unexplainable biological meanings [3]. Moreover, in comparison with the number of genes, we can only collect a small number of samples at present because of the high expense. In fact, there are only a small number of genes which are related or informative to a tumor. Therefore, informative gene selection to a tumor is often used as a preprocessing technique in the tumor diagnosis or classification.

* Corresponding author.

D. Liu et al. (Eds.): ISNN 2007, Part I, LNCS 4491, pp. 1227–1236, 2007.

Actually, informative gene selection (to a tumor), i.e., finding the genes that are discriminative between the normal and tumorous phenotypes, has been studied extensively in the past several years. Typically, informative genes are selected by ranking genes according to a kind of criterion, such as t, F, rank sum and χ^2 test statistics [4],[5],[6], [7]. Generally, these traditional methods just select the top k genes (k is a fixed positive integer). In this way, there again appears a serious problem that informative genes selected through individual gene evaluations are often highly correlated, which also leads to a low prediction accuracy of the diagnosis system.

To maintain a high prediction accuracy of the diagnosis system, we should find a set of uncorrelated or independent but still highly informative genes. In order to do so, many researchers have made the redundancy analysis on genes for selecting the independent informative gene set for tumor diagnosis [8],[9],[3]. However, many of these approaches are too sensitive to the order of genes according to their individual ranks such that too many genes are eliminated and thus some useful information may be lost. On the other hand, some valuable information can also be discovered by evaluating the classification capability of combinations of genes [10]. To this end, we established a post-filtering gene selection algorithm to select informative genes of a tumor with a microarray dataset based on redundancy and multi-gene analysis [11].

To further improve the efficiency of informative gene set selection, we can consider the structure of genes expressed by the rows of the matrix of the microarray dataset. That is, these genes consists of some different functional clusters to the tumor. If we can get these clusters, we can implement the post-filtering gene selection algorithm on each cluster to select the typical and independent genes. In this way, we can get a compact set of informative genes which is efficient for the tumor diagnosis.

Based on the above idea, this paper further proposes a new approach to the informative gene set selection. Since the Distance Sensitive Rival Penalized Competitive Learning (DSRPCL) algorithm [12], as a generalization of the original rival penalized competitive learning algorithm [13], can allocate an appropriate number of clusters for an input dataset automatically, we utilize it to classify the genes into a number of functional clusters. Then, we use the post-filtering gene selection algorithm on each cluster to select the typical and independent informative genes. Finally, we get the compact set of informative genes for tumor diagnosis.

In the sequel, we introduce the DSRPCL algorithm for discovering the functional gene clusters in Section 2. Section 3 describes the post-filtering gene selection algorithm on each gene cluster for typical and independent informative gene selection. The experiment results of the proposed informative gene set selection method as well as their comparisons are presented in Section 4. Finally, we conclude the paper in Section 5.

2 The DSRPCL Algorithm for Functional Gene Clusters

Given a microarray dataset $(x_i^\mu)_{N\times d}$ of N genes and d samples, we let it be $S = \{X^\mu\}_{\mu=1}^N$, where $X^\mu = [x_1^\mu, x_2^\mu, \cdots, x_d^\mu]^T$ represents the μ-th gene through its expression levels over all the d samples. Suppose that X^μ is just an input to a simple competitive learning network, which only has one layer of units. Initially there are n units with the weight vector $W_i = [w_{i1}, w_{i2}, \cdots, w_{id}]^T$ respectively for the i-th unit. All the weight vectors can be represented by a big vector $W = \text{vec}[W_1, W_2, \cdots, W_n]$. For each input X^μ, the basic idea of the DSRPCL algorithm is that not only the weight vector of the winner unit is modified to adapt to the input, but also the weight vectors of the rivals or losers are punished to keep away from the input. As a weight vector diverges to infinity, the corresponding cluster becomes empty and can be canceled. As a result, we can automatically obtain the number of gene clusters as well as their centers, or "representative genes", of these clusters assuming n is larger than the true number of the actual gene clusters. As a result, the genes are automatically divided into several functional clusters by classifying each gene into the cluster whose center is closest to it.

Table 1. The DSRPCL algorithm and its variants

1 Randomly initialize the vector $W_1^{(0)}, \cdots, W_n^{(0)}$, and let $T = 0$.

2 Update W_i with a learning rate η $(0 \le \eta \le 1)$:

1) Batch DSRPCL:

$$\Delta W_i = -\eta \frac{\partial E(W)}{\partial W_i} = \begin{cases} \eta \sum_\mu (X^\mu - W_i), & \text{if } i = c(\mu), \\ -\eta \sum_\mu \|X^\mu - W_i\|^{-P-2}(X^\mu - W_i), & \text{otherwise.} \end{cases}$$

2) DSRPCL1:

$$\Delta W_i = \begin{cases} \eta (X^\mu - W_i), & \text{if } i = c(\mu), \\ -\eta \|X^\mu - W_i\|^{-P-2}(X^\mu - W_i), & \text{otherwise.} \end{cases}$$

3) DSRPCL2:

$$\Delta W_i = \begin{cases} \eta (X^\mu - W_i), & \text{if } i = c(\mu), \\ -\eta \|X^\mu - W_i\|^{-P-2}(X^\mu - W_i), & \text{if } i = r(\mu), \\ 0, & \text{otherwise.} \end{cases}$$

4) SARPCL:

a) Let $\lambda = e^{(-k_1 T - k_0)}$, $\eta = \eta_0/(c_1\, T + c_0)$ and $t = 0$.

b) Randomly select X^μ from $S = \{X^1, \ldots, X^N\}$, and take $\xi \sim$Uniform$[0, 1]$.

c) $$\Delta W_i = \begin{cases} \eta (X^\mu - W_i), & \text{if } i = c(\mu), \\ -\eta \|X^\mu - W_i\|^{-P-2}(X^\mu - W_i), & \text{otherwise.} \end{cases}$$

If $\xi \le \lambda$, let $\Delta W_i = -\Delta W_i$.

d) If $t < M$, let $t = t + 1$ and return to STEP b).

e) If $\lambda < \varepsilon$, stop.

3 If $|E(W)^{(T+1)} - E(W)^{(T)}| > \varepsilon_1$, let $T = T + 1$, and return to STEP 2; otherwise, stop.

Mathematically, the DSRPCL algorithm tries to minimize the following cost function:

$$E(W) = E_1(W) + E_2(W)$$
$$= \frac{1}{2}\sum_{\mu} \|X^{\mu} - W_{c(\mu)}\|^2 + \frac{1}{P}\sum_{\mu, i\neq c(\mu)} \|X^{\mu} - W_i\|^{-P}, \quad (1)$$

where $c(\mu)$ is the index of the winner unit for the μ-th gene. That is, $W_{c(\mu)}$ is the nearest weight vector for X^{μ}. P is a positive constant. Ma and Wang[12] obtained the derivatives of $E(W)$ with respect to w_{ij} as follows:

$$\frac{\partial E(W)}{\partial w_{ij}} = -\sum_{\mu} \delta_{i,c(\mu)}(x_j^{\mu} - w_{ij}) + \sum_{\mu}(1 - \delta_{i,c(\mu)})\|X^{\mu} - W_i\|^{-P-2}(x_j^{\mu} - w_{ij}) \quad (2)$$

where $\delta_{i,j}$ is the Kronecker function. The DSRPCL algorithm is just a gradient descent algorithm based on the above derivatives of the cost function $E(W)$. Table 1 gives the details of the DSRPCL algorithm and its variants, where we denote it as the batch DSRPCL algorithm, the DSRPCL1 algorithm is the adaptive DSRPCL algorithm, and the DSRPCL2 algorithm modifies only the rival weight vector (i.e., the second winner) so that $E_2(W)$ is only affected by the largest term with $r(\mu)$. The other variant of the DSRPCL algorithm is the simulated annealing rival penalized competitive learning (SARPCL) by applying the simulated annealing mechanism to the DSRPCL1 algorithm. The stopping threshold value ε is a pre-fixed small positive number. k_0, k_1, c_0 and c_1 are positive constant numbers which can be selected by experience.

According to the properties of the DSRPCL algorithm shown in [12], when n is selected to be large enough, the DSRPCL algorithm can lead to a number of functional gene clusters from which we can detect the typical and independent informative genes more efficiently.

3 The Post-filtering Gene Selection Algorithm for Discovering the Independent Informative Genes

In order to attain a compact set of independent and informative genes to a tumor, we can remove the redundant informative genes through the post-filtering gene selection algorithm proposed in [11]. Initially, genes are ranked by the goodness of classification or diagnosis with the individual gene expression profile on the samples, which is measured through a statistical test under the null hypothesis that no difference exists between the gene expression profiles of tumorous and normal samples. The first gene will certainly be selected, but two types of redundant genes are identified and dealt with differently from the second gene to the last one. Let A be the expression profile of a selected gene and B be that of another candidate gene. Firstly, B will be abandoned if it is highly correlated with A because no much valuable classification information is added. Secondly, B is possibly redundant if the p-value of the test on the combination of genes B and A is greater than that of A alone. Otherwise, B will be chosen as a new informative gene. Furthermore,

Table 2. The post-filtering gene selection algorithm

0. Let iteration number $i = 0$, and the output gene set $O = \emptyset$,.
Order the gene sets:
1. If $i == 0$, let $S^i = \{X^{\mu_1}\}_{\mu_1=1}^{N_1}$, N_1 is the initial number of genes; otherwise, $S^i = \{S^i_{\mu_i}\}_{\mu_i=1}^{N_i}$, where $S^i_{\mu_i} = \{R_l^{i-1}, R_j^{i-1}\}$, $i > 1$, and $l, j = 1, \cdots, \text{length}(R^{i-1})$, satisfying $S_l^i \cap S_j^i = \emptyset$ and $T(S_l^i) \leq T(S_j^i)$ for $i \geq 1$, $\forall\, l < j$, and $l,\ j = 1, \cdots, N_i$.
Redundancy analysis:
2. Let $r = 0$, $\mu_i = 1$.
3. If $T(S^i_{\mu_i}) \geq \alpha$, let $r = r + 1$, $R^i_r = S^i_{\mu_i}$; else
   1) let $j = \mu_i + 1$.
   2) if $\text{corr}(S^i_{\mu_i}, S^i_j) > \beta$, let $S^i = S^i - S^i_j$;
      elseif $T(S^i_{\mu_i}) < T(S^i_{\mu_i}, S^i_j)$, let $r = r + 1$, $R^i_r = S^i_j$
   3) if $j < \text{length}(S^i)$, let $j = j + 1$, go to STEP 2); otherwise, stop.
4. If $\mu_i < \text{length}(S^i)$, let $\mu_i = \mu_i + 1$, go to STEP 3.
5. $O = O \cup S^i$.
6. If $i < M$, let $i = i + 1$, got to STEP 1.

the possibly redundant genes are combined into subsets and evaluated again from the view of multi-genes in a similar way with the individual genes.

Based on the above ideas, the post-filtering gene selection algorithm can be descried in Table 2. The ith iteration of the algorithm evaluates the classification capability of 2^{i-1} genes and removes the redundant ones. S^i represents an ordered list of subsets of 2^{i-1} genes according to p-values of a kind of statistical test denoted by $T(\cdot)$. In the redundancy analysis, an element in S^i will be removed if its correlation coefficient with other selected gene subsets are greater than β. Furthermore, an element in S^i will be moved to R^i if it attains a p-value less than α or that of the combination with some already selected gene subsets. The $(i+1)^{th}$ iteration will evaluate larger gene subsets by combining two smaller ones from R^i. If α is relatively large, we tend to select genes with good classification by individual evaluations. On the other hand, if α is too small, we tend to select genes with good classification by multi-gene analysis. We can stop the algorithm when no more informative genes can be found in an iteration. Alternatively, for simplicity, the algorithm stops when it reaches a user-specified maximum number of iterations M.

As we implement the post-filtering gene selection algorithm on each of the functional gene clusters obtained via the DSRPCL algorithm, we can get a set of typical independent informative genes from different aspects with a high speed since the number of genes in each cluster is generally much decreased.

4 Experimental Results and Comparisons

We tested the effectiveness of our proposed gene set selection method on the colon cancer dataset[1] through the support vector machine. This dataset contains the expression profiles of 2000 genes in 22 normal tissues and 40 tumor tissues.

[1] Retrieved from `http://microarray.princeton.edu/oncology/database.html`

Before our experiments, we normalized the gene expression profiles with zero mean and unit variance in order to eliminate the possible noise. For a tumor diagnosis system, i.e., a binary classifier, we took its prediction accuracy as the evaluation metric of this method. We used the radial basis functions (RBF's) as the kernels to build SVMs with a MATLAB toolbox called $OSUSVM3.0$ [2]. There were two parameters γ and C to be selected. We took a grid search procedure from 16×16 pairs of γ and C ($\gamma,\ C = 2^{-7}, 2^{-6}, \cdots, 2^{8}$), and chose the values optimizing the performance of SVMs on the training dataset [14].

Table 3. Results of clustering on the colon dataset

clustering method	n	K	the number of genes in each cluster
Batch DSRPCL	12	6	560, 413, 186, 36, 317, 488
DSRPCL1	10	6	178, 412,484, 351, 366,209
DSRPCL2	10	6	353, 409,40, 518, 193, 487
SARPCL	10	6	153, 55, 421, 248, 470,653

The first step was to cluster gene profiles using the DSRPCL algorithm and its variants, which stopped when the difference of the cost functions $E(W)$ between two successive steps was less than a threshold value 1.e-3. In addition, the selection of parameters was important in the DSRPCL algorithm and its variants. If P was too small, the power of de-learning might become so strong that the result of clustering was wrong. P was usually selected around 0.15 [12].

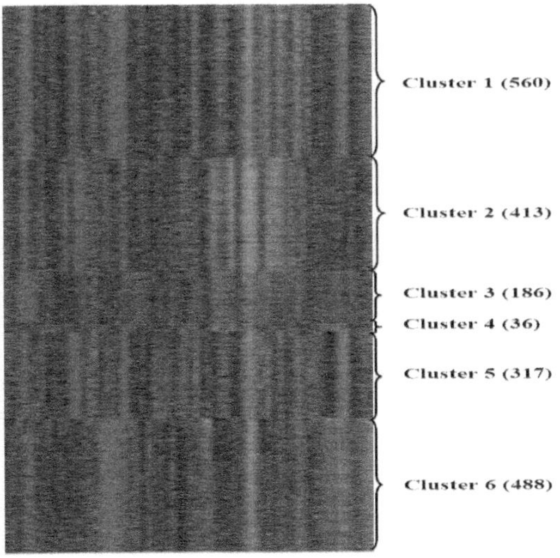

Fig. 1. The clustering result of Batch DSRPCL on the colon dataset

[2] Available on http://eewww.eng.ohio-state.edu/ ~ maj /osu_svm

With the initial number of clusters n, Table 3 showed the final number of clusters K and the number of genes in each cluster after clustering by DSRPCL and its variants on the colon dataset. For the SARPCL algorithm, we set $t = 500$, $k_1 = 0.005$, $k_0 = 1.200$, $c_1 = 0.015$ based on experiments on a small simulated dataset. Both DSRPCL and it variants automatically clustered genes into 6 clusters. Figure 1 illustrated the result of Batch DSRPCL clustering through the program *TreeView* [3], in which every row represented a gene profile and every column a sample. In this experiment, we applied the hierarchical clustering method to samples before clustering genes for the convenience of visual investigation. The positive expression was shown in red, and the negative expression was shown in green. Visually, 6 clusters distinguished themselves with others well, which might imply some biological significance.

Table 4. The LOOCV results of several methods of informative gene selection (Numbersu in brackets showed the number of selected informative genes)

clustering method	α	0.001	0.005	0.01	0.05
Traditional method	i=0	90.3% (60)	91.9% (137)	91.9% (188)	88.7% (389)
Post-filtering	i=1	91.9% (7)	91.9% (10)	91.9% (10)	91.9% (13)
	i=2	93.5% (9)	95.2% (19)	95.2% (22)	95.2% (34)
	i=3	93.5% (10)	96.8% (23)	98.4% (28)	95.2% (49)
Batch DSRPCL	i=1	93.5% (11)	100% (13)	100% (14)	96.8% (19)
	i=2	95.2% (65)	93.5% (59)	95.2% (56)	95.2% (67)
	i=3	96.8% (145)	96.8% (147)	95.2% (148)	96.8% (135)
DSRPCL1	i=1	95.2% (15)	98.4% (17)	96.8% (17)	93.5% (23)
	i=2	93.5% (57)	95.2% (59)	96.8% (59)	95.2% (67)
	i=3	96.8% (137)	96.8% (119)	96.8% (123)	96.8% (127)
DSRPCL2	i=1	93.5% (13)	95.2% (16)	96.8% (17)	96.8% (26)
	i=2	93.5% (65)	96.8% (58)	96.8% (57)	93.5% (70)
	i=3	96.8% (145)	96.8% (122)	96.8% (121)	96.8% (130)
SARPCL	i=1	96.8% (14)	98.4% (13)	98.4% (16)	95.2% (20)
	i=2	95.2% (58)	95.2% (59)	95.2% (64)	93.5% (72)
	i=3	95.2% (134)	95.2% (147)	95.2% (132)	93.5% (140)

The second step was to apply the post-filtering informative gene selection algorithm to each gene cluster. For each cluster, we conducted the leave-one-out cross-validation (LOOCV) experiments [15]. The classifier was successively learned on $d - 1$ samples and tested on the remaining one. We applied the gene set selection in each cross-validation trial on the training samples of that trial. The construction of a classifier was restricted to the selected informative genes using the training data. Finally, we computed the average prediction accuracy

[3] Which can be downloaded from `http://rana.lbl.gov`

and the number of informative genes for the d results as our evaluation result. Here, we used 2-sample rank sum test for individual genes and 2-sample Hotelling T^2 test as the multivariate statistical test on multiple genes, and chose $\beta = 0.6$ [11]. The prediction accuracies of several methods were presented in Table 4. The traditional method selected genes only through rank sum test [6] without further filtering ($i = 0$). Post-filtering chose genes of more informative by applying the post-filtering algorithm to those selected by the traditional method. From Table 4, while the best prediction accuracy achieved by the post-filtering algorithm was 98.4% using 28 genes, the prediction accuracy could achieve 100% by the Batch DSRPCL algorithm using only 13 genes. Moreover, we noticed the post-filtering method often needed several iterations to reach a higher prediction accuracy, but using the DSRPCL algorithm and its variants usually achieved the highest prediction accuracy in the first iteration because the clustering process provided some useful information for finding the informative genes.To compare with the traditional clustering methods, we used the profile clustering program *Cluster*[4] written by Eisen to cluster genes on the colon dataset by using these methods. Here, we set $\alpha = 0.005$, $\beta = 0.6$, and used rank sum test. By letting the number of cluster be 6, K-means clustering method completed the clustering after 35 iterations. The numbers of genes in every cluster were 75, 1042, 444, 28, 356 and 55, respectively. By using the hierarchical clustering method, when the genes were divided into 6 clusters the numbers of genes in every cluster were 4, 293, 341, 471, 5 and 886, respectively. From the results in Table 5, it can be found that the DSRPCL algorithm and its variants as the unsupervised clustering methods were better than the traditional clustering methods. Particularly, the DSRPCL methods not only needed not specify the number of clusters, but also led to better prediction accuracies.

Table 5. The LOOCV results of post-filtering algorithm using different clustering methods (Numbers in brackets showed the number of selected informative genes)

	hierarchical	K-means	Batch DSRPCL	DSRPCL1	DSRPCL2	SARPCL
i=1	87.1% (16)	90.3% (19)	100% (13)	98.4% (17)	95.2% (16)	98.4% (13)
i=2	96.8% (82)	95.2% (73)	93.5% (59)	95.2% (59)	96.8% (58)	95.2% (59)
i=3	95.2% (166)	96.8% (153)	96.8% (147)	96.8% (119)	96.8% (122)	95.2% (147)

5 Conclusions

Traditional informative gene set selection methods are most simple and fast, but selected genes may be highly correlated and redundant, which strongly influences the prediction or diagnosis accuracy of a tumor. The post-filtering gene selection algorithm tries to overcome this redundancy problem. However, it becomes very slow if the number of genes is very large. Moreover, it is also difficult to explain

[4] Which can be downloaded from `http://rana.lbl.gov`

these selected informative genes biologically. By utilizing the DSRPCL algorithm and its variants to cluster genes automatically, we can divide genes into several functional clusters without specifying the number of clusters in advance. In such a way, the post-filtering gene selection algorithm performs more effectively on each cluster due to the low computing complexity. Of course, the clustering process itself is time-consuming, but it can result in some relatively uncorrelated gene clusters which not only benefit the explainable biological meaning but also aid selecting most unique and uncorrelated genes. All in all, the proposed informative gene set selection based on DSRPCL and redundancy analysis can achieve higher diagnosis accuracy with a smaller number of informative genes and might reduce the computing complexity and benefit the biological explanation.

Acknowledgements

This work was supported by the Natural Science Foundation of China for Project 60471054.

References

1. Dudoit D.S., Fridyand J., Speed T. P.: Comparison of Discrimination Methods for the Classification of Tumor Using Gene Expression Data. Journal of American Statistical Association **97** (2002) 77-87
2. Guyon, I., Weston, J., Barnhill, S., Vapnik, V.: Gene Selection for Cancer Classification Using Support Vector Machine. Machine Learning **46** (2002) 389-422
3. Yu, L., Liu, H.: Redundancy Based Feature Selection for Microarray Data. Proceedings of the Tenth ACM Conference on Knowledge Discovery and Data Mining (SIGKDD'04) (2004) 737-742
4. Golub T.R., Slonim D.K., Tamayo P., et al.: Molecular Classification of Cancer: Class Discovery and Class Prediction by Gene Expression Monitoring. Science **286** (1999) 531-537
5. Ding C.: Analysis of Gene Expression Profiles: Class Discovery and Leaf Ordering. Proceedings of the 6th Annual International Conference on Computational Molecular Biology (RECOMB'02) (2002) 127-136
6. Deng L., Ma J., Pei J.: Rank Sum Method for Related Gene Selection and Its Application to Tumor Diagnosis. Chinese Science Bulletin **49** (2004) 1652-1657
7. Luo, J., Ma, J.: A Multi-Population X-2 Test Approach to Informative Gene Selection. Lecture Notes in Computer Science **3578** (2005) 406-413
8. Koller, D., Sahami, M.: Toward Optimal Feature Selection. Proceedings of the 13th International Conference on Machine Learning (ICML'96) (1996) 284-292
9. Xing, E.P., Jordan, M.I., Karp R.M.: Feature Selection for High-Dimensional Genomic Microarray Data. Proceedings of the 18th International Conference of Machine Learning (ICML'01) (2001) 601-608
10. Bo, T., Jonassen, I.: New Feature Subset Selection Procedures for Classification of Expression Profiles. Genome Biology **3** (2002) RESEARCH0017.1–0017.11
11. Wang, L., Ma, J.: A Post-Filtering Gene Selection Algorithm Based on Redundancy and Multi-Gene Analysis. International Journal of Information Technology **11** (2005) 36-44

12. Ma, J., Wang, T.: A cost-Function Approach to Rival Penalized Competitive Learning (RPCL). IEEE Transactions on Systems, Man and Cybernetics, Part B: Cybernetics **36** (2006) 722-737
13. Xu, L., Krzyzak, A., Oja, E.: Rival Penalized Competitive Learning for Clustering Analysis, RBF Net, and Curve Detection. IEEE Transactions on Neural Networks **4** (1993) 636-649
14. Hsu, C., Chang, C., Lin, C.: A Practical Guide to Support Vector Classification. National Taiwan University. Department of Computer Science and Information Engineering, given in the web: www.csie.ntu.edu.tw/~cjlin/papers/guide/guide.pdf.
15. Ben-Dor, A., Friedman, N., Yakhini, Z.: Scoring Genes for Relevance. Agilent Technical Report no.AGL-2000-13 (2000)

Incremental Learning and Its Application to Bushing Condition Monitoring

Christina B. Vilakazi and Tshilidzi Marwala

School of Electrical and Information Engineering
University of the Witwatersrand
Private Bag 3, Wits, 2050, Johannesburg, South Africa
c.vilakazi@gmail.com,t.marwala@ee.wits.ac.za

Abstract. The problem of fault diagnosis of electrical machine has been an ongoing research in power systems. Many machine learning tools have been applied to this problem using static machine learning structures such as neural network, support vector machine that are unable to accommodate new information as it becomes available into their existing models. This paper presents a new method to bushing fault condition monitoring using fuzzy ARTMAP(FAM). FAM is introduced for bushing condition monitoring because it has the ability to incrementally learn information as it becomes available. An ensemble of classifiers is used to improve the classification accuracy of the systems. The testing results show that FAM ensemble gave an accuracy of 98.5%. Furthermore, the results show that fuzzy ARTMAP can update its knowledge in an incremental fashion without forgetting previously learned information.

1 Introduction

Most of transformer failures are linked to bushing faults, which results in lengthy downtime that have economic consequences. Hence, there is a high demand for a cost effective and automated condition monitoring that can detect faults as early as possible. Early diagnosis and fault identification is an important activity for maximizing a plant's lifetime, operational costs and levels of safety. Incipient faults in bushing transformers are either electrical or thermal in nature and can degrade the oil and cellulose insulation leading to the formation of dissolved gases. These faults can be detected and monitored using dissolved gas-in-oil analysis (DGA). DGA has gained worldwide acceptance as a diagnostic method for the detection of oil-filled equipment internal faults [1]. Fault gases are produced by degradation of the transformer oil and solid insulating materials such as paper, pressboard and bushing board, which are all made of cellulose. The rate of cellulose and oil degradation is significantly increased in the presence of a fault inside the bushing. In the past years, various fault diagnosis techniques have been proposed, including the conventional key gas method, ratio method [2]. Recently computational intelligence methods, such as expert system [3], fuzzy logic [4] and artificial neural network [5][6] have been employed for bushing condition monitoring.

D. Liu et al. (Eds.): ISNN 2007, Part I, LNCS 4491, pp. 1237–1246, 2007.

Many artificial intelligence based methods for fault diagnosis rely heavily on adequate and representative set of training. However, in real-life applications it is often common that the available data is incomplete, inaccurate and changing. It is also often common that the training data becomes available only in small batches and that some new classes only appear in subsequent data collection stages. Hence, there is a need to update the classifier in an incremental fashion without compromising on the classification performance of previous data.The paper aims to improve the classification accuracy of a single fuzzy ARTMAP (FAM) by using an ensemble of fuzzy ARTMAP. Furthermore, the incremental capability of the FAM is implemented for bushing fault diagnosis.

2 Background

2.1 Dissolve Gas-In-Oil Analysis (DGA)

DGA is the most commonly used diagnostic technique for oil-filled machines such as transformers and bushings. DGA is used to detect oil breakdown, moisture presence and partial discharge activity. Fault gases are produced by degradation of transformer and bushing oil and solid insulation, such as paper and pressboard, which are all made of cellulose. The gases produced from the transformer and bushing operation can be listed as follows [6]:

- Hydrocarbons gases and hydrogen: methane, ethane, ethylene, acetylene and hydrogen.
- Carbon oxide: carbon monoxide and carbon dioxide.
- Naturally occuring gases: nitrogen and oxygen.

The symptoms of faults are classified into four main groups corona, low energy discharge, high energy discharge and thermal. The quantity and types of gases reflect the nature and extent of the stressed mechanism in the bushing.

2.2 Ensemble of Classifiers

Ensemble approaches typically aim at improving the classifier accuracy on a complex classification problem through divide-and-conquer approach. Perrone and Copper [7] proved mathematically that an ensemble method can be used to improve the classification ability of a single classifier. The success of an ensemble method depends on two factors; a pool of diverse individual classifier to be fused and a proper decision fusion method. A proper classifier team should be robust and generate the best fusion performance. It should also be optimal so that it can reduce the time for calculation while improving the classification accuracy. Various classifier selection techniques such as Q static, generalized diversity and agreement have been used successfully for classifier selection [8][9]. Many researchers have found that the dependency among classifier can affect the fusion results. Goebel et al. [10] recommend an effective method for classifier selection based on calculating the correlation degree of n classifiers. This method is used to select the classifiers to be fused. This correlation degree is given by [10];

$$\rho_n = \frac{nN^F}{N - N^F - N^R + nN^F}, \tag{1}$$

where N^F is the number of samples which are misclassified by all classifier N^R represent the samples which were classified correctly by all classifiers. N is the total number of experiments example. Generally, smaller correlation degree can lead to better performance of classifier fusion because independent classifiers can give more effective information. A number of decision fusion techniques exists; some of these include majority and weighted majority voting, trained combiner fusion, median, min and max. In this study, majority voting decision fusion scheme is used. This scheme considers each of the predictions produced by the available classifiers as a vote.

2.3 Overview of the Fuzzy ARTMAP

Fuzzy ARTMAP is a neural network architecture based on adaptive resonance theory that is capable of supervised learning of arbitrary mapping of clusters in the input space and their associated labels. The key feature of this type of network architecture is that it is capable of fast, online, incremental learning, classification and prediction [11]. The architecture of the FAM is shown in Fig. 1. The FAM shown in Fig. 1 consists of fuzzy ART modules, ART_a and ART_b, that are linked together via an inter-ART module, F^{ab}, called a map field. During supervised learning, ART_a receives an input in the complement code form $\underline{I}^t = \underline{A}^t = (\underline{a}^T, \underline{a}^{c^T})$ and ART_b receives an input also in the complement code $\underline{I}^t = \underline{B}^t = (\underline{b}^T, \underline{b}^{c^T})$. Note that each component in $\underline{I}$ is in the interval [0,1] and complement coding is a normalization rule that preserves amplitude information. For example, $\underline{I}^t = (\underline{a}^T, \underline{a}^{c^T}) = (a_1, \ldots, a_n, a_1^c, \ldots, a_n^c) =$

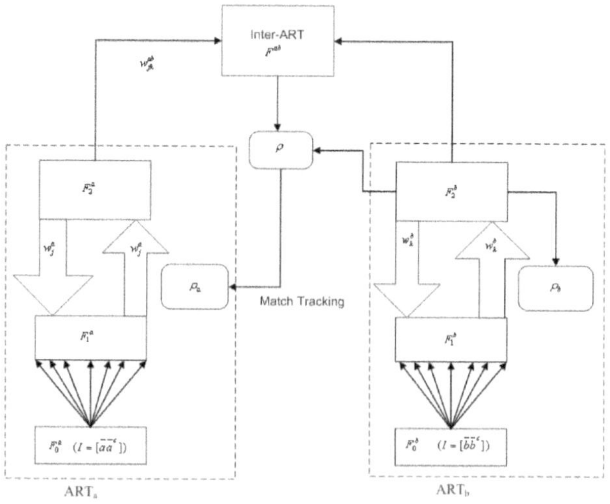

Fig. 1. Architecture of the fuzzy ARTMAP

$(a_1, \ldots, a_n), 1 - a_1, \ldots, 1 - a_n(a = 0.4, \ldots, 0.8, 0.6, \ldots, 0.2)^T$. The map field is used to form predictive associations between categories and to realize the match-tracking rule. It ensures autonomous system operation in real time and works by increasing the vigilance parameter ρ_a of ART_a. Parameter ρ_a calibrates the minimum confidence that ART_a must have in a recognition category, or hypothesis, activated by an input A in order for ART_a to accept that category, rather than search for a better one through an automatically controlled process of hypothesis testing. The lower the value of ρ_a is the larger the number of categories is. A predictive failure of at ART_b increases ρ_a by the minimum amount needed to at ART_a, using a mechanism called match tracking. Hypothesis testing leads to the selection of a new ART_a category, which focuses attention on a new cluster of A that is better able to predict B. Owing to these mechanisms, FAM systems becomes one of a rapidly growing family of incremental learning pattern recognition systems.

2.4 Incremental Learning

An algorithm can be called an incremental learning algorithm if it is able to learn additional information while retaining previously learned knowledge. It should also be able to accommodate the new classes that may be introduced by the incoming data [12]. The FAM fulfills all these criteria. Analysis of the biological learning has led to the understanding that incremental learning can be of three types [13];

- Structure Incremental Learning: The structure of the neural network is changed during learning.
- Learning Parameter Incremental Learning: Is where learning parameters are adapted during learning.
- Data Incremental Learning: The data set or its complexity is increased in stages during learning.

The incremental learning in the FAM is a combination of structure and data incremental learning. As data of different classes is added to the system or more data on existing classes are added, the structure of the system adapts creating a larger number of categories in which to map the output labels. Learning new information without requiring access to previously used data, however, raises 'stability-plasticity dilemma'. This dilemma indicates that a completely stable classifier maintains the knowledge from previously seen data, but fails to adjust in order to learn new information, while a completely plastic classifier is capable of learning new data but lose prior knowledge.

3 Proposed System

The proposed system is implemented using the fuzzy ARTMAP. The overview of the proposed system is shown in Fig. 2. The population of classifiers is used to

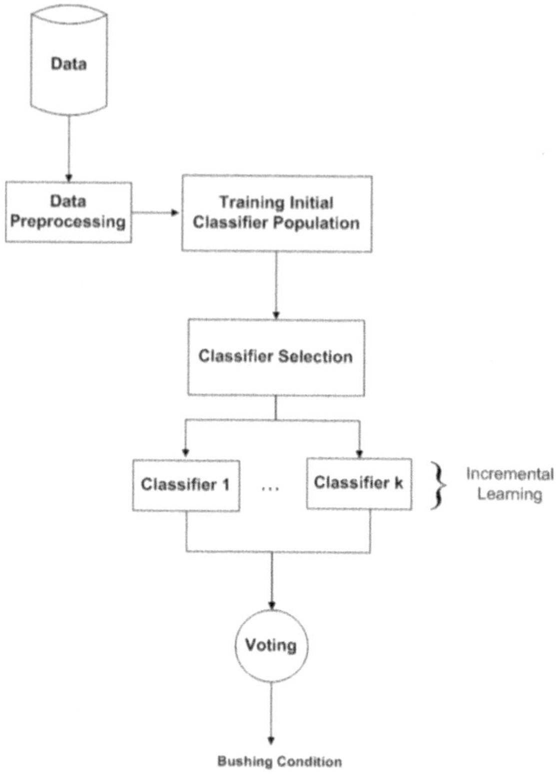

Fig. 2. Block Diagram of the proposed system

introduce classification diversity. A classifier selection process is executed using the correlation measure. As a result, an optimal team of classifier is formed to improve classification accuracy. After the classifier selection, majority voting fusion algorithm is employed for the final decision.

3.1 Data Transformation

The mentioned variables undergo a min-max normalization. The normalization is requirement for the FAM, since the FAM complement coding assumed the data is normalized. The equation for min-max normalization for a single feature is given by

$$x_{norm} = \frac{x - x_{min}}{x_{max} - x_{min}}. \tag{2}$$

Where, x_{min} and x_{max} are the minimum and maximum value for that feature from the data, respectively. The performance of the trained classifier are evaluated using the standard classification accuracy;

$$accuracy = \frac{N_C}{N_T} \times 100\%. \tag{3}$$

Where, N_C is the number of correct classification and N_T is the total number of data points in the data sets.

4 Experimental Results

The available data was divided into three that contains three classes only data set; training, validation and testing data sets. The validation data set is used to evaluate the performance of each classifiers in the initial population and this performance is used to select the classifiers for the ensemble. The testing data set is used to evaluate the performance of the system on unseen data. The first experiment will compare the performance of fuzzy ARTMAP with multi-layer perceptron (MLP), support vector machine (SVM), Extension Neural Network (ENN). The second experiment will show that an ensemble of classifier improve the classification accuracy compared to a single classifier. The last experiment evaluates the incremental capability of the fuzzy ARTMAP for bushing condition monitoring using two new classes.

4.1 Comparison of Classifiers

This first experiment aims to compare the performance of the fuzzy ARTMAP in terms of classification accuracy and time with MLP, SVM and ENN.

ENN is a new pattern classification system based on concepts from neural networks and extension theory [14]. The extension theory uses a novel distance measurement for classification processes. An ENN with a learning rate of 0.356 was used.

SVM is a machine learning algorithm based on the statistical learning theory. The kernel function is important for SVM classifier and the most popular kernel functions are the linear, polynomial, Gaussian RBF and sigmoid. In this experiment, an Gaussian RBF with 50 gaussian centers is used.

MLP is a supervised neural network that uses backpropagation. The input data vector is presented to the network input layer while the output layer is presented with the actual output. The network is refined through a process of error backpropagation, where the resultant error between the actual and target output is minimized. An MLP with 15 hidden layer neurons was used in this study.

Table 1. Comparison of classifiers for bushing fault diagnosis

Classifier	Validation Accuracy(%)	Testing Accuracy (%)	Time (s)
MLP	98.00	95.87	0.2031
SVM	97.70	95.63	30.7031
ENN	99.00	98.50	0.0156
fuzzy ARTMAP	98.50	97.50	0.5687

4.2 Creation of Ensemble

The creation of an ensemble of classifiers follows a series of steps. First, an initial population of twelve classifiers with different permutations of the training data are created. This permutation is needed in order to create diversity of the classifier being added since FAM learns in an instance-based fashion, which makes the order in which the training patterns are received an important factor. From the created population, the best performing classifier of the team is selected based on the classification accuracy. Then the correlation degree between the best classifier and all the members of the population is calculated using Eq. 1. From the unselected classifiers, the classifier with low correlation is selected for fusion. This is repeated until all the unselected classifiers are selected. Then fuse together the selected classifiers using the majority voting to form the final decision. The selection results for different numbers of classifiers are shown in Table 2. Fig. 3 shows the effect of classifier selection and shows that the optimal results is achieved with six classifiers.

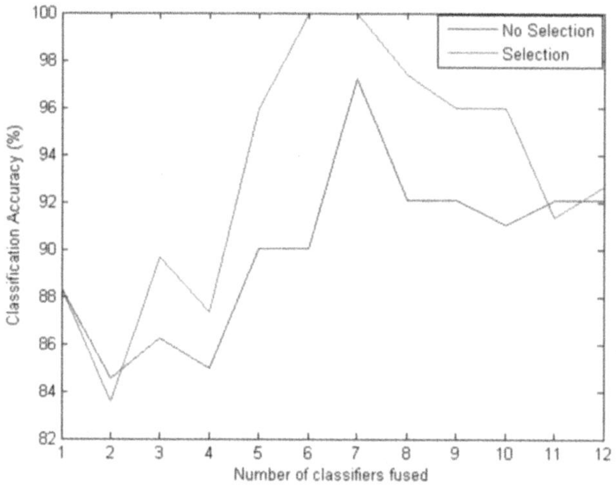

Fig. 3. The effect of classifier selection

Table 3 shows the classification accuracy for both the best classifier and the ensemble. The table shows that the ensemble of FAM performs better that the best FAM classifier.

4.3 Incremental Learning

Any condition monitoring system that has to be used online must have incremental learning capability for it to be rendered useful. As mentioned earlier an incremental learning algorithm must accommodate both new data and new

Table 2. Results of the fuzzy ARTMAP classifiers

Classifier	Validation Accuracy(%)	Correlation (ρ)
1	98.5	Best Classifier
2	97.5	0.624
3	97.5	0.6319
4	97.5	0.6239
5	95.0	0.6310
6	95.0	0.6296
7	95.0	0.6359
8	92.5	0.6377
9	92.5	0.6143
10	92.5	0.6178
11	92.5	0.6340
12	92.5	0.6476

Table 3. Comparison of Classification Accuracy

Classifier	Validation Accuracy(%)	Testing Accuracy (%)
Best Classifier	98.5	95.2
Ensemble	100.0	98.5

classes that may be present in the new data without compromising the classification accuracy of the previously learned data. Carpenter *et al.* [11] showed that the fuzzy ARTMAP is able to accommodate these types of data.

In this paper, we demonstrate the incremental ability of the fuzzy ARTMAP for new classes. Using the existing model of the classifier in the ensemble, each classifier is trained on data of the fourth new class. Experimentation is performed on independent test set for this class for both the best classifier and ensemble of the classifiers. Both the best classifier and the ensemble gave a classification accuracy of 100%. Experimentation of the system on the testing data set of the initial three classes is performed, this is to determine how addition of new information affects previously learned information. The classifier accuracy achieved is 94.88% for the best classifier and 98% for the ensemble. This simple experiment shows that the system is able to learn new classes, while still preserving existing knowledge of the system. The small change in the accuracy of the three classes is due to the tradeoff between stability and plasticity of the classifier during training. It is should be note that the classification performance of fuzzy ARTMAP is always 100% on training data, since according to the ARTMAP learning algorithm convergence is achieved only when all training data are correctly classified. Furthermore, once a pattern is learned, a particular cluster is assigned to it, and future training does not alter this clustering. The learning ability of the system on a further fifth class is shown in Fig. 4. The graph shows that as more training examples are added the ability of the system to correctly classify the data increases as shown by the increase in classification accuracy. The classification accuracy of the best classifier and the ensemble is 95% and

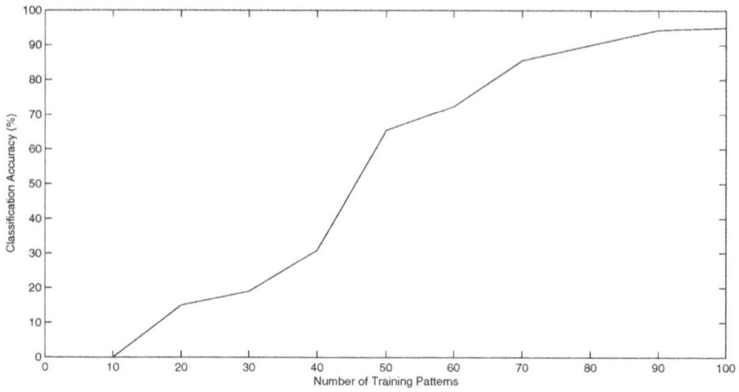

Fig. 4. Incremental learning of the best classifier on class 5

96.33%, respectively. The best classifier gave a classification accuracy of 90.20% while the ensemble gave a classification accuracy of 89.67% on the original test data with three classes. Again, the change in the accuracy of the three classes is due to the tradeoff between stability and plasticity of the classifier during training.

5 Discussion and Conclusion

The fuzzy ARTMAP gave slightly better results than the MLP and SVM however ENN slightly outperform the fuzzy ARTMAP. The good results of FAM might be due to the fact that the structure of the fuzzy ARTMAP adapts, creating a larger number of categories in which to map the output labels as the training data becomes available. The ensemble of classifiers gave an improvement on the classification accuracy from a single fuzzy ARTMAP. In the If new information has to be added to the classifier for the batch learning this involve discarding the existing classifier and training a new classifier using accumulated data. Another method involve modifying the weights of the classifier using the misclassified instances only. However, these methods suffer from catastrophic forgetting and they require access to old data and require optimization of the training parameters. In the incremental learning system, the old system can learn new information on-line without requiring access to old data or forgetting previously learned information.

References

1. Saha, T.K.: Review of Modern Diagnostic Techniques for Assessing Insulation Condition in Aged Transformers. IEEE Trans. Electrical Insulation **10** (2003) 903-917
2. Rogers R.R.: IEEE and IEC codes to Interpret Faults in Transformers using Gas in Oil Analysis IEEE Trans. Electrical Insulation **13** (1978) 349-354

3. Lin, C.E, Ling, J.M., Huang, C.L.: An Expert System for Transformer Fault Diagnosis using Dissolved Gas Analysis. IEEE Trans. Power Delivery **8** (1993) 231-238
4. Mofizul, S.I., Wu, T., Ledwich, G.: A Novel Fuzzy Logic Approach to Transformer Fault Diagnosis. IEEE Trans. Dielectric and Electricrical Insulation **7** (2000) 177-186
5. Bhattacharyya, S.K., Smith, R.E., Haskew T.A.: A Neural Network Approach to Transformer Fault Diagnosis using Dissolved Gas Analysis Data. In Proceedings of North American Power Symposium **12** (1993) 125-129
6. Dhlamini, S.M., Marwala, T.: Using SVM, RBF and MLP for bushings. In Proceedings of IEEE Africon (2004) 613-617
7. Perrone, M.P., Cooper, L.N.: Neural Networks for Speech and Image Processing, When Networks Disagree: Ensemble Methods for Hybrid Neural Networks, Chapman Hall (1993).
8. Petrakos, M., Benediktsson, J.A., Kannellopoulos, I.: The Effect of Classifier Agreement on the Accuracy of the Combined Classifier in Decision Level Fusion. IEEE Trans. Geoscience and Remote Sensing **39** (2001) 2539-2546
9. Kuncheva, L.I., Whitaker, C.J., Shipp, C.A., Duin, R.P.W.: Limits of the Majority Vote Accuray in Classifier Fusion. Pattern Analysis and Applications **6** (2003) 22-31
10. Goebel, K., Yan, W.Z., Cheetham, W.: A Method to Calculate Classifiers Correlation for Decision Fusion. In Proceeding of Decision and Control (2002) 135-140
11. Carpenter, G.A., Grossberg, S., Markuzon, N., Reynolds, J.H., Rosen, D.B.: Fuzzy ARTMAP: A Neural Network Architecture for Incremental Supervised Learning of Analog Multidimensional Maps. IEEE Trans. Neural Networks **3** (1992) 698-713
12. Polikar, R., Udpa, L., Udpa, A.S., Hanovar, V.: Learn++: An Incremental Learning Algorithm for Supervised Neural Networks. IEEE Trans. Systems, Man and Cybernetics **31** (2001) 497-508
13. Chalup, S.K.: Incremental Learning in Biological and Machine Learning Systems. Int. J. Neural Systems **12** (2002) 90-127
14. Wang, Z., Zhang, Y., Li, C., Liu, Y.: ANN Based Transformer Fault Diagnosis. In Proceedings of American Power Conference **11** (1997) 428-432

Approximation Property of Weighted Wavelet Neural Networks

Shou-Song Hu, Xia Hou, and Jun-Feng Zhang

College of Automatic Engineering
Nanjing University of Aeronautics and Astronautics
Nanjing 210016, China
djhx@263.net
http://www.nuaa.edu.cn

Abstract. A new weighted wavelet neural network is presented, And the approximation capability of such weighted wavelet neural network is also studied based on the traits of Lebesgue partition, the operator theory and the topology structure of the relatively compact set in Hilbert space. The simulation results indicate that the weighted wavelet neural network is a uniformed approximator, which can approximates the nonlinear function in compact set by arbitrary precision.

1 Introduction

Function approximation is a fundamental subject of artificial neural networks. The selection of node functions is crucial for the approximation property and the convergence rate of a network. In neural networks, two types of activation functions are commonly used: global, as in Back-propagation networks (BPN), and local, as in radial basis function networks (RBFN). They have different approximation properties, and given enough nodes, both networks are capable of approximating any continuous function with arbitrary accuracy [1,2,3]. However, due to its multilayered structure and the greedy nature of the BP algorithm in BPN, the training processes often settle in undesirable local minima. The RBFN performs better in learning functions with local variations and discontinuities due to the introduction of two parameters in its node functions-the center and the variance. However, the two parameters and the number of the hidden nodes have to be determined by training because there is not any theoretical guidance in choosing them.

Recently, with the development of the wavelet theory, some properties of wavelets and associated scaling functions have been successfully used in the approximation of target functions by neural networks. Wavelet neural networks are a novel powerful class of neural networks that incorporate the most important advantages of multiresolution analysis. Such networks preserve all the features of common neural networks, but also have rigorous mathematical foundations based on wavelet theory. Since it was proposed, the wavelet neural network has been studied by a sequence of papers [4,5,6] and applied to the fault diagnosis

D. Liu et al. (Eds.): ISNN 2007, Part I, LNCS 4491, pp. 1247–1254, 2007.

[7,8], signal processing [9] and time series prediction [10]. Unlike the RBFN, the hierarchy and the thresholds of the wavelet network, as well as the number of the hidden nodes are closely related by the theory of wavelets. Therefore, less computation is required in wavelet neural network designing and training.

As an extension of wavelets, a weighted wavelet can preserve all the advantages the wavelet has. Furthermore, it can simultaneously have several properties very useful in practical applications such as orthogonality, regularity, and compact support, also even if the weighted scale function make a jump, which is impossible for a scalar wavelet. Based on the considerations above, a model of weighted neural network is proposed and its universal approximation properties are proved in this paper. Both theoretical and experimental results show that the weighted neural network has some advantages over the wavelet network, especially in the approximation of jump functions.

The remainder of this paper is organized as follows. Section 2 gives some preliminaries and the weighted scaling function is also studied in this section. Section 3 presents the main results of weighted wavelet networks as a uniform approximator. In section 4, weighted wavelet networks is applied to nonlinear function approximation. Section 5 presents the conclusions of this work.

2 Preliminaries

The basic idea of wavelets and multiresolution analysis is to use the dyadic translates and dilates of one function as a basis of L_2. Wavelets are often constructed based on the Fourier transform, as the translation and dilation on the frequency domain acts as the algebra operations, so these wavelets are also called as algebraic wavelets. Classical examples of compactly supported algebraic wavelets are Daubechies orthogonal wavelets [11] and spline wavelets [12].

Algebraic wavelets are biorthogonal with respect to the L_2 inner product space. Consider an integrable positive function $\omega(x)$ of the space L_2, defined the weighted inner product of the space

$$L_2 : \langle f, g\rangle_\omega = \int_{-\infty}^{+\infty} \omega(x) f(x)\overline{g(x)}dx,$$

the associated norm is $\|f, g\| = \sqrt{\langle f, g\rangle_\omega}$. Define a multiresolution analysis as a sequence of closed subspaces V_j such that:

1. $V_j \subset V_{j+1}$;
2. $\bigcup_{j=-\infty}^{+\infty} V_j$ is dense in the space L_2 and $\bigcap_{j=-\infty}^{+\infty} V_j = \{0\}$;
3. Scaling functions $\varphi_{j,k}$ exist so that $\{\varphi_{j,k}\}_k$ is a Riesz basis of V_j.

These imply that for every scaling function $\varphi_{j,k}$, the coefficients $\{h_{j,k,l}\}$ exist so that $\varphi_{j,k} = \sum_l h_{j,k,l}\varphi_{j+1,2k+l}$. Each scaling function satisfies a different refinement relation. The dual multiresolution analysis consists of spaces $\widetilde{V}_j$ with bases generated by dual scaling functions $\widetilde{\varphi}_{j,k}$ that are biorthogonal with the scaling functions $\langle \varphi_{j,k}, \widetilde{\varphi}_{j,k'} \rangle_\omega = \delta_{k-k'}$. Note that the coefficients $h_{j,k,l}$ of the refinement relation 1 can be written as the following by the weighted inner product: $h_{j,k,l} = \langle \varphi_{j,k}, \widetilde{\varphi}_{j+1,2k+l} \rangle_\omega$. The weighted scaling functions satisfy the following properties:

Support: $\operatorname{supp} \varphi_{j,k} = a^{-j}[-N+1+k, N+k]$;
Unitary: $\int_{-\infty}^{+\infty} \omega(x)\varphi_{j,k}(x)dx = 1$;
Biorthogonal: $\langle \varphi_{j,k}, \widetilde{\varphi}_{j,k'} \rangle_\omega = \delta_{k-k'}$;
Vanishing moment: $\sum_k M_{j,k}^p \varphi_{j,k} = x^p$, where $M_{j,k}^p = \langle x^p, \widetilde{\varphi}_{j,k} \rangle_\omega$ $(0 \le p < N)$.

Then the weighted wavelet and dual wavelet are chosen as

$$\begin{aligned} \Psi_{j,k} &= \varphi_{j+1,ak} - \varphi_{j+1,ak+1} + \varphi_{j+1,ak+2} + \cdots + (-1)^{a-1}\varphi_{j+1,ak+a-1} \\ &= \sum_{l=ak}^{ak+a-1} (-1)^l \varphi_{j+1,l} \end{aligned}$$

and $\widetilde{\Psi}_{j,k} = \sum_l g_{j,k,l}\widetilde{\varphi}_{j+1,2k+l}$, where $g_{j,k,l} = (-1)^l h_{j,k+[l/a],1-l}$.

So the weighted wavelet function satisfied the following characteristics:

Orthogonal: $\langle \widetilde{\varphi}_{j,k}, \Psi_{j,k'} \rangle_\omega = 0$;
Orthogonal 2: $\langle \varphi_{j,k}, \widetilde{\Psi}_{j,k} \rangle_\omega = 0$;
Biorhogonal: $\langle \Psi_{j,k}, \widetilde{\Psi}_{j',k'} \rangle_\omega = \delta_{j-j'}\delta_{k-k'}$;
Decay: $\int_{-\infty}^{+\infty} \omega(x)\Psi_{j,k}(x)dx = 0$;
Vanishing moment: $\int_{-\infty}^{+\infty} \omega(x)x^p\widetilde{\Psi}_{j,k}(x)dx = 0 (0 \le p < N)$;
Compact support: $\operatorname{supp} \Psi_{j,k} = \operatorname{sup} p \widetilde{\Psi}_{j,k} = a^{-j}[-D+K, D+K+1)$.

The normal wavelet functions can be looked as the special weighted wavelet functions with the weighted function $\omega(x) = 1$.

3 Main Results

The essential of the weighted wavelet neural network virtually is the parameters adjustment can be conducted automatically, that is to say, the weighted wavelet neural network measures the position and direction (circum-rotation) of the hyperplane continuously, and so the adjustment of the model space partition domain. Many nerve cells formed the hyperplane in space, the space is divided into many regions. When the weighted wavelet function is used as activation, the space can be separated into many hyper-cone surfaces on account of the distributing of samples in the high dimensional space. The process of the nonlinear function approximation will determine which hyper-cones the input samples belong to, i.e. the nonlinear function can be considered as the set composed by these hyper-cones. Such-and-such, any nonlinear function will be matched to the nearest hyper-cone, so the capabilities of the wavelet neural network for approximation and pattern recognition is just hereinto.

The point set D denotes as the n dimension space, and $C(D)$ is defined as the linear space that all of bounded real continuous function in D. The symbol Ψ, satisfying $\int |\Psi(x)|dx = 1$, shows the integral weighted wavelet of $C(D)$. For any positive number $P > 0$, B^p represents the hyper cube $[-p, p]^n$. For any positive number $q > 0$ and a multiple index $\sum = (r_1, r_2, \cdots, r_n) \in N^n$, the sign

$$L = \{(I_k, t_k)\}_{k \in \sum} = \{([x_{i-1}, x_i], t_i) : x_{i-1} \leq t_i \leq x_i, i = 1, \cdots, n\}$$

represents the partition of $C(D)$ constituted by n intervals I_k according to the index $\sum$ and the center t_k in L, which can be called as Lebesgue partition [13]. The diameter $d(I_k)$ of the area composed of the n interval is defined as $d(I_k) = \sup_{x,y \in I_k} \|x - y\|$, then the norm of the partition L is described as $|L| = \max_{k \in \sum} d(I_k)$. The cubage V_k of the partition L is $V_k = \prod_{i=1}^{n} V_{k_i}$, where V_{k_i} is the length of I_k circumvolves with the ith coordinate.

The following conclusion is obtained:

Theorem. The weighted wavelet neural networks hidden cells output function $\Psi(x)$ has compact support and satisfies $\int |\Psi(x)|dx = 1$, the set $A \subset C(R^n)$ is relatively compact set, ε is a given positive number, then there exists positive numbers σ, δ and some continued function $\widetilde{f}$ of an n elements nonlinear function $f(x)$ such that for any $q > 1$ and the Lebesgue partition L satisfied $|L| < \delta$ in B^q, then

$$\Big| f(x) - \sum_{k \in \sum} \phi_k(f) \Psi[(a - t_k)/\sigma] \Big| < \varepsilon$$

where $\phi_k(f) = \frac{1}{\sigma^n} \widetilde{f}(t_k) V_k$, there exists a weighted wavelet neural network such that for any $\varepsilon > 0$, the continuous nonlinear function is approximated by this network.

Proof. According to *Ascoli−Arzelâ* theory [14], A is uniformly bounded compact set , then for any continuous function $f(x)$ in A, there exists positive number $h > 0$ such that $|f(x)| < h$, $\forall x \in B^q$, $f \in A$. Let the function $f(x)$ be continual extension in R^n, and let $\widetilde{A} = \{\widetilde{f} = f, f \in B^q;\ \widetilde{f} = 0, otherwise\}$, obviously

$$|\widetilde{f}(x)| < h, \quad \forall x \in R^n, \widetilde{f} \in \widetilde{A} \tag{1}$$

Let $q_0 > 0$, because the weighted wavelet function Ψ is compactly supported and integrable, thus

$$\int_{R^n \setminus B^{q_0}} |\Psi(x)|dx < \frac{\varepsilon}{8h} \tag{2}$$

As the compact set $\widetilde{A}$ is equicontinuous after the extension, so there exists positive number $\delta > 0$, for any $x, y \in R^n$ and any function $\widetilde{f} \in \widetilde{A}$,

$$\|x - y\| < \delta \Rightarrow |\widetilde{f}(x) - \widetilde{f}(y)| < \frac{\varepsilon}{4\|\Psi\|_1} \tag{3}$$

where $|\Psi\|_1 = \int |\Psi(x)|dx$.

Let a positive number $\sigma > 0$ such that $|\sigma x\| < \delta$ for all $x \in C(R^n)$, let

$$\Psi_\sigma(x) = \frac{1}{\sigma^n}\Psi(\frac{x}{\sigma}), \quad x \in R^n \tag{4}$$

Let $a \in B^1$, definite the following transform:

$$(\Psi_\sigma * \widetilde{f})(a) = \int_{R_n} \widetilde{f}(a-x)\Psi_\sigma(x)dx = \int_{R_n} \widetilde{f}(a-\sigma x)\Psi(x)dx$$

For any $\widetilde{f} \in \widetilde{A}$, according to (1) (2) (3),

$$\begin{aligned}
\left|(\Psi * \widetilde{f})(a) - \widetilde{f}(a)\right| &\leq \int_{R_n} \left|\widetilde{f}(a-\sigma x) - \widetilde{f}(a)\right| \cdot \left|\Psi(x)\right| dx \\
&\leq \int_{B^{q_0}} \left|\widetilde{f}(a-\sigma x) - \widetilde{f}(a)\right| \cdot \left|\Psi(x)\right| dx \\
&\quad + \int_{R^n \setminus B^{q_0}} \left|\widetilde{f}(a-\sigma x) - \widetilde{f}(a)\right| \cdot \left|\Psi(x)\right| dx \\
&< \frac{\varepsilon}{4\|\Psi\|_1} \int_{B^{q_0}} \left|\Psi(x)\right| dx + 2h \int_{R^n \setminus B^{q_0}} \left|\Psi(x)\right| dx \\
&< \frac{\varepsilon}{4} + 2h \cdot \frac{\varepsilon}{8h} = \frac{\varepsilon}{2}
\end{aligned} \tag{5}$$

According to the definition of $\widetilde{f}$, $\widetilde{f}(x) = 0$ always holds, for all $x \in R^n \setminus B^q$, hence $(\Psi_\sigma * \widetilde{f})(a) = \int_{B^q} \Psi_\sigma(a-x)\widetilde{f}(x)dx$. Define the mapping $T_{f,a}(x) : B^q \to R$ as

$$\begin{aligned}
T_{f,a}(x) &= \Psi_\sigma(a-x)\widetilde{f}(x) \\
\Psi_\sigma * \widetilde{f}(a) &= \int_{B^q} T_{f,a}(x)dx
\end{aligned} \tag{6}$$

Let $T = \{T_{f,a}\} : f \in A, a \in B^q$, the points which are mapped by the function in relatively compact set constitute a relatively compact set, and $\widetilde{A}$ and the set $\{\Psi_\sigma(a-x) : a \in B^1\}$ are also relatively compact sets, hence T is a relatively compact set, too. There must exist a positive number δ such that for any one partition L satisfied $|L| < \delta$ we have

$$\left|\int_{b^q} T_{f,a}(x)dx - \sum_{k \in \sum} T_{f,a}(t_k)V_k\right| < \frac{\varepsilon}{2} \tag{7}$$

for any $T_{f,a} \in T$ and any $t_k \in I_k$.

The following inequality can be obtained from the inequalities (5) (6) (7) for any $a \in B^1$:

$$\left|\widetilde{f}(x) - \sum_{k \in \sum} T_{f,a}(t_k)V_k\right| < \varepsilon, \quad \widetilde{f} \in \widetilde{A}$$

Because $\widetilde{f}(a) = f(a)$ for all $a \in B^1$, so as long put

$$\phi_k(f) = \frac{1}{\sigma^n}\widetilde{f}(t_k)V_k, \quad k \in \sum$$

i.e.

$$\left|\widetilde{f}(x) - \sum_{k\in\sum} \phi_k(f)\Psi_\sigma(a-x)\right| < \varepsilon,$$

hence

$$\left|f(x) - \sum_{k\in\sum} \phi_k(f)\Psi_\sigma(a-x)\right| < \varepsilon$$

4 Simulation

The approximation capability of the weighted wavelet neural networks for nonlinear function has been analyzed. Then the following nonlinear function is considered at the application of the weighted wavelet neural network:

$$f(x) = 1/2[\cos(2x + 100\varphi) + \sin(3x + 3\varphi)]$$

In this simulation, Haar wavelet is chosen as the weighted wavelet activation function

$$h(x) = \begin{cases} +1 & : \quad 0 \le x < 0 < 1/2 \\ -1 & : \quad 1/2 < x \le 1 \\ 0 & : \quad otherwise \end{cases}$$

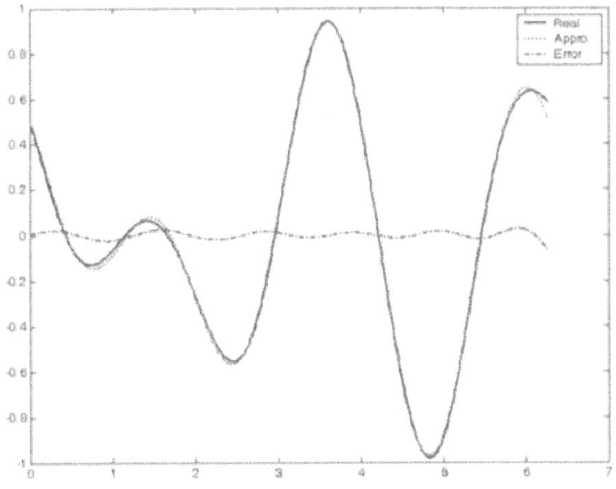

Fig. 1. $\varphi = 1$, real, approximation and error

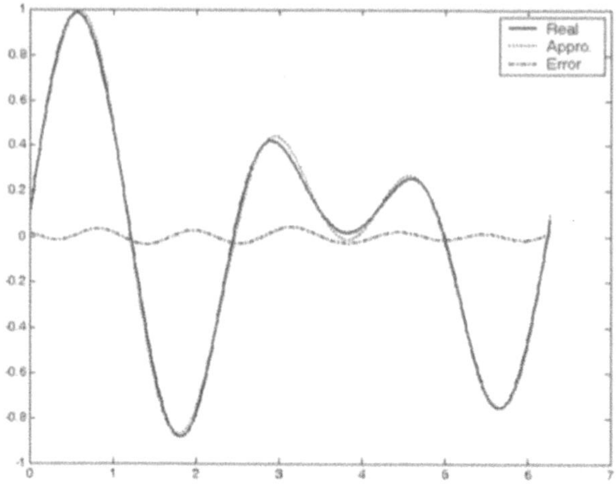

Fig. 2. $\varphi = 2$, real, approximation and error

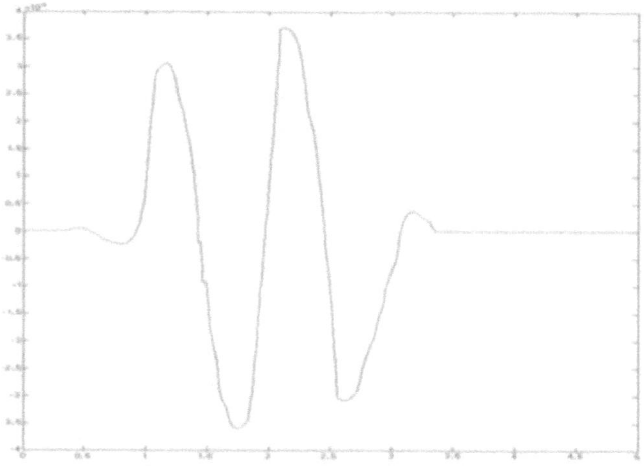

Fig. 3. Weighted wavelet function

For any given phase φ, the weighted wavelet neural networks can trace nonlinear function accurately, i.e. different phase φ does not affect the approximation capability of the weighted wavelet neural networks. (In figure 1 and figure 2, all error values is no more than 9% of the smallest error.)

It is the other case in the weighted l wavelet function is obvious figure by figure 3 so the conclusion can be drawn that the wavelet neural network is uniformed approximator [15].

5 Conclusion

The application of the weighted wavelet neural network is widely in practice, this paper pays more attention on the analysis for the approximation capability of the weighted wavelet neural network based on Lebesgue partition in Hilbert space, operator theory and relatively compact. The simulation results indicate that the weighted wavelet neural network is a uniformed approximator.

Acknowledgments. Supported by the National Natural Science Key Foundation of China (60234010), Aeronautic Science Foundation (05E52031) of China.

References

1. Funahashi, K.: On the Approximate Realization of Continuous Mappings by Neural Networks. Neural Networks **1** (1989) 183-192
2. Cybenko, G.: Approximation by Superpositions of a Sigmoidal Function. Mathematics of Control, Signals and Systems **1** (1989) 303-314
3. Hornik, K.: Approximation Capabilities of Multiplayer Feedforward Networks. Neural networks **1** (1991) 251-257
4. Pati, Y.C., Krishnaprasad, P. S.: Discrete Affine Wavelet Transforms for Analysis and Synthesis of Feedforward Neural Networks. Advances in Neural Information Processing Systems **1** (1991) 743-749
5. Zhang, Q.H., Benveniste, A.: Wavelet Networks. IEEE Trans. Neural Networks **1** (1992) 889-898
6. Zhang, Q.H.: Using Wavelet Network in Nonparametric Estimation. IEEE Trans. Neural Networks **1** (1997) 227-236
7. Oussar, Y., Rivals, I., Personnaz, L., Dreyfus, G.: Training Wavelet Networks for Nonlinear Dynamic Input-output Modeling. Neurocomputing **1** (1998) 173-188
8. Xu, J.X., Tan, Y.: Nonlinear Adaptive Wavelet Control Using Constructive Wavelet Networks. In: Proceeding of American Control Conference. Arlington, VA (2001)
9. Parasuraman, K., Elshorbagy, A.: Wavelet Networks: An Alternative to Classical Neural Networks. In: Proceedings of International Joint Conference on Neural Networks. Montreal, Canada (2005)
10. Tang, X.Y., Zhang, Y.L.: Pattern Recognition for Fault Based on BP Wavelet Network. Computer Engineering **1** (2003) 94-96
11. Daubechies, I.: Orthonormal Bases of Compactly Supported Wavelets. Comm. Pure. Appl. Math. **1** (1988) 909-996
12. Chui, C.K., Wang, J.Z.: A General Framework of Compactly Supported Splines and Wavelets. J. Approx. Theory **1** (1992) 263-304
13. Krishnaswami, A.: A Fundamental Invariant in the Theory of Partitions. Topics in Number Theory. Kluwer. Acad. Publ. Dordrecht (1999)
14. Bartsch, R.: Ascoli-Arzela-theory Based on Continuous Convergence in an (almost) Non-Hausdorff Setting. Categorical Topology **1** (1996) 221-240
15. Ying, H.: Sufficient Conditions on Uniform Approximation of Multivariate Functions by General Takagi-Sugeno Fuzzy Systems with Linear Rule Consequent. IEEE Trans. Sys., Man. and Cyber. **1** (1998) 515-520

Estimation of State Variables in Semiautogenous Mills by Means of a Neural Moving Horizon State Estimator*

Karina Carvajal[1] and Gonzalo Acuña[2]

[1] Facultad de Ingeniería, Universidad de Atacama.
Copayapu 485, Copiapó, Chile
[2] Facultad de Ingeniería, Universidad de Santiago de Chile, USACH
Avda. Ecuador 3659, Santiago, Chile
kcarvajal@diicc.uda.cl

Abstract. A method of moving horizon state estimation (MHSE) including a recurrent neural network as the dynamic model is used as an estimator of the filling level of the mill for a semiautogenous ore grinding process. The results are compared to those of a simple neural network acting as an estimator. They show the advantages of the Neural-MHSE, especially concerning robustness under large perturbations of the state variables (index of agreement > 0.9), which would favor its application to industrial scale processes.

1 Introduction

In semiautogenous (SAG) grinding of ores the optimum operating conditions of the mills are strongly dependent on the correct determination of the relevant state variables of the process. Unfortunately, the prevailing conditions in a SAG mill make it difficult to measure these variables on line and in real time, something that is particularly critical in the case of the filling level state variable of the mill.

Software sensors have proved to be powerful tools for determining state variables that cannot be measured directly [1]. A software sensor is an algorithm for on-line estimation of relevant unmeasured process variables. This kind of on-line estimation algorithms can use any model: phenomenological, statistical, artificial intelligence based, or even a combination of them. In general, for these kinds of sensors it is necessary to have appropriate dynamic models that account for the evolution of the relevant state variables. Many of the techniques used for the implementation of sensors require rather precise descriptive models of the process which in the industrial case are difficult to achieve. Additionally, they are time consuming, highly sensitive to calibration and fine tuning, and often do not consider realist models representing the disturbances, own of a industrial process [2].

The moving horizon state estimator (MHSE) is an approach of software sensors in which a dynamic estimation problem is converted into a nonlinear static optimization over a time horizon and is presented as an alternative for carrying out estimation

* This work thanks the partial financial support of Fondecyt under project 1040208.

D. Liu et al. (Eds.): ISNN 2007, Part I, LNCS 4491, pp. 1255–1264, 2007.

techniques in industrial scale processes [2]. In this paper, the dynamic model included in the MHSE is an external recurrent neural network that allows approximating the evolution of certain state variables of the nonlinear dynamic process of SAG grinding. In this way the capacity of neural networks to approximate nonlinear complex functions with an arbitrary precision is used [3].

This paper is divided into the following sections: a brief description of the SAG grinding process; the used neural model; a definition of MHSE; the used methodology; and the reached results and final conclusions.

2 Description of the Semiautogenous Grinding Process

The objective of the concentration processes in mining is to recover the particles of valuable species (copper, gold, silver, etc.) that are found in mineralized rocks. The concentration process is divided into three steps: crushing, grinding and flotation.

In the grinding process, the size of the particles from the crushing continues to be reduced to obtain a maximum granulometry of 180 micrometers (0.18 mm). The grinding process is carried out using large rotary equipment or cylindrical mills in two different ways: conventional or SAG grinding.

In SAG mills the grinding occurs by the falling action of the ore from a height close to the diameter of the mill. The SAG involves the addition of metallic grinding media to the mill, whose volumetric filling level varies from 4% to 14% of the mill's volume, as show figure 1. The ore is received directly from the primary crushing in a size of about 8 inches. This material is reduced in size under the action of the same mineralized material and under the action of numerous 5-inch diameter steel balls [4].

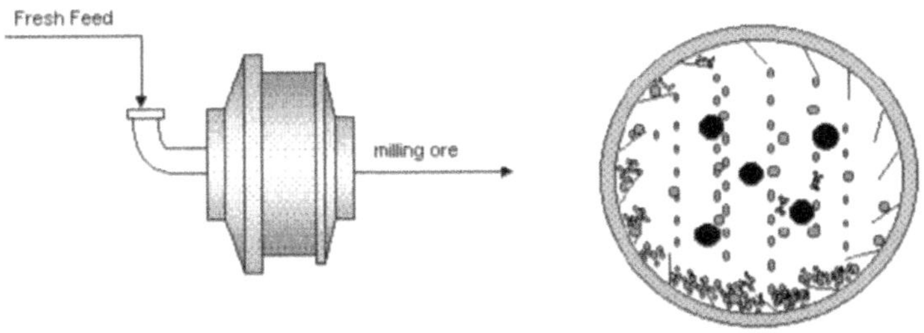

Fig. 1. SAG Mill Scheme – External and Internal look

In the operation of these mills the aim is to work under conditions that imply the maximum installed power consumption, but this means working under unstable operating conditions because an increase in the filling level of the mill beyond the point of maximum consumption leads to an overfilling condition (see figure 2).

Furthermore, the maximum power value that can be consumed by a SAG mill is not constant and depends mainly on the internal load density, the size distribution of the feed, and the condition of the lining. The filling level of the inner load that corresponds

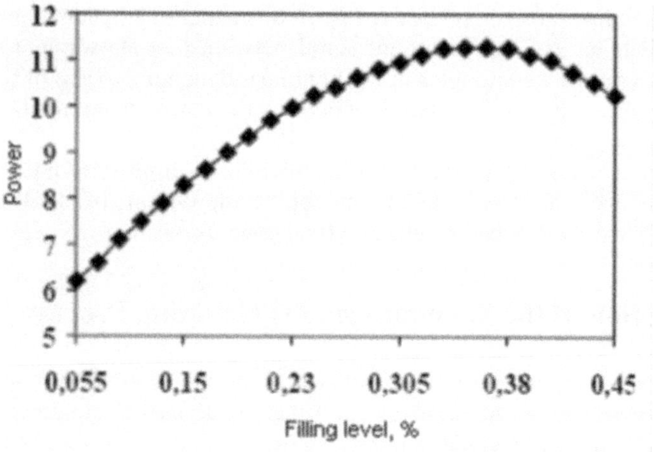

Fig. 2. Typical variation of power consumption versus filling level of SAG mill

to the maximum power consumption is related to the filling level with grinding media and the motion of the inner load. For that reason the operators of the SAG circuit must try to conjugate these factors in order to achieve first the stabilization of the operation, and then try to improve it [4]. That is why it is important for them to have reliable and timely information on the filling level of the mill.

3 Modeling by Means of Neural Networks

In the SAG grinding process to be modeled two state variables are considered: the filling level, which is the variable that it is desired to estimate on line and in real time– and the proportion of fine granulometry in the mill's feed. Data on the filling level are available indirectly from the torque variable produced by the motion of the load in the mill.

The pressure on the mill's shaft bearings, closely related to the filling level, is considered as an output together with the fresh ore feed flow to the mill. On-line and real-time measurements are available for filling level, fines and bearing pressure variables.

To carry out the modeling of this dynamic process the canonical form of a recurrent neural network was used, because it has been shown that any of these networks can be transformed into their canonical form consisting of a static neural network with appropriate external recurrences, therefore feasible to be trained using the well-known backpropagation algorithm [5]. After several trainings, it was decided to use a network consisting of an input layer (with four inputs: fresh feed, bearing pressure, filling level and fines at time t), a hidden layer with nine neurons, and an output layer (with four outputs: fresh feed, bearing pressure, filling level and fines at time t+1). Sigmoid logarithmic transfer function was used in all the layers.

The data used in this paper were derived from several days of operation of a SAG mill at the industrial level. After preprocessing the data, nearly 6000 points were chosen. Of those points 3500 were used for training, 1500 were used for validation to avoid overtraining, and the remainder to test the model. The optimization algorithm used was the quasi-Newton BFGS included in Matlab. The data were normalized between 0 and 1. 1% noise was added to the fresh feed and bearings pressure variables. The architecture of the neural network used is shown in Fig. 3.

Several algorithms of training were proved, like Levenberg-Marquardt backpropagation algorithm (trainlm in Matlab), Gradient Descent with adaptive learning rate backpropagation (traingda in Matlab), Gradient descent backpropagation (traingd in Matlab) and BFGS quasi-Newton backpropagation (trainbfg in Matlab).

Trainbfg was chosen because this algorithm performed better than the others. Trainbfg is a network training function that updates weight and bias values according to the BFGS quasi-Newton method.

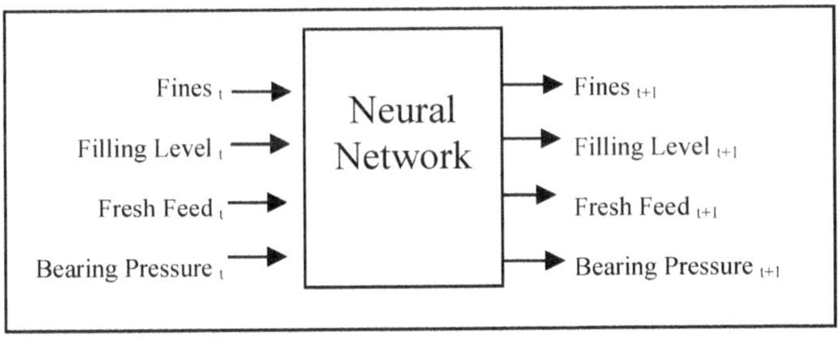

Fig. 3. Neural network architecture for modelling the evolution of the grinding process. Fines and Filling Level are considered as state variables while Fresh Feed and Bearing Pressure are the measured output variables.

4 Moving Horizon State Estimator

In general, the classic estimators assume a precise model of the system, an assumption which, unable to be satisfied in most real applications, turns into an error in the estimation that often increases with time. Furthermore, and due to the difficulty of including into the above methods some restrictions typical of industrial settings, such as infrequent sampling, compliance with conservation laws, and inequalities and the need to recalibrate every now and then the estimation from adequate measurements, an estimation technique was developed adapted from concepts derived from predictive control methods: the so-called moving horizon state estimator, MHSE [2], [6].

In short, the MHSE renders a dynamic state estimation problem into a static optimization problem in which the value of an initial vector at the beginning of a horizon is sought in order to minimize the difference between the system's real output and its evolution given by the model and its initial conditions.

A central aspect of this method is the minimization of an objective function or criterion consisting of the sum of the estimation errors of the output over a certain time horizon. That is, the MHSE looks for the state vector, at the beginning of the horizon that minimizes the chosen criterion.

Formally, if we have the nonlinear model of a system given by

$$\dot{x}(t) = f\left(x(t), u(t)\right). \\ y(t) = g(x(t)). \tag{1}$$

The MHSE tries to determine the vector, at the beginning of the horizon that minimizes a criterion, for example of the type

$$J = \frac{1}{2}\sum_{i=t_{sh}}^{t_k}\left[Y_i - \hat{Y}_i\right]^2. \tag{2}$$

where Y is the output over the horizon (lh) under consideration, and $\hat{Y}$ is the estimation of the output over that horizon. In this case the variables used for the minimization are the pressure on the bearings and the fresh feed (on-line measurable variables).

In the case of complex nonlinear systems the problem of the optimization to determine the optimum state vector at the start of the horizon cannot be solved analytically, and for that reason numerical techniques must be used. They can be grouped roughly into classical or heuristic techniques, among which those that are locally or globally convergent can be pointed out [7]. In this paper a local optimization technique was used.

5 Neural Moving Horizon State Estimator

To develop a Neural MHSE, the phenomenological model is replaced by the previously trained neural network, including a feedback of the state variables (Fig. 4). This dynamic neural network model allows an estimation of the value of the state variables at the end of the horizon from their optimum initial value provided by the moving horizon state estimation method.

The MHSE algorithm used is the following:

1. Initialization of the state vector: x_{sh}.
2. Search for the optimum initial state vector: x_{sh}* (by means of some optimization method).
3. Calculation of the final solution of the horizon using a dynamic model of the process: x_k.
4. Return to step 2 to carry out the calculation of x_{k+1} before moving the horizon one step forward.

A scheme of this algorithm is showed in Fig. 5. The used optimization method is that included in the Matlab fmincon function, which finds the minimum value of a nonlinear multivariate function, subjet to restrictions.

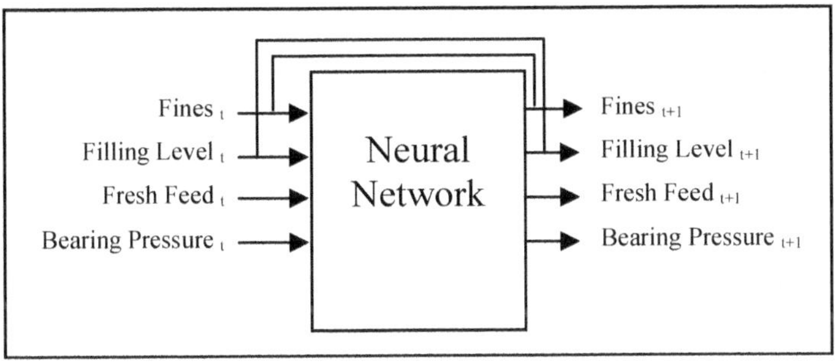

Fig. 4. Neural Network Estimator for the Fines and Filling Level variables. This neural network is included in the MHSE for producing the Neural MHSE estimator.

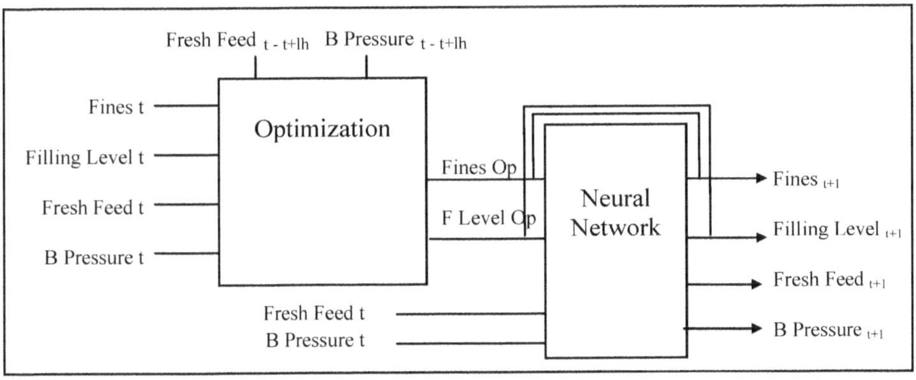

Fig. 5. Neural Moving Horizon State Estimator

6 Analysis and Discussion of Results

To evaluate the estimator, 800 new data points were used. 1% noise and a perturbation of the filling level and fines variables were incorporated into these data to test the robustness of the method.

The Neural MHSE (Fig 5) is compared to a Neural Network (Fig 4) acting as an estimator. The estimation error is quantified using the following error indices: root mean square error, standard residual error and Index of Agreement [8]. The Index of Agreement (IA) indicates the degree of fit of the estimated and the simulated values of a given variable (equation 3); a value above 0.9 indicates a good estimation. The Root Mean Square Error (RMS) is the square root of the mean square error (equation 4) and the Residual Standard Deviation (RSD) is the standard deviation of the residuals (equation 5). RMS and RSD values close to zero also indicate goodness-of-fit [1].

$$IA = 1 - \frac{\sum_{i=1}^{n} (o_i - p_i)^2}{\sum_{i=1}^{n} (|o_i'| - |p_i'|)^2} \tag{3}$$

$$RMS = \sqrt{\frac{\sum_{i=1}^{n} (o_i - p_i)^2}{\sum_{i=1}^{n} o_i^2}} \tag{4}$$

$$RSD = \sqrt{\frac{\sum_{i=1}^{n} (o_i - p_i)^2}{N}} \tag{5}$$

where o_i are observed values, p_i are estimated values, $p_i' = p_i - o_m$ and $o_i' = o_i - o_m$, where o_m is the mean value of observed values.

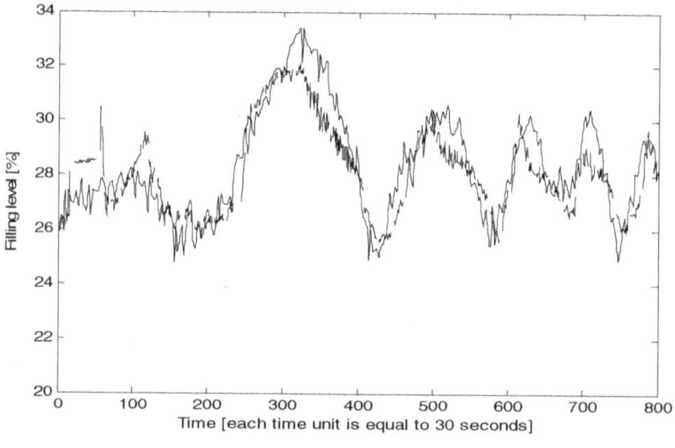

Fig. 6. Neural Network Estimator Output for Filling Level (dashed line) with perturbation in filling level. The real data are represented by continuous line.

Figures 6, 7, 8 and 9 show the results obtained by applying a perturbation to the filling level and fines variables during approximately 20 time units. Table 1 shows the result of the indices obtained by means of the Neural Moving Horizon Estimation Method and the Neural Network acting as estimator for tests with the perturbation.

From these results it is seen that the Neural MHSE achieves an IA above 90% for the fines variable and the filling level variable, which is the state variable that it is desired to be estimated on line and in real time. It is also seen that, in the presence of perturbations, both estimators differ, being the Neural MHSE more robust.

Table 1. IA, RMS and RSD for the Neural MHSE and the Neural Network acting as an estimator

Indices	Neural Network Estimator		Neural MHSE (lh = 10)	
	Filling Level	Fines	Filling Level	Fines
IA	0.9226	0.8398	0.9284	0.9541
RMS	0.1275	0.3849	0.1248	0.2019
RSD	0.0780	0.1718	0.0901	0.0764

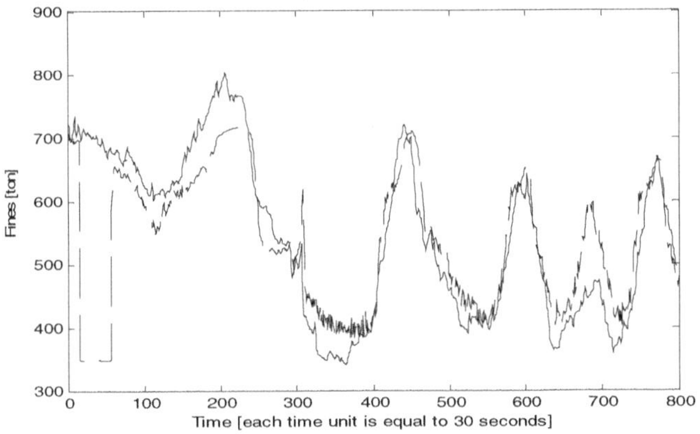

Fig. 7. Neural Network Estimator Output for Fines (dashed line) with perturbation in filling level and fines. The real data are represented by continuous line.

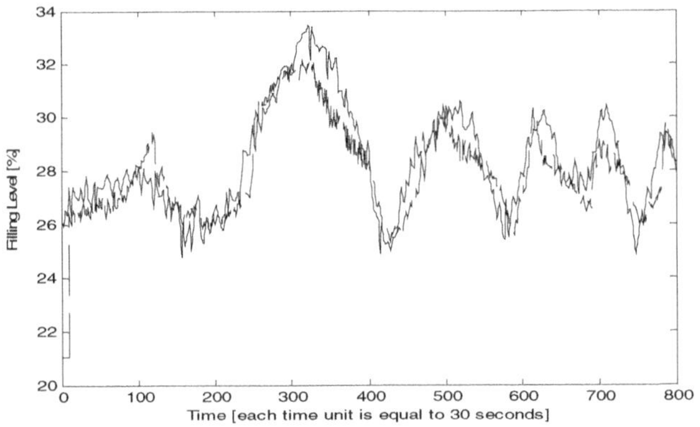

Fig. 8. Neural MHSE Output for Filling Level (dashed line), with perturbation in filling level and fines. The real data are represented by continuous line.

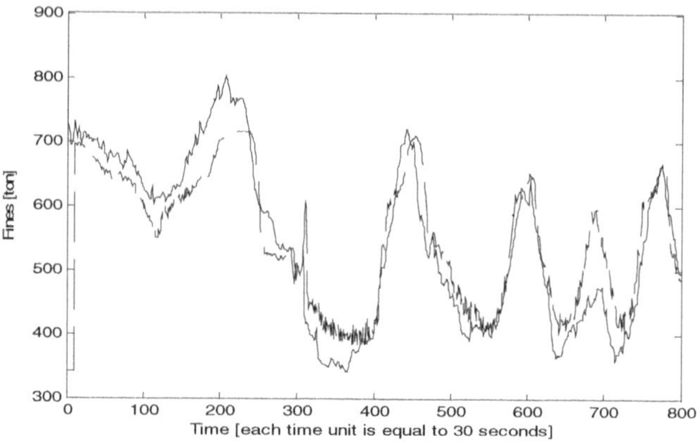

Fig. 9. Neural MHSE Output for Fines (dashed line), with perturbation in filling level and fines. The real data are represented by continuous line.

7 Conclusions

This paper presents a neural moving horizon state estimator applied to a semiautogenous grinding process in order to estimate the fines and filling level variables.

For modeling the process a static network with external recurrences was used, allowing a dynamic system to be modeled. The training algorithm used is the classical backpropagation method with a second order quasi-Newton optimization method and sigmoid logarithmic transfer function method, requiring the data from a previous time t to predict the output at a time t+1.

The results delivered by the estimator are very satisfactory in view of the Index of Agreement (> 0.9) and an RMS (< 0.1), pointing to an estimation very close to real values. The robustness of the Neural MHSE method after adding perturbations to the variables involved in the process should be stressed.

It must be pointed out that this work shows the possibility of complementing the moving horizon state estimation method, of increasing use in industrial processes, with neural networks established as dynamic models, reinforcing the advantages of both methodologies in an extremely important application in our country.

References

1. Vaněk, M., Hrnčiřík, P., Vovsík, J., Náhlík, J.: On-line Estimation of Biomass Concentration Using A Neural Network and Information about Metabolic State. Bioproc. and Biosys. Eng. **27** (2004) 9-15
2. Valdés-González, H., Flaus, J.M., Acuña, G.: Moving Horizon State Estimation with Global Convergence Using Interval Techniques: Application to Biotechnological Processes. Journal of Process Control **13**(4) (2003) 325-336
3. Hornik, K., Stinchcombe, M., White, H.: Multilayer Feedforward Networks Are Universal Approximators. Neural Networks **2** (1989) 359-366

4. Magne, L., Valderrama, W., Pontt, J.: Visión Conceptualy Estado de la Tecnología en Molienda Semiautógena. Workshop SAG'99. Viña del Mar, Chile (1999)
5. Nerrand, O., Roussel-Ragot, P., Personnaz, L., Dreyfuz, G., Marcos, S.: Neural Networks and Non-linear Adaptative Filtering: Unifying Concepts and New Algorithms. Neural Computation **5** (1993) 165-199
6. Allgöwer, F., Badgwell, T., Qin, J., Rawlings, J., Wright, S.: Nonlinear Predictive Control and Moving Horizon Estimation – An Introductory Overview. Advances in Control Highlights of ECC'99, Paul M. Frank (Ed.), Springer-Verlag, Chapter 12, (1999) 391-449
7. Cherrualt, Y. : Optimisation: méthodes locales et globales, Presses Universitaires de France (PUF), Collections mathématiques, France, (1999)
8. Robeson, S.M., Steyn, D.G.: Evaluation and Comparison of Statistical Forecast Models for Daily Maximum Ozone Concentrations. Atmos. Environ., 24B (1990) 303-312

A New Adaptive Neural Network Model for Financial Data Mining

Shuxiang Xu[1] and Ming Zhang [2]

[1] School of Computing, University of Tasmania, Locked Bag 1359, Launceston, Tasmania 7250, Australia
Shuxiang.Xu@utas.edu.au
[2] Department of Physics, Computer Science & Engineering, Christopher Newport University, Newport News, VA 23606, USA
mzhang@pcs.cnu.ed

Abstract. Data Mining is an analytic process designed to explore data (usually large amounts of data - typically business or market related) in search of consistent patterns and/or systematic relationships between variables, and then to validate the findings by applying the detected patterns to new subsets of data. One of the most commonly used techniques in data mining, Artificial Neural Networks provide non-linear predictive models that learn through training and resemble biological neural networks in structure. This paper deals with a new adaptive neural network model: a feed-forward higher order neural network with a new activation function called neuron-adaptive activation function. Experiments with function approximation and stock market movement analysis have been conducted to justify the new adaptive neural network model. Experimental results have revealed that the new adaptive neural network model presents several advantages over traditional neuron-fixed feed-forward networks such as much reduced network size, faster learning, and more promising financial analysis.

1 Introduction

Data Mining is an analytic process designed to explore data (usually large amounts of data - typically business or market related) in search of consistent patterns and/or systematic relationships between variables, and then to validate the findings by applying the detected patterns to new subsets of data. The ultimate goal of data mining is prediction - and predictive data mining is the most common type of data mining and one that has the most direct business applications. The process of data mining usually consists of three stages: (1) the initial exploration, (2) model building or pattern identification with validation/verification, and (3) deployment. Data mining tools predict future trends and behaviours, allowing businesses to make proactive, knowledge-driven decisions. Data mining tools can answer business questions that traditionally were too time-consuming to resolve. They scour databases for hidden patterns, finding predictive information that experts may miss because it lies outside their expectations. One of the most commonly used techniques in data mining,

D. Liu et al. (Eds.): ISNN 2007, Part I, LNCS 4491, pp. 1265–1273, 2007.

Artificial Neural Network (ANN) technology offers highly accurate predictive models that can be applied across a large number of different types of financial problems [1, 2, 3].

ANN's are computer programs implementing sophisticated pattern detection and machine learning algorithms to build predictive models from large historical databases. When designing an ANN model, one of the most important decisions to make is the selection of the neuron activation function. The traditional activation functions are the sigmoid function, generalized sigmoid functions, the radial basis function, non-sigmoid functions, and so on. One common characteristic of these activation functions is that they are all fixed and can not be adjusted to adapt to different problems. Activation function is critical as the behaviour and performance of an ANN model largely depends on it. So far there have been limited studies with emphasis on setting a few free parameters in the neuron activation function (a neuron activation function with a few free parameters is called a *neuron-adaptive activation function* in this paper). ANN's with such activation function seem to provide better fitting properties than classical architectures with fixed activation function neurons. Zhang *et al* [4] established a higher order ANN with adaptive activation functions for automating financial data modelling. Vecci *et al* [5] studied the properties of an ANN which is able to adapt its activation function by varying the control points of a Catmull-Rom cubic spline. Their simulations confirmed that the special learning mechanism allows to use the network's free parameters in a very effective way. In Chen and Chang [6], real variables *a* (gain) and *b* (slope) in the generalised sigmoid activation function were adjusted during learning process. A comparison with classical ANN's to model static and dynamical systems was reported, showing that an adaptive sigmoid (i.e., a sigmoid with free parameters) leads to an improved data modelling. In Campolucci *et al* [7], an adaptive activation function built as a piecewise approximation with suitable cubic splines had arbitrary shape and allowed to reduce the overall size of the neural networks, trading connection complexity with activation function complexity. Other authors such as Hu *et al* [8], Yamada *et al* [9] also studied the properties of neural networks with adaptive activation functions.

Higher-Order Neural Networks (HONN's) (Lee, Doolen, et al [10]) are networks in which the net input to a computational neuron is a weighted sum of products of its inputs. Such neuron is called a Higher-order Processing Unit (HPU) (Lippmann, [11]). It was known that HONN's can implement invariant pattern recognition (Psaltis et al [12], Reid et al [13], Wood et al [14]). Giles et al [15] showed that HONN's have impressive computational, storage and learning capabilities. In Redding et al [16], HONN's were proved to be at least as powerful as any other Feed-forward Neural Network (FNN) architecture when the order of the networks is the same. Kosmatopoulos et al [17] studied the approximation and learning properties of one class of recurrent HONN's and applied these architectures to the identification of dynamical systems. Identification schemes based on higher order network architectures were designed and analysed. Thimm et al [18] proposed a suitable initialization method for HONN's and compared this method to weight initialization techniques for FNN's. A large number of experiments were performed which leaded to the proposal of a suitable initialization approach for HONN's.

In this paper we established an ANN model with a neuron-adaptive activation function. We proposed an empirical justification of a neuron-adaptive activation function with 4 free parameters. Following the definition of the neuron-adaptive activation function, we conducted experiments with function approximation and stock market movement analysis to exhibit the advantages of ANN model with our neuron-adaptive activation function over traditional ANN's with fixed activation functions.

2 A Neuron-Adaptive Activation Function

Definition 2.1. A *Neuron-adaptive Activation Function (NAF)* is defined as:

$$\Psi(x) = A1 \cdot e^{-B1 \cdot x^2} + \frac{A2}{1 + e^{-B2 \cdot x}}, \tag{2.1}$$

where A1, B1, A2, B2 are real variables which will be adjusted (as well as weights) during training.

In our experiments (sections 4 and 5) we used a learning algorithm that is based on steepest descent rule [19] to adjust free parameters in (2.1) as well as connection weights between neurons. Basically our algorithm is not far from traditional back propagation algorithm, however, as parameters in (2.1) can be adjusted, it provides more flexibility and better approximation and simulation ability for our neural network model.

3 The Learning Algorithm

We use the following notations:

$I_{i,k}(u)$	the input or internal state of the ith neuron in the kth layer
$w_{i,j,k}$	the weight that connects the jth neuron in layer $k-1$ and the ith neuron in layer k
$O_{i,k}(u)$	the value of output from the ith neuron in layer k
$A1, B1, A2, B2$	adjustable variables in activation function
$\theta_{i,k}$	the threshold value of the ith neuron in the kth layer
$d_j(u)$	the jth desired output value
β	learning rate
m	total number of output layer neurons
l	total number of network layers
r	the iteration number
η	momentum

The input to the ith neuron in the kth layer is:

$$input_{i,k} = \sum_j \left(w_{i,j,k} O_{j,k-1}\right) + \prod_j out_{j,k-1} - \theta_{i,k}, \tag{3.1}$$

$$O_{i,k}(u) = \Psi\left(I_{i,k}(u)\right) = A1_{i,k} \cdot e^{-B1_{i,k} \cdot I_{i,k}(u)} + \frac{A2_{i,k}}{1 + e^{-B2_{i,k} \cdot I_{i,k}(u)}}. \tag{3.2}$$

To train our neural network an energy function

$$E = \frac{1}{2} \sum_{j=1}^{m} \left(d_j(u) - O_{j,l}(u)\right)^2 \tag{3.3}$$

is adopted, which is the sum of the squared errors between the actual network output and the desired output for all input patterns. In (3.3), m is the total number of output layer neurons, l is the total number of constructed network layers (here $l = 3$). The aim of learning is undoubtedly to minimize the energy function by adjusting the weights associated with various interconnections, and the variables in the activation function. This can be fulfilled by using a variation of the steepest descent gradient rule [10] expressed as follows:

$$w_{i,j,k}^{(r)} = \eta w_{i,j,k}^{(r-1)} + \beta \frac{\partial E}{\partial w_{i,j,k}}, \tag{3.4}$$

$$\theta_{i,k}^{(r)} = \eta \theta_{i,k}^{(r-1)} + \beta \frac{\partial E}{\partial \theta_{i,k}}, \tag{3.5}$$

$$A1_{i,k}^{(r)} = \eta A1_{i,k}^{(r-1)} + \beta \frac{\partial E}{\partial A1_{i,k}}, \tag{3.6}$$

$$B1_{i,k}^{(r)} = \eta B1_{i,k}^{(r-1)} + \beta \frac{\partial E}{\partial B1_{i,k}}, \tag{3.7}$$

$$A2_{i,k}^{(r)} = \eta A2_{i,k}^{(r-1)} + \beta \frac{\partial E}{\partial A2_{i,k}}, \tag{3.8}$$

$$B2_{i,k}^{(r)} = \eta B2_{i,k}^{(r-1)} + \beta \frac{\partial E}{\partial B2_{i,k}}. \tag{3.9}$$

To derive the gradient information of E with respect to each adjustable parameter in equations (3.4)-(3.9), we define

$$\frac{\partial E}{\partial I_{i,k}(u)} = \zeta_{i,k}, \tag{3.10}$$

$$\frac{\partial E}{\partial O_{i,k}(u)} = \xi_{i,k}. \tag{3.11}$$

Now, from equations (3.2), (3.3), (3.10) and (3.11), we have the partial derivatives of E with respect to adjustable parameters as follows:

$$\frac{\partial E}{\partial w_{i,j,k}} = \frac{\partial E}{\partial I_{i,k}(u)} \frac{\partial I_{i,k}(u)}{\partial w_{i,j,k}} = \zeta_{i,k} O_{j,k-1}(u), \tag{3.12}$$

$$\frac{\partial E}{\partial \theta_{i,k}} = \frac{\partial E}{\partial I_{i,k}(u)} \frac{\partial I_{i,k}(u)}{\partial \theta_{i,k}} = -\zeta_{i,k}, \tag{3.13}$$

$$\frac{\partial E}{\partial A1_{i,k}} = \frac{\partial E}{\partial O_{i,k}} \frac{\partial O_{i,k}}{\partial A1_{i,k}} = \xi_{i,k} e^{-B1_{i,k} \cdot I_{i,k}}, \tag{3.14}$$

$$\begin{aligned}\frac{\partial E}{\partial B1_{i,k}} &= \frac{\partial E}{\partial O_{i,k}} \frac{\partial O_{i,k}}{\partial B1_{i,k}} \\ &= -\xi_{i,k} \cdot A1_{i,k} \cdot I_{i,k} \cdot e^{-B1_{i,k} \cdot I_{i,k}},\end{aligned} \tag{3.15}$$

$$\frac{\partial E}{\partial A2_{i,k}} = \frac{\partial E}{\partial O_{i,k}} \frac{\partial O_{i,k}}{\partial A2_{i,k}} = \xi_{i,k} \cdot \frac{1}{1+e^{-B2_{i,k} \cdot I_{i,k}}}, \tag{3.16}$$

$$\frac{\partial E}{\partial B2_{i,k}} = \frac{\partial E}{\partial O_{i,k}(u)} \frac{\partial O_{i,k}(u)}{\partial B2_{i,k}} = \xi_{i,k} \cdot \frac{A2_{i,k} \cdot I_{i,k}(u) \cdot e^{-B2_{i,k} \cdot I_{i,k}(u)}}{\left(1+e^{-B2_{i,k} \cdot I_{i,k}(u)}\right)^2}. \tag{3.17}$$

And for (3.10) and (3.11) the following equations can be computed:

$$\zeta_{i,k} = \frac{\partial E}{\partial I_{i,k}(u)} = \frac{\partial E}{\partial O_{i,k}(u)} \frac{\partial O_{i,k}(u)}{\partial I_{i,k}(u)} = \xi_{i,k} \cdot \frac{\partial O_{i,k}(u)}{\partial I_{i,k}(u)}, \tag{3.18}$$

while

$$\frac{\partial O_{i,k}(u)}{\partial I_{i,k}(u)} = A1_{i,k} \cdot B1_{i,k} \cdot e^{-B1_{i,k} \cdot I_{i,k}(u)} + \frac{A2_{i,k} \cdot B2_{i,k} \cdot e^{-B2_{i,k} \cdot I_{i,k}(u)}}{\left(1+e^{-B2_{i,k} \cdot I_{i,k}(u)}\right)^2}, \tag{3.19}$$

and

$$\xi_{i,k} = \begin{cases} \sum_j \zeta_{j,k+1} w_{j,i,k+1}, & \text{if } 1 \le k < l; \\ O_{i,l}(u) - d_i(u), & \text{if } k = l. \end{cases} \tag{3.20}$$

All the training examples are presented cyclically until all parameters are stabilized, i.e., until the energy function E for the entire training set is acceptably low and the network converges.

4 Experiments to Justify NAF (2.1)

In this section we try to justify, empirically, the definition of our NAF (2.1). We show why we used a combination of the following two elementary functions as our activation function:

$$e^{-x^2}, \ \frac{1}{1+\mathrm{e}^{-x}}.$$

Our first experiment was to construct an ANN with NAF (2.1) to approximate a function of one variable:

$$f(x) = \sin x(x+1) + \cos(x^2 - 1) - 2 \quad x \in [0, 2\pi], \tag{4.1}$$

after training the learned NAF became:

$$\Psi(x) = 0.01e^{-2.11x^2} + \frac{3.39}{1+e^{-1.33x}}, \tag{4.2}$$

$$x, y \in [0, \ 2\pi]$$

$$(A1 = 0.02, B1 = 3.08, A2 = 1.37, B2 = 3.15).$$

Note that in the above equation, the coefficient $A1$ for the elementary function e^{-x^2} is a small real number 0.01, while the coefficient $A2$ for the other elementary function $\frac{1}{1+e^{-x}}$ is 3.39. As these coefficients are *learned* during training, we inferred that, for approximating the specific function (4.1), the elementary function $\frac{1}{1+e^{-x}}$ plays more important roles than e^{-x^2}. (This is also why we call our neuron activation function *adaptive*.) We compared our adaptive ANN with standard ANN (with sigmoid activation function), and the experimental results are displayed in Table 4.1 (RMS: Root-Mean-Squared). Table 4.1 clearly demonstrates the advantages of NAF (2.1) over traditional sigmoid function with regard to network size, training speed and simulation error.

Table 4.1. Adaptive ANN and standard ANN to approximate function (4.1) (HL: Hidden Layer)

Neural Network	No. HL	HL Nodes	Epoch	RMS
Adaptive ANN	1	2	1,000	0.001198
Standard ANN	1	2	5,000	0.834219
Standard ANN	1	5	5,000	0.087769
Standard ANN	1	11	5,000	0.009132
Standard ANN	1	12	5,000	0.004379

Next, we approximated the following function of two variables:

$$f(x, y) = \sin(x + y^2) + x(y+3) - 1, \qquad x, y \in [0, 3]. \tag{4.3}$$

In this example, we constructed an ANN with 2 input neurones, 1 hidden layer, and 1 output neurone. The learned NAF after training became:

$$\Psi(x) = 4.45e^{-1.44x^2} + \frac{0.001}{1+e^{-8.32x}}, \tag{4.4}$$

$$x, y \in [0, 3]$$

$$(A1 = 2.45, B1 = 2.18, A2 = 0.01, B2 = 4.89).$$

In this activation function (4.4), the learned coefficients *A1* and *A2* (for the two elementary functions) are 4.45 and 0.001, respectively. Similarly, this time we inferred that to approximate function (4.3), e^{-x^2} plays more important roles than $\frac{1}{1+e^{-x}}$. Table 4.2 demonstrates the advantages of NAF (2.1) over traditional sigmoid function with regard to network size, training speed and simulation error.

Table 4.2. Adaptive ANN with NAF and standard ANN to approximate function (4.3) (HL: Hidden Layer)

Neural Network	No. HL	HL Nodes	Epoch	RMS Error
Adaptive ANN	1	2	2,000	0.001011
Standard ANN	1	2	5,000	0.908875
Standard ANN	1	5	5,000	0.082567
Standard ANN	1	10	5,000	0.008129
Standard ANN	1	15	5,000	0.005998

5 Adaptive ANN Model for Financial Analysis

We used ANN with NAF to simulate the Commonwealth Bank of Australia share prices data. The economic data were downloaded from the Commonwealth Bank of Australia official web site (www.commbank.com.au). For our experiments at this stage we only used 254 data points (Nov 2000 to Nov 2001). The following NAF was obtained after training:

$$\Psi(x) = 4.99e^{-3.88x^2} + \frac{3.28}{1+e^{-4.19x}}, \tag{5.1}$$

$$(A1 = 8.37, B1 = 4.25, A2 = 10.08, B2 = 3.13).$$

In this situation, we inferred that for simulating this specific financial data set, e^{-x^2} and $\frac{1}{1+e^{-x}}$ are of equal importance. The detailed comparison between our adaptive ANN and traditional standard ANN for this example is illustrated in Table 5.1, Figure 5.1, and Figure 5.2.

Table 5.1. Adaptive ANN and Standard ANN to Simulate Share Prices (HL: Hidden Layer)

Neural Network	No. HL	HL Nodes	Epoch	RMS Error
Adaptive ANN	1	5	5,000	0.021099
Standard ANN	1	5	12,000	0.932928
Standard ANN	1	11	12,000	0.864908
Standard ANN	1	14	12,000	0.088726
Standard ANN	1	18	12,000	0.060934

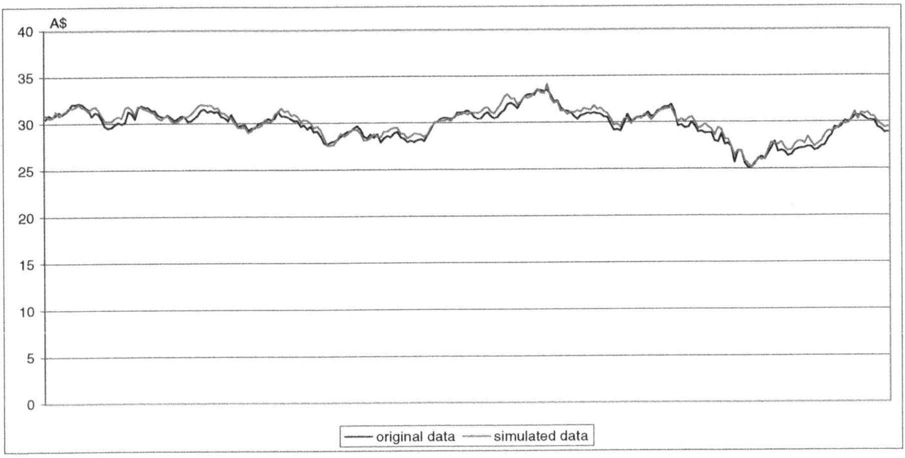

Fig. 5.1. Adaptive ANN with NAF to Simulate the Commonwealth Bank Share Prices (*Nov 2000 – Nov 2001*)

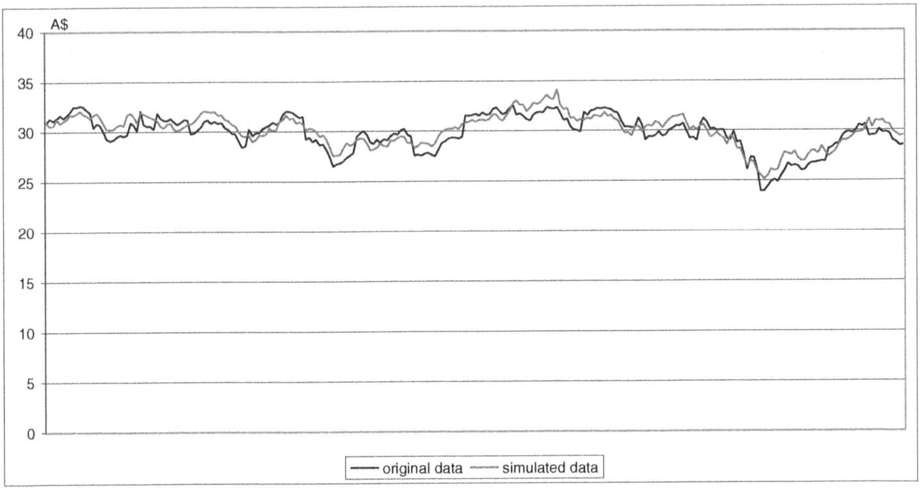

Fig. 5.2. Standard ANN to simulate the Commonwealth Bank Share Prices (Nov 2000 – Nov 2001)

6 Conclusions

In summary, by reviewing experiments described above and many other experiments that were not depicted in this paper, we concluded that we have created an adaptive higher order ANN model which is superior to existing standard ANN for the purpose of data mining. Our experiments exposed the advantages of our adaptive higher order ANN with NAF over traditional ANN such as increased training speed, much reduced network size and simulation error. Our next step will be to justify theoretically the proposed neuron-adaptive activation function for adaptive ANN model, and to explore the generalisation ability of ANN with NAF in financial forecasting.

References

1. Adriaans, P., Zantinge, D.:Data Mining. Addison-Wesley (1996)
2. Sarker, Ruhul A., Abbass, Hussein A., Newton, Charles S.: Data mining : A Heuristic Approach. Idea Group Pub./Information Science Publishing (2002)
3. Han., J., Kamber, M.: Data Mining : Concepts and Techniques. Morgan Kaufmann Publishers (2001)
4. Zhang, M., Xu, S., Fulcher, J.: Neuron-adaptive Higher Order Neural Network Models for Automated Financial Data Modelling. IEEE Transactions on Neural Networks **13** (1) (2002)
5. Vecci, L., Piazza, F., Uncini, A.: Learning and Approximation Capabilities of Adaptive Spline Activation Function Neural Networks. Neural Networks **11** (1998) 259-270
6. Chen, C.T., Chang, W.D.: A Feedforward Neural Network with Function Shape Autotuning. Neural Networks **9** (4) (1996) 627-641
7. Campolucci, P., Capparelli, F., Guarnieri, S., Piazza, F., Uncini, A.: Neural Networks with Adaptive Spline Activation Function. Proceedings of IEEE MELECON 96, Bari, Italy, (1996) 1442-1445
8. Hu, Z., Shao, H.: The Study of Neural Network Adaptive Control Systems. Control and Decision **7** (1992) 361-366
9. Yamada, T., Yabuta, T.: Remarks on a Neural Network Controller Which Uses an Auto-tuning Method for Nonlinear Functions. IJCANN **2** (1992) 775-780
10. Lee, Y.C., Doolen, G., Chen, H., Sun, G., Maxwell, T., Lee, H., Giles, C.L.: Machine Learning Using a Higher Order Correlation Network. Physica D: Nonlinear Phenomena **22** (1986) 276-306
11. Lippman, R.P.: Pattern Classification Using Neural Networks. IEEE Commun. Mag. **27** (1989) 47-64
12. Psaltis, D., Park, C.H., Hong, J.: Higher Order Associative Memories and Their Optical Implementations. Neural Networks **1** (1988) 149-163
13. Reid, M.B., Spirkovska, L., Ochoa, E.: Simultaneous Position, Scale, Rotation Invariant Pattern Classification Using Third-order Neural Networks. Int. J. Neural Networks **1** (1989) 154-159
14. Wood, J., Shawe-Taylor, J.: A Unifying Framework for Invariant Pattern Recognition. Pattern Recognition Letters **17** (1996) 1415-1422
15. Giles, C.L., Maxwell, T.: Learning, Invariance, and Generalization in Higher Order Neural Networks. Applied Optics **26** (23) (1987) 4972-4978
16. Redding, N.J., Kowalczyk, A., Downs, T.: Constructive Higher-order Network Algorithm That Is Polynomial Time. Neural Networks **6** (1993) 997-1010
17. Kosmatopoulos, E.B., Polycarpou, M.M., Christodoulou, M.A., Ioannou, P.A.: High-order Neural Network Structures for Identification of Dynamical Systems. IEEE Transactions on Neural Networks **6** (2) (1995) 422-431
18. Thimm, G., Fiesler, E.: High-order and Multilayer Perceptron Initialization. IEEE Transactions on Neural Networks **8** (2) (1997) 349-359
19. Rumelhart, D.E., McClelland, J.L.: Parallel Distributed Computing: Exploration in the Microstructure of Cognition. Cambridge, MA: MIT Press (1986)

A Comparison of Four Data Mining Models: Bayes, Neural Network, SVM and Decision Trees in Identifying Syndromes in Coronary Heart Disease

Jianxin Chen[1], Yanwei Xing[2], Guangcheng Xi[1], Jing Chen[1], Jianqiang Yi[1], Dongbin Zhao[1], and Jie Wang[2]

[1] Key Laboratory of Complex Systems and Intelligence Science
Institute of Automation, Chinese Academy of Sciences
100080, Beijing, China
{jianxin.chen,guangcheng.xi}@ia.ac.cn
[2] Guanganmen Hospital,Chinese Academy of Chinese Medical Science
100053, Beijing, China

Abstract. Coronary heart disease (CHD) is a serious disease causing more and more morbidity and mortality. Combining western medicine and Traditional Chinese Medicine (TCM) to heal CHD becomes especially necessary for medical society today. Since western medicine faces some problems, like high cost and more side effects. TCM can be a complementary alternative to overcome these defects. Identification of what syndrome a CHD patient caught has been a challenging issue for medical society because the core of TCM is syndrome. In this paper, we carry out a large-scale clinical epidemiology to collect data with 1069 cases, each of which must be a CHD instance but may be diagnosed as different syndromes. We take blood stasis syndrome (frequency is 69%) as an example, employ four distinct kinds of data mining algorithms: Bayesian model; Neural Network; Support vector machine and Decision trees to classify the data and compare their performance. The results indicated that neural network is the best identifier with 88.6% accuracy on the holdout samples. The next is support vector machine with 82.5% accuracy, a slight higher than Bayesian model with 82.0% counterpart. The decision tree performs the worst, only 80.4%. We conclude that in identifying syndromes in CHD, neural network can provide a best insight to clinical application.

1 Introduction

Coronary heart disease (CHD) is one of leading causes of morbidity and mortality in China [1]. Each year, about 1 million have heart attacks and 1 million die of CHD-related causes. Furthermore, it is on the rise and has become a true pandemic that respects no borders. Despite so many advance techniques and research achievements are emerging and have been applied into cure of CHD by modern medicine, death rate of CHD is still growing in China. This may

D. Liu et al. (Eds.): ISNN 2007, Part I, LNCS 4491, pp. 1274–1279, 2007.

be because cure by western medicine costs CHD patients so much that a lot of people can not afford. Moreover, side effects caused during taking western medicine treatments are great. Therefore, in China, most of CHD patients take Traditional Chinese Medicine (TCM) as a complementary alternative to heal CHD for its low cost and no side effects.

TCM has been always regarded as a key component in five thousand years of Chinese civilization history. TCM, whose core is syndrome, or 'Zheng' in Chinese, is on her way to modernization, aim-ing to be accepted, like western medicine, as sciences [2],[3]. The kernel of TCM is syndrome. Every herbal is prescribed in accord with syndromes, among which blood stasis syndrome is a ubiquitous and important subject, especially in arterial vascular diseases. Clinical data driven statistical research is becoming a common complement for biological or molecular research for CHD and the resulting achievements contribute significantly to medical society [4]. Predicting and diagnosing syndromes the CHD patients catch is one of most critical and challenging tasks for TCM society. It is the combination of the series effects of CHD, the potential benefits of the research outcomes and the purpose for further understanding the essential of CHD that motivate us to carry out a large-scale clinical epidemiology survey on CHD and apply data mining techniques to discovery knowledge in the surveyed data. The data mining algorithms subsume four kind of popular supervised classification algorithms: Bayesian method, Neural network, Support vec-tor machine (SVM), and decision tree.

Under the background of supervised classification problem, data mining algorithms mainly comprise of four broadly used kinds: Bayesian method, neural network, support vector machine and decision trees. Each kind is developed quickly and usually combines with each other to solve some hard problems [5]. Bayesian network (BN) is chosen from Bayesian method to perform classification here. Multilayer perceptron (MLP) with back propagation learning algorithm is selected from neural networks for its higher performance in doing classification than other algorithm, such as radial basis function, recurrent neural network. We used a well-known support vector machine algorithm, Platt's SMO Algorithm [6], as a representative of SVM classification since it can process both categorical and numerical variables. For decision tree kind, Quinlan's C4.5 algorithm [7] is employed for performing tasks in this paper.

2 Clinical Data Collection

We select 80 symptoms that are closely related to CHD and frequently appeared in the literatures concerning on CHD. Since TCM and western share the same symptoms' name, making the further understanding and processing of data more creditable. In the survey, the data set was recruited from 5 clinical centers located in two provinces (Beijing and Henan) from the same demographic area and at the same time from June 2005 to October 2006, where total 1069 patients who suffer from CHD were surveyed. Inclusion criteria contain four items:

1. Each case must accord with CHD's diagnosis criterions, which are instituted by American College of Cardiology (ACC) together with American Heart Association (AHA) in 2002.
2. Each case is verified by Coronary Artery Angiography as at least one branch of coronary artery main branch's diameter stenosis is larger than 70% or coronary artery left diameter stenosis is greater than 50%.
3. Each case must be attached with an informed consent signed by each patient.
4. Each patient must age more than 35.

Alternatively, exclusion criteria have two items:

1. Any patient catches ST-segment elevation acute myocardial infarction will be excluded.
2. Any patient also suffers from inter-current serious diseases such as liver or kidney's disease will be excluded.

Each symptom has four levels: none, light, middle, severe. Each case is diagnosed as a syndrome by experienced TCM experts. Each symptom is considered as an attribute, the diagnosed syndrome is taken as a response. Among 1069 cases, the blood stasis syndrome is 717 cases, occupying about 67% of whole data.

3 Classification Models

New data always needs already existed algorithms to test its performance. We employed four types of classification algorithms: Bayesian model, neural networks, SVM and decision trees. These models were jigged for inclusion in this research due to their popularity in the recently published documents. The following is a brief introduction to the four classification algorithms and the parameter setting of each model.

3.1 Bayesian Network

A Bayesian network (BN) is a graphical model that encodes probabilistic relationships among attributes of interest. Several advances have been made to ameliorate Bayesian network to fit all kinds of realistic problems [8]. We select stimulated annealing as method for searching network structures. Estimator is BayesNetEstimator, which is the base class for estimating the conditional probability tables of a Bayesian network once the structure has been learned.

3.2 Multilayerperceptron

Multilayer perceptron (MLP) are feed forward neural networks trained with the stan-dard back propagation algorithm. The number of hidden layers is equal to (number of attributes + number of classes) / 2, here is 41.Learning rate is 0.3.

3.3 Support Vector Machine

The SVM is a state-of-the-art maximum margin classification algorithm rooted in statistical learning theory [11],[12]. SVM performs classification tasks by maximizing the margin separating both classes while minimizing the classification errors. We used sequential minimal optimization algorithm to train the SVM here.

3.4 Decision Trees C4.5

As the name implies, this algorithm recursively separates observations in branches to construct a tree for the purpose of improving the prediction accuracy. In doing so, they use mathematical algorithms information gain to identify a variable and corresponding threshold for the variable that splits the input observation into two or more subgroups. This step is repeated at each leaf node until the complete tree is constructed. This step is repeated at each leaf node until the complete tree is constructed Confidence factor is set as 0.01. The minimum number of instances per leaf is 2.

4 Performance Evaluation and Results

4.1 Performance Measures

We employed three hackneyed performance measures: accuracy, sensitivity and specificity. A distinguished confusion matrix is obtained to calculate the three measures. Confusion matrix is a matrix representation of the classification results. the upper left cell denotes the number of samples classifies as true while they were true (i.e., TP), and lower right cell denotes the number of samples classified as false while they were actually false (i.e., TF). The other two cells (lower left cell and upper right cell) denote the number of samples misclassified. Specifically, the lower left cell denoting the number of samples classified as false while they actually were true (i.e., FN), and the upper right cell denoting the number of samples classified as true while they actually were false (i.e., FP).Once the confusion matrixes were constructed, the accuracy, sensitivity and specificity are easily calculated as: sensitivity = TP/(TP + FN); specificity = TN/(TN + FP). Accuracy = (TP + TN)/(TP + FP + TN + FN);10-fold cross validation is used here to minimize the bias produced by random sampling of the training and test data samples. Extensive tests on numerous data sets, with different learning strategies, have shown that 10 is about the right number of folds to get the best estimate of error, and there is also some theoretical evidence that backs this up [9],[10].

4.2 Results

Every model was evaluated based on the three measures discussed above (classification accuracy, sensitivity and specificity). The results were achieved using average value of 10 fold cross-validation for each algorithm. As shown in Fig. 1, Bayesian model (BN) achieved classification accuracy of 0.82 with a sensitivity of 0.87 and a specificity of 0.72. The SVM achieved classification accuracy of 0.825 with a sensitivity of 0.88 and a specificity of 0.716. The decision trees (C4.5) achieved a classification accuracy of 0.8920 with a sensitivity of 0.9017 and a specificity of 0.8786. However, the neural network model (MLP) preformed the best of the four models evaluated. MLP achieved a classification accuracy of 0.8920 with a sensitivity of 0.9017 and a specificity of 0.8786. Why neural

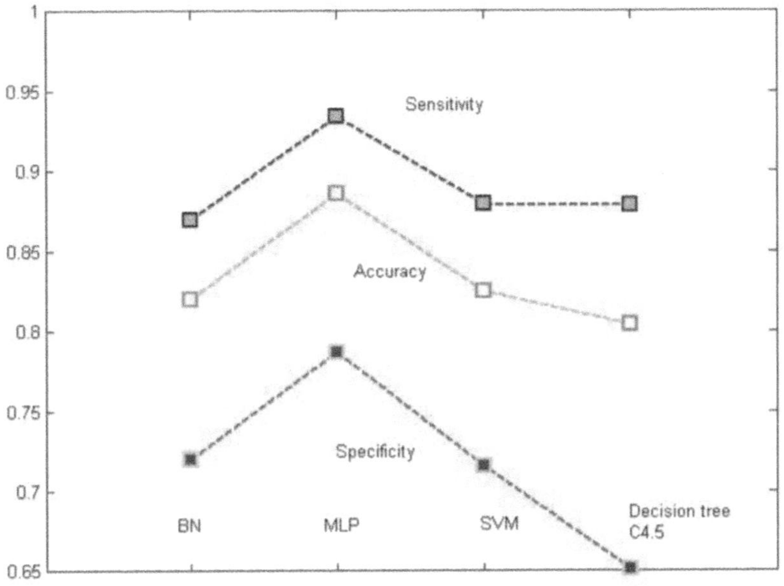

Fig. 1. Four models' three performance measures -Sensitivity, Specificity and Accuracy. Neural network is with best performance.

network (MLP) performs best? We know that syndrome is combination of symptoms. Indeed, syndrome is a diagnosed concept produced by mean of mapping symptoms to TCM expert's brains. So syndrome is identified by human brain and neural network is considered as best modeler of human brain. Furthermore, neural network can approximate arbitrary mapping, while syndrome is a mapping of symptoms. These two reasons may explain why neural network is with best performance.

5 Conclusion

In this paper, we employ four kinds of popular data mining models to perform classification task in identifying syndrome (blood stasis syndrome here) in CHD. The data was recruited from 5 clinical hospitals with whole 1069 cases. We used

10-fold cross validation to compute confusion matrix of each model and then calculate the three performance measures-sensitivity, specificity and accuracy to evaluate four kinds of models. We found that the Bayesian model (BN) achieved classification accuracy of 0.82 with a sensitivity of 0.87 and a specificity of 0.72. The SVM achieved classification accuracy of 0.825 with a sensitivity of 0.88 and a specificity of 0.716. The decision trees (C4.5) achieved a classification accuracy of 0.8920 with a sensitivity of 0.9017 and a specificity of 0.8786. However, the neural network model (MLP) preformed the best of the four models evaluated. MLP achieved a classification accuracy of 0.8920 with a sensitivity of 0.9017 and a specificity of 0.8786. We also explain why neural network perform best. The results showed here make clinical application more accessible, which will provide great advance in healing CHD.

Acknowledgments

The work has been supported by 973 Program under grant No. (2003CB517106 and 2003CB517103) and NSFC Projects under Grant No. 60621001, China.

References

1. World Health Organization.: World Health statistics Annual. Geneva, Switzerland,World Health Organization (2006)
2. Normile, D.: The New Face of Traditional Chinese Medicine. Science **299** (2003) 188-190
3. Xue, T.H., Roy, R.: Studying Traditional Chinese Medicine. Science **300** (2003) 740-741
4. Wang, Y.b., Zhang, W.L., etc.: VKORC1 Haplotypes are Associated with Arterial Vascular Diseases (Stroke, Coronary Heart Disease, and Aortic Dissection). Circulation **113** (2006) 1615-1621
5. Brudzewski, K., Osowski, S., Markiewicz, T.: Classification of Milk by Means of an Electronic Nose and SVM Neural Network. Sensors and Actuators B **98** (2004) 291-298
6. Keerth, S., Shevade, K., etc.: Improvements to Platt's SMO Algorithm for SVM Classifier Design. Neural Computation **13** (2001) 637-649
7. Quinlan, R.: C4.5: Programs for Machine Learning. Morgan Kaufmann Publishers, San Mateo, CA (1993)
8. Huang, C.L., Shih, H.C., Chao, C.Y.: Semantic Analysis of Soccer Video Using Dynamic Bayesian Network. IEEE. Transactions on Multimedia **8**(2006) 749-760
9. Witten, I.H., FrankMichalewicz, E.Z.: Data Mining: Practical Machine Learning Tools and Techniques. 2nd edn, Morgan Kaufmann, San Francisco (2005)
10. Delen, D., Walker, G., Kadam, A.: Predicting Breast Cancer Survivability. Artif. Intell. Med **34** (2005) 113-127
11. Vapnik, K.: Statistical Learning Theory. Wiley, New York (1998)
12. Graf, A., Wichmann, F., Bulthoff, H., etc.: Classification of Faces in Man and Machine. Neural Computation **18** (2006) 143-165

A Concept Lattice-Based Kernel Method for Mining Knowledge in an M-Commerce System

Qiudan Li, Chunheng Wang, Guanggang Geng, and Ruwei Dai

Key Laboratory of Complex Systems and Intelligence Science,
Institute of Automation, Chinese Academy of Sciences, Beijing
{qiudan.li,chunheng.wang,guanggang.geng,ruwei.dai}@ia.ac.cn

Abstract. With the vast amount of mobile user information available today, mining knowledge of mobile users is getting more and more important for a mobile commerce (M-commerce) system. Vector space model (VSM) is one of the most popular methods to achieve the above goal. Unfortunately, it can not identify the latent information in the user feature space, which decreases the quality of personalized services. In this paper, we present a concept-lattice based kernel method for mining the hidden user knowledge. The main idea is to employ concept lattice for constructing item proximity matrix, and then embed it into a kernel function, which transforms the original user feature space into a user concept space, and at last, perform personalized services in the user concept space. The experimental results demonstrate that our method is more encouraging than VSM.

1 Introduction

The advances in wireless technologies and the rapid growth of mobile users have led to the emergence of mobile commerce (m-commerce). M-commerce is defined as the transactions of commodities, services, or information over the Internet through the use of mobile handheld devices [1]. However, due to the inherent limitations of mobile devices and wireless network, personalization is the key to the success of M-commerce. Personalization in M-commerce is a matching process between mobile users and contents, according to the specific context information. It can be implemented by combining the user knowledge, content knowledge and context knowledge together. Therefore, mining the knowledge of similar mobile users from user profiles is getting more and more important for a mobile commerce (M-commerce) system.

Vector space model (VSM) is one of the most popular methods to achieve the above goal, where a mobile user is represented by the products the user has bought. Unfortunately, it can not identify the latent information in the user feature space, which decreases the quality of personalized services.

Kernel-based learning methods (Kms) are a state-of-the-art class of learning algorithms. They are modular systems, formed by a general purpose learning module and by a kernel, which defines the mapping into the concept space. The mapping corresponds to a matrix when it is linear. Kms have the advantage of accessing feature

D. Liu et al. (Eds.): ISNN 2007, Part I, LNCS 4491, pp. 1280–1285, 2007.

spaces that would be either too expensive or too complicated to represent [2]. Concept lattice, which is a concept analysis technique, provides an efficient way to mine the latent information among various items.

Inspired by the good ideas in [2]-[3],[5], in this paper, we present a concept-lattice based kernel method for mining the hidden user knowledge. The main idea is to employ concept lattice for constructing item proximity matrix, and then embed it into a kernel function, which transforms the original user feature space into a user concept space, and at last, perform personalized services in the use concept space. The proposed method takes good advantages of the kernel method and concept lattice theory, namely, the concept modeling and computation abilities without knowing the detailed representation of concept user space, which more adapts to the mobile commerce environment. The experimental results demonstrate that our method is more encouraging than VSM.

The rest of the paper is organized as follows: Section2 reviews related works in kernel methods, concept lattice theory and m-commerce; section3 describes the proposed concept-lattice based kernel method; section4 provides experimental results; section5 concludes the paper and points out some future works.

2 Related Works

In this section, we briefly present some of the research works related to kernel methods, concept lattice, and m-commerce applications.

Kernel methods provide a natural framework for pattern analysis in text, the similarity measure can often be obtained by borrowing methods from other research fields. [2] has studied the problem of introducing semantic information into a kernel based learning method by latent semantic indexing (LSI). [3]has introduced the Vector Space family of kernel methods. Besides the information retrieval fields, kernel methods have also been successfully applied to tasks including classification and regression.

The theory of concept lattice proposed by prof. Rudolf Wille is a powerful technology for content processing [4]. It has been applied to many research areas, such as information retrieval, knowledge representation and knowledge discovery, logic and AI, etc [4]-[7]. An excellent overview of the concept lattice can be found in [7]. However, their application in m-commerce systems has not been fully exploited. [8] has adopted the concept lattice to solve the scalability problem in collaborative filtering systems.

A four-level framework for m-commerce has been proposed in [9], which consists of m-commerce applications, user infrastructure, middleware and network infrastructure levels. The framework can be used to the design of a mobile commerce application, it is a foundation of our m-commerce system.

Compared with the existing works, in this paper, we focus on the user modeling of the m-commerce system, namely, exploring the potentials of concept lattices and kernel methods for m-commerce applications, to provide the personalized services from a new angle.

3 Proposed Concept Lattice-Based Kernel Method

In this section, we describe the proposed concept lattice-based kernel method for mining the hidden user knowledge. In an m-commerce application, mining the similar knowledge is to identify the nearest neighbors of active mobile user by analyzing the similarity relations between the user and the history mobile users. Vector space model (VSM) can be used to achieve this goal, where a mobile user is represented by the products the user has bought. Unfortunately, it can not identify the latent information in the user feature space, which decreases the quality of personalized services. The primary motivation behind the proposed method is the fact that users who have bought similar products may be also similar, therefore, identifying the latent information among products can help to mine the similar knowledge from user profiles accurately. Kernel method provides a way to map the original user feature space to the user concept space, then, the hidden latent information can be identified in the obtained concept space. The core problem of the kernel method is the construction mechanism of the product item proximity matrix. Inspired by the excellent works done in information retrieval areas [2],[3],[5], in this paper, we adopt the theory of concept lattice to solve the key problem.

The proposed method consists of three distinct components. The first component constructs the item proximity matrix that captures the latent information among various products items. The second component embeds it into a kernel function, which transforms the original user feature space into a user concept space. The third component applies the concept lattice-based kernel to perform personalized services. The details on these components are presented in the remainder of this section. The following formal descriptions on kernel and concept lattice are from [2],[4],respectively.

3.1 Building the Concept Lattice-Based Kernel

In this paper, the user feature space F is a subset of the vector space R^n. Let u, v be any two feature vectors. A kernel corresponds to dot products in user concept space C via a map φ, which can be defined as follows:

$$\varphi : F \to C,$$
$$u \mapsto \varphi(u),$$

that is

$$k(u,v) = <\varphi(u), \varphi(v)>.$$

Given a kernel, an associated user concept space can be constructed. We consider the simplest case, that is, φ is a linear transformation. Therefore, it can be written as:

$$\varphi(u) = Pu.$$

Where P is any appropriately shaped matrix. In our application problem, the matrix P is called item proximity matrix, whose entries encode the amount of latent relations between products of the collection.

For the VSM, P is equal to the identity matrix I. In this paper, the goal is to construct the more reasonable matrix P by the theory of concept lattice.

The matrix P can be constructed by the following steps:

First, we represent the user feature space F as an item context, which is a triple (I,U,R), where I is a set of items, U is a set of users, $R \subseteq I \times U$, written as iRu, meaning the user u has bought the product item i.

Then, a concept lattice can be constructed from the item context in an incremental way. In the hasse diagram-based representation, each node is a concept including extent and intent, there exists order relations among these concepts.

Definition1 Concept, extent and intent: Let (I,U,R) be the context, for $A \subseteq I$, the set of users that commonly has bought all the items in A is defined as $A' = \{u \in U | iRu \ for\ all\ i \in A\}$. For $B \subseteq U$, the set of items commonly have been bought by users in B is defined as $B' = \{i \in I | uRi \ for\ all\ u \in B\}$, A concept is a pair (A, B), where $A' = B$, $B' = A$. A and B are called the extent and the intent of the concept, respectively.

Definition2 Order relation, concept lattice: Let $C(I,U,R)$ be the set of concepts of context (I,U,R), and let (A_1,B_1), $(A_2,B_2) \in C(I,U,R)$, (A_1,B_1) is the sub-concept of (A_2,B_2) or (A_2,B_2) is the super-concept of (A_1,B_1), written as $(A_1,B_1) \le (A_2,B_2)$, if and only if $A_1 \subseteq A_2$ ($B_1 \supseteq B_2$). "$\le$" is an ordered relation of concepts. The partially ordered set $(C(I,U,R);\le)$ is a concept lattice.

Finally, we will identify the item node from the obtained item concept lattice. The definition of Item Node is given as follows.

Definition3 Item node: A concept (A, B) is called an item node of item s if s belongs to A and B contains all the attributes of s.

The distance between two item nodes i, j is defined as the length of the shortest path between the two nodes, written as $d(i, j)$. Here the shortest path means that there does not exist any other concept between any two adjacent item nodes of the path. Moreover, the path should not contain the node, of which extent or intent is empty. A similarity metric of item node i and j can be defined as: $S(i, j) = 1/(d(i, j)+1)$. Based on the induced similarity metric, the item proximity matrix is constructed.

For a $n\times1$ vector u_a and a $n\times t$ matrix U_h containing t history mobile users and n product items, by using the theory of concept lattice, we obtain a $n\times n$ matrix P, and next, we get a $1\times t$ vector q by multiplying u_a', P with U_h, which contains the similarity between the active mobile user and history mobile users. By setting the threshold and number of nearest neighbors, at last, we can get the specified neighbors.

From the above process, we can notice the proposed method has the following advantages: 1) the concept lattice theory has been used to construct the kernel function according to the m-commerce application problems, which can mine implicit relations among user feature spaces; 2) the method takes good advantage of the kernel method, namely, we need not know the explicit representation in user concept space, which more adapts to the m-commerce application, since the relations among the product items are complex; 3) the method can be easily extended, we can construct more complex kernels based on the proposed method.

3.2 Applying the Method for Personalized Services

For the found nearest neighbors of the active mobile user, we scan through the products that have been bought by these neighbors and have not been bought by the active mobile user, then, calculate the frequency count of the products, and finally, get the Top-N products items with largest count values.

4 Implementation of the Proposed Method in an M-Commerce System

Providing personalized services is important to the success of m-commerce system. The Personalization Engine Module consisting of user model, content model and context model is the core of the system. Under specific context information including time, location, weather, activity, etc, a personalized service can be derived by mining the knowledge of similar mobile users. We implement the proposed method in the system, it is one of the components in User Model.

The data set comes from users' rating data stored in Data Service Layer of the system. It reflects users' preferences on restaurants, namely, whether or not the user likes the restaurant under some context information. We randomly select 180 mobile users, 100 restaurants, and total 2160 ratings from the database. Each user has rated 10 or more restaurants. We divide the data set into 80% training set and 20% test set.

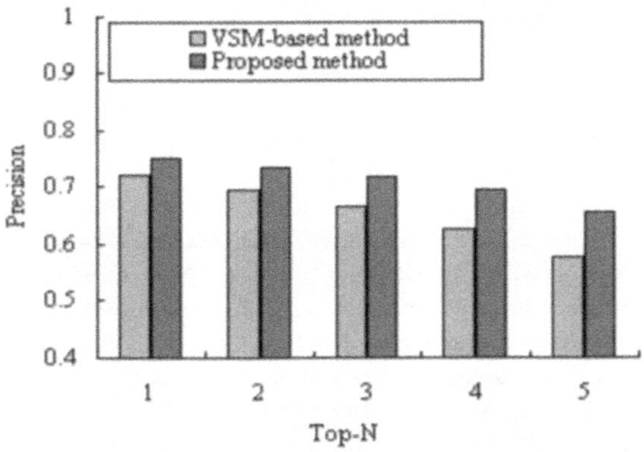

Fig. 1. Comparison of two methods

We report the results in terms of precision, the common measure used in information retrieval research. It is defined as follows [10]:

$$Precision = \frac{size\ of\ hit\ set}{size\ of\ Top-N\ set}.$$

The comparison of the proposed method and VSM-based method is shown in Fig.1. The results are average precisions on test set. In the experiment, we use $N = 1 \sim 5$ as the number of items be provided by the two methods.

As can be seen from the results, in terms of precision, the proposed method outperformed VSM-based method. The results confirmed that the proposed method can identify the latent information among user feature space, and mine the hidden knowledge from the user profiles.

5 Conclusions and Future Works

In this paper, we presented a novel concept-lattice based kernel method, and applied the method to mine the hidden user knowledge for an m-commerce system. The experimental results demonstrate that our method is more encouraging than VSM. The proposed method sets up a useful linkage between the theory of concept lattice and the machine learning method.

Future research will involve the following issues: firstly, to further investigate the more efficient method for constructing the item proximity matrix; secondly, to build the more reasonable kernel function based on the proposed method, which will make the concept space satisfy the practical needs better.

References

1. Shi, N.S.: Mobile Commerce Applications. Hershey, PA: Idea Group Pub (2004)
2. Cristianini, N., Shawe-taylor, J., Lodhi, H.: Latent Semantic Kernels. Journal of Intelligent Information Systems **18** (2002) 127-152
3. Shaw-Taylor, J., Cristianini, N.: Kernel Methods for Pattern Analysis. Cambridge, UK: Cambridge University Press (2004)
4. Carpineto, C., Romano, G.: Concept Data Analysis: Theory and Applications. John Wiley& Sons (2004)
5. Carpineto, C., Romano, G.: Order-Theoretical Ranking. Journal of the American Society for Information Science **51** (2000) 587-601
6. Godin, R., Gecsei, J., Pichet, C.: Design of a Browsing Interface for Information Retrieval. In Proceedings of the 12th International Conference on Research and Development in Information Retrieval (ACM SIGIR'89), Cambridge, MA: ACM (1989) 32-39
7. Uta Priss.: Formal Concept Analysis in Information Science, Annual Review of Information Science and Technology, ASIST **40** (2006) 521-543
8. du Boucher-Ryan, P., Bridge, D.: Collaborative Recommending using Formal Concept Analysis. Knowledge-Based Systems **19** (2006) 309-315
9. Varshney, U., Vetter, R.: Mobile Commerce: Framework, Applications and Networking Support. Mobile Networks and Applications **7** (2002) 185-198
10. Sarwar, B., Karypis, G., Konstan, J., Riedl, J.: Analysis of Recommendation Algorithms for E-Commerce. In Proceedings of ACM E-Commerce (2000) 158-167

A Novel Data Mining Method for Network Anomaly Detection Based on Transductive Scheme

Yang Li[1,2], Binxing Fang[1], and Li Guo[1]

[1] Institute of Computing Technology, Chinese Academy of Sciences, Beijing China, 100080
[2] Graduate School of Chinese Academy of Sciences, Beijing China, 100080
liyang@software.ict.ac.cn

Abstract. Network anomaly detection has been a hot topic in the past years. However, high false alarm rate, difficulties in obtaining exact clean data for the modeling of normal patterns and the deterioration of detection rate because of "unclean" training set always make it not as good as we expect. Therefore, we propose a novel data mining method for network anomaly detection in this paper. Experimental results on the well-known KDD Cup 1999 dataset demonstrate it can effectively detect anomalies with high true positives, low false positives as well as with high confidence than the state-of-the-art anomaly detection methods. Furthermore, even provided with not purely "clean" data (unclean data), the proposed method is still robust and effective.

1 Introduction

As an important branch of intrusion detection field, anomaly detection has been an active area of research in network security since it was originally proposed by Denning [1]. A lot of data mining methods have been proposed for this hotspot [2]. Anomaly detection algorithms have the advantage over misuse detection that they can detect new types of intrusions as deviations from normal usage. In this problem, given a set of normal data to train from, and given a new piece of test data, the goal of the intrusion detection algorithm is to determine whether the test data belong to "normal" or to an anomalous behavior. However, anomaly detection methods suffer from a high rate of false alarms. This occurs primarily because previously unseen (yet legitimate) system behaviors are also recognized as anomalies, and hence flagged as potential intrusions. Moreover, if the training set is contaminated by the "noisy" data, the detection performance of anomaly detection methods would deteriorate sharply.

In this paper, we present a novel data mining method for network anomaly detection. It is based on TCM-KNN (Transductive Confidence Machines for K-Nearest Neighbors) algorithm, which is successfully applied to pattern recognition, fraud detection and outlier detection [6], [7]. The most distinguished characteristic for it is that it need not construct a classifier as the traditional data mining methods and is immune to the effect of "noisy" data in training dataset, therefore, it has better detection performance than the traditional anomaly detection methods in practice. A series of experiments on the well-known KDD Cup 1999 dataset demonstrate our method has higher detection rate (also named true positive rates) and lower false alarm (also

D. Liu et al. (Eds.): ISNN 2007, Part I, LNCS 4491, pp. 1286–1292, 2007.

named false positive rates) than the state-of-the-art anomaly detection methods. Furthermore, it holds good performance even inferred by the "noisy" data in training set.

2 TCM-KNN Algorithm

Transductive Confidence Machines (TCM) introduced the computation of the confidence using algorithmic randomness theory [5]. Unlike traditional methods in data mining, transduction can offer measures of reliability to individual points, and uses very broad assumptions except for the iid assumption (the training as well as new (unlabelled) points are independently and identically distributed). There exists a universal method of finding regularities in data sequences. This p-value serves as a measure of how well the data supports or not a null hypothesis (the point belongs to a certain class). The smaller the p-value, the greater the evidence against the null hypothesis (i.e., the point is an outlier with respect to the current available classes). Users of transduction as a test of confidence have approximated a universal test for randomness (which is in its general form, non-computable) by using a p-value function called strangeness measure [4]. The general idea is that the strangeness measure corresponds to the uncertainty of the point being measured with respect to all the other labeled points of a class.

Imagine we have a intrusion detection training set $\{(x_{1,} y_1),...,(x_{n,} y_n)\}$, of n elements, where $X_i = \{x_i^1, x_i^2,..., x_i^n\}$ is the set of feature values (such as the connection duration time, the packet length, etc.) extracted from the raw network packet (or network flow such as TCP flow) for point i and y_i is the classification for point i, taking values from a finite set of possible classifications (such as normal, DoS attack, Probe attack, etc.), which we identify as $\{1,2,3,...,c\}$. We also have a test set of s points similar to the ones in the training set, our goal is to assign to every test point one of the possible classifications. For every classification we also want to give some confidence measures.

In this paper, we combine K-Nearest Neighbors (KNN) algorithm with TCM for TCM-KNN algorithm and it is noted that TCM can be combined with any other data mining methods such as SVM. We denote the sorted sequence (in ascending order) of the distances of point i from the other points with the same classification y as D_i^y. Also, D_{ij}^y will stand for the jth shortest distance in this sequence and D_i^{-y} for the sorted sequence of distances containing points with classification different from y. We assign to every point a measure called the individual strangeness measure. This measure defines the strangeness of the point in relation to the rest of the points. In our case the strangeness measure for a point i with label y is defined as

$$\alpha_{iy} = \frac{\sum_{j=1}^{k} D_{ij}^{y}}{\sum_{j=1}^{k} D_{ij}^{-y}}, \tag{1}$$

where k is the number of neighbors used. Thus, our measure for strangeness is the ratio of the sum of the k nearest distances from the same class to the sum of the k nearest distances from all other classes. This is a natural measure to use, as the strangeness of a point increases when the distance from the points of the same class becomes bigger or when the distance from the other classes becomes smaller.

Provided with the definition of strangeness, we will use equation (2) to compute the p-value as follows:

$$p(\alpha_t) = \frac{\#\{i : \alpha_i \geq \alpha_t\}}{n+1}, \tag{2}$$

where # denotes the cardinality of the set, which is computed as the number of elements in finite set. α_t is the strangeness value for the test point (assuming there is only one test point, or that the test points are processed one at a time), is a valid randomness test in the iid case. The proof takes advantage of the fact that since our distribution is iid, all permutations of a sequence have the same probability of occuring. If we have a sequence $\{\alpha_1, \alpha_2, ..., \alpha_m\}$ and a new element α_t is introduced then α_t can take any place in the new (sorted) sequence with the same probability, as all permutations of the new sequence are equiprobable. Thus, the probability that α_t is among the j largest occurs with probability of at most $\frac{j}{n+1}$.

3 Anomaly Detection Framework Based on TCM-KNN Algorithm

In standard TCM-KNN, we are always sure that the point we are examining belongs to one of the classes. However, in anomaly detection, we need not assign a point constructed from the network packets to a certain class, we only attempt to pinpoint the point in question is normal or abnormal.

Therefore, we propose to use a modified definition of α as follows:

$$\alpha_{iy} = \sum_{j=1}^{k} D_{ij}^{y} . \tag{3}$$

This new definition will make the strangeness value of a point far away from the class considerably larger than the strangeness of points already inside the class. With respect to our anomaly detection task, there are no classes available, then the above test can be administered to the data as a whole (it all belonged to one class - normal). Therefore, it only requires a single α_i per point (as opposed to computing one per class), and the τ used directly reflects the confidence level ($1-\tau$) is required.

The process of our new simplified TCM-KNN algorithm for anomaly detection is depicted in Figure 1:

Figure 2 shows us an anomaly detection framework based on TCM-KNN algorithm and it clearly illustrates how to apply the proposed method to the realistic anomaly detection scenario.

```
Parameters: k (the nearest neighbors to be used), m (size of training
dataset), τ (preset threshold), r (instance to be determined)
for i = 1 to m {
calculate D_i^y according to equation (1) for each one in training dataset
and store;
calculate strangeness α according to equation (3) for each one in
training dataset and store;
        }
calculate the strangeness for r according to equation (3);
calculate the p-values for r according to equation (2);
if (p≤τ)
determine r as anomaly with confidence 1−τ and return;
else
claim r is normal with confidence 1−τ and return;
```

Fig. 1. Pseudocode of the TCM-KNN algorithm for anomaly detection

The framework includes two phases: training phase and detection phase. In the first phase, three important jobs should be considered:

a) Data collection for modeling: representative data for normal network behaviors should be collected for our method to modeling. It is worth noting here that as anomaly detection, attack data is no need for us to collect.

b) Feature selection & vectorlization: to meet the requirement of TCM-KNN which mainly depends on the distance calculation based on vectors, feature selection and vectorlization work should be employed. For instances, the duration time of a TCP connection, the ratio between the number of SYN packets, etc. might be selected for the features. They are mostly the same as those in KDD Cup 1999 dataset whose connections meta information have been extracted as 41 features.

c) Modeling by TCM-KNN algorithm: for the last step, TCM-KNN algorithm introduced in this paper then calculates the strangeness and p-value for each instance in the training dataset as discussed in Figure 1, thus to construct the anomaly detection engine.

For the detection phase, all the real-time data collected from the network also should be preprocessed to vectors according to the selected features having been acquired in training phase, then would be directed to the anomaly detection engine based on TCM-KNN, benign or malicious traffic would be determined.

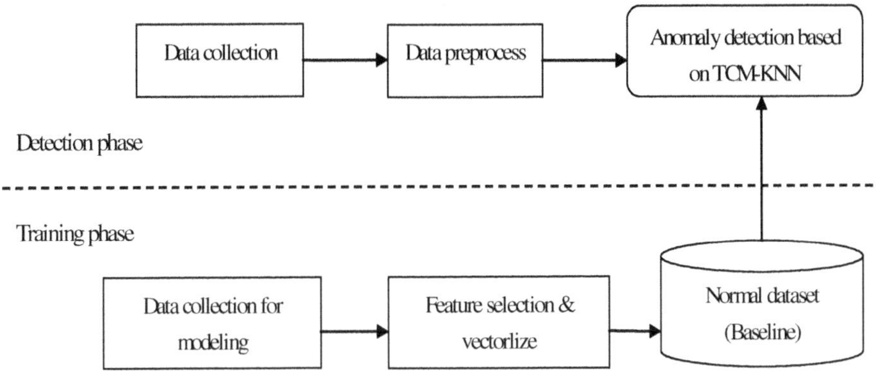

Fig. 2. Anomaly Detection Framework Based on TCM-KNN

4 Experimental Results

4.1 Dataset and Preprocess

In our experiments, we select the well-known KDD Cup 1999 dataset (KDD 99) [8] as our test dataset. It includes connections information summarized from the original TCP dump files. A connection is a sequence of TCP packets starting and ending at some well defined times, between which data flows to and from a source IP address to a target IP address under some well defined protocol. Each connection is labeled as either normal, or as an attack, with exactly one specific attack type. Each connection record consists of about 100 bytes. The attacks contain 24 different types of attacks that are broadly categorized in four groups such as Probes, DoS (Denial of Service), U2R (User to Root) and R2L (Remote to Local).

Before beginning our experiments, we preprocessed the dataset. First, we normalized the dataset. For the numerical data, in order to avoid one attribute will dominate another attribute, they were normalized by replacing each attribute value with its distance to the mean of all the values for that attribute in the instance space. For discrete or categorical data, we represent a discrete value by its frequency. That is, discrete values of similar frequency are close to each other, but values of very different frequency are far apart. As a result, discrete attributes are transformed to continuous attributes.

4.2 Experimental Results

In the contrast experiments between TCM-KNN and the most distinguished anomaly detection methods proposed by authors in [3], we used the sampled "noisy" dataset for training and test (it includes 2048 normal instances and 1870 attack instances). We adopted tenfold cross-validation approach to make the experiment. For the unsupervised anomaly detection algorithms, we set their parameters as the same in [3] for the convenience of comparison. For our TCM-KNN, k is set 50 and τ 0.05 (therefore, the confidence level is 0.95). Figure 3 shows the comparison results of them. It is clear that our method demonstrates higher TP and especially the lower FP than the other three methods.

Moreover, we also use both "clean" dataset and "unclean" dataset for training, to test the adaptive performance of our TCM-KNN algorithm. The result is depicted in Table 1. It clearly shows that a little difference can be observed when we use the two types of training dataset. It strongly demonstrates the proposed TCM-KNN method can be a good candidate for anomaly detection in realistic network environment than the other three methods, because acquiring purely "clean" dataset for training is often impossible and the relatively "unclean" dataset is reasonable. Therefore, a robust detection performance in such a "noisy" network environment is a necessity for anomaly detection method. The results demonstrate our TCM-KNN method has such a good performance.

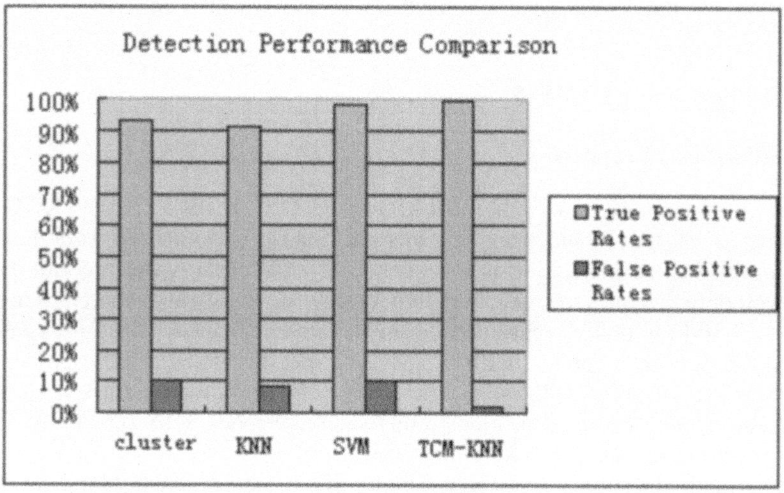

Fig. 3. Detection performance comparison results between TCM-KNN and the other three distinguished anomaly detection methods. The left bar for each method denotes TP (true positive rates) and the right one denotes FP (false positive rates).

Table 1. Running results using both "clean" and "unclean" training dataset

	clean dataset	unclean dataset
TP	99.44%	99.42%
FP	1.74%	2.37%

5 Conclusions and Future Work

In this paper, we propose a novel anomaly detection method based on TCM-KNN data mining algorithm. Experimental results demonstrate its effectiveness and advantages over traditional unsupervised anomaly detection methods. As our preliminary work, a lot of work should be improved in the future. Among them, how to reduce the computational cost of TCM-KNN is the most important one. Data reduction and feature selection will be focused around and thereafter the real application of TCM-KNN for anomaly detection would be carried out.

References

1. Denning, D.E.: An Intrusion Detection Model. IEEE Transactions on Software Engineering (1987) 222-232
2. Lee, W., Stolfo, S.J.: Data Mining Approaches for Intrusion Detection. Proceedings of the 1998 USENIX Security Symposium (1998)
3. Eskin, E., Arnold, A., Prerau, M., Portnoy, L., Stolfo, S.J.: A Geometric Framework for Unsupervised Anomaly Detection: Detecting Intrusions in Unlabeled Data. In D. Barbara and S. Jajodia (editors), Applications of Data Mining in Computer Security, Kluwer (2002)
4. Gammerman, A., Vovk, V.: Prediction algorithms and confidence measure based on algorithmic randomness theory. Theoretical Computer Science (2002) 209-217
5. Li, M., Vitanyi, P.: Introduction to Kolmogorov Complexity and its Applications. 2nd Edition, Springer Verlag (1997)
6. Proedru, K., Nouretdinov, I., Vovk, V., Gammerman, A.: Transductive confidence machine for pattern recognition. Proc. 13th European conference on Machine Learning (2002) 381-390
7. Daniel Barbará, Carlotta Domeniconi, James P. Rogers: Detecting outliers using transduction and statistical testing. In: Proceedings of the 12th ACM SIGKDD international conference on Knowledge discovery and data mining. USA (2006) 55-64
8. Knowledge discovery in databases DARPA archive. Task Description. http://www.kdd.ics.uci.edu/databases/kddcup99/task.html

Handling Missing Data from Heteroskedastic and Nonstationary Data

Fulufhelo V. Nelwamondo and Tshilidzi Marwala

School of Electrical and Information Engineering,
University of the Witwatersrand
Private Bag 3, Wits, 2050, South Africa
{f.nelwamondo,t.marwala}@ee.wits.ac.za

Abstract. This paper presents a computational intelligence approach for predicting missing data in the presence of concept drift using an ensemble of multi-layered feed forward neural networks. An algorithm that detects concept drift by measuring heteroskedasticity is proposed. Six instances prior to the occurrence of missing data are used to approximate the missing values. The algorithm is applied to simulated time series data sets resembling non-stationary data from a sensor. Results show that the prediction of missing data in non-stationary time series data is possible but is still a challenge. For one test, up to 78% of the data could be predicted within 10% tolerance range of accuracy.

1 Introduction

The problem of missing data has intensively been researched but remains largely unresolved. One of the reasons for this is that the complexity of approximating missing variables is highly dependent on the problem domain. This complexity furthermore increases when data is missing in an online application where data has to be used as soon as it is available. A challenging aspect of the missing data problem is when data is missing from a time series that exhibit nonstationarity. Most learning techniques and algorithms that have been developed thus far assume that data will continuously be available. Furthermore, they assume that data conform to a stationary distribution.

There are many nonstationary quantities in nature that fluctuate with time. Common examples are the stock market, weather, heartbeats, seismic waves and animal populations. There are some engineering and measurement systems that are dedicated at measuring nonstationary quantities. Such instruments are not immune to failures. Computational Intelligence (CI) approaches have previously been proposed for non-stationary data such as the stock market prediction, but the volatility of the data makes the problem complex. The 2003 Nobel Prize Laureates in Economics, Granger [1] and Engle [2] had an excellent contribution to nonlinear data. Granger showed that the traditional statistical methods could be misleading if applied to variables that wander over time without returning to some long-run resting point [1]. Engle on the other hand had a ground-breaking discovery of Autoregressive Conditional Heteroskedasticity

D. Liu et al. (Eds.): ISNN 2007, Part I, LNCS 4491, pp. 1293–1302, 2007.

(ARCH), a method to analyse unpredictable movements in financial market prices and also applicable in risk assessment [2]. Many techniques for solving missing data have been developed and discussed at length in literature [3]. However, no attempt has yet been made to approximate missing data in strictly non-stationary processes, where concepts change with time. The challenge with missing data problems in this application is that the approximation process must be complete before the next sample is taken. Moreover, more than one technique may be required to approximate the missing data due to drifting of concepts. As a result, the computation time, amount of memory required and the model complexity may grow indefinitely as new data continually arrive [4].

This paper challenges the above mentioned problems by proposing a computational intelligence technique that uses an ensemble of of multi-layer perceptons feedfoward neural networks. In the proposed algorithm, concept drift is detected by measuring heteroskedasticity. The proposed technique learns new concepts incrementally. The resulting state of knowledge is then used in predicting the missing values. A brief definition of nonstationary and chaotic systems will be discussed, followed by a discussion on the principle of concept drift and its detection techniques. A formal algorithm of approximating missing data in the presence of concept drift will then be presented, followed by the empirical evaluation and results.

2 Definition of Nonstationary and Chaotic Systems

Nonstationarity is a common property to many macroeconomic and financial time series data [5]. Nonstationarity means that a variable has no clear tendency to return to a constant value or a linear trend. Often, stationarity is defined in terms of the mean and the auto-covariance. If data is randomly sampled and is found to have a constant mean, and the auto-covariance is a function that depends only on the distance in placement, the data is considered to be stationary, or more formally, wide-sense stationary. Chaotic systems on the other hand are nonstationary systems that are highly sensitive to initial conditions. This makes such systems very difficult to predict. The reason for this difficulty is that *concepts drift* before reaching a particular observation.

Some work has been done for dealing with missing data in nonstationary time series [6]. In most cases, attempts are made to first make the data stationary using deferencing procedures [6]. In cases of missing data, applying the differencing techniques proposed in literature is not feasible for solving the problem of missing data [7]. An interesting method was proposed by Stefanakos and Anthnassoulis [6], which operates by completing missing values at the level of uncorrelated residuals after removing any systematic trends such as periodic components. Their method [6], is complex and only works well with seasonal data. It therefore would not be precise in cases where the concept being predicted or learned changes with time.

3 The Principle of Concept Drift

The principle of concept drift implies that the concept about which data is obtained may shift from time to time. Predicting values of a rapidly drifting concept is not possible if the concept changes each time step without restriction [4]. The rate of concept drift is defined as the probability that the target function disagrees over two successive examples [8]. There are two types of concept drifts that have been reported in literature. These types are categorized by the rate of the drift and are referred to as *sudden* and *gradual* concept drift.

One drawback of concept drift is that for a high volume of non-stationary data streams, the time it takes to predict may grow indefinitely [4]. In all cases of concept drift, incremental methods that continuously revise and refine the approximation model need to be devised and these methods need to incorporate new data as they arrive. This can be achieved by continually using recent data while not forgetting past data. However, in some cases, past data might be invalid and may need to be forgotten [4]. Harries and Sammut [9] have developed an off-line method for partitioning data streams into a set of time-dependent conceptual clusters. Their approach was, however, aimed at detecting concept drift in off-line systems. This work looks at a technique of detecting concept drift in an on-line application.

4 Concept Drift Detection Using Heteroskedasticity

Techniques of detecting concept drift are quite essential in time series data. The biggest challenge to this task is due to data being collected over time. Ways of detecting concept drift may vary in accordance to the pattern at which the concept is drifting. In most cases, the use of a window, where old examples are forgotten has proved to be sufficient [4][8]. Known examples of window based algorithms include Time-Window Forgetting, FLORA and FRANN [10]. A cyclically drifting concept exhibits a tendency to return to previously visited states. However, there are many algorithms such as STAGGER [11] and FLORA 3 [12] that have been developed to handle cyclic concept drift. In this kind of drift, old examples need not be forgotten as they may reappear at a later stage. An effective missing data estimator must be able to track such changes and to quickly adapt to them. In light of this challenge, this work proposes the use of heteroskedasticity as a means of detecting concept drift.

Heteroskedasticity occurs when the variables in a sequence have differing variances. Heteroskedasticity can arise in a variety of ways such as changes in behaviours of data under different conditions. Heteroskedasticity has been modelled as Autoregressive Conditional Heteroskedasticity (ARCH) or Generalized Autoregressive Conditional Heteroskedasticity (GARCH). Only recently, a new model has been developed and this model is Non-stationary Non-linear Heteroskedasticity (NNH) that assumes stochastic volatility [13]. For a volatile NNH model, we consider the sample auto correlations of the squared processes of obtaining the data from the sensor. The sample autocorrelations are defined as [13]:

$$R^2_{nk} = \frac{\sum_{k+1}^{n}(y_t^2 - \overline{y}_n^2)(y_{t-k}^2 - \overline{y}_n^2)}{\sum_{t=1}^{n}(y_t^2 - \overline{y}_n^2)^2} \tag{1}$$

where $\overline{y}_n^2$ denotes the sample mean of y_t^2 and $y_t = \sigma_t \epsilon_t$. We assume ϵ to be independent identically distributed (iid) (0,1) and is updated using filtration γ denoting information available at time t whereas σ on the other hand is adapted to γ_{t-1}.

The NNH model therefore specifies the conditional heteroskedasticity as a function of some explanatory variables, completely in parallel with the conventional approach. This work considers an aspect of NNH that the variable affecting the conditional heteroskedasticity is non-stationary and typically follows a random walk [13].

5 Missing Data Approximation in the Presence of Concept Drift

This section will present an algorithm used in this investigation.

5.1 Learning and Forgetting

The most common technique for learning is through the use of a *window* and this technique only trusts the most recent examples. Examples are added to the window as they arrive and oldest ones removed from the window. In the simplest case, the window will be of a fixed size, however, adaptive windows have also been reported in the literature [10]. It is very important to choose the window size very well as small windows will help in fast adaptation to new concepts, meanwhile, bigger windows will offer a good generalization. In this case, the choice of the window size is a compromise between fast adaptability and good generalization [14].

The idea of forgetting an example has been criticized for weakening of the existing description items [10]. This kind of forgetting also assumes that only the latest examples are relevant, which might not be the case. Helmbold and Long [8] have however shown that it is sufficient to use a fixed number of previous examples. An algorithm that removes inconsistent examples more efficiently will manage to track concept sequences that change more rapidly [10]. In this work, the window is chosen such that it is not too narrow to accommodate a sufficient number of examples. Again, the window size is chosen to avoid slowing down the reaction to concept drift.

5.2 Algorithm Description

The proposed algorithm uses an ensemble of regressors and avoids discarding old knowledge by discarding old networks. Instead, networks are stored and ranked according to a particular concept. The algorithm is divided into three sections namely training, validation and testing as described below.

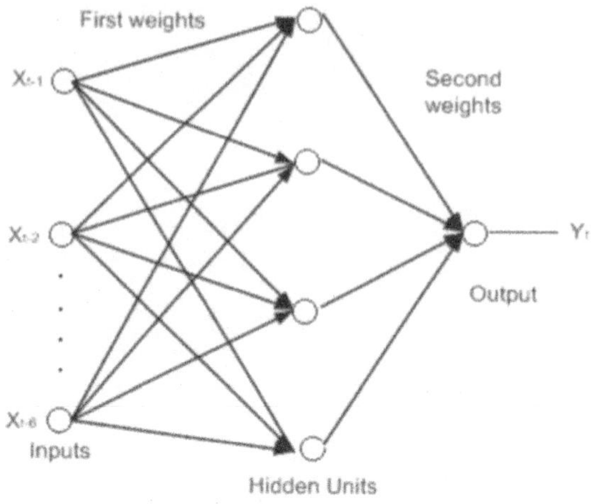

Fig. 1. MLP network used to predict missing data. Six previous data samples are used to predict the current missing value

Training: Batch learning is initially used. In this training, each missing datum is predicted using the past i instances where $i = 6$ in the application of this work. This implies that the window size is fixed at six samples. While sliding this window through the data, the heteroskedasticity of each window is calculated. All vectors are then grouped according to their heteroskedasticity range. This process result in disordering the sequence of the data. An ensemble of neural networks to predict data for a particular heteroskedasticity is then trained. An MLP neural network as shown in Fig. 1 is used.

In the entire range of heteroskedasticity [0, 1], a subrange of length 0.05 was used and various neural networks were trained. This practice leads to 20 subranges and as a result, 20 trained neural networks. In a case where data does not have a heteroskedastiity value of a particular subrange, no network will be trained for such a subrange. In this paper, each network was assigned to a subrange and was optimized for such range. The objective here was to have at least one neural network designed for each individual subrange. However, this does not imply that only one network will be used in a particular subrange as we need to add diversity to the system. The next step is validation.

Validation: All networks created in the training section above are subjected to a validation set containing all the groupings of data. Each regressor is tested on all groups and weights are assigned accordingly. The weight is assigned using the weighted majority scheme given as [15]:

$$\alpha_k = \frac{1 - E_k}{\sum_{j=1}^{N}(1 - E_k)} \tag{2}$$

Table 1. An illustration of how weights are assigned to each neural network after validation

Network	range 1	range 2	range 3	range 4	...	range 20
1	$\alpha_1^{(1)}$	$\alpha_2^{(1)}$	$\alpha_3^{(1)}$	$\alpha_4^{(1)}$	...	$\alpha_{20}^{(1)}$
2	$\alpha_1^{(2)}$	$\alpha_2^{(2)}$	$\alpha_3^{(2)}$	$\alpha_4^{(2)}$	...	$\alpha_{20}^{(2)}$
3	$\alpha_1^{(3)}$	$\alpha_2^{(3)}$	$\alpha_3^{(3)}$	$\alpha_4^{(3)}$	...	$\alpha_{20}^{(3)}$
$\vdots$	$\vdots$	$\vdots$	$\vdots$	$\vdots$	$\ddots$	$\vdots$
20	$\alpha_1^{(20)}$	$\alpha_2^{(20)}$	$\alpha_3^{(20)}$	$\alpha_4^{(20)}$	...	$\alpha_{20}^{(20)}$

where E_k is the estimate of model k's error on the validation set. This leads to each network having 20 weights, forming a weight vector as shown in Table 1.

Testing: When missing data are detected, i instances before the missing data are used to create a vector of instances. The heteroskedasticity of this vector is evaluated. From all the networks created, only those networks that have high weights assigned to them in the validation set for the same range of heteroskedasticity are chosen. In this application, all available networks are used and the missing values are approximated as shown below

$$f(x) = y \equiv \sum_{k=1}^{k=N} \alpha_i f_k(x) \tag{3}$$

where α is the weight assigned during the validation stage when no data were missing and N is the total number of neural networks used. For a given network, the weights are normalised such that $\sum_{i=1}^{i=N} \alpha_i \approx 1$.

After *enough* new instances have been sampled, the training process is repeated. The next section reports on the evaluation of the algorithm on two data sets.

6 Empirical Evaluation

6.1 Simulated Data Set

Case study 1: Firstly the algorithm proposed in Section 5.2 above is evaluated on the time series data produced by numerical simulation. A sequence of uncorrelated Gaussian random variables is generated with zero mean and variance of 0.108 as done by Stephanos and Anthanassoulis [6]. In this study, data is simulated as if it is coming from a sensor that measures some variable that exhibit nonstationary characteristics. The data is made to show some cyclic behavior that simulates a cyclic concept drift. Fig. 2(A) shows a sample of the simulated data.

Case study 2: The second test data was created using the Dow Jones stock market data. The stock market is well known for being difficult to predict as it exhibits nonstationarity. The opening price of the Dow Jones stock is also simulated as some data collected from some sensor and sampled at a constant interval. A sample of this data is shown in Fig. 2(B).

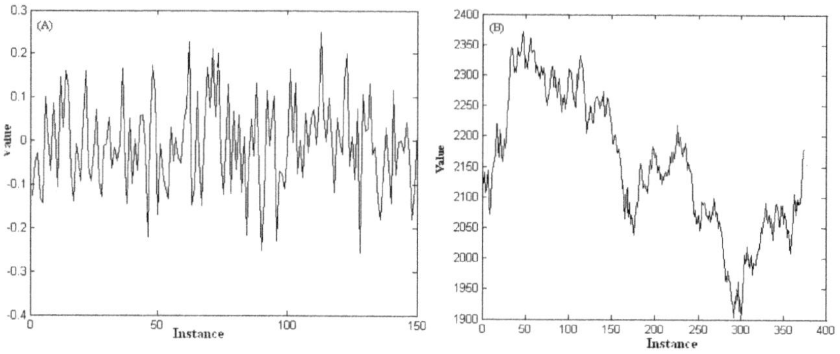

Fig. 2. Sample data with (A) cyclic concept drift and (B) gradual and sudden concept drift

The relative performance of the algorithm is measured by how close the prediction is to the actual data. The results obtained with this data are summarised in the next section.

7 Experimental Results

Firstly the effect of the number of regressors on the error was evaluated. Mean Square Error was computed for each prediction and results are presented in figure 3. Performance in terms of accuracy is shown in Fig.4 for the study in case 1 and this figure evaluates predictability of missing sensor values within 10% tolerance.

It can be seen from both figures that as the number of estimators is increased, the Mean Squared Error reduces. However, there is a point at which increasing the number of regressors does not significantly affect the error. For each case study, the algorithm was tested with 500 missing points and the same algorithm was used to estimate the missing values. Performance was calculated based on how many missing points are estimated within a given percentage tolerance as follows:

$$Accuracy = \frac{n_\tau}{N} \times 100\% \tag{4}$$

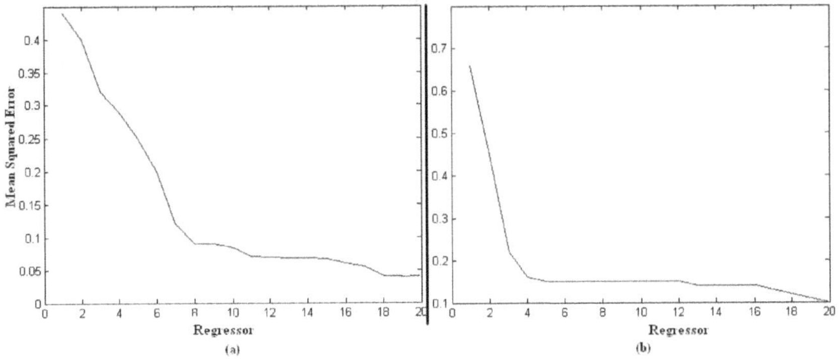

Fig. 3. Effect of the number of regressors on the Mean Square Error for (a) the simulated data and (b) the real data of the from the stock market

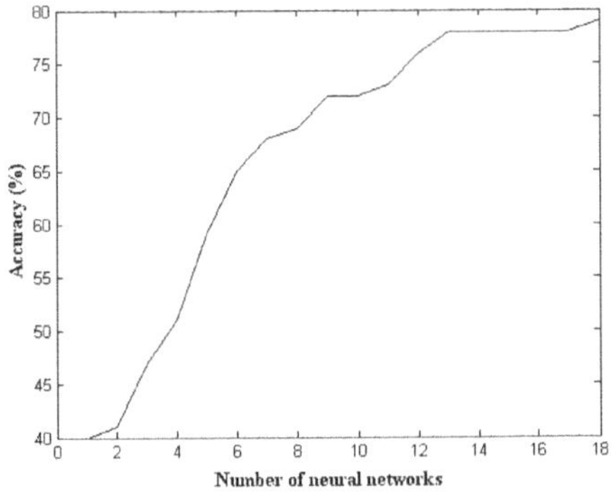

Fig. 4. Effect of the number of regressors on the prediction accuracy for Case study 1

where n_τ is the number of predictions within a 10% tolerance and N is the total number of instances being evaluated. Results are summarised in Table 2. In addition, the correlation coefficients between the missing data and the estimated data are computed and the results are shown in Table 2.

Table 2. Results obtained from both case studies and the Correlation coefficients between the estimated and the actual values for prediction within 10%

Case study	Estimate within 10%	Estimate within 5%	Corr Coefficient
1	78%	41%	0.78
2	46%	25%	0.26

Results in Table 2 show that prediction of missing data when there is a large concept drift is not very accurate. It can be seen that there is poor correlation between the estimated data and the actual data for Case study 2. This point was was further investigated in this study, paying particular attention to the data set of Case study 2. The best results obtained in estimation of missing data in that case are shown in Fig. 5. It was observed that there is a time lag of approximately two instances. Pan *et al* [16] also found a lag in their stock market prediction. Findings in this paper show that this lag is responsible for the poor correlation coefficient reported in Table 2. Results obtained here give some light to the use of heteroskedasticity as a measure of concept drift.

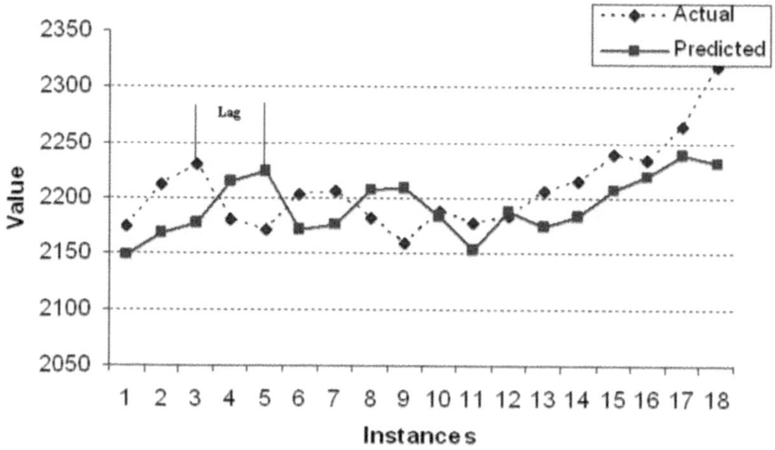

Fig. 5. Best results obtained with the data set of case study 2

8 Discussion and Conclusion

This paper presented an algorithm to approximate missing data in non- stationary time series that may also possess concept drift. An ensemble of estimators has been used and the final output was computed using the weighted approach of combining regression machines. Results show that the predictability increases as the number of neural networks used is increased. This is seemingly caused by the concept drift. The concept simply drifts from a region of one mastered by one neural network to a region mastered by another. The approach used in this work can therefore work as it covers the entire range of possible outcomes. However, the major disadvantage is that data covering the entire range needs to be there in the beginning. Furthermore, an ensemble with a large number of neural network slows down the prediction which then reduces the usability of the method in fast sampled data. This method can be computationally expensive as it requires a large memory to store all the networks and their assigned weights.

References

1. Granger, C.W.J.: Time series analysis, cointegration and applications. Nobel Price Lecture (2003) 360–366
2. Engle, R.: Autoregressive conditional heteroskedasticity with estimates of the variance of uk inflation. Econometrica **50** (1982) 987–1008
3. Little, R.J.A., Rubin, D.B.: Statistical Analysis with Missing Data. Wiley, New York (1987)
4. Last, M.: Online classification of nonstationary data streams. Intelligent Data Analysis **6** (2002) 129–147
5. Engle, R.F.: Time-series econometrics: Cointegration and autoregressive conditional heteroskedasticity. Advanced Information on the Bank of Sweden Prize in Economic Sciences in Memory of Alfred Nobel (2003) 1–30
6. Stefanakos, C., Athanassoulis, G.A.: A unified methodology for analysis, completion and simulation of nonstationary time series with missing values, with application to wave data. Applied Ocean Research **23** (2001) 207–220
7. Ljung, G.M.: A note on the estimation of missing values in time series. Commun Statist **18**(2) (1989) 459–465
8. Helmbold, D.P., Long, P.M.: Tracking drifting concepts using random examples. In: Proceedings of the Fourth Annual Workshop on Computational Learning Theory, AAAI (1991) 13–23
9. Harries, M., Sammut, C.: Extracting hidden context. Machine Learning **32** (1988) 101–126
10. Kubat, M., Widmer, G.: Learning in the presence of concept drift and hidden contexts. Machine Learning **23** (1996) 69–101
11. Schlimmer, J.C., Granger, R.H.: Incremental learning from noisy data. Machine Learning **1** (1986) 317–354
12. Widmer, G., Kubat, M.: Effective learning in dynamic environments by explicit context tracking (1993)
13. Park, J.Y.: Nonlinear nonstationary heteroskedasticity. Journal of Econometrics **110** (2002) 383–415
14. Scholz, M., Klinkenberg, R.: An ensemble classifier for drifting concepts. In: Proceedings of the Second International Workshop on Knowledge Discovery in Data Streams, Porto, Portugal, ECML (2005)
15. Merz, C.J.: Using correspondence analysis to combine classifiers. Machine Learning (1997) 1–26
16. Pan, H., Tilakaratne, C., Yearwood, J.: Predicting australian stock market index using neural networks, exploiting dynamical swings and inter-market influences. Journal of Research and Practice in Information Technology **37** (2005) 43–55

A Novel Feature Vector Using Complex HRRP for Radar Target Recognition*

Lan Du, Hongwei Liu, Zheng Bao, and Feng Chen

National Lab. of Radar Signal Processing, Xidian University,
Xi'an, Shaanxi 710071, China
dulan@mail.xidian.edu.cn

Abstract. Radar high-resolution range profile (HRRP) has received intensive attention from the radar automatic recognition (RATR) community. Since the initial phase of a complex HRRP is strongly sensitive to target position variation, which is referred to as the initial phase sensitivity, only the amplitude information in the complex HRRP, what is called the real HRRP, is used for RATR. This paper proposes a novel feature extraction method for the complex HRRP. The extracted complex feature vector contains the difference phase information between range cells but no initial phase information in the complex HRRP. The recognition algorithms, frame-template-database establishment methods and preprocessing methods used in the real HRRP-based RATR can also be applied to the proposed complex feature vector-based RATR. The recognition experiments based on measured data show that the proposed complex feature vector can obtain better recognition performance than the real HRRP if only the cell interval parameters are proper.

1 Introduction

A high-resolution range profile (HRRP) is the amplitude of the coherent summations of the complex time returns from target scatterers in each range cell, which represents the projection of the complex returned echoes from the target scattering centers onto the radar line-of-sight (LOS). It contains target structure signatures, such as target size, scatterer distribution, etc., thereby radar HRRP target recognition has received intensive attention from the radar automatic target recognition (RATR) community [1~5]. According to the definition of HRRP given in the above literatures, an HRRP is a real vector, which is the amplitude of complex returned echo vector in baseband. Obviously, the phase information in the complex returned echo is not applied to recognition.

Similar to the scatterer distribution information contained in the real HRRP, the phase information contained in the complex HRRP should also be valuable for RATR. However, the initial phase of a complex HRRP, which is a function of the ratio of target distance to radar wavelength, is very sensitive to the variation of target's radial

* This work was partially supported by the National Science Foundation of China (NO.60402039) the National Defense Advanced Research Foundation of China (NO.51307060601).

D. Liu et al. (Eds.): ISNN 2007, Part I, LNCS 4491, pp. 1303–1309, 2007.

distance. Until now there has been little work in the field of the complex HRRP-based RATR. The literature [1] directly applied the adaptive Gaussian classifier (AGC), which can be used in the real HRRP-based RATR [3], to the complex HRRP-based RATR under the condition of using simulated turntable data. For simulated turntable data, the targets are assumed to only rotate around the turntable center in the simulation and the initial phase sensitivity problem need not be dealt with. However, in applications, at least the test samples are measured during the target's movement.

For high resolution radar system, we can get a complex echo vector consisting of the discrete echo components from different range cells. If the complex echo from one range cell is multiplied by the conjugate complex echo from another range cell, the initial phase will be counteracted and the corresponding difference phase, which is independent of the target position, will remain. By this property, a complex HRRP with some range cells' delay multiplied by the conjugate vector of the original complex HRRP will form a new complex feature vector with the difference phase information in the complex HRRP. And considering the influence of amplitude property on the recognition, each product term, i.e. each components in the above conjugate product vector, should be divided by a corresponding conversion factor. The recognition experiments based on measured data show that the proposed complex feature vector can obtain better recognition results than the real HRRP in the minimum Euclidean distance classifier and the AGC if only the cell interval parameters are proper.

2 Characteristic Comparison Between Real HRRP and Complex HRRP

As stated in the literatures [2,3,5], several issues have to be considered when real HRRP is applied to RATR, i.e. the target-aspect, time-shift and amplitude-scale sensitivities of real HRRP. Another important characteristic of real HRRP is that the average profile of a real HRRP frame can reduce the target-aspect sensitivity of the real HRRPs in a large extent [2].

Since the amplitude profile of a complex HRRP is the corresponding real HRRP, complex HRRP is also sensitive to the variation of target aspect, time shift and amplitude scale. For the target-aspect sensitivity, a set of complex HRRPs from a target-aspect sector without scatterers' motion through range cells (MTRC), is defined to be a complex HRRP frame, which can also be represented by a frame template. For the time-shift sensitivity, the time-shift compensation approaches in the training phase and in the test phase used in real HRRP-based RATR are also suitable for complex HRRP-based RATR. For the amplitude-scale sensitivity, complex HRRP should also be amplitude normalized before being used in RATR.

As discussed in Section 1, the initial phase of a complex HRRP is sensitive to the target distance variation. Similar to the time-shift sensitivity, there are also two kinds of initial phase sensitivity in RATR using complex HRRP. a) In the training phase, the initial phases of each training frame should be calibrated, which can be achieved by using the autofocus techniques in ISAR imaging. b) In the test phase, before a test sample is matched with a training frame, its initial phase should be calibrated with the training samples' initial phases. Obviously, similar to the time-shift compensation in the test phase, the autofocus techniques in ISAR imaging are not suitable for this process, since the test sample and the training frame may be from different targets. Actually, it is hard to deal with the initial phase sensitivity of complex HRRP in the test phase of complex HRRP-based RATR.

Because the average profile of a real HRRP frame can represent the stable scatterer auto-term (SAT) profile, the average profile of each frame can be used as the corresponding frame template in real HRRP-based RATR [2]. Then can the average profile of a complex HRRP frame be used as the corresponding frame template in complex HRRP-based RATR? Actually, after being amplitude normalized, the zero-Doppler slice in an ISAR image is identical with the average profile of a motion compensated complex HRRP frame, because the ISAR image is the Fourier transformation of the complex HRRP frame along the range cell dimension and its zero-Doppler channel output is theoretically identical with the average profile. The average profile of a real HRRP frame can represent the projection of the target's scatterer distribution onto the radial range dimension, while the average complex profile of a complex HRRP frame only represents the zero-Doppler slice of the target's scatterer distribution. Therefore, the average complex profile cannot be used as the corresponding frame template for complex HRRP-based RATR. Moreover, because the focus location in an ISAR image varies with the autofocus approach, the average complex profile of a complex HRRP frame is not exclusive. This further demonstrates that the average complex profile of a complex HRRP frame is not the suitable frame template for RATR using complex HRRPs.

Therefore, the recognition algorithms and frame-template-database establishment methods used in the real HRRP-based RATR can not be directly applied to the complex HRRP-based RATR.

3 A Novel Feature Extraction Method Using the Difference Phase Information in the Complex HRRP

3.1 The Fundamental Idea

Suppose there are radial displacement and little uniform rotation for a target. If the transmitted signal is $s(t)e^{j\omega_c t}$ with ω_c denoting the carrier angular frequency and the returned complex HRRP is denoted as a discrete complex vector, the mth complex returned echo from the nth range cell ($n=1,2,\cdots,N$) in baseband is

$$\begin{aligned} x_n(t,m) &= \sum_{i=1}^{L_n} \sigma_{ni} s\left(t-\frac{2R_{ni}(m)}{C}\right) e^{-j[\frac{4\pi}{\lambda}R_{ni}(m)]} \approx s\left(t-\frac{2R(m)}{c}\right)\sum_{i=1}^{L_n}\sigma_{ni} e^{-j\left\{\frac{4\pi}{\lambda}\left[R(m)+\Delta r_{ni}(m)\right]\right\}} \\ &= s\left(t-\frac{2R(m)}{c}\right) e^{j\psi(m)} \sum_{i=1}^{L_n} \sigma_{ni} e^{j\phi_{ni}(m)} \end{aligned} \tag{1}$$

where L_n is the number of target scatterers in the nth range cell, σ_{ni} is the strength of the ith scatterer in the nth range cell, $R_{ni}(m)$ denotes the radial distance between the radar and the ith scatterer in the nth range cell in the mth returned echo, $R(m)$ denotes the radial distance between the target reference center in the mth returned echo and the radar, $\Delta r_{ni}(m)$ represents the radial displacement of the ith scatterer in the nth range cell in the mth returned echo, and $\psi(m) = -\frac{4\pi}{\lambda} R(m)$ denotes the initial phase of the mth returned echo. Without loss of generality, we assume that the transmitted signal $s(t)$ is a rectangular and narrow pulse with unit amplitude. After the time-shift compensation for the complex envelope, Eq. (1) can be written as

$$x_n(m) = e^{j\psi(m)} \sum_{i=1}^{L_n} \sigma_{ni} e^{j\phi_{ni}(m)} \tag{2}$$

Then the mth complex HRRP is

$$\begin{aligned} \tilde{\boldsymbol{x}}(m) &= \left[x_1(m), x_2(m), \cdots, x_n(m), \cdots, x_N(m)\right]^T \\ &= e^{j\psi(m)} \left[\sum_{i=1}^{L_1} \sigma_{1i} e^{j\phi_{1i}(m)}, \sum_{i=1}^{L_2} \sigma_{2i} e^{j\phi_{2i}(m)}, \cdots, \sum_{i=1}^{L_n} \sigma_{ni} e^{j\phi_{ni}(m)}, \cdots, \sum_{i=1}^{L_N} \sigma_{Ni} e^{j\phi_{Ni}(m)}\right]^T \end{aligned} \tag{3}$$

Obviously, the complex HRRP varies with the initial phase. According to Eqs. (2) and (3), for the complex echoes from all range cells in a complex HRRP, their initial phases are same. If the complex echo from the n_1 th range cell is multiplied by the conjugate complex echo from the n_2 th ($n_1 \neq n_2$) range cell, then

$$\begin{aligned} y_{n_1 n_2}(m) &= x_{n_1}(m) \cdot x_{n_2}^*(m) = \left[e^{j\psi(m)} \sum_{k=1}^{L_{n_1}} \sigma_{n_1 k} e^{j\phi_{n_1 k}(m)}\right] \cdot \left[e^{-j\psi(m)} \sum_{l=1}^{L_{n_2}} \sigma_{n_2 l} e^{j\phi_{n_2 l}(m)}\right] \\ &= \sum_{k=1}^{L_{n_1}} \sum_{l=1}^{L_{n_2}} \sigma_{n_1 k} \sigma_{n_2 l} \exp\left\{-j\frac{4\pi}{\lambda}\left[\Delta r_{n_1 k}(m) - \Delta r_{n_2 l}(m)\right]\right\} \end{aligned} \tag{4}$$

Obviously, the product term $y_{n_1 n_2}(m)$ is independent of the initial phase. Let $L_{n_1 n_2} = L_{n_1} \cdot L_{n_2}$, $\sigma_{n_1 n_2 i} = \sigma_{n_1 k} \cdot \sigma_{n_2 l}$, $\Delta r_{n_1 n_2 i}(m) = \Delta r_{n_1 k}(m) - \Delta r_{n_2 l}(m)$. Then

$$y_{n_1 n_2}(m) = \sum_{i=1}^{L_{n_1 n_2}} \sigma_{n_1 n_2 i} \exp\left[-j\frac{4\pi}{\lambda} \Delta r_{n_1 n_2 i}(m)\right] = \sum_{i=1}^{L_{n_1 n_2}} \sigma_{n_1 n_2 i} e^{j\phi_{n_1 n_2 i}(m)} \tag{5}$$

The product term $y_{n_1 n_2}(m)$ can also be regarded as the complex echo from a suppositional range cell in a special kind of complex HRRP, which is the vector product of the complex HRRP with J ($n_2 - n_1 = J$) range cells' delay and the conjugate vector of the original complex HRRP. If x_{ni} and y_{ni} represent the initial cross and radial coordinates of the ith scatterer in the nth range cell, then

$$\begin{aligned} y_{n_1 n_2}(m) &= \sum_{k=1}^{L_{n_1}} \sum_{l=1}^{L_{n_2}} \sigma_{n_1 k} \sigma_{n_2 l} \exp\left\{-j\frac{4\pi}{\lambda}\left\{\begin{matrix}\left(y_{n_1 k} - y_{n_2 l}\right) \cdot \left[\cos(m \cdot \Delta\phi) - 1\right] \\ +\left(x_{n_1 k} - x_{n_2 l}\right)\sin(m \cdot \Delta\phi)\end{matrix}\right\}\right\} \\ &\overset{\substack{y_{n_1 k} \approx y_{n_1} \\ y_{n_2 k} \approx y_{n_2}}}{\approx} \sum_{k=1}^{L_{n_1}} \sum_{l=1}^{L_{n_2}} \sigma_{n_1 k} \sigma_{n_2 l} \exp\left\{-j\frac{4\pi}{\lambda}\left\{\begin{matrix}\left(y_{n_1} - y_{n_2}\right) \cdot \left[\cos(m \cdot \Delta\phi) - 1\right] \\ +\left(x_{n_1 k} - x_{n_2 l}\right)\sin(m \cdot \Delta\phi)\end{matrix}\right\}\right\} \\ &= \sum_{k=1}^{L_{n_1}} \sum_{l=1}^{L_{n_2}} \sigma_{n_1 k} \sigma_{n_2 l} \left\{\begin{matrix}\cos\left[\theta_{n_1 n_2}(m) + \varphi_{n_1 k n_2 l}(m)\right] \\ +j\sin\left[\theta_{n_1 n_2}(m) + \varphi_{n_1 k n_2 l}(m)\right]\end{matrix}\right\} \end{aligned} \tag{6}$$

where $\Delta\phi$ denotes the target-aspect difference between two neighboring echoes. When J is small, $\theta_{n_1n_2}(m) \ll \varphi_{n_1kn_2l}(m)$ and $\theta_{n_1n_2}(m)$ can be ignored. Then

$$y_{n_1n_2}(m) \approx \sum_{k=1}^{L_{n_1}}\sum_{l=1}^{L_{n_2}} \sigma_{n_1k}\sigma_{n_2l}\exp\left[j\varphi_{n_1kn_2l}(m)\right] = \sum_{k=1}^{L_{n_1}}\sum_{l=1}^{L_{n_2}} \sigma_{n_1k}\sigma_{n_2l}\left\{\cos\left[\varphi_{n_1kn_2l}(m)\right] + j\sin\left[\varphi_{n_1kn_2l}(m)\right]\right\} \quad (7)$$

The product term $y_{n_1n_2}(m)$ in Eq. (7) depends on the cross coordinates' differences between the scatterers in the n_1 th and the n_2 th range cells. Then the physical mechanisms of the product term are similar to those of the power term of the real HRRP [2]. The above feature extraction method uses the conjugate product between the complex echoes from two range cells with J intervals (J is small) to counteract the initial phase, yet the feature vector is in power form, and because of the amplitudes of the two complex echoes being similar, the amplitude profile of the product vector can increase the amplitude's dynamic range of the original complex HRRP, reduce the contribution of weak scatterers to the similarity measurement, and enlarge the speckle effect. According to the literatures [3~5], this is unfavorable for the recognition from the viewpoint of using the HRRP's amplitude property. In order to convert the conjugate product vector into the one in amplitude form, we propose a novel feature vector

$$\begin{aligned} z_J(m) &= \left[z_{J1}(m), z_{J2}(m), \cdots, z_{Jn}(m), \cdots, z_{J(N-J)}(m)\right]^T \\ &= \left[\frac{x_1(m)\cdot x_{1+J}^*(m)}{\sqrt{|x_1(m)|_2^2 + |x_{1+J}^*(m)|_2^2}}, \frac{x_2(m)\cdot x_{2+J}^*(m)}{\sqrt{|x_2(m)|_2^2 + |x_{2+J}^*(m)|_2^2}}, \cdots, \right. \\ &\qquad \left. \frac{x_n(m)\cdot x_{n+J}^*(m)}{\sqrt{|x_n(m)|_2^2 + |x_{n+J}^*(m)|_2^2}}, \cdots, \frac{x_{N-J}(m)\cdot x_N^*(m)}{\sqrt{|x_{N-J}(m)|_2^2 + |x_N^*(m)|_2^2}}\right]^T \end{aligned} \quad (8)$$

For both large echoes, their product term is divided by a large conversion factor, while for both small echoes, their product term is divided by a small conversion factor, therefore, such feature extraction method can reduce the amplitude's dynamic range of the product vector and the amplitude property of the complex feature vector is more favorable for recognition. The complex feature vector in Eq. (8) is referred to as the complex feature vector with difference phases in this paper. According to what discussed above, such complex feature vector can be regarded as a complex echo vector consisting of the discrete echo components from the suppositional range cells. The amplitude property of such feature vector is more favorable for recognition. If the complex feature vectors extracted from all training and test samples by our feature extraction method are applied to RATR, due to containing no initial phases of the complex HRRPs, the recognition algorithms, frame-template-database establishment methods and preprocessing methods used in the real HRRP-based RATR can also be applied to the proposed complex feature vector-based RATR.

3.2 The Recognition Algorithm

According to the idea in Section 3.1, the complex feature vector with difference phases is the product vector divided by the corresponding conversion factors, therefore,

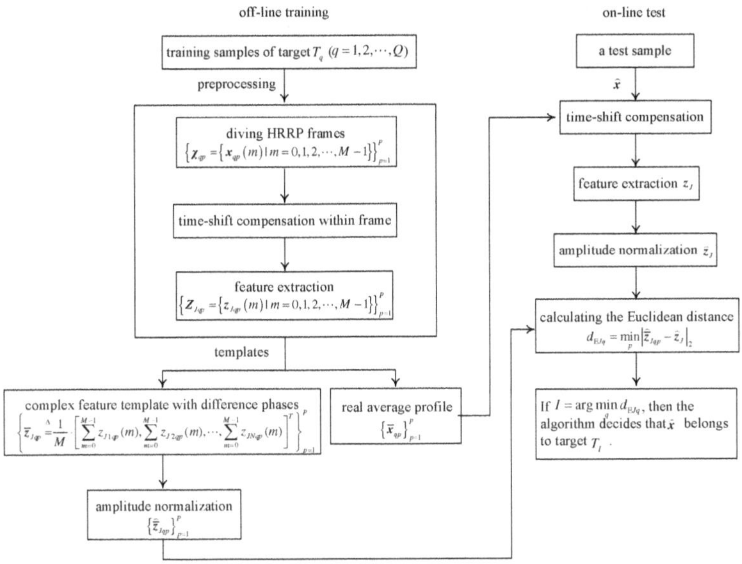

Fig. 1. The flow chart of the minimum Euclidean distance classifier using the proposed featureextraction method

the range cell order of a complex HRRP will influence the feature extraction. However, due to the time-shift sensitivity of an HRRP, which has been discussed in Section 2, the range cell order of an HRRP varies with the range window, and all of the samples used in the training and test phases should be time-shift compensated. To solve the time-shift sensitivity, all of the complex HRRPs used in the training and test phases are time-shift compensated by the slide correlation processing in our recognition algorithm, thus the complex feature vectors with difference phases extracted from the time-shift compensated complex HRRPs are independent of the time shift and can be used to classifiers directly. Fig. 1 takes the minimum Euclidean distance classifier as an example to show how to apply the feature extraction method proposed in this paper to classifiers.

4 Experimental Results

The results presented in this paper are based on real airplane data, which are measured by radar with the center frequency of 5520MHz and the bandwidth of 400MHz. Training data cover almost all of the target-aspect angles of test data, but their elevation angles are different.

Fig. 2 (a) and (b) shows the recognition rates of the proposed complex feature vectors with $J=1,2,3,4,5$ and the real HRRP versus the power parameter α in the minimum Euclidean distance classifier [2,4] and the AGC [1,3]. The recognition experiments based on measured data show that the proposed feature vector can obtain better recognition results than the real HRRP in the minimum Euclidean distance classifier and the AGC if only the cell interval parameters are proper.

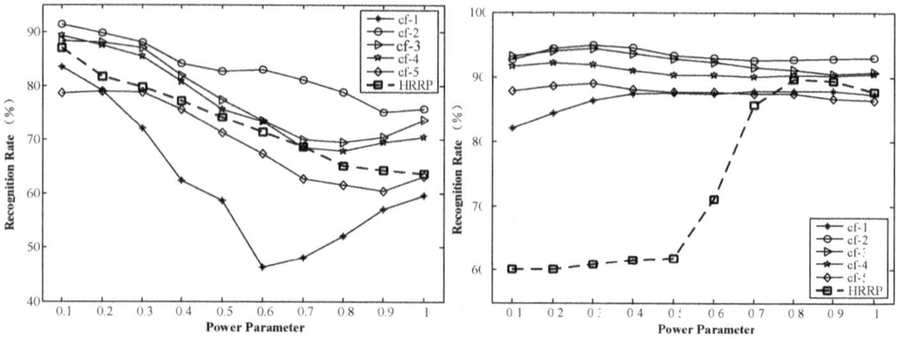

(a) The minimum Euclidean distance classifier (b) The AGC

Fig. 2. The average recognition rates of the real HRRP and the complex feature vectors with $J = 1,2,3,4,5$ versus the power parameter in the minimum Euclidean distance classifier and the AGC

5 Conclusion

A novel feature extraction method which can counteract the initial phase and leave the difference phase information in the complex HRRP is proposed in this paper. The recognition experiments based on measured data show that the proposed feature extraction method can effectively use the difference phase information in the complex HRRP and improve the recognition performance.

References

[1] Jacobs, S.P.: Automatic Target Recognition using High-Resolution Radar Range Profiles, Ph. D. Dissertation, Washington University, 1999.

[2] Du, L., Liu, H.W., Bao, Z.: Radar HRRP Target Recognition Based on Higher-Order Spectra, IEEE Trans. Signal Process **53 (7)** (2005) 2359-2368

[3] Du, L., Liu, H.W., Bao, Z., Zhang, J.Y.: A Two-Distribution Compounded Statistical Model for Radar HRRP Target Recognition, IEEE Trans. Signal Process **54 (6)** (2006) 2226-2238.

[4] Van der Heiden, R., Groen, F.C.A.: The Box-Cox Metric for Nearest Neighbor Classification Improvement, Pattern Recognition **30 (2)** (1997) 273-279.

[5] Liu, H.W., Bao, Z.: Radar HRR Profiles Recognition Based on SVM with Power-Transformed-Correlation Kernel, Springer LNCS, Berlin: Springer-Verlag, Aug. (2004) 531-536

A Probabilistic Approach to Feature Selection for Multi-class Text Categorization

Ke Wu[1], Bao-Liang Lu[1,*], Masao Uchiyama[2], and Hitoshi Isahara[2]

[1]Department of Computer Science and Engineering
Shanghai Jiao Tong University
800 Dong Chuan Rd., Shanghai 200240, China
{wuke,bllu}@sjtu.edu.cn
[2]Knowledge Creating Communication Research Center,
National Institute of Information and Communications Technology,
3-5 Hilaridai, Seika-cho, Soraku-gun, Kyoto, 619-0289 Japan
{mutiyama,isahara}@nict.go.jp

Abstract. In this paper, we propose a probabilistic approach to feature selection for multi-class text categorization. Specifically, we regard document class and occurrence of each feature as events, calculate the probability of occurrence of each feature by the theorem on the total probability and utilize the values as a ranking criterion. Experiments on Reuters-2000 collection show that the proposed method can yield better performance than information gain and χ-square, which are two well-known feature selection methods.

1 Introduction

Text categorization is a process of assigning a text document into some predefined categories. Many information retrieval applications[1], such as filtering, routing or searching for relevant information can benefit from the text categorization research. However, a major characteristic, or difficulty of text categorization problem is the high dimensionality of the feature space. Dimension reduction techniques can be applied to handle the problem. They have attracted much recently since effective feature reduction can improve the prediction performance and learning efficiency, provide faster predictors possibly requesting less information on the original data, reduce complexity of the learned results, and save more storage space.

Dimension reduction techniques can typically be grouped into two categories, which are feature extraction(FE) and feature selection(FS). The traditional FE algorithms reduce the dimension of data by linear algebra transformations while FS algorithms reduce the dimension of data by directly selecting features from

* To whome correspondence should be addressed. This work was supported in part by the National Natural Science Foundation of China under the grants NSFC 60375022 and NSFC 60473040, and the Microsoft Laboratory for Intelligent Computing and Intelligent Systems of Shanghai Jiao Tong University.

D. Liu et al. (Eds.): ISNN 2007, Part I, LNCS 4491, pp. 1310–1317, 2007.

the original vectors. Although FE algorithms have been proved to be very effective for dimension reduction, the high dimension of data sets in text domain fails FE algorithms due to their expensive computational cost. Therefore FS algorithms are more popular in text domain.

In text categorization, FS algorithms are typically performed by assigning a score to each term and keeping some number of terms with the highest scores while discarding the rest. Numerous feature selection metrics have been proposed, e.g. information gain(IG), odds ratio, χ-square(CHI), document frequency (DF) , mutual information(MI) and SVM-based featurez selection [3], etc. These metrics have been extensively examined in binary classification and most have been extended to multi-class. However, SVM-based feature selection for multi-class text classification still has not been investigated although SVM-based feature selection yields better performance than some well-known feature selection metrics, at the same level of a feature set size. In this paper, we extend this metric to a multi-class classification case in the text domain and compared our proposed metric to two well-known feature selection measures, i.e., IG and CHI in the multi-class case.

The remainder of the paper is organized as follows. In section 2, we describe the multi-class learning algorithms and feature selection methods to be used in our experiments. In section 3, we introduce SVM-based feature selection method in the binary classification and extend it to multi-class classification. In section 4, the experimental results on Reuters-2000 collection are presented and analyzed. In Section 5, we conclude the paper.

2 Multi-class Classification and Feature Selection

2.1 Multi-class Classifiers

Naïve Bayes(NB). The multinomial model as described in [6] is used. The basic idea is based on the assumption that the probability of each word event in a document is independent of the word's context and position in the document and word event in a document conforms to a multinomial distribution. The predicted class for document d is the one that maximizes the posterior probability that $P(c|d) \propto P(c)\prod_t P(t|c)^{tf(t,d)}$, where $P(c)$ is the prior probability that a document belongs to class c, $P(t|c)$ is the probability that a word t occurs given class c and $tf(t,d)$ is the number of occurrences of word t in a document d.

k-Nearest-Neighbor(k-NN). Its basic idea is that it classifies a new sample into a predefined class based on a local vote by its k-nearest neighbors. In k-NN algorithm, classification is delayed until a new sample arrives. All samples in training set correspond to points in an n-dimensional Euclidean space and usually Euclidean distance is used to calculate the nearest neighbors of a new sample. If most of those k-nearest neighbors of a new sample are in class c, then assign the sample to c.

Rocchio [4]. It constructs a prototype vector for each category using both the centroid of positive training samples and the centroid of negative training samples. The prototype vector is calculated as follows:

$$\boldsymbol{c_j} = \alpha \frac{1}{|C_j|} \sum_{\boldsymbol{d} \in C_j} \frac{\boldsymbol{d}}{\|\boldsymbol{d}\|} - \beta \frac{1}{|D - C_j|} \sum_{\boldsymbol{d} \in D - C_j} \frac{\boldsymbol{d}}{\|\boldsymbol{d}\|}, \tag{1}$$

where α and β are parameters that adjust the relative impact of positive and negative training samples, respectively. When classifying a new document, Rocchio classifier computes either the dot product or cosine value between the new document and the prototype vector of each class, and then assigns the new document into the class with the highest dot product or cosine value.

2.2 Feature Selection Methods

Information Gain [2]. This feature ranking criterion is based on information theory. It measures the number of bits of information obtained for category prediction by knowing the presence or absence of a term in a document. Let $\{c_i\}_{i=1}^{m}$ denote the set of categories in the target space. The information gain of term t is defined as follows:

$$\begin{aligned} IG(t) &= -\sum_{i=1}^{m} P(c_i) \log P(c_i) \\ &+ P(t) \sum_{i=1}^{m} P(c_i|t) \log P(c_i|t) + P(\bar{t}) \sum_{i=1}^{m} P(c_i|\bar{t}) \log P(c_i|\bar{t}). \end{aligned} \tag{2}$$

Chi-square(CHI). CHI [2,7] measures the lack of independence between t and c_i and can be compared to the chi-square distribution with one degree of freedom to judge extremeness. It is defined as follows:

$$\chi^2(t, c_i) = \frac{N[P(t, c_i)P(\bar{t}, \bar{c_i}) - P(t, \bar{c_i})P(\bar{t}, c_i)]^2}{P(t)P(\bar{t})P(c_i)P(\bar{c_i})}, \tag{3}$$

where N is the total number of documents. For each category, we caculate the χ^2 statistic between each unique term in training data set and the corresponding category, and then we combine the category-specific scores of each term into two scores as follows:

$$\chi^2_{avg}(t) = \sum_{i=1}^{m} P(c_i)\chi^2(t, c_i), \tag{4}$$

$$\chi^2_{max}(t) = \max_{i=1}^{m}\{\chi^2(t, c_i)\}. \tag{5}$$

In this paper, we use $\chi^2_{avg}(t)$ for multi-class text classification.

3 A Probabilistic Feature Selection Approach

In the linear case of binary classification, the output of a trained SVM can be expressed as:

$$F(x) = sign(\boldsymbol{w}^T \cdot \boldsymbol{x} + b) = sign(\sum\nolimits_i w_i x_i + b). \tag{6}$$

From (6), we can see that a feature i with the weight w_i close to 0 has a smaller effect on the prediction than features with large absolute values of w_i. If the classifier performs well, the input feature subset with the largest weights should correspond to the most informative features [9]. As a result, $|w_i|$ is an evidence for feature ranking. This method was introduced by Brank *et al.* [3] for binary text categorization in 2002.

On the other hand, how to extend svm-based binary feature selection into multi-class feature selection remains unresolved in text categorization. In this Section, we apply a probabilistic approach to implement the extension of svm-based binary feature selection.

Intuitively, we try to extract features that all classes regard as important from candidate features and rank them higher. To the end, we apply a similar method to one-versus-all multi-class SVMs [12] to decompose a multi-class problem into a series of two-class sub-problems and combine these results in a probabilistic way.

More specifically, we first construct k two-class classifiers, one for each class, where k is the number of classes. The ith SVM is trained with all the samples from the ith class against all the samples from the rest classes and thus k decision functions are generated. Consequently, the ith decision function $F_i(x)$ is used as a binary classification sub-model criterion for discriminating the ith class from the all other classes. On the other hand, we regard document class and occurrence of each feature as events, calculate the probability of occurrence of each feature by the theorem on the total probability and utilize the values as a ranking criterion.

Assume that there are sure event E and impossible event $\varnothing$. Let E_i indicates the event that the ith class is true. According to probability theory, events E_1, E_2, $\ldots$, E_k constitute a sample space S. S corresponds to sure event E, where $E = E_1 \cup E_2 \cup \ldots \cup E_k$ and $E_i \cap E_j = \varnothing, i \neq j$. $P(E_i)$ is the prior probability that the ith class is true. Define a random event F as a event that a feature is selected as a discriminative feature. Let f_m denote the mth feature and let $P(F = f_j|E_i)$ denote the conditional probability of the event that f_j is selected as a discriminative feature given that E_i has occurred. When event E_i occurs, the ith binary classification sub-model is effective for determining the final classification result.Under the feature ranking criterion $R^{(i)}$of ith sub-model, we can derive $P(F = f_j|E_i)$ by the following equation.

$$P(F = f_j|E_i) = \frac{r_j^{(i)}}{\sum_{j=1}^{n} r_j^{(i)}}. \tag{7}$$

According to the theorem on the total probability, $P(F = f_j)$ can be derived from $P(F = f_j|E_i)$ and $P(E_i)$.

$$P(F = f_j) = \sum_{i=1}^{k} P(F = f_j|E_i)P(E_i). \tag{8}$$

The above probability can be exploited as a feature selection metric for multi-class classification.

Table 1. Accuracy rates of k-NN classifiers for various feature set sizes

Dimensionality	IG	CHI	PSVM
50	41.00±0.41	44.75±0.39	**50.30**±0.38
100	38.89±0.40	41.00±0.36	**56.63**±0.34
200	43.02±0.50	42.80±0.52	**61.92**±0.41
300	43.62±0.48	44.80±0.52	**63.44**±0.54
400	43.91±0.49	44.66±0.50	**63.33**±0.51
500	44.20±0.49	44.91±0.50	**63.25**±0.54
800	45.28±0.53	45.65±0.55	**64.62**±0.51
1000	45.96±0.47	46.44±0.54	**65.06**±0.51
2000	47.08±0.47	47.30±0.49	**64.94**±0.42
3000	47.57±0.47	47.72±0.47	**53.10**±0.46
4000	47.69±0.46	47.84±0.47	**53.31**±0.46
5000	47.89±0.43	**47.93**±0.44	47.64±0.46
8000	**48.26**±0.42	48.22±0.44	48.11±0.40
10000	48.37±0.43	**48.38**±0.42	48.20±0.40
50000	**48.63**±0.41	48.62±0.42	48.55±0.41
100000	**48.63**±0.41	**48.63**±0.41	48.58±0.42
159300	48.63±0.41	48.63±0.41	48.63±0.41

4 Experiments

In the paper, the Reuters-2000 collection[1] is used to conduct all experiments. It includes a total of 806,791 documents, with news stories covering the period from 20 Aug 1996 through 19 Aug 1997. We divided this time interval into a training period, which includes all the 504,468 documents dated 14 April 1997 or earlier, and test period, consisting of the remaining 302,323 documents. We used the same 16 categories that were selected in [3]. These statistics for the selected subset of 16 categories approximately follows the distribution for all 103 categories. The selected set of categories includes: godd, c313, gpol, ghea, c15, e121, gobit, m14, m143, e13, e21, gspo, e132, c183, e142, and c13. A document may belong to one or more categories, but we simply think that a document belongs to one category.

In our experiments, data are documents from the above 16 categories. More specifically, the training data set contains 282,010 document and the test data set consists of 175,807 documents was divided into 10 parts with the approximately equal size. Two state-of-the-art feature selection methods for text categorization, i.e. IG and CHI as our baseline, are investigated on the data set. We ignored the case of the word surface form and removed words according to a standard stop list containing 523 stop-of-words and words that occur less than 4 times from the corpus. Consequently, we used the bag-of-words model to represent documents containing a vocabulary of 159,300. Additionally, the normalized TF-IDF score was used to weight features. In addition, we respectively applied the above three

[1] http://about.reuters.com/researchandstandards/corpus/

Table 2. Accuracy rates of the NB classifiers for various feature set sizes

Dimensionality	IG	CHI	PSVM
50	**56.04**±0.56	53.80±0.48	53.58±0.42
100	62.62±0.44	**65.25**±0.43	60.71±0.37
200	69.75±0.31	69.88±0.40	**70.41**±0.20
300	71.27±0.27	71.62±0.43	**73.20**±0.15
400	72.61±0.36	72.99±0.28	**75.70**±0.17
500	73.32±0.35	73.34±0.35	**76.37**±0.24
800	74.95±0.28	74.75±0.32	**77.71**±0.19
1000	75.59±0.33	75.74±0.34	**78.08**±0.24
2000	77.23±0.28	77.00±0.27	**78.94**±0.25
3000	78.38±0.31	78.34±0.27	**79.38**±0.25
4000	78.95±0.29	78.96±0.29	**79.49**±0.28
5000	79.22±0.29	79.27±0.28	**79.32**±0.26
8000	79.59±0.25	79.75±0.25	**79.79**±0.22
10000	79.84±0.25	**80.05**±0.21	79.97±0.20
50000	81.50±0.24	**81.63**±0.24	81.26±0.19
100000	82.33±0.27	**82.37**±0.28	82.04±0.24
159300	82.29±0.22	82.29±0.22	82.29±0.22

Table 3. Accuracy rates of the Rocchio classifiers for various feature set sizes

Dimensionality	IG	CHI	PSVM
50	40.86±0.32	43.20±0.37	**52.28**±0.41
100	51.01±0.38	53.47±0.43	**56.69**±0.39
200	58.53±0.40	59.96±0.32	**62.52**±0.31
300	62.98±0.47	61.87±0.39	**63.51**±0.25
400	64.40±0.39	63.87±0.40	**66.56**±0.21
500	65.40±0.34	63.60±0.43	**67.74**±0.24
800	68.07±0.37	67.06±0.34	**69.26**±0.24
1000	69.29±0.35	69.44±0.29	**70.11**±0.28
2000	71.52±0.35	70.87±0.35	**72.17**±0.41
3000	72.79±0.40	72.40±0.36	**73.27**±0.37
4000	73.45±0.37	73.06±0.38	**73.69**±0.38
5000	73.76±0.35	73.57±0.39	**73.96**±0.36
8000	74.38±0.36	74.13±0.35	**74.60**±0.35
10000	74.64±0.35	74.35±0.39	**74.70**±0.36
50000	**75.17**±0.35	75.08±0.34	75.13±0.34
100000	**75.20**±0.34	75.19±0.35	75.19±0.34
159300	75.21±0.35	75.21±0.35	75.21±0.35

feature selection metrics to three well-known multi-class classifiers, that is , kNN, NB and Rocchio.

We use LibSVM[13] for obtaining the weight of each feature. Three typical classifiers, k-NN, NB and Rocchio, are from Rainbow toolkit and default parameters were used in the experiments. The experimental results are shown in Tables 1 through 3, where IG denotes information gain metric, CHI denotes Chi-square metric, and PSVM denotes SVM-based probabilistic criterion metric. The first column indicates the feature set size in the experiments and the last three columns indicate the correct rates and their scale in percentage.

From Tables 1 through 3, we can observe that our metric can perform better than both IG and CHI. On the one hand, for three feature selection metrics, there is a comparable performance when the size of feature set becomes large. On the other hand, the proposed metric performs better than two existing metrics when the size of feature selection remain small. In Table 1, there is amazingly better performance for the proposed metric than IG and CHI until the size of feature set reaches 5000. In Tables 2 and 3, PSVM has better performance than that of IG and CHI in almost all cases. The results in Table 1 indicate that our proposed metric can make more relevant features have high feature rank, since k-NN classifier is sensitive to irrelevant features. In addition, the results in Table 2 indicate that our metric can obtain more informative features for all classes, since NB classifier has a preference to positive features for each class. Also, our metric can suppress some features with high rank, which is obtained by a statistical bias, since tf-idf scheme in Rocchio classifier has a strong relation with data.

5 Conclusion

We have presented a novel feature selection metric for multi-class classification. The empirical results of our study indicate that the proposed method has better performance than IG and CHI. Although IG and CHI are two state-of-the-art feature selection metrics, they simplely consider the relation between the single feature and some classes, and they easily suffer from the data bias. The proposed probabilistic approach can effectively avoid the case. However, it could be affected by the noise samples on support boundary. Therefore, in future work, we will investigate effect of the noise samples on support boundary on feature selection.

Acknowledgment

The authors thank Zhi-Gang Fan for his valuable advice.

References

1. Baeza-Yates, R., Ribeiro-Neto, B.: Modern Information Retrival. Addison-Wesley (1999)
2. Yang, Y., Pedersen, J.O.: A Comparative Study on Feature Selection in Text Categorization. Proc. Of the 14th Int. Conf. on Machine Learning (1997) 412-420
3. Brank, J., Grobelnik, M., Milic-Frayling, N., Mladenic, D.: Feature Selection Using Support Vector Machines. In Proc. 3rd Int. Conf. on Data Mining Methods and Databases for Engineering, Finance, and Other Fields (2002)
4. Ittner, D.J., Lewis, D.D., Ahn, D.D.: Text Categorization of Low Quality Images. In Symposium on Document Analysis and Information Retrieval,Las Vegas (1995) 301-315
5. Fan, Z.G., Lu, B.L.: Fast Recognition of Multi-View Faces with Feature Selection. 10th IEEE International Conference on Computer Vision (2005)76-81
6. McCallum, A., Nigam, K.: A Comparision of Event Models for Naive Bayes Text Classification. AAAI-98 Workshop on Learning for Text Categorization (1998)
7. Sebastiani, F.: Machine Learning in Automated Text Categorization. ACM computing Surveys **34** (1) (2002) 1-47
8. Vapnik, V.: The Nature of Statistical Learning Theory. Springer-Verlag, New York (2000)
9. Guyon, I., Weston, J., Barnhill, S., Vapnik, V.: Gene Selection for Cancer Classification Using Support Vector Machines. Machine Learning **46** (2002) 389-422
10. Fan, Z.G., Wang, K.A., Lu, B.L.: Feature Selection for Fast Image Classification with Support Vector Machines. Proc. ICONIP 2004, LNCS, **3316** (2004) 711-720
11. Heisele, B., Serre, T., Prentice, S., Poggio, T.: Hierarchical Classification Andfeature Reduction for Fast Face Detection with Support Vector Machines. Pattern Recognition **36**(2003) 2007-2017
12. Rifkin, R., Klautau, A.: In Defense of One-Vs-All Classification. Journal of Machine Learning Research **5** (2004) 101-141
13. Chang, C.C., Lin, C.J.: LIBSVM: A Library for Support Vector Machines [EB/OL]. http://www.csie.ntu.edu.tw/ cjlin/libsvm (2001)

Zero-Crossing-Based Feature Extraction for Voice Command Systems Using Neck-Microphones

Sang Kyoon Park[1], Rhee Man Kil[1], Young-Giu Jung[2], and Mun-Sung Han[2]

[1] Division of Applied Mathematics
Korea Advanced Institute of Science and Technology
373-1 Guseong-dong, Yuseong-gu, Daejeon 305-701, Korea
scalphunter@kaist.ac.kr, rmkil@kaist.ac.kr
[2] Smart Interface Research Team
Electronics and Telecommunications Research Institute
161 Gajeong-dong, Yuseong-gu, Daejeon 305-700, Korea
reraj@etri.re.kr, msh@etri.re.kr

Abstract. This paper presents zero-crossing-based feature extraction for the speech recognition using neck-microphones. One of the solutions in noise-robust speech recognition is using neck-microphones which are not affected by the environmental noises. However, neck-microphones distort the original voice signals significantly since they only capture the vibrations of vocal tracts. In this context, we consider a new method of enhancing speech features of neck-microphone signals using zero-crossings. Furthermore, for the improvement of zero-crossing features, we consider to use the statistics of two adjacent zero-crossing intervals, that is, the statistics of two samples referred to as the second order statistics. Through the simulation for speech recognition using the neck-microphone voice command system, we have shown that the suggested method provides the better performance than other approaches using conventional speech features.

1 Introduction

One of the difficult but promising research fields in the automatic speech recognition (ASR) is the speech recognition in noisy and/or reverberant environments. Especially, the noise problem is the biggest obstacle which hinders vendor from producing a commercial level speech related products. One of the solutions for the noise problem is using a neck-microphone which is not affected by the environmental noises. A neck-microphone is a sensing device which captures the vibration of human vocal tract by contacting to the speaker's neck. This type of mechanism is especially useful in noisy environments since the device doesn't get any propagated signal from other sound sources. Nowadays, many neck-microphones are used in factories or military operations. Although the neck microphone has its own merits of the robustness to environmental noises, the

D. Liu et al. (Eds.): ISNN 2007, Part I, LNCS 4491, pp. 1318–1326, 2007.

shortcomings also exist. Due to its sensing mechanism, the device cannot receive the high frequency components of the voice accurately. There are also a lot of distortions to one's voice so that speaker's identity or exact pronunciation could be lost. In this context, we consider a new methodology for enhancing speech features extracted from neck-microphone signals using zero-crossings.

Zero-crossings have been used to find noise-robust speech features[1,2,3]. An example of such use is the zero-crossing peak-amplitudes (ZCPAs)[3] proposed by Kim et al. They considered a model of the neural transduction of acoustic signals based on two parallel mechanisms of auditory nerve fibers: rate and temporal representations. They demonstrated that the auditory model based on zero-crossing features is more robust in noisy environments than other popularly used feature extraction methods, such as LPCC or MFCC. This is mainly due to the dominant frequency principle[4,5] which states that the number of zero-crossings per unit time is close to two times the frequency of the dominant signal when one exists. From this observation, we propose a new method of extracting speech features using zero-crossings from neck-microphone signals. In our method, we also consider to use the concept of gray level co-occurrence matrix (GLCM)[6,7,8,9] which has been usually used for image processing like texture analysis. In this paper, the concept of GLCM is applied to find the suitable zero-crossing features for speech recognition.

To show the validity of our method, the experiments of speech recognition using neck-microphones for various schemes were conducted. For the purpose of comparison, a neck-microphone ASR systems using LPCC, MFCC, and the suggested zero-crossing-based speech features, were used for the recognition of isolated speech commands. The database of speech commands is composed of about 5200 Korean voice samples captured by the neck-microphone. In these experiments, two types of classifiers, namely the time-delayed neural network (TDNN) and hidden Markov model (HMM) classifiers were used. Through the simulation for speech recognition using the neck-microphone voice command system, we have shown the suggested method provides the better performance than other approaches using LPCC or MFCC in both classifiers.

2 Zero-Crossing Features for Neck-Microphone Signals

A Neck-microphone is a sensing device which captures the vibration of human vocal tract. Due to its sensing mechanism, there are also a lot of distortions to one's voice. For example, the recorded speech sample which is the pronunciation of Korean digit "sah-m" (meaning three) using the ordinary and neck microphones is shown in Figures 1-(a) and (b) respectively. As shown in these figure, a neck-microphone has no problem in receiving a low frequency components, but even the visual comparison shows the ability of microphones to tell between high and low frequency components: one can hardly see the difference between first "sah-" part from 'm' part in Figure 1-(b) while the ordinary microphone recording has the obvious difference between them as shown in Figure 1-(a).

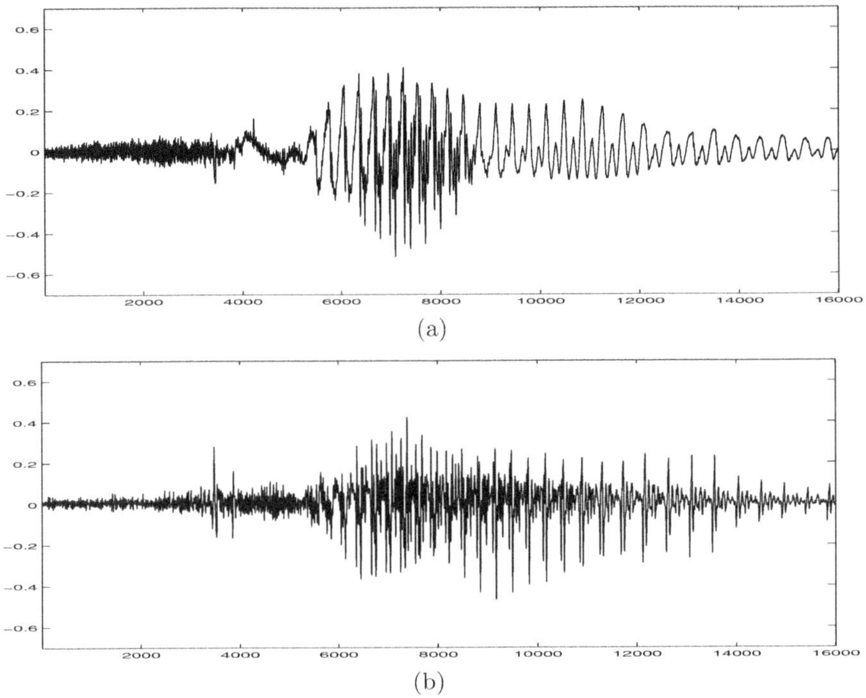

Fig. 1. Korean speech "sah-m": (a) and (b) represent the recorded waveform using the ordinary and neck microphones respectively

To enhance the speech features of the speech samples using neck-microphones, we consider to use zero-crossings since the zero-crossing based features such as the ZCPA[3] have shown the effectiveness in recognizing heavily corrupted speech samples in noisy environments. The ZCPA coding of a speech signal was motivated by auditory signal processing. In this coding, a synchronous neural firing is represented by the upward-going zero-crossing event of the signal at the output of each bandpass filter. Each peak-amplitude between successive zero-crossings is detected and used to simulate the neuronal firing rate. To explain the ZCPA coding, let us denote $x_i(t)$ as the output signal of the ith channel of the filter-bank. Suppose there are N (upward) zero-crossings, and zero-crossing times are represented by t_n, $n = 1, 2, \cdots, N$ satisfying $x_i(t_n) = 0$. Then, the ZCPA coded signal $\tilde{x}_i(t)$ is represented by

$$\tilde{x}_i(t) = \begin{cases} \max\limits_{t_{n-1} < t \leqslant t_n} x_i(t), & \text{if } t = t_n, \\ 0, & \text{otherwise.} \end{cases} \quad (1)$$

for $n = 1, 2, \cdots, N$ where t_0 represents the starting time of $x_i(t)$. The concept of ZCPA coding is illustrated in Figure 2. The ZCPA features are represented by the histogram of frequency bins which are inverses of two adjacent zero-crossing intervals. This histogram can be regarded as a pseudo-spectrum for an original

Fig. 2. The ZCPA coding: the signal is represented by the time at which upward zero-crossing occurs and the peak amplitude within the zero-crossing interval

speech signal. The obtained ZCPA feature can be modified by manipulating the time or spectral domain afterwards. More detailed theoretical aspects and application results were described in [3]. Some post processing approaches of the ZCPA features which can be considered in this work are RASTA filtering[10], peak mean subtraction (PMS), and temporal masking[11].

3 Frequency Level Co-occurrence Matrix (FLCM) for Zero-Crossing Features

For the improvement of zero-crossing features, we consider to use the statistics of two adjacent zero-crossing intervals, that is, the statistics of two samples referred to as the second order statistics. Note that we usually consider the first order statistics, that is, the statistics of individual samples such as mean or variance of samples. The methods using the second order statistics have been used in the feature extraction of images for texture classification[6,7,8,9]. In our work, we will consider to use the second order statistics to improve the features extracted from the neck-microphone signals. First, let us consider the output signal $x_i(t)$ from the ith channels of the filter-bank. Then, there are zero-crossing points t_n, $n = 1, 2, \cdots, N$. Here, the zero-crossing intervals τ_n, $n = 2, 3, \cdots, N$ are defined by

$$\tau_n = t_n - t_{n-1}, \tag{2}$$

as illustrated in Figure 3. Then, we can consider the frequency of t_n as

$$freq(t_n) = 1/\tau_n. \tag{3}$$

For our convenience, we consider the bark scale frequency bins, $freq\_bin(k)$, $k = 1, 2, \cdots, B$. Then, the transition probability from the frequency bin k to frequency bin l can be described by

$$T(k,l) = Pr\{freq(t_{n-1}) \in freq\_bin(k) \; and \; freq(t_n) \in freq\_bin(l)\}. \tag{4}$$

Here, we define the matrix **T** in which the (k,l)th element is given by (4). This matrix represents the co-occurrence matrix of the frequency transition of the signal. For instance, if a signal is pure sine wave, then the matrix **T** would have only one element on the diagonal line of the matrix. If a signal has smooth transitions in frequencies or mostly periodic components, then the elements around the diagonal line in **T** have larger values than other elements in **T**. From the defined matrix **T**, we can extract some useful features from a speech signal. In the case of texture analysis of image processing[6,7], we can consider the features of energy, entropy, contrast, inverse difference momentum, etc. In our case, we consider a new feature to detect the voice activity in the neck-microphone signals. Here, let us describe the following procedure for determining the frequency level co-occurrence matrix (FLCM) **C** from the neck-microphones:

1. Take a sample $x_i(t),\ i = 1, 2, \cdots, M(=$ number of channels$)$ from each channel of a filter-bank.
2. For each channel, the ZCPA signal $\tilde{x}_i(t)$ and the zero-crossing points $t_n,\ i = 1, 2, \cdots, N(=$ mumer of zero-crossings$)$ are obtained.
3. Construct the co-occurrence matrix **C** of zero-crossings:
   - for $i = 1, 2, \cdots, M$ do
     - $C_i(k,l) = 0,$ for $k, l = 1, 2, \cdots, B(=$ number of frequency bins$)$.
     - for $n = 2, 3, \cdots, N$ do
       if $freq(t_{n-1}) \in freq\_bin(k)$ *and* $freq(t_n) \in freq\_bin(l)$, then
       $C_i(k,l) = C_i(k,l) + \tilde{x}_i(t_{n-1})$.
   - $C(k,l) = \sum_{i=1}^{M} C_i(k,l)$.

In the step 3 of this procedure, we consider the co-occurrence matrix weighted by the peak amplitude of the signal since the peak amplitude of the signal is also important to detect the speech features in noisy (or corrupted) environments. The suggested co-occurrence matrix $C(k,l)$ can be easily applied to construct

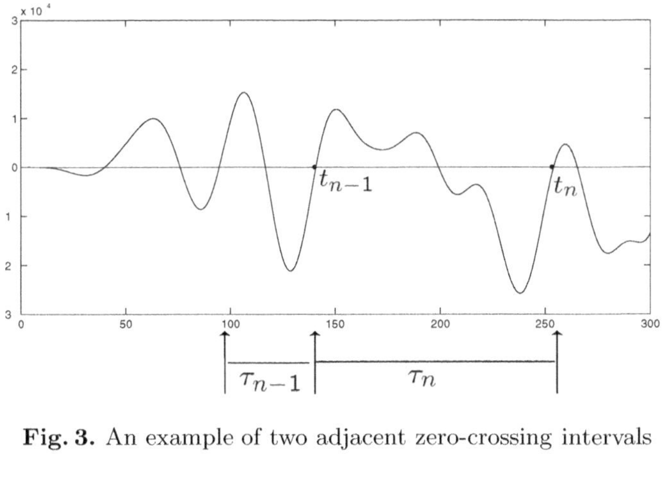

Fig. 3. An example of two adjacent zero-crossing intervals

the same pseudo-spectrum using by the ZCPA features. For instance, the kth bin of the pseudo-spectrum $S(k)$ using the ZCPA feature can be represented by

$$S(k) = \sum_{l=1}^{B} C(k,l). \tag{5}$$

In this work, the main objective is to retrieve a useful statistics from the FLCM. Since the FLCM contains the frequency transition information, there are many possible features of the transition characteristics of the signal. Among them, we consider the periodicity to detect the voice activity in the neck-microphone signal. Here, the periodicity is defined by

$$P = \frac{\sum_{k=1}^{B} C(k,k)}{E}, \tag{6}$$

where E represents the energy defined by

$$E = \sum_{k=1}^{B} \sum_{l=1}^{B} C(k,l). \tag{7}$$

These two features are simple but it can contribute to the improvement of the speech recognition rate which will be explained in the next section.

4 Simulation for Speech Recognition Using Neck-Microphones

To show the effectiveness of our approach, we performed the experiments of speaker-independent speech recognition using two types of classifiers, they are TDNN[12] and HMM[13] classifiers. For the benchmark data, we used the Korean speech samples of 35 voice commands composed of a set of 4200 training samples spoken by one group of 100 males and also a set of 900 test samples spoken by another group of 50 males. In these experiments, the 16 channel FIR bandpass filters of the 100th order were used as a filter-bank. The center frequencies of bandpass filters were between 200 Hz and 4000 Hz and set from 1 to 16 in Bark scale which can be obtained by

$$\text{Bark\_scale}(f) = \frac{26.81}{1 + (1960/f)} - 0.53,$$

where f represents a frequency in Hz. The ZCPA features[3] were extracted from the channels of a filter-bank and applied to the nonlinear compression of

$$C(x) = \log(1 + 20x),$$

where x represents the value of ZCPA histogram.

For the speech recognition using TDNNs, we need some pre-processing for using TDNN as a recognizer. The endpoint detection (EPD) should be performed before extracting a ZCPA feature is required. Unlike HMMs, we cannot directly put an input vector of arbitrary length to TDNN since the input layer of a TDNN is fixed. In this work, energy based EPD[14,15,16,17] was used. The time normalization is also essentially needed to match output feature vectors' dimension to the number of input nodes of TDNN. This is like dynamic time warping to a constant line. Here, the window size was changed according to the center frequency of a channel. An inverse relation between a window size and a center frequency holds since every frame of output should have similar number of zero crossings. In this simulation, we used the window size of $15/f_c$ where f_c represents the center frequency of a channel.

We performed the simulation for speech recognition using MFCC, MFCC with RASTA filtering[10], ZCPA, ZCPA with RASTA filtering, and ZCPA plus periodicity defined in (6) with RASTA filtering. Here, the TDNNs with 64 frame inputs, 2 hidden layers of 16 units, and the output layer of 4 units (64-16-16-4) were used. The simulation results are listed in Table 1. These simulation results showed us that 1) ZCPA-based features outperformed MFCC-based features, 2) RASTA filtering was effective in improving the recognition rate, and 3) the periodicity feature helped to improve the performance of ZCPA features.

We also performed the simulation of speech recognition using HMMs. In these experiments, we consider the speech recognition using LPCC, MFCC, and the suggested ZCPA features. For each feature, energy (E), delta (D), and acceleration (A) were added. In the ZCPA case, the energy defined by (7) was used. The simulation results are listed in Table 2. These simulation results showed us that ZCPA-based features outperformed LPCC-based or MFCC-based features.

In summary, the ZCPA-based feature were effective in the speech recognition using neck-microphones compared to other features. The FLCM features were also helpful to improve the performance of ZCPA features especially when we use TDNN classifiers. Note that these performance of speech recognition is not dependent upon environmental noises since neck-microphone signals are only captured by the vibration of speaker's vocal tract and not affected by propagated signals from other sound sources. This makes the neck-microphone-based command system useful in severely noisy environments compared to the ordinary microphone-based command system in which the recognition rate for this type

Table 1. Simulation results for speech recognition using TDNNs

Feature Type	Recognition Rate(%)
MFCC	60.7
MFCC with RASTA	65.6
ZCPA	82.2
ZCPA with RASTA	85.3
ZCPA + P with RASTA	90.8

Table 2. Simulation results for speech recognition using HMMs

Feature Type	Recognition Rate(%)
LPCC with (E, D, A)	83.4
MFCC with (E, D, A)	86.2
ZCPA with (E, D, A)	91.3

of task is usually more than 95 percent in clean environments and significantly reduced in severely noisy environments.[1]

5 Conclusion

In this paper, we presented the zero-crossing-based features for speech recognition using neck-microphones. Through the simulation for speech recognition using the neck-microphone voice command system, we have shown that suggested method provides the better performance than other approaches using conventional features such as LPCC or MFCC. We also showed that the features using the second order statistics such as the FLCM features were helpful to improve the performance of ZCPA features. The possibility of using the second order statistics in speech recognition is interesting and the further investigation of this method remains as subjects for future research.

References

1. Kay, S., Sudhaker, R.: A Zero Crossing-Based Spectrum Analyzer. IEEE Transactions on Acoustics, Speech, and Signal Processing **34** (1) (1986) 96-104
2. Sreenivas, T., Niederjohn, R.: Zero-Crossing Based Spectral Analysis and Svd Spectral Analysis for Formant Frequency Estimation in Noise. IEEE Transactions on Signal Processing **40** (2) (1992) 282-293
3. Kim, D., Lee, S., Kil, R.M.: Auditory Processing of Speech Signals for Robust Speech Recognition in Real-World Noisy Environments. IEEE Transactions on Speech and Audio Processing **7** (1) (1999) 55-69
4. Blachman, N.: Zero-Crossing Rate for the Sum of Two Sinusoids or a Signal Plus Noise. IEEE Transactions on Information Theory (1975) 671-675
5. Kedem, B.: Time series analysis by higher order crossings. IEEE Press (1994)
6. Haralick, R.M., Shanmugam, K., Dinstein, I.: Texture Features for Image Classification. IEEE Transactions on Systems, Man and Cybernetics **3** (6) (1973) 610-621
7. Davis, L.S., Johns, S.A., Aggarwal, J.K.: Texture Analysis Using Generalized Co-Occurrence Matrices. IEEE Transactions on Pattern Recognition and Machine Intelligence **1** (3) (1979) 251-259

[1] The recognition rate for the speech signals mixed with white Gaussian noises is usually reduced to less than 40 percent of the performance in clean environments when the signal to noise ratio is less than 5 dB.

8. Clausi, D.A., Jernigan, M.E.: A Fast Method to Determine Cooccurrence Texture Features Using a Linked List Implementation. Remote Sensing of Environment (1996) 506-509
9. Clausi, D.A., Zhao, Y.: Rapid Extraction of Image Texture by Co-Occurrence Using a Hybrid Data Structure. Computers and Geosciences **28** (6) (2002) 763-774
10. Hermansky, H.: Rasta Processing of Speech. IEEE Transactions on Speech and Audio Processing **2** (4) (1994) 578-589
11. Ghulam, M., Fukuda, T., Horikawa, J., Nitta, T.: A Noise-Robust Feature-Extraction Method Based on Pitch-Synchronous Zcpa for Asr. In Proceesings of INTERSPEECH-ICSLP **1** (2004) 133-136
12. Hanazawa, T., Hinton, G., Shikano, K., Waibel, A., Lang, K.: Phonem Recognition Using Time Delay Neural Networks. IEEE Transactions on Acoustics, Speech, and Signal Processing **37** (1) (1989) 328-339
13. Young, S., Kershaw, D., Odell, J., Ollason, D., Valtchev, V., Woodland, P.: HTK Book. Microsoft Corporation (2000)
14. Rabiner, L.: A Tutorial on Hidden Markov Models and Selected Applications in Speech Recognition. Proceedings of the IEEE **77** (2)(1989) 257-286
15. Rabiner, L., Sambur, M.: An Algorithm for Determining the Endpoints of Isolated Utterances. The Bell System Technical Journal **54** (2) (1975) 297-315
16. Savoji, M.H.: Endpointing of Speech Signals. Speech Communication **8** (1) (1989) 46-60
17. Mak, B., Junqua, J., Reaves, B.: A Robust Algorithm for Word Boundary Detection in the Presence of Noise. IEEE Transactions on Speech and Audio Processing **2** (3) (1997) 406-412

Memetic Algorithms for Feature Selection on Microarray Data

Zexuan Zhu[1,2] and Yew-Soon Ong[1]

[1] Division of Information Systems, School of Computer Engineering, Nanyang Technological University, Nanyang Avenue, Singapore 639798
[2] Bioinformatics Research Centre, Nanyang Technological University, Research TechnoPlaza, 50 Nanyang Drive, Singapore 637553

Abstract. In this paper, we present two novel memetic algorithms (MAs) for gene selection. Both are synergies of Genetic Algorithm (wrapper methods) and local search methods (filter methods) under a memetic framework. In particular, the first MA is a Wrapper-Filter Feature Selection Algorithm (WFFSA) fine-tunes the population of genetic algorithm (GA) solutions by adding or deleting features based on univariate feature filter ranking method. The second MA approach, Markov Blanket-Embedded Genetic Algorithm (MBEGA), fine-tunes the population of solutions by adding relevant features, removing redundant and/or irrelevant features using Markov blanket. Our empirical studies on synthetic and real world microarray dataset suggest that both memetic approaches select more suitable gene subset than the basic GA and at the same time outperforms GA in terms of classification predictions. While the classification accuracies between WFFSA and MBEGA are not significantly statistically different on most of the datasets considered, MBEGA is observed to converge to more compact gene subsets than WFFSA.

1 Introduction

Over the recent years, microarray technology has attracted increasing interest in many academic communities and industries. One major application of microarray technology lies in cancer classification. Thus far, a significant amount of new discoveries have been made and new bio-markers for various cancer have been detected from microarray data. However, due to the nature of the microarray gene expression data, cancer classification has remained a great challenge to computer scientists.

Microarray data is characterized with thousands of genes but with only a small number of samples available for analysis. This makes learning from microarray data an arduous task under the effect of curse of dimensionality. Furthermore, microarray data often contains many irrelevant and redundant features, which affect the speed and accuracy of most learning algorithms. Therefore, feature selection (also commonly known as gene selection in the context of microarray) is widely used to address these problems.

In general, feature selection methods can be categorized into filter and wrapper methods [1]. Filter methods assess the merits of features according to their

D. Liu et al. (Eds.): ISNN 2007, Part I, LNCS 4491, pp. 1327–1335, 2007.

individual relevance or discriminative power with respect to the target classes. Since these methods do not involve the induction algorithm, they are relatively inexpensive to compute. Wrapper methods, on the contrary, use the induction algorithm itself to evaluate the candidate feature subsets. They select feature subsets more suitable for the induction algorithm, generally at the expense of a higher computational time when compared to filter methods. One key issue of wrapper method is how to search the space of feature subsets. On microarray data, as the number of genes (features) are typically very large, most of existing search methods (e.g., complete search, heuristic search, and random search) face the problems of intractable computational time.

Genetic Algorithm (GA) has well known ability to produce high quality solution within tractable time even on complex problems [3,4,5]. It has been naturally used for gene selection and has shown promising performance in dealing with microarray data [2]. Unfortunately, due to the inherent nature of GA, it often takes a long time to locate the local optimum in a region of convergence and may sometimes not find the optimum with sufficient precision. One way to solve this problem is to hybridize GA with some memetic operators (also known as local search operators) [6,7,8] which are capable of fine-tuning and improving the solutions generated by the GA more precise and efficient. This form of evolutionary algorithms are often referred to as Memetic algorithms (MAs) [6,7].

In this paper, we present a comparison study on two MAs we have recently proposed for gene selection [8,9] on synthetic and real microarray data. Both are synergies of the Genetic Algorithm (wrapper methods) and local search methods (filter methods) under a memetic framework. In particular, the Wrapper-Filter Feature Selection Algorithm (WFFSA) [8] fine-tunes the population of GA solutions by adding or deleting features based on univariate feature filter ranking method. The second MA approach, Markov Blanket-Embedded Genetic Algorithm (MBEGA) [9], fine-tunes the GA population by adding the relevant features, removing the redundant and/or irrelevant features using Markov blanket technique.

The rest of this paper is organized as follows. Section 2 presents the two MAs proposed for gene selection. Section 3 presents the numerical results obtained from our empirical study on synthetic and real microarray datasets. Analysis of the numerical results and some discussions are also presented in the section. Finally, Section 4 concludes this paper.

2 System and Methodology

In this section, we describe the memetic framework of WFFSA and MBEGA. The local search method or filter method or otherwise known also as meme used in the WFFSA and MBEGA are then described in subsection 2.1.

The proposed Memetic framework for gene selection is briefly outlined in Figure 1. At the start of the MA search, a population of potential gene subset solution is generated randomly with each chromosome encoding a candidate gene subset. In the present work, each chromosome is composed of a bit string of length equal to the total number of features or genes in the problem of interest. Using binary encoding, a bit of '1' ('0') implies the corresponding gene is selected (excluded). The fitness of each chromosome is then obtained using an objective function based on the induction algorithm:

$$Fitness(c) = J(S_c) \tag{1}$$

where S_c denotes the selected gene subset encoded in a chromosome c, and the gene selection objective function, $J(S_c)$, provides a measure on the classification error for the given gene subset S_c. In this study, we use support vector machine (SVM) as the classifier since it has shown superior performance over other methods on microarray data. Further, to reduce the computational cost incurred, the leave-one-out error of SVM is estimated using the radius margin bound [10]. When two chromosomes are found having similar fitness (i.e. for a misclassification error of less than one data instance, the difference between their fitness is less than a small value of $\varepsilon = 1/n$, where n is the number of instances), the chromosome with a smaller number of selected genes is given greater chance of surviving to the next generation.

Memetic Algorithm for Gene Selection
BEGIN

1. **Initialize:** Randomly generate an initial population of feature subsets encoded with binary string.
2. **While**(*not converged or computational budget is not exhausted*)
3. Evaluate fitness of all feature subsets in the population based on $J(S_c)$.
4. Select the elite chromosome c_b to undergo local search.
5. Replace c_b with improved chromosome c_b'' using Lamarckian learning.
6. Perform evolutionary operators based on restrictive selection, crossover, and mutation.
7. **End While**

END

Fig. 1. Memetic algorithm for gene selection

In each MA generation, the elite chromosome, i.e., the one with the best fitness value, then undergoes a local search procedure in the spirit of Lamarckian learning [6]. Subsequently, the population of chromosome then undergoes evolutionary operators that includes linear ranking selection, and our proposed restrictive crossover and restrictive mutation [8] operators with elitism. To accelerate the evolutionary search, a constraint on the maximum number of '1' bits,m, in each chromosome is imposed.

2.1 Local Search

The local search procedure proposed is a recipe of two heuristic operators, namely *Add* and *Del*. For a given selected gene subset encoded in chromosome

c, we define $\mathbf{X}$ and $\mathbf{Y}$ as the subsets of selected and excluded genes encoded in c, respectively. An *Add* operator inserts genes from $\mathbf{Y}$ into $\mathbf{X}$, while a *Del* operator removes existing genes from $\mathbf{X}$ to $\mathbf{Y}$. The important question is which gene to add or delete from a given chromosome that encodes potential gene subset. Here, we consider two possible scheme for adding or deleting genes in WFFSA and MBEGA.

1. **Filter Ranking (WFFSA):** All features are ranked using a filter method. In this study the ReliefF [11] is considered. *Add* operator selects a feature from $\mathbf{Y}$ using the linear ranking selection method described in [12], and moves it to $\mathbf{X}$. *Del* selects a feature from $\mathbf{X}$ also using linear ranking selection and moves it to $\mathbf{Y}$. The outline for *Add* and *Del* operators are provided in Figures 2 and 3, respectively.
2. **Markov Blanket [13] (MBEGA):** Here, both the *Add* and *Del* operators selects a feature from $\mathbf{Y}$ using also the linear ranking selection approach. However, MBEGA differs in the use of the *C-correlation* measure [14] instead of ReliefF in WFFSA for ranking of features (see Figure 2 for the details). Further for a given X_i, MBEGA proceeds to remove all other features in $\mathbf{X}$ that have been covered by X_i using the approximate Markov blanket[1] [14]. If a feature X_j has a Markov blanket given by X_i, this suggests that X_j gives no additional useful information beyond X_i on class C. Hence, X_j may be considered as redundant and could be safely removed. If there is no feature in the approximate Markov blanket of X_i, the operator then tries to delete X_i itself. The detailed procedure for *Del* operator is outlined in Figure 4.

***Add* Operator:**
BEGIN

1. Rank the features in $\mathbf{Y}$ in descending order based on ReliefF in WFFSA while the *C-correlation* measure in MBEGA.
2. Select a feature Y_i in $\mathbf{Y}$ using linear ranking selection [12] such that the higher the quality of a feature in $\mathbf{Y}$, the more likely it will be selected to move to $\mathbf{X}$.
3. Add Y_i to $\mathbf{X}$.

END

Fig. 2. *Add* operator

It is possible to quantify the computational complexity of the two local operators based on the search range L, which defines the maximum numbers of *Add* and *Del* operations. Therefore, with L possible *Add* and L possible *Del* operations, there are a total of L^2 possible combinations of *Add* and *Del* operations that may be applied on a chromosome during local learning. Since our

[1] Here we use the approximate Markov blanket [14] instead of a complete Markov blanket [13] to reduce the computational expense involved.

***Del* Operator in WFFSA:**
BEGIN

1. Rank the features in **X** in ascending order using ReliefF.
2. Select a feature X_i in **X** using linear ranking selection [12] such that the lower the quality of a feature in **X**, the more likely it will be selected to move to **Y**.
3. Remove X_i to **Y**.

END

Fig. 3. *Del* operator in WFFSA

***Del* Operator in MBEGA:**
BEGIN

1. Rank the features in **X** in descending order based on *C-correlation* measure.
2. Select a feature X_i in **X** using linear ranking selection [12] such that the higher the *C-correlation* value of a feature in **X**, the more likely it will be selected.
3. Eliminate all features in $\mathbf{X} - \{X_i\}$ which are in the approximate Markov blanket of X_i. If no feature is eliminated, remove X_i itself.

END

Fig. 4. *Del* operator in MBEGA

previous study in [8] suggests $L = 4$ and the improvement first strategy gives better search performances than several others scheme considered in WFFSA, we use such a configurations in the present comparison study between WFFSA and MBEGA for gene selection. The details of the local search learning procedure used to improve only the elite chromosome of each GA search generation is outlined in Figure 5.

3 Empirical Study

In this section, we investigate the performances of WFFSA and MBEGA on both synthetic and real world microarray data. Results of basic GA (i.e., without local search) are also presented for comparison. In our empirical study, WFFSA, MBEGA and GA use the same parameter configurations with population size = 50, crossover probability = 0.6, and mutation rate = 0.5. The stopping criteria for all three algorithms are defined by a convergence to the global optimal or a maximum computational budget of 2000 fitness functional calls is reached. It is worth noting that the fitness function calls made to $J(S_c)$ in the local search are also included as part of the total fitness function calls for fair comparison to the GA. The maximum number of bit '1' in the chromosome m is set to 50.

Local Search
BEGIN

Select the elite chromosome c_b to undergo memetic operations.
For $l = 1$ **to** L^2
    Generate a unique random pair $\{a, d\}$ where $0 \leq a, d < L$.
    Apply a times *Add* on c_b to generate a new chromosome c_b'.
    Apply d times *Del* on c_b' to generate a new chromosome c_b''.
    Calculate fitness of modified chromosome c_b'' based on $J(S_c)$.
    If c_b'' is better than c_b either on fitness or number of features
        Replace the genotype c_b with c_b'' and stop memetic operation.
    End If
End For

END

Fig. 5. Local search procedure

To prevent overfitting, we consider a balanced .632+ external bootstrap [15] for estimating the prediction accuracy of a given gene subset. At each bootstrap, a training set is sampled with replacement from the original dataset, and the test data is formed by unsampled instances. Note that $J(S_c)$ uses only the training data while the prediction accuracy of a feature subset is evaluated based on the unseen test data. The external bootstrap is repeated 30 times for each dataset and the average results are reported.

3.1 Synthetic Data

We begin our study of the proposed MAs on synthetic microarray data since the true optimal gene subset is known beforehand. Here, the two-class synthetic data used is composed of 4030 genes and 80 samples with 40 samples for each class label. The centroid of these two classes are located at (-1,-1,-1) and (1,1,1). Three groups of relevant genes are generated from a multivariate normal distribution, with 10 genes in each group. All these relevant genes are generated using variance 1 and a mean vector $[\mu_1, \mu_2, \ldots, \mu_{80}]$, of $\mu_1 = \ldots = \mu_{40} = -1$ and $\mu_{41} = \ldots = \mu_{80} = 1$. The correlation between intra-group genes is 0.9, whereas the correlation between inter-group genes is 0. Hence, genes in the same group are redundant with each other and the optimal gene subset to separate the two classes consists of any 3 relevant genes from different groups. Another 4000 irrelevant genes are added to the data. Among these 4000 genes, 2000 are drawn from a normal distribution of N(0,1) and the other 2000 genes are sampled with a uniform distribution of U[-1,1].

The results of the feature selection by GA, WFFSA, and MBEGA are tabulated in Table 1. Both the WFFSA and MBEGA outperform GA in terms of classification accuracy, showing lower test error rates in Table 1 than the latter. MBEGA obtains the lowest test error rates among all three methods. Both

Table 1. Feature selection by each algorithm on synthetic data

Algorithm	GA	WFFSA	MBEGA
Test Error	0.0701	0.0250	**0.0202**
#Selected Groups	2.6	**3**	**3**
#Selected Genes	34.1	35.5	**9.7**
#Selected Relevant Genes	3.2	23.2	8.2
#Selected Redundant Genes	0.6	20.2	5.2
#Selected Irrelevant Genes	30.9	12.3	1.5

#Selected Genes = #Selected Relevant Genes + #Selected Irrelevant Genes;
#Selected Relevant Genes = #Selected Groups + #Selected Redundant Genes.

WFFSA and MBEGA also select more compact feature subset than GA. They consistently select relevant genes found in all the 3 groups, while GA fails to do so and converges to subsets that contain irrelevant genes at the end of the computational budget of 2000 fitness function calls. Among all, MBEGA selects the smallest subset of relevant genes with the use of Markov blanket for local learning. Since the filter ranking procedure in WFFSA considers individual gene contributions during local learning, it cannot identifying the interactions between the genes. Therefore, unlike MBEGA, WFFSA is incapable of removing the redundant genes.

3.2 Microarray Data

In this section, we consider some real world microarray datasets having significantly large number of features (genes). In particular, 6 publicly available datasets (i.e., Colon Cancer, Central Nervous System, Leukemia, Breast Cancer, Lung Cancer, and Ovarian Cancer) in [16] are considered in the present study. In Table 2, the average test error and average number of selected genes of each feature selection algorithm (across 30 runs using the .632+ bootstraps) on the eleven datasets are reported. In gene selection, evaluation of the objective function based on SVM and local learning takes up the overwhelming bulk of the computation. To speedup the gene selection process, we considered the Grid problem solving environment developed in [17,18]. By doing so, a sub-linear improvement in the gene selection efficiency can be achieved via parallelism since gene subsets and local search in an EA generation can be conducted independently and simultaneously across multiple compute nodes.

It is consistent with the earlier results in section 3.1 on synthetic data that both WFFSA and MBEGA obtain lower test error rates than GA (see Table 2). This suggests the local searches in both MAs have successfully helped to fine-tune the GA more accurately and efficiently. As a result, smaller subset of important genes that generates improved classification accuracies are found for the datasets. Further, the results in Table 2 also suggests that WFFSA and MBEGA are competitive to each other in terms of classification accuracy. Both outperforms each other on three out of the eleven datasets considered. Nevertheless, it is

Table 2. Performance of feature selection algorithms on microarray datasets

		GA	WFFSA	MBEGA
Colon	*err*	0.1899	0.1523	**0.1434**
(2000 × 60)	$\overline{\lvert S_c \rvert}$	23.3	**23.1**	24.5
Central Nervous System	*err*	0.317	**0.277**	0.2779
(7129 × 60)	$\overline{\lvert S_c \rvert}$	24.1	22.1	**20.5**
Leukemia	*err*	0.0769	**0.0313**	0.0411
(7129 × 72)	$\overline{\lvert S_c \rvert}$	25.2	29.6	**12.8**
Breast	*err*	0.3119	0.2582	**0.1926**
(24481 × 97)	$\overline{\lvert S_c \rvert}$	22.1	27.8	**14.5**
Lung	*err*	0.0193	**0.0088**	0.0104
(12533 × 181)	$\overline{\lvert S_c \rvert}$	24.4	28.6	**14.1**
Ovarian	*err*	0.0057	0.0050	**0.0048**
(15154 × 253)	$\overline{\lvert S_c \rvert}$	23.3	18.7	**9.0**

err: test error; $\overline{|S_c|}$: average number of selected genes.

worth highlighting that once again MBEGA converges to more compact gene subsets than WFFSA by eliminating the redundant genes that exists.

4 Conclusions

In this paper, two MAs feature selection algorithms WFFSA and MBEGA are investigated and compared using synthetic and real world microarray data. With the inclusion of local learning based on filter and/or Markov blanket guided *Add/Del* operators in the basic GA, both MAs have been observed to converges to solutions with more compact feature/gene subsets at improved classification performance and efficiency. The prediction accuracy of both MAs described are not significantly different when experimented on synthetic and microarray data. Nevertheless, MBEGA generally leads to a lower redundant features in the final solution than WFFSA. For diagnostic purpose in clinical practice, a smallest subset of genes would generally be preferred if good predictive performance is maintained. Therefore, MBEGA would be more suitable for gene selection.

Acknowledgement

This work has been funded in part under the A*STAR SERC Grant No. 052 015 0024 administered through the National Grid Office.

References

1. Kohavi, R., John, G.H.: Wrapper for Feature Subset Selection. Artificial Intelligence, **97(1-2)** (1997) 273-324
2. Ong, Y.S., Keane, A.J.: A Domain Knowledge Based Search Advisor for Design Problem Solving Environments. Engineering Applications of Artificial Intelligence, **15(1)** (2002) 105-116

3. Lim, M.H., Yu, Y., Omatu, S.: Extensive Testing of a Hybrid Genetic Algorithm for Solving Quadratic Assignment Problems. Computational Optimization and Applications **23** (2002) 47-64
4. Ong, Y.S., Nair, P.B., Lum, K.Y.: Max-min Surrogate-assisted Evolutionary Algorithm for Robust Aerodynamic Design. IEEE Trans. on Evolutionary Computation **10(4)** (2006) 392-404
5. Wahde, M., Szallasi, Z.: A Survey of Methods for Classification of Gene Expression Data Using Evolutionary Algorithms. Expert Review of Molecular Diagnostic **6(1)** (2006) 101-110
6. Ong, Y.S., Keane, A.J.: Meta-Lamarckian in Memetic Algorithm. IEEE Trans. on Evolutionary Computation **8(2)** (2004) 99-110
7. Ong, Y.S., Lim, M.H., Zhu, N., Wong, K.W.: Classification of Adaptive Memetic Algorithms: A Comparative Study. IEEE Transactions on Systems, Man and Cybernetics-Part B **36(1)** (2006) 141-152
8. Zhu, Z., Ong, Y.S., Dash, M.: Wrapper-filter Feature Selection Algorithm Using a Memetic Framework. IEEE Transactions On Systems, Man and Cybernetics-Part B, accepted 2006.
9. Zhu, Z., Ong, Y.S., Dash, M.: Markov Blanket-embedded Genetic Algorithm for Gene Selection. Pattern Recognition, submitted 2006.
10. Vapnik, V.: Statistical Learning Theory. Wiley (1998)
11. Robnic-Sikonja, M., Kononenko, I.: Theoretical and Empirical Analysis of ReliefF and RReliefF. Machine Learning **53(1-2)** (2003) 23-69
12. Baker, J.E.: Adaptive Selection Methods for Genetic Algorithms. In Proc. Int'l Conf. Genetic Algorithm and Their Applications (1985) 101-111
13. Koller, D., Sahami, M.: Toward Optimal Feature Selection. 13th International Conference on Machine Learning. Morgan Kaufmann, Bari, Italy (1996)
14. Yu, L., Liu, H.: Efficient Feature Selection via Analysis of Relevance and Redundancy. Journal of Machine Learning Research **5** (2004) 1205-1224
15. Braga-Neto, U.M., Dougherty, E. R.: Is Cross-validation Valid for Small-sample Microarray Classification, Bioinformatics **20(3)** (2004) 374-380
16. Li, J., Liu, H.: Kent Ridge Biomedical Data Set Repository, Http://sdmc-lit.org.sg/GEDatasets (2002)
17. Salahuddin, M., Hung, T., Soh, H., Sulaiman, E., Ong, Y.S., Lee, B.S., Ren, Y.: Grid-based PSE for Engineering of Materials (GPEM), CCGrid 2007, submitted.
18. Lim, D., Ong, Y.S., Jin, Y., Sendhoff, B., Lee, B.S.: Efficient Hierarchical Parallel Genetic Algorithms Using Grid Computing. Future Generation Computer Systems: The International Journal of Grid Computing: Theory, Methods and Applications, doi:10.1016/j.future.2006.10.008.

Feature Bispectra and RBF Based FM Signal Recognition

Yuchun Huang, Zailu Huang, Benxiong Huang, and Shuhua Xu

Dept. of Electronics and Information Engineering,
Huazhong Univ. of Sci. & Tech.(HUST), Wuhan, 430074, P.R. China
supermiracle@163.com

Abstract. Automatic communication signal (e.g., FM signal) classification and identification focus on finding the fine feature contained in the almost approximate noisy communication signal comprehensively identifying the the same or different version of transmitters in modern electronic warfare. Direct use of HOS becomes unavailable for on-line application because of its huge computation time and memory space especially in the case of high frequency FM signal. This paper presents a novel view to improve the HOS analysis efficiency by sub-sampling while preserving the noise-contaminated fine feature and eliminating the random Gaussian noise. FM signal-related feature bispectra are also introduced to translate the 2-D feature matching pattern to a 1-D one applicable for an optimal adaptive k-means iterative RBF classifier. Computer simulations show that this novel feature bispectra outperform AIB and SB in terms of computation time and recognition rate for on-line steady FM signal recognition.

1 Introduction

Modern communication signal processing is focusing on the fine feature contained in the communication signal and trying to identify the individual or the like signal transmitting device apart from the traditional requirement for reliability and efficiency in modulation and demodulation, which is quite difficult and complicated but of fundamental significance to modern electronic warfare, target tracking, antispy and secure communication. FM signal, one of the most common and widely used communication signals, can be generated by the same version of a manufacturer or by different versions of various manufacturers, but bears approximate modulation parameters. The intrinsic little variance of transmitter in the performance of oscillator, the nonlinearity of the amplifier, the interaction of other circuit components are bound with the slight change in the output performance parameters of FM signal, such as carrier frequency, modulation index, harmonic components, parasite modulation amplitude, etc. In practice, any detectable, unique, steady feature mentioned can act as the characteristic ID. However, real-world FM signals almost suffers stochastic noises including modulation interference, propagation interference, device electronic noise, sampling noise and unwanted affection from near electromagnetic components. These noises reduce

D. Liu et al. (Eds.): ISNN 2007, Part I, LNCS 4491, pp. 1336–1345, 2007.

SNR and have different properties (additive or multiplicative, Gaussian or non-Gaussian), so the discriminant ability and accuracy of the parameters are greatly limited. Fortunately lots of experiments show that most of these parameters can be the characteristic fine feature of the transmitter. Since these cover a variety of topics, some are ignored and only the basic, main features are discussed in the simplest case.

Sub-sampling technique in time domain has higher SNR by deserting some samples seriously affected by noise while maintaining the source entropy according to the accordant degree of samples. The problem is how to determine the sub-sampling principle and the number of samples large enough to ensure the lowest limit of source coding. The period gram and its variants, modern spectral estimation are famous for measuring signal parameters, but second-order measures contain no phase information, cannot identify nonlinear phase coupling, and are more vulnerable to noise [2]. Higher order spectra are powerful tools in non-Gaussian and non-linear signal processing. They remain both amplitude and phase information of a signal, suppress additive colored Gaussian noise of unknown power spectrum, extract information due to deviations from Gaussianity, detect and characterize nonlinear properties in signals as well as identify nonlinear feature. In recent years, HOS found successful application in many cases of feature extraction, such as HRRP, sonar signal, speaker recognition, fault feature, image matching, etc. Actually HOS signal processing has become one of the kernels of modern signal processing.

However, HOS spectra (hereafter discuss the most common bispectrum only) are multidimensional function. Direct use of them results in a 2-D feature matching score that is not practicable for on-line automatic signal classification and identification. To overcome this difficulty, many types of 1-D slices, integrated bispectra were developed, including horizontal/diagonal slices, axially integrated bispectra(AIB), radially integrated bispectra (RIB), circularly integrated bispectra (CIB)[1]. All these methods transformed 2-D pattern of bispectra to 1-D one, their difference lies in the integrated paths or slices selection. Obviously, any slices bispectra or integrated bispectra is only an approximate subset of the discrete bispectra of a sequence, cannot obviate the redundant trivial or even baneful bifrequency points (f1, f2) for signal recognition. Selected Bispectra(SB) make use of Fisher's class separability as discriminant measure to the key bispectra with maximum interclass separability as feature vectors. But calculation of SB becomes huge as the number of versions of FM signals increases.

The novel view presented in this paper is aiming at the efficiency of bispectra computation and feature extraction. Though all the sampling data contain information, some are redundant and not necessarily the key data for fine feature in the case of FM signal. Usually FM signal has a relatively high carrier frequency (f_c) at tens of KHz (or more) and the modulated voice signal runs at the range of 300–3400Hz. To extract the fine feature and obviate the stochastic voice, more attention must be paid to the frequency components between f_c and $f_c + 300$ for a real FM signal to detect the nonlinearity and modulation parameter of the transmitter. Actually some FM transmitting devices do produce pilot noise-squelching signal outband of voice. Accordingly, it is not the optimal choice to make bispectra analysis on the basis of total sampling data. If fc of FM signal can be coarsely estimated, the number of data for bispectra analysis can be greatly reduced by sub-sampling. The simple and effective method for the estimation of carrier frequency of narrow-band signal is by means of Hilbert transform. Real bandlimited FM signal can then be shifted to a relatively low frequency range. Low frequency thus results in the low sampling rate consistent with preservation of the fine feature. The sub-sampling principle depends on the f_c. Since some prior information (f_c) of FM signal is obtained, the bifrequency feature selection of bispectra can also make use of it to choose the significant feature bispectra as the input of RBF classifier. On account of the complexity of real FM signal, this paper deals with different versions of FM signal with approximate f_c in the simplest case.

2 Feature Bispectra and RBF Classifier

2.1 FM Signal

For simplicity, let consider the sinusoid FM signal(most pilot noise-squelching signal can be attributed to this type) defined by

$$\begin{aligned} S_{FM}(t) &= A\cos\left[2\pi f_c t + K_F \int A_m \cos(2\pi f_m t)\, dt\right] \\ &= A\cos\left[2\pi f_c t + \beta_{FM}\sin(2\pi f_m t)\right] . \end{aligned} \tag{1}$$

where β_{FM} is the modulation index and represents the phase deviation of FM signal.

Expanding (1) into products of sines and cosines, we get

$$S_{FM}(t) = A \sum_{n=-\infty}^{\infty} J_n(\beta_{FM}) \cos\left[2\pi (f_c + n\, f_m)\, t\right] . \tag{2}$$

where $J_n(\beta_{FM})$ is the n-order bessel coefficient.Provided $\beta_{FM} > n$, the modulation index β_{FM} can be coarsely estimated from

$$\beta_{FM} = \frac{2n\, J_n(\beta_{FM})}{J_{n-1}(\beta_{FM}) + J_{n+1}(\beta_{FM})} . \tag{3}$$

2.2 Coarse Frequency Estimation

Consider a real signal $x(t)$ with a time-varying phase in additive noise

$$y(t) = x(t) + \eta(t) . \tag{4}$$

where narrowband signal $x(t) = A\cos[2\pi f_c t + \theta(t)]$, $\eta(t)$ is real white Gaussian noise with variance σ_η^2. The analytical signal associated with $y(t)$ is given by

$$z(t) = y((t) + j\, H[y(t)] = A\exp\{j[\theta(t) + 2\pi f_c t]\} + \epsilon(t) . \tag{5}$$

where H represents the Hilbert transform and is a complex-valued white Gaussian noise with i.i.d. real and imaginary part with total variance $\sigma_\epsilon^2 = 2\,\sigma_\eta^2$.

The Instantaneous Frequency(IF) is defined by

$$f(t) = f_c + \frac{1}{2\pi}\frac{d\theta(t)}{dt} . \tag{6}$$

So the carrier frequency f_c can be estimated by Least Square Error linear fitting method shown in (7)

$$\sum \left[\text{unwrap}\left(\text{angle}\left(z\left[n\right]\right)\right) - \left(2\pi f_c n + \varphi\right)\right]^2 = min . \tag{7}$$

where unwrap$\left(\text{angle}\left(z\left[n\right]\right)\right)$ is the unwrapped continuous phase of discretized sequence of $z(t)$. Obviously,this is a coarse IF estimation method.This method can also estimate the modulation index β_{FM} of sinusoid signal.

2.3 Subsampling for Bispectra Data

Bispectrum is defined as 2-D Fourier transform of the third order cumulant.Let $\{x[n], n = 0, 1, \ldots, N-1\}$ be a discrete time signal,its discrete Fourier transform $\{X[k]\}$ is defined as

$$X[k] = \frac{1}{N}\sum_{n=0}^{N-1} x[n]\, e^{-j\,k\frac{2\pi}{N}n}, \qquad k = 0, 1, \ldots, N-1 . \tag{8}$$

The power spectrum of $\{x[n]\}$ is defined as

$$P[k] = \frac{1}{N} X[k]\, X^*[k] . \tag{9}$$

and its bispectrum can be defined as

$$B[k1, k2] = \frac{1}{N} X[k1]\, X[k2]\, X^*[k1+k2] . \tag{10}$$

The normalized bispectrum bicoherence can be defined as

$$BIC[k1, k2] = \frac{|B[k1, k2]|}{\sqrt{P[k1]\,P[k2]\,P[k1+k2]}} . \tag{11}$$

So $BIC[k1, k2]$ is a representation of the degree of Quadratic Phase Coupling (QPC).

By the symmetry properties, the bispectra of a real FM signal (2), many discrete peaks in the 3-D graph shown Fig. 1, are uniquely defined by its Principal Domain(PD): $0 \le k2 \le k1 \le k1 + k2 \le \pi$, provided that there is no bispectral aliasing. It can be seen from Fig.1 that the main bispectra components of FM signal concentrate at the center of the bifrequency plane and that there are large 'margin bispectra' area around. Obviously, the 'margin bispectra' contribute little to the FM signal recognition. So the detailed 'margin bispectra' can be

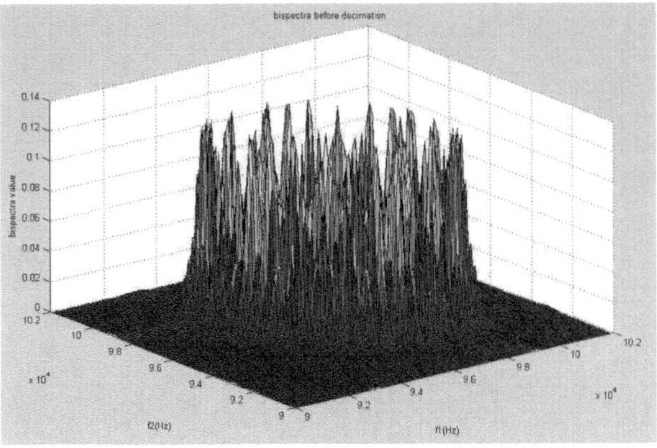

Fig. 1. Bispectra before decimation

suppressed someway. By the Carson formula below, the bandwidth of a FM signal (2) can be defined as

$$B_{FM} = 2\left(\beta_{FM} + 1\right) f_m \ . \tag{12}$$

where indicates that components outside $[f_c - B_{FM}/2, f_c + B_{FM}/2]$ are small enough to be ignored as shown in the 'margin bispectra' of Fig.1.

Based on the fact that a FM signal (2) is approximately band-passed, high carrier frequency can be transformed to zero by some shift operation in frequency domain which means certain decimation in time domain for a discrete sequence. The new sampling rate F_t can be estimated by

$$\begin{cases} F_t > B_{FM} \\ \mathrm{mod}\ (f_c, F_t) \approx 0 \end{cases} \tag{13}$$

to avert possible large aliasing distortion in frequency domain. Note that: 1) the new sampling rate F_t can be far below the Nyquist frequency, so it is sub-sampling and inevitable to introduce little acceptable distortion in frequency domain; 2) From (10) is also sub-sampling in bispectra domain. But by sub-sampling the 'margin bispectra' can be greatly suppressed, as shown in Fig.2. Moreover the reduction of sampling data can improve the efficiency of bispectra calculation and loss little accuracy for FM signal recognition; 3) In practice, the estimation of β_{FM} and f_c may bring some error. To mask their little fluctuation, F_t can be relaxed to cover the bandwidth of almost all observed, approximate FM signals by enhancing $(\beta_{FM} + 1)$ or f_m.

2.4 Feature Bispectra Selection

Although sub-sampling mentioned above reduces the computation complexity of bispectra, M-point bispectra still results in a M × M 2-D feature vector which is a heavy burden for the ensuing RBF classifier. The common method

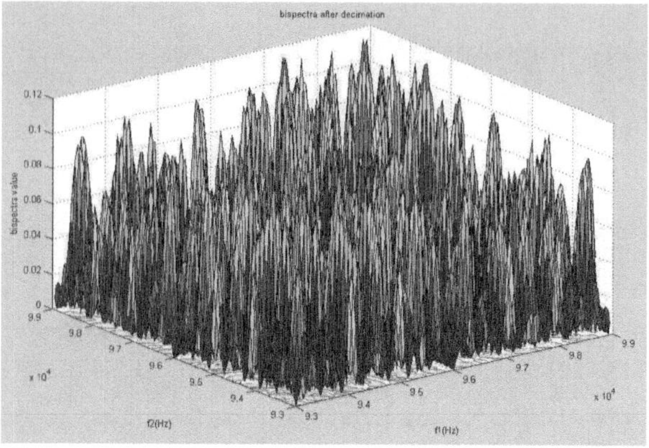

Fig. 2. Bispectra after decimation

to this problem is integrated bispectra or selected bispectra. Bispectrum slicing is also useful in many cases. But the problem is how to select the type and position of bispectra slices. Based on the bandpass property of FM signal and the symmetry property of bispectra, a significant bispectra region of sub-sampled FM signal(named 'feature bispectra')can be determined by

$$\begin{cases} 0 \le f_1 \le \frac{B_{FM}}{2} \\ -\frac{B_{FM}}{2} \le f_2 \le \frac{B_{FM}}{2} \end{cases} \quad \text{or} \quad \begin{cases} 0 \le k_1 \le \frac{B_{FM} M}{2F_t} \\ -\frac{B_{FM} M}{2F_t} \le k_2 \le \frac{B_{FM} M}{2F_t}. \end{cases} \tag{14}$$

Where M is the sampling points in frequency. The bispectra contents in both the first and the fourth quadrant are selected because only the first quadrant or the triangular PD is not sufficient in many applications [3] to get a better understanding of the quadratic coupling nonlinear signal. Though feature bispectra may contain redundant information, it is easy to understand and extract the feature vector of FM signal. When the bandwidth of FM signal is small, feature bispectra can greatly reduce the size of RBF classifier which is of significance to on-line FM signal recognition.

2.5 RBF Classifier

Automatic signal classification and identification can be subdivided into three phases: data acquisition, feature extraction, target classification and identification. Once the salient feature bispectra vector is extracted, the next problem is to compare it with those vectors of known FM signal versions stored in the database to identify the version of the collected target. This is a problem of pattern match in high dimensional space.

RBF neural network is nonlinear and belongs to a class of neural network model in which the activation of a hidden unit is determined by the distance between the input feature vector and a known one. RBF can make any arbitrary mapping between the input and output patterns [4]. Its learning speed, generalization, adaptation capability and intrinsic multi-class resolution are suitable for FM signal classification and recognition.

One crucial problem for training the RBF Network is the proper selection of hidden neurons and their parameters (center, width). Chen et al proposed the orthogonal least squares (OLS) procedure for choosing the centers one by one with the fixed common width from the input vectors [5]. Supervised learning schemes, such as gradient descent[4], all samples-based BP[6], growing and pruning RBF[7] and etc., are however very time- consuming. One form of overcoming this difficulty is to design the network by employing such unsupervised clustering techniques as SOM [9], k-means and its variants [8], evolutionary clustering [10]. Here an optimal adaptive k-means iteration algorithm is developed for the feature bispectra classification. This sequential learning algorithm offers not only on-line all parameters adaptability but robustness to center initialization. It creates neurons one at a time, if necessary. At each iteration the normalized input feature that results in lowering the network error of a K neurons RBF the most, is drawn and the similarity measure between the newly selected input feature and the kth cluster center is checked:

$$d\left(x, c_k\right) = \left\| x - c_k \right\|^2 / \upsilon_k, \qquad k = 1, \ldots, K \,. \tag{15}$$

If the similarity $d\left(x, c_k\right)$ falls beneath the threshold, the newly input x is used to create a hidden neuron; else is assigned to the cluster c_{nr} with highest similarity to x and only the parameters of cluster c_{nr} update as follows:

$$\begin{aligned} c_{nr}^{t+1} &= c_{nr}^{t} + \eta^{t}\left(x^{t} - c_{nr}^{t}\right), \\ \hat{\upsilon}_{nr}^{t+1} &= \alpha \hat{\upsilon}_{nr}^{t} + (1-\alpha)\left\| x^{t} - c_{nr}^{t} \right\|^{2} . \end{aligned} \tag{16}$$

Where the superscript and subscript denote the iteration step and the cluster index respectively, is a constant very close to 1 that defines the estimation precision. The initial width value $\hat{\upsilon}_k^0$ of all clusters is set to the same small number, which allows the fast disappearance of initialization effects. The learning rate η^t at iteration t is defined as:

$$\eta^{t} = \ln K + \sum_{i=1}^{K} \hat{\upsilon}_{i,norm}^{t} \ln \hat{\upsilon}_{i,norm}^{t} \,. \tag{17}$$

where $\hat{\upsilon}_{i,norm}^{t} = \hat{\upsilon}_i^t / \sum \hat{\upsilon}_j^t$ is the normalized width of ith cluster.

Finally the error of the new network is checked, and if low enough iteration is finished. Otherwise the above procedure is repeated until the error goal is met, or the maximum number of neurons is reached.

Obviously the proposed algorithm improves the traditional k-means by dynamic modification of the learning rate. The learning rate is derived from the difference between the quality of the current clusters and of the target ones. The entropy of normalized width is employed in (17) to ensure that the learning rate increases when the current clusters are far from the target ones and otherwise reduces.

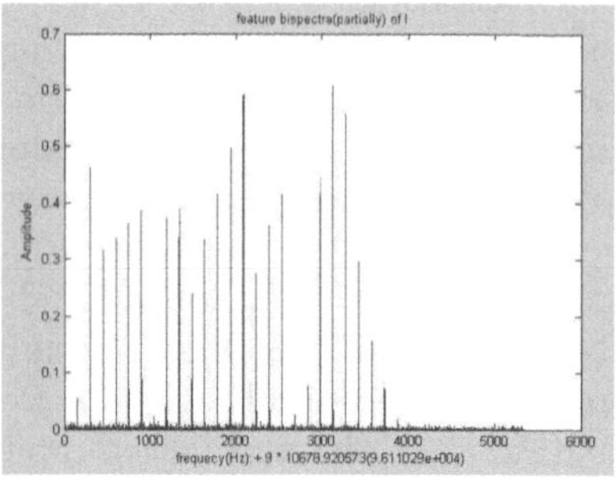

(a) partial feature bispectra graph of FM signal I

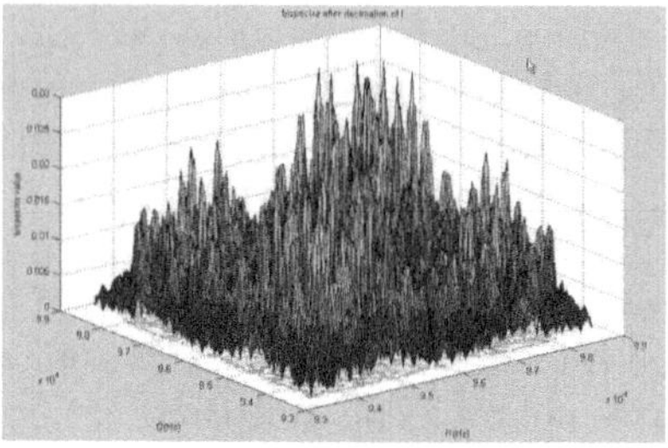

(b) 3-D bispectra graph of FM signal I

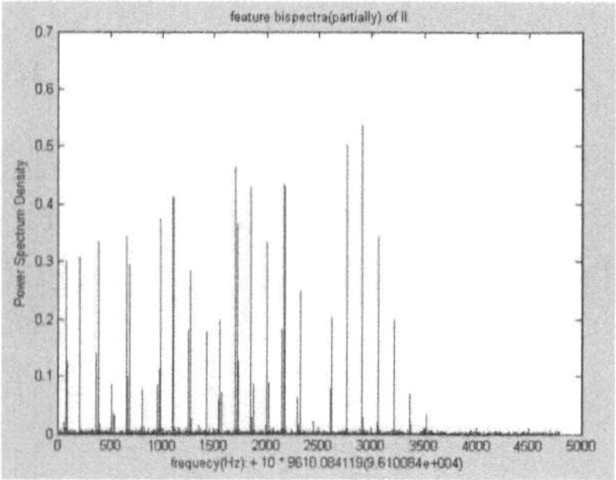

(c) partial feature bispectra graph of FM signal II

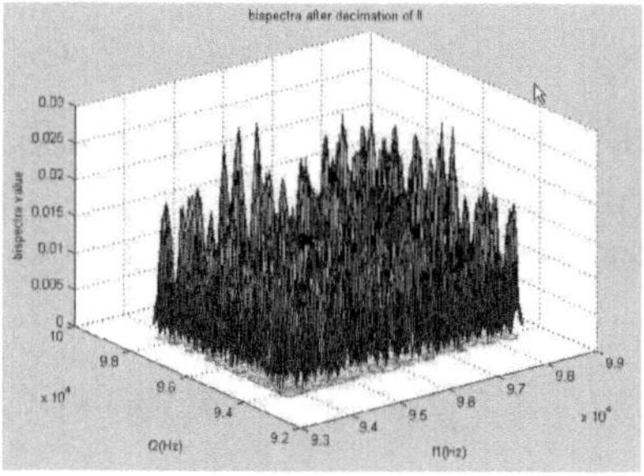

(d) 3-D bispectra graph of FM signal II

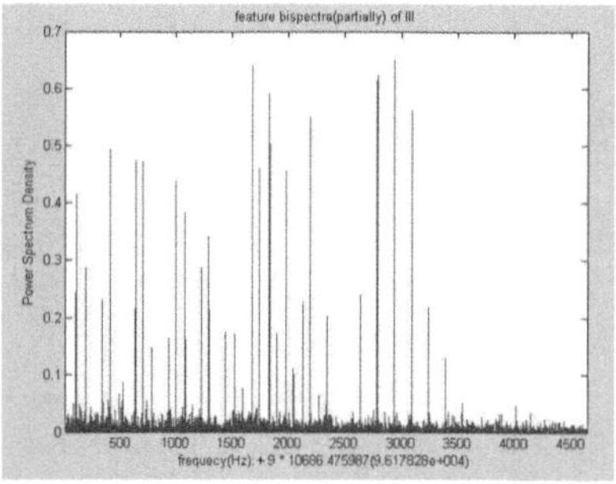

(e) partial feature bispectra graph of FM signal III

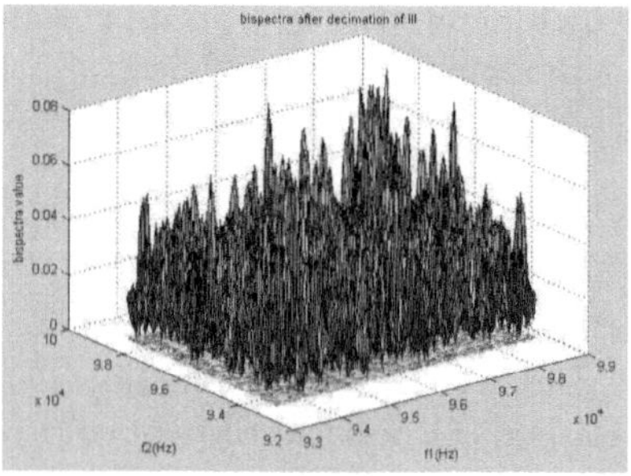

(f) 3-D bispectra graph of FM signal III

Fig. 3. Feature bispectra(partially) and 3-D bispectra graph of three versions of FM signal

3 Simulation

Feature bispectra is tested for three versions of FM signals in the presence of additive Gaussian white noise. The basic signal and modulation parameters of typical one of each version are shown in Table 1.[1]

Table 1. Parameters of three typical FM signal

Parameters	Signal I	Signal II	Signal III
A	1	1	1
Fm(Hz)	150	150	150
f_c(Hz)	96114.8991	96107.2513	96158.0324
f_{c}_est(Hz)	96110.285	96100.842	96178.285
f_s(Hz)	8*96K	8*96K	8*96K
β_{FM}	19.3297	17.7985	17.3982
β_{FM}_est	17.0118	15.7568	8.2218
Time-Len(s)	0.6	0.6	0.6
SNR(dB)	-3	-3	-9

where there is little overlapping in the parameters of three typical FM signals.

The 3-D bispectra and partial feature bispectra graph of the three FM signals are figured out in Fig.3. The left column (Fig.3(a), Fig.3(c), Fig.3(e)) represents part of the feature bispectra and the right column (Fig.3(b), Fig.3(d), Fig.3(f)) is the 3-D bispectra graph.

These figures show: 1) the three versions of FM signal are separable by feature bispectra; 2) the position of peaks changes with the carrier frequency; 3)the width of peaks and amplitude at same frequency change with the modulation index.

So the feature bispectra act as a comprehensive representation of the FM signal. Moreover, the intrinsic ability of feature bispectra is to find the quadratic coupling nonlinearity in the real-world FM signal (ignored in simulation for simplicity).

Table 2. Comparison of Performance of three Bispectrum Method

Method	AIB	SB	Feature Bispectra
Recognition rate	79.6%	82.4%	80.8%
Recognition time(s)	19.2	17.3	9.2
Training time(s)	19.7	22.5	9.8

[1] f_{c}_est and β_{FM}_est are the mean estimation of carrier frequency,modulation index respectively.

As far as the calculation time is concerned, feature bispectra outweigh AIB and SB at the same recognition rate level, which presents a window to the successful application of bispectra for FM signal analysis and recognition (Table 2).[2]

4 Conclusion

FM signal-related feature bispectra are a comprehensive representation of the fine features of its transmitter. They are easy to extract and provide considerable savings of computer time and memory space while preserving a high recognition rate. So feature bispectra and RBF- based classifier are applicable for on-line FM signal recognition. More experiments yet should be carried out to perfect the fine feature family of FM signals in any complex cases.

References

1. Zhang, X.: A New Feature Vector using Selected Bispectrum for Signal Classification with Application in Radar Target Recognition. IEEE Trans. Signal Processing **49** (2001) 1875-1885
2. Marple, S.L: A Tutorial Overview of Modern Spectral Estimation. International Conference on Acoustic, Speech, and Signal Processing **4** (1989) 2152-2157
3. Hinich, M.J.: On the Principal Domain of the Discrete Bispectrum of a Stationary Signal. IEEE Trans. Signal Processing **43** (1995) 2130-2134
4. Haykin, S.: Neural Networks-A Comprehensive Foundation. Prentice Hall, 2nd edition (1999)
5. Chen, S., Cowan, C.F.N, Grant, P.M.: Orthogonal Least Squares Learning Algorithm for Radial Basis Function Networks. IEEE Trans. Neural Networks **2** (1991) 302-309
6. Bors, A. G., Gabbouj, M.: Minimal Topology for a Radial Basis Functions Neural Network for Pattern Classification. Dig. Signal Process. **4** (1994) 173-188
7. Mao, K.Z.: RBF Neural Network Center Selection based on Fisher Ratio Class Separability Measure. IEEE Trans. Neural Network **13** (2002) 1211-1217
8. Sing, J.K.: Improved K-Means Algorithm in the Design of RBF Neural Network. Decision, Identification& Estimation in Image Processing, TENCON (2003) 841-845
9. Chang, C., Fu, S.: Image Classification Using a Module RBF Neural Network. ICIC (2006) 270-273
10. Ma, P.C.H.: An Evolutionary Clustering Algorithm for Gene Expression Microarray Data Analysis. IEEE Trans. Evolutionary Computation **10** (2006) 196-314

[2] This is the average results of 500 times random simulations.The time value is just for reference and changes with the computation environment.

A Rotated Image Matching Method Based on CISD*

Bojiao Sun and Donghua Zhou

Department of Automation, Tsinghua University,
100084 Beijing, China
sunbj03@mails.tsinghua.edu.cn,
zdh@mail.tsinghua.edu.cn

Abstract. In the image registration process, there always exists rotation transformation. The ordinary methods such as NCC (Normalized Cross Correlation Algorithm), SD (Square Difference Algorithm), SSDA (Sequential Similarity Detection Algorithm), are not suitable for rotated image registration. In this paper, a method based on circular template, intensity distribution and SD is proposed for rotation image registration. Through the CPs (Control Points) got by the proposed method, transformation model and least square method, the rotation parameters are obtained. Experimental results verify its effectiveness. Compared with the existing feature-based approaches, it is easier to obtain CPs and needs no salient objects.

1 Introduction

Image registration plays an important role in many fields, such as image-aided navigation system, etc. Its purpose is to establish coordinate relation between images. The reference image (pre-store and pre-treatment image) and sensed image (acquired image timely from the sensor) are usually needed to match.

Because the images are acquired at different conditions such as different platform or different time, etc, there usually exist transformations such as rotation, translation, etc, which makes the image registration difficult. The original methods, such as NCC (Normalized Cross Correlation Algorithm) [1], SSDA (Sequential Similarity Detection Algorithm) [2], SD (Square Difference Algorithm) [3], are not suitable for rotation image registration. A new method called MI (Mutual Information) [4],[5] is still not suitable for this situation. So rotated image registration becomes a difficult problem.

To solve this problem, Goshtasby used circular template and invariant moment criteria for matching [6]. Since it needs to compute seven invariant moments, it is time-consuming. In this paper, a method based on circular template, intensity distribution and SD method is proposed. Through the CPs (Control Points) got

* This work was supported by NSFC (Grant No. 60574084) and the national 973 program of China (Grant No. 2002CB312200).

D. Liu et al. (Eds.): ISNN 2007, Part I, LNCS 4491, pp. 1346–1352, 2007.

by the proposed method and transformation model, the rotation parameters are obtained. Experimental results verify its effectiveness.

This paper is organized as follows. The introduction is given above. Section 2 presents the details of the algorithm. Some experiments are given in section 3. Some discussions and conclusions are presented in section 4.

2 Algorithm

2.1 Basic Principle

In image registration process, template is always square or rectangle. If the image rotates, the contents of the template will change. So the ordinary methods, especially area-based methods, are not suitable for its registration. Circular template has been tried in image registration. Because rotation makes the corresponding relation of point to point between reference image and sensed image change, the ordinary methods still don't work. Experimental results verified our statement.

It is found that although the corresponding relation changes after rotation, the contents of the template at the same area will not change when circular template is used. On this premise, a suitable character should be found to treat this situation. Perhaps the intensity proportion of different gray-level or the intensity distribution in the circular template is a good character. At the same time, as entropy is the good way to represent the intensity distribution, it can be used as a criterion in image registration.

According to this idea, circular template and entropy are used for image matching. However, we are surprised that the experimental results are not right. The reason is that after rotation, the entropy of circular template changes due to interpolation effect and entropy is not robust to compare intensity distribution. To solve this problem, square difference algorithm is further used to compare the intensity distribution, and experimental results verify its effectiveness.

2.2 Algorithm

Suppose A is the reference image and B is the sensed image with some rotation to the reference image. In order to match the rotated images, a square template B_i is chosen from the sensed image. The area in its inscribed circle B_{ic} is regarded as the circular template and the proportion $p_{kB_{ic}}$ of different gray level k in B_{ic} is calculated. Then we need to find the circular template A_{jc} with the same size of B_{ic} in A and calculate the proportion $p_{kA_{jc}}$ of different gray level k in A_{jc}. The square difference method (SD) is used to compare $p_{kB_{ic}}$ with $p_{kA_{jc}}$.

$$D_{B_{ic}A_{jc}} = \sum_{k=1}^{K} \left(p_{kB_{ic}} - p_{kA_{jc}}\right)^2 \qquad k = 1, \cdots, K\,. \tag{1}$$

Where K is total number of gray levels and $D_{B_{ic}A_{jc}}$ is intensity distribution difference of B_{ic} and A_{jc}.

Global search strategy is used until all possible positions are tried. A_{jc} with smallest $D_{B_{ic}A_{jc}}$ is regarded as the matching template. The center of A_{jc} is the matching point of the center of B_{ic}.

The coordinate relation between A and B is obtained according to transformation method. In the matching process, at least two matching points are needed. To increase the robustness of this method, least square method and more control points are used. The transformation model is detailed in section 2.3.

2.3 Estimation of Transformation Parameters

Suppose that the transformation between image A and B is a rigid transformation composed of rotation, translation and scale with the scaling factor 1. (x, y) and (x', y') are matching points in the reference image and the sensed image, respectively. The coordinate relation between the reference image and the sensed image can be expressed as follows [7]:

$$\begin{pmatrix} X' \\ Y' \end{pmatrix} = s \cdot \begin{pmatrix} \cos\theta & -\sin\theta \\ \sin\theta & \cos\theta \end{pmatrix} \begin{pmatrix} X \\ Y \end{pmatrix} + \begin{pmatrix} d_x \\ d_y \end{pmatrix} . \tag{2}$$

Where θ is the angle that the sensed image rotates relevant to the reference image, d_x and d_y are, respectively, the horizontal translation and the vertical translation, s is the scaling factor.

Let $cos\theta = \mu$, $sin\theta = \nu$ and $s = 1$, then formula (2) can be rewritten as:

$$\begin{aligned} \mu \cdot X - \nu \cdot Y + d_x &= X' \\ \nu \cdot X + \mu \cdot Y + d_y &= Y' \end{aligned} . \tag{3}$$

In order to increase the robustness of this method, more CPs and least square method are used. Let

$$\mathbf{C} = \begin{pmatrix} X_1 & -Y_1 & 1 & 0 \\ Y_1 & X_1 & 0 & 1 \\ \vdots & \vdots & \vdots & \vdots \\ X_m & -Y_m & 1 & 0 \\ Y_m & X_m & 0 & 1 \end{pmatrix} \quad \mathbf{Z} = \begin{pmatrix} \mu \\ \nu \\ d_x \\ d_y \end{pmatrix} \quad \mathbf{D} = \begin{pmatrix} X_1' \\ Y_1' \\ \vdots \\ X_m' \\ Y_m' \end{pmatrix} . \tag{4}$$

Then the transformation parameters can be derived from the formula (4).

$$\mathbf{Z} = (\mathbf{C}^{\mathbf{T}}\mathbf{C})^{-1}\mathbf{C}^{\mathbf{T}}\mathbf{D} . \tag{5}$$

3 Experimental Results

To verify the effectiveness of this approach, many experiments using circular template, intensity distribution and SD method have been done. In order to express it concisely, circular template, intensity distribution and SD method is denoted by CISD. Two examples are given as follows.

Example 1. The reference image is an optical aerial image with the size of 250×250. It is shown in Fig. 1 (a). The sensed image is obtained from Fig. 1 (a) after rotation of 75° anticlockwise and is shown in Fig. 1 (b). The selected control points (CPs) are shown in Fig. 1 (b). Fig. 1 (a) shows the matched points in the reference image using CISD method.

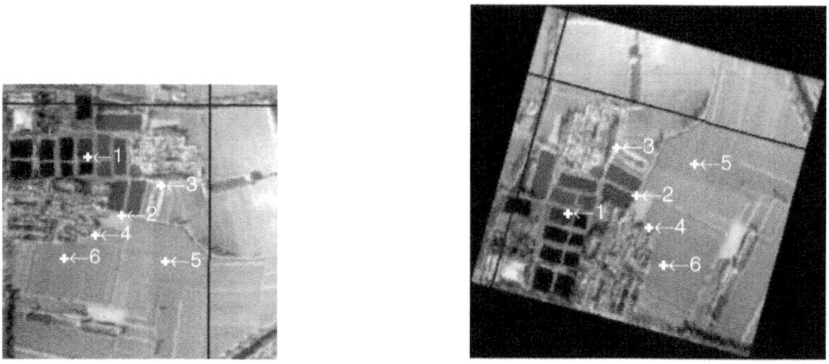

(a) (b)

Fig. 1. Experimental results of example 1. (a) The reference image (optical aerial image) and matching points got by CISD method. (b) The sensed image obtained from Fig. 1 (a) after rotation of 75° anticlockwise and its CPs.

Table 1 shows the CPs of the reference image and the sensed image, and also the transformation parameters of example 1.

Table 1. CPs and transformation parameters of example 1

	Sensed image	Reference image (CISD)		
	B_{io} (CPs of B)	A_{io} (CPs of A)	Matching position	Errors
1	(85, 183)	(81, 65)	(81, 65)	(0, 0)
2	(144, 167)	(112, 118)	(112, 118)	(0, 0)
3	(127, 125)	(148, 91)	(148, 90)	(0, -1)
4	(155, 195)	(88, 136)	(88, 136)	(0, 0)
5	(195, 139)	(152, 160)	(152, 160)	(0, 0)
6	(168, 228)	(60, 157)	(59, 157)	(-1, 0)
	matching rate		100%	
μ =0.2605	ν =0.9631	θ =74.8647	d_x =1.3375	d_y =-244.0470

The transformation parameters are derived from the formula (5). So the coordinate between reference image and sensed image is:

$$\begin{pmatrix} X' \\ Y' \end{pmatrix} = \begin{pmatrix} 0.2605 & -0.9631 \\ 0.9631 & 0.2605 \end{pmatrix} \begin{pmatrix} X \\ Y \end{pmatrix} + \begin{pmatrix} 1.3375 \\ -244.0470 \end{pmatrix} . \quad (6)$$

The rotated angle using proposed method is 74.8647 degree. Its angle error is 0.1353 degree.

Example 2. The reference image is a middle infrared aerial image with the size of 250×250. It is shown in Fig. 2 (a). The sensed image is obtained from Fig. 2 (a) after rotation of 220° clockwise and is shown in Fig. 2 (b). The selected control points (CPs) are shown in Fig. 2 (b). Fig. 2 (a) shows the matched points in the reference image using CISD method.

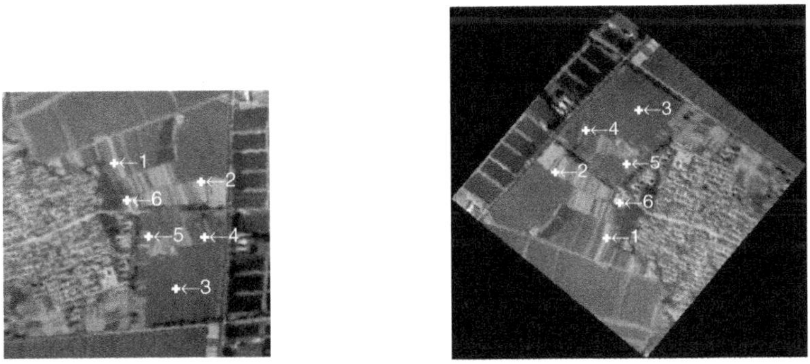

(a) (b)

Fig. 2. The experimental results of example 2. (a) The reference image (middle infrared aerial image) and matching points got by CISD method. (b) The sensed image obtained from Fig. 2 (a) after rotation of 220° clockwise and its CPs.

Table 2 shows the CPs of the reference image and the sensed image, and also the transformation parameters of example 2.

According to formula (5), the coordinate between reference image and sensed image is:

$$\begin{pmatrix} X' \\ Y' \end{pmatrix} = \begin{pmatrix} -0.7688 & -0.9400 \\ 0.9400 & -0.7688 \end{pmatrix} \begin{pmatrix} X \\ Y \end{pmatrix} + \begin{pmatrix} 193.6757 \\ -355.0277 \end{pmatrix} . \quad (7)$$

The rotated angle using the proposed method is 219.7763. Its angle error is 0.2237 degree.

Through observing the corresponding points in the reference image, calculating transformation model and comparing the obtained angle with the true rotated angle, it is found that the matching points got by CISD method are correct. The experimental results illustrate that the proposed method gives an effective way to deal with the rotated image registration.

Table 2. CPs and transformation parameters of example 2

	Sensed image	Reference image (CISD)		
	B_{io} (CPs of B)	A_{io} (CPs of A)	Matching position	Errors
1	(155, 233)	(108, 70)	(108, 69)	(0, -1)
2	(104, 167)	(189, 88)	(189, 87)	(0, -1)
3	(187, 105)	(165, 188)	(165, 188)	(0, 0)
4	(135, 125)	(192, 140)	(192, 139)	(0, -1)
5	(175, 159)	(140, 139)	(140, 139)	(0, 0)
6	(168, 198)	(120, 105)	(120, 104)	(0, -1)
	matching rate		100%	
μ =-0.7688	ν =0.9400	θ =219.7763	d_x =193.6757	d_y =-355.0277

4 Conclusions and Discussions

The images are often acquired at different time or different platform, etc, so there usually exists rotation, which makes image matching very difficult.

When the template is square or rectangle, the contents of rotated image is different from that of original image. If the template is circular, the situation changes. No matter how big scale the image rotates, the contents of it don't change except for the interpolation effect. A suitable invariant character can be used to describe the character of this area. On the contrary, the ordinary method such as NCC (Normalized Cross Correlation method), SD (Square Difference method) cannot work, because the corresponding relation changes when the image rotates.

It is found that intensity distribution of circular template doesn't change when rotation exists. Entropy is tried for rotated image registration, but it fails. The reason is that it is not robust to interpolation.

In this paper, a new algorithm based on circular template, intensity distribution and SD method, is proposed for image registration with rotation and translation. In order to get more CPs and reduce the computational time at the same time, several small subimages are chosen in the sensed image. Using the proposed method, several corresponding matching CPs are obtained in the reference image. The transformation parameters between the reference image and the sensed image can be calculated using transformation model and the least square method.

Compared with the existing feature-based approaches, it is easier to obtain CPs by the proposed approach and it needs no salient objects. The effectiveness of the proposed approach is verified through experiments.

References

1. Pratt, W.: Digital Image Processing. Wiley New York (1991)
2. Barnea, D.I., Silverman, H.F.: A Class of Algorithms for Fast Digital Image Registration. IEEE Trans. Computing **21** (1972) 179-186

3. Zitova, B., Flusser, J.: Image Registration Methods: A Survey. Image and Vision Computing **21** (2003) 977-1000
4. Viola, P., Wells, W.M.: Alignment by Maximization of Mutual Information. Int. J. Computer Vision **24** (1997) 137-154
5. Maes, F., Vandermeulen, D., Suetens, P.: Medical Image Registration Using Mutual Information. Proceedings of the IEEE **91** (2003) 1699-1722
6. Goshtasby, A.: Template Matching in Rotated Images. IEEE Trans. Pattern Analysis and Machine Intelligence **7** (1985) 338-344
7. Li, H., Manjunath, B.S., Mitra, S.K.: A Contour-based Approach to Multisensor Image Registration. IEEE Trans. Image Processing **4** (1995) 320-334

Author Index

Lecture Notes in Computer Science

For information about Vols. 1–4380

please contact your bookseller or Springer

Vol. 4510: P. Van Hentenryck, L. Wolsey (Eds.), Integration of AI and OR Techniques in Constraint Programming for Combinatorial Optimization Problems. X, 391 pages. 2007.

Vol. 4506: D. Zeng, I. Gotham, K. Komatsu, C. Lynch, M. Thurmond, D. Madigan, B. Lober, J. Kvach, H. Chen (Eds.), Intelligence and Security Informatics: Biosurveillance. XI, 234 pages. 2007.

Vol. 4504: J. Huang, R. Kowalczyk, Z. Maamar, D. Martin, I. Müller, S. Stoutenburg, K.P. Sycara (Eds.), Service-Oriented Computing: Agents, Semantics, and Engineering. X, 175 pages. 2007.

Vol. 4493: D. Liu, S. Fei, Z. Hou, H. Zhang, C. Sun (Eds.), Advances in Neural Networks – ISNN 2007, Part III. XXVI, 1215 pages. 2007.

Vol. 4492: D. Liu, S. Fei, Z. Hou, H. Zhang, C. Sun (Eds.), Advances in Neural Networks – ISNN 2007, Part II. XXVII, 1321 pages. 2007.

Vol. 4491: D. Liu, S. Fei, Z.-G. Hou, H. Zhang, C. Sun (Eds.), Advances in Neural Networks – ISNN 2007, Part I. LIV, 1365 pages. 2007.

Vol. 4486: M. Bernardo, J. Hillston (Eds.), Formal Methods for Performance Evaluation. VII, 469 pages. 2007.

Vol. 4483: C. Baral, G. Brewka, J. Schlipf (Eds.), Logic Programming and Nonmonotonic Reasoning. IX, 327 pages. 2007. (Sublibrary LNAI).

Vol. 4482: A. An, J. Stefanowski, S. Ramanna, C.J. Butz, W. Pedrycz, G. Wang (Eds.), Rough Sets, Fuzzy Sets, Data Mining and Granular Computing. XIV, 585 pages. 2007. (Sublibrary LNAI).

Vol. 4481: J.T. Yao, P. Lingras, W.-Z. Wu, M. Szczuka, N.J. Cercone, D. Ślęzak (Eds.), Rough Sets and Knowledge Technology. XIV, 576 pages. 2007. (Sublibrary LNAI).

Vol. 4480: A. LaMarca, M. Langheinrich, K.N. Truong (Eds.), Pervasive Computing. XIII, 369 pages. 2007.

Vol. 4479: I.F. Akyildiz, R. Sivakumar, E. Ekici, J.C.d. Oliveira, J. McNair (Eds.), NETWORKING 2007. Ad Hoc and Sensor Networks, Wireless Networks, Next Generation Internet. XXVII, 1252 pages. 2007.

Vol. 4472: M. Haindl, J. Kittler, F. Roli (Eds.), Multiple Classifier Systems. XI, 524 pages. 2007.

Vol. 4471: P. Cesar, K. Chorianopoulos, J.F. Jensen (Eds.), Interactive TV: a Shared Experience. XIII, 236 pages. 2007.

Vol. 4470: Q. Wang, D. Pfahl, D.M. Raffo (Eds.), Software Process Change – Meeting the Challenge. XI, 346 pages. 2007.

Vol. 4464: E. Dawson, D.S. Wong (Eds.), Information Security Practice and Experience. XIII, 361 pages. 2007.

Vol. 4463: I. Măndoiu, A. Zelikovsky (Eds.), Bioinformatics Research and Applications. XV, 653 pages. 2007. (Sublibrary LNBI).

Vol. 4462: D. Sauveron, K. Markantonakis, A. Bilas, J.-J. Quisquater (Eds.), Information Security Theory and Practices. XII, 255 pages. 2007.

Vol. 4459: C. Cérin, K.-C. Li (Eds.), Advances in Grid and Pervasive Computing. XVI, 759 pages. 2007.

Vol. 4453: T. Speed, H. Huang (Eds.), Research in Computational Molecular Biology. XVI, 550 pages. 2007. (Sublibrary LNBI).

Vol. 4452: M. Fasli, O. Shehory (Eds.), Agent-Mediated Electronic Commerce. VIII, 249 pages. 2007. (Sublibrary LNAI).

Vol. 4450: T. Okamoto, X. Wang (Eds.), Public Key Cryptography – PKC 2007. XIII, 491 pages. 2007.

Vol. 4448: M. Giacobini et al. (Ed.), Applications of Evolutionary Computing. XXIII, 755 pages. 2007.

Vol. 4447: E. Marchiori, J.H. Moore, J.C. Rajapakse (Eds.), Evolutionary Computation,Machine Learning and Data Mining in Bioinformatics. XI, 302 pages. 2007.

Vol. 4446: C. Cotta, J. van Hemert (Eds.), Evolutionary Computation in Combinatorial Optimization. XII, 241 pages. 2007.

Vol. 4445: M. Ebner, M. O'Neill, A. Ekárt, L. Vanneschi, A.I. Esparcia-Alcázar (Eds.), Genetic Programming. XI, 382 pages. 2007.

Vol. 4444: T. Reps, M. Sagiv, J. Bauer (Eds.), Program Analysis and Compilation, Theory and Practice. X, 361 pages. 2007.

Vol. 4443: R. Kotagiri, P.R. Krishna, M. Mohania, E. Nantajeewarawat (Eds.), Advances in Databases: Concepts, Systems and Applications. XXI, 1126 pages. 2007.

Vol. 4440: B. Liblit, Cooperative Bug Isolation. XV, 101 pages. 2007.

Vol. 4439: W. Abramowicz (Ed.), Business Information Systems. XV, 654 pages. 2007.

Vol. 4438: L. Maicher, A. Sigel, L.M. Garshol (Eds.), Leveraging the Semantics of Topic Maps. X, 257 pages. 2007. (Sublibrary LNAI).

Vol. 4433: E. Şahin, W.M. Spears, A.F.T. Winfield (Eds.), Swarm Robotics. XII, 221 pages. 2007.

Vol. 4432: B. Beliczynski, A. Dzielinski, M. Iwanowski, B. Ribeiro (Eds.), Adaptive and Natural Computing Algorithms, Part II. XXVI, 761 pages. 2007.

Vol. 4431: B. Beliczynski, A. Dzielinski, M. Iwanowski, B. Ribeiro (Eds.), Adaptive and Natural Computing Algorithms, Part I. XXV, 851 pages. 2007.

Vol. 4430: C.C. Yang, D. Zeng, M. Chau, K. Chang, Q. Yang, X. Cheng, J. Wang, F.-Y. Wang, H. Chen (Eds.), Intelligence and Security Informatics. XII, 330 pages. 2007.

Vol. 4429: R. Lu, J.H. Siekmann, C. Ullrich (Eds.), Cognitive Systems. X, 161 pages. 2007. (Sublibrary LNAI).

Vol. 4427: S. Uhlig, K. Papagiannaki, O. Bonaventure (Eds.), Passive and Active Network Measurement. XI, 274 pages. 2007.

Vol. 4426: Z.-H. Zhou, H. Li, Q. Yang (Eds.), Advances in Knowledge Discovery and Data Mining. XXV, 1161 pages. 2007. (Sublibrary LNAI).

Vol. 4425: G. Amati, C. Carpineto, G. Romano (Eds.), Advances in Information Retrieval. XIX, 759 pages. 2007.

Vol. 4424: O. Grumberg, M. Huth (Eds.), Tools and Algorithms for the Construction and Analysis of Systems. XX, 738 pages. 2007.

Vol. 4423: H. Seidl (Ed.), Foundations of Software Science and Computational Structures. XVI, 379 pages. 2007.

Vol. 4422: M.B. Dwyer, A. Lopes (Eds.), Fundamental Approaches to Software Engineering. XV, 440 pages. 2007.

Vol. 4421: R. De Nicola (Ed.), Programming Languages and Systems. XVII, 538 pages. 2007.

Vol. 4420: S. Krishnamurthi, M. Odersky (Eds.), Compiler Construction. XIV, 233 pages. 2007.

Vol. 4419: P.C. Diniz, E. Marques, K. Bertels, M.M. Fernandes, J.M.P. Cardoso (Eds.), Reconfigurable Computing: Architectures, Tools and Applications. XIV, 391 pages. 2007.

Vol. 4418: A. Gagalowicz, W. Philips (Eds.), Computer Vision/Computer Graphics Collaboration Techniques. XV, 620 pages. 2007.

Vol. 4416: A. Bemporad, A. Bicchi, G. Buttazzo (Eds.), Hybrid Systems: Computation and Control. XVII, 797 pages. 2007.

Vol. 4415: P. Lukowicz, L. Thiele, G. Tröster (Eds.), Architecture of Computing Systems - ARCS 2007. X, 297 pages. 2007.

Vol. 4414: S. Hochreiter, R. Wagner (Eds.), Bioinformatics Research and Development. XVI, 482 pages. 2007. (Sublibrary LNBI).

Vol. 4412: F. Stajano, H.J. Kim, J.-S. Chae, S.-D. Kim (Eds.), Ubiquitous Convergence Technology. XI, 302 pages. 2007.

Vol. 4411: R.H. Bordini, M. Dastani, J. Dix, A.E.F. Seghrouchni (Eds.), Programming Multi-Agent Systems. XIV, 249 pages. 2007. (Sublibrary LNAI).

Vol. 4410: A. Branco (Ed.), Anaphora: Analysis, Algorithms and Applications. X, 191 pages. 2007. (Sublibrary LNAI).

Vol. 4409: J.L. Fiadeiro, P.-Y. Schobbens (Eds.), Recent Trends in Algebraic Development Techniques. VII, 171 pages. 2007.

Vol. 4407: G. Puebla (Ed.), Logic-Based Program Synthesis and Transformation. VIII, 237 pages. 2007.

Vol. 4406: W. De Meuter (Ed.), Advances in Smalltalk. VII, 157 pages. 2007.

Vol. 4405: L. Padgham, F. Zambonelli (Eds.), Agent-Oriented Software Engineering VII. XII, 225 pages. 2007.

Vol. 4403: S. Obayashi, K. Deb, C. Poloni, T. Hiroyasu, T. Murata (Eds.), Evolutionary Multi-Criterion Optimization. XIX, 954 pages. 2007.

Vol. 4401: N. Guelfi, D. Buchs (Eds.), Rapid Integration of Software Engineering Techniques. IX, 177 pages. 2007.

Vol. 4400: J.F. Peters, A. Skowron, V.W. Marek, E. Orłowska, R. Słowiński, W. Ziarko (Eds.), Transactions on Rough Sets VII, Part II. X, 381 pages. 2007.

Vol. 4399: T. Kovacs, X. Llorà, K. Takadama, P.L. Lanzi, W. Stolzmann, S.W. Wilson (Eds.), Learning Classifier Systems. XII, 345 pages. 2007. (Sublibrary LNAI).

Vol. 4398: S. Marchand-Maillet, E. Bruno, A. Nürnberger, M. Detyniecki (Eds.), Adaptive Multimedia Retrieval: User, Context, and Feedback. XI, 269 pages. 2007.

Vol. 4397: C. Stephanidis, M. Pieper (Eds.), Universal Access in Ambient Intelligence Environments. XV, 467 pages. 2007.

Vol. 4396: J. García-Vidal, L. Cerdà-Alabern (Eds.), Wireless Systems and Mobility in Next Generation Internet. IX, 271 pages. 2007.

Vol. 4395: M. Daydé, J.M.L.M. Palma, Á.L.G.A. Coutinho, E. Pacitti, J.C. Lopes (Eds.), High Performance Computing for Computational Science - VECPAR 2006. XXIV, 721 pages. 2007.

Vol. 4394: A. Gelbukh (Ed.), Computational Linguistics and Intelligent Text Processing. XVI, 648 pages. 2007.

Vol. 4393: W. Thomas, P. Weil (Eds.), STACS 2007. XVIII, 708 pages. 2007.

Vol. 4392: S.P. Vadhan (Ed.), Theory of Cryptography. XI, 595 pages. 2007.

Vol. 4391: Y. Stylianou, M. Faundez-Zanuy, A. Esposito (Eds.), Progress in Nonlinear Speech Processing. XII, 269 pages. 2007.

Vol. 4390: S.O. Kuznetsov, S. Schmidt (Eds.), Formal Concept Analysis. X, 329 pages. 2007. (Sublibrary LNAI).

Vol. 4389: D. Weyns, H.V.D. Parunak, F. Michel (Eds.), Environments for Multi-Agent Systems III. X, 273 pages. 2007. (Sublibrary LNAI).

Vol. 4385: K. Coninx, K. Luyten, K.A. Schneider (Eds.), Task Models and Diagrams for Users Interface Design. XI, 355 pages. 2007.

Vol. 4384: T. Washio, K. Satoh, H. Takeda, A. Inokuchi (Eds.), New Frontiers in Artificial Intelligence. IX, 401 pages. 2007. (Sublibrary LNAI).

Vol. 4383: E. Bin, A. Ziv, S. Ur (Eds.), Hardware and Software, Verification and Testing. XII, 235 pages. 2007.

Vol. 4381: J. Akiyama, W.Y.C. Chen, M. Kano, X. Li, Q. Yu (Eds.), Discrete Geometry, Combinatorics and Graph Theory. XI, 289 pages. 2007.

GPSR Compliance

The European Union's (EU) General Product Safety Regulation (GPSR) is a set of rules that requires consumer products to be safe and our obligations to ensure this.

If you have any concerns about our products, you can contact us on ProductSafety@springernature.com

In case Publisher is established outside the EU, the EU authorized representative is:

Springer Nature Customer Service Center GmbH
Europaplatz 3
69115 Heidelberg, Germany

Batch number: 10761593

Printed by Printforce, the Netherlands